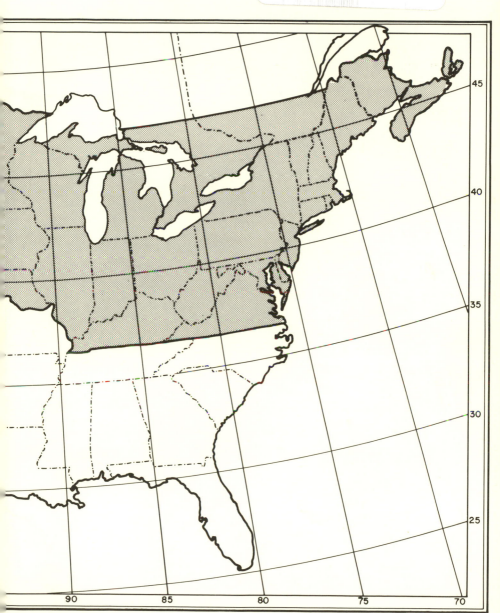

Manual of Vascular Plants of Northeastern United States and Adjacent Canada
Second Edition

Manual of Vascular Plants of Northeastern United States and Adjacent Canada
Second Edition

THE LATE HENRY A. GLEASON, Ph.D.
Head Curator, Emeritus

and

ARTHUR CRONQUIST, Ph.D.
Senior Scientist

The New York Botanical Garden

The New York Botanical Garden
Bronx, New York 10458, USA

The New York Botanical Garden
Southern Blvd. at 200th St.
Bronx, New York 10458, USA

Copyright © 1991, by
New York Botanical Garden

Library of Congress Cataloging-in-Publication Data

Gleason, Henry A. (Henry Allan), 1882–
    Manual of vascular plants of northeastern United States and
adjacent Canada / Henry A. Gleason and Arthur Cronquist. — 2nd ed.
    p.   cm.
    Includes bibliographical references and index.
    ISBN 0-89327-365-1
    1. Botany—Northeastern States. 2. Botany—Canada, Eastern.
3. Plants—Identification.   I. Cronquist, Arthur.   II. Title.
QK117.049   1991
582.0974—dc 20   91-23110
                                                    CIP

ISBN No. 0-89327-365-1

Preface

This manual is intended to facilitate the identification of vascular plants growing wild in the northeastern United States and adjacent Canada. The part of the United States covered extends from the Atlantic Ocean west to the western boundaries of Minnesota, Iowa, northern Missouri, and southern Illinois, south to the southern boundaries of Virginia, Kentucky, and Illinois, and south to the Missouri River in Missouri. In Canada the area covered includes Nova Scotia, Prince Edward Island, New Brunswick, and the parts of Quebec and Ontario lying south of the 47th parallel of latitude.

This second edition of the manual is based on the first edition (1963), which in turn was based on the New Britton and Brown Illustrated Flora (1952), by H. A. Gleason and collaborators. Gleason himself wrote the treatment for more than three-fourths of the species of his flora, and I did a little more than half of the remainder, comprising the family Compositae (Asteraceae) except the Vernonieae. I now take full responsibility for the entire contents of the manual.

In general, I retain the previous treatment unless recent literature or consultation with other botanists or my own field experience suggests the need for change. Perhaps the most drastic changes are in the treatments of *Panicum* and the stemless blue violets. For *Panicum* I have drawn heavily on the treatment by Gould & Clark (1978) of the *Dichanthelium* group. The treatment of the stemless blue violets reflects my own observations in the field over several decades.

It seems appropriate to abandon the historically useful but now moribund Englerian arrangement of the families and orders of angiosperms, in favor of a more modern arrangement taken from the second (1988) edition of my own *Evolution and Classification of Flowering Plants.*

In addition to the artificial key to families (based on that of Gleason), a synoptical arrangement in key form is presented for the families, orders, and subclasses of dicotyledons and monocotyledons. Inasmuch as these orders and subclasses are not amenable to mutually exclusive phenetic characterization, the formal treatment of angiosperms in this text goes directly from the class to the family. The higher taxa of gymnosperms and vascular cryptogams, in contrast, are phenetically meaningful and are therefore formally presented in the text.

So many botanists have helped me with one or another part of the manuscript that some of the contributions may have escaped my memory. I recall receiving assistance from Susan G. Aiken, Ihsan Al-Shehbaz, Gregory J. Anderson, George W. Argus, T. M. Barkley, Mary E. Barkworth, Rupert C. Barneby, Jerry M. Baskin, David M. Bates, Randall J. Bayer, Joseph M. Beitel, Brian M. Boom, Ralph E. Brooks, Leo P. Bruederle, Judith M. Canne, Lorraine Caruso, Henrietta L. Chambers, Steven E. Clemants, Tom S. Cooperrider, Stephen Darbyshire, Douglas R. Dewey, Laurence J. Dorr, Richard H. Eyde, Robert B. Faden, David E. Fairbrothers, John D. Freeman, Fred C. Galle, Robert K. Godfrey, James W. Grimes, Jon J. Hamer, James W. Hardin, Robert R. Haynes, Robert Hill, Steven R. Hill, Robert Hilton, Peter C. Hoch, Noel H. Holmgren, Christopher J. Humphries, David Hunt, Hugh H. Iltis, Duane Isely, David M. Johnson, Almut G. Jones, Robert B. Kaul, Khani, Joseph H. Kirkbride Jr., Robert R. Kowal, Robert Kral, Eric E. Lamont, Lieu Fah-Seong, James L. Luteyn, Ronald L. McGregor, John McNeill, John T. Mickel, Richard S. Mitchell, Michael Nee, Eugene C. Ogden, Gerald B. Ownbey, Chongwook Park, Thomas S. Patrick, C. Thomas Philbrick, Richard W. Pohl, Duncan M. Porter, R. K. Rabeler, Peter H. Raven, James L. Reveal, Anton A. Reznicek, Bruce K. Riggs, Reed C. Rollins, Paul E. Rothrock, John T. Semple, Terry L. Sharik, Chung-Fu Shen, Charles J. Sheviak, Ernest Small, David Snyder, Otto T. Solbrig, David M. Spooner, Lisa M. Standley, G. Ledyard Stebbins, O. A. Stevens, Ronald L. Stuckey, William Wayt Thomas, Gordon C. Tucker, Leonard Uttal, Sam P. Vander Kloet, David P. Vernon, Karl A. Vincent, Edward G. Voss, Warren H. Wagner, Robert S. Wallace, Thomas Wieboldt, John J. Wurdack and Peter F. Zika. The typing has been done in large part by Lucy F. Kluska, but also in considerable part by my wife, Mabel A. Cronquist, who has been a devoted assistant during the many years of preparation. To all, named and unnamed, my thanks. None of the people who helped me is responsible for any of the opinions presented, which are my own.

ABBREVIATIONS

avg	average, averaging
ca	circa, about
cal	calyx
cm	centimeter, centimeters
cor	corolla, corollas
cosmop	cosmopolitan
cult.	cultivated, cultivation
dm	decimeter, decimeters
fl, fls, fld	flower, flowers, flowered
fr, frs	fruit, fruits
infl, infls	inflorescence, inflorescences
invol, invols	involucre, involucres
lf, lvs	leaf, leaves
lfl, lfls	leaflet, leaflets
m	meter, meters
mm	millimeter, millimeters
pet	petal, petals
pl, pls	plant, plants
±	more or less
reg.	region, regions
sep	sepal, sepals
sp., spp.	species (sing.), species (pl.)
ssp., sspp.	subspecies (sing.), subspecies (pl.)
tep	tepal, tepals
var., vars.	variety, varieties

SOME OF THE MORE FREQUENTLY CITED BOTANICAL AUTHORS

(Abbreviations of the names of other authors are in general compatible with the Kew List)

Adans. Michel Adanson, 1727–1806
Aiton William Aiton, 1731–1793
Anderss. Nils Johan Andersson, 1821–1880
Andrz. Antoni Lukianowicz Andrzejowski, 1785–1868
Arnott George Arnott Walker Arnott, 1799–1868
Ashe William Willard Ashe, 1872–1932
L. H. Bailey Liberty Hyde Bailey, 1858–1954
Baillon Henri Ernest Baillon, 1827–1895
Barneby Rupert Charles Barneby, 1911–
Bartram William Bartram, 1739–1823
P. Beauv. Ambroise Marie Françoise Joseph Palisot de Beauvois, 1752–1820
Benth. George Bentham, 1800–1884
Bernh. Johann Jacob Bernhardi, 1774–1850
Besser Wilibald Swibert Joseph Gottlieb von Besser, 1784–1842
E. Bickn. Eugene Pintard Bicknell, 1859–1925
M. Bieb. Friedrich August Marschall von Bieberstein, 1768–1826
Boiss. Pierre Edmond Boissier, 1810–1885
B. Boivin Joseph Robert Bernard Boivin, 1916–1985
Bonpl. Aimé Jacques Alexandre Bonpland, 1773–1858
F. Boott Francis Boott, 1792–1863
W. Boott William Boott, 1805–1887
Brainerd Ezra Brainerd, 1844–1924
Britton Nathaniel Lord Britton, 1859–1934
R. Br. Robert Brown, 1773–1858
BSP. Nathaniel Lord Britton, 1859–1934; Emerson Ellick Sterns, 1846–1926; Justus Ferdinand Poggenburg, 1840–1893
Cass. Alexandre Henri Gabriel Cassini, 1781–1832
Cham. Ludolf Adalbert von Chamisso, 1781–1838
Chapman Alvin Wentworth Chapman, 1809–1899
Cronq. Arthur Cronquist, 1919-
A. DC. Alphonse Louis Pierre Pyramus de Candolle, 1806–1893
DC. Augustin Pyramus de Candolle, 1778–1841
Decne. Joseph Decaisne, 1807–1882
Desf. Réné Louiche Desfontaines, 1750–1833
D. Don David Don, 1799–1841
G. Don George Don, 1798–1856
Douglas David Douglas, 1798–1834
Duchesne Antoine Nicolas Duchesne, 1747–1827
Dunal Michel Félix Dunal, 1789–1852
Durand Élie Magloire Durand, 1794–1873
D. C. Eat. Daniel Cady Eaton, 1834–1895
Ehrh. Friedrich Ehrhart, 1742–1795
Elliott Stephen Elliott, 1771–1830
Engelm. George Engelmann, 1809–1884
Engler Heinrich Gustav Adolf Engler, 1845–1930
Farw. Oliver Atkins Farwell, 1867–1944
Fern. Merritt Lyndon Fernald, 1873–1950
Fischer Friedrich Ernst Ludwig von Fischer, 1782–1854
Forssk. Pehr Forsskål, 1732–1763
Fries Elias Magnus Fries, 1794–1878
Gaertner Joseph Gaertner, 1732–1791
Gaertner f. Carl Friedrich von Gaertner, 1772–1850
Gleason Henry Allan Gleason, 1882–1975
J. F. Gmelin Johann Friedrich Gmelin, 1748–1804
J. G. Gmelin Johann Georg Gmelin, 1709–1755
S. G. Gmelin Samuel Gottlieb Gmelin, 1745–1774
A. Gray Asa Gray, 1810–1888
S. F. Gray Samuel Frederick Gray, 1766–1828

Greene Edward Lee Greene, 1843–1915
Griseb. August Heinrich Rudolf Grisebach, 1814–1879
H. & A. William Jackson Hooker, 1785–1865, and George Arnott Walker Arnott, 1799–1868
HBK. Friedrich Wilhelm Heinrich Alexander von Humboldt, 1769–1858; Aimé Jacques Alexandre Bon-
  pland, 1773–1858; Carl Sigismund Kunth, 1788–1850
Heller Amos Arthur Heller, 1867–1944
Hemsley William Botting Hemsley, 1843–1924
Hook. William Jackson Hooker, 1785–1865
Hook. f. Joseph Dalton Hooker, 1817–1911 (son of William)
Hornem. Jens Wilken Hornemann, 1770–1841
Jacq. Nikolaus Joseph von Jacquin, 1727–1817
Johnston Ivan Murray Johnston, 1898–1960
M. E. Jones Marcus Eugene Jones, 1852–1934
Juss. Antoine Laurent de Jussieu, 1748–1836
Kunth Carl Sigismund Kunth, 1788–1850
Kuntze Carl Ernst Otto Kuntze, 1843–1907
L. Carl Linnaeus, 1707–1778
L. f. Carl von Linné, 1741–1783 (son of Carl Linnaeus)
Lam. Jean Baptiste Antoine Pierre de Monnet de Lamarck, 1744–1829
Ledeb. Carl Friedrich von Ledebour, 1785–1851
Lehm. Johann Georg Christian Lehmann, 1792–1860
Less. Christian Friedrich Lessing, 1809–1862
Lindl. John Lindley, 1799–1865
Link Johann Heinrich Friedrich Link, 1767–1851
Mackenzie Kenneth Kent Mackenzie, 1877–1934
Marshall Humphry Marshall, 1722–1801
Medikus Friedrich Casimir Medikus, 1736–1808
Meissn. Carl Daniel Friedrich Meissner, 1800–1874
C. A. Meyer Carl Anton Andreevic von Meyer, 1795–1855
Michx. André Michaux, 1746–1802
Miller Philip Miller, 1691–1771
Moench Conrad Moench, 1744–1805
Moq. Christian Horace Bénédict Alfred Moquin-Tandon, 1804–1863
Morton Conrad Vernon Morton, 1905–1972
F. Muell. Ferdinand Jacob Heinrich von Mueller, 1825–1896
Muell.-Arg. Jean Müller of Aargau, 1828–1896
Muhl. Gotthilf Henry Ernest Muhlenberg, 1753–1815
Munz Philip Alexander Munz, 1892–1974
Murray Johan Andreas Murray, 1740–1791
Nash George Valentine Nash, 1864–1921
Nees Christian Gottfried Daniel Nees von Esenbeck, 1776–1858
A. Nels. Aven Nelson, 1859–1952
Nutt. Thomas Nuttall, 1786–1859
Olney Stephen Thayer Olney, 1812–1878
Pallas Peter Simon von Pallas, 1741–1811
E. J. Palmer Ernest Jesse Palmer, 1875–1962
Pavón José Antonio Pavón, 1750–1844
Pennell Francis Whittier Pennell, 1886–1952
Pers. Christiaan Hendrik Persoon, 1761–1836
Planchon Jules Émile Planchon, 1823–1888
Poggenburg Justus Ferdinand Poggenburg, 1840–1893
Poiret Jean Louis Marie Poiret, 1755–1834
C. Presl Carel Borowag Presl, 1794–1852
Pursh Frederick Traugott Pursh, 1774–1820
Raf. Constantine Samuel Rafinesque-Schmaltz, 1783–1840
R. & S. Johann Jacob Roemer, 1673–1819; Josef August Schultes, 1773–1831
Reichenb. Heinrich Gottlieb Ludwig Reichenbach, 1793–1879
Rich. Louis Claude Marie Richard, 1754–1821
Richardson John Richardson, 1787–1865
B. L. Robinson Benjamin Lincoln Robinson, 1864–1935
Roemer Johann Jacob Roemer, 1763–1819
Ruíz Hipólito Ruíz Lopez, 1754–1815
Rupr. Franz Josef Ruprecht, 1814–1870
Rydb. Per Axel Rydberg, 1860–1931
Sarg. Charles Sprague Sargent, 1841–1927
Schlecht. Diederich Franz Leonhard von Schlechtendahl, 1794–1866
Schultes Josef August Schultes, 1773–1831
Schultz-Bip. Carl Heinrich Schultz of Zweibrücken (Bipontinus), 1805–1867
Scop. Giovanni Antonio Scopoli, 1723–1788

Scribn. Frank Lamson Scribner, 1851–1938
Small John Kunkel Small, 1869–1938
J. E. Smith James Edward Smith, 1759–1828
Spach Édouard Spach, 1801–1879
Sprengel Curt Polycarp Joachim Sprengel, 1766–1833
Standley Paul Carpenter Standley, 1884–1963
Sterns Emerson Ellick Sterns, 1846–1926
Steudel Ernst Gottlieb von Steudel, 1783–1856
Swartz Olof Peter Swartz, 1760–1818
T. & G. John Torrey, 1796–1873; Asa Gray, 1810–1888
Torr. John Torrey, 1796–1873
Thunb. Carl Peter Thunberg, 1743–1828
Trel. William Trelease, 1857–1945
Underw. Lucien Marcus Underwood, 1853–1907
Vict. Frère Alexandre Marie-Victorin (Conrad Kirouac), 1885–1944
Wahl Herbert Alexander Wahl, 1900–1975
Wahlenb. Georg (Göran) Wahlenberg, 1780–1851
Walter Thomas Walter, 1740–1789
S. Wats. Sereno Watson, 1826–1892
Wherry Edgar Theodore Wherry, 1885–1982
Wieg. Karl McKay Wiegand, 1873–1942
Willd. Carl Ludwig von Willdenow, 1765–1812
A. Wood Alphonso Wood, 1810–1881
Wormsk. Morten Wormskiöld, 1783–1845
Zucc. Joseph Gerhard Zuccarini, 1797–1848

GLOSSARY

The definitions in the glossary are principally for usage within the vascular plants. Many of the terms also have a broader meaning, or a different meaning in some other group.

a-, ab- Latin prefix, meaning not, or different from, or away from, or without.

abaxial On the side away from the axis, or turned away from the axis. (Compare *adaxial.*)

abortive Hardly or imperfectly developed, usually implying that development begins normally but stops short of normal maturity. (Compare *vestigial.*)

acaulescent Literally, without a stem; in common botanical usage, with the leafy part of the stem so short that the leaves are all clustered in a basal rosette.

accrescent Increasing in size with age, as for example a calyx that continues to enlarge after anthesis.

accumbent cotyledons Cotyledons with one edge against the radicle. (Compare *incumbent cotyledons.*)

achene The most generalized type of dry, indehiscent fruit, usually one-seeded, lacking the specialized features that mark a caryopsis, nut, samara, or utricle.

achlamydeous Without a perianth.

acrodromous venation Venation with one or more pairs of prominent lateral veins arching outward and forward from at or near the base, often then converging distally.

acropetal Near the tip or distal end, as opposed to near the base or proximal end; proceeding from the proximal toward the distal end. (Compare *basipetal.*)

acroscopic Toward the distal as opposed to the proximal end. (Compare *basiscopic.*)

actinomorphic Radially symmetrical, as in a regular flower.

aculeate Prickly; beset with prickles.

acuminate Gradually and concavely tapering to a narrow tip or sharp point. (Compare *acute, attenuate.*)

acute Sharp-pointed (as to shape, but not necessarily as to texture), with more or less straight distal margins that form an angle of less than 90 degrees at the tip.

ad- Latin prefix, meaning to or toward.

adaxial On the side toward the axis, or turned toward the axis. (Compare *abaxial.*)

-adelphous Greek combining form, indicating an arrangement or union in groups; as, monadelphous, in one group, diadelphous, in two groups, etc.

adherent Sticking together. Like adnate, the term is applied only to unlike parts, but it usually implies a less firm or less perfect union than adnate. (Compare *coherent.*)

adnate Grown together, or attached; applied only to unlike parts, as stipules adnate to the petiole, or stamens adnate to the corolla. (Compare *connate.*)

adventitious Arising from mature nonmeristematic tissues, especially if such a development would not ordinarily be expected. Any root that originates from a stem is considered to be adventitious.

adventive Introduced but not naturalized, or only locally established.

aestival Appearing or blooming in or pertaining to the summer.

aestivation The arrangement of flower parts in the bud, especially the position of the petals (or sepals) with respect to each other.

aggregate fruit A fruit derived from two or more pistils of a single flower; e.g., a raspberry. An aggregate fruit that also has some other flower part(s) in addition to the pistils is said to be aggregate accessory; e.g., blackberries and strawberries, in which the receptacle is an integral part of the mature fruit.

alate Winged.

alkaloid Any of a large, chemically diverse group of nitrogenous, pharmacologically active ring-compounds produced by plants.

alloploid, allopolyploid Polyploid with the genomes derived from two or more distinct species. (Compare *autoploid.*)

allopatric Occupying different geographic regions. (Compare *sympatric.*)

alternate Situated singly at each node, as the leaves on a stem; situated regularly between organs of another kind, as stamens alternate with the petals.

alveolate Honeycombed, with deep, angular cavities (alveoli) separated by thin partitions.

ament See catkin.

amphibious Living both on land and in water, especially if growth continues on land after the water has receded or evaporated.

amphitropous ovule An ovule with the body half-inverted, so that the funiculus is attached near the middle, and the micropyle points at right angles to the funiculus; hemianatropous.

amplexicaul Clasping the stem, as the base of some kinds of leaves.

ampliate Enlarged, dilated.

an- Green prefix, equivalent to Latin a-, ab-.

anastomosing Rejoining (after branching) to form a network.

anatropous ovule An ovule with the body fully inverted, so that the micropyle is basal, adjoining the funiculus.

andro-dioecious Producing male flowers on one plant and perfect flowers on another.

androecium All of the stamens of a flower, considered collectively.

androgynophore A stalk arising from the receptacle, on which the andoecium and the gynoecium of some kinds of flowers are borne.

androgynous spike (in *Carex*) A spike with both staminate and pistillate flowers, the staminate above the pistillate. (Compare *gynaecandrous spike.*)

anemophilous Pollinated by wind.

angiosperm A member of a group of plants (Magnoliophyta) characterized by having the ovules enclosed in an ovary.

annual Yearly. A plant that germinates, flowers, and sets seed during a single growing season, in the northern hemisphere during a single calendar year. A winter annual germinates in the fall and fruits the following spring or summer.

annular In the form of a ring, or marked by transverse rings.

annulus A little ring; the partial or complete ring of specialized, thick-walled cells that encircles the sporangium of typical ferns.

ante- Latin combining form, meaning before or in front of.

anterior Literally, in front. The anterior side of a flower is the side toward the subtending bract, rather than the side toward the axis of the inflorescence; thus the lower lip of a bilabiate flower is the anterior lip. (Compare *posterior.*)

anther The part of a stamen that bears the pollen, consisting of one or usually two pollen-sacs and a connecting layer between them.

anthesis The period during which a flower is fully expanded and functional, ready to shed or receive pollen.

anthocyanin A chemical class of water-soluble pigments, ranging in color from blue or violet through purple to crimson, often found in the central vacuole of a cell, especially in petals.

anthoxanthin   A group of pigments closely allied to anthocyanin, but ranging in color from yellow to orange or orange-red.

antrorse   Directed forward or upward; opposite of *retrorse*.

apetalous   Without petals.

aphyllopodic   With the lowest leaves reduced to scales, so that the first foliage leaves are well above the base of the plant, as in some species of *Carex*. (Compare *phyllopodic*.)

apiculate   Ending abruptly in a small, usually sharp point.

apocarpous   With the carpels separate from each other (or with only one carpel). (Compare *syncarpous*.)

apogamous   Having the most common type of apomixis, in which no gametes are produced, and successive generations have the same chromosome number. In vascular cryptogams, bypassing meiosis and fertilization, so that the spores have the same chromosome number as mature plants.

apomixis   The setting of seed without fertilization.

apotropous ovule   An ovule which (if erect) has the raphe ventral (between the partition and the body of the ovule), or which (if pendulous) has the raphe dorsal. (Compare *epitropous ovule*.)

appressed   Lying close to (pressed against) an organ, as hairs appressed to a leaf or leaves appressed to a stem.

arachnoid   Provided with a cobwebby, usually sparse pubescence of relatively long, soft, tangled hairs.

arborescent   Tree-like, or becoming a tree, or with nearly the size and form of a tree.

arcuate   Curved into an arc or a circle, without regard to direction.

areola, areole   A small, clearly bounded area on a surface, as an area bounded by the veinlets of a leaf.

areolate   Marked with areolae.

argenteous   Silvery in color. (Compare *canescent, cinereous*.)

aril   A specialized, usually fleshy outgrowth from the funiculus that covers or is attached to the mature seed; more loosely, any appendage or thickening of the seed-coat.

arillate   Provided with an aril.

aristate   Tipped with an awn or bristle.

aristulate   Diminutive of aristate.

articulate, articulated   Jointed, with a predetermined point of natural separation.

ascending   Growing obliquely upward (stems); directed obliquely forward in respect to the organ to which they are attached (parts of a plant).

asepalous   Without sepals.

assurgent   Ascending (stems).

atomate, atomiferous, atomiferous-glandular   Bearing scattered sessile or subsessile glands.

atro-   Latin prefix, meaning dark or blackish.

attenuate   Tapering very gradually to a very slender tip; more extreme than acute or acuminate.

auricle   A small projecting lobe or appendage at the base of an organ.

auriculate   Provided with one or more auricles.

austral   Southern; opposite of *boreal*.

autoploid, autopolyploid   Polyploid with the genomes all derived from the same species. (Compare *alloploid*.)

autotrophic   Nutritionally independent, making its own food from raw materials obtained more or less directly from the substrate. The term is usually interpreted to include mycorrhizal as well as nonmycorrhizal plants, so long as they are photosynthetic.

awn   A slender, usually terminal or dorsal bristle.

axil   The position (not a structure) between the axis and a lateral organ such as a leaf; the point of the angle between a stem and a leaf.

axile placenta   A placenta along the central axis (or along the vertical midline of the septum) of an ovary with two or more locules.

axillary   Located in or arising from an axil.

axis   A portion of a plant from which a series of lateral organs or branches arise, as the axis of an inflorescence. (Compare *rachis*.)

baccate   Resembling or having the structure of a berry.

banner   The upper, usually enlarged petal of a papilionaceous flower; the standard.

barbate   Bearded with long, stiff hairs.

barbed   Bearing short, firm, retrorse points.

barbellate   Diminutive of barbate or barbed.

basifixed   Attached by the base. (Compare *dorsifixed*.)

basipetal   Developing or proceeding from the distal end toward the base. (Compare *acropetal*.)

basiscopic   Toward the basal or proximal (as opposed to the distal) end. (Compare *acroscopic*.)

beak   A comparatively short and stout terminal appendage on a thickened organ, as a seed or a fruit; not used for a flat organ such as a leaf.

bearded   Bearing a tuft or ring of rather long hairs.

berry   The most generalized type of fleshy fruit, derived from a single pistil, fleshy throughout, and containing usually several or many seeds; more loosely, any pulpy or juicy fruit.

betalain   A chemical class of nitrogenous, water-soluble pigments, consisting of betacyanins (blue or violet to purple or red) and betaxanthins (yellow to orange or orange-red).

bi-   Latin prefix, meaning two.

biennial   Living two years only and blooming the second year.

bifid   More or less deeply cleft from the tip into two usually equal parts.

bilabiate   Two-lipped.

bipinnate   Twice pinnate, the primary pinnae again pinnate.

bivalvate   Opening by two valves.

blade   The expanded, terminal portion of a flat organ such as a leaf, petal, or sepal, in contrast to the narrowed basal portion.

bloom   A waxy powder covering a surface, making it glaucous or pruinose.

boreal   Northern; opposite of *austral*.

brachiate   Having opposite, widely spreading branches.

bract   A specialized leaf, from the axil of which a flower or flower-stalk arises; more loosely, any more or less reduced or modified leaf associated with a flower or an inflorescence, but not a part of the flower itself; sometimes applied also to a specialized leaf subtending an inflorescence; in conifers, one of the primary appendages of the cone axis, in the axils of which the ovuliferous scales are borne.

bracteal   Having the form or position of a bract.

bracteate, bracted   Provided with bracts.

bracteole, bractlet   Diminutive of bract; strictly, a bract that is borne on a petiole instead of subtending it.

bud   An undeveloped leafy shoot, or an undeveloped flower. Vegetative buds are often enclosed by reduced, specialized leaves called bud-scales.

bulb A short, vertical, underground shoot that has modified leaves or thickened leaf-bases prominently developed as food-storage organs. (Compare *corm.*)

bulbil, bulblet Diminutive of bulb; one of the small new bulbs arising around the parent bulb; a bulb-like structure produced by some plants in the axils of leaves or in place of flowers.

bullate Covered with rounded projections resembling unbroken blisters.

caducous Falling off very early. (Compare *deciduous.*)

caespitose See cespitose.

callus A firm thickening; the firm, thickened base of the lemma in many grasses.

calyculate Provided with a set of small bracts around the calyx, suggesting an outer calyx; in the Asteraceae, provided with a set of distinctive small bracts around the base of the involucre.

calyptra A deciduous lid or cap investing or terminating a structure.

calyx All of the sepals of a flower, collectively.

cambium A lateral meristem; specifically, the vascular or intrafascicular cambium, which produces xylem internally and phloem externally.

campanulate Bell-shaped, usually descriptive of a corolla or calyx.

campylodromous venation Venation with one or more pairs of prominent lateral veins from at or near the base, at least the outer ones basally downcurved before arching forward.

campylotropous ovule An ovule with the body distorted by unequal growth, so that the micropyle and the adjacent funiculus appear to lie along one side of the ovule instead of at the base.

cancellate With a fine, regular, ladderlike, reticulate pattern.

canescent Pale or gray, because of a fine, close, whitish pubescence. (Compare *argenteous, cinereous.*)

capillary With the form of a hair.

capitate Headlike, or in a head.

capsule A dry, dehiscent fruit composed of more than one carpel.

carina A keel.

carinal Pertaining to the keel; in *Equisetum,* pertaining to the longitudinal ridges of the stem.

carinate Keeled.

carotene A chemical group of fat-soluble, yellow to orange or red, long-chain hydrocarbons.

carotenoid pigment Any carotene or xanthophyll.

carpel The fertile leaf (megasporophyll) of an angiosperm, which bears the ovules. A pistil is composed of one or more carpels.

carpophore The part of the receptacle that in some kinds of flowers is prolonged between the carpels as a central axis, e.g., in the Apiaceae. In *Silene,* the stalk of the ovary, to which the filaments and petals may be basally adnate.

cartilaginous Tough and firm, but somewhat flexible, like cartilage.

caruncle An excrescence near the hilum of some seeds; a sort of aril.

caryopsis The fruit of a member of the grass family, typically dry indehiscent, and with the seed-coat of the single seed adnate to the pericarp.

castaneous Chestnut-colored; dark reddish-brown.

catkin A dense, bracteate spike or raceme with a nonfleshy axis bearing many small, naked or apetalous flowers; an ament.

caudate With a tail-like basal or terminal appendage.

caudex   A short, more or less vertical, often woody, persistent stem at or just beneath the surface of the ground, serving as the perennating organ from which new aerial stems arise each year. A caudex may surmount a taproot, or it may produce adventitious roots like a rhizome.

caulescent   With an obvious, leafy stem that has visible internodes. (Compare *acaulescent*.)

cauline   On or pertaining to a stem. Cauline leaves are attached to the stem distinctly above the ground, in contrast to basal leaves.

cell   As used in taxonomic description, the locule of an ovary or the theca of an anther, i.e., a cavity or compartment within which special bodies are produced; in general biological usage, an organized unit of protoplasm, bounded by a membrane or a cell wall, or the wall itself, after the death of the protoplasm.

cernuous   Nodding, or tending to nod.

cespitose   Growing in dense, low tufts.

chaff   Thin, dry scales; the receptacular bracts of some Asteraceae.

chartaceous   Papery in texture.

chlorophyll   The characteristic green pigment of plants, an essential enzyme in photosynthesis.

choripetalous   Composed of or characterized by separate petals.

chorisepalous   Composed of or characterized by separate sepals.

ciliate   With a fringe of marginal hairs.

ciliolate   Diminutive of ciliate.

cinereous   Ashy in color, usually because of a covering of short hairs; somewhat darker than canescent, and less shiny than argenteous.

circinate   Coiled from the tip downward, with the apex as the center, like the fiddlehead of a fern.

circum-   Latin prefix, meaning around, as around a circle.

circumboreal   Occurring all the way around the north pole, in both Eurasia and America.

circumscissile   Dehiscing by an encircling transverse line, so that the top comes off as a lid or cap.

cladophyll   A branch or stem that has the form and function of a leaf; same as phylloclade.

clasping   Partly surrounding another organ at the base.

clathrate   Lattice-like.

clavate   Shaped like a club or a baseball bat, being thicker toward the distal end than at the base.

clavellate   Diminutive of clavate; slightly thickened at the distal end.

claw   The narrow, basal portion of some sepals and petals, in contrast to the broader blade.

cleft   Cut about half-way to the midrib or base, or a little deeper; deeply lobed. There is no sharp distinction between lobed, cleft, and parted, which in general apply to progressively deeper divisions.

cleistogamous flower   A self-pollinating flower that remains closed, setting seed without opening.

climbing   Growing more or less erect without fully supporting its own weight, instead leaning, scrambling, twining, or attaching onto some other structure such as a tree or wall.

clone   A group of individuals originating by vegetative multiplication from a single individual.

coalescent   Running together so as to form a single unit.

cochlear Coiled like a snail-shell.

coetaneous With the flowers developing at the same time as the leaves. (Compare *precocious, serotinous.*)

coherent Sticking together, but not organically united. The term applies only to like parts. (Compare *adherent, connate, connivent.*)

collar The outer side of a grass (or sedge) leaf at the juncture of the blade and sheath.

collateral Side by side.

colleter A mucilage-secreting stout trichome found on stipules or bud-leaves of some angiosperms.

colonial Forming colonies. The term is used chiefly for plants with underground connections between separate aerial stems.

column A firm group of united filaments, as in many Malvaceae; in orchids, the united filaments and style, collectively.

coma A tuft of (usually long and soft) hairs, especially on a seed.

commissure The face by which two carpels cohere. The term is usually applied only to schizocarps, as in the Apiaceae.

comose Provided with a coma.

complete flower A flower with calyx, corolla, androecium, and gynoecium.

compound leaf A leaf with two or more distinct leaflets.

compound ovary or pistil An ovary or pistil composed of more than one carpel.

concolored, concolorous Of uniform color, both sides or all parts colored alike.

conduplicate Folded together lengthwise, usually into two equal, more or less appressed halves.

cone A cluster of sporophylls or ovuliferous scales on an axis; a strobilus.

connate Grown together or mutually attached. The term is properly applied only to like organs, as filaments connate into a tube or leaves connate around a stem. (Compare *adnate.*)

connective The tissue of an anther that connects the pollen-sacs.

connivent Converging or coming together, but not organically united.

contorted (aestivation) Same as convolute aestivation.

convolute (aestivation) Arranged so that one edge of each petal (or sepal) is covered, and the other exposed.

cordate Shaped like a stylized heart, with the notch at the base. (Compare *obcordate.*)

cordilleran Relating to or occurring in the system of mountain ranges that covers much of western North America.

coriaceous Leathery in texture.

corm A short, vertical, underground stem that is thickened as a perennating food-storage organ, without prominently thickened leaves. (Compare *bulb.*)

corniculate Bearing little horns or crests, especially near the summit.

corolla All of the petals of a flower collectively.

corona A set of petal-like structures or appendages between the corolla and the androecium, derived by modification of the corolla or of the androecium.

coroniform Crown-shaped.

corymb Strictly, a simple, racemose inflorescence that is flat-topped or round-topped because the outer pedicels are progressively longer than the inner; loosely, any inflorescence having the form of a corymb.

corymbiform Having the form but not necessarily the structure of a corymb.

corymbose In a corymb.

costa A prominent rib or vein; in ferns, the midrib of a pinna or pinnule.

costate   Longitudinally ribbed.

cotyledon   A leaf of the embronic plant within a seed.

creeping   Growing along (or beneath) the surface of the ground and rooting at intervals, usually at the nodes.

crenate   Toothed along the margin, the teeth rounded. (Compare *dentate, serrate.*)

crenulate   Diminutive of crenate.

crested   Provided with an elevated, often complex appendage or rib on the summit or back.

crisped, crispate   Irregularly curled or crooked so as to be distinctly 3-dimensional rather than plane (said of hairs or leaf-margins).

cruciform   Cross-shaped.

cucullate   Hooded or hood-shaped, or provided with a separate hood or cucullus.

culm   The aerial stem of a grass or sedge.

cultigen   A kind of plant that has originated in cultivation.

cuneate   Wedge-shaped or triangular, with the narrow end at the point of attachment.

cupuliform   Cup-shaped.

cusp   An abrupt, sharp, often rigid point.

cuspidate   Tipped with a cusp.

cuticle   The waxy layer covering the epidermis of a leaf or stem.

cyathium   The pseudanthium (false flower) of a *Euphorbia,* consisting of a cupulate involucre (often with petaloid appendages) which contains a central naked pistillate flower (consisting of a single pistil) surrounded by several naked staminate flowers (consisting each of a single stamen).

cyclic   Occurring in definite cycles or whorls.

cyme   A broad class of inflorescences characterized by having the terminal flower bloom first, commonly also with the terminal flower of each branch blooming before the others on that branch.

cymule   Diminutive of cyme; often applied to the ultimate dichasia or monochasia of which many compound cymes are composed.

deciduous   Falling after completion of the normal function. A deciduous tree is one that normally loses its leaves at the approach of winter or the dormant season. (Compare *caducous, persistent.*)

declined   Curved downward.

decompound   Repeatedly (and often irregularly) compound into numerous leaflets.

decumbent   With a prostrate or curved base and an erect or ascending tip.

decurrent   With an adnate wing or margin extending down the stem or axis below the point of insertion.

decussate   Arranged oppositely, with each pair set at right angles to the pair above and the pair below.

deflexed   Bent downward.

dehiscent   Opening at maturity, releasing or exposing the contents.

deliquescent   With the central axis melting away irregularly into a series of smaller branches. (Compare *excurrent.*) Term also applied to flowers that collapse and become slimy after anthesis.

deltoid, deltate   Shaped more or less like an equilateral triangle, with one of the sides as the base.

dendritic   Branched in treelike fashion, as the hairs of some Brassicaceae.

dentate   Provided with spreading, pointed teeth. (Compare *serrate, crenate.*)

denticulate   Diminutive of dentate.

depauperate   Small or poorly developed because of unfavorable environment.

determinate inflorescence   An inflorescence in which the terminal flower blooms first, stopping the elongation of the main axis; in a strictly determinate inflorescence the terminal flower of each branch likewise blooms first. A cymose inflorescence. (Compare *indeterminate inflorescence.*)

di-   Greek prefix, meaning two.

diadelphous   Arranged in two fascicles.

dichasial cyme   A cyme that is repeatedly branched in dichasial fashion.

dichasium   A 3-flowered cymule in which the development of the terminal flower is followed by that of the two opposite or subopposite lateral flowers.

dichlamydeous   Having two kinds of perianth-members, i.e., sepals and petals.

dichotomous   Forking more or less regularly into two branches of similar size.

didymous   Developing in pairs, or composed of two similar but almost separate parts.

didynamous   With 4 stamens in 2 unequal pairs, as in many Lamiaceae.

digitate   Same as palmate.

dimidiate   Appearing as if one half of an otherwise symmetrical structure were wanting.

dimorphic   Of two forms.

dioecious   Producing male and female flowers (or other reproductive structures) on different individuals.

diploid   Having two full chromosome-complements per cell.

diplostemonous   With two cycles of stamens.

disarticulating   Separating at a predetermined joint.

disciform   With the form of a disk. In the Asteraceae, with the central flowers of a head perfect (or functionally staminate) and the marginal ones pistillate but without a ligule.

discoid   Resembling a disk. In the Asteraceae, with the flowers of a head all tubular and perfect (or functionally staminate).

disk   An outgrowth from the receptacle of a flower, surrounding the base of the ovary; often derived by reduction and modification of the innermost set of stamens; in the Asteraceae, the central part of the head, composed of tubular flowers.

distal   Toward or at the tip or far end. (Compare *proximal.*)

distichous   In two vertical rows or ranks.

distinct (as applied to plant organs)   Not connate with similar organs. (Compare *free.*)

dithecal   Having or composed of two thecae.

diurnal   Pertaining to the daytime. Diurnal flowers are open during the day.

divaricate   Widely spreading from the axis or rachis.

divided   Cut into distinct parts, as a leaf that is cut to the midrib or the base.

dolabriform   Pick shaped. (See *malpighiaceous hairs.*)

dorsal   Pertaining to or located on the back. (Compare *ventral.*)

dorsifixed   Attached by the back. (Compare *basifixed, versatile.*)

dorsiventral   Flattened, with the two flattened sides unlike; having a back side and a belly side.

dorsoventral   From the front to the back, as opposed to from one side to the other; also sometimes used as a synonym of dorsiventral.

drupe   A fleshy fruit with a firm endocarp that permanently encloses the usually solitary seed, or with a portion of the endocarp separately enclosing each of two or more seeds.

drupelet   Diminutive of drupe.

dyad   A set or group of two.

e-, ex-   Latin prefix, meaning without, or from, or away from.

echinate   Beset with prickles.

ecotype An infraspecific population adapted to a particular habitat or set of similar habitats.

elaiosome An oily appendage or excrescence, especially on seeds, offering food to ants.

ellipsoid Elliptic in long-section and circular in cross-section (applied only to 3-dimensional bodies).

elliptic With approximately the shape of a geometrical ellipse (applied only to flat bodies).

emarginate With a small notch at the tip.

embryo The young plant in a seed; the young sporophyte, before it has begun to take on its mature form and size.

endemic Confined to a particular geographic area.

endocarp The inner part of the pericarp.

endosperm Food storage tissue of a seed, derived from the triple-fusion nucleus of the embryo-sac.

ensiform Sword-shaped.

entire With a continuous unbroken margin, not toothed or cut.

entomophilous Pollinated by insects.

epappose Without a pappus.

ephemeral Lasting for only a short time. An ephemeral flower lasts only about a day or less. An ephemeral annual is one that completes its growth in spring or early summer and survives the dry period as seeds.

epi- Greek prefix, meaning upon, often used to mean outer or outermost.

epicalyx A set of bracts, so closely placed beneath the calyx as to suggest an additional, outer calyx.

epigaeous germination Germination in which the cotyledons are brought above the ground. (Compare *hypogaeous germination.*)

epigynous With the perianth and stamens attached at the top of the ovary, rather than beneath it, i.e., with the ovary inferior. (Compare *hypogynous, perigynous.*)

epipetalous Attached to the petals or corolla.

epiphyte A plant without connection to the soil, growing upon another plant, but not deriving its food or water from it. (Compare *parasite.*)

epitropous ovule An ovule which (if erect) has the raphe dorsal, or which (if pendulous) has the raphe ventral (between the partition and the body of the ovule). (Compare *apotropous ovule.*)

eprophyllate In *Juncus,* without prophylls.

equitant Astride, as if riding. The term is used for flat, conduplicate, unifacial leaves, the margins of the blade connate to form an edge facing the stem; such leaves often enfold each other in two ranks, as in *Iris.*

erose With an irregular margin, as if gnawed.

eu- Greek prefix, meaning true or real or typical.

eusporangiate fern A fern with sporangia that develop each from a group of sporangial initials, have a jacket layer more than one cell thick at maturity, do not have a highly specialized mechanism of dehiscence, and contain a large and indefinite number of spores. (Compare *leptosporangiate fern.*)

even-pinnate Pinnately compound, but lacking a terminal leaflet, so that typically there is an even number of leaflets. (Compare *odd-pinnate.*)

evergreen Remaining green throughout the winter.

ex- Same as e-.

excurrent With a continuing central axis, from which lateral branches arise. (Compare *deliquescent.*)

exfoliating   Peeling off in layers.

exocarp   The outermost layer of the pericarp.

explanate   Spreading out flat.

exsert, exserted   Projecting out beyond an envelope, as stamens from a corolla; opposite of *included.*

exstipulate   Without stipules.

extra-   Latin prefix, meaning outside of.

extrorse   Turned toward the outside, or facing outward. Opposite of *introrse.*

falcate   Sickle-shaped, or curved like a hawk's beak.

farinose   Covered with a meal-like powder.

fascicle   A small, close bundle or cluster.

fastigiate (branches)   Close together, nearly parallel, and usually erect.

fenestrate   With one or more window-like openings.

ferrugineous   Rust-colored.

fertile   Capable of normal reproductive functions, as a fertile stamen produces pollen, a fertile pistil produces ovules, and a fertile flower normally produces fruit, although it may lack stamens. (Opposite of *sterile.*)

fibrillose   Composed of or breaking down into fibers or fibrils.

-fid   Suffix meaning deeply cut.

filament   The stalk of a stamen, i.e., the part that supports the anther.

filiform   Very slender, thread-like.

fimbriate   Unevenly fringed with slender projections along the margin.

fimbrillate   Diminutive of fimbriate.

fistulose, fistulous   Hollow, lacking pith, especially if somewhat inflated.

flabellate   Fan-shaped.

flabelliform   Flabellate, or nearly flabellate.

flaccid   Weak and lax, hardly if at all capable of supporting its own weight.

flagelliform   Very slender and elongate, with the form of a flagellum.

fleshy   Thick and juicy; succulent.

floccose   Covered with very long, soft, fine, loosely spreading and more or less tangled hairs; like tomentose, but looser and more open.

floret   A little flower; an individual flower of a definite cluster, as in the head of a composite or the spikelet of a grass.

floricane   The flowering (second year) stem of a *Rubus.* (Compare *primocane.*)

foliaceous   Leaf-like in flatness, color, and texture.

-foliolate   Suffix indicating the number of leaflets in a compound leaf, e.g., trifoliolate, with 3 leaflets.

follicle   A fruit, derived from a single carpel, which dehisces along the seed-bearing suture at maturity.

fornix   One of a set of small, often arch-shaped scales or appendages in the tube or throat of a corolla of some kinds of plants, as in many borages.

fovea   A little pit.

free (as applied to plant organs)   Not attached to another kind of organ, as stamens free from the corolla. (Compare *adnate, distinct.*)

free-central placenta   A placenta consisting of a free-standing column or projection from the base of a compound, unilocular ovary.

frond   The leaf of a fern; by extension, any relatively large leaf.

fruit   A ripened ovary, together with any other structures that ripen with it and form a unit with it.

frutescent   Becoming shrubby.

fruticose   Shrubby.

fugacious   Falling or disappearing early, usually applied to parts of a flower.

fulvous   Tawny, dull yellow.

funiculus   The stalk of an ovule.

furcate   Forked: often in combining forms, e.g., trifurcate, three-forked.

fuscous   Gray-brown, especially if rather dark.

fusiform   Spindle-shaped, thickest near the middle and tapering to both ends.

galea   The strongly concave or helmet-like upper lip of certain bilabiate corollas, as in *Castilleja.*

galeate   Having the form of a galea, or provided with a galea.

gametophyte   The generation that has *n* chromosomes and produces gametes as reproductive bodies. In angiosperms the female gametophyte is the embryo-sac and the male gametophyte is the pollen-grain.

gamo-   Greek prefix, meaning (in this context) connate, as gamopetalous or gamosepalous.

geminate   In pairs.

gemma   A small, vegetative bud that can become detached from the plant and grow into a new plant.

geniculate   Abruptly bent or twisted.

gibbous   Abruptly swollen on one side, commonly near the base.

glabrate   Nearly glabrous, or becoming glabrous.

glabrescent   Becoming glabrous or nearly so.

glabrous   Smooth, without hairs (trichomes) or glands.

gladiate   Sword-shaped, either straight or somewhat curved.

gland   A protuberance, appendage, or depression on the surface of an organ (or on the end of a trichome) which produces a sticky or greasy, viscous substance.

glandular   Provided with glands, or functioning as a gland.

glaucescent   Becoming glaucous.

glaucous   Covered with a fine, waxy, removable powder that imparts a whitish or bluish cast to the surface, as a prune or a cabbage-leaf.

globose   More or less spherical.

glochidia   Hairs or hair-like outgrowths with retrorse barbs at the tip.

glomerate   Densely compacted in clusters or heads.

glomerule   A small, compact, head-like cyme; any dense, small cluster.

glume   One of a pair of bracts, found at the base of a grass spikelet, which do not subtend flowers. (Compare *lemma, palea.*)

glutinous   Provided with a firm, somewhat sticky covering or component.

-gonous   Greek suffix, meaning angled.

gymnosperm   A member of a group of plants (Pinophyta) characterized by having ovules that are not enclosed in an ovary.

gynaecandrous spike (in *Carex*)   A spike with both staminate and pistillate flowers, the staminate below the pistillate. (Compare *androgynous spike.*)

gynandrous   With the stamens adnate to the pistil, as in orchids.

gynobase   An enlargement or prolongation of the receptacle of some flowers, as in the Boraginaceae and Lamiaceae.

gynobasic style   A style that is attached directly to the gynobase, as well as to the individual carpels or nutlets.

gyno-dioecious   Producing female flowers on one plant and perfect flowers on another.

gynoecium   All the carpels of a flower, collectively.

gynophore A central stalk in some flowers, bearing the gynoecium.
habit The general appearance or manner of growth of a plant; e.g., an annual habit, a branching habit, a deciduous habit, a shrubby habit.
habitat The environmental conditions or kind of place in which a plant grows.
hair An epidermal appendage, usually slender, sometimes branched, not stiff or stout enough to be called a spine or prickle, not flattened as a scale.
halophyte A plant adapted to growth in salty soil.
hamate Hooked at the tip.
haploid Having only one complement of chromosomes per cell. (Compare *diploid*.)
hastate Shaped like an arrowhead, but with the basal lobes more divergent. (Compare *sagittate*.)
haustorium A specialized structure through which a parasite extracts nourishment from its host.
head An inflorescence of sessile or subsessile flowers crowded closely together at the tip of a peduncle. Unless otherwise specified, a head is presumed to belong to the racemose (indeterminate) group of inflorescences.
-hedral Greek suffix, relating to surfaces; e.g., a tetrahedral spore has 4 surfaces.
helicoid cyme A sympodial cyme with the apparent main axis curved in more or less of a helix, because the successive lateral branches that make up the axis all arise on the same side. (Compare *scorpioid cyme*.)
hemi- Greek prefix, meaning half.
herb A plant, either annual, biennial, or perennial, with the stems dying back to the ground at the end of the growing season.
herbaceous Adjectival form of herb; also, leaf-like in color or texture, or not woody.
hermaphrodite With both sexes together in the same individual, or in the same flower.
hetero- Greek prefix, meaning unlike, or of differing sorts.
heteromorphic Of two or more distinct forms or types.
heterosporous Producing two different kinds of spores (usually of very unequal size), one of which gives rise to male gametophytes, the other to female; opposite of homosporous.
heterostylic With styles of different (usually two) lengths in flowers of different individuals, some surpassing, others surpassed by the stamens.
heterotrophic Parasitic or saprobic, as opposed to autotrophic.
hexa- Greek prefix, meaning six.
hexaploid Having six complements of chromosomes per cell.
hilum The scar of the seed at its point of attachment.
hirsute Pubescent with rather coarse or stiff but not pungent, often bent or curved hairs; coarser than villous, but less firm and sharp than hispid.
hirsutulous Diminutive of hirsute.
hirtellous Diminutive of hirsute.
hispid Pubescent with coarse and firm, often pungent hairs. (Compare *hirsute*.)
hispidulous Diminutive of hispid.
homo- Greek prefix, meaning alike, or all of the same sort.
homosporous Producing only one kind of spore, which gives rise to a gametophyte that produces both antheridia and archegonia; opposite of heterosporous.
hyaline Thin and translucent or transparent.
hydathode An epidermal structure (often ending a vein) that exudes water.
hydrophyte A plant adapted to life in the water.
hygroscopic Having the property of taking up and holding moisture from the air, sometimes swelling and shrinking, or changing in position, according to the humidity.

hypanthium A ring or cup around the ovary, formed either by marginal expansion of the receptacle or by the union of the lower parts of the calyx, corolla, and androecium; when the petals and stamens appear to arise from the calyx-tube, that part of the apparent calyx-tube which is below the attachment of the petals is the hypanthium.

hypo- Greek prefix, meaning beneath.

hypogaeous germination Germination in which the cotyledons remain beneath the ground. (Compare *epigaeous germination*.)

hypogynous With the perianth and stamens attached directly to the receptacle; more generally, beneath the gynoecium. (Compare *epigynous, perigynous*.)

-iferous Latin combining form, indicating carrying or bearing.

imbricate Arranged in a tight spiral, so that the outermost member has both edges exposed, and at least the innermost member has both edges covered; more loosely, in a shingled arrangement.

imperfect flower A flower that has stamens but not pistil(s), or pistil(s) but not stamens, regardless of what other flower parts may be present or absent. (Compare *perfect flower, complete flower, incomplete flower*.)

in- Latin prefix, meaning, in different contexts, not, in, or into.

incised Rather deeply and sharply (and often irregularly) cut.

included Contained within an envelope, not projecting beyond it; opposite of exserted.

incomplete flower A flower that lacks one or more of the kinds of structures found in a complete flower, q.v.

incumbent cotyledons Cotyledons with the back of one of them against the radicle. (Compare *accumbent cotyledons*.)

indehiscent Not dehiscent, remaining closed at maturity.

indeterminate inflorescence An inflorescence that blooms from the base upwards, so that theoretically it could continue to elongate indefinitely. (Compare *determinate inflorescence*.)

indument The epidermal appendages of a plant or an organ considered collectively; same as vestiture. The ramentum of many ferns is not considered part of the indument.

induplicate Valvate, with the margins infolded.

indurate, indurated Hardened.

indusium An epidermal outgrowth or reflexed and modified leaf-margin that covers the sori of many kinds of ferns; more generally, any cupulate structure subtending something else.

inferior ovary An ovary with the other floral parts (calyx, corolla, and androecium) attached to its summit; an epigynous flower has an inferior ovary.

inflorescence A flower-cluster of a plant; the arrangement of the flowers on the axis.

infra- Latin prefix, meaning beneath or within or less than; opposite of supra-.

infundibuliform Funnelform.

innocuous Harmless, hence unarmed or spineless.

innovation An offset at the base of a stem.

inserted Attached to, referring to the point of origin of an organ.

inter- Latin prefix, meaning between or among.

intercalary meristem A meristem, separated from the apical meristem, that produces primary tissues; e.g., the meristem at the base of the leaf-blade in grasses.

internode The part of a stem between two successive nodes.

intra- Latin prefix, meaning within.

intrastaminal Within (as opposed to outside of) the androecium.

introrse Directed or turned inward. Opposite of *extrorse*.

involucel Diminutive of involucre; an involucre of the second order.

involucre Any structure that surrounds the base of another structure; in angiosperms usually applied to a set of bracts beneath an inflorescence.

involute Rolled inward, so that the lower side of an organ is exposed and the upper concealed. (Compare *revolute.*)

irregular flower A flower in which the petals (or less often the sepals) are dissimilar in form or orientation.

isomerous With the same number of parts as something else; a flower that has isomerous stamens has the same number of stamens as sepals or petals.

jaculator A modified funiculus that expels the seeds in the Acanthaceae.

keel A sharp or conspicuous longitudinal ridge; also the two partly united lower petals of many Fabaceae.

labellum Lip; the odd petal of an orchid.

labiate Lipped; usually in compounds such as bilabiate, i.e., two-lipped, referring to a sympetalous corolla with upper and lower sets of lobes.

lacerate Torn, or with an irregularly jagged margin.

laciniate Cut into narrow and usually unequal segments.

lactiferous Bearing a milky latex.

lacuna An empty space or gap in the midst of tissue.

lacustrine Pertaining to or growing around lakes.

laminar Thin and flat, as in a leaf-blade; pertaining to the leaf-surface, as opposed to the margins.

laminar placentation An arrangement of the ovules on the ventral surface (rather than the margins) of the carpel.

laminar sori Sori borne on the lower surface of the leaf, removed from the margins.

lanate Woolly.

lanceolate Lance-shaped; as used in recent American botanical literature, much longer than wide, widest below the middle, and tapering to both ends (or rounded to the base); like ovate, but narrower.

lanuginous Woolly.

lanulose Diminutive of lanate.

latex A colorless to more often white, yellow, or reddish liquid, produced by some plants, characterized by the presence of colloidal particles of terpenes dispersed in water.

laticifer A tube containing latex.

leaflet An ultimate unit of a compound leaf.

leaf-trace A vascular bundle, from the point where it leaves the stele to the point where it enters the leaf.

legume The fruit of a member of the Fabales (Fabaceae, Caesalpiniaceae, Mimosaceae), composed of a single carpel, typically dry, several-seeded, and dehiscing down both sutures.

lemma One of the pair of bracts (lemma and palea) that generally subtend the individual flowers in grass spikelets; the lemma has its back at the outside of the spikelet, whereas the palea has its back to the rachilla. Some spikelets also have one or more sterile lemmas, with similar orientation but without a palea or flower, as if they were extra glumes.

lenticel A slightly raised area in the bark of a stem or root, consisting of loosely arranged, nearly or quite unsuberized cells.

lenticular Shaped like a double-convex lens.

lepidote Scaly; covered with small scales.

leptosporangiate fern   A fern with sporangia that develop from a single sporangial initial, have a jacket-layer only one cell thick at maturity, have a specialized annulus governing dehiscence, and usually contain a relatively small and definite number of spores. (Compare *eusporangiate fern.*)

liana   A climbing, woody vine.

ligneous   Woody.

ligulate   Having a ligule. In the Asteraceae, a ligulate head is composed solely of ligulate flowers.

ligule   Literally, a little tongue. Term applied to the flattened part of the ray-corolla in the Asteraceae, and to the appendage on the adaxial side of the leaf at the junction of blade and sheath in many Poaceae and some Cyperaceae.

limb   The expanded part of a sympetalous corolla above the throat; the expanded part of any petal or leaf.

linear   Line-shaped; very long and narrow, with essentially parallel sides.

lip   A projection or expansion of something; one of the two segments or sets of lobes of a bilabiate (two-lipped) corolla or calyx; the odd petal of an orchid.

litoral, littoral   Along or pertaining to shores.

livid   Pale lead-colored.

lobe   A projecting segment of an organ, too large to be called a tooth or an auricle, but with the adjoining sinuses usually extending less than half-way to the base or midline.

locule   A seed-cavity (chamber) in an ovary or fruit; a compartment in any container.

loculicidal   Dehiscing along the midrib or outer median line of each locule, i.e., "through" the locules. (Compare *septicidal, poricidal.*)

lodicule   One of the tiny scales that may represent a vestigial perianth in grasses.

loment   A legume composed of one-seeded joints.

lunate   Crescent-shaped.

lyrate   Pinnatifid, with the terminal lobe evidently the largest and usually rounded.

lysigenous   Originating by dissolution or degeneration of cells or tissue. (Compare *schizogenous.*)

macro-   Latin prefix, equivalent to Greek mega-, q.v.

maculate   Spotted.

malpighiaceous hairs   Pick-shaped hairs, attached more or less toward the middle or at some point along the length, rather than at the end. Same as *dolabriform hairs.*

marcescent   Withering and persistent, as petals and stamens in some kinds of flowers.

marginal placenta   A placenta along the suture of the ovary of a simple pistil. (Compare *parietal placenta.*)

maritime   Pertaining to the ocean or the seacoast; growing near the ocean, under the influence of salt water.

matutinal   Of the morning; opening or functioning in the morning.

mega-   Greek prefix, meaning large, or by extension, female.

megasporangium   A sporangium that produces one or more megaspores.

megaspore   A spore that gives rise to a female gametophyte. Megaspores are usually much larger than corresponding microspores, which give rise to male gametophytes. (Compare *microspore.*)

megasporophyll   A sporophyll that bears or subtends one or more megasporangia.

membranaceous, membranous   Thin and flexible, like a membrane, as in an ordinary leaf, in contrast to *chartaceous, coriaceous,* or *succulent.*

mericarp   An individual carpel of a *schizocarp,* q.v.

-merous   Greek suffix, referring to the number of members of a set (as in a cycle of floral organs), usually with a numerical prefix, as trimerous, pentamerous, etc. Such terms often refer to the perianth only.

mesic Moist; neither very wet nor very dry.

mesocarp The central layer of pericarp that is divided into three layers. (Compare *endocarp, exocarp.*)

mesophyte A plant adapted to growth under ordinary moisture conditions; intermediate between hydrophyte and xerophyte.

micro- Greek prefix, meaning small, or by extension, male.

micropyle The opening through the integument(s) of an ovule to the nucellus.

microsporangium A sporangium that produces microspores.

microspore A spore that gives rise to a male gametophyte. (Compare *megaspore.*)

microsporophyll A sporophyll that bears or subtends one or more microsporangia.

mixed bud A bud that gives rise to both flowers and leaves, i.e., to a leafy branch with one or more flowers.

mixed inflorescence An inflorescence with cymose and racemose components, the sequence of flowering neither strictly determinate nor strictly indeterminate.

mixed panicle An inflorescence of paniculate appearance, but of mixed cymose and racemose components.

monad A cell or small unitary structure, detached from other such structures.

monadelphous stamens Stamens with the filaments or anthers all connate, usually forming a tube.

moniliform Literally, necklace-like; constricted at regular intervals.

mono- Greek prefix, meaning one.

monocarpic Blooming only once and then dying; usually applied only to perennials.

monocephalous With a single head.

monochasium A 2-flowered cymule, with a terminal flower and a single lateral flower.

monochlamydeous Having only one kind of perianth-member, forming a single cycle.

monoecious With unisexual flowers, both types borne on the same individual.

monopodial With the branches or appendages arising from a simple axis. (Compare *sympodial.*)

monothecal Having or composed of a single theca.

mucro A short, sharp, slender point.

mucronate Tipped with a mucro.

mucronulate Diminutive of mucronate.

multi- Latin prefix, meaning many.

multicipital Literally, many-headed; often applied to a caudex or rootcrown from which several or many stems arise.

multiple fruit A fruit derived from several or many flowers, as a pineapple or mulberry.

muricate Beset with small, sharp projections.

muriculate Diminutive of muricate.

muticous Blunt or innocuous, without a spine.

mycorrhiza A symbiotic association of a fungus and the root of a vascular plant; by extension, other symbiotic associations of fungi with higher plants.

mycotrophic Evidently modified as a result of mycorrhizal association; dependent on mycorrhizal association.

naked Lacking various organs or appendages, in contrast to other forms in which these organs or appendages are present; e.g., a naked flower lacks a perianth.

napiform Turnip-shaped.

naturalized Thoroughly established, but originally coming from another area.

naviculate Boat-shaped.

nectary A structure that produces nectar, usually but not always in association with a flower.

nerve A prominent longitudinal vein of a leaf or some other organ.

net-veined With the veins forming a network. Term traditionally applied to pinnate and palmate venation collectively, as opposed to parallel venation.

neutral flower A flower that has neither stamens nor pistil.

nigrescent Blackish, or becoming black.

node A place on a stem where a leaf is (or has been) attached.

nodose Knobby, used especially in describing roots.

nodulose Diminutive of nodose.

nucellus The megasporangial wall of a seed plant, which typically encloses the female gametophyte; the part of an ovule next within the integuments.

nude Same as naked.

nut A relatively large, dry, indehiscent fruit with a hard wall, usually containing only one seed.

nutlet Diminutive of nut; a very thick-walled achene; one of the dry, indehiscent, one-seeded half-carpels of the Boraginaceae and Lamiaceae.

ob- Latin prefix, meaning in a reverse direction.

obcordate Like cordate, but with the notch at the tip instead of at the base.

obdiplostemonous With two cycles of stamens, the outer cycle opposite the petals.

oblanceolate Like lanceolate, but broadest above the middle and tapering to the base.

oblate Shorter from base to tip than across the middle; usually applied to 3-dimensional objects.

oblong Shaped more or less like a geometrical rectangle (other than a square).

obovate Like ovate, but larger toward the distal end.

obsolete So much reduced as to be scarcely detectable or entirely suppressed. The term implies that the structure was better developed in ancestral forms. (Compare *abortive, rudimentary, vestigial*.)

obtuse Blunt, with the sides coming together at an angle of more than 90 degrees.

ochroleucous Yellowish-white.

ocrea, ochrea A sheath around the stem just above the base of the leaf, derived from stipules, as in many Polygonaceae.

ocreola, ocreole Diminutive of ocrea; usually applied to the minute sheaths in the inflorescence of Polygonaceae.

odd-pinnate Pinnately compound and with a terminal leaflet, so that typically there is an odd number of leaflets. (Compare *even-pinnate*.)

offset A short, usually prostrate shoot, primarily propagative in function, originating near the ground-level at the base of another shoot; as applied to arrangements, nearly but not quite opposite.

operculate Having an operculum.

operculum A little lid; the deciduous cap of a circumscissile fruit or other organ.

opposite Situated directly across from each other at the same node or level, as the leaves or leaflets of some plants; situated directly in front of (on the same radius as) another organ, as stamens opposite the petals.

orbicular Essentially circular in outline; applied to flat structures.

orifice A mouth or other opening.

orthotropous ovule A straight (unbent) ovule with the funiculus at one end and the micropyle at the other.

oval   Broadly elliptic.

ovary   The structure that encloses the ovules of angiosperms; the expanded, basal part of a pistil, containing the ovules.

ovate   Shaped like a long-section through a hen's egg, with the larger end toward the base. (Term applied to plane surfaces; compare *ovoid, obovate.*)

ovoid   Shaped like a hen's egg. (Term applied to solid objects; compare *ovate.*)

ovule   A young or undeveloped seed; the megasporangium, plus the enclosing integuments, of a seed-plant. For ovule-types, see anatropous, campylotropous, and orthotropous ovules.

palate   A raised part of the lower lip of a corolla, constricting or closing the throat.

palea   One of the pair of bracts (lemma and palea) that generally subtend the individual flowers in grass spikelets; the palea has its back to the rachilla. (Compare *lemma.*)

paleaceous   Chaffy.

palmate   With three or more lobes or nerves or leaflets or branches arising from a common point.

palmatifid   More or less deeply cut in palmate fashion.

pandurate   Fiddle-shaped.

panicle   A branching indeterminate inflorescence, usually broadest near the base and tapering upwards. The term is often loosely used to apply also to mixed panicles, q. v.

paniculate   Arranged in a panicle.

paniculiform   With the form, but not necessarily the structure of a panicle.

pannose   Densely and closely tomentose, the tomentum forming a felt-like layer.

papilionaceous flower   A flower having the structure typical of the Fabaceae, with a banner petal, two wing petals, and two partly connate keel petals.

papillate, papillose   Covered with papillae, i.e., with short, rounded, blunt projections.

pappus   The modified calyx crowning the ovary (and achene) of the Asteraceae, consisting variously of hairs, scales, bristles, or a mixture of these.

parallel-veined   With several or many more or less parallel main veins, the network of smaller veins not obvious. (Compare *net-veined.*)

parasite   A plant that derives its food or water chiefly from another plant to which it is attached. (Compare *epiphyte.*)

parenchyma   A tissue composed of relatively unspecialized, usually thin-walled cells.

parietal placenta   A placenta along the walls or on the intruded partial partitions of a compound, unilocular ovary. (Compare *marginal placenta.*)

parted   Deeply cut, usually more than half way to the midvein or base.

pectinate   Comb-like, with a single row of narrow spreading appendages (or hairs) of uniform size, like the teeth on a comb.

pedate   Palmately divided or parted, with the lateral segments 2-cleft.

pedicel   The stalk of a single flower in an inflorescence.

peduncle   The stalk of an inflorescence or of a solitary flower.

pedunculate   Borne on a peduncle.

pellucid   Transparent or translucent.

peltate   Shield-shaped, attached by the lower surface instead of by the base or margin.

pendulous   Hanging or drooping.

penicillate   With a tuft of short hairs at the end.

penta-   Greek prefix, meaning five.

pentadelphous   Connate into five groups.

pentamerous   With five parts of a kind.

pepo   A fleshy fruit with a hard, inseparable rind, and without a pit or core; e.g., cucumber, melon, and other Cucurbitaceae.

perennial   A plant that lives more than two years.

perfect flower   A flower with both an androecium and a gynoecium, whether or not it has a calyx and corolla. (Compare *imperfect flower.*)

perfoliate leaf   A leaf with the basal margins connate around the stem, so that the stem appears to pass through the leaf.

perianth   All of the sepals and petals (or tepals) of a flower, collectively.

pericarp   The matured ovary wall of a fruit.

perigynium   The special bract that encloses the achene of a *Carex.*

perigynous   Having the perianth and stamens united into or borne on a basal saucer or cup (the hypanthium) distinct from the ovary; more generally, around the base of the gynoecium, as a perigynous disk. (Compare *epigynous, hypogynous.*)

perisperm   Food storage tissue in the seed, derived from the nucellus.

persistent   Remaining attached after the normal function has been completed.

petal   A member of the inner set of floral leaves, usually colored or white and serving to attract pollinators.

petaloid   Petal-like, especially in color and texture.

petaliferous   Having petals.

petiolate   Having a petiole.

petiole   A leaf-stalk.

petiolule   The stalk of a leaflet of a compound leaf.

-phile   Greek combining form, indicating loving; e.g., calciphile plants do best in calcareous soil.

-phobe   Greek combining form, indicating hating or avoiding; e.g., calciphobe plants do not grow well in calcareous soil.

phreatophyte   A plant that characteristically has the major part of its root-system in soil permanently saturated with water, although there is no permanent standing water above the ground.

phyllary   An involucral bract of the Asteraceae.

phylloclade   A portion of the stem having the general form and function of a leaf; same as cladophyll.

phyllode   A more or less expanded, bladeless petiole.

phyllopodic   With the lowest leaves well developed instead of reduced to scales. (Compare *aphyllopodic.*)

pilose   Rather sparsely beset with straight spreading hairs.

pilosulous   Diminutive of pilose.

pin   The type of heterostylic flower that has a relatively long style and short stamens. (Compare *thrum.*)

pinna   One of the primary lateral divisions of a pinnately compound leaf.

pinnate   With two rows of lateral branches or appendages, or parts along an axis, like the barbs on a feather.

pinnatifid   More or less deeply cut in pinnate fashion.

pinnatilobate   With pinnately arranged lobes.

pinnipalmate   Intermediate between pinnate and palmate, as the venation of some leaves in which the first pair of lateral veins are much larger than the others.

pinnule   Diminutive of pinna; an ultimate leaflet of a compound leaf that is two or more times compound.

pistil   The female organ of a flower, composed of one or more carpels, and ordinarily differentiated into ovary, style, and stigma.

pistillate flower   A flower with one or more pistils, but no stamens.

pith   A central, parenchymatous tissue in a stem (or root) surrounded by the central cylinder of vascular tissue (stele).

placenta   The tissue of the ovary to which the ovules are attached. For placentation types, see axile, free-central, marginal, and parietal placentas.

plagiotropic   Growing horizontally or at an oblique angle to the substrate.

plicate   Folded like a fan.

-plinerved   With the principal lateral veins arising from the midvein distinctly above the base of the blade. The term is commonly used with a numerical prefix, as triplinerved, indicating 3 main veins, the 2 lateral ones arising from the central one.

-ploid   Pseudo-Greek suffix, used with a numerical prefix to indicate the number of chromosome-complements in each cell, as haploid, diploid, triploid, etc.

plumose   Feathery; applied to a slender organ, such as a style or pappus-bristle, with a dense long pubescence, or with pinnately arranged lateral bristles.

plumule   The part of the embryo of a seed that gives rise to a shoot; it lies above (and in dicotyledons between) the cotyledon(s). Same as epicotyl.

plurilocular   With several or many locules.

pod   Any kind of dry, dehiscent fruit.

pollen   The mass of young male gametophytes (pollen-grains) of a seed plant, at the stage when they are released from the anther or microsporangium.

pollination   In angiosperms, the transfer of pollen from the anther to the stigma; in gymnosperms, from the microsporangium to the micropyle.

pollinium   A coherent cluster of many pollen-grains, transported as a unit during pollination, as in many Orchidaceae and Asclepiadaceae.

poly-   Greek prefix, meaning many.

polyandrous   Having many stamens.

polycarpellate, polycarpous, polycarpic   Having many carpels.

polygamo-dioecious   Nearly dioecious, but with some of the flowers perfect.

polygamo-monoecious   Nearly monoecious, but with some of the flowers perfect.

polygamous   With intermingled perfect and unisexual flowers.

polymorphic   Occurring in several different forms.

polypetalous   With the petals separate from each other. (Compare *sympetalous.*)

polyploid   With three or more chromosome-complements in each cell.

polysepalous   With the sepals separate from each other.

polystemonous   With many stamens.

polystichous   Arranged in several longitudinal rows.

pome   A fruit with a core, such as an apple; technically, a fleshy fruit derived from a compound, inferior ovary, with a papery or bony endocarp and usually several seeds.

poricidal   Opening by pores.

porrect   Directed outward and forward.

posterior   Literally, behind. The posterior side of a flower is the side toward the axis of the inflorescence, rather than the side toward the subtending bract; thus the upper lip of a bilabiate flower is the posterior lip. (Compare *anterior.*)

praemorse   Ending abruptly, as if bitten off.

precocious   Developing very early; with the flowers developing before the leaves. (Compare *coetaneous, serotinous.*)

prickle   A sharp outgrowth from the epidermis or bark. (Compare *spine, thorn.*)

primocane   The first-year cane of a *Rubus*; in the second year the canes produce flowers and are called floricanes.

procumbent   Prostrate or trailing, but not rooting at the nodes.

prophyll   One of the pair of bracteoles at the base of a flower, as in some species of *Juncus*.

prostrate   Flat on the ground.

proterandrous, protandrous   Liberating pollen before the stigma is receptive.

proterogynous, protogynous   With the stigma receptive before the pollen is liberated.

protostele   A stele with a solid core of xylem (or at least without leaf-gaps), lacking a pith.

proximal   Toward or at the base or the near end. (Compare *distal.*)

pruinose   Strongly glaucous, like a prune.

pseudanthium   A compact inflorescence with small individual flowers, the whole simulating a single flower.

pseudo-   Greek prefix, meaning false.

puberulent, puberulous   Minutely pubescent, usually with rather soft, curled hairs.

pubescent   Bearing hairs (trichomes) of any sort.

pulverulent   Appearing powdery, as if dusted over.

pulvinate   Cushion-like.

pulvinus   The swollen base of a petiole or petiolule, which may (as in many legumes) govern the attitude of the leaf or leaflet.

punctate   Dotted, usually with small pits that may be translucent or glandular.

puncticulate   Diminutive of punctate.

pungent   Firmly sharp-pointed, especially if thereby rendered uncomfortable to the touch or grasp; as applied to odor, sharp or acrid or penetrating.

pustular, pustulate, pustulose   With little blisters or pustules.

pyrene   A small, bony unit, derived from the endocarp of a drupe or pome, and containing a seed. Such a fruit often has two or more pyrenes.

pyriform   Pear-shaped.

raceme   A more or less elongate inflorescence with pedicellate flowers arising in acropetal sequence (from the bottom up) from an unbranched central axis. (Compare *corymb, spike, panicle.*)

racemiform   Having the form but not necessarily the structure of a raceme.

racemose inflorescence   A broad class of inflorescences characterized by flowering in acropetal sequence. (Compare *cyme.*)

rachilla   Diminutive of rachis; the axis of a spikelet in the grasses and some sedges.

rachis   A main axis, such as that of a compound leaf.

radiate head   In the Asteraceae, a head with the marginal flowers ligulate and the central ones tubular. (Compare *discoid head, ligulate head.*)

radical   Pertaining to the root; radical leaves are basal leaves that seem to arise from the root-crown.

rame   As used in the Poaceae, the flowering stem of a grass.

rameal   Pertaining to or located on a branch.

ramentum   The scaly appendages or outgrowths from the epidermis on the stem and leaves of many ferns, differing from trichomes in being distinctly flattened and more than one cell wide.

raphe   The part of the funiculus that is permanently adnate to the integument of the ovule, commonly visible as a line or ridge on the mature seed coat.

raphide   A needle-shaped crystal in a plant cell.

ray The ligule or ligule-bearing flower in the Asteraceae; one of the branches of an umbel; etc.

receptacle The basal part of a flower, representing the end of a stem (or pedicel), to which the other flower-parts are attached; in the Asteraceae, the end of the peduncle, to which the flowers of the head are attached.

recurved Curved backward.

reflexed Bent backward, more abruptly or fully than *recurved*.

regular flower A flower in which the members of each circle of parts (at least the sepals and petals) are similar in size, shape, and orientation.

reniform Kidney-shaped.

repand With a shallowly sinuate or slightly wavy margin.

repent Creeping.

replum A frame-like placenta that persists on the pedicel after the sides (valves) of the fruit have fallen, as in the Brassicaceae and some Capparaceae.

resupinate Upside-down; inverted by twisting of the pedicel, as the flowers of most orchids.

reticulate Forming a network, as the veins of a leaf.

retrorse Directed backward or downward; opposite of *antrorse*.

retuse With a small terminal notch in an otherwise rounded or blunt tip.

revolute Rolled outward or backward, so that the upper side of an organ is exposed and the lower concealed. (Compare *involute*.)

rhizoid A structure of rootlike form and function, but of simple anatomy, lacking xylem and phloem.

rhizomatous Bearing rhizomes.

rhizome A creeping underground stem.

rib One of the main longitudinal veins of a leaf or other organ.

riparian Pertaining to or growing along stream-banks.

rootstock Same as rhizome.

rosette A cluster of leaves or other organs arranged in a circle or disk, often in a basal position.

rostellate Diminutive of rostrate.

rostellum A little beak; in orchids a slender extension from the upper edge of the stigma.

rostrate Beaked, with a short, stout terminal appendage.

rosulate Arranged in a little rosette.

rotate Wheel-shaped or saucer-shaped; flat and circular in outline, as a sympetalous corolla with spreading lobes and without a tubular basal part.

rotund Round or rounded.

ruderal Weedy, growing in waste or disturbed places.

rudimentary Poorly developed, but reflecting an early evolutionary stage rather than evolutionary reduction. (Compare *vestigial*.)

rufous Reddish, especially reddish-brown, like the breast of an American robin.

rugose Wrinkled.

rugulose Diminutive of rugose.

runcinate Sharply cleft or pinnatifid, with backward-pointing segments.

runner A slender stolon; a long, slender, prostrate stem rooting at the nodes and tip.

saccate Furnished with or in the shape of a sac or pouch.

sagittate Arrowhead-shaped, with the basal lobes more or less in line with the body. (Compare *hastate*.)

salverform With a slender tube and abruptly spreading limb, as the corolla of a phlox.

samara   An indehiscent, winged fruit.

sanguineous   Blood-red.

saprobe, saprobic   Same as saprophyte and saprophytic.

saprophyte   A plant (or fungus) that lives on dead organic matter, neither parasitic nor making its own food. The term is often loosely extended to include nongreen myco-trophic plants that get their food from the symbiotic fungal associate.

scaberulous, scabrellate   Diminutive of scabrous.

scabridulous   Diminutive of scabrous.

scabrous, scabrid   Rough to the touch, due to the structure of the epidermis or to the presence of short, stiff hairs.

scalariform   Ladder-like, or with ladder-like markings.

scale   Any small, thin or flat structure.

scandent   Climbing.

scape   A leafless (or merely bracteate) peduncle arising from the ground-level in acau-lescent plants.

scapiform   Scape-like, but not strictly leafless.

scapose   With the flowers on a scape.

scarious   Thin, dry, and chaffy in texture, not green.

scattered   Irregularly arranged.

schizocarp   A fruit that splits into separate carpels at maturity, as in the Apiaceae.

schizogenous   Arising by splitting or separation of tissue, as a schizogenous duct. (Compare *lysigenous*.)

sclerenchyma   A strengthening (but not conducting) tissue composed of thick-walled, lignified cells that are nearly or quite without living contents at maturity.

scorpioid cyme   A sympodial cyme with a zigzag rachis, the successive lateral branches that make up the rachis arising on different sides. The term has often been incorrectly used in place of helicoid cyme.

scurfy   Beset with small, bran-like scales.

secund   With the flowers or branches all on one side of the axis (often by twisting of the pedicels).

semi-   Latin prefix, meaning half.

sepal   A member of the outermost set of floral leaves, typically green or greenish and more or less leafy in texture.

sepaloid   Sepal-like, especially in color and texture.

septate   Provided with one or more partitions (septa).

septicidal   Splitting through the septa, so that the carpels are separated.

septifragal   Dehiscent, with the valves breaking away from the septa.

septum   A partition; in an ovary, a partition formed by the connate walls of adjacent carpels.

sericeous   Silky, from the presence of long, slender, soft, more or less appressed hairs.

serotinous   Literally, late; with the flowers developing after the leaves have fully expanded. (Compare *coetaneous, precocious.*)

serrate   Toothed along the margin with sharp, forward-pointing teeth. (Compare *dentate, crenate.*)

serrulate   Diminutive of serrate.

sessile   Attached directly by the base, without a stalk.

seta   A bristle.

setaceous   Bristle-like.

setose   Beset with bristles.

setulose Diminutive of setose.

sheath An organ that partly or wholly surrounds another organ, such as the sheath of a grass leaf, which surrounds the stem.

sigmoid Doubly curved, like the letter s.

siliceous Containing or composed of silica.

silicle A fruit like a silique, but not much (if at all) longer than wide.

silique An elongate capsule in which the two valves are deciduous from the persistent, frame-like replum to which the seeds are attached. The silique and silicle are the characteristic fruits of the Brassicaceae.

simple leaf A leaf with the blade all in one piece (although it may be deeply cleft), not compound.

simple pistil A pistil composed of a single carpel.

sinuate With a strongly wavy margin.

sinus The cleft or recess between two lobes or segments of an expanded organ such as a leaf.

sobol A slender, shortly rhizome-like stem, originating from an ordinary axillary bud at or just below the ground-level, and turning upward distally to become an ordinary aerial stem.

sordid Of a dull, dingy, or dirty hue.

sorus A cluster of sporangia, as in ferns.

spadix A spike with small, crowded flowers on a thickened, fleshy axis.

spathe A large, usually solitary bract subtending and often enclosing an inflorescence; the term is used only in the monocotyledons.

spatulate Shaped like a spatula, rounded above and narrowed to the base.

species The smallest groups that are consistently and persistently distinct, and distinguishable by ordinary means.

sperm A motile or transported gamete that can fuse with an egg to form a zygote; the male gamete.

spicate Arranged in a spike.

spiciform With the form, but not necessarily the structure of a spike.

spike A more or less elongate inflorescence of the racemose (indeterminate) type, with sessile or subsessile flowers. (Compare *head, raceme.*)

spikelet Literally, a small spike; in grasses and many sedges, one of the ultimate flower-clusters, each consisting of 1–many flowers plus their subtending bracts.

spine A firm, slender, sharp-pointed structure, representing a modified leaf or stipule; more loosely, a structure having the appearance of a true spine. (Compare *prickle, thorn.*)

spinule Diminutive of spine, and not necessarily indicating a modified leaf.

spinulose Provided with spinules, or with a spinule at the tip.

sporangiophore A stalk bearing sporangia, as in *Equisetum.*

sporangium A case or container for spores.

spore A one-celled reproductive structure other than a gamete or zygote, in vascular plants always representing the first cell of the gametophyte generation.

sporocarp A specialized body (not obviously a leaf or cone-scale) within which sporangia are borne.

sporophyll A leaf (often more or less modified) that bears or subtends one or more sporangia.

sporophyte The generation that has $2n$ chromosomes and produces spores as reproductive bodies. In angiosperms the megaspore is retained in the ovule and develops into the embryo-sac, and the microspores develop into the pollen-grains; thus the sporophyte seems to be the whole plant, and the gametophyte a mere stage in reproduction.

spur   A hollow appendage of the corolla or calyx.

squamella   A small scale.

squarrose   Abruptly spreading or recurved at some point above the base.

stamen   The microsporophyll of an angiosperm; the male organ of a flower, consisting of an anther and usually a filament.

staminate flower   A flower with one or more stamens, but no pistil.

staminode   A modified stamen that does not produce pollen.

standard   The uppermost petal of a papilionaceous flower; the banner.

stele   The primary vascular structure of a stem or root, together with any other tissues (such as the pith) that may be enclosed.

stellate   Star-shaped. Stellate hairs have several or many branches from the base.

sterile   Unproductive or infertile, as a flower that does not produce seeds, or a stamen that does not produce pollen.

stigma   The part of the pistil that is receptive to pollen.

stipe   The stalk of a structure, without regard to its morphological nature. The term is usually applied only where more precise terms such as petiole, pedicel, or peduncle cannot be used, as the stipe of an ovary.

stipellule   Diminutive of stipule; a stipule-like body at the base of a leaflet.

stipitate   Borne on a stipe.

stipulate   Provided with stipules.

stipule   One of a pair of basal appendages found in association with the leaves of many species.

stolon   An elongate, creeping stem on the surface of the ground. The term is sometimes loosely applied also to slender rhizomes near the surface of the ground.

stoloniferous   Producing stolons.

stoma, stomate   A special kind of intercellular space in epidermal tissue, bounded by a pair of guard-cells which, under certain conditions, close off the space by changing shape.

stramineous   Straw-colored.

striate   Marked with fine, more or less parallel lines.

strict   Very straight and upright, not at all lax or spreading.

strigillose   Diminutive of strigose.

strigose   Provided with straight, appressed hairs, usually all pointing in more or less the same direction.

strigulose   Diminutive of strigose.

strobilus   A cluster of sporophylls on an axis; a cone.

strophiole   An appendage at the hilum of certain seeds.

strumose   Covered with cushion-like swellings.

style   The slender stalk that typically connects the stigma(s) to the ovary.

stylopodium   An enlargement or disk-like expansion at the base of the style, as in the Apiaceae.

sub-   Latin prefix, meaning under, or almost, or not quite.

suberose   Corky in texture.

subulate   Awl-shaped.

succulent   Fleshy and juicy; more specifically, a plant that accumulates reserves of water in its fleshy stems or leaves, due to the high proportion of hydrophilic colloids in the cell sap.

suffrutescent Half-shrubby, or somewhat shrubby, or dying back to a persistent, woody base.

suffruticose Becoming somewhat woody; suffrutescent.

sulcate Marked with longitudinal grooves.

super-, supra- Latin combining form meaning above, or upon, or more than; opposite of infra-.

superior ovary An ovary that is attached to the receptacle above the level of attachment of the other flower parts. A flower with a superior ovary may be either hypogynous or perigynous.

surcurrent Extending upward; said of a pinnule whose base extends upward and forms a wing along the rachis; opposite of decurrent.

suture A seam or line of fusion; usually applied to the vertical lines along which a fruit may dehisce.

syconium The specialized, multiple fruit of a fig.

symbiosis A close physical association between two different kinds of organisms, typically with benefit to both.

sympatric Occupying the same geographic region. (Compare *allopatric*.)

sympetalous With the petals connate, at least toward the base. Same as gamopetalous.

sympodial With the apparent main axis consisting of a series of short axillary branches.

syn-, sym- Greek prefix, meaning united.

syncarpous With united carpels. (Compare *apocarpous*.)

syngenesious With connate stamens or anthers.

synsepalous With the connate sepals; same as gamosepalous.

taproot The main central root, from which smaller branch roots originate.

taxon (pl. taxa) Any taxonomic entity, of whatever rank.

taxonomy A study aimed at producing a system of classification of organisms that best reflects the totality of their similarities and differences; a classification produced by such a study.

tendril A slender, coiling or twining organ (representing a modified stem or leaf or part thereof) by which a climbing plant grasps its support.

tepal A sepal or petal, or a member of an undifferentiated perianth.

terete Cylindrical (usually solid), round in cross-section.

ternate Borne in threes.

testa The seed-coat, derived from the integuments of the ovule.

tetra- Greek prefix, meaning four.

tetrad A mutually attached set of 4, forming a single unit.

tetradinous Occurring in tetrads.

tetradynamous stamens A set of four long and two (outer) short stamens, as in many Brassicaceae.

tetramerous With four parts of a kind.

tetraploid With four chromosome-complements in each cell.

thalloid Resembling or consisting of a thallus.

thallus A plant body that is not clearly differentiated into roots, stems, and leaves.

theca A compartment or locule, as in an anther.

thorn A stiff, woody, modified stem with a sharp point; more loosely, any structure that resembles a true thorn. (Compare *prickle, spine*.)

throat The orifice of a sympetalous corolla or gamosepalous calyx, or the somewhat expanded part between the proper tube and the limb; in grasses, the upper margins of the sheath.

thrum A type of heterostylic flower that has a relatively short style and long stamens. (Compare *pin.*)

thyrse An elongate, narrow, mixed panicle, consisting of a series of racemosely arranged small cymes.

thyrsoid Resembling or having the quality of a thyrse.

tomentose Covered with tangled or matted, woolly hairs.

tomentulose Diminutive of tomentose.

tomentum An indument of crooked, matted or tangled hairs.

torose Alternately contracted and expanded.

torulose Diminutive of torose.

torus The receptacle of a flower or of a head.

tracheid The most characteristic cell type in xylem, long, slender, and tapering at the ends, with a lignified secondary wall and a definite lumen, but without living contents at maturity.

trailing Prostrate but not rooting.

translator A structure connecting the pollinia of adjacent anthers in the Asclepiadaceae.

tri-, triplo- Latin or Greek prefix, meaning three.

trichome Any hair-like outgrowth from the epidermis.

trichotomous Forking in threes.

trifoliate, trifoliolate With three leaves, with three leaflets.

trigonous With three angles (applied to solid bodies).

trilocular With three locules.

trimerous With three parts of a kind.

trimorphic Of three forms.

trinerved Three-nerved, ordinarily with all three nerves arising directly from the base. (Compare *triplinerved.*)

triplinerved Three-nerved, with the two lateral nerves arising distinctly above the base as branches from the midnerve. (Compare *trinerved.*)

triquetrous With three sharp or projecting angles (applied to solid bodies).

truncate With the apex (or base) transversely straight or nearly so, as if cut off.

tuber A thickened part of a rhizome, usually at the end, serving in food-storage and often also in reproduction. The term is also sometimes loosely applied to tuberous roots.

tubercle A small swelling or projection, usually distinct in color or texture from the organ on which it is borne, as the tubercle on the achene of *Eleocharis.*

tuberous Thickened like a tuber.

tunicate Covered or provided with sheathing leaf-bases that form concentric circles when viewed in cross-section, as the bulb of an onion.

turbinate Top-shaped.

turgid Swollen; expanded by internal pressure.

turion A small, bulb-like offset, as for example in some species of *Epilobium.*

twining Growing in a spiral, usually around the stem of some other plant that serves as a support.

umbel A racemose inflorescence with greatly abbreviated axis and elongate pedicels; in a compound umbel the primary branches are again umbellately branched at the tip.

umbellate Arranged in umbels.

umbellet One of the ultimate umbellate clusters of a compound umbel.

umbelliform With the form, but not necessarily the structure of an umbel. The term is applied especially to umbel-like inflorescences that are condensed cymes rather than condensed racemes.

umbo A blunt or rounded elevation or protuberance on the end or side of a solid organ, as on the scales of many pine-cones.

umbonate Provided with an umbo.

uncinate Hooked at the summit.

undulate Wavy-margined.

uni- Latin prefix, meaning one.

unifacial leaf A leaf with the margins joined, so that only the morphologically abaxial surface is exposed, as in *Iris* and some spp. of **Juncus**.

unilocular With a single locule.

unisexual flower A flower with an androecium or a gynoecium, but not both.

urceolate Urn-shaped or pitcher-like, contracted at or just below the mouth.

utricle A small, thin-walled, one-seeded, more or less inflated fruit.

valvate Opening by valves; arranged with the margins of the petals (or sepals) adjacent throughout their lengths, without overlapping. (Compare *imbricate*.)

valve One of the portions of the ovary wall into which a capsule separates at maturity; in anthers opening by pores, the portion of the anther-wall covering the pore.

vascular Pertaining to conduction. Vascular plants are those which have xylem and phloem; a vascular bundle is a strand of xylem and phloem and associated tissues.

vein A vascular bundle, especially if externally visible, as in a leaf.

velum The membranous flap that partly covers the sporangium of *Isoetes*.

velutinous Velvety; covered with fine, soft, short, spreading pubescence.

ventral Pertaining to or located on the front or belly side. The adaxial surface of a leaf or carpel is considered to be the ventral side, and the seed-bearing suture of an ordinary carpel is the ventral suture. (Compare *dorsal*.)

ventricose Inflated or swelling out on one side only, or unequally, as the corolla of many species of *Penstemon*.

vernation The arrangement of leaves in the bud.

verrucose Warty; covered with wart-like projections.

versatile anther An anther attached near the middle (instead of at the end), and therefore capable of swinging freely on the filament. (Compare *dorsifixed, basifixed*.)

verticil A whorl of leaves or flowers.

verticillaster A false whorl, composed of a pair of nearly sessile cymes in the axils of opposite leaves or bracts, as in many Lamiaceae.

verticillate Arranged in whorls.

vespertine Pertaining to or opening or functional in the evening.

vestigial Poorly developed, as a result of evolutionary reduction from a better developed state. (Compare *rudimentary*.)

vestiture Same as indument.

villosulous Diminutive of villous.

villous, villose Pubescent with long, soft, often bent or curved but not matted hairs. (Compare *hirsute, sericeous, tomentose*.)

virgate Wand-like; slender, straight, and erect.

viscid Sticky or greasy.

viscidulous Diminutive of viscid.

viviparous Sprouting or germinating on the parent plant, as the bulbils in the inflorescence of some plants.

weed A plant that aggressively colonizes disturbed habitats or places where it is not wanted.

whorl A ring of 3 or more similar structures radiating from a node or common point.

wing A thin, flat extension or projection from the side or tip of a structure; one of the two lateral petals in a papilionaceous flower.

xero- Greek prefix, meaning dry.

xeromorphic Having the form or appearance of a xerophyte, usually with some obvious adaptation to reduce transpiration or survive desiccation.

xerophyte A plant adapted to life in dry places.

zygomorphic flower A bilaterally symmetrical flower, capable of being bisected by only one plane into similar halves. Most irregular flowers are zygomorphic.

GENERAL KEYS

The keys that follow are intended solely to facilitate the identification of plants. They are based, as far as practicable, on the most easily observed characters and consequently often omit any mention of the more important technical characters upon which the classification of plants is chiefly based. Also, since most families of large or even moderate size include plants of diverse aspect and superficially diverse structure, the same family may often be reached through two or more routes in the key.

The key is arranged in eighteen sections, the first of which is a general section through which a plant may be referred to one of the following sections where it may be traced further. This arrangement is used partly to avoid the waste of space incurred by progressively greater indentation of the key couplets toward the right-hand margin and partly to bring the alternative or contrasting characters nearer to each other so that they can be more readily compared. The student will therefore trace his plant first through Section 1, General Section, which will refer him directly to certain families or to one of the following sections. If he is so referred, he will then turn to the proper section, trace the plant through it, and again be referred to a family or frequently to a genus in the text.

Certain plants of our flora may be easily misinterpreted by the student, and the keys are so arranged that some such plants may be traced to their proper family even if an error has been made. Thus the monocotyledonous *Arisaema* may be identified, even if it has been mistaken for a dicotyledon; the dicotyledonous *Eryngium,* with distinctly parallel-veined leaves, may be traced also through the monocotyledons. This is not done to encourage careless observation, but to help the beginner with some excusable errors. In three families and in one genus of a fourth family the flowers are so highly modified in structure that they may not be understood by the beginning student. In the Orchidaceae and Asclepiadaceae the stamens, styles, and stigmas are of unusual structure and attachment and are not easily recognized. In the Asteraceae the individual flowers are small, often dimorphic, and aggregated into heads simulating a single flower. In the genus *Euphorbia* the apparent flower is a cup-shaped involucre subtending several staminate flowers, each consisting of a single stamen only, and one pistillate flower, consisting of a single pistil only. In the Lemnaceae the plant is not differentiated into stem and leaf, and the flowers are minute and rarely observed. These five groups are keyed in the General Section. Since they include many common and conspicuous plants, students are advised to become familiar with them and thereafter, in identifying other plants through the General Section, to pass over the steps of the key that lead to them.

In seven genera of submersed aquatic plants the flowers are minute and easily overlooked, or rarely produced, or of peculiar structure or arrangement. These plants will therefore be identified with difficulty. They are *Ceratophyllum,* with whorled, dissected leaves; *Podostemum,* with alternate, forked or dissected leaves; *Vallisneria,* with greatly elongate basal leaves, in fresh water, *Zostera,* with elongate alternate leaves, in salt water; *Elodea,* with whorled, small, narrow, simple leaves; and *Zannichellia* and *Najas,* with opposite, lanceolate to filiform leaves.

SECTION 1—GENERAL SECTION

1 Vascular cryptogams, not producing seeds or fls, both generations physiologically independent at maturity (the gametophyte sometimes depending wholly on its stored food, but not remaining organically attached to the sporophyte); spores the typical disseminules; ferns and fern-like plants .... Section 2, p. xlii

1 Seed plants, the gametophyte never becoming physiologically independent, the young sporophyte as well as the female gametophyte retained within the old sporophyte (the visible plant) for some time; seeds the typical disseminules (includes also some angiosperms with the fls replaced by or altered into bulblets or tufts of minute lvs, as noted in Section 5).

2 Gymnosperms, not producing true fls; ovules borne on the surface of a scale or partly embedded in a fleshy disk, never enclosed in an ovary; ovuliferous scales commonly grouped to form

cones; style and stigma none; perianth none; trees or shrubs with narrow, needle-like or scale-
like, mostly evergreen lvs (Division Pinophyta) .... Section 3, p. xliii
2 Angiosperms, typically producing true fls, but these in some members small or greatly reduced or
rarely produced, or the fls replaced by or altered into bulblets or tufts of minute lvs; ovules
borne within a generally closed ovary; stigma(s) present, usually elevated above the ovary on
a style; perianth present or none; plants woody or herbaceous, of various aspect, often with
broad flat lvs, but sometimes with narrow, scale-like lvs suggesting those of gymnosperms
(Division Magnoliophyta).
  3 Plants epiphytic or parasitic, growing upon or attached to the stems of other plants and completely
without connection to the soil, at least at the time of flowering .... Section 4, p. xliii
  3 Plants rooting in the soil or growing in water (includes some root-parasites and nongreen mycotrophs as well as ordinary green plants).
    4 Plants of unusual or highly specialized structure, as discussed in the introduction to the key.
      5 Plants small, floating on quiet water or stranded on shore, thalloid, without distinction of
stem and leaf, with one or few roots produced from the lower surface, or wholly without
roots; fls minute, only rarely observed. (Compare also Salviniaceae, in Section 2.) .... Lemnaceae, p. 650
      5 Plants otherwise, of more ordinary vegetative form and structure, not thalloid.
        6 Plants with small inflorescences closely resembling individual fls, each composed of a
cup-shaped invol bearing around its margin 1–5 glands, these sometimes with projecting points or petal-like appendages; from each invol protrude few to several
stamens and one stipitate 3-lobed ovary with 3 or more styles .... *Euphorbia,* p. 335
        6 Plants otherwise; inflorescences various, sometimes suggesting individual fls, but then
of very different structure from *Euphorbia,* the individual fls readily recognizable
on close inspection.
          7 Plants with the fls closely aggregated on a common receptacle into a head subtended
by an invol, the latter often suggesting a cal and the whole head often casually
resembling a single fl; cor sympetalous, either tubular and ± regular or tubular
at the base only and strap-shaped or unequally cleft above (or lacking in the
pistillate fls of a few genera); stamens connate (seldom merely connivent) by their
anthers into a tube through which the style protrudes; ovary inferior, the cor (if
present) arising from its summit; cal none, but its place often occupied by a ring
of hairs or awns or scales; fl-heads in general resembling those of a Dandelion,
Daisy, Goldenrod, or Thistle, but varying in size and color .... Asteraceae,[1] p. 518
          7 Plants otherwise, with more ordinary inflorescences; stamens, style and stigma greatly
modified from the ordinary form and scarcely recognizable as such, all joined to
form a special structure in or near the center of the fl.
            8 Fls irregular, the lower (rarely the upper) pet differing from the other two in size or
shape or color; ovary inferior, appearing below the perianth .... Orchidaceae, p. 849
            8 Fls regular, the 5 pet alike; ovary superior, located above the attachment of the
perianth but usually concealed .... Asclepiadaceae, p. 394
    4 Plants without any of the syndromes of special features described above.
      9 Plants producing inflorescences only, the fls modified into or replaced by bulblets or tufts
of minute lvs .... Section 5, p. xliv
      9 Plants typically bearing fls[2] at the proper season and in due course producing fruits and
seeds.
        10 Monocotyledons; lvs usually parallel-veined, often sheathing at the base; parts of the
fl (especially the perianth-members) commonly in threes or sixes; seed with a single
cotyledon (or none); vascular bundles scattered in the stem (Class Liliopsida[3]) ..... Section 6, p. xliv
        10 Dicotyledons; lvs usually net-veined; parts of the fl (at least the perianth members)
commonly in fours or fives; seed with two cotyledons; vascular bundles usually
arranged in a ring surrounding a central pith (or cavity) (Class Magnoliopsida).

---

[1] Through superficial observation a few other plants with fls in involucrate heads are occasionally mistaken for Asteraceae.
Most of these may be excluded because of their superior ovary or separate pet or well developed cal. The Dipsacaceae, with
separate stamens, may be distinguished from those few Asteraceae with merely connivent stamens by having only 4 (instead
of 5) stamens.

[2] Some plants of this group produce fls so small or so concealed that they are rarely observed; a few produce fls but have
lost the ability to produce seeds, as *Hemerocallis fulva.*

[3] The chief differences between the monocotyledons and dicotyledons are in the structure of the seed and in the arrangement
and structure of the vascular bundles of the stem. The structure and appearance of the flowers and the venation of the leaves
are also usually distinctive, but all these characters are subject to exception. Nevertheless, the general habit and aspect of the
plants are so characteristic that every one of them is ordinarily referred to the proper place without hesitation. Certain other
points about the two groups may also assist in identification, but they apply only to plants of our region and may lead to error
if used elsewhere. All angiospermous trees, shrubs, and woody vines in our region are dicotyledons, except *Arundinaria,* a tall
grass with jointed stems and grass-like lvs sheathing at the base, and *Smilax,* vines climbing by tendrils arising from stipules.
All plants with grass-like lvs sheathing the stem are monocotyledons. The great majority of monocotyledons have alternate
lvs. All plants with four colored (or white) pet are dicotyledons, except *Maianthemum,* which has parallel-veined lily-like lvs.
All plants with 5 or 7 or more pet are dicotyledons. All plants with 5 or 7 stamens are dicotyledons, except a few aquatic
herbs. All plants with more than 7 stamens *and* a single ovary are dicotyledons, except *Limnobium.* Most plants with irregular
fls are dicotyledons; among the monocotyledons irregular fls occur in *Commelina, Pontederia, Thalia,* and *Canna,* all with
distinctly parallel-veined lvs, and in the Orchidaceae.

SECTION 2—THE VASCULAR CRYPTOGAMS

1 Plants free-floating at or near the surface of the water; lvs tiny .... Salviniaceae, p. 30
1 Plants rooted in the substrate; lvs variously small to large.
   2 Stems conspicuously jointed, bearing at each node a whorl of small, narrow, scale-like lvs united
      at the base; sporangia borne under the caps of peltate sporangiophores that are aggregated to
      form a terminal cone .... Equisetaceae, p. 7
   2 Stems not evidently jointed; lvs various, only seldom small and whorled; terminal cones wanting
      or quite different structure from the above.
      3 Sporangia borne in the axils of vegetative or ± modified lvs that are often aggregated into a
         terminal cone, or embedded in the adaxial side of the lf toward its base; lvs always narrow
         or small, and with a single unbranched midvein.
         4 Plants homosporous; lvs without a ligule .... Lycopodiaceae, p. 1
         4 Plants heterosporous; lvs ligulate (i.e., with a small adaxial appendage near the base).
            5 Lvs and sporophylls small, not over 5 mm; stems ± elongate and branched .... Selaginellaceae, p. 4
            5 Lvs and sporophylls elongate, generally at least 5 cm; stem very short, corm-like, unbranched
               .... Isoetaceae, p. 5
      3 Sporangia variously arranged, but not axillary to lvs; no terminal cone; lvs with a branching
         vein-system, small and simple to more often large and compound.
         6 Lvs 4-foliolate, resembling a 4-lvd clover; aquatic or amphibious plants .... Marsileaceae, p. 30
         6 Lvs otherwise, not clover-like; plants terrestrial or sometimes emergent from shallow water.
            7 Plants eusporangiate, the sporangia thick-walled, 2-valved, without an annulus, borne on
               a special sporophore that projects from the upper side of the lf, not forming sori, and
               without an indusium; spores numerous, commonly more than 1000; lvs sometimes
               bent over in bud, but not circinate .... Ophioglossaceae, p.. 9
            7 Plants leptosporangiate, the sporangia thin-walled, with a well developed or sometimes
               only poorly developed annulus; spores not more than 512; lvs circinate in bud.
               8 Sporangia not borne in sori; annulus poorly developed, consisting of an apical or api-
                  colateral group of thick-walled cells; fertile lvs, or portions of lvs, obviously unlike
                  the sterile.
                  9 Sporangia borne in clusters on modified short lf-segments, each with 256–512 spores;
                     large, coarse ferns, not climbing .... Osmundaceae, p. 12
                  9 Sporangia in 2 rows on each fertile segment, each with 128–256 spores; large, climbing
                     ferns or small, acaulescent ferns .... Schizaeaceae, p. 13
               8 Sporangia borne in sori; annulus well developed, forming a complete ring or a ring
                  interrupted by the stalk of the sporangium; fertile lvs, or portions of lvs, similar to
                  or unlike the sterile.
                  10 Blade filmy, only one cell thick; annulus complete, oblique; spores up to 512 .... Hymenophyllaceae, p. 13

10 Blade of normal structure, several or many cells thick; annulus interrupted by the
 stalk; spores (32)64.
 11 Rhizome and lvs with hairs only, lacking scales; sori marginal or submarginal,
  protected by the reflexed lf-margin (or parts of it); spores tetrahedral .. Dennstaedtiaceae, p. 14
 11 Rhizome (and often the lvs) ± scaly (the scales sometimes narrow, but distinctly
  flattened), often also hairy.
  12 Indusium formed by the reflexed lf-margin, or none; petiole with a single
   vascular bundle, always slender, very often dark and wiry; spores tet-
   rahedral . . . . Adiantaceae, p. 15
  12 Indusium present or absent, if present then not formed by the reflexed lf-
   margin; petiole with 2 or more vascular bundles (at least toward the
   base), sometimes slender and dark as in the Adiantaceae, but often
   stouter and paler; spores bilateral.
   13 Indusium none, the sori naked; lvs evergreen, only once cleft, the pinnae
    sessile and broad-based, or the segments confluent . . . . Polypodiaceae, p. 14
   13 Indusium present, or if wanting, then the lvs not as above.
    14 Lvs strongly dimorphic, the sori hidden by the inrolled margins of
     the strongly modified fertile pinnules . . . . Onocleaceae, p. 28
    14 Lvs monomorphic or sometimes dimorphic, but the sori clearly
     visible.
     15 Sori in two rows, one row on each side of the midvein of the
      pinna or segment, protected by an elongate, flap-like in-
      dusium that opens toward the midvein . . . . Blechnaceae, p. 29
     15 Sori otherwise, the indusium, if elongate and flap-like, opening
      toward the margin or tip of the pinna or segment . . . . Aspleniaceae, p. 18

SECTION 3—GYMNOSPERMS

1 Seeds borne singly or paired at the ends of short lateral branches, provided with a fleshy (aril-like)
 red covering, not borne in cones; lvs alternate (i.e., spirally arranged), flat, sharp, narrowed to a
 shortly subpetiolar base . . . . Taxaceae, p. 32
1 Seeds borne in small to large cones that have several or many cone-scales attached to the axis of the
 cone, without an aril or aril-like covering; lvs various, but not as above in all respects.
 2 Cone-scales and lvs opposite or whorled in 3's; cones in ours small, not more than about 1 cm
  long and thick, sometimes berry-like . . . . Cupressaceae, p. 36
 2 Cone-scales spirally imbricate; lvs alternate, or scattered, or (in *Pinus*) in bundles of 2–5, or (in
  Larix) numerous and clustered on short spur-branches, never definitely opposite or whorled;
  cones usually somewhat larger.
  3 Cones with a bract behind (beneath) each cone-scale, the cone-scale longer or shorter than its
   subtending bract; lvs persistent and plants evergreen, except in *Larix* . . . . Pinaceae, p. 32
  3 Cones without bracts; branchlets in ours deciduous (with the lvs) in autumn . . . . Taxodiaceae, p. 36

SECTION 4—EPIPHYTES AND
BRANCH-PARASITES

1 Stems slender, twining, or looping from one host-plant to another; lvs none; plants yellow, cream-
 color, brownish or nearly white . . . . Cuscutaceae, p. 410
1 Stems not twining.
 2 Lvs well developed, green (*Phoradendron*) . . . . Viscaceae, p. 327
 2 Lvs minute or essentially lacking.
  3 Stem 2 cm long or less, attached to the branches of spruce, tamarack, or pine; northern states
   only (*Arceuthobium*) . . . . Viscaceae, p. 327
  3 Stem elongate, filiform, rootless, pendent from the branches of trees; Md. and s. . . . . Bromeliaceae, p. 820

SECTION 5—PLANTS PRODUCING INFLORESCENCES ONLY, THE FLOWERS MODIFIED INTO BULBLETS OR TUFTS OF MINUTE LEAVES

1 Monocotyledons, with linear, terete, or grass-like lvs.
  2 Lvs with a strong odor of garlic or onion (*Allium*) .... Liliaceae, p. 823
  2 Lvs without a strong odor.
    3 Lf-blades cross-septate (*Juncus*) .... Juncaceae, p. 658
    3 Lf-blades not cross-septate.
      4 Grasses, with the lf-sheath split opposite the blade.
        5 Plants bulbous at the base (*Poa bulbosa*) .... Poaceae, p. 737
        5 Plants not bulbous at the base (*Festuca rubra*) .... Poaceae, p. 737
      4 Sedges, with closed lf-sheath (a few spp. of *Cyperus*) .... Cyperaceae, p. 667
1 Dicotyledons, with lvs lanceolate or broader.
  6 Lvs toothed, at least toward the apex (*Saxifraga foliolosa*) .... Saxifragaceae, p. 231
  6 Lvs entire (*Polygonum viviparum*) .... Polygonaceae, p. 128

SECTION 6—MONOCOTYLEDONS

In our territory, all Monocotyledons are herbaceous except the Agavaceae, *Arundinaria,* and some spp. of *Smilax.* All of these can be reached through this key and through Section 7. The family Orchidaceae is reached directly through Section 1.

1 Perianth none, or chaffy, scale-like, or bristle-like, never petal-like in color or texture or size.
  2 Fls in the axils of chaffy or husk-like scales and usually entirely or largely concealed by them, or the stamens and styles protruding at anthesis; perianth none, or represented by bristles or minute scales only; fls regularly arranged in spikes or spikelets or heads of uniform size and structure.
    3 Plants with linear basal lvs only, each scape bearing a single terminal button-like head ..... Eriocaulaceae, p. 656
    3 Plants otherwise, *either* with cauline lvs, *or* with sheathing scales only, *or* with several to many spikes, spikelets, or heads of fls on each stem.
      4 Lf-sheaths generally split lengthwise on the side opposite the blade; lvs usually 2-ranked; stems rounded or flat, never triangular in cross-section, usually hollow; anthers appearing versatile .... Poaceae, p. 737
      4 Lf-sheaths continuous around the stem or becoming ruptured only in age; lvs usually 3-ranked, or reduced to sheathing scales only; stems often triangular in cross-section, usually with a pith; anthers basifixed .... Cyperaceae, p. 667
  2 Fls *either* not in the axils of chaffy bracts, *or,* if subtended by bracts, then equaling or exceeding them and not concealed.
    5 Plants aquatic, with floating or wholly submersed lvs (or stranded on shore at low water); fls submersed or floating or barely raised above the water-surface.
      6 Fls axillary and solitary or very few together, often very inconspicuous and easily overlooked.
        7 Lvs all or chiefly alternate, only the uppermost sometimes opposite.
          8 Plants of fresh or rarely brackish water; sep 4, subtending 4 stamens and 4 superior ovaries .... Potamogetonaceae, p. 639
          8 Plants of salt or brackish water, or of alkaline or saline water inland; perianth none; stamens 1 or 2; ovaries 1–4.
            9 Lvs ca 0.5 mm wide; fr exsert, the ovaries stipitate .... Ruppiaceae, p. 645
            9 Lvs 3–10 mm wide; fr not exsert, merely exposed by the spreading of the lf-sheath .... Zosteraceae, p. 647
        7 Lvs all opposite.
          10 Ovary 1, with 2–4 stigmas; fr solitary; lvs 1–3 cm .... Najadaceae, p. 645
          10 Ovaries usually 4, each with its own stigma; frs 2–4; lvs 3–10 cm .... Zannichelliaceae, p. 646
      6 Fls in spikes or heads.
        11 Fls perfect, the spikes or heads all alike or essentially so.
          12 Lower fls in each bright yellow spike with 6 tep and 6 stamens (*Orontium*) .... Araceae, p. 648
          12 Lower fls all with 4 sep and 4 stamens; spikes not bright yellow .... Potamogetonaceae, p. 639
        11 Fls unisexual, in dense globose heads, the lower heads pistillate, the upper ones staminate .... Sparganiaceae, p. 818

  5 Plants of land or shallow water, the lvs (under normal conditions) and the fls completely emersed.
      13 Fls individually minute, in spikes or heads, each spike or head surrounded or subtended by
            a single large, white or colored spathe; lvs never linear or grass-like. (*Orontium,* some-
            times emersed, has a bright yellow spike without a conspicuous spathe.) .... Araceae, p. 648
      13 Fls or fl-clusters not subtended or surrounded by a large spathe; lvs linear or nearly so,
            sometimes grass-like.
          14 Infl a dense elongate spike.
              15 Spike terminal and erect.
                 16 Spike uniform from base to apex; fls perfect .... Juncaginaceae, p. 639
                 16 Spike differentiated into a basal pistillate portion and a more slender, terminal,
                      staminate portion .... Typhaceae, p. 820
              15 Spike apparently lateral .... Acoraceae, p. 647
          14 Infl of subglobose heads, or racemes, or loose open clusters.
              17 Fls unisexual, the lower heads pistillate .... Sparganiaceae, p. 818
              17 Fls all perfect and alike.
                 18 Ovary 1; fr a loculicidal 3-valved capsule .... Juncaceae, p. 658
                 18 Ovaries 3 or 6, connivent but separating at maturity .... Juncaginaceae, p. 639
1 Perianth present, at least the inner segments (and usually all segments) petal-like in color or texture
    or size.
  19 Fls unisexual; plants monoecious or dioecious.
      20 Perianth differentiated into 3 green or greenish sep and 3 white or pinkish pet; stamens almost
            invariably more than 6 .... Alismataceae, p. 632
      20 Perianth-segments alike or essentially so.
          21 Plants aquatic; stamens 3–12 .... Hydrocharitaceae. p. 636
          21 Plants terrestrial; stamens 3–6.
              22 Ovary inferior; twining herbaceous vines .... Dioscoreaceae, p. 844
              22 Ovary superior; plants not twining, though sometimes climbing by tendrils.
                 23 Lvs relatively broad and net-veined; fls in small umbels; plants (except one sp.)
                      with stipular tendrils .... Smilacaceae, p. 843
                 23 Lvs narrow and parallel-veined; fls not in umbels; no tendrils .... Liliaceae, p. 823
  19 Fls perfect.
      24 Ovary or ovaries superior.
          25 Ovaries several to many in each fl.
              26 Pistils 3; lvs partly alternate, partly basal .... Scheuchzeriaceae, p. 638
              26 Pistils more than 3, often numerous; lvs all basal.
                 27 Perianth of 3 green sep and 3 white or pinkish pet; fr an achene; fls in panicles
                    or umbels .... Alismataceae, p. 632
                 27 Perianth of 6 pink tep, the 3 outer smaller than the inner and more deeply colored;
                    fr a follicle; fls in an umbel .... Butomaceae, p. 631
        25 Ovary one in each fl.
          28 Fls irregular.
              29 Sep green; pet separate, colored, the lower one much smaller than the two (usually
                 blue) upper ones; fertile stamens 3 (*Commelina*) .... Commelinaceae, p. 654
             29 Sep and pet both colored, united at base; stamens 6; aquatic or marsh plants Pontederiaceae, p. 821
          28 Fls regular or nearly so.
             30 Perianth differentiated into cal (usually green) and colored cor.
               31 Stamens 3, or rarely 2.
                 32 Lvs linear, all basal; cor yellow; fls in heads at the top of a scape .... Xyridaceae, p. 652
                 32 Lvs lanceolate, cauline; cor bluish-purple; fls not in heads (*Murdannia*)
                     .... Commelinaceae, p. 654
              31 Stamens 6.
                 33 Lvs alternate or basal, lanceolate to linear; fls in umbels (*Tradescantia*)
                   .... Commelinaceae, p. 654
                 33 Lvs in a single whorl of (normally) 3; fl solitary, terminal (*Trillium*) ...... Liliaceae, p. 823
             30 Perianth not differentiated into cal and cor, its divisions essentially similar in
               texture and color.
              34 Stamens 3; tep 6, united below into a slender tube; aquatic or mud plants Pontederiaceae, p. 821
              34 Stamens 6 or in one genus 4, as many as the segments of the perianth.
                 35 Infl and perianth white-woolly .... Haemodoraceae, p. 823
                 35 Infl and perianth not white-woolly.
                   36 Lvs perennial, stiff, spine-tipped; tep large, mostly 4–7 cm (*Yucca*)   Agavaceae, p. 842
                 36 Lvs annual, or if perennial (*Xerophyllum*), then the tep under 1 cm
                    .... Liliaceae, p. 823
      24 Ovary inferior, appearing below the perianth.
          37 Aquatic plants with submersed or floating lvs .... Hydrocharitaceae, p. 636
          37 Terrestrial or marsh or bog plants.
             38 Twining vines (perfect fls very rarely produced) .... Dioscoreaceae, p. 844
             38 Plants not twining.
               39 Infl conspicuously white-woolly .... Haemodoraceae, p. 823
               39 Infl not white-woolly.
                 40 Stamens 3.
                 41 Stamens opposite the outer members of the perianth; lvs developed;
                   plants of wide distribution .... Iridaceae, p. 845

41 Stamens opposite the inner members of the perianth; lvs represented by
   minute scales; plants in our range known only from se. Va. and near
   the head of Lake Michigan .... Burmanniaceae, p. 849
40 Stamens 5 or 6.
42 Stamens 5 (*Eryngium,* a dicotyledon with parallel-veined lvs, keyed here
   for convenience) .... Apiaceae, p. 364
42 Stamens 6.
43 Lvs perennial, very stiff and firm; fls numerous in an elongate spike
   (*Agave*) .... Agavaceae, p. 842
43 Lvs annual, softer; infl various .... Liliaceae, p. 823

SECTION 7—DICOTYLEDONOUS TREES, SHRUBS, AND WOODY VINES

A few woody monocotyledons are included here as well as in Section 6

1 Plants with coarse, firm, narrow, evergreen lvs clustered on a short, stout stem; stamens 6; tep 6, more than 1.5 cm .... Agavaceae, p. 842
1 Plants of different habit; fls not *at once* with 6 stamens and 6 tep more than 1.5 cm.
  2 Stems thick and fleshy, flattened and jointed, or subglobose to short-cylindric, always spiny or spinulose; fls large, with numerous pet and
    stamens; foliage lvs reduced to functionless scales or lacking .... Cactaceae, p. 94
  2 Stems of more ordinary structure and proportions.
    3 Lvs reduced to alternate subulate spines; fls yellow, irregular (*Ulex*) .... Fabaceae, p. 273
    3 Lvs present at the proper season and functional, although in some spp. small or scale-like.
      4 Lvs and lf-scars opposite or whorled.
        5 Lvs still unexpanded at anthesis.
          6 Perianth of both cal and cor.
            7 Pet separate; fls mostly unisexual; stamens commonly 8; ovary superior, 2-lobed .... Aceraceae, p. 351
            7 Pet united; fls perfect; stamens 5; ovary inferior (*Lonicera*) .... Caprifoliaceae, p. 507
          6 Perianth of a single series of parts, or lacking.
            8 In the staminate or perfect fls:
              9 Stamens 2–4, usually 2 (*Fraxinus, Forestiera*) .... Oleaceae, p. 462
              9 Stamens 5–10, usually 8.
                10 Sep or cal-lobes 4, spreading, yellowish; shrubs (*Shepherdia*) .... Elaeagnaceae, p. 306
                10 Sep or cal-lobes commonly 5, erect, often red; trees .... Aceraceae, p. 351
            8 In the pistillate fls:
              11 Hypanthium well developed, bearing the sep at its margin, its orifice nearly or quite closed by the disk (*Shepherdia*) ....
                 .... Elaeagnaceae, p. 306
              11 Hypanthium none.
                12 Ovary prominently 2-lobed .... Aceraceae, p. 351
                12 Ovary not lobed (*Fraxinus, Forestiera*) .... Oleaceae, p. 462
        5 Lvs partly or fully developed at anthesis.
          13 Lvs compound.
            14 Plants climbing or trailing.
              15 Cor none; cal regular; sep 4, petaloid; stamens numerous (*Clematis*) .... Ranunculaceae, p. 47
              15 Cor well developed, irregular, 5-lobed; stamens 4 (*Bignonia, Campsis*) .... Bignoniaceae, p. 492
            14 Plants neither climbing nor trailing.
              16 Pet well developed and conspicuous.
                17 Pet united to form a gamopetalous cor; shrubs.
                  18 Stamens 4; lvs palmately compound (*Vitex*) .... Verbenaceae, p. 428
                  18 Stamens 5; lvs pinnately compound (*Sambucus*) .... Caprifoliaceae, p. 507
                17 Pet separate; shrubs or trees.
                  19 Lfls 3; fls in drooping racemes; stamens 5 .... Staphyleaceae, p. 349
                  19 Lfls 5–7; fls in erect racemes or panicles; stamens 6–8 .... Hippocastanaceae, p. 350
              16 Pet none.
                20 Stamens commonly 8; ovary 2-lobed .... Aceraceae, p. 351
                20 Stamens 2–4; ovary not lobed (*Fraxinus*) .... Oleaceae, p. 462
          13 Lvs simple.
            21 Perianth composed of a single series of parts (usually termed the cal) or with the cal and cor not differentiated.
              22 Lvs palmately lobed .... Aceraceae, p. 351
              22 Lvs entire, not lobed.
                23 Perianth of numerous brownish-red segments; fls perfect .... Calycanthaceae, p. 40
                23 Perianth-segments 4 or 5, mostly greenish; fls unisexual.
                  24 Lvs smooth or merely puberulent (*Buckleya, Nestronia*) .... Santalaceae, p. 325
                  24 Lvs silvery, scurfy-lepidote (*Shepherdia*) .... Elaeagnaceae, p. 306
            21 Perianth composed of both cal and cor, the cal inconspicuous in some spp.
              25 Stamens more numerous than the segments or lobes of the cor.
                26 Pet united into a gamopetalous cor .... Ericaceae, p. 201
                26 Pet distinct.
                  27 Stamens 10 or fewer.
                    28 Lvs palmately lobed .... Aceraceae, p. 351
                    28 Lvs pinnately lobed or without lobes.
                      29 Lvs serrate or pinnately lobed (*Hydrangea*) .... Hydrangeaceae, p. 225
                      29 Lvs entire (*Decodon*) .... Lythraceae, p. 309
                  27 Stamens more than 10.
                    30 Fls white.
                      31 Ovary 1, at least partly inferior (*Decumaria, Philadelphus*) .... Hydrangeaceae, p. 225
                      31 Ovaries 4, superior (*Rhodotypus*) .... Rosaceae, p. 238

30 Fls yellow.
    32 Lvs punctate with translucent dots; styles 2–5, separate, or somewhat united at base .... Clusiaceae, p. 141
    32 Lvs not punctate; style 1 .... Cistaceae, p. 154
25 Stamens as many as the pet or cor-lobes, or fewer.
  33 Cor of separate pet.
    34 Fls in terminal heads or cymes (*Cornus*) .... Cornaceae, p. 323
    34 Fls axillary or in axillary clusters.
      35 Stamens 2; cor-lobes 4, elongate-linear, pure white (*Chionanthus*) .... Oleaceae, p. 462
      35 Stamens 4 or more, as many as the pet; pet never elongate or pure white.
        36 Style short, not lobed; stamens alternate with the pet .... Celastraceae, p. 328
        36 Style lobed or cleft; stamens opposite the pet (*Rhamnus*) .... Rhamnaceae, p. 340
  33 Cor gamopetalous.
    37 Ovary inferior.
      38 Fls numerous in dense globose heads; lvs entire (*Cephalanthus*) .... Rubiaceae, p. 501
      38 Fls not in dense globose heads; lvs entire, toothed, or lobed .... Caprifoliaceae, p. 507
    37 Ovary superior.
      39 Cor distinctly irregular.
        40 Trees with broadly cordate lvs .... Bignoniaceae, p. 492
        40 Low shrubs with small, oblong or ovate lvs (*Conradina, Cunila, Thymus*) .... Lamiaceae, p. 432
      39 Cor regular.
        41 Ovaries 2; stems elongate, twining or trailing.
          42 Cor yellow or blue (varying to whitish), salverform, without corona (*Trachelospermum, Vinca*) ..
          .... Apocynaceae, p. 393
          42 Cor dark red, rotate; corona rotate, bearing 10 long appendages (*Periploca*) .... Asclepiadaceae, p. 394
        41 Ovary 1; stems (except *Gelsemium*) erect or spreading.
          43 Stamens 2; cor-lobes 4 .... Oleaceae, p. 462
          43 Stamens 4 or 5; cor-lobes 4 or 5.
            44 Mature lvs less than 2 cm; dwarf arctic-alpine shrubs.
              45 Cal closely subtended by 3 conspicuous bracts (*Diapensia*) .... Diapensiaceae, p. 217
              45 Cal not bracted at base (*Loiseleuria*) .... Ericaceae, p. 201
            44 Mature lvs much larger; plants southern, or escaped from cult.
              46 Stamens 4.
                47 Ovary bilocular, with numerous ovules; fr capsular, with numerous seeds ....
                .... Buddlejaceae, p. 462
                47 Ovary with 4 uniovulate chambers; fr separating into 4 nutlets (*Callicarpa*) ....
                .... Verbenaceae, p. 428
              46 Stamens 5 (*Gelsemium*) .... Loganiaceae, p. 385
4 Lvs and lf-scars alternate on the stem.
  48 Plants of the Grass Family, with hollow, jointed, bamboo-like stems and elongate, parallel-veined lvs sheathing the stem at base
    (*Arundinaria*) .... Poaceae, p. 737
  48 Plants not grass-like.
    49 Plants of the Aster Family, with numerous small fls closely aggregated in small heads subtended by an involucre (see more
      detailed discussion in Section 1) (*Artemisia, Baccharis*) .... Asteraceae, p. 518
    49 Plants not of the Aster family
      50 Plants regularly or frequently dioecious, an individual bearing either staminate or pistillate fls but not both.
        51 Plants climbing, or in the absence of support sometimes trailing.
          52 Plants producing tendrils.
            53 Stems almost always prickly; lvs or their lobes entire; pet and sep together 6; tendrils arising from stipules ..
            .... Smilacaceae, p. 843
            53 Stems unarmed; lvs or their lobes dentate, or the lvs compound; pet 4 or 5 (quickly deciduous); tendrils arising
            from the stem opposite the lvs .... Vitaceae, p. 342
          52 Plants not producing tendrils.
            54 Plants climbing by adventitious roots; lvs trifoliolate .... Anacardiaceae, p. 353
            54 Plants twining; lvs simple.
              55 Stamens 5; pistil 1; lvs pinnately veined (*Celastrus*) .... Celastraceae, p. 328
              55 Stamens 6 or more; pistils 3 or more; lvs palmately veined .... Menispermaceae, p. 65
        51 Plants erect, ascending, or nearly prostrate, but not climbing.
          56 Fls in catkins or catkin-like clusters, these globose to ovoid or elongate; cor none; cal small and inconspicuous or
            none.
            57 Juice milky; cal present but minute .... Moraceae, p. 74
            57 Juice not milky; cal none.
              58 In the pistillate fls:
                59 Ovules many; trees or shrubs; twigs and bracts not resinous-dotted (some trees have resinous buds) .
                .... Salicaceae, p. 166
                59 Ovule solitary; shrubs; twigs and bracts resinous-dotted .... Myricaceae, p. 79
              58 In the staminate fls:
                60 Twigs and bracts densely resinous-dotted .... Myricaceae, p. 79
                60 Twigs and bracts not resinous-dotted .... Salicaceae, p. 166
        56 Fls not in catkins or catkin-like structures; either cal or cor or both almost invariably present; individual fls in
          many spp. large or conspicuous.
          61 Lvs compound, present at anthesis.
            62 Lvs bipinnate (or some of them only once pinnate in *Gleditsia*); ovary of a single carpel, ripening into a
            large, flat pod .... Caesalpiniaceae, p. 271
            62 Lvs once pinnate, or merely trifoliolate; carpels several.
              63 Lvs beset with translucent dots; stems often prickly .... Rutaceae, p. 356
             63 Lvs without translucent dots; stems unarmed.
               64 Lfls entire except for one or more coarse teeth near the base, each tooth with a large gland beneath;
                carpels united at the axis only, each with a single ovule, maturing into separate samaras ...
                .... Simaroubaceae, p. 355
              64 Lfls otherwise, lacking the glands described above; carpels firmly united into a compound ovary
                that ripens into a single fr.
                65 Trees; fls in large, open, terminal panicles; fr an inflated, partly 3-locular capsule ....
                .... Sapindaceae, p. 350
                65 Shrubs (sometimes arborescent); infl otherwise, sometimes large and terminal, but then con-
                gested; fr a small, berry-like drupe .... Anacardiaceae, p. 353
          61 Lvs simple, or absent at anthesis.

66 Lvs only 3–8 mm, revolute, present at anthesis; pet none .... Empetraceae, p. 200
66 Lvs much larger at maturity, or absent at anthesis.
  67 In the pistillate fls:
    68 Perianth not differentiated into cal and cor, or lacking.
      69 Perianth-members 6, yellow; trees or shrubs with fragrant wood (*Lindera, Sassafras*) ....
        .... Lauraceae, p. 41
      69 Perianth-members 3–5, or lacking.
        70 Style *and* stigma one.
          71 Style very short; stigma almost sessile .... Aquifoliaceae, p. 329
          71 Style elongate, curved or coiled at the summit (*Nyssa*) .... Cornaceae, p. 323
        70 Style divided above; stigmas 2–4.
          72 Lvs very inequilateral at base (*Celtis*) .... Ulmaceae, p. 71
          72 Lvs essentially equilateral .... Rhamnaceae, p. 340
    68 Perianth differentiated into cal and cor, the cal sometimes small and inconspicuous.
      73 Fls in a loose or dense terminal panicle .... Anacardiaceae, p. 353
      73 Fls axillary or in axillary clusters.
        74 Cor small, not over 8 mm wide; style 1; pet distinct or nearly so.
          75 Style very short; stigma almost sessile .... Aquifoliaceae, p. 329
          75 Style otherwise.
            76 Style elongate, coiled or curved at the summit (*Nyssa*) .... Cornaceae, p. 323
            76 Style divided above the middle .... Rhamnaceae, p. 340
        74 Cor 15–40 mm wide if flattened out; styles 2–6, usually 4; cor gamopetalous ....
          .... Ebenaceae, p. 218
  67 In the staminate fls:
    77 Stamens more numerous than the sep or the pet, or the perianth lacking.
      78 Perianth minute or none; stamens commonly 12 (*Nyssa*) .... Cornaceae, p. 323
      78 Perianth well developed and conspicuous.
        79 Perianth yellow, of 6 petal-like segments; stamens 9; trees or shrubs with fragrant wood
          (*Lindera, Sassafras*) .... Lauraceae, p. 41
        79 Cal 4-parted; cor gamopetalous; stamens commonly 16 .... Ebenaceae, p. 218
    77 Stamens as many as the cal-lobes, as many as the pet if they are present.
      80 Fls in a loose or dense terminal panicle .... Anacardiaceae, p. 353
      80 Fls axillary or in axillary clusters.
        81 Stamens alternate with the sep, opposite the pet if they are present ..... Rhamnaceae, p. 340
        81 Stamens opposite the sep, alternate with the pet (if present).
          82 Lvs very inequilateral at base (*Celtis*) .... Ulmaceae, p. 71
          82 Lvs essentially equilateral .... Aquifoliaceae, p. 329
50 Plants not dioecious, the fls either perfect or unisexual; if unisexual, both staminate and pistillate fls borne on the same
  plant, but sometimes in infls of different size or structure. (In *Corylus* the pistillate fls are in small catkins resembling
  lf-buds; they may be recognized at anthesis by the protruding red styles.)
  83 Fls (or some of them, especially the staminate) in catkins or catkin-like clusters or dense globose heads, always unisexual
    and individually small and inconspicuous.
    84 Staminate fls in ellipsoid to elongate-cylindric catkins.
      85 Pistillate fls solitary or in small clusters.
        86 Lvs pinnately compound .... Juglandaceae, p. 77
        86 Lvs simple, but in some spp. deeply lobed .... Fagaceae, p. 81
      85 Pistillate fls in catkins, heads, or cone-like structures.
        87 Juice milky; cal present .... Moraceae, p. 74
        87 Juice not milky; cal lacking.
          88 Pistillate fls 2 or 3 behind each bract .... Betulaceae, p. 87
          88 Pistillate fls solitary behind each bract .... Myricaceae, p. 79
    84 Staminate fls in dense globose heads.
      89 Lvs pinnately veined, with a single midvein (*Fagus*) .... Fagaceae, p. 81
      89 Lvs palmately veined, with 3–5 principal veins.
        90 Lvs star-shaped, with 5 triangular, finely serrate lobes; perianth none (*Liquidambar*) . Hamamelidaceae, p. 70
        90 Lvs not star-shaped, their margins coarsely and sharply toothed; perianth present but minute ....
          .... Platanaceae, p. 70
  83 Fls not in catkins or dense globose heads, in many (not all) spp. perfect or individually large and conspicuous.
    91 Perianth none, or of a single series of parts, or not differentiated into a distinct cal and cor.
      92 Lvs compound.
        93 Stems climbing .... Lardizabalaceae, p. 65
        93 Stems not climbing.
          94 Plants armed with stout prickles (*Aralia*) .... Araliaceae, p. 362
          94 Plants unarmed.
            95 Tree; lfls 11–41 .... Simaroubaceae, p. 355
            95 Shrub 3–5 dm; lfls commonly 5 (*Xanthorhiza*) .... Ranunculaceae, p. 47
      92 Lvs simple, or lacking at the time of anthesis.
        96 Stamens more numerous than the members or lobes of the perianth, or the perianth nearly lacking.
          97 Climbing vines.
             98 Stamens 6; cal-lobes 1 or 3 (*Aristolochia*) .... Aristolochiaceae, p. 42
            98 Stamens 8; cal-lobes 5 (*Brunnichia*) .... Polygonaceae, p. 128
          97 Erect shrubs or trees.
             99 Lvs palmately lobed; stamens ca 15; tep 5 .... Sterculiaceae, p. 146
            99 Lvs entire or merely toothed.
               100 Pistils numerous; stamens numerous; tep 9–12 (*Magnolia*) .... Magnoliaceae, p. 38
               100 Pistil one; stamens 12 or fewer; perianth-lobes 4–6.
                  101 Perianth minute (*Nyssa*) .... Cornaceae, p. 323
                  101 Perianth well developed.
                    102 Perianth-lobes 4 (in some spp. obscure); stamens 8.
                      103 Lvs glabrous beneath, or still undeveloped at anthesis ... Thymelaeaceae, p. 312
                      103 Lvs silvery-lepidote beneath .... Elaeagnaceae, p. 306
                  102 Perianth-segments 6; stamens 9 or 12 .... Lauraceae, p. 41
        96 Stamens as many as the lobes or divisions of the perianth.
          104 Style one, simple or branched above.
            105 Climbing or trailing vines.
              106 Climbing by tendrils; fls not in umbels .... Vitaceae, p. 342

106 Climbing by adventitious roots; fls in umbels (*Hedera*) .... Araliaceae, p. 362
105 Shrubs or small trees.
  107 Fls in terminal infls.
    108 Fls in a terminal peduncled raceme (*Pyrularia*) .... Santalaceae, p. 325
    108 Fls in a freely branched terminal cyme (*Cornus*) .... Cornaceae, p. 323
  107 Fls in lateral or axillary clusters.
    109 Style unbranched, with a single stigma.
      110 Style very short; stigma almost sessile; foliage not scurfy-lepidote ....
             .... Aquifoliaceae, p. 329
      110 Style elongate; foliage scurfy-lepidote (*Elaeagnus*) .... Elaeagnaceae, p. 306
    109 Style 2–4-lobed; stigmas 2–4 .... Rhamnaceae, p. 340
104 Styles more than one.
  111 Styles 2.
    112 Stems prickly (*Oplopanax*) .... Araliaceae, p. 362
    112 Stems unarmed .... Ulmaceae, p. 71
  111 Styles 3 .... Buxaceae, p. 331
91 Perianth present and clearly differentiated into cal and cor.
 113 Ovaries 3 to many, distinct or nearly so; stamens more than 10.
  114 Sep 5; pet 5; lvs simple or compound .... Rosaceae, p. 238
  114 Sep 3; pet 6 or 9; lvs simple.
    115 Stems ringed at each node by a stipular scar; fls white to greenish-yellow .... Magnoliaceae, p. 38
    115 Stems not ringed at the nodes; fls lurid purple or dark red .... Annonaceae, p. 40
 113 Ovary one (the styles and stigmas may be more than one); *Ailanthus*, in the Simaroubaceae, with up to 10
    stamens and with weakly united carpels that separate in fr, is keyed here.
  116 Cor irregular, or the pet only one.
    117 Stamens 5 or 8.
      118 Stamens 5; pet united below; lvs simple (*Rhododendron*) .... Ericaceae, p. 201
      118 Stamens 8; pets distinct; lvs pinnately compound .... Sapindaceae, p. 350
    117 Stamens 10.
      119 Pet one, blue; fls in dense spike-like racemes; lvs pinnately compound (*Amorpha*) ....
            .... Fabaceae, p. 273
      119 Pet or cor-lobes 5; lvs and infl various.
        120 Upper 3 pet separate or nearly so; lower 2 pet ± united and concealing the stamens.
          121 The large upper median pet in front of the lateral ones; lvs (commonly absent at
              anthesis) simple, broadly rotund-cordate (*Cercis*) ...... Caesalpiniaceae, p. 271
          121 The large upper median pet behind the lateral ones; lvs usually compound, seldom
              simple .... Fabaceae, p. 273
        120 Upper 3 pet united nearly to their summit; lower 2 pet separate nearly to the base;
          stamens not concealed; lvs simple (*Rhododendron*) .... Ericaceae, p. 201
  116 Cor regular or essentially so.
    122 Cor gamopetalous.
      123 Styles or long style-branches 4 or 5.
        124 Cor-lobs 4; stamens 4–16; lvs entire .... Ebenaceae, p. 218
        124 Cor-lobes 5 or 6; stamens very numerous.
          125 Cal-lobes valvate; bractlets linear, about as long as the cal; lvs lobed (*Hibiscus*)
            .... Malvaceae, p. 147
          125 Cal-lobes imbricate; bractlets at base of the cal minute or none; lvs finely serrulate
            .... Theaceae, p. 140
      123 Style one; stigma one.
        126 Stamens more numerous than the cor-lobes.
          127 Trees with bipinnately compound lvs (*Albizia*) .... Mimosaceae, p. 270
          127 Shrubs with simple lvs.
            128 Stamens distinct at base .... Ericaceae, p. 201
            128 Stamens monadelphous or pentadelphous toward the base.
              129 Stamens pentadelphous .... Symplocaceae, p. 219
              129 Stamens monadelphous.
                130 Styles separate; fls 6–10 cm wide .... Theaceae, p. 140
                130 Style one; fls smaller .... Styracaceae, p. 219
        126 Stamens as many as the cor-lobes.
          131 Stamens nearly or quite free from the cor.
            132 Style very short; stigma nearly or quite sessile .... Aquifoliaceae, p. 329
            132 Style well developed .... Ericaceae, p. 201
          131 Stamens inserted on cor-tube.
            133 Low, matted or prostrate plants; mature lvs 5–15 mm .... Diapensiaceae, p. 217
            133 Erect or straggling or climbing shrubs; lvs larger.
              134 Stamens opposite the cor-lobes; petaloid staminodia also present ....
                .... Sapotaceae p. 218
              134 Stamens alternate with the cor-lobes; no staminodia.
                135 Stamens 4 (*Callicarpa*, *Clerodendron*) .... Verbenaceae, p. 428
                135 Stamens 5 (*Lycium*, *Solanum*) .... Solanaceae, p. 400
    122 Cor of separate pet.
      136 Ovary actually or apparently inferior.
        137 Stamens twice as many as the pet, or more numerous.
          138 Style one (*Vaccinium*) .... Ericaceae, p. 201
          138 Styles 2–5 .... Rosaceae, p. 238
        137 Stamens just as many as the pet.
          139 Pet 4.
            140 Fls white, in terminal cymes (*Cornus*) .... Cornaceae, p. 323
            140 Fls yellow, in small axillary clusters, opening in autumn, winter, or earliest
              spring (*Hamamelis*) .... Hamamelidaceae, p. 70
         139 Pet 5.
            141 Fls in racemes or small corymbiform clusters (*Ribes*) ... Grossulariaceae, p. 226
            141 Fls in numerous umbels .... Araliaceae, p. 362
      136 Ovary superior.
        142 Mature lvs scale-like or cylindric, less than 1 cm.
          143 Fls racemose or paniculate, 4–5-merous; lvs scale-like .... Tamaricaceae, p. 163

167 Lvs glabrous or puberulent, not revolute.
  168 Racemes lateral .... Cyrillaceae, p. 199
  168 Racemes terminal (*Itea*) ....
            .... Grossulariaceae, p. 226
166 Fls axillary, or in axillary clusters, or in terminal
  or lateral panicles.
  169 Stamens opposite the pet; style 3-cleft or
    3-lobed (*Ceanothus, Rhamnus*) ......
             .... Rhamnaceae, p. 340
  169 Stamens alternate with the pet; style very short,
    the stigma almost sessile ....
           .... Aquifoliaceae, p. 329

SECTION 8 – HERBACEOUS DICOTYLEDONS WITH UNISEXUAL FLOWERS

Some plants that also have perfect fls are included here for convenience. If perfect fls are found, it is recommended that such plants be traced through a different section of the key.

1 Lvs lacking or reduced to small scales.
  2 Plants without green color, growing under beech trees (*Epifagus*) .... Orobanchaceae, p. 488
  2 Plants with chlorophyll, not growing under beech trees.
    3 Plants with jointed stems, growing in salt marshes or saline soil (*Salicornia*) .... Chenopodiaceae, p. 95
    3 Plants with normal (not jointed) stems, growing in fresh water or rarely on muddy shores .... Haloragaceae, p. 307
1 Lvs not reduced to scales.
  4 Lvs dissected into numerous narrow or filiform segments; plants aquatic or sometimes on shores.
    5 Lvs pinnately dissected .... Haloragaceae, p. 307
    5 Lvs palmately dissected .... Ceratophyllaceae, p. 46
  4 Lvs not dissected.
    6 Lvs compound.
      7 Lvs 3-foliolate or palmately compound.
        8 Fls in a dense head on a fleshy spadix, wholly or partly concealed by a large unilateral
          spathe (*Arisaema, Pinellia,* monocotyledons keyed here for convenience) .... Araceae, p. 648
        8 Fls in umbels or compound umbels or in spikes or panicles.
          9 Fls in umbels or compound umbels.
            10 Lvs alternate or basal (*Sanicula*) .... Apiaceae, p. 364
            10 Lvs cauline, in a single whorl (*Panax*) .... Araliaceae, p. 362
          9 Fls in spikes or panicles.
            11 Perianth conspicuous; stamens numerous; pistils several or many (*Clematis*) .... Ranunculaceae, p. 47
            11 Perianth minute; stamens 5; pistil one (*Cannabis*) .... Cannabaceae, p. 73
      7 Lvs pinnately compound, or ternately or pinnately decompound.
        12 Fls in one to many umbels (*Aralia*) .... Araliaceae, p. 362
        12 Fls not in umbels.
          13 Fls in globose heads or short subglobose spikes.
            14 Lvs once pinnately compound (*Sanguisorba*) .... Rosaceae, p. 238
            14 Lvs twice pinnately compound .... Mimosaceae, p. 270
          13 Fls solitary or in panicles.
            15 Cauline lvs opposite.
              16 Stamens numerous; pistils several to many (*Clematis*) .... Ranunculaceae, p. 47
              16 Stamens 3; pistil one with one style .... Valerianaceae, p. 515
            15 Cauline lvs alternate.
              17 Plants climbing (*Cardiospermum*) .... Sapindaceae, p. 350
              17 Plants erect or essentially so.
                18 Fls either with cal or cor but not both, or the perianth wanting (*Thalic-*
                  trum) .... Ranunculaceae, p. 47
              18 Fls with cal and cor (both small).
                19 Stamens 10; pistils usually 2 (*Astilbe*) .... Saxifragaceae, p. 231
                19 Stamens 15 or more; pistils usually 3 or 4 (*Aruncus*) .... Rosaceae, p. 238
    6 Lvs simple.
      20 Lvs all basal.
        21 Fls in short or elongate spikes .... Plantaginaceae, p. 459
        21 Fls in small open panicles (*Rumex*) .... Polygonaceae, p. 128
      20 Lvs all or mainly cauline.
        22 Lvs opposite or whorled.
          23 Foliage densely stellate-pubescent or lepidote (*Croton, Crotonopsis*) .... Euphorbiaceae, p. 332
          23 Foliage glabrous to pubescent, but not densely stellate or lepidote.
            24 Fls solitary; plants of mud, swamps, or shallow water.
              25 Lvs in whorls of 6–12 .... Hippuridaceae, p. 458

```
    25 Lvs opposite.
        26 Perianth none; stamen one; styles 2 ............................ Callitrichaceae, p. 458
        26 Perianth present, irregular, falling quickly; stamens 2; style 1 (Hemian-
            thus) ................................................................. Scrophulariaceae, p. 465
 24 Fls in axillary or terminal clusters.
    27 Infls axillary.
        28 Principal lvs evidently lobed (Humulus) ......................... Cannabaceae, p. 73
        28 Lvs entire or merely toothed.
            29 Lvs entire or essentially so.
                30 Spikes distinctly peduncled ........................... Plantaginaceae, p. 459
                30 Heads sessile or nearly so (Alternanthera) ............. Amaranthaceae, p. 104
            29 Lvs distinctly serrate.
                31 Styles 2; stamens 8–20 (Mercurialis) ................... Euphorbiaceae, p. 332
                31 Style one; stamens 4 or 5 ............................... Urticaceae, p. 75
    27 Infls terminal.
        32 Pet lacking (Iresine) ........................................ Amaranthaceae, p. 104
        32 Pet present, white or colored.
            33 Stamens 3; style one ..................................... Valerianaceae, p. 515
            33 Stamens 10; styles mostly 5 (Silene) ..................... Caryophyllaceae, p. 112
22 Lvs alternate.
    34 Foliage densely stellate-tomentose or lepidote (Croton, Crotonopsis) .......... Euphorbiaceae, p. 332
    34 Foliage glabrous or pubescent, but not densely stellate or lepidote.
        35 Cal and cor both present, the latter usually white or colored.
            36 Plants twining; sep, pet, and stamens each 6; pistils 3 or 6 (Cocculus) .. Menispermaceae, p. 65
            36 Plants not twining.
                37 Plants climbing by tendrils or occasionally trailing; stamens 3; pistil 1;
                    ovary inferior ......................................... Cucurbitaceae, p. 164
                37 Plants erect or spreading or creeping; stamens more than 6; pistils 4 or
                    more, separate or connivent.
                    38 Lvs sessile or nearly so, fleshy; stamens 8 or 10; pistils 4 or 5 (Sedum)
                        ..................................................... Crassulaceae, p. 229
                    38 Lvs distinctly petioled; pistils and stamens more numerous.
                        39 Fls in a terminal panicle (Napaea) ................ Malvaceae, p. 147
                        39 Fls solitary (Dalibarda, Rubus) .................. Rosaceae, p. 238
        35 Cal present or absent; cor always lacking; cal petal-like in only two spp.
            40 Fls in small axillary clusters, always individually very small.
                41 Key for pistillate fls:
                    42 Style one, unbranched (Parietaria) .................. Urticaceae, p. 75
                    42 Styles 2 or 3.
                        43 Styles 3, each of them 2-lobed or -branched (Acalypha, Phyl-
                            lanthus) ...................................... Euphorbiaceae, p. 332
                        43 Styles 2 or 3, unbranched.
                            44 Sep and bracts acute, scarious ................ Amaranthaceae, p. 104
                            44 Sep (often lacking) and bracts herbaceous (Atriplex, Che-
                                nopodium) ................................. Chenopodiaceae, p. 95
                41 Key for staminate fls:
                    45 Fls or fl-clusters subtended and often exceeded by bracts.
                        46 Bracts foliaceous, lobed or cleft (Acalypha) ............... Euphorbiaceae, p. 332
                        46 Bracts entire.
                            47 Sep and bracts scarious, acute ................ Amaranthaceae, p. 104
                            47 Sep and bracts herbaceous (Parietaria) ........ Urticaceae, p. 75
                    45 Fls or fl-clusters not bracteate.
                        48 Lvs with stipules (Phyllanthus) ................... Euphorbiaceae, p. 332
                        48 Lvs without stipules (Atriplex) .................. Chenopodiaceae, p. 95
            40 Fls in spikes, racemes, or panicles, usually terminal, but in some spp. op-
                posite the lvs or basal.
                49 Infls arising on the lower part of the stem below the lvs .............. Buxaceae, p. 331
                49 Infls terminal or lateral, with or above the lvs.
                    50 Sep petal-like in size or color.
                        51 Lvs entire; stem glabrous .......................... Phytolaccaceae, p. 92
                        51 Lvs lobed; stem spiny (Cnidoscolus) ................ Euphorbiaceae, p. 332
                    50 Sep minute or none, green or pale, not petal-like.
                        52 Divisions of the perianth 6, in two cycles (Rumex) ........ Polygonaceae, p. 128
                        52 Divisions of the perianth 5, or fewer, or the perianth lacking.
                            53 Sep acute, scarious, mingled with similar acute scarious
                                bracts ................................... Amaranthaceae, p. 104
                            53 Sep not acute, or not scarious, or not mingled with scarious
                                bracts.
                                54 Key for pistillate fls or fr:
                                    55 Ovary 3-locular; fr a 3-locular capsule with 3 or 6
                                        seeds .......................... Euphorbiaceae, p. 332
                                    55 Ovary unilocular; fr a one-seeded utricle ...... Chenopodiaceae, p. 95
                                54 Key for staminate fls:
                                    56 Stamens numerous; lvs very large, rotund in gen-
```

eral outline, palmately veined and lobed (*Ricinus*) .... Euphorbiaceae, p. 332

56 Stamens as many as the divisions of the cal, or fewer; lvs not as above.

57 Sep more or less united into a 2–5-lobed cal Euphorbiaceae, p. 332

57 Sep separate .... Chenopodiaceae, p. 95

SECTION 9—HERBACEOUS DICOTYLEDONS WITH PERFECT FLOWERS LACKING A PERIANTH

1 Aquatic plants, either submersed or growing on wet shores after recession of the water.
  2 Stamens 4–8 .... Haloragaceae, p. 307
  2 Stamen one .... Podostemonaceae, p. 307
1 Plants of land or shallow water; if in water, the stems erect and rising above the surface.
  3 Lvs entire or merely toothed.
    4 Lvs alternate, cordate, entire .... Saururaceae, p. 42
    4 Lvs whorled, linear or narrowly lanceolate .... Hippuridaceae, p. 458
  3 Lvs deeply lobed or compound .... Ranunculaceae, p. 47

SECTION 10—HERBACEOUS DICOTYLEDONS WITH PERFECT FLOWERS, PERIANTH IN ONE CYCLE, AND INFERIOR OVARY

1 Stamens more numerous than the lobes or members of the perianth.
  2 Lobes of the perianth 1 or 3; stamens 6 or 12 .... Aristolochiaceae, p. 42
  2 Lobes of the perianth 2 or 4; stamens 4–8, or numerous.
    3 Stamens numerous; plants of moist or dry soil; stipules conspicuous (*Sanguisorba*) .... Rosaceae, p. 238
    3 Stamens 4–8; plants of water, muddy shores, or wet soil; stipules minute or none.
      4 Lvs linear or pinnately lobed .... Haloragaceae, p. 307
      4 Lvs ovate or rotund (*Chrysosplenium*) .... Saxifragaceae, p. 231
1 Stamens as many as or fewer than the lobes or members of the perianth.
  5 Lvs opposite or whorled.
    6 Fls in dense terminal heads.
      7 Heads subtended by (usually 4) large white bracts (*Cornus*) .... Cornaceae, p. 323
      7 Heads subtended by green bracts, or not bracteate .... Dipsacaceae, p. 516
    6 Fls not in heads.
      8 Lvs whorled.
        9 Fls in umbels; lvs in a single whorl (*Panax*) .... Araliaceae, p. 362
        9 Fls in cymes or cymose panicles or short axillary infls; lvs in several or many whorls (*Asperula, Galium*) .... Rubiaceae, p. 501
      8 Lvs opposite.
        10 Fls axillary, few or solitary; stamens 4.
          11 Style one (*Ludwigia*) .... Onagraceae, p. 313
          11 Styles 2 (*Chrysosplenium*) .... Saxifragaceae, p. 231
        10 Fls in terminal corymbiform clusters or panicles; stamens 3 .... Valerianaceae, p. 515
  5 Lvs alternate or basal.
    12 Stamens 4 or fewer; division of the perianth 3 or 4.
      13 Lvs with large conspicuous stipules (*Alchemilla, Sanguisorba*) .... Rosaceae, p. 238
      13 Lvs without stipules.
        14 Divisions of the perianth 3; stamens 3 (*Proserpinaca*) .... Haloragaceae, p. 307
        14 Divisions of the perianth 4; stamens 4.
          15 Styles 2; lvs oval to rotund, crenate (*Chrysosplenium*) .... Saxifragaceae, p. 231
          15 Style one; lvs lanceolate to linear, entire or nearly so (*Ludwigia*) .... Onagraceae, p. 313
    12 Stamens 5; divisions or lobes of the perianth 5.
      16 Fls in heads or umbels; lvs usually (but not always) compound or deeply lobed.
        17 Styles 2 .... Apiaceae, p. 364
        17 Styles 5 (*Aralia*) .... Araliaceae, p. 362
      16 Fls in terminal or axillary cymules, or solitary and pediceled in the axils; lvs entire or nearly so (*Comandra, Geocaulon*) .... Santalaceae, p. 325

SECTION 11—HERBACEOUS DICOTYLEDONS WITH PERFECT FLOWERS, PERIANTH IN ONE CIRCLE,[4] AND SUPERIOR OVARY

1 Ovaries more than one in each fl, sometimes united up to the middle, but distinct in the upper half.
  2 Lvs with conspicuous stipules (*Alchemilla, Sanguisorba*) .... Rosaceae, p. 238
  2 Lvs without stipules.
    3 Ovaries united nearly to the middle (*Penthorum*) .... Saxifragaceae, p. 231
    3 Ovaries essentially distinct.
      4 Lvs toothed or cleft or lobed or compound, or if simple and entire, then the fls not in racemes
        .... Ranunculaceae, p. 47
      4 Lvs simple and entire; fls in simple racemes .... Phytolaccaceae, p. 92
1 Ovary one in each fl (the styles or stigmas may be distinct).
  5 Stamens more than twice as many as the parts or lobes of the perianth.
    6 Lvs modified into hollow pitchers, all basal .... Sarraceniaceae, p. 152
    6 Lvs of normal structure, simple or compound.
      7 Perianth small and inconspicuous, pale or greenish.
        8 Lvs compound, alternate; fls in terminal racemes (*Actaea, Cimicifuga*) .... Ranunculaceae, p. 47
        8 Lvs simple, opposite or whorled; fls axillary .... Aizoaceae, p. 93
      7 Perianth well developed, brightly colored.
        9 Lvs entire.
          10 Lvs large, deeply cordate at base; aquatic or mud plants .... Nymphaeaceae, p. 44
          10 Lvs linear, terete, succulent; land plants (*Talinum*) .... Portulacaceae, p. 109
        9 Lvs compound, dissected, lobed, or spiny-toothed.
          11 Parts of the perianth 5; juice not milky or colored.
            12 Lvs dissected; fls solitary, individually large (*Nigella*) .... Ranunculaceae, p. 47
            12 Lvs bipinnately compound; fls individually small, in dense heads (*Mimosa*) .... Mimosaceae, p. 270
          11 Parts of the perianth 4 or 8; juice milky or colored .... Papaveraceae, p. 66
  5 Stamens up to twice as many as the lobes or parts of the perianth, but not more.
    13 Styles 2 or more.
      14 Lvs reduced to small scales; succulent plants with jointed stems, in salt marshes or saline
        soil (*Salicornia*) .... Chenopodiaceae, p. 95
      14 Lvs not scale-like.
        15 Lvs basal or crowded near the base.
          16 Lvs broad, palmately veined and lobed (*Heuchera*) .... Saxifragaceae, p. 231
          16 Lvs linear, terete, succulent (*Talinum*) .... Portulacaceae, p. 109
        15 Lvs all or chiefly cauline.
          17 Lvs opposite or whorled.
            18 Lvs crenate or very shallowly lobed (*Chrysosplenium*) .... Saxifragaceae, p. 231
            18 Lvs entire.
              19 Ovary and capsule unilocular .... Caryophyllaceae, p. 112
              19 Ovary and capsule 3–5-locular .... Aizoaceae, p. 93
          17 Lvs alternate.
           20 Stamens 10; styles mostly 10 .... Phytolaccaceae, p. 92
           20 Stamens fewer than 10; styles 2–5.
             21 Stipules sheathing the stem above the base of each lf; fr an achene ..... Polygonaceae, p. 128
             21 . Stipules none.
              22 Stamens as many as the tep, or fewer .... Chenopodiaceae, p. 95
              22 Stamens more numerous than the tep.
                23 Stamens less than twice as many as the tep (*Eriogonum*) ...... Polygonaceae, p. 128
                23 Stamens twice as many as the tep (*Chrysosplenium*) .... Saxifragaceae, p. 231
    13 Style one or none (two or more stigmas may be discernible).
      24 Stamens more numerous than the lobes or divisions of the perianth.
        25 Plants white to yellow, pink, or red; lvs reduced to scales; fls or fl-clusters terminal
          (*Monotropa*) .... Monotropaceae, p. 215
        25 Plants with green color in the foliage lvs.
          26 Divisions of the perianth 5.
            27 Lvs compound.
              28 Lvs 3-foliolate (cleistogamous fls of *Amphicarpa, Lespedeza*) .... Fabaceae, p. 273
              28 Lvs bipinnately compound (*Mimosa*) .... Mimosaceae. p. 270
            27 Lvs simple.
              29 Parts of the perianth distinctly unequal, two of them smaller than the other
                3 (apetalous plants of *Helianthemum*) .... Cistaceae, p. 154

---

[4] In several families or genera keyed in this section two circles of perianth, the cal and the cor, are present, but one falls early and may easily be overlooked. Such plants are also keyed under other sections.

SECTION 12—HERBACEOUS DICOTYLEDONS WITH PERFECT FLOWERS; CAL AND COR BOTH PRESENT; OVARIES TWO OR MORE IN EACH FLOWER

5 Fls irregular.
  6 Stamens numerous; lvs compound or deeply divided (*Aconitum, Delphinium*) .... Ranunculaceae, p. 47
  6 Stamens 5; lvs serrate or lobed (*Heuchera*) .... Saxifragaceae, p. 231
5 Fls regular.
  7 Sep and pet each 3.
    8 Aquatic plants with entire floating lvs or submersed dissected lvs .... Cabombaceae, p. 45
    8 Land or mud plants; lvs otherwise.
      9 Land plants; lvs alternate, pinnately divided .... Limnanthaceae. p. 361
      9 Mud plants; lvs opposite, entire (*Crassula*) .... Crassulaceae, p. 229
  7 Sep or pet more than 3.
    10 Sepal-like bracts 3; petal-like sep 5–12; lvs all basal, 3–5-lobed (*Hepatica*) .... Ranunculaceae, p. 47
    10 Sep and pet each 4 or more.
      11 Pet united for part of their length into a tubular or salverform cor.
        12 Lvs opposite .... Apocynaceae, p. 393
        12 Lvs alternate.
          13 Plants erect; fls in terminal clusters (*Amsonia*) .... Apocynaceae, p. 393
          13 Plants trailing; fls solitary on axillary peduncules (*Dichondra*) .... Convolvulaceae, p. 406
      11 Pet separate.
        14 Lvs rotund, centrally peltate; fls solitary, 12–25 cm wide .... Nelumbonaceae, p. 44
        14 Lvs not centrally peltate; fls smaller.
          15 Lvs succulent, fleshy .... Crassulaceae, p. 229
          15 Lvs not succulent.
            16 Hypanthium none; sep separate to the base .... Ranunculaceae, p. 47
            16 Hypanthium present, appearing like a cal-tube, saucer-shaped or cup-shaped,
                bearing the sep and pet at its margin.
              17 Pistils as many as the pet, or more numerous .... Rosaceae, p. 238
              17 Pistils fewer than the pet.
                18 Lvs simple, sometimes shallowly lobed .... Saxifragaceae, p. 231
              18 Lvs compound.
                19 Lvs trifoliolate or once-pinnately compound; fls yellow (*Agri-*
                  monia, Waldsteinia) .... Rosaceae, p. 238
                19 Lvs ternately decompound; fls white or nearly so.
                  20 Stamens 10; carpels separating only in fr (*Astilbe*) ...... Saxifragaceae, p. 231
                  20 Stamens 15 or more; carpels distinct from the beginning
                    (*Aruncus*) .... Rosaceae, p. 238

SECTION 13—HERBACEOUS DICOTYLEDONS WITH PERFECT FLOWERS; CAL AND COR BOTH PRESENT; OVARY ONE IN EACH FLOWER, INFERIOR

1 Stamens more numerous than the pet.
  2 Stamens twice as many as the pet or sep, commonly 8 or 10, rarely 12.
    3 Style 1, sometimes elongate, sometimes short or almost wanting.
      4 Land plants; cor evident, white yellow, pink, or purple .... Onagraceae, p. 313
      4 Mud or water plants.
        5 Pet bright yellow, 1 cm or more (*Ludwigia*) .... Onagraceae, p. 313
        5 Pet minute, whitish or greenish (*Myriophyllum*) .... Haloragaceae, p. 307
    3 Styles 2 or more.
      6 Styles 2 .... Saxifragaceae, p. 231
      6 Styles 3 or more.
        7 Lvs simple; sep 2, commonly caducous .... Portulacaceae, p. 109
        7 Lvs ternately compound; cal with 3 or more teeth .... Adoxaceae, p. 514
  2 Stamens more than twice as many as the sep or pet, 20 or more.
    8 Lvs of normal structure, alternate .... Loasaceae, p. 165
    8 Lvs reduced to functionless scales or absent; stems thick and fleshy, spiny .... Cactaceae, p. 94
1 Stamens as many as the pet or corolla-lobes, or fewer.
  9 Pet separate.
    10 Pet 2; stamens 2 (*Circaea*) .... Onagraceae, p. 313
    10 Pet 4 or 5; stamens 4 or 5.
      11 Pet 4.
        12 Fls in dense heads subtended by large white bracts; land plants (*Cornus*) .... Cornaceae, p. 323
        12 Fls never in heads and never subtended by conspicuous white bracts; aquatic or marsh
        plants.
          13 Submersed lvs or lf-like organs pinnately dissected into narrow segments.
            14 Floating lvs rhombic or deltoid; pet white, conspicuous .... Trapaceae, p. 313

SECTION 14—HERBACEOUS DICOTYLEDONS WITH PERFECT FLOWERS; CAL AND COR BOTH PRESENT; FLOWERS IRREGULAR, WITH ONE SUPERIOR OVARY AND WITH STAMENS MORE NUMEROUS THAN THE PET OR DIVISIONS OF THE COR

SECTION 15—HERBACEOUS DICOTYLEDONS WITH PERFECT FLOWERS; CAL AND COR BOTH PRESENT; FLOWERS REGULAR, WITH ONE SUPERIOR OVARY AND WITH STAMENS MORE NUMEROUS THAN THE PET OR DIVISIONS OF THE COR

24 Lvs simple, either opposite or whorled.
  25 Fls minute, sessile in the axils; lvs opposite (*Crassula*) .... Crassulaceae, p. 229
  25 Fls large, solitary, terminal; lvs in a single whorl of 3 (*Trillium*, a monocotyledon
     with net-veined lvs, keyed here for convenience) .... Liliaceae, p. 823
23 Pet 4 or more.
  26 Sep 6 or more; pet 6 or more.
    27 Lvs a single opposite pair, peltate, bearing a solitary fl between them (*Podo-
     phyllum*) .... Berberidaceae, p. 63
    27 Lvs more than 2; fls normally several on each plant.
      28 Styles as many as the pet; hypanthium scarcely developed; lvs thick and
       fleshy (*Sempervivum*) .... Crassulaceae, p. 229
      28 Style one; hypanthium tubular, bearing the sep and pet at or near its margin;
       lvs thin .... Lythraceae, p. 309
  26 Sep 4 or 5; pet 4 or 5.
    29 Lvs compound, or divided nearly or quite to the base.
      30 Lvs opposite.
       31 Fls yellow; lvs pinnately compound .... Zygophyllaceae, p. 357
       31 Fls anthocyanic or nearly white; lvs palmately divided or dissected ... Geraniaceae, p. 359
      30 Lvs alternate;
       32 Styles 5; lvs trifoliolate .... Oxalidaceae, p. 357
       32 Style one.
        33 Principal lvs twice pinnatifid, but not truly compound (*Ruta*) ...... Rutaceae, p. 356
        33 Principal lvs truly compound.
         34 Lvs palmately compound .... Capparaceae, p. 174
         34 Lvs pinnately or bipinnately compound
          35 Fls axillary or racemose .... Caesalpiniaceae, p. 271
          35 Fls in dense axillary heads .... Mimosaceae, p. 270
    29 Lvs simple, entire or merely serrate to shallowly lobed.
      36 Style one.
       37 Hypanthium cup-shaped or tubular, bearing the sep and pet at or near
        its margin.
        38 Anthers opening longitudinally .... Lythraceae, p. 309
        38 Anthers opening by terminal pores .... Melastomataceae, p. 322
       37 Hypanthium none.
        39 Sep essentially alike in size and shape.
         40 Stamens free from the cor; filaments separate, all bearing an-
          thers .... Pyrolaceae, p. 213
         40 Stamens inserted on the cor; filaments monadelphous, only 5
          of them bearing anthers (*Galax*) .... Diapensiaceae, p. 217
        39 Sep obviously dissimilar in width, or two pairs of them partly united
         .... Cistaceae, p. 154
      36 Styles 2 or more.
       41 Ovary lobed, each lobe bearing a style.
        42 Styles 2; lvs all or chiefly basal, the cauline none, or few and small,
         or a single pair only .... Saxifragaceae, p. 231
        42 Styles 4 or 5; lvs all or chiefly cauline .... Crassulaceae, p. 229
       41 Ovary not lobed, the styles arising together from its summit.
        43 Lvs serrate; fls minute, sessile or nearly so in the axils (*Bergia*) ... Elatinaceae, p. 140
        43 Lvs entire or only very minutely serrulate; fls otherwise.
         44 Fls yellow .... Clusiaceae, p. 141
         44 Fls white or anthocyanic, never yellow .... Caryophyllaceae, p. 112

SECTION 16—HERBACEOUS DICOTYLEDONS WITH PERFECT FLOWERS; CAL AND COR BOTH PRESENT; FLOWERS WITH SEPARATE PET, ONE SUPERIOR OVARY, AND STAMENS JUST AS MANY AS THE PET OR FEWER

1 Lvs compound or dissected.
  2 Fls solitary on lfless scapes.
    3 Cor irregular, spurred; pet 5 (*Viola*) .... Violaceae, p. 157
    3 Cor regular; pet 8 (*Jeffersonia*) .... Berberidaceae, p. 63
  2 Fls borne on leafy stems.

   4 Fls in loose, open, peduncled clusters.
      5 Pet 6; stamens 6; lvs thrice ternately compound, with separate lfls (*Caulophyllum*) .... Berberidaceae, p. 63
      5 Pet 5; stamens 5; lvs merely dissected .... Geraniaceae, p. 359
   4 Fls in other sorts of infls.
      6 Lvs once pinnate.
         7 Fls sessile or nearly so in the lf-axils, yellow .... Caesalpiniaceae, p. 271
         7 Fls in dense terminal spikes, anthocyanic or rarely white (*Dalea*) .... Fabaceae, p. 273
      6 Lvs twice pinnate; fls in dense, peduncled, axillary heads .... Mimosaceae, p. 270
1 Lvs entire to deeply lobed.
  8 Lvs opposite.
    9 Sep 2 or 3.
      10 Sep 2; pet 3 or 5 .... Portulacaceae, p. 109
      10 Sep and pet each 2, or each 3 (*Elatine*) .... Elatinaceae, p. 140
    9 Sep and pet each 4–6 (or in *Sabatia* up to 12).
      11 Lvs deeply palmately lobed (*Geranium*) .... Geraniaceae, p. 359
      11 Lvs entire to serrate.
        12 Style one.
          13 Hypanthium well developed, cup-shaped to tubular, bearing the sep and pet at or
             near its margin .... Lythraceae, p. 309
          13 Hypanthium none.
            14 Stamens alternate with the divisions of the cor .... Gentianaceae, p. 386
            14 Stamens opposite the divisions of the cor .... Primulaceae, p. 220
        12 Styles 2–5.
          15 Lvs glandular-serrate (*Bergia*) .... Elatinaceae, p. 140
          15 Lvs entire.
            16 Ovary and capsule 4–5-locular .... Linaceae, p. 345
            16 Ovary and capsule unilocular.
              17 Fls yellow (*Hypericum*) .... Clusiaceae, p. 141
              17 Fls white to pink, lilac, or red.
                18 Pet separate to the base .... Caryophyllaceae, p. 112
                18 Pet united at base (*Sabatia*) .... Gentianaceae, p. 386
  8 Lvs alternate or basal.
    19 Lvs shallowly to deeply *palmately* lobed.
      20 Fls irregular, one pet prolonged at base into a spur .... Violaceae, p. 157
      20 Fls regular.
        21 Fls with a conspicuously fringed corona as well as a cal and cor; climbing or trailing
          vines .... Passifloraceae, p. 163
        21 Fls without a corona; not vines.
          22 Style one; pet 6; lvs peltate (*Diphylleia*) .... Berberidaceae, p. 63
          22 Styles 2(3); pet 5; lvs not peltate .... Saxifragaceae, p. 231
    19 Lvs entire, serrate, crenate, or *pinnately* lobed.
      23 Styles 2 or more.
        24 Lvs all basal or nearly basal.
          25 Lvs glabrous or nearly so .... Plumbaginaceae, p. 139
          25 Lvs densely covered all around with conspicuous stipitate glands .... Droseraceae, p. 153
        24 Lvs cauline, not stipitate-glandular .... Linaceae, p. 345
      23 Style one or none.
        26 Hypanthium tubular, resembling a cal-tube, bearing the sep, pet, and stamens at or
          near its margin (*Lythrum*) .... Lythraceae, p. 309
        26 Hypanthium not developed, or very short and not tubular.
          27 Fls irregular, the pet dissimilar in size or shape or one pet prolonged into a spur
            .... Violaceae, p. 157
          27 Fls regular; spur lacking.
            28 Pet and sep each 4 (*Coronopus, Lepidium*) .... Brassicaceae, p. 175
            28 Pet and sep each 5.
              29 Lvs pinnately lobed (*Erodium*) .... Geraniaceae, p. 359
              29 Lvs entire or serrate.
                30 Fls terminal, or in a terminal infl.
                  31 Fls solitary, terminal (*Parnassia*) .... Saxifragaceae, p. 231
                  31 Fls in a terminal umbel or raceme (*Dodecatheon, Lysimachia*) .. Primulaceae, p. 220
                30 Fls axillary (*Cubelium*; a short spur is present but is easily overlooked)
                  .... Violaceae, p. 157

SECTION 17—HERBACEOUS DICOTYLEDONS WITH PERFECT FLOWERS; CAL AND COR BOTH PRESENT; FLOWERS REGULAR AND GAMOPETALOUS, WITH ONE SUPERIOR OVARY; STAMENS JUST AS MANY AS THE LOBES OF THE COR

26 Aquatic plants with floating lvs or with submersed dissected lvs.
  27 Lvs floating, broad, entire; fls arising apparently from the petiole just below its summit
     (*Nymphoides*) ....Menyanthaceae, p. 412
  27 Lvs submersed, dissected; fls in whorls on an emersed, floating, jointed axis (*Hottonia*)
     .... Primulaceae, p. 220
26 Land or marsh plants, without floating or submersed dissected lvs.
  28 Lvs basal; infl terminating a scape.
    29 Fls 4-merous, in spikes or heads; cor scarious .... Plantaginaceae, p. 459
    29 Fls 5-merous; cor of normal petaloid texture, not scarious.
      30 Style one.
        31 Fls in umbels (*Androsace, Dodecatheon, Primula*) .... Primulaceae, p. 220
        31 Fls in a slender spike (*Galax*) .... Diapensiaceae, p. 217
      30 Styles 5 or rarely 3 .... Plumbaginaceae, p. 139
  28 Lvs all or chiefly cauline.
    32 Ovary deeply lobed, appearing like 2 or 4 separate ovaries.
      33 Lvs opposite (*Mentha, Pycnanthemum*) .... Lamiaceae, p. 432
      33 Lvs alternate.
        34 Styles 2; ovary 2-lobed; stems creeping (*Dichondra*) .... Convolvulaceae, p. 406
        34 Style one; ovary 4-lobed .... Boraginaceae, p. 419
    32 Ovary one, not conspicuously lobed or divided.
      35 Lvs opposite or whorled (those of the infl sometimes alternate).
        36 Fls in dense heads or dense short spikes; cor 4-lobed or 4-divided.
          37 Cor scarious; lvs linear .... Plantaginaceae, p. 459
          37 Cor of normal petaloid texture; lvs broader than linear (*Phyla*) ...... Verbenaceae, p. 428
        36 Fls not in dense heads, but in some plants in short, crowded racemes; cor
          4–12-lobed or -divided.
          38 Stamens distinctly opposite the cor-lobes .... Primulaceae, p. 220
          38 Stamens alternate with the cor-lobes (in some plants inserted so near
            the base of the cor that the position is not readily ascertained).
            39 Stamens inserted in the sinuses between the cor-lobes; matted or
              prostrate half-shrubs, with the lvs less than 15 mm .... Diapensiaceae, p. 217
            39 Stamens inserted on the tube of the cor or at its base.
              40 Lvs connected at base by a stipular line.
                41 Cor 5-lobed .... Loganiaceae, p. 385
                41 Cor 4-lobed .... Buddlejaceae, p. 462
              40 Lvs with no trace of stipules.
                42 Lobes of the cor 5.
                  43 Stigmas 3 .... Polemoniaceae, p. 413
                  43 Stigma one, either capitate or 2-lobed.
                    44 Ovary unilocular; lvs entire .... Gentianaceae, p. 386
                    44 Ovary 2-locular or rarely 4-locular; lvs almost al-
                      ways ± dentate or angled .... Solanaceae, p. 400
                42 Lobes of the cor 4 or 6–12 .... Gentianaceae, p. 386
      35 Lvs alternate.
        45 Lvs conspicuously lobed, dissected, or compound (not merely hastate or
          sagittate).
          46 Twining vines, or sometimes merely trailing if a support is lacking.
            47 Cor funnelform (*Ipomoea, Quamoclit*) .... Convolvulaceae, p. 406
            47 Cor rotate (*Solanum*) .... Solanaceae, p. 400
          46 Erect or spreading or prostrate plants, never twining.
            48 Cor rotate or saucer-shaped.
              49 Anthers all, or all but one, connivent around the style; pet entire
                (*Lycopersicon, Solanum*) .... Solanaceae, p. 400
              49 Anthers separate; pet fringed (*Phacelia*) ....Hydrophyllaceae, p. 416
            48 Cor campanulate to funnelform, tubular, or salverform.
              50 Lvs 3-foliolate ....Menyanthaceae, p. 412
              50 Lvs variously lobed, dissected, or compound, but not 3-folio-
                late.
                51 Ovary 3-locular; stigmas or style-branches 3 .... Polemoniaceae, p. 413
                51 Ovary unilocular; stigmas or style-branches 2 ....Hydrophyllaceae, p. 416
        45 Lvs entire or toothed to only shallowly lobed.
          52 Lvs reduced to small scales; fls 4-merous (*Bartonia*) .... Gentianaceae, p. 386
          52 Lvs small or large, but not scale-like; fls mostly 5-merous.
           53 Fls or fl-clusters axillary or lateral.
            54 Fls sessile or nearly so, solitary in the axils of the lvs.
              55 Lvs densely pubescent; cor ca 1 cm wide (*Evolvulus*) .. Convolvulaceae, p. 406
              55 Lvs glabrous; cor ca 1 mm wide (*Centunculus*) .... Primulaceae, p. 220
             54 Fls pediceled and solitary, or in axillary or lateral clusters.
              56 Cor 5-lobed to below the middle of the limb (*Hydrolea*) .
                ....Hydrophyllaceae, p. 416
              56 Cor-limb entire, or merely angled, or very shallowly lobed.
                57 Stigmas 2–4 .... Convolvulaceae, p. 406
                57 Stigma one.

58 Twining vines.
   59 Fls with a large, funnelform or salverform cor
       and separate stamens .... Convolvulaceae, p. 406
   59 Fls with a rotate cor and connivent stamens
       (*Solanum*) .... Solanaceae, p. 400
58 Not twining; stems spreading or prostrate to erect
   .... Solanaceae, p. 400
53 Fls or fl clusters terminal.
  60 Fls solitary and terminal.
    61 Cor 15 mm or less; fls terminating short branches.
      62 Sep linear to narrowly lanceolate, acuminate; erect herb
        of western states (*Heliotropium*) .... Boraginaceae, p. 419
      62 Sep oblanceolate to obovate, blunt; matted half-shrubs
        of the coastal plain or northeastern mountains ... Diapensiaceae, p. 217
    61 Cor 7 cm or more; fls in the forks of the stem (*Datura*) .... Solanaceae, p. 400
  60 Fls several to many in terminal clusters.
    63 Infls paniculate or cymose.
      64 Styles 2 (*Hydrolea*) .... Hydrophyllaceae, p. 416
      64 Style one.
        65 Principal lvs up to 6 cm, linear or narrowly lan-
         ceolate (*Collomia*) .... Polemoniaceae, p. 413
        65 Principal lvs much longer than 6 cm and propor-
         tionately much wider (*Nicotiana*) .... Solanaceae, p. 400
    63 Infl a raceme, spike, or umbel.
      66 Cor rotate or saucer-shaped.
        67 Anthers separate; filaments, or some of them, hairy
         (*Verbascum*) .... Scrophulariaceae, p. 465
        67 Anthers connivent; filaments short and smooth
         (*Solanum*) .... Solanaceae, p. 400
      66 Cor campanulate to funnelform or salverform.
        68 Fls ca 3 cm wide, in a bracted secund apparent
         raceme or spike (*Hyoscyamus*) .... Solanaceae, p. 400
        68 Fls less than 1 cm wide.
         69 Fls 3–6 mm wide, in a bractless spike-like heli-
          coid cyme (*Heliotropium*) .... Boraginaceae, p. 419
         69 Fls ca 2 mm wide, in an open raceme (*Samolus*)
          .... Primulaceae, p. 220

SECTION 18—HERBACEOUS DICOTYLEDONS WITH PERFECT FLOWERS AND ONE SUPERIOR OVARY; CAL AND COR BOTH PRESENT; FLOWERS GAMOPETALOUS, EITHER DISTINCTLY IRREGULAR OR WITH STAMENS FEWER THAN THE LOBES OF THE COR

1 Anther-bearing stamens 5.
  2 Ovary deeply 4-lobed around the base of the central style (*Echium, Lycopsis*) .... Boraginaceae, p. 419
  2 Ovary not lobed.
    3 Cor rotate or saucer-shaped (*Verbascum*) .... Scrophulariaceae, p. 465
    3 Cor narrowly campanulate to funnelform (*Hyocyamus*) .... Solanaceae, p. 400
1 Anther-bearing stamens 2 or 4, rarely 3 (one or more sterile filaments or staminodes may also be
  present).
  4 Cor prolonged at base into a spur or sac.
    5 Cal deeply 5-parted.
      6 Lvs all or mostly cauline .... Scrophulariaceae, p. 465
      6 Lvs all basal; scapes one-fld (*Pinguicula*) .... Lentibulariaceae, p. 494
    5 Cal deeply 2-parted, apparently of 2 sep (*Utricularia*) .... Lentibulariaceae, p. 494
  4 Cor not spurred or saccate at base.
    7 Plants without green color, the lvs reduced to functionless scales .... Orobanchaceae, p. 488
    7 Plants with green color in the lvs and stems.
      8 Lvs alternate or basal.

9 Stamens 2.
   10 Plants with leafy stem or bracteate scape (*Besseya, Veronica*) ....Scrophulariaceae, p. 465
   10 Plants with lfless scapes bearing a slender or stout or interrupted spike .... Plantaginaceae, p. 459
9 Stamens 4.
   11 Fls solitary and terminal, or solitary in the axils.
      12 Fls solitary and terminal on lfless scapes or basal peduncles (*Limosella*) ......Scrophulariaceae, p. 465
      12 Fls axillary .... Pedaliaceae, p. 491
   11 Fls in terminal racemes or spikes, often conspicuously bracted.
      13 Cal split to the base on the lower median line; cor 3–5 cm long and wide .... Pedaliaceae, p. 491
      13 Cal 4-lobed, or divided vertically into lateral halves, or equally 5-parted; cor smaller
         ....Scrophulariaceae, p. 465
8 Lvs opposite or whorled.
   14 Ovary deeply 4-lobed, appearing like 4 separate ovaries around the base of the central
      style; plants usually with square stems and often aromatic when crushed .... Lamiaceae, p. 432
   14 Ovary not conspicuously 4-lobed; plants only seldom square-stemmed and aromatic.
      15 Stamens 2.
         16 Cor scarious .... Plantaginaceae, p. 459
         16 Cor petaloid in texture.
            17 Fls in terminal racemes or spikes, or solitary or paired in the axils of the lvs
               ....Scrophulariaceae, p. 465
            17 Fls in axillary racemes or spikes.
               18 Cor almost regular, with a 4-lobed limb ....Scrophulariaceae, p. 465
               18 Cor distinctly bilabiate (*Dicliptera, Justicia*) .... Acanthaceae, p. 490
      15 Stamens 4.
         19 Alternative key, based principally on technical characters.
            20 Ovary unilocular, with a single ovule, the fr an achene (*Phryma*) .... Verbenaceae, p. 428
            20 Ovary 2–4-locular, or with many ovules in a single locule.
               21 Fr splitting at maturity into 2–4 one-seeded nutlets.
                  22 Cor strongly bilabiate; fls not in axillary heads or spikes .... Lamiaceae, p. 432
                  22 Cor weakly bilabiate, or the fls in axillary heads or spikes .... Verbenaceae, p. 428
               21 Fr capsular.
                  23 Fr an elongate (1–2 dm) prominently 2-beaked capsule (*Proboscidea*)
                     .... Pedaliaceae, p. 491
                  23 Fr shorter (ca 1 cm or less), not conspicuously beaked.
                     24 Capsule elastically dehiscent to the base; seeds borne on hooked
                        projections from the placenta .... Acanthaceae, p. 490
                     24 Capsule not elastically dehiscent, seldom splitting to below the
                        middle; seeds not borne on hooks ....Scrophulariaceae, p. 465
         19 Alternative key, based principally on superficial characters.
            25 Cor distinctly and strongly bilabiate or very irregular.
               26 Fls in dense, peduncled, axillary heads or short-cylindric axillary spikes
                  (*Phyla*) .... Verbenaceae, p. 428
               26 Fls in racemes or panicles or in terminal spikes, or solitary in the axils.
                  27 Upper lip of the cor very small or apparently lacking (*Ajuga, Teu-
                     crium*) .... Lamiaceae, p. 432
                  27 Upper lip of the cor well developed.
                     28 Upper lip of the cor apparently composed of 4 lobes, the lower
                        lip of one lobe only (*Isanthus, Trichostema*) .... Lamiaceae, p. 432
                     28 Upper lip of 2 lobes (in some plants ± united), the lower lip
                        3-lobed.
                        29 Cal very irregular, the 3 upper lobes subulate, the 2 lower
                           short and broad (*Phryma*) .... Verbenaceae, p. 428
                        29 Cal regular or irregular, 2-lobed, 4-lobed, or 5-lobed, but
                           never with subulate upper lobes ....Scrophulariaceae, p. 465
            25 Cor nearly regular and about equally lobed, the upper lip sometimes slightly
               smaller than the lower, or the cor ± oblique across the limb.
               30 Cor salverform, with a slender tube of almost uniform diameter.
                  31 Stamens inserted near the middle of the cor-tube (*Verbena*) ...... Verbenaceae, p. 428
                  31 Stamens inserted near the base of the cor-tube (*Buchnera*) ......Scrophulariaceae, p. 465
               30 Cor funnelform or campanulate, with a broad tube widened upward.
                  32 Cal 5-lobed, split to the base along the lower median line; cor 3–5
                     cm long and wide .... Pedaliaceae, p. 491
                  32 Cal 4-lobed, or about equally 5-lobed, never split to the base on the
                     lower side only.
                     33 Cal-lobes 5, narrow or setaceous, much exceeding the short tube;
                        fls anthocyanic.
                        34 Stem roughly pubescent with retrorse hairs; lvs often with 1
                           or 2 small lobes near the base (*Tomanthera*) ......Scrophulariaceae, p. 465
                        34 Stems glabrous or minutely pubescent or villous; lvs never
                           lobed (*Ruellia*) .... Acanthaceae, p. 490
                     33 Cal-lobes 4, or short and relatively broad, or fls yellow .....Scrophulariaceae, p. 465

SYNOPTICAL ARRANGEMENT OF THE SUBCLASSES OF MAGNOLIOPSIDA

1 Plants relatively archaic, the flowers typically apocarpous, always polypetalous or apetalous (but sometimes synsepalous) and generally with an evident perianth, usually with numerous (sometimes laminar or ribbon-shaped) stamens initiated in centripetal sequence, the pollen-grains mostly binucleate and often uniaperturate or of uniaperturate-derived type; ovules bitegmic and crassinucellar; seeds very often with a tiny embryo and copious endosperm, but sometimes with a larger embryo and reduced or no endosperm; cotyledons occasionally more than 2; plants very often accumulating benzyl-isoquinoline or aporphine alkaloids, but without betalains, iridoid compounds, or mustard-oils, and only seldom strongly tanniferous .... I. Magnoliidae.
1 Plants more advanced in one or more respects than the Magnoliidae; pollen-grains triaperturate or of a triaperturate-derived type, cotyledons not more than 2; stamens not laminar, generally with well defined filament and anther; plants only rarely producing benzyl-isoquinoline or aporphine alkaloids, but often with other kinds of alkaloids, or tannins, or betalains, or mustard-oils, or iridoid compounds.
  2 Flowers more or less strongly reduced and often unisexual, the perianth poorly developed or wanting, the flowers often borne in catkins, but never forming bisexual pseudanthia, and never with numerous seeds on parietal placentas; pollen-grains often porate and with a granular rather than columellar infratectal structure, but also often of ordinary type .... II. Hamamelidae.
  2 Flowers usually more or less well developed and with an evident perianth, but if not so, then usually either grouped into bisexual pseudanthia or with numerous seeds on parietal placentas, only rarely with all the characters of the Hamamelidae, as listed above; pollen-grains of various architecture, but rarely if ever both porate and with a granular infrastructure.
    3 Flowers polypetalous or less often apetalous or sympetalous; if sympetalous then usually either with more stamens than corolla-lobes, or with bitegmic or crassinucellar ovules; ovules only rather seldom with an integumentary tapetum; carpels 1–many, distinct or more often united to form a compound pistil; plants often tanniferous or with betalains or mustard-oils.
      4 Stamens, when numerous, usually initiated in centrifugal (seldom centripetal) sequence; placentation various, often parietal or free-central or basal, but also often axile; species with few stamens and axile placentation usually either bearing several or many ovules per locule, or with a sympetalous corolla, or both.
        5 Plants usually either with betalains, or with free-central to basal placentation (in a compound ovary), or both, but lacking both mustard-oils and iridoid compounds, and tanniferous only in the two smaller orders; pollen-grains trinucleate or seldom binucleate; ovules bitegmic, crassinucellar, most often campylotropous or amphitropous; plants most commonly herbaceous or nearly so, the woody species usually with anomalous secondary growth or otherwise anomalous stem-structure; petals distinct or wanting, except in the Plumbaginales; in the largest order (Caryophyllales) the sieve-tubes with a unique sort of P-type plastid, and the seeds usually with perisperm instead of endosperm .. III. Caryophyllidae.
        5 Plants without betalains, and the placentation only rarely (except in the Primulales) free-central or basal; plants often with mustard-oils or iridoid compounds or tannins; pollen-grains usually binucleate (notable exception: Brassicaceae); ovules various, but seldom campylotropous or amphitropous except in the Capparales; plants variously woody or herbaceous, many species ordinary trees; petals distinct or less often connate to form a sympetalous corolla, seldom wanting; seeds only seldom with perisperm; plastids of the sieve-tubes usually of S-type, but in any case not as in the Caryophyllales .... IV. Dilleniidae.
      4 Stamens, when numerous, usually initiated in centripetal (seldom centrifugal) sequence; flowers seldom with parietal placentation (notable exception: many Saxifragaceae) and also seldom (except in parasitic species) with free-central or basal placentation in a unilocular, compound ovary, but very often (especially in species with few stamens) with 2–several locules that have only 1 or 2 ovules in each; flowers polypetalous or less often apetalous, only rarely sympetalous; plants often tanniferous, and sometimes with iridoid compounds, but only rarely with mustard-oils and never with betalains .... V. Rosidae.
    3 Flowers sympetalous (rarely polypetalous or apetalous); stamens generally isomerous with the corolla-lobes or fewer, never opposite the lobes; ovules unitegmic and tenuinucellar, very often with an integumentary tapetum; carpels most commonly 2, occasionally 3–5 or more; plants only seldom tanniferous, and never with betalains or mustard-oils, but often with iridoid compounds or various other sorts of repellants .... VI. Asteridae.

Synoptical Arrangement of the Orders and Families of Magnoliidae as Represented in our Flora

1 Plants with ethereal oil cells in the parenchymatous tissues; pollen uniaperturate or derived from a uniaperturate type, not triaperturate.
  2 Plants woody, ours trees or shrubs; fls with an evident perianth that may or may not be differentiated into sep and pet.

      3 Fls hypogynous; pollen in ours uniaperturate; nodes trilacunar or multilacunar .... Order MAGNOLIALES.
         4 Lvs with evident, deciduous stipules; nodes multilacunar .... Magnoliaceae.
         4 Lvs without stipules; nodes trilacunar .... Annonaceae.
      3 Fls perigynous; pollen inaperturate or biaperturate; nodes unilacunar, with 2 traces .... Order LAURALES.
         5 Lvs opposite; anthers opening by longitudinal slits; pistils several or many .... Calycanthaceae.
         5 Lvs alternate; anthers opening by uplifting valves; pistil solitary .... Lauraceae.
   2 Plants herbaceous, or if woody then forming vines.
      6 Perianth wanting; pollen uniaperturate; seeds with scanty endosperm and abundant perisperm
         .... Order PIPERALES, Saururaceae.
      6 Perianth of well developed, connate (often petaloid) sep; pollen inaperturate; seeds with abundant
         endosperm and no perisperm .... Order ARISTOLOCHIALES, Aristolochiaceae.
1 Plants without ethereal oil cells; pollen (except most Nymphaeales) of triaperturate or triaperturate-
    derived type.
   7 Plants aquatic, and lacking vessels in the shoot; placentation laminar or apical .... Order NYMPHAEALES.
      8 Plants rooted to the substrate; fls long-pedunculate and reaching or exserted from the surface of
        the water, perfect, with evident pet; carpels 2–many; ovules bitegmic, anatropous; lvs (except
        some Cabombaceae) alternate, long-petiolate, with floating or emergent, cordate or peltate
        blade.
         9 Carpels individually embedded in the enlarged, obconic receptacle; ovule 1; seeds with large
            embryo, no perisperm, and virtually no endosperm; pollen triaperturate .... Nelumbonaceae.
         9 Carpels not embedded in the receptacle; ovules 2–many; seeds with small embryo, some
            endosperm, and abundant perisperm; pollen uniaperturate or inaperturate.
           10 Plants acaulescent, the lvs all simple and arising directly from the rhizome, with a floating
              blade; pet 8–many; carpels ± firmly united into a compound, plurilocular ovary;
              ovules and seeds numerous .... Nymphaeaceae.
           10 Plants with long, slender, leafy, distally floating stems in addition to the rhizomes, with
              or without floating lf-blades, often some or all of the lvs submersed and dissected;
              pet 3; carpels distinct; ovules and seeds 2 or 3 .... Cabombaceae.
      8 Plants rootless, free-floating, submersed; fls inconspicuous, sessile, without pet; carpel 1; ovule
        solitary, unitegmic, orthotropous; lvs all sessile, whorled, dissected .... Ceratophyllaceae.
   7 Plants terrestrial or occasionally aquatic, generally with vessels in the shoot; placentation of diverse
    types, often marginal.
     11 Gynoecium mostly apocarpous or of a single carpel; sep usually more than 2; plants without
       protopine .... Order RANUNCULALES.
       12 Fls perfect; herbs or less often woody plants.
         13 Carpels usually 2 or more and distinct, seldom solitary or weakly united; stamens
            usually numerous, spirally arranged; anthers opening by longitudinal slits .... Ranunculaceae.
         13 Carpel solitary; stamens as many as and opposite the pet; anthers usually opening by
            2 uplifting valves .... Berberidaceae.
       12 Fls unisexual; twining woody vines.
         14 Lvs compound; embryo small, straight; endosperm copious; ovules numerous .... Lardizabalaceae.
         14 Lvs simple; embryo rather large, very often curved or even coiled; endosperm copious
            to often scanty or none; ovule solitary .... Menispermaceae.
     11 Gynoecium syncarpous, often of 2 carpels; sep mostly 2, sometimes 3; plants commonly
       producing protopine (an isoquinoline alkaloid) .... Order PAPAVERALES.
       15 Sep fully enclosing the bud; fls regular; stamens distinct, numerous; plants producing milky
         or colored latex, often in articulated laticifers .... Papaveraceae.
       15 Sep small, not enclosing the developing bud; fls strongly irregular; stamens 6, diadelphous;
         plants without latex or laticifers, but commonly with elongate secretory cells .... Fumariaceae.

Synoptical Arrangement of the Orders and Families of Hamamelidae as Represented in our Flora

1 Pistillate (or perfect) fls either producing more than one fr, or producing a single dehiscent fr .....
   .... Order HAMAMELIDALES.
   2 Carpels several, distinct; fr indehiscent; connective apically enlarged and peltate .... Platanaceae.
   2 Carpels 2, united; fr dehiscent; connective sometimes prolonged, but not enlarged and peltate ..... Hamamelidaceae.
1 Pistillate (or perfect) fls producing a single indehiscent fr .... Order URTICALES.
   3 Fls not in aments; ovary unilocular; cal often present, though small.
      4 Ovary with 2 styles; ovule apical, pendulous, anatropous or amphitropous.
         5 Plants with laticifers and milky juice; ours all woody .... Moraceae.
         5 Plants without laticifers and without milky juice.
           6 Woody plants; leaves entire or merely toothed .... Ulmaceae.
           6 Herbs (sometimes vines); lvs evidently lobed or palmately compound .... Cannabaceae.
      4 Ovary with a single style; ovule basal or nearly so, erect, orthotropous; ours herbs .... Urticaceae.
   3 Fls, or at least the staminate ones, in aments; ovary with 1–several locules; cal mostly wanting or
    very much reduced; woody plants.
      7 Ovule solitary, orthotropous or hemitropous; aromatic plants with resinous-dotted lvs.
        8 Lvs pinnately compound; ovary bilocular below and unilocular above, the ovule borne at the
        summit of the partial partition .... Order JUGLANDALES, Juglandaceae.

8 Lvs simple (though sometimes pinnately lobed); ovary fully unilocular, with a basal ovule .
.... Order MYRICALES, Myricaceae.
7 Ovules 2 or more, anatropous; plants not aromatic, or not strongly so, the lvs not resinous-
dotted; ovary 2- to several-locular at least below, often unilocular above .... Order FAGALES.
9 Carpels 3 or sometimes 6; pistillate fls not in aments; frs generally subtended or enclosed
individually or in small groups by a characteristic multibracteate cupule; pollen triaperturate .... Fagaceae.
9 Carpels 2; pistillate fls borne in aments; fr without a cupule, but often subtended or enclosed
by a foliaceous hull derived from 2 or 3 bracts; pollen 2–7-porate .... Betulaceae.

SYNOPTICAL ARRANGEMENT OF THE ORDERS AND FAMILIES OF
CARYOPHYLLIDAE AS REPRESENTED IN OUR FLORA

1 Ovules 1–many, campylotropous or amphitropous; seeds with a peripheral, usually curved or annular
embryo bordering or surrounding a ± abundant perisperm, or the perisperm sometimes scanty
or wanting; sieve-tubes with a characteristic type of plastid that has a subperipheral ring of
proteinaceous filaments; most families producing betalains but not anthocyanins; plants not
tanniferous; perianth, ovary, and placentation diverse, but the perianth unlike that of the next
two orders .... Order CARYOPHYLLALES.
2 Gynoecium either of 1–many distinct carpels, each with a single ovule, or of 2 or more carpels
united to form a compound ovary with as many locules and ovules as carpels.
3 Sep distinct, not petaloid; carpels several; infl racemose; lvs alternate .... Phytolaccaceae.
3 Sep united to form a distally lobed tube that commonly simulates a sympetalous cor and is
sometimes closely subtended by sepaloid bracts; carpel solitary; infl cymose or head-like;
lvs opposite .... Nyctaginaceae.
2 Gynoecium otherwise, always of 2 or more carpels united to form a compound ovary, the ovary
either unilocular or with as many locules as carpels, in the latter case with more than one ovule
per carpel.
4 Fls perigynous to epigynous; tep and stamens few and cyclic to often numerous; plants ±
succulent.
5 Unarmed lf-succulents; ovary superior to inferior, usually plurilocular .... Aizoaceae.
5 Spiny stem-succulents; ovary inferior, unilocular .... Cactaceae.
4 Fls hypogynous; tep and stamens mostly few and cyclic, but sometimes more numerous in
Portulacaceae; plants succulent or not.
6 Perianth monochlamydeous, commonly small and inconspicuous, or sometimes wanting;
ovules mostly solitary on a basal placenta in a unilocular ovary.
7 Perianth mostly green or greenish and ± herbaceous; filaments distinct .... Chenopodiaceae.
7 Perianth generally dry and scarious or membranous; filaments distinct or often connate
below .... Amaranthaceae.
6 Perianth generally dichlamydeous (except notably the Molluginaceae, with axile placentation
and usually several or many ovules).
8 Sep typically 2, but sometimes more numerous; stamens most commonly as many as and
opposite the pet, seldom more numerous; plants (as in the preceding 6 families) producing betalains but not anthocyanins .... Portulacaceae.
8 Sep 4 or 5; stamens up to twice as many as the pet (or sep), not at once of the same number
as and opposite the pet; plants producing anthocyanins but not betalains.
9 Ovary with 2–5 locules and axile placentation; pet small and inconspicuous or more
often wanting .... Molluginaceae.
9 Ovary unilocular, with central or basal placentation; pet usually well developed, but
sometimes small or wanting .... Caryophyllaceae.
1 Ovule solitary on a basal placenta, anatropous or orthotropous; seeds without perisperm, the straight
or curved, peripheral or embedded embryo commonly associated with ± abundant endosperm;
sieve-tube plastids without protein crystalloids; plants tanniferous and producing anthocyanins
but not betalains.
10 Perianth not clearly differentiated into cal and cor, the 2–6 tep either in a single whorl or more
often in two ± similar sets of (2)3 each; carpels (2)3; stamens 2–9, but only seldom 5, generally
in 2 or 3 whorls; ovule orthotropous; lvs often with conspicuous sheathing stipules ....
.... Order POLYGONALES, Polygonaceae.
10 Perianth differentiated into a pentamerous, gamosepalous cal and a pentamerous, generally gamopetalous cor; carpels 5; stamens 5, opposite the pet (or cor-lobes); ovules anatropous; lvs
exstipulate .... Order PLUMBAGINALES, Plumbaginaceae.

SYNOPTICAL ARRANGEMENT OF THE ORDERS AND FAMILIES OF
DILLENIIDAE AS REPRESENTED IN OUR FLORA

1 Fls mostly polypetalous, or sometimes apetalous, only seldom (Cucurbitaceae) sympetalous.
2 Plants insectivorous .... Order NEPENTHALES.

   3 Lvs modified to form pitchers; ovary plurilocular, with axile placentation; style solitary, expanded
      above and peltate or umbrella-like .... Sarraceniaceae.
   3 Lvs not forming pitchers; ovary unilocular, with parietal or basal placentation; styles 3, bipartite
      to the base .... Droseraceae.
  2 Plants not insectivorous.
    4 Fls much reduced, unisexul, without an evident perianth, borne in catkins; plants woody ....
       .... Order SALICALES, Salicaceae.
    4 Fls normally developed, with an evident perianth, not in catkins; plants woody or herbaceous.
     5 Placentation mostly axile, only seldom parietal.
       6 Sep imbricate; indument not stellate or lepidote; seeds lacking cyclopropenyl fatty acids ... Order THEALES.
        7 Lvs alternate .... Theaceae.
        7 Lvs opposite or whorled.
         8 Lvs stipulate; no secretory cavities or canals .... Elatinaceae.
         8 Lvs exstipulate; well developed secretory cavities or canals present in most or all
           organs .... Clusiaceae.
       6 Sep valvate; indument stellate or lepidote; seeds containing cyclopropenyl fatty acids ... Order MALVALES.
        9 Anthers tetrasporangiate and dithecal; epicalyx wanting; filaments distinct or connate in
         groups (or the anthers sessile); woody plants.
          10 Carpels firmly united in fl and fr; filaments distinct or often united into 5 antepetalous
            bundles .... Tiliaceae.
          10 Carpels only weakly united in fl, separating in fr; anthers in our genus sessile .... Sterculiaceae.
        9 Anthers bisporangiate and monothecal; fls very often with an epicalyx; filaments all
         connate into a tube; mostly herbs, a few woody plants .... Malvaceae.
     5 Placentation mostly parietal.
      11 Plants without mustard-oil and myrosin-cells; carpels most often 3 .... Order VIOLALES.
       12 Ovary superior.
        13 Fls hypogynous, and without a corona; not vines.
         14 Stamens mostly 8–50, if only 5, then the pet 3 .... Cistaceae.
         14 Stamens and pet each 4 or 5.
          15 Lvs stipulate, of normal proportions; style solitary; ours herbs .... Violaceae.
          15 Lvs exstipulate, much reduced and scale-like; styles distinct or merely
            connate at the base; woody plants .... Tamaricaceae.
        13 Fls evidently perigynous, and with an extrastaminal corona; often (including our
         spp.) vines .... Passifloraceae.
       12 Ovary inferior.
        16 Fls unisexual; stamens 2–5, often apparently 3; cor mostly sympetalous; mostly
         tendriliferous vines .... Cucurbitaceae.
        16 Fls perfect; stamens 10–numerous; cor of separate pet; ours not vines .... Loasaceae.
      11 Plants with mustard-oil and myrosin-cells .... Order CAPPARALES.
       17 Stigma 1, entire or slightly lobed, usually borne on a style; pet mostly 4; carpels 2;
        ovary closed, the ovules borne on a replum.
        18 Ovary and fr unilocular; stamens 4–many, but never tetradynamous; fls generally
         with an evident gynophore or androgynophore .... Capparaceae.
        18 Ovary and fr bilocular; stamens typically 6 and tetradynamous, rarely fewer; fls
         without a gynophore .... Brassicaceae.
       17 Stigmas as many as the 3–4 carpels, sessile and well separated around the distal margins
        of the open ovary; pet mostly (4)5–8 .... Resedaceae.
1 Fls mostly sympetalous (except mainly some Ericales, these with specialized embryological features).
  19 Fls with most or all of the features of the ericoid embryological syndrome (ovules unitegmic,
    tenuinucellar; endosperm-development cellular; endosperm-haustoria formed at both ends
    of the embryo-sac; testa a single layer of cells, or even wanting; style hollow, the cavity fluted
    in alignment with the locules of the ovary; anthers with a fibrous layer in the wall, becoming
    inverted during ontogeny, so that the morphological base appears to be apical, dehiscing by
    apparently terminal pores; anther-tapetum of the glandular type, with multinucleate cells;
    pollen in tetrads); filaments very often attached directly to the receptacle; plants strongly
    mycotrophic, and often producing iridoid compounds and other secondary metabolites of
    restricted occurrence .... Order ERICALES.
  20 Embryo normally developed, with 2 cotyledons; plants woody, always chlorophyllous.
    21 Pollen in monads; pet, when present, distinct or only very shortly connate at the base.
     22 Ovules 1–3 per locule; intrastaminal nectary-disk present; fr indehiscent; seed-coat
      none .... Cyrillaceae.
     22 Ovules numerous; nectary-disk wanting; fr a loculicidal capsule; seed-coat a single
      layer of cells .... Clethraceae.
    21 Pollen mostly in tetrads, but if in monads then the corolla evidently sympetalous; pet,
     when present, very often united to form a sympetalous cor.
     23 Perianth weakly or scarcely differentiated into cal and cor, consisting of 3–6 separate,
      distinct members arranged in 1 or 2 cycles; ovules solitary in each locule; many
      or all of the fls generally unisexual .... Empetraceae.
     23 Perianth clearly differentiated into cal and cor, usually with more than 3 (typically
      5) members in each of 2 cycles; the cor sympetalous or seldom polypetalous .... Ericaceae.
  20 Embryo very small and undifferentiated, without cotyledons; herbs or half-shrubs.
    24 Plants usually with green lvs; anthers opening by seemingly apical pores; pollen usually
     in tetrads; pet separate .... Pyrolaceae.

24 Plants without chlorophyll, the lvs reduced to scales; anthers opening by longitudinal slits, or sometimes by seemingly apical pores; pollen in monads; pet separate or united .... Monotropaceae.
19 Fls with less than half of the features of the ericoid embryological syndrome; filaments nearly always attached to the cor-tube; plants not strongly mycotrophic, without iridoid compounds, and without the special secondary metabolites of the Ericales.
25 Placentation mostly axile, less often parietal; functional stamens various, very often more numerous than the cor-lobes or alternate with them.
26 Perennial herbs or low half-shrubs; functional stamens as many as and alternate with the cor-lobes ....Order DIAPENSIALES, Diapensiaceae.
26 Trees or shrubs; functional stamens either more numerous than the cor-lobes or as many as the cor-lobes and then opposite them .... Order EBENALES.
27 Plants with a well developed latex-system; pubescence of 2-armed (malpighian) hairs .... Sapotaceae.
27 Plants without a latex-system; pubescence not of 2-armed hairs.
28 Fls unisexual; style deeply cleft .... Ebenaceae.
28 Fls perfect; style undivided.
29 Pubescence of stellate hairs or peltate scales; stamens all in a single series; anthers linear .... Styracaceae.
29 Pubescence of simple hairs or none; stamens grouped into 5 fascicles; anthers short and broad .... Symplocaceae.
25 Placentation free-central or basal in a unilocular ovary; functional stamens as many as and opposite the cor-lobes .... Order PRIMULALES, Primulaceae.

SYNOPTICAL ARRANGEMENT OF THE ORDERS AND FAMILIES OF ROSIDAE AS REPRESENTED IN OUR FLORA

1 Fls mostly of relatively primitive structure, commonly either apocarpous (with 1–many carpels), or with ± numerous stamens, often with both of these features; ovules solitary to very often several or many per carpel; plants never with internal phloem, never parasitic, and never highly modified aquatics.
2 Gynoecium mostly of 2–many carpels, only seldom (notably in the subfamily Prunoideae of the Rosaceae) of a single carpel, and then the plants not with the special features of the Fabales or Proteales .... Order ROSALES.
3 Seeds mostly with ± well developed endosperm.
4 Plants woody.
5 Lvs opposite .... Hydrangeaceae.
5 Lvs alternate .... Grossulariaceae.
4 Plants herbaceous.
6 Plants succulent; carpels as many as the pet, distinct or united only at the base; fls hypogynous or nearly so .... Crassulaceae.
6 Plants not succulent; carpels fewer than the pet, often 2, variously distinct or united; fls almost hypogynous in *Parnassia,* otherwise perigynous or partly epigynous .... Saxifragaceae.
3 Seeds mostly with very scanty or no endosperm (endosperm well developed in *Physocarpus*) .... Rosaceae.
2 Gynoecium of a single carpel.
7 Pet present; perianth most commonly 5-merous; stamens most commonly 10; lvs mostly compound, mostly stipulate and with a basal pulvinus; fr of various types, but most commonly dry and dehiscent along both sutures, with 2 or more seeds .... Order FABALES.
8 Fls hypogynous or nearly so; cor regular; pet mostly valvate, often cônnate below to form a tube; filaments often colored and long-exserted to form the conspicuous part of the infl; lvs bipinnately compound .... Mimosaceae.
8 Fls ± distinctly perigynous; cor usually ± strongly irregular; pet imbricate, distinct or only the 2 lower ones connate; filaments generally not forming the conspicuous part of the infl; lvs once or twice compound, seldom simple.
9 Cor not papilionaceous; adaxial (upper) pet borne internally to the lateral ones and often smaller than them; sep mostly distinct; filaments distinct or variously connate, but not forming a sheath around the pistil .... Caesalpiniaceae.
9 Cor mostly papilionaceous, the adaxial pet (banner) borne externally to the others and generally the largest, folded along the middle line so as to embrace the others in bud; 2 lateral pet (wings) similar inter se and mostly distinct; 2 lower pet innermost, similar inter se, mostly connate distally to form a keel enfolding the androecium and gynoecium; sep mostly connate below to form a tube (beyond the hypanthial base), stamens usually connate by their filaments to form a sheath around the pistil .... Fabaceae.
7 Pet wanting; cal 4-merous; stamens 4 or 8; lvs simple, exstipulate, without a pulvinus; fr drupelike .... Order PROTEALES, Elaeagnaceae.
1 Fls more advanced in one or another respect, mostly syncarpous, typically with not more than twice as many stamens as sep or pet, very often with only 1 or 2 ovules per carpel; plants sometimes (Myrtales) with internal phloem, sometimes (Santalales) parasitic, sometimes (Podostemales) highly modified aquatics.

10 Stem with internal phloem; fls usually strongly perigynous or epigynous, often tetramerous .... Order MYRTALES.
 11 Ovary superior.
    12 Ovules several or many per carpel; ovary 2–several-locular; fr capsular .... Lythraceae.
    12 Ovule solitary in the pseudomonomerous ovary; fr indehiscent .... Thymelaeaceae.
 11 Ovary half to more often fully inferior.
    13 Fr indehiscent, horned, 1-seeded; plants aquatic .... Trapaceae.
    13 Fr not horned, usually (including all our aquatic spp.) dehiscent and with several or many
        seeds.
       14 Anthers opening by longitudinal slits; connective without appendages; lvs pinnately
           veined .... Onagraceae.
       14 Anthers opening by terminal pores; connective appendaged near the base; lvs with
           3–several prominent longitudinal veins .... Melastomataceae.
10 Stem without internal phloem; fls variously hypogynous to perigynous or epigynous, often pen-
    tamerous.
 15 Plants parasitic or hemiparasitic; gynoecium very often unusual in the number or arrangement
     of ovules, which are often much reduced in structure .... Order SANTALALES.
    16 Plants terrestrial, partial root-parasites; ovules pendulous from the summit of the free-
        central placenta .... Santalaceae.
    16 Plants aerial, attached to the branches of their host; ovules embedded in and scarcely
        differentiated from the placenta .... Viscaceae.
 15 Plants fully autotrophic.
    17 Fls generally ± reduced (not very much so in some Euphorbiaceae) and often unisexual,
        the perianth mostly poorly developed or wanting; styles generally distinct or only
        basally connate.
       18 Endosperm wanting; fls perfect, hypogynous or nude, with numerous ovules; highly
           modified aquatics ....Order PODOSTEMALES, Podostemonaceae.
       18 Endosperm well developed; fls with only 1 or 2 ovules per locule, otherwise very
           diverse in structure; habit various.
          19 Fls epigynous, not grouped into pseudanthia; plants mostly aquatic, very often
              with dissected lvs ....Order HALORAGALES, Haloragaceae.
          19 Fls hypogynous or naked, often grouped into pseudanthia; usually not aqua-
              tic ....:....................... Order EUPHORBIALES.
             20 Ovules apotropous, the raphe dorsal; styles borne on or forming continuations
                 of the apical margin of the ovary .... Buxaceae.
             20 Ovules epitropous, the raphe ventral; ovary with a single deeply cleft common
                 style, or with the styles or stigmas clustered at the top .... Euphorbiaceae.
    17 Fls of ordinary type, not much reduced, very often with a nectary-disk.
       21 Lvs mostly simple and entire or merely toothed, only seldom (as in many Vitaceae)
           compound.
          22 Fls epigynous .... Order CORNALES, Cornaceae.
          22 Fls hypogynous, or sometimes half-epigynous or somewhat perigynous.
             23 Fls regular or nearly so; anthers opening by longitudinal slits.
                24 Stamens distinct; fls usually with a solitary (sometimes deeply cleft) style,
                    or the stigma(s) sessile.
                   25 Stamens alternate with the pet .... Order CELASTRALES.
                      26 Ovules erect and basal-axile .... Celastraceae.
                      26 Ovules pendulous, apical-axile .... Aquifoliaceae.
                   25 Stamens opposite the pet .... Order RHAMNALES.
                      27 Fls perigynous; fr drupaceous; ovules solitary in each locule .... Rhamnaceae.
                      27 Fls hypogynous; fr baccate; ovules paired in each locule .... Vitaceae.
                24 Stamens connate by their filaments toward the base; styles distinct or
                    sometimes connate below .... Order LINALES, Linaceae.
             23 Fls ± strongly irregular, and commonly with poricidal anthers ....
                .... Order POLYGALALES, Polygalaceae.
       21 Lvs (except Balsaminaceae) mostly compound or conspicuously lobed or cleft.
          28 Ovary superior.
             29 Plants woody .... Order SAPINDALES.
                30 Lvs stipulate; ovules often more than 2 per locule.
                   21 Stamens bicyclic or tricyclic; ovules epitropous; lvs without a ter-
                       minal lfl .... Zygophyllaceae.
                   31 Stamens unicyclic; ovules apotropous; lvs with a terminal lfl .... Staphyleaceae.
                30 Lvs exstipulate; ovules mostly 1 or 2 per locule.
                   32 Disk mostly extrastaminal or wanting; ovules apotropous; stamens
                       very often 8.
                      33 Lvs alternate .... Sapindaceae.
                      33 Lvs opposite.
                         34 Fls evidently irregular; fr a (2)3(4)-carpellate, usually 1-seed-
                             ed capsule; lvs palmately compound .... Hippocastanaceae.
                         34 Fls regular; fr a double samara; lvs in most spp. simple and
                             palmately lobed .... Aceraceae.
                   32 Disk mostly intrastaminal, annular or sometimes modified into a
                       gynophore; stamens only seldom 8.

35 Ovules apotropous, solitary in the single locule; plants strongly resinous .... Anacardiaceae.
35 Ovules epitropous; locules 2 or more; plants not strongly resinous.
    36 Lvs glandular punctate; parenchymatous tissues with scattered secretory cavities containing aromatic ethereal oils .... Rutaceae.
    36 Lvs not glandular-punctate; plants often with scattered secretory cells, but without secretory cavities.
        37 Stamens connate by their filaments; seeds with well developed endosperm .... Meliaceae.
        37 Stamens distinct; seeds without endosperm .... Simaroubaceae.
29 Plants herbaceous .... Order GERANIALES.
    38 Fls apparently regular and not obviously spurred; stamens mostly twice as many as the pet, sometimes some of them staminodial; lvs compound or deeply cleft.
    39 Ovary with distinct terminal styles; fr a loculicidal capsule .... Oxalidaceae.
    39 Ovary with a single gynobasic style; fr of 1-seeded mericarps.
        40 Ovules apotropous; plants producing mustard-oil .... Limnanthaceae.
        40 Ovules epitropous; plants without mustard-oil .... Geraniaceae.
    38 Fls strongly irregular, one of the sep petaloid and with a conspicuous free spur; stamens 5; lvs simple, merely toothed .... Balsaminaceae.
28 Ovary inferior .... Order ARALIALES.
    41 Carpels 2–5; fr a berry or berry-like drupe; fls commonly in umbels or heads that are often grouped into various sorts of compound infls, but not forming regular compound umbels .... Araliaceae.
    41 Carpels 2; fr a dry schizocarp; fls very often in compound umbels .... Apiaceae.

SYNOPTICAL ARRANGEMENT OF THE ORDERS AND FAMILIES OF ASTERIDAE AS REPRESENTED IN OUR FLORA

1 Fls much reduced, nearly or quite without a perianth, either unisexual or perfect and epigynous; stamen solitary; plants mostly aquatic .... Order CALLITRICHALES.
  2 Gynoecium seemingly of a single carpel, the ovary unilocular, with a single style and a single terminal ovule; fls perfect and epigynous, or some or all of them unisexual; fr an achene or drupelet .... Hippuridaceae.
  2 Gynoecium bicarpellate, with 2 styles or a deeply bifid style; ovary compartmented into 4 locelli, each with a single pendulous, axile ovule; fls unisexual; fr of 4 separating nutlets .... Callitrichaceae.
1 Fls usually with a ± well developed perianth (typically a sympetalous cor and a cal or pappus), but if not, then differing in other respects from the foregoing group; plants terrestrial or seldom aquatic.
  3 Ovary superior, or less than half inferior.
    4 Plants nearly always with opposite or whorled lvs, and nearly always with internal phloem; fls mostly regular or nearly so and with as many stamens as cor-lobes .... Order GENTIANALES.
      5 Ovary one, the carpels fully united except for the often distinct or lobed stigmas, or seldom the style more deeply cleft; style not notably thickened and modified distally; plants without a latex system, and without cardiotonic glycosides.
        6 Lvs with interpetiolar stipules, or these sometimes reduced to mere interpetiolar lines; ovary bilocular, with axile placentas .... Loganiaceae.
        6 Lvs exstipulate; ovary unilocular, with parietal (often intruded) placentas .... Gentianaceae.
      5 Ovaries 2, the carpels united only by the style or style-head; style thickened and modified at the tip; plants with a well developed latex-system, and commonly producing cardiotonic glycosides.
        7 Androecium without translators, the pollen not forming pollinia; carpels united by part or all of the style below the thickened style-head .... Apocynaceae.
        7 Androecium provided with translators, the pollen coherent to form pollinia; fls with a corona; carpels united only by the thickened style-head .... Asclepiadaceae.
    4 Plants only rarely at once with opposite (or whorled) lvs and internal phloem, and then with irregular fls that have fewer stamens than cor-lobes.
      8 Ovary generally consisting of 2 biovulate carpels, each with 2 uniovulate lobes or locelli, rarely (as in *Phryma* of the Verbenaceae) more reduced; fr typically consisting of separate, closed half-carpels (nutlets); plants without internal phloem .... Order LAMIALES.
        9 Lvs mostly alternate; fls mostly regular or nearly so and with as many stamens as cor-lobes; stems not square .... Boraginaceae.
        9 Lvs opposite; fls ± irregular and with 2–4 stamens; young stems commonly square.
          10 Style terminal or nearly so, the ovary only shallowly or not at all lobed at the tip; plants seldom aromatic .... Verbenaceae.
          10 Style commonly gynobasic (as in most Boraginaceae), uniting the otherwise essentially distinct lobes of the ovary, or seldom the ovary lobed only part way (one-third or more) to the base; plants commonly aromatic .... Lamiaceae.

8 Ovary consisting of 2–several carpels with 2–many ovules each, but the carpels only rarely
   divided into uniovulate segments; frs diverse, most commonly capsular or baccate, not
   consisting of half-carpellary nutlets.

  11 Cor scarious, persistent, regular; fls mostly anemophilous, generally tetramerous as to cal,
   cor, and androecium; lvs phyllodial, ± parallel-veined, very often all basal . . . .
   . . . . Order PLANTAGINALES, Plantaginaceae.

  11 Cor otherwise; fls mostly entomophilous or ornithophilous, variously pentamerous or
   tetramerous or otherwise, with isomerous or anisomerous stamens; lvs various in
   form and structure, sometimes much reduced, but not phyllodial and parallel-veined,
   and only seldom all basal. . . . . Order SOLANALES.

    12 Fls regular or nearly so and with as many functional stamens as cor-lobes, typically
   pentamerous; plants very often producing alkaloids, but only seldom iridoid
   compounds, and without orobanchin.

      13 Twining stem-parasites, not rooted in the ground at maturity, nearly or quite
   without chlorophyll; embryo without well differentiated cotyledons . . . . Cuscutaceae.

      13 Autotrophic plants, rooted in the ground; embryo with evident cotyledons.

        14 Plants aquatic, and producing iridoid compounds; stem often with scattered
   vascular bundles; ovules numerous on parietal placentas . . . . Menyanthaceae.

        14 Plants terrestrial (sometimes of wet places) without iridoid compounds; other
   characters not combined as in Menyanthaceae.

          15 Stem with internal phloem; plants commonly with alkaloids; placentation
   axile to basal.

            16 Ovules ± numerous; carpels, when (as usually) 2, obliquely oriented,
   neither median nor collateral; plants without latex; cotyledons
   not plicate; style simple, the stigma only shortly or scarcely lobed
   . . . . Solanaceae.

            16 Ovules 2 per carpel; carpels, when (as usually) 2, median (one anterior,
   the other posterior); plants with latex-canals or latex-cells; cotyledons plicate; style simple or often ± deeply cleft, or the styles
   distinct . . . . Convolvulaceae.

          15 Stem without internal phloem; plants without alkaloids.

            17 Carpels 3; stigmas 3; placentation axile . . . . Polemoniaceae.

            17 Carpels 2; stigmas 2 or only one; placentation parietal, seldom axile
   . . . . Hydrophyllaceae.

    12 Fls mostly irregular and with fewer functional stamens than cor-lobes, or sometimes
   with regular, tetramerous cor and 2 or 4 stamens; plants commonly producing
   orobanchin and iridoid compounds, but only seldom alkaloids . . . . . . Order SCROPHULARIALES.

      18 Insectivorous herbs, often aquatic; placentation free-central . . . . Lentibulariaceae.

      18 Plants not insectivorous, and only occasionally aquatic; placentation various, but
   not free-central.

        19 Cor mostly 4-lobed and regular or nearly so (wanting in some Oleaceae);
   mostly woody plants with opposite or whorled lvs, rarely herbs.

          20 Stamens 4; ovules ± numerous in each locule . . . . Buddlejaceae.

          20 Stamens 2(–4); ovules 2 in each locule (more in *Forsythia*) . . . . Oleaceae.

        19 Cor mostly 5-lobed and/or ± strongly irregular.

          21 Plants parasitic, without chlorophyll; embryo minute, undifferentiated;
   lvs reduced and alternate; placentation parietal . . . . Orobanchaceae.

          21 Plants partly or fully autotrophic, producing chlorophyll; embryo with 2
   cotyledons; lvs ± well developed, opposite or alternate; placentation
   axile.

            22 Herbs.

              23 Fr explosively dehiscent, the seeds with an enlarged and spe-
   cialized funiculus that serves as a jaculator . . . . Acanthaceae.

              23 Fr indehiscent or dehiscent, but not explosively so, the funiculus
   of ordinary type.

                24 Plants with specialized mucilaginous hairs, the herbage com-
   monly slimy; seeds with scanty or no endosperm . . . . Pedaliaceae.

                24 Plants without specialized mucilaginous hairs; seeds with ±
   well developed endosperm . . . . Scrophulariaceae.

            22 Trees or woody vines . . . . Bignoniaceae.

3 Ovary inferior, or at least half inferior.

  25 Fls borne in involucrate, centripetally flowering heads; anthers connate (or connivent) into a
   tube around the style, which pushes out the pollen; ovary unilocular (though bicarpellate),
   with a solitary basal ovule . . . . Order ASTERALES, Asteraceae.

  25 Fls borne in various sorts of infls, but if in involucrate heads then the heads basically cymose
   in structure; fls with or often without a specialized pollen-presentation mechanism similar
   to that of the Asterales; ovary with 1–several locules and 1–many ovules in each locule
   (or some locules empty).

    26 Lvs alternate; stamens free from the cor, or attached at the base of the tube; fls with a ±
   well developed pollen-presentation mechanism resembling that of the Asterales; herbs,
   characteristically storing carbohydrate as inulin . . . . Order CAMPANULALES, Campanulaceae.

    26 Lvs opposite or whorled; stamens attached to the cor-tube, usually well above the base; fls
   without the sort of pollen-presentation mechanism described above; plants variously
   woody or herbaceous, without inulin.

27 Stipules usually present and interpetiolar, bearing colleters on the inner surface (some-
    times reduced to interpetiolar lines or enlarged into lvs); cor regular and with
    isomerous stamens .... Order RUBIALES, Rubiaceae.
27 Stipules typically none, when present usually small and adnate (at least basally) to the
    petiole, in any case without colleters; cor regular or more often irregular, the
    stamens as many as or often fewer or more numerous than the cor-lobes ..... Order DIPSACALES.
    28 Plants mostly woody, seldom herbaceous; stamens mostly as many as the cor-
        lobes, seldom fewer; ovules often more than one per locule .... Caprifoliaceae.
    28 Plants herbaceous; stamens seldom of the same number as the cor-lobes; ovules
        not more than one per locule.
        29 Stamens twice as many as the cor-lobes, paired at the sinuses, each with only
            a single pollen-sac .... Adoxaceae.
        29 Stamens as many as or fewer than the cor-lobes, of normal structure, each
            with 2 pollen-sacs.
            30 Fls individually enclosed or subtended by a ± cupulate epicalyx or in-
                volucel, borne in compact, involucrate, basically cymose heads; ovary
                basically bicarpellate, but pseudomonomerous and strictly unilocular
                .... Dipsacaceae.
            30 Fls without an epicalyx or involucel, borne in various sorts of infls, but
                not in involucrate heads; ovary basically tricarpellate, but two of the
                carpels empty and often vestigial .... Valerianaceae.

Synoptical Arrangement of the Subclasses, Orders, and Families of LILIOPSIDA, as Represented in our Flora

1 Plants either with apocarpous (sometimes monocarpous) fls, or ± aquatic, or very often both, always
    herbaceous, but never thalloid; vascular system generally not strongly lignified, often much re-
    duced, the vessels confined to the roots, or wanting; endosperm wanting; pollen trinucleate ...
    .... Subclass ALISMATIDAE.
    2 Perianth differentiated into evident sep and pet; fls often bracteate.
        3 Fls hypogynous; carpels distinct or only basally connate .... Order ALISMATALES.
            4 Ovules numerous, scattered over the inner surface of the carpel; pollen-grains monosulcate;
                embryo straight; lvs linear, not differentiated into blade and petiole .... Butomaceae.
            4 Ovule solitary, ventral-basal; pollen-grains pantoporate; embryo horseshoe-shaped; lvs typi-
                cally but not always with an expanded blade .... Alismataceae.
        3 Fls epigynous, with a compound ovary and parietal or nearly basal or a modified sort of laminar
            placentation .... Order HYDROCHARITALES, Hydrocharitaceae.
    2 Perianth, when present, not differentiated into sep and pet; bracts wanting, except in Scheuchzeria-
        ceae .... Order NAJADALES.
        5 Emergent marsh-plants with a terminal spike or raceme.
            6 Ovules 2, basal-marginal; carpels essentially distinct; infl with a bract subtending each pedicel
                .... Scheuchzeriaceae.
            6 Ovules solitary in each carpel, the carpels attached along their inner margin to an erect, elongate
                axis; infl bractless .... Juncaginaceae.
        5 Plants immersed or with floating lvs, only the infl often emergent.
            7 Pollen-grains globose or isobilateral; plants of fresh or alkaline or brackish water (*Ruppia*
                seldom marine).
                8 Fls perfect.
                    9 Tep 4; stamens 4; fruiting carpels sessile; ovule ventromarginal, generally near the base;
                        pollen-grains globose .... Potamogetonaceae.
                    9 Tep none; stamens 2; fruiting carpels long-stipitate; ovule pendulous from the apex;
                        pollen-grains of a unique isobilateral type .... Ruppiaceae.
                8 Fls unisexual.
                    10 Pistil solitary, forming a unilocular ovary surmounted by 2–4 elongate stigmas; ovule
                        basal, erect .... Najadaceae.
                    10 Pistils several, each with a style and stigma; ovule ventral-apical, pendulous .... Zannichelliaceae.
            7 Pollen-grains thread-like; plants marine; stamen solitary; gynoecium a single unilocular, bi-
                carpellate pistil .... Zosteraceae.
1 Plants with syncarpous (seldom pseudomonomerous) fls, usually terrestrial (or epiphytic), much less
    often aquatic, although sometimes not only aquatic but also thalloid and free-floating; vascular
    system, endosperm, and pollen various, but not combined as in the Alismatidae.
    11 Small, free-floating, thalloid aquatics, the minute fls rarely produced and only 2–4 in an infl.
        (Family extracted in this synopsis from the Order Arales in the subclass Arecidae.) .... Lemnaceae.
    11 Plants neither free-floating nor thalloid; fls either larger or more numerous.
        12 Infl (in our order) a spadix .... Subclass ARECIDAE, order ARALES.
            13 Lvs ensiform, unifacial; ovary 2–3-locular; anthers introrse .... Acoraceae.
            13 Lvs with a ± expanded, bifacial blade; ovary (in ours) unilocular; anthers extrorse .... Araceae.
        12 Infl otherwise.
            14 Nectar and nectaries mostly wanting (sometimes present at the tips of the tep); perianth
                in the more archaic families trimerous and with well differentiated sep and pet, in

the more advanced families reduced and chaffy and often not obviously trimerous, or wanting, the families with reduced perianth typically adapted to wind-pollination; ovary always superior (or nude); vessels generally present in all vegetative organs; endosperm largely or wholly starchy, commonly mealy and with compound starch-grains . . . . Subclass COMMELINIDAE.

15 Fls perfect, and with ± showy pet that are well differentiated from the sep, adapted to pollination by insects . . . . Order COMMELINALES.

  16 Lvs mostly or all basal, with an open sheath often not well differentiated from the blade; infl a terminal head or dense spike with the fls sessile in the axils of firm, spirally arranged, closely imbricate bracts . . . . Xyridaceae.

  16 Lvs well distributed along the stem, with closed sheath; fls pedicellate in more open infls . . . . Commelinaceae.

15 Fls perfect or unisexual, without showy pet, the perianth (when present) sometimes in 2 series, but dry and chaffy; fls (except some Eriocaulaceae) adapted to wind-pollination or self-pollination.

  17 Fls aggregated into dense, pseudanthial, involucrate heads; lvs all basal, without a well differentiated sheath; ovary plurilocular, with 1 ovule per locule . . . . Order ERIOCAULALES, Eriocaulaceae.

  17 Fls in various sorts of infls, but not in pseudanthial heads; lvs variously cauline or basal, but in either case with a well developed sheath; ovary mostly either unilocular or with more than one ovule per locule (or both).

    18 Fr capsular; ovary with 1–3 locules and 3–many ovules; fls with an evident (though small) biseriate, chaffy perianth . . . . Order JUNCALES, Juncaceae.

    18 Fr indehiscent; ovary unilocular, with a solitary ovule, or rarely plurilocular with one ovule per locule; perianth present or not.

      19 Ovule(s) anatropous, pendulous from near the summit of the ovary; plants aquatic or semiaquatic . . . . Order TYPHALES.

        20 Infl of dense, globose heads; perianth of pistillate fls of 2–6 small tep; frs sessile or nearly so, distributed by water or animals . . . . Sparganiaceae.

        20 Infl a dense, elongate, cylindric spike; perianth of pistillate fls of numerous capillary bristles; frs long-stipitate, distributed by wind . . . . Typhaceae.

      19 Ovules at once anatropous and pendulous from near the summit of the ovary; plants terrestrial to occasionally semiaquatic . . . . Order CYPERALES.

        21 Fls spirally or less often distichously arranged on the axis of the spike or spikelet, usually each fl seemingly or actually subtended by only a single scale, without an evident scale between the fl and the axis; seed-coat generally free from the pericarp; lf-sheath usually closed; stem usually solid, very often triangular; carpels 3 or less often 2; embryo embedded in the endosperm; pollen-grains borne in pseudomonads . . . . Cyperaceae.

        21 Fls distichously arranged on the axis of the spikelet (or only one per spikelet), each fl ordinarily subtended by a pair of scales (lemma and palea), the palea inserted between the fl and the axis; seed-coat usually adnate to the pericarp; lf-sheath usually open; stem usually hollow, never triangular; carpels 2, seldom 3; embryo peripheral to the endosperm; pollen-grains in true monads . . . . Poaceae.

14 Nectar and nectaries (often septal nectaries) generally present; perianth generally well developed, not reduced and chaffy, the fls typically adapted to pollination by insects or other animals; ovary superior to inferior; vessels most often confined to the roots.

22 Sep usually well differentiated from the pet; endosperm typically starchy and mealy, with compound starch-grains; ours rootless, pendulous epiphytes with filiform lvs . . . . Subclass ZINGIBERIDAE, Order BROMELIALES, Bromeliaceae.

22 Sep usually petaloid in form and texture, only rarely green and herbaceous; endosperm when present typically very hard, with reserves of hemicellulose, protein, and oil, less commonly starch, but not mealy in texture, the starch-grains when present typically simple rather than compound; not pendulous, rootless epiphytes . . . . Subclass LILIIDAE.

  23 Seeds of ordinary number and structure, with well developed embryo and endosperm; plants not obviously mycotrophic; septal nectaries very often present; ovary superior or inferior . . . . Order LILIALES.

    24 Food-reserves of the seed consisting mainly or wholly of starch; endosperm never very hard; stomates paracytic, varying to seldom almost anomocytic.

      25 Aquatic or semiaquatic plants; lvs mostly with a distinct (sometimes inflated) petiole and an expanded, bifacial blade with curved-convergent main veins and evident cross-veins; infl glabrous, or at least not conspicuously long-hairy . . . . Pontederiaceae.

      25 Terrestrial geophytes; lvs mostly linear and parallel-veined, with an equitant, unifacial blade; infl conspicuously long-hairy . . . . Haemodoraceae.

    24 Food-reserves of the seed consisting mainly or wholly of hemicellulose, protein, and oil, with little or no starch; endosperm usually very hard; stomates most commonly anomocytic.

      26 Lvs mostly narrow, parallel-veined, and without a distinct petiole, the blade sessile or with a basal sheath, sometimes broader and even

net-veined (as in *Trillium*), but only seldom with a broad, net-veined
blade on a distinct petiole.

    27 Habit lilioid, i.e., the plants geophytes with soft, annual (rarely pe-
rennial) lvs, the stem herbaceous and without secondary growth.

        28 Stamens as many as the tep (typically 6); ovary superior or less
often inferior .... Liliaceae.

        28 Stamens 3, opposite the outer tep; ovary inferior .... Iridaceae.

    27 Habit agavoid or yuccoid, i.e., the plants coarse, often shrubby or
arborescent xerophytes with firm or succulent, mostly perennial
lvs, the stem commonly with secondary growth of monocotyledonous type .... Agavaceae.

  26 Lvs mostly with a well defined petiole and a broad, net-veined blade.

    29 Ovary superior .... Smilacaceae.

    29 Ovary inferior .... Dioscoreaceae.

23 Seeds very numerous and tiny, with minute, usually undifferentiated embryo and
very little or no endosperm; plants strongly mycotrophic, sometimes without
chlorophyll; nectaries diverse, but only seldom septal; ovary inferior .... Order ORCHIDALES.

    30 Stamens 3 or 6, symmetrically arranged and free from the style; pollen-grains
not cohering in pollinia .... Burmanniaceae.

    30 Stamens 1 or 2, adnate to the style to form a column; pollen grains ± coherent
in pollinia .... Orchidaceae.

DIVISION LYCOPODIOPHYTA

Vascular cryptograms; perennial herbs; stem unbranched or more often dichotomously to monopodially branched, but without axillary buds, essentially protostelic, but the stele sometimes dissected; lvs mostly alternate or opposite, seldom in pseudowhorls, narrow or very small or both, with a single unbranched midvein; sporangia axillary to the sporophylls, or adnate to the adaxial base of the sporophyll; sporophylls resembling vegetative lvs or more often ± modified, often aggregated into a terminal strobilus (cone); gametophyte tiny, thalloid, photosynthetic or mycorhizal, or depending on its stored food. Two classes.

CLASS *LYCOPODIOPSIDA*

Lvs eligulate, small, in ours never as much as 2 cm; spores numerous in each sporangium, all alike; sperms biflagellate. A single order, Lycopodiales.

FAMILY **LYCOPODIACEAE**, the Clubmoss Family

The only family of the order. Two genera, the other (*Phylloglossum*) an Australian monotype with clustered basal lvs and a solitary cone on a naked scape.

1. LYCOPODIUM L. Clubmoss. Leafy-stemmed perennial herbs, often evergreen. 400, cosmop. (*Diphasium, Huperzia, Lepidotis, Lycopodiella, Pseudolycopodiella.*)

Species in a given section or subgenus are cytologically similar and commonly form ± sterile but vegetatively vigorous hybrids that often reproduce effectively by rhizomes or gemmae, sometimes forming considerable populations. Some of these hybrids are: *L.* × *brucei* (Cranfill) Lellinger = *L. appressum* × *L. prostratum*; *L.* × *buttersii* Abbe = *L. lucidulum* × *L. selago*; *L.* × *copelandii* Eiger = *L. alopecuroides* × *L. appressum*; *L.* × *habereri* House = *L. digitatum* × *L. tristachyum*; *L.* × *issleri* (Rouy) Lawalrée = *L. alpinum* × *L. complanatum*; *L.* × *zeilleri* (Rouy) Beitel = *L. complanatum* × *L. tristachyum.*

1 Sporophylls borne on the stem in zones alternating with zones of sterile lvs, or all the lvs becoming fertile; no terminal cones (*Huperzia, Selago*).
  2 Lvs with stomates on the upper (as well as the lower) surface, erect or spreading, equal or only slightly unequal, mostly 3–8 × 0.6–1.2 mm, broadest near or below the middle.
    3 Lvs erect or spreading, equal, mostly 3–5 × 0.6–0.8 mm, broadest at or near the base, entire; northern, extending s. only in the higher mts. . . . . 1. *L. selago.*
    3 Lvs spreading, slightly unequal, mostly 5–8 × 0.8–1.2 mm, broadest near the middle and obscurely toothed above; Pa. to Minn., s. to N.C., Ala., and Mo. 2. *L. porophilum.*
  2 Lvs without stomates on the upper surface, spreading, zonally unequal so that the shoot looks ragged, the larger ones mostly 7–12 × 1.2–2+ mm, usually broadest above the middle and toothed above the middle; widespread with us . . . . 3. *L. lucidulum.*
1 Sporophylls aggregated into evident terminal cones.
  4 Mature sporophylls green, similar to the vegetative lvs in shape; sporangia subglobose (*Lepidotis, Lycopodiella*).
    5 Lvs and sporophylls evidently toothed or ciliate; mainly coastal plain.
      6 Lateral stems proximally arching, only distally prostrate and rooting . . . . 4. *L. alopecuroides.*
      6 Lateral stems wholly prostrate, rooting distally . . . . 5. *L. prostratum.*
    5 Lvs (at least of the erect stem) entire or nearly so, or the sporophylls sometimes with a single pair of teeth near the widened base.
      7 Plants dwarf, up to ca 1 dm; lvs ± spreading; boreal and montane . . . . 6. *L. inundatum.*
      7 Plants taller, mostly 1–3.5 dm; lvs on erect stems ± appressed; mainly coastal plain . . . . . 7. *L. appressum.*
  4 Mature sporophylls yellowish, broader and shorter than the lvs; sporangia (except in no. 8) reniform.
    8 Lvs in 6–10 ranks, entire to ciliate or denticulate.
      9 Lvs merely acute or tipped with a short, sharp spinule; cones solitary.
        10 Prostrate stems dorsiventral, creeping, with 2 rows of large lateral lvs and 4 ranks of

smaller medial lvs; cones on long, erect, remotely lvd peduncles (*Pseudolyco-podiella*) .... 8. *L. carolinianum.*
10 Prostrate stems radially ± symmetrical, the lvs nearly uniform; cones sessile at the ends of densely leafy stems (*Lycopodium, p.p.*).
  11 Prostrate stems superficial, leafy; erect stems simple or with few long branches; lvs 5–11 mm .... 9. *L. annotinum.*
  11 Prostrate stems deeply subterranean; erect stems freely branched; lvs 2.5–5 mm .... 10. *L. obscurum.*
9 Lvs with long hair-like (sometimes deciduous) tips; cones 1–several per peduncle (*Lycopodium p.p.*) .... 11. *L. clavatum.*
8 Lvs in 4 or 5 ranks (except sometimes on the peduncles), entire (*Diphasium, Diphasiastrum*).
12 Vegetative branches appearing leafy, the stems largely hidden by the lvs; cones sessile or short-pedunculate.
  13 Lvs mostly 5-ranked, adnate less than half their length; branchlets not flattened ...... 13. *L. sitchense.*
  13 Lvs mostly 4-ranked on the sterile branchlets, the lateral ones adnate at least half their length; branchlets obviously flattened.
    14 Lvs of the lower rank tending to have a short, stout, ascending-spreading, sub-petiolar base and an erect or incurved blade .... 12. *L. alpinum.*
    14 Lvs of the lower rank appressed .... 14. *L. sabinaefolium.*
12 Vegetative branches appearing as flattened, wing-margined, remotely scaly stems, the free tips of the lvs up to 2 mm; peduncles well defined.
  15 Lvs of the ventral rank 1–1.5 mm; branchlets mostly 1.2–2 mm wide, with a bluish-green cast beneath .... 15. *L. tristachyum.*
  15 Lvs of the ventral rank up to 1 mm; branchlets mostly 2–3 mm wide, bright green or bright yellow-green beneath.
    16 Aerial branches with annual constrictions; cones without a sterile tip .... 16. *L. complanatum.*
    16 Aerial branches without such constrictions; cones usually with an evident slender sterile tip .... 17. *L. digitatum.*

1. Lycopodium selago L. Fir-c. Horizontal stems short; erect or ascending stems 6–20 cm, 2–3 mm thick (excl. lvs); lvs in ca 8 ranks, subulate or triangular-subulate, broadest at or near the base, gradually acuminate, entire, glossy, 3–5 × 0.6–0.8 mm, often with axillary gemmae; sporophylls resembling the vegetative lvs, borne in zones at the beginning of each season's growth (or all the lvs eventually becoming fertile); sporangia broadly reniform, 1–1.5 mm wide; spores 32–38 microns in diameter, the sides concave, the commissural faces smooth, the outer face papillate, the commissures not in furrows; high polyploid, ca $2n=268$. (*Huperzia s.*) The circumboreal var. *selago,* with ± spreading lvs, occurs in moist, often rocky and wooded places s. to Mass., N.Y., Wis., Minn., and in the mts. (but not at the highest elevs.) to N.C. and Tenn. Alpine and subalpine plants of exposed sites from Nf. to N.Y., and s. irregularly in the highest mts. to N.C. and Tenn., with more tufted, short stems and erect-appressed lvs, have been associated (perhaps incorrectly) with the European var. *appressum* Desv.

2. Lycopodium porophilum Lloyd and Underw. Rock-c. Much like *L. selago* var. *selago,* but with looser, shortly creeping base and with slight, inconspicuous zonation of longer and shorter lvs (approaching but not matching no. 3 in both of these regards); lvs spreading, mostly 5–8 × 0.8–1.2 mm, broadest near the middle and slightly toothed above; high polyploid. Acid, rocky (usually sandstone) cliffs and ledges; Pa. to Minn., s. to N.C., Ala., and Mo.

3. Lycopodium lucidulum Michx. Shining c. Stems lax, with ± elongate, rooting, leafy base and ascending branches to 25 cm, 1.5–2 mm thick (excl. lvs); lvs mostly 6-ranked, spreading or reflexed, shining, oblanceolate, usually toothed above the middle, lacking stomates on the upper surface, unequal in length so that the shoot looks ragged, the early lvs of the season the shortest and commonly subtending sporangia, the later ones mostly 7–12(–15) × 1.2–2+ mm; spores 20–30 microns in diameter; otherwise much like no. 1; high polyploid, $2n=134$. Wet woods and rocks in acid soil; Nf. to Ont. and Minn., s. to S.C., n. Ga., n. Ala., and nw. Ark. (*Huperzia l.*)

4. Lycopodium alopecuroides L. Foxtail-c. Sterile stems proximally arching, then recurved to the ground and distally prostrate and rooting, mostly 2–3+ mm thick (excl. lvs), the lvs many-ranked, twisted into a ± erect position so that the upper side of the stem looks densely spreading-leafy and the lower side nearly bare, all linear-subulate, 6–7 × 0.6–1 mm, strongly ciliate-denticulate; fertile stems erect, 1–2.5 dm, the densely crowded lvs to 11 mm; cones sessile, 2–6 cm, 1–2 cm thick; sporophylls green, linear-subulate, 6–11 mm, widely spreading, not much enlarged at the base, strongly ciliate to above the middle; sporogia globose, 0.8–1 mm thick; spores as in no. 6; $2n=156$. Bogs and moist banks in acid soil; Mass. to Fl. and Tex., chiefly on the coastal plain, but disjunct in the Cumberland mts.

5. Lycopodium prostratum Harper. Featherstem-c. Sterile stems prostrate throughout (though not rooting proximally), mostly 1–1.5(–2) mm thick (excl. lvs), many of the lvs tending to spread laterally from the axis, parallel to the ground; cones 4–8 cm; otherwise much like no. 4. Open, moist or wet places, and open wet coniferous woods, mainly on the coastal plain; N.C. to Fla. and Tex., and up the Mississippi embayment to w. Ky.

6. Lycopodium inundatum L. Bog-c. Sterile stems ± prostrate, irregularly rooting, flattened, the lvs of the lower side twisted into an ascending position; lvs 8–10-ranked, linear-subulate, mostly entire, 5–7 × 0.5–0.7 mm; fertile stems few, erect, to 1 dm, 1 mm thick; sporophylls ascending or spreading, crowded into a sessile terminal cone 1.5–5 cm × 6–12 mm, all except the lower abruptly expanded near the base and usually with a single pair of sharp teeth near the expansion, otherwise like the lvs; sporangia subglobose, 1–1.3 mm wide; spores 43 microns or more in diameter, the sides convex, the outer face wavy-reticulate, the commissural faces papillose, the commissures sunken in deep furrows; $2n=156$. Acid soil of bogs, shores, and meadows, often in seasonally inundated sites; circumboreal, s. to Md., sw. Va., O., Wis., Minn., and Calif.

7. Lycopodium appressum (Chapman) Lloyd & Underw.   Southern c. Like a larger version of no. 6; lvs 5–8 × 1–1.2 mm, sparingly toothed; fertile stems 1–3.5 dm, 1.5–5 mm thick, with appressed lvs; cones slender, 2.5–7.5+ cm, 3–8 mm thick, with ascending to appressed sporophylls; $2n=156$. Acid soil of bogs, shores, and meadows; mainly on the coastal plain from se. N.H. to Fla. and Tex. and up the Mississippi embayment to w. Ky.; irregularly disjunct in the mts. of N.C. and Tenn. and in sw. Mich. (*L. inundatum* var. *bigelovii*.)

8. Lycopodium carolinianum L.   Slender c. Prostrate stems creeping and rooting, to 10 cm, branched, ca 1 mm thick, dorsiventral, with 6-ranked lvs, the 2 lower (seemingly lateral) ranks spreading, lance-ovate, 3–7 × 1.3–2.1 mm, acuminate, entire, the 4 upper ranks erect, much smaller, lanceolate, 2.5–5 × 0.7–1.2 mm, attenuate; erect stems 1–3 dm, terete, ca 1 mm thick, the lvs scattered, equal, bract-like, linear-subulate, 3–5 mm, ascending or appressed; cones solitary, 2–12 cm, 5–7 mm thick; sporophylls yellowish-green, broad-based, with subulate spreading tip; sporangia subglobose, 1–1.3 mm wide; spores with smooth commissural faces, otherwise as in no. 6; $2n=156$. Bogs and sandy pine-barrens in acid soil; Mass. and L.I. to Fla. and Tex., mainly on the coastal plain, and interruptedly pantropical.

9. Lycopodium annotinum L.   Stiff c. Prostrate stems superficial, elongate, simple or sparsely branched, 1.5–2.5 mm thick, sparsely leafy, the lvs of the lower ranks twisted; erect stems simple or 1–2-forked, 0.5–3 dm, the lvs 5–11 mm, linear-subulate or linear-oblanceolate, spinulose-tipped, 8-ranked, subverticillate, with 4 lvs per whorl; cones terminal, sessile, 1.5–4 cm, 4.5–7.5 mm thick; sporophylls yellowish, broad-based, acuminate, the margins hyaline and erose; sporangia reniform, 1.5 mm wide; spores 28–36 microns in diameter, the sides convex, the commissural faces smooth, the outer open-reticulate, the commissures not in furrows; $2n=68$. Acid, moist to dry soil, often in coniferous woods; circumboreal, s. to N.J. and Minn., and in the mts. to Va. and W.Va.

10. Lycopodium obscurum L.   Princess-pine, ground-pine. Prostrate stems deeply subterranean, elongate, branched; erect stems scattered, simple below, freely branched above, to 3 dm, the main axis 2 mm thick, the branchlets terete or flattened; lvs numerous, shining, 6–8-ranked, linear-subulate, all alike or ± dimorphic, 4.6 × 0.8–1.2 mm, entire, firm, sharp-pointed; cones terminating some of the branches; sessile, 1–4 cm, 5–7 mm thick; sporophylls yellowish, broad-based, abruptly acuminate, with hyaline, inconspicuously erose margins; sporangia reniform, 1.5–2 mm wide; spores with the comissural faces more finely reticulate than the outer, otherwise as in no. 9; $2n=68$. Moist woods and bog-margins in acid soil; boreal Amer. and Asia, s. in e. U.S. to n. Ga., Ala., Ind., and Mo.

11. Lycopodium clavatum L.   Running pine. Horizontal stems superficial, elongate, arching and rooting at intervals, 2–3 mm thick, not dorsiventral, the lvs curved upwards; erect stems branched, to 2.5 dm; lvs mostly 10-ranked, uniform, linear-subulate, 4–7 × 0.5–0.8 mm, slightly or scarcely toothed, attenuate to a hyaline hair-like tip 2–3 mm; cones 1.5–11 cm, 3–7 mm thick, 1–6 on remotely bracteate peduncles 1.5–15 cm; sporophylls yellowish, broad-based, abruptly acuminate, with scarious erose margins; sporangia reniform, ca 1.5 mm wide; spores 28–36 microns in diameter, the sides convex, the commissural and outer faces angular-reticulate, the commissures not in furrows; $2n=68$. Open dry woods and rocky places in acid soil; circumboreal, s. in e. U.S. to N.C., W.Va., Mich., and Io.

12. Lycopodium alpinum L.   Alpine c. Much like no. 13, but the lvs of the ultimate vegetative branchlets strictly 4-ranked, decussately opposite, subtrimorphic; every other pair of lvs conspicuously decurrent on the stem as a pair of flanges, the flange continuous with one margin of the lf, so that the stem appears somewhat flattened or wing-margined with the lvs of the lateral rows twisted; free tips of these lateral lvs mostly 1.5–2 mm; dorsal and ventral lvs distinctly different from the lateral ones and also differrent inter se, those of the upper side subulate, straight, loosely appressed or closely ascending, 2–3 mm, shortly decurrent, those of the lower side (which appears to be concave because of the way the lateral lvs are twisted) tending to have a short, stout, ascending-spreading subpetiolar base (this often laterally compressed) and an erect (parallel to the stem) or incurved blade 2–3 mm and lanceolate or lance-ovate; $2n=46$. Cold woods and subalpine meadows and rocky slopes; circumboreal, s. to Keweenaw Co., Mich. and the Gaspé penins. of Que.

13. Lycopodium sitchense Rupr.   Sitka c. Much like no. 14, but the sterile branchlets not at all flattened or dorsiventral, the lvs uniform, mostly 5-ranked, subulate, 3–4 × 0.5–0.7 mm, adnate much less than half their length; $2n=46$. Exposed, subalpine habitats; e. Asia and n. N. Amer., s. in our range to Me., N.H., and Ont.

14. Lycopodium sabinaefolium Willd.   Juniper-c. Horizontal stems superficial to slightly subterranean, elongate, rooting at intervals, with distant, yellowish bract-lvs; erect stems dichotomously branched, to 2 dm, the sterile branches flattened and dorsiventral at least distally, 1.5 mm wide, their lvs in 4 ranks, linear-subulate, sharp-pointed, entire, those of the dorsal and ventral ranks appressed, shortly adnate, those of the lateral ranks slightly larger, adnate half their length, with spreading, apically incurved tip; lvs with stomates on both sides; cones 1.5–2.5 cm, 2.5–4 mm thick, solitary or paired on branching peduncles with slightly reduced, appressed, 6-ranked lvs; sporophylls yellowish, broad-based, abruptly narrowed upwards, with scarious erose margins; sporangia and spores as in no. 11; $2n=46$. Subalpine woods and meadows; Lab. to Ont., s. to N.Y., Pa., and Mich. Thought to have originated as a hybrid of nos. 13 and 15, but now stable and fertile.

15. Lycopodium tristachyum Pursh.   Wiry ground-cedar. Much like no. 16; horizontal stems deeply subterranean; erect stems repeatedly and rather irregularly branched, the principal branches commonly at several successive levels, subdichotomous, the ultimate branchlets 1.2–2 mm wide (incl. lvs), appearing narrower than in no. 16, somewhat bluish-glaucous beneath; lvs of the ventral rank 1–1.5 mm, the tip sometimes reaching the base of the next lf; cones mostly 3–4; sporophylls and sporangia a little smaller than in no. 16; $2n=46$. Coniferous woods and sandy places; Nf. and Lab. to Man., s. to Md., Va., Ind., and Minn., and in the mts. to S.C., Ga., and Ala.

16. Lycopodium complanatum L.   Northern ground-cedar. Horizontal stems elongate, mostly shallowly subterranean, 1–2 mm thick, with distant, reduced lvs; erect stems to 3 dm, branched, the branches indeterminate, with annual constrictions, the branchlets obviously flattened, 2–3 mm wide (incl. lvs) shiny, bright green or somewhat yellowish-green; lvs 4-ranked, entire, adnate over half their length, the upper appressed, the lateral having a short, deltoid free part with slender, spreading or incurved tip, the lower much reduced, seldom 1 mm long; cones 1–2.5 cm, 3–5 mm wide; without a sterile tip, 1–4+ on forked, remotely bracteate peduncles; sporophylls yellowish, broad-based, with scarious erose margins; sporangia reniform, 1.4–1.7 mm wide; spores as in no. 11; $2n=46$. Woods and rocky slopes in acid soil; circumboreal, s. in our range to Pa. and Minn.

17. Lycopodium digitatum Dillen.   Southern ground-cedar. Horizontal stems surficial or nearly so; erect stems more regularly branched, the branches determinate, without annual constrictions, the branch-systems flattened and fan-shaped; cones 2–4+ cm, commonly (not always) with a slender sterile tip several mm, the typically 4 peduncles tending to be subumbellately clustered; otherwise much like no. 16; $2n=46$. Dry open woods or meadows; Nf. and Que. to Man. and Minn., s. to Ga., Tenn., and La. (*L. dillenianum; L. flabelliforme; L. complanatum* var. *f.*)

Class *ISOETOPSIDA*

Lvs ligulate (i.e., with a small ventral appendage a little above the base), either very small or elongate and narrow; plants heterosporous; female gametophyte largely enclosed in the megaspore wall, which merely breaks open at one end. Two orders.

Order SELAGINELLALES

Stems slender, subdichotomously branched, erect, arching, or prostrate; roots adventitious, often borne on specialized lfless descending rhizophores; lvs sessile, small (ours under 5 mm), with a central unbranched vascular bundle, spiral or in 4 ranks and then dimorphic, entire, with a small membranous ligule near the axil; sporangia bivalvate, unilocular; spores tetrahedral, the megaspores normally 4 to a sporangium, yellowish or white, with a thick sculptured wall, the microspores numerous, minute, mostly red or yellow, spinulose; gametophytes minute, largely contained within the spore wall, lacking chlorophyll; sperms biflagellate.

Family **SELAGINELLACEAE**, the Selaginella Family

The only family of the order, with only the genus *Selaginella*.

1. SELAGINELLA P. Beauv., nom. conserv.   Spikemoss. Terrestrial evergreen herbs superficially resembling mosses. 500, cosmop.

1 Lvs spirally arranged, all alike.
  2 Lvs thin, toothed, loose, not bristle-tipped; sporophylls many-ranked .... 1. *S. selaginoides.*
  2 Lvs thick, ciliolate, appressed, bristle-tipped; sporophylls 4-ranked .... 2. *S. rupestris.*
1 Lvs 4-ranked, dimorphic, the ovate lateral ones larger than the median ones .... 3. *S. apoda.*

1. Selaginella selaginoides (L.) Link. Northern s. Sterile stems prostrate, mostly 2–5 cm, without rhizophores; fertile stems erect, 6–10(–20) cm, 0.4–0.5 mm thick (excl. lvs); lvs spirally arranged, ascending or spreading, thin, lanceolate, remotely spinulose-toothed, acuminate, not bristle-tipped, up to 4 × 1 mm; cones subcylindric, mostly 1.5–3 cm, to 5 mm wide, the sporophylls mostly 10-ranked, larger and more prominently spinulose than the lvs; megaspores pale yellowish, 0.5 mm wide, with low tubercles on the commissural faces; $2n=18$. Open rocks and wet banks; circumboreal, s. to N.S., n. Mich., n. Minn., and Nev.

2. Selaginella rupestris (L.) Spring. Rock-s. Stems much branched and with numerous rhizophores, 1 mm thick (excl. lvs), the creeping and the erect (to 6 cm) branches similar except that the latter terminate in tetragonous cones 1–1.5 cm × 1 mm; lvs spirally arranged, appressed, thick and firm, subulate, ciliolate, 1.5–2 × 0.25–4 mm, tipped by a scabrous slender white bristle 1–1.5 mm; sporophylls 4-ranked, evidently broader than the lvs; megaspores yellowish, 0.5 mm wide, faintly reticulate or rugose on all faces, often only 1 or 2 per sporangium, and the plants then usually apogamous and lacking microsporangia; $2n=18$. Rocky or sandy, mostly acid soil; s. Greenl. to ne. Alta., s. to n. Ga. and Okla.

3. Selaginella apoda (L.) Spring. Meadow-s. Plants delicate, the stems 0.2–0.4 mm thick, branched, ± prostrate with assurgent tips; lvs thin, 4-ranked, the lateral ones spreading, ovate, 1–2 × 0.7–1 mm, acute, obscurely ciliolate-dentate, the dorsal and ventral ones appressed, up to 1 × 0.6 mm, acute to long-acuminate; cones 1–2 cm, the sporophylls 4-ranked, slightly larger than the marginal lvs; megaspores white, 0.3–0.5 mm wide, the outer face reticulate; $2n=18$. Meadows and streambanks, esp. in calcareous soil; s. Que. to Wis., s. to Fla. and Tex. (*S. eclipes.*)

ORDER ISOETALES

Stem a fleshy, flattened, 2–3-lobed corm mostly 1–2.5 cm wide, covered above by the expanded lf-bases and below by roots originating near the groove between the lobes; lvs spirally arranged, narrow and elongate, 5–60 cm, entire, hollow, divided into 4 longitudinal cavities with scattered cross-walls, and often with 4 or more peripheral fibrous strands in addition to the central fibrovascular bundle, bearing a small ligule on the inner face above the embedded basal sporangium, which is partly or wholly covered by the velum, a fold of lf-tissue; outer lvs commonly megasporophylls, the inner microsporophylls; sporangia thin-walled, irregularly divided by transverse partitions; megaspores fairly numerous, tetrahedral with a large outer face, usually whitish, often roughened; microspores very numerous, bilateral, grayish; gametophytes minute, largely contained in the spore walls, lacking chlorophyll; sperms multiflagellate.

FAMILY **ISOETACEAE**, the Quillwort Family

The only family of the order, containing only the genus *Isoetes*.

1. ISOETES L. Quillwort, Merlin's grass. Perennial, mostly aquatic or amphibious herbs with linear, grass-like lvs; $x=11$. 100+, cosmop. Hybrids are not infrequent. Some such are *I.* × *eatonii* Dodge, = *I. engelmannii* × *E. echinospora*; *I.* × *foveolata* A. A. Eaton, = *I. riparia* × *I. tuckermannii*; and *I.* × *harveyi* A. A. Eaton = *I. lacustris* × *I. tuckermannii*; *I.* × *hickeyi* W. Taylor & Luebke = *I. lacustris* × *I. echinospora*.

1 Megaspores densely spinulose; boreal .... 6. *I. echinospora.*
1 Megaspores otherwise, variously ornamented (but not spinulose) or nearly smooth.
  2 Plant amphibious or even terrestrial; lvs with abundant stomates and (except in *I. riparia*) usually
     with 4 or more peripheral fibrous strands.
    3 Lvs usually with several peripheral strands.
      4 Sporangia pale, not spotted; megaspores raised-reticulate; widespread .... 1. *I. engelmannii.*
      4 Sporangia brown-spotted, or wholly brown; megaspores tuberculate or with short low ridges
        to nearly smooth.

5 Lvs usually black at base, and with a pale line down the middle of the adaxial surface;
   megaspores mostly 0.25–0.45 mm wide; Mississippi drainage .... 2. *I. melanopoda.*
5 Lvs usually pale or brown at base, and without such a line; megaspores mostly 0.4–0.65
   mm wide.
   6 Sporangia 6–10(–14) mm; plants terrestrial, calciphile; Mississippi drainage; midsouthern
      .... 3. *I. butleri.*
   6 Sporangia 3–5 mm; plants amphibious, not calciphile; Va. to S.C. .... 4. *I. virginica.*
3 Lvs without peripheral strands; megaspores with crowded tubercles and rough crests; Atlantic   5. *I. riparia.*
2 Plants normally submersed, only rarely amphibious; lvs with few or no stomates and without
   peripheral strands; mainly northern.
   7 Megaposres avg >0.6 mm wide; lvs fleshy-firm, dark green; usually in water 1–3 m deep .... 7. *I. lacustris.*
   7 Megaspores avg <0.6 mm wide; lvs soft and flaccid, yellow-green to bright green; usually in
      water less than 1 m deep .... 8. *I. tuckermanii.*

1. **Isoetes engelmannii** A. Braun.   Engelmann's q. Lvs 6–35(–50) cm × 0.6–2 mm, erect, not twisted, with abundant stomates and usually 4 peripheral strands; sporangia 6–13 mm, pale, one-third to two-thirds covered by the velum; megaspores 0.45–0.6 mm wide, raised-reticulate; $2n=22$. Amphibious, submersed in spring, usually later emersed; N.H. to Fla., w. to Ind., Ill., se. Mo., and Miss.

2. **Isoetes melanopoda** Gay & Durieu.   Black-footed q. Lvs 7–40 cm × 0.3–1.2 mm, erect, not or scarcely twisted, with abundant stomates and usually 4 peripheral strands, usually blackish at base, and with a pale line down the middle of the adaxial side; sporangia (5–)8–15(–20) mm, finely brown-dotted at maturity; one-fifth to two-thirds covered by the velum; megaspores 0.25–0.45 mm wide, with tubercles or short low ridges; $2n=22$. Amphibious, mostly in shallow temporary ponds, or nearly terrestrial; Minn. and S.D. to e. Tex. and Ala., e. occasionally to Ill., Tenn., nw. Ga., and reputedly s. N.J.

3. **Isoetes butleri** Engelm.   Butler's q. Often ± dioecious (unique among our spp.); lvs 6–20 cm × 0.3–0.7 mm, erect, twisted, whitish or pale reddish-brown toward the base, with abundant stomates and 4 or more peripheral strands; sporangia oblong, 6–10(–14) mm, up to one-half covered by the velum, marked with fine brown lines at maturity; megaspores (0.4-)-0.5–0.65 mm wide, low-tuberculate or nearly smooth; $2n=22$. Terrestrial on thin, seasonally wet soil, mostly over limestone; sc. Ky. to nw. Ga., w. to Ark. and e. Kans.

4. **Isoetes virginica** N.E. Pfeiffer.   Virginia q. Lvs 15–30 cm × 0.5 mm, recurved, somewhat twisted, commonly brown toward the base, with abundant stomates and 4 or more peripheral strands; sporangia broadly oblong, 3–5 mm, brown at maturity, one-fourth covered by the velum; megaspores 0.4–0.5 mm wide, provided with short, low ridges. Amphibious on clay soils in temporary pools, often over sandstone; rare and sporadic from Va. to S.C., mainly on the Piedmont.

5. **Isoetes riparia** Engelm.   Riverbank-q. Lvs mostly 10–20 cm × 0.5–1.5 mm, ± erect, somewhat twisted, bright green to somewhat yellow-green, whitish toward the base, with stomates but without peripheral strands; sporangia 3–8 mm, pale or brown-dotted to wholly brown at maturity, up to one-third covered by the velum; megaspores 0.35–0.65+ mm wide; beset with crowded tubercles and rough crests; $2n=44$; perhaps an alloploid derived from nos. 1 and 6. Amphibious (seldom in water over 0.6 m deep) in ponds, streams, and fresh or slightly brackish rivers, rooted in gravel or mud; s. Que., se. Ont., N. Engl., and e. N.Y. to W.Va. and S.C. (*I. canadensis; I. dodgei; I. saccharata.*)

6. **Isoetes echinospora** Durieu.   Spiny-spored q. Lvs mostly 5–15 cm × 0.5–1.5 mm, erect or sometimes spreading, soft, gradually tapering from the base to a long, very slender tip, often stomatiferous toward the tip, without peripheral strands; sporangia 4–7 mm, usually brown-dotted at maturity, in ours usually at least half covered by the velum; megaspores 0.25–0.6 mm wide, densely short-spiny, the spines sharp to blunt or even truncate, sometimes some of them bifid; $2n=22$. Submersed or amphibious in ponds, lakes, and slow streams, rooted in sand or mud; common, circumboreal, s. to N.J., Pa., O., Mich., Minn., Colo., and Calif. The Amer. plants may be weakly distinguished as var. *muricata* (Durieu) Engelm. (*I. muricata; I. braunii,* a preoccupied name; *I. setacea,* misapplied.)

7. **Isoetes lacustris** L.   Deep-water q. Lvs mostly 5–10(–20) cm × 0.7–2 mm, erect to often recurved, relatively coarse and firm, carrying their width well toward the acute tip, fleshy and brittle, twisted, dark green, generally lacking both stomates and peripheral strands; sporangia 3–5 mm, pale, only rarely spotted, up to one-half covered by the velum; megaspores mostly 0.5–0.8 (avg >0.6) mm wide, beset with jagged crests or high ridges that may form an irregular reticulum, varying to rarely (*I. hieroglyphica*) with a loose reticulum of low rounded ridges or with scattered low mounds; $2n=110$. Submersed (0.5–)1–3+ m in cold ponds, lakes, or streams, rooted in gravel; circumboreal, s. to N.J., Mich., Wis., Minn., and Colo.; disjunct in the mts. of Va. and Tenn. (*I. macrospora.*)

8. **Isoetes tuckermanii** A. Braun.   Tuckerman's q. Lvs mostly 5–20+ cm × 0.25–1.5 mm, erect to often recurved, thin, soft and flaccid, twisted or not, yellowish-green to bright green, with few or no stomates and without peripheral strands; sporangia 2–5 mm, pale or uniformly tan, seldom brown-spotted, up to one-third covered by the velum; megaspores mostly 0.4–0.65 (avg <0.6) mm wide, commonly high-crested as in no. 7, but varying to rarely (*I. acadiensis*) with a loose reticulum of low, rounded ridges; $2n=44$. Rather shallowly (to 1 m) submersed in ponds or lakes, less often in rivers, rooted in gravel, sand, or mud, rarely amphibious; N.S. and P.E.I. to Que., s. to N.Y. and Md.

DIVISION EQUISETOPHYTA

Vascular cryptogams; stems longitudinally ribbed and grooved (with a vascular bundle beneath each rib), jointed, with a persistent meristem at the base of each internode; lvs whorled; sporangiophores closely grouped in successive whorls to form a terminal cone; spores numerous, minute, all alike in appearance; gametophyte tiny, thalloid, photosynthetic. A single living class (Equisetopsida), with a single living order (Equisetales), family (Equisetaceae) and genus.

FAMILY **EQUISETACEAE**, the Horsetail Family

Plants without secondary thickening; ribs of the stem offset at the nodes; stem ordinarily with a large central cavity and two additional sets of longitudinal cavities; the *vallecular cavities,* which traverse the cortex beneath the grooves of the stem, and the small *carinal cavities,* which are adjacent and internal to the vascular bundles (on alternate radii with the vallecular cavities); stomates in regular longitudinal bands in the grooves of the stem; lvs small, scale-like, often not photosynthetic, connate to form a sheath at each node, each lf continuous with a ridge of the internode below; branches, if present, usually whorled, alternating with the lvs, arising from small superficial primordia at the nodes and breaking through the base of the sheath; sporangiophores peltate, each with 5–10 elongate sporangia under a polygonal cap; spores each with 4 spirally wound bands (elaters) with enlarged tip; sperms multiflagellate.

1. EQUISETUM L.   Horsetails and scouring rushes. Rhizomatous colonial perennials with annual or perennial green stems, the epidermis with silicified cell walls and often with regularly arranged silicified projections, so that the surface is minutely rough to the touch. $2n=216$ or nearly so. 15, cosmop.

Sterile but vegetatively vigorous and persistent hybrids are often produced between species of the same subgenus. Some of these are: *E.* × *ferrissii* Clute = *E. hyemale* × *E. laevigatum*; *E.* × *litorale* Kuhlewein = *E. arvense* × *E. fluviatile*; *E.* × *nelsonii* (A. A. Eaton) Schaffner = *E. laevigatum* × *E. variegatum*; *E.* × *trachyodon* A. Braun = *E. hyemale* × *E. variegatum*.

1 Stems evergreen, mostly simple, at least not with regularly whorled branches. Subg. *Hippochaete.* Scouring rushes.
  2 Stems 1–3 dm, 3–12-ridged; central cavity up to ⅓ the diameter of the stem.
    3 Central cavity wanting; vallecular cavities 3 . . . . 1. *E. scirpodes.*
    3 Central cavity evident, mostly larger than the 5–12 vallecular ones . . . . 2. *E. variegatum.*
  2 Stems 2–15+ dm, 16–50-ridged; central cavity more than half the diameter of the stem.
    4 Cones distinctly apiculate; sheaths usually with an inframedial or basal black band as well
      as an apical one, the teeth deciduous or often persistent . . . . 3. *E. hyemale.*
    4 Cones blunt or inconspicuously apiculate; sheaths black-banded at the tip only, the teeth
      promptly deciduous. Forms of . . . . 4. *E. laevigatum.*
1 Stem annual, simple or often with regularly whorled branches. Subg. *Equisetum.* Horsetails.
  5 Fertile and sterile stems alike, green; fr summer.
    6 Stem shallowly 9–32-grooved, the central cavity more than half the diameter of the stem.
      7 Teeth of the sheaths promptly deciduous; stems normally simple . . . . 4. *E. laevigatum.*
      7 Teeth of the sheaths persistent; stems simple or often branched . . . . 5. *E. fluviatile.*
    6 Stem deeply 5–10-sulcate, the central cavity less than ⅓ the diameter of the stem, scarcely
      larger than the vallecular cavities . . . . 6. *E. palustre.*
  5 Fertile and sterile stems unlike; fr spring.
    8 Fertile stems becoming green and branched, ± persistent.
      9 Teeth of the sheaths free or nearly so, blackish with pale margins . . . . 7. *E. pratense.*
      9 Teeth cohering in several broad brown lobes . . . . 8. *E. sylvaticum.*
    8 Fertile stems whitish to brownish, unbranched, soon withering.
      10 Less robust (measurements in descr.); widespread . . . . 9. *E. arvense.*
      10 More robust (measurements in descr.); n. Mich . . . . 10. *E. telmateia.*

1. **Equisetum scirpoides** Michx.   Dwarf s. r. Stems all alike, evergreen, prostrate or ascending, often bent or contorted, 0.7–2.5 dm, 0.5–1 mm thick, unbranched or with a few long branches, with 3 primary ridges, the ridges tuberculate, broadly and deeply concave, so that the stem appears 6-ridged; central cavity none, vallecular cavities 3, large; stomates in 2 rows in each principal furrow; sheaths 3–4 mm, flaring, with a broad black band above a usually green base, the 3 scarious-margined teeth each with a subulate, often deciduous tip; cones small, 3–5 mm, subsessile, apiculate. Moist, often swampy places, especially in coniferous woods; circumboreal, in Amer. s. to Conn., N.Y., s. Ill., Io., S.D., and Wash.

2. **Equisetum variegatum** Schleicher.   Variegated s. r. Stems all alike, evergreen, ascending or erect, 1–3 dm, 1–2.5 mm thick, branched near the base, 5–12-ridged, the ridges shallowly furrowed and with 2 rows of siliceous tubercles; central cavity ¼–⅓ the diameter of the stem, mosty larger than the 5–12 vallecular ones, stomates in 2 rows in each principal furrow; sheaths mostly 2–5 mm, poorly marked at the base, slightly flared upwards, green, with an apical blackish band, the body of each tooth 1–2 mm with a black or blackish, 2-ridged midstripe and conspicuous white-hyaline margins, rather abruptly contracted to a hair-like deciduous tip 0.5–1 mm; cones small, scarcely 1 cm, subsessile, apiculate. Wet thickets, bogs, and sandy shores; circumboreal, in Amer. s. to Pa., Ill., Minn., and Colo. Ours is var. *variegatum*.

3. **Equisetum hyemale** L.   Common s.r. Stems all alike, evergreen, mostly unbranched, ± erect, 2–15+ dm, 4–14 mm thick, with mostly 16–50 broad, rounded, roughened ridges; central cavity at least ¾ the diameter of the stem, much larger than the small vallecular ones; stomates in 2 rows in each furrow; sheaths mostly 5–15 mm, black-banded at the tip and usually also at or near the base, the dark, scarious-margined, basally connate teeth articulate to the sheath but tending to be irregularly subpersistent; cones short-pedun-culate, 1–2.5 cm, evidently apiculate. Streambanks and other moist or wet places; circumboreal, s. in Amer. to Fla., Calif., and C. Amer. Amer. pls are var. *affine* (Engelm.) A. A. Eaton (*E. affine; E. prealtum; E. robustum.*)

4. **Equisetum laevigatum** A. Braun.   Smooth s. r. Stems all alike, annual (reputedly sometimes perennial toward the southwest), mostly unbranched, 3–10 dm, 3–8 mm thick, with mostly (10–)16–32 ridges, these smooth or commonly with inconspicuous, low, transverse wrinkle-ridges; central cavity mostly ⅔–¾ the diameter of the stem, much larger than the small vallecular ones; stomates in 2 lines in each furrow; sheaths mostly 7–15 mm, mainly green, black-banded at the apex only, or the lowest ones sometimes black also at the base or throughout, the dark, scarious-margined teeth promptly and regularly deciduous; cones short-pedunculate, 1–2.5 cm, blunt or often inconspicuously short-apiculate. Open wet places; s. B.C. to Baja Calif., e. to s. Ont., w. N.Y., O., w. Va., Ill., Okla., and Tex. (*E. kansanum.*)

5. **Equisetum fluviatile** L.   Water-h. Stems annual, all alike, to 1 m or more, shallowly 9–25-ridged, the ridges smooth or nearly so, the stomates in a single broad band in each furrow; central cavity commonly more than ¾ the diameter of the stem, the vallecular cavities commonly lacking except near the base; sheaths green, 4–9 mm, with persistent, narrow, sharp, black or blackish teeth 1.5–3+ mm, these not or scarcely hyaline-margined; branches none to often numerous and whorled at the middle and upper nodes, 4–6-angled, simple, the first internode slightly shorter than the associated stem-sheath; cone pedunculate, 1–2 cm, de-ciduous. In shallow water, along muddy shores, and in marshes and bogs; circumboreal, in Amer. s. to Pa., Ill., Io., Nebr., and Wash.

6. **Equisetum palustre** L.   Marsh-h. Stems annual, all alike, 2–8 dm, deeply 5–10-sulcate, smooth to the touch but with minute transverse ridge-wrinkles on the angles, the stomates in a single broad band in each furrow, the central cavity small, less than ⅓ the diameter of the stem, about the size of the vallecular cavities; sheaths green, 5–10 mm, rather loose, with persistent teeth, these 3–7 mm, black or dark brown at least in part, with evident, often rather broad, pale and hyaline margins; branches few and irregular to numerous and whorled at the middle and upper nodes, 5–6-angled, simple, the first internode shorter than the associated stem-sheath; cone pedunculate, 1–3.5 cm, blunt, deciduous. Streambanks, wet meadows, and marshes; cir-cumboreal, in Amer. s. to Pa., Ill., N.D., and Wash.

7. **Equisetum pratense** Ehrh.   Meadow-h. Stems annual, dimorphic, the sterile ones 2–5 dm, 1–3 mm thick, smoothish toward the base, othewise beset with high, blunt, siliceous tubercles or very short, high, transverse ridge-crests on the (6–)10–18 ridges, with rather small vallecular cavities and a fairly large central cavity commonly ⅓–½ the diameter of the stem, the stomates in 2 broad bands in each furrow; sheaths 2–6 cm, green, the teeth 1–2 mm, persistent, only basally connate, with pale, hyaline margins and a firmer dark midstripe; branches regularly whorled at the middle and upper nodes, solid, mostly 3-angled, simple, the first internode up to as long as the associated stem-sheath; fertile stems with simple branches, otherwise as in *E. sylvaticum*. Streambanks and moist woods; circumboreal, s. in our range to N.J. and ne. Io.

8. **Equisetum sylvaticum** L.   Woodland-h. Stems annual, dimorphic, the sterile ones 3–7 dm, 1.5–3 mm thick, mostly 10–18-ridged, each ridge with 2 rows of sharp hooked spinules, the stomates in 2 bands in the furrows, the central cavity larger than the vallecular ones and most more than half the diameter of the stem, the sheaths 1–2 cm, basally green, disatlly brown, with persistent, irregularly connate brown teeth; branches regularly whorled, solid, 4–5-angled, commonly again branched; the first internode commonly longer than the associated sheath of the main stem; fertile stems subprecocious, at first simple and pale, later producing whorls of green, mostly compound branches and often themselves becoming green; cones long-peduncled, 1.5–3 cm, deciduous, not apiculate. Cool moist woods; circumboreal, in Amer. s. to Md., W.Va., Ky., and Io.

9 **Equisetum arvense** L.   Common or field-h. Stems annual, dimorphic, the sterile ones 1.5–6(–10) dm, 1.5–5 mm thick, (4–)10–14-ridged, with well developed vallecular cavities and small central cavity ca ¼ the

diameter of the stem; the stomates in 2 broad bands in the furrows, the sheath 5–10 mm, with persistent, brown, free or partly connate teeth 1.5–2 mm; branches regularly whorled at the middle and upper nodes, 3–4-angled, solid, sometimes again branched, the first internode longer than the associated stem-sheath; fertile stems whitish to brownish, precocious and soon withering, to 3 dm, simple, to 8 mm thick, their sheaths 14–20 mm, with large, partly connate teeth 5–9 mm; cones long-pedunculate, 0.5–3.5 cm, not apiculate. Cosmopolitan, somewhat weedy; in moist to moderately dry habitats throughout our range.

10. Equisetum telmateia Ehrh.   Giant h. Resembling no. 9, but much more robust; sterile stems 5–30 dm, 0.5–2 cm thick, with more numerous (commonly 20–40) but lower ridges, their sheaths mostly 1–2.5 cm, with slender, attenuate teeth 3–8 mm, the branches 4–6-angled, fertile stems 2.5–6 dm, 1–2.5 cm thick, their sheaths 2–5 cm, with 20–30 prominent teeth connate in groups of 2–4; cone 4–10 cm. Moist low places; Europe and N. Afr.; Alaska panhandle to s. Calif, formerly (but perhaps not now) in n. Mich. The Amer. pls have been segregated as var. *braunii* (Milde) Milde.

DIVISION POLYPODIOPHYTA Ferns

Vascular cryptogams; ours perennial herbs; stem unbranched to more often dichoto-mously or monopodially branched; stele usually with lf-gaps; lvs mostly alternate, typically large and compound or cleft and with a branching vascular system, but sometimes ± reduced; sporangia borne on vegetative or modified lvs, often at the margins or on the lower surface; sporophylls sometimes modified into specialized sporocarps, but never aggregated into a terminal strobilus; gametophyte small, thalloid, photosynthetic or less often mycorhizal, or depending on its stored food; sperms multiflagellate. A single class, Polypodiopsida.

ORDER OPHIOGLOSSALES

Plants homosporous, eusporangiate; sporangia numerous, relatively large (commonly ca 1 mm thick), thick-walled, 2-valved, without an annulus, stalked to sessile or even embedded in the axis of the simple or branched sporophore, not indusiate; spores very numerous, >1000; gametophytes small, subterranean, mycorhizal. A single family.

FAMILY OPHIOGLOSSACEAE, the Adder's tongue Family

Plants herbaceous, soft and rather fleshy, lacking sclerenchymatous tissue, with a short, simple, erect underground stem, mycorhizal roots, and a single lf, or sometimes 2 or more lvs; lvs petiolate (the petiole called a common stalk), the common stalk and the stalk of the sporophore continuous as a stem-like axis on which the vegetative portion of the blade (hereinafter simply called the blade) appears to be borne laterally; petiole enlarged at the base and partly or wholly enclosing a bud; blade with a branching vein-system, entire to pinnately (or ternate-pinnately) compound or dissected. 3/60.

1 Lf-blades mostly lobed or compound; veins free, forked; sporangia short-stalked .... 1. *Botrychium*.
1 Lf-blades simple and entire; veins anastomosing; sporangia sunken and coherent in 2 rows .... 2. *Ophioglossum*.

1. BOTRYCHIUM Swartz.   Grape-fern. Lvs glabrous or sometimes with simple hairs; blades mostly pinnately or ternate-pinnately or subpalmately compound or dissected, seldom simple, the veins forked and free; sporophore 1–several times pinnate, the nu-merous sporangia short-stalked and free, spores 1500–2000. 30+, cosmop. Depauperate or juvenile plants of related spp. are scarcely to be distinguished; spp. 5–10 are morpho-logically confluent, but said to be genetically distinct.

1 Blade small and not much dissected, seldom over 7 cm wide, the pinnae entire to pinnatifid; blade
    glabrous. (*Botrychium*)

2 Pinnae mostly obovate to flabellate, rounded or subtruncate above and entire to sometimes distally few-lobed; blade and sporophore ± erect in bud.
  3 Lowest pair of pinnae evidently the largest (at least in well developed plants); blade often not reaching the base of the sporophore . . . . 1. *B. simplex.*
  3 Lowest pair of pinnae not obviously larger or different from the next pair above; blade ca = or mostly longer than the stalk of the sporophore . . . . 2. *B. lunaria.*
2 Pinnae mostly lobed to pinnatifid; obtuse to acuminate; at least the tip of the blade reflexed in bud.
  4 Blade oblong to ovate in outline and usually stalked; apex of the blade reflexed and clasping the erect sporophore in bud . . . . 3. *B. matricariaefolium.*
  4 Blade deltoid and usually sessile; blade and sporophore reflexed in bud . . . . 4. *B. lanceolatum.*
1 Blade larger and more dissected, often over 7 cm wide, the pinnae mostly 1–3 times again compound; blade slightly hairy, at least in bud; blade and sporophore reflexed in bud.
  5 Blade with a long stalk attached well below the middle of the apparent stem, evergreen. (*Sceptridium*)
    6 Blade regularly dissected, the terminal segments not notably different from the others.
      7 Blade very finely dissected, its ultimate segments linear or narrowly oblong, with mostly 1–2 veinlets . . . . 8. *B. dissectum.*
      7 Blade less dissected, with broader segments that have more veinlets.
        8 Ultimate segments of the blade somewhat flabellate . . . . 10. *B. jenmanii.*
        8 Ultimate segments not flabellate.
          9 Ultimate segments rounded and obscurely toothed or entire . . . . 5. *B. mutifidum.*
          9 Ultimate segments acute and coarsely and irregularly toothed or cleft . . . . 6. *B. rugulosum.*
    6 Blade unequally dissected, the terminal segments, and the terminal segments of the lateral pinnae, tending to be elongate and not much cleft.
    10 Ultimate lf-segments mostly obtuse or rounded; blade remaining green . . . . 7. *B. oneidense.*
    10 Ultimate segments mostly acute to acuminate; lvs often turning bronzy in autumn.
      11 Blade mostly tri- or quadripinnate, the pinnules rounded or subtruncate at base . . . . . . 8. *B. dissectum.*
      11 Blade mostly bipinnate, the pinnules ± cuneate at base . . . . 9. *B. biternatum.*
  5 Blade sessile, attached near the middle of the apparent stem, deciduous (*Osmundopteris*) . . . . . . 11. *B. virginianum.*

1. **Botrychium simplex** E. Hitchc. Little g.-f. Plants 3–16 cm, wholly glabrous; blade and sporophore ± erect in bud; blade supramedial or more often inframedial, on a 5–20 mm stalk, mostly 1–2(–4) × 0.5–1.5(–3) cm, often not reaching the base of the sporophore, simple or more often with (1)2–3(4) pairs of subopposite pinnae, these mostly obovate, basally cuneate, rounded above, entire or sometimes lobulate or the basal ones even somewhat pinnatifid, the basal pair larger than the next pair above at least in well developed plants; upper pinnae tending to be confluent; 2*n*=90. Open, marshy places, meadows, and edges of woodland ponds; circumboreal, s. to N.H., Pa., Io., and Calif. May, June. Ours is var. *simplex.*

2. **Botrychium lunaria** (L.) Swartz. Moonwort. Plants 3–18+ cm, wholly glabrous; blade and sporophore ± erect in bud; blade bright green in life, sessile or on a short stalk to 5 mm, borne near the middle of the apparent stem and seldom surpassed by it, 1.5–7 × 0.7–3(–3.5) cm, distinctly pinnate, with (1)3–6(7) pairs of similar pinnae, these sessile, crowded and often overlapping, dichotomously veined, without a midrib, flat, broadly flabellate, broader than long, the proximal margin forming an obviously retrorse angle with the rachis, the lowest pair of pinnae not evidently different from the next pair; sporophore and its stalk each 0.5–7 cm, subequal or either one the longer. Open fields and meadows, or sandy or gravelly streambanks; circumboreal and also scattered in the S. Hemisphere, in our range s. to N.Y., Mich., and Minn. June–Aug.

The foregoing description applies to the widespread, nomenclaturally typical, diploid (2*n*=90) phase of the species. It passes freely into a diploid shade form, often called var. *onondagense* (Underw.) House, with the blade on an evident stalk generally over 5 mm, the pinnae more remote, often somewhat spoon-shaped rather than flat, and not notably wider than long, the proximal margin diverging widely from the axis. A distinctive dwarf diploid ally of var. *onondagense,* often with persistent gametophytes, occurs in mature deciduous forests in Mich., Wis., and Minn., and has been called *B. mormo* W. H. Wagner. A more yellow-green tetraploid (2*n*=180), morphologically scarcely separable from diploid var. *onondagense,* is also widespread in the N. Amer. portion of the range of *B. lunaria,* and has been distinguished as *B. minganense* Vict. A hexaploid much like the *minganense* phase has been described as *B. campestre* W. H. Wagner.

3. **Botrychium matricariaefolium** A. Braun. Daisy-lvd g.-f. Plants 5–28 cm, wholly glabrous; bud with the blade bent over at the tip and clasping the erect sporophore; blade supramedial on the apparent stem, deciduous, with a 2–15 mm stalk, submembranous, oblong or ovate, mostly 1.5–7(–9) × 0.4–7(–7) cm, the pinnae 2–7 pairs, subopposite, obtuse or acutish, obtusely sublobate or in well developed plants again pinnatifid with obtusely lobed segments; sporophore 3–12+ cm, pinnate or bipinnate, its stalk 1–4 cm; 2*n*=180. Thickets and rich woods in subacid soil; circumboreal, in Amer. from Nf. to Alta., s. to N.J., Md., Va., N.C., Tenn., n. O., Wis., S.D., and Colo.; Patagonia. June–Aug. Ours is var. *matricariaefolium.* (*B. neglectum; B. oblanceum; B. pseudopinnatum; B. ramosum* of auth.)

4. **Botrychium lanceolatum** (S. G. Gmelin) Angström. Lance-lvd g.-f. Plants 6–40 cm, wholly glabrous; blade and sporophore reflexed in bud; blade supramedial, dark green, deciduous, sessile or with a short stalk to 6 mm, deltoid, mostly 1–6 × 1–9 cm, with 2–5 pairs of ± pinnatifid pinnae, the lowest pair of pinnae much the largest, the pinnae and pinnules mostly longer than wide; sporophore 1.5–9 cm, mostly bipinnate,

its stalk mostly (0.5–)1–3 cm, surpassed by the blade; $2n=90$. Circumboreal, s. in Amer. to Va., W.Va., O., Minn., Colo., and Wash. July–Aug. Var. *lanceolatum,* a stout, fleshy plant with the lobes of the pinnae 1–5 mm wide, broad-based, and apically rounded, is a subarctic and boreal plant of mt. slopes and meadows, s. to n. Me. (*B. hesperium.*) Var. *angustisegmentum* Pease & Moore, a laxer and more membranous plant with the lobes of the pinnae 1–2.5 mm wide, rather distant, narrow at the base and apically acute, occurs in moist, shady woods and margins of swamps from Nf. to Ont. and Wis., s. to Va., W.Va., and O.

5. Botrychium multifidum (S. G. Gmelin) Rupr.   Leathery g.-f. Plants 10–45 cm; bud pilose, its blade and sporophore reflexed; blade inframedial or nearly basal on an evident stalk, evergreen, rather fleshy or coriaceous, sparingly pilose, 3–15 × 3–20 cm, plane in life, 2–4 times ternate-pinnately compound, the pinnae and lesser segments often crowded, the ultimate segments less than twice as long as wide, obtuse or rounded, subentire to crenate or dentate; sporophore long-stalked, bipinnate or tripinnate; $2n=90$. Meadows and open woods; circumboreal, in Amer. s. to N.C., Ill., Io., and Calif. Mostly Aug.–Sept. (*B., matricariae; B. silaifolium.*)

6. Botrychium rugulosum W. H. Wagner.   Blade convex and rugulose in life; ultimate segments angular, acute and more coarsely and irregularly toothed or cleft; otherwise much like no. 5; $2n=90$. Woods, moist pastures, and swampy places; N.H., Conn., and s. Que. to Wis. and Minn. Rare.

7. Botrychium oneidense (Gilbert) House.   Blunt-lobed g.-f. Much like the less-dissected phase of no. 8, but the persistently green, 2–3 times pinnate blade with mostly obtuse or rounded ultimate segments, the major segments tending to be ovate; $2n=90$. Rich, moist woods and swamps; N.B. to Ont. and Minn., s. to N.C. and Ind. Aug.–Nov.

8. Botrychium dissectum Spreng.   Lace-frond g.-f. Plants 10–30+ cm; bud pilose, its blade and sporophore reflexed; blade inframedial or nearly basal on an evident 3–7 cm stalk, somewhat coriaceous, persistent, often reddish when first expanding and tending to turn bronzy in autumn, mostly 3–4 times pinnate but highly variable, sometimes finely divided throughout, more often unequally dissected with the terminal segment of the principal divisions tending to be more elongate and less divided than the other segments; primary and secondary segments often crowded; ultimate segments mostly acute or acuminate, but with rounded or subtruncate base, fertile spike 2–10 cm, long-stalked; $2n=90$. In a wide range of habitats; N.S. and N.B. to Wis. and se. Minn., s. to Fla. and La. Aug.–Nov. (*B. d.* var. *obliquum,* the more common phase with less strongly dissected lvs.)

9. Botrychium biternatum (Savigny) Underw.   Sparse-lobed g.-f. Blade not strongly dissected, mostly bipinnate, the terminal segment of the principal divisions tending to be more elongate and less divided than the other segments; major lateral segments tending to be oblong; ultimate segments ± cuneate at base; otherwise much like the less-dissected forms of no. 8. $2n=90$. Woods, swamps, and old fields; se. U.S., n. to Md., Ind., and Mo. Sept.–Nov. (*B. dissectum* var. *tenuifolium.*)

10. Botrychium jenmanii Underw.   Alabama g.-f. Blade 3–4 times ternate-pinnate, with short, broad, flabellate, distally rounded and finely toothed ultimate segments; otherwise much like no. 5; $2n=180$. Open woods, old fields, and pastures; W.I., n. to Va., Tenn., and Ala. Sept.–Nov. Rare.

11. Botrychium virginianum (L.) Swartz.   Rattlesnake-fern. Plants 20–75 cm; bud becoming exposed on one side, its blade and sporophore both reflexed; blade medial or somewhat supramedial, sessile, thin, deciduous, sparingly pilose, deltoid, mostly 7–20 × 10–30 cm, 2–4 times ternate-pinnately compound, the pinnules and segments decurrent, the ultimate segments acutely toothed; sporophore bipinnate or tripinnate, mostly 6–15 cm, on a stalk 7–20 cm; $2n=184$. Woods and moist, open places; Nf. and Lab. to Alas., s. to Fla. and Calif.; also in s. Mex. and irregularly in Eurasia. Spring and early summer.

2. OPHIOGLOSSUM L.   Adder's-tongue. Lvs glabrous; blade simple and entire, mostly elliptic, net-veined; sporophore unbranched, the sporangia in 2 rows, ± embedded and coherent. 30, cosmop.

1 Blade acute, apiculate; principal areoles enclosing smaller ones .... 1. *O. engelmannii.*
1 Blade rounded to acutish, not apiculate; principal areoles not enclosing smaller ones .... 2. *O. vulgatum.*

1. Ophioglossum engelmannii Prantl.   Bases of old lvs persisting on the rhizomes; lvs often 2 or more, the blade usually somewhat trough-shaped in life, acute at each end, apiculate, the veins forming areoles that enclose many smaller areoles and also some free veinlets; otherwise much like *O. vulgatum.* Woods, pastures, and ledges, mostly in calcareous soils; nw. Va. to s. Ill. and e. Kans., s. to Fla., Ariz., and Mex. Spr.–early summer.

2. Ophioglossum vulgatum L.   Plants 7–35 cm; stipe (1.5–)6–13(–19) cm; lvs mostly solitary; blade flat, elliptic-oblong to ovate, rounded to obtuse or subacute, but not apiculate, mostly 4–8 × (1–)2–4 cm, sessile or short-stalked; venation regularly areolate, the areoles all small with mostly free included veinlets, not enclosing secondary areoles; sporophore 1–5 cm × 2.5–4 mm on a stalk mostly 7–14 cm; sporangia 0.6–1.2 mm thick; $2n=960$–1320. Circumboreal, in Amer. from N.S. to Wash., s. to Fla. and Tex. Two vars. with us:

Var. *pseudopodum* (S. F. Blake) Farw.   Northern a.-t. Blade widest at the middle, tapering to the base, pale green, herbaceous; basal sheath membranous and ephemeral; spores avg 50–60 microns; grassy marsh-edges and ditches; northern, mostly n. of the glacial boundary, but s. in the mts. to s. Va. (*O. pusillum.*)

Var. *pycnostichum* Fern.   Southern a.-t. Blade widest below the middle, more abruptly rounded to the base, dark green, firm; basal sheath leathery and ± persistent; spores avg 35–45 microns; low woods and flood plains, southern, mostly s. of the glacial boundary, s. to Fla. and Mex. (*O. pycnostichum.*)

Order POLYPODIALES

Plants homosporous, leptosporangiate, the sporangia thin-walled and with a ± definite annulus, borne on vegetative or modified lvs, often ± covered by an indusium, but not enclosed in a sporocarp; terrestrial (sometimes epiphytic) or occasionally aquatic plants, ours herbaceous, the lvs (often called fronds) circinate in bud, arising from a rhizome or caudex, the petiole (often called a stipe) often stem-like in aspect. (Filicales.)

Family OSMUNDACEAE, the Royal Fern Family

Rhizome stout, sclerenchymatous, covered with persistent lf-bases and fibrous roots but lacking scales, lvs erect, coarse, compound, scaleless, ± hairy, the long petiole with stipule-like wings at the base; blade divided into fertile and sterile parts, or some lvs sterile and others fertile, the fertile lvs or parts without expanded green tissue; sporangia all maturing at once, borne in clusters on short lf-segments, not in sori, rather large, pear-shaped with a short stout stalk, the walls 2 cells thick, bivalvate; annulus poorly developed, consisting of an apical or apicolateral cluster or short row of thickened cells; spores 256–512, green; gametophyte terrestrial, fleshy, elongate, dark green. 3/18.

1. OSMUNDA L.   Coarse ferns with pinnate-pinnatifid or bipinnate lvs. 10, cosmop.

1 Lvs bipinnate, the sterile pinnules serrulate; fertile part of the blade terminal . . . . 1. *O. regalis.*
1 Lvs pinnate-pinnatifid, the sterile segments entire.
　2 Fertile and sterile lvs distinct . . . . 2. *O. cinnamomea.*
　2 Fertile pinnae inserted near the middle of vegetative lvs . . . . 3. *O. claytoniana.*

1. **Osmunda regalis** L.   Royal fern. Lvs mostly 5–18 dm; petiole glabrous; blade bipinnate, broadly ovate, to 5.5 dm wide; pinnae 5–7 pairs, subopposite, the lowest slightly reduced; pinnules alternate, 7–10 to a side, oblong, to 7 × 2.3 cm, sessile or nearly so, rounded and oblique at the base, obtuse, serrulate, finely and closely veined; fertile pinnae borne at the ends of some of the blades, several pairs, the larger segments oblong, 6–11 × 2–3 mm, at first greenish, eventually brownish; 2n=44. Swamps and moist places, mostly in acid soil; circumboreal, in Amer. from Nf. to Sask., s. to Fla., Tex., and trop. Amer. Spring–early summer. The Amer. plant is var. *spectabilis* (Willd.) A. Gray.

2. **Osmunda cinnamomea** L.   Cinnamon-fern. Lvs clustered, 6–12(–16) dm; petiole densely woolly when young; blade pinnate-pinnatifid, dimorphic, the sterile ones oblong, to 3 dm wide, relatively firm, shining green, gradually tapering to the tip, as also the individual pinnae; pinnae oblong-lanceolate, sessile, 15–25 pairs, alternate or subopposite, to 15 × 3 cm, with a tuft of brownish hairs at the base, deeply pinnatifid, the segments 15–20 on a side, alternate, broadly oblong, obtuse or subacute, usually entire, laxly ciliate, the veins 9–12 pairs per segment; fertile lvs vernal, central, surrounded by the larger sterile outer ones, soon withering brown, essentially bipinnate, the rachis long-woolly, the pinnae 2–4 cm × 8–15 mm; sporangia cinnamon-brown; 2n=44. Swamps, streambanks, and other moist places in subacid soil; Lab. to Minn., s. to Fla., Tex., N.M., and trop. Amer.; e. Asia. Ours is var. *cinnamomea.*

3. **Osmunda claytoniana** L.   Interrupted fern. Much like no. 2; lvs soft, dull blue-green, foreshortened and blunt at the tip, as also the individual pinnae; rachis soon glabrate; segments of the sterile pinnae slightly hairy but not ciliate; outer lvs usually sterile, the inner larger and with 1–5 pairs of fertile pinnae near the middle of the blade; fertile pinnae to 6 × 2 cm, much smaller than the vegetative ones above and below; sporangia dark brown; 2n=44. Open, moist woods and margins of swamps, in subacid or neutral soil; Nf. to Ont. and Minn., s. to Ga., Tenn., and Ark.; e. Asia. Ours is var. *claytoniana.* A hybrid with no. 1 is *O. × ruggii* Tryon.

FAMILY **SCHIZAEACEAE**, the Curly Grass Family

Rhizomes lacking scales but with septate hairs; lvs dimorphic or divided into fertile and sterile parts, the fertile segments narrow and obviously different from the sterile; venation typically dichotomous and free; sporangia in 2 rows on each fertile segment, usually solitary (not in sori), large, ovoid or pyriform, ± curved, thin-walled, marginal in origin but apparently on the lower surface, ± covered either by recurved margins or by outgrowths from the lf-tissue, without a true indusium, vertically dehiscent, the annulus an apical group of thick-walled cells; spores 128–256; terrestrial plants, scarcely fern-like in appearance. 5/150.

1 Lvs climbing, 4 dm or more; rhizomes long-creeping .... 1. *Lygodium.*
1 Lvs not climbing, 1.5 dm or less; rhizome erect .... 2. *Schizaea.*

1. **LYGODIUM** Swartz, nom. conserv. Climbing fern. Rhizome creeping, slender, dichotomously branched; lvs distichous, remote, sympodially developed, one branch of each dichotomy elongate, the other short and bearing 2 terminal palmately lobed lfls, the whole lf thus resembling a stem with lvs; sterile pinnae basal, the fertile apical, with reduced lf-tissue; sporangia each borne on a short vein, covered by a membranous indusium-like outgrowth from the lf-surface, short-stalked, strongly curved, the annulus appearing lateral; spores tetrahedral; gametophyte thalloid. 3, warm reg.

1. **Lygodium palmatum** (Bernh.) Swartz. Rhizome black, wiry; lvs to 3 m, the slender rachis flexuous; lfls long-stalked (1–2 cm), 2–4 × 3–6 cm, deeply cordate, palmately lobed, the lobes mostly 6, to 2.5 × 1.2 cm, rounded above, entire; fertile pinnae several times dichotomously branched, the numerous ultimate segments 3–5 × 1.5 mm; sporangia 6–10 per segment; $2n=60$. Moist thickets and woods in acid soil; irregularly from s. N.H. and e. N.Y. to O. and sw. Mich., s. to Fla. and Miss.

2. **SCHIZAEA** J. E. Smith, nom. conserv. Curly grass. Rhizome erect; lvs borne spirally, circinate in vernation, the sterile simple and linear, the fertile with several apical pairs of segments; sporangia sessile, protected by the incurved, long-ciliate lf-margin, erect or slightly bent, the annulus terminal; spores bilateral; gametophyte filamentous. 30, ± cosmop.

1. **Schizaea pusilla** Pursh. Rhizome short and slender, densely rooting, mostly simple; lvs numerous, glabrous, the sterile 2–6 cm × 0.3–0.4 mm, entire, curled, without a petiole; fertile lvs erect, 8–12 cm, the fertile segments pinnate, with 4–7 pairs of pinnae 2–3.5 × 0.4–0.6 mm; sporangia 8–14 per segment; $2n=206$. On hummocks in bogs or wet grassy places, in acid soil; irregularly from Nf. and N.S. to L.I., N.J. and Del.; reported from the Bruce Peninsula, Ont.; rare and local.

FAMILY **HYMENOPHYLLACEAE**, the Filmy Fern Famly

Rhizome scaleless, delicate, branching; lvs circinate in bud, the blade only 1 cell thick, lacking stomates, glabrous, with reduced vascular tissue; sori marginal, terminal on the veins, with a valvate or tubular indusium, usually sessile or sunken in the lf-tissue; sporangia borne basipetally on an elongate receptacle, sessile or subsessile, with an oblique complete annulus; spores 32–512; gametophytes monoecious, thalloid or filamentous, long-lived, photosynthetic; small, delicate, chiefly epiphytic ferns. 6/500.

1. **TRICHOMANES** L. Bristle-fern. Sori marginal, in tubular indusial pockets, the sporangia on a slender, bristle-like axis that may be exserted from the indusium. 275, warm reg.

1. **Trichomanes boschianum** Sturm. Appalachian b.-f. Lvs scattered on the delicate, long-creeping rhizome; petiole to 7 cm, margined nearly to the bsae with a green herbaceous wing, sparingly septate-pilose; blade light green, ovate to lanceolate, to 19 × 4 cm, bi- or tripinnatifid, the rachis broadly winged, sparingly

pilose, the pinnae 6–10 pairs, rhombic, the pinnules 2–4 pairs, obcuneate, the ultimate segments 1-veined; sori green, terminal on lateral veins at the base of the lf-lobes, 1–4 to a pinnule, 2 mm; $2n$=72, 144. Rockhouses and moist, overhanging cliffs, usually on sandstone; Ky. and s. O. to s. Ill., s. to w. S.C., Ga., Ala., and Ark., the filamentous gametophytes of this or other spp. n. to Vt. and N.H.

FAMILY **POLYPODIACEAE**, the Polypody Family

Sori laminar, borne in 1–several rows on each side of the costa, without an indusium; sporangia thin-walled and with a well developed vertical annulus interrupted by the stalk; spores (32)64, bilateral; rhizome scaly; petiole eventually abscising to leave a short projection on the rhizome; blade simple or once pinnate, glabrous or with short hairs or few to many scales; veins free or often in part reticulate. 45/350+. Family in the past often broadly defined to include the Adiantaceae, Aspleniaceae, Blechnaceae, Dennstaedtiaceae, and Onocleaceae.

1. **POLYPODIUM** L. Polypody. Lvs scattered along the branched, widely creeping rhizome, firm, evergreen, deeply lobed to once pinnate with sessile, broad-based pinnae, the veins free or areolate only along the costae; petiole with 3 vascular bundles below, these uniting distally; sori rotund or elliptic, borne subterminally on the vein-forks. 150+, cosmop.

1 Petiole and lower surface of the blade densely scaly; lf-segments entire . . . . 1. *P. polypodioides.*
1 Petiole and lower surface of the blade scaleless, lf-segments denticulate . . . . 2. *P. virginianum.*

   1. **Polypodium polypodioides** (L.) Watt.   Resurrection-fern. Rhizome 1.3–1.8 mm thick, its scales narrowly subulate, 2–3 mm, numerous, appressed, fimbriate, with a dark median stripe; lvs 6–25 cm, the petiole 1–9 cm, densely appressed-scaly; blades oblong, 2–6 cm wide, acute, the segments 8–20 pairs, broadly linear or linear-spatulate, 2–5 mm wide, entire, apically rounded, green and glabrous above, densely covered beneath with persistent peltate scales; veins obscure, mostly once-forked, some anastomosing; sori in deep excavations near the margin, without intermingled hairs; $2n$=74. On rocks and trees, mostly in subacid soil; se. Md. to Fla., w. to Ill., Mo., Okla., and Tex.; tropical Amer. Plants of the U.S. are var. *michauxianum* Weatherby.

   2. **Polypodium virginianum** L.   Common p. Rhizome 1.5–3 mm thick, the scales 2.5–5 mm, basally cordate, apically filiform-attenuate, entire, concolorous, light brown; lvs 8–40 cm, the petiole mostly 3–15 cm, scaly only at the very base; blade narrowly oblong, mostly 3–6 cm wide, narrowed to an acuminate tip, the rachis sparsely scaly beneath; segments mostly 12–20 pairs, 3–7 mm wide, slightly denticulate, apically obtuse, glabrous; veins evident, free, mostly twice-forked, terminating in hydathodes; sori superficial, borne close to the midvein of the segments, the sporangia intermingled with long glandular golden hairs; $2n$=74, 111, 148. On rocks and banks in circumneutral soil, or seldom on tree-trunks; Nf. to Yukon, s. to Ga., Ala., Ark., and S.D. Perhaps equally well treated as var. *virginianum* of a broadly defined, circumboreal *P. vulgare* L.

FAMILY **DENNSTAEDTIACEAE**, the Bracken Family

Sori marginal or appearing submarginal, protected both by the inrolled lf-margin (or portions of it) and by an extrorsely oriented flap of tissue on the ventral lf-surface just within the margin; sporangia thin-walled and with a well developed vertical annulus interrupted by the stalk; spores (32)64, tetrahedral; rhizome and lvs with hairs only (or glabrous), lacking scales; petiole with a single gutter-shaped bundle, or with several bundles that unite distally to form a gutter-shaped bundle. 16/350.

1 Sori globular, individually indusiate; ultimate lf-segments incised . . . . 1. *Dennstaedtia.*
1 Sori linear, confluent, collectively covered by the infolded lf-margin; ultimate lf-segments entire . . . . 2. *Pteridium.*

   1. **DENNSTAEDTIA** Bernh.   Petiole with a single gutter-shaped bundle; lvs soft, membranous, their veins free, apically thickened, not reaching the margin; sori marginal at the vein-tips, the thin, whitish, cupulate indusium formed partly by a recurved modified tooth of the lf-margin; sporangia few; delicate ferns with elongate, branched rhizomes and

fragrant, distant, erect, deciduous compound lvs, the rhizome, petiole and blade with septate hairs. 70, mainly trop.

1. Dennstaedtia punctilobula (Michx.) Moore.   Hay-scented fern. Lvs mostly 4–8 dm, the petiole elongate, to 2 dm, basally castaneous, the fragrant blade lanceolate or lance-ovate, to 25 cm wide, bipinnate-pinnatifid; pinnae 17–25 pairs, alternate or subopposite, short-petiolulate, lanceolate, to 13 × 4.5 cm, the pinnules subsessile, 14–25 pairs, deeply pinnatisect, the segments typically 6–8 pairs, ovate, with 2–3 oblique acute teeth on each side; rachises, costae, and costules with short, spreading, glandular hairs; $2n=68$. Mostly in open woods or clearings or on rocky slopes; Nf. and N.S. to Ont. and Minn., s. to Ga., Ala., and Ark. (*Dicksonia* p.)

2. PTERIDIUM Gled., nom. conserv.   Bracken fern. Petiole coarse, ± erect and stem-like, with several vascular bundles that unite distally to form a gutter-shaped bundle; blade ternate-pinnately 2–3 times compound, the ultimate segments pinnatifid but otherwise entire; pinnae opposite or nearly so; pinnules alternate, their segments oblong to linear with revolute margins; veins "free" but connected by a submarginal vascular strand on which the contiguous and confluent linear sori are borne, these protected by the recurved, modified lf-margin and a minute or often nearly obsolete hyaline inner indusial flap of tissue; coarse ferns with widely creeping rhizomes and coriaceous, deciduous lvs, the rhizome and often also the lvs provided with septate hairs. 1–several, by interpretation, cosmop.

1. Pteridium aquilinum (L.) Kuhn.   Lvs mostly (3–)5–15(–20) dm overall, the erect petiole mostly somewhat shorter than the more spreading blade; $2n=104$. Many habitats; widespread at least in the N. Hemisphere. Three var. with us:

a Lvs ± densely short-hairy, the reflexed indusial margin also ciliate; pinnules nearly at right angles to
    the costa; cordilleran region of w. N. Amer., and locally in n. Mich., s. Ont. and s. Que. .... var. *pubescens* Underw.
a Lvs glabrous or only slightly (rarely more densely) hairy; pinnules at an oblique angle to the costa.
  b Longest entire lf-segment or part of a segment ca 4 times as long as wide, the terminal segments
    mostly 5–8 mm wide; margins of the ultimate segments commonly somewhat hairy; throughout
    our range, and also in Eurasia .... var. *latiusculum* (Desv.) Underw.
  b Longest entire lf-segment or part of a segment ca 9 times as long as wide, the terminal segments
    mostly 2–4.5 mm wide; margins of the ultimate segments nearly or quite glabrous; se. U.S.,
    n. to Mo., s. Ind., O., and on the coastal plain to Cape Cod, Mass. .... var. *pseudocaudatum* (Clute) Heller.

FAMILY **ADIANTACEAE**, the Maidenhair Fern Family

Sori mostly submarginal and usually covered when young by the reflexed margins of the lf-segments (or the margins virtually flat and unmodified); sporangia thin-walled and with a well developed vertical annulus interrupted by the stalk; spores (32)64, tetrahedral; rhizome scaly; petiole slender, with a single vascular bundle, often eventually breaking off well above the ground; blade (1)2–several times compound (usually pinnately so), with free veins. 35/500+.

Thalloid, irregularly shaped gametophytes of a sp. of *Vittaria* occur in the s. Appalachian mts., n. to N.Y. and O.

1 Sori short, distinct, borne on the veins on the underside of the individual reflexed marginal flaps of
    the ultimate lf segments .... 1. *Adiantum.*
1 Sori soon confluent as a submarginal band, ± covered by the usually inflexed margin of the lf-segments.
  2 Fertile lvs with notably narrow and elongate ultimate segments, these 1–3(–5) mm wide.
    3 Fertile segments not mucronate; petiole pale or greenish at least distally .... 2. *Cryptogramma.*
    3 Fertile segments cartilaginous-mucronate; petiole castaneous .... 3. *Aspidotis.*
  2 Fertile lvs with relatively or absolutely broader ultimate segments.
    4 Blade with copious white powder beneath; w. Missouri .... 5. *Notholaena.*
    4 Blade not powdery; widespread.
      5 Lvs strongly dissected, the ultimate segments generally well under 1 cm, often evidently hairy
        beneath (glabrous in one of our spp.) .... 4. *Cheilanthes.*
      5 Lvs less dissected, at least the larger ultimate segments 1 cm or more, glabrous or only sparsely
        and inconspicuously hairy beneath .... 6. *Pellaea.*

1. ADIANTUM L.   Maidenhair fern. Lvs shining, resistant to wetting, deciduous in our spp.; petiole slender and wiry, dark and polished; blade usually glabrous, 1–several times pinnate or partly dichotomous, the ultimate segments petiolulate, mostly broad and ± flabellate; proper indusium none, but the reflexed, modified marginal flaps of the fertile segments providing a false indusium; sori borne singly on the underside of the individual indusial flaps; graceful woodland ferns with rather short, scaly rhizomes. 200, cosmop.

1. **Adiantum pedatum** L.   Northern m. Rhizome short-creeping, 3–5 mm thick, its scales light brown, 3–6 × 1–2 mm; lvs few or solitary, with an erect, purple or purplish-black petiole 1–6 dm, the blade shorter and set nearly at right angles to the petiole, mostly 1–4 dm long and about as wide, or wider, reniform-orbicular when well developed, subequally forked at the top of the petiole into 2 recurved-spreading rachises that bear 2–several progressively shorter pinnae along the outside of the curve; larger pinnae with mostly 15–35 alternate, short-stalked, widely spreading pinnules on each side of the costa; pinnules 12–22 × 5–9 mm, obliquely oblong, the midrib near the lower margin, which is entire and straight or concave, the upper (distal) margin more convex, lobulate, and bearing the oblong indusial flaps; $2n=58$, 116. Moist woods and streamsides, in circumneutral soil, and on serpentine slopes at scattered stations from Nf. to Vt.; Nf. to Alas. and adj. Asia, s. to Ga., La., Okla., and Calif. Most e. Amer. plants belong to the nomenclaturally typical phase of the sp.; serpentine plants from both e. and w. N. Amer. tend to have fewer and smaller pinnae and have been segregated as subsp. *calderi* Cody.

2. **Adiantum capillus-veneris** L.   Southern m. Rhizome short-creeping, 1.5–2.5 mm thick, its slender dark brown scales 1.5–3 mm, scarcely 0.5 mm wide; lvs scattered, lax, mostly 1–4(–5) dm overall; petiole purplish-black, nearly or fully as long as the blade; blade (once or) twice (or in part thrice) pinnate, mostly 2–4 times as long as wide; pinnae alternate, few, mostly 5–12 on each side; ultimate segments petiolulate, subflabellate or rhombic, deeply parted with dentate lobes, lacking a midrib; fertile pinnules slightly smaller than the sterile; $2n=60$. Moist calcareous rocks and banks; Va., Ky., Mo., S.D., and Utah, s. to the tropics, and in warmer parts of the Old World; sometimes escaped farther north.

2. CRYPTOGRAMMA R. Br.   Rock-brake. Lvs small, evergreen or deciduous, dimorphic; petiole weak, glabrous or merely hairy, scaly only at the base, stramineous to greenish at least distally; blades glabrous, bipinnate to tripinnate, the sterile ones spreading, membranous, with largely green rachis, the ultimate segments dentate or crenate, the veins not reaching the margin; fertile lvs more erect and surpassing the sterile ones, with narrow, entire ultimate segments, the revolute margin modified into a false indusium covering the sori, which are borne on the vein-tips. 2–4, N. Hemisphere.

1   Lvs densely clustered on a short, much-branched rhizome . . . . 1. *C. crispa.*
1   Lvs scattered on a more elongate, slender rhizome . . . . 2. *C. stelleri.*

1. **Cryptogramma crispa** (L.) R. Br.   Parsley-fern. Rhizome short and compactly branched, sclerified, beset with old petiole-bases, the scales 4–6 × 1–1.5 mm, with firm dark midstripe; lvs numerous, tufted, evergreen, firm, with stramineous to greenish petioles, those of the sterile lvs 3–15 cm, those of the fertile lvs longer (to 20+ cm) and nearly or fully equaling the sterile blades; sterile blades 2–11 × 1.5–6 cm, ca thrice pinnate, with ± rounded, toothed ultimate segments to 8 × 4 mm; fertile blades 4–14 × 1.5–5 cm, 2–3 times pinnate, the ultimate segments linear or linear-oblong, 4–12 × 1–2 mm, the margins broadly reflexed but not much modified, tending to open out as the sporangia mature; sori eventually covering the whole lower surface of the pinnule; $2n=60$ (Amer. pls). Rock crevices; Que. to Alas., s. to n. Mich., ne. Minn., and N.M.; Eurasia. The Amer. pls are var. *acrostichoides* (R. Br.) C. B. Clarke. (*C. a.*)

2. **Cryptogramma stelleri** (S. G. Gmelin) Prantl.   Slender r.-b. Rhizomes slender, soft, creeping, pilose and scaly, the pale brown, hyaline-reticulate scales not over 2 × 0.5 mm; lvs scattered, thin and soft, deciduous; petioles castaneous or dark purple basally, usually greenish or stramineous distally, 2.5–15 cm, the fertile ones commonly longer than the sterile; sterile blades 3–8 × 2–5 cm, 2–3 times pinnate with confluent, flabellate to ovate, distally toothed pinnules 5–15 × 3–10 mm; fertile blades with longer and narrower, sometimes linear pinnules up to 2 cm × 3(–5) mm, the rather narrowly reflexed, evidently hyaline-scarious indusial margin more strongly modified than in *C. crispa*; $2n=60$. Calcareous rocks and moist shady slopes, circumboreal, in Amer. from Nf. to Alas., s. to n. N.J. (formerly), n. Pa., W.Va., Ill., Io., Utah, and Wash.

3. ASPIDOTIS Copeland.   Lvs glabrous, evergreen, shiny, often dimorphic; petiole slender and wiry, mostly brown; blade 2–4 times pinnate, the ultimate segments ± elongate, distantly toothed, with free veins; margins of the fertile segments reflexed and abruptly white-scarious (in ours), forming a well defined common false indusium for the submarginal sori; small, mesophytic rock-ferns with a short rhizome beset with firm, narrow, dark brown scales. 3 Amer., 1 Afr.

1. Aspidotis densa (Brackenr.) Lellinger. Pod-fern. Lvs clustered on the much-branched rhizome, usually all alike and fertile, the sterile lvs, when present, few and smaller than the fertile and with somewhat smaller, relatively broader, sharply toothed or incised segments; petioles castaneous, 4–18 cm, commonly longer than the blade; fertile blade 2.5–7 cm, a third to fully as wide, often deltoid, tripinnate, the rachis proximally dark like the petiole, distally greenish; pinnae mostly 4–7 offset pairs, the lowest ones the largest; ultimate segments numerous and crowded, ± confluent, linear or elliptic-linear, 3–12 cm × 1–2 mm, tapering to a cartilaginous-mucronate tip; sori set on the indusial flap near its line of junction with the unmodified part of the reflexed margin; $2n=60$. Cliff-crevices and moist, rocky slopes; cordilleran region of w. U.S., and locally on serpentine in s. Ont. and Que. (*Cryptogramma d.; Pellaea d.; Cheilanthes siliquosa.*)

4. CHEILANTHES Swartz, nom. conserv. Lip-fern. Lvs glabrous or more often evidently hairy or scaly, evergreen, uniform; petiole slender and wiry, dark and shining; lvs 2–4 times pinnate, the ultimate segments small, with clavate-tipped free veins; sori borne on the vein-ends just within the margins of the pinnules, often confluent, the point of attachment covered by the reflexed margins, which form a ± continuous, unmodified or partly scarious false indusium, but the mature sporangia often conspicuously exserted; small, ± xeromorphic rock-ferns, the rhizome beset with slender, brown to blackish scales. 100+, widespread, mostly New World.

1 Petiole and rachis hairy or glabrous, but without scales.
  2 Lvs evidently tomentose or villous-hirsute beneath.
    3 Lvs woolly beneath; rhizome short, multicipital, the lvs tufted .... 1. *C. feei.*
    3 Lvs villous-hirsute beneath; rhizome shortly creeping, the lvs subdistant .... 2. *C. lanosa.*
  2 Lvs glabrous or nearly so beneath .... 3. *C. alabamensis.*
1 Petiole and rachis with flattened white scales mixed with jointed hairs.
  4 Scales linear-subulate, mainly confined to the petiole and rachis .... 4. *C. tomentosa.*
  4 Scales lanceolate, borne on the costae as well as on the petiole and rachis .... 5. *C. eatonii.*

1. Cheilanthes feei Moore. Slender l.-f. Rhizome short, multicipital, its scales linear-subulate, 3–6 × 0.5 mm, orange-brown with black midstripe; lvs tufted, 5–20 cm, the petiole to 10 cm, equaling or shorter than the blade, dark purplish brown, long-hairy, not scaly; blade linear-oblong, 3–13 × 1.5–4 cm, 3–4 times pinnate, loosely and copiously villous-tomentose (usually tawny or rusty) beneath; pinnae 6–12 pairs, spaced out below, petiolulate, ovate to ovate-oblong; pinnules 3–6 pairs, their ultimate segments 2–3 pairs, commonly only 1–1.5 mm, the margins loosely inrolled but unmodified and not covering the mature sporangia, which often spread out over the lower surface; $2n=87$ (triploid). Calcareous bluffs and rocks; Wis. and Ill. to Alta. and B.C., s. to Ark., Tex., and Ariz.; disjunct in Va.

2. Cheilanthes lanosa (Michx.) D. C. Eat. Woolly l.-f. Rhizome rather shortly creeping, its scales lance-linear, 2–3 mm × ca. 0.3 mm, brown with a dark midstripe in age; lvs somewhat scattered, 1–3 dm, the 2–8 cm petiole much shorter than the blade, purplish, hirsute but not scaly; blade lance-linear, 2–5 cm wide, bipinnate-pinnatifid to subtripinnate, green and sparsely hairy above, villous-hirsute beneath with shining, whitish, jointed hairs; pinnae 12–20 pairs, ovate, petiolulate; pinnules 7–10 pairs, their ultimate segments ovate, obtuse, decurrent, entire, the slightly and irregularly recurved (but otherwise unmodified) margin scarcely covering the few sori; $2n=60$. Cliffs and shale outcrops, mostly in subacid soil; Conn. and N.Y. to Wis. and Minn., s. to Ga. and Tex.

3. Cheilanthes alabamensis (Buckley) Kuntze. Alabama l.-f. Rhizome rather shortly creeping, its numerous scales linear-subulate, 2–3 mm × 0.25 mm or less, orange-brown, concolorous; lvs not closely tufted, 11–30 cm, the 2–9 cm petiole equaling or shorter than the blade, black, hairy on the upper side, not scaly; blade lance-linear, 2–4.5 cm wide, bipinnate to bipinnate-pinnatifid, glabrous or with a few scattered septate hairs; pinnae 18–25 pairs; pinnules oblong or linear-oblong, 2–8 mm, acute, entire or auriculate at the upper base or pinnatifid with 1–3 pairs of segments, the margins inrolled and with a modified, indusial edge; $2n=87$ (triploid). Calcareous bluffs and rocks; sw. Va. to Tenn., and Mo., s. and w. to Ga., Tex., Ariz., and Mex.

4. Cheilanthes tomentosa Link. Woolly l.-f. Rhizome short, erect, multicipital, its scales linear, 5–6 mm, brown with a dark central band; lvs tufted, 15–40 cm, the 5–11 cm petiole evidently shorter than the blade, hirsute-lanuginose and with intermingled linear-subulate white scales up to ca 0.1 mm wide; blade lance-linear or lanceolate, 1.5–4 cm wide, tripinnate, green and sparsely villous above, densely rusty-woolly beneath, tripinnate, the rachis woolly and with intermingled slender, hair-like scales like those of the petiole, but these scarcely or not at all extending to the costae and lower surface; pinnae 15–25 pairs, lance-ovate; pinnules 6–11 pairs; ultimate segments obovate or suborbicular, typically 1–2 times as long as wide, decurrent, the margins inrolled and with a pale, scarious edge; $2n=87$ (triploid). Rock ledges in subacid soil; w. Va., W.Va., and Ky., s. and w. to Ga., Ark., Tex., and Mex.; reported from Pa. (*C. lanosa* sensu Fernald, misapplied.)

5. Cheilanthes eatonii Baker. Eaton's l.-f. Much like no. 4, but more scaly, the scales lanceolate, mostly 0.2–1 mm wide, borne on the rachis and costae (especially the lower side) and to some extent on the lower surface of the pinnules, where intermingled with the rusty tomentum; terminal segment of the pinnae and pinnules more elongate, typically 2–3 times as long as wide; spores 32; $2n=87$ (triploid). Cliff-crevices and talus-slopes; mts. of Va. and W.Va.; Okla. and Tex. to Utah and Ariz. (*C. castanea,* the phase with the upper lf-surface thinly hairy to glabrate, the only phase in our range.)

5. NOTHOLAENA R. Br.   Cloak-fern.   Much like *Cheilanthes* and *Pellaea,* and not sharply distinct from either; axes of the pinnae and pinnules in most spp. (including ours) upcurled when dry, exposing the lower lf-surface; margins of the lf-segments flat to revolute, scarcely or not at all modified. 75, mostly dry New World.

1. **Notholaena dealbata** (Pursh) Kunze.   Powdery c.-f. Rhizome-scales brown, concolorous, 2–5 mm; lvs tufted, evergreen, 7–15 cm, the slender petiole 3–9 cm, reddish-brown, glabrous; blades monomorphic, deltoid-pentagonal, tripinnate or tripinnate-pinnatifid, up to 5 × 4 cm, the rachis glabrous; pinnae opposite, petiolulate, deltoid-ovate, ca 4 pairs, the lowest the largest, the upper smaller and simple, sessile or adnate; ultimate segments spatulate (terminal flabellate), sessile, 2–3.5 mm, cuneate at base, often crenate at the apex, pruinose above, copiously white-powdery beneath, the margin only slightly recurved and scarcely modified; sporangia large, mostly solitary; 2n=54. Calcareous cliffs; Mo. and Nebr. to Okla. and c. Tex. (*Cheilanthes d.; Pellaea d.*)

6. PELLAEA Link, nom. conserv.   Cliff-brake.   Lvs firm, evergreen, glabrous or sparsely hairy, 1–4 times pinnate, with free veins, the pinnae and often also the pinnules narrowed to a minutely or evidently petiolate base; ultimate segments folded downward along the midrib under water-stress, concealing the lower surface; petioles wiry, often breaking off well above the ground to leave a persistent base; sori borne on the vein-ends just within the margins of the lf, often confluent, the point of attachment covered by the reflexed margins, which form a continuous false indusium, but the mature sporangia often conspicuously exserted; small, ± xeromorphic ferns, many spp. (including ours) with short, branched rhizomes forming a multicipital caudex that is so densely beset with long, narrow, brown scales as to appear brown-woolly. 85, New World, S. Afr., and New Zealand.

1 Lvs monomorphic, lax; petiole and rachis glabrous or slightly hairy . . . . 1. *P. glabella.*
1 Lvs slightly dimorphic, stiffly erect; petiole and rachis strongly hairy . . . . 2. *P. atropurpurea.*

1. **Pellaea glabella** Mettenius.   Smooth c.-b. Similar to *P. atropurpurea,* but with monomorphic lvs mostly 10–20 cm, the pinnae somewhat decurrent, gradually departing at acute angles to the rachis; basal pinnae with 3–7 segments, less than twice as long as the upper; longest ultimate segments mostly 1–2 cm; petiole and rachis glabrous or slightly hairy; scales of the rhizome discrete, uniformly rust-brown, lustrous, the tips not entangled; margins of the fertile segments usually strongly reflexed, leaving but few sporangia exposed; ours mostly tetraploid (2n=116) and apogamous, the sporangia with 32 spores. Calcareous cliffs or bluffs; Vt. to Va. and Tenn., w. to Minn., Kans., and Tex., and in the western cordillera. Our pls are var. *glabella.* (*P. atropurpurea* var. *bushii.*)

2. **Pellaea atropurpurea** (L.) Link.   Purple c.-b. Rhizome-scales 3–6 mm, appressed, matted, dull tawny or rusty, the young tips tan and entangled; lvs slightly dimorphic, stiffly erect, mostly 20–40 cm; petiole shorter than the blade, purplish-brown, obviously hairy (as also the rachis) especially on the upper side; pinnae 5–11 pairs, petiolulate, the sterile ones ovate to broadly oblong, the basal mostly at least twice as long as the upper and with 3–15 segments, the ultimate segments mostly 1–7.5 × 0.5–1 cm; fertile segments narrower than the sterile, linear-oblong, acute, mucronate, the irregularly revolute margin opening and commonly leaving many sporangia exposed, eventually becoming flat; 2n=87, an apogamous triploid with 32 spores per sporangium. Calcareous rocks and open woods; s. Que., Vt., and R.I. to Minn. and Wyo., s. to Fla., Ariz., and Mex.; disjunct in Sask., Alta., and B.C. Thought to have originated by hybridiziation of *P. glabella* with another undetermined sp.

FAMILY **ASPLENIACEAE**, the Spleenwort Family

Sori borne on the lower surface of the lf, provided with an indusium unrelated to the lf-margin, or without an indusium; sporangia thin-walled and with a well developed vertical annulus interrupted by the stalk; spores (32)64, bilateral; rhizome scaly; lvs simple to more often pinnatifid or innately compound or decompound, with free or seldom net venation, all about alike or only slightly dimorphic; petiole with 2–several vascular bundles, slender or often coarse. 50 +/3000+. (Aspidiaceae.)

1 Indusium elongate, flap-like, with a linear attachment.
  2 Lvs soft, deciduous; indusia (especially the upper) often crossing veins .... 3. *Athyrium.*
  2 Lvs firm, evergreen; indusia not crossing veins.
    3 Lvs simple, entire or inconspicuously toothed.
      4 Many of the veins of the lf anastomosing; lvs rooting at the slender tip (*A. rhizophyllum*)
      .... 1. *Asplenium.*
      4 Veins all free; lvs not rooting.
        5 Blade cordate at the base; indusia numerous in pairs facing each other along the veins
        .... 2. *Phyllitis.*
        5 Blade tapering at the base; indusia only 2, each borne just within the lf-margin and
          subtending a compound sorus running the length of the blade (*A. septentrionale*)
          .... 1. *Asplenium.*
    3 Lvs pinnatifid to tripinnate, or sometimes merely forked .... 1. *Asplenium.*
1 Indusium short, attached at a point (or underneath the sorus), or absent.
  6 Indusium inferior, i.e., borne under the sorus and surrounding the sporangia, deeply lacerate into
    broad or filamentous segments, often inconspicuous .... 5. *Woodsia.*
  6 Indusium attached at one side of the sorus, or peltate and centrally attached, or wanting.
    7 Indusium hood-like, its free tip arched over the sorus and commonly thrown back as the
      sorus expands .... 4. *Cystopteris.*
    7 Indusium not hood-like.
      8 Indusium peltate, with a central stalk, opening all around; lvs evergreen .... 9. *Polystichum.*
      8 Indusium attached laterally, or absent; lvs deciduous or (spp. of *Dryopteris*) evergreen.
        9 Lf-segments ciliate; acicular unicellular hairs present on the rachis above .... 7. *Thelypteris.*
        9 Lf-segments not ciliate; acicular hairs wanting.
          10 Indusium none; petiolar bundles 2; rhizome long and slender .... 6. *Gymnocarpium.*
          10 Indusium present; bundles 3–7; rhizome short and stout .... 8. *Dryopteris.*

1. ASPLENIUM L. Spleenwort. Petiole mostly wiry, green to black, sparsely scaly, otherwise glabrous, with 2 bundles below, these united above into a single bundle mostly X-shaped in section; lvs clustered, the blade simple and entire to thrice pinnate; veins free except in *A. rhizophyllum*; sori on the lower lf-surface, ± elongate, the indusia attached at one side, all facing distally, thin and hyaline, glabrous, entire; mostly ± evergreen ferns with short-creeping to suberect rhizomes and membranous to coriaceous lvs. 600+, mostly warm reg. (*Camptosorus.*)

Our spp. produce ± sterile hybrids and also enter into a large polyploid complex. The following names are believed to apply to sterile hybrids, as indicated:

A. × *alternifolium* Wulf. = *A. septentrionale* × *trichomanes*
A. × *clermontiae* Syme = *A. ruta-muraria* × *trichomanes*
A. × *ebenoides* R. R. Scott = *A. platyneuron* × *rhizophyllum*
A. × *gravesii* Maxon = *A. bradleyi* × *pinnatifidum*
A. × *herb-wagneri* Taylor & Mohlenb. = *A. pinnatifidum* × *trichomanes*
A. × *inexpectatum* (E. L. Braun) Morton = *A. rhizophyllum* × *ruta-muraria*
A. × *kentuckiense* McCoy = *A. pinnatifidum* × *platyneuron*
A. × *shawneense* (Moran) H. E. Ballard = *A. rhizophyllum* × *trichomanes*
A. × *trudellii* Wherry = *A. montanum* × *pinnatifidum*
A. × *virginicum* Maxon = *A. platyneuron* × *trichomanes*
A. × *wherryi* D. M. Smith = *A. bradleyi* × *montanum*

1 Lvs merely toothed or entire, or sometimes irregularly forked.
  2 Lvs irregularly few-forked, or linear and grass-like, not rooting .... 10. *A. septentrionale.*
  2 Lvs neither forking nor grass-like, tapering to a flagelliform rooting tip .... 9. *A. rhizophyllum.*
1 Lvs pinnatilobate or 1–3 times pinnate.
  3 Lvs merely pinnatilobate, the segments (except sometimes the lowest ones) broad-based and con-
    fluent .... 8. *A. pinnatifidum.*
  3 Lvs 1–3 times pinnate with stalked or narrow-based and sessile pinnae.
    4 Lvs 2–3 times pinnate, at least toward the base.
      5 Petiole wholly green; pinnae evidently stalked .... 1. *A. ruta-muraria.*
      5 Petiole brown at least toward the base; pinnae sessile or inconspicuously short-stalked.
        6 Petiole distally green; blade mostly lanceolate, not parallel-sided .... 2. *A. montanum.*
        6 Petiole brown at least toward the base; blade ± oblong, almost parallel-sided .... 3. *A. bradleyi.*
    4 Lvs merely once pinnate, the pinnae mostly crenate or entire.
      7 Rachis purple-black or dark brown.
        8 Lvs slightly dimorphic, the sterile ones prostrate or arching and much shorter than the erect
          fertile ones .... 4. *A. platyneuron.*
        8 Lvs monomorphic.
          9 Pinnae oblong, usually more than twice as long as wide .... 5. *A. resiliens.*
          9 Pinnae nearly oval, not much longer than wide .... 6. *A. trichomanes.*
      7 Rachis green .... 7. *A. viride.*

1. Asplenium ruta-muraria L. Wall-rue. Rhizome 1.5–2 mm thick, the lanceolate scales 1.5–4 mm; lvs numerous, spreading, 4–17 cm, the petiole to 7 cm, green blade ovate or rhombic, firm, bipinnate or bipinnate-pinnatifid below, merely pinnate above, the rachis green; pinnae alternate, ca 6 pairs, long-petiolulate (petiolule 2–5 mm), the basal ones with 3–6 flabellate or rhombic, finely toothed and sometimes deeply lobed pinnules to 12 mm; $2n=144$. Rock-crevices and ledges, mostly on limestone; interruptedly circumboreal, in Amer. from Mass. and Vt. to Mich., s. to Ala., Ill., Mo., and Ark. (*A. cryptolepis,* the Amer. pls.)

2. Asplenium montanum Willd. Mt.-s. Rhizome ca 1 mm thick, the numerous lanceolate scales 3–4 mm; lvs numerous, spreading, 5–20 cm, the petiole 2–10 cm, dark brown toward the base, green distally; blade deltoid-ovate to deltoid-lanceolate, to 7(–10) cm wide, firm, pinnate-pinnatifid to bipinnate-pinnatifid, the rachis green; pinnae 5–12 pairs, alternate, shortly (to 2 mm) petiolulate or sessile, deltoid, with up to 5 pairs of segments, these oblong to spatulate, coarsely toothed or incised; $2n=72$. Cliff-crevices in noncalcareous rocks; Mass., s. Vt., and s. N.Y. to O. and Ky., s. to Ga. and Ala.

3. Asplenium bradleyi D. C. Eat. Bradley's s. Rhizome 1–1.5 mm thick, the numerous narrowly lanceolate scales blackish, 2–4.5 mm; lvs numerous, spreading, 7–25 cm, the petiole brown, to 10 cm; blade linear-oblong, to 20 × 4.5 cm, pinnate-pinnatifid to bipinnate, the rachis brown proximally, green for the distal third or half, bearing scattered slender scales; pinnae 8–15 pairs, alternate or the basal subopposite, short-stalked to oblong-lanceolate, sometimes with only one free, superior basal segment (otherwise incised-serrate) or pinnate throughout with 5+ pairs of segments, these obtuse, sharply serrate. Rare, on noncalcareous cliffs and rocks or acid soil; s. N.Y. and N.J. to Ga. and Ala., w. to O., Mo., and Okla. Name applied to both the sterile hybrid ($2n=72$) between spp. 2 and 4, and to the fertile alloploid ($2n=144$) derived from it.

4. Asplenium platyneuron (L.) Oakes. Ebony-s. Rhizome 2–3.5 mm thick, the few linear-subulate black scales 2.5–4 mm; lvs subdimorphic, 5–50 cm, with dark brown petiole and rachis; sterile blades spreading, 5–10 cm, very shortly petiolate, pinnate, with 15–22 pairs of sessile, alternate, crowded, oblong pinnae 7–10 × 3.5–4.5 mm, these rounded and remotely serrulate, tending to be auriculate at the upper base; fertile lvs few, erect, to 4.5 dm × 6 cm, short-petiolate, the blade linear-oblong, acute, narrowed basally, pinnate, the 30–50 pairs of alternate pinnae less crowded, more linear-oblong (except the reduced, more deltoid lowermost ones), to 3.5 cm × 6 mm, more prominently auriculate at the upper base and sometimes at the lower base as well, mostly serrate, seldom more deeply cleft; $2n=72$. Woods, banks, and rocks in circumneutral soil; Que. and Ont. to se. Minn., Kans., and Colo., s. to Fla., Tex., and Ariz.

5. Asplenium resiliens Kunze. Black-stalked s. Rhizome 1–2 mm thick, the few linear-subulate dark brown scales 2.5–3.5 mm; lvs numerous, erect, uniform, 5–30 cm, the petiole to 7.5 cm, dark brown or purplish-black; blade linear-oblong, to 25 × 2.5 cm, pinnate throughout, acute above, narrowed below, thin, the straight, rather thick rachis almost wholly dark, with a few septate hairs but not scales; pinnae 14–30 pairs, mostly opposite, slightly deflexed, mostly 8–12 × 3–4 mm, mostly oblong, auriculate at the upper (and sometimes also the lower) base, rounded, slightly or scarcely crenulate, subsessile; spores 32 per sporangium; $2n=108$, an apogamous triploid. Crevices of calcareous rocks, or in circumneutral soil; s. Pa. to Ill., Mo., and Kans., s. to trop. Amer.

6. Asplenium trichomanes L. Maidenhair-s. Rhizome 1.5–2 mm thick, the lanceolate brown scales 1.5–2 mm; lvs numerous, 5–25 cm, the purple-brown petiole to 6 cm; blades uniform, linear, pinnate throughout, to 19 × 1.7 cm, acute to attenuate above, usually not narrowed below, thin, dark green, the thin, flexible rachis almost wholly dark; pinnae 9–20 pairs, mostly opposite, oval to round oblong, oblique with the upper side the larger, but not auriculate, 4–8 × 3–6 mm, mostly crenate, subsessile; $2n=72, 144$. On shaded, mostly calcareous rocks; interruptedly circumboreal, and nearly throughout our range, s. to Ga. and Tex.

7. Asplenium viride Hudson. Green s. Rhizome 1 mm thick, the scales few, 3–4 mm; lvs numerous, 5–15 cm, the petiiole to 4 cm, brownish at base, green distally; blade thin, not evergreen, linear-oblong, pinnate throughout, up to 1.5 cm wide, acute, scarcely narrowed toward the base, the rachis green and flattened; pinnae 9–16 pairs, subopposite at base, alternate upwardly, sessile, 3–9 × 2–4 mm, obtuse, cuneate at base, bluntly toothed, not auriculate; $2n=72$. Shaded limestone rocks; circumboreal, in Amer. s. to N.S., n. Vt., n. N.Y., Wis., Colo., and Wash.

8. Asplenium pinnatifidum Muhl. Lobed s. Rhizome 1 mm thick, its scales 4–6 mm; lvs spreading, 1–3 dm, the petiole to 12 cm, brown basally or throughout; blade firm, lance-linear, to 2.5 cm wide, deeply pinnatifid or pinnate at the base, merely lobed above, the caudate-attenuate tip wavy or entire; rachis green; pinnae or lobed to 7 pairs or more, the basal ones mostly subrotund to ovate, bluntly toothed; $2n=144$, believed to be an alloploid from spp. 2 and 9. Crevices in noncalcareous rocks, and in acid soil; N.J. and Pa. to Ga., w. to Ill., Mo., and Okla.

9. Asplenium rhizophyllum L. Walking fern. Rhizome ca 2 mm thick, the scales lance-subulate, entire, dark brown, iridescent; petiole flattened, winged, green except at the dark brown scaly base; blade simple, 5–30 cm, undulate-margined, generally cordate at base with rounded auricles, to 3 cm wide at base, tapering gradually to a flagelliform rooting tip; veins anastomosing, the areoles mostly in 2 rows, the outer veinlets free; $2n=72$. Mostly on rocks, esp. limestone; Que. and Ont. to Minn., s. to Ga. and Okla. (*Camptosorus r.*)

10. Asplenium septentrionale (L.) Hoffm. Forked s.; grass-fern. Plants forming large, very dense tufts, with numerous, crowded, slender lvs 5–15 cm, grass-like in aspect; petiole greenish, very slender, ca 0.5 mm thick, ± erect, generally longer than the blade; blades irregularly forked, with a few narrow segments (or some unforked), each segment 1–2 cm × 1–2 mm, entire or with a few long, slender teeth (reduced pinnules), and with a single compound sorus (subtended by a ± continuous indusium just within the margin on each side) running its whole length; $2n=144$. Interruptedly circumboreal, in Amer. mostly cordilleran, but disjunct on shale in Monroe and Hardy cos., W.Va.

2. PHYLLITIS Hill. Hart's tongue.

Petiole coarse, scaly, with 2 curved bundles at the base; lvs entire; veins pinnate, free, twice forked, ending short of the margin in linear hydathodes; sori linear, paired, one on the anterior branch of a vein, the other on the posterior branch of the next higher vein, the membranous, entire indusia facing each other and overlapping when young; evergreen terrestrial ferns with erect scaly rhizomes and numerous firm, glossy green, elongate, simple lvs. 8, widespread.

1. **Phyllitis scolopendrium** (L.) Newman. Rhizome-scales lance-ovate, 5–6 × 1–2 mm, glabrous, iridescent; petiole 4–12 cm × 1–2 mm, sulcate above, densely scaly when young; blade up to 35 × 4 cm, acute or acutish, cordate-auriculate at base, sparsely scaly beneath, especially along the midrib; sori up to 2 cm; $2n=144$. On limestone rocks and in sink-holes; rare and local; N.B., Bruce Peninsula of Ont., upper penins. of Mich., c. N.Y., and e. Tenn. and n. Ala. Our native plants are var. *americana* Fern. The Eurasian var. *scolopendrium,* diploid ($2n=72$), somewhat larger, and differing in a number of minor details, is established here and there as an escape from cult.

3. ATHYRIUM Roth.

Petiole stramineous, stout, with 2 bundles at the base, these uniting upwardly into a single trough-shaped bundle; blade pinnate to tripinnate, sparingly scaly, otherwise glabrous or sparsely hairy; veins free, simple or forked, reaching the margin; indusia borne on the anterior side of the vein, often recurved and crossing the vein and thus hooked or horseshoe-shaped, rarely back to back, subentire or toothed, often ciliate; large wood-ferns with thick, creeping to suberect rhizomes and thin, light green, deciduous lvs, the rhizome and often also the petioles with conspicuous, entire, fibrous scales; rhizome in our spp. beset with old swollen, starch-filled petiole-bases. 150+, cosmop.

1 Blade merely pinnate, the pinnae linear and subentire .... 1. *A. pycnocarpon.*
1 Blade merely pinnate-pinnatifid to tripinnate-pinnatifid.
  2 Blade merely pinnate-pinnatifid; veins simple; sori straight or nearly so .... 2. *A. thelypterioides.*
  2 Blade mostly bipinnate or even tripinnate, the lower segments of the pinnae mostly discrete,
    sometimes even stalked; veins forking; sori often hooked .... 3. *A. filix-femina*

1. **Athyrium pycnocarpon** (Sprengel) Tidestrom. Glade-fern. Rhizome ± creeping; lvs scattered, 6–11 dm, the blade longer than the petiole, 11–20 cm wide, gradually tapering above, only slightly reduced below, glabrous except for a few septate brown hairs on the rachis, costae, and veins; pinnae 20–30 pairs, subdimorphic, the fertile one slightly narrower than the sterile; sterile pinnae narrowly lance-linear, 7–12 cm × 10–13 mm, subsessile and rounded or subtruncate at the base, gradually tapering to the base, subentire or slightly toothed, the veins simple or forked, tending to run to the sinuses; indusia glabrous, narrow, elongate, straight or nearly so, following the anterior vein-branch and often reaching from near the costa to near the margin; $2n=80$. Cool woods and talus slopes in circumneutral soil; s. Que. to Ont. and Minn., s. to Fla. and La. (*Asplenium p.; Diplazium p.; Homalosorus p.*)

2. **Athyrium thelypterioides** (Michx.) Desv. Silvery glade-fern. Rhizome short-creeping, the lvs at the front end, 3.5–11 dm, the blade longer than the petiole, 13–22 cm wide, sharply acuminate, slightly reduced below, pinnate-pinnatifid, the rachis septate-hairy and sparsely scaly, pinnae 15–20 pairs below the pinnatifid tip, deeply pinnatifid, the costal wing 0.5–1 mm wide, the segments oblong, barely to strongly toothed, 6–13 × 3–6 mm; veins simple or forked, directed into the teeth; indusia 3–7 pairs per segment, light brown, firm, elongate, mostly 1.5–2 mm, straight or nearly so, seldom crossing the veins; $2n=80$. Rich moist woods; N.S. and N.B. to Ont., and Minn., s. to Ga., La., and Okla. (*Asplenium acrostichoides; Diplazium t.*)

3. **Athyrium filix-femina** (L.) Roth. Lady-fern. Rhizome short-creeping to suberect, the lvs clustered near its tip, 4–10 dm; blade 10–35 cm wide, acuminate, slightly reduced below, sparsely scaly, otherwise glabrous or minutely glandular, mostly bipinnate or bipinnate-pinnatifid; pinnae 20–30 pairs below the pinnatifid tip, lance-linear, subsessile, attenuate-acuminate, the pinnules mostly serrate to deeply parted, obtuse to acuminate; veins forking, directed into the teeth; indusia dark brown, thin, ciliate (at least when young) mostly short and ± curved or hooked, commonly crossing the veins; $2n=80$. Moist woods, meadows, and streambanks; cosmopolitan and highly variable, two vars. in our area:

Var. **michauxii** Mettenius. Petiole commonly ca half as long as the blade, the basal scales often blackish; blade widest near the middle, the fourth or fifth pair of pinnae the largest; cilia of the indusia eglandular; spores yellowish to brownish. Nf. and Que. to Ont. and S.D., s. to Md., Va., Ky., and Mo. (*A. angustum.*)

Var. **asplenioides** (Michx.) Farw. Petiole nearly or fully as long as the blade, the basal scales pale; blade widest below the middle, the second or third pair of pinnae the largest; cilia of the indusium gland-tipped; spores dark brown or blackish. Fla. to e. Tex., n. to R.I., Pa., s. Ind., Mo., Okla., and e. Kans. (*A. asplenioides.*)

4. CYSTOPTERIS Bernh., nom. conserv.   Bladder-fern. Petiole slender, with 2 bundles, scaly only at the base, much shorter than the blade in our spp.; blade mostly 2–4 times pinnate or ternate-pinnate; veins forking, free, reaching the margin, slightly thickened at the tip; sori rotund, each borne on a veinlet that continues to the margin; indusium hyaline, basiscopic (i.e., attached under the proximal margin of the sorus on the entering veinlet), ± hood-like, its free tip arched over the sorus (towards the distal margin of the pinnule) and commonly thrown back as the sorus expands, sometimes tiny and caducous; delicate woodland ferns with deciduous lvs, the petiole-bases not persistent. 10, cosmop.

Our spp. of *Cystopteris* enter into a polyploid complex that compromises the essential differences among the species. *C. bulbifera* and *C. protrusa* are diploid, $2n=84$. *C. fragilis* is tetraploid ($2n=168$) in our area, but is considered to include some hexaploids and octoploids in the Old World. Two other established polyploids combine the features of *C. bulbifera* with those of *C. protrusa* (in one case) or *C. fragilis* (in the other). *C. tennesseensis* Shaver (tetraploid) may be an alloploid derived from *C. bulbifera* and *C. protrusa,* and *C. laurentiana* (Weatherby) Blasdell may be an alloploid derived from *C. bulbifera* and *C. fragilis* var. *fragilis.* Both of these apparent alloploids usually have some poorly developed or abortive bulbils along the lf-rachis, and they tend to have some minute glands on the indusia. In *C. tennesseensis* the blade is widest at the base, as in *C. bulbifera,* whereas in *C. laurentiana* it is widest above the base, as in *C. fragilis. C. laurentiana* is northern, occurring from Nf. and Que. to Wis. and n. Minn., with a disjunct station in Centre Co., Pa. *C. tennesseensis* is more southern, occurring from Va. and Ky. to Kans. and Okla. There are also some ± sterile polyploids in this complex. *C.* × *illinoensis* R. C. Moran (triploid, $2n=126$) may reflect hybridization of *C. bulbifera* with *C. fragilis* var. *mackayi,* and *C.* × *wagneri* R.C. Moran (tetraploid, $2n=168$) may reflect hybridization of *C. tennesseensis* with *C. fragilis* var. *mackayi.*

1 Lowest pair of pinnae a little shorter than (or merely equaling) the next pair; no bulbils.
2 Rhizome short, densely scaly, the growing point not produced beyond the clustered lvs .... 1. *C. fragilis*
2 Rhizome long-creeping, sparsely scaly but densely villous, the growing point produced 2–4 cm
    beyond the often more scattered lvs .... 2. *C. protrusa*
1 Lowest pair of pinnae a little longer than the next pair; rachis commonly bulbiliferous .... 3. *C. bulbifera*

1. Cystopteris fragilis (L.) Bernh.   Brittle b.-f. Lvs glabrous or inconspicuously glandular, in a small cluster on a short, densely scaly rhizome, 0.5–3.5(–4) dm, the dark brown (or distally greenish) petiole shorter than or up to as long as the blade; blade mostly 3.5–20(–25) × 1.5–8 cm, (2–)2.5–4(–5) times as long as wide, 2–3 times pinnate, with 8–18 pairs of pinnae (the members of a pair often offset, especially the distal pairs), the lowest pair equaling or generally a little smaller than the next pair, the upper ones gradually reduced so that the blade tapers into a narrow tip; veins typically directed to the teeth of the pinnules, but the tooth sometimes apically slightly indented so that the vein ends in a small sinus; indusium pale, not glandular; $2n=188$ in ours. Moist, mostly wooded slopes and ledges in circumneutral soil; circumboreal, s. in our range to Va. and Mo. We have 2 vars. with widely overlapping distribution:

Var. **fragilis.** Basal pinnules broad-based and mostly sessile or subsessile; indusium up to 1 mm, usually cleft at the apex; relatively northern, s. to Mich., Vt., and in the mts. to N.C. (*C. dickieana.*)

Var. **mackayi** Lawson. Lvs less cleft, the pinnules more remote, the lowermost pinnules only slightly lobed and mostly cuneate at the base; indusium ca. 0.5 mm, or nearly obsolete, usually entire; a little more southern, from N.S. and N.B. and s. Que. to Minn. and Nebr., s. to Va., s. Ill., and Mo., and in the mts. to n. Ga. (*C. tenuis.*)

2. Cystopteris protrusa (Weatherby) Blasdell.   Lowland b.-f. Rhizome long-creeping, sparsely scaly but densely villous, the growing point produced 2–4 cm beyond the often more scattered lvs; petiole greenish or stramineous except often at the very base; minor axes of the blade with scattered glands; veins typically running into marginal teeth; lowermost pinnules mostly cuneate at the base, and often short-petiolulate; indusium ca 0.5 mm, usually entire; $2n=84$; otherwise much like no. 1. Abundant in mesic woods; N.Y. and s. Ont. to s. Minn., s. to Ga., Ala., La., and e. Kans. (*C. fragilis* var. *p.*)

3. Cystopteris bulbifera (L.) Bernh.   Bulblet b.-f. Larger and coarser than no. 1, the lvs 3–8(–15) dm, the blade 6–9 cm wide, with (15–)20–30 pairs of pinnae, the lowest pair a little larger than the next pair, the reduced upper pinnae more numerous than in no. 1, so that the lf tapers more gradually to the tip; rachis commonly with 1 or more well developed bulbils on the lower side beyond the middle, these up to 4 or 5 mm; costae with scattered pale short-stalked glands; veins mostly ending in the smaller sinuses (reflecting an exaggeration of the tendency seen in some specimens of no. 1 for the principal teeth to be apically indented); indusium brownish, rounded or truncate at the tip, bearing scattered small, pale, short-stalked glands; $2n=84$; otherwise much like no. 1.

5. WOODSIA R. Br.   Cliff-fern. Petiole slender and wiry, shorter than the blade, scaly at least at the base, with 2 bundles, these uniting distally; blade pinnate to subtripinnate, scaly or scaleless, often hairy or glandular, the veins forking and free, not reaching the margin; sori borne on the veins, often confluent in age, rotund but with an elongate receptacle; indusium inferior, borne below and ± surrounding the sporangia, consisting variously of septate hairs around a minute basal disk, or of a small sac that soon ruptures irregularly to form a small basal disk with spreading, unequal segments; small to medium-sized ferns of rocky places, with tufted lvs and persistent, elongate petiole-bases on a short rhizome that is covered with yellowish to dark brown scales. 20–25, cosmop.

1 Petioles jointed below the middle, their persistent bases all of similar length; blade not glandular; indusium a minute disk with long, septate marginal hairs.
  2 Lvs glabrous and scaleless above the base; blade up to 1.5 cm wide .... 1. *W. glabella.*
  2 Lvs hairy, the rachis and petiole also scaly; blade 1–3.5 cm wide.
    3 Segments of the pinnae mostly 2–3 pairs, not ciliate; indusial hairs few .... 2. *W. alpina.*
    3 Segments of the pinnae mostly 4–7 pairs, ciliate, indusial hairs many .... 3. *W. ilvensis.*
1 Petioles not jointed, the persistent bases of unequal length; blade often glandular; indusium somewhat expanded at the base, its segments 2–several cells wide at the base.
  4 Rachis remotely scaly; indusial lobes broad; lvs without long septate hairs .... 4. *W. obtusa.*
  4 Rachis not scaly.
    5 Blade glandular and with long, glandless, septate hairs; indusial lobes mostly broad, lacerate, with filamentous tips .... 5. *W. scopulina.*
    5 Blade glabrous or merely glandular; indusial lobes narrow and hair-like above the base .... 6. *W. oregana.*

1. Woodsia glabella R. Br.   Smooth c.-f. Rhizome scales lanceolate, $3–4 \times 1–1.5$ mm, brown, concolorous, denticulate, not ciliate; lvs 5–16 cm; petiole stramineous, glabrous, scaly only at the base, articulate below the middle; blade linear, pale green, 8–14 mm wide, slightly narrowed below, glabrous and scaleless, pinnate, with green rachis; pinnae sessile, 8–14 pairs, suborbicular to deltoid, $5–9 \times 5–6$ mm, rounded to acute, trilobed or pinnatifid with 2–3 pairs of entire or crenate segments; indusium a minute disk with a ring of numerous long, brown, septate marginal hairs; $2n=78$. Calcareous rocks; circumboreal in Amer from Nf. to Alas., s. to N.S., Mass., n. N.H., n. Vt., N.Y. (Adirondack and Catskill Mts.), Ont. and B.C.

2. Woodsia alpina (Bolton) S. F. Gray.   Alpine c.-f. Rhizome-scales lanceolate or lance-ovate, $4–6 \times 1–2$ mm, brown, denticulate, sparsely fimbriate; lvs 6–15 cm; petiole bright brown, sparingly scaly and hairy, articulate below the middle; blade linear, 1–2 cm wide, slightly narrowed below, pinnate-pinnatifid, sparsely hairy on the rachis and lower surface, the rachis also sparsely scaly; pinnae sessile, 8–15 pairs, ovate, obtuse, deeply pinnatifid, the segments 2–3(4) pairs, oblong to suborbicular, the lower sublobate, not ciliate; indusium a small disk with a few long septate marginal hairs; $2n=156$, thought to be an alloploid derived from nos. 1 and 3. Rock crevices in cool sites; circumboreal, s. to N.S., Vt., N.Y., Mich., and Minn. A sterile triploid hybrid with no. 3 is *W.* × *gracilis* (Lawson) Butters.

3. Woodsia ilvensis (L.) R. Br.)   Rusty c.-f. Rhizome-scales abundant, brown, concolorous, lanceolate, $4–6 \times 0.7–1$ mm, attenuate, denticulate, sparsely fimbriate-ciliate; lvs 5–22 cm; petiole shining brown, articulate below the middle, scaly and hirsute; blade oblong-lanceolate, 2–3.5 cm wide, slightly narrowed below, pinnate-pinnatifid, the rachis hairy and scaly; pinnae 10–16 pairs, firm, narrowly oblong, sessile, dark green, deeply pinnatifid, the (2–)4–7 pairs of segments oblong, rounded, sparingly long-ciliate, the larger ones irregularly lobed, the costae with pale, linear scales and long septate hairs; indusium a minute disk with a marginal fringe of long hairs; $2n=82$. Rock-crevices, mostly on acid substrate; circumboreal, s. to B.C., Sask., nw. Io., n. Ill., O., Pa., and in the mts. to N.C. *W.* × *abbeae* Butters is a sterile diploid hybrid with *W. oregana.*

4. Woodsia obtusa (Sprengel) Torr.   Blunt c.-f. Rhizome-scales few, lance-linear, $2–3.5 \times 0.3–0.6$ mm, attenuate, sparingly glandular-fimbriate, with a dark central stripe when mature; lvs 25–55 cm; petiole stramineous, not articulate, sparingly glandular-hairy and scaly when young; blade lanceolate, 3.5–10 cm wide, slightly narrowed below, bipinnate-pinnatifid, the rachis glandular-hairy and sparingly scaly; pinnae 13–18 pairs, rather remote, subsessile, ovate or lance-ovate, sparsely glandular-hairy on the costae, costules, and veins; pinnules 6–13 pairs, oblong, rounded, deeply pinnatifid with 4–6 pairs of rounded, crenate segments; indusial segments 4–6, several cells wide, glandular-margined, incurved and covering the sporangia when young; $2n=152–156$. Shady ledges and loose talus-slopes, mostly in circumneutral soil; Me. to Minn. and e. Nebr., s. to Fla. and Tex.

5. Woodsia scopulina D. C. Eat.   Mt. c.-f. Rhizome-scales lanceolate, 4–5 mm, hair-pointed, subdenticulate, sparsely ciliate, with an irregular black midstripe; lvs 18–40 cm, evidently stipitate-glandular (the glands clavate) and sparsely to copiously beset with ± elongate, flattened, septate white hairs at least on the rachis and the lower surface of the blade; petiole brown, hairy, scaly only at the base, not articulate; blade lance-linear, 3–7 cm wide, acute, slightly narrowed below, bipinnate, the rachis not scaly; pinnae 9–25 pairs, deltoid-ovate to lanceolate, acuminate, sessile, scaleless, the pinnules 7–10 pairs, oblong, obtuse, crenate to subpinnatifid; indusial segments 3–6, incurved, several cells wide at least below, deeply lacerate; $2n=76$. Cliffs and rocky slopes; cordilleran region from Alta. and s. Alas. to Calif. and N.M., e. irregularly to Que.; isolated in the Ozark Mts. and in the s. Appalachian Mts. from Va. and W.Va. to N.C. and Tenn. (*W. appalachiana*, the Appalachian and Ozarkian plants.) *W.* × *maxonii* Tryon is a triploid sterile hybrid with tetraploid *W. oregana*.

6. **Woodsia oregana** D. C. Eat. Western c.-f. Rhizome-scales light brown, often lacking a dark stripe, lance-linear, seldom over 0.5 mm wide, subentire; lvs 7–25 cm; petiole dark brown toward the scaly base, not articulate, without long hairs; blade lanceolate, 1–5 cm wide, acuminate, slightly narrowed below, sub-bipinnate to subtripinnate, glabrous to glandular (the glands cylindric), but without long hairs; pinnae 7–17 pairs, deltoid to lanceolate, sessile, mostly bipinnatifid with 5–7 pairs of segments, or merely pinnatifid with toothed segments; indusial segments several, narrow, 2–3 cells wide at the base, commonly uniseriate distally; $2n=76$, 152. Rock crevices; cordilleran region from B.C. and Alta. to Calif. and N.M., and irregularly e. to Okla., Wis., Sask., Que., w. N.Y., and Vt. Strongly glandular plants from the Great Lakes region (esp. in Minn. and Wis.) have been called var. *cathcartiana* (B. L. Robinson) Morton. At least some of these are tetraploid, in contrast to the diploid, less glandular or glabrous var. *oregana*.

6. **GYMNOCARPIUM** Newman. Oak-fern. Petiole stramineous with a dark base, slender, longer than the blade, sparsely scaly at the base, with two vascular bundles free to the apex; blade deltoid or pentagonal, 2–3 times pinnate; lower pinnae distinctly the largest, articulate at the base, long-petiolulate and inequilateral; veins free, forking, reaching the margin; sori on the anterior vein-branches; indusium none; rather small, delicate ferns with long, branched, scaly rhizomes and scattered, thin, deciduous lvs, the old petiole-bases not persisting. 6, N. Amer., Eurasia, and the Pacific Islands.

1 Lvs nearly or quite eglandular . . . . 1. *G. dryopteris.*
1 Lvs densely and finely glandular . . . . 2. *G. robertianum.*

1. **Gymnocarpium dryopteris** (L.) Newman. Oak-f. Petiole 10–30 cm; blade yellow-green, scaleless, deltoid-pentagonal, to 18 × 25 cm, bipinnate-pinnatifid to tripinnate-pinnatifid, glabrous or occasionally slightly glandular especially along the rachis, pinnae several pairs, opposite, the members of the lowest pair deltoid and each nearly as long as the rest of the blade, with a petiolule up to 2.5 cm long, distinctly asymmetrical, the lowest basiscopic segment evidently the largest, mostly a third as long as the main rachis or longer; ultimate segments oblong, 8–18 mm, obtuse, the margins crenate to subpinnatifid, not recurved; veins 2–6 pairs, mostly simple; $2n=160$ (ours). Cool woods and talus-slopes; circumboreal, s. to Va., N.C., O., Ill., Io., and Ariz. A vegetatively reproducing hybrid with the next is *G.* × *heterosporum* W. H. Wagner.

2. **Gymnocarpium robertianum** (Hoffman) Newman. Limestone oak-f. Much like no. 1; blade green, densely and finely glandular (as also the petiole), a little narrower, deltoid, to 14 × 16 cm; segments of the basal pinnae ovate-oblong, obtuse, sessile, the lowest basal one mostly less than a fourth as long as the main rachis; ultimate segments to 9 mm, with entire to crenate, mostly recurved margins; $2n=160$ (ours). On calcareous rocks; circumboreal, s. to Pa., Mich., Io., and Ida. (*G. jessoense; Dryopteris r.; Phegopteris r.*)

7. **THELYPTERIS** Schmidel, nom. conserv. Petiole stramineous, with 2 bundles that unite below the blade; lvs bipinnatifid or pinnate-pinnatifid, sparingly or not at all scaly, pubescent with acicular, unicellular hairs, at least on the upper side of the rachis, often also on the costae; rachis grooved, the marginal ridges continuous, elevated above the insertion of the costae; veins free, reaching the margins, running into the teeth (if present); sori intramarginal, each borne on a vein that continues to the margin; indusium when present small and inconspicuous, hyaline, often ciliate and/or glandular, reniform or horseshoe-shaped, attached at the sinus on the basipetal side of the sorus, or in some spp. wholly wanting; medium-sized wood-ferns with scattered, thin, light green, deciduous lvs and long, slender, blackish, sparingly (seldom densely) scaly rhizomes. 500+, cosmop.

1 Blades triangular, broadest at the base; indusium none.
 2 Wings of the lf-rachis not extending down to the lowest pair of pinnae . . . . 1. *T. phegopteris.*
 2 Wings of the lf-rachis extending down to the lowest pair of pinnae . . . . 2. *T. hexagonoptera.*
1 Blades narrowly lanceolate, barely to strongly narrowed at the base; indusium present.
 3 Blade strongly narrowed toward the base, the lowest pinnae rudimentary . . . . 3. *T. noveboracensis.*
 3 Blade only slightly narrowed at the base, the lowest pinnae well developed

4 Veins all simple; indusium glandular-margined .... 4. *T. simulata.*
4 Veins, or many of them, forked; indusium ciliate or rarely glabrous .... 5. *T. palustris.*

1. **Thelypteris phegopteris** (L.) Slosson.   Narrow or northern beech-fern.   Lvs scattered on long, slender, densely hairy and scaly rhizomes, to 5 dm, the pilose and scaly petiole mostly longer than the blade; blade 10–25 cm, mostly 1–2 times as long as wide, ciliate-margined, pinnate-pinnatifid, pinnae mostly 10–15 opposite or offset pairs below the pinnatifid apex, to 3 cm wide, approximate, broadly sessile and narrowly confluent along the rachis, only the lowest pair a little removed and often retrorsely divergent, commonly a little shorter than the second pair; segments of the pinnae oblong, obtuse, entire or the largest crenate to subpinnatifid; veins simple or once forked; rachis, costae and veins hairy beneath, the rachis also with abundant brown, lance-ovate scales; indusium none; $2n=90$. Moist cliffs and woods; circumboreal, s. in Amer. to N.C., Tenn., Io., and Oreg. (*Dryopteris p.; Phegopteris polypodioides.*)

2. **Thelypteris hexagonoptera** (Michx.) Weatherby.   Broad or southern beech-fern.   Much like no. 1; rhizome sparingly hairy and densely scaly; lvs 3–8 dm; petiole hairy and sparsely scaly below, smooth above; blade to 4 dm, mostly a little wider than long, the lowest pair of pinnae the largest, widely spreading, the rachis winged throughout; pinnae up to 7 cm wide; veins simple or rarely a few forked; scales of the rachis few, whitish, lance-triangular; $2n=60$. Moist woods and thickets; Que. and Me. to Ont. and Minn., s. to n. Fla. and Tex. (*Phegopteris h.; Dryopteris h.*)

3. **Thelypteris noveboracensis** (L.) Nieuwl.   New York fern.   Rhizome with a few pale, appressed, lance-ovate scales, otherwise glabrous; lvs to 9 dm, the petiole dark and scaly at the base, sparsely hairy, 12–20 cm, much shorter than the blade; blade narrowly lance-elliptic, acuminate, to 15 cm wide, pinnate-pinnatifid, gradually narrowed below, the lower pinnae reduced to mere rudiments; pinnae many pairs, sessile, lance-linear, to 15 mm wide, with linear-oblong, obtusish, subentire segments; rachis, costa, and veins hairy, without scales; veins simple or rarely a few forked; indusia pale, long-ciliate; $2n=54$. Mixed woods and swamp-margins; Nf. to Ont. and se. Wis., s. to Ga., Ala., and Ark. (*Aspidium n.; Dryopteris n.*)

4. **Thelypteris simulata** (Davenp.) Nieuwl.   Massachusetts f.   Resembling no. 5; petioles of the sterile blades to 2 dm, of the fertile to 3.5 dm, mostly a little shorter than the blades; blades a little narrower, to 5 × 1.5 dm; costae and veins only sparsely or scarcely hairy; veins simple, those of the sterile segments 4–5 pairs, of the fertile ones 6–7 pairs; indusia margined with minute glands; $2n=128$. Swamps and moist woods, in acid soil; N.S. to Va. and W.Va.; disjunct in the driftless area of Wis. (*Aspidium s.; Dryopteris s.*)

5. **Thelypteris palustris** Schott.   Marsh-fern.   Rhizome with a few appressed, ovate scales, otherwise glabrous; lvs subdimorphic; petioles black at the base, glabrous, scaleless, the sterile to 3.5 dm, shorter or longer than the blades, the fertile to 7 dm, longer than the blades; blade lanceolate or lance-oblong, to 4 × 2 dm, pinnate-pinnatifid, acuminate, only slightly narrowed at the base; pinnae lance-linear, subsessile, the sterile to 2 cm wide, with obtuse, entire, broadly oval segments to 10 × 5.5 mm, the fertile segments oblong, to 4 mm wide, with incurved margins; costae scarcely scaly, these and the costules hairy; veins 6–8 pairs per segment, those of the sterile blades once forked, those of the fertile blades simple or the basal ones mostly forked; indusia ciliate or rarely glabrous; $2n=70$. Marshes and bog-margins; widespread in the N. Hemisphere, and throughout our range. (*Aspidium thelypteris; Dryopteris thelypteris.*) Our plants are var. *pubescens* (Lawson) Fern.

8. **DRYOPTERIS** Adans., nom. conserv.   Wood-fern, shield-fern. Petiole stout, stramineous or green, with 3–7 free bundles, beset with chaffy scales, especially toward the base; blade pinnate-pinnatifid to tripinnate-pinnatifid, sometimes beset with flattened or hair-like, multicellular scales, otherwise glabrous or merely glandular, the minor axes decurrent on the major ones, the ridges on the upper side of the rachis not raised above the level of insertion of the costae; veins free, mostly ending a little short of the margin in elongate hydathodes; sori roundish, intramaginal (or nearly marginal) on the veins; indusium well developed, persistent, reniform or horseshoe-shaped, glabrous or sometimes glandular; mostly medium-sized to rather large mesophytic ferns with dark green, often evergreen lvs, the rhizomes short, stout, erect or shortly creeping, scaly and commonly covered with old petiole-bases, the lvs tufted, often in a vase-like configuration; $x=41$, at least for European and temperate American spp. 150, cosmop.

Our spp. produce frequent sterile hybrids and also enter into a polyploid complex. The following names are believed to apply to hybrids, as indicated:

D. × *algonquinensis* D. Britton = *D. fragrans* × *marginalis.*
D. × *benedictii* Wherry = *D. carthusiana* × *clintoniana.*
D. × *boottii* (Tuckerman) Underw. = *D. cristata* × *intermedia.*
D. × *burgessii* Boivin = *D. clintoniana* × *marginalis.*
D. × *dowellii* (Farw.) Wherry = *D. clintoniana* × *intermedia.*
D. × *leedsii* Wherry = *D. celsa* × *marginalis.*

D. × *neowherryi* W. H. Wagner = *D. goldiana* × *marginalis.*
D. × *pittsfordensis* Slosson = *D. carthusiana* × *marginalis.*
D. × *separabilis* Small = *D. celsa* × *intermedia.*
D. × *slossoniae* Wherry = *D. cristata* × *marginalis.*
D. × *triploidea* Wherry = *D. carthusiana* × *intermedia.*
D. × *uliginosa* Druce = *D. carthusiana* × *cristata.*

1 Lvs less dissected, mostly pinnate-pinnatifid to bipinnate.
  2 Lvs small, to 3.5 dm overall, to 5.5 cm wide, the ultimate segments to 5 mm .... 1. *D. fragrans.*
  2 Lvs larger, mostly 3–11 dm, over 5.5 cm wide, the larger segments well over 5 mm.
    3 Sori submarginal .... 2. *D. marginalis.*
    3 Sori laminar, borne well within the margins.
      4 Lvs evidently dimorphic, the deciduous fertile ones larger and more erect than the ev-
        ergreen sterile ones; fertile pinnae ± louvered in life .... 3. *D. cristata.*
      4 Lvs only slightly (in no. 4) or not at all dimorphic; fertile pinnae lying in the plane of the
        blade.
        5 Ramentum mainly restricted to the petiole, scanty or wanting from the rachis and
          costae; lvs coarsely dissected, with mostly 10–18 pairs of pinnae, the larger pinnae
          with 10–20 pairs of mostly broad-based segments, these often over 1.5 cm.
          6 Sori borne close to the midvein of the segment .... 5. *D. goldiana.*
          6 Sori borne midway between the midvein and the margin.
            7 Lvs seldom over 2 dm wide; basal pinnae relatively broad-based and trian-
              gular, mostly 1.5–2.5 times as long as wide; northern .... 4. *D. clintoniana.*
            7 Lvs mostly 2–3 dm wide; basal pinnae not notably broad-based, mostly 3–4
              times as long as wide; mostly southern .... 6. *D. celsa.*
        5 Ramentum abundant on the rachis and costae as well as on the petiole; lvs more
          finely dissected, with mostly 20–30 pairs of pinnae, the larger pinnae with (15–)20+
          pairs of segments, the larger segments (0.7–)1–1.5(–2) cm, many of them ±
          narrow-based; northern .... 7. *D. filix-mas.*
1 Lvs more dissected, bipinnate-pinnatifid to tripinnate-pinnatifid.
  8 First lower pinnule of the basal pinna short, no longer than the adjacent pinnule, not obviously
    longer or wider than the first upper pinnule; petiole, rachis, costae, and indusia bearing
    minute stalked glands .... 8. *D. intermedia.*
  8 First lower pinnule of the basal pinna longer than the adjacent pinnule, and notably longer and
    wider than the first upper pinnule; no glands, except sometimes on the indusium.
    9 Basal pinnules on the upper and lower sides of the lowest pinna subopposite, rarely more
      than 4 mm apart, the lower one up to twice (thrice) as long as the upper one; lvs deciduous
      .... 9. *D. carthusiana.*
    9 Basal pinnules on the upper and lower sides of the lowest pinna distinctly offset, 4–15 mm
      apart, the lower one mostly (2–)3–5 times as long as the upper one.
      10 Lvs rather widely spreading .... 10. *D. campyloptera.*
      10 Lvs rather closely ascending .... 11. *D. expansa.*

1. Dryopteris fragrans (L.) Schott. Fragrant w.-f. Plants *Woodsia*-like in aspect; rhizomes ± erect; lvs evergreen (and long-persistent after death), firm, fragrant, glandular and scaly; petiole short, to 7 cm; blade lance-linear, pinnate-pinnatifid, to 30 × 5.5 cm, acuminate, gradually narrowed below, the lowest pinnae very small; pinnae numerous and crowded, mostly 15–25 pairs, linear-oblong, to 27 × 8 mm, with 6–10 pairs of oblong, obtuse, crenate segments to 5 mm, not spinulose; veins obscure, 2–3 pairs per segment; sori large, often only 1–4 per segment; indusia glandular-margined; $2n{=}82$. Cliff-crevices, often on limestone; interruptedly circumboreal, s. to N.Y., Ont., n. Mich., Wis., and Minn.

2. Dryopteris marginalis (L.) A. Gray. Marginal w.-f. Rhizome ascending to erect; lvs firm, evergreen; petiole 1–2 dm, with a tuft of slender, concolorous, pale reddish-brown scales at the base and smaller, scattered scales above; blade lance-oblong, (2–)3–4.5 × 1–2.5 dm, broadest just above the base, glabrous but with a few hair-like scales, pinnate-pinnatifid to bipinnate, with 15–20 pairs of pinnae, the larger pinnae with mostly 10–15(–18) pairs of segments, the larger segments mostly (0.7–)1–2(–2.5) cm, with 6–9 pairs of veins; sori submarginal, near the sinuses between the teeth; $2n{=}82$. Woods and talus-slopes; Nf. to S.C. and Ga., w. to s. Ont., Minn., Kans., and Okla.

3. Dryopteris cristata (L.) A. Gray. Crested w.-f. Rhizome horizontal and short-creeping to somewhat ascending; lvs somewhat dimorphic, the deciduous fertile ones larger and more erect than the evergreen sterile ones; petiolar scales light brown, concolorous; fertile lvs mostly 3.5–8 dm, with mostly 10–25 pairs of pinnae, these 5–9 cm, to 2.5(–4) cm wide, the segments up to 20 × 8 mm, often broadly confluent, the lowest pinnae reduced, broadly triangular, up to about twice as long as wide; fertile pinnae twisted at an angle to the plane of the blade in life; sori midway between the midvein and the margin; sterile lvs half to three-fourths as long as the fertile; $2n{=}164$. Marshes and wet woods; Nf. to Sask. and B.C., s. to N.C., Tenn., Io., Nebr., and Ida.; Eur.

4. Dryopteris clintoniana (D. C. Eat.) Dowell. Clinton's w.-f. Sterile and fertile lvs similar or only slightly different, evergreen, to 13 dm; petiole to 3.5 dm, scaly at the base, the scales ovate, ca. 1 cm, concolorous or with darker brown center; blade lanceolate, commonly 1.5–2 dm wide, pinnate-pinnatifid, ± acuminate;

pinnae mostly 10–18 pairs, the larger ones mostly 7–11 × 2–4 cm, short-petiolulate, lance-oblong, acuminate, the 10–15 pairs of segments oblong, 10–23 × 6–9 mm, obtuse, incurved-serrate or biserrate, the teeth subspinulose; veins 5–7 pairs per segment; basal pinnae relatively short and broad-based, mostly 1.5–2.5 times as long as wide; $2n=246$, thought to be an alloploid of nos. 3 and 5. Swamps and wet woods; Me. and Que. to Wis., s. to N.Y., n. N.J., n. Pa., n. O., and n. Ind.

5. **Dryopteris goldiana** (Hook.) A. Gray. Goldie's w.-f. Rhizome mostly ascending or erect; lvs semi-evergreen, 7.5–12 dm, the petiole to 4 dm, densely scaly at the base, the subentire scales 1–2 cm, lance-attenuate, tending to be dark brown medially and paler marginally; blade ovate, 2–3.5 dm wide, abruptly acuminate, not much narrowed below, pinnate-pinnatifid, eglandular, the rachis and costae with some pale narrow scales; pinnae 12–16 pairs below the pinnatifid tip, oblong-lanceolate, mostly 3–4.5 cm wide, sub-sessile, the larger ones with 15–20 pairs of segments, the larger segments 15–30 × 6–9 mm, incurved-serrate or -biserrate, the teeth spinulose; lower basal segment of the lowest pinna shorter than the second segment; veins mostly 7–10 pairs per segment, mostly twice-forked, sori near the midvein of the segments; $2n=82$. Moist woods in rich circumneutral soil; N.B. to s. Ont. and Minn., s. to S.C., n. Ga., Tenn., and Io. A rare hybrid with *Polystichum lonchitis* in Ont. has been named *Dryostichum × simpsonii* W. H. Wagner.

6. **Dryopteris celsa** (W. Palmer) Small. Log-fern. Rhizome horizontal, short-creeping; lvs to 12 dm, deciduous; petiole 3–5 dm, with large, ± bicolored scales (dark medially, with paler margins) and small, almost hair-like, pale brown scales; blade ovate-oblong, 3.5–6(–8.5) dm, (1.5–)2–3 dm wide, not much tapered at the base, gradually tapering to the tip, pinnate-pinnatifid with up to 18 pairs of pinnae, the larger pinnae with ca 12–15 pairs of distantly serrulate, scarcely spinulose segments, the larger segments mostly 1.5–2.5 cm; a few of the basal segments of the larger pinnae contracted below; basal pinna with the first segments shorter than the adjacent ones; sori borne midway between the midvein and the margins; $2n=164$, thought to be an alloploid derived from no. 5 and the more southern *D. ludoviciana* (Kunze) Small. On rotting logs and in humus-rich soil in moist or wet places; N.J. and se. Pa. to Va., S.C., and n. Ga., w. to s. Ill., e. Mo., and Tex.; disjunct in w. N.Y. and sw. Mich. Rare and local.

7. **Dryopteris filix-mas** L. Male fern. Rhizome ascending or erect; lvs mostly deciduous, appearing more delicate and finely divided than in spp. 5 and 6; petiole a third as long as the blade, its numerous scales pale, concolored, the broad, toothed ones intermingled with narrow, hair-like ones; blade to 10+ dm, mostly 1.5–3 dm wide; pinnate-pinnatifid with numerous (mostly 20–30 pairs) of pinnae to 15 × 4 cm, the largest pinnae near or a little below the middle and with (15)20+ pairs of segments, many of the segments evidently contracted at the base; larger ultimate segments (0.7–)1–1.5(–2) cm; ramentum ± plentiful on the rachis and costae, sometimes on the lower surface of the blade, chiefly of narrow, elongate, hair-like scales, that of the rachis sometimes partly of more evident (but still narrow) scales; $2n=164$. Thickets, moist woods, and streambanks; circumboreal, extending s. to Vt., n. Mich., ne. Ill., S.D., and Mex.

8. **Dryopteris intermedia** (Muhl.) A. Gray. Fancy w.-f. Rhizome ± ascending or erect; lvs evergreen; petiole, rachis, costae, and indusia bearing minute stalked glands; petiole 1–3 dm, its broad, pale brown scales with an irregular darker area; blade 2–5 × 1–2 dm, broadest just above the obtuse base, abruptly tapered below the acuminate tip, bipinnate-pinnatifid to tripinnate-pinnatifid, the lowest pinnae usually with the lowest pair of pinnules shorter than the next pair, the lowest basal pinnule only slightly or not at all larger than the subopposite acroscopic one, the ultimate segments finely spinulose-toothed; sori midway between the midvein and the margins of the ultimate segments; $2n=82$. Moist woods and swamps; Nf. to Ga., w. to Minn. and Ark. (*D. austriaca* var. *i.; D. spinulosa* var. *i.*)

9. Dryopteris carthusiana (Villars) H. P. Fuchs. Toothed w.-f. Rhizome horizontal, short-creeping; lvs deciduous, essentially glabrous except for the chaffy, concolorous pale brown scales that beset at least the lower part of the petiole and sometimes also part of the rachis; petiole mostly a quarter to a third as long as the blade; blade 2–5+ dm, a third to half as wide, broadest a little below the middle, bipinnate-pinnatifid to tripinnate, with mostly 10–15 pairs of pinnae below the pinnatifid tip; pinnae (at least the lower ones) mostly oblique, the basal pinnule of the lower side of the lowest pinna longer than the one next to it and up to twice (thrice) as long as the subopposite upper pinnule; ultimate segments finely spinulose-toothed; sori midway between the midvein and the margins; $2n=164$. Moist or wet woods and swamps; interruptedly circumboreal, in Amer. s. to S.C., Ark., and Wash. (*D. spinulosa; D. austriaca* var. *s.*)

10. **Dryopteris campyloptera** Clarkson. Mt. w.-f. Rhizome horizontal, short-creeping; lvs deciduous, rather widely ascending-spreading, the petiole commonly about as long as the blade, beset with light brown, concolorous scales; blade mostly 3–5 × 2–3 dm, triangular to pentagonal, tripinnate-pinnatifid, with ca 15–20 pairs of pinnae; first two pinnules of the basal pinna offset by 4–12 mm, the basiscopic one evidently wider than the acroscopic one and (2–)3–5 times as long; ultimate segments finely spinulose-toothed; sori midway between the midvein and the margins; indusium sometimes with some stipitate glands, the lf otherwise glandless; rhizome horizontal, short-creeping; $2n=164$, usually thought to be an alloploid of spp. 8 and 11 but hardly to be distinguished morphologically from the latter. Moist woods; Lab. and Nf. to s. Que., w. Mass., N.Y., and Pa., and s. in the mts. to N.C. and Tenn. (*D. spinulosa* var. *americana; D. austriaca,* misapplied.)

11. **Dryopteris expansa** (C. Presl) Fraser-Jenkins & Jermy. Northern w.-f. Much like no. 10; rhizome often more ascending or suberect; lvs more upright, the petiole half to fully as long as the blade; scales with a dark center; $2n=82$. Moist woods and rocky slopes; interruptedly circumboreal, in Amer. s. to Que., s. Ont., n. Mich., nw. Minn., and Calif. (*D. dilatata.*)

9. POLYSTICHUM Roth, nom. conserv.   Petioles stout, with 4 or 5 vascular bundles, much shorter than the blades, beset with persistent, brownish, notably dimorphic scales, some broad and often toothed or cleft, others small, appressed, hair-like; blade scaly especially on the rachis and costae, otherwise glabrous or occasionally glandular, pinnate to subtripinnate, the principal pinnae subsessile on a narrow, very shortly petiolulate base, the uppermost ones reduced and confluent; pinnae or pinnules spinulose-toothed, mostly subfalcate; veins obscure, free, mostly once-forked; sori borne on the veins in one or more rows on each side of the midrib; indusium peltate, rotund, persistent; ferns with stout, scaly, erect or short-creeping rhizomes and clustered, firm, evergreen lvs. 135, cosmop.

1 Pinnae with a superior basal auricle, but not pinnatifid.
  2 Blade not reduced at the base; fertile pinnae smaller than the sterile .... 1. *P. acrostichoides.*
  2 Blade gradually narrowed below; fertile and sterile pinnae similar .... 2. *P. lonchitis.*
1 Pinnae pinnately parted, the larger ones with 9 or more pairs of pinnules .... 3. *P. braunii.*

    1. Polystichum acrostichoides (Michx.) Schott.   Christmas-fern. Rhizome short-creeping; lvs 3–7.5 dm, the green petiole a fourth to two-fifths as long as the blade; blade pinnate, 5–12 cm wide, lance-linear, acuminate, not reduced at base; pinnae alternate, 20–35 pairs, linear-oblong, acute, sharply auricled at the upper base, dark green and scaleless above, paler and with hair-like scales beneath; fertile upper pinnae abruptly smaller; sori commonly in 2 rows on each side of the midrib; indusium thick, entire, blackish wnen dry; $2n$=82. Woods and open thickets; N.S. to Wis., s. to Fla., Tex., and Mex. A hybrid with no. 2 is *P.* × *hagenahii* Cody.

    2. Polystichum lonchitis (L.) Roth.   Northern holly-fern. Rhizome erect or ascending; lvs 2–6 dm, the petiole very short; blade rigid, lance-linear, 3–7 cm wide, pinnate, gradually narrowed below; pinnae alternate, 25–40 pairs, lanceolate (the reduced lower ones subtriangular), acuminate, sharply auricled at the upper base, scaly beneath, darker green and scaleless above; fertile pinnae like the sterile; sori in a single row on each side of the midrib; indusium lacerate; $2n$=82. Cool shaded rocks and hillsides, mostly on limestone; circumboreal, s. to N.S., Que., Ont., Mich., and Calif.

    3. Polystichum braunii (Spenner) Fée.   Braun's holly-fern. Rhizome erect, lvs 3–9 dm; petiole 13–18 cm, densely scaly; blade lanceolate, 7–17 cm wide, acuminate, reduced below, essentially bipinnate; pinnae 30–40 pairs, acuminate, alternate above, the lower subopposite; pinnules 9–15 pairs on the larger pinnae, ovate-deltoid to ovate-oblong, acute, spinulose-denticulate with incurved teeth, the veins of both surfaces bearing linear scales; sori in one row on each side of the midrib; indusium pale, sublobate; $2n$=164. Upland woods and rock ledges in circumneutral soil; circumpolar, s. to Mass., N.Y., Pa., Mich., Wis., and n. Minn. The Amer. plants differ slightly from the Eurasian ones and have been segregated as var. *purshii* Fern. A hybrid with no. 1 is *P.* × *potteri* Barrington.

FAMILY **ONOCLEACEAE**, the Sensitive Fern Family

    Lvs strongly dimorphic; fertile lvs much smaller than the sterile, erect, brown to black at maturity, bipinnate with free venation; sori hidden by the inrolled margins of the pinnules, and also provided with a hyaline, extrorse, sometimes obscurely or poorly developed indusium, usually several to a pinnule, borne at the ends of the veins, round at first, later often confluent, each with an elongate receptacle; sporangia thin-walled and with a well developed vertical annulus interrupted by the stalk; spores 64, bilateral, winged, green; sterile lvs pinnate (or merely pinnatifid) with wavy-margined to deeply pinnatifid pinnae and free or anastomosing veins; petiole stout, stramineous with a dark, densely scaly base, its 2 vascular bundles wholly distinct. 3/5.

1 Sterile lvs with free venation .... 1. *Matteuccia.*
1 Sterile lvs with net venation .... 2. *Onoclea.*

    1. MATTEUCCIA Todaro.   Ostrich-fern. Coarse ferns with a stout erect crown bearing a dense spiral of large, deciduous sterile lvs and later in the season some smaller fertile ones; sterile lvs pinnate with pinnatifid pinnae, the lowermost pinnae diminutive, the venation free; fertile lvs firm, persistent throughout the winter, with spreading-ascending, linear to moniliform pinnae; indusium hyaline, hood-like, extrorse. 3, the other 2 Asian.

1. **Matteuccia struthiopteris** (L.) Todaro.   Clonal, with deep-seated long-creeping black rhizomes as well as erect leafy crowns; sterile lvs 0.5–2(–3) m, short-petiolate; blade ± hairy along the rachis and costae, oblanceolate, 1.5–5 dm wide, pinnae numerous, alternate, gradually reduced toward the base of the blade, more abruptly so upwards, long-acuminate, deeply pinnatifid with 20+ pairs of segments, these pinnately open-veined, oblong, obtuse, with slightly revolute, crenulate margin; fertile lvs produced from midsummer to early fall, not over 7 dm, the numerous pinnae 2–6 cm × 2–4 mm, obtuse, their numerous, crowded pinnules hooded; $2n=80$. Swamps and moist woods in circumneutral soil; circumboreal, in Amer. s. to Va., Mo., S.D., and B.C. (*M. pensylvanica; Pteretis nodulosa.*)

2. **ONOCLEA** L.   Sensitive fern. Ferns with long-creeping, branching, sparsely scaly rhizomes, the lvs arising from the top half of the rhizome; sterile lvs deciduous, very sensitive to frost, ± deeply pinnatifid (truly pinnate at the base) with wavy-margined to pinnatilobate segments, the venation evidently reticulate; fertile lvs produced from midsummer to early fall, firm, persistent throughout the winter, the pinnules inrolled and globular; indusium hood-like, obscurely extrorse. Monotypic.

1. **Onoclea sensibilis** L.   Sterile lvs to 1 m, the blade 18–40 × 15–35 cm, with a broadly winged rachis except near the base, the segments 8–12 pairs, opposite, 1.5–5 cm wide, with scattered white hairs on the midrib and veins beneath, glabrous above; fertile lvs smaller, the blade to 17 cm, the globular pinnules 3–4 mm wide; $2n=74$. Swamps, open woods, and meadows in neutral to slightly acid soil; Nf. and Lab. to Fla., w. to Minn., Ark., Tex., and irregularly to Colo.; e. Asia.

Family **BLECHNACEAE**, the Deer-Fern Family

Sori borne on the lower surface of the lf, elongate, in a single row on each side of the midvein of the pinna or segment, protected by an elongate, flap-like indusium that opens toward the midvein; sporangia thin-walled and with a well developed vertical annulus interrupted by the stalk; spores 64, bilateral; rhizome scaly; petiole with 2–several vascular bundles that may unite distally; lvs pinnate or pinnate-pinnatifid, with free or net venation, often ± dimorphic. 8/250.

1. **WOODWARDIA** J. E. Smith.   Chain-fern. Petiole about equaling the blade, with 2–several bundles; lvs monomorphic or dimorphic, pinnatifid to pinnate-pinnatifid, the fertile ones, when differentiated, with linear segments; blade sparsely scaly and glandular on the rachis and costae, otherwise glabrous; veins partly anastomosing, forming one or more series of areoles along the midvein of the pinnae or segment; sori superficial (our spp.) or often set in pits, oblong to linear, spanning 2 or more areoles, arranged in a chain-like row on each side of the costa; large bog-ferns with coarse, long-creeping rhizomes and scattered, erect, deciduous lvs. 10, mainly N. Hemisphere.

1 Lvs monomorphic; pinnae 15–20 pairs, distinct, pinnatifid . . . . 1. *W. virginica.*
1 Lvs dimorphic; pinnae 7–10 pairs, the sterile ± confluent, merely serrulate . . . . 2. *W. areolata.*

1. **Woodwardia virginica** (L.) J. E. Smith.   Virginia c.-f. Rhizome 6–10 mm thick; lvs monomorphic, to 14 dm, the petiole purplish-brown, shining, the blade lance-oblong, to 5 × 2.5 dm, acuminate, pinnate-pinnatifid, the pinnae 15–20 pairs, pinnatifid with a rachis 1–1.5 mm wide, the segments 15–20 pairs, ovate to broady oblong, with cartilaginous margins; veins forming a single series of areoles along the costa and costules, otherwise free, indusia thin, soon recurved, those along the costules only 1–1.5 mm, those along the costae larger; $2n=70$. Swamps and moist places in acid soil; N.S. to Fla., w. to Ont., Mich. Ill., and Tex.; Bermuda. (*Anchistea v.*)

2. **Woodwardia areolata** (L.) Moore.   Netted c.-f. Rhizome 2–3.5 mm thick; lvs dimorphic, 3–8 dm, the sterile ones with stramineous petiole and deltoid-ovate blade 10–17 cm wide, deeply pinnatifid or nearly pinnate at the base, the segments 7–10 pairs, 1–2 cm wide, finely serrulate, net-veined with several series of areoles, but the veins free at the margin; fertile lvs surpassing the sterile, with shining, purple-black petiole, the segments narrowly linear, with a single series of areoles along the costa; indusia linear, 4–8 mm, opening tardily and not reflexed in age; $2n=70$. Swamps and wet woods in acid soil; N.S. to n. Fla., w. to Mich., Mo., and e. Tex., commonest on the coastal plain. (*Lorinseria a.*)

Order MARSILEALES

Plants heterosporous; sori enclosed within a sporocarp, which represents a modified, folded pinna with connate margins, each sorus bearing both micro- and megasporangia; sporangia thin-walled, with vestigial or no annulus; aquatic or amphibious plants, rooted in the substrate. A single family.

Family MARSILEACEAE, the Water-clover Family

Stem a creeping, superficial rhizome; lvs alternate, in 2 rows, erect, filiform or with a long petiole and 2 or 4 approximate lfls; sporocarps firm, compressed-ovate to globose, slender-pedunculate or subsessile, the peduncle basally adnate to the petiole; sori 2–many in each sporocarp; spores tetrahedral, the microspores 16–64, minute, the megaspores larger, solitary in each megasporangium; gametophytes minute, nongreen, the female largely contained within the megaspore. 3/50.

1. MARSILEA L.   Water-clover. Lvs long-petiolate, the blade mostly floating or emergent, with 4 approximate lfls (2 juxtaposed pairs) regularly spreading from the petiole-tip and suggesting a 4-lvd clover; lfls dichotomously and reticulately veined, and with an intramarginal connecting vein; sporocarp longitudinally ± bilocular and with many transverse partitions, finally dehiscent by 2 valves, somewhat compressed, ± ovate or elliptic, in many spp. (including ours) attached laterally to the distal part of the peduncle, which forms a raphe that may project as an "inferior" tooth; a "superior" tooth often also present on the upper side of the sporocarp a little above the base; sori several or many, each with central megasporangia and lateral microsporangia on an elongate receptacle. 45, cosmop.

1 Rhizome rooting between as well as at the nodes; lfls glabrous or nearly so; widespread .... 1. *M. quadrifolia.*
1 Rhizome rooting only at the nodes; lfls appressed-hairy; western .... 2. *M. vestita.*

1. Marsilea quadrifolia L.   European w.-c. Rhizome rooting between as well as the nodes; petioles to 30 cm; lfls glabrous or nearly so, those of emergent lvs 7–30 mm long and wide, those of floating lvs avg a little larger; peduncle attached 1–12 mm above the base of the petiole, 3–20 mm, simple or often branched 1–4 mm above its base to bear 2 or 3 sporocarps; sporocarp 4–5.5 × 3–4 mm, 2.5 mm thick, hairy when young, soon glabrate, the inferior tooth mostly wanting, the superior one wanting or up to 0.2 mm; sori 10–17 per sporocarp, with ca 12 microsporangia and 3–7 megasporangia per sorus. In quiet shallow water; native of Europe, locally escaped and established in our range.

2. Marsilea vestita Hook. & Grev.   Hairy w.-c. Rhizome rooting only at the nodes; lfls appressed-hairy on both sides, the emergent ones 5–20 × 5–15 mm, spatulate to flabellate, slightly oblique, at least the inner margin usually concave; peduncle 2–25 mm, usually unbranched, attached at or up to 3 mm above the base of the petiole; sporocarp 3.5–7.5 × 3–6.5 mm, the inferior tooth 0.3–0.6 mm, the superior one 0.4–1.2 mm and often hooked; sori 14–22 per sporocarp, with 27–64 microsporangia and (2–)9–15 megasporangia per sorus. Cordilleran region of w. U.S. and adj. Can. and Mex., e. to Minn., Io., and La. (*M. mucronata.*) Ours is the widespread var. *vestita.*

Order SALVINIALES

Plants heterosporous; microsporangia and megasporangia borne in separate sori, each sorus completely enclosed by its basally attached indusium, which forms a sort of sporocarp; sporangia thin-walled, with vestigial or no annulus; free-floating aquatic plants (seldom on wet soil), with or without roots. A single family.

Family SALVINIACEAE, the Floating Fern Family

Stem filiform, branched; lvs small, crowded, alternate or whorled, simple or bilobed or dissected, sessile or nearly so; sporocarps soft and thin-walled, each containing a single sorus; male sorus with ± numerous microsporangia; female sorus with a single megasporangium containing a single megaspore; gametophytes minute, usually nongreen, largely contained within the spore-wall. 2/15.

1. **AZOLLA** Lam. Mosquito-fern. Delicate, moss-like plants, subdichotomously branched above every third lf; roots slender, simple, inconspicuous; lvs alternate, minute, sessile, unequally bilobed, one lobe submersed and serving as a float, the other smaller, emersed, and containing a colony of a symbiotic blue-green alga, *Anabaena azollae*; sporocarps borne on the first leaf of a branch, paired at the tip of a short, bipartite secondary lobe originating from the submersed lobe near its base, appearing almost axillary; microspores 32 or 64, aggregated into 4 or more massulae, each massula with its own thin wall, which bears numerous slender appendages (glochidia) with terminal, retrorse hooks; megaspore with a hemispheric or broadly bell-shaped basal part and a sharply marked, complex, conical terminal part that eventually shows 3 basally spreading, apically confluent, longitudinal valves. 6, cosmop.

1 Glochidia without cross-walls; plants mostly 5–10 mm wide .... 1. *A. caroliniana.*
1 Glochidia with several cross-walls; plants mostly 10–15 mm wide .... 2. *A. mexicana.*

1. **Azolla caroliniana** Willd. Eastern m.-f. Plants rarely fruiting, minute, mostly 5–10 mm wide, dichotomously branched almost throughout; upper lf-lobes 0.5–0.6 mm, much smaller than the lower, not much imbricate; massulae ca 0.3 mm, the glochidia without cross-walls; megaspores unknown in U.S.; 2*n*=48. Floating in still water; Mass. and N.Y. to Fla. and La. on the coastal plain, and up the Mississippi embayment to w. Ky. and (?) s. Ill.; W. Ind. Rare.

2. **Azolla mexicana** Presl. Mexican m.-f. Plants seldom fruiting, a little larger than no. 1, mostly 10–15 mm wide, ± pinnately branched when well developed, dichotomous only at the periphery; upper lf-lobes 0.7–0.9 mm, not much smaller than the lower, ± strongly imbricate; microsporocarps mostly a little over 1 mm thick; massulae 0.2–0.3 mm; glochidia with several cross-walls; megaspores 0.4–0.5 mm, the rounded base strongly pitted. Floating in still water; widespread in w. U.S., s. to n. S. Amer., and e. to Wis., Ill., and Ark. Rare.

DIVISION PINOPHYTA Gymnosperms

Woody plants with seeds, the gametophyte generation reduced to a set of stages in the reproduction of the sporophyte; microsporophylls aggregated on an axis, forming a small, simple or compound strobilus (the male cone); ovules produced in simple or compound female cones, or sometimes singly, or in pairs on slender stalks, in any case exposed to the air at the time of pollination and receiving pollen at the micropyle, not enclosed in an ovary; female gametophyte multicellular, with 500 or more cells, usually producing more or less definite archegonia; food reserves of the seed stored in the body of the female gametophyte; no triple fusion cell produced. Three subdivisions, the Pinicae, Cycadicae, and Gneticae. Only the Pinicae are represented in the wild in our range.

SUBDIVISION *PINICAE*

Small to large, usually excurrently branched trees, or sometimes deliquescently branched shrubs, in any case with small pith and well developed secondary growth, the secondary wood generally containing tracheids and resin canals (mucilage cavities in *Ginkgo*) but never with vessels; leaves simple, alternate, opposite, or ternate (or in clusters on specialized short spurs), in ours slender and needle-like to less often reduced and scale-like (larger and fan-shaped in *Ginkgo*), with a single midvein or with a dichotomously branching vein system, but never net-veined; male cones simple, the microsporophylls attached directly to the central axis; female cones (when present) generally compound, i.e., the ovules borne on structures (the ovuliferous scales) axillary to the primary appendages of the cone-axis; cotyledons usually several, seldom only 2 or 3. Two classes, the Pinopsida (as principally described above) and Ginkgoöpsida. The class Ginkgoöpsida has only a single living species, *Ginkgo biloba* L., which is often cultivated in our range but not known to escape.

CLASS *PINOPSIDA*

Leaves entire or slightly toothed, never fan-shaped, in ours always narrow and elongate (needle-like) or reduced and scale-like; wood with resin ducts. Two living orders, the Taxales and Pinales.

ORDER TAXALES

Seeds borne singly, terminal or subterminal on short, bracteate, axillary shoots, not in cones, provided with an aril-like fleshy covering at maturity; male cones small, with spirally arranged, peltate microsporophylls; lvs spirally arranged, linear, twisted at the petiolar base so as to form flat sprays; mostly dioecious evergreen trees and shrubs. A single family, Taxaceae.

FAMILY TAXACEAE, the Yew Family

Characters of the order Taxales. 3/15.

1. TAXUS L.   Yew. Fleshy covering of the seed bright orange-red at maturity, globular-cupuliform, open at the end. 10, widespread, mainly N. Temp.

1. **Taxus canadensis** Marshall.   American y.; ground-hemlock. Straggling shrub with ascending or rarely erect stems to 2 m; lvs 1–2 cm × 1–2 mm, abruptly narrowed to a sharp point, attached by the shortly petiolar base to a decurrent sterigma on the twig; scales of winter-buds keeled, ± acute; fleshy seed ca 5 mm; $2n=24$. Coniferous woods and bogs; Nf. and Lab. to Minn. and se. Man., s. to Va., Ky., and Io.
   T. baccata L., the English yew, *T. chinensis* (Pilger) Rehder, the Chinese yew, and *T. cuspidata* Siebold & Zucc., the Japanese yew, occasionally escape locally from cult.

ORDER PINALES, Conifers

Ovules and seeds borne in more or less definite cones, these compound, the primary cone scales sometimes exserted beyond the ovuliferous scales and appearing as bracts. The order consists of the 3 following families and the Araucariaceae, the latter in the Southern Hemisphere. Except as otherwise specified, the word cone in the following text refers to the female cone.

FAMILY PINACEAE, the Pine Family

Female cones with several or many spirally arranged ovuliferous scales, each scale subtended by a short or elongate bract and bearing 2 ovules adaxially near the base; mature cones dry and often woody, not at all berry-like; seeds mostly winged; male cones with spirally arranged microsporophylls; monoecious trees or shrubs, lvs spirally arranged, or sometimes fascicled, linear or needle-like, never scale-like. 10/150.

1 Dwarf branches not produced; lvs all alternate.
  2 Cones erect; cone-scales deciduous at maturity; lvs attached directly to the twig, leaving a smooth,
     round scar on falling .... 1. *Abies.*
  2 Cones pendulous; cone-scales persistent; lvs attached to persistent, elevated or projecting sterigmata.
    3 Lvs quadrangular, sessile on the sterigmata; resin-ducts 2 in each lf .... 2. *Picea.*
    3 Lvs flat, short-petiolate on the sterigmata; resin-duct 1 in each lf .... 3. *Tsuga.*
1 Dwarf branches produced, bearing clusters of lvs.
  4 Lvs deciduous, numerous in each cluster, also scattered on long twigs .... 4. *Larix.*
  4 Lvs evergreen, all in clusters of 2–5 .... 5. *Pinus.*

1. ABIES Miller. Fir. Male cones ovoid or short-cylindric, pendulous on short stalks from the axils of lvs of the preceding year, subtended by conspicuous bud-scales; female cones erect, cylindric, typically confined to the upper part of the tree, the scales individually deciduous at maturity, so that the cone rarely falls intact; bracts longer than the ovuliferous scales at anthesis, but often surpassed and hidden at maturity; evergreen trees, the lvs attached directly to the twigs and leaving a smooth, round scar on falling; $2n=24$. 40, N. Temp.

1 Bracts of mature cones hidden or slightly exserted, not reflexed .... 1. *A. balsamea.*
1 Bracts of mature cones strongly exserted, reflexed .... 2. *A. fraseri.*

1. Abies balsamea (L.) Miller. Balsam f. Spire-topped tree to 25 m, the gray bark eventually becoming scaly; twigs minutely hairy; lvs 12–25 mm; blunt or minutely notched, twisted at the base and tending to lie in one plane; cones 5–10 cm, with broadly rounded, appressed scales; bracts stipitate, the blade quadrate to rotund, mucronate to aristate, included or occasionally shortly exserted, but only rarely exceeding the subtended scale, the exserted part, when present, sometimes spreading, but not reflexed. Moist woods and swamps; Nf. and Lab. to the Mackenzie Valley, s. to N.Y., Mich., Minn., and in the mts. to W.Va. and n. Va.

2. Abies fraseri (Pursh) Poiret. Fraser-f.; she-balsam. Much like no. 1, but not so evidently spire-topped; lvs often conspicuously notched at the tip; cone-scales nearly reniform, bracts strongly exserted, exceeding the subtended scale, reflexed, retuse and aristate at the erose summit. High elev. in the Blue Ridge Mts.; sw. Va. to N.C. and Tenn.

2. PICEA A. Dietr. Spruce. Male and female cones terminating twigs of the previous year's growth; female cones pendulous, with persistent scales much exceeding the bracts; evergreen trees, the quadrangular lvs jointed at the base to a short, projecting sterigma that persists on the denuded branch; resin-ducts 2 in each lf; $2n = 24$. 30+, N. Temp.
  P. abies (L.) Karsten, the Norway s., with cones 10–15 cm, occasionally escapes locally from cult.

1 Twigs and bud-scales glabrous .... 1. *P. glauca.*
1 Twigs and lower-bud scales pubescent.
  2 Lvs gray-green, glaucous; cones long-persistent .... 2. *P. mariana.*
  2 Lvs yellow-green and shiny, not glaucous; cones soon deciduous .... ?............ 3. *P. rubens.*

1. Picea glauca (Moench) Voss. White s. Tree to 25 m, the crown slender but not notably spire-like; twigs glabrous; lvs 8–18 mm, markedly glaucous, especially when young; winter buds obtuse, with obtuse, glabrous scales; cones ellipsoid to short-cylindric, 3.5–5 cm, deciduous, pale brown, the scales thin and flexible at the entire margin. Mostly in rich, moist soil; Nf. and Lab. to Alas., s. to n. N.Y., Mich., Minn., S.D., and B.C.

2. Picea mariana (Miller) BSP. Black s. Tree to 10(–25) m, with a narrow but scarcely spire-like crown, the branches tending to droop; twigs pubescent with crooked or glandular, cylindric hairs; lvs 6–18 mm, radially spreading, straight, often blunt, strongly glaucous, grayish-green; winter buds acute, gray-brown, the lowest scales pubescent and subulate; cones dark purple before maturity, dull grayish-brown when ripe, ovoid, 1.5–3.5 cm, persistent for many years; cone-scales firmly attached to the rigid axis, with thin, brittle, often somewhat pointed, usually dentate or erose margin, often with a bucktooth projection. Usually in sphagnum bogs, in our range, but sometimes on subalpine slopes, as in the White Mts. of N.H.; Nf. and Lab. to Alas., s. to n. N.J., Pa., Mich., Wis., and B.C.

3. Picea rubens Sarg. Red s. Tree to 30 m, with a relatively broad and pyramidal or conic crown, the branches longer than in no. 2 and tending to be horizontally spreading, with upturned tip; twigs pubescent as in no. 2, but the hairs more steeple-shaped; lvs 10–30 mm, sharply pointed, shiny yellow-green, not glaucous, those on the upper side forwardly ascending or erect, the others all laterally ascending-spreading or tending to be upcurved, those on the lower side twisted at the base; winter buds acute, chestnut-brown, the lowest scales pubescent and subulate; cones green or purplish-green before maturity, brown or reddish-brown when ripe, oblong-ovoid, 3–4.5 cm, mostly falling the first winter after maturity, more fragile than in no. 2, the cone-scales easily detached from the axis, but with firmer, more rounded, often entire margin. Rocky woods and hillsides, esp. in the mts.; Que. and Ont. to Pa. and N.J., and s. in the higher mts. to N.C. and Tenn. (*P. australis; P. rubra.*) Hybridizes extensively with no. 2.

3. **TSUGA** Carrière.   Hemlock. Male cones axillary near the tip of the twigs of the previous season, peduncled, globose; female cones terminal on lateral branches of the previous season, pendulous on a short peduncle, the persistent scales much larger than the minute bracts; evergreen trees, the flat lvs jointed at the base to minute, persistent sterigmata; resin-duct 1 in each lf; $2n=24$. 14, temp. Asia and N. Amer.

1 Cones 12–20 mm, their principal scales round-obovate .... 1. *T. canadensis.*
1 Cones 25–35 mm, their principal scales oblong-ovate .... 2. *T. caroliniana.*

   1. Tsuga canadensis (L.) Carrière.   Eastern h. Tree to 30 m, the leader nodding; twigs pubescent; lvs 8–15 mm, blunt, usually minutely spinulose on the margin, twisted at the petiolar base to form flat sprays; cones thickly ellipsoid, 12–20 mm, scales persistently erect or ascending, the exposed portion of the middle ones distinctly broader than long. Moist soil, esp. on rocky ridges and hillsides; N.S. to Mich., Wis., and occasionally Minn., s. to N.J., Del., O., and Ind., and in the mts. to Ga. and Ala.
   2. Tsuga caroliniana Engelm.   Carolina-h. Similar to no. 1; lvs entire, averaging slightly larger, more diversely oriented, so that the sprays are not so flat; cones 25–35 mm, the scales widely spreading at maturity, oblong-ovate, the exposed portion of the middle ones at least as long as wide. Ravines and rocky hillsides in the Blue Ridge Prov., rarely on the Piedmont; sw. Va. to n. Ga.

4. **LARIX** Miller.   Larch. Male and female cones sessile or subsessile on leafless dwarf branches of previous year's growth, globose or subglobose; lvs deciduous, needle-like; branches of 2 sorts, (1) long shoots of the current year, with scattered lvs, and (2) dwarf branches produced laterally on the previous year's growth, slowly elongating for many years and producing crowded clusters of numerous lvs. 10, mainly N. Temp.

   1. Larix laricina (Duroi) K. Koch.   Tamarack. Tree to 20 m; lvs soft, very slender, 10–25 mm, light green; cones with 10–18 glabrous scales, red at first, pale brown and 1–2 cm when ripe. Mostly in bogs; Nf. and Lab. to Alas., s. to n. N.J., W.Va., O., n. Ill., and Minn.

5. **PINUS** L.   Pine. Male cones in fascicles at the base of the current year's growth; female cones woody, maturing at the end of the second or third season and often long-persistent; apophysis (exposed portion of the mature cone-scale) marked by a transverse line or ridge interrupted in the center by a linear to round or quadrate, often elevated or spine-bearing umbo, or the umbo sometimes terminal; evergreen trees or shrubs with dimorphic branches and lvs, the long branches (except in seedlings) bearing only scale-lvs, from the axils of which early appear very short dwarf branches, each bearing a cluster of (1)2–5 needle-like lvs, the cluster surrounded at the base by a bundle-sheath of 1 or more membranous scale-lvs; dwarf branches eventually (after 2–several years) deciduous with the lvs; evergreen trees or shrubs; $2n=24$. Nearly 100, N. Hemisphere.

1 Lvs in clusters of 5, each lf with 1 fibrovascular bundle; bundle-sheath deciduous; umbo terminal;
    wood soft, not very resinous (Sect. *Strobus*) .... 1. *P. strobus.*
1 Lvs in clusters of 2 or 3, each lf with 2 fibrovascular bundles; bundle-sheath persistent; umbo lateral;
    wood hard, strongly resinous (Sect. *Pinus*).
  2 Umbo without a terminal spine, or with a minute, reflexed, appressed spine; lvs in 2's.
    3 Lvs 9–16 cm .... 2. *P. resinosa.*
    3 Lvs 2–7 cm.
      4 Cone reflexed; larger branches orange-brown; lvs 3–7 cm .... 3. *P. sylvestris.*
      4 Cone ± erect; branches not orange-brown; lvs 2–3.5 cm .... 4. *P. banksiana.*
  2 Umbo with a terminal spine or prickle.
    5 Lvs mostly 14–50 cm, in 3's.
      6 Cones 7 cm or more, a third or half as thick.
        7 Cones 15–25 cm; lvs 20–30(50) cm .... 5. *P. palustris.*
        7 Cones 7–13 cm; lvs 14–25 cm .... 6. *P. taeda.*
      6 Cones 4–6 cm, nearly as thick .... 7. *P. serotina.*

5 Lvs mostly 3–12 cm, in 2's or 3's.
  8 Cone-scales with a triangular, spreading, often upcurved spine 3–8 mm .... 8. *P. pungens.*
  8 Cone-scales with a slender or triangular spine 1–3 mm.
    9 Lvs about 2(1.5–3) mm wide, usually in 3's, 7–12 cm; twigs brown, not glaucous .... 9. *P. rigida.*
    9 Lvs about 1(0.7–1.5) mm wide.
      10 Lvs in 2's, 4–7 cm; twigs purplish-brown, or glaucous and purplish-gray ....10. *P. virginiana.*
      10 Lvs in 2's or 3's, 7–12 cm; twigs pale, glaucous .... 11. *P. echinata.*

1. **Pinus strobus** L.   White p. Tall tree, to 70 m, with long, irregular branches; bark becoming thick, dark, and furrowed; wood pale, soft, not very resinous; lvs very slender, in 5's, mostly persisting 2 years, pale green and glaucous, 8–13 cm, with 1 fibrovascular bundle; bundle-sheath deciduous; cones commonly borne near the tips of the longer branches, cylindric, often bent, 10–15 cm, the apophysis not thickened, the umbo resinous and terminal, unarmed; seed (wing included) 2–3 cm. Many habitats, esp. in fertile or well drained, sandy soil; Nf. to Minn. and se. Man., s. to Del., n. Ga., Ky., and Io.

2. **Pinus resinosa** Aiton.   Red p.; Norway-p. Tree to 40 m; bark becoming light red-brown, with scaly plates; winter buds ca 1.5 cm, the scales red-brown with white-fringed margins, not very resinous; lvs in 2's, persisting 4–5 years, dark green, 9–16 cm, snapping cleanly when bent; cones spreading, conic-ovoid, 4–8 cm, the apophysis thickened, dark smooth and spineless; seeds 1.5–2 cm. Dry, sandy or rocky soil; Nf. and N.S. to se. Man., s. to Conn., Pa., n. Ill., and Minn., and in the mts. to W.Va.
    P. nigra Arnold, Austrian pine, may escape locally from cult. It has dark bark and stiff lvs that do not break cleanly when bent.

3. **Pinus sylvestris** L.   Scotch p. Tree to 30 m; bark of the larger branches and a segment of the main trunk conspicuously orange-brown and appearing blistered; lvs in 2's, bluish-green, usually twisted, 3–7 cm × ca 1.5 mm; cones yellow-brown, soon reflexed, short-ovoid to oblong, often bent, 3–6 cm, the apophysis thickened, the umbo scarcely elevated, spineless. Native of Europe, occasionally escaped from cult. in our range.

4. **Pinus banksiana** Lambert.   Jack-p. Usually a small tree, or sometimes to 20 m, with spreading branches; winter buds less than 1 cm, pale cinnamon-brown, very resinous; lvs in 2's, persisting 2–3 years, usually curved, 2–3.5 cm × 1–1.5 mm; cones erect or nearly so, long-persistent, typically remaining closed for several or many years or until fire, usually curved and asymmetrical, conic, yellowish-brown, 3–5 cm, the apophysis thickened, the umbo not elevated, spineless or with a minute, reflexed spine; seeds ca 1.5 cm. Dry or sterile, sandy or rocky soil; Que. to Me., Vt. and n. N.Y., w. to nw. Ind., Minn., and the Mackenzie Valley.

5. **Pinus palustris** Miller.   Longleaf-p. Tall tree, to 30 m; bark becoming orange-brown, with rough, scaly plates; twigs very stout; winter buds large, up to 2.5 cm, covered with silvery-white, fringed scales; lvs in 3's, closely crowded at the ends of the branches, persistent 2 years, light green, soft, often drooping, 20–30(–50) cm; cones divergent, often pendent, conic-oblong, 15–25 × 5–6 cm, opening at maturity and soon falling; apophysis thick and prominently ridged, the umbo elevated, with a small, reflexed spine; seeds ca 4 cm. Moist or dry, sandy soil, largely on the coastal plain, less commonly on the piedmont; se. Va. to Fla. and Tex. (*P. australis* Michx. f., an illegitimate substitute name.)

6. **Pinus taeda** L.   Loblolly-p. Tree to 30 m; bark becoming red-brown, with scaly plates; terminal buds smaller than in no. 5, scarcely 1.5 cm, their scales red-brown, differing from no. 7 in not being very resinous; lvs in 3's, persistent 3 years, slender, dark green, soft and flexible, 14–25 cm; cones divergent, 7–13 cm, opening at maturity and falling the next year; apophysis thick, the umbo elevated into a triangular projection with a very short, spreading or reflexed spine; seeds ca 2.5 cm. Moist, sandy soil, chiefly on the coastal plain and (southward) on the piedmont; Cape May, N.J., to Fla. and Tex., and n. in the interior to Tenn. and Ark.

7. **Pinus serotina** Michx.   Pond-p. Tree to 15(25) m; terminal buds very resinous; lvs in 3's, dark green, flexible, 15–25 cm; cones divergent or somewhat reflexed, globose-ovoid, 4–6 cm, usually persistent and remaining closed for several years; apophysis thickened, the umbo conic, with a straight or reflexed spine ca 1 mm; seeds ca 2.5 cm. Swamps and wet soil, chiefly on the coastal plain; Fla. to se. Md. and s. Del., and reported from s. N.J.

8. **Pinus pungens** Lambert.   Table-mountain p. Small, often crooked tree to 10(20) m, with widely spreading branches; lvs in 2's or rarely some of them in 3's, dark bluish-green, stiff, sharp, usually twisted, mostly 4–9 cm × 1.5–2 mm; cones spreading or deflexed, conic-ovoid, mostly 6–9 cm long-persistent, often remaining closed for several years or until fire; apophysis thick, strongly elevated at the center, the umbo pyramidally projecting, with a very stout, spreading or upwardly curved spine 3–8 mm; seeds ca 2 cm; male cones reddish-purple. Dry or rocky soil, chiefly in the mts.; Pa. and adj. N.J. to n. Ga. Readily recognizable by its strongly armed cones.

9. **Pinus rigida** Miller.   Pitch-p. Tree to 20 m, with spreading, irregular branches; bark dark and very rough; twigs brown; terminal buds ca 1.5 cm, their scales chestnut-brown, fringed, resinous; lvs mostly in 3's, persisting 2–3 years, stiff, dark green, 7–12 cm × ca 2 mm; cones divergent, conic-ovoid, 4–7 cm, long-persistent but generally opening at maturity; apophysis thickened and somewhat elevated, the umbo elevated and with a slender spine 1–3 mm; seeds ca 1.5 cm. Dry, rocky or sandy soil; s. Me. to s. Que. and s. Ont., s. to n. Ga., and with outlying stations in c. and w. Ky.; dominant on the pine-barrens of N.J., but seldom on the coastal plain farther south.

10. Pinus virginiana Miller. Scrub-p.; Virginia-p. Small tree, to 10 m, or seldom to 30 m, with long, horizontal branches; bark becoming dark red-brown; twigs purplish-brown to purplish-gray; winter buds less than 1 cm, very resinous, pale cinnamon-brown; lvs in 2's, persistent 3–4 years, dark green, soft and flexible, 4–8 cm × ca 1 mm; cones dark red-brown, oblong-conic, 4–7 cm, opening at maturity but persisting for several years; apophysis elevated, the umbo strongly raised, with a straight, slender spine 1–3 mm; seeds 1.5–2 cm. Dry, sandy or sterile soil; s. N.Y. to s. Ind., s. to Ga., Ala., and ne. Miss.

11. Pinus echinata Miller. Yellow or shortleaf p. Tree to 35 m; bark at first scaly and nearly black, later becoming red-brown and scaly-plated; twigs pale and glaucous; winter buds less than 1 cm, red-brown to gray-brown, not strongly resinous; lvs in 2's or 3's, persistent 2–4 years, soft, dark green, 7–12 cm × ca 1 mm; cones divergent, ovoid to ovoid-conic, 4–6 cm, opening at maturity but persisting several years; apophysis somewhat elevated, the umbo low-pyramidal, with a stout or slender, recurved or spreading spine 1–2 mm; seeds ca 2 cm. Dry, sandy or rocky soil; s. N.Y. to n. Fla., w. to s. O., s. Mo., e. Okla., and e. Tex.

FAMILY TAXODIACEAE, the Bald Cypress Family

Female cones woody, with spirally arranged ovuliferous scales, these peltate, apparently without subtending bracts; seeds winged; male cones with spirally arranged, peltate microsporophylls; monoecious, evergreen or (ours) deciduous trees; leaves spirally arranged, needle-like or scale-like. 10/16.

1. TAXODIUM Rich. Male cones numerous, in drooping terminal panicles, the microsporophylls mostly 6–8; female cones on branches of the previous year, long-persistent, subglobose, the several or many scales irregularly quadrangular above, each with 2 ovules; deciduous trees with dimorphic branches, some persistent and forming axillary buds, others deciduous in autumn with their lvs. 2, the other Mexican.

1. Taxodium distichum (L.) Rich. Bald cypress. Tree to 40 m, the trunk concavely thickened below and conspicuously buttressed at least in mature trees; bark reddish to light brown, with shallow furrows, becoming flaky; leafy branchlets mostly spreading horizontally; lvs narrowly linear, 5–15 mm, divergent, appearing 2-ranked on the branchlets, the free portion of the lf basally contracted and twisted; panicles ca 1 dm, initiated in the autumn, developing in the spring; cones pendulous at the ends of the twigs, 2–3 cm thick, with dull, rugose scales; 2n=22. Brownwater riverine and lacustrine margins and swamps, occasionally in brackish water, chiefly on the coastal plain; Del. (and locally and perhaps only recently n. to L.I.) to Fla. and Tex., n. in the Mississippi Valley to s. Ind. We have only the widespread var. *distichum*, as here described. Var. *imbricarium* (Nutt.) Croom (*T. ascendens*), the pond-cypress, extends n. only to N.C.

FAMILY CUPRESSACEAE, the Cypress Family

Female cones small, the scales opposite or ternate, apparently without subtending bracts, woody and distinct, or often ± coalescent to form a berry-like or drupe-like cone at maturity; microsporophylls of male cones few, opposite; monoecious or dioecious evergreen trees or shrubs; lvs opposite or whorled, crowded and often overlapping, needle-like or more often scale-like, sometimes dimorphic. 15/140.

1 Cone woody or leathery, the scales eventually separating; monoecious; seeds ± winged.
  2 Cone globose, the scales peltate, not overlapping .... 1. *Chamaecyparis.*
  2 Cone oblong-ovoid, the scales overlapping .... 2. *Thuja.*
1 Cone berry-like or drupe-like, somewhat fleshy, usually colored, remaining closed; usually dioecious;
  seeds wingless .... 3. *Juniperus.*

1. CHAMAECYPARIS Spach. Female cones subglobose, woody, dehiscent, with 3–6 pairs of ovuliferous scales each bearing 2–5 ovules, the scales becoming peltate and polygonal, pointed or elevated in the middle; seeds winged; microsporophylls few; monoecious trees with small, appressed, scale-like, decussately opposite lvs, the lateral ones folded and upper and lower flat or rounded, often with a resin-gland; branchlets numerous from the axils of the lateral lvs, forming ± flattened sprays; 2n=22. 6, N. Temp.

1. **Chamaecyparis thyoides** (L.) BSP.   Atlantic white cedar. Slender tree to 25 m; ultimate branchlets ca 1 mm wide; lvs imbricate, ovate, 2–4 mm, pointed, usually glandular, persisting and enlarging on the older branches; cones 6–8 mm, the scales ca 6, pointed in the center. Swamps and bogs, chiefly on the coastal plain; s. Me. and c. N.H. to Fla. and Miss.

2. THUJA L.   Arbor Vitae.

Female cones oblong-ovoid, leathery or woody, with 4–6 pairs of overlapping scales, the upper and lower scales sterile, the middle ones each with 2–3 ovules; seeds winged; microsporophylls few; monoecious trees with shreddy bark and small, appressed, imbricate, scale-like, decussately opposite lvs, the lateral ones folded, the upper and lower flat and with a prominent resin-gland, branchlets numerous from the axils of the lateral lvs, forming soft, flat sprays. 6, N. Temp.

1. **Thuja occidentalis** L.   Northern white cedar. Conical tree to 20 m, with widely spreading branches; ultimate branchlets very soft and flat, 1–2 mm wide; lvs closely imbricate, broadly ovate to rotund, 2–4 mm, glandular, persisting and enlarging on the older branches; cone ca 1 cm, the outer scales nearly as long as the inner; $2n=22$. Moist or wet soil, often in swamps; Que. and N.S. to Hudson Bay, s. to N.J., O., n. Ind. and Ill., Wis., Minn., and in the mts. to N.C. and Tenn.

3. JUNIPERUS L.   Juniper.

Female cones indehiscent, berry-like or drupe-like, usually colored (commonly bluish), the several scales coalescent and somewhat fleshy at maturity; seeds 1–10 per cone, wingless; male cones with numerous microsporophylls; dioecious or occasionally monoecious shrubs or small trees with scale-like or needle-like lvs opposite or whorled in 3's; $2n=22$. 60, mainly N. Temp.

1. **Juniperus communis** L.   Common j. Lvs in whorls of 3, jointed at the base, crowded, linear, loose or spreading, pungent, 6–18 mm, with a median white stripe above; cones axillary, bluish or black, 6–13 mm thick, mostly 3-seeded. Circumboreal, s. in Amer. to Pa., Wis., Minn., the Rocky Mt. states, and irregularly to S.C. and s. Ind. Nearly all of our plants belong to the var. *depressa* Pursh, the common Amer. phase of the sp.; decumbent but not prostrate, forming flat-topped circular patches mostly 0.5–2 m tall and several m wide; lvs spreading or ascending, mostly straight, (6)10–18 mm, the white stripe narrower than each margin; mostly in dry, rocky, or otherwise poor soil; occasional more arborescent specimens, with a ± definite central axis, have been mistaken for the arborescent Eurasian var. *communis*, which has the white stripe on the lvs at least as broad as each margin. The high-northern, circumboreal var. *montana* Aiton reaches only the ne. margin of our range, as on the coast of N.S. and the coast and mts. of n. Me.; it is prostrate or nearly so and trailing, with oblong, abruptly short-pointed, incurved and imbricate lvs 6–10 mm, these with a broad white stripe, evidently wider than each margin (var. *alpina*; var. *saxatilis*, misapplied; *J. sibirica; J. nana*).

2. **Juniperus virginiana** L.   Eastern red cedar. Shrub or tree with a dense crown, to 20 m; juvenile lvs subulate, pungent, 5–7 mm, spreading or ascending; lvs of adult branches scale-like, appressed, ovate or lance-ovate, 2–4 mm, obtuse or subacute, convex on the back; cones terminal on short, straight peduncles, subglobose, blue-glaucous, 5–7 mm thick; seeds 1 or 2, pitted toward the base. In a variety of soils, esp. in dry, calcareous sites; s. Me. and s. Que. to N.D., s. to Ga., nw. Fla., and Tex. Two weakly distinguished vars.: Var. *virginiana,* occurring from Va. to s. Mo. and southward, has a relatively broad, ovoid crown with widely spreading branches, and the seeds are strongly pitted. Var. *crebra* Fern., the more northern phase, has a narrowly spire-shaped crown with distinctly ascending branches, and the seeds are only obscurely pitted.

3. **Juniperus horizontalis** Moench.   Creeping j. Branches closely prostrate, often much elongate, with numerous erect branches 1–3 dm; lvs mostly scale-like and appressed, varying from ovate and 1–2 mm to oblong and up to 4 mm; cones on short, recurved peduncles, blue, 5–8 mm thick; seeds mostly 3–5, not pitted. Rocky or sandy places; nearly transcontinental in Can., from Nf., Lab., and Que to Yukon and n. B.C. and disjunct in Alas., more scattered s. to Me. (esp. along the coast), N.H., N.Y., Mich. (notably along the shores of the Great Lakes), Wis., Minn., and Wyo. Hybridizes with *J. virginiana.*

DIVISION MAGNOLIOPHYTA
Flowering Plants

Woody or herbaceous plants with seeds, the gametophyte generation reduced to a set of stages in the reproduction of the sporophyte; sexual reproductive structures grouped into specialized short shoots called fls, each fl typically with sporophylls (stamens and carpels) collectively surrounded by two sets of floral envelopes (cal and cor) that together constitute the perianth, one or both perianth-sets sometimes reduced or wanting, or the two sets sometimes not differentiated from each other; stamens (microsporophylls) forming a spiral or one or more cycles next inward from the perianth (or adnate to it), typically with an elongate, slender filament (forming a basal stalk) and two distal, parallel pollen-sacs that (together with their connecting tissue) make up the anther; carpels (megasporophylls) 1–many, comprising the innermost set of floral lvs, folded together and sealed individually to form as many simple pistils, or collectively to form a compound pistil, the pistil(s) in either case with an enlarged, basal (or stipitate) part (the ovary) that contains the ovules, and a terminal or lateral part (the stigma or stigmas) that is receptive to pollen and is sessile on the ovary or more often elevated above it on a slender style; pollen grains (male gametophytes) 2–3-nucleate, germinating on the stigma and producing a pollen-tube that penetrates the tissue of the ovary and typically reaches the few-nucleate embryo-sac (female gametophyte) through the micropyle; sperms two in each pollen-tube, one fertilizing the egg in the embryo-sac, the other fusing with 2 nuclei (or the product of fusion of 2 nuclei) of the embryo-sac to form the endosperm-nucleus, thus effecting double fertilization; xylem generally containing vessels. Two classes, the Magnoliopsida and Liliopsida.

Class *MAGNOLIOPSIDA* Dicotyledons

Embryo mostly with 2 cotyledons; floral parts, when of definite number, typically in sets of 5, less often 4, seldom 3 (carpels often fewer); pollen mostly of triaperturate or triaperturate-derived type; herbs or woody plants, generally with an intrafascicular cambium, usually with vessels in the stem and leaves as well as in the roots, the vascular bundles in herbaceous forms usually borne in a ring that encloses a pith; leaves mostly pinnately or palmately net-veined. 6 subclasses, 64 orders, and 324 families.

Family **MAGNOLIACEAE**, the Magnolia Family

Fls perfect, regular, hypogynous; cal and cor scarcely differentiated, the perianth generally composed of 3 or more whorls, each of 3 free members, the outer whorl arbitrarily considered to be the cal; stamens numerous, distinct, spirally arranged; pollen monosulcate; pistils ± numerous, spirally arranged, inserted on an expanded or elongate receptacle, usually free and distinct at anthesis, but sometimes more or less concrescent in fr; ovules mostly 2; seeds with tiny, dicotyledonous embryo and copious endosperm; woody pls with alternate, stipulate, simple, entire or lobed lvs and usually large, solitary, terminal fls. 13/225.

1 Lvs entire or with 2 small basal lobes; fr dehiscent, exposing the seeds ....1. *Magnolia.*
1 Lvs mostly 4-lobed, broadly retuse; fr indehiscent, samaroid .... 2. *Liriodendron.*

1. MAGNOLIA L. Magnolia. Perianth of 9–15 similar or scarcely differentiated members, in 3–5 whorls; anthers introrse; pistils many, on an elongate receptacle, ripening into a cone of coherent, extrorsely dehiscent frs; seeds red or pink, remaining attached by a threadlike funiculus for some time after dehiscence; trees or shrubs with large lvs and fls. 80, trop. to warm-temp.

1 Lvs acute to broadly rounded at the base.
  2 Pet 3–5 cm; lvs strongly glaucous beneath .... 1. *M. virginiana.*
  2 Pet 5–14 cm; lvs green or rusty-pubescent beneath.
    3 Lvs evergreen, thick and leathery, ± rusty beneath .... 2. *M. grandiflora.*
    3 Lvs deciduous, thinner, green beneath.
      4 Lvs scattered, broadest near or below the middle; pet 5–8 cm, greenish-yellow .... 3. *M. acuminata.*
      4 Lvs clustered at the ends of the twigs, broadest well above the middle; pet 8–14 cm, white ...... 4. *M. tripetala.*
1 Lvs distinctly cordate at the base.
  5 Pet white, 7–11 cm; lvs green and glabrous beneath .... 5. *M. fraseri.*
  5 Pet white with a basal rosy spot, 12–18 cm; lvs glaucous and finely hairy beneath .... 6. *M. macrophylla.*

1. Magnolia virginiana L.   Sweet bay. Tall shrub or slender tree to 20 m; terminal winter bud 1–2.5 cm, finely silky; lvs aromatic, scattered, leathery, evergreen in the south, oblong to elliptic or oblanceolate, 8–15 cm, a third or half as wide, obtuse, basally acute to rounded, the lower surface glaucous and finely hairy; fls fragrant, white, subglobose, ca 5 cm thick; pet 9–12, leathery, obovate, concave, 3–5 cm; fr cone ellipsoid, 3–5 cm, the follicles glabrous; $2n=38$. Wet woods and margins of swamps; Fla. to Tex., n. to Pa., N.J., Long I., and e. Mass., chiefly on CP. May–July.

2. Magnolia grandiflora L.   Bull-bay, southern m. Handsome tree to 30 m; winter terminal bud 1–2.5 cm, evidently hairy; lvs scattered or somewhat clustered toward the ends of the twigs, leathery, evergreen, elliptic, 12–25 cm, a third or half as wide, acute to obtuse, basally acute to rounded-obtuse, the lower surface conspicuously rusty with a ± persistent, thin, villous tomentum; fls fragrant, creamy-white, cup-shaped, ca 10 cm thick; pet 6–12, obovate or spatulate, broadly rounded distally, concave, 7–13 cm; fr cone ellipsoid, 6–10 cm, the follicles woolly-villous; $2n=$ca. 114. Rich woods and moist bottomlands; CP from se. Va. to Fla. and Tex. Apr.–June.

3. Magnolia acuminata (L.) L.   Cucumber-tree. Slender tree to 25 m; winter terminal bud 1–2.5 cm, white-hairy; lvs scattered, deciduous, broadly oblong or elliptic, 10–25 cm, half to two-thirds as wide, abruptly acuminate, basally acute to rounded or truncate, the lower surface green and minutely hairy; fls greenish-yellow; pet 6, ascending, 5–8 cm, oblanceolate or oblong-spatulate, abruptly narrowed to the tip; styles filiform, promptly deciduous; fr-cone ellipsoid, 5–8 cm, the follicles glabrous; $2n=76$. Rich woods; s. and w. N.Y. and s. Ont. to s. Mo. and e. Okla., s. to Ga., w. Fla., and La. May, June.

4. Magnolia tripetala (L.) L.   Umbrella-tree. Small tree to 10 m; winter terminal bud 2–3 cm, glabrous; lf scars with minute resin glands; lvs closely clustered beneath the fls, deciduous, thin, lance-obovate, at anthesis 1.5–2 dm, at maturity to 5 dm, abruptly and sharply acuminate, tapering from the middle to an acute base, the lower surface green, finely and sparsely hairy; fls white, malodorous; pet 6–9, oblanceolate to narrowly obovate, spreading, 8–14 cm; style thick, persisting as a beak on the glabrous follicle; fr-cone ellipsoid-cylindric, 7–12 cm; $2n=38$. Rich woods; Ga. to Ark., n. to s. Pa., W.Va., O., Ky., and e. Mo. May.

5. Magnolia fraseri Walter.   Umbrella-tree, mountain-m. Small tree to 15 m; twigs glabrous; winter terminal bud 3–5 cm, glabrous; lvs ± crowded beneath the fls, deciduous, thin, rhombic-ovate, at anthesis 1–2 dm, at maturity (2)2.5–4 dm, ca half as wide, acute to obtuse, concavely narrowed from above the middle to the base, where prolonged into short, rounded basal lobes, green and glabrous on both sides; fls white, fragrant; pet 6–9, elliptic, 7–10 × 3–4 cm, somewhat clawed; style thick, persisting as a beak on the glabrous follicle; fr-cone ellipsoid, 5–10 cm; $2n=38$. Rich woods; mts. of w. Va. to e. Ky., s. to n. Ga. and Ala.; locally naturalized in Md. May, June.

6. Magnolia macrophylla Michx.   Umbrella-tree, bigleaf m. Tree to 15 m; twigs tomentose; winter terminal bud 3–5 cm, tomentose; lvs deciduous, thin, obovate or pandurate, very large, mostly 5–10 dm, acute or obtuse, concavely narrowed from above the middle to a broad base that is prolonged into short, broadly rounded basal lobes, the lower surface strongly glaucous and finely hairy; fls fragrant; pet 6, elliptic to elliptic-obovate, 12–18 × 4–9 cm, white with a rosy basal spot; style thick, persisting as a beak on the tomentose follicle; fr-cone globose or ovoid, 5–10 cm thick; $2n=38$. Rich woods, chiefly in the mts.; Appalachian region from Va., Ky., and s. O. to n. Ga., w. to Ark. and La. May, June.

2. LIRIODENDRON L.   Tulip-tree, yellow poplar. Sep 3, soon reflexed; pet 6, erect, broadly ovate, forming a cup-shaped cor; anthers extrorse; pistils many, on an elongate receptacle, rpiening into a cone of dry, samaroid, eventually deciduous frs; deciduous trees with simple, 4-lobed lvs and large but inconspicuously colored fls. 2, the other in China.

1. Liriodendron tulipifera L.   Sturdy tree to 60 m; lvs long-petioled, 8–14 cm long and about as wide, broadly retuse; fls solitary at the ends of the branches; sep pale green; pet 4–5 cm, greenish-yellow, with a large basal orange blotch within; frs narrow and elongate, 3–4 cm; $2n=38$. Rich woods; Vt. to s. Mich., s. Ill., and se. Mo., s. to Fla. and La., often cult. May, June.

FAMILY **ANNONACEAE**, the Custard-apple Family

Fls perfect, regular, hypogynous; sep 3, distinct or basally connate; pet 6, in 2 whorls; stamens numerous, distinct, spirally arranged; filament very short; connective often prolonged beyond the extrorse anther; pollen uniaperturate to biaperturate or inaperturate; pistils usually ± numerous on a flat to conic or subglobose receptacle, free and distinct; ovules 1–many; fr commonly of separate, fleshy, indehiscent, stipitate, berry-like carpels, or the carpels sometimes coalescent to form an aggregate fleshy fr, seldom dry and dehiscent or indehiscent; seeds large, with small, dicotyledonous embryo and abundant, ruminate endosperm, woody pls with alternate, exstipulate, simple, entire lvs. 130/2300.

1. **ASIMINA** Adans.   Pawpaw. Sep soon deciduous; outer pet imbricate, inner smaller, valvate; pollen in tetrads, uniaperturate; pistils (1–)3–15, usually only 1–3 maturing; ovules several or many; fr fleshy, aromatic; shrubs or small trees with deciduous lvs. 10, e. U.S.

1 Fls relatively large, 3–4 cm wide; lvs mostly 15–35 cm at maturity . . . . 1. *A. triloba.*
1 Fls smaller, less than 2 cm wide; lvs mostly 6–15(–20) cm at maturity . . . . 2. *A. parviflora.*

1. **Asimina triloba** (L.) Dunal.   Usually shrubby, sometimes a tree to 10 m; young twigs hairy; lvs oblong-obovate, becoming 15–35 cm after anthesis, abruptly short-acuminate, gradually narrowed below; petiole 5–10 mm; fls from wood of the previous year, lurid purple, 3–4 cm wide, the sep 8–12 mm long, the outer pet 1.5–2.5 cm, broadly ovate, outcurved, the inner ovate, nearly erect, and nectariferous at the base; receptacle hemispheric or subglobose; peduncles hairy, (10–)15–20(–25) mm at anthesis; frs solitary or few, 6–15 × 3–4 cm, yellowish-brown, rounded at the tip; seeds flattened, 2–2.5 cm; $2n=18$. Rich, damp woods; w. N.Y. and s. Ont. to s. Mich. and e. Nebr., s. to Fla. and Tex. Apr, May.

2. **Asimina parviflora** (Michx.) Dunal.   Much like no. 1; shrub to 4 m; lvs 6–15(–20) cm after anthesis; fls 12–15 mm wide, the sep 4–7 mm long, the outer pet 1–1.3 cm; peduncles 3–8 mm at anthesis; frs 2–6 cm; seeds 1–1.5 cm; $2n=18$. Pine or oak woods on the coastal plain and less frequently on the piedmont; se. Va. to n. Fla., w. to e. Tex. Apr.

FAMILY **CALYCANTHACEAE**, the Strawberry-shrub Family

Fls perfect, regular, perigynous, the hypanthium urceolate or campanulate; perianth of numerous (15–30) more or less petaloid tepals spirally arranged on the outside of the hypanthium; stamens 5–20, spirally arranged at the top of the hypanthium; filaments short or none; connective prolonged beyond the extrorse anther; 10–25 staminodes present internal to the stamens; pollen-grains dicolpate; pistils 5–35, distinct, spirally arranged within the hypanthium cup; style elongate, filiform, with decurrent stigma; ovules 2, but one abortive; achenes enclosed in the enlarged, fleshy hypanthium; embryo with 2 spirally twisted cotyledons; endosperm none; aromatic shrubs with opposite, simple, petiolate, entire, exstipulate lvs and solitary fls short-pedicelled at the end of short leafy branches. 3/5.

1. **CALYCANTHUS** L., nom. conserv.   Carolina-allspice, strawberry-shrub. Hypanthium urceolate, with a narrow mouth; tepals slender; stamens 10–20, ribbon-shaped. 2, U.S.

1. **Calycanthus floridus** L.   Shrub 1–3 m; twigs and petioles glabrous or nearly so; lvs ovate to elliptic or oblong, 5–15 × 2–6 cm, glabrous beneath or sparsely hairy especially when young; fls aromatic, 4–7 cm wide; tepals brown-purple, linear-oblong ; pistils 10–35; fr (pseudocarp) 2–6 × 1–3 cm; $2n=22$. Rich mt. woods; s. N.Y., s. Pa. and s. O. to Ga., nw. Fla., and se. Miss. June, July. Our plant is var. *glaucus* (Willd.) T. & G. (*C. f.* var. *laevigatus; C. fertilis.*) The more hairy var. *floridus* is more southern.

FAMILY **LAURACEAE**, the Laurel Family

Fls perfect or unisexual, regular, perigynous; perianth of (4)6 sepaloid members in 2 cycles; stamens 3–12, most often 9, the filament often with a pair of basilateral nectariferous appendages, the innermost set of 3 stamens often reduced to staminodes; anthers opening by 2 or 4 uplifting valves; pollen inaperturate or sometimes monosulcate; ovary superior, unilocular and apparently unicarpellate, with a single pendulous ovule; fr a 1-seeded berry or drupe; seeds with large, straight, dicotyledonous embryo and without endosperm; aromatic trees and shrubs with alternate, simple, exstipulate, entire or lobed leaves and small, usually yellowish fls. 30–50/2000+.

1 Lvs evergreen; fls perfect .... 1. *Persea.*
1 Lvs deciduous; fls unisexual, the stamens or pistils reduced and nonfunctional.
   2 Infl terminal, peduncled; fr blue; lvs palmately veined, often lobed .... 2. *Sassafras.*
   2 Infls lateral, sessile; fr red; lvs pinnately veined, not lobed .... 3. *Lindera.*

1. PERSEA Miller, nom. conserv.   Fls perfect; inner series of tepals evidently larger than the outer; fertile stamens 9, in 3 series, the filaments of the inner series each with a pair of basilateral glands; fourth series of stamens represented by staminodes; fr subtended by the persistent, somewhat accrescent tepals; evergreen trees or tall shrubs. 200, mainly trop.

1. Persea palustris (Raf.) Sarg.   Swamp-bay. Tall shrub or tree to 10(–15) m, with thin, gray, fissured bark; twigs densely rusty-villous-tomentose; lvs lanceolate to oblong or oblanceolate, 10–15 cm, evidently and loosely villous or villous-tomentose beneath, but often glabrescent except for the usually persistently hairy midvein; petiole 1–2 cm; infls of axillary, compound dichasia, with numerous fls, the peduncle mostly 2–7 cm long and longer than the petiole; inner tepals 4–5 mm; anthers 4-locular; filaments puberulent; fr black, subglobose to ellipsoid, ca 1 cm; 2*n*=24. Swamps and wet woods on the coastal plain; Md. to Fla., Tex., and Bahama I. May, June (*P. pubescens; P. borbonia* var. *pubescens.*)
   The closely allied but less hairy *P. borbonia* (L.) Sprengel, red bay, with the lvs thinly and closely rufous-strigulose beneath, and with short peduncles scarcely surpassing the petioles, has been attributed to our range but appears to be wholly more southern.

2. SASSAFRAS Nees.   Sassafras. Dioecious; tep 6, persistent; stamens 9, the 3 inner each with a pair of stalked glands at the base; anthers introrse, 4-locular; pistillate fls with 6 short staminodia; pedicels apically expanded at maturity, forming, together with the persistent, accrescent perianth, a cup-like base below the fr; deciduous trees or shrubs. 2, the other Chinese.

1. Sassafras albidum (Nutt.) Nees.   Shrub or tree to 30 m; bark becoming deeply furrowed; lvs long-petioled, silky beneath at least when young, with 3 main veins, highly variable in form, ovate and entire to 2–3(–5)-lobed, often asymmetrical; fls greenish-yellow, produced at the branch-tips of the previous year, expanding with the lvs of the season; peduncles and pedicels at first short, later red and to 1 cm; fr blue, ellipsoid, 1 cm; 2*n*=48. Disturbed woods, thickets, roadsides, and old fields; s. Me. to Mich. and Mo., s. to Fla. and e. Tex. Apr., May. (*S. variifolium.*)

3. LINDERA Thunb., nom. conserv.   Spice-bush. Mostly dioecious; tep 6, deciduous; stamens 9, the 3 inner each with a pair of basilateral glands; anthers introrse, 2-locular; pedicels short, slender to maturity; trees or shrubs. 100, mainly Asian.

1. Lindera benzoin (L.) Blume.   Much-branched shrub to 5 m; lvs obovate to oblong or elliptic, 6–12 cm, pointed at both ends, pinnately veined, the uppermost lvs on each twig commonly the largest; petioles 5–12 mm; fls yellow, 6–7 mm wide, appearing before the lvs in dense clusters ca 2 cm thick from the nodes of last year's stems; fr short-stalked, red, ellipsoid, 6–10 mm; 2*n*=24. Abundant in rich, moist woods; s. Me. to Mich., s. to Fla. and Tex. Mar.–May. (*Benzoin aestivale.*) The northern var. *benzoin,* with the lvs and twigs glabrous, extends s. to Va. and Mo., and in the mts. to Ga. The more southern var. *pubescens* (E. J. Palmer & Steyerm.) Rehder, with the lower lf surfaces and young twigs hairy, extends n. to se. Va., s. O., sw. Mich., and Mo.

FAMILY **SAURURACEAE**, the Lizard's tail Family

Fls perfect, hypogynous, without perianth; stamens 6 or 8 in 2 whorls, or 3 in a single whorl; pollen monosulcate; carpels 3–5, these conduplicate and distinct above the connate base in *Saururus*, united into a unilocular compound ovary with parietal placentation in the other genera; ovules orthotropous, (1)2–4 and laminal-lateral in *Saururus*, 6–10 on each placenta in the other genera; seeds with small, dicotyledonous embryo, scanty endosperm, and abundant, starchy perisperm; aromatic perennial herbs with small fls in dense, bracteate spikes or racemes. 5/7.

1. **SAURURUS** L. Lizard's tail. Stamens 6–8; carpels united at the base only, each with a short, outcurved style, stigmatic along the inner side; fr subglobose, indehiscent, strongly rugose; herbs with jointed stems, broad, alternate lvs, and white fls in slender spikes. 2, the other Asian.

1. **Saururus cernuus** L. Branched, 5–12 dm; lvs with long, basally sheathing petiole, the blade cordate-ovate, 6–15 cm; spikes 1 or 2, peduncled, terminal but often surpassed by axillary branches, 6–15 cm, nodding at the tip before anthesis; filaments white, 3–4 mm, much surpassing the pistils; fr 2–3 mm thick; $2n=22$. Swamps and marshes; s. N. Engl. and s. Que. to Minn., s. to Fla. and Tex. June–Aug.

FAMILY **ARISTOLOCHIACEAE**, the Birthwort Family

Fls perfect, mostly trimerous, epigynous or sometimes merely perigynous; cal gamosepalous and often ± corolloid, regular and 3-lobed, or irregular (often S-shaped or pipe-shaped) and then 3-lobed or 1-lobed or lobeless; pet minute or none; stamens mostly 6 or 12, often adnate to the style or connivent around it; anthers extrorse; pollen inaperturate; ovary superior to more often inferior, with (4–)6 united carpels and as many locules, or the partitions sometimes not reaching the center, the numerous ovules accordingly axile or parietal; fr mostly capsular; seeds with very small, dicotyledonous embryo and abundant, fleshy endosperm; aromatic perennial herbs, or more often twining woody vines, with alternate, simple, entire or sometimes trilobed, mostly exstipulate leaves and large, axillary (in ours) fls, these commonly red-purple or brown and often fetid. 8–10/600.

1 Stamens 12; cal regular.
  2 Ovary superior to half-inferior; styles separate, each prolonged beyond the extrorse stigma into 2
     horns; lvs perennial .... 1. *Hexastylis.*
  2 Ovary wholly inferior; styles united; stigma capitate, 6-lobed; lvs annual .... 2. *Asarum.*
1 Stamens 6; cal either irregularly lobed or with strongly bent tube .... 3. *Aristolochia.*

1. **HEXASTYLIS** Raf. Heart-leaf, little brown jugs. Cal regular, tubular, 3-lobed, glabrous outside; pet none; stamens 12, with very short or no filament, appressed to the styles; connective only shortly (if at all) prolonged; ovary superior to half-inferior, 6-locular; styles 6, erect, distinct, each prolonged beyond the conspicuous, extrorse stigma into 1 or 2 erect horns; fr a fleshy capsule, enclosed by the accrescent cal, the two rupturing together at maturity; perennial, evergreen herbs from short or long rhizomes, bearing annually a petiolate, leathery, broadly cordate or hastate, glabrous lf, the old lvs persistent; fls solitary, on a short peduncle from the axil of a bract, greenish or purplish outside, dark red-purple inside; seeds carunculate. Genus often included in *Asarum*. 4, se. U.S., possibly also some in China.

1 Lvs mostly round-cordate; style-extension above the stigma merely notched at the tip.
  2 Fls smaller, the cal-tube 0.7–1.5 cm long, the lobes 5–9 mm wide at base .... 1. *H. virginica.*
  2 Fls larger, the cal-tube 1.5–4 cm long, the lobes 10–20 mm wide at base .... 1. *H. shuttleworthii.*
1 Lvs mostly triangular-hastate; style extension bifid to the stigma .... 3. *H. arifolia.*

1. **Hexastylis virginica** (L.) Small. Rhizome short and freely branched; lvs cordate-orbicular, 5–10 cm, commonly broadest above the summit of the petiole, the basal lobes broadly rounded, the sinuses broad to

narrow or even closed; cal-tube 0.7–1.5 cm, cylindric to slightly constricted above or slightly flaring, the lobes erect to spreading 3–10 mm, 5–9 mm wide at base; ovary superior to half inferior; style-extension merely notched at the tip; $2n=26$. Moist or dry woods; s. Md. to W.Va. and e. Ky., s. to se. Va., Ga., and Ala. Apr., May. (*H. contracta, H. heterophylla, H. memmingeri, H. naniflora.*)

2. **Hexastylis shuttleworthii** (Britten & Baker) Small. Much like no. 1, sometimes producing slender creeping rhizomes as well as short branching ones; cal tube 1.5–4 cm, often ± constricted above, the spreading lobes 6–17 mm, 10–20 mm wide at base; ovary wholly superior. Rich woods; Va. and W.Va. to Ga. and Ala. Apr., May. (*H. lewisii, H. macranthum.*)

3. **Hexastylis arifolia** (Michx.) Small. Lvs mostly triangular in outline, 5–15(–20) cm long and about as wide, mostly broadest below the summit of the petiole, with comparatively wide and shallow sinuses, or varying even in the same colony to cordate-ovate; cal urceolate, the tube 1.2–2.8 cm, constricted above, the erect to spreading lobes 2–8 mm, about as wide at base as long; style-extension bifid to the stigma; $2n=26$. Moist or dry woods; s. Va. to se. Ky., s. to n. Fla., Ala., and se. La. Apr., May. Var. *arifolia,* the s. and e. phase of the species, in our area restricted to se. Va., has ± spreading cal-lobes 3–8 mm long and wide. Var. *ruthii* (Ashe) Blomquist, the nw. phase of the species, in our range restricted to sw. Va. and se. Ky., has erect or slightly spreading cal-lobes 2–5 mm long and wide.

2. **ASARUM** L. Wild ginger. Cal regular, tubular below, deeply 3-lobed, hairy outside; pet none or vestigial; stamens 12, with well developed filament, closely appressed to the styles; connective prolonged as a subulate tip; ovary inferior, 6-locular; styles coherent in a column, expanded at the tip into a 6-lobed stigma; fr capsular, opening irregularly; seeds large, ovoid, wrinkled, carunculate; herbs, the elongate rhizome producing annually a pair of petiolate, broad, membranous, hairy lvs, these deciduous at the end of the season; fls solitary, red-brown, short-peduncled, arising between the pair of lvs. 90, N. Hemisphere.

1. **Asarum canadense** L. Plants hairy, especially the petioles and cal; lvs cordate-rotund to cordate-reniform, mostly 8–12 cm wide at anthesis, larger at maturity; peduncle stout, 2–5 cm; fls 2–4 cm; calyx tube short, erect, the lobes spreading or reflexed; $2n=26$. Rich woods, usually in colonies; N.B. and Que. to Ont. and Minn., s. to N.C., n. Ala., and n. La. Apr., May. Var. *canadense,* with essentially the range of the species, has the cal-lobes spreading from the base, curved upward beyond the middle, and gradually narrowed into a revolute-margined slender tip 0.5–2 cm. (*A. acuminatum,* the extreme form.) Var. *reflexum* (E. Bickn.) B. L. Robinson, from N.Y. to Mich. and Io., and southward, commoner in the Middle West, has the cal-lobes reflexed, often closely appressed to the ovary, triangular, abruptly contracted into a short, tubular, 2–4 mm tip. (*A. ambiguum,* a form intermediate to var. *canadense.*)

3. **ARISTOLOCHIA** L. Cal tubular, irregular, S-shaped or pipe-shaped and 3-lobed at the summit, or straight and with a single unilateral terminal lobe; pet none; stamens in ours 6; anthers sessile, adnate to the style; ovary inferior, 6-locular (in ours); style short and stout; stigma broad, with 6 (in ours) lobes or angles; fr a septicidal, lobed or angled capsule, with numerous compressed, horizontal seeds; herbs or woody vines, usually with broad, alternate, palmately veined lvs, and with small or large fls solitary or fascicled in the axils. 500, cosmop.

1 Erect, ascending, or reclining herbs.
  2 Perianth S-shaped, 3-lobed at the tip; fls nearly basal . . . . 1. *A. serpentaria.*
  2 Perianth nearly straight, with 1 terminal lobe; fls in axillary fascicles . . . . 2. *A. clematitis.*
1 Twining, woody vines.
  3 Petioles and perianth densely hairy; peduncles bractless . . . . 3. *A. tomentosa.*
  3 Petioles and perianth glabrous or nearly so; peduncles with a leafy bract . . . . 4. *A. macrophylla.*

1. **Aristolochia serpentaria** L. Virginia-snakeroot. Stems erect from a rhizome, to 6 dm, thinly hairy; lvs ovate to oblong or nearly linear, acuminate, truncate to sagittate or cordate at base, 6–12 cm; lower lvs reduced to scales; fls solitary on slender, scaly peduncles from the lowest nodes; perianth S-shaped, enlarged at base and summit, 1–2 cm, the spreading limb madder-purple, unequally or indistinctly 3-lobed; $2n=28$. Moist or dry upland woods; Conn. and s. N.Y. to s. Mich., Ill. and Mo., s. to Fla. and Tex. May, June.

2. **Aristolochia clematitis** L. Birthwort. Stems ascending or reclining, 5–10 dm; lvs broadly cordate, 5–10 cm long and wide, blunt to subacute, pale beneath, the broadly rounded basal auricles somewhat incurved below the rounded sinuses; fls in small axillary fascicles, the yellow perianth nearly straight, 3 cm, enlarged around the ovary, terminating in an ovate lobe on one side; $2n=14$. Native of s. Europe, occasionally escaped from cult. in our range. Summer.

3. **Aristolochia tomentosa** Sims. Pipe-vine. High-climbing woody twiner; lvs broadly round-cordate, 1–2 dm long and wide, softly hairy beneath, peduncles axillary, solitary or paired, 2–4 cm, densely tomentose, bractless; perianth bent, 4 cm, densely tomentose, nearly closed at the throat, the limb dark madder-purple, spreading to reflexed, subequally 3-lobed; fr cylindric, 6–8 cm; $2n=28$. Wet, alluvial woods; s. Ind. to se. Kans., s. to Ga., nw. Fla., and Tex.; locally intr. in Mass. June.

4. **Aristolochia macrophylla** Lam. Dutchman's pipe. High-climbing woody twiner; lvs broadly round-cordate, 1–4 dm long and wide, glabrous, or minutely hairy beneath; peduncles solitary or few together in the axils, glabrous, very slender, bearing near the middle a cordate-clasping, foliaceous bract 1–2 cm wide; perianth glabrous, strongly bent, 4 cm, much dilated at base, gradually narrowed to the throat, the limb dark madder-purple, 2.5 cm wide, subequally 3-lobed; fr cylindric, 5–8 cm. Rich mt. woods; Appalachian region from s. Pa. to n. Ga.; cult. and escaped northward. June. (*A. durior*, probably misapplied.)

Family **NELUMBONACEAE**, the Lotus-Lily Family

Fls perfect, regular, hypogynous; tep numerous, spirally arranged, a few outer ones green and sepaloid, the others petaloid; stamens numerous (ca 200) spirally arranged; pollen tricolpate; pistils numerous (ca 10–40), individually sunken in (but free from) the enlarged, obconic, spongy receptacle, only the sessile stigma protruding; ovule solitary, pendulous; fr of separate, hard-walled nuts loose in the cavities of the accrescent receptacle; seed solitary, filled by the large, dicotyledonous embryo; pls aquatic; lvs arising directly from the rhizome, alternate, long-petiolate, with large, circular, centrally peltate blade, some blades floating, others raised above the water; fls elevated above the water on long, stout peduncles arising from the rhizome. 1/2.

1. **NELUMBO** Adans. Lotus-lily. Characters of the family.

1. **Nelumbo lutea** (Willd.) Pers. American l.-l., water-chinquapin. Lvs 3–7 dm wide; peduncles rising to 1 m above the water, the solitary, pale yellow fl 15–25 cm wide; mature receptacle 1 dm wide, flat on top, the acorn-like frs 1 cm thick; $2n=16$. In quiet water; irregularly from N.Y. and s. Ont. to Minn., s. to Fla., Tex., and W.I.; N. Engl., where probably only intr. Summer. (?*N. pentapetala*.)

N. nucifera Gaertner, the Oriental sacred lotus, with pink or white fls, rarely escapes from cult.

Family **NYMPHAEACEAE**, the Water-Lily Family

Fls perfect, regular, hypogynous to perigynous or epigynous; sep 4–5, sometimes petaloid; pet 8–many, distinct in ours, often passing into the stamens; stamens numerous, spirally arranged; filament not strongly differentiated from the connective and anther, the elongate pollen sacs embedded in the adaxial side of the long, flat, usually narrow staminal blade; pollen monosulcate; gynoecium of 5–35 carpels ± firmly united into a compound, plurilocular, superior to inferior ovary that is apically constricted beneath an expanded disk bearing as many radiant stigmatic lines as the number of locules; ovules numerous, laminar, fr an irregularly dehiscent, spongy to leathery berry; seeds small, with small, dicotyledonous embryo, rather scanty endosperm, and copious perisperm; pls aquatic; lvs arising directly from the rhizome, alternate, long-petiolate, with cordate or hastate to peltate, floating to sometimes emergent or wholly submerged blade; fls borne singly at or just above the surface of the water on long, stout peduncles arising from the rhizome. 5/50.

1 Fls yellow, often tinged with red or purple or green; sep petaloid; pet inconspicuous .... 1. *Nuphar.*
1 Fls white or rarely pink; sep green; pet conspicuous .... 2. *Nymphaea.*

1. **NUPHAR** J. E. Smith, nom. conserv. Yellow water-lily, spatterdock. Sep mostly 6, sometimes only 5, concave, imbricate, petaloid, forming a subglobose or deeply saucer-shaped yellow fl; pet many, much smaller than the sep, usually shorter than the numerous, hypogynous stamens; stigmatic disk with 6–24 radiating lines; lf-blades basally cordate to hastate; petioles and peduncles with numerous minute air-cavities; $2n=34$. (*Nymphaea* of auth., misapplied.) 15, N. Temp. and cold.

1. Nuphar microphylla (Pers.) Fern. Lvs both submersed and floating, the latter mostly 5–10 cm, the sinus usually more than half as long as the midvein; petioles flattened on the upper side; fls 1.5–2 cm thick; sep usually 5, yellow within; anthers 1.5–3 mm, shorter than the filaments; stigmatic disk red, 3–6 mm wide at anthesis, scarcely enlarged in fr, 6–10-rayed; fr 15 mm, with constricted neck; $2n=34$. Ponds; Nf. to N.J. and Pa., w. to n. Mich., n. Wis., and ne. Minn. Summer. (*N. lutea* subsp. *pumila*.) A hybrid with *N. variegata* has been called *N.* × *rubrodisca* Morong.

2. Nuphar variegata Durand. Floating lvs 10–25 cm, two-thirds as wide, usually rounded above, the narrow basal sinus seldom over half as long as the midvein; submersed lvs few or none; petioles flattened on the upper side and narrowly winged; fls (2.5)3.5–5 cm thick, sep mostly 6, usually reddish within or on the basal half; anthers 4–7 mm, longer than the filaments; stigmatic disk green or greenish, 1 cm wide, mostly 10–15-rayed; fr ovoid, 2–4.5 cm, only slightly constricted below the disk; $2n=34$. Ponds; Nf. to B.C., s. to Del., n. O., Io., Kans., and Ida. Summer. (*N. fraterna*.)

3. Nuphar advena (Aiton) Aiton f. Floating and submersed lvs rarely present; emersed lvs mostly 2–4 dm, on stout petioles, the subacute basal lobes separated by a broadly triangular sinus; petioles terete, or oval in cross section; fls 3–5 cm thick; sep 6, yellow within or rarely suffused with purple; anthers 3–7 mm, longer than the filaments; stigmatic disk greenish, 10–15 mm wide at anthesis, mostly 14–18-rayed; fr broadly ovoid, 4 cm, mostly thicker than long. Sluggish streams, less often in ponds; s. Me. to Fla. and Cuba, w. to Wis., Mo., Kans., Tex., and n. Mex. Summer. (*N. fluviatilis; N. macrophylla; N. puteorum*.)

4. Nuphar sagittifolia (Walter) Pursh. Lvs both floating and submerged, oblong-lanceolate, 15–40 × 5–10 cm, the basal sinus only 2–5 cm; fls 2.5–3 cm thick; anthers 3–4 mm; stigmatic disk 8–15 mm wide at anthesis, 10–14-rayed. Tidal reaches of streams; Va. to S.C. Summer. A hybrid with *N. advena* has been called *N.* × *interfluitans* Fern.

2. NYMPHAEA L., nom. conserv. Water-lily. Sep 4, green; pet numerous, white or pink, inserted on the ovary; stamens numerous, inserted on the ovary; inner stamens with linear filament and long anther, the outer with progressively widened, petal-like filament and shorter anther; stigmatic disk with 6–30 radiating lines; lf-blades floating, rotund or elliptic with a basal sinus; petioles and peduncles with 4 large air-passages. (*Castalia*.) 30, widespread.

1. Nymphaea odorata Aiton. Rhizome with or without knotty tubers; lf-blades 1–3 dm wide, rotund, with a narrow sinus, often purple or red beneath; fls ± fragrant, open from early morning to noon or midafternoon, mostly 7–20 cm wide; sep 3–10 cm, the receptacle circular at the level of their insertion; pet 17–25, about equaling the sep, white or rarely pink, narrowly elliptic-oblong to spatulate, tapering to a subacute or rounded tip; stamens mostly 40–100; stigmatic lines 10–25; sep closely inflexed over the top of the fr; seeds 2–4 mm; $2n=56, 84$. Quiet waters; Nf. to Man., n. Minn., and n. Mich., s. on or near the coastal plain to Fla. and Tex. Summer. Var. *odorata*, with nearly the range of the species, has flat lvs, and the sep are seldom over 7 cm. (*N. tuberosa; C. minor* is the small-flowered extreme.) Var. *gigantea* Tricker, restricted to the coastal plain from Del. to Fla. and Tex., has the lf-margin upturned, and the fls average larger, the sep up to 10 cm. (*C. lekophylla*.)

    N. alba L., a related European sp., is casually intr. in N. Engl. It has the petioles crowded on the rhizome, the blades deep purple beneath, the fls open nearly all day, and the pet rounded above.

2. Nymphaea tetragona Georgi. Pygmy w.-l. Rhizome erect, the petioles and peduncles arising near the tip; lf-blades commonly 7–12 cm long, two-thirds to three-fourths as wide, with a fairly open sinus; fls white, inodorous, relatively small, mostly 4–8 cm wide, opening in the afternoon; sep 2–3 cm, the receptacle quadrangular at the level of their insertion; pet 8–17; stamens mostly 20–40; stigmatic lines 6–9; sep becoming stiffly erect, not arched over the fr. Cold ponds and lakes; Me. and Ont. to n. Mich. and n. Wash., and northward; also in Eurasia. Summer. (*C. leibergii*.)

Family CABOMBACEAE, the Water-Shield Family

Fls perfect, regular, hypogynous; sep and pet mostly 3(4) each and all ± petaloid, free and distinct or nearly so; stamens 3–18, differentiated into a slightly flattened filament and fairly conventional anther; pollen monosulcate or sometimes trichotomosulcate; pistils 2–18, distinct, each tapering into a short style; ovules 1–3, laminar along or near the dorsal suture; fr small, leathery, indehiscent, achene-like or follicle-like; seeds small, with small, dicotyledonous embryo, scanty endosperm, and copious perisperm; caulescent aquatic herbs with creeping rhizomes and elongate, distally floating, leafy stems; lvs alternate, long-petiolate, with floating, centrally peltate blade, or most or all of them opposite or whorled, submerged, short-petiolate, and much dissected; fls subaerial, solitary and long-pedunculate from the upper nodes. 2/8.

1 Lvs all with floating, entire blade; stamens 12–18 .... 1. *Brasenia.*
1 Principal lvs submersed, palmately dissected; stamens 3–6 .... 2. *Cabomba.*

1. **BRASENIA** Schreber.   Water-shield. Sep somewhat petaloid, colored within; stamens 12–18, or reputedly up to 36; filaments of 2 lengths; anthers introrse, pistils 4–8; submerged parts covered with a mucilaginous jelly; lvs alternate, mostly crowded near the summit, long-petioled, with floating, centrally peltate blade. Monotypic.

1. **Brasenia schreberi** J. F. Gmelin.   Stem to 2 m; lf-blades elliptic, 4–12 cm, half as wide; fls dull purple, on stout peduncles to 15 cm; sep and pet persistent, the latter 12–16 mm, the former somewhat shorter; 2n=80. Quiet water; N.S. and e. Que. to Minn., s. to Fla. and Tex.; also on the Pacific slope, and in tropical Amer., and the Old World. Summer.

2. **CABOMBA** Aublet.   Sep and pet similar, both petaloid; stamens 3–6; anthers extrorse; pistils 3–6; principal lvs opposite, palmately finely dissected, submersed; some small, alternate lvs with floating, entire, peltate blade usually also present. 7, New World.

1. **Cabomba caroliniana** A. Gray.   Fanwort. Stem to 2 m, branched; submersed lvs rotund, 2–5 cm wide, on petioles 1–3 cm; floating lvs few and small, linear-elliptic, 6–20 mm, commonly slightly constricted at the middle, often bifid at one end; peduncles 3–10 cm; sep and pet 6–12 mm, white with yellow base, the pet auriculate near the base; stamens 6; 2n=24. Ponds and quiet streams; N.J. to O., s. Mich., and Mo., s. to Fla. and Tex., and occasionally intr. farther north, as in Mass. and N.Y.

FAMILY **CERATOPHYLLACEAE**, the Hornwort Family

Fls small and inconspicuous, axillary, hydrogamous, hypogynous, unisexual, solitary in the axils, the male generally above the female; sep tiny, 8–15 in a single whorl, connate at the base; pet none; stamens mostly 10–20, distinct, spirally arranged on the flat receptacle, filament short and broad; pollen sacs extrorse on the thickened, somewhat laminar connective, which is prolonged into a pair of prominent points; pollen inaperturate; pistil solitary, simple, tapering into a filiform style; ovule solitary, suspended; fr an achene; seed with thin testa, lacking both perisperm and endosperm; embryo with 2 thickened cotyledons; plumule well developed, with several embryonic lvs; radicle vestigial; submersed, aquatic, rootless perennial herbs with elongate, branching stems, and whorled, dichotomously dissected lvs. 1/6.

1. **CERATOPHYLLUM** L.   Hornwort; coontail. The only genus of the family.

1 Achenes with 2 spines near the base; lf-segments conspicuously serrate on one side .... 1. *C. demersum.*
1 Achenes with basal and also several marginal spines; lf-segments entire or nearly so .... 2. *C. echinatum.*

1. **Ceratophyllum demersum** L.   Stems elongate, freely branched, forming large masses; lvs in whorls of 5–12, rather stiff, variable, 1–3 cm, mostly once or twice forked, the ultimate segments linear, flat, 0.5 mm wide, but nearly doubled in width at the broad, antrorse teeth; achene ellipsoid, 4–6 mm, with 2 basal spines; first 2 pairs of plumular lvs unbranched; 2n=24. Abundant in quiet water; widespread from s. Can. to S. Amer., and widely distributed in the Old World.

2. Ceratophyllum echinatum A. Gray. Much like no. 1; lvs more flaccid, the larger ones mostly 3(4) times forked, the ultimate segments commonly narrower, entire or with a few small, very slender teeth; achenes with several unequal marginal spines as well as 2 basal ones, conspicuously tuberculate on the surface; all plumular lvs forked. Quiet water; N.S. and N.B. to B.C., s. to Fla. and trop. Amer.; less often collected than no. 1.

FAMILY **RANUNCULACEAE**, the Buttercup Family

Fls hypogynous, regular (except *Aconitum* and *Delphinium*), with all parts free and distinct (pistils connate in *Nigella*); sep usually imbricate, in some genera small and soon deciduous, in others petal-like; pet present or absent; stamens usually numerous, often some of them modified into nectariferous staminodes, these when larger and more showy called pet; pistils 1–many, simple; stigma lateral or terminal; fr an achene, utricle, follicle, or berry; seeds with copious endosperm and small, dicotyledonous embryo, or sometimes with larger embryo and little or no endosperm; mostly herbs, seldom shrubs or woody vines, the lvs usually alternate, sometimes opposite or whorled, or all basal, often cleft or compound. 50/2000.

1 Ovules 2–many in each ovary; fr a follicle or berry.
  2 Lvs simple, variously entire or merely toothed to deeply palmatifid, but not compound and not finely dissected into narrow segments.
    3 Fls regular.
      4 Fr a follicle; rhizomes and roots not yellow; fls 1–many; ovules often more than 2.
        5 Staminodes and pet none (though the sep are petaloid) .... 1. *Caltha.*
        5 Staminodes well developed (though not petaloid).
          6 Staminodes narrowly oblong, flat; native .... 2. *Trollius.*
          6 Staminodes hollow and somewhat 2-lipped; introduced .... 3. *Helleborus.*
      4 Fr a berry; rhizomes yellow internally; fls solitary; ovules 2 .... 21. *Hydrastis.*
    3 Fls strongly irregular.
      7 Posterior sep strongly arched or hooded, scarcely spurred; pet not spurred .... 8. *Aconitum.*
      7 Posterior sep prolonged into a long spur; upper pet spurred .... 9. *Delphinium.*
  2 Lvs compound, with distinct lfls, or finely dissected into narrow segments.
    8 Lvs finely dissected into narrow segments; ovaries and follicles connate .... 4. *Nigella.*
    8 Lvs with distinct lfls; ovaries and follicles distinct.
      9 Shrub .... 22. *Xanthorhiza.*
      9 Herbs.
        10 Sep small, soon deciduous; fls small and numerous, deriving their color and size principally from the numerous stamens; lvs large, 2–3 times compound.
          11 Fr a follicle; stigma borne on a definite style .... 6. *Cimicifuga.*
          11 Fr a berry; stigma essentially sessile .... 7. *Actaea.*
        10 Sep well developed, petaloid; fls conspicuous, mostly few; lvs diverse.
          12 Staminodes developed into conspicuous pet, long-spurred at the base; sep in ours seldom white; lvs 2–3 times compound .... 16. *Aquilegia.*
          12 Staminodes wanting, or small and inconspicuous, neither petaloid nor spurred; sep white or nearly so.
            13 Staminodes none; lvs 1–3 times compound .... 17. *Isopyrum.*
            13 Staminodes present, clavate, nectariferous in the hollow summit; lvs trifoliolate .... 20. *Coptis.*
1 Ovule one in each ovary; fr an achene (a dehiscent utricle in *Trautvetteria*).
  14 Lvs linear, all basal; sep short-spurred at the base; fls small, solitary .... 14. *Myosurus.*
  14 Lvs either distinctly broader, or not all basal, or usually both; sep not spurred.
    15 Pet ± well developed, yellow or white to orange or red; cauline lvs alternate.
      16 Pet with a nectariferous spot or scale at the base .... 13. *Ranunculus.*
      16 Pet unappendaged .... 15. *Adonis.*
    15 Pet wanting, or represented by inconspicuous staminodes.
      17 Lvs alternate; sep small, inconspicuous, usually soon deciduous.
        18 Fr indehiscent; lvs ternately decompound .... 19. *Thalictrum*
        18 Fr dehiscent; lvs simple, but deeply lobed .... 5. *Trautvetteria.*
      17 Lvs opposite or whorled (the stem sometimes naked except for the involucral lvs just beneath the fl); sep well developed and petaloid.
        19 Sep valvate in bud, usually 4 .... 12. *Clematis.*
        19 Sep imbricate in bud, mostly 5 or more.
          20 Lvs merely lobed (or the involucral ones entire) to deeply incised or parted, but not truly compound; stigma minute, unilateral on the style.
            21 Cauline leaves resembling the basal, remote from the peduncled fls .... 10. *Anemone.*
            21 Cauline lvs represented only by 3 involucral lvs closely subtending the fl .... 11. *Hepatica.*
          20 Lvs ternately compound, with distinct lfls; stigma sessile, capitate .... 18. *Anemonella.*

1. CALTHA L. Marsh-marigold, cowslip. Sep 5–9, petaloid; pet none; stamens many; staminodes none; pistils 4–many, the style very short; follicles compressed to subterete; low perennial poisonous herbs of wet places, with alternate, petiolate, entire or merely toothed lvs and peduncled, axillary or terminal fls. 15, circumboreal.

1 Fls bright yellow, 1.5–4 cm wide; pistils 4–15; widespread .... 1. *C. palustris.*
1 Fls pink or white, 1 cm wide; pistils 20–40; northwestern .... 2. *C. natans.*

1. Caltha palustris L. Stems hollow, 2–6 dm, branched above; basal lvs long-petioled, the broadly cordate-rotund blade usually with a deep and narrow sinus; cauline lvs with progressively shorter petiole and wider sinus; fls bright yellow, 1.5–4 cm wide; stamens 50–120; anthers linear-oblong or lance-oblong, 2 mm; follicles 4–15, 10–15 mm; 2*n*=32. Wet woods and meadows, swamps, bogs, and shallow water; circumboreal, s. to Va., W.Va., Ind., Ill., and Io., and in the mts. to N.C. and Tenn. Apr.–June. Forms with the basal leaves somewhat flabellate, with a relatively broad and open sinus, do not appear to be taxonomically separable. (var. *flabellifolia.*)

2. Caltha natans Pallas. Stems slender, floating; lvs reniform to broadly cordate, 2–5 cm; fls pink or white, 1 cm wide; stamens 12–25; anthers subrotund, 0.5 mm; follicles 20–40, in a dense head, 4–5 mm; 2*n*=32. Ponds or slowly flowing streams; circumboreal, s. to n. Wis. and n. Minn. July, Aug.

2. TROLLIUS L. Globe-fl. Sep 5 or more, large, petaloid; pet none; staminodes numerous, shorter than the stamens, nectariferous at the base; stamens very numerous; filaments slender; anthers linear; pistils several to many; style subulate; follicles thin-walled; perennial herbs with alternate, palmately cleft lvs and large, solitary, terminal fls. 15, the others Eurasian.

1. Trollius laxus Salisb. Stem 1–5 dm; basal lvs long-petioled, cauline 1–3, chiefly above the middle, sessile or subsessile and often approximate; blades subrotund, 3–5-parted with 3-lobed and coarsely toothed, cuneate-obovate segments; sep 5–7, greenish-yellow to bright yellow or cream, elliptic, 15–20 mm, at first incurved-ascending; staminodes oblong, 3–5 mm; follicles 1 cm; 2*n*=32. Swamps, wet woods and wet meadows; Conn. to Del., Pa., and O. Apr.–June.

3. HELLEBORUS L. Sep 5, large, petaloid; pet none; stamens numerous, the outer 8–10 modified into staminodes, short, upcurved, basally clawed, distally clavate or obovoid, hollow and somewhat 2-lipped; pistils mostly 3 or 4, with erect, slender style; fr follicular; poisonous perennial herbs with alternate, palmately cleft (or the upper entire) lvs and large white, green, or purple fls terminating the stem and branches. 20, Eurasia.

1. Helleborus viridis L. Green hellebore. Plants 2–4 dm; basal lvs long-petioled, divided into 7–11 sharply toothed segments; cauline smaller, with shorter petiole and fewer segments; bracts divided and somewhat leaf-like; fls green, 4–6 cm wide, on peduncles 2–5 cm; follicles connate at the base; 2*n*=32. Native of Europe, ± established here and there in our range. Earliest spring.

H. niger L., called Christmas rose because it flowers so early, is widely cultivated and may escape. It has entire bracts and white to pale purple fls.

4. NIGELLA L. Fennel-fl. Sep 5, petaloid; pet none; staminodes 5 or more; short, clawed, distally 2-lobed; stamens many; pistils 5–10, connate to about the middle, prolonged above into subulate styles; fr of connate, beaked follicles; annuals with finely dissected lvs and large, regular, terminal fls. 15+, Medit. region and w. Asia.

1. Nigella damascena L. Love-in-a-mist. Plant 3–6 dm; fls bluish, 3–4 cm wide, closely subtended by an involucre of dissected lvs; staminodes villous; 2*n*=12. Native of s. Europe, occasionally escaped from cult. in our range.

5. TRAUTVETTERIA Fischer & C. A. Meyer. False bugbane. Sep 4 or 5, broadly obovate, caducous; pet and staminodes none; stamens very numerous, the filaments widened distally and often wider than the short, oblong anthers; pistils several, small; styles short, subulate; fr a 4-nerved, thin-walled, 1-seeded, dehiscent utricle; stout perennials with large, alternate, palmately divided lvs and broadly rounded, corymbiform clusters of small white fls. Monotypic.

1. Trautvetteria caroliniensis (Walter) Vail. Plants 5–10 dm, hairy in the infl; lvs broadly reniform, eventually to 3 dm wide, deeply 5–11-lobed, the lobes narrowly to broadly cuneate, irregularly toothed or cleft; basal lvs long-petioled, the cauline much smaller and sessile or subsessile; infl long-peduncled; filaments to 1 cm; fr semi-ellipsoid to obovoid, with a short, persistent style. Mt. woods, especially in damp or wet soil; Pa. and Ky. to Ga.; also on prairies in s. Ind. and w. Ill., and on limestone bluffs in s. Mo.; w. N. Amer.; Japan. June, July. Ours is var. *caroliniensis.*

6. CIMICIFUGA Wernisch.   Bugbane. Sep 4–5, petaloid, very convex, caducous; pet none; stamens very numerous, with long, white filaments and rounded or obovoid anthers, a few of the outermost usually transformed into expanded, 2-horned staminodes; pistils 1 or more; follicles several-seeded; tall perennial poisonous herbs with few, large, commonly basally disposed, ternately or pinnately compound or decompound lvs; broad lfls, and erect, elongate, many-fld racemes or panicles. 20, N. Hemisphere.

1 Ovary 1 (very rarely 2 or 3), sessile; style stout; follicle firm-walled.
  2 Staminodes present; principal lvs with mostly (15)20 or more lfls .... 1. *C. racemosa.*
  2 Staminodes wanting; principal lvs with mostly 3–9(15) lfls .... 2. *C. rubifolia.*
1 Ovaries 3 or more, stipitate; style subulate; follicles papery .... 3. *C. americana.*

1. Cimicifuga racemosa (L.) Nutt.   Black snakeroot; black cohosh. Plant 1–2.5 m; lvs ternate and then pinnately once or twice compound with mostly (15–)20 or more rotund to oblong, coarsely and sharply toothed or cleft lfls seldom over 1 dm, cuneate to cordate at base; infl usually branched, slender, 2–8 dm; pedicels not bracteolate; sep 5 mm; stamens 8–10 mm; staminodes oblanceolate or oblong, 2–3 mm; caducous; follicles firm-walled, transversely veined, 6–9 mm, the style becoming lateral and recurved; seeds rough but not chaffy; 2*n*=16. Moist or dry woods; Mass. to N.Y., O., Ind., and Mo., s. to S.C., Va., and Tenn. June–Aug.

2. Cimicifuga rubifolia Kearney.   Appalachian b. Plant seldom over 1.5 m; lvs with mostly 3–9(–15) lfls, typically ternate with the lateral segments but not the terminal one again ternate, the lfls relatively large, mostly 1–2.5 dm long and wide, cordate at the base; pedicels bibracteolate at the base (in addition to the small subtending bract); staminodes none; follicle 8–15 mm; seeds covered with chaffy scales; otherwise much like no. 1. Cool mt. woods, uncommon; sw. Va. to N.C. and Tenn.; disjunct in w. Ky. and s. Ill. (Aug.) Sept. (*C. racemosa* var. *cordifolia,* misapplied.)

3. Cimicifuga americana Michx.   American b., summer cohosh. Plant 1–1.5 m; lvs and infl as in no. 1; staminodes broadly oval, petaloid, saccate and nectariferous at the base within; ovaries mostly 3 or 5(8) tipped with a short, subulate style, short-stipitate, the stipe elongating to 5–8 mm; follicles fragile, 10–15 mm with subterminal, erect or ascending style; seeds densely covered with chaffy scales. Mt. woods; s. Pa. to N.C. and Tenn. Aug., Sept.

7. ACTAEA L.   Baneberry. Sep 3–5, obovate, petaloid, caducous; pet 4–10, deciduous, spatulate to obovate, clawed; stamens numerous, the elongate filaments mostly widened upwards; anthers oval; pistil 1; stigma broad, sessile, bilobed; fr 1 several-seeded berry; poisonous, odoriferous perennial herbs with large, ternately 2–3 times compound lvs, sharply toothed lfls, and small, white fls in a dense, long-peduncled, terminal raceme. 5, circumboreal.

1 Fr red, rarely white; fruiting pedicels slender; ovary wider than the stigma .... 1. *A. rubra.*
1 Fr white, rarely red; fruiting pedicels very stout; stigma wider than the ovary .... 2. *A. alba.*

1. Actaea rubra (Aiton) Willd.   Red b. Much like no. 2; lfls commonly hairy at least on the veins beneath; pet spatulate to obovate; ovary wider than the stigma, narrowed above to a short beak; fruiting pedicels slender, 0.4–0.7 mm thick; fr red, rarely white; 2*n*=16. Rich woods; circumboreal, in Amer. from Lab. and Nf. to Alas., s. to Conn., n. N.J., n. Ind., Io., and Ariz. May–July. (*A. eburnea; A. erythrocarpa; A. spicata* var. *rubra; A. alba* of some authors, misapplied.)

2. Actaea alba (L.) Miller.   Doll's eyes. Stem 4–8 dm; lfls ovate or ovate-oblong, usually glabrous beneath; raceme short at anthesis, with congested fls, later elongate; pet 2.5–4 mm, mostly spatulate; stigma sessile, wider than the ovary; fruiting pedicels red, 1–2 mm thick; berries ellipsoid-globose, 1 cm, white, rarely red, with persistent stigma; 2*n*=16. Rich woods; e. Que. to Ont. and Minn., s. to Ga., La., and Okla. May–June. (*A. pachypoda.*)

8. ACONITUM L.   Monkshood, aconite. Fls irregular; calyx corolloid, with unequal sep, the upper one (called the helmet) the largest, strongly arched or hooded, its tip prolonged forward and downward into a short beak; upper 2 pet concealed under the helmet, clawed, nectariferous at the tip; lower 3 pet vestigial or wanting; stamens numerous; pistils 3–5, the fr follicular; poisonous perennial herbs with broad, palmately cleft lvs and showy, mostly blue or white fls in a terminal raceme or panicle. 50+, Eurasia and N. Amer.

1 Fls pale to deep blue (varying to seldom white); stem from a thickened, turnip-like root that is renewed each year.
  2 Lvs deeply cleft, but not to the base, the segments much broader than linear; native.
    3 Helmet rounded and dome-like, about as high as long, the hood scarcely differentiated from the beak; n. of the glacial boundary . . . . 1. *A. noveboracense.*
    3 Helmet mostly somewhat higher than long, the hood prolonged upwards and more or less evidently differentiated from the projecting beak; mostly s. of the glacial boundary . . . . 2. *A. uncinatum.*
  2 Lvs divided to the base, with linear-oblong segments; escaped from cult. . . . . 3. *A. napellus.*
1 Fls white or ochroleucous; stem from fascicled, slender roots . . . . 4. *A. reclinatum.*

    1. Aconitum noveboracense A. Gray. New York m. Stem erect, to 1(–1.5) m, from a thickened, turnip-like root that is renewed each year, the new root alongside the old one; lvs rather few, deeply 5–7-cleft into cuneate-obovate segments, these deeply, sharply, and repeated incised; infl usually simple, its axis and pedicels with straight, spreading hairs 1 mm, fls few, dark purple to blue (varying to seldom white); helmet ca 14–17 mm long (measured to the tip of the beak) and about as high (or not quite so high), rounded and domelike, the beak scarcely differentiated from the hood; $2n=16$. Streambanks, talus, and cold woods; c. and ec. N.Y., ne. O., and in and near the driftless area of Wis. and Io. July–Aug. Perhaps properly to be subordinated to the cordilleran *A. columbianum* Nutt.

    2. Aconitum uncinatum L. Southern m. Stem slender, weak, often leaning on other plants, arising from a turnip-like root as in no.1, but the new root well separated from the old one by a slender stalk; cauline lvs numerous, deeply 3–5-cleft into narrow or broadly rhomboid segments; infl short, few-fld, often glabrous as in no. 1; fls blue; helmet 15–22 mm long, mostly higher than long, the hood somewhat prolonged upward and ± evidently differentiated from the projecting beak. Rich woods; s. Pa. to O. and s. Ind., s. to N.C., w. S.C., n. and w. Ga., and c. Tenn. Aug.–Oct. Two vars.:

Upper lvs 3-lobed, the lobes broad and only shallowly toothed, with broad teeth; rachis glabrous, pedicels usually pubescent only at the tip, but sometimes pubescent throughout; virtually throughout the range of the sp., but chiefly at low elev. . . . . var. *uncinatum.*
Upper lvs mostly 5-lobed, more cuneate at the base, often trifid or deeply serrate with acuminate teeth; pedicels and usually also the rachis of the infl pubescent; morphologically somewhat intermediate between var. *uncinatum* and the wholly more northern *A. noveboracense*; chiefly in the mts., from sw. Pa. and w. Md. to Tenn. and N.C. and S.C., seldom farther e. in Va. and N.C. (var. *acutidens*) . . . . var. *muticum* DC.

    3. Aconitum napellus L. Garden aconite. Stem stout and erect, to 12 dm, from a turnip-like root; lvs divided to the base, the divisions deeply cleft, usually to below the middle, into numerous linear-oblong segments; inflorescence elongate, with crowded blue to variegated or white fls; rachis and pedicels nearly glabrous, but with remote, incurved hairs. Native of Europe, rarely escaped from cult. in N. Engl. July–Sept.

    4. Aconitum reclinatum A. Gray. White m. Stem arising from fascicled, slender roots, slender and weak, sometimes erect and to 1 m, sometimes longer and reclining or scrambling; cauline lvs many, deeply(3)5-cleft into cuneate-obovate segments, these sharply and coarsely incised above; infl elongate, usually branched, the axis and pedicels with short, incurved hairs; fls white or ochroleucous; helmet about twice as high as long, produced above into a conic or subcylindric hood. Rich woods; mts. and upper Piedmont from sw. Pa. to W.Va., Va., N.C., and reputedly n. Ga. June–Sept.

    9. DELPHINIUM L. Larkspur. Fls irregular; cal corolloid, with unequal sep, the upper one prolonged backward into a spur usually equaling or surpassing the blade; pet mostly 4, the two upper very inequilateral, each with a long spur extending into the spurred sep, the two lower clawed, abruptly deflexed at the middle, often bifid; pet united into 1 in our introduced spp., with 1 spur; stamens numerous; pistils 1–5, the fr follicular; annual or perennial poisonous herbs with basal or cauline, palmately cleft lvs and a terminal, simple or branched raceme of blue to pink or white fls. (*Consolida.*) 200, widespread.

1 Plants perennial; pistils 3 or seldom 5 (*Delphinium* proper).
  2 Lower pet not bifid; roots tuberous; follicles divergent . . . . 1. *D. tricorne.*
  2 Lower pet bifid; roots not tuberous; follicles erect.
    3 Stem below the infl pubescent; seeds covered with striolate scales.
      4 Fls usually blue; seeds obscurely winged, with appressed scales . . . . 2. *D. carolinianum.*
      4 Fls white or very pale; seeds wingless, with projecting scales . . . . 3. *D. virescens.*

3 Stem below the infl glabrous; seeds shining, without scales .... 4. *D. exaltatum.*
1 Plants annual; pistil 1; introduced (segregate genus *Consolida*) .... 5. *D. ambiguum.*

1. Delphinium tricorne Michx. Dwarf l. Plants 2–6 dm, from a cluster of short, tuberous roots; lvs few, mostly at or near the base, deeply divided into several oblong-linear or cuneate segments; raceme 8–20 cm, softly short-villous, the pedicels 1–3 cm, the fls few, blue or sometimes white; spur 12–18 mm; lateral sep 11–15 mm, lower pet not bifid; follicles 10–15 mm, divergent; seeds smooth; triangular. Rich, moist woods; Pa. to s. Minn., s. to N.C., Ga., Ala., and Okla. Apr., May.

2. Delphinium carolinianum Walter. Carolina l. Stem 5–10(–15) dm, pubescent throughout, glandular above; lvs well distributed along the stem, deeply dissected into linear segments seldom over 2 mm wide, the upper petioles notably short or wanting; basal lvs soon deciduous, mostly wanting at anthesis; racemes elongate, often branched, virgate, the lower pedicels to 2 cm; fls mostly blue (except for the usually white upper pet), rarely white; spur 13–18 mm; lower pet bifid, bearded; follicles 10–15+ mm; seeds obscurely 3-winged, the sides covered with appressed scales; 2*n*=16 (32). Dry woods, prairies, and sand-hills; n. Fla. to e. Tex., n. to N.C., Ky., Ill., Io., and w. Kans. May–July. (*D. azureum; D. nortonianum.*)
D. elatum L., the candle-larkspur, a tall, stout, European sp. with a long, dense infl, is widely cult. and locally adventive. It has pubescent stems, scarcely cleft lower pet, and cellular-reticulate seeds.

3. Delphinium virescens Nutt. Prairie l. Stem 4–12 dm, finely and often densely pubescent, often or usually glandular; lvs chiefly basal or below the middle of the stem, deeply dissected into linear segments mostly 2–5 mm wide; raceme elongate, virgate, the lower pedicels seldom over 1.5 cm; fls white, greenish-white, or bluish-white; spur 10–14 mm; upper pair of pet sometimes yellow, lower pet bifid, bearded; follicles 15–20 mm; seeds wingless, densely covered with projecting scales; 2*n*=16. Prairies and plains; N.D., s. Man., Minn., and w. Wis., s. to Ill., La., Tex., Colo., and N.M., and e. occasionally into c. Ky. and nw. Ga. May–July. (*D. penardii.*) Perhaps better treated as a var. of *D. carolinianum.*

4. Delphinium exaltatum Aiton. Tall l. Stem to 2 m, smooth; lvs large, pale beneath, divided into a few cuneate to lanceolate segments 1–2.5 cm wide, these with 1–4 coarse, sharp lobes above the middle; raceme elongate, often compound, to 3 dm, the pedicels and often the axis pubescent; lowest pedicels 1–2.5 cm; fls blue or purple; spur 10–15 mm; lateral sep 9–11 mm; seeds triangular, wrinkled, otherwise smooth and shining. Rich woods; Pa. to N.C., w. to O., Tenn., and s. Mo. July, Aug.

5. Delphinium ambiguum L. Rocket l. Branching annual, 3–7 dm; lvs dissected into narrowly linear segments; fls 2–4 cm wide, variously blue, violet, pink, or white, in an elongate raceme; pet 2, united, with a single spur; ovary 1, densely hairy; follicle hairy, mostly 12–18 mm (including the style). Native of Europe, occasionally escaped from cult. Summer. (*D. ajacis,* misapplied; *Consolida a.*)
D. consolida L., Forking l. a similar sp., sometimes also escapes. It has fewer fls in a short-corymbiform raceme, glabrous pistil, and follicles 8–14 mm. (*Consolida regalis.*)

10. ANEMONE L. Anemone, windflower. Sep 4–20, petaloid; pet none; pistils numerous, in a subglobose to cylindric head, pubescent; style short or elongate; stigma minute; achenes flattened, clavate, or fusiform, tipped with the persistent style; perennials from a rhizome or caudex, with palmately deeply divided basal lvs and an erect stem with a whorl of 3 or more involucral lvs subtending one or more elongate peduncles; fls medium-sized to large, white to blue or red or greenish. (*Pulsatilla.*) 100+, N. Hemisphere.

1 Style at maturity 2–4 cm, plumose; sep 25–35 mm .... 9. *A. patens.*
1 Style at maturity 0.4 cm or less; sep less than 25 mm.
  2 Achenes thinly pubescent or strigose, in a subglobose head; plants rhizomatous.
    3 Involucral lvs distinctly petioled.
      4 Lateral segments of the involucral lvs usually deeply incised on the outer margin; central
        segment usually broadest and serrate only above the middle .... 1. *A. quinquefolia.*
      4 Lateral segments of the involucral lvs usually serrate on the outer margin; central segment
        usually broadest at or below the middle and serrate to below the middle .... 2. *A. lancifolia.*
    3 Involucral lvs sessile .... 3. *A. canadensis.*
  2 Achenes concealed by the long, dense, cottony pubescence: head mostly ovoid to cylindric.
    5 Involucral lvs sessile or nearly so.
      6 Sep mostly 10–20; heads narrowly ovoid to cylindric, 17–25 mm; rhizome tuber-like or with
        a series of small tubers .... 4. *A. caroliniana.*
      6 Sep mostly 5–8; heads subglobose or broadly ovoid, 6–20 mm; not tuberous.
        7 Plants with slender rhizomes; lf-segments rarely cleft to the middle .... 5. *A. parviflora.*
        7 Plants with a caudex; lf-segments again cleft to below the middle .... 6. *A. multifida.*
    5 Involucral lvs distinctly petioled; plants with a caudex.
      8 Head of achenes cylindric, to 1 cm thick; plants typically with more than 3 involucral lvs,
        and with the branches above the involucre naked .... 7. *A. cylindrica.*
      8 Head more ovoid or ellipsoid, mostly over 1 cm thick; plants with 3 involucral lvs, the
        branches often each with an additional pair of involucel lvs .... 8. *A. virginiana.*

1. Anemone quinquefolia L.   Wood-a.   Delicate herb mostly 1–2 dm from a slender rhizome; basal lf solitary, long-petioled; lfls 3–5, rhombic to cuneate-obovate, coarsely and unevenly toothed or incised, chiefly above the middle; involucral lvs similar but smaller, the lateral lfls commonly incised on the outer margin, the middle one 2–5 cm, broadest distinctly above the middle, cuneate and entire from the middle to the base; sep mostly 5, white, or reddish beneath, 10–22 mm; achenes fusiform, 3–4.5 mm, short-hirsute; $2n=32$. Moist woods; Que. to Man., s. to Md., O., n. Ind., and ne. Io., and in the mts. to n. Ga. Apr.–June. Eastern plants (w. to Ont. and O.) with the stem glabrous below the involucre, are var. *quinquefolia*. More western plants, with the stem spreading-villous below the involucre, are var. *bifolia* Farw. (var. *interior*.)

2. Anemone lancifolia Pursh.   Lance-lvd a.   Much like *A. quinquefolia,* and not sharply distinct, averaging a little larger, mostly 1.5–3 dm; lfls of the involucre rhombic-ovate to elliptic, usually 3.5–7 cm, the lateral ones coarsely toothed nearly to the rounded base along the outer margin, rarely incised, the central lfl broadest at or below the middle and serrate to below the middle. Rich mt. woods; s. Pa. to Ga. Apr., May. (*A. trifolia* of auth., misapplied.) Perhaps better treated as *A. quinquefolia* var. *lancifolia* (Pursh) Fosberg. Very small (merely depauperate?) plants from the southern Appalachians, smaller in all dimensions than *A. quinquefolia* and *A. lancifolia,* are rarely observed and have been described as *A. minima* DC. Their proper status remains to be determined.

3. Anemone canadensis L.   Canadian a.   Rhizomatous perennial, 2–8 dm; basal lvs long-petioled, rotund, deeply 3-parted, the segments more or less cleft and sharply toothed; involucral lvs similar but sessile; peduncles 1–3, the lateral ones commonly with a small secondary involucre; sep 5, white, mostly 12–20 mm long; achenes flat, suborbicular, 3–5 mm, strigose; style 2–5 mm, strigose; $2n=14$. Sandy shores, damp prairies, and wet meadows, abundant; e. Que. to Alta., s. to Md., W.Va., Ky., Mo., and N.M. May–Aug.

4. Anemone caroliniana Walter.   Prairie-a.   Plants 1–4 dm, from a small, globular tuber that produces 1– several delicate succulent rhizomes; basal lvs deeply 3-parted, the segments deeply and irregularly incised into few or several acute divisions; involucral lvs borne below the middle, sessile, smaller than but otherwise much like the basal; peduncle solitary, villous; sep 10–20, white to rose or purple, 10–22 mm, narrowly oblong; fruiting head narrowly ± ellipsoid, 13–20 mm; achenes densely woolly, the erect style ca = the body and projecting from the wool; $2n=16$. Dry prairies and barrens; S.D. to Tex., e. to Wis., Ind., La., and occasionally to Ga., S.C., and N.C. Apr., May.

5. Anemone parviflora Michx.   Small-fld a.   Plant 1–3 dm, mostly villous, from slender rhizomes; basal lvs long-petioled, deeply 3-parted into broadly cuneate-obovate segments, these crenately lobed or incised to above the middle into broad, usually obtuse divisions; involucral lvs sessile, otherwise much like the basal; peduncle solitary; sep 5–6, white or bluish, broadly oblong to oval or elliptic, 8–15 mm; fruiting head short-ovoid, 1 cm or less; achenes densely woolly; style short, glabrous; $2n=16$. Wet, rocky ledges, usually in calcareous soil; Lab. to Alas., s. to Que., n. Minn., Colo., and Oreg.; n. Asia. June–Aug.

6. Anemone multifida Poiret.   Cut-lvd a.   Plants 1–6 dm, from a stout, often branched caudex, usually villous; basal lvs long-petioled, deeply 3-parted, the segments deeply incised or lobed into acute, linear-oblong divisions; involucral lvs similar but subsessile; peduncles 1–3, the lateral ones often with a small secondary involucre; sep 5–9, white, yellowish, or red, 5–15 mm; fruiting head subglobose; achenes 2–3.5 mm, densely woolly; style short, glabrous; $2n=32$. Shores and rocky banks in calcareous soil; Nf. to Alas., s. to Me., Vt., n. N.Y., n. Mich., S.D., N.M., and Calif.; s. S. Amer. May, June. Our plants, sometimes distinguished as var. *hudsoniana* DC., are scarcely separable from the South American var. *multifida* sens. strict.

7. Anemone cylindrica A. Gray.   Long-headed a., thimbleweed.   Plants 3–10 dm, from a short caudex; basal and involucral lvs similar, the former several, the latter 3–10 and commonly twice as many as the mostly naked peduncles, both types petiolate, broad, deeply 3–5-parted into basally narrow and cuneate segments that are again few-toothed or cleft above the middle; peduncles mostly 2–6, 1–3 dm; sep 5(6), greenish-white, 8–12 mm long; styles crimson in life; fruiting head dense, cylindric, 2–4.5 cm, scarcely 1 cm thick; achenes densely woolly, 2 mm, the style 0.5–1 mm, subglabrous, seldom strongly spreading; $2n=16$. Dry, open woods and prairies; Me. to B.C., s. to N.J., O., Mo., and Ariz. June–Aug.

8. Anemone virginiana L.   Tall a., thimbleweed.   Much like *A. cylindrica,* averaging coarser and more loosely hairy; primary lf-segments rhombic-ovate, with the margins toothed or incised to below the middle and ± convex toward the base; involucral lvs 3; peduncles mostly 1–3, some of them often with a 2-lvd secondary involucre; sep greenish-white or less often pure white; seldom red, 7–13(16) mm; fruiting head ovoid, 1.5–3 cm; achenes densely woolly; style 1–1.5 mm, short-hairy, often ± spreading; $2n=$ 16. Dry or open woods; e. Que. to N.D., s. to Ga., Ala., and Ark. June–Aug. Some plants from the n. part of the range diverge toward *A. cylindrica* in their somewhat narrower heads and in having the lf-segments more cuneate below. These have been segregated as var. *alba* A. Wood or *A. riparia* Fern., but they do not appear to be sharply separable from *A. virginiana.*

9. Anemone patens L.   Pasque-fl, prairie-smoke.   Plants 1–4 dm, from a short, branched caudex, villous throughout; basal lvs long-petioled, reniform in outline, the segments deeply and repeatedly incised into narrow, acute divisions; involucral lvs similar but smaller and sessile; sep 5–7, blue or purple to white, elliptic to oblong, 25–35 mm, villous on the back; achenes clavate, 3–4 mm; style plumose, flexuous, 2–4 cm in fr; $2n=16$. Dry prairies and barrens; Eurasia, and from Alas. to Wash. and Tex., e. to Wis. and n. Ill. Apr., May. The Amer. plants are probably referable to the var. *multifida* Pritzel, described from Siberia. (*A. ludoviciana; A. nuttalliana; Pulsatilla patens.*)

11. HEPATICA Miller. Hepatica. Sep 5–12, often 6, petaloid; pet and staminodes none; pistils numerous; ovary tapering to a short style; achenes conic to fusiform, hairy; perennial herbs with basal, simple, lobed lvs and several 1-fld scapes bearing a calyx-like involucre of 3 entire bracts just beneath the fls. 4, Europe and N. Amer. Our 2 spp. closely allied and also close to European *H. nobilis* Miller.

1 Lvs lobed to near the middle, the lobes blunt or rounded .... 1. *H. americana.*
1 Lvs lobed to well below the middle, the lobes acute .... 2. *H. acutiloba.*

1. Hepatica americana (DC.) Ker Gawler. Round-lobed h. Lvs densely long-pilose when young, especially beneath, but glabrate in age, persistent and green until the following spring, 3-lobed, the lobes broadly obtuse or rounded, the terminal one often wider than long; length of the lf blade about 2 (to 2.5) times the distance from the summit of the petiole to the sinuses; scapes 5–15 cm, villous, as also the petioles; bracts obtuse, nearly as long as the sep; fls bluish to white or pink, 12–25 mm wide, without nectar; 2*n*=14. Dry or moist but upland woods; Que. and N.S. to Minn. and Man., s. to Ga., Tenn., and Mo. Mar., Apr. (*H. triloba* and *H. hepatica,* misapplied; *H. nobilis* var. *obtusa.*)

2. Hepatica acutiloba DC. Sharp-lobed h. Much like *H. americana* and hybridizing, but maintaining its identity; lvs 3-lobed or occasionally 5–7-lobed, deeply cordate at the base, the lobes broad, acute; length of the lf blade about 3 times the distance from the summit of the petiole to the sinuses; bracts acute, about equaling the sep. Similar habitats; Que. to Minn., s. to Ga., Ala., and Mo. Mar., Apr. (*H. nobilis* var. *acuta.*)

12. CLEMATIS L. Clematis. Sep petaloid, valvate in bud, commonly 4, sometimes more; pet none; staminodes present or absent; pistils numerous; fr a flattened achene, terminated by the elongate, persistent style; erect perennial herbs, or herbaceous or somewhat woody vines climbing by the prehensile lf-rachis, with opposite, simple or compound lvs and solitary or panicled, medium-sized to large fls, usually dioecious; *x*=8. (*Atragene, Viorna.*) 300, cosmop. Species of the section *Viorna* are called Leather-flower or Curly heads.

1 Sep thin, spreading or ascending; plants climbing or scrambling; lvs compound.
  2 Sep white to yellow or green; staminodes none (*Clematis*).
    3 Fls numerous, paniculate, 2–3 cm wide, white (*Flammula*).
      4 Fls perfect; anthers 1.5–3 mm; intr .... 1. *C. terniflora.*
      4 Fls mostly polygamo-dioecious; anthers less than 1.5 mm; native.
        5 Lvs trifoliolate .... 2. *C. virginiana.*
        5 Lvs biternate, or pinnately 5-foliolate .... 3. *C. catesbyana.*
    3 Fls solitary or in groups of 3, 3–8 cm wide (*Viticella*).
      6 Lfls coarsely toothed; fls 3–5 cm wide, yellow and green .... 4. *C. orientalis.*
      6 Lfls mostly entire; fls 5–8 cm wide, creamy-white .... 5. *C. florida.*
  2 Sep reddish-violet; staminodes present (*Atragene*) .... 6. *C. occidentalis.*
1 Sep thick and firm, connivent below, free above, forming an urn-shaped cal (*Viorna*).
  7 Plants climbing or scrambling; lvs or many of them compound and evidently petiolate.
    8 Sep 3–5 cm, the margins above the middle broad, thin, ± crisped .... 7. *C. crispa.*
    8 Sep 1.5–2.5 cm, not crisp-margined distally.
      9 Lower surface of the lvs green, not glaucous, sometimes hairy; nodes and rachis-joints ± pubescent; sep hairy on the back.
        10 Style glabrous or nearly so beyond the middle; lvs thick and firm, strongly reticulate above .... 8. *C. pitcheri.*
        10 Style at first villous, later plumose throughout; lvs thin, not strongly reticulate above
        .... 9. *C. viorna.*
      9 Lower surface of the lvs glabrous and glaucous, as also the nodes and rachis-joints; sep generally glabrous on the back.
        11 Lvs thick and firm, strongly reticulate above; western sp. .... 10. *C. versicolor.*
        11 Lvs thin, not strongly reticulate above; chiefly eastern and southern spp.
          12 Plants climbing vines; most or all of the lvs compound, often with 6–10 lfls .. 11. *C. glaucophylla.*
          12 Plants only weakly or scarcely climbing; lvs simple, or the upper sometimes compound with 2–6 lfls .... 12. *C. addisonii.*
  7 Plants erect or nearly so; lvs mostly simple and sessile or subsessile.
    13 Persistent style long-plumose throughout its length; eastern spp.
      14 Lvs glaucous beneath, the later ones often compound; sep glabrous outside .... 12. *C. addisonii.*
      14 Lvs not glaucous, all simple; sep pubescent outside.
        15 Hairs of the summit of the achene and base of the style erect or ascending.
          16 Lower lf-surfaces glabrous or nearly so; mature styles mostly 2–3(–3.5) cm .... 13. *C. viticaulis.*
          16 Lower lf-surfaces generally evidently hairy; mature styles mostly (3–)3.5–6 cm..... 14. *C. ochroleuca.*
        15 Hairs at the summit of the achene and base of the style spreading or reflexed.

17 Lvs persistently woolly beneath .... 15. *C. coactilis.*
17 Lvs glabrous or nearly so beneath .... 16. *C. albicoma.*
13 Persistent style glabrous distally; western sp. .... 17. *C. fremontii.*

1. Clematis terniflora DC.   Yam-lvd c. Stem climbing, 2–3 m; herbage glabrous; lfls mostly 5, ± coriaceous, triangular-ovate to broadly ovate, obtuse to acuminate, entire or rarely with an entire lobe, rounded to cordate at the base, the terminal one long-stalked; fls numerous, white, 2–3 cm wide; sep densely pubescent along the margins and on the back; achenes glabrous or minutely appressed-hairy, 8 mm; mature style plumose, 2–3 cm; $2n=16$, 48, 64. Native of Japan, commonly cult. and often escaped. July–Sept. (*C. dioscoreifolia; C. maximowicziana; C. paniculata* Thunb., not J. F. Gmelin.)

2. Clematis virginiana L.   Virgin's bower. Stems climbing, 2–3(–5) m, the lower part persistent and becoming woody; lvs mostly trifoliolate (uppermost sometimes simple), the lfls 2–10 cm, ovate to subcordate, acuminate, usually coarsely mucronate-toothed, glabrous or sparsely (seldom more copiously) hairy beneath; mostly polygamo-dioecious; panicles many, on axillary peduncles mostly 2–6 cm; sep white, oval to oblong-spatulate, 10–15 mm, hairy at least on the back; achenes many in a subglobose head, 4 mm, pubescent, light to dark brown or greenish-brown; mature style flexuous, strongly plumose, 2–4 cm; $2n=16$. Moist soil; N.S. and e. Que. to Man., s to Ga. and La. July–Sept.

3. Clematis catesbyana Pursh.   Much like *C. virginiana,* but the lvs 5-foliolate or biternate, lfls 3–6 cm, usually 3-lobed or entire; peduncles mostly 3.5–8 cm; sep linear to obovate; achenes reddish or purplish-brown to dark blackish-purple. Moist soil; Va. to Fla., w. to Ky. and Ark. July, Aug. (*C. micrantha; C. ligusticifolia,* a more western sp., the name sometimes misapplied here.)

4. Clematis orientalis L.   Woody climber with small (2.5 cm), sharply and coarsely toothed lfls; fls 1–3 in the axils, yellow and green, 3–5 cm wide; $2n=16$. Native of Japan, naturalized in w. U.S., and sometimes escaped in our range.

5. Clematis florida Thunb.   Woody climber with usually twice ternate lfls; lfls ovate-lanceolate, entire or with a few teeth; fls creamy-white, 5–8 cm wide. Native of China, seldom escaped from cult. in our range.

6. Clematis occidentalis (Hornem.) DC.   Purple c. Subglabrous trailer or climber, to 2 m; lfls 3, long-stalked, ovate, entire to crenate or lobed; fls solitary, mostly axillary; sep reddish-violet, lance-ovate, 3–5 cm, softly villous; outer filaments progressively broader and with smaller anthers; the outermost often completely sterile and prominently veined; achenes villous, in a dense, globular head; mature styles flexuous, plumose, 3–5 cm; $2n=16$. Rocky woods; e. Que. to Man., s. to N.J., Del., O., nw. Ill., and ne. Io., and in the mts. to N.C. May, June. (*C. verticillaris, Atragene americana.*)

C. viticella L., a Eurasian climber with blue to purple or rosy fls 3–5 cm wide, lacking staminodes, and with a glabrous style, occasionally escaped from cult.

7. Clematis crispa L.   Climber; lvs pinnate; lfls mostly 2–4 pairs, lanceolate to ovate, acuminate, entire or rarely 2–3-lobed, glabrous, the lower lfls on each lf much the larger; cal urceolate; sep 3–4.5 cm, connivent about half-length, the body narrowly lanceolate, dilated above the middle into broad, thin, undulate or crisped margins; mature style 2–3 cm, with numerous ascending hairs rarely over 1.5 mm; $2n=16$. Swamps and wet woods on the coastal plain and piedmont; se. Va. to Fla. and Tex., n. in the interior to s. Ill. May–Aug.

8. Clematis pitcheri T. & G.   Stems climbing or scrambling, generally somewhat hairy at least at the nodes; lfls usually 3–5 pairs, lance-ovate to cordate-ovate, entire to 2–3-lobed or even deeply parted, firm, prominently reticulate on both sides, subglabrous except for the generally somewhat hairy rachis-joints, or seldom evidently hairy beneath, not glaucous; cal urceolate; sep 1.2–2.5 cm, lance-ovate, acute or shortly acuminate, tomentose on the margins, minutely hairy or glabrate on the back; style at anthesis glabrous, at maturity 2–3 cm, firm, somewhat hairy below, essentially glabrous above the middle; $2n=16$. Dry or moist woods; Ind. to Nebr., s. to Tenn., Ark., Tex., N.M., and ne. Mex. June–Aug.

9. Clematis viorna L.   Stem climbing, generally somewhat hairy at least at the nodes; principal lvs with 2–4 pairs of lfls, some of the lfls often trifoliolate; lfls lanceolate to ovate, entire or 2–3-lobed, thin, not prominently reticulate, generally hairy beneath and at the rachis-joints, not glaucous; cal urceolate; sep ovate or lance-ovate, 1.5–2.5 cm, caudate-acuminate, thinly hairy on the back, densely tomentose at the margins; style at anthesis hirsute, at maturity 3–5 cm, densely plumose throughout; $2n=16$. Moist woods and thickets; Pa. to Ill. and Mo., s. to Ga. and Miss. (*C. gattingeri; Viorna viorna; V. flaccida,* a more hairy form.)

10. Clematis versicolor Small.   Stems climbing; herbage glabrous, glaucous; lfls mostly 4 pairs, firm, strongly reticulate on both sides, ovate-oblong to cordate-ovate, often 2–3-lobed; cal urceolate; sep ovate-oblong, 1.5–2.5 cm, short-acuminate, bluish-lavender to reddish-purple, or whitish distally, densely tomentose on the margins, otherwise glabrous; style at anthesis densely hairy, at maturity 4–5 cm, densely long-plumose throughout; $2n=16$. Dry, calcareous woods; chiefly Ozarkian, from s. Mo., Okla., and Ark. to Ky. and Tenn. June, July.

11. Clematis glaucophylla Small.   Stems climbing; herbage glabrous and glaucous; lvs mostly compound, but simple ones often also present, the latter ovate to broadly cordate-ovate, subsessile, the former with 4–5 pairs of lance-ovate to cordate-ovate lfls, the lower pairs sometimes 3-lobed, the upper pairs progressively smaller, all thin, not strongly reticulate; cal urceolate; sep lance-ovate, 1.5–2.5 cm, shortly caudate acuminate, deep cherry red, glabrous on the back, tomentose on the margins; mature style 4–5 cm, densely plumose; $2n=16$. Moist woods and along streams; N.C. to Ky. and Okla., s. to Fla. and Miss.; rare. June, July. (*Viorna glaucophylla.*)

12. Clematis addisonii Britton. Stem at first erect, but often soon becoming procumbent or scrambling, up to 1 m long; herbage glabrous and glaucous; lvs thin, not strongly reticulate; the simple ones broadly ovate, subsessile, obtuse and mucronate, the compound ones (when present) appearing later, especially on the branches, with 2(3) pairs of round-ovate lfls, the lowest pair much exceeding the upper; cal urceolate; sep narrowly ovate, 1.5–2.5 cm, glabrous on the back, tomentose on the margins distally; mature style 2.5–3.5 cm, strongly plumose; $2n=16$. Dry limestone hills; w. Va. Apr.–June, July.

13. Clematis viticaulis Steele. Much like *C. ochroleuca*; stem less hairy; branches usually overtopping the central axis; lvs dark green, glabrous or nearly so beneath, smaller, mostly 4–7.5 × 2–4.5 cm; mature peduncles 1–5 cm, shorter than the subtending lvs; sep only sparsely hairy outside; mature styles 2–3(–3.5) cm; $2n=16$. Shale barrens in Bath and Rockbridge cos., Va. Apr.–June.

14. Clematis ochroleuca Aiton. Stems erect, 2–6 dm, simple or branched, ± villous-tomentose, the branches seldom overtopping the central axis; lvs subsessile, ovate, to 12 cm, half as wide (or a little wider), rather light green, softly hairy beneath when young, glabrescent in age; peduncles elongating to 5–20 cm at maturity, regularly exceeding the uppermost lvs; fls solitary and terminal, cal urceolate; sep 2–3.5 cm, densely sericeous outside; base of the achenes appressed-puberulent with short hairs, these becoming longer distally, but always erect or ascending, as are also the lower hairs of the style; mature style (3–)3.5–6 cm, densely plumose; $2n=16$. In woods and disturbed open places; s. N.Y. to Ga., mostly on the piedmont. May, June.

15. Clematis coactilis (Fern.) Keener. Much like *C. albicoma*, but more strongly pubescent, the stems densely sericeous, the rather light green lvs densely woolly or sericeous beneath; branches seldom overtopping the central axis; lvs averaging a little larger, the better developed ones mostly 6–11 × 4–9 cm; mature styles mostly 3–4.5 cm; $2n=16$. Apr.–June. Local on shale barrens in w. Va.

16. Clematis albicoma Wherry. Stem erect, 2–5 dm, simple to more often much branched, the branches usually overtopping the central axis; lvs subsessile, dark green, lance-ovate or ovate, glabrous or slightly hairy along the veins beneath when young, the larger ones mostly 4–8 × 1.5–5 cm; mature peduncles 2–10 cm; cal urceolate; sep 1.5–3 cm, densely hairy outside; achenes densely hairy, the hairs of the distal portion and the base of the style reflexed or widely divergent; mature style mostly 2–4 cm, densely plumose; $2n=16$. Apr.–June. Local on shale barrens in w. Va. and e. W.Va.

17. Clematis fremontii S. Wats. Stems erect, 2–6 dm, simple or branched, villous or glabrate; lvs subsessile, lance-elliptic to broadly ovate, firm, reticulate, glabrescent in age, the larger ones mostly 7–13 × 4–9 cm; mature peduncles mostly 3–6 cm, shorter than the uppermost lvs; cal urceolate; sep lanceolate, long-acuminate, densely tomentose on the margins, otherwise glabrous or nearly so; achenes densely tomentose; mature style mostly 1.5–3 cm, tomentose (not plumose) below, glabrous above the middle; $2n=16$. Barrens and prairies; e. and s. Mo.; disjunct in c. Kans. and adj. Neb. Apr., May.

13. RANUNCULUS L.

13. RANUNCULUS L. Buttercup, crowfoot, spearwort. Fls regular; sep green or yellowish, 3, 4, or more commonly 5, rarely more; pet mostly 5, sometimes fewer or more, each with a nectariferous pit or scale on the upper side at the base; stamens mostly numerous, rarely as few as 5; pistils numerous in a globose, ovoid, or cylindric head; ovule ordinarily 1, erect or ascending; fr an achene; annual or perennial herbs with alternate, entire to much dissected lvs and yellow, white, or rarely red fls; juice acrid, poisonous. (*Batrachium, Ceratocephalus, Ficaria.*) 250+, ± cosmop.

1 Pet white, or suffused with yellow at the base, plants aquatic. (*Batrachium.*)
  2 Lvs all floating and shallowly lobed .... 26. *R. hederaceus.*
  2 Lvs all submersed, dissected into filiform or narrowly linear segments.
    3 Lvs soft, mostly collapsing when withdrawn from water, the free petiole about equaling the
      stipular part; achenes 10–20(–40), the beak 0.2–0.5 mm; pedicels not recurved .... 27. *R. trichophyllus.*
    3 Lvs firmer, seldom collapsing when withdrawn from water, the free petiole none or much
      shorter than the stipular part.
      4 Achene-beak 0.2–0.5 mm; achenes 30–45(–80); mature pedicels recurved .... 28. *R. subrigidus.*
      4 Achene-beak 0.7–1.5 mm; achenes (–7)15–25; pedicels not recurved .... 29. *R. longirostris.*
1 Pet yellow; plants aquatic or more often terrestrial.
  5 Sep normally 3, rarely 4.
    6 Lvs deeply 3-parted; fls 1 cm wide. (*Coptidium.*) .... 30. *R. lapponicus.*
    6 Lvs entire or sinuately lobed; fls 2.5–4 cm wide. (*Ficaria.*) .... 31. *R. ficaria.*
  5 Sep usually 5, rarely more.
    7 Lvs (both cauline and basal) entire to merely denticulate or crenate.
      8 Principal lvs cordate-ovate to ovate or reniform. (*Cyrtorrhyncha.*) .... 24. *R. cymbalaria.*
      8 Principal (or all) lvs linear to elliptic or lanceolate, acute or tapering at the base.
        9 Plants perennial; achenes 1.4–2 mm.
          10 Sep 5–7 mm; chief cauline lvs 1–3 cm wide; achene-beak 1–1.3 mm .... 17. *R. ambigens.*
          10 Sep 2–4 mm; cauline lvs up to 1 cm wide; achene-beak 0.3–0.5 mm .... 18. *R. flammula.*
        9 Plants annual; achenes 0.7–1.2 mm.
          11 Pet 5, 6–9 mm, much exceeding the sep .... 19. *R. texensis.*
          11 Pet 1–5, 1–2 mm, about equaling the sep .... 20. *R. pusillus.*
    7 Lvs (all or in part) deeply lobed, dissected, or compound.

12 Plant a small scapose annual with dissected lvs (*Ceratocephalus.*) .... 25. *R. testiculatus.*
12 Plants otherwise, either perennial, or leafy-stemmed, or both.
  13 Basal and cauline lvs distinctly unlike, some or all of the former merely crenate, the
    latter sessile or subsessile and deeply parted; achenes turgid, without a sharp or
    wing-like differentiated border.
    14 Pet 4–10 mm, wider and longer than the sep.
      15 Basal lvs truncate or rounded or cuneate at base, commonly longer than
        wide; achenes flattened at base .... 12. *R. rhomboideus.*
      15 Basal lvs cordate or subcordate at base, commonly wider than long; achenes
        turgid throughout .... 13. *R. harveyi.*
    14 Pet 1.5–3 mm, shorter and narrower than the sep.
      16 Beak of the achene 0.2 mm.
        17 Lvs and young stems glabrous or nearly so, or the stems puberulent
          above; basal lvs with ± cordate base .... 14. *R. abortivus.*
        17 Lvs and young stems ± villous; basal lvs seldom at all cordate .... 15. *R. micranthus.*
      16 Beak of the achene 0.7–1 mm, curved or hooked .... 16. *R. allegheniensis.*
  13 Basal and cauline lvs similar in general pattern, the latter merely smaller, or with
    fewer segments or shorter petioles; achenes various.
    18 Achenes turgid, without a sharp or wing-like differentiated border.
      19 Pet 4–14 mm; achenes 1.2–2.5 mm, beaked; plants mostly aquatic.
        20 Margin of achenes below the middle thickened and corky; pet mostly 6–
          14 mm .... 21. *R. flabellaris.*
        20 Margin of achenes rounded but not tumid; pet mostly 4–8 mm .... 22. *R. gmelinii.*
      19 Pet mostly 2–4 mm; achenes 0.8–1.2 mm, nearly beakless; plants terrestrial
        or at most semi-aquatic .... 23. *R. sceleratus.*
    18 Achenes flattened, with a sharp, differentiated border paralleled by a nerve on
    each face.
      21 Achenes papillate or spiny; introduced, weedy spp.
        22 Pet 5–10 mm; spines, if present, not hooked.
          23 Achenes 20–40, 2–3 mm, merely papillate; beak 0.2–0.5 mm .... 9. *R. sardous.*
          23 Achenes 4–9, 4.5–7 mm, strongly spiny, beak 2–3 mm .... 10. *R. arvensis.*
        22 Pet 1–3 mm, achenes 1.5–3 mm, with slender, hooked spines .... 11. *R. parviflorus.*
      21 Achenes smooth or merely pubescent; native and introduced spp.
        24 Pet 7–15 mm; anthers 1.2 mm or longer (or a little shorter in *R. acris*).
          25 Style short, outcurved, stigmatose along the upper (inner) side (vis-
            ible at 10 × magnification); introduced weeds.
            26 Terminal segment of the principal lvs stalked.
              27 Stem usually creeping, rarely erect, not bulbous .... 1. *R. repens.*
              27 Stem erect, bulbously thickened at the base .... 2. *R. bulbosus.*
            26 Terminal segment of the principal lvs sessile, usually connected
              to the lateral segments by green tissue .... 3. *R. acris.*
          25 Style elongate, nearly straight, stigmatose only at the tip; native, not
            weedy.
            28 Principal lvs as wide as or wider than long; mature receptacle
              ellipsoid or clavate .... 7. *R. hispidus.*
            28 Principal lvs mostly longer than wide; mature receptacle conic
              .... 8. *R. fascicularis.*
        24 Pet 2–5 mm; anthers 0.7–1 mm (or a little longer in *R. recurvatus*).
          29 Achene-beak straight or somewhat curved, not hooked; terminal
            segment or all segments of the principal lvs stalked.
            30 Pet equaling or longer than the sep .... 4. *R. macounii.*
            30 Pet distinctly shorter than the sep .... 5. *R. pensylvanicus.*
          29 Achene-beak strongly hooked; terminal segment or lobe of principal
            cauline lvs sessile, usually connected to the lateral ones by green
            tissue .... 6. *R. recurvatus.*

1. Ranunculus repens L.   Creeping b. Hirsute to strigose or subglabrous perennial, mostly creeping, rarely ascending or erect; lvs petioled, 3-parted, the terminal segment stalked, all segments broadly obovate to subrotund, cleft or lobed, sharply toothed; pet 8–15 mm, two-thirds as wide; anthers 1–2 mm; achenes broadly and obliquely ovate, 2.5–3.5 mm, sharply but narrowly margined, the beak triangular, usually curved, 0.8–1.5 mm; $2n$ mostly= 32. Native of Europe, naturalized in fields, lawns, roadsides, and wet meadows. May–July. Robust, subglabrate plants have been called var. *glabratus* DC., and double-flowered ones have been called var. *degeneratus* Schur or var. *pleniflorus* Fern.

2. Ranunculus bulbosus L.   Bulbous b. Hirsute, cormose perennial, erect, 2–6 dm, the stems bulbously thickened at the base, basal lvs long-petioled, 3-parted, the terminal segment stalked, all segments broadly obovate to subrotund, deeply cleft and lobed; cauline lvs few, much smaller and less divided; fls long-pedicellate; pet broadly obovate, 8–14 mm; anthers 2–3 mm, rarely less; achenes broadly and obliquely obovate, 2.5–3.5 mm, smooth, much thicker apically than basally, distinctly margined; beak stout, outcurved, 0.6–1 mm; $2n$=16. Native of Europe, naturalized in fields, meadows, and lawns. Apr.–July.

3. Ranunculus acris L.   Common or meadow-b., blister-plant. Spreading-hairy perennial, slender, eventually to 1 m, leafy mostly below the middle, the long branches with few, widely separated, much reduced lvs; lvs reniform, deeply 3-cleft, the broadly cuneate-obovate segments incised or cleft into oblong or linear lobes; pet broadly obovate, often retuse, 8–16 mm, twice as long as the sep; achenes in a subglobose head, flat, margined, broadly and obliquely obovate, 2–3 mm, the beak 0.4–1 mm; $2n$=14, 28, 29, 32. Native of Europe, widely introduced in N. Amer. as a weed in fields and meadows and along roadsides. May–Sept. Double-flowered forms occur.

4. Ranunculus macounii Britton. Marsh-c. Much like no. 5, but sometimes glabrous, the stems widely branched and sometimes rooting from the lower nodes; pet 3.5–5 mm, equaling or commonly longer than the sep; fruiting heads mostly 7–12 mm, ovoid-cylindric; achenes 2.7–3.3 mm; beak triangular-subulate, 0.8–1.8 mm; $2n$=32. Marshes and wet meadows; Lab. to Alas. and e. Asia, s. to Que., n. Mich., Io., and Ariz. June–Aug.

5. Ranunculus pensylvanicus L.f. Bristly c. Spreading-hairy annual or perennial, erect, simple or branched, 3–7 dm; basal lvs soon withering, cauline few, petioled; lf-segments 3, the terminal one stalked, all deeply 3-lobed, the lobes toothed or incised; fls few, short-pedicellate; pet obovate, 2–4 mm, shorter than the reflexed sep; anthers 0.7–1 mm; achenes in a cylindric or ovoid-cylindric head mostly 10–15 mm, subrotund, 1.8–2.7 mm, sharply margined; beak flat-subulate, 0.6–1.4 mm, straight or curved; $2n$=16. Marshes, ditchbanks, and wet meadows; Nf. to Alas. and e. Asia, s. to N.J., Pa., O., n. Ind., n. Ill., Minn., and Ariz. July, Aug.

6. Ranunculus recurvatus Poiret. Hooked c. Stems 2–7 dm, sparsely hirsute, with few branches and fls; lvs all petioled except the uppermost, broadly reniform or rotund, 3-cleft to below the middle, ± pilose; pet pale yellow, 3.5–6 mm, narrowly oblong-obovate, about equaling the sep; achenes very flat, nearly rotund, 2 mm, sharply margined; beak firm, hooked or coiled, 1 mm; $2n$=32. Moist or dry woods; Que. and Me. to n. Minn., s. to Ga., Miss., and Okla. Apr.–June.

7. Ranunculus hispidus Michx. Hispid b. Fibrous-rooted perennial 1.5–9 dm, sometimes with some somewhat thickened but elongate (over 5 cm) roots; rhizome short, regenerated in part each year, only the oldest portion withering; basal lvs the largest, the blade at least as wide as long, mostly 3-lobed or trifoliolate, the acute or acuminate lobes or segments (especially the terminal one) variously incised, lobed, or merely toothed; pet 5–8(–10), widest above the middle, 8–14 × 3–10 mm, equaling or up to twice as long as the sep; receptacle ± clavate or ellipsoid above the broad staminal zone; body of the achenes subrotund or obovate, 2–3.5 mm, the nerves bordering the marginal keel tending to be raised so that the margin is ± tricarinate, the keel sometimes winged; beak straight or nearly so, 1.8–3 mm; $2n$=32, 64. N.S. and s. Que. to se. Sask., s. to Fla. and e. Tex. Apr., May. Three vars., with broadly overlapping ranges:

a Plants erect, neither stoloniferous nor becoming repent, typically growing in well drained uplands; sep usually spreading; achenes narrowly margined; $2n$=32; N.Y. to Ga., w. to Ill., Mo., se. Kans., and e. Okla. (*R. marilandicus.*) .... var. *hispidus*.
a Plants lax and rank-growing, either stoloniferous or becoming repent, typically in lowland, swampy habitats.
  b Achenes narrowly margined; sep spreading; plants becoming repent, some of the stems eventually arched back to the ground and rooting at the nodes; $2n$=64; more northern than the other vars., from N.S. and s. Que. to se. Sask., s. to N.J., n. Va., Ill., and Mo. (*R. caricetorum; R. septentrionalis,* misapplied.) .... var. *caricetorum* (Greene) T. Duncan.
  b Achenes very broadly margined; sep reflexed; plants becoming stoloniferous but not repent; $2n$=32; N.Y. to Minn., s. to Fla. and e. Tex. (*R. carolinianus; R. nitidus; R. septentrionalis.*) .... var. *nitidus* (Elliott) T. Duncan.

8. Ranunculus fascicularis Muhl. Thick-root b. Perennial 1–2 dm or at maturity 3 dm tall, erect or ascending, not becoming repent; pubescence mostly appressed; rhizome short, regenerated each year; some roots very slender, others usually becoming thickened in late season and up to 5 cm long; lvs mostly basal, the blade ovate in outline, longer than wide, the terminal segment stalked, all segments deeply lobed and the lobes generally incised or coarsely crenate; cauline lvs 1–3, smaller, sessile or nearly so, less divided; fls long-pedicellate; pet 5–7(–10), widest near or below the middle, 8–14 × 3.5–5.5 mm; anthers 1.3–1.8 mm; receptacle ± conic above the broad staminal zone; achenes rotund, the body 2–3.5 mm, sharply margined, but the adjacent lateral nerves not much raised; beak slender, straight or nearly so, 1.5–3 mm; $2n$=32. Prairies and dry woods; Mass. to s. Ont. and Minn., s. to N.J., Md., W.Va., Tenn., La., and Tex. Apr., May.

9. Ranunculus sardous Crantz. Annual, usually several-stemmed, 1–6 dm, much branched above; principal lvs mostly clustered near the base, 3-parted, the terminal segment stalked, all segments broadly triangular-obovate, deeply lobed and incised; cauline lvs much smaller, with fewer and narrower segments; fls numerous, long-pedicellate; pet 5–8(–10) mm; receptacle hairy in fr; achenes 30–40, broadly obovate to subrotund, the margin smooth and sharply differentiated, the sides papillate; beak 0.2–0.5 mm; $2n$=16, 48. Native of Europe, found here and there in our range as a weed. May–July.

10. Ranunculus arvensis L. Corn-c., hunger-weed. Annual, 2–4 dm; lvs deeply divided into many linear and entire or narrowly cuneate-oblong and incised segments; fls long-pedicellate; pet obovate, 5–8 mm; receptacle hairy in fr; achenes 4–9, broadly and obliquely obovate, 4.5–7 mm, the sides and margins with numerous slender spines to 2.5 mm; beak subulate, 2–3 mm; $2n$=32. Native of Europe, occasionally found as a weed in our range, esp. along the Atlantic coast. Apr.–Aug.

11. Ranunculus parviflorus L. Stickseed-c. Annual, 1–3 dm, branched from the base, softly villous; lvs cordate-rotund, palmately 3–5-lobed, the lobes incised and sharply toothed; pedicels short and surpassed by the lvs at first, later elongating; pet 1–3 mm; receptacle glabrous; achenes subrotund, 1.5–3 mm, the margin nearly or quite smooth, the sides with many short, slender, hooked spines from swollen bases; beak flat, triangular, 0.5–1 mm; 2n=14, 28. Native of Europe, found here and there in our range as a weed. Apr.–July.

12. Ranunculus rhomboideus Goldie. Prairie-b. Perennial, 1–2 dm at anthesis, spreading-hairy; basal lvs ovate-oblong to broadly ovate, 1–5 cm, long-petioled, crenate mostly above the middle, tapering to rounded at the base; cauline lvs sessile or subsessile, cleft into a few linear segments; fls few–several; pet oblong-elliptic, 5–9 mm, much surpassing the villous sep; achenes in a globose head, obliquely obovate, 2–2.8 mm, flattened below, turgid above; beak very short; 2n=16. Dry, open woods and prairies; Mich. and w. Ont. to Io., S.D., and Sask. Apr., May. (*R. ovalis.*)

13. Ranunculus harveyi (A. Gray) Britton. Ozark b. Glabrous or very sparsely hairy, 2–4 dm, usually much branched; basal lvs broadly round-ovate to reniform, 2–3 cm, wider than long, crenate, rarely 3-lobed, cordate or subcordate at base; cauline lvs divided to the base into 3–5 linear or oblanceolate segments; pet spatulate-elliptic, 6–8 mm, twice as long as the sep; achenes in a globose head, dull brown, obovate to subrotund, turgid, 1.3–1.7 mm, the straight, slender beak 0.3–0.5 mm. Rocky, open woods; s. Ill. and s. Mo. to Ala. and Ark. Apr.

14. Ranunculus abortivus L. Small-fld c. Stems erect, branched above, 2–5 dm, glabrous or sometimes puberulent above; basal lvs reniform to rotund, ± cordate at the base, chiefly merely crenate, but one or more of them often variously lobed or divided; cauline lvs sessile or subsessile, usually deeply 3–5-divided, the segments varying from broadly linear and entire to oblanceolate or even obovate and irregularly toothed or incised; pet 2–3 mm, rhombic, shorter than the sep; achenes in a short-ovoid head on a villous receptacle, broadly obovate, turgid, 1.5 mm, the beak very short; 2n=16. Moist or dry woods, abundant and variable; Lab. to Alas., s. to Fla., Tex., and Colo. Apr.–June. The var. *eucyclus* Fern., with slender, flexuous stems and suborbicular, deeply cordate basal lvs with narrow sinus, occurs chiefly from N.Y. and N. Engl. to Nf. and Que. Var. *abortivus,* stouter and more erect, not markedly flexuous, with more reniform basal lvs with a broad sinus, is mostly but not entirely more southern than var. *eucyclus.* Other described vars. are of doubtful significance.

15. Ranunculus micranthus Nutt. Small-fld c. Much like no. 16; some of the roots conspicuously thickened; lvs sparsely villous with straight hairs 1–2 mm, the basal broadly ovate in outline, broadly cuneate to rounded or truncate at base, or a few occasionally subcordate; 2n=16. Dry or moist woods; Mass. to N.C. and W.Va.; s. O. to Tenn., w. to S.D. and Okla. Apr., May.

16. Ranunculus alleghiensis Britton. Allegheny c. Plants erect, glabrous or subglabrous, branched above, 2–5 dm; basal lvs reniform or broadly flabellate, truncate to subcordate at base, crenate or some variously lobed or parted; cauline lvs sessile or subsessile, 3–5-parted, the lobes linear, varying to cuneate-obovate, the broader ones lobed or incised; pet 1.5 mm, shorter than the reflexed sep; achenes in a globose head, turgid, 1.4–2 mm, the beak firm, strongly curved or hooked, 0.7–1 mm. Moist or dry woods; chiefly in the mts.; Mass. to O., N.C., and Tenn. Apr., May.

17. Ranunculus ambigens S. Wats. Water-plantain s. Stems hollow, rooting from the lower nodes, to 1 m, ascending; cauline lvs with dilated, sheathing petiole and well defined, lanceolate blade to 15 × 3 cm, usually denticulate; pet 5–10 mm, equaling or a little exceeding the sep; achenes 1.5–2 mm, the horizontal beak 1–1.3 mm. Swamps, marshes, and shores, mostly in heavy soil; Me. to Minn., s. to Va., Tenn., and La. June–Sept. (*R. obtusiusculus* Raf., probably misapplied.)

18. Ranunculus flammula L. Creeping s. Lvs entire or slightly toothed; pet obovate, mostly 3–7 mm, about twice as long as the sep; achenes 1.3–1.7 mm, the erect or ascending beak 0.2–0.5 mm. 2n=32. Sandy or muddy shores; circumboreal, s. to Mass., Pa., Mich., Minn., and the western cordillera. July–Sept. Our commonest phase is the circumboreal var. *filiformis* (Michx.) DC., slender, prostrate, and creeping, 1–5 dm long, sending up at each node a few lvs and a naked or few-lvd, 1-fld stem 3–15 cm, the lvs filiform or linear, with scarcely expanded blade to 1.5 mm wide; 2n=32. (*R. reptans.*) The very similar var. *ovalis* (Bigel.) L. Benson, with expanded blades mostly 1.5–7 mm wide, is sporadic with us but common farther west. The chiefly Eurasian var. *flammula,* in our range restricted to N.S., is stouter, with ascending or reclining stems 1–5 dm, sometimes rooting at the lower nodes but not creeping, branched above and with several fls, and with the lvs mostly 3–10 mm wide.

19. Ranunculus texensis Engelm. Coastal-plain s. Annual; stems weak and slender, 2–8 dm, sometimes reclining, sometimes rooting at the lower nodes and even becoming colonial, much branched at least distally; lvs petioled, the basal cordate to cuneate, the principal cauline ones linear-elliptic to lanceolate, 5–15 mm wide, obtuse, entire or sparsely dentate; pet 5.5–9 mm, much exceeding the sep; stamens ca 25; achenes 0.7–1.1 mm, minutely pitted, very minutely beaked. Marshes and muddy ground on the coastal plain; s. Del. to Tex., n. in the interior to s. Ill.and Ind. May, June. (*R. laxicaulis; R. subcordatus; R. oblongifolius,* misapplied.)

20. Ranunculus pusillus Poiret. Low s. Annual; stems weak, ascending or erect, 1–5 dm, branched from the base, many-fld; lower lvs long-petioled, ovate or lance-ovate, 1–4 cm, basally obtuse or rounded; cauline lvs progressively narrower, most of them linear, scarcely differentiated into blade and petiole; pet 1–5, 2 mm, equaling or a little shorter than the sep; stamens 10 or fewer; achenes 1 mm, very shortly beaked. Ditches, muddy ground, and shallow water, mostly on the coastal plain; s. N.Y. to Fla. and Tex., n. in the interior to Mo., Ind., and O. May, June.

21. Ranunculus flabellaris Raf. Yellow water-c. Submersed lvs ternately decompound with linear segments 1–2 mm wide; emersed lvs (when present) broadly reniform, 3-parted, the segments 3-cleft, the division in turn 3-lobed; fls on long, stout peduncles; sep 4–8 mm; pet 6–14 mm; achenes plump, 1.3–2.5 mm, the sides rugose, the margin corky-thickened around the base and the ventral side below the middle; beak 0.7–1.5 mm; 2n=32. Quiet water and muddy shores, seldom wholly emersed; Me. to B.C., s. to Va., Ky., Ind., Ill., and La. Apr.–June. (*R. delphinifolius.*)

22. Ranunculus gmelinii DC. Yellow water-c. Much like no. 21; segments of the submersed lvs mostly 2–4 mm wide; emersed lvs pentagonal, the principal segments again lobed and crenately toothed; sep 4–6 mm; pet mostly 4–8 mm; achenes nearly or quite smooth on the sides, the margins rounded but not tumid or conspicuously corky; beak 0.4–0.7 mm; 2n=16, 32. Shallow water and muddy shores, seldom wholly emersed; boreal Amer. and Asia, s. to Me., Mich., Io., N.D., and Colo. July, Aug. Our plants as here described are mostly glabrous or nearly so and represent the var. *hookeri* (D. Don) L. Benson, widespread in Amer. The var. *gmelinii,* tending to be hirsute, and with smaller fls (sep 2.5–3 mm, pet 3.5–4 mm) occurs chiefly in Alas. and nw. Can., but extends to n. Minn.

23. Ranunculus sceleratus L. Cursed c. Stems erect, 2–6 dm, stout, hollow, glabrous, branched above and many-fld; basal and lower cauline lvs reniform, deeply 3-parted, the segments again cleft or lobed; upper lvs much smaller, commonly of 3 linear-oblong segments; pet 2–3 mm, shorter than the sep; achenes very numerous in a short-cylindric head, turgid, 0.8–1.2 mm, marginless, and very nearly beakless; 2n=32. Marshes, ditchbanks, and swampy meadows; circumboreal, in Amer. s. to Va., Mo., N.M., and Calif. Apr.–Aug. Our common phase, with minute, irregular transverse ridges on the central part of each face of the achene, is the widespread var. *sceleratus.* The chiefly western Amer. var. *multifidus* Nutt., with the basal lvs usually deeply twice cleft, and with the achene-surfaces smooth except for a circle of minute pin-prick depressions on each face, extends e. to Minn. and Io.

24. Ranunculus cymbalaria Pursh. Seaside c. Glabrous, stoloniferous perennial, 5–15 cm tall; lvs mostly basal; petioles 2–5 cm; blades cordate-ovate to ovate or reniform, 5–25 mm, rounded above, usually cordate below, ± crenate; fls few; pet 3–5 mm, scarcely surpassing the sep; achenes longitudinally nerved, 1.5–2 mm, numerous in a short-cylindric head to 12 mm; 2n=16. May–Oct. In mud, especially in brackish or alkaline places; irregularly distributed throughout much of N. Amer., except se. U.S.; also in Asia. Our plants are var. *cymbalaria.*

25. Ranunculus testiculatus Crantz. Bur-buttercup. Small, scapose annual to 1 dm, thinly silky-tomentose; lvs 1–4 cm, ternate or biternate with narrow segments; sep greenish, ovate-lanceolate, persistent, accrescent to 5–6 mm; pet narrow, 4–8 mm, pale yellow, fading to whitish with pink veins; achenes (25–)35–70, widely spreading, forming a cylindric or ovoid bur (1–)1.5–2.5 cm, individually 5–7 mm, 3-locellar, the central chamber with a single seed, the 2 empty lateral chambers bulged-inflated on the upper side, the abrupt stiff beak longer than the body. Vernal ephemeral Eurasian weed, now widespread in w. U.S. and occasionally intr. with us, as in Io. and O. (*Ceratocephalus t.*)

26. Ranunculus hederaceus L. Ivy-lvd c. Stems creeping; submersed lvs none; floating lvs long-petioled, shallowly 3–5-lobed; fls white, 5–7 mm wide; receptacle glabrous; 2n=16. Shallow water and wet shores; coastal plain from se. Pa. to S.C.; also Nf. and Europe. May–Aug.

27. Ranunculus trichophyllus Chaix. White water-c. Stems submersed, elongate, branched; lvs submersed, ternately or binately dissected into filiform segments, soft and collapsing when withdrawn from water, 2–3 cm, the free petiole mostly about equaling the glabrous stipular part; fls at the water-surface, white, 1–1.5 cm wide; stamens mostly 10–25, achenes 10–20(–40), with a minute beak only 0.2–0.5 mm; 2n=32. Ponds and slow streams; circumboreal, but not at the highest latitudes; 2n=32. Ponds and slow streams; circumboreal, but not at the highest latitudes; in Amer. s. to N.J., Pa., Minn., and the western cordillera. June–Aug. We have two vars. Var. *trichophyllus,* with essentially the range of the sp., has the receptacle hirsutulous, with tufted hairs. (*R. aquatilis* var. *capillaceus.*) Var. *calvescens* W. Drew, from Que. to Minn., and Pa., chiefly in N. Engl., has the receptacle glabrous or with a few scattered hairs.

28. Ranunculus subrigidus W. Drew. White water-c. Resembling nos. 27 and 29, and intergrading with no. 29; lvs of no. 29; achenes of no. 27, but more numerous, mostly 30–45(–80); stamens mostly 5–10; mature pedicels recurved; 2n=16. Ponds and slow streams; irregularly from Que. to B.C., and s. to Del., Mich., Minn., S.D., Tex., and Mex. June–Sept. (*R. circinatus* var. *s.*) True *R. circinatus* Sibth. is here interpreted as being strictly Eurasian.

29. Ranunculus longirostris Godron. White water-c. Stems submersed, elongate, branched; lvs submersed, ternately or binately dissected into filiform segments, relatively firm, not collapsing when withdrawn from water, 1–2 cm, the free petiole none or much shorter than the expanded, pubescent stipular part; fls at the water-surface, white, 1–1.5 cm wide; stamens 10–20; achenes (7–)15–25, with a straight, slender beak 0.7–1.5 mm. Ponds and slow streams; sw. Que. to Del., Tenn., and Ala., w. to Sask., Ida., Nev. and Tex. May–July.

30. Ranunculus lapponicus L. Lappland b. Rhizomes elongate, bearing at each node a solitary stem 1–2 dm and 1(–2) basal lvs; lvs petioled, reniform, deeply 3-parted, the flabellate-obovate segments coarsely crenate to shallowly lobed; cauline lf 1 or none, smaller, short-petioled; fls 8–12 mm wide; sep 3; achene-body oblong, 2–3 mm, turgid below, flattened above; beak slender, sharply curved or hooked; 2n=16. Wet soil; circumboreal, is in Amer. to n. Me., n. Minn., and B.C. June, July.

31. Ranunculus ficaria L. Lesser celandine. Perennial from a cluster of tuberous roots, 1–3 dm, with a few short internodes below and 1 or few long peduncles; lvs long-petiolate, cordate to oblong-cordate, blunt, entire or sinuately toothed, often with axillary bulbils; lower lvs apparently opposite; sep 3–4, green; pet 8–

12, yellow, narrowly obovate, mostly 10–18 mm; achenes beakless; $2n=16, 24, 32$. Native of Eurasia, casually escaped from cult. in our range.

14. MYOSURUS L.　Mouse-tail. Fls regular; sep 5, spurred at the base; pet 5, or none, long-clawed, the scarcely dilated blade nectariferous at the summit; stamens 5–20; pistils numerous, on a long, slender receptacle; achenes quadrate or hexagonal with a large basal scar, areolate on the sides, truncate and ridged on the summit; small annuals with linear lvs and 1-fld scapes. 6–7, widespread.

1. Myosurus minimus L.　Scape 3–15 cm; lvs linear or nearly so, the blade scarcely differentiated; sep yellowish-green, the blade 3(–5)-nerved, narrowly oblong, 3–5 mm, the spur 1.5–2 mm; pet about equaling the sep; spike of pistils 5–10 mm, elongating to as much as 6 cm in fr; achenes 2.5 mm long, 1 mm wide, shortly beaked; $2n=16$. Moist soil; circumboreal, in Am. chiefly western, extending e. to s. Ont., Ky., and even Va. Apr.–June.

15. ADONIS L.　Pheasant's eye. Much like *Ranunculus,* but without the nectariferous scale or pit at the base of the pet; cauline lvs alternate, sessile or nearly so, dissected into numerous narrowly linear segments; fls solitary, terminating the stem or its branches; sep 5, green, shorter than the 5–20 conspicuous pet; ovule and seed suspended. 20, temp. Eurasia.

1 Pet 10–20, bright yellow; perennial . . . . 1. *A. vernalis.*
1 Pet 5–8, red with dark basal spot; annual . . . . 2. *A. annua.*

1. Adonis vernalis L.　Spring-adonis. Perennial, 1–4 dm; pet 10–20, mostly bright yellow, 2–3.5 cm; anthers yellow; achenes evidently hairy, with a short, hooked beak; $2n=16$. Native of Europe, occasionally escaped in our range, esp. toward the south. Late summer.

2. Adonis annua L.　Fall-adonis. Annual, 3–7 dm; pet 5–8, red with a dark basal spot, 8–15 mm; anthers blackish-purple; achenes glabrous, with a straight beak; $2n=32$. Native of Eurasia, occasionally escaped in our range, esp. toward the south. Late summer.

16. AQUILEGIA L.　Columbine. Fls regular, but complex in structure; sep 5, short-clawed; pet 5, the blade prolonged backward from the base below the fl into an elongate, hollow spur; stamens numerous, often ± connivent, the innermost staminodial; follicles mostly 5, erect; perennial herbs from a stout, caudex-like rhizome, with petiolate, 2–3 times ternately compound, alternate lvs and few but showy fls. 70, N. Hemisphere.

1 Fls red and yellow . . . . 1. *A. canadensis.*
1 Fls blue or purple to white or pink . . . . 2. *A. vulgaris.*

1. Aquilegia canadensis L.　Canada-c. Stems 3–20 dm; basal lvs large, long-petiolate, cauline reduced upwards; lfls broadly obovate to subrotund, crenately toothed or lobed; fls nodding, 3–4 cm; sep red; pet with yellow blade and nearly straight red spur, the spur 1.5–2.5 cm; fls rarely wholly yellow or salmon-colored; stamens projecting in a column; $2n=14$. Dry woods, rocky cliffs and ledges, and even peat bogs; N.S. to Sask., s. to Fla. and Tex. Apr.–June. (*A. coccinea; A. latiuscula.*)

2. Aquilegia vulgaris L.　European c. Foliage as in no. 1; fls nodding, blue, varying to purple, white, or pink, (2.5–)3–4(–5) cm long and wide, stamens not longer than the sep; spurs short and thick, strongly incurved; follicle-beak 5–10 mm; $2n=14$. Native of Eurasia, escaped from cult. and locally established esp. in the cooler parts of our range. May–July.

A. brevistyla Hook., a western species, has been reported from Minn. It has ascending or nodding fls 1.5–2.5 cm, with the yellowish-white sep a little shorter than the blue or purplish pet, and the follicle-beak is 2–5 mm.

17. ISOPYRUM L.　Sep mostly 5, petaloid; pet none; staminodes in ours none; stamens numerous, the filaments narrowly clavate; pistils few; fr follicular; perennial herbs with basal and alternate cauline, 1–3-ternately compound lvs and 1–few white fls terminating the stem or on axillary peduncles. 25, N. Temp.

1. Isopyrum biternatum (Raf.) T. & G.　Roots with many small, tuber-like thickenings; stems slender, 1–4 dm; basal lvs long-petioled, 2–3-ternate; cauline lvs smaller, less petiolate and less dissected; lfls broadly obovate, 3-lobed; fls white, 1.5–2 cm wide; follicles mostly 4, semi-ovoid, compressed, divergent; $2n=14$. Moist woods; w. N.Y. and s. Ont. to Minn., s. to the piedmont of Va. and the Carolinas, and to w. Fla. and Ark. Apr., May. (*Enemion b.*)

18. ANEMONELLA Spach. Rue-anemone.

Sep 5–10, petaloid; pet and staminodes none; pistils several, usually 8–12; achenes fusiform, saliently several-ribbed, tipped with the persistent, sessile, capitate stigma; perennial, glabrous herb from a small cluster of fusiform tubers, bearing a few long-petioled, 2–3-ternately compound basal lvs, an erect stem, and a few-fld umbel subtended by 2 or 3 virtually opposite or whorled, sessile, ternately compound involucral lvs. Monotypic.

1. **Anemonella thalictroides** (L.) Spach. Stems slender, 1–2(–3) dm; lf-segments broadly ovate to subrotund, 1–3 cm, 3-toothed distally, rounded to subcordate below; fls without nectar; sep white to pale pink-purple, 10–15 mm; $2n=14$, 42. Dry or moist woods; N.H. to Minn. and Kan., s. to Fla., Miss., and Ark. Apr., May. (*Thalictrum t.*)

19. THALICTRUM L. Meadow-rue.

Fls perfect or more often unisexual; plants hermaphrodite, dioecious, or polygamo-dioecious; sep 4–5, green or petaloid, soon deciduous; pet none; filaments elongate, capillary to clavate or much dilated; ovaries several, 1-seeded; style very short; stigma elongate; achenes usually conspicuously veined or ribbed, sometimes stipitate; perennial herbs with ternately decompound lvs and most ample terminal panicles of greenish, yellowish, purplish, or white fls, the color due principally to the stamens; polyploid series based on $x=7$. 50+, mostly N. Temp.

1 Fls perfect; plants of w. Va. and e. Ky. to Ala. and Ga.
  2 Achene concave on the upper margin, ca twice as long as the stipe .... 1. *T. clavatum.*
  2 Achene straight on the upper margin, about as long as the stipe .... 2. *T. mirabile.*
1 Fls all or mostly unisexual; plants of various distribution.
  3 Lfls all or predominantly with 4 or more teeth or lobes, usually distinguishable as 3 lobes each
      with 1–3 teeth.
    4 Cauline lf subtending the lowest flowering branch long-petioled; fl early spring .... 3. *T. dioicum.*
    4 Cauline lf subtending the lowest flowering branch sessile or nearly so, the free petiole none or
      no longer than the stipular dilation; fl late spring or summer.
      5 Achenes more convex on the lower side than the upper; boreal sp. .... 4. *T. venulosum.*
      5 Achenes more convex on the upper side than the lower; s. Appalachian spp.
        6 Plant caudiciferous; achene 2.5–4 mm; terminal lfl mostly longer than wide .... 5. *T. coriaceum.*
        6 Plant rhizomatous; achene 4–6 mm; terminal lfl mostly wider than long .... 6. *T. steeleanum.*
  3 Lfls all or nearly all entire or with 2 or 3 entire lobes, seldom with a few additional teeth.
    7 Lfls finely pubescent beneath with capitate-glandular hairs, rarely glabrous; filaments capillary
      or slightly dilated; anthers 1.8–2.7 mm .... 8. *T. revolutum.*
    7 Lfls glabrous or merely puberulent beneath, not glandular.
      8 Lfls predominantly entire, glabrous, rarely over 15 mm; filaments slightly dilated; anthers
        0.7–1.2 mm; Va. and s. .... 7. *T. macrostylum.*
      8 Lfls predominantly 3-lobed, usually puberulent beneath, mostly over 15 mm.
        9 Anthers 0.8–1.5 mm; filaments strongly dilated above, constricted beneath the anthers;
          achene narrowed to a stipitate base .... 9. *T. pubescens.*
        9 Anthers 1.5–3.5 mm; filaments slightly or scarcely dilated above; achene obtuse at the
          sessile or subsessile base .... 10. *T. dasycarpum.*

1. **Thalictrum clavatum** DC. Mountain m.-r. Stem 3–7 dm, from a cluster of thickened roots, often little exceeding the basal lf; lfls 2–3 cm long and wide, pale beneath; panicle widely branched, few-fld, with elongate, slender pedicels; fls perfect; filaments white, generally more than 3 mm, strongly dilated above; anthers minute; achenes falcate, concave on the upper margin, strongly convex on the lower, 4–5.5 mm, conspicuously 3-nerved on each face; stipe 1.5–3 mm. Moist woods and cliffs in and near the mts.; Va. and W.Va. to e. Ky., s. to S.C. and Ga. May–July.

2. **Thalictrum mirabile** Small. Much like *T. clavatum,* and perhaps not specifically distinct, but more lax and delicate; filaments generally less than 3 mm; achenes 2.5–4 mm, straight on the upper margin, convex on the lower; stipe 2.5–3.5 mm, nearly equaling the achene. Wet sandstone cliffs; e. Ky. and Tenn. to Ala. June.

3. **Thalictrum dioicum** L. Early m.-r. Plants dioecious, 3–7 dm at anthesis; stem arising from a subterranean, fibrous-rooted crown; lvs glabrous, thin, flat-margined; petioles all elongate, the uppermost 3–6 cm and subtending the infl; stipules of upper lvs broadly ovate to semicircular or semilunate, mostly much wider than long; filaments and anthers yellow or greenish-yellow; achenes sessile or subsessile, 4 mm, strongly ribbed, straight and essentially symmetrical; $2n=28$, 42. Abundant in moist woods; Que. to Man., s. to S.C., Ga., Ala., and Mo. Fl with or before the expansion of the lvs on deciduous trees.

4. Thalictrum venulosum Trelease. Northern m.-r. Stems glabrous or nearly so, arising 3–10+ dm from the upturned end of a rhizome; lvs usually at least sparsely glandular beneath, but less so than in no. 8, those below the infl petioled, those subtending the panicle-branches sessile; panicle with ascending branches; body of the achene 4–6 mm, its lower margin more convex than the upper; 2*n*=42. Rocky or gravelly soil, often along shores; Lab. to B.C., s. to Vt., n. N.Y., n. Mich., n. Wis., n. Minn., Wyo., and Oreg. June, July. Our plants have been distinguished on dubious grounds from typical cordilleran *T. venulosum* as var. *confine* (Fern.) Boivin. (*T. confine.*)

5. Thalictrum coriaceum (Britton) Small. Stems to 1 m, from a stout caudex; upper lvs sessile, the terminal lfl in each set of 3 commonly triangular-obovate, longer than wide, narrowed to the base; panicle very loosely and widely branched; filaments slender, colored; anthers 2.5–3.5 mm; achenes short-stipitate, turgid, ovoid, 2.5–4 mm long and half as wide, the upper side more convex than the lower; style persistent, usually deflexed; 2*n*=70, 140. Rich woods in and near the mts.; Md., Va., W.Va., Ky., Tenn., N.C., and Ga. Late May, June.

6. Thalictrum steeleanum B. Boivin. Stems 1–1.5 m, from a rhizome; upper lvs sessile, the terminal lfl in each set of 3 broadly rounded to subcordate at the base and broader than long; panicle to 3 dm, loosely branched; filaments slender, colored; anthers 3–4.5 mm; achenes short-stipitate, ellipsoid, 4–6 mm long and a third as thick, the upper side more convex than the lower; style persistent, deflexed. Moist, alluvial thickets; s. Pa. and Md. to n. Va. and W.Va. Late May, June.

7. Thalictrum macrostylum Small & Heller. Small-lvd m.-r. Stem slender, ca 1(–1.5) m; lfls firm and conspicuously reticulate-veiny beneath, narrowly revolute, ovate to oblong or obovate, entire or some of them with 2 or 3 lobes, glabrous, 5–15 mm; panicle large, very loosely branched; sep 1–2 mm; filaments slightly dilated, narrower than the anther; anthers 0.7–1.2 mm; stigma 0.6–1.7 mm; achene 3 mm, with low, inconspicuous ribs. Rich woods and meadows; s. Va. to S.C., w. to Ala. and Miss. June.

8. Thalictrum revolutum DC. Skunk m.-r. Plants stout, odorous, 1–2 m; lfls variable in shape and size, mostly subcoriaceous and prominently reticulate-veiny beneath, narrowly revolute, finely pubescent beneath with short, capitate-glandular hairs (rarely glabrous and glaucous); filaments capillary or slightly dilated above; anthers 1.8–2.7 mm; stigmas 2–3.5 mm; achenes ellipsoid or lanceolate, 4–6 mm, usually minutely glandular-hairy; 2*n*=ca 134. Dry woods and prairies; Mass. to s. Ont., O., Ill., and Mo., s. to Fla., Ala., and Ark. June–July.

9. Thalictrum pubescens Pursh. Tall m.-r. Stems mostly 1–3 m, arising from a short crown; cauline lvs sessile; lfls glabrous to more minutely puberulent beneath, mostly 3-lobed and more than 15 mm; infl ± rounded above; sep 2–3.5 mm, ± elliptic and rounded; filaments white, dilated distally, often wider than the anther, constricted at the tip; anthers usually narrowly obovoid, blunt, 0.8–1.5 mm; stigmas 0.5–2 mm, fiddlehead-curved; achenes somewhat glandular, drying dark, 3–5 mm, a third to half as wide, narrowed below to a short stipe 0.3–0.5 mm; lowest achenes much reflexed, the head of achenes therefore subglobose; 2*n*=84, 154. Rich woods, wet meadows, and streambanks; Lab. and Que. to Ont., s. to s. N.C., Tenn., and Ind. June–Aug., often later than other spp. (*T. polygamum.*)

10. Thalictrum dasycarpum Fischer & Avé-Lall. Purple m.-r. Stems ca 1(–2) m; cauline lvs sessile; lfls mostly 3-lobed and more than 15 mm, distinctly puberulent beneath, or in western plants occasionally glabrous; infl ± pyramidal and pointed; sep 3–5 mm, ± lanceolate and acute; filaments white, capillary or slightly dilated above, not constricted at the tip; anthers linear, 1.5–3.5 mm, sharply apiculate; stigmas ± straight, 2–4.5 mm (mostly over 3 mm in fr); achenes 4–6 mm, obtuse at the sessile or subsessile base; outermost achenes widely spreading or somewhat deflexed, the head of achenes therefore ± hemispherical; 2*n*=168. Wet meadows, shores, and streambanks; sw. Ont. to Alta. and Wash., s. to O., Ind., Mo., Okla., and Ariz. June, July. The glabrous phase has been called var. *hypoglaucum* (Rydb.) Boivin.

20. COPTIS Salisb. Goldthread. Sep 4–7, petaloid; pet none; staminodes 4–7, smaller than the sep, clavate, nectariferous in the hollow summit; stamens numerous, with broad anthers; pistils 3–7, stipitate even at anthesis, narrowed to a short, slender, persistent style; fr follicular; perennial herbs with basal, evergreen, ternately compound lvs and erect, lfless or minutely bracteate scapes with 1–few white fls. 10, N. Temp. and Arctic.

1. Coptis trifolia (L.) Salisb. Rhizomes slender, bright yellow; lfls 3, subsessile, crenate or lobulate with mostly mucronate teeth, the middle lfl cuneate-obovate, the others inequilateral, broadly rounded on the lower side; peduncle 1-fld, 5–15 cm; fls white, 12–16 mm wide; follicles (style included) 8–13 mm; 2*n*=18. Damp, mossy woods and bogs; Greenl. to Alas. and e. Asia, s. to N.J., N.C., n. Ind., Io., and B.C. The e. Amer. plants are segregated on minor characters as var. *groenlandica* (Oeder) Fassett.

21. HYDRASTIS L. Golden Seal. Sep 3, petaloid, falling when the fl opens; pet none; stamens and pistils numerous; styles very short; ovules 2; berries dark red, 1–2-seeded, forming a head; rhizomes knotty, yellow, containing berberine; perennial herbs with alternate lvs and a solitary, terminal fl. 2, the other Asian.

1. **Hydrastis canadensis** L. Stem 2–5 dm, hairy, bearing usually one basal lf and 2 cauline lvs near the top; blades broadly cordate-rotund, 5-lobed, palmately veined, 3–10 cm wide at anthesis, later sometimes to 25 cm, the lobes incised, doubly serrate, short-acuminate; peduncle hairy, ca 1 cm; filaments 5–8 mm; $2n=26$. Deep, rich woods; Vt. to Mich. and Minn., s. to N.C., Tenn., and Ark. Apr., May.

22. **XANTHORHIZA** Marshall. Yellow-root. Sep 5, small, spreading; pet none, staminodia 5, clawed, the nectariferous part transversely oblong, 2-lobed; stamens 5–10; pistils mostly 5–10; style subulate; ovules 2; fr an obliquely oblong follicle, with a single distal seed (or 2 seeds); low shrub with yellow wood, bearing a crowded cluster of pinnately compound lvs and several simple racemes or raceme-like panicles of small, purple-brown fls. Monotypic.

1. **Xanthorhiza simplicissima** Marshall. Stems 3–5 dm; lvs long-petioled; lfls mostly 5, lance-ovate to broadly ovate, acute or acuminate, toothed to deeply cleft; infls 5–12 cm; sep lanceolate, broadly clawed, 3 mm; stamens and staminodes 1 mm; follicles thin-walled, pale brown, sparsely hairy; $2n=36$. Moist woods, chiefly in the mts.; s. N.Y. to Pa. and Ky., s. to S.C., w. Fla., and Ala. Apr. (*Zanthorhiza apiifolia.*)

FAMILY **BERBERIDACEAE,** the Barberry Family

Fls hypogynous, regular, perfect, all parts free and distinct, the cal, cor, and androecium often bicyclic; sep 4 or 6, sometimes early deciduous, in some genera petaloid; pet as many as or more than the sep, sometimes represented only by staminodial nectaries; stamens as many as the pet and opposite them, with anthers opening by uplifting valves, or (in *Podophyllum*) twice as many and opening by longitudinal slits; ovary 1, 1-celled and seemingly monocarpous, tapering to a sessile stigma; ovules 1–many; fr a berry or capsule (see also *Caulophyllum*); seeds with small, or slender and elongate, dicotyledonous embryo and copious endosperm, often arillate; herbs or shrubs with alternate or basal (see also *Podophyllum*), simple to compound leaves usually dilated at the base or stipulate, and solitary, racemose or cymose fls. 13/650.

1 Smooth perennial herbs.
  2 Fls solitary.
    3 Flowering stem with a pair of opposite lvs below the infl .... 1. *Podophyllum.*
    3 Flowering stem strictly scapose, the lvs all basal .... 3. *Jeffersonia.*
  2 Fls racemose to cymose or paniculate.
    4 Lvs simple, lobed .... 2. *Diphylleia.*
    4 Lvs ternately compound .... 4. *Caulophyllum.*
1 Spiny shrubs .... 5. *Berberis.*

1. **PODOPHYLLUM** L. May-apple; mandrake. Sep 6, biseriate, falling early; pet 6 or 9, 2- or 3-seriate; stamens twice as many as the pet; anthers opening by longitudinal slits; ovary thick-ovoid, with a large, sessile stigma; fr a large, ovoid, fleshy, many-seeded berry; perennial herbs from a rhizome, usually colonial, the flowering stem with a pair of lvs and a short-peduncled, solitary terminal fl. 10, N. Hemisphere.

1. **Podophyllum peltatum** L. Radical lvs scattered, with a stout, erect, stem-like petiole and peltate, deeply lobed blade to 3 or 4 dm wide; stem erect, 3–5 dm, with a pair of petiolate, half-round to cordate or unequally peltate, similarly lobed but mostly smaller lvs; peduncle nodding, short, fl white (pink), 3–5 cm, wide; fr yellow, 4–5 cm, the pulp edible when ripe; $2n=12$. Seeds and herbage poisonous. Moist, preferably open woods, Que. to Minn., s. to Fla. and Tex. May.

2. **DIPHYLLEIA** Michx. Umbrella-lf. Sep 6, biseriate, ovate-elliptic; stamens 6; ovary ellipsoid; stigma sessile; fr a few-seeded berry; smooth perennial herbs with a thick rhizome, the flowering stem with 2 alternate lvs peltate near the margin, the basal lvs scattered, erect, with centrally peltate blade; fls in a pedunculate cyme, white. 3, the other 2 in e. Asia.

1. Diphylleia cymosa Michx.   Basal lvs long-petioled, the blade 2–5 dm wide, 2-cleft, radially many-lobed and sharply dentate; cauline lvs a little smaller, similarly cleft; cyme 5–10 cm wide; fls many, white, 15 mm wide; fr blue, 1 cm, on red pedicels 1–3 cm; $2n=12$. Cool mt. woods; Blue Ridge prov. from Va. to Ga. May, June.

3. JEFFERSONIA Barton.   Twin-lf; rheumatism root. Sep usually 4, uniseriate, early deciduous; pet usually 8, biseriate; ovary ovoid, with many ovules and a broad, sessile stigma; fr capsular, obovoid or pyriform, opening in the distal half by a horizontal cleft extending half way around it, the top forming a lid; smooth perennial herbs with a few basal lvs deeply divided into 2 obliquely half-ovate segments, the fls solitary on a naked scape. 2, the other in Manchuria.

1. Jeffersonia diphylla (L.) Pers.   Scape 1–2 dm at anthesis; petioles at first shorter than the scapes, later elongating to 2–5 dm; blades immature at anthesis, eventually 8–15 cm; fls white, 1–3 cm wide; fr 2–3 cm; $2n=12$. Rich woods, preferring calcareous soil; w. N. Y. and s. Ont. to se. Minn., s. to Md. and Ala. Apr., May.

4. CAULOPHYLLUM Michx.   Blue cohosh. Sep 6, biseriate, petaloid, subtended by 3 or 4 sepal-like bracts; pet 6, biseriate, represented only by small, stipitate-flabellate, gland-like nectaries opposite and much shorter than the sep; stamens 6; ovary asymmetrical, tapering above and with a minute apicolateral stigma; poisonous smooth perennial herbs, the erect stem bearing above the middle a single large, sessile, triternate lf, simulating 3 biternate lvs, and another smaller lf just below the infl; ovary soon ruptured by the enlarging seeds, which ripen fully exposed on short, stout stalks and resemble drupes. 2, the other in e. Asia.

1. Caulophyllum thalictroides (L.) Michx.   Erect, 3–8 dm, glaucous when young; lfls obovate-oblong, 2–5-lobed above the middle, 5–8 cm at maturity; infl a small, panicle-like or raceme-like cyme, 3–6 cm; fls yellowish-green or greenish-purple, the slender, spreading, somewhat twisted sep 1 cm; seeds bitter, dark blue, 5–8 mm, on stalks nearly as long; $2n=16$. Rich, moist woods; N.B. to Ont. and Man., s. to S.C., Ala., and Mo. Apr., May. Some authors recognize 2 largely sympatric vars. Var. *thalictroides* has variously yellow, purple or green sep mostly 4–5 mm. Var. *giganteum* Farwell blooms 2 weeks earlier and has fewer fls larger in all parts, the consistently purple sep 6–8 mm. (*C. g.*)

5. BERBERIS L.   Barberry. Sep 6, biseriate, petaloid, subtended by 2 or 3 small bracts; pet 6, biseriate, usually smaller than the sep and each with 2 basal glands; stamens 6; ovary with 1–few ovules; fr a berry; spiny shrubs; primary lvs of the shoot modified into simple or often trifid spines, bearing axillary fascicles of small foliage lvs; fls yellow, in racemes, or contracted into umbel-like clusters, or solitary; fr in ours red. 600, widespread.

1 Lvs entire; fls in small clusters or solitary . . . . 1. *B. thunbergii.*
1 Lvs spinulose-toothed; fls racemose.
  2 Pet notched at the tip; racemes mostly 5–10-fld . . . . 2. *B. canadensis.*
  2 Pet entire; racemes mostly 10–20-fld . . . . 3. *B. vulgaris.*

1. Berberis thunbergii DC.   Japanese b. Densely and divaricately branched, to 2 m; spines usually simple; lvs obovate to spatulate, mostly obtuse, entire, short-petiolate; fls solitary or in umbel-like clusters of 2–4, 8 mm wide; fr 1 cm; $2n=28$. Native of Japan, frequently escaped from cult. along roadsides and in thickets in our range. May.

2. Berberis canadensis Miller.   American b. Erect, rather sparsely branched, 1–2 m; younger twigs mostly purple, brown, or reddish; lvs obovate to spatulate, 2–6 cm, coarsely spinulose-dentate, the veinlets indistinct beneath; racemes 2–4 cm, with 5–10 fls; pet notched at the tip; fr nearly 1 cm; $2n=28$. Rocky woods; mts. of Va. and W.Va. to Ga. and Ind. May.

3. Berberis vulgaris L.   Common or European b. Freely branched, to 3 m, the twigs gray or yellowish-gray; lvs obovate or obovate-oblong, 2–5 cm, finely spinulose-denticulate, the veinlets prominently reticulate beneath; racemes 3–6 cm, with 10–20 fls on pedicles 5–10 mm; pet entire; fr 1 cm, tart; $2n=28$. Native of Europe, formerly much planted and escaped, now largely purposefully exterminated as the alternate host of the common "stem" rust of wheat. May, June.

FAMILY **LARDIZABALACEAE,** the Lardizabala Family

Fls unisexual, regular, hypogynous, trimerous; sep 3 or 6; pet none, or represented by nectaries; stamens 3 or 6; ovaries 3, distinct; ovules numerous; mostly woody climbers with alternate, exstipulate, palmately compound lvs. 8/30.

Akebia quinata (Houtt.) Decne., a native of e. Asia, sometimes escapes from cult. It is a high-climbing woody twiner, with 5 retuse lfls and fragrant, purplish-brown fls in small, axillary racemes, the lower pistillate, 2–3 cm wide, the upper staminate and smaller; $2n=32$.

FAMILY **MENISPERMACEAE,** the Moonseed Family

Fls small, unisexual, mostly trimerous, regular, hypogynous; tepals not much differentiated, generally alternating in 4 cycles, the 2 outer (cal) exceeding the 2 inner (cor), or cor none; stamens 6–many, distinct or connate; pistils typically 3, sometimes fewer or more, distinct, simple; ovule 1, suspended; fr a drupe; seeds with a large, dicotyledonous embryo and often little or no endosperm; twining, woody vines with simple, alternate, exstipulate, palmately veined lvs and small, white or greenish fls in cymes or cymose panicles or seeming racemes. 70/400.

1 Lvs peltate near the margin; pet 6–9; stamens 12–24 .... 1. *Menispermum.*
1 Lvs not peltate.
  2 Pet 6; stamens 6; stigma entire; fr red .... 2. *Cocculus.*
  2 Pet none; stamens 12; stigma radially lobed; fr black .... 3. *Calycocarpum.*

1. MENISPERMUM L.   Moonseed. Sep (4–)6(–8), longer than the 6–9 pet; stamens 12–24; pistils (2–)3(–4); drupe subglobose, developing obliquely, the terminal stigmatic scar far to one side at maturity; stone flattened, thickened into 3 rough ridges over most of its margin; woody twiners, with broad lvs peltate near the margin. 2, the other in e. Asia.

1. Menispermum canadense L.   Climbing 2–5 m; lvs slender-petioled, broadly ovate to suborbicular, 10–15 cm long and wide, shallowly 3-lobed to entire; fr bluish-black, 6–10 mm, poisonous; $2n=52$. Moist woods and thickets; w. Que. and w. N. Engl. to Man., s. to Ga. and Okla. June, July.

2. COCCULUS DC., nom. conserv.   Sep, pet, and stamens each 6; pistils 3 or 6, with entire stigma; fr flattened on both sides, the stigmatic scar lateral; stone thickened and rough on the margin for three-fourths of its circumference, somewhat excavated on the sides; woody twiners. 10, warm reg.

1. Cocculus carolinus (L.) DC.   Carolina m.; coral beads. Slender climber to 2–3 m; lvs variable, deltoid to ovate, 5–15 cm, entire to 3-lobed, cordate to truncate at base, acute to retuse at the mucronate tip, usually hairy beneath; staminate infls 1 dm, pistillate much shorter; fls 3–4 mm wide; fr red, 5–8 mm; $2n=78$. Moist woods and thickets; Va. to Fla. and Tex., n. in the interior to s. Ind. and Mo. June–Aug. (*Epibaterium c.*)

3. CALYCOCARPUM (Nutt.) Spach.   Cupseed. Sep 6; pet none; stamens 12; pistils 3, narrowed to the summit and there radially dilated into a lobed stigma; fr ellipsoid, flattened on one side; stone deeply excavated on one side, its margin somewhat incurved and erose; woody twiner. Monotypic.

1. Calycocarpum lyonii (Pursh) A. Gray.   Climbing to the tree-tops; lvs 1–2 dm long, somewhat wider, 3–7-lobed, braodly cordate at base, the lobes acuminate; panicles slender, 1–2 dm; fls numerous, 5 mm wide; fr black, 15–25 mm. Moist woods and swamps; s. Ind. to s. Mo., s. to Fla., La., and Okla. May, June.

FAMILY **PAPAVERACEAE**, the Poppy Family

Fls regular, perfect, hypogynous; sep 2 or 3, fully enclosing the bud, caducous; pet 4 or more (rarely none), distinct, showy; stamens numerous; nectaries none; pistil of 2, less often 3 or 4, seldom (as in *Argemone* and *Papaver*) more numerous carpels, united to form a compound, typically unilocular ovary; ovules numerous on parietal placentas, but these sometimes (as in *Papaver*) deeply intruded as partial partitions, or the ovary seldom (as in *Glaucium*) fully partitioned into 2 locules; fr typically a unilocular capsule, dehiscent by partial or complete abscission of elongate valves alternating with the placentas, which form a persistent replum, or the valves sometimes (as in *Argemone* and *Papaver*) reduced and subapical, so that the capsule opens by a ring of pores; capsule bilocular with deciduous valves in *Glaucium*; seeds with copious endosperm and small, dicotyledonous embyro, sometimes arillate; herbs or shrubs with milky or colored juice, mostly alternate lvs, and usually large fls. 25/200.

Three spp., occasionally found as garden escapes, not included in the keys below, are: *Eschscholzia californica* Cham., California poppy; lvs ternately dissected into linear segments; fls long-peduncled, yellow or orange, 3–6 cm wide. *Macleaya cordata* (Willd.) R. Br., Plume poppy; perennial, to 2.5 m, with large, pinnately lobed lvs and large panicles; sep 2, caducous; pet none. *Platystemon californicus* Benth., Cream-cups; annual, 1–3 dm, with chiefly opposite, linear or lance-linear, entire lvs, long-peduncled, cream-yellow fls 3 cm wide, and fr of several separating carpels that fragment transversely.

1 Pet 8 or more; fl solitary on a scape; lf 1, basal . . . . 1. *Sanguinaria.*
1 Pet 4 or rarely 6; cauline lvs present.
  2 Foliage not spiny.
    3 Fls yellow.
      4 Style well developed, about 1 cm; fr ellipsoid, bristly . . . . 2. *Stylophorum.*
      4 Style very short or none; fr elongate-linear, glabrous.
        5 Fls in small, umbel-like infls; pet 1 cm; fr 2–5 cm, unilocular . . . . 3. *Chelidonium.*
        5 Fls solitary; pet 2.5–5 cm; fr 15–30 cm, bilocular . . . . 4. *Glaucium.*
    3 Fls red or purple to white . . . . 5. *Papaver.*
  2 Foliage spiny . . . . 6. *Argemone.*

1. **SANGUINARIA** L. Bloodroot. Sep 2; pet 8, or up to 16; ovary narrowed above to a short style terminated by a capitate, 2-lobed stigma; placentas 2; capsule fusiform, crowned by the persistent style, dehiscent from the base upwards by elongate valves; seeds arillate; perennial herb with red juice, from a stout rhizome that sends up a single, lobed lf and a large, white fl on a scape. Monotypic.

    1. **Sanguinaria canadensis** L. Lf orbicular in outline, sometimes 2 dm wide at maturity, 3–9-lobed, the lobes undulate to coarsely toothed; scape 5–15 cm at anthesis; fls without nectar, white (pink), 2–5 cm wide; 4 pet usually longer than the others and the fls quadrangular in outline; fr 3–5 cm; $2n=18$. Rich woods; N.S. to Ont. and Man., s. to Fla. and Okla. Apr.

2. **STYLOPHORUM** Nutt. Celandine-poppy. Sep 2; pet 4, yellow; ovary ellipsoid, abruptly narrowed to an elongate style; stigma 2–4-lobed; placentas 2–4; capsule dehiscent from the top downwards by elongate valves; seeds arillate; perennial, rhizomatous herbs with saffron-colored juice, deeply lobed, mostly basal lvs, and a few-fld, terminal, umbelliform infl. 2, the other in e. Asia.

    1. **Stylophorum diphyllum** (Michx.) Nutt. Stem 3–5 dm at anthesis; basal lvs several, long-petioled, thin, broadly oblong to ovate, pinnately divided almost or quite to the midvein into 5–7 oblong or obovate, obtusely lobed or toothed segments; cauline lvs 1 pair, smaller than the basal; buds on erect pedicels 2–5 cm; sep hairy; pet 2–3 cm; style 1 cm; fr ellipsoid or thick-fusiform, bristly, 2–3 cm; $2n=20$. Rich, moist woods; w. Pa. to s. Mich. and Wis., s. to Tenn. and Ark. Apr.–May.

3. **CHELIDONIUM** L. Celandine. Sep 2; pet 4, yellow; ovary narrowly cylindric, glabrous, narrowed above into a very short style with 2-lobed stigma; fr smooth, dehiscent

from the base upwards by elongate valves; seeds arillate; biennial with saffron-colored juice, deeply lobed lvs, and few-fld umbelliform infls. Monotypic.

1. Chelidonium majus L.   Branched, 3–8 dm; cauline lvs several, alternate, much like those of *Stylophorum diphyllum*; sep glabrous; pet 1 cm; fr 3–5 cm; $2n=12$. Eurasian sp., well established in moist soil from Que. to Io., s. to Ga. and Mo. Apr.–Sept.

4. GLAUCIUM Miller.   Horned poppy.
Sep 2; pet 4, yellow; ovary cylindric, bilocular; stigmas sessile, 2-lobed; fr elongate, usually curved, bilocular, dehiscent from the top downward by elongate valves; herbs with saffron-colored juice, large, pinnately lobed lvs, and large, yellow fls terminating the branches or on leafy-bracted axillary peduncles. 20, Eurasia.

1. Glaucium flavum Crantz.   Stout biennial or perennial 3–9 dm; lvs firm, hairy, the lowest petioled, the others sessile or nearly so, ovate or oblong, irregularly pinnatifid; fls 5–9 cm wide, their short peduncles strongly thickened at maturity; fr 15–30 cm, 5 mm thick; $2n=12, 24$. Native of Europe, established in waste places, especially near the coast, from Mass. to Va., and occasionally inland to Mich. June. (*G. glaucium.*)

5. PAPAVER L.   Poppy.
Sep 2; pet normally 4, white or colored, our spp. never yellow; ovary of 4–many carpels, the placentas deeply intruded as partial partitions; stigmas as many as the carpels, sessile and radiating on a disk terminating the ovary; fr opening by small valves just below the margin of the stigmatic disk; seeds not arillate; herbs with milky or colored juice and large, usually long-peduncled fls terminating the stem and branches. 100, cosmop.

Three cult. spp. that occasionally escape in our range are included in the key but not described.

1 Cauline lvs cordate-clasping.
  2 Buds 2–4 cm; fr 2.5–6± cm .... 1. *P. somniferum.*
  2 Buds 1–2 cm; fr up to 2 cm; tulip-p. .... *P. glaucum* Boiss. & Hausskn.
1 Cauline lvs not clasping.
  3 Fr narrow, ca twice as long as wide; stigmatic rays 4–9(10).
    4 Perennial; anthers yellow; fr strongly ribbed; Morocco-p. .... *P. atlanticum* (J. Ball) Cosson.
    4 Annual; anthers purple; fr not strongly ribbed .... 2. *P. dubium.*
  3 Fr broad, scarcely longer than thick; stigmatic rays (5–)8–15.
    5 Robust perennial; pet 4–8 cm; buds 2–3.5 cm; Oriental p. .... *P. orientale* L.
    5 Slender annual (biennial); pet 1.5–4 cm; buds 0.5–2 cm .... 3. *P. rhoeas.*

1. Papaver somniferum L.   Opium-p. Stout annual to 1 m, glaucous and mainly glabrous except for the distally hispid peduncles; lvs sessile, coarsely toothed or shallowly lobed, with cordate-clasping base; buds 2–4 cm; pet 3–6 cm, purple or red to white; fr glabrous, 2.5–6+ cm, subglobose or broadly obovoid; stigmatic rays 8–15; $2n=22$. Native of the Mediterranean region, modified in cult., often escaped in our range. June–Sept.

2. Papaver dubium L.   Sparingly branched, ± hispid annual 3–6 dm; lvs pinnately divided, the pinnae toothed to deeply incised; peduncles appressed-hairy above; buds under 2 cm; pet 1.5–3(–3.5) cm, red to pink or red-orange, often with a basal dark spot; anthers purple; fr glabrous, 1.5–2.2 cm, obconic or narrowly obovoid, not strongly ribbed; stigmatic rays 5–9; $2n=42$. Native of Europe, sparingly intr. or escaped in waste places in our range. May–Aug.

3. Papaver rhoeas L.   Shirley-p.; corn-p. Sparingly branched annual (biennial) to 1 m, ± hispid, the peduncles spreading-hispid throughout; lvs pinnately divided, the pinnae usually lobed or incised; buds 0.5–2 cm, pet 1.5–4 cm, variously red, purple, pink, white, or streaked, often with a basal dark spot; fr glabrous, ovoid to subglobose, 1–2 cm; stigmatic rays 8–15, typically 10; $2n=14$. Native of Eurasia and N. Afr., widely escaped in our range, but seldom abundant. May–Sept.

6. ARGEMONE L.   Prickly poppy.
Stems, lvs, and sep spiny; sep generally 3; pet generally 6; ovary unilocular, with 4–6 parietal placentas; style short or none; stigma dilated, 4–6-radiate; capsule spiny, the upper third opening by valves; coarse herbs with thistle-like foliage and large fls terminating the branches; juice in our spp. yellow. 20, New World.

1 Pet yellow or cream; stamens 30–50 .... 1. *A. mexicana.*
1 Pet white, varying to pink; stamens 100 or more .... 2. *A. albiflora.*

    1. Argemone mexicana L. Mexican poppy. Annual, mostly 2.5–8 dm; lvs usually blotched with paler green; fls closely subtended by the upper lvs; pet yellow or cream, 1.5–3.5 cm; stamens 30–50; $2n=28$. Native of tropical Amer., occasionally escaped from cult. as far n. as Mass. May–Sept.

    2. Argemone albiflora Hornem. White p. p. Annual or biennial, 4–10(15) dm; lvs not blotched; fls usually closely subtended by the upper lvs; pet white or rarely suffused with pink, 2–4 cm; stamens 100 or more; $2n=28$. Dry, open, often sandy places, especially in disturbed sites; se. U.S., occasionally escaped from cult. n. to Ill. and Conn. May–Aug. (*A. alba* and *A. intermedia,* misapplied.)

FAMILY **FUMARIACEAE**, the Fumitory Family

    Fls perfect, hypogynous, strongly irregular, often bilaterally symmetrical; sep 2, bract-like, often ± peltate, not enclosing the developing buds, commonly caducous; outer pet 2, connivent below, distinct above, one or both spurred or saccate at the base; inner pet 2, slender below, ± dilated at the summit and connivent over the stigma; stamens 6, the elongate filaments connate in 2 sets opposite the outer pet, each set bearing a median bilocular and 2 lateral unilocular anthers, and usually with a nectary-spur at the base of the filaments; ovary of 2 carpels, unilocular, stigma 2-lobed; fr typically a unilocular capsule, dehiscent by partial or complete abscission of 2 elongate valves alternating with the parietal placentas, which form a persistent replum, or sometimes dry and indehiscent; seeds with copious endosperm and small, dicotyledonous embryo, generally arillate; glabrous herbs with alternate, ± dissected lvs, watery juice, and variously cymose to racemose, paniculate, or solitary fls. 19/400.

1 Corolla bilaterally symmetrical, the 2 outer pet spurred or saccate at base.
  2 Plants acaulescent, the lvs all basal .... 1. *Dicentra.*
  2 Plants climbing; leaves cauline .... 2. *Adlumia.*
1 Corolla asymmetrical, one outer pet spurred or saccate at base.
  3 Ovary elongate; fr an elongate capsule with a persistent style .... 3. *Corydalis.*
  3 Ovary subglobose; fr subglobose, indehiscent, the style deciduous .... 4. *Fumaria.*

    1. DICENTRA Bernh., nom. conserv. Sep minute; cor ovate or cordate, bilaterally symmetrical; pet weakly united, the outer 2 large, saccate or spurred at base, spreading or ascending at the summit, the inner much narrower, connate toward the dilated and ornamented apex, the claws separate; ovary slender, tapering to the long style; stigma with 2 crests or horns; fr generally capsular, with a persistent replum; handsome perennials with showy fls mostly in racemes or mixed panicles; in subg. *Dicentra* (including all our spp.) the lvs are all basal and the infl terminates a scape. (*Bicuculla.*) 15–20, N. Amer. and e. Asia.

1 Fls in racemes; rootstock covered with bulblets.
  2 Fls cordate-triangular, the spurs divergent; bulblets white or pink .... 1. *D. cucullaria.*
  2 Fls cordate-ovate, the spurs rounded; bulblets yellow .... 2. *D. canadensis.*
1 Fls in mixed panicles; rootstock without bulblets .... 3. *D. eximia.*

    1. Dicentra cucullaria (L.) Bernh. Dutchman's breeches. Rhizome very short, covered with small, white or pink, tear-shaped bulblets; lvs typically 2 per scape, somewhat yellow-green, not glaucous, long-petioled, broadly triangular in outline, decompound, the ultimate segments linear or narrowly oblong; scapes 1–3 dm, bearing a terminal raceme of nodding white fls suffused with yellow at the summit; cor 15–20 mm, its spurs subacute, divergent; nectary-spurs mostly 2–3 mm; $2n=32$. Rich woods; N.S. and e. Que. to n. Minn., s. to Ga., Ark., and Kans.; disjunct in Wash., Oreg., and Ida. Apr.–June.

    2. Dicentra canadensis (Goldie) Walp. Squirrel-corn. Much like no. 1; bulblets fewer and about twice as big, pea-shaped, yellow; lvs typically 1 per scape, blue-green, glaucous; cor narrowly ovate, its spurs short, broadly rounded, scarcely divergent; nectary-spurs 0.5–1 mm; $2n=64$. Rich woods; s. Me. and s. Que. to s. Minn., s. to N.C., Tenn., and Mo. Apr., May.

    3. Dicentra eximia (Ker Gawler) Torr. Wild bleeding heart. Lvs and scapes from a short, stout, scaly rhizome, without bulblets; lvs eventually to 4 dm (petiole included), ternately dissected into lanceolate or oblong, coarsely toothed segments; scape 2–5 dm, bearing a mixed panicle of nodding pink fls; cor narrowly ovoid, 18–25 mm, cordate at base, the spurs short and rounded; tips of the outer pet divergent, 5–6 mm; nectary-spurs virtually obsolete; $2n=16$. Dry or moist mt. woods; chiefly s. Appalachian sp., from N.C. and Tenn. to W.Va., Pa., N.J., and s. and w. N.Y. May–Aug.

2. ADLUMIA Raf., nom. conserv.

Allegheny-vine. Sep scale-like; cor bilaterally symmetrical, narrowly flattened-ovoid, subcordate at base; pet all connate about half their length, the outer constricted above to form an ovate appendage, the inner narrow, dilated at the summit into a transversely oval appendage; cor persistent after anthesis with little change of color, becoming spongy, enclosing the slender, 2-valved capsule; handsome biennial vine, climbing by the upper part of the rachis of the pinnate-ternatively decompound lvs; fls pearly-pink, in axillary panicles. 2, the other in Korea.

1. **Adlumia fungosa** (Aiton) Greene. Acaulescent during the first year, with several ascending, non-prehensile, decompound lvs; climbing the second year to 3 m, with slender, elongate stems and large, delicate, prehensile lvs, their rachis elongate and the uppermost lfls greatly reduced; $2n=32$. Woods, chiefly in the mts.; Que. to Wis., s. to Del., N.C., Tenn., and Ind. June–Sept.

3. CORYDALIS Vent., nom. conserv.

Corydalis. Fls asymmetrical; sep small, appressed; pet elongate, ± connivent, the upper one of the 2 outer ones spurred at the base, at the apex somewhat dilated and keeled or winged, the lower one narrower, similarly dilated and keeled, but not spurred at the base (sometimes gibbous), the 2 inner (lateral) pet narrower, similarly dilated and keeled or winged, and coherent over the stigma; capsule slender, often torulose, 2-valved, with a persistent replum and a slender, persistent style; seeds black, shining, minutely low-tuberculate; herbs with cauline, alternate, bipinnately dissected lvs and short, bracteate racemes of short-pediceled fls. (*Capnoides.*) 300, mainly N. Temp.

1 Fls rose to pink-purple or greenish-purple, with yellow tip; stem erect .... 1. *C. sempervirens.*
1 Fls yellow; stems erect to prostrate.
  2 Outer pet folded at the summit into a median, hollow keel not ornamented by a projecting wing; spurred pet 13–16 mm; cleistogamy rare .... 2. *C. aurea.*
  2 Outer pet with a prominent, flat wing along the median line near the summit; cleistogamy common.
    3 Spurred pet 11–15 mm, the spur 4–6 mm; fruiting pedicels short, 2–5 mm .... 3. *C. micrantha.*
    3 Spurred pet 7–9 mm, the spur 1–2 mm; fruiting pedicels elongate, (5–)10–15 mm .... 4. *C. flavula.*

1. **Corydalis sempervirens** (L.) Pers. Tall c. Erect, glaucous biennial 3–8(–10) dm, much-branched at least above; lower lvs petioled, the upper subsessile; fls in small panicles at the ends of the branches; bracts minute, lanceolate, cor (10–)12–17 mm, incl. the 2.5–5 mm spur; sep broadly ovate, 2–4 mm; fr ± erect, 2.5–4(–5) cm; seeds ca 1 mm wide, with an obtuse margin; $2n=16$. Dry or rocky woods; Nf. to Alas., s. to N.Y., Pa., n. Ind., Minn., Mont., and B.C., and along the mts. May–Sept.

2. **Corydalis aurea** Willd. Golden c. Prostrate-ascending annual or biennial, 2–5 dm, the stems sympodial; racemes dense, 1–3 cm, often surpassed by the upper lvs; bracts lanceolate; sep broadly ovate, 1.5–2 mm, erose; cor bright yellow, 13–16 mm, incl. the 4–5 mm spur; outer pet folded distally along the median line into a conspicuous but wingless keel; fr smooth, spreading or drooping, 1.5–2.5 cm; seeds ca 2 mm wide, with a narrow ring-margin; $2n=16$. Rocky banks or sandy soil; Que. to Alas., s. to Pa., Mich., n. Ill., and Minn., and widespread in w. U.S. May–July. Our plants are var. *aurea*; var. *occidentalis* Engelm., often attr. to our range (as *C. montana* Engelm.) is more southwestern.

3. **Corydalis micrantha** (Engelm.) A. Gray. Slender c. Erect or ascending annual, 1–3 dm, glaucous or nearly green; racemes usually well surpassing the lvs; bracts narrowly lanceolate; cor pale yellow, 11–15 mm, incl. the 4–6 mm spur; crest of the upper pet low, entire; frs ± erect, 1–1.5(–2) cm; often some or all fls cleistogamous, these 1–5 in a raceme, inconspicuous, often with smaller fr; seeds ca 1.5 mm wide, without a ring-margin; $2n=16$. Moist, especially sandy soil; Ill. to s. Minn. and S.D., s. to the Gulf, and n. along the Atlantic coast to N.C. Our plants are var. *micrantha*. Vars. *australis* and *texensis* are more southern.

4. **Corydalis flavula** (Raf.) DC. Short-spurred c. Green or glaucous annual 1–3 dm, at first erect, often becoming prostrate, the stem sympodial; racemes barely if at all exceeding the lvs; bracts ovate, foliaceous, 4–8 mm; cor pale yellow, 7–9 mm, incl. the 1–2 mm spur; crest of the upper pet usually toothed; frs spreading or drooping, 1.5–2 cm, on pedicels commonly (0.5–)1–1.5 cm; fls sometimes cleistogamous; seeds ca 2 mm wide, with a narrow, acute ring-margin; $2n=16$. Moist soil; e. N.Y. to s. Ont., s. Mich. and S.D. s. to N.C., La., and Okla. Apr., May.

4. FUMARIA L.   Fumitory. Fls asymmetrical; sep small, closely appressed; pet elongate, ± connivent, the upper one of the 2 outer ones spurred at the base, the 2 outer dilated at the summit, the 2 inner (lateral) coherent at the summit over the stigma; fr 1-seeded, indehiscent, globose or nearly so; style deciduous; mostly annuals with the aspect of *Corydalis*. 50, Old World.

1. **Fumaria officinalis L.**   Lax but rather robust, diffusely branched, 2–8 dm; racemes dense, many-fld, 2–4 cm; cor 8 mm, the tube red-purple, the summit dark red; fr 2.5 mm. Native of Europe, intr. in waste ground here and there in our range; $2n=32$. May–Aug.

FAMILY PLATANACEAE, the Plane-tree Family

Plants anemophilous, monoecious, the numerous, tiny fls in unisexual, spherical heads; sep 3–4(–7), distinct or basally connate; staminate fls with tiny pet alternating with the sep, and with a single set of stamens opposite the sep; filaments very short; anthers elongate, the connective prolonged into a peltate appendage; pistillate fls generally without pet, often with small staminodia, and with (3–)5–8(–9) distinct carpels commonly in 2–3 whorls; stigma decurrent along the recurved, linear style and encroaching on the slender ovary; ovule solitary, pendulous, ± orthotropous; fr a spherical head of linear achenes, each achene subtended by many long hairs; seeds with slender, dicotyledonous embryo and scanty endosperm; trees with alternate, stipulate, mostly palmately veined lvs; winter buds at first concealed by the hollow base of the petiole. 1/6-7, most spp. N. Amer., but 2 Asian.

1. PLATANUS L. Characters of the family.

1. **Platanus occidentalis L.**   Sycamore, buttonwood. Tree to 50 m, the trunk to 3 m thick, with pale greenish-gray to nearly white, exfoliating bark; lvs mostly reniform in outline, broadly 3–5 lobed, sharply serrate; stipules foliaceous, toothed or lobed; heads solitary, long-peduncled, 2.5–3 cm thick in fr; achenes 7–8 mm, nearly equaled by the subtending hairs; $2n=42$. Moist or wet, alluvial soil; sw. Me. to s. Mich. and se. Minn., s. to Fla. and Tex.
   P. × hybrida Brot., the London plane, a fertile putative hybrid of *P. occidentalis* and *P. orientalis* L., is widely planted as a street-tree and occasionally escapes. It typically has 2 heads per infl, and the lvs are often more truncate or broadly cuneate at the base; $2n=42$. (*P × acerifolia*.)

FAMILY HAMAMELIDACEAE, the Witch Hazel Family

Fls regular, perfect or unisexual, usually partly or wholly epigynous, 4–5-merous; cal small or none; pet present or absent; stamens commonly 4–5, sometimes ± numerous; carpels 2 in our genera, united into a compound, partly inferior, bilocular ovary with 1–several ovules per locule; fr a woody capsule, dehiscent across the top; seeds with large, dicotyledonous embryo and ± well developed endosperm; shrubs or trees with simple, alternate lvs and deciduous stipules. 26/100±.

1 Pet none, fls in dense heads; lvs palmately lobed . . . . 1. *Liquidambar*
1 Pet present, evident, linear; fls in small, axillary clusters; lvs toothed . . . . 2. *Hamamelis*

1. LIQUIDAMBAR L. Sweet gum, red gum. Monoecious, the fls unisexual, in dense heads, without perianth, but mingled with small scales; staminate heads ovoid, in a terminal raceme; stamens numerous; pistillate heads solitary, long-peduncled; pistillate fls consisting of a bilocular ovary with 2 long sytles; ovules orthotropous, several in each locule; frs capsular, beaked by the elongate styles, concrescent below into a globose head, each ovary with 1–2 seeds; trees with palmately veined and lobed lvs. 6, mainly N Hemisphere.

1. Liquidambar styraciflua. Tall tree, the smaller branches often with broad, corky wings; lvs long-petioled, rotund to reniform in outline, deeply 5-lobed, the lobes divergent, triangular, finely serrate; staminate infl 5–10 cm; mature pistillate heads brown, 3–4 cm thick including the indurate styles; $2n=32$. Moist or wet woods; Conn. to s. O., s. Ill. and Okla., s. to Fla. and Guatemala. May.

2. HAMAMELIS L.

Witch-hazel. Fls perfect, 4-merous, with linear pet and small, triangular sep; stamens 4, opposite and much shorter than the sep, alternating with scale-like staminodes; styles 2, short; ovule 1 per cell, suspended, anatropous; capsule half-inferior, pubescent; tall shrubs with toothed lvs, the fls in short-peduncled axillary clusters. 6, N. Amer. and e. Asia.

1. Hamamelis virginiana L. Shrub to 5 m; lvs broadly obovate or obovate–oblong, 5–15 cm, with several or many rounded teeth, inequilateral at the broadly rounded or subcordate base, often stellate beneath; pet spreading, 1.5–2 cm, yellow or seldom reddish; sep dull yellowish-brown within; fr ovoid or thickly ellipsoid, 1–1.5 cm, the hypanthium often bearing the persistent sep; $2n=24$. Moist woods; Que. and N.S. to n. Mich. and Minn., s. to Fla. and Tex. Oct., Nov., the fr ripe a year later.

Family ULMACEAE, the Elm Family

Fls regular or nearly so, hypogynous or perigynous, perfect or unisexual; sep (2–)5(–9), distinct or often connate below; pet none, stamens erect in bud, as many as and opposite the cal members, rarely 2 or 3 times as many; ovary superior, bicarpellate, unilocular; ovule solitary, suspended, anatropous or amphitropous; styles 2, stigmatic along the inner side; fr a samara or a small drupe or nut; seeds with dicotyledonous embryo and scanty or no endosperm; woody plants; lvs alternate, simple, distichous, often oblique at base; fls small and inconspicuous, anemophilous, in small cymes or racemes or fascicles, or solitary and axillary. 18/150+.

```
1 Fls perfect; fr a samara; venation strictly pinnate ............................................. 1. Ulmus.
1 Some of the fls staminate; fr a drupe or drupe-like nut, not winged.
    2 Ovary and fr stipitate, the fr strongly roughened; venation pinnate ................... 2. Planera.
    2 Ovary and fr sessile in the cal, smooth; leaves ± 3-veined at base ................... 3. Celtis.
```

1. ULMUS L.

Elm. Fls perfect, in short racemes or fascicles; cal campanulate, 4–9-lobed or cleft; anthers extrorse; styles short; fr a flat, 1-seeded samara, usually shortly stipitate and often surmounted by the persistent and sometimes enlarged styles; embryo straight; trees with pinnately veined, inequilateral, usually doubly serrate lvs; the small fls appear from buds resembling but a little larger than the lf-buds. 20, mainly N. Hemisphere.

Three cult. Old-World spp. that occasionally escape are included in the key but not described.

```
1 Fls appearing before the lvs in the spring; cal campanulate, merely lobed.
    2 Fls in fascicles; samara-wing glabrous except for the sometimes ciliate margins.
        3 Fascicles loose, the pedicels to 2 cm; fr ciliate; lvs not very rough ................. 1. U. americana.
        3 Fascicles dense, the fls subsessile; fr not ciliate.
            4 Native tree of characteristic arching form; lvs 10–20 cm, very rough above; fr pubescent over
              the seed; stamens 5–9 ............................................................. 2. U. rubra.
            4 Cult. trees, occasionally escaped, not arching; lvs mostly 2–10 cm; fr glabrous; stamens mostly
              3–5.
                5 Lvs nearly symmetrical, simply serrate (or nearly so), smooth above; Siberian or Chinese
                  .e. ............................................................................. U. pumila L.
                5 Lvs strongly asymmetrical at base, doubly serrate, rough above.
                    6 Base of the longer side of the lf forming a rounded lobe ± overlapping or concealing the
                      short petiole; seed central; Scotch e. ........................................ U. glabra Hudson.
                    6 Base of the lf not overlapping or concealing the petiole; seed above the middle of the fr;
                      English e. ..................................................................... U. procera Salisb.
    2 Fls in racemes; samara-wing pubescent on the sides, also ciliate-margined.
        7 Racemes 2–4 cm; fr 1.5–2 cm; well grown lvs 8–14 cm ............................... 3. U. thomasii.
        7 Racemes 1 cm or less; fr 1 cm; well grown lvs 4–7 cm ............................... 4. U. alata.
1 Fls and fr appearing in the autumn; cal divided to the base ............................... 5. U. serotina.
```

1. Ulmus americana L.   White or American e. Graceful, arching tree to 40 m, the twigs smooth or short-hairy; lf-buds glabrous or minutely puberulent; lvs ovate-oblong to somewhat obovate, mostly 8–14 cm, glabrous to scabrellous above; fls fascicled, on unequal pedicels to 2 cm; cal usually oblique, lobed; stamens 5–9; stigmas white; fr elliptic, 1 cm, densely ciliate, the sides glabrous and strongly reticulate; $2n=56$. Usually in moist, fertile soil; N.S. and Que. to Sask., s. to Fla. and Tex.

2. Ulmus rubra Muhl.   Slippery or red e. Tree to 20 m, with the shape of no. 1, the twigs scabrous-pubescent; winter-buds densely covered with red-brown hairs; lvs oblong to obovate, thick and stiff, usually 10–20 cm, very rough above; fls fascicled, subsessile; stamens 5–9; stigmas pink; fr suborbicular, 1.5–2 cm, pubescent over the seed, otherwise glabrous, scarcely reticulate; $2n=28$. Moist woods; s. Me. and s. Que. to e. N.D., s. to Fla. and Tex. (*U. fulva.*)

3. Ulmus thomasii Sarg.   Rock-e. Tree to 30 m, the thinly hairy twigs often becoming irregularly winged with 2 or more plates of cork after their second year; buds thinly hairy; lvs oblong to obovate, 8–14 cm, distinctly cordate on one side at base, glabrous above; fls in slender racemes to 4 cm; cal merely lobed; fr elliptic, 1.5–2.3 cm, pubescent on the sides, the stipe 1 mm; $2n=28$. Rich upland woods; sw. Que. to Minn., s. to N.J., W.Va., w. Tenn., and n. Ark. (*U. racemosa* Thomas.)

4. Ulmus alata Michx.   Winged e. Tree to 15 m, the thinly hairy twigs commonly developing 2 opposite plates of cork, beginning the first or second season; buds glabrous or finely pubescent; lvs oblong or somewhat obovate, mostly 4–7 cm, less than half as wide, scarcely cordate at base, glabrous to scabrellous above; fls in short racemes seldom 1 cm; fr narrowly ovate, 1 cm incl. the 2 mm stipe, ciliate, pubescent on the sides; $2n=28$. Moist or dry uplands; se. Va. to s. Ind., s. Ill., and Mo., s. to Fla. and Tex.

5. Ulmus serotina Sarg.   September e. Tree to 20 m; lvs oblong to oval or oblong-obovate, usually 6–10 cm, glabrous above; fls in short racemes, appearing in the fall; cal divided to the base into oblanceolate segments; fr elliptic, long-ciliate, 1–1.5 cm, the stipe 2 mm, the incurved teeth 2 mm; $2n=28$. Upland woods; irregularly from Ky. and s. Ill. to nw. Ga., Ala., Miss., Ark., and e. Okla.

2. PLANERA J. F. Gmelin.   Planer-tree. Polygamo-monoecious, the staminate fls in small fascicles from winter-buds, the fertile ones on short, slender pedicels from the axils of lvs of the season, sometimes lacking stamens; cal campanulate, with 4–5 rounded lobes; anthers extrorse; ovary short-stipitate; styles stout, divergently recurved; fr ellipsoid, drupe-like; embryo straight; woody plants with pinnately veined lvs. Monotypic.

1. Planera aquatica (Walter) J. F. Gmelin.   Tall shrub or small tree to 10 m; lvs lance-ovate to oblong or ovate, usually 3–7 cm, rounded at base, glabrous to minutely pubescent; staminate fls barely protruding from the bud-scales; pistillate fls 1–3 from each axil; fr conspicuously stipitate above the persistent cal, the endocarp thin and papery, the exocarp raised into longitudinal ridges bearing soft, finger-like projecting processes 1–3 mm. Swamps; Fla. to Tex., n. to w. Ky., s. Ill. and s. Mo. Fls with the lvs, April.

3. CELTIS L.   Hackberry, sugarberry. Monoecious or polygamous; staminate fls in small clusters near the base of the twigs of the season; fertile fls solitary or sometimes paired, from the upper axils of the same twigs, often lacking stamens; cal deeply 5-parted; anthers introrse; style very short, with elongate, recurved-divergent stigmas; fr a drupe with thin, ± sweet pulp and a hard stone; embryo curved; woody pls with entire or serrate lvs 3-nerved from a ± oblique base; fls ephemeral, borne in early spring. 70, mainly N. Hemisphere.

1 Lvs lanceolate, ± long-acuminate, entire or rarely with a few teeth; style promptly deciduous .... 1. *C. laevigata.*
1 Lvs either distinctly broader than lanceolate, or evidently serrate, or both; style ± persistent.
    2 Lvs evidently serrate to well below the middle; fruiting pedicels surpassing the petioles .... 2. *C. occidentalis.*
    2 Lvs entire, or with a few scattered teeth above the middle; fruiting pedicels about as along as the
        petioles .... 3. *C. tenuifolia.*

1. Celtis laevigata Willd.   Southern h. Tree to 30 m; lvs yellowish-green, lanceolate, 4–8 cm, ± long-acuminate, entire or seldom with a few long teeth, broadly cuneate to rounded at base, glabrous or nearly so except for the ciliate margins; style promptly deciduous after flowering; fr dark orange-red, 5–9 mm, the pedicel 3–6 mm, equaling or shorter than the subtending petiole; $2n=20$. Wet grounds; Fla. to Tex. and ne. Mex., n. to s. Mo., s. Ind., and s. Va. (*C. mississippiensis; C. smallii.*)

2. Celtis occidentalis L.   Northern h. Commonly 6–15(35) m, tending to form a symmetrical, openly branched tree; lvs dark green above, paler beneath, lance-ovate to broadly ovate or deltoid, those of the fertile branches 6–12 cm, conspicuously serrate, abruptly acuminate to very long-acuminate, oblique at the base, one side (seldom both) generally ± strongly cordate, varying from thin and smooth to firm, impressed-veiny and scabrous above, and pubescent beneath, as in the next sp.; major areoles (between the primary lateral veins) usually 5–8 on each side; style tardily deciduous, ± persistent on the developing fr; fr ellipsoid to subglobose, 7–13 mm, dark red to nearly black, sweet and tasty, on a pedicel (7–)10–20(–25) mm that usually notably surpasses the subtending petiole; stone conspicuously pitted. Usually in moist, rich soil, often on floodplains; s. Que. to s. Man., s. to Va., Ark., and Okla., and locally to N.C., Ga., Ala., and Miss. Highly variable, but not clearly divisible into vars. (*C. canina; C. crassifolia.*)

3. **Celtis tenuifolia** Nutt. Dwarf h. Shrub or small, irregularly and compactly branched tree to 5 m (rarely more); lvs firm, dark green above, paler beneath, ovate or broadly ovate to deltoid, mostly 3–6 cm, entire or with a few low teeth above the middle, scabrous and impressed-veiny above, pubescent beneath and on the petiole, only slightly cordate at the oblique to nearly symmetrical base; major areoles mostly 3–5 on each side, style tardily deciduous; fr subglobose, 5–9 mm, salmon-colored, insipid, on a pedicel 3–6 mm that is up to about as long as the subtending petiole; stone shallowly and obscurely pitted. Rocky hills and barrens, sometimes on dunes; Fla. to Tex., n. to N.J., Ind., Mo., and Okla., and locally to s. Mich. and s. Ont. (*C. georgiana; C. occidentalis* var. *pumila.*)

Family **CANNABACEAE**, the Indian Hemp Family

Fls unisexual, small and inconspicuous, the male with 5 erect stamens on short or very short filaments opposite the 5 sep, the female with a short, ± entire, membranous cal enclosing the ovary, or the cal often much reduced in *Cannabis*; pet none; ovary unilocular, with a single pendulous, anatropous ovule; style short, with 2 elongate, filiform stigmas; fr a small nut or an achene, in *Humulus* invested by the persistent cal; seeds with curved or coiled, dicotyledonous embryo and scanty endosperm; erect or twining herbs without milky juice; lvs opposite at least below, palmately lobed or compound (except sometimes the upper); infls axillary to the upper (often reduced) lvs, the male relatively loose, branched, and many-fld, the female more compact and few-fld. 2/4.

1 Stems twining, retrorsely spinulose; principal lvs merely lobed .... 1. *Humulus.*
1 Stems erect, not spinulose; principal lvs palmately compound .... 2. *Cannabis.*

1. **HUMULUS** L. Hops. Dioecious; pistillate fls in short spikes, paired, each pair subtended by a foliaceous bract; ovary enclosed by the membranous, lobeless cal; fr an achene enclosed in the persistent cal and covered by the accrescent bracts; twining, herbaceous, perennial vines with rough stems and broad, opposite, usually lobed lvs. 3, N. Temp.

1 Principal lvs 3-lobed; bracts of the pistillate spikes entire; perennial .... 1. *H. lupulus.*
1 Principal lvs 5–9-lobed; bracts of the pistillate spikes spinulose-ciliate; annual .... 2. *H. japonicus.*

1. **Humulus lupulus** L. Hops. Rhizomatous perennial vine to 10 m; principal lvs as broad as long, cordate at base, 3-lobed to below the middle, the upper lvs often broadly ovate and lobeless; petiole shorter than the blade; staminate infls 5–15 cm; pistillate spikes 1 cm, becoming 3–6 cm, the persistent, accrescent bracts entire and mostly blunt, stramineous, very glandular at base; 2n=20. Moist soil; N.S. to Man., Mont., and Calif., s. to N.C., Ark., and Ariz., and widely distributed in the Old World. July, Aug. 5 geographically significant vars., 3 in our range.

1 Plants less hairy, the central lobe of most lvs with fewer than 20 hairs per cm along the midrib beneath, and with fewer than 25 glands per 10 sq. mm between the veins beneath; native to Europe, cult. and escaped in our range .... var. *lupulus.*
1 Plants more hairy, the central lobe of the lvs with more than 20 hairs per cm along the midrib beneath, and with more than 25 glands per 10 sq. mm between the veins; native.
　2 Lvs with hairs between the veins beneath, and with more than 100 hairs per cm of length along the midrib beneath; smaller lvs usually unilobed, midwestern .... var. *pubescens* E. Small.
　2 Lvs lacking hairs between the veins beneath, and with fewer than 100 hairs per cm of length along the midrib beneath; smaller lvs mostly 3-lobed; widespread e. of the Rocky Mts. (*H. americanus?*) .... var. *lupuloides* E. Small.

2.. **Humulus japonicus** Siebold & Zucc. Japanese h. Weedy annual vine; stem and lvs much rougher than in no. 1; petioles commonly longer than the 5–9-lobed blades; uppermost lvs usually 3–5-lobed; mature pistillate spikes to 2 cm, their bracts relatively small, under 1 cm, densely spinulose-ciliate, dull green, conspicuously and abruptly acuminate, scarcely glandular, releasing the ripe achenes; 2n=32 (pistillate), 34 (staminate). Native of e. Asia, escaped from cult. from N. Engl. to N.C. and Mo. July–Oct.

2. CANNABIS L. Hemp, Indian Hemp. Ordinarily dioecious; pistillate fls in small clusters on short, leafy branches from the upper axils, each fl semi-enclosed by a closely subtending, abruptly acuminate bract; cal short and barely lobed, or sometimes longer but hyaline and becoming adnate to the pericarp; fr a thick-lenticular achene, with a closely subtending accrescent bract; erect annual herbs; lvs petiolate, the principal ones palmately compound with 5–9 lfls, commonly opposite, the upper smaller, usually alternate, with fewer lfls or undivided. Monotypic.

1. **Cannabis sativa** L. Stem 1–3 m; lfls lance-linear to lance-elliptic, pubescent, 5–15 cm, toothed; achene in ours mostly 3.8–4.5 mm. Native of Eurasia, cult. in many parts of the world, and found here and there in our range. Late summer. Complex sp., differentiated under the influence of cult. into a northern subsp. *sativa,* used for production of fiber (hemp) from the bast, and a southern subsp. *indica* (Lam.) E. Small & Cronq., harvested for an intoxicating, hallucinogenic drug in the epidermal resin glands. The crude flowering or fruiting material is marijuana; the extracted resin is hashish. Plants growing wild in our range are all or nearly all descended from plants formerly cult. for fiber, and the resin has a relatively small percentage of intoxicant. They may be referred to subsp. *sativa* var. *sativa.* Our plants appear to be in process of reversion toward the wild-adapted var. *spontanea* Vavilov (*C. ruderalis*), marked by its relatively small size, and more especially by its smaller (less than 3.8 mm), irregularly blotched achenes with a variable period of dormancy. Parallel vars. adapted to growth in cult. or in the wild exist in subsp. *indica.*

FAMILY **MORACEAE**, the Mulberry Family

Fls unisexual; cal of (2–)4–5(6) distinct or ± connate sep; pet none; stamens usually as many as and opposite the sep, rarely fewer or only one, often incurved in bud; ovary superior to inferior, bicarpellate, or one carpel ± reduced or even suppressed, the ovary bilocular or much more often unilocular; style or style-branches stigmatic along the inner surface, generally 2, sometimes one of them ± reduced or even wholly suppressed; ovules solitary in each locule (or one locule empty), generally pendulous and anatropous to hemitropous or campylotropous; frs generally fleshy, often connate (or ripening collectively with the receptacle) to form a syncarp; seeds with straight or more often curved embryo, the cotyledons often unequal; endosperm usually present and fatty; woody (rarely herbaceous) plants with milky juice, opposite or alternate, simple or compound lvs and small or minute fls crowded in dense clusters or heads. 40/1000.

1 Lvs entire, pinnately veined; stems usually thorny .... 1. *Maclura.*
1 Lvs toothed or lobed, palmately veined; stems unarmed.
  2 Style undivided; pistillate catkins and fr globose .... 2. *Broussonetia.*
  2 Style deeply divided; pistillate catkins and fr short-cylindric .... 3. *Morus.*

1. MACLURA Nutt., nom. conserv. Osage-orange. Dioecious; staminate fls numerous in loose, globose to oblong, peduncled axillary heads, with 4-parted cal and 4 stamens; pistillate fls coherent in dense, globose, axillary heads, the 4-lobed cal closely surrounding the ovary, the style filiform and elongate; fr a hard, globose syncarp, composed of the enlarged common receptacle and calyces, completely concealing the achenes; trees, usually thorny, with alternate, entire lvs. 11, mainly warm reg.

1. **Maclura pomifera** (Raf.) C. K. Schneider. Tree to 20 m; lvs lance-ovate, 6–12 cm, acuminate, petiolate; thorns stout, 1–2 cm; fr yellowish-green, 6–12 cm thick, the surface convoluted. Rich, moist soil; Ark., Okla., and Tex., occasionally adventive or persistent after cult. with us. May, June. (*Toxylon p.*)

2. BROUSSONETIA L'Hér., nom. conserv. Paper-mulberry. Dioecious; staminate fls in cylindric catkins, with deeply 4-lobed cal and 4 stamens; pistillate fls in dense, globose heads, with 4-lobed cal and one long, exserted style; fr globose, dense, consisting chiefly of the accrescent calyces, from which the orange achenes protrude; trees or shrubs with serrate or lobed, alternate lvs. 7, e. Asia, Polynesia.

1. **Broussonetia papyrifera** (L.) Vent. Tall shrub or small tree with smooth bark and pubescent twigs; lvs firm, broadly ovate, serrate and often lobed, scabrous above, densely pubescent beneath; staminate catkins

slender, 5 cm; pistillate catkins 1–2 cm thick; fr 2–3 cm thick; $2n=26$. Native of China and Japan, occasionally escaped from cult. from N.Y. to Mo. and s. Apr., May. (*Papyrius p.*)

3. MORUS L. Mulberry. Monoecious or dioecious; fls in cylindric catkins, the staminate longer and more loosely flowered than the pistillate; cal deeply 4-parted; stamens 4; style deeply 2-parted; fr a short-cylindric, edible syncarp resembling a blackberry, composed of juicy, accrescent but scarcely coherent calyces, each enclosing a small, seed-like achene, with the remains of the style protruding; trees with alternate, serrate or lobed (often mitten-shaped), palmately veined lvs. 10, widespread.

1 Lvs pubescent beneath, scabrous or glabrous above .... 1. *M. rubra.*
1 Lvs glabrous, or sparsely pubescent along the larger veins beneath .... 2. *M. alba.*

1. Morus rubra L. Red m. Forest tree to 20 m, with dark, scaly bark; lvs thin, broadly ovate, obovate, or subrotund, coarsely serrate, sometimes mitten-shaped or even 2–4-lobed, abruptly acuminate into a conspicuous point to 4 cm, glabrous or scabrous above, softly pubescent beneath; fr dark purple, 2–3 cm; $2n=28$. Rich woods, often on flood-plains; s. Vt. to se. Minn. and e. Neb., s. to Fla. and Tex. Fr late June, July.

2. Morus alba L. White m. Tree to 15 m; lvs glabrous or nearly so, or with white, spreading hairs along the veins beneath, acute or short-acuminate, otherwise as in no. 1; fr white, pink, or pale purple to nearly black; $2n=28$. Native of e. Asia, often escaped from cult. in our range.

Family URTICACEAE, The Nettle Family

Fls unisexual or rarely perfect; cal 3–5-lobed or parted, usually more deeply so in the staminate fls, or the sep distinct; pet none; stamens as many as the cal-lobes and opposite them, inflexed in bud, elastically reflexed when the pollen is shed; ovary superior, unilocular, with a single erect, orthotropous ovule from the base; style 1; fr usually an achene, often enclosed in an accrescent cal; seeds with straight, dicotyledonous embryo and usually a thin endosperm; herbs (ours) with opposite or alternate, simple, usually (except in *Parietaria*) stipulate lvs and axillary (in ours), simple or branched clusters of inconspicuous small fls. 45/700.

1 Plants with obvious stinging hairs.
  2 Lvs opposite .... 1. *Urtica.*
  2 Lvs alternate .... 2. *Laportea.*
1 Pls without stinging hairs, or these minute and sparse.
  3 Lvs normally opposite.
    4 Fls in axillary spikes; achene completely enclosed by the cal .... 3. *Boehmeria.*
    4 Fls in axillary panicles or glomerules; achene exceeding the cal .... 4. *Pilea.*
  3 Lvs alternate .... 5. *Parietaria.*

1. URTICA L. Nettle, stinging nettle. Dioecious or monoecious; male fls with 4 subequal sep, 4 stamens, and vestigial ovary; sep of the female fls unequal, the outer 2 small and inconspicuous; stigma capitate-tufted; fr a lenticular achene enclosed by the 2 inner sep; herbs with opposite lvs, ± beset with stinging bristles; fls minute, in panicles, spikes, or head-like clusters from the upper axils. 25, widespread.

1 Rhizomatous perennial; stipules 5–15 mm, erect; infls usually surpassing the petioles .... 1. *U. dioica.*
1 Annual; stipules 1–3 mm, spreading or deflexed; infls usually shorter.
  2 Upper lvs reduced; lf-teeth commonly blunt, with convex sides .... 2. *U. chamaedryoides.*
  2 Upper lvs scarcely reduced; lf-teeth commonly sharp, with straight sides .... 3. *U. urens.*

1. Urtica dioica L. Erect, rhizomatous perennial, usually simple, to 2 m; lvs serrate, acute or acuminate, 5–15 cm; stipules lance-linear, 5–15 mm; infls branched, many-fld, commonly exceeding the subtending petioles; fruiting cal 2 mm, pubescent; achene ovate, 1.5 mm. Nearly cosmopolitan, either as a native or an intr. sp., and highly variable. June–Sept. Var. *dioica*, native to Europe but well established in our range, is dioecious, weak-stemmed, and rather densely hairy, with mostly broadly ovate, cordate-based lvs with stinging hairs on both sides, and with the teeth commonly 5–6 mm deep; $2n=52$. Var. *procera* (Muhl.) Wedd., the native phase, is usually monoecious, stouter and more sparsely hairy, with ovate to lance-oblong lvs that are seldom cordate at the base, with the stinging hairs commonly confined to the lower surface, and with the teeth averaging 2–3.5 mm deep; $2n=26$ in ours. (*U. gracilis; U. viridis.*)

2. Urtica chamaedryoides Pursh. Fibrous-rooted or weakly taprooted annual, 3–8 cm, branched from the base, subglabrous except for a few stinging hairs; stipules oblong, 1–3 mm; lvs crenate-serrate, the lower broadly ovate to subrotund, the upper progressively narrower and smaller; fl-clusters subglobose to short-spicate, shorter than the petioles; achenes ovate, 1–1.5 mm, less than 1 mm wide. Moist woods and thickets; W.Va., s. O., and Ky. to Mo. and Okla., s. to Ala. and Mex. Spring and summer.

3. Urtica urens L. Taprooted annual, simple or branched, 2–5 dm, with numerous stinging hairs; lvs long-petioled, elliptic to broadly ovate and blunt or rounded in outline, deeply incised-serrate, the teeth triangular, acute, upper lvs usually larger than the lower; stipules oblong, 1–3 mm, fl-clusters oblong, often shorter than the petioles; achenes triangular, 1.5–2.5 × 1–1.5 mm; $2n=24, 26, 52$. Native of Europe, widely distributed elsewhere as a weed, and occasionally found in our range. May–Sept.

2. LAPORTEA Gaudich., nom. conserv. Monoecious; staminate fls with 5 distinct sep and 5 stamens; sep of the pistillate fls 4, the outer pair much smaller, or 3 or 2 by abortion; style elongate, pubescent, persistent; fr an oblique achene, much surpassing the 2 persistent sep; herbs, shrubs, or trees with small fls, often in large clusters. 25, widespread.

1. Laportea canadensis (L.) Wedd. Fibrous-rooted, often rhizomatous perennial herb, 5–10 dm, often flexuous; lvs alternate, broadly ovate, 8–15 cm, acuminate, coarsely serrate, hairy; staminate fls in cymes from the lower axils, seldom surpassing the petioles; pistillate fls in loose, elongate, spreading, divaricately branched cymes from the upper axils, usually much surpassing the petioles; style 3–4 mm, flat, D-shaped, with reflexed style; $2n=26$. Rich, moist woods; N.S. to Man., s. to Ga. and Okla. July, Aug (*Urticastrum divaricatum.*)

3. BOEHMERIA Jacq. Monoecious or dioecious; male fls minute, with 4-parted cal and 4 stamens; female fls with tubular, flattened to ovoid cal minutely 2–4-toothed at the summit; ovary enclosed by the cal; style filiform, exsert; fr an achene surrounded by the accrescent cal, the latter narrowly 2 winged; style persistent; herbs, shrubs, or small trees with opposite or alternate lvs and axillary infls. 100, warm reg.

1. Boehmeria cylindrica (L.) Swartz. False nettle. Perennial, usually dioecious herb of nettle-like aspect, but without stinging hairs; stems erect, usually simple, 4–10 dm; lvs opposite, long-petioled, ovate to lance-ovate, acute or acuminate, coarsely serrate, 3-nerved; fls minute, in small, capitate clusters along an un-branched axis, forming interrupted or continuous spikes from the upper axils; fr ovate, minutely winged. Moist or wet soil; Que. and Ont. to Minn., s. to Fla. and N.M. July, Aug. (*B. drummondiana*, the sun-form, with smaller, firmer, more hairy lvs.)

4. PILEA Lindl., nom. conserv. Clearweed. Monoecious or dioecious; cal of the male fls deeply 4-parted; stamens 4; cal of the female fls deeply 3-parted, the segments often unequal or gibbous or hooded; staminodes minute, scale-like; stigma sessile; fr a com-pressed, thin-walled achene, loosely subtended by the persistent cal; herbs with opposite lvs, inconspicuous, connate stipules, and axillary cymes of small, greenish fls, the numerous cystoliths appearing as minute, whitish lines on the surface of the lf in herbarium-speci-men; blades with scattered, minute stinging hairs. 200, mainly trop.

1 Mature achenes green to yellowish, usually marked with purple on the sides . . . . 1. *P. pumila.*
1 Mature achenes dark purple or olivaceous, with narrow pale margin . . . . 2. *P. fontana.*

1. Pilea pumila (L.) A. Gray. Annual, very smooth and pellucid, 1–5 dm; lvs long-petioled, shining, ovate, 3–12 cm, serrate or crenate-serrate, broadly cuneate to rounded at base, usually glabrous, 3-nerved from the base, the nerves narrowly winged; terminal lf-tooth elongate; cymes 5–30 mm, from the middle and upper axils; cal-lobes equal or nearly so; achenes pale green, usually with slightly raised purple spots, 1.3–2 mm, 55–70% as wide as long; seed smooth. Moist, rich, shaded soil, often in colonies; Que. to Minn. s. to Fla., La., and Okla. July–Sept. (*Adicea p.*) Northeastern plants have the lvs sharply cuneate at base with 11 or fewer rounded teeth on each side. Toward the s. and w. the lf-base tends to be more rounded and the teeth sharper and more numerous; this poorly defined phase has been called var. *deamii* (Lunell) Fern.

2. Pilea fontana (Lunell) Rydb. Much like no. 1, the herbage less pellucid; achenes dark purple or dark olivaceous to nearly black, pale-margined, irregularly elevated on the sides, broadly ovate, 1.3–2 mm, 70

85% as wide; seed tuberculate. Wet, boggy or springy soil; P.E.I. to Minn. and N.D., s. to Va., Ind., and Nebr.; Fla. July–Sept. (*P. opaca.*)

5. PARIETARIA L. Pellitory. Monoecious or polygamous; staminate fls with deeply 4-parted cal and 4 stamens; cal of the pistillate fls tubular at base, 4-lobed; ovary compressed, ovoid; stigma subsessile; fr an achene, loosely enclosed by the accrescent cal; herbs with alternate, entire, exstipulate lvs and short, axillary clusters of few fls subtended and exceeded by green bracts. 30, widespread.

1. **Parietaria pensylvanica** Muhl. Pubescent annual, erect or ascending, 1–4 dm; lvs petioled, thin, lanceolate, 3–8 cm, acuminate, entire, 3-nerved from above the cuneate base, obscurely scabrellate above; fls from the middle and upper axils; bracts lance-linear, 5 mm; achenes smooth, shining, ovoid, 1 mm, somewhat 2-edged; $2n=14$. Dry woods and moist banks, sometimes weedy; Me. to B.C., s. to Va., Ala., Tex., and Nev. June–Sept.

P. nummularia Small, a prostrate or ascending small plant with reniform or suborbicular to rhombic-ovate leaves 0.5–1 cm wide, native to se. U.S., has been reported as a weed from Wilmington, Del.

Reports of *P. floridana* Nutt., another southern sp., with the lvs 3-nerved from the truncate or rounded base, may be based on small plants of *P. pensylvanica.*

Family **JUGLANDACEAE**, the Walnut Family

Fls tiny, unisexual, borne in unisexual (seldom bisexual) catkins, the bracts, bractlets, and tepals variously connate and adnate and subject to diverse morphological interpretations,; staminate catkins commonly (including our genera) elongate, drooping, borne on the twigs of the previous year, or at the base of the current year's growth; staminate fls solitary in the axils of the bracts of the catkin, each with usually 4 tepals adnate to the 2 bractlets, with which they form a common structure, or the tepals (or both tepals and bractlets) obsolete; stamens 2–many; filaments short; pistillate fls mostly in short, few-fld catkins terminating the twigs of the season, or solitary and terminal, each fl subtended by a variously developed set of ± connate bracteoles and a single primary bract, these collectively often forming a minute, perianth-like involucre; ovary inferior, with 1–4 calyx-teeth toward the summit (or the calyx-teeth obsolete, as in *Carya*), mostly bilocular (or 4-locular by the intrusion of "false" partitions) below, unilocular above, the partition not reaching the top of the cavity; styles 2, distinct or basally connate, often with plumose stigma; ovule solitary, erect, orthotropous, borne at the summit of the partition; fr a rather large, sometimes samaroid nut, or ± drupe-like, the softer, fibrous husk (representing the ripened involucre) sometimes splitting to release the hard nut; embryo large, dicotyledonous; endosperm none; aromatic, anemophilous trees or shrubs with mostly alternate, pinnately compound, exstipulate lvs (ours deciduous) that are provided with small, resinous, peltate, basally embedded gland-scales. 7–8/60.

1 Pith transversely partitioned; husk of the fr indehiscent; median lateral lfls the largest .... 1. *Juglans.*
1 Pith not partitioned; husk dehiscent in 4 valves; terminal lfls usually the largest .... 2. *Carya.*

1. JUGLANS L. Staminate catkins protruding from the buds in autumn, elongating in spring, densely fld, pendulous; bract adherent to the "perianth," except at its summit; "perianth" spreading, 3–6-lobed, with 8–40 stamens on its upper side; anthers glabrous; pistillate fls in short spikes terminating the branches, closely subtended by a 3-lobed, cup-shaped involucre that ripens with the fr to form a husk; perianth minute, 4-parted; husk indehiscent, clammy-glandular; nut indehiscent but ± distinctly 2-valved; trees with partitioned pith and odd-pinnate lvs, the median-lateral lfls the largest, inequilateral or falcate; lfls with conduplicate vernation. 20, widespread.

J. regia L., the English walnut, rarely escapes from cult. It has 7–9 glabrous, entire lfls.

1 Pith dark chocolate-brown; bark with smooth ridges; fr oblong-ovoid, somewhat pointed .... 1. *J. cinerea.*
1 Pith tan to creamy; bark with very rough ridges; fr mostly subglobose .... 2. *J. nigra.*

1. Juglans cinerea L. Butternut. Tree to 30 m; bark grayish-brown, with smooth ridges; pith dark brown; a dense pad of short hairs often present long the upper margin of old lf-scars; lfls 11–17, oblong-lanceolate, acuminate; pubescence, especially of the lower lf-surface, largely or wholly of stellate, few-rayed hairs; fr ovoid-oblong, 4–7 cm, somewhat pointed; nut ovoid to short-cylindric, longer than thick, very rough, marked with 2 or 4 obscure longitudinal ridges. Rich, moist soil; N.B. to Minn., s. to S.C., Ga., and Ark. (*Wallia c.*)

2. Juglans nigra L. Black walnut. Tree to 40 m; bark nearly black, with rough ridges; pith rather light brown; upper margin of lf-scars glabrous; lfls 11–23, oblong-ovate, acuminate; stellate hairs few or none; fr subglobose, not pointed, 5–8 cm thick; nut commonly subglobose and slightly flattened, very rough, rather distinctly 2-valved. Rich, moist soil; Vt. to Minn. and S.D., s. to Ga., the Fla. panhandle, and Tex. (*Wallia n.*)

2. CARYA Nutt., nom. conserv. Hickory. Staminate catkins slender, elongate, borne in peduncled groups of 3 at the summit of the previous year's growth or at the base of that of the current year; "perianth" 2–3-lobed, closely subtended and usually surpassed by the bract; stamens 3–10, commonly 4; anthers pilose; pistillate fls solitary or in spikes of 2–10, terminating the branches, each closely subtended by a cup-shaped, 4 lobed, perianth-like involucre that ripens with the fr to form a husk; tepals obsolete, husk ± dehiscent into 4 valves, releasing the hard shelled nut; trees with hard, heavy wood, continuous pith, and odd-pinnate lvs, the 3 terminal lfls usually the largest; lfls with involute vernation. (*Hicoria.*) 15, e. N. Amer. and e. Asia.

1 Scales of the terminal bud 4–6, valvate; fr with narrowly winged sutures.
  2 Nut (and fr) elongate, cylindric; kernel edible . . . . 1. *C. illinoinensis.*
  2 Nut (and fr) flattened, as broad as long, angular; kernel bitter.
    3 Winter-buds bright orange-yellow; husk splitting to the middle; widespread . . . . 2. *C. cordiformis.*
    3 Winter-buds brown; husk splitting to the base; southern . . . . 3. *C. aquatica.*
1 Scales of the terminal bud more than 6, imbricate; fr usually wingless.
  4 Fr (except in no. 6) 3.5–7 cm, the husk mostly 4–8 mm thick, splitting to the base; terminal buds
    10–25 mm, bearing some straight, appressed hairs.
    5 Lvs permanently soft-hairy beneath; lfls mostly 7 or 9.
      6 Nuts strongly compressed, 3–6 cm, ± wedge-shaped at base . . . . 4. *C. laciniosa.*
      6 Nuts terete or slightly compressed, 1.5–3 cm, rounded at base . . . . 5. *C. tomentosa.*
    5 Lvs at maturity glabrous or nearly so beneath, the hairs mostly limited to the larger veins
      and vein-axils; lfls mostly 5 or 7.
      7 Nut prominently angled from base to summit; lfls most often 5.
        8 Fr 2.5–3.5 cm; terminal lfl oblanceolate, a fourth as wide as long . . . . 6. *C. carolinae-septentrionalis.*
        8 Fr 3.5–5 cm; terminal lfl obovate, a third to half as wide as long . . . . 7. *C. ovata.*
      7 Nut angled only above the middle; lfls most often 7 . . . . 8. *C. ovalis.*
  4 Fr 2–3.5 (–4) cm, the husk 2–4 mm thick; terminal buds rarely over 10 mm, resinous, without
    straight, appressed hairs except for marginal cilia.
    9 Mature bark separating into long plates; nut compressed, angled above; petiole, rachis, and
      midvein nearly or quite glabrous at maturity . . . . 8. *C. ovalis.*
    9 Mature bark rough, furrowed; nut (except in *C. pallida*) only slightly or not at all compressed,
      not strongly angled.
      10 Outer bud scales sparsely resinous-dotted; husk splitting to the middle or not at all; kernel
        usually astringent, scarcely edible; widespared . . . . 9. *C. glabra.*
      10 Outer bud scales covered with small, yellowish scales; husk splitting to the base; kernel
        edible; southern.
        11 Petiole, rachis, and midvein nearly or quite glabrous at maturity; nut light brown,
          scarcely angled . . . . 10. *C. texana.*
        11 Petiole, rachis, and midvein beneath copiously, loosely, and ± persistently hairy; nut
          white, prominently angled . . . . 11. *C. pallida.*

1. Carya illinoinensis (Wangenh.) K. Koch. Pecan. Bark deeply furrowed; lfls 11–17, oblong-lanceolate, the lateral conspicuously falcate, the terminal commonly on a stalk 2–4 mm; frs in spikes of 3–10, ellipsoid or cylindric, 3–5 cm, narrowly winged to the base; nut ellipsoid or cylindric, terete, 2.5–4 cm, brown, smooth, short-pointed; kernel edible, each half barely notched at the tip; $2n=32$. Wet alluvial forests; sw. O. to Io. and e. Kans., s. to Ala., Tex., and N. Mex. (*C. pecan.*) *C.* × *brownii* Sarg. is a hybrid with *C. cordiformis*; *C.* × *nussbaumeri* Sarg., with *C. laciniosa*; *C.* × *lecontei* Little., with *C. aquatica*; and *C.* × *schneckii* Sarg., with *C. tomentosa.*

2. Carya cordiformis (Wangenh.) K. Koch. Bitternut-h. Bark scaly; winter-buds bright orange-yellow; lfls mostly (5)7–9(11), the lateral lanceolate to lance-ovate or elliptic, the terminal commonly long-cuneate at base and sessile or nearly so; staminate catkins clustered just beneath the new growth on the wood of the previous year's twigs; fr obovoid to subglobose, often somewhat flattened, 2.5–3.5 cm, winged chiefly above the middle, splitting to about the middle; nut subglobose to obovoid, 1.5–3 cm, at least two-thirds as thick obscurely angled, otherwise smooth, tipped with a slender, persistent point; kernel bitter; cotyledons bifid $2n=32$. Woods; s. Me. and sw. Que. to Minn. and e. Neb., s. to Fla. and Tex. *C.* × *laneyi* Sarg. is a hybric with *C. ovata.*

3. **Carya aquatica** (Michx. f.) Nutt. Water-h. Bark separating into long scales; winter-buds brown; lfls (7)9–11(–15), lanceolate, the terminal long-cuneate at base and sessile or nearly so; frs 1–4, strongly flattened, ovoid or obovoid, 2.5–3.5 cm, narrowly winged, splitting to the base at maturity; nut obovoid to nearly discoid, 1.5–3 cm, half as thick as wide, strongly wrinkled, short-pointed; kernel bitter; cotyledons bifid; $2n=32$. Swamps and flood-plains; se. Va. to Fla. and Tex., and n. inland to s. Ill. and se. Mo.

4. **Carya laciniosa** (Michx. f.) Loudon. Shellbark-h. Bark as in no. 7; twigs and branchlets notably stout; buds large; lfls mostly 7–9, permanently pubescent beneath with fascicled hairs, the terminal oblanceolate to obovate, the lateral very inequilateral at base; fr ellipsoid to subglobose, 3.5–7 cm, its husk 6–12 mm thick, eventually splitting to the base; nut strongly compressed, 3–6 cm, prominently angled, somewhat wedge-shaped at base, very thick-walled; $2n=32$. Flood plains; N.Y. and s. Ont. to Io. and e. Kans., s. to N.C., Ga., Miss., and e. Okla.

5. **Carya tomentosa** (Poiret) Nutt. Mockernut-h. Bark gray, becoming dark gray and deeply furrowed in age, not separating into plates; lfls mostly 7 or 9, persistently pubescent beneath with fascicled hairs, the terminal oblanceolate ot obovate; fr ellipsoid to obovoid, rarely ovoid, 3.5–5 cm, its stony husk 2–6 mm thick, eventually splitting to the base; nut 1.5–3 cm, slightly compressed to subterete, ± angled, subglobose or ovoid to obovoid, rounded at base, thick-walled; kernel edible; $2n=64$. Dry woods; Mass. to Ind., se. to Io. and Mo., s. to Fla. and Tex. (*C. alba*.)

6. **Carya carolinae-septentrionalis** (Ashe) Engler & Graebner. Carolina shagbark-h. Much like no. 7; twigs very slender, with a small black bud, terminal lfl lanceolate or oblanceolate, a fifth to a fourth as wide as long, little different from the adjacent ones; fr and nut only two-thirds as large as in no. 7; $2n=32$. Upland, often calcareous woods; s. Va. to Tenn. and Ga. Perhaps better treated as *C. ovata* var. *australis* (Ashe) Little.

7. **Carya ovata** (Miller) K. Koch. Shagbark-h. Bark light gray, soon separating into long plates; twigs stout, with a large gray bud; lfls 5, or 7 on sprouts, pubescent beneath when young, soon glabrate except for subapical tufts of hairs on the teeth, the terminal obovate, much larger and proportionately wider than the lateral; fr subglobose to broadly obovoid, mostly 3.5–5 cm; husk 3–12 mm thick, eventually splitting to the base; nut compressed, 2–3 cm, two-thirds as wide or wider, rounded at base, usually sharp-pointed; kernel sweet and edible; $2n=32$. Rich, moist soil; s. Me. and s. Que. to se. Minn. and w. Neb., s. to Ga. and e. Tex.

8. **Carya ovalis** (Wangenh.) Sarg. Sweet pignut-h., red h. Bark light or dark gray, eventually separating into fairly long plates; buds glabrous to pubescent, resinous, usually with some slender, appressed hairs; petioles and rachis soon glabrate; lfls usually 7(5), soon glabrous beneath except for scattered hairs on the larger veins and in the vein-axils; terminal lfl oblanceolate to narrowly obovate; fr subglobose, 2–4.5 cm, the husk 2–5 mm thick, eventually splitting to the base; nut pale, 4-angled above the middle; kernel sweet and edible; $2n=32$. Upland woods; Mass. to Wis., s. to Ga., Miss., and Mo. (*C. microcarpa; Hicoria borealis.*) Highly variable, but scarcely separable into significant vars.

9. **Carya glabra** (Miller) Sweet. Pignut-h. Bark light or dark gray, on younger trunks tight and patterned, becoming finely furrowed and checked, later rough and deeply furrowed; outer bud-scales sparsely resinous-dotted, otherwise glabrous; petiole, rachis, and lower lf-surface soon glabrous, or pubescent only on the larger veins and in the vein-axils; lfls usually 5(7), the terminal one broadly oblanceolate to obovate; fr subglobose to obovoid, 2–3.5(–5) cm; husk 2–6 mm thick, eventually dehiscent to the middle, sometimes only along one or two sutures; nut 2–3 cm, rather thin-shelled, pale buff, slightly 4-angled or smooth, round or compressed; kernel astringent, scarcely edible, varying to bland; $2n=64$. Upland woods; s. Vt. and se. N.H. to Mich., se. Io., and e. Kans., s. to Fla. and e. Tex.

10. **Carya texana** Buckley. Ozark-h., black h. Bark nearly black, rough, deeply furrowed; buds 5–8 mm, densely covered with resinous, yellowish scales; petiole, rachis, and lower lf-surface densely rusty-pubescent when young, at maturity subglabrous, or pubescent only on the main veins and in the vein-axils; lfls (5)7, pale beneath, the terminal lance-obovate, commonly under 10 cm; fr obovoid, 2–4 cm, densely resinous, eventually splitting to the base; nut pale brown, broadly ovoid or subglobose, scarcely angled, ± compressed, kernal edible; $2n=64$. Dry upland woods; s. Ind. to Mo., s. to La. and Tex. (*C. buckleyi.*) Our plants are usually segregated as var. *arkansana* (Sarg.) Little.

11. **Carya pallida** (Ashe) Engler & Graebner. Pale h. Bark rough, deeply furrowed; buds 5–8 mm, densely covered with resinous, yellowish scales; petiole, rachis, and lower side of the principal veins copiously, loosely, and ± persistently pubescent with fascicled hairs; lfls 7(9), pale beneath, densely scurfy when young, nearly glabrous on the actual surface at maturity; fr ellipsoid to obovoid or subglobose, 2–4 cm, densely resinous, eventually splitting to the base along some of the sutures; nut white, compressed, nearly round, prominently angled; kernel edible. Moist or dry soil; Fla. to La., n. to Ark., s. Ill., Ky., and Va., and on the coastal plain to s. N.J.

Family **MYRICACEAE**, The Bayberry Family

Fls tiny, borne in unisexual catkins, without perianth, commonly one fl (or pseudanthial partial infl) in the axil of each bract; staminate fls commonly (including our spp.) with 2–8 (often 4) stamens; ovary bicarpellate, unilocular, the 2 short styles distinct or united at base, each with a linear-elongate stigma; ovule solitary, basal, erect, orthotropous; fr an achene or small nut or a small, thin-skinned drupe, sometimes enveloped by persistent, accrescent bracteoles; seeds with a straight, dicotyledonous embryo, nearly or quite without endosperm; aromatic, anemophilous shrubs or small trees with alternate, simple, pinnately veined lvs that characteristically bear 2 sorts of trichomes, one sort elongate, white, slender, and unicellular, the other sort multicellular, with a short, ± embedded stalk and an expanded, glandular or resinous, yellow or brownish head. 3/50.

1 Lvs pinnately lobed, stipulate; bracteoles forming a bur around the frs .... 1. *Comptonia.*
1 Lvs entire or merely toothed, exstipulate; bracteoles not forming a bur .... 2. *Myrica.*

1. COMPTONIA L'Hér. Sweet fern. Pistillate fl initially subtended by a primary bract and bracteoles, but an indefinite number (often 4) of linear-subulate secondary bracteoles arising adaxially to each of the latter, ± coalescent below, accrescent, elongating and persistent so that the pistillate infl forms a bur; ovary glabrous, without papillae; achene smooth, hard; rhizomatous, aromatic shrubs with deciduous, pinnately lobed, stipulate lvs. Monotypic.

1. Comptonia peregrina (L.) J. M. Coulter. Dioecious or seldom monoecious, much branched shrub to 1.5 m; terminal bud wanting; lvs linear-oblong, 6–12 × 0.8–1.5 cm, deeply pinnately lobed; staminate catkins clustered near the tips of the twigs, arising directly from the distal axillary buds, cylindric, mostly 1.5–4 cm, nodding, their bracts ± quadrangular, acuminate; anthers formed in the fall; pistillate catkins subglobose, at maturity forming a bur 1–2 cm thick; achene ellipsoid-conic or subcylindric, 3–5 mm; 2*n*=32. Dry, often barren and sandy soil; N.Y. to N.C., w. S.C., and n. Ga., w. to Sask., Minn., Ill., and Tenn. Mar.–June. (*Myrica asplenifolia.*)

2. MYRICA L. Pistillate fl subtended by a primary bract and 2–6 bracteoles, the latter persistent or deciduous, often resembling tiny sep, not forming a bur; ovary glabrous, or often covered with waxy papillae, sometimes also hairy, ripening into an achene or drupelet; aromatic shrubs or small trees, ours all dioecious, with entire or merely toothed, exstipulate lvs. (*Cerothamnus, Gale, Morella.*) Spp. 2–4 are closely related and not always sharply distinct. Nearly 50, widespread.

1 Pistillate bractlets 2, accrescent, thickened, sandwiching the flattened achene .... 1. *M. gale.*
1 Pistillate bractlets 4–6, remaining small and inconspicuous; fr a waxy, subglobose drupelet.
　2 Principal lvs mostly 1.5–3 cm wide (or more) and 2.5–4 times as long, often deciduous.
　　3 Ovary and fr hairy as well as waxy; lvs definitely deciduous .... 2. *M. pensylvanica.*
　　3 Ovary and fr glabrous or nearly so; lvs subpersistent .... 3. *M. heterophylla.*
　2 Principal lvs mostly 0.8–1.5(–2) cm wide and 4–5 times as long, evergreen .... 4. *M. cerifera.*

1. Myrica gale L. Sweet gale. Freely branched shrub to 1.5 m; lvs deciduous, oblanceolate, 3–6 cm; terminal bud wanting; distal axillary buds floral, the catkins in anthesis before the lvs expand, the staminate cylindric, 1–2 cm, with depressed-triangular bracts, the pistillate ovoid, 8–10 mm, with subrotund bracts; anthers formed in autumn; lvs bractlets 2, persistent, accrescent, becoming much-thickened, clasping and about equaling the flattened, ovate, beaked, glabrous and resinous-dotted achene, 2.5–3 mm, the tips divergent; 2*n*=48, 80. Swamps and shores; circumboreal, in Amer. s. to N.J., Pa., Mich., Minn., and Oreg., and (disjunct?) in N.C. (*Gale palustris.*)

2. Myrica pensylvanica Mirbel. Northern bayberry. Bushy shrub 0.5–2 m; lvs deciduous, broadly ob-lanceolate or elliptic, mostly 4–8 × 1.5–3 cm and 2.5–4 times as long as wide, obtuse or rounded and minutely apiculate, entire or with a few low teeth toward the tip, generally (as also the twigs) with some glandless white hairs, in addition to the resinous glands, at least when young, the glands sometimes of 2 sorts as in no. 4, sometimes not; terminal bud present; outer bud scales glabrous, ciliate, broadly rounded distally; staminate catkins produced below the leafy branches in May or June, cylindric, 6–15 mm, with broadly quadrate bracts; anthers formed in the spring; pistillate catkins slender, 5–10 mm, with ovate bracts; bracteoles 4–6, ± persistent but remaining small and inconspicuous; ovary densely hairy as well as papillate; frs solitary or few in a cluster, subglobose, 3.5–5 mm, covered with a thick layer of white wax that masks the underlying papillae, and also ± densely short-hairy; 2*n*=16. Dry hills and shores, especially near the coast, from Nf. to N.C., and less commonly inland to O. and s. Ont. (*Morella p.; Myrica* and *Cerothamnus caroliniensis* of authors, perhaps not of Miller.)

3. Myrica heterophylla Raf. Southern bayberry. Much like no. 2; lvs sometimes larger, up to 12 × 5 cm, subpersistent, evergreen southward, generally eventually deciduous with us, punctate on the lower side only, sometimes virtually without eglandular hairs, the glands of the lower surface all shining and about alike; outer bud-scales usually ciliate-margined and hairy on the back, merely obtuse; ovary and fr nearly or quite without trichomes, the fr rarely more than 4 mm; fl in April or May with us, earlier southward; $2n=16$. Swamps and moist, low ground on the coastal plain; N.J. to Fla., Ark., and Tex. (*Morella h.; ? Myrica caroliniensis* Miller.)

4. Myrica cerifera L. Wax-myrtle. Plants sometimes low (5–10 dm) and colonial by rhizomes, sometimes taller and arborescent, up to 10 m or more; lvs evergreen, typically oblanceolate, varying to oblong or lanceolate, mostly 3–9 × 0.6–1.5 cm and 4–5 times as long as wide, usually acute, entire or with a few sharp teeth above the middle, finely punctate on both sides, nearly or quite without eglandular hairs, as also the twigs; glands on the lower lf-surface ± persistent and evidently of 2 sorts, some shining, golden-yellow, others dull brownish and less resinous; staminate catkins produced in the axils of old lvs in April or May (earlier southward), ovoid, 6–12 mm, with obovate-rotund bracts; pistillate catkins linear, 5–10 mm, loose, with lance-ovate bracts, bracteoles 4–6, ± persistent but remaining small and inconspicuous; ovary glabrous; frs 2–3.5 mm, with a thick, waxy coating, glabrous; $2n=16$. Moist or wet, sandy soil on the coastal plain; N.J. to Fla., Tex., Ark., and Mex.; W.I. (*Cerothamnus c., Morella c.; Myrica pusilla.*)

FAMILY **FAGACEAE**, the Beech Family

Fls tiny, inconspicuous, unisexual (the plants monoecious) or some of them perfect; staminate fls borne in catkins or heads, with 4–7 (most commonly 6) minute, scale-like tepals, these distinct or connate below, or sometimes almost obsolete; stamens (4–)6–12 (–40), on slender filaments; pistillate fls 1–7(–15) together at the base of the staminate catkins or from separate axils, individually or collectively subtended by an involucre that develops into a cupule; ovary inferior, of (2)3 or sometimes 6(7–12) carpels, with 3–7 minute tepals around the summit, or these obsolete; styles as many as the carpels; locules as many as the carpels, but the septa sometimes not reaching the summit of the ovarian cavity; ovules 2 in each locule, pendulous, anatropous; fr generally a nut with a stony to leathery pericarp, subtended at the base or more or less enclosed individually or collectively by an accrescent cupule or hull that appears to be composed of numerous, imbricate, concrescent bracts; seed solitary, with large, straight, dicotyledonous embryo and without endosperm; anemophilous or sometimes secondarily entomophilous trees or shrubs with alternate, simple, entire to deeply lobed, pinnately veined leaves and deciduous stipules. 6–8/800.

1 Staminate fls in peduncled heads; nuts sharply triangular . . . . 1. *Fagus.*
1 Staminate fls in slender catkins; nuts rounded or ± flattened on 1 or 2 sides.
  2 Pistillate fls 2–4 per involucre; mature involucre prickly, enclosing 1–3 nuts . . . . 2. *Castanea.*
  2 Pistillate fls solitary; mature involucre of concrescent scales, covering the base or rarely all of the
    single nut . . . . 3. *Quercus.*

1. FAGUS L. Beech. Staminate fls in small heads on drooping peduncles, the bracts caducous, the stamens 8–16; pistillate fls usually paired on a short peduncle, concealed by the numerous subulate bracts; ovary 3-locular; fr a sharply 3-angled nut; 1-seeded, borne normally in pairs within the accrescent, 4-valved involucre; trees with straight-veined lvs, a vein running to each tooth; fls appearing with the lvs, the staminate from the lower axils, the pistillate from the upper. 10, N. Temp.

1. Fagus grandifolia Ehrh. American b. Tree to 30 m, with smooth, gray bark; lvs short-petioled, ovate to oblong-obovate, densely silky when young, later glabrous above, usually remaining silky at least on the midvein beneath; winter-buds elongate, slender, sharp, mostly 1.5–2.5 cm. N.S. to n. Fla., w. to Wis., e. Ill., se. Mo., e. Okla. and e. Tex. Var. *grandifolia*, red b., northern, occurring on rich upland soils throughout most of our range, usually has sharply serrate lvs, and the prickles of the fr are 4–10 mm, erect to spreading or recurved. Var. *caroliniana* (Loudon) Fern. & Rehder, white b., more southern, occurring on moist or wet lowland soils especially on and near the coastal plain, n. to Mass., s. O., Ill., and Mo., has usually more markedly acuminate, often merely denticulate lvs, and the prickles of the fr are 1–3(–4) mm, mostly abruptly reflexed from near the base. A third phase, sometimes called gray b., growing at the margins of the spruce-fir forest, s. even to N.C., differs from the other two in that the segments of the cupule do not cover the nutlets at maturity; unlike the others, it tolerates calcareous soil.

2. CASTANEA Miller. Staminate fls in slender, elongate axillary catkins, several together in the axil of a minute, caducous bract, the campanulate cal 6-parted, the stamens

12–20; pistillate fls borne on the base of the staminate catkins or from separate axils, 2–4 within each ovoid, prickly involucre; ovary usually 6-locular; frs solitary or 2–3 (rarely more) together within the accrescent, long-spined, 2–4-valved involucre; trees or shrubs with straight-veined lvs, the fls appearing after the lvs, at least sometimes partly entomophilous. 10, N. Temp.

1 Tree; mature lvs essentially glabrous beneath; nuts usually 2 or 3 together .... 1. *C. dentata.*
1 Shrub 2–5 m; lvs gray-tomentose beneath; nuts usually solitary .... 2. *C. pumila.*

1. Castanea dentata (Marshall) Borkh. Chestnut. Tree to 30 m; lvs oblong-lanceolate, acuminate, coarsely and sharply serrate with ascending or incurved teeth, glabrous or nearly so, short-petioled; staminate catkins to 20 cm; mature involucres 5–6 mm thick, with very numerous spines 1 cm or more, the nuts 1.5–2 cm, usually 2 or 3 together, flattened on one or two sides; $2n=24$. Original range from s. Me. to se. Mich., s. to Del., Ky., and s. Ill., and along the mts. to Ala., usually in acid upland soils; now nearly exterminated by blight.

C. *mollissima* Blume, the Chinese chestnut, with the lvs tomentose beneath, occasionally escapes from cult.

2. Castanea pumila (L.) Miller Chinquapin. Shrub 2–5 m, often colonial, or occasionally a small tree; lvs lance-oblong to oblong-obovate, to 15 cm, or longer on rapidly growing shoots, sharply serrate with ascending or salient teeth to merely bristle-toothed, gray-tomentose beneath; staminate spikes to 15 cm; involucres 2–3.5 cm thick, usually several or many in a large, compact head or spike, the nuts solitary, ovoid, terete, 1–1.5 cm. Dry or moist acid soil; s. N.J. to Ky. and Ark., s. to Fla. and Tex. Ours is var. *pumila. C. × neglecta* Dode is a hybrid with no. 1.

3. QUERCUS L. Oak. Staminate fls in slender, naked catkins, the bracts caducous or none, the cal divided to the base into 6(3–7) segments; stamens 3–12; pistillate fls solitary or in small spikes, each subtended by a bract and surrounded by an involucre of many scales; ovary 3-locular; nut not subtended and partly enclosed by the accrescent involucre; trees or shrubs, the fls appearing before the lvs; $2n=24$. 400+, cosmop.

Our oaks fall into 2 subgenera: in *Lepidobalanus,* the white oaks (spp. 1–12), the fr matures at the end of the first season, the stigmas are sessile or nearly so, the abortive ovules lie at the base of the seed, and the lvs or their lobes are not bristle-tipped. In *Erythrobalanus,* the red and black oaks (spp. 13–28), the fr matures during the second year, the styles are elongate, the abortive ovules lie at the top of the seed, and in most spp. the lvs or their lobes are bristle-tipped.

Most or all of the spp. within each subgenus can and frequently do hybridize, but natural intersubgeneric hybrids are unknown. The following names are believed to apply to hybrids as indicated:

Q. × *asheana* Little = *cinerea* × *laevis*
Q. × *atlantica* Ashe = *cinerea* × *laurifolia*
Q. × *beadlei* Trel. = *alba* × *michauxii*
Q. × *bebbiana* C. Schneider = *alba* × *macrocarpa*
Q. × *benderi* Baenitz = *rubra* × *coccinea*
Q. × *blufftonensis* Trel. = *falcata* × *laevis*
Q. × *brittonii* W. T. Davis = *ilicifolia* × *marilandica*
Q. × *bushii* Sarg. = *marilandica* × *velutina*
Q. × *byarsi* Sudw. = *macrocarpa* × *michauxii*
Q. × *caduca* Trel. = *cinerea* × *nigra*
Q. × *cocksii* Sarg. = *laurifolia* × *velutina*
Q. × *comptoniae* Sarg. = *lyrata* × *virginiana*
Q. × *cravenensis* Little = *cinerea* × *marilandica*
Q. × *deamii* Trel. = *macrocarpa* × *muehlenbergii*
Q. × *demarei* Ashe = *nigra* × *velutina*
Q. × *dubia* Ashe = *phellos* × *velutina* or *laevis*
Q. × *egglestonii* Trel. = *imbricaria* × *shumardii*
Q. × *exacta* Trel. = *imbricaria* × *palustris*
Q. × *faxonii* Trel. = *alba* × *prinoides*
Q. × *fernaldii* Trel. = *rubra* × *ilicifolia*
Q. × *fernowii* Trel. = *alba* × *stellata*
Q. × *filialis* Little = *phellos* × *velutina*
Q. × *garlandensis* E. J. Palmer = *falcata* × *nigra*

Q. × *giffordii* Trel. = *ilicifolia* × *phellos*
Q. × *guadelupensis* Sarg. = *macrocarpa* × *stellata*
Q. × *hawkinsiae* Sudw. = *rubra* × *velutina*
Q. × *heterophylla* Michx. = *rubra* × *phellos*
Q. × *hillii* Trel. = *macrocarpa* × *muehlenbergii*
Q. × *humidicola* E. J. Palmer = *bicolor* × *lyrata*
Q. × *incomita* E. J. Palmer = *falcata* × *marilandica*
Q. × *jackiana* Schneid. = *alba* × *bicolor*
Q. × *joorii* Trel. = *falcata* × *shumardii*
Q. × *leana* Nutt. = *imbricaria* × *velutina*
Q. × *lowellii* Sarg. = *rubra* × *ilicifolia*
Q. × *ludoviciana* Sarg. = *falcata* × *phellos*
Q. × *mellichampii* Trel. = *laevis* × *laurifolia*
Q. × *mutabilis* E. J. Palmer & Steyermark = *palustris* × *shumardii*
Q. × *neopalmeri* Sudw. = *nigra* × *shumardii*
Q. × *palaeolithicola* Trel. = *ellipsoidalis* × *velutina*
Q. × *podophylla* Trel. = *cinerea* × *velutina*
Q. × *porteri* Trel. = *rubra* × *velutina*
Q. × *rehderi* Trel. = *ilicifolia* × *velutina*
Q. × *richteri* Baenitz. = *rubra* × *palustris*
Q. × *robbinsii* Trel. = *coccinea* × *ilicifolia*
Q. × *rudkinii* Britton = *marilandica* × *phellos*
Q. × *runcinata* (A. DC.) Engelm. = *rubra* × *imbricaria*
Q. × *sargentii* Rehder = *prinus* × *robur*
Q. × *saulii* C. Schneider = *alba* × *prinus*
Q. × *schuettei* Trel. = *bicolor* × *macrocarpa*
Q. × *shirlingii* Bush = *imbricaria* × *shumardii*
Q. × *stelloides* Palm. = *prinoides* × *stellata*
Q. × *sterilis* Trel. = *marilandica* × *nigra*
Q. × *sterretti* Trel. = *lyrata* × *stellata*
Q. × *subfalcata* Trel. = *falcata* × *phellos*
Q. × *subintegra* Trel. = *cinerea* × *falcata*
Q. × *substellata* Trel. = *bicolor* × *stellata*
Q. × *tridentata* Engelm. = *imbricaria* × *marilandica*
Q. × *vaga* E. J. Palmer & Steyerm. = *palustris* × *velutina*
Q. × *willdenowiana* Zabel = *falcata* × *velutina*
Q. × *walteriana* Ashe = *laevis* × *nigra*

1 Lvs (or most of them) entire or merely sinuate.
  2 Lvs linear-oblong to elliptic or slightly obovate, broadest near the middle, often more than 2.5 times as long as wide.
    3 Lvs permanently pubescent on the general surface beneath.
      4 Lvs not bristle-tipped, the cartilaginous margin continuous around the tip.
        5 Lvs plane, finely and tightly stellate beneath (see descr.) .... 11. *Q. virginiana*.
        5 Lvs revolute-margined, more coarsely and loosely stellate-tomentose beneath .... 12. *Q. geminata*.
      4 Lvs bristle-tipped, the bristle, if broken off, leaving a truncate scar.
        6 Veinlets elevated on the upper surface; lvs mostly 10–15 × 3.5–6 cm .... 13. *Q. imbricaria*.
        6 Veinlets depressed on the upper surface; lvs mostly 5–10 × 1.5–3 cm .... 14. *Q. cinerea*.
    3 Lvs soon glabrescent beneath, except sometimes for the vein-axils or midvein.
      7 Lvs dull green beneath, 4–6 times as long as wide .... 15. *Q. phellos*.
      7 Lvs glossy-green beneath, usually 2.5–4 times as long as wide.
        8 Lvs blunt-tipped; moist lowlands .... 16. *Q. laurifolia*.
        8 Lvs sharply acute or spinulose-apiculate; well-drained uplands .... 17. *Q. hemisphaerica*.
  2 Lvs of a broadly obovate or cuneate-spatulate type, 1–2.5 times as long as wide.
    9 Lvs glabrous on the general surface beneath, cuneate at base .... 18. *Q. nigra*.
    9 Lvs rusty-puberulent beneath, rounded or subcordate at base .... 19. *Q. marilandica*.
1 Lvs prominently toothed or lobed.
  10 Lobes or teeth of the lvs rounded to acute, but never bristle-tipped.
    11 Lvs distinctly lobed, the lobes 1–5(6) on each side.
      12 Pubescence of the lower lf-surface none or essentially none at maturity.
        13 Lvs pale and glaucous beneath; nut 1.5–2.5 cm .... 1. *Q. alba*.
        13 Lvs dull green beneath; nut 1–1.5 cm .... 18. *Q. nigra*.
      12 Pubescence of the lower lf-surface persistent, perceptible to the touch.
        14 Lower scales of the acorn-cup flat or barely concave; lvs pubescent beneath with erect, few-branched hairs only.
          15 Lf-lobes constricted at base; lf-pubescence of sessile hairs .... 2. *Q. stellata*.
          15 Lf-lobes not constricted; lf-pubescence of stipitate hairs .... 3. *Q. margaretta*.
        14 Lower scales of the acorn-cup strongly arched or galeate, the cup appearing tuberculate; lvs usually pubescent beneath with very fine, horizontally spreading, many-branched hairs, as well as commonly with some erect, few branched hairs.

16 Inner cup-scales acute or obtuse .... 4. *Q. lyrata.*
16 Inner cup-scales caudate-tipped .... 5. *Q. macrocarpa.*
11 Lvs coarsely toothed, with 3–14 teeth on each side, the sinuses extending less than one-third
of the way to the midvein.
  17 Acorns on peduncles 2–7 cm and exceeding the petioles; lower lf-surface with densely
    stellate hairs and longer, erect, few-branched hairs .... 6. *Q. bicolor.*
  17 Acorns sessile or on peduncles not exceeding the adjacent petioles.
    18 Pubescence of the lower lf-surface wholly of erect, few-branched hairs .... 7. *Q. michauxii.*
    18 Pubescence of the lower lf-surface all or chiefly of short, widely spreading, stellate
    hairs, these many to very few.
      19 Lf-teeth rounded; cup 2–2.5 cm wide; lvs sparsely hairy beneath .... 8. *Q. prinus.*
      19 Lf-teeth ending in a projecting callous papilla; cup 1–1.5 cm wide; lvs usually
      completely covered by tomentum beneath.
        20 Tall tree; lvs with mostly 9–14 pairs of lateral veins .... 9. *Q. muehlenbergii.*
        20 Colonial shrub 1–3(–5) m; lvs with mostly 5–8 pairs of lateral veins .... 10. *Q. prinoides.*
10 Lobes of the lvs bristle-tipped.
  21 Mature lvs stellate-tomentose or -tomentulose on the general surface beneath.
    22 Stellate hairs minute, just within the range of a 10× lens.
      23 Lvs small, mostly 5–10 cm, with broadly triangular lobes; shrub or small tree to 5
      m .... 20. *Q. ilicifolia.*
      23 Lvs larger, mostly 10–20 cm, often with long, narrow lobes; tall trees.
        24 Terminal lf-lobe relatively short and broad, often tapering from the base, not
        falcate .... 21. *Q. pagoda.*
        24 Terminal lf-lobe elongate and narrow, often somewhat falcate .... 22. *Q. falcata.*
    22 Stellate hairs conspicuous, their structure easily discernible at 10×.
      25 Lvs broadly obovate, 3–5-lobed at the apex, the other lobes small or none ...... 19. *Q. marilandica.*
      25 Lvs oblong, elliptic, or slightly obovate, the lateral lobes the largest .... 24. *Q. velutina.*
  21 Mature lvs glabrous beneath, or hairy on the midvein or in vein-axils.
    26 Lvs broadly obovate or cuneate-spatulate, widened and shallowly 3-lobed distally.
      27 Lvs rounded to subcordate at base, minutely rusty-hairy beneath .... 19. *Q. marilandica.*
      27 Lvs broadly or narrowly cuneate at base, glabrous beneath .... 18. *Q. nigra.*
    26 Lvs oblong, elliptic, or slightly obovate, the median lateral lobes the largest.
      28 Petioles of well grown lvs 0.5–1.5 cm; inner cup-scale inflexed .... 23. *Q. laevis.*
      28 Petioles of well grown lvs 2.5–7 cm; cup-scales not inflexed.
        29 Terminal buds 4-angled, 7–10 mm, densely hairy; inner cup-scales loose, forming
        a fringe .... 24. *Q. velutina.*
        29 Terminal buds ovoid, 3–5(–7) mm, their scales glabrous or ciliate; innermost
        cup-scales almost wholly covered.
          30 Lvs shallowly lobed, the length of a principal lateral lobe, measured from
          the center of the sinus above to its tip, rarely more than twice the
          distance from the center of the sinus to the midvein .... 25. *Q. rubra.*
          30 Lvs deeply lobed, the length of a median lateral lobe, measured as above,
          generally 3–6 times the distance from sinus to midvein.
            31 Acorn-cup shallowly to deeply saucer-shaped, covering a quarter to a
            third of the nut.
              32 Cup 10–16 mm wide; nut 10–13 mm .... 26. *Q. palustris.*
              32 Cup 16–25 mm wide; nut 15–28 mm .... 27. *Q. shumardii.*
            31 Acorn-cup turbinate, covering at least a third and usually half of the
            nut.
              33 Cup-scales pubescent and dull; inner bark yellowish or orange .... 28. *Q. ellipsoidalis.*
              33 Cup-scales partly glabrous, shining; inner bark reddish .... 29. *Q. coccinea.*

1. **Quercus alba** L.   White oak.   Tall tree with light gray, coarsely flaky, shallowly furrowed bark and widely spreading branches; twigs soon glabrescent; lvs obovate or oblong-obovate, cuneate at base, thinly floccose beneath at first (the pubescence can be rolled off with one's thumb), glabrous or nearly so at maturity, pale beneath, the lobes 3 or 4(5) pairs, ascending, oblong to ovate, rounded or rarely acute; acorns sessile or on pedicels to 4 cm, the cup deeply saucer-shaped, pubescent within, covering a fourth to a third of the nut; nut ovoid to cylindric-ovoid, 1.5–2.5 cm. Upland woods; Me. to Mich. and Minn., s. to n. Fla. and e. Tex. Extreme forms in which the lobes are scarcely more than large teeth may be distinguished from the chestnut-oaks by the lack of pubescence.
Q. robur L., the English oak, differing from *Q. alba,* among other respects, in its basally cordate-auriculate lvs, occasionally escapes from cult.

2. **Quercus stellata** Wangenh.   Post-o.   Small or large tree with thick, rough, deeply furrowed bark; lvs 9–15 × 5–10 cm, obovate or often cruciform, thick, variable, usually with a few large, rounded lobes, the main pair of lateral lobes usually constricted at base and with truncate to retuse tip, pubescent on both sides with erect, few-branched hairs, those of the upper surface often deciduous just above the base; petioles and twigs persistently pubescent; acorns sessile or nearly so, the cup hemispheric or pyriform, covering half the nut, its scales flat or nearly so; nut ovoid, 1–1.5 cm. Dry upland woods and barrens; se. Mass. and s. N.Y. to O., Ind., and s. Io., s. to Fla. and Tex., common along our s. border.

3. **Quercus margaretta** Ashe. Sand post-o. Shrub or small tree to 10 m, spreading underground and becoming colonial; lvs oval or obovate, 5–10 × 2–5 cm, soon glabrescent above, thinly but softly pubescent beneath with stipitate, few-branched hairs; lateral lobes 1–2(3) pairs, each lobe entire, oval, not constricted at base, rounded at the summit, or the terminal lobe occasionally 3–5-toothed; fr as in no. 2. Dry, especially sandy soil; coastal plain from se. Va. to Fla. and La. (*Q. stellata* var. *m.*)

4. **Quercus lyrata** Walter. Overcup-o. Tree to 30 m, with thick, gray bark, the twigs soon glabrescent; lvs narrowly obovate, densely gray-tomentose beneath with horizontally spreading, many-branched stellate hairs, above which project a few erect, few-branched hairs, or with only the latter type of hair; lateral lf-lobes 2–4 pairs, oval, rounded to subacute, rarely emarginate, the lower usually much smaller and separated from the upper by often elongate sinuses; acorns 1.5–2.5 cm, about as thick; cup covering half to all the conic-ovoid nut, the lower scales strongly concave. Wet woods and swamps on the coastal plain; s. N.J. to Fla. and Tex., n. in the interior to s. Ind., s. and w. Ill., and Mo.

5. **Quercus macrocarpa** Michx. Bur-o. Low shrub to tall tree, the latter with rough, deeply furrowed bark; lvs large, oblong to obovate, cuneate at base, pale beneath with a close, fine stellate pubescence, with 4–7 pairs of blunt or acute lateral lobes, a pair of sinuses near the middle usually deeper than the others; winter-buds pubescent; fr sessile or stoutly short-peduncled, the cup covering a third to more often two-thirds or even all the nut, pubescent within, the marginal scales acuminate into slender awns forming a terminal fringe; nut depressed-ovoid to narrowly ovoid, 1–4 cm thick. Var. *macrocarpa*, a tree to 50 m, occurs in moist woods and alluvial flood-plains from N.B. and Que. to Ont. and s. Man., s. to Va., La., and Tex. Var. *depressa* (Nutt.) Engelm. (*Q. mandanensis*), a shrub up to 5 m, with acorns of minimal size that have a relatively short marginal fringe, occurs on dry uplands and bluffs from w. Minn. and Io. to Sask., Mont., and Wyo.

6. **Quercus bicolor** Willd. Swamp white O. Tree to 30 m, with thick, dark, furrowed and flat-ridged bark; lvs obovate, cuneate to a rounded or acute base, with 6–10 pairs of coarse, irregular, rounded teeth, pale or whitened beneath with pubescence of both short, horizontally spreading and erect, few-branched stellate hairs, rather velvety to the touch; winter-buds glabrous or nearly so; peduncles (2–)4–7 cm, usually with 2 acorns; cup hemispheric, covering about half the nut, the lower scales strongly convex on the back, the upper scales long-acuminate or even caudate; nut broadly ovoid, 1.5–2.5 cm. Flood-plains and other poorly drained sites; Que. and Me. to s. Mich. and c. Minn., s. to N.C., Tenn., and n. Ark.

7. **Quercus michauxii** Nutt. Swamp chestnut-o. Tree to 30 m, with light gray bark like no. 1; petioles to 3 cm, persistently pubescent beneath, glabrescent above; blades obovate, 7–11 cm wide, with 10–15 pairs of ovate or triangular, rather uniform, acute to rounded, usually ascending teeth, pubescent beneath with erect, few-branched hairs; acorns 2.5–3.5 cm, the cup covering half the ovoid or conic-ovoid nut, 2–3 cm wide, its large scales free and when dry often spreading. Low or wet soil, especially alluvial flood-plains; coastal plain from N.J. to Fla. and Tex., n. in the interior to Mo. and s. Ind., and at scattered stations in Ky. and Tenn. (*Q. prinus*, misapplied.)

8. **Quercus prinus** L. Rock chestnut o. Tree to 30 m, with thick, deeply furrowed, dark bark; lvs obovate to oblong or even ovate, cuneate to rounded at base, the lower surface green or grayish, thinly pubescent with flatly appressed, spreading, stellate hairs, each margin with 10–15 rather regular, oval, obtuse teeth; acorns 2.5–3.5 cm, the cup hemispheric, 2 cm wide, its comparatively few scales almost completely con-crescent, covering a third to half the nut. Dry or moist, upland or rocky woods, chiefly in the Appalachian region and its foothills; Me. to n. Ga., extending to the coast as far s. as Va.,and w. to s. Ill. and n. Miss.; disjunct in s. Mich. (*Q. montana.*) The identity of the Linnean type of *Q. prinus* is debatable. We here follow current usage.

9. **Quercus muehlenbergii** Engelm. Yellow o. Tall tree (to 25 m) with rather thin, gray, flaky bark; lvs densely pubescent or tomentose beneath with fine, horizontally spreading, grayish hairs, lanceolate to narrowly oblong, oblanceolate or obovate, 10–20 cm, broadly cuneate to rounded at base, with 9–14 veins on each side, the veins running straight to the teeth and nearly parallel, the teeth commonly sharp, ascending, and often incurved, each ending in a minute, papilliform projection; acorns sessile or nearly so, 1–2 cm, the cup 1–2 cm wide, its very numerous small scales free at the tip; nut ovoid. In good, chiefly calcareous soils; Vt. to se. Minn. and w. Nebr., s. to n. Fla., Ala., and Tex. (*Q. prinoides* var. *acuminata.*)

10. **Quercus prinoides** Willd. Chinquapin-o. Much like no. 9 and hybridizing with it, but a colonial shrub 1–3(–5) m; lvs oblong-obovate, 4–10 × 2–6 cm, usually cuneate at base, with 5–8 lateral veins and as many low teeth on each side. Dry, rocky slopes and barrens, often near the coast, preferably in calcareous soil; Mass to N.C., w. to n. Ind., s. Mich., and Okla.

11. **Quercus virginiana** Miller. Live o. Large, evergreen tree (to 20 m), with widely spreading branches that often support *Tillandsia*, or smaller in difficult habitats and only shrubby (such plants often called var. *minima* Sarg.); bark furrowed and cross-checked into small plates; lvs firm, flat, narrowly elliptic to oblong, mostly 4–8 × 1–2 cm, blunt, entire, cuneate to obtuse at the base, glabrous above, closely and tightly cinereous with minute stellate hairs beneath (these scarcely distinguishable at 20×, sometimes even at 30×), less obviously veiny than no. 12; acorns solitary or paired, 1.5–2.5 cm, the cup turbinate, 8–15 mm, its scales acute, closely appressed. Dry or moist soil on the coastal plain; se. Va. to Fla., Tex., and ne. Mex.

12. **Quercus geminata** Small. Sand live o. Much like no. 11, often smaller; lvs revolute-margined, the upper surface often rounded rather than plane; pubescence of the lower lf-surface coarse and looser, the individual stellae readily visible at 20×, often even at 10×; main veins impressed on the upper lf-surface,

elevated on the lower; acorns commonly paired. Dry or moist soil on the coastal plain; se. Va. to Fla. and Miss. (*Q. virginiana* var. *maritima,* the dwarf phase.)

13. Quercus imbricaria Michx. Shingle-o.; jack-o. Medium-sized tree; bark with flat gray ridges and shallow furrows; twigs soon glabrescent; lvs dark green, shining, firm, oblong or lance-oblong, occasionally oblong-obovate, 10–17 × 3.5–7 cm, bristle-tipped, entire, glabrous above, softly and loosely pubescent beneath; a crown of dry tan or brown leaves remaining well into winter and even into early spring; acorn 12–16 mm, the cup 6–8 mm, with relatively few and broad scales. Dry upland soils; e. Pa. to s. Mich., O., and Kans., s. to N.C., Ga., and Ark.

14. Quercus cinerea Michx. Blue-jack o. Small tree, to 15 m; twigs of the season cinereous-tomentulose; lvs with a dull blue-green cast, especially beneath, firm, elliptic to oblong or narrowly obovate-oblong, 5–10 × 1–3 cm, obtuse or rounded and bristle-tipped, entire, broadly cuneate to rounded at base, permanently cinerous-tomentulose beneath with minute, stellate hairs; veinlets of the upper lf-surface minutely depressed; acorn 1–1.5 cm, the cup saucer-shaped, covering a third of the nut, its scales ovate-oblong, obtuse. Dunes and dry pine-barrens; se. Va. to Fla. and Tex., mostly on the coastal plain, and n. to Okla. (? *Q. incana* Bartram, perhaps misapplied.)

15. Quercus phellos L. Willow-o. Tall shrub, or tree to 25 m; twigs of the season reddish-brown, glabrous; lvs shiny-green above, thin, oblong-linear to lanceolate or oblanceolate, 6–12 × 1–2(–2.5) cm, bristle-tipped, entire, usually cuneate at base, glabrous above with finely reticulate, elevated veinlets, dull (and somewhat paler) green beneath and glabrous or with the stellate pubescence persistent especially in the vein-axils or along the midvein; acorns 1–1.5 cm, the cup shallowly saucer-shaped, its scales closely appressed, pubescent except at the margin. Swamps and moist soil; s. N.Y. to Fla. and Tex., chiefly on the coastal plain, n. in the interior to s. Ill.

16. Quercus laurifolia Michx. Laurel-o. Tree to 25 m; twigs glabrous; reddish-tinted; lvs usually yellowish or green when young, only seldom bronzy-red at maturity coriaceous, glossy beneath, narrowly oblong to lanceolate, elliptic, rhombic, or narrowly obovate, typically 5.5–9 × 1.5–3 cm, rounded-obtuse at the tip, tapering to a narrow, acute base, consistently entire, essentially glabrous except usually for some persistent tomentum in the vein-axils beneath, obscurely reticulate or with the reticulum more prominent on the lower surface than the upper, subpersistent, but turning brown in the fall and deciduous before spring; petiole 2–5 mm; acorns 1.5 cm, the cup depressed-hemispheric. Moist or wet soil, chiefly on the coastal plain; se. Va. to Fla., s. Ark., and e. Tex. (*Q. obtusa.*)

17. Quercus hemisphaerica Bartram. Darlington-o. Much like no. 16; twigs grayish; lvs bronzy-red and translucent when young, at maturity typically 3.5–5 × 1–2 cm, seldom larger, sharply acute (and often shortly bristle-pointed) at the tip, obtuse or merely acutish at the base, glabrous, evidently reticulate-veiny with the reticulum more prominent above, persistent and green throughout the winter and into early spring, sometimes some of them with a few teeth; petiole 1–2 mm. Well drained soil in uplands; coastal plain from se. Va. to Fla. and se. Tex.

18. Quercus nigra L. Water-o. Tree to 25 m; bark gray to black, smooth when young, roughened with scaly ridges in age; twigs soon glabrous; buds thinly pubescent, 3–6 mm; lvs cuneate-obovate or cuneate-spatulate and abruptly widened above, entire or shallowly and irregularly 2–5-lobed, 4–10 × 1.5–5 cm, scarcely bristle-tipped, long-cuneate to the base, glabrous and dull green on both sides, or with some stellate hairs on the vein-axils; acorns 1–1.5 cm, the cup saucer-shaped, with numerous narrow, closely appressed scales, covering a third of the nut. Usually in damp or wet soil; Cape May Co., N.J. to Fla. and Tex., chiefly on and near the coastal plain, and n. in the interior to Mo.

19. Quercus marilandica Muenchh. Black-jack o. Small or medium-sized tree with deeply checked bark and thinly pubescent twigs; buds densely pubescent, 5–8 mm; lvs broadly obovate or triangular-obovate, mostly (6–)10–20(–30) cm and half to fully as wide, shallowly and broadly 3-lobed across the summit, rarely with additional lobes below, sometimes nearly or quite entire, rounded or subcordate at base, brownish-green beneath with minute, closely appressed hairs, rarely softly stellate-tomentose; petioles mostly 0.5–1.5 cm; acorns 1.5–2 cm, the cup turbinate, enclosing half the nut, its scales relatively few and large, sericeous. Dry or sterile, especially sandy soil; s. N.Y. to s. Io., s. to Fla. and Tex. The vegetatively persistent, often locally abundant hybrid with *Q. ilicifolia* is *Q.* × *brittonii* W. T. Davis.

20. Quercus ilicifolia Wangenh. Bear-o. Shrub or small tree to 5 m; twigs of the season thinly hairy or glabrous; lvs oblong or oblong-obovate, 5–10 cm, half or two-thirds as wide, typically with 5 short, broadly triangular lobes, occasionally with an extra pair of small lobes or with toothed lobes, closely and finely gray-tomentulose beneath; petioles 1–3 cm, glabrescent; acorns ovoid, 1.2–2 cm, the cup turbinate or deeply saucer-shaped, covering a third to half the nut, its scales relatively few and large, closely appressed. Rocky, sandy or sterile soil; s. Me. to c. N.Y., O., W.Va., and N.C.

21. Quercus pagoda Raf. Cherrybark-o. Large tree, to 30(–35) m; bark of middle-sized trunks with numerous small, flat plates, much as in *Prunus serotina,* at full maturity finely multiridged and -furrowed; lvs more regularly and uniformly lobed than in no. 22, the middle lobes usually with short, widely spreading upper edge and longer, more tapering lower edge, the terminal lobe relatively short and broad, often tapering from the base, not falcate; otherwise much like no. 22. Typically in bottomland hardwood forests; s. N.J. to Ga., nw. Fla., and e. Tex. mainly on the coastal plain, and up the Mississippi embayment to w. Ky. and s. Ind. (*Q. falcata* var. *pagodaefolia.*)

22. Quercus falcata Michx.   Southern red o; Spanish o. Large tree, to 30 m; bark dark, thick, the rough ridges separated by deep, narrow furrows; twigs of the season persistently pubescent; lvs highly variable, commonly with 1–4 pairs of lateral lobes, these and especially the terminal lobe elongate, 2–5 times as long as wide, the lateral ones rather narrowly triangular and long-tapering, the terminal one more oblong and tending to be somewhat falcate; lvs persistently cinereous-tomentose beneath and on the long (2–5 cm) petiole; acorns 1–1.5 cm, about a third covered by the deeply saucer-shaped cup. Dry or sandy soil in the e., more often in wet soil in the w.; N.J. and sw. Pa. to Fla. and Tex., chiefly on the coastal plain, n. in the interior to O., Ind., and Mo. (*Q. rubra,* misapplied; *Q. triloba.*)

23. Quercus laevis Walter.   Turkey-o. Tree to 20 m; lvs obovate, on petioles 0.5–1.5 cm, pubescent in the vein-axils beneath, otherwise glabrous, usually cuneate at base, deeply lobed, the lateral lobes elongate, oblong or slightly broadened distally, with 2–3 bristle-tipped teeth at the tip; acorn 2–2.5 cm, the cup turbinate or deeply saucer-shaped, 2–2.5 cm wide, covering half or a third of the nut, the marginal scales abruptly inflexed. Dry, especially sandy soil, chiefly on the coastal plain; se. Va. to Fla. and La. (*Q. catesbaei.*)

24. Quercus velutina Lam.   Black o. Tree to 40 m, with very dark, rough bark, the inner bark yellow or orange; lvs glossy, shallowly or deeply lobed, pubescent in the vein-axils beneath, otherwise glabrous or persistently and loosely stellate over the lower surface and along the midvein above, the petioles and twigs of the season commonly thinly hairy; buds 4-angled, densely pubescent, 7–10 mm; acorns 1.5–2 cm, the cup turbinate, covering half the nut, wholly hairy inside, the scales relatively few and large, pubescent, the uppermost loose, prominently projecting and forming a marginal fringe. Usually in dry or sterile upland soil and on dunes; s. Me. to Mich. and se. Minn., s. to Fla. and Tex.

25. Quercus rubra L.   Northern red o. Tree to 50 m; bark smoothish and with a reddish-purple cast until the trunk is 2–3 dm thick, eventually developing broad, shallow furrows between the narrow, flat, gray ridges; inner bark reddish or red-brown; young twigs glabrous, dark reddish-brown; lvs dull green, 10–20 cm, soon glabrous throughout or often with small tufts of persistent hairs on the vein-axils beneath, 7–11-lobed, the lobes roughly triangular, broadest at base, bristle-tipped and usually with a few lateral teeth, little if at all longer than the width of the central body of the blade; crown-lvs often more deeply cleft than the lower lvs as just described; acorn to 3 cm, its cup shallow, saucer-shaped, 2–3 cm wide and covering a fifth or a fourth of the nut, or sometimes the cup somewhat narrower and deeper and enclosing a third of the somewhat smaller (2–2.5 cm) nut; acorn-cup with a ring of hairs inside around the scar. (*Q. borealis.*)

26. Quercus palustris Muenchh.   Pin-o. Tree to 30 m, the lower branches ± persistent, slender, widely spreading and usually slightly declined; bark smooth, or with shallow furrows and flat ridges; mature twigs glabrous; lvs shining, paler beneath, glabrous except for the conspicuous tufts of stellate hairs in the vein-axils beneath, often cuneate at the base and decurrent on the petiole; lf-lobes 2 or 3 pairs, much longer than the width of the central body of the blade, often widened distally and toothed, bristle-tipped; upper edge of lf-lobes ± perpendicular to the midvein; secondary lf-lobes fewer than in no. 27; winter-buds relatively small, acute, subglabrous; acorn-cup saucer-shaped, 1–2.5 cm wide, covering a fourth to a third of the nut, with very small, puberulent scales; nut 10–14 mm. Mostly in low, periodically flooded places, but susceptible to cult. on upland sites; Mass. and Vt. to Mich., Io., and Kans., s. to N.C., Tenn., and Okla.

27. Quercus shumardii Buckley.   Shumard o. Tall tree with close, deeply furrowed bark; twigs soon glabrescent; lvs smooth at maturity except for conspicuous tufts of stellate hairs in the vein-axils beneath, deeply (5)7-lobed, the sinuses commonly elliptic and extending three-fourths the distance to the midvein, the lobes oblong or usually widened distally, each with several conspicuous teeth or small lobes above; acorn-cup flatly saucer-shaped, 1–2.5 cm wide, with closely imbricate, minutely puberulent scales, covering a third or a fourth of the 1.5–3 cm nut. Chiefly in moist, lowland soil; Pa. and Va. to Ind., s. Mich., Mo., and Kans., s. to Fla. and Tex.

28. Quercus ellipsoidalis E. J. Hill.   Northern pin-o. Middle-sized tree; twigs soon glabrescent; lvs glabrous except for small tufts of stellate hairs in the vein-axils beneath; lateral lobes 2–3 pairs, separated by rounded sinuses, usually extending more than half way to the midvein, usually widened and several-toothed distally; acorn-cup turbinate, 9–14 mm wide, smooth inside, with closely appressed, puberulent scales, covering a third of the 12–20 mm nut. Dry upland soil; s. Mich. and adj. O. to Minn. and sw. Ont. and Mo. Possibly better included in no. 29.

29. Quercus coccinea Muenchh.   Scarlet o. Tree to 25 m, the twigs soon glabrescent; inner bark red; lvs glabrous or with minute tufts of stellate hairs in the vein-axils beneath, deeply 7–9-lobed, the lobes oblong or more often somewhat broadened distally, each with a few bristle-tipped teeth or shallow lobes, the sinuses rounded, the upper often much wider than the lower; upper edge of the lf-lobes forwardly arcuate; acorn-cup-turbinate, 16–22 mm wide, smooth inside, covering about half the 1.5–2 cm nut; scales relatively large, closely appressed, soon losing most of their pubescence, the cup becoming glossy. Dry upland soil, commoner ne.; sw. Me. to Ga., w. to s. Mich., Mo., and Miss.

FAMILY **BETULACEAE**, the Birch Family

Fls tiny, inconspicuous, unisexual (the plants monoecious), in unisexual catkins of rather different appearance, the male pendulous and more or less elongate, with a flexuous axis, the female pendulous or erect, shorter, firm, often woody; each bract of the catkin subtending 1–3 fls; cal of 1–6 minute, scale-like sep, or obsolete; pet none; stamens commonly as many as and opposite the sep, or (when the sep are obsolete) sometimes up to as many as 18; filaments very short, distinct or basally connate; anthers often deeply divided from the summit so that the pollen sacs are ± distinct; ovary inferior or nude, bicarpellate, with 2 styles, bilocular below, unilocular above, with 1–2 pendulous, anatropous ovules in each locule; fr a nut or more commonly a 2-winged samara, without a basal cupule, but sometimes with a foliaceous hull derived from 2 or 3 involucral bracts; seed solitary, with a large, dicotyledonous embryo and little or no endosperm; anemophilous trees or shrubs with alternate, simple, pinnately straight-veined, usually toothed lvs and deciduous stipules. 6/120.

1 Scales of the pistillate catkins soon deciduous; involucre greatly accrescent, enclosing or subtending
    a single nut or nutlet; half-anthers pilose at the tip.
   2 Fr 1–1.5 cm, closely enveloped by the involucre; pistillate catkins compact, head-like . . . . 1. *Corylus.*
   2 Fr 0.5 cm, not closely enveloped by the involucre; pistillate catkins more elongate.
      3 Involucre a closed, ovoid sac, inflated in fr; bark rough, flaky . . . . 2. *Ostrya.*
      3 Involucre expanded, open, 2–3-lobed; bark smooth . . . . 3. *Carpinus.*
1 Scales of the pistillate catkin accrescent, each subtending 2 or 3 flat frs; involucre none; anthers or
    half-anthers glabrous.
   4 Pistillate scales thin, deciduous with or soon after the frs; half-anthers 4, monothecal . . . . 4. *Betula.*
   4 Pistillate scales thick, becoming indurate, long-persistent, anthers 4, dithecal . . . . 5. *Alnus.*

1. CORYLUS L.   Hazel-nut. Staminate catkins emerging in autumn, reaching anthesis in early spring, each scale subtending a pair of small bractlets and a single naked fl with 4 stamens, the filaments deeply bipartite, each segment bearing a half-anther; pistillate catkins small, ovoid, the few, closely imbricate scales concealing the fls except for the elongate stigmas, each fl subtending by a minute bract and 2 bractlets, these greatly accrescent at maturity and enclosing the hard-shelled, edible nut; shrubs or small trees with doubly serrate lvs, colonial by roots. 10+, N. Temp.

1 Mature involucre 1.5–3 cm, with laciniate lobes; young twigs mostly glandular . . . . 1. *C. americana.*
1 Mature involucre 4–7 cm, with a long, subterete beak; twigs glandless . . . . 2. *C. cornuta.*

1. Corylus americana Walter.   American h.-n. Shrub 1–3 m, the young twigs and petioles ± pubescent (hairs red when young) and normally with stout, stipitate glands; lvs broadly ovate to obovate, finely doubly serrate, broadly rounded to cordate at base, paler and ± pubescent beneath; staminate catkins on short woody peduncles or branchlets; mature involucre pubescent but not bristly, 1.5–3 cm, the broad, laciniate bracts commonly separate above the nut, at least on one side; nut compressed, 1–1.5 cm, usually wider than long; 2*n*=22. Dry or moist woods and thickets; Me. to Sask., s. to Ga., La., and Okla.

2. Corylus cornuta Marshall.   Beaked h. Shrub 1–3 m, the young twigs villous at first; lvs oblong or oblong-obovate, short-acuminate, coarsely and doubly serrate, broadly rounded to subcordate at base, pale green beneath and ± pubescent, especially on the veins and in the vein-axils; catkins sessile or nearly so; involucre 4–7 cm, usually densely bristly below, prolonged beyond the nut into a long, slender beak cut at the summit into narrowly triangular lobes; nut short-ovoid, scarcely compressed, 1–1.5 cm; 2*n*=22. Moist woods and thickets; Nf. to B.C., s. to N.J., Pa., O., Mo., Oreg., and in the mts. to n. Ga. (*C. rostrata.*)

2. OSTRYA Scop., nom. conserv.   Hop-hornbeam, ironwood. Staminate catkins dense, the spirally arranged, depressed-ovate scales each terminating in a sharp point and subtending a cluster of several stamens without perianth; filaments short, divided at the summit, each branch bearing an apically pilose half-anther; pistillate catkins slender, loosely fld, the ovate, hairy bracts caducous, subtending 2 fls each enclosed within an ovoid pouch composed of united bract and bractlets, the cal minute; bracts accrescent and inflated in fr, forming a hops-like strobilus; fr a flattened-ovoid nutlet; trees or tall shrubs with rough, flaky bark, the catkins opening with the lvs in the spring. 10, mainly N. Temp.

1. Ostrya virginiana (Miller) K. Koch.   Tree to 20 m; twigs and petioles at first pilose and sometimes glandular, later glabrescent; lvs narrowly to broadly oblong or ovate, short-acuminate, sharply and often doubly serrate; fruiting catkins short-cylindric, 3–5 cm, each inflated sac flattened-ovoid 1–3 cm, bristly at base; nutlet 5 mm; 2*n*=16. Moist or dry woods and banks; N.S. to Man., s. to Fla. and Tex.

3. CARPINUS L.   Staminate catkins pendulous, the scales ovate, each bearing a single naked fl composed of several stamens; filaments short, each divided at the summit and bearing 2 apically pilose half-anthers; pistillate catkins slender, somewhat shorter, with ovate, deciduous scales; pistillate fls in pairs, each subtended by a minute bract adnate at base to 2 minute bractlets, the cal also minute; bracts and bractlets accrescent in fr; fr a small, ribbed nutlet; trees or shrubs, fl in early spring. 30, N. Temp.

1. Carpinus caroliniana Walter.   Hornbeam, blue beech, musclewood, ironwood. Tall shrub or small tree to 10 m, with flattened trunk and smooth, muscular-looking, blue-gray or ashy-gray bark; lvs oblong to oblong-ovate, 5–12 cm, acute or shortly acuminate, sharply and often doubly serrate; fruiting catkins ovoid to short-cylindric, 2–5 cm, the bracts (1.5–)2–3(–4) cm, halberd-shaped, with 1 or 2 divergent basal lobes, entire or with a few coarse teeth especially along one side of the middle lobe. Moist woods; N.S. to Minn., s. to Fla. and Tex. Most of our plants belong to the var. *virginiana* (Marshall) Fern., with the lvs beset with conspicuous dark glands beneath. Along the s. margin of our range this passes into var. *caroliniana* of se. U.S., lacking the dark glands of the often smaller and less toothed lvs, and with blunter, less toothed bracts.

4. BETULA L.   Birch. Staminate catkins elongate, the scales ovate to rotund, each subtending 3 fls; cal minute, 4-parted or usually 2-parted, one lobe much exceeding the other; stamens 2, each short filament divided near the summit and the anther-sacs separate; pistillate catkins ovoid to cylindric, the scales closely imbricate, generally 3-lobed, relatively thin, at maturity herbaceous or coriaceous, deciduous with or soon after the frs, the catkin disintegrating; fr ordinarily a small samara, the elliptic body with 2 membranous lateral wings and terminated by the short, persistent styles, or sometimes the wings virtually obsolete; trees or shrubs, the bark often separable into thin layers; the catkins appear in the fall and come into anthesis early in the spring (midsummer in *B. michauxii*). 50, N. Hemisphere.

The taxonomy of *Betula* is complicated by extensive hybridization and introgression, yet the species maintain their identity over large areas of sympatry. Many of the hybrids have been given names. *B.* × *borealis* Spach is *B. papyrifera* × *B. pumila*; *B.* × *caerulea* Blanchard and *B.* × *caerulea-grandis* Blanchard are *B. populifolia* × *B. papyrifera* var. *cordifolia*; *B.* × *minor* (Tuckerman) Fern. is *B. glandulosa* var. *glandulosa* a dwarf, alpine form of *B. papyrifera*, perhaps locally alloploid; *B.* × *purpusii* C. K. Schneider is *B. alleghaniensis* × *B. pumila*; *B.* × *murrayana* Barnes & Dancik is an octoploid backcross of *B.* × *purpusii* with *B. alleghaniensis*; and *B.* × *sandbergii* Britton is *B. papyrifera* × *B. glandulosa* var. *glandulifera*.

1 Large shrubs or usually trees; bark usually separable in layers; lvs and frs various.
  2 Bark yellowish-gray to dark reddish brown or blackish-brown; samara up to about twice as wide
      as the body, not (or only slightly) retuse; fruiting catkins (except in no. 3) sessile or nearly so.
    3 Lvs rounded to subcordate at base; crushed twigs with the flavor of wintergreen.
      4 Bark yellowish-gray, finely shaggy; pistillate scales pubescent, 6–13 mm .... 1. *B. alleghaniensis*
      4 Bark dark reddish-brown, smooth, eventually becoming rough with large, thick, loose-edged
        plates; pistillate scales glabrous, 5–7 mm .... 2. *B. lenta*.
    3 Lvs cuneate at base; not aromatic .... 3. *B. nigra*.
  2 Bark whitish; samara mostly more than twice as wide as the body, retuse at the summit; fruiting
      catkins peduncled.
    5 Lvs pubescent in the vein-axils beneath .... 4. *B. papyrifera*.
    5 Lvs glabrous beneath .... 5. *B. populifolia*.
1 Shrubs with close bark; lvs 0.5–3(–5) cm, with 2–6 pairs of lateral veins; samara less than twice as
    wide as the body, scarcely retuse.
  6 Pistillate scales 3-lobed; fr definitely though narrowly winged; lvs mostly 1–3(–5) cm.
    7 Resin-glands none .... 6. *B. pumila*.
    7 Resin-glands copiously scattered on the young stems and lvs .... 7. *B. glandulosa*.
  6 Pistillate scales oblong, entire; fr wingless; lvs 0.5–1 cm .... 8. *B. michauxii*.

1. Betula alleghaniensis Britton.   Yellow b. Tree to 30 m, the lustrous, yellowish-gray bark exfoliating in thin plates, appearing finely shaggy, eventually becoming dark and roughened in large trees; crushed twigs with the flavor of wintergreen; lvs lance-ovate to ovate, ovate-oblong or somewhat obovate, 6–10 cm, short-acuminate, coarsely and sharply toothed, rounded or subcordate at base, at maturity softly pubescent on the veins beneath, especially in the vein-axils, often paired on short spur-branches; lateral veins mostly 9–12 pairs; fruiting catkins sessile or nearly so, ovoid to short-cylindric, 2–3 cm; scales 6–13 mm, pubescent and ciliate, with ascending or divergent, oblong lobes; frs 2.5–4.5 mm, round-obovoid to subrotund or oblate, the body 1.5–2.5 mm wide, more than half as wide as the whole fr; $2n=84$. Moist, chiefly seral woods; Nf.

to se. Man., s. to Del., Pa., O., n. Ind., Wis., Minn., and occasionally Io., and along the mts. to n. Ga. (*B. lutea.*)

2. Betula lenta L.   Sweet or cherry b. Tree to 25 m, the smooth, reddish-brown bark exfoliating in thin layers, eventually becoming rough, with large, thickened, loose-edged plates on large trees; crushed twigs with a strong flavor of wintergreen; lvs ovate or ovate-oblong, 5–10 cm, acute or short-acuminate, very finely and sharply serrate, rounded to subcordate at base, appressed-villous on the veins beneath; lateral veins 9–12 pairs, impressed above; fruiting catkins sessile or nearly so, short-cylindric to ellipsoid or somewhat obovoid, 2–3 cm; scales glabrous, 5–7 mm, the middle lobe somewhat prolonged, the oblong lateral lobes divergent or ascending; frs broadly triangular-obovoid, 2.4–3 mm and about as wide, the body half as wide; $2n=28$. Moist woods; s. Me. and s. Que. to Del., O., and Ky., and in the mts. to Ga. and Ala. (*B. uber,* a striking Mendelian variant in Smyth Co., Va., with rounded lvs 2–4 cm, and 3–6 pairs of lateral veins.)

3. Betula nigra L.   River or red b. Tree to 30 m, the brown bark exfoliating in thin layers, scaly-roughened on larger trunks; lvs ovate-oblong or deltoid-ovate, 4–8 cm, acute, sharply double-serrate above, entire toward the cuneate base, tomentose beneath when young, soon glabrescent on the surface but remaining softly villous on the veins beneath and tomentose on the petiole; lateral veins 6–10 pairs; fruiting catkins on peduncles 5–8 mm, cylindric, 1.5–3 cm; scales pubescent, 6–8 mm, parted above the middle into 3 oblong lobes; frs pubescent, depressed-ovate, 3–4 × 4–7 mm, the body 2.5–4 mm wide. Swamps and flood-plain forests; N.H. to Fla., w. to s. O., s. Mich., se. Minn., e. Kans. and Tex.

4. Betula papyrifera Marshall.   White or paper-b. Usually a small to middle-sized tree, occasionally to 30 m, often slightly leaning rather than strictly erect; bark white or nearly so, with horizontal (often semilunate) black marks about the branches, easily separable into thin layers, the peeled plates showing salmon-pink on the inside; lvs ovate, 5–10 cm, acuminate, sharply serrate or doubly serrate, cuneate to rounded at base, glabrous above, very sparsely pubescent beneath, usually only along the veins or in the vein-axils; fruiting catkins 3–5 cm; scales 3.9–6.2 mm, two-thirds to fully as wide, the lateral lobes broadly falcate-obovate, divergent, the middle lobe tapering; frs elliptic, deeply retuse, broadly winged, 1.8–3.4 × 2.7–5 mm, the body 0.9–1.5 mm wide; mostly polyploid, often $2n=70$. Seral in moist or dry soil after fire or other disturbance; Lab. to Alas., s. to N.J., W.Va., n. Ind., and ne. Io. Becoming very dwarf at and above timberline in the White Mts. of N.H. and perhaps elsewhere. Most of our plants are var. *papyrifera*, as described above. The well marked but wholly confluent var. *cordifolia* is separately described.

Var. *cordifolia* (Regel) Fern.   Mt. Paper-b. More strictly erect and more branching below; lvs ovate to broadly ovate, short-acuminate, doubly serrate, cordate or subcordate (seldom merely truncate) at base, ± pubescent beneath; fruiting catkins longer, avg 5 cm; pistillate scales 5.6–8.7 × 4–5.4 mm, the lateral lobes ascending, scarcely dilated, the middle lobe somewhat prolonged and oblong; frs 4.1–6.7 mm wide, the body 1.2–2 mm wide; $2n=48$ (42, 56). Nf. and e. Que. to n. N.Y.; high Blue Ridge of N.C.; n. Minn. (*B. cordifolia.*)

B. alba L. (*B. pubescens* Ehrh.), the Old World counterpart of *B. papyrifera,* occasionally escapes from cult. It is more stiffly erect, with evidently resinous buds and mostly smaller lvs (3–5 cm on the fertile shoots) and pistillate catkins (1.5–3 cm). In Eurasia it is largely confined to wet habitats, whereas *B. pendula* grows in ordinary well drained soil.

5. Betula populifolia Marshall.   Gray b. Like a smaller and shorter-lived version of no. 4, seldom over 10 m tall, the trunks commonly clustered and erect-ascending, seldom over 1.5 dm thick; bark more dead-white with proportionately more black marking; lvs smaller, broadest near the often subtruncate base, and concavely tapering to a long, slender point, doubly serrate except on the base, glabrous beneath; fruiting catkins 1.3–3 cm; scales densely short-hairy on both sides; obtriangular in outline, 3–4.2 × 3–4.5 mm, the broad, falcate-obovate lateral lobes widely spreading, much larger than the terminal lobe and partly covering it; frs oblate, deeply retuse, broadly winged, 2.2–3 mm wide, the body 0.6–1 mm wide. Mostly diploid, $2n=28$. Seral in upland woods and old fields; N.S. to s. Que., s. N.J. and Pa., with outlying stations in n. Va., s. Ont., n. O., and ne. Ind.

B. pendula Roth, the European weeping birch, occasionally escapes from cult. and would key to *B. populifolia.* It has drooping twigs; middle-sized trunks are beset with vertical, diamond-shaped, rough black marks that eventually coalesce.

6. Betula pumila L.   Swamp-b. Erect, branching shrub 1–4 m, with close, brown bark; lvs firm, obovate to broadly ovate or orbicular, 2–3 cm, obtuse to broadly rounded above, coarsely dentate, rounded to broadly cuneate at base, softly pubescent on both sides when young, glabrescent with age, the hairs mostly 1–2 mm; lateral veins 3–6 pairs, the veinlets conspicuously raised-reticulate; fruiting catkins 1–2 cm, on peduncles 5–10 mm; scales 2.8–5.5 mm, about as wide, the lateral lobes widely divergent, constricted at base; frs depressed-obovate to suborbicular or oblate, truncate or rounded above, 1.6–2.8 × 1.8–3.7 mm, the body half as wide as the whole fr; $2n=56$. In bogs, often forming large colonies; Nf. and Que. to s. Ont. and Mich., s. to N.J., Md., c. O., and n. Ind.

7. Betula glandulosa Michx.   Dwarf b. Depressed or erect shrubs to 2 m, with close brown bark, the twigs and young lvs conspicuously dotted with resin-glands; lvs small, firm, obovate, varying to flabellate-obovate or rotund, roundly toothed, with mostly 3–5 pairs of lateral veins; petioles very minutely puberulent; fruiting catkins 1–2.5 cm; scales 2.5–4 mm, the lateral lobes ascending, or often widely divergent; frs broadly ovate to obovate, 1.2–2.5 × 1.5–2.3 mm, the body 1–1.5 mm wide. Bogs and wet alpine slopes and summits; widespread in boreal N. Amer., s. to the higher mts. of Me., N.H., and N.Y., and also to n. Ind., Minn., and the w. cordillera. Our ne. mt. plants, $2n=28$, with small lvs 1–2(3) cm that lack long hairs, are var. *glandulosa.* The more w., less alpine plants, $2n=56$ as in *B. pumila,* with the lvs avg larger (2–3 or even 5 cm) and ±

pilose when young, are var. *glandulifera* (Regel) Gleason, on ecological and cytologic grounds perhaps better called *B. pumila* var. *glandulifera* Regel, but morphologically more like *B. glandulosa.*

8. Betula michauxii Spach.   Newfoundland dwarf b. Small, rhizomatous shrub 1–6 dm; stems simple or branched, minutely cinereous-tomentulose; lvs broadly flabelliform-obovate, 0.5–1 cm, glabrous, shining, firm, coarsely reticulate, incised around the rounded to subtruncate summit, imbricate-ascending along the branches; fruiting catkins 0.5–1 cm, sessile on very short spur-branches; bracts oblong or oblong-ovate, entire or nearly so, narrower than the wingless, ovate to subrotund frs, tightly appressed, often with spreading tip. Bogs and open barren-lands; Lab. and Nf., and local in N.S. (*B. terrae-novae.)*

5. ALNUS Miller.   Alder. Staminate catkins pendulous, usually clustered, each bract subtending 3 fls with minute, 4-parted cal and 4 stamens with short, undivided filaments; pistillate catkins becoming woody, long-persistent, short, ovoid to ellipsoid or short-cylindric, each bract subtending 2 fls without cal; bracts of the pistillate catkins adnate to the bractlets, the compound unit cuneate, rounded or truncate and lobed at the summit; fr a small achene or samara, crowned with the 2 short, persistent styles and surrounded by a thin margin or membranous wing; trees or shrubs with 3-angled pith and broad, ovate to obovate, deciduous lvs. 30, mostly N. Hemisphere.

1 Winter buds sessile, covered by imbricate, unequal scales; frs broadly winged .... 1. *A. viridis.*
2 Winter buds evidently stalked, covered by 2 or 3 equal scales; frs merely margined.
  2 Pistillate catkins relatively small, mostly 1–1.5(2) cm, subsessile or short-pedunculate, often closely
    clustered; lvs with mostly 8–14 principal veins on a side.
    3 Lvs doubly serrate (or almost lobulate), with teeth of irregular size .... 2. *A. incana.*
    3 Lvs mostly serrulate with fine, almost regular teeth .... 3. *A. serrulata.*
  2 Pistillate catkins larger, mostly 1.5–3 cm, mostly evidently pedunculate; lvs with mostly 5–8 prin-
    cipal veins on a side.
    4 Fl late summer or fall; lvs ± pointed; local on Delmarva peninsula .... 4. *A. maritima.*
    4 Fl spring; lvs broadly rounded (or retuse) above; sparingly intr. .... 5. *A. glutinosa.*

1. Alnus viridis (Villars) Lam.   Green or mt. a. Tall colonial shrub; younger parts ± glutinous; winter buds virtually sessile, acuminate, with ca 5 imbricate scales; lf-bearing stems usually with both long shoots and short spurs, the latter bearing the lvs (unique among our spp. in this regard, the other spp. usually not forming both long and short shoots); lvs glutinous, especially beneath and when young, broadly elliptic to ovate, broadly obtuse to rounded or subcordate at base, with mostly 6–9 main veins on each side, green and shiny (not glaucous) beneath, sometimes sparsely velvety-rusty, finely and sharply toothed, or almost laciniate; catkins appearing with the lvs, the pistillate ones in infls with 2–3 lvs at the base, at maturity 1–1.5(2) cm, evidently slender-pedunculate; filaments 0.5–1 mm; fr 2–3 mm, the body only 1–2 mm wide and surrounded by a broad, pale membranous wing; $2n=28$. Bogs, shores, and cold woods; nearly circumboreal, in Am. s. to Mass., N.Y., Wis., Minn., Ida., and Calif., and at disjunct stations in Pa. and on Roan Mt. (N.C.–Tenn.). Boreal Amer. and e. Amer. pls, as here described, are var. *crispa* (Michx.) House. (*A. alnobetula; A. crispa; A. mollis,* the pubescent extreme.)

2. Alnus incana (L.) Moench.   Speckled a. Tall colonial shrub or small tree; winter buds evidently stalked, with 2–3 equal scales; lvs oval, elliptic, or ovate, broadest near or below the middle, obtuse to short-acuminate, sharply and irregularly double-serrate or almost lobulate, broadly obtuse to rounded at base, with mostly 8–14 principal veins on a side, soon becoming pale and glaucous beneath, ± pubescent, especially on the veins, not glutinous; staminate catkins mature before the lvs begin to open; anthers virtually sessile; fruiting catkins mostly sessile or only short-pedunculate, crowded in small, naked infls, 1–1.5(2) cm, seldom much over 1 cm thick; frs 2–3.5 mm, narrowly coriaceous-winged or merely thin-margined; $2n=28$. Wet soil; circumboreal, s. to Md., W.Va., Ill., and the w. cordillera. E. Amer. pls, as here described, are var. *americana* Regel. (*A. rugosa; A. incana* ssp. *rugosa.)*

3. Alnus serrulata (Aiton) Willd.   Smooth a. Much like no. 2; lvs elliptic to obovate, tending to be broadest above the middle, obtuse to rounded, serrate with very fine, sharp, nearly regular teeth, obtuse to cuneate at base, beneath green and glabrous, or thinly pubescent on the veins; $2n=28$. Wet soil; N.S. and s. Que. to n. Fla., w. to n. Ind., s. Ill., Mo., Okla., and w. Tex.

4. Alnus maritima (Marshall) Nutt.   Brook-a. Tall shrub or small tree; lvs oblong to obovate, obtuse to short-acuminate, serrate with low, distant, ascending or incurved teeth, broadly cuneate at base, dull green and glabrous beneath; principal lateral veins mostly 5–8 on a side, mostly curved-ascending, often not reaching the margins; fruiting catkins evidently pedunculate, not leafy-bracteate, relatively large, 1.5–3 cm long, often 1.5 cm thick, with very broad scales; fr ovate to obovate, 3–4 mm, ⅔ as wide; catkins in anthesis in late summer or fall (unique among our spp.); $2n=28$. Edges of ponds and small streams, often in standing water; Delmarva peninsula; disjunct in s.c. Okla., where said to have been intr. from Delmarva by relocated Indians. (*A. metaporina.)*

5. Alnus glutinosa (L.) Gaertner.   Black a. Tree, with young parts very glutinous; lvs broadly oval to obovate or subrotund, broadly rounded above or retuse, broadly cuneate to rounded at base, finely serrate, with 5–8 principal veins on a side; fruiting catkins very glutinous, evidently pedunculate in axillary infls, 1.5–2.5 cm, 10–13 mm thick; frs elliptic or suborbicular, 2.5–3.5 mm, narrowly bordered; $2n=28$. Native of Eurasia and n. Afr., planted especially along roadsides and occasionally escaped in our range. (*A. alnus; A. vulgaris.*)

FAMILY PHYTOLACCACEAE, the Pokeweed Family

Fls rather small, perfect or sometimes unisexual, regular, hypogynous; sep 4 or 5; pet generally none; stamens typically twice as many as the sep, sometimes fewer or more numerous; gynoecium of 1–several carpels, these separate or ± connate into a ring, each with a single ovule, style, and stigma; seeds with peripheral, dicotyledonous embryo curved around a ± abundant perisperm; herbs or sometimes woody plants (then with anomalous secondary growth), with alternate, entire lvs, producing betalains but not anthocyanins, the fls most commonly in racemes or spikes. 18/125.

1. **PHYTOLACCA** L. Pokeweed, pokeberry. Fls perfect or unisexual (the pls then dioecious), racemose, with 5 sep, 5–30 stamens, and a ring of 5–15 separate or ± connate carpels; herbs or woody pls. 35, mostly warm regions.

1. **Phytolacca americana** L. Coarse, glabrous, perennial herb to 3 m, branching above; lvs lance-oblong to ovate, 1–3 dm, on petioles 1–5 cm; racemes 1–2 dm, pedunculate, nodding in fr; fls perfect, greenish-white or pinkish, 6 mm wide; stamens and carpels usually each 10; carpels united except for the styles; berry depressed-globose, juicy, dark purple, 1 cm thick; $2n=36$. Fields, fence-rows, and damp woods; Me. to Minn., s. to the Gulf of Mexico. July–Sept. (*P. decandra.*)

P. acinosa Roxb., of China and Japan, is reported to be established as a weed in Wis. It resembles *P. americana* in aspect, but has 8 stamens and 8 distinct carpels. The carpels ripen into distinct frs, each with a thin dry pericarp and a large black seed, the set of frs collectively no larger than a berry of *P. americana*; $2n=36$. (*P. esculenta.*)

FAMILY NYCTAGINACEAE, the Four-O'clock Family

Fls perfect (in ours) or unisexual, hypogynous; cor none; cal gamosepalous, campanulate or funnelform, often corolloid, its base closely investing the ovary; stamens 1–many (1–5 in ours); filaments slender, exsert, usually unequal; gynoecium a single carpel with a long, slender style and a single basal ovule; fr a utricle or achene or nut, commonly enclosed in the persistent (and often indurated) base of the cal-tube, the collective structure called an anthocarp; seed with large, dicotyledonous, straight or more often curved, peripheral embryo and abundant or scanty perisperm; herbs or sometimes woody plants, commonly with anomalous secondary growth, producing betalains but not anthocyanins; lvs opposite or rarely alternate, simple, entire, exstipulate; 1–several fls often enclosed by a cal-like involucre, the cor-like cal and cal-like involucre easily mistaken for a fl with cal and cor. 30/300.

1. **MIRABILIS** L. Umbrella-wort. Cal corolloid, funnelform, ± 5-lobed; stamens 3–5; stigma capitate; anthocarp indurated, 5-angled or -ribbed; involucre well developed, 5-lobed, subtending 1–5 fls; perennial herbs (sometimes woody at the base) with opposite lvs and terminal, cymose infls. 60, New World.

1 Cal-tube above the ovary 1–2 mm; involucres 3–5-fld. (*Oxybaphus.*)
  2 Lvs ± ovate, with cordate or truncate base; infl not glandular .... 1. *M. nyctaginea.*
  2 Lvs linear to lanceolate, tapering to the base; infl ± glandular-pubescent.
    3 Lvs lanceolate, the principal ones mostly 1 cm wide or more; stem pubescent.
      4 Stem hirsute (at least at the nodes) with hairs 1–2 mm .... 2. *M. hirsuta.*
      4 Stem with short (0.5 mm) incurved hairs in 2 strips on each internode .... 3. *M. albida.*
    3 Lvs linear, seldom over 5 mm wide; stem glabrous or nearly so below the infl .... 4. *M. linearis.*
1 Cal-tube above the ovary 2–3 cm; involucres 1-fld. (*Mirabilis* proper.) .... 5. *M. jalapa.*

1. **Mirabilis nyctaginea** (Michx.) MacMillan.   Heart-lvd u.-w. Stem nearly smooth, branched above, to 1.5 m; lvs ovate-oblong to deltoid-ovate, acute, with cordate or truncate base, glabrous or nearly so, on petioles 1–3 cm; invol saucer-shaped, 1 cm wide, densely ciliate, accrescent in fr; cal pinkish-purple, 1 cm; anthocarp narrowly obovoid, densely hairy, rough on the sides and on the 5 prominent ribs; 2n=58. Dry soil; Man. to Ill. and La., w. to Mont. and Colo.; now established in waste places e. to Mass. and Va. May–Aug. (*Allionia n.; Oxybaphus n.; O. floribunda.*)

2. **Mirabilis hirsuta** (Pursh) MacMillan.   Hairy u.-w. Stems erect or decumbent, to 1 m, ± hirsute, especially about the nodes, with spreading hairs 1–2 mm, becoming glandular-hairy in the infl; lvs lance-linear to lance-ovate, the larger commonly 1–2 cm wide; invol 5 mm, glandular-hairy, becoming 1–2 cm; cal pink, 1 cm; anthocarp narrowly obovoid, 4–5 mm, hairy, rugose on the sides and ridges; 2n=58. Dry prairies, hills, and barrens; Wis. to Sask., s. to Mo., Tex., and Ariz. Summer. (*Allionia h.; A. pilosa; Oxybaphus h.*)

3. **Mirabilis albida** (Walter) Heimerl.   Pale u.-w. Stems erect or decumbent, 2–10 dm; each internode below the infl with 2 longitudinal strips of incurved hairs mostly less than 0.5 mm; infl glandular-hairy; invol 4 mm, becoming 1.5–3 cm wide at maturity; cal pink, 7–10 mm; anthocarp narrowly obovoid, 5 mm, thinly hairy, coarsely tuberculate on the sides and ridges. Dry prairies, sandhills, and barrens; S.C. to Tenn., Mo. (and reputedly Io.), and Kans., s. to La. and Tex. Summer (*Oxybaphus a.; Allionia a.; A. decumbens.*)

4. **Mirabilis linearis** (Pursh) Heimerl.   Narrow-lvd u.-w. Stems erect or decumbent, 3–10 dm, glaucous and commonly glabrous below the usually glandular-hairy infl; lvs linear, without differentiated petiole, rarely over 5 mm wide, glaucous, thick and succulent, becoming very wrinkled when dry, invol pubescent, 5 mm, becoming 1–2 cm wide; cal pale pink to purple, 1 cm; anthocarp narrowly obovoid, 5–7 mm, with rugose sides and smooth ridges; 2n=52. Dry or sandy soil, prairies, and barrens; N.D. (and reputedly Minn.) to Mont., s. to Tex., Ariz., and n. Mex.; intr. in Mo., and rarely adventive in e. U.S. Summer. (*Allionia l.; Oxybaphus l.*)

5. **Mirabilis jalapa** L.   Four-o'clock. Perennial from a thick, tuberous root, but grown as an annual; lvs ovate, petioled; invol cal-like, of 5 united bracts, deeply 5-lobed, 1-fld, 6–8 mm at anthesis; cal cor-like, showy, opening in late afternoon, white, yellow, red, purple, or variegated, funnelform, with long, slender tube 2–3 cm, the limb 2–3.5 cm wide; 2n=58. Native of tropical Amer., widely cult., and occasionally escaping in our range.

Family **AIZOACEAE**, the Fig Marigold Family

Fls perfect (in ours) or seldom unisexual, perigynous to epigynous; sep commonly 5; pet typically numerous, in 1–6 cycles, mostly linear, or sometimes (including both of our spp.) wanting; stamens 5-merous, sometimes basally connate into groups; gynoecium of 2–5 or more numerous carpels united to form a compound ovary with distinct (or distally distinct) styles, or sometimes apparently reduced to a single carpel; ovary generally with as many locules as carpels; ovules solitary to usually ± numerous in each locule, on axile, basal, apical, or parietal placentas, campylotropous to almost anatropous; fr most commonly a loculicidal capsule, but in ours circumscissile, often ± included in the persistent cal; seeds with an elongate, dicotyledonous, peripheral embryo curved around the abundant perisperm; succulent herbs (ours) or less commonly shrubs, very often with anomalous secondary growth, producing betalains but not anthocyanins; lvs opposite or alternate, simple and generally entire, exstipulate, or sometimes with interpetiolar stipules; fls in rather small cymose infls, or solitary in the axils. 60/2500.

Tetragonia tetragonioides (Pallas) Kuntze, New Zealand Spinach, is cult. and occasionally escaped or persistent. It is a succulent annual with alternate, deltoid-ovate lvs, inconspicuous axillary fls, a half-inferior ovary with a single pendulous ovule in each of the several locules, and small, hard, obconic, indehiscent, 1-seeded frs crowned with 2–5 short horns. (*T. expansa.*)

1 Styles and locules 3–5; seeds numerous; lvs without stipules . . . . 1. *Sesuvium.*
1 Style and locule 1 (in ours) or 2; seeds few; lvs with interpetiolar stipules . . . . 2. *Trianthema.*

1. **SESUVIUM** L.   Sea-purslane. Fls perfect, perigynous; sep 5, spreading at anthesis, persistent, each usually with a dorsal appendage near the tip; pet none; stamens 5–many; ovary 3–5-locular, with 3–5 styles and many ovules; capsule circumscissile; seeds arillate; succulent, fibrous-rooted herbs with opposite, exstipulate lvs, the small fls solitary or few in the axils, sessile or nearly so. 8, warm reg.

1. Sesuvium maritimum (Walter) BSP.   Annual, glabrous and succulent throughout, prostrate to ascending, 1–3 dm, freely branched; lvs oblanceolate to broadly spatulate, 1–3 cm, the members of a pair commonly unequal; sep 2–3 mm including the short appendage; stamens 5; fr ovoid, 4 mm; seeds numerous, smooth, under 1 mm. Sea-beaches; N.Y. to Tex. and W.I.; scattered inland in Kans. and Okla. July–Sept.

2. TRIANTHEMA L.   Fls perfect, perigynous; sep 5, each usually with a dorsal appendage near the tip; pet none; stamens mostly 5–10; ovary unilocular or bilocular, with accordingly 1 or 2 styles; ovules few; capsule circumscissile; seeds arillate; succulent, shortly taprooted herbs with opposite lvs basally connate into a short sheath with interpetiolar stipules; fls small, solitary or few in the axils. 20, warm reg.

1. Trianthema portulacastrum L.   Perennial, glabrous or sparsely hairy, diffusely branched from the base, decumbent or ascending, to 1 m; lvs round-obovate, 1–4 cm, slender-petiolate, the members of a pair somewhat unequal; fls sessile, partly concealed in the petiolar sheath; ovary unilocular, style 1; membranous hypanthium and small cal pushed off by growth of the fr; fr 4–5 mm; asymmetrically crested around the broad top; seeds 1–4, rough, 2 mm. Widespread tropical weed, extending n. to Okla. and Mo., and rarely about our Atlantic ports. May–Sept. Reputedly poisonous to livestock.

FAMILY **CACTACEAE**, the Cactus Family

Fls epigynous, mostly regular and perfect; tepals generally numerous and spirally arranged, all showy and petaloid, or often the outer ones more sepaloid, but not sharply differentiated into two types, all united below to form a hypanthium; stamens numerous, arising spirally or in groups from the hypanthium; gynoecium of 3–many carpels united to form a compound, inferior, unilocular ovary with a single style and as many radiating stigmas as carpels; ovules numerous on parietal placentas, mostly campylotropous; fr a dry, pulpy or juicy berry, seeds with straight or more often curved, dicotyledonous embryo and abundant to scanty or no perisperm; spiny (seldom unarmed) stem-succulents, producing betalains but not anthocyanins, usually with scarcely developed lvs, the spines commonly restricted to regularly arranged areoles (the modified nodes) that usually also have clusters of small, detachable bristles (glochids); fls sessile, mostly large and solitary at the areoles. 30+/1000+.

1 Stems branched, jointed .... 1. *Opuntia.*
1 Stems subglobose, not jointed .... 2. *Coryphantha.*

1. OPUNTIA Miller.   Prickly pear. Stems branched and jointed, the joints cylindric to flattened; spines and glochids arising from the areoles, or the plants virtually spineless; fls borne within the areoles near the tips of joints of the previous year; pet and sep rotate from the summit of the scarcely prolonged hypanthium; stamens shorter than the pet; seeds wingless. 150+, New World.

1 Spines solitary or occasionally paired, borne at only a few areoles .... 1. *O. humifusa.*
1 Spine usually several together, borne at most areoles.
  2 Joints strongly flattened, mostly 6–10 cm, not easily detached .... 2. *O. macrorhiza.*
  2 Joints turgid, not much flattened, 2–5 cm, easily detached .... 3. *O. fragilis.*

1. Opuntia humifusa (Raf.) Raf.   Eastern p. p. Prostrate or spreading, forming large mats; roots mostly fibrous, only seldom tuberous-thickened; joints of the stem flattened, oblong to suborbicular, 4–12 cm at maturity; areoles commonly 10–25 mm apart, spineless, or some of them with 1, rarely 2 spines 1.5–3 cm; fls 4–8 cm wide, yellow, often with a red center; outer sep subulate to lanceolate; fr red or purple, 2.5–5 cm, edible; seeds discoid, with an indurate, regular margin. On rocks, shores, sand-dunes, or sandy prairies; e. Mass. to s. Ont. and s. Minn., s. to Fla. and e. Tex. June, July. (*O. calcicola; O. compressa; O. opuntia; O. pollardii; O. rafinesquei; O. vulgaris,* misapplied.)

2. Opuntia macrorhiza Engelm.   Plains p. p. Prostrate and spreading, forming mats to 1.5 m wide; usually some roots tuberous-thickened; joints of the stem mostly 6–10 × 5–7 cm, flat, orbicular to obovate; areoles on full-grown joints 15–30 mm apart, with (1–)3–6 spines 3.5–5.5 cm; fls yellow with reddish center, 5–7 cm wide; outer sep ovate, acute to acuminate; fr reddish-purple, 2.5–4 cm, juicy; seeds discoid with a rough, irregular, corky margin. Prairies and plains; S.D. to Ariz. and Tex., e. occasionally to Wis., s. Mich., Ill., and reputedly O. and Ky. May–July. (*O. mesacantha; O. tortispina,* misapplied.)

O. polyacantha Haw., a similar western sp. with 6–10 spines per areole and dry frs, has been reported e. to Wis. and Mo.

3. Opuntia fragilis (Nutt.) Haw. Little p. p. Prostrate or spreading, forming dense mats to 5 dm wide; joints orbicular to obovate, very turgid, 2–5 × 1–2.5 cm, easily detached; areoles crowded, commonly 3–6 mm apart, usually coarsely white-woolly, all or nearly all armed with (1–)3–7 strongly barbed spines 1.2–2.5 cm; fls yellow to greenish, 4–5 cm wide; fr fleshy and greenish to reddish when young, dry and tan at maturity, inedible, 1.2–1.5 cm; seeds discoid, with an irregular margin; 2*n*=66. Dry prairies and plains; Ill. and Wis. and n. Mich. to B.C., n. Tex., and Ariz. May–July.

2. CORYPHANTHA (Engelm.) Lemaire. Stems cylindric to globose, solitary or forming clumps, not jointed, covered with cylindric to conic tubercles, each of these with a terminal areole bearing radial and usually also central spines but lacking glochids; fls arising individually on new growth of the season at the base of an adaxial groove descending from the areole; hypanthium funnelform-prolonged above the ovary; fr thin-walled, fleshy; seeds small, dark, wingless. 30+, N. Amer.

1 Central spines wanting or only 1 or 2; fls yellow to yellowish-red .... 1. *C. missouriensis.*
1 Central spines mostly 4; fls dark purplish-pink .... 2. *C. vivipara.*

1. Coryphantha missouriensis (Sweet) Britton & Rose. Stems solitary or in clumps, 5–10 cm; tubercles elongate, 12–15 mm; areoles very woolly; spines pubescent, at first yellowish, becoming dark gray; central spines 0–1(2), the radial ones 10–20, spreading, straight, 1–2 cm; fls 2.5 cm, yellow to yellowish-red; fr red, 1–2 cm; seeds black, 1 mm; 2*n*=44. Dry plains and hillsides, often among grasses; N.D. to Io., Mo., and Ark., w. to Mont. and Ariz. June, July. Ours is var. *missouriensis.* The var. *caespitosa* (Engelm.) L. Benson, with larger fls (ca 5 cm) and seeds (2.5 mm), may also be sought at our w. border.

2. Coryphantha vivipara (Nutt.) Britton & Rose. Stems solitary or in clumps, to 30 cm; tubercles elongate, 5–20 mm; areoles moderately woolly; central spines mostly 4, 1 pointing downward, 1–2 cm, red or basally white; radial spines 12–20, 1 cm, white; fls 4 cm, dark purplish-pink; fr green 1.2–2.5 cm; seeds brown, 1.5–2 mm; 2*n*=22. Dry, grassy plains; w. Minn. to Kans. and Okla., w. to Alta. and Ariz. May–Aug. (*Mammillaria v.; Neomammillaria v.*) Ours is var. *vivipara.*

FAMILY **CHENOPODIACEAE**, the Goosefoot Family

Fls small, perfect or less often unisexual; sep (1)5(6), distinct or basally connate, rarely wanting, greenish and ± herbaceous, or rarely somewhat scarious or even hyaline; pet none; stamens as many as and opposite the sep, or seldom fewer; gynoecium of 2–3(5) carpels united to form a compound, unilocular ovary with distinct or connate styles; ovary superior or seldom (as in *Beta*) half-inferior; ovule solitary, basal, ± campylotropous; fr a utricle or a small nut, indehiscent or seldom with irregular or circumscissile dehiscence, the pericarp thin and sometimes adherent to the seed, often subtended by the persistent cal or by persistent bracteoles, sometimes multiple; seeds mostly lenticular, with elongate, dicotyledonous, annular or spirally twisted embryo peripheral to the well developed perisperm, or seldom without perisperm; herbs or shrubs, producing betalains but not anthocyanins, with simple, exstipulate, alternate or seldom opposite (sometimes much reduced) lvs; fls 1–many and glomerate in the axils or in bracteate or ebracteate spikes, panicles, or cymes, the bracts and bracteoles (if present) mostly blunt and herbaceous, seldom scarious. 100/1500.

1 Lvs represented only by small, opposite, scarcely projecting scales .... 9. *Salicornia.*
1 Lvs more normally developed, all or chiefly alternate.
  2 Fr free from the cal and exposed (though sometimes hidden by the bract to which it is axillary);
    cal of 1–3 sep.
    3 Lvs linear, entire; sep 1–3, scarious, minute .... 7. *Corispermum.*
    3 Lvs, at least the main ones, with 2 large salient teeth; sep 1, herbaceous .... 6. *Monolepis.*
  2 Fr generally surrounded and ± enclosed either by the persistent cal or by the paired bracteoles.
    4 Fr enclosed and concealed by a pair of rhombic, rounded, or deltoid bracteoles .... 8. *Atriplex.*
    4 Fr surrounded or enclosed by the persistent cal.
      5 Foliage lvs all entire and linear to narrowly lanceolate.
        6 Sep at maturity membranous or scarious, erect or nearly so .... 5. *Axyris.*
        6 Sep at maturity herbaceous or coriaceous, ± incurved over the fr.
          7 Fruiting sep each with a flat, horizontal dorsal wing, or transversely keeled.

```
8 Lvs unarmed .........................................................................  4. Kochia.
8 Lvs, especially those of the infl, spiny-tipped .......................................11. Salsola.
  7 Fruiting sep concave or cucullate or carinate or spiny on the back.
  9 Plants glabrous or white-mealy or glandular-puberulent.
    10 Fls few to several in glomerules, these forming a terminal panicle ..............  1. Chenopodium.
    10 Fls sessile in groups of 3 in the axils of the lvs ...................................  10 Suaeda.
  9 Plants villous, especially about the infl ................................................  3. Bassia.
5 Foliage lvs either toothed or lobed or broader than lanceolate.
 11 Fruiting cal horizontally winged ....................................................  2. Cycloloma.
 11 Fruiting cal unchanged in form, its lobes flat or rounded to carinate or cucullate .........  1. Chenopodium.
```

1. CHENOPODIUM L. Goosefoot. Fls perfect (seldom some of them pistillate); cal persistent, mostly 2–5-parted (most commonly 5-, less often 3-), the short, usually blunt segments commonly incurved over the fr (cal only shallowly lobed in one sp.); stamens 1–5, typically isomerous with the cal-lobes; styles 2(–5); fr laterally compressed (the seed erect) or more often flattened across the top (the seed horizontal), thin-walled, the pericarp often adherent to the ± lenticular seed; embryo annular; ours herbs (most spp. annual) with entire or toothed to ± deeply lobed lvs and small, greenish to reddish fls, these in most spp. sessile in glomerules (the glomerules either axillary or in terminal spike-like or panicle-like infls), but in other spp. in compact cymes that may collectively form a thyrse, or otherwise disposed. (*Blitum, Roubieva.*) 100+, cosmop.

```
1 Mature cal obovoid-urceolate, shallowly-toothed, raised-reticulate. (Roubieva.) ....................  1. C. multifidum.
1 Mature cal lobed to the middle or below, not veiny or reticulate.
  2 Ultimate branches of the diffuse small infls appearing as slender spines ........................ 11. C. aristatum.
  2 Plants unarmed.
    3 Foliage beset with resinous glands or glandular hairs.
      4 Fls in dense glomerules disposed in short or elongate spike-like infls.
        5 Perianth smooth to minutely puberulent; seeds all or mostly horizontal ..............  2. C. ambrosioides.
        5 Perianth densely yellowish-glandular; seeds erect ....................................  3. C. pumilio.
      4 Fls in a slender, thyrsoid infl of small lateral cymes ....................................  4. C. botrys.
    3 Foliage not glandular.
      6 Seeds (or most of them) erect. (Blitum.)
        7 Perennial; styles enlarged in fr, 1–1.5 mm; sep mostly 4 or 5 ......................  5. C. bonus-henricus.
        7 Annual; styles not enlarged, not evident in fr; sep of most fls 3.
          8 Lvs green on both sides.
            9 Glomerules of fls relatively few and large, 5–15 mm thick at maturity and becoming
              fleshy ....................................................................................  7. C. capitatum.
            9 Glomerules smaller and more numerous, up to ca 5 mm thick at maturity, not be-
              coming fleshy ..........................................................................  6. C. rubrum.
          8 Lvs densely white-mealy beneath ..........................................................  8. C. glaucum.
      6 Seeds horizontal; sep consistently 5.
        10 Sep at maturity rounded on the back in conformity with the fr, the midvein flat or only
           slightly raised.
           11 Lvs densely white-mealy beneath; plant very fetid ............................  9. C. vulvaria.
           11 Lvs green on both sides, or sometimes slightly mealy when young; plants not strongly
              odorous.
              12 Lvs entire or nearly so .................................................  10. C. polyspermum.
              12 Lvs sharply and coarsely toothed.
                 13 Seed and fr rounded on the margin ..................................  12. C. urbicum.
                 13 Seed and fr keeled or sharply angled around the margin.
                    14 Lf-teeth 1–4 per side; seed shining, 1.5–2.5 mm wide ............. 13. C. gigantospermum.
                    14 Lf-teeth several per side; seed dull, 1.2–1.5 mm wide ................... 14. C. murale.
        10 Sep at maturity not in full conformity with the fr, either carinate, or cucullate, or strongly
           elevated in the middle, the cal thus ± pentagonal or almost star-shaped.
           15 Lvs relatively broad, lanceolate to broadly ovate, rhombic, or deltoid; secondary veins
              developed.
              16 Pericarp papery, brittle when dry, loosely enclosing the seed ...............  15. C. standleyanum.
              16 Pericarp very thin and membranous, when dry ± closely adherent to the seed
                 and usually scarcely separable from it.
                 17 Pericarp smooth or only obscurely roughened as seen at 10× or even 20× ........ 16. C. album.
                 17 Pericarp evidently roughened and cellular-reticulate as seen at 10× (or in any
                    case at 20×) .............................................................. 17. C. berlandieri.
           15 Lvs narrower, lanceolate to linear, mostly entire, 1- or 3-nerved, without evident
              secondary veins.
              18 Lvs 3-nerved; pericarp loose, readily separable from the seed ................. 18. C. pratericola.
              18 Lvs 1-nerved; pericarp firmly adherent to the seed ...................... 19. C. leptophyllum.
```

1. Chenopodium multifidum L. Prostrate or spreading, odoriferous perennial; stems to 6 dm, villous-puberulent; lvs numerous, obscurely puberulent, finely resinous-glandular beneath, subsessile, 1–4 cm, narrowly oblong, deeply and irregularly pinnatilobate, the upper much reduced; fls in small axillary glomerules; cal urceolate, shallowly 5-lobed, at maturity 1.5–2.5 mm, obovoid, raised-reticulate-veiny, and loosely but completely enclosing the fr; stamens 5; styles 2–5, basally connate; pericarp thin, free, seed erect, obovate, thick-lenticular, smooth, 1 mm wide; $2n=32$. Native of S. Amer., intr. in waste places in our range, esp. southward. (*Roubieva m.*)

2. Chenopodium ambrosioides L. Mexican tea, wormseed. Malodorous annual or short-lived perennial, erect, to 1(–2) m, with ascending branches; lvs beset with minute yellow glands, the lower blades lanceolate to ovate, to 12 cm, acute, deeply sinuate-pinnatifid to merely serrate, cuneate at base, the upper progressively less toothed or entire; fls sessile in small glomerules that are disposed in slender, somewhat elongate spikes, these in turn forming terminal panicles; cal 5-lobed, not obviously glandular; styles 3–4; seeds horizontal or some of them erect, thick-lenticular, not margined, 0.6–1 mm wide, dark brown, shining; $2n=16, 32, 64$. Native of tropical Amer., established as a weed in gardens, roadsides, and waste places n. to Me. and Wis.

3. Chenopodium pumilio R. Br. Annual; stems spreading or prostrate, 2–4 dm; lvs densely beset with large yellow glands beneath, oblong or lanceolate, 1–3 cm with 2–4 coarse teeth per side; glomerules small, aggregated into small axillary spikes to 10 mm; cal deeply 5 cleft, beset with large yellow glands; seeds erect, 0.5–0.6 mm wide, lenticular with a raised margin; $2n=16$. Native of Australia, naturalized in waste places in our range. (*C. carinatum.*)

4. Chenopodium botrys L. Jerusalem-oak, feather-geranium. Annual 2–6 dm, pubescent with short glandular hairs, strongly but not unpleasantly aromatic; lvs oblong to ovate, the lower to 8 cm, sinuate-pinnatifid, the upper much smaller, pinnatilobate to entire; fls in numerous small lateral cymes along the main axes, forming long, slender, terminal infls; cal deeply 5-lobed, glandular-hairy, the segments rounded on the back; seeds horizontal or some of them erect, dull blackish, 0.6–0.8 mm wide, thick-lenticular, not margined; $2n=18$. Native of Europe, established as a weed in waste places in our range.

C. graveolens Willd., a native of sw. U.S. and trop. Amer., with the cal beset with yellow glands, the segments at maturity developing a conical tubercle on the back near the top, is rarely adventive in our range.

5. Chenopodium bonus-henricus L. Good King Henry. Erect or ascending, thick-rooted perennial to 7 dm; lvs numerous, long-petioled, broadly triangular-hastate, to 12 cm, entire or undulate; fls in small glomerules aggregated into a terminal, branching, paniculiform infl; sep mostly 4 or 5, erose; styles 2(3), persistent and conspicuous, 1–1.5 mm in fr; seeds all or mostly erect, black, shining, lenticular with a raised margin, 1.5–2 mm wide; $2n=36$. Native of Europe, occasionally found as a weed in our range.

6. Chenopodium rubrum L. Alkali-blite. Annual, branched from the base, prostrate to erect, 1–8 dm, the foliage and fls commonly tinged with red or becoming red at maturity; lvs commonly rhombic-ovate or oblong, with a conspicuous lateral tooth on each side, cuneate below the teeth, varying to entire or several-toothed, green on both sides; fls in ± numerous glomerules, these up to ca. 5 mm thick at maturity, in small plants chiefly axillary, in larger ones aggregated into terminal paniculiform infls; sep mostly 3; seeds erect, shiny dark brown, lenticular, 0.6–1 mm wide; $2n=36$. Salt marshes and brackish soil, Nf. to N.J.; more abundant inland, from Ind. and Io. to Wash. and Calif., and occasionally adventive elsewhere. Tall, erect plants with lvs 4–10 cm and large infls are typical *C. rubrum*. Prostrate plants with small lvs and only axillary glomerules have been segregated on insufficient grounds as *C. humile* Hook. or *C. rubrum* var. *humile* (Hook.) S. Wats.

7. Chenopodium capitatum (L.) Aschers. Strawberry-blite. Erect or ascending annual 2–6 dm, branched from the base; lower petioles often much exceeding the blades, the upper much shorter; lvs triangular or triangular-hastate, to 10 cm, acute, broadly truncate to an acute base, above the lateral angles entire to coarsely sinuate-dentate; fls in a limited number of globose clusters 5–10 mm thick at anthesis (to 15 mm in fr) these forming a terminal lfless spike, or some also in the upper axils; cal deeply 3-parted, the segments oblong to obovate, concave, at maturity commonly enlarged, fleshy, deliquescent-coalescent, and bright red; seeds erect, dull black, 0.7–0.8 mm wide, narrowly margined; $2n=18$. Woodland clearings, often following a fire, and along roadsides in waste places; circumboreal, in our range s. to N.J., Ind., and Mo. (*Blitum c.*)

C. foliosum (Moench) Aschers., with the glomerules of fls all subtended by reduced lvs and with the seeds channeled around the periphery, is occasionally adventive (from Eurasia) in our range.

8. Chenopodium glaucum L. Oak-lvd g. Annual, usually branched from the base, prostrate to erect, 1–4 dm; lvs lanceolate to oblong or ovate, 1–4 cm, cuneate at base, entire or merely undulate, or with 2–4 low teeth per side, densely white-mealy beneath, especially when young; fls in small glomerules aggregated into usually interrupted spiciform clusters equaling or shorter than the lvs, or also in a short terminal panicle; sep mostly 3(4); seeds dark brown, shining, lenticular, 0.6–1 mm wide, loose in the pericarp, those of the lateral fls of the cymules erect, that of the terminal fl in the cymule horizontal; $2n=28$. Native of Europe, occasionally found as a weed in our range, especially northward. (*C. salinum.*)

9. Chenopodium vulvaria L. Stinking g. Fetid annual, strongly white-mealy, erect or ascending, 1–5 dm, widely branched from the base; lvs broadly ovate, 1–3 cm and about as wide, entire, rounded or broadly cuneate to the petiole, densely white-mealy beneath, glabrescent above; infls few, of small glomerules in compact leafy spikes, chiefly from the upper axils; sep 5, white-mealy, rounded on the back; seeds horizontal, dull black, thick-lenticular, loose in the pericarp, 1 mm wide; $2n=18$. Native of Eurasia, found here and there in our range as a weed.

10. Chenopodium polyspermum L. Erect or ascending annual to 1 m; stems 4-angled, lvs green, ovate to elliptic, 2–7 dm, obtuse or subacute, entire or with an obscure tooth at the widest part, rounded or cuneate to a short petiole; infls axillary, loose, cymosely much branched, many-fld, shorter than or equaling the lvs; cal smooth, its 5 apiculate segments rounded on the back; seeds horizontal, shiny-black, 1 mm wide; $2n=18$. Native of Eurasia, occasionally found as a weed in our range.

11. Chenopodium aristatum L. Branching, glabrous or granular-scaberulous annual to 3 dm, turning red in autumn; lvs linear or narrowly elliptic, entire, to 6 cm × 8 mm; fls individually solitary in the forks of the diffusely and slenderly branched axillary dichasial cymes, the ultimate branches appearing as slender spines 1–4 mm; sep 5; seeds horizontal to erect, dull blackish, 0.6–0.7 mm wide. Native to c. and e. Asia, locally established as a weed in Mich.

12. Chenopodium urbicum L. City-g. Erect annual 3–10 dm; lvs on short or long petioles, triangular to rhombic-ovate or lanceolate, acute, sharply and coarsely toothed, truncate or cuneate at base, shining above, usually slightly white-mealy beneath when young, to 8 × 5 cm; cal smooth or white-mealy, its 5 segments rounded on the back; seeds horizontal, shiny dark brown, rounded at the margin, 1.0–1.2 mm wide; $2n=18$, 36. Native of Europe, sparingly intr. in waste places in our range.

13. Chenopodium gigantospermum Aellen. Maple-lvd g. Erect, bright green annual to 1.5 m; lvs long-petioled, broadly ovate to deltoid, 5–20 cm, acute or acuminate, truncate to rounded or cordate at base, bearing on each side 1–4 large, salient teeth separated by broadly rounded sinuses; infl a loose, sparsely fld, terminal panicle of short, interrupted spikes, the branches often white-mealy; cal sparsely or not at all mealy; seeds horizontal, loosely or tightly enclosed in the readily separable pericarp, shiny-black, obscurely striolate, 1.5–2.5 mm wide, with a bluntly keeled margin; $2n=36$. Disturbed ground and moist woods; Que. to Yukon, s. to Va., Ky., Ark., Tex., and Calif. Closely allied to the European diploid *C. hybridum.*

14. Chenopodium murale L. Nettle-lvd g. Erect, branching annual to 8 dm; lvs long-petioled, deltoid to rhombic-ovate, acute or acuminate, sharply and coarsely toothed, truncate to broadly cuneate at base, sometimes a little white-mealy, to 8 × 6 cm; infls mostly axillary, divergent, branched; cal ± white-mealy, its 5 segments inconspicuously keeled on the back; seeds adnate to the pericarp, horizontal, dull black, sharply margined, 1.2–1.5 mm wide; $2n=18$. Native of Europe, intr. in waste places in our range.

15. Chenopodium standleyanum Aellen. Woodland-g. Erect or arching annual to 1(–2) m; lvs thin, green or sparsely mealy, lanceolate to rarely ovate, to 8 cm, acute, entire or the larger with a few low teeth, acute or cuneate at base; fls single to few in small glomerules, these forming short, interrupted spikes that are often grouped into a loose, slender, sometimes nodding panicle; cal ± white-mealy, scarcely covering the fr; pericarp smooth, papery, fragile, easily separable; seeds horizontal, black, shining, ca 1 mm wide, smooth to faintly striolate, longitudinally striolate over the radicle; $2n=18$. Dry, open woods; widespread in e. U.S. and adj. Can., from s. Que. to Fla., and w. to S.D. and Tex. (*C. boscianum,* misapplied.)

16. Chenopodium album L. Lamb's quarters, pigweed. Erect annual, much branched and to 1 m when well developed, the stems only seldom pigmented at the nodes, the lvs and infl often turning reddish late in the season; lvs generally ± white-mealy, rhombic-ovate to lanceolate, 3–10 cm, broadly cuneate at base, the larger (lower) ones mostly 1.5–2+ times as long as wide and almost always toothed; fls in dense glomerules, forming interrupted or continuous spikes that are grouped into a terminal paniculiform infl; cal ± white-mealy, its segments ± strongly carinate or cucullate, covering the fr, or sometimes looser and exposing the fr.; style divided to the base, wholly deciduous; pericarp very thin and delicate, closely adherent to the seed and usually scarcely separable from it, smooth or only obscurely roughened when viewed at 10× or even 20×, uniformly black or blackish; seeds horizontal, black, shining, mostly 1.0–1.5 mm wide, usually marked with a faint radial furrow, otherwise smooth or nearly so; $2n=54$. June–Oct. Polymorphic European weed, now widely intr. in N. Amer. and elsewhere. Sometimes divided into an indefinite number of infraspecific taxa or specific segregates. (*C. glaucophyllum; C. lanceolatum,* a relatively narrow-lvd form.) In our range the two following segregates are particularly noteworthy and probably merit some sort of taxonomic recognition:

C. opulifolium Schrader. Differing from characteristic *C. album* mainly in its wider lvs, the larger ones nearly as wide as long.

C. missouriense Aellen. Lvs mostly wider than in typical *C. album,* the lower ones up to 1.5 times as long as wide; stem consistently purple at the nodes; infl less compact, more delicate and flexuous; seeds avg smaller, 0.9–1.2 mm wide; fl only in late season, mostly Sept. The name *C. album* var. *missouriense* (Aellen) Bassett & Crompton is available.

17. Chenopodium berlandieri Moq. Pitseed g. Much like *C. album,* but the pericarp evidently roughened and cellular-reticulate when viewed at 10× (or in any case at 20×); a minute (0.1 mm) undivided style-base persistent on the fr (at least in var. *zschackei*); $2n=36$. Common native Amer. weed, widespread in our range and s. into Mex. Three ± distinguishable vars. with us:

a Pericarp with a light yellowish area around the style, where it does not fit closely to the seed; sep broadly wing-keeled; herbage white-mealy to a varying degree, as in *C. album*; seeds mostly 1.0–1.5 mm wide; mainly in c. U.S., w. of the Appalachian Mts. .... var. *zschackei* (Murr) Murr.
a Pericarp wholly black or blackish, without a light area around the style; style-base less prominent than in var. *zschackei,* or wanting; sep less strongly keeled; herbage only thinly mealy, distinctly greener than in characteristic plants of var. *zschackei* and *C. album.*
b Seeds mostly 1.5–2.3 mm wide; plants when well developed relatively robust, often 1 or even 2 m

tall, with large, drooping infls and with ovate or rhombic lvs 6–10(–15) cm, but smaller plants are of a size with the next var.; widespread in our range. (*C. bushianum; C. paganum,* mis-applied.) .... var. *bushianum* (Aellen) Cronq.

  b Seeds mostly 1.3–1.7 mm wide; plants relatively small, seldom over 5 dm, the infls small and erect, the lvs mostly lanceolate or lance-elliptic, often less than 4 cm, seldom over 6 cm; mainly coastal or coastal plain, often on sea-beaches; also about the s. end of Lake Michigan. (*C. macrocalycium.*) .... var. *macrocalycium* (Aellen) Cronq.

18. Chenopodium pratericola Rydb. Erect annual to 8 dm with ascending branches; lvs erect or ascending, oblong to lanceolate or lance-ovate, entire or the larger ones often few-toothed or subhastate, at least the larger ones 3-nerved from the base, but without any apparent secondary veins, mostly 2–4 cm × 4–15 mm, 3–5 times as long as wide; infl white-mealy, of many small glomerules in short, terminal or subterminal, erect or ascending spikes, forming a slender paniculiform infl; sep 5, carinate when ripe, tending to spread and expose the fr; pericarp loose, freely separable from the seed; seeds horizontal, black and shining, mostly 1.0–1.4 mm wide; $2n=18$. In dry, open places in w. and c. U.S., and extending e. in woodlands to the Appalachian region and s. Que. (*C. foggii; C. desiccatum* var. *leptophylloides,* misapplied.)

19. Chenopodium leptophyllum Nutt. Similar in aspect to no. 18, but the lvs narrower, linear or lance-linear, entire, somewhat fleshy, 1-nerved, mostly 2–4 cm × 2–4 mm; pericarp firmly attached to the seed, seed ca 1 mm wide or less. $2n=18$. Widespread in dry ground in w. U.S., and occasionally found in disturbed sites toward the w. margin of our range. (*C. pallescens; C. subglabrum.*)

2. CYCLOLOMA Moq. Winged pigweed. Fls perfect or pistillate; cal persistent, 5-lobed to the middle, the segments usually keeled, incurved over the ovary; stamens 5; ovary flattened, discoid; styles (2)3; cal at maturity developing below its lobes a continuous, horizontal, erose or toothed, membranous wing; fr plano-convex, horizontal, black, pu-berulent, the pericarp free from the seed; embryo annular; densely branched annual tum-bleweed with pale green lvs and slender, few-fld, terminal, interrupted spikes of minute green fls subtended by minute bracts; fr often turning purple. Monotypic.

  1. Cycloloma atriplicifolium (Sprengel) J. M. Coulter. Plants 1–8 dm, pubescent when young; lvs soon deciduous, lanceolate, coarsely and irregularly sinuate-toothed, the lower to 8 cm, the upper progressively reduced; all terminal branchlets floriferous, forming spikes 2–6 cm; fls sessile; mature cal 3–4 mm wide; seeds 1.5 mm; $2n=36$. Dry or sandy ground, often as a weed; Man. to Ind., Ark., and Tex., w. to Utah and N.M., and occasionally adventive eastward.

3. BASSIA All. Fls perfect, solitary or in short spikes from the upper axils, forming terminal spikes or spike-like panicles; sep 5, pubescent, incurved in fl, persistent and incurved over the fr, and developing dorsal appendages; stamens 5; ovary depressed-globose; styles filiform; seed horizontal, the embryo annular; shrubs or annual herbs with much-branched stems, numerous small, hairy lvs, and minute fls, villous at least in the infl. 20, Old World.

1 Three sep bearing a short dorsal process in fr .... 1. *B. hirsuta.*
1 All sep bearing a slender hooked spine at maturity .... 2. *B. hyssopifolia.*

  1. Bassia hirsuta (L.) Aschers. Bushy-branched annual 2–4 dm, turning pink in autumn; lvs fleshy and subterete, pubescent, linear-oblong, 1–1.5(2) cm; spikes paniculiform, usually with a flexuous axis; cal 1 mm; fr 3 × 2 mm; one lower and 2 upper sep bearing a short, stout, dorsal process; seed 1.5 mm wide; $2n=18$. Native of Europe, now abundant on beaches and salt marshes from Mass. to Va.

  2. Bassia hyssopifolia (Pallas) Kuntze. Five-hook bassia. Much like no. 1; lvs flat, to 2(–3.5) cm; spikes slender, terete, 4 mm thick, with straight axis; all sep with a slender, hooked dorsal spine; $2n=18$. Native of Eurasia, intr. especially along the seashore from Mass. to N.J., and also in w. U.S.

4. KOCHIA Roth. Fls perfect or pistillate; cal minute, subglobose, 5-lobed; stamens 5, exsert; ovary depressed, styles 2–3, elongate, filiform; cal at maturity enclosing the fr, each sep bearing a horizontal wing; pericarp thin, free; seed horizontal, flat; embryo annular; annual or perennial herbs or shrubs with narrow lvs and small fls sessile in the axils of bracts, forming dense, axillary or terminal spikes. 30+, mainly Old World.

  1. Kochia scoparia (L.) Schrader. Summer-cypress. Bushy-branched, erect annual to 1 m, usually villous above; lvs linear to narrowly lanceolate, sessile, + pubescent; spikes 5–10 mm, villous; bracteal lvs 3–10 mm; fls solitary or paired; cal at anthesis 0.3 mm long, at maturity 2.5 mm wide, star-shaped, each sep incurved over the fr and bearing a short dorsal wing; styles villous, to 2 mm; seed 1.5 mm wide; $2n=18$. Native of Europe, naturalized in w. U.S. and occasionally adventive from the w. or escaped from cult. in our range. Western plants are mostly more villous and have short lvs 1–3 cm with spreading cilia to 2 mm. Cult. and escaped plants, sometimes called var. *culta* Farw., are less villous or glabrous and have inconspicuously ciliate lvs 3–5 cm; they turn bright red in the autumn.

5. AXYRIS L. Pistillate fls in short, bracted, panicled spikes terminating the branches; cal of 3 or 4 hyaline sep; ovary compressed, at maturity minutely 2-winged at the summit; styles 2, filiform; staminate fls several to many in a short terminal spike, or solitary or few at the top of the pistillate spikes; sep and stamens 3–5; seed flattened, obovate, closely invested by the very thin pericarp, the embryo horseshoe-shaped, the radicle basal; annual herbs with alternate, entire lvs. 7, USSR and Korea.

1. Axyris amaranthoides L. Erect, 2–8 dm, with numerous short, ascending branches; lvs short-petioled, lanceolate, stellate-hairy beneath, to 8 cm, progressively reduced above; pistillate panicles commonly in long, naked peduncles; bracts of the pistillate fls in 3's, the middle one the largest; pistillate cal 2 mm, stellate-hairy; staminate spikes 2–10 mm, the fls minute; fr brown, 2–3 mm; $2n=18$. Native of Siberia, becoming a weed in our range, especially northward.

6. MONOLEPIS Schrader. Fls perfect; cal of one herbaceous abaxial sep; stamen 1, between the sep and the ovary; ovary compressed; styles 2, very short; pericarp free from the seed, but apparently adherent when dried, cellular-reticulate; seed flat, vertical, acute-margined; embryo annular; branched annuals with small, entire or few-toothed lvs and dense clusters of small sessile fls in the upper axils, forming a leafy terminal spike. 3, w. N. Amer.

1. Monolepis nuttalliana (Schultes) Greene. Poverty weed. Winter annual; stems slightly succulent, much branched, spreading or ascending, 1–5 dm; lvs lanceolate to ovate, to 5 cm (petiole included), entire or variously toothed, commonly with a single large salient tooth on each side near the middle, cuneate to the base; fls numerous, much shorter than the lvs; sep green, spatulate or obovate, 1–2 mm. Dry, often alkaline soil: Man. to w. Mo., Tex., and n. Mex., w. to the Pacific, and occasionally adventive eastward.

7. CORISPERMUM L. Bugseed. Fls perfect; cal minute, scarious, of a single adaxial sep, or sometimes also with 2 smaller abaxial sep; stamens 1–3, exsert; ovary compressed, erect, exceeding the cal; styles 2, short; fr indurate, indehiscent, planoconvex, erect, angled or narrowly margined or winged; pericarp closely adherent to the erect seed; embryo annular; much-branched herbs with linear or subulate lvs, the fls solitary in the axils of the bracts, forming a continuous or interrupted terminal spike. Late summer, fall. 50, Eurasia.

1 Fr merely angled, not winged .... 1. *C. orientale.*
1 Fr conspicuously (though rather narrowly) wing-margined.
  2 Spikes dense, continuous except at the very base; bracts strongly overlapping and concealing the
    axis and frs .... 2. *C. hyssopifolium.*
  2 Spikes slender, interrupted; bracts not overlapping, exposing the axis, the lower narrower than the
    frs, the upper scarcely longer than the frs .... 3. *C. nitidum.*

1. Corispermum orientale Lam. Much like no. 2, but with merely angled frs 2–3 mm; infl ± densely pubescent, esp. when young; bracts overlapping, but exposing the axis along the lower part of the infl; $2n=18$. Native of Eurasia, now established as a weed in much of the U.S., esp. in sandy soil and along lake-shores. (*C. emarginatum; C. villosum.*)

2. Corispermum hyssopifolium L. Slender, much-branched, 1–6 dm, often pubescent esp. when young and in the infl; lvs linear, 1–6 cm × 1–3 mm, often early deciduous; spikes densely fld, 2–8 cm × 4–8 mm; bracts ovate, 4–10 mm, long-acuminate, concealing the frs and the axis of the infl; frs elliptic-obovate, 3–4.5 mm, with a pale, firm wing to 0.5 mm wide; $2n=18$. Native of Eurasia, now well established as a weed over much of the U.S., esp. in sandy soil and along lake-shores.

3. Corispermum nitidum Kit. Much like no. 2; spikes loosely fld, to 10 cm, 3–4 mm thick; bracts ovate to linear-oblong, 2–8 mm, scarcely imbricate, exposing the axis, the lower narrower than the fr, the upper only scarcely longer than the fr; fr 2–3.5 mm, with a narrow, pale wing; $2n=18$. Native of Eurasia, now well established as a weed over much of the U.S., esp. in sandy soil and along lake-shores.

8. ATRIPLEX L.   Monoecious; staminate fls with 3–5 sep and as many stamens; pistillate fls all or mostly enclosed individually by 2 broad bracteoles, without perianth except in no. 2; ovary wall very thin; styles 2, filiform; embryo annular, the radicle variously inferior (near the base, and directed downward or horizontally) to median (lateral, directed upwards) or superior (near the summit, and directed upwards); annual (incl. all our spp.) or perennial herbs or shrubs; fls minute, sessile or short-pedicellate in glomerules or borne singly in the axils or in terminal spikes or panicles; lvs usually opposite below and alternate above, ± mealy at least when young, sometimes later green and glabrate. Spp. 3–6 form a morphologically confluent, mostly self-pollinated polyploid series, whose members are sharply distinct at some but not all points of contact. 100+, widespread.

1 Bracteoles enclosing the fr at maturity herbaceous or spongy to somewhat succulent, easily separable
      or eventually separating to the base.
    2 Fruiting bracteoles toothed around the summit; radicle superior .... 2. *A. arenaria.*
    2 Fruiting bracteoles entire or with low teeth below the middle; radicle inferior to sometimes median.
       3 Fruiting bracteoles rotund-ovate to suborbicular, obtuse, conspicuously reticulate-veiny; seed
            apparently central .... 1. *A. hortensis.*
       3 Fruiting bracteoles triangular or triangular-hastate to rhomboidal, not conspicuously veined; ±
            acute; seed basal.
          4 Plant a widespread ruderal weed, only casually coastal; bracteoles not at all spongy-thickened
            .... 3. *A. patula.*
          4 Plants maritime, or of saline places inland, not especially weedy; bracteoles usually ± spongy-
            thickened toward the base.
             5 Infl leafy-bracteate throughout .... 4. *A. glabriuscula.*
             5 Infl leafy-bracteate only at the base, or not at all.
                6 Principal lvs triangular, hastate, or rhomboidal, with sharp angles or lobes .... 5. *A. hastata.*
                6 Principal lvs linear to lanceolate or oblong, entire or merely toothed .... 6. *A. littoralis.*
1 Bracteoles enclosing the fr at maturity hard or bony in texture over the seed.
    7 Bracteoles cuneate-rotund in outline; radicle superior; western, native .... 7. *A. argentea.*
    7 Bracteoles broadly triangular or rhomboid; radicle inferior.
       8 Plants prostrate, maritime; fruiting bracteoles very finely reticulate .... 8. *A. laciniata.*
       8 Plants erect or spreading; veinlets of the bracteoles dark and conspicuous, many of them with
            free ends; widespread weed of waste places .... 9. *A. rosea.*

1. Atriplex hortensis L.   Orache.   Erect or decumbent, to 2 m; lvs becoming green, ovate to broadly deltoid, to 15 cm; staminate and pistillate fls in separate spikes or mingled, forming a large, paniculate infl; some pistillate fls with a horizontal seed, 4–5-lobed persistent cal, and no bracteoles, but most with vertical seed, no perianth, and 2 persistent bracteoles (as in other spp. of *Atriplex*); fruiting bracteoles thin, separate, to 15 mm, rotund-ovate or suborbicular, broadly obtuse, entire or denticulate, conspicuously reticulate; seed apparently central, its radicle inferior; 2*n*=18. Native of Asia, cult. as a pot herb, and rare as a waif in our range. (*A. nitens.*)

2. Atriplex arenaria Nutt.   Seabeach o.   Erect or prostrate, 2–5 dm; lvs silvery-scurfy, broadly oblong to obovate, entire or nearly so, 1–3 cm, tapering to the base; staminate spikes slender, 1–2 cm; pistillate fls in small, axillary glomerules, or intermingled with the lower staminate ones; fruiting bracteoles cuneate-rotund, 4–6 mm, somewhat wider, with 4–7 triangular teeth along the rounded summit, crested on each side of the center near the base with 2–4 projections to 2.5 mm; radicle superior; 2*n*=18. Sandy seashores; N.H. to Fla. and Tex. Perhaps better treated as a var. of the more tropical *A. pentandra* (Jacq.) Standley, but the proper nomenclatural combination not yet made.

3. Atriplex patula L.   Spearscale.   Plants to 1(–1.5) m, erect and widely branched, or sometimes ± prostrate; lvs green especially on the upper side, the principal ones 4–12 cm, lanceolate to rhombic-hastate with a cuneate base; infls numerous, interrupted-spiciform, more continuous upwards, leafy-bracteate only near the base; fruiting bracteoles thin, foliaceous, joined near the base or almost to the middle, obscurely veined or only the midvein prominent, mostly 3–7 mm, ovate to more often rhombic-triangular or triangular-hastate, acute to acuminate, entire or with 1 or 2 short lateral teeth, smooth or with irregular appendages on the back; seeds dimorphic, some brown, 2.5–3.5 mm wide, others black, smaller, 1.2–3 mm wide; radicle ± inferior; mainly or wholly tetraploid on *x*=9. Widespread ruderal weed, only casually maritime, intr. from Eurasia. (*A. acadiensis.*)

4. Atriplex glabriuscula Edmondston.   Plants to 1 m, prostrate to sometimes erect; lvs green, the principal ones mostly 4–10 cm, usually deltoid-hastate and irregularly toothed above the lobes; infls of loose glomerules forming a terminal thyrse, leafy-bracteate throughout; fruiting bracteoles spongy-thickened toward the base, obscurely veined, mostly 5–13 mm, united nearly to the middle, ovate-triangular to rhombic-triangular, abruptly acuminate, the lateral angles rounded, with a few small teeth or entire, the back often roughened; brown seeds 2.5–4 mm wide; black seeds smaller, 1.5–2.5 mm wide, or wanting; radicle median, ascending; dipoid on *x*=9. Sea beaches; Nf. and Que. to R.I., and widespread on the Atlantic coast of Europe, whence probably intr. (*A. patula* ssp. *g.*)

5. **Atriplex hastata** L.   Plants erect and to 1 m, or prostrate; lvs green, the principal ones 2–10 × 2–9 cm, hastate to triangular or rhomboidal, with sharp basal angles or lobes; infls lfless except at base; fruiting bracteoles foliaceous and somewhat (but usually not strongly) spongy-thickened toward the base, joined near the base, obscurely to ± evidently veined, 3–10 mm, triangular-hastate to triangular ovate, the lateral angles ± rounded, entire or toothed; seeds dimorphic; brown seeds 1.5–3 mm wide; black seeds harder, 1–2 mm wide; radicle inferior; diploid on x=9. On sea beaches and in salt marshes or saline soil inland; widespread in the N. Hemisphere, but in N. Amer. perhaps only intr. (*A. patula* var. *h.; A. triangularis; A. franktonii.*)

6. **Atriplex littoralis** L.   Erect, to 1.5 m; lvs green, the principal ones 2–12 cm, mostly linear to lanceolate or oblong and entire or merely toothed, obtuse to acuminate, narrowed to a short petiole; infls interrupted below, lfless above the base; fruiting bracteoles joined towards the ± spongy-thickened base, obscurely to evidently veined, 3–7 mm, broadly triangular to ovate, ± acute, usually with 2(–several) evident tubercles on the back; brown seeds 2–3 mm wide; black seeds 1.5–2 mm wide; radicle inferior; $2n$=18. Sea beaches, and also commonly inland in saline habitats; widespread in N. Temperate regions. (*A. patula* ssp. *l.*) Most of our coastal plants are reputedly diploid, and may be intr. Inland and some coastal plants are reputedly hexaploid and probably native. (*A. subspicata.*)

7. **Atriplex argentea** Nutt.   Silverscale.   Much-branched herb to 8 dm, silvery-scurfy throughout; lvs lanceolate to ovate or deltoid, 1–3 cm, often as wide, fls axillary and in terminal spikes, the staminate and pistillate mingled or separate; fruiting bracteoles cuneate-rotund, 4–7 mm long and wide, united to beyond the middle, irregularly toothed across the broadly rounded summit, the central part indurate, the tubercles few or none, to 2 mm; radicle superior; $2n$=36. Widespread in dry or alkaline soils in w. U.S. and adj. Can., occasionally intr. e. to Mich., Mo., and reputedly O.

8. **Atriplex laciniata** L.   Spreading or prostrate, to 5 dm; lvs ovate to orbicular, 2–4 cm, scurfy on both sides, with Kranz anatomy; fls axillary or in short terminal spikes; fruiting bracteoles rhombic, 5–8 mm long and wide, entire, 3-nerved, very finely and obscurely reticulate, the central part indurate, the tubercles few or none; seeds light brown, 3.5–4 mm wide; radicle inferior. Seashores about the Gulf of St. Lawrence, in N.B., P.E.I., N.S., and Que.; also on the Atlantic coast of Europe; perhaps only intr. with us. (*A. sabulosa.*)

9. **Atriplex rosea** L.   Red orache.   Erect, usually much-branched, to 2 m; lvs lanceolate to rhombic-ovate, to 7 cm, deeply sinuate-dentate, thinly gray-scurfy; staminate fls in the uppermost axils and in terminal spikes less than 1 cm; pistillate fls axillary; fruiting bracteoles broadly ovate to rhombic, 4–10 mm long and wide, conspicuously dentate, united to the middle, the central part indurate; tubercles none to several; veins conspicuous, many with free tips; $2n$=18. Native of Eurasia, established as a weed in alkaline soil in w. U.S. and occasionally found in waste places in our range.

9. SALICORNIA L.   Glasswort, samphire.

Fls perfect or pistillate, in groups of 3 and deeply sunken in pits in the axis of a fleshy spike above the axils of scale-like bracts; cal obpyramidal, fleshy, not lobed, opening by a small terminal slit only, becoming spongy at maturity; stamens 1 or 2, the anthers exsert; ovary laterally compressed; styles 2, short; fr enclosed in the cal, the pericarp none; annual or perennial succulent herbs with jointed, usually freely branched stems, the lvs reduced to minute, opposite, scarcely projecting scales; uppermost scales crowded, forming a cylindric terminal spike. All spp. vary greatly in height and branching; all branches normally terminate in a spike, the fls appearing in late summer. The plants often turn bright red in autumn. (*Sarcocornia.*) 25, widespread.

1 Rhizomatous perennial; the 3 fls of a cluster inserted at the same level .... 1. *S. virginica.*
1 Annual; 2 lateral fls nearly contiguous below the base of the central one.
  2 Scales below the spikes mucronate; spike mostly 4.5–6 mm thick .... 2. *S. bigelovii.*
  2 Scales below the spikes rounded to merely acutish; spike mostly 1.5–4.5 mm thick.
    3 Joints of the spike mostly evidently longer than thick; widespread sp. .... 3. *S. europaea.*
    3 Joints of the spike about as thick as or thicker than long; western sp. .... 4. *S. rubra.*

1. **Salicornia virginica** L.   Perennial g.   Rhizomatous perennial; main stems hard or woody, prostrate, rooting, forming mats and emitting erect or ascending flowering stems 1–3 dm; spikes 1–4 cm × 2.5–3 mm, the joints 2.5–3 mm; central fl cuneate-obovate, truncate across the top, scarcely surpassing the obliquely ovate lateral ones; seeds 0.7–1 mm. Salt marshes; Mass. to Tex.; Alas. to Calif., W.I., Europe, and n. Afr. (*S. ambigua; Sarcocornia perennis.*)

2. **Salicornia bigelovii** Torr.   Dwarf g.   Erect annual, mostly 1–4 dm; scales below the spikes mucronate; spikes 2–10 cm, the joints 2–3.5 × 4.5–6 mm; lateral fls nearly contiguous below the lateral angles of the larger central one; seeds 1–1.5 mm. Salt marshes; N.S. to Tex.; also in Calif., Mex., and W.I.

3. **Salicornia europaea** L.   Samphire.   Annual, 1–5 dm, erect or the lower branches prostrate; scales below the spike obtuse or rounded; spikes erect or ascending, 2–6 × 1.5–3(–4.5) mm, the joints mostly 2–4 mm; fls broadly obovate, the central one much surpassing the lateral ones but reaching only to 0.5–1 mm below the node above; seeds 1.3–2 mm; $2n$=18, 36. The commonest sp. of salt marshes; Que. to Fla., and in saline soil inland to Mich.; also Alas. to Calif., W.I., and widespread in the Old World. (*S. herbacea, S. prostrata.*)

4. **Salicornia rubra** A. Nels.   Western g. Slender annual 1–2.5 dm; scales below the spike acutish, those of the spikes blunt or rounded; spikes very numerous, 1–5 cm, the joints 2–2.5 × 2–3 mm; central fl broadly rounded above the cuneate base, reaching nearly or quite to the node above, the lateral ones obliquely ovate; seeds 1–1.2 mm; $2n=18$. Saline soil; w. Minn. to Sask. and w. Kans., w. to B.C. and Nev.

10. SUAEDA Forsskål.   Sea-blite.

Fls small, perfect or unisexual; cal at anthesis deeply 5-lobed, at maturity unchanged except in size, or some or all lobes becoming carinate, cucullate, or winged; stamens 5; ovary depressed or flattened; styles 2–5, short; fr completely enclosed by the cal; pericarp thin and delicate, usually loose; seed commonly horizontal, occasionally vertical; embryo spirally coiled; perisperm scanty or none; herbs (our spp. all annual) with very numerous, linear lvs, the upper reduced to bracts, forming terminal spikes and bearing in the axils the solitary or clustered (commonly 3) fls. (*Dondia*.) The taxonomy of the genus is in need of careful reconsideration. 50, widespread.

1 Sep at maturity essentially equal, rounded on the back or keeled, not hooded.
  2 Sep rounded on the back or obscurely keeled.
    3 Seeds 1.5–2 mm wide; lvs mostly 10–30+ mm .... 1. *S. maritima*.
    3 Seeds 1–1.5 mm wide; lvs 3–15 mm .... 2. *S. richii*.
  2 Sep narrowly keeled .... 3. *S. linearis*.
1 Sep at maturity distinctly unequal, 1–3 of them more prominently hooded than the others.
  4 One or 2 sep sharply pointed; coastal salt marshes .... 4. *S. americana*.
  4 All sep sharply pointed; alkaline or saline soil mainly in the interior .... 5. *S. calceoliformis*.

1. **Suaeda maritima** (L.) Dumort.   White s.-b. Prostrate to erect, 1–5 dm, more robust than no. 2; lvs green or more often glaucous, acute, plano-convex, to 3(5) cm on the primary axis; lvs the lower bracts like the lvs, the upper gradually reduced to 5 mm; cal at maturity 3 mm wide; sep equal or nearly so, rounded at the tip, rounded on the back or obscurely keeled; seeds 1.5–2 mm wide; $2n=36$. Coastal salt-marshes; intr. from Que. to Va., Fla. and La.; widespread in Eurasia. (*Dondia m.*)

2. **Suaeda richii** Fern.   Procumbent, forming mats up to 5 dm wide; lvs dark green, linear to linear-oblong, blunt, rounded on both sides, to 15 mm; spikes 2–10 cm; bracts shorter (3–5 mm) and broader; cal at maturity 2 mm wide, the sep equal or nearly so, rounded on the back and at the tip; seeds 1–1.5 mm wide; $2n=36$. Coastal salt-marshes; N.S. and se. Nf. to Me. and ne. Mass.

3. **Suaeda linearis** (Elliott) Moq.   Southern s.-b. Erect or ascending; 2–8 dm, usually much-branched; lvs dark green, plano-convex, the primary ones to 4 cm, the rameal ones progressively shorter; spikes often elongate, usually dense; bracts 4–7 mm; cal at maturity 2 mm wide, sep narrowly keeled; seeds 1–1.5 mm wide; $2n=54$. Coastal salt-marshes; Me. to Tex. and W.I. (*Dondia l.*)

4. **Suaeda americana** (Pers.) Fern.   Northeastern s.-b. Stems procumbent at base, 2–3 dm, the flowering tips ascending; lvs green, eventually red, plano-convex, to 2 cm; spikes 5–10 cm, rather loose, the bracts 5–12 mm; cal at maturity 2–2.5 mm wide, irregular, 1–2 of the sep sharply pointed, more strongly hooded than the others; seeds 1–1.5 mm wide. Coastal salt-marshes; Que. and N.S. to N.J. (*Dondia a.*)

5. **Suaeda calceoliformis** (Hook.) Moq.   Plains s.-b. Erect to decumbent, 1–8 dm; lvs green or glaucous, plano-convex, widest at the base, the primary ones to 4 cm, those of the branches much shorter; spikes usually dense; bracts lance-ovate, 2–6 mm; evidently broader than the foliage lvs; cal at maturity 1.5–2 mm, irregular, 1–3 sep sharply hooded, the others less so or merely carinate; seeds 1–1.5 mm wide; tetraploid, $2n=36, 54$. Saline or alkaline soil; B.C. and s. Yukon to Calif., e. to Sask., w. Minn., and Tex., and occasionally intr. eastward, as in n. Ill. and s. Mich. (*S. depressa* and *Dondia depressa*, misapplied; *S. erecta*.) *S. rolandii* Bassett & Crompton, from N.B., N.S., and Que., may be distinct; lvs not wide-based; seeds dimorphic, 1.5–2 mm; $2n=90$.

11. SALSOLA L.

Fls perfect; cal deeply 5-lobed, the segments at maturity incurved over the fr, transversely carinate or winged, the tips connivent and erect; stamens usually 5; styles 2; seed thick, horizontal; embryo spirally coiled; perisperm none; ours branching, annual herbs with linear, succulent, ± spinulose-tipped lvs, the fls solitary or few in the axils of the shorter and spinier upper lvs (bracts), each subtended by a pair of bracteoles. 50+, widespread.

1 Sep at anthesis soft, muticous, the midvein obscure; inland tumbleweeds.
  2 Bracts of the infl ± appressed; embryo annular .... 1. *S. collina*.
  2 Bracts of the infl evidently spreading; embryo closely coiled in several layers .... 2. *S. tragus*.
1 Sep stiff, with an evident midvein excurrent as a sharp point; maritime plants .... 3. *S. kali*.

1. **Salsola collina** Pallas. Katune. Young plants with a strong, erect, monopodial stem, later becoming more bushy-branched; leaves somewhat dilated at the base; bracts ± appressed, imbricate, short, almost subulate, exceeding the bracteoles; sep often connate with the bracteoles, wingless or with a small, erose wing; embryo annular; 2n=18, otherwise much like *S. tragus*. Native of Siberia, well established as a weed on the Great Plains, and e. occasionally to Minn., Io., Ill., and even Vt.

2. **Salsola tragus** L. Russian thistle. Glabrous or hispidulous, freely and diffusely branched, becoming rigid, forming a loose ball 2–10 dm that eventually breaks off at the ground and becomes a tumbleweed; lvs narrowly linear, 1–3 (the lower to 6) cm, reduced upwards and passing into the divergent, spinescent, ± setulose-serrulate, basally scarious-margined bracts; bracteoles 2.5–3.5 mm, also pungent; sep at anthesis soft, muticous, the midvein obscure; fruiting cal 3–10 mm wide, the lower fls often with the sep merely carinate, the others with the sep developing a prominently veined wing on the back; embryo closely coiled in several layers; 2n=36. Native to Eurasia, now widespread as a weed, abundant in the drier parts of w. N. Amer., less commonly in our range. (*S. kali* var. *tenuifolia*; *S. iberica*; *S. pestifer*; *S. ruthenica*.)

3. **Salsola kali** L. Saltwort. Chaffy-hispid (rarely glabrous), to 4 dm, with long, prostrate or ascending branches from the base; lvs rather firm, linear subulate, to 3 cm, 1–2 mm wide, reduced upwards and passing into the divergent, spinescent bracts; bracteoles spinescent, longer than the fl, deflexed in fr; sep stiff, with a distinct midvein shortly exserted as a subulate spine-tip, in fr wingless or with a deeply and irregularly cleft or pectinate wing, the whole 4–6(7) mm wide; embryo closely coiled in several layers; 2n=36. Sea-beaches; Nf. to La., and on the coast of w. Europe.

FAMILY **AMARANTHACEAE**, the Amaranth Family

Fls small, hypogynous or nearly so, generally regular, perfect or less often unisexual; sep mostly 3–5, seldom only 1 or 2 or even wanting, generally dry and scarious or membranous, distinct or ± connate at the base; pet none; stamens generally as many as the sep and opposite them, seldom fewer; filaments distinct or more often connate at the base into a tube, the tube often produced into teeth or lobes between the anthers, in extreme cases (e.g., *Froelichia*) the androecium simulating a small, sympetalous cor with filaments attached at the sinuses; gynoecium of 2–3(4) carpels united to form a compound, unilocular ovary with a single, often evidently lobed style; ovule usually solitary and basal, or in a few genera (e.g., *Celosia*) the ovules several on a basal or short, free-central placenta; ovules ± distinctly campylotropous; fr a utricle or a small nut, or a circumscissile, 1-seeded capsule (pyxis), often subtended or ± enclosed by the persistent cal; seeds lenticular, with an elongate, dicotyledonous embryo surrounding the abundant perisperm; herbs, seldom climbers or shrubs, producing betalains but not anthocyanins, with simple, exstipulate, alternate or opposite lvs; fls solitary to more often in cymose or variously compound infls, often subtended by scarious or membranous (sometimes conspicuously pigmented) bracts or bracteoles. 65/900.

1 Lvs alternate; anthers tetrasporangiate and becoming bilocular.
  2 Ovule and seed solitary; style very short, with elongate stigmatic branches; fls unisexual . . . . 1. *Amaranthus*.
  2 Ovules and seeds 2 or more; style elongate; stigma capitate; fls perfect . . . . 2. *Celosia*.
1 Lvs opposite; anthers bisporangiate and becoming unilocular.
  3 Fls perfect, in dense heads or spikes, or in axillary glomerules.
    4 Sep distinct or nearly so.
      5 Infls becoming elongate-spicate; anthers tetrasporangiate and bilocular . . . . 3. *Achyranthes*.
      5 Infls persistently short and dense; anthers bisporangiate and unilocular . . . . 4. *Alternanthera*.
    4 Sep united into a tube that becomes indurate in fr . . . . 5. *Froelichia*.
  3 Fls unisexual, in loose pyramidal panicles; plants dioecious . . . . 6. *Iresine*.

1. **AMARANTHUS** L. Amaranth. Monoecious or dioecious; fls small, each subtended by bracts and bracteoles; cal of 1–5 scarious or membranous sep separate to the base, often aristate, or wanting from the pistillate fls; stamens 1–5, with short filament and linear-oblong anther; ovary short and broad, compressed; style short or virtually none; stigmas (2)3(5), slender, pubescent; ovule 1; fr a thin-walled to coriaceous utricle, indehiscent or bursting irregularly, or commonly circumscissile at the middle, crowned by the persistent stigmas; seed flattened or lenticular, round to obovate; prostrate to erect annuals, usually much branched, with alternate, petiolate, entire or sinuate lvs; fls in small axillary

clusters, or aggregated into axillary or terminal, simple or panicled spike-like thyrses. 50, widespread. (*Acnida*.) Spp. 8–13 are ± connected by intermediates of hybrid origin.

1 Plants monoecious (the staminate and pistillate fls intermingled or in separate infls).
  2 Fls all or chiefly in small axillary clusters, sometimes also forming a small terminal panicle.
    3 Fr circumscissile at the middle; seeds suborbicular.
      4 Plants prostrate; bracts about equaling the cal . . . . 1. *A. blitoides.*
      4 Plants bushy-branched tumbleweeds; bracts 2–3 times as long as the cal . . . . 2. *A. albus.*
    3 Fr indehiscent; seeds (except in no. 6) obovate or elliptic.
      5 Sep of the pistillate fls 4 or 5; lvs emarginate or not.
        6 Axis of the short infls not much thickened; sep equaling or a little shorter than the smooth or rugulose fr.
          7 Sep mostly 2–4 mm; seeds 2–2.5 mm . . . . 3. *A. pumilus.*
          7 Sep mostly 1–1.5 mm; seeds 0.8–1 mm . . . . 4. *A. crispus.*
        6 Axis of the short infls strongly thickened (to 1 mm); sep longer than the papillate fr . . . . 5. *A. crassipes.*
      5 Sep of the pistillate fls 2 or 3; lvs conspicuously emarginate . . . . 6. *A. blitum.*
  2 Fls chiefly in elongate, terminal, spike-like thyrses or compound panicles; axillary panicles or clusters of much smaller size may also be present.
    8 Stem bearing a pair of spines at the base of most lvs . . . . 7. *A. spinosus.*
    8 Stems without spines.
      9 Fr circumscissile at the middle.
        10 Cultivated ornamentals, sometimes casually escaped or persistent; infl generally very large and showy, bright red or bright yellow or bright green; fr equaling or exceeding the bract and sep.
          11 Infl stiffly erect; style branches thickened at the base; bract equaling the spreading style-branches . . . . 10. *A. hypochondriacus.*
          11 Infl lax, nodding; style branches slender at the base; bract surpassed by the style branches.
            12 Sep straight, the inner oblong, acutish, the shorter ones less than 2 mm, about half as long as the utricle; style-branches erect . . . . 12. *A. cruentus.*
            12 Sep outcurved above, the inner spatulate and obtuse or emarginate, all over 2 mm and nearly equaling the utricle; style-branches spreading . . . . 13. *A. caudatus.*
        10 Weeds; infl generally smaller and dull; fr surpassed by the bract and longer sep.
          13 Sep obtuse or emarginate, somewhat outcurved above . . . . 8. *A. retroflexus.*
          13 Sep acute, straight or nearly so.
            14 Style-branches spreading from the base; infl stiff, unbranched or with a few long branches . . . . 9. *A. powellii.*
            14 Style-branches erect; infl lax, with many crowded, short lateral branches . . . . 11. *A. hybridus.*
      9 Fr indehiscent.
        15 Fr smooth, flat-ellipsoid; sep of pistillate fls 2 . . . . 14. *A. deflexus.*
        15 Fr rugose when dry, flat-ovoid; pistillate sep 3 . . . . 15. *A. viridis.*
1 Plants dioecious.
  16 Fls pistillate.
    17 Sep regularly present, at least 1 mm, with distinct midvein.
      18 Sep 5, at least the inner ones spatulate; style-branches 2(3).
        19 Outer sep 2–2.5 mm, obtuse, merely apiculate . . . . 16. *A. arenicola.*
        19 Outer sep 3–4 mm, acute, the midvein excurrent as a rigid point . . . . 17. *A. palmeri.*
      18 Sep 1 or 2, lanceolate or linear; style-branches 3 or 4 . . . . 18. *A. rudis.*
    17 Sep none, or rarely 1 or 2 and vestigial, then less than 1 mm and without visible midvein; style-branches 3 or 4.
      20 Seeds ca 1 mm; fr 1.5–2 mm; larger lvs usually ovate or lanceolate . . . . 19. *A. tuberculatus.*
      20 Seeds 2–3 mm; fr 2.5–4 mm; lvs usually narrowly lanceolate to linear . . . . 20. *A. cannabinus.*
  16 Fls staminate.
    21 Outer sep with heavy midvein, often definitely longer than the inner; bracts with heavy midrib, mostly 1.5–2 mm.
      22 Outer sep mostly 2–3 mm, obtuse, apiculate, but the dark midvein not excurrent . . . . 16. *A. arenicola.*
      22 Outer sep mostly 3–4 mm, acuminate, the midrib excurrent as a rigid spine.
        23 Bract 4–6 mm, equaling or surpassing the outer sep . . . . 17. *A. palmeri.*
        23 Bract ca 1.5–2 mm, definitely shorter than the outer sep . . . . 18. *A. rudis.*
    21 Outer sep with more slender midvein, not appreciably longer than the inner; bracts mostly with slender midvein, up to ca 1.5 mm.
      24 Bract ca 1–1.5 mm, the midrib conspicuously excurrent . . . . 19. *A. tuberculatus.*
      24 Bract ca 1 mm or less, the midrib scarcely excurrent . . . . 20. *A. cannabinus.*

    1. Amaranthus blitoides S. Wats. Monoecious; stems prostrate, much branched, 2–6 dm; lvs numerous, pale green, oblong to obovate, 1–2 cm, obtuse or rounded, attenuate to a long petiole, often crowded near the branch-tips; fls in short, dense, axillary clusters; bracts about equaling the sep, acuminate, scarcely aristate; sep 4–5, with conspicuous, branching green veins; those of the pistillate fls ovate to oblong, straight, acuminate, unequal; stamens 3; style-branches short, recurved; fr thick-lenticular, 2–2.5 mm, about equaling the longest sep, smooth or nearly so, circumscissile at the middle; seed suborbicular, 1.4–1.7 mm; $2n=32$. Native of w. U.S., intr. in our range and common as a weed except in the extreme northeast. (*A. graecizans,* misapplied.)

2. Amaranthus albus L. Tumbleweed. Monoecious; bushy-branched, to 1 m high and wide, the stems whitish; lvs of flowering branches elliptic to oblong or obovate, 0.5–3 mm, pale green, obtuse or rounded, attenuate to a long petiole; early lvs often to 8 cm; fls in short, dense, axillary clusters; bracts rigid, subulate, 2–3 times as long as the fls; sep 3, with simple midvein, those of the pistillate fls straight, acutish, unequal, the longest nearly equaling the fr; stamens 3; style-branches very short, erect; fr lenticular, 1.3–1.7 mm, rugulose when dry, circumscissile at the middle; seed suborbicular, 0.7–1 mm; 2n=32. Fields and waste ground; native to the prairies and plains of c. N. Amer., now a weed throughout our range and elsewhere. (*A. graecizans,* misapplied.)

3. Amaranthus pumilus Raf. Monoecious; stems fleshy, prostrate or decumbent, branched from the base, 1–4 dm; lvs fleshy, round-obovate, 1–2 cm, ± retuse at the broadly rounded summit, abruptly narrowed to a 5–10 mm petiole; fls in dense axillary clusters; stamens 5; sep of pistillate fls 5, oblong-oblanceolate, unequal, cucullate, 2–4 mm, twice as long as the lanceolate bracts; fr fleshy, indehiscent, thick-fusiform, 4–5 mm, smooth or rugulose; seed elliptic, 2–2.5 mm. Sea-beaches; Mass. to N.C.; rare.

4. Amaranthus crispus (Lesp. & Théven) N. Terrac. Monoecious; stem prostrate, much-branched, 2–6 dm, minutely pubescent; lvs very many, ovate-oblong, 0.5–1.5 cm, acute to rounded, minutely cuspidate, crisped on the margin; infls axillary, 3–10 mm; sep of pistillate fls 4 or 5, spatulate-obovate, 1–1.5 mm, minutely aristate; fr flattened-obovoid, thin-walled, indehiscent, minutely roughened, 1.5–2 mm; seed obovate, reddish-brown, 1 mm. Native of Argentina, occasionally found in waste places in our range, especially about seaports.

5. Amaranthus crassipes Schlechtendahl. Monoecious; prostrate or decumbent, 2–6 dm; lvs long-petioled, obovate or elliptic, 1–4 cm, rounded above; infls axillary, the short axes thickened (to 1 mm) and indurate; bracts minute; sep of pistillate fls 5, spatulate, 2–2.5 mm; fr compressed-obovoid, 1.4–1.8 mm, papillate above the middle, coriaceous, indehiscent; seed broadly obovate, 1–1.4 mm. Native of tropical Amer., rarely found about seaports in our range.

6. Amaranthus blitum L. Monoecious; stems prostrate to erect, to 6 dm; lvs rhombic-ovate to obovate, 1–3 cm, broadly retuse; petiole about equaling the blade; fls chiefly in axillary clusters, also in short (1–2 cm) terminal panicles; sep of pistillate fls 2, oblong, 1.5–2 mm, at the margins of the fr, a narrower and shorter third one often present at one side; fr broadly elliptic, 1.7–2.2 mm, thin-walled, indehiscent, smooth; seed suborbicular, 1–1.2 mm; 2n=34. An Old-World weed of unknown origin; occasionally found about seaports in our range, and at scattered stations in disturbed sites inland. (*A .lividus.*)

7. Amaranthus spinosus L. Spiny a. Monoecious; stem erect, branched, to 1 m, bearing at most nodes a pair of divergent spines 5–10 mm; lvs lance-ovate to ovate, 3–6 cm, narrowed to an obtuse, mucronate tip, broadly cuneate to the long petiole; spikes numerous, 5–15 cm, 6–10 mm thick, the terminal often chiefly or wholly staminate, the basal part and the axillary clusters mostly pistillate; sep of the pistillate fls 5, oblong, 1–1.5 mm; fr 1.5–2 mm, indehiscent or bursting irregularly, the terminal part spongy and roughened; seed suborbicular, 0.7–1 mm; 2n=32, 34. A pantropical weed, probably originally from the New World, extending n. in our range to N.Y., Pa., Ind., and Mo., seldom farther n.

8. Amaranthus retroflexus L. Redroot, rough pigweed. Monoecious; stem stout, erect, usually branched, to 2 m; plant scurfy-villous in and for some distance below the infl; lvs long-petioled, ovate or rhombic-ovate, usually hairy beneath at least along the veins, up to 10(15) cm in well developed plants; principal infl a terminal panicle of several or many short, densely crowded, obtuse spike-like thyrses, the whole 5–20 cm; similar but smaller panicles produced from the upper axils; bracts rigid, subulate, 4–8 mm, much surpassing the cal; sep 5, those of the pistillate fls oblong-lanceolate, slightly outcurved above, rounded or truncate or emarginate at the tip, and often minutely mucronate, with simple midvein, at maturity 3–4 mm and much-surpassing the fr; stamens 5; style-branches fairly long, erect or slightly recurved; fr compressed, 1.5–2 mm, circumscissile at the middle, the upper part rugulose; seed round-obovate, 1–1.2 mm; 2n=34. Native of tropical Amer., now a common weed throughout most of the U.S. and also in the Old World.

9. Amaranthus powellii S. Wats. Monoecious, to 2 m, freely branched, puberulent to villous in the infl, otherwise glabrous or puberulent; lvs long-petioled, lance-ovate to deltoid-elliptic, to 10(15) cm in well developed plants, glabrous or nearly so; infl terminal, stiff, dense and spike-like, unbranched or with a few widely spaced long branches, dull greenish, not showy; bracts ca 5 mm, much surpassing the sep and frs, with a very thick, excurrent midrib; sep 3–5, with simple midvein, those of the pistillate fls sharply acute, nearly straight, unequal, 2–3(–3.5) mm, the longer (outer) ones generally surpassing the fr; stamens as many as the sep; style-branches elongate, recurved from the thickened base; fr slightly rugose, circumscissile; seed dark brown, suborbicular, 1–1.3 mm; 2n=34. Native to open habitats in w. N. and S. Amer., now a common weed in w. U.S., and becoming established in our range.

10. Amaranthus hypochondriacus L. Prince's feather. Cult. ornamental and grain amaranth, apparently derived from *A. powellii;* infl large and showy, stiffly erect, mostly bright red or bright yellow in life, or sometimes bright green; bracts with rather strong midrib, about equaling the style-branches, these spreading from the thickened base; sep all over 2 mm, nearly equaling the fr, all lanceolate, acute; seeds of ornamental forms generally dark brown, of grain forms mostly pale ivory. Amer. cultigen, originally cult. for grain, now much more commonly so in the Old World than in the New; widely cult. as an ornamental, and occasionally escaping or persistent, but scarcely established in our range.

11. Amaranthus hybridus L. Smooth pigweed. Monoecious, to 2 m, usually freely branched, scurfy-villous at least in the infl; lvs long-petioled, ovate or rhombic-ovate or lanceolate, mostly acutish, the larger ones on well developed plants often 15 cm or more; infl dull greenish or slightly reddish, not showy, terminal, lax, generally with many short, crowded lateral branches and often with nodding tip; bracts mostly 3–4 mm, slightly surpassing the sep and fr, with fairly thick, long-excurrent midrib; sep 5, those of the pistillate fls mostly 1.5–2 mm, about equaling the fr, with simple midvein, straight, acute; stamens 5; style-branches rather short, erect; fr rugose, circumscissile; seed dark brown, ca 1 mm; $2n=32$. Originally native to tropical Amer., now a cosmopolitan weed.

12. Amaranthus cruentus L. Red a. Cultigen apparently derived from *A. hybridus*; infl large and showy, mostly bright red in life, lax, at least the terminal thyrse generally nodding or drooping; bract of the pistillate fls with slender midrib, not surpassing the fr; sep 5, straight, unequal, the shorter (inner) ones less than 2 mm, oblong, acutish, ca half as long as the fr; style-branches erect; seeds in ornamental forms dark brown; $2n=32$, 34. Originally cult. for grain and dyestuff, now relictual as a grain-plant, but widely cult. as an ornamental and pot-herb, and occasionally escaping or persistent, but scarcely established in our range. (*A. hybridus* ssp. *c*.)

13. Amaranthus caudatus L. Love-lies-bleeding. Cult. ornamental and grain-amaranth, apparently derived from the S. Amer. *A. quitensis* HBK.; monecious, stout, erect, to 2 m; lvs long-petioled, elliptic to rhombic-ovate, to 2 dm; infl large and showy, generally bright red in life, composed of several to many drooping, spike-like thyrses, the terminal one 1–3 dm, much longer than the lateral; bract of the pistillate fls with slender midrib, not surpassing the fr; sep 5, outcurved above, the inner spatulate and obtuse or emarginate, all over 2 mm and nearly equaling the fr; style-branches slender, spreading from the base; seed most commonly pale ivory, and when fresh flushed with red around the rim, less commonly dark brown; $2n=32$, 34. Amer. cultigen, originally cult. for grain, now relictual as a grain-plant, but widely cult. as an ornamental, occasionally escaping or persistent, but scarcely established in our range.

14. Amaranthus deflexus L. Monoecious; prostrate or ascending, 2–5 dm; lvs lanceolate to broadly ovate, 1–3 cm; petiole about equaling the blade; terminal panicle dense, narrowly ovoid, 2–5 cm; smaller clusters also present in the upper axils; bracts shorter than the fr; sep of the pistillate fls 2, oblanceolate, two-thirds as long as the fr, acute; fr smooth, flat-ellipsoid, 2.5 mm, half as wide, indehiscent; seed obovate, shining, 1.2 mm; $2n=34$. Probably native to tropical Amer., occasionally found in waste places in our range, especially along the coast.

15. Amaranthus viridis L. Green a. Monoecious; erect, to 1 m; lvs broadly ovate or rhombic-ovate, 3–7 cm, often retuse, acute or rounded at base; thyrses few or several, the lateral ascending, not much shorter than the terminal, forming a panicle 1–2 dm; bracts much shorter than the fls; sep of the pistillate fls 3, oblanceolate, shorter than the fr, acute; fr compressed-obovoid, 1.5 mm, very rugose when dry, indehiscent; seed orbicular, sharp-edged, 1 mm; $2n=34$. Probably native to tropical Amer., now a pantropic weed and occasionally adventive in our range.

16. Amaranthus arenicola I. M. Johnst. Sandhill-a. Dioecious; erect, to 2 m; lvs long-petioled, oblong to lanceolate, 2–8 cm; terminal thyrse 1–4 dm, 1 cm thick, nearly continuous, the lateral ones shorter; bracts 1.5–2.5 mm, mostly shorter than the sep, with heavy, not or scarcely excurrent midvein; male fls with 5 subequal sep, these 2–3 mm, acute, apiculate, but the dark midvein not excurrent; female fls with 5 recurved-spatulate sep, the outer 2–2.5 mm, the inner 1.5–2 mm, all emarginate or obtuse, but the outer apiculate; fr 1.5 mm, circumscissile at the middle; style-branches usually 2; seed 1–1.2 mm, round, dark reddish-brown; $2n=32$. Dry, commonly sandy places; w. Mo. to N.D., Mont., and Tex., and intr. in disturbed soil at scattered stations e. to the Atlantic. (*A. torreyi*, misapplied.)

17. Amaranthus palmeri S. Wats. Careless weed. Dioecious; erect, 3–10(–20) dm; lvs long-petioled, rhombic-ovate, or rhombic-lanceolate, 3–10 cm; terminal thyrse to 5 dm, 1–1.5 cm thick, the lateral ones shorter or none; bracts mostly 4–6 mm, with heavy, spinosely excurrent midvein; male fls with 5 unequal sep; the outer 3.5–4 mm, acuminate, with conspicuous, long-excurrent midvein, the inner 2.5–3 mm, obtuse or emarginate; female fls with 5 recurved-spatulate sep, the outer 3–4 mm, acute, with the midvein excurrent as a rigid point, the inner 2–2.5 mm, emarginate; fr 1.5–2 mm, circumscissile at the middle; style-branches 2(3); seed 1–1.3 mm, dark reddish-brown; $2n=32$, 34. Dry soil; s. Calif. to Okla., Kans., Neb., and La., s. to Mex., and at scattered stations in our range as a weed.

18. Amaranthus rudis Sauer. Dioecious; stout, erect, 5–20 dm; lvs long-petioled, rhombic-oblong to oval, to 10 cm, the upper narrow and reduced; staminate infl a single long, terminal, spike-like thyrse or of numerous short, panicled branches; pistillate thyrses usually elongate and very slender, single or numerous in a panicle; bracts mostly 1.5–2.5 mm, with heavy, excurrent midrib; male fls with 5 unequal sep, the outer ca 3 mm, acuminate, with conspicuous, excurrent midvein, the inner 2.5 mm, obtuse or emarginate; female fls with 1 or 2 sep, the shorter one vestigial, the other ca 2 mm, narrowly lanceolate, acuminate, with heavy, excurrent midvein; fr 1.5 mm, circumscissile at the middle, thin, rugose; style-branches 3 or 4; seeds plump, suborbicular, 1 mm; $2n=32$. Sandy fields and waste places, chiefly in moist soil; Wis. and N.D. to Tex. and Ala., and at scattered stations eastward as an adventive. (*Acnida* and *Amaranthus tamariscinus*, misapplied.)

19. Amaranthus tuberculatus (Moq.) Sauer. Dioecious; prostrate to erect, to 2 m; lvs and infls highly variable; bracts 1–1.5 mm, with slender, strongly excurrent midvein; male fls with 5 subequal sep 2.5–3 mm, the outer acuminate, the inner obtuse or emarginate, the midvein not excurrent; female fls usually without perianth, occasionally with 1 or 2 vestigial sep under 1 mm; fr 1.5–2 mm, indehiscent or bursting irregularly, smooth or roughened; style branches 3 or 4; seeds 0.8–1 mm, lenticular to obovoid, dark reddish-brown;

$2n=32$. Margins of fresh water; Vt. to N.D., s. to N.J., O., Tenn., La. and Kans. (*A. ambiens; Acnida altissima; Acnida subnuda.*)

20. **Amaranthus cannabinus** (L.) Sauer. Dioecious; stout, erect, 1–2.5 m, lvs narrowly lanceolate, to 15 × 4 cm, the upper shorter and narrower, all acuminate to an obtuse tip; fls in large terminal panicles composed of numerous slender, spike-like thyrses, the staminate commonly naked, the pistillate usually leafy; bracts 1–2 mm, the midrib slender in males, heavier in females, scarcely excurrent; male fls with 5 subequal sep 2.5–3 mm, the outer acute, the inner emarginate, the midvein not excurrent; female fls usually without perianth, rarely with 1 or 2 irregular, vestigial sep less than 1 mm; fr 2.5–4 mm, indehiscent, fleshy, 3–5-ribbed, often rugose; style-branches 3–5; seed 2–3 mm, very flat, commonly obovate, dark reddish-brown. Salt and brackish marshes along the coast; Me. to Fla. (*Acnida c.*)

2. **CELOSIA** L. Fls perfect; sep 5, distinct, erect, scarious, striate; stamens 5, connate at base; ovary subglobose; ovules 2–several; style none to elongate; stigmas capitate or subulate; utricle usually circumscissile at the middle; seeds 2–several, lenticular; annual or perennial herbs, or shrubs in the tropics, with alternate, petiolate, entire or lobed lvs; fls in terminal or axillary spikes or axillary glomerules; commonly white, silvery, or colored. 60, widespread.

1. **Celosia argentea** L. Erect, glabrous annual to 1 m; lvs lanceolate to nearly linear, 8–15 cm; spikes dense, terminating the stem and in the upper axils, 2–15 cm; sep lance-oblong, 6–8 mm, in wild plants silvery, in cult. plants also pink, yellow, or red; style indurate and exserted at maturity; $2n=36$, 72. Native of tropical Amer., occasionally escaped from cult. in our range.

C. cristata L., or *C. argentea* var. *cristata* (L.) Kuntze, the Cockscomb, rarely escapes from cult. in our range. It is a cultigen derived from *C. argentea*, with cristate, fan-shaped, or distorted spikes in a variety of colors.

3. **ACHYRANTHES** L. Fls perfect; sep 4–5, distinct, erect, firm-chartaceous; stamens connate below to form a short tube with (4)5 alternating filaments and as many alternating sterile lobes; anthers tetrasporangiate and bilocular; style solitary, with a capitate stigma; ovule one; utricle membranous, indehiscent; herbs with opposite, petiolate, entire lvs and greenish fls in elongate, terminal and axillary spikes. 100, Old World.

1. **Achyranthes japonica** (Miq.) Nakai. Perennial 7–15 dm; lvs elliptic, 3–13 × 1.5–7 cm, acute or acuminate, tapering to a petiole up to 3.5 cm, short-hairy above and on the veins beneath; spikes at first dense, elongating to as much as 2 dm and then more open; fls divergent at anthesis, sharply reflexed in fr, each subtended by a membranous bract 2 mm and 2 rigid, subulate-spinose bracteoles 2–4 mm, each bracteole with 2 basal, suborbicular, membranous auricles, the bracteoles and fl falling as a unit; sep 5, lance-linear, 4–5 mm, acuminate; sterile lobes of the androecium short, broad, subentire; fr oblong, 2.5 mm, its slender style 1 mm. Wooded, annually flooded river-banks; native of e. Asia, intr. in Ky. and W.Va. Late summer.

4. **ALTERNANTHERA** Forsskål. Fls perfect; sep 5, distinct, unequal, white or greenish; filaments united at least below, the 3–5 antheriferous filaments alternating with 3–5 sterile ones; anthers bisporangiate and unilocular, style solitary, with a capitate stigma; ovule one; utricle membranous, indehiscent; annual or perennial herbs or shrubs with opposite, petiolate or sessile, entire or obscurely denticulate lvs; fls white or greenish, in terminal heads or short spikes. 170, warm reg.

1. **Alternanthera philoxeroides** (Mart.) Griseb. Alligator-weed. Stout perennial herb 3–10 dm, ascending or decumbent, villous at least in the axils; lvs subsessile, linear-elliptic to elliptic or obovate, 3.5–11 × 0.5–2 cm; peduncles axillary or terminal, simple, 1–5 cm, pilose; heads globose, 1.5 cm thick, white; bracts a fourth as long as the sep, broadly ovate, acute or acuminate, glabrous; sep ovate-oblong, 6 mm, acute or acutish, subchartaceous; staminodes surpassing the anthers; style elongate, the stigma entire. Wet places; native to tropical Amer., intr. as a weed in the coastal states from se. Va. to Fla. and Tex.

Tidestromia lanuginosa (Nutt.) Standley, a branching, prostrate or decumbent annual native to w. U.S., with broadly ovate to orbicular lvs 1–3 cm subtending small, axillary glomerules of small fls, is occasionally found in waste places along our w. border.

Gomphrena globosa L., the Globe amaranth, a branching annual to 1 m, with terminal, colored heads 2–2.5 cm thick, and with very woolly fls, occasionally escapes from cult.

5. **FROELICHIA** Moench. Cottonweed. Fls perfect, each subtended by a scarious bract and 2 bractlets; cal tubular, becoming flask-shaped or conic and indurate in fr,

shortly 5-lobed, densely woolly; stamens 5, the filaments united into a tube equaling the cal and suggesting a sympetalous cor, the 5 oblong, bisporangiate and unilocular anthers alternating with 5 ligulate lobes; style slender; stigma capitate; ovule 1; fr a membranous, indehiscent utricle, included in the indurated cal-tube; ours annuals with narrow, opposite lvs and woolly terminal spikes. 20, New World.

1 Mature cal flask-shaped, symmetrical; larger lvs mostly 1–2.5 cm wide .... 1. *F. floridana.*
1 Mature cal obliquely conic; larger lvs mostly 0.5–1 cm wide .... 2. *F. gracilis.*

1. Froelichia floridana (Nutt.) Moq.   Common c. Erect, (3)5–20 dm, canescent or thinly tomentose, the stem stout, tending to be quadrangular; larger lvs mostly 5–12 × 1–2.5 cm; upper internodes progressively longer; spikes terminating elongate peduncles, often branched, eventually to 10 cm, the fls closely set in a 5-ranked spiral, averaging ca 15–20 fls per cm; bractlets rotund, scarious, much shorter than the cal; mature cal soon deciduous, flask-shaped, symmetrical, mostly (4)5–5.5 mm, densely woolly, with 2 lateral entire to dentate or erose wings; seeds 1.6–1.8 mm; 2n=ca 78. Dry, especially open, sandy soil; coastal states from s. Del. and e. Md. to La.; disjunct inland from Ind. and Wis. to s. Ill. and Mo., and from S.D. to Tex. July–Sept. There are 3 vars., 2 in our range. Coastal plain plants, var. *floridana,* are robust, up to 2 m, with the lowest internode of the infl often 1–2 dm; the peduncles are lanate with hairs up to 2 mm, and the larger lvs are mostly elliptic-lanceolate and broadest near or below the middle. Plants of the upper Mississippi Valley region, var. *campestris* (Small) Fern., average smaller, seldom over 1 m, with the lowest internode of the infl seldom over 1 dm; the peduncles are more closely hairy, with hairs mostly under 0.5 mm, and the larger lvs are more elliptic-oblanceolate, broadest above the middle, and obtuse or rounded at the tip.

2. Froelichia gracilis (Hook.) Moq.   Slender c. Slender, erect to nearly prostrate, commonly branched from near the base, 2–7 dm; lvs mostly below the middle of the plant, linear to narrowly lanceolate, the larger ones up to ca 8 × 1 cm; spikes 1–3 cm; fls in a 3-rowed spiral, avg 6–10 per cm; mature cal obliquely conic, mostly 3.5–4 mm; seeds 1.2–1.4 mm; otherwise much as in no. 1; 2n=54. Dry soil; Ind. to Io. and Colo., s. to Ark. and n. Mex.; adventive especially along railroad-tracks e. to s. Ont., Mass., and S.C. July–Sept.

6. IRESINE P. Browne.   Bloodleaf. Fls perfect or unisexual, the plants then variously polygamous to monoecious to dioecious; cal deeply 5-parted; stamens 5; filaments connate at base; anthers bisporangiate and unilocular, ovary compressed; style very short; stigmas usually filiform; ovule 1; utricle compressed, membranous, indehiscent; herbs or shrubs with opposite, petiolate lvs and small or minute fls in spikes or aggregated into panicles. 70, irregular.

1. Iresine rhizomatosa Standley.   Dioecious, rhizomatous perennial 5–15 dm, pilose at the nodes; lvs thin, bright green, lance-ovate, acuminate, entire, to 14 cm, a third to half as wide; panicles terminal and from the upper axils, pyramidal, 1–3 dm, with innumerable tiny fls; bracts and bracteoles ovate, shorter than the cal; sep lance-ovate, 1.2–1.5 mm, those of the pistillate fls subtended by long (eventually 3–5 mm) hairs; fr rotund, 2–2.5 mm. Damp woods; Md. to Va. and Ala., w. to Kans. and Tex. Aug.–Oct. (*I. paniculata,* misapplied.)

Family **PORTULACACEAE**, the Purslane Family

Fls mostly perfect, hypogynous or (*Portulaca*) semi-epigynous, regular or seldom (spp. of *Montia*) slightly irregular; sep 2, or seldom (spp. of *Lewisia*) up to 9, persistent or seldom (*Talinum*) deciduous; pet (2–)4–6 or seldom (spp. of *Lewisia*) more numerous (up to ca 18), distinct or sometimes basally connate, mostly imbricate, often ephemeral; stamens most commonly as many as and opposite the pet, or sometimes (as in *Portulaca*) more numerous and then sometimes grouped into bundles; filaments free or sometimes basally adnate to the pet or to the short cor-tube; ovary unilocular (often showing vestiges of partitions at the base) mostly composed of 2–3 (up to 9 in *Portulaca*) carpels with as many distinct styles, or seldom (as in *Talinum*) with a single (then usually lobed or cleft) style; ovules (1)2–many on a free-central or basal placenta; fr usually a loculicidal or circumscissile capsule; seeds commonly lenticular; embryo slender, dicotyledonous, peripheral, ± curved around the abundant perisperm; herbs (all ours) or rarely shrubs, producing betalains but not anthocyanins, often somewhat succulent; lvs simple, entire, opposite or alternate (or all basal); stipules scarious, or modified into tufts of hair, or wanting; fls solitary or more often in various sorts of cymose or racemose or head-like infls. 20/500.

1 Ovary partly inferior; capsule circumscissile; fls sessile or nearly so .... 1. *Portulaca.*
1 Ovary superior; capsule opening lengthwise; fls pediceled.
  2 Fls solitary or racemose; lvs flat (though sometimes narrow).
    3 Perennial from a corm or rounded, tuberous root; cauline lvs only 2 .... 2. *Claytonia.*
    3 Annual, or perennial from a rhizome or fibrous root system; cauline lvs more than 2 .... 3. *Montia.*
  2 Fls in long-peduncled terminal cymes; lvs terete or nearly so .... 4. *Talinum.*

1. PORTULACA L.   Sep 2; pet 4–6, commonly 5 (more in double-fld forms); stamens 6–many; ovary partly inferior; styles or style-branches (2)3–several; capsule circumscissile near the middle; seeds many; ours succulent annuals with mostly alternate cauline lvs, the uppermost crowded and forming an involucre to the fls, these sessile or nearly so, solitary or glomerate at the ends of the stem and branches. 100+, mostly warm reg.

1 Lvs flat, spatulate to obovate; plants glabrous or nearly so; fls yellow.
  2 Seeds with low, blunt tubercles; stamens mostly 6–10; widespread sp. .... 1. *P. oleracea.*
  2 Seeds with sharp, conic tubercles; stamens 7–19; southwestern sp. .... 2. *P. retusa.*
1 Lvs subterete, linear; plants densely hairy at the nodes; fls variously colored .... 3. *P. grandiflora.*

1. Portulaca oleracea L.   Common purslane. Stem prostrate, usually purplish-red, glabrous, repeatedly branched, forming large mats, edible when young; lvs edible, succulent but flat, spatulate to obovate-cuneate, 1–3 cm, commonly rounded at the summit, the cauline alternate or occasionally opposite; fls solitary or in small terminal glomerules, sessile, self-compatible, yellow, 5–10 mm wide; sep ± acute; stamens mostly 6–10; style-branches 4–6; seeds with low, blunt tubercles, requiring light for germination; polyploid series on $x=9$. A cosmopolitan weed, probably originally native to s. Asia. All summer.

2. Portulaca retusa Engelm.   Much like *P. oleracea,* and perhaps better considered to be a phase of it; lvs tending to be retuse at the summit; calyx-lobes obtuse in bud; stamens mostly 7–19; style-branches mostly 3–4; seeds covered with sharp, conic tubercles. Disturbed sites; Utah and Ariz. to w. Mo. and reputedly Io. and Minn. (? *P. neglecta.*)

3. Portulaca grandiflora Hook.   Moss-rose. Stems diffusely branched, ascending or widely spreading, 2–4 dm, hairy at the nodes; lvs linear, subterete, 1–3 cm, the cauline alternate, the terminal crowded, forming an involucre intermingled with long hairs; fls white or various brilliant shades of red or yellow, 2–4 cm wide; stamens numerous, ca 40 or more; style-branches 5–9; $2n=18, 36$. Native of Argentina, occasionally escaped from cult. All summer.

P. mundula I. M. Johnston, occurring from the Ozark region to n. Mex., probably does not reach our range. It is smaller in all respects than *P. grandiflora*; the fls have ca 10–15(–30) stamens, and red-purple pet 5–8 mm.

2. CLAYTONIA L.   Spring-beauty. Sep 2, herbaceous, ovate, persistent in fr; pet 5, oval or elliptic, spreading; stamens 5, opposite the pet; ovules 6; style with 3 short lobes; capsule ovoid, opening by inrolling valves; perennial herbs from a rounded corm (our spp.) or fleshy taproot, with one or few lvs from the base and a single opposite pair on the stem below the loose terminal raceme; our spp. have delicate glabrous stems 1–3 dm from the corm, lengthening during and after anthesis, the 5–15 long-pediceled fls with pink-veined, otherwise white or pale pinkish (yellow) pet 1–1.5 cm. 10, N. Amer.

1 Lvs (petiole included) 3–8 times as long as wide, the petiole evident .... 1. *C. caroliniana.*
1 Lvs mostly at least 8 times as long as wide, the petiole often scarcely differentiated .... 2. *C. virginica.*

1. Claytonia caroliniana Michx.   Cauline lvs, including the petiole, mostly 3–6(–11) cm, the blade usually 10–15(–30) mm wide, tending to be diamond-shaped or ovate or lance-ovate, acute at the base and clearly distinguished from the petiole; $2n=16$–38, mostly 24 in our range and 16 in the s. Appalachians. A rare yellow form occurs in Md. Cool woods, sometimes with no. 2 and rarely hybridizing with it; N.S. to Minn., and s. along the mts. to N.C., Tenn., and n. Ga. Early spring, disappearing by early summer.

2. Claytonia virginica L.   Cauline lvs mostly at least 8 times as long as wide and long-tapering to the base, the blade sessile or merging gradually into the short, poorly differentiated petiole, rarely less than 7 cm (petiole included); $2n=12$–72+. Rich woods and fields; N.S. to Minn., s. to Ga., La., and Tex. Early spring, disappearing by midsummer. Two weakly defined vars.: var. *virginica* (*C. media, C. robusta*), with lvs mostly 5–10(–20) mm wide, and with chromosome numbers based eventually on $x=8$, is the more n. phase, and var. *acutiflora* DC. (*C. simsii*), with the lvs mostly 1–4 mm wide and with chromosome numbers based eventually on $x=6$, is the more s. phase, but both vars. are well represented in much of our range.

Yellow-fld plants related to var. *virginica* are locally and sporadically distributed in N.J., Pa., and Md. The oldest name for these is *C. virginica* f. *lutea* R. J. Davis, based on plants from Md. These and the Pa.

plants occur in mixed and intergrading populations with typical *C. virginica.* The N.J. plants, at least, occur in wetter, shadier habitats than is typical of *C. virginica,* and last throughout much of the summer. These have been called f. *hammondiae* Kalmbacher. Further study is in order.

3. MONTIA L.   Sep 2, herbaceous, persistent; pet 3 or 5, often slightly unequal, sometimes connate below; stamens as many as and opposite the pet, or some members of the cycle missing; ovules and style-branches 3; fr a 3-valved capsule with 1–3 seeds; fibrous-rooted annual or perennial herbs without basal lvs, ours with several pairs of opposite lvs, the fls in axillary or terminal racemes, or solitary in the axils, or in small axillary cymules. 50, mostly N. Amer.

1 Plants perennial, with slender rhizomes and stolons; stamens 5; pet 5–8 mm .... 1. *M. chamissoi.*
1 Plants annual or seldom perennial, without rhizomes; stamens 3; pet ca 1.5 mm .... 2. *M. fontana.*

1. Montia chamissoi (Ledeb.) Greene.   Erect perennial 5–20 cm, colonial from long, slender rhizomes and long, slender, subnaked stolons, both of these often with bulblet-like offsets; lvs spatulate or obovate, 2–5 cm; fls mostly 3–10 in lax, terminal and axillary racemes, on pedicels 1–3 cm; sep 2–3 mm; pet 5, white or pinkish, 5–8 mm; stamens 5. Wet, often springy or boggy habitats; widespread in the w. cordillera; disjunct in ne. Io. and se. Minn.; also along the upper Delaware R. in Pa. June, July.

2. Montia fontana L.   Blinks. Lax, branching annual or seldom perennial, 1–3 dm, often rooting from the nodes; lvs oblanceolate or spatulate or obovate, 0.5–1.3 cm; pedicels solitary or paired in the axils and terminal, to 1 cm, or the fls in 2–4-fld cymules from the upper axils; pet 5, white, ca 1.5 mm, often unequal, united at the base into a short tube split on one side; stamens 3. Springs, brooks, or wet soil, often ± aquatic; irregularly cosmopolitan, with 4 vars. differing in the ornamentation of the seed-coat. June–Sept. Our plants, with the seed-coat shiny and nearly smooth, belong to the chiefly boreal var. *fontana,* occurring s. to N.S. and Me. (*M. lamprosperma; M. rivularis,* a ± perennial phase.)

4. TALINUM Adans.   Fame-fl. Sep 2, caducous; pet 5 (rarely more), distinct, fugacious; stamens 4–45, distinct; filaments slender; ovary ovoid; styles partly or wholly united; capsule 1-celled, many-seeded; seeds cochleate, flattened, minutely roughened; glabrous, succulent, annual or perennial herbs with conspicuous fls in long-peduncled bracteate cymes. Our spp. have a short taproot that in age becomes multicipital, producing from each crown a short stem that has numerous crowded, succulent, terete lvs 3–8 cm, and is prolonged above into a slender scape cymosely branched above and bracteate at the nodes; bracts triangular, 2–5 mm, prolonged backward at base; fls pinkish to deep rosy-red, open for only a few hours in full sunshine; midveins of the capsule-valves often persistent, as well as the placentas; seeds gray over black, 1 mm wide, minutely roughened. 30+, mostly warm N. Amer.

1 Stamens 4–8 .... 1. *T. parviflorum.*
1 Stamens 10–45.
  2 Pet 5–10 mm; stamens 10–25.
    3 Stigma nearly capitate; seeds minutely roughened .... 2. *T. teretifolium.*
    3 Stigmas 3, linear; seeds minutely roughened and strongly wrinkled .... 3. *T. rugospermum.*
  2 Pet 10–16 mm; stamens 25–45 .... 4. *T. calycinum.*

1. Talinum parviflorum Nutt.   Prairie f.-fl. Infl 8–15 cm; pet 4–7 mm, pale or dull pink; stamens 4–8, commonly 5 or 6, stigma capitate; fr 3–6 mm; $2n=24, 48$. Thin, acidic soil overlying rocks; Minn. (and reputedly sw. Wis.) to S.D. and Colo., s. and w. to Ark., Tex., and Ariz.; disjunct in c. Ala. June, July. Fls open from noon to late afternoon.

2. Talinum teretifolium Pursh.   Appalachian f.-fl. Infl 1–3 dm; pet 5–10 mm, pink, stamens 10–25; anthers oblong, 1 mm; stigma capitate; fr 5–6 mm; seed minutely roughened; $2n=48$. Thin soil overlying rocks, especially serpentine; Del., Pa., and W.Va., s. to Ga. and Ala. All summer. Fls open in early afternoon only.

3. Talinum rugospermum Holzinger.   Sand f.-fl. Infl 1–2 dm; pet 6–8 mm; stamens 10–25; anthers subglobose, 0.5 mm; style divided a fourth to a third of its length into linear stigmas; fr 4–5 mm; seeds minutely roughened and strongly wrinkled. Thin soil overlying sandstone, and on sand prairies; nw. Ind. to Minn., Neb., and Kans., and reputedly in ne. Tex. July, Aug. Fls open in late afternoon only.

4. Talinum calycinum Engelm.   Rockpink f.-fl. Infl 1–3 dm; sep prolonged at base like the bracts; pet 10–16 mm, deep rose-pink to rose-red; stamens 20–45; stigma capitate; fr 6–8 mm. Barrens and rocky ledges; s. Ill. and s. Mo. to Neb., Kans., and Tex. May, June. Fls open only in early to mid-afternoon.

FAMILY **MOLLUGINACEAE**, the Carpet-weed Family

Fls perfect or rarely unisexual, hypogynous; sep (4)5, persistent, mostly distinct; pet small or more often wanting; stamens (2)3–10, or sometimes numerous, the filaments distinct or basally connate; gynoecium of mostly 2–5 carpels united to form a compound, 2–5-locular ovary with separate styles (style solitary in *Glinus*) and axile placentation; ovules 1–many in each locule, ± campylotropous; fr dry, opening loculicidally or by transverse slits, or rarely indehiscent, commonly surrounded by the persistent cal; seeds with an elongate, dicotyledonous, peripheral embryo curved around the abundant perisperm; mostly herbs, only slightly or not at all succulent, often with anomalous secondary growth, producing anthocyanin but not betalains; lvs alternate, opposite, or whorled, simple and entire, exstipulate or with small, deciduous stipules; fls in cymose, often loose and open infls, or borne singly in the axils, commonly small and inconspicuous. 13/100.

1. **MOLLUGO** L. Carpet-weed. Fls perfect; sep 5, distinct; stamens 3–10; ovary superior, 3–5-locular; styles 3–5; ovules many; fr a thin-walled loculicidal capsule; seeds without an aril; herbs with opposite or whorled lvs and pedicellate, axillary, white or greenish fls. 20, warm reg.

1. **Mollugo verticillata** L. Glabrous annual, repeatedly forked, forming mats to 4 dm wide; lvs in whorls of 3–8, narrowly to broadly oblanceolate, 1–3 cm, tapering to a short petiolar base; fls 2–5 from each node, on pedicels 5–15 mm, pale green or white, 4–5 mm wide; stamens 3 or 4; fr ovoid, 3 mm; seeds numerous, arcuate-ridged; $2n=64$. Apparently native to tropical Amer., but now a common weed in moist soil and dunes nearly throughout temperate N. Amer. June–Sept.

Glinus lotoides L. is intr. from Africa into s. U.S., n. to c. Mo. It is a low annual, with forking, stellate-villous stems, small, rounded, opposite lvs, and clustered small fls with 3–5 stamens, 3–5 locules, a single short style, and tuberculate seeds.

FAMILY **CARYOPHYLLACEAE**, the Pink Family

Fls mostly perfect, usually regular (4)5-merous as to the cal and cor; pet generally distinct, sometimes of very ordinary nature, sometimes bifid from the tip, sometimes with a long basal claw, and then often with variously developed appendages ventrally at the juncture of claw and blade; stamens commonly 5 or 10, seldom only 1–4; filaments variously hypogynous, free and distinct, or basally adnate to the pet to form a short or ± elongate tube that may or may not be adnate to a gynophore, or inserted at the edge of a glandular disk surrounding the ovary, or even (*Scleranthus*) borne on a hypanthium; ovary superior, mostly 2–5-carpellary, typically unilocular but sometimes with partitions at the base, rarely (as in spp. of *Silene*) plurilocular for much of its length; styles distinct or ± united; ovules 1–many on a central (or proximally axile) placenta, the placenta often continuous and attached to the top as well as the base of the ovary, but sometimes free at the top, or reduced to a basal nubbin; fr most commonly a capsule dehiscent by as many or twice as many valves or apical teeth as there are styles, less often indehiscent and utricular, and then sometimes enclosed in the persistent, indurated cal or hypanthium; seeds commonly lenticular; embryo slender, dictoyledonous, excentric, commonly peripheral and curved around the abundant perisperm; herbs (all ours) or rarely shrubs, generally producing anthocyanins, but not betalains, the stems commonly swollen at the nodes; lvs opposite or rarely (as in *Corrigiola*) alternate, simple, entire, often connected by a transverse line at the base, provided with scarious or hyaline stipules in the Paronychioideae, otherwise exstipulate; fls commonly in dichasial cymes, or sometimes solitary. 75/2000.

1 Lys with evident, scarious or hyaline stipules. (Paronychioideae.)
  2 Fr capsular, with several or many seeds; pet present.
    3 Styles and valves of the fr mostly 5; lvs appearing whorled . . . . 9. *Spergula.*
    3 Styles and valves of the fr mostly 3; lvs opposite . . . . 10. *Spergularia.*
  2 Fr a 1-seeded utricle; pet none.

    4 Stipules of each nodal pair (between the lf-bases) distinct, long-pointed, evidently longer than
       wide .... 11. *Paronychia.*
    4 Stipules of each nodal pair united to form a fringed scale about as broad as high .... 12. *Herniaria.*
1 Lvs without stipules (but the reduced cauline lvs with a few fringing bristles toward the base in
   Stipulicida).
    5 Sep distinct, or connate only at the base (but borne on a cupulate hypanthium in *Scleranthus*).
     (Alsinoideae.)
      6 Fr a 1-seeded utricle; pet none; fls perigynous, with a cupulate hypanthium resembling a cal-
       tube .... 8. *Scleranthus.*
      6 Fr capsular, with several or many seeds; pet present or sometimes absent; fls essentially hypog-
       ynous.
        7 Pet entire or merely emarginate or erose-fimbriate at the tip, or sometimes wanting.
         8 Plants fleshy; staminal disk conspicuous; seeds 3 mm or more; maritime .... 3. *Honckenya.*
         8 Plants not fleshy; staminal disk inconspicuous or none; seeds less than 2 mm; most spp.
          not maritime.
          9 Styles 4 or 5, as many as the sep .... 1. *Sagina.*
          9 Styles or style-branches 3, fewer than the sep.
           10 Styles 3, distinct; sep pointed or blunt, not notched; widespread.
            11 Infl ± cymose or racemiform, or of solitary flowers, but not umbelliform; pet
              entire or emarginate .... 2. *Arenaria.*
            11 Infl umbelliform; pet erose-fimbriate at the tip .... 4. *Holosteum.*
           10 Style 1, trifid; 3 of the sep notched at the tip; se. Va. and s. .... 5. *Stipulicida.*
        7 Pet ± deeply bifid, or seldom wanting.
         12 Fr cylindric, opening by apical teeth; styles in our spp. mostly 5 (3 in *C. nutans*, 4 in *C.
         diffusum*) .... 6. *Cerastium.*
         12 Fr ovoid or oblong, opening by valves to about the middle; styles mostly 3 (5 in *S.
         aquatica*) .... 7. *Stellaria.*
    5 Sep connate to form a well developed, toothed or lobed tube. (Silenoideae.)
     13 Styles 3–5.
      14 Styles mostly 5 (in some spp. occasionally 4 but predominantly 5).
       15 Cal-lobes much longer than the tube; pet not appendaged .... 13. *Agrostemma.*
       15 Cal-lobes much shorter than the tube; pet appendaged.
        16 Cal neither glandular nor inflated .... 14. *Lychnis.*
        16 Cal glandular, often inflated in fr .... 15. *Silene.*
      14 Styles mostly 3 (in some spp. occasionally 4 but predominantly 3) .... 15. *Silene.*
     13 Styles 2.
      17 Cal ebracteate.
       18 Fls 2 cm long, or longer.
        19 Cal tubular, 20-nerved; pet appendaged .... 16. *Saponaria.*
        19 Cal ovoid, strongly wing-angled, 5-costate; pet not appendaged .... 17. *Vaccaria.*
       18 Fls 1 cm long, or less .... 18. *Gypsophila.*
      17 Cal subtended by 1–3 pairs of bracts.
       20 Cal 20–40-nerved .... 19. *Dianthus.*
       20 Cal 5-nerved or 15-nerved .... 20. *Petrorhagia.*

1. SAGINA L.   Pearlwort. Fls solitary and terminal or axillary, or in terminal cymes; sep mostly 5 or 4; pet as many as the sep, white, entire or emarginate, or none; stamens as many as and opposite the sep, or twice as many; styles as many as and alternate with the sep; valves of the capsule opposite the sep; seeds numerous, minute; low, matted, delicate annuals or perennials with opposite, basally connate, narrow (often linear-subulate) lvs, lacking stipules. 15, mostly cold-temp. N. Hemisphere.

1 Pet obviously longer than the sep, commonly nearly twice as long .... 1. *S. nodosa.*
1 Pet up to about as long as the sep, or none.
  2 Perennial, often with basal rosettes or sterile shoots; sep 4 (seldom 5 on some fls), spreading in fr.
    3 Lvs mucronate or shortly aristate, 1 mm wide or less; intr., weedy .... 2. *S. procumbens.*
    3 Lvs merely acutish, 1–2 mm wide; local in Ky. and Tenn. .... 3. *S. fontinalis.*
  2 Slender annual without persistent basal rosettes or sterile shoots; sep 5(4), erect-appressed in fr .... 4. *S. decumbens.*

    1. Sagina nodosa (L.) Fenzl.   Obviously perennial, decumbent to erect, 5–15 cm, with conspicuous tufts of basal lvs; lvs linear-subulate, 5–20 cm, triquetrous, the tip mucronate, boat-shaped, the upper cauline lvs reduced; short sterile shoots replacing some or all fls; fls terminating the slender branches, or on slender pedicels 4–12 mm in a branching infl; sep 5 (or 4 on some fls), broadly ovate or oblong, 2 mm, erect-appressed at maturity and about equaling the 5-valved fr; pet nearly twice as long as the sep; stamens 10; seeds 0.5 mm, plump, obliquely subreniform or nearly globose; minutely tessellate-tuberculate in lines or ridges, scarcely grooved. Polyploid series based on $x=11$. Rocky or sandy soil, moist shores, or beaches; circumboreal, s. in Amer. to Mass., Lake Superior (incl. Keweenaw Point), and Alta. Summer. The widespread native Amer. phase, glabrous throughout, or with only the pedicels and calyx-bases glandular-hairy, is var.

borealis (Crow) Cronq. The European var. *nodosa,* with the stems, pedicels, calyx-bases, and often the lf-margins glandular-hairy, is intr. along the coast from Nf. and Que. to Mass.

2. Sagina procumbens L. Glabrous, branching perennial, prostrate to ascending, the stems 2–10 cm; lvs linear-subulate, 3–10(–15) mm, up to about 1 mm wide, mucronate or shortly aristate, sometimes minutely ciliate, frequently subtending short shoots or lf-fascicles; fls solitary or few at the tips of the stem and branches, or some of them on axillary pedicels, often nodding after anthesis, but finally erect; sep 4(5), 2–2.5 mm, spreading after maturity; pet shorter than the sep, or none; stamens as many as the sep; fr about equaling the sep, commonly 2–3 mm, flattened, obliquely triangular, sulcate along the 2 dorsal angles, very finely roughened; 2*n*=22. Moist soil and rocky places, often a weed in paths or pavements; circumboreal, s. in Amer. to Md., Mich., and Kans. Summer.

3. Sagina fontinalis Short & Peter. Kentucky p. Delicate, glabrous perennial; stems ascending, 10–15 cm; lvs narrowly oblanceolate to linear, 1–2 cm × 1–2 mm, acutish, somewhat fleshy, fls solitary, tetramerous, appearing lateral; lvs reduced upwards, becoming bract-like but not scarious; seeds 0.6 mm, orbicular-reniform, shining, minutely papillate. Wet cliffs; Lexington, Ky., and Nashville, Tenn. (*Alsine f.; Arenaria f.; Stellaria f.*)

4. Sagina decumbens (Ell.) T. & G. Slender, branching annual, ascending or decumbent, 3–10(–15) cm, often with short sterile axillary shoots; fls terminating the slender branches, commonly some also on slender, axillary glabrous pedicels 10–15 mm; sep 5(4), 1.5–2 mm, often purple-tipped or -margined; pet up to ca as long as the sep, or wanting; stamens as many as the sep, or up to twice as many; fr 2–3 mm, ellipsoid, evidently surpassing the erect-appressed sep; seeds light brown, 0.25–0.3 mm, flattened, obliquely triangular, sulcate along the 2 dorsal angles, very minutely glandular-tuberculate or essentially smooth; 2*n*=36. Wet places or dry, sandy soil; Mass. to Ill., s. to Fla. and Tex. Mar.–May. The apetalous form has been called var. *smithii* (A. Gray) S. Wats. *S. occidentalis,* of Wash., Oreg., and Calif. is scarcely different from *S. decumbens.*

S. japonica (Sw.) Ohwi, from Japan and China, is casually intr. in N.Y., Conn., and Mass., and may be expected elsewhere. Much like *S. decumbens* in aspect, it differs in its slightly succulent lvs, glandular-hairy (at least distally) pedicels, shorter, more globose fr, and dark brown, evidently muricate, plump seeds without a dorsal groove.

2. ARENARIA L.   Sandwort. Fls solitary or in terminal cymes; sep 5; pet 5, white, entire to emarginate; stamens normally 10; styles (2)3(–5); ovary unilocular; ovules numerous; primary valves of the capsule 3, entire or 2-cleft; seeds ± reniform; low annual or perennial herbs without stipules. (*Minuartia, Moehringia, Sabulina.*) 250, widespread.

1 Lvs relatively broad, with expanded blade; fr dehiscent by 6 valves.
  2 Rhizomatous, non-cespitose perennials; seeds strophiolate.
    3 Sep 3–6 mm, acutish to acuminate, often longer than the pet; lvs mostly ± acute .... 1. *A. macrophylla.*
    3 Sep 2–3 mm, obtuse to sometimes acutish, much shorter than the pet; lvs mostly ± obtuse ...... 2. *A. lateriflora.*
  2 Cespitose perennials, or taprooted annuals; seeds not strophiolate.
    4 Perennial; lvs lanceolate to oblanceolate, mostly 1.5–3 cm .... 3. *A. lanuginosa.*
    4 Annual; lvs ovate, less than 1 cm .... 4. *A. serpyllifolia.*
1 Lvs narrow, linear or linear-subulate, without expanded blade; fr with 3 valves.
  5 Sep acute, prominently ribbed.
    6 Primary lvs with secondary lvs fascicled in their axils .... 5. *A. stricta.*
    6 Primary lvs only present.
      7 Short-lived perennial (sometimes annual) mostly 2–10 cm; lvs 5–8 mm .... 6. *A. rubella.*
      7 Annual, mostly 1–3 dm; lvs 1–2 cm; southern .... 7. *A. patula.*
  5 Sep obtuse (or seldom acutish), nerveless or obscurely nerved.
    8 Lvs rigid, linear-subulate; plants obviously perennial .... 8. *A. caroliniana.*
    8 Lvs soft, linear, plane; delicate perennial or annual .... 9. *A. groenlandica.*

1. Arenaria macrophylla Hook. Much like no. 2; lvs mostly 2–5 cm × 3–8 mm, lanceolate, elliptic, or oblanceolate, obtusish to more often acute or acuminate; sep 3–6 mm, broadly lanceolate, acutish to acuminate, from a little shorter to often evidently longer than the pet; fr shorter than the sep, globose-oblong, sometimes dehiscing only to the middle; seed 1.5 mm, with a conspicuous strophiole; 2*n*=48. Woods, stony slopes, and plains; Lab. and Que. to Vt., Mass., n. Mich., and Wis.; and in the w. cordillera. May–Aug. (*Moehringia m.*)

2. Arenaria lateriflora L. Colonial by rhizomes, the stem leafy, retrorsely puberulent; lvs 1–3 cm × 5–10 mm, usually ovate to elliptic-oblong and obtuse, sometimes narrower and acutish; cymes 1–5-fld, the peduncle 1–3 cm, the pedicels 5–15 mm, usually puberulent; sep 2–3 mm, ovate or obovate, obtuse or acutish, faintly 3–5-nerved, glabrous, scarious-margined; pet 4–6 mm; fr 3–5 mm, the 6 valves separating nearly to the base; seed 1–1.4 mm, smooth and glossy, with a small strophiole; 2*n*=48. Woods or sometimes open places; circumboreal, s. in Amer. to Pa., Mo., and s. Calif. May–July. Highly variable. (*Moehringia l.*)

3. Arenaria lanuginosa (Michx.) Rohrb. Lax perennial; stem retrorsely puberulent, to 5 dm; lvs narrowly lance-elliptic or elliptic-oblanceolate, 15–30 × 2–8 mm, ± pustulate, ciliate at base; pedicels axillary, slender, mostly 1.5–4 cm, ascending to reflexed, puberulent; sep lanceolate, 3–4 mm, acute or acutish, 1-nerved, glabrous, often pustulate, sometimes ciliolate at base; pet minute or usually none; seed 0.6–0.8 mm, flattened, somewhat keeled dorsally, smooth, shining, black or dark reddish-black; $2n$=44. Open woodlands and rocky places or damp thickets; se. Va. to Ark., s. to tropical Amer. Our plants are var. *lanuginosa*.

4. Arenaria serpyllifolia L. Thyme lf-s. Diffuse, delicate or wiry, puberulent annual 5–30 cm; lvs usually 8–10 pairs, 3–8 × 3–5 mm, ovate, acute, sparsely scabrid-puberulent, 3–5-nerved, mostly much shorter than the internodes; infl short or extending to the middle of the stem; bracts leafy, pedicels slender, 4–8 mm; sep 2.5–4 mm, lance-ovate, acuminate, 3–5-nerved, somewhat carinate, scarious-margined, scabrid-puberulent or often glandular; pet usually shorter than the sep; fr ovoid-conic, ± exceeding the sep, dehiscent to an uncertain depth by 6 teeth or valves; seed plump, 0.4–0.6 mm, gray-black or reddish-brown, tesselate-tuberculate; tetraploid on x=10. Native of Eurasia, now found throughout most of temperate N. Amer. as a weed in sandy or stony places. May–Aug. A delicate, diploid phase with relatively small lvs, fls and frs, less common in our range than typical *A. serpyllifolia*, is var. *tenuior* Mert. & Koch. (*A. leptoclados*.)

5. Arenaria stricta Michx. Diffuse annual or perennial, glabrous or sometimes hairy, commonly with numerous short sterile shoots on prostrate branches; stems 1–4 dm, decumbent or erect, leafy for one-third to two-thirds their length, usually with short, leafy, fascicle-like sterile axillary shoots; primary lvs 8–30 mm, subulate-setaceous, somewhat involute, 3-nerved; infl open, often extending to the middle of the stem; pedicels slender, 5–20 mm; sep 3.5–6.5 mm, broadly lanceolate, acute, scarious-margined, (1)3-nerved; pet 5–8 mm, entire; fr equaling or a little longer than the sep, the 3 valves dehiscent to the middle or beyond; seed 0.8–1.5 mm, brown-black, low-tuberculate; $2n$=22–30. Rocky or gravelly, often calcareous soil. (*Minuartia s.*) Var. *stricta*, perennial or annual, mostly lax and diffuse, the lvs mostly 1.5–3 cm and the stem leafy to above the middle, the pet usually longer than the sep, is common and variable in our range, from s. Ont. and N.Y. to Minn., s. to Va. and Ark. June–July. Var. *litorea* (Fern.) B. Boivin, often cespitose, leafy below the middle, the lvs 5–15 mm, and the pet equaling or mostly shorter than the sep, occurs from Nf. to Minn. and Alas. and B.C. July–Sept. (*A. litorea; A. dawsonensis; Sabulina d.*) Var. *texana* B. L. Robinson, rigidly cespitose, usually only the lower third of the 5–20 cm stem leafy, the lvs 5–20 mm, the pet usually longer than the rigid sep, occurs from Mo. to Neb. and Tex., and reputedly e. to O.

6. Arenaria rubella (Wahlenb.) J. E. Smith. Short-lived perennial or sometimes annual, compact or loose, 2–10(–15) cm, often decumbent, leafy mostly below the middle, glabrous to glandular-hairy; lvs 5–8 × 0.3–0.8 mm, obtuse or acutish, prominently 3-nerved, puberulent or sometimes glabrous; cymes usually with several fls; pedicels slender, 5–12 mm; sep 3–4.5 mm, lanceolate, acute, strongly 3-nerved; pet usually shorter than the sep; fr membranous, the 3 valves obtuse, entire or slightly emarginate, dehiscing to below the middle; seeds 0.4–0.7 mm, tuberculate; $2n$=24, 26. Moist, rocky or gravelly places; circumboreal, s. in Amer. to Que., n. Vt., Ont., and N.M. June–Aug. (*Minuartia r.; Sabulina propinqua.*)

7. Arenaria patula Michx. Annual, usually much-branched at base, glabrous or minutely glandular-puberulent; stems slender, 1–2 dm, decumbent or erect; lvs mostly cauline, several pairs, 1–2 cm × 0.5–1 mm; infl open, often extending to below the middle; pedicels 15–30 mm; sep 4.5–6 mm, narrowly lanceolate, acute; pet white, 5–8 mm; fr equaling or shorter than the sep, the 3 valves dehiscent to the middle; seeds gray-brown, low-tuberculate. Apr.–June. (*Minuartia p.*) Var. *patula*, occurring in rocky soil, barrens, and meadows, from Ind. and Minn. to Va., Ala., and Tex., has 5-nerved sep, and the seeds are 0.5–0.7 mm. Var. *robusta* (Steyerm.) Maguire, occurring on limestone slopes in Mo., Ky., and Tenn., has 3-nerved sep, and the seeds are 0.7–0.9 mm.

8. Arenaria caroliniana Walter. Perennial from a taproot and woody caudex, erect or ascending, 1–2(3) dm, the stem very leafy below, densely glandular-puberulent and minutely scabrous, or seldom glabrous; lvs firm, linear-subulate, 3–10 mm, glabrous, eventually 1–3-nerved, triangular in section, abruptly acutish or obtuse; pedicels 1–6 cm, capillary, ascending or erect; sep ovate, obtuse, 3 mm, faintly 1–3-nerved; pet 5–8 mm; fr ovoid, the 3 valves separating nearly to the base; seeds plump, 1.2 mm, reddish-brown, tuberculate. Sandy soil and pine-barrens on the coastal plain; N.Y. to Fla. May–July. (*A. squarrosa; Minuartia c.; Sabulina c.*)

9. Arenaria groenlandica (Retz.) Sprengel. Delicate, glabrous annual (perennial?); lvs linear or linear-oblanceolate, 5–20 mm, obtuse, rather soft and fleshy; fls few to many, the infl sometimes extending to below the middle; pedicels 5–25 mm; sep 3–3.5 mm, oblong to oblong-ovate, obtuse or acutish, faintly 1-nerved; pet 5–10 mm, conspicuously exceeding the sep; fr broadly conic, the 3 valves separating to the base, obtuse, slightly notched; seed 0.7–0.8 mm, minutely tuberculate, reddish-brown. $2n$=20. (*Minuartia g.*) Var. *groenlandica* is a mat-forming plant 5–10(–15) cm, the cymes 3–7-fld, the sep 4–5.5 mm, the pet 6–10 mm, occurring in rocky places from Greenland to N.S. and e. Que. and s. in the higher mts. to Me., N.H., Vt., and N.Y., and in N.C. (*Sabulina g.*) Var. *glabra* (Michx.) Fern. is 1–2 dm, seldom matted, with simple or forking stems, the cymes mostly 9–15-fld, the sep 3–4 mm, the pet 4–6(–8) mm, occurring in the foothills and lower mts. from Me. and N.H. to S.C. and Tenn. (*Minuartia g.; Sabulina g.*) The closely related *A. cumberlandensis* Wofford & Kral, from rock-houses in n. Tenn., may be expected in our range in Ky. It differs from *A. groenlandica* var. *glabra* in its slightly wider and softer, more veiny leaves and in having only 1–3 flowers per stem.

3. HONCKENYA Ehrh. Fls solitary in the forks of the stem or in leafy terminal cymes, usually functionally unisexual; sep 5; pet 5; stamens normally 10; staminal disk conspicuous; styles (2)3–5(6); ovary usually 3–5-locular, but the capsule unilocular; seeds few, large, pyriform, basally rostrate; fleshy, glabrous perennials with entire, exstipulate lvs. 2, mostly cool N. Hemisphere.

1. **Honckenya peploides** (L.) Ehrh. Colonial by rhizomes and runners, forming dense colonies often 1–2 m wide; stems 1–5 dm, simple to much-branched; lvs 1–2 cm, elliptic or elliptic-ovate to obovate or sometimes lanceolate or oblanceolate; sep ovate or lance-ovate, 3–7 mm; pet in male fls ca equaling the sep, in female fls rarely over 2 mm; stamens in male fls about equaling the pet, in female fls reduced or abortive; fr depressed, 4–8 mm long, 5–12 mm thick; seeds 3–5 mm, reddish-brown; $2n=48$–70. Sea-beaches and sand-dunes; circumboreal, s. on the Atlantic coast to Va. June–July. Our plants are var. *robusta* (Fern.) House, with little-branched, few-fld stems 2–5 dm, the sep 4–5 mm. (*Arenaria p.*)

4. HOLOSTEUM L. Jagged chickweed. Fls in terminal, pedunculate, cymose umbels; sep 5; pet 5, irregularly erose-fimbriate at the tip; stamens 3–10; styles 3; capsule thin-walled, ± cylindric, many-seeded, opening by 6 valves or apical teeth; seeds compressed, rough, attached to the inner face; annual or biennial herbs without stipules. 6, temp. Eurasia.

1. **Holosteum umbellatum** L. Annual, somewhat glaucous; stems tufted, unbranched, 1–3 dm, stipitate-glandular especially near the middle, generally glabrous at top and bottom; lvs stipitate-glandular around the margins, the basal ones tufted, oblong to oblanceolate, 1–2.5 cm, often short-petiolate, the cauline few and sessile; peduncle much surpassing the lvs, with 3–15 fls on slender pedicels that elongate to 1.5–3 cm in fr; sep ca 3 mm at anthesis, somewhat accrescent; pet generally a little longer than the sep; stamens 3–5, alternipetalous; capsule well surpassing the cal, its valves ± strongly outrolled from the tip; seeds 1 mm; $2n=20$. A weed of waste places, native to Eurasia, widely intr. in the U.S. and nearly throughout our range. Apr–May.

5. STIPULICIDA Michx. Fls small, in condensed, few-fld, minutely bracteate fascicles at the branch-tips; sep 5, 2 merely blunt, 3 with sharply marked, expanded, white-hyaline scarious, rounded and distally notched tip; pet 5, white, clawed, minutely toothed or entire, subpersistent, barely surpassing the sep; stamens 3–5; style trifid above a short but evident undivided basal portion; capsule 3-valved; seeds few, on a short free-central placenta; short-lived, taprooted, slender, rigidly and angularly (often subdichotomously) branched, glabrous plants with small lvs in an overwintering basal rosette, the cauline lvs reduced to tiny scales at the nodes. (Monotypic, se. U.S.)

1. **Stipulicida setacea** Michx. Mostly 1–2 dm, with 1–many wiry stems from the base; basal lvs 0.5–2 cm long overall, with long, slender petiole and short, abruptly expanded blade 1.5–4 mm wide; rosette beset with slender, chaffy scales amongst the lvs; nodal scale-lvs 1–2 mm, commonly with a few fringing bristles toward the base; sep 1–1.5 mm, barely surpassed by the fr. Sandhills and dry sandy pine woods on the coastal plain; se. Va. to Fla. and La. Apr.–Aug. Ours is the widespread var. *setacea*. (*S. filiformis*.)

6. CERASTIUM L. Mouse-ear chickweed. Fls in terminal cymes, or sometimes solitary; sep (4)5; pet (4)5, retuse to bifid or seldom entire, or occasionally wanting; stamens (4–)10; styles (3–)5; capsule usually surpassing the sep, cylindric, membranous, often curved, dehiscent by (6–)10 short apical teeth; seeds numerous, obovate-reniform, dorsally grooved, papillate-tuberculate; low annual or perennial herbs with rather small, opposite, exstipulate lvs. 100, widespread.

1 Plants ± distinctly perennial.
  2 Pet much longer than the sep, commonly 2–3 times as long .... 1. *C. arvense.*
  2 Pet about as long as the sep .... 2. *C. vulgatum.*
1 Plants annual or winter-annual.
  3 Sep (and commonly also the bracts) provided with long, forward-pointing, eglandular hairs that
     protrude well beyond the tip.
    4 Fls in dense clusters, the pedicels mostly shorter than the sep .... 3. *C. viscosum.*
    4 Fls in open dichasia, the pedicels mostly longer than the sep .... 4. *C. brachypetalum.*
  3 Sep and bracts without such hairs.
    5 Styles mostly 3 or 4 and capsule teeth 6 or 8 (unlike all our other spp.).

6 Styles 3; capsule-teeth 6; sep 5; pet 5; stamens 10 .... 5. *C. dubium.*
6 Styles mostly 4 and capsule-teeth 8; sep, pet, and stamens mostly 4 .... 6. *C. diffusum.*
5 Styles 5 and capsule-teeth 10.
  7 Bracts, at least the upper, with ± evident hyaline-scarious distal margins and tip; lvs relatively
    small, up to 1.5 cm; stamens 5(–10).
    8 Pet equaling or slightly surpassing the sep, cleft 1–1.5 mm .... 7. *C. pumilum.*
    8 Pet shorter than the sep, cleft less than 1 mm .... 8. *C. semidecandrum.*
  7 Bracts wholly herbaceous; lvs larger, mostly 1.5–5 cm; stamens (5–)10 .... 9. *C. nutans.*

1. **Cerastium arvense** L. Field-c.; starry grasswort. Glabrous to densely villous and often glandular perennial, ascending or erect, 1.5–4(–6) dm, branched and often matted at the base; lvs linear to narrowly ovate, 2–7 cm × 1–15 mm, usually subtending conspicuous axillary fascicles or short shoots; sep lanceolate, acute, 5–8 mm; pet conspicuous, white, bilobed, 2–3 times as long as the sep; polyploid series based on $x=18$ (or 9?). Rocky, gravelly, or sandy places, especially in calcareous or magnesian soils, often a weed in abandoned fields and meadows; throughout most of the north-temperate and boreal regions. Apr.–Aug. Highly variable, and often segregated into numerous confluent spp. or vars. (*C. campestre; C. oreophilum; C. strictum; C. velutinum.*) An extremely villous phase on serpentine in se. Pa. and ne. Md. is var. *villosissimum* Pennell.
    Cerastium tomentosum L., with similarly large fls but differing in its tomentose herbage, occasionally escapes from cult., as does the very similar *C. biebersteinii* DC.

2. **Cerastium vulgatum** L. Short-lived perennial, perhaps sometimes blooming the first year, viscid-puberulent, 1.5–5 dm, tending to sprawl and root at the nodes, and with short, often matted, basal sterile branches; lvs ovate or lance-ovate, 1–2 cm × 3–12 mm, obtuse or acutish, the lower oblanceolate or spatulate; infl often at first compact, but becoming more open in age, the mature pedicels mostly 5–12 mm; sep 4.5–6 mm, lance-oblong, scarious-margined, without the long distal hairs of *C. viscosum*; pet about equaling the sep, evidently bifid, commonly to 1 mm or more; fr cylindric, often curved, 8–10 × 2–3 mm; polyploid series based on $x=18$ (or 9?). Native of Eurasia, now established as a weed over most of N. Amer., often in lawns. Apr.–Oct. (*C. fontanum.*)

3. **Cerastium viscosum** L. Clammy c. Viscid-pubescent annual or winter-annual 0.5–3 dm; lvs mostly 1–2.5 cm × 5–15 mm, ovate to obovate or the lower spatulate, obtuse or rounded to acutish; infl compact, or becoming more open in age and then with discrete glomerules of fls; pedicels 1–5 mm, mostly shorter than the sep; bracts wholly herbaceous; sep 4–5 mm, lanceolate, acute, somewhat scarious-margined, the sep and commonly also the bracts provided with long, forward-pointing, eglandular hairs that protrude well beyond the tip, and sometimes glandular as well; pet a little shorter than or about equaling the sep, evidently bifid, commonly to 1 mm or more, or sometimes none; stamens 10; fr 6–8(–10) × 1.5–2 mm, sometimes upcurved; $2n=72$. Native of Eurasia, now a cosmopolitan weed, and found ± throughout our range, especially southward. Apr.–July. (*C. glomeratum.*)

4. **Cerastium brachypetalum** Pers. Similar to *C. viscosum,* but with a more open, dichasially branched infl, the pedicels mostly longer than the sep; bracts conspicuously long-hairy, not glandular; pet often shorter than the sep; stamens up to 10. Polyploid series probably based on $x=9$ or 18. Native of Eurasia, now widely but irregularly intr. in our range, especially southward. Apr., May.

5. **Cerastium dubium** (Bast.) O. Schwarz. Lax, finely glandular-puberulent annual 1–3 dm; lvs 1–3 cm × 1–2 mm; fls in lax, open dichasia, the pedicels to 15 mm; bracts wholly herbaceous; sep 5, 3.5–6 mm; pet 5, about equaling or a little longer than the sep, shortly cleft; styles 3; fr 6-toothed; $2n=36, 38$. European weed, casually intr. with us, as in Ill. (*C. anomalum.*)

6. **Cerastium diffusum** Pers. Glandular-puberulant annual to 3 dm; lvs 0.5–2 cm, the lower oblanceolate to spatulate, the upper ovate to elliptic; bracts usually herbaceous; pedicels much longer than the sep; fls mostly 4(5)-merous; sep 4–7 mm; pet shorter than the sep; stamens 4(5); fr with 8(10) teeth; $2n=72$. European weed, casually intr. with us, as in Ill. Ours is var. *diffusum.*

7. **Cerastium pumilum** Curtis. Much like no. 8; scarious margins and tips of the bracts shorter and less conspicuous; pedicels ± erect; pet equaling or slightly surpassing the sep; notched 1–1.5 mm; seeds minutely papillate; $2n=72+$. European weed, casually intr. with us, as in N.J., Md., Va., Ill., Mich., and Ont. Ours is var. *pumilum.*

8. **Cerastium semidecandrum** L. Viscid-pubescent annual 0.5–2 dm; lvs relatively small, mostly 0.5–1(1.5) cm, the basal oblanceolate, the cauline ovate to broadly elliptic; infl often compact as in *C. viscosum,* varying to moderately open and dichasial, the pedicels up to a little longer than the sep, usually deflexed in fr; bracts conspicuously scarious-margined and -tipped, the distal portion of the upper ones generally wholly scarious for ca 1 mm or more; sep 3–5 mm, lanceolate, acute, stipitate-glandular, with few or no eglandular hairs, scarious-margined; pet shorter than the sep, only shallowly notched, generally to a depth of less than 0.5 mm; stamens 5 or sometimes 10; capsule 4.5–7 mm, less than twice as long as the sep; seeds smooth or nearly so; $2n=36$. Native of Eurasia, now intr as a weed in much of our range, but not so ubiquitous as *C. viscosum.*

9. **Cerastium nutans** Raf. Viscid-pubescent annual 1–4.5 dm, mostly softer and leafier than our intr spp; lvs narrowly lance-oblong to oblanceolate, 1.5–5(–8) cm × 5–10(–13) mm, acutish; infl ± open, loosely cymose; bracts wholly herbaceous; sep 3.5–5 mm, relatively broad, thin, and blunt, thinly to fairly copiously provided with short to fairly long, sometimes gland-tipped hairs that do not surpass the sep-tip; pet from a

little shorter to evidently longer than the sep; bifid to a depth of mostly 1–2 mm, or seldom wanting; stamens 10 or seldom (especially in apetalous forms) only 5; fr 8–15 mm, straight or curved, mostly 2–3 times as long as the sep. $2n=34$–36. Woods and open places, sometimes in disturbed sites, but not so weedy as our intr spp; N.S. to Mack., s. to Ga., Ariz., and Oreg. Apr.–July. (*C. brachypodum,* the arbitrarily delimited, geographically coextensive phase with pedicels mostly 0.5–1.5 cm, instead of 1.5–4 cm as in typical material.)

7. **STELLARIA** L.   Chickweed (the broad-lvd spp); Stitchwort (the narrow-lvd spp); Starwort. Fls solitary in the forks of the stem, or in terminal cymes; sep 5; pet 5, bifid, often deeply so, or lacking; stamens mostly 10, sometimes fewer; styles mostly 3(4) but 5 in *S. aquatica;* ovules numerous; capsule ordinarily dehiscent by twice as many valves as there are styles, but in *S. aquatica* with 5 apically notched or shortly bifid valves; low annual or perennial herbs with mostly rather small, opposite, exstipulate lvs; x most commonly$=13$. (*Alsine, Myosoton.*) 100+, widespread, esp. N. Temp.

1 Styles normally 5; lvs 2–8 cm, ovate or lance-ovate; cymes open, leafy .... 1. *S. aquatica.*
1 Styles normally 3(4); lvs and infl various.
  2 Lvs relatively broad, either ovate to obovate or broadly elliptic in shape, or mostly more than 1
    cm wide, or both.
    3 Plants essentially glabrous; lvs 3–15 mm, sessile; wet, mostly brackish places .... 11. *S. humifusa.*
    3 Plants ± hairy, at least in lines or strips on the upper part of the stem and on the pedicels; lvs
      often larger.
      4 Intr, weedy annual; lf-blades 1–3 cm, the middle and lower petiolate .... 2. *S. media.*
      4 Native, sylvan perennial; lf-blades 2–9 cm, sessile or less often petiolate .... 3. *S. pubera.*
  2 Lvs narrower, mostly linear, lanceolate, or narrowly elliptic, seldom as much as 1 cm wide.
    5 Fls in branched, often dichotomous cymes.
      6 Pet conspicuously surpassing the sep.
        7 Bracts herbaceous; seeds 2–2.5 mm .... 4. *S. holostea.*
        7 Bracts scarious; seeds 0.7–1.6 mm.
          8 Infl open, divaricately branched; pedicels spreading or reflexed.
            9 Fls mostly few; sep 3.5–4.5 mm, weakly 3-nerved; seeds virtually smooth .... 7. *S. longifolia.*
            9 Fls mostly many; sep 4.5–5.5 mm, strongly 3-nerved; seeds tuberculate .... 8. *S. graminea.*
          8 Infl less open; pedicels ascending or erect.
            10 Sep 5–8 mm, acute or acuminate; seeds 1.4–1.6 mm .... 5. *S. palustris.*
            10 Sep 4–4.5 mm, obtuse or acutish; seeds not over 1 mm .... 6. *S. longipes.*
      6 Pet shorter than the sep, or lacking.
        11 Seeds strongly papillate; cymes axillary, few-fld, all the fls subtended by tiny, scarious
          bracts .... 9. *S. alsine.*
        11 Seeds nearly smooth; cymes terminal, few–many-fld, the lower bracts herbaceous, often
          leafy .... 10. *S. borealis.*
    5 Fls usually solitary in the forks of the stem, appearing lateral or axillary or terminal.
      12 Stems mostly 25–50 cm; seeds nearly smooth .... 10. *S. borealis.*
      12 Stems up to 20 cm; seeds obviously sculptured.
        13 Sep equaling or a little longer than the fr; mostly maritime .... 11. *S. humifusa.*
        13 Sep distinctly shorter than the fr; mostly inland, seldom maritime .... 12. *S. crassifolia.*

1. **Stellaria aquatica** (L.) Scop.   Giant c. Decumbent perennial, rhizomatous and rooting at the lower nodes; stem glandular-hairy, to 8 dm; lvs 2–8 × 1–4 cm, ovate or lance-ovate, acute, mostly sessile, the lowermost short-petiolate; fls in open, leafy, terminal cymes; pedicels 1–3 cm, eventually reflexed; sep lance-ovate, acute to obtusish, glandular-hairy, 5–6 mm (to 9.5 mm in fr); pet much exceeding the sep; styles normally 5; fr ovoid, dehiscent by 5 valves, each valve notched or shortly 2-cleft at the tip; seeds 0.7–0.8 mm, orbicular-reniform, strongly papillate; $2n=28$. Native of Europe, intr. as a weed in wet and waste places; Que. and Ont. to Minn., s. to N.C., Mo., and Kans. May–Nov. (*Alsine a.; Myosoton a.*)

2. **Stellaria media** (L.) Villars.   Common c. Annual, the weak, branched stems to 4 dm, puberulent in 1 or 2 broad lines; lvs mostly 1–3 cm, ovate to obovate, the upper sessile, the lower with progressively longer petiole, often ciliate toward the base or on the petiole, otherwise mostly glabrous; fls solitary or in small, terminal, leafy cymes; sep lance-oblong, 3.5–6 mm, obtuse or acutish, ± hairy and pustulate; pet shorter than the sep; stamens 3–5(10); fr ovoid, usually deflexed; seeds 0.9–1.2 mm, suborbicular, bluntly papillate, so that the margin looks wavy; $2n$ mostly$=40$–44. A highly variable weed of waste places, cult. areas, meadows, and woodlands, intr. from the Old World, but sometimes appearing like a native. Early spring–late fall. (*Alsine m.*)
    S. pallida (Dumort.) Piré, the lesser c., has been reported from Mich. and will doubtless be found elsewhere in our range. Much like *S. media* in aspect, but with pale, yellowish-green foliage and usually cleistogamous and apetalous fls; cal 3 mm or less, often with a basal red band; stamens 1–3(–5); ripe frs erect; seeds up to ca 0.8 mm, acutely papillate, so that the margin looks prickly; $2n=22$. Native to Europe.

3. **Stellaria pubera** Michx.   Star c. Erect or ascending perennial 1.5–4 dm, thinly spreading-hairy, the stem often pubescent in 1 or 2 lines; early floriferous shoots followed by taller, more vigorous, chiefly vegetative ones; lvs elliptic to lanceolate or oblanceolate or ovate, 2–9 × 1–3 cm, acute; fls in leafy, open, terminal

cymes; pedicels 1–3 cm; pet and the ovoid fr shorter than the sep; seeds 1.7–2 mm, asymmetrically reniform, coarsely sulcate-papillate. Woods; N.J. to Ill., s. to Fla. and Ala. Apr.–June. Var. *pubera* (2*n*=30), with all but the lowermost lvs sessile, and with obtuse to acute sep mostly 4–6 mm that are usually puberulent on the back, occurs mainly in and e. of the Appalachian Mts., but may be found anywhere in the range of the sp. as a whole. Var. *silvatica* (Beguinot) Weatherby (2*n*=60), with the middle and lower lvs narrowed to a winged petiole 1–2 cm, and with the lance-acuminate sep 7–10 mm and glabrous except for the ciliate margins, occurs chiefly in and w. of the Appalachian Mts. If the var. *silvatica* is treated as a distinct sp. it apparently takes the name *S. corei* Shinners. *S. silvatica* (Béguinot) Maguire is preoccupied, and *Alsine tennesseensis* properly applies to a form of var. *pubera* instead of var. *silvatica*.

4. Stellaria holostea L. Easter-bell. Rhizomatous perennial, 1.5–4.5 dm, the 4-angled stems glaucous and sometimes sparsely hispidulous; lvs narrowly lanceolate, sessile, 2–8 cm × 2–10 mm, hispidulous on the midrib and margins, otherwise sparsely puberulent and glabrate; infl open, the ciliate bracts herbaceous; pedicels pubescent, slender, 2–7 cm; sep broadly lanceolate, 6–9 mm, acute, glabrous, obscurely 5–11-veined; pet conspicuously exceeding the sep, notched less than half-way to the base; seeds 2–2.5 mm, obliquely reniform, coarsely papillate; 2*n*=26. Native of Eurasia, often escaped from cult. in our range. Apr.–June. (*Alsine h.*)

5. Stellaria palustris Retz. Meadow-starwort. Much like no. 4, but less robust and wholly glabrous, the lvs 1.5–4 cm × 1.5–4 mm; bracts scarious; sep narrowly lanceolate, 5–8 mm, distinctly 3-veined, shorter than the pet, these bifid almost to the base; seeds 1.4–1.6 mm, coarsely rugose-tuberculate; 2*n*=ca. 130. Native of Europe, established in grassy places and along shores locally in Que. (*S. glauca.*)

6. Stellaria longipes Goldie. Long-stalked stitchwort. Rhizomatous perennial 0.5–2(3) dm, glabrous and often glaucous; lvs 1–4 cm, lance-linear or linear, widest at or near the base; cymes dichotomous, terminal or appearing lateral; bracts scarious; pedicels ascending to erect, 1–3(6) cm; sep 4–4.5 mm, lanceolate or lance-ovate, obtuse or acutish; pet surpassing the sep; fr ovoid-conic, conspicuously longer than the sep, dark purple when ripe; seeds oblong to oval, 0.75–1 mm, obscurely sculptured; 2*n*=26–107, most often 52, 78, or 104. Grassy places along streams or in moist, gravelly or sandy areas; circumboreal, s. in Amer. to N.S., N.Y., Ind., Minn., N.M., and Calif. (*Alsine l.*)

7. Stellaria longifolia Muhl. Long-lvd stitchwort. Perennial; stems weak. 1.5–4.5 dm, glabrous or sometimes scabrous on the 4 prominent angles; lvs linear or lanceolate to narrowly elliptic or oblanceolate, 2–5 cm × 2–6 mm, sometimes ciliate at base; infl terminal or pseudolateral, seldom over 10 cm, divaricately branched, few-fld; bracts scarious, sometimes ciliolate; pedicels slender, spreading or reflexed; sep lanceolate or lance-ovate, acute, weakly 3-nerved, scarious-margined, seldom ciliolate, 3.5–4.5 mm in fr; pet exceeding the sep; fr somewhat or much surpassing the sep, stramineous or dark, short- to elongate-conic; seeds oblong, 0.7–1 mm; obscurely sculptured, virtually smooth; 2*n*=26. Moist, grassy places and damp woods; circumboreal, s. in Amer. at least to Va., Mo., Kans., Ariz., and Calif. (*Alsine l.*) Hybridizes with no. 10.

8. Stellaria graminea L. Common stitchwort. Perennial; stems weak, 3–5 dm, glabrous or sometimes scabrous on the 4 prominent angles; lvs 1.5–5 cm × 1.5–7 mm, linear to lance-linear, often ciliate at base; infl terminal, diffuse, often extending half-way to the base; bracts scarious, ciliolate at the base; pedicels slender, spreading or reflexed; sep lanceolate, acute or acuminate, strongly 3-nerved, 4.5–5.5 mm in fr, the margins scarious, commonly ciliolate at least at base; pet shorter than to barely surpassing the sep, bifid almost to the base; seeds 0.8–1.2 mm, oblong-subreniform or orbicular-subreniform, coarsely rugose-tuberculate; 2*n*=26, 39. Native of Europe, intr. in grassy places, fields, roadsides, etc., from Nf. and Que. to Minn., s. to S.C. and Kans. May–July. (*Alsine g.*)

9. Stellaria alsine Grimm. Bog-stitchwort. Slender perennial with decumbent or ascending, glabrous, angled stems, often rooting at the nodes; lvs mostly 1.5–3 cm × 1.5–6(–10) mm, elliptic to oblanceolate or linear-oblong, acute, the upper sessile, the lower petiolate, glabrous except for the basally ciliate margins; cymes axillary, few-fld; bracts tiny, narrowly lanceolate, scarious; pedicels slender, clavate at the summit; fls with a definite short hypanthium; sep 2.5–3.5 mm, narrowly lanceolate, acute; pet shorter than the sep; fr ovoid, ± exceeding the sep; seeds 0.5–0.7 mm, obovate, somewhat caudate, strongly papillate or rugose-papillate; 2*n*=24, 26. Cold marshes, streams, and springs; Nf. and Que. to Md. and W.Va., and occasionally elsewhere in our range; widespread in Eurasia and perhaps only intr. with us, but appearing native. May–Aug. (*Alsine uliginosa.*)

10. Stellaria borealis Bigelow. Northern stitchwort. Rhizomatous perennial; stems weak, much-branched, angled, glabrous to scabrous, to 5 dm; lvs lanceolate to lance-linear or oblong, 1–4(–6) cm × 2–8 mm, usually narrowed at the base, the margins often ciliate or scabrous; fls (1–)5–50+ in a lax terminal cyme, the lower bracts herbaceous and often foliaceous, the uppermost often reduced and scarious; pedicels 1–4 cm, erect or arching, seldom reflexed; sep lanceolate, 2–3.5(–4.5) mm in fr; pet shorter than the sep, or usually lacking; styles 1–2 mm, at first erect, later spreading; fr rather firm, ovoid, surpassing the sep, often dark; seeds oblong-obovate, 0.7–0.9 mm, obscurely sculptured; 2*n*=52. Moist, usually shaded places; Greenl. and Lab. to Alas., s. to N.Y., n. W.Va., Mich., Wis., Minn., Colo., and Oreg. May–Aug. Ours is the widespread var. *borealis*. (*Alsine b.; S. calycantha,* misapplied.)

11. Stellaria humifusa Rottb. Salt-marsh stitchwort. Much like no. 12; more strongly fleshy; lvs averaging smaller and often relatively wider, mostly 3–10(–15) × 1.5–4(5) mm; sep at anthesis mostly (3)3.5–5 mm, at maturity equaling or a little longer than the fr; 2*n*=26. Salt marshes and other wet, subsaline habitats along or near the seacoast; circumboreal, s. in Amer. to Me. and Oreg. Summer. (*Alsine h.*)

12. Stellaria crassifolia Ehrh. Fleshy stitchwort. Low, matted or erect, somewhat fleshy, glabrous perennial 1–2 dm; lvs soft, lanceolate or lance-elliptic, 1–2 cm × 1–2.5 mm, usually shorter than the internodes; fls mostly solitary, usually nodding; pedicels subcapillary, 1–2 cm; sep at anthesis mostly 2–3(3.3) mm, lanceolate, obtuse or acutish; pet surpassing the sep; fr ovoid, distinctly surpassing the sep; seeds 0.8 mm, oblong-orbicular, obviously (but not strongly) concentrically rugose-tuberculate; 2n=26. Streamsides, marshes, and other cold, wet places, mainly inland, but sometimes encroaching into the habitat of no. 11; circumboreal, s. in Amer. to the Gaspé peninsula, Mich., Minn., Colo., and Calif. June–Aug. (*Alsine c.*)

8. SCLERANTHUS L. Knawel. Fls in compact, terminal and axillary cymose clusters, perfect, perigynous, the cupulate hypanthium resembling a cal-tube; sep 5; pet none; stamens 1–10; ovary ovoid; styles 2, distinct; ovule one on a basal placenta; hypanthium becoming indurate in fr, crowned by the persistent sep and enclosing the membranous utricle; seed obovoid, beaked at the micropylar end; low herbs with diffusely forking stems and opposite, exstipulate lvs connate at base. 10, Old World.

1. Scleranthus annuus L. Annual k. Low, diffuse, spreading, glabrous or puberulent annual or biennial to 15 cm; lvs linear, the larger ones 5–25 mm; fls sessile or subsessile; sep equaling or longer than the hypanthium, lanceolate, ± acute, with a very narrow scarious border (ca 0.1 mm wide) near the tip; stamens usually 5–10; 2n=22, 44. A weed in fields, roadsides, and waste places; native of Eurasia, established in our range from Que. to Wis., s. to S.C. and Mo. All summer.

S. perennis L., perennial k., another Eurasian sp., is established in s.c. Wis. and may be expected elsewhere in our range. It differs in its perennial habit and its obtuse or rounded sep with a conspicuous white-scarious border 0.3–0.5 mm wide near the tip.

9. SPERGULA L. Spurrey. Infl terminal, dichotomously branched, the pedicels reflexed; sep 5; pet 5; stamens 10 or 5; styles normally 5; valves of the capsule normally 5, opposite the sep; seeds compressed or globose, acutely margined or winged; succulent annuals with whorled lvs and scarious stipules. 5, Eurasia.

1 Seed subglobose, merely margined or narrowly winged; lvs channelled beneath .... 1. *S. arvensis.*
1 Seed compressed, broadly winged; lvs not channelled .... 2. *S. morisonii.*

1. Spergula arvensis L. Simple or much-branched annual to 40 cm, sparingly (seldom more copiously) glandular-pubescent; lvs 2–5 cm, narrowly linear or subulate, channelled beneath, clustered at the nodes in two opposite sets of 6–8, appearing verticillate; stipules small but evident; sep ovate, 2–3 mm, obtuse, glandular-puberulent; pet white, obovate, obtuse, shorter to longer than the sep; stamens 10 or sometimes 5; fr broadly ovoid, surpassing the sep; seed 1–1.5 mm, blackish, subglobose, minutely roughened, conspicuously white-papillate and white-margined or narrowly winged; 2n=18. Native of Europe, widespread as a weed in cult. ground and waste places; N.S. to Alas., s. to Fla. and Calif. May–Aug. (*S. sativa.*)

2. Spergula morisonii Boreau. Much like no. 1, but the lvs not channelled and the seeds compressed, with a brownish, striate wing barely or scarcely half as wide as the body; pet ovate, contiguous or slightly overlapping, obtuse; 2n=18. European weed, occasionally intr. in our coastal states and fully established at least in s. N.J.

S. pentandra L., another European weed, is only casual in our coastal states. It has lanceolate, acute, non-contiguous pet, and seeds with a shining, white-transparent wing more than half as wide as the body.

10. SPERGULARIA (Pers.) J. & C. Presl. Sand-spurrey. Fls in divaricately branched terminal cymes; sep 5, pet 5; stamens 2–10; styles normally 3; capsules normally 3-valved, dehiscent to the base; seeds orbicular-reniform or obovoid, globose or compressed; low, mostly succulent herbs with linear or setaceous, opposite lvs, often with secondary lf-fascicles; stipules pale, scarious, connate. (*Tissa.*) 40, widespread.

S. diandra (Guss.) Boiss., a European sp. with black, wingless seeds 0.4–0.5 mm, has been reported from the coast of Mass. All our other spp. have brown seeds.

1 Stamens 6–10.
  2 Seeds 0.6–1.1 mm, smooth or nearly so, ± winged; pet white (pink); fleshy halophytes .... 1. *S. media.*
  2 Seeds 0.4–0.6 mm, papillate, wingless; pet pink; scarcely fleshy, and not halophytic .... 2. *S. rubra.*
1 Stamens 2–5.
  3 Seeds 0.8–1.4 mm, minutely reticulate, not papillate; lvs not mucronate .... 3. *S. canadensis.*
  3 Seeds 0.6–0.8 mm, smooth or glandular-papillate; lvs short-mucronate .... 4. *S. marina.*

1. **Spergularia media** (L.) C. Presl.   Erect or prostrate, much-branched annual or perennial to 40 cm, glabrous or sparsely glandular-puberulent above; lvs 1–5 cm × 0.8–2 mm, with short or no mucro; stipules deltoid, 1.5–6 mm; sep narrowly ovate, 3–6 mm; pet white (seldom pink), shorter than the sep; stamens (7–)9–10; fr 5–8 mm; seed 0.6–1.1 mm, smooth or minutely sculptured, the marginal wings 0.1–0.4 mm wide, entire or erose; 2n=18, 36. Native of Europe, intr. in salt flats and marshes in coastal and c. N.Y., and along salted highways in O., Mich., and Ill.; to be expected elsewhere in our range. (*S. maritima.*)

2. **Spergularia rubra** (L.) J. & C. Presl.   Roadside s.-s. Simple or much-branched annual or short-lived perennial, prostrate or ascending, 5–30 cm, glabrous or sparsely glandular-hairy above; lvs linear-filiform, mucronate, scarcely fleshy, 3.5–35 × 0.4–1.2 mm; stipules triangular-acuminate, 2.5–5 mm; sep 3.5–5 mm, lanceolate; pet consistently pink, shorter than the sep; stamens (6–)10; fr 3.5–5 mm; seed 0.4–0.6 mm, papillate, wingless; 2n=36, 54. Native of Europe, intr. as a weed in sandy or gravelly soil from Nf. to Mich., Wis., and B.C., s. to Md., Ala., N.M., and Calif. May–Sept. (*Tissa r.*)

3. **Spergularia canadensis** (Pers.) D. Don.   Northern s.-s. Prostrate or decumbent, much-branched annual to 25 cm, glabrous below the infl; lvs 6–45 × 0.6–2 mm, blunt, fleshy; stipules deltoid, 1–2.8 mm; sep ovate, 2.2–3.2 mm, obtuse, glabrous; pet white or pink, equaling or shorter than the sep; stamens 2–5; fr 3.5–5 mm; seed 0.8–1.4 mm, minutely reticulate or ridged, with or without an erose wing. Muddy or sandy tidal areas; Nf. and Que. to N.Y.; Alas. to Calif. July–Sept. (*Tissa c.*)

4. **Spergularia marina** (L.) Griseb.   Salt-marsh s.-s. Simple or much-branched annual, erect or prostrate, to 35 cm, glabrous to glandular-puberulent; lvs 5–40 × 0.5–1.5 mm, fleshy, short-mucronate; stipules deltoid, 2–4 mm; sep ovate, 2.5–5 mm, obtuse; pet white or pink, ovate, shorter than the sep; stamens 2–5; fr equaling or surpassing the cal; seed 0.6–0.8 mm, smooth or glandular-papillate, wingless or occasionally with an erose wing to 0.4 mm wide; 2n=36. Native of Eurasia, probably only intr. in Amer.; coastal brackish and saline marshes, and flat and alkaline areas inland; salted highways in O. and Mich.; Que. to B.C., s. to Fla. and Baja Calif. June–Sept. (*S. leiosperma; S. salina; Tissa m.*)

11. PARONYCHIA Miller.   Whitlow-wort.

Fls perfect in loose to dense, evidently bracteate cymes or in the forks of the stem, hypogynous or nearly so; sep distinct or united at base, ± strongly cucullate; pet none; stamens usually 5, inserted at the base of the cal, sometimes alternating with as many minute teeth or setiform staminodes; ovary maturing as an ovoid to globose or obovoid, membranous utricle; style 2-parted; ovule one on a basal placenta; seed smooth, reddish-black; annual or perennial herbs with small, opposite, entire lvs and conspicuous hyaline stipules. (*Anychia, Anychiastrum.*) 40+, cosmop.

Corrigiola littoralis L., a slender, low, European annual with *alternate,* narrowly oblanceolate lvs, conspicuous, caudate stipules, congested cymose infl, and small, pentamerous fls with 3 sessile stigmas, is occasionally found near the coast, formerly often as a ballast weed, but is not certainly established with us.

1 Plants perennial.
  2 Cal 3–6 mm, the sep obviously short-awned.
    3 Fls in dense cymes, concealed by the silvery hyaline bracts; cal conspicuously appressed-hairy .... 1. *P. argyrocoma.*
    3 Fls in repeatedly forked cymes, not concealed by bracts; cal glabrous or minutely puberulent ...... 2. *P. virginica.*
  2 Cal 1–1.5 mm, the sep merely cuspidate .... 3. *P. riparia.*
1 Plants annual.
  4 Cal 2–3 mm; stems puberulent .... 4. *P. fastigiata.*
  4 Cal 1–1.5 mm; stems glabrous .... 5. *P. canadensis.*

1. **Paronychia argyrocoma** (Michx.) Nutt.   Silver w.-w., silverling. Taprooted perennial with numerous retrorsely silky-hairy, simple or forked stems 5–30 cm, forming mats or tufts; lvs 1–3 cm, linear or lance-linear, acute, apiculate, glabrous to silky; stipules silvery-hyaline, lanceolate, acute; fls sessile in dense cymes, subtended and mostly concealed by the conspicuous silvery bracts; cal 3.5–6 mm, appressed-hairy; sep connate at base, hooded above, with a hirsute awn 0.5–1.5 mm; fr membranous, pubescent at the top. Rocky slopes, ridges, or ledges, usually high in the mts; Me., N.H., and Mass. to Va., W.Va., N.C., Tenn., and Ga. July–Sept.

2. **Paronychia virginica** Sprengel.   Appalachian w.-w. Taprooted perennial with numerous glabrous or rough-puberulent, prostrate to erect stems 1–4 dm; lvs narrowly linear, 2–3 cm × 0.4–0.8 mm, mucronate; stipules entire, hyaline, lance-attenuate, often connate at base; fls numerous in repeatedly forking cymes, sessile, the alternate bracts foliar and provided with a pair of hyaline stipules; sep 3–4 mm; lance-linear, 3-nerved, the awn glabrous, thick, divergent, usually under 1 mm. Open or wooded places, crevices and ledges or rocky places, usually at low altitudes. July–Oct. Our plants, occurring in w. Md., W.Va., and w. Va., are usually low and matted, with numerous sterile shoots, the stems completely glabrous or sometimes minutely puberulent above; these are var. *virginica.* (*P. dichotoma* Nutt., not DC.) Another var. occurs in Okla. and Tex.

3. Paronychia riparia Chapm.   Southern w.-w. Taprooted perennial with minutely puberulent or glabrate, wiry, prostrate or sometimes erect stems 2–6(–12) dm; lvs broadly to narrowly elliptic or elliptic-oblanceolate, 10–25 × 2–6 mm, acute, minutely serrulate, glabrous or minutely puberulent; flowering branches delicate, tending to be cyme-bearing on one side, less often from alternate sides of successive nodes, or the infl forking; fls subsessile; cal 1–1.5 mm; sep oblong-ovate, cucullate, minutely cuspidate, ciliolate on the margins, otherwise glabrous or scantily puberulent between the 3 rather strong nerves. Sandy soils of the coastal plain; se. Va. to c. Fla. July–Oct. (*P. baldwinii* ssp. *r.*)

4. Paryonchia fastigiata (Raf.) Fern.   Forked chickweed. Annual with forked, erect or diffusely spreading, puberulent stems 5–25 cm; lvs of the primary branches narrowly lance-elliptic to oblanceolate, 5–20 mm, acute to obtuse, frequently white-punctate; foliar bracts extending through the repeatedly forking cymes; cal 2–3 mm, glabrous or scantily puberulent; sep lance-linear, 1–3-nerved, very often corrugated, the tip of the hooded apex abruptly apiculate or short-awned; styles united below; fr glabrous, included or barely exserted. Dry woods or openings; Mass. to Wis. and e. Minn., s. to Fla. and Tex. July–Sept. Var. *fastigiata*, with the range of the sp., is erect to low and diffuse, usually reddish or brownish at maturity, the lvs 1–2 cm, usually serrulate, the stipular bracts lanceolate, equaling or shorter than the cal, the style much shorter than the ovary. Var. *paleacea* Fern., occurring more or less throughout the range of the sp., differs in its usually greenish color and in its lance-attenuate stipular bracts equaling or mostly surpassing the cal. Var. *nuttallii* (Small) Fern., apparently local in se. Pa., differs from var. *fastigiata* in having sep with a stout white awn 0.2 mm. Var. *pumila* (A. Wood) Fern., occurring from Pa. and O. to Ga. and Ala., mainly on shale-barrens, is low, diffusely and usually horizontally branched, the lvs 7–12 mm and entire, the stipules lance-ovate, ciliolate, the style as long as the ovary. (*Anychiastrum montanum.*)

5. Paronychia canadensis (L.) A. Wood.   Forked chickweed. Glabrous annual with slender, erect, forking stems and almost capillary ultimate branches; lvs 5–30 × 2–8 mm, elliptic to oval, thin, entire, usually punctate; infl diffuse; cal 1–1.5 mm; sep oblong-ovate, scarious-margined, 1-nerved, flat, the apiculate hood very short; fr exceeding the sep, obovate-spheroid, granular at the tip; styles short, free nearly to the base, recurved; plants not highly variable. Sandy soil and open places; N.H. to Minn., s. to Va., Ga., Ala., Mo., and Kans. June–Sept. (*Anychia c.*)

12. HERNIARIA L.   Much like *Paronychia,* but the stipules between adjacent lf-bases united to form a fringed scale about as broad as long; bracts inconspicuous; stigmas 2, short, virtually sessile. 35, Old World.

1. Herniaria glabra L.   Freely branched, mat-forming, subglabrous annual or perennial; lvs numerous, subsessile, 2–6 × 1–3 mm or some smaller; interpetiolar stipules 0.5–1 mm; fls congested in small cymose axillary (or seemingly lf-opposed) clusters (the clusters sometimes forming short axillary branches), 4–5-parted, 0.5–1 mm; stamens 2–5; staminodes 4–6; $2n=18$. Native of Europe, casually intr. here and there in our range.
  H. incana Lam., a strong perennial with densely white-hairy lvs up to 12 × 3 mm, and fls ca 2 mm, has been reported as an introduction (from Europe) on Assateague I. in Md.
  H. hirsuta L., a slender annual otherwise resembling *H. incana,* has been collected (as an introduction from Europe) in Md.

13. AGROSTEMMA L.   Corn-cockle. Cal-lobes 5, much longer than the coarsely 10-ribbed tube; pet 5, without auricles or appendages; stamens 10; styles (4)5; capsule dehiscent by (4)5 ascending teeth; annual with opposite, entire, exstipulate lvs. 2, Eurasia.

1. Agrostemma githago L.   Stem to 1 m, thinly hairy; lvs linear or lanceolate, 8–12 cm × 5–10 mm; fls reddish, solitary at the ends of the branches, on pedicels to 2 dm; cal-tube 12–18 mm, elliptic-ovoid, the lobes 2–4 cm, lance-linear, acute or acuminate; pet 2–3 cm, oblanceolate, retuse; fr 14–18 mm; $2n=48$. Native of Europe, widely established as a weed of grainfields and waste places, most abundant in the n. part of our range. July–Sept.

14. LYCHNIS L.   Fls mostly perfect; cal-tube 10-ribbed, with 5 short teeth, variously glabrous or hairy but never glandular and never inflated; pet 5, the claw narrow, usually expanded upwards into conspicuous lateral auricles at the juncture with the blade, and also generally with a pair of appendages on the inner surface at the juncture; stamens 10; styles 5, or in occasional individuals sometimes 4 or 6; capsules unilocular, or seldom largely 5-locular, dehiscent by (4)5(6) entire teeth (as many teeth as styles); seeds reniform, muricate to tuberculate; perennial herbs with opposite, entire, exstipulate lvs; $x=12$. 12, temp. Eurasia.

1 Plants densely tomentose; cal-lobes twisted .... 1. *L. coronaria.*
1 Plants variously pubescent, but not tomentose; cal-lobes plane.

2 Cauline lvs numerous, mostly 10–20 pairs, broad, mostly 2–5 cm wide . . . . 2. *L. chalcedonica.*
2 Cauline lvs few, mostly 2–5 pairs, narrow, less than 1.5 cm wide.
  3 Pet deeply cleft; fr not partitioned . . . . 3. *L. flos-cuculi.*
  3 Pet entire or merely emarginate; fr partitioned to above the middle . . . . 4. *L. viscaria.*

1. Lychnis coronaria (L.) Desr. Mullein-pink. Gray-tomentose perennial 4–8 dm, the stout stems rarely branched; basal lvs 5–10 × 1–3 cm, the cauline 5–10 pairs, usually smaller; fls few, the pedicels 5–10 mm; cal 12–15 mm, the narrowly lanceolate lobes 4–7 mm, connivent, twisted; pet crimson, 2–3 cm, without auricles, the appendages narrowly lanceolate, 1.5–2.5 mm, the blade broadly obovate, entire or emarginate, 10–15 mm; fr 12–16 mm; 2*n*=24. Native of Europe, often escaped from cult. in our range. June–Aug.

2. Lychnis chalcedonica L. Scarlet lychnis, Maltese cross. Perennial, 3–6 dm; stem hirsute; basal lvs spatulate or oblanceolate to lanceolate; cauline lvs 10–20 pairs, lanceolate or lance-ovate, 5–12 × 2–5 cm, acute, sparsely hairy or glabrate, serrulate-ciliate; infl terminal, ± capitate; fls numerous, red (white); cal 12–17 mm at maturity, the 10 coarse ribs strigose-hirsute, the lobes 2.5–3.5 mm; pet 14–18 mm, the claw ciliate, the appendages tubular, 2–3 mm, the blade 7–9 mm, deeply bilobed; ovary on a stipe 4–6 mm; fr 1 cm; 2*n*=24, 48. Native of Asia, occasionally escaped from cult. in our range. June–Sept.

3. Lychnis flos-cuculi L. Ragged robin. Perennial, 3–8 dm; stem thinly strigose above; cauline lvs mostly 4–5 pairs, lanceolate or lance-oblong, 5–10 cm × 8–12 mm, acute, sessile or the lowest petiolate; infl openly paniculate, much branched; fls rose-purple or sometimes white; cal 6–10 mm, campanulate, glabrous except for the densely ciliate short lobes; pet 15–20 mm, without auricles, the appendages 2–3 mm, dissected, the blade 8–12 mm, deeply 2-lobed, the lobes oblong-linear, with a pair of slender lateral teeth, or dissected; fr 5–7 mm; 2*n*=24. Native of Europe, frequently escaped from cult. in our range. July–Sept.

4. Lychnis viscaria L. German catchfly. Perennial, 2–8 dm; viscid in the condensed, narrow infl; lvs mostly basal, narrow, acuminate; cal glabrous or nearly so, ca 1 cm; pet 12–15 mm, the appendages conspicuous, the blade dark red, pink, or seldom white, entire or emarginate; fr 6–8 mm, partitioned to above the middle; 2*n*=24. Native of Europe, sometimes escaped from cult. and occasionally persisting as a weed in our range. May–July. (*Viscaria vulgaris.*)

15. SILENE L. Catchfly, campion. Fls perfect or sometimes unisexual; cal-tube (5–)10–30-veined, with 5 short teeth, variously hairy or glandular or glabrous, sometimes inflated; pet 5, the claw narrow, usually expanded above into evident auricles at the juncture with the blade, and usually with a pair of appendages on the inner surface at the juncture; stamens 10, styles 3, less often 4 or 5, rarely more; ovary usually stipitate, the stamens and pet often adnate to the stipe; capsule unilocular or more or less completely plurilocular, dehiscent by twice as many teeth as the number of styles, or less often by only as many teeth as styles; seeds reniform to globose, roughened; annual to perennial herbs with opposite (seldom whorled), entire, exstipulate lvs. 400, mostly N. Temp.

1 Styles mostly 5.
  2 Cal much-inflated, at least in age; cauline lvs mostly 1–4 cm wide; weeds.
    3 Fls white, opening in the evening, odorous . . . . 1. *S. latifolia.*
    3 Fls red-purple or bright pink, opening in the morning, inodorous . . . . 2. *S. dioica.*
  2 Cal not inflated; cauline lvs less than 1 cm wide; native, western . . . . 4. *S. drummondii.*
1 Styles mostly 3.
  4 Plants perennial.
    5 Fls bright red.
      6 Blade of the pet 2-lobed or 2-cleft; cauline lvs 2–8 pairs.
        7 Cauline lvs 2–4 pairs, mostly oblanceolate; carpophore 2–4 mm . . . . 5. *S. virginica.*
        7 Cauline lvs 5–8 pairs, broadly lanceolate to subrotund; carpophore 6–8 mm . . . . 6. *S. rotundifolia.*
      6 Blade of the pet entire or irregularly denticulate; cauline lvs 10–20 pairs . . . . 7. *S. regia.*
    5 Fls white, pink, or purple, never bright red.
      8 Pet dichotomously several times cleft, or fimbriate; auricles lacking.
        9 Middle cauline lvs opposite; pet usually 8-cleft, not lanate at base . . . . 8. *S. ovata.*
        9 Middle cauline lvs in whorls of 4; pet fimbriate, lanate at base . . . . 9. *S. stellata.*
      8 Pet entire or bilobed, the lobes sometimes with a small lateral tooth.
        10 Plants mostly 2–8 dm tall.
          11 Lvs mainly or wholly cauline, mostly 1–3.5 cm wide.
            12 Fr 3-locular, appendages of the pet minute or none; intr., weedy . . . . 10. *S. vulgaris.*
            12 Fr essentially unilocular; appendages 1–1.6 mm; native, not weedy . . . . 11. *S. nivea.*
          11 Lvs basally disposed, the cauline ones seldom over 1 cm wide . . . . 12. *S. nutans.*
        10 Plants 2 dm or less.
          13 Plants 8–20 cm, not pulvinate; cal 15–22 mm, pubescent . . . . 13. *S. caroliniana.*
          13 Plants 3–6 cm, pulvinate-cespitose; cal 3–10 mm, glabrous . . . . 14. *S. acaulis.*
  4 Plants annual.
    14 Cal about 30-veined . . . . 19. *S. conica.*
    14 Cal 10-veined.

15 Plants glabrous or nearly so.
  16 Carpophore 7–8 mm; cal 13–17 mm .... 15. *S. armeria.*
  16 Carpophore 1 mm; cal 4–10 mm .... 16. *S. antirrhina.*
15 Plants densely pubescent throughout.
  17 Pet deeply bilobed; appendages usually as wide as long.
    18 Cal in fr 10–15 mm, not inflated, the veins hirsute .... 17. *S. dichotoma.*
    18 Cal in fr 25 mm or more, inflated, the veins glandular .... 3. *S. noctiflora.*
  17 Pet entire or sometimes slightly toothed; appendages linear .... 18. *S. gallica.*

1. **Silene latifolia** Poiret. White campion, white cockle. Dioecious annual to more often biennial or short-lived perennial from a stout root, 4–12 dm, usually hairy, becoming glandular above; cauline lvs as many as 10 pairs, lanceolate to broadly elliptic, 3–10 × 1–4 cm, acute, 3–5-nerved; infl usually much branched; fls white, odorous, opening in the evening; cal in male fls 1.5–2 cm, 10-nerved, in female 2–3 cm, 20-nerved and becoming much inflated; pet 2–4 cm, the claw exsert, auriculate, the appendages 1–1.5 mm, erose, the blade deeply bilobed; styles mostly 5; fr 10–15 mm, dehiscent by 10 erect or spreading teeth; seeds gray, with blunt tubercles; $2n=24$. Native of Europe, established as a common weed throughout much of N. Amer. May–Sept. (*S. pratensis; S. alba,* a preoccupied name; *Lychnis a.*)

2. **Silene dioica** (L.) Clairv. Red campion, red cockle. Much like no. 1 and hybridizing with it; consistently perennial; stem nearly or quite eglandular; fls red-purple or bright pink, inodorous, opening in the morning; claws hardly auriculate; appendages of the pet 1 mm; fr with recurved teeth; seeds reddish-black, with acute tubercles; $2n=24$. Native of Europe, now widespread as an occasional weed in n. U.S. and s. Can. May–Sept. (*Lychnis d.; Melandrium d.*)

3. **Silene noctiflora** L. Sticky cockle, catchfly. Annual, 2–8 dm, densely and coarsely hirsute below, viscid with abundant glandular hairs above; lvs lance-ovate to elliptic-oblanceolate, 5–12 × 2–4 cm, the basal petiolate, the cauline narrower, sessile, infl loosely branched; fls mostly perfect; cal 1.5 cm at anthesis, inflated and 2.5–3 cm in fr, the 10 nerves glandular and freely anastomosing, the lance-linear lobes 5–9 mm; pet blade 7–10 mm long, deeply 2-lobed, pink above, yellowish beneath, inrolled during the day, opening in the evening; auricles 1–1.5 mm, the broad appendages 0.5–1.5 mm, entire or erose; carpophore 1–3 mm; styles 3; fr 3-locular, opening by 6 teeth; seeds 0.8–1 mm, rugose-papillate; $2n=24$. Native of Europe, widely distributed as a weed in most of the U.S. and s. Can. July–Sept.

4. **Silene drummondii** Hook. Perennial, 2–5 dm, retrorse-puberulent, becoming glandular and often hirsute in the infl; lvs mostly basal, the lanceolate or elliptic to oblanceolate blade 3–10 cm × 5–12 mm, petiolate; cauline lvs remote, 2–4 pairs, much reduced, usually linear; fls white or pinkish, perfect, 1–several in a loose, narrow infl; cal narrowly tubular, 1–1.5 cm in fl, enlarged but not inflated in fr, 10-nerved, glandular; pet included in the cal or the blade shortly exserted, the claw flaring upwards and usually broader than the blade, often auriculate, the blade 1–3 mm, retuse or shallowly lobed, the appendages less than 0.5 mm; styles (4)5; fr unilocular, opening by (4)5 teeth; $2n=24$, 48. Dry hillsides and plains; B.C. to Ariz., e. to Sask., N.D., S.D., Neb., and Clay Co., Minn. June–Aug. (*Lychnis d.*)

5. **Silene virginica** L. Fire-pink. Short-lived perennial, 2–8 dm, at least the stems glandular and puberulent; basal lvs oblanceolate or spatulate, 4–10 cm × 8–18 mm, petiolate; cauline lvs 2–4 pairs, sessile or nearly so, to 15 or even 30 cm and 3 cm wide; infl open, 7–11-fld, leafy-bracteate; cal broadly tubular, 18–22 mm; pet crimson, the tubular appendages 3 mm, the linear-oblong blade 15–22 mm, bilobed; carpophore 2(–4) mm; styles 3, capsule dehiscent by 6 teeth; $2n=48$. Rich woods or open woodlands and rocky slopes; N.J. and w. N.Y. to s. Ont. and se. Mich., s. to Ga. and Okla. May–Sept.

6. **Silene rotundifolia** Nutt. Taprooted perennial, 2–7 dm, the stems weak, freely branched, thinly glandular-pilose; cauline lvs 5–8 pairs, sessile or the lower short-petiolate, broadly lanceolate to elliptic-obovate or suborbicular, 3–10 × 2–7 cm, often ciliate; infl open; fls few; cal broadly tubular, 2–2.5 cm, thinly pilose and viscid-puberulent; pet with ciliate claw, the auricles 1–2 mm, the appendages 1–1.5 mm, the blade crimson, glandular-ciliate, deeply bifid into lanceolate segments; carpophore 6–8 mm; styles 3; capsule dehiscent by 6 teeth; $2n=48$. Exposed cliffs and banks; W.Va. and s. O. to Ala. and Ga. Summer.

7. **Silene regia** Sims. Wild pink. Taprooted perennial, 5–16 dm, glabrous to glandular-puberulent; cauline lvs 10–20 pairs, lanceolate to broadly ovate, 4–12 × 1.5–7 cm, infl narrow, compound, 1.5–3 dm, leafy-bracteate; cal tubular, 1.8–2.5 cm, glandular; pet crimson, the auricles prominent, the tubular appendages 2–4 mm, the narrowly oblong or elliptic, entire or irregularly denticulate blade 12–20 × 3–5 mm; carpophore 3–5 mm; styles 3; capsule dehiscent by 6 teeth; $2n=48$. Prairies and open woods; O. to e. Mo., s. to Ala. and Ga. July.

8. **Silene ovata** Pursh. Coarse perennial, 3–15 dm, puberulent or hirsutulous; lvs sessile, ovate to broadly lanceolate, 6–12 × 3–8 cm, acute or acuminate; infl open-paniculate, 1–5 dm; cal tubular, 6–8 mm, becoming inflated and 10–12 mm in fr; pet white, without auricles, the appendages minute, the blade 7–9 mm, deeply dichotomously dissected with usually 8 linear segments; carpophore 2–2.5 mm, styles 3; capsule dehiscent by 6 teeth; $2n=48$. Woods; Va. and Ky. to Ga., Ala., and Ark. Summer.

9. **Silene stellata** (L.) Aiton f. Starry campion. Perennial with several simple, puberulent stems 3–12 dm; cauline lvs mostly in whorls of 4, lance-linear to lance-ovate, 3–10 cm, to 4 cm wide, acuminate; infl loosely paniculate; cal campanulate, 5–12 mm, finely puberulent; pet white, 8–11 mm, lanate at base, the auricles and appendages none, the blade fimbriately 8–12-lobed; carpophore 2–3 mm; styles 3; capsule dehiscent by

6 teeth; $2n=48$. Woods. June–Aug. Var. *stellata,* with essentially glabrous pedicels and cal, occurs from Conn. to O. and Nebr., s. to Ga. and Tex. Var. *scabrella* (Nieuwl.) Palmer & Steyerm., with densely crisp-puberulent pedicels and cal, occurs from O. and Mich. to N.D., s. to Tenn. and Tex.

10. **Silene vulgaris** (Moench) Garcke. Bladder-campion. Robust perennial, 2–8 dm, glabrous and glaucous, or seldom hirsutulous; lvs mainly or wholly cauline, lance-ovate to oblanceolate, 3–8 × 1–3 cm, abruptly acuminate, sometimes ciliolate, the cauline often clasping; infl open; cal papery, inflated and umbilicate, 1 cm in fl, accrescent to 2 cm in fr, with 20 equal main veins and also reticulate-veiny; pet white, without auricles, the appendages minute or none, the blade 3.5–6 mm, deeply bilobed; carpophore 2–3 mm; styles 3; fruit trilocular, opening by 6 teeth; $2n=24$. Native of Europe, occasionally found as a weed in waste places throughout most of temperate N. Amer. Summer. (*S. cucubalus; S. latifolia* (Mill.) Britt. & Rend.))

S. cserei Baumg., a related biennial sp. of se. Europe, is also occasionally intr. in our range. It has broadly ovate lvs and a smaller, less inflated cal, only 10–12 mm in fr, with 10 long and 10 short veins and few or no anastomosing lateral veins.

11. **Silene nivea** (Nutt.) Otth. White campion. Rhizomatous perennial, 2–3 dm, glabrous or puberulent; lvs mainly or wholly cauline, sessile or short-petiolate, lanceolate or lance-oblong, 5–10 × 1–3.5 cm, long-acuminate; fls few, mostly axillary; cal tubular-campanulate, 1.5 cm, umbilicate, glabrous or hirsute; pet white, without auricles, the appendages oblong, 1–1.6 mm, entire or erose, the blade cuneate, 8 mm, 2-lobed or 2-cleft; styles 3; fr unilocular, opening by 6 teeth; $2n=48$. Woods; N.J., Pa., Md., and Va. to Tenn., Minn., S.D., Neb., and Mo. June, July. (*S. alba* Muhl.)

12. **Silene nutans** L. Perennial, 3–5 dm; basal lvs narrowly spatulate, 4–10 cm × 5–15 mm; cauline lvs 1–3 pairs, mostly smaller, petiolate or subsessile, oblong to lanceolate; infl open, compound; cal tubular-conic, 7–10 mm, glandular-puberulent; fls nodding, later erect; pet pink or white, the auricles inconspicuous, the appendages linear, 2 mm (rarely obsolete), the blade 5 mm, lobed half-length; styles 3; fr trilocular below, opening by 6 teeth; $2n=24$. Native of Europe, occasionally escaped from cult. in the e. part of our range. June–Sept.

13. **Silene caroliniana** Walter. Wild pink. Taprooted perennial with slender simple stems 8–20 cm, usually glandular-hairy; lvs oblanceolate 5–12 cm × 12–30 mm, the basal petiolate, the cauline 2–3 pairs, often sessile; infl short, dense, with usually 5–13 fls; cal tubular, 15–22 mm; pet white or dark pink, without auricles, the appendages oblong, entire, 1.5–2 mm long, the blade cuneate-obovate, 8–15 mm, entire or retuse, usually irregularly crenate; seeds 1.3–1.5 mm, prominently inflated, tuberculate; $2n=48$. Var. *pensylvanica* (Michx.) Fern., with broadly tubular, thinly to densely pubescent but eglandular cal, and with the claws usually about equaling the cal, occurs in rocky, usually calcareous woods from s. O. and Ky. to Mo. and n. Ala.; Apr, May. Var. *caroliniana* is more southern.

14. **Silene acaulis** L. Moss-campion. Pulvinate-cespitose, taprooted perennial, 3–6° cm, glabrous except the scabrid or ciliolate margins of the lvs; lvs crowded, imbricate, 4–15 × 0.8–1.5 mm, lanceolate, sessile; fls subsessile, or solitary on peduncles 2–2.5 cm; plants dioecious, with the staminate fls shorter than the pistillate, or the fls sometimes perfect; cal 3.5–7 mm, glabrous; pet purple or lavender, seldom white, 8–12 mm, cuneate-oblong or oblanceolate with little distinction between claw and blade, the blade entire to deeply 2-lobed, the auricles none, the appendages minute or none; fr 3-locular, barely exsert; $2n=24$. Circumboreal, s. in e. N. Amer. to alpine summits of N.H. Our plants may be distinguished from the European typical phase on minor characters as var. *exscapa* (All.) DC.

15. **Silene armeria** L. Sweet William catchfly. Glabrous or rarely sparsely puberulent annual 1–7 dm, the stem sometimes with slightly glutinous zones below some of the upper nodes; lvs sessile and ± clasping (basal more spatulate), 2–5 × 0.5–1.5 cm, rather numerous; infl open or compact, the ultimate cymes congested; cal tubular-clavate, much constricted below, umbilicate, 13–17 mm, 10-nerved; pet pink or lavender, auricles none, appendages linear, 2–3 mm, the blade 4–7 mm, obovate, truncate to emarginate; carpophore 7–8 mm; fr almost wholly 3-locular; seeds 0.6 mm wide, rugose; $2n=24$. Native of Europe, formerly much cult., and found as a casual weed in our range. June, July.

16. **Silene antirrhina** L. Annual, 2–8 dm, often puberulent below, glabrous above except usually for some glutinous bands below the upper nodes; basal lvs oblanceolate to spatulate; cauline lvs oblanceolate to linear, 3–6 cm × 9–12 mm, the margins ciliate near the base; infl open; fls mostly ± numerous; cal 4–10 mm, 10-nerved; pet white or pink, equaling or exceeding the cal, 2-lobed or obsolete, the appendages minute or none; carpophore 1 mm; fr 4–10 mm, 3-locular; seeds 0.5–0.8 mm wide, papillate; $2n=24$. Sandy soil and waste places, through the U.S. and s. Can., apparently native. Summer.

17. **Silene dichotoma** Ehrh. Strongly hirsute annual 3–8 dm; lvs lanceolate to oblanceolate, 3–8 cm × 3–35 mm, the lower usually ciliate-petiolate, the upper sessile; infl usually one or more times dichotomous, with leafy-bracteate, monochasial, raceme-like branches; fls mostly perfect; cal narrowly tubular, 10–15 mm, the 10 green nerves hirsute; pet white to reddish, without auricles, the appendages truncate, 0.2 mm, the blade rhombic-cuneate, 5–9 mm, deeply 2-lobed; stamens exsert (or vestigial); carpophore 2–4 mm; fr 3-locular; seeds 1–1.3 mm wide, finely rugose; $2n=24$. Native of Eurasia, widespread as a weed in the U.S. Summer.

18. **Silene gallica** L. Annual 1–4 dm, conspicuously hirsute (or the lvs merely puberulent), and glandular above; basal lvs few, oblanceolate to spatulate, the cauline narrower, 1.5–4 cm × 2–8(–15) mm; fls ± erect, in a leafy-bracteate, raceme-like, secund monochasial cyme; cal 6–9 mm and tubular at anthesis, later inflated but scarcely accrescent, 10-nerved, glandular and hairy; pet white to pink, slightly longer than the cal, entire

or sometimes slightly toothed, the appendages linear and entire; carpophore to 1 mm; fr 3-locular except at the tip; seeds 1 mm wide, finely corrugate-rugose; $2n=24$. Native of Eurasia, occasionally intr. in our range. Apr–July. (*S. quinquevulnera,* with a crimson spot on the pet; *S. anglica,* misapplied.)

S. pendula L., a Mediterranean sp. with horizontal or nodding fls, cal 12–18 mm in fr, the pet notched, occasionally escapes from cult.

19. Silene conica L. Erect annual 2–5 dm, the stems puberulent below, glandular above; lvs linear to narrowly lanceolate or oblanceolate, 2–5 cm × 1.5–6 mm; fls 5–13; pedicels 1–3 cm; cal ca 30-nerved, eventually 12–17 × 5–7 mm, the lobes attenuate, 5–7 mm; pet white to reddish, conspicuously surpassing the cal, the blade 3–6 mm, shallowly 2-lobed, the appendages 1–2 mm, deeply 2-lobed; fr 8–10 mm, 3-locular at the base; seeds 0.6–0.9 mm wide; $2n=20$. Native of Eurasia, sparingly intr. as a weed of waste places mainly along our Atlantic coast, but also inland to Mich. May–July.

S. conoidea L., a related Eurasian sp., with the cal becoming 2–3 cm and much inflated, the pet-blade 8–12 mm, and the seeds 1.2–1.5 mm wide, has been collected in Del.

16. SAPONARIA L. Soapwort, bouncing Bet. Infl ± congested or of open cymes; cal cylindric, membranous, 20-nerved; pet with distinct claw and blade, the appendages conspicuous, subulate; stamens 10; styles 2(3); capsule dehiscent by 4(–6) teeth; seeds rotund-reniform, uniformly reticulate; perennial herbs. 30, temp. Eurasia.

1. Saponaria officinalis L. Colonial by rhizomes, erect, 4–8 dm, mostly glabrous; lvs 7–10 × 2–4 cm, elliptic to elliptic-ovate or lance-elliptic, acute; infl congested and subcapitate to open and oblong-pyramidal, to 15 cm, the primary bracts leafy, the ultimate ones scarious; fls fragrant, often double; cal 1.5–2.5 cm, the lobes triangular-attenuate, the tube often becoming deeply bilobed; pet white or pinkish, the auricles none, the appendages conspicuous, the blade 8–15 mm, oblong to oblong-obovate; stamens exsert; $2n=28$. Native of the Old World; formerly cult. and now established as a weed of roadsides and waste places or along railways throughout most of temperate N. Amer. Summer.

17. VACCARIA Wolf. Cow-herb. Infl a loose, open, paniculate cyme; cal strongly 5-ribbed, sharply wing-angled; pet pink or rose, without auricles or appendages; stamens 10; styles 2(3); capsule firm, dehiscent by 4(6) ascending teeth; seeds large, globose; much branched annuals. 4, mainly temp. Eurasia.

1. Vaccaria hispanica (Miller) Rauschert. Taprooted annual, 2–6 dm, branched above, glabrous and glaucous; cauline lvs 5–10 × 2–4 cm, lanceolate to lance-ovate, acute, clasping or the lower connate; cal 12–17 mm, ovoid or flask-shaped; pet 18–22 mm, pink, the blade 6–8 mm, obovate, retuse; stamens exsert; fr 6–8 mm; seeds 2–2.6 mm, minutely tuberculate, reddish-brown to black; $2n=30$. Native of Europe, widely distributed as a weed through temperate N. Amer. (*V. pyramidata, V. segetalis, V. vaccaria, Saponaria vaccaria.*)

18. GYPSOPHILA L. Cymes dichasially branched, often diffuse; bracts scarious; cal short, campanulate to turbinate, 5-nerved, with scarious commissures, ebracteate; pet scarcely differentiated into blade and claw, without auricles or appendages; stamens 10; styles 2; capsule globose to oblong, opening by 4 ascending teeth; seeds compressed, subreniform; annual or perennial herbs, none native. 125, mainly temp. Eurasia.

1 Plants evidently perennial.
  2 Larger lvs broad-based, clasping, mostly 1–2 cm wide .... 1. *G. scorzonerifolia.*
  2 Lvs narrowed to the base, not clasping, up to 1 cm wide .... 2. *G. paniculata.*
1 Plants annual .... 3. *G. muralis.*

1. Gypsophila scorzonerifolia Ser. Coarse, diffusely branched perennial, mostly 5–10 dm, glabrous and glaucous below, glandular-hairy above; lvs oblong-lanceolate to ovate, mostly 2–10 × 1–2 cm, obtuse or abruptly acute, clasping at the base; pedicels and cal glandular; cal 2.5–4 mm; pet 4–6 mm, pale purplish-pink or white, drying darker; capsule globose. Native to the lower Volga region in the U.S.S.R., and occasionally escapes in our range, as in Mich. and Ind. Summer.

G. perfoliata L., another European sp., is also occasionally intr. in our range. It is much like *G. scorzonerifolia,* but the cal and pedicels are glabrous, and the lvs are sometimes wider, up to 3.5 cm.

2. Gypsophila paniculata L. Baby's breath. Diffusely branched perennial, mostly 4–10 dm, glabrous and glaucous, or the stem somewhat scabrous below; lvs lanceolate, acuminate, 2–7 cm × 2.5–10 mm, not clasping; pedicels and cal glabrous, the cal 1.5–2 mm; pet 2–4 mm, white or pale reddish; capsule globose; $2n=34$. Native of Europe, cult. and occasionally escaped in our range. Summer.

3. **Gypsophila muralis** L.   Cushion baby's breath. Diffusely branched, slender annual, 5–40 cm, glabrous above, often puberulent below; lvs 5–15(–25) × 1–2(–3) mm, linear or nearly so, acute; pedicels 1–2 cm, spreading or ascending from the axils of all but the lower lvs; cal 2.5–4 mm, ± obconic, its lobes ciliate-margined; pet oblanceolate, 6–10 mm, emarginate, pink or white; fr oblong, slightly surpassing the cal; ovules 24–36; 2n=34. Native of Eurasia, found as a weed here and there in the n. part of our range. Summer.

   G. elegans M. Bieb., another annual sp. native to se. Europe and Asia Minor, occasionally escapes from cult. in our range. It is larger than *G. muralis,* commonly 2–5 dm, with larger lvs mostly 2–4 cm × 3–10 mm, a broader, more campanulate cal, shorter, more globose capsule, and 12–24 ovules.

19. DIANTHUS L.   Pink. Fls solitary or in open or capitate cymes; cal subtended by 1–3 pairs of bracts, cylindric, with 20 or more nerves; pet 5, without auricles or appendages; stamens 10; styles 2; capsule dehiscent by 4 teeth; seeds disciform, concave-convex, the hilum at the center of the concave surface; embryo straight; annual to perennial herbs. 300, Old World.

1 Fls solitary or few, long-stalked.
  2 Pet-blade 5–10 mm, merely toothed .... 1. *D. deltoides.*
  2 Pet-blade 12–18 mm, fringed-cleft .... 2. *D. plumarius.*
1 Fls clustered in 1 or more head-like cymes.
  3 Perennial; lvs mostly 1–2 cm wide; cal glabrous .... 3. *D. barbatus.*
  3 Annual or biennial; lvs mostly well under 1 cm wide; cal villous-puberulent .... 4. *D. armeria.*

   1. **Dianthus deltoides** L.   Maiden-p. Perennial, 1–4 dm from slender rhizomes, glabrous or hispidulous; basal lvs oblanceolate, acute, 1.5–3 cm × 1.5–3 mm; cauline lvs 5–10 pairs, lance-linear, acute, 2–4 cm; fls scattered on pedicels 1–4 cm; cal 12–18 mm, 30–40-nerved; pet-blade red-purple, lavender, or white, 5–10 mm, toothed around the end; fr equaling the cal; 2n=30. Native of Europe, occasionally escaped from cult, mainly in the n. part of our range. May–Aug.

   2. **Dianthus plumarius** L.   Garden-p. Loosely multicipital perennial, 1–3 dm, glabrous and often glaucous; lvs linear, 2–8 cm × 1.5–3 mm, 3-nerved, the 2–8 pairs of cauline ones shorter than the basal; fls 1–5 on pedicels 1–3 cm; cal 15–30 mm, ca 40-nerved, the lobes 4–5 mm, pet-blade red to white, 12–18 mm, fringed-cleft to near the middle; fr surpassing the cal; 2n=30, 90. Native of Europe, occasionally escaped from cult. mainly in the n. part of our range. May–Aug.

   3. **Dianthus barbatus** L.   Sweet William. Stout, glabrous perennial 3–6 dm; cauline lvs 5–10 pairs, lanceolate to oblanceolate, mostly 6–10 × 1–2 cm, acute or acuminate, the basal ones wider; infl a many-fld head with narrow, leafy bracts; cal glabrous, 15–18 mm, ca 40-nerved; pet-blade whitish to dark red, 5–10 mm, broad, toothed around the broad summit; fr 1 cm. 2n=30. Native of Eurasia, occasionally escaped from cult. especially in the n. part of our range. June–Aug.

   D. carthusianorum L., cluster-head p., perennial and largely glabrous like no. 3, but with linear lvs, puberulent calyx-lobes, and consistently dark crimson pet, is locally escaped in our range, as in Mich. and N.H.

   4. **Dianthus armeria** L.   Deptford-p. Annual or biennial, 2–6 dm, the stem usually strigose below the nodes; basal lvs numerous, narrowly oblanceolate; cauline lvs 5–10 pairs, linear to lanceolate, 3–8 cm × 2–8 mm, pubescent; fls in congested, 3–9-fld cymes, often surpassed by the slender, erect bracts; cal villous-puberulent, 12–20 mm, (20–)25-nerved; pet-blade elliptic-oblanceolate, 4–5 mm, dentate, pink or rose, dotted with white; fr equaling the cal; 2n=30. Native of Europe, established as a weed from Que. and Ont. to B.C., s. to Fla. and Ark. May–July.

20. PETRORHAGIA (Seringe) Link.   Fls individually terminating the branches, or often clustered into head-like, conspicuously involucrate cymes terminating the branches; cal often (including our spp.) subtended by 1–3 pairs of bracts; cal-tube 5-nerved, generally with membranous, veinless commissures beneath the sinuses (unlike *Dianthus*); pet without auricles or appendages; stamens 10; styles 2; capsule dehiscent by 4 teeth; seeds as in *Dianthus.* 25, mainly Mediterranean.

1 Fls solitary (–3) at the branch-tips, the cal well surpassing the subtending bracts .... 1. *P. saxifraga.*
1 Fls in conspicuously involucrate, head-like terminal cymes, the invol about equaling the fls .... 2. *P. prolifera.*

   1. **Petrorhagia saxifraga** (L.) Link.   Caespitose perennial 0.5–4 dm, slender and freely branched; at least the lower internodes finely scabrous; lvs linear, 0.5–2(–3) cm, to 1(–2) mm wide, 1-veined, scabro-ciliolate toward the base; fls solitary (–3) at the branch-tips, closely subtended by 2(4) scarious bracts, these evidently shorter than the 5-veined, 3–6 mm cal; pet-blade short, pink or white, 3-veined, retuse or obcordate; seeds tuberculate; 2n=30, 60. Eurasian sp., cult. and sporadically escaped in our range, sometimes as a weed of lawns or roadsides. June–Oct. (*Tunica s.*)

2. **Petrorhagia prolifera** (L.) P. W. Ball & Heywood.   Childing pink. Annual to 6 dm, simple or sparingly branched; internodes mostly glabrous; lvs linear or linear-oblong, 1–6 cm, to 2(–3) mm wide, 3-veined, scabrous-margined; fls in (1–)3–7-fld capitate cymes terminating the stem and branches, these infls closely subtended by 2–3 pairs of large, scarious, brown, broadly ovate-oblong or obovate bracts forming an invol mostly 1–1.5 cm about equaling the fls, the fls also individually subtended by a pair of bracts; cal cylindric, 10–13 mm, 15-veined (5 veins the stronger); pet-blade short, pink, 1-veined, shallowly obcordate; seeds reticulate; $2n=30$. European sp., cult. and sporadically escaped into disturbed sites in our range. May–Sept. (*Tunica p.*)

Family **POLYGONACEAE**, the Smartweed Family

Fls relatively small, perfect or unisexual; tep 2–6, basally connate into a minute or evident floral tube, green and herbaceous to often colored and ± petaloid, most often in 2 similar or slightly dissimilar cycles of 3, but not clearly differentiated into sep and pet, sometimes 5 and spirally uniseriate; stamens 2–9, rarely more, most commonly 3 + 3, but not infrequently 8 (notably in spp. of *Polygonum*); filaments distinct or basally connate; ovary superior, unilocular, (2)3(4)-carpellate, with distinct or basally united styles; ovule solitary on a basal placenta; fr an achene or small nut, lenticular or often trigonous, sometimes closely subtended by the persistent, sometimes accrescent tepals, or enclosed in a fleshy hypanthium; seeds with copious endosperm; embryo straight or curved; commonly excentric or peripheral (seldom centric), with 2 cotyledons; herbs (almost all ours) or less often woody plants, sometimes climbing or twining, producing anthocyanins but not betalains; lvs mostly alternate, seldom opposite or whorled, simple and usually entire, seldom cleft, sometimes articulate at base; stipules commonly well developed and connate into a usually scarious or hyaline (often bilobed or fringed) sheath (ocrea) around the stem, or (notably in *Eriogonum*) sometimes much reduced or wanting; fls borne in various sorts of infls, often in small, involucrate fascicles, often individually subtended by a persistent ocreola, evidently articulate to the pedicel, and often with a distinctly stipitate base above the articulation. 30/1000.

1 Fls in small clusters surrounded by a toothed or lobed involucre; stamens 9; lvs exstipulate . . . .   1. *Eriogonum.*
1 Fls not in involucrate clusters; stamens usually 8 or fewer; lvs stipulate.
  2 Achenes with 2 or 3 broad, flat wings, much exceeding the perianth . . . .   2. *Oxyria.*
  2 Achenes lenticular or trigonous, not winged, often ± enclosed by the perianth.
    3 Tep 6, in 2 series of 3, those of the inner series notably enlarged in fr . . . .   3. *Rumex.*
    3 Tep mostly 5 in a single spiral, usually subequal in fr.
      4 Perianth at maturity less than 1 cm, wingless, or with a keel or median wing on all 3 outer
         tep; plants without tendrils, only a few spp. climbing.
        5 Pedicels jointed near the base; fls in terminal racemes . . . .   4. *Polygonella.*
        5 Pedicels jointed at or near the summit; fls in various sorts of infls.
          6 Achene wholly or in considerable part enclosed by the ± accrescent perianth; embryo
             at one angle of the achene; habit various . . . .   5. *Polygonum.*
          6 Achene exserted for most of its length; embryo within the endosperm; annual with
             triangular-hastate lvs and an open, corymbiform infl . . . .   6. *Fagopyrum.*
      4 Perianth at maturity more than 2 cm, coriaceous, with one wing extending down the pedicel
         to the joint; climbing, woody, tendriliferous vines . . . .   7. *Brunnichia.*

1. **ERIOGONUM** Michx.   Fls perfect, in small clusters surrounded by a cup-like, 2–8-toothed or -lobed involucre and commonly mingled with filiform bractlets; tep 6, usually somewhat united at the base, biseriate; stamens 9, inserted on the base of the tep; filaments slender, usually exsert; styles 3, filiform, distinct; stigmas capitellate; fr a trigonous achene; annual or perennial herbs, or less often shrubs, with alternate, opposite, or whorled lvs, lacking stipules. 200, N. Amer.

1 Lvs mainly or all cauline, all alternate . . . .   1. *E. longifolium.*
1 Lvs basal and cauline, the cauline ones whorled . . . .   2. *E. allenii.*

1. **Eriogonum longifolium** Nutt.   Coarse, taprooted perennial mostly 1–2 m; stem woolly-villous, freely branched above; lvs ± numerous, alternate, white-tomentose beneath, glabrous above, lanceolate or lance-elliptic, broadly sessile or with an expanded, chartaceous, short-petiolar base, the larger ones mostly 10–15

× 1.5–3 cm; invols numerous, 3–4 mm, closely woolly-villous, shortly lobed; perianth closely woolly-villous outside, tapering to a shortly stipitate base; tep all alike, yellowish inside; filaments glabrous. Ours in cedar-glades; mainly Ozarkian, from Mo. and Kans. to La. and Tex., but irregularly e. to Ky., Tenn., Ala., and Fla. June–Oct. Our plants, as here described, belong to var *harperi* (Goodman) Reveal, local in Ky. (Christian Co.), Tenn., and Ala. (*E. harperi.*)

2. Eriogonum allenii S. Wats. Tomentose perennial 3–5 dm, erect from a thick, woody root, branched above; lvs glabrescent above, the basal ones long-petioled, with ovate or ovate-oblong blade 5–15 × 2–7 cm, the cauline ones in whorls of 3–5, short-petiolate, oval or oblong, progressively reduced; infl broad and rather flat-topped, with many invols, these 6–8 mm, lobed to the middle; perianth tapering to a shortly stipitate base; tep bright yellow, broad, obtuse, 3 mm, the outer glabrous, the inner inconspicuously hairy below; filaments villous toward the base. Shale-barrens; w. Va. and adj. W.Va. July, Aug.

2. OXYRIA Hill. Mountain sorrel. Fls perfect; tep 2 + 2, the inner broader; stamens 6, about equaling the tep; ovary flattened, with 2 short, divergent styles, the stigmas tufted; fr a flat achene surrounded by a broad, radially striate wing, cordate at base, subtended by the persistent, slightly enlarged tep, the 2 outer spreading, linear-oblong, the inner appressed, obovate-spatulate; perennial glabrous herbs with a stout, multicipital root, bearing several basal and few cauline lvs on long petioles, the blade broadly cordate to reniform; fls minute, in branching panicles; foliage acid. Monotypic.

1. Oxyria digyna (L.) Hill. Stem 1–4 dm; lvs 3–5 cm long and wide, palmately veined; fls 1 mm; fr orbicular, 3–4 mm, turning red; 2n=14. Moist, rocky slopes and ledges; circumboreal, s. in our range to N.S. and n. N.H. Summer.

Rheum rhabarbicum L., Rhubarb, a coarse perennial with large basal lvs, stout pertioles, 6 tep, 6 stamens, and 3-winged achenes, occasionally escapes from cult. (*R. rhaponticum*, misapplied.)

3. RUMEX L. Dock, sorrel. Fls perfect or unisexual; tep 3 ± 3, the outer usually narrower at anthesis; stamens 6, on short filaments; ovary trigonous; styles 3, spreading or deflexed over the angles of the ovary, each with a branched, stellate stigma; achene trigonous, closely invested by the accrescent inner tep; our spp. herbs, usually coarse; lvs subtending by sheathing stipules; fls small, greenish or suffused with red, in small verticils aggregated into a compound infl. At maturity the inner tep are called valves; often the midrib of a valve enlarges into an evident protuberance called a grain or tubercle. Our spp. all fl in summer and fr in late summer and fall. 200, widespread.

1 Lvs, or some of them, generally hastate or sagittate; mostly dioecious; foliage acid to the taste.
  2 Valves about equaling the achene and closely investing it . . . . 1. *R. acetosella.*
  2 Valves expanded into broad, reticulate wings much exceeding the achene.
    3 One or more valves with a conspicuous basal grain; lvs mainly sagittate . . . . 2. *R. acetosa.*
    3 Valves without grains; lvs mainly hastate (or entire) . . . . 3. *R. hastatulus.*
1 Lvs not hastate or sagittate; fls mostly or all perfect; lvs seldom acid.
  4 Valves without grains.
    5 Mature valves less than 1 cm wide.
      6 Pedicel of the fr with a visibly swollen joint near or below the middle.
        7 Lower lvs 2–4 times as long as wide . . . . 4. *R. longifolius.*
        7 Lower lvs 2–3 dm long and wide, broadly cordate at base . . . . 5. *R. alpinus.*
      6 Pedicel of the fr without a visibly swollen joint . . . . 7. *R. occidentalis.*
    5 Mature valves 2–3 cm wide . . . . 6. *R. venosus.*
  4 Valves (or at least one of them) with a prominent grain.
    8 Valves little if at all wider than the face of the achene, 2–3 mm long . . . . 18. *R. conglomeratus.*
    8 Valves notably wider than the achene, the sides enlarged into projecting wings appressed to the
      corresponding wings of other valves, the fr thus 3-winged.
      9 Margins of the valves entire to dentate or undulate.
        10 Base of the grain distinctly above the base of the valve; pedicel without a visibly swollen
           joint . . . . 8. *R. orbiculatus.*
        10 Base of the grain even with the base of the valve, or projecting below it; pedicel with a
           visibly swollen joint.
          11 Fr with one grain, this much less than half as long as the valve . . . . 9. *R. patientia.*
          11 Fr with 1–3 grains, the larger ones at least half as long as the valve.
            12 Pedicels 2–5 times as long as the fr; grains 3, projecting below the valves . . . . . . 10. *R. verticillatus.*
            12 Pedicels seldom more than twice as long as the fr; grains 1–3, not projecting below
              the valves.
              13 Lvs crisp-margined; grains two-thirds as wide as long . . . . 11. *R. crispus.*
              13 Lvs flat; grains up to half as wide as long.
                14 Valve evidently longer and much more than twice as wide as the grain.

15 Valves broadly rotund-ovate, very obtuse; grain 1(−3) .... 12. *R. altissimus*
15 Valves deltoid-ovate, subacute; grains 3 .... 13. *R. salicifolius*.
14 Valve not much longer than and about twice as wide as the grain .... 14. *R. pallidus*.
9 Margins of the valves with a few long, slender, spinose or bristly teeth.
16 Well developed grain 1, perennials.
17 Verticils all separate; pedicels about equaling the fr .... 15. *R. pulcher*.
17 Upper verticils contiguous; pedicels much longer than the fr .... 16. *R. obtusifolius*.
16 Well developed grains 3; annuals .... 17. *R. maritimus*.

1. **Rumex acetosella** L. Red s. Dioecious perennial 1–4 dm from slender creeping roots; lvs variable, usually 3-lobed, the terminal lobe narrowly elliptic to oblong, the lateral much smaller, triangular, divergent; lf-base below the lobes truncate to long-cuneate; infl sometimes half the length of the shoot; pedicel jointed next to the fl; outer tep lanceolate; inner tep in male fls 1.5–2 mm, obovate, in female broadly ovate; achene 1.5 mm, shiny golden brown, closely invested by the valves; 2*n*=14, 28, 42. Fields, lawns, and waste places, in acid soils; native of Eurasia, naturalized throughout most of N. Amer., and highly variable.

2. **Rumex acetosa** L. Green s. Dioecious perennial 3–10 dm, the stems 1–few from a stout root, usually simple to the infl; lvs oblong, all or chiefly sagittate, the lower long-petioled, the upper subsessile; infl 1–2 dm, usually lfless, open, the branches mostly simple; pedicel jointed at about midlength; tep of staminate fls 2–3 mm, the outer oblong, the inner obovate; outer tep of pistillate fls triangular-ovate, nearly 2 mm, soon reflexed; valves thin, broadly round-cordate, 4–6 mm long and wide; reticulate-veiny, the midrib of at least one of them dilated at the base into an evident grain or tubercle; achene dark brown, 2–2.5 mm; 2*n*=14 (♀), 15 (♂). Native of Eurasia, in our range an occasional weed from Conn. and Pa. northward.
The closely related Eurasian sp. *R. thyrsiflorus* Fingerh., with a denser infl, the primary branches again branched, and with the lf-lobes often a little more divergent, has been reported in n. Mich. and on the s. side of the St. Lawrence R. in Que.

3. **Rumex hastatulus** Baldwin. Wild s. Dioecious perennial from a stout, woody root, 4–10 dm, branched from the base; basal lvs numerous, cauline lvs linear to hastately 3-lobed, the terminal lobe usually linear-oblong, the basal lobes commonly widely divergent, often unequal; infl 1–3 dm; staminate tep 1.5 mm, the outer obovate-oblong, the inner broadly obovate; outer tep of pistillate flowers reflexed in fr, barely 1 mm; valves broadly round-cordate, 2.5–3.5 mm, reticulate-veiny; achenes brown, 1.5 mm; 2*n*=26 (♀), 27 (♂). Sandy soil of the coastal plain; Mass. to Fla. and Tex., and n. in the Miss. Valley to Kans. and s. Ill.

4. **Rumex longifolius** DC. Yard-d. Coarse perennial to 1.5 m; lower lvs narrowly oblong, broadest near the middle, tapering to an acute base; mature pedicels visibly jointed near the base; valves rotund to subreniform, 4–6 × 5–7 mm, entire or toothed, reticulate-veiny, without grains; 2*n*=40, 60. Native of n. Europe, sparingly intr. in our range, mainly toward the north. (*R. domesticus*.)

5. **Rumex alpinus** L. Stout, rhizomatous perennial to 12 dm; lower lvs subrotund, 2–3 dm long and wide, broadly rounded above, broadly cordate at base; cauline lvs much smaller; mature pedicels visibly jointed near the base; valves cordate-ovate, 3–5 mm, without grains; 2*n*=20. Native of Europe, sparingly naturalized in the ne. part of our range.

6. **Rumex venosus** Pursh. Veiny d. Glabrous, rhizomatous perennial 2–6 dm; lvs ovate to oblong or lanceolate, flat, entire, 4–12 cm; infl very dense in fr; pedicels weakly jointed near midlength; valves rose-color, 1.5–2 × 2–3 cm, without grains; 2*n*=40. Dry soil in waste places; native of w. U.S., rarely adventive in our area.

7. **Rumex occidentalis** S. Wats. Western d. Stout, taprooted perennial to 2 m; lower lvs 10–30 cm, oblong or linear-oblong, abruptly cordate or subcordate at the base, often crisp-margined; infl of crowded glomerules, at maturity forming a continuous slender panicle to 5 dm; pedicels in fr 5–10 mm, not visibly jointed, but disjointing at a fourth their length; valves cordate-ovate to cordate-triangular, 4–7 mm, about as wide, entire, conspicuously reticulate, without grains; achene 3.5–4.5 mm; 2*n*=ca 140, 160, 200. Wet soil; N.S. and c. Me. to Nf.; widespread in the w. Cordillera. (*R. fenestratus*.)

8. **Rumex orbiculatus** A. Gray. Great-water-d. Stout perennial to 2.5 m, simple to the infl; lvs mostly flat, lanceolate, acute to rounded at base; infl to 5 dm, the several ascending branches mostly subtended by reduced lvs; mature pedicels obscurely jointed near the base; valves subrotund, 5–8 mm, about as wide, basally broadly truncate; grains 3, narrowly lance-acuminate, half as long as the valve, the base distinctly above the base of the valve; 2*n*=60, 160. Swamps and shallow water; Nf. and Que. to N.D., s. to N.J., Ind., and Nebr. (*R. britannica,* misapplied.)

9. **Rumex patientia** L. Patience-d. Stout, taprooted, glabrous perennial to 2 m, simple to the infl; lvs pale green, basally disposed, somewhat crisped, oblong to oblanceolate, to 15 cm wide, the larger (lower) truncate to subcordate at base, the smaller acute; lateral veins of the lvs making an angle of 45–60° with the midrib; infl to 5 dm, with stout, ascending branches subtended by reduced lvs, otherwise leafless, the verticils eventually contiguous; pedicels 5–10 mm, visibly jointed near the base; valves broadly rounded, 6–9 mm, very blunt, deeply cordate, subentire to crenate-undulate; grain 1, up to a third as long as the valve; 2*n*=60. Native of Europe, occasionally found in waste places in our range. Hybridizes with nos. 11 and 16.
Two other closely related ruderal European spp., rarely adventive with us, have small grains, as in *R. patientia,* but on all 3 valves, and the valves are usually ± toothed; the lateral veins of the lvs make an angle of 60–90° with the midrib. *R. cristatus* DC. has red-brown valves with numerous irregular teeth ca 1 mm,

and the lvs are smooth beneath. *R. kerneri* Borbas has dark brown valves with smaller, sometimes indistinct teeth, and the lvs are minutely papillose-scabrid beneath.

10. Rumex verticillatus L. Water-d., swamp-d. Stout, taprooted, glabrous perennial to 1.5 m, with many short axillary branches; lvs flat, lanceolate or lance-linear, narrowed to the base; infl 2–4 dm, with usually few ascending leafless branches, the verticils becoming contiguous in the upper half, pedicels straight, slender, 10–15 mm, deflexed, jointed near the base; valves broadly triangular-ovate, 3.5–6 mm, nearly or fully as wide, or even a little wider than long, obtuse, truncate at base, very thick toward the center; grains 3, lanceolate, two-thirds as long as the valve, the blunt base projecting 0.5 mm below the valve; $2n=60$. Swamps and wet lowland woods; Que. and Ont. to Wis. and Kans., s. to Fla. and Tex. (*R. floridanus.*)

11. Rumex crispus L. Curly d. Stout, taprooted perennial to 1.5 m, simple to the infl; lvs strongly crisped, the larger commonly rounded to subcordate at the base; infl large, with many ascending or erect branches, becoming compact at maturity, with some linear lvs intermingled; pedicels flexuous, 5–10 mm, spreading-decurved, jointed near the base; valves thin, broadly ovate, 4–5 mm long and wide, entire or nearly so, obtuse, truncate or subcordate at base; grains 3, often unequal, the larger broadly ovoid, very turgid, rounded at both ends, half as long as the valve; $2n=60$. Native of Europe, found as a weed of roadsides, fields, and waste ground throughout the U.S. and s. Can. (*R. elongatus.*) Hybridizes with nos. 9 and 16.

R. stenophyllus Ledeb., a related European halophyte rarely adventive with us, will key here. It differs in its often less strongly crisped lvs and especially in the numerous small but distinct teeth along the margins of the valves.

12. Rumex altissimus A. Wood. Pale d. Stout perennial to 1 m, often branched; lvs all cauline, pale, flat, lanceolate or more often lance-ovate, acute or acuminate, obtuse or rarely rounded at base; infl loose, 1–3 dm, with usually few short ascending branches; pedicels 3–5 mm, jointed just above the base; valves firm, subrotund, 4–6 mm long and wide, broadly truncate at base; grains 1–3, when plural often unequal, the larger lanceolate, acute, half to two-thirds as long as the valve; $2n=20$. Swamps and wet soil; N.H. to Mich., Minn., and Colo., s. to Va., Ga., Tex., and Ariz.

13. Rumex salicifolius J. A. Weinm. Taprooted, glabrous perennial to 1 m, often branched; lvs flat, pale green or glaucous, narrowly lanceolate, long-tapering at both ends; infl 1–3 dm, the few ascending branches usually subtended by linear lvs; pedicels 2–4 mm, jointed near the base; valves thick, triangular, 3–6 mm long and wide, acute or subacute; grains 3, lanceolate, acute, half to two-thirds as long as the valve; $2n=20$ (ours), 40. Moist, often brackish or saline soil; Que. to B.C., s. to N.Y., Pa., Ky., Mo., and Mex. (*R. mexicanus,* a southwestern tetraploid.) Our plants, as here described, are sometimes segregated as var. *triangulivalvis* (Danser) Hickman. (*R. t.*)

14. Rumex pallidus Bigelow. Seabeach-d. Branching perennial from a white root, ascending, or decumbent at base, 2–7 dm; lvs flat, glaucous, narrowly lanceolate, acute to long-cuneate at base; infl 1–2 dm, with numerous widely divergent branches; pedicels 2–4 mm, jointed just above the base; valves broadly ovate, 3–4 × 2–3 mm, obtuse, broadly truncate at base, the margins scarcely projecting; grains 3, turgid, nearly as long as the valves and more than half as wide; $2n=20$. Sandy or rocky beaches and coastal swamps; Nf. and Que. to L.I. Appears to integrade with no. 13.

15. Rumex pulcher L. Fiddle-d. Slender, branching perennial 2–8 dm from a taproot; lvs basally disposed, the lower commonly constricted just above the cordate base, hairy on the petiole and veins beneath; infl very large, with many divergent spike-like branches, the glomerules separate, mostly subtended by reduced lvs; pedicels stout, conspicuously jointed, about as long as the fr; valves triangular-ovate, 4–5 mm, conspicuously reticulate, each with 2–6 spinose teeth on each margin below the middle; fully developed grain 1, verrucose, the other valves bearing imperfect small grains; $2n=20$. Native of Europe, intr. in waste places from N.Y. to Fla., Okla., and Tex., and on the Pacific coast.

16. Rumex obtusifolius L. Bitter d. Stout perennial to 12 dm from a large tap or branched root, usually single-stemmed and simple to the infl; lower lvs cordate at base, broadly oblong to ovate, to 15 cm wide, the upper much smaller; infl freely branched, the lower verticils commonly separate and leafy-bracted; pedicels conspicuously longer than the fr, jointed below the middle; valves triangular-ovate, 3.5–5 mm, with 2–4 spinose teeth on each margin, one bearing a large, minutely wrinkled grain, the others merely with a slightly thickened midrib; $2n=50$. Native of Europe, commonly naturalized in waste ground, especially in moist soil, from Que. and N.S. to B.C., s. to Fla. and Ariz.

17. Rumex maritimus L. Golden d. Fibrous-rooted, hollow-stemmed annual to 8 dm, often bushy-branched; lvs narrowly lanceolate to oblong-linear, the larger commonly truncate to subcordate at base, the margins often crisped; panicle large and freely branched, many of the verticils subtended by small lvs; pedicels jointed near the base, twice as long as the fr; valves triangular-ovate, 2–3 mm, with 2 or 3 divergent marginal bristles on each side, the free tip triangular; grains 3; $2n=40$. Shores, streambanks, and wet ground, avoiding acid soils; widespread in Eurasia, and irregularly in N. and S. Amer. Typical *R. maritimus* is European. We have 2 vars. Var. *fueginus* (Philippi) Dusén, widespread in both N. and S. Amer., has narrowly lanceolate grains, 0.3–0.4 mm wide and tapering to the summit, and the valves have relatively long marginal bristles, up to ca. 2.5 mm. Var. *persicarioides* (L.) R. Mitchell, occurring in coastal marsh areas from L.I. and Mass. to Que., and from n. Calif. to B.C., has broader, ellipsoid grains, 0.4–0.6 mm wide and rounded to the summit, nearly concealing the body of the valve, which has shorter marginal bristles, 0.5–1.5 mm. (*R. persicarioides,* but the name often misapplied to var. *fueginus.*)

18. Rumex conglomeratus Murray. Slender, glabrous, taprooted perennial to 1 m, mostly single-stemmed, lower lvs cordate or subcordate at base; panicle forming as much as half the shoot, its branches long and often divergent, the middle and lower fascicles of fls usually subtended by lvs; pedicels 1–2 mm, jointed near the base; valves oblong-ovate, 2–3 mm, 1 mm wide, the margins entire or rarely denticulate, not projecting as wings; grains commonly 3, at maturity nearly as wide as the valve; $2n=20$. Native of Europe, naturalized in waste places here and there in our range.

Rumex sanguineus L., with the valves 3–4 mm, only one bearing a subglobose grain, is probably only a waif with us.

4. POLYGONELLA Michx. Jointweed. Fls perfect or unisexual; tep 5, petaloid, persistent and ± accrescent; stamens 8 (5 + 3), the filaments sometimes dimorphic; ovary trigonous; styles 3, distinct; stigmas capitate; fr a smooth, sharply trigonous achene, subtended and loosely invested by the perianth; embryo slender, straight or nearly so; taprooted herbs or subshrubs of somewhat heathlike aspect, commonly freely branched, with small lvs and white to red fls in compound racemes, each fl solitary in the axil of a sheathing bract; lvs jointed to the summit of the ocrea; pedicels articulated just above the base. 10, N. Amer.

1 Suffrutescent perennial; outer 2 tep soon becoming reflexed . . . . 1. *P. polygama.*
1 Annual; tep all remaining loosely appressed . . . . 2. *P. articulata.*

1. Polygonella polygama (Vent.) Engelm. & A. Gray. Slender, somewhat woody perennial 2–6 dm, often freely branched; lvs flat, mostly narrowly cuneate, 5–20(–35) × (1–)2–4(–6) mm, with hyaline margins at least distally; infl of few–many racemes 1–3 cm; pedicels straight, divergent, 1 mm at anthesis; some fls pistillate, others functionally staminate; tep white to pink, ovate, 1–1.5 mm, the 2 outer becoming reflexed during anthesis, persistent but scarcely enlarging, the 3 inner remaining loosely appressed, in pistillate fls enlarging to 2–3 mm in fr; achene 1.5–2 mm. $2n=28$. Dry, sandy soil and pine barrens on the coastal plain; se. Va. to Fla. and Tex. Sept.

2. Polygonella articulata (L.) Meissner. Slender, wiry annual 1–5 dm, branched at least above; lvs narrow, revolute, 5–20 mm, 1 mm wide or less, not hyaline-margined; racemes erect or ascending, 2–3.5 cm; pedicels 2–3 mm, slender, decurved; fls perfect; tep white or greenish to pink and red, 1.5–2 mm, larger in fr, the 2 outer obovate, keeled above, the 3 inner elliptic, all remaining loosely appressed; achene 2–2.5 mm; $2n=32$. Dry, acid sands; Me. and s. Que. to ne. N.C., mostly on the coastal plain; shores of the Great Lakes, and inland dunes to s. Ill., Io. and Minn. July, Oct. (*Delopyrum a.*)

5. POLYGONUM L. Fls perfect or unisexual; pedicels jointed at or near the summit; tep (4)5(6), connate below; stamens 3–8; ovary flattened or trigonous, with 2–3 minute, usually capitate stigmas, the styles usually short or obsolete, separate or united; fr a lenticular or trigonous achene, wholly or in large part enclosed by the persistent perianth; embryo at one angle of the achene; annual or perennial herbs (seldom shrubs), commonly with well developed ocreae; fls borne in axillary clusters or terminal (often also axillary) spike-like racemes, or in terminal heads or panicles. Most of the several sections have been taken by some authors as distinct genera. Our spp. mostly flower from early or midsummer into fall. 200, widespread, mainly temp.

1 Fls in small clusters axillary to ordinary or ± reduced lvs; lvs jointed at base . . . . Sect. *Polygonum.*
1 Fls in terminal (or also axillary) spikes or racemes or panicles or heads; lvs not jointed.
  2 Outer tep not keeled or winged at maturity.
    3 Lvs chiefly basal; stem simple, with a single short dense raceme . . . . Sect. *Bistorta.*
    3 Lvs chiefly cauline; stem often branched, commonly with several infls.
      4 Styles short, withering in fr; tep commonly 5, but sometimes only 4.
        5 Stems glabrous or pubescent.
          6 Fls subtended by ocreolae, in spike-like racemes . . . . Sect. *Persicaria.*
          6 Fls subtended by chaffy bracts, in dense heads . . . . Sect. *Cephalopilon.*
        5 Stems armed with reflexed prickles on the angles . . . . Sect. *Echinocaulon.*
      4 Styles elongate, persistent, becoming indurate; tep 4 . . . . Sect. *Tovara.*
  2 Outer tep evidently keeled or winged at maturity.
    7 Stems commonly twining or trailing, rarely erect and only a few dm tall . . . . Sect. *Tiniaria.*
    7 Stems stout, erect, freely branched, commonly 1–2 m . . . . Sect. *Pleuropterus.*

SECTION POLYGONUM (Avicularia). Knotweed, Knotgrass

Our spp. annual herbs with branched (often repeatedly branched) stems; lvs small, narrowed to the base and jointed to a short, petiole-like part of the ocrea; fls small, solitary or few and short-pedicellate to subsessile from the upper (or upper and middle) ocreae; tep mostly 5, ± connate at least below, somewhat accrescent; stamens (3–)5(–8); styles 3, distinct or partly united; achenes mostly or all trigonous, often unequally so, often some of those produced in late season larger and of different shape and texture from the earlier ones; $x=10$. Plants with the fls axillary to ± ordinary lvs are said to be homophyllous; those with ± evidently terminal infls, the fls axillary to ± much reduced lvs or bracts, are said to be heterophyllous, but the distinction is not always clear. Spp. 1 & 2, here for convenience keyed with sect. *Polygonum,* are now often referred to sect. *Duravia,* a group not otherwise represented with us.

1 Pedicels abruptly recurved at the summit; fls and fr deflexed .... 1. *P. douglasii.*
1 Pedicels straight or nearly so; fls and fr ascending.
  2 Lvs plicate with 2 longitudinal folds, minutely spinulose-serrulate .... 2. *P. tenue.*
  2 Lvs flat or revolute, entire or obscurely erose.
    3 Perianth ± bottle-shaped, constricted above the achene.
      4 Fruiting perianth divided ca ⅓ length; plants homophyllous .... 3. *P. achoreum.*
      4 Fruiting perianth divided ca ¾ length; plants heterophyllous .... 4. *P. erectum.*
    3 Perianth otherwise, not constricted above the achene.
      5 Achenes glossy; tep at maturity notably white or pink (at least at the margins) and loosely
        spreading above; maritime plants.
        6 Ocreae of the lower nodes very conspicuous, 7–10 mm, mostly 8–16-nerved .... 5. *P. glaucum.*
        6 Ocreae smaller, seldom as much as 7 mm, mostly 3–7-nerved .... 6. *P. oxyspermum.*
      5 Achenes dull, or if glossy then the tep wholly yellow-green and appressed to the achene at
        maturity; widespread, maritime or inland plants.
        7 Outer 3 tep flat, equaling or shorter than the inner 2.
          8 Plants heterophyllous; sprawling agrarian weed .... 7. *P. aviculare.*
          8 Plants homophyllous.
            9 Lvs mostly 2–5 times as long as wide; prostrate street and dooryard weed .... 8. *P. arenastrum.*
            9 Lvs mostly 5–9 times as long as wide; usually ± ascending weed .... 9. *P. neglectum.*
        7 Outer 3 tep cucullate, distinctly surpassing the inner 2 in fr and ± concealing them.
          10 Achenes abruptly short-beaked, finely and uniformly granular; maritime .... 10. *P. fowleri.*
          10 Achenes neither beaked nor uniformly granular; widespread, sometimes maritime.
            11 Plants prostrate; lvs mostly 2–4 times as long as wide .... 11. *P. buxiforme.*
            11 Plants ± ascending or erect; lvs mostly 4–12 times as long as wide.
              12 Fls evidently exserted from the ocreae, on pedicels mostly 2–3.5 mm; plants
                ± strongly heterophyllous .... 12. *P. ramosissimum.*
              12 Fls scarcely exserted, on pedicels less than 2 mm, nearly homophyllous .... 13. *P. prolificum.*

SECTION BISTORTA. Bistort

Alpine perennials from a short, thick rhizome, with chiefly basal lvs, a simple stem, and a single terminal spike-like or head-like raceme.

1 Raceme slender, 0.5–1 cm thick, the lower fls usually replaced by bulblets .... 14. *P. viviparum.*
1 Raceme stout, 1–2 cm thick; bulblets none .... 15. *P. bistorta.*

SECTION PERSICARIA. Smartweed

Annual or perennial herbs with linear to ovate, sessile or short-petioled lvs not jointed at base; fls in terminal, spike-like racemes, subtended by ocreolae, or a few from the upper axils; pedicels jointed at the summit; tep (4)5(6), united at base, somewhat accrescent; stamens (3–)8; styles 2 or 3; achenes dark brown or black, lenticular or trigonous.

1 Perennials from rhizomes or stolons.
  2 Racemes terminal, solitary or paired .... 16. *P. amphibium.*
  2 Racemes terminal and axillary, usually several or numerous.
    3 Ocreae entire or nearly so, not bristly-fringed .... 17. *P. densiflorum.*
    3 Ocreae fringed around the summit with bristles.
      4 Tube and lobes of the perianth evidently glandular-punctate.

5 Racemes not much interrupted, the ocreolae commonly overlapping; lvs mostly 2–3.5(–
   4.5) cm wide .... 18. *P. robustius.*
5 Racemes much interrupted, especially below; lvs seldom over 2 cm wide .... 23. *P. punctatum.*
4 Perianth not glandular-punctate .... 19. *P. hydropiperoides.*
1 Taprooted annuals.
   6 Ocreae entire or merely lacerate, not bristly-fringed.
      7 Tep 4(5), the outer ones strongly 3-nerved in fr, each nerve ending in an anchor-shaped fork ... 20. *P. lapathifolium.*
      7 Tep 5, with inconspicuous, irregularly forked nerves .... 21. *P. pensylvanicum.*
   6 Ocreae fringed around the summit with bristles.
      8 Tube and lobes of the perianth evidently glandular-punctate.
         9 Tep 4(5); achene dull; ocreolae commonly overlapping .... 22. *P. hydropiper.*
         9 Tep 5; achene shining; ocreolae (especially the lower) more remote .... 23. *P. punctatum.*
      8 Perianth not evidently glandular-punctate.
         10 Lvs lance-ovate to linear, seldom as much as 5 cm wide.
            11 Peduncles and upper stem stipitate-glandular .... 24. *P. careyi.*
            11 Peduncles and stems not glandular.
               12 Ocreolae long-ciliate, the cilia 2–3.5 mm .... 25. *P. cespitosum.*
               12 Ocreolae entire, or with a few short cilia up to 1 mm .... 26. *P. persicaria.*
         10 Lvs broadly ovate to cordate, mostly 5–10 cm wide .... 27. *P. orientale.*

Section Cephalopilon

Herbs, differing from section *Persicaria* mainly in the dense, capitate inflorescence, with the flowers subtended (and often surpassed) by chaffy bracts.

One sp. in our range .... 28. *P. nepalense.*

Section Echinocaulon..Tear-thumb

Annual or perennial, branching herbs, the slender stems with long internodes armed with reflexed prickles,
reclining on other plants; lvs broad.

1 Ocreae scarious or hyaline, not foliaceous-expanded.
   2 Lvs tapering to the base .... 29. *P. bungeanum.*
   2 Lvs sagittate or hastate.
      3 Lvs sagittate; achenes trigonous .... 30. *P. sagittatum.*
      3 Lvs hastate; achenes lenticular .... 31. *P. arifolium.*
1 Ocreae, especially of the upper lvs, conspicuously foliaceous-expanded .... 32. *P. perfoliatum.*

Section Tovara. Jumpseed, Virginia knotweed

Erect herbs, perennial from a rhizome, with broad lvs and elongate, terminal, slender, much interrupted
racemes; tep 4; styles 2, indurated and reflexed in fr.

One sp. in our range .... 33. *P. virginianum.*

Section Tiniaria

Annual or perennial herbs, commonly twining or trailing, rarely erect; stigmas minute, capitellate; outer
perianth lobes keeled or winged at maturity, the wing usually extending along the pedicel as far as the joint.

1 Mature perianth scarcely surpassing the achene, the outer lobes keeled or barely winged.
   2 Ocreae smooth; achenes dull black; styles united .... 34. *P. convolvulus.*
   2 Ocreae reflexed-bristly at base; achenes shining; styles divergent .... 35. *P. cilinode.*
1 Mature perianth much surpassing the achene, the outer lobes strongly winged ...... 36. *P. scandens.*

Section Pleuropterus

Tall, stout, erect, perennial herbs with broad lvs and large, panicled infls; stigmas minute, fringed; outer
perianth-lobes conspicuously winged at maturity.

1 Lvs broadly truncate at base; fls functionally unisexual, the plants dioecious . . . . 37. *P. cuspidatum.*
1 Lvs cordate at base; fls perfect . . . . 38. *P. sachalinense.*

1. Polygonum douglasii Greene. Slender annual, 2–6 dm, with numerous ascending branches; lvs linear-oblong to narrowly lanceolate, 2–5 cm × 2–8 mm, subulate-tipped, flat, often somewhat revolute; fls remote, only 1–3 per ocrea; pedicels 2–3 mm, exsert and soon reflexed; perianth 3–4 mm, cleft nearly to the base; achene black, shiny, 3–4 mm; $2n=40$. Open, rocky or gravelly soil; Que. and n. N. Engl. to ne. Minn. and Io.; widespread in w. U.S.

2. Polygonum tenue Michx. Slender annual, 1–4 dm, with ascending or erect branches; lvs linear, 1–3 cm, subulate-tipped, minutely spinulose-serrulate, plicate in 2 folds near the midvein; fls remote, mostly only 1 per ocrea; perianth deeply cleft, the oblong segments connivent over the achene; achene black, sharply trigonous, 2.5–4 mm; $2n=20, 30, 32$. Dry, chiefly acid soils; Me. to Minn., s. to Ga. and Tex.

3. Polygonum achoreum S. F. Blake. Annual to 5(–7) dm, homophyllous, erect at first, but often becoming prostrate with upturned branches; lvs bluish-green, oval to elliptic with rounded tip, mostly 10–35 × 3–15 mm; ocreae 4–10 mm, becoming brown and lacerate; perianth yellow-green, 2.5–3.7 mm, somewhat bottle-shaped and constricted above the achene, cleft only ca ⅓ its length, the 3 outer lobes narrow, curved, strongly cucullate; achenes 2.4–3.3 mm, yellow-green to tan, uniformly papillose; late-season achenes exsert, to 5 mm, olivaceous, smooth; $2n=40, 60$. A weed in waste places, widespread in our region, w. to the Rocky Mts.

4. Polygonum erectum L. Much like no. 3, but somewhat heterophyllous (more so in age), with green or bright yellow-green lvs, the larger ones mostly 25–60 × 10–30 mm; ocreae hyaline-silvery, entire to slightly lacerate; perianth cleft ca ¾ its length; achenes light to dark brown, striate-papillose. A weed in waste places, widespread (but not common) in the U.S. and s. Can.

5. Polygonum glaucum Nutt. Seaside knotweed. Annual to 3(–5) dm, homophyllous, diffusely branched, prostrate or nearly so, commonly with upturned branch-tips, very pale and glaucous; lvs lanceolate or narrowly elliptic, 1–3 cm × 2–8 mm; ocreae conspicuous, mostly 7–10 mm (at least on the lower nodes), silvery-scarious distally, the lower half becoming brown, with 8–16 usually scabrellous nerves; pedicels included in the ocreae; mature perianth 3–4 mm, divided nearly to the base, the lobes loosely spreading, obovate-oblong, obtuse or broadly rounded above, with green midrib and white to pink margins; achene 3–4 mm, exsert, glossy, brown to blackish; $2n=40$. Sandy beaches; Mass. to Ga. (*P. maritimum,* misapplied.)

6. Polygonum oxyspermum C. A. Meyer & Bunge. Much like no. 5, often greener; ocreae to 5(–7) mm, brown, 3–7-nerved, lacerate; mature perianth 3–5 mm; achenes 3.5–5 mm (to 6.5 mm in late fls); $2n=40$. Damp beaches; Atlantic coast from N.S. and N.B. to Nf., and in nw. Europe, perhaps only intr. in Amer. (*P. acadiense; P. raii.*)

7. Polygonum aviculare L. Freely branched, sprawling to suberect, heterophyllous annual, to 1 or reputedly 2 dm; early lvs lanceolate or lance-ovate, 2.5–6 cm × 4–15 mm, the later ones only a third as large; ocreae 4–8 mm, hyaline, becoming lacerate; mature perianth 2.5–4 mm, divided to well below the middle, its segments subequal, with white to pink margins and flat tip; achenes 2.2–3.2 mm, dark brown, striate-papillose, included or barely exserted; $2n=40, 60$. Cosmopolitan weed of cult. fields and waste places.

8. Polygonum arenastrum Boreau. Dooryard knotweed. Freely branching, homophyllous annual, prostrate and commonly mat-forming; lvs bluish-green, broadly elliptic to oval-oblong, obtuse or rounded, 5–20 × 3–9 mm, often deciduous in late season; ocreae 3–6 mm, bifid, eventually lacerate; mature perianth 1.8–2.6 mm, divided to about the middle, the segments with white to pink margins and flat, obtuse or rounded tip; achenes 1.5–2.3 mm, shortly included to shortly exserted, dark brown, striate-papillate, late season ones to 4 mm, olivaceous, smooth; $2n=40$. Abundant weed of dooryards, sidewalks, and streets, native of Europe, now widespread in temperate N. Amer. (*P. aviculare,* misapplied.)

9. Polygonum neglectum Besser. Prostrate to ascending annual, much like no. 8, but more sparsely branched, and with narrower, linear oblong to lance-linear lvs 8–30 mm; perianth avg a bit larger, to 3 mm, divided half to ¾ length; achenes avg larger, 2–3 mm; $2n=40, 60$. A weed of roadsides, vacant lots, and pastures, seldom on beaches and dunes; Eurasia and much of N. Amer., doubtfully native. (*P. franktonii,* of maritime and freshwater beaches, dunes, and shores in se. Can.)

10. Polygonum fowleri B. L. Robinson. Prostrate to ascending, homophyllous annual to 5 dm, with relatively few, divergent branches; lvs elliptic-oblong to oblanceolate, 1–4 cm × 3–10 mm; ocreae seldom over 4 mm, lacerate, brown, the pedicels included or barely exsert; mature perianth 3–5 mm, cleft nearly to the base, its segments oblong, obtuse, cucullate (at least the outer) with white or pink margins, after anthesis appressed to the achene; achenes 2.8–4 mm, broadly short-beaked distally, finely granular-roughened, included (to 5 mm and exsert in late season); $2n=40, 60$. Sea-beaches and margins of salt-marshes; Lab. and Nf. to Me., and on the n. Pacific coast. (*P. allocarpum.*)

11. Polygonum buxiforme Small. Prostrate, freely branched and mat-forming, homophyllous annual; lvs oblong to oblanceolate, 3–20 × 1–8 mm; ocreae to 5 mm, hyaline-silvery distally; pedicels barely exsert; mature perianth 2–3 mm, divided to below the middle, the undivided part asymmetric, ventricose on one side, the segments greenish with white or pinkish margins; outer 3 segments cucullate; achenes 2–2.8 mm, included, dark brown, minutely striate-papillose, one side broader; late-season frs exsert, to 4.5 mm, olivaceous, smooth; $2n=60$. Packed, nondrifting sands, borders of marshes and dunes, and sandy soils, both maritime and inland; rare but widespread in the U.S. and s. Can.

12. Polygonum ramosissimum Michx.   Freely branched, mostly erect or ascending annual 3–10 dm (or reputedly to 20 dm), ± strongly heterophyllous, the rameal lvs abruptly smaller than those of the main stem; lower internodes to 5 cm, the upper progressively shorter; lvs linear to narrowly lance-elliptic, 1–6 cm × 2–5(–10) mm, obtuse or acute, flat; fls 1–3 and exsert from the upper ocreae on pedicels 2–3.5 mm, forming racemes to 15 cm, mature perianth 3–4 mm, deeply cleft, the 3 outer tep cucullate and notably exceeding the inner 2, especially in fr; achenes smooth and ± shiny, the earlier ones blackish, 2–3.5 mm, thickly ovoid, and included in the perianth, but some of the later ones often paler, narrowly ovoid, 4–5 mm, and exsert; $2n=60$. Unstable, usually moist, sometimes saline habitats, sometimes in coastal marshes, but also widespread inland; Me. and Que. to Wash., s. to Del., Ill., and N.M. (*P. atlanticum; P. exsertum; P. triangulum.*)

13. Polygonum prolificum (Small) B. L. Robinson.   Much like no. 12, and intergrading with it; plants nearly homophyllous; lvs rounded or obtuse at the tip, conspicuously veined, becoming rugose-veiny when dry; fls not so strongly restricted to the stem-tips, scarcely exsert, on pedicels less than 2 mm; $2n=60$. Brackish shores and marshes along the coast, from Me. and Va.; less commonly and irregularly w. to Oreg., Calif., and N.M.

14. Polygonum viviparum L.   Erect perennial 1–4 dm from a short, stout rhizome, the stem simple and few-lvd; lowest lvs linear-oblong, 3–8 × 0.5–1.5 cm, long-petioled, the others progressively reduced and nearly linear; raceme 1, slender, 3–6 × 0.5–1 cm, the lower fls usually replaced by bulblets; pedicels jointed at the summit; tep white or pink, obtuse, the inner the wider; stamens long-exserted; styles 3, longer than the ovary; achene trigonous; $2n=66–132$. Moist, usually calcareous soil; circumboreal, s. in our range to n. Minn., n. Mich., and the higher mts. of N.E. Late June–Aug. (*Bistorta v.*)

15. Polygonum bistorta L.   Perennial to 1 m, the rhizome commonly contorted; basal lvs oblong-ovate or triangular-ovate, rounded to a broadly winged petiole; raceme stout, 5–9 × 1–2 cm, floriferous to the base; otherwise much like no. 14; $2n=24$, 48. Native of Eurasia, rarely escaped from cult. in our range. May–Aug.

16. Polygonum amphibium L.   Water smartweed. Rhizomatous, functionally dioecious perennial, aquatic or amphibious or terrestrial, the terrestrial forms erect or ascending, up to 1(–2) m, with hairy, typically lance-acuminate lvs, the more strongly aquatic forms with floating branches (and often stolons) and lvs, the lvs then glabrous, typically elliptic and obtuse or rounded; ocreae with or without a bristly fringe, in aerial forms often with a flared collar; racemes stout, dense, mostly 2–15 × 1–2 cm, usually borne singly or in pairs at the tips of major branches; tep 5, pink or red, bluntly rounded above, united below, 4–5 mm at anthesis, to 7 mm in fr, becoming veiny and enclosing the achene, this dark, shiny, doubly convex, 2.5–3 mm; $2n=66, 96, 98$. Widespread in the N. Hemisphere, and in S. Afr. Our members of the species may be organized into 2 copiously intergrading ecotypic vars. The aquatic-adapted extreme, var. *stipulaceum* N. Coleman, flowers only when in water; its infls are up to 4 cm, and the lvs are typically glabrous, floating, and elliptic. (*P. natans, P. hartwrightii.*) The terrestrial-adapted extreme, var. *emersum* Michx., grows on land or is strongly emergent from shallow water, and does not produce floating leaves or shoots; its infls are 4–15 cm, the lvs are hairy and acuminate, and the peduncles are glandular-hairy. (*P. coccineum, P. muhlenbergii.*) Var. *amphibium* is strictly Eurasian. (*Persicaria a.*)

17. Polygonum densiflorum Meissner.   Rhizomatous perennial to 1.5 m, often rooting from the swollen nodes; lvs lanceolate, acuminate, mostly 10–25 × 2–5 cm; ocreae smooth, entire or nearly so, not fringed; peduncles strigulose; racemes slender, to 10 cm; ocreolae obconic, truncate, scarcely oblique, commonly overlapping; perianth white, 2.5–3 mm, enlarging to 3–4 mm, conspicuously surpassing the fr, this black, 2–2.5 mm, 70–100% as wide, strongly biconvex. Swamps and shallow water of the coastal plain; s. N.J. to Fla. and Tex., n. in the interior to s. Mo.; W.I. and S. Amer. (*Persicaria portoricensis.*)

Polygonum polystachyum Wallich, the Kashmir plume, is a Himalayan sp. that is reported to spread occasionally from cult. in the ne. part of our range. It would key to *P. densiflorum,* from which it differs in its showy, open-paniculate infls, often wider lvs (to 8 cm wide) that tend to be narrowly truncate or cordate-auriculate at base, and in other respects. It is more nearly related to nos. 37 and 38, but the tepals are not keeled or winged and are scarcely accrescent.

18. Polygonum robustius (Small) Fern.   Coarse smartweed. Robust perennial from a stout rhizome, to 1 m or more; stems often 1 cm thick; lvs lanceolate, 2–3.5(–4.5) cm wide; ocreae fringed with bristles; racemes not much interrupted, the ocreolae commonly overlapping, usually not fringed; pedicels at maturity exserted 2–4 mm; perianth white, glandular-punctate, 3–3.5 mm; achenes black, shining, trigonous, 2.7–3.5 mm. Wet soil and shallow water on the coastal plain or near the coast; N.S. to Fla. and tropical Amer.

19. Polygonum hydropiperoides Michx.   False water-pepper. Rhizomatous perennial, erect or ascending, to 1 m; lvs lanceolate to linear; ocreae always strigose and ciliate; peduncles usually strigose, especially above; racemes slender, erect, often interrupted below; ocreolae separate or scarcely overlapping, obconic, usually short-ciliate; perianth white, greenish, or pink, conspicuously exsert; achenes ovoid, black, smooth and shining, trigonous with rounded angles and concave sides, 1.8–3 mm; $2n=20,40$. Wet soil, beaches, marshes, and shallow water; Que. to B.C., s. to S. Amer. Highly variable, and divisible into 3 vars.:

1 Hairs of the ocreae (not the cilia) adnate at base, erect and appressed; lvs glabrous or somewhat strigose; $2n=40$.

2 Mature achenes included in the perianth; fls pink (white); plate-glands wanting; widespread .... var. *hydropiperoides*.
2 Mature achenes slightly exsert from the incurved tepal-tips; fls creamy or greenish; large, pale,
   plate-shaped glands present on the lower lf-surfaces and sometimes also on the tep; chiefly on
   the coastal plain. (*P. o.; P. o.* var. *adenocalyx.*) .... var. *opelousanum* (Riddell) Stone.
1 Hairs of the ocreae long and usually spreading, enlarged at base, not adnate; plants coarse and stout,
   on the average hairier and with wider lvs than the other vars.; fls ovoid, greenish-creamy; $2n=20$;
   chiefly along the Gulf coast, but extending into our range, mainly in forms transitional to var.
   hydropiperoides, along the coastal plain (*P. s.*) .... var. *setaceum* (Baldw.) Gleason.

20. Polygonum lapathifolium L.   Dock-leaved smartweed. Erect annual to 15 dm, or occasionally prostrate; lvs variable, commonly lanceolate and acuminate, often tomentose beneath; ocreae entire or merely lacerate, not fringed-ciliate; summit of the peduncle and axis of the infl glabrous to densely glandular; racemes numerous, nodding, 1–5 cm; fls or ocreolae ovate; perianth rose, white, or green, 3–4 mm at maturity, 4(5)-lobed to below the middle, the outer lobes becoming obovate, strongly 3-nerved, each nerve divided at the summit into 2 recurved branches; achene lenticular, flat or concave on both sides, 1.7–3.2 mm, 75–90% as wide; $2n=22$. Probably both native and intr., chiefly in moist soil, nearly throughout temperate N. Amer., and in the Old World. (*P. scabrum, P. tomentosum.*)

21. Polygonum pensylvanicum L.   Pennsylvania smartweed. Erect, branching annual to 2 m, or seldom prostrate; lvs lanceolate, acuminate; ocreae becoming lacerate, but not fringed-ciliate; racemes numerous, erect, commonly cylindric, blunt, 1.5–3 cm, ocreolae ovate, entire or minutely ciliate; perianth rose or white, scarcely exceeding the achene, 5-lobed to below the middle, the segments with several inconspicuous, branched and anastomosing veins; achenes lenticular, concave on both sides, 2.6–3.4 mm, 85–100% as wide; $2n=44$. Fields and waste ground, especially in moist, rich soil; N.S. and Que. to Minn. and S.D., s. to Fla. and Tex. Highly variable, but scarcely divisible into significant vars. Heterostylic southern plants, reaching our range in s. Ill., have been segregated as *P. bicorne* Raf. (*P. longistylum* Small), but the populational relationship of these plants to the widespread, homostylic *P. pensylvanicum* remains to be elucidated.

22. Polygonum hydropiper L.   Water-pepper. Erect or spreading peppery-tasting annual to 6 dm, simple or branched, essentially glabrous, often reddish; lvs narrowly lanceolate to lance-ovate, sometimes strigulose on the veins; ocreae glabrous or strigose, short-ciliate, those from the middle of the stem upward distended by concealed cleistogamous fls; racemes slender but usually continuous, commonly arched or nodding at the summit; ocreolae glabrous, short-ciliate, obliquely truncate, commonly overlapping; fls greenish; tep 4(5), glandular-punctate, usually white-margined; achenes lenticular or more often obtusely trigonous, dark brown to black, dull, 2.2–3.3 mm; $2n=20, 22$. Native of Europe, commonly found in wet soil from Que. to B.C., s. to Ala. and Calif. (*Persicaria h.*)

23. Polygonum punctatum Elliott.   Dotted smartweed. Slender annual or perennial to 1 m, simple or branched, erect or ascending; lvs narrowly lanceolate or elliptic, glabrous, to 20 × 2 cm; ocreae glabrous or strigose; racemes slender, erect or arched, to 10 cm, much interrupted, especially below, where the internodes may be 2–3 cm; fls greenish; tep 5, 2 mm; pedicels exsert 1–2 mm at maturity; achenes smooth and shining, lenticular or trigonous, 2.4–3 mm, two-thirds as wide; $2n=44$. Wet soil, open swamps, and shallow water; Que. to Fla., and w. to the Pacific; also in tropical Amer. (*Persicaria p.; Polygonum acre.*) Two geographically coextensive vars. of doubtful significance:

1 Rhizomatous perennial; infls ± contiguous, relatively uniform in size and terminating leafy branches;
   achenes predominantly trigonous .... var. *punctatum*.
1 Mostly taprooted annuals; some infls notably elongate, interrupted, branch-like, from lower nodes;
   achenes predominantly lenticular (var. *leptostachyum*) .... var. *confertiflorum* (Meissner) Fassett.

24. Polygonum careyi Olney.   Erect, branching, glandular-pubescent annual to 12 dm; lvs lanceolate; ocreae spreading-hirsute and bristly-fringed; infl loosely branched; peduncles stipitate-glandular; racemes cylindric, drooping, 3–6 cm; perianth pink or rose, 2.5–3 mm; stamens 5(–8); achenes black, smooth and shining, ovoid or commonly obovoid, 1.8–2.4 × 1.5–2 mm, thickly biconvex. Moist or wet ground, fields, roadsides, and meadows; Me. to Wis. (and reputedly Minn.), s. to Del., Ind., and Ill. (*Persicaria c.*)

25. Polygonum cespitosum Blume.   Freely branched, soon decumbent annual to 1 m, glabrous or partly strigose; lvs thin, dark green, lanceolate to elliptic or oblanceolate; ocreae strigulose or glabrous, ciliate with bristles 5–10 mm, racemes dense, 2–4 cm × 5 mm; ocreolae overlapping, their cilia mostly 2–3.5 mm, often equaling or surpassing the fls; achenes black, smooth and shining, trigonous, 2–2.5 mm; $2n=22, 24$. Native of e. Asia, now a common weed, especially in moist soil, in much of our range, especially eastward. (*Persicaria longiseta.*) Most or all of our plants represent the var. *longisetum* (De Bruyn) Stewart. Var. *cespitosum*, with shorter bristles on the ocreae and ocreolae, may also be expected.

26. Polygonum persicaria L.   Lady's thumb. Erect or ascending annual to 8 dm, often much branched, glabrous or nearly so; lvs narrowly lanceolate, often with a dark blotch; ocreae membranous, strigillose, short-ciliate; spikes dense, straight, cylindric, 1–4 cm × 7–12 mm; ocreolae obliquely truncate, entire or with a few short cilia to 1 mm, pedicels mostly included; perianth pink or rose, 2–3.5 mm; achenes lenticular or occasionally trigonous, black, smooth and shining, broadly ovate, 1.8–3 mm, averaging four-fifths as wide; $2n=44$. Native of Europe, now a common weed throughout our range, s. to Fla. and Tex., and w. to Alaska and Calif. (*P. dubium, P. maculosum; P. puritanorum; Persicaria persicaria.*)

27. Polygonum orientale L. Prince's feather. Erect, branching, pubescent annual to 2.5 m; lvs ovate, broadly rounded at base; ocreae villous, ciliate, often with a spreading, herbaceous collar; infl large; peduncles densely hairy; racemes commonly drooping, dense, cylindric, 3–8 cm, fls rose to crimson; achenes flat, lenticular, 2.8–3.5 mm, about as wide, beakless, smooth and shining; $2n=22$, 24. Native of India, often escaped in waste places near gardens. (*Persicaria o.*)

28. Polygonum nepalense Meissner. Lax annual to 4 dm; lvs 1–5 cm, often splashed with red, glabrous or often bristly-hirsute beneath, ovate or deltoid, truncate or abruptly contracted to the winged petiole, which tends to be expanded and clasping at base; ocreae often bristly below; peduncles often stipitate-glandular; infl capitate, the greenish-white bracts equaling or exceeding the 8–30 fls; perianth greenish-white or antho-cyanic, 2 mm at anthesis, accrescent to 3 mm and strongly surpassing the achene; achenes lenticular, red-brown, reticulate-ridged, elliptic, 1.5–2 mm; $2n=48$. Moist, shady places; native of Asia, sometimes grown as a ground cover, and occasionally escaped into relatively stable habitats in N.Y., Conn., and Pa.

Polygonum capitatum Buch.-Ham., a related sp., is reported to escape occasionally from cult. It is a rhizomatous perennial with short-petiolate or subsessile lvs.

29. Polygonum bungeanum Turcz. Erect, taprooted annual to 7 dm, with the aspect of sect. *Persicaria*; stem beset with scattered short, stout, recurved prickles, otherwise largely glabrous except for the conspic-uously stipitate-glandular branches of the infl; lvs lanceolate or narrowly elliptic, 5–15 × 1–2.5 cm, tapering to the short petiole, rough-strigose along the midrib and main veins at least beneath, otherwise largely glabrous; ocrea coarsely antrorse-strigose, but with few or no cilia around the top; racemes lax, rather short; mature perianth 3.5–4 mm, constricted at the base; achenes plump, dark, dull, 2.5–3 mm long and wide. Native of e. Asia, locally intr. as a weed in Minn., and to be expected elsewhere.

30. Polygonum sagittatum L. Arrow-lvd tearthumb. Slender annual 1–2 m, reclining on other plants, or erect when young, the stems 4-angled, reflexed-prickly as are usually also the petioles and peduncles and the midrib beneath, lvs sometimes pubescent on margins, lanceolate to elliptic, 2–10 cm, to 2.5 cm wide, sagittate at base, the lobes directed downwards; infl of short, head-like, long-peduncled terminal and axillary racemes seldom over 1 cm; perianth pink to white or green; style 1.5 mm, trifid to the middle; achene trigonous, 2.2–3 mm; $2n=40$. Marshes and wet meadows; Nf. and Que. to Sask., s. to Ga. and Tex. (*Tracaulon s.*)

31. Polygonum arifolium L. Halberd-lvd tearthumb. Much like no. 30; lvs hastate, to 20 × 15 cm, finely stellate beneath, the basal lobes triangular, divergent; peduncles commonly glandular above; perianth 2.5 mm, pink; styles 2, 0.5 mm; achene lenticular, 3.5–5 mm. Marshes, swamps, and wet meadows; N.B. to Minn., s. to Ga. and Mo. Aug–Sept. (*Tracaulon a.*)

32. Polygonum perfoliatum L. Mile-a-minute. Annual, clambering and with reflexed-prickly stems and petioles as in the preceding 2 spp., climbing to several m; lvs thin, glabrous deltoid to hastate-cordate, shortly peltate at base, 3–8 cm, as wide as or wider than long, palmately or pinnipalmately veined; ocreae foliaceous-expanded; upper lvs represented only by the conspicuous, perfoliate ocreae; racemes spike-like, 1–2 cm, few-fld; stigmas 3; perianth 3–5 mm, persistent, thickening to form a fleshy, berry-like, iridescent-blue covering over the rotund-trigonous achene; $2n=24$. Native of e. Asia, intr. and established in Pa., Md., and W.Va.; to be expected to spread.

33. Polygonum virginianum L. Jumpseed. Erect, rhizomatous perennial 5–10 dm; lvs lanceolate to ovate, to 15 cm, acuminate, rough-hairy to glabrous; petioles to 2 cm; ocreae pubescent and long-ciliate; racemes very slender, terminal, 1–4 dm, the ocreolae well separated below, becoming contiguous or overlapping upwards, 1–3-fld; pedicels divergent, jointed at the summit; tep 4, greenish-white or suffused with pink, 2.5 mm, the 2 lateral ones external and somewhat smaller than the median ones, scarcely changed in fr; achene lenticular, ovate, 4 mm; styles 2, persistent, indurate, deflexed, connivent, hooked at the tip; $2n=44$. Moist woods; N.H. to Minn. and Nebr., s. to Fla. and Tex. (*Tovara v.*)

34. Polygonum convolvulus L. Black bindweed. Trailing or twining annual to 1 m; stem, petioles, and often the lf-veins scabrellate in lines; lvs sagittate to triangular-cordate, ocreae smooth; racemes interrupted, naked or with a few small lvs at base, 2–6 cm; fls in clusters of 3–6; pedicels 1–2 mm, jointed above; perianth 1.5–2 mm, green outside, white inside; achene dull black, 3–4 mm, closely invested but not exceeded by the perianth; outer tep often with narrowly winged midrib; styles united; $2n=20$, 40. Roadsides, railways, and waste ground throughout our range; native of Europe. May–Oct. (*Bilderdykia c., Tiniaria c.*)

35. Polygonum cilinode Michx. Fringed bindweed. Twining or trailing to occasionally erect perennial, to 2 m, pubescent to subglabrous; lvs ovate to triangular-ovate with deeply cordate base; ocreae very oblique, reflexed-bristly at base; racemes long-peduncled, mostly branched, 4–10 cm, the small fl-clusters remote; fls white, 1.5–2 mm; achene very glossy, black, closely invested but scarcely surpassed by the perianth, the 3 outer tep rarely narrowly winged; styles separate, divergent; $2n=20$, 40. Dry woods and thickets. Nf. and Que. to n. Minn., s. to Pa. and s. Mich., and in the mts. to N.C. (*Bilderdykia c.; Tiniaria c.*)

36. Polygonum scandens L. False buckwheat. Twining perennial to 5 m, the stem sharply angled, often scabrellate on the angles; lvs oblong-ovate or broadly cordate, acuminate, deeply cordate at base; ocreae very oblique, smooth, racemes from most upper axils, usually unbranched, leafless or nearly so, 5–11 cm, inter-rupted; pedicels winged above the joint; perianth 1.5–2.5 mm, white, achene very glossy black, 3–5 mm, closely invested and much exceeded by the perianth; outer 3 tep at maturity with broadly winged midrib, the whole fr, measured from the joint in the pedicel, 7–15 mm; $2n=20$, and some higher number. Moist woods, thickets, and roadsides; Que. to N.D., s. to Fla. and Tex. (*Bilderdykia s.; Tiniaria s.*) Three vars.:

1 Perianth 10–15 mm long at maturity (measured from the joint), with flat or crisped (occasionally
   crenate) wings; native, widespread in our range .... var. *scandens.*
1 Perianth 7–10 mm long at maturity.
  2 Wings of the perianth flat, mostly entire, well developed; native of Europe, occasionally intr. in
    our range, especially eastward. (*P. dumetorum.*) .... var. *dumetorum* (L.) Gleason.
  2 Wings of the perianth conspicuously and irregularly crenate or lacerate, relatively poorly developed,
    especially distally; native from N.E. to Io., s. to Fla. and Tex. (*P. cristatum.*) ....
    .... var. *cristatum* (Engelm. & A. Gray) Gleason.

Polygonum aubertii L. Henry, the Silver lace vine, occasionally escapes from cult. It is much like *P. scandens,* but is more woody and has a profusion of openly branched, paniculate, creamy-white infls.

37. **Polygonum cuspidatum** Sieb. & Zucc.   Japanese knotwood, Mexican bamboo. Stout, rhizomatous perennial, 1–3 m, stems terete, becoming striate-ridged; lvs broadly ovate, 8–15 × 5–12 cm, abruptly acuminate, broadly truncate at base, the basal angles prominent; racemes numerous from the upper axils, often branched, forming a series of panicles 8–15 cm; fls functionally unisexual, the plants dioecious, perianth white or greenish-white; outer tep narrowly winged along the midrib; styles 3; achenes trigonous, 3 mm, invested by the enlarged perianth, the outer tep broadly winged; $2n$=44, 88. Native of Japan, escaped from cult. and well established in our range. (*Pleuropterus zuccarinii; Reynoutria japonica.*)

38. **Polygonum sachalinense** F. W. Schmidt.   Giant knotwood. Much like no. 37; stems angular; lvs avg larger, to 30 cm, tending to be more gradually acute, ovate with cordate base, the basal lobes broadly rounded; fls perfect; $2n$=44. Native of e. Asia, rarely escaped from cult. in our range. (*Reynoutria s.*)

6. **FAGOPYRUM** Miller.   Fls perfect; tep 5, petaloid, unequal in width, connate toward the base; stamens 8; ovary 3-angled; styles 3, with capitellate stigma; fr a trigonous achene, much exceeding the persistent but scarcely enlarged cal; embryo within the endosperm, with broad, often folded cotyledons; annual herbs with alternate, petiolate, broadly triangular lvs and obliquely truncate ocreae; fls white or pink, pediceled in fascicles disposed in racemes arising from the axils of short bracts. 15, temp Eurasia.

1. **Fagopyrum esculentum** Moench.   Buckwheat. Annual, 2–6 dm, the stem pubescent in lines above; lvs broadly triangular-hastate, the lower long-petioled; fl-clusters usually crowded and compact to form a terminal, corymbiform infl; tep elliptic obtuse, 2–3 mm, achene smooth and shining, 5–7 mm, with smooth, entire angles, much exceeding the tep. $2n$=16. Commonly escaped from cult., but not long persistent. June–Sept. (*F. sagittatum.*)
   Fagopyrum tataricum (L.) Gaertner, India wheat, seldom escapes from cult. Its racemes are more scattered along the stem, and the achenes are smaller and dull, with irregularly protuberant angles.

7. **BRUNNICHIA** Banks.   Fls perfect; tep 5, connate at base; stamens 8; styles 3, with 2-lobed stigma; fr an achene enclosed by the accrescent, coriaceous perianth; base of perianth and its pedicel above the joint winged on one side, the whole resembling a samara; vines, climbing by tendrils terminating short, few-lvd lateral branches, with alternate, petiolate lvs; fls in small fascicles spicate on a long axis; ocreae none; stipules represented by a narrow pubescent ring around the stem. Monotypic.

1. **Brunnichia cirrhosa** Gaertner.   Buckwheat-vine. High-climbing woody vine; lvs ovate or ovate-oblong, to 15 cm, acute or acuminate, truncate or subcordate at base; lower spikes axillary, the upper forming a loose, leafless panicle; pedicels 5–10 mm, jointed at the lower third, dilated above; perianth 3–4 mm at anthesis; fr brown, 2.5–3.5 cm × 4–8 mm, corky, the coriaceous tep connivent; achene 8–10 mm, roundly trigonous below, sharply so toward the acuminate tip. $2n$=48. Wet woods and margins of swamps on the coastal plain; s. Ill. and Mo. to S.C., Fla., and Tex. July. (? *B. ovata* (Walter) Shinners, an older but insufficiently identified name.)

FAMILY **PLUMBAGINACEAE**, the Leadwort Family

Fls regular, perfect, hypogynous, pentamerous throughout; cal dry and ± scarious, persistent, the tube often conspicuously ribbed and with membranous intervals, the lobes often showy and somewhat petaloid; corolla sympetalous or seldom of essentially distinct, clawed pet, often persistent; stamens as many as and opposite the corolla-lobes (or pet);

ovary unilocular, with a solitary basal ovule on a long funiculus; fr dry, indehiscent or circumscissile; mostly herbs or low shrubs with alternate (or all basal), simple, entire lvs, the herbage with characteristic scattered, often depressed chalk-glands; fls in panicles or racemes or cymose heads. 12/400.

1. **LIMONIUM** Miller.   Sea-lavender. Cal tubular to obconic or funnelform, scarious except at base, shallowly 5-lobed, the lobes sometimes alternating with shorter teeth; pet 5, nearly distinct, long-clawed; stamens 5, each attached to the base of a pet; styles 5; fr a utricle; scapose perennial herbs with basal lvs and large branching infls, the fls in clusters of 1–3, enveloped at base by 2–3 scarious bracts. 200+, cosmop.

1. **Limonium carolinianum** (Walter) Britton.   Taprooted, mostly 2–7 dm tall; lvs firm, glabrous, the blade mostly elliptic to obovate, spatulate, or oblanceolate and 1–4(–7) cm wide; infl ± widely branched above the middle, with a few scale-like bracts; fls in small, sessile, secund clusters; cal scarcely surpassing the bracts, mostly (4–)5–6(–7) mm including the narrow, white, ca 1 mm lobes, glabrous or in ours more often ± strigose, especially on the nerves below; cor lavender, about equaling or a little shorter than the cal. Salt marshes along the coast; Lab. to ne. Mex. July–Sept. (*L. nashii*.)

FAMILY **THEACEAE**, the Tea Family

Fls regular, mostly perfect and hypogynous or nearly so, usually 5-merous; sep imbricate, distinct or often basally connate; pet distinct or only basally connate, in ours large and imbricate; stamens mostly numerous, the filaments often united at the base into a ring or into 5 bundles opposite and adnate to the base of the pet; ovary (2–)3–5(–10)-locular; styles or style-branches (or stigmas) as many as the locules; ovules (1)2 to more often several or many in each locule; fr in ours a loculicidal capsule; trees or shrubs with alternate, simple, exstipulate lvs. 40/600, warm reg.

1. **STEWARTIA** L.   Pet 5(6), white, barely connate at base, pubescent beneath; stamens numerous, the filaments united at base into a narrow ring adnate to the base of the corolla; ovary 5-locular; capsule woody, with 1–4 seeds per locule, splitting from the top down; sep persistent into fr; shrubs or trees with deciduous lvs and large white fls short-pedicelled or subsessile in the axils. 6, the other 4 in e. Asia.

1 Styles distinct; seeds dull, thin-edged or narrowly winged .... 1. *S. ovata.*
1 Styles united, only the stigma 5-lobed; seeds shining, plump .... 2. *S. malachodendron.*

1. **Stewartia ovata** (Cav.) Weatherby.   Mountain Stewartia. Shrub to 5 m, lvs elliptic to oblong-ovate, 6–15 cm, distinctly acuminate, obscurely to conspicuously serrulate, broadly acute to rounded below, hairy beneath; fls subsessile, 7–10 cm wide; filaments purple or white; styles 5; fr ovoid, 2 cm, sharply 5-angled, pointed, villous; seeds dark, dull, flattish, thin-edged or (especially at one end) narrowly winged; $2n=30$. Mt. woods; e. Ky. to n. Va. and n. Ala. June. (*S. pentagyna; Malachodendron p.*)

2. **Stewartia malachodendron** L.   Silky S. Shrub to 6 m; lvs elliptic to obovate-oblong, 4–10 cm, acute or abruptly acuminate at both ends, obscurely serrulate, finely hairy beneath; fls subsessile, 6–9 cm wide; filaments purple; style thick, much shorter than the stamens, with a 5-lobed stigma; fr depressed-globose, 1.5 cm; seeds shining brown, plump, with angular (not thin or winged) margins; $2n=30$. Moist woods on the coastal plain, seldom more inland; e. Va. to Fla. and La.

FAMILY **ELATINACEAE**, the Waterwort Family

Fls regular, perfect, hypogynous, 2–5-merous, pet small, stamens in 1 or 2 series; ovary 2–5-locular, with axile or basal-axile placentas; styles 2–5; fr a many-seeded septifragal capsule; seeds without endosperm; herbs, mostly small, of wet soil or shallow water, with simple, opposite (or whorled), stipulate, entire or toothed lvs and small, axillary, solitary or cymose fls. 2/40, widespread.

1 Fls 5-merous; plants glandular-puberulent .... 1. *Bergia.*
1 Fls (in our spp.) 2–3-merous; plants glabrous .... 2. *Elatine.*

1. BERGIA L. Fls 5-merous; sep distinct; stamens 5 or 10; ovary 5-locular, the placentas typically extending to its summit; fr globose or ovoid; seeds numerous, minute, cylindric or clavate, obscurely areolate; small marsh-herbs with axillary fls and firm-walled frs. 20+, widespread.

1. Bergia texana (Hook.) Seub. Branched, diffuse or ascending annual 1–3 dm, glandular-puberulent; lvs elliptic-oblong to oblong-spatulate, 1–3 cm, narrowed to the base, glandular-denticulate; stipules lanceolate, glandular-ciliate; fls 1–3 in the axils on pedicels 1–3 mm; sep lance-acuminate, 2.5–3 mm, scarious-margined, rough on the keeled midnerve; pet oblong, shorter than the sep; stamens usually 5; fr globose, 2 mm; seeds 0.3–0.5 mm. Muddy or sandy shores; s. Ill. and Kans. to Wash. and Calif.

2. ELATINE L. Waterwort. Sep and pet each 2–4; stamens as many (in our spp.) or twice as many as the sep; ovary 2–4-locular, the placentas axile or basal-axile; seeds several per cell, cylindric, minutely areolate; small aquatic annuals with inconspicuous axillary fls and thin-walled frs. 20, widespread.

1 Seeds clearly axile, attached to the placenta at different levels; fls mostly 3-merous .... 1. *E. triandra.*
1 Seeds basal-axile; fls mostly 2-merous .... 2. *E. minima.*

1. Elatine triandra Schkuhr. Creeping or floating annual; lvs linear to obovate, 3–8 mm; fls mostly 3-merous; seeds axile, set at different levels in the fr, marked with transversely elongate, 6-sided areoles, the angular ends of the areoles dovetailed with those of adjacent rows, the longitudinal ridges appearing narrow or broken; 2n=ca 40. Three vars.: Var. *triandra* has linear to spatulate, often retuse lvs, and seeds with 16–26 areoles in each longitudinal row; Wis. and N.D. to Wash. and Calif., also in Eurasia, and intr. in Me. Var. *americana* (Pursh) Fassett has obovate, rarely retuse lvs, otherwise like var. *triandra*; muddy tidal shores, Que. and N.B. to Va., and also in Mo. Var. *brachysperma* (A. Gray) Fassett has obovate lvs, and seeds with 9–15 areoles in each longitudinal row; s. O. and c. Ill. to w. U.S.

2. Elatine minima (Nutt.) Fischer & C. A. Meyer. Annual, creeping and forming small mats on mud, with branches to 5 cm; lvs oblong to narrowly obovate, rarely over 4 mm; fls usually 2-merous; seeds basal-axile, erect, all terminating at about the same level, marked with narrowly elliptic areoles, the ends of the areoles not dovetailed with those of adjacent rows, and the longitudinal strips between the rows appearing distinct and straight under a hand-lens. Nf. to N.Y., w. to Ont. and Minn., s. along the coast to Md. and Va.

FAMILY CLUSIACEAE, the Mangosteen Family

Fls perfect, hypogynous, regular (at least as to the cor), polypetalous, the perianth mostly 4–5-merous; stamens typically numerous and centrifugal, varying to only 5, the elongate filaments sometimes basally connate; ovary of 2–5 carpels, unilocular or partly or wholly 2–5-locular, the placentas accordingly parietal or axile; styles as many as the carpels, sometimes united below; fr a many-seeded capsule, usually septicidal; herbs or shrubs with simple, opposite, exstipulate, entire lvs ordinarily beset with translucent internal glands and often also superficially punctate; fls cymose (or solitary), usually yellow. Our plants all belong to the subfamily Hypericoideae, as principally described above. 50/1200, trop. and N. Temp.

1 Pet yellow (or coppery); fls without hypogynous glands .... 1. *Hypericum.*
1 Pet pink or flesh-colored; 3 hypogynous glands alternating with 3 fascicles of 3 stamens each .... 2. *Triadenum.*

1. HYPERICUM L. St. John's-wort. Perianth 4–5-merous; sep persistent, often unequal, sometimes only 2 fully developed; pet convolute, typically yellow, occasionally more orange or coppery; stamens 5 to more often ± numerous, distinct or ± connate at base, often into 3 or 5 definite bundles; no hypogynous glands; ovary 2–5-carpellate, unilocular or partly or wholly plurilocular, often with intruded partial partitions; styles persistent, often connate below; fr capsular; seeds short-cylindric, areolate; herbs or shrubs with cymose (rarely solitary) fls. (*Ascyrum, Sarothra.*) 400, mainly N. Temp.

1 Perianth mostly 4-merous; shrubs or subshrubs (*Ascyrum*).
  2 Styles mostly 3(4); lvs elliptic-oblong, 7–15 mm wide .... 1. *H. stans.*
  2 Styles mostly 2; lvs linear to oblanceolate, mostly 1.5–6 mm wide.
    3 Plants erect, mostly single-stemmed, mostly (3–)5–12(–15) dm .... 2. *H. hypericoides.*
    3 Plants decumbent, many-stemmed, seldom over 3 dm .... 3. *H. stragulum.*
1 Perianth mostly 5-merous; herbs or shrubs.
  4 Shrubs; fr 3–5-locular, or with 3–5 strongly intruded placentas.
    5 Styles and locules mostly 5, occasionally only 4, rarely only 3 in no. 5.
      6 Capsule not lobed, or only slightly lobed; fls seldom more than 7 .... 4. *H. kalmianum.*
      6 Capsule deeply lobed; fls usually 15 or more .... 5. *H. lobocarpum.*
    5 Styles mostly 3, seldom 4.
      7 Lvs and sep articulate at base; infls of 1–many fls, appearing leafy.
        8 Fls (1–)3–many per infl; fr with intruded placentas that meet in the center but do not join.
          9 Fr up to 6 × 3 mm; fls 7–many per infl .... 6. *H. densiflorum.*
          9 Fr at least 7 × 3.5 mm; fls (1–)3–7 per infl .... 7. *H. prolificum.*
        8 Fls terminal and solitary (–3); placentas in fr intruded but not meeting .... 8. *H. frondosum.*
      7 Lvs and sep not articulate; infls many-fld, appearing naked .... 9. *H. nudiflorum.*
  4 Herbs (some spp. with strictly unilocular fr are woody at base).
    10 Styles united or closely connivent below, persistent as a single straight beak on the fr and
        eventually split by its dehiscence; stigmas 3(4), minute; stamens many.
      11 Plants of wet places, rhizomatous, with scattered stems.
        12 Sep oblanceolate to narrowly obovate; lvs plane, 2–3 times as long as wide .... 10. *H. ellipticum.*
        12 Sep lanceolate to ovate; lvs revolute, 4–6 times as long as wide .... 11. *H. adpressum.*
      11 Plants of dry places, the stems commonly clustered, somewhat woody below.
        13 Sep equal or nearly so, 2.5–5 mm; stamens 45–85 .... 12. *H. sphaerocarpum.*
        13 Sep very unequal, the outer pair the larger, 7–12 mm; stamens 120–200 .... 13. *H. dolabriforme.*
    10 Styles separate to the base or nearly so, often divergent; fr not beaked; stigmas capitate; stamens
        few to many.
      14 Styles 5; fr 5-locular .... 14. *H. pyramidatum.*
      14 Styles 3(4); fr unilocular or trilocular (4-locular).
        15 Fr wholly 3(4)-locular; stamens connate at base into 3 or 5 fascicles; cor marcescent.
          16 Branches sharply ridged below each lf; seeds 1–1.3 mm .... 15. *H. perforatum.*
          16 Branches terete or nearly so; seeds less than 1 mm.
            17 Fr 4–7 mm, conspicuously glandular; pet black-dotted; widespread.
              18 Fls small; sep 2.5–4 mm; pet 4–7 mm; styles 2–4 mm .... 16. *H. punctatum.*
              18 Fls larger; sep 5–7 mm; pet 8–12 mm; styles 6–10 mm .... 17. *H. pseudomaculatum.*
            17 Fr 8–11 mm, essentially glandless; pet not markedly black-dotted; sw. Va. 18. *H. mitchellianum.*
        15 Fr unilocular; stamens not connate into fascicles; nor in most spp. deciduous.
          19 Foliage pubescent; sep ciliate .... 19. *H. setosum.*
          19 Foliage glabrous.
            20 Stamens 50–80; styles 2–4 mm .... 20. *H. denticulatum.*
            20 Stamens 5–22; styles ca 1 mm or less.
              21 Infl obviously cymose, each node generally with a terminal fl and 2 opposite
                lateral branches; lvs commonly 3–7-nerved from the base.
                22 Fr ellipsoid, rounded above; sep broadest near the middle.
                  23 Bracts subulate; sep acute, about equaling the fr .... 21. *H. mutilum.*
                  23 Bracts elliptic, leaf-like; sep obtuse, much shorter than the fr .... 22. *H. boreale.*
                22 Fr ovoid or conic, narrowed more gradually above than below; sep
                  broadest well below the middle.
                  24 Lvs deltoid-lanceolate to deltoid-ovate, broadly rounded or sub-
                    cordate at base; outer nerves diverging at about right angles
                    to the midrib .... 23. *H. gymnanthum.*
                  24 Lvs linear to lanceolate or oblanceolate, narrowed or rounded at
                    base, 1–7-nerved, the outer nerves strongly ascending from
                    the base.
                    25 Lvs 5–7-nerved, broadly obtuse or rounded at base .... 24. *H. majus.*
                    25 Lvs 1–3-nerved, tapering to the base .... 25. *H. canadense.*
              21 Infl apparently racemose, the branches often alternate, bearing a series of
                solitary axillary fls; lvs linear or subulate, strictly 1-nerved (*Sarothra*).
              26 Lvs linear, 5–20 mm; fr but little longer than the sep .... 26. *H. drummondii.*
                26 Lvs subulate, appressed, 1–4 mm; fr 2–3 times as long as the sep .. 27. *H. gentianoides.*

1. Hypericum stans (Michx.) P. Adams & Robson. St. Peter's-wort. Erect shrub 3–8(–10) dm; lvs firm, articulate at the slightly clasping base, elliptic-oblong, 1.5–3.5 cm × 7–15 mm, rounded above; fls distinctly pediceled (3–10 mm) above the subulate or lanceolate bracteoles; sep 4, the 2 outer broadly round-ovate or subrotund, 10–18 mm, subcordate at base, 6–7-nerved, the 2 inner narrowly lanceolate, two-thirds as long; pet 4, yellow, obliquely obovate, 12–18 mm; stamens numerous, the filaments distinct or weakly connate at base; styles 3(4), very short; fr unilocular, exserted from the persistent cal. Sandy soil; L.I. and N.J. to Fla., w. to s. Ky., e. Okla., and e. Tex. July, Aug. (*Ascyrum s.*)

2. Hypericum hypericoides (L.) Crantz. St. Andrew's cross. Shrub (3–)5–12(–15) dm, generally with a single main stem below and freely branched above; lvs firm, articulate at the narrow base, obtuse or rounded above, 1–3 cm × 1.5–6 mm, variable in shape but typically linear-elliptic to linear-oblong and broadest near the middle; bracteoles subulate, borne within 1 mm of the cal; 2 outer sep mostly 6–11 × 3–12 mm, rounded at base, obscurely 3–5-nerved, the 2 inner much smaller or even obsolete; pet 4, yellow, narrowly oblong-elliptic, 8–11 mm; stamens numerous, the filaments distinct or weakly connate at base; styles 2, very short; fr unilocular, enclosed by or somewhat exserted from the persistent cal. Dry or moist, sandy or rocky soil; N.J. and Md. to Ky., s. Mo., and Okla., s. to Fla., Tex., W.I., and C. Amer. June–Aug. (*Ascyrum h.*)

3. Hypericum stragulum P. Adams & Robson. Decumbent shrub, with several prostrate stems giving rise to numerous erect branches mostly 1–3 dm, forming low, compact mats or mounds 3–4 dm wide, lvs generally uniform in size and shape, typically oblanceolate and broadest above the middle; otherwise much like no. 2. July, Aug. Dry, rocky slopes and moist, rich woods; Nantucket, L.I., and N.J. to Va., N.C., and n. Ga., w. to s. O., se. Kans., e. Okla., and e. Tex. July, Aug. (*Ascyrum hypericoides* var. *multicaule.*)

4. Hypericum kalmianum L. Branching shrub to 1 m; lvs firm, often revolute, articulate at the base, linear to narrowly oblong, 2–4 cm × 3–8 mm; fls 2–3.5 cm wide, in small cymes of 3–7 terminating the branches; stamens very many, distinct; styles (4)5; fr (4)5-locular, narrowly ovoid, 7–10 mm, the style-base persistent as a slender beak; 2*n*=18. Dunes and rocky shores about Lakes Erie, Huron and Michigan; inland in c. Wis. and along the Ottawa R. in Que. Late June–Aug.

5. Hypericum lobocarpum Gattinger. Infl a compound dichasium of usually 15 or more fls, with similar dichasia in the upper 2–4 lf-axils, producing a panicle-like aspect; styles (3–)5; fr mostly 5.5–8 mm, deeply sulcate or lobed, quasi- (3-) 5-locular; otherwise much like no. 6; 2*n*=18. Moist bottom-lands and banks of streams and ditches; se. Mo., s. Ill., and w. Tenn. to Miss., La., e. Tex., and se. Okla. June–Aug. (*H. densiflorum* var. *l.*)

6. Hypericum densiflorum Pursh. Shrub to 2(–3) m, much-branched above; lvs articulate at the base, linear to narrowly elliptic or commonly oblanceolate, 2–4(–5) cm × 5–8 mm; fls 7–many in a large, compound, dichasial cyme, 10–15 mm wide; styles 3(4), connate at the base and forming a beak on the fr; fr mostly 4.5–6 mm, up to 3 mm thick, shallowly or obscurely sulcate, quasi- 3(4)-locular, the partitions meeting in the center but not joined. Wet meadows and moist bottomlands, less often on rocky slopes, coastal plain from N.J. to S.C., and in the mts. from sw. Pa. to n. Ga. and c. Ala. July–Sept.

7. Hypericum prolificum L. Diffusely branched shrub to 2 m, the twigs sharply 2-edged; lvs articulate at the base, linear to oblong or narrowly elliptic, 3–6 cm × 4–15 mm, usually abruptly narrowed to a short petiole; infl a small terminal cyme of mostly 3–7 fls, often with additional cymules from the upper axils; pet 7–10 mm; styles 3(4), connate at the base and forming a beak on the fr; fr mostly 7–14 mm, quasi- 3(4)-locular, the partitions meeting in the center but not joined; 2*n*=18. Many habitats, from swamp-margins to cliffs and woods; N.Y. to s. Mich. and Minn., s. to Ga. and La. July–Sept. (*H. spathulatum.*)

8. Hypericum frondosum Michx. Branching shrub 5–10 dm; lvs articulate at the base, narrowly oblong to elliptic or lance-ovate, 3–6 × 1–1.5 cm; fls solitary or 3 at the ends of the branches, subsessile, or rarely 1 or 2 in the upper axils; sep foliaceous, broadly elliptic or ovate, the larger 10–15 mm; pet 10–20 mm; styles mostly 3(4); fr ovoid conic, 10–13 mm, thick-walled, the placentas intruded but not meeting. Cedar glades, river-bluffs, cliffs, and rocky hills; Tenn. and Ky. to s. Ind., s. Ga., and e. Tex., and occasionally escaped elsewhere. Late May–July. (*H. aureum.*)

9. Hypericum nudiflorum Michx. Branching shrub 5–20 dm; lvs linear-oblong to narrowly elliptic, 3–6 cm, not articulate; fls numerous in open, terminal cymes with lance-subulate bracts only 2–3 mm; sep about equal, narrowly oblong, 3–4 mm; pet 6–10 mm; styles 3(4); fr ovoid, 4–5.6 mm, with evidently intruded placentas; 2*n*=18. Wet soil; se. Va. to Tenn., s. to n. Fla. and s. Miss. June–Aug.

10. Hypericum ellipticum Hook. Rhizomatous herb 2–5 dm, usually simple to the infl; lvs elliptic, 1–3(–4) cm, a third to half as wide, obtuse or rounded at both ends or somewhat tapering below, not revolute; cymes with few–many fls; bracts linear or lanceolate; sep oblanceolate to narrowly obovate, to 6 mm; pet 5–7 mm; stigmas 3(4), minute, fr unilocular (the placentas slightly intruded), ovoid, 5–6 mm, rounded to a short beak; 2*n* reported as 16, 18. Wet shores and marshes; Nf. and N.S. to w. Ont., s. to Conn., N.Y., Mich., Minn., and at higher altitudes to W.Va. and N.C. July–Aug.

11. Hypericum adpressum Barton. Rhizomatous herb 3–8 dm, coarser than no. 10, usually simple to the infl; lvs many, linear-oblong to narrowly elliptic, 3–6 cm × 5–10 mm, revolute, acute or subacute, tapering to the base; infl usually with many fls; bracts minute, subulate; sep lanceolate to ovate, 2–7 mm; pet 6–8 mm; stigmas 3(4), minute; fr ovoid, 4–6 mm, gradually narrowed to the beak, unilocular but with intruded placentas; 2*n*=18. Marshes, shores, and wet meadows; e. Mass. to Ga. and Tenn., and occasionally inland to Ind. and Ill. July, Aug.

12. Hypericum sphaerocarpum Michx. Erect perennial 3–7 dm, somewhat woody below, often rhizomatous but the stems clustered; lvs linear-oblong to narrowly elliptic, 3–7 cm × 5–15 mm, with or without evident lateral veins, acute or obtuse, tapering to the base; infl much-branched and usually compact; bracts lanceolate; sep lanceolate to ovate, equal or nearly so, 2.5–5 mm; pet 5–9 mm; stamens 45–85; stigmas 3(4), minute; fr globose to ovoid, 5–7 mm, slender-beaked, strictly unilocular; seeds 2–2.7 mm. Upland woods, prairies, and barrens; O. and Ky. to Io., Kans., and Okla., s. to Ala. and Ark. June–Aug. (*H. cistifolium,* misapplied.)

13. Hypericum dolabriforme Vent. Spreading or ascending perennial 2–5 dm, somewhat woody below, often rhizomatous but the stems clustered; lvs linear to linear-oblong, 2–4 cm, without evident lateral veins; cymes compact and few-fld; sep very unequal, the outer pair ovate or broadly lanceolate, 7–12 mm, the others smaller; pet 9–13 mm; stamens 120–200; stigmas 3(4), minute; fr ovoid, 5–8 mm; seeds 1.5–1.8 mm. Limestone outcrops, cedar-glades, etc.; Ky. and s. Ind. to Ga. June–Aug.

14. Hypericum pyramidatum Aiton. Branched perennial herb 7–15 dm; lvs lanceolate to elliptic, 4–10 cm, acute or obtuse, sessile and sometimes clasping; fls few, chiefly solitary at the ends of the branches, 4–6 cm wide; stamens very numerous, united at base into 5 sets; styles 5, united below but not persistent; stigmas capitate; fr ovoid, 15–30 mm, 5-locular. Moist soil; Que. to Minn., s. to Pa., Ind., and Kans. June–Aug. (*H. ascyron*, misapplied.)

15. Hypericum perforatum L. Common St. John's-wort. Perennial, 4–8 dm, with numerous very leafy decussate branches, these sharply ridged below the base of each lf; lvs sessile, linear-oblong, 2–4 cm on the main axis, half as large on the branches; fls numerous, forming a large, rounded or flattened compound cyme; sep narrowly lanceolate, acuminate, 4–6 mm, with few or no black glands; pet oblong, 8–12 mm, black-dotted near the margin; stamens many, fascicled; stigmas capitate; fr 3-locular; seeds 1–1.3 mm; $2n=32$. Native of Europe, abundant as a weed in fields, meadows, and along roads throughout much of the U.S. and s. Can. June–Sept.

16. Hypericum punctatum Lam. Spotted St. John's-wort. Erect perennial 5–10 dm, with few branches below the infl; lvs oblong-elliptic to narrowly oblong, the larger 4–6 cm and over 1 cm wide, blunt or even retuse; infl usually small, crowded; fls short-pediceled, 8–15 mm wide; sep heavily dotted and lined with black, ovate-oblong, obtuse or broadly acute, 2.5–4 mm; pet 4–7 mm, copiously beset with amber glands; seeds under 1 mm; $2n=16$. Moist or dry soil, fields, and open woods; Que. to Minn., s. to Fla., Miss., and Okla. June–Aug. (*H. subpetiolatum.*)

17. Hypericum pseudomaculatum Bush. Erect perennial 4–8 dm, branched above; lvs sessile, lanceolate or narrowly elliptic to lance-ovate, the lower obtuse or subacute, the upper acute; infl many-fld, less compact than no. 16; fls 15–25 mm wide, sep lanceolate, acute or acuminate, 5–7 mm, conspicuously black-dotted; pet 8–12 mm, conspicuously dotted; stamens many, fascicled; styles 6–10 mm, seldom persistent; stigmas capitate; fr 5–7 mm, glandular as in no. 16; $2n=16$. Moist or dry soil; Ill. to Okla., s. to Fla. and Tex. June–Aug.

18. Hypericum mitchellianum Rydb. Sparingly branched perennial 4–8 dm; lvs sessile and somewhat clasping; ovate-oblong, the main ones 4–6 cm, half as wide, obtuse; fls comparatively few, in small, crowded cymes; sep 6–10 mm, lance-acuminate, marked with long black lines, but seldom dotted; pet 6–12 mm, scarcely dotted; stamens many, fascicled; styles 3–5 mm; fr 3-locular, ovoid, 8–11 mm, essentially glandless; $2n=16$. Moist slopes at high altitudes in the mts.; sw. Va., w. N.C., and e. Tenn. July.

19. Hypericum setosum L. Pubescent throughout; stems 3–8 dm, simple or virgately branched above; lvs numerous, erect, lanceolate to ovate, 5–15 mm, gradually reduced upwards; sep ciliate; pet 5–7 mm; stamens 20–40; styles 3(4), 1.5–2 mm; otherwise much like no. 20; $2n=12$. Bogs and wet soil on the coastal plain; se. Va. to Fla. and La. June–Sept.

20. Hypericum denticulatum Walter. Coppery St. John's-wort. Glabrous perennial herb 2–7 dm; cymes small and few-fld to large and much-branched; bracts lanceolate or subulate; cal erect and somewhat ac-crescent; pet 5–10 mm; coppery-yellow, persistent and inflexed around the fr; stamens mostly 50–80, some-what connate at the base; but not fascicled; fr ovoid, 3–5 mm, unilocular; styles 3, distinct, persistent, 2–4 mm; $2n=24, 48$. Wet woods, bogs, and marshes. July–Sept. Var. *denticulatum*, occurring chiefly on the coastal plain from N.J. to Ga. has elliptic to subrotund lvs 8–20 mm, one-third to two-thirds as wide, typically appressed and shorter than the internodes; the stem is usually simple and virgate, with a small, few-fld cyme, and the sep are oblanceolate to obovate, varying to sometimes lanceolate or ovate. Var. *acutifolium* (Elliott) S. F. Blake, is more inland (in our range), occurring from Md., W.Va., s. O., s. Ind., and s. Ill., s. to La. and n. Fla.; it has narrower lvs commonly 2–3(–4) cm and a fifth to a third as wide, typically ascending or spreading, and seldom shorter than the internodes; the stem is usually more branched above, with an open, many-fld cyme, and the sep are lanceolate to narrowly ovate. (*H. acutifolium, H. virgatum.*)

21. Hypericum mutilum L. Erect annual or perennial 1–8 dm, usually much-branched above; lvs lanceolate to elliptic or ovate, 1–4 cm, obtuse to subacute, broadly obtuse to rounded at base, 3–5-nerved; infl commonly widely much-branched; bracteal lvs subulate, 1–4 mm; sep linear-oblong, acute, about equaling the fr; pet 1.5–2.5 mm; stamens 5–16; styles 3, under 1 mm; fr green, ellipsoid, unilocular, 2–3.5 mm; seeds 0.4–0.7 mm; $2n=16$. Wet soil; Nf. and Que. to Man., s. to Fla. and Tex. July–Sept. (*H. parviflorum.*)

22. Hypericum boreale (Britton) E. Bickn. Perennial from slender rhizomes, 1–4 dm, much-branched above; lvs sessile, elliptic, oblong, or oval, obtuse or rounded at both ends 3–5-nerved, the larger ones 1–2 cm, half as wide or a third as wide; bracts like the cauline lvs but smaller; sep blunt, evidently shorter than the ellipsoid fr, this 3–5 mm, deep purple, unilocular; pet 2.5–3 mm; stamens 8–15; styles 3(4), under 1 mm; seeds 0.6–0.75 mm; $2n=16$. Wet, sandy or mucky soil; Nf. and Que. to w. Ont., s. to Va., O., Ind., and n. Ill. July–Sept.

23. Hypericum gymnanthum Engelm. & A. Gray. Erect annual 2–7 dm, simple or seldom branched above; lvs deltoid-lanceolate or deltoid-ovate, broadly rounded or subcordate at the closely sessile base, 1.5–3 cm, 5–7-nerved; infl loose and open with long internodes, the cymes raised above the uppermost foliage lvs on a long peduncle; bracts subulate; sep lanceolate, acuminate, 3.5–5 mm; pet narrow, 3–3.5 mm; stamens 10–14; styles 3(4), under 1 mm; fr lanceolate or narrowly ovoid, 3–5 mm; seeds 0.5 mm, the longitudinal ridges barely visible at 10×; $2n=16$. Moist or wet, usually sandy soil, chiefly on the coastal plain; s. N.J. to Fla. and Tex., and at scattered more inland stations to Pa., O., Ind., Ill., and Mo.

24. Hypericum majus (A. Gray) Britton. Much like no. 25, but more distinctly perennial from a rhizome, and avg taller and with more compact infl; lvs commonly ascending, mostly lanceolate to narrowly oblong or narrowly elliptic, 2–4 cm × 3–9 mm, 5–7-nerved, broadly obtuse or rounded at the weakly clasping base, the basal margins of a pair of lvs meeting around the stem; sep lanceolate, acute, 4.5–6.5 mm; pet 3.5–4 mm; stamens 14–21; styles 3(4), up to 1 mm; fr maroon, ellipsoid-conic, 5–7 mm; seeds 0.5–0.7 mm; $2n=16$. Wet meadows and shores; Nf. and Que. to B.C., s. to N.J., O., Ill., Kans., and Colo. July–Sept. Depauperate plants may approach no. 25.

25. Hypericum canadense L. Annual, or perennating by short, leafy stolons; stems slender, erect, 1–6 dm; lvs linear to narrowly oblanceolate, 1–4 cm × 1–4(–5) mm, obtuse, tapering to the narrow base, 1-nerved or weakly 3-nerved; infl compact to more often loose, usually with fewer fls than no. 21; bracts subulate; sep lanceolate, acute or acuminate, 3–5 mm; pet 2.5–3 mm; stamens 12–22; styles 3(4), under 1 mm; fr purple, mostly conic, 4–6 mm; seeds 0.5 mm; $2n=16$. Sandy or muddy shores or wet meadows; Nf. and Que. to Minn., s. to Ga. and Ala., more common eastward. July–Sept. Hybridizes with several other spp; one such apparent hybrid, approaching no. 21 or 22, has been called *H.* × *dissimulatum* E. Bickn.

26. Hypericum drummondii (Grev. & Hook.) T. & G. Nits-and-lice. Much-branched annual 2–8 dm, the branches usually alternate; lvs numerous, erect or ascending, linear, 5–20 mm; fls mostly solitary in the axils, short-pediceled; sep lance-linear, 3–6 mm, often surpassing the 3–5 mm pet; stamens 10–22; styles 3, up to 1 mm; fr ovoid, 4–6 mm; seeds light to dark brown, 1 mm, coarsely and conspicuously rugose and areolate; $2n=24$. Dry soil; Md. to O., Ill., Io., and se. Kans., s. to Fla. and Tex. July–Sept. (*Sarothra d.*)

27. Hypericum gentianoides (L.) BSP. Orange-grass. Erect annual 1–5 dm, with numerous slender, usually opposite branches; lvs appressed, scale-like, 1–4 mm; fls subsessile, chiefly solitary in the axils; sep lance-linear, 1.5–2.5 mm; pet 2–4 mm; stamens mostly 5–10; styles 3, 1 mm; fr slender-conic, 4–7 mm; seeds 0.4–0.8 mm, pale brown, obscurely areolate; $2n=24$. Sterile, especially sandy soil; Me. to Ont. and Minn., s. to Fla. and Tex. June–Sept. (*Sarothra g.*)

2. TRIADENUM Raf. Marsh St. John's-wort. Perianth 5-merous; pet imbricate, flesh-color or pinkish; stamens 9, the filaments connate into 3 fascicles of 3, alternating with 3 conspicuous hypogynous glands; ovary wholly 3-locular; styles separate; stigmas capitate; perennial, glabrous marsh-herbs with small fls in axillary and terminal cymules. 10, e. N. Amer. and e. Asia.

1 Lvs with translucent glands, dark-punctate beneath.
  2 Lvs sessile, cordate or subcordate at base.
    3 Sep at maturity 5–8 mm, acute or acuminate; styles 2–3 mm .... 1. *T. virginicum*.
    3 Sep at maturity 3–5 mm, obtuse or rounded; styles 0.5–1.5 mm .... 2. *T. fraseri*.
  2 Lvs narrowed to the base, usually petiolate .... 3. *T. walteri*.
1 Lvs without translucent glands or superficial dots .... 4. *T. tubulosum*.

1. Triadenum virginicum (L.) Raf. Erect, rhizomatous perennial, 3–6 dm; lvs 3–6 cm, oblong to ovate-oblong or elliptic, 2–3(–4) times as long as wide, rounded or retuse above, cordate or subcordate at the sessile base, dark-punctate beneath, dotted with translucent glands; sep lanceolate, acute or acuminate, 5–8 mm; pet 8–10 mm; fr trilocular, cylindric, 8–12 mm, gradually tapering to the 2–3 mm styles; $2n=38$. Bogs, marshes, and wet shores; N.S. to Fla. and Miss., mostly near the coast, or inland at low altitudes across N.Y. to s. Ont., O., and the s. end of Lake Michigan. July, Aug. (*Hypericum v.*)

2. Triadenum fraseri (Spach) Gleason. Much like no. 1; sep elliptic or oblong, obtuse or rounded, 3–5 mm; pet 5–8 mm; fr ovoid or cylindric, 7–12 mm, rather abruptly narrowed to the 0.5–1.5 mm styles. Bogs, marshes, and wet shores; Nf. and Que. to Minn., s. to Conn., N.Y., O., n. Ind., and Neb., and in the mts. to W.Va. July, Aug. (*Hypericum f.*)

3. Triadenum walteri (S. G. Gmelin) Gleason. Erect, rhizomatous perennial to 1 m, simple or branched above; lvs translucent-dotted, lance-elliptic to oblong or oblong-elliptic, 5–12 × 1–3 cm, dark-punctate beneath, obtuse or rounded above, narrowed below to a petiole 3–15 mm; cymes terminal and from the upper axils, on peduncles 1–4 mm; pet 5–7 mm; fr 7–12 mm; styles 1 mm. Swamps and marshes on the coastal plain; Md. to Fla. and Tex., n. in the interior to s. Ind. and s. Mo. July–Sept. (*Hypericum w.; H. petiolatum* Walter, not L.)

4. Triadenum tubulosum (Walter) Gleason. Erect, rhizomatous perennial 6–10 dm, branched above; lvs thin, lacking translucent glands or superficial dots, oblong or elliptic-oblong to elliptic-oblanceolate, 7–13 × 2–5 cm, obtuse or rounded above, usually tapering to an acute to truncate base; cymes few-fld, on peduncles to 12 mm; sep at maturity narrowly oblong, acute, 4–6 mm; pet 5–7 mm; fr 9–11 mm; styles 1 mm. Wooded swamps on the coastal plain; se. Va. to La., n. in the interior to s. Mo. and s. O. July–Sept. (*T. longifolium; Hypericum t.*)

FAMILY **TILIACEAE**, the Linden Family

Fls regular, perfect or sometimes unisexual, hypogynous, mostly 4-merous, sometimes with an epicalyx; sep valvate, usually distinct; pet distinct, usually conspicuous; stamens (10–) ± numerous, distinct or the filaments sometimes basally connate into 5 or 10 groups, often some of them staminodial; anthers bilocular; ovules (1)2–several in each of the 2–several locules of the ovary, the placentation generally axile; style 1, with a capitate or lobed stigma; fr capsular or indehiscent; trees or shrubs, rarely herbs, often stellate-hairy, with alternate, simple, stipulate, usually palmately or pinnipalmately veined lvs, the fls in various sort of cymose infls. 50/450, mainly warm reg.

1. **TILIA** L.   Basswood, linden.   Fls 5-merous; stamens numerous, distinct or often united into 5 antepetalous bundles, in our native spp. the innermost member of each group modified into an oblanceolate or spatulate, petaloid staminode; style dilated into a shallowly lobed stigma; ovary tomentose, with 2 ovules in each of the 5 locules; fr tomentose, indehiscent, nutlike, 1–2-seeded; trees with broad, petiolate, palmately veined lvs very oblique at the base, and fragrant, white or ochroleucous fls in axillary cymes, the long peduncle adnate about to the middle of a characteristic narrow, elongate, short-petioled, foliaceous bract. 50, mainly N. Temp.

1. **Tilia americana** L.   Basswood, American linden. Tree to 40 m; lvs broadly ovate to subrotund, mostly 7–15 cm, sharply serrate, cordate or truncate at base; pet narrowly oblong to oblanceolate, 7–12 mm, tapering to the base; fr subglobose, 6–8 mm; 2n=82. Rich woods; N.B. to Man., s. to Fla. and Tex.; disjunct in Mex. July. Four vars., two with us. The mainly northern var. *americana,* American b., extending to s. to nw. N.C. and e. Okla., has the lvs green beneath and glabrous to sparsely stellate-hairy (or more densely so when young), commonly with conspicuous tufts of hairs in the vein-axils beneath. (*T. neglecta,* an introgressant toward the next var.) The mainly s. Appalachian var. *heterophylla* (Vent.) Loudon, white b., extending n. to sw. Pa. and s. O., w. to s. Ill., with outliers n. and w., is more hairy, the lvs permanently covered beneath with fine white or brown stellate hairs. (*T. heterophylla; T. michauxii; T. monticola; T. pubescens.*)
   Two European spp. with mostly smaller (5–10 cm) lvs and without staminodes escape locally from cult. *T. platyphyllos* Scop. has strongly 5-ribbed frs, whereas *T. cordata* Miller has the frs only weakly or scarcely ribbed.
   T. petiolaris DC., pendent silver l., of w. Asia, rarely escapes from cult. It has pendulous twigs, the young ones tomentulose, and suborbicular lvs white-felted beneath.

FAMILY **STERCULIACEAE**, the Cacao Family

Much like the Tiliaceae; fls mostly 5-merous; sep or cal-lobes valvate; pet commonly small, or often wanting; stamens often seated on an androgynophore, typically with the filaments all connate into a tube around the gynoecium, but in our genus the anthers sessile on the androgynophore; carpels mostly 5, scarcely to evidently united (sometimes only by their styles), or even wholly distinct; mostly woody plants, with alternate, stipulate lvs, often stellate-hairy. 65/1000, mostly warm reg.

1. **FIRMIANA** Marsili.   Fls regular, functionally unisexual; cal-tube cupuliform and with a dense band of hairs and a nectariferous disk surrounding the prominent, exserted androgynophore; pet none; stamens ca 15, with sessile anthers, in staminate fls forming a globular clump concealing the pistillodes, in pistillate fls well formed but sterile, closely subtending the ovary; carpels at first connate (even the styles), later separating; follicles stipitate, papery, 2–4-seeded; small, monoecious or subdioecious trees with deciduous, palmately lobed lvs and caducous stipules. 15, Old World trop.

1. **Firmiana simplex** (L.) Wight.   Chinese parasol-tree. Tree to 15 m; lvs 8–40 cm long and wide, stellate-hairy on both sides at first, later ± glabrate especially above, the 3–5 lobes sharp-pointed; fls in large mixed panicles to 4 dm; cal-lobes 5, narrow, firm, yellow or white inside, densely stellate outside, spreading-reflexed, to 10 mm, much longer than the tube; follicles thin and flat, to 12 × 5 cm, densely stellate; seeds pea-sized; 2n=50. Native of se. Asia, cult. in se. U.S. n. to Va. and Md., and occasionally escaped. June, July.

Family **MALVACEAE**, the Mallow Family

Fls regular, hypogynous, 5-merous, usually perfect, often with an epicalyx (its members here called bractlets); sep distinct or often connate at base, valvate; pet distinct, often adnate to the base of the filament-tube, convolute or sometimes imbricate; stamens mostly numerous, the filaments all connate into a tube for most of their length; anthers unilocular; carpels (1)2–many (often 5), either loosely coherent in a ring around the base of the single style and then separating at maturity, or wholly united into a compound ovary; style with as many long or short terminal branches as carpels; ovules 1–many per carpel; fr a loculicidal capsule, or of separating, dehiscent or indehiscent mericarps; herbs, seldom shrubs or small trees, often stellate-hairy, with alternate, stipulate, simple and entire to more or less dissected, generally palmately veined lvs. 75/1000+ cosmop.

1 Carpels 5, united into a compound ovary; fr a loculicidal capsule; stamen-column bearing anthers
    along the sides, 5-toothed at the summit (Hibisceae).
  2 Style-branches short, the stigmas nearly sessile; bracts 3, broad .... 1. *Gossypium.*
  2 Style-branches elongate; bracts 6–many, linear.
    3 Locules of the fr several-seeded; pet yellow, or over 4 cm, or the plant woody .... 2. *Hibiscus.*
    3 Locules 1-seeded; ours an herb with pink pet 2–4 cm .... 3. *Kosteletzkya.*
1 Carpels 5–many, loosely united in a ring, separating at maturity; anthers at the summit of the stamen-
    column, sometimes also a few along the sides (Malveae).
  4 Ovules and seeds 2 or more per carpel.
    5 Carpels beakless; our sp. with large pink-purple pet 2.5–3 cm .... 4. *Iliamna.*
    5 Carpels beaked at the tip; pet either yellow or less than 1 cm.
      6 Bractlets none; seeds 3 or more in each carpel .... 5. *Abutilon.*
      6 Bractlets 3; seeds 2 per carpel, separated by a transverse septum .... 6. *Modiola.*
  4 Ovules and seeds 1 per carpel.
    7 Style-branches filiform, stigmatose along the inner side.
      8 Fls perfect; epicalyx usually present (wanting in 2 spp. of *Callirhoe*).
        9 Bractlets 6–9, united at base .... 7. *Althaea.*
        9 Bractlets 3 or fewer, or none.
          10 Pet obcordate; intr. .... 8. *Malva.*
          10 Pet truncate; native .... 9. *Callirhoe.*
      8 Fls unisexual, the plants dioecious; epicalyx none .... 10. *Napaea.*
    7 Style-branches terminating in truncate or capitate stigmas.
      11 Bractlets 3 .... 11. *Malvastrum.*
      11 Bractlets none.
        12 Carpels at maturity beakless (in our spp.); lvs (ours) deeply parted .... 12. *Sphaeralcea.*
        12 Carpels at maturity prominently beaked; lvs merely toothed or lobed.
          13 Lateral walls of the mature carpels persistent; fls yellow or white .... 13. *Sida.*
          13 Lateral walls of the mature carpels evanescent; fls pale blue .... 14. *Anoda.*

1. GOSSYPIUM L. Cotton. Bractlets of the epicalyx 3, large, foliaceous, cordate, ± incised; stamen-tube bearing anthers along the side but not at the summit; style undivided, with 5 subsessile stigmas; ovary 3–5-locular, ripening into a loculicidal capsule; seeds covered with long, usually white fibers. 20, warm reg.

1. Gossypium hirsutum L. Upland c. Lvs orbicular in outline, with 3–5 entire triangular lobes, cordate at base; fls white, the pet 5–7 cm; $2n=52$. Native of C. Amer., cult. in s. U.S. and often appearing in our range as a waif along railways from the south.

2. HIBISCUS L. Bractlets of the epicalyx numerous, commonly about 12, linear; stamen-column with anthers along the sides; style-column prolonged, bearing 5 short branches with capitate stigmas; cor large; ovary 5-locular; ovules several in each locule; fr a subglobose to ovoid or prismatic, loculicidal capsule subtended or enclosed by the persistent, accrescent cal; seeds glabrous or roughened or hairy; annual or perennial herbs or shrubs, mostly with large fls. 200+, mostly warm reg.

1 Stem woody .... 1. *H. syriacus.*
1 Stem herbaceous.
  2 Lvs merely serrate or lobed; pet 6 cm or more, the cor funnelform.

3 Cal and lower lf-surface glabrous or very nearly so .... 2. *H. laevis.*
3 Cal and lower lf-surface densely but finely stellate .... 3. *H. moscheutos.*
2 Lvs divided nearly to the base; pet to 4 cm; cor widely spreading .... 4. *H. trionum.*

1. **Hibiscus syriacus** L. Rose of Sharon. Branching shrub to 6 m; lvs ovate in outline, usually 3-lobed, coarsely serrate or dentate, glabrous; cal densely but finely stellate; fr 4–6 cm, white or variously pink, red, blue, or violet; fr prismatic, pointed, densely stellate-tomentose; seeds long-hairy on the margins; $2n=80$–92. Native of e. Asia, persistent after cult. and occasionally escaped. July–Sept.

2. **Hibiscus laevis** All. Smooth rose-mallow. Glabrous perennial 1–2 m; lvs triangular in outline, the basal lobes, if developed, divergent, the terminal lobe long-acuminate, 2–6 times as long as the body of the lf; bractlets tapering to a filiform point; pet pink with darker center, 6–8 cm; seeds pubescent; $2n=38$. Marshes and shallow water, on or near the coastal plain; s. Pa. to Fla. and Tex., and n. in the interior to O. and Minn. Aug.–Sept. (*H. militaris.*)

3. **Hibiscus moscheutos** L. Rose-mallow. Perennial 1–2 m, the stems canescent above; lvs lanceolate to ovate, occasionally shallowly lobed beyond the middle, glabrous or soon glabrate above, stellate-canescent beneath; peduncles jointed near or above the middle, often bearing a ± well developed lf; involucral bracts usually not ciliate, cor 10–20 cm wide, pink or white, often with a red or purple center; fr rounded or tapering to the often short-beaked summit, glabrous except the sutures; seeds glabrous; $2n=38$. Marshes along the coast and inland; Mass. and N.Y. to O., s. Wis., and Mo., s. to the Gulf, and disjunct in Calif.; sometimes escaped elsewhere. Most of our plants are var. *moscheutos,* as described above. (*H. palustris.*) The mostly more southern or western var. *occidentalis* Torr., with the frs and both sides of the lvs pubescent, and the bracts usually ciliate, enters our range in s. Ill., w. Ky., and adj. Ind. (*H. lasiocarpus.*)

4. **Hibiscus trionum** L. Flower of an hour. Branching, hairy annual 3–5 dm; lvs long-petioled, deeply 3 parted, the segments oblong to obovate, coarsely serrate or lobed; cal conspicuously veined, hispid on the nerves with spreading simple hairs from swollen bases; pet 1.5–4 cm, pale yellow with purple base, expanded only a few hours; mature cal inflated, 5-angled, enclosing the hirsute fr; seeds finely verrucose; $2n=28, 56$. Native of s. Europe; fields, roadsides, and waste places from N.S. to Minn., s. to Fla. and Tex. July–Sept.

3. **KOSTELETZKYA** C. Presl, nom. conserv. Carpels 1-seeded; capsule eventually breaking into 5 deciduous segments, each composed of halves of adjacent carpels; seeds smooth; otherwise as in *Hibiscus.* 30, N. Amer., Afr., Madag.

1. **Kosteletzkya virginica** (L.) C. Presl. Seashore-mallow, Perennial to 1 m, rather roughly stellate; lvs triangular-ovate in outline, 6–15 cm, often hastate with 2 divergent lobes; fls axillary or in a leafy-bracteate terminal panicle; bractlets 6–10, linear, 3–9 mm; pet pink, obovate, 2–4 cm; capsule depressed-globose, 5-locular, hispid over the summit, 1 cm thick. Salt or brackish marshes; L.I. to Fla., Tex., and W.I. Aug.–Sept. Var. *virginica,* occurring from Va. s., has the hairs of the fr mostly 1.2–2.4 mm. Var. *aquilonia* Fern., occurring from N.C. n., has the hairs of the fr rarely over 1 mm, and the plants tend to have smaller fls, shorter bractlets, and less densely hairy frs than var. *virginica.*

4. **ILIAMNA** Greene. Bractlets of the epicalyx 3, narrowly linear, persistent; pet large; stamen-column bearing anthers at the summit; carpels numerous, 2–4-seeded, at maturity rounded at the summit, dehiscent to the base, thin-walled and plane on the sides, densely hirsute on the back with long, erect, simple hairs; branching perennial herbs with large, thin, palmately lobed, maple-like lvs and handsome fls solitary or fascicled in the upper axils. 7, U.S.

1. **Iliamna remota** Greene. Stellate-hairy perennial 1–2.5 m; lvs 1–2 dm long and wide, 5–7-lobed; pedicels 1 cm; cal-lobes triangular-ovate, 10–15 mm; pet pink-purple, 2.5–3 cm; mature carpels oblong, 8–10 mm, the thin sides glabrous, reticulate-veined near the dorsal margin; seeds finely hairy. On an island in the Kankakee R., Ill. (the original station), and near the summit of Peters Mt., Va.; local along railroads in Ind. and Va. June, July. (*Sphaeralcea r.; Phymosia r.; Iliamna corei,* the Va. plant.)

5. **ABUTILON** Miller. Epicalyx wanting; stamen-column with anthers at the top; carpels 5–many; ovules 3–9 per carpel; styles 5–many, slender, stigmatic at the top; mature carpels dehiscent across the top, rounded or beaked at the summit, eventually falling from the axis; herbs or shrubs, usually pubescent, with broad, cordate, angular or lobed lvs and axillary, usually yellow fls. 100+, warm reg.

1. **Abutilon theophrasti** Medikus. Velvet-leaf; pie marker. Stout branching annual 1–1.5 m, softly hairy throughout with few-armed stellate hairs; lvs cordate, 10–15 cm long and wide, acuminate, entire or obscurely toothed, on petioles about as long; peduncles joined above the middle, elongating to 2–3 cm at maturity; fls

yellow, 1.5–2.5 cm wide; head of fr 2–3 cm wide; carpels commonly 10–15, densely hairy, with conspicuous horizontally spreading beaks; $2n=42$. Native of s. Asia, established as a weed in fields and waste places nearly throughout our range but more abundant southward. July–Oct. (*A. abutilon*.)

6. MODIOLA Moench. Bractlets of the epicalyx 3, linear to obovate; carpels numerous, at maturity hispid and 2-horned at the summit, dehiscent, with thin lateral walls, 2-seeded, the seeds separated by a triangular partial septum intruded from the outer wall; annual or biennial with divided lvs and small red-purple fls. Monotypic.

1. **Modiola caroliniana** (L.) G. Don. Bristly mallow. Stems prostrate or ascending, often rooting at the nodes; lvs broadly ovate to subrotund, 3–5-cleft and irregularly incised; fls on elongate axillary pedicels; pet 3–5 mm; $2n=18$. Moist or dry soil and waste places; tropical Amer., n. as a native to S.C. and reputedly Va., and occasionally farther as an adventive.

7. ALTHAEA L. Bractlets of the epicalyx 6–9, connate at base; stamen-column with anthers at the summit; carpels numerous, usually 15 or more, at maturity falling away from the central axis and indehiscent; style-branches as many as the carpels, slender and elongate, stigmatic along the inner side; pubescent perennial herbs with conspicuous fls in the upper axils. 12, temp. Eurasia.

1. **Althaea officinalis** L. Marsh-mallow. Erect, branched, 5–12 dm; lvs ovate, 5–10 cm, coarsely serrate and commonly shallowly 3-lobed, velvety; fls several in peduncled clusters from the upper axils, pink, 3 cm wide; bractlets narrowly lanceolate; $2n=42$. Native of Europe; naturalized in salt marshes from Mass. to Va., and locally inland. July–Sept.
Althaea rosea L., hollyhock, occasionally escapes. It is mostly unbranched and 1.5–3 m, with cordate-orbicular lvs 1–3 dm wide and large fls commonly 10 cm wide.

8. MALVA L. Mallow. Bractlets of the epicalyx 3, linear to obovate; stamen-column bearing anthers at the summit; pet truncate to (in our spp.) obcordate; carpels 10–20, beakless, one-seeded, indehiscent, the sides incompletely covering the seed; styles slender, stigmatic along the inner side; annual to perennial herbs with broad, toothed to lobed or parted lvs, the fls solitary or fascicled in the axils. 25, temp. Eurasia.

1 Upper lvs 5–7-divided to below the middle; pet 2–3.5 cm, erect perennials.
  2 Bractlets linear to narrowly oblanceolate, essentially glabrous on the back .... 1. *M. moschata*.
  2 Bractlets ovate to obovate, densely stellate-hairy on the back .... 2. *M. alcea*.
1 Upper lvs very rarely lobed as far as the middle; pet to 2.5 cm; prostrate to erect annuals or biennials.
  3 Bractlets oblong-ovate; pet red-purple, 2–2.5 cm .... 3. *M. sylvestris*.
  3 Bractlets linear or narrowly lanceolate; pet white or slightly tinged with pink or purple, less than
    2 cm.
    4 Plants erect, to 2 m; fls subsessile in axillary fascicles .... 4. *M. verticillata*.
    4 Plants depressed, usually branched from the base; fls evidently pedicellate.
      5 Pet twice as long as the sep; mature carpels rounded and not rugose on the back .... 5. *M. neglecta*.
      5 Pet about equaling or only slightly surpassing the sep; mature carpels flat and rugose-reticulate
        on the back.
        6 Cal only slightly accrescent, not reticulate-veiny; fruiting pedicels mostly more than 10 mm
          .... 6. *M. rotundifolia*.
        6 Cal strongly accrescent and reticulate-veiny in fr; pedicels mostly less than 10 mm .... 7. *M. parviflora*.

1. **Malva moschata** L. Musk-m. Erect perennial, 4–10 dm, rough-hairy; lvs orbicular in outline, 5–7-parted, the segments of the upper lvs again deeply pinnatifid; fls partly solitary on long pedicels from the upper axils, but chiefly crowded in terminal clusters; bractlets linear to narrowly lanceolate or oblanceolate, ciliate, glabrous or nearly so on the back; pet white to pale purple, triangular, 2–3.5 cm; mature carpels densely hairy, rounded on the back, not rugose; $2n=42$. Native of Europe, escaped from cult., along roadsides and in waste places; Que. and N.S. to B.C., s. to Va. and Mo., more common northward. June–Sept.

2. **Malva alcea** L. Vervain-m. Much like no. 1; stem distinctly stellate-hairy; primary segments of the upper lvs merely shallowly lobed or coarsely toothed above the middle; bractlets oblong to ovate or obovate, densely stellate on the back, as also the cal; mature carpels glabrous or sparsely pubescent, keeled on the back; $2n=84$. Native of Europe, occasionally escaped from cult., especially in ne. U.S. June–Sept.

3. **Malva sylvestris** L. High m. Erect biennial, 4–10 dm, hirtellous to glabrate; lvs orbicular or reniform in outline, shallowly 3–7-lobed, the lobes broadly rounded, serrate; petioles hairy only or chiefly in a single line on the upper side; bractlets oblong to ovate or obovate; fls fascicled in the upper axils, on pedicels to 5

cm; pet red-purple, 2–2.5 cm; mature carpels rugose-reticulate on the back, glabrous or sparsely hairy; $2n=42$. Native of Eurasia, escaped from cult. here and there in our range. June–Aug.

4. Malva verticillata L.   Whorled m.   Erect annual to 2 m, with large, long-petioled lvs mostly (6–)8–15(–20) cm wide, reniform, with 5–7 rounded, crenate lobes, the upper on progressively shorter petioles; fls subsessile in axillary fascicles; pet 6 mm, white or bluish-white; fr as in no. 6; $2n=$ca 84. Probably native of e. Asia, escaped from cult. here and there in our range. July–Sept. One form has the lvs crisp-margined.

5. Malva neglecta Wallr.   Common m.; cheeses.   Prostrate to ascending annual or biennial to 1 m, usually branched from the base; lvs long-petioled, orbicular or reniform, 2–6 cm wide, shallowly 5–9-lobed, crenate, with cordate or subcordate base; fls fascicled in the axils, on pedicels to 3 cm; bractlets narrow; pet obcordate, 6–12 mm, white or slightly tinted with pink or purple, twice as long as the sep; mature carpels usually 12–15, rounded on the back and usually finely hairy, not rugose or reticulate, the lateral faces not radially veined, the whole ring of carpels with a crenate outline, the depressed central portion of the head a third as wide as the head; $2n=42$. Native of Eurasia and n. Afr., now abundant as a weed in gardens and waste places throughout temperate N. Amer. May–Oct.

6. Malva rotundifolia L.   Dwarf m.   Much like no. 5 in aspect; pedicels often shorter, but generally at least 10 mm in fr; pet mostly 4–5 mm, slightly if at all longer than the sep; mature carpels 8–11, commonly 10, glabrous or hairy, conspicuously rugose-reticulate on the back and radially veined on the lateral faces, the margins sharply angled but not winged, the whole head of carpels approximately circular in outline; central depressed area one-fifth the diameter of the head; $2n=42$. Native of Europe; widely established as a weed in w. and s. U.S., and local with us. (*M. pusilla.*)

7. Malva parviflora L.   Little m.   Much like nos. 5 and 6; pedicels mostly less than 10 mm, even in fr; cal strongly accrescent and reticulate-veiny in fr; fr as in no. 6, but the margins of the carpels narrowly winged; $2n=42$. Native of the Mediterranean region, widespread as a weed in c. and w. U.S., and local with us.

9. CALLIRHOE Nutt.   Poppy-mallow. Bractlets of the epicalyx 3 or none; each cal-segment with 3 strong nerves; stamen-column often bearing anthers along the sides as well as at the top; pet broadly truncate; carpels 10–20, rounded or short-beaked at the tip, one seeded, sometimes eventually dehiscent at the tip; styles slender, elongate, stigmatic along the inner side; perennial herbs from a thick (often turnip-like) or fusiform root (our spp.), or annual from a slender taproot; lvs triangular to rotund in outline, crenate to deeply divided; fls large, purple, red-purple, or occasionally pink or white. 9, N. Amer.

1. Callirhoe triangulata (Leavenw.) A. Gray.   Clustered p.-m.   Stems ascending, 4–10 dm; herbage (including the cal) harshly stellate-hairy; principal lvs chiefly triangular with broadly truncate to cordate base, crenate (seldom some of them cleft), the lower notably long-petiolate; reduced upper lvs often much narrower and/or few-cleft; fls short-pedicellate, several and crowded at the ends of more elongate axillary and terminal peduncles; bractlets 3–8 mm, spatulate or obovate; cal-lobes deltoid-ovate, 2–5+ mm, the tips connivent in bud into a short beak; pet 16–32 mm, red with a white basal spot, entire or slightly erose; mature carpels hairy, thin-walled, not rugose; $2n=30$. Sandy prairies; s. Wis. and ne. Io. to Ill., w. Ind., and se. Mo.; irregularly from Miss. to Ga. and N.C. July–Sept.

2. Callirhoe involucrata (T. & G.) A. Gray.   Purple p.-m.   Stems ascending, to 10(–20) dm; herbage rough-pubescent to hirsute, often with simple hairs, those around the base of the cal stiff, spreading, 1–3 mm; lvs orbicular in general outline, deeply 5–7-lobed, the principal divisions oblanceolate to cuneate-obovate, lobed or incised; stipules mostly 5–15 mm; peduncles (pedicels) mainly axillary, 3–10 cm, 1-fld; bractlets lanceolate or linear, 8–17 mm; gynodioecious; cal-lobes lanceolate or lance-ovate, 9–16 mm, with separate, loose, free tips 4.5–10+ mm, pet red or purple with a white basal spot, 19–32 mm (9–20 in functionally pistillate fls); mature carpels strongly rugose, strigose at least distally; $2n$ mostly=30, 60. Open places, esp. dry upland prairies; Ill. to s. Minn. and w. Wyo., s. to Ark., Tex., and n. Mex.; adventive in Pa. May–Oct. Our plants belong to the nominal var. *involucrata,* as here described.

3. Callirhoe bushii Fern.   Bush's p.-m.   Stems ± erect, 5–14 dm; herbage pubescent with mainly simple and spreading bristly hairs, varying to merely strigose or subglabrous; lvs suborbicular to ovate in outline, 5–7-cleft, with oblong or ovate, coarsely toothed segments; stipules ovate, 8–16+ mm; peduncles (pedicels) mainly axillary, to 15 cm, 1-fld; fls all perfect; bractlets lanceolate or ovate, 10–17 × 3–6 mm, the slender tips connivent in bud to form an evident beak; pet red or pale red, 20–32 mm, erose-denticulate across the top; mature carpels rugose-reticulate; $2n=56$. Glades, rocky woods, and disturbed sites; Mo. and se. Kans. to n. Ark. and e. Okla., barely entering our range in Mo. May–Aug.

4. Callirhoe alcaeoides (Michx.) A. Gray.   Pale p.-m. Stems curved-ascending, 1.5–6 dm, covered with 4-rayed appressed hairs, as also the petioles, peduncles, and lower lf-surfaces; basal and lower lvs cordate to triangular or ovate, crenate-lobulate or often some or all of them ± lobed or cleft, the other lvs few, smaller, and generally deeply cleft; stipules 5–10 mm, persistent; peduncles (pedicels) axillary to reduced upper lvs or bracts, 1-fld; gynodioecious, the functionally pistillate fls with smaller sep and pet and fewer, nonfunctional anthers; epicalyx wanting; cal hispid-strigose; sep-tips connivent in bud, forming a short beak; pet white to pink or mauve, erose or fimbriate across the top, mostly 15–25 mm (7–17 mm in pistillate fls); mature carpels indurate, reticulate-rugose, copiously strigose; $2n=28$. Prairies and roadsides; Ill. to Tenn., w. to Nebr., Kans., and Tex., more common westward. May–July.

5. Callirhoe digitata Nutt.   Finger p.-m. Erect, 5–20 dm, the stem glaucous and glabrous or nearly so; lvs glabrous to rather thinly strigose, palmately deeply cleft into 5–10 finger-like segments that may be again cleft; stipules slender, 6–8 mm, caducous; infl an open panicle, the fls long-pedicellate, all perfect; epicalyx wanting; cal, carpels, and usually the pedicels glabrous or nearly so; sep-tips connivent in bud, forming a short beak; pet red to purple with a basal white spot, 18–31 mm, erose-denticulate to fimbriate across the top; mature carpels indurate, rugose-reticulate; $2n=28$. Limestone glades and rocky prairies; Ozarkian, mainly in sw. Mo., nw. Ark., and adj. Kans. and Okla., but locally intr. along railroads in Clay Co., Mo. May–Sept.

10. NAPAEA L.   Glade-mallow. Dioecious; epicalyx none; cal-lobes ovate or triangular, shorter than the tube; pet white; stamen-column bearing anthers at the summit in the staminate fls, in the pistillate fls reduced to a short sheath surrounding the ovary and bearing sessile anthers in a distal ring; carpels ca 10, 1-seeded, at maturity becoming rugose on the back and sides, irregularly dehiscent, beakless or obscurely apiculate; tall perennial herb. Monotypic.

1. Napaea dioica L.   Stems 1–2 m, branched above; lvs orbicular in outline, 1–3 dm, deeply 5–9-lobed or parted, the lobes coarsely toothed or incised; fls very many in a large terminal panicle; cal 5–8 mm, sparsely pilosulous with simple hairs or glabrous; pet of staminate fls 5–9 mm, of pistillate fls much smaller; mature carpels 5 mm, ribbed on the back. $2n=$ca 30. Moist alluvial woods; Va. to Pa., w. to Wis. and Io., rarely escaped elsewhere. June–Aug.

11. MALVASTRUM A. Gray, nom. conserv.   Bractlets of the epicalyx 3, narrowly linear; cal-lobes broadly ovate, persistent; stamen-column bearing anthers at the summit; pet yellow; carpels 5–many, separating at maturity and in our sp. dehiscent, 1-seeded, at maturity conforming with the shape of the seed and not much larger than it, the distal and basal parts not differentiated; annual or perennial herbs with lanceolate to ovate, serrate but not lobed lvs, and small, short-pediceled fls solitary or fascicled in the upper axils. 14, New World.

1. Malvastrum hispidum (Pursh) Hochr.   Erect, branching annual 2–5 dm, closely stellate, the hairs of the stem chiefly 4-armed, those of the lvs and cals often only 2–3-armed; lvs linear-oblong or oblong to lance-elliptic, seldom over 1 cm wide, remotely toothed; fls solitary in the axils; sep at maturity valvate-connivent, their margins turned outward, forming a 5-winged cal; pet yellow, 3–4.5 mm; mature carpels 5–6, thin-walled, beakless, minutely hairy; $2n=36$. Dry ground, prairies, and barrens; Ky. and Tenn. to Io., Kans., and Okla. June–Sept. (*M. angustum; Sphaeralcea angusta; Sidopsis hispida.*)

12. SPHAERALCEA A. St.-Hil.   Scarlet mallow. Bractlets of the epicalyx 0–3, usually caducous; stamen-column bearing anthers at the summit; pet usually emarginate; carpels 10–15, densely stellate on the back, at maturity differentiated into a lower, indehiscent, fertile part comprising at least half of the whole, reticulate on the sides and commonly tuberculate on the back, and a terminal, sterile, dehiscent part with plane sides; herbs with crenate or dentate to deeply parted lvs and a terminal raceme or thyrse of yellowish or brick-red fls. 60, New World and S. Afr.

1. Sphaeralcea coccinea (Pursh) Rydb.   Densely stellate perennial 1–2 dm from deep-seated running roots, the stems often clustered on a stout caudex; lvs 2–5 cm, deeply 3-parted, the divisions variously lobed; fls in a compact, leafy raceme; pet 1–2 cm, rusty-red; mature carpels densely stellate, 1-seeded, reticulate on the sides, tuberculate on the back; $2n=10$. Dry prairies and plains; w. Io. and w. Minn. to Man. and w. Tex., w. to the Rocky Mts. May, June. (*Malvastrum c.*)

13. SIDA L. Epicalyx wanting (in our spp.); cal 5-lobed, persistent and enclosing the fr, the ten costae leading alternately to the lobes and sinuses; stamen-column bearing anthers at the top; carpels 5–15, each with a single pendulous ovule; mature mericarps each with a lower, indehiscent, 1-seeded portion and an upper, dehiscent portion that is often tipped with 1 or 2 erect or connivent or inflexed beaks; upper and lower parts of the mericarp set apart dorsally by a shoulder that is an extension and joining of 2 lateral ribs; herbs and small shrubs with mostly small fls solitary or fascicled in the axils or forming a terminal panicle. 150+, warm reg.

1 Lvs ovate to linear, crenate or subentire; pet yellow.
  2 Carpels 5, each 2-beaked at maturity; fls fascicled in the axils .... 1. *S. spinosa.*
  2 Carpels mostly 10; fls solitary in the axils.
    3 Mature carpels with a single beak (this split into 2 by dehiscence); pedicels visibly jointed above
      the middle; annual .... 2. *S. rhombifolia.*
    3 Mature carpels with 2 beaks; pedicels not jointed; perennial .... 3. *S. elliottii.*
1 Lvs broadly ovate to rotund in outline, deeply lobed; pet white .... 4. *S. hermaphrodita.*

    1. Sida spinosa L. Prickly s. Branching annual 3–6 dm, commonly with a short spine-like process at the base of each lf; lvs petiolate, 2–4 cm, crenate, lance-ovate to oblong or elliptic, basally subtruncate to subcordate; fls fascicled in the axils; pedicels 2–12 mm, the longer ones jointed near or above the middle; cal thinly stellate; pet pale yellow, 4–6 mm, carpels 5, each tipped with 2 erect, hispidulous beaks; $2n=14$, 28. Fields, roadsides, and waste places; pantropical, extending n. occasionally as far as Mass. and Mich. July–Oct.

    2. Sida rhombifolia L. Arrowleaf s. Branching annual (in our range) to 1 m; lvs oblong to ovate or rhombic, 2–6 cm, crenate chiefly beyond the middle; cal thinly stellate; pet pale yellow, 4–8 mm; carpels commonly 10, each tipped by a single erect, subulate beak grooved longitudinally and split into two at dehiscence; $2n=14$, 28. A weed in waste places and along roads; pantropical, n. occasionally to se. Va., and as a ballast-waif farther n. All summer.

    3. Sida elliottii T. & G. Coastal plain s. Branching perennial to 1 m; lvs linear to narrowly oblong, 2–6 cm × 3–8 mm, firm, sharply serrate to subentire; fls solitary in the axils; pedicels jointless, 5–25 mm, cal thinly stellate, the midrib usually pilosulous; pet bright yellow, 10–15 mm; carpels commonly 10, each tipped by 2 slender, erect, puberulent beaks. Dry or moist soil; se. Va. to Tenn. and se. Mo., s. to Fla. and Tex. June–Aug. (*S. inflexa.*)

    4. Sida hermaphrodita (L.) Rusby. Virginia s. Perennial, 1–2(–4) m, stellate-hairy when young; lvs broadly ovate to rotund in outline, 1–2 dm, deeply 3–7-lobed, the lobes lanceolate, irregularly serrate, the middle one the longest; fls in a terminal cymose panicle; cal thinly to densely stellate; pet white, 6–10 mm; carpels commonly 10, at maturity acuminate into a slender, erect beak 3 mm; $2n=28$. Loose, sandy, often rocky soil in unstable habitats along rivers and on flood-plains; c. Pa. to Md., Va., W.Va., e. Tenn., s. O., and apparently disjunct in nw. O., ne. Ind., and s. Mich., where perhaps adventive. On the Atlantic slope mainly confined to the Potomac and Susquehanna watersheds.

14. ANODA Cav. Much like *Sida,* but the lateral walls separating the carpels eventually obliterated. 10, warm New World.

    1. Anoda crista (L.) Schlechtendahl. Spurred anoda. Erect, branching, hairy annual to 1 m; lvs triangular-ovate or hastate, 5–10 cm, fls axillary, long-pediceled, pale blue, 1 cm wide; mature cal saucer-shaped, 2 cm wide, much exceeding the 8–12 carpels. Sw. U.S. and tropical Amer., occasionally adventive toward the s. part of our range.

FAMILY **SARRACENIACEAE**, the Pitcher-plant Family

Fls perfect; sep 5; pet 5 or none; stamens (10–) numerous, with short filament and introrse anthers; ovary (3)5-locular, with numerous ovules on axile (or distally intruded-parietal) placentas; style 2, subentire or branched above; fr a loculicidal capsule; seeds numerous, small, with a minute, dicotyledonous embryo and copious endosperm; perennial, insectivorous herbs with hollow, pitcher-shaped lvs and large, nodding fls solitary on a scape or in few-fld racemes. 3/15, New World.

    1. SARRACENIA L. Pitcher-plant. Sep 5, broad and spreading; pet 5, incurved, soon deciduous; stamens many; ovary large, subglobose; style slender at base, above extended into 5 rays connected by tissue and forming a 5-angled or -lobed, umbrella-shaped body with the minute stigmas beneath it at the angles; fls solitary on scapes; lvs bearing a broad or narrow wing along the adaxial side and prolonged beyond the pitcher on the abaxial side into a broad hood; $x=13$. 8, e. N. Amer.

1 Lvs curved-ascending from the base, broadly winged; pet mostly dark red-purple .... 1. *S. purpurea.*
1 Lvs slender, erect, narrowly winged; pet yellow .... 2. *S. flava.*

1. Sarracenia purpurea L. Lvs 1–2 dm, obovoid, the wing semi-elliptic to semi-oblanceolate, usually 1–5 cm wide, the hood sessile by a very broad base; scapes 3–5 dm; fls 5–7 cm wide, the pet dark red-purple or rarely yellowish; $2n=26$. Sphagnum-bogs, sandy or marly shores, and other wet places; Lab. and Nf. to n. Fla. and s. La., restricted to the coastal states southward, but far inland n. of the glacial boundary, to Man., Minn., n. Ill., Ind., and O. May–Aug. Two rather weak geographic vars.: var. *purpurea,* northern, s. to Md., Del., and N.J., with pitchers more than 3 times as long as wide, usually glabrous outside, the pet deep maroon to occasionally red; and var. *venosa* (Raf.) Fern., southern, n. to Md., Del. and N.J., with pitchers less than 3 times as long as wide, usually bristly outside, the pet mostly bright red to pink. (*S. venosa.*)

2. Sarracenia flava L. Lvs erect, 3–8 dm, slender, gradually widened distally and widest at the summit, the wing seldom as much as 1 cm wide, the hood depressed-ovate and raised on a stipe-like base; scapes to 1 m; fls 8–10 cm wide, the pet yellow; $2n=26$. Sandy bogs and wet pine-lands; se. Va. to n. Fla. and s. Ala. May. A hybrid with no. 1 is *S.* × *catesbaei* Elliott.

FAMILY **DROSERACEAE**, the Sundew Family

Fls regular, perfect, hypogynous, usually 5-merous; pet distinct, convolute; cal, cor, and stamens withering persistent; anthers versatile, extrorse; ovary unilocular with (3–) many ovules on parietal placentas or on an essentially basal placenta; styles distinct and often also deeply bifid, or (*Dionaea*) united to form a common style; fr usually a loculicidal capsule; seeds 3–many, ± spindle-shaped, with copious endosperm and a short, dicotyledonous embryo; annual to perennial insectivorous herbs of bogs and marshes, commonly acaulescent and with a rosette of basal lvs, these basically alternate (rarely whorled), often circinate in bud, the blade either modified as an active trap or more often provided with irritable, mucilage-tipped tentacle-hairs. 4/100, widespread.

1. DROSERA L. Sundew. Pet white, pink, or purple, broadened distally; stamens 4–8, as many as the pet; styles usually 3, bipartite to the base; ovules in 2–5 rows on each placenta; capsule usually 3-valved; testa loose; lf-blades filiform to peltate; stipules scarious, variously fringed or divided, or rarely wanting; infl a sympodial cyme, nodding at the undeveloped tip. 100, widespread.

1 Lf-blades filiform, not distinct from the petiole .... 1. *D. filiformis.*
1 Lf-blades expanded, distinct from the petiole.
  2 Lf-blades suborbicular, broader than long .... 2. *D. rotundifolia.*
  2 Lf-blades linear to spatulate or cuneate, longer than broad.
    3 Stipules adnate to the petiole.
      4 Lf-blades elongate-spatulate; seeds fusiform, 1–1.5 mm .... 3. *D. anglica.*
      4 Lf-blades linear; seeds rhomboidal, 0.5–0.8 mm .... 4. *D. linearis.*
    3 Stipules nearly free, or lacking.
      5 Scapes glabrous; stipules conspicuous.
        6 Seeds 0.7–1 mm, irregularly papillate; pet 4–5 mm, white .... 5. *D. intermedia.*
        6 Seeds 0.4–0.5 mm, with papillate-corrugate ridges; pet 6–7 mm, pink, varying to almost
          white .... 6. *D. capillaris.*
      5 Scape glandular-hairy; stipules absent or much-reduced; seeds 0.3–0.4 mm, pitted .... 7. *D. brevifolia.*

1. Drosera filiformis Raf. Lvs erect, 8–25 cm, covered with long, purple, glandular hairs, the blade not differentiated from the petiole; stipules adnate, fimbriate on the marginis, forming matted brown wool at the base of the lvs; scape 6–22 cm, with 4–16 fls 1 cm wide; sep glandular hairy, 4–7 mm; pet purple, broadly ovate, erose at the tip; seeds black, 0.5–0.8 mm, ellipsoid, abruptly constricted near the ends, coarsely crateriform-pitted, the pits in 16–20 rows; $2n=20$. Along the coast; Mass. and R.I. to Del.; S.C. to Fla. and La. July–Sept.

2. Drosera rotundifolia L. Petioles 1.5–5 cm, flat, glandular-hairy; lf-blades 4–10 mm, broader than long, much shorter than the petiole; stipules 4–6 mm, adnate, fimbriate along the upper half, scape glabrous, 7–

35 cm, with 3–15 fls 4–7 mm wide; sep 4–5 mm; pet white to pink, spatulate, longer than the sep; seeds light brown, lustrous, 1–1.5 mm, sigmoid-fusiform, finely and regularly striate longitudinally; $2n=20$. Common in bogs and swamps; circumboreal, in Amer. s. to S.C., Ga., Tenn., Ill., and Calif. June–Sept.

3. Drosera anglica Hudson. Petioles 3–7 cm, glabrous or sparsely glandular-hairy; lf-blades narrowly obovate to elongate-spatulate, 15–35 × 3–4 mm; stipules 5 mm, adnate, fimbriate along the upper half; scape glabrous, 6–25 cm, with 1–9 fls 6–7 mm wide; sep minutely glandular-denticulate, 5–6 mm; pet white, spatulate, 6 mm; seeds black, sigmoid-fusiform, 1–1.5 mm, longitudinally striate-areolate; $2n=40$. Peaty and boggy places; circumboreal, s. in Amer. to Que., n. Me., Bruce Co., Ont., n. Minn., n. Mich., n. Wis., and Calif. June–Aug. (*D. longifolia,* in part.) Perhaps originally an amphiploid of nos. 2 and 4.

4. Drosera linearis Goldie. Petioles 3–7 cm, flat, glabrous, lf-blades linear, 2–5 cm × 2 mm; stipules 5 mm, adnate, marginally fimbriate; scape glabrous, 6–13 cm, with 1–4 fls 6–8 mm wide; sep minutely glandular-denticulate, 4–5 mm; pet white, obovate, 6 mm; seeds black, oblong-obovoid, rhomboidal, 0.5–0.8 mm, densely and irregularly crateriform-pitted; $2n=20$. Bogs and wet shores, especially in calcareous soils; se. Can., s. to Me., Mich., Wis., and Minn., and w. irregularly through s. Can. to B.C. and nw. Mont. July–Aug.

5. Drosera intermedia Hayne. Stem 1–8 cm, with lvs in a rosette or also at intervals for several cm; petioles 2–5 cm, glabrous; lf-blades oblong-spatulate to obovate, 8–20 cm × 4–5 mm, with long glandular hairs on the upper side; stipules adnate for 1 mm, then breaking into several setaceous segments 2–5 mm; sep 3–4 mm; pet 4–5 mm, white; seeds reddish-brown, oblong, 0.7–1 mm, blunt at the ends, densely and irregularly covered with long papillae; $2n=20$. In wet places and shallow water; interruptedly cicumboreal, in Amer. from Nf. to Fla., w. to Minn., Ill., Tenn., and Tex. July, Aug. (*D. longifolia,* in part.) Hybridizes with no. 2.

6. Drosera capillaris Poiret. Petioles 0.6–4 cm, sparsely glandular-hairy; lf-blades broadly spatulate, 5–10 × 3–5 mm, usually shorter than the petiole; stipules free, or adnate the first 1 mm, then breaking into numerous setaceous segments 3–5 mm; scape glabrous, 4–20 cm, with 2–20 fls; sep 3–4 mm; pet 6–7 mm, pink, varying to almost white; seeds brown, ellipsoid to oblong-ovoid, 0.4–0.5 mm, asymmetric, coarsely papillose-corrugated in 14–16 ridges; $2n=20$. Common in wet low places, chiefly on or near the coastal plain; Va. to Fla. and Tex., and in Tenn.; also in W.I., Mex., and n. S. Amer. May–Aug.

7. Drosera brevifolia Pursh. Petioles 5–10 mm, dilated, glabrous; lf-blades cuneate, 4–10 mm, usually longer than the petiole; stipules absent or reduced to one or 2 minute, setaceous segments; scape 4–9 cm, with 1–8 fls; sep glandular-hairy, 2.5–3.5 mm, pet rose to white, obovate, 4–8 mm; seeds black, 0.3–0.4 mm, obovoid-oblong, crateriform-pitted, with 10–12 rows of pits; $2n=20$. Common in wet low places, chiefly on or near the coastal plain, from Va. to Fla. and Tex., and in Ky. and Tenn.; also in S. Amer. Mar–May. (*D. annua; D. leucantha; D. maritima.*)

FAMILY **CISTACEAE**, the Rock-rose Family

Fls hypogynous, perfect, regular except the cal; sep typically 5, the 2 outer much smaller or narrower than the 3 inner, and sometimes adnate to them; pet 5 or sometimes 3, distinct, often fugacious, or lacking in some fls; stamens irregular in number, often numerous, the filaments distinct; ovary superior, unilocular, with 3(–10) parietal (often intruded) placentas; ovules few to many; style 1, or the stigma(s) sessile; fr a loculicidal capsule, usually separating to the base and ± enclosed by the persistent cal; seeds with abundant endosperm and a dicotyledonous, usually curved or coiled or folded embryo; herbs or shrubs with simple, entire, alternate, opposite, or verticillate lvs and cymose fls. 8/200, widespread.

1 Pet (in petaliferous fls) 5, yellow, conspicuous.
  2 Lvs lanceolate or oblong; style very short, not over ca 1 mm .... 1. *Helianthemum.*
  2 Lvs narrowly linear or scale-like; style slender, ca 2 mm .... 2. *Hudsonia.*
1 Pet 3, dark red, minute; style none .... 3. *Lechea.*

1. HELIANTHEMUM Miller. Frostweed. Sep 5, but the outer 2 much narrower (and often shorter) than the inner and often partly or largely adnate to them; pet 5 in the first fls of the season, convolute in bud, yellow, fugacious, the later fls in our spp. mostly cleistogamous and apetalous; stamens 10–50 or more in open fls, (3)4–6(–8) in cleistogamous ones; style in ours shorter than the ovary, up to ca 1 mm with a capitate stigma, or the stigma virtually sessile; placentas membranous-intruded; embryo curved into a hook, or ring or ± circinate; perennial herbs (ours), or small shrubs with well developed but relatively small and narrow, alternate (ours), short-petiolate or virtually sessile lvs and stellate-hairy herbage, the frs glabrous. (*Crocanthemum.*) 100, widespread.

1 Petaliferous fls solitary (2) at the top of the stem (later overtopped by lateral branches); capsules of
   cleistogamous fls with 5–14 seeds.
  2 Stem at maturity with strongly ascending branches and branchlets .... 1. *H. canadense.*
  2 Stem at maturity with widely divergent branches and branchlets .... 2. *H. dumosum.*
1 Petaliferous fls in infls of 2–10+ at the top of the stem (sometimes later overtopped by lateral branches);
   capsules of cleistogamous fls with 1–2(3) seeds.
  3 Stems ± clustered on a branching caudex; lvs cuneate at base .... 3. *H. bicknellii.*
  3 Stems scattered on a creeping rhizome; lvs attenuate at base .... 4. *H. propinquum.*

    1. Helianthemum canadense (L.) Michx. Stems ± clustered and ± erect, at first anthesis simple or nearly
so and mostly 1.5–3 dm, with a single (2) terminal open fl, later much-branched, to 3–6 dm, with strongly
ascending branches and branchlets overtopping the main axis and bearing small terminal and axillary glomerules of small, cleistogamous fls; lvs of the main stem mostly 2–3(–4) cm × 5–8 mm, elliptic-oblanceolate
or narrowly elliptic, the lower side densely stellate-pubescent, the upper greener, its shining surface visible
through the sparse covering of mixed stellate and longer (1mm) simple hairs, but these lvs deciduous, and
those of the branches smaller and with fewer or no long hairs; cal of open fls 6–9 mm, with evident long,
simple hairs as well as smaller stellate ones, the outer sep well developed, more than half as long as the inner;
pet 8–15 mm; cal of cleistogamous fls merely stellate, the outer sep largely adnate to the inner, with a tiny
free point; fr of open fls 6–7 mm, with 30–45 seeds, of cleistogamous fls 2–3 mm, with 5–10 seeds; seeds
papillate; $2n=20$. Dry, sandy soil or upland woods; Me. and N.S. to Minn., s. to Ga., Tenn., and Mo. Late
May–June. (*Crocanthemum c.; H. majus.*)

    2. Helianthemum dumosum (E. Bickn.) Fern. Resembling no. 1, but lower, often only 1 dm at first
anthesis, becoming diffusely branched and up to 3(–4) dm, with widely spreading branches and branchlets,
the central axis often obscured; upper lf-surface more densely stellate than in no. 1 (but less densely so than
the lower surface) and often with fewer or no long hairs; some fls transitional between the early, open ones
and the later, cleistogamous ones, which are typically solitary or paired at the tips and forks of the branches;
fr of open fls 4–7 mm, with 16–50 seeds, of cleistogamous ones 3–4.5 mm, with 8–14 seeds. Dry, sandy soil
and barrens; Mass. to L.I., blooming earlier than the other spp.

    3. Helianthemum bicknellii Fern. Stems ± clustered on a loosely branched caudex, mostly 2–5 dm, at
first anthesis simple or nearly so and with a terminal infl of (2–)6–10(–18) open, petaliferous fls, later with
numerous erect or appressed branches and branchlets bearing axillary glomerules of cleistogamous fls, but
these later branches not much if at all overtopping the main axis; lvs linear-oblong to narrowly elliptic,
cuneate at base, stellate-pubescent (more densely so beneath), without longer hairs, those of the main axis
2–3 cm, the others smaller; cal of open fls 5–7 mm, stellate-pubescent, without longer hairs, the outer sep
nearly as long as the inner; pet 8–12 mm; outer sep of cleistogamous fls with an evident, linear free tip 0.5–
1+ mm, fr of open fls 3.5–5 mm, with 12–26 seeds, of cleistogamous fls 1.5–2 mm, with 1–2(3) seeds; seeds
indistinctly reticulate. Dry, usually sandy soil; Me. to Minn. and s. Man., s. to n. Ga., Ark., and Colo. June–
July, 2–3 weeks later than no. 1. (*H. majus* and *Crocanthemum majus*, misapplied; *C. bicknellii.*)

    4. Helianthemum propinquum E. Bickn. Stems scattered on a creeping rhizome, 1–3 dm; lvs linear-
spatulate to oblong-linear, attenuate at base, those of the main axis 1–3 cm; petaliferous fls in a terminal
cymose cluster of 2–6, later often surpassed by the lateral branches; outer sep evidently shorter than the
inner, the free tip of those of the cleistogamous fls reduced to a mere knob 0.2–0.5 mm; frs of open fls with
12–15 seeds; otherwise much as in n. 3. Dry, sandy soil; Mass. and se. N.H. to Md., and disjunct in the mts.
of N.C. and Tenn. Blooming earlier than no. 3, and later than no. 1.

    2. HUDSONIA L. False heather. Sep basically 5, but the 2 outer each connate with
an inner one, which appears as a subulate lobe above the middle; sep spreading at anthesis,
later connivent and twisted and enclosing the fr; pet 5, convolute in bud, yellow, fugaceous,
but often drying persistent; stamens 8–30(8–20 in our spp.); placentas nerve-like, biovulate;
style slender, ca 2 mm, with a minute stigma; seeds 1–3; embryo coiled into a closed
hook; much branched shrubs or subshrubs, forming dense mats or mounds or low bushes
as much as 1 m across, with small, alternate, linear or scale-like lvs and numerous fls,
each solitary and axillary or terminating a short axillary branch. 2, N. Amer.

1 Pedicels 5–15 mm; lvs 3–6 mm, linear or subulate, loose or spreading, green, sparingly villous .... 1. *H. ericoides.*
1 Pedicels 0–3 mm; lvs 1–3 mm; lance-ovate, closely appressed, usually densely tomentose .... 2. *H. tomentosa.*

    1. Hudsonia ericoides L. Lvs linear subulate, 3–6 mm, loosely erect to somewhat spreading, green,
sparingly villous; pedicels 5–15 mm; frs cylindric, loosely persistent after maturity; $2n=20$. On acid rocks
from Me. to Nf., and locally abundant in pine barrens from N.H. to Del.; S.C. May, June.

    2. Hudsonia tomentosa Nutt. Lvs lance-ovate or triangular, ± scale-like, 1–3 mm, closely imbricate and
appressed to the twigs except on vigorous shoots, densely white-tomentose (less so on vigorous shoots);
pedicels up to 3 mm; frs cylindric, promptly deciduous; $2n=20$. Locally abundant on and about coastal dunes
from Me. to N.C., and inland in sandy habitats from Lab. and N.E. to W.Va., w. to Wis., Minn., Man., and

s. Mack. May–July. A hybrid with no. 1 is *H.* × *spectabilis* Morse. The names *H. intermedia* and *H. tomentosa* var. *intermedia,* sometimes misapplied to the hybrid, are based on specimens of *H. tomentosa* from the Adirondack Mts.

3. LECHEA L.   Pinweed. Sep 5, the 2 outer linear or lanceolate, the 3 inner broadly ovate to obovate, concave in conformity to the capsule; pet 3, mostly smaller than the sep, imbricate in bud, reddish, marcescent; stamens mostly (3–)5–15(–25); style none; stigmas 3, plumose, sessile; ovules 2 on each side of the 3 intruded, expanded and shield-like, parietal placentas; fr 3-valved, maturing 1–6 seeds, largely or wholly enclosed by the persistent cal; embryo straight or curved; perennial (seldom biennial) herbs with few or solitary, erect stems, small, alternate to sometimes opposite or whorled, 1-nerved, entire, sessile or short-petiolate lvs, and large, finely leafy panicles of very numerous minute fls in mid- or late summer, and producing basal shoots with numerous crowded lvs late in the season. 20, N. Amer.

1 Outer sep about as long as or longer than the inner.
  2 Hairs of the stem spreading; inner sep keeled, glabrous except sometimes the keel . . . . 1. *L. mucronata.*
  2 Hairs of the stem appressed or strongly ascending; inner sep hairy, scarcely keeled.
    3 Mature cal with an indurate, shiny, yellowish, obconic to subcylindric base 0.5 mm, contrasting
        with the softer upper portion; fruiting pedicels averaging 2+ mm . . . . 4. *L. racemulosa.*
    3 Mature cal not so differentiated; fruiting pedicels averaging less than 2 mm.
      4 Lvs very narrow, the cauline ones averaging 10 (+) times as long as wide . . . . 2. *L. tenuifolia.*
      4 Lvs wider, the cauline ones averaging 5 times (or less) as long as wide . . . . 3. *L. minor.*
1 Outer sep not reaching much if at all beyond the middle of the inner.
  5 Fr and cal (measured from the basal articulation) narrowly obovoid, twice as long as thick (before
      dehiscence) . . . . 4. *L. racemulosa.*
  5 Fr and cal globose to broadly ovoid, notably less than twice as long as thick.
    6 Herbage canescent with copious appressed hairs; lower lf-surface ± pubescent.
      7 Flowering branches spreading at 45° or more; lvs of basal shoots elliptic or oblong, 5–12 mm,
          a third or a fourth as wide; mainly Atlantic seaboard . . . . 5. *L. maritima.*
      7 Flowering branches diverging at 30° or less; lvs of basal shoots narrowly lanceolate, 3–5 mm;
          Middle West and Ont. . . . . 6. *L. stricta.*
    6 Herbage green; lvs hairy beneath only on the midrib and margin, or glabrous.
      8 Fruiting calyx obovoid, acute at base; seeds 1–3(4), flattened-ovoid . . . . 7. *L. pulchella.*
      8 Fruiting calyx subglobose, blunt at base; seeds 4–6, shaped like a section of an orange . . . . 8. *L. intermedia.*

1. Lechea mucronata Raf.   Stems 2–8 dm, spreading-villous; lvs villous beneath on the margins and midvein, otherwise glabrous or nearly so, often some of them whorled, those of the basal shoots ovate-elliptic, to 15 mm, the cauline ones lanceolate to oblanceolate or elliptic, 1–3 cm; fls densely aggregated on short lateral branches; inner sep very concave, smooth except for the broadly linear, roughened, sometimes sparsely hairy keel; outer sep rough, sometimes pilose, about equaling the inner; fr subglobose, about equaling the cal. Fields and open woods in dry or sandy soil; N.H. to Mich., Ill., Mo., and Okla., s. to Fla. and Tex. (*L. villosa.*)

2. Lechea tenuifolia Michx.   Stems 1–3 dm, with a few erect or ascending hairs; lvs of basal shoots linear, crowded, 3–6 mm; cauline lvs linear or filiform, to 2 cm, usually under 1(–1.5) mm wide and at least 10 times as long as wide, sparsely hairy only beneath, soon deciduous; panicle occupying half the plant or more, its numerous branches often racemiform and secund; inner sep concave, with a conspicuous midvein but not keeled, hairy over most or all of the surface; outer sep linear, usually distinctly exceeding the inner; fr broadly ovoid to subglobose, usually slightly shorter than and enclosed by the cal. Dry soil, upland woods, and barrens; s. Me. to S.C., chiefly near the coast; Minn. to Ind., s. and sw. to Miss. and Tex. A more robust and more hairy form, chiefly to the w. and sw. of our range, but also in nw. Ill., has been segregated as var. *occidentalis* Hodgdon.

3. Lechea minor L.   Stems 2–5 dm, with erect or strongly ascending hairs; lvs sparsely pilose beneath on margins and midvein, those of the basal shoots commonly opposite or whorled, elliptic-ovate, 3–6 mm, the cauline lanceolate to narrowly oblong, mostly avg not more than 5 times as long as wide, short-petiolate, often whorled; fls in loose, irregular clusters or short racemes; inner sep rounded on the back, with very narrow midrib, scarcely keeled, ± pubescent over the surface; outer sep green, similarly hairy, equaling or usually exceeding the inner; fr subglobose to ellipsoid, about equaling or slightly surpassing the cal, the top exposed. Dry soil, usually in sunny places; Mass. and Vt. to s. Ont. and n. Ind., s. to Fla. and La.

4. Lechea racemulosa Michx.   Stems 2–4 dm, thinly appressed-hairy, lvs usually glabrous above, pilose on the margins and on the midrib beneath, those of the basal shoots oblong or elliptic, 4–6 mm, a third as wide, often whorled, the cauline ones narrowly lanceolate or oblanceolate, 5–8 times as long as wide; panicle usually occupying half the plant, with numerous ascending branches, the fls in short, racemiform clusters on pedicels 1–3 (avg 2 or more) mm at maturity; outer sep from distinctly shorter than to about as long as the inner; fruiting cal with a differentiated basal part, as indicated in the key; seeds 1–3. Dry soil; se. N.Y. to O. and s. Ind., s. to Ga. and Ala.; also nw. Ind. and s. Mo.

5. **Lechea maritima** Leggett. Plants canescent; stems erect or ascending, 1–4 dm; lvs canescent across the surface beneath, those of the basal shoots mostly whorled, elliptic, 5–12 mm, 3–4 times as long as wide, the cauline ones narrowly lanceolate or oblanceolate, to 2 cm; panicle forming half to two-thirds of the plant, with divergent branches; pedicels very hairy; sep hairy, about equaling or slightly exceeding the subglobose or broadly obovoid fr; seeds 2–5(6). Sandy soil near the sea; N.B. to Va.; also inland in N. Engl. The Va. plants have been segregated on minor differences as var. *virginica* Hodgdon, and those from N.B. as var. *subcylindrica* Hodgdon.

6. **Lechea stricta** Leggett. Plants canescent; stems commonly procumbent for 1–5 cm, thence abruptly erect, 1–4 dm; lvs canescent across the surface beneath, those of the basal shoots crowded but not conspicuously whorled, 3–5 mm, narrowly lanceolate, the cauline ones narrowly lanceolate to oblanceolate, to 15 mm; panicle forming a third to half the plant, with closely ascending branches, the ultimate branchlets crowded and racemiform; pedicels equaling or surpassing the cal, very hairy; sep densely hairy, the inner 3 about equaling the subglobose fr; seeds 3 or 4. Dry, sandy woods, prairies, and shores; s. Ont.; nw. Ind. to Minn. and Nebr.

7. **Lechea pulchella** Raf. Stems 2–8 dm, thinly appressed-hairy; lvs of basal shoots whorled or scattered, narrowly lanceolate to oblanceolate, 4–8 mm, sparsely hairy beneath on the midrib and margin; lvs similar but larger, to 2 cm; panicle occupying half the plant, the primary branches diverging at 45–60°, the ultimate branches with fls in crowded or racemiform clusters; pedicels usually shorter than the cal; outer sep short; fr subglobose or broadly ellipsoid, often distinctly surpassing the obovoid, basally acute cal; seeds 1–3(4) flattened-ovoid, smooth, dark brown. Dry soil; Mass. to s. Ont. and nw. Ind., s. to Del., n. Va., and O., and along the coastal plain to Fla. and La. (*L. leggettii.*)

8. **Lechea intermedia** Leggett. Stem 2–6 dm, thinly appressed-hairy; lvs of basal shoots oblong-lanceolate to narrowly elliptic, 3–7 mm, sparsely pilose on the midrib and margin beneath, or glabrous; cauline lvs linear-oblong, with a few hairs on the midrib beneath, or glabrous; panicle occupying a third or half the plant, narrowly cylindric, the lateral branches seldom over 5 cm; pedicels equaling or surpassing the sep; cal subglobose, with obtuse to abruptly rounded base; outer sep short; fr subglobose to slightly ovoid or obovoid, barely exceeding the sep; seeds 4–6, shaped like a section of an orange, pale brown, partly and irregularly invested with a gray membrane. Dry, sterile or sandy soil; Cape Breton I. to Ont. and Minn., s. to Pa., O., Ill., and in the mts. to Va.; also Nebr., N.D., and Sask. (*L. juniperina.*)

FAMILY **VIOLACEAE**, the Violet Family

Fls hypogynous, perfect, in ours (tribe Violeae) strongly irregular; sep 5, commonly persistent; pet 5, distinct, the lowest one in ours prolonged behind into a spur; stamens 5, the filaments very short, distinct or ± connate, the anthers commonly connivent around the ovary, often the 2 lower ones (Violeae) or all of them (the woody genera) with a gland-like or spur-like nectary on the back; connective often prolonged into a membranous appendage; ovary superior, unilocular, with mostly 3 parietal placentas; style solitary, often distally enlarged or otherwise modified, the stigma simple or lobed, often oblique; ovules (1–) several or many on each placenta; fr a loculicidal capsule (ours) or sometimes a berry; seeds with straight embryo and flat cotyledons embedded in the copious, oily endosperm, often arillate; herbs (ours) or less often woody plants; lvs alternate or seldom opposite or even whorled, stipulate, simple and entire or toothed to sometimes lobed or dissected, but without well defined lfls, the fls axillary, often nodding, or in racemes, heads, or panicles. 16/800, cosmop.

1 Sep without auricles; stamens connate into a sheath, none spurred .... 1. *Hybanthus.*
1 Sep with basal auricles; stamens merely connivent, the 2 lower ones spurred .... 2. *Viola.*

1. **HYBANTHUS** Jacq., nom. conserv. Green violet. Sep subequal, without auricles; pet small, subequal (in ours), emarginate, recurved at the tip, the lower one wider than the others, gibbous or spurred at base, and more strongly notched at the tip; stamens (in ours) connate into a sheath surrounding the pistil and bearing a bilobed gland at the base; style hooked at the tip; fr a 3-valved capsule; leafy-stemmed perennial herbs (ours) or shrubs with alternate (seldom opposite or even whorled) lvs and inconspicuous, greenish-white fls pedicellate in the axils. (*Cubelium.*) 150, mostly warm reg.

1. **Hybanthus concolor** (T. Forster) Sprengel. Stems to 1 m, solitary or clustered, from a crown of fibrous roots; herbage ± hairy; lvs broadly elliptic to ovate-oblong, 7–16 cm, abruptly acuminate, tapering to slender petioles 1–2 cm; fls greenish-white, 4–5 mm, solitary or several in the axils on strongly recurved peduncles jointed beyond the middle; sep linear, nearly as long as the pet; fr oblong-ellipsoid, 1.5–2cm; seeds subglobose,

5 mm; $2n=48$. Rich woods and ravines; Vt. and Ont. to Mich. and Kans., s. to N.C., Ga., and Ark. May, June. (*Cubelium c.*)

Hybanthus verticillatus (Ortega) Baillon, a southwestern sp., closely approaches our range in e. Kans. and may be sought in Mo. It is smaller, seldom over 4 dm, with narrow, subsessile, alternate to casually opposite or even verticillate lvs mostly well under 1 cm wide, and with smaller fls, frs, and seeds.

2. **VIOLA** L. Violet. Sep usually with posterior auricles; pet somewhat unequal, the 2 lateral ones often bearded internally at the base, the lower one usually with a basal spur or sac and sometimes bearded at the throat; stamens 5, the 2 lowermost bearing appendages that extend into the spur of the lower pet; ours herbs with axillary or basal 1-fld peduncles bearing 2 small bracts near the middle. Most spp. produce normal petaliferous fls in spring, and very fertile cleistogamous fls in summer. All spp. with cyanic (rather than yellow) fls produce occasional white-fld forms. Hybrids between closely related spp. are common, and spp. 3–8 form an intergrading polyploid complex. 400, mainly N. Temp.

1 Plants essentially acaulescent.
  2 Fls bright yellow .... 1. *V. rotundifolia.*
  2 Fls various shades of blue or violet to purple, varying to white.
    3 Petaliferous fls with prominently protruded stamens; cleistogamous fls none .... 2. *V. pedata.*
    3 Petaliferous fls with stamens not protruded; cleistogamous fls well developed.
      4 Rhizome thick and often short; stolons wanting; lateral pet bearded.
        5 Blades of the principal lvs relatively narrow, mostly narrowly ovate or narrower, generally at least 1.5 times as long as wide, subtruncate to shallowly cordate at base, sometimes all of them merely crenate-serrate, sometimes some of them with notably larger and more spreading teeth at base or with small basal lobes .... 3. *V. sagittata.*
        5 Blades of the principal lvs wider, mostly less than 1.5 times as long as wide, variously truncate to often strongly cordate at base, and varying from merely toothed to strongly lobed or palmately dissected.
          6 Lvs all merely crenate-serrate, cordate at base.
            7 Lf-blades held at an angle to the strongly ascending to suberect petioles; plants of moist or wet places.
              8 Fls relatively narrow, the lateral pet directed forward at an angle; cordilleran sp., rare in cold calcareous bogs in the n. part of our range .... 6. *V. nephrophylla.*
              8 Fls wide open, almost pansy-like in form, the lateral pet spreading; common and widespread spp.
                9 Hairs of the beard of the lateral pet short, generally well under 1 mm, many or all of them strongly clavate or with a conspicuous expanded knob at the end .... 4. *V. cucullata.*
                9 Hairs of the beard of the lateral pet longer, commonly some of them 1 mm or more, only slightly or not at all expanded distally .... 5. *V. sororia.*
            7 Lvs mostly flat on the ground, or nearly so, forming a rosette; plants of dry or well drained soil; mostly southern .... 7. *V. villosa.*
          6 Lvs, or some of them (at least some of the ones produced in the summer) usually lobed or cleft, or at least prominently and sharply toothed near the base, the blade ascending or erect in alignment with the petiole; plants typically of well drained soil .... 8. *V. palmata.*
      4 Rhizome slender and elongate; plants stoloniferous (except in nos. 11 and 12); lateral pet bearded or beardless.
        10 Style ending in a small, down-pointed hook; intr., locally escaped .... 9. *V. odorata.*
        10 Style dilated upwards, cuspidate, with a conic beak on the lower side; native.
          11 Pet mostly lilac or lavender.
            12 Spur short, ca 2 mm; stolons present .... 10. *V. palustris.*
            12 Spur large and blunt, 4–7 mm; stolons lacking .... 11. *V. selkirkii.*
          11 Pet white, the lower with brown-purple lines toward the base.
            13 Lvs relatively broad, up to ca 1.2 times as long as wide, evidently cordate at base.
              14 Lvs reniform, with shallow basal sinus, obviously wider than long; no stolons .... 12. *V. renifolia.*
              14 Lvs cordate-ovate to cordate-orbicular, with a deep basal sinus; plants stoloniferous.
                15 The 2 upper pet in life strongly reflexed and somewhat twisted; fr ± purple .... 13. *V. blanda.*
                15 The 2 upper pet not (or only slightly) recurved, not twisted; fr green ... 14. *V. macloskeyi.*
            13 Lvs narrower, generally at least 1.5 times as long as wide, only slightly or not at all cordate at base.
              16 Lvs mostly 1.5–2.5 times as long as wide .... 15. *V. primulifolia.*
              16 Lvs mostly more than 3 times as long as wide .... 16. *V. lanceolata.*
1 Plants with leafy stems (though sometimes almost acaulescent at first flowering), the fls axillary.
  17 Stipules entire or only shallowly toothed, at least the lower ones somewhat scarious.
    18 Fls yellow, the pet commonly with brown-purple veins near the base, sometimes purplish-tinged outside.

19 Lvs relatively narrow, mostly (2–)2.5–6 times as long as wide, subentire, mostly tapering
    to the base; prairie sp. .... 17. *V. nuttallii.*
19 Lvs broader, seldom as much as twice as long as wide; spp. of forested regions.
  20 Lf-blades, or some of them, often 3-parted; unlobed lvs, when present, with truncate
     to broadly tapering base .... 18. *V. tripartita.*
  20 Lf-blades merely toothed, with cordate or hastate base.
    21 Lvs hastate or narrowly cordate; rhizome long, slender, white .... 19. *V. hastata.*
    21 Lvs broadly cordate; rhizome short, stout, brown or yellowish .... 20. *V. canadensis.*
17 Stipules pectinately toothed or fringed (varying to rarely entire in *V. adunca*), not at all scarious.
  22 Stipules neither leaf-like nor lyrate-pinnatifid; style slender, not capitate; perennials.
    23 Lateral pet bearded; spur mostly 3–6 mm; style-tip bent.
      24 Fls creamy-white; lvs finely toothed; sep ciliate .... 22. *V. striata.*
      24 Fls lavender to violet (except albinos); lvs coarsely toothed; sep not ciliate.
        25 Stems erect to prostrate, but not as in no. 25; northern.
          26 Stipules lance-linear, bristly-toothed less than half their length; lvs subtrun-
            cate to subcordate at base .... 23. *V. adunca.*
          26 Stipules broadly lanceolate, bristly-toothed usually over half their length; lvs
            with ± strongly cordate base .... 24. *V. conspersa.*
        25 Stems becoming prostrate and terminating in a rosette that bears petaliferous fls
          the next year; southern .... 25. *V. walteri.*
    23 Lateral pet beardless; spur 7–12 mm; style-tip straight .... 26. *V. rostrata.*
  22 Stipules lf-like and lyrate-pinnatifid; style much enlarged upwards; annuals, except sometimes
    no. 28.
    27 Pet of open fls much longer than the sep.
      28 Middle lobe of the stipules linear to linear-spatulate, not toothed or crenate; mid-
        and late-season fls cleistogamous .... 27. *V. rafinesquii.*
      28 Middle lobe of the stipules spatulate or oblong-spatulate, distinctly toothed or crenate;
        fls all open (i.e., chasmogamous) .... 28. *V. tricolor.*
    27 Pet shorter than or merely equaling the sep; fls all open .... 29. *V. arvensis.*

1. **Viola rotundifolia** Michx.   Round-lvd yellow v. Acaulescent; rhizome elongate, covered with persistent lf-bases; lvs at anthesis broadly cordate-ovate, 2–3 cm, nearly as wide, finely undulate-crenate, their petioles with crisp hairs often extending onto the lower part of the blade; blades enlarging in summer to 5–12 cm, orbicular-ovate, with strongly cordate base and overlapping basal lobes; fls developing with the lvs, on rather short peduncles, bright yellow, the 2 lateral pet bearded, the 3 lower with brown veins; style clavate and abruptly capitate; cleistogamous fls on short, lfless stolons, appearing racemose; fr ellipsoid, 5–8 mm; seeds ivory; $2n=12$. Deep, rich, usually coniferous woods; se. Que. and Me. to sw. Ont., s. to Pa., N.J., Del., and along the mts. to S.C. and Ga. Apr., May.

2. **Viola pedata** L.   Bird's-foot v. Acaulescent, with an erect rootstock, not stoloniferous, glabrous or nearly so; principal lvs 3-parted, the lateral segments again 3–5-cleft into linear or lanceolate divisions often with 2–4 teeth near the tip; fls 2–4 cm wide; pet beardless, all lilac-purple, or less commonly the upper 2 dark violet; tips of the large, orange stamens conspicuously exsert; cleistogamous fls none; style clavate, beakless, obliquely concave at the top, the stigma within a small protuberance near the center of the cavity; seeds coppery; $2n=56$. Dry fields and open woods; Me. to Minn., s. to n. Fla. and e. Tex. May, June.

3. **Viola sagittata** Aiton.   Arrowhead-v. Acaulescent, glabrous or hairy; principal lvs relatively narrow, mostly ovate or narrower, generally at least 1.5 times as long as wide, subtruncate to shallowly cordate at base, sometimes all merely crenate-serrate, sometimes some of them with notably larger and more spreading teeth at the base or with small basal lobes; fls 2–2.5 cm wide, violet-purple, the 3 lower pet bearded at base; style, as in the following 5 spp., dilated upwards in a vertical plane, capitate, with a conic beak on the lower side, the stigma within the tip of the beak; cleistogamous fls on erect or ascending peduncles; fr ovoid; seeds brown; $2n=54$. Dry to moist, open woods, clearings, and meadows, less often along streams, often in drier, sunnier habitats than no. 5; Me. and N.S. to Minn., s. to Fla. and La. Apr.–June. (*V. fimbriatula; V. emarginata,* the name applied to apparent hybrids with other stemless blue spp.)

4. **Viola cucullata** Aiton.   Blue marsh-v. Glabrous, acaulescent; rhizome short, oblique or horizontal, often branching to form colonies; lvs all merely crenate-serrate, cordate at base, the blade typically held at an angle to the strongly ascending petiole, ovate to reniform at anthesis, the summer-lvs up to 10 cm wide, all except the earliest acute or shortly acuminate; fls on long, slender peduncles, typically elevated well above the lvs; pet light violet to sometimes white, usually with a dark-veined eye, the 2 lateral pet densely bearded with short hairs generally well under 1 mm, many or all of the hairs strongly clavate or with a conspicuously expanded knob at the end; spurred pet glabrous, somewhat shorter than the lateral ones; cleistogamous fls long and slender, on long, erect peduncles, the auricles nearly as long as the sep; fr ovoid-cylindric, only slightly longer than the sep; seeds nearly black; $2n=54$. Bogs, swamps, and other wet or moist places in shade or sun, not weedy; Nf. to Ont. and Minn., s. to N.C., Ga., Tenn., and e. Ark. Apr.–July.

5. **Viola sororia** Willd.   Dooryard-v. Glabrous to often evidently spreading-hairy, acaulescent; rhizome short, stout, oblique or horizontal, sometimes branching to form colonies; lvs all merely crenate-serrate, cordate at base, the blade typically held at an angle to the strongly ascending petiole, ovate to orbicular or reniform, acute to often obtuse or rounded above, the summer-lvs to 13 cm wide; fls seldom elevated much above the lvs, 2–3 cm wide, relatively broad and open, appearing flat-faced; pet relatively wide, deep violet

to lavender or white, the 2 lateral ones densely bearded with relatively long hairs (many of them ca. 1 mm or more) that are only slightly or not at all expanded upwards; spurred pet glabrous or sometimes bearded, nearly or fully as long as the lateral ones; cleistogamous fls stout, on horizontal to erect peduncles; fr thickly ellipsoid, evidently surpassing the sep; seeds dark brown; $2n=54$. Moist or wet places, sometimes in disturbed sites and somewhat weedy; Nf. and Que. to B.C., s. to Fla., Tex., and Calif. Apr.–June. (*V. affinis,* a form with the spurred pet bearded; *V. papilionacea,* a subglabrous form; *V. priceana,* the Confederate violet, with pale gray-blue fls and a conspicuous violet eye; *V. novae-angliae,* a form transitional to no. 3; *V. floridana; V. langloisii; V. latiuscula; V. missouriensis; V. pratincola; V. septentrionalis.*)

6. **Viola nephrophylla** Greene. Lvs glabrous, relatively small, the later ones not notably larger than the earlier ones, seldom as much as 7 cm wide; fls often elevated somewhat above the lvs, narrower and more elongate than in no. 5, the relatively narrow lateral pet directed forward at an angle, tapering to an almost clawed base; spurred pet bearded or not; $2n=54$; otherwise much like no. 5. Mainly in the w. cordillera, but extending e. in cold, wet places (often with a high pH) to ne. U.S. (Mich., Wis., n. N.Y., and New Engl.) and se. Can. May–July.

7. **Viola villosa** Walter. Southern woolly v. Acaulescent, with a short, oblique to erect rootstock, downy throughout, varying to glabrous except for the appressed-hairy upper surface of the lvs; lvs tending to be evergreen, often color-patterned, flat on the ground or nearly so, spreading out to form a rosette, relatively small, the blade 1.5–7.5 cm, broadly cordate-ovate to cordate-orbicular or reniform-cordate, usually obtuse or rounded above, commonly with a narrow basal sinus, all merely crenate or subentire; stipules entire to fimbriate; fls short-pedunculate, violet, 1.5–2 cm wide, the pet relatively narrow and mostly directed forward, so that the fls seldom appear fully open, the 3 lower pet bearded, the spur large and globose; cleistogamous fls on prostrate to erect peduncles; frs ovoid or ovoid-cylindric; seeds brown; $2n=54$. Dry or well drained, often sandy soil in open woods and clearings; se. U.S. from Fla. to Tex., n. to Va., Tenn., and s. Ind., and occasionally to N.Y. and s. Conn. Apr., May. (*V. hirsutula,* the form with the lvs glabrous beneath.)

8. **Viola palmata** L. Wood-v. Acaulescent, with a short, oblique to erect rootstock, evidently spreading-hairy to virtually glabrous; lvs ascending or erect, the blade ± aligned with the petiole, the early ones often cordate and merely toothed as in no. 5, but at least the later ones (or all of them) variously lobed or dissected, or at least with large, sharp, spreading teeth toward the base, often trilobed, with broad, merely toothed central lobe and narrower, more cleft lateral lobes; fls about equaling or somewhat surpassing the early lvs, often surpassed by the later ones, deep to pale violet, varying to streaked or white, mostly 2–3 cm wide, the 3 lower pet bearded, or the spurred one seldom glabrous; cleistogamous fls on prostrate to erect peduncles; frs ovoid to ellipsoid, evidently surpassing the sep; seeds brown; $2n=54$. Well drained soil in woods, clearings, and glades; Me. to Minn., s. to Fla. and Tex. Apr., May. The common phase in most of our range, as principally described above, is var. *palmata.* (*V. brittoniana* and *V. septemloba,* with strongly dissected lvs; *V. chalcosperma, V. esculenta, V. lovelliana, V. pectinata, V. stoneana, V. triloba, V. viarum.*) Westward this passes into var. *pedatifida* (G. Don) Cronq. (*V. pedatifida, V. egglestonii*), mainly of the prairies from Alta. to Okla., e. to Man., Mich., Ind., and the cedar glades of Ky. and Tenn.; lvs more strongly dissected, typically 3-parted, each division again cleft into linear lobes, these often again cut into 2–4 segments, the earliest and latest lvs almost as much cut as the others. Extreme plants of var. *palmata,* from well outside the area of var. *pedatifida,* would pass as the latter var. in the absence of geographic data.

9. **Viola odorata** L. Sweet v. Finely hairy, with a long-creeping rhizome and spreading by numerous leafy stolons rooting at the nodes; lvs broadly ovate to orbicular, 2–6 cm, rounded or obtuse at the tip, evenly crenate, cordate at base, peduncles equaling or shorter than the lvs; fls very fragrant, typically deep violet, varying to white, the lateral pet usually bearded; style recurved-hooked at the tip; sep ciliate; ovary and fr hairy; cleistogamous fls on recurving peduncles, with broadly ovoid, hairy, purplish fr; seeds large, ochroleucous; $2n=20$. Native of Europe, cult. in many forms and often escaped, especially about dwellings. Apr., May.

10. **Viola palustris** L. Northern marsh-v. Delicate, glabrous plants from a long, slender rhizome, also stoloniferous; lvs thin, 2.5–3.5 cm wide, orbicular to reniform, crenulate, cordate or subcordate at base, the basal sinus wider than in no. 11; fls 1.5 cm wide; pet pale lilac or lavender (or nearly white) with purple veins, the lateral ones sparsely bearded, the spur 2 mm; style dilated upwards in a vertical plane, with a conic beak on the lower side, the stigma within the tip of the beak, as in the next 6 spp.; frs 5–6 mm, green, ellipsoid; seeds buff; $2n=24, 48$. Wet or moist places; circumboreal, in Amer. from Lab. to Alas., s. to the alpine regions of Me. and N.H., Manitou Isl., Mich., and in the West to Colo. and Calif. June–Aug. (*V. epipsila,* the diploid, including the Michigan plants.)

11. **Viola selkirkii** Pursh. Great-spurred v. Rather delicate plants from a long, slender, rhizome, without · stolons; lvs glabrous beneath, minutely spreading-hairy above, 1.5–3 cm at anthesis, later larger, crenate, broadly ovate-cordate, with a narrow basal sinus and converging or overlapping lobes; fls numerous, 1.5 cm wide; sep eciliate; pet pale violet, beardless, the spur large and blunt, 4–7 mm; frs small, stoutly ellipsoid or globose, 4–6 mm; seeds buff; $2n=24$. Cold mt. woods and shady ravines; circumboreal, s. in Amer. to n. Pa., n. O., Wis., Minn., S.D., and B.C. May–July.

12. **Viola renifolia** A. Gray. Kidney-lvd v. Rhizome elongate, becoming stout and scaly; stolons wanting; lvs reniform to reniform-cordate, distantly crenate-serrate, rounded above or shortly blunt-tipped, variously hairy on both sides, or on the lower surface only, or wholly glabrous; fls 7–10 mm wide; sep narrowly lanceolate; pet white, all beardless or bearded, the 3 lower with brown-purple veins near the base, the spur short and rounded; frs ellipsoid, 4–5 mm, the cleistogamous ones on horizontal to erect peduncles; seeds

brown with darker marks; $2n=24$. Arbor-vitae swamps and cold woods; Nf. to B.C., s. to Mass., Conn., N.Y., Mich., Minn., and Colo. May–July.

13. Viola blanda Willd. Sweet white v. Acaulescent, with slender creeping rhizomes and later in the season with well developed leafy or naked stolons, glabrous to commonly with some short, stiff, white hairs on the upper surface of the basal lobes of the lvs; petioles and peduncles usually red-tinged; lvs broadly cordate-ovate to cordate-orbicular, dark green with a satiny sheen, acute or acutish, the basal sinus narrow (or the lobes overlapping), avg larger than in no. 14, the largest ones 2–4+ cm wide at anthesis; fls often fragrant, on peduncles shorter than or surpassing the lvs; pet white, the 3 lower with brown-purple veins near the base, all beardless or nearly so, the 2 upper usually linear-oblong and reflexed-twisted, the 2 lateral pointing forward, so that the fl does not look flat-faced; cleistogamous fls on prostrate peduncles, with purplish, ovoid frs 4–6 mm; seeds dark brown, acute at base; $2n=48$. Cool ravines and moist, shady slopes in deep humus, usually under evergreens; s. Me. to N.J., O., and s. Ind., and in the mts. to N.C., Tenn., and n. Ga. Apr., May. (*V. incognita.*)

14. Viola macloskeyi F. Lloyd. Wild white v. Acaulescent, with slender creeping rhizomes, and freely stoloniferous, glabrous or variously hairy; petioles and peduncles usually greenish; lvs evidently crenate, broadly cordate-ovate to reniform, commonly 1–3 cm wide at anthesis, later to 8 cm, rounded to blunt to acutish above, the basal sinus more open than in no. 13; fls fragrant or inodorous, on erect peduncles equaling or rising above the lvs; pet white, the 3 lower with brown-purple veins near the base, the 2 upper narrowly obovate, only slightly (or not at all) recurved, not twisted, the 2 lateral ones beardless or scantily bearded, straighter and more spreading than in no. 13; cleistogamous fls on prostrate or erect peduncles, with green, ellipsoid or ovoid frs 4–6 mm; seeds brown or nearly black; $2n=24$. Springy land, and along cold streams, often in shallow water; Lab. and Nf. to B.C., s. to Del., O., Ind., Io., and in the mts. to S.C., Tenn., n. Ga., and Calif. Apr.–July. Our plants, as here described, are var. *pallens* (Banks) C. L. Hitchc., in contrast to the cordilleran var. *macloskeyi,* with smaller, less toothed lvs. (*V. pallens.*)

15. Viola primulifolia L. Primrose-lvd v. Acaulescent, with slender creeping rhizomes, and producing leafy stolons after anthesis; lvs variously hairy or glabrous, oblong to ovate or obovate, finely crenate, obtuse or rounded at the tip, rounded to truncate or cuneate at base, mostly 1.5–2.5 times as long as wide; peduncles shorter to longer than the lvs; pet white, the 3 lower with brown-purple veins near the base, the 2 lateral ones beardless or slightly bearded; cleistogamous fls on short, erect peduncles; frs oblong-ellipsoid, 7–10 mm; seeds reddish-brown to black; $2n=24$. Moist, open meadows and stream-banks, especially in sandy soil; N.B. to Fla., w. to Ind., Tenn., Okla., and Tex., more common eastward. Apr.–June.

16. Viola lanceolata L. Strap-lvd v. Acaulescent, glabrous or somewhat hairy, with slender creeping rhizomes, and in summer producing many creeping stolons with cleistogamous fls; lvs narrow, mostly more than 3 times as long as wide, tapering to the base, crenulate to serrate; pet white, the 2 upper linear-oblong or broadly oblong to obovate, the 3 lower with brown-purple veins near the base, all beardless; cleistogamous fls on erect peduncles; frs green, ellipsoid, 5–8 mm; seeds brown; $2n=24$. Wet, open places, often along streams and ponds, especially in sandy soil; Me. and N.B. to Fla., w. to w. Ont., Minn., e. Okla., and e. Tex. Apr.–June. Two vars.: Var. *lanceolata,* with lanceolate lvs mostly 3.5–6 times as long as wide, is chiefly northern, and widespread in our range. Var. *vittata* (Greene) Weath. & Griscom, with narrower lvs mostly 6–16 times as long as wide, is more southern, occurring chiefly on the coastal plain from Va. to Fla. and Tex., with a disjunct area at the s. end of Lake Michigan. (*V. vittata.*)

17. Viola nuttalli Pursh. Yellow prairie-v. Finely hairy to subglabrous, with numerous ascending stems to 1(–1.5) dm from a stout, deep-seated rhizome; stems bearing lvs and fls from near the base; lf-blades mostly lanceolate or lance-elliptic, varying to sometimes lance-ovate, 2–6 cm, tapering (seldom more abruptly contracted) to petioles nearly or quite as long, mostly (2–)2.5–8 times as long as wide; fls 1–1.5 cm wide, the pet yellow, with brown-purple lines near the base and often purplish-tinged outside, the lateral pet beardless or slightly bearded; style-head bearded; frs subglobose or ellipsoid, 6–8 mm; $2n=24$. Prairies and plains; widespread in the w. cordillera and e. onto the Great Plains, reaching Kans., Nebr., S.D., and w. Minn. Apr.–July. The plains plants, as here described, are var. *nuttallii.* Several other vars., differing in ploidy-level and in form, size, and vestiture of the lvs, occur farther w.

18. Viola tripartita Elliott. Three-parted v. Plants 1.5–4.5 dm from a short rhizome, usually puberulent or glandular-puberulent above; lvs 2–4(–7) near the top of the stem, 3–10 cm, often some (sometimes all) of them ovate to lance-ovate with long tip and truncate to broadly tapering base, usually some or all ± deeply 3-cleft, the terminal segment the largest; fls on slender, erect, axillary peduncles; sep lance-linear, ciliolate; pet bright yellow with brown-purple lines toward the base, the lateral ones bearded; style with a capitate, bearded summit; frs stoutly ellipsoid, 8–10 mm; seeds pale. Rich woods; sw. Pa. and s. O. to N.C., S.C., Ga., Ala., and ne. Miss. Apr., May.

19. Viola hastata Michx. Spear-lvd v. Plants 0.5–2.5 dm from a long, slender, branching, brittle, white rhizome, glabrous or slightly puberulent; lvs 2–4 near the top of the stem, the blades halberd-shaped to oblong-cordate, 2–8 cm at anthesis, larger later, dark green, usually with irregular silvery blotches above, shallowly but saliently serrate; occasionally one long-petioled, reniform-cordate radical lf accompanies the flowering stem, and others may arise from the rhizome; fls on slender, erect, axillary peduncles; sep lance-linear, with scarious, eciliate margins; pet bright yellow with brown-purple lines toward the base, purple-tinged on the back; style bearded at the capitate summit; frs ellipsoid, 6–8 mm; seeds pale brown; $2n=12$. Rich woods; Pa. and O. to S.C., n. Ga., and e. Tenn. Apr., May.

20. Viola pubescens Aiton. Yellow forest-v. Plants 1–4.5 dm from a short, stout, densely scaly, light brown or yellowish rhizome, softly hairy throughout, varying to essentially glabrous; lvs 2–4 near the top, often accompanied by 1–several long-petiolate, reniform-cordate radical lvs; cauline lvs orbicular-ovate to broadly ovate, short-pointed, crenate-dentate, with cordate or truncate-decurrent base, 4–10 cm, usually a little wider than long; stipules broadly ovate; stem often with 1 or 2 stipule-like sheathing bracts well above the base; fls on slender, axillary peduncles; pet clear yellow with brown-purple veins near the base, the lateral ones bearded; style bearded at the capitate summit; frs 10–12 mm, woolly or glabrous; seeds pale brown; $2n=12$. Rich woods, or sometimes in meadows; N.S. to Man. and N.D., s. to S.C., Ga., La., and Tex. Apr.–June. (*V. pensylvanica; V. eriocarpa,* the more glabrate phase.)

21. Viola canadensis L. Tall white v. Glabrous or minutely to less often evidently hairy; stems 2–4 dm; basal lvs well developed, long-petioled, ± cordate at base, often relatively wider than the cauline ones; cauline lvs several, the lower widely spaced, the upper larger and more crowded, mostly cordate and 5–10 cm; stipules lance-acuminate, entire, whitish and subscarious; fls slender-pedunculate from the upper axils; pet white inside, with a yellow base, the 3 lower ones with purplish lines toward the base, the lateral ones bearded, all (but especially the upper pair) ± purplish-tinted on the outside and sometimes less strongly so on the inside; style bearded at the capitate summit; frs ellipsoid-globose; $2n=24$. Moist woods, Nf. to Alas. and B.C., Ala., Ark., and Ariz. Mostly Apr.–July. Most of our plants belong to var. *canadensis,* widespread in e. U.S. and adj. Can., with a short, stout rhizome, short or no pubescence, and lvs usually longer than wide. The chiefly western var. *rugulosa* (Greene) C. L. Hitchcock, colonial by long stolons or superficial rhizomes, usually evidently hairy, and with the lvs often wider than long, extends e. to Wis. and Io. and is disjunct in the mts. of sw. Va., nw. N.C., and e. Tenn. (*V. rugulosa.*)

22. Viola striata Aiton. Creamy v. Plants glabrous or nearly so, clustered, 6–30 cm, reclining or erect from a short rhizome; lower cauline and basal lvs orbicular-ovate and blunt, the upper cordate-ovate and acute, with many small teeth, all plicate-rugulose; stipules large, foliaceous, lanceolate to lance-oblong, pointed, fimbriate-toothed; fls very numerous, rising well above the lvs; sep fimbriate-ciliate, lance-linear, with pronounced basal auricles; pet cream-color or ivory, usually with brown-purple veins near the base, but not washed with anthocyanins either front or back, the lateral ones strongly bearded, the spur thick and well-developed, 3–4 mm; style slender, the tip bent; cleistogamous fls on leafy branches 2–15 cm from the lf-axils in summer; frs ellipsoid-globose, 4–5 mm; seeds pale brown; $2n=20$. Usually in sun or light shade, along ditches and streams, often weedy; Mass. to s. Wis. and reputedly Minn., s. to Ga., Ark., and e. Okla. Apr.–June. A hybrid with no. 26 is *V.* × *brauniae* Glover & Cooperrider. One with no. 24 is *V.* × *eclipses* H. Ballard.

23. Viola adunca J. E. Smith. Hook-spurred v. Tufted on short or elongate rhizomes, at first anthesis erect, 2–8 cm, and nearly acaulescent, later to 15 cm and more spreading or prostrate, with evidently leafy stems; lvs darker green and thicker than in no. 24, ovate to suborbicular, obtuse, crenulate, subtruncate to subcordate at base, mostly 1–2.5(–3) cm wide; stipules lance-linear, with fimbriate-incised teeth above the middle; peduncles elongate; sep narrowly lanceolate, not ciliate; pet violet, the lateral ones bearded; spur 4–6 mm, straight and blunt, or often tapering to a short, incurved point; style slender, bent at the tip; frs 4–5 mm, ellipsoid; seeds dark brown; $2n=20, 40$. Sandy or gravelly, moist or sometimes dry soil; Greenl. and Lab. to Alas., s. to N.Y., Mich., Minn., and Calif. May–July. The widespread and highly variable var. *adunca* is ± puberulent. (*V. arenaria; V. subvestita.*) Var. *minor* (Hook.) Fern., more boreal or in our range more alpine, is essentially glabrous. (*V. labradorica.*)

24. Viola conspersa Reichb. American dog-v. Glabrous; stems beginning to fl when only 1–2 cm, becoming up to 20 cm in summer, clustered on an oblique, occasionally branched rhizome; basal lvs thin, pale green, glabrous or nearly so, reniform to orbicular, rounded or blunt, with a rather open sinus, or cordate; stem-lvs round-cordate, obtuse, mostly (1–)1.5–4(–5) cm wide; stipules broadly lanceolate, fimbriate-toothed at least above; fls surpassing the lvs; sep narrowly lanceolate, entire; pet light blue-violet, with darker veins, the 2 lateral ones bearded; spur 4–5 mm; style slender, with bent tip; frs ellipsoid, 4–5 mm; seeds light brown; $2n=20$. Woods and meadows; N.S. and Que. to Minn. and e. N.D., s. to Pa. and O., and in the mts. to S.C., Tenn., and Ala. May–July.

25. Viola walteri House. Walter's v. Finely puberulent throughout; stems several, at first upright, later prostrate and with cleistogamous fls, and terminating in a rosette that bears petaliferous fls the next year; lvs orbicular, 2–3 × 3–4 cm, blunt, cordate at base, often purplish beneath especially on the veins; stipules sparsely fimbriate-toothed; fls on long, slender peduncles arising from the basal rosette and later from the lower axils of the elongating stems; pet light violet with darker veins, the lateral ones bearded, the spur 3–4 mm; style slender, bent at the tip, or with a terminal cusp at right angles to the axis; frs 4–5 mm, ellipsoid-globose; seeds light brown; $2n=20$. Dry or moist woods and ledges; s. O. to Va., s. to Fla., Ark., and Tex. Apr., May.

26. Viola rostrata Pursh. Long-spurred v. Essentially glabrous, the stems erect or spreading from a branched, oblique, short rhizome, 5–12 cm at anthesis, later 15–25 cm, lvs ovate, often broadly so, 2–4 cm, cordate at base, all but the lowermost acute; stipules lanceolate, fimbriate-toothed above the middle; fls elevated above the lvs; pet light violet, with darker veins forming a pronounced eye, all beardless; spur 7–12 mm, slender, slightly upcurved at the tip; style slender, the tip straight, smooth; fr stoutly ellipsoid, 5–6 mm; seeds light yellow-brown; $2n=20$. Shady slopes and woodlands, usually in deep humus; Me. and Que. to Mich. and e. Wis., s. to Ga. and Ala. Apr.–June. Hybridizes with nos. 22 and 24.

27. **Viola rafinesquii** Greene. Wild pansy. Slender, glabrous or subglabrous annual 5–40 cm, often branched from the base, the internodes usually exceeding the lvs; lvs almost entire, variable in shape, the lower ± orbicular, the upper spatulate to broadly oblanceolate; stipules conspicuous, foliaceous, often ciliate, pectinately cut, the middle lobe elongate-spatulate and entire or nearly so; fls long-peduncled, the early ones generally chasmogamous; sep ciliate, half as long as the deep blue or bluish-white to cream-colored ("ivory") pet, these ca 1 cm or nearly 1 cm; style much enlarged above, with a globose, hollow tip and a wide orifice on the lower side; mid- and late-season (rarely all) fls cleistogamous, with small or no pet; frs oblong-ellipsoid, 4–6 mm; seeds pale brown; 2*n*=34. Fields, open woods, roadsides, and grassy places; Mass. and N.Y. to Nebr. and Colo., s. to S.C., Ga., and Tex. (*V. kitaibeliana* var. *rafinesquii.*)

28. **Viola tricolor** L. Johnny-jump-up. Glabrous or hairy annual or short-lived perennial to 3 dm, often branched from the base; lvs all crenate, the lowest ones oribicular or cordate, the upper oblong to elliptic and basally cuneate; stipules foliaceous, laciniate to lyrate-pinnatifid, the middle lobe oblanceolate, distinctly toothed; internodes usually shorter than the lvs; fls 1.5–2.5 cm wide, the sep two-thirds as long as the pet; pet variously colored, the upper usually darker than the lower; style as in no. 27; cleistogamous fls wanting; frs ellipsoid or oblong-ellipsoid, 6–10 mm; seeds dark brown; 2*n*=26. Native of Europe, extensively cult. and casually escaped in our range.

Viola × *wittrockiana* Gams, the garden pansy, originating by hybridization between *V. tricolor* and other spp. such as *V. lutea* Hudson, and selected for large fls (5–10 cm wide) in diverse colors, occasionally escapes or persists briefly after cult.

29. **Viola arvensis** Murray. European field-pansy. Pubescent to glabrate annual to 30 cm, often branched from the base, the stem relatively robust, with short, reflexed hairs on the angles; lvs variable, the lowest orbicular to ovate, the upper oblong to elliptic or narrowly elliptic, all crenate; stipules foliaceous, laciniate into 5–9 slender segments and usually 1 lflike segment often nearly as large as the main blade; fls 1–1.5 cm long, scarcely 1 cm wide; pet shorter than or barely equaling the broadly lanceolate sep, all pale yellow ("ivory") or occasionally with purplish tips; spur short; style as in no. 27; cleistogamous fls wanting; frs globose to ellipsoid, 5–10 mm; seeds brown; 2*n*=34. Native of Europe, now established as a weed in or near fields or along roadsides throughout our range and s. and w. Apr.–Sept.

FAMILY **TAMARICACEAE**, the Tamarisk Family

Fls regular, perfect, hypogynous, 4–5-merous; sep distinct or nearly so; stamens mostly as many (in our spp.) or twice as many as the pet, seated on or beneath a hypogynous nectary-disk; ovary of (2)3–4(5) carpels, unilocular, but sometimes with ± deeply intruded partial partitions, the styles distinct or merely connate at base; ovules on parietal or basal placentas; fr a capsule; seeds several or many, minute, provided with long hairs that usually form a tuft at one end; shrubs or small trees with small, alternate, estipulate, mostly sessile, commonly subulate or scale-like lvs with embedded, multicellular, salt-excreting glands, and with small fls solitary or more often in slender, scaly-bracteate racemes, spikes, or panicles. 4/100, Old World.

1. **TAMARIX** L. Tamarisk. Stamens 4–5; pet often long-persistent; ovary conic; placentas basal or nearly so; styles 3–4, short-clavate or obovate; fr conic, separating completely into 3 or 4 valves; deciduous shrubs or trees with small, scale-like, often imbricate lvs and numerous small fls in racemes or spikes borne on the old wood or aggregated into a terminal panicle. 50, Old World.

Two spp. occasionally escape from cult. in our range, but are scarcely members of our established flora. *T. parviflora* DC. has 4-merous fls, with the filaments seated on the 4 lobes of the disk. *T. chinensis* Lour. has 5-merous fls, with the filaments arising from (or just outside) the sinuses of the 5-lobed disk (the disk lobes often emarginate). (*T. ramosissima; T. pentandra,* an illegitimate name.)

FAMILY **PASSIFLORACEAE**, the Passion-flower Family

Fls regular, usually perfect, perigynous with a well-developed, saucer-shaped to tubular hypanthium; sep (3–)5(–8), persistent, alternating with the pet and attached with them to the margin of the hypanthium, which also bears a corona of varied and often complex structure; stamens (4)5(–many), mostly alternate with the pet, inserted near the center of the hypanthium, or raised on an androgynophore; ovary superior, often on a gynophore, unilocular, with (2)3(–5) parietal placentas and numerous ovules; styles mostly distinct or connate only at base; fr a capsule or berry; seeds generally arillate; mostly vines climbing by axillary tendrils, with alternate, commonly lobed lvs, usually with small stipules; our spp. perennial. 16/650, mostly warm reg.

1. PASSIFLORA L.   Passion-flower. Fls 5-merous, perfect; stamens 5, monadelphous around the gynophore; corona a double or triple fringe; styles 3, elongate, with capitate or clavate stigma; fr a berry; herbaceous or woody vines. 400, mostly warm reg.

1 Lvs deeply 3-lobed, the lobes serrulate; fls 4–6 cm wide; fr 5 cm .... 1. *F. incarnata*.
1 Lvs shallowly 3-lobed, the lobes entire; fls 2–2.5 cm wide; fr 1 cm .... 2. *P. lutea*.

1. Passiflora incarnata L.   Maypops. Stems climbing or trailing to 8 m, pubescent; petioles pubescent, glandular at or near the summit; lvs deeply 3-lobed, puberulent beneath, truncate or rounded to a small cuneate base, the lobes lance-ovate, constricted at base, acuminate, serrulate; peduncle elongate, with glandular-serrate bracts above; fls 4–6 cm wide, the pet and sep white, the outer corona 2 cm, purple or pink, the inner much shorter; berry edible, yellow, ellipsoid, 5 cm; 2*n*=18, 36. Fields, roadsides, thickets, and open woods; Va. and sw. Pa. to s. O., s. Ill. and Okla., s. to Fla. and Tex. June–Aug.

2. Passiflora lutea L.   Climbing or trailing to 3 m; petioles not glandular; lvs truncate to subcordate at base, with margins meeting across the petiole, 3-lobed, the lobes semi-orbicular or depressed-ovate, entire, obtuse or rounded; fls greenish-yellow, 2–2.5 cm wide; exterior corona yellow; fr purple, 1 cm thick. 2*n*=24, 84. Moist soil; all summer. Var. *lutea,* with the cal, petioles, and young stems pilose, occurs from Pa. and W.Va. to Fla., Tenn., and Ala. Var. *glabriflora* Fern., glabrous, occurs from O. to Mo. and Okla., s. to Ala. and Tex.

FAMILY **CUCURBITACEAE**, the Gourd Family

Monoecious or dioecious, the fls unisexual, epigynous, regular, gamopetalous or polypetalous; cal (3–)5(–6)-lobed, sometimes to the base; stamens attached to the hypanthium or seldom around the summit of the ovary, basically 5, unithecal, and alternate with the corolla-lobes, but in nearly all spp. showing some degree of displacement, reduction, and fusion, often apparently 3, 2 double with dithecal (tetrasporangiate) anthers, and one single, with a unithecal (bisporangiate) anther; ovary mostly of 3 carpels, unilocular with intruded, often much-enlarged parietal placentas, or these seldom joined in the center and the ovary thus plurilocular; style mostly solitary, with 1–3(–5) usually bilobed stigmas or stigma-lobes; fr usually a berry or pepo, less often a capsule; seeds (1–)many, large, commonly compressed, with oily embryo and 2 large, flat cotyledons; herbaceous or softly woody plants, trailing or more often climbing by spirally coiled tendrils, with mostly white or yellow or greenish fls and simple, alternate, exstipulate, often lobed lvs. 90/700, mainly warm reg.

1 Fls yellow, 5–10 cm wide or larger; fr globose, 5–10 cm .... 1. *Cucurbita*.
1 Fls white or greenish to yellow, well under 5 cm wide; fr less than 5 cm thick.
  2 Filaments distinct; ovary and fr smooth or merely hairy .... 2. *Melothria*.
  2 Filaments connate into a column; ovary and fr prickly or spiny.
    3 Staminate cor 6-lobed; fr inflated, dehiscent, 4-seeded .... 3. *Echinocystis*.
    3 Staminate cor 5-lobed; fr not inflated, indehiscent, 1-seeded .... 4. *Sicyos*.

1. CUCURBITA L.   Monoecious; sep 5; cor yellow, campanulate, the 5 lobes recurved at the tip; stamens apparently 3, the filaments distinct, the anthers connivent and contorted; style 1, with 3 divided stigmas; fr a pepo; mostly trailing, annual or perennial herbs with forked tendrils, large, lobed or angled to merely toothed lvs, large yellow fls solitary in the axils, and large, firm-walled, many-seeded frs. 15, New World.

1. Cucurbita foetidissima HBK.   Wild pumpkin. Stems rough, trailing from a thick perennial root, often to several m; herbage malodorous; lvs triangular-ovate, 1–2 dm, irregularly and finely toothed and often angularly lobed, rough on both sides; fls yellow, campanulate, 5–10 cm long and wide; fr subglobose, 5–10 cm, greenish-orange, smooth. $2n$=40, 42. Dry soil; Mo. to Tex. and Calif., and intr. along railways farther e. Summer. (*Pepo f.*)

2. MELOTHRIA L.   Melonette. Monoecious; cal campanulate, 5-dentate; cor small, campanulate, 5-lobed; stamens apparently 3, the short filaments distinct, the straight, oblong anthers distinct or barely connivent; hypanthium greatly constricted above the 3-locular ovary; ovules numerous; style short, with 3 slender stigmas; fr pulpy, with numerous horizontal seeds; climbing herbs, with mostly undivided tendrils, the very small fls solitary or in small clusters. 10, New World.

1. **Melothria pendula** L.   Creeping cucumber. Slender, glabrous vine 1–2 m; lvs orbicular in outline, cordate at base, shallowly or deeply 5-lobed, 3–7 cm; fls few, the pistillate solitary, slender-peduncled, 8 mm wide, the staminate smaller, 2–6 in a short raceme terminating a slender peduncle; fr ovoid, green, 1 cm. Woods; s. Va., s. Ind., and s. Mo. to Fla. and n. Mex. June–Aug.
   Thladiantha dubia Bunge, a dioecious Asiatic vine with cordate-ovate lvs, axillary yellow fls 1.5–2.5 cm, and with strongly ribbed and cross-ribbed ellipsoid fr 5 cm, occasionally escapes in s. Can. and n. N. Engl.

3. ECHINOCYSTIS T. & G., nom. conserv.   Monoecious; fls 6-merous, the cal small, the rotate cor deeply lobed; stamens united by their filaments into a column, the anthers connivent, nearly straight; ovary in ours bilocular with 2 erect ovules in each locule; style with a broad, lobed stigma; fr bladdery-inflated, at length dry, dehiscent by 2 pores at the summit; herbs with branched tendrils, angular or lobed lvs, and small white or greenish fls, the staminate in long racemes, the pistillate short-peduncled, solitary or in small clusters. Monotypic.

1. **Echinocystis lobata** (Michx.) T. & G.   Balsam-apple, wild cucumber. High-climbing annual; lvs long-petioled, orbicular in outline, with (3–)5(–7) sharp, triangular lobes; staminate fls in long erect racemes, the cor 8–10 mm wide, with lanceolate lobes, the pistillate few or solitary, short-peduncled from the same axils; fr ovoid, 3–5 cm, green, weakly prickly; $2n$=32. Moist ground and thickets, throughout our range and s. to Fla. and Tex.; often cult. July–Sept. (*Micrampelis l.*)

4. SICYOS L.   Monoecious; fls 5-merous; cal small, 5-toothed; cor rotate, 5-lobed; anthers and filaments united into a column; ovary unilocular, with 1 suspended ovule; stigmas 3; fr dry, indehiscent, not inflated, covered by prickly bristles; annual vines with branched tendrils, broad, angular or lobed lvs, and small, white or greenish fls, the pistillate in small capitate clusters, the staminate in corymbiform racemes. 15, mainly trop.

1. **Sicyos angulatus** L.   Bur-cucumber. Climbing sometimes to several m; lvs orbicular in outline, shallowly 3–5-lobed, usually with a deep basal sinus, the lobes denticulate, acuminate; pistillate peduncles eventually 5–8 cm, the staminate usually longer; staminate cor 8–10 mm wide, lobed to the middle; fr ovoid, 1.5 cm, hairy and spiny; $2n$=24. Damp soil; Me. and Que. to Minn., s. to Fla. and Ariz. July, Aug.

FAMILY **LOASACEAE**, the Loasa Family

Fls perfect, regular, epigynous; sep (4)5(–7), persistent; pet (4)5(–7) or 10 by development of petaloid staminodes, induplicate-valvate, distinct or sometimes basally adnate to the stamen-tube, or even forming a sympetalous, lobed cor; stamens 10 to more often numerous, distinct or basally connate into a short tube or into antepetalous bundles, some of them often modified into petaloid or scale-like or nectariferous staminodia; ovary of 3–5(–7) carpels, inferior, unilocular, with parietal, often ± deeply intruded placentas; style solitary; ovules 1–many on each placenta; fr generally a capsule; seeds mostly with abundant, oily endosperm; erect or climbing herbs or shrubs, usually with an indument of bristly, barbed, hooked, or stinging hairs; lvs exstipulate, simple but often lobed. 14/2000+, New World.

1. MENTZELIA L.   Hypanthium ovoid to cylindric or obconic; sep 5; pet 5–10; stamens 10–200, distinct or connate into fascicles; ovules 2–many on 3 parietal placentas;

style 3-lobed; fr a capsule; annual or perennial herbs or shrubs, armed with stiff, barbed hairs, with alternate, serrate to pinnatifid lvs and 1–many often large, white to yellow or orange fls terminating the branches. (*Nuttallia*.) 50, New World.

1 Pet 5, 0.8–1.2 cm .... 1. *M. oligosperma.*
1 Pet 10, 5–7 cm .... 2. *M. decapetala.*

1. Mentzelia oligosperma Nutt. Stickleaf. Perennial with stramineous, freely branched stems to 1 m; lvs short-petioled, or the upper sessile, lanceolate to ovate, coarsely and sinuately toothed, the larger 3–6 cm; pet 5, golden, 8–12 mm; stamens ca 20; fr cylindric, 1 cm; $2n=20$, 22. Rocky hills and barrens; Mo. to S.D. and Colo., s. to Tex. June–Aug.

2. Mentzelia decapetala (Pursh) Urban & Gilg. Sand-lily, evening star. Stout, sparingly branched biennial 2–5 dm; lvs sessile or short-petioled, narrowly lanceolate or oblong, sharply cleft or pinnatifid, the larger 6–15 cm; pet 10, pale yellow to nearly white, 5–7 cm; stamens 100–200; fr 3–5 cm; $2n=22$. Dry prairies and plains; N.D. to Alta. and Ida., s. to w. Io., Tex., and n. Mex.; adventive in Ill. July–Sep. (*Nuttallia d.*)

FAMILY SALICACEAE, the Willow Family

Dioecious; fls in catkins (aments), without perianth, each fl subtended by a small bract (commonly called a scale) and either provided with one or 2 enlarged basal glands (*Salix*) or subtended by an obliquely cup-shaped disk (*Populus*); staminate fls consisting of (1)2–many stamens; pistillate fls consisting of a single unilocular ovary with 2–4 carpels and as many stigmas, the stigmas sometimes bifid or otherwise cleft, with or without a common style; placentation parietal or sometimes basal; fruit a 2–4-valved capsule; seeds usually ± numerous, small, without endosperm, covered with long white hairs and adapted to distribution by wind; trees or shrubs, typically fl in early spring; lvs alternate, simple, entire or toothed, usually stipulate, but the stipules often deciduous. 2/300+, widespread.

1 Fls individually subtended by an obliquely cup-shaped disk, but without glands; if-buds with several
  scales; stamens 5–many; trees with pendulous catkins .... 1. *Populus.*
1 Fls lacking a disk but provided with one or 2 enlarged basal glands; lf-buds with one scale; stamens
  1–8; trees or shrubs, with erect or pendulous catkins .... 2. *Salix.*

1. POPULUS L. Catkins drooping, appearing before the lvs, their scales toothed, lobed, fimbriate, or densely ciliate; each fl set on a cupulate, commonly oblique disk that may be homologous with the "glands" of *Salix* fls; stamens 5–80, on short filaments; styles 2–4-valved, trees or tall shrubs with soft, light wood, mostly ovate to deltoid, deciduous lvs, scaly, often viscid winter-buds, and elongate catkins that mature before the lvs are fully expanded in the spring. 40, widespread.

KEY BASED PRIMARILY ON FLOWERS AND WINTER BUDS

1 Scales of the catkins merely dentate or with 3–7 linear-triangular, long-ciliate lobes; stamens 5–12;
  stigmas 2, slender, each deeply bifid (Sect. *Populus*).
  2 Scales of the staminate fls merely dentate; twigs and terminal bud tomentose .... 1. *P. alba.*
  2 Scales of the staminate fls deeply 3–7-lobed; twigs and terminal bud glabrous or somewhat hairy,
    but not tomentose.
    3 Terminal bud dull brown, pubescent .... 2. *P. grandidentata.*
    3 Terminal bud shining, glabrous or nearly so .... 3. *P. tremuloides.*
1 Scales of the catkins deeply and conspicuously fimbriate; stamens 12–80; stigmas 2–4, broadly dilated.
  4 Stigmas elevated on distinct, elongate styles; terminal bud not viscid; stamens 12–20 (Sect. *Leu-
    coides*) .... 4 *P. heterophylla.*
  4 Stigmas sessile or nearly so; terminal bud viscid; stamens (15–)20–80.
    5 Stigmas 2 (sometimes 3 in nos. 6 and 8); stamens (15–)20–30 (to 40 in no. 6).
      6 Native tree of wet woods and shores northward (Sect. *Tacamahacca*) .... 5. *P. balsamifera.*
      6 Introduced trees, always near cultivation (Sect. *Aigeiros*, in part, except no. 6, an intersectional
        hybrid).
        7 Trees mostly of narrowly columnar form, but sometimes broader; floral disk 1–2 mm wide;
          stigmas 2; ovules and seeds 4–8 per placenta .... 7. *P. nigra.*
        7 Trees of broader form, not columnar; floral disk 2–4 mm wide; stigmas 2 or 3; ovules and
          seeds 6–14(–20) per placenta.

8 Stamens 25–40; petioles only weakly compressed .... 6. *P.* × *jackii.*
8 Stamens (15–)20–30; petioles strongly compressed .... 8. *P.* × *canadensis.*
5 Stigmas 3 or 4; stamens (30–)40–80. (Sect. *Aigeiros,* in part) .... 9. *P. deltoides.*

KEY BASED PRIMARILY ON LEAVES

1 Petioles distinctly flattened laterally.
 2 Lvs with a definite translucent border; teeth (under a lens) terminating in a callous incurved point
  in no. 9, less so in no. 8, more spreading in no. 7; buds viscid; fr ovoid, its pedicel 3–10 mm;
  rough-barked trees.
  3 Lvs relatively coarsely toothed, the largest teeth (1.5–)2–5 mm deep, many or all of the lvs with
   2–5 prominent glands at the base of the blade on the upper side; capsules 3–4-valved; native
   on floodplains and along streams .... 9. *P. deltoides.*
  3 Lvs more finely toothed, the largest teeth 1–2 mm deep, few or none of the lvs glandular at base;
   capsules 2-valved (2–3-valved in no. 8); intr. trees, always near cultivation.
   4 Trees mostly of narrowly columnar form, but sometimes broader; largest teeth of lvs ca 1
    mm deep; lvs of short shoots generally ± cuneate at base; lvs of long shoots mostly
    broader than long .... 7. *P. nigra.*
   4 Trees of broader form, never columnar; largest teeth of lvs mostly 1.5–2 mm deep; lvs of
    short shoots generally rounded at base; lvs of long shoots about as broad as long ...... 8. *P.* × *canadensis.*
 2 Lvs without a translucent border; teeth (under a lens) mostly rounded at summmit, not incurved;
  buds not or scarcely viscid; fr slenderly conic, its pedicel 1–2 mm; smooth-barked (only tardily
  rough-barked) trees.
  5 Lvs of short shoots with more than 15 teeth per side, the teeth small, not over ca 1 mm deep;
   lvs glabrous .... 3. *P. tremuloides.*
  5 Lvs of short shoots with fewer than 15 teeth per side, the teeth larger, the largest ones 1.5–6 mm
   deep; lvs tomentose-puberulent when young, especially beneath .... 2. *P. grandidentata.*
1 Petioles terete or nearly so, or flattened at the summit only (weakly flattened throughout in no. 6),
 usually shallowly channeled above.
 6 Mature lvs evidently tomentose beneath, 3–5-lobed, or coarsely and irregularly serrate; fruiting
  pedicels 1–2 mm .... 1. *P. alba.*
 6 Mature lvs glabrous beneath, or thinly pubescent only on the main veins, finely and regularly
  toothed.
  7 Lvs obtuse or rounded above, cordate at base, permanently pubescent on the veins beneath and
   at the very base of the blade above; fruiting pedicels 10–15 mm .... 4. *P. heterophylla.*
  7 Lvs acute or acuminate, with cuneate to truncate or subcordate base, glabrous on both sides;
   fruiting pedicels 2–3 mm.
   8 Native tree of wet woods and shores northward; lvs of short shoots mostly ovate or lance-
    ovate and acute; capsules 2-valved, with 15–22 seeds per placenta .... 5. *P. balsamifera.*
   8 Intr. hybrid cultigen, always near cultivation; lvs of short shoots mostly deltoid-ovate and
    acuminate; capsules 2–3-valved, with 7–14(–20) seeds per placenta .... 6. *P.* × *jackii.*

1. Populus alba L.   White or silver poplar.   Trees with whitish-gray bark and mostly widely spreading branches; terminal bud and young twigs tomentose; lvs white-tomentose beneath, palmately 3–7-lobed on long shoots, ovate and irregularly dentate on short ones; stamens 6–10; stigmas 2, bifid, filiform; fr narrowly ovoid, the pedicel 1–2 mm; seeds (1)2(3) per placenta; 2*n*=38, 57. Native of Eurasia, commonly planted, and occasionally escaped. A columnar form is cv. 'Bolleana' (*P. bolleana.*)
 Populus × *canescens* (Aiton) J. E. Smith, the gray poplar, is a hybrid of *P. alba* and the Eurasian aspen, *P. tremula* L. Occasionally planted and rarely escaped, it differs most obviously from *P. alba* in having the lvs merely coarsely toothed (none lobed), and glabrescent in age.

2. Populus grandidentata Michx.   Big-toothed aspen. Tree with light greenish-gray bark (darker than no. 3), becoming dark brown and furrowed in age; petioles strongly flattened; terminal buds dull brown, finely hairy; lvs tomentose-puberulent when young, especially beneath, later glabrous, dark green above, pale and glaucous beneath, often with a pair of basal glands, mostly 5–12 cm, the lowest veins strongly ascending as in no. 3; lvs of short shoots coarsely and remotely toothed, with mostly 5–12 large, projecting teeth (the largest ones 1.5–6 mm deep) on each side; lvs of long shoots often with smaller and more numerous teeth; scales of catkins shallowly cleft into 5–7 lance-triangular lobes; stamens 5–12; stigmas 2, linear, bifid; frs slenderly conic, 3–5 mm, on pedicels 1–2 mm, with 5–8 seeds per placenta; 2*n*=38. Upland woods, often in drier soil than no. 3; N.S. to Minn., s. to N.C., Tenn., and n. Mo., abundant northward. A hybrid with no. 3 is *P.* × *barnesii* W. H. Wagner, or *P.* × *smithii* B. Boivin.

3. Populus tremuloides Michx.   Aspen, quaking a. Tree with light grayish-green smooth bark, becoming dark and furrowed in age; petioles strongly flattened; terminal buds brown, shining, glabrous or nearly so; lvs glabrous, dark green above, light green and slightly glaucous beneath, without basal glands, 3–10 cm, rotund-ovate to reniform-cordate, shortly acuminate to merely apiculate, the lowest lateral veins strongly ascending; lvs of short shoots finely and regularly toothed, with mostly 18–30 teeth up to 1 mm deep on each side, those of the long shoots often with even smaller and more numerous teeth, or subentire; scales of the catkins cleft to below the middle into 3–5 lance-triangular lobes; stamens, stigmas, and frs much as in no. 2; 2*n*=38, 57, 76. Moist upland woods and streamsides, often on cut-over or burned land; Lab. to Alas., s. to N.J., Va., Tenn., Mo., and Mex.

4. Populus heterophylla L.   Swamp-cottonwood. Tall tree, with shaggy, dull brown bark; buds usually ± pubescent, not or scarcely glutinous; petioles terete, or slightly flattened only near the summit; lvs broadly ovate, 12–20 cm, obtuse or rounded, finely crenate-serrate with incurved teeth, cordate at base, tomentose when young, becoming glabrous above except at the very base of the blade, glabrous below except along the larger veins; scales of the catkins long-ciliate; stamens 12–20; stigmas 2 or 3, broadly dilated, each on a conspicuous style; frs ovoid, 2–3-valved, 7–12 mm, on pedicels 10–15 mm, forming a loose raceme. Wet low woods and swamps; sw. Conn. to Fla. and La., n. in the interior to s. Mich.

5. Populus balsamifera L.   Balsam-poplar. Tall tree with dark gray furrowed bark and glabrous or hairy twigs; terminal buds glabrous, viscid; petioles ± terete; lvs lance-ovate to lance-ovate, acute or short-acuminate, finely serrate or subentire, cuneate to rounded or subcordate at base, glabrous, dark green above, white and glaucous (often streaked with orange resin) beneath; scales of the catkins long-ciliate; stamens 20–30; stigmas 2, nearly sessile, broadly dilated; frs ovoid, 2-valved, 5–8 mm, crowded on short pedicels, forming a compact, spike-like raceme; seeds 15–22 per placenta; 2*n*=38. Wet woods, river-banks, and shores; Lab. to Alas., s. to Conn., n. Pa., n. Ind., n. Io., Nebr., and Colo. (*P. tacamahacca.*)

6. Populus × jackii Sarg.   Balm of Gilead. Rare natural hybrid of nos. 5 and 9, cult. as a single pistillate clone. (*P. gileadensis; P. balsamum-gileadense; P. candicans,* misapplied.)

7. Populus nigra L.   Black poplar. Tall tree with dull gray branches and dark, furrowed bark on older trunks; petioles distinctly flattened; winter buds resinous; lvs pubescent when young, finely toothed, the teeth up to ca 1 mm deep, more spreading than in no. 9, only slightly or not at all incurved; lvs of short shoots rhomboid-ovate, with cuneate base, of long shoots broadly deltoid-ovate, truncate at base, abruptly pointed; floral disk 1–2 mm wide, stamens 20–30; stigmas 2, broadly dilated; frs ovoid 7–9 mm, twice as long as the pedicels, with 4–8 seeds per placenta; 2*n*=38. Native of Eurasia, frequently planted, and occasionally escaped. The Lombardy poplar is a common horticultural form with erect branches, forming a long, columnar crown. (*P. italica; P. nigra* var. *italica; P. dilatata.*)

8. Populus × canadensis Moench.   Carolina poplar. A frequently planted hybrid between no. 9 and the columnar form of no. 7, commonly with a crown of intermediate shape, the lvs also intermediate, usually without basal glands; floral disk 2–3.5 mm wide; capsules 2–3-valved, with 6–11 seeds per placenta; 2*n*=38. (*P. × eugenei, P. canadensis* var. *eugenei,* the staminate clone most frequently distributed as Carolina poplar.)

9. Populus deltoides Marshall.   Cottonwood. Tall tree with ± spreading branches and a broad crown, the bark becoming deeply furrowed and nearly black; terminal buds glabrous, viscid; petioles strongly flattened; lvs glabrous, typically broadly deltoid, 8–14 cm, nearly as wide, short-acuminate, broadly truncate or subcordate at base, serrate with incurved, callous-tipped teeth (the largest teeth mostly 2–5 mm deep), generally bearing 2–5 prominent glands on the upper side at the very base; lowest lateral veins of the lvs widely spreading; scales of the catkins fringed; floral disk 1.5–4 mm wide; stamens (30–)40–80; stigmas 3 or 4, broadly dilated; frs ovoid, 3–4-valved, 6–10 mm, on pedicels 3–10 mm, forming a loose raceme; seeds (3–)7–10(–18) per placenta; 2*n*=38. Low woods and moist prairies and river-banks; Que. and N. Engl. to Fla. and Tex., and w. to the base of the Rocky Mts. Var. *deltoides,* the eastern cottonwood, widespread in our range, has gray or reddish-brown twigs, glabrous winter-buds, and mostly 3–5 glands at the base of the lvs, the lvs of the short shoots with mostly 15–30 teeth per side. The wholly confluent var. *occidentalis* Rydb., the Great Plains cottonwood, forming gallery forests along the rivers across the plains, and encroaching into the westerly Great Lakes region, has yellowish twigs, finely pubescent winter buds, mostly only 2 basal glands on the lvs, and mostly only 5–15 teeth on each side of the often more deltoid-ovate and more long-acuminate lvs. (*P. sargentii; P. deltoides* ssp. *monilifera.*)

2. SALIX L.   Willow. Catkins mostly ascending or spreading, seldom drooping, developing before, with, or less often after the lvs, their scales mostly entire, sometimes erose or shallowly toothed at the tip; fls with a single short and broad to slender and elongate ventral protuberance called a gland, or the staminate fls of some spp. also with a dorsal gland; stamens typically 2, seldom only 1 (by fusion of 2), or in some spp. 3–8; stigmas 2, entire or more often bifid; capsules 2-valved; shrubs or less often trees, sometimes depressed and mat-forming; winter-buds covered by a single nonresinous scale. 300+, mainly N. Temp. Taxonomy complicated by hybridization.

8 Stipules usually minute and caducous or obsolete; lvs and twigs glabrous almost from
   the first; petioles 6–20 mm; northern . . . . .2. *S. amygdaloides.*
7 Lvs not glaucous beneath, or seldom faintly so . . . . 3. *S. nigra.*
6 Buds blunt; margins of the scale fused.
 9 Lvs notably long-acuminate . . . . 4. *S. lucida.*
 9 Lvs acute or merely short-acuminate.
  10 Scales copiously long-hairy nearly or quite throughout . . . . 5. *S. serrissima.*
  10. Scales long-hairy chiefly near the base, largely glabrous distally . . . . 6. *S. pentandra.*
5 Stamens 2.
 11 Lvs entire . . . . 15. *S. pedicellaris.*
 11 Lvs toothed.
  12. Scales pale, commonly yellowish, often soon deciduous.
   13 Alien trees, cult. and sparingly escaped in our range.
    14 Twigs mostly erect or spreading, only seldom pendulous; frs 3–5.5 mm.
     15 Lvs and twigs ± sericeous; lvs with 7–10 teeth per cm of margin . . . . 7. *S. alba.*
     15 Lvs and twigs glabrous or nearly so, at least at maturity; lvs with 4–6 teeth
      per cm of margin . . . . 8. *S. fragilis.*
    14 Twigs pendulous; frs 1.5–2.5 mm . . . . 9. *S. babylonica.*
   13 Native many-stemmed colonial shrub to 3(–5) m . . . . 10. *S. exigua.*
  12 Scales dark, commonly dark brown or blackish, generally more persistent.
   16 Young vegetative organs glabrous; lvs with a balsamic fragrance . . . . 16. *S. pyrifolia.*
   16 Young vegetative organs ± pubescent; lvs not balsamic.
    17 Peduncles 0.3–1 cm, bracteate; catkins 2–4 cm; style 0.5 mm . . . . 17. *S. eriocephala.*
    17 Peduncles 1–3 cm, leafy; catkins (2–)3–8 cm; style 0.7–1.5 mm.
     18 Lvs glaucous beneath, the surfaces soon glabrate; pedicels 1–3 mm . . . . 18. *S. myricoides.*
     18 Lvs green beneath, usually permanently lanate; pedicels 0.5–1 mm . . . . 19. *S. cordata.*
4 Frs pubescent; scales dark (except in nos. 10 and 11); stamens 2 (or only one).
 19 Scales pale, yellowish (sometimes reddish at the tip).
  20 Catkins with the lvs; lvs seldom over 1.5 cm wide, commonly ca 10 times as long as wide
   . . . . 10. *S. exigua.*
  20 Catkins subprecocious; lvs 1.5–3 cm wide, less than 5 times as long as wide . . . . 11. *S. bebbiana.*
 19 Scales dark, commonly dark brown or blackish.
  21 Catkins precocious, or sometimes only subprecocious, sessile or short-pedunculate.
   22 Stamens partly or wholly connate into one; many of the lvs subopposite . . . . 30. *S. purpurea.*
   22 Stamens 2, wholly distinct; lvs all alternate.
    23 Twigs notably glaucous . . . . 21. *S. pellita.*
    23 Twigs not glaucous.
     24 Frs 3–5 mm; lvs closely serrate . . . . 22. *S. sericea.*
     24 Frs 6–10 mm; lvs entire or toothed, but not closely serrate.
      25 Frs subsessile, the pedicels mostly less than 1 mm; lvs glabrous by the
       time they are fully expanded; style mostly 1–1.5 mm . . . . 20. *S. planifolia.*
      25 Frs on pedicels 1–3 mm; lvs often more persistently hairy; style less than
       1 mm.
       26 Lvs usually soon glabrous or only sparsely hairy beneath, seldom more
        persistently hairy; catkins larger (see descr.); shrub or small tree
        2–5(–7) m . . . . 23. *S. discolor.*
       26 Lvs ± persistently hairy beneath; catkins smaller; shrubs 0.5–3 m.
        27 Lvs larger, 3–10(–15) × 1–2(–3) cm, stipulate . . . . 24. *S. humilis.*
        27 Lvs smaller, 2–5 cm × 7–10(–12) mm, exstipulate . . . . 25. *S. occidentalis.*
  21 Catkins developing with or after the lvs, on short, mostly leafy peduncles (virtually sessile
   in no. 29).
   28 Low shrubs, rarely as high as 2 m.
    29 Twigs puberulent; lvs sericeous to glabrate beneath; frs 2–4 mm, on pedicels 1–2
     mm . . . . 26. *S. argyrocarpa.*
    29 Twigs and usually also the lvs evidently tomentose; frs 5–8 mm, subsessile . . . . 27. *S. candida.*
   28 Taller shrubs or trees, mostly 2–7 m.
    30 Lvs becoming glabrous (or subglabrous) and glaucous beneath; native . . . . 28. *S. petiolaris.*
    30 Lvs permanently sericeous beneath; alien sp., seldom escaped . . . . 29. *S. viminalis.*

 1. Salix caroliniana Michx. Carolina-w. Shrub or tree to 10 m, or larger westward; buds conic, sharp-tipped, the short (3–7 mm) scale with free, overlapping margins; petioles and twigs yellowish to dark brown, usually white-hairy the first year, the petiole with glandular dots or processes at the summit; stipules well developed and persistent, 7–15 mm, broadly reniform, serrulate; lvs spreading, lance-linear to lanceolate (or the young ones often oblanceolate and obtuse), 7–15 × 0.7–3 cm, mostly 5–10 times as long as wide, long-acuminate, closely serrulate, dark green or yellowish-green above, densely glaucous beneath, often hairy, especially along the midrib, the areoles minute beneath; catkins with the lvs, 3–11 cm, on lax, leafy peduncles 2–5(–7) cm; scales yellowish, villous, deciduous; stamens (4–)6(–8); fr narrowly lance-ovate, 3–6 mm, granular-roughened; pedicels 1.5–5 mm; style 0.1–0.2 mm. Floodplains and other moist or wet low places; Del., Md., and the Potomac and Ohio valleys (to Pittsburgh and s. Ind.), w. to e. Kans. and Okla., s. to Cuba and Guat. (*S. ambigua; S. longipes; S. wardii.*)

 2. Salix amygdaloides Andersson. Peach-lf-w. Shrub or tree 3–20 m, with 1–4 often leaning trunks to 4 dm thick, the bark fissured; lvs and twigs a little hairy at the very beginning, but soon glabrous; twigs yellowish to reddish-brown, tending to droop; buds sharp-pointed, the short (2–5 mm) scale with free, overlapping margins; stipules typically minute and caducous or obsolete, rarely well developed; lvs thin, lanceolate to lance-ovate, 5–10(–15) × 1–3.5 cm, 3–6 times as long as wide, abruptly long-acuminate, closely serrate, yellowish-green above, strongly glaucous and minutely areolate beneath; petioles 6–20 mm, often twisted, with or without small glands at the end; catkins with the lvs, linear, lax, the staminate ones 4–7 cm, the pistillate becoming 5–20 cm; scales yellowish, villous especially within, deciduous; stamens 3–7, typically 5; fr lanceolate, 4–7 mm, smooth; pedicels 1.5–3 mm; style 0.2–0.4 mm; $2n=38$. Floodplains and other moist or wet low places; Vt., N.H., ne. N.Y. and adj. Que., w. to se. B.C., and s. to Pa., Ky., Ark., and Ariz. An apparent hybrid with no. 3 is *S.* × *glatfelteri* C. K. Schneider.

3. Salix nigra Marshall.   Black w. Shrub or tree to 20 m or more, with 1–4 often leaning trunks to 5 dm thick; twigs slender, yellowish-brown to dark reddish-brown, often hairy when young; buds sharp-pointed, the scale with overlapping free margins; stipules usually well developed, up to 12 mm, semicordate, acute, serrulate, often deciduous; lvs commonly pendulous, lance-linear or narrowly lanceolate, 6–10(–12) × 1–1.5(–2) cm, long-acuminate, often falcate, finely serrulate, the upturned lateral veins forming a continuous marginal vein, deep green above, paler beneath (but not or scarcely glaucous) and with very small areoles, often hairy when young; petioles 4–10 mm, usually hairy, commonly glandular at the tip; catkins with the lvs, slender, lax, 2.5–7 cm, on leafy peduncles 1–3 cm; scales yellowish, villous, deciduous; stamens (3–)6(7); fr ovoid-conic, 3–5 mm, glabrous; pedicels 0.5–1.5 mm; style 0.2 mm; $2n=38$. Abundant in alluvial soil along streams and in meadows; s. N.B. to c. Minn., s. to Fla. and Tex., and w. across s. U.S. to Calif., s. into Mex. Ours is var. *nigra*. (*S. marginata*.)

4. Salix lucida Muhl.   Shining w. Shrub or small tree to 6 m; bark brown, smooth; twigs and young lvs usually with caducous, reddish-brown hairs, otherwise glabrous, the twigs and buds shining, chestnut-brown; buds blunt, the scale with fused margins (as in spp. 5–30); stipules well developed and persistent, 2–5 mm, reniform to semilunate, glandular along the margin; lvs lanceolate to lance-ovate, 5–15 × 1.5–4 cm, 3–5 times as long as wide, mostly abruptly long-acuminate or attenuate, sharply glandular-serrate, shining green above, paler green beneath; petioles 5–15 mm, glandular at the tip; catkins with the lvs, 2–5 cm, on lateral leafy peduncles 1–2.5 cm, the scales yellow, thinly hairy, deciduous; stamens 3–6; fr lanceolate, 4.5–7 mm, glabrous; pedicels 0.5–1 mm; style 0.5–0.8 mm; $2n=76$. Moist or wet low ground; Nf. and Lab. to Sask., s. to Del., W.Va., Io., and Nebr.; Roanoke Co., Va. Plants with the twigs and lower lf-surfaces ± persistently hairy are occasionally found in the n. part of our range (var. *intonsa* Fern.). A hybrid with no. 7 is *S.* × *jesupii* Fern.

5. Salix serissima (L. H. Bailey) Fern.   Autumn-w. Shrub 1–4 m, with olive-brown branches, the twigs and bud-scales yellow-brown, glabrous, shining; stipules minute or none; lvs glabrous from the first, reddish when young, firm, lanceolate to lance-elliptic or lance-oblong, 5–10 × 1–2.5(–3.5) cm, finely glandular-serrate, acute to short-acuminate, shining dark green above, subglaucous beneath; petioles glandular above at the tip; catkins with or somewhat after the lvs, the staminate 1.5–3(–4) cm, the pistillate stout, 1.5–3.5(–5) cm at maturity, on lateral leafy peduncles 1–3.5 cm; scales pale yellow, white-pilose nearly or quite throughout, those of the pistillate catkins deciduous; stamens 3–7; fr narrowly conic, 7–10 mm glabrous, on pedicels 1–2 mm; style up to 1 mm; $2n=76$. Swamps and bogs; Nf. to Alta., s. to n. N.Y., n. Ind., c. Minn., and Colo. Late May–June, the fr maturing in late summer or fall.

6. Salix pentandra L.   Bay-lvd w. Much like no. 5, but the scales hairy only at the base outside, glabrous or nearly so distally; shrub or small tree to 7 m; staminate catkins mostly 3–6+ cm, the pistillate 3.5–7 cm; stamens 4–9, mostly 5; fr 5–8 mm, on pedicels 0.5–1.5 mm; $2n=76$. Native of Europe, cult. and occasionally escaped mainly in the n. part of our range; May–June, the fr maturing in mid- and late summer.

7. Salix alba L.   White w. Tree to 25 m; twigs spreading or sometimes pendulous, greenish to yellowish, long-sericeous, somewhat brittle at base; stipules minute and caducous; lvs narrowly lanceolate to lanceolate, 4–10(–15) × 1–2.5 cm, acuminate, with 7–10 glandular teeth per cm of margin, glaucous beneath, ± silvery-silky on both sides but especially beneath; petioles sericeous, 3–6 mm, those on vigorous young shoots glandular at the tip; catkins with the lvs, 3–6 cm, on peduncles 1–4 cm with 2–4 small lvs; scales as in no. 8; stamens mostly 2; fr ovoid-conic, 3–5 mm, glabrous, sessile or subsessile; style 0.2–0.4 mm; $2n=76$. Native of Europe, sparingly escaped from cult. in our range. (*S. a.* var. *vitellina*, golden w., a long-established cultivar with bright yellow twigs.) Hybridizes freely with nos. 8 and 9.

8. Salix fragilis L.   Crack-w. Tree to 20 m, the trunk to 1 m thick; twigs stout, ascending or spreading, greenish to dark red, glabrous or slightly hairy, very brittle at base; stipules small, semi-cordate, caducous; lvs narrowly lanceolate to lanceolate, 7–12(–15) × 1.5–3.5 cm, more coarsely and remotely serrate than in no. 7, with 4–6 glandular teeth per cm of margin, dark green above, ± glaucous beneath, glabrous or nearly so at maturity; petioles 7–20 mm, glabrous, glandular above at the tip; catkins with the lvs, 4–8 cm, on peduncles 1–3(–5) cm with 2–5 small lvs, the scales greenish-yellow, crisp-villous, deciduous; stamens 2; frs narrowly conic, 4–5.5 mm, glabrous, on a pedicel 0.5–1 mm; styles 0.3–1 mm; $2n=38, 76, 114$. Native of Europe, often escaped from cult. in our range.

9. Salix babylonica L.   Weeping w. Tree to 12 m, the cult. and escaped forms with long, slender, pendulous, olivaceous or yellow-brown, glabrous or subglabrous, basally brittle twigs; stipules lanceolate, 2–7 mm, or none; lvs lance-linear, 8–12 × 0.5–1.5 cm, mostly falcate, long-acuminate, finely but unevenly spinulose-serrulate, glabrous, yellowish-green above, ± glaucous beneath; petioles 7–12 mm, tomentose, often glandular above at the tip; catkins with the lvs, 1–2.5 (staminate to 4) cm, on peduncles 0.5–1.5 cm with 2–4 tiny lvs;

scales pale yellow, crinkly-hairy, ± persistent; stamens 2; frs narrowly ovoid, 1.5–2.5 mm, sessile; style almost obsolete; stigmas brush-like (unique among our spp.); $2n = 76$. Native of China, sparingly escaped from cult. in our range.

10. Salix exigua Nutt.   Sandbar-w. Many-stemmed colonial shrub 1.5–3(–5) m, the slender, brown to reddish-brown, leafy twigs sericeous at least when young, often soon glabrous; stipules minute and caducous (rarely to 3 mm) or obsolete; lvs linear or broadly linear to linear-oblanceolate, 5–14 cm × 5–12(–18) mm, acute or acuminate at both ends, subsessile, remotely and irregularly spinulose-denticulate, green on both sides but paler beneath, often sericeous when young, later glabrous (or the new lvs produced after insect-attack more densely and permanently sericeous); catkins with the lvs, 1–3 together on very short lateral branches from axillary buds of the previous year, and often again in midsummer, terminating leafy lateral branches 3–10 cm; scales pale, yellowish, soon deciduous; stamens 2; pistillate catkins lax, 2–6(–8) cm; frs narrowly lanceolate, 5–8 mm; pedicels 0.5–1.5 mm; style virtually obsolete; $2n=38$. Common on sandbars, mudbars, and moist alluvial soil; widespread in the U.S., s. Can., and n. Mex. Plants from e. of the Rocky Mts., as here described, constitute the subsp. *interior* (Rowlee) Cronq. (*S. interior; S. longifolia* Muhl., not Lam.), which has 2 vars. All or nearly all of our plants belong to the var. *angustissima* (Andersson) Reveal & Broome, which has the ovaries hairy at least when young, although the frs are often glabrate. The more n. or nw. var. *pedicellata* (Andersson) Cronq., with the ovaries and frs glabrous from the first, may be sought along the w. margin of our range.

11. Salix bebbiana Sarg.   Beaked w. Shrub or small tree 2–5 m, with one or few stems; twigs slender, divaricate, brownish, hairy to glabrate; stipules mostly small or none; lvs elliptic to broadly rhombic-oblanceolate or obovate-oval, acute or abruptly short-acuminate, 4–8(–10) × 1.5–4) cm, subentire to undulate-crenate, glaucous and rugose-reticulate beneath, ± gray-hairy on both sides, but sometimes eventually glabrate; catkins subprecocious, the staminate small, subsessile, the pistillate 2–7 cm, very lax, on bracteate peduncles 5–20 mm; scales 1–2 mm, greenish-yellow with reddish tip, pilose; stamens 2; frs lanceolate-rostrate, 7–10 mm, finely hairy; pedicels 3–6 mm, finely hairy; style 0.1–0.4 mm; $2n=38$. Moist or wet places; Nf. to Alas., s. to N.J., Md., Ill., S.D., and Ariz. (*S. rostrata* Richardson, not Thuill.)

12. Salix herbacea L.   Herb-like w. Slender, subherbaceous plant, creeping in moss and humus, glabrous throughout, except the sometimes sparsely hairy scales; twigs chestnut, subascending; buds ovate, 1.5–2.5 mm; lvs oval to suborbicular or even wider than long, 1–3 × 1–2.5 cm, crenulate-dentate, green on both sides, reticulate beneath; catkins after the lvs, 10–17 mm, with 5–12(–16) fls, terminating short, 2–4-lvd twigs; scales light to dark brown; stamens 2, the filaments distinct; staminate fls biglandular; frs glabrous, subsessile, lanceolate, 3–6 mm; style 0.5 mm; $2n=38$. Grassy, sandy, or rocky places; circumboreal, s. in N. Amer. to Mt. Katahdin, Me., Mt. Washington, N.Y., and Essex Co., N.Y. June–Sept. A hybrid with no. 13 has been called *S.* × *peasei* Fern.

13. Salix uva-ursi Pursh.   Bearberry-w. Prostrate, matted alpine shrub with stoutish, leafy, brown twigs; buds ovate, 1–2 mm, brown, glabrous, stipules mostly none; petioles 2–4 mm; lvs narrowly to broadly elliptic, or obovate, 5–20(–25) × 2–7(–10) mm, ± glandular-crenate, acute at base, green, shining, and reticulate above, glaucous and strongly reticulate beneath, glabrous or sometimes silky-villous when young, sometimes persistent and marcescent; catkins with or shortly after the lvs, 1–4 cm, on leafy peduncles 0.5–2 cm, scales brown with blackish tip, 1–1.5 mm, villous; stamen 1 (by cohesion) or rarely 2; staminate fls biglandular; frs more numerous than in no. 12, conic, 3–4 mm, subsessile, glabrous; style 0.7–1.5 mm; $2n=38$. Exposed, rocky places; Greenl. and Baffin Isl., s. to the mts. of Me., N.H. and N.Y. June–Aug.

14. Salix arctophila Cockerell.   Trailing shrub; vegetative parts all glabrous; twigs yellowish-brown, sometimes pruinose; buds 3–8 mm; petioles 5–12(–24) mm; lvs elliptic-obovate to obovate (the apical ones rarely obovate-oval and subcordate), 2–4(–5) × 1.5–3 cm, acute to apiculate, shallowly crenate-denticulate toward the tip, dark green and shining above, glaucous beneath, prominently and slenderly raised-reticulate on both sides; catkins after the lvs, 3–9 cm, rather lax, erect on leafy peduncles 1–4 cm; scales blackish, long-villous; stamens 2; staminate fls biglandular; fr slender, 6.5–8 mm, thinly silky; pedicels 0.5–1 mm; style 1–1.5 mm; $2n=76$. Marshy, mossy, low places in the mts., or sometimes in drier areas; Mt. Katahdin, Me., n. to Greenl. and Mack. (*S. groenlandica*.)

15. Salix pedicellaris Pursh.   Bog-w. Slender, sparingly branched shrub (2–)4–10(–15) dm, with dark, glabrous twigs; stipules none; petioles short, mostly 2–7 mm; lf-blades at first silky, but very soon glabrous, entire, narrowly or broadly elliptic to oblanceolate or obovate; (2–)3.5–6(–8) × 0.5–2(–3) cm, entire, often slightly revolute, green above, commonly glabrous and glaucous beneath, finely raised-reticulate on both sides, with parallel primary lateral veins; catkins with the lvs, 1–3 cm, on leafy peduncles 2–4(–5) cm; scales 1 mm, yellowish-brown, hairy within, at least near the tip, the hairs surpassing the scale, otherwise generally glabrous; stamens 2; frs narrowly lanceolate, 4–7 mm; pedicels 2–3 mm; style 0.1–0.3 mm; $2n=38$, 57, 76. Sphagnum-bogs and swamps; Nf. to Yukon and B.C., s. to N.Y., N.J., n. Io., and Oreg.

16. Salix pyrifolia Andersson.   Balsam-w. Shrub to 5 m, occasionally arborescent; twigs glabrous, at first yellowish, becoming red and shining; bud-scale 2–8 mm, red; stipules wanting, or ovate and 1–2 mm; petioles 0.8–2 cm; lvs reddish while unfolding, thin and translucent at first, ovate to lanceolate or lance-oblong, 4–10(–13) × 2–4(–5) cm, acute to abruptly acuminate, crenate-serrate with often glandular teeth, rounded to cordate at base, deep green above, glaucous and finely reticulate beneath, glabrous, with a balsamic fragrance; catkins with or after the lvs, 2–6(–8) cm, lax, the staminate on bracteate peduncles 2–8 mm, the pistillate on leafy peduncles 1–3 cm; scales 2–2.5 mm, reddish-brown, white-villous, becoming glabrate outside;

stamens 2; frs lanceolate-rostrate, 6–8 mm, glabrous; pedicels 2.5–3.5 mm; style 0.3–0.5 mm; $2n=38$. Moist ot wet or swampy ground; Lab. and Que. to B.C., s. to N.S., n. N. Engl., n. N.Y., and n. Minn. Fls mid-May–Aug. (*S. balsamifera.*)

17. Salix eriocephala Michx.   Diamond-w. Shrub 2–4(–6) m, or sometimes arborescent; twigs, bud-scales, and petioles typically reddish-brown to dark brown, hairy when young; stipules semi-ovate to subreniform, 5–10(–15) mm, acutish, persistent; petioles 5–14 mm; lvs pubescent and reddish or purplish when young, later glabrous, typically oblong-lanceolate, serrulate, 7–12(–15) × 1.5–3(–4) cm, short-acuminate, rounded to subcordate at base, dark green above, ± glaucous beneath, rigid and veiny in age, catkins with or a little before the lvs, 2–5 cm, subsessile or on bracteate peduncles 3–10 mm; scales 1–1.5 mm, dark brown, crinkly-hairy; stamens 2; frs lance-ovate, 4–6 mm, glabrous; pedicels 1–2 mm; style ca 0.5 mm; $2n=38$. Along watercourses and in lowlands; transcontinental, from Nf. and Que. to Va., w. to Yukon, B.C., and Calif.; disjunct in Ala., w. Ga., and w. Fla. (*S. cordata* Muhl., not Michx.; *S. gracilis; S. lutea; S. missouriensis; S. rigida.*) Complex sp., probably consisting of several vars., but the limits of these not yet clearly understood.

18. Salix myricoides (Muhl.) J. Carey.   Bluelf-w. Shrub 1–4 m; twigs yellowish to dark brown, hairy when young; stipules semi-ovate, 5–10 mm; petioles 5–12 mm; lvs lanceolate or lance-ovate to ovate-oval or obovate-oval, 4–9(–12) × 1.5–4.5(–6) cm, thickish, acute to short-acuminate, glandularly crenate-serrate, mostly rounded or cordate at base, dark green above, strongly glaucous beneath, sometimes sparsely hairy when young; catkins with or shortly before the lvs, 2–8 cm, on leafy peduncles 0.5–1.5 cm; scales 1.5–2 mm, brownish-black, long-villous; stamens 2; frs narrowly lanceolate, 5–8 mm, glabrous; pedicels 1–3 mm; style 0.7–1.5 mm. Sandy shores, calcareous slopes, and sometimes swamps; Nf. and e. Que. to n. Me. and the Great Lakes region, s. to Ill. and Wis., and n. to Hudson Bay. (*S. glaucophylla* Bebb, not Besser; *S. glaucophylloides.*) Plants with more hairy lvs and twigs have been called var. *albovestita* (C. Ball) Dorn. Depauperate plants in sterile sand chiefly along the Great Lakes, to 2 m, with more coriaceous and veiny lvs 2.5–5 × 0.5–2 cm, may constitute another var., but the proper combination (using the epithet *brevifolia*) has not yet been made.

19. Salix cordata Michx.   Dune-w. Shrub (1–)2–3 m, the vegetative parts all ± pubescent, the young twigs densely gray-tomentose; stipules cordate-ovate or subreniform, 6–15 mm; petioles 4–8 mm, somewhat clasping; lvs lance-ovate to broadly ovate, 4–6(–8) × 1.5–3(–4) cm, abruptly acuminate, glandular-serrate, rounded or cordate at base, green on both sides, ± lanate, strongly nerved beneath; catkins with the lvs, 5–8 cm, on 3–5-lvd peduncles 1–2.5 cm; scales brown, densely long-villous; stamens 2; frs lanceolate, 5–8 mm, glabrous; pedicels 0.5–1 mm; style 0.7–1.5 mm; $2n=44$. Sandy and alluvial shores, or often on dunes; Nf. and e. Que. to N.S., n. Me., and w. around the Great Lakes to Ill., Wis., and n. Mich., n. to Hudson Bay. (*S. adenophylla; S. syrticola.*) Plants with the twigs, petioles and blades (except the midrib) ± glabrate or glabrous have been called var. *abrasa* Fern.

20. Salix planifolia Pursh.   Erect shrub 1–3 m; twigs stoutish, divaricate, chestnut to red-purple or brown, puberulent at first, soon glabrous and shining; stipules small or more often none; petioles 4–10 mm; lvs rather crowded, narrowly elliptic or elliptic-oblanceolate or narrowly obovate, 3–6(–8) × 1–2.5(–3.5) cm, acute at both ends or the largest obtuse at base, entire or sometimes ± crenate-serrulate, glabrous and glossy above, sparsely sericeous to glabrous and glaucescent beneath, the primary lateral veins numerous, closely spaced; catkins precocious (reputedly sometimes coetaneous—not provided for in the key), sessile, 2.5–5 cm; scales 2–3 mm, black, long-villous; stamens 2; frs lanceolate, 4–7 mm, finely sericeous, subsessile, the pedicels only 0.1–0.5 mm; style 1–1.5 mm; $2n=76$. Wet meadows, lake-shores, streambanks, and rocky places; Lab. to Yukon, s. to the mts. of Me., N.H., and Vt., and to n. Minn. and the w. cordillera. (*S. phylicifolia* ssp. *planifolia.*)

21. Salix pellita Andersson.   Satiny w. Shrub or small tree 3–5 m; twigs brittle, yellowish to olive-brown or reddish-brown, thinly hairy or glabrous, soon becoming glaucous; stipules none; petioles to 10 mm; lvs lance-linear to lanceolate or linear-oblanceolate, 4–10(–13) cm × 8–15(–22) mm, acuminate, subrevolute, entire to undulate-crenate, glabrate and with subimpressed veins above, glaucescent and densely satiny-silky beneath, becoming glabrate; primary lateral veins numerous, closely parallel, diverging at a wide angle; catkins precocious, sessile or on bracteate peduncles to 1 cm, the staminate rarely produced in our area, the pistillate 2–5 cm; scales 1–2 mm, blackish, long-villous; frs lanceolate, 4–6 mm, sessile or subsessile, silky; style 0.8–1.2 mm; $2n=38, 76$. Alluvial or gravelly riverbanks, shores, and swamps; Nf. and Lab. to s. Sask., s. to N.S. and the n. parts of Me., N.H., Vt., and Mich. Closely allied to the cordilleran *S. drummondiana* Barratt.

22. Salix sericea Marshall.   Silky w. Shrub or small tree, to 4 m; twigs brittle at the base, slender, brown, densely gray- to brown-tomentose when young, later glabrate except often at the nodes; stipules lanceolate to semi-ovate, 3–10 mm, often deciduous; petioles slender, 5–10 mm, puberulent; lvs narrowly lanceolate to lanceolate, 6–10(–14) × 1–2.5(–3) cm, acuminate, cuneate at base, closely glandular-serrate, dark green and puberulent to glabrous above, glaucous and densely to thinly short-sericeous beneath; catkins ± precocious, 1–4 cm, subsessile or on narrow-bracted peduncles to 1 cm; scales 1 mm, blackish, villous; stamens 2; frs ovoid-oblong, 3–5 mm, blunt, short-sericeous; pedicels 0.5–1(–1.5) mm; style 0.1–0.3 mm. Moist, rocky ground, often in or near running water; N.S. and N.B. to Wis. and e. Io., s. to S.C., n. Ga. and n. Ark. (*S. coactilis.*) A possible hybrid with no. 29 is *S. × subsericea* (Andersson) C. K. Schneider.

23. Salix discolor Muhl.   Pussy-w. Few-stemmed shrub or small tree 2–5(–7) m; twigs rather stout, reddish to dark brown, hairy when young, usually later glabrate; buds large, to 1 cm; stipules small to large, rounded to semi-ovate; lvs mostly elliptic or elliptic-oblanceolate, 4–8(–10) × 1.5–3.5(–5) cm, acute or shortly acuminate, subentire to undulate-crenate, with flat margins, at maturity generally dark green and glabrous (seldom puberulent) above, the lower lf-surface becoming glaucous, generally reddish-strigillose when young, sometimes more persistently so, varying to loosely and persistently hairy; catkins precocious, sessile or nearly so, bractless or few-bracted, stout, and staminate 2–4 cm, ornamental, the pistillate 4–8(10) cm in fr; scales 1.5–2.5 mm, dark brown, long-villous; stamens 2; frs lanceolate, beaked, 7–10(–12) mm, densely gray-hairy; pedicels 1.5–3 mm; style 0.5–0.7 mm, the stigmas at least as long; $2n=76$, 95, 114. Common in swamps and wet ground; Nf. to Alta., s. to Del., Ky., Mo., S.D., and Mont. $S. \times conifera$ Wangenh. is probably a hybrid with no. 24.

S. caprea L., the goat or florist's willow, a European sp. with obovate to suborbicular lvs 5–13 × 2.5–5.5(–8) cm, apiculate to shortly acute, ± crenate-serrate and tomentose, with irregularly revolute margins, otherwise much like no. 23, furnishes the pussy-willows of florists for Easter-decorations. It occasionally escapes about nurseries.

S. cinerea L., the gray florist's willow, with more tomentose, narrower lvs than *S. caprea*, acute at both ends, is similarly used and rarely escapes. The long prominent ridges on decorticated wood are distinctive.

24. Salix humilis Marshall. Upland-w. Colonial shrub 1–3 m; twigs flexible, yellowish to brown, velutinous-tomentose, or eventually glabrate; stipules lanceolate, often deciduous; lvs oblanceolate to narrowly obovate, 3–10(–15) × 1–2(–3) cm, acute or abruptly short-acuminate, somewhat revolute-margined, entire or sparingly undulate-crenate, dark green (and often puberulent) above, glaucous, somewhat rugose, and ± gray-tomentose beneath (sometimes also with some reddish hairs), becoming glabrate; petioles 3–7 mm; catkins precocious, sessile or subsessile, often recurved, the staminate 1–2 cm, the pistillate 1.5–4 cm at maturity; scales 1.5–2 mm, blackish, long-villous; stamens 2; frs narrowly lanceolate-rostrate, 6–10 mm, gray-hairy; pedicels 1–2 mm; style 0.2–0.4 mm; $2n=38$, 76. Scattered but common in open woods, dry barrens, and mesic or wet prairies; Nf. and s. Que. to N.D., s. to Fla. and Tex.

25. Salix occidentalis Walter. Much like no. 24, but less common and smaller in all respects; shrub to 1 m, with exstipulate, narrowly oblanceolate lvs mostly 2–5 cm × 7–10(–12) mm; petioles 0.5–3 mm; staminate catkins 0.5–1.2 cm, the pistillate ones 1–2 cm at maturity. Dry open places, often in more exposed habitats than no. 24; Me. to Minn., s. to S.C., Ga., La., and Okla. (*S. tristis; S. humilis* var. *microphylla.*).

26. Salix argyrocarpa Andersson. Silvery w. Bushy shrub 5–17 dm, often depressed at base; twigs short, leafy, full of lf-scars, reddish-brown, puberulent when young; stipules none, or minute, and caducous; petioles to 6 mm; lvs narrowly elliptic-oblanceolate, 2–5 cm, to 1 cm wide, acute, subrevolute and subentire to closely crenulate, dark green with impressed veins above, glaucescent and ± long-sericeous to glabrate beneath, the primary lateral veins numerous, wide-angled, closely parallel; catkins with the lvs, 1–2.5 cm, on leafy peduncles 0.5–1.2 cm; staminate fls biglandular; scales 1–2 mm, brownish-black, thinly villous; frs lance-ovate, 2–4 mm, densely short-sericeous; pedicels 1–2 mm; style 0.5 mm; $2n=76$. Moist ravines and alpine slopes; mts. of n. Vt., n. N.H., and n. Me. to Nf. and Lab. A hybrid with no. 20 is $S. \times grayi$ C. K. Schneider.

27. Salix candida Fluegge. Sage-lvd w. Freely branched low shrub to 1(–1.5) m; branches divaricate, brownish, glabrous, full of lf-scars; twigs densely white-tomentose; stipules lanceolate, glandular; lvs linear-oblong, oblong, or narrowly to rarely broadly oblanceolate, 4–8(–12) × 0.7–1.5(–2.5) cm, mostly acute at both ends, revolute, entire to glandular-crenate, dull and thinly tomentose with sunken veins above, densely white-tomentose beneath; catkins with the lvs, 1–5 cm, on leafy-bracted peduncles 0.5–2 cm; scales 1.5–2.5 mm, brown, persistent, woolly-villous; stamens 2; frs lanceolate, 5–8 mm, subsessile, white-tomentose; style 1–1.5 mm; $2n=38$. Calcareous wetlands in glaciated areas; Lab. to Alas., s. to N.J., Pa., O., Ind., Ill., Io., and Colo. An extreme form with most of the tomentum of the lvs soon deciduous, the lvs glaucous beneath, is f. *denudata* (Andersson) Rouleau. A hybrid with no. 17 is $S. \times rubella$ Bebb; one with no. 28 is $S. \times clarkei$ Bebb.

28. Salix petiolaris J. E. Smith. Meadow-w. Clumpy shrub to few-stemmed tree (northward), (1–)2–7 m; twigs slender, yellowish and puberulent to dark brown and glabrous; stipules minute or none; petioles slender, 5–15 mm; lvs narrowly lanceolate or narrowly oblanceolate, 4–10(–15) × 0.8–2(–3) cm, acute to acuminate, subentire to closely glandular-serrulate, densely sericeous on both sides when unfolding from the bud, later dark green and glabrous above, glabrous and glaucous beneath (rarely more persistently hairy); catkins with the lvs, 1–3.5 cm, lax, subsessile or on bracteate peduncles 1–2 cm; scales 1.5–2 mm, brown, thinly villous; stamens 2; frs lanceolate-rostrate, 4–8 mm, thinly sericeous; pedicels 2.5–5 mm; style 0.1–0.5 mm; $2n=38$. Moist meadows, stream-banks, and lake-shores; N.B. to N.J. and Pa., w. to Alta., n. Nebr., and Colo. Low shrubs with small lvs 3–6 cm and subentire have been called var. *angustifolia* Andersson (*S. gracilis*), but the 3 characters are not well correlated.

29. Salix viminalis L. Silky osier. Few-stemmed shrub or tree to 15 m; twigs slender, yellowish to reddish-brown, puberulent to glabrous; stipules mostly wanting, or lance-linear and up to 9 mm; petioles to 1 cm; lvs lance-linear, 5–12(–17) cm × 5–15(–20) mm, long-acuminate, revolute, entire or shallowly undulate, dull green and puberulent with impressed veins above, densely short-sericeous with yellow midrib beneath, the primary lateral veins numerous, wide-angled; catkins with the lvs, subsessile, 2–6 cm, mostly naked; scales 1.5–2.5 mm, black, thinly villous; stamens 2; frs lanceolate, 4–6 mm, subsessile, densely short-sericeous; style 0.7–1.2 mm; $2n=38$. Native of Eurasia, occasionally escaped from cult. in our range. A hybrid with *S. caprea* is $S. \times smithiana$ Willd.

30. Salix purpurea L. Basket w.; purple osier. Many-stemmed shrub 1–2.5 m; twigs slender, greenish-yellow or rarely purplish, glabrous; stipules none; lvs short-petioled, mostly linear-oblanceolate or spatulate or linear, 4–7(–10) cm × 7–14(18) mm, entire below, irregularly serrate toward the tip, finely raised-reticulate on both sides, with a purplish cast, glaucescent beneath, glabrous, at least some of those towards the ends of the twigs subopposite; catkins precocious, 2–3.5 cm × 5–8 mm, sessile, bracteate, sub-opposite; scales broadly obovate, blackish, becoming glabrate; stamens 2 but the filments and often the anthers united; frs ovoid-conic, 3–4 mm, obtuse, sessile, short-hairy; style 0.1–0.2 mm; $2n=38$. Native of Europe, sparingly escaped from cult. in our range.

FAMILY CAPPARACEAE, the Caper Family

Fls typically (including our spp.) borne in terminal, bracteate racemes, hypogynous or nearly so, usually perfect, regular or somewhat irregular; receptacle usually ± prolonged into a gynophore (the ovary appearing stipitate) or androgynophore; sep mostly 4, distinct or basally connate; pet 4, often unequal, commonly long-clawed; stamens mostly 6–many, never tetradynamous; filaments ± elongate, exsert; ovary bicarpellate, unilocular with 1 style; ovules ± numerous on 2 parietal placentas; fr in ours an elongate capsule, dehiscent by 2 valves, with a persistent, frame-like replum; seeds mostly reniform or reniform-orbicular, with little or no endosperm; plants producing mustard-oils; ours annuals with alternate, trifoliolate or palmately compound lvs. 45/800, mainly warm reg.

1 Pet entire; fr long-stipitate (above the pedicel); stamens 6 .... 1. *Cleome.*
1 Pet retuse or laciniate; fr sessile or short-stipitate; stamens 6–many .... 2. *Polanisia.*

1. CLEOME L.   Pet equal or nealry so, entire; receptacle bearing an adaxial, usually rounded or pointed, mostly greenish or yellowish gland between the cor and the androe-cium; style mostly (incl our spp.) very short; frs usually long-stipitate, reticulately veined, opening from the base upwards, the valves deciduous; annuals with palmately compound or trifoliolate lvs, simple bracts, and handsome fls in elongating terminal racemes. 200, mainly warm reg.

In addition to the following spp., the Middle-Eastern *C. ornithopodioides* L. is adventive on chrome ore piles in Ky. It is slender, to 5 dm, with strongly glandular stem, trifoliolate lvs (the lfls to 1.5 cm) and small fls (the pet only 1–2 mm) and frs (1.5–3.5 cm).

1 Lfls 3; stem glabrous, or sparsely and finely villous when young .... 1. *C. serrulata.*
1 Lfls of the principal lvs commonly 5–7; stem viscid-pubescent .... 2. *C. hassleriana.*

1. Cleome serrulata Pursh.   Stinking clover. Stems 5–8 dm, branched above, pale, glabrous or nearly so; petioles not spiny; lfls 3, narrowly lanceolate, 3–6 cm, sharply acuminate or cuspidate, entire, glabrous at maturity, sometimes finely and sparsely villosulous when young; bracts crowded, lanceolate, short-petioled; pet subequally disposed, pink (white), oblong, 1 cm, the blade narrowed to the base and much exceeding the claw; fr torulose, 3–5 cm, glabrous, the stipe 1.5–2 cm, about equaling the pedicel; $2n$ reputedly $=32$, 34, 60. Prairies; Ind. to Mo. and Minn., w. to Sask., Wash., Calif., and Tex. July, Aug.

2. Cleome hassleriana Chodat.   Spider-fl. Stems to 1.5 m, branched above, viscid-pubescent; petioles commonly with a pair of short spines at the base; lfls of the main lvs 5–7, oblanceolate, finely hairy and often finely serrulate, the central one 6–10 cm; pet all oriented toward one side of the fl, white to pink, the blade obovate, 1 cm, abruptly narrowed to a claw ca half as long; fr 4–10 cm, about equaling the stipe, this nearly twice as long as the pedicel; $2n=20$. Native of trop. Amer., occasionally escaped from cult. in our range, especially southward. June–Oct. (*C. spinosa,* misapplied.)

2. POLANISIA Raf.   Clammy-weed. Pet unequal (the adaxial pair the longer), long-clawed, emarginate to deeply erose or laciniate; stamens 6–many, unequal; receptacle bearing a prominent, fleshy, apically concave (seldom truncate), brightly colored, adaxial gland between the cor and androecium; stamens 6–27, unequal, the outer shorter; style short to elongate; fr sessile or short-stipitate, erect, reticulate-veiny, stipitate-glandular, the valves persistent; malodorous, viscid-hairy annuals with trifoliolate lvs and terminal racemes of white or ochroleucous to pinkish fls. 4, N. Amer. (*Cristatella.*)

1 Lfls 5–20 mm wide; frs 5–10 mm wide; bracts simple .... 1. *P. dodecandra.*
1 Lfls 0.5–4 mm wide; frs 3–4 mm wide; bracts trifoliolate .... 2. *P. jamesii.*

1. Polanisia dodecandra (L.) DC. Stems 2–6+ dm; petioles about as long as the blades; lfls ovate or obovate to lance-elliptic, 1.5–6+ × 0.5–2 cm; fls numerous (15–30 open fls and buds); bracts simple; pet white to pink, irregularly truncate to deeeply emarginate; stamens (7–)10–27; nectary solid, 1–2 × 0.5–1.5 mm, with bright orange or orange-red top, inconspicuous in fr; style 3–40 mm, slender, withering and deciduous in fr; frs oblong to narrowly fusiform, flattened but somewhat inflated, mostly 3–6 cm × 5–10 mm, on a stipe 2–6 mm, the valves separating apically for ⅓ to ⅔ their length; seeds mostly 15–40+, 1.7–3 mm; 2n=20. Sandy or gravelly places, especially along streams; N.H. and s. Que. to Alta. and Wash., s. to Md., Tenn., Ark., Tex., and n. Mex. July–Sept. Four well marked vars., 2 in our range: Var. *dodecandra,* the more northeastern phase, common in our range, has the larger pet 3.5–6.5(–8) mm, and the longest stamens 4–10(–14) mm. (*P. graveolens.*) Var. *trachysperma* (T. & G.) Iltis, more western, extends e. to Minn. and Mo. and is occasionally intr. eastward; it has larger fls, the larger pet (7–)8–13(–16) mm, and the longest stamens (9–)12–30 mm, usually much exceeding the pet. (*P. t.*)

2. Polanisia jamesii (T. & G.) Iltis. Stems 1–3(–4) dm; petioles usually shorter than the blades; lfls linear or linear-elliptic, 0.5–4 cm × 0.5–4 mm; fls few (5–9 open fls and buds); bracts trifoliolate; pet white to ochroleucous, strongly dimorphic, the larger pair 4–5 mm, emarginate and shallowly laciniate with 3–6 segments, the smaller 2–3 mm and more deeply laciniate; stamens 6–9; nectary 2–3.5 × 0.5 mm, tubular, yellow, persistent and drying pink-purple; style 0.5–2 mm, persistent in fr; frs linear-fusiform, slightly inflated, 1.5–3 cm × 3–4 mm, dehiscing nearly to the base; seeds 4–18, 1.7–1.8 mm; 2n=20. Sandy prairies and plains; Great Plains (n. as far as S.D.) e. to Wis. and Ill. (*Cristatella j.*)

Family **BRASSICACEAE**, the Mustard Family

Fls typically borne in terminal, elongating, mostly bractless racemes (these occasionally with a few long branches), or rarely terminal and solitary on a scape, hypogynous, perfect, regular or nearly so, only seldom (as in *Lunaria*) with an evident gynophore; sep 4; pet 4, diagonal to the sep, commonly with an elongate claw and an abruptly spreading blade, collectively forming a cross, or seldom wanting; androecium tetradynamous, with 6 stamens, the 2 outer shorter than the 4 inner, or seldom the stamens fewer than 6; ovary bicarpellate, nearly always divided into 2 locules by a thin, unvasculated septum connecting the 2 parietal placentas, the placental frame forming a persistent replum; ovules (1–) several or many on each placenta, in 2 rows separated by the partition; style solitary, often short (or the stigma sessile); fr dry and usually dehiscent, generally a silique (elongate) or a silicle (short), the valves falling away from the persistent replum at maturity, or seldom the fr indehiscent and sometimes transversely jointed; endosperm very scanty or none; plants producing mustard-oils; herbs (all ours) or rarely shrubs, with alternate (rarely opposite), simple to often pinnately ± dissected lvs, only seldom with definite lfls. (Cruciferae.) The following strictly artificial key avoids the more technical characters of the staminal glands, stigma, and embryo. 340/3350, mostly temp reg.

1 Fr transversely distinctly segmented into 2 unlike members, one of these indehiscent . . . . Group I.
1 Fr not transversely segmented into unlike members, though usually beaked by the persistent style and
  sometimes constricted between the seeds.
  2 Fr flattened at right angles to the septum, the septum therefore much narrower than the width of
    the fr, its position indicated by a nerve or stricture along the median line of both faces; fr never
    more than 4 times as long as wide . . . . Group II.
  2 Fr not flattened, or flattened parallel to the septum, the septum therefore as wide as the fr; fr short
    or long.
    3 Fr very thin and flat, more than 6 mm wide, on a gynophore more than 10 mm . . . . Group III.
    3 Fr not flattened, or if flattened then less than 6 mm wide, usually sessile.
      4 Fr relatively short and broad, not over 3 times as long as wide . . . . Group IV.
      4 Fr relatively long and slender, 3–100 times as long as wide.
        5 Pet none . . . . Group V.
        5 Pet present.
          6 Fls yellow . . . . Group VI.
          6 Fls white, pink, blue, or purple . . . . Group VII.

Group I

1 Upper member of the fr a flat or angled beak, usually shorter than the lower member.
  2 Seeds in 2 rows in each locule; pet with dark veins; beak seedless . . . . 5. *Eruca.*
  2 Seeds in 1 row in each locule; pet uniformly colored; beak usually with 1–3 seeds.

3 Valves with 3–5 prominent veins; sep spreading to reflexed .... 2. *Sinapis.*
3 Valves with only a prominent midvein; sep usually erect to ascending .... 1. *Brassica.*
1 Upper member of the fr distinctly longer or thicker than the lower.
  4 Upper member many times longer than the inconspicuous lower member .... 6. *Raphanus.*
  4 Upper member lanceolate to subglobose, not more than 4 times as long as the lower one.
    5 Upper member subglobose, not corky; pet yellow; style elongate; lvs hirsute .... 7. *Rapistrum.*
    5 Upper member lanceolate to ovoid, corky; pet white or anthocyanic; style absent; lvs glabrous
      .... 8. *Cakile.*

Group II

1 Pet unlike, the outer pair much larger than the inner pair; median (paired) filaments with a basal
   appendage .... 15. *Teesdalia.*
1 Pet essentially equal, or none; filaments unappendaged.
  2 Lvs all basal, subulate, entire; plants mostly aquatic .... 17. *Subularia.*
  2 Lvs all or partly cauline, flat; plants terrestrial.
    3 Pet bright yellow; fr samaroid, indehiscent .... 13. *Isatis.*
    3 Pet white or none; fr not samaroid, usually dehiscent.
      4 Racemes lateral, spreading or drooping; fr wrinkled or tuberculate .... 11. *Coronopus.*
      4 Racemes all or chiefly terminal, erect or nearly so; fr smooth.
        5 Seeds 1 or 2 in each fr.
          6 Fr dehiscent, flat or somewhat distended over the seed, winged or sharply keeled at the
             margin .... 10. *Lepidium.*
          6 Fr indehiscent; didymous, each valve separately distended over the seed, obtuse at the
             margin .... 12. *Cardaria.*
        5 Seeds several in each fr.
          7 Fr oval to orbicular, winged laterally and distally, rounded or obtuse at base; plants
             glabrous or with simple hairs .... 14. *Thlaspi.*
          7 Fr obtriangular or triangular-obcordate, wingless, acute at base; plants with forked or
             stellate hairs .... 16. *Capsella.*

Group III

A single genus .... 21. *Lunaria.*

Group IV

1 Fr indehiscent, often woody, or spongy or corky in texture, commonly asymmetrical, or appendaged,
   or verrucose, or pitted, containing 1–4 seeds.
  2 Fls white .... 12. *Cardaria.*
  2 Fls yellow.
    3 Cauline lvs auriculate-clasping at base.
      4 Fr conic-obovoid; mature pedicels appressed, 2–4 mm; plant glabrous .... 18. *Myagrum.*
      4 Fr subglobose, pitted; mature pedicels spreading, to 1 cm; plant hairy .... 19. *Neslia.*
    3 Cauline lvs narrowed to the base .... 20. *Bunias.*
1 Fr dehiscent, the walls thin or papery or membranous or herbaceous, symmetrical or slightly curved;
   seeds often numerous.
  5 Pet none .... 25. *Draba.*
  5 Pet present.
    6 Pet yellow.
      7 Cauline lvs bipinnate; fr narrowly clavate .... 43 *Descurainia.*
      7 Cauline lvs entire to once pinnatifid.
        8 Fr pyriform to globose or ovoid, or short-cylindric, nearly as thick as wide.
          9 Pubescence of simple hairs or none .... 32. *Rorippa.*
          9 Pubescence of stellate hairs.
            10 Cauline lvs sessile, auriculate at base .... 42. *Camelina.*
            10 Cauline lvs tapering to the base, not auriculate .... 26. *Lesquerella.*
        8 Fr distinctly flattened.
          11 Fr oblong, 2–3 times as long as wide .... 25. *Draba.*
          11 Fr broadly oblong to orbicular, less than twice as long as wide .... 22. *Alyssum.*
    6 Pet white.
      12 Pet bifid.
        13 Plants scapose, to 1.5 dm; fr glabrous; style obsolete .... 25. *Draba.*
        13 Plants with leafy stems to 7 dm; fr stellate-hairy; style elongate .... 24. *Berteroa.*
      12 Pet rounded or slightly emarginate at the summit.
        14 Fr ovoid to obovoid, nearly as thick as wide .... 27. *Armoracia.*
        14 Fr orbicular to elliptic or oblong, distinctly flattened.

15 Plant with 2-pronged hairs lying longitudinally on the stem and lvs .... 23. *Lobularia.*
15 Plant with stellate hairs, or rarely glabrous.
  16 Fr orbicular or nearly so, minutely retuse, with thin or wing-like margin .... 21. *Alyssum.*
  16 Fr distinctly longer than wide, obtuse to acute .... 25. *Draba.*

Group V

1 Principal cauline lvs simple .... 29. *Cardamine.*
1 Principal cauline lvs bipinnatifid .... 43. *Descurainia.*

Group VI

1 Plants ± pubescent with 2-pronged or stellate or branched hairs.
  2 Lvs bipinnate or bipinnatifid .... 43. *Descurainia.*
  2 Lvs entire or toothed.
    3 Fr not more than 5 times as long as wide, distinctly flattened .... 25. *Draba.*
    3 Fr 7 to many times as long as wide, terete or 4-angled .... 37. *Erysimum.*
1 Plants pubescent with simple hairs, or glabrous.
  4 Lvs strictly entire, the upper cordate-clasping .... 9. *Conringia.*
  4 Lvs toothed to pinnatifid.
    5 Seeds in 2 rows in each locule.
      6 Beak of the fr flat, 5–12 mm long, as wide as the body .... 5. *Eruca.*
      6 Beak of the fr slender, 4 mm long or less.
        7 Fls 2–5 mm wide .... 32. *Rorippa.*
        7 Fls 10 mm wide or more .... 4. *Diplotaxis.*
    5 Seeds in one row in each locule.
      8 Cauline lvs auriculate or cordate-clasping at base .... 33. *Barbarea.*
      8 Cauline lvs not auriculate or clasping.
        9 Seeds subglobose .... 1. *Brassica.*
        9 Seeds ovoid to oblong.
          10 Fr-valve 3-nerved; fls ebracteate .... 39. *Sisymbrium.*
          10 Fr-valve 1-nerved; lower fls usually bracteate .... 3. *Erucastrum.*

Group VII

1 At least some trichomes branched.
  2 Fr distinctly flattened .... 31. *Arabis.*
  2 Fr terete or quadrangular.
    3 Cauline lvs auriculate-clasping or cordate-clasping at the sessile base; fr 5–10 cm .... 31. *Arabis.*
    3 Cauline lvs not clasping at the base, sessile or petioled.
      4 Fls large, the pet 2–2.5 cm; fr 5–10 cm .... 36. *Hesperis.*
      4 Fls small, the pet 2–4 mm; fr 1–2.5 cm.
        5 Plant annual from a slender taproot .... 41. *Arabidopsis.*
        5 Plant perennial, with a branched caudex .... 40. *Braya.*
1 Trichomes unbranched or absent.
  6 Early fls solitary on long scapes; small winter-annuals .... 28. *Leavenworthia.*
  6 All fls in racemes; habit various.
    7 Stem and pedicels stipitate-glandular; fr with a beak 4–15 mm .... 35. *Chorispora.*
    7 Plant not glandular; fr beakless or short-beaked.
      8 Fr flattened parallel to the septum.
        9 Fr less than 5 times as long as wide .... 25. *Draba.*
        9 Fr more than 5 times as long as wide.
          10 Replum narrowly winged; fr dehiscing elastically, the valves becoming coiled .... 29. *Cardamine.*
          10 Replum not winged; fr not dehiscing elastically, the valves not coiling.
            11 Cauline lvs all pinnatisect or pinnate .... 30. *Sibara.*
            11 Cauline lvs entire or merely toothed .... 31. *Arabis.*
      8 Fr terete or quadrangular.
        12 Upper lvs entire, obtuse, cordate-clasping, glabrous; fr 8–12 cm .... 9. *Conringia.*
        12 Lvs otherwise; fr mostly well under 8 cm.
          13 Cauline lvs deltoid, coarsely toothed; plants in life with the odor of garlic .... 38. *Alliaria.*
          13 Cauline lvs distinctly otherwise; plants not smelling of garlic.
            14 Cauline lvs petiolate, not clasping; plants aquatic, with freely rooting, submersed
                or partly floating stem .... 32. *Rorippa.*
            14 Cauline lvs petiolate or sessile, auriculate-clasping, dentate, acuminate; plants
                terrestrial, or if aquatic then neither submersed nor floating .... 34. *Iodanthus.*

1. BRASSICA L. Mustard. Sep usually erect to ascending, saccate at base; pet yellow (our spp.), obovate, clawed; staminal glands 4, rounded, evident; anthers oblong; ovary subcylindric, scarcely narrowed to the short style; ovules several or many in one row in each locule; stigma capitate; fr subterete or angled, ± elongate, often torulose, terminated by a sterile, slender, terete to angular beak, this nerveless or 1-nerved on each side; valves with a prominent midnerve, the other nerves much weaker, scarcely parallel, often anastomosing; cotyledons conduplicate; seeds large, subglobose, in one row in each locule; coarse, weedy annuals and biennials, at least the lower lvs pinnatifid, the fls usually conspicuous. Dwarf specimens of all our spp. occur under unfavorable conditions and are not provided for in the descriptions. 35, Old World.

1 Upper lvs petiolate, not clasping at base.
  2 Mature frs and pedicels ascending; fr terete or nearly so .... 1. *B. juncea.*
  2 Mature frs and pedicels erect or appressed; fr ± 4-angled .... 2. *B. nigra.*
1 Upper lvs sessile and clasping .... 3. *B. rapa.*

1. Brassica juncea (L.) Czernj. Brown, Chinese, or Indian m. Glabrous, often glaucous annual 3–10 dm; lower lvs to 2 dm, pinnatifid and dentate, the upper progressively reduced, short-petioled or sessile; fls 12–15 mm wide; mature pedicels ascending, 10–15 mm; frs ascending, subterete, 1.5–4 cm; seeds 2 mm, conspicuously and evenly reticulate; $2n=36$. Native of Asia, established as a weed throughout our range and w. to the Pacific. June–Oct.

2. Brassica nigra L. Black m. Annual to 15 dm, rough-hairy below, glabrous and glaucous above; lvs all petioled, ovate to obovate, the lower commonly lobed, the upper merely toothed; fls 1 cm wide; mature pedicels erect, 3–4 mm; frs erect, 1–2 cm, smooth, quadrangular because of the stout midnerves of the valves; seeds 1.5–2 mm, minutely rough-reticulate; $2n=16$. European weed, naturalized in fields and waste places throughout the U.S. May–Aug.

3. Brassica rapa L. Annual to 8 dm, green and sparsely setose-hispid, at least along the lf-margins and the midrib beneath; lower lvs petioled, pinnately ± lobed; upper lvs oblong to lanceolate, dentate or entire, sessile and clasping; fls 1 cm wide; mature pedicels widely ascending, 1–2 cm; frs ascending to suberect, terete or nearly so, 3–5 cm, the slender beak 8–15 mm; seeds 1.5–2 mm, minutely roughened; $2n=20$.
    B. rapa as here defined includes the field-weed (often called *B. campestris* L.) and a number of cultivars derived from it, notably turnip and bird's rape. The related sp. *B. oleracea* L., glabrous and glaucous, $2n=18$, includes a host of cultivars (cabbage, cauliflower, broccoli, Brussels sprouts, etc.) derived from a wild European form that is often segregated as *B. sylvestris* (L.) Miller. Rape and rutabaga are referred to *B. napus* L. ($2n=38$), thought to be an alloploid of the other two spp. Infraspecific nomenclature here is tangled.

2. SINAPIS L. Much like *Brassica,* but the valves of the fr with 3–5 prominent veins; beak 3-nerved on each side, often flattened and 2-edged (ensiform), often with 1–3 seeds; sep spreading to reflexed. 7, Mediterranean.

1 Fr glabrous or rarely bristly, 2 mm thick; beak a third as long to nearly as long as the body; seeds 7–
  13 .... 1. *S. arvensis.*
1 Fr bristly, 4 mm thick; beak 1–2 times as long as the body; seeds 4–8 .... 2. *S. alba.*

1. Sinapis arvensis L. Charlock. Rough-hairy to subglabrous annual 2–8 dm; lvs obovate, the lower coarsely toothed or sometimes lobed, the upper progressively reduced, merely toothed; fls 1.5 cm wide; mature pedicels ascending, 5 mm; frs ascending, linear, subterete, the body 1–2 cm × 1.5–2.5 mm, smooth or rarely bristly; beak flattened-quadrangular, commonly half as long as the body; seeds 7–13, smooth, 1–1.5 mm; $2n=18$. A European weed, now widespread in fields, gardens, and waste places in the U.S. May–July. (*Brassica arvensis; B. kaber.*)

2. Sinapis alba L. White mustard. Rough-hairy annual 3–7 dm; lvs obovate in outline, the lower to 2 dm, lyrate, the upper progressively reduced, less lobed or merely toothed; fls 1.5 cm wide; mature pedicels divergent, 6–15 mm; frs divergent or ascending, commonly bristly at least when young, the body 10–15 mm, the valves much distended over the 4–8 seeds; beak 1–2 cm, flat, often curved; seeds smooth, 2 mm; $2n=24$. A European weed, now found occasionally in fields and waste places in our range, sometimes cult. for its seeds. (*Brassica alba; B. hirta.*)

3. ERUCASTRUM C. Presl. Sep erect to spreading, somewhat cuccullate at the tip; pet yellow, spatulate; each pair of long stamens subtended by a short, pyramidal gland; a short gland between the ovary and each short stamen; ovary cylindric; ovules numerous; style very short; stigma capitate; fr elongate, 4-angled, conspicuously beaked, the valves sharply 1-nerved; seeds ovoid or oblong, in 1 row in each locule; herbs with pinnatifid lvs, the pubescence of simple hairs or none. 17, mainly Mediterranean.

1. Erucastrum gallicum (Willd.) O. E. Schulz. Dog-mustard. Annual or biennial 3–6 dm, resembling *Brassica*; basal and lower lvs oblanceolate, to 15 cm, sparsely hairy, deeply pinnatifid, the segments crenately or angularly dentate, the terminal one much the largest; cauline lvs progressively reduced; racemes bracteate below; pedicels slender, ascending, 6–10 mm; sep erect; pet 4–8 mm, pale yellow or whitish; fr usually upcurved, the body 2–4 cm, the beak 3 mm; $2n=30$. Waste places; a European weed, now widespread in the U.S. and s. Can. May–Sept.

4. DIPLOTAXIS DC. Sep erect or ascending, not saccate; pet yellow or white, gradually narrowed to the claw; short stamens subtended by a hemispheric or reniform gland; each pair of long stamens subtended by a short, prismatic gland; ovary linear, with many ovules; style scarcely differentiated; stigma capitate; fr linear, slender, somewhat compressed, shortly beaked, the valves with a prominent midnerve; seeds small, smooth, in 2 rows in each locule; herbs with usually toothed or pinnately lobed lvs and middle-sized fls; pubescence of simple hairs or none. 25, Europe, Mediterranean.

1 Lvs mostly basal or nearly so; fr not stipitate .... 1. *D. muralis.*
1 Lvs mostly cauline; fr on a stipe 1–2 mm .... 2. *D. tenuifolia.*

1. Diplotaxis muralis (L.) DC. Sand-rocket; stinking wall-rocket. Glabrous or sparsely hairy annual (to perennial), erect or decumbent, 2–5 dm, branched from base; lvs chiefly at or near the base, oblanceolate, long-attenuate below, toothed to sometimes pinnatifid; lower pedicels becoming remote and 1–1.5 cm; pet yellow; frs ascending, 2–4 cm × 2 mm, not stipitate; $2n=42$. Native of Europe, naturalized in waste places, especially in sandy soil, in much of the n. part of our range and westward. May–Sept.

2. Diplotaxis tenuifolia (L.) DC. Slimleaf w.-r. Glabrous or subglabrous, taprooted perennial (but sometimes blooming the first year), erect, 3–8 dm; lvs mostly cauline but below the middle of the stem, somewhat fleshy and glaucous, foetid when crushed, oblanceolate to narrowly oblong, toothed to more often deeply pinnatifid; lower pedicels becoming remote and 2–3 cm; pet yellow; frs ascending, 2–5 cm, the base of the valves elevated 1–2 mm above the scars of the perianth; $2n=22$. Native of Europe, established as a weed in waste places here and there in our range, especially northward, and in w. U.S. May–Sept.

D. erucoides (L.) DC., annual or nearly so, with white, violet-veined pet, is a casual ballast-weed with us.

5. ERUCA Miller. Rocket-salad. Sep erect, the inner somewhat saccate at base; pet spatulate, the blade tapering gradually to the long slender claw, staminal glands minute; ovary cylindric, gradually tapering into the elongate style; ovules numerous; stigma minute; fr slightly flattened, several-seeded, tipped with a prominent flat beak; valves conspicuously 1-nerved; seeds in 2 rows in each locule; herbs with branched stems and large, mostly pinnately or lyrately lobed lvs; pubescence sparse, of simple hairs. 4, Mediterranean and e. Afr.

1. Eruca sativa Miller. Garden-rocket. Annual, 2–8 dm; lvs irregularly lyrate, the terminal segment the largest; lower lvs petiolate, the upper sessile; sep caducous, the inner not cucullate; pet white to sordid or ochroleucous, with a few conspicuous purple veins, 1.5–2 cm, the claw surpassing the cal; anthers oblong; frs erect on short, stout pedicels, the body 12–25 × 3–6 mm, glabrous, the narrowly triangular beak 5–12 mm; $2n=22$. Native of Europe, where cult. for salad; intr. or escaped here and there in our range. May, June. Sometimes subordinated or submerged in *E. vesicaria* (L.) Cav., a Mediterranean plant with the sep subpersistent and all cucullate, and with acutish anthers.

6. RAPHANUS L. Sep erect, obtuse, somewhat saccate at base; pet large, yellow, pink-purple or white, broadly obovate, abruptly narrowed to the claw; ovary cylindric; style elongate, scarcely narrower than the ovary; stigma capitate; fr long-beaked, transversely divided into 2 members, the lower small and usually seedless or abortive, the upper indehiscent, its 2–10 large seeds in a single row, separated by constrictions or spongy cross-partitions; coarse annual or biennial herbs, with at least the lower lvs pinnatifid; pubescence of simple hairs. 3, w. Eurasia, Mediterranean.

1. **Raphanus raphanistrum** L.   Jointed charlock.   Coarse annual from a stout taproot, 3–8 dm, usually sparsely hispid; lower lvs obovate-oblong, 1–2 dm, pinnatifid into 5–15 oblong segments, the lower segments very small, the upper progressively larger; upper lvs reduced and often entire; pet yellow, turning white, 10–15 mm; mature pedicels ascending, 8–15 mm; fr nearly cylindric when fresh, when dry becoming prominently several-ribbed and constricted between the 4–10 seeds, the body 2–4 cm, the beak 1–3 cm, the lower member 1–2 mm, its upper margin indicated by a line or projecting rim extending half way to all around the fr; $2n=18$. Native of Eurasia, established as a weed in fields and waste places and along roadsides throughout most of the U.S. and s. Can. June–Aug.

   R. sativus L., the radish, with more thickened taproot, pink-purple to white pet, and spongy fr not constricted between the seeds, the lower member usually obsolete, occasionally escapes from cult. but apparently does not long persist.

7. **RAPISTRUM** Crantz, nom. conserv.   Sep ascending, cucullate at the tip; pet yellow, narrowly obovate, gradually narrowed to the claw; short stamens subtended by a flat gland, long stamens by a short-conic gland; anthers oblong; ovary cylindric, the style about as thick, the stigma truncate; fr short, terete, coriaceous, tipped by the elongate, persistent style, transversely divided into 2 members, the lower cylindric to obovoid, dehiscent by 2 valves, bearing 1–3 suspended seeds, the upper larger, subglobose, indehiscent, obtusely 5-ribbed and usually verrucose, containing one erect seed; hairs simple. 2, w. Eurasia, Mediterranean.

   1. **Rapistrum rugosum** (L.) All.   Annual, 2–10 dm, hispid at least below; lvs mostly basal, oblanceolate to obovate, pinnatifid to coarsely dentate; cauline lvs few, smaller, merely dentate; pet 6–10 mm; mature pedicels stout, erect, 2–4 mm; lower member of the fr 2–3 mm, upper member 3–4 mm; beak 1–4 mm; $2n=16$. Native of Eurasia, rarely adventive in waste places with us. All summer.

8. **CAKILE** Miller.   Sea-rocket.   Pet obovate, pink or purplish to white; short stamens subtended by a minute gland, long stamens separated by a larger gland; anthers ovate-oblong; ovary cylindric, with imperfect septum; ovules 2–4; style not differentiated; stigma truncate; fr indehiscent, corky when dry, transversely divided into 2 dissimilar joints, the lower persistent, 1-seeded or seedless, the upper eventually deciduous, fertile, usually 1-seeded, ovoid or lance-ovoid, 2- or 4-angled or ± 8-angled (2 sutures, and 3 nerves for each valve); seed suspended in the lower joint, erect in the upper; coarse, succulent annuals of coastal sands, much-branched, with spreading or ascending stems and oblanceolate to obovate, dentate to pinnatifid lvs. 7, widespread.

   1. **Cakile edentula** (Bigelow) Hook.   Bushy-branched, 1–8 dm; lvs spatulate and sinuately toothed, varying to pinnatilobate or subentire; fls pale purple to white, 5 mm wide; pedicels at maturity indurate, 2–8 mm; lower joint of fr turbinate-cylindric, 5–12 mm, or sterile and much-reduced, the upper joint 4-angled or obscurely 8-angled, always evidently larger than the lower, with a somewhat flattened beak, and conspicuously narrowed below to the ariculation; $2n=18$. Coastal sands; Atlantic coast from s. Lab. to N.C., and along the shores of the Great Lakes; intr. on the Pacific coast and in Australia. Maritime plants and some of these from the Great Lakes (especially about the s. end of Lake Michigan) have the upper segment of the fr plump, 5–9 mm wide, and short-beaked, the beak shorter than the seed-bearing part; these are var. *edentula*. Most of the plants from the Great Lakes have the upper segment of the fr slender, 3–5 mm wide, and long-beaked, the beak as long as or longer than the seed-bearing part; these are var. *lacustris* Fern.

   C. maritima Scop., a European sp. with deeply pinnatifid lvs, and with the lower segment of the fr dilated into projecting teeth at the joint, is occasionally adventive about Atlantic seaports and appears to be established on beaches in the Chesapeake Bay region.

9. **CONRINGIA** Heister.   Hare's-ear mustard.   Sep erect, saccate at base; pet long-clawed, bright yellow to nearly white; short stamens subtended by a U-shaped gland; ovary cylindric, gradually tapering to the short style; ovules many; stigma capitate; fr elongate, slender, many-seeded, terete or 4-angled, tapering to a short beak; seeds in 1 row in each locule, oblong, granular-roughened; annual or biennial glabrous herbs, often glaucous; lvs entire, at least the upper cordate-clasping. 6, w. Eurasia, Mediterranean.

   1. **Conringia orientalis** (L.) Andrz.   Erect annual to 8 dm; lvs pale, the lower narrowed to the base, the upper oval, elliptic, or oblong, broadly rounded above, cordate-clasping, entire; pet ochroleucous, narrowly obovate, 10–12 mm; pedicels and frs widely divergent, the former 1–1.5 cm, the latter 8–12 cm × 2–3 mm,

4-angled; $2n=14$. Native of Eurasia, occasionally found as a weed in our range, and more common in nw. U.S.

10. LEPIDIUM L. Pepperweed, peppergrass. Sep oblong to rotund; pet small, white (yellow), linear to spatulate, or often lacking; stamens 6, or by abortion 4 or 2; anthers small, oval; ovary flat; ovules 1 per locule, suspended; style short or none; stigma capitate; fr a silicle, flattened contrary to the narrow septum, thin or somewhat distended over the seeds, ovate to orbicular or obovate, often winged, commonly retuse, tipped by the persistent style or stigma; herbs with linear to elliptic, entire to pinnatifid lvs. 175, cosmop.

1 Cauline lvs (or some of them) auriculate or cordate-clasping at the sessile base.
  2 Lvs all entire or merely toothed, or the lower shallowly lobed; pet white . . . . 1. *L. campestre.*
  2 Lower lvs pinnately dissected; pet yellow . . . . 2. *L. perfoliatum.*
1 Cauline lvs not auriculate or clasping.
  3 Fr notched at the tip; annual or biennial, mostly 1–5 dm.
    4 Fr 2–4 mm; pedicels spreading or ascending; stamens 2.
      5 Cotyledons accumbent (the edges against the radicle); pet well developed . . . . 3. *L. virginicum.*
      5 Cotyledons incumbent (the back of one against the radicle); pet short or none.
        6 Fr obovate-obcordate; plants not foetid; lower lvs coarsely toothed to pinnately lobed . . . . 4. *L. densiflorum.*
        6 Fr ovate to elliptic; plants foetid; lower lvs bipinnately lobed . . . . 5. *L. ruderale.*
    4 Frs 5–7 mm; pedicels closely ascending or suberect; stamens 6 . . . . 6. *L. sativum.*
  3 Fr not notched at the tip; rhizomatous perennial, mostly 1–2 m . . . . 7. *L. latifolium.*

1. **Lepidium campestre** (L.) R. Br. Field-cress. Densely short-hairy annual or biennial 2–5 dm; basal lvs elongate, oblanceolate, entire to shallowly lobed; cauline lvs erect or ascending, lanceolate to narrowly oblong, 2–4 cm, entire or denticulate, sessile, clasping by acute auricles; racemes dense, to 15 cm, the mature pedicels divergent, 4–8 mm; pet white, 2–2.5 mm; stamens 6; anthers yellow; frs oblong-ovate, 5–6 mm, broadly winged above, the short style barely or scarcely exsert; $2n=16$. Native of Europe, established as a weed of fields, roadsides, and waste places through our range and westward. May, June.
L. heterophyllum (DC.) Benth., a European perennial with purple anthers and more evidently (though shortly) exserted style, is sporadic in N. Engl., N.Y., and N.J.

2. **Lepidium perfoliatum** L. Clasping pepperweed. Erect, sparsely hairy annual 2–5 dm, branched above; basal and lower cauline lvs pinnately dissected into linear segments; upper cauline lvs broadly ovate to subrotund, entire, deeply cordate, the rounded auricles about as long as the rest of the blade and usually overlapping; sep firm, broadly oval, 1–1.3 mm; pet narrowly spatulate, yellow, 1–1.5 mm; stamens 6; fr rhombic-elliptic, 4 × 3 mm, obscurely retuse; style about as long as the notch; $2n=16$. Native of Europe, well established as a weed in w. U.S., e. to Mich. and O. and occasionally elsewhere in our range. May, June.

3. **Lepidium virginicum** L. Poor-man's pepper. Erect annual or biennial 1–5 dm; basal lvs oblanceolate, sharply toothed to pinnatifid or even bipinnatifid; upper lvs smaller, oblanceolate to linear, dentate to entire, acute, narrowed to the base; racemes numerous, to 1 dm; pet equaling to twice as long as the sep; stamens 2; fr broadly elliptic to orbicular, widest at or below the middle, 2.5–4 × 2–3.5 mm, narrowly winged across the tip; style included in the notch; $2n=32$. Dry or moist soil, fields, gardens, roadsides, and waste places; Nf. to Fla., w. to the Pacific states. Our plants, with accumbent cotyledons, are var. *virginicum.*

4. **Lepidium densiflorum** Schrader. Prairie-pepperweed. Thinly short-hairy annual or biennial, 2–5 dm; basal lvs 4–7 cm, coarsely dentate to pinnately lobed; cauline lvs shorter, linear or narrowly oblanceolate, mostly entire, sharply acute; mature racemes erect, 5–10 cm, with 9–15 frs per cm; pet none, or shorter than the sep, linear to narrowly spatulate; stamens 2; fr 2–3.3 mm, nearly or fully as wide, obovate-obcordate, generally broadest a little above the middle, narrowly winged above; stigma included in the notch; cotyledons incumbent; $2n=32$. Dry or moist soil, waste grounds, roadsides, and pastures; throughout our range and w. to the Rocky Mts. May, June. (*L. apetalum,* misapplied; *L. neglectum.*)

5. **Lepidium ruderale** L. Stinking pepperweed. Foetid, minutely hairy annual or biennial, 2–5 dm; basal lvs commonly bipinnately lobed; lower cauline lvs usually pinnatifid, the upper entire or dentate, rounded or obtuse at the tip; stamens 2; mature racemes loose, with 6–10 frs per cm; fr ovate to elliptic, generally broadest near or below the middle, 2–3 mm, three-fourths as wide, wingless; stigma included in the notch; seeds wingless; $2n=16$. Native of Europe, intr. in waste places here and there in our range.

6. **Lepidium sativum** L. Garden-cress. Glabrous, ± glaucous annual 2–4 dm; lvs pinnately dissected into a few linear, oblong, or oblanceolate segments; fls 2 mm wide; stamens 6; fr elliptic-oval, 5–7 mm, two-thirds as wide, deeply notched; style half as long to nearly as long as the notch; mature pedicels erect or closely ascending, 2–4 mm; $2n=16, 24$. Native probably of w. Asia, escaped from cult. especially in the ne. part of our range.

7. **Lepidium latifolium** L. Perennial pepperweed. Rhizomatous perennial to 2 m; herbage glabrous or nearly so; lower lvs long-petioled, to 30 × 8 cm; cauline lvs lanceolate to narrowly ovate, entire to dentate, the main ones 1–4 cm wide; infl paniculately branched, with many short racemes that do not elongate;

stamens 6; fr sparsely hirsutulous, suborbicular, 2 mm, wingless, rounded at both ends, not notched; stigma subsessile; $2n=24$. Native of s. Europe and w. Asia, established along the coast from Mass. to L.I. and at widely scattered stations elsewhere.

11. CORONOPUS Zinn, nom. conserv. Wart-cress. Sep oval, scarious-margined; pet minute, white, spatulate, scarcely exceeding the cal; stamens 6 or by abortion only 2, the short ones flanked on one side by a minute gland; ovary short, with one ovule per locule; style very short or none; stigma capitate; fr short, compressed contrary to the narrow septum, indehiscent, but the mericarps eventually separating from the septum; herbs, glabrous or with simple hairs, and with short axillary racemes. 10, widespread.

1. **Coronopus didymus** (L.) J. E. Smith. Swine-cress. Foetid annual or biennial, thinly hairy, much-branched, spreading or ascending, 1–4 dm; lvs oblong, 1–3 cm (including the short petiole), pinnatifid, the segments entire or with a few deep teeth; racemes rather loose, 1–3 cm, the mature pedicels slender, 2 mm; fr didymous, distended over the seeds, coarsely wrinkled on the surface, 1.7–3 mm, cordate at base, evidently notched at summit, the style and stigma included in the notch; $2n=32$. Native of S. Amer., established as a weed throughout our range and elsewhere, especially southward. Summer. (*Carara d.*)

C. squamatus (Forssk.) Aschers., a Eurasian sp., is occasionally adventive near the coast. It is glabrous and has very dense racemes with apiculate, conspicuously ridged or reticulate or warty frs; $2n=32$. (*C. procumbens.*)

12. CARDARIA Desv. White top; hoary cress. Sep (at least the outer) blunt; pet white, equal, slightly longer to twice as long as the sep, with narrow claw and obovate, retuse blade; short stamens separated by a pair of minute glands; a minute gland also between each pair of long stamens; seeds 1 per locule; style evident, 1–2 mm, with a capitate stigma; fr an indehiscent silicle, subglobose and evidently inflated to somewhat flattened contrary to the partition, the base obtuse to rounded or cordate, the summit acutish to rounded or truncate, only rarely emarginate; cotyledons incumbent, the back of one against the radicle; perennial herbs, glabrous to canescent with short, simple hairs, the lvs entire or merely toothed; $x=8$. 3, Eurasia.

Calepina irregularis (Asso) Thellung, a glabrous European annual or biennial with a lax raceme, unequal pet, and rugose frs, is casually intr. in Va. and N.C. The cauline lvs are clasping, with acute, patent auricles.

1 Sep and fr glabrous or nearly so, the fr somewhat compressed.
  2 Fr ± cordate at base, usually constricted at the septum . . . . 1. *C. draba.*
  2 Fr obtuse to rounded or subtruncate at base, generally not constricted at the septum . . . . : 2. *C. chalepensis.*
1 Sep and fr short-hairy, the fr evidently inflated, not compressed . . . . 3. *C. pubescens.*

1. **Cardaria draba** (L.) Desv. Heart-podded h.c. Erect perennial to 6 dm, vigorously spreading by creeping roots, short-hairy below, less so or glabrate upwards, the sep and fr glabrous or very nearly so; cauline lvs oblong to ovate or obovate, ascending or erect, sessile, auriculate, irregularly toothed or entire, up to 9 × 4 cm; racemes numerous, dense, commonly forming a white top across the summit of the pl; mature pedicels 10–15 mm; sep 2–2.5 mm; fr 2.5–3 × 3–5 mm, somewhat inflated but usually constricted at the narrow partition, cordate or subcordate at base; $2n=32, 64$. Fields, roadsides, and waste places; native of the Middle East and e. Mediterranean region, now a serious weed in w. U.S. and adj. Can., and occasionally found in our range. May–July. (*Lepidium d.*)

2. **Cardaria chalepensis** (L.) Hand.-Mazz. Lens-podded h.c. Very much like no. 1, and hybridizing with it, but the fr 2.5–6 × 4–6 mm, rounded to obovate or transversely elliptic, not or scarcely constricted at the septum, the base obtuse to rounded or subtruncate; $2n=48, 64, 80$. Native to the Middle East, now intr. in N. Amer. in the same kinds of places as no. 1.

3. **Cardaria pubescens** (C. A. Meyer) Jarmolenko. Globe-podded h.c. Much like the previous 2 spp., but the pubescence extending to the sep and frs; fr inflated, obovoid to subglobose, 3–4.5 × 2.5–4.5 mm; mature pedicels shorter, up to 10 mm, but the racemes often more elongate at maturity; $2n = 16$. Native of the Middle East and c. Asia, now intr. in N. Amer. in the same kinds of places as no. 1. (*Hymenophysa p.*)

13. ISATIS L. Woad. Sep divergent, somewhat unequal; pet yellow, narrowly obovate, exceeding the sep; short stamens flanked and subtended by a pair of confluent glands; each pair of long stamens subtended by a single gland; ovary flat, with a single locule and ovule; stigma sessile, 2-lobed; fr indehiscent and somewhat samaroid, compressed contrary to the normal position of the replum, with strongly flattened valves; seed single and median, replacing the replum; herbs, usually tall, with leafy, branching stems. 60, Eurasia, n. Afr.

1. **Isatis tinctoria** L. Biennial, usually single-stemmed, 5–12 dm, glabrous and glaucous; basal lvs oblanceolate, long-petioled; cauline lvs sessile, lanceolate to oblong, mostly 2–10 cm, entire or nearly so, auriculate-clasping; infl of numerous racemes, forming a large terminal panicle; fls 6 mm wide; frs drooping on short, slender pedicels, oblong to obovate, mostly 8–15 × 2.5–6 mm, thickened and corky in the center; $2n=28$. Native of Europe, where traditionally cult. as the source of a blue dye; occasionally found as a weed in our range. May–July.

14. **THLASPI** L. Penny-cress. Sep ascending; pet white (in ours) to pink or blue, spatulate to obovate; short filaments flanked on each side by a semicircular gland; filaments slender, with ovate anthers; ovary ellipsoid to obovoid, somewhat flattened; style very short; frs orbicular to obovate or obcordate, strongly flattened contrary to the septum, distended over the seeds, keeled or winged at the margin, retuse at the summit; seeds normally 4 or more per locule; herbs, usually glabrous, with auriculate lvs. 75, mostly Eurasia.

1 Fr 10–17 mm, the distal notch deeper than wide .... 1. *T. arvense.*
1 Fr 4–7 mm, the distal notch mostly wider than deep .... 2. *T. perfoliatum.*

1. **Thlaspi arvense** L. Field p.-c. Glabrous annual or winter-annual 1–5 dm; cauline lvs sessile, oblong to lanceolate, entire or few-toothed, bearing at base 2 narrow, divergent, acute auricles 1–5 mm; fls 3 mm wide, the pet twice as long as the sep; mature pedicels ascending; frs orbicular to broadly elliptic, 10–17 mm, winged all around, deeply (2–3 mm) notched; seeds concentrically ridged-striate, numerous in each locule; $2n=14$. Native of Europe, commonly established in waste places throughout much of N. Amer. Apr.–June.

2. **Thlaspi perfoliatum** L. Thoroughwort p.-c. Annual 1–3 dm; cauline lvs sessile, ovate or ovate-oblong, 1–3 cm, entire or with a few teeth, the basal auricles blunt or rounded; pet 2–3 mm; mature pedicels horizontally spreading; frs obovate, 4–7 mm; convex below, the wing gradually widening to the summit, the wide open notch ca 0.5–1 mm deep and 1–2 mm wide; seeds smooth, 2–4 per locule; $2n=14, 28, 42, 70$. Native of Eurasia, now found in fields, roadsides, and waste places here and there in our range. Apr., May.

15. **TEESDALIA** R. Br. Shepherd's cress. Sep ascending; pet short, white; stamens 4–6, the median (paired) filaments with a prominent adaxial scale-appendage at base; ovary flattened, suborbicular, with 2 ovules per locule; style very short, or none; fr broadly ovate, flattened contrary to the septum, retuse, keeled at the margin and narrowly winged distally; annuals with the aspect of *Lepidium,* glabrous or with short, simple hairs; lvs tufted at the base, the short, erect or ascending stems leafless or few-leaved. 2, Europe, Mediterranean.

1. **Teesdalia nudicaulis** (L.) R. Br. Pls glabrous or occasionally short-hairy; basal lvs spatulate in outline, 2–5 cm, lyrate-pinnatifid with rounded or obtuse lobes; cauline lvs none or much reduced; stems 0.5–2 dm; fls 2 mm wide; pet unequal, the 2 outer (abaxial) ones well surpassing the sep, the 2 inner half or two-thirds as long, about equaling the sep; stamens 6; fr 3–3.5 mm; style ca 0.1–0.2 mm; $2n=36$. Native of Europe, now found here and there in the U.S. as a casual weed. Apr.–June.

Iberis amara L. and *I. umbellata* L., spp. of Candytuft, occasionally escape. They would key to *Teesdalia* because of their dissimilar pet and short frs flattened contrary to the septum, but they have well developed cauline lvs, longer (5–9 mm) frs, and larger (5–10 mm wide) fls. In *I. amara* the lvs have a few large teeth and the infl elongates after anthesis; in *I. umbellata* the lvs are entire and the infl remains short, forming a dense, umbel-like cluster in fr.

16. **CAPSELLA** Medikus, nom. conserv. Shepherd's purse. Sep short-oblong, ascending; pet white, obovate; short stamens flanked by a pair of minute glands; ovary flat, obovate; ovules 6–many per locule; style very short; frs strongly flattened at right angles to the septum, triangular-obcordate, truncate to broadly notched at the summit, commonly narrowly winged distally; annual or biennial herbs, pubescent with stellate hairs, with a tuft of basal lvs and a few reduced cauline lvs on the branching stems. Monotypic.

1. **Capsella bursa-pastoris** (L.) Medikus. Sparingly branched annual or winter-annual 1–6 dm; basal lvs oblong, 5–10 cm, pinnatilobate; cauline lvs much smaller, lanceolate to linear, entire or denticulate, auriculate;

racemes becoming much-elongate; pedicels at maturity divaricate, 1–2 cm; sep 2 mm; pet 1.5–4 mm; fr triangular-obcordate, with straight to slightly concave or slightly convex margins, 4–8 mm; 2n=16, 32. Probably native to s. Europe, now a cosmopolitan and ubiquitous weed. (*C. gracilis; C. rubella;* etc.)

17. SUBULARIA L.   Awlwort. Sep elliptic-ovate; pet white, narrowly oblong-ovate; glands forming a continuous ring around the base of the ovary within the stamens; ovary broadly elliptic, with 2–7 ovules per locule; stigma sessile; fr elliptic or oblong, somewhat flattened contrary to the septum, few-seeded; small plants with a basal rosette of subulate or linear lvs and a short, loose, few-fld raceme of minute fls. 2, the other African.

1. **Subularia aquatica** L.   Glabrous annual; lvs narrowly subulate, 1–5 cm; scape 2–10 cm, with 2–10 rather remote fls; sep ± persistent in fr; frs 2–4 mm, short-stipitate, elliptic to broadly obovate, on ascending pedicels; 2n=28. Submerged along the margins of fresh-water ponds and lakes, less often on muddy shores; circumboreal, s. in Amer. to Me., Vt., Minn., Wyo., and Calif. June–Aug. Amer. pls are var. *americana* (G. A. Mulligan & Calder) B. Boivin.

18. MYAGRUM L.   Sep erect, the outer oblong, the inner ovate; pet yellow, narrowly spatulate; glands confluent, forming a ring around the base of the stamens; ovary sub-cylindric, with 1 ovule per locule; style elongate; fr indehiscent, corky, conic-obovoid, flattened, truncate, tipped with the stout 4-angled style, enclosing the single seed in the center near the base, bearing on each side distally an inflated hollow appendage; herbs with sessile, clasping lvs and elongate racemes. Monotypic.

1. **Myagrum perfoliatum** L.   Glabrous and glaucous annual 3–5 dm; cauline lvs narrowly oblong; obtuse, entire or sparsely denticulate; fls sessile or subsessile, 2 mm wide; mature pedicels appressed, thickened distally, 2–4 mm; fr 5–7 mm; 2n=14. Native of Eurasia, established as a weed here and there in our range.

19. NESLIA Desv., nom. conserv.   Sep oblong, obtuse; pet yellow, spatulate, gradually tapering into the claw; each short stamen flanked by a pair of fleshy, U-shaped glands, the arms with short lateral processes; ovules 2 per locule; fr firm, thick-walled, indehiscent, very slightly compressed, reticulate and pitted; branching annuals with minute fls and clasping sessile lvs. 2, Mediterranean and Near East.

1. **Neslia paniculata** (L.) Desv.   Ball-mustard. Much-branched, to 8 dm, pubescent with branched or 2-pronged hairs; cauline lvs oblong-lanceolate, 3–6 cm, acute, entire or nearly so; fls 1.5 mm wide; mature pedicels slender, divergent, to 1 cm; frs 2 × 2.5 mm; 2n=14. Native of the Near East, found as an occasional weed here and there in our range. June, July.

20. BUNIAS L.   Sep oblong, ascending; pet yellow (in ours), narrowly oblong, grad-ually narrowed to the base; short stamens surrounded at base by an annular gland; ovary narrowly ovoid, with 1–2 ovules per locule; fr hard, indehiscent, tapering above into the short style; seeds 1–4, the fr often constricted between them; septum indurate; coarse herbs with large, dandelion-like lvs. 6, Mediterranean, Asia.

1. **Bunias orientalis** L.   Glabrous or sparsely glandular biennial or short-lived perennial to 1 m; lvs oblong, acuminate, the lower to 2 dm, ± pinnatifid or coarsely dentate, often with 2 large salient teeth near the base, the upper lvs progressively reduced, none clasping; mature pedicels to 10–15 mm; pet 4–8 mm; fr obliquely ovoid, 5–10 mm, coarsely and irregularly verrucose; 2n=14. Native of s. Europe, becoming established as a weed here and there in our range. May–July.
  B. erucago L., a hispid and glandular European annual or biennial, is locally intr. in Pa. and Va. It has pet 8–13 mm and 4-winged frs 10–12 mm.

21. LUNARIA L.   Sep narrow, saccate at base; pet large, purple, with broad blade and narrow claw; glands large, annular; pistil elevated on a short gynophore that elongates greatly in fr; ovules 2–3 per locule; fr very broad and flat, reticulately veined, the silvery septum horizontally striolate; seeds reniform; sparsely hairy herbs with cordate or deltoid-cordate, coarsely dentate, acuminate lvs, those toward the base opposite or nearly so. 3, Europe.

1. Lunaria annua L. Honesty. Annual to 1 m; cauline lvs short-petioled or sessile; fls 2 cm wide; frs broadly elliptic, rounded at both ends, 3.5–5 cm, two-thirds as wide; $2n=30$. Native of se. Europe, occasionally escaped from cult. May, June.

L. rediviva L., perennial honesty, money-plant, a perennial with the cauline lvs all petioled, and with frs 5–8 × 1.5–2.5 cm, acute at both ends, also occasionally escapes.

22. ALYSSUM L. Pet yellow, varying to white, gradually narrowed to the base; short filaments flanked by a single gland on each side; anthers short and blunt; ovary short, compressed; ovules 1–8 per locule; style slender but short; fr elliptic to orbicular, flattened parallel to the septum, each locule with a few seeds; herbs with entire or serrate lvs; ± pubescent with stellate hairs. 170, mainly Eurasia.

1. Alyssum alyssoides (L.) L. Annual, 5–25 cm, stellate-hairy throughout; lvs oblanceolate, 6–15 mm, entire, obtuse; fls pale yellow or nearly white, 2 mm wide; pet narrowly oblong; frs on widely divergent pedicels, orbicular, 3–4 mm, flat at the margin, convex toward the center; seeds 2 per locule; $2n=32$. Native of Europe, abundant as a weed in waste places nearly throughout our range, and to w. U.S. May, June. (*Clypeola a.*)

A. petraeum Ard., a short-lived perennial with larger, bright yellow fls and glabrous, elliptic frs, is reported to be established (as an escape) in Oneida Co., N.Y. (*Aurinia p.*) *A. saxatile* L., a matted perennial with yellow fls and glabrous frs, also sometimes escapes.

23. LOBULARIA Desv., nom. conserv. Pet white or lavender purple, with slender claw and broadly rounded blade; short filaments flanked on each side by a pair of glands of different size; anthers short and blunt; ovary ovoid, with 1–8 ovules per locule; style very short; fr orbicular to ovate, flattened parallel to the septum; style persistent; herbs with entire lvs, ± pubescent with 2-pronged hairs longitudinally appressed to the stem and branches. 4, mostly Mediterranean.

1. Lobularia maritima (L.) Desv. Sweet alyssum. Lax, diffusely branched annual to perennial, 1–3 dm, finely strigose; lvs linear to linear-oblanceolate, 2–5 cm, acute; fls fragrant, 4 mm wide; fr flattened with convex sides, elliptic, 3 × 2 mm, very sparsely hairy; ovules 1 per locule; $2n=22, 24$. Native of s. Europe, often escaped from cult. in our range. All summer. (*Koniga m.*)

24. BERTEROA DC. Sep ascending; pet white or sometimes yellow, deeply 2-lobed, gradually narrowed to the base; short filaments flanked on each side by a short semicircular gland; anthers oblong; ovary ellipsoid, with 2–6 ovules per locule; style elongate, persistent; fr elliptic or oval, flattened parallel to the septum, few-seeded; herbs with chiefly entire lvs, ± pubescent with stellate hairs. 5, temp. Eurasia.

1. Berteroa incana (L.) DC. Hoary alyssum. Stellate-canescent annual to perennial, stiffly erect, usually branched above, to 7 dm; lvs oblanceolate, (1–)2–5 cm, acute, entire; fls white, 3 mm wide; fr elliptic, plump, thinly hairy, 5–8 × 3–4 mm, convex over the whole surface; seeds 3–6 per locule; $2n=16$. Native of Europe, established as a weed throughout much of our range, especially northward.

B. mutabilis (Vent.) DC., with strongly flattened frs 4.5–6 mm wide, is rarely adventive with us.

25. DRABA L. Sep ascending or erect, blunt; pet yellow or white, rounded, emarginate, or rarely bifid, narrowed below to a claw, or in some spp. reduced or wanting; anthers short, oval or oblong; ovary ovoid, with 2–many ovules per locule; style short or none; stigma capitate; glands various; fr a silicle, seldom 5 times as long as wide; herbs with entire to dentate lvs, the hairs simple to branched or stellate. 350, mostly cool N. Hemisphere, 65 in S. Amer.

1 Pls annual or winter-annual or biennial from a slender root.
  2 Annual or winter-annual; cauline lvs not very numerous; widespread, often weedy spp.
    3 Pet bifid nearly to the middle (unique among our spp.); lvs all basal . . . . 1. *D. verna.*
    3 Pet distally rounded or emarginate; one or more of the lvs generally cauline.
      4 Frs small, only 2.5–4 × 0.8–1.4 mm . . . . 2. *D. brachycarpa.*
      4 Frs larger, mostly 5–20 × 1.5–2.5 mm.
        5 Rachis and pedicels (and usually also the frs) glabrous.
          6 Pedicels short, less than two-thirds as long as the frs; pet white . . . . 3. *D. reptans.*
          6 Pedicels as long as or generally longer than the frs; pet yellow . . . . 4. *D. nemorosa.*
        5 Rachis, pedicels, and frs pubescent . . . . 5. *D. cuneifolia.*

2 Biennial; cauline lvs very numerous and crowded; boreal, not weedy .... 6. *D. incana.*
1 Pls perennial, often with a multicipital caudex.
  7 Style short, not over 1 mm, or virtually obsolete; northern spp.
    8 Fr densely and evenly stellate-hairy .... 7. *D. cana.*
    8 Fr glabrous, or seldom sparsely stellate.
      9 Basal lvs regularly and evenly pubescent with symmetrically branched stellate hairs.
        10 Fr usually ± twisted, a fourth or a fifth as wide as long; style 0.3–0.8 mm .... 8. *D. arabisans.*
        10 Fr straight, a fourth to two-fifths as wide as long; style 0.1–0.3 mm .... 9. *D. glabella.*
      9 Basal lvs loosely and unevenly pubescent with irregularly branched hairs .... 10. *D. norvegica.*
  7 Style well developed, 1.5–3 mm; southern sp. with stellate-hairy fr .... 11. *D. ramosissima.*

1. Draba verna L. Whitlow-grass. Annual or winter-annual with very slender scapes 5–20 cm; lvs crowded in a basal rosette, oblanceolate to spatulate or obovate-oblong, 1–2 cm, with simple and branched hairs; pedicels ascending, the lowest to 3 cm in fr; fls white, 2–3 mm wide; pet bifid nearly to the middle; fr 4–10 mm, glabrous; $2n=14$–64. Native of Eurasia, naturalized throughout much of N. Amer. Apr., May. (*Erophila v.*) Complex sp., often divided into several intraspecific taxa of doubtful significance.

2. Draba brachycarpa Nutt. Annual or winter-annual to 2 dm, usually much-branched; basal lvs few and evanescent, elliptic to obovate; cauline lvs narrower, the uppermost nearly linear; racemes all elongate at maturity, 2–6 cm; pedicels ascending, glabrous to minutely stellate, usually 2–3 mm; pet white or yellow, to 3 mm, or none; frs oblong-elliptic, 2.5–4 × 0.8–1.4 mm, glabrous; $2n=16$, 24. Dry woods and fields; Va. to Ind. and Kans., s. to Fla. and Tex. Mar.–May.

3. Draba reptans (Lam.) Fern. Simple or basally branched annual or winter-annual 5–20 cm; basal lvs oblanceolate to spatulate or obovate, 1–3 cm, blunt, entire, pubescent with mostly simple hairs above, stellate beneath; cauline lvs few and near the base of the stem; pet white, to 5 mm, or none; racemes at maturity congested, the glabrous axis 5–20 mm, rarely more; pedicels glabrous, ascending, mostly a third to half as long as the fr; fr ascending or erect, 1–2 cm, glabrous or strigulose; $2n=16$, 30, 32. Dry, sterile or sandy soil; Mass. and R.I. to s. Ont., Minn., and Wash., s. to N.C., Ga., and Calif. Apr., May. Our pls are var. *reptans.* (*D. caroliniana; D. micrantha.*)

4. Draba nemorosa L. Annual or winter-annual to 3 dm; basal lvs lance-ovate to oval, elliptic, or obovate, 1–2.5 cm; cauline lvs similar, few to several, all below the middle of the stem, often dentate, pubescent with simple and branched hairs; pet yellow, turning white in age, 2 mm; mature raceme loose and elongate, to 2 dm, glabrous, with widely spreading or ascending glabrous pedicels to 3 cm; frs ascending to erect, glabrous or hirsutulous, linear-oblong, 5–10 mm; $2n=16$. Dry soil, prairies, and hillsides; widespread in Eurasia and much of w. N. Amer., extending e. to Lake Superior and locally in Mich. and sw. Que. (*D. lutea.*)

5. Draba cuneifolia Nutt. Annual or winter-annual, 1–2.5 dm, simple or branched at the base; basal lvs oblanceolate to narrowly obovate, 1–3 cm, coarsely dentate, rough-hairy; cauline lvs few and only near the base; pet white, to 4 mm, or none; mature racemes 5–10 cm, the axis and divaricate pedicels pubescent; pedicels half as long to equaling the fr; frs widely spreading, narrowly elliptic, 6–11 mm, blunt, minutely strigulose; $2n=32$. Dry woods and barrens; Ill., Ky., and Mo., w. and s. to Calif. and n. Mex.; Fla. May.

6. Draba incana L. Biennial, producing the first year a rosette that elongates the second year into a leafy flowering stem, the cauline lvs usually very numerous, not strongly contrasting with the basal, gradually reduced upwards, the foliage forming a narrow cone; basal lvs narrowly oblanceolate, to 3 cm × 5 mm, the cauline becoming ovate toward the top of the stem, rounded at base, often with a few salient teeth; stems 1–4 dm, simple or branched above; pet white, 3–5 mm; frs crowded and closely ascending on short pedicels, oblong, elliptic, or lanceolate, glabrous to stellate, 5–10 mm, acute; persistent style 0.1–0.3 mm; $2n=32$. Various habitats; Greenl. to Hudson Bay and N.B.; Isle Royale, Mich., n. Eurasia.

7. Draba cana Rydb. Perennial with dense basal rosettes, the numerous lvs oblanceolate or spatulate, densely stellate-hairy, commonly less than 3 cm; cauline lvs (3–)5–8(–12), lanceolate to ovate, 5–20 mm; stems 1–3 dm; raceme elongating, the lower fls remote, axillary, only the upper crowded; pet white, 4 mm; frs closely ascending on short pedicels, oblong to lance-oblong, broader at the base than at the tip, often twisted, 5–12 × 1.5–2 mm, stellate-hairy; persistent style 0.2–0.5 mm; $2n=32$. Rocky ledges and gravelly or rocky soil; Siberia and boreal Amer., s. in our range to N.B., the mts. of n. N. Engl., n. Mich., and e. Wis. May–July. (*D. lanceolata* and *D. stylaris,* both misapplied.)

8. Draba arabisans Michx. Perennial, 1–4 dm, simple to freely branched above; basal lvs narrowly oblanceolate to spatulate, to 6 cm, often sharply toothed, uniformly stellate-hairy; cauline lvs few to several, oblong to obovate, narrowed or acute at base, often dentate; mature racemes loose, to 10 cm, commonly glabrous or nearly so; pet yellow in bud, white at anthesis, 4–6 mm; frs glabrous, lanceolate to narrowly oblong, 7–12 mm long, a fourth or a fifth as wide, usually soon twisted, or in depauperate specimens shorter, ovate, and straight; persistent style 0.3–0.8 mm; $2n=96$. Rocks and cliffs; Nf. to w. Ont., s. to n. N. Engl., N.Y., and n. Mich. May, June.

9. Draba glabella Pursh. Much like no. 8; cauline lvs mostly obtuse or broadly rounded at base; frs glabrous or sparsely hirtellous, oblong or lance-oblong, 7–11 × 2–3.5 mm, mostly one-fourth to two-fifths as wide as long, straight or nearly so; persistent style 0.1–0.3 mm, or virtually obsolete; $2n=64$, 80. Rocky or gravelly places and cliff-crevices; circumboreal, s. in our range to N.B., n. N.Y., and Isle Royale. May–July.

10. **Draba norvegica** Gunnerus. Perennial; basal lvs oblanceolate to obovate, 1–2 cm; cauline lvs 1–5, ovate, 5–10 mm, rounded at base, the upper often subtending a fl; stems 1–2 dm; mature racemes dense, forming a third to half the total height of the pl; pedicels and frs ascending; pet white; frs oblong, elliptic, or lance-oblong, mostly 7–9 × 2.2–3.2 mm, veiny, glabrous or seldom sparingly stellate-puberulent; persistent style 0.1–0.3 mm, or virtually obsolete; 2n=48. Rocky ledges and cliffs; e. and ne. Can. and adj. U.S.; n. Europe; known in our range only from Cape Breton Isl. and Cook Co., Minn.

11. **Draba ramosissima** Desv. Mat-forming perennial with a branching caudex; basal lvs oblanceolate, 1–5 cm, tapering to a long slender base; cauline lvs scattered, lanceolate to oblanceolate, both types sharply acute and with a few sharp, slender, divergent teeth; flowering stems 1–4 dm, thinly hairy, often branched above to form an eventually open and rather diffuse infl; pet white, 6–7 mm; mature pedicels thinly stellate, rather widely ascending, to 15 mm; frs elliptic to narrowly oblong, 4–10 mm, often twisted, thinly stellate; persistent style 1.5–3 mm; 2n=16. Dry mt. woods; Md. and Va. to W.Va., e. Ky., N.C., and Tenn. Apr., May.

26. **LESQUERELLA** S. Wats. Sep erect to spreading, obtuse; pet yellow (in ours), obovate to narrowly spatulate; glands at base of the small stamens 2, often confluent or appendaged; ovary subglobose, often short-stipitate; ovules 1–15 per locule; style slender, ± elongate, usually persistent; stigma capitate; fr short, usually inflated, globose to ellipsoid or obovoid, or sometimes ± flattened; herbs ± canescent with stellate hairs or scales, and with entire or toothed to seldom pinnatifid lvs. 90, N. and S. Amer., Siberia.

L. gracilis (Hook.) S. Wats., an annual sp. of Texas, is adventive in waste places in Ill. & Mo.

1 Frs 2–3 mm, on straight, divaricate pedicels; Tenn., Ky., and s. Ind. .... 1. *L. globosa.*
1 Frs (3–)3.5–5.5(–6) mm, on recurved pedicels; more northern or western .... 2. *L. ludoviciana.*

1. **Lesquerella globosa** (Desv.) S. Wats. Slender, taprooted perennial to 5 dm, erect and often branched near the base, copiously leafy to the base of the infl; basal lvs mostly 2.5–5 × 0.5–1.5 cm, obovate to oblanceolate, short-petiolate, entire or merely sinuate-toothed to sometimes pinnatifid; cauline lvs smaller, oblanceolate to linear, often erect and appressed; racemes with up to 50 fls, elongating to 0.5–1 dm at maturity; pet 4–7 mm; ovules 2 per locule; mature pedicels straight, widely spreading or horizontal, 7–15 mm; frs subglobose, 2–3 mm, at first loosely stellate, in age glabrate; septum wrinkled, much larger than the replum; style 2–3.5 mm; 2n=14. Calcareous rocks and barrens; c. Tenn., nc. Ky., and s. Ind. Apr.–May.

2. **Lesquerella ludoviciana** (Nutt.) S. Wats. Taprooted perennial with several erect or ascending stems 1–4 dm; basal lvs narrow, to 10 × 1 cm, entire or shallowly toothed, the cauline progressively smaller and not very numerous; racemes elongating to as much as 1.5 dm; pet 6–10 mm; ovules (2–)4–6(–8) per locule; pedicels elongating to 1–2 cm and soon becoming recurved; frs subglobose or obovoid, (3–)3.5–5.5(–6) mm, copiously and loosely stellate; style 3–5 mm; 2n=10, 30. Prairies, plains, and barrens; w. Ill. and e. Minn.; N.D. to Kans., w. to Mont., n. Ariz., and s. Nev. Apr.–Aug. (*L. argentea,* a preoccupied name.)

27. **ARMORACIA** Gaertn., Meyer, & Scherb., nom. conserv. Sep ascending, elliptic to obovate; pet white, obovate, gradually narrowed to the claw; short stamens subtended by a U-shaped gland; long stamens subtended by a small conic gland; anthers linear-oblong; ovary ovoid to ellipsoid; ovules numerous; style slender, stigma large, capitate; frs inflated, obovoid or ellipsoid, tipped with the slender style and conspicuous stigma; valves with an inconspicuous midnerve, otherwise nerveless; glabrous herbs. 4, Europe to Siberia.

1 Pl aquatic, with lax, submersed or prostrate stems; fr unilocular .... 1. *A. lacustris.*
1 Pl terrestrial (often in wet soil), with tall erect stems; fr bilocular .... 2. *A. rusticana.*

1. **Armoracia lacustris** (A. Gray) Al-Shehbaz & V. Bates. Lake-cress. Fibrous-rooted perennial; stems commonly submersed; submersed lvs pinnately dissected into numerous filiform segments; emersed lvs, if any, lanceolate to narrowly oblong, 3–7 cm, dentate; pet 6–8 mm; mature pedicels divaricate, 1 cm; frs rarely maturing, ellipsoid or somewhat obovoid, 5–8 mm, unilocular; persistent style slender, 2–4 mm. In quiet water, or also on muddy shores; Que. to Minn., s. to Fla. and Tex. May–Aug. (*A. aquatica; Neobeckia a.; Radicula a.*)

2. **Armoracia rusticana** Gaertn., Meyer, and Scherb. Horse-radish. Erect perennial to 1 m from thick roots; lower lvs long-petioled, the blade oblong, 1–3 dm, cordate at base; upper lvs smaller, short-petioled to sessile, lanceolate; racemes several, terminal and from the upper axils; pet 6–8 mm; mature pedicels ascending, 8–12 mm; frs obovoid, bilocular, to 6 mm, or falling early; style 0.3 mm, with broad, persistent

stigma; seeds rarely maturing; $2n=32$. Native of se. Europe and w. Asia; commonly cult. and widely escaped into moist soil. May–July. (*A. armoracia; A. lapathifolia; Radicula armoracia.*)

28. LEAVENWORTHIA Torr. Sep obtuse, erect to spreading; pet white to anthocyanic or yellow, narrowly spatulate-obovate to lingulate, retuse to truncate; ovary cylindric, with a short, stout, scarcely differentiated style and capitate stigma; frs typically linear to oblong and compressed parallel to the septum, but varying to terete or subglobose; valves reticulately veined, the midnerve evident only near the base; seeds in a single row, flattened, evidently winged to virtually wingless; racemes arising from the base of the pl, either reduced to a single long-pedicellate fl (seemingly on a scape), or with a short rachis scarcely surpassing the basal lvs and bearing a few long pedicels; small, glabrous winterannuals of thin, vernally wet, summer-dry soil in calcareous cedar-glades, and now also in disturbed habitats, as along roadsides, blooming in early spring, the earliest lvs simple, the others lyrate-pinnatifid. 7, c. and sc. U.S.A.

1 Pet emarginate, 6–10 mm; terminal lf-lobe much larger than the distal lateral ones.
  2 Frs flattened, not torulose; seeds evidently winged .... 1. *L. exigua.*
  2 Frs torulose; seeds virtually wingless .... 2. *L. torulosa.*
1 Pet truncate, 5–7 mm; terminal lf-lobe only slightly larger than the distal lateral ones .... 3. *L. uniflora.*

    1. Leavenworthia exigua Rollins. Principal lvs 1.5–5 cm, the lobes shallowly toothed, the terminal lobe markedly larger than the lateral ones; sep green; fls white with a light yellow eye, the pet 6–9 mm, shallowly emarginate; frs strongly flattened, obtuse above and below, oblong, 1–2 cm × 3.5–5.5 mm, the persistent style 2–3 mm; seeds 4–10, evidently winged, irregularly orbicular, 3–4 mm, the areolae of the reticulum much smaller over the embryo than on the wing; radicle straight; $2n=22$. Ky. to Ga. and Ala. Ours is var. *laciniata* Rollins.

    2. Leavenworthia torulosa A. Gray. Much like no. 1 in aspect; sep pale lavender to pale greenish; fls white to lavender, often with a yellow eye, the pet 6–10 mm, emarginate; frs linear, strongly torulose even when young, 1.5–3 cm × 2.5–4 mm, nearly as thick as wide, the persistent style 2.5–5 mm; seeds nearly wingless, longer than wide, 2.5–3.5 mm, the areolae of the reticulum nearly uniform throughout; radicle slightly bent; $2n=30$. S. Ky. to s. Tenn.

    3. Leavenworthia uniflora (Michx.) Britton. Principal lvs 2–10 cm × 5–10 mm, deeply pinnatifid with up to 9 pairs of acutely toothed lobes, gradually increasing in size from the petiole towards the tip, the terminal lobe only slightly larger than the adjacent lateral pair; sep at first greenish, later purplish; pet white, narrowly lingulate, truncate, usually erect, 5–7 mm, frs thick and somewhat fleshy, 1.5–3 cm × 3–5 mm, nearly as thick as wide, the persistent style 1.5–3 mm; seeds broadly winged, suborbicular, 3–4 mm, the areolae smaller over the embryo than on the wing; radicle strongly bent; $2n=30$. Fl before the other spp., and becoming somewhat weedy. Sw. O. to nw. Ga. and nw. Ark.

29. CARDAMINE L. Bitter-cress; toothwort. Sep erect, obtuse; pet white to pink or purple, or rarely none; stamens in some spp. fewer than 6; short stamens subtended by a semicircular gland; anthers oblong; ovary cylindric, gradually tapering to the short, slender style; stigma truncate or nearly so; fr linear, terete or quadrangular, straight, many-seeded, tipped with the persistent style, which in the *Dentaria* group commonly forms an evident, slender beak 6–8 mm; persistent frame of the replum relatively broad or slightly wingmargined; seeds in a single row in each locule, flattened, wingless; annual to perennial herbs with simple to pinnately or palmately compound (even biternate) lvs; pubescence none or of simple hairs. 200, cosmop. (*Dentaria.*)

    Diplotaxis erucoides (q.v.) might be sought here, but has an evidently emarginate stigma, and 2 rows of seeds in each locule of the fr.

1 Principal lvs palmately 3–5-cleft or compound; pet large, ca 10–20 mm. (*Dentaria*; toothwort.)
  2 Rhizome not segmented, or with alternately enlarged and slightly constricted regions; axis of the
    stem above the cauline lvs glabrous, rarely hairy.
    3 Rhizome of uniform diameter; cauline lvs mostly 2; hairs of the margins of the lfls 0.1 mm,
      appressed .... 1. *C. diphylla.*
    3 Rhizome with alternately enlarged and slightly constricted regions; cauline lvs typically 3; hairs
      on margins of lfls 0.2–0.3 mm, spreading .... 2. *C. × maxima.*
  2 Rhizome definitely formed of segments joined by fragile connectives.
    4 Axis of the stem above the cauline lvs hairy, rarely glabrous; hairs of the margins of the lfls and
      upper stem 0.2–0.3 mm, spreading .... 3. *C. concatenata.*
    4 Axis of the stem above the cauline lvs glabrous, rarely hairy.

5 Cauline lvs biternate, the segments entire; lf-margins without trichomes .... 4. *C. dissecta.*
5 Cauline lvs ternate, the segments toothed; hairs of the margins of the lfls 0.1 mm, appressed
.... 5. *C. angustata.*
1 Principal lvs simple to pinnately compound (sometimes pinnately trifoliolate); pet variously large or small. (*Cardamine*; bitter cress.)
  6 Principal cauline lvs simple, or with 1 or 2 minute lateral lobes only.
    7 Pedicels rapidly elongating, those of the oldest fls commonly 1 cm, at maturity to 4 cm; pet 5–16 mm.
      8 Stems erect from a short, tuberlike rhizome; pet 7–16 mm.
        9 Sep green, turning yellow in age; pet white, rarely pink .... 6. *C. rhomboidea.*
        9 Sep purple, turning brown in age; pet pink to purple, rarely white .... 7. *C. douglassii.*
      8 Stems decumbent and stoloniferous at base; pet 5–8 mm .... 8. *C. rotundifolia.*
    7 Pedicels 1–3 mm at anthesis, at maturity seldom over 5(–10) mm; pet less than 5 mm, or none.
      10 Pet none; axis elongating after anthesis; frs divergent .... 9. *C. longii.*
      10 Pet 3–5 mm; axis scarcely elongating; frs crowded and erect .... 10. *C. bellidifolia.*
  6 Principal cauline lvs, or most of them, with at least 2 well developed lateral lobes.
    11 Pet 6–13 mm, the fls 6–16 mm wide; pls perennial.
      12 Cauline lvs with mostly 3 lfls; s. Appalachian region .... 11. *C. clematitis.*
      12 Cauline lvs with numerous (mostly 7–17) lfls; more northern .... 12. *C. pratensis.*
    11 Pet 1.5–3 mm, or none, the fls mostly 3–4 mm wide; pls annual or biennial.
      13 Basal lvs few, or absent at anthesis; cauline lvs mostly 4–10(–20).
        14 Cauline lvs mostly 4–8 cm, their lateral lfls relatively broad, commonly oval or broadly oblong, generally decurrent on the rachis .... 13. *C. pensylvanica.*
        14 Cauline lvs mostly 2–4 cm, with narrower, non-decurrent lateral segments .... 14. *C. parviflora.*
      13 Basal lvs numerous and persistent; cauline lvs commonly 2–5 .... 15. *C. hirsuta.*

1. **Cardamine diphylla** (Michx.) A. Wood. Broad-lvd t. Rhizome continuous; stem glabrous, 2–4 dm; cauline lvs commonly 2, opposite or nearly so; lfls 3, coarsely crenate-toothed, half as wide as long, the terminal rhombic-ovate, the lateral obliquely ovate, obtuse to rounded at base on the broad outer side; basal lvs similar to the cauline; hairs of the lf-margins appressed, very short, ca 0.1 mm; sep 5–8 mm; pet 11–17 mm, dull purplish; frs 2–4 cm; 2*n*=96. Rich woods; Que. and N.B. to Minn., s. to N.C., Ga., and Ala. Apr., May. (*Dentaria d.; D. incisa,* a form with narrower, laciniately toothed lfls.)

2. **Cardamine × maxima** A. Wood. Three-lvd t. Rhizome with alternately enlarged and slightly constricted regions; stems 1.5–4 dm, glabrous; basal lvs similar to the cauline, usually present at anthesis; cauline lvs typically 3 and distinctly alternate, ternate, the segments petiolulate, lance-ovate to rhombic-ovate, crenate-toothed, the lateral asymmetrical, or with a lateral lobe, or bifid nearly to the base; hairs of the lf-margins spreading, 0.2–0.3 mm; fls and frs much as in no. 1; 2*n*=208, perhaps other nos. Rich woods; Me. and N.B. to Mich. and Wis., s. to W.Va. and Ky. Apr., May. (*Dentaria m.*) Said to be a sterile hybrid between nos. 1 and 3.

3. **Cardamine concatenata** (Michx.) O. Schwarz. Five-parted t. Rhizome constricted at intervals, the segments 2–3 cm; stems 2–4 dm, shortly spreading-hairy above; basal and cauline lvs similar, the former usually absent at anthesis, the latter typically in a whorl of 3 above the middle of the stem, deeply 3-parted or 3-foliolate, the segments linear or lanceolate, subentire to laciniate-toothed, the lateral segments often, the terminal segment sometimes deeply bifid, the whole lf appearing 5-parted; sep 5–8 mm; pet pale lavender, 12–19 mm; frs 2–4 cm; 2*n*=128, 240, 256. Moist rich woods; Me. and s. Que. to Minn., s. to Fla., La., and Okla. Apr., May, a little earlier than nos. 1 and 5. (*Dentaria laciniata.*) Some possible hybrids with no. 5 have been called *Dentaria anomala* Eames.

4. **Cardamine dissecta** (Leavenw.) Al-Shehbaz. Dissected t. Rhizome constricted at intervals, the segments elongate; stems 2–4 dm, glabrous; basal lvs usually present at anthesis, resembling the cauline; cauline lvs 2, opposite or nearly so, biternate into linear entire segments 2–4 mm wide, these sometimes with lateral linear lobes; sep 4–6 mm; pet 10–14 mm, anthocyanic; frs 2–4 cm; 2*n*=64, 112. Rich moist woods; c. O. and s. Ind. to n. Ga. and n. Ala. Mar., Apr. (*Dentaria furcata; D. multifida.*)

5. **Cardamine angustata** O. E. Schulz. Appalachian t. Rhizome constricted at intervals, stems 2–4 dm, glabrous or rarely a little hairy; basal lvs like those of no. 1, with broad, coarsely crenate-toothed segments; cauline lvs 2(3), opposite or nearly so, their segments narrowly lanceolate, sharply toothed to entire, as in no. 3; sep 5–8 mm; pet 10–15 mm, faintly pinkish-lavender; frs 2–4 cm; 2*n*=128. Rich woods; N.J. to Ind., s. to Ga. and Miss. Apr., May. (*Dentaria heterophylla.*) A very similar pl from Conn., with the segments of the cauline lvs confluent at base, has been called *Dentaria incisifolia* Eames.

6. **Cardamine rhomboidea** (Pers.) DC. Spring-cress. Stems borne singly or few together from a short, stout, tuber-like rhizome, barely subterranean rhizome, simple or branched above, 2–4.5 dm up to the first fl, the raceme elongating in anthesis so that the whole pl may reach 6 dm; lvs all simple, the lower long-petioled, the 4–8 cauline ones progressively less so or sessile; basal blades rotund or cordate, mostly about as wide as long, only rarely purplish beneath, generally deciduous before anthesis; cauline blades narrowly oblong to ovate; herbage sparsely to copiously beset with very short hairs up to 0.15(0.2) mm; sep green, turning yellow after anthesis; pet white (rarely pink), 7–16 mm; pedicels at maturity ascending to divergent, the lower to 4 cm; fr slender, 1.5–2.5 cm, often abortive, its beak 2–4.5 mm; 2*n*=64, 80, 96. Moist or wet woods or shallow water; Que. to Minn., S.D., and Kans., s. to Fla. and Tex. Apr.–June. (*C. bulbosa.*)

7. **Cardamine douglassii** Britton.　Pink spring-cress. Much like no. 6, and hybridizing with it, but mostly shorter (1–2.5 dm up to the first fl), with consistently longer hairs (these 0.2–0.8 mm), pink or purple (rarely white) pet, and purple sep that turn brown in age; rhizome sometimes superficial and becoming green; lvs avg shorter, the basal ones usually purplish beneath and tending to be longer than wide; cauline lvs mostly 3–5; 2n=64, 96, 144. Moist woods; N.H. to s. Minn., s. to Va., Tenn., and Mo. Apr., May, cresting 2–3 weeks before no. 6.

8. **Cardamine rotundifolia** Michx.　Trailing b.-c. Stems weak, decumbent or ascending from a long, slender rhizome, emitting stolons with small lvs; larger lvs often with 1 or 2 small lateral lobes on the petiole; cauline lvs mostly petiolate, ovate to rotund, rounded to subcordate at base; pet white, 6–8 mm; raceme and pedicels elongating after anthesis, the pedicels to 1–2 cm; frs divergent or ascending, 1–2 cm, the beak 2.5–3 mm. Wet soil, swamps, or running water; N.Y. and O. to N.C. and Ky. May, June.

9. **Cardamine longii** Fern.　Salt-marsh b.-c. Stems weak and diffuse from fibrous roots, 1–3 dm; lvs all petioled, subrotund to reniform, rounded or cordate at base; raceme poorly developed, elongating after anthesis; pet none; mature pedicels 1–3(–6) mm; frs ascending, 5–12 mm, the beak 0.5–1 mm. Borders of salt-marshes; Me. to Va. Summer.

10. **Cardamine bellidifolia** L.　Northern b.-c. Glabrous perennial with densely tufted stems 3–10 cm; lvs long-petioled, oval or elliptic, 5–10 mm; pet 3–5 mm; axis of the raceme scarcely elongating, the frs 2 cm, erect, crowded, on pedicels 3–5(–10) mm, the beak 1–1.5 mm; 2n=16. Circumboreal, s. in our range to the higher mts. of Me. and N.H. Summer.

11. **Cardamine clematitis** Shuttlew.　Appalachian b.-c. Perennial from a slender rhizome, decumbent or ascending, 2–4 dm; basal and cauline lvs nearly alike, but the cauline gradually smaller and with shorter petioles, trifoliolate or the middle and upper often simple; petioles with a pair of small (0.5–4 mm) auricles at the base, or sometimes exauriculate; lfls obovate or rotund to broadly ovate or reniform, entire or commonly shallowly few-lobed, the terminal usually 2–5 cm long and wide, each lobe mucronate; pet white, 6–8 mm; mature pedicels ascending, 5–12 mm; frs erect or ascending, 2–4 cm, the beak 2–3 mm. Wet places in the mts.; Va. and s. W.Va. to S.C., Ga. and Tenn. Apr.–June. Plants without petiolar auricles are sometimes treated as a distinct sp., *C. flagellifera* O. E. Schulz (*C. hugeri.*)

12. **Cardamine pratensis** L.　Cuckoo-fl. Erect perennial 2–5 dm; basal lvs long-petioled, with 3–8 broadly oval to rotund or reniform lfls 5–20 mm, the terminal the largest; lowest cauline lvs sometimes like the basal, the others all shorter, on shorter petioles, the mostly 7–17 lfls oval to oblong or linear; pet obovate, 8–15 mm; mature pedicels ascending, 8–15 mm; frs suberect, 2.5–4 cm, the beak 1–2 mm; 2n=16–96. Circumboreal, with 2 vars. in our range. Var. *pratensis,* a native of Eurasia, is intr. in moist soil of roadsides, lawns, and meadows from Nf. to N.Y. and N.J.; it has pink fls, the terminal lfl of the basal lvs is angularly dentate, and the lateral lfls of the cauline lvs tend to be sessile; 2n=<56. Apr.–July. Var. *palustris* Wimmer & Graebner is native in bogs and swamps throughout most of Can., extending s. to N.J., O., n. Ind., and Minn.; it has white fls, the terminal lfl of the basal lvs is entire or nearly so, and the lateral lfls of the cauline lvs are commonly contracted to a short petiole; 2n=>56. May, June. (*C. palustris.*)

13. **Cardamine pensylvanica** Muhl.　Pennsylvania b.-c. Erect or spreading annual or biennial to 6 dm; stem usually hispidulous at least below, seldom glabrous; terminal lfl of basal lvs obovate to orbicular, 1–2 cm wide; cauline lvs commonly 4–8 cm, the terminal segment cuneate-obovate, the lateral smaller, broadly oblong to oval, 3–8 mm, usually decurrent on the rachis, entire or often with a few teeth or shallow lobes; pet white, 1.5–4 mm; mature pedicels ascending; frs ascending, 2–3 cm, the stylar beak 0.5–2 mm; 2n=32, 64. Swamps and wet woods; Nf. and Que. to Minn. and B.C., s. to Fla. and Calif. Apr.–June. Small pls of drier habitats approach *C. parviflora.*

　　　C. impatiens L., a weedy European annual or biennial with lvs sagittate-auriculate at base; 2n=16, is adventive in our coastal states, and w. to Mich.

14. **Cardamine parviflora** L.　Dry-land b.-c. Annual or biennial with glabrous, mostly solitary stems 1–3 dm; terminal lfl of basal lvs oblong to broadly cuneate-obovate; cauline lvs commonly 2–4 cm, with 3–6 pairs of segments, the terminal segment linear to cuneate-oblong, entire or toothed, the lateral not much smaller, linear or linear-spatulate or narrowly oblong, usually 1–3 mm wide and entire, not decurrent; fls and fr much like no. 13, but the stylar beak not over 1 mm; 2n=16. Usually in dry soil; Que. and Nf. to Minn., s. to Ga. and Ark.; also in the Old World. The Amer. pls are var. *arenicola* (Britton) O. E. Schulz.

15. **Cardamine hirsuta** L.　Hoary b.-c. Annual with glabrous stems 1–4 dm; basal lvs numerous and conspicuous in comparison with the few small stem-lvs, the pl appearing subscapose; terminal lfl of basal lvs rotund to reniform, entire to shallowly few-lobed; petioles of the cauline lvs ciliate at least at base; lateral segments of the cauline lvs gradually narrower toward the top of the stem, obovate-oblong to linear-spatulate, often with 1 or 2 teeth, commonly hirsutulous above; pet white, 2–3 mm; pedicels narrowly ascending; frs erect, 1.5–2.5 cm; 2n=16. Widely distributed in the Old World; naturalized in moist, especially sandy soil from s. N.Y. to Ill. and Ala. Mar., Apr.

30. **SIBARA** Greene.　Resembling *Cardamine* and *Arabis*; sep erect; pet small, white to pink or purple; nectar-glands small, associated with individual stamens, or obsolete; anthers oblong; fr linear, flattened parallel to the septum or subterete, the valves nerveless or faintly 1-nerved at the base; replum narrow, not margined; seeds several and uniseriate

in each locule, narrowly winged (in our sp.) or wingless; annual or biennial herbs, glabrous or sparsely pubescent below with simple or branched hairs; lvs pinnatifid. 10, N. Amer., mainly western.

1. Sibara virginica (L.) Rollins. Resembling *Cardamine parviflora* in aspect; stiff, ascending or decumbent, 1–4 dm, often branched from the usually hirsute base, the hairs simple or bifurcate; basal lvs oblong, 4–8 cm, deeply pinnatifid or lyrate-pinnatifid into oblong divergent segments; cauline lvs similarly pinnatifid, or the uppermost lance-linear and entire; racemes compact; pet 2–3 mm, white or pinkish; frs ascending (often more strongly so than the short pedicels), 1.5–2.5 cm × 1.5–2 mm, scarcely beaked; seeds suborbicular, narrowly winged, 1.2–1.5 mm wide; $2n=16$. Woods, fields, and roadsides; Va. to Ind. and Kans., s. to Fla. and Tex. Mar.–May. (*Arabis v.*)

31. ARABIS L. Rock-cress. Sep erect or spreading, the outer sometimes saccate at base; pet white, yellowish, or pink, narrowly spatulate to oblong or obovate; anthers oblong; ovary cylindric; ovules numerous; style short and scarcely differentiated; stigma truncate; fr linear, elongate, flat or subterete, the valves often with a midvein or reticulately veined; seeds flattened, often winged; annual to perennial herbs, the basal lvs petioled, usually simple, the cauline smaller and usually sessile; pubescence variable but usually present. 180, N. Hemisphere.

1 Pedicels and frs at maturity erect or nearly so.
  2 Frs terete or nearly so, 0.8–1.3 mm wide .... 1. *A. glabra.*
  2 Frs flat.
    3 Seeds in 1 row in each locule; frs 0.7–1.1 mm wide, stem hairy below .... 2. *A. hirsuta.*
    3 Seeds in 2 rows; frs 1.5–2.5 mm wide; stem commonly glabrous .... 3. *A. drummondii.*
1 Pedicels and frs at maturity ascending, spreading, or reflexed.
  4 Cauline lvs not auricled or sagittate at base.
    5 Frs 2–4.5 cm, ascending .... 4. *A. lyrata.*
    5 Frs (4)5–11 cm, widely spreading, recurved, or reflexed.
      6 Cauline lvs hairy; frs 1.8–3.2 mm wide .... 5. *A. canadensis.*
      6 Cauline lvs glabrous; frs. 1.2–2 mm wide.
        7 Plants simple or few-branched, the infl a raceme; cal 3–5 mm; fl late Apr.–early June .... 6. *A. laevigata.*
        7 Plants freely branched, the infl a panicle; cal 2–3 mm; fl mid-July–Aug. .... 7. *A. serotina.*
  4 Cauline lvs auricled or sagittate at base.
    8 Frs and pedicels ascending, spreading, or somewhat decurved.
      9 Cauline lvs (except possibly the very lowest) glabrous on both sides.
        10 Basal lvs glabrous or sparsely pubescent with simple hairs only.
          11 Pet 3–5 mm, equaling or slightly exceeding the sep .... 6. *A. laevigata.*
          11 Pet 5–8 mm, nearly or fully twice as long as the sep .... 8. *A. missouriensis.*
        10 Basal lvs stellate-hairy on both sides .... 9. *A. divaricarpa.*
      9 Cauline lvs distinctly pubescent.
        12 Pubescence on the upper side of the cauline lvs none or of simple hairs.
          13 Pet 6–9 mm; mature pedicels 6–18 mm; seeds winged.
            14 Frs 6–9 cm; cauline lvs chiefly entire .... 8. *A. missouriensis.*
            14 Frs 2.5–4 cm; cauline lvs chiefly coarsely serrate .... 10. *A. patens.*
          13 Pet 2–3 mm, barely exceeding the sep; mature pedicels 2–4 mm; seeds wingless .... 11. *A. shortii.*
        12 Pubescence of both sides of the cauline lvs of stellate or forked hairs .... 12. *A. perstellata.*
    8 Frs pendent; mature pedicels abruptly reflexed .... 13. *A. holboellii.*

1. Arabis glabra (L.) Bernh. Tower-mustard. Biennial or short-lived perennial 3–15 dm; stem commonly hirsute below with mostly simple (sometimes bifurcate) hairs, glabrous and glaucous above; lower lvs commonly with Y-shaped hairs; cauline lvs overlapping below, more remote above, all lanceolate or lance-oblong, auriculate-clasping, usually glabrous and glaucous; pet 3–6 mm; style 1–1.5 mm; mature pedicels erect, 7–16 mm; frs erect, subterete, 5–9 cm × 0.8–1.3 mm, the valves with prominent midnerve reaching nearly or quite to the tip; seeds in 2 rows in each locule, angular, very narrowly winged; $2n=12$. Usually in dry soil; circumboreal, in Amer. s. to N.C., Ark., Kans., N.M., and Calif. May, June.

2. Arabis hirsuta (L.) Scop. Much like no. 1, 2–8 dm; cauline lvs oblong to lance-linear, 1–3 cm, at least the lower pubescent; style 0.5–1 mm; frs flat, 3–5 cm × 0.7–1.1 mm, the valves 1-nerved to or somewhat beyond the middle; seeds in 1 row, the wing evident or lacking; $2n=32$. Circumboreal, with 3 vars. in Amer., as follows:

1 Pet 5–9 mm; stem hirsute near the base, otherwise mostly glabrous; sw. Wis., otherwise confined to
    the western cordilleran region .... var. *glabrata* T. & G.
1 Pet 3–5 mm.
  2 Stem and lvs ± spreading-hirsute; woods, banks, and ledges, chiefly in calcareous soil; Que. to
    Alas., s. to Pa., n. Ga., Ind., Ark. and Ariz. .... var. *pycnocarpa* (M. Hopkins) Rollins.

2 Stem strigose with malpighiaceous hairs; lvs glabrous or nearly so; woods and hillsides in calcareous regions, O. to Ill., s. to La. and Ark. .... var. *adpressipilis* (M. Hopkins) Rollins.

3. Arabis drummondii A. Gray. Erect biennial 3–9 dm, glabrous or with a few 2-pronged hairs below, often glaucous; cauline lvs sessile, lanceolate to narrowly oblong, 2–8 cm, acute, entire or with a few teeth, auriculate at base or the uppermost acute; pet 5–9 mm; mature pedicels erect, 10–15 mm; frs straight, erect, flat, 4–7 cm × 1.5–2.5 mm, the seeds in 2 rows; $2n=14$, 20. Various habitats; Lab. and Nf. to B.C., s. to Del., O., Io., and Ariz. May–Aug.

4. Arabis lyrata L. Erect or ascending biennial or perennial, 1–4 dm, branched from the usually hirsute base; basal lvs spatulate, 2–4 cm, entire to pinnately lobed; cauline lvs linear to narrowly spatulate, narrowed to the base, the lowest sometimes with a few short teeth or lobes; pet 3–8 mm; mature pedicels widely ascending, 6–15 mm, the fr continuing about the same direction as the pedicel, 2–4.5 cm × 1 mm, the valves 1-nerved to beyond the middle; seeds oblong to elliptic, wingless, 1 mm; $2n=16$, 32. Dry woods and fields, especially in sandy soil; Vt. to Alas., s. to Va., Ky., and Mo., and in the mts. to Ga.; also in e. Asia. May–July.

5. Arabis canadensis L. Sickle-pod. Stout biennial to 1 m, sparsely hairy at base; cauline lvs lanceolate to elliptic or oblong, 3–10 cm, narrowed to the base, chiefly sharply and remotely dentate, ± hairy; pet 3–5 mm; pedicels erect in bud, widely divergent at anthesis, in fr reflexed and 7–12 mm; frs pendulous, flat, ± curved, 7–11 cm × 2–3.5 mm, prominently veined; seeds in 1 row, 3 mm long, as wide as the valve, the wing 0.5–1 mm wide; $2n=14$. Woods; Me. to Minn., s. to Ga. and Tex. May–July.

6. Arabis laevigata (Muhl.) Poiret. Biennial to 1 m, simple or few-branched, glabrous and glaucous throughout, except the sparsely hirsutulous, spatulate basal lvs; cauline lvs (averaging ca 13 below the first fl) narrowly lanceolate, 5–15 cm, serrate to entire, usually sagittate at the sessile base; cal 3–5 mm; pet white, 3–6 mm, equaling to a fourth longer than the sep; pedicels at maturity widely spreading, 7–12 mm; frs widely spreading to somewhat decurved, arcuate, linear, flat, 5–10 cm × 1.2–2 mm, nerveless or faintly nerved to the middle; seeds in 1 row, oblong, narrowly winged; $2n=14$. Woods and hillsides; Que. to S.D., s. to Ga. and Okla. Apr.–June. Var. *burkii* Porter, with nonsagittate cauline lvs, occurs from Pa. and Md. to Va. and W.Va., sometimes on shale-barrens.

7. Arabis serotina E. Steele. Freely branched, the infl a panicle with numerous ± pedulous fls; lvs without auricles, but often even more deciduous; cal short, mostly 2–3 mm; frs 4–8 cm, straight or nearly so; $2n=14$; otherwise much like no. 6. Shale-barrens of Va. and W.Va. July–Aug, well after no. 6.

8. Arabis missouriensis Greene. Biennial, 2–5 dm, glabrous to sparsely hairy; basal lvs persistent, dentate to pinnatifid; cauline lvs numerous (averaging 25 below the first fl), erect, lanceolate to linear, the lower commonly with a few sharp teeth, the upper mostly entire; pet creamy-white, 6–8 mm, nearly or fully twice as long as the sep; mature racemes lax and elongate, with ascending pedicels 5–10 mm; frs at first erect, becoming widely divergent and commonly arcuate, 5–9 cm × 1.2–2 mm, flat, the valves 1-nerved to or beyond the middle; seeds in 1 row, quadrate-oblong, 1.5–2 mm, conspicuously winged. Moist or dry, rocky or sandy woods and hills; Me. to N.Y. and N.J.; Ind., Wis., and Mich. to Mo., Ark. and Okla. May–July. (*A. viridis.*)

9. Arabis divaricarpa A. Nels. Erect biennial to 1 m, glabrous above the base; basal lvs oblanceolate, finely stellate on both sides; cauline lvs lance-linear to oblong, erect or nearly so, 2–5 cm, auriculate, entire, glabrous; pet pinkish or white, 5–8 mm; mature pedicels spreading, 6–12 mm; frs at first erect, soon spreading, linear, straight or nearly so, 3–9 cm × 1.2–2.2 mm, the valves 1-nerved to or beyond the middle; seeds in 1 or 2 rows; $2n=14$, 20, 21, 28. Sandy or rocky soil; Que. to Yukon, s. to N.H., Wis., n. Io., and Calif. June, July. (*A. brachycarpa; A. confinis.*)

10. Arabis patens Sulliv. Biennial or short-lived perennial 3–6 dm, the herbage hirsute with simple hairs, or also some forked ones on the stem; cauline lvs lanceolate or lance-oblong to lance-ovate, 2–5 cm, usually rounded to the broadly clasping base, coarsely serrate chiefly below the middle, or the upper lvs entire, hirsute with simple hairs on both sides; pet white, 6–9 mm; pedicels at maturity widely ascending, glabrous, 10–16 mm; frs widely ascending or spreading, linear, flat, 2.5–4 cm × 1 mm, the valves sharply 1-nerved usually to the tip; seeds oblong, narrowly winged. Moist rocky woods; Pa. to Ind., s. to w. N.C., Ala., and e. Tenn. May.

11. Arabis shortii (Fern.) Gleason. Biennial (sometimes perennial?) 2–5 dm, the stem branched at base, often decumbent, thinly pubescent with mostly forked hairs; cauline lvs oblanceolate to oblong-obovate, 2–5 cm, even the uppermost usually sharply toothed, somewhat narrowed to the clasping base, pubescent with simple hairs above, stellate ones beneath; pet white, a little exceeding the sep; pedicels at anthesis hirsute, at maturity 2–4 mm and usually glabrous; fr widely spreading, flat, 1.5–3 cm × 1 mm, often minutely hairy, the valves nerveless or 1-nerved near the base; seeds in 1 row, wingless. Rich moist woods; N.Y. to Minn., s. to Va. and Ark. Apr., May. (*A. dentata.*)

12. Arabis perstellata L. Braun. Much like no. 11, but perennial, and with the lvs closely and softly stellate-hairy on both sides, also with longer simple hairs on the upper side; pet pink, 3–5 mm; pedicels to 8 mm, fr more densely hairy; $2n=14$. Wooded hillsides in e. Ky. Apr.

13. Arabis holboellii Hornem. Biennial or short-lived perennial 2–9 dm, densely pubescent with dendritically branched hairs, those of the stem appressed; basal lvs oblanceolate, 2–5 cm, the cauline lance-linear to oblong, 1–4 cm, usually erect, auriculate, mostly revolute-margined; pet pink to white, 7–10 mm; mature pedicels abruptly deflexed at base, 6–15 mm; frs pendent, 3–8 cm × 1–2 mm, the valves 1-nerved below the middle; seeds in 1 row, suborbicular, 1 mm; $2n=14$, 21, 28, 35, 42. Dry rocky or gravelly soil; Greenl. to Alas., s. to Que., n. Mich., Colo., and Calif. June, July. Our plants, as here described, are var. *retrofracta* (Graham) Rydb., a chiefly cordilleran phase that also occurs in n. Mich.

32. RORIPPA Scop. Sep ascending; pet yellow or white, spatulate to obovate, small, often scarcely exceeding the sep, gradually narrowed to the claw; short stamens flanked at base by a pair of minute glands, or these confluent into an annular gland; long stamens separated by a short conic gland; ovary cylindric; style very short; stigma capitate or shortly 2-lobed; ovules numerous; fr subterete, subglobose to cylindric, with short style and persistent stigma, the valves thin, with obscure midnerve; seeds numerous, minute, in most spp in 2 rows in each locule; annual to perennial herbs with toothed to pinnatifid lvs, the pubescence of simple hairs or none, 80, cosmop.

1 Fls white; plants rooting from the lower nodes .... 1. *R. nasturtium-aquaticum.*
1 Fls yellow (rarely white); plants usually not rooting from the lower nodes.
  2 Pls perennial, with creeping roots or rhizomes; pet somewhat surpassing the sep.
    3 Frs about equaling or evidently longer than the pedicels, often more than 6 mm long.
      4 Lvs evidently lobed or pinnatifid, and generally auriculate at base.
        5 Stems decumbent or ascending; lateral lf-segments entire or with a few low teeth; beak of
          the fr 1.3–2.5 mm .... 2. *R. sinuata.*
        5 Stems ± erect; lateral lf-segments commonly sharply toothed; beak 0.5–1 mm .... 3. *R. sylvestris.*
      4 Lvs merely toothed, neither lobed nor auriculate .... 4. *R. indica.*
    3 Frs shorter than the pedicels, up to about 6 mm.
      6 Frs cylindric, 2–6 mm, 2–4 times as long as wide; style up to half as long as the fr; stem
        hollow; cauline lvs usually not auricled .... 5. *R. amphibia.*
      6 Frs globose, 1–2 mm, as wide as long; style as long as the fr; cauline lvs auricled .... 6. *R. austriaca.*
  2 Pls annual or biennial, taprooted; pet up to about as long as the sep, or none.
    7 Well grown mature pedicels mostly 3–7 mm; fr mostly 0.5–1.5 times as long as the pedicel .... 7. *R. palustris.*
    7 Well grown mature pedicels rarely over 3 mm; fr (1–)1.5–4 times as long as the pedicel.
      8 Seeds ca 20–70; pet presnt .... 8. *R. tenerrima.*
      8 Seeds ca 100–200; pet often wanting .... 9. *R. sessiliflora.*

1. Rorippa nasturtium-aquaticum (L.) Hayek. Water-cress. Perennial; stems submersed or partly floating, or prostrate on mud, freely rooting; lvs of 3–9 obtuse segments, the lateral ovate to rotund, the terminal much larger; usually rotund; fls 5 mm wide, the pet twice as long as the sep; mature pedicels divaricate, 8–15 mm; frs slender, 1–2.5 cm, the beak 1 mm; seeds coarsely reticulate; $2n=32$, 48, 64. Native of Eurasia, now widely established in clear quiet water throughout the U.S. and s. Can. All summer. (*Radicula n.; Sisymbrium n.; Nasturtium microphyllum; N. officinale.*) Some botanists now restrict the name *R. nasturtium-aquaticum* to the "diploid" race ($2n=32$), and call the tetraploids ($2n=64$) *R. microphylla* (Boenn.) Hylander, regarding the plants with $2n=48$ as vegetatively reproducing hybrids. The two types differ as follows:

1 Seeds in 2 rows in each locule, with ca 25–50(–60) polygonal depressions on each face; $2n=32$ ..... *R. nasturtium-aquaticum.*
1 Seeds in 1 row in each locule, with ca 100–150(–175) polygonal depressions on each face; $2n=64$ .... *R. microphylla.*

2. Rorippa sinuata (Nutt.) A. S. Hitchc. Western yellow-cress. Perennial 1–4 dm, usually with several ascending branches from near the base, the stem and lower surface (or at least the midrib) of the lvs sparsely to densely beset with small, vesicular, hemispherical trichomes that may be collapsed and scale-like in dried specimens; lvs all cauline, 3–8 × 0.5–1.5 cm, oblanceolate or oblong, auriculate-clasping or not, pinnatifid, the segments commonly 5–7 pairs, spreading or ascending, usually obtuse, entire or with a few teeth, separated by rounded sinuses; lateral sep somewhat saccate at base; pet light yellow, 3.5–5.5 mm, surpassing the sep; mature pedicels slender, ascending to recurved, 5–9 mm; frs subterete, 5–12 × 1.5–2 mm, straight or upcurved, often beset with vesicular trichomes, abruptly narrowed to a slender stylar beak 1.3–2.5 mm; $2n=16$. Moist, especially sandy soil, fields, and shores; Ill. to Alta. and Wash., s. to Ark., Tex., and Calif., and occasionally adventive eastward. Apr.–July. (*Radicula s.*)

3. Rorippa sylvestris (L.) Besser. Creeping yellow-cress. Erect perennial 2–6 dm, usually branched above, glabrous or sparingly hirsute on the lower part of the stem; basal rosettes present in young plants; cauline lvs slender-petiolate, oblong, deeply pinnatifid or divided into numerous oblong or obovate segments, the lower lvs to 2 dm, more than 2 cm wide, the upper progressively reduced, the segments mostly acute, sharply or angularly toothed or in depauperate pls subentire; pet 4–5 mm, twice as long as the sep; mature pedicels widely divergent or somewhat deflexed, often arcuate, 5–10 mm; frs nearly straight, 10–15 × 1.5 mm, the style 0.5–1 mm; seeds often wanting; $2n=32$, 40, 48. Native of Eurasia, intr. in wet soil nearly throughout our range. May–Aug. (*Radicula s.*)

4. Rorippa indica (L.) Hiern.   Indian yellow-cress.   Low perennial 1–3 dm, essentially glabrous throughout; lvs mainly cauline, the principal ones petiolate, with lance-elliptic to lance-rhombic, merely toothed blade up to 7 × 2 cm; pet 3 mm, slightly surpassing the sep; mature pedicels widely ascending; frs cylindric, upcurved, 1 cm, with a short, thick stylar beak 0.5(–1) mm. Native of se. Asia, sparingly intr. as a weed in N.Y. and elsewhere. May, June.

5. Rorippa amphibia (L.) Besser.   Erect or ascending perennial to 1 dm; stems hollow, often rooting below; lvs lanceolate to oblanceolate, to 12 cm, the lower or submersed ones commonly laciniate to pectinately parted, the upper narrower and merely dentate; pet about twice as long as the sep; mature pedicels 8–15 mm, horizontally spreading or somewhat deflexed; frs ellipsoid, straight, the body 3–6 mm, the stylar beak 1–2 mm; $2n=16, 32$. Native of Eurasia, established in swamps and shallow water from Que. to Me. and Mass. and perhaps elsewhere. June, July.

     R. × *prostrata* (Bergeret) Schinz & Thell., another European plant, is sparingly intr. in some of our coastal states. It is intermediate between nos. 3 and 5, and is thought to be an unstable segregating series of hybrid origin. It differs from no. 5 in its solid stems and more pinnatifid lvs, and from no. 3 in its smaller frs (5–6 mm) and sessile or subsessile, less cleft (or subentire) upper lvs.

6. Rorippa austriaca (Crantz) Besser.   Field yellow-cress. Perennial to 6 dm, from rather thick and fleshy creeping roots; lvs oblanceolate, 3–10 cm, mostly obtuse, dentate to subentire, gradually tapering to an auriculate base; pet slightly longer than the sep; mature pedicels ascending, 7–15 mm; frs globose, 1–2 mm, about as long as wide, the replum circular in outline, often not maturing seeds; style about as long as the fr; $2n=16$. Native of Europe, sparingly established as a weed in fields here and there in our range. June, July.

7. Rorippa palustris (L.) Besser.   Common yellow-cress. Taprooted annual or biennial, erect, to 1 m, the taller pls commonly much-branched; lvs lanceolate to oblong-obovate, at least the lower pinnatifid; pet 1.5–2.5 mm, about equaling the sep, mature pedicels widely spreading, 3–7 mm; frs 2.5–9 mm, up to twice as long as the pedicels; middle cauline lvs generally distinctly winged toward the base and auriculate, so that the lf appears ± sessile; $2n=32$. (*R. islandica*, misapplied.) *R. palustris* is circumboreal, with 3 vars. in our range.

1   Lvs hirsute on the lower surface; stems hirsute usually up to the terminal raceme; widespread mainly
     n. of the glacial boundary in our range, and in the w. cordillera .... var. *hispida* (Desv.) Rydb.
1   Lvs glabrous on the lower surface; stems glabrous or sparsely hirsute below.
   2   Pls mostly 1–4 dm, with slender stems mostly under 3 mm thick; lvs thin, the upper as well as the
       lower often pinnatifid; frs 4–9 mm, not constricted in the middle; native of Eurasia, occasionally
       intr. in our range, mainly in the coastal states and w. to the Great Lakes .... var. *palustris*.
   2   Pls mostly 4–10 dm, with stout stems mostly over 3 mm thick; lvs thick, the upper ones generally
       merely toothed; frs 2.5–5 mm, constricted at the middle; widespread in c. and e. U.S. and adj.
       Can. .... var. *fernaldiana* (Butters & Abbe) Stuckey.

8. Rorippa tenerrima Greene.   Delicate, freely branched, glabrous or subglabrous annual; stems 1–2 dm, decumbent to prostrate, much-branched from the base; basal and cauline lvs short-petiolate, neither clasping nor auriculate, oblong to oblanceolate or spatulate, mostly 2–6 × 1–1.5 cm, usually lyrate-pinnatifid; pet ca 1 mm or less, a little shorter than the sep; mature pedicels 1.5–3.5 mm, ascending; frs cylindric, straight or incurved, sometimes constricted at the middle, mostly 3–8 × 1–1.5 mm; seeds ca 20–70. Wet, disturbed places; widely scattered in w. U.S. and adj. Can. and Mex., e. occasionally to St. Louis. (*R. obtusa*, misapplied; *R. truncata*.)

9. Rorippa sessiliflora (Nutt.) A. S. Hitchc.   Southern yellow-cress. Glabrous annual 2–4 dm, normally erect and branched; lower lvs oblong-oblanceolate, coarsely crenate and often ± pinnatifid below the middle, the upper progressively smaller, petiolate, oblanceolate to ovate, entire to crenate; sep strongly saccate; pet wanting; stamens 3–6; mature pedicels widely divergent, 0.5–2 mm; frs spreading or ascending, thick (often over 2 mm), often slightly falcate, 7–10 mm, with numerous (commonly 150–200) seeds, these tiny (to 0.5 mm), deeply pitted, appearing rough at 10×; style up to 0.4 mm, or none; $2n=16$. Wet soil and shallow water; Fla. to Tex., n. to D.C., O., Ind., s. Minn., and Neb. May–Oct. (*Radicula s.*)

     R. teres (Michx.) Stuckey, a sp. chiefly of the coastal plain of s. U.S. and s. to Mex., extends up the Mississippi embayment occasionally as far as St. Louis. It would key to *R. sessiliflora*, from which it differs in having short but evident pet 1–2 mm, fewer (mostly 100–150) seeds, an often longer style (up to 1.5 mm), and in often being provided with vesicular trichomes.

33. BARBAREA R. Br., nom. conserv.   Winter-cress.

Sep oblong, ascending; pet yellow, spatulate to obovate; short stamens partly surrounded at base by a semicircular gland; long stamens separated by a short erect gland; anthers oblong; ovary cylindric, narrowed to a slender style; ovules several per locule; stigma truncate; frs linear, terete or obscurely 4-angled, tipped by the persistent short style, the valves with strong midnerve and anastomosing intermediate nerves; biennials or short-lived perennials, glabrous or with a few simple hairs, the basal lvs mostly pinnatifid with a large terminal lobe and 2–several smaller lateral ones, the cauline smaller, entire to pinnatifid. (*Campe.*) 20, N. Temp.

1 Basal lvs with 1–4 pairs of lateral lobes (or simple); frs mostly 0.5–4(–5) cm, on pedicels up to 1 mm
thick.
2 Beak of the fr 2–3 mm; pet 6–8 mm; widespread intr. sp. . . . . 1. *B. vulgaris.*
2 Beak of the fr 0.5–1.5(–2) mm; pet 3–5 mm; native northern sp. . . . . 2. *B. orthoceras.*
1 Basal lvs with 4–10 pairs of lateral lobes; frs 4.5–7 cm, on pedicels 1.2–1.8 mm thick . . . . 3. *B. verna.*

1. **Barbarea vulgaris** R. Br. Yellow rocket. Dark green, erect, branched above, 2–8 dm, bitter; basal lvs petiolate, with 1–4 pairs of small, elliptic to ovate lateral lobes and a large, ovate to rotund terminal one; cauline lvs progressively reduced, the upper sessile and generally lobed rather than pinnatifid, or the uppermost entire or merely toothed; fls crowded at anthesis; pet 6–8 mm; mature pedicels 3–6 × 0.5 mm; frs erect and appressed to ascending or spreading, 1.5–3 cm, the beak 2–3 mm; 2*n*=16. Native of Eurasia, now widely naturalized as a weed in wet meadows and in damp soil of fields, roadsides, and gardens. Apr.–June. (*B. barbarea; Campe b.; B. stricta; Campe s.*)

2. **Barbarea orthoceras** Ledeb. Northern w.-c. Stems simple or branched above, 3–8 dm; basal lvs simple or with 1–4 pairs of lateral lobes; middle and upper lvs deeply toothed or lyrately lobed, often with 3 or 4 pairs of lfls; pet 3–5 mm; mature pedicels clavate-thickened, to 1 mm; frs erect to spreading, 2–4 cm, the beak 0.5–1.5(–2) mm; 2*n*=16. Swamps and wet woods; circumboreal, s. in Amer. to Me., N.H., Ont., and Calif. June–July. (*B. americana.*)

3. **Barbarea verna** (Miller) Aschers. Early w.-c. Erect, 3–8 dm, peppery; lower lvs oblong, to 2 dm, with 4–10 pairs of ovate, repand lateral lobes; cauline lvs smaller, with 3–8 pairs of oblong to linear, mostly entire lateral lobes; pet 6–8 mm; mature pedicels 3.5–7 × 1.2–1.8 mm; frs widely ascending, straight or nearly so, 4.5–7 cm, the beak 1–2.5 mm; 2*n*=16. Native of Eurasia, naturalized in damp soil, fields, roadsides, etc. from Nf. to Wash., s. to Fla. and Calif. Apr.–June. (*Campe v.*)

34. IODANTHUS T. & G. Sep erect, the inner somewhat saccate at base; pet pale violet to nearly white, the blade triangular-obovate, gradually narrowed to the exsert claw; short stamens surrounded at base by an annular gland; anthers linear-oblong; ovary cylindric; style short, scarcely differentiated; stigma capitate; frs linear, terete or nearly so, short-beaked, slightly constricted below the seeds; valves with an inconspicuous midnerve; seeds numerous, in 1 row. 4, N. Amer.

1. **Iodanthus pinnatifidus** (Michx.) Steudel. Purple rocket. Perennial to 1 m, simple below the infl, glabrous or nearly so; lvs thin, lanceolate to elliptic or oblong, acute or acuminate, commonly sharply dentate, often laciniate, rarely double-dentate or merely crenate, tapering to a petiole-like base, frequently auriculate around the stem, the larger or lower often pinnatifid at base with 1–4 pairs of small segments; racemes elongating before anthesis; sep obtuse, 3–5 mm; pet 10–13 mm; pedicels 6–10 mm; frs slender, widely divergent, 2–4 cm, the valves covered with minute transparent papillae. Moist or wet alluvial woods; w. Pa. and W.Va. to Ill., Minn., Tenn., Kans., Ark., and Okla. May, June.

35. CHORISPORA R. Br., nom. conserv. Sep erect, saccate at base; pet with slender claw evidently longer than the blade; ovary linear, gradually tapering into the slender style; ovules numerous; frs elongate, slightly constricted at short intervals, transversely septate at the constrictions, prolonged into the indurate, subulate style, at maturity breaking transversely into 2-seeded joints; herbs with purple, white or yellow fls. 13, Mediterranean and Near East.

1. **Chorispora tenella** (Pallas) DC. Blue mustard. Widely branched annual 2–4 dm; stem, axis of the infl, and pedicels sparsely beset with stipitate glands; lvs lanceolate to oblanceolate, 3–8 cm, with a few low teeth; sep 7 mm; pet blue-purple, ca 12 mm; pedicels 2–5 mm; frs ascending, the body 2–3 cm, the beak 1–2 cm; 2*n*=14. Native of c. Asia, established as a weed in w. U.S. and occasionally adventive in our range. May–July.

Malcolmia maritima R. Br., an Old World sp. habitally similar to *C. tenella,* but eglandular and without the cross-partitions in the scarcely beaked fr, is adventive on L.I.

36. HESPERIS L. Dame's rocket. Sep erect, the outer narrow, crested to near the summit, the inner broad, saccate at base; pet purple to white, the blade obovate, the claw surpassing the cal; short stamens subtended by a U-shaped gland; long stamens glandless; anthers linear-oblong; ovary cylindric; style scarcely narrower; stigma 2-lobed, decurrent along the style; ovules numerous; frs widely spreading on stout pedicels, linear, terete or nearly so, scarcely dehiscent; stigma persistent; valves indistinctly 1-nerved; seeds large (3–4 mm), angularly fusiform, in one row; tall biennial or perennial herbs with showy fls; pubescence of both simple and branched hairs. 25, Eurasia.

1. Hesperis matronalis L.   Erect biennial or perennial 5–10 dm, often branched above; lvs lanceolate or deltoid-lanceolate, short-petioled or sessile, remotely and sharply denticulate, pubescent above with simple hairs, below chiefly with branched hairs; sep ± pubescent, fls fragrant; pet purple, varying to pink or white, 2–2.5 cm; frs 5–10 cm, somewhat constricted between the seeds; 2n=16, 24, 26, 28, 32. An old-fashioned ornamental, escaped along roads and in open woods and moist bottomlands; Que. and N.S. to Mich. and Io., s. to Ga. and Ky. May, June.

37. ERYSIMUM L.   Sep erect or connivent, often saccate at base, the outer often minutely appendaged at apex; pet in our spp. yellow to orange, obovate or spatulate, abruptly or gradually narrowed to a long claw; short stamens subtended by a semicircular or annular gland; each pair of long stamens also subtended by a gland; anthers linear-oblong; ovary linear-cylindric, hairy; ovules numerous; style very short; stigma capitate, 2-lobed; frs elongate, ± 4-angled, hairy, the valves with a prominent midnerve; seeds in one row; herbs with narrow, entire to pinnatifid lvs, ± pubescent with appressed, 2–5(–7)-pronged hairs. All our spp. are ± densely pubescent on the stem, lvs, sep, and fr, and even on the back of the pet, especially at the base of the blade. 200, N. Temp.

1 Pet 15–25 mm; well grown frs 6–10 cm .... 1. *E. asperum.*
1 Pet 10 mm or less.
  2 Sep 2–3.5 mm; pet 3.5–5.5 mm; frs 1.2–3 cm .... 2. *E. cheiranthoides.*
  2 Sep 4.5–7 mm; pet 6–10 mm.
    3 Frs widely spreading, 5–12 cm; anthers 1 mm .... 3. *E. repandum.*
    3 Frs erect or nearly so, 1.5–5 cm; anthers (1.5–)2–2.5 mm.
      4 Some of the hairs of the upper lf-surface 4-pronged (use 20×) .... 4. *E. hieraciifolium.*
      4 None of the hairs 4-pronged .... 5. *E. inconspicuum.*

1. Erysimum asperum (Nutt.) DC.   Western wallflower. Erect biennial or short-lived perennial, 2–10 dm; lvs linear to oblanceolate, entire or with a few low teeth; sep 1 cm; pet bright yellow to orange-yellow, 1.5–2.5 cm, the blade half as long as the claw; mature racemes much-elongate, with stout, divergent pedicels 7–15 mm; frs ± spreading, 4–10 cm; 2n=36. Prairies, sand-hills, and open woods; O. to Minn. and Mo., and widespread in w. U.S., rarely adventive farther e. along railroads. May, June. (*Cheirinia a.; Erysimum arkansanum.*) Our plants, representing the Great Plains phase of the sp., are var. *asperum.*

2. Erysimum cheiranthoides L.   Wormseed-mustard. Erect, often simple annual 2–10 dm; lvs linear to oblanceolate, entire or barely sinuate; sep 2–3.5 mm; pet bright yellow, 3.5–5.5 mm; mature racemes elongate, with straight rachis and slender, widely divergent pedicels (6–)8–12(–14) mm; frs ascending to erect, (1.2–)1.5–2.5(–3) cm; 2n=16. Usually in wet soil, but also a weed in fields and roadsides; originally European, but now circumboreal, s. in Amer. to N.C., Ark., Colo., and Oreg. June–Aug. (*Cheirinia c.*)

3. Erysimum repandum L.   Treacle-mustard; bushy wallflower. Annual, 2–4 dm, often much-branched; lvs linear to narrowly lanceolate or oblanceolate, the smaller often entire, the larger commonly conspicuously sinuate-dentate to pinnatifid; sep densely stellate, 4.5–5.5 mm; pet pale yellow, 6–10 mm; anthers 1 mm; mature racemes elongate, the rachis often zigzag, the pedicels very thick, divergent, 2–8 mm; frs widely spreading, 5–12 cm; 2n=14, 16. Native of Europe, found as an occasional weed in waste places in our range and westward. (*Cheirinia r.*)

4. Erysimum hieraciifolium L.   Tall wormseed-mustard. Biennial or perennial, much like no. 5, usually but not always with somewhat broader lvs, these up to 1 or even 1.5 cm wide and sometimes evidently toothed; hairs of the upper lf-surface mainly 3-pronged, with a considerable admixture of 4-pronged ones; 2n=32, 40. European weed, now found from N.S. to Sask., Mich., and Wis. June, July.

5. Erysimum inconspicuum (S. Wats.) MacMillan.   Erect, often simple perennial 3–8 dm; lvs mostly erect or ascending, linear to oblanceolate, entire or obscurely and remotely sinuate-dentate, the cauline rarely over 5 mm wide; pubescence of the upper lf-surface of mixed 2-pronged and some 3-pronged hairs; sep densely stellate, 5–7 mm; pet pale yellow, 6–10 mm; anthers (1.5–)2–2.5 mm; mature racemes elongate, the stout, ascending pedicels 3–9 mm; frs erect or nearly so, 1.5–4 cm; 2n=54. Dry soil of prairies, plains, and upland woods; Alas. to Nev., e. to Man., Minn., and n. Mich., and occasionally adventive eastward. May–Aug. (*E. parviflorum; Cheirinia inconspicua; C. syrticola.*)

38. ALLIARIA Heister.   Garlic-mustard. Pet white, spatulate, gradually narrowed to the claw; short stamens surrounded at base by an annular gland; each pair of long stamens separated by a trigonous gland; filaments flattened; anthers oval; ovary cylindric, with few-several ovules per locule; style not differentiated; stigma capitate; frs linear, elongate, 4-angled, each valve with a conspicuous midnerve and smaller lateral veins; seeds in one row, nearly cylindric, longitudinally striate; erect biennials or perennial with broad-leaves, and with the odor of garlic. 2, Eurasia.

1. Alliaria petiolata (Bieb.) Cavara & Grande. Biennial to 1 m, simple or little-branched, glabrous or with a few simple hairs; lower lvs reniform, the others deltoid, 3–6 cm long and wide, acute, coarsely toothed; pet 5–6 mm; mature pedicels stout, 5 mm; frs widely divergent, 4–6 cm; seeds black, 3 mm; $2n=42$. Native of Europe, now found as a weed in gardens and moist woods throughout most of our range. May, June. (*A. alliaria; A. officinalis.*)

39. SISYMBRIUM L. Sep obtuse, ascending; pet small, yellow, obovate to spatulate, gradually narrowed to the claw; glands of the short stamens usually annular; filaments slender; anthers oblong; ovary cylindric; style short, scarcely differentiated; stigma capitate; ovules numerous; frs elongate, linear or subulate, terete or slightly quadrangular, tipped with the minute, persistent style; valves 3-nerved, with conspicuous midnerve and thinner lateral nerves; seeds in one row, oblong, smooth or nearly so; ours annuals or winter-annuals, ± pubescent with simple hairs, at least the lower lvs deeply pinnatifid. 90, widespread.

1 Frs subulate, closely appressed to the rachis; pedicels erect, 2–3 mm .... 1. *S. officinale.*
1 Frs linear, widely spreading; pedicels divergent, 5–20 mm.
  2 Pedicels at maturity nearly or quite as thick as the fr; frs 5–10 cm .... 2. *S. altissimum.*
  2 Pedicels at maturity slender; frs 1–5 cm.
    3 Pet 5–8 mm; sep 3–4 mm; infl elongate, the young frs scarcely or not at all overtopping the fls
    .... 3. *S. loeselii.*
    3 Pet 3–4 mm; sep 2–3 mm; infl very contracted, the young frs distinctly overtopping the fls and
    buds .... 4. *S. irio.*

1. Sisymbrium officinale (L.) Scop. Hedge-mustard. Erect, 3–8 dm; lower lvs petioled, deeply pinnatifid, the segments oblong to ovate or the terminal one rotund, angularly toothed, upper lvs sessile or nearly so and often entire; pet light yellow, 3–4 mm; racemes stiffly erect, the mature pedicels closely appressed, 2–3 mm, thickened above; frs closely appressed, subulate, 8–15 mm, 1–1.5 mm wide at base; $2n=14$. Native of Eurasia, commonly established as a weed throughout most of the U.S. May–Sept.

2. Sisymbrium altissimum L. Tumbling mustard, Jim-Hill m. Erect, to 1 m, commonly simple below and much-branched above, glabrous or sparsely pilose; lvs petioled, pinnatifid, the lower with 5–8 pairs of linear and entire to lanceolate and serrate segments, the upper small, with fewer segments; pet pale yellow, 6–8 mm; pedicels nearly or quite as thick as the fr, ascending, 5–10 mm; frs ascending or spreading, slender, 5–10 cm × 1–1.5 mm; $2n=14$. Native of Eurasia, established as a weed of fields and waste places throughout most of the U.S. and adj. Can., especially northward. June–Aug. (*Norta a.*)

3. Sisymbrium loeselii L. Tall hedge-mustard. Erect, 5–12 dm, the stem generally reflexed-hirsute below; lower lvs lyrate-pinnatifid and dentate, usually hirsute, the lateral segments triangular to ovate, acute, spreading or deflexed, the terminal one larger, triangular; infl elongating progressively, the fls and buds overtopping the frs; sep 3–4 mm; pet lemon-yellow, 5–8 mm; anthers ca 1.5 mm; mature pedicels divergent, 5–20 mm; frs ascending to erect, slender, terete, 1.5–4 cm; $2n=14$. Native of se. Europe and w. Asia, occasionally found in our range, and becoming a weed in w. U.S. June–Aug.

4. Sisymbrium irio L. London-rocket. Erect, to 6 dm, spreading-hairy below; lower lvs deeply pinnatifid with oblong to ovate, entire to dentate or angularly lobed segments, the upper lvs with fewer and smaller lobes; infl very contracted, the young frs elongating rapidly and projecting beyond the fls; pet 3–4 mm; anthers ca 0.7 mm; pedicels slender, divergent, 5–10 mm; frs linear, 2.5–5 cm; $2n=14, 28, 42, 56$. Native of Eurasia, occasionally found in our range, and becoming a weed in the Pacific states. May, June. (*Norta i.*)

40. BRAYA Sternb. & Hoppe. Sep somewhat cucullate; pet white or suffused with purple, oblong-obovate; short stamens subtended by a U-shaped gland; filaments flat; anthers cordate-ovate; ovary cylindric, with 2–several ovules per locule; style very short; stigma capitate; fr linear, often torulose, the valves with prominent midnerve and obscure secondary nerves, the septum composed of thick-walled elongate cells and appearing transversely striolate at 20×; herbs with simple or branched hairs. 20, circumboreal.

1. Braya humilis (C. A. Meyer) B. L. Robinson. Perennial from a stout caudex; stems clustered, 1–2 dm, sparsely hairy; lvs linear-oblanceolate, 1–3 cm; with a few sharp teeth near the tip or entire; ± pubescent with 2–4-pronged hairs; mature racemes 4–10 cm, with strongly ascending pedicels 3–6 mm; pet 3–4 mm; frs 10–25 × 1 mm, hairy at least when young; seeds in 1 row; $2n=28, 40, 42, 56, 64$. Limestone cliffs; Willoughby Mt., Vt.; n. Mich.; widespread in boreal Amer. and Siberia.

41. ARABIDOPSIS Heynh.   Sep oblong, obtuse; pet white, spatulate, gradually narrowed to the base; filaments capillary; anthers ovate; short stamens subtended by a minute semicircular gland; each pair of long stamens separated by a minute gland; ovary cylindric, ovules numerous; style undifferentiated; stigma truncate; fr linear, subterete, the valves with conspicuous midnerve, the cells of the septum vertically elongate, herbs, ± pubescent with branched hairs. 15, N. Hemisphere.

1. **Arabidopsis thaliana** (L.) Heynh.   Mouse-ear cress.   Annual or winter-annual 1–4 dm; lvs chiefly in a basal rosette, oblong to spatulate, 1–5 cm, stellate-hairy; cauline lvs smaller, linear to narrowly oblong; sep pilose; pet 2–3 mm; mature racemes very open, the widely divergent pedicels 5–10 mm; frs divergent or ascending, 1–2 cm × 1 mm; 2n=10. Native of Europe, naturalized in fields and waste places, especially in sandy soil, throughout most of the U.S. Apr.–June. (*Sisymbrium t.*)

42. CAMELINA Crantz.   False flax.   Sep erect, obtuse, the outer slightly saccate at base; pet yellow, spatulate; filaments linear; anthers ovate; short stamens flanked at base by a pair of semicircular glands; ovules 4–12 per cell; style slender, persistent; stigma capitate; fr obovoid or pyriform, firm-walled, slightly flattened parallel to the septum, somewhat keeled along the sutures, narrowed to the short-stipitate base; annuals or winter-annuals with simple and branched hairs; stems branched above; basal lvs narrowly spatulate, often repand; cauline lvs linear to lanceolate, clasping by a sagittate-auriculate base. 6, Eurasia.

1 Herbage rough-hairy, the short stellate hairs exceeded by simple hairs 1–2 mm . . . . 1. *C. microcarpa.*
1 Herbage sparsely hairy or glabrate, the few simple hairs not longer than the stellae . . . . 2. *C. sativa.*

1. **Camelina microcarpa** Andrz.   Small-seed f. f.   Erect, 3–7 dm, rough-hairy; lvs entire or remotely denticulate; frs erect, 2.5–5 mm, obscurely rugulose, on pedicels 4–9 mm; seeds ca 1 mm; 2n=40. Fields and waste places, usually in sandy soil; native of the Old World, established as a casual weed throughout most of the U.S. and s. Can. Apr.–June.
2. **Camelina sativa** (L.) Crantz.   Large-seed f. f.   Much like no. 1; stem and lvs glabrous to sparsely pubescent, the simple hairs not projecting beyond the stellate ones; frs 7–9 mm, inconspicuously veiny, on pedicels 10–21 mm; seeds 1–2 mm; 2n=40. Native of the Old World, established as a casual weed in n. U.S. and adj. Can., less common with us than no. 1.
    C. alyssum (Miller) Thell., a European sp. with lobed or deeply toothed lvs, lax racemes, and depressed-globose frs 6–10 mm, is casually intr. in s. Can. (*C. dentata.*)

43. DESCURAINIA Webb & Berth., nom. conserv.   Sep ovate, obtuse; pet yellow, small, obovate or spatulate; anthers ovate or oblong; staminal glands minute or none; ovary cylindric; style very short, as thick as the ovary; stigma capitate; ovules numerous; frs linear or clavate, terete or slightly quadrangular, tipped with the very short persistent style; valves with a prominent midnerve; seeds elliptic or oblong; annual or biennial herbs; ± pubescent, some or all of the hairs branched; lvs 1–3 pinnate with numerous small segments. 40, mostly temp. New World. (*Sophia.*)

1 Frs clavate, 1–2 mm wide, 3.5–8 times as long as wide; seeds ± in 2 rows; infl usually glandular . . . . 1. *D. pinnata.*
1 Frs linear, 0.5–1.3 mm wide, 6–30 times as long as wide; seeds in 1 row; infl not glandular.
    2 Lvs mostly once-pinnate, with incised lfls; frs 7–12 mm; seeds 4–10 per locule . . . . 2. *D. richardsonii.*
    2 Lvs mostly 2–3-pinnate; frs 15–25 mm; seeds (12–)15–25 per locule . . . . 3. *D. sophia.*

1. **Descurainia pinnata** (Walter) Britton.   Tansy-mustard.   Erect, 2–7 dm; lvs oblong to oblanceolate, the lower the larger, bipinnate, or pinnate with deeply pinnatifid segments, the upper progressively reduced and less dissected; fls 2–4 mm wide; mature racemes elongate, to 3 dm, with widely divergent pedicels 5–20 mm; frs narrowly clavate, 5–13 × 1–2 mm; seeds in 2 rows in each locule, at least near the middle; 2n=14, 28, 42. Dry, open or sparsely wooded places; throughout most of N. Amer. Two vars. in our range. Var. *pinnata,* occurring chiefly on the coastal plain from se. Va. to Tex., has canescent, nonglandular herbage (sometimes glandular in the infl), and spreading or ascending frs 5–10 (avg 7) mm. (*Sisymbrium canescens.*) Var. *brachycarpa* (Richardson) Fern., a chiefly northern plant extending s. to N. Engl., W.Va., Tenn., and Tex., has glandular (sometimes also hairy) herbage (incl. the infl) and ascending to erect frs 6–13 (avg 9) mm. (*Sophia p.; S. incisa; S. intermedia; S. millefolia.*)

2. **Descurainia richardsonii** (Sweet) O. E. Schulz. Gray t.-m. Erect, to 1 m, ± canescent, lvs lanceolate to oblanceolate, pinnate with entire to lobed or deeply pinnatifid segments; fls 3 mm wide; mature racemes dense, the frs overlapping, the pedicels 3–6 mm, narrowly ascending; frs appressed to the rachis, linear, 5–12 × 0.7–1.3 mm; septum without veins; seeds in each locule 4–10, in one row; $2n=14, 28, 42$. Throughout w. N. Amer., from nw. Can. to n. Mex., e. to Minn., Ont., and e. Que. June, July. Our pls, as here described, are var. *richardsonii*. (*Sophia richardsoniana; S. hartwegiana; Sisymbrium incisum.*)

3. **Descurainia sophia** (L.) Webb. Herb-sophia, t.-m. Erect, 3–8 dm, canescent throughout; lvs ovate to obovate or the upper narrower, 2–3 times pinnatifid into linear or narrowly oblanceolate segments; fls 3 mm wide; mature racemes loose, with widely ascending pedicels 8–14 mm; frs narrowly linear, 15–25 × 0.5–1 mm; septum with longitudinal veins; seeds in each locule (12–)15–25, in one row; $2n=28$. Native of Eurasia, established as a weed in fields and waste places throughout most of the U.S. and s. Can. May–July. (*Sophia s.; S. multifida; Sisymbrium sophia.*)

FAMILY RESEDACEAE, the Mignonette Family

Fls hypogynous, perfect, irregular and with an evident gynophore or androgynophore; sep 4–8, sometimes unequal; pet mostly 4–8, valvate, distinct, unequal, the innermost one the largest and usually conspicuously fringed-appendiculate on the upper part of the back; stamens 3–50+; carpels (2)3–6(7), ordinarily united to form a superior, compound, unilocular ovary that is open at the top and has small, sessile, well separated stigmas around the rim, the ovule borne on parietal placentas; fr usually a gaping capsule; seeds mostly reniform, with little or no endosperm; mostly herbs, producing mustard oil; lvs alternate, entire to deeply pinnatifid; fls small, in elongating racemes or spikes. 6/70.

1. **RESEDA** L. Sep and pet each 4–8, commonly 5 or 6, the sep nearly equal, the pet unequal, at least the larger generally consisting of a rounded to oblong, flattened or concave base, bearing on its back one or more projecting appendages (seldom the pet merely fringed-cleft); upper pet the largest, the lateral and lower progressively smaller, with simpler appendages; stamens 10–25, subtended on the upper side by a large, rounded disk; carpels mostly 3 or 4; ours all herbs with dense terminal racemes of whitish to greenish or yellow fls ca 6 mm wide, in summer. 55, Eurasia, n. Afr.

1 Fls greenish-white; carpels 4; lvs rather regularly pinnatifid .... 1. *R. alba.*
1 Fls greenish-yellow; carpels 3; lvs irregularly pinnatifid, with few segments, or entire .... 2. *R. lutea.*

1. **Reseda alba** L. Erect, taprooted annual or perennial to 8 dm, glabrous and somewhat glaucous; lvs pinnately parted into several linear-oblong, evenly distributed, acute segments; fls greenish-white, fragrant; pet 5 or 6, commonly with 3 distinct or basally connate appendages, the central one the smallest; filaments persistent into fr; ovary and fr usually with 4 apical lobes; $2n=20$. Waste places; Me. and O. to Del.; intr. from the Mediterranean region.

2. **Reseda lutea** L. Erect or ascending, diffusely branched, taprooted biennial or perennial to 8 dm, glabrous; lvs irregularly pinnatifid, with a few narrow, entire or cleft lateral segments chiefly near or above the middle; fls greenish-yellow; pet usually 6, with appendages as in no. 1; filaments deciduous; ovary and fr usually with 3 apical lobes; $2n=48$. Waste places, N. Engl. to Io., Mo., and Md.; intr. from Europe.

R. odorata L., Mignonette, with chiefly undivided lvs, and very fragrant fls on pedicels twice as long as the cal, the 6 pet greenish or greenish-yellow, with numerous filiform-clavate appendages, occasionally escapes from cult.

R. luteola L., the Dyer's rocket, or yellow-weed, with entire lvs, short pedicels, and 4 pet that are merely toothed or cleft, is occasionally adventive from Europe.

FAMILY CYRILLACEAE, the Cyrilla Family

Fls perfect, hypogynous, regular or nearly so, mostly 5-merous; sep imbricate, persistent; pet distinct or nearly so; stamens as many as and alternate with the pet, or twice as many; anthers opening by longitudinal slits or apical pores; pollen in monads; intrastaminal nectary-disk present; ovary superior, of 2–5 carpels, with as many locules; style short or nearly suppressed, shortly lobed or entire; ovules 1–3 per locule, pendulous from near the tip; fr indehiscent; seed-coat wanting; embryo slender, straight, with small, slightly expanded cotyledons, embedded in the copious endosperm; glabrous shrubs or small trees with alternate, simple, exstipulate, entire lvs and small fls in racemes. 3/14.

1. **CYRILLA** Garden. Pedicels with a pair of persistent bractlets; pet thickened and glandular at and below the middle; stamens 5, alternate with the pet; filaments terete; anthers opening longitudinally, dorsoversatile, with separate basal lobes; ovary 2–3-locular; ovules 3 per locule; fr sulcate. Monotypic.

1. **Cyrilla racemiflora** L. Leatherwood; ti-ti. Shrub or small tree to 10(–20) m; lvs leathery, persistent through much of the winter, oblanceolate to obovate, 5–10 cm, the petiole 5–8 mm; racemes several, borne just below the leafy branches of the season, spreading, slender, 6–15 cm, floriferous to the base; bracts subulate, equaling or longer than the pedicels; pet white, oblanceolate, 3 mm, slightly surpassing the stamens; fr gray, round-ovoid to depressed-globose, 2–3 mm, excluding the persistent short, 2–3-lobed style; $2n=40$. Swamps and wet woods on the coastal plain; se. Va. to Tex., s. into n. S. Amer. June.

FAMILY **CLETHRACEAE**, the Clethra Family

Fls perfect, hypogynous, regular, mostly 5-merous; sep persistent, connate below, the lobes imbricate; pet imbricate, distinct or nearly so; stamens bicyclic, free or barely adnate to the base of the pet; anthers becoming inverted in ontogeny, deeply sagittate, the free tips of the pollen-sacs appearing apical, each opening by a slit-like distal pore; pollen in monads; ovary superior, trilocular, often nectariferous at the base; ovules numerous on axile placentas; fr a loculicidal capsule; seeds numerous, often winged, with copious endosperm and a short, dicotyledonous embryo; seed-coat a single layer of cells; shrubs or small trees, usually stellate-hairy, with simple, exstipulate lvs and small, white (or pinkish) fls in racemes or panicles. 1/65.

1. **CLETHRA** L. White alder; sweet pepper-bush. Characters of the family.

1 Filaments glabrous; lvs obovate-oblong, obtuse or subacute .... 1. *C. alnifolia.*
1 Filaments hirsute; lvs oblong or elliptic, acuminate .... 2. *C. acuminata.*

1. **Clethra alnifolia** L. Coast w. a. Shrub to 3 m; lvs obovate-oblong, 5–10 cm, obtuse or subacute, finely and sharply serrate distally, mostly entire below the middle, tapering to a petiole 0.5–1(–2) cm, racemes erect, 5–15(–20) cm, densely short-hairy as also the pedicels (2–5 mm) and cal; pet oblong-obovate, 5–7 mm, glabrous; filaments glabrous; style strigillose; frs ± ascending, pubescent, 3 mm thick; $2n=32$. Swamps and moist woods, mostly near the coast; N.S. and Me. to Fla. and Tex. July, Aug.

2. **Clethra acuminata** Michx. Mountain w. a. Shrub to 6 m; lvs oblong to elliptic, mostly 10–20 cm, half as wide, finely serrate over most of their length, acuminate, the base acute or rounded, sparsely villosulous beneath, on villous petioles 1.5–4.5 cm; racemes 8–15 cm, tomentose, as also the pedicels and cal; pet 6–8 mm, pilose internally at base; filaments hirsute; style glabrous, frs ± pendulous, densely villous, 3–4 mm thick. Rich woods in the mts.; sw. Pa. to e. Ky. and w. Va., s. to Tenn. and ne. Ga. July, Aug.

FAMILY **EMPETRACEAE**, the Crowberry Family

Fls small, regular, hypogynous, perfect or often unisexual; perianth of 3–6 tepals, these ± evidently biseriate and differentiated into 2 or 3 imbricate sep and as many imbricate pet, or 3–4 and all about alike in a single series; stamens (2)3(4), alternate with the pet when the pet are differentiated from the sep; anthers becoming inverted in ontogeny, opening by longitudinal slits; pollen in tetrads; no nectary-disk; ovary superior, 2–9-locular, with a short, cleft style; ovules 1 per locule, basal-axile; fr a juicy or dry drupe with 2–9 stones; seeds with thin coat and straight, elongate, dicotyledonous embryo embedded in copious endosperm; small, evergreen, ericoid shrubs or shrublets; lvs small, alternate or subverticillate, simple, with a basal pulvinus, revolute, with a deep furrow beneath. 3/5.

1 Fls axillary; stigmas 6–9 .... 1. *Empetrum.*
1 Fls in small terminal heads; stigmas 2–5, commonly 3 .... 2. *Corema.*

1. EMPETRUM L.   Crowberry.   Fls unisexual or (ours) often perfect; sep, pet, and stamens each 4; filaments elongate; ovary depressed-globose, 6–9-locular; style very short, with 6–9 expanded, spreading, toothed or lobed stigmas; fr a drupe with 6–9 stones, the style persistent; dwarf, much-branched shrubs with greenish or purplish fls subsessile in the axils; x=13. 2, mainly cool N. Hemisphere.

1 Youngest parts of the stems minutely stipitate-glandular; fr black .... 1. *E. nigrum.*
1 Youngest parts of the stems tomentose or villous, not glandular; fr red or purple .... 1. *E. rubrum.*

1. Empetrum nigrum L.   C. Stem bushy-branched, diffusely spreading, to 4 dm long, minutely stipitate-glandular when young; lvs numerous, crowded, on the older parts of the stem widely spreading or reflexed, narrowly elliptic or linear-oblong, 4–8 mm; fr 6–9 mm, black, often glaucous. On rocks; circumboreal, s. to the mts. of N. Engl. and N.Y. (and to e. L.I.), n. Mich., and n. Minn. Summer. Typical *E. nigrum* is a European diploid with unisexual fls, the plants dioecious. Our plants belong to the circumboral tetraploid var. *hermaphroditum* (Hagerup) T. Soerensen, with perfect or sometimes partly unisexual fls.

2. Empetrum rubrum L.   Red c. Much like no. 1; stems villous to tomentose when young, not glandular; fr red or purple. The typical form is antarctic and mostly dioecious. N. Amer. plants are subsp. *eamesii* (Fern. & Wieg.) R. Good, tetraploid and with mostly perfect (or polygamous) fls. Our ssp. has two vars.: Var. *eamesii* (Fern. & Wieg.) Cronq. has bright red frs and short (2.5–4 mm), broad, crowded, erect or ascending lvs; it occurs from N.S. to Nf. (*E. eamesii.*) Var. *atropurpureum* (Fern. & Wieg.) R. Good has dark red or purplish frs and longer (4–6 mm), linear, ascending or soon divergent, scarcely crowded lvs; it occurs from e. Que. to the higher mts. of Me. and N.H., Vt., and N.Y., and in ne. Minn. and on Isle Royale, Mich.

2. COREMA D. Don.   Broom-crowberry. Dioecious; sep 3–4; pet none; stamens 2; ovary 2–5-locular; style slender, with divergent, subulate stigmas, deciduous; drupe globose, dry, usually 3-seeded; bushy shrubs with numerous crowded, alternate or often subverticillate lvs and small, sessile, terminal heads of purplish fls. 2, the other in Iberia and the Azores.

1. Corema conradii (Torr.) Torr.   Much-branched, to 5 dm tall and 2 m wide; lvs linear, 3–6 mm; heads few fld, each fl subtended by 1 or more sepal-like bracts; filaments purple, 5 mm; anthers brown-purple; drupe 1.5 mm. Usually in sandy soil, mostly near the coast, but also in Ulster Co., N.Y.; Nf. to N.J. Apr.–May. Best developed northward.

FAMILY **ERICACEAE**, the Heath Family

Fls regular or somewhat irregular, hypogynous to sometimes epigynous, mostly perfect and 4–5-merous; cor gamopetalous or much less often polypetalous, the lobes convolute or imbricate; stamens free from the cor or attached to its very base, as many or more often twice as many as the pet, rarely intermediate in number; anthers becoming inverted during ontogeny, so that the morphological base is apical, opening by pores at the apparent apex, or by more or less elongate slits extending downward from the apex, or rarely opening longitudinally their whole length, the apparent tip often drawn out into a pair of slender (sometimes connate) tubules (awns), and the body sometimes with 2 or 4 evident, slender appendages (spurs); pollen generally in tetrahedral tetrads; ovary superior to sometimes inferior, mostly 4–5-locular (10-locellar in *Gaylussacia*); placentation axile; ovules usually several or numerous in each locule (1 per locule in *Arctostaphylos,* 1 per locellus in *Gaylussacia*); fr capsular, often with a persistent axis, or sometimes a berry or drupe; seeds (5) ± numerous, sometimes winged, with copious endosperm and a straight (often very short) dicotyledonous embryo; seed-coat a single layer of cells; shrubs, small trees, or lianas, seldom almost herbaceous, with simple, exstipulate lvs (these often small and firm, with revolute margins, and then said to be ericoid, but also often quite ordinary) and small (often urceolate) to large fls typically in bracteate racemes or corymbs, but sometimes solitary and terminal or axillary. Genera 1–7 belong to the tribe Rhododendreae, 8–18 to the Andromedeae, 19 to the Arbuteae, 20 and 21 to the Ericeae, and 22 and 23 to the Vaccinieae. 125/3500.

1 Ovary superior; fr except in 3 genera (nos. 17–19) capsular and dehiscent.
  2 Pet all separate.
    3 Fruit 5-locular; anthers with terminal pores; lvs woolly beneath .... 1. *Ledum.*
    3 Fruit 2–3-locular; anthers opening longitudinally; lvs glabrous .... 2. *Leiophyllum.*
  2 Pet united at least toward the base (rarely 3 well joined and 2 distinct).
    4 Fls 4-merous; fr 4-locular.
      5 Lvs alternate; native plants.
        6 Deciduous shrub 1–2 m; fr a thick-walled septicidal capsule .... 3. *Menziesia.*
        6 Prostrate, evergreen shrub; fr berry-like ....17. *Gaultheria.*
      5 Lvs opposite or whorled; introduced plants.
        7 Lvs opposite, 2–4 mm .... 20. *Calluna.*
        7 Lvs whorled, ours 3–10 mm .... 21. *Erica.*
    4 Fls 5-merous; fr mostly 5-locular (2–3-locular in *Loiseleuria*).
      8 Lvs 2–10 mm; dwarf, arctic-alpine shrubs with fls ca 1 cm or less.
        9 Anthers spurless; fr septicidal; lvs 4–10 mm, loose or spreading.
          10 Cor campanulate; anthers opening longitudinally; lvs opposite .... 6. *Loiseleuria.*
          10 Cor constricted above; anthers with terminal pores; lvs alternate .... 7. *Phyllodoce.*
        9 Anthers spurred; fr loculicidal; lvs 2–4 mm, erect, imbricate in rows .... 8. *Harrimanella.*
      8 Lvs mostly over 10 mm; plants not dwarf arctic-alpine shrubs, except one sp. of *Rhododendron*
        with the fls well over 1 cm.
        11 Cor salverform to campanulate or rotate, not constricted above.
          12 Prostrate shrubs; cal closely subtended by 2 large bracts; fr berry-like .... 18. *Epigaea.*
          12 Erect or ascending shrubs; cal not subtended by bracts; fr capsular.
            13 Cor campanulate to funnelform.
              14 Cor 1 cm, with short lobes wider than long; fr loculicidal .... 16. *Zenobia.*
              14 Cor well over 1 cm, with well developed lobes; fr septicidal .... 4. *Rhododendron.*
            13 Cor rotate or saucer-shaped .... 5. *Kalmia.*
        11 Cor tubular to ovoid, urceolate, or globose, ± constricted at the mouth.
          15 Pedicels without bracteoles, or with bracteoles at the very base.
            16 Fls in short terminal infls; anthers spurred.
              17 Fr a drupe with 5 pyrenes; lvs flat, glabrous .... 19. *Arctostaphylos.*
              17 Fr capsular; lvs revolute, white-hairy beneath .... 9. *Andromeda.*
            16 Fls in lateral infls; anthers with or without spurs.
              18 Fls in umbelliform lateral clusters that may be borne in the axils of the lvs
                or may be distributed along lfless branches to form a compound, race-
                miform to paniculiform infl; anthers spurred .... 10. *Lyonia.*
              18 Fls in lateral racemes; anthers spurless .... 11. *Leucothoe.*
          15 Pedicels with 2 bracteoles near or above the middle or at the tip.
            19 Fls in lateral or terminal racemes.
              20 Pedicels bracteolate at the summit; lvs deciduous; anthers awned .... 12. *Eubotrys.*
              20 Pedicels bracteolate near the middle; anthers not awned.
                21 Lvs evergreen; anthers spurred; capsule globose .... 13. *Pieris.*
                21 Lvs deciduous; anthers without spurs; capsule oblong-ovoid .... 14. *Oxydendrum.*
            19 Fls solitary in the axils of foliage lvs.
              22 Lower lf-surface and cal lepidote; fr a loculicidal capsule .... 15. *Chamaedaphne.*
              22 Lower lf-surface and cal glabrous; fr berry-like ....17. *Gaultheria.*
1 Ovary largely or wholly inferior; fr indehiscent, ± fleshy.
  23 Erect shrubs; fls 4–5-merous.
    24 Ovary 4–5-locular; ovules and seeds ± numerous .... 22. *Vaccinium.*
    24 Ovary 10-locellar; ovules and seeds 10 .... 23. *Gaylussacia.*
  23 Trailing shrubs; fls 4-merous.
    25 Lvs glabrous; ovary wholly inferior; cal not becoming fleshy; berry not with a wintergreen
      flavor .... 22. *Vaccinium.*
    25 Lvs beset with short setae; ovary only partly inferior; cal becoming fleshy; fr with a wintergreen
      flavor ....17. *Gaultheria.*

1. LEDUM L.   Fls 5-merous; cal-lobes very short; pet 5, separate, spreading; stamens 5–10; filaments slender; anthers short, dorsifixed, with 2 terminal pores; ovary puberulent, 5-locular; style elongate; capsule septicidal; erect or diffuse, freely branched shrubs with thick, evergreen, alternate, sessile or subsessile lvs and long-pedicellate fls in a crowded (sometimes umbelliform) terminal corymb. 5, boreal.

  1. Ledum groenlandicum Oeder.   Labrador-tea.   Shrub to 1 m with densely villous twigs; lvs lanceolate to narrowly elliptic, 2–5 cm, revolute, dark green above and sparsely villous when young, densely villous-tomentulose beneath; pedicels puberulent, 1–2 cm; fls white, 1 cm wide; stamens 5–7; fr slenderly ovoid, 5–

7 mm, the style persistent; $2n=26$. Bogs and wet shores, or sometimes on rocky alpine slopes; Greenl. to Alas., s. to Mass., n. N.J., Pa., Mich., Wis., and Minn. June, July.

2. **LEIOPHYLLUM** Hedwig f.   Sand-myrtle.   Fls 5-merous; cal deeply divided; cor subrotate, of 5 separate pet; stamens mostly 10, the slender filaments longer than the pet, the minute, subglobose anthers opening longitudinally; ovary 2–3(–5)-locular; style about equaling the filaments; stigma not expanded; capsule ovoid, septicidal, subtended by the persistent cal; low, freely branched shrubs with small, leathery, persistent lvs and numerous white or pinkish fls from the axils of the crowded upper lvs, forming an umbelliform cluster. Monotypic.

1. **Leiophyllum buxifolium** (Bergius) Elliot.   Widely branched shrub 1–10 dm; lvs alternate to opposite, ovate to oblong, 6–12 mm; pedicels 5–10 mm, glabrous or stipitate-glandular; pet 3–4 mm; anthers purple; fr 3–4 mm; $2n=24$. Dry, sandy pine-barrens, and on outcrops of acid rocks in the mts.; coastal plain of N.J., N.C., and S.C., and at a few way-stations inland to the higher mts. (mainly Smoky and Blue Ridge) of N.C., S.C., Ga., Tenn., and Ky. (*Dendrium b.; D. hugeri.*)

3. **MENZIESIA** J. E. Smith.   Fls 4-merous; cal saucer-shaped, shallowly lobed; cor urceolate-campanulate, shallowly lobed; stamens 8, included; anthers linear, opening by 2 short, pore-like terminal slits; ovary 4-locular; style included; capsule short, thick-walled, septicidal; shrubs with alternate, deciduous, short-petioled lvs and umbelliform or cor-ymbiform clusters of fls from terminal buds of the previous year. 7, N. Temp.

1. **Menziesia pilosa** (Michx.) Juss.   Minnie-bush.   Shrub 1–2 m, with densely hairy twigs; petioles 2–4 mm; lvs thin, elliptic to oblong-obovate, 1.5–4 cm, with a yellowish callous point, pilose above, finely hairy beneath and with a few narrow lacerate or cleft scales on the midvein; fls 2–several, on stipitate-glandular pedicels 1–3 cm; sep glandular-ciliate; cor greenish-red to ochroleucous, 8–10 mm, its erect lobes 1–2 mm; fr minutely glandular, 6 mm. Mt. woods; Pa. to Ga. May, June.

4. **RHODODENDRON** L.   Cal 5-lobed, much shorter than the cor; cor campanulate to funnelform, conspicuous, regularly or commonly irregularly 5-lobed; stamens mostly 5 or 10, with elongate filaments and short anthers with 2 terminal pores; ovary superior, 5-locular; style elongate; capsule septicidal; shrubs or trees with alternate, deciduous or evergreen lvs and showy fls in terminal racemes or umbel-like clusters from scaly buds. Our spp. $2n=26$, except as noted. (*Azalea.*) 850, mainly N. Hemisphere.

1 Lvs thick, leathery, evergreen, strictly entire (*Rhododendron*).
  2 Lvs 5–20 cm; large shrubs or small trees.
    3 Cal-lobes mostly 2–4 mm; pedicels and ovary stipitate-glandular .... 1. *R. maximum.*
    3 Cal-lobes mostly under 1 mm; pedicels and ovary hirsutulous .... 2. *R. catawbiense.*
  2 Lvs 1–1.5 cm; shrub to 3 dm .... 3. *R. lapponicum.*
1 Lvs thin, deciduous, ciliate.
  4 Cor regular or somewhat irregular, funnelform, not deeply divided; stamens 5 (*Azalea*).
    5 Fls anthocyanic, ranging from white to pink or pale purple.
      6 Fls appearing in the spring with or before the lvs.
        7 Cor-lobes about as long as the tube; widespread spp.
          8 Ovary hirsute, not glandular; lvs mostly glabrous beneath except for appressed setae
            along the midvein; fl-buds glabrous .... 4. *R. periclymenoides.*
          8 Ovary stipitate-glandular; lvs permanently soft-hairy beneath; fl-buds pubescent ..... 7. *R. prinophyllum.*
        7 Cor-lobes conspicuously shorter than the tube; spp. of the coastal plain.
          9 Blooming plants over 5 dm, usually not colonial; bud-scales canescent .... 6. *R. canescens.*
          9 Blooming plants early to 5 dm, strongly colonial; buds mostly glabrous .... 7. *R. atlanticum.*
      6 Fls appearing in later spring or early summer when the lvs are full-grown.
        10 Sep under 1 mm; exsert part of the filaments (beyond the cor-tube) equaling or shorter
          than the tube; pedicels puberulent and glandular.
          11 Cor-tube less than twice as long as the lobes; Me. to O. and Fla. .... 8. *R. viscosum.*
          11 Cor-tube at least twice as long as the lobes; se. Va. and s. .... 9. *R. serrulatum.*
        10 Sep 1.6–3(–5) mm; exsert part of the filaments longer than the cor-tube; pedicels glandular,
          not puberulent .... 10. *R. arborescens.*
     5 Fls xanthic, ranging from yellow to orange, vermilion, or scarlet.
      12 Fls appearing with the lvs in May .... 11. *R. calendulaceum.*
      12 Fls appearing in late June or July when the lvs are well grown .... 12. *R. bakeri.*
  4 Cor irregular, divided nearly to the base, the upper segment shallowly 3-lobed, the lower 2 segments
    linear or oblong; stamens 10 (*Rhodora*) .... 13. *R. canadense.*

1. **Rhododendron maximum** L.   White laurel; rose-bay. Shrub or small tree to 10 m; lvs leathery, evergreen, oblong-obovate, 1–2 dm, abruptly narrowed to an acute tip, more gradually narrowed and ± cuneate at base, glabrous above, obscurely scurfy-tomentose beneath; fls numerous in an umbel-like infl, rose to white, spotted with yellow or orange within, 3.5–4 cm wide; pedicels and ovary stipitate-glandular; sep oblong to ovate or subrotund, 2–4 mm; fr 10–15 mm. Moist or wet woods, often in dense colonies; N.S. to s. Ont. and O., s. especially in the mts. to Ga. and Ala. June, July.

2. **Rhododendron catawbiense** Michx.   Red laurel; rose-bay. Shrub or small tree to 6 m; lvs leathery, evergreen, oval or elliptic, 5–15 cm, broadest near the middle, rounded to subacute at both ends, glabrous; fls lilac-purple, somewhat larger than in no. 1; pedicels and ovary hirsutulous, not glandular; sep broadly triangular or semicircular, under 1 mm; fr 20–25 mm. Mt. woods; Va. to Ky., s. to Ga. and Ala. May, June. *R. minus* Michx., a mostly more southern sp., has been reported (perhaps as in introduction) from Bell Co., Ky. It has short cal-lobes as in no. 2, but the pedicels, ovary, and lower surface of the lvs are lepidote with small, sessile, glandular, pale to reddish-brown scales.

3. **Rhododendron lapponicum** (L.) Wahlenb.   Lappland rose-bay. Freely branched low shrub 1–3 dm, sometimes depressed and forming loose mats; lvs leathery, evergreen, elliptic, 10–15 mm, densely lepidote, especially beneath; fls few in a terminal, umbel-like cluster, bright purple, campanulate, 1.5 cm wide; stamens 5–10; fr 5 mm; $2n=26$, 52. Circumboreal, s. to the higher mts. of N. Engl. and N.Y., and rarely to c. Wis. June.

4. **Rhododendron periclymenoides** (Michx.) Shinn.   Pinkster-fl. Colonial shrub to 2 m, branched above; winter bud scales glabrous; lvs oblong to narrowly obovate, not fully expanded at anthesis, later 5–10 cm, essentially glabrous except for the marginal cilia and a row of appressed setae along the midrib beneath; fls fragrant to nearly inodorous; pedicels and cor-tube hirsute or rarely glandular; sep mostly under 1 mm, shorter than their cilia; cor pink to white, 4–5 cm, the lobes tending to be arcuate-reflexed and to have loosely revolute margins, so that they look narrow and the limb is open, in contrast to no. 5, with flatter lobes and a fuller limb; ovary densely strigose, fr sparingly so. Moist or dry woods and bogs; Vt. and Mass. to c. N.Y., s. to S.C. and Tenn.; s. Ill. Late Apr., May. (*R. nudiflorum,* an illegitimate name; *Azalea n.*)

5. **Rhododendron prinophyllum** (Small) Millais.   Rosebud-azalea. Much like no. 4, sometimes to 5 m, usually not colonial; winter bud scales pubescent; lvs softly and permanently hairy beneath, with or without setae on the midvein; fls very fragrant; pedicels, cor-tube, and ovary stipitate-glandular; cor bright pink to sometimes nearly white, flat-faced, 4–5 cm wide, the tube about as long as the lobes; stamens ca twice as long as the cor-tube; fr sparsely stipitate-glandular. Moist or dry woods, especially in the mts.; Me. and s. Que. to Va. and Ky.; s. Ill., s. Mo., Ark., and e. Okla. May. (*R. roseum; R. canescens,* misapplied; *Azalea p.; A. rosea,* an illegitimate name.)

6. **Rhododendron canescens** (Michx.) Sweet.   Piedmont-azalea. Tall shrub, sometimes to 5 m, usually not colonial; lvs felty-pubescent beneath; pedicels and cor-tube densely and finely pubescent as well as stipitate-glandular; cor white to more often pink, the tube longer than the lobes, rather abruptly expanded to the limb; stamens ca 3 times as long as the cor-tube; ovary densely strigose, fr sparsely so, not glandular. Moist woods; chiefly on the piedmont and coastal plain; Del. to Fla. and Tex. Apr., early May.

7. **Rhododendron atlanticum** (Ashe) Rehder.   Coastal azalea. Vigorously colonial from stout rhizomes, sending up short-lived flowering branches mostly 3–5 dm; winter-buds glabrous; mature lvs narrowly obovate, 3–5 cm, glabrous or minutely stipitate-glandular, often glaucous beneath, but not setose; fls in long-stalked clusters, with or before the lvs; pedicels and cor-tube glandular; cal-lobes 1.5–3 mm; cor white or flushed with red, 4 cm wide. Sandy soil on the coastal plain; s. N.J. to Tex. Apr., May. (*Azalea a.*)

8. **Rhododendron viscosum** (L.) Torr.   Swamp-azalea. Shrub 1–2 mm, often colonial, much branched and with ± setose-pubescent young twigs; lvs 3–6 cm, setose-hairy, not glandular, oblanceolate to obovate, acute to rounded and mucronulate; bud-scales broadly rounded; fls spicy-fragrant; pedicels and cor-tube stipitate-glandular and minutely hairy; cor white (pink), 2–4 cm wide, the tube 2–3 cm, the lobes 1.5–2 cm; style projecting 1–5(–10) mm beyond the anthers at anthesis, usually puberulent for 1 cm or more at the base. Wet woods and swamps; Me. to O. and Fla. June, July. (*Azalea v.*)

9. **Rhododendron serrulatum** (Small) Millais.   Hammock sweet azalea. Tall shrub or small tree to 7 m; setae of the young twigs often more appressed than in no. 8, or the twigs merely puberulent, lvs 3–6 cm, narrowly obovate, acute or obtuse; bud-scales, especially the inner, acute and aristate; pedicels and cor-tube stipitate-glandular and minutely hairy; cor white, 2.5–3 cm wide, the tube 2.5–3 cm, at least twice as long as the lobes; style at anthesis projecting 1 cm beyond the anthers, usually glabrous, or puberulent only at the very base. Wet woods on the coastal plain; se. Va. to Fla. and La. June, July. (*Azalea s.*) Perhaps better treated as *R. viscosum* var. *serrulatum* (Small) Ahles.

10. **Rhododendron arborescens** (Pursh) Torr.   Smooth or sweet azalea. Tall shrub or small tree to 6 m, not colonial, with glabrous twigs and bud-scales; lvs oblanceolate or rarely obovate, 5–8 cm, obscurely hairy on the midvein above and beneath, otherwise glabrous; fls fragrant; pedicels and cor-tube stipitate-glandular, not otherwise hairy; sep 1.5–3(–5) mm; cor white, occasionally flushed with red, often with a yellow blotch on the upper lobe, 4–5 cm wide, the tube 2.5–3 cm; exsert pair of filaments somewhat longer than the tube; stamens and style red, the style ca 1 cm longer than the stamens. Upland woods, especially in the mts.; Pa. to Ky., s. to Ga. and Ala. June, July. (*Azalea a.*)

11. **Rhododendron calendulaceum** (Michx.) Torr.   Flame-azalea. Much branched upright shrub to 3 m, not colonial; lvs oblong to obovate or oblanceolate, not yet full-grown at anthesis, later 5–10 cm, half as

wide; winter-buds glabrous; fls in clusters of 5–15, scarcely fragrant, yellow to orange and red, 4–5 cm wide, the tube usually glandular-setose; upper lobe of the cor only slightly wider than the others; sep oblong-ovate, glandular-ciliate, 2–3 mm; $2n=52$. Mt. woods; Pa. and s. O. to Ga. and Ala. May. (*Azalea c.*) South of our range enters into a polyploid complex with no. 12.

12. Rhododendron bakeri (Lemon & McKay) Hume.   Cumberland azalea.   Branching shrub to 3 m, ± colonial; lvs narrowly obovate, 3–6 cm, less than half as wide, full-grown at anthesis; fls several in a short raceme, orange-red to red, 4 cm wide, the upper lobe of the cor much wider than the others and marked with a large orange spot; sep round-ovate, glandular-ciliate, scarcely 1 mm; $2n=26$. Mt. woods; Cumberland Plateau region from e. Ky. and w. Va. to e. Tenn., n. Ga., and n. Ala. Late June, July. (*R. cumberlandense.*)

13. Rhododendron canadense (L.) Torr.   Rhodora.   Branching shrub to 1 m; lvs deciduous, narrowly oblong to elliptic, 2–5 cm, ciliate, finely hairy, especially beneath; fls appearing with or before the lvs, few and subsessile in a terminal umbel-like cluster; cor rose-purple to white, 2 cm, divided almost to the base into 3 segments, the upper segment erect, shallowly 3-lobed, the 2 lower narrow, spreading; $2n=52$. Bogs and wet woods; Nf. and Que. to Ont., s. to e. Pa. and n. N.J. (*Rhodora c.*)

5. KALMIA L.   Laurel. Fls 5-merous; cal deeply lobed; cor subrotate, shallowly lobed, 10-saccate about halfway from base to margin; stamens 10, the short anthers at first fitting into the cor-sacs, the filaments later springing inward; anthers opening by 2 large apical slits; ovary 5-locular; style elongate; stigma truncate; capsule ovoid to depressed-globose, indented at the top, septicidal; evergreen shrubs with conspicuous fls. 6, N. Amer.

<pre>
1 Fl-clusters lateral .... 1. K. angustifolia.
1 Fl-clusters terminal.
  2 Lvs opposite, subsessile, 1–4 cm, whitened beneath with fine hairs .... 2. K. polifolia.
  2 Lvs mostly alternate, petioled, 5–10 cm, glabrous .... 3. K. latifolia.
</pre>

1. Kalmia angustifolia L.   Sheep-l.   Branching shrub to 1 m; lvs firm, most of them ternate or opposite, short-petioled, ellipitic or oblong, 3–5 cm, a fourth or half as wide, dull or somewhat bluish-green, glabrous or obscurely hairy above, paler and glabrous or with scattered stipitate glands beneath; fls several or many in short-peduncled, umbelliform or corymbiform racemes from the axils of last year's lvs, on pedicels to 2 cm, these and the bracteoles and cal stipitate-glandular as well as finely puberulent; cor 6–12 mm wide, reddish-purple to deep pink; fr depressed-globose, 3–5 mm thick; $2n=24$. Mostly in acid soils; Nf. and Lab. to Va., and inland to Mich. and s. Ont. May–July.

The var. *angustifolia*, as principally described above, gives way at the s. to var. *carolina* (Small) Fern., Carolina sheep-laurel, occurring at the margins of bogs and swamps on the coastal plain from se. Va. to S.C. and in the mts. from sw. Va. to Tenn. and N. Ga. Var. *carolina* is sometimes taller, to 1.5 m; lvs softly and permanently velvety-canescent beneath and with scattered stipitate glands; cal and bracteoles finely velvety-canescent, not glandular. Apr., May. (*K. c.*)

2. Kalmia polifolia Wangenh.   Swamp-l.   Sparingly branched shrub to 1 m, with sharply 2-edged twigs; lvs opposite, subsessile, leathery, linear to lanceolate, 1–4 cm, revolute, whitened beneath with very fine hairs; fls from the upper axils, each subtended by a short folded bract, on pedicels 1–3 cm, forming a terminal corymbiform cluster; sep broadly ovate, imbricate, 3 mm; cor rose-purple, 10–16 mm wide; fr subglobose, 6–7 mm; $2n=48$. Bogs; Lab. to Alas., s. to n. N.J., Mich., Minn., and Calif. May, June.

3. Kalmia latifolia L.   Mt.-l.   Shrub or small tree 2–3(–10) m, often forming dense thickets; petioles 1–2 cm; lvs all or mostly alternate, leathery, elliptic or lanceolate, 5–10 cm, usually acute at both ends, rich green, glabrous; fls numerous in terminal corymbs, on pedicels 1–4 cm; sep lance-oblong, 3 mm, not imbricate; cor white to rose with purple markings, 2–2.5 cm wide; fr depressed-globose, 6–8 mm wide; $2n=24$. Woods, chiefly in sandy or rocky, acid soils; se. Me. to Ga., w. to s. O., s. Ind., Miss., w. Fla., and se. La.

6. LOISELEURIA Desvaux, nom. conserv.   Alpine azalea. Fls 5-merous; cal deeply lobed; cor campanulate, lobed to about the middle; stamens 5, included; anthers subglobose, opening longitudinally; pistil as long as the stamens; stigma capitate; ovary 2–3-locular; capsule ovoid, subtended by the persistent cal, septicidal, each of the 2–3 valves eventually 2-cleft. Monotypic.

1. Loiseleuria procumbens (L.) Desvaux.   Woody, diffusely branched, depressed or decumbent, forming bushy clusters rarely over 1 dm; petioles 1–2 mm; lvs opposite, leathery, evergreen, narrowly elliptic, 5–8 mm, tomentose beneath between the broad midvein and the strongly revolute margins; fls short-pediceled from the axils of crowded upper lvs, forming a small umbelliform cluster; cor pink or white, 3–4 mm; fr rounded-ovoid, 4 mm; $2n=24$. Circumboreal, s. to the higher mt. summits of Me., N.H., and N.Y. June–Aug. (*Chamaecistus p.*)

7. **PHYLLODOCE** Salisb. Fls mostly 5-merous; cal deeply lobed, persistent; cor ovoid and constricted at the throat to campanulate, shallowly to deeply lobed; stamens twice as many as the cor-lobes; anthers narrowly oblong, shortly 2-lobed above, each lobe with a terminal pore; ovary mostly 5-locular; style stout, about equaling the stamens; stigma capitate; capsule septicidal; low or depressed shrubs with small evergreen lvs and one to few fls on erect pedicels from the axils of the upper lvs, forming a small umbelliform cluster. 7, N. Temp. and Arctic.

1. **Phyllodoce caerulea** (L.) Bab. Mountain-heath. Diffusely branched from a woody root, 5–15 cm; lvs alternate, crowded in many rows, narrowly oblong, 4–10 mm; minutely serrulate; petiole 1 mm, arising from a swollen sterigma; pedicels 1–2 cm, becoming 3–4 cm, minutely stipitate-glandular; sep lanceolate, 4 mm; cor ovoid, constricted at the throat, purple, 1 cm, the rounded lobes 1 mm; fr 4 mm, minutely glandular; $2n=24$. Circumboreal, s. to alpine summits in Me. and N.H. July, Aug.

8. **HARRIMANELLA** Coville. Fls 5-merous; cal deeply 5-parted, persistent; cor campanulate, lobed to the middle or beyond; style ovoid to conic, constricted at base; stigma minute, truncate; filaments short and flat; anthers approximate to the pistil at the basal constriction of the style; anthers subglobose, opening by 2 terminal pores, bearing 2 very slender, spreading or deflexed, dorsal spurs; capsule subglobose, loculicidal; dwarf, matted shrubs of moss-like aspect, with numerous crowded, linear, flat, evergreen lvs erect and loosely imbricate in several rows, the fls solitary, nodding, terminal, white or pink. 2, circumboreal.

1. **Harrimanella hypnoides** (L.) Coville. Stems 2–20 cm; lvs 2–4 mm; pedicels erect, 5–20 mm; cor nearly 1 cm; $2n=32, 48$. Circumboreal, s. to alpine summits of Me., and (?) N.Y., and to the n. shore of Lake Superior. July, Aug. (*Cassiope h.*)

9. **ANDROMEDA** L. Bog-rosemary. Fls 5-merous; cal saucer-shaped, deeply lobed; cor urceolate, the broad, short lobes spreading or recurved; stamens half as long as the cor-tube; filaments narrowly triangular, pubescent; anthers ovate, opening by 2 terminal pores, each pollen-sac tipped with a slender, outwardly divergent spur; ovary 5-locular, depressed-globose; style columnar, about equaling the cor-tube; stigma subcapitate; capsule loculicidal, globose or nearly so; low shrubs with alternate, evergreen, narrow lvs conspicuously whitened beneath when young, the fls small, white or pink, in terminal, umbelliform corymbs. 2, circumboreal.

1. **Andromeda glaucophylla** Link. Shrub to 5 dm; lvs leathery, linear to narrowly elliptic, 2–5 cm, entire, revolute (often completely so), whitened beneath when young by a dense indument of fine erect hairs; fls often pink, on recurved pedicels seldom over 8 mm; cor 5–6 mm thick; $2n=48$. Acid bogs; Nf. and Lab. to Sask., s. to N.J., W.Va., n. Ind., and Minn. May, June.

10. **LYONIA** Nutt., nom. conserv. Fls 5-merous; cal saucer-shaped to campanulate, deeply divided, the lobes valvate; cor globose to tubular or ovoid, the lobes short; stamens well included; filaments basally dilated, geniculate above; anthers ovate, dorsifixed, spurred but awnless, opening by 2 terminal pores; style columnar to fusiform; stigma truncate; capsule globose to truncate-ovoid, loculicidal, the sutures thickened and appearing as 5 rounded ribs; deciduous or evergreen shrubs with alternate, entire or minutely serrulate lvs; fls white to rose, on long pedicels in umbelliform lateral clusters that may be borne in the axils of the lvs or may be distributed along lfless branches to form a compound, racemiform to paniculiform infl. 35, N. Hemisphere.

1 Cor globose-ovoid, 3–5 mm; sep 1–1.5 mm .... 1. *L. ligustrina.*
1 Cor tubular, 7–13 mm; sep 3–10 mm.
  2 Lvs deciduous; umbelliform clusters of fls not subtended by lvs .... 2. *L. mariana.*
  2 Lvs evergreen, bearing the umbelliform clusters of fls in their axils .... 3. *L. lucida.*

1. **Lyonia ligustrina** (L.) DC. Male-berry. Deciduous shrub to 4 m; lvs thin, mostly obovate or oblanceolate, 3–7 cm, acute or abruptly acuminate, very minutely serrulate; petioles 2–4 mm; umbelliform clusters

mostly 2–8-fld, very numerous, forming a panicle at the end of the branches of the previous year; sep broadly triangular, 1–1.5 mm; cor globose or broadly ovoid, 3–5 mm; stamens with a pair of short spurs near the junction of the anther and filament; fr subglobose, 4 mm thick. Swamps and wet soil; Me. to N.Y., O., Ky. and Ark., s. to Fla. and Tex. June, July. (*Xolisma l.; Arsenococcus l.*) Foliage, bracts, and pubescence variable.

   2. **Lyonia mariana** (L.) D. Don.   Staggerbush. Deciduous shrubs to 2 m; lvs elliptic to oblong, 3–6 cm, commonly obtuse, entire, hairy on the veins beneath; petioles 2–4 mm; umbelliform clusters not subtended by lvs; pedicels 1–2 cm, bracteolate at the very base; sep narrowly oblong, 4–10 mm, deciduous in winter; cor 9–13 mm; stamens with a pair of ± well developed spurs near the junction of the anther and filament; fr truncate-ovoid, 5 mm thick and usually longer. $2n=24$. Moist, sandy soil, especially near the coast; R.I. to e. Pa., Tenn., s. Mo., Tex., and Fla. (*Neopieris m.*)

   3. **Lyonia lucida** (Lam.) K. Koch.   Evergreen swamp-fetterbush. Evergreen shrub to 2 m; lvs leathery, glossy, elliptic to narrowly obovate, 3–7 cm, acute or abruptly acuminate, entire, with a distinct intramarginal vein; petioles 2–5 mm; sep oblong, 3–5 mm, persistent; cor 7–9 mm; stamens with a pair of ± well developed spurs near the junction of the anther and filament; fr subglobose or depressed, 5 mm thick; $2n=24$. Wet woods; se. Va. to Fla. and La. May. (*L. nitida; Neopieris n.; Desmothamnus l.*)

11. LEUCOTHOE D. Don.

11. LEUCOTHOE D. Don.   Dog-laurel. Fls 5-merous; cal campanulate to saucer-shaped, persistent, the short lobes imbricate at base; cor tubular, constricted at the throat, the short, ovate lobes outcurved; stamens included, much shorter than the cor-tube; filaments flat, very narrowly triangular; anthers oblong, dorsifixed, spurless and virtually awnless, the pollen-sacs separate above and each opening by a terminal pore; ovary 5-locular; style about equaling the cor-tube; stigma capitate, 5-lobed; capsule depressed-globose, ± retuse, loculicidal; seeds flattened, angular, shining, reticulate; evergreen shrubs with alternate, commonly spinulose-serrulate lvs and dense axillary racemes of white fls, each pedicel 2-bracteolate at base, subtended and about equaled by an ovate bract. 40, N. Amer. and e. Asia.

1 Sep broadly ovate, obtuse or rounded; racemes 2–4(–5) cm .... 1. *L. axillaris.*
1 Sep lance-ovate, acute or subacute; racemes 4–10 cm .... 2. *L. walteri.*

   1. **Leucothoe axillaris** (Lam.) D. Don.   Swamp d.-l. Shrub 1–2 m; lvs leathery, lanceolate or oblong to oblanceolate or elliptic, 8–15 cm, acute to abruptly short-acuminate, entire to spinulose-serrulate; petioles 5–10 mm; racemes sessile, dense, 2–4(–5) cm, with 8–30 fls; sep broadly ovate, obtuse or rounded, 2 mm; cor 6–8 mm; filaments usually spreading-hairy as well as papillate; seeds angular. Wet woods on the coastal plain; se. Va. to Fla. and La. Apr., May.

   2. **Leucothoe walteri** (Willd.) Melvin.   Mountain d.-l. Much like no. 1; lvs long-acuminate, commonly spinulose-serrulate; petioles 10–15 mm; racemes 4–10 cm, with 20–60 fls; sep lance-ovate, acute or subacute, 2 mm; cor 5–7 mm; filaments papillate, usually not hairy; seeds elliptic or oblong, shining; $2n=22$. Along streams in the mts.; Va. to N.C., Tenn., Ga. and Ala.; disjunct in Baltimore Co., Md. May. (*L. editorum; L. fontanesiana; L. catesbaei,* misapplied.)

12. EUBOTRYS Nutt.

12. EUBOTRYS Nutt.   Fetterbush. Fls 5-merous; cal campanulate to saucer-shaped, persistent, the lobes imbricate at base; cor tubular, constricted at the throat, the short, rounded lobes spreading or recurved; stamens included, much shorter than the cor-tube; filaments glabrous, flat, narrowly triangular; anthers oblong or obovoid, dorsifixed, the pollen-sacs separate in the distal half, awned, opening by a terminal pore; style slender; stigma truncate, fr depressed-globose, loculicidal; seeds not reticulate; deciduous shrubs with thin, alternate lvs and elongate racemes of white fls terminating the branches of the previous year, the bracts soon deciduous; each pedicel with 2 bracteoles just beneath the cal. 2, N. Amer.

1 Racemes commonly straight, erect or ascending; sep lanceolate; anthers 4-awned .... 1. *E. racemosa.*
1 Racemes commonly curved or recurved; sep ovate; anthers 2-awned .... 2. *E. recurva.*

   1. **Eubotrys racemosa** (L.) Nutt.   Deciduous swamp-f. Shrub 1–4 m; lvs oblong to oblanceolate or obovate, 3–8 cm, acute or short-acuminate, finely serrulate; petioles 1–3 mm; racemes secund, 3 cm or more, mostly terminating lfless branches of the previous year, soon surpassed by leafy shoots; pedicels 2–3 mm, thickened distally; sep lanceolate, acute, 2–3 mm; cor 7–9 mm; anthers 4-awned, fr 4 mm thick, not lobed; seeds shaped like a section of an orange. Woods and barrens in moist acid soil, chiefly on the coastal plain; e. Mass. to Fla. and La. May, June. (*Leucothoe r.*)

2. **Eubotrys recurva** (Buckley) Britton. Deciduous mt.-f. Much like no. 1; racemes curved or recurved, 5–12 cm; sep ovate, acute, 2.5–3 mm; cor 7–9 mm; anthers 2-awned; fr 5–6 mm thick, 5-lobed; seeds oblanceolate, flat, winged. Moist woods in the mts.; w. Va. and se. Ky. to ne. Ga. May. (*Leucothoe r.*)

13. **PIERIS** D. Don. Fls 5-merous, cal persistent, campanulate, the sep leathery, valvate at base; cor ovoid, much-constricted above, 5-saccate at base alternately with the sep, sharply 5-angled in bud; stamens included, much shorter than the cor; filaments flat, sometimes geniculate, glabrous; anthers oblong, opening by 2 terminal pores, bearing 2 short, outwardly deflexed spurs; ovary 5-lobed; style columnar; stigma truncate; capsule globose, not retuse, not thickened at the sutures; seeds few in each locule, narrowly oblong, finely cellular-reticulate; shrubs with alternate, evergreen lvs and white fls in terminal panicles. 10, N. Amer., e. Asia.

1. **Pieris floribunda** (Pursh) Benth. & Hook. Evergreen mt.-fetterbush. Shrub to 2 m; lvs leathery, lanceolate to lance-elliptic or oblong, 3–7 cm, entire or minutely serrulate, ciliate, acuminate; petioles 5–10 cm; fls solitary at the nodes of an elongate, bracteate raceme, the short pedicel 2-bracteolate near the middle, the few to several racemes each 3–6 cm, forming a terminal panicle; sep triangular, 2–3 mm; cor 5–6 mm; fr 5–6 mm; seeds 2.5–3 mm. Mt. woods; Va. and W.Va. to Ga. May. (*Andromeda f.*)

14. **OXYDENDRUM** DC. Sourwood. Fls 5-merous; cal deeply parted, the lobes imbricate, spreading or reflexed at anthesis, persistent; cor conic-ovoid, much-constricted above, the tube canescent, the short lobes spreading or recurved; stamens 10, included; filaments flat, tapering, puberulent; anthers linear, opening by short slits above the middle, each of the 2 pollen-sacs prolonged into an erect tube as long as the body; style slender; stigma truncate; capsule oblong-ovoid, 5-angled, loculicidal; tree or tall shrub with alternate, deciduous lvs and numerous white fls secund in several elongate, lfless racemes that form a terminal panicle. Monotypic.

1. **Oxydendrum arboreum** (L.) DC. Tree to 20 m, beginning to fl at 2–3 m; lvs oblong, elliptic, or lance-ovate, 10–15 cm, acuminate, entire or serrulate; petioles 1–1.5 cm; racemes 4–7 at end of each branch, widely spreading, 5–15 cm, persistent; pedicels 3–8 mm, minutely bracteolate near the middle; cor 6–7 mm; fr 5–7 mm; 2*n*=24. Woods, Pa. to s. Ind., s. to Fla. and La. June, July.

15. **CHAMAEDAPHNE** Moench, nom conserv. Leatherleaf. Fls 5-merous; cal deeply parted, persistent, closely subtended by 2 bractlets, the lobes imbricate; cor nearly cylindric, slightly narrowed at the throat; stamens 10, included; filaments flat; anthers oblong, each of the 2 pollen-sacs prolonged into an erect tube as long as the body and opening by a terminal or subterminal pore; ovary 5-locular, subtended by a 10-lobed disk; style elongate; stigma truncate; capsule depressed-globose, loculicidal; evergreen shrub with alternate lvs and numerous nodding, axillary fls forming a leafy raceme; stem and lower lf-surface and cal brown-lepidote. (*Cassandra.*) Monotypic.

1. **Chamaedaphne calyculata** (L.) Moench. Branching shrub to 1.5 m; petioles 1–3 mm; lvs oblong or elliptic or rarely obovate, 1.5–5 cm, minutely crenulate; pedicels 2–5 mm; cor white, 6–7 mm; 2*n*=22. In bogs, often forming dense thickets; circumboreal, s. to N.J., O., n. Ind., n. Ill., and (chiefly in the mts.) to N.C. Apr.–June.

16. **ZENOBIA** D. Don. Fls 5-merous; cal saucer-shaped, the lobes valvate; cor campanulate, the broad, short lobes outcurved; stamens included; filaments slender; anthers lanceolate, dorsifixed, the pollen-sacs united throughout, each opening by a terminal pore and with 2 erect slender awns; ovary subglobose; style about equaling the cor-tube; stigma truncate; capsule subglobose, 5-lobed, loculicidal; seeds minutely rugose; shrubs with deciduous, veiny, crenate-serrate lvs and white or pink fls on long pedicels in umbelliform clusters from the nodes of a lfless terminal axis, forming an elongate panicle. Monotypic.

1. **Zenobia pulverulenta** (Bartram) Pollard. Shrub 1–2 m; lvs mostly broadly oval or elliptic, rounded or occasionally acute at both ends, dull green above, usually glaucous beneath; infls 1–4 dm, pedicels 1–3 cm; fls fragrant; cor 9–10 mm; fr 6–7 mm thick. Wet woods on the coastal plain; se. Va. to S.C. May. (*Z. cassinefolia.*)

17. GAULTHERIA L. Fls 4–5-merous; cal closely subtended by 2 bracteoles, campanulate to saucer-shaped, deeply divided; cor tubular to campanulate, shallowly lobed; stamens included; filaments short, flat; anthers oblong, the pollen-sacs separate or nearly so, each tipped by 2 erect awns; ovary 4–5-locular, wholly or partly superior; style short, columnar; stigma truncate; capsule thin-walled, completely enclosed in the fleshy, white or colored, accrescent cal, forming a dry or mealy berry-like fr with a characteristic flavor; erect to creeping shrubs with alternate, persistent lvs and usually white fls in racemes or panicles or (in our spp.) solitary in or just above the axils. 150, widespread.

1 Leafy stems suberect; fls 5-merous; ovary superior; berries red .... 1. *G. procumbens.*
1 Leafy stems prostrate; fls 4-merous; ovary partly inferior; berries white .... 1. *G. hispidula.*

1. Gaultheria procumbens L. Wintergreen; checkerberry. Leafy stems suberect from a horizontal rhizome, 1–2 dm, with a few lvs crowded near the top; petioles 2–5 mm; blades elliptic or oblong to rarely subrotund, 2–5 cm, entire or crenulate, glabrous; fls 5-merous, on nodding pedicels 5–10 mm; cor barrel-shaped, 7–10 mm, the rounded lobes 1 mm; fr bright red, 7–10 mm; $2n=44, 88$. Dry or moist woods in acid soil; Nf. to Man., s. to Va., Ky., n. Ind., Minn., and in the mts. to Ga. and Ala. July, Aug.

2. Gaultheria hispidula (L.) Muhl. Creeping snowberry; moxie. Stems prostrate, 2–4 dm, very leafy, bristly especially when young; lvs short-petioled, broadly elliptic to subrotund, 5–10 mm, smooth above, sparsely bristly beneath; fls few, 4-merous, on recurved pedicels 1 mm; cal campanulate; cor campanulate, 2–5 mm, the rounded lobes a third as long as the tube; filaments obovate; ovary partly inferior; fr white, 5–10 mm; $2n=24$. Bogs and wet woods, often on decaying logs; Nf. and Lab. to B.C., s. to N.J., Pa., Mich., and Minn., and in the mts. to N.C. May, June. (*Chiogenes h.*)

18. EPIGAEA L. Trailing arbutus. Plants functionally dioecious, the fls 5-merous, either pistillate, with vestigial stamens, or apparently perfect but with unexpanded stigmas and functionally staminate; each fl closely subtended by 2 ovate bracts nearly or quite as long as the cal; sep distinct to the base, strongly imbricate; cor salverform, the thick tube densely pubescent within, the lobes oblong-ovate; stamens included; filaments slender, hairy below; anthers linear, awnless, opening longitudinally; ovary villous above; style slender; stigma 5-lobed; capsule fleshy, depressed-globose, at length 5-valved, subtended by the persistent cal and bracts, white-pulpy within; prostrate, creeping, evergreen shrub, with leathery, alternate lvs and conspicuous, pink to white, fragrant fls in short, crowded, terminal and axillary spikes. 3, the others in Japan, Caucasus, and Asia Minor.

1. Epigaea repens L. Stems branched, 2–4 dm, hirsute; lvs ovate or oblong, 2–10 cm, entire, rounded or cordate at base, ± pilose, especially when young; petiole pubescent, half as long as the blade; spikes 2–5 cm; bracts ciliate; cor-tube 8–15 mm, the lobes 6–8 mm. Sandy or rocky, acid soil; Nf. and Que. to Sask., s. to Fla., Miss., and Io. Apr., May.

19. ARCTOSTAPHYLOS Adans. nom. conserv. Fls 5-merous; cal saucer-shaped, the sep imbricate, distinct to the base; cor ovoid, the rounded lobes spreading or recurved; stamens 10; filaments pubescent, much-dilated below, much shorter than the cor; anthers subglobose, opening by 2 terminal pores and bearing 2 deflexed terminal awns; ovary 5-locular, conic-ovoid, subtended by a 10-lobed disk; style columnar; stigma capitate; ovules 1 per locule; fr a fleshy or mealy drupe with 5 bony nutlets; freely branched shrubs (ours low and creeping) with persistent, alternate lvs and white to pink fls in short, few-fld, terminal racemes. 45, mainly w. N. Amer.

1 Lvs entire, evergreen; fr red; nutlets partly or wholly concrescent .... 1. *A. uva-ursi.*
1 Lvs serrulate, marcescent; fr black; nutlets separate .... 2. *A. alpina.*

1. Arctostaphylos uva-ursi (L.) Sprengel. Bearberry. Prostrate shrub, forming mats to 1 m wide; lvs leathery, evergreen, oblanceolate to oblong-obovate, 1–3 cm, obtuse or rounded, entire, tapering to the base; sep broadly ovate, 1.5 mm; cor commonly white or tinged with pink, 4–6 mm; fr bright red, dry or mealy, inedible, 6–10 mm, the 5 nutlets partly or usually wholly concrescent. Sandy or rocky soil. May, June. (*Uva-ursi uva-ursi.*) Circumboreal, in N. Amer. from Lab. to Alas., s. to Va., n. Ind., Ill., N.M., and Calif. Variable, especially in pubescence, but not taxonomically divisible, the differences reflecting combinations of genetic and environmental factors.

2. **Arctostaphylos alpina** (L.) Sprengel. Alpine bearberry. Stems tufted or prostrate, the branches 1–2 dm tall; lvs withering at the end of the season and persistent, thin, oblanceolate to spatulate-obovate, 2–5 cm (petiole included), acute to rounded, serrulate, rugose above, reticulate beneath; cor 4–5 mm; fr black, juicy, 6–10 mm. Greenl. to Alas., s. to the higher mts. of Me. and N.H. May, June. (*Mairania a.; Arctous a.*)

20. **CALLUNA** Salisb. Heather. Fls 4-merous, each closely subtended by 3 pairs of decussate bracts simulating a cal; sep distinct, petaloid, scarious, much exceeding the tubular, deeply 4-lobed cor; stamens 8, included, connivent; anthers lance-subulate, each with 2 short basal spurs, the pollen sacs distinct for more than half their length above, opening longitudinally nearly to the base; stigma deeply 4-lobed; capsule 4-valved, enclosed by the persistent perianth; branched shrubs with small, opposite, decussate lvs and small, white to pink or rose-purple fls in ± secund racemes. Monotypic.

1. **Calluna vulgaris** (L.) Hull. Ascending shrub to 1 m; lvs lance-oblong; 2–4 mm, sessile, the larger auriculate; cal 3–4 mm; $2n=16$. Widespread in Europe; naturalized in sandy places along the coast from Nf. to N.J. and rarely inland. July–Sept.

21. **ERICA.** L. Heath. Fls 4-merous; sep distinct; cor ovoid to campanulate, shallowly lobed; stamens 8; anthers with or without appendages; capsule subglobose, enclosed by the persistent cor; branching shrubs with narrow, whorled, subsessile lvs and pink to purple fls. 600+, Old World. Our spp. all intr. from Europe.

1 Cor ovoid; anthers included; sep linear to lanceolate; fls in terminal racemes.
  2 Lvs and cal conspicuously glandular-ciliate; lf-surface canescent .... 1. *E. tetralix.*
  2 Lvs and cal glabrous .... 2. *E. cinerea.*
1 Cor campanulate; anthers exsert; sep broadly ovate; fls in 2-fld axillary racemes .... 3. *E. vagans.*

1. **Erica tetralix** L. Cross-lvd h. Shrub to 4 dm; lvs narrowly oblong, at least the younger ones canescent, glandular-ciliate, as also the sep and bracts; cor ovoid, 7–8 mm; anthers with long basal appendages; $2n=24$. Locally from Me. to N.J. and in W.Va. July–Sept.

2. **Erica cinerea** L. Twisted h. Shrub to 4 dm; lvs linear, essentially glabrous, as also the sep and bracts; cor ovoid, 6–7 mm; anthers with short, broad basal appendages; $2n=24$. Nantucket Isl., Mass. July, Aug.

3. **Erica vagans** L. Cornish h. Shrub to 5 dm; lvs narrowly linear, glabrous; sep glabrous, broadly ovate; cor campanulate, 3–4 mm; anthers exsert, without appendages; $2n=24$. Nantucket Isl., Mass. Summer.

22. **VACCINIUM** L. Fls 4–5-merous, epigynous (seldom only half-epigynous); cal spreading or appressed; cor tubular or ovoid to campanulate, shallowly to deeply lobed; stamens 8 or 10, commonly included; anthers with 2 pollen-sacs, each sac opening by a pore at the end of a terminal tubule; style slender, usually surpassing the stamens; fr a many-seeded berry, commonly 4–5-locular, or sometimes 8–10-locellar; evergreen or deciduous shrubs or small trees with entire or serrulate lvs and solitary or racemose fls terminal or in the axils. (*Cyanococcus, Oxycoccus, Polycodium.*) 400+, widespread. Spp. 9–15 form a difficult complex.

1 Cor shallowly lobed or toothed, the 4–5 lobes conspicuously shorter than the tube.
  2 Anthers with a pair of conspicuous spurs in addition to the terminal tubules.
    3 Cor open-campanulate, open in bud; stamens exsert (*Polycodium*) .... 1. *V. stamineum.*
    3 Cor tubular to ovoid, obovate, or urceolate, ± constricted at the summit, closed in bud; stamens
      included.
      4 Southern plants; fr dry, inedible, 10-locellar (*Batodendron*) .... 2. *V. arboreum.*
      4 Northern plants; fr juicy, edible, 4–5-locular (*Vaccinium*).
        5 Fls mostly 4-merous, 1–4 from the axils of bud-scales; lvs entire .... 3. *V. uliginosum.*
        5 Fls mostly 5-merous, solitary in the lf-axils; lvs mostly toothed.
          6 Lvs aristate-serrulate throughout their length, or at least in the upper half.
            7 Plants mostly 1–2 dm; lvs mostly 1–3 cm .... 4. *V. cespitosum.*
            7 Plants mostly 5–10(–15) dm; lvs mostly 3–6 cm .... 5. *V. membranaceum.*
          6 Lvs entire, or aristate-serrulate only below the middle .... 6. *V. ovalifolium.*
  2 Anthers not spurred.
    8 Cor campanulate; lvs leathery, evergreen, rarely 2 cm.
      9 Fls 4-merous, in short terminal clusters; northern sp. (*Vitis-idaea*) .... 7. *V. vitis-idaea.*
      9 Fls 5-merous, in lateral clusters; southern sp. (*Herpothamnus*) .... 8. *V. crassifolium.*

8 Cor tubular to urceolate, constricted above; lvs deciduous, very often over 2 cm; fls jointed to
    the pedicel (unlike all other sections) (*Cyanococcus*).
10 Lowbush-blueberries, rarely as much as 8 dm, extensively colonial.
  11 Lower lf-surfaces stipitate-glandular .... 9. *V. tenellum.*
  11 Lower lf-surfaces not glandular (though the margin may be glandular-puberulent).
    12 Plants very dwarf, rarely 1 dm; high-northern .... 10. *V. boreale.*
    12 Plants taller, usually over 1 dm; widespread.
      13 Lvs serrulate.
        14 Lvs relatively small, seldom more than 3 × 1.5 cm, green or much less
          often glaucous, glabrous or nearly so .... 11. *V. angustifolium.*
        14 Lvs larger, the larger ones generally 3–5 cm long and 1.5 cm wide or more,
          evidently glaucous beneath, glabrous or sometimes ± evidently hairy;
          occasional forms of .... 12. *V. pallidum.*
      13 Lvs entire or nearly so.
        15 Lvs glaucous beneath, glabrous or sometimes somewhat hairy, the larger
          ones generally 3–5 cm long and 1.5 cm wide or more .... 12. *V. pallidum.*
        15 Lvs not glaucous, evidently hairy at least when young, seldom more than
          3 × 1.5 cm .... 13. *V. myrtilloides.*
10 Highbush-blueberries, mostly 1–3(–5) m, typically crown-forming.
  16 Lvs mostly 3–8 × 1.5–4 cm, entire or sometimes serrulate; widespread .... 14. *V. corymbosum.*
  16 Lvs mostly 1.5–3 × 0.8–1.5 cm, serrulate; s. coastal plain .... 15. *V. elliottii.*
1 Cor 4-cleft to below the middle, the lobes recurved at anthesis.
17 Stem erect; lvs deciduous, serrulate, ciliate (*Oxycoccoides*) .... 16. *V. erythrocarpum.*
17 Stem trailing; lvs evergreen, entire (*Oxycoccus*).
  18 Pedicels bearing above the middle a pair of green, lf-like bracteoles 2–4 mm .... 17. *V. macrocarpon.*
  18 Pedicels bearing near or usually below the middle a pair of red, scale-like bracteoles to 0.5–
    1.5(–2) mm .... 18. *V. oxycoccos.*

1. **Vaccinium stamineum** L.   Deerberry. Deciduous shrub to 1.5 m, glabrous to densely hairy; lvs thin, oblong or elliptic to obovate, 3–10 cm, acute or acuminate, entire, petioles 2–3 mm; fls in racemes or panicles on specialized branches subtended by leafy bracts like the foliage lvs but usually much smaller; sep half-orbicular to ovate or deltoid, 1–2.5 mm; cor obconic, open even in bud, greenish or suffused with purple, eventually 4–6 mm, 5-lobed nearly to the middle; stamens exsert, the conspicuous terminal tubules long-exsert, much longer than the dorsal spurs; style prematurely exsert; fr yellowish or greenish to blue, to 1 cm, palatable to more often scarcely edible; 2*n*=24. Dry woods; Me. to Ind. and Mo., s. to Fla., Tex., and Mex. May, June. (*Polycodium s.; P. candicans; V. caesium; V. melanocarpum; V. neglectum.*) Highly variable, some of the numerous segregates perhaps worthy of taxonomic notice.

2. **Vaccinium arboreum** Marshall.   Farkleberry. Divergently branched shrub to 5 m; lvs firm, deciduous, oblanceolate to obovate, 2–5 cm, entire, narrowed to a short petiole, shining above, dull and sometimes puberulent beneath; fls few in a short raceme or solitary, on pedicels 6–10 mm; cor white, obovoid, 4.5–5.5 mm, constricted at the throat; stamens included; anthers with 2 divergent dorsal spurs 1–1.2 mm and erect terminal tubules 1.5 mm; style slightly exsert; fr black, inedible. Dry, sandy or rocky woods; se. Va. to Ky., s. Ind., Mo., and se. Kans., s. to Fla. and Tex. May, June. (*Batodendron a.*)

3. **Vaccinium uliginosum** L.   Bog-bilberry. Low, diffuse or matted, stoloniferous shrubs; lvs deciduous, firm, elliptic to oblong-obovate, blunt, entire; fls all or mostly 4-merous, in clusters of 1–4 from the axils of bud-scales, on pedicels 2–5 mm; sep ovate or triangular, 1 mm; cor pink, ovoid; anthers spurred; fr sweet, dark blue or black; *x*=12. Circumboreal, s. to n. N. Engl., n. N.Y., and n. Mich. June, July. Most of our plants belong to the circumboreal tetraploid var. *alpinum* Bigelow, often growing in bogs or muskeg, 1–3 dm, with lvs 10–25 mm and densely hairy on both sides, the frs 4–6 mm. (*V. pubescens.*) In alpine sites we also have the diploid var. *gaultherioides* (Bigelow) Bigelow, occurring mainly from Greenl. and e. Can. to the high mts. of N. Engl., dwarf and tufted or matted, the glabrous or subglabrous lvs 5–10 mm and the fr 2–4 mm. (*V. microphyllum.*)

4. **Vaccinium cespitosum** Michx.   Dwarf bilberry. Branched, 1–2 dm, colonial by superficial rhizomes; lvs deciduous, firm, 1–3 cm, oblanceolate or cuneate to obovate, obtuse or rounded above, tapering to the base, finely aristate-serrulate at least in the upper half; fls solitary in the axils of the lower lvs of the season, on decurved pedicels 3 mm; sep very short and broad; cor 5 mm, usually pink; anthers spurred; fr blue, 6–8 mm; 2*n*=24. Circumboreal, s. to n. N. Engl., n. N.Y., Wis., and n. Minn. June, July.

5. **Vaccinium membranaceum** Douglas.   Mt.-bilberry. Branched, 5–10(–15) dm, usually ± colonial; lvs thin, deciduous, ovate-oblong to oblong, 3–6 cm, acute, finely aristate-serrulate throughout with incurved teeth, rounded at base, glabrous; petioles 2–3 mm; fls solitary in the lower axils of the season, nodding on pedicels 1–2 mm; sep very short and broad or nearly obsolete; cor subglobose, 5–6 mm; anthers spurred; fr purple or black, 8–10 mm. Moist woods; Bruce Penins., Ont.; Upper Penins., Mich.; Alta. and B.C. to n. Calif. and n. Utah; Ariz. May, June.

6. **Vaccinium ovalifolium** J. E. Smith.   Tall bilberry. Bushy-branched shrub to 1.5 m, crown-forming or seldom colonial; lvs thin, deciduous, elliptic or oval, 2–5 cm, obtuse or rounded at both ends, entire or aristate-serrulate only below the middle, glabrous, paler green beneath, on petioles 1–2 mm; fls solitary in the lower axils of the season, on spreading or decurved pedicels 4–8 mm; cal-lobes very short or nearly

obsolete; cor 7–10 mm; much-constricted at the throat; anthers spurred; fr purplish-black, nearly 1 cm. Moist woods; Nf. to N.S. and e. Que.; n. Mich. and adj. Ont.; Black Hills; Alas. to Ida. and Oreg. May.

7. **Vaccinium vitis-idaea** L.   Lingonberry, mt.-cranberry, partridge-berry. Diffusely branched or prostrate shrub to 2 dm; lvs leathery, evergreen, subsessile, elliptic to obovate-oblong, 8–18 mm, rounded at both ends, sparsely dotted with erect black glands beneath; fls few in small terminal clusters, each on a short, glandular pedicel axillary to a bud-scale; sep glandular-ciliate; cor campanulate, 5–7 mm, 4-lobed nearly to the middle; stamens 8, included; fr red, edible, nearly 1 cm; 2n=24. In bogs and on rocks; boreal N. Amer. and Eurasia, s. to the higher mts. of N. Engl., the shore of Me., and to Minn. and B.C. (*Vitis-idaea vitis-idaea; Vitis-idaea punctata.*)

8. **Vaccinium crassifolium** Andrews.   Creeping blueberry. Roots woody, tuberous-thickened at intervals; stems trailing, to 1 m; lvs leathery, evergreen, elliptic, 5–15 mm, glabrous, remotely serrulate, much-thickened at the margin; fls in small lateral clusters much shorter than the subtending lvs; cor pink, 4 mm; fr purple-black; 2n=24. Pine-barrens and pocosins; se. Va. to S.C. and adj. Ga. Apr., May. (*Herpothamnus c.*)

9. **Vaccinium tenellum** Aiton.   Southern blueberry. Colonial from rhizomes, the aerial stems 2–4 dm; young twigs densely pubescent; lvs deciduous, oblanceolate to oblong-obovate, 1.5–3 cm, usually acute, finely ciliate-serrulate, tapering to the base, stipitate-glandular beneath; fls appearing with the lvs, in umbelliform racemes; cor pink to red, ellipsoid-conic, 5–7 mm; fr black, 5–8 mm; 2n=24. Sandy woods and barrens on the coastal plain; se. Va. to Ga. Apr. (*Cyanococcus c.*)

10. **Vaccinium boreale** I. V. Hall & Aald.   Sweet hurts. Dwarf, rarely to 1 dm, with superficial rhizomes, forming small, dense colonies as part of the ground-level vegetation; lvs deciduous, narrowly elliptic, 1–2 cm × 2–6 mm, sharply serrulate, green and glabrous or nearly so; cal and pedicel glaucous; cor cylindric, 3–4 mm, white; fr blue-glaucous, 3–5 mm; 2n=24. Alpine meadows and exposed, rocky sites; Nf., Lab., and Que., s. to mt.-tops of Me., n. N.H., n. Vt., and n. N.Y. June, July.

11. **Vaccinium angustifolium** Aiton.   Common lowbush-blueberry. Shrubs 1–6 dm, extensively colonial; lvs deciduous, ± elliptic, mostly (1–)1.5–3 cm, a third to half as wide, sharply serrulate, green and glabrous (or with a few hairs along the veins) to less often glaucous; cal and pedicel often glaucous; cor cylindric to urceolate, 4–6 mm; fr blue-glaucous or less often black, 5–10 mm; 2n=48. Moist or dry, sandy or rocky soil, often in burned-over sites; Lab. and Nf. to Man., s. to N.J., Pa., Ill., and Minn., and in the mts. to N.C. (*V. brittonii; V. lamarckii; Cyanococcus a.*)

12. **Vaccinium pallidum** Aiton.   Hillside-blueberry. Colonial shrubs 2–8(–10) dm; bud-scales rounded above; lvs ovate to broadly elliptic, entire or seldom serrulate, the larger ones mostly (2.5–)3–5 cm, half as wide, glaucous beneath, the others sometimes somewhat hairy; cal and pedicel glaucous; cor cylindric to urceolate, 4–8 mm; fr 5–8 mm, blue-glaucous or rarely black, 2n=24, or seldom 48. Dry upland woods and old fields; Me. to Minn., s. to Ga., Ala., Ark., and e. Okla. Apr.–June. (*V. altomontanum; V. vacillans; Cyanococcus liparis; C. subcordatus; C. v.*) Morphologically intermediate between nos. 11 and 13, but apparently distinctive.

13. **Vaccinium myrtilloides** Michx.   Velvetleaf-blueberry. Colonial shrubs 2–5 dm; bud-scales sharply acute; lvs thin and soft, elliptic to lance-elliptic, 1–3(–4) cm, half or a third as wide, entire, softly hairy beneath, not glaucous; cor broadly cylindric, 4–5 mm, white or tinged with pink; fr blue-glaucous, 4–7 mm; 2n=24. Moist or dry soil and bogs; Lab. to B.C., s. to Pa., Ind., ne. Io., and Minn., and in the mts. to Va. and W.Va. May–July. (*V. canadense; Cyanococcus c.*)

14. **Vaccinium corymbosum** L.   Highbush-blueberry. Crown-forming shrubs (0.5)1–3(–5) m; lvs ovate to narrowly elliptic, deciduous, 3–8 × 1.5–4 cm, variously serrulate or often entire, hairy (rarely glandular) or glabrous and sometimes glaucous; cor white or greenish-white to pink, cylindric to urceolate cylindric, 5–10 mm; fr blue-glaucous to black, 5–12 mm; 2n=24, 48, 72, ours mostly 48. Open swamps and bogs, or sometimes in upland woods or in old fields, from the coastal plain to the mt.-tops; Me. and N.S. to Fla., w. to Mich., n. Ill., Ky., Ark., e. Okla., and e. Tex. Apr.–July. (*V. atrococcum; V. australe; V. caesariense; V. constablei; V. marianum; V. simulatum; Cyanococcus atrococcus; C. margarettae; C. s.*) Highly variable, not readily sorted into recognizable taxa. *V. caesariense* Mackenzie, a diploid with glabrous, glaucous, entire lvs, glabrous or subglabrous twigs, and blue-glaucous fr, is ± distinguishable from the variable remainder of the group, diploid or often polyploid, with variously entire or serrulate, glabrous or hairy lvs and twigs, and blue or often black fr.

15. **Vaccinium elliottii** Chapman.   Southern highbush-blueberry. Lvs mostly 1.5–3 × 0.8–1.5 cm, serrulate; cor narrowly urceolate, some shade of pink; otherwise much like no. 14; 2n=24. Open flatwoods and ravines, and dry uplands along river-valleys, only seldom in swampy places; se. Va. to Fla. and La. on the coastal plain and piedmont. Fl early, usually before the lvs.

16. **Vaccinium erythrocarpum** Michx.   Southern mt.-cranberry, dingleberry. Erect, branching shrub to 2 m; lvs thin, deciduous, subsessile, lanceolate to lance-ovate, 3–6 cm, minutely serrulate, ciliate, acuminate; pedicels solitary in the axils, nodding, not bracteate, 8–15 mm; cor 1 cm, pink or red to white, cleft to below the middle, the 4 narrow lobes revolute; anthers 8–11 mm, the tubules about equaling the thecae; fr red to deep purple or black, 1(–1.5) cm, usually insipid. Mt. forests; Va. and W.Va. to n. Ga. and c. Tenn. June–July. (*Oxycoccus e.; Hugeria e.*)

17. **Vaccinium macrocarpon** Aiton.   Cranberry. Evergreen trailing shrub with very slender stems; lvs leathery, subsessile, elliptic-oblong, 5–15 mm (the largest usually > 1 cm), rounded at the tip, flat or scarcely revolute; fls 2–6 in a cluster, solitary in the axils of the reduced lower lvs of a normal branch; pedicels 1–3

cm; bracteoles above the middle of the pedicel, green, 2–4 × 1–2 mm; cor white to pink, 1 cm wide, deeply 4-cleft, the lobes strongly reflexed; styles 5–7 mm; fr red, 1–1.5 cm, tart; $2n=24$. Bogs; Nf. to Man., s. to Va., O., and n. Ill., and locally in the mts. to N.C. and Tenn. June–Aug. (*Oxycoccus m.*)

18. Vaccinium oxycoccos L. Small cranberry. Much like no. 17; lvs flat to strongly revolute, ovate or elliptic-ovate, 2–10 mm, commonly acute; fls 1–4 from the axils of the uppermost reduced lvs, forming a terminal cluster; bracteoles red, scale-like, 0.5–1.5(–2) × <1 mm, borne somewhat above to considerably below the middle of the pedicel; style 3–4 mm; fr 6–12 mm; $2n=24, 48, 72$. Bogs; circumboreal, s. to N.J., Pa., O., n. Ind., and Minn. May–July. (*Oxycoccus o.; O. palustris; O. quadripetalus.*)

23. GAYLUSSACIA HBK., nom. conserv. Huckleberry. Fls 5-merous; hypanthium short, obconic; cal 5-lobed, persistent; cor tubular or conic, scarcely if at all constricted above, the short lobes erect or outcurved; stamens 10, included; anthers with 2 pollen-sacs, each sac prolonged into a terminal tubule; ovary inferior, 10-locellar, ripening into a fleshy drupe with 10 seed-like nutlets, edible but of poor quality; freely branched shrubs, often colonial, commonly conspicuously glandular, with small, alternate, deciduous or evergreen lvs and white or pinkish fls in lateral, bracteate racemes. 50, New World.

1 Lvs entire, deciduous, glandular at least on the lower side.
  2 Sep, pedicels, bracts, and lvs with sessile glands, not pubescent.
    3 Racemes commonly exceeding the lvs; lvs glandular only beneath .... 1. *G. frondosa.*
    3 Racemes commonly shorter than the lvs; lvs glandular on both sides .... 2. *G. baccata.*
  2 Sep, pedicels, bracts, and lvs stipitate-glandular, the sep, pedicels, and bracts also rather densely
    pubescent .... 3. *G. dumosa.*
1 Lvs serrulate, leathery, evergreen, not glandular .... 4. *G. brachycera.*

1. Gaylussacia frondosa (L.) T. & G. Dangleberry. widely branched, to 2 m; lvs deciduous, elliptic or oblong to oblong-obovate, 3–6 cm, beset with minute, sessile, resinous glands beneath; racemes slender, loose, usually surpassing the lvs; bracts linear-oblong to elliptic, deciduous; pedicels usually much longer than the fls; pubescence of the infl and fls sparse or none, the glands sessile, if present, and the sep not ciliate; cor 3–4.5 mm, its tube two-thirds as thick; fr dark blue, glaucous, 7–8 mm. Moist woods and thickets, mostly near the coast; N.H. to Fla. and Ala., and sometimes inland to O., W. Va., and Tenn. May, June. (*Decachaena f.*)

2. Gaylussacia baccata (Wangenh.) K. Koch. Black h. Much-branched shrub to 1 m; lvs deciduous, elliptic to oblong or oblanceolate, 2–5 cm, entire, resinous-glandular on both sides; racemes short, rarely surpassing the lvs; pedicels commonly shorter than the fls; bracts oblong to linear, deciduous; pubescence of the infl and cal usually copious, the glands sessile and commonly numerous, the sep not ciliate; cor 4–6 mm, its tube half as thick; fr black, 6–8 mm; $2n=24$. Dry, sandy or rocky soil, woods, and thickets; Nf. and Que. to Ont. and Man., s. to Ga., Ala., and Mo. May, June. (*Decachaena b.*)

3. Gaylussacia dumosa (Andr.) T. & G. Dwarf h. Colonial by long, woody rhizomes, sending up leafy branches 2–5 dm; lvs deciduous, oblanceolate to obovate, 2–4 cm, entire, inconspicuously stipitate-glandular; racemes rather dense; bracts oval or oblong, foliaceous, persistent, 5–12 mm; pedicels, hypanthium, and cal pubescent and copiously stipitate-glandular; sep glandular-ciliate; cor 5–9 mm; fr black, hairy, 5–8 mm. Usually in wet, sandy soil or in bogs, on or near the coastal plain; Nf. to Fla. and La. May, June. (*Lasiococcus d.*) Var. *bigeloviana* Fern., from Nf. to Va., has the bracts conspicuously glandular and the fls 7–9 mm. Var. *dumosa*, from L.I. to La., has thinner bracts, often shining, sparsely glandular to glabrous on the surface, the fls 5–7 mm.

4. Gaylussacia brachycera (Michx.) A. Gray. Box-h. Plants not glandular; stems 2–4 dm, bearing a conspicuous ridge below the base of each lf; colonial by long, woody rhizomes; lvs leathery, evergreen, elliptic, 12–22 mm, finely serrulate, glabrous; racemes few-fld, subsessile, shorter than the subtending lvs; cor 5 mm; fr dark blue, edible. Wooded hillsides, chiefly in the mts.; Md. and Del. to Pa., Ky., and e. Tenn.; very local except in W.Va. May, June. (*Buxella b.*)

Family **PYROLACEAE**, the Shinleaf Family

Fls regular, perfect, hypogynous, (4)5-merous; sep distinct or shortly connate at base; pet distinct, imbricate; stamens twice as many as the pet, distinct, free from the pet; anthers becoming inverted during ontogeny, so that the morphological base is apical, opening by 2 pores at the apparent apex, sometimes produced into short tubes beneath the pores; pollen usually in tetrads; ovary imperfectly 5-locular, the deeply intruded placentas not joined in the center; stigma 5-lobed; ovules numerous; fr a loculicidal capsule; seeds numerous; seed-coat a single layer of cells; embryo tiny, undifferentiated, embedded in the endosperm; strongly mycotrophic, glabrous, perennial herbs or half-shrubs; lvs alternate to sometimes opposite or subverticillate, persistent through the winter except in *Moneses,* all basal or nearly so except in *Chimaphila,* rarely poorly developed or even suppressed; fls in racemes, corymbs, or umbels, or solitary and terminal. 3–4/45.

1 Half-shrubs with cauline lvs; fls umbellate or corymbose . . . . 1. *Chimaphila.*
1 Herbs with chiefly basal (rarely no) lvs; fls racemose or solitary.
  2 Fls racemose; fr opening from the base upward . . . . 2. *Pyrola.*
  2 Fl solitary and terminal; fr opening from the top downward . . . . 3. *Moneses.*

1. CHIMAPHILA Pursh.

Pet 5, spreading; filaments strongly dilated below the slender tip; anthers plump, the pollen-sacs flattened-oblong, attached at the middle, rounded at the base, connate below the filament, separate and tapering above, opening by a wide terminal pore; style very short, nearly immersed in the depressed top of the ovary; capsule depressed-globose, loculicidal from the top down; low, perennial, evergreen half-shrubs from a creeping rhizome, with thickish, denticulate, opposite or often subverticillate, wholly cauline lvs and terminal, long-peduncled, few-fld umbels or corymbs of white or pink fls. 4–5/N. Hemisphere.

1 Lvs oblanceolate, tapering to the base, not marked with white . . . . 1. *C. umbellata.*
1 Lvs lanceolate, obtuse or rounded at base, striped with white . . . . 2. *C. maculata.*

1. Chimaphila umbellata (L.) Barton.   Prince's pine; pipsissewa. Stems spreading, the fertile branches erect, 1–3 dm; lvs oblanceolate, 3–6 cm, acute or mucronate, sharply toothed, especially above, nearly entire below, tapering to a short petiole; fls 4–8, corymbose or subumbellate, 10–15 mm wide; dilated part of the filaments ciliolate; $2n=26$. Dry woods, especially in sandy soil. June–Aug. Circumboreal, with 2 vars. in Amer. Var. *cisatlantica* S. F. Blake (*C. corymbosa*), occurring from Que. and N.S. to Minn., s. to Va., W.Va., and n. Ind., has relatively thin, veiny lvs and mostly recurved pedicels. Var. *occidentalis* (Rydb.) S. F. Blake (*C. occidentalis*), from n. Mich. to B.C., s. to Colo. and Calif., has thicker, less veiny lvs and often ascending pedicels.

2. Chimaphila maculata (L.) Pursh.   Spotted wintergreen. Flowering branches 1–2 dm tall; lvs 2–7 cm, lanceolate, striped with white along the midvein and to a lesser extent along the primary lateral veins, remotely and sharply toothed, acute to rounded at base, short-petioled; fls (1)2–5 and umbellate, 12–18 mm wide, dilated part of the filaments villous. Dry woods, especially in sandy soil; Me. and N.H. to Mich., s. to S.C., Ga. and Ala. June–Aug., after no. 1.

2. PYROLA L.

Shinleaf, wintergreen. Pet 5; filaments elongate, slender; anthers subapically attached, opening by terminal pores, the pollen sacs often prolonged above into short tubes; style ± elongate; capsule opening from the base upward; rhizomatous herbs, the few broad, petiolate lvs nearly basal, persistent throughout the winter (rarely the lvs suppressed); fls in an erect, terminal, long-peduncled raceme. 40/N. Temp.

1 Style and filaments straight or nearly so, the stamens connivent around the style.
  2 Style exserted at anthesis; raceme secund . . . . 1. *P. secunda.*
  2 Style not exserted at anthesis; raceme not secund . . . . 2. *P. minor.*
1 Style strongly declined at anthesis; filaments upcurved, the anthers not surrounding the style.
  3 Sep (measured above the sinuses) about as broad as long, or broader.
    4 Lvs mostly 1–3 cm (rarely suppressed); anther-tubes abruptly narrowed from the pollen-sacs and
       evidently prolonged; sep broadly ovate, obtuse or subacute . . . . 3. *P. chlorantha.*
    4 Lvs mostly 3–7 cm; anther-tubes scarcely narrowed from the pollen-sacs, very short; sep tri-
       angular, shortly acuminate . . . . 4. *P. elliptica.*
  3 Sep distinctly longer than broad.
    5 Pet white; sep oblong or ovate-oblong, not overlapping at base . . . . 5. *P. rotundifolia.*
    5 Pet pink or pinkish; sep essentially triangular, slightly overlapping at base . . . . 6. *P. asarifolia.*

1. Pyrola secunda L.   One-sided s. Lvs elliptic to broadly ovate or subrotund, 1.5–4 cm, obtuse or rounded, entire to crenate-serrate, obtuse or rounded at base, often separated by conspicuous internodes; scape 1–2 dm; raceme crowded, secund; sep semi-orbicular to ovate, 0.5–1 mm; pet white or greenish, 5 mm; anthers 1.5–2 mm, truncate, opening by large pores, rounded at base; style elongate, exsert at anthesis; $2n=38$. Moist woods and mossy bogs; circumboreal, in Amer. s. to N.J., Md., Ind., Minn., and N.M. Differs from other spp. of *Pyrola* in having a 10-lobed, hypogynous disk, basally bituberculate pet, and simple pollen grains (not in tetrads), and perhaps best considered to form a distinct genus, *Orthilia secunda* (L.) House. (*Ramischia s.*)

2. Pyrola minor L.   Little s. Lvs elliptic or round-oblong, 2–4 cm, rounded to truncate at both ends; scape 5–15 cm, with a loose raceme of 5–15 nodding white fls; pet 5 mm; anthers very plump, 1 mm or less, broadly truncate to 2 wide terminal pores; style at anthesis ca 2 mm, about equaling the pet; 2*n*=46. Moist woods, circumboreal, in Amer. s. to the higher mts. of N. Engl. and n. Mich., and in the w. cordillera. (*Braxilia m.; Erxlebenia m.*)

3. Pyrola chlorantha Swartz.   Lvs 1–3(–4) cm, often shorter than the petiole, obovate to broadly elliptic or subrotund, rounded to truncate at the summit, rounded to broadly cuneate at base but scarcely decurrent, seldom ± reduced or even wholly suppressed; scapes 1–2.5 dm, seldom with more than 10 fls; sep broadly ovate-triangular, broader than long, obtuse or subacute; pet 4–9 mm, white, ± veined with green; anthertubes abruptly narrowed from the pollen sacs and of different color, prolonged 0.6–0.8 mm and usually separate, often curved; style declined; 2*n*=46. Dry woods; circumboreal, in Amer. s. to Md., W.Va., Ind., and the w. cordillera. June–Aug. (*P. virens; P. oxypetala.*)

4. Pyrola elliptica Nutt.   Elliptic s. Lvs 3–7 cm, commonly longer than the petiole, broadly elliptic or oblong to somewhat obovate, subacute to rounded above, acute to rounded at base and always decurrent on the petiole; scapes 1.5–3 dm, often with more than 10 fls; sep triangular, about as broad as long, very shortly acuminate; pet 5–9 mm, white, ± veined with green; anther-tubes scarcely narrowed or differentiated from the pollen-sacs, prolonged only 0.2–0.3 mm and connivent; style declined. Dry upland woods; Nf. and Que. to Minn. and B.C., s. to Del., W.VA., and Io. June–Aug.

5. Pyrola rotundifolia L.   Rounded s. Lvs firm, mostly broadly elliptic to subrotund, 2.5–7 cm, broadly rounded above, rounded to truncate or short-cuneate at base, and always somewhat decurrent on the petiole; scapes 1.5–3 dm, usually with 1 or 2 scale-lvs; sep oblong or ovate-oblong, 3–4 mm, nearly twice as long as wide, erose or undulate, not overlapping at base; pet white, 8–10 mm; anthers 3–3.5 mm, minutely cuspidate at base, abruptly narrowed into very short tubes above; style declined; 2*n*=46. Dry or moist woods and bogs; circumboreal, s. in Amer. to N.C., Ky., Ind., and Minn. July, Aug. Our plants are nearly all var. *americana* (Sweet) Fern., as described above. The more northern, circumboreal var. *rotundifolia*, smaller in all parts, the sep 1.5–3 mm, the pet 5–7 mm, the anthers 2–2.5 mm, reaches our range in N.S.

6. Pyrola asarifolia Michx.   Pink s. Lvs mostly 3–6 cm; scapes 1.5–3 dm, with usually 1–3 scale-lvs; sep triangular, 2–3 mm, acute, entire, widest just above the base and therefore slightly overlapping; pet 5–7 mm, pink to pale purple; anthers cuspidate at base, abruptly narrowed above into very short tubes; style declined. Moist woods and bogs; Nf. to Alas., s. to N.Y., n. Ind., Minn. and N.M. July, Aug. (*P. uliginosa.*)

3. MONESES Salisb.   One-fld shinleaf. Pet 5; filaments slender, slightly widened near the base; anthers subapically attached, the pollen-sacs narrowed above the filament and each with an apical pore; ovary subglobose, concave at the top; style well developed; stigma peltate, with 5 erect marginal lobes; capsule subglobose, opening from the top down; delicate perennial from a very slender rhizome, with 1–4 sets of deciduous, opposite or ternate lvs near the base; fls white, solitary, terminal, nodding, long-peduncled. Monotypic.

1. Moneses uniflora (L.) A. Gray.   Plants 3–10 cm; lvs suborbicular, 1–2 cm, finely toothed or subentire; petioles 5–10 mm; fls fragrant, 12–20 mm wide; 2*n*=26. Damp woods and bogs; circumboreal, s. in Amer. to Conn., N.Y., Mich., Minn. and N.M. July, Aug. (*Pyrola u.*)

FAMILY **MONOTROPACEAE**, the Indian Pipe Family

Fls regular, perfect, hypogynous, (3)4–5(–8)-merous; sep distinct or sometimes shortly connate, imbricate, or rarely obsolete; pet distinct or connate into a lobed tube, imbricate or convolute, commonly of about the same color as the stem, rarely wanting; stamens 6–10(–14), mostly twice as many as the sep or pet, distinct or shortly connate at base, attached directly to the receptacle; anthers often becoming inverted during ontogeny, so that the morphological base is apical, opening lengthwise or sometimes by terminal pores or by slits across the broad tips of the pollen-sacs, sometimes awned or appendaged at the morphological base; pollen in monads, but often forming a coherent mass; ovary (4)5(6)-locular with axile placentation, or unilocular with strongly intruded, parietal placentas; style mostly short, with a capitate or peltate, often lobed stigma; fr a capsule or berry; seeds numerous and tiny; seed-coat a single layer of cells; embryo minute, undifferentiated, embedded in the well developed to sometimes very scanty endosperm; strongly mycotrophic perennial herbs without chlorophyll, variously white to pink, red, purple, yellow or brown in color, with much-reduced, scale-like, alternate lvs, the fls solitary or in a bracteate raceme (or a raceme of cymules). 10/12.

1 Pet separate; fr erect .... 1. *Monotropa.*
1 Pet united to form a campanulate or ureceolate cor; fr nodding.
  2 Cor campanulate; anthers awnless; fr a berry .... 2. *Monotropsis.*
  2 Cor globose-ovoid; anthers awned; fr a capsule .... 3. *Pterospora.*

1. **MONOTROPA** L.   Pet 4–6(–8), distinct, all or some saccate at base, the cor urceolate or broadly tubular; sep 0–5; stamens 8 or 10, (–14), on slender, hairy filaments, alternately longer and shorter; anthers transverse, opening by 1 or 2 clefts across the top; ovary 4–5(6)-locular; style short, thick; stigma broad, peltate, umbilicate; capsule ovoid to subglobose, loculicidal, opening from the top downward; white, yellow, pink or red plants, turning black in drying, with erect stems beset with small scale-lvs, nodding fls the same color as the stem, and erect frs. 2, N. Hemisphere.

1 Fls solitary .... 1. *M. uniflora.*
1 Fls few to many, racemose .... 2. *M. hypopithys.*

    1. **Monotropa uniflora** L.   Indian Pipe. Stems 1–2 dm, solitary or several, waxy-white, rarely pink or red; fl solitary, nodding, odorless, 10–17 mm; sep often none; pet broadly oblong, slightly widened distally; anthers opening by 2 clefts across the top; style longer than the ovary; stigma glabrous. Rich woods; Nf. to B.C., s. to Fla., Calif., and Colombia; also e. Asia. June–Aug.

    2. **Monotropa hypopithys** L.   Pinesap. Stems 1–3 dm, often clustered, yellow, tawny, pink, or red, ± pubescent; raceme dense, at first nodding, erect at anthesis; fls 8–18 mm, the lower usually 4-merous, the terminal often larger and 5-merous; sep lanceolate, erect, unlike the basally saccate pet; anthers opening by a single cleft into 2 very unequal valves; style shorter than the ovary; stigma ± villous at the margin; 2n=16, 48. Moist or dry woods, usually in acid soil; interruptedly circumboreal, but not at high latitudes, in Amer. s. to Fla. and Mex. (*Hypopitys americana.*)

2. **MONOTROPSIS** Schwein.   Sweet pinesap. Cor gamopetalous, cupulate, 5-cleft; sep 5, erect; stamens 10; filaments slender, glabrous; anthers fitting under the stigma, subglobose, unappendaged, each pollen-sac opening by a long slit on one side; ovary glabrous, with deeply intruded parietal placentas; style short; stigma capitate; fr a berry; purple or brownish-purple plants, the simple stems beset with numerous ovate, scarious, crowded and overlapping scale-lvs; raceme at first nodding, erect at anthesis. Monotypic.

    1. **Monotropsis odorata** Elliott.   Stems 5–12 cm; fls pinkish, 1 cm, with the odor of violets, nearly equaled by the ovate bracts, the pedicel often with 2 or 3 bracteoles; sep lanceolate; cor rather fleshy, persistent, its lobes ovate-oblong, shorter than the tube. Dry woods; Del. and Me. to Ky., s. to S.C. and Ala. Feb.–Apr.

3. **PTEROSPORA** Nutt.   Pine-drops. Cor gamopetalous, urceolate, 5-lobed, persistent; stamens 10; filaments flattened, elongate; anthers thick-triangular, somewhat inflexed but not strongly inverted, each pollen-sac opening by a large lateral slit and with a retrorse, filiform appendage at the morphological base; ovary 5-locular; style columnar, short; stigma fleshy, capitate, shallowly 5-lobed; capsule depressed-globose, deeply umbilicate, loculicidal, opening from the morphological base toward the distal end; seeds with a broad terminal wing; pink to reddish or purplish (or eventually brownish) plants with small scale-lvs and long racemes of nodding fls and frs. Monotypic.

    1. **Pterospora andromedea** Nutt.   Stems erect, simple, 3–10 dm, glandular-hairy, with numerous scale-lvs especially below; fls numerous in a terminal raceme 1–3 dm; cor 6–7 mm, white to more often reddish, its spreading-recurved lobes 1–2 mm; 2n=16. Coniferous woods, chiefly in dry soil; P.E.I. to N.Y. and Pa., w. to s. Ont. and Mich.; widespread in the w. cordillera, from s. Can. to Mex. June–Aug.

FAMILY **DIAPENSIACEAE**, the Diapensia Family

Fls regular, perfect, hypogynous, 5-merous except the 3-locular ovary; cor sympetalous, with imbricate or convolute lobes, or (*Galax*) the pet virtually distinct; functional stamens 5, attached to the cor-tube alternate with the lobes; 5 staminodes often alternate with the stamens, and then sometimes connivent or connate with them to form a tube; anthers often inflexed, but not becoming inverted, opening by longitudinal or transverse slits, sometimes awned at the base; pollen in monads; placentation axile; fr a loculicidal capsule; seeds several or numerous, small, with copious endosperm around a straight or slightly curved, dicotyledonous embryo; evergreen herbs or subshrubs with alternate or opposite (often all basal), simple, exstipulate lvs and small or medium-sized, white or pink, solitary or racemose fls. 6/18.

1 Subshrubs with cauline lvs; staminodes wanting.
  2 Fls few, pedunculate; pollen-sacs divergent; arctic-alpine .... 1. *Diapensia.*
  2 Fls numerous, sessile; pollen-sacs parallel; southern .... 2. *Pyxidanthera.*
1 Scapose herbs with basal lvs; staminodes present.
  3 Style none; fls racemose, 3–5 mm; stamens and staminodes monadelphous .... 3. *Galax.*
  3 Style filiform; fls (in ours) solitary, 15–25 mm; stamens distinct .... 4. *Shortia.*

1. DIAPENSIA L. Cal deeply 5-parted; cor campanulate, the 5 lobes broadly obovate, about equaling the tube; stamens 5, adnate to the cor up to the sinuses; filaments broad and flat; pollen-sacs 2, transversely divergent, opening by longitudinal slits; style slender, elongate; stigma truncate; fr subglobose; dwarf, branched and matted, evergreen subshrubs with mostly opposite lvs; fls white, each closely subtended by 3 bracts, solitary on a terminal peduncle. 4, the others Asian.

  1. Diapensia lapponica L. Stems matted, 5–10 cm; lvs crowded and imbricate, spatulate, 6–15 mm; peduncles 1–4 cm; bracts ovate; cal-lobes obovate-oblong; cor 7–10 mm; fr 5 mm; 2n=12. Circumboreal, s. to alpine summits in Me., N.H., Vt., and N.Y. June.

2. PYXIDANTHERA Michx. Pyxie; flowering moss. Sep distinct; cor broadly campanulate, the obovate lobes exceeding the tube; stamens 5, adnate to the cor up to the sinuses; filaments broad and flat, much shorter than the cor-lobes; anthers inflexed, the 2 pollen-sacs parallel, acute, opening by transverse slits; style slender; stigma rounded; branched, prostrate, evergreen subshrubs with small, alternate or fascicled, crowded lvs and numerous sessile fls solitary at the ends of short branches. Monotypic.

  1. Pyxidanthera barbulata Michx. Forming mats up to 1 m wide; lvs oblanceolate, 3–8 mm, sharply acute, usually hairy toward the base; fls commonly very numerous, white to rosy, 5–8 mm wide; fr globose, 2 mm; 2n=12. Sandy pine-barrens; L.I.; N.J.; se. Va. to S.C. Apr., May.

3. GALAX L. Sep distinct; cor cleft nearly to the base; stamens and staminodes monadelphous, the tube adnate to the cor at the base, free above, 10-lobed at the top, the lobes opposite the sep bearing anthers; anthers inflexed, with a single transversely dehiscent pollen-sac; style none; stigma broad; fr ovoid; evergreen perennial herbs with long-petioled basal lvs and erect scapes bearing a long, spike-like raceme of small white fls. Monotypic.

  1. Galax aphylla L. Plants malodorous; lvs orbicular to broadly ovate, 4–15 cm wide, serrate, deeply cordate at base, on petioles 8–20 cm; scapes 2–4 dm, floriferous in the terminal 5–10 cm; fls 3–5 mm; fr 2–3 mm; 2n=12, 24. Moist or dry woods, chiefly in the mts.; w. Md. to Ky., s. to Ga. and Ala., and extending to the coastal plain in N.C. and s. Va. June, July.

4. SHORTIA T. & G. Cal deeply cleft; cor open-campanulate, cleft to the middle or below; stamens distinct; 5 small, scale-like staminodia incurved over the ovary; anthers inflexed, with 2 pollen-sacs; style filiform; stigma obscurely 3-lobed; capsule ovoid; evergreen, rhizomatous, perennial herbs with long-petioled basal lvs and erect scapes bearing (in our sp.) a rather large, solitary fl. 8, the others e. Asian.

  1. **Shortia galacifolia** T. & G.   Lvs orbicular to broadly elliptic or subcordate, 2–7 cm, toothed, on petioles 3–14 cm; scapes 1–2 dm; fls nodding; cor white, rarely pale pink or blue, 1.5–2.5 cm, the broadly rounded lobes conspicuously erose; fr 5–7 mm, surpassed by the persistent, slightly accrescent cal; $2n=12$. Shady banks along streams, and in moist, wooded ravines; scattered localities from Amherst Co., Va. to S.C. and Ga. Mar.–Apr. (*Sherwoodia g.*)

FAMILY SAPOTACEAE, the Sapodilla Family

Fls regular, perfect, hypogynous, sympetalous, usually 5-merous, the sep distinct or nearly so, imbricate, the cor-lobes imbricate and often with secondary lobes or appendages; stamens attached to the cor-tube, in 1–3 whorls, often some of them reduced to staminodes, at least one anther-bearing whorl opposite the cor-lobes; ovary with a single axile or axile-basal ovule in each of the 5(2–14+) locules; fr fleshy, indehiscent; seeds large, commonly with a large, excavated scar of attachment; trees or shrubs with milky juice and with simple, mostly alternate, mostly entire, usually exstipulate lvs and mostly small fls borne (singly or) in small cymose clusters at the nodes. 50+/1100.

  1. **BUMELIA** Swartz, nom. conserv.   Southern buckthorn. Fls 5-merous throughout; cor-lobes with a petaloid appendage on each side; stamens 5, opposite the cor-lobes; staminodes petaloid, alternate with the cor-lobes; shrubs or small trees, often thorny, with alternate lvs (often clustered on dwarf lateral branches) and sessile lateral cymose umbels of small, white or greenish fls. 23, New World.

1 Mature lvs glabrous or nearly so, the hairs, if any, fine and appressed .... 1. *B. lycioides.*
1 Mature lvs loosely woolly-villous beneath .... 2. *B. lanuginosa.*

  1. **Bumelia lycioides** (L.) Pers.   Smooth s. b. Shrub or small tree to 10 m; lvs elliptic to elliptic-oblanceolate, 4–13 × 1–4.5 cm, reticulate-veiny, silvery-strigulose beneath when young, soon glabrate; fls 10–60 in a cluster, the pedicels 4–10 mm, glabrous or nearly so; cor 3–4 mm, the tube shorter than or equaling the lobes; fr ellipsoid to obovoid or subglobose, 7–13 mm; seeds 6–9 mm. Low woods, thickets, and river-banks; Fla. to e. Tex., n. to s. Va., s. Ind., and s. Mo. June–July.
  2. **Bumelia lanuginosa** (Michx.) Pers.   Woolly s. b. Much like no. 1; lvs oblanceolate to sometimes obovate or elliptic, the larger ones mostly 5–10 × 1.5–3.5 cm, loosely woolly-villous with persistent gray hairs beneath; pedicels often hairy; $2n=24$. Dry or rocky woods; Fla. to Tex. and n. Mex., n. to Mo. and Kans. June, July. Our plant, as here described, is var. *oblongifolia* (Nutt.) R. Clark.

FAMILY EBENACEAE, the Ebony Family

Dioecious or polygamo-dioecious; cal deeply lobed, persistent and often accrescent in fr; cor sympetalous; stamens 2–4 times as many as the cor-lobes, hypogynous or attached to the base of the cor-tube in staminate fls, in the pistillate fls none or fewer and usually sterile; ovary plurilocular, with mostly 3–8 carpels and as many or twice as many chambers; style deeply cleft, or the styles distinct; ovules apical-axile, 1 per chamber; fr a berry; seeds with thin testa and abundant, hard endosperm; cotyledons flat; trees or shrubs with alternate, entire, exstipulate lvs. 5/450.

  1. **DIOSPYROS** L.   Persimmon. Dioecious or nearly so, the pistillate fls commonly with nonfunctional stamens; fls usually 4-merous; cal deeply lobed; cor urceolate to rotate, with spreading or recurved lobes, those of the staminate and the fertile fls usually different in size; stamens 4–many, usually 16, hypogynous or inserted at the base of the cor; ovary usually 4-locular, or 8-locellar. 400, mainly warm reg.

  1. **Diospyros virginiana** L.   Tree to 15 m, sometimes colonial, with deeply furrowed and cross-checked bark; lvs oval or oblong, 8–15 cm, shortly acuminate, on petioles 1–2 cm; fls greenish-yellow, the staminate

solitary or 2–3 together, 1 cm, with usually 16 linear, acuminate anthers about as long as the cor-tube, the pistillate solitary, 1.5–2 cm, with larger cal and usually 8 sterile anthers; ovary mostly 8-locellar; fr yellowish-brown, 2–4 cm thick, edible after frost; $2n=60$, 90. Mostly in dry woods; se. Conn. and s. N.Y. to Fla., w. to Io., Kans., and Tex. Var. *virginiana,* chiefly e. of the Miss. R., has glabrous twigs with thin lvs acute or cuneate at base and glabrous or nearly so beneath. Var. *pubescens* (Pursh) Dippel, chiefly w. of the Miss. R., has hirsutulous twigs with firm lvs broadly cuneate to rounded or subcordate at base and usually thinly but permanently hairy beneath.

FAMILY **STYRACACEAE**, the Storax Family

Fls regular, perfect, sympetalous, hypogynous to epigynous, 4–5-merous; cal in our spp. adherent to the ovary, its lobes short; stamens mostly twice as many as the cor-lobes, all in one series, with linear anthers; ovary 3–6-locular, with 1–many (mostly 4–6) ovules per locule, the placentation axile; style 1; fr mostly dry; seeds with abundant, oily endosperm; cotyledons broad; trees or shrubs, stellate-hairy to glabrous, with simple, alternate, exstipulate lvs. 10/150.

1 Fls 5-merous; lower half of the cal adherent to the ovary; fr not winged .... 1. *Styrax.*
1 Fls 4-merous; cal almost completely adherent to the ovary; fr winged .... 2. *Halesia.*

1. **STYRAX** L. Storax. Fls 5-merous; cal obconic, its lower half adherent to the ovary; cor-lobes equaling or (ours) longer than the tube, eventually spreading or recurved; stamens 10, inserted at the base of the cor, monadelphous at base; ovary 3-locular below, unilocular above; ovules few in each locule; fr dry, crustaceous, usually 1-seeded, often irregularly dehiscent, its lower third adherent to the persistent cal. 120, widespread.

1 Lvs narrowly elliptic, glabrous or nearly so; cor-lobes 10–12 mm .... 1. *S. americanus.*
1 Lvs broader, commonly obovate, tomentose beneath, cor-lobes 15–22 mm .... 2. *S. grandifolius.*

1. **Styrax americanus** Lam. American snowbell. Branched shrub to 4 m; lvs narrowly elliptic, 3–10 cm, acuminate, usually serrulate distally, paler beneath but glabrous or nearly so; fls on short, lateral, leafy branches, a few solitary or paired in the axils and 2–4 in short, terminal, semi-drooping racemes; cal 3–4 mm; cor-lobes 10–12 mm, longer than the tube; fr subglobose or obovoid, 6–8 mm thick; $2n=16$. Swamps and wet woods, chiefly on the coastal plain; se. Va. to Fla. and La., n. in the Miss. valley to s. Ind., and at scattered stations in s. Pa., Ky., s. O., and n. Ind. May, June. Our plants are var. *americanus.* A more hairy var. (*pulverulentus*) occurs southward.

2. **Styrax grandifolius** Aiton. Big-leaf snowbell. Shrub or small tree to 9 m; lvs broadly obovate to broadly elliptic or sometimes broadly ovate, 5–20 cm, acute or shortly acuminate, distantly toothed to entire, pale and ± stellate-tomentose beneath; racemes longer and more drooping than in no. 1, with 5–20 fls; cal 4–5 mm, canescent; cor-lobes 15–22 mm; fr subglobose, 7–9 mm thick; $2n=32$. Woods, often in drier sites than no. 1; Va. to c. Fla., w. to Ky., s. Ill., s. Mo., Ark., and e. Tex. Apr.

2. **HALESIA** Ellis. Silver-bell tree. Fls 4-merous; cal narrowly obconic, adherent to the ovary, only the very short lobes free; cor broadly campanulate; stamens commonly 8(16), inserted near the base of the cor-tube, monadelphous below; ovary inferior; ovules 4 per locule; fr dry and indehiscent, the oblong body conspicuously winged; fls white, in umbel-like lateral clusters. 6, se. U.S. and e. Asia.

1. **Halesia tetraptera** Ellis. Shrub or small tree to 10(–30) m; lvs oblong or elliptic to oblong-obovate, acute or acuminate, finely serrulate, becoming 10–15 cm after anthesis; fls in clusters of 2–5, on slender pedicels 1–2 cm; cor opening before maturity, at anthesis 2–2.5 cm; fr 2.5–3.5 cm, 4-winged; $2n=24$. Rich, moist woods; w. Va. to s. O. and s. Ill., s. to Fla. and e. Tex. Apr., May. (?*H. carolina* L., an earlier name, perhaps misapplied.)

FAMILY **SYMPLOCACEAE**, the Sweet-lf Family

Fls regular, 5-merous, mostly perfect, epigynous or half-epigynous, the campanulate cal adherent to the ovary as far as the sinuses; cor nearly polypetalous, the lobes connate only at base; stamens 12–many, attached to the base of the cor-tube, usually in more than one series, or in fascicles alternate with the cor-lobes; anthers short and broad; ovary 2–5-locular, each locule with 2–4 ovules on the axile placenta; style 1; fr indehiscent, dry or ± fleshy, most commonly a single-stoned drupe; endosperm well developed; cotyledons very short; trees or shrubs with simple, alternate, exstipulate, often sweet-tasting lvs. 1/300+.

1. SYMPLOCOS Jacq.   Sweet-lf; horse-sugar. Characters of the family. 300+, mostly warm regions.

1. Symplocos tinctoria (L.) L'Hér.   Shrub to 5 m; lvs elliptic to commonly oblanceolate, to 15 cm, acute or short-acuminate, entire, narrowed to the base, glabrous or nearly so at maturity; fls sessile in lateral clusters on old wood, usually precocious; cor yellow, the lobes 7–9 mm; stamens in 5 clusters alternating with the cor-lobes, each cluster with 6–10 filaments united at base and about as long as the cor; fr slenderly ellipsoid, dry, 6–10 mm. Woods, chiefly on the coastal plain; Del. to Fla. and La., n. in the Miss. valley to Ark. and Tenn. Apr., May.

FAMILY **PRIMULACEAE**; the Primrose Family

Fls regular, perfect, hypogynous (half-epigynous in *Samolus*), usually 5-merous; cor (absent in *Glaux*) sympetalous; stamens as many as the cor-lobes and opposite them (free in *Glaux*); ovary unilocular, with several or many ovules on a free-central placenta; style 1; stigma capitate or truncate; fr a capsule; seeds with copious endosperm; herbs with simple (pinnatifid in *Hottonia*), exstipulate lvs. 30/1000.

1 Lvs all or chiefly cauline.
  2 Lvs deeply pinnatifid into linear segments; plants aquatic . . . . 7. *Hottonia*.
  2 Lvs entire or nearly so; plant terrestrial or amphibious.
    3 Lvs all or chiefly opposite or whorled.
      4 Lvs in a single terminal whorl; fls commonly 7-merous . . . . 2. *Trientalis*.
      4 Lvs opposite or in several whorls; fls 5–6-merous.
        5 Fls yellow (sometimes pale) . . . . 1. *Lysimachia*.
        5 Fls white, pink, red, or blue.
          6 Fls nearly sessile; cor none . . . . 3. *Glaux*.
          6 Fls long-pediceled; cor present . . . . 4. *Anagallis*.
    3 Lvs all or chiefly alternate; fls never yellow.
      7 Fls nearly sessile in the axils . . . . 5. *Centunculus*.
      7 Fls in racemes.
        8 Cal-lobes much shorter than the tube . . . . 6. *Samolus*.
        8 Cal divided nearly to the base . . . . 1. *Lysimachia*.
1 Lvs in a basal rosette; infl a bracteate umbel terminating a scape.
  9 Cor-lobes spreading or ascending.
    10 Cor-tube longer than the cal; style well developed, slender . . . . 8. *Primula*.
    10 Cor-tube equaling or shorter than the cal; style minute . . . . 9. *Androsace*.
  9 Cor-lobes sharply reflexed from the base . . . . 10. *Dodecatheon*.

1. LYSIMACHIA L.   Loosestrife. Fls 5(6)-merous in our spp.; cal spreading, parted nearly or quite to the base; cor-tube very short, the limb broadly campanulate to rotate; stamens 5(6), distinct or monadelphous, sometimes alternating with small staminodes; perennial herbs with opposite or whorled, or rarely alternate lvs and yellow or rarely white fls, often marked with dots or lines, in axillary or terminal racemes or panicles or solitary in the axils. (*Steironema; Naumbergia.*) 150, cosmop. Our first 6 species are self-incompatible and produce frequent hybrids.

L. clethroides Duby, differing from all our other spp. in its alternate lvs and white fls, occasionally escapes from cult., but does not appear to be established; it is native to e. Asia.

1 Staminodes alternating with the stamens; filaments free; lvs not punctate. (*Steironema, Seleucia.*)
  2 Principal lvs ovate or lance-ovate, broadest not far from the base, abruptly narrowed to the well defined petiole, generally 1.5 cm wide or more (or sometimes a little narrower in *L. radicans*, marked by its reclining habit).

3 Plants erect.
  4 Plants with long, slender, creeping rhizomes; petiole prominently ciliate throughout its length
    .... 1. *L. ciliata.*
  4 Plants with a short, freely rooting rhizome that may be scarcely more than a root-crown;
    petiole glabrous, or ciliate mainly below the middle, the distal part glabrous or with a
    few scattered cilia.
    5 Petiole not ciliate, or ciliate only at the very base; lvs mostly 1.5–3.5 cm wide; dry upland
      woods .... 2. *L. tonsa.*
    5 Petiole ± ciliate below the middle, often with a few scattered cilia distally; lvs rarely over
      2 cm wide; wet places .... 4. *L. hybrida.*
3 Plants soon becoming prostrate, often rooting at the nodes .... 3. *L. radicans.*
2 Principal lvs narrowly lanceolate to linear, seldom as much as 1.5 cm wide, gradually narrowed to
  a long, tapering base, often without a clearly defined petiole.
  6 Lvs pinnately veined, the margins scabrellous or even ciliate, not revolute.
    7 Plants with a short, stout, freely rooting rhizome .... 4. *L. hybrida.*
    7 Plants with long, slender, stoloniform rhizomes .... 5. *L. lanceolata.*
  6 Lvs with obscure or obsolete lateral veins, the margins smooth and revolute .... 6. *L. quadriflora.*
1 Staminodes none; lvs punctate.
  8 Filaments separate; cor-lobes linear, much shorter than the stamens. (*Naumbergia.*) .... 7. *L. thyrsiflora.*
  8 Filaments connate below; cor-lobes broader, at least as long as the stamens. (*Lysimachia.*)
    9 Pet not dotted or lined with black or red; plants hairy; naturalized spp.
      10 Cal-lobes 3–5 mm, red-margined; cor-lobes entire .... 8. *L. vulgaris.*
      10 Cal-lobes 5–8 mm, not red-margined; cor-lobes glandular-ciliate .... 9. .... *L. punctata.*
    9 Pet dotted or lined with black or dark red; plants glabrous or hairy; native and intr. spp.
      11 Fls all axillary.
        12 Stems prostrate; cor-lobes 10–15 mm .... 10. *L. nummularia.*
        12 Stems erect; cor-lobes 6–8 mm .... 11. *L. quadrifolia.*
      11 Fls all or mostly in a terminal raceme.
        13 Fls all, or all except the very lowest, subtended by greatly reduced linear bracts .... 12. *L. terrestris.*
        13 Lower fls from the axils of normal foliage-lvs .... 13. *L.* × *producta.*

1. Lysimachia ciliata L. Fringed l. Erect, 4–13 dm, rhizomatous; lvs ovate to lanceolate, 4–14 × 1.5–6 cm, the base rounded (or even subcordate) or obtuse to cuneate, the petiole 0.5–5 cm, evidently ciliate its whole length; fls solitary in the upper axils, on pedicels 1.5–6.5 cm; cal-lobes 3.5–9 × 1.5–4.5 mm, with 3–5 parallel often reddish-brown veins; cor-lobes 5–12 × 3–9 mm; stamens distinct; staminodes linear or subulate; $2n= 34, 92, 96, 100, 108, 112$. Moist or wet lot ground; in shade or sun; N.S. and Que. to B.C. and Alas., s. to Ga., La., N.M., and Oreg. May–Sept. (*Steironema c.*)

2. Lysimachia tonsa (Wood) Knuth. Appalachian l. Much like no. 1, but the rhizome very short and freely rooting, scarcely more than a mere crown; lvs mostly ovate, the larger 4–8 × 1.5–3.5 cm, the long petiole not ciliate, or ciliate only at the very base; $2n=34, 102$. Dry, upland woods, mostly in the mts.; Ky. and w. Va. to Ga. and n. Ala. Late May–Aug. (*Steironema intermedium* Kearney, not Janka.)

3. Lysimachia radicans Hook. Trailing l. Rhizome short and freely rooting; stems slender, to 1 m, laxly and diffusely branched, soon becoming prostrate or spreading and often rooting; nodes usually with a fringe of short hairs; lvs lanceolate or lance-ovate, the larger ones 4–10 × 1–2.5 cm, scabrous on the margin, broadly acute to rounded at base, on long, eciliate petioles; cal-lobes and cor-lobes 3–5 mm; stamens distinct; staminodes triangular-subulate, obtuse; fr equaling or exceeding the cal; $2n=34$. Swamps and wet soil, usually in shade; e. and c. Va.; Mo. and s. Ill. to e. Tex. and w. Fla. June–Aug. (*Steironema r.*)

4. Lysimachia hybrida Michx. Mississippi-valley l. Erect from a short, stout, freely rooting rhizome, to 1 m, the lateral branches usually numerous and exceeding the subtending lvs; principal lvs lance-linear to narrowly oblong or lance-ovate, 4–10 × 1–2 cm, flat, green beneath, tapering or more rounded to the base, the petiole ciliate mainly below the middle, sometimes with a few scattered cilia distally; cal-lobes thin, 4–-8 m, the midnerve and usually also the 2 or 4 lateral ones visible under a hand-lens; cor-lobes 6–10 × 4–10 mm; stamens distinct; staminodes triangular-ovate to subulate; fr 4–6.5 mm thick; $2n=34$. Sloughs, wet woods, and wet prairies; Me. and s. Que. to Fla., irregularly w. to Alta., Wash., and Ariz., commonest along the Atlantic seaboard from Me. to Md., and in the upper Miss. Valley. Late June–Aug. (*Steironema h.*)

5. Lysimachia lanceolata Walter. Lance-lvd l. Erect from long, slender, stoloniform, superficial rhizomes, 2–9 dm, the lateral branches scarcely exceeding the subtending lvs; lower cauline lvs ovate to obovate, petioled; principal lvs linear to lanceolate or narrowly oblong, to 15 × 1.5 cm, often folded along the midrib, paler beneath, scabrellous on the margin or ciliate near the base, gradually tapering below, often with little or no distinction of petiole and blade; cal-lobes firm, 4–7 mm, the nerves, especially the lateral ones, obscure or hidden; cor-lobes 5–10 × 4–6 mm; fr 2–4.5 mm thick; $2n=34$. Moist or wet woods and prairies; N.J. and Pa. to n. Fla., w. to Wis., Io., Mo., and e. Tex. June–Aug. (*Steironema l.; S. heterophyllum.*)

6. Lysimachia quadriflora Sims. Smooth l. Erect, 3–10 dm from a short rhizome, usually with short branches from the upper nodes; lvs firm, usually ascending, linear, 3–8 cm × 2–7 mm, gradually tapering to the sessile base, the margins smooth and revolute or with a few cilia at the very base, the lateral veins obscure or obsolete; fls almost entirely in terminal clusters on the stem and branches; cal-lobes lanceolate, 4–6 mm; cor-lobes 7–12 × 5–9 mm; stamens distinct; staminodes triangular to subulate; $2n=34$. Moist or wet soil, especially on prairies; Mass. to Ga., w. to Man. and Ark., chiefly n. of the Ohio R. and w. of the Appalachians, and in the Ozark region of Mo. July, Aug. (*Steironema q.*)

7. Lysimachia thyrsiflora L.   Swamp-l. Erect, 3–7 dm, from long, stoloniform rhizomes; lvs punctate, narrowly lanceolate to linear, 5–12 cm; infl a few dense, short racemes 1–3 cm on spreading peduncles 2–4 cm arising from the middle axils; fls mostly 6-merous; cal-lobes narrowly lanceolate, 2.5 mm; cor-lobes linear, 4–5 mm, marked with black; stamens separate, erect, nearly twice as long as the pet; $2n=54$. Swamps; circumboreal, s. to N.J., O. and Mo. (*Naumbergia t.*)

8. Lysimachia vulgaris L.   Garden-l. Stems erect, to 1 m, from long, stoloniform rhizomes, ± soft-hairy; lvs whorled or opposite, punctate, lanceolate to lance-ovate, 7–12 cm, soft-hairy beneath; infl a terminal raceme and a series of peduncled short racemes or panicles from the upper axils; cal-lobes 2.5–4.5 mm, red-margined; cor-lobes 8–12 mm, entire; $2n=56, 84$. Native of Eurasia, occasionally escaped from cult. in our range.

9. Lysimachia punctata L.   Spotted l. Erect, to 1 m, rarely branched, pubescent, spreading by stoloniform rhizomes; lvs chiefly in whorls of 3 or 4, occasionally only opposite, punctate, lanceolate, 5–10 cm; fls in axillary whorls, usually more numerous than the subtending lvs, the uppermost whorls with smaller lvs and shorter internodes; pedicels 1–2 cm; cal-lobes lance-linear, 5–8 mm; cor-lobes 12–16 mm, glandular-ciliolate on the margins. $2n=30$. Native of Eurasia, now an occasional weed in our range, especially northward.

10. Lysimachia nummularia L.   Moneywort. Creeping, often forming mats; lvs opposite, punctate, short-petioled, broadly quadrate to subrotund, 1–2.5 cm; fls solitary in the axils, the pedicels about equaling the lvs; cal-lobes foliaceous, triangular-ovate, 8 mm; cor-lobes 10–15 mm, dotted with dark red; $2n=32, 36, 43, 45$. Native of Europe, escaped from cult. into moist places nearly throughout our range, often weedy. June–Aug.

11. Lysimachia quadrifolia L.   Whorled l. Erect from long, stoloniform rhizomes, 3–9 dm, the stem glabrous or sparsely hairy, rarely branched; lvs in whorls of (3)4(–7), punctate, narrowly or broadly lanceolate, 5–10 cm, hairy beneath, widely spreading; fls axillary on spreading pedicels 2–5 cm, the upper half or two-thirds of the plant floriferous; cal-lobes lance-oblong, 2–4 mm; pet 6–8 mm, oblong or elliptic, yellow with dark lines; $2n=84$. Moist or dry upland soil, chiefly in open woods; Me. to S.C., w. to Wis., e. Minn., Ill., Ky., and Ala.

12. Lysimachia terrestris (L.) BSP.   Bulbil-l. Erect from long, stoloniform rhizomes, 4–8 dm, the stem glabrous, often branched but usually with a single raceme; lvs opposite, punctate, narrowly lanceolate, the principal ones 5–10 cm; elongate bulblets develop in the axils late in the season; raceme erect, 1–3 dm, many-fld, its bracts linear or subulate, 3–8(–10) mm; pedicels 8–15 mm; cor-lobes narrowly elliptic, 5–7 mm, yellow, marked with dark lines; $2n=84$. Open swamps and wet soil; Nf. to S.C., w. to Minn., Sask., Io., and Tenn. June–Aug. A vigorous sterile hybrid with no. 7, called *L.* × *commixta* Fern., has numerous peduncled racemes from the middle axils, as well as a terminal raceme; it propagates vegetatively, and sometimes occurs in the absence of the parents.

13. Lysimachia × producta (A. Gray) Fern.   A ± fertile hybrid derived from nos. 11 and 12, more nearly resembling no. 12, but with the bracts of the well developed terminal raceme larger, the lower ones often foliaceous, pedicels to 3 cm; principal lvs often whorled. Moist soil; Me. to Md. and less commonly to Va., N.C., Ky., and Wis. June, July.

2. TRIENTALIS L.   Starfl.

Fls ordinarily 7-merous; cal deeply divided into nearly separate lance-linear sep; cor rotate, with very short tube and lanceolate to ovate lobes; stamens at the base of the cor, the slender filaments connected by a membranous ring; capsule 5-valved, many-seeded; low herbs, perennial by a slender rhizome, bearing a few lvs all or mostly clustered in a whorl at the top, from the axils of which appear one or more white fls on long, slender pedicels. 3, N. Temp.

1. Trientalis borealis Raf.   Stem 1–2 dm, usually with a small scale-lf near the middle and a whorl of lanceolate, acuminate lvs 4–10 cm at the top; pedicels 1–few, 2–5 cm; sep lance-linear; cor 8–14 mm wide, its lobes slenderly acuminate and mucronate; $2n=96$. Rich woods and bogs; Lab. and Nf. to Alta., s. to Pa., n. O., n. Ill., and Minn.; also on the coastal plain from Mass. to Va., and irregularly s. in the Appalachians to n. Ga. May, June. (*T. americana.*)

3. GLAUX L.   Sea-milkwort.

Cal campanulate, petaloid, cleft to below the middle into 5 oblong-obovate, ± spreading, obtuse, persistent lobes; cor none; stamens hypogynous, shorter than the cal and alternate with its lobes; capsule broadly ovoid, 5-valved, somewhat beaked, few-seeded; small, succulent, glabrous perennial herbs with opposite, sessile lvs and axillary, subsessile, white to red fls. Monotypic.

1. Glaux maritima L.   Erect and simple to diffusely branched and often prostrate, 5–30 cm; lvs linear to broadly oblong or oval, 5–25 × 1.5–10 mm; cal 4–5 mm; $2n=30$. Moist or dry saline soil; circumboreal, in Amer. s. to Va. and N.M.

4. ANAGALLIS L. Pimpernel. Sep distinct; cor rotate, deeply 5-parted, the lobes convolute in bud; stamens inserted near the base of the cor-tube; filaments hairy; capsule circumscissile near the middle; annual or perennial herbs with opposite, entire lvs and small, axillary, pediceled fls.25, widespread.

1. **Anagallis arvensis** L. Annual, usually much-branched, diffuse to erect, 1–3 dm; stem 4-angled; lvs sessile, decussate, elliptic to ovate, 1–2 cm; fls solitary in the axils; pedicels slender, ascending, in fr recurved; sep 3–4 mm; cor equaling or somewhat surpassing the cal, open only in fair weather; $2n=40$. Native of Eurasia, now a weed in temperate regions almost throughout the world, sometimes along salted highways. The common var. *arvensis* has scarlet or brick-red to sometimes white fls on pedicels that usually exceed the lvs. Var. *caerulea* (Schreber) Gren. & Godron, rare in our range, mostly more southern than var. *arvensis* in Europe, has blue fls on pedicels usually shorter than the lvs.

5. CENTUNCULUS L. Chaffweed. Fls ephemeral, 4(5)-merous; sep narrowly lanceolate; cor marcescent over the capsule; shorter than the cal, the tube broad, with anthers at its throat, the lobes lance-ovate, spreading, soon becoming erect and connivent; style very short; capsule subglobose, circumscissile; low annual with chiefly alternate lvs and minute pink fls nearly sessile in the axils. Monotypic.

1. **Centunculus minimus** L. Erect or ascending, 4–12 cm; lvs subsessile, oblong or obovate, 5–10 mm; fls in most axils; sep 2 mm; cor shorter than the cal; $2n=22$. Moist or wet soil; irregularly cosmopolitan, especially in warmer regions, n. along the coast to N.S. and inland to O. and Minn. Apr.–Sept. (*Anagallis m.*)

6. SAMOLUS L. Water-pimpernel. Cal campanulate, its broadly triangular lobes shorter than the tube; cor deeply 5-lobed, with a minute, scale-like staminode in each sinus; stamens about equaling the cor-tube and inserted near its base; ovary ca two thirds inferior; style very short or obsolete; stigma capitellate; capsule globose, dehiscent by 5 valves as far as the base of the cal-lobes; perennial herbs with entire lvs and small white fls in terminal (or in part axillary) racemes on slender pedicels bracteolate near the middle. 10, cosmop.

1. **Samolus floribundus** HBK. Stems 1–3 dm, branched at least above; lvs alternate and in a basal tuft, spatulate to obovate, obtuse or rounded, 2–5(–10) cm (including the petiole); racemes 3–15 cm; pedicels widely spreading, the lower usually 10–15 mm; cal-tube 1.5 mm; fls 2–3 mm wide; fr 2.5 mm. Brackish shores along the coast, and muddy banks of streams and ditches inland; tropical Amer., n. to N.B., s. Mich., and B.C. May–Sept. (*S. parviflorus.*)

7. HOTTONIA L. Featherfoil. Sep 5, elongate, imbricate in bud; cor (in ours) tubular or somewhat salverform, shorter than the cal, 5-lobed to near the middle; stamens very short, inserted near the base of the cor; style slender, included; capsule depressed-globose; aquatic herbs, the submersed lvs deeply pinnatifid into linear segments, the stem bearing a terminal umbel of racemes, the peduncle and axis of each inflated, constricted at the joints, and bearing whorls of bracteate fls. 2, the other Eurasian.

1. **Hottonia inflata** Elliott. Stems ordinarily submersed, to 5 dm; lvs oblong, 2–6 cm; peduncles several, partly emersed, the successive segments progressively shorter; fls 3–10 at each node, sessile or on pedicels to 15 mm, each subtended by a linear bract; sep linear, 4–10 mm; cor 4–5 mm, white; $2n=22$. Quiet shallow water or occasionally in wet soil, chiefly on the coastal plain; Me. to Fla. and La., n. in the Miss. valley to s. Ind., and irregularly to O. and W.Va. May–Aug.

8. PRIMULA L. Primrose. Cal campanulate to tubular, 5-lobed, persistent; cor funnelform or salverform, the tube longer than the cal (in our spp.), the lobes retuse or 2-lobed; stamens included, borne on the cor-tube; filaments very short; capsule 5-valved at the tip; scapose perennials, the often large and showy fls in a terminal umbel or head, or in successive whorls on the lfless scape, and subtended by small or leafy bracts. 200, N. Hemisphere.

In addition to the following spp., *P. veris* L., the English cowslip, with yellow fls and somewhat inflated cal, occasionally escapes in the n. part of our range.

1 Bracts of the umbel 3–6 mm; cal 3–6 mm .... 1. *P. mistassinica.*
1 Bracts of the umbel 6–14 mm; cal 6–9 mm .... 2. *P. laurentiana.*

1. Primula mistassinica Michx.   Mistassini p. Lvs oblanceolate or spatulate, 2–7 cm, denticulate or entire, long-tapering to the base, glabrous but the lower side often densely farinose with a white or yellow powder that may also appear on the cal; scapes to 2.5 dm; bracts subulate or narrowly lanceolate, 3–6 mm, slightly thickened but not gibbous or saccate at base; fls 1–10; cal 3–6 mm, lobed to about the middle; cor-tube yellow, the limb anthocyanic but generally pale, seldom white, 1–2 cm wide; $2n=18$. Rocks, cliffs, and gravelly shores; Lab. to Me. and n. N.Y., w. to n. Ill., Mich., Wis., and Alas. May, June. (*P. intercedens.*)

2. Primula laurentiana Fern.   Laurentide p. Much like no. 1, but larger, the lvs to 10 cm, the scape to 4 dm, the fls sometimes as many as 15; bracts lance-linear, 6–14 mm, distinctly saccate at base; cal 6–9 mm; cor 8–13 mm wide; $2n=54, 72$. Wet rocks; Lab. to N.S. and Me. June, July.

9. ANDROSACE L.   Cal-tube obconic, equaling or exceeding the triangular lobes; cor salverform, constricted at the throat; stamens included; style very short; stigma capitate; capsule 5-valved, enclosed in the persistent cal; small winter-annuals with a rosette of basal lvs and one to several scapes bearing terminal, involucrate umbels of very small fls. 100, N. Temp. and Arctic.

1 Bracts of the umbel oblanceolate to oblong-obovate, 3–8 mm .... 1. *A. occidentalis.*
1 Bracts of the umbel narrowly lanceolate to subulate, 1–3 mm .... 2. *A. septentrionalis.*

1. Androsace occidentalis Pursh.   Western A. Lvs oblanceolate to oblong or oblong-obovate, up to 3 cm, entire to denticulate, grayish-puberulent with simple hairs; scapes 1–many, pubescent with branched hairs; umbels 3–10-fld, the bracts 3–8 mm, resembling the lvs in shape, less than 4 times as long as wide; cal 4–5 mm, carinate below each lobe, the expanded, greenish, lance-ovate, pubescent lobes about equaling the nearly glabrous tube; cor white, scarcely exceeding the cal-tube, the limb 2.5 mm wide; $2n=20$. Dry, usually sandy soil; Ind. to Man., w. to B.C. and Ariz. Apr.

2. Androsace septentrionalis L.   Northern A. Lvs oblanceolate, 1–3 cm, toothed or entire, sparsely to densely hairy with simple or forked hairs; scapes usually many, to 10+ cm, usually ± densely hairy with fine, branched hairs; umbels 3–25-fld, the bracts lanceolate or subulate, 1–3 mm, at least 4 times as long as wide; cal 2.5–4 mm, strongly carinate, the tube evidently longer than the ± deltoid lobes; cor white, the tube shortly exsert, the limb 3–4 mm wide; $2n=20$. Circumboreal and polymorphic, extending s. to Que., Minn., and Ariz. May, June. (*A. puberulenta.*)

10. DODECATHEON L.   Shooting star. Cal deeply parted, persistent; cor cleft almost to the base, its lobes reflexed; filaments very short, ± connate; anthers elongate, erect, connivent; style about equaling the anthers; capsule opening by 5 short terminal valves; glabrous perennials with a basal rosette of lvs, erect scapes, and a terminal, bracteate umbel of handsome nodding fls on ascending or erect pedicels; fr erect. 13, N. Amer.

1 Wall of the capsule thick and firm, not flexible .... 1. *D. meadia.*
1 Wall of the capsule thin, flexible .... 2. *D. radicatum.*

1. Dodecatheon meadia L.   Eastern s.s. Lvs oblong to oblanceolate or rarely ovate, 6–20 cm, usually tinged or marked with red at the base; scapes 2–6 dm; fls few to many; cor-lobes narrow, white to lavender or lilac, rarely magenta, 1–2.5 cm; capsules dark reddish-brown, 8–18 mm, thickest near the base, firm-walled. Moist or dry woods and prairies; Md. to Ga. and Ala., w. to s. Wis., se. Minn., Io., Okla. and Tex. May, June. Three vars.:

a Lvs gradually tapered into the petiole.
  b Plants robust; fls 4–25; cal-lobes 4–8 mm; anthers 7–10 mm; fr 7–17 mm; $2n=88$; range of the sp.
  .... var. *meadia.*
  b Plants slender; fls 1–14; cal-lobes 3–5 mm; anthers 4–7 mm; fr 7–10 mm; chiefly Ozarkian, from
    s. Mo. to Okla. and Tex., e. occasionally to sw. Va. .... var. *brachycarpum* (Small) Fassett.
a Lvs cordate at the base or abruptly narrowed to the petiole; $2n=44$; shaded cliffs s. of the glacial
    boundary in s. Ill. and w. Ky. .... var. *frenchii* Vasey.

2. Dodecatheon radicatum Greene.   Western s.s. Much like no. 1, but usually smaller and blooming 2 weeks later; lvs not tinged with red toward the base; cor-lobes mostly deep rose-purple, rarely white, averaging smaller than in no. 1; fr thickest near the middle, the walls thin and fragile; $2n=44$. Moist hillsides; driftless area of sw. Wis. and adj. Minn. to nw. Ill.; irregularly from e. Mo. to Pa., s. of the glacial border; throughout the w. cordillera. Apr.–June. Our plants are var. *radicatum.* (*D. amethystinum.*)

FAMILY **HYDRANGEACEAE**, the Hydrangea Family

Fls borne in complex, basically cymose infls, perfect and regular (or the marginal ones often neutral and irregular, with enlarged, petaloid sep), mostly half to fully epigynous with an only shortly or scarcely prolonged hypanthium, but hypogynous or nearly so in some extralimital small genera; cal-lobes 4–5(–12); pet 4–5(–12), distinct; stamens mostly 8–many; filaments distinct or only slightly connate at base; carpels (2)3–5(–12), united to form a compound ovary, the styles in most genera distinct or shortly connate below, often stigmatic for much of their length; ovules mostly several or many on each of the axile or intruded parietal placentas; fr a capsule, or seldom a berry; seeds with a straight, linear, dicotyledonous embryo embedded in the endosperm; mostly shrubs or woody vines or small trees with opposite (rarely whorled), exstipulate, simple (sometimes lobed) lvs. 17/170.

1 Stamens 20–40.
  2 Pet 4(5); erect shrubs .... 1. *Philadelphus.*
  2 Pet 7–10; slender, woody vines .... 2. *Decumaria.*
1 Stamens 8 or 10 .... 3. *Hydrangea.*

1. PHILADELPHUS L. Mock-orange. Fls perfect, 4(5)-merous; sep valvate, persistent, usually tomentose within; pet large, white; stamens 20 or more; ovary inferior, mostly 4-locular with numerous ovules; styles in our spp. elongate, united to or beyond the middle, the stigmas separate and obovoid or united and linear; capsule indurate, obovoid, loculicidal; shrubs with simple, opposite, entire or toothed lvs and conspicuous fls in terminal cymes or short racemes. 50, widespread.

1 Fls 5 or more in a determinate raceme.
  2 Bark of the season's growth brown, exfoliating the second year .... 1. *P. coronarius.*
  2 Bark of the season's growth gray, not exfoliating later .... 2. *P. pubescens.*
1 Fls in cymules of 3, or solitary.
  3 Styles separate above; stigmas thick; lvs glabrous or nearly so beneath .... 3. *P. inodorus.*
  3 Styles united; stigmas linear, connate; lvs softly pubescent beneath .... 4. *P. hirsutus.*

1. Philadelphus coronarius L. European m.-o. Shrub to 4 m; bark of young twigs brown, exfoliating the second year; lvs short-petioled, ovate or ovate-oblong, 4–8 cm, acuminate, remotely dentate, pubescent in and near the axils of the 3 main veins beneath, otherwise glabrous; fls in terminal racemes of 5–7, 3 cm wide, fragrant; pedicels, hypanthium, and outer side of sep glabrous; $2n=26$. Native of Europe, often escaped from cult. in our range. June, July.
    P. tomentosus Wallich, a similar Himalayan sp. with the lvs softly hairy beneath, also sometimes escapes.

2. Philadelphus pubescens Loisel. Ozark m.-o. Stiffly branched shrub 1–3 m; bark of young twigs gray, not exfoliating; lvs ovate-elliptic, 5–8 cm, acuminate, glabrous above, the lower surface (as also the pedicels, hypanthium, and outer side of the sep) usually soft-hairy, but varying to glabrous; fls in racemes of 5–7, 2.5 cm wide, scarcely or only slightly fragrant; $2n=26$. On limestone river-bluffs; Ky. and Tenn. to Ala., Mo., s. Ill., Okla., and e. Tex. May, June. (*P. intectus,* the glabrous extreme.)

3. Philadelphus inodorus L. Appalachian m.-o. Arching shrub 1–3 m; bark of young twigs brown, exfoliating the second year; lvs ovate to oblong, 4–8 cm, toothed or entire, acuminate, rounded to acute at base, glabrous, or very sparsely hairy beneath; fls in cymules of 3 or solitary, 3.5–5.5 cm wide; hypanthium, pedicels, and outer side of sep glabrous; styles separate above; seeds long-tailed; $2n=26$. Streambanks and moist hillsides or cliffs, chiefly in the mts.; e. Pa., Va. and Tenn., Ga. and Ala., sometimes escaped from cult. northward. May, June. (*P. grandiflorus.*)

4. Philadelphus hirsutus Nutt. Cumberland m.-o. Shrub 1–2 m; bark of young twigs brown, exfoliating and turning gray the second year; lvs lance-ovate to broadly ovate, 3–6 cm, acuminate, finely and sharply denticulate, obtuse or rounded at the base, puberulent or glabrescent above, softly pubescent beneath; pedicels, hypanthium, and outer side of sep pubescent; fls 2 cm wide, solitary or in cymules of 3; styles fully united, even the extrorse, linear stigmas connate; seeds tailless; $2n=26$. Cliffs, river-bluffs, and rocky banks; Ky. to N.C., Ga., and Ala. May, June.

2. DECUMARIA L. Fls perfect, epigynous, 7–10-merous; sep small; pet oblong, spreading; stamens 20–30; ovary 7–10-locular; style short and thick, with 7–10-lobed stigma; capsule turbinate, strongly many-ribbed, its summit depressed-conic, crowned with the persistent style, splitting longitudinally between the ribs; slender woody vines, climbing by aerial rootlets, with opposite lvs and terminal corymbiform clusters of small white fls. 2, the other in China.

1. Decumaria barbara L. Climbing hydrangea. High-climbing vine; lvs petioled, ovate to ovate-oblong, glabrous and shining above, glabrous or minutely hairy beneath; infl 4–10 cm wide, fragrant; capsule 6–10 mm; $2n=28$. Swamps and wet woods on the coastal plain, and also more inland southward; se. Va. to Fla. and La. May, June.

3. HYDRANGEA L. Hydrangea. Fls ± epigynous, 4–5-merous except the 2–4(5)-locular ovary; fertile fls small, with minute sep, short pet, 8 or 10 stamens, and 2–4(5) short, stout styles; sterile fls consisting only of a large, white or colored, spreading cal of 3 or 4 sep; fr a capsule, consisting of the enlarged, strongly ribbed hypanthium enclosing the ripened ovary, tipped by the persistent styles and dehiscing between them; shrubs or woody vines with opposite (rarely whorled) lvs and numerous small fls in crowded cymes; fls all perfect, or the marginal sterile, or all sterile. 80, widespread.

1 Infl flat-topped or broadly convex; native .... 1. *H. arborescens.*
1 Infl pyramidal or ovoid; cult. and rarely escaped .... 2. *H. paniculata.*

1. Hydrangea arborescens L. American h. Straggling shrub 1–3 m; lvs ovate-oblong to broadly round-ovate, acuminate, serrate, subcordate to acute at base, glabrous above; infl flat-topped or broadly convex, 5–10 cm wide, fertile throughout or with some marginal fls sterile or rarely wholly sterile (the wholly sterile form often cult.); pet 5, white, 3 mm; stamens 10; ovary inferior, with 2(3) carpels; sep of the sterile fls (3)4, subrotund, nearly 1 cm, white, reticulate-veined; seeds elliptic, with blunt ends; $2n=36$. Dry or moist, often rocky woods and hillsides; s. N.Y. to O., Mo., and Okla., s. to Ga., La., and Ark. June, July. The widespread var. *arborescens* has the lvs glabrous or with only a few scattered hairs along the midrib beneath. The well marked var. *discolor* Ser. (var. *deamii*) occurring from c. Ind. and Ill. to N.C., Ga., and Okla., has the lvs ± densely pubescent beneath with minutely tuberculate hairs. The var. *discolor* might perhaps equally well be treated as a distinct, closely allied species, *H. cinerea* Small.

2. Hydrangea paniculata Siebold. Straggling shrub to 5 m; lvs ovate to elliptic; panicles pyramidal or ovoid, 15–25 cm; pet 5; stamens 10; carpels 2–4; ovary only half-inferior; sterile sep 4, 15 mm, white changing to greenish or bluish in age; seeds linear, minutely caudate; $2n=36$, 72. Native of e. Asia, sometimes escaped from cult. in our range. Aug., Sept. A form with all the fls sterile is called peegee hydrangea.

H. quercifolia Bartram, oak-lvd h., a sp. of s. U.S. with similar infl but with sharply 5–7-lobed lvs, sometimes also escapes from cult.

Deutzia scabra Thunb., a shrub with opposite lvs, regular, perfect (sometimes double) fls, 10 stamens, and 3–5 styles, rarely escapes from cult.

FAMILY **GROSSULARIACEAE**, the Gooseberry Family

Fls regular or nearly so, mostly perfect, perigynous to more often epigynous with a prolonged, saucer-shaped to tubular hypanthium, the (3–)5(–9) persistent sep appearing as lobes on the hypanthium or sometimes forming a cal-tube that extends beyond the hypanthium, sometimes ± petaloid and more showy than the proper pet, these sometimes wanting; stamens mostly as many as and opposite the sep, but a second, functional or staminodial set sometimes present alternate with the sep; carpels 2–3(–7) generally united to form a compound, superior to usually ± inferior ovary, this plurilocular with axile placentas, or unilocular with ± intruded parietal placentas; fr a capsule or berry with numerous, often arillate seeds; endosperm copious to scanty; cotyledons 2; shrubs or sometimes trees with mostly alternate, simple (but often deeply cleft), exstipulate or seldom (as in *Itea*) stipulate lvs, the fls mostly in racemes, less often in panicles or small umbels or corymbs or solitary and axillary. 25/300+.

1 Ovary superior; lvs pinnately veined, serrulate .... 1. *Itea.*
1 Ovary inferior; lvs palmately veined and lobed .... 2. *Ribes.*

1. ITEA L.   Virginia willow.   Hypanthium small, broadly cupulate; sep 5, small, erect; pet 5, narrowly oblong; stamens 5; filaments pubescent; ovary superior, bilocular, longitudinally 2-grooved, hairy; style grooved; stigma capitate; fr an indurate capsule, at first 2-grooved, tipped with the persistent style, eventually septicidal to the summit of the style; shrubs with chambered pith, superposed axillary buds, pinnately veined, serrulate lvs and slender, terminal racemes of small white fls. 20, e. U.S., e. Asia, w. Malaysia.

1. Itea virginica L.   Shrub to 3 m, with hairy twigs and infl; lvs elliptic to oblong-lanceolate, 3–8 cm, acute or acuminate, acute at base, glabrous or nearly so; stipules tiny, slender, caducous; racemes 4–15 cm; pedicels 2–3 mm; sep-tips eventually deciduous; pet 5–6 mm; capsules ± reflexed, slender, 7–10 mm including the style; $2n=22$. Swamps and wet woods, chiefly on the coastal plain; s. N.J. and e. Pa. to Fla. and La., n. in the Miss. valley to s. Ill. May, June.

2. RIBES L.   Hypanthium prolonged beyond the inferior ovary, saucer-shaped to tubular; sep longer than the small pet; sep, pet, and stamens (4)5, the stamens opposite the sep and (except in *R. lacustre*) usually inserted at the level of the sinuses; ovary unilocular with 2 parietal placentas; style usually bifid; fr a many seeded berry; shrubs with palmately veined and lobed lvs (many of them fascicled on short lateral branches) and small fls in small clusters or racemes. (*Grossularia, Limnobotrya.*) 150, N. Temp. and Andes.

1 Fls solitary or in small corymbiform clusters of 2–4; pedicels not articulate at the summit; hypanthium
    above the ovary ± elongate, campanulate to obconic or tubular; plants usually spiny. (Goose-
    berries.)
  2 Stamens at full anthesis evidently longer than the sep.
    3 Stamens 9–12 mm; hypanthium white; western . . . . 1. *R. missouriense.*
    3 Stamens 6–8 mm; hypanthium purplish-green; eastern . . . . 2. *R. rotundifolium.*
  2 Stamens at full anthesis ca equaling or shorter than the sep.
    4 Hypanthium above the ovary campanulate to cupulate above a rounded base.
      5 Stamens as long as the pet, shorter than the oblong sep; native . . . . 3. *R. cynosbati.*
      5 Stamens as long as the round-obovate sep; rarely escaped . . . . 4. *R. uva-crispi.*
    4 Hypanthium above the ovary obconic or cylindric.
      6 Stamens ca equaling the pet, shorter than the sep . . . . 5. *R. oxyacanthoides.*
      6 Stamens well surpassing the pet, ca equaling the sep . . . . 6. *R. hirtellum.*
1 Fls racemose; pedicels articulate at the summit at maturity; hypanthium above the ovary saucer-
    shaped to tubular; plants (except *R. Lacustre*) unarmed. (Currants.)
  7 Ovary and fr bristly with glandular hairs.
    8 Stems spiny; pet broadly flabellate to half-orbicular . . . . 7. *R. lacustre.*
    8 Stems unarmed; pet cuneate-obovate . . . . 8. *R. glandulosum.*
  7 Ovary and fr not bristly (sessile resinous glands may be present).
    9 Lvs dotted with shining resinous glands beneath; fr black.
      10 Pedicels much longer than the minute bracts; cal densely hairy outside.
        11 Hypanthium cupulate or short-campanulate; racemes drooping; intr. . . . . 9. *R. nigrum.*
        11 Hypanthium saucer-shaped; racemes erect or ascending native . . . . 10. *R. hudsonianum.*
      10 Pedicels much shorter than the conspicuous bracts; cal not very hairy . . . . 11. *R. americanum.*
    9 Lvs without glands beneath.
      12 Fls yellowish-green to dull purple; hypanthium saucer-shaped; fr red.
        13 Anther-sacs almost adjacent; lateral lf-lobes directed forward; native . . . . 12. *R. triste.*
        13 Anther-sacs separated by a connective as broad as the sac; lateral lf-lobes spreading;
          cult. and sometimes escaped . . . . 13. *R. sativum.*
      12 Fls golden-yellow; hypanthium tubular; fr black or rarely yellow . . . . 14. *R. odoratum.*

1. Ribes missouriense Nutt.   Missouri gooseberry. Nodal spines stout; lvs 1.5–3.5 cm, rotund in outline, the 2 principal sinuses extending nearly to the middle, softly hairy at least beneath, or glabrate in age; peduncles puberulent or glandular to subglabrous, 1–2 cm, usually much exceeding the 5–13 mm pedicels; floral bracts ciliate with glandular and eglandular hairs; ovary and fr glabrous; hypanthium 1.5–2.5 mm, cupulate, white or creamy; sep oblanceolate, 5–7.5 mm, reflexed; pet cuneate-obovate, white (or pinkish in age), 2–3.5 mm; stamens 9–12 mm; style 10–14 mm, bifid; fr red to purple; $2n=16$. Moist or dry upland woods; Ind. to Minn. and Tenn., w. to Mont. and Kans. May.

2. Ribes rotundifolium Michx.   Appalachian gooseberry. Nodal spines mostly short and slender, or wanting; lvs 1.5–3.5 cm long and wide, broadly obovate to rotund in outline, glabrous or minutely villosulous on both sides; peduncles glabrous, 5–10 mm, the pedicels 4–7 mm; floral bracts glabrous or sparsely glandular; ovary and fr glabrous; hypanthium subcylindric, mostly purplish or greenish-purple; sep oblanceolate, 4–5 mm; pet white, spatulate-obovate, 2–2.5 mm; stamens 6–8 mm; style 6–8 mm, bifid; fr pale purple. Rocky upland woods; Mass. and N.Y. to w. N.C. and e. Tenn. Apr., May.

3. **Ribes cynosbati** L.   Dogberry. Internodes prickly or smooth; nodal spines (0)1–2(3), 5–15 mm; lvs 2–5 cm long and wide, truncate to cordate at base, pubescent, usually not glandular; peduncles 7–25 mm, with 1–3(4) fls on pedicels 5–16 mm, the frs thus held well away from the stems; bracts glandular-ciliate; hypanthium campanulate from a rounded base, 3--4.5 mm (above the ovary); sep broadly oblong, 2.5–4 mm, soon reflexed; pet obovate, subtruncate, 1–2.5 mm; stamens equaling the pet; style equaling the sep, mostly undivided; ovary with stalked glands that become stiff prickles on the greenish to pale red fr; $2n=16$. Moist woods; Que. to n. Minn. and e. N.D., s. to n. Ga., sw. Ala., Ark., and e. Okla. May, June.

4. **Ribes uva-crispi** L.   Garden-gooseberry. Stems with stout nodal spines, often also bristly; lvs to 6 cm wide; hypanthium broadly campanulate above the ovary, 3 mm long and wide, pubescent; sep pubescent, spreading or reflexed, round-obovate, 3.5–4 mm, longer than the hypanthium; pet 2–2.5 mm long and wide; stamens and style nearly or quite equaling the sep; fr pubescent and usually also with some gland-tipped bristles; $2n=16$. Naive of the Old World, occasionally escaped from cult. in our range. (*R. grossularia; Grossularia reclinata.*)

5. **Ribes oxyacanthoides** L.   Northern gooseberry. Nodal and internodal spines commonly present; lvs 1.5–3 × 2–3.5 cm, glandular and hairy at least beneath, broadly cuneate to truncate (rarely subcordate) at base; peduncles and pedicels each 2–6 mm; bracts finely glandular-ciliate; ovary and hypanthium glabrous; hypanthium above the ovary obconic, 2.5–4 mm; sep glabrous, oblong, blunt, 2.5–4 mm; pet obovate, 2–3 mm; stamens ca equaling to slightly exceeding the pet, but shorter than the sep; fr smooth, greenish-purple; $2n=16$. Moist woods; n. Mich. to Hudson Bay, w. to Alas., Wyo., and Nev. May, June. Ours is var. *oxycanthoides*. (*R. setosum*, a western phase.)

6. **Ribes hirtellum** Michx.   Much like no. 5, and not always sharply distinct; nodal and internodal spines few and weak, or often absent; lvs softly hairy to glabrous, not glandular, often cuneate at base; bracts villous-ciliate; ovary and hypanthium glabrous; fls slightly smaller; stamens and style ca twice as long as the pet, just equaling the sep; $2n=16$. Rocky woods and cliffs; Nf. and Que. to Alta., s. to N.Y., Pa., n. Ind., ne. Ill., and Nebr. May, June. (*R. huronense.*) Variable, and sometimes divided into several vars. on the basis of pubescence and lf-shape.

7. **Ribes lacustre** (Pers.) Poiret.   Spiny swamp-currant. Stems very spiny on the internodes, with clusters of longer nodal spines; petioles glandular and hairy; blades deeply 3–5-lobed and coarsely toothed, glabrous or nearly so; racemes loosely spreading or drooping; rachis, pedicels, ovary and fr glandular-bristly; hypanthium above the ovary broadly cupulate, 1 mm, glabrous; sep broadly cuneate-obovate, 2 mm, wider than long; pet flabellate to half-orbicular, 1.5 mm; stamens inserted just above the ovary, equaling the pet; style bifid to the middle; fr dark purple or black; $2n=16$. Swamps and wet woods; Lab. to Alas., s. to Mass., Pa., Mich., Minn., and the western cordillera. May, June. (*Limnobotrya l.*)

8. **Ribes glandulosum** Grauer.   Skunk-currant. Stems reclining or sprawling, unarmed; lvs 5(7)-lobed to about the middle, serrate or double serrate, glabrous or sparsely villosulous on the veins beneath; racemes erect or ascending, glandular-hairy at least on the pedicels, the ovary and fr glandular-bristly; hypanthium above the ovary deeply saucer-shaped, 1 mm, glabrous; sep oblong-ovate from rounded sinuses, 2–2.5 mm; pet cuneate-obovate, 1–1.5 mm; style bifid nearly to the base; fr dark red; $2n=16$. Swamps and wet woods; Nf. to Mack. and B.C., s. to Me., Vt., Mich., Minn., and in the mts. to N.C. May, June. (*R. prostratum.*)

9. **Ribes nigrum** L.   Garden black currant. Erect, unarmed shrub; lvs 3–5-lobed to about the middle, dotted with resinous glands beneath; racemes drooping; pedicels 2–8 mm, much exceeding the minute, ovate bracts; hypanthium above the ovary short-campanulate; sep greenish-purple within; ovary commonly with sessile resinous glands; fr black; $2n=16$. Native of Eurasia, occasionally escaped from cult.

10. **Ribes hudsonianum** Richardson.   Western black currant. Stem erect or ascending, unarmed; lvs shallowly to deeply cordate, 5-lobed, gland-dotted beneath and sometimes also minutely hairy; racemes erect or ascending; pedicels 3–7 mm, much exceeding the minute bracts; hypanthium above the ovary saucer-shaped, 1 mm; sep white or nearly so, oblong-elliptic, 3.5–4.5 mm, densely hairy outside; pet cuneate-oblong, 1.5–3 mm, obscurely 3-lobed; ovary usually with a few sessile, resinous glands, otherwise glabrous; fr black. Swamps and wet woods; n. Ont. to Alas., s. to Mich., Io., Wyo., and B.C. May, June.

11. **Ribes americanum** Miller.   Eastern black currant. Stems erect, unarmed; lvs broadly truncate to shallowly cordate at base, 3–5-lobed, gland-dotted and sparsely villosulous beneath; racemes drooping, many fld; pedicels 0–2 mm, much exceeded by the conspicuous lanceolate bracts; hypanthium above the ovary narrowly campanulate, 3.5–4.5 mm; sep greenish-white or yellowish, oblong-ovate, 4.5–5 mm, glabrous or very sparsely villosulous; pet oblong, very blunt, 2.5–3 mm; fr black, glabrous; $2n=16$. Moist woods; N.S. to Man. and Alta., s. to Del., W.Va., Ind., Io., Nebr., and Colo. Apr., May. (*R. floridanum.*)

12. **Ribes triste** Pallas.   Swamp red currant. Unarmed straggling shrub; lvs glabrous to softly hairy beneath, broadly truncate to shallowly cordate at base, mostly 5-lobed, the lobes toothed from sinus to tip, the two middle sinuses the deepest, the lateral lobes direct forward; racemes drooping, the axis and pedicels (1–4 mm) often with short-stipitate glands; hypanthium above the ovary saucer-shaped, less than 1 mm; sep greenish-purple, broadly rhombic-ovate, 2 mm; pet cuneate, 1 mm, truncate or notched; fr red; $2n=16$. Bogs and wet woods; Nf. to Alas., s. to N.J., Mich., Wis., Minn., and Alta.; n. Asia. June, July.

13. **Ribes sativum** Syme.   Garden red currant. Erect, unarmed shrub; lvs mostly 5-lobed, the lateral lobe widely spreading; infl not glandular; anther-sacs widely separated by a broad connective; fr red; $2n=16$. Native of Eurasia, occasionally escaping from cult. in our range.

14. **Ribes odoratum** H. H. Wendl. Buffalo-currant. Erect, unarmed shrub; lvs broadly cuneate to truncate at base, ciliate, finely puberulent to glabrate beneath, deeply 3(5)-lobed, the lobes entire in their lower half, often few-toothed above; fls fragrant, golden-yellow; hypanthium above the ovary tubular, 11–15 mm; sep oblong-ovate, 5–6.5 mm, broadly rounded above; pet 2.5–3.5 mm, erose above; fr black or rarely yellow; $2n=16$. Cliffs and rocky hillsides; Minn. to Ark., w. and s. to S.D., Colo., and Tex.; widely cult. and often escaped farther e. Apr.–June.

FAMILY **CRASSULACEAE**, the Stonecrop Family

Fls regular, hypogynous or nearly so, (3)4–5-merous or occasionally more, usually perfect; sep, pet, and pistils normally of the same number, the stamens as many or twice as many; sep distinct or ± united at base; pet in our genera distinct or very shortly connate below; carpels generally distinct or nearly so and ripening into follicles; a small nectariferous appendage borne externally on each carpel near the base; succulent herbs or shrubs with fleshy, simple lvs and usually cymose fls. 25/900.

1 Stamens twice as many as the pet (or fewer in *Sedum* × *erythrostictum*, a coarse perennial).
  2 Fls 4–5(–9)-merous; lvs not at once flat and forming dense, monocarpic rosettes .... 1. *Sedum.*
  2 Fls 6–40-merous; lvs flat and forming dense, monocarpic rosettes .... 2. *Sempervivum.*
1 Stamens as many as the pet; ours small annuals with 3–4-merous fls .... 3. *Crassula.*

1. **SEDUM** L. Stonecrop. Fls mostly 4–5(–9)-merous; pet distinct or united only at the very base; stamens usually twice as many as the pet, the antepetalous ones usually adnate at base to the pet; carpels distinct or nearly so, tapering to the short, stout subulate style; fr follicular; seeds numerous; succulent herbs or half-shrubs with thick or terete, alternate, opposite, or whorled lvs and small to middle-sized, yellow or white to anthocyanic fls. In addition to the following native and intr. spp., a number of spp. are cultivated and may escape locally. 300, mainly N. Temp.

1 Fls mostly or all unisexual, the plants essentially dioecious .... 1. *S. rosea.*
1 Fls mostly or all perfect.
  2 Pet white to light or deep pink or purple.
    3 Infl repeatedly branched, not secund; lvs toothed or entire; fls mostly 5-merous.
      4 Lvs flat, relatively large, mostly 1.5–8 × 0.5–3.5 cm, toothed or entire.
        5 Lvs mostly 3–8 × 1–3.5 cm; stems tufted on a coarse-rooted caudex or crown.
          6 Pet pale; sep under 4 (avg 2.5) mm; lvs not much reduced upwards.
            7 Carpels 5, producing seeds .... 2. *S. telephioides.*
            7 Carpels 0–5, not producing seeds .... 3. *S.* × *erythrostictum.*
          6 Pet deep pink; sep over 5 (avg 6) mm; lvs strongly reduced upwards .... 4. *S. purpureum.*
        5 Lvs mostly 1.5–3 × 0.5–2 cm; fibrous-rooted, matforming, with ± erect fl stems .... 5. *S. spurium.*
      4 Lvs subterete, 0.6–1.5 cm, well under 0.5 cm wide, entire .... 6. *S. album.*
    3 Infl forking at base into several spreading, secund, sympodial cymes; lvs entire.
      8 Infl and upper part of stem glandular-hairy; fls (4–)6(–8)-merous .... 7. *S. hispanicum.*
      8 Infl and stem glabrous; fls 4(–7)-merous.
        9 Lvs, or many of them, whorled in 3s (rarely only opposite) .... 8. *S. ternatum.*
        9 Lvs alternate.
          10 Lvs subterete, the cauline ones mostly with a sagittate basal spur .... 9. *S. pulchellum.*
          10 Lvs somewhat flattened, with a short, truncate basal spur .... 10. *S. glaucophyllum.*
  2 Pet yellow.
    11 Lvs whorled in 3s .... 11. *S. sarmentosum.*
    11 Lvs alternate.
      12 Fls mostly 5-merous; follicles divergent.
        13 Lvs relatively broad, ± ovoid .... 12. *S. acre.*
        13 Lvs narrow, ± linear .... 13. *S. sexangulare.*
      12 Fls (5–)7(–9)-merous; follicles erect .... 14. *S. reflexum.*

1. **Sedum rosea** (L.) Scop. Roseroot. Dioecious perennial from a thick, scaly caudex; stems axillary to the caudex-scales, erect, 1–4 dm, lvs sessile, often somewhat glaucous, oblanceolate or obovate, 2–4 cm, entire or toothed toward the acute summit; infl compact, repeatedly branched, 1–5 cm wide; fls 4-merous or sometimes 5-merous; rarely 3-merous; pet oblanceolate to elliptic-oblong, 2–4.5 mm; follicles connivent, with short, spreading beaks. Cliffs and ledges; circumboreal, s. in the w. cordillera to Calif. and N.M., along the Atlantic coast to N.S. and Me., and disjunct inland in N.H., N.Y., Pa., N.C. to Tenn., and Minn. May–July. (*Rhodiola r.*) A complex, variable sp., here considered to consist of 2 major geographic phases that are

morphologically ill-defined but cytologically apparently distinct. Var. *rosea,* with $2n=22$, occurs in Atlantic N. Amer. and n. Eurasia; it has narrow pet (ca 1 mm wide or less) that are usually yellow or yellowish, sometimes red toward the tip. Var. *integrifolium* (Raf.) A. Berger, with $2n=36$, is widespread in the w. cordillera, with outlying stations in the driftless area of se. Minn. and in c. N.Y. (where it adjoins the range of var. *rosea*); it has slightly wider pet (mostly 1–1.8 mm wide, but sometimes narrower) that are usually red or reddish, but sometimes largely yellow. (*S. integrifolium; S. rosea* var. *leedyi.*)

2. **Sedum telephioides** Michx.   Rather coarse-rooted perennial from a stout caudex, the stems tufted, 2–6 dm, simple to the infl; lvs alternate or opposite, ± elliptic, narrowed to the base, flat, 3–8 × 1.5–3.5 cm, entire or often coarsely few-toothed, sometimes glaucous, not much reduced upwards; infl broadly rounded, much-branched, not secund, 4–10 cm wide, the branchlets narrowly winged; fls mostly 5-merous; sep ca 2.5 mm; pet white to light pink, 4.5–6 mm; nectaries white or pale yellow, ca 1.4 times as long as wide; ovaries white, pale green, or pink; frs short-stipitate, erect with divergent beaks; $2n=24$, seldom 48. Dry, rocky places in the mts. from s. Pa. to N.C.; disjunct in s. Ind. and Ill. and w. Ky., and rarely escaped elsewhere. Aug.–Oct.

3. **Sedum × erythrostictum** Miq.   Much like no. 2 in aspect; pet white with green midnerve; nectaries yellow; stamens 0–10; carpels 0–5, pink, infertile. Sterile triploid ($3x=36$) from Japan, perhaps originating as a hybrid of *S. spectabile* Boreau and *S. viridescens* Nakai, occasionally escaped from cult. in our range. (*S. × alboroseum.*)

4. **Sedum purpureum** (L.) J. A. Schultes.   Live forever. Much like no. 2; perennial from clusters of tuberous-thickened, carrot-like white roots; lvs alternate, or opposite on the branches, sessile, elliptic-oblong, dull green or only slightly glaucous, evidently reduced upwards; sep over 5 (avg 6) mm; pet deep pink, almost red; nectaries yellow, ca 1.7 times as long as wide; ovaries pink, seldom setting seed; triploid, $3x=36$. Native of Europe, escaped from cult. and well established in disturbed, moist sites in our range, especially northward. Aug. Perhaps better treated as *S. telephium* var. *purpureum* L., but in our range forming a distinctive, sharply delimited unit.

5. **Sedum spurium** M. Bieb.   Fibrous-rooted perennial with branched, creeping stems and more erect branches (or stem-tips) 1–2 dm; lvs opposite, thick, flat, obovate or obovate-oblong, cuneate to a sessile or shortly petiolar base, 1.5–3 × 0.5–2 cm, with prominently glandular margins, coarsely crenate above; infl small, compactly branched; fls mostly 5-merous; pet white to pink or reddish-purple, 8–11 mm, spreading above an erect base; frs ± erect. Native of the Caucasus, sometimes escaped from cult. July, Aug.

6. **Sedum album** L.   Fibrous-root perennial from creeping stems, forming mats with more erect flowering stems 1–2 dm; lvs alternate, subterete, 0.6–1.5 cm, obtuse, scarcely spurred at base; infl convex, rather dense, repeatedly branched; fls mostly 5-merous; pet white, 3–4 mm; frs pink, erect; $2n=68$, 136. Native of Eurasia, sometimes escaped from cult. June, July.

7. **Sedum hispanicum** L.   Slender, fibrous-rooted annual to sometimes biennial or perennial, 5–15 cm, often freely branched, generally glandular-hairy upwards; lvs alternate, linear, subterete, 0.5–1.8 cm, glaucous; fls (4–)6(–8)-merous, numerous in unilateral, sympodial cymes; pet 4–7 mm, white with pink or green midvein; frs widely divergent; $2n=40$. Native of se. Europe, occasionally escaped from cult. May–July.

8. **Sedum ternatum** Michx.   Fibrous-rooted perennial from creeping stems, sending up usually a single flowering stem 1–2 dm and several short, leafy, sterile shoots; lvs green, those of the sterile shoots and the lower ones of the fertile shoots mostly in whorls of 3(4), obovate, 1–2 cm long, often 1 cm wide, entire, cuneate to the base; upper lvs of fertile stems opposite or sometimes alternate, oblanceolate to nearly linear; infl of 2–4 divergent, secund, sympodial cymes; fls mostly 4-merous; pet white, widely spreading, 5–9 mm; filaments only very shortly adnate to the pet; frs divaricate; $2n=16$, 24, 32, 48. Rocks, cliffs and woods; N.J. to n. Ga., w. to Io. and Ark., and occasionally escaped n. to N. Engl., N.Y., and s. Mich. May, June.

9. **Sedum pulchellum** Michx.   Mostly biennial from a cluster of fibrous roots, less often annual or perennial; stems simple or often branched at the base, 1–3 dm; lvs numerous, alternate, entire, linear or somewhat clavate, ± terete, 1–3 cm, the basal spur (at least on the flowering stems) commonly sagittate; infl of 3–7 widely divergent or recurved, secund, sympodial cymes up to 1 dm; fls mostly 4(–7)-merous; pet pinkish-white to pale pink or purple, ± spreading, 4–8 mm; filaments only very shortly adnate to the pet; frs divaricate; $2n=22$, 44, 66. Moist rocks and cliffs, Ky. and Tenn. to Mo., Okla., and Tex. May, June.

10. **Sedum glaucophyllum** R. T. Clausen.   Fibrous-rooted perennial, the stems tufted, spreading or prostrate, branched, with more erect flowering shoots to 1 dm; lvs blue-green or pale green, alternate, entire, somewhat flattened (generally at least twice as wide as thick), oblanceolate to spatulate, narrowed to a petiole-like base, or linear upwards, with a short, truncate basal spur, divergent, becoming recurved; infl of 2–4 short, widely divergent or recurved, secund, sympodial cymes; fls mostly 4(5)-merous; pet white, widely spreading, 5–8 mm; filaments only very shortly adnate to the pet; frs divaricate; $2n=28$, less often 44, 56. Rocky places, mainly in the mts.; Md., Va., W.Va., and nw. N.C. Apr–June. Closely allied to *S. nevii* A. Gray, which has thicker lvs, more closely striate seeds, and $2n=12$, and is confined to the s. end of the s. Appalachian Mts., beyond our range.

11. **Sedum sarmentosum** Bunge.   Fibrous-rooted perennial with long, creeping stems, forming loose mats, and with decumbent or erect flowering shoots ca 1 dm; lvs mostly in whorls of 3, thick but flattened, oblanceolate-elliptic to lance-ovate, 2–3 cm; infl of a few divergent, branched, sympodial cymes; fls mostly 5-merous; pet yellow, widely spreading, lance-linear, 5–8 mm, connate for ca 0.5 mm; filaments basally

adnate to the minute cor-tube; carpels erect in fl, becoming divergent, sterile; $2n=72$. Native of China, frequently escaped in our range. Summer.

12. Sedum acre L. Golden carpet. Fibrous-rooted evergreen perennial from creeping stems, forming mats; flowering stems 5–10 cm, the lower part commonly clothed with persistent dead lvs; lvs crowded, imbricate, ovoid, terete, 2–5 mm, blunt; infl of 2 (–several) short, branched, sympodial cymes; fls 5-merous; pet yellow, 6–8 mm, spreading, connate for ca 0.5 mm; filaments basally adnate to the minute cor-tube; carpels erect in fl, divergent in fr; $2n=80$. Native of Eurasia, cult. in many forms, and often escaped in our range, especially northward. June, July.

13. Sedum sexangulare L. Much like no. 12; lvs linear, subterete, blunt, 3–6 mm, alternate, crowded to form 5 or 6 ranks; infl of ca 3 divergent, sympodial cymes; pet 3–5 mm; $2n=74$. Native of Europe, escaped from cult. in N.H. and Vt. June, July.

14. Sedum reflexum L. Fibrous-rooted perennial with much-branched, prostrate stems, forming loose mats, giving rise to many short sterile shoots and a few flowering shoots 1.5–3.5 dm; lvs lance-linear, terete or subterete, apiculate, 5–12 × 1–2.5 mm; infl nodding and subglobose in bud, concave in fr; fls (5–)7(–9)-merous; pet yellow, spreading, 5–7 mm; frs erect; $2n=108$. Native of Europe and n. Afr., occasionally escaped from cult. in our range. Summer. (*S. rupestre,* misapplied.)

2. SEMPERVIVUM L. Fls 6–40-merous; pet distinct or united only at base; stamens twice as many as the pet, the filaments usually pubescent at least at base; carpels distinct or nearly so, the ovaries pubescent, the styles outcurved; fr follicular; seeds numerous; succulent perennials with dense, monocarpic rosettes of broad, flat, alternate, entire, ciliate, sessile lvs, acaulescent in the vegetative phase, and reproducing vegetatively by axillary stolons or seldom by division of the rosettes. 25, Eurasia.

1. Sempervivum tectorum L. Hen-and-chickens. Plants with axillary stolons; rosette-lvs obovate or oblong-lanceolate, with a pungent mucro, narrowed to the base, ciliate, otherwise glabrous; flowering stems 3–6 dm; cauline lvs narrowly obovate to oblong, glandular-hairy; infl very pubescent, freely branched into several spreading cymes; fls mostly 12–16-merous; pet pink-purple, 1 cm, spreading, entire; $2n=36$, 72. Native of Europe, sometimes escaped from cult. in our range, esp. in N. Engl. July–Sept.

S. heuffelii Schott, another European sp., is reported to be established locally in Wis. It has 6-merous fls with suberect, pale yellow, glandular-ciliate pet, and it lacks stolons.

3. CRASSULA L. Fls (3)4–5(–9)-merous; pet distinct or united only at base; stamens as many as and alternate with the pet, adnate to the base of the sep; carpels distinct or nearly so, the style short or even wanting, the stigma terminal; fr follicular; seeds (1–) several or many; succulent herbs or shrubs with thick, opposite lvs and mostly small fls. 250, cosmop.

1. Crassula aquatica (L.) Schönl. Pygmy-weed. Branching annual, rooting at the base, 2–10 cm; lvs linear, 3–6(–8) mm, spreading, connate around the stem; fls subsessile, solitary in the axils, mostly 4-merous, ca 1 mm wide, the pet white, erect or slightly spreading; seeds numerous; $2n=42$. Muddy shores near the coast, usually in the intertidal zone; Nf. and Que. to Md.; Minn.; also La. to Tex., and on the Pacific coast and in Eurasia. Late summer. (*Tillaea a.*)

FAMILY SAXIFRAGACEAE, the Saxifrage Family

Fls regular or sometimes somewhat irregular, perfect or seldom some or all of them unisexual, ± perigynous (almost hypogynous in *Parnassia*) to epigynous, the (3)4–5(–10) sep commonly appearing as lobes of the hypanthium; pet typically as many as the sep, often clawed, sometimes cleft or dissected, well developed or often relatively small and inconspicuous (and then sometimes irregularly developed and fewer than the sep), or even wanting; stamens bicyclic or often unicyclic, one set sometimes staminodial; carpels 2–5(–7), usually ± connate at least below to form a compound, distally often ± deeply lobed ovary, each lobe commonly prolonged into a stylar beak with a usually capitate stigma (in *Parnassia* the stigmas sessile atop the ovary); fr dry, dehiscent, most often septicidal or dehiscent along the ventral sutures of the carpels above their level of union; seeds generally ± numerous, with a straight, dicotyledonous embryo embedded in the endosperm; herbs or seldom subshrubs with alternate (sometimes all basal) or less often opposite, simple or compound, exstipulate or inconspicuously stipulate lvs. 40/700.

1 Lvs ternately decompound .... 1. *Astilbe.*
1 Lvs simple, though often deeply cleft.
  2 Carpels 5(–7); stamens 10 .... 2. *Penthorum.*
  2 Carpels 2–4; stamens 4–10.
    3 Pet present; fls 5-merous or very rarely 4-merous.
      4 Functional stamens twice as many as the pet.
        5 Pet entire or nearly so.
          6 Carpels in fl and fr equal in size .... 3. *Saxifraga.*
          6 Carpels in fl and fr very unequal .... 4. *Tiarella.*
        5 Pet deeply pinnatifid or fimbriate .... 5. *Mitella.*
      4 Functional stamens as many as and alternate with the pet.
        7 Lvs toothed or lobed; fls in panicles or cymes; staminodes wanting.
          8 Ovary bilocular, with axile placentation.
            9 Pet marcescent-persistent; seeds narrowly winged; lvs in ours shallowly lobed .... 6. *Sullivantia.*
            9 Pet deciduous; seeds not winged; lvs in ours cleft to the middle .... 7. *Boykinia.*
          8 Ovary unilocular, with 2 parietal placentas .... 8. *Heuchera.*
        7 Lvs entire; fls solitary on scape-like stems; staminodes opposite the pet .... 9. *Parnassia.*
    3 Pet none; fls 4-merous .... 10. *Chrysoplenium.*

1. **ASTILBE** Buch.-Ham.   Polygamo-dioecious; fls regular, 5-merous; hypanthium saucer-shaped; pet narrow; stamens 10; ovary superior; carpels 2(3), ± connate (except the styles) and with axile placentas, separating into erect follicles at maturity; tall perennial herbs with large, alternate, decompound lvs and large panicles of small, white or yellowish fls. 25, N. Amer., e. Asia.

    1. **Astilbe biternata** (Vent.) Britton.   False goat's-beard. Stems clustered on an enlarged caudex, 1–2 m, glandular hairy above; lvs petioled, 2 or 3 times ternately compound, the lfls sessile or petiolulate, oblong to ovate, rounded to subcordate at base, acuminate, sharply serrate; pet narrowly oblanceolate, 2 mm, often none in the pistillate fls; follicles connivent, 4 mm; 2*n*=28. Moist, north-facing woods in the mts., Va. and W.Va. to Ky., s. to Ga. June. Habitually much like *Aruncus.*
    A. japonica (Morren & Decne.) A. Gray, a smaller sp. (rarely to 1 m), with larger fls and more cuneate-based lfls, rarely escapes from cult.

2. **PENTHORUM** L.   Ditch-stonecrop. Fls perfect; sep 5(–7); pet small or more often wanting; stamens 10; carpels 5(–7), united nearly to the middle, abruptly narrowed to the short styles, divergent at maturity, forming an angularly 5(–7)-lobed, 5(–7)-locular capsule with horn-shaped lobes, eventually opening by abscission of the lobes; each carpel with a pluriovulate ventral (marginal) placenta just above the united part of the ovary; perennial rhizomatous herbs with erect stems, alternate, serrulate lvs, and a terminal 2–4-branched cyme of secund greenish fls. Sometimes referred to the Crassulaceae. 3, e. N. Amer., e. Asia.

    1. **Penthorum sedoides** L.   Erect, 2–7 dm, simple or branched above, glabrous below, stipitate-glandular in the infl; lvs lanceolate to narrowly elliptic, 5–10 cm, sharply serrate, acuminate at both ends; cymes 2–8 cm; sep oblong-lanceolate; capsule 5–6 mm wide; 2*n*=18. Marshes and muddy soil; Me. to Ont. and Minn., s. to Fla. and Tex. July–Sept.

3. **SAXIFRAGA** L.   Saxifrage. Fls perfect, regular or slightly irregular, 5-merous; hypanthium usually adnate to the base of the carpels, sometimes essentially free; pet mostly narrow; stamens 10; carpels mostly 2, connate at least at the base, with axile placentas, ripening into a bilobed capsule or nearly distinct follicles; perennial herbs with the principal lvs usually in basal rosettes or crowded on basal shoots, the flowering stem commonly with few or no lvs, the fls mostly in branched, cymose or cymose-paniculate infls, rarely solitary and terminal. (*Antiphylla, Chondrosea, Hydatica, Leptasea, Micranthes.*) 350, Arctic, N. Temp., Andes.

1 Cauline lvs present below the base of the infl.
  2 Principal lvs rotund to reniform in outline, 3–5(–7)-lobed, long-petioled.
    3 Bulblets present in the axils of the upper lvs; pet 6–10 mm .... 1. *S. cernua.*
    3 Bulblets none; pet 3–5 mm .... 2. *S. rivularis.*
  2 Principal lvs narrow, entire or toothed or terminally lobed, sessile or nearly so.
    4 Lvs with conspicuous terminal teeth or lobes .... 3. *S. tricuspidata.*
    4 Lvs entire, or serrate along the lateral margins.

5 Pet mostly purple; lvs opposite .... 4. *S. oppositifolia.*
5 Pet yellow or white; lvs alternate.
  6 Lvs linear oblong, entire, often ciliate; fls yellow .... 5. *S. aizoides.*
  6 Lvs obovate-oblong, sharply but finely serrate; fls white .... 6. *S. aizoon.*
1 Cauline lvs none below the infl, the branches commonly subtended by minute lvs.
  7 Normal fl 1, terminal, the others replaced by tufts of minute lvs .... 7. *S. foliolosa.*
  7 Normal fls several to many.
    8 Filaments clavate; lvs sharply and saliently toothed.
      9 Lf-blades 3–8 times as long as wide .... 8. *S. micranthidifolia.*
      9 Lf-blades not more than about twice as long as wide .... 9. *S. caroliniana.*
    8 Filaments subulate or filiform; lvs toothed or entire.
      10 Fls irregular; 3 pet clawed and auriculate at the base of the blade .... 12 *S. michauxii.*
      10 Fls regular; pet neither clawed nor auriculate.
        11 Sep erect to ascending or spreading at anthesis.
          12 Hypanthium entirely free from the carpels .... 10. *S. careyana.*
          12 Hypanthium adnate to the base of the carpels .... 11. *S. virginiensis.*
        11 Sep reflexed at anthesis .... 13. *S. pensylvanica.*

1. **Saxifraga cernua** L. Leafy bulbil-s. Rhizomatous; stems solitary or few, 8–20 cm; lower lvs long-petioled, the blade reniform in outline, 10–25 mm wide, 3–7-lobed; upper lvs progressively smaller, shorter-petioled, and fewer-lobed, the uppermost ovate, entire, some with small axillary bulblets; fl solitary, terminal, white; pet 6–10 mm; $2n=36$–72. Circumboreal, s. to the Gaspé Penins. of Que., Mt. Washington (N.H.), n. Minn., and Colo. July.

2. **Saxifraga rivularis** L. Alpine-brook s. Rhizomatous; stems tufted, 2–10 cm; lower lvs long-petioled, the blade reniform in outline, 4–13 mm wide, 3–5(–7)-lobed; upper lvs on shorter petioles, smaller, less cleft or entire, without bulblets; fls 1–4, long-pediceled, white; pet 3–5 mm; $2n=26$, 52, 56. Circumboreal, s. to e. Que., the White Mts., of N.H., and Mont. July.

3. **Saxifraga tricuspidata** Rottb. Three-toothed s. Stems 1–2 dm from a branching caudex; lvs of the basal shoots crowded, narrowly oblong to obovate-oblong, 1–2 cm, sessile, ciliate, terminating in 3 erect, sharp, stiff teeth; cauline lvs few, remote, 3–10 mm, often entire; cyme convex, 2–5 cm wide; pet white, 6–7 mm; $2n=26$. Rocky places; circumboreal, s. to Lab., n. Mich., and B.C. July, Aug. (*Leptasea t.*)

4. **Saxifraga oppositifolia** L. Purple alpine s. Densely matted from a branching caudex, the flowering stems 2–10 cm; lower lvs closely imbricate, opposite, obovate to spatulate, 3–6 mm, ciliate; cauline lvs similar, few; fl solitary, erect, usually purple, 1 cm wide; $2n=26$, 52. Circumboreal, s. in rocky places to e. Que., n. Vt., n. N.Y., and Wyo. June, July.

5. **Saxifraga aizoides** L. Yellow alpine s. Matted from a branching caudex, the erect flowering stems 5–20 cm; lvs thick, linear-oblong, 1–2 cm × 2–4 mm, sessile, entire, sometimes ciliate, those of the flowering stems fewer but much like the others; cymes usually several-fld, often elongate; fls 5–8 mm wide, yellow, usually dotted with orange; $2n=26$. Circumboreal, s. to Vt., w. N.Y., n. Mich., and B.C. June–Aug. (*Leptasia a.*)

6. **Saxifraga aizoon** Jacq. White alpine s. Perennial, 1–3 dm, with dense basal rosettes; basal lvs stiff, sessile, obovate-oblong to narrowly oblong-spatulate, 1–3 cm, ciliate at base, sharply but finely serrate, each tooth with a minute, lime-encrusted pore; cymes several-fld, often elongate; pet white, 3–4 mm; $2n=$ 28. Circumboreal, s. to N.S., N.H., Vt., Mich., and Minn. July. (*Chondrosea a.*)

7. **Saxifraga foliolosa** R. Br. Naked bulbil-s. Stems 5–20 cm from a slender caudex; basal lvs in a small rosette, oblanceolate to narrowly obovate, essentially sessile, 1–2 cm, ciliate below, few-toothed above; cauline lvs none; infl sparsely branched, most of the fls replaced by axillary bulbils, usually with a single terminal fl; pet white, clawed, 4–5 mm, subequal; $2n=56$, 64. Circumboreal, s. to Mt. Katahdin, Me. (*Hydatica f.; S. stellaris var. comosa.*)

8. **Saxifraga micranthidifolia** (Haw.) Steudel. Lettuce-s. Stems 3–8 dm from a short, stout rhizome, branched above, with a large, open infl; blades of the basal lvs oblanceolate to oblong, 3–8 times as long as wide, mostly 8–20 cm, gradually tapering to a winged petiole, sharply and saliently toothed, each tooth terminating in a minute, slender mucro; fls 6 mm wide; sep reflexed; pet oval, white, with a yellow spot near the base; filaments clavate; seeds not spinulose; $2n=22$. Wet cliffs and mt. brooks; e. Pa. to Va., W.Va., and N.C. May, June. (*Micranthes m.*)

9. **Saxifraga caroliniana** A. Gray. Carolina-s. Stems 2–5 dm from a short, thick rhizome, branched above, with a large, open, paniculiform infl; blades of the basal lvs broadly oblong to ovate or subrotund, 5–10 × 3–7 cm, abruptly narrowed into a broad, winged petiole of about equal length, coarsely and saliently toothed, each tooth blunt or truncate under a lens; cauline lvs none; bracteal lvs lanceolate, 5–10 mm; sep soon reflexed; pet ovate, clawed, 3–4 mm, white with yellow spots below the middle; filaments clavate; seeds not spinulose. Rocky woods in the mts.; Va. and N.C. June. (*Micranthes c.*)

10. **Saxifraga careyana** A. Gray. Carey's s. Habitally much like no. 9, avg a little smaller; hypanthium free from the carpels; sep erect or eventually ± spreading; pet more nearly oblong or elliptic, scarcely clawed, less prominently spotted; filaments filiform-subulate. Moist, rocky places; mts. of s. Va., N.C., and Tenn. May–July. (*Micranthes c.*)

11. **Saxifraga virginiensis** Michx.   Early s. Flowering stems solitary or few together, 1 dm at first flowering, later to 4 dm; basal lvs oblong or ovate, 2–5 cm, entire or serrate, obtuse, cuneately narrowed to a margined petiole; infl branched, at first compact, later lax and open; hypanthium adnate to the base of the carpels; sep spreading or ascending; pet spatulate to obovate, 4–6 mm, white; seeds minutely papillate in longitudinal rows; $2n=20$ + several B, 28. Moist or dry open woods and rock-ledges; N.B. to Man., s. to Ga., La., and Okla. Apr., May. (*Micranthes v.*) Many aberrant forms have been described.

12. **Saxifraga michauxii** Britton.   Claw-s. Stems 1–5 dm from a short, stout rhizome, also with offsets, branched above, with a large, open infl; blades of basal lvs elliptic, oblong, or lanceolate, very coarsely and saliently serrate with 4–8 teeth on each side, gradually narrowed below to a broad petiole; bracteal lvs similar but much smaller; fls irregular, the 3 larger pet 3–4 mm, clawed, the ovate blade spotted with yellow and auriculate at base, the 2 smaller pet oblanceolate, unspotted; seeds spinulose in 2 rows. Moist or wet ledges and rocky woods in the mts.; Va. to Ky., s. to Ga. June–Oct. (*Hydatica petiolaris; Saxifraga leucanthemifolia* Michx., not LaPey.)

13. **Saxifraga pensylvanica** L.   Swamp-s. Erect, 3–10 dm; lvs all basal, entire to glandular-denticulate or obscurely serrulate, sparsely hairy, the larger ones 10–20 cm; infl at first compact, later usually lax and elongate; sep soon reflexed; pet greenish-white or purple, 2–3 mm, 1-nerved or 3-nerved, narrowly lanceolate to lance-ovate; body of the follicles erect, connivent at least to the middle, the styles erect or divergent; diploids and polyploids up to 14x, based on $x=14$. Wet meadows and bogs; Me. to Minn., s. to Va. and Mo. May, June. (*Micranthes p.; S. forbesii.*)

4. **TIARELLA** L.   Foam-fl. Fls perfect, 5-merous, regular or nearly so, barely perigynous; hypanthium small, campanulate; pet clawed, linear to elliptic; stamens 10; carpels 2, unequal, united below into a superior, unilocular ovary, distinct above; placentas parietal, near the base; rhizomatous herbs with broad, palmately veined, lobed basal lvs and an erect, usually lfless stem with a bracteate raceme or panicle of white fls. 1 Asia, 5 N. Amer.

1. **Tiarella cordifolia** L.   Plants erect, 1–3.5 dm, usually with long stolons; basal lvs broadly cordate-ovate to subrotund, shallowly 3–5-lobed, crenate, sparsely hairy; raceme at first crowded, later to 1 dm; pedicels 5–10 mm; sep elliptic-obovate, 2–3.5 mm, blunt; pet 3–5 mm; fr thin-walled, the larger carpel 10 mm, the smaller one often only half as long; $2n=14$. Rich woods; N.S. to Ont. and e. Wis., s. to Ga. and Ala. Spring. Var. *collina* Wherry (*T. wherryi* Lakela), without stolons, occurring n. as far as Ky. and Va., is of doubtful status.

5. **MITELLA** L.   Bishop's cap; mitrewort. Fls perfect, regular, perigynous, 5-merous; hypanthium turbinate to saucer-shaped, adnate to the base of the carpels; pet narrow, deeply pinnatifid or fimbriate to entire; stamens 10 (in ours) or 5; carpels 2, connate into a 2-beaked, unilocular, superior (to half-inferior) ovary with low-parietal placentas; rhizomatous herbs with basal or alternate, palmately veined lvs from the rhizome, the flowering stems lfless or nearly so, with a terminal bracteate raceme of small white, greenish, or purple fls. 12, N. Amer., e. Asia.

1 Lvs on the flowering stem a single opposite pair; fls white; pet 2 mm .... 1. *M. diphylla.*
1 Lvs on the flowering stem one or none; fls yellowish-green; pet 3–5 mm .... 2. *M. nuda.*

1. **Mitella diphylla** L.   Two-lvd m. Stems 1–4 dm, sparsely hairy below, glandular-puberulent above; basal lvs long-petioled, ovate-rotund with cordate base, shallowly 3–5-lobed, crenate, hairy; cauline lvs 2, mostly sessile, smaller, mostly 3-lobed; raceme 5–15 cm; pedicels 1–2 mm; fls white; pet 2 mm, deeply fimbriate-pinnatifid, with obliquely ascending segments; seeds few, black, shining, 1–1.5 mm; $2n=14$. Rich woods; Que. to Minn., s. to Va., Ga., and Mo. May, June. An apparent hybrid with *M. nuda* is *M.* × *intermedia* Bruhin.

2. **Mitella nuda** L.   Naked m. Stems 0.5–2 dm, hairy below and glandular throughout; basal lvs rotund to reniform with cordate base, obscurely lobed, crenate, hairy; cauline lf 1 or none, ovate, sessile, few-toothed; racemes 2–10 cm; pedicels 1–5 mm; fls yellowish-green; pet 3–5 mm, very deeply pectinate-fimbriate, with divergent segments; seeds few, black, shining, 1 mm; $2n=28$. Bogs and wet woods, usually in moss; Lab. to Mack., s. to Pa., Mich., Minn., and Mont.; e. Asia. May, June.

6. **SULLIVANTIA** T. & G.   Fls perfect, regular, partly epigynous with a prolonged, deeply cupulate hypanthium bearing the sep, pet, and stamens at its margin; pet obovate or spatulate, marcescent-persistent; stamens 5, opposite and much shorter than the sep; carpels 2, connate below to form a bilocular ovary with numerous ovules on axile placentas,

tapering above into the prominent distinct stylar beaks that eventually dehisce ventrally; stigmas capitate; capsule largely contained within the persistent hypanthium; seeds narrowly winged; fibrous-rooted perennial herbs with largely basal, long-petioled, ± stipulate, reniform or reniform-orbicular, palmately veined, shallowly lobulate and toothed lvs and cymose panicles of small white fls. 3, N. Amer.

1. **Sullivantia sullivantii** (T. & G.) Britton. Lvs round-reniform, 3–7 cm wide, sparsely pubescent to subglabrous, with numerous shallow lobes each terminating in 2 or 3 short teeth; infl lax, 2–4 dm, with ascending branches, conspicuously glandular; pet 3 mm; $2n=14$. Moist shaded cliffs; s. O. and e. Ky. to Mo., nw. Ill., sw. Wis., ne. Io., and se. Minn., in unglaciated areas. June, July. (*S. renifolia*, the more western plants.)

7. **BOYKINIA** Nutt., nom. conserv. Fls perfect, regular, ± epigynous with a prolonged, deeply cupulate or campanulate hypanthium bearing the sep, pet, and stamens at its margin; pet obovate or spatulate, deciduous; stamens 5, opposite and shorter than the sep; carpels 2, connate below to form a bilocular ovary with numerous ovules on axile placentas, tapering above into the 2 short, distinct stylar beaks that eventually dehisce ventrally; stigmas capitate; capsule largely contained within the persistent hypanthium; seeds wingless; rhizomatous perennial herbs with basal and alternate, long-petioled, ± stipulate, palmately lobed lvs and panicled cymes of small white fls. 7, N. Amer.

1. **Boykinia aconitifolia** Nutt. Aconite-saxifrage. Stems 3–6 dm, glandular-hairy above; basal lvs broadly reniform, 5–10 cm wide, 5–9-lobed to the middle, the lobes laciniately toothed; cymes at first compact, later elongating and racemiform; hypanthium glandular-hairy; pet obovate, short-clawed, 3–5 mm; $2n=12$. Moist mt. woods; Va., W.Va., and Ky. to Ga. and Ala. June, July. (*Therofon a.*)

8. **HEUCHERA** L. Alum-root. Fls perfect, perigynous, 5-merous, regular or obliquely irregular; hypanthium saucer-shaped to tubular, adnate to the lower part of the carpels; sep green or petaloid; pet small, linear to narrowly obovate, short-clawed, or occasionally absent; stamens 5, opposite the sep; carpels 2, united below into a unilocular ovary with 2 parietal placentas; capsule 2-beaked; herbs from a short rhizome or caudex, with long-petioled, palmately lobed and toothed basal lvs, a naked or rarely few-lvd stem, and a panicle of small fls. 35, N. Amer.

1 Free portion of the hypanthium above its divergence from the ovary (measured to the sinuses) shorter
    than the adnate portion.
  2 Hypanthium ± villous with long white hairs; fls (measured to the end of the sep) 1.5–4 mm; mostly
      on cliffs and ledges; fl summer.
    3 Lf-teeth ± acute; seeds muricate .... 1. *H. villosa.*
    3 Lf-teeth broadly rounded to a mucronate summit; seeds nearly smooth .... 2. *H. parviflora.*
  2 Hypanthium glandular-puberulent; fls 3–8 mm; mostly dry woods or prairies; fl spring.
    4 Petioles minutely glandular-puberulent, glabrous, or with a few scattered hairs.
      5 Pet entire or shortly toothed; free part of hypanthium 0.5–1.5 mm .... 3. *H. americana.*
      5 Pet fimbriate; free part of hypanthium 1.5–2 mm .... 4. *H. × hispida.*
    4 Petioles villous with long, spreading hairs .... 5. *H. × hirsuticaulis.*
1 Free portion of the hypanthium longer than the adnate portion.
  6 Petioles spreading-hirsute; Mich. to Ind. and Ark. and westward.
    7 Free portion of the hypanthium up to 2 mm; stamens exserted .... 5. *H. × hirsuticaulis.*
    7 Free portion of the hypanthium 2–6 mm; stamens included to exserted .... 6. *H. richardsonii.*
  6 Petioles glabrous or merely puberulent; Pa. to Ky. and southward.
    8 Style exserted (beyond the sep-tips), or included by less than 1 mm; sep ± spreading .... 7. *H. pubescens.*
    8 Style included by 1.5 mm or more; sep inflexed, closing the fl .... 8. *H. longiflora.*

1. **Heuchera villosa** Michx. Maple-lvd a.-r. Plants 2–8 dm; scape with a few small, scale-like bracts; lvs angular, with triangular, acute lobes and teeth, rough-hairy on both sides, or mainly on the lower side or only on the veins beneath; petioles and stems ± spreading-villous with multicellular trichomes; infl congested at anthesis, with pedicels up to 3 mm, later more open; fls regular, 1.5–3(–3.5) mm at anthesis (measured to the sep-tips); hypanthium glandular-villous, the free part up to 0.5 mm, shorter than the adnate part, white or pink, usually with green-tipped sep; pet white to pink, linear, entire, often spirally coiled, much exceeding the sep; stamens exserted; seeds sharply muricate; $2n=14$. Moist, shaded ledges and cliffs, especially on acid rocks; irregularly distributed from the mts. of Va. and W.Va. to S.C. and Tenn., w. to s. Ind., Mo., and Ark.; rare in w. Pa. and w. N.Y. Late June–Sept. Ours is var. *villosa*. Var. *arkansana* (Rydb.) E. B. Smith is Ozarkian.

2. **Heuchera parviflora** Bartling. Small-fld a.-r. Plants 1–4.5 dm; scape with a few small, scale-like bracts; lvs rounded, with shallow, broadly rounded lobes and with the teeth broadly rounded to a short mucro, more softly pubescent than in no. 1; infl open, the pedicels 3–17 mm at anthesis; fls as in no. 1, but the white (pink) pet narrowly oblanceolate, reflexed, seldom coiled; seeds merely inconspicuously ridged; $2n=14$. A rare plant on moist, shaded ledges and cliffs, on sandstone or limestone; W.Va. to N.C. and Ala., w. to s. Ind., s. Ill., Mo., and Ark. July–Sept. Most of our plants are var. *parviflora*, with spreading-villous stems and petioles and somewhat villous lvs (the upper surface often scantily so), and with the bracts subtending the floral branches toothed and often foliaceous. The Ozarkian var. *puberula* (Mackenzie & Bush) E. Wells, with shorter, denser, more glandular pubescence on the petioles, scapes, and lvs, and with the bracts subtending the floral branches entire and scale-like, extends e. occasionally to c. Ky. (*H. puberula.*)

3. **Heuchera americana** L. Common a.-r. Plants 4–14 dm; scape with a few small, scale-like bracts; petioles glabrous or minutely glandular-puberulent, seldom with a few longer spreading hairs; lvs often white-mottled, variously hairy or glabrous, 5–9-lobed with lobes of various depths, often pentagonal in outline, the teeth rounded to acute, mucronate; stem glandular in the infl, otherwise glabrous; fls only slightly oblique, 3–7 mm (to the sep-tips), the free part of the hypanthium 0.5–1.5 mm; stamens and style long-exserted; pet spatulate, greenish, white, or pink, entire or with small teeth; $2n=14$. Dry upland woods; Conn. to Ga., w. to s. Ont., s. Mich., Ind., s. Ill., and e. Okla. Apr.–June (July).

4. **Heuchera × hispida** Pursh. Hybrid swarm of spp. 3 and 7, for convenience designated with a binomial; scape with up to 5 lflike bracts; petioles glabrous or minutely glandular; lvs 7–9-lobed, the lobes mostly broadly rounded, the teeth rounded or obtuse to the mucronate apex; stem glandular in the infl, otherwise glabrous or nearly so; fls distinctly oblique, 6–8 mm above, 4.5–6 mm below; free part of the hypanthium 1.5–2 mm; pet purple, longer and wider than the sep, with fimbriate margins; stamens and style exserted. Mt. woods; Va., W.Va., and N.C.

5. **Heuchera × hirsuticaulis** (Wheelock) Rydb. Hybrid swarm of spp. 3 and 6, for convenience designated with a binomial; scape with small, scaly bracts; petioles conspicuously spreading-hirsute; fls from small and nearly regular to larger and strongly oblique; free part of the hypanthium 1–2 mm; pet greenish, white, or pink, narrower than the sep, subentire or finely toothed; stamens and style exserted. Dry uplands woods and prairies, largely supplanting the original parents in the area where their ranges overlap, from upper Mich. to e. Ind., s. and w. to Tenn., Ark., Kans., Nebr., and Minn.

6. **Heuchera richardsonii** R. Br. Prairie a.-r. Plants 2–9 dm; scape with a few small, scale-like bracts; petioles strongly spreading-hirsute; lvs broadly cordate-ovate, glabrous or nearly so above, hairy at least on the veins beneath, shallowly 7–9-lobed, each lobe with 3–5 rounded to acute, mucronate lobules; stems ± hirsute, becoming glandular above; infl relatively narrow and congested; fls very oblique, 7–10 mm on the upper side, half as long below, with wide sinuses between the sep; free part of the hypanthium 2–6 mm; pet mostly green or greenish-white (pink), 2–4 mm, narrowly spatulate, finely toothed; stamens shortly included to long-exserted; style shortly included to shortly exserted; $2n= 14, 28$. Prairies and dry woods; s.c. Mich. and nw. Ind. to nw. Ark. and ne. Okla., w. and n. to Colo. and Alta. May–July.

7. **Heuchera pubescens** Pursh. Appalachian a.-r. Plants 3–9 dm; scape with up to 5 lf-like bracts; petioles and stems glabrous or glandular-puberulent; lvs round-cordate, conspicuously 5–7-lobed, the teeth broadly rounded to acute, mucronate; bracts of the infl somewhat scarious; fls oblique, 7–11 mm on the upper side, ⅔–¾ as long on the lower, the free part of the hypanthium 2–6 mm; sep somewhat spreading, rounded, separated by narrow sinuses; pet white to pink or purple, 2.5–5 mm, somewhat inflexed, broadly spatulate, generally fimbriate-margined; stamens and style typically barely exserted, varying to barely included or evidently exserted; $2n=14$. Upland woods in the Ridge and Valley geological province, from s. Pa. and w. Md. to Va. and W.Va. May, June.

8. **Heuchera longiflora** Rydb. Closed-fld a.-r. Plants 3–9 dm; scape with small and scale-like or larger and leaf-like bracts; petioles glabrous or minutely glandular-puberulent; lvs round-cordate, glabrous or nearly so, or short-hairy on the veins beneath, often white-mottled, 5–9-lobed, the teeth very broad and flat or barely rounded, mucronate; stems short-hirsute; infl lax, with herbaceous bracts; fls horizontal, distinctly oblique, 7–9 mm on the upper side, ¾ as long on the lower; sep inflexed, closing the fl, rounded, 2–3.5 mm, separated by narrow sinuses; pet white to pink or purple, inflexed, spatulate, 2–5 mm, often fimbriate; stamens evidently included to barely exserted; styles included 1.5–3 mm (from the cal-tip) at anthesis; $2n=14$. Rich woods on limestone substrate; mts. of e. Ky., s. W.Va., and sw. Va. to N.C. and Ala. May, June.

9. **PARNASSIA** L. Grass of Parnassus. Fls perfect; hypanthium short; sep, pet, and stamens each 5; staminodia 5, opposite the pet, each consisting of 3 or more sterile stamens ± connate at base and separate above, their anthers reduced to glands; ovary superior or partly inferior, unilocular; ovules numerous on 4 parietal placentas; stigmas 4, nearly sessile, commissural; capsule 4-valved, loculicidal; seeds numerous, oblong, angular, cellular-reticulate; glabrous perennial herbs, the lvs entire, palmately veined, mostly long-petioled in a basal rosette, or a single sessile one on the erect, 1-fld scape; fls erect, the pet white, conspicuously veined. 50, N. Temp.

1 Staminodes 3-parted, not dilated at base; sep with firm center and subscarious margins.
  2 Staminodes 4–9 mm, equaling or shorter than the stamens.
    3 Blades of the lower lvs longer than wide, obtuse or subcordate at base .... 1. *P. glauca.*
    3 Blades of the lower lvs wider than long, reniform .... 2. *P. asarifolia.*
  2 Staminodes 10–16 mm, conspicuously longer than the stamens .... 3. *P. grandifolia.*
1 Staminodes 5–many-cleft, dilated at base; sep foliaceous, not scarious-margined.
  4 Staminodes with (5)7(9) filaments; lvs all narrowed to the base .... 4. *P. parviflora.*
  4 Staminodes with 9–many filaments; lvs broadly rounded or cordate at base .... 5. *P. palustris.*

1. **Parnassia glauca** Raf. American g.-o.-p. Plants 2–4 dm; blades of basal lvs stiff and leathery, ovate to round-ovate, 3–5 × 1.5–4.5 cm, blunt to subacute, at base broadly rounded to subcordate, decurrent on the upper part of the petiole; cauline lf usually present, at or well below the middle, sessile or short-petioled, the blade like the basal ones but smaller; pet sessile, 10–18 mm, with several strong veins, the central 5 usually unbranched; staminodes white, 4–7 mm, a little shorter than the stamens, 3-parted for ⅗ or ⅘ their length; 2*n*=32. Calcareous bogs, shores, and wet meadows; Nf. and Que. to Sask., s. to N.J., Pa., Ind., Io., and S.D. Aug., Sept. (*P. americana; P. caroliniana,* misapplied.)

2. **Parnassia asarifolia** Vent. Kidney-lvd g.-o.-p. Plants 2–4 dm; blades of basal lvs thin and flexible, reniform, 3–5 cm wide, wider than long; cauline lf sessile near the middle of the stem, like the basal but smaller; pet narrowed to a basal claw, 12–18 mm, with 11–15 radiating veins; staminodes white, 7–9 mm, a little shorter than the stamens, 3-parted for ⅗ or ⅘ their length; 2*n*=32. Streambanks and springy or boggy soil, chiefly in the mts.; Va. and W.Va. to Ga., and w. to Ark. and e. Tex. Aug.–Oct.

3. **Parnassia grandifolia** DC. Big-lvd g.-o.-p. Plants 2–4 dm; blades of basal lvs broadly round-ovate, 5–8 × 3–7 cm, obtuse to broadly rounded above, usually cordate or subcordate below; cauline lf sessile or subsessile near the middle of the stem, like the basal but smaller; pet sessile, 15–22 mm, with 7–9 veins, the outer pair of veins emitting numerous lateral branches; anthers reddish; staminodes white, 10–16 mm, evidently longer than the stamens, 3-parted for ⅘ or ⁹⁄₁₀ their length, the segments knobbed at the tip; ovary greenish; 2*n*=32. Wet, calcareous soil in the mts.; Va. and W.Va. to n. Fla., and w. to Ark., s. Mo., and e. Tex.; reputedly in N.J. Sept., Oct.

4. **Parnassia parviflora** DC. Small-fld g.-o-p. Plants 1–3 dm; blades of basal lvs ovate to oblong, 1–3.5 cm, ½ to ¾ as wide, narrowed at the base and decurrent on the petiole; cauline lf usually present, erect, lanceolate to ovate, 3–15 mm, narrowed to a sessile base; pet 5–13-nerved, 6–10 mm, ⅛ to ½ longer than the green sep; staminodes green or yellow, 3.5–5 mm, dilated to an obovate or reverse-triangular scale with (5)7(9) short filaments on the distal margin; 2*n*=36. Calcareous shores and wet meadows; Lab. to B.C., s. to Que., Ont., Mich., Wis., S.D., and N.M. June–Aug. Perhaps better as *P. palustris* var. *parviflora* (DC) B. Boivin.

5. **Parnassia palustris** L. Arctic g.-o.-p. Plants 2–4 dm; blades of basal lvs broadly ovate to subrotund, 1–3 cm, ¾ to as wide as long, at base broadly rounded to more often cordate, not decurrent; cauline lf as large as the basal, broadly ovate, sessile and cordate-clasping; pet 5–17-nerved, 9–13 mm, half longer to twice as long as the lance-triangular green sep; staminodes green or yellow, 5–9 mm, dilated to an obovate or reverse-triangular scale with 9–23 short, slender filaments on the distal margin; 2*n*=18, 27, 36, 54. Calcareous shores and wet meadows; circumboreal, s. in Amer. to Nf., Mich., Minn., Wyo., and B.C.

10. CHRYSOSPLENIUM L.

10. CHRYSOSPLENIUM L. Fls perfect, regular, perigynous, 4-merous; sep spreading; pet none; stamens 4–8, inserted in the notches of an 8-lobed disk that nearly fills the center of the fl; filaments very short; carpels 2, united below into a unilocular ovary with 2 parietal placentas; styles protruding through the center of the disk, divergently curved; capsule 2-lobed, dehiscent across the top; inconspicuous, branching perennials with small, simple lvs, the fls small, terminal or in small terminal cymes, or appearing axillary by proliferation of the stem. 40, cool reg.

1 Fls solitary; lvs entire or obscurely and irregularly toothed, the lower opposite .... 1. *C. americanum.*
1 Fls in small cymules; lvs with 5–7 conspicuous rounded teeth, all alternate .... 2. *C. iowense.*

1. **Chrysosplenium americanum** Schwein. Stems decumbent, branched, 5–20 cm; lvs all short-petioled, the lower opposite, the upper alternate, ovate to rotund, 5–15 mm, entire or obscurely and irregularly toothed; fls solitary at the end of the branches, 4–5 mm wide, the ovate sep greenish-yellow or greenish-red or green marked with brown-purple, the disk red or brown-purple or green; stamens mostly 8; 2*n*=24. Springy or muddy soil, usually in shade; Que. and Ont. to Sask., s. to Va., Ky., and Ind. Apr.–June.

2. **Chrysosplenium iowense** Rydb. Stem arising from the end of last year's rhizome; lvs all alternate, reniform, conspicuously crenate-toothed, the lower long-petioled; fls 3–5 together in small, nearly sessile, terminal cymules; sep bright yellow; stamens 5–8; 2*n*=96, ± 120. Wet, mossy slopes in ne. Io., and widely distributed in the Arctic. May–July.

FAMILY **ROSACEAE**, the Rose Family

Fls commonly regular and perfect, perigynous or epigynous, usually (4)5-merous as to the sep and pet, or the pet seldom wanting; hypanthium saucer-shaped to cup-shaped or urceolate; stamens (1–)10–many, typically in multiples of 5, often 20 in all; pistils 1– many, superior and usually distinct, or (in subfamily Maloideae) united into a compound, inferior ovary with axile placentas and distinct styles; ovules 1–several; fr variously an achene, follicle, drupe, or pome, or a set of coherent drupelets, with or without an enlarged and modified hypanthium or receptacle; endosperm mostly wanting or nearly so, but copious in *Physocarpus*; herbs or woody plants with mostly alternate, simple or compound, stipulate or seldom (as in *Spiraea*) exstipulate lvs and commonly with conspicuous fls. 100/3000.

Several spp. not keyed and described in the following text occasionally escape from cult. Among these are *Exochorda racemosa* (Lindl.) Rehder, a simple-lvd, white-fld shrub with a superior ovary of 5 weakly united carpels, ripening into a winged capsule; *Rhodotypos scandens* (Thunb.) Makino, a simple-lvd, white-fld shrub differing from all our other Rosaceae in its opposite lvs; and *Kerria japonica* (L.) DC. a shrub with simple lvs and bright yellow fls.

SYNOPSIS OF THE SUBFAMILIES AND TRIBES IN OUR REGION

1 Carpels 1–many, superior, distinct (or becoming connate in fr).
  2 Carpels 2–many, rarely only 1 and then not ripening into a drupe. (Subfamily *Rosoideae*).
    3 Fr a follicle; carpels (1)2–5. (Genera 1–5) .... Tribe Spiraeeae.
    3 Fr an achene or drupelet; carpels (1–few–) many.
      4 Hypanthium slightly concave to flat, convex, or elongate.
        5 Ovule 1, or ovules 2 and superposed (Genera 6–13) .... Tribe Potentilleae.
        5 Ovules 2, collateral. (Genera 14–15) .... Tribe Rubeae.
      4 Hypanthium excavate, completely enclosing the ovary or ovaries.
        6 Carpels 1–4; hypanthium at maturity ± indurate. (Genera 16–19) .... Tribe Sanguisorbeae.
        6 Carpels numerous; hypanthium at maturity fleshy. (Genus 20) .... Tribe Roseae.
  2 Carpel 1; fr a drupe. (Subfamily Prunoideae; Genus 21) .... Tribe Pruneae.
1 Carpels 2–5, united into a compound, inferior ovary; fr a pome. (Subfamily *Pomoideae*; Genera 22–
  26) .... Tribe Pomeae.

ARTIFICIAL KEY TO THE GENERA

1 Lvs simple.
  2 Herbs.
    3 Pet none; stamens 4; hypanthium permanently enclosing the 1(2–3) ovaries .... 16. *Alchemilla*.
    3 Pet present; stamens many; hypanthium not enclosing the 5–many ovaries.
      4 Pistils 5–10; fr dry; plants scapose; lvs serrate .... 15. *Dalibarda*.
      4 Pistils many, concrescent into an aggregate, fleshy fr; plants leafy-stemmed; lvs lobed .... 14. *Rubus*.
  2 Trees, shrubs, or woody vines.
    5 Ovary 1.
      6 Style 1; ovary superior; fr a drupe .... 21. *Prunus*.
      6 Styles 2–5; ovary inferior; fr a pome (sometimes small and berry-like).
        7 Ovary and fr 10-locular by the intrusion of a false septum between the ovules of each
          primary locule; most spp. with fls in racemes; unarmed .... 26. *Amelanchier*.
        7 Ovary and fr 2–5-locular; fls not in racemes.
          8 Styles distinct; actual carpels within the pericarp very hard and bony, seed-like; mostly
            thorny trees and arborescent shrubs ....25. *Crataegus*.
          8 Styles in most spp. connate at the base; actual carpels leathery or papery, easily opened
            to expose the seeds within.
            9 Trees or arborescent shrubs, most spp. thorny; fr mostly well over 1 cm thick .... 22. *Pyrus*.
            9 Shrubs to 3 m, unarmed; fr to 1 cm thick .... 23. *Aronia*.
    5 Ovaries 2–many.
      10 Ovaries 2–5; fr of follicles; unarmed shrubs.
        11 Stipules or stipular scars present; lvs ± lobed; endosperm present .... 1. *Physocarpus*.
        11 Stipules none; lvs entire or serrate; endosperm wanting .... 2. *Spiraea*.
      10 Ovaries numerous, concrescent into an aggregate fleshy fr; prickly or unarmed shrubs .... 14. *Rubus*.
1 Lvs compound or dissected, or deeply trifid and again 3-lobed.
  12 Stamen 1, lvs deeply trifid and again 3-lobed; small annuals .... 17. *Aphanes*.
  12 Stamens 5–many; lvs compound or dissected; habit various.

13 Lvs trifoliolate or palmately once compound; plants, except *Rubus* and one sp. of *Potentilla,*
    herbs.
  14 Ovaries 2–10.
    15 Pet white to pink; fr of follicles .... 5. *Porteranthus.*
    15 Pet yellow; fr of achenes.
      16 Stamens numerous; lfls with numerous teeth .... 10. *Waldsteinia.*
      16 Stamens 5; lfls merely 3–5-toothed at the summit .... 8. *Sibbaldia.*
  14 Ovaries numerous.
    17 Styles filiform, elongate; lower, middle and upper lvs conspicuously different in shape
        .... 13. *Geum.*
    17 Styles short and inconspicuous; lvs all nearly alike, except in size.
      18 Bractlets present, alternating with the sep; ovaries ripening into achenes, but the
          receptacle sometimes fleshy; plants unarmed.
        19 Bractlets broadly 3-toothed at the summit; fls yellow; receptacle fleshy, in-
            edible .... 7. *Duchesnea.*
        19 Bractlets entire, or obscurely toothed on the lateral margins only.
          20 Receptacle dry; fls yellow, seldom white or red-purple .... 11. *Potentilla.*
          20 Receptacle fleshy, edible at maturity; fls white (pink) .... 6. *Fragaria.*
      18 Bractlets none; ovaries ripening into coherent drupelets; plants usually prickly .... 14. *Rubus.*
13 Lvs pinnately compound, or ternately 2–several times compound or dissected.
  21 Lvs ternately 2–several times compound or dissected.
    22 Stamens numerous; carpels 3–5, ripening into follicles; pls 1–2 m .... 3. *Aruncus.*
    22 Stamens 5, carpels mostly 5–10, ripening into achenes; pls 1–3 dm .... 9. *Chamaerhodos.*
  21 Lvs pinnately once compound.
    23 Herbs.
      24 Pet none; sep 4; infl a dense spike or head; pistils 1 or 2 .... 19. *Sanguisorba.*
      24 Pet 5; sep 5; infl more open.
        25 Hypanthium turbinate, armed with hooked prickles; pistils 2 .... 18. *Agrimonia.*
        25 Hypanthium flat to saucer-shaped or hemispheric, unarmed; pistils 5 or more.
          26 Pistils 5–15, ripening into a whorl of follicle-like achenes .... 12. *Filipendula.*
          26 Pistils numerous, ripening into a head of achenes.
            27 Styles short and inconspicuous .... 11. *Potentilla.*
            27 Styles elongate and either jointed or plumose .... 13. *Geum.*
    23 Trees, shrubs, or woody vines.
      28 Carpels 2–5.
        29 Carpels 5, distinct, superior, becoming dry and dehiscent in fr .... 4. *Sorbaria.*
        29 Carpels 2–4, connate below and half-inferior; fr a small pome .... 24. *Sorbus.*
      28 Carpels ± numerous, distinct, or ± concrescent in fr.
        30 Unarmed shrub with yellow fls and closely crowded, narrow lfls 1–2 cm .... 11. *Potentilla.*
        30 Plants distinctly otherwise, usually prickly, only seldom with yellow fls, and
            usually with larger lfls.
          31 Hypanthium globose to urceolate, with a constricted orifice, concealing
              the ovaries and achenes .... 20. *Rosa.*
          31 Hypanthium flat to hemispheric, the ovaries and the usually concrescent
              drupelets exposed .... 14. *Rubus.*

1. PHYSOCARPUS Maxim., nom. conserv.   Ninebark. Hypanthium shallowly cupulate; sep 5, triangular, persistent; pet 5, suborbicular, spreading; stamens 20–40; pistils
(1–)3–5, sometimes weakly united below; style elongate; stigma capitate; ovules few; fr a
firm-walled inflated follicle, dehiscent on both sutures; seeds hard, shining, with endosperm; shrubs with stipulate, simple, usually lobed lvs and corymbose white fls. 10+, N.
Amer., Asia.

  1. Physocarpus opulifolius (L.) Maxim.   Shrub to 3 m, the bark exfoliating in narrow strips; petioles 1–
3 cm; lvs ovate to obovate, the principal ones ± 3-lobed, irregularly serrate, broadly cuneate to truncate at
base; pedicels 1–2 cm; sep 1.5–2.5 mm; fls 7–10 mm wide; fr 5–10 mm, glabrous or stellate-pubescent;
$2n=18$. Moist, sandy or rocky soil, especially along streambanks and shores; Que. to N.D. and Colo., s. to
N.C., Tenn., and Ark., also escaped from cult. May–July. The form with stellate fr, commoner toward the
west, has been called var. *intermedius* (Rydb.) B. L. Robinson. (*Opulaster opulifolius; O. australis.*)

  2. SPIRAEA L.   Spiraea. Hypanthium cupulate or turbinate; sep 5, pet 5, small,
spreading; stamens 15–many; pistils mostly 5; styles terminal; ovules 2–several; fr a firm
follicle, dehiscent on the ventral suture; endosperm none; shrubs with exstipulate, simple
lvs and white to pink or purple fls in terminal or lateral infls. 70, N. Hemisphere.

1 Infls simple, of unbranched umbels or corymbs; intr. .... 1. *S. prunifolia.*
1 Infls compound, branched.
  2 Infl flattened or merely convex, corymbiform, the floriferous part wider than long.

3 Lvs acute to obtuse or rounded; fls white, seldom pink; native.
  4 Lvs up to about twice as long as wide, obtuse to subcordate at base .... 2. *S. betulifolia.*
  4 Lvs more than twice as long as wide, cuneate at base .... 3. *S. virginiana.*
3 Lvs long-acuminate; fls mostly pink; intr. .... 4. *S. japonica.*
2 Infl more elongate, paniculiform, mostly longer than wide.
  5 Lvs glabrous or nearly so; sep spreading; follicles glabrous .... 5. *S. alba.*
  5 Lvs densely tomentose beneath; sep reflexed after anthesis; follicles pubescent .... 6. *S. tomentosa.*

    1. Spiraea prunifolia Siebold & Zucc.   Shrub 2–3 m; lvs ovate or ovate-oblong, short-petioled, 2–4 cm, finely serrulate; umbels axillary, sessile or nearly so, with 3–6 white, commonly double fls 8–10 mm wide on pubescent pedicels 1–2 cm. Native of e. Asia, often escaped in our range. May.
    Three other cult. spp. that would key to *S. prunifolia* occasionally escape. *S. thunbergii* Siebold has lance-linear, sparsely serrate lvs and sessile or subsessile umbels with 3–6 fls on glabrous pedicels 1–1.5 cm. *S. chamaedryfolia* L. has ovate or ovate-oblong lvs with numerous sharp teeth, and peduncled dense corymbs of fls with exserted stamens, terminating leafy branches. *S. vanhouttei* (C. Briot) Zabel, the Bridal wreath, has rhombic-ovate to obovate lvs few-toothed above, and peduncled corymbs of fls with short stamens, terminating short leafy branches.
    2. Spiraea betulifolia Pallas.   Birch-lvd s. Simple or sparsely branched shrub to 1 m; lvs broadly ovate or oblong to obovate, 4–7 cm long and usually at least half as wide, obtuse or rounded at the tip, rounded to broadly cuneate at the base, irregularly toothed (tending to be doubly serrate) above the middle, glabrous or sparsely villosulous beneath; infl terminal, compound, broadly dome-shaped, 3–10 cm wide, the branches and pedicels thinly hairy to glabrous; hypanthium glabrous or sparsely villous without, white-strigose to glabrate within; fls white to seldom pink, 4–5 mm wide; sep 0.5–0.8 mm; fr glabrous, 2–3 mm; 2*n*=54. Mt. woods; w. Md., Va., and W.Va. to Ala. June, July. Our plants, as here described, are var. *corymbosa* (Raf.) Maxim. Other disjunct vars. occur in w. N. Amer. and e. Asia.
    3. Spiraea virginiana Britton.   Appalachian s. Simple or sparsely branched shrub to 1 m; lvs oblong-oblanceolate or oblanceolate, 3–6 cm × 10–18 mm, acute and mucronate, entire or with a few low teeth near the tip, somewhat glaucous beneath; infl terminal, compound, short and broad, 3–8 cm wide, the branches glabrous or villous; hypanthium glaucous; fls white, 5–6 mm wide; sep 0.8–1.1 mm; fr 1.5 mm. Damp, rocky banks in the mts.; W.Va. and Va. to Tenn., N.C., and n. Ga.
    4. Spiraea japonica L.f.   Japanese s. Shrub to 1.5 m; lvs lanceolate or lance-ovate, 8–15 × 2.5–5 cm, long-acuminate, sharply toothed, acute at base; infl terminal, compound, short and broad, 5–20 cm wide; pedicels and hypanthium finely pubescent; fls mostly pink, 5 mm wide; fr 2.5–3 mm; 2*n*=18. Native of Japan, often escaped from cult. June, July.
    5. Spiraea alba Duroi.   Meadowsweet. Shrub to 2 m; lvs 3–7 cm, mostly broadest above the middle, toothed, glabrous or nearly so; infl terminal, compound, paniculiform; fls white, seldom pinkish, 4–7 mm wide; sep spreading; fr glabrous; 2*n*=36. Nf. and Que. to Alta., s. to N.C., Mo., and S.D. June–Aug. Two copiously intergrading vars., widely overlapping geographically as well as morphologically:
    Var. *alba.* Twigs dull brown or yellow-brown; lvs relatively narrow, oblanceolate, 3–4+ times as long as wide, finely and sharply toothed; infl evidently puberulent, as also the hypanthium and the usually obtuse sep. Wet meadows, swamps, and shores; relatively western, from Alta. and N.D. and S.D. to w. Que., s. to Ind., Mo., and N.C.
    Var. *latifolia* (Aiton) Dippel. Twigs purple-brown or red-brown; lvs broader, broadly oblanceolate to oblong or obovate, mostly 2–3 times as long as wide, more coarsely and bluntly toothed; infl glabrous or nearly so; hypanthium glabrous; sep acute. Moist or dry, often upland or rocky soil, old fields, and meadows; relatively eastern, from e. Nf. and e. Que. to N.C., w. to Mich. (*S.l.*)
    S. salicifolia L., a similar Eurasian sp. that sometimes escapes from cult., has pink fls, and the lvs are broadest below the middle.
    6. Spiraea tomentosa L.   Hardhack. Simple or sparsely branched shrub to 12 dm; lvs ovate to oblong or lanceolate, 3–5 cm, beneath white or rufous with a dense tomentum and prominently veined; infl terminal, branched, elongate, 5–15 cm; hypanthium tomentose; sep reflexed after anthesis; fls pink or rarely white, 3–4 mm wide; fr pubescent; 2*n*=24. Swamps and wet meadows; N.S. and N.B. to Que. and Minn., s. to N.C., Tenn., and Ark. July–Sept. Var. *tomentosa,* chiefly eastern, has 11–20 fls per cm of axis; the ill-defined, mostly western var. *rosea* (Raf.) Fern. has 6–10 fls per cm of axis.

    3. ARUNCUS L.   Dioecious; fls 5-merous; hypanthium saucer-shaped; sep triangular; pet spatulate to obovate; stamens 15 or more, unequal, vestigial in pistillate fls; carpels 3(4–5), distinct, erect, with divergent styles, capitate stigmas, and several ovules, vestigial in staminate fls; follicles oblong, turgid, ventrally dehiscent; tall, shortly rhizomatous perennial herbs with alternate, exstipulate, 2–3-ternately compound lvs and small white fls in numerous racemes aggregated into a large terminal panicle. Much like *Astilbe* in aspect, but sharply distinct on floral characters. 10+, N. Temp.

    1. Aruncus dioicus (Walter) Fern.   Goat's beard. Erect, 1–2 m; lvs to 5 dm; lfls lance-ovate or ovate-oblong to broadly ovate, 5–15 cm, acuminate, doubly serrate, rounded to cordate at base, the lateral usually oblique, sometimes 2-lobed; infl 1–3 dm; pet ca 1 mm or less; fr reflexed, 2 mm, with persistent style; $2n=18$. Rich, north-facing woods; Pa. to O., Ind., and Io., s. to N.C., Ala., and Ark. May, June. The poorly defined, chiefly western var. *pubescens* (Rydb.) Fern. has subcylindric frs, and the lvs are usually or always pubescent beneath, in contrast to the chiefly more eastern var. *dioicus,* with semi-ovoid frs (strongly convex on the back) and with the lvs glabrous or pubescent beneath. (*A. alleghheniensis.*)

    A. sylvester Kostel., a European sp. occasionally escaped from cult., has frs 2.5–3.5 mm, the style deciduous.

4. SORBARIA (Ser.) A. Braun, nom. conserv.

Hypanthium broadly cup-shaped; sep 5, soon reflexed; pet 5, elliptic-ovate; stamens numerous; carpels 5, distinct, with elongate, clavate style, capitate stigma, and several ovules; follicles dehiscent on both sutures; shrubs with stipulate, pinnately compound lvs and small white fls in large panicles. 10, N. Amer., Asia.

    1. Sorbaria sorbifolia (L.) A. Braun. Shrub 1–2 m, the younger parts with a flocculent, deciduous, stellate tomentum; stipules lanceolate, 1 cm; lfls lanceolate, 3–7 cm, acuminate, doubly serrate; infl 1–3 dm, with ascending branches; sep 1.3–1.5 mm, often erose; pet 2.5–3 mm; stamens to 8 mm; pistils pubescent; $2n=36$. Native of e. Asia, sometimes found along roadsides and fence-rows as an escape from cult. July. (*Schizonotus s.*)

5. PORTERANTHUS Britton.

Hypanthium narrowly campanulate, herbaceous; sep 5, erect; pet 5, spreading, linear or narrowly oblanceolate; stamens 10 or more, usually ca 20, scarcely exserted; carpels 5, distinct, with filiform style, capitate stigma, and few ovules; follicles soon rupturing the hypanthium, dehiscent ventrally and to some extent dorsally; seeds large, flattened; perennial rhizomatous herbs with subsessile, stipulate, trifoliolate lvs and a few terminal or subterminal, long-pedicellate, white or pale pink fls, the hypanthium often purple or red. (*Gillenia,* a preoccupied name.)

1 Stipules ovate, nearly as wide as long, laciniate-toothed .... 1. *P. stipulatus.*
1 Stipules narrowly linear, entire or ciliate .... 2. *P. trifoliatus.*

    1. Porteranthus stipulatus (Muhl.) Britton. Midwestern Indian-physic. Stems 4–10 dm, thinly hairy; stipules persistent, broadly ovate, 1–3 cm, laciniate-toothed; lfls lanceolate, 5–8 cm, tapering to both ends, sharply serrate, or some of those on the lower lvs often pinnatifid with narrow, toothed segments; hypanthium 4.5–5.5 mm; pet 10–13 mm; follicles 6–8 mm, glabrous or nearly so; $2n=18$. Dry or moist upland woods; w. N.Y. to O., Ill., and Kans., s. to Ga. and Tex. June, July. (*Gillenia s.*)

    2. Porteranthus trifoliatus (L.) Britton. Mountain Indian-physic. Stems 5–10 dm, glabrous or sparsely hairy; stipules narrowly linear, 5–10 mm, soon deciduous; lfls lanceolate to oblong to oblanceolate, 4–10 cm, sharply and irregularly serrate, acute to tapering at both ends; lower lvs similar to the upper; hypanthium 5.5–8 mm; pet 12–22 mm; frs 5.5–8 mm, hairy; $2n=18$. Dry or moist upland woods, chiefly in the mts.; s. Ont. to Del., N.C., and Ga., w. to e. O., e. Ky., and Ala., and irregularly to s. Mich., s. Ill., and s. Mo. May, June. (*Gillenia t.*)

6. FRAGARIA L.

Strawberry. Hypanthium saucer-shaped; sep 5, alternating with foliaceous bracts of nearly equal size; pet 5, white, obovate to subrotund; stamens numerous; with short filaments; pistils numerous, inserted on a prolongation of the receptacle; style slender, lateral; fr of numerous achenes on a greatly enlarged, juicy, edible red receptacle, subtended by the persistent calyx and bracts; perennial herbs, usually spreading freely by runners, with basal, 3-foliolate, serrate lvs and scape-like peduncles bearing few to several fls. 30, N. Temp., S. Amer.

1 Achenes superficial on the mature receptacle; terminal tooth of the lfls usually surpassing the adjacent
    lateral ones; pet mostly 5–7 mm .... 1. *F. vesca.*
1 Achenes set in pits on the mature receptacle; terminal tooth of the lfls usually surpassed by the adjacent
    lateral ones; pet mostly 7–10 mm .... 2. *F. virginiana.*

    1. Fragaria vesca L. Thin-lvd wild s. Lfls sessile or nearly so, ovate to obovate, ± silky beneath, bright green or somewhat yellowish-green, thin, often bulging on the upper side between the principal lateral veins, these diverging from the midrib at an angle of ca 45°, the teeth sharp and rather divergent, the terminal tooth more than half as wide as the adjacent lateral ones and (equaling or) projecting beyond them; peduncles at anthesis usually shorter than the lvs, later surpassing them; pedicels unequal and commonly proliferous, the infl eventually racemiform or paniculiform; pet 5–7 mm; $2n=14$. Europe and N. Amer. Apr.–June. The chiefly European var. *vesca*, with ovoid or subglobose fr and with the long hairs of the petioles and peduncles widely spreading or even retrorse, is widely intr. in our range, and thought to be native in se. Can. Var. *americana* Porter, native from Nf. to Man., s. to Va., Ind., and Neb., has slenderly ovoid or ellipsoid fr, and the long hairs of the petioles and peduncles are ascending or appressed.

2. **Fragaria virginiana** Duchesne. Thick-lvd wild s. Lfls petiolulate, glabrate to sericeous beneath, dark green or somewhat blue-green, thick and firm, flat, the principal lateral veins divergent from the midrib at an angle of ca 30°, the teeth blunter and less divergent than in no. 1, the terminal tooth commonly less than half as wide as the adjacent lateral ones and surpassed by them; infl with as many as 12 fls on pedicels of about equal length, forming a corymbiform cluster mostly shorter than the lvs, even at maturity; pet 7–10 mm; fr 1–1.5 cm thick; $2n=56$. Throughout much of the U.S. and s. Can. Apr.–June. Variable in orientation of pubescence, shape of the fr and other characters, but the several varieties or segregate species often recognized in our range need to be re-examined. (*F. australis; F. canadensis; F. grayana; F. terrae-novae.*) *F. ananassa* Duchesne, the cultivated strawberry, thought to be derived from hybrids of *F. virginiana* and *F. chiloensis* Duchesne, sometimes escapes. The fr, even in wild forms, is usually more than 1.5 cm thick, and the pet are 10–15 mm; the lfls are larger and firmer than in our wild spp. and they tend to be evergreen.

7. **DUCHESNEA** J. E. Smith. Hypanthium saucer-shaped; sep 5, alternating with and exceeded by large, foliaceous, 3-toothed bracts; pet 5, yellow, cuneate-oblong, truncate or slightly retuse; stamens 20–25; filaments filiform, pistils numerous; style filiform, lateral; receptacle becoming fleshy and red at maturity, but not juicy or edible, bearing numerous superficial achenes; perennial herbs from a short rhizome, bearing several long-petiolate, 3-foliolate basal lvs and emitting slender stolons with scattered lvs, the solitary 1-fld peduncles arising from the nodes. 2, Asia.

1. **Duchesnea indica** (Andrews) Focke. Indian strawberry. Lfls ovate or elliptic, 2–4 cm, crenate, sparsely strigose beneath; peduncles 3–10 cm; fls 14–18 mm wide; fr resembling a strawberry but insipid, 1 cm thick; $2n=42, 84$. Native of Asia, established in moist waste places here and there in our range. Apr.–Aug.

8. **SIBBALDIA** L. Hypanthium saucer-shaped; sep 5, oblong, acute, reticulately veined, alternating with linear-oblong bractlets; pet 5, yellow, minute, spatulate; stamens 5; pistils 5–10, rarely more; ovule 1, ascending; style filiform, lateral; fr a small, smooth achene; perennial herbs with long-petioled, 3-foliolate lvs and small fls in cymes. 6–7, all but ours Asiatic.

1. **Sibbaldia procumbens** L. Flowering stems 5–10 cm, from a multicipital caudex, lfless or few-lvd below the congested cymes; basal lvs overtopping the stem; lfls cuneate to obovate, 1–3 cm, sparsely pilose, 3–5-toothed at the tip; lfls longer than the bractlets, twice as long as the pet; $2n=14$. Circumboreal, s. to alpine regions of e. Que. and n. N.H., and to Colo. and Calif. June, July.

9. **CHAMAERHODOS** Bunge. Hypanthium saucer-shaped; sep 5; bractlets none; pet 5, cuneate-obovate; stamens 5, opposite the pet, with short filaments; ovaries 5–10, rarely more; ovule ascending; style nearly basal, slender, jointed below; fr an achene; taprooted perennial herbs with ternately decompound cauline lvs and ample bracteate cymes of numerous small, white or purplish fls. 5, all but ours strictly Siberian.

1. **Chamaerhodos erecta** (L.) Bunge. Glandular-pubescent and hirsute, short-lived perennial (seldom biennial) 1–3 dm, branched at least above; lvs dissected into blunt, linear oblong segments, the lower lvs long-petioled, the upper short-petioled to sessile and less divided or entire; hypanthium and lanceolate sep (these 2 mm) hispid; pet white, barely exceeding the sep. Dry plains and open hills; Alas. to Colo. and Utah, e. to w. Minn., and disjunct at Keweenaw Point, Mich.; c. and ne. Asia. June, July. (*C. nuttallii.*)

10. **WALDSTEINIA** Willd. Barren strawberry. Hypanthium obconic, its mouth contracted by a conspicuous disk; sep triangular; bractlets usually wanting; pet yellow; stamens numerous, the slender filaments erect and persistent; carpels 2–6, free and distinct, inserted on a hairy receptacle, enclosed by the hypanthium except for the elongate, terminal, protruding, deciduous styles; ovule 1, basal; fr an achene; perennial rhizomatous herbs with the aspect of *Fragaria,* having 3-foliolate basal lvs and a naked or bracteate peduncle bearing a cyme of small fls. 6, N. Temp.

1. Waldsteinia fragarioides (Michx.) Tratt. Petioles elongate; lfls broadly cuneate-obovate, serrate and usually shallowly lobed, the lateral ones asymmetrical; peduncles about equaling the lvs, with few to several fls. Moist or dry woods; Me. and w. Que. to Minn., s. to Pa., Ind., and in the mts. to Ga. and Ala. Apr., May. The widespread var. *fragarioides,* with obovate to broadly elliptic pet mostly 5–10 mm long and more than half as wide, obtuse or rounded and evidently exceeding the sep, tends to give way in the s. Appalachian mts. (at least southward) to var. *parviflora* (Small) Fern., with lance-elliptic or narrowly elliptic pet mostly 2.5–5 mm long, less than half as wide, often acute, and shorter than to barely exceeding the sep. (*W. parviflora; W. doniana,* a garden plant probably of var. *fragarioides,* but the name often used for var. *parviflora.*)

11. POTENTILLA L. Cinquefoil or five-fingers, names applied to the spp. with 5 lfls. Hypanthium saucer-shaped to hemispheric; sep (4)5, alternating with as many, usually somewhat narrower or shorter, foliaceous bractlets; pet (4)5, diverse in size and shape; stamens 5–many, commonly 20; pistils inserted on a prolongation of the receptacle, (10–) numerous; ovaries short; ovule 1; style slender, terminal or lateral or nearly basal, articulated with the ovary; fr a head of achenes, often partly enclosed by the persistent, accrescent cal; herbs or small shrubs, with compound, stipulate lvs and small or medium-sized, yellow or seldom white or red-purple fls. $x=7$; many spp are polyploid, or apomictic, or both. 200, N. Hemisphere.

1 Pet red-purple; semiaquatic near-shrubs with creeping base .... 23. *P. palustris.*
1 Pet yellow to white; habit and habitat various.
  2 Shrubs; lfls 5–7, 1–2 cm .... 22. *P. fruticosa.*
  2 Herbs or occasionally subshrubs, the latter with fewer or much longer lfls.
    3 Plants dwarf, cespitose, arctic-alpine, less than 1 dm; lfls 3; fls yellow .... 12. *P. robbinsiana.*
    3 Plants taller, or with elongate, prostrate stems, not arctic-alpine except sometimes in no. 20,
      which has white fls; lfls 3–many.
      4 Fls solitary on naked pedicels from the nodes of the usually creeping stem.
        5 Lvs pinnately compound, tomentose beneath.
          6 Achenes as thick as wide, deeply furrowed; lvs beneath densely sericeous, with long
            straight hairs overlying the tomentum .... 18. *P. anserina.*
          6 Achenes somewhat flattened, rounded on the back; lvs beneath tomentose on the surface,
            appressed-hairy on the veins .... 19. *P. pacifica.*
        5 Lvs palmately compound, hirsutulous to strigose or sericeous beneath.
          7 Fls yellow; principal lvs 5-foliolate, the upper sometimes 3–4-foliolate.
            8 Fls 5-merous as to the sep and pet.
              9 Fls 10–15 mm wide; anthers 0.6–1.0 mm.
                10 Lowest fl usually in the axil of the second well developed cauline lf .... 1. *P. simplex.*
                10 Lowest fl usually in the axil of the first well developed cauline lf, or even from
                  the axil of a scarcely developed early lf .... 2. *P. canadensis.*
              9 Fls 18–25 mm wide; anthers 1.3–2.0 mm .... 3. *P. reptans.*
            8 Many or all of the fls 4-merous as to the sep and pet .... 4. *P. anglica.*
          7 Fls white; lvs 3-foliolate .... 5. *P. sterilis.*
      4 Fls few to many, cymose.
        11 Lower surfaces of the lvs with straight hairs or none, never tomentose.
          12 Principal lvs below the infl 3-foliolate
            13 Fls white.
              14 Lfls with several teeth, chiefly in the distal half .... 5. *P. sterilis.*
              14 Lfls 3(–5)-toothed at the summit, otherwise entire .... 20. *P. tridentata.*
            13 Fls yellow.
              15 Pet and sep subequal; stamens (15–)20; achenes usually ridged .... 6. *P. norvegica.*
              15 Pet much shorter than the sep; stamens 5–10(–15); achenes smooth .... 7. *P. rivalis.*
          12 Principal lvs below the infl 5–several-foliolate.
           16 Lvs pinnately compound, with mostly 7 or more lfls.
            17 Style terminal; pet yellow; achenes with a corky ventral appendage .... 8. *P. paradoxa.*
            17 Style deeply lateral; pet ochroleucous to white; achenes not appendaged .... 21. *P. arguta.*
          16 Lvs palmately compound, or closely pinnate but with only 5 lfls.
           18 Anthers 1.0–1.5 mm; plants erect, simple to the infl .... 9. *P. recta.*
           18 Anthers 0.2–0.7 mm; plants freely branched, often not erect.
            19 Pet and sep subequal; stamens 20; anthers 0.5 mm .... 11. *P. intermedia.*
            19 Pet half as long as the sep; stamens 5 or 10; anthers 0.2 mm .... 7. *P. rivalis.*
        11 Lower surfaces of the lvs tomentose, with or without straight hairs also.
         20 Lvs palmately compound.
           21 Infl much-branched, leafy; cauline lvs well developed; anthers 0.5 mm; intr.
            22 Lvs distinctly tomentose beneath, the surface usually concealed .... 10. *P. argentea.*
            22 Lvs obscurely and thinly tomentose beneath, the pubescence chiefly of straight
              hairs .... 11. *P. intermedia.*
          21 Infl sparingly branched, inconspicuously bracteate; cauline lvs ± reduced; anthers
            1 mm; native, western.

23 Lfls divided more than half way to the midrib .... 13. *P. flabelliformis.*
23 Lfls merely toothed .... 14. *P. pulcherrima.*
20 Lvs pinnately compound, the lfls remote or sometimes crowded and almost palmate.
24 Style filiform throughout, 1.5–2 mm; stipules entire or with low teeth.
25 Lfls green above, crowded, the lf almost palmate .... 14. *P. pulcherrima.*
25 Lfls densely white-tomentose or sericeous above, well separated, the lf clearly pinnate.
26 Lvs generally tomentose above, with few or no straight hairs; lfls commonly irregular in size .... 15. *P. effusa.*
26 Lvs generally sericeous above, the numerous straight hairs overlying the tomentum; upper lfls progressively larger than the lower .... 16. *P. hippiana.*
24 Style thickened and glandular at base, 1 mm; stipules of the cauline lvs commonly laciniate; lvs glabrous to sericeous above .... 17. *P. pensylvanica.*

1. **Potentilla simplex** Michx.   Old-field f.-f. Stems and basal lvs from a short rhizome to 8 cm, the stems slender, with long internodes, villous to glabrate, at first erect and often 2–3 dm tall at first flowering, but soon widely ascending, eventually arching to the ground and producing a terminal tuber; lfls 5, oblanceolate to elliptic or obovate, to 7 cm, evidently toothed; fls yellow, 10–15 mm wide, solitary on slender axillary pedicels, the lowest one usually in the axil of the second well developed cauline lf; anthers 0.6–1.0 mm. Dry woods and fields; Nf. to Minn., s. to Ala. and Tex. Apr.–June.

2. **Potentilla canadensis** L.   Running f.-f. Much like no. 1, but lower and mostly even more slender, the rhizome very short, praemorse, the stems at first anthesis only 5–15 cm, erect or ascending, soon greatly elongating and prostrate, often rooting at the nodes; lowest fl usually in the axil of the first well developed cauline lf, sometimes even from the axil of a scarcely developed lower cauline lf. Dry woods and fields; P.E.I., N.S., and N.B., s. in the coastal states to Ga., inland to s. Ont., s. O., and e. Tenn. Apr.–June. (*P. pumila.*) Not sharply set off from no. 1, with which it sometimes grows, but forming a distinctive taxon of more limited distribution. Most of what has passed as *P. canadensis* is properly *P. simplex.*

3. **Potentilla reptans** L.   Creeping f.-f. Stems prostrate and creeping, to 1 m; principal lvs long-petioled; lfls 5 or 7, obovate or oblong-obovate, the median 2–5 cm, the lateral progressively shorter, all crenate; fls solitary (2) from the axils, long-pedicellate, yellow, 18–25 mm wide; bractlets often exceeding the sep; anthers 1.3–2.0 mm; 2*n*=28. Native of Eurasia, intr. in lawns, roadsides, and waste places; N.S. to O. and Va. June–Aug.

4. **Potentilla anglica** Laichard.   Trailing f.-f. Stems diffusely branched, trailing or later ascending; lower lvs often 5-foliolate, the upper 3–4-foliolate; lfls cuneate-obovate, 1–2 cm, with 5–9 sharp, antrorse teeth near the tip; fls yellow, 12–18 mm wide, solitary on slender axillary pedicels, many or all of them 4-merous; 2*n*=56. Native of Europe, intr. locally in Nf., N.S., Pa., and on the Pacific coast. May–Sept. (*P. procumbens.*)

5. **Potentilla sterilis** (L.) Garcke.   Strawberry-potentilla. Strawberry-like plant with basal trifoliolate lvs and long stolons bearing much smaller lvs; lfls broadly elliptic to obovate, to 5 cm, deeply toothed, the terminal tooth shorter and narrower than the adjacent lateral ones; peduncles 1–3-fld, with bract-like lvs subtending the branches; fls white, 10–15 mm wide; style high-lateral; achene villous at the scar of the style; 2*n*=28. Native of Eurasia and possibly Nf.; rarely intr. in e. U.S. May.

6. **Potentilla norvegica** L.   Strawberry-weed. Stout, leafy annual to short-lived perennial, the stem mostly branched and many-fld, hirsute below, subtomentose above; lfls 3, elliptic to broadly obovate, to 8 cm, crenate; fls yellow, nearly 1 cm wide, the pet nearly as long as the sep; bractlets and sep lance-ovate, subequal at anthesis, the sep accrescent in fr and to 15 mm; stamens usually 20; anthers 0.3 mm; achenes flattened, 1 mm, usually marked with curved longitudinal ridges; 2*n*=42, 56, 70. In a wide variety of habitats; circumboreal, s. in Amer. to S.C., Tenn., Tex., and Calif. June–Aug. The native American plants tend to differ slightly from the European ones, and have been segregated as *P. monspeliensis* L. or *P. norvegica* var. *hirsuta* (Michx.) T. & G., but no clear distinction can be made, and the Eurasian phase has also apparently been intr. into our range.

7. **Potentilla rivalis** Nutt.   Brook-f.-f. Mostly annual or biennial, freely branched, ascending to erect, 4–8 dm, glabrate to softly pubescent, with a diffusely branched leafy infl; lfls 3–5, pinnate or palmate, narrowly obovate-oblong, 2–5 cm, usually blunt, strongly toothed; fls 4–5 mm wide; bractlets and sep subequal, twice as long as the pale yellow pet; stamens 5 or 10; anthers 0.2 mm; style terminal; achenes smooth, 0.7 mm. Streambanks and damp soil; Minn. to Ill., w. to B.C. and Calif., occasionally adventive further e. Summer. (*P. millegrana; P. pentandra.*)

8. **Potentilla paradoxa** Nutt.   Diffuse potentilla. Annual to short-lived perennial, diffusely branched, spreading or ascending, 2–5 dm, glabrate to hairy; principal lvs pinnately compound; lfls 7–11, oblong to cuneate-obovate, 1–3 cm, crenate to sharply serrate; infl much-branched, leafy, many of the lvs pinnate, others 3-foliolate; fls yellow, 5–7 mm wide; bractlets, sep and pet subequal; stamens 10–20; style terminal; achenes smooth, the inner face with a corky thickening nearly as large as the body. Moist or wet soil; Man. to Ill. and La., w. to Wash. and N.M. Summer. (*P. nicolletii.*)

9. **Potentilla recta** L.   Sulphur-f.-f. Perennial; stems erect, 4–8 dm, simple to the infl, pubescent; lvs palmately compound, the lower long-petioled, with 5–7 lfls, the upper smaller, with shorter or no petiole

and only 3 lfls; lfls radially divergent, narrowly oblanceolate, deeply toothed; infl flattened, with many fls; sep and bractlets lance-ovate, subequal; pet sulphur-yellow, 1 cm; anthers 1.0–1.5 mm; style terminal; achenes striate with low curved ridges; $2n=42$. Native of Europe, found as a frequent weed in dry soil and waste places throughout our range. June–Aug. (*P. sulphurea.*)

10. Potentilla argentea L.. Silvery f.-f. Perennial, freely branched, depressed or ascending, 1–5 dm; lvs digitate, the larger with 5 lfls, these linear-oblanceolate to narrowly obovate, 1.5–3(–5) cm, silvery-tomentose beneath, long-cuneate and entire at base, bearing above the middle 2–4 oblong or linear teeth with revolute margins; fls yellow, 7–10 mm wide; pet and sep subequal; anthers 0.5 mm; achenes nearly smooth; $2n=14$, 28, 42. Native of Eurasia, found as an occasional weed in dry soil and waste places in our range. June–Sept.
P. inclinata Villars (*P. canescens*), a European sp. rarely intr. in our range, differs from *P. argentea* most notably in having some long, spreading, simple hairs on the stem, and also some long, simple hairs on the lf-veins beneath, in addition to the tomentum.

11. Potentilla intermedia L. Perennial from a stout root, erect or decumbent, leafy, 3–7 dm; lvs digitate, the larger with 5 lfls, these oblanceolate to obovate, 3–5 cm, deeply and often irregularly serrate above the entire cuneate base, villous or sericeous beneath, with or more often without a thin tomentum; infl much-branched; fls numerous, 8–10 mm wide; pet yellow, about equaling the sep; stamens 20; cal and bractlets accrescent, to 15–20 mm; achenes wrinkled, with conspicuous longitudinal ridges; $2n=28$, 42, 56. Native of Eurasia, found as an occasional weed in dry soil and waste places in our range. June–Sept.

12. Potentilla robbinsiana Oakes. White-Mt. potentilla. Plants dwarf, cespitose, the basal or near-basal lvs forming a tuft 2–4 cm; lfls 3, digitate, cuneate-obovate, long-pilose, especially beneath, with 2–4 deep teeth on each side; fls yellow, 6 mm wide; bractlets elliptic, 2 mm, equaling the oblong sep; pet obcordate, barely surpassing the sep. Rare and local, in rocky places at high elev. in the White Mts. of N.H.; formerly also on Mt. Mansfield in Vt., and reputedly still present at an undisclosed location in Vt. June.

13. Potentilla flabelliformis Lehm. Comb-f.-f. Perennial from a stout caudex, erect, 4–8 dm; basal lvs long-petioled, palmately compound, with 5–7 lfls, the cauline much smaller, often with 3-lfls, becoming bract-like in the infl; lfls oblanceolate to narrowly obovate, densely tomentose beneath, green and often glabrous above, cleft well over half way to the midrib into narrow segments; infl an open cyme with ascending branches; pet yellow, 6–10 mm; anthers 1 mm; $2n=56–70$. Prairies, rocky banks, and dry woods; s. B.C. to Oreg., e. to Sask. and Mont.; vicinity of Keweenaw Point, Mich. July, Aug. (*P. gracilis,* misapplied.)

14. Potentilla pulcherrima Lehm. Much like *P. flabelliformis,* but the lfls shallowly and obtusely serrate, and sometimes closely crowded-pinnate instead of strictly palmate; $2n=56–108$. Similar habitats; s. B.C. to Nev., e. to Man., ne. Minn., and S.D.

15. Potentilla effusa Douglas. Great-Plains potentilla. Perennial from a thick, woody caudex, 2–4 dm; lvs pinnately compound, white-tomentose on both sides, the basal elongate with 5–11 lfls, the cauline few and much smaller, the uppermost 2–4 cm; infl an open cyme with long, ascending branches; sep 4–6 mm, usually much exceeding the lanceolate bracts; pet yellow, barely surpassing the sep; style filiform, 1.5–2 mm. Dry plains; Minn. to Sask. and N.M. June–Aug.

16. Potentilla hippiana Lehm. Woolly potentilla. Much like *P. effusa*; lfls more sericeous than tomentose above, and increasing regularly in size toward the summit of each lf; sep 5–7 mm, little exceeding the bractlets, slightly shorter than the pet; $2n=42–98$. Dry hills, prairies, and plains; Keweenaw Point, Mich.; w. Minn. to Alta. and N.M. June–Aug.

17. Potentilla pensylvanica L. Perennial from a stout taproot and often a branched caudex, decumbent to erect, 2–8 dm, the stem thinly to densely tomentose; lvs pinnately compound, glabrous to sericeous above, subglabrous to white-tomentose beneath, the lower long-petioled, the cauline progressively reduced and less petioled or subsessile; lfls 5–13 (+), or 3 in smaller lvs, the upper 3 approximate and often confluent, oblanceolate to obovate, deeply toothed or cleft; stipules of the cauline lvs commonly laciniate; infl usually a compact cyme, seldom over 5 cm wide; pet yellow; style 1 mm, terminal, thickened and glandular at base; $2n=14$, 28, 56. A polymorphic sp. of n. and w. N. Amer. and much of temperate and boreal Eurasia, often divided into a number of spp. or infraspecific taxa. Our plants can be sorted into 3 morphologically confluent but cytologically apparently distinctive vars.:

a Lfls mostly 5–7, crowded, subpalmately disposed, cleft nearly to the midrib; fls mostly 1 cm wide or less; bractlets and pet shorter than the sep; octoploid; widespread in N. Amer., the common phase with us, extending s. to Me., N.H., Ind., and Minn. (*P.p.* var. *glabrata; P. litoralis; P. pectinata.*)
.... var. *bipinnatifida* (Douglas) T. & G.
a Lfls mostly (5) 7–13 (+), not so crowded, evidently pinnately disposed; fls mostly 1 cm wide or more.
  b Lfls cleft half way to the midrib, or often more deeply cleft; bractlets about as long as the sep; pet generally longer than the sep; plants erect or decumbent at base, up to 5(–7) dm; tetraploid; widespread in N. Amer. but more common westward, in our range rare e. of Minn. (*P. atrovirens*) .... var. *pensylvanica.*
  b Lfls cleft about half way to the midrib; bractlets distinctly longer than the sep; pet about as long as or shorter than the sep; plants erect, 2–4 dm; diploid; dry, sandy soil and stabilized dunes on the northern prairies of c. N. Amer., from w. Minn. and w. Man. to Alta. (*P. finitima; P. atrovirens,* misapplied.) .... var. *arida* B. Boivin.

18. Potentilla anserina L.   Silver-weed. Perennial, at first acaulescent, with tufted basal lvs, soon emitting long stolons that root and produce similar but smaller clusters of lvs at the nodes; lvs oblanceolate, to 3 dm, pinnately compound with numerous lfls often alternating with other much smaller ones; lfls oblanceolate or narrowly elliptic, to 4 cm, increasing in size distally, sharply toothed, tomentose beneath and also silvery-sericeous with long appressed hairs, glabrous to sericeous above; stolons, pedicels and leaf-axes generally villous, the hairs often spreading; fls yellow, 1.5–2.5 cm wide, solitary on slender pedicels from the nodes of the stolons and sometimes from the original plant; bractlets often toothed; style lateral; achenes 2.5 mm, about as thick, deeply furrowed on the summit and back; 2n=28, 35, 42. Moist or wet, open places; circumboreal, s. to N.Y., n. Ind., Io., and N.M. May–Sept. (*Argentina anserina*.)

19. Potentilla pacifica Howell.   Coastal silver-weed. Much like *P. anserina*; stolons, pedicels, and leaf axes glabrous or sparsely appressed-hairy; lfls averaging slightly smaller, gray-tomentose beneath on the surface, sericeous on the veins only; achenes somewhat flattened laterally, 1.5–2 mm, rounded or narrowly ridged on the back, not furrowed; 2n=28. Wet, sandy seacoasts, commonly in brackish soil; Lab. to L.I., and the Pacific coast of N. Amer. (*Argentina p.*)

20. Potentilla tridentata Sol.   Mountain white potentilla. Perennial from a branching caudex, woody at the base, 1–3 dm; lvs mostly near the base; lfls 3, palmate, firm, oblong-lanceolate, (1–)1.5–2.5(–3) cm, entire below, shallowly 3(–5)-toothed at the truncate summit, glabrous above, obscurely strigose beneath; fls several in a flattened cyme, white, 1–1.5 cm wide; bractlets lanceolate, somewhat shorter than the acute, triangular-ovate sep; ovary and achenes villous; style lateral; 2n=14, 28. Rocks and gravelly shores, sometimes at high elev.; Greenl. to Mack., s. to Conn., Mich., Io., and in the mts. to Ga. June–Aug. (*Sibbaldiopsis t.*)

21. Potentilla arguta Pursh.   Tall potentilla. Perennial from a stout rhizome or branching caudex, ± viscid-pubescent throughout with brownish hairs; stems erect, 3–10 dm, simple up to the narrow, crowded infl; lvs pinnately compound, the basal long-petioled; lfls 7–11, or only 5 in the uppermost lvs, increasing in size distally, to 7 cm, often alternating with minute folioles; fls ochroleucous or nearly white, 12–18 mm wide; sep ovate, much longer than the lanceolate bractlets, nearly as long as the pet; style deeply lateral; achenes obovoid, 1 mm, finely striate; 2n=14. Dry woods and prairies; e. Que. to Mack., s. to D.C., Ind., Mo., and Ariz. June, July. Ours is var *arguta*. (*Drymocallis agrimonioides*.)

22. Potentilla fruticosa L.   Shrubby five-fingers. Bushy-branched shrub to 1 m; lvs numerous, short-petioled, pinnately compound with 5–7 crowded, narrow, entire, often revolute lfls 1–2 cm, the 3 terminal ones often confluent; fls solitary or few at the ends of the branches, yellow, 2–3 cm wide; bractlets lanceolate, as long as but much narrower than the ovate, acuminate sep; ovaries and achenes villous; style lateral; 2n=14, 28. Wet meadows, bogs, and shores, especially in calcareous soil; circumboreal, s. to N.J., n. Ill., S.D., and Ariz., and reported from Tenn. June–Sept. (*Dasiphora f.*) Our plants, and many of the Eurasian ones, are diploid and hermaphrodite, and have been segregated as *P. floribunda* Pursh. Typical *P. fruticosa*, of Siberia and n. Europe, is tetraploid and functionally dioecious.

23. Potentilla palustris (L.) Scop.   Marsh-potentilla. Stems coarse, somewhat woody below, reddish-brown, 2–6 dm, decumbent or ascending from a long rhizome, rooting below; lvs long-petioled, pinnately compound; lfls 5–7 (the upper 3 adjacent), narrowly oblong to elliptic, 5–10 × 1–3 cm, sharply toothed, glaucous beneath; infl leafy, few-fld; fls red-purple, 2 cm wide; bractlets narrowly lanceolate, much shorter than the acuminate sep; pet half as long as the sep; style lateral; achenes smooth, attached to the enlarged, spongy receptacle; 2n=28, 35, 42, 62–64. Swamps, bogs, and streambanks; circumboreal, s. to N.J., O., Io., and Calif. June–Aug. (*Comarum p.*)

12. FILIPENDULA Miller.   Hypanthium cupulate; fls 5–7-merous; sep ovate or oblong, reflexed; pet spreading; stamens 20–40, inserted at various levels in 10 longitudinal rows opposite the perianth-members; pistils 5–15, in a circle, the ovaries erect; styles clavate, outcurved; stigmas large, capitate; frs 1-seeded, resembling a follicle in appearance but indehiscent; perennial, rhizomatous herbs with stipulate, pinnately compound lvs (some tiny lfls commonly intermingled with the principal ones) and large panicles of white to pink fls, the lateral branches of the panicle surpassing the axis. 10, Temp. Eurasia and N. Amer.

1 Lvs with 2–5 pairs of well developed lateral lfls.
  2 Lateral lfls 3–5-lobed; fls pink; fr straight .... 1. *F. rubra.*
  2 Lateral lfls usually merely coarsely toothed; fls white; fr twisted .... 2. *F. ulmaria.*
1 Lvs with (8–)10–25 pairs of well developed lateral lfls .... 3. *F. vulgaris.*

1. Filipendula rubra (Hill) B. L. Robinson.   Queen of the prairie. Pls glabrous, 1–2 m; terminal lfl reniform in outline, to 2 dm wide, deeply cleft into 5–9 lanceolate or oblong, serrate segments; lateral lfls 2–5 pairs, sessile, to 1 dm, shallowly to deeply 3–5-lobed; infl 1–2 dm wide; sep broadly round-ovate; pet mostly 5, rich pink, 2–4 mm; frs glabrous, erect, oblong, 6–8 mm. Low woods and wet prairies or meadows; N.Y. to Minn., s. to N.C. and Ky., and sometimes escaped from cult. elsewhere. June, July.

2. Filipendula ulmaria (L.) Maxim.   Queen of the meadow. Pls 1–2 m; terminal lfl rotund in outline, 6–15 cm wide, deeply divided into 3–5-ovate or ovate-oblong, serrate segments; lateral lfls 2–5 pairs, 2–8 cm, oblong to ovate, coarsely and sharply doubly serrate, seldom shallowly lobed; sep ovate-triangular, obtuse, pet 5(6) white, 2–5 mm; frs glabrous, 3–4 mm, tightly and spirally imbricated, each going through about a half-turn; $2n=14$, 16, 24. Native of Eurasia, occasionally escaped from cult., especially northeastward. June–Aug. Var. *ulmaria* has the lfls closely tomentose beneath; var. *denudata* (J. & C. Presl) Beck has the lfls green and sparsely hairy beneath with long straight hairs.

3. Filipendula vulgaris Moench.   Dropwort. Sparsely hairy or glabrate, to 8 dm; lfls 10–25 pairs, 1–2 cm, ovate-oblong; pet usually 6, white, 5–9 mm; frs straight, pubescent; $2n=14$, 16. Native of Eurasia, occasionally escaped from cult. in e. U.S. May–July. (*F. hexapetala.*)

13. GEUM L.   Avens. Hypanthium turbinate or hemispheric; sep commonly imbricate; bractlets linear to oblong, or wanting; pet elliptic to obovate; stamens 10–many; ovaries numerous on an elongate, cylindric receptacle; style filiform, elongate; achenes tipped by the long, persistent style; fibrous-rooted perennial herbs with variable foliage, the lower lvs pinnate, the middle smaller and commonly trifoliolate, the upper often simple; fls solitary to many in terminal corymbs; pet white to yellow, or suffused with pink or purple. (*Sieversia.*) 50, mostly N. Hemisphere.

1 Style jointed near or above the middle, the basal part hooked at the tip and becoming indurate, the
   terminal part usually deciduous; dispersal as stick-tights.
  2 Head of achenes evidently stipitate above the cal; terminal segment of the style attached to the
     very tip of the basal segment; bractlets none. (*Stylipus.*) . . . . 1. *G. vernum.*
  2 Head of achenes sessile or nearly so; terminal segment of the style attached laterally to the basal
     segment; bractlets usually present. (*Geum.*)
    3 Sep green or greenish, reflexed at anthesis; pet spreading; fls erect.
      4 Pet white or ochroleucous.
        5 Pedicels minutely velvety-puberulent, with or without a few longer hairs; receptacle densely
          hirsute among the achenes.
          6 Pet pure white, longer or only slightly shorter than the sep . . . . 2. *G. canadense.*
          6 Pet ochroleucous, much shorter than the sep . . . . 3. *G. virginianum.*
        5 Pedicels hirsute; receptacle glabrous or inconspicuously short-hairy . . . . 4. *G. laciniatum.*
      4 Pet golden yellow.
        7 Terminal segment of the basal lvs rotund to reniform, cordate or truncate at base, much
          larger than the lateral segments . . . . 5. *G. macrophyllum.*
        7 Terminal and principal lateral segments of the basal lvs essentially alike in size and shape,
          narrowly to broadly cuneate at base . . . . 6. *G. aleppicum.*
    3 Sep purple or crimson, erect or ascending at anthesis; pet erect or ascending; fls ± nodding . . . . 7. *G. rivale.*
1 Style not obviously jointed, in fr elongate and plumose at least below; dispersal by wind. (*Sieversia.*)
  8 Lateral lfls of the basal lvs well developed, gradually increasing in size distally, but not very different
     from the terminal one . . . . 8. *G. triflorum.*
  8 Lateral lfls of the basal lvs poorly developed or none . . . . 9. *G. peckii.*

1. Geum vernum (Raf.) T. & G.   Spring avens. Stems erect or ascending, 3–6 dm; some basal lvs long-petioled, simple or nearly so, others smaller and pinnate; principal cauline lvs pinnate or some trifoliolate, with several toothed and often deeply cleft divisions, the upper and lower cauline lvs reduced; sep triangular, reflexed; bractlets none; pet yellow or white, 1–2 mm; head of achenes on a stipe 1–2 mm; achenes 2–3 mm, minutely appressed-puberulent; terminal segment of the style 0.7 mm, inserted on the very tip of the basal segment; $2n=42$. Rich woods; s. Ont. to s. Wis., Io., and Kan., s. to Tenn. and nw. Okla.; also c. and s. N.Y., N.J., and Pa. Apr, May.

2. Geum canadense Jacq.   White a. Stems slender, 4–10 dm, glabrous or sparsely pubescent below, becoming minutely but densely velvety-puberulent above and on the pedicels, often also with a few scattered long hairs; basal lvs long-petioled, with mostly 3 obovate lfls; cauline lvs mostly shorter-petioled, with more oblong-lanceolate to rhombic lfls, the uppermost simple and subsessile; pet white, nearly as long as the sep, or longer; head of frs obovoid, 10–15 mm; receptacle densely bristly, the hairs protruding among the ovaries at anthesis but shorter than the achenes, these hairy at least above, 2.5–3.5 mm excluding the style; $2n=42$. Dry or moist woods; N.S. to Minn. and N.D., s. to Ga. and Tex. May, June. (*G. camporum.*)

3. Geum virginianum L.   Cream-colored a. Much like no. 2; lvs larger and more coarsely toothed, the terminal lfl usually much larger than the lateral ones, often deeply incised; stem hirsute below; pet ochroleucous, 2–3 mm, half or two-thirds as long as the sep; $2n=42$. Moist upland woods; Mass. and N.Y. to Ind., s. to N.C. and Tenn. June, July. (*G. flavum.*)

4. Geum laciniatum Murray.   Rough a. Stem 4–10 dm, hirsutulous with mostly deflexed hairs; lower lvs long-petioled, pinnately compound, the segments pinnately lobed and incised; upper lvs 3-foliolate with cuneate-obovate segments, or merely 3-lobed; pedicels minutely puberulent and densely hirsute, the hairs commonly 1–1.5 mm long; sep triangular, 4–10 mm; pet white, 3–5 mm; head of achenes subglobose, 12–

18 mm; receptacle glabrous or with a few short bristles; achenes 3–5 mm, excluding the style, glabrous or sparsely hirsute; $2n=42$. In moist or wet soil; N.S. to s. Ont. and Mich., s. to N.J., w. Va., W.Va., Ky., and Mo. May, June.

5. **Geum macrophyllum** Willd. Big-lvd a. Stem to 1 m, hirsute, especially below; basal lvs long-petioled, the terminal segment rotund to reniform, 5–12 cm wide, often 3-lobed, the lateral few, much smaller, interspersed with several to many minute lfls; upper lvs short-petioled to sessile, deeply 3-lobed or 3-foliolate; pet yellow, 4–7 mm, somewhat exceeding the sep; achenes minutely pubescent throughout, hirsute about the summit; basal segment of the style minutely glandular, the terminal segment minutely pubescent; $2n=42$. Moist woods and rocky ledges, sometimes weedy; Nf. and Lab. to Alas. and e. Asia, s. to Me., Vt., N.Y., Mich., Minn., and Calif. May–July. Var. *perincisum* (Rydb.) Raup, occurring from n. Mich. w. and n., has relatively strongly dissected lvs, the terminal lfl of the middle cauline ones 3-cleft nearly to the base with the lobes again cleft or laciniate. (*G. oregonense.*) The widespread var. *macrophyllum* has less dissected lvs, the terminal lfl of the middle cauline ones 3-cleft to about the middle with the lobes merely toothed.

6. **Geum aleppicum** Jacq. Yellow a. Stems stout, to 1 m, very hirsute, especially below; lvs highly variable, but their lfls or lobes mostly obovate or oblong, distinctly cuneate at base; lower lvs on long petioles, the principal lfls often mingled with minute ones; lfls mostly incised-serrate; pet yellow, about equaling or slightly exceeding the sep; achenes numerous, ca 200 or more, divaricately hispid about the summit, otherwise glabrous or appressed-hairy; basal segment of the style often hirsute below; terminal segment conspicuously short-hirsute; receptacle glabrous or merely short-hispid; $2n=42$. Swamps and wet meadows; interruptedly circumboreal. Amer. plants, the var. *strictum* (Aiton) Fern., as here described, occur from Nf. and Que. to Yukon, s. to N.J., Ind., Ill., Io., and N.M. May–July.

Geum urbanum L., a Eurasian sp. with somewhat shorter pet, fewer (mostly less than 100) achenes, and only very minutely pubescent terminal segment of the style, is adventive as a weed in e. U.S. May, June.

7. **Geum rivale** L. Water-a. Stems 3–6 dm from a short, stout rhizome, sparsely hirsutulous; basal lvs to 3 dm, with 3(5) principal lfls, the terminal one broadly obovate or subrotund, serrate, ± 3-lobed, the others adjacent, narrowly obovate; additional lateral lfls remote and irregular; cauline lvs much smaller, reduced above, variously toothed to divided; fls several, campanulate, nodding, the pedicels eventually elongate; sep purple, ascending or erect, 7–10 mm; bractlets linear, 3–4 mm; pet yellowish, suffused with purple and purple-veined, usually a little shorter than the sep; style at anthesis jointed near the middle, the distal segment plumose, to 8 mm at maturity; achenes spreading, with decurved styles; $2n=42$. Swamps and wet meadows; Nf. and Que. to Alta., s. to N.J., Pa., Ind., Mich., and Calif. May–July. *G. pulchrum* Fern., with clear yellow pet conspicuously exceeding the sep, has been thought to be a hybrid with *G. macrophyllum,* but may be merely a form of *G. rivale.*

8. **Geum triflorum** Pursh. Prairie smoke. Stems 2–4 dm from a short, stout rhizome, hairy throughout; basal lvs 1–2 dm, oblanceolate, pinnately compound; lfls 7–17, the lateral progressively larger toward the lf-tip, cuneate-oblong, to 5 cm, irregularly laciniate or lobed, the terminal one similar but wider, often confluent with the upper lateral ones and scarcely larger than them; cauline lvs few and small, laciniate; peduncles eventually to 1 dm; fls nodding, broad-based, narrowed distally in life; sep shorter than the linear bractlets; pet purplish, 8–12 mm, about equaling the bractlets, suberect; mature styles 3–5 cm, strongly plumose except the terminal 2–5 mm; $2n=42$. Dry woods and prairies; w. N.Y. to Minn. and Io., w. to B.C. and Calif. May, June. Our plants are var. *triflorum.* (*Sieversia t.*)

9. **Geum peckii** Pursh. White-mt.-a. Stems 1.5–4 dm from a coarse rhizome, nearly smooth below, more hairy above; basal lvs with a rounded-reniform terminal lfl 5–10 cm wide, shallowly 5–7-lobed, sharply and irregularly serrate, truncate to subcordate at base, and 0–6 much smaller, laciniate lateral lfls rarely over 1 cm; cauline lvs resembling the lateral lfls of the basal lvs; sep triangular, twice as long as the bractlets; pet yellow, 8–15 mm, spreading; mature styles 8–10 mm, hairy about half their length; $2n=42$. Damp slopes and alpine meadows in the higher mts. of Me. and N.H. July, Aug. (*Sieversia p.*)

14. **RUBUS** L. Bramble. Hypanthium small, flat to hemispheric; sep usually 5, valvate, spreading to reflexed, commonly with a shortly caudate tip; bractlets none; pet as many as the sep, erect or spreading, spatulate to obovate or elliptic; stamens numerous; pistils numerous, inserted on a convex to conic receptacle that often elongates in fr; ovules 2, collateral, only one maturing; style filiform or clavate; fr a cluster of drupelets, falling together (or sometimes separately), the receptacle falling with the cluster of drupelets or remaining attached to the pedicel; shrubs or less often perennial herbs, very often prickly, with simple or more commonly compound, serrate or lobed lvs and small to large, perfect or unisexual, white to pink or red fls; infl determinate, but commonly with the aspect of a raceme or corymb or panicle; $x=7$. 200+, cosmop.

In subgenera *Rubus* and *Idaeobatus* the plant sends up from a perennial base a series of biennial stems. During their first year these are termed *primocanes*; they are usually unbranched and normally do not fl. During their second year they are known as *floricanes*; they increase no more in length, but emit a number of short lateral branches with a few lvs and usually a terminal fl or infl. The lvs of the primocane are compound; those of the floricane are often partly simple, and often of a different shape.

Most of the spp. of subg. *Rubus* grow in disturbed habitats, often representing an early stage in plant succession. Different species may grow intermingled in such places. Some of the same species also grow in more stable habitats, with some sorting out of species by habitat.

The taxonomy of *Rubus* is complicated by hybridization, polyploidy, and apomixis. The subgenus *Rubus* (blackberries) is particularly difficult, and the conservative treatment here presented is subject to extensive change when a proper biosystematic study can be made. Some of the many names listed in synonymy should probably be transferred to the partial list of putative hybrids, hybrid segregates, and local populations of hybrid origin presented below.

R. × *aculiferus* = *allegheniensis* × *setosus*.
R. × *adjacens* = *hispidus* × *setosus*.
R. × *alter* = *hispidus* × *setosus*.
R. × *arcuans* = *recurvicaulis* × *setosus*.
R. × *bicknellii* = *recurvicaulis* × *setosus*.
R. × *biformispinus* = *allegheniensis* × *hispidus*.
R. × *blanchardianus* = *hispidus* × *setosus*.
R. × *electus* = *allegheniensis* × *hispidus*.
R. × *glandicaulis* = *allegheniensis* × *setosus*.
R. × *harmonicus* = *hispidus* × *setosus*.
R. × *jacens* = *hispidus* × *setosus*.
R. × *jactus* = *allegheniensis* × *hispidus*.
R. × *laevior* = *allegheniensis* × *hispidus*.
R. × *licitus* = *allegheniensis* × *pensilvanicus*.
R. × *mainensis* = *flagellaris* × *hispidus*.
R. × *miscix* = *pensilvanicus* × *setosus*.

R. × *montpelierensis* = *allegheniensis* × *setosus*.
R. × *neglectus* = *idaeus* × *occidentalis*.
R. × *peculiaris* = *pensilvanicus* × *setosus*.
R. × *permixtus* = *allegheniensis* × *hispidus*.
R. × *pudens* = *hispidus* × *setosus*.
R. × *ravus* = *allegheniensis* × *setosus*.
R. × *rixosus* = *hispidus* × *setosus*.
R. × *rosa* = *allegheniensis* × *pensilvanicus*.
R. × *sanfordii* = *allegheniensis* × *hispidus*.
R. × *sceleratus* = *allegheniensis* × *setosus*.
R. × *severus* = *recurvicaulis* × *setosus*.
R. × *tholiformis* = *hispidus* × *setosus*.
R. × *trifrons* = *hispidus* × *setosus*.
R. × *vigoratus* = *hispidus* × *setosus*.
R. × *viridifrons* = *hispidus* × *setosus*.

1 Principal foliage-lvs simple, merely lobed and toothed.
  2 Lf-lobes broadly triangular, acute. (Subg. *Anoplobatus*.)
    3 Fls rose-purple (rarely white); fr rather dry; widespread in our range .... 1. *R. odoratus*.
    3 Fls white; fr juicy; cordilleran sp, local in Minn., Mich., and Ont. .... 2. *R. parviflorus*.
  2 Lf-lobes broadly rounded.
    4 Stems herbaceous, unarmed, 1–3 dm; dioecious; pet much surpassing the sep. (Subg. *Chamae-*
      morus.) .... 3. *R. chamaemorus*.
    4 Stems more woody, taller, to 2 m; fls perfect; pet not surpassing the sep .... 7. *R. idaeus*.
1 Principal foliage-lvs compound (reduced simple lvs may be present in the infl).
  5 Stems herbaceous or nearly so, not well differentiated into primocanes and floricanes, unarmed, or
    occasionally with a very few weak bristles; stipules oblanceolate to obovate. (Subg. *Cylactis*.)
    6 Pet mostly white or greenish-white, 4–8 mm; flowering stems 1.5–5 dm; lfls 4–8 cm .... 4. *R. pubescens*.
    6 Pet pink, (8–)10–16 mm; flowering stems 0.5–1.5 dm; lfls 1–3.5 cm .... 5. *R. acaulis*.
  5 Stems woody, biennial, persisting throughout the winter and bearing fls the second year, usually ±
    prickly or bristly; stipules linear or setaceous.
    7 Fr separating from the receptacle, which is persistent on the pedicel. (Subg. *Idaeobatus*; rasp-
      berries.)
      8 Pet equaling or shorter than the sep; floricane-lvs 3-foliolate or simple.
        9 Infl few-fld, corymbiform to umbelliform or short-racemiform; cal gray-tomentose and
          commonly minutely glandular-pubescent.
          10 Pedicels with stout curved prickles, not glandular; fr usually black .... 6. *R. occidentalis*.
          10 Pedicels with straight weak bristles, also finely glandular; fr usually red .... 7. *R. idaeus*.
        9 Infl many-fld and paniculiform; cal and pedicels densely shaggy with long, glandular, reddish
          or purple hairs .... 8. *R. phoenicolasius*.
      8 Pet surpassing the sep; floricane-lvs pinnately 5–9-foliolate .... 9. *R. illecebrosus*.
    7 Fr separating from the stem with the central receptacle included. (Subg. *Rubus* or *Eubatus*;
      blackberries; the trailing spp. are also called dewberries.)
      11 Coarse, heavily armed, scrambling climbers (or sprawling in the open); infl cymose-panic-
        ulate, with ± numerous fls; branches and pedicels of the infl armed with firm, flat
        prickles; intr. spp.
        12 Lfls laciniate, or compound and again cleft, green beneath .... 23. *R. laciniatus*.
        12 Lfls merely toothed, grayish-tomentose beneath.
          13 Prickles falcate; pet white or slightly pinkish; stems canescent toward the tip .... 24. *R. discolor*.
          13 Prickles straight or nearly so; pet pale pink to red; stems essentially glabrous .... 25. *R. bifrons*.
      11 Plants of various habit, but not as above; infl in most spp. racemiform or by reduction the
        fls solitary; pedicels in most spp. unarmed; native, except no. 12.
        14 Primocanes prostrate or low-arching, generally rooting at least at the tip.
          15 Armature largely or wholly of slender, small-based prickles or stiff or glandular
            bristles (broad-based prickles may also be present).

16 Bristles of the stem usually numerous, glandular; fls normally solitary on each
  branch of the floricane; southern sp. .... 10. *R. trivialis.*
16 Bristles of the stem not glandular (minute glandular hairs may be present among
  the bristles); infl racemiform; widespread sp .... 11. *R. hispidus.*
15 Armature wholly of stout, stiff, commonly hooked prickles with ± expanded base;
  bristles none, but minute glandular hairs may be present.
  17 Young primocanes glaucous; some lower nodes of the racemiform or panicu-
    liform infl bearing 2 or 3 fls or bearing a short branch with 2 or more fls...... 12. *R. caesius.*
  17 Young primocanes not glaucous; lower nodes of the infl 1-fld.
    18 Infl racemiform, the few or several fls subtended by stipules only, or the
      lowest one in the axil of an expanded lf .... 13. *R. recurvicaulis.*
    18 Infl leafy, the 1–several fls all, or all except one of the two terminal fls (when
      two are present), subtended by simple or trifoliolate lvs, or the fl solitary
      and terminal.
      19 Terminal lfl of the trifoliolate floricane-lvs ± ovate, with broadly round-
        ed to subcordate base and sharply acute to long-acuminate tip ..... 14. *R. flagellaris.*
      19 Terminal lfl of the trifoliolate floricane-lvs oblong to oblanceolate or
        obovate, with narrowly to broadly cuneate base and obtuse to very
        abruptly and inconspicuously acuminate tip .... 15. *R. enslenii.*
14 Primocanes normally erect or ascending, not rooting.
  20 Armature largely or wholly of stiff bristles or slender, small-based prickles .... 16. *R. setosus.*
  20 Armature wholly of stout, stiff, often hooked prickles with ± expanded base.
    21 Glandular hairs abundant on the pedicels and often elsewhere.
      22 Terminal lfl of primocane-lvs rarely over three-fifths as wide as long ... 17. *R. allegheniensis.*
      22 Terminal lfl of primocane-lvs three-fourths as wide as long .... 18. *R. orarius.*
    21 Glandular hairs wanting, or sometimes a very few present on the pedicels.
      23 Lfls lanceolate to oblong or ovate, not gray or white beneath.
        24 Terminal lfl of primocane-lvs at least half as wide as long, the base
          broadly obtuse to rounded or cordate.
          25 Lvs glabrous beneath, or nearly so; plants usually only sparsely
            armed, the prickles mostly few, small, and weak .... 19. *R. canadensis.*
          25 Lvs softly pubescent beneath; plants usually more strongly armed
            .... 20. *R. pensilvanicus.*
        24 Terminal lfl of primocane-lvs less than half as wide as long, generally
          cuneate at base .... 21. *R. argutus.*
      23 Lfls distinctly oblanceolate to obovate, broadest well above the middle,
        densely and closely white- or gray-tomentose beneath .... 22. *R. cuneifolius.*

1. Rubus odoratus L. Flowering raspberry. Widely branched unarmed shrub 1–2 m, becoming densely
and coarsely glandular above; lvs subrotund to triangular or reniform in outline, 1–2 dm wide, (3)5-lobed a
half or a third their length, the lobes triangular, acute, irregularly serrate; fls rose-purple (white), in a loose,
open, widely branched cyme; sep with ± dense purple elongate gland-tipped hairs or bristles; pet obovate,
1.5–2.5 cm; fr depressed, 1 cm thick, dryish and rather insipid, the drupelets tending to fall separately; 2*n*=14.
Moist, shady places and margins of woods; N.S. and Me. to Mich., s. to N.C. and Tenn. June–Aug. (*Rubacer o.*)

2. Rubus parviflorus Nutt. Thimbleberry. Unarmed shrub 1–2 m, the younger parts, petioles, and pedicels
usually stipitate-glandular; lvs rotund to reniform in outline, 1–2 dm wide, shallowly ca. 5-lobed, the lobes
serrate, acute; fls few, white, in a long-pedunculate cymose cluster; sep with orange-yellow very short gland-
tipped hairs; pet elliptic-ovate, 1.5–2 cm; fr juicy, edible, red, 1.5 cm thick, pubescent, falling as a unit from
the persistent receptacle; 2*n*=14. Open woods and thickets; Bruce Peninsula, Ont., to n. Mich and n. Minn.;
S.D.; widespread in the western cordillera. May–July. (*Rubacer p.*)

3. Rubus chamaemorus L. Cloudberry, baked-apple-berry. Stems arising 1–3 dm from a creeping rhizome,
essentially herbaceous, erect, unbranched, often flexuous, unarmed, the lowest nodes bearing stipules only;
lvs commonly 2 or 3, long-petioled, simple, rotund to reniform in outline, the largest 4–9 cm wide, shallowly
5–7-lobed, serrate; fls solitary, terminal, long-peduncled, white, 2–3 cm wide, unisexual; sep becoming reddish;
fr edible, orange to red, 1.5–2 cm thick, quickly deciduous from the dry receptacle; 2*n*=56. Bogs and wet
mt. slopes; circumboreal, extending s. to N.B., Me., and N.H. June, July.

4. Rubus pubescens Raf. Dwarf raspberry. Horizontal stems long-creeping at or near the surface of the
soil, ± leafy when aerial; upright stems arising near the base or along the length of the horizontal ones,
herbaceous, unarmed, rarely with a few bristles, 1.5–5 dm, bearing 2–5 long-petiolate 3-foliate lvs with
oblanceolate stipules; lfls rhombic-ovate to obovate, 4–8 cm, acute or acuminate, sharply toothed especially
beyond the middle, tending to be entire along the cuneate lower portion; peduncle terminal, 1–3-fld, occasionally 1 or 2 additional fls arising from the axils; pedicels usually with a few stipitate glands; pet white,
seldom pink, 4–8 mm; fr dark red, 5–10 mm thick, only tardily separating from the spongy receptacle; 2*n*=14,
28. Damp woods and bogs; Lab. to Yukon and B.C., s. to N.H., W.Va., Ind., Colo., and Wash. May–July.
(*R. triflorus.*)

5. Rubus acaulis Michx. Perennial herb from slender, wiry rhizomes, 5–25 cm; lvs 2 or 3; lfls 3, the
terminal short-petiolulate, broadly obovate, 1–3.5 cm, cuneate and entire below, deeply serrate and obtuse
to often broadly rounded above, the lateral similar but asymmetrical, often with a broad lateral lobe; fls

solitary; pedicels and cal glandless; sep narrowly triangular, caudate, to 1 cm; pet pink, (8–)10–16 mm; fr subglobose, 1 cm thick, only tardily deciduous from the receptacle; $2n=14$. Cold bogs and wet meadows; Lab. to Alas., s. to e. Que., n. Minn., and Colo. June–Aug. Closely related to the chiefly Eurasian *R. arcticus* L., and possibly better treated as *R. arcticus* var. *grandiflorus* Ledeb.

6. Rubus occidentalis L. Black raspberry, black-cap. Stems erect or ascending, or sometimes arching and rooting at the tip, not glandular, strongly glaucous at least the first year, sparsely beset with stout, straight or hooked prickles with expanded base, as are also the petioles and especially the pedicels; lfls mostly 3, or 5 on the primocanes, the intermediate pair then adjacent to the lower; uppermost lvs of the floricane often simple; terminal lfl broadly ovate, rounded or subcordate at base, sharply, deeply and irregularly serrate; lower lfls similar but smaller and narrower; all lfls thinly gray-tomentose beneath; fls 3–7 in a dense, umbelliform cyme, often also 1 or 2 from the upper axils; pet white, shorter than the sep, narrowly obovate, at first erect, soon deciduous; fr purple-black (yellowish), with narrow belts of white tomentum between the drupelets, 1 cm thick, separating as a unit from the persistent receptacle; $2n=14$. Dry or moist woods, fields, and thickets; Que. to N.D. and e. Colo., s. to Ga. and Ark. May, June. Also cult. in many cultivars. Hybridizes with no. 7.

7. Rubus idaeus L. Red raspberry. Stems to 2 m, erect, spreading, or decumbent, sparsely to densely armed with slender-based prickles and stiff, glandular bristles; primocane lfls 3 or 5, ovate to lanceolate, acuminate, sharply serrate mostly above the middle, softly gray-pubescent beneath; intermediate pair of lfls, when present, inserted close to the terminal lfl; infl an umbelliform cyme of 2–5 fls, often also with 1 or 2 solitary fls from the upper axils; pedicels stipitate-glandular and ± bristly, the bristles exceeding the glandular hairs and gland-tipped when young; sep soon reflexed; pet white or greenish-white, spatulate to obovate, erect, shorter than the sep; fr red (yellow), 1–1.5 cm thick, separating as a unit from the persistent receptacle; $2n=14$, seldom more. Dry or moist woods, fields, and roadsides; Nf. to Alas., s. to Pa., Ind., Io., and Ariz. May–July. The Amer. and e. Asian plants, as here described, constitute the var. *strigosus* (Michx.) Maxim., in contrast to the eglandular Eurasian var. *idaeus,* which is cult. in many cultivars. At the subspecific level our plants would be subsp. *melanolasius* Focke. (*R. strigosus.*)

8. Rubus phoenicolasius Maxim. Wineberry. Stems arching, to 2 m, armed with a few slender prickles, densely shaggy with long (3–5 mm), purple, glandular hairs, as also the petioles and infl; lfls 3, densely white-tomentose beneath, the terminal one broadly ovate, abruptly short-acuminate, rounded at base, the lateral similar but much smaller; infl a many-fld cymose panicle; sep glandular-hairy; pet white, narrowly ovate, erect, much shorter than the sep; fr red, 1 cm thick, separating as a unit from the persistent receptacle; $2n=14$. Native of e. Asia, well established as an escape from cult. in our range, especially along the Atlantic coast.

9. Rubus illecebrosus Focke. Strawberry-raspberry. Stems sparsely beset with wide-based prickles, otherwise glabrous and glandless; lvs pinnately compound; lfls 5–9, lanceolate, 7–10 cm, sessile, serrate; cor white, 3–4 cm wide; pet spreading, broadly obovate; fr red, to 3 cm, separating as a unit from the persistent receptacle; $2n=14$. Native of Japan, escaped from cult. here and there in our range.

10. Rubus trivialis Michx. Coastal-plain dewberry. Primocanes trailing, rooting at least at the tip, armed with a few short, stout, ± recurved prickles with expanded base, also hispid with many reddish glandular bristles; primocane lvs ± evergreen, 5-foliolate, on long, prickly petioles; lateral lfls sessile or subsessile; terminal lfl on a prickly and hispid petiolule a fourth to a third as long as the blade, mostly elliptic or oblong, a third to half as wide as long, serrate; flowering branches (excluding the peduncle) rarely over 1 dm, often only 1–3 cm, with a few small 3-foliolate lvs and commonly 1 (rarely 2) peduncles much surpassing the lvs. Chiefly in dry or sandy soil; Md. (?) and se. Va. to Fla. and Tex., n. in the interior to Mo. Apr., May.

11. Rubus hispidus L. Swamp-dewberry. Primocanes trailing to low-arched, usually rooting at the tip, armed with slender, weak, straight or barely curved, slightly reflexed prickles commonly 2–4 mm, these usually with small base, sometimes with an expanded base to 5 mm, numerous to almost wanting; shorter glandular hairs usually also present; primocane lvs 3- or 5-foliolate, firm and usually shining, the terminal lfl mostly short-petiolate, oblong-ovate to obovate, broadest above the middle; infl distinctly racemiform, commonly with several or many fls; fr sour. In a wide range of habitats, even in peat-bogs. Abundant nearly throughout our range; Que. and N.S. to Wis., s. to N.C. and Mo. June–Aug. (*R. ambigens; R. compos; R. cubitans; R. davisiorum; R. distinctus; R. eflagellaris; R. emeritus; R. fassettii; R. furtivus; R. huttonii; R. kalamazoensis; R. missouricus; R. multiformis; R. novanglicus; R. paganus; R. paludivagus; R. parlinii; R. plus; R. porteri; R. rosendahlii; R. rowleei; R. schoolcraftianus; R. spiculosus; R. tardatus; R. vagulus; R. vegrandis; R. vigil; R. zaplutus.*)

12. Rubus caesius L. European dewberry. Primocanes ascending, soon arched and rooting at the tip, sparsely beset with stout hooked prickles with expanded base, not bristly, glaucous when young; primocane lvs 3-foliolate, the lfls broadly ovate, subcordate at base, softly pubescent beneath; infl subpaniculiform, some of its axillary branches with 2 or more fls. Native of Eurasia, occasionally escaped from cult.

13. Rubus recurvicaulis Blanchard. Blanchard's dewberry. Infl racemiform, with few or several fls mostly subtended by stipules only, or the lowest one axillary to an expanded, petiolate lf; pedicels tending to be ascending or spreading rather than erect; otherwise much like *R. flagellaris,* but often rising a little higher above the ground. In pockets of soil on ledges and cliffs, as well as in disturbed, often sandy habitats; N.S. and Que. to Minn., s. to Md., W.Va., and Ind., most abundant northward. June–Aug. (*R. adenocaulis; R. arenicola; R. arundelanus; R. boyntonii; R. bracteoliferus; R. complex; R. conabilis; R. cordifrons; R. folioflorus; R. fraternalis; R. grandidens; R. grimesii; R. licens; R. minnesotanus; R. noveboracus; R. pauper; R. perpauper; R. pityophilus; R. plicatifolius; R. satis; R. setospinosus; R. tantulus; R. vagus.*)

14. Rubus flagellaris Willd. Northern dewberry. Primocanes prostrate or low-arched, normally rooting at least at the tip, never bristly, ± armed with small, stout, curved, somewhat hooked prickles with expanded base; foliage and pubescence various, but lfls of the compound floricane lvs regularly of an ovate type, and sharply acute or acuminate, broadly rounded to the base or even subcordate, widest distinctly below the middle; fls occasionally solitary and terminal, more often 2–5, each pedicel, except possibly that of one of the two terminal fls, subtended by a 3-foliolate lf or by a simple lf with expanded blade, the pedicels elongate and suberect. Often in pockets of soil on ledges and cliffs, as well as in disturbed habitats. Chiefly northern, from e. Can. to Minn., but extending s. to Ga. and Ark. May, June. (*R. baileyanus; R. curtipes; R. foliaceus; R. invisus; R. jaysmithii; R. maltei; R. meracus; R. occidualis; R. plexus; R. profusiflorus; R. redundans; R. roribaccus; R. temerarius.*)

15. Rubus enslenii Tratt. Southern dewberry. Primocanes slender, prostrate or low-arching, normally rooting at least at the tip; armature of small, slender to stout, recurved or reflexed prickles from an expanded base, the prickles never numerous or large and in some forms nearly lacking; lvs relatively small, dull, glabrous or nearly so; terminal lfl of primocane lvs oblong to oblanceolate or obovate, usually with ± straight sides below the middle; floricane lfls distinctly cuneate toward the base, oblanceolate or obovate, mostly obtuse or very abruptly short-acuminate; infl few-fld, leafy, most of the pedicels subtended by simple or trifoliolate lvs, or often reduced to a single terminal fl. Most commonly in open oak-hickory woods; generally southern, but extending n. to s. Me., c. N.H., O., sw. Mich., and s. Wis. Apr.–June. (*R. akermanii; R. cathartium; R. celer; R. centralis; R. clarus; R. connixus; R. decor; R. depavitus; R. felix; R. hypolasius; R. imperiorum; R. indianensis; R. iniens; R. injunctus; R. kentuckiensis; R. leviculus; R. longipes; R. michiganensis; R. nefrens; R. obvius; R. particularis; R. pernagaeus; R. rosagnetis; R. scambens; R. sewardianus; R. subinnoxius; R. tetricus; R. whartoniae.*)

16. Rubus setosus Bigel. Bristly blackberry. Primocanes erect, ascending, or arched, rising to about 1 m or less, not rooting, beset with stiff bristles or slender, often soft prickles, the latter straight or nearly so, spreading or somewhat reflexed, mostly slender to the base, often glandular when young; primocane lvs 3- or 5-foliolate; petiole commonly as bristly as the stem; lfls usually glabrous or nearly so, but sometimes evidently hairy beneath, even velvety; infl few- to many-fld, racemiform or corymbiform, or sometimes reduced and leafy; pedicels usually glandular; fr of mediocre quality. Typically in moist or wet, low ground, but also in old fields and other disturbed habitats; Que. to Wis., s. to Va. and Ill. June–Aug. (*R. angustifoliatus; R. ascendens; R. beatus; R. benneri; R. clausenii; R. discretus; R. dissensus; R. elegantulus; R. groutianus; R. hispidoides; R. jejunus; R. lawrencei; R. mediocris; R. nocivus; R. racemiger; R. regionalis; R. rotundior; R. schneideri; R. semisetosus; R. uniformis; R. univocus; R. vermontanus; R. wheeleri.*)

17. Rubus allegheniensis T. C. Porter. Common blackberry. Stems 0.5–2(–3) m, mostly erect or nearly so, the young primocanes often sparsely glandular; primocane lvs softly pubescent beneath; terminal lfl typically ovate-oblong, varying to ovate, 1–2 dm, widest near or below the middle, long-acuminate, finely and sharply serrate, rounded to truncate or subcordate at base; lateral lfls usually 4, smaller; armature of the stem of nearly straight prickles spreading at right angles or barely reflexed, much flattened at the long base; prickles of the petioles, pedicels, and midveins similar but prominently hooked; infl racemiform, commonly elongate and many-fld, the lower 1 or 2(3) fls subtended by lvs, the others by stipules only; pedicels tomentose and glandular; fls 2 cm wide; sep acute to more commonly short-caudate; pet cuneate and separate at base. Our commonest tall blackberry, occurring in a wide variety of mostly disturbed habitats, from N.S. and Que. to Minn., s. almost throughout our range and along the mts. to N.C. and Tenn. May–July. (*R. abbrevians; R. alumnus; R. attractus; R. concameratus; R. flavinanus; R. frondisentis; R. inclinis; R. nuperus; R. ortivus; R. paulus; R. perinvisus; R. pugnax; R. reravus; R. saltuensis.*)

18. Rubus orarius Blanchard. Much like *R. allegheniensis*; primocane lfls mostly 3, or 5 with the intermediate pair subsessile or poorly developed; terminal lfl of primocane lvs 8–15 cm, broadly ovate to subrotund, at least three-fourths as wide as long, commonly short-acuminate or abruptly acuminate; sharply or coarsely or irregularly serrate, broadly rounded to subcordate at base; infl racemiform, but not so long and many-fld as in no. 17. Que. to Wis., s. to Md., W.Va., and Mo. Perhaps actually a series of hybrids of *R. allegheniensis* with *R. pensilvanicus* and/or other spp. The name *R. orarius* may not properly apply, but is here retained for historical continuity with Gleason's treatment while we await a proper biosystematic study of the group.

19. Rubus canadensis L. Smooth blackberry. Much like *R. pensilvanicus*, which might perhaps better be merged with it, but usually only sparsely armed, with few, small, and weak prickles, and with glabrous or subglabrous lvs. Often in open deciduous forest or on mt. ledges, as well as in disturbed habitats; Nf. and Que. to Minn., s. through N.Y. and along the mts. to n. Ga. June, July (*R. amicalis; R. kennedyanus; R. multilicius.*)

20. Rubus pensilvanicus Poiret. Pennsylvania-blackberry. Stems usually stout, 1–3 m; armature of the primocanes of straight, spreading, or slightly reflexed prickles from an expanded base; shorter hooked prickles usually present on the petioles, often on the petiolules, and occasionally on the midveins and on the axis of the raceme, rarely in much weaker form on the pedicels; primocane lvs softly pubescent beneath, 3- or 5-foliolate; terminal lfl broadly ovate, 6–12 cm, half to nine-tenths as wide, broadest near or above to often well below the middle, distinctly acuminate, broadly rounded or subcordate at base, very coarsely and irregularly serrate to doubly serrate; floricane lvs usually with elliptic to rhombic or obovate lfls coarsely toothed above the middle, many of them simple and broadly ovate or subrotund; raceme usually short and compact, few-fld, well surpassing the lvs, varying to loose and open with long-pedicellate fls. Me. and N.H.

to Minn., s. to Va., W.Va., Ind., and Mo. May–July. (*R. acer; R. amnicola; R. associus; R. avipes; R. barbarus; R. bellobatus; R. bractealis; R. brainerdii; R. bushii; R. cardianus; R. condensiflorus; R. congruus; R. cupressorum; R. defectionis; R. dissitiflorus; R. facetus; R. floricomus; R. floridus; R. frondosus; R. independens; R. insulanus; R. latifoliolus; R. libratus; R. multispinus; R. ostryifolius; R. pauxillus; R. perfoliosus; R. philadelphicus; R. praepes; R. pratensis; R. recurvans; R. subsolanus; R. wisconsinensis.*)

21. Rubus argutus Link. Southern blackberry. Stems erect, to 2 m, glabrous; armature of the stems of straight, spreading or slightly reflexed prickles 3–8 mm from a flattened base; petioles, flowering branches, and occasionally the pedicels and midveins commonly armed with smaller, distinctly hooked prickles; primocane lvs 5-foliolate, the terminal lfl long-stalked, commonly 8–12 cm, the intermediate pair conspicuously stalked and the lfls not overlapping; lfls all about alike in shape, oblong or oblong-oblanceolate, typically less than half as wide as long, sharply or often coarsely and irregularly serrate, widest above the middle, thence sharply acuminate to the tip, below the middle ± cuneate, with the sides usually conspicuously straight or straightish but serrate, acute to obtuse or rounded at base, softly pubescent beneath; pedicels villous, glandless or with a few scattered glands; fls in short, loose, open racemiform cymes. Chiefly a southern sp., from Md., Ky., and Mo. to Fla., and Tex., but n. occasionally to Mass. May, June. (*R. blakei; R. fatuus; R. jugosus; R. louisianus.*)

22. Rubus cuneifolius Pursh. Sand-blackberry. Stems erect or nearly so, to 1 m, heavily armed with stout, straight or hooked prickles from a widely expanded base; primocane lvs 3(5)-foliolate, on a strongly armed petiole; terminal lfl seldom over 5 cm, rhombic to obovate, widest well above the middle, truncate to rounded or broadly obtuse, short-apiculate, or rarely merely acute, cuneate from the broadest part with nearly straight margins to the base, simply serrate in the distal half with rather broad low teeth, entire in the basal half, densely white- or gray-tomentose beneath; infl of usually 1–3 fls on ascending pedicels, all or mostly subtended by lvs with expanded blade. Dry, sandy soil, chiefly on the coastal plain; Conn. and L.I. to Fla. and Ala.; disjunct in Hillsboro Co., N.H. May–July. (*R. longii; R. sejunctus.*)

23. Rubus laciniatus Willd. Evergreen blackberry. Stems coarse, scrambling to several m, strongly armed with flattened, recurved prickles; lvs evergreen, primarily (3)5-foliolate, the lfls laciniately cleft or again compound and cleft, the lower side evidently hairy but green; fls numerous in cymose panicles; cal-lobes usually prickly; pet pinkish to sometimes white, usually trifid; fr ± elongate, 1–1.5 cm thick; $2n=28$. European cultigen, sparingly naturalized from Mass. to Minn. and southward. June–Aug.

24. Rubus discolor Weihe & Nees. Himalayan blackberry. Stems coarse, scrambling to several m, armed with strong, flattened, curved prickles; primocanes canescent toward the tip; lvs tardily deciduous, grayish-tomentose beneath, bright green and ± glabrous above, those of the primocanes mostly 5-foliolate; lfls broadly ovate to oblong-obovate, usually abruptly short-acuminate, 6–12 cm, sharply toothed; infl cymose-paniculate, often thyrsoid, with strong, flat prickles to 1 cm; cal canescent-tomentose, its unarmed lobes without long appendages; pet white to sometimes pinkish; fr subglobose to slightly elongate, 1.5 cm thick; $2n=28$. European sp., occasionally escaped from cult. in the s. part of our range. June–Aug. (*R. procerus.*)

25. Rubus bifrons Vest. Much like no. 24; stems glabrous or nearly so; prickles of stem and infl long, strong, subulate, straight, spreading; lfls 3 or 5; pet pale pink to red; $2n=28$. European sp., occasionally escaped from cult. in our range.

15. DALIBARDA L. Fls of two types, petaliferous and apetalous; hypanthium saucer-shaped; sep 5 or 6, without bractlets; pet 5, white, elliptic; stamens numerous, with long slender filaments; pistils 5–10; ovaries villous; ovules 2; style terminal, slender, elongate; fr a dry, achene-like, 1-seeded drupe; perennial herbs with slender creeping stems, simple cordate lvs, and solitary, white, long-peduncled fls. Monotypic.

1. **Dalibarda repens** L. Dewdrop. Stems very slender; petioles downy; lvs cordate, 3–5 cm, sparsely hairy; peduncles of the petaliferous, usually sterile fls 5–10 cm, of the fertile apetalous fls 2–5 cm and often recurved; hypanthium densely hairy; sep unequal, usually 3-toothed at the tip; pet 4–8 mm; frs few, 3–4 mm. Swamps and moist woods; Que. and N.S. to Mich. and Minn., s. to N.J. and O. June–Sept.

16. ALCHEMILLA L. Hypanthium ellipsoid to turbinate or campanulate, permanently enclosing the ovary or ovaries; sep mostly 4, spreading or ascending; bractlets sometimes present; pet none; stamens 4(5), inserted on the outer margin of the disk; pistil commonly 1(2–3); style basal; fr an achene; rhizomatous perennial herbs with small clusters of fls arranged in openly branched, cymose infls. 100+, widespread.

1. **Alchemilla vulgaris** L. Lady's mantle. Rhizome very stout; stems 2–6 dm, mostly simple below the infl; lvs reniform, 3–10 cm wide, shallowly and roundly 5–9-lobed, the lobes sharply serrate; hypanthium campanulate, 1.5 mm, its throat nearly closed by the disk; sep 1.5 mm, alternating with smaller bractlets; $2n=64$–$100+$. Widely distributed in Eurasia, and intr. in New England, N.Y., and se. Can. May–Aug. (*A. pratensis.*) Variable, and often divided into numerous apomictic microspecies. Among these, *A. filicaulis*

Buser, *A. glabra* Neyg., *A. gracilis* Opiz, *A. monticola* Opiz, *A. venosa* Juz. and *A. xanthochlora* Rothm., have been ascribed to our range.

17. APHANES L. Fls in small, condensed, lf-opposed cymes that are partly enclosed by the conspicuous, foliaceous, connate stipules, stamen 1(2), inserted on the inner margin of the disk; small, taprooted annuals with deeply dissected lvs; otherwise like *Alchemilla,* and sometimes included therein. 20, widespread.

> **1. Aphanes microcarpa** (Boiss. & Reuter) Rothm. Diffusely branched, hairy, seldom over 10 cm; free petiole 1 mm; blade subrotund, 3–5 mm, deeply trifid, each segment deeply 3-toothed; stipules with several oblong lobes almost as long as the undivided part; hypanthium scarcely 1 mm at anthesis, under 2 mm in fr; sep connivent; $2n=16$. Native of s. Europe, intr. into lawns and waste places from L.I. to Ala. Apr.–June. (*Alchemilla m.*)
> *A. arvensis* L. has been reported from N.S. It is coarser than *A. microcarpa,* to 30 cm, with blades up to 1 cm, the lobes of the stipules triangular, about half as long as the undivided part, the fruiting hypanthium 2+ mm, and the sep somewhat spreading; $2n=48$.

18. AGRIMONIA L. Agrimony, harvest-lice. Hypanthium obconic to hemispheric, armed above with hooked bristles, becoming indurate and 10-grooved at maturity; sep spreading at anthesis, later incurved at base and connivent above, forming a beak on the fr; pet 5, yellow, small; stamens 5–15; ovaries 2, concealed within the hypanthium, ripening into achenes; style terminal; perennial herbs from stout rhizomes, the 3–many large lfls of the pinnately compound lvs interspersed with much smaller ones; stipules foliaceous, usually deeply toothed or laciniate; fls in a long, interrupted, spike-like raceme, the short peduncle subtended by a laciniate bract, the very short pedicels by a pair of 3-lobed bractlets. All our spp. fl in July and Aug. 15, N. Temp.

1 Hypanthium glabrous or glandular, or sparsely strigose only in the furrows; native.
  2 Axis of the infl conspicuously glandular, sometimes also hairy.
    3 Larger lvs with 3–9 principal lfls; axis of the infl with long hairs or none.
      4 Axis with scattered long spreading hairs; mature hypanthium 3–5 mm . . . . 1. *A. gryposepala.*
      4 Axis usually merely glandular; mature hypanthium 2–2.5 mm . . . . 2. *A. rostellata.*
    3 Larger lvs with 11–23 principal lfls; axis of the infl finely and densely short-hairy, sometimes
      also with long spreading hairs . . . . 3. *A. parviflora.*
  2 Axis of the infl eglandular, finely and often densely short-hairy.
    5 Lvs conspicuously glandular beneath; roots not tuberous-thickened . . . . 4. *A. striata.*
    5 Lvs either eglandular beneath or with a very few glands mostly concealed by the dense velvety
      pubescence; roots tuberous-thickened.
      6 Principal lvs with at least 5 well developed lfls . . . . 5. *A. pubescens.*
      6 Principal lvs with 3 well developed lfls . . . . .6. *A. microcarpa.*
1 Hypanthium evidently hirsute on the ribs as well as in the furrows; intr. . . . . 7. *A. eupatoria.*

    1. Agrimonia gryposepala Wallr. Common a. Roots fibrous; stem stout, 5–15 dm, glandular and ± long-hirsute; stipules large and foliaceous, semi-cordate, 1–2 cm wide; principal lfls of the larger lvs 5–9, lance-ovate to elliptic or obovate, coarsely and often bluntly serrate, glabrous or nearly so above, beneath conspicuously glandular, sparsely hirsute chiefly or wholly on the veins; axis of the infl glandular, also hirsute with long spreading hairs; pedicels hirsute; hypanthium glandular only, or with a few short stiff hairs below, 3–5 mm at maturity, expanded above; bristles numerous in several rows; $2n=56$. Moist or dry open woods; Me. to Ont. and Mont., s. to N.J., Ind., Kans., and in the mts. to Tenn. and N.C.; also in Calif. and N.M.

    2. Agrimonia rostellata Wallr. Woodland a. Roots sometimes tuberous-thickened; stem slender, to 1 m; stipules lanceolate to semi-ovate, rarely over 1 cm wide; principal lfls of the larger lvs 3–9, thin, oblong-obovate or obovate, very coarsely and usually bluntly serrate; conspicuously glandular beneath, otherwise glabrous or with short hairs on the veins beneath; axis of the infl glandular, seldom with a few spreading bristles; hypanthium 2–2.5 mm at maturity, hemispheric or broadly campanulate, glandular, not hairy. Moist rich woods; Conn. to Ind. and Kans., s. to S.C., Ga., La., and Okla.

    3. Agrimonia parviflora Aiton. Southern a. Roots fibrous; stems stout, to 12 dm, densely hirsute and also finely short-hairy below; principal lfls of the larger lvs 11–23, lanceolate, sharply serrate, glandular beneath and sparingly pubescent, especially on the veins; axis of the infl glandular, finely pubescent with short, mostly ascending hairs, with or without long spreading hairs; pedicels very short, erect, at maturity abruptly deflexed at the bracteoles; hypanthium turbinate, 3 mm, glandular, sometimes with a few stiff hairs below; outer bristles much shorter than the inner. Damp woods; Conn. and N.Y. to s. Mich., s. Wis., and S.D., s. to Hispaniola and Mex.

    4. Agrimonia striata Michx. Roadside-a. Roots fibrous; stems stout, to 1 m or more, hirsute below, pubescent and glandular above; stipules lanceolate to semi-ovate, 1–2 cm; principal lfls of the larger lvs 7–

11, the upper 5 commonly directed forwards, lance-ovate, coarsely serrate, glabrous or nearly so above, gland-dotted beneath and sparsely hairy, especially on the veins; axis of the infl densely pubescent with ascending hairs, commonly also with some long flexuous hairs; fls densely crowded; pedicels short, the 3-cleft bractlet commonly surpassing the hypanthium; mature hypanthium reflexed, turbinate, 4–5 mm, minutely strigose in the deep furrows; $2n=56$. Dry or moist woods; Que. and N.S. to e. B.C., s. to R.I., N.Y., Mich., Io., and S.D., and in the mts. to W.Va. and Ariz.

5. Agrimonia pubescens Wallr. Downy a. Roots tuberous-thickened; stems stout, to 1 m or more, densely short-hairy and ± hirsute; stipules lanceolate to semi-ovate; principal lfls of the larger lvs 5–13, lanceolate to elliptic or narrowly obovate, coarsely serrate, glabrous, scabrellous, or sparsely pubescent above, velvety-pubescent beneath; axis of the infl densely short-pubescent; mature hypanthium campanulate, 2.5–3 mm, minutely strigose in the deep furrows, also ± beset with short stiff ascending hairs, especially below. Dry, open woods; Me. to Mich. and se. S.D., s. to N.C., Ga., and Okla.

6. Agrimonia microcarpa Wallr. Low a. Roots tuberous-thickened; stems usually simple, 3–6 dm, hirsute below; lvs mostly clustered below, the larger with only 3 well developed lfls, often with a much smaller pair below them and a few minute lfls interspersed; petiole long-hirsute; lfls oblong-obovate, usually rounded above, regularly serrate, velvety-pubescent beneath; mature hypanthium turbinate, 2–2.5 × 2.5–3 mm, minutely strigose in the furrows. Woodlands; N.J. and Pa. to Fla. and e. Tex. (*A. platycarpa* Wallr., a form with larger lvs better distributed along the stem.)

7. Agrimonia eupatoria L. Medicinal a. Lvs softly and densely pubescent or sometimes subtomentose beneath, with long stiff hairs on the veins; axis of the infl glandular or not, ± hirsute with long spreading hairs; hypanthium hirsute throughout with long, spreading or ascending hairs; pet 4–5 mm; $2n=28$. Formerly used in folk-medicine and as a source of a golden yellow dye. Native of Eurasia, occasionally intr. in our range.

19. SANGUISORBA L.   Burnet. Fls 4-merous, perfect or unisexual; hypanthium urceolate, contracted at the mouth, ± 4-angled, not prickly; sep petaloid; pet none; stamens (2)4 or many; pistils 1 or 2; ovaries included in the hypanthium; ovule 1, suspended; style terminal; fr an achene, enclosed by the indurate hypanthium; perennial herbs from a thick rhizome, or annuals, the lvs pinnately compound with serrate to pinnatifid lfls, the fls small, borne in dense spikes or heads. 25, N. Amer., Eurasia.

1 Pistil and achene 1; stamens 4; principal lfls usually well over 2 cm.
  2 Stamens much longer than the white sep; spikes 3–12 cm .... 1. *S. canadensis.*
  2 Stamens equaling or shorter than the purple sep; heads 1–3 cm .... 2. *S. officinalis.*
1 Pistils and achenes 2; stamens numerous; lfls seldom over 2 cm .... 3. *S. minor.*

1. Sanguisorba canadensis L.   American b. Perennial from a thick rhizome; stem erect, usually simple below, to 15 dm; lower lvs to 5 dm (petiole included), the upper progressively reduced; stipules foliaceous; lfls 7–15, petiolulate, ovate to oblong or elliptic, 3–8 cm, blunt, sharply serrate; spikes 1–several, long-pedunculate, cylindric, 3–12 cm; fls all perfect; hypanthium not roughened between the wings; sep white, elliptic, spreading, 2–3 mm; $2n=56$. Marshes, wet meadows, and damp prairies; Nf. and Lab. to Man., s. to N.J., Pa., O., and Ind., and in the mts. to W.Va., Ky. and N.C. July–Sept.

2. Sanguisorba officinalis L.   Great b. Much like no. 1; stems to 1 m; spikes thickly ellipsoid to short-cylindric, 1–3 cm; stamens equaling or shorter than the purple-brown sep, with filiform red filaments; $2n=28$, 56. Native of Eurasia, rarely escaped from cult.

3. Sanguisorba minor Scop.   Salad-b. Perennial from a caudex-like rhizome; stems 2–7 dm; lower lvs numerous and well developed, with usually 7–17 lfls, the upper lvs progressively reduced; lfls ovate to rotund, 5–20 mm, with 3–7 sharp teeth on each side; heads several on elongate peduncles, short-ovoid to globose, 8–20 mm; fls subtended by ciliate bracts, the lower ones perfect or staminate, the upper pistillate; sep green or brown, 2.5–5 mm, the inner pair broader than the outer; stamens numerous, the long filaments drooping; pistils 2; mature hypanthium 5 mm, very rough between the wings; achenes 2, semi-ovoid; $2n=28$, 56. Native of Eurasia, established as a weed along roadsides and in fields and waste places here and there in our range. May, June. (*Poterium sanguisorba.*)

20. ROSA L.   Rose. Hypanthium globose to urceolate, with a constricted orifice; sep usually long-attenuate or prolonged into a foliaceous tip, often persistent in fr; pet large, spreading at anthesis, pink to red, white, or yellow; stamens very numerous, inserted near the orifice of the hypanthium on relatively short filaments; ovaries mostly numerous, inserted on the bottom or also on the sides of the hypanthium; styles usually barely exserted, distinct or ± united; fr a bony achene; mature hypanthium, called a hip, commonly colored and pulpy or fleshy; shrubs or woody vines, usually prickly; lvs pinnately compound with 3–11 serrate lfls, the stipules commonly large and adnate to the petiole;

x=7. Genus beset with hybridization and polyploidy. 100+, mainly N. Temp. The application of the names *R. spinosissima* and *R. cinnamomea* is now so confused that both have been abandoned in Flora Europaea. We do likewise.

1 Stipules free from the petiole for most of their length; intr. climber . . . . 1. *R. bracteata.*
1 Stipules adnate to the petiole for more than half their length.
  2 Styles united into a column, protruding from the orifice of the hypanthium.
    3 Lfls 3 or 5, 3–10 cm; pet 2–3 cm, usually pink; native . . . . 2. *R. setigera.*
    3 Lfls (5)7 or 9, 1–3 cm; pet 1–2 cm, usually white; intr.
      4 Stipules pectinately toothed and glandular-ciliate; styles glabrous . . . . 3. *R. multiflora.*
      4 Stipules merely dentate; styles pubescent . . . . 4. *R. wichuraiana.*
  2 Styles distinct, only slightly if at all exserted, usually only the stigmas protruding.
    5 Fls solitary at the branch-tips, the pedicel not subtended by a bract; intr.
      6 Lfls 3–7, 2–6 × 2–3 cm; pet (2.5–)3–4.5 cm, deep pink . . . . 5. *R. gallica.*
      6 Lfls 7–11, 0.5–2 × 0.5–1 cm; pet 1–2(–2.5) cm, white or ochroleucous, rarely pink . . . . 6. *R. pimpinellifolia.*
    5 Fls solitary or corymbose; when solitary, the pedicel bracteate near its base.
      7 Sep unlike in size and shape, the outer pinnatifid with several lanceolate segments; orifice of
        the hypanthium 1 mm wide; styles shortly exserted; intr.
        8 Lower surface of the lvs evidently stipitate-glandular.
          9 Styles pubescent; sep erect and persistent in fr . . . . 7. *R. eglanteria.*
          9 Styles glabrous; sep soon deciduous . . . . 8. *R. micrantha.*
        8 Lower surface of the lvs glabrous or nearly so . . . . 9. *R. canina.*
      7 Sep all entire, or some with 1–4 linear, scarcely foliaceous appendages arising near the base;
        orifice of the hypanthium 2–4 mm wide, closed and covered by the heads of the stigmas.
        10 Twigs of the season tomentose; pet mostly 3–5 cm; intr. . . . . 10. *R. rugosa.*
        10 Twigs glabrous or inconspicuously hairy; pet mostly 2–3 cm; native, except no. 19.
          11 Hypanthium and pedicel ± stipitate-glandular; sep after anthesis usually spreading
            or reflexed, eventually deciduous.
            12 Lfls finely toothed, the teeth near the widest part of the lfl avg ca 0.5 mm high,
              measured along the distal margin.
              13 Infrastipular prickles present at many nodes, stout, broad-based, decurved;
                internodal prickles none or very few . . . . 11. *R. palustris.*
              13 Infrastipular prickles scarcely differing from the numerous internodal ones,
                all straight and slender . . . . 12. *R. nitida.*
            12 Lfls coarsely toothed, the teeth near the widest part of the lfl avg ca 1 mm high,
               measured along the distal margin.
              14 Infrastipular prickles stout, often decurved, flattened toward the base; inter-
                nodal prickles commonly lacking . . . . 13. *R. virginiana.*
              14 Infrastipular prickles slender, straight, terete; internodal prickles commonly
                 numerous . . . . 14. *R. carolina.*
          11 Hypanthium and pedicel generally glabrous; sep long-persistent, usually ± erect after
            anthesis and often connivent.
            15 Infrastipular prickles none or not differentiated from the internodal ones.
             16 Fls borne at the top of the stems of the season and often also on lateral
              branches from stems of the previous year; lfls ordinarily 9 or 11; inter-
              nodes always prickly . . . . 15. *R. arkansana.*
            16 Fls borne only on lateral branches from stems of the previous year; lfls or-
              dinarily 5 or 7.
              17 Floral bracts and upper stipules glabrous or pubescent, or also glandular-
                dentate; stems typically unarmed, or with slender prickles on only
                the lower internodes . . . . 16. *R. blanda.*
             17 Floral bracts and upper stipules conspicuously glandular; stems densely
                prickly on most of the internodes . . . . 17. *R. acicularis.*
           15 Infrastipular prickles present.
            18 Infrastipular prickles straight, slender; fls not double; native . . . . 18. *R. woodsii.*
            18 Infrastipular prickles curved, broad-based; fls often double; intr. . . . . 19. *R. majalis.*

1. Rosa bracteata Wendl. McCartney-r. Stems climbing to 6 m, with broad-based, paired, recurved infrastipular prickles, when young densely pubescent and covered with stiff stipitate glands; stipules lanceolate, adnate only at base; petioles 1 cm or less; lvs evergreen; lfls 5–9, shining above, tomentose beneath, elliptic to obovate, scarcely serrate; fls solitary, subtended by large bracts; sep ovate, long-acuminate, densely tomentose without and at base within; pet white, 2–3 cm; styles connate, exserted; 2*n*=14. Native of China, cult. in the South and established from Va. and Tenn. to Fla. and Tex.

2. Rosa setigera Michx. Climbing prairie-r. Stems climbing 2–4 m, glabrous, sparsely armed with stout, decurved, basally flattened prickles, or occasionally unarmed; stipules very narrow, glandular-ciliate, the free tip lanceolate, spreading; petioles stipitate-glandular, sometimes sparsely prickly; lfls 3 or 5, firm, glossy dark green, glabrous and impressed-veiny above, paler and glabrous to subtomentose beneath, lanceolate to oblong-ovate, 3–10 cm, the lateral subsessile, the terminal long-stalked; pedicels, hypanthium, and sep stipitate-glandular; dioecious; fls numerous; sep soon reflexed, deciduous; pet usually pink, 2–3 cm; styles connate, protruding from the hypanthium; hips small, brown-green; $2n=14$. Thickets and fence-rows; sw. Ont. and O. to Io. and Kans., s. to w. N.C., Ga., La., and Tex. (*R. rubifolia.*)

3. **Rosa multiflora** Thunb.   Multiflora-r. Vigorously colonial; stems climbing or scrambling to 3 m; stipules conspicuously pectinate-serrate and glandular-ciliate; lfls 5–11, to 2.5 cm, elliptic to obovate; infl many-fld; sep often caudate-tipped, often with some slender lateral lobes, eventually deciduous; pet white (pink) 1–1.5(–2) cm; styles connate, exserted 1.5–3 mm, glabrous; $2n=14$. Native of e. Asia, often escaped from cult.

4. **Rosa wichuraiana** Crépin.   Memorial r. Much like no. 3, but with smaller, rounder, firmer lfls, jagged-dentate stipules, fewer fls, and pubescent styles; semi-evergreen southward; $2n=14$. Native of e. Asia, often planted along highways.

5. **Rosa gallica** L.   French r. Erect colonial shrub to 1(–1.5) m; stem with stout, hooked prickles and numerous bristles, the latter usually gland-tipped; lf-rachis with scattered short prickles; lfls 3–7, leathery, rugose above, sparsely hairy beneath, 2–6 × 2–3 cm, broadly elliptic to ovate, rounded to subcordate at base; fls mostly solitary at the branch-tips, often double; hypanthium and pedicels glandular-setose; sep deflexed and deciduous after anthesis, the outer usually pinnatifid; pet typically deep pink, (2.5–)3–4.5 cm; $2n=28$. Native of Europe, rarely escaped from cult.

Rosa centifolia L., the cabbage-r., also occasionally escapes. It differs from *R. gallica* in its stouter, taller habit, coarser prickles, thinner, longer lfls, and nodding fls.

6. **Rosa pimpinellifolia** L.   Scotch r. Erect colonial shrub to 1 m, the stems armed with numerous slender, straight prickles mixed with stiff bristles; lfls 7–11, subglabrous, 0.5–1.5(–2) × 0.5–1 cm; fls solitary at the branch-tips, bractless; pedicel glabrous; hypanthium wide-mouthed, glabrous; pet 1–2(–2.5) cm, white or ochroleucous (sometimes more yellowish at base), seldom pink; hips purplish-black; $2n=28$. Native of Eurasia, occasionally escaped from cult. (*R. spinosissima,* in part.)

7. **Rosa eglanteria** L.   Eglantine; sweetbrier. Stems erect, 1–3 m, armed with stout, unequal flattened, hooked prickles, often with some acicles (slender, straight prickles) and glandular setae intermingled, especially on the flower-branches; foliage sweetly aromatic, with an apple-like odor; lfls 5–9, firm, 1–2.5 cm, broadly elliptic to ovate or subrotund, doubly serrate with gland-tipped teeth, ± pubescent and coarsely stipitate-glandular beneath or on both sides; fls 1–3; pedicels and hypanthium stipitate-glandular or glandular-pubescent, the hypanthium not much thickened around the orifice; pet 1–2 cm, pink; sep glandular-ciliate, erect or spreading, ± pubescent in fr, some or all of them pinnately lobed; styles short-hairy, shortly exserted; $2n=35$. Native of Europe, often escaped from cult. (*R. rubiginosa.*)

8. **Rosa micrantha** J. E. Smith.   Eglantine, sweetbrier. Much like no. 7; stems more arching, to 3.5 m; prickles subequal; sep soon deciduous; style glabrous; $2n=28$, 35, 42. Native of Europe, frequently escaped in our range.

Rosa tomentosa J. E. Smith, another European sp., is reported to be established on Prince Edward Island. It is much like no. 8, but the lvs have a resinous odor and are more tomentose beneath, and the stem often has straighter prickles.

9. **Rosa canina** L.   Dog-r. Arching shrub 1–3 m, the stems armed with stout, flattened, hooked prickles; stipules of the principal lvs linear, those of the uppermost lvs dilated, to 4 mm wide; lfls 5 or 7, ovate-elliptic, 2–4 cm, glabrous or with a few deciduous glands on the main veins beneath; pedicels and hypanthium usually glabrous; wall of the hypanthium conspicuously thickened around the very narrow orifice; fls 1–4, long-pedicellate; sep soon reflexed, then deciduous; pet 1.5–2.5(–3) cm, pink to white; styles shortly exserted; $2n=35$. Native of Europe, escaped from cult. here and there in our range.

10. **Rosa rugosa** Thunb.   Japanese r. Stems 1–2 m, densely prickly, the infrastipular prickles larger, decurved; young parts of the stem, young prickles, and prickle-bases densely tomentose; lfls (5)7 or 9, dark green and rugose above, 2.5–5 cm; pedicels glandular-bristly and pubescent; hypanthium smooth or nearly so; pet 3–5 cm; hips dark red, mostly depressed-globose and 2–3 cm thick, crowned with the persistent, erect sep; $2n=14$. Native of e. Asia, frequently escaped from cult., often maritime.

11. **Rosa palustris** Marshall.   Swamp-r. Much-branched shrub to 2 m; infrastipular prickles stout, decurved, 3–6 mm, the flattened base at least half as long; internodal prickles none; stipules very narrow, scarcely widened distally; lf-rachis softly pubescent; lfls commonly 7, narrowly elliptic to oblanceolate, softly pubescent on the midrib beneath, finely serrate, the teeth usually less than 2 mm wide and avg 0.5 mm high; fls solitary or in small corymbs; pedicel and hypanthium stipitate-glandular; sep soon spreading or reflexed, then deciduous; pet pink, 2–3 cm; hips red, only 7–12 mm thick; $2n=14$. Swamps, marshes, and streambanks; N.S. to Minn., s. to the Gulf of Mexico. Evidently hybridizes with *R. blanda,* and probably to a lesser extent with *R. carolina.*

12. **Rosa nitida** Willd.   New England r. Slender colonial shrub to 1 m; infrastipular and internodal prickles scarcely differentiated, the latter usually very abundant, all straight and slender, rising from a slightly flattened base rarely more than a fifth as long as the prickle; stipules usually widened distally, conspicuously glandular-ciliate or glandular-dentate; lf-rachis and lower lf surface commonly glabrous, sometimes sparsely villosulous; lfls 7 or usually 9, narrowly elliptic to oblong, finely serrate, the teeth mostly distinctly antrorse and avg ca 0.5 mm high; fls 1–3; pedicels and hypanthium stipitate-glandular; sep, pet and hips as in no. 11; $2n=14$. Moist or wet soil; Nf. and s. Que. to Conn.

13. **Rosa virginiana** Miller.   Virginia-r. Stout and branched, to 2 m, scarcely colonial; infrastipular prickles stout, usually somewhat decurved, the flattened base tending to be more than half as long as the prickle;

internodal prickles few and similar to the others, or none; stipules usually glandular-dentate, somewhat dilated above, the free part lanceolate to semiovate; lfls usually 7 or 9, glossy, oblong to oval or ovate, often more than half as wide as long, coarsely toothed, the teeth avg ca 1 mm high; fls solitary or few together on branches from old wood; pedicel and hypanthium stipitate-glandular; sep often conspicuously elongate, to 3 cm, with lanceolate foliaceous tip, soon spreading or reflexed, then deciduous; pet pink, 2–3 cm; hips red, 10–15 mm thick; $2n=28$. Moist or dry soil; Nf. to Pa. and Va., and inland irregularly to Mo. Hybridizes with *R. carolina*, *R. nitida*, and *R. palustris*. (*R. lucida*.)

14. **Rosa carolina** L.   Pasture-r. Colonial; stems simple or little-branched, rarely over 1 m, usually copiously armed with internodal prickles, these all about alike in form, but variable in form, straight, slender, and terete to the base, which is mostly less than half as long as the body; infrastipular prickles scarcely differentiated; stipules glandular-dentate to entire; lfls 3–7, oblong to oval or subrotund, often more than half as wide as long, coarsely toothed, in some forms glandular on the margin; fls usually borne singly on stems of the season; pedicel and hypanthium stipitate-glandular; sep attenuate into a linear tip, or rarely with a narrow foliaceous appendage, soon spreading or reflexed, then deciduous; pet pink, 2–3 cm; hips red, 8–12 mm thick; $2n=28$. Uplands woods, dunes, and prairies; Me. to Minn., s. to Fla. and Tex. (*R. housei; R. humilis; R. lyonii; R. obovata*.) Hybridizes extensively with *R. arkansana* (forming *R.* × *rudiuscula*), and *R. virginiana*, and perhaps less often with *R. acicularis* and *R. palustris*.

15. **Rosa arkansana** T. C. Porter.   Dwarf prairie-r. Colonial; only half-shrubby; stems under 1 m, usually densely prickly; prickles slender, straight, unequal, the infrastipular and internodal ones essentially alike; stipules pubescent, usually entire, or glandular-dentate toward the tip; lfls (7)9 or 11, 1–4 cm, firm, obovate or obovate-oblong, sharply serrate, very often pubescent beneath; fls corymbose, terminating the nearly herbaceous stems of the season and often also on short lateral branches from older stems; hypanthium and pedicel usually glabrous or nearly so; sep persistent, often becoming erect and connivent; pet pink (white) to deep rose, 1.5–3 cm; hips purplish or red, 10–15 mm thick; $2n=28$. Prairies and plains, or in open or brushy sites eastward; N.Y. to Alta., s. to D.C., Ind., Mo., Tex., and Colo. (*R. conjuncta; R. pratincola; R. suffulta*.)

16. **Rosa blanda** Aiton.   Smooth r. Colonial; stems to 1.5 or 2 m, unarmed or with few to many slender prickles toward the base, these not extending onto the flowering branches; stipules entire to glandular-dentate; lfls mostly 5 or 7, narrowly elliptic-oblong to oval or obovate, to 5 cm, coarsely toothed, especially above the middle; fls solitary or corymbose on lateral branches from stems of the previous year; pedicels and hypanthium glabrous; sep persistent, usually erect and connivent after anthesis; pet pink, 2–3 cm; hips red, 8–15 mm thick; $2n=14$. Dry woods, hills, prairies, and dunes; Que. to Man., s. to N.Y., Pa., Ind., and Mo. (*R. subblanda*.) The stipules, rachis, and lower lf-surface are usually softly hairy or tomentulose. A striking form found along the shores of the Great Lakes, with glabrous stipules and rachis, and small, firm, glabrous lfls, may be worthy of varietal recognition. *R. johannensis* Fern., occurring from Que. to Me. and n. N.Y., may represent another var., or possibly it may reflect hybridization with another sp. The lvs are glabrous or nearly so, and the persistent sep are reflexed.

17. **Rosa acicularis** Lindley.   Bristly r. Colonial; stems to 1(–2) m, usually densely beset with straight, slender, unequal prickles, even on the flowering lateral branches; stipules pubescent, glandular on the margin, when young densely covered with short-stipitate glands, as also the bracts; rachis usually pubescent and glandular; lfls 5 or 7(9), elliptic to ovate or obovate, 1.5–4.5(–8) cm, usually doubly serrate, often glandular; fls usually solitary on lateral branches from stems of the previous year; hypanthium and pedicel glabrous, or the pedicel seldom stipitate-glandular; pet pink or deep rose, 1.5–3 cm; sep persistent, becoming erect and connivent; hips 1–2 cm thick, dark blue or purplish; $2n=42$. Upland woods, hills, and rocky banks; N.B. and Que. to Alas. and Eurasia, s. to W.Va., Mich., nw. Ill., Io., and N.M. The Amer. plants, as here described, are var. *bourgeauiana* Crépin, or subsp. *sayi* (Schwein.) W. H. Lewis. (*R. bourgeauiana; R. sayi*.)

18. **Rosa woodsii** Lindley.   Western r. Colonial; stems to 1 m, appearing stiff and with crowded lvs, provided with straight or somewhat curved, slender infrastipular prickles 3–5 mm, and often with other stout or weak prickles as well; stipules rarely to 15 mm, densely stipitate-glandular, also glandular along the margin; rachis glandular, often also bristly; lfls 5 or 7, mostly 1–2 cm, elliptic or oval, sparsely glandular beneath, the teeth glandular on the longer margin; fls commonly corymbose on lateral branches from stems of the previous year; pedicel and hypanthium generally glabrous; sep persistent, often becoming erect; pet pink, 1.5–2.5 cm; hips red, mostly 6–12 mm thick; $2n=14$. Prairies and plains; Minn. to Mo., w. to Mack., e. Wash., e. Oreg., s. Calif., and n. Mex. Our plants, as here described, are var. *woodsii*. (*R. fendleri; R. macounii*.) The taller and laxer var. *ultramontana* (S. Wats.) Jepson, with larger lfls, is cordilleran. *R. woodsii* hybridizes with *R. blanda*.

19. **Rosa majalis** Herrm.   Cinnamon-r. Colonial; stems 1–2 m, armed with curved, broad-based infrastipular prickles and often also with a few internodal prickles; lfls 5 or 7, sparsely pubescent above, densely so beneath, as also the rachis; pedicel glabrous, short, often not surpassing the bract; hypanthium wide-mouthed, glabrous; fls often double; pet dark rose-color, 2–3 cm; hips red; $2n=14$. Native of Eurasia, often escaped from cult. (*R. cinnamomea*, misapplied.)

21. PRUNUS L.

21. PRUNUS L. Hypanthium cup-shaped, obconic, or urceolate; sep spreading or reflexed, usually soon deciduous; pet 5, white to pink or red, elliptic to obovate, spreading; stamens ca 20; pistil 1, simple, 2-ovulate, inserted at the bottom of the hypanthium and bearing a terminal style; fr a 1-seeded drupe, the exocarp fleshy or juicy, the endocarp (stone) hard; trees or shrubs with simple, serrate lvs, very often with a pair of large glands at the summit of the petiole, the fls conspicuous, umbellate or solitary from axillary buds or short lateral branches, or racemose and terminal; bark commonly with conspicuous horizontal lenticels, relatively smooth, or breaking up into smooth platelets. (*Amygdalus, Cerasus, Padus*.) 200, mainly N. Temp.

1 Fls and fr sessile or subsessile; ovary and fr velvety or tomentose; intr. (*Amygdalus*).
  2 Fls pink to dark pink; stone pitted; lvs lanceolate or lance-oblong . . . . 1. *P. persica.*
  2 Fls white or slightly pinkish; stone smooth; lvs broadly ovate to subrotund . . . . 2. *P. armeniaca.*
1 Fls and fr pedicellate; ovary (or at least the fr) glabrous.
  3 Cherries; stone globose or subglobose; fls in most spp. either in terminal racemes or in small clusters
    that are subtended by leafy green bracts arising from the same bud as the fls (except nos. 6
    and 7, which have the sep glabrous above).
    4 Infl distinctly racemose, with 20 or more fls, the pedicels much shorter than the axis; lvs below
      the raceme of ordinary size (*Padus*).
      5 Hypanthium glabrous within; pet ca 4 mm; stone smooth; native.
        6 Sep entire or inconspicuously glandular-erose, persistent under the fr; lf-teeth very short,
          appressed or incurved . . . . 3. *P. serotina.*
        6 Sep conspicuously glandular-erose, soon deciduous; lf-teeth triangular, salient or ascending
          . . . . 4. *P. virginiana.*
      5 Hypanthium pubescent within; pet 6–10 mm; stone sculptured; sep and lf-margins of no. 4;
        intr. . . . . 5. *P. padus.*
    4 Infl of few-fld umbel-like clusters, or of short racemes with not more than about 12 fls, the
      pedicels then much longer than the axis (*Cerasus*).
      7 Clusters of fls bractless or bracted only by the bud-scales.
        8 Sep glandular-serrate; fr nearly black; teeth at the middle of the ± cartilaginous-thickened
          lf-margin 1–4 mm apart, less than 0.5 mm high . . . . 6. *P. pumila.*
        8 Sep glandless; fr bright red; teeth at the middle of the unthickened lf-margin 0.5–1 mm
          apart, irregular in size . . . . 7. *P. pensylvanica.*
      7 Clusters of fls and frs ± leafy-bracted at the base.
        9 Pedicels arising from a common axis, the infl thus a short raceme . . . . 8. *P. mahaleb.*
        9 Pedicels wholly separate, the infl thus umbel-like.
          10 Sep entire; pet 9–15 mm; lvs 5–15 cm; small to middle-sized trees.
            11 Lvs persistently hairy beneath; proximal margin of the lf-teeth not much longer
              than the distal margin, the gland therefore well removed from the sinus . . . . 9. *P. avium.*
            11 Lvs soon glabrous; proximal margin of the lf-teeth very much longer than the distal
              margin, the gland therefore near the sinus . . . . 10. *P. cerasus.*
          10 Sep glandular on the margin; pet 5–7 mm; lvs 3–5 cm; bushy shrub . . . . 11. *P. fruticosa.*
  3 Plums; stone flattened to turgid, ± 2-edged; fls in small clusters (or solitary) that are not leafy-
    bracteate (although other buds from the same axil may produce lvs); sep pubescent or tomentose
    inside, at least at the base, except in no 20.
    12 Lf-teeth glandless, acute or acuminate (sometimes with a callous point); sep without marginal
      glands.
      13 Lvs ± obovate-oblong, abruptly acuminate; fr red or purple-red; pet mostly 8–15 mm.
        14 Sep glabrous on the lower side; petiole usually not glandular at the summit; lvs cuneate
          to obtuse at base; commonly colonial; widespread . . . . 12. *P. americana.*
        14 Sep pubescent on the lower side; petiole usually glandular at the summit; lvs ± broadly
          rounded at base; not colonial; southwestern . . . . 13. *P. mexicana.*
      13 Lvs lanceolate to subrotund, broadest near or below the middle, obtuse to acute or gradually
        acuminate; fr dark purple to black; pet 4–8 mm; eastern.
        15 Pedicels and lower surface of mature lvs distinctly pubescent; Atlantic seaboard sp. . . . . 14. *P. maritima.*
        15 Pedicels and mature lvs glabrous or nearly so; Alleghenian sp. . . . . 15. *P. alleghaniensis.*
    12 Lf-teeth gland-tipped (at maturity sometimes with only a minute scar marking the former
      presence of a gland); sep, except in no 16, with marginal glands.
      16 Fls mostly 4 or 5 per cluster, occasionally fewer; fr red to yellow; native.
        17 Lvs 3–6 cm, usually folded lengthwise; sep without marginal glands . . . . 16. *P. angustifolia.*
        17 Lvs mostly 5–13 cm, not folded; sep with marginal glands.
          18 Lvs less than half as wide as long; pet 4–7 mm.
            19 Fls opening before the lvs have expanded; lf-teeth very low and almost point-
              less, the gland next to the sinus . . . . 17. *P. munsoniana.*
            19 Fls opening when the lvs are half-grown; lf-teeth triangular, ascending, with a
              terminal gland . . . . 18. *P. hortulana.*
          18 Lvs more than half as wide as long; pet 10–15 mm . . . . 19. *P. nigra.*
      16 Fls commonly solitary or paired, rarely 3; fr dark blue to black; intr. 20. *P. domestica.*

1. Prunus persica (L.) Batsch. Peach. Small tree to 10 m; lvs lanceolate or lance-oblong, 8–15 cm; fls solitary, sessile or subsessile, usually pink and ca 3 cm wide, sometimes much smaller and nearly red, appearing in early spring before the lvs; fr subglobose, velvety, to 10 cm thick in cult. plants, usually much smaller and hard in wild plants; stone compressed, deeply pitted; $2n=16$. Native of China, occasionally escaped from cult. especially southward.

2. Prunus armeniaca L.   Apricot.   Spreading tree to 10 m; lvs broadly ovate to subrotund, abruptly acuminate, 5–10 cm, nearly as wide; fls solitary, sessile or nearly so, white or pale pinkish, 2–3 cm wide, appearing in early spring; fr subglobose, very thinly velvety, 3–5 cm thick; stone compressed, smooth; $2n=16$. Native of w. Asia, long in cult. and rarely adventive in our range.

3. Prunus serotina Ehrh.   Wild black cherry.   Tree to 25 m; bark aromatic, breaking up into small plates and appearing scaly-roughened; lvs firm, lanceolate to oblong or oblanceolate, 6–12 cm, with mostly 15 or more pairs of inconspicuous lateral veins, acuminate at the tip, acute or obtuse at base, finely incurved-serrate; racemes terminating leafy twigs of the current season, 8–15 cm; pedicels 3–6 mm; sep oblong or triangular, 1–1.5 mm, entire or sparsely glandular-erose, persistent under the fr, pet white, 4 mm, with subrotund blade; fr dark purple or black, 1 cm thick, edible when fully ripe; $2n=32$. Formerly a forest tree, now abundant as a weed-tree of roadsides, waste land, and forest-margins; N.S. to N.D. and sw. Ont., s. to Fla., Ariz., and Guatemala. May.

4. Prunus virginiana L.   Choke-cherry.   Tall shrub or small tree to 10 m; lvs thin, oblong to obovate, 5–12 cm, with mostly 8–11 pairs of fairly conspicuous lateral veins, obtuse to acute or abruptly short-acuminate above, obtuse to rounded at base, sharply serrate with slender ascending teeth; racemes terminating leafy twigs of the season, 6–15 cm; pedicels 5–8 mm; sep broadly triangular to semi-circular, 1–1.5 mm, conspicuously glandular-erose, deciduous soon after anthesis; pet white, 4 mm, with subrotund blade; fr dark red or nearly black, 8–10 thick, astringent but edible; $2n=12$. In a wide variety of habitats, from rocky hills and dunes to borders of swamps; Nf. to B.C., s. to N.C., Tex., and Calif. May, June. Ours is var. *virginiana*.

5. Prunus padus L.   European bird-cherry.   Tall shrub or small tree to 15 m; bark foetid; lvs oblong-obovate, 5–10 cm, short-acuminate, finely and sharply toothed, racemes loose and often drooping, 8–15 cm; pedicels 10–15 mm; hypanthium pubescent within; sep ovate-oblong, 2 mm, deeply erose; pet white, elliptic, 6–10 mm; fr inedible, nearly black, 6–8 mm; stone sculptured; $2n=32$. Native of Europe, now rarely cult. and occasionally established in woods, thickets, and roadsides.

6. Prunus pumila L.   Sand-cherry.   Low, diffusely branched, decumbent or prostrate shrubs, to 1(–3) m; lvs narrowly to broadly oblanceolate, varying to oblong, 4–10 cm, obtuse or acute, the margin firm or cartilaginous, finely and remotely glandular-serrate, long-cuneate to acute at base, glabrous, often glaucous beneath; fls in clusters of 2–4, on pedicels 4–12 mm; sep glandular-serrulate; pet elliptic to round-obovate, 4–8 mm; fr nearly black, subglobose, 1–1.5 cm thick, edible; $2n=16$. Four fairly well marked vars. may be recognized:

a Plants ± erect, often diffusely branched, 3–10(–30) dm.
   b Lvs oblanceolate, 10–18 mm wide and 3–6 times as long, long-tapering at base; sand dunes and
     sandy soil, especially on the shores of the Great Lakes; also inland in the Lake states .... var. *pumila*.
   b Lvs oblong or oblong-obovate, mostly 2–3 cm wide and 2–3 times as long; dry or rocky woods;
     Me. to Man., s. to Del., N.C., Ind., and n. Ill. (*P. susquehanae*) ....var. *cuneata* (Raf.) L. H. Bailey.
a Plants depressed, mostly 1–3(–6) dm; lvs oblanceolate, often narrowly so, to 2 cm wide.
   c Stems prostrate and mat-forming, with ascending or erect short branches; gravelly or sandy beaches
     and shores, especially in calcareous regions; e. Que. and N.B. to Ont., Mass., n. N.Y., n. N.J.,
     and e. Pa.; rep. from Wis. .... var. *depressa* (Pursh) Bean.
   c Stems merely decumbent or ascending, hardly mat-forming; dry sandy hills and rocky slopes; Minn.
     and nw. Io. to Kans., w. to N.D., Wyo., and Colo. (*P. besseyi*) .... var. *besseyi* (Bailey) Gleason.

7. Prunus pensylvanica L.f.   Pin-cherry.   Slender shrub or small tree to 15 m; bark smooth; lvs lanceolate or oblong-lanceolate to rarely oblong or obovate, 6–12 cm, commonly less than half as wide, long-acuminate, obtuse or rounded at base, glabrous from the beginning, or initially hairy only along the midrib beneath, finely and irregularly serrate, the gland near the sinus; petiole usually glandular at the summit; fls in umbel-like clusters of 2–5, on pedicels 1–2 cm; sep glabrous; pet white, 5–7 mm, villous-pubescent on the back near the base; fr red, juicy, acid, 6 mm thick; stone subglobose; $2n=16$, 32. Dry or moist woods and forest-clearings, often abundant after fires; Nf. and Lab. to Mack. and B.C., s. to Pa. and Minn., irregularly in the mts. to N.C. and Tenn., and elsewhere to Ill., Io., and Colo. Apr., May.

8. Prunus mahaleb L.   Mahaleb-cherry.   Shrub or tree to 10 m, with widely spreading branches; young twigs glandular-puberulent; lvs broadly ovate to orbicular, 3–7 cm, more than half as wide, abruptly short-acuminate, rounded to subcordate at base, finely serrate, the gland near the sinus; corymbs on short, leafy-bracted branches, 4–10-fld, with a distinct axis; pedicels 1–2 cm; cal and hypanthium glabrous; pet white, glabrous, 5–8 mm; fr dark red to black, bitter, 6 mm thick; stone subglobose; $2n=16$. Native of Europe, occasionally escaped from cult. in our range.

9. Prunus avium L.   Sweet cherry.   Cylindric or pyramidal tree to 20 m; lvs ovate-oblong to obovate, 8–15 cm, short-acuminate, persistently long-hairy beneath, chiefly along the main veins, coarsely and often doubly serrate, the teeth salient or ascending, blunt or retuse, bearing the gland at the summit; pedicels 2–5 cm; sep glabrous; pet 9–15 mm, glabrous; fr globose, 1.5–2.5 cm; stone subglobose; $2n=16$. Native of Eurasia, often escaped from cult. in our range, even appearing like a native.

10. Prunus cerasus L.   Sour cherry, pie-cherry.   Tree to 10 m, usually with broadly rounded crown; lvs ovate-elliptic to obovate, 5–9 cm, acute, soon glabrous beneath, serrate or doubly serrate, the teeth usually wider than high, rounded, with a gland near the sinus; fls several in a sessile umbel, on pedicels 2–3 cm; sep glabrous; pet 10–15 mm; fr globose, 1.5–2 cm, bright red, tart; stone subglobose; $2n=32$. A cultigen thought to be derived by hybridization of *P. avium* with *P. fruticosa*; occasionally escaped from cult. in our range.

11. Prunus fruticosa Pallas.   Ground-cherry. Bushy shrub to ca 1 m; lvs oblong-obovate, 3–5 cm, obtuse, finely serrate, the gland near the sinus; fls in small leafy-bracted clusters; sep glandular on the margin; pet 5–7 mm; fr globose, dark red, 1 cm; $2n=32$. Native of Eurasia, occasionally escaped from cult. in our range.

12. Prunus americana Marshall.   Wild plum. Shrub or small tree to 8 m, spreading from the roots and forming thickets; lvs obovate to obovate-oblong, usually somewhat hairy beneath, 6–10 cm, abruptly and sharply acuminate, acute to obtuse at base, sharply and coarsely (often doubly) serrate, the teeth spreading or antrorse, acuminate, 1–2 mm, glandless but tending to have a callous point; petiole mostly glandless; fls 2--4 in an umbel; pet white, 10–15 mm; sep often pubescent on the upper side, often distally toothed, glandless or with a few very obscure glands; fr red to yellow, glaucous, 2–3 cm thick; stone compressed; $2n=16$. Moist woods, roadsides and fence-rows; N.H. to Man. and Mont., s. to n. Fla. and Okla. May, June.

13. Prunus mexicana S. Wats.   Bigtree-plum. Tree to 12 m, not sprouting from the roots; lvs firm, oblong to oblong-obovate, 6–10 cm, abruptly acuminate, broadly rounded at base, coarsely and usually doubly serrate with broad, acuminate, ovate-triangular teeth 1–2 mm, slightly pubescent and somewhat rugose above, permanently soft-pubescent and ± reticulate-veiny beneath; petiole usually with a pair of distal glands; pedicels usually glabrous, 8–20 mm; hypanthium glabrous or sparsely pubescent; sep narrowly oblong, 2–3 mm, pubescent on both sides; pet white, 8–10 mm; fr purplish-red, glaucous, globose or ellipsoid, 2–3 cm long; stone turgid, obovoid to subglobose. Rocky open woods; Mo. and s. Ill. to Ala. and Tex., and n. Mex., with outlying stations to O., Ky. and S.D. Apr. (*P. americana* var. *lanata*.)

14. Prunus maritima Marshall.   Beach-plum. Straggling shrub 1–2.5 m, rarely arborescent and to 4 m; lvs firm, oblong or elliptic, 4–7 cm, half as wide, obtuse or acute, broadly cuneate to obtuse at base, finely serrate with broadly triangular to semicircular, abruptly acuminate teeth, rugose above, finely and rather roughly hairy beneath; pedicels 5–10 mm, finely hairy; hypanthium pubescent; sep oblong, 2.5 mm, finely hairy; pet obovate, 5–6 mm, cuneate at base; fr glaucous, purplish-black to sometimes red or yellow, subglobose, 1.5 cm; stone turgid, truncate at base; $2n=16$. Dunes and sandy soil near the coast; N.B. to Md., occasionally extending some distance inland; c. Mich. May. (*P. gravesii*, a single unusual clone-colony in Conn., with small, subrotund lvs.)

15. Prunus alleghaniensis T. C. Porter.   Allegheny-plum. Straggling, often thorny, colonial shrub or low tree to 4 m; lvs scarcely out of the bud at anthesis, at maturity lanceolate to oblong, 5–8 cm, gradually acute or acuminate above, acute to obtuse at base, glabrescent, sharply serrate, the teeth triangular, acuminate, 0.5–1 mm on the distal side, pedicels 5–10 mm, glabrous; hypanthium minutely puberulent or nearly glabrous; sep oblong-ovate, 2 mm, pubescent on both sides but the back often only inconspicuously so; pet white, 6–8 mm; fr dark purple, subglobose, 12 mm thick; stone turgid. Dry, rocky woods, sometimes on shale-barrens; Pa. and w. Md. to W.Va., with outlying stations to Conn., w. Va., and e. Tenn. Rare. Apr., May; before the lvs.

16. Prunus angustifolia Marshall.   Chickasaw-plum. Much-branched, often thorny shrub or small tree to 5 m, forming thickets; lvs scarcely developed at anthesis, at maturity trough-shaped, lanceolate to narrowly elliptic or oblanceolate, 3–6 cm, acute or short-acuminate, smooth and shining above, very finely serrulate, the teeth 12–20 per cm, bearing a gland near the sinus; pedicels (3–8 mm), hypanthium, and outer side of the sep smooth or nearly so, the inner side of the sep hairy especially near the base; pet 4–6 mm; fr subglobose, red, not glaucous, 1–1.5 cm thick; stone turgid. Usually in sandy or sterile soil, open woods, thickets, and fence-rows; chiefly from Va. to Fla., w. to Kans. and Tex., but also irregularly to s. N.J., s. O., Ind., Ill., and Neb. Apr.

17. Prunus munsoniana Wight & Hedrick.   Wildgoose-plum. Small tree; lvs lanceolate or oblong-lanceolate, 5–10 cm, gradually acuminate, obtuse to rounded at base, hairy beneath at least along the midvein, very finely serrate, the teeth projecting less than 0.5 mm and bearing a gland next to the sinus; pedicels glabrous, 8–12 mm; sep glandular on the margin, pubescent above, often also below; pet white, 4–7 mm; fr red, 2 cm; $2n=16$. Moist woods and thickets; chiefly in Mo. and e. Kans., but irregularly to O., Ky., Tenn., Miss., La., and Tex. Apr., before the lvs have expanded.

18. Prunus hortulana L. H. Bailey.   Hortulan plum. Tree to 10 m; lvs oblong-lanceolate to oblong, 8–13 cm, long-acuminate, obtuse to rounded at base, sparsely hairy beneath or eventually subglabrous, finely serrate with triangular ascending teeth bearing a terminal gland; pedicels glabrous, 8–15 mm; sep glandular on the margin, pubescent above, at least at base, glabrous below; pet 5–7 mm; fr red, varying to yellow, 2–3 cm; $2n=16$. Moist woods and thickets; chiefly in Mo. and Ill., but also irregularly to s. O., sw. Va., Tenn., n. Ark., ne. Okla., and e. Kans. Apr., May, the fls opening when the lvs are half-grown.

19. Prunus nigra Aiton.   Canada-plum. Small tree to 10 m; lvs obovate to broadly oblong-obovate, 7–12 cm, abruptly acuminate, broadly cuneate to rounded or subcordate at base, coarsely and often doubly serrate with irregular, triangular-ovate, rather blunt, gland-tipped teeth often 2 mm high, ± hairy beneath, at least in the vein-axils; fls in clusters of 3 or 4, on reddish pedicels 1–2 cm; sep glandular on the margin, pubescent above, glabrous beneath; pet 10–15 mm; fr ellipsoid, red, varying to yellow, 2–3 cm; $2n=16$. Moist woods and thickets; N.S. (intr.) and Me. to Man., s. to Conn., N.Y., n. O., Ind., Ill., and ne. Io. May.

20. Prunus domestica L.   Much-branched shrub or small tree, usually with thorny branches; lvs oblanceolate to obovate, 4–7 cm, obtuse to broadly rounded above, narrowed to rounded at base, hairy beneath, sometimes eventually glabrous; fls only 1 or 2(3) in a cluster, 2 cm wide; fr dark blue to black, 2–3 cm, the stone subglobose, scarcely keeled, adherent to the flesh; $2n=48$. A European cultigen, thought to be derived by hybridization of *P. cerasifera* with *P. spinosa,* occasionally escaped from cult. along roadsides and fence-rows in our range. Most of our plants are ssp. *institia* (L.) C.K. Schneider, the bullace or Damson-plum, as described above. (*P. institia.*) Less often we have ssp. *domestica,* the common plum, more arborescent, generally unarmed, with larger fr 4–7 cm, and a compressed, keeled stone that often separates from the flesh. *P. spinosa* L., the blackthorn or sloe, is also sometimes found in our range. It is smaller and spinier than *P. domestica,* with smaller lvs (2–4 cm), and smaller, inedible fr 1–1.5 cm.

22. PYRUS L. Apple, crab-apple, pear. Hypanthium globose to obovoid; sep 5, ascending to spreading or recurved; pet 5, pink to white, elliptic to obovate, short-clawed; stamens 15–50, shorter than the pet; ovary inferior, 3–5-locular, the styles separate or connate below; fr a fleshy pome, each locule normally with 2 seeds; trees or shrubs, sometimes thorny, with simple, alternate, toothed or lobed lvs and large fls in the spring in simple umbels or umbel-like clusters on dwarf lateral branches (fr-spurs); $x=17$. Our native spp. hybridize with each other and with *P. malus,* and many specimens are hard to identify. Many cult. crab-apples are of hybrid origin. 60, N. Temp.

Chaenomeles speciosa (Sweet) Nakai, the Japanese quince, rarely escapes from cult. It is a thorny shrub to 2(–3) m, with lustrous, oblong or elliptic, finely serrate lvs 4–8 cm, and red, pink, or white fls 4–5 cm wide. (*C. lagenaria.*)

1 Lvs involute or convolute in bud, never lobed; not thorny; cult. and escaped.
  2 Styles separate to the base; mouth of the hypanthium closed around the styles; lvs involute in bud;
    anthers red; fr with stone-cells .... 1. *P. communis.*
  2 Styles connate at base; mouth of the hypanthium open; lvs convolute in bud; anthers yellow; fr
    without stone-cells .... 2. *P. malus.*
1 Lvs folded in bud, often lobed; stems tending to be thorny; native, except no. 3.
  3 Cal deciduous; styles 3–4; fr scarcely 1 cm; intr. .... 3. *P. sieboldii.*
  3 Cal persistent; styles 5; fr 1.5–3.5 cm; native.
    4 Hypanthium and pedicels glabrous or sparsely villous; lvs usually glabrous.
      5 Lvs broadly ovate to broadly lanceolate, sharply acute or acuminate .... 4. *P. coronaria.*
      5 Lvs lanceolate to elliptic or oblong, obtuse or subacute .... 5. *P. angustifolia.*
    4 Hypanthium and pedicels tomentose; lvs permanently pubescent beneath .... 6. *P. ioensis.*

1. Pyrus communis L. Pear. Pyramidal tree to 15 m; lvs elliptic to subrotund, finely and rather bluntly serrulate, involute in bud, glabrous at maturity; fls white, 2.5–3 cm wide; hypanthium nearly closed by a ring of tissue at the summit, surrounding the free styles; anthers red; cal persistent; fr pyriform to obovoid, 6–15 cm, containing stone-cells; $2n=34$. Eurasian cultigen, occasionally escaped.

P. pyrifolia (Burm. f.) Nakai, the Chinese pear, differing in its more sharply serrate lvs, slightly larger fls, and subglobose fr, seldom escapes in our range.

2. Pyrus malus L. Apple. Widely spreading tree to 15 m; lvs elliptic to ovate, finely serrate, convolute in bud, permanently pubescent beneath; fls white, tinged with pink, 3 cm wide; hypanthium densely tomentose, open at the mouth; anthers yellow; cal persistent; fr globose or ovoid-globose, 6–12 cm; $2n=34$. Eurasian cultigen, occasionally escaped. Probably hybridizes with native crab-apple.

P. prunifolia Willd., the Chinese crab, and *P. baccata* L., the Siberian crab, occasionally escape from cult. Both differ from *P. malus* in their glabrous or subglabrous hypanthium, pedicels, and lvs, and small, tart fr. *P. prunifolia* has pubescent twigs, persistent cal, and fr ca 2 cm. *P. baccata* has glabrous twigs, deciduous twigs, deciduous cal, and fr ca 1 cm.

3. Pyrus sieboldii Regel. Toringo crab. Small tree with spreading branches; twigs pubescent; lvs pubescent on both sides, later glabrous above, ovate or elliptic, acuminate, 2.5–6 cm, serrate, those of the long shoots also 3(5)-lobed; pedicels and cal pubescent; fls 2 cm wide, pink or very deep rose in bud, finally nearly white; styles 3–4 (unique among our spp.); fr 6–8 mm, with deciduous cal; $2n=34, 58, 85$. Native of Japan, escaped and becoming established in Mass.

4. Pyrus coronaria L. Sweet crab. Tall shrub or low tree with thorny branches, rarely to 10 m; lvs ovate to triangular-ovate to sometimes broadly lanceolate, acute or acuminate, mostly with rounded to cordate base, 4–10 cm, glabrous beneath or sparsely villous when young, serrate, or commonly with a few triangular lobes near the widest part; hypanthium and cal glabrous or sparsely villous; fls at first pink, fading white; anthers pink or salmon-color; fr subglobose, greenish, 2–3 cm; $2n=68$. Woods and thickets; N.Y. to Mich., Ill., and occasionally Mo., s. irregularly to n. Ga., n. Ala., and ne. Ark. (*Malus bracteata; M. coronaria; M. glaucescens; M. lancifolia.*) Hybrids of this with no. 5 or with *P. malus* are called *P.* × *platycarpa* L. H. Bailey.

5. Pyrus angustifolia Aiton. Southern crab. Much like no. 4; lvs firm, oblong to elliptic, 3–7 cm, obtuse to subacute, tapering to the base, glabrous or sparsely pubescent beneath when young, those of the fl-spurs unlobed, those of rapidly growing shoots often with several shallow lobes; $2n=34, 68$. Woods and thickets; mainly from N.C. to n. Fla. and w. to La., but occasionally to N.J., s. O., w. Ky., Ark., and e. Tex. (*Malus a.*)

6. Pyrus ioensis (A. Wood) L. Bailey. Prairie crab. Tall shrub or low tree to 10 m, usually thorny; twigs tomentose; lvs firm, ovate-oblong to broadly elliptic, 6–10 cm, coarsely serrate and usually also shallowly lobed, acute or short-acuminate, usually persistently pubescent beneath; hypanthium and pedicels densely tomentose; fls 3.5–4 cm wide; anthers pink or salmon-color; fr subglobose, green, 2–3 cm; $2n=34$. Woods and thickets; n. Ind., s. Wis., and se. Minn. to S.D., Neb., Tex., and La.; disjunct ? in Mich. (*Malus i.*) A hybrid with *P. malus* is *P.* × *soulardii* L. H. Bailey.

23. ARONIA Medikus, nom. conserv. Chokeberry. Hypanthium broadly obconic; sep 5, spreading at anthesis, usually glandular on the margin; pet 5, broadly obovate or rotund, short-clawed, spreading; stamens usually 20; ovary densely woolly on the summit; styles 5, connate at base, long-persistent; fr a small, red to black pome; shrubs with simple, alternate, glandular-serrate lvs bearing a row of slender glands along the midvein on the upper side, and with rounded or flattened compound clusters of small white fls. Often submerged (along with *Sorbus*) in the closely allied genus *Pyrus*. Only our 2 spp.

In addition to the two following diploid, sexual spp., there is a series of ± self-perpetuating hybrids and hybrid-descendants, partly apomictic, possibly in part tetraploid, with dark purple or purple-black frs and with the herbage usually less pubescent than in *A. arbutifolia,* often ± glabrate; the lvs do not turn red. Such plants may occur with or without one or both of the parental species. This variable, partly stabilized population, which appears to be augmented by continuing hybridization, may be called *A. prunifolia* (Marshall) Rehder. (*A. atropurpurea; Pyrus floribunda.*)

1 Frs bright red; lower surfaces of the lvs generally ± densely hairy .... 1. *A. arbutifolia.*
1 Frs inky black; lvs generally glabrous or nearly so .... 2. *A. melanocarpa.*

1. Aronia arbutifolia (L.) Elliott. Red c. Rhizomatous, colonial shrub (0.2–)1–2(–4) m; twigs and pedicels rather densely villous or tomentose, varying to rarely glabrous; petioles 3–10 mm; lvs elliptic or oblong-lanceolate to obovate, 3–7 cm, acute to short-acuminate, ± densely tomentose beneath, varying to rarely glabrous (the glabrous plants sometimes occurring well beyond the range of no. 2), turning bright red in the fall; sep usually conspicuously glandular; pet 4–6 mm; fr bright red, 4–10 mm thick, maturing late and persisting into winter; $2n=34$. Bogs, swamps, and wet woods, less often in dry soil; Nf. to Fla. and Tex., southward on and near the coastal plain, but also in the mts., extending to Ky. and W.Va. Apr.–July, fr Sept.–Nov. (*Pyrus a.*)

2. Aronia melanocarpa (Michx.) Elliott. Black c. Much like no. 1, but the twigs, pedicels, and lower surfaces of the lvs generally glabrous or nearly so, varying to seldom somewhat more evidently hairy; lvs not turning bright red; fr black, ripening earlier and soon withering, but sometimes persistent; $2n=34$. Similar habitats; Nf. and s. Lab. to n. Ga. and Ala., w. to e. Minn., ne. Io., and se. Mo., widely overlapping the range of no. 1, especially northward, but wanting from the coastal plain and adjacent piedmont. May–July, fr Aug.–Oct. (*Pyrus m.*)

24. SORBUS L. Mountain-ash. Hypanthium obconic; sep triangular, ascending; pet 5, obovate to orbicular, spreading; stamens 15–20; pistils in our spp. 2–4, half-inferior, ± separate above, the styles distinct; fr a small orange-red pome, each locule with 1 or 2 elongate, flattened seeds; trees or shrubs with odd-pinnate lvs (lfls in ours 11–17) and numerous white fls in branched, round or flattened clusters. Often submerged in *Pyrus*. 100, N. Temp.

1 Twigs, lvs, branches of infl, and hypanthium ± glabrous; winter-buds glutinous; native.
  2 Lfls ± long-acuminate, mostly 3–5 times as long as wide; fr 4–7 mm thick .... 1. *S. americana.*
  2 Lfls abruptly short-acuminate or merely acute, 2–3(–3.5) times as long as wide; fr 7–10 mm thick
  .... 2. *S. decora.*
1 Twigs, lower surface of the lfls, branches of the infl, and hypanthium all ± densely white-villous;
  winter-buds white-villous, not glutinous; intr. .... 3. *S. aucuparia.*

1. Sorbus americana Marshall. American m.-a. Shrub or tree to 10(–15) m; twigs glabrous or nearly so; winter-buds glutinous, with glabrous or sparsely ciliate scales; lfls lanceolate to narrowly oblong, gradually acuminate, 5–9 cm, mostly 3–5 times as long as wide, sharply serrate, paler and usually glabrous beneath; infl 6–15 cm wide; hypanthium and sep glabrous; pet obovate, 3–4 mm, narrowed to the base, equaling or longer than the stamens; fr 4–7 mm thick; $2n=34$. In moist or wet soil; Nf. to Minn., s. to Pa. and n. Ill., and in the mts. to n. Ga. May, June. (*Pyrus a.*)

2. Sorbus decora (Sarg.) C. K. Schneider. Showy m.-a. Shrub or tree to 10 m; winter-buds glutinous, the principal scales glutinous on the back, the inner usually conspicuously brown-ciliate; lfls oblong or oblong-elliptic, abruptly short-acuminate or merely acute, 3.5–8 cm, (2–)2.5–3(3.5) times as long as wide, sharply serrate, paler and glabrous or sparsely pilose beneath; infl 6–15 cm wide; hypanthium glabrous or very sparsely pilose; pet orbicular, 4–5 mm, about equaling the stamens; fr 7–10 mm thick; $2n=68$. Moist or dry, often rocky soil; Lab. to Minn., s. to Conn., N.Y., n. O., n. Ind., and ne. Io. May, June. (*Pyrus d.; Sorbus subvestita.*)

3. Sorbus aucuparia L. European m.-a.; rowan-tree. Tree to 10 m, the younger parts ± white-villous, not glutinous; lfls oblong, 3–5 cm, acute to obtuse, sharply or bluntly serrate; pet orbicular, 4–5 mm, about equaling the stamens; fr ca 1 cm thick; $2n=34$. Native of Europe, often escaped from cult. and even appearing as native in our range. (*Pyrus a.*)

25. CRATAEGUS L. Hawthorn. Fls perfect, regular, in compound or nearly simple cymes, or rarely solitary; pet 5, white, rarely pink or red; sep 5, entire or glandular-serrate; stamens 5–25, arranged alternately in 1–5 rows; ovary inferior or free at the tip, the 1–5 carpels with as many persistent free styles; fr a small, globose to ovoid or pyriform pome, usually red or reddish, sometimes yellow, greenish, or blue-black, with 1–5 bony, usually 1-seeded nutlets; small trees or shrubs with usually thorny and often flexuous branches; lvs simple, deciduous, toothed and often also lobed, those of the flowering branches (the floral lvs) often unlike those of strictly vegetative branches; $x=16$, 17. Our spp. all fl in the spring and fr in the fall. 100+, N. Temp. and Andes.

The spp. of *Crataegus* characteristically occur in disturbed sites or seral communities, such as pastures, edges of the forest, open second-growth woods, and thickets along streams. It has not been considered necessary to repeat these comments on the habitat for each species in the following text.

The taxonomy of *Crataegus* is complicated by hybridization, polyploidy, aneuploidy, and apomixis. Several hundred species have been described from our range alone. A proper understanding of the group will require a thorough biosystematic study; the treatment here presented is rough and more than ordinarily tentative. Some names suspected or believed to apply to hybrids or plants of hybrid origin are listed before the key to accepted species.

C. × *anomala* Sarg. = *C. coccinea* × *mollis.*
C. × *apiomorpha* Sarg. = *C. coccinea* × *flabellata.*
C. × *ardua* Sarg. = *C. punctata* × *succulenta.*
C. × *aulica* Sarg. = *C. coccinea* × *dilatata.*
C. × *celsa* Sarg. = *C. punctata* × *succulenta.*
C. × *chadsfordiana* Sarg. = *C. pruinosa* × *succulenta.*
C. × *collicola* Ashe = *C. crus-galli* × *punctata.*
C. × *corusca* Sarg. = *C. coccinea* × *mollis.*
C. × *danielsii* Palmer = *C. crus-galli* × *punctata.*
C. × *densiflora* Sarg. = *C. flabellata* × *rotundifolia.*
C. × *desueta* Sarg. = *C. brainerdii* × *punctata.*
C. × *disperma* Ashe = *C. crus-galli* × *punctata.*
C. × *divida* Sarg. = *C. brainerdii* × *succulenta.*
C. × *durobrivensis* = *C. pruinosa* × *punctata.*
C. × *ellwangeriana* Sarg. = *C. coccinea* × *mollis.*
C. × *formosa* Sarg. = *C. flabellata* × *pruinosa.*
C. × *fretalis* Sarg. = *C. coccinea* × *flabellata.*
C. × *glareosa* Ashe = *C. pruinosa* × *succulenta.*
C. × *haemocarpa* Ashe = *C. flabellata* × *pruinosa.*
C. × *harryi* Sarg. = *C. brainerdii* × *punctata.*
C. × *hirtiflora* Sarg. = *C. mollis* × *punctata.*
C. × *ideae* Sarg. = *C. brainerdii* × *rotundifolia.*
C. × *illecebrosa* Sarg. = *C. coccinea* × *coccinioides.*
C. × *immanis* Ashe = *C. pruinosa* × *rotundifolia.*
C. × *improvisa* Sarg. = *C. brainerdii* × *coccinea.*
C. × *incaedua* Sarg. = *C. calpodendron* × *punctata.*
C. × *integriloba* Sarg. = *C. punctata* × *succulenta.*
C. × *kellermanii* Sarg. = *C. pruinosa* × *punctata.*
C. × *kelloggii* Sarg. = *C. mollis* × *rotundifolia.*
C. × *kennedyi* Sarg. = *C. brainerdii* × *pruinosa.*
C. × *kingstonensis* Sarg. = *C. brainerdii* × *coccinioides.*
C. × *knieskerniana* Sarg. = *C. coccinea* × *flabellata.*
C. × *laetifica* Sarg. = *C. crus-galli* × *succulenta.*

C. × *laneyi* Sarg. = *C. brainerdii* × *succulenta.*
C. × *laurentiana* Sarg. = *C. rotundifolia* × *succulenta.*
C. × *lecta* Sarg. = *C. mollis* × *pruinosa.*
C. × *lettermanii* Sarg. = *C. mollis* × *punctata.*
C. × *littoralis* Sarg. = *C. intricata* × *pruinosa.*
C. × *locuples* Sarg. = *C. mollis* × *pruinosa.*
C. × *lucorum* Sarg. = *C. coccinea* × *flabellata.*
C. × *mansfieldensis* Sarg. = *C. coccinea* × *punctata.*
C. × *media* Sarg. = *C. flabellata* × *pruinosa.*
C. × *membranacea* Sarg. = *C. pruinosa* × *succulenta.*
C. × *menandiana* Sarg. = *C. punctata* × *succulenta.*
C. × *merita* Sarg. = *C. coccinea* × *flabellata.*
C. × *neobaxteri* Sarg. = *C. punctata* × *rotundifolia.*
C. × *nitida* (Engelm) Sarg. = *C. crus-galli* × *viridis.*
C. × *nitidula* Sarg. = *C. punctata* × *rotundifolia.*
C. × *nuda* Sarg. = *C. crus-galli* × *succulenta.*
C. × *pausiaca* Ashe = *C. crus-galli* × *punctata.*
C. × *persimilis* Sarg. = *C. crus-galli* × *succulenta.*
C. × *pilosa* Sarg. = *C. intricata* × *pruinosa.*
C. × *puberis* Sarg. = *C. flabellata* × *punctata.*
C. × *putata* Sarg. = *C. pruinosa* × *succulenta.*
C. × *randiana* Sarg. = *C. brainerdii* × *flabellata.*
C. × *rotundata* Sarg. = *C. pruinosa* × *rotundifolia.*
C. × *rubrocarnea* Sarg. = *C. brainerdii* × *flabellata.*
C. × *shirleyensis* Sarg. = *C. brainerdii* × *punctata.*
C. × *silvestris* Sarg. = *C. pruinosa* × *punctata.*
C. × *spatiosa* Sarg. = *C. pruinosa* × *succulenta.*
C. × *vailiae* Britton = *C. calpodendron* × *uniflora.*
C. × *vittata* Ashe = *C. coccinea* × *flabellata.*
C. × *websteri* Sarg. = *C. brainerdii* × *calpodendron.*
C. × *whitakeri* Sarg. = *C. calpodendron* × *mollis.*
C. × *xanthophylla* Sarg. = *C. coccinea* × *flabellata.*

1 Veins of the lvs running to the sinuses as well as to the points of the lobes; fls 0.7–1.3 cm wide.
  2 Styles and nutlets mostly 3–5; native, mostly southern.
    3 Cal tending to be deciduous from the fr; lvs acute or acuminate, often cordate at base, 2–5 cm
      wide .... 1. *C. phaenopyrum.*
    3 Cal persistent on the fr; lvs rounded to acutish at the tip, not cordate, often less than 2 cm wide.
      4 Infl glabrous; lvs glabrous, the floral ones mostly narrowly obovate, 0.6–1.5 cm wide, not
        much lobed .... 2. *C. spathulata.*
      4 Infl villous; lvs densely hairy when young, the floral ones mostly deltoid or broadly ovate,
        evidently lobed, usually over 1.5 cm wide .... 3. *C. marshallii.*
  2 Style and nutlet usually 1, seldom 2 or 3; escaped from cult. .... 4. *C. monogyna.*
1 Veins of the lvs running only to the points of the lobes or of the larger teeth; fls often larger.
  5 Fr purplish-black; thorns 1–2.5 cm; nutlets pitted on the inner side .... 5. *C. douglasii.*
  5 Fr red to yellow or green; thorns often longer; nutlets with or without pits.
    6 Nutlets mostly 2–3, rounded at the ends, with a deep pit on the inner side.
      7 Young twigs glabrous to nearly so; lvs 3–6 × 2–5 cm .... 6. *C. succulenta.*
      7 Young twigs tomentose or villous; lvs 5–9 × 4–8 cm .... 7. *C. calpodendron.*
    6 Nutlets not pitted (sometimes slightly pitted in *C. brainerdii* but then 2–5 and acute).
      8 Fls solitary or rarely 2–3 together; shrub to 2 m .... 8. *C. uniflora.*
      8 Fls in compound (rarely simple) cymes; plants often arborescent.
        9 Floral lvs mostly narrowed to an acute or acutish to attenuate base.
          10 Floral lvs lobeless or nearly so, mostly obovate to oblong-elliptic.
            11 Lvs glossy above; nutlets usually 1 or 2(–5); fr remaining hard and dry.
              12 Lvs, twigs and infl glabrous or nearly so; widespread .... 9. *C. crus-galli.*
              12 Lvs, twigs and infl hairy at least when young; southern .... 10. *C. berberifolia.*
            11 Lvs dull above; nutlets 3–5; fr often becoming mellow or succulent.
              13 Lvs thin, the veins obscure above; fr 0.5–0.8 cm thick .... 11. *C. viridis.*
              13 Lvs thick and firm, the veins impressed above; fr 0.8–1.5 cm thick.
                14 Lvs larger, the floral ones mostly 2.5–6 cm; widespread .... 12. *C. punctata.*
                14 Lvs smaller, the floral ones mostly 1.5–3 cm; southern .... 13. *C. michauxii.*
          10 Floral lvs usually lobed, mostly elliptic or rhombic-ovate or rarely obovate.
            15 Petiole and base of blade conspicuously glandular; fr (especially in no. 13) usually
              yellowish or bronze-green, remaining hard and dry.
              16 Floral lvs usually not over 3 cm, highly variable, sometimes lobed above the
                middle; southern .... 13. *C. michauxii.*
              16 Floral lvs mostly 3–5 cm, more uniform and lobed nearly to the base, seldom
                nearly lobeless; widespread .... 14. *C. intricata.*
            15 Petiole and blade eglandular or nearly so; fr red or rarely yellow, often becoming
              mellow or succulent; mostly northern.
              17 Lvs rounded to obtuse or acutish at the tip; nutlets not pitted .... 15. *C. chrysocarpa.*
              17 Lvs usually abruptly acuminate; nutlets tending to have a shallow pit on the
                inner surface .... 16. *C. brainerdii.*
        9 Floral lvs mostly broad at the rounded or truncate base, usually lobed or incised.
          18 Plants relatively strongly pubescent, the fr hairy at least at the ends, the mature lvs
            hairy beneath at least along the veins, the infl tomentose; filaments much shorter
            than the pet .... 17. *C. mollis.*
          18 Plants less pubescent or glabrous, the fr glabrous or villous, the infl glabrous or villous, the mature
            lvs generally glabrous; filaments nearly as long as the pet.
            19 Fls large, mostly 2.2–2.6 cm wide; vegetative lvs often cordate.
              20 Lvs essentially glabrous from the first, crisped at the edges .... 18. *C. coccinioides.*
              20 Lvs short-hairy at least on the upper side when young, not crisped .... 19. *C. dilatata.*
            19 Fls smaller, not over 2 cm wide; vegetative lvs rounded to truncate or subcordate
            at base.
              21 Sep usually deeply glandular-serrate .... 20. *C. coccinea.*
              21 Sep entire or sometimes slightly glandular-serrate.
                22 Fruiting cal prominent and usually elevated; young lvs usually glabrous ..... 21. *C. pruinosa.*
                22 Fruiting cal small and closely sessile; young lvs with short hairs above ..... 22. *C. flabellata.*

1. **Crataegus phaenopyrum** (L.f.) Medikus. Washington-thorn. Tree to 12 m; lvs mostly ovate or cordate, acuminate, dark green and glossy above, glabrous or nearly so, 2–6 × 2–5 cm, coarsely and irregularly serrate and with 1–4 pairs of lateral lobes, the basal pair of lobes the largest, the others very shallow, the veins running to the sinuses as well as to the points of the lobes; petiole 1–2 cm, eglandular; fls 1–1.3 cm wide, in compound glabrous cymes; sep short-deltoid, acute or apiculate, tending to fall from the fr; fr 4–5 mm thick, bright scarlet; nutlets 3–5. Both native and as an escape from cult.; Pa. to n. Fla., w. to Ill. and Mo.

2. **Crataegus spathulata** Michx. Littlehip-h. Arborescent shrub or more often a tree to 8 m; lvs narrowly obovate, 0.5–3 × 0.6–1.5 cm, glabrous, dark green and glossy above, attenuate at the base into a short winged petiole, crenate or crenately lobed only near the tip, the veins running to the sinuses as well as to the points of the lobes; fls 0.7–1 cm wide, in compound glabrous cymes; fr 4–6 mm thick, red, with a broad low persistent cal, thin dry flesh, and 3–5 nutlets. Usually in moist or fertile soil; s. Va. to Fla. and Tex., and n. in the Mississippi Valley to s. Mo. and s. Ill.

3. **Crataegus marshallii** Eggl. Parsley-h. Young twigs villous, later glabrate; lvs broadly ovate to deltoid, 0.5–5 cm long and often as wide or wider, deeply and laciniately lobed, the lobes coarsely toothed or again cleft near the tip, villous above and thickly gray-pubescent beneath when young, glabrous or nearly so at

maturity; petiole villous when young, often equaling or longer than the blade; fls 1–1.2 cm wide, in villous compound cymes; fr 4–8 mm thick; plants otherwise much like no. 2. S. Va. to Fla. and Tex., n. in the Mississippi Valley to se. Mo. and s. Ill.

4. Crataegus monogyna Jacq. Oneseed-h. Arborescent shrub or tree to 12 m; thorns 1–2 cm, lateral on the twigs and also terminating short stout branches; lvs ovate to rhombic or obovate, cuneate to truncate at base, glabrous, or sometimes slightly hairy beneath when young, dark green, deeply cleft with 3–5(7) lobes, the veins running to the sinuses as well as to the points of the lobes; petioles from ¼ to fully as long as the blade; fls 0.9–1.2 cm wide, in glabrous or slightly villous compound cymes; sep deltoid, entire; fr red, 5–8 mm thick, with thin flesh and usually a single nutlet. Native of Europe and w. Asia, often escaped from cult. in our range. (*C. oxyacantha,* a rejected name.)

C. laevigata (Poiret) DC., English h., is also rarely found as an escape. It has 2–3 styles and nutlets, and less deeply cut lvs; red-fld forms are common. (*C. oxyacantha,* misapplied.)

5. Crataegus douglasii Lindl. Black h. Tree to 12 m, or sometimes an arborescent shrub; twigs glabrous, unarmed or with scattered stout thorns 1–2.5 cm; lvs oblong-obovate to elliptic, often broadly so, mostly 2–4 × 1.5–3 cm and with 2–4 pairs of small, shallow, often irregular lateral lobes, or larger and more deeply cleft on vegetative shoots, firm, dark green and glossy above, essentially glabrous; fls 1–1.3 cm wide, in mostly 5–12-fld glabrous compound cymes; fr 0.8–1 cm thick, dark wine-color to black when fully ripe, succulent; nutlets 3–5, with a large pit on the inner surface. Local near Lake Superior and Lake Huron in ne. Minn., n. Mich., and Ont.; widespread in the w. cordillera.

6. Crataegus succulenta Schrader. Fleshy h. Tree to 8 m, or sometimes an arborescent shrub, the twigs glabrous or slightly hairy when young, armed with stout thorns 3–4.5 cm; lvs firm, the upper side often strigose when young, dark green, glossy, and glabrous at maturity, the lower side glabrous to ± persistently hairy, the blade elliptic or rhombic to ovate or oblong-obovate, shallowly lobed especially above the middle, mostly 3–6 × 2–5 cm; petiole 1–2 cm, usually eglandular; fls 1–1.7 cm wide, in glabrous to villous compound cymes; sep glandular-serrate, becoming reflexed and often eventually deciduous; fr bright red, 0.7–1.2 cm thick; nutlets mostly 2–3(4), rounded at the ends, with a large, deep pit on the inner face. Usually in dry or rocky ground; N. Engl. and se. Can. to Pa. and in the mts. to N.C. and Tenn., w. to Man., Mo., Neb., and w. U.S. (*C. laxiflora.*)

7. Crataegus calpodendron (Ehrh.) Medikus. Pear-h. Much like no. 6, but more prominently hairy and with larger lvs; twigs villous or tomentose when young; lvs 5–9 × 4–8 cm, short-villous above and usually also beneath; petiole 1–1.5 cm; infl ± villous or tomentose. S. Ont. and N.Y. to Ga. and Ala., w. to Minn., Kans., and Tex. (*C. fontanesiana; C. whittakeri.*)

8. Crataegus uniflora Muenchh. Oneflower-h. Shrub 1–2 m with slender, spreading branches; twigs villous when young; lvs obovate, 1.5–3 × 0.8–2 cm, or larger and sometimes broadly elliptic or suborbicular on vegetative shoots, cuneate and often glandular at base, crenate or serrate, firm, scabrate and shining above, pubescent at least on the veins beneath; petiole stout, 2–5 mm, pubescent; fls 1–1.6 cm wide, solitary or rarely 2 together on short-pubescent pedicels; sep foliaceous, pectinate or deeply glandular-serrate; fr 0.8–1.2 cm thick, greenish-yellow or dull red, with a broad, slightly elevated cal and thin hard flesh; nutlets usually 5. Usually in sandy or rocky ground; L.I. and Pa. to Fla., w. to O., Mo., and e. Tex.

9. Crataegus crus-galli L. Cockspur-thorn. Arborescent shrub or more often a tree to 10 m, often with a broadly rounded or depressed top and widely spreading branches; twigs glabrous or nearly so; lvs glabrous or nearly so, dark green and glossy above, the floral ones mostly obovate or narrowly obovate to sometimes elliptic, 2–6 × 1–3.5 cm, only obscurely if at all lobed, those on vegetative shoots more often 5–9 × 3–8 cm and sometimes more evidently lobed; petiole 0.3–2 cm; fls 0.8–1.8 cm wide, in glabrous or subglabrous, usually lax, compound cymes; sep lanceolate or lance-linear, entire or occasionally somewhat serrate; fr 0.6–1.5 cm thick, green to dull or bright red, remaining rather hard and dry; nutlets usually 1 or 2(3–5); 2n=51, 68. Que. to Fla., w. to Minn., Kans., and Tex. (*C. acutifolia; C. canbyi; C. hannibalensis; C. ohioensis; C. permixta; C. prunifolia; C. pyracanthoides; C. regalis; C. schizophylla; C. tantula; C. vallicola.*) Typical *C. crus-galli* has 10 usually white anthers; plants with 20 pink anthers have been segregated as *C. tenax* Ashe (*C. fontanesiana,* misapplied).

10. Crataegus berberifolia T. & G. Barberry-h. Much like no. 9, but the lvs, twigs, and infl evidently pubescent at least when young, often persistently so, the lvs becoming glossy above. S. Ill. and w. Ky. to Ga., La., se. Kans., and Tex. (*C. engelmannii; C. fecunda; C. paradoxa.*)

11. Crataegus viridis L. Green h. Tree to 12 m; thorns few or none; lvs variable and often asymmetric, thin, yellowish-green, dull, glabrous except for tufts or lines of persistent tomentum along the veins beneath, the floral ones obovate to rhombic, elliptic, oblong-ovate, or lanceolate, commonly toothed to near the base and not evidently lobed, 2.5–6 × 1.5–3 cm, the vegetative ones more often ovate to suborbicular and usually lobed at least toward the base; fls 1.2–1.5 cm wide, in glabrous compound cymes; sep lanceolate, entire, usually eventually deciduous; fr red to orange-red or rarely yellow, 5–8 mm thick; nutlets 3–5. Del. to Fla. and Tex., and n. in the Mississippi Valley to sw. Ind., Mo., se. Kans., and Okla.

12. Crataegus punctata Jacq. Dotted h. Arborescent shrub or usually a small tree to 10 m; twigs glabrous, usually with thorns 4–6 cm; lvs thick and firm, with impressed veins above, dull, glabrous or short-hairy on one or both surfaces, the floral ones mostly obovate, 2.5–6 × 1.5–4.5 cm, lobeless or only obscurely and shallowly lobed, the vegetative ones mostly larger, sometimes more rhombic or suborbicular, and often ± lobed; fls 1–2 cm wide, in villous or glabrous compound cymes; sep lanceolate or lance-linear, entire or glandular-serrate; fr 0.8–1.5 cm thick, dull or sometimes bright red to yellow; nutlets 3–5; $2n=34$. Nf., Que., and N. Engl. to S.C. and Ga., w. to Minn., Io., and Okla. (*C. collina; C. peoriensis; C. recedens; C. suborbiculata; C. verruculosa.*)

13. **Crataegus michauxii** Pers. Yellow h. Shrub or small tree mostly 1–4 m, the zig-zag twigs sometimes villous when young; thorns short; lvs cuneate-spatulate to rhombic-elliptic, obtuse or pointed at the tip, narrowed below to the short, glandular petiole, 1.5–3 × 1–2.5 cm, or the vegetative ones larger, to 5 × 4 cm, glandular or glandular-serrate on the margins, lobeless or shallowly 3-lobed above, yellow-green, firm, the veins distinctly impressed above; fls 1.6–1.8 cm wide, in slightly villous, 2–4-fld cymes; sep lanceolate, finely glandular-serrate; fr yellow to reddish, 0.9–1.3 cm thick, with a prominent cal, thick dry flesh, and 3–5 nutlets. Often in sandy or gravelly soil; s. Va. (at least reputedly) to Fla. and Miss. (*C. flava*, misapplied.)

14. **Crataegus intricata** Lange. Biltmore-h. Irregularly and often intricately branched shrub 1–3 m, or a small tree 3–8 m, with few or many slender thorns 2.5–5 cm, the twigs glabrous or villous; lvs thin but firm, ovate to elliptic, rhombic, or even suborbicular, 2.5–6 × 1.5–5 cm (or the vegetative ones larger), usually with several pairs of shallow lateral lobes, glabrous to short-villous; petiole 1–3 cm, evidently glandular; fls 1.3–2 cm wide, 4–10 in simple or compound, lax, glabrous or villous cymes, these with numerous bracts conspicuously and copiously stipitate on the margins; sep narrow or broad, glandular-serrate or entire; fr green or yellow to orange or red, 0.7–1.5 cm thick, with thin dry flesh and mostly 3–5 nutlets. N. Engl. and s. Ont. to N.C. and Ala., w. to Mich. and Ark. (*C. biltmoreana; C. boyntonii; C. foetida; C. fortunata; C. neobushii; C. rubella; C. stonei.*)

15. **Crataegus chrysocarpa** Ashe. Fireberry-h. Arborescent shrub or small tree to 7 m, with glabrous or villous, thorny twigs; lvs glabrous to strigose or short-villous on one or both sides when young, firm, dull or sometimes shiny, the veins tending to be impressed on the upper surface, the blade elliptic or suborbicular to rhombic or broadly obovate or ovate, 2–9 × 1.5–6 cm, rounded to obtuse or acutish at the tip, generally with several pairs of rather shallow lateral lobes; petiole ¼ to ⅔ as long as the blade, glabrous or slightly glandular; fls 1.2–2 cm wide, in glabrous or slightly villous compound cymes; fr dark or bright red to rarely yellow, tending to become mellow or succulent, 0.8–1.5 cm thick, with 3–5 nutlets. Nf. and Que. to N.Y., Pa. and Va., w. to Man. and Mo., and in the Rocky Mts. (*C. brunetiana; C. dodgei; C. faxonii; C. irrasa; C. jackii; C. jonesiae; C. margaretta; C. mercerensis; C. oakesiana; C. rotundifolia* Moench, not Lam.)

16. **Crataegus brainerdii** Sarg. Brainerd-h. Arborescent shrub or small tree to 8 m, with glabrous, usually thorny twigs; lvs glabrous, or often strigose-scaberulous above especially when young, mostly elliptic to ovate, abruptly acuminate, with shallow lateral lobes, 3–7 × 2–5 cm; petiole up to half as long as the blade, glabrous or sometimes slightly glandular; fls 1.3–1.8 cm wide, in glabrous compound cymes; fr 0.8–1.5 cm thick, red, tending to become mellow; nutlets 2–5, acute, tending to be shallowly pitted on the inner face. Que. and N. Engl. to N.Y. and in the mts. to N.C., w. to Mich. (*C. coleae; C. dunbarii; C. eatoniana; C. macauleyae; C. pinguis.*)

17. **Crataegus mollis** (T. & G.) Scheele. Downy h. Tree to 12 m, with a broad, rounded top of widely spreading branches, or sometimes an arborescent shrub, sparingly thorny or nearly thornless; young twigs villous; lvs densely short-hairy above and ± tomentose beneath, becoming glabrous above and more thinly hairy beneath at maturity, highly variable in shape, 3–10 × 2.5–8 cm, usually lobed, rather deeply so on the vegetative shoots; fls 1.6–2.3 cm wide, in tomentose compound cymes; filaments much shorter than the pet; sep coarsely glandular-serrate, tomentose on both sides; fr red, often with pale dots, hairy at least near the ends, 0.9–1.6 cm thick, becoming mellow; nutlets 3–5. Commonest in limestone-regions; N. Engl. and se. Can. to Ala., w. to Minn., Kans., and Okla. (*C. arnoldiana; C. canadensis; C. submollis.*)

18. **Crataegus coccinioides** Ashe. Kansas-h. Arborescent shrub or more often a small tree to 7 m; twigs slender, armed with stout, dark purple thorns 3–5 cm; lvs broadly ovate or deltoid, truncate or subcordate or rarely rounded at base, sharply and deeply serrate and divided into 4–5 pairs of triangular lateral-lobes, 4–6 × 3.5–5 cm, or larger on vegetative shoots, essentially glabrous from the first, thin but firm, yellow-green, crisped at the edges when mature; fls ca 2.5 cm wide, in mostly 4–7-fld glabrous cymes; sep laciniately glandular-serrate; fr bright red, 1.3–1.7 cm thick, succulent, with a very broad, nearly sessile cal; nutlets 5. Mostly in limestone regions; s. Ill. to se. Kans. and n. Ark.

19. **Crataegus dilatata** Sarg. Broadleaf-h. Much like no. 18, but the lvs short-hairy at least on the upper side when young or even at maturity, not crisped, and more likely to be cordate. Que. and s. Ont. to N. Engl. and N.Y.

20. **Crataegus coccinea** L. Scarlet h. Arborescent shrub or tree to 10 m, the top tending to be compact and conical; twigs glabrous, or villous when young; lvs glabrous, or often roughly short-hairy above at least when young, broad-based, mostly ovate to suborbicular, with 3–5 pairs of lateral lobes, mostly 3–7 × 2.5–6 cm; fls 1.3–2 cm wide, in glabrous or villous compound cymes; petiole ½–⅘ as long as the blade, often glandular; sep glandular-serrate, usually deeply so; fr 0.7–1.3 cm thick, bright red and often succulent, with 3–5 nutlets. N. Engl. and se. Can. to Del. and W.Va. (and in the mts. to N.C.), w. to Minn., Ill., and Ky. (*C. confragosa; C. hillii; C. holmesiana; C. pedicellata; C. pennsylvanica; C. pringlei; C. putnamiana; C. tortilis.*)

21. **Crataegus pruinosa** (Wendl.) K. Koch. Frosted h. Arborescent shrub or small tree to 8 m, with intricately thorny branches; lvs firm, mostly rounded or truncate to subcordate at base, rather broadly ovate or deltoid to rhombic or elliptic, with shallow lateral lobes, mostly 2.5–6 × 2–5 cm, usually glabrous from the first; petiole ⅓ to nearly as long as the blade; fls 1.2–2 cm wide, in small, nearly simple, glabrous cymes; sep narrow, entire or nearly so; fr red or sometimes greenish, mostly 1–1.5 cm thick, with prominent, usually

elevated cal; nutlets (2)3–5; 2n=51, 68. Nf. to N.C., w. to Wis. and Okla. (*C. compacta; C. crawfordiana; C. disjuncta; C. franklinensis; C. gattingeri; C. gaudens; C. jesupii; C. leiophylla; C. mackenzii; C. milleri; C. platycarpa; C. porteri; C. rugosa; C. virella.*)

22. Crataegus flabellata (Bosc) K. Koch. Fanleaf-h. Arborescent shrub or small tree to 8 m; twigs glabrous, lvs thin to rather firm, slightly to evidently short-hairy above when young, and sometimes also villous along the veins beneath, essentially glabrous at maturity, broadly ovate to deltoid or broadly elliptic, rounded or obtuse to truncate or subcordate at the base, usually with shallow lateral lobes, 2.5–6 × 2–5 cm or sometimes larger on vegetative shoots; fls 1.3–2 cm wide, in glabrous or somewhat villous compound cymes; sep narrow, generally entire or nearly so; fr red, 0.8–1.5 cm thick, palatable, usually becoming mellow or succulent; nutlets 3–5. N. Engl. and se. Can. to Ga., w. to Minn., Ill., and La. (*C. basilica; C. beata; C. beckwithiae; C. brumalis; C. compta; C. ferrissii; C. filipes; C. gravis; C. iracunda; C. iterata; C. lemingtonensis; C. levis; C. macrosperma; C. populnea; C. schuettei; C. silvicola; C. stolonifera.*)

26. AMELANCHIER Medikus. Serviceberry, shadbush, juneberry. Hypanthium obconic, campanulate, or saucer-shaped; sep 5, erect to reflexed, persistent; pet 5, white; stamens usually 20, shorter than the pet; ovary inferior, 5-locular in our spp., the 5 styles distinct or partly connate; ovules 2 per locule, soon separated by a partition growing in from the back of each carpel, the 10-seeded pome therefore apparently 10-locular; unarmed shrubs or trees with simple, alternate, serrate lvs and medium-sized fls in rather short, often leafy-bracteate racemes (rarely reduced to a single fl) terminating the branches of the season and opening with or before the lvs; x=17. 20, N. Temp.

Genus beset with hybridization, polyploidy, and apomixis (as shown by the frequent apparently fertile triploids). Plants called *A. intermedia* Spach probably reflect hybridization between *A. canadensis* and *A. arborea* or *A. laevis*. Plants called *A. interior* Nielsen, found mainly in Minn., Wis., e. Io., and n. Mich., may be a hybrid swarm or a set of segregating polyploids derived from *A. laevis* and *A. sanguinea*. Plants called *A. wiegandii* Nielsen may reflect hybridization between *A. sanguinea* and *A. arborea* or *A. laevis*. Plants called *A. nantucketensis* Bicknell may be hybrids between *A. canadensis* and *A. spicata*. Other hybrids also occur and may be to some degree self-perpetuating.

1 Fls commonly 1–4(5) together, long-pedicellate (1–3 cm) in leafy-bracted fascicles that do not have
  the aspect of a raceme; summit of ovary densely tomentose, tapering into the styles; lvs imbricate
  in bud; northern sp. .... 1. *A. bartramiana.*
1 Fls several to many in racemes; ovary rounded to nearly flat on top, glabrous or tomentose; lvs
  conduplicate in bud.
  2 Ovary tomentose at the summit, some of the tomentum often persisting into fr.
    3 Lf-teeth mostly less than twice as many as the relatively strong, straight or slightly curved, simple
      or once-forked lateral veins, which commonly extend into the teeth; mostly bushy, clumped
      shrubs or small trees.
      4 Pedicels relatively short, the lower mostly 5–10 mm at anthesis; western sp., reaching our
        range in Io. and Minn. .... 2. *A. alnifolia.*
      4 Pedicels longer, the lower mostly more than 10 mm at anthesis; widespread northeastern sp.
        .... 3. *A. sanguinea.*
    3 Lf-teeth mostly more than twice as many as the relatively weak, usually curved lateral veins,
      which branch and anastomose and fade out near the margin; stoloniferous colonial shrubs
      to 1.5 m.
      5 Lvs tomentose beneath at anthesis .... 4. *A. spicata.*
      5 Lvs glabrous or nearly so beneath at anthesis .... 5. *A. fernaldii.*
  2 Ovary usually glabrous or nearly so at the summit; lf-margins and venation as in spp. 4 and 5.
    6 Pet relatively short, 4–12 mm; racemes firm, erect or ascending; fruiting sep erect to spreading
      or loosely recurved.
      7 Stoloniferous colonial shrub to 1.5 m; pedicels at anthesis 2–6 mm, the longest becoming 3–
        12 mm in fr; lvs scarcely exserted from the bud at anthesis; mainly on the coastal plain
        .... 6. *A. obovalis.*
      7 Bushy, clumped shrub or small tree mostly 2–5 m; pedicels longer, at anthesis mostly 5–15
        mm, the longest becoming 12–20 mm in fr; lvs at anthesis exserted from the bud but still
        folded and scarcely half-grown; mainly in the coastal states, but not limited to the coastal
        plain .... 7. *A. canadensis.*
    6 Pet relatively long, (10–)12–22 mm; racemes lax, drooping in rain or soon after the twigs are
      cut; fruiting sep tightly reflexed; widespread spp.
      8 Lvs at anthesis mostly much less than half-grown, tomentose beneath; pedicels relatively
        short, mostly not over 2 cm even in fr; fr dryish, insipid .... 8. *A. arborea.*
      8 Lvs at anthesis about half-grown, glabrous beneath, and with a coppery red cast (the latter
        feature shared only with sp. no. 1); pedicels longer, the lower becoming 2.5–5 cm in fr;
        fr juicy, sweet .... 9. *A. laevis.*

1. Amelanchier bartramiana (Tausch) Roemer. Mountain serviceberry. Shrub to 2 m, often in clumps; lvs imbricate in bud, short-petiolate, oblong-elliptic or somewhat obovate, often mucronate, tapering to the base, sharply serrate, glabrous, ± coppery when young but less so than in no. 9; petiole 3–10(–15) mm; fls solitary or more often 2–4(5) together on glabrous pedicels 1–3 cm, 1 terminal, the others from adjacent lf-axils; sep triangular-acuminate, ascending or spreading, persistently tomentose on the upper side; pet elliptic, 6–10 mm, often more than half as wide; ovary densely tomentose at the summit, tapering into the styles; fr dark purple, 1 cm, longer than thick, often some of the seeds abortive; sexual. Moist woods, swamps, and streambanks; Lab. to James Bay and Man., s. to N. Engl., n. Mich., and n. Minn., and at higher altitudes in the mts. to Pa. and W.Va. May–Aug.

2. Amelanchier alnifolia Nutt. Western serviceberry, saskatoon. Clustered shrub or small tree mostly 2–5 m; lvs soon glabrous or nearly so, broadly elliptic to quadrate-rotund, to 6 cm, mostly 1–1.5 times as long as wide, coarsely toothed and strongly veined as in no. 3, often toothed only distally or across the broadly rounded or subtruncate summit; fls in short racemes, these mostly less than 2.5 cm in fl, to 3.5 cm in fr; pedicels relatively short, the lowest mostly 5–10 mm at anthesis; sep spreading to recurved, 1.5–3.5 mm; pet mostly 10–20 mm; ovary tomentose at the summit; fr subglobose, ca 1 cm, dark purplish, sweet and tasty. Thickets, borders of woods, and banks of streams; widespread in the w. cordillera and n. Great Plains, reaching our range in w. Minn. and nw. Io., and reputedly with outlying stations in Ont. and even w. Que. May.

3. Amelanchier sanguinea (Pursh) DC. New England serviceberry. Erect or straggling shrub or small tree to 3(–6) m, usually with several clumped stems; lvs at anthesis half-grown and tomentose beneath, eventually glabrous, oblong to subrotund or quadrate, to 7 cm, mostly 1–1.5 times as long as wide, rather coarsely toothed (often only above the middle), with mostly 3–5 teeth per cm of margin, seldom more than 20 per side; lateral veins more than half as many as the teeth, parallel, straight or slightly curved, simple or once (rarely twice) forked, each vein or fork running to a tooth; racemes loose and open, the lower pedicels mostly (1–)1.5–3(–4) cm; sep recurved from near the middle after anthesis; ovary tomentose at the summit; mostly polyploid. Mostly in open woods; Me. and s. Que. to Minn., s. to N.Y., n. N.J., Mich., and Io., and irregularly in the mts. to N.C. and Tenn. May, June. Var. *sanguinea*, with the range of the species, has pedicels up to 3 cm, the pedicels and sep usually retaining their tomentum until after anthesis; the hypanthium is cupulate, 4–7 mm wide, and the pet are mostly 10–18 mm. (*A. humilis; A. huronensis.*) Var. *grandiflora* (Wieg.) Rehder, restricted to c. N.Y., has the pedicels and hypanthium soon glabrescent, the hypanthium saucer-shaped and 7–9 mm wide, the pedicels 2.5–4 cm, and the pet 15–22 mm. (*A. amabilis.*)

4. Amelanchier spicata (Lam.) K. Koch. Dwarf serviceberry. Stoloniferous colonial shrub 3–10(–15) dm; lvs a quarter to half-grown at anthesis and then densely tomentose beneath, at maturity glabrous and pale beneath, ovate to oblong, elliptic, or obovate-oblong, usually 2–5 cm, acute to obtuse or rounded, finely and sharply toothed, the lateral veins curved forward, branched and anastomosing near the margin, not definitely prolonged into the teeth, the teeth almost always more than twice as many as the veins; racemes short and dense; pedicels thinly pubescent, the lowest 7–15 mm; sep soon recurved or reflexed from near the middle; pet mostly 5–10 mm, half as wide; ovary tomentose at the summit; mostly polyploid. Dry woods, old fields, and rocky banks; Que. and Me. to Minn., s. to N.Y., Mich., Io., and in the mts. to N.C. May, June. (*A. mucronata; A. stolonifera.*) Occasional plants of no. 8 with the ovary ± tomentose on top will key here but are tall, non-colonial shrubs.

5. Amelanchier fernaldii Wieg. St. Lawrence serviceberry. Straggling stoloniferous shrub to 1 m; lvs thinly tomentose when young, at anthesis almost full grown and glabrous, otherwise much like those of no. 4; racemes lax, the pedicels 1–3 cm at anthesis, to 4 cm in fr, glabrous or nearly so; hypanthium glabrous or nearly so; sep lanceolate, erect or spreading at anthesis. Limestone hills and barrens; Nf. to Que., N.S., and n. Me., and reputedly to n. Mich. and n. Minn. June, July. (*A. gaspensis.*)

6. Amelanchier obovalis (Michx.) Ashe. Coastal plain serviceberry. Stoloniferous colonial shrub 3–15 dm; lvs oblong to broadly elliptic or subrotund, 3–5 cm, acute to broadly rounded, finely and sharply serrate chiefly above the middle, the teeth more than twice as many as the veins, which branch and anastomose toward the margin; lvs thinly villous above and densely tomentose beneath when young, soon glabrescent and pale, at anthesis scarcely exserted beyond the bud-scales; fls in dense, spike-like, scarcely pedunculate erect racemes; pedicels villous-tomentose, 2–6 mm, the lowest becoming 3–12 mm in fr; hypanthium tomentose; sep spreading or ascending, 1.5–2.5 mm; pet obovate, 4–8 mm, half as wide; ovary glabrous at the summit; diploid and polyploid. Open woods, pine-barrens, and streamsides on the coastal plain; N.J. and e. Pa. to S.C. Apr.

7. Amelanchier canadensis (L.) Medikus. Shadbush; eastern serviceberry. Bushy shrub of alder-like aspect, usually in clumps, or a small tree to 8 m; lvs elliptic or elliptic-oblong to somewhat obovate, to 8 cm, rounded to subacute at the tip, often mucronate, broadly cuneate to rounded at base, finely and often obscurely serrate, the blunt teeth at least twice as many as the veins, which branch and anastomose toward the margin; lvs at anthesis ± exserted from the bud but still folded and densely tomentose beneath, glabrescent at maturity; racemes erect, short, dense; pedicels tomentose, at anthesis mostly 5–15 mm, the lowest becoming 12–20 mm in fr; hypanthium tomentose externally; sep 2–4 mm, soon spreading or somewhat recurved, but not strongly reflexed; pet usually oblong-obovate, 5–12 mm; ovary glabrous at the summit; fr dark purple to black; mostly diploid, sometimes polyploid. Swamps and shores and moist woods; Nf. to Miss., inland to c. N.H. and w.c. N.Y., but southward chiefly on the coastal plain. Apr.–June. (*A. lucida.*)

8. **Amelanchier arborea** (Michx. f.) Fern. Downy serviceberry. Much like no. 9, seldom only 2 m tall; lvs typically oblong-obovate, much less than half-grown at anthesis, densely pubescent beneath when young, not (or scarcely) coppery, usually glabrous or nearly so at maturity; pedicels shorter, mostly not over 2 cm even in fr, often silky; sep 2–3 mm; pet (10–)12–15+ mm; ovary glabrous at the summit, varying to occasionally somewhat tomentose; fr dark red-purple, tending to be dry and insipid; mostly diploid. Dry or rocky upland woods; Me. and N.B. and s. Que. to sw. Ont. and Minn., s. to Ga., nw. Fla., La., and e. Okla. (*A. arborea* var. *cordifolia*.)

9. **Amelanchier laevis** Wieg. Smooth serviceberry. Tall, erect shrub or slender tree to 15 m; lvs elliptic to ovate-oblong or ovate, to 8 cm, shortly acuminate (or merely abruptly acute), finely and sharply serrate nearly to the rounded or subcordate base, the teeth more than 20 per side, and more than twice as many as the lateral veins, which curve and anastomose before reaching the margins; lvs at anthesis half-grown and glabrous or nearly so, with a prominent coppery-red cast that disappears before maturity; racemes lax, drooping in rain or soon after the twigs are cut; pedicels glabrous, 1–3 cm, the lower becoming 2.5–5 cm in fr; hypanthium glabrous externally; sep 3–4 mm, strongly reflexed after anthesis, tightly so in fr; pet slender, oblong to narrowly cuneate, mostly 15–22 mm; ovary glabrous on top; fr dark purple-red to black, sweet and juicy; mostly diploid, at least partly apomictic. Dry or moist upland woods; Nf. to Ont. and Minn., s. to Md., Ind., and Io., and in the mts. to Ga. and Ala. Apr.–June, a bit before no. 8

FAMILY **MIMOSACEAE**, the Mimosa Family

Fls regular, hypogynous or nearly so, mostly perfect and 5-merous; sep commonly united to form a tube with valvate lobes, or sometimes much-reduced; pet distinct or often connate below to form a tube, valvate; stamens as many or more often twice as many as the pet, or numerous, the anthers small, the filaments distinct or connate below, commonly colored and exserted to form the most conspicuous part of the infl; pistil 1, simple; fr a legume; seed-coat commonly with a horseshoe-shaped groove on the flat surface near the margin; embryo large, generally straight; endosperm in most genera wanting or very scanty; herbs or more often woody plants with alternate, stipulate, usually bipinnately compound lvs and numerous small fls in heads, spikes, or dense racemes. 50–60/3000.

1 Tree or tall shrub; stamens very numerous; intr. .... 1. *Albizia.*
1 Herbs; stamens 5–10; native.
  2 Stems prickly; stamens 10, rose-purple .... 2. *Mimosa.*
  2 Stems unarmed; stamens in ours 5, white or greenish .... 3. *Desmanthus.*

1. **ALBIZIA** Durazz. Cal campanulate, gamosepalous, 5-toothed; corolla funnelform, gamopetalous, deeply 5-lobed; stamens numerous, long-exserted, the capillary filaments connate below; fr broadly linear, flat, several-seeded; unarmed trees or shrubs with bipinnately compound lvs and pink or white fls in heads. 150, mainly tropical.

1. **Albizia julibrissin** Durazz. "Mimosa"; silk-tree. Widely spreading, soft-wooded tree to 10 m; lvs 2–5 dm; pinnae 4–12 pairs, 1–2 dm; lfls 20–30 pairs, oblong, 10–15 mm, mucronate, very inequilateral; fls pink, in long-peduncled heads forming a terminal raceme; cal 2–2.5 mm; cor 5–8 mm; stamens 15 mm; fr 10–15 × 2 cm; $2n=36$. Native of tropical Asia, hardy as far n. as s. N.Y., and often escaped from cult. southward. June–Aug.

2. **MIMOSA** L. Fls perfect or the lower ones staminate; cal very small; cor gamopetalous, funnelform, 5-lobed; stamens mostly 10, with long, slender filaments; fr prickly, oblong to linear, obtusely tetragonal, the sutures (collectively forming a replum) about as wide as the persistent valves and separating from them at maturity; sprawling perennial herbs, armed with numerous stout hooked prickles; lvs bipinnate, sensitive, with small lfls; fls rose-purple, numerous in axillary peduncled heads. Nearly 600, New World. Description here based entirely on the small series Quadrivalves of section Batocaulon, forming the traditional genus *Schrankia*.

1. **Mimosa quadrivalvis** L. Sensitive brier. Stems branched, arched or decumbent, to 1 m long, strongly ribbed, armed with stout hooked prickles; lvs 6–15 cm, the rachis and rachillae prickly; pinnae 3–5(–7) pairs, 1–5 cm; lfls 8–11(–14) pairs, oblong or elliptic, often shortly cuspidate, 1.5–9 mm, the veins few, obscure to

very prominent and reticulate beneath; heads long-peduncled, 1.5–2.5 cm thick; fr linear, strongly ribbed, 4–8(–12) cm, prickly on the ribs, the slender beak 0.5–2 cm; $2n=36$. Dry, sterile or sandy soil; widespread in tropical and warm-temperate parts of the New World, in the U.S. n. to w. Va., Ky., w. Io., and S.D. May–Sept. Some 11 ± geographic vars., 2 in our range. Var. *nuttallii* (DC.) Beard ex Barneby, w. of the Mississippi R., has the lateral veins of the lfls prominently raised beneath; its heads are 1.5–2.5 cm thick (7–11 mm without the filaments); and the stems and pods are not hairy. (*Schrankia nuttallii; S. uncinata,* misapplied; *Leptoglottis n.*) Var. *angustata* (T. & G.) Barneby, mainly e. of the Mississippi R. and Gulf coastal, has the lateral veins of the lfls obscure; its heads are 1–1.5 cm thick (5–7.5 mm without the filaments); and the stems and pods are often finely puberulent. (*Schrankia microphylla; Leptoglottis m.; Moronga m.*)

3. **DESMANTHUS** Willd. Cal gamosepalous, campanulate, 5-toothed; pet 5, distinct or slightly connate at the very base; stamens 5 (in ours) or 10, usually long-exserted; fr oblong to linear, smooth; seeds few–several; unarmed perennial herbs or shrubs with bipinnate lvs, with a prominent gland on the upper side of the rachis between the 2 lowest pinnae (and sometimes between the next pair); fls white or greenish-white, in long-peduncled axillary heads. 25, New World.

1. **Desmanthus illinoensis** (Michx.) MacMillan. Bundleflower. Taprooted herb with clustered stems to 1 m or more; stipules filiform, 6–10 mm; lvs bipinnate, 5–10 cm, nearly sessile; pinnae 6–12 pairs, 2–4 cm; lfls 20–30 pairs, 1.5–3 mm, often ciliate; peduncles 2–6 cm, ascending; pet 2 mm; stamens 5; frs strongly curved or somewhat twisted together into a dense subglobose head, thin, 1–2.5 cm × 4–7 mm; seeds 3–5 mm, nearly as wide; $2n=28$. Moist or dry soil, riverbanks, prairies, and pastures; O. to N.D., s. to Fla. and N.M., and occasionally intr. elsewhere. All summer. (*Acuan i.*)

FAMILY **CAESALPINIACEAE**, the Caesalpinia Family

Fls perfect or sometimes unisexual, slightly to evidently perigynous; sep mostly 5, distinct or nearly so, or the 2 upper ones ± connate, imbricate or rarely valvate, sometimes reduced; cor of (0–)5 distinct, imbricate, similar or more often ± dissimilar pet, but not papilionaceous, the uppermost pet generally internal to the 2 upper lateral ones and often smaller than them, stamens (1–)10(–numerous), distinct or with the filaments variously connate, but not usually forming a sheath around the pistil, and not long-exserted, all alike or variously unlike, sometimes some of them staminodial; pistil 1, simple; fr a legume, sometimes indehiscent; embryo large, straight, endosperm mostly wanting or very scanty; herbs or more often woody plants, with alternate, stipulate, pinnately or less often bipinnately compound lvs (seldom unifoliolate or simple) that typically have a pulvinus at the base of the lf and at the base of each lfl, and with small to often large and showy fls mostly in racemes or spikes. 150/2000.

1 Trees or tall shrubs; anthers versatile, opening lengthwise.
  2 Lvs simple, preceded by the rich pink fls . . . . 1. *Cercis.*
  2 Lvs compound; fls white to greenish or yellow, appearing with or after the lvs.
    3 Lfls ca 1 cm wide or less; stems usually thorny, fls in compact racemes . . . . 2. *Gleditsia.*
    3 Lfls mostly 1.5–4 cm wide; stems unarmed; fls in open panicles . . . . 3. *Gymnocladus.*
1 Herbs; anthers basifixed, opening by an apical pore or short slit.
  4 Stamens all with normal anthers; pod elastically dehiscent . . . . 4. *Chamaecrista.*
  4 Upper 3 stamens sterile; pod indehiscent or inertly dehiscent . . . . 5. *Senna.*

1. **CERCIS** L. Fls perfect; hypanthium short, hemispheric; cal irregular, 5-toothed; cor irregular, almost papilionaceous, but the upper petal the smallest and internal to the 2 upper lateral ones, the lower petals the largest, distinct; stamens 10, distinct; anthers versatile, longitudinally dehiscent; pods thin and flat, narrowly oblong, rather tardily dehiscent, several-seeded, strongly margined along the upper suture; trees or shrubs with simple, cordate, entire lvs and sessile fascicles of pedicellate fls borne on the stems of the previous year and opening before the lvs. 6, North Temperate.

1. **Cercis canadensis** L. Redbud. Widely branched tree to 12 m; lvs broadly cordate, 5–12 cm, usually wider than long, obtuse to shortly acuminate, glabrous or sparsely hairy beneath; fls magenta-pink, 1 cm, in small clusters on slender pedicels 6–12 mm; pods pointed at both ends, 6–10 × 1–1.5 cm; $2n=14$. Moist woods; Conn. and s. N.Y. to s. Mich., Io., and e. Neb., s. to Fla. and n. Mex. Our plants are var. *canadensis*.

2. GLEDITSIA L. Plants polygamous, with reduced fls in separate staminate and fertile infls appearing with the lvs from spur-branches; hypanthium short-campanulate; perianth regular, pubescent, of 2 similar yellowish-green series of 3–5 members each, the pet a little larger than the sep; stamens mostly 5–8; anthers versatile, longitudinally dehiscent; pods large, flat, oval to elongate, woody and scarcely (or very tardily) dehiscent; trees with branched thorns (some cultivars unarmed), usually with 2 kinds of lvs, those of the long shoots bipinnate, those of the spurs mostly only once pinnate. 14, widespread.

1 Pod elongate, normally many-seeded; widespread .... 1. *G. triacanthos.*
1 Pod short, oval, 1(–3)-seeded; southern .... 2. *G. aquatica.*

1. Gleditsia triacanthos L. Honey-locust. Tree to 20 or even 40 m; lvs with 9–14 pairs of lfls or 4–7 pairs of pinnae; lfls oblong-lanceolate, obscurely crenate, 2–4 cm on pinnate lvs, 1–2 cm on bipinnate ones; petioles pubescent; staminate racemes 3–7 cm, densely many-fld; fertile racemes loose, with fewer, more evidently pedicellate, pistillate or perfect fls; pods 15–40 × 3–4 cm, dark brown, firm, pubescent when young, the seeds ca 2 cm apart and separated by sweetish pulp; $2n=28$. Rich moist woods; Pa. to Tenn. and w. Fla., w. to s. Minn., se. S.D. and Tex., and widely cult. elsewhere. May. *G.* × *texana* Sargent, with pods 10–15 cm, lacking pulp, is a rare hybrid with no. 2.

2. Gleditsia aquatica Marshall. Water-l. Much like no. 1; petioles glabrous; pods elliptic or ovate, 3–5 cm, pointed at both ends, glabrous, with 1(–3) seeds, lacking pulp; $2n=28$. Swamps and low woods, mainly along larger rivers, tolerant of flooding; S.C. to Fla. and Tex., n. in the Mississippi Valley to s. Ind. and s. Ill. Apr., May.

3. GYMNOCLADUS Lam. Plants polygamo-dioecious; fls regular, perfect or unisexual; hypanthium tubular; sep and pet each 5, nearly alike, inserted in a single series; stamens 10, distinct, alternately long and short; anthers versatile, longitudinally dehiscent; pods broadly oblong, woody, indehiscent, flat but thick, with a few large seeds separated by pulp; trees, without small twigs but with very large, bipinnately compound lvs; fls in terminal panicles. 1 e. U.S., 4 e. Asia.

1. Gymnocladus dioica (L.) K. Koch. Kentucky coffee-tree. Tree to 30 m; lvs to nearly 1 m, with 3–6 pairs of pinnae, each with several pairs of ovate, abruptly acuminate, short-petiolulate lfls mostly 3–8 × 1.5–4 cm, or the lowest pinnae replaced by single lfls; fls greenish-white, softly pubescent, in terminal panicles 6–20 cm; hypanthium 10–15 mm; pet and sep oblong or oblanceolate, 8–10 mm, exceeding the stamens; pods 8–25 × 3–5 cm; seeds thick and hard, 10–15 mm wide and long; $2n=28$. Rich moist woods, seldom abundant; N.Y. to s. Minn. and se. S.D., s. to w. Va., Tenn., Ark., and Okla., mainly midwestern. May.

4. CHAMAECRISTA (L.) Moench. Fls perfect, 5-merous; pedicels 2-bracteolate near or above the middle; hypanthium very short; pet yellow, sometimes marked with red near the claw, often unequal, the upper one internal to the adjacent lateral ones; stamens 5 or 10, all with anthers; anthers basifixed, longer than the short filaments, opening by an apical pore or short slit; pods compressed, not segmented, elastically dehiscent, the valves coiling; seeds in our spp. quadrate, faintly marked with shallow pits in rows; lvs evenly once-pinnate, generally with one or more large glands on the petiole or rachis, in our spp. somewhat sensitive to the touch; our spp. herbs. (Often included in *Cassia*.) 250, widespread.

1 Stamens 10, longest anthers commonly ca 8 mm; pet 1–2 cm, subequal .... 1. *C. fasciculata.*
1 Stamens 5; anthers to about 3 mm; pet very unequal, the longest 6–8 mm .... 2. *C. nictitans.*

1. Chamaecrista fasciculata (Michx.) Greene. Partridge-pea; locust-weed. Annual; stems erect, ascending or rarely spreading, pubescent; petiolar gland saucer-shaped, sessile or nearly so; lfls 5–18 pairs, oblong, usually 1–2 cm, acute to obtuse, mucronate; fls 1–6 in short, axillary, bracteate racemes; pedicels 1–2 cm; pet subequal, 1–2 cm, 4 of them red-marked at base; stamens 10, very unequal; anthers linear, several times longer than their filaments, the two largest commonly 8 mm; pods linear-oblong, 3–6 cm, flat, straight; $2n=16$. In a wide variety of ± open, often disturbed habitats; Mass. to s. Minn., s. to Fla. and Mex., commoner southward. July–Sept. (*C. chamaecrista; Cassia f.*)

2. Chamaecrista nictitans (L.) Moench. Wild sensitive plant. Annual; stems erect, 1–5 dm, glabrous to villous; petiolar gland saucer-shaped, short-stalked; lfls 7–20 pairs, oblong, 6–15 mm, aristate, usually glabrous; fls solitary or 2–3 together, on pedicels 2–4 mm; pet very unequal, one 6–8 mm and nearly or fully twice as long as the others; stamens 5, unequal, filaments very short; anthers 1.5–3 mm; pods oblong, straight, flat, 2–4 cm × 3–6 mm; $2n=16$. Dry, especially sandy soil, upland woods, dunes, and disturbed habitats; Mass. and s. Vt. to N.Y., O., Mo., and Kans., s. to tropical Amer. July–Sept. (*C. procumbens; Cassia n.*)

5. **SENNA** Miller. Fls perfect, ± irregular; pedicels without bracteoles; hypanthium very short; sep 5; pet 5, yellow, often unequal; stamens mostly 10 but unlike, the lowest (abaxial) ones fully developed, the upper ones progressively reduced, in our spp. the 3 upper ones without normal anthers; anthers basifixed, at least half as long as the filaments, opening by an apical pore or short slit; pods terete or quadrangular to compressed, often transversely septate, indehiscent or inertly dehiscent; lvs evenly once-pinnate, generally with one or more large glands on the petiole or rachis; our spp. all herbs. (Often included in *Cassia*.) 240, widespread.

1 Lfls oblong to elliptic or ovate, obtuse or acuminate; pod flattened, with some evidence of transverse segmentation; gland near the base of the petiole.
  2 Lfls mostly 6–10 pairs, obtuse or acute; fls several in each axil, racemose.
    3 Petiolar gland clavate to obovoid above a short stipe; ovary villous; joints of the pod about as long as wide .... 1. *S. hebecarpa.*
    3 Petiolar gland short-cylindric or rounded or dome-shaped; ovary appressed-hairy; joints of the pod about twice as wide as long .... 2. *S. marilandica.*
  2 Lfls 3–6 pairs, mostly sharply acuminate; fls 1–3 in each axil .... 3. *S. occidentalis.*
1 Lfls, especially the terminal pair, obovate and broadly rounded above; pod scarcely flattened, with no evidence of transverse segmentation; gland between the lowest pair of lfls .... 4. *S. obtusifolia.*

1. **Senna hebecarpa** (Fern.) Irwin & Barneby. Northern wild senna. Erect perennial 0.5–2 m, glabrous or villous above; stipules subsetaceous; petiolar gland clavate to obovoid, constricted at base into a short stipe; lfls commonly 6–10 pairs, oblong or elliptic, 2–5 cm, acute or obtuse, mucronate; infl of several axillary, many-fld racemes, forming a terminal panicle; buds nodding; sep unequal; pet 10–15 mm, slightly dissimilar; filaments about equaling the anthers; ovary densely villous; pods 7–12 cm × 5–9 mm, tardily dehiscent, sparsely villous, the joints nearly square; seeds nearly as wide as long, flat with a depressed center. Moist open woods, roadsides, and streambanks; Mass. and s. N.H. to s. Wis., s. to N.C., e. Tenn., and Ill. July, Aug. (*Cassia h.; C. marilandica* and *Ditremexa m.* of authors, misapplied.)

2. **Senna marilandica** (L.) Link. Southern w. s. Erect perennial, 1–2 m, glabrous or sparsely villous above; stipules lance-linear; petiolar gland short-cylindric or rounded or dome-shaped; lfls commonly 4–8 pairs, oblong or elliptic, 2–5 cm, acute or obtuse, mucronate; infl of several axillary racemes, less floriferous than no. 1; buds nodding; sep unequal; pet 10–15 mm, slightly dissimilar; ovary appressed-hairy; pods glabrous or with scattered minute incurved hairs, tardily dehiscent, 6–10 cm × 8–11 mm, the joints about twice as wide as long; seeds nearly twice as long as wide, plump, with depressed center; $2n=28$. Moist open woods and streambanks; s. N.Y. to s. Wis. and e. Neb., s. to Fla. and Tex. July, Aug. (*Cassia m.; C. medsgeri; Ditremexa medsgeri.*)

3. **Senna occidentalis** (L.) Link. Coffee-s. Annual (in our range), 3–10 dm, malodorous, bearing in each upper axil a single fl or 2–3 fls in a raceme with a very short common axis; gland depressed or dome-shaped, near the base of the petiole; lfls 3–6 pairs, ovate to eliptic or lance-ovate, 3–8 cm, acuminate; buds nodding; sep unequal; pet 10–15 mm, slightly dissimilar; pods straight or slightly curved, flattened, 7–14 cm × 6–9 mm, tardily dehiscent, glabrous or sparsely hairy, weakly segmented; seeds 4–5 mm, flattened-obovoid with depressed center; $2n=28$. Native of the Old World tropics, widely naturalized in the warmer parts of the New World, extending n. to N.C. and Ark., and rarely in our range as an adventive. Late summer. (*Cassia o.; Ditremexa o.*)

4. **Senna obtusifolia** (L.) Irwin & Barneby. Sickle-pod. Annual to 1 m, malodorous; fls solitary or paired in the upper axils, forming a terminal infl when young; petiolar gland between he two lowest lfls; lfls 2–3 pairs, elliptic to obovate, the terminal pair always obovate, 4–7 cm, the others smaller; buds nodding; pet 10–15 mm, slightly dissimilar; pods divaricate, ± strongly curved, dehiscent, 10–18 cm × 3–6 mm, scarcely flattened, not segmented; seeds thick, shining, obliquely truncate at both ends, marked with a diagonal stripe on each side; $2n=28$. Moist woods; pantropical (probably originally American), extending n. to Va., Ky., and Mo., and occasionally farther as an adventive. July–Sept. (*Cassia o.; C. tora,* in part.)

FAMILY **FABACEAE**, the Pea or Bean Family

Fls perfect, slightly to evidently perigynous; sep mostly 5, ± connate into a lobed, often bilabiate tube; cor typically papilionaceous and consisting of 5 pet, the uppermost one (the *standard* or *banner*) external to the others and usually the largest; rarely, as in *Centrosema* and *Clitoria,* the fl resupinate, with the banner lowermost; 2 lateral pet, called *wings,* similar inter se and distinct or sometimes lightly adherent in a small area to the keel; 2 lowermost pet innermost, similar inter se, mostly connate distally to form a *keel* enfolding the stamens and pistil; seldom (*Amorpha*) only the banner present, the other pet suppressed; stamens (5–)10, most often 9 of the filaments connate into an open sheath around the pistil, the 10th (uppermost) one partly or wholly separate from the other 9 (stamens then said to be *diadelphous*), or less often all the stamens connate by their filaments to form a closed (or adaxially split) sheath (stamens then said to be *monadelphous*), or (*Aeschynomene*) the stamens diadelphous and 5 + 5, the sheath cleft above and below, or the filaments sometimes all distinct; pistil 1, simple; fr a legume, typically dry and dehiscent on both sutures, but sometimes indehiscent and even jointed, rarely fleshy; embryo large, the short hypocotyl-radicle typically bent back against the thickened cotyledons; endosperm mostly wanting or very scanty; herbs or less often woody plants, with alternate, stipulate, pinnately or less often palmately compound (or trifoliolate), seldom unifoliolate or simple lvs that typically have a basal pulvinus; fls borne in racemes or spikes or heads or seldom in panicles, often large and showy. 440/12,000.

1 Stamens wholly distinct.
  2 Lvs pinnately compound; tree with fls in panicles (Tribe Sophoreae.) . . . . 1. *Cladrastis.*
  2 Lvs digitately 3-foliolate; perennial herbs, fls in racemes. (Tribe Thermopsideae.)
    3 Pod turgid or inflated; fls variously colored . . . . 2. *Baptisia.*
    3 Pod flattened; fls yellow . . . . 3. *Thermopsis.*
1 Stamens monadelphous (sometimes only at the base) or diadelphous.
  4 Anthers alternately of 2 forms, 5 subglobose and versatile, 5 oblong or linear and basifixed; filaments monadelphous; lvs, if pinnate, with not more than 4 lfls; fls (except in *Lupinus* and *Ononis*) mostly yellow or orange-yellow.
    5 Pod dehiscent, 2–several-seeded.
      6 Lfls entire (sometimes reduced or wanting); fls not red.
        7 Filament-tube closed; fr flattened, not inflated. (Tribe Genisteae.)
          8 Shrubs; lvs simple or trifoliolate or reduced.
            9 Plants unarmed; fls in terminal racemes.
              10 Keel-pet straight beneath; style curved abruptly near the tip; fls in ours 1–1.5 cm
                . . . . 4. *Genista.*
              10 Keel-pet curved beneath; style curved throughout its length; fls in ours 2–2.5 cm
                . . . . 5. *Cytisus.*
            9 Plants spiny; fls axillary to spines . . . . 6. *Ulex.*
        8 Herbs; lvs palmately compound . . . . 7. *Lupinus.*
        7 Filament-tube split on the upper side; fr inflated. (Tribe Crotalarieae.) . . . . 8. *Crotalaria.*
      6 Lfls (or lvs) serrate; fls pale red; depressed, suffruticose, ± spiny plants. (Tribe Trifolieae, in small part.) . . . . 25. *Ononis.*
    5 Pod indehiscent, 1-seeded or sometimes 2–3-seeded. (Tribe Aeschynomeneae, in large part.)
      11 Lfls 3 . . . . 14. *Stylosanthes.*
      11 Lfls normally 4.
        12 Stipules adnate to the petiole, not spurred; fr unarmed, maturing underground . . . . 15. *Arachis.*
        12 Stipules free from the petiole, spurred at base; fr spiny, aerial . . . . 16. *Zornia.*
  4 Anthers essentially uniform in size and shape (except in *Orbexilum* and *Pediomelum*); filaments diadelphous, except in a few genera; lvs and fls various.
    13 Lfls serrulate, 3; herbs, sometimes lax or trailing, but not twining. (Tribe Trifolieae, in large part.)
      14 Cor withering and persistent . . . . 26. *Trifolium.*
      14 Cor deciduous after anthesis.
        15 Fr straight, nearly or quite symmetrical; racemes elongate . . . . 27. *Melilotus.*
        15 Fr asymmetric, ± curved or coiled; fls in heads or head-like racemes . . . . 28. *Medicago.*
    13 Lfls entire, except often in *Vicia,* which has more than 3 lfls; habit various.
      16 Filaments all or alternately dilated at the summit.
        17 Fr not jointed. (Tribe Loteae.)
          18 Fr elongate, dehiscent; fls, if in heads, subtended by entire bracts . . . . 21. *Lotus.*
          18 Fr short, scarcely protruding from the cal, tardily if at all dehiscent; heads subtended by 3-cleft bracts . . . . 22. *Anthyllis.*
        17 Fr jointed. (Tribe Coronilleae.)
          19. Fr terete or angled; cal bilabiate; ours perennial . . . . 23. *Coronilla.*
          19 Fr flattened; cal regular or nearly so; ours annual . . . . 24. *Ornithopus.*
      16 Filaments all slender.
        20 Herbage ± gland-dotted; fr not regularly dehiscent into 2 valves; herbs, not twining, or sometimes shrubs.
          21 Fr spiny, few-seeded; cor papilionaceous; lvs pinnate. (Tribe Galegeae, in small part.) . . . . 20. *Glycyrrhiza.*
          21 Fr unarmed, 1–2-seeded; lvs and fls various.
            22 Lvs pinnately (3)5–many-foliolate; infls terminal or lf-opposed. (Tribe Amorpheae.)

23 Pet 5 (but 4 of them sometimes resembling petaloid staminodes); stamens
      5–10 .... 35. *Dalea.*
23 Pet 1 (the banner); stamens 10, all with anthers, not petaloid .... 36. *Amorpha.*
22 Lvs digitately 3–5-foliolate, or pinnately 3-foliolate, or some of them unifo-
      liolate; infls axillary. (Tribe Psoraleeae.)
      24 Lvs pinnately trifoliolate, the terminal lfl long-stalked .... 37. *Orbexilum.*
      24 Lvs digitately compound, all lfls sessile or on petiolules of about equal
            length (or the upper lvs unifoliolate).
            25 Cal gibbous in fl, inflating or enlarging slightly in fr; fr ± circumscissile
                  .... 38. *Pediomelum.*
            25 Cal tubular-campanulate, not inflating or enlarging in fr, but flaring
                  back and tearing along a lateral sinus; fr indehiscent .... 39. *Psoralidium.*
20 Herbage not gland-dotted; fr and habit various.
      26 Fr indehiscent and 1-seeded, or 2–several-seeded and transversely septate into
            1-seeded joints that eventually disarticulate.
            27 Filament-tube cleft on the upper and lower side; stamens 5 + 5. (Tribe Aes-
                  chynomeneae, in small part) .... 13. *Aeschynomene.*
            27 Filament-tube cleft on the upper side; stamens 9 + 1.
                  28 Lfls numerous, commonly 11–31. (Tribe *Hedysareae.*)
                        29 Fr with (1)2–several joints, visibly constricted between the joints ...... 31. *Hedysarum.*
                        29 Fr with a single 1-seeded joint .... 32. *Onobrychis.*
                  28 Lfls 3. (Tribe Desmodieae.)
                        30 Fr with (1)2–several joints, visibly contracted between the joints ...... 33. *Desmodium.*
                        30 Fr with a single 1-seeded joint .... 34. *Lespedeza.*
      26 Fr dehiscent or indehiscent, normally with 2 or more seeds (1–2-seeded and de-
            hiscent in *Rhynchosia*), not jointed (transversely septate in *Sesbania,* but not
            disarticulating).
            31 Lvs even-pinnate, the terminal lfl represented by a tendril or bristle (tendril
                  wanting in *Vicia faba,* which has 2 or 3 pairs of large lfls). (Tribe Vicieae.)
                  32 Style bearded all around at the tip (like a bottle-brush) or on the outer
                        (abaxial) side only, or glabrous .... 29. *Vicia.*
                  32 Style bearded along the inner (adaxial) side toward the tip (like a tooth-
                        brush) .... 30. *Lathyrus.*
            31 Lvs odd-pinnate (except in *Sesbania,* which has numerous lfls) or trifoliolate,
                  without tendrils.
                  33 Ovary without a disk; lfls mostly 7–70; plants (except *Wisteria*) not twining.
                        34 Racemes terminal; if also lateral, then opposite the lvs. (Tribe Tephro-
                              sieae.)
                              35 Herbs; lfls with numerous straight parallel primary lateral veins .... 9. *Tephrosia.*
                              35 Woody twiners; lfls net-veined .... 10. *Wisteria.*
                        34 Racemes axillary.
                              36 Lvs even-pinnate, with mostly 30–70 lfls; fr transversely septate.
                                    (Tribe Sesbanieae.) .... 12. *Sesbania.*
                              36 Lvs odd-pinnate, with mostly 7–31 lfls; fr not septate.
                                    37 Trees or shrubs.
                                          38 Fr flat; fls pink to white. (Tribe Robinieae.) .... 11. *Robinia.*
                                          38 Fr inflated; fls yellow. (Tribe Galegeae, in small part.) .... 17. *Colutea.*
                                    37 Herbs. (Tribe Galegeae, in principal part.)
                                          39 Keel obtuse to acute, not beaked .... 18. *Astragalus.*
                                          39 Keel abruptly short-beaked .... 19. *Oxytropis.*
                  33 Ovary surrounded by a disk at the base; lfls mostly 3(5–7 in *Apios,* some-
                        times only one in *Rhynchosia*); most spp. twining or trailing herbs.
                        (Tribe Phaseoleae.)
                        40 Style bearded; keel-pet strongly incurved to coiled.
                              41 Standard about as long as the wings and keel; fls resupinate.
                                    42 Cal-lobes exceeding the tube, nearly equal; style bearded only
                                          about the stigma .... 40. *Centrosema.*
                                    42 Cal-lobes shorter than the tube, unequal, the upper 2 partly
                                          connate; style bearded along one side .... 41. *Clitoria.*
                              41 Standard little if at all longer than the wings and keel; fls not
                                    resupinate.
                                    43 Cal-lobes 5; keel-pet coiled .... 42. *Phaseolus.*
                                    43 Cal-lobes 4 by fusion of the upper 2; keel-pet merely incurved.
                                          44 Fls in elongate racemes; intr. .... 43. *Dolichos.*
                                          44 Fls in long-peduncled heads; native .... 44. *Strophostyles.*
                        40 Style glabrous; keel-pet nearly straight to ± incurved, not coiled.
                              45 Ovules and seeds 1 or 2; cal-lobes 5, the upper 2 connate more
                                    than half their length, but distinct above; fls orange or yellow
                                    .... 45. *Rhynchosia.*
                              45 Ovules and seeds more than 2 (except in apetalous fls of *Amphi-
                                    carpaea*); cal otherwise; fls rarely wholly yellow or orange.
                                    46 Cal with only one well developed lobe, the upper 4 much
                                          reduced or obsolete .... 46. *Apios.*
                                    46 Cal 4-lobed by fusion of the upper 2.

47   Fls of 2 kinds, petaliferous and apetalous, the petaliferous
      ones without bracteoles .... 47. *Amphicarpaea.*
47   Fls all petaliferous; cal with a pair of bracteoles at base.
   48   Axis of the infl glandular-thickened at the base of each
       pedicel.
     49   Pods wingless; fls in ours 2–2.5 cm; intr. .... 48. *Pueraria.*
     49   Pods narrowly 2-winged along the upper suture;
       fls in ours 1.2–1.5 cm; native .... 49. *Dioclea.*
   48   Axis of the infl not glandular-thickened .... 50. *Galactia.*

1. CLADRASTIS Raf.   Yellow-wood. Cal short, campanulate, 5-lobed; cor large, the pet clawed; stamens 10, distinct; pods flat, thin, linear, short-stipitate, several-seeded; trees with exstipulate odd-pinnate lvs and conspicuous white fls in ample panicles. 1 e. U.S., 5 e. Asia.

     1. Cladrastis lutea (Michx. f.) K. Koch. Smooth-barked tree to 15 m; lfls 6–9, alternate, broadly ovate to obovate or the terminal one rhomboid, 6–13 cm; panicle hanging, 1–3 dm; standard reflexed, the claw 2–3 mm, the blade broadly rounded, 1.5 cm; wings and keel-pet all distinct, nearly alike, their claws 6–8 mm, their blades oblong, 16–20 mm, cordate at base; ovary pubescent; pods 7.5–10 cm; $2n=28$. Rich woods and rocky bluffs; irregularly from w. Va. and w. N.C. to s. O., s. Ind., Ala., Miss., s. Mo., Ark. and Okla., and sometimes escaped from cult. farther n. May, early June. (? *C. kentukea, Sophora kentukea,* a possibly older name of doubtful application.)

     2. BAPTISIA Vent.   Wild or false indigo. Cal bilabiate, the upper lip entire to 2-lobed, the lower evidently 3-lobed; standard reniform to suborbicular, its sides usually reflexed; wings and keel subequal, straight, oblong; stamens 10, distinct; ovary usually shorter than the style; pods stipitate, papery to woody, globose to cylindric or thick-lenticular, terminating in a curved beak; perennial herbs from thick rhizomes, the stem usually much branched, with conspicuous white, yellow, or violet fls in terminal racemes; lvs digitately 3-foliolate, commonly blackening in drying; $x=9$. Hybrids frequent. 17, U.S.

. 1   Fls wholly or largely dull violet-blue .... 1. *B. australis.*
1   Fls white, cream-color, or yellow.
   2   Bracts 1–3 cm, persistent, foliaceous, reticulately veined .... 2. *B. bracteata.*
   2   Bracts less than 1 cm, usually soon deciduous, not obviously reticulate.
     3   Fls white, or with a tinge of blue on the standard.
       4   Fls 18–25 mm; pods thick, drooping, the stipe twice as long as the cal .... 3. *B. lactea.*
       4   Fls 12–18 mm; pods cylindric, erect, the stipe scarcely exceeding the cal .... 4. *B. alba.*
     3   Fls yellow.
       5   Fls 8–16 mm; pods 8–15 mm; lfls 1–3(–4) cm; racemes very numerous .... 5. *B. tinctoria.*
       5   Fls 20–28 mm; pods 20–30 mm; lfls 4–8 cm; racemes solitary or few .... 6. *B. cinerea.*

     1. Baptisia australis (L.) R. Br.   Blue false i. Plants to 15 dm, glabrous, with one or few racemes to 4 dm; petioles 5–18 mm; stipules or some of them persistent until anthesis, lanceolate or lance-ovate, 8–20 mm; lfls oblanceolate to obovate, the larger ones 4–8 cm; bracts caducous; pedicels 5–15 mm; fls 20–27 mm; cal 8–10 mm; pods 3–5 cm, the stipe slightly exceeding the cal; $2n=18$. Moist, usually rocky or gravelly soil, woods, and prairies; s. N.Y. to N.C. and n. Ga., w. to Neb. and Tex., and intr. in Vt. May, June. Var. *australis,* as here described, is replaced w. of the Mississippi R. by var. *minor* (Lehm.) Fern., with smaller lfls, larger fls, and more inflated and more evidently stipitate pods. (*B. minor.*) A hybrid of var. *minor* with no. 2 is *B.* × *bicolor* Greenman & Larisey.

     2. Baptisia bracteata Elliott.   Plains w. i. Plants 4–8 dm, pubescent throughout; stipules persistent, foliaceous, lanceolate, those of the lower lvs to 4 cm, the upper gradually reduced; petioles 2–5 mm; lfls oblanceolate, the larger ones to 8 cm; raceme usually solitary, ± declined, 1–2 dm, secund; bracts persistent, lanceolate and oblong, 1.5–4 cm; pedicels 1.5–4 cm; fls pale yellow to nearly white, 20–28 mm; pods ellipsoid, 3–5 cm, pubescent, tapering to a conspicuous beak. Prairies and open upland woods; s. Mich. to se. Minn. and w. Neb., s. to Miss. and Tex.; e. Ala. to S.C. and w. N.C. May, June. Our plants are var. *glabrescens* (Larisey) Isely. (*B. leucophaea.*) The more southeastern var. *bracteata* has longer petioles, commonly 4–10 mm.

     3. Baptisia lactea (Raf.) Thieret.   Milky w. i. Plants 1–2 m, widely branched, glabrous and glaucous; stipules lanceolate, 5–10 mm, some usually persisting until anthesis; petioles 6–12 mm; lfls narrowly obovate to oblanceolate, 3–6 cm; racemes solitary or few, erect, 2–6 dm; bracts lance-ovate, caducous; pedicels 4–10 mm; fls white or with a tinge of purple on the standard, 18–25 mm; cal 8 mm, densely pubescent within; pods black, drooping, ellipsoid-oblong, firm and thick-walled, 2.5–4 cm × 8–12 mm, abruptly narrowed to a short beak; stipe well exserted; $2n=18$. Prairies and open upland woods; O. and s. Mich. to Minn., Neb., and Tex., and e. in the coastal states to Fla. and N.C.; locally (intr.?) in w. N.Y. June, July. Ours is var. *lactea.* (*B. leucantha.*)

4. Baptisia alba (L.) R. Br. White w. i. Plants to 1.5 m, much branched, glabrous or nearly so; stipules minute, caducous; petioles 5–10 mm; lfls oblanceolate or oblong-lanceolate, 2–4 cm × 8–15 mm; racemes solitary or few, erect, 2–5 dm; bracts 5 mm, caducous; pedicels 3–6 mm; fls white, 12–18 mm; cal 7 mm; pods erect, cylindric, brown, 2–3 cm × 6–8 mm; stipe scarcely exceeding the cal. Dry, sandy woods; Va. (Delmarva Penins.) to Ga., w. Fla., Ala., and e. Tenn. May.

5. Baptisia tinctoria (L.) R. Br. Yellow w. i. Plants to 1 m, widely branched; stipules setaceous, deciduous; petioles 1–3 mm; lfls obovate from a cuneate base, 1–3(–4) cm, rounded to slightly retuse at the tip; racemes very numerous, terminating most of the branches; bracts setaceous, caducous; pedicels 3–6 mm; fls yellow, 8–13(–16) mm; cal glabrous; pods thick-lenticular to subglobose, 8–15 × 6–8 mm; stipe 5–10 mm; $2n=18$. Dry sterile or sandy soil; s. Me. to Ga. and Tenn., w. occasionally to s. Ont., Mich., Ind., Ill., and (only intr.?) to Wis. and Minn. June, July. A hybrid with no. 3 is *B.* × *deamii* Larisey; one with no. 4 is *B.* × *pinetorum* Larisey.

6. Baptisia cinerea (Raf.) Fern. & Schubert. Carolina w. i. Plants 3–6 dm, ± pubescent, with a few widely spreading branches; stipules lanceolate, those of the lower lvs usually persistent; petioles very stout, 3–6 mm; lfls subcoriaceous, elliptic to oblanceolate, 4–8 cm; racemes usually solitary, 1–2 dm; fls yellow, 20–28 mm; pods ellipsoid, firm-walled, 2–3 cm, tapering to the beak, the stipe to 10 mm. Dry sandy soil on the coastal plain; se. Va. to S.C. May. (*B. villosa* of auth., misapplied.)

3. THERMOPSIS R. Br.

Buckbean. Cal bilabiate, the upper lip shallowly lobed or merely notched, the lower deeply 3-lobed; pet nearly equal in length, the standard suborbicular, the wings and keel oblong; stamens 10, distinct; ovary sessile or short-stipitate, longer than the style; pod linear, flat, several-seeded; perennial, usually rhizomatous herbs with 3-foliolate lvs and ± foliaceous stipules, bearing yellow fls in terminal racemes. 25, N. Amer., ne. Asia.

1. Thermopsis mollis (Michx.) M. A. Curtis. Piedmont-b. Stems erect, 4–10 dm, finely appressed-hairy; petioles 1–2.5 cm; stipules lanceolate, 1–2 cm; lfls rhombic elliptic, 4–8 cm; racemes mostly loose, 6–15 cm; bracts lanceolate; cal finely hairy, its lower lobes triangular, acuminate; cor 15 mm; pods 5–7 cm, irregularly spreading, hairy or glabrate. Woods, typically on the piedmont, sometimes in the mts. especially southward; s. Va. to Ga. May

T. villosa (Walter) Fern. & Schubert, the Blue-Ridge b. occurs mainly in N.C., Tenn., and n. Ga. and is locally escaped northward. It has broad, clasping stipules, mostly more than 1 cm wide, and villous, often denser infl and pods, the pods erect or strongly ascending. (*T. caroliniana.*)

4. GENISTA L.

Cal bilabiate, the upper lip 2-toothed, the lower 3-toothed or -lobed; standard ovate; wings obliquely obovate or oblong, auriculate on one side; keel-pet oblong, straight beneath; stamens 10, monadelphous into a closed tube adherent to the claws of the wings and keel; style curved abruptly near the tip; small shrubs with simple or 3-foliolate lvs and yellow fls in terminal racemes or panicles. 90, Old World.

1. Genista tinctoria L. Dyer's greenweed. Slender shrub to 1(–2) m; lvs simple, sessile, elliptic to lanceolate, 1–3(–5) cm × 3–10(–15) mm; racemes erect, 3–6 cm, leafy-bracteate, often several and forming a terminal panicle; fls 10–15 mm; pods 1.5–3 cm × 5 mm; $2n=48$. Native of Europe, escaped from cult. in dry sterile soil from Me. and n. N.Y. to Va. June–Sept.

5. CYTISUS L.

Cal broadly cup-shaped, rather shallowly bilabiate, each lip minutely toothed; standard suborbicular, reflexed; wings elliptic, auriculate on one side at base; keel-pet curved beneath; stamens 10, monadelphous in a closed tube; filaments alternately long with minute anthers and short with linear ones; style curved throughout its length; pods elongate, flat, several-seeded; unarmed shrubs with green twigs, 1–3-foliolate lvs (or lfless), minute or no stipules, and conspicuous yellow (white) fls. 33, Old World.

1. Cytisus scoparius (L.) Link. Scotch broom. Bushy shrub to 2 m, with stiff, slender, green, 5-angled branches; lvs small, mostly 3-foliolate, or the upper 1-foliolate, the petiole 2–8 mm, the lfls obovate, 5–10 mm; fls 20–25 mm, solitary or paired in the upper axils, forming long terminal racemes; pods 3–5 cm, villous along the sutures; $2n=46$. Native of Europe, well established in dry sandy soil as an escape from cult., from N.S. to Va., and on the Pacific coast. May, June.

6. ULEX L.

Cal bilabiate, cleft nearly to the base, both lips ovate, the upper minutely 2-toothed, the lower 3-toothed; standard broadly obovate; wings and keel oblong-obovate, obtuse, the keel hairy along the lower margin; stamens 10, monadelphous in a closed tube, alternately long with globose anthers and short with linear ones; pods short, few-seeded, almost enclosed by the persistent cal; shrubs without normal lvs except in the juvenile stage, the petioles reduced to linear-subulate spines; fls bright yellow, solitary or paired in the axils of the spines. 20, Old World.

1. **Ulex europaeus** L. Gorse. Much branched shrub to 2 m, all branches ending in spiny tips; spine-tipped petioles 5–15 mm, numerous, crowded; fls short-pedicelled along the smaller branches, forming terminal panicles; each fl subtended by 2 ovate bracteoles 2–3 mm; cal loosely villous, 12–16 mm; pet subequal, 15–20 mm; pods 2–4-seeded, 1–2 cm, villous-tomentose, subtended by the persistent cal; $2n=96$. Native of Europe, escaped and established in sandy soil and waste places along the coast, from Mass. to Va. June.

7. **LUPINUS** L. Lupine. Cal deeply bilabiate, the upper lip 2-toothed, the lower entire or 3-lobed; standard suborbicular with strongly reflexed sides; wings united toward the summit; keel-pet strongly convex on the lower side, prolonged into a beak-like apex; stamens 10, monadelphous into a closed tube for about half their length; anthers alternately linear or minute and subglobose; pods oblong, flattened, with 2–several seeds; herbs with palmately compound, 1–many-foliolate lvs and handsome white, yellow, pink, or blue fls in terminal racemes or spikes. 200, widespread.

1. **Lupinus perennis** L. Sundial-l. Perennial; stems ± erect, 2–6 dm, thinly pubescent; petioles 2–6 cm; lfls 7–11, oblanceolate, 2–5(–6) cm; racemes erect, arising above the lvs, 1–2(–3) dm, the fls numerous, blue varying to pink or white; lower lip of the cal entire, 6–8 mm, the upper lip half as long; standard 12–16 mm, half as wide; pod pubescent, 3–5 cm; $2n=48, 96$. Dry, open woods and clearings; s. Me. to Fla., w. to Minn. and Ind. May, June.

Two w. Amer. spp. are casually escaped from cult. in n. New Engl. and adj. Can. *L. polyphyllus* Lindley differs in being taller and coarser, to 1 m or more, with 11–17 lfls to 13 cm, and infls 2–4 dm. *L. nootkatensis* Donn has villous herbage and large fls 1.5–2 cm in racemes 1–3 dm.

8. **CROTALARIA** L. Rattlebox. Cal obscurely bilabiate, the upper lip less deeply cleft than the lower and with broader lobes; standard suborbicular, short-clawed; wings not auriculate; keel-pet connivent on both margins, strongly convex on the lower side; stamens 10, monadelphous below the middle, the sheath cleft on the upper side; filaments alternately long with subglobose anthers, and short with linear anthers; distal part of the style usually with 1 or 2 lines of hairs; pods subglobose to cylindric or ellipsoid, inflated; seeds 2–many; annual or perennial herbs, or shrubs in the tropics, with simple (in all our spp.) or trifoliolate lvs and usually yellow fls in racemes. 600, mostly warm reg.

1 Standard 6–15 mm, shorter than to slightly longer than the cal; lvs elliptic to lanceolate or linear, broadest near or below the middle; infls lf-opposed.
  2 Stems ± erect; stipules mostly well developed and decurrent, the stem thus winged; typical lvs more than 3 times as long as wide.
    3 Stem ± spreading-hairy; upper lf-surface hairy .... 1. *C. sagittalis.*
    3 Stem closely appressed-hairy; upper lf-surface glabrous .... 2. *C. purshii.*
  2 Stems ± spreading; stipules usually reduced; typical lvs 2–3 times as long as wide .... 3. *C. rotundifolia.*
1 Standard 15–25 mm, greatly exceeding the cal; lvs oblanceolate to obovate from a cuneate base, widest near the summit; infls terminal and axillary .... 4. *C. spectabilis.*

1. **Crotalaria sagittalis** L. Weedy r. Annual in our range; stems ± erect, simple or branched above, 1–4 dm, with spreading or loosely ascending hairs; stipules, at least on the upper lvs, decurrent on the stem; lvs pubescent on both sides, sessile or nearly so, lanceolate to linear, 3–8 cm, to 1.5 cm wide; racemes lf-opposed, 2–4-fld, cal villous, to 9 mm; lower sep lance-linear; standard 6–12 mm; pods oblong, 1.5–3 cm, inflated, nearly sessile in the cal; $2n=32$. Dry open soil and waste land, usually as a weed in our range and probably only intr. except along our s. border; Mass. and Vt. to Minn. and Nebr., s. to tropical Amer. June–Sept.

2. **Crotalaria purshii** DC. Coastal plain r. Perennial from a thick woody root; stems erect, branched from the base, 2–5 dm, with closely appressed hairs; lvs finely appressed-hairy on the lower surface, glabrous on the upper, averaging narrower than in no. 1, the upper usually linear and seldom over 5 mm wide, the lower often narrowly lanceolate. Dry sandy soil; se. Va. to n. Fla. and e. La., mainly on the coastal plain. Summer.

3. **Crotalaria rotundifolia** J. F. Gmelin. Low r. Perennial from a slender root; stems spreading or decumbent, 1–4 dm; stipules ± reduced or wanting; racemes erect, open, few-fld. Dry, open woods and sandy

soil; se. Va. to Fla., La., and tropical Amer. Summer. Our plants are var. *vulgaris* Windler, with ± spreading pubescence and relatively broad lvs mostly 2–3 times as long as wide. (*C. angulata*, probably misapplied.)

4. Crotalaria spectabilis Roth. Showy r. stems erect, to 1 m; principal lvs narrowly obovate from a cuneate base, to 15 cm; racemes terminal and axillary, elongate and many-fld, the central one to 3 dm; bracts ovate, 8–12 mm; cal glabrous; standard 15–25 mm. Native of the Old World tropics, widely naturalized in tropical regions; se. Va., Mo., and scattered through s. U.S. Late summer.

9. TEPHROSIA Pers., nom. conserv. Cal-tube hemispheric, slightly oblique, the lobes lanceolate, exceeding the tube; standard subrotund, short-clawed, sericeous on the back; wings and keel connivent, broadly auriculate on the upper side above the short claw, the wings obovate-oblong, the keel semicircular; stamens 10, the upper one at least partly free; style in our spp. bearded along the inner side; fr linear, several-seeded, dehiscent; our spp. perennial herbs from long roots, with odd-pinnate lvs, no stipellules, and medium-sized yellow or purple fls in terminal or lf-opposed racemes; lfls in our spp. with numerous straight parallel lateral veins. 400+, mostly warm reg.

1 Uppermost stamen connate with the staminal tube about ⅓ its length; stem monopodial, not much
   branched, the racemes all obviously terminal; widespread .... 1. *T. virginiana.*
1 Uppermost stamen free; stem sympodial, some of the racemes opposite the lvs; se. U.S. .... 2. *T. spicata.*

1. Tephrosia virginiana (L.) Pers. Goat's rue. Stems 1–several, ± erect, 2–7 dm, unbranched or with a few weak axillary branches; herbage densely villous or sericeous to sparsely strigillose or partly glabrous; lfls mostly (9–)15–25(–31), elliptic or linear-oblong, 1–3 cm; racemes terminal to the main stem and sometimes to axillary branches; short-pedunculate or sessile, compact, mostly 4–8 cm; pedicels 4–10(–17) mm; bracts setaceous, deciduous; fls 1.5–2 cm, usually bicolored, the standard yellow to ochroleucous, the wings and keel pink or pale purple; fr 3.5–5.5 cm, strigose to villous; 2*n*=22. Old fields, open woods, and dunes, often in sandy soil; s. N.H. to Fla., w. to Wis., se. Minn., Kans., and Tex. June, July (*Cracca v.*)

2. Tephrosia spicata (Walter) T. & G. Southern g. r. Stems 1–many, decumbent to erect, 3–6 dm, flexuous, sympodially branched; herbage strigose to hirsute or partly glabrous, the pubescence often rusty; lfls mostly 9–17, elliptic to oblong or obovate, 1–2.5 cm; racemes usually several, terminal and opposite the lvs at the upper nodes, long-pedunculate, at first compact, soon becoming interrupted by elongation of the axis; bracts large, lanceolate, persistent, stipulate; fls 12–17 mm, nearly white at first, becoming pink or red, and drying purple; fr 3–5 cm, spreading hairy to occasionally subglabrous; 2*n*=22. Sandy fields, open woods, and barrens; s. Del. to Fla., e. to se. Ky., s. Tenn., and La. June, July. (*Cracca s.*)

10. WISTERIA Nutt., nom. conserv. Wisteria. Cal-tube hemispheric to campanulate, bilabiate; cal-lobes unequal, the 2 upper short or completely connate, the lowest one often elongate; standard reflexed, rotund, with 2 basal callosities; wings obliquely obovate, clawed, auriculate at base of the blade; keel upcurved, auriculate at base; stamens 10, diadelphous; fr linear, flattened, dehiscent, many-seeded; woody twining vines with odd-pinnate lvs and handsome blue fls in large racemes at the end of short branches, appearing before the lvs are fully expanded. (*Kraunhia.*) 6, e. Asia, e. N. Amer.

1 Ovary and fr glabrous; wings with an elongate narrow auricle parallel to the claw.
  2 Upper lip of the cal less than half as long as the tube; infl mostly 4–15 cm .... 1. *W. frutescens.*
  2 Upper lip of the cal nearly or fully as long as the tube; infl mostly 15–30 cm .... 2. *W. macrostachya.*
1 Ovary and fr pubescent; wings with a short divergent auricle .... 3. *W. floribunda.*

1. Wisteria frutescens (L.) Poiret. Atlantic w. High-climbing vine; lfls 9–15, ovate-oblong or lance-ovate, 4–6 cm, acuminate, obscurely hairy beneath; racemes compact, 4–15 cm; pedicels 4–10 mm; axis, pedicels, and cal finely pubescent and also with some short-stipitate, oblong or clavate glands; fls blue-purple, 1.5–2 cm; cal campanulate, the upper 2 lobes broadly triangular to almost completely connate, forming an upper lip up to about a third as long as the tube; lower lobes lanceolate, the median slightly the longer; blade of the wing-pet prolonged backward into a linear auricle about as long as the claw and ± parallel to it; fr glabrous; 2*n*=16. Moist or wet woods and river-banks; Va. to Fla. and Tex., n. in the interior to Ark. Late Apr., May.

2. Wisteria macrostachya Nutt. Mississippi w. Much like no. 1; infl often more copiously glandular; racemes mostly 1.5–3 dm; upper cal-lobes almost completely connate, forming an upper lip nearly or quite as long as the tube; lower lobes lanceolate, the median much the larger; 2*n*=16, 48. Alluvial forests; s. Ind. to s. Mo., s. to La., and occasionally escaped elsewhere, as in Kans. May.

3. Wisteria floribunda (Willd.) DC. Japanese w. Lfls mostly 13–19, lanceolate to ovate; pubescence of the infl of simple hairs only; cal hemispheric; racemes 2–5 dm; fls 1.5–2 cm, opening sequentially from base to tip of the infl; auricle of the wing-pet short, triangular, divergent from the claw; fr velvety; $2n=16$. Native of Japan, occasionally escaped from cult. in our range, especially southward, and in se. U.S.

W. sinensis (Sims) Sweet, the Chinese w., with 7–13 lfls and shorter (1.5–2 dm) racemes of slightly larger (2.5 cm) fls opening almost simultaneously, also escapes in our range, especially southward.

11. ROBINIA L. Locust. Cal-tube hemispheric or broadly campanulate, bilabiate, the lower 3 lobes subequal, the upper 2 connate toward the base; cor large; standard suborbicular, ± reflexed; wings and keel long-clawed with a rounded lobe at the base of the blade, the wings obliquely obovate, the keel strongly upcurved; stamens 10, diadelphous, the uppermost one connate with the others at its middle, the staminal sheath villous at base; fr elongate, flat, many-seeded (or some ovules abortive), trees or shrubs with odd-pinnate lvs, stipules setaceous or modified into spines, and showy white, pink, or purple fls in axillary racemes. 4, N. Amer.

1 Fls white; ovary and fr glabrous; twigs neither hispid nor glandular; trees .... 1. *R. pseudoacacia.*
1 Fls pink or rose; ovary and fr hispid; twigs hispid or glandular; shrubs or small trees.
  2 Twigs merely glandular-viscid; tall shrub or small tree; lfls 13–25 .... 2. *R. viscosa.*
  2 Twigs hispid-setose; shrub to 2(–3) m; lfls 7–13 .... 3. *R. hispida.*

1. **Robinia pseudoacacia** L. Black l. Tree to 25 m, the twigs and peduncles puberulent, becoming subglabrous; stipules commonly modified into spines 1 cm; lfls 7–19, oval or elliptic, 2–4 cm; racemes drooping, many-fld, 1–2 dm; fls white, fragrant, 1.5–2.5 cm; cal finely hairy, the upper lip truncate or broadly notched; ovary glabrous; fr 5–10 cm; $2n=20$. Native from Pa. and s. Ind. to Okla., s. to Ga. and Ala., and often escaped from cult. n. to N.S. and Que. May, June.

2. **Robinia viscosa** Vent. Clammy l.; rose-acacia. Large shrub or tree to 5(–12) m, the twigs and peduncles viscid with numerous large, sessile or subsessile glands; stipules usually setaceous; lfls 13–25, lance-ovate to oval; fls crowded in short suberect racemes, pink with a yellow spot on the standard, 2.5 cm; cal finely hairy, the upper lip cleft into 2 sharp triangular lobes shorter than the body of the lip; fr hispid with stipitate-glandular hairs; $2n=20$. Native in mt. woods from Pa. to Ala., but found along roadsides as an escape from cult. nearly throughout our range. Ours is var. *viscosa.* June.

3. **Robinia hispida** L. Bristly l.; rose-acacia. Rhizomatous shrub 1–2(–3) m; stems, peduncles, and cal densely to sparsely hispid with glandular hairs 2–5 mm, those of the stem tending to persist and become indurate; lfls 7–13, ovate-oblong to subrotund, 3–6 cm; racemes usually 3–10-fld; fls rose or pink-purple, 2.5–3 cm; ovary densely glandular; fr densely hispid. Mt. woods; Va. and Ky. to Ga. and Ala., sometimes escaped from cult. as far n. as Me. and Minn. June, July. The common var. *hispida* is triploid ($2n=30$) and fruits poorly or not at all (the frs often formed, but nearly or quite seedless); the rare var. *fertilis* (Ashe) R. T. Clausen, probably not native with us, is diploid and fertile ($2n=20$). Var. *fertilis* is being grown as ground cover on mine spoil etc. in Pa. Vigorous subglabrous triploids, native from sw. Va. to N.C. and n. Ga., sometimes cult. elsewhere, have been called var. *rosea* Pursh; an origin through hybridization with *R. pseudoacacia* has been suggested.

12. SESBANIA Scop., nom. conserv. Cal-tube campanulate to hemispheric, with short, equal, triangular lobes; standard rotund to reniform, short-clawed; wings short-clawed, obliquely oblong-obovate, obscurely auriculate; keel-pet oblanceolate, long-clawed, strongly upcurved, auriculate at base; filaments 10, diadelphous, the tube gibbous at base; fr compressed, linear, the ± numerous seeds separated by transverse partitions; herbs or shrubs with even-pinnate lvs of numerous lfls and short, bracteolate, axillary racemes of yellow fls. 50, warm reg.

1. **Sesbania exaltata** (Raf.) Cory. Sesban. Rank, glabrous annual to 4 m; lfls 30–70, narrowly oblong, 1–3 cm; fls 2–6 in short racemes, yellow mottled with purple, 1.5–2 cm; fr 10–20 cm × 3–4 mm, subtended by the persistent cal and tipped with the long, slender style; $2n=12$. Alluvial soil; tropical Amer. to Fla., Tex., and s. Mo., and rarely adventive along the coast in our range. (*S. macrocarpa.*)

13. AESCHYNOMENE L. Joint-vetch. Cal deeply bilabiate, the tube short, the 2 lips divergent, foliaceous, shallowly lobed; pet subequal, clawed; standard suborbicular to ovate; wings narrowly obovate; keel boat-shaped, obtuse; stamens diadelphous, the filament-tube cleft above and below into 2 groups of 5 stamens each; fr stipitate, elongate, indehiscent, transversely segmented and eventually separating into reticulately veined, 1-seeded joints, herbs (in the tropics shrubs) with pinnately compound lvs and mostly yellow fls in axillary racemes. 150, mostly warm reg.

1. Aeschynomene virginica (L.) BSP.   Northern j.-v. Stems erect, branched, to 1 m or more, sparsely pustulate-hairy above; lvs odd-pinnate, the lfls very numerous, oblong, 1–2 cm; racemes often leafy; fls 1–6, yellow with red veins, 10–15 mm, subtended by ovate, sessile, serrate bracts; fr 2–7 cm, on a stipe 1.5–2 cm; joints 4–10, each 5–8 × 5–6 mm, sparsely pustulate-hairy; $2n=40$. Sandy or muddy river-banks and fresh to brackish tidal shores; s. N.J. to S.C. Aug., Sept.

14. STYLOSANTHES Swartz.   Pencil-fl. Hypanthium elongate, pedicel-like; cal-lobes unequal, the lowest triangular, acute, larger than the 4 ovate-oblong obtuse upper, the 2 uppermost connate about half-length; standard broadly obovate to orbicular; wings shorter than the standard, oblong-obovate, clawed, biauriculate at the base of the blade; keel curved upward, about equaling the wings; stamens 10, monadelphous, the anthers alternately oblong and subglobose; joints of the fr 2, the lowest generally sterile and stipe-like, the upper fertile, flattened, tipped with the hooked persistent style; perennial herbs with pinnately trifoliolate lvs and small fls in short leafy spikes; stipules adnate to the petiole and connate into a tube around the stem, subulate-tipped. 25, mostly warm reg.

1. Stylosanthes biflora (L.) BSP.   Stems wiry, branched from the base, depressed to ascending or erect, 1–5 dm, with or without spreading setae; lfls narrow, 1–4 cm, 3–6 times as long as wide, acute and subulate-tipped to merely mucronate; spikes short; bracteal lvs usually 1-foliolate, the blade and stipular base often setose-ciliate as well as puberulent; fls 2–6 in the spike, orange-yellow (white), the standard 7–9 mm; fr 4–5 mm, thinly hairy, obliquely ovate, reticulate-veiny; $2n=20$. Dry or rocky woods and barrens; s. N.Y. to O., s. Ill., and Kans., s. to Fla. and La. June–Aug. (*S. riparia.*)

15. ARACHIS L.   Peanut. Hypanthium slender, greatly elongate; cal distinctly bilabiate; cor yellow; standard much longer than the other pet; stamens 10, monadelphous; anthers alternately oblong and globose; fr developing underground, 2–3-seeded, turgid, indehiscent, slightly constricted between the seeds; perennial herbs with normally 4-foliolate, evenly pinnate lvs, large stipules adnate to the petiole, and small fls. 20+, S. Amer.

1. Arachis hypogaea L.   Stem soon decumbent; lfls oblong-obovate to obovate, 3–6 cm; fls in the lower axils; standard orbicular, 12–16 mm; pedicel elongating and curved down after anthesis, pushing the fr into the soil; $2n=40$. Native of S. Amer., cult. in se. U.S. and occasionally escaped along our s. border. Summer.

16. ZORNIA J. F. Gmelin.   Cal irregular, membranous, the 2 upper lobes connate most of their length, the lateral ones much shorter, the lowest one lanceolate, about equaling the upper; standard broadly obovate to suborbicular, much exceeding the other pet; wings short-clawed, the blade half-orbicular, minutely auriculate; keel boat-shaped, obtuse; stamens 10, monadelphous, the tube closed; anthers alternately oblong and globose; fr transversely segmented and eventually separating into 1-seeded indehiscent joints; herbs with palmately compound 2- or 4-foliolate lvs and mostly small fls in interrupted terminal spikes, each fl subtended by a pair of large bracts. 80, warm reg.

1. Zornia bracteata (Walter) J. F. Gmelin.   Stems slender, prostrate, much branched, 2–6 dm; stipules free from the petiole, lanceolate, prolonged below into a short auricle; lfls commonly 4, 1–3 cm, lance-linear to obovate; spikes terminal, with 3–10 fls 1–2 cm apart; bracts 7–10 mm, broadly oval, erect, auriculate; fls yellow, 1 cm; joints of the pod 2–4, bristly, 3–4 mm; $2n=20$. Dry sandy soil; se. Va. to Fla. and Mex. Summer.

17. COLUTEA L.   Cal-tube hemispheric, slightly oblique, the lobes triangular-acuminate, subequal; standard rotund to reniform; wings and keel-pet strongly auriculate at base, the wings falcate, oblong, the keel upcurved; stamens 10, diadelphous, the uppermost one wholly free; ovary elongate; style hairy distally, hooked at the summit, shrubs with odd-pinnate lvs and medium-sized yellow fls in axillary racemes. 25, Old World.

1. Colutea arborescens L.   Bladder-senna. Shrub 2–4 m; lfls 7–13, oval to obovate, 1.5–2.5 cm; fls 2–8 in the raceme, yellow marked with red, 2 cm; fr membranous, ovoid or fusiform, 5–8 × 3 cm, indehiscent; $2n=16$. Native of the Mediterranean region, rarely escaped from cult. in our range.

18. ASTRAGALUS L.   Milk-vetch. Cal-tube campanulate to cylindric, often oblique; cal-lobes triangular to linear-subulate; standard obovate-oblong to rotund, usually surpassing the obtuse or emarginate wings; keel obtuse to acute, not beaked; stamens 10, diadelphous; fr highly variable, usually dehiscent; ours perennial herbs with numerous pinnate lfls and anthocyanic to ochroleucous or white fls in axillary racemes. (*Atelophragma, Batidophaca, Geoprumnon, Holophacos, Homalobus, Phaca, Pisophaca, Tium, Xylophacos.*) 1500+, widespread.

1 Fr unilocular, although one or both sutures may be ± inflexed or intruded.
  2 Ovary and pod strictly glabrous.
    3 Pod obviously stipitate, drooping; cor white or ochroleucous.
      4 Fr compressed, both sutures acute, the body 1–1.5 cm; cor scarcely 1 cm .... 1. *A. tenellus.*
      4 Fr triquetrous, the body 2–3 cm; cor 1.5–2 cm .... 2. *A. racemosus.*
    3 Pod sessile, erect or spreading.
      5 Fr bladdery-inflated; fls whitish; stems generally 3–9 dm and ± erect .... 3. *A. neglectus.*
      5 Fr narrow, not inflated; fls typically anthocyanic; stems low, 1–3 dm .... 4. *A. distortus.*
  2 Ovary and pod evidently (though sometimes sparingly) pubescent.
    6 Hairs malpighian (both ends free); stem (excluding peduncles) mostly under 1(–1.5) dm; fr sessile,
      ± spreading, leathery.
      7 Fls ochroleucous, 8–14 mm; cal-tube 3–3.5 mm, the lobes nearly as long .... 5. *A. lotiflorus.*
      7 Fls purple, 15–25 mm; cal-tube 6–10 mm, the lobes much shorter .... 6. *A. missouriensis.*
    6 Hairs basifixed; stems usually well over 1 dm; fr otherwise.
      8 Fr firm, subterete, not or scarcely sulcate; western dryland sp. .... 7. *A. flexuosus.*
      8 Fr. thin-walled, ± trigonous; mesophytic, chiefly northern spp.
        9 Stems clustered on a surficial root-crown; fr not or but slightly sulcate.
          10 Fr 13–25 mm, on a stipe 2–5 mm; fls 9–12 mm .... 10. *A. robbinsii.*
          10 Fr 5–13 mm, essentially sessile; fls 6–8 mm .... 11. *A. eucosmus.*
        9 Stems scattered on subterranean caudex-branches; fr deeply sulcate .... 12. *A. alpinus.*
1 Fr bilocular by intrusion of the sutures to form a complete partition.
  11 Racemes usually much shorter than the subtending lf; intr.
    12 Fr stipitate, linear-ellipsoid, 3–4 cm .... 13. *A. glycyphyllos.*
    12 Fr subsessile, broadly ovoid or subglobose, 0.6–1.4 cm .... 14. *A. cicer.*
  11 Racemes at least equaling the subtending lf; native.
    13 Pods ± erect, dehiscent at the tip, with normal walls; stipules connate.
      14 Fls erect or strongly ascending at anthesis, ochrolecuous or purple; fr sulcate.
        15 Hairs basifixed; fr loosely hirsute; stems scattered, from a subterranean source .... 15. *A. agrestis.*
        15 Hairs malpighian; fr finely strigose; stems clustered at the surface .... 16. *A. adsurgens.*
      14 Fls spreading or reflexed at anthesis, commonly ochroleucous; fr subterete .... 17. *A. canadensis.*
    13 Pods spreading or somewhat ascending, indehiscent or very tardily dehiscent, the valves thick
      and fleshy, becoming spongy in age; stipules free.
      16 Stems spreading-hairy; cal gibbous-saccate at base .... 8. *A. tennesseensis.*
      16 Stems strigose; cal not or only slightly gibbous .... 9. *A. crassicarpus.*

1. **Astragalus tenellus** Pursh. Pulse m.-v. Thinly strigose or subglabrous, leafy perennial, 2–5 dm, the stems branched, clustered on a stout root or branched crown; stipules connate; lfls 11–21, oblong to elliptic or obovate, 1–3 cm, strigose beneath; racemes numerous, short-peduncled, loose, 2–6 cm, with 4–20 ochroleucous fls 6–9 mm; cal-tube 2–2.5 mm, the subulate lobes 1–2.5 mm; fr drooping, compressed, oblong, glabrous, 1–1.5 cm, on a stipe to 7 mm; 2*n*=24. Prairies, hillsides, and open woods; Minn. to Yukon, s. to N.M. June. (*Homalobus t.; H. stipitatus.*)

2. **Astragalus racemosus** Pursh. Alkali m.-v. Stems clustered on a stout taproot and caudex, 2–6 dm, strigillose; lfls 15–31, narrow, 1–3.5 cm, strigillose beneath; racemes 5–10 cm, long-peduncled, with 20–70 nodding, ochroleucous fls 15–20 mm; cal-tube 5–6 mm, the subulate lobes 1.5–3 mm; fr drooping, linear-oblong, 2–3 cm, glabrous, triquetrous, the upper suture acute, the lower side flattened or broadly sulcate; stipe 3.5–7 mm; 2*n*=24. Badlands and gullies, often in seleniferous soil; Sask., N.D., and S.D. to Utah and n. Mex., and reported from w. Minn. May, June. Ours is var. *racemosus*. (*Tium r.*)

3. **Astragalus neglectus** (T. & G.) Sheldon. Cooper m.-v. Stems clustered on a taproot, ± erect, 3–9 dm, branching, usually hollow, glabrous or nearly so, amply leafy; stipules free; lfls 11–23, oblong to elliptic or obovate, 1–3 cm, strigose beneath; racemes several, each scarcely surpassing the subtending lf, with 10–20 white or chloroleucous fls 11–15 mm; cal-tube cylindric, 4–5 mm, the lobes half as long; fr erect, sessile, straight, fusiform or obliquely ovoid, inflated, 1.5–2 cm × 8–18 mm, glabrous, the sutures scarcely sulcate; 2*n*=22. River-banks and lake-shores, especially on limestone; irregularly from c. N.Y. and se. Ont. to Sask.; N.D., and S. D.; Grant Co., W.Va., and Alleghany Co., Va. June. (*A. cooperi; Phaca n.*)

4. **Astragalus distortus** T. & G. Ozark m.-v. Stems clustered on a stout taproot, 1–3 dm, ascending or spreading, sparsely strigulose, or glabrous below; stipules free; lfls 13–25, elliptic to obovate, 3–11 mm, sparsely strigose beneath; peduncles 5–14 cm; racemes short, with 10–20 anthocyanic to seldom white fls 10–16 mm; cal-tube 2–4 mm, the triangular lobes half as long; fr ascending, subsessile, broadly linear to ovate-elliptic, arcuate, 1.5–2.5 cm × 3.5–7 mm, turgid but not inflated, glabrous, sulcate on both sutures. Shale-barrens of W.Va., Md., and w. Va.; dry prairies, barrens, and open woods from Ill. and Io. to Miss., La., and Tex. May–July. Ours is var. *distortus*. (*Holophacos d.*)

5. **Astragalus lotiflorus** Hook. Lotus m.-v. Stems clustered on a stout taproot, up to ca 1 dm; stipules free; lfls 7–17, oblong to elliptic or oblanceolate, 5–20 mm, densely canescent with malpighian hairs beneath,

less so or subglabrous above; infls of chasmogamous or cleistogamous fls, the former type slender-pedunculate, subcapitate but elongating in age, with 5–17 ochroleucous fls 8–14 mm; cal-tube 3–3.5 mm, the lobes nearly or quite as long; fr spreading or ascending, sessile, fusiform, 1.5–3.5 cm × 5–8 mm, densely short-hairy, long-pointed, the upper margin nearly straight, the lower convex, not or scarcely sulcate; $2n=26$. Sandy prairies, dunes, and hills, Minn. and Io. to se. B.C., s. to Tex. and N.M. May, June. (*Batidophaca l.; B. cretacea.*) Some or all infls on many plants produce only cleistogamous fls 3–5 mm in short subsessile racemes.

6. **Astragalus missouriensis** Nutt. Nuttall m.-v. Stems clustered on a stout taproot, ± prostrate, up to 1(–1.5) dm, much overtopped by the lvs and peduncles; stipules free; lfls 9–17, elliptic or elliptic, 5–13 mm, densely canescent with malpighian hairs; peduncles 4–10 cm; racemes 2–4 cm, remaining short, with 5–15 purple fls 15–25 mm, these appearing narrower than in no. 5; cal-tube oblique, 6–10 mm, the subulate lobes much shorter; fr spreading, oblong, 1.5–2.5 cm × 5–9 mm, densely strigose, subterete, sessile; $2n=22$. Dry hillsides, bluffs, and prairies; w. Io. and w. Minn. to s. Alta., Tex., and N.M. May, June. Ours is var. *missouriensis.*

7. **Astragalus flexuosus** Douglas. Pliant m.-v. Stems scattered along the branches of a subterranean caudex, ± ascending, 2–6 dm, branched, thinly canescent, the hairs basifixed; stipules ± free; lfls 11–25, linear to oblong or oblanceolate, 5–15 mm, strigose beneath; racemes 2–3 cm at anthesis, later elongating, with 12–25 white or purplish fls 8–11 mm; cal-tube 3–4 mm, the lobes triangular, 0.5–1.5 mm; fr spreading or pendulous, 1.2–2.5 cm × 3–5 mm, linear-oblong, firm, subterete, scarcely or not sulcate, minutely strigose, sessile or nearly so; $2n=22$. Sandy prairies and bluffs; w. Minn. to Sask., se. B.C., N.M., and Ariz. Ours is var. *flexuosus.* (*Pisophaca f.; P. elongata.*)

8. **Astragalus tennesseensis** A. Gray. Tennessee m.-v. Stems clustered on a stout taproot, ± ascending, 1–5 dm, leafy, softly spreading-villous; lower stipules membranous, broadly ovate, 7–20 mm; lfls 23–33, narrowly oblong to elliptic, 1–2 cm, glabrous above, long-hairy beneath; cal-tube 6–11 mm, saccate-gibbous at base, villous; cor ochroleucous, 1.5–2 cm; fr ascending, bilocular, indehiscent, 2.5–4 cm × 7–13 mm, curved, acuminate, thick-walled, thinly long-hairy, wrinkled when dry, pithy within. Cedar glades of Tenn. and n. Ala.; upper Illinois and Rocky rivers in n.c. Ill. Apr.–June. (*Geoprumnon t.*)

9. **Astragalus crassicarpus** Nutt. Ground-plum; prairie-plum. Stems clustered on a stout taproot and caudex, 1–6 dm, decumbent with ascending tip, strigose; stipules 3–10 mm; lfls 15–29, oblanceolate to elliptic, 8–15 mm, appressed-hairy on both sides or glabrous above; fls 17–25 mm; cal-tube 6–8 mm; fr subglobose, abruptly pointed, indehiscent, glabrous, bilocular, the walls 2–5 mm thick at maturity; $2n=22$. Prairies and plains; Apr., May. Var. *crassicarpus,* with purple fls, with the pubescence of the cal appressed or ascending, partly or wholly of black hairs, and with the fr 1.5–2.5 cm, occurs from Wis. to Alta., s. to n. Mo., Okla., Tex., and N.M. (*Geoprumnon c.; Astragalus caryocarpus.*) Var. *trichocalyx* (Nutt.) Barneby, with ochroleucous fls (sometimes purple-tipped), with the cal densely villous-tomentulose, and with frs 2.5–3 cm, occurs from c. Ill. to Mo., Ark., Okla., and Tex. (*Geoprumnon t.; Astragalus mexicanus,* misapplied.) Three other vars. are more western.

10. **Astragalus robbinsii** (Oakes) A. Gray. Robbins m.-v. Stems clustered on the surficial crown of a taproot, 1–5 dm, glabrous or thinly strigose; lower stipules large and veiny, the upper smaller and spreading or deflexed; lfls 7–17, oblong or elliptic, 1–2 cm, glabrous on both sides or sparsely strigillose beneath; racemes at anthesis compact, 2–3 cm, later elongating; cal-tube black-hairy, 3–3.5 mm, the lobes subulate, half as long; wings and keel progressively shorter than the standard; fr black-hairy, reflexed, oblong or slightly curved, obscurely and bluntly trigonous, flattened or depressed on the lower side but hardly sulcate, the lower suture intruded to form a narrow partial partition; stipe 2–5 mm; $2n=16, 32$. Local on shelving limestone along rivers or in the mts.; Nf. and Lab. to N.S. and Vt., and widespread in the w. cordillera. (*Atelophragma r.*) Var. *robbinsii,* with relatively small whitish fls ca 9 mm, and with obscurely strigose fr 10–15 mm, formerly occurred near Burlington, Vt., but is now thought to be extinct. Var. *minor* (Hook.) Barneby, widespread in the w. cordillera, and occurring from Lab. and N.S. to Vt., has the lfls strigose beneath, purple fls 10–12 mm, and conspicuously strigillose fr 13–25 mm, short-acuminate but scarcely beaked. (Var. *blakei.*) Var. *jesupii* Eggleston, occurring along the Connecticut R. in Vt. and N.H., differs from var. *minor* in that the lfls are glabrous or nearly so beneath, and the thinly strigulose fr ends in a slender beak 1.5–3 mm. Other vars. occur to the n. and w.

11. **Astragalus eucosmus** B. L. Robinson. Elegant m.-v. Much like no. 10, but the fls 6–8 mm, the fr sessile, shorter and relatively broader, 5–13 × 3–5 mm, often more flattened; $2n=16, 32$. Rocky slopes or gravel-bars; Lab. to Alas., s. to Nf. and N. Me., and in the w. to Colo. June, July. (*Atelophragma elegans.*)

12. **Astragalus alpinus** L. Alpine m.-v. Stems glabrous or nearly so, decumbent, 1–5 dm, scattered on a freely branching, subterranean caudex; stipules connate; lfls mostly 15–25, narrowly oblong to oval, 1–2 cm, sparsely hairy on both sides or glabrous above; racemes long-peduncled, 2–4 cm, loosely few-fld, elongating with maturity; fls nodding, purple or purplish, 10–13 mm; cal-tube broad, 2 mm, the triangular lobes 1 mm; keel slightly to evidently surpassing the wings, equaling or a little shorter than the standard; fr pendulous, lance-oblong, 8–13 mm, ± falcate, pubescent with mixed black and white hairs, evidently trigonous, deeply sulcate on the lower side, the suture also intruded to form a partial partition; stipe 2–4 mm; $2n=16, 32+$. Circumboreal, extending s. on gravelly river-banks and lake-shores to Nf., Me., Vt., and Wis., and in the w. to Colo. June–Aug. Typical var. *alpinus,* with the cal and fr densely pubescent with loosely ascending hairs ca 0.5 ± mm, occurs in Bayfield Co., Wis., and is widespread far beyond our range. (*Atelophragma a.*) Var. *brunetianus* Fern., with the hairs of the pod shorter (0.2–0.4 mm) and usually appressed, occurs from e. Vt. to Que. and Nf. (*A. labradoricus.*)

13. Astragalus glycyphyllos L.   Licorice m.-v. Coarse perennial from a stout taproot; stems decumbent or procumbent, 3–10 dm; stipules foliaceous; lfls 9–15, oval, 1–4.5 cm, glabrous or nearly so, peduncles 2–4(–8) cm, shorter than the subtending lf; cor dingy ochroleucous, 11–14 mm; fr incurved-ascending, bilocular, glabrous, linear, 3–4 cm, on a stipe 2–4 mm; $2n=16$. Native of Europe, rarely established in fields and waste places, notably in Ind. and n. N.Y. May–Aug.

14. Astragalus cicer L.   Chick-pea m.-v. Diffuse, leafy perennial with prostrate or weakly ascending stems 2.5–7 dm; stipules of middle and upper lvs connate below or not, lanceolate, narrowly pointed, usually deflexed; lfls 17–29(–31), lance-elliptic or oblong, 0.5–3.5 cm, pubescent at least beneath; peduncles 4–11 cm, shorter than the subtending lf; cor ochroleucous, 12–16 mm; fr ascending or spreading, bilocular, hairy, subsessile, firm-walled, broadly ovoid or subglobose, 6–14 × 5–8(–10) mm. Native of Europe, casually intr. in disturbed moist places in w. U.S., and rarely established with us, as in Ingham Co., Mich.

15. Astragalus agrestis Douglas.   Field m.-v. Stems slender, 1–3 dm, scattered on slender rhizomes or arising from a subterranean root-crown, glabrous or sparsely strigose; stipules of the middle and upper lvs connate below, with greenish free blade; lfls 13–21, linear-oblong to lanceolate, sparsely hairy on both sides, the hairs basifixed; racemes dense, 2–4 cm, long-peduncled; fls strongly ascending or erect at anthesis, purple, 15–22 mm; cal ± hirsute, mostly with mixed black and white hairs, the subulate lobes half as long as the tube; fr erect, bilocular, obliquely ovoid, 7–10 mm, densely long-white-hirsute, deeply sulcate on the lower margin; $2n=16$. Prairies, moist meadows, and stream-banks; Minn. and Io. to Yukon, s. to Kans., N.M., and Calif.; Eurasia. May. (*A. goniatus; A. dasyglottis.*)

16. Astragalus adsurgens Pallas.   Standing m.-v. Much like no 15, but with pick-shaped (malpighian) hairs, sometimes more densely pubescent, and with the stems arising from a surficial crown or caudex surmounting the taproot; stipules soon becoming pale and scarious; fls purple to ochroleucous, 12–20 mm; fr densely strigulose; $2n=32$. Prairies, plains, and foothills, often in rocky or gravelly places; Minn., w. Ont., and n. Io. to N.M., Oreg., and Alas., and widespread in Asia. June. Ours is var. *robustior* Hook. (*A. chandonnettii; A. striatus.*)

17. Astragalus canadensis L.   Canada m.-v. Erect, rhizomatous, usually robust perennial to 15 dm; stipules connate, lanceolate to deltoid, 3–18 mm; lfls 15–35, oblong or elliptic, 1–4 cm, strigose beneath with malpighian hairs; racemes long-peduncled, 5–15 cm, with spreading or somewhat reflexed, white or ochroleucous fls 12–17 mm; pods numerous, crowded, erect, 10–15 mm, bilocular, subterete with the ventral suture keeled, glabrous, or seldom short-hairy; $2n=16$. Open woods, river-banks, and shores, usually in moist soil; throughout most of the U.S. and s. Can.; Siberia. June–Aug.

19. OXYTROPIS DC., nom. conserv.   Very much like *Astragalus*, except for the abruptly short-beaked keel; fr sessile, ovoid to cylindric, sulcate on the upper side and partly or wholly 2-locular by intrusion of the sutures; perennial herbs, acaulescent or nearly so, the lvs and peduncles crowded on a multicipital caudex surmounting a taproot; stipules in our spp. adnate to the petiole; fls purple to ochroleucous, in dense spikes or heads on long axillary scape-like peduncles; our spp. becoming 1–4 dm. 300, N. Temp.

1 Lfls mostly whorled, commonly in 4's; herbage densely long-villous .... 1. *O. splendens.*
1 Lfls opposite or scattered; pubescence various, but mostly not densely long-villous.
  2 Herbage strigose-canescent with malpighian hairs; fr leathery .... 2. *O. lambertii.*
  2 Herbage glandular or silky-pilose, the hairs basifixed; pods thinner.
    3 Herbage, bracts, and cal not viscid; bracts densely hairy on the back .... 3. *O. campestris.*
    3 Herbage, bracts, and cal glandular-viscid, the glands most readily visible on the cal; bracts
      glabrous or nearly so on the back (except for the glands) .... 4. *O. viscida.*

1. Oxytropis splendens Douglas.   Showy loco-weed. Densely long-villous throughout; lfls narrowly lanceolate, sharply acuminate, 1–2 cm, mostly in 4's; spikes 3–8 cm; bracts linear, mostly longer than the fls; fls red-purple, drying violet, 13–18 mm; fr ovoid, densely villous, 10–17 mm, short-beaked; $2n=16$. Prairies and plains; Minn. and w. Ont. to Alas., Alta., and n. N.M. July.

2. Oxytropis lambertii Pursh.   Purple l.-w. Lfls 9–19, linear to narrowly oblong, 1–2.5 cm, thinly strigose-canescent with malpighian hairs; spikes 4–10 cm; cal densely villous; fls purple (white), 1.5–2 cm; fr cylindric, very firm, sometimes soon glabrous, the body 8–15 mm, the beak 3–7 mm; $2n=48$. Dry prairies and plains; Minn. and s. Man. to nw. Mo., Okla., and Tex., w. to Mont., Utah, and Ariz. May, June. Ours is var. *lambertii.*

3. Oxytropis campestris (L.) DC.   Herbage, especially the lvs, densely villous-sericeous at first with long, lax but not strongly spreading hairs, later more thinly hairy or partly glabrous; lfls 15–31, lance-linear to lance-ovate, 5–25 mm; spikes dense, 2–4 cm, much longer in fr; fls purple, 1.5–2 cm; fr lance-ovoid, rather chartaceous, not rigid, sparsely or moderately pubescent with loose or subappressed hairs, 1.5–2.5 cm, including the beak; $2n=32$, 48. Rocks, cliffs, and gravelly shores; circumboreal, in Amer. s. to N.S., Me., Wis., Colo., and Oreg. July. The foregoing description is based primarily on var. *johannensis* Fern., which occurs in N.S., n. Me., and s. of the St. Lawrence R. in Que. Var. *chartacea* (Fassett) Barneby, from n. Wis., tends to have slightly looser and more persistent pubescence, and has smaller pods only 8–15 mm. (*O. chartacea.*)

4. Oxytropis viscida Nutt.   Plants pilose-hirsute (often thinly so) and ± viscid throughout, especially on the cal, with wart-like glands, staining paper yellowish in drying; lfls (19–)25–39, oblong-ovate to lance-linear, 5–17 mm; spikes 2–3 cm; fls in ours purple, 12–20 mm; fr ovoid to oblong, rather chartaceous, not rigid, short-hairy and usually verrucose-glandular, the body mostly 8–15 mm, the beak to 5 mm; $2n=32$. Mostly in alpine and subalpine habitats in the cordilleran region from Alas. and Yukon to Calif. and Colo., e. across most of subarctic Amer., and isolated in the Gaspé region and along the international boundary in ne. Minn. and adj. Ont., where it grows on slate cliffs. June. Ours is var. *viscida* (*O. ixodes*).

20. GLYCYRRHIZA L.   Licorice.   Cal bilabiate, the tube campanulate, oblique, the upper 2 lobes united part-way; pet usually acute, the standard oblong-obovate, tapering to the base, the wings and keel shorter, oblong, clawed; stamens 10, diadelphous for about half their length; anthers alternately large and small; ovary cylindric; style smooth, nearly straight; fr indehiscent or only tardily dehiscent, slightly flattened to somewhat inflated, glandular or spiny, few-seeded; gland-dotted perennial herbs with odd-pinnate lvs and dense axillary racemes of white, yellow, or blue fls. 20, ± cosmop.

1. Glycyrrhiza lepidota Pursh.   Wild l.   Colonial by creeping roots; stems to 1 m, the younger parts and the lower side of the lfls dotted with minute glands; lfls 11–19, oblong or lanceolate, those of the main axis 2.5–5 cm, the others smaller and narrower; racemes erect, dense, shorter than the subtending lvs; fls pale yellow, 12–15 mm; fr 1.5 cm, brown, densely beset with hooked prickles; $2n=16$. Moist prairies; Minn. to Alta. and Wash., s. to Ark., Tex., and Calif. and intr. in disturbed sites farther east. May, June. Our native plant, var. *lepidota,* has minutely pubescent stems; the far-western var. *glutinosa* (Nutt.) S. Wats., with the stem stipitate-glandular, is adventive in Va.

21. LOTUS L.   Cal campanulate or obconic, the elongate teeth subequal; pet clawed; standard obovate, not auriculate, its claw generally with thickened, infolded margins; keel-pet usually beaked; stamens 10, diadelphous; some or all of the filaments dilated at the tip; ovary sessile; pod several-seeded, dehiscent, oblong to linear, terete (in our spp.), flattened, or 4-angled; herbaceous or suffrutescent plants with pinnately compound, 1–5-foliolate lvs and solitary or umbellate fls. (*Acmispon, Hosackia.*) 100, mainly N. Temp.

1 Lfls 5, or apparently so; stipules none; fls mostly in compact umbels.
  2 Relatively robust, tetraploid, wide-lvd plants (details in descr.) .... 1. *L. corniculatus.*
  2 Relatively slender, diploid, narrow-lvd plants .... 2. *L. tenuis.*
1 Lfls 3; stipules represented by glands; fls solitary .... 3. *L. purshianus.*

1. Lotus corniculatus L.   Birdsfoot-trefoil.   Taprooted, usually ± glabrous perennial with prostrate to suberect stems to 6 dm; lvs subsessile, 5-foliolate, the lower pair of lfls evidently removed from the 3 crowded terminal lfls and resembling foliaceous stipules; lfls 5–15 mm, elliptic to oblanceolate, mostly 1.5–2.5 times as long as wide; fls mostly 4–8 together, in long-peduncled, head-like umbels from the upper axils; pedicels 1–3 mm; cal-tube 2.8–3.5 mm; cor mostly 10–16 mm, bright yellow, becoming orange and marked with brick-red; filaments unequal, the 5 longer dilated at the tip; fr 1.5–3.5 cm; $2n=24$. Native of Europe, cult. as a forage-crop, and widely established in meadows and disturbed habitats in the U.S., common in our range. June–Aug.

2. Lotus tenuis Waldst. & Kit.   B.-t.   Much like no. 1, and not always sharply distinct exomorphically, but less robust and typically in drier, more difficult habitats; lfls mostly linear or lance-linear and 2.5–5+ times as long as wide; fls mostly (1)2–5 together; cal-tube 1.8–2.8 mm; cor 7–10(–12) mm, yellow, drying orange; $2n=12$. Native of Europe, widely intr. in U.S., but more common in arid regions. June–Aug. (*L. corniculatus* var. *tenuifolius.*)

3. Lotus purshianus (Benth.) Clements & Clements.   Spanish clover.   Annual, ± erect, branching, 2–6 dm, lvs subsessile; stipules represented by minute glands; lfls 3, 1–2 cm, the terminal one stalked; fls solitary in the upper axils; pedicels about equaling the lvs, bearing a terminal bract as long as the fl; cor 5–8 mm, at first ochroleucous, turning pink or salmon, the keel yellow-tipped; filaments about equal, all dilated at the tip; fr spreading or deflexed, 2–3 cm; $2n=14$. Various habitats. May–Sept. Two vars.: Var. *purshianus,* usually pubescent, and with the lfls mostly 2–3.5 times as long as wide, occurs from the Pacific states and adj. Can. and Mex., e. to w. Minn., w. Mo., and Tex., and is occasionally adventive farther e. (*Hosackia americana.*) Var. *helleri* (Britton) Isely, usually glabrate or only slightly pubescent, and with the lfls mostly 4–5 times as long as wide, occurs in a disjunct area on the piedmont from s. Va. to n. Ga. (*L. helleri; Hosackia h.; Acmispon h.*)

22. ANTHYLLIS L.   Cal ellipsoid, becoming inflated, bilabiate with oblique orifice; pet long-clawed, the standard broadly obovate, auriculate at base, the wings and keel connivent; stamens 10, monadelphous nearly to the summit, the free tips of the filaments dilated, wider than the anthers; ovary stipitate; style with a tenuous basal part and an indurate distal part; fr enclosed by the cal, tardily or not at all dehiscent, 1–few-seeded; herbs or shrubs with odd-pinnate or 1-foliolate lvs and white, purple, or yellow fls in apparently terminal, usually pedunculate, head-like umbels. 20+, Mediterranean.

1. **Anthyllis vulneraria** L. Kidney-vetch. Annual or short-lived perennial, with several ascending or suberect stems to 5 dm; lvs mostly crowded below, the first ones long-petiolate and unifoliolate, obovate to lanceolate, 1–5 cm, the others mostly with 5–11 oblong to oblanceolate lfls; head 3–4 cm thick, subtended by 3-cleft bracts; fls 1.5 cm, yellow; cal hirsute, 1 cm; $2n=12$. Native of Europe, casually intr. here and there in our range.

23. **CORONILLA** L. Cal-tube campanulate to hemispheric, bilabiate, the broad lower lip 3-toothed, the narrow upper one shallowly cleft; pet subequal, clawed, the standard orbicular, the wings ovate-oblong, the keel-pet upcurved, lanceolate, acute; stamens diadelphous; fr few–several-seeded, transversely jointed; herbs or shrubs with 1–3-foliolate or odd-pinnate lvs and small or medium-sized fls in long-peduncled axillary umbels. 20, Europe, Mediterranean.

1. **Coronilla varia** L. Crown-vetch. Loosely ascending perennial 3–10 dm; lvs sessile, 6–15 cm; lfls 11–25, oblong to obovate, 1–2 cm; umbels 10–20-fld; fls 10–15 mm, pinkish; fr linear, 2–6 cm, 4-angled, with 3–7 joints; $2n=24$. Native of the Mediterranean region, widely planted along highways in our range, and frequently established elsewhere. May–Sept.

24. **ORNITHOPUS** L. Cal-tube much longer than the lanceolate teeth; standard obovate to orbicular; keel obtuse, shorter than the oblong wings; stamens 10, diadelphous; fr terete or compressed, transversely segmented and usually constricted between the segments, eventually breaking at the constrictions; herbs with odd-pinnate lvs of numerous lfls; fls white to pink or red, in small heads. 6, mainly Mediterranean.

1. **Ornithopus sativus** Brot. Serradela. Erect or ascending, pubescent annual 2–7 dm; lvs 4–7 cm; lfls 9–18 pairs, 4–8 mm; peduncles 2–4 cm, from the upper axils; heads 2–6-fld, subtended by a pinnate leafy bract; fls white or pink, 5–9 mm; fr with 3–5 strongly nerved oval joints each 3–4 mm, the terminal one beaked; $2n=14$. Native of the Mediterranean region, established on Long Island. June–Aug.

25. **ONONIS** L. Rest-harrow. Cal regular, deeply 5-cleft; standard orbicular; wings obovate-oblong, two-thirds as long as the standard; keel usually beaked and upcurved; stamens 10, monadelphous into a closed tube, the filaments alternately long and short; pod small, oblong or ovate, few-seeded, subtended or enclosed by the persistent cal; herbs or shrubs with stipules adnate to the petiole, serrate lfls, and solitary or paired axillary fls. 75, mainly Mediterranean.

1. **Ononis campestris** G. Koch. Suffruticose, ± glandular-pubescent plants, prostrate or creeping, 3–6 dm, the branches often ending in weak thorns; stipules about as long as the petiole; lfls 3, or only 1 when subtending fls, elliptic-oblong to oval, 1–2 cm; fls pale red, 1–2 cm; $2n=30$. Native of w. Europe, occasionally adventive in waste places about eastern cities. May–Sept. (*O. spinosa,* misapplied.)

26. **TRIFOLIUM** L. Clover. Cal-tube campanulate to tubular, sometimes oblique, the lobes setaceous to triangular, often unequal; pet all separate (spp. 15–17) or ± united into a tube, usually withering and persistent after anthesis; standard ovate to oblong or obovate, often folded about the wings or with only its summit outwardly curved; stamens 10, diadelphous, filaments dilated below the anthers; fr short, straight, often included in the persistent cal, dehiscent or indehiscent, 1–6-seeded; herbs with 3-foliolate, serrulate lvs and small fls in heads, spikes, or head-like racemes or umbels. In the absence of fr, our spp. may be distinguished from *Medicago* by their strongly bilabiate cal. 250, widespread.

1 Fls white to pink, red, purple, or crimson, never yellow.
  2 Fls essentially sessile in the heads.
    3 Cal-tube glabrous or uniformly pubescent, not gibbous.
      4 Individual fls 10–20 mm.

5 Stipules obtuse or abruptly narrowed to a short point, the free part shorter than the adnate
part.
  6 Heads globose to broadly round-ovoid, short-pedúncled or sessile; stipules abruptly
    narrowed to a short awn; lfls broadest near the middle .... 1. *T. pratense*.
  6 Heads usually ovoid or cylindric, long-peduncled; stipules obovate, obtuse; lfls broadly
    ovate .... 2. *T. incarnatum*.
5 Stipules gradually tapering to a long slender free tip longer than the adnate part.
  7 Lfls broadest at or below the middle; stem without deflexed hairs .... 3. *T. medium*.
  7 Lfls broadest well above the middle; stem with ± deflexed hairs .... 4. *T. hirtum*.
4 Individual fls 5–7 mm.
  8 Heads peduncled; sep exceeding the cor .... 5. *T. arvense*.
  8 Heads sessile; sep shorter than the cor .... 6. *T. striatum*.
3 Cal-tube much more densely pubescent on one side, gibbous or soon becoming so, at maturity
greatly inflated.
  9 Creeping perennial; fls in normal position, the standard uppermost .... 7. *T. fragiferum*.
  9 Ascending annual; fls inverted, the standard lowermost .... 8. *T. resupinatum*.
2 Fls evidently pedicellate in the heads, the pedicels commonly 2 mm or more.
  10 Lobes of the cal less than twice as long as the tube; intr., weedy spp.
    11 Stems creeping, rooting .... 9. *T. repens*.
    11 Stems ascending, not rooting .... 10. *T. hybridum*.
  10 Lobes of the cal 2–4 times as long as the tube; native spp.
    12 Fls 5–6 mm; wings acute .... 11. *T. carolinianum*.
    12 Fls 8–12 mm; wings obtuse.
      13. Plants stoloniferous, perennial, essentially glabrous; lfls broadly obovate .... 12. *T. stoloniferum*.
      13 Plants not stoloniferous; lfls hairy or less often glabrous; lfls of narrower shape.
        14 Lfls ovate to narrowly obovate; stems erect or ascending; annual or biennial .... 13. *T. reflexum*.
        14 Lfls linear to oblanceolate; stems prostrate; perennial .... 14. *T. virginicum*.
1 Fls yellow, or turning brown after anthesis.
  15 Lvs palmately trifoliolate, the lfls all sessile or nearly so: stipules about equaling the petiole .... 15. *T. aureum*.
  15 Lvs pinnately trifoliolate, the terminal lfl distinctly stalked; stipules about half as long as the
    petiole.
    16 Fls 3.5–5 mm; lf-rachis 1–3 mm .... 16. *T. campestre*.
    16 Fls 2.5–3.5 mm; lf-rachis scarcely 1 mm .... 17. *T. dubium*.

1. **Trifolium pratense** L. Red c. Short-lived perennial, ascending or suberect, to 8 dm, the stem appressed-hairy; stipules oblong, the free part short, abruptly narrowed to a short awn; lower lvs long-petioled, upper short-petioled to sessile; heads sessile or on peduncles to 2 cm, globose to round-ovoid; fls 13–20 mm; cal glabrous to sparsely pilose, the tube 3–4 mm, the lobes setaceous, one 4–7 mm, four 2–5 mm; cor magenta, varying to nearly white; standard obovate-oblong, equaling or a little longer than the oblong obtuse wings; $2n=14$. Native of Europe, widely cult. and escaped in fields and roadsides nearly throughout temperate N. Amer. May–Aug.

2. **Trifolium incarnatum** L. Crimson c. Annual, 3–8 dm, the stem appressed-hairy; stipules blunt, ± erose and usually red or purple around the summit; lfls broadly obovate, tapering to the base, broadly rounded or truncate above; heads ovoid or cylindric, 3–5 cm, on long, erect peduncles; fls 10–15 mm; cal-tube densely villous, 3 mm, the linear-subulate lobes subequal, 5–7 mm; cor crimson, straight, the standard oblanceolate, obtuse; $2n=14$. Native of Europe, found in old fields, waste places, and roadsides as an escape from cult. May–Aug.

3. **Trifolium medium** L. Zigzag c. Biennial or perennial, ascending, 2–5 dm, the stems sparsely appressed-pubescent; stipules tapering to a long, lance-linear tip; lfls elliptic or oblong, widest at or below the middle; heads subglobose, sessile or on peduncles to 3 cm; fls 14–20 mm; cal-tube glabrous or sparsely hirsute, conspicuously pubescent within at the throat, 3.5–4.5 mm, the lobes narrow, spreading hairy, one 4–6 mm, four 2–5 mm; pet purple, the standard obovate-oblong, about as long as the wings; $2n=ca$ 70–84+. Native of Europe, sometimes escaped from cult. in e. U.S. and N.S. June–Aug.

4. **Trifolium hirtum** All. Annual, 1–4 dm, the stem softly pubescent with somewhat deflexed hairs; stipules tapering to an elongate linear tip; lfls obovate, heads closely sessile, subglobose; fls 12–15 mm; cal densely hirsute with ascending hairs, the lobes setaceous, 4–6 mm, exceeding the tube; cor purple, the standard straight, very narrow, acute, much exceeding the wings; $2n=10$. Native of s. Europe, intr. as a weed in Va. Summer.

5. **Trifolium arvense** L. Rabbit-foot c. Annual, erect, freely branched, 1–4 dm, the stems soft-hairy; petioles 4–10 mm, exceeded by the subulate tips of the stipules; lfls narrowly oblanceolate, 1–2 cm, serrulate only at the tip; heads densely fld, ovoid to cylindric, 1–3 cm, on peduncles 1–3 cm; cal densely villous with hairs to 2 mm, the tube 1.5–2 mm, the setaceous lobes 3–5 mm, surpassing and partly concealing the small, whitish or pinkish cor; $2n=14$. Native of Eurasia and n. Afr., found as a casual weed of roadsides and waste places, especially in light, sandy soil, throughout much of the U.S. and s. Can. May–Sept.

6. **Trifolium striatum** L. Annual or biennial, decumbent or ascending, 1–3 dm; lfls obovate-oblong to obovate, 8–15 mm; heads ovoid, 10–15 mm, sessile in the upper axils and appearing terminal; fls 5–7 mm; cal densely pubescent with ascending hairs, the tube 2–3 mm, the subulate teeth 0.7–1.5 mm, the lowest one slightly the longer; pet bright pink, with darker veins; $2n=14$. Native of Europe, locally established as a weed along the eastern seaboard, notably at Cape Cod, Mass. June–Aug.

7. Trifolium fragiferum L. Strawberry-c. Habitally like *T. repens*; peduncles becoming 8–15 cm; heads globose or ovoid, 10–15 mm thick at anthesis; fls 6–7 mm; cal bilabiate, the upper lip 2–3 mm, half longer than the lower and much more villous, the lobes narrowly linear, 0.7–1 mm; cor rosy; cal at maturity becoming reddish, greatly enlarged, strongly reticulate, and gibbous on the upper side, the lower lip scarcely altered; $2n=16$. Native of Eurasia and n. Afr., becoming established as a lawn-weed here and there in our range. Summer.

8. Trifolium resupinatum L. Persian c. Ascending or decumbent annual; peduncles 2–5 cm; cal at maturity resembling that of no. 7; cor inverted, the standard toward the lower side of the head; $2n=16$. Native of Europe, intr. in grass-land and established at widely scattered stations in our range. Early summer.

9. Trifolium repens L. White c. Perennial; stems creeping, sending up long-petioled lvs and long-peduncled heads; stipules connate beyond the level of adnation to the petiole to form a pale, tubular-amplexicaul sheath, but the tips separate; lfls broadly elliptic to obovate, rounded to retuse at the summit, 1–2 cm; fls 7–11 mm, distinctly pedicellate; cal glabrous, the tube 1.8–3 mm, the nerves leading to the acute sinuses sharply defined and ending in a purple spot; cal-lobes narrowly triangular, acuminate, unequal, the longest about equaling the tube; cor white or tinged with pink, the standard elliptic-obovate, rounded at the tip, exceeding the obtuse wings; $2n=32$ (64). Native of Eurasia, commonly planted and escaped in lawns and roadsides throughout most of temperate N. Amer. All summer.

10. Trifolium hybridum L. Alsike c. Ascending perennial 3–8 dm; stipules lance-ovate, their tips free beyond the level of adnation to the petiole; lfls oval to elliptic, broadly rounded to retuse at the summit; heads numerous, globose, on peduncles 2–8 cm; fls 7–10 mm, distinctly pedicellate; cal glabrous, the nerves leading to the obtuse sinuses faint or obsolete; cal-lobes linear-subulate, not very unequal, 1.7–2.5 mm, slightly exceeding the tube; cor white and pink, turning brown after anthesis, the standard obovate-oblong, 2 mm longer than the obtuse wings; $2n=16$. Native of Eurasia, escaped from cult. throughout temperate N. Amer., commoner northward. Summer.

11. Trifolium carolinianum Michx. Carolina c. Perennial; stems tufted, branched from the base, 1–3 dm, prostrate to erect; lfls obovate, 8–15 mm, glabrous or sparsely pubescent; peduncles, pedicels and cal hirsutulous; heads globose, on peduncles to 12 cm; fls 5–6 mm; cal-tube 1 mm, the lobes unequal, the upper 2 the largest, 2.5–3.2 mm, narrowly triangular and conspicuously 3-nerved; cor purplish, the standard broadly ovate, longer than the sharply acute wings. Dry fields, open woods, and waste land; s. Va. to Fla. and Tex., n. in the interior to Mo. and Kans. Spring and summer.

12. Trifolium stoloniferum Muhl. Running buffalo-c. Stoloniferous perennial, essentially glabrous; flowering stems assurgent, 1–3 dm; lvs all long-petioled and arising from the ground level (either from the central crown or from stolons) except for a short-petioled, opposite pair subtending the fl-head; lfls broadly obovate, 2–3 cm long and wide, commonly shallowly retuse; stipules green, leafy, lanceolate to ovate, 1–2 cm; peduncles 3–8 cm; fls 9–12 mm; cal-tube 1.5–2.5 mm, the lobes subequal, 3–4.5 mm; cor white, often with pink-purple veins, the obovate standard conspicuously exceeding the obtuse wings; seeds 1.7–2 mm; $2n=16$, 32. Moist, well drained disturbed woods associated with streams; rare; W.Va and O. to e. Kans. and n. Ark. Apr.–July.

13. Trifolium reflexum L. Annual buffalo-c. Biennial or winter-annual, forming an overwintering rosette, variously hairy or glabrous, erect or ascending 2–5 dm, the stems branched from the base, not stoloniferous; lvs with progressively shorter petioles upwards; lfls ovate, oval, elliptic, or narrowly obovate, 2–3 cm, much longer than wide, conspicuously serrate, often marked with purple bands; peduncles 2–8 cm; heads globose; fls 8–12 mm, on pedicels to 1 cm; cal-tube 1.2–1.7 mm, the lobes linear, 3–7 mm; standard red or white, broadly elliptic-obovate, exceeding the pink or white obtuse wings; seeds 1.2–1.5 mm; $2n=16$. Upland woods and prairies; w. N.Y. and s. Ont. to Io. and Kans., s. to Fla. and Tex. May–July.

14. Trifolium virginicum Small. Shale-barren c. Perennial from a stout taproot, not stoloniferous; stems prostrate, pubescent, 1–2 dm; lfls linear to oblanceolate, 2–6 cm, 3–7 times as long as wide, thinly silky beneath; peduncles thinly hairy, 4–10 cm, pedicels hairy; fls 10–12 mm; cal hirsutulous, the tube 1.5–2 mm, the lobes linear, 3–4.5 mm; cor nearly white, the standard broadly obovate-oblong, obtuse, exceeding the obtuse wings. Shale-barrens; s. Pa., Md., w. Va., and W.Va. May, June.

15. Trifolium aureum Pollich. Palmate hop-c. Annual or biennial, much-branched, mostly erect, 2–5 dm, the stem appressed-hairy; petioles 5–12 mm, about equaling the lance-oblong stipules; lfls all sessile or nearly so, oblanceolate to obovate-oblong, 1–2 cm; heads short-cylindric, 1–2 cm, on peduncles 1–4 cm from the upper axils; pedicels 0.5 mm; fls 5–7 mm; cal strongly 2-lipped, glabrous, the tube 5-nerved, 1 mm, the lobes lance-linear, the lower 1.2–1.8 mm, the upper half as long; cor yellow, the standard obovate, conspicuously striate-sulcate in age, usually serrulate, the wings dilated, concave and somewhat spreading at the summit; $2n=16$. Native of Eurasia, established as a weed along roadsides and in waste places from Nf. to B.C., to S.C. and Ark. May–Sept. (*T. agrarium,* perhaps misapplied.)

16. Trifolium campestre Schreber. Pinnate hop-c. Annual, much-branched, 1–4 dm, the stem pubescent; petioles 8–12 mm, twice as long as the obliquely ovate stipules; lfls oblong-obovate, 8–15 mm, the terminal one on a stalk 1–3 mm; heads globose to short-cylindric, 8–15 mm, compact, with usually 20–30 fls; fls 3.5–5 mm; cal as in no. 15; cor yellow, the standard obovate, with 5 conspicuous diagonal veins on each side, much exceeding the spoon-shaped, slightly divergent wings; $2n=14$. Native of Eurasia and n. Afr., established as a weed along roadsides and in waste places throughout much of N. Amer. May–Sept. (*T. procumbens,* perhaps misapplied.)

17. Trifolium dubium Sibth. Little hop-c. Much like no. 16; stalk of the terminal lfl scarcely 1 mm; heads smaller, 5–10 mm, commonly with 5–15 fls each 2.5–3.5 mm; standard inconspicuously veined; $2n=28$, 32. Native of Europe, established as a weed in waste ground throughout much of N. Amer. Summer.

27. MELILOTUS Miller. Sweet clover. Cal eventually deciduous, the tube campanulate, the lobes subequal, subulate to lanceolate, acute or acuminate; pet separate; standard oblong to obovate, usually longer than the others; wings and keel coherent, but the wings not spurred on the surface and with only a short basilateral lobe; stamens 10, diadelphous; ovary short, sessile or somewhat stipitate, but the style ± elongate; pod ovate or rotund, slightly compressed to subglobose, 1–4-seeded, usually indehiscent; annual or biennial (seldom short-lived perennial) herbs with 3-foliolate, serrulate lvs, the terminal lfl stalked; fls white or yellow, small, numerous in elongate peduncled racemes from the upper axils, with a repeating trip-mechanism; stipules partly adnate to the petiole. 20, Old World.

```
1 Fls white  ............................................................................................  1. M. alba.
1 Fls yellow.
   2 Pods glabrous.
      3 Fls 4.5–7 mm, on decurved pedicels 1.5–2 mm; fr 3–5 mm  ...........................  2. M. officinalis.
      3 Fls 2–3 mm, on ascending pedicels 0.5–0.8 mm; fr 1.5–3 mm  .......................  3. M. indica.
   2 Pods minutely pubescent, the hairs visible under a 10× lens  ....................  4. M. altissima.
```

1. Melilotus alba Medikus. White s.c. Erect, taprooted biennial (annual) 1–3 m; lfls lanceolate or oblanceolate to narrowly oblong or rarely ovate, 1–2.5 cm; racemes numerous, 5–20 cm (peduncle included); pedicels 1–2 mm, fls white, 3.5–5(–6) mm, the banner distinctly longer than the wings; ovules 2–4; fr 2.5–3 mm, reticulate-veiny; seeds 1 or 2; $2n=16$. Native of Eurasia, established as a weed along roadsides and in waste places, especially in calcareous soil, from N.S. to B.C., s. to Mex. and W.I. Summer and Fall.

2. Melilotus officinalis (L.) Pallas. Yellow s.c. Ascending or erect, taprooted biennial (annual) 0.5–1.5(–2) m; lfls oblanceolate to obovate, 1–2.5 cm; racemes 5–15 cm (peduncle included); pedicels 1.5–2 mm, decurved; fls yellow, 4.5–7 mm, the wings about as long as the banner; cal-teeth narrowly lanceolate, subulate; ovules 4–7; fr glabrous, 3–5 mm, strongly cross-ribbed; seeds usually 1 or 2; $2n=16$. Native of Eurasia, established as a weed of waste places throughout our range, w. to the Pacific, not common southward. Summer.

3. Melilotus indica (L.) All. Ascending or erect annual 2–6 dm; racemes very densely fld, 3–8 cm; pedicels at anthesis 0.5–0.8 mm, ascending; fls yellow, 2–3 mm; cal-teeth lance-oblong, obtuse; fr 1.5–3 mm, strongly reticulate-veiny; otherwise as no. 2; $2n=16$. Native of the Mediterranean region, now a cosmopolitan weed, abundant in the s. and Pacific states, and adventive in our range. Summer.

4. Melilotus altissima Thuill. Much like no. 2, but sometimes short-lived perennial, and with pubescent, reticulate-veined, mostly 2-seeded frs 5–6 mm; $2n=16$. Native of Eurasia, adventive here and there in our range.

28. MEDICAGO L. Cal-tube campanulate, the 5 lobes subequal; standard obovate or oblong, longer than the oblong erect wings, these with a prominent spur on the surface and a large retrorse basilateral lobe; keel blunt, shorter than and adherent to the wings; stamens 10, diadelphous, all free from the cor; style short; pod straight to more often ± coiled, usually indehiscent, 1–several-seeded; herbs with 3-foliolate, serrulate lvs, the terminal lfl stalked; fls discharging pollen by an explosive trip-mechanism, small, yellow or blue, in axillary heads or short head-like racemes. 80, Old World, mainly Eurasia.

```
1 Plants deep-rooted, long-lived perennials; fls 6–12 mm  .............................................  1. M. sativa.
1 Plants annual or biennial; fls yellow, 2–4 mm.
   2 Pod reniform, unarmed  ...............................................................  2. M. lupulina.
   2 Pods closely coiled, spiny.
      3 Stem glabrous or nearly so; stipules deeply incised  .....................................  3. M. polymorpha.
      3 Stem softly and densely pubescent; stipules entire or minutely toothed  .........................  4. M. minima.
```

1. Medicago sativa L. Alfalfa. Deep-rooted perennial with slender stems up to 1 m; stipules lance-ovate, toothed; lfls oblanceolate or oblong to narrowly obovate, 1.5–3 cm, toothed at the summit; peduncles erect, about equaling the subtending lvs; heads subglobose to short-cylindric; fls 6–12 mm, on pedicels 2–3 mm; cal-lobes lance-linear, 1–3 mm, about equaling the tube; pod finely hairy; $2n=16$, 32. June–Sept.

A highly diversified species, with pronounced but confluent variation. The two subspecies characterized below are connected by a hybrid swarm called ssp. × *varia* (T. Martyn) Arcangeli. Additional variants may be characterized in more detailed treatments.

Ssp. sativa. Alfalfa, blue a. Decumbent or loosely erect, to 1 m; fls blue-violet; pods coiled into 1.5–4 complete turns. Native of the Caspian area, widely cult. and escaped in temperate regions, found throughout the U.S. and s. Can.

Ssp. falcata (L.) Arcangeli. Yellow a. Laxer than ssp. *sativa*, often ± prostrate; fls yellow; pods falcate (or nearly straight) or coiled up to 1.5 turns. Native of Siberia and n. Europe, now widely cult. and escaped, but not so common as ssp. *sativa*.

2. Medicago lupulina L. Black medick. Annual or biennial with prostrate or ascending stems to 8 dm; stipules lanceolate, entire or toothed; lfls elliptic to obovate, 1–2 cm; peduncles much exceeding the subtending lvs; heads globose to short-cylindric, to 1 cm; fls yellow, 10–50 per head, 2–4 mm; pods nearly black, 2–3 mm, reniform, unarmed, 1-seeded, the conspicuous veins tending to be longitudinal; $2n=16$. Native of Eurasia, common as a weed in our range and elsewhere. May–Sept.

3. Medicago polymorpha L. Smooth bur-clover. Annual; stems prostrate to ascending, branched from the base, usually glabrous or nearly so, 2–5 dm; stipules deeply incised; lfls triangular-obovate, 6–15 mm, two-thirds as wide, obtuse to retuse, glabrous, or obscurely villosulous beneath; peduncles about equaling the subtending petioles; fls yellow, mostly 2–5(–8) per head, 2.5–4 mm; fr 2–5 times spirally coiled, 4–6 mm wide excluding the spines, these 2–3 mm, arising in 2 rows from an elevated ridge, minutely hooked at the tip; $2n=14$. Native of the Mediterranean region, now established as a weed in s. U.S. and the Pacific states, and occasionally adventive in our range. Summer. (*M. hispida.*)

4. Medicago minima (L.) Bartal. Downy bur-clover. Much like no. 3, but softly pubescent throughout; stipules entire or minutely toothed; lvs often smaller and shorter-petiolate; peduncles surpassed by the subtending lvs; $2n=16$. Native of Eurasia, established as a weed in sandy soil from se. Va. to Fla. and Tex., and occasionally adventive farther n. May–Sept.

29. VICIA L. Vetch. Cal regular or irregular, often gibbous; standard with a broad claw overlapping the wings, its blade obovate to subrotund; wings oblong or narrowly obovate, adherent to and usually surpassing the keel; stamens 10, diadelphous, the tube ending obliquely; style pubescent all around at the summit, or sometimes on the outer (abaxial) side only, otherwise glabrous; fr flat to terete, dehiscent, in some spp. transversely septate; seeds 2–many; herbs with usually small stipules and once pinnate lvs, the terminal lfl generally represented by a tendril; stem never winged; fls in axillary racemes or few-fld axillary clusters. 140, widespread.

1 Infl of solitary or paired, sessile or subsessile, axillary fls or few-fld, subsessile, axillary racemes; annual, except no. 5; style pubescent on the outer (abaxial) side near the tip. (*Vicia*, except spp. 6 & 7, *Fava*.)
  2 Tendrils present, at least on the upper lvs; fr mostly 1.5–7 cm.
    3 Cal regular or nearly so, the lobes of similar shape and not very unequal, more than half as long as the tube.
      4 Fls 10–30 mm; cal 8–17 mm; tendrils mostly branched.
        5 Fls violet to white, without yellow markings.
          6 Fls 18–30 mm; pod pale brown at maturity .... 1. *V. sativa.*
          6 Fls 10–18 mm; pod nearly black at maturity .... 2. *V. angustifolia.*
        5 Fls dull yellow, or yellow marked with violet .... 3. *V. grandiflora.*
      4 Fls 5–8 mm; cal 4–6 mm; tendrils all simple .... 4. *V. lathyroides.*
    3 Cal distinctly irregular, the upper 2 lobes much shorter than the lower and less than half as long as the tube.
      7 Lfls commonly 4–9 pairs, 1–2.5 cm; perennial .... 5. *V. sepium.*
      7 Lfls 1–3 pairs 3–6 cm; annual .... 6. *V. narbonensis.*
  2 Tendrils none, fr mostly 8–20 cm .... 7. *V. faba.*
1 Infl of distinctly peduncled racemes or long-peduncled fls; annual to perennial; style pubescent all around near the tip, or (No. 8) glabrous.
  8 Fls 3–7 mm, white or whitish, solitary or in racemes of 2–8; annual. (*Ervum.*)
    9 Cal-lobes equal or nearly so; fr mostly 2-seeded, hirsute .... 8. *V. hirsuta.*
    9 Cal-lobes distinctly unequal; fr mostly 4-seeded, glabrous .... 9. *V. tetrasperma.*
  8 Fls over 7 mm, blue or white, often more than 8 in the raceme. (*Cracca.*)
    10 Cal very gibbous at base, the pedicel appearing ventral; mostly annual or biennial.
      11 Pubescence of the raceme of long spreading hairs .... 10. *V. villosa.*
      11 Pubescence of the raceme of short appressed or incurved hairs .... 11. *V. dasycarpa.*
    10 Cal only slightly gibbous at base, the pedicel appearing basal; perennial.
      12 Stipules all or mostly sharply serrate; fls 15–27 mm .... 12. *V. americana.*
      12 Stipules entire; fls 8–13 mm.
        13 Fls white (keel blue-tipped), in loose racemes; cal-lobes subequal .... 13. *V. caroliniana.*
        13 Fls blue (rarely white), in dense racemes, cal-lobes very unequal .... 14. *V. cracca.*

1. Vicia sativa L. Common v. Slender annual to 1 m, erect to ascending, often climbing; lfls 4–8 pairs, oblong to elliptic or even obovate, or sometimes linear, 1–3.5 cm; stipules half-sagittate, often sharply serrate, with a glandular spot beneath; fls mostly paired in the upper axils, subsessile, violet or purple, rarely white,

18–30 mm; cal-tube campanulate, 5–7 mm, its lobes linear-subulate, the upper 3–7 mm, the lower 4–9 mm; fr flattened, light brown, 2.5–7 cm, 4–12-seeded; $2n=12$. Native of s. Europe, widely cult. in numerous races, and often escaped in our range. July–Sept.

2. Vicia angustifolia L.   Narrow-lvd v.   Much like no. 1; lfls 3–5 pairs, linear to narrowly elliptic, seldom broader; stipules with or without a glandular spot beneath; fls 12–18 mm; cal-tube 4–6 mm, the lobes 3–6 mm; fr nearly black; $2n=12, 14$. Native of Europe, established in fields, roadsides, and waste places throughout our range. Perhaps better treated as *V. sativa* var. *angustifolia* (L.) Ehrh.

3. Vicia grandiflora Scop.   Bigflower-v.   Ascending annual to 6 dm; lfls 3–7 pairs, 1–2 cm; fls usually paired (–4) in the upper axils, 25–35 mm, dull yellow in aspect, or the standard marked with violet, the keel commonly dark purple; cal-tube 6–9 mm, the linear-subulate subequal lobes half to two-thirds as long; style with a conspicuous tuft of hairs just behind the tip on the outer (abaxial) side, glabrous on the other side; $2n=14$. Native of sw. Europe and w. Asia, occasionally intr. in our range from N.Y. to Va. and Ky., and in se. U.S.

4. Vicia lathyroides L.   Spring-v.   Weak annual, 5–25 cm, branched at base; lfls 2–4 pairs, to 1.5 cm; tendrils simple; stipules entire; fls 1 or 2 in the axils, blue or violet, 5–8 mm, subsessile; cal 4–6 mm; fr 1.5–3 cm, 6–12-seeded; $2n = 12$. Native of the Mediterranean region, locally intr. in our range.

5. Vicia sepium L.   Hedge-v.   Erect or climbing perennial to 10 dm; stipules small, half-sagittate or ovate; lfls 4–9 pairs, ovate to ovate-oblong, entire, 1–2.5 cm; fls 2–6 in subsessile axillary clusters, blue-violet to white, 8–15 mm; cal distinctly irregular, the tube 4–5 mm, brown-villous, the lobes abruptly narrowed above a broadly triangular base into a slender tip, the lower half as long as the tube, the upper much shorter; fr 2–3.5 cm × 5–8 mm, 3–7-seeded, with smooth sutures; $2n=14$. Native of Europe, intr. here and there in our range, especially northward.

6. Vicia narbonensis L.   Narbonne-v.   Coarse, erect annual 3–6 dm with 4-angled stems, large semi-cordate stipules and ample lvs with tendrils; lfls 1–3 pairs, obovate to elliptic, serrate or entire, 3–6 cm; fls 1 or 2 in the axils, blue or violet, 15–25 mm; cal distinctly bilabiate; fr 4–7 cm × 10–15 mm, 4–6-seeded, with pectinate-fringed sutures; $2n=14$. Native of the Mediterranean region, established as a weed near Washington, D.C.

7. Vicia faba L.   Broad bean.   Stout erect annual 5–20 dm with 4-angled stems and no tendrils; lfls 2 or 3 pairs, large, ovate or elliptic, 4–10 cm; fls 20–30 mm, mostly white, veined with violet, the wings with a large violet spot; fr 8–20 × 1–3 cm, with 2–4 large seeds; $2n=14$. Long cult. in Europe, occasionally cult. in Amer. and rarely escaped.

8. Vicia hirsuta (L.) S. F. Gray.   Tiny v.   Slender annual, climbing or decumbent, 3–6 dm; lfls usually 6–8 pairs, seldom fewer, linear to narrowly elliptic, 5–15 mm; peduncles 1–3 cm, with 3–8 whitish fls 3–4 mm; cal-lobes subulate, subequal, 1–1.5 mm, about equaling the tube; style not hairy, only the capitate stigma papillate; fr flattened, hirsute, 6–10 mm, normally 2-seeded; $2n=14$. Native of Europe, intr. in fields, roadsides, and waste places throughout most of the U.S. and s. Can. May–Aug.

9. Vicia tetrasperma (L.) Moench.   Four-seed v.   Slender annual, climbing or decumbent, 3–6 dm; lfls mostly 2–5 pairs, seldom more, linear-oblong to oblanceolate or narrowly elliptic, 1–2 cm; peduncles 1–3 cm, with 1 or 2(–4) light purple to white fls 3–7 mm; lowest cal-lobe linear, about equaling the tube, the upper much shorter and triangular; style minutely hairy all around near the tip; fr flat, glabrous, with (3)4(–6) seeds; $2n=14$. Native of Eurasia, intr. in fields and waste places from Que. to Wis., s. to Fla. and Tex. May–Aug.

10. Vicia villosa Roth.   Hairy v.   Annual or biennial to 1 m, the stems spreading-villous, especially upwards; lfls usually 5–10 pairs, narrowly oblong to lance-linear, 1–2.5 cm; racemes long-pedunculed, dense, secund, with 10–40 fls; cal irregular, villous, the tube 2.3–4 mm, very gibbous, the pedicel apparently ventral; upper lobes linear-triangular, 0.8–1.5 mm; lateral and lower lobes linear above a triangular base, the lowest 2–5 mm, long-villous; cor slender, 12–20 mm, the spreading blade of the standard less than half as long as the claw; fr 2–3 cm, glabrous; $2n=14$. Native of Europe, intr. in fields, roadsides, and waste places throughout most of the U.S. and s. Can. June–Aug.

11. Vicia dasycarpa Tenore.   Woolly-pod v.   Much like no. 10; stem and lvs appressed-hairy to glabrate; racemes mostly 5–15-fld; cal-lobes a little shorter, scarcely villous; pod ± pubescent, varying to glabrous; $2n=14$. Native of s. Europe, intr. in fields, roadsides and waste places at scattered stations from Me. to Mont., s. to Ga. and Calif. May–Aug. (*V. villosa* ssp. *eriocarpa* and var. *glabrescens*.)

12. Vicia americana Muhl.   American v.   Trailing or climbing perennial to 1 m; stipules all or mostly sharply serrate; lfls usually 4–8 pairs, 1.5–3 cm; racemes shorter than the subtending lvs, loose, with 2–9 blue-purple fls 15–27 mm; cal-tube very oblique, 3.5–5.5 mm, the upper lobes to 1 mm, the lower 1.2–3 mm; fr 2.5–3.5 cm; $2n=14, 28$. Moist woods; c. N.Y. and s. Ont., s. in the mts. to w. Va., w. to Minn. and Mo., and widely distributed in w. N. Amer. May–July. Most of our plants are the widespread var. *americana*, with elliptic or oblong lfls 5–12 mm wide, the mucronate summit otherwise obtuse to broadly rounded, the numerous lateral veins leaving the midrib at an angle of about 45 degrees and branching and anastomosing before reaching the margin. The var. *minor* Hook. (var. *angustifolia*, *V. sparsifolia*) grows chiefly on the Great Plains and westward, but sometimes occurs within our western border; it has thick, narrow, densely puberulent lfls with prominent, essentially unbranched lateral veins that leave the midrib at a very narrow angle. Occasional plants in our range with truncate or emarginate lfls sharply toothed toward the summit, otherwise like var. *americana*, resemble or actually represent the otherwise chiefly cordilleran var. *truncata* (Nutt.) Brewer. (*V. truncata.*)

13. Vicia caroliniana Walter. Pale v. Trailing or climbing perennial to 1 m; lfls usually 5–10 pairs, elliptic to oblong-lanceolate, 1–2 cm, rounded to obscurely retuse at the tip, mucronate; stipules lanceolate or half-sagittate, entire; racemes (peduncle included) 6–10 cm, loose, with 7–20 white fls 8–12 mm; cal nearly regular, 3 mm or less, the lobes essentially equal, broadly triangular, under 1 mm; fr 1.5–3 cm; 2*n*=14. Moist woods and thickets; N.Y. to Wis., s. to Fla. and Tex. May, June.

14. Vicia cracca L. Bird-v. Trailing or climbing perennial to 1 m; lfls usually 5–11 pairs, linear to narrowly oblong, 1–3 cm, with few lateral veins leaving the midrib at a very narrow angle; stipules entire; racemes long-peduncled, dense, secund, equaling or exceeding the subtending lf, with 20–50 crowded blue (white) fls 8–13 mm; cal-tube oblique, gibbous at base, 2–3 mm, the upper lobes broadly triangular, 0.3–0.7 mm, the lowest linear-triangular, 1.3–2.5 mm; blade of the standard about as long as the claw; 2*n*=14, 28. Fields, roadsides, and meadows; native of Eurasia and possibly ne. N. Amer., widely naturalized in n. U.S. and s. Can., especially eastward, and found essentially throughout our range. June–Aug.

V. tenuifolia Roth, much like *V. cracca*, but commonly with more elongate peduncles and with 10–25 fls 12–16 mm, the blade of the standard twice as long as the claw, is rarely intr. in our range.

30. LATHYRUS L. Vetchling; wild pea. Cal regular or irregular, usually gibbous at base; standard broadly obovate to rotund, the fls typically appearing wider in proportion to length than in *Vicia*; wings narrowly to broadly obovate, lightly coherent with the upcurved keel; stamens 10, diadelphous, the tube ± truncate at the end; style usually bent at nearly a right angle, bearded along the inner (adaxial) side distally; fr flat to terete with 2–many seeds, dehiscent; herbs, the stems often winged, the lvs pinnately compound, usually (incl. all our spp.) ending in a tendril; stipules often notably large; fls red-purple to white or yellow, in racemes. 150, widespread.

1 Lfls 2 or more pairs; cal irregular; native spp.
  2 Stipules essentially symmetrical, basally attached, with 2 basal lobes .... 1. *L. maritimus.*
  2 Stipules distinctly asymmetrical, laterally attached (i.e., with only one basal lobe).
    3 Stipules semicordate (the basal lobe broadly rounded); fls ochroleucous .... 2. *L. ochroleucus.*
    3 Stipules semisagittate (the basal lobe sharp); fls red-purple (white).
      4 Racemes with (5-)10–20 fls; lfls 8–12; stems wingless; dry woods .... 3. *L. venosus.*
      4 Racemes with 2–6(–9) fls; lfls 4–8(10); stems often winged; moist places .... 4. *L. palustris.*
1 Lfls a single pair; cal regular or irregular; intr. spp.
  5 Stem wingless.
    6 Fls yellow .... 5. *L. pratensis.*
    6 Fls red-purple .... 6. *L. tuberosus.*
  5 Stem winged.
    7 Annual; fr hairy; cal-lobes essentially alike in size and shape .... 7. *L. hirsutus.*
    7 Perennial; fr glabrous; cal-lobes unequal, the lower twice as long as the upper .... 8. *L. latifolius.*

1. Lathyrus maritimus (L.) Bigelow. Beach-p. Stout, rhizomatous perennial, decumbent to suberect, to 1 m; stipules foliaceous, broadly ovate, 1.5–4 × 1–2.5 cm, symmetrical, with 2 ± well developed basal lobes; lfls 3–6 pairs, somewhat fleshy, oblong to obovate, 3–5 cm, half as wide; peduncles equaling or shorter than the subtending lvs, with usually 5–10 purple fls (1.2–)1.5–2.5(–3) cm; cal irregular, the lowest lobe lance-linear, nearly twice as long as the triangular upper ones; 2*n*=24. Sea-beaches and lake-shores; circumboreal, in our range s. on the Atlantic coast to N.J., on the shores of Lake Champlain, Oneida Lake, N.Y., and the Great Lakes. June–Aug. (*L. japonicus.*) Glabrous plants, in our range abundant on the Great Lakes and occasional along the seacoast, are var. *glaber* (Ser.) Eames. Plants pubescent at least on the cal, pedicels, and lower lf-surface, abundant on the coast and very rare on the Great Lakes, are var. *pellitus* (Fern.) Gleason.

2. Lathyrus ochroleucus Hook. White p. Glabrous rhizomatous perennial, to 8 dm; stipules semicordate, 1.5–3 cm, the broadly rounded basal lobe often irregularly toothed; lfls 3–5 pairs, thin, elliptic to lance-ovate or obovate, 2.5–5 cm, racemes shorter than the subtending lf, with 5–10 ochroleucous fls 12–18 mm; cal irregular, the upper lobes triangular, a fourth to half as long as the lower, the lateral lobes broadest distinctly above the base; 2*n*=14. Dry upland woods and thickets; Vt. to Man. and nw. Can., s. to Pa., Ill., Io., and Wash. May–July.

3. Lathyrus venosus Muhl. Forest-p. Stout, rhizomatous perennial to 1 m; stipules narrow, semisagittate; lfls 8–12, elliptic, 3–6 cm; peduncles stout, shorter than to equaling the subtending lf, with a dense raceme of purple fls 12–20 mm; cal distinctly oblique, the lower teeth much exceeding the upper but shorter than the tube; 2*n*=28. Woods and thickets. June, July. Var. *venosus*, with the cal and herbage glabrous or nearly so, or the stems slightly hairy, occurs mostly in the mts. from e. Pa. and adj. N.J. to e. W.Va. and Ga. Var. *intonsus* Butters & St. John, with the cal and herbage sparsely to densely hairy, occurs from s. Ont. to Minn. and Sask., s. to Ind. and Mo., as well as in the range of var. *venosus* and at one station in e. Que.

4. Lathyrus palustris L. Marsh-p. Slender rhizomatous climbing perennial, to 1 m, the stem often winged; stipules semisagittate, often serrate; lfls 4–8(10), linear to elliptic, 2–8 cm × 3–20 mm; peduncles about equaling the subtending lf, with a raceme of 2–6(–9) red-purple (whitish) fls 12–20 mm; cal distinctly oblique, the lower lobes much longer than the upper; $2n=42$. Wet meadows, swamps, shores, and wet woods; circumboreal, in Amer. s. to N.J., Pa., O., Ind., Mo., Colo., and Calif. June, July. Var. *palustris,* with winged stems and usually 6–8 lfls, occurs throughout the range of the sp. Typical var. *palustris* is glabrous or nearly so; a pubescent form, called var. *pilosus* (Cham.) Ledeb., has the same distribution but is more common along the seacoast; it is only doubtfully significant. Forms with narrow and broad lfls occur in both typical var. *palustris* and the hairy phase. The well marked but still not wholly distinct var. *myrtifolius* (Muhl.) A. Gray, consistently glabrous, with wingless (but sometimes acutely ridged) stems and usually 4(6) lfls that tend to be shorter (2–4 cm), broader, and more rounded than in var. *palustris,* occurs chiefly from N.Y. to Wis. and Ill., especially about the Great Lakes.

5. Lathyrus pratensis L.   Meadow-p. Rhizomatous perennial, 3–8 dm, the stems angled but not winged; stipules lanceolate, 1–3 cm, about equaling the wingless petiole, with or without small basal lobes; lfls 2, lance-linear to narrowly elliptic, 2–4 cm; racemes long-peduncled, with 2–8 yellow fls 12–18 mm; $2n=7-42$. Native of Eurasia, intr. in waste places here and there in our range.

6. Lathyrus tuberosus L.   Tuberous vetchling. Rhizomatous perennial with small tubers; stems 2-angled, wingless, to 8 dm; stipules lanceolate, 5–12 mm, with a single short basal lobe; lfls 2, oblanceolate to elliptic, 1.5–3 cm; peduncles 5–8 cm, bearing a crowded raceme of 2–10 red-purple fls 12–16 mm, the standard broader than long; $2n=14$. Native of Europe and w. Asia, intr. at a few scattered stations chiefly toward the n. part of our range.

7. Lathyrus hirsutus L.   Annual; stem winged, often branched from the base, to 1 m; stipules lanceolate, with a short basal lobe; lfls lance-linear, 2–5 cm; peduncles 3–10 cm, with 1–3 violet to pink or white fls 1–1.5 cm; cal-lobes ovate or lance-ovate, subequal, 2.5–3.5 mm, acuminate, about equaling the tube; fr 2–5 cm, papillose-hirsute, 5–10-seeded; $2n=14$. Native of Europe, intr. in se. Va. and at scattered stations in s. U.S.

L. odoratus L., the sweet pea, a wing-stemmed annual with fragrant fls 2–4 cm, occasionally escapes, but apparently does not persist.

8. Lathyrus latifolius L.   Everlasting p. Rhizomatous perennial; stems trailing or climbing to 2 m, broadly winged, 5–10 mm wide; stipules ovate or lanceolate, with a basal lobe, 2.5–4 cm; petiole broadly winged; lfls 2, lanceolate to elliptic, 4–8 × 1–3 cm; peduncles 10–20 cm, with a raceme of 4–10 anthocyanic or white fls 1.5–3 cm; cal-lobes very unequal; fr 6–10 cm, 10–15-seeded; $2n=14$. Native of s. Europe, tending to escape from cult. esp. toward the s. part of our range. June–Aug.

L. sylvestris L., with more narrowly winged stem and petioles, lance-linear stipules 1–3 cm, fls 13–20 mm; and fr 4–7 cm, otherwise much like no. 8, also occasionally escapes in our range.

31. HEDYSARUM L.   Cal-tube broadly campanulate; cal-lobes triangular, equal or nearly so; standard narrowly obovate, tapering to the base; wings oblong to nearly linear, usually shorter than the keel, the blade auriculate at base; keel-pet obtuse, semirhombic-ovate; stamens diadelphous; ovary elongate, fr indehiscent, transversely segmented and ultimately separating into joints; perennial herbs with odd-pinnate lvs and elongate, peduncled, axillary racemes. 100, mostly boreal and N. Temp.

1. Hedysarum alpinum L.   Stems mostly several, 3–7 dm; stipules connate opposite the lf-base; lfls 15–21, oblong to elliptic or lanceolate, those of the principal lvs 15–30 × 5–10 mm; racemes secund, 5–10 cm, on peduncles 8–15 cm; fls 12–16 mm, purple, seldom white; keel surpassing the standard and wings; fr with 8 (1)2–5 flat joints each 5–9 × 3.5–6 mm; $2n=14$. River-banks, hills, and ledges, in moist soil; Siberia and boreal N. Amer., s. to Me., n. Vt., n. Mich., Man., Wyo., and B.C. June–Aug. Our plants are var. *americanum* Michx. (*H. boreale,* misapplied.)

32. ONOBRYCHIS Miller.   Cal campanulate, nearly regular, the teeth slender; standard obcordate to obovate, about equaling the keel and much longer than the wings; stamens diadelphous, the upper stamen united with the others only near the middle; fr indehiscent, compressed, ovate to subrotund, with a distinct, usually toothed or spiny margin; 1–2-seeded; herbs with scarious stipules, odd-pinnate lvs with numerous lfls, and long-peduncled, axillary narrow racemes. 100+, Eur. and Medit. to c. Asia.

1. Onobrychis viciaefolia Scop.   Sainfoin. Perennial, 3–6 dm, sparingly branched; lfls 11–29, oblanceolate to oblong, 1–3 cm, peduncles from the upper axils, stout, 1–3 dm; racemes dense, 5–10 cm; bracts brownish, lanceolate; fls rosy, 10–14 mm; cal-lobes linear-setaceous, 3–5 mm, usually ciliate or long-hairy; fr obliquely rotund-ovate, 5–8 mm, strongly and coarsely reticulate-veined, the broadly rounded lower suture beset with short stout spines or coarse teeth; $2n=28$. Native of s. Europe, occasionally intr. in our range. June, July.

33. DESMODIUM Desv., nom. conserv.   Tick-trefoil. Cal ± bilabiate, the upper 2 lobes connate for all or most of their length, the lower 3 separate; standard oblong to suborbicular, narrowed at base; wings oblong; keel nearly straight; stamens 10, usually diadelphous; fr indehiscent, stipitate, transversely segmented and eventually separating into 1-seeded joints (articles), ± beset with hooked hairs; ours all perennial herbs with ± petioled (rarely subsessile), pinnately 3-foliolate lvs and purple or pink or less often white fls in elongate racemes or panicles. (*Meibomia.*) 300, mainly warm reg.

1 Stamens monadelphous; cal-lobes less than half as long as the tube; stipe much exceeding the persistent remains of the stamens; upper suture of the fr glabrous.
  2 Infl borne on a lfless stem arising from the base; pedicels 1–2 cm .... 1. *D. nudiflorum.*
  2 Infl terminating the leafy stem; pedicels 2–12 mm.
    3 Lvs usually clustered just beneath the infl, this 3–8 dm; fls purple (white) .... 2. *D. glutinosum.*
    3 Lvs scattered; infl 1–2 dm; fls white .... 3. *D. pauciflorum.*
1 Stamens diadelphous; cal-lobes more than half as long as the tube; stipe shorter than the remains of the stamens except sometimes in no. 21; upper suture pubescent with hooked hairs, except commonly in no. 21.
  4 Stems prostrate; lfls regularly more than half as wide as long.
    5 Articles glabrous or nearly so on the sides, generally twisted; fls ochroleucous .... 4. *D. ochroleucum.*
    5 Articles pubescent on the sides, not twisted; fls anthocyanic.
      6 Avg well grown lvs with the petiole 2–5 cm and the terminal lfl 3–7 cm.
        7 Lfls mostly suborbicular or broadly obovate; stipules ovate-acuminate, obliquely cordate, 5–12 mm; widespread sp. .... 5. *D. rotundifolium.*
        7 Lfls (at least the terminal one) more ovate or rhombic; stipules obliquely ovate to lance-acuminate, 4–8 mm; sp. of limited range .... 19. *D. humifusum.*
      6 Avg well grown lvs with petiole 1–2 cm and terminal lfl 1.5–3 cm .... 6. *D. lineatum.*
  4 Stems erect or strongly ascending; lfls various.
    8 Terminal lfl not more than 1.5 cm wide, nor more than a fourth as wide as long.
      9 Lfls thick, strongly reticulate; articles 1–3, the lower margin rounded.
        10 Lvs subsessile; lfls pubescent beneath; widespread sp. .... 7. *D. sessilifolium.*
        10 Lvs evidently petiolate; lfls glabrous beneath, or nearly so; coastal plain.
          11 Articles flat or concave on the upper margin; mostly in dry places .... 8. *D. strictum.*
          11 Articles convex on the upper margin; mostly in moist or wet places .... 9. *D. tenuifolium.*
      9 Lfls thin, not strongly reticulate; articles 3–6, sub-triangular .... 17. *D. paniculatum.*
    8 Terminal lfl broader, in well grown lvs more than 1.5 cm wide or more than a fourth as wide as long.
      12 Lvs pubescent beneath with hooked hairs (these visible under a 10× lens).
        13 Petiole much longer than the stalk of the terminal lfl; stipules 6–15 mm, ± persistent.
          14 Articles semi-rhomboid, 7–13 mm; fls pinkish, becoming green; pedicels 6–14 mm .... 10. *D. canescens.*
          14 Articles rounded on both margins, 4.5–7 mm; fls white; pedicels 12–23 mm .... 11. *D. illinoense.*
        13 Petiole about as long as the stalk of the terminal lfl; stipules 2–4 mm, caducous .... 12. *D. fernaldii.*
      12 Lvs glabrous or pubescent beneath, but without hooked hairs.
        15 Lower margin of the articles gradually curved, nearly semicircular.
          16 Fls 8–13 mm; cal 3.5–7 mm; stipe of the fr 2–6 (avg 3–4) mm.
            17 Terminal lfl much longer than the petiole and petiolule together; lvs appressed-hairy beneath .... 13. *D. canadense.*
            17 Terminal lfl only slightly if at all longer than the petiole and petiolule together; lvs softly velvety-villous beneath .... 20. *D. viridiflorum.*
          16 Fls 4–6 mm; cal 2–3 mm; stipe of the fr 1–3 (avg 1.5–2) mm.
            18 Lateral lfls distinctly longer than the petiole; lfls ± pubescent.
              19 Terminal lfl of avg lvs 3–7 cm, usually less than half as wide .... 14. *D. rigidum.*
              19 Terminal lfl of avg lvs 1.5–2.5 cm, usually more than half as wide .... 15. *D. ciliare.*
            18 Lateral lfls about as long as the petiole; lfls glabrous or nearly so .... 16. *D. marilandicum.*
        15 Lower margin of the articles abruptly curved near the middle, the article somewhat triangular or semi-rhomboid in outline.
        20 Articles 4.5–9 mm, two-thirds as wide; stipules 2–8 mm, mostly caducous.
          21 Lfls oblong-lanceolate, commonly 3–8 times as long as wide .... 17. *D. paniculatum.*
          21 Lfls broader, seldom more than 2.5(–3) times as long as wide.
            22 Lvs pubescent beneath; upper suture of the articles densely hairy.
              23 Lvs pubescent beneath but not densely velvety-villous .... 18. *D. glabellum.*
              23 Lvs densely velvety-villous beneath .... 20. *D. viridiflorum.*
            22 Lvs glabrous beneath; upper suture nearly or quite glabrous .... 21. *D. laevigatum.*
        20 Articles 8–11 mm, half as wide; stipules 8–17 mm, usually persistent .... 22. *D. cuspidatum.*

1. Desmodium nudiflorum (L.) DC. Naked t.-t. Stem normally forked at base, one branch sterile, 1–3 dm, with lvs crowded at the summit, the other fertile, usually lfless, ascending, 4–10 dm; stipules deciduous; lateral lfls ovate to ovate-oblong, the terminal one elliptic to ovate, 4–10 cm; fls 6–8 mm, pink to purple (white); pedicels 1–2 cm; stamens monadelphous; cal-lobes short; stipe glabrous, 10–18 mm; articles 2–4, semi-obovate, 6–10 mm, glabrous on the upper margin. Rich woods; Me. to Minn., s. to Fla. and Tex. July–Sept.

2. Desmodium glutinosum (Muhl.) A. Wood. Cluster-lf t.-t. Stem solitary, erect, 1–4 dm, generally bearing near the summit several long-petioled lvs and prolonged into a terminal panicle 3–8 dm; stipules narrow, persistent; lateral lfls asymmetrically obovate, acuminate; terminal lfl round-ovate, 7–15 cm, long-acuminate; fls 6–8 mm, purple (white); pedicels 3–8 mm; stamens monadelphous; cal-lobes short; stipe glabrous, 4–12 mm; articles (1)2–3(4), triangular or semi-obovate, 8–12 mm; glabrous on the upper margin. Rich woods; N.S. to Sask., s. to Fla. and Mex. July. (*D. acuminatum; D. grandiflorum,* misapplied.)

3. **Desmodium pauciflorum** (Nutt.) DC.   Few-fld t.-t. Stem solitary, decumbent to erect, 2–7 dm, often branched; lvs scattered, the lowest near the base of the stem; stipules narrow, caducous; lateral lfls ovate or oblong-ovate; terminal lfl rhombic-ovate, 3–8 cm, mostly longer than wide, acute; panicle 1–2 dm; fls rather few, white, 4–6 mm, on pedicels 6–12 mm; cal-lobes short; stamens monadelphous; stipe 4–11 mm, finely uncinate, seldom glabrous; articles 1–3, triangular, 9–12 × 5.5–8 mm, glabrous on the upper margin. Rich woods; N.Y. to O. and Io., s. to Fla. and Tex. July, Aug.

4. **Desmodium ochroleucum** M. A. Curtis.   White t.-t. Stems prostrate, to 1 m, branched, pubescent; principal petioles 2.5–5 cm; stipules lance-ovate, 6–9 mm; lfls ovate to rounded-ovate or rhomboid, pubescent to subglabrous, the terminal one 4–7 cm; racemes axillary and terminal, simple or nearly so, the bracts caducous; fls ochroleucous, 8 mm, on pedicels 12–20 mm; stipe 2–4 mm; articles 2–4, usually twisted, 7–9 × 5.5–6.5 mm, pubescent on the margins, glabrous or nearly so on the strongly reticulate sides. Dry woods, especially in sandy soil; N.J. to N.C. and Ga., w. to se. Mo. July, Aug.

5. **Desmodium rotundifolium** DC.   Round-lvd t.-t. Stems prostrate, to 1.5 m, branched, densely villous to subglabrous; principal petioles 3–5 cm; stipules ovate, persistent, 8–12 mm; lfls mostly suborbicular or broadly obovate, ± villous, the terminal one 3–7 cm; racemes axillary and terminal, very open, simple or nearly so, to 3 dm, with ovate bracts; fls anthocyanic, 8–11 mm, on pedicels 6–13 mm; stipe 3–5 mm; articles 3–6, 5–7.5 × 3–5 mm, pubescent on the sides and margins, not twisted. Dry woods and barrens; Vt. and Mass. to s. Mich., s. to Ga., Mo., and La. July–Sept. (*Meibomia michauxii.*)

6. **Desmodium lineatum** DC.   Matted t.-t. Stems prostrate, to 7 dm, branched and often mat-forming, densely villous to subglabrous; petioles mostly 1–2 cm; stipules lanceolate, 2–5 mm; lfls broadly ovate to elliptic to rotund, 1–3 cm, rounded at both ends; panicles terminal or from the upper axils, to 4 dm; fls 4–5 mm, purple; stipe 1.5–2.5 mm; articles 2–4, 4–6 × 3–4 mm. Dry upland woods and barrens, especially in sandy soil on the coastal plain; se. Md. to Fla. and Tex. July, Aug. (*Meibomia arenicola.*)

7. **Desmodium sessilifolium** (Torr.) T. & G.   Sessile-lvd t.-t. Stems 1–few, erect, mostly simple, 6–15 dm, finely hairy; petioles 1–5 mm, shorter than the stalk of the terminal lfl; stipules lanceolate, 3–5 mm, caducous; lfls thick, oblong to linear-oblong, mostly 3–7 cm × 6–12 mm, rather densely hairy beneath as well as reticulate-veiny; infl large, with several strict, ascending branches 1–3 dm; fls mostly pinkish to lavender, 5 mm, on pedicels 1.5–4.5 mm; cal 2–3 mm; stipe 1–3 mm; articles 1–3, convex above, rounded below, 4–6.5 × 3–4.5 mm. Dry, sandy soil; R.I. to s. Ont., Mich., and Kans., s. to N.C., Miss., and Tex., commoner westward. July, Aug.

8. **Desmodium strictum** (Pursh) DC.   Pine-barren t.-t. Stems 1–few, erect, to 12 dm, minutely hairy; petioles much exceeding the stalk of the terminal lfl; lfls thick, linear, 3–7 cm, mostly 6–10 times as long as wide, glabrous or nearly so, strongly reticulate-veiny; infl 2–4 dm, with several ascending branches; fls 3–5 mm, on pedicels 6–11 mm; cal 1.5–2 mm; stipe 1–2 mm; articles 1–2(3), 4–6 × 3–4.5 mm, rounded on the lower margin, flat or slightly concave on the upper. Pine barrens on the coastal plain; N.J. to Fla. and La. July, Aug.

9. **Desmodium tenuifolium** T. & G.   Narrow-lvd t.-t. Very much like no. 8; articles 3.5–5 × 2.5–3.5 mm, convex on the upper margin. Moist savannas, margins of pocosins, and shallow ditches on the coastal plain; se. Va. to Fla. and La.

10. **Desmodium canescens** (L.) DC.   Hoary t.-t. Stem arising from a long taproot, erect, 6–15 dm, often much-branched, spreading-hairy as well as uncinate-puberulent (as also the infl); stipules ovate, persistent, 6–12 mm; petioles nearly as long as the lateral lfls; lfls thin, ovate or lance-ovate, finely uncinate-puberulent beneath, the terminal lfl 6–12 cm; panicle of several racemes, fls 10 mm, pinkish, becoming green (white), on pedicels 6–14 mm; cal 5 mm; stipe 2–5 mm; articles 4–6, semi-rhomboidal, 7–13 mm. Fields and woods; Mass. to Wis. and Neb., s. to Fla. and Tex.

11. **Desmodium illinoense** A. Gray.   Prairie t.-t. Stem erect, 1–2 m, pubescent with hooked hairs, commonly with a single long terminal raceme or few-branched panicle; stipules ovate, acuminate, 10–15 mm; petioles much longer than the stalk of the terminal lfl; lfls lance-ovate, rough on both sides with hooked hairs, strongly reticulate beneath, the terminal one 6–10 cm; axis of the infl softly spreading-hairy as well as commonly with some hooked hairs; fls 8–10 mm, white, on pedicels 12–23 mm; stipe 2.5–5 mm; articles 2–5, 4.5–7 mm, with rounded margins. Rich prairie soil; s. Ont. to Wis. and Neb., s. to O., Okla., and n. Tex. July, Aug.

12. **Desmodium fernaldii** Schubert.   Fernald's t.-t. Stem erect, to 13 dm, uncinate-puberulent, as also the axis of the infl; stipules lance-subulate, 2–4 mm, caducous; petiole about as long as the stalk of the terminal lfl; lfls ovate or rhombic-ovate, subglabrous above, strongly reticulate and minutely uncinate-puberulent on the veins beneath; fls 8–10 mm; stipe 2.5–5 mm; articles 2–5, 6–8 mm. Dry, sandy soil; se. Va. to Fla. and Tex. July, Aug. (*D. rhombifolium,* misapplied.)

13. **Desmodium canadense** (L.) DC.   Canadian t.-t. Stem erect, branched above, to 2 m, uncinate-puberulent and spreading-hairy at least above, stipules linear-subulate, to 8 mm; petioles 2–20 mm, the petiole and lf-rachis together a sixth to a half as long as the terminal lfl; lfls oblong or lance-oblong, appressed-hairy

beneath, the terminal one 5–9 cm; racemes densely fld, with conspicuous lance-ovate bracts; fls 10–13 mm, on pedicels 5–8 mm; cal 5–7 mm; stipe 2–4 mm; articles 3–5, 5–6.5 mm, the lower margin rounded. Moist soil, thickets, and river-banks; Que. and N.S. to Alta., s. to Md., w. Va., Mo., and Okla., common. July, Aug.

14. Desmodium rigidum (Elliott) DC.   Stiff t.-t. Stems erect, to 12 dm, usually simple to the infl, densely uncinate-puberulent, and with a few long hairs near the nodes; stipules narrow, 2–5 mm, caducous, petioles much shorter than the lateral lfls; lfls oblong to ovate, firm, scaberulous above, minutely villous and strongly reticulate beneath, the terminal one 2.5–7 cm; fls 4–6 mm, on pedicels 6–17 mm; cal 2.5–3 mm; stipe 1.5–3 mm; articles 1–3, obliquely obovate, 3.5–5.5 mm, the lower margin gradually curved. Dry woods and thickets; Mass. to s. Mich. and Ind., s. to Fla., La., and Tex. July, Aug. (?*D. obtusum.*)

15. Desmodium ciliare (Muhl.) DC.   Little-lf t.-t. Stem(s) slender, erect, with few to many long, spreading hairs, 4–10 dm, with a large terminal panicle; stipules narrow, 2–4 mm, caducous; petioles spreading-hairy, 2–10 mm; lfls oblong to oval, very blunt, pilose on both sides, varying to subglabrous; fls 4–5 mm, on pedicels 4–9 mm; cal 2 mm; stipe 1–2 mm; articles 1–3, obliquely obovate, 3.5–5 × 3–4 mm, the lower margin gradually curved. Dry, often sandy soil; Mass. to Ind., Mo., and se. Kans., s. to Fla., Cuba, and Tex. July, Aug. (*D. obtusum,* probably misapplied.)

16. Desmodium marilandicum (L.) DC.   Maryland t.-t. Stem(s) slender, erect, 6–12 dm, glabrous or nearly so; stipules slender, 2–6 mm, caducous; petioles slender, about equaling the lateral lfls; lfls thin, ovate or oval, 1.5–3.5 cm, glabrous, or inconspicuously hairy beneath; fls 4–6 mm, on pedicels 8–19 mm; cal 2–3 mm; stipe 1.5–2.5 mm; articles 1–3, somewhat obovate, 4–5 × 3–3.5 mm. Dry upland woods and thickets; Mass. to Mich., and Mo., s. to Fla. and Tex. July, Aug.

17. Desmodium paniculatum (L.) DC.   Stem(s) slender, erect, 6–12 dm, usually branched above, usually glabrous or nearly so; stipules narrow, 2–4 mm, caducous, petioles mostly 1.5–5 cm; lfls thin, lanceolate or oblong, mostly 3–8 times as long as wide, glabrous or somewhat appressed-hairy, to 10 × 2.5 cm; infl large and branching, 1–4 dm; fls 6–8 mm, on pedicels 4–11 mm; cal 2.5–3.5 mm; stipe 2.5–3.5 mm; articles 3–6, ± triangular, 5–7.5 × 3.5–4.5 mm. Dry woods; s. Me. to s. Ont., Mich., and Neb., s. to Fla. and Tex. July, Aug. The broader-lvd forms grade into no. 18.

18. Desmodium glabellum (Michx.) DC.   Stem(s) 5–12 dm, ± erect, usually much-branched above, ± densely uncinate-puberulent, or (in the phase called *D. dillenii* Darlington or *D. perplexum* Schubert) provided with longer spreading hairs, or with both sorts of pubescence; stipules slender, 4–8 mm; petioles mostly 2–5 cm; lfls ± pubescent especially beneath, the terminal one commonly oblong-ovate, 4–9 cm, a third to two-thirds (avg half) as wide; fls 7–10 mm; cal 4 mm; stipe 3–5 mm; articles 2–4, 5–9 × 3.5–5 mm; 2n=22. Dry upland woods; Me. to Wis. and Neb., s. to Fla. and Tex. July, Aug.

19. Desmodium humifusum (Muhl.) Beck.   Stems prostrate, to 1 m or more, uncinate-puberulent and with longer spreading hairs; stipules ovate- to lance-acuminate, 4–8 mm; petioles 2.5–5 cm; lfls appressed-hairy on both sides and sometimes with a few hooked hairs, the terminal one ovate or rhombic, 5–7 × 3–5 cm; racemes axillary and terminal, branching; fls 7–9 mm, on pedicels to 9 mm; articles 3 or 4, 6–8 × 4–5 mm, deltoid (the upper suture straight) or rhombic (both sutures angled). Dry sandy woods; Mass. to Md.; s. Mo. July, Aug. (*D. glabellum,* misapplied.)

20. Desmodium viridiflorum (L.) DC.   Velvety t.-t. Stem erect, 8–18 dm, finely pubescent; stipules lanceolate to ovate, 4–6 mm, often brick-red; terminal lfl broadly ovate to rhombic or deltoid-ovate, 5–10 cm, half to three-fourths as wide, sparsely short-hairy above, densely velvety-villous beneath; infl 2–5 dm; fls 6–9 mm long, pink, turning green, on pedicels 3–8 mm; cal 3.5–5 mm; stipe 2.5–6 mm; articles 3–5, 5–7 mm, the lower margin rounded or angled. Dry woods; s. N.Y. to Ind., Mo., and Okla., s. to Fla. and Tex. July, Aug. (*D. nuttallii,* the form with subtriangular articles.)

21. Desmodium laevigatum (Nutt.) DC.   Smooth t.-t. Stem erect, 5–12 dm, glabrous or nearly so below the much-branched infl; stipules lance-subulate, 5–8 mm; lfls thin, ovate, glabrous, glaucous beneath, the terminal one 4–8 cm, half to three-fourths as wide; fls 8–10 mm, deep rose or purple, on pedicels 10–20 mm; cal 3.5–5 mm; stipe 3.5–6.5 mm; articles 2–5 4.5–8 mm, glabrous or nearly so on the upper suture, the lower suture abruptly curved near the middle. Dry upland woods; s. N.Y. to Ind. and Mo., s. to Fla. and Tex. July, Aug.

22. Desmodium cuspidatum (Muhl.) Loudon.   Big t.-t. Stem stout, erect, leafy, to 2 m; stipules lanceolate, 8–17 mm, usually persistent; principal petioles 5–8 cm, flattened distally; lfls thin, bright green above, paler beneath, the lateral lance-ovate, the terminal ovate, 6–12 cm, sharply acuminate; infl a simple raceme or sparsely branching panicle; fls 6–12 mm, pink, on pedicels 3–7 mm; cal 4–5 mm; stipe 2–5 mm; articles 3–7, rhomboidal, 8–11 mm. Dry upland woods and thickets; July, Aug. The var. *cuspidatum* (*D. bracteosum, D. grandiflorum*), occurring from Mass. and Vt. to Mich. and Wis., s. to Ga. and Okla., has the stem, lvs, stipules, and cal glabrous or nearly so. The var. *longifolium* (T. & G.) Schubert, occurring from O. to Minn. and Nebr., s. to Ga. and ne. Tex., is ± hairy.

34. LESPEDEZA Michx.   Lespedeza. Cal-tube short, the narrow lobes subequal, or the 2 upper partly connate; standard suborbicular to oblong-ovate, short-clawed; wings oblong, straight, clawed and auriculate, connivent with the keel; keel-pet straight, blunt, obliquely obovate; stamens 10, equal, diadelphous; fr oval to elliptic, indehiscent, 1-seeded;

herbs or shrubs with numerous small trifoliolate lvs and spicate, racemose or solitary, purple to ochroleucous fls in sessile or peduncled axillary infls, each fl subtended at base by 2 or 4 small bractlets; stipellules wanting; many spp. have both petaliferous and apetalous fls with similar or different frs. 25, widespread.
The following names are believed to be based on hybrids:

L. × *acuticarpa* Mackenzie & Bush = *L. violacea* × *virginica.*
L. × *brittonii* E. Bickn. = *L. procumbens* × *virginica.*
L. × *hirta* var. *appressipilis* S. F. Blake = *L. angustifolia* × *hirta.*
L. × *longifolia* DC. = *L. capitata* × *hirta.*
L. × *manniana* Mackenzie & Bush = *L. capitata* × *violacea.*
L. × *neglecta* Mackenzie & Bush = *L. stuevii* × *virginica.*
L. × *nuttallii* Darl. = *L. hirta* × *intermedia.*
L. × *oblongifolia* (Britton) W. Stone = *L. angustifolia* × *hirta.*
L. × *simulata* Mackenzie & Bush = *L. capitata* × *virginica.*

1 Shrubs or half-shrubs 1–3 m, escaped from cult.; fls purple; stipules slender but persistent.
  2 Racemes lax, tending to droop; cal-lobes evidently longer than the tube .... 1. *L. thunbergii.*
  2 Racemes compact, erect or strongly ascending; cal-lobes up to about as long as the tube .... 2. *L. bicolor.*
1 Herbs up to 1.5 m, the stems sometimes ± woody but then of annual duration.
  3 Perennial; stipules subulate or setaceous, soon withering or deciduous; cal-lobes acuminate; native, except no. 13.
    4 Fls violet or purple, in axillary racemes; apetalous fls commonly numerous, in sessile or subsessile axillary clusters.
      5 Plants trailing or procumbent.
        6 Pubescence of the stem and petioles appressed .... 3. *L. repens.*
        6 Pubescence of the stem and petioles spreading .... 4. *L. procumbens.*
      5 Plants ascending to erect.
        7 Peduncles of the petaliferous racemes evidently longer than their subtending lvs; keel 1–2 mm longer than the wings .... 5. *L. violacea.*
        7 Peduncles shorter than their subtending lvs; keel equaling or shorter than the wings.
          8 Lvs strigose on both sides, or glabrous above; upper 2 cal-lobes connate at least half their length.
            9 Lfls oblong or linear-oblong, 3–7 mm wide, usually 4–6 times as long .... 6. *L. virginica.*
            9 Lfls elliptic, 6–20 mm wide, 1.5–3 times as long .... 7. *L. intermedia.*
          8 Lower surface of the lvs densely velvety; upper 2 cal-lobes not connate, or only shortly so .... 8. *L. stuevei.*
    4 Fls ochroleucous, with or without a purple spot; apetalous fls few or none, when present mingled with the petaliferous ones.
      10 Fls in spikes or heads; cal-lobes all separate; wings exceeding the keel.
        11 Infl of globose to short-ovoid heads, short-pedunculate; cal-lobes 6–10 mm .... 9. *L. capitata.*
        11 Infls spicate, the peduncle usually longer than the infl; cal-lobes 3–7 mm.
          12 Lfls oval to obovate, half to three-fourths as wide as long .... 10. *L. hirta.*
          12 Lfls linear to narrowly oblong, up to a third as wide as long.
            13 Spikes dense, continuous; fls (incl. cal) 6–9 mm; coastal plain .... 11. *L. angustifolia.*
            13 Spikes loose, interrupted; fls 4–6 mm; Middle West .... 12. *L. leptostachya.*
      10 Fls solitary or in axillary clusters of 2 or 3; upper cal-lobes connate about half their length; wings and keel equal .... 13. *L. cuneata.*
  3 Annual; stipules lance-ovate, scarious, striate, persistent; cal-lobes blunt; intr.
    14 Stem retrorsely appressed-pubescent .... 14. *L. striata.*
    14 Stem antrorsely appressed-pubescent .... 15. *L. stipulacea.*

1. **Lespedeza thunbergii** (DC.) Nakai. Shrub or half-shrub 1–3 m; stipules slender but persistent; lfls elliptic or lance-elliptic, 1–4 cm, glabrous above, paler and strigose beneath; fls showy, purple, 12–17 mm, borne in numerous lax, axillary racemes; cal-lobes slender, long-pointed, evidently longer than the tube, the 2 upper connate most of their length; $2n=22$. Native of China and Japan, rarely escaped from cult. in our range. Sept., Oct.

2. **Lespedeza bicolor** Turcz. Much like no. 1; lfls avg wider; fls smaller, 9–13 mm, borne in numerous strongly ascending racemes that tend to form a sort of terminal, leafy panicle; cal-lobes more lance-triangular, not so long-pointed, not obviously longer than the tube; the 2 upper generally not so long-connate; $2n=18$. Native of Japan, occasionally escaped from cult. in se. U.S., n. to Va. July–Oct.

3. **Lespedeza repens** (L.) Barton. Smooth trailing l. Stems clustered, trailing or procumbent, to 1 m, finely appressed-hairy to glabrate; petioles shorter than the lfls; lfls mostly 1–2 cm, half as wide, appressed-hairy on both sides or glabrous above; peduncles much exceeding the subtending lvs, bearing a loose raceme of 3–8(–12) purple fls 5–8 mm; cal-lobes 1–2 mm, the upper two largely connate; pet about equal; fr finely appressed-hairy; $2n=20$. Dry woods and fields; Conn. to Wis. and Kans., s. to n. Fla. and Tex. June–Sept.

4. Lespedeza procumbens Michx. Downy trailing l. Much like no. 3, but the stem, petioles, lower lf-surfaces, and fr sparsely to densely villous with spreading hairs; racemes with mostly 8–12 fls; $2n=20$. Dry upland woods; N.H. to Wis., s. to Ga., w. Fla., and Tex. Aug., Sept.

5. Lespedeza violacea (L.) Pers. Violet l. Stems erect or ascending, usually much-branched, 4–8 dm, glabrous to sparsely pubescent with erect or ascending hairs; petioles often nearly as long as the lfls; lfls elliptic, 1–4 cm, half as wide, appressed-hairy beneath, glabrous or nearly so above; racemes few-fld, much exceeding the subtending lvs, fls purple, 6–10 mm, the wings shorter than the keel; cal-lobes 1.5–3 mm, the 2 uppermost connate beyond the middle; $2n=20$. Dry upland woods; Mass. and Vt. to Wis., s. to Ga., Kans., and Tex., commonest in the Midwest. Aug., Sept. (*L. prairea.*).

6. Lespedeza virginica (L.) Britton. Virginia l. Stems 1–few, erect or ascending, simple or virgately branched above, strigose, 3–10(–15) dm; lvs usually erect or strongly ascending; lfls linear to narrowly oblong, 3–5(–7) mm wide, (3–)4–6 times as long, short-strigose on both sides, or rarely glabrous above; fls purple, 6–8 mm, the apetalous ones chiefly along the middle of the stem in small, subsessile clusters, the others in short, few-fld racemes from the upper axils, on peduncles shorter than the lvs, forming a crowded, leafy infl; cal-lobes of the petaliferous fls 1.7–3 mm, the 2 uppermost connate to the middle or beyond; fr thinly and inconspicuously strigose; $2n=20$. Dry upland woods; Mass. and Vt. to Wis., s. to n. Fla., Kans., and Tex. Aug., Sept.

7. Lespedeza intermedia (S. Wats.) Britton. Wand-l. Stems seldom as much as 1 m, usually virgately branched above; lvs generally spreading; lfls elliptic, 6–20 mm wide, 1.5–3 times as long, strigose beneath, usually glabrous above, otherwise much like no. 6; $2n=20$. Dry upland woods; Vt. to Mich. and Okla., s. to Fla. and Tex. Aug., Sept. (*L. frutescens,* misapplied.)

8. Lespedeza stuevei Nutt. Velvety l. Stems erect, simple or virgately branched above, to 1(–1.5) m, softly spreading-hairy; lvs spreading, short-petioled; lfls ovate-oblong to elliptic or obovate, spreading-hairy beneath, ascending-hairy above, those of the main stem 1.5–3.5 cm; apetalous fls numerous in sessile axillary clusters; petaliferous fls purple, 6–8 mm, in dense racemes in the upper axils, on peduncles shorter than the subtending lvs; cal-lobes 1.7–3.5 mm, the 2 uppermost connate less than half their length; fr densely villous; $2n=20$. Dry upland woods and barrens; Mass. to n. Fla., w. to Tex., and n. inland to s. Ind., Mo., and Kans. Aug., Sept.

9. Lespedeza capitata Michx. Bush-clover. Stems ± erect, 6–15 dm, simple or branched above, densely villous; petioles 2–5 mm, shorter than the stalk of the terminal lfl; lfls to 4.5 × 1.8 cm, variable in shape and pubescence, usually appressed-hairy on both sides or glabrous above; heads numerous, forming a thyrsoid infl, subglobose to short-ovoid, 12–25 mm, with very many, densely crowded fls; peduncles usually shorter than the subtending lvs and rarely longer than the heads; fls 8–12 mm, ochroleucous; cal-lobes villous, 6–10 mm; fr pubescent, conspicuously shorter than the cal; $2n=20$. Open dry woods, sand-dunes, and prairies; Me. and s. Que. to Minn. and S.D., s. to Ga., w. Fla., and Tex. July–Sept. (*L. c.* var. *stenophylla,* the narrow-lvd extreme; *L. velutina,* the form with spreading-hairy lvs.)

10. Lespedeza hirta (L.) Hornem. Hairy l. Stems erect, to 15 dm, variously hairy; principal petioles mostly 5–20 mm; lfls oval or elliptic to somewhat ovate or obovate, 1.5–4 cm, half to three-fourths as wide, variously hairy, or glabrous above; spikes numerous, short-cylindric, 1–3 cm, usually on peduncles equaling or surpassing the subtending lvs; fls ochroleucous, 6–9 mm; cal-lobes densely hairy, 3.5–7 mm; fr elliptic, pointed at both ends, about as long as the cal; $2n=20$. Abundant in dry soil; Me. to Fla., w. to Wis., Ill., Mo., Okla., and Tex. Late summer. Two vars.: var. *hirta,* with the upper surface of the lfls glabrous, strigose, or pilose, but not silvery, and with the stems villous to pilose, has nearly the range of the sp., but is rare within the range of var. *curtissii.* Var. *curtissii* (Clewell) Cronq., with the lfls densely silvery-strigillose and the stems appressed-hairy or densely short-pilose, occurs on the s. coastal plain from s. Va. to Fla. and Ala.

11. Lespedeza angustifolia (Pursh) Elliott. Narrow-lvd l. Stems erect, simple or branched above, 5–10 dm, finely pubescent with usually appressed hairs; petioles 1–5(–10) mm; lfls oblong to linear, 2–6 cm × 3–7 mm, glabrous or minutely hairy above, thinly to densely sericeous (or subvelutinous) beneath, or with spreading hairs along the midvein; spikes cylindric, 1–3 cm, ± densely fld, on ascending peduncles 1–5 cm; fls 6–9 mm, ochroleucous; cal-lobes densely hairy, 4–6.5 mm; fr densely hairy, nearly or quite as long as the cal-lobes; $2n=20$. Dry or moist, acid soils, chiefly on the coastal plain; e. Mass. to Fla. and La.; c. Tenn. Late summer.

12. Lespedeza leptostachya Engelm. Prairie l. Stems ± erect, nearly simple, 5–10 dm, sericeous; petioles 4–10 mm; lfls narrowly oblong, 2–4 cm × 3–7 mm, obtuse and mucronate, sparsely appressed-pubescent above, sericeous beneath; spikes slender, loose and interrupted, 2–3 cm, on peduncles 1–2 cm; fls 4–6 mm, ochroleucous, the cor about equaling the cal; fr densely villous, equaling the cal. Dry prairies; n. Ill., s. Wis., n. Io., and s. (rarely c.) Minn. Late summer.

13. Lespedeza cuneata (Dum. Cours.) G. Don. Chinese l. Stems erect, somewhat woody, to 1.5 dm, with numerous long virgate branches, very leafy; petioles 2–5 mm; lfls ascending or erect, linear-cuneate, 10–25 mm, truncate and mucronate at the tip, glabrous above, sericeous beneath; branches sulcate, hairy on the angles; fls mostly solitary (2–3) in the upper axils, white marked with purple or pink, 7–9 mm, the wings and keel about equal; cal sericeous, the lobes lance-subulate, 3 mm, the 2 upper connate about half their length; fr oval, 3 mm; $2n=20$. Native of e. Asia, cult. in se. U.S. and now planted and established along roadsides and in disturbed sites in the s. part of our range, n. to N.J., L.I., and even Mich.

14. Lespedeza striata (Thunb.) Hook. & Arnott. Japanese clover. Annual, erect or diffuse, much branched, 1–4 dm; stems sparsely pubescent with retrorsely curved or appressed hairs; petioles 1–2 mm; lfls oblong-obovate, 1–2 cm, a third as wide; stipules lance-ovate, brown, scarious, many-nerved, persistent, 4–6 mm; fls pink or purple, 1–3 in the upper axils, sessile or on pedicels to 4 mm, the petaliferous ones 6–7 mm, mingled with the apetalous ones; cal-tube 1.5–2 mm, its lobes subequal, oblong, reticulate, about as long as the tube; fr obovate, acute, scarcely reticulate, 3–4 mm; 2*n*=22. Fields and upland woods, mostly in dry soil; N.J. and Pa. to Ind. and Kans., s. to the Gulf; native of e. Asia. July, Aug.

15. Lespedeza stipulacea Maxim. Korean clover. Much like no. 14; stems 1–6 dm, sparsely pubescent with ascending or erect appressed hairs; petioles 2–10 mm; lfls broadly obovate, two-thirds as wide as long, or wider; spikes in the upper axils, dense, leafy, 1–2 cm; cal-tube 1 mm, its lobes ovate, equaling the tube; fr oval or obovate, 3 mm, strongly reticulate; 2*n*=20. Native of e. Asia, intr. as a forage plant, and becoming naturalized as a weed from N.J. and Va. to Ind. and Mo. and southward. Summer.

35. DALEA Lucanus., nom. conserv. Cal campanulate to tubular or ellipsoid, strongly to obscurely 10-ribbed and often angled, the lobes in our spp. equal or subequal; pet all long-clawed, only the banner borne on the hypanthium, the other 4 set in sockets on the staminal tube or between the filaments at its summit; stamens 5–10, all with anthers; ovary short, with 2 ovules, only one fertile; style elongate; fr indehiscent, enclosed by the persistent cal; glandular-punctate herbs (ours) or shrubs with spirally twisted hairs (sometimes glabrous except for the mouth of the cal), small, odd-pinnate lvs and terminal bracteate spikes or heads of numerous small fls, *x*=7. (*Kuhnistera, Parosela, Petalostemum.*) *Petalostemum*, often held as distinct, can be seen as an integral part of *Dalea* when the whole genus is considered. 160, New World.

1 Fls evidently papilionaceous, with ± conventional banner, wings, and keel-pet, although the wings and keel-pet are borne on the sides of the staminal sheath; stamens 9 or 10, all with anthers, the free filaments much shorter than the staminal sheath.
  2 Blades of the keel-pet connate by one margin to form a conventional keel enfolding the anthers; spikes loose, each fl enfolded by a bract that eventually falls with it; perennial. (Subg. *Parosela*.)
    .... 1. *D. enneandra.*
  2 Blades of the keel-pet distinct; spikes dense, with closely overlapping fls; bracts caducous; annual.
    (Subg. *Dalea*, sect. *Dalea*) .... 2. *D. leporina.*
1 Fls not evidently papilionaceous, the wings and keel-pet borne at the summit of the staminal column, alternating with the filaments of the 5 stamens, and appearing as petaloid staminodes; free filaments about equaling the staminal sheath. (Subg. *Dalea*, sect. *Kuhnistera*, incl. *Petalostemum*.)
  3 Principal lvs with mostly 19–31 lfls, herbage glabrous .... 5. *D. foliosa.*
  3 Principal lvs with mostly 3–13 lfls, or (*D. villosa*) the lfls more numerous and the herbage then densely soft-hairy.
    4 Cal densely sericeous or villous; fls anthocyanic, seldom white.
      5 Principal lvs with mostly (9)11–17(–21) lfls .... 3. *D. villosa.*
      5 Principal lvs with (3)5(7) lfls .... 4. *D. purpurea.*
    4 Cal-tube glabrous or minutely pubescent, the lobes appearing ciliate; fls white.
      6 Spikes few, cylindric, the principal ones more than 1.5 cm long and more than twice as long as thick .... 6. *D. candida.*
      6 Heads ± numerous, to 1.5 cm, globose or short-ovoid, less than twice as long as thick .... 7. *D. multiflora.*

1. Dalea enneandra Nutt. Sailpod dalea. Perennial 6–12 dm, glabrous up to the silky calyces, freely branched into a diffuse panicle of loose spikes; lfls mostly 5–13, glaucous, 4–11 mm, narrow; spikes 5–30-fld, the fls ± 2-ranked; bracts broadly ovate to obovate or truncate, enfolding and eventually falling with the cal; fls of subg. *Parosela*, white, 6–8 mm; cal-tube 3–3.5 mm, the triangular-aristate teeth 3.5–4.5 mm, very plumose, eventually spreading; stamens 9; fr 3–4 mm; 2*n*=14. Dry prairies and hillsides; w. Io. and w. Mo. to Mont., Colo., and N.M. July, Aug. (*Parosela e.*)

2. Dalea leporina (Aiton) Bullock. Hare's-foot d. Erect, leafy annual 3–10(–15) dm, simple to much branched, glabrous below the infl; lfls mostly 15–35+, oblong to oblanceolate or obovate, 3–12 mm; spikes erect, dense, 1.5–8 cm; bracts caducous, ovate to lance-acuminate or caudate; fls of sect. *Dalea*, white or anthocyanic, 5–7 mm; cal very villous, its subulate lobes mostly shorter than the tube, erect in fr; stamens 9 or 10; anthers shortly exserted; fr 2.5–3 mm; 2*n*=14. Prairies and open woods, or in disturbed sites; widespread in Latin Amer., n. to N.D., Minn., and Ind., and occasionally adventive farther e. June–Aug. (*D. alopecuroides; Parosela a.*)

3. Dalea villosa (Nutt.) Sprengel. Downy prairie-clover. Coarse, densely leafy, softly villous perennial 2–5 dm (or the stems a little longer and diffuse); lfls mostly (9)11–17(–21), rather narrowly elliptic or oblanceolate, often ± folded, 5–10 mm, the terminal one smaller than the adjacent ones; spikes fairly dense but not cone-like, becoming 3–12 cm; bracts slender, setaceous-tipped, deciduous; fls of sect. *Kuhnistera*, pale purple or pinkish, seldom white, 5–6 mm; cal densely spreading-hairy, not bracteolate; 2*n*=14. Sandy prairies; Wis. to Sask., Mont., Colo., ne. N.M., and Okla. July, Aug. Ours is var. *villosa*. (*Petalostemum v.*)

4. Dalea purpurea Vent. Purple p.-c. Perennial, 2–10 dm, glabrous or thinly hairy to tomentulose; lvs numerous; often with axillary fascicles; lfls (3)5(7), narrow, involute when dry, 8–25 mm; spikes scarcely elevated above the lvs, very dense and cone-like, becoming cylindric and 2–7 cm; bracts, except the lowest, mostly short and papery, the oblanceolate body villous above, tapering to a dark, subulate, glabrous or shortly hairy tip; fls of sect. *Kuhnistera,* 5–7 mm, rose-purple; cal densely villous or sericeous, bluntly pentagonal, not prominently ribbed, not bracteolate; $2n=14$. Dry prairies and open glades; Ind. to Man. and Alta., s. to Ky., Ark., Tex., and N.M. June, July. Ours is var. *purpurea.* (*Petalostemum p.*)

5. Dalea foliosa (A. Gray) Barneby. Cedar-glade p.-c. Glabrous perennial 2–8 dm; lfls of the principal lvs mostly 19–31, oblong to elliptic, mostly flat, 5–10 mm; spikes very dense, ovoid, becoming cylindric and 2–5 cm; bracts slender, caudate-acuminate, surpassing the cal, which is flanked at the base by a pair of bracteoles; fls of sect. *Kuhnistera,* 5–6 mm, rose-purple (white); cal-tube 10-angled, glabrous, the short teeth densely hairy within; $2n=14$. Cedar glades and calcareous barrens; c. Tenn. to n. Ala., and formerly also on bluffs and stony flats along the upper Illinois R. and tributaries in ne. Ill. July. (*Petalostemum f.*)

6. Dalea candida Michx. White p.-c. Erect, glabrous perennial 3–10 dm; lfls 5–9, narrow to fairly broad, flat or loosely folded, 10–35 mm; spikes 1–few, the terminal one becoming cylindric, 1.5–6 cm, more than twice as long as thick; bracts (except the lowest) surpassing the cal-lobes in bud and at least equaling them in fl, the oblanceolate body tapering to a caudate-acuminate tip; cal flanked at base by a pair of bracteoles; fls of sect. *Kuhnistera,* 4–6 mm, white; cal-tube sharply 10-ribbed, glabrous to shortly villous, the lobes finely ciliate; $2n=14$. Dry prairies and dry upland woods; Wis. and Ind. to Tenn. and Ala., w. to Alta., Ariz., and n. Mex. June, July. Most of our plants belong to var. *candida,* as here described. (*Petalostemum c.*) The more western var. *oligophylla* (Torr.) Shinners (*P. occidentale*), with laxer spike (the axis usually partly visible in pressed specimens), somewhat smaller lfls (seldom over 20 mm), often pubescent cal-tube, and more often diffuse stems, barely reaches our range in w. Io. and w. Minn. June, July.

7. Dalea multiflora (Nutt.) Shinners. Rounded-headed p.-c. Much like no. 6; larger cauline lvs with 7–13 lfls; heads ± numerous, subglobose or shortly ovoid to 1.5 cm, less than twice as long as thick. Dry hills and prairies; w. Io. and w. Mo. to se. Neb., Kans., Tex., and n. Mex. July, Aug. (*Petalostemum m.*).

36. AMORPHA L. Cal-tube obconic; cal-lobes half-orbicular to lanceolate, equal or the lowest one slightly larger, standard obovate or broad cuneate, ± folded about the base of the stamens; wings and keel none; stamens 10, exsert; filaments united only near the base; ovary short, 2-ovuled; fr oblong, often curved, slightly compressed, indehiscent, 1–2-seeded; shrubs with pinnately compound lvs, numerous penninerved lfls, and small anthocyanic (white) fls in dense spike-like racemes from the upper axils and terminal. 15, N. Amer.

1 Shrubs to 1 m, lfls mostly 0.5–2 cm; cal-lobes at least half as long as the tube.
  2 Lvs only inconspicuously or scarcely punctate, usually canescent; racemes usually clustered in groups
    of 5–20+ .... 1. *A. canescens.*
  2 Lvs conspicuously punctate, glabrous or nearly so; racemes solitary .... 2. *A. nana.*
1 Shrubs mostly 1–4 m; lfls mostly 2–6 cm; cal-lobes (at least the upper 4) much shorter.
  3 Lvs shining above, blackening in drying; cal with inconspicuous or no glands .... 3. *A. nitens.*
  3 Lvs not shining, not blackening in drying; cal-tube strongly glandular .... 4. *A. fruticosa.*

1. Amorpha canescens Pursh. Lead-plant. Shrub 5–10 dm, erect or ascending, simple or sparingly branched, ordinarily closely canescent throughout; petioles 1–3 mm; lfls crowded, 13–20 pairs, 1–2 cm, inconspicuously or scarcely punctate; racemes usually clustered in groups of 5–20+, the terminal one 5–15 cm, the others shorter; cal-lobes lanceolate, subequal, mostly 1–1.5 mm, half to two-thirds as long as the tube; banner 4.5–6 mm; fr canescent, 4 mm; $2n=20$. Dry prairies and sandy open woods; s. Mich. to Minn. and Sask., s. to Ill., Mo., Ark., Tex., and N.M. June, July. The rare hybrid with no. 4 is *A. × notha* E. J. Palmer.

2. Amorpha nana Nutt. Smooth l.-p. Shrub 3–6(–9) dm, glabrous or inconspicuously strigillose; petioles 4–8 mm; lfls 6–15 pairs, 5–15 mm, conspicuously punctate beneath; racemes 3–7 cm, solitary at the branch-tips; fr 5 mm, strongly glandular, not hairy; otherwise much like no. 1. Dry prairies; Io. and Minn. to Sask., Kans., Colo., and N.M. June, July. (*A. microphylla.*)

3. Amorpha nitens Boynton. Shining false indigo. Foliage blackening in drying; lfls oblong-ovate, 2–6 cm, glabrous to sparsely pubescent, thin, shining above; petiolules glabrous or sometimes sparsely short-hairy; cal-tube eglandular or with a few very small, inconspicuous, slightly elevated punctate glands; upper margin of the pod nearly straight; otherwise much like no. 4. Thickets and streambanks; s. Ill. and w. Ky. to Ga., La., and e. Okla. May, June.

4. Amorpha fruticosa L. False indigo. Branching shrub to 4 m; foliage not blackening in drying; petioles 2–5 cm; lfls 4–10+ pairs, green but not shining, 2–4 cm, usually sparsely short-hairy beneath; petiolules almost always pubescent; racemes (1)2–several, 6–20 cm; cal-tube 2–3 mm, the upper 4 lobes 0.5 mm, broadly triangular to half-orbicular, the lowest one somewhat longer and narrower; banner 5–6 mm; fr 5–9 × 2–4.5 mm, strongly glandular, the upper margin usually strongly bulged upward; $2n=40$. Moist woods and stream-banks; N.H. to Minn. and s. Sask., s. to Fla., Tex., s. Calif., and n. Mex. May, June. Variable, but not clearly divisible. (*A. croceolanata,* a southern phase with loose, often tawny or orange pubescence.)

37. ORBEXILUM Raf.   Cal campanulate, sometimes slightly oblique at base, but not gibbous, its lobes equal or the lower (especially the median one) the longest; standard obovate to rotund, usually clawed at base; wings about equaling the standard but longer than the keel; stamens 10(9), usually diadelphous, the upper median filament wholly or mostly free (or suppressed); ovule normally 1; fr short, ± flattened, indehiscent, in ours strongly cross-wrinkled and exserted from the cal; usually ± glandular-punctate herbs with pinnately 3-foliolate lvs, the terminal lfl long-stalked; fls in long-peduncled axillary spikes or racemes, usually blue; $x=11$. (Often included in *Psoralea*.) 8, N. Amer.

1 Principal lfls 2.5–5 cm wide; fr ovate, acuminate; plants rhizomatous . . . . 1. *O. onobrychis.*
1 Principal lfls under 2 cm wide; fr suborbicular; plants taprooted . . . . 2. *O. pedunculatum.*

1. Orbexilum onobrychis (Nutt.) Rydb.   Rhizomatous perennial, simple or sparingly branched, erect, to 1.5 m; lfls bright green, lanceolate or lance-ovate, 5–10 cm, half as wide, acuminate, thinly pubescent beneath; stipules and bracts subulate; peduncles nearly equaling the subtending lvs; racemes 3–8 cm; fls blue, varying to white, 6–7 mm; cal minutely pubescent to glabrate, 2–3 mm, slightly oblique, the short, broad lobes subequal; fr ovate, acuminate, 8–10 mm. Open woods and moist prairies; O. and Ky. to Io. and Mo.; also mt. woods, w. Va. to e. Tenn. and w. S.C. June, July. (*Psoralea o.*)

O. stipulatum (T. & G.) Rydb., eglandular, with diffuse stems, elliptic-ovate lfls, ovate stipules, and short-pedunculate head-like racemes with ovate bracts, originally on limestone-rocks along the Ohio R. in Ky. and Ind., appears to be extinct.

2. Orbexilum pedunculatum (Miller) Rydb.   Sampson's snakeroot. Perennial from a fusiform taproot, often branched from the base, erect, 3–8 dm; lfls narrowly oblong or elliptic to lance-linear, 5–7 cm, gradually narrowed to an obtuse tip, glabrous to thinly pubescent; peduncles ascending, much exceeding the subtending lvs; racemes 2–6 cm; fls blue, 6–7 mm, subtended by deciduous bracts; cal ± hirsute, the tube 1.5–2 mm, the lobes triangular, pointed, the upper about equaling the tube, the lower longer; fr suborbicular, 4–5 mm; $2n=22$. Dry or moist open woods and prairies. May, June. (*Psoralea p.*) Two well marked vars., intergradient s. of our range:

Var. *pedunculatum.* Floral bracts glandless or nearly so, ovate, gradually narrowed to a long-acuminate tip, deciduous long before the corolla is full-grown; cal glandless; lfls 4–7 times as long as wide. O. to Ill. and Kans., s. to Ga. and Tex.

Var. *gracile* (Chapman) Grimes. Floral bracts densely glandular-punctate, broadly ovate to subrotund, abruptly narrowed to a caudate tip, persistent until the fl is nearly mature; cal glandular-punctate; lfls 3–5 times as long as wide. Mainly coastal plain, Va. to Fla. (*Psoralea psoralioides.*)

38. PEDIOMELUM Rydb.   Much like *Orbexilum,* but the lvs digitately 3–5-foliolate, or the upper 1-foliolate; cal-tube gibbous at base on the upper side; standard broadly oblanceolate to obovate or suborbicular; fr opening by transverse rupture, not strongly cross-ribbed; $x=11$. (Often included in *Psoralea*.) 21, N. Amer.

1 Plants with a fusiform to globose, tuberous-thickened, often deep-seated, edible root; fr with a sword-
    shaped beak as long as or longer than the body, the upper part of the body falling off with the
    beak; fls 11–20 mm.
  2 Lvs subsessile, 3-foliolate or the upper 1-foliolate; fls 11–14 mm . . . . 1. *P. canescens.*
  2 Lvs long-petiolate, 5-foliolate; fls 15–20 mm.
    3 Plants strigose; petioles mostly 2–3 cm . . . . 2. *P. cuspidatum.*
    3 Plants conspicuously spreading-hairy; fr hirsute; main petioles (3)4–10 cm . . . . 3. *P. esculentum.*
1 Plants with woody rhizomes; fls 4–10 mm . . . . 4. *P. argophyllum.*

1. Pediomelum canescens (Michx.) Rydb.   Three-lfl prairie-turnip. Much branched, 3–10 dm from a tuberous-thickened root; petioles under 1 cm, equaling or shorter than the well developed petiolules; lower lvs 3-foliolate, the upper often 1-foliolate; lfls elliptic to obovate, 2–5 cm, blunt or rounded, gray-villous beneath; fls blue, 11–14 mm; cal very gibbous at base, thinly villous, glandular-punctate, the tube 3–4 mm, the 4 upper lobes 3.5–4.5 mm, the lowest one 5–6 mm. Dry, sandy soil; se. Va. to Fla. and Ala. (*Psoralea c.*)

2. Pediomelum cuspidatum (Pursh) Rydb.   Strigose p.-t. Much branched, thinly strigose, 3–6 dm from a tuberous-thickened root; petioles mostly 2–3 cm; lfls commonly 5, narrowly obovate or oblong-obovate, 2–4 cm; spikes dense, 2–6 cm; fls blue, 15–20 mm, sessile or nearly so; cal gibbous, densely glandular, otherwise glabrous or sparsely hirsute, the tube 5–6 mm, the lowest lobe much the longest, lanceolate, 8–10 mm; fr glandular, short-beaked. Prairies and plains; Minn. to Mont., s. to Ark. and Tex. June. (*Psoralea c.*)

3. Pediomelum esculentum (Pursh) Rydb.   Shaggy p.-t. Much branched, 1–4 dm from a tuberous-thickened root, conspicuously spreading-hairy; petioles of at least the larger lvs mostly (3–)4–10 cm; lfls 5, oblong to oblanceolate or narrowly obovate, 2–6 cm; spikes dense, leafy-bracteate, 3–8 cm; fls blue, sessile or nearly so, 15–20 mm; cal-tube gibbous, 5–6.5 mm, glabrous or sparsely hirsute, the lobes subequal, lance-linear,

7-10 mm, ciliate; fr hirsute, the flat beak evidently longer than the body; 2n=22. Dry prairies and plains; Minn. and Wis. to Mo. and Tex., w. to Alta. and N.M. May–July. (*Psoralea e.*)

4. Pediomelum argophyllum (Pursh) Grimes.   Silvery scurf-pea. Much branched, densely white-sericeous throughout, eglandular, 3–6 dm, rhizomatous; petioles 1–3+ cm; lfls 3 or 5, narrowly elliptic to oblanceolate or narrowly ovate, 2–5 cm; spikes 2–8 cm; interrupted, with 1–5 fascicles of dark blue, sessile fls 8–10 mm; cal-tube 2–3 mm; lowest cal-lobe lance-linear, 4.5–7 mm, twice as long as the upper 4; fr densely silky; 2n=22. Dry prairies and plains; Minn. and Wis. to Mo., w. to Sask., Mont., and N.M. June–Aug. (*Psoralea a.; Psoralidium a.*)

39. PSORALIDIUM Rydb.   Resembling *Pediomelum*, but the cal flaring back and tearing along a lateral sinus in fr; fr indehiscent, densely glandular, rotund to ovate or shortly elliptic in profile; x=11. (Often included in *Psoralea*.) 3, N. Amer.

1 Fls mostly blue, in loose, interrupted racemes surpassing the lvs, fr ovoid . . . . 1. *P. tenuiflorum.*
1 Fls white or the keel blue, in short, congested racemes often surpassed by the lvs; fr subglobose . . . . . 2. *P. lanceolatum.*

1. Psoralidium tenuiflorum (Pursh) Rydb.   Gray scurf-pea. Erect, much branched, to 1 m, gray-strigose, rhizomatous; petioles 0.4–1 cm; lfls 3 or 5, narrowly elliptic to oblanceolate, subglabrous above, mostly 2–4 cm × 5–8 mm; racemes loose and open, 4–10 cm, on elongate peduncles exceeding the lvs; pedicels 1.5–3 mm; fls 1–4 per node, blue (white), 5–8 mm; cal hirsute and glandular-punctate, the tube 1.5–2 mm, the lobes narrowly triangular, acute, the lowest 1.2–1.8 mm, the other 4 shorter; fr flattened-ovoid, 7–8 mm; 2n=22. Dry prairies and plains; Mont. to Utah and Ariz., e. to Minn., Ind., and Mo. May, June. (*Psoralea t.; Psoralea floribunda.*)

2. Psoralidium lanceolatum (Pursh) Rydb.   Much like no. 1; petioles mostly 1–2 cm; racemes short, congested, often surpassed by the lvs; fls white or the keel blue; cal-teeth subequal; fr subglobose, 4–5 mm; 2n=22. Dry, sandy soil; Wash. to Calif., e. to N.D., w. Io., and Okla. May–Sept. (*Psoralea l.*)

40. CENTROSEMA (DC.) Benth. nom. conserv.   Spurred butterfly-pea. Fls resupinate, the banner lowermost; cal nearly regular, the tube hemispheric, the lobes subequal; standard rotund, much exceeding the other pet, with a minute median spur at the base; wings oblong, obtuse; keel-pet broadly semi-obovate, strongly incurved; stamens 10, diadelphous, the anthers all alike; style glabrous, strongly incurved to the pilose stigma; fr elongate, many-seeded, with a longitudinal ridge near each margin, tightly coiled after dehiscence; twining herbs with pinnately compound lvs of 3 stipellate lfls, and with solitary or clustered axillary fls. 45, New World.

1. Centrosema virginianum (L.) Benth.   Climbing or trailing perennial to 1 m, minutely pubescent; stipules lanceolate, 2–4 mm, striate; lfls oblong to elliptic or ovate, 3–6 cm; fls 1–4 in a short-peduncled raceme, violet, 2–3 cm; bractlets ovate, striate, 8–12 mm, much surpassing the cal-tube; cal-lobes linear, 6–12 mm; fr flat, 7–12 cm × 4 mm, the persistent style 2–2.5 cm; 2n=18. Dry sandy woods and barrens; s. N.J. to Fla., w. to Ky., Ark., and Tex. (*Bradburya v.*)

41. CLITORIA L.   Butterfly-pea. Fls resupinate, the banner lowermost; cal somewhat gibbous at base, irregular, the 2 upper lobes partly connate, shorter than the lower; standard rotund-obovate, retuse, narrowed to the base, much exceeding the other pet; keel-pet strongly incurved but not twisted, shorter than the wings and coherent with them to the middle; stamens 10, monadelphous below; anthers all alike; style strongly incurved, pubescent on the concave side; fr stipitate, linear-oblong, flattened, dehiscent, several-seeded; twining herbs or shrubs with pinnately 3–9-foliolate, stipellate lvs and large, bracteolate fls solitary or in small clusters on axillary peduncles. 40, warm reg.

1. Clitoria mariana L.   Perennial, glabrous, twining herb to 1 m; stipules lance-ovate, striate, 2–4 mm; stipellules and bracteoles similar but smaller; lfls 3, ovate, ovate-oblong, or oblong-lanceolate, 3–6 cm, obtuse; peduncle shorter than the petiole of the subtending lf, bearing ordinarily 1 or 2 pale blue or pink-lavender fls 4–6 cm, each fl closely subtended by a pair of bracteoles; cal-tube 10–14 mm, exceeding the triangular-ovate teeth; fr 3–6 cm, tipped by the persistent style. Dry upland woods and barrens; L.I. and N.J. to s. O., s. Ind., Mo., and Okla., s. to Fla. and Tex. June, July. (*Martiusia m.*)

42. PHASEOLUS L.   Bean. Cal-tube hemispheric, slightly oblique; lobes 5, subequal, shorter than the tube; standard rotund, its sides folded over the other pet; wings semi-obovate, coherent with the keel; keel-pet linear-oblong, at the summit incurved and spirally coiled; stamens 10, diadelphous; style bearded along the upper side, spirally coiled and thickened beyond the middle; fr terete or flattened, with few–several seeds; twining or erect herbaceous perennials with pinnately trifoliolate lvs, small few-nerved stipules and bracts, and small to medium-sized fls each closely subtended by a pair of minute bracteoles. 200+, mainly warm New World.

1. **Phaseolus polystachios** (L.) BSP. Wild b. Twining perennial often to 3 or 4 m; lfls broadly ovate or rhombic-ovate, softly pubescent beneath, often glabrous above, the terminal one 3–8 cm, the others usually asymmetric; racemes slender, to 15 cm; fls purple, 10–12 mm; upper cal-lobes semi-orbicular, the others broadly triangular; keel coiled in about one and a half complete turns; fr flat, 3–6 cm, coiled after dehiscence; seeds 2 mm; $2n=22$. Moist woods and thickets; s. Me. to O., Ill. and Mo., s. to Fla. and Tex. July, Aug.
Two cult. spp., *P. vulgaris* L., the common bean, and *P. coccineus* L., the scarlet runner, occasionally escape but apparently never persist.

43. **DOLICHOS** L. Cal-tube hemispheric, slightly oblique, 4-lobed by fusion of the 2 uppermost; upper lobe broadly triangular, barely emarginate; lower 3 lobes lance-triangular; standard oblate, broader than long, with 2 prominent keel-like appendages at the base within; wings long-clawed, the obovate blade upcurved; keel-pet narrowly oblong, bent upward at the middle at about a right angle; stamens 10, diadelphous; style bearded along the upper side; fr oblong, flat, few-seeded; herbs with large, 3-foliolate lvs and axillary racemes of medium sized, anthocyanic (white) fls. 60, Old World.

1. **Dolichos lablab** L. Hyacinth bean; Egyptian b. Perennial; stems twining, 1–3 m; lfls broadly rhombic-ovate; racemes to 3 dm; fls purple or white, 1.5–2.5 cm; fr 3–8 × 1.5–2 cm; $2n=20, 22, 24$. Native of Asia, rarely escaped from cult. as far n. as N.Y.

44. **STROPHOSTYLES** Elliott. Woolly bean. Cal irregular, 4-lobed through fusion of the 2 upper lobes; lowest lobe the longest, much exceeding the tube; standard orbicular or broadly round-ovate, its sides at base folded over the other pet; wings much shorter than the keel, oblong, upcurved; keel-pet broadest near the middle, abruptly contracted and strongly curved upward into a beak; stamens 10, diadelphous; style bearded along the upper side; fr subterete, elongate, coiled after dehiscence, with several woolly seeds; twining or trailing herbs with pinnately trifoliolate, stipellate lvs and long-peduncled heads of small pink-purple to white fls, each closely subtended by a pair of striate bracteoles. 3, e. N. Amer.

1 Cal-tube and bracteoles glabrous or with a few appressed hairs; fls 8–14 mm.
  2 Annual; bracteoles lanceolate, acute, at least as long as the cal-tube .... 1. *S. helvola.*
  2 Perennial; bracteoles ovate or oblong, blunt, up to half as long as the cal-tube .... 2. *S. umbellata.*
1 Cal-tube and bracteoles densely hirsute; fls 5–8 mm .... 3. *S. leiosperma.*

1. **Strophostyles helvola** (L.) Elliott. Annual w.b. Annual twiner to 1 m; lfls ovate to rarely oblong, 2–6 cm, sparsely pilose at least beneath, some of them commonly with a lateral lobe on one or both sides, or with concave lateral margins; peduncles 0.5–3 dm; fls few–several, 8–14 mm, pink-purple, turning greenish; cal-tube glabrous or with a few appressed hairs; bracteoles lanceolate, acute, usually glabrous, extending at least to the sinuses of the cal; fr 4–9 cm, sparsely appressed-hairy; seeds 5–10 mm, persistently woolly; $2n=22$. Dry sandy soil, often on cinders; Que. to Minn. and S.D., s. to Fla. and Tex. (*S. missouriensis.*)

2. **Strophostyles umbellata** (Muhl.) Britton. Perennial w.b. Much like no. 1; perennial; lfls 2–5 cm, never lobed, but occasionally with barely concave lateral margins, pilose beneath, glabrous or minutely short-hairy above; bracteoles ovate or oblong, blunt, up to half as long as the cal-tube; seeds 3–5 mm, rarely longer; $2n= 22$. Dry, sandy, upland woods; s. N.Y. to Fla. and Tex., chiefly on the coastal plain, n. in the interior to s. Ind., Mo., and Okla.

3. **Strophostyles leiosperma** (T. & G.) Piper. Small-fld w.b. Annual; lfls narrowly to broadly oblong or lanceolate, 3–5 cm, pilose on both sides, more densely so beneath, with stiff hairs 1–2 mm; cal and bracteoles hirsute, the bracteoles lanceolate, varying in length; fr 2–4 cm, spreading-hairy; seeds 2.5–3 mm, the pubescence easily detached; otherwise much like no. 1. Dry or moist sandy soil, upland woods, dunes and shores; O. to Wis. and N.D., s. to Ala. and Tex. (*S. pauciflora.*)

45. **RHYNCHOSIA** Lour. Cal-tube short, obconic, nearly regular, the lobes about equal, lanceolate, much exceeding the tube, the 2 upper connate part-way; pet about equal

in length; standard broadly obovate, narrowed to the base; wings oblong, blunt; keel-pet oblong, somewhat incurved; stamens 10, diadelphous; ovary and fr pubescent; style glabrous, fr flat, dehiscent, 1–2-seeded; trailing, twining, or erect perennial herbs with 1-foliolate or pinnately 3-foliolate lvs and lacking stipellules; fls small, yellow, in axillary racemes, without bracteoles. (*Dolicholus.*) 200, warm reg.

1 Stems erect; lfls densely tomentose beneath; cal-lobes 4–6 mm .... 1. *R. tomentosa.*
1 Stems prostrate or twining-climbing; lfls sparsely hairy beneath; cal-lobes 7–10 mm .... 2. *R. difformis.*

1. Rhynchosia tomentosa (L.) H. & A. Stems erect, 3–8 dm, closely pubescent all around with short, ± ascending hairs; lfls obovate-oblong or oblong, 3–7 cm, half to two-thirds as wide, pubescent above (often densely so), densely gray-tomentose beneath; racemes 2–3 cm, the fls densely crowded in the distal half; cal-lobes 4–6 mm, the upper 2 separate about half their length; fr hirsutulous with short hairs 0.5–1 mm from papillose bases, beneath them densely villosulous (scarcely visible at 10×); $2n= 22$. Dry, sandy woods and barrens, chiefly on the coastal plain; Del. to Ky., Tenn., Fla. and La. July–Sept. Ours is var. *tomentosa.* (*R. erecta; Dolicholus e.*)

2. Rhynchosia difformis (Elliott) DC. Twining vine, prostrate or climbing to 4 m, densely pubescent on the angles with spreading or somewhat reflexed hairs, thinly pubescent on the sides; terminal lfl broadly elliptic, rhombic, or subrotund, 2–5 cm, three-fourths as wide to wider than long, sparsely pubescent on both sides; lowest lvs often 1-foliolate; cal-lobes 7–10 mm, the upper 2 separate a third to half their length; fr pubescent with slender straight hairs over a minute puberulence; $2n=22$. Dry sandy woods and barrens on the coastal plain; se. Va. to Fla. and Tex. June, July. (*R. tomentosa,* misapplied.)

46. APIOS Fabr. Ground-nut. Cal-tube broadly hemispheric, the 4 upper lobes very short or obsolete, the lowest one lance-triangular, about equaling the tube; standard rotund, soon reflexed, obscurely auricled near the base; wings curved-oblong-obovate, soon deflexed below the keel, short-clawed, auricled at the base of the blade; keel-pet strongly curved, horseshoe-shaped, blunt, broadest at the truncate base; stamens 10, diadelphous; style glabrous, curved with the keel; fr linear, several-seeded, coiled after dehiscence; perennial twining herbs from slender rhizomes bearing tuberous thickenings, with pinnately compound lvs of 5–7 obscurely stipellate lfls and axillary peduncled racemes, the nodes bearing prominent callous thickenings; each fl closely subtended by a pair of caducous, linear, 1-nerved bractlets, 10, e. Asia and e. U.S.

1 Standard brown-purple, rounded or retuse at the summit .... 1. *A. americana.*
1 Standard greenish-white, the red-tinged summit prolonged into a spongy appendage .... 2. *A. priceana.*

1. Apios americana Medikus. Common g.-n. Rhizomes with 2 or more tubers; stipules and stipellules minute, caducous; lfls of the principal lvs 5–7, lanceolate to ovate, 4–6 cm, glabrous to short-hairy beneath; racemes dense (or looser southward), with solitary or paired brown-purple fls 10–13 mm; standard broader than long, usually retuse; fr 5–10 cm; $2n=22$, and perhaps 44. Moist woods; Que. to Minn. and S.D., s. to Fla. and Tex. July, Aug. (*A. tuberosa; Glycine apios.*)

2. Apios priceana B. L. Robinson. Kentucky g.-n. Much like no. 1, the tubers fewer and larger; fls greenish-white, suffused with rose toward the tip, 17–22 mm; standard terminating in a spongy protuberance 4 mm. Rocky woods; Ky. and reputedly Tenn. Aug. Rare and little-known.

47. AMPHICARPAEA Elliott. Hog-peanut. Cal slightly irregular, the tube short-cylindric, the lobes 4 through fusion of the upper 2; standard obovate, narrowed to the base, sometimes auricled below the middle; wings and keel slightly shorter, with elongate slender claws exceeding the blade, the wings oblong, auricled at the base of the blade, the keel-pet narrowly obovate, nearly straight, not auricled; stamens 10, diadelphous; anthers all alike; style glabrous; fr flat, oblong, pointed at both ends, mostly 3-seeded; plants producing, in addition to the typical fls and frs as described above, some nearly or wholly apetalous fls on long, filiform branches at the base of the stem, these developing small, 1-seeded, often subterranean frs; twining herbs with pinnately compound lvs of 3 stipellate lfls and small purplish to white fls in axillary racemes or panicles, each pedicel subtended by a striate bract; bractlets none. (*Falcata.*) 3, Afr., Asia, N. Amer.

1. Amphicarpaea bracteata (L.) Fern. Annual to 1.5 m; stipules and stipellules ovate or lanceolate, striate; lfls 2–8 cm, ovate or rhombic-ovate, broadly rounded at base; infls peduncled from many of the axils, with broadly obtuse to subtruncate bracts and 2–many pale purple to whitish fls 12–18 mm; cal-lobes 1–2 mm; frs 1.5–4 cm, strigose at least on the sutures; $2n=20$. Highly variable especially in pubescence, but scarcely divisible into vars. Abundant in woods and thickets; Que. and N.S. to Man. and Mont., s. to Fla. and Tex. Aug., Sept. (*A. monoica; A. pitcheri.*)

48. PUERARIA DC. Cal 4-lobed through fusion of the upper 2, the lowest lobe the longest, the two lateral ones the shortest, all exceeding the tube; standard round-obovate; keel-pet narrowly oblong, nearly straight, auriculate at base; stamens 10, the uppermost one coherent with the others only near the middle; style elongate, glabrous; fr flat, several-seeded, hirsute; tall, woody or half-woody twiners with ovate stipules, large, pinnately trifoliolate lvs, and reddish-purple fls in dense axillary racemes or panicles. 12, e. Asia.

    1. **Pueraria lobata** (Willd.) Ohwi. Kudzu-vine. Stems 10–30 m, the younger parts hairy; lfls broadly ovate to subrotund, 10–15 cm, often 2–3-lobed, hairy beneath; infls 1–2 dm, the axis and pedicels densely sericeous; fls 2–2.5 cm; lowest cal-lobe 8–12 mm; standard yellow at base; fr 4–5 cm; $2n=22, 24$. Native of Japan, cult. for forage and erosion-control in se. U.S., escaped and extending n. into the s. part of our range, sometimes even to Mass., N.Y., and O. Late summer. (*P. thunbergiana.*)

49. DIOCLEA HBK. Cal-tube nearly regular, the lobes 4 by fusion of the upper 2, the lowest lobe exceeding the lateral ones; standard broadly obovate, auricled at base; wings obovate-oblong; keel-pet semi-elliptic, nearly straight, stamens 10, the uppermost one coherent with the others for half its length; fr oblong, flattened, few-seeded, dehiscent, narrowly 2-winged along the upper suture; twining woody or half-woody vines with large, pinnately trifoliolate lvs, minute stipules, and small blue or purple fls borne several at each node on an elongate axis and each closely subtended by a pair of bracteoles. 30, mainly trop. Amer.

    1. **Dioclea multiflora** (T. & G.) C. Mohr. Stems high-climbing, finely retrorse-hairy; lfls broadly ovate to rotund, 5–15 cm, sparsely hairy beneath; peduncles 3–6 cm, bearing a crowded raceme 2–5 cm; fls 12–15 mm; fr 5 cm. Alluvial woods; w. Ky. to Ga. and La. June.

50. GALACTIA P. Browne. Milk-pea. Cal obconic, nearly regular, 4-lobed by fusion of the upper 2, the lobes nearly equal, somewhat exceeding the tube; standard broadly obovate, narrowed to a short claw, obscurely auricled; wings oblong, blunt, auricled at base above the slender claw, slightly shorter than the straight, semi-obovate, short-clawed keel-pet; stamens 10, diadelphous; ovary pubescent; style glabrous, fr narrowly oblong, flat, few-seeded, twisted after dehiscence; twining or trailing perennial herbs with pinnately trifoliolate lvs, minute stipules and bracts, and short racemes; fls small, purplish, often 2 or 3 at a node, each closely subtended by a pair of minute bracteoles. 50, mostly warm New World.

    The proper application of the names *G. regularis* and *G. volubilis* is in dispute, and the traditional usage of *G. regularis,* in particular, may well be incorrect. We here follow the tradition, rather than accepting newer interpretations that are also debatable.

1 Stems mostly prostrate and straight or twining only at the tip; fls 12–18 mm; pubescence closely
     appressed or none .... 1. *G. regularis.*
1 Stems mostly twining-climbing; fls 7–14 mm.
  2 Pubescence appressed; fls mostly 10–14 mm .... 2. *G. macreei.*
  2 Pubescence spreading; fls mostly 7–10 mm .... 3. *G. volubilis.*

    1. **Galactia regularis** (L.) BSP. Trailing m.-p. Stems mostly prostrate and straight or twining only at the tip, retrorsely strigillose or glabrate; lfls elliptic to ovate-oblong or oblong-lanceolate, strigillose or glabrate beneath, glabrous to strigillose-scaberulous above, the terminal one 2–5 cm; racemes few-fld, 3–6(–10) cm, the fls nearly adjacent, 12–18 mm; cal-tube glabrous or strigose, the lobes 3.5–7 mm; frs 3–5 cm, strigose. Dry sandy woods and barrens, chiefly on the coastal plain; s. N.Y. to Fla. and Miss. July, Aug. (? *G. glabella.*)
    2. **Galactia macreei** M. A. Curtis. Macree's m.-p. Resembling no. 1 in pubescence, and no. 3 in habit and infl. the fls of intermediate size; $2n=20$. Damp or wet thickets, low woods, and pond-margins; coastal plain from se. Va. to Fla. and Tex. July, Aug. (*G. volubilis,* probably misapplied.)
    3. **Galactia volubilis** (L.) Britton. Hairy m.-p. Stems twining, usually climbing, to 1.5 m; stem, cal, and at least the lower surface of the lfls loosely spreading-hairy; lfls commonly ovate or ovate-oblong, varying to narrowly oblong; racemes often 1 dm, loose, the fls well separated by internodes often 1–3 cm; fls 7–10

mm; cal-lobes 2–4 mm; fr 2–4 cm, shortly and densely spreading-hairy; $2n=20$. Dry upland woods and barrens; s. N.Y. to s. O., Mo., and Kans., s. to Fla. and Tex. July, Aug. (*G. regularis,* probably misapplied.) Plants occurring mainly in the Mississippi Valley region, with the lfls loosely hairy on the upper as well as the lower surface, may warrant recognition as var. *mississippiensis* Vail. (*G. mississippiensis.*)

FAMILY ELAEAGNACEAE, the Oleaster Family

Fls in racemes or small umbels or solitary in the axils, perfect or unisexual, regular (2)4(6)-merous, strongly perigynous, apetalous; hypanthium in perfect and pistillate fls ± tubular, constricted above the ovary, in staminate fls cupulate to almost flat; sep appearing as lobes on the hypanthium, valvate, often somewhat petaloid; stamens at the throat of the hypanthium, as many as (and alternate with) or twice as many as the sep; filaments very short; ovary unilocular, with a single basal ovule and a long, slender style; fr drupe-like or berry-like, the dry achene enveloped by (but free from) the persistent, fleshy or mealy base of the hypanthium, which very often has a bony inner layer; seed with a straight embryo, 2 expanded, thickened cotyledons, and little or no endosperm; woody plants, clothed with a lepitode or stellate indument, the lvs simple, entire, penniveined, exstipulate. 3/50.

1 Lvs opposite; stamens 8; fls unisexual, the plants dioecious . . . . 1. *Shepherdia.*
1 Lvs alternate; stamens 4; fls perfect or some of them staminate . . . . 2. *Elaeagnus.*

1. SHEPHERDIA Nutt., nom. conserv. Dioecious; fls in small clusters on twigs of the previous year, the pistillate ones usually few; sep greenish-yellow within; fls with an 8-lobed disk at the summit of the hypanthium; stamens 8, alternating with the lobes of the disk; pistillate fls with the hypanthium constricted at the summit; fr ± berry-like, without a well developed stone; lvs opposite. (*Lepargyrea.*) 3, N. Amer.

1 Lvs green and nearly glabrous above . . . . 1. *S. canadensis.*
1 Lvs silvery-lepidote above (as well as beneath) . . . . 2. *S. argentea.*

1. Shepherdia canadensis (L.) Nutt. Rabbit-berry. Unarmed deciduous shrub 1–3 m; lvs ovate to narrowly elliptic or lanceolate, the larger ones 3–7 × 1.5–4 cm, obtuse to subcordate at base, green and nearly glabrous above, densely whitish-lepidote-stellate and with scattered brownish scales beneath; fls in last year's naked axils; staminate fls with ovate, spreading sep 1.5–3 mm; pistillate fls with sep 1.5–3 mm, the mouth of the hypanthium occluded by pubescence; fr red or sometimes yellow, bitter, ellipsoid, 6–9 mm; $2n=22$. Various habitats, often on calcareous substrate; Nf. to Alas., s. to N.Y., n. Ind., S.D., and Ariz. Apr., May, with the lvs.

2. Shepherdia argentea Nutt. Buffalo-berry. Widely branched shrub or small tree to 6 m, colonial by roots; branches stiff, often shortly thorn-tipped; lvs silvery-lepidote on both sides, oblong to lanceolate, 2–5 × 0.5–1.5(–2.5) cm, narrowed to the base; fls a little smaller than in no. 1; fr scarlet, 6–9 mm, edible but sour; $2n=26$. River-banks and canyons; Minn. and Io. to B.C. and N.M. Apr., May, with the lvs.

2. ELAEAGNUS L. Fls perfect or unisexual, in small lateral clusters on twigs of the current year; stamens 4, scarcely exsert; hypanthium tubular-turbinate above the globose, investing base, lobed not more than ca half-length, deciduous from the top of the developing fr; fr drupe-like, the persistent hypanthium-base rather dry and mealy, with a hard, bony, striate-fluted inner layer; lvs alternate. 45, Eurasia, N. Am.

1 Lvs silvery-lepidote above (as well as beneath).
　2 Lvs 1.5–3 times as long as wide; nectary scarcely developed; native . . . . 1. *E. commutata.*
　2 Lvs 3–8 times as long as wide, rarely wider; nectary shortly tubular; escaped . . . . 2. *E. angustifolia.*
1 Lvs soon green and glabrescent above; escaped . . . . 3. *E. umbellata.*

1. Elaeagnus commutata Bernh. Silver-berry. Shrub 2–5 m, colonial by roots, the young twigs brown-lepidote; lvs ovate-oblong to obovate-oblong, 2–7 cm, silvery-lepidote on both sides; fls 12–15 mm, densely silvery; sep spreading, 2.5–4 mm, yellow within; disk scarcely developed; fr 1 cm, ellipsoid, densely silvery; $2n=28$. River-banks and moist hillsides; Que. to Alas., s. to Minn., N.D., and Utah. June. (*E. argentea.*)

2. **Elaeagnus angustifolia** L. Russian olive; oleaster. Small, usually thorny tree to 10 m, the young twigs densely silvery; lvs lanceolate, 3–10 cm, silvery on both sides; fls as in no. 1, except the style surrounded by a short, tubular disk just above the constriction of the hypanthium; fr 1 cm, yellow, with silvery scales; $2n=28$. Native of Eurasia, planted for ornament and often escaped in c. and w. U.S., rarely eastward. June, July.

3. **Elaeagnus umbellata** Thunb. Shrubby tree to 5 m; lvs soon green and glabrescent above; hypanthium-tube ca twice as long as the sep; fr red, finely dotted with pale scales, juicy and edible, 6–8 mm, on pedicels ca 1 cm; $2n=28$. Native of e. Asia, becoming common in much of our range. May, June.

E. multiflora Thunb., another e. Asian sp., also occasionally escapes. It differs from no. 3 in its short hypanthium, ca equaling the sep, and in its larger fr, 1–1.5 cm on pedicels 1.5–2.5 cm. It is greener than *E. umbellata*, and fls a little earlier.

Family **PODOSTEMACEAE**, the River-weed Family

Fls perfect, much reduced, with superior or naked ovary, the sep vestigial or wanting; pet none; stamens in 1–several cycles; ovary mostly bicarpellate and bilocular, with axile placentation and distinct or basally connate styles; fr a septicidal capsule with numerous tiny seeds; embryo straight, with 2 cotyledons; endosperm none; submerged or partly floating aquatics of fast rivers with a stony bed, producing aerial fls and frs at times of low water; plants vegetatively highly modified, commonly ± thalloid. 40/200+.

1. **PODOSTEMUM** Michx. River-weed. Fls naked, borne singly in sessile, axillary, tubular spathes; stamens 2(–7), on one side of the ovary, their filaments united more than half length; staminodes 2, short, linear; ovary short-stipulate, the stipe elongating and elevating the 10-ribbed capsule on an evident pedicel; valves of the fr 2, one persistent, the other deciduous. 17, mainly tropical New World.

1. **Podostemum ceratophyllum** Michx. Thread-foot. Stems attached to stones by fleshy disks, 2–10 cm, usually branched; lvs alternate, dilated into a pair of stipule-like appendages at base, 1–10 cm, very narrow and entire or commonly ± forked into numerous narrow segments; spathe 2–3 mm, pedicels eventually to 1 cm; fr 2.5–3 mm. In rapidly flowing water; N.S. and N.B. and Me. to Ga., w. to s. Ont., W.Va., Ky., Okla., and La.; W.I., C. Amer.

Family **HALORAGACEAE**, the Water-milfoil Family

Fls epigynous, regular, bibracteolate, 3- or 4-merous, mostly small and anemophilous; sep valvate, persistent; stamens as many or twice as many as the sep, with short filament and large anther; ovary with 3 or 4 carpels and locules and distinct, feathery styles, each locule with a solitary, pendulous ovule; fr small, nut-like or drupaceous, sometimes separating into mericarps; embryo straight, cylindric, dicotyledonous, embedded in the soft endosperm; exstipulate herbs (some extralimital genera woody), often aquatic. 8/100.

1 Fls 4-merous; emersed lvs (except in *M. aquaticum*) bract-like, much reduced .... 1. *Myriophyllum.*
1 Fls 3-merous; emersed lvs foliaceous .... 2. *Proserpinaca.*

1. **MYRIOPHYLLUM** L. Water-milfoil. Fls unisexual, 4-merous, small, sessile in the axils of bracteal or foliage lvs, often aggregated into an emersed terminal spike; cal of 4 teeth, or obsolete; pet present or not; stamens 8 or 4 (even in one sp.); fr hard, ± 4-lobed, eventually splitting into 4 mericarps; vegetatively plastic herbs, submersed in quiet water or rooting on muddy shores, usually with once-pinnately dissected lvs. 20+, cosmop.

1 Lvs and bracts entire, or the lvs wanting .... 1. *M. tenellum.*
1 Lvs pinnately divided, with filiform segments.
  2 Foliage lvs partly or wholly alternate, subopposite, or irregularly scattered.
    3 Mature mericarps distinctly tuberculate on the back.
      4 Fls in emersed spikes .... 2. *M. pinnatum.*
      4 Fls in the axils of submersed foliage lvs .... 3. *M. farwellii.*

3 Mature mericarps rounded on the back, smooth or barely rugulose .... 4. *M. humile.*
2 Foliage lvs all distinctly whorled.
  5 Emergent lvs mostly 2.5–3.5 cm; petioles 5–7 mm; dioecious, ours pistillate .... 5. *M. aquaticum.*
  5 Emergent lvs under 2 cm; petioles 0–2 mm; monoecious, the upper fls staminate.
    6 Bracts and fls, except sometimes the lowermost, alternate .... 6. *M. alterniflorum.*
    6 Bracts and fls whorled.
      7 Bracts subtending the staminate fls cleft or deeply laciniate-toothed .... 7. *M. verticillatum.*
      7 Bracts subtending the staminate fls entire to sharply serrulate.
        8 Bracts subtending the staminate fls entire, not exceeding the fls.
          9 Lf-segments numerous, mostly 12–20 per side; no turions .... 8. *M. spicatum.*
          9 Lf-segments fewer, mostly 5–12 per side; propagating by turions .... 9. *M. sibiricum.*
        8 Bracts sharply serrulate, much exceeding the fls .... 10. *M. heterophyllum.*

1. **Myriophyllum tenellum** Bigelow.   Flowering stems very slender, mostly erect and unbranched, lfless or with a few remote scale-lvs only; spikes emersed, 2–5 cm; bracts all or chiefly alternate, oblong to obovate, entire, shorter than to slightly exceeding the fls; mericarps rounded on the back and at base, minutely granular, 1 mm. $2n=14$. Nf. and Que. to N.J., w. to w. Ont. and Minn.

2. **Myriophyllum pinnatum** (Walter) BSP.   Lvs partly or largely alternate or scattered, 1–3 cm, with capillary segments; emersed lvs sometimes developing and resembling the bracts, these chiefly whorled, much exceeding the fls, to 18 mm, with a flat rachis to 1 mm wide and on each side 3–5 ascending teeth 1–2 mm; spikes emersed; fr 2 mm, deeply lobed, each mericarp bearing a flat or concave longitudinal ridge with sharply tuberculate margins. Mass. to O., Io., and S.D., s. to Ga. and Tex., more abundant southward. (*M. scabratum.*)

3. **Myriophyllum farwellii** Morong.   Plants wholly submerged, producing terminal turions; lvs 1–3 cm, with filiform segments, all or many of the lvs alternate, subopposite, or irregularly scattered; fls axillary to foliage lvs; stamens 4; fr 2–2.5 mm, each mericarp with 2 low, denticulate or tuberculate, longitudinal ridges. N.S. and Que. to N.Y., w. to Minn.

4. **Myriophyllum humile** (Raf.) Morong.   Submersed lvs chiefly alternate, subopposite, or irregularly scattered; whorled lvs also present; no turions; fls axillary to submersed foliage lvs or in short emersed spikes; bracts mostly alternate, longer than the fls, irregularly pinnately parted, varying to linear and entire; mericarps 0.7–1.2 mm, obliquely truncate, rounded on the back, smooth or nearly so. N.S. and Vt. to Md., w. locally to Ill. and Minn.

5. **Myriophyllum aquaticum** (Vell. Conc.) Verdc.   Parrot's feather. Stems stout, very leafy; lvs all whorled, 2–5 cm, with 10–18 segments on each side, the lower segments much reduced, the rachis to 1 mm wide; petioles 5–7 mm; fls axillary to scarcely modified lvs in emergent infls; fr 1.5–2 mm, minutely granular; pls dioecious, only the pistillate, white fls known in N. Amer. Native of S. Amer., commonly cult. in aquaria and often escaped in s. U.S., n. to N.Y., W.Va., and Mo. (*M. brasiliense.*)

6. **Myriophyllum alterniflorum** DC.   Stems very slender; no turions; most internodes under 1 cm; lvs whorled, up to 1(–3) cm, often shorter than the internodes; spikes emersed, 2–5 cm; bracts mostly alternate, lance-linear, entire or minutely callous-serrulate, shorter than or about equaling the fls, or the very lowest pinnatisect and exceeding the fls; mericarps 1.5 mm, truncate, rounded on the back and at base, minutely granular-roughened; $2n=14$. Greenl. to Nf., Mass., N.Y., and Ont.; Europe.

7. **Myriophyllum verticillatum** L.   Much like no. 9; lvs with 9–17 segments per side; floral bracts less strongly reduced and modified, those subtending the pistillate fls (1–)2–several times as long as the fl, sometimes to 1 or even 1.5 cm, even those subtending the staminate fls evidently cleft or laciniate-toothed; pistillate fls with persistent sep mostly 0.5–1 mm; staminate fls with yellow-green pet; stamens 8, less often 4; turions abscised from the stem in late fall or early winter, dark yellow-green, their lvs less modified than in no. 9; $2n=28$. In acid to circumneutral waters to 3 m deep; circumboreal, s. to Mass., N.Y., Ind., Ill., Nebr., and Utah. (*M. hippuroides,* misapplied to the phase with 4 stamens.)

8. **Myriophyllum spicatum** L.   European w.-m. Much like no. 9; stem thickened below the infl to almost double its width farther down, curved to lie parallel with the water-surface, usually branching freely near the water-level; shoot-tip more tassel-like, with internodes mostly 1–3 cm; no turions; lvs with more numerous (mostly 12–20) long, straight segments per side; lower bracts often pectinate and often somewhat exceeding the frs; mericarps 2–3 mm; $2n=42$. European sp., widely intr. in N. Amer., tolerant of nitrogenous pollution.

9. **Myriophyllum sibiricum** Komarov.   Common w.-m. Stems loosely erect in the water, often whitish, not much branched above, not much thickened below the infl; shoot-tip knob-like, with crowded lvs, the internodes mostly ca 1 cm or less; lvs whorled, 1–5 cm, with mostly 5–12 slender, widely arcuate segments to a side; spikes emersed, mostly 4–10 cm, with whorled bracts and fls, the fls at the upper nodes pistillate, at the upper staminate; lower bracts mostly serrate or somewhat pectinate-toothed, up to about as long as the frs, the upper (subtending the staminate fls) entire and oblong-spatulate or ovate-oblong; pistillate fls without perianth, or the sep vestigial and under 0.5 mm; staminate fls with small pinkish pet; stamens 8; mericarps ca 2 mm, smooth or nearly so; plants in late season developing axillary (and often terminal) dark green turions 2+ cm, these eventually separating from the parent plant by decay, the new plants with a characteristically curved or U-shaped stem-base; $2n=42$. In circumneutral to alkaline water; Lab. to Alas., s. to Md., Va., W.Va., Ill., Mo., N.M., and Calif.; also in n. Eurasia. (*M. exalbescens.*)

10. Myriophyllum heterophyllum Michx. Stem stout, to 3 mm thick; internodes seldom over 1.5 cm; lvs whorled, (1.5–)2–4 cm, pinnatisect into capillary segments; spikes emersed, 5–15 (–40) cm; bracts whorled, persistent and eventually reflexed, lanceolate to oblong or even obovate, sharply serrulate, 4–18 mm (but fairly uniform on each plant) stamens 4; fr subglobose, 1–1.5 mm, each mericarp rounded or more commonly with 2 low keels on the back, scabrous to obscurely tuberculate, with an ascending-recurved beak. N.Y. to w. Ont. and N.D., s. to Fla. and Tex., more abundant southward.

2. PROSERPINACA L. Mermaid-weed. Fls perfect, 3-merous; sep triangular, persistent and somewhat accrescent; pet none; stamens 3; ovary 3-locular; styles 3, subulate, stigmatose beyond the middle; fr indehiscent, bony, 3-seeded and ± triangular; amphibious or marsh herbs, perennial by a rhizome-like base; stem beset with short, dark spicules; lvs all alternate, at least the submersed ones, when present, deeply pinnatisect; fls small, purplish or greenish, solitary or 2 or 3 together, sessile in the axils of emersed lvs. 2, N. Amer.

1. Prosperpinaca palustris L. Common m.-w. Stems decumbent and prostrate or rooting at base, often colonial, the floriferous branches erect, 1–4 dm; submersed lvs, if present, 2–4 cm, ovate or broadly oblong, deeply pinnatisect into narrow segments; emersed lvs linear-oblong to linear-oblanceolate, 2–6(–8) cm, serrate, tapering to the base; fr ovoid-pyramidal, 2–5 mm long and wide. Swamps, marshes, wet shores, and shallow water; Que. and N.S. to w. Ont. and Minn., s. to Cuba and Guatemala. July, Aug. The widespread var. *crebra* Fern & Griscom has the fr 2–4 mm wide, acutely angled but not winged, the sides nearly flat. Var. *palustris* (*P. platycarpa*), on the coastal plain from Mass. to La., has the frs 3.5–5 mm wide, sharply angled, or narrowly winged above or throughout, with concave sides. Var. *amblyogona* Fern., chiefly Ozarkian, extending e. to Ind., has frs 2–4 mm wide, plump, obtusely angled with convex sides.

2. Prosperpinaca pectinata Lam. Coastal plain m.-w. Emersed lvs ovate-oblong in outline, 1.5–3 cm, deeply pinnatisect, the rachis scarcely wider than the 6–12 pairs of lateral segments; fr 2.5–4 mm wide, irregularly ridged, obtusely angled with nearly flat sides; otherwise much like no. 1. Sandy bogs of the coastal plain; N.S. to Fla. and La. June–Aug.

P. intermedia Mackenzie, found on the coastal plain from N.S. to Va., often with the 2 preceding spp., may be a hybrid between them. Its emersed lvs are shallowly pectinate-pinnatifid, with a flat rachis 2–3 mm wide and ascending segments 1–4 mm; the fr is as in *P. palustris* var. *crebra.*

Family **LYTHRACEAE**, the Loosestrife Family

Fls perfect, regular or irregular, strongly perigynous, 4–8-merous; pet borne at the top of the hypanthium or within its tube, crumpled in bud, or seldom wanting; stamens as many or more often twice as many as the pet (or sep), rarely numerous, attached within the hypanthium, the ± elongate filaments often alternately unequal; ovary superior, often subtended by a hypogynous disk, plurilocular (sometimes incompletely so, the partitions not reaching the summit); style 1, with capitate stigma, or the stigma rarely sessile; ovules ± numerous; fr dry, usually capsular; seed with a straight embryo, 2 expanded cotyledons, and no endosperm; lvs simple, entire, exstipulate, opposite or less often alternate or whorled. 25/nearly 500, mostly in warm reg.

In addition to the following genera and spp., *Lagerstroemia indica* L., the crape-myrtle, occasionally escapes toward the s. part of our range. It is a tall shrub with large panicles of purple fls, the 6 pet long-clawed, the stamens ca 40.

1. DECODON J. F. Gmelin. Water-willow. Fls 4–5(–7)-merous, trimorphic, the styles and stamens of 3 possible lengths, any one fl with 8 or 10 stamens alternately of 2 lengths; hypanthium hemispheric or broadly campanulate; sep short, triangular, alternating with longer, linear appendages; pet conspicuous, narrowed at the base; disk none; capsule subglobose, 3(–5)-locular, loculicidal. Monotypic.

1. **Decodon verticillatus** (L.) Elliott. Perennial, woody below, with slender stems 1–3 m, usually arched and rooting at the tip; lvs opposite or more often in whorls of 3 or 4, short-petiolate, lanceolate, 5–15 × 1–4 cm; fls in dense cymes in the upper axils, the narrow pet pink-purple, 10–15 mm; fr 5 mm thick; 2n=32. Swamps and still water-courses. July–Sept. Var. *verticillatus*, with the pedicels and lower lf-surface ± tomentulose, occurs mostly near the coast from Me. to Fla. and La., and in the Mississippi Valley to Ind. and Mo. Var. *laevigatus* T. & G., with glabrous pedicels and lvs, is mostly inland, from N.S. and Que. to Minn., s. to e. Tenn.

2. ROTALA L. Fls (3)4(–6)-merous; hypanthium campanulate to globose or urceolate, with (as in our sp.) or without intersepalar appendages; pet present or absent; stamens 4(–6); ovary imperfectly (3)4(–6)-locular; disk wanting; capsule septicidal, (2)3–4-valved, transversely minutely and densely striate; glabrous annuals of wet places, with small, inconspicuous, bibracteolate fls solitary in the axils. 45, mainly Old World tropics.

1. **Rotala ramosior** (L.) Koehne. Tooth-cup. Erect or prostrate, 1–4 dm, often freely branched; lvs opposite or whorled, linear-spatulate, 1–3(–5) cm, mostly obtuse; sep alternating with triangular appendages of nearly the same size; pet white or pink, minute, caducous; fr 3–5 mm; 2n=32. In mud or wet soil; near the coast from Mass. to Fla. and Tex.; inland from O. and Mich. to Minn., Ark., and Okla.; also on the Pacific coast, in Mexico, and trop. Amer.

3. AMMANNIA L. Fls typically 4-merous; hypanthium globose or campanulate, usually with small intersepalar appendages; sep short; pet present or not, equaling or barely surpassing the sep; stamens 4(8); ovary imperfectly 2–4-locular; disk wanting; capsule rupturing irregularly, not striate; glabrous annuals of wet places, with narrow, sessile, opposite lvs and inconspicuous, bibracteolate fls in small axillary cymes of (1–)3–15. 25, mainly tropical.

1 Style 1.5–3 mm, exserted from the hypanthium at least in fr.
  2 Fr mostly 2–3.5 mm thick, shortly surpassing the cal; peduncles and pedicels evident, slender ...... 1. *A. auriculata.*
  2 Fr mostly 3.5–6 mm thick, not surpassing the cal.
    3 Pet deep rose-purple; anthers deep yellow; peduncles and pedicels often evident .... 2. *A. coccinea.*
    3 Pet pale lavender; anthers light yellow; peduncles and pedicels virtually wanting .... 3. *A. robusta.*
1 Style 0.5 mm, included in the hypanthium even in fr .... 4. *A. latifolia.*

1. **Ammannia auriculata** Willd. Redstem. Erect, 1–3 dm, pyramidally branched, with short ascending branches beginning above the base; lvs membranous, linear or lance-linear, 1.5–5 cm, cuneate to a barely auriculate sessile base; cymes 3–15-fld, on slender axillary peduncles 4–10 mm; pedicels 1–3 mm; fls small, the hypanthium at anthesis mostly 1.5–2.5 mm, the caducous pet mostly 1–2 mm, deep rose-purple; anthers 4, yellow; sep triangular, shortly surpassed by the fr, this 2–3.5 mm thick; style persistent, 1.5–3 mm; ovary and fruit 4-locular; 2n=32. Borders of ponds, and similar wet places; N.D. to Tex., and irregularly pantropical, perhaps not entering our range, most of our plants so identified in the past being extremes of no. 2. July–Oct.

2. **Ammannia coccinea** Rottb. Tooth-cup. Amphidiploid of nos. 1 and 3; stems often freely branched; cymes mostly with 3 or more fls, subsessile or on a short stout peduncle to 4+ mm; pedicels short and stout, or none; hypanthium mostly 2–3.5 mm; pet 1.7–2.5+ mm, deep rose-purple; anthers deep yellow; fr 3.5–4.5 mm thick; 2n=66. N.J. and Pa. to N.D., s. to trop. Amer.; Calif. July–Sept.

3. **Ammannia robusta** Heer & Regel. Sessile t.-c. Stems 2–10 dm, freely branched when well developed, the branching tending to begin at the base, with the lowest pair of branches initially decumbent and often nearly as long as the main stem; lvs rather thick and fleshy, linear or narrowly oblong, to 10 × 1.5 cm, usually much smaller, ± clasping at the truncate or auriculate, often dilated base; fls 1–3 per axil, virtually sessile, the stout peduncle and pedicels up to ca 1 mm, or wanting; fls a little larger than in no. 1, the hypanthium mostly 2.5–4.5 mm at anthesis, the caducous pet 2–3+ mm, pale lavender in life; anthers 4 or 8, light yellow; hypanthium about equaling and largely enclosing the fr, but open at the top, ovary and fruit 4-locular; the persistent style 1.5–3 mm, exserted; fr globose, 3.5–5+ mm thick; 2n=34. Edges of ponds, grassy swales, and similar wet places; N.J.; O. to N.D. and Wash., s. to C. Amer. July–Sept. Often confused with no. 2 in the past.

4. Ammannia latifolia L. Stem stout, 2–5 dm, simple or branched near the base; lvs spatulate to linear-oblong or lanceolate, mostly 2–5 cm, the lower cuneate, the upper truncate to cordate-auriculate at base; cymules sessile or nearly so; sep obtuse, often minutely mucronate; pet pink, 1 mm, caducous, or wanting; style only 0.5 mm, included in the hypanthium even in fr; $2n=48$. Wet ground on the coastal plain; N.J. to Fla. and Tex., s. to trop. Amer. July–Sept. (*A. koehnei; A. teres.*)

4. DIDIPLIS Raf. Water-purslane. Fls 4-merous; hypanthium campanulate; sep triangular; intersepalar appendages wanting; pet none; stamens (2–)4, included; ovary imperfectly 2-locular; style very short or none; disk wanting; capsule globose, indehiscent; annual amphibious herb with narrow, opposite lvs and minute greenish fls solitary in the axils. Monotypic.

1. Didiplis diandra (Nutt.) A. Wood Submersed or rooting in mud; stems weak, branched, 1–4 dm; lvs numerous, the submersed ones linear, 1–2.5 cm, sessile and truncate at the base, the emersed ones shorter, narrowly elliptic, tapering at the base. Minn. and Wis. to Kans. and O., s. to Va., Fla., and Tex. Summer. (*Peplis d.*)

5. LYTHRUM L. Loosestrife. Fls 6-merous, often dimorphic or trimorphic as to the style and stamens; hypanthium cylindric or narrowly campanulate, 8–12-nerved; sep short, alternating with appendages in the sinuses; pet red-purple to white; stamens (4–)6 or 12; ovary bilocular, with or without a disk; style slender; capsule septicidal or septifragal, enclosed by the persistent hypanthium; herbs or shrubs, usually virgately branched, with opposite, alternate, or whorled lvs and numerous bibracteolate fls sessile in the axils or in small cymes aggregated into a terminal spike-like thyrse. 30, ± cosmopolitan.

1 Fls numerous in a terminal spike-like thyrse; stamens mostly 12 .... 1. *L. salicaria.*
1 Fls solitary or paired in the axils of the lvs; stamens mostly (4–)6.
  2 Perennial; fls heterostylic.
    3 Hypogynous disk present; many lvs alternate, mostly longer than the internodes .... 2. *L. alatum.*
    3 Hypogynous disk absent; lvs chiefly opposite, mostly shorter than the internodes .... 3. *L. lineare.*
  2 Annual; fls all alike; hypogynous disk absent .... 4. *L. hyssopifolia.*

1. Lythrum salicaria L. Purple l. Stout, erect perennial herb 5–15 dm, glabrous or more often pubescent or, especially upwards, even subtomentose; lvs opposite or ternate, sessile, lanceolate to nearly linear, 3–10 cm, the larger ones cordulate at base; spikes 1–4 dm, interspersed with foliaceous, lanceolate to ovate bracts; fls trimorphic as to the style and stamens; appendages of the hypanthium linear, fully twice as long as the sep; pet red-purple, 7–12 mm; stamens mostly 12, alternately longer and shorter; disk wanting; $2n=30$, 50, 60. Native of Eurasia; in wet places from Nf. and Que. to N.D., s. to Va., Mo., and Kans., and occasionally w. to the Pacific. July–Sept.

L. virgatum L., much like *L. salicaria* but glabrous throughout, with the lvs narrowed to the base, and the appendages about equaling the sep, is locally established as an escape from cult. in Mass. and N.E.

2. Lythrum alatum Pursh. Winged l. Erect, glabrous, virgately branched perennial 4–8 dm, the stem 4-angled; lvs thick and firm, sessile, linear-oblong to lance-ovate, usually only the lowest opposite, those below the branches to 4 cm, broadest below the middle or near the truncate to broadly rounded or subcordate base, those of the branches crowded, proportionately narrower, obtuse to rounded at the base; fls solitary in most of the upper axils, heterostylic, either the stamens or the style exserted; pedicels 1–3 mm; hypanthium 4–7 mm, sharply but narrowly 12-winged; intersepalar appendages twice as long as the sep; pet purple, obovate, 2–6 mm; disk present; $2n=20$. Moist or wet soil, especially on prairies; Me. to N.D. and se. Wyo., s. to Fla. and Tex., and locally elsewhere, perhaps as an introduction. June–Sept. Most of our plants are var. *alatum,* as described above. (*L. dacotanum.*) Var. *lanceolatum* (Elliott) T. & G., more robust, up to 12 dm, with lanceolate lvs narrowed to the base, is chiefly more southern, barely reaching our range in se. Va.

3. Lythrum lineare L. Narrow-lvd l. Erect, glabrous perennial 4–15 dm, much branched above; lvs mostly opposite, narrowed to a sessile base, mostly shorter than the internodes, the lower linear-oblong, 1–4 cm, to 4 mm wide, the upper smaller, those subtending the fls linear, mostly 8–15 mm; fls axillary, heterostylic as in no. 2, short-pedicellate; hypanthium 4 mm; appendages triangular, about equaling the sep; pet pale purple to white, 3–4 mm; $2n=20$. Brackish marshes on the coastal plain; N.J. to Fla. and La. July–Sept.

4. Lythrum hyssopifolia L. Annual l. Pale green annual 1–6 dm; stem 4-angled; lvs linear to narrowly oblong, mostly alternate, 1–2 cm, a few often elliptic and to 6 mm wide; fls pale purple to white, sessile, solitary or paired; hypanthium half as long as the subtending lf; stamens and pet 4–6, the pet cuneate, 2–3 mm. $2n=20$. Mostly in wet soil, especially around salt-marshes; Me. to N.J. and e. Pa.; s. O.; also on the Pacific coast, in S. Amer., and widely distributed in the Old World. June–Sept.

6. CUPHEA P. Browne. Fls irregular, 6-merous; hypanthium tubular, prominently 12-ribbed, swollen or spurred at the base on one side, oblique at the mouth; stamens usually 11 or 12, alternately unequal; pet unequal; ovary with a curved gland at its base, imperfectly 2-locular; style slender; capsule included in the hypanthium, opening along the upper side, the hypanthium also splitting along the upper side; herbs or shrubs with opposite lvs and solitary or racemose fls. 200+, mainly tropical and subtropical Amer.

1. **Cuphea viscosissima** Jacq. Blue wax-weed. Erect, weedy annual, 1.5–6 dm, sparingly branched, very viscid-hairy throughout; lvs long-petioled, lanceolate to ovate, 2–5 cm; fls solitary or paired in the upper axils, short-pedicellate; hypanthium nearly 1 cm; pet red-purple, clawed, the upper 2 the largest, ca. 8 mm; $2n=12$. Dry soil; N.H. to n. Ill. and Kans., s. to Ga. and La., commoner southward. July–Sept. (*C. petiolata*, a preoccupied name.)

FAMILY **THYMELAEACEAE**, the Mezereum Family

Fls perfect or sometimes unisexual, usually regular, mostly 4–5-merous, perigynous with a campanulate to tubular hypanthium on which the usually ± petaloid sep appear as lobes, or the sep scarcely developed; pet small or none; stamens typically bicyclic, in our genera 8; ovary superior, commonly subtended by a hypogynous disk, in our genera unilocular, with one well developed carpel and a second vestigial one; style one, often excentric, sometimes very short; ovule solitary, pendulous; fr mostly indehiscent; seed with a straight embryo, 2 expanded cotyledons, and little or no endosperm; highly poisonous woody plants or seldom herbs; lvs simple, entire, exstipulate, alternate in our genera. 50/500, cosmop.

1 Fr drupaceous, exposed at maturity; larger lvs over 3 cm; shrubs.
  2 Sep well developed, petaloid; stamens and style included . . . . 1. *Daphne.*
  2 Sep scarcely developed; stamens and style exsert . . . . 2. *Dirca.*
1 Fr dry, enclosed in the persistent hypanthium; lvs under 2 cm; ours annual . . . . 3. *Thymelaea.*

1. **DAPHNE** L. Fls perfect; hypanthium tubular, not constricted above the ovary, bearing 8 short included stamens near the middle; sep 4, spreading, petaloid; pet none; style very short, with a large, capitate stigma; fr a drupe; shrubs with conspicuous fls in axillary or terminal clusters. 70, Old World.

1. **Daphne mezereum** L. Mezereum. Shrub to 1 m; lvs deciduous, oblanceolate to oblong, 4–8 cm, tapering to the base; fls appearing in early spring in sessile lateral clusters of 2–4; hypanthium 7–9 mm, pubescent; sep ovate, pink to purple or white, 5 mm; fr red or yellow, to 1 cm; $2n=18$. Native of Eurasia, well established as an escape from cult. from N.S. nd Nf. to N.Y. and O.

2. **DIRCA** L. Leatherwood. Fls perfect; hypanthium narrowly funnelform, slightly constricted above the ovary, thence differentiated into a cylindric tube and slightly spreading limb, bearing 8 exsert stamens at the summit of the tube; sep minute; pet none; style slender, exsert, exceeding the stamens; fr a drupe; shrubs with jointed twigs and very tough bark, bearing lateral clusters of 2–4 small, subsessile, pale yellow fls subtended by hairy bud-scales in early spring before the lvs appear. 2, N. Amer.

1. **Dirca palustris** L. Shrub 1–2 m; lvs broadly oblong-obovate or obovate to ovate, 5–8 cm, on petioles 2–5 mm; fls 7–10 mm, the stamens protruding 3 mm; fr ellipsoid, 12–15 mm, pale yellowish-green, quickly deciduous. Rich, moist woods; N.S. and s. Que. to Minn., s. to Fla., Ala., Ark., and Okla.

3. **THYMELAEA** Miller. Hypanthium urceolate or tubular, persistent about the dry, indehiscent fr; sep 4; pet none; stamens 8, included; style slightly eccentric; lvs small, sessile; most spp. dioecious or polygamous evergreen shrubs, but ours an annual herb with perfect fls. 20, Medit., temp. Asia.

1. **Thymelaea passerina** (L.) Cosson & J. Germain. Taprooted annual to 6 dm, simple or with long, ascending branches; herbage glabrous or nearly so; lvs small, narrow, to 15 × 2.5 mm, with a jointed, cartilaginous base, fls 1–3 in the axils, 2–3 mm; hypanthium short-hairy, shortly 4-lobed, scarcely corolloid; fr ovoid, 2 mm, with thin pericarp appressed to the hard, black seed; $2n=40$. Disturbed places; Eurasian sp., intr. in O., Ill., Wis., Io., Neb. and Kans. Apr.–Sept.

Family **TRAPACEAE**, the Water-chestnut Family

Fls perfect, regular, tetramerous; sep valvate, joined at the base to form a short hypanthial tube, 2 or all 4 of them persistent and indurate-accrescent in fr as horn-like or spine-like projections; stamens alternate with the pet; ovary bilocular, partly inferior, the lower part sunken in the receptacle, surrounded at the base by a cupular disk, becoming almost wholly inferior in fr; style subulate, with a capitate stigma, each locule with a solitary, pendulous, apical-axile ovule, but one locule and its ovule aborting after anthesis, so that the fr is unilocular and 1-seeded; fr indehiscent, with a thin, evanescent, fleshy exocarp and a persistent, stony endocarp; seed with 2 very unequal cotyledons, one large, starchy, and retained within the fr, the other small and scale-like, growing out (along with the radicle and plumule) through the terminal pore left by the fall of the style; endosperm wanting; aquatic annuals, rooted in the substrate but often breaking off and becoming free-floating; submersed part of the stem with elongate internodes and opposite or subopposite or ternate, elongate, filiform-dissected lf-like organs; aerial part of the stem short, bearing densely crowded, rosulate lvs with small, cleft, deciduous stipules, the blade ± rhombic, the elongate petiole with a float near midlength. 1/15, trop. and subtrop. Africa and Eurasia. (Hydrocaryaceae.)

1. TRAPA L.   Water-nut; water-chestnut. Characters of the family.

1. Trapa natans L.   Stem mainly submersed, to 1 m or more; blades of aerial lvs 2–5 cm wide, sharply serrate above; fls axillary, short-peduncled; pet white, 8 mm; fr 3–4 cm wide, with 4 stout horns; $2n$=ca 36, ca 48. Native of Eurasia, established and becoming abundant in the Hudson, Potomac, and Mohawk rivers, and locally elsewhere in our range. June–Sept.

Family **ONAGRACEAE**, the Evening-primrose Family

Fls mostly perfect, epigynous and often with the hypanthium prolonged beyond the ovary, regular or occasionally somewhat irregular, (2–)4(–7)-merous; sep valvate, often reflexed at anthesis; pet distinct or seldom wanting; stamens as many or more often twice as many as the sep, often alternately unequal; pollen-grains tending to cohere in large groups; carpels generally as many as the sep, united to form a compound ovary with axile (or intruded-parietal) placentation, or rarely the ovary pseudomonomerous; style with a capitate or discoid to 4-lobed stigma; ovules (1–) several or numerous in each locule; fr a capsule, less often a small nut or a berry; embryo-sac monosporic, 4-nucleate; endosperm initially diploid, wanting at maturity; cotyledons 2; herbs (ours) or less often woody plants, with simple and entire or toothed to sometimes lobed or lyrate-pinnatifid lvs; stipules in most genera wanting or small and caducous. 17/675, mostly New World, esp. w. U.S.

1 Fls 4-merous or sometimes 5(6)-merous; lvs opposite or more often alternate.
  2 Fr capsular, many-seeded; fls regular or nearly so.
    3 Sep persistent on the fr; no free hypanthium; pet yellow or none .... 1. *Ludwigia.*
    3 Sep deciduous after anthesis; hypanthium often prolonged beyond the ovary.
      4 Seeds with a coma (tuft of hairs) at the free end; hypanthium only shortly or not at all prolonged
        beyond the ovary; pet white to pink or purple .... 2. *Epilobium.*
      4 Seeds without a coma; hypanthium often elongate; pet in most spp. yellow.
        5 Stigma evidently 4-cleft, the slender lobes stigmatic all around .... 3. *Oenothera.*
        5 Stigma ± peltate, the receptive surface above-within .... 4. *Calylophus.*
  2 Fr indehiscent, 1–few-seeded; pet white to pink or red, often unilaterally disposed .... 5. *Gaura.*
1 Fls dimerous; lvs strictly opposite; pet tiny, white .... 6. *Circaea.*

1. LUDWIGIA L.   Water-primrose, primrose-willow. Fls 4– or 5(–7)-merous; hypanthium not prolonged beyond the summit of the ovary; sep persistent; pet yellow (in our spp.) or white, soon deciduous, or often wanting; stamens as many or twice as many as the sep; ovary cylindric to obconic or prismatic, often angled or winged; style short, with capitate or lobed stigma; ovules numerous; capsule dehiscent longitudinally or by a terminal pore; seeds when pluriseriate in each locule naked and with an evident raphe, when uniseriate surrounded by endocarp from which they may be separated easily or only with difficulty; herbs (ours) or shrubs, mostly of wet places (sometimes aquatic), with alternate or opposite lvs and small or medium-sized diurnal fls sessile or short-pedicellate in the upper axils (in ours) or in terminal spikes or heads. (*Jussiaea.*) 75, mainly trop. Amer.

1 Stamens 8 or 10(12), twice as many as the sep.
  2 Fls 5(6)-merous; fr mostly 2–4 cm; seeds uniseriate in each locule.
    3 Seeds loose in the horseshoe-shaped endocarp; plants erect, without rhizomes .... 1. *L. leptocarpa.*
    3 Seeds adnate to and almost completely covered by the endocarp; plants rhizomatous or with a
      creeping base or partly floating.
      4 Anthers 2.5–3.5 mm; pet mostly 1.5–2.5 cm; flowering stems ± erect .... 2. *L. uruguayensis.*
      4 Anthers 1–1.7 mm; pet mostly 1–1.5 cm; flowering stems ± creeping or floating .... 3. *L. peploides.*
  2 Fls 4-merous; fr 1–2 cm; seeds pluriseriate in each locule .... 4. *L. decurrens.*
1 Stamens 4, as many as the sep; fr up to ca 1 cm; seeds pluriseriate.
  5 Lvs alternate; flowering stems erect or ascending.
    6 Fls definitely pedicellate; pet present and ± conspicuous.
      7 Lvs tapering to the base; pet about equaling the sep .... 5. *L. alternifolia.*
      7 Lvs rounded at the sessile base; pet surpassing the sep .... 6. *L. hirtella.*
    6 Fls sessile or very nearly so; pet (except *L. linearis*) minute or none.
      8 Capsule nearly or fully as thick as long, subspherical or cubic-spherical.
        9 Bracteoles 2–5 mm, borne at or often above the base of the ovary.
          10 Herbage and fr glabrous or nearly so.
            11 Capsule sharply 4-angled or narrowly winged .... 7. *L. alata.*
            11 Capsule roundly or scarcely 4-angled, or shallowly grooved .... 8. *L. polycarpa.*
          10 Herbage and fr evidently hairy .... 9. *L. pilosa.*
        9 Bracteoles less than 1 mm, at or below the base of the ovary, or obsolete .... 10. *L. sphaerocarpa.*
      8 Capsule cylindric or obconic, about two or three times as long as thick.
        12 Pet present; lvs linear .... 11. *L. linearis.*
        12 Pet wanting; lvs lanceolate .... 12. *L. glandulosa.*
  5 Lvs opposite; flowering stems prostrate and creeping, or partly floating.
    13 Fls evidently pedicellate; pet present but small .... 13. *L. brevipes.*
    13 Fls sessile; pet none .... 14. *L. palustris.*

1. Ludwigia leptocarpa (Nutt.) H. Hara. Water-willow. Freely branched, ± fibrous-rooted annual or perennial 3–20 dm, softly spreading-hairy throughout, or the lvs subglabrous; lvs alternate, lanceolate or narrowly elliptic or elliptic-oblanceolate, mostly 5–15 × 1–3 cm, tapering to a sessile or shortly petiolar base; fls 5(6)-merous, evidently short-pedicellate; ovary slender, angular, 10–15 mm; pet 5–10 mm; stamens 10(12); anthers 1 mm; fr subcylindric, 10–12-nerved, 3–4 cm; seeds uniseriate in each locule, loose in the horseshoe-shaped endocarp; $2n=32$. Swamps, ditches, and pond-margins; tropical Amer., n. to se. U.S., and in the Mississippi Valley to w. Ky. June–Sept. (*Jussiaea l.*)

2. Ludwigia uruguayensis (Cambess.) H. Hara. Showy w.-p. Rhizomatous perennial, often some of the stems slender, floating, and glabrous, but fls produced only on erect or ascending, usually soft-hairy stems 3–6 dm; lvs alternate, those of the flowering stems lanceolate to oblanceolate, mostly 3–6 cm, on petioles seldom over 1 cm, generally hairy; pet mostly 15–25 mm; anthers 2.5–3.5 mm; otherwise much as in no. 3. Swamps and ponds at a few stations in our range from s. N.Y. southward; N.C. to Tex., and s. to Argentina. Summer. (*Jussiaea u.; J. michauxiana.*)

3. Ludwigia peploides (HBK.) Raven. Creeping w.-p. Stems prostrate or floating or ascending at the tip, rooting at the nodes, glabrous or very sparsely villosulous; lvs alternate, lanceolate to oblanceolate or obovate, narrowed below to a petiole 2–4 cm; fls 5-merous; ovary at anthesis cylindric, 7–12 mm, its pedicel 1–6 mm; pet mostly 1–1.5 cm; stamens 10; anthers 1–1.7 mm; fr subcylindric, 10-nerved, 2–4 cm, its pedicel 3–8 cm; seeds uniseriate in each locule, individually enclosed by the endocarp, the seed and endocarp together 1 mm, subtruncate at the end opposite the funiculus; $2n=16$. Ponds and swamps; s. Ind. to Kans., s. to La. and Tex., thence e. to N.C. and intr. about Phila. and probably elsewhere; pantropical. Our plants are var. *glabrescens* (Kuntze) Shinners. (*Jussiaea repens*, misapplied.)

4. Ludwigia decurrens Walter. Wingstem w.-p. Erect, freely branching glabrous annual to 2 m, the stem 4-angled and commonly 4-winged by the decurrent lf-bases; lvs alternate, lanceolate to linear, sessile or nearly so, the main ones mostly 5–12 × 1–3.5 cm; pedicels 1–5 mm; fls tetramerous; ovary at anthesis obconic, 4-angled, 8–10 mm; pet 8–12 mm; stamens 8; fr slenderly pyramidal, 4-angled or narrowly 4-winged, 1–2 cm; seeds multiseriate in each locule, ellipsoid, 0.3–0.4 mm; $2n=16$. Swamps and shallow water, chiefly on the coastal plain; Md. to Fla. and Tex., n. in the interior to s. Ind. and s. Mo., and s. to Argentina. July–Sept. (*Jussiaea d.*)

5. Ludwigia alternifolia L. Square-pod w.-p. Erect, often freely branched perennial 4–12 dm from a cluster of fleshy-fibrous roots; lvs alternate, lanceolate or lance-linear, acute or acuminate, 5–10 cm, tapering to the sessile or subpetiolar base; pedicels 3–5 mm, with 2 lanceolate bracteoles near the top; fls 4-merous; pet about equaling the sep; stamens 4; fr 5–6 mm, glabrous or hairy, nearly cubic, with rounded base, slightly wing-angled, opening by a terminal pore, and eventually loculicidal; pedicels up to about as long as the fr; $2n=16$. Swamps and wet soil; Mass. and s. Ont. to Io. and s. Neb., s. to Fla. and Tex. July, Aug. Ozarkian plants, extending e. to s. Ind., tend to be evidently short-hairy, and have been distinguished from the more eastern or northern, essentially glabrous plants as var. *pubescens* E. J. Palmer & Steyerm.

6. **Ludwigia hirtella** Raf.   Rafinesque's w.-p. Erect perennial 4–10 dm from a cluster of fleshy-fibrous roots; stem spreading-hirsute; lvs alternate, erect or ascending, lanceolate or lance-linear, obtuse or acutish, 2–5 cm, rounded at the sessile base, densely pubescent on both sides; pedicels 3–8 mm, bracteolate near the top; fls 4-merous; pet 10–15 mm, exceeding the erect to spreading sep; fr spreading-hirsute, equaling or shorter than the pedicel, otherwise as in no. 5. Swamps on the coastal plain; s. N.J. to Fla., Ark., and Tex. June–Sept.

L. **virgata** Michx., another coastal plain sp., has been reported from se. Va. It differs from no. 6 in its subglabrous herbage and soon reflexed sep.

7. **Ludwigia alata** Elliott.   Wing-cup w.-p. Fr broadly cupulate, 3–4 mm, sharply 4-angled or narrowly 4-winged; $2n=48$; otherwise much like no. 8. Swamps and bogs on the coastal plain; se. Va. to Fla. and Miss. July–Nov.

8. **Ludwigia polycarpa** Short & Peter.   Top-pod w.-p. Perennial 2–10 dm from a stoloniferous base, usually much branched above, the stems 4-angled, glabrous; principal lvs lanceolate or lance-linear, 4–12 cm, acuminate, tapering to a sessile or obscurely petiolar base, glabrous except on the margin; fls 4-merous, sessile in elongate, interrupted, leafy spikes; pet minute and greenish or none; stamens 4; fr short-cylindric or commonly widened above, 4–7 × 3–5 mm, roundly 4-sided or shallowly grooved, glabrous, its well developed bracteoles narrow, inserted at or often well above the base; $2n=32$. Swamps, marshes, and wet prairies; Mass. and Conn.; locally in Pa. and Va.; s. Ont. to Minn., s. to Ky., s. Ill, s. Mo., and e. Kans. July–Sept.

9. **Ludwigia pilosa** Walter.   Hairy w.-p. Perennial 5–12 dm from a stoloniferous base, usually much branched above, softly spreading-hairy throughout; lvs lanceolate, the larger 3–10 cm; fls 4-merous, sessile or nearly so, tending to be crowded toward the end of short lateral branches, subtended by much reduced, often ± elliptic lvs; pet none; fr 3–4 mm long and thick, not angled, the well developed bracteoles inserted above its base; $2n=32$. Wet ground; on the coastal plain from se. Va. to Fla. and Tex. June–Aug. (L. ravenii.)

10. **Ludwigia sphaerocarpa** Elliott.   Round-pod w.-p. Perennial to 1 m from a stoloniferous base, commonly much branched above, glabrous or ± pubescent; lvs lanceolate to linear, the larger commonly 5–10 cm, tapering to the base, those subtending the fls much smaller, fls tetramerous, subsessile, separated by distinct internodes; pet none; fr subglobose, finely pubescent, 2.5–4 mm, not angled; bracteoles attached at the base of or below the fr, very rarely half as long as the fr, often minute or obsolete; $2n=32$. Swamps and wet soil; Mass. to Fla. and Tex.; also in Ind. and Mich. July–Sept. Variable, and often divided into several vars.

11. **Ludwigia linearis** Walter.   Narrow-lvd w.-p. Perennial 4–10 dm from a cluster of fleshy-fibrous roots, often also stoloniferous, glabrous to often minutely granular-pulverulent throughout; lvs linear, 2–6 cm × 1–5 mm; fls 4-merous; sep 3–4 mm; pet 3–5 mm; anthers 1 mm; fr subcylindric, somewhat widened above, obscurely 4-sided, 6–9 mm; $2n=16$. Wet soil, muddy shores, or shallow waters; s. N.J. to Fla., Ark., and Tex., chiefly on the coastal plain. July–Sept.

12. **Ludwigia glandulosa** Walter.   Small-fld w.-p. Stoloniferous perennial, the basal part of the stem prostrate and rooting, the rest ± erect, 3–10 dm, glabrous or nearly so, freely and slenderly branched; cauline lvs lanceolate, 3–10 cm × 5–15 mm, tapering to a short petiole or petiolar base, with conspicuous secondary veins; fls sessile, 4-merous; sep 1.5–2 mm; pet none; anthers ca 0.3 mm; fr glabrous or minutely puberulent, cylindric or obscurely 4-grooved, 4–8 mm; $2n=32$. Wet or muddy soil or shallow water; coastal plain from se. Va. to Fla. and Tex., n. in the interior to s. Ind. and Kans. July–Sept.

13. **Ludwigia brevipes** (Long) Eames.   Coastal plain water-purslane. Stems prostrate and creeping, or partly floating; lvs 1–3 cm, spatulate or oblanceolate, tapering to the base but scarcely petiolate; pedicels 5–15 mm; sep 4–6 mm; pet yellow, equaling or shorter than the sep; fr clavate, to 10 mm. Shallow water and wet shores at widely scattered stations along the coastal plain from N.J. to S.C.

14. **Ludwigia palustris** (L.) Elliott.   Common water-purslane. Stems flaccid, prostrate and creeping, or partly floating; lvs 0.5–3 cm, lanceolate to ovate, abruptly narrowed to a petiole about as long; fls minute, sessile, 4-merous, apetalous; fr oblong-obovoid, 3–4 mm, roundly 4-sided, the wall much thickened and corky below each sep, very thin below each sinus and revealing the outline of the seeds when dry; bracteoles minute or wanting; $2n=16$. Creeping on mud and partly floating in shallow water; widespread in the n. hemisphere, and throughout our range.

2. **EPILOBIUM** L.   Willow-herb. Fls 4-merous; hypanthium usually prolonged a little beyond the summit of the elongate, slender ovary and simulating a cal-tube; sep ovate to lanceolate, deciduous from the fr; pet white to pink or purple; stamens 8, often unequal; style short; stigma clavate, ovoid, or 4-cleft; capsule linear, elongate, 4-valved; seeds numerous, small, bearing a terminal coma of hairs; annual or usually perennial herbs, rarely subshrubs, with alternate or opposite, linear to ovate lvs; fls many to few in elongate or short terminal racemes, or even solitary. 200, N. and S. Temp. and Arctic.

1 Taprooted annual; epidermis of the stem exfoliating below .... 12. *E. brachycarpum.*
1 Fibrous-rooted, often rhizomatous perennials; epidermis not exfoliating.
  2 Stigma 4-cleft.
    3 Pet entire; lvs subentire, glabrous or nearly so; native .... 1. *E. angustifolium.*

3 Pet notched at the tip; lvs toothed, evidently hairy; introduced.
  4 Pet 10–15 mm, shallowly notched; lvs somewhat clasping . . . . 2. *E. hirsutum.*
  4 Pet 4–9 mm, deeply notched; lvs not clasping . . . . 3. *E. parviflorum.*
2 Stigma entire.
  5 Lvs entire or nearly so, revolute, mostly well under 1 cm wide; hairs of the stem not in decurrent
     lines.
    6 Pubescence spreading . . . . 4. *E. strictum.*
    6 Pubescence appressed.
      7 Lvs glabrous or nearly so on the upper surface . . . . 5. *E. palustre.*
      7 Lvs evidently (though finely) pubescent across the upper surface . . . . 6. *E. leptophyllum.*
  5 Lvs usually toothed, flat, often 1 cm wide or more; stem generally with lines of pubescence
     decurrent from the lf-bases.
    8 Plants overwintering by turions . . . . 7. *E. glandulosum.*
    8 Plants without turions.
      9 Stems erect, mostly 3–10 dm and freely branched above; plants perennating by persistent
        basal rosettes, not soboliferous.
        10 Coma nearly white; buds obtuse or rounded; seeds broadly short-beaked . . . . 8. *E. ciliatum.*
        10 Coma brown; mature fl-buds with projecting sep-tips; seeds beakless . . . . 9. *E. coloratum.*
      9 Stems decumbent to erect, mostly 0.5–4 dm tall and simple above the base; plants sobolifer-
        ous.
        11 Frs mostly 3.5–5.5+ cm; plants avg larger and more erect than no. 11, to 4(5) dm; lvs
          1.5–4.5 cm . . . . 10. *E. hornemannii.*
        11 Frs mostly 2–3.5(-4) cm; plants mostly 0.5–1.5 dm; lvs 1–2 cm . . . . 11. *E. alpinum.*

1. **Epilobium angustifolium** L. Fireweed; great w.-h. Perennial from coarse, running, rhizome-like roots, erect, usually simple-stemmed, 1–3 m, puberulent in the infl, often otherwise glabrous or nearly so; lvs alternate, numerous and crowded, lanceolate or lance-linear, to 1.5(–2) dm, sessile or subsessile; racemes terminal, elongate; fls many, pink-purple (white); pet 1–2 cm, short-clawed; hypanthium not prolonged beyond the ovary; style basally hairy; stigma deeply 4-cleft; fr 3–8 cm; 2*n*=36, 72, 108. Many habitats, especially moist soils rich in humus, often abundant after fires; circumboreal, s. to N.J., O., n. Ill., Nebr., and N.M. June–Sept. (*Chamaenerion a.; C. spicatum.*) Ours represent the tetraploid var. *canescens* A. Wood, in contrast to the more northern, diploid var. *angustifolium.*

2. **Epilobium hirsutum** L. Hairy w.-h. Rhizomatous perennial; stems erect, much branched, 5–20 dm, softly and densely villous above; lvs mostly opposite (but the bracts alternate), hairy on both sides, lanceolate or oblong and somewhat clasping, the larger ones 5–12 × 1–3 cm, sharply serrulate, with evident lateral veins; pedicels 1 cm or less; pet red-purple, 10–15 mm, shallowly notched; hypanthium shortly prolonged beyond the ovary; stigma 4-cleft; fr 5–8 cm; coma nearly white; 2*n*=36. Native of Eurasia and n. Afr., intr. in moist or wet soil, especially in disturbed sites, from s. Me. and s. Que. to Mass. and Md., w. to n. O., Mich., and ne. Ill.

3. **Epilobium parviflorum** Schreber. Small-fld hairy w.-h. Much like no. 2, but smaller, to 8 dm, the lvs 2–8 cm, sessile or subsessile but not clasping, the upper commonly alternate; pet 4–9 mm, deeply notched. European sp., intr. into wet places in Mich. and s. Ont. July–Sept.

4. **Epilobium strictum** Muhl. Northeastern w.-h. Perennial from slender rhizomes, erect, simple or spar-ingly branched above, 3–6 dm; lvs lanceolate or lance-linear, entire or nearly so, revolute, 2–4 cm × 3–8 mm, the uppermost or all but the lowermost ones alternate; pubescence of the younger stems, upper lf-surface, pedicels and fr soft and usually dense, composed of straight, spreading or somewhat ascending hairs; pet notched, pink, 5–8 mm; coma pale dingy brown. Bogs and swamps; Que. to Minn., s. to Va., O., and n. Ill. (*E. densum; E. molle* Torr., not Lam.)

5. **Epilobium palustre** L. Marsh w.-h. Perennial from very slender, nearly naked stolons, each ending in a turion from which the stem of the next year arises; stems slender, simple or branched, 1–5(–8) dm, canescently strigose-puberulent in the infl, but the middle and lower part usually more sparsely strigillose-puberulent, often in vertical strips; lvs all opposite, or the upper more often alternate, sessile or nearly so, sometimes with axillary fascicles, broadly linear to more often lanceolate or lance-linear, 2–7 cm × 2–15 mm, entire or nearly so, often revolute, not very hairy, the lower side commonly puberulent along the midrib but otherwise glabrous or nearly so, the upper side glabrous or with a few scattered hairs esp. along the midrib; infl usually nodding in bud, but soon erect; fls short-pedicellate in the upper axils, the pedicels elongating to 0.5–2.5 cm in fr; pet white to pink or lilac, 4–6 mm, evidently notched; 2*n*=36. Wet low ground; circumboreal, s. in Amer. to Pa., Wis., Colo., and Nev. July, Aug. (*E. oliganthum.*)

6. **Epilobium leptophyllum** Raf. American marsh w.-h. Much like no. 5, but more hairy, avg more robust, and with consistently narrow lvs; pls 2–10 dm, often freely branched and with numerous lvs; stems canescently strigillose-puberulent nearly or quite throughout; lvs more often with axillary fascicles, mostly 1.5–7 cm × 1–7 mm, revolute, evidently strigose-puberulent across the whole upper surface and along the midrib (or across the whole surface) beneath; 2*n*=36. Wet meadows, bogs, and marshy ground; Nf. to Mack. and B.C., s. to N.C., Kans., and Calif. July, Aug. (*E. nesophilum.*)

7. **Epilobium glandulosum** Lehm. Northern w.-h. Much like no. 8, but the stem arising from a large, subterranean turion; pls often simple or only sparsely branched even when well developed; lvs avg wider, mostly ovate or narrowly ovate; infl tending to be leafy and narrow, not much branched; pet mostly (5–)6–

10 mm, rose-purple to rarely white; seeds 1–1.5 mm, similarly ridged, or seldom merely papillate; $2n=36$. Wet places, typically in cooler, more stable habitats than no. 8; Nf. to Alas. and Asia, s. to N.S., N.B., Vt., n. N.Y., n. Mich., Wis., Minn., and in the w. to N.M. and Calif. June–Aug. (*E. boreale; E. leptocarpum; E. steckerianum; E. ciliatum* var. *g.*) Hybridizes with no. 8.

8. Epilobium ciliatum Raf. American w.-h. Perennating by persistent basal rosettes, without turions or long rhizomes; stems erect, mostly solitary, freely branched when well developed, to 1 or even 1.5 m, puberulent at least above, usually in decurrent lines, sometimes even sericeous, and often also glandular in the infl; lvs opposite, or the upper offset or alternate, lanceolate or lance-ovate, mostly 3–12 × 0.5–3.5 cm, serrulate, the teeth somewhat remote (usually 2–5 per cm of margin) or small and obscure; fls numerous; sep 2–6 mm, not projecting in bud; pet 2–6 mm, white (pink), notched; fr 4–10 cm, on a ± evident pedicel 2–15+ mm; seeds numerous, ca 1 mm, broadly short-beaked, longitudinally finely ribbed (evidently so at 20×); coma nearly white; $2n=36$. Wet places, often in unstable habitats, variable and often somewhat weedy; Nf. and Lab. to Alas., s. to Va., w. N.C., Ind., Io., Calif., Tex., Mex., and C. Amer.; Chile and Arg. June–Aug. Ours is the widespread var. *ciliatum.* (*E. adenocaulon; E. americanum; E. perplexans.*)

9. Epilobium coloratum Biehler. Eastern w.-h. Much like no. 8; erect, to 1 m, usually freely branched, ± pubescent, especially above, with short, incurved hairs, the pubescence of the stem tending to be in lines; lvs chiefly opposite, narrowly lanceolate or oblong-lanceolate, the larger often 1 dm, sharply and irregularly serrate with numerous (4–8 per cm of margin) teeth; fls numerous; cal-teeth free at the tip in bud, projecting; pet 4–5 mm, pink or white; fr 3–5 cm; seeds abruptly rounded and beakless at the summit; coma brown; $2n=36$. Wet soil; Me. to Minn., s. to N.C., Ala., Ark., and Tex. July–Oct. A hybrid with no 8 has been called *E.* × *wisconsinense* Ugent.

10. Epilobium hornemannii Reichb. Hornemann's w.-h. Much like no. 11, but coarser and more nearly erect, to 4(5) dm, the stem usually not sigmoid, the lvs 1.5–4.5 cm, entire or often toothed, the erect frs 3.5–5.5(–7.5) cm, slender, often only ca 1 mm thick, or up to 1.5 mm; $2n=36$. Streambanks and other wet places, with us at upper elev. in the mts.; circumboreal, s. to Me., n. N.H., n. N.Y., and the Rocky Mts. June–Aug. Two ecogeographically coextensive vars. Var. *hornemannii* has anthocyanic pet 5–8 mm; var. *lactiflorum* (Hausskn.) D. Löve, has white to ochroleucous pet 2.5–5 mm. (*E. lactiflorum.*)

11. Epilobium alpinum L. Alpine w.-h. Soboliferous perennial, without turions; stems loosely clustered, scaly and rhizome-like at the base, 0.5–1.5 dm, often forming small, loose mats or cushions, tending to be sigmoidally curved rather than strictly erect, the lower part puberulent in decurrent lines, the upper part usually more generally puberulent and commonly somewhat glandular; lvs sessile or short-petiolate, often crowded, entire or obscurely few-toothed, mostly 1–2 cm, blunt; infl nodding to erect; fls few, axillary, the pedicels to 2 cm in fr; pet anthocyanic, 3.5–6 mm; fr 2–3.5(–4) cm, glabrous or sparsely glandular-hairy; seeds finely cellular-roughened or papillate (at 20×); coma readily deciduous; $2n=36$. Wet places, sometimes on talus, with us at upper elev. in the mts.; circumboreal, s. to Me., n. N.H., n. N.Y., and the Rocky Mts. June–Aug. We have only the widespread var. *alpinum,* with slender frs ca 1 mm (seldom to 1.5 mm) thick, and seeds mostly 0.7–1.3 mm. (*E. anagallidifolium.*)

12. Epilobium brachycarpum C. Presl. Annual w.-h. Taprooted annual 2–10+ dm, freely branched at least when well developed, glabrous below, often glandular-puberulent above, the stem with an exfoliating outer layer; lvs alternate nearly throughout, linear to narrowly elliptic or lanceolate, scarcely veiny, entire or irregularly toothed, with a firm brownish apical gland, the principal ones of the main stem 2–7 cm × 2–8 mm, often soon deciduous; fascicles of reduced lvs commonly present in the main axils; racemes terminating the numerous branches, lax, few-fld; pedicels filiform, 5–20 mm; pet pink or white, 2–6 mm; stigma with 4 very short, erect lobes; fr 1.5–3 cm; seeds 1.5–2.5 mm; coma deciduous; $2n=24$. Dry, open places; widespread in w. U.S., and occasionally adventive farther e., perhaps native in w. Que. and the Bruce Peninsula of Ont. Variable, with many minor named forms. (*E. paniculatum.*)

3. OENOTHERA L. Evening-primrose. Fls 4-merous, solitary in the axils or forming a terminal raceme or spike, commonly vespertine and ephemeral, but matutinal or diurnal and lasting several days in spp. 13–17; hypanthium usually slenderly tubular and prolonged well beyond the ovary (at least 3 mm in all but one of our spp.), deciduous from the fr; sep at first connate, often with terminal or subterminal appendages, splitting at anthesis and reflexed; pet in most of our spp. yellow, in two white or pink, often large and showy; stamens 8, equal or alternately unequal; anthers versatile (sometimes obscurely so in small fls); sporogenous tissue continuous in each microsporangium; ovary 10-locular; ovules numerous; style ± elongate; stigma with 4 slender lobes stigmatic all around; capsule prismatic to terete or winged, linear to ovoid or obovoid, in some spp. deeply sulcate and strongly angled; herbs with alternate, mostly narrow lvs. (*Kneiffia, Hartmannia, Lavauxia, Megapterium, Meriolix, Peniophyllum, Raimannia.*) 80, New World.

---

1 Ovary terete or nearly so; fr terete, obscurely and roundly 4-angled, or sulcate; stamens equal (except in no. 10).

2 Fls yellow; fl-buds erect (or the whole stem nodding at the tip).

3 Fr thickest near the base; seeds horizontal, very angular, not pitted. (*Oenothera.*)
  4 Sep-appendages separate from the base, subterminal, the actual end of the sep represented by
    a distinct ridge or small lobe.
    5 Pet 2–4 cm; anthers 8–12 mm; shale-barrens .... 1. *O. argillicola.*
    5 Pet 1–2 cm; anthers 4–7 mm; widespread .... 2. *O. parviflora.*
  4 Sep-appendages closely connivent at base, strictly terminal.
    6 Pet 1–2.5 cm; style mostly less than 2 cm; sep 1–2.5 cm .... 3. *O. biennis.*
    6 Pet 3–5 cm; style 2–6 cm; sep 2–6 cm .... 4. *O. glazioviana.*
3 Fr essentially linear; seeds ascending, obovoid to fusiform or ellipsoid, ± pitted, not sharply
  angled. (*Raimannia.*)
  7 Hypanthium-tube barely strigose; fls numerous in dense terminal spikes.
    8 Fls larger, the sep 1.5–2.5 cm, the pet 1.5–2.5(–3) cm; normal diploid .... 5. *O. rhombipetala.*
    8 Fls smaller, the sep 0.5–1.5 cm, the pet 0.5–1.5(–2) cm; complex heterozygote .... 6. *O. clelandii.*
  7 Hypanthium-tube hirsute; fls few in the upper axils.
    9 Lvs entire or obscurely sinuate-denticulate; coastal plain .... 7. *O. humifusa.*
    9 Lvs deeply sinuate-dentate to pinnatifid; widespread .... 8. *O. laciniata.*
2 Fls white to pink; buds nodding.
  10 Stamens equal; ovary and hypanthium minutely spreading-puberulent. (*Anogra.*) .... 9. *O. nuttalii.*
  10 Stamens alternately unequal; ovary and hypanthium strigose. (*Hartmannia.*) .... 10. *O. speciosa.*
1 Ovary 4-angled or narrowly 4-winged, at least above; fr sharply 4-angled or 4-winged; stamens, at
  least in spp. 13–17, alternately unequal.
  11 Lvs runcinate-pinnatifid, all basal; fr 4-winged. (*Lavauxia.*) .... 11. *O. triloba.*
  11 Lvs entire or merely toothed; fr and habit various.
    12 Pet 5–7 cm; wings of the fr 1–2.5 cm wide; fls vespertine. (*Megapterium*) .... 12. *O. macrocarpa.*
    12 Pet up to 3 cm; fr 4-angled and narrowly 4-winged; fls diurnal. (*Kneiffia.*)
      13 Fls larger; pet (10–)15–30 mm; style 10–20 mm; anthers 4–8 mm.
        14 Infl and fr with predominantly or exclusively gland-tipped hairs (or glabrous) .... 13. *O. tetragona.*
        14 Infl and fr with predominantly or exclusively eglandular hairs.
          15 Inconspicuously short-hairy or subglabrous; fr evidently clavate; free sep-tips
            mostly 1 mm or less .... 14. *O. fruticosa.*
          15 Conspicuously pilose-hirsute with hairs mostly 1–3 mm; fr only slightly or scarcely
            clavate; free sep-tips mostly 1–3 mm .... 15. *O. pilosella.*
      13 Fls smaller; pet 3–10 mm; style 1–10 mm; anthers 0.5–2.5 mm.
        16 Perennial; cauline lvs well over 1 mm wide; bracts 5+ mm .... 16. *O. perennis.*
        16 Annual; cauline lvs not over 1 mm wide; bracts up to 2 mm .... 17. *O. linifolia.*

1. Oenothera argillicola Mackenzie. Shale-barren e.-p. Biennial or short-lived perennial to 1.5 m, glabrous or nearly so throughout, or somewhat hairy in the infl, or the lvs minutely puberulent; lvs narrow, the basal narrowly oblanceolate, to 30 × 2 cm, the cauline varying to narrowly lanceolate or lance-elliptic, seldom over 15 × 1.5 cm; infls spicate and ± leafy-bracteate, the tip nodding; hypanthium 3.5–4.5 cm, the appendages 3–7 mm, subterminal as in no. 2; pet 2–4 cm; anthers 8–12 mm; style 2–3.5 cm; fr 2–3 cm, tapering upwards, half as thick at the top as at the base; seeds 1–1.5 mm, prismatic-angled; 2*n*=14. Shale-barrens; n. Pa. to w. Va. and e. W.Va. July–Sept.

2. Oenothera parviflora L. Small-fld e.-p. Resembling no. 3 and often mistaken for it; infl-tips nodding; hypanthium 1.5–4 cm; sep 1–2 cm, the free tips 1–3 mm, subterminal, the actual end of the sep represented by a distinct transverse ridge or a small lobe; pet 1–2 cm; anthers 4–7 mm; style 0.5–2.5 cm; fr 1.5–4 cm, only slightly tapering; seeds 1.8–2.2 mm; 2*n*=14, a complex heterozygote. Disturbed open places, abundant especially northward; Nf. to Va. and W.Va., w. to Minn. and Ark. July–Oct. Highly variable. Closely and minutely canescent-strigulose plants, without gland-tipped hairs, occur mainly from Mass. to Va. and have been called var. *oakesiana* (Robbins) Fern. (*O. oakesiana*), in contrast to var. *parviflora,* widespread but more common northward, with intermingled long and short hairs, often some of them gland-tipped, varying to nearly glabrous. (*O. angustissima; O. cruciata.*)

3. Oenothera biennis L. Common e.-p. Biennial or short-lived perennial, 0.5–2 m; lvs lanceolate to oblong, mostly 1–2 dm, acute or acuminate, entire to repand-dentate, often crisped on the margin, sessile or short-petioled, glabrous to canescent, fls in stiff, terminal, simple or branched, ± leafy-bracteate spikes; hypanthium 2–5 cm; sep 1–3 cm, their appendages terminal, 1–5 mm, connivent for most of their length before anthesis; pet yellow, 1–2.5 cm; anthers 4–7 mm; fr 1.5–4 cm, stout, tapering upwards; seeds 1.2–1.8 mm, angular-prismatic; 2*n*=14, a complex heterozygote. Fields, roadsides, prairies, and waste places, throughout most of the U.S. and s. Can. July–Sept. More or less divisible into 3 vars.:

1 Plants ± densely grayish-strigose to grayish-hirsute; infl with few or no gland-tipped hairs; the common
  phase in w. U.S., seldom found e. of Ill. (*O. rydbergii; O. strigosa.*) .... var. *canescens* T. & G.
1 Plants greener, more sparsely pubescent to subglabrous; infl with some gland-tipped hairs.
  2 Pet mostly 1–2 cm; sep-tips mostly 1–3 mm; the common phase in most of e. U.S. .... var. *biennis.*
  2 Pet mostly 2–2.5 cm; sep-tips mostly 3–4 mm; mostly above 800 m in the s. Appalachian Mts.,
    from s. Pa. to Va. and N.C. .... var. *austromontana* (Munz) Cronq.

4. Oenothera glazioviana Micheli. Garden e.-p. Much like *O. biennis* var. *biennis,* but with broader cauline lvs and larger fls, the sep 2–6 cm, with tips 5–8 mm, the pet 3–5 cm, the styles 2–6 cm; 2*n*=14, a complex heterozygote. Cultigen, widely but sparingly escaped. (*O. erythrosepala; O. lamarckiana,* misapplied.)

5. Oenothera rhombipetala Nutt. Longspike e.-p. Much like no. 6, but an outcrossing diploid with larger fls, the hypanthium 2–3.5 cm, the sep 1.5–2.5 cm, the pet 1.5–2.5(3) cm, often rhombic-obovate. Fields and prairies in sandy soil; Great Plains, from S.D. to Okla. and Tex., e. occasionally at least as far as se. Minn. June–Sept. (*Raimannia r.*)

6. Oenothera clelandii Dietrich, Raven & W. L. Wagner. Erect or ascending biennial 4–10 dm; lvs linear to narrowly lanceolate or linear-oblong, 3–8 cm, entire or remotely and obscurely denticulate, short-hairy on both sides; fls autogamous, numerous, crowded in a terminal, leafy-bracteate spike 1–3 dm; hypanthium 1.5–3 cm, sparsely strigose; sep linear, 0.5-1.5 cm, reflexed separately or ± connivent; pet yellow, 0.5–1.5(–2) cm, broadly elliptic or ovate, usually acute; fr linear, 10–18 mm, usually curved; seeds dark brown, obovoid or fusiform, obscurely pitted; 2*n*=14, a complex heterozygote. Fields and prairies in sandy soil; Mich. to Minn., Io., Ill., and Ind., and irregularly to Ark., Ky., N.J., and N.Y. June–Sept. (*O. rhombipetala,* in part.)

7. Oenothera humifusa Nutt. Spreading e.-p. Perennial; stems rather woody, commonly branched from the base, spreading or ascending, 2–5 dm; lvs oblong to oblanceolate, 1–2(–3) cm, entire or shallowly sinuate-denticulate, densely canescent-strigose; fls usually few, sessile in the upper axils, not forming a distinct spike, or occasionally in a spike 1–2 dm; hypanthium and ovary hirsute, the hypanthium 2–3 cm; sep 8–12 mm, separately reflexed; pet 1 cm; fr linear, 2–3 cm, usually curved; seeds dark brown, obovoid or fusiform, obscurely pitted; 2*n*=14. Sand-dunes and beaches along the coast; N.J. to La. June, July. (*Raimannia h.*) Possibly hybridizes with no. 8.

8. Oenothera laciniata Hill. Cut-lf e.-p. Annual, simple or branched from the base, decumbent to erect, 1–4(–8) dm; lvs oblanceolate to oblong or lanceolate, 3–8 cm, tapering to the base, prominently sinuate-dentate to pinnatifid, green, sparsely puberulent to glabrous above; fls few, sessile in the axils of the scarcely reduced upper lvs, not forming a distinct spike; hypanthium hirsute, 1.5–3.5 cm; sep 6–12 mm, reflexed separately or connivent; pet 5–18 mm; fr linear, straight or curved, 1.5–3.5 cm; seeds thick-ellipsoid, pale brown, conspicuously pitted; 2*n*=14. Dry, usually sandy soil; Me. to S.D., s. to S. Amer. May–Oct. Our plants, as here described, are var. *laciniata.*

9. Oenothera nuttallii Sweet. White-stemmed e.-p. Perennial, spreading by creeping roots; stem erect, 3–8 dm, with white bark exfoliating below, glabrous below, minutely glandular-puberulent in the infl; lvs linear to linear-oblong, 2–8 cm, acute, entire or nearly so, tapering to the base, glabrous above, puberulent with incurved hairs beneath; fls sessile in the upper axils, malodorous, nodding in bud; hypanthium 2.5–4 cm; sep at anthesis connivent, unilaterally reflexed or spreading; pet white, turning pink, 1.5–2.5 cm; stamens equal; fr linear-oblong, 1–3 cm, minutely glandular, shallowly 4-sulcate; seeds in 1 row in each locule; 2*n*=14, 28. Prairies and plains; Wis. to Alta., s. to Kans. & Colo. June, July. (*Anogra n.*)

10. Oenothera speciosa Nutt. White e.-p. Perennial by creeping roots (sometimes annual?), ascending to erect, 1–6 dm; lvs linear to lanceolate or oblanceolate, 3–8 cm, gradually tapering to a petiolar base, entire to runcinate-pinnatifid, the basal half commonly more coarsely toothed than the upper; fls few, sessile in the upper axils, erect, buds nodding, long-acuminate; hypanthium slender, 1–2 cm, strigose; sep lance-linear, reflexed at anthesis, 1.5–2.5 cm, usually long connivent; pet obcordate, white or pink, 2–4 cm; stamens alternately unequal; fr narrowly obovoid or clavate, 10–18 mm, 4-sided, roundly 8-ribbed and deeply sulcate, tapering to both ends, sterile below; 2*n*=14. Dry, open places; Mo. and Kans. to Tex., and intr. eastward. May, June. (*Hartmannia s.*)

11. Oenothera triloba Nutt. Stemless e.-p. Winter-annual to biennial or short-lived perennial from a stout taproot, nearly stemless; lvs oblanceolate, 1–2 dm, runcinate-pinnatifid, often nearly to the midvein; fls axillary, the earliest sessile, the later ones on stout pedicels to 1 cm; hypanthium slender, 3–10 cm; sep lanceolate, 10–18 mm, reflexed, usually unilaterally so and connivent; pet obovate, 1–2.5 cm, pale yellow, drying whitish; stamens alternately unequal; fr hard, woody, long-persistent, ellipsoid or obovoid, 1–2 cm, the 4 wings each with a lateral tooth or angle. Dry woods, barrens, and prairies; Ky. and Tenn. to Kans. and Tex. May, June. (*Lavauxia t.*)

12. Oenothera macrocarpa Nutt. Wing-fruit e.-p. Perennial, subacaulescent or to 5 dm, decumbent to erect, silky-strigose; lvs linear-lanceolate, 5–10 cm, acute or acuminate, entire or nearly so, narrowed to a long petiolar base; fls few, subsessile in the axils; hypanthium slender, 7–10 cm; sep lance-linear, 2–4 cm, spreading unilaterally at anthesis and persistently connivent; pet yellow, obovate, 5–7 cm; fr 4–7 cm, slender, 4-sided, the angles bearing semi-elliptic wings 1–2.5 cm wide. Dry rocky barrens in calcareous soil; s. Ill. and Mo. to Nebr. and Colo., s. to Tex. May–July. (*O. missouriensis; Megapterium missouriense.*)

13. Oenothera tetragona Roth. Northern sundrops. Much like no. 14, and not always sharply distinct; usually only rather sparsely hairy, varying to virtually glabrous (especially below the infl), and sometimes glaucous; most or all of the hairs of the infl and fr minutely gland-tipped; lvs ovate to nearly linear, avg wider than in no. 14, entire or often remotely dentate, the larger ones 3–10 cm; infl and fls erect; sep-tips wholly connivent or free for 1–2 mm; pet (1–)1.5–3 cm; fr mostly 6–15 mm, thickest near or a little above the middle; 2*n*=28. Fields and open woods, at higher elev. than the next where their ranges overlap; N.S. and Que. to Mich., s. to n. Ga., ne. Ala., and Mo. May–Aug. (*O. glauca; O. fruticosa* ssp. *g.; Kneiffia g.; K. hybrida; K. latifolia; K. t.*)

14. Oenothera fruticosa L. Southern sundrops. Fibrous-rooted perennial to 1 m, simple or branched above, variously strigose or (especially upwards) spreading-hairy (but the hairs mostly shorter than in no. 15), or nearly glabrous below the infl, eglandular, or with only a few of the hairs gland-tipped; lvs narrowly elliptic to ovate or nearly linear, usually entire or nearly so, seldom over 6 cm; fls diurnal, out-crossing, several in a compact infl that only seldom nods at the tip; hypanthium 0.5–1.5 cm; sep (0.5–)1–2 cm, the tips short, usually free for only ca 1 mm or less; pet (1–)1.5–2.5 cm, notched at the summit; filaments 5–15 mm, alternately unequal; anthers 4–7 mm; style 12–18 mm; stigmas held above the anthers at anthesis; fr strongly tetragonal or narrowly 4-winged, evidently clavate or even obpyramidal, the body mostly 4–10 mm, tapering to a short or usually ± elongate pedicel-like base; $2n$=28, 42, 56. Meadows, fields, and open woods, often in disturbed habitats; Mass. to Fla., w. to Ind., Mo., Okla., and La. June–Aug. (*O. linearis; O. longipedicellata; Kneiffia allenii; K. f.*)

15. Oenothera pilosella Raf.   Midwestern sundrops. Fibrous-rooted, shortly rhizomatous perennial to 8 dm, simple or branched above, pilose-hirsute, especially above, with hairs 1–3 mm (seldom subglabrous); lvs lanceolate or lance-elliptic, 3–10 × 0.5–2.5 cm, mostly sessile, acute or acuminate, remotely denticulate or dentate or entire; infl compact, many-fld, the buds erect; fls diurnal, out-crossing, hypanthium (1–)1.5–2.5 cm; sep 1–2 cm, their free tips 1–3 mm; pet (1–)1.5–2.5 cm, notched at the summit; filaments alternately unequal; anthers 4–8 mm; style 10–20 mm; fr elliptic or linear-elliptic to rather narrowly clavate, the body 8–15 mm, tapering to a sessile or shortly pedicel-like base; $2n$=56. Moist soil, wet meadows, fields, and open woods; Io. and s. Wis. to La., e. to s. Ont., w. N.Y., and w. W.Va., and occasionally escaped as far e. as Vt., Mass., and Va. May–July. (*O. pratensis; Kneiffia p.; O. fruticosa* var. *hirsuta.*)

16. Oenothera perennis L.   Little sundrops. Fibrous-rooted perennial (1–)2–6 dm, usually simple; principal lvs oblanceolate to elliptic, (1–)3–6 cm, obtuse, narrowed to a petiolar base, the bracteal shorter and proportionately narrower, but still well developed and generally exceeding the ovary; infl nodding, the axis straightening during anthesis, the fls becoming erect and opening singly, autogamous, diurnal; hypanthium 3–10 mm; pet 5–10 mm, obcordately notched; filaments 3–4 mm, alternately unequal; anthers 1.5–2.5 mm; style 3–10 mm, about equaling the anthers; fr ellipsoid-clavate or oblong, glandular puberulent (or glabrate in age), the body 4–10 mm, tapering to a short pedicel-like base; $2n$=14, a complex heterozygote. Moist or dry soil, fields, meadows, and open woods; Nf., and Que. to s. Man., s. to Va., and Mo., and in the mts. to N.C. and S.C. June–Aug. (*O. pumila; Kneiffia pumila; K. perennis.*)

17. Oenothera linifolia Nutt.   Thread-lf sundrops. Slender, taprooted annual 1–5 dm, branched above, sparsely strigose or glandular-puberulent to subglabrous; basal lvs ovate to obovate or narrowly elliptic, 1–2(–4) cm, short-petiolate; cauline lvs numerous, crowded, linear-filiform, 1–4 cm, less than 1 mm wide; fls diurnal, autogamous, in well developed plants numerous, forming a long terminal raceme, individually subtended by deltoid bracts 1–2 mm; hypanthium 1–3 mm; pet 3–5 mm; filaments short, alternately unequal; anthers 0.5–1 mm; style 1–4 mm; fr sessile or short-stipitate, 4–7(–10) mm, ellipsoid-rhomboid, with tapering base, strongly 4-angled but scarcely winged; $2n$=14. Rock ledges and sandy barrens; Mo. and se. Kans. to e. Tex., e. to w. Fla., Ga., Tenn., Ky., and rarely N.C. Apr.–June. (*Peniophyllum l.*)

4. CALYLOPHUS Spach.   Much like *Oenothera* (and often included therein), but the stigma peltate, entire or often very shallowly and broadly 4-lobed, the receptive surface above-within; sporogenous tissue divided into packets in each microsporangium. (*Meriolix, Salpingia.*) 6, w. N. Amer.

1. Calylophus serrulatus (Nutt.) Raven.   Perennial to 6 dm, usually with clustered stems, often somewhat woody below, generally strigose-canescent at least above; lvs numerous, linear to narrowly oblong or oblanceolate, 2–6 cm, entire or more often sharply serrulate; fls sessile in the upper axils, matutinal, autogamous; hypanthium funnelform, 3–12 mm, 4-angled, ± pubescent, sometimes only on the angles, sep triangular-ovate, with sharply raised midrib, reflexed at anthesis; pet yellow, 5–12 mm; stamens 8, alternately unequal; style only shortly or scarcely surpassing the hypanthium; stigma discoid, slightly 4-lobed; fr linear, canescent, roundly 4-angled, 1.5–3 cm; seeds 1.5 mm, sharply angled; $2n$=14, a complex heterozygote. Dry prairies and plains; Wis., w. Mich., and Minn. to Alta., s. to Ill., Ark., Tex., Ariz., and Chih. June, July. (*Oenothera s.; Meriolix s.; M. intermedia.*)

5. GAURA L.   Fls 4-merous; hypanthium tubular, usually about as long as the ovary, deciduous from the fr; sep reflexed; pet short-clawed, ephemeral, white, becoming pink or red, in most spp. ± unilaterally disposed; stamens 8, usually with a small scale internally at the base of each filament; sporogenous tissue divided into packets in each microsporangium; ovary initially 4-locular, with 1(2) ovules per locule, later ± unilocular by disintegration of the partitions; stigma 4-cleft; fr small, hard, indehiscent, 1–few-seeded; herbs with alternate, entire to toothed or in part lyrate lvs and long, terminal spikes or racemes of small fls. 20, N. Amer.

1 Strong perennials with the stems clustered on a coarse root or caudex.
  2 Fr on a distinct, slender pedicel (0.5–)1–3 mm long and scarcely 0.5 mm thick . . . . 1. *G. filipes.*
  2 Fr sessile, but abruptly contracted below the middle into a very stout, thick-terete, stipe-like base
        more than 1 mm thick . . . . 2. *G. coccinea.*
1 Coarse annuals or biennials, usually single-stemmed; fr tapering to a sessile base.
  3 Fls larger, the sep and pet 6+ mm, the anthers 2–5 mm . . . . 3. *G. biennis.*
  3 Fls smaller, the sep and pet up to 3.5 mm, the anthers 0.5–1 mm . . . . 4. *G. parviflora.*

1. Gaura filipes Spach. Threadstalk gaura. Clumped perennial 6–15(–20) dm, with numerous, crowded lvs in the lower ⅓–⅔ and long, subnaked, wand-like, branching infls, finely strigose throughout, varying to subglabrous or shortly spreading-hairy, with only the ovary, hypanthium, and sep strigose; lvs linear or linear-oblanceolate to lance-elliptic, entire or with a few coarse teeth, mostly 2–7 cm × 2–15 mm; bracts 1–2 mm, caducous; fls shortly slender-pedicellate, the pedicel at maturity of the fr (0.5–)1–3 mm, less than 0.5 mm thick; sep 5–12 mm, separately reflexed; pet 5–10 mm, turning pink; fr finely strigillose, 5–10 mm, fusiform, sharply 4-angled especially upwards; seeds 1 or 2; $2n=14$. Dry woods and fields; Ky. and s. Ind. to S.C., Fla., and La. Aug., Sept. (*G. michauxii.*)

G. villosa Torr., of dunes and sandy flats on the Great Plains, is said to be locally naturalized in N.J. It is coarser than *G. filipes*, somewhat woody toward the base, and densely villous or villosulous at least below.

2. Gaura coccinea Pursh. Scarlet g. Clumped perennial 2–5(10) dm, spreading by roots, strigose or partly villous, varying to subglabrous; lvs crowded, lanceolate to narrowly oblong, mostly 1–4 cm and entire or coarsely few-toothed, or the lower larger and more cleft, but deciduous; spikes 5–15 cm, densely fld and elongating with age, often lax or nodding at the tip; bracts 2–5 mm; sep 5–9 mm, separately reflexed; pet 3–7 mm, becoming orange-red or deep maroon; fr strigose-puberulent, sessile, the terminal part 3–5 mm, strongly 4-angled and pyramidal (but with concave sides), abruptly contracted to a stout, 8-ribbed, subterete, stipe-like base mostly 1.5–2 mm long and nearly or fully as thick; seeds (1–)3 or 4; $2n=14, 28, 42, 56$. Dry prairies and plains; s. Sask. and w. Minn. to w. Mo. and Tex., w. to B.C. and Calif., and s. to Mex.; occasionally intr. eastward to N.Y. May–Aug.

3. Gaura biennis L. Biennial g. Coarse, erect biennial or winter-annual, branched above, ± hairy, and glandular at least in the infl; lvs lanceolate or oblong-lanceolate, to 12 cm, narrowed to the base, remotely or obscurely denticulate or sinuate; spikes several, many-fld; sep 6–18 mm, reflexed in pairs; pet 6–15 mm, turning pink or red; anthers 2–5 mm; fr short-hairy, 5–9 mm, 4-sided, with a prominent thick rib on each angle and a slender one on each face, fusiform, tapering to a sessile or subsessile base; $2n=14$. Open places, often in disturbed habitats; Mass. to Wis. and Neb., s. to N.C., Ala., and Tex. July–Oct. Two cytologically distinctive but hybridizing and morphologically confluent vars.:

var. *biennis*. Plants to 2 m; stem and infl ± densely villous, never strigose; seeds 3–6; self-compatible, often self-pollinated, complex heterozygote with about half the pollen stainable; mainly e. of the Mississippi R., extending also into Io.

var. *pitcheri* T. & G. Plants to 4 m; stem and infl strigose to partly villous or hirtellous; seeds 2–4; self-incompatible, with 90+ percent stainable pollen; mainly w. of the Mississippi R., but also common throughout Ill. (*G. longiflora.*)

4. Gaura parviflora Douglas. Small-fld g. Coarse, erect, taprooted, mostly single-stemmed annual or biennial to 2 m, branched above, glandular and softly spreading-villosulous; lvs lance-ovate to lanceolate or oblong, obtuse at base, remotely denticulate, to 10 cm; spikes much elongate, many-fld; fls autogamous, virtually regular; sep 2–3.5 mm, reflexed separately; pet 1.5–3 mm, turning pink; anthers 0.5–1 mm; fr glabrous or seldom short-hairy, 5–10 mm, fusiform to lanceolate, tapering to a sessile base, with a rounded rib on each of the 4 angles and a slender rib on each face; seeds 3–4; $2n=14$. Weed of fields, pastures, streamsides, etc.; Ind. to Wash., s. to Tex. and n. Mex., and occasionally adventive eastward. May–Aug.

6. CIRCAEA L. Enchanter's nightshade. Fls 2-merous; hypanthium short, tubular, deciduous after anthesis; pet white (or pinkish in age), obcordate or deeply notched; stamens 2; ovary 1- or 2-locular, with a solitary ovule in each locule; stigma shallowly bilobed or capitate; fr small, indehiscent, obovoid or pyriform, usually slightly compressed, beset with ± hooked bristles; rhizomatous perennial herbs with opposite, petiolate lvs and small fls paniculately disposed in terminal, bracteate racemes; $x=11$. 7, N. Temp. and Arctic.

1 Open fls ± well spaced; fr bilocular .... 1. *C. lutetiana.*
1 Open fls clustered at the top of the raceme; fr unilocular .... 2. *C. alpina.*

1. Circaea lutetiana L. Common e. n. Erect, to 7(–10) dm from slender rhizomes; lvs oblong-ovate, 6–12 cm, usually not more than about half as wide, acuminate, very shallowly sinuate-denticulate, rounded or barely subcordate at base; petioles subterete; infls many-fld, to 2 dm, elongating early so that the open fls as well as the frs are well spaced; pedicels filiform, spreading, finally reflexed, 3–6(–10) mm, glandular-pubescent; sep 2–2.5 mm, glabrous or glandular-pubescent; pet 2.5–4 mm, bilobed less than half-length; anthers 0.7–1 mm; stigma shallowly bilobed; disk prominent, 0.2–0.4 mm high; fr 3.5–5 mm, equally bilocular, each half normally with 3 large and 2 small rounded ridges separated by narrow furrows; $2n=22$. Moist woods, interruptedly circumboreal, in Amer. from N.S. and s. Que. to s. Man., s. to Ga., La., and Okla. June–Aug. Ours is var. *canadensis* L. (*C. quadrisulcata*, properly an Asian var.)

A sterile but vegetatively vigorous hybrid between our spp. is vegetatively intermediate and has unequally bilocular frs; the infl elongates early as in no. 1. The name *C.* × *intermedia* Ehrh. was based on the European hybrid between *C. lutetiana* var. *lutetiana* and *C. alpina* var. *alpina*.

2. Circaea alpina L. Alpine e. n. Rhizomes tuberous-thickened at the end; stems weak and soft, 1–3 dm; lvs thin, ovate or deltoid-ovate, 2–6 cm, usually more than half as wide, acute, sharply and coarsely undulate-dentate, broadly truncate to more commonly cordate at base; petioles thin and ± flat above, and with a conspicuous thin median wing beneath; racemes rarely with more than 15 fls, elongating to 1 dm in fr, the fls at anthesis all clustered at the still unelongated top; pedicels 2–6 mm, glabrous, spreading or slightly deflexed in fr; sep 1–2 mm, white or pinkish, glabrous; pet 1–2.5 mm, bilobed to about the middle or less; anthers 0.2–0.3 mm; disk nearly obsolete; stigma capitate or inconspicuously bilobed; fr 2–3 mm, unilocular and 1-seeded, not ribbed or furrowed; $2n=22$. Moist or wet woods and mossy bogs; circumboreal, s. to N.Y., Ind., Io., and S.D., and in the mts. to N.C., Tenn., and N.M. June–Aug. Ours is var. *alpina*.

FAMILY MELASTOMATACEAE, the Melastome Family

Fls commonly showy, mostly perfect, partly or wholly epigynous (or strongly perigynous with the hypanthium enveloping but free from the ovary), mostly 4–5-merous, regular except for the stamens; sep valvate; pet convolute, distinct; stamens usually twice as many as the pet; filaments generally inflexed in bud (each anther sometimes occupying a cavity between the hypanthium and the ovary), commonly twisted at anthesis so as to bring all the anthers to one side of the fl; anthers mostly opening by terminal pores; connective commonly provided with one or more appendages of diverse form; nectaries mostly wanting; ovary mostly 3–5-locular, with a terminal style and capitate or punctate stigma; ovules ± numerous, mostly axile; fr a loculicidal capsule or a berry; embryo with 2 commonly unequal cotyledons; endosperm wanting; lvs simple, usually opposite and entire, mostly exstipulate, commonly with 3–9 prominent veins, the lateral ones arching and subparallel. 200/4000, mostly of warm reg., esp. S. Amer.

1. **RHEXIA** L.   Meadow-pitchers; meadow-beauty. Fls 4-merous; hypanthium tubular at anthesis, becoming urceolate in fr, the basal part distended by the developing capsule, the terminal part persisting as a tubular or flaring neck crowned by the persistent sep; lower part of the hypanthium weakly adnate to the lower part of the ovary, easily separated from it, and becoming free at maturity; anthers unilocular at anthesis, each opening by a single terminal pore, in most spp. (not *R. petiolata*) elongate-linear, curved, and prolonged at the base 0.5–1 mm below and internal to the filament-tip; connective with a short, fragile, spur-like appendage externally at the base; fr a capsule; seeds cochlear; perennial herbs with handsome lavender-purple to white fls in terminal cymes, appearing in late summer; lvs sessile or short-petioled, mostly 3-veined, rather remotely setose-ciliate (sometimes subserrulately so); herbage often beset with long, gland-tipped bristly hairs. 12, e. N. Amer. Spp. 3–6 frequently hybridize when at the same level of ploidy.

1 Anthers 1–2 mm, nearly straight; pet ascending .... 1. *R. petiolata.*
1 Anthers 5–11 mm, evidently curved; pet spreading.
   2 Sep widely (almost horizontally) flaring, tapering to a distinct awn-tip .... 2. *R. aristosa.*
   2 Sep more ascending (at least below), acute to acuminate, but not awned.
     3 Stem subequally quadrangular, the faces almost flat, the angles sharp or winged.
       4 Angles of the stem obviously (though rather narrowly) winged .... 3. *R. virginica.*
       4 Angles of the stem scarcely or not at all winged .... 4. *R. interior.*
     3 Stem very unequally quadrangular, one pair of opposing faces broad, dark green, convex, the
         other much narrower, paler, concave.
       5 Body of the hypanthium ± setose; pet 12–18 mm; anthers 5–8 mm; widespread .... 5. *R. mariana.*
       5 Body of the hypanthium glabrous; pet 18–25 mm; anthers 8–11 mm; coastal plain .... 6. *R. nashii.*

1. **Rhexia petiolata** Walter.   Short-stemmed m.-p. Stems 2–6 dm from a short, somewhat woody caudex, 4-angled, essentially glabrous; lvs erect or ascending, subsessile, ovate or lance-ovate, 1–2 cm, sometimes bristly above; hypanthium glabrous or nearly so, strongly constricted between the globose base and the short, abruptly flared neck, 5–7(–9) mm at maturity; sep triangular, ciliate, shortly awn-tipped; pet 10–20 mm, glabrous, ascending; anthers oblong, nearly straight, 1–2 mm, scarcely prolonged at the base; self-compatible, the other spp. all self-sterile and with widely spreading pet; $2n=22$. Acid bogs on the coastal plain; se. Va. to Fla. and Tex. (*R. ciliosa.*)

2. **Rhexia aristosa** Britton.   Bristly m.-p. Stems 2–7 dm from a usually tuberous-thickened base, simple to bushy with ascending branches, prominently 4-angled or even somewhat winged, bristly-setose at the nodes, otherwise glabrous; lvs erect or ascending, sessile, lanceolate or lance-linear, 2–3.5 cm × 3–7 mm, sometimes setose above; hypanthium densely long-spreading-setose near the rim, otherwise glabrous or sparsely setose, at maturity 7–10 mm, the neck up to about as long as the ovoid base; sep widely flaring from the base, tapering to an awn-tip 0.5–1.5 mm; pet dull lavender, 10–20 mm, the outer surface often appressed-

bristly especially toward the margins; anthers 5–8 mm; $2n=22$. Local in sandy or peaty acid bogs on the coastal plain; N.J. to Ga. and Ala.

3. **Rhexia virginica** L.   Wing-stem m.-p. Stems 2–10 dm from an often tuberous-thickened, otherwise fibrous-rooted base, subequally quadrangular and evidently winged, usually ± bristly at the nodes, otherwise bristly to glabrous; lvs ovate to lance-ovate, 2–7 cm, usually a third to half as wide, sometimes bristly on one or both sides; hypanthium smooth or bristly, 7–10 mm at maturity, the neck mostly shorter than the body; sep narrow, acute or acuminate, 2–4 mm; pet rose-lavender, 15–20 mm, often bristly on the back; anthers 5–7 mm; $2n=22$, 44. Moist, open places; N.S. to Wis., s. to Fla. and Tex.

4. **Rhexia interior** Pennell.   Showy m.-p. Stems 2–10 dm from creeping rhizomes, subequally quadrangular but scarcely winged, bristly-setose; hypanthium sparsely to copiously bristly, 10–13 mm at maturity; pet bright rose-lavender, 18–25 mm, generally with some bristles on the back; otherwise much like no. 5; $2n=44$. Moist or wet open places. Two geographically discrete but otherwise scarcely distinguishable segments: Plants of the Atlantic Coastal Plain, from s. N.J. to S.C., have been segregated as *R. ventricosa* Fern. & Griscom. More inland plants, occurring from s. Ind. to c. Ala., w. to se. Kans., e. Okla., and n. La., are considered to be *R. interior* proper.

5. **Rhexia mariana** L.   Dull m.-p. Weedy, often freely branched perennial 2–10 dm from shallow, slender, stoloniform rhizomes; herbage and hypanthium evidently spreading-setose; stems inequably and often obscurely quadrangular, the 4 raised lines delimiting 2 narrow, flat to concave sides and 2 broader, often rounded ones; principal lvs lance-elliptic to lance-linear or lance-ovate, 2–5(–8) cm, seldom more than a third as wide; hypanthium mostly 6–10(11) mm at maturity, the neck as long as or longer than the body; pet 12–18 mm, dull lavender to white; generally glabrous; anthers 5–8 mm; $2n=22$, 44. Moist, open places, often in disturbed habitats; Mass. to Fla., w. to s. Ind., s. Mo., and Tex.

6. **Rhexia nashii** Small.   Nash's m.-p. Like a large-fld form of no. 5, the hypanthium (9)10–15(–20) mm at maturity, the pet mostly dull lavender to rose-purple and 18–25 mm; anthers 8–11 mm; rhizomes sometimes tuberiferous; stems sometimes to 15 dm; hypanthium commonly with some bristles around the rim, otherwise glabrous or nearly so; pet with some glandular bristles on the back; $2n=44$, 66. In wetter sites than no. 5; coastal plain from se. Va. to Fla. and La. (*R. mariana* var. *purpurea*.) Intergrades with no. 5.

FAMILY **CORNACEAE**, the Dogwood Family

Fls perfect or sometimes unisexual, regular, epigynous, 4- or less often 5-merous; cal represented by teeth or a minute rim around the summit of the ovary, or obsolete; pet distinct, valvate or imbricate; stamens as many as and alternate with the pet, or up to 15 and ± bicyclic; ovary inferior, 1–4(–9)-locular; style terminal, with a capitate or lobed stigma (or the styles distinct); ovules solitary in each locule, apical or apical-axile or apical-parietal, pendulous; fr most commonly a drupe with a single 1–5-locular, usually longitudinally grooved stone; embryo dicotyledonous, embedded in the copious, oily endosperm; woody plants (seldom herbs) with simple, exstipulate, opposite or less often alternate lvs; fls individually small, borne in various kinds of often cymose-capitate infls, the latter sometimes subtended by a whorl of large, petaloid bracts. 15/110+, cosmop. but mostly N. Temp.

1 Lvs opposite or apparently whorled (alternate in only one sp.); stamens and pet each 4; fls perfect .... 1. *Cornus*.
1 Lvs alternate; stamens ca 10; pet minute; fls ± unisexual .... *Nyssa*.

1. **CORNUS** L.   Dogwood. Fls mostly perfect, 4-merous; sep usually minute; pet small, valvate; style well developed; stigma capitate; fr a drupe, the stone bilocular (except in *C. canadensis*) but often only 1-seeded; fls in cymose or cymose-paniculate infls or, by reduction of the branches, in umbelliform or head-like clusters. (*Chamaepericlymenum, Svida, Cynoxylon.*) 50+, widespread but mainly N. Temp.

Named hybrids include *C.* × *acadiensis* Fern. (*sericea* × *alternifolia*), *C.* × *arnoldiana* Rehder (*racemosa* × *amomum* var. *schuetzeana*), and *C.* × *slavinii* Rehder (*sericea* × *rugosa*). Others include *C. amomum* × *C. stricta*, *C. drummondii* × *C. stricta*, *C. drummondii* × *C. amomum* var. *schuetzeana*, *C. drummondii* × *C. racemosa*, and *C. racemosa* × *C. stricta*.

1 Fls in open cymes; bracts none or minute; mature drupes blue to white.
  2 Lvs alternate, though tending to be clustered toward the branch-tips. (*Mesomora*.) .... 1. *C. alternifolia*.
  2 Lvs strictly opposite. (*Svida*.)
    3 Style at anthesis conspicuously dilated in the distal 1 mm; sep relatively large, 1–2 mm; pith
        brown; fr blue .... 2. *C. amomum*.

3 Style of essentially uniform diameter throughout, or gradually dilated upwards; sep smaller, rarely 1 mm; pith and fr various.
  4 Twigs bright red; stone dark brown with yellow stripes; fr white .... 3. *C. sericea.*
  4 Twigs not bright red; stone flesh-colored, not striped; fr blue or white.
    5 Well grown lvs with 6–8 lateral veins on each side; fr blue .... 4. *C. rugosa.*
    5 Lvs with 3 or 4(5) lateral veins on each side.
      6 Lvs scabrous-strigose above, strigose or more loosely short-hairy beneath; lateral veins mostly arising in the lower half of the blade; fr white (light blue) .... 5. *C. drummondii.*
      6 Lvs thinly strigose to subglabrous on both sides, but smooth to the touch; lateral veins more evenly spaced.
        7 Twigs maroon; infl flat to convex; fr light blue; southern .... 6. *C. stricta.*
        7 Twigs tan; infl convex to often pyramidal; fr white; northern .... 7. *C. racemosa.*
1 Fls in a dense, head-like, pseudanthial cluster, subtended by 4 large, usually white, petal-like bracts; mature drupes bright red.
  8 Herbs 1–2 dm from a woody rhizome. (*Chamaepericlymenum.*)
    9 Lateral veins arising from the midvein in the basal third of the lf .... 8. *C. canadensis.*
    9 Lateral veins arising from or very near the base of the lf .... 9. *C. suecica.*
  8 Tree or tall shrub. (*Cynoxylon.*) .... 10. *C. florida.*

1. **Cornus alternifolia** L.f. Pagoda-d. Shrub or small tree to 6 m, the twigs with white pith; lvs alternate, thin, ovate to oblong or obovate, 5–10 cm, conspicuously acuminate, pale green and minutely strigillose beneath, the lateral veins 4 or 5 on each side; petioles 8–50 mm, even on the same twig; infl usually hemispheric; sep very short or obsolete; fr blue; stone with a deep pit at the summit; $2n=20$. Rich woods and thickets; Nf. and N.S. to Minn., s. to Fla., Ala., and Ark. May–July. (*Svida a.*)

2. **Cornus amomum** Miller Knob-styled d. Solitary shrub 1–3(–5) m; young twigs loosely pubescent, later ± glabrate and becoming maroon, the pith dark brown; old bark fissuring; infl a flat to somewhat rounded cyme; sep 1–2 mm; distal 1 mm of style abruptly thickened to twice the diameter below; fr blue, 6–9 mm; stone strongly ridged; $2n=22$. Moist or wet woods and along streams; Me. and Que. to Minn., s. to Ga., Ark., and Okla. May–July. Two well marked geographic vars.:
Var. *amomum.* Lvs ovate to broadly elliptic, commonly 6–12 cm, 1.5–2.2 times as long as wide, subtruncate to rounded at base, with 4–6 veins on a side, the lower surface usually greenish and nonpapillose and provided with both appressed and loosely spreading, whitish or very often rusty hairs especially along the veins. Chiefly southeastern, from Miss., Ala., and Ga. n. to s. Ind., s. O., Pa., N.J., and sometimes N.Y. and Mass.
Var. *schuetzeana* (C. A. Meyer) Rickett. Lvs 6–9 cm, 2.2–4 times as long as wide, cuneate at base, with 3–5 veins on a side, the lower surface usually whitish-papillose and with mainly or wholly appressed, whitish to only occasionally rusty pubescence. Chiefly northern and western, from Me. and s. Que. to Minn., s. to N.Y., Ky., Ind., Ill., Ark., and Okla. (*C. obliqua; C. purpusii.*)

3. **Cornus sericea** L. Red osier-d. Shrub 1–3 m, often forming dense thickets; twigs bright red; pith large and white; lvs lanceolate to elliptic or ovate, mostly 5–10 cm and a third to two-thirds as wide, acute to gradually acuminate at the tip, acute to broadly rounded at base, distinctly whitened beneath; lateral veins in well grown lvs 5–7 to a side; infl flat or slight convex; fr white (reputedly sometimes blue), 7–9 mm; stone brownish-black, with 7–9 vertical yellow stripes; $2n=22$. Streambanks and moist woods; Nf. to Alas., s. to Pa., Ind., Ill., and n. Mex. May–Aug. Pubescence of the lower lf-surface typically sparse and strictly appressed, that of the twigs minute and mostly appressed. (*C. baileyi,* a form common about the Great Lakes and in s. Ont., with the pubescence partly spreading; *C. interior,* a chiefly western form, also found e. to Mich. and Ind., with the infl and twigs tomentose.) (*C. stolonifera.*)

4. **Cornus rugosa** Lam. Round-lvd d. Shrub 1–4 m, commonly more tree-like in form than no. 3, often with a single main stem from the base; twigs glabrous or nearly so, light or yellowish-green, often shaded or mottled with red; pith white; lvs ovate to rotund, mostly 7–12 cm, abruptly acuminate, broadly cuneate or usually rounded at base, minutely scaberulous-strigose above, softly and loosely white-hairy to merely strigose beneath with hairs 0.5–1 mm; lateral veins 6–8 on a side; infl flat or slightly convex; fr light blue, 6 mm; $2n=22$. Moist or dry, sandy or rocky soil, typically in better-drained sites than no. 3; Que. to n. Ont. and Man., s. to N.J., Pa., n. O., n. Ind. and Io., and in the mts. to Va. May–July. (*C. circinata; Svida r.*)

5. **Cornus drummondii** C. A. Meyer. Rough-lvd d. Shrubs or occasionally small trees to 6 m; twigs scabrous, olive to pinkish-brown, with white or tan pith; old bark gray, sometimes fissuring; lvs lance-ovate to elliptic or broadly ovate, mostly 5–8 cm, commonly abruptly acuminate, broadly cuneate to subcordate at base, scabrous-strigose above and evidently rough to the touch, minutely papillose-whitened beneath and with appressed or spreading hairs often 0.5 mm; lateral veins 3 or 4(5) on a side, tending to arise from the lower half of the blade; infl flat or convex; fr white (light blue), 4–8 mm, on red pedicels; $2n=22$. Wet woods and streambanks; O. and s. Mich. to s. Wis., Io., and Neb., s. to Ga. and Tex. May, June. (*C. priceae.*)

6. **Cornus stricta** Lam. Southern swamp-d. Much like no. 7; shrub 1–3 m; twigs maroon-brown or maroon-black, becoming gray or gray-brown; old bark fissured; lvs green on both sides, scarcely paler beneath; infl flat-topped to more often convex; fr pale blue (white). Swamps and river-banks, in our range confined to the coastal plain; s. Del. to Fla. and Tex., and n. inland to Tenn., s. Ind., s. Ill., se. Mo., Ark., and s. Okla. May, June. (? *C. foemina,* a doubtful name; *Svida s.*)

7. **Cornus racemosa** Lam. Northern swamp-d. Shrub 1–5 m, often forming thickets; twigs glabrous, at first green, soon becoming tan and eventually gray-brown; old bark mostly smooth and gray; pith white (tan);

lvs lanceolate to elliptic, mostly 4–8 cm, a third to half as wide, abruptly acuminate, cuneate at base, often papillose-whitened beneath, sparsely strigose to glabrous on both sides, with 3 or 4(5) lateral veins to a side; infls often very numerous, convex to often pyramidal and paniculiform; fr at first leaden, becoming white (light blue), 5–8 mm, on reddish pedicels; $2n=22$. Moist soil, woods, thickets, roadsides, and streambanks; Me. and s. Que. to s. Man., s. to Va., s. Ill., and Mo., adjoining but only slightly overlapping the range of no. 6; the two might well be considered vars. of one sp. (*C. paniculata;* ? *C. foemina,* a doubtful name.)

8. Cornus canadensis L. Dwarf cornel; bunchberry. Colonial from woody rhizomes; stems erect, 1–2 dm, with a cluster of 4–6 apparently whorled lvs at the summit and below them 1 or 2 pairs of scales or reduced lvs; lvs short-petioled, lanceolate to oblanceolate or often obovate, 4–8 cm, acute at both ends; lateral veins 2 or 3 pairs, arising from the midvein below the middle; fl-cluster solitary and pseudanthial, on a peduncle 1–3 cm; bracts 4, white (purplish), lance-ovate to broadly ovate, acute or acuminate, 1–2 cm; fls greenish or yellowish to purplish; frs red, globose, 8 mm; stone smooth, often unilocular; $2n=44$. Moist acid woods and bogs; Greenl. to Alas. and e. Asia, s. to N.J., Pa., Ind., and Minn., and in the mts. to Md., W.Va., Va. and Calif. June, July. (*Chamaepericlymenum c.*)

9. Cornus suecica L. Swedish cornel. Much like no. 8; stems 5–15 cm, regularly bearing a few pairs of cauline lvs as well as the terminal cluster or pair and often emitting lateral branches from their axils; lvs sessile; lateral veins usually 2 pairs, arising from or very near the base of the lf; bracts white, 1 cm; fls dark purple; stone lightly grooved, bilocular; $2n=22$. Wet woods and rocks; circumboreal, s. in Amer. to Nf., N.S., and Que. (*Chamaepericlymenum s.*)

10. Cornus florida L. Flowering d. Widely branched small tree (or large shrub) to 10 m, the bark becoming closely and deeply checked; lvs ovate to elliptic or obovate, mostly 6–10 cm and half as wide, abruptly acuminate, pale beneath, strigillose on both sides; bracts 4, white (pink), obcordate, notched at the tip, 3–6 cm; fls yellowish, 20–30 in the cluster; fr red, ellipsoid, 10–15 mm; $2n=22$. Woods; Me. to s. Ont., Mich., Ill., Mo., and Okla., s. to Fla. and ne. Mex. May, June, before the lvs are fully grown. (*Cynoxylon f.*)

2. NYSSA L. Sour gum. Dioecious or polygamo-dioecious; infls (or solitary fls) axillary, pedunculate; staminate fls in short racemes or heads, with minute sep, 5–10 small, imbricate pet, and 8–15 stamens much longer than the pet in 2 ± definite whorls; "perfect" (but functionally pistillate) fls solitary or in infls of 2–4+ and sessile (in ours) on slender peduncles, the sep, when present, usually 5, the pet 5–8, the stamens fewer than in the staminate fls and about equaling the pet, or wanting; ovary unilocular and with a single style and ovule, seldom with a more or less well developed second locule and style; fr a drupe, crowned by the persistent disk, each locule opening apically by a triangular valve at the time of germination; deciduous trees or large shrubs with alternate lvs. 5, N. Amer. and se. Asia.

1 Staminate fls pedicellate; pistillate fls (1)2 or more per peduncle; fr 1–1.5 cm .... 1. *N. sylvatica.*
1 Staminate fls sessile; pistillate fls solitary; fr 2–3 cm .... 2. *N. aquatica.*

1. Nyssa sylvatica Marshall. Black tupelo; black gum. Tree to 30 m; lvs often crowded distally except on rapidly growing shoots, elliptic to obovate, 4–15 cm, usually abruptly short-acuminate to an obtuse or acute tip, entire or rarely with a few coarse teeth, broadly cuneate at base; staminate fls in an umbel or umbel-like raceme, the peduncles 1–3 cm, the pedicels 1–5 mm; fertile fls 2–4, sessile at the end of a peduncle that elongates to usually 3–6 cm, the 5 sep small but evident; fr blue-black, ovoid to globular, 1–1.5 cm; stone with ca 10 shallow grooves separated by low, rounded ridges; $2n=44$; fl May, June. The var. *sylvatica,* as thus described, is a common constituent of moist forests from s. Me. to Fla., w. to se. Wis., s. and s. Ill., Mo., and Tex. The well marked var. *biflora* (Walter) Sargent, swamp blackgum, grows in wet places, often in deep fresh-water swamps, on the coastal plain from Del. to Fla. and La. It differs in its narrower, more oblanceolate, blunter lvs, shorter fruiting peduncles 1–3 cm, and usually only 2 pistillate fls together; the trunk becomes much swollen at the base when the tree grows in standing water. (*N. biflora.*)

2. Nyssa aquatica L. Cotton-gum, water-tupelo. Tree to 30 m; lvs ovate to elliptic or rarely obovate, 10–15 cm, half as wide, conspicuously acuminate, mucronate at the tip, entire or more often with several large, salient teeth, broadly cuneate to cordate at base, distinctly glaucous-green beneath; staminate fls several, crowded and sessile on a peduncle 1–2 cm; pistillate fl solitary on a peduncle 2–3 cm, its cal obsolete; fr purple, ellipsoid or ovoid, 2–3 cm, constricted to the apical disk; stone with 8–10 sharp, narrow wings. Deep fresh-water swamps on the coastal plain, often with *Taxodium;* Va. to Fla. and e. Tex., n. in the Mississippi embayment to s. Mo. and s. Ill. May. (*N. uniflora.*)

Family SANTALACEAE, the Sandalwood Family

Fls perfect or often unisexual, regular, commonly (and in all our spp.) epigynous, monochlamydeous, the tep usually 4 or 5, distinct or forming a valvately lobed, often fleshy hypanthial cup or tube; stamens as many as and opposite the tep; nectary-disk seated atop the ovary or lining the hypanthium; ovary (2)3(–5)-carpellate, unilocular or partitioned only at the base, the erect, free-central placenta bearing 1–4 ovules pendulous from the summit; style with a capitate or lobed stigma; embryology often unusual; fr a nut or drupe with a solitary seed lacking a testa; embryo straight, dicotyledonous, embedded in the copious endosperm; green root-parasites with exstipulate, simple, entire (sometimes reduced) lvs and small fls in diverse sorts of infls. 35/400, cosmop.

1 Lvs alternate or scattered.
  2 Herbs.
    3 Fls in compact, terminal, cymose-paniculiform infls; hypanthium evident .... 1. *Comandra.*
    3 Fls in axillary cymules; hypanthium wanting .... 2. *Geocaulon.*
  2 Shrubs; fls in terminal racemes .... 3. *Pyrularia.*
1 Lvs opposite; shrubs.
    4 Fls axillary, or in small axillary infls .... 4. *Nestronia.*
    4 Fls terminal, or in small terminal infls .... 5. *Buckleya.*

1. **COMANDRA** Nutt. Bastard toad-flax. Fls perfect, epigynous; hypanthium well developed, lined with a nectary that is shallowly lobed at the summit alternate with the stamens; tep mostly 5, appearing as lobes on the hypanthium, each bearing a tuft of hairs behind and ± adherent to the stamen; ovary unilocular; placental column geniculate near midlength; stigma capitate; ovules mostly 3; fr a dry drupe, subglobose or with a short neck at the top, surmounted by the persistent sep; glabrous low perennial herbs, colonial by rhizomes, with rather small, alternate or scattered, subsessile or short-petiolate lvs and terminal cymose-paniculiform infls of small fls. 2, the other in Eur.

  1. **Comandra umbellata** (L.) Nutt. Stems 1–4 dm, simple or branched; lvs 2–5 cm; tepals whitish, about equaling the hypanthium; $2n=28$, 52 (?). Prairies, shores, and upland woods; widespread in N. Amer., from Nf. and Que. to B.C., s. to Ga., Ark., and N.M. May–July. Most of our plants belong to var. *umbellata*, the characteristic phase of e. N. Amer., with lanceolate to elliptic or ovate, ± evidently reticulate-veiny lvs green on both sides or somewhat paler beneath but not glaucous, the tep mostly 2–3 mm long and the fr mostly 4–6 mm thick. (*C. richardsiana; C. u.* var. *decumbens.*) The well marked, more western var. *pallida* (A. DC) M. E. Jones barely reaches our range, as in w. Minn. It has thicker, ± glaucous, often narrower lvs without evident lateral veins, longer, relatively narrower tep mostly 3–4 mm, and larger fr mostly 6–10 mm thick. (*C. pallida.*)

2. **GEOCAULON** Fern. Fls mostly in axillary cymules of 3, the 2 lateral often functionally staminate and soon deciduous, the central one perfect; tep distinct, spreading; nectary-disk nearly flat, covering the top of the ovary; style very short, conic; drupe juicy; otherwise much like *Comandra.* Monotypic.

  1. **Geocaulon lividum** (Richardson) Fern. Stems 1–3 dm; lvs elliptic to oblong or narrowly ovate, 1–4 cm, blunt or rounded, reticulate-veiny; fls in shortly slender-peduncled dichasia from a few upper axils, these soon becoming middle axils by continued growth of the axis; tep greenish-purple, triangular, 1.5 mm; filaments very short and broad, scarcely exceeding the lobes of the disk; fr orange-yellow to scarlet, 5–10 mm thick; $2n=26$. Sphagnum-bogs and wet coniferous woods; Lab. and Nf. to Alas. and B.C., s. to Me., N.H., n. N.Y., and n. Mich. June–Aug. (*Comandra l.*)

3. **PYRULARIA** Michx. Subdioecious; staminate fls with short, broadly obconic hypanthium; tep usually 5; disk prolonged into as many short, glandular lobes, bearing stamens in the sinuses, the pistil infertile; fertile fls with longer hypanthium and smaller anthers on shorter filaments, the style short and stout, the stigma capitate; fr a pyriform to subglobose drupe; trees or shrubs with thin, alternate lvs and small greenish fls in peduncled terminal racemes; $2n=38$. 4, the others in e. Asia.

  1. **Pyrularia pubera** Michx. Buffalo-nut; oil-nut. Much branched shrub to 5 m, parasitic on the roots of woody plants; lvs thin, petiolate, ovate-oblong to obovate, 5–15 cm, ± acuminate, acute to rounded at base, puberulent when young; staminate racemes 3–8 cm, many-fld; fertile racemes shorter and fewer-fld; tep triangular-ovate, acute, 2 mm, bearing a tuft of hairs behind and ± adherent to each stamen; fr 1.5–3 cm, with thin exocarp and large stone; endosperm very oily. Oak-hickory forest in the mts.; w. Va. to s. Pa. and e. Ky., s. to Ga. May.

4. NESTRONIA Raf. Dioecious, or some fls perfect; staminate fls with obconic hypanthium, the 4 tep each with a tuft of hairs behind and ± adherent to the stamen; pistillate fls with campanulate hypanthium, 4 tep, and shortly conic, 4-lobed style, often also with stamens; fr an ellipsoid, 1-seeded drupe; shrubs with glabrous twigs and foliage, opposite lvs, and small greenish fls, the staminate in slender-pedunculate axillary umbels, the pistillate solitary in the axils. Monotypic.

1. **Nestronia umbellula** Raf. Colonial shrub to 1 m, parasitic on the roots of pines and deciduous trees and shrubs; lvs short-petioled, lance-ovate, 3–7 cm, acute at both ends or rounded at base, glaucous beneath; staminate umbels 3–10 fld, shorter than the subtending lvs; hypanthium smooth, the staminate 2–3 mm, the pistillate 5–6 mm (ovary included); tep spreading, 2.5–3 mm, puberulent at the margin; fr yellowish-green, 10–18 mm. Moist or more often dry woods in the piedmont region; s. Va. to Ala.; rare. May.

5. BUCKLEYA Torr. Dioecious; staminate fls with flat, spreading hypanthium, 4–5-lobed cal, and conspicuous, flat, 4–5-lobed nectary disk bearing the stamens in its sinuses, the filaments curved-ascending; pistillate hypanthium slenderly obconic, bearing 4 foliaceous bracts at its summit, the 4 tep erect; fr an ellipsoid yellowish-green drupe; shrubs with puberulent twigs and foliage, opposite lvs, and small greenish fls, the staminate in small terminal umbels, the pistillate solitary and terminal. 4, the others in China and Japan.

1. **Buckleya distichophylla** Torr. Shrub to 1.5 m, parasitic on the roots of *Tsuga* and other hosts, with fan-like ascending branches, the branchlets simulating a single compound lf; lvs lanceolate, acuminate, cuneate to the subsessile base, puberulent especially on the margin and midvein, increasing in size on each branchlet from base to summit, to 7 cm; staminate umbels 3–10-fld, subsessile, the ovate tep 2.5–3 mm; hypanthium (and ovary) of the pistillate fls 7–8 mm, the bracts at its summit lanceolate, 7–9 mm, the sep triangular, erect, 2 mm; fr 2 cm, the bracts persistent nearly to maturity. Mt. woods at low elev.; s. Va., N.C., and Tenn.; very rare. May.

FAMILY **VISCACEAE**, the Christmas-mistletoe Family

Fls unisexual, regular, epigynous, monochlamydeous; tep valvate, small, seldom more than 4, often reduced to points around the rim of the ovary; stamens as many as and opposite the tep, the filaments often very short (or wanting); ovary 3–4-carpellate, unilocular or solid, containing a massive, placenta-like body nearly or quite filling the locule and including 2 bisporic embryo-sacs; stigma small, sessile or on a very short style; embryology complex and unusual; fr a shining, sometimes explosive berry; seed solitary, without a testa, surrounded or capped at one end by viscid tissue; embryo rather large, dicotyledonous, embedded in the starchy green endosperm; ± green, brittle shrublets hemiparasitic on the branches of trees, often dichasially branched, usually with nodal constrictions; lvs opposite, exstipulate, simple, entire, often reduced to scales; fls small, mostly borne in spike-like, often branching infls. (Loranthaceae, in part.) 7–8/350, cosmop., especially trop.

1 Lvs flat, 2–6 cm; stems to 4 dm .... 1. *Phoradendron.*
1 Lvs minute, scale-like; stems to 3 cm .... 2. *Arceuthobium.*

1. PHORADENDRON Nutt. Monoecious or dioecious; perianth deeply 3-lobed, subglobose; anthers 3, sessile on the base of the perianth-lobes, opening transversely; ovary ovoid; fr a subglobose to ovoid, sessile berry; small shrubs, parasitic on trees, with well developed (in most spp.), opposite, coriaceous lvs and small fls in short, axillary spikes. 200, New World.

1. **Phoradendron serotinum** (Raf.) M.C. Johnston. American Christmas-mistletoe. Freely branched, the stems thick, brittle, 2–4 dm; internodes 1.5–5 cm; lvs oblong to obovate, 2–6 cm, blunt or rounded, cuneate at base; dioecious; spikes solitary or few together, 1–5 cm; staminate infls of 3–6 segments, each segment with 15–60 fls; pistillate infls of 3–5 segments, each segment with 4–11 fls; fr 4–5 mm, white; $2n=28$. On many spp. of trees; N.J. to s. O., s. Ind., and s. Mo., s. to Fla. and Tex. Oct.–Dec. (*P. flavescens; P. leucarpum,* the spelling a correctable orthographic error for *leucocarpum; not P. leucocarpum* Patschofsky.)

2. ARCEUTHOBIUM M. Bieb. Dwarf m. Dioecious; fls solitary or few in the axils of scale-like, connate, decussate lvs; staminate perianth 2–5-lobed; anther sessile at the middle of each lobe, opening transversely; pistillate perianth 2-lobed; berry on a short, recurved pedicel. 28, mainly of N. Amer.

1. Arceuthobium pusillum Peck. Eastern d. m. Stem 0.5–1.5(–3) cm, simple or with few short branches, greenish-brown; lvs semiorbicular, 1 mm wide; fls freely produced, resembling short lateral branches until expanded; fr 2 mm, on a pedicel about equalling the subtending lf. Nf. and Que. to Minn. and Sask., s. to n. N.J., Pa., and Mich., chiefly on black spruce, less often on white spruce, occasionally on other conifers. June, July.

FAMILY **CELASTRACEAE**, the Staff-tree Family

Fls perfect or less often unisexual, regular, 4–5-merous, mostly hypogynous, commonly rather small and not very showy; pet distinct, mostly imbricate; stamens alternate with the pet, seated outside or upon a nectary-disk; ovary superior but sometimes ± embedded in the disk, 2–5-locular, with 2–several axile, apotropous, mostly erect or ascending ovules per locule; style generally short, sometimes wanting; stigma capitate or lobed; fr in ours a loculicidal capsule; seeds commonly covered by a well developed aril; embryo with a short radicle and 2 large, flat cotyledons, embedded in the copious, oily endosperm; woody plants with simple lvs and minute or no stipules. 50/800, cosmop.

1 Lvs alternate; stems twining; fls commonly functionally unisexual .... 1. *Celastrus.*
1 Lvs opposite; stems not twining; fls perfect.
  2 Ovary 3–5-locular; aril orange to red; lvs deciduous or evergreen .... 2. *Euonymus.*
  2 Ovary 2-locular; aril whitish; lvs evergreen .... 3. *Paxistima.*

1. CELASTRUS L. Bittersweet. Dioecious or polygamo-dioecious; fls 5-merous, the staminate with 5 stamens about as long as the pet, inserted on the margin of the cup-shaped disk, and with a vestigial pistil, the pistillate with vestigial stamens and a well developed ovary, stout columnar style, and 3-lobed stigma; ovules 2 per locule, collateral; fr 3-valved, each valve covering 1 or 2 seeds completely enclosed in a fleshy red aril; woody twiners with deciduous, alternate, serrulate lvs and small whitish or greenish fls. 30, mostly trop.

1 Fls numerous in terminal panicles; lvs mostly ca twice as long as wide .... 1. *C. scandens.*
1 Fls usually 2 or 3 in short axillary cymes; lvs broader .... 2. *C. orbiculatus.*

1. Celastrus scandens L. American b. Climbing to several m; lvs elliptic to oblong or ovate, acuminate, 5–10 cm; panicles terminal, 3–8 cm; frs orange, several in a cluster, nearly 1 cm; seeds ellipsoid, 6 mm, covered by the bright red aril; 2n=46. Road-sides and thickets, usually in rich soil; Que. and Ont. to Man. and Wyo., s. to w. S.C., n. Ga., e. Ala., La., and Tex. May, June.

2. Celastrus orbiculatus Thunb. Oriental b. Much like no. 1; lvs suborbicular to broadly oblong-obovate; fls few in small axillary cymes much shorter than the subtending lvs; 2n=46. Native of e. Asia, established as an escape in open woods and thickets here and there from Conn. s.

2. EUONYMUS L. Fls perfect, 4–6-merous, small, solitary or cymose in the axils; pet widely spreading; stamens very short, inserted at the margin of the broad disk, in which the 3–5-locular ovary is largely immersed; ovules 2–6 per locule, ± ascending; stigma sessile, 3–5-lobed; fr colored, capsular, often 3–5-lobed, with 1–6 seeds per locule, each seed covered by a red or orange aril; shrubs or small trees, prostrate to erect, sometimes climbing by adventitious roots; twigs often 4-angled, sometimes winged; lvs opposite, deciduous in our native spp. 200+, cosmop. Several cult. spp. that only occasionally escape are included in the key, but not numbered or described.

1 Fls 4-merous; fr smooth (drying rugose).
  2 Twigs with 2–4 conspicuous corky wings; lvs subsessile, turning bright red in autumn, then deciduous
  .... 1. *E. alatus.*

2 Twigs not winged; lvs ± petiolate, not turning bright red.
  3 Frs subglobose; petioles short, ca 5 mm; locally escaped from cult.
    4 Lvs persistent; fls in dense cymes, in June or July.
      5 Erect shrub or small tree; Japanese spindle-tree .... *E. japonicus* Thunb.
      5 Trailing or climbing shrub; Chinese spindle-tree .... *E. fortunei* (Turcz.) Hand.-Mazz.
    4 Lvs tardily deciduous; fls in loose cymes in Aug. and Sept. .... *E. kiautschovicus* Loes.
  3 Frs deeply lobed; petioles long, (5–)10–20 mm.
    6 Lvs finely hairy beneath; fls brownish-purple; aril bright red .... 2. *E. atropurpureus.*
    6 Lvs glabrous beneath; fls greenish-white; aril orange .... 3. *E. europaeus.*
1 Fls 5-merous; frs strongly tuberculate; petioles 1–5 mm.
  7 Erect shrub with stiffly divergent branches; uppermost lvs lanceolate to ovate .... 4. *E. americanus.*
  7 Trailing shrub, or with a few ascending branches; uppermost lvs obovate .... 5. *E. obovatus.*

1. Euonymus alatus (Thunb.) Siebold. Winged burning bush. Shrub to 2.5 m; twigs with 2–4 conspicuous corky wings; lvs subsessile, elliptic to obovate, to 6(–8) cm, sharply serrulate, turning bright red in autumn, then deciduous; fls 4-merous, green, 6–8 mm wide; fr smooth, purplish; aril orange. Native of e. Asia, widely cult. and locally escaped and established in our range. May, June.

2. Euonymus atropurpureus Jacq. Wahoo. Erect shrub to 6 m, vigorously colonial by rhizomes; petioles 1–2 cm; lvs lance to lance-ovate, 6–12 cm, acuminate, finely serrulate, acute at base, finely hairy beneath, turning pale yellow in autumn, then deciduous; fls few–many in divergently branched cymes 2–4 cm wide on axillary peduncles 2–5 cm, 4-merous, brownish-purple, 6–8 mm wide; fr red, smooth, 1.5 cm thick, 4-lobed, sometimes 1–3 of the lobes reduced; aril bright red. Moist woods; N.Y. to N.D., s. to Fla. and Tex. June.

3. Euonymus europaeus L. European spindle-tree. Shrub to 6 m, much like no. 2; lvs glabrous beneath; fls few, rarely more than 7, in long-peduncled axillary cymes, greenish-white, 6–10 mm wide; aril orange; $2n=64$. Native of Europe, escaped from cult. throughout most of our range, especially eastward. May, June.

4. Euonymus americanus L. Strawberry-bush; American burning bush. Erect shrub to 2 m, with stiffly divergent 4-angled twigs; lvs firm, on petioles 1–3 mm, lanceolate to elliptic or ovate, 3–7 cm, acute to acuminate, turning red or reddish in autumn, then deciduous; fls 5-merous, solitary or in cymes of 2–3 on axillary peduncles 1–3 cm, 10–12 mm wide, greenish-purple; pet narrowed to a very short claw; fr 3–5-lobed, 1.5 cm thick, crimson when ripe, strongly tuberculate; aril orange to scarlet; $2n=64$. Moist woods; s. N.Y. to s. O. and se. Mo., s. to Fla. and Tex. May, June.

5. Euonymus obovatus Nutt. Running s.-b. Prostrate shrub, the slender stems taking root, often sending up a few branches rarely to 3 dm; lvs thin, on petioles 3–5 mm, the terminal ones obovate, 3–6 cm, the lower usually somewhat narrower and smaller; fls 5-merous, greenish-purple, axillary, solitary or in cymes of 2–5, 6–8 mm wide; pet scarcely narrowed at base; fr usually 3-lobed, strongly tuberculate, 1.5 cm thick. Rich moist woods and hillsides; w. N.Y. to Mich., s. to W.Va., Ky., and Mo. May.

3. PAXISTIMA Raf. Mt.-lover. Fls perfect, 4-merous; stamens short; ovary bilocular, sunken in the center of the disk; capsule 2-valved; seeds 1 or 2; aril lacerate, whitish; small evergreen shrubs with opposite lvs and small axillary fls. 2, the other in w. U.S. (*Pachistima* and *Pachystima*, variant spellings.)

1. Paxistima canbyi A. Gray. Stems decumbent and rooting below, the distal part erect or ascending, branched, 2–4 dm; lvs numerous, leathery, sessile, linear-oblong, 1–2 cm, obtuse, revolute, entire or serrulate; pedicels solitary or few from some of the upper axils, very slender, 6–10 mm; pet green, 2.5 mm. Rocky woods in the mts.; w. Va. to s. O. and e. Ky. Apr., May.

FAMILY **AQUIFOLIACEAE**, the Holly Family

Fls unisexual or sometimes perfect, regular, hypogynous, 4–8-merous, mostly in axillary infls; pet imbricate, very shortly connate at base (*Ilex*) or distinct; stamens usually alternate with the pet, commonly adnate to the base of the cor in *Ilex;* nectary-disk wanting; ovary mostly 4–8-locular, with a single pendulous, apical-axile, apotropous ovule in each locule; style short or none; stigma lobed or capitate; fr a drupe with usually as many stones as carpels (or fewer by abortion); seeds with a very small, dicotyledonous embryo and abundant, oily endosperm; shrubs or trees with simple, mostly alternate lvs and minute or no stipules. 4 genera, all but about 20 spp. belonging to *Ilex.*

1 Pet obovate, shortly connate at base; stamens inserted on the cor-tube .... 1. *Ilex.*
1 Pet linear-oblong, distinct; stamens free and distinct .... 2. *Nemopanthus.*

1. ILEX L. Holly. Polygamo-dioecious; fls 4–8-merous, unisexual or sometimes perfect, the pistillate usually bearing stamens with small anthers, the staminate often with a vestigial pistil; cal small but evident, persistent under the fr; pet obovate, shortly connate at base; stamens alternate with the pet, barely adnate to the cor-tube; ovary 4–8-locular; style very short and stout, or none; fr black or red (yellow); trees or shrubs with simple, alternate lvs and axillary clusters of small, white or greenish fls. 300–400, cosmop.

1 Lvs coriaceous, evergreen (as shown by their persistence on year-old branches).
  2 Fls 4-merous; fr red (yellow); lvs not punctate beneath.
    3 Lvs spine-tipped and commonly with some salient, spine-tipped teeth .... 1. *I. opaca.*
    3 Lvs innocuous, merely crenulate ....2. *I. vomitoria.*
  2 Fls 6–8-merous; fr black; lvs punctate beneath.
    4 Lvs remotely crenulate above the middle, with a few curved-appressed teeth .... 3. *I. glabra.*
    4 Lvs with a few small, subspinulosely salient teeth above the middle, or entire .... 4. *I. coriacea.*
1 Lvs thin, deciduous.
  5 Sep ciliate, even at maturity of the fr.
    6 Pedicels shorter than the subtending petioles; lf-teeth evident.
      7 Pet entire or very minutely erose; nutlets smooth on the back .... 5. *I. verticillata.*
      7 Pet ciliate; nutlets grooved on the back .... 7. *I. montana.*
    6 Pedicels (or pedicels and peduncles together) longer than the subtending petioles; lf-teeth obscure;
      nutlets grooved on the back .... 8. *I. amelanchier.*
  5 Sep entire.
    8 Fls mostly 4–5-merous; lvs obtuse to rounded or acutish; nutlets grooved on the back .... 9. *I. decidua.*
    8 Fls mostly 6–8-merous; lvs distinctly acuminate; nutlets smooth on the back .... 6. *I. laevigata.*

1. Ilex opaca Aiton. American holly. Bushy shrub or small tree to 15 m; lvs evergreen, coriaceous, elliptic to oblong or ovate, 5–10 cm, tipped with a stout spine, and usually with 2–several strongly salient, spine-tipped teeth on each side; fls mostly 4-merous, the staminate in peduncled clusters; fr 8–10 mm, bright red (yellow); nutlets grooved on the back; $2n=36$. Sandy soil near the coast from Me. to Md.; widely distributed in the s. states from Va. to Ky. and s. Mo., s. to Fla. and Tex. May, June.

2. Ilex vomitoria Aiton. Yaupon. Shrub or small tree to 8 m; lvs evergreen coriaceous, oval to oblong or elliptic, 2–4 cm, obtuse at both ends, glabrous, finely crenate, the apex of each tooth closely appressed to the sinus; fls mostly 4-merous, the staminate in short-peduncled cymes, the pistillate cymes subsessile; fr red, 5 mm; nutlets grooved on the back; $2n=40$. Swamps and wet woods on the coastal plain; se. Va. to Fla., Ark., and Tex. May.

3. Ilex glabra (L.) A. Gray. Inkberry. Shrub 1–3 m with finely puberulent twigs; lvs coriaceous, evergreen, oblanceolate, 2–5 cm, a fourth to a third as wide, obtuse, cuneate at base, punctate beneath, bearing 1–3 minute appressed teeth on each side near the tip; fls mostly 6–8-merous, the staminate in small, pedicellate clusters, the pistillate mostly solitary; fr black, 4–5 mm, firm, dry, persistent; nutlets smooth on the back. Bogs, swamps, and wet woods on the coastal plain; N.S. to Fla. and La. June, July.

4. Ilex coriacea (Pursh) Chapman. Gallberry. Shrub to 5 m; twigs glabrous or viscid-puberulent; lvs coriaceous, evergreen, oblanceolate, 3.5–8 cm, acute or obtuse, minutely mucronate, tapering to the base, puberulent along the midvein above, otherwise glabrous, entire or with a few low salient teeth usually in the upper third of the blade; fls mostly 6–8-merous, the staminate numerous in dense fascicles, the pistillate few or soltiary; fr black, soft, deciduous at maturity; nutlets smooth on the back. Wet woods; se. Va. to Fla. and La. May, June. (*I. lucida,* misapplied.)

5. Ilex verticillata (L) A. Gray. Winterberry. Shrub to 5 m; lvs round-obovate to elliptic or lance-oblong, usually acuminate, with numerous appressed or salient teeth; staminate fls 4–6-merous, several in each fascicle; pistillate fls 5–8-merous, solitary or few together; sep ciliate; fr bright red, varying to yellow, 7 mm thick; nutlets smooth on the back; $2n=36$. Swamps and wet woods. May, June. Four vars.:

a Lvs 1–2.5 times as long as wide, generally broadest above the middle.
  b Lvs about twice as long as wide, commonly thin, 4–8 cm, distinctly acuminate, glabrous beneath
    or only sparsely pubescent on the main veins; staminate cymes distinctly peduncled; Nf. and
    Que. to Ont. and Mich., s. to Md., W.Va., and Ind. (var. *tenuifolia; I. bronxensis*) .... var. *verticillata.*
  b Lvs 1.2–1.7 times as long as wide, commonly firm and only 2–4 cm, rounded above and barely
    apiculate, ± pubescent on the surface beneath; staminate cymes subsessile, densely many-fld;
    Mass. to Mich. .... var. *cyclophylla* Robinson.
a Lvs commonly 3–4 times as long as wide, broadest near or below the middle.
  c Principal lvs at least 2 cm wide, thinly to densely pubescent on the veins and surface beneath; the
    common form of the sp. from N.J. to O. and Mo., s. to Ga. and Miss., but also n. to Me., Ont.,
    and Minn. .... var. *padifolia* (Willd.) T. & G.
  c Lvs numerous and crowded, less than 2 cm wide, sharply serrate, glabrous beneath or sparsely
    pubescent on the veins; wet sands near the sea; N.S. to L.I. .... var. *fastigiata* (Bickn.) Fern.

6. Ilex laevigata (Pursh) A. Gray. Smooth winterberry. Shrub to 3 m, with glabrous twigs; lvs elliptic to oblanceolate or obovate, 4–8 cm, usually distinctly acuminate, serrulate over most of their length with appressed or salient teeth, acute or cuneate or base, usually sparsely villous on the veins beneath; fls 6–8-merous, 1 or 2 in an axil, the staminate on slender pedicels 8–16 mm, the pistillate fls and frs on short pedicels 3–5 mm; sep entire; fr orange-red (yellow), 8 mm; nutlets smooth on the back. Swamps and wet woods; Me. to N.Y. and S.C., chiefly e. of the mts. June.

7. **Ilex montana** (T. & G.) A. Gray. Biglf-holly. Shrub or small tree to 10 m; lvs ovate or lance-ovate, 6–12 cm, broadest near the middle, conspicuously acuminate, rather sharply serrate with ascending or salient teeth; pedicels shorter than the subtending petioles; fls 4–5(6)-merous, the sep and pet ciliate; fr red, 6 mm thick; nutlets grooved on the back; 2n=40. Mt. woods; May, June. Var. *montana* (*I. monticola*), occurring from w. Mass. to N.C., has the lvs usually at least twice as long as wide, glabrous beneath or only sparsely pubescent (especially along the veins), the sep glabrous except for the ciliate margins. Var. *mollis* (A. Gray) Britton, occurring from Ky. to Ga. and Ala. has the lvs commonly less than twice as long as wide, velvety-pubescent beneath, and the cal villous. (*I. beadlei.*)

8. **Ilex amelanchier** M. A. Curtis. Sarvis-holly. Tall shrub; lvs elliptic to oblong, 5–10 cm, acute or subacuminate, apparently entire, but actually minutely appressed-serrulate, fls 4–5-merous, the staminate in pedunculate clusters, the pistillate solitary on a pedicel equaling or surpassing the subtending petiole; sep ciliate; fr red, nutlets grooved on the back. Swamps and wet woods on the coastal plain; se Va. to Fla. and La.; rare.

9. **Ilex decidua** Walter. Possum-haw. Shrub or small tree to 10 m, with glabrous twigs; lvs oblanceolate to narrowly obovate, 3–7 cm, rounded or obtuse at the tip, finely crenulate, narrowly cuneate at base, glabrous above, pubescent at least along the midvein beneath; fls in small peduncled fascicles, often on dwarf branches, 4–5-merous; sep entire; fr red; nutlets grooved on the back; 2n=40. Swamps and wet woods, chiefly on the coastal plain; D.C. to Fla. and Tex., n. in the interior to s. Ind., s. Ill., and s. Mo. May. (*I. longipes.*)

2. **NEMOPANTHUS** Raf.   Mt.-holly. Much like *Ilex,* but the pet narrow and distinct, the stamens free from the pet, and the sep minute or obsolete (or those of the pistillate fls more evident but deciduous as the fr matures); fls 4–5-merous, 1(–3) in the axils, notably long-pedicellate; fr red (yellow). Only 2 spp.

1 Lvs acuminate, evidently glandular-serrulate, the larger ones mostly 5–10 cm; local .... 1. *N. collinus.*
1 Lvs obtuse or acute and mucronate, mostly 3–5(–6) cm, entire or nearly so; widespread .... 2. *N. mucronatus.*

1. **Nemopanthus collinus** (Alexander) R. C. Clark.   Appalachian mt.-holly. Lvs acuminate, evidently glandular-serrulate, the larger ones mostly 5–10 cm, only rather loosely and inconspicuously reticulate-veiny; petioles greenish or stramineous; sep minute but evident at 10×, subpersistent on the developing fr; anthers 0.8–1.7 mm; nutlets 5–6 mm, dorsally and usually laterally ribbed; otherwise much like no. 2. Swamps, bogs, and streamsides in the mts. of W.Va., Va., and N.C.; rare and local. May, June. (*Ilex longipes,* misapplied.)

2. **Nemopanthus mucronatus** (L.) Trel.   Common mt.-holly. Branching shrub to 3 m; petioles 5–12 mm, often reddish; lvs elliptic, 2–5(–6) cm, obtuse or rounded to acute and mucronate at the tip, entire or very minutely and remotely toothed, rounded at base, evidently and closely reticulate-veiny; pedicels 1–2 cm, or even 3 cm in fr; cal virtually obsolete; anthers 0.3–0.8 mm; pet yellowish, 2 mm; fr red (yellowish), 6 mm thick; nutlets 4–5 mm, dorsally smooth or ribbed, laterally smooth; 2n=40. Moist or wet places; Nf. and Que. to Minn., s. to W.Va. and Ind. March–June.

Family **BUXACEAE,** the Boxwood Family

Monoecious or dioecious, the fls small, regular, hypogynous, apetalous and with separate or no sep; stamens 4–30, distinct; disk wanting; ovary of (2)3 carpels, with as many biovulate locules that may be partitioned into uniovulate locelli; styles distinct, persistent, borne on or forming continuations of the apical margin of the ovary, the stigma decurrent along the inner side and commonly with a median furrow; ovules apical-axile, pendulous, apotropous; fr capsular or fleshy; seeds black and shining, commonly with a straight, axile, dicotyledonous embryo and abundant oily endosperm, usually carunculate; evergreen or seldom deciduous woody plants or (*Pachysandra*) suffrutescent herbs with simple, opposite or less often alternate, leathery, exstipulate lvs, the fls commonly in spikes or dense racemes or heads. 5/60, cosmop.

1. **PACHYSANDRA** Michx.   Staminate fls with 4 sep, stamens 4, opposite the sep, long-exsert, with very stout filaments; pistillate fls with 4 or more sep; ovary bi- or tricarpellate, with 4 or 6 uniovulate locelli; fr capsular or baccate, lobed; suffrutescent plants with long rhizomes, the ± evergreen, alternate lvs mostly crowded toward the

summit of the erect or ascending stems, the fls in spikes, a few at the base pistillate, the upper all staminate. 4, the others in e. Asia.

1. **Pachysandra procumbens** Michx. Allegheny-spurge. Stems 1.5–4 dm; lvs broadly ovate to rotund, 3–8 cm, with a few coarse teeth above, abruptly narrowed below to a long petiole; spikes 1–few, lateral from near the base, 5–10 cm; fr a densely pubescent, 3-lobed, basally circumscissile capsule 1.5 cm. Rich woods; Ky., W.Va., and w. N.C. to nw. Fla. and s. La. Apr., May.
P. terminalis Siebold & Zucc., with narrower lvs, terminal infls, and bicarpellate, baccate white fr 6–7 mm, sometimes escapes.

FAMILY **EUPHORBIACEAE**, the Spurge Family

Fls regular, hypogynous, nearly always unisexual; perianth generally small and inconspicuous, seldom (as in staminate fls of *Cnidoscolus*) larger and showy, the sep distinct or connate below, in 1 or 2 cycles, or wanting; stamens 1–many, distinct or variously connate; a nectary-disk often present; ovary of (2)3(4) carpels and as many locules, with 1 or 2 pendulous, apical-axile, epitropous ovules per locule, rarely (*Crotonopsis*) unilocular but apparently with 2 styles; styles distinct, or united below, sometimes branched or fringed; fr typically a capsular schizocarp, but sometimes of diverse other types; seeds commonly carunculate; embryo with 2 ± expanded cotyledons; plants often with milky or colored juice; lvs mostly simple and alternate, but sometimes compound or opposite or whorled; stipules usually present, sometimes reduced to glands. 300/7500, cosmop.

1 Perianth none; staminate fls of a single stamen; pistillate fls of a single pistil; one central pistillate fl and few–many staminate fls arising within a cupulate involucre and simulating a perfect fl with a usually stipitate pistil; juice milky . . . . 10. *Euphorbia.*
1 Perianth present in staminate or pistillate fls or both; fls clearly unisexual, the 2 types differing in position, or size, or character of the bracts; juice not milky except in *Cnidoscolus.*
  2 Ovules and seeds 2 per locule; fls in ours 2–4 in the axils; lvs entire . . . . 1. *Phyllanthus.*
  2 Ovules and seeds 1 per locule; fls otherwise arranged; lvs in many spp. toothed.
    3 Principal lvs conspicuously lobed.
      4 Lvs peltate; stem glabrous; juice not milky . . . . 2. *Ricinus.*
      4 Lvs not peltate; stem armed with stinging hairs; juice milky . . . . 3. *Cnidoscolus.*
    3 Principal lvs entire or merely toothed.
      5 Pubescence of branched or stellate hairs or scales.
        6 Fr 2–3-locular, dehiscent; hairs of the lower lf-surface strictly stellate . . . . 4. *Croton.*
        6 Fr unilocular, indehiscent; hairs of the lower lf-surface stellate but with the branches united at base to form a flat scale . . . . 5. *Crotonopsis.*
      5 Pubescence of simple hairs or none.
        7 Styles undivided.
          8 Herbage glabrous; infls terminal . . . . 6. *Stillingia.*
          8 Herbage pubescent; infls mostly lf-opposed or on short axillary branches . . . . 7. *Tragia.*
        7 Styles laciniate or pinnatifid or irregularly branched; some or all infls axillary.
          9 Lvs alternate; carpels 3 (sometimes 1 or 2 infertile) . . . . 8. *Acalypha.*
          9 Lvs opposite; carpels 2 . . . . 9. *Mercurialis.*

1. **PHYLLANTHUS** L. Monoecious; fls minute, in small axillary cymules; cal 4–6-lobed; pet none; staminate fls with mostly 3–5 stamens, an evident extrastaminal disk, and reduced or no pistil; ovary globose, the 3 short styles ± bifid; seeds 2 per locule, without a caruncle; herbs or shrubs with numerous simple, alternate, entire, stipulate lvs. 750, mainly tropical.

1. **Phyllanthus caroliniensis** Walter. Glabrous annual, usually erect, 1–4 dm, with many ascending branches in 2 ranks; lvs 2-ranked on the stems, subsessile, oblong-obovate, 1–2 cm, entire, rounded above; fls 2–4 in the axils of the lvs, one staminate, 1–3 pistillate, 1–2 mm wide, on pedicels 0.5 mm; sep oblong-spatulate, usually 6; stamens 3, on very short filaments; capsule subglobose, 1.5–2 mm thick; seeds sharply angled, 1 mm; 2*n*=36. Moist, usually sandy soil, low woods, and meadows; Pa. to Ill. and Mo., s. to trop. Amer. July–Sept. Ours is var. *caroliniensis.*

2. **RICINUS** L. Monoecious; stamens very numerous, with repeatedly branching filaments; styles 3, united at base, each bifid and plumose; capsule large, 3-lobed, covered with soft prickles; seeds 2, carunculate. Monotypic.

1. Ricinus communis L. Castor-bean. Tall, stout, glabrous annual (perennial and arborescent in warmer climates) with large (1–4 dm wide), peltate, subrotund, palmately 6–11-lobed lvs and large racemes of fls; seeds 1 cm, violently poisonous; $2n=20$. Rarely adventive in waste places with us.

3. CNIDOSCOLUS Pohl. Monoecious; pet none; staminate cal gamosepalous, petaloid, white, with 5 imbricate lobes; disk annular, extrastaminal; stamens 8–10+, at least the inner connate below; pistillate cal small and caducous; ovary trilocular; styles several times bifid or laciniate; seeds 3, carunculate; herbs or woody plants with septate pith and milky juice, armed with stinging hairs; lvs simple, alternate, coarsely toothed or lobed; fls in terminal cymes, the staminate above the few pistillate ones. 50, New World.

1. **Cnidoscolus stimulosus** (Michx.) Engelm. & A. Gray. Bull-nettle. Deep-rooted, 1–few-stemmed perennial to 6 dm, armed throughout with stiff hairs 3–6 mm, these least abundant on the lvs; lvs long-petioled, subrotund in outline, deeply 3–5-lobed and often again cleft; cymes few-fld; cal of staminate fls salverform, the limb 14–25 mm wide; stamens 10, unequal; fr prismatic; seeds subcylindric, 6–8 mm; $2n=36$. Dry, sandy soil, chiefly on the coastal plain; se. Va. to Fla. and Tex. June, July. (*Jatropha s.*)

4. CROTON L. Monoecious or dioecious; cal 5–7-lobed; pet mostly as many as and about equalling the sep in staminate fls, but reduced or wanting in pistillate fls; stamens (3–)8–20+, inflexed in bud; anthers elongate; disk present; ovary (2)3-locular, with 1 ovule per locule; styles distinct or nearly so, bifid or dissected; fr capsular; seeds (1–)3, carunculate; herbs, shrubs, or trees with stellate pubescence, the small fls borne chiefly in condensed, spike-like, terminal infls. 600+, cosmop.

In addition to the following spp., the southwestern sp. *C. lindheimerianus* Scheele has been reported from Ill. It is a diffusely branched, monoecious annual with entire lvs and 3 styles, each bifid to the base.

1 Lvs toothed, and with 1 or 2 glands at the petiole-tip; styles 3, bifid . . . . 1. *C. glandulosus.*
1 Lvs entire or merely undulate, not glandular; styles otherwise.
  2 Plants monoecious.
    3 Styles 2, each bifid; ovary 2-locular; fr 1-seeded . . . . 2. *C. monanthogynus.*
    3 Styles 3, each 2–3 times dichotomous; ovary 3-locular; seeds normally 3 . . . . 3. *C. capitatus.*
  2 Plants dioecious; styles 3, each cleft to the base into 4 or more branches . . . . 4. *C. texensis.*

1. **Croton glandulosus** L. Tooth-lvd croton. Annual, 2–6 dm, roughly stellate-hairy; lvs narrowly oblong to oblong-ovate, 3–7 cm, coarsely serrate, with 1 or 2 large glands at the summit of the petiole; staminate fls with 4 sep, 4 pet, and 7–9 stamens; pistillate fls with 5 sep and no pet, each of the 3 styles bifid nearly to the base; $2n=16$. Dry or sandy soil; widespread in trop. and subtrop. Amer., n. to Va., Ind., Io., and Neb., and adventive farther north. Ours have been called var. *septentrionalis* Muell. Arg.

2. **Croton monanthogynus** Michx. Prairie-tea; one-seed croton. Annual, 2–6 dm, widely dichotomously branched, the lowest branches often 3–5 together; lvs ovate or ovate-oblong, 1–4 cm, entire, blunt or apiculate, broadly rounded to truncate at base, silvery-green and rather densely stellate; infls less than 1 cm; staminate fls minute, with 3–5 sep and pet and 3–8 stamens; pistillate fls with 5 sep and no pet, becoming reflexed; ovary bilocular, one locule later abortive; styles 2, deeply bifid; seed 1; $2n=16$. Dry or sterile soil; O. to Io., Neb., and Colo., s. to Fla., Tex., and n. Mex.; occasionally adventive farther n. or e.

3. **Croton capitatus** Michx. Woolly croton. Annual, to 1 m, sparingly and somewhat umbellately branched above; lvs narrowly triangular to oblong, lanceolate, or ovate-oblong, 4–10 cm, the principal ones with the blade rounded at base and less than twice as long as the petiole; fls in dense terminal infls 1–3 cm; staminate fls with 5 sep, 5 pet, and 10–14 stamens; pistillate fls apetalous, the 6–10 sep accrescent to 1 cm; styles 3, each 2–3 times dichotomous, the stigmas thus 12–24; $2n=20$. Dry, usually sandy soil and waste places; O. to Ind. and Neb., s. to Fla. and Tex. Ours are var. *capitatus.*

4. **Croton texensis** (Klotzsch) Muell. Arg. Texas croton. Dioecious, apetalous annual; stems to 1 m, sparingly branched chiefly above; lvs oblong-linear to lance-oblong, 2–8 cm, obtuse or abruptly acute, rounded to obtuse at base; staminate plants commonly with very numerous fls, the sep 5, the stamens 8–12; pistillate plants usually stouter, with fewer fls, the styles 3, each divided nearly to the base into 4 or more branches; fr 3-locular. Dry prairies; S.D. and Wyo. to Okla., Tex., and Ariz., and often adventive in the w. part of our range, especially along railroads.

5. CROTONOPSIS Michx. Monoecious; staminate fls at the summit of short spikes, minute, with 5 sep, 5 pet, and 5 initially inflexed stamens; pistillate fls at the base of the spikes, with 3–5-cleft cal and no pet; ovary with a single locule and 2 bifid styles; ovule solitary; fr lenticular, indehiscent, lepidote; seed without a caruncle; annual with the stem, lower lf-surface, sep and ovary coated with appressed round scales; lvs exstipulate, alternate, or subwhorled beneath the branches. Only 2 spp.

1 Hairs of the upper lf-surface with overlapping arms to 1 mm; scales of the fr with broad central disk
   and very short, appressed, radiating arms .... 1. *C. elliptica*.
1 Hairs of the upper lf-surface with very short (to 0.3 mm) non-overlapping arms; scales of the fr with
   minute central disk and longer radiating arms .... 2. *C. linearis*.

    1. **Crotonopsis elliptica** Willd. Stems 1–4 dm, slender, usually repeatedly branched; lvs lance-linear to lance-ovate, 1–3 cm, softly stellate-hairy above, the longer arms to 1 mm; frs only 1 or 2 per spike, with a median vein of each side, the lepidote hairs with broad central disk and very short appressed arms. In sand, sandy soil, or rocky barrens; Conn. to e. Pa., s. and w. to Fla. and Tex., and n. in the interior to s. Ill., Mo., and se. Kans. Late summer.
    2. **Crotonopsis linearis** Michx. Much like no. 1; lvs linear to lance-linear, 1.5–4 cm, the stellate hairs of the upper surface small, with non-overlapping arms to 0.3 mm; spikes slender, 1–2 cm, with 3–6 pistillate fls; fr not veined, its lepidote hairs with a minute central axis and conspicuous radiating, often ascending arms. Dry sand or sandy soil on the coastal plain; S.C. to Fla. and Tex., n. in the interior to s. Ill. and se. Mo., and disjunct in e. Io. and nw. Ill. Late summer.

    6. **STILLINGIA** Garden. Monoecious, apetalous; no disk; staminate fls with 2-lobed cal and 2 stamens, the filaments connate below; pistillate fls with 3 sep; ovary with 3 locules and ovules; styles united at base, not lobed; capsule 3-seeded, the base of each valve persisting after the fertile portion has fallen; caruncle sunken in a ventral depression of the seed; glabrous shrubs or perennial herbs with simple, alternate, glandular-serrate lvs and terminal spikes of small fls. 25+, mainly New World.

    1. **Stillingia sylvatica** L. Queen's delight. Herb 5–10 dm, often emitting a whorl of branches from the base of the primary spike; lvs 3–8 cm; lance-ovate, elliptic, or oblanceolate, serrate with incurved glandular teeth; spikes stout, erect, 5–10 cm, the staminate portion eventually deciduous; staminate fls clustered in the axils of minute bracts, each bract with 2 conspicuous, saucer-shaped glands; fr glabrous, 1 cm. Dry woods; se. Va. to Fla., Tex., and N.M., n. in the interior to Kans. May, June.

    7. **TRAGIA** L. Monoecious; cal 3–6-lobed; pet none; disk small or none; stamens 2–5+; ovary with 3 locules and ovules; styles connate at base, recurved-spreading above, undivided; capsule 3-lobed; seeds without a caruncle; perennial herbs or half-shrubs, often twining, ± pubescent, often with some stinging hairs; infls slender, racemiform, lf-opposed or terminating leafy-bracteate axillary branches, with a few pistillate fls below and many staminate ones above. 100+, mostly tropical.

1 Lvs ± acute at base; stamens 2 .... 1. *T. urens*.
1 Lvs cordate at base; stamens 3 .... 2. *T. cordata*.

    1. **Tragia urens** L. Narrow-lvd tragia. Stems erect, usually branched above, 2–4 dm, thinly and softly pubescent; lvs sessile or nearly so, linear to narrowly oblong, oblanceolate or elliptic, ± acute at base, entire or merely undulate, or with a few sinuate teeth distally; racemes to 10 cm, many of them on leafy-bracteate axillary branches, the very numerous staminate fls with 4 sep and 2 stamens; fr pubescent, 7–9 mm thick; $2n=44$. Dry sandy soil; se. Va. to Fla. and Tex. May, June.
    2. **Tragia cordata** Michx. Heart-lvd tragia. Stems erect or often twining and to 1.5 m, hirsute; lvs long-petioled, ovate or triangular-ovate, 5–10 cm, more than half as wide, acuminate, coarsely serrate, cordate at base; racemes numerous, lf-opposed, 3–6 cm; staminate fls with 3 sep and 3 stamens; staminate pedicel exceeding the bract, but deciduous from near the base; pistillate fls few, the cal 5-lobed; fr bristly, 10–15 mm thick. Rich soil of rocky woods or hillsides; s. Ind. to s. Mo., s. to Fla. and La. June–Aug. (*T. macrocarpa.*)

    8. **ACALYPHA** L. Copperleaf. Monoecious; pet and disk none; staminate fls with 4 valvate sep and 4–8 stamens, the filaments slender, the anthers elongate and often curved or coiled; pistillate fls with 3(–5) sep; styles slender, elongate, irregularly branched or lacerate into filiform segments; capsule (1–)3-locular; seeds with a caruncle; herbs, shrubs, or trees with alternate, petiolate, stipulate, usually serrate lvs and minute fls in axillary or terminal, often conspicuous, racemiform condensed cymose infls, each fl or fascicle of

fls subtended by a bract; our spp. annual herbs blooming from July to Oct., with staminate fls only 0.5 mm wide and aggregated into axillary heads or short spikes, the pistillate fls with initially small but rapidly accrescent bracts that equal or exceed the mature pubescent frs. 400, mostly New World.

1 Pistillate fls all or chiefly in terminal spikes, the staminate fls in axillary clusters .... 1. *A. ostryaefolia.*
1 Pistillate and staminate fls all in axillary infls, the same axis commonly with staminate fls above and
   pistillate below.
  2 Bracts 5–9-lobed; lf-blades not much longer than the petioles.
    3 Fr 3-seeded; seeds 1–2 mm .... 2. *A. rhomboidea.*
    3 Fr 2-seeded; seeds 2–3 mm .... 3. *A. deamii.*
  2 Bracts (9)10–15-lobed; lf-blades much longer than the petioles.
    4 Lobes of the bracts linear to oblong; bracts usually eglandular .... 4. *A. virginica.*
    4 Lobes of the bracts ovate to deltoid; bracts usually glandular .... 5. *A. gracilens.*

1. Acalypha ostryaefolia Riddell. Rough-pod c. Erect, 3–6 dm, branched above, finely hairy; lvs widely spreading or often dropping, thin, ovate-oblong to broadly ovate, abruptly short-acuminate, finely and regularly toothed, rounded to cordate at base; pistillate fls in terminal spikes 3–8 cm, also a few in the axillary staminate clusters; bracts tending to embrace the fr, cleft to about the middle into 13–17 linear segments; fr deeply 3-lobed, 3–4 mm thick, beset with slender projections and also ± pubescent. A weed in grain-fields, gardens, and waste places; N.J. to Ind. and Neb., s. to Fla. and Tex.

2. Acalypha rhomboidea Raf. Rhombic c. Stems erect, simple or branched, 2–6 dm, glabrous or puberulent in lines or throughout with incurved hairs, often with a few spreading hairs as well; petioles slender, divaricate, those of the larger lvs regularly more than half as long as the blades, these spreading, lanceolate to ovate, with a tendency to be rhombic; pistillate bracts 5–9-lobed, usually stipitate-glandular (visible at 10×) but without long hairs; staminate spikes scarcely exceeding the bracts; seeds 3, 1–2 mm. Dry or moist soil of open woods, roadsides, waste places, and gardens: Que. to N.D., s. to Fla. and Tex. Perhaps better treated as *A. virginica* var. *rhomboidea* (Raf.) Cooperrider.

    A. australis L., an east-Asian sp., has been reported as locally established at a number of stations in New York City. It differs from *A. rhomboidea* in its cordate, merely crenate bracts beneath the pistillate fls.

3. Acalypha deamii (Weath.) Ahles. Two-seed c. Much like no. 2, but the fr with 2 seeds 2–3 mm; petioles spreading, the blade abruptly drooping; 2n=40. Mesic sites at scattered localities from s. O. and s. Ind. to Tenn. and Ark. (*A. rhomboidea* var. *d.*)

4. Acalypha virginica L. Virginia c. Stem erect, usually branched, 2–6 dm, pubescent with incurved hairs, often also ± hirsute; petioles commonly much exceeding the bracts that they subtend and a third to half as long as the blades, or shorter in depauperate plants; blades lance-ovate, shallowly crenate; pistillate bracts 8–14 mm, deeply cleft into (9)10–15 oblong to linear acute segments, usually with some long, spreading hairs; staminate spikes equalling or slightly surpassing the bracts; 2n=40. Dry or moist open woods, fields and roadsides; Me. to Ind. and se. S.D., s. to Fla. and Tex.

5. Acalypha gracilens A. Gray. Short-stalk c. Stem erect, simple or branched from near the base, 2–5 dm, pubescent with incurved hairs; petioles about a fourth as long as the blades and usually shorter than the bracts they subtend; blades nearly linear to oblong or narrowly elliptic, 2–5 cm, obscurely crenate to entire; pistillate bracts usually arcuate-recurved, 5–10 mm, shallowly lobed into 9–15 deltoid or ovate, usually glandular segments; staminate spike mostly 5–15 mm, usually conspicuously exceeding the bracts; seeds 3 (except in var. *monococca*), 1.5–2 mm; 2n=40. Moist or dry, sandy soil, open woods, fields, and meadows; Me. to Wis., s. to Fla. and Tex. Many southern plants (n. to s. Ill.) have more elongate staminate spikes, to 3 or 4 cm, and have been distinguished as var. *fraseri* (Muell.-Arg.) Weatherby. The Ozarkian var. *monococca* Engelm., differing from var. *gracilens* in its one-seeded fruits, extends n. to Mo. and s. Ill. (*A. monococca.*)

9. MERCURIALIS L. Mercury. Dioecious; pet and disk none; sep 3; stamens 8–20, with short anthers; pistillate fls with 2 elongate staminodia and a bilocular, biovulate ovary; styles separate, laciniate-fringed; seeds 2, delicately carunculate; herbs with opposite, punctate, petiolate, stipulate lvs, the staminate fls in peduncled axillary spike-like infls, the pistillate in short axillary clusters. 7 or 8, Old World.

1. Mercurialis annua L. Herb-mercury. Spreading or ascending annual, 1–5 dm; lvs lanceolate to lance-ovate, 3–6 cm, serrate; staminate spikes exceeding the lvs; fr hirsute; 2n=16 and multiples. Native of Europe, w. Asia, and n. Afr., escaped or adventive in waste places here and there in our range and s. May–Sept.

10. EUPHORBIA L. Spurge. Fls greatly reduced, the staminate consisting of a single stamen, the pistillate of a single pistil; several staminate fls surrounding one pistillate fl inserted at the base of a cupulate involucre, the whole constituting an infl termed a cyathium but simulating a single fl with a stipitate ovary; involucre 4–5-lobed, usually bearing glands in the sinuses; these in some spp. with white or colored petaloid margin; ovary 3-locular and 3-ovulate; styles 3, bifid; capsule 3-lobed, 3-seeded; herbs, shrubs, or trees of diverse form, with milky, often highly acrid juice. The several subgenera and even some of the sections have often been taken as distinct genera, but the boundaries are difficult to draw. 1500, cosmop.

1 Glands of the involucre without petaloid appendages; many or all of the lvs alternate except in nos. 6 and 14; stipules absent or vestigial.
  2 Glands of the involucre regularly 4, oval to crescent-shaped, never bilabiate; stem terminated by a proliferating umbel of 3–many rays subtended by a whorl of lvs. (Subg. *Esula.*) . . . . Group I.
  2 Glands of the involucre usually 1, rarely more, tangentially bilabiate; stem not terminated by an umbel subtended by a whorl of lvs. (Subg. *Poinsettia.*) . . . . Group II.
1 Glands of the involucre with small or conspicuous, white or rarely colored, petaloid appendages, or the appendages scarcely developed in some spp. with strictly opposite lvs; stipules present.
  3 Glands and petaloid appendages 5; lvs opposite or much more often alternate or partly whorled, equilateral at base; plants, except no. 18, ± erect; involucres not truly axillary, usually distinctly terminal; plants annual or perennial. (Subg. *Agaloma.*) . . . . Group III.
  3 Glands and petaloid appendages 4; lvs all opposite, oblique at base; plant mostly prostrate to ascending (suberect in nos. 25 and 29); involucres borne singly or in small fascicles in the axils of many of the lvs or also in terminal clusters; our spp. annual. (Subg. *Chamaesyce.*) . . . . Group IV.

Group I. Subgenus *Esula*

1 Glands of the involucre transversely oval or elliptic or subrotund. (Sect. *Tithymalus.*)
  2 Seeds smooth or very obscurely reticulate.
    3 Cauline lvs 5–10 cm, entire; involucres 3 mm; perennial . . . . 1. *E. purpurea.*
    3 Cauline lvs 1–5, finely serrulate; involucres 1.5 mm; annual.
      4 Involucres glabrous outside; styles serrate, bifid a third to half length . . . . 2. *E. obtusata.*
      4 Involucres almost always villous outside; styles united at base, each branch bifid a fifth its length . . . . 3. *E. platyphyllos.*
  2 Seeds distinctly areolate; annual.
    5 Ovary and fr smooth; involucres 2 mm . . . . 4. *E. helioscopia.*
    5 Ovary and fr strongly roughened; involucres scarcely 1 mm . . . . 5. *E. spathulata.*
1 Glands of the involucre reniform or crescent-shaped with the concave side outward.
  6 Cauline lvs strictly opposite and decussate; seeds 4–6 mm. (Sect. *Lathyris.*) . . . . 6. *E. lathyrus.*
  6 Cauline lvs all or mainly alternate; seeds 1–3 mm. (sect. *Esula.*)
    7 Rays of the umbel mostly 3–5; seeds not smooth; annual, except sometimes no. 7.
      8 Seeds pitted.
        9 Seeds pitted about uniformly on both inner and outer faces.
          10 Pits small, punctiform or circular, uniformly scattered . . . . 7. *E. commutata.*
          10 Pits large, transversely elongate, forming 4 vertical rows . . . . 8. *E. falcata.*
        9 Seeds pitted only on the outer face, the inner face with 2 longitudinal furrows . . . . 9. *E. peplus.*
      8 Seeds tuberculate . . . . 10. *E. exigua.*
    7 Rays of the umbel usually more than 5; seeds smooth; colonial perennials.
      11 Principal cauline lvs 1–2 cm wide, distinctly pinnately veined . . . . 11. *E. lucida.*
      11 Principal cauline lvs mostly less than 1 cm wide, essentially 1-nerved.
        12 Principal lvs 3–8 cm × 4–8 mm; plants 3–7 dm . . . . 12. *E. esula.*
        12 Principal lvs 1–3 cm × 1–3 mm; plants 1.5–3(–4) dm . . . . 13. *E. cyparissias.*

Group II. Subgenus *Poinsettia*

1 Principal cauline lvs all or mostly opposite; herbage ± pubescent . . . . 14. *E. dentata.*
1 Principal cauline lvs all or mostly alternate; herbage glabrous or nearly so . . . . 15. *E. cyathophora.*

Group III. Subgenus *Agaloma*

1 Lvs obtuse to retuse; plants perennial. (Sect. *Tithymalopsis.*)
  2 Appendages white (rarely green), conspicuous, much exceeding the gland and commonly as long as or longer than wide.
    3 Stems branched from the middle or above, with several to many well developed, alternate lvs below the primary branches . . . . 16. *E. corollata.*
    3 Stems branched from the base, the foliage lvs all on the branches, opposite or whorled . . . . 17. *E. marilandica.*
  2 Appendages very short or apparently none, no longer (radially) than the gland.
    4 Appendages green; stem branched from the base, the lvs all on the branches and opposite, not ciliate . . . . 18. *E. ipecacuanhae.*

    4 Appendages white; stem simple below and bearing there a few alternate, much reduced or scale-
      like, soon deciduous lvs; well developed lvs ciliate .... 19. *E. mercurialina*.
  1 Lvs terminating in a sharp point; plants annual.
    5 Lvs all opposite, narrow, mostly well under 1 cm wide. (Sect. *Zygophyllidium*.) .... 20. *E. hexagona*.
    5 Lvs below the infl all alternate, broad, mostly well over 1 cm wide. (Sect. *Petaloma*.) .... 21. *E. marginata*.

GROUP IV. Subgenus *Chamaesyce*

  1 Young stem and lvs glabrous.
    2 Lvs strictly entire; seeds smooth.
      3 Stipules separate.
        4 Plants of shores and dunes of the Atlantic and the Great Lakes; appendages (measured radially)
          no longer than the glands, or lacking.
          5 Seeds 2–2.6 mm; widely distributed .... 22. *E. polygonifolia*.
          5 Seeds 1.5–1.9 mm; se. Va. and southward .... 23. *E. ammannioides*.
        4 Plants of inland sites, from Indiana w.; seeds 1.3–2 mm.
          6 Appendages minute, rarely longer than wide; stems prostrate, branched from the base .... 24. *E. geyeri*.
          6 Appendages conspicuous, ovate or ovate-oblong; stems erect or ascending, first branched
            well above the base .... 25. *E. missurica*.
      3 Stipules united into a triangular, scale-like, often lobed or fringed structure .... 26. *E. serpens*.
    2 Lvs serrulate (at 10×), at least around the tip.
      7 Seeds smooth to obscurely roughened or pitted, not transversely ridged .... 27. *E. serpyllifolia*.
      7 Seeds with 3 or 4 evident transverse ridges .... 28. *E. glyptosperma*.
  1 Young stems pubescent, at least in a line on each side; lvs usually pubescent, at least on one side
    when young, tending to be glabrescent in age.
    8 Ovary and capsule glabrous.
      9 Stems ascending or suberect, mostly simple below, puberulent when young .... 29. *E. nutans*.
      9 Stems prostrate or widely spreading, branched from the base, spreading-hirsute .... 30. *E. vermiculata*.
    8 Ovary and capsule pubescent.
      10 Involucre cleft on one side a fourth to a third its length; lvs mostly ± oblong and 2–3 times
         as long as wide; styles 0.3–0.4 mm .... 31. *E. maculata*.
      10 Involucre cleft on one side fully half its length, lvs mostly obovate-oblong or obovate and less
         than twice as long as wide; styles 0.5–0.7 mm .... 32. *E. humistrata*.

1. Euphorbia purpurea (Raf.) Fern.   Purple s. Stout perennial to 1 m from a short, stout rhizome; cauline lvs elliptic to narrowly lance-oblong, 5–10 × 1–3 cm, obtuse, entire, often sparsely villous beneath; lvs subtending the umbel shorter and broader, commonly ovate, those of the umbel depressed-ovate to reniform; rays of the primary umbel 5–8, rarely fewer; involucres 3 mm, glabrous, often purplish above; fr 6–8 mm, rough, with small irregular projections; seeds plump, subglobose, 3–4 mm, smooth, mottled brown. Dry or moist woods, rare; N.J. and Pa. to W.Va. and N.C. May, June. (*E. darlingtonii; Tithymalus d.; Galarhoeus d.*)

2. Euphorbia obtusata Pursh.   Woodland s. Glabrous annual, 2–7 dm, simple to the umbel or with a few lateral branches; cauline lvs oblong-oblanceolate, 1.5–4 cm, very obtuse, finely serrulate, subcordate and somewhat clasping at the sessile base; lvs subtending the primary umbel shorter and broader, those of the umbel broadly ovate; rays of the primary umbel usually 3; involucres 1.5 mm, strictly glabrous outside; styles separate to the base, each bifid a third to half its length; fr 3–3.5 mm, verrucose; seeds thickly lenticular, 2–2.5 mm and nearly as wide, smooth or obscurely reticulate. Woods; Pa. to Ind. and Io., s. to S.C. and Tex. Apr., May. (*Tithymalus o.; Galarhoeus o.*)

3. Euphorbia platyphyllos L.   Annual, 3–7 dm, commonly with many branches below the umbel; cauline lvs oblanceolate, 2–5 cm, mostly acute, finely serrulate, sessile or nearly so, tapering to a narrow or subcordate base, often sparsely villous beneath; lvs subtending the umbel shorter and broader, those of the umbel depressed-ovate to subrotund; rays of the primary umbel commonly 5; involucres 1.5 mm, almost always finely villous; styles united at base, each branch bifid a fifth its length; fr 3 mm; $2n=28$. Native of Europe; intr. in waste places here and there in our range. June–Sept. (*Tithymalus p.; Galarhoeus p.*)

4. Euphorbia helioscopia L.   Wartweed. Annual, smooth or nearly so, 2–5 dm, the upper internodes progressively longer; cauline lvs spatulate, 1.5–5 cm, very blunt or retuse, finely and sharply serrulate; lvs subtending the primary umbel somewhat oblique, broadly elliptic to obovate; rays of the primary umbel 5, on well grown plants repeatedly branched; involucres 2 mm; fr smooth, 3 mm; seeds ovoid, 2–2.5 mm, conspicuously areolate; $2n=42$. Native of Europe; intr. abundantly from e. Kan. to w. Ont., n. Mich., and N.Y., and occasionally elsewhere. Apr.–Sept. (*Tithymalus h.; Galarhoeus h.*)

5. Euphorbia spathulata Lam.   Prairie-s. Glabrous annual, 1–6 dm, often branched from the base; cauline lvs oblong-spatulate, 1–4 cm, obtuse, finely serrulate, the upper truncate or even subcordate at the sessile base; lvs subtending the umbel similar; lvs of the umbel shorter and relatively wider, broadly oblong to ovate or deltoid-ovate; rays of the primary umbel commonly 3, repeatedly dichotomous; involucre delicate, 0.8 mm; ovary and fr beset with numerous conic to cylindric processes, the fr 2.5 mm; seeds thick-lenticular, 1.5–1.8 mm, marked with a reticulum of fine, sharp, low ridges. Prairies, barrens, and rocky hills; Minn. and Mo. to Mont. and Tex., w. to the Pacific. May, June. (*E. dictyosperma; Tithymalus arkansanus; T. missouriensis; Galarhoeus a.; G. m.*)

6. **Euphorbia lathyrus** L. Caper- or myrtle-s. Annual or biennial, glabrous and glaucous, to 1 or even 2 m, usually simple to the umbel; cauline lvs opposite, linear or lance-linear, sessile at the truncate or subcordate base, those subtending the umbel similar, usually 4, those of the umbel lance-ovate to deltoid-ovate; involucres 4 mm; fr 1 cm, thick-walled, tardily dehiscent; seeds 4–6 mm, very plump, wrinkled and carunculate; $2n=20$. Native of the Mediterranean region, escaped from cult. here and there in our range. May–Sept. (*Tithymalus l.; Galarhoeus l.*)

7. **Euphorbia commutata** Engelm. Tinted s. Glabrous, 2–4 dm, often branched from the base, the decumbent stems often emitting new shoots the next year; young stems and lvs commonly somewhat red-tinted; cauline lvs numerous; oblanceolate to obovate, or the upper oval to ovate, commonly sessile; lvs subtending the umbel broadly elliptic to ovate or subcordate, those of the umbel broadly triangular-reniform, tending to be connate; rays of the primary umbel usually 3; fr smooth, 3 mm; seeds pale gray, 1.5–2 mm, deeply and uniformly pitted on both faces; $2n=28$. Moist woods and shaded hillsides; Pa. to s. Ont., Mich., and Minn., s. to Fla. and Tex. Apr.–June. (*Tithymalus c.; Galarhoeus c.*) Southern plants, extending n. to Va., Ky., and Mo., with all cauline lvs oblanceolate and long (5–12 mm)-petioled, have been segregated as var. **erecta** Norton.

8. **Euphorbia falcata** L. Annual, 1–4 dm, usually with numerous alternate branches; cauline lvs oblanceolate, 1–2 cm, acute, caducous; lvs subtending the umbel similar; lvs of the umbel round-ovate or reniform, 5–10 mm, conspicuously mucronate; rays of the umbel commonly 3, each repeatedly dichotomous; seeds brown beneath a gray coating, deeply sculptured on both sides with several transverse furrows; $2n=16$. Native of the Mediterranean region, becoming established in waste places and along roads from Pa. and O. to Va. and W.Va. July–Sept. (*Tithymalus f.*)

9. **Euphorbia peplus** L. Petty s. Annual, 1–3 dm, usually much branched; cauline lvs oblong-obovate or subrotund, usually petiolate, 1–2 cm, rounded to retuse at the summit, those subtending the umbel broadly ovate to obovate, those of the umbel smaller, ovate; rays of the umbel 3, repeatedly dichotomous; fr with 2 low longitudinal keels on each valve; seeds 1.5 mm, marked on the outer face with 4 rows of (2)3 or 4 large pits, on the inner face with 2 deep longitudinal furrows; $2n=16$. Native of Eurasia, widely intr. in N. Amer. from N.B. to Alas., s. to Fla. and Calif., but seldom abundant. Summer. (*Tithymalus p.; Galarhoeus p.*)

10. **Euphorbia exigua** L. Annual, 1–3 dm, often branched from the base; cauline lvs linear or linear-oblong, 1–2.5 cm, sessile, those subtending the umbel narrowly lanceolate, truncate or subcordate at base, those of the umbel similar but progressively smaller; rays of the umbel 3–5; seeds gray, 1–1.5 mm, finely and rather sharply tuberculate; $2n=16$, 24. Native of Europe and w. Asia, sporadically intr. and established from e. Can. to W.Va. and O. Summer.

11. **Euphorbia lucida** Waldst. & Kit. Shining s. Rhizomatous perennial to 1 m; cauline lvs lance-oblong or oblong, 5–10 × 1–2 cm, sessile and often somewhat clasping, shining above, distinctly pinnately veined, the evident lateral veins leaving the midrib at angles of 45–90°; lvs subtending the umbel shorter, ovate; lvs of the umbel broadly round-deltoid; involucres 2.5 mm; fr 3.5–4 mm; seeds 2.5–3 mm; $2n=36$. Native of Europe, rarely established as a weed in our range. (*E. agraria,* misapplied; *Tithymalus l.; Galarhoeus l.*)

12. **Euphorbia esula** L. Leafy s. Strong-rooted, vigorously colonial perennial, erect, 3–7 dm, glabrous, usually with numerous alternate flowering branches below the umbel; cauline lvs linear to lance-linear or narrowly oblanceolate, mostly 3–8 cm × 3–8 mm, obtuse to mucronate, essentially 1-nerved, the lateral veins very obscure, leaving the midrib at an angle of 15–35°; lvs subtending the umbel shorter and broader, lanceolate to ovate, those of the umbel opposite, broadly cordate or reniform; rays of the primary umbel mostly 7–15; fr 3–3.5 mm, finely granular; seeds ellipsoid, 2–2.5 mm; $2n=20$, 60, 64. Native of Eurasia, widely established in N. Amer. from New England to the Pacific, s. to Md., Ind., Io., and Colo. Summer. (*Tithymalus e.; Galarhoeus e.; Euphorbia intercedens; E. podperae; E. virgata,* at least as applied to N. Amer. plants.)

13. **Euphorbia cyparissias** L. Cypress-s. Colonial by rhizomes, loosely erect, 1.5–3(–4) dm; cauline lvs very numerous, crowded, linear, 1–3 cm × 1–3 mm, 1-nerved, those subtending the umbel similar, those of the umbel broadly cordate; rays of the umbel usually 10 or more; fr 3 mm, seldom produced, slightly granular-roughened; seeds plump, smooth, 1.5–2 mm; $2n=20$, 40. Native of Eurasia, commonly established in our range in roadsides, waste ground, and cemeteries; Me. to Wash., s. to Va. and Ark. Apr.–July. (*Tithymalus c.; Galarhoeus c.*)

14. **Euphorbia dentata** Michx. Toothed s. Erect annual, 2–6 dm, often branched, with hairy herbage; lvs all or mostly opposite, petiolate, linear to ovate, coarsely toothed to subentire; infl congested, mingled with reduced green lvs; involucres 2–3 mm, with fimbriate lobes and a conspicuous, fleshy, flattened-obconic, tangentially bilabiate gland; styles bifid half their length, or deeper; fr smooth, 5 mm thick; seeds ovoid, rough-tuberculate, 2.5–3 mm, usually carunculate; $2n=14(?)$, 28, 56. Dry soil; Ill. and Wis. to Wyo. and Mex., and established as a weed on roadsides and waste places, especially in cindery soil, e to Mass., N.Y. and Va. July–Sept. (*Poinsettia d.*)

15. **Euphorbia cyathophora** Murray. Fire-on-the-mountain. Erect annual, to 1 m, usually branched, glabrous or nearly so (or the lvs seldom sparsely pilose beneath); lvs mainly or all alternate (or the uppermost opposite), variable, even on the same plant, from linear to broadly oblong or ovate or pandurate, entire or serrate to lobed, the upper mostly lobed and blotched with red or white at base; involucral gland solitary, bilabiate, cupulate, usually wider than high; fr smooth, 6–8 mm thick; seeds tuberculate, 3–3.5 mm, without a caruncle; $2n=28$, 56. Moist soil, often in shade; trop. Amer and s. U.S., cult. and escaped n. to Va., Wis., and Minn. Summer. (*E. heterophylla* and *Poinsettia h.,* misapplied; *Poinsettia c.*)

16. Euphorbia corollata L. Flowering s. Erect perennial from a deep root, glabrous to villous, 3–10 dm, usually simple below, umbellately or paniculately branched above; lvs not ciliate, the cauline ones alternate, linear to elliptic, 3–6 cm, those subtending the primary branches similar, whorled, those of the infl smaller and often opposite; involucres numerous, forming a corymbiform or paniculiform cyme to 3 dm wide; pedicels, except a few lower ones, less than 1 cm; appendages white (green), conspicuous, ovate or oblong to obovate or rhombic, commonly longer than wide, 1.5–4 mm; seeds ovoid; 2n=28. Dry woods and old fields in the e., in the w. abundant on prairies; Mass. and N.H. to Minn., s. to Fla. and Tex. June–Sept. (*Tithymalopsis c.*) Two confluent geographic vars:

Var. *corollata*. Stems usually 3–10 together, 2.5–5 mm thick at base; nodes below the umbel (25–)35–60(–75); infl neither conspicuously leafy nor separated into distinct groups of involucres; involucres mostly 6–8+ mm wide across the appendages; the more widespread and n. phase, extending s. throughout Va. and to w. N.C., n. Ga., Ala., and Tex.

Var. *paniculata* Boiss. Stems 1–3 together, mostly 1.5–3 mm thick at base; nodes below the infl 20–30; infl often either conspicuously leafy or tending to consist of distinct, compact groups of involucres; involucres mostly 4–6 mm wide across the appendages; se. phase, from Va. and s. Md. to n. Fla., w. to Tenn. and Miss. (*E. apocynifolia; E. pubentissima; E. zinniiflora.*)

17. Euphorbia marilandica Greene. More or less intermediate between nos. 16 and 18, resembling no. 18 early in the season, and no. 16 later; stems dichotomously or trichotomously branched from the base or near it, beginning to bloom when only 1 dm, later growing to 5 dm; lvs lance-linear to oval; involucres long-pediceled from the forks of the stem; appendages white, orbicular to rhombic. Probably not a true species, but instead either a series of hybrids or unusual individuals of 16 and 18. Sandy soil; Md.; reported from s. N.J. to S.C. (*Tithymalopsis m.; E. arundelana.*)

18. Euphorbia ipecacuanhae L. Ipecac-s. Stems several from a stout vertical root, repeatedly dichotomous (or trichotomous) from the base upward, eventually to 3 dm, spreading to ascending; lvs eciliate, opposite, red or green, variable, even on the same plant, from linear to broadly ovate or obovate, the early ones small, the later ones to 5 cm, involucres solitary and long-pediceled from each node, terminal but often apparently lateral by suppression of one branch; appendages scarcely developed, consisting of a minute green margin on the gland; seeds angular. In sand, chiefly on the coastal plain; L.I. to Ga. Apr.–May (July). (*Tithymalopsis i.*)

19. Euphorbia mercurialina Michx. Cumberland-s. Perennial from a stout root, 2–4 dm, cymosely branched above the middle; lvs below the first branches few, sessile, oval to rotund, to 1 cm, or scale-like; lvs subtending the branches opposite or in 3's, those of the branches opposite, all ciliate, narrowly ovate, 3–6 cm, obtuse or subacute, rounded at base, on petioles 2–6 mm; involucres few, terminating the axis at each fork, on pedicels 1–3(–5) cm; appendages forming a narrow, undulate-marginal white rim on the glands. Rich soil on wooded slopes of ravines; Cumberland Plateau region from s. Ky. to N. Ala. and nw. Ga. Apr.–June. (*Tithymalopsis m.*)

20. Euphorbia hexagona Nutt. Six-angled s. Erect annual, 1–5 dm, with numerous opposite, very slender branches, but usually maintaining a central axis; principal lvs opposite, narrowly elliptic or oblong, 3–5 cm, equilateral at base, those of the branches shorter, often narrowly linear or filiform; involucres mostly unisexual (some bisexual), minutely pediceled in the upper axils, 2–3 mm, pubescent; appendages 5; fr eventually exsert well beyond the involucre on a pedicel 2–4 mm; seeds gray, nearly cylindric, 2.5 mm, minutely tuberculate. Prairies, plains, and barrens; Minn. to Mont., s. to Mo., Tex., and N.M. July–Sept. (*Zygophyllidium h.*)

21. Euphorbia marginata Pursh. Snow-on-the-mountain. Annual; stems erect, 3–8 dm, softly villous, especially above; cauline lvs alternate, sessile, broadly ovate to elliptic or obovate-oblong, 4–10 cm; lvs subtending the 3-rayed infl whorled or sometimes merely opposite, those within infl smaller, margined with white or wholly white; cymes crowded; involucres hairy, the 5 lobes deeply fimbriate, the 5 appendages white, conspicuous, reniform to broadly ovate; fr 3-lobed, 6–7 mm thick; seeds ovoid, 4 mm, tuberculate; 2n=56. Prairies and plains; Minn. to Mo., w. to Mont. and N.M., and escaped from cult. elsewhere. Summer. (*Dichrophyllum m.; Lepadena m.*)

22. Euphorbia polygonifolia L. Seaside-s. Prostrate, usually divergently branched or forming mats, glabrous throughout; lvs opposite, linear-oblong to lance-oblong, mostly 8–15 mm, entire, slightly inequilateral at base; appendages very small or none; fr 3–3.5 mm; seeds plump, smooth, compressed-ovoid, gray, 2–2.6 mm. Sand dunes and sandy beaches of the Atlantic from Que. to Ga., and of the Great Lakes, probably excluding Lake Superior. July–Oct. (*Chamaesyce p.*)

23. Euphorbia ammannioides HBK. Southern seaside-s. Usually prostrate, repeatedly branched, 1–3 dm, forming mats, glabrous throughout; lvs opposite, narrowly to broadly oblong, mostly 5–12 mm, entire, truncate or shallowly retuse; appendages small or none; fr 2–2.3 mm; seeds scarcely compressed, smooth, gray mottled with brown, 1.5–1.9 mm. Sandy beaches on or near the coast; se. Va. to Fla., Tex., and trop. Amer. June–Sept. (*Chamaesyce ingallsii.*)

24. Euphorbia geyeri Engelm. Dune-s. Glabrous annual, branched from the base, usually prostrate, 1–3 dm; lvs oblong to broadly elliptic, 5–10 mm, entire, broadly rounded to shallowly retuse at the apex; appendages inconspicuous, short-ovate, rarely more than twice as long as the gland (measured radially); stamens 5–17 per involucre; fr 2 mm; seeds not compressed, smooth, roundly 3-angled, 1.3–1.6 mm. Inland sandy prairies and dunes; Ind. to N.D. and e. Mont., s. to Okla., Tex., and N.M. July–Oct. (*Chamaesyce g.*)

25. Euphorbia missurica Raf. Missouri s. Glabrous annual; stems decumbent to suberect, 1–6 dm, repeatedly forked; lvs linear to oblong, 1–3 cm, entire, obtuse to rounded or retuse; appendages ovate to oblong, conspicuous, more than twice as long as the glands; stamens 29–48 per involucre; fr strongly 3-lobed, 2–2.5 mm; seeds not compressed, smooth, roundly 3-angled, mottled white and brown, 1.5–2 mm. Rocky or sandy soil; Minn. to Mont., s. to Mo., Tex., and N.M. Summer. Var. *missurica* (*E. zygophylloides; Chamaesyce z.; C. nuttallii*), with scarcely angled seeds, capillary smaller branches, and capillary peduncles to 11 mm, occurs from Mo. to Tex. Var. *intermedia* (Engelm.) Wheeler (*E. petaloidea; Chamaesyce p.*) with definitely angled seeds, and stouter branches and peduncles, the latter to 4 mm, occurs from Minn. to Mont. and Tex.

26. Euphorbia serpens HBK.   Round-lvd s.   Glabrous annual; stems prostrate, freely branched, 1–4 dm, the internodes of the branches very short and the plants thus very leafy; lvs broadly oblong to suborbicular, 2–7 mm, entire; stipules united into a scale-like structure often lobed or fringed at the tip; appendages very small; fr 3-angled, 1–1.5 mm; seeds quadrangular, smooth, 1 mm. Moist alluvial soil; s. Ont. to Mont., s. to Tenn., Fla., Ariz., and trop. Amer., and intr. here and there eastward. July–Oct. (*Chamaesyce s.*)

27. Euphorbia serpyllifolia Pers.   Thyme-lvd s.   Much like no. 28; stamens 5–12; seeds smooth or obscurely roughened or pitted. Dry, rocky soil; n. Mich., Minn., and Io., to B.C., Calif. and Mex.; and occasionally intr. eastward. July–Oct. (*Chamaesyce s.*)

28. Euphorbia glyptosperma Engelm.   Ridge-seed s.   Glabrous annual; stems mostly prostrate, freely branched, 1–3 dm, often forming mats; lvs narrowly to broadly oblong or ovate, 4–15 mm, strongly inequilateral, minutely serrulate, especially on the rounded summit, varying to entire; appendages short but evident; stamens 1–5 (typically 4) per involucre; fr depressed-ovoid, sharply 3-angled, 1.5–2 mm; seeds 1–1.3 mm, sharply 4-angled, marked with several conspicuous transverse ridges. Dry sandy soil; Que. and N.B. to B.C., s. to N.Y., Ind., Mo., and Ariz. June–Oct. (*Chamaesyce g.*)

29. Euphorbia nutans Lagasca.   Eyebane. Annual; stems to 8 dm, obliquely ascending at least in the upper half, the lower half often erect, the younger parts puberulent, often in a single longitudinal strip, with usually incurved hairs to 0.3 mm, the older parts glabrous or nearly so; lvs opposite, oblong or oblong-ovate, 1–3.5 cm, about a third as wide, serrulate, usually conspicuously inequilateral; fr 2–2.5 mm, strongly 3-lobed, glabrous; seeds gray or pale brown, 1–1.5 mm; 2n=12. Dry or moist soil; N.H. to Mich. and N.D., s. to Fla. and Tex.; abundant as a weed in lawns and gardens and intr. elsewhere in the world. June–Oct. (*E. maculata,* misapplied; *E. preslii; Chamaesyce rafinesquii*)

30. Euphorbia vermiculata Raf.   Hairy s.   Annual; stems prostrate to ascending, to 4 dm, hirsute ± uniformly from base to tip with spreading hairs 0.5–1.5 mm; lvs obliquely ovate-oblong to ovate, 0.5–2 cm, usually half to two-thirds as wide, serrate; fr 1.5–2 mm, strongly 3-lobed, glabrous; seeds gray or pale brown, 1–1.3 mm. Fields, roadsides, and waste ground; N.S. to Mich., s. to N.J., O., and Ind.; also Ariz., and N.M. July–Oct. (*E. hirsuta; Chamaesyce p.*)

31. Euphorbia maculata L.   Milk-purslane; spotted s.   Annual; stems prostrate or nearly so, to 4 dm, often forming circular mats, sparsely to densely villous or villous-puberulent; lvs dark green, often with a red spot, oblong or ovate-oblong to linear-oblong, mostly 5–15 mm and a third to half as wide, often widest below the middle; involucre cleft on one side a fourth to a third of its length; ovary and fr strigose, the fr 1.5 mm; styles 0.3–0.4 mm, bifid a fourth to a third their length; seeds quadrangular, 1 mm, nearly smooth or with a few inconspicuous transverse ridges, becoming mucilaginous when wet; 2n=28. Very abundant as a weed in lawns, gardens, and waste places, also in meadows and open woods; Que. to N.D., s. to Fla. and Tex., and intr. as a weed elsewhere in the world. June–Sept. (*E. supina; Chamaesyce m.*)

E. prostrata Aiton, much like no. 31, but with the lvs more variable in outline, chiefly less than 1 cm, with the styles bifid to the base or nearly so, and with the sharply quadrangular seeds marked with 3 or 4 conspicuous transverse ridges, is rarely adventive (from s. U.S.) and probably not persistent in our range. (*E. chamaesyce,* misapplied.)

E. stictospora Engel.,a chiefly sw. sp. with virtually entire styles, and with mottled, punctate-pitted seeds depressed-punctate at base and sharply acute at the tip, barely encroaches into our range in w. Io.

E. hirta L., decumbent to erect, to 6 dm, usually conspicuously hairy, with oblong or lance-oblong lvs 1–4 cm and the fls numerous in dense, axillary and terminal, often peduncled clusters, is widespread in trop. and warm-temp. regions, including s. U.S., but only rarely adventive and probably not persistent with us.

32. Euphorbia humistrata Engelm.   Much like no. 31; sometimes rooting at the lower nodes; lvs pale green, chiefly obovate to obovate-oblong, or narrower on the smaller branches, mostly 5–15 mm and more than half as wide, commonly widest above the middle, inconspicuously serrulate; involucre cleft on one side for fully half its length (this the most dependable distinction from no. 31); styles 0.5–0.7 mm; seeds smooth or nearly so. Moist alluvial soil; O. to Kans., s. to Ala. and Tex., and occasionally adventive farther e. Aug.–Oct. (*Chamaesyce h.*)

FAMILY **RHAMNACEAE**, the Buckthorn Family

Fls regular, 4- or 5-merous, perfect or sometimes unisexual, rather shortly perigynous; pet distinct, often clawed, rarely wanting; stamens as many as and alternate with the sep, opposite and often enfolded by the concave or hooded pet; ovary sessile on the intrastaminal nectary-disk or ± immersed in it (thus sometimes seemingly inferior), 2–3(–5)-locular, with a single erect, basal ovule per locule; style terminal, undivided to often deeply cleft; fr a drupe or sometimes a capsule or schizocarp; embryo large, with 2 cotyledons; endosperm usually scanty or wanting; woody plants with simple lvs and small fls. 55/900, cosmop.

1 Erect or spreading trees or shrubs.
  2 Lvs 3-nerved from the base; fls numerous; pet present, long-clawed .... 1. *Ceanothus*.
  2 Lvs not 3-nerved; fls few; pet, when present, with short or no claw .... 2. *Rhamnus*.
1 Climbing woody vines .... 3. *Berchemia*.

1. CEANOTHUS L. Sep inflexed, at length deciduous above the hypanthium; pet long-clawed, in bud incurved with the infolded sides of the broad blade clasping the anther, at anthesis spreading or deflexed between the sep; stamens at anthesis free and exsert; ovary 3-angled and 3-locular, immersed in the disk; style 3-lobed; fr a 3-lobed capsule-like drupe subtended by the persistent hypanthium, the exocarp thin, dark, coriaceous, the endocarp eventually loculicidally dehiscent; shrubs with alternate, 3-nerved, glandular-serrate lvs and numerous small white (in our spp.) to blue fls in sessile or short-peduncled umbels aggregated into terminal or axillary pedunculate panicles. 55, N. Amer.

1 Fls in axillary or terminal infls borne on the branches of the current season.
  2 Infls terminating axillary peduncles; lvs mostly ovate .... 1. *C. americanus*.
  2 Infls terminating leafy shoots of the season; lvs mostly oblong or elliptic .... 2. *C. herbaceus*.
1 Fls in infls borne below the leafy twigs on stems of the previous year .... 3. *C. sanguineus*.

    1. Ceanothus americanus L. New Jersey tea; redroot. Shrub to 1 m, often freely branched; lvs narrowly to broadly ovate, 3–8 cm, usually more than half as wide, obtuse or acute, broadly cuneate to rounded or subcordate at base, the lateral nerves often naked for 1–3 mm at base; infls on axillary peduncles, the lower peduncles progressively longer and up to 2 dm, each panicle short-cylindric to ovoid, occasionally branched, often subtended by 1–3 reduced lvs, the individual umbels of the panicle usually separated by distinct internodes; fr depressed-obovoid, 5–6 mm; 2*n*=24. Upland woods, prairies, and barrens; Que. to Minn., s. to Fla. and Tex. June, July. Var. *americanus*, occurring mostly in and e. of the mts., has the leaves sparsely pilose only along the nerves beneath. Var. *pitcheri* T. & G., occurring w. of the mts., has the lvs pubescent across the surface beneath. Var. *intermedius* T. & G. occurring on the coastal plain from Va. s., differs from var. *americanus* in its very much branched habit, more numerous infls, and small lvs commonly 2–4 cm.

    2. Ceanothus herbaceus Raf. Prairie-redroot. Bushy shrub up to 1 m; lvs typically oblong to elliptic, varying to lance-oblong or oblanceolate-oblong, 2–6 × 1–2 cm, obtuse or subacute, the lateral nerves never naked and often arising unevenly 1–3 mm above the base of the lf; panicles several to many, terminating the leafy branches of the season, on peduncles rarely to 5 cm, hemispheric to short-ovoid, the component umbels closely set; fr 4–5 mm. Sandy or rocky soil, prairies, and plains; a few stations in Vt., N.Y., and Que.; Mich. to Minn. and e. Mont., s. to nw. Ind., Ark., and Tex. May, June. (*c. ovatus; C. pubescens*.)

    3. Ceanothus sanguineus Pursh. Wild lilac. Shrub 1–3 m; lvs ovate-elliptic, 4–7 cm, rounded at both ends; infls borne on branches of the previous year, ascending, 8–15 cm, leafless, floriferous for half their length or more, the component umbels separated by internodes to 2 cm. Rocky hills, Keweenaw Co., Mich.; Mont. to B.C. and Calif. June.

2. RHAMNUS L. Buckthorn. Fls 4–5-merous, perfect or unisexual, the staminate with vestigial pistil, the pistillate with shorter or smaller stamens; pet lacking (in one sp.) or present, small, often obcordate with the sides folded about the stamen; ovary 2–4-locular; fr a drupe with 2–4 1-seeded stones; shrubs or trees with alternate or opposite, pinnately veined, usually toothed lvs and green or greenish-white fls solitary or umbellate in the axils of a few of the lower lvs of the current season. 100, widespread.

1 Fls perfect, 5-merous; style undivided; winter-buds naked, hairy; germination hypogeal. (*Frangula*.)
  2 Lvs entire except for a few marginal glands above, commonly more than half as wide as long;
    pedicels and hypanthium glabrous or nearly so .... 1. *R. frangula*.
  2 Lvs minutely serrulate, commonly a third as wide as long; pedicels puberulent .... 2. *R. caroliniana*.
1 Fls functionally unisexual, 4- or 5-merous; style cleft a third its length, or more; winter-buds scaly;
  germination epigeal. (*Rhamnus*.)
  3 Sep and stamens each 5; pet none; stones of the drupe 3 .... 3. *R. alnifolia*.
  3 Sep, pet, and stamens each 4; stones of the drupe mostly 2 or 4 (not 3).
    4 Lvs mostly at least half as wide as long, with (2)3(4) lateral veins on each side .... 6. *R. cathartica*.
    4 Lvs narrower, or with more numerous veins, or usually both.
      5 Lvs alternate; petioles mostly less than 1 cm; native .... 4. *R. lanceolata*.
      5 Many of the lvs opposite or subopposite; petioles mostly 1–2 cm; locally intr. .... 5. *R. citrifolia*.

    1. Rhamnus frangula L. European alder-b. Shrub to 7 m; lvs all or mostly alternate, oblong or usually obovate-oblong, 5–8 cm, commonly more than half as wide, acute to abruptly short-acuminate, entire or with a few marginal glands near the tip; umbels sessile, with 2–8 fls, appearing after the lvs; pedicels glabrous or subglabrous, mostly unequal, 3–10 mm; fls perfect, 5-merous; pet broadly obovate, scarcely clawed, cleft at the tip, 1–1.4 mm; styles united; fr red, turning nearly black, with 2–3 stones; $2n=20$–26. Native of Eurasia, escaped from cult. from Que. and N.S. to N.J., Ind., and Ky., especially in wet soil. May, June. (*Frangula alnus.*)

2. **Rhamnus caroliniana** Walter.   Carolina b. Usually a tall shrub, occasionally a tree to 10 m; lvs alternate, oblong or elliptic, 6–12 cm, a third (to half) as wide, usually widest above the middle, acute or subacuminate, broadly obtuse to a rounded base, minutely and often remotely serrulate; umbels sessile or on peduncles to 1 cm, appearing after the lvs; pedicels 3–6 mm, evidently puberulent at anthesis, as also the hypanthium; fls (1–)3–8, perfect, 5-merous; pet broadly obcordate, 1–1.3 mm; styles united; fr red, turning black, with 3 stones. Moist woods and alluvial soil, chiefly on the coastal plain, but also inland; se. Va. to s. O. and s. Mo., s. to Fla. and Tex. May, June. Var. *caroliniana,* chiefly e. of the Ozark region, has the lvs villosulous beneath when young, but soon glabrate. Var. *mollis* Fern., with the lvs persistently soft-pubescent beneath, is Ozarkian, extending w. and s. to Tex. and Miss., and e. to s. Ind.

3. **Rhamnus alnifolia** L'Hér.   American alder-b. Shrub to 1(1.5) m; lvs alternate, lance-oblong to elliptic or lance-ovate, obtuse to acuminate, crenate-serrate; petioles 5–15 mm; umbels sessile, 1–3-fld; fls appearing with the lvs, 5-merous; hypanthium saucer-shaped; sep 1.5–2 mm; pet none; styles 1 mm, connate a third their length; stamens very short in the pistillate fls; fr black, 6–8 mm; stones 3. Swamps and bogs; Nf. and Que. to B.C., s. to N.J., Pa., O., n Ind., n. Ill., Io., and Calif.; also in Tenn.

4. **Rhamnus lanceolata** Pursh.   Lance-lvd b. Shrub 1–2 m; lvs alternate, lanceolate to elliptic, 3–8 cm at maturity, usually less than half as wide, short-acuminate, finely serrulate with incurved teeth, the somewhat upcurved lateral veins several (generally more than 4) on each side; petiole mostly 4–10 mm; fls appearing with the lvs, 4-merous, the staminate mostly 2–3 per axil, the pistillate mostly solitary; pet broadly obcordate, 1–1.2 mm in the staminate fls, 0.5–0.8 mm in the pistillate; styles 2, connate two-thirds their length; fr black; stones 2. Moist, especially calcareous soil. Var. *lanceolata,* with the lvs and young branches densely and softly pubescent at anthesis, the lvs permanently pubescent beneath, occurs in mt. woods from s. Pa. to Ala. Var. *glabrata* Gleason, with the lvs and young branches at anthesis nearly or quite glabrous, occurs from O. and Ky. to S.D., Kans., and Ark.

5. **Rhamnus citrifolia** (Weston) W. Hess & Stearn.   Much like no. 4, but the lvs, or many of them, opposite or subopposite, on petioles mostly 1–2 cm; pistillate fls 1–3 per axil. Native of ne. Asia, sparingly intr. into our range. (*R. davurica.*)

6. **Rhamnus cathartica** L.   Common b. Functionally dioecious shrub or small tree to 6 m, some of the branches usually ending in short thorns; lvs mostly opposite or subopposite, or some of them alternate, broadly elliptic, oblong, or elliptic-obovate, 3–6 cm, at least half as wide, the lateral veins (2)3(4) on each side, strongly upcurved; petiole a third to two-thirds as long as the blade; fls appearing with the lvs, 4-merous; sep 2–3 mm; pet erect, lanceolate, 1–1.3 mm in staminate fls, 0.6 mm in the pistillate; style 4-fid half its length; fr black, 5–6 mm thick, commonly with 4 stones; $2n=24$. Native of Eurasia, escaped from cult. at many places in our range.

3. **BERCHEMIA** Necker, nom. conserv.   Supple-jack. Fls perfect, 5-merous, perigynous with a large annular disk; sep triangular; pet oblong, acute, as long as but narrower than the sep, their margins involute around the filaments; ovary 2-locular; style short, undivided; fr a drupe with thin, leathery exocarp and bilocular stone; twining woody vines with alternate lvs and small greenish-white fls in terminal and axillary panicles. 20+, mostly Old World trop.

1. **Berchemia scandens** (Hill) K. Koch.   Stems high-climbing, tough and flexible; lvs petioled, ovate, oval, or oblong, 3–6 cm, acute to rounded, entire or undulate, glabrous, with several conspicuous parallel ascending veins; panicles 1–5 cm; sep and pet 2 mm; fr oval or oblong, slightly compressed, 6–8 mm. Swamps and wet woods, chiefly on the coastal plain; se. Va. to Fla. and Tex., n. in the interior to s. Ill. May.

FAMILY **VITACEAE**, the Grape Family

Fls regular, hypogynous, perfect or unisexual, mostly 4–5-merous; cal small, often scarcely lobed, or nearly obsolete; pet small, valvate in bud; stamens as many as and opposite the pet; ovary often somewhat sunken in the nectary disk, bilocular, with 2 erect, apotropous, nearly basal ovules per locule; style simple, with a discoid or capitate stigma; seeds with a conspicuous abaxial chalazal knot and 2 deep adaxial grooves; embryo small, with 2 cotyledons, embedded in the copious endosperm; mostly woody vines with lf-opposed tendrils and infls, the lvs stipulate, simple or compound, often palmately lobed or veined. 11/700, mainly trop. and subtrop.

1 Pet separate and spreading at anthesis; lvs simple or compound.
2 Disk cup-shaped with a free margin; lvs simple or bipinnate .... 1. *Ampelopsis.*

2 Disk wholly adnate to the ovary; lvs simple or with 3–5(7) lfls .... 2. *Parthenocissus.*
1 Pet connate above, the cor deciduous in one piece before anthesis; lvs simple .... 3. *Vitis.*

1. **AMPELOPSIS** Michx.   Fls 5-merous, perfect or unisexual; cal saucer-shaped, scarious, shallowly lobed; pet separate and spreading at anthesis; stamens erect and well developed in staminate fls, shorter and probably functionless in the pistillate; pistil surrounded at base by a cup-shaped, entire or lobulate disk, reduced and sterile in staminate fls; fr a berry with thin flesh and 2–4 seeds; shrubs or woody vines, with or without tendrils opposite the lvs, and with small greenish fls in cymose clusters. 20, N. Amer. and Asia.

1 Lvs simple, sometimes lobed.
    2 Lvs merely toothed, or some of them slightly lobed; young twigs glabrous .... 1. *A. cordata.*
    2 Lvs, or some of them, evidently lobed; young twigs hairy .... 2. *A. brevipedunculata.*
1 Lvs bipinnately compound .... 3. *A. arborea.*

1. Ampelopsis cordata Michx.   Raccoon-grape. High-climbing vine; tendrils few or none on the flowering branches; lvs deltoid-ovate, 6–12 cm, sharply serrate-dentate, broadly truncate to shallowly cordate at base, glabrous or nearly so except at the summit of the long petiole, the terminal tooth acuminate; infls long-peduncled, arising opposite the lvs, repeatedly forked, glabrous; fls 4 mm wide, the pet glabrous; fr bluish, 7–10 mm; 2*n*=40. Alluvial woods, chiefly on or near the coastal plain; s. O. to Io. and e. Neb., s. to S.C. and Fla. June. (*Cissus ampelopsis.*)

2. Ampelopsis brevipedunculata (Maxim.) Trautv.   Porcelain-berry. Much like no. 1, but more pubescent, the young twigs, the petioles, and the lower side of the lvs (at least the main veins) rather coarsely ± hairy; lvs, or some of them, evidently (not necessarily very deeply) 3-lobed, varying to deeply 5-lobed with some of the lobes again cleft; fr hard, bright blue to marbled or white; 2*n*=20. Native of ne. Asia, cult. in the U.S. and now well established as an escape from N. Engl. to N.C., w. to Ont. and Mich. July, Aug.

3. Ampelopsis arborea (L.) Koehne.   Pepper-vine. High-climbing vine but often also bushy, with few or no tendrils; lvs bipinnate or partly tripinnate, 1–2 dm, broadly triangular, the lfls ovate to rhombic, 2–5 cm, on short petiolules or subsessile, deeply toothed with a few large triangular teeth; infls much shorter than the lvs, each of the principal forks with a compact, subumbellate cluster of fls; fr dark purple to black; 2*n*=40. Wet or alluvial soil, chiefly on or near the coastal plain; se. Va. to Fla., Tex., and n. Mex., and n. in the interior to s. Ill. June, July. (*Cissus a.*)

2. **PARTHENOCISSUS** Planchon, nom. conserv.   Fls 5-merous, perfect or unisexual; disk wholly adnate to the ovary; pet separate and spreading at anthesis; stamens short, erect; berries with thin flesh and 1–4 seeds; woody vines, trailing or climbing by tendrils, with palmately lobed or compound lvs and small fls in infls borne opposite the lvs or aggregated into terminal clusters. 15, N. Amer. and Asia.

1 Lvs all palmately compound, mostly 5-foliolate.
    2 Tendrils many-branched, with numerous adhesive disks; infl with a central axis .... 1. *P. quinquefolia.*
    2 Tendrils few-branched, without adhesive disks; infl dichotomous .... 2. *P. vitacea.*
1 Most of the lvs simple and 3-lobed .... 3. *P. tricuspidata.*

1. Parthenocissus quinquefolia (L.) Planchon.   Virginia-creeper; woodbine. High climbing vine, adhering to its support by numerous adhesive disks developing at the ends of the much branched tendrils; lvs long-petioled, the lfls dull green above, paler and slightly glaucous beneath, elliptic to obovate, 6–12 cm, abruptly acuminate, sharply serrate chiefly beyond the middle, cuneate to the base, sessile or on petiolules to 15 mm; infls terminal and from the upper axils, forming panicles usually longer than wide, with a well marked zigzag sympodial axis and divergent branches, the fls in terminal unbelliform clusters; frs nearly black, 6 mm; 2*n*=40. Moist soil; Me. to O., Io., and Nebr., s. to Fla. and Tex. June. (*Psedera q.; P. hirsuta.*)

2. Parthenocissus vitacea (Knerr) A. Hitchc.   Grape-woodbine. Much like no 1, tendrils few-branched, almost invariably without adhesive disks; lfls more strongly toothed, glossy-green above, glabrous to thinly pubescent beneath; infl forked at the summit of the peduncle, the 2(3) branches divergent, each producing a broad, rounded cluster of fls that may be again dichotomous; 2*n*=40. Moist soil; Que. and n. N. Engl. to N.J., w. to Man. and Ind., thence to sw. Wyo., s. Utah, Ariz., and Tex. June. (*P. inserta,* perhaps misapplied; *Psedera vitacea.*)

3. Parthenocissus tricuspidata (Siebold & Zucc.) Planchon.   Boston-Ivy. High-climbing and freely branched; tendrils with preformed adhesive disks; lvs mostly or all cordate and 3-lobed, but often a few of them trifoliolate. Native of China and Japan, locally escaped from cult. here and there in our range.

3. **VITIS** L.   Grape.   Fls 5-merous, actually or functionally unisexual, the plants subdioecious; cal obsolete or nearly so; pet separate at the base, but cohering at the summit

and early deciduous; sterile fls with 5 erect stamens on long filaments and a vestigial pistil; fertile fls with a well developed pistil and 5 short, reflexed, functionless stamens; a 5-lobed nectary disk present at the base of the ovary; fr a juicy berry; seeds ovoid, 4, or fewer by abortion; simple-lvd woody vines with (rarely without) lf-opposed tendrils, the fls mostly cymose-paniculate, with the pedicels very often umbellately clustered. 60, N. Hemisphere.

1 Pith continuous through the nodes; bark tight, not shredding .... 1. *V. rotundifolia.*
1 Pith interrupted by a diaphragm at each node; bark becoming shreddy.
  2 Lvs pubescent beneath (at least when young) with long cobwebby hairs that tend to lie parallel to
    the surface, sometimes completely concealing it; simple straight hairs may also occur on the
    veins.
    3 Tendrils and/or panicles occasionally or regularly produced from 3 or more successive nodes.
      4 Lvs permanently densely tomentose beneath, the surface concealed .... 2. *V. labrusca.*
      4 Lvs at maturity merely pubescent or slightly cobwebby on the veins beneath .... 3. *V. novae-angliae.*
    3 Tendrils and/or panicles not produced at more than 2 successive nodes.
      5 Petioles and branches of the season, at anthesis or later, glabrous or very sparsely pilose .... 4. *V. aestivalis.*
      5 Petioles and branches of the season persistently pubescent.
        6 Hairs of the petioles and twigs very short (to 0.5 mm), numerous and spreading .... 5. *V. cinerea.*
        6 Hairs of the petioles and twigs cobwebby and ± appressed .... 6. *V. baileyana.*
  2 Lvs pubescent beneath when young, especially on the veins, with short, rather straight and ±
    spreading hairs, glabrous at maturity except in the vein-axils.
    7 Tendrils or panicles produced opposite most lvs, commonly lacking opposite each third lf; lvs
      cordate to ovate or triangular.
      8 Branches of the current season green to gray or brown.
        9 Diaphragm interrupting the pith at the lowest nodes of the flowering branches about 2 mm
          thick; berries black, not glaucous .... 7. *V. vulpina.*
        9 Diaphragm up to 1 mm thick; berries distinctly glaucous .... 8. *V. riparia.*
      8 Branches of the current season bright purplish-red .... 9. *V. palmata.*
    7 Tendrils none, or produced only opposite the uppermost lvs; lvs mostly reniform .... 10. *V. rupestris.*

    1. Vitis rotundifolia Michx. Muscadine-g.; scuppernong-g. High-climbing vine with tight bark, the young branches with many lenticels; pith continuous through the nodes; lvs firm, glossy on both sides, rotund to cordate-ovate, 6–12 cm, often somewhat wider, rarely slightly lobed, coarsely and irregularly serrate, sub-glabrous at maturity except for tufts of hair in the vein-axils, the basal sinus 90° or wider, with entire margins; panicles short (2–5 cm), densely fld; fr few in a subglobose cluster, 1–2 cm, thick-skinned; $2n=40$. Moist soil; Del. to Ky. and Tex.

    2. Vitis labrusca L. Fox-grape. High-climbing vine; pith interrupted at the nodes by a diaphragm; tendrils or infls from 3 or more successive nodes; lvs firm, round-cordate, 1–2 dm, usually shallowly 3-lobed, shallowly serrate, the lower surface persistently covered by a dense tomentum, this reddish or rusty when young, sometimes later fading to ashy-gray; peduncles and twigs eventually glabrous; panicles ovoid, 4–8 cm; fr dark red to nearly black, 1–2 cm; $2n=38$. Woods, roadsides, and thickets; Me. to s. Mich., s. to S.C. and Tenn. Ancestor of the Concord and many other cult. grapes, including many hybrids, these collectively called *V. labruscana* L. H. Bailey and occasionally found as waifs.

    3. Vitis novae-angliae Fern. New-England g. In some ways intermediate between nos. 2 and 8; tendrils and/or infls often produced from 3 successive nodes; young lvs distinctly tomentose, becoming nearly glabrous in age, the teeth comparatively short and wide; frs black, very glaucous, 12–17 mm. Me. to Vt., Conn. and N.J. A doubtful sp., most abundant in the region where *V. labrusca* and *V. riparia* are the only other native grapes.

    4. Vitis aestivalis Michx. Summer-g.; pigeon-g. High-climbing vine, the pith interrupted by nodal dia-phragms; twigs terete; tendrils or infls lacking opposite each 3rd lf; lvs broadly cordate-ovate to subrotund, with narrow (45°) to broad basal sinus, usually shallowly to deeply 3–5-lobed, when young covered with a reddish or rusty cobwebby tomentum on both sides, sometimes also with straight hairs on the veins, glaucous and persistently ± floccose-tomentose beneath; petioles and stems glabrous or sparsely pilose after disap-pearance of the tomentum; infl usually long (5–15 cm) and slender; fr dark purple or black, 5–10 mm; $2n=38$. Moist or dry soil, open forests, roadsides, and thickets; Mass. to Ont. and s. Minn., s. to Ala. In var. *aestivalis,* occurring over most of the range of the sp., the persistent tomentum tends to obscure the glaucescence of the lower surface of the lvs. In var. *argentifolia* (Munson) Fern. (*V. bicolor*), occurring chiefly inland and in the n. part of the range, the tomentum is more completely deciduous, leaving the lower surface of the lf distinctly blue-green or silvery.

    5. Vitis cinerea Engelm. Graybark-g.; pigeon-g. High-climbing vine, the pith interrupted by nodal dia-phragms; twigs angular; tendrils or infls lacking opposite each 3rd lf; lvs subrotund to broadly ovate, 1–2 dm, shallowly 3-lobed or lobeless, irregularly serrate, when young densely covered beneath and partly above with cobwebby tomentum, as also the petioles and branches, the tomentum largely disappearing above but persistent and floccose beneath; twigs and petioles, after disappearance of the tomentum, densely short-hairy; infls 1–2 dm, slender; fr nearly black, 4–6 mm; $2n=38$. Moist alluvial soil. Var. *cinerea,* with the tomentum of the lvs ashy-gray, occurs from c. Ind. to s. Wis. and e. Kans., s. to Fla. and Tex., and has been reported from se. Va. Var. *floridana* Munson, with rufescent tomentum, occurs from se. Va. to Fla. (*V. simpsonii.*)

6. **Vitis baileyana** Munson. Possum-g. High-climbing vine, the angled branches at first cobwebby or tomentose, later glabrescent; lvs mostly small, 6–10 cm, lobeless or with 3 short lobes, thinly cobwebby-pubescent, especially along the veins; frs in close, dense clusters, black, 4–7 mm. Woods and thickets; w. V. to e. Ky., s. to Ga. and Ala. A problematical sp., combining the general lf-characters of no. 7 with the pubescence of no. 4.

7. **Vitis vulpina** L. Frost-g.; winter-g.; chicken-g. High-climbing vine, the pith interrupted by nodal diaphragms 2–5 mm thick; twigs terete; lvs rotund in outline, 5–15 cm, lobeless or slightly 3-lobed (the lobes ± divergent), coarsely serrate (the teeth mostly wider than high, with one or both sides convex), pubescence beneath when young and persistently so in the vein-axils; infls long and slender, loosely fld, 10–15 cm; fr black, not glaucous, 5–10 nm; raphe carried over the end of the seed as a distinct raised ridge; $2n=38$. Woods and thickets; N.J. to s. Wis. and Neb., s. to Fla. and Tex. (*V. cordifolia.*)

8. **Vitis riparia** Michx. Frost-g.; river-bank g. High-climbing vine, the pith interrupted by diaphragms 0.5–1 (or at maturity 2) mm thick; lvs rotund, 1–2 dm, usually 3-lobed (the lobes mostly pointing forward) coarsely and sharply serrate (the teeth commonly longer than wide, with one or both sides concave), pubescent beneath when young, permanently so in the vein-axils and sometimes along the veins; panicles 5–10 cm; fr black, heavily glaucous, 6–12 mm; $2n=38$. Moist woods, roadsides, and thickets; N.B. and Que. to Mont., s. to Va., Tenn., Ark., and Tex. (*V. vulpina,* misapplied.) Perhaps properly to be submerged in *V. vulpina.*

9. **Vitis palmata** M. Vahl. Red, cat-, or catbird-g. High-climbing vine with bright purplish-red branches; lvs triangular-ovate to subrotund, 8–14 cm, 3–5-lobed, often to below the middle, sparsely pubescent beneath on the veins and in the vein-axils, the terminal lobe in 3-lobed lvs triangular and prolonged to an acuminate tip, in 5-lobed lvs constricted at base to rounded sinuses; infls 8–15 cm; fr black, slightly or not at all glaucous. Swamps; s. Ind., c. Ill., and Mo. to La. and Tex.

10. **Vitis rupestris** Scheele. Sand- or sugar-g. Rarely climbing, the stems prostrate or ascending or merely reclining on bushes; tendrils commonly none, or a few produced opposite the uppermost lvs only; lvs firm, paler beneath, often ± folded, reniform to depressed-ovate, 5–10 cm, somewhat wider, unlobed, coarsely toothed, with a broad basal sinus, glabrous on both sides or sparsely hairy along the larger veins; infls small, rarely to 10 cm; fr 6–10 mm, black; $2n=38$. Dry hills and rocks; s. Ill. to Okla., s. to Tenn., Ark., and Tex.; also se. Pa., where perhaps only escaped.

FAMILY **LINACEAE**, the Flax Family

Fls regular, hypogynous, perfect, (4)5-merous; sep imbricate, distinct or nearly so; pet convolute, distinct, clawed, usually caducous; small nectary-glands commonly present external to the stamens; stamens as many as and alternate with the pet, sometimes alternating also with filamentous or tooth-like staminodes; filaments expanded below and connate into a tube; carpels (2)3–5, united to form a compound, plurilocular ovary; styles distinct or connate below; ovules 2 in each primary locule, apical-axile, pendulous, collateral, commonly separated by an often incomplete "false" septum; fr a septicidal capsule; seeds with a straight, dicotyledonous embryo and scanty endosperm; mostly (incl. all our spp.) herbs, with simple, entire, typically narrow and sessile lvs and basically cymose infls. 6/220, cosmop.

1 Fls 5-merous; sep entire or merely ciliate or toothed along the margins .... 1. *Linum.*
1 Fls 4-merous; sep 3-lobed across the broad summit .... 2. *Radiola.*

1. **LINUM** L.   Flax. Fls 5-merous throughout; sep lanceolate to ovate, 1–5-nerved, often ciliate; stamens in some spp. alternating with minute, tooth-like staminodes; capsule partly or completely 10-locellar; herbs with rather small, sessile, often narrow lvs and usually several to many fls in an evidently cymose to racemiform or paniculiform infl. 200, cosmop. (*Cathartolinum.*)

1 Pet blue; fr 5–10 mm thick.
  2 Perennial; sep entire; lvs 1-nerved, or obscurely 3-nerved at base .... 1. *L. perenne.*
  2 Annual; inner sep ciliolate; lvs 3-nerved more than half-length .... 2. *L. usitatissimum.*
1 Pet yellow or white; fr 2–5 mm thick.
  3 Pet white; staminodes present; annual .... 3. *L. catharticum.*
  3 Pet yellow; staminodes wanting; annual or perennial.
    4 Styles partly united; annual.
     5 Styles united only near the base; sep persistent .... 4. *L. sulcatum.*
     5 Styles united to above the middle; sep deciduous in fr .... 5. *L. rigidum.*
    4 Styles wholly distinct; normally perennial.
     6 Fr turbinate or pyriform or ovoid or ovoid-globose, narrowed to the summit.

```
7 False septa clearly incomplete; septa ciliate ......................................... 6. L. intercursum.
7 False septa virtually complete; septa not ciliate ..................................... 7. L. floridanum.
6 Fr depressed-globose.
  8 Margins of the inner sep with some prominent, stalked glands ........................... 8. L. medium.
  8 Margins of the sep glandless or with tiny sessile glands.
    9 Branches terete or nearly so; infl broad and ± corymbiform ...................... 9. L. virginianum.
    9 Branches striate-angled; infl ± elongate and paniculiform ........................... 10. L. striatum.
```

1. Linum perenne L.   Wild blue f. Perennial, 3–7 dm, the stems mostly clustered; lvs erect, very numerous, linear, 1–3 cm, 1-nerved, or obscurely 3-nerved at base; fls heterostylic; sep ovate, 5–7 mm, obtuse and minutely mucronate, entire; pet blue, 12–23 mm; staminodes subulate or tooth-like, 1 mm; stigmas ellipsoid-capitate; fr 5–9 mm, ovoid-globose, scarcely beaked, the 10 segments acute or shortly acuminate; false septa incomplete, long-ciliate; $2n=18$. Native of Europe, occasionally intr. in disturbed habitats in our range, as in Wis. and Mich., where it has been mistaken for the closely allied homostylic cordilleran sp. *L. lewisii* Pursh. May–July.

2. Linum usitatissimum L.   Common f. Annual; stems 1–several, erect, to 1 m; lvs lance-linear, 3-nerved, 1.5–3.5 cm; sep 7–9 at maturity, the inner minutely ciliate on the scarious margin; pet blue, 10–15 mm; staminodes minute, tooth-like; stigmas linear-clavate; fr 6–10 mm, subglobose and abruptly short-beaked, tending to be only tardily and incompletely dehiscent, the mericarps acuminate; false septa very incomplete, long-ciliate; $2n=30$. European cultigen, often escaped or adventive in fields and roadsides in U.S. and s. Can. Summer. (*L. humile.*)

3. Linum catharticum L.   White f. Glabrous annual 1–3 dm, simple below, branched above into a dichasial cyme; lvs all or mostly opposite, oblong to obovate, 3–15 mm; pedicels very slender, terete, the lowest 10–25 mm, the upper shorter; sep glandular-ciliate; pet white with a yellow claw, 4–8 mm; staminodia tiny, triangular; fr ovoid-globose, 2–2.5 mm, fragile; false septa incomplete, long-ciliate; $2n=16$. Native of Europe; intr. in disturbed sites from Nf. and N.S. to N.J. and Pa., w. to n. Mich. July–Sept. (*L. pratense; Cathartolinum c.*)

4. Linum sulcatum Riddell.   Grooved yellow f. Glabrous annual 2–8 dm, branched above, the paniculiform infl with ascending, racemiform branches; pedicels up to 3(–5) mm; lvs narrow, all or mainly alternate, 1–2.5 cm, with a pair of minute dark glands at base; sep long-acuminate, 3–6 mm, all conspicuously glandular-toothed, persistent; pet yellow, 5–10 mm; styles 2–4.5 mm, united for 0.3–1.5 mm at base; fr ovoid-globose, rounded to acute at the summit, 2.5–3.5 mm, the sharp-pointed mericarps cohering basally in pairs; false septa incomplete, long-ciliate; $2n=30$. Dry, often sandy soil, prairies, and upland woods; Mass., Vt., and s. Que. to Man., s. to Ga. and Tex., more common westward. June, July. (*Cathartolinum s.*)

5. Linum rigidum Pursh.   Stiffstem yellow f. Annual 2–5 dm, usually simple below and with ascending, racemiform branches above, the pedicels up to about 1 cm, or longer in fr; stem sparsely puberulent or scabrous below, otherwise glabrous; lvs narrow, alternate, 1–3 cm, sometimes with a pair of minute dark glands at base; sep 4.5–7 mm, short-awned, all glandular-toothed, deciduous as the fr matures; pet yellow, 10–17 mm; styles 5–9 mm, united to above the middle; fr ovoid, 4–5 mm, dehiscing through the essentially complete, non-ciliate false septa into 5 2-seeded segments, the members of a pair united at base by a triangular cartilaginous plate, the true septa not separating; $2n=30$. Prairies and plains; Minn. to Mo. and Tex., w. to Alta., Colo., and N.M. Ours is var. *rigidum.* (*Cathartolinum r.*)

6. Linum intercursum E. Bickn.   Bicknell's yellow f. Glabrous perennial 2–8 dm, simple below the infl or few-branched at base; lvs mostly 25–50 below the infl, narrowly elliptic to oblanceolate, 1–2.5 cm × 1.5–5 mm, opposite below, usually alternate above; infl of slender, rather stiffly ascending-spreading branches; pedicels 1–4 mm; sep lanceolate, acute, 2–3 mm, the inner usually (and the outer seldom) sparsely glandular-toothed; pet yellow, 4–7 mm; fr turbinate, ± pointed, 2–3 mm, readily dehiscing into 10 sharp-pointed mericarps; exposed portion of the fr anthocyanic; false septa evidently incomplete, glabrous; septa sparsely but evidently ciliate; $2n=36$. Open oak or pine woods and open places on the coastal plain from e. Mass. to N.C., thence inland to Tenn. and Ala.; isolated in nw. Ind. June, July. (*Cathartolinum i.*)

7. Linum floridanum (Planchon) Trel.   Coastal plain yellow f. Much like no. 6, but with more numerous (mostly 50–150) lvs below the infl, these narrower, linear-oblanceolate or oblanceolate; infl mostly smaller; pet 6–10 mm; fr ovoid or ovoid-globose or pyriform, scarcely pointed; false septa nearly complete, glabrous, as also the true septa; $2n=36$. Moist or dry woods and pine-barrens on the coastal plain; se. Va. to Fla. and La., n. in the Mississippi Valley to s. Ill.; isolated near D.C. June, July. (*Cathartolinum f.*)

8. Linum medium (Planchon) Britton.   Common yellow f. Glabrous perennial (rarely annual) 2–7 dm, simple below the infl; lvs mostly 20–70(–100) below the infl, narrowly lanceolate to oblanceolate, the lower (or nearly all) opposite, the larger ones 1–2.5 cm × 2–5 mm, gradually reduced above; infl with ± elongate, rather stiffly ascending-spreading branches, the pedicels up to 3(–5) mm; sep lanceolate, acute, the outer entire, 2–3.5 mm, the inner somewhat shorter, broader, and evidently gland-toothed; pet yellow, 4–8 mm; fr ± persistent, depressed-globose, 2 mm, commonly suffused with purple above; false septa virtually complete, glabrous, as also the septa. Dry upland woods and beaches, s. Me. to Fla., w. to s. Ont., Mich., Ill., Io., e. Kans., and e. Tex. June, July. Var. *medium*, found about Lake Erie and w. Lake Ontario, in s. Ont. and rarely adj. U.S., is tetraploid ($2n=72$) and has relatively thick, blunt lvs 3–5 mm wide; the fr tends to split at the summit only. Var. *texanum* (Planchon) Fern., occupying the rest of the range of the species, is diploid and has thinner, more pointed, narrower lvs mostly 1.5–3.5 mm wide; the fr splits completely into 10 deciduous mericarps. (*Cathartolinum m.*)

9. Linum virginianum L.   Virginia yellow f. Glabrous perennial 2–7 dm; stems 1–several from the base, simple below the infl; lvs mostly 10–40 below the infl, elliptic to oblanceolate or obovate, 1.5–2.5 cm × 3–6 mm, the lower opposite, the upper often alternate; stem with a single narrow wing decurrent from the base of each lf but not extending to the lf below; infl open-corymbiform, typically with a few rather long, slender, ascending-spreading branches; pedicels 2–8(–10) mm, terete or nearly so; sep lance-ovate, the outer acute, entire, 2–4 mm, the inner broader, often with a few tiny sessile glands along the margins; pet yellow, 3.5–5.5 mm; fr depressed-globose, scarcely 2 mm, readily splitting into 10 mericarps, shattering and falling; false septa nearly complete, glabrous; true septa usually sparsely ciliate; $2n=36$. Upland woods; Mass. and s. Mich. to s. Mich., n. Ill., and Mo., s. to Ga. and Ala., southward mostly avoiding the coastal plain. June, July. (*Cathartolinum v.*)

10. Linum striatum Walter.   Ridgestem yellow f. Glabrous perennial 3–9 dm; stems 1–several from the base, erect-ascending, simple below the infl, prominently ridged-striate above; lvs mostly 25–50 below the infl, thin, elliptic to oblanceolate or obovate, the larger ones 1.5–3.5 cm × 4–10 mm, the lower opposite, the upper often alternate; infl ± elongate-paniculiform, with slender, spreading branches; pedicels sharply angular, 1–4 mm; sep lanceolate to ovate, acute, 1.5–3 mm, entire or the inner with a few tiny marginal glands; pet yellow, 3–5 mm; fr depressed-globose, scarcely 2 mm, pale, splitting freely into 10 mericarps, shattering and falling; false septa nearly complete, glabrous, as also the true septa; $2n=36$. Damp or wet woods, swamps, and bogs; Mass. to n. Fla., w. to Mich., se. Mo., and e. Tex. June, July. (*Cathartolinum s.*)

2. RADIOLA Hill.   Much like *Linum,* differing mainly in its 4-merous fls, apically cleft sep, and diminutive, freely branching habit. Monotypic.

1. Radiola linoides Roth.   Diminutive annual 3–10 cm, dichotomously or dichasially branched; lvs opposite, ovate to oblong, 2–3 mm; sep 1 mm, 3-cleft ca ¼ or ⅓ length across the broad summit, the lobes narrow, pointed; pet white, 1 mm; fr globose, 1 mm, soon splitting into 4 valves each partly divided by an incomplete false septum; $2n=18$. Native of Eurasia and n. Afr., established in N.S. July, Aug. (*Millegrana radiola.*)

FAMILY **POLYGALACEAE**, the Milkwort Family

Fls perfect, mostly hypogynous, ± strongly irregular; sep typically 5, the 2 inner often much the larger; pet 3(–5), typically adnate to the filaments to form a common tube; stamens 3–10, commonly 8 in 2 sets of 4; anthers basifixed, usually opening by apical or subapical pores or short slits; intrastaminal nectary-disk sometimes present, or modified into an adaxial gland; ovary superior, 2–5(–8)-locular with a terminal, usually curved, often unequally bilobed style; ovules solitary in each locule, apical-axile, pendulous, epitropous; fr most often a loculicidal capsule; seeds often hairy, and often with an evident micropylar aril; embryo straight, with 2 cotyledons; herbs or woody plants with simple, entire, alternate to seldom opposite or whorled lvs and terminal or axillary spikes, racemes, or panicles of fls. (*Galypola, Pilostaxis.*) 12/750, cosmop.

1. POLYGALA L.   Milkwort. Sep 5, the 3 outer small, the 2 inner (lateral) ones, called wings, much larger and often petaloid; pet 3, all ± united below with the filament-tube, the 2 upper ones similar, the lower one keel-shaped or boat-shaped with a fringe-like crest (in our spp.); stamens (6)8, the filaments monadelphous into a sheath split along the upper side; ovary bilocular; seed usually with an aril; ours herbs with alternate or verticillate lvs and small (except in the first sp.) fls in open to dense and spike-like or head-like racemes. 500, cosmop.

1 Fls large, 13–19 mm; stamens 6; well developed lvs few, mostly near the summit .... 1. *P. paucifolia.*
1 Fls smaller, rarely over 10 mm; stamens 8, lvs well distributed, or mostly near the base.
  2 Fls yellow or orange in life, sep decurrent on the pedicels as narrow wings.
    3 Racemes numerous in terminal compound cymes.
      4 Basal lvs elliptic to spatulate or obovate; wings acuminate; seeds hairy .... 2. *P. ramosa.*
      4 Basal lvs linear or lance-linear; wings abruptly cuspidate; seeds glabrous .... 3. *P. cymosa.*
    3 Racemes few, solitary at the end of leafy branches; lf-shape of no. 2 .... 4. *P. lutea.*
  2 Fls rose-purple to white or greenish, never yellow or orange; sep not decurrent.
    5 Cor 7–10 mm, more than twice as long as the wings; spike very dense .... 5. *P. incarnata.*
    5 Cor less than 5 mm, about equaling to shorter than the wings.
      6 Plants biennial or perennial, the stems arising from a thick, knotty root, or several stems arising from the same base.
        7 Fls white, in dense, spike-like racemes; cleistogamous fls none.
          8 Cauline lvs (3–)5–30 mm wide; basal and lower lvs much reduced .... 6. *P. senega.*

8 Cauline lvs 1–2 mm wide; basal lvs much like the cauline .... 7. *P. alba.*
7 Fls rose-purple to white, in loose racemes; subterranean cleistogamous fls produced .... 8. *P. polygama.*
6 Plants annual, the slender stems solitary from a small taproot.
9 Racemes slender, distinctly tapering to the apex.
10 Racemes elongate, becoming interrupted below through persistence of the frs on the
  lengthening axis; wings about as long as the fr .... 9. *P. ambigua.*
10 Racemes shorter, remaining continuous and appearing truncate at base through prompt
  falling of frs; wings shorter than the fr .... 10. *P. verticillata.*
9 Racemes capitate, ovoid, or short-cylindric, usually blunt at the tip.
11 Lvs all alternate; wings glabrous.
12 Wings about twice as long as the cor .... 11. *P. sanguinea.*
12 Wings about equaling the cor.
13 Bracts deciduous from the axis after anthesis .... 12. *P. mariana.*
13 Bracts long-persistent on the axis.
14 Wings 3–5 mm; pedicels 1.5–2.5 mm; racemes 8–13 mm thick .... 13 *P. curtissii.*
14 Wings 2–2.5 mm; pedicels 0.5–1.5 mm; racemes 5–6 mm thick .... 14. *P. nuttallii.*
11 Lvs all or chiefly whorled; wings minutely roughened on the inner side.
15 Peduncles mostly 2–8 cm; wings 1.5–2.5 mm wide, not cuspidate .... 15. *P. brevifolia.*
15 Peduncles very short, seldom 1 cm; wings 3–4 mm wide, cuspidate .... 16. *P. cruciata.*

1. **Polygala paucifolia** Willd. Flowering wintergreen. Colonial perennial with slender rhizomes emergent from small tubers; stems 8–15 cm, with several scattered scale-lvs 2–8 mm, and near the summit 3–6 elliptic to oval lvs 1.5–4 cm; a few cleistogamous fls usually present at the base; chasmogamous fls 1–4, each on a long pedicel from the very short terminal peduncle, rose-purple to white, 13–19 mm, the cor about equaling the obovate wings; stamens 6; stigma obscurely bilobed; fr suborbicular, 6 mm, notched at the tip; seed with a 3-lobed aril. Moist rich woods; Que. and N.B. to Sask., s. to Conn., N.Y., Wis., and in the mts. to Ga. May, June. (*Triclisperma p.*)

2. **Polygala ramosa** Elliott. Low pine-barren milkwort. Simple or branched annual, 1–4 dm; basal lvs narrowly elliptic to spatulate or obovate, obtuse, 1–3 cm, generally withering after anthesis, the cauline ones progressively narrower, becoming linear-oblong or linear; infl 5–15 cm wide, the racemes numerous, rather loose, 8–10 mm thick; pedicels 1–2 mm; fls yellow, drying dark green; wings acuminate, 3 mm, about equaling the cor; fr suborbicular; seeds finely pubescent; 2*n*=64. Pine-barren swamps on the coastal plain; s. N.J. to Fla. and Tex. June–Aug. (*Pilostaxis r.*)

3. **Polygala cymosa** Walter. Tall pine-barren m. Biennial, 4–11 dm, simple to the infl; basal lvs grass-like, linear or nearly so, acuminate, 3–12 cm, persistent in a rosette, the cauline similar but shorter; infl much branched, 6–20 cm wide, the numerous racemes 1–3, 1 cm thick, pedicels 1–2 mm; fls yellow, drying green; wings abruptly cuspidate, 2–4 mm, surpassing the cor; keel inflexed at the tip; fr didymous; seeds glabrous; 2*n*=64, 68. Acid sandy marshes and wet pine-barrens; Del. to Fla. and La. June–Aug. (*Pilostaxis c.*)

4. **Polygala lutea** L. Orange m. Biennial or perennial; stems often clustered, simple or branched, 1–4 dm; basal lvs oblong-obovate or obovate, 2–5 cm, obtuse or rounded, the cauline progressively narrower, becoming oblanceolate, 1.5–3 cm; stems and branches terminating in a lfless peduncle 3–10 cm, with a dense terminal head-like raceme 1–3 cm × 10–18 mm; fls orange, drying yellow; wings abruptly cuspidate, 5–7 mm, equaling the cor; fr obovate; seeds short-hairy; aril nearly as long as the seed; 2*n*=64, 68. Sandy acid swamps and bogs of the coastal plain; L.I. to Fla. and La. June–Aug. (*Pilostaxis l.*)

5. **Polygala incarnata** L. Pink m. Annual; stems slender, glaucous, simple or sparingly branched, 2–6 dm; lvs alternate, erect or ascending, linear, 5–12 mm; racemes dense, 1–4 cm × 10–15 mm; fls pale rose-purple, 7–10 mm, the cor more than twice as long as the wings, promptly deciduous after anthesis; seeds hairy, 2 mm, with a cellular aril 1 mm. Dry soil, upland woods, barrens, and prairies; L.I. to Mich., Wis., Io., and Kan., s. to Fla. and Tex. June–Aug. (*Galypola i.*)

6. **Polygala senega** L. Seneca-snakeroot. Perennial; stems mostly clustered, 1–5 dm, usually unbranched, minutely puberulent; lvs alternate, the lowest reduced or scale-like, the others lance-linear or wider, 3–8 cm × 3–30 mm; racemes dense, 1.5–4 cm × 5–8 mm, the peduncle 1–3 cm; fls white; wings broadly elliptic, 3–3.5 mm, exceeding the cor; fr suborbicular; seeds hairy, 2–3 mm, the aril nearly or quite as long. Dry or moist woods and prairies, often in calcareous soil; N.B. to Alta., s. to Ga. and Ark. May, June.

7. **Polygala alba** Nutt. White m. Taprooted perennial; stems several to many, usually simple, 2–4 dm, glabrous; lower lvs narrowly oblanceolate, most cauline lvs linear, 1–2 mm wide; racemes long-pedunculate, slender, 3–8 cm × 5–8 mm, tapering to the tip, often interrupted below; fls white or greenish-white; wings elliptic, 2.5–4 mm, somewhat exceeding the cor; fr elliptic, twice as long as wide; seeds 2–2.5 mm, the aril up to half as long; 2*n*=24, ca. 72, 104–108. Prairies, rocky slopes, and barrens; N.D. to Tex., w. to Mont. and Ariz.; rep. from Minn. May–Aug.

8. **Polygala polygama** Walter. Bitter m. Biennial; stems clustered, decumbent, glabrous, 1–2.5 dm, simple at anthesis, later sparingly branched; lowest lvs spatulate to obovate, 1 cm; cauline lvs linear-oblanceolate to oblong-oblanceolate, 1–3 cm × 2–7 mm, obtuse to subacute; racemes loose and open, 2–10 cm; fls rose-purple to white; wings obovate, 4–6 mm, exceeding the cor; cleistogamous fls secund in slender subterranean racemes; 2*n*=56. Dry, usually sandy soil; Me. to Mich. and Minn., s. to N.J., w. Va., O., Ind., and Io.; also along the coastal plain from e. Va. to Fla. and Tex. The northern plants have a slightly denser raceme than the southern and are often distinguished as var. *obtusata* Chodat.

9. **Polygala ambigua** Nutt. Loose m. Slender annual 1–4 dm, branched above; lvs alternate or scattered or the lowest occasionally in whorls of 2–5, linear, 1–2 cm × 1–2(–3) mm; peduncles 2–8 cm; bracts deciduous; racemes very slender, the whole axis to 10 cm, the floriferous or fructiferous part 1–5 cm, becoming interrupted at base; pedicels 0.5 mm; fls white or greenish-white; wings about equaling the fr. Open woods and fields, especially in sandy soil; Me. to n. N.H., Mich., and Mo., s. to Ga., Ala., and Okla. June–Oct. Wings usually 1.3–1.7 mm; plants from S.C. to se. Va., and from Mo., Ark., and Okla., with wings 2–2.6 mm, have been called *P. verticillata* var. *dolichoptera.* (*P. verticillata* var. *a.*)

10. **Polygala verticillata** L. Whorled m. Erect, divergently branched annual, 1–4 dm; lvs linear to linear-oblong, 1–2(–3) cm × 1–3 mm, at least the lower (sometimes all) in whorls of 2–5; lower branches of the infl usually opposite or whorled; peduncle elongate; racemes continuous, conic or cylindric-conic, the floriferous and fructiferous part 6–15 mmm, appearing truncate at base, the whole axis to 4 cm; fls white, greenish, or occasionally pinkish, the lower dropping promptly as the fr matures; wings half to two-thirds as long as the fr; $2n=34$. Moist, sandy soil, grasslands, and woods. July–Oct. Var. *isocycla* Fern., occurring nearly throughout the range of the sp. (Vt. to Man., s. to Fla. and Tex.) has finely pubescent seeds, short pedicels a fourth to a third as long as the fr, greenish-white sep, narrow, dense racemes on peduncles 0.5–4 cm, the plant with widely spreading branches. Var. *verticillata,* occurring from Me. to Mich., s. to Tenn., has hirsute seeds, pedicels a third to half as long as the fr, often purplish sep, wider, looser racemes on peduncles 2–7 cm, and the branches are ascending.

11. **Polygala sanguinea** L. Blood-m. Erect annual 1–4 dm, simple or branched above; lvs linear or narrowly elliptic, 1–4 cm × 1.5 mm; racemes sessile or short-peduncled, very dense, head-like, rounded or short-cylindric, 1 cm thick, the floriferous portion 1–2 cm, the whole axis to 4 cm; fls rose-purple, white, or greenish; wings oval, 3–5 mm, or longer in fr, blunt, with conspicuous midvein; cor half as long as the wings; seed pyriform, the 2 linear lobes of the aril extending beyond the middle. Fields, meadows, and open woods; N.S. to Minn., s. to S.C. and La. July–Sept. (*P. viridescens.*)

12. **Polygala mariana** Miller. Maryland-m. Erect annual, simple or branched above, 1–4 dm; lvs linear, or the lowest narrowly spatulate, 1–2 cm × 1–2 mm; peduncles well developed; bracts falling with the fls; racemes short-ovoid or cylindric, very blunt, to 3 cm but bearing attached fls only in the uppermost 1 cm; fls pink or rose; wings obovate, 2.5–3.5 mm, slightly exceeding the cor; aril a fourth or a third as long as the seed; $2n=34$. Dry or sandy soil on the coastal plain; s. N.J. to Fla. and Tex.; also inland in Tenn. and rep. from Ky. June–Sept. (*P. harperi.*)

13. **Polygala curtissii** A. Gray. Appalachian m. Erect annual, simple or branched above, 1–4 dm; lvs linear to linear-oblong or narrowly oblanceolate, 1–2 cm × 1–4 mm; peduncles well developed; bracts lanceolate, persisting after anthesis on the elongating axis; racemes dense, cylindric, blunt or apiculate (by the young buds), 1–2 cm × 8–13 mm, or the whole axis to 4 cm; fls rose-purple; wings obovate-oblong, tapering to the base, 3–5 mm, a third as wide, exceeding the cor; aril a third as long as the hairy seed; $2n=40$. Dry sandy soil; Del. to O., s. to S.C. and Miss. June–Sept.

14. **Polygala nuttallii** T. & G. Nuttall's m. Much like no. 13, but smaller and more slender, to 3 dm; lvs 5–15 mm, 1 mm wide; racemes 5–6 mm thick; wings only 2–2.5 mm, about equaling the cor; $2n=46$. Dry sandy soil and barrens; e. Mass. to Ga. July–Sept.

15. **Polygala brevifolia** Nutt. Shortleaf-m. Much like no. 16, the stems more slender, with longer branches; racemes elevated on a distinct peduncle, the one terminating the main axis usually 1–3 cm; fls pale rose-purple; wings ovate, 3–3.5 × 1.5–2.5 mm, minutely apiculate or obtuse. Sandy swamps on the coastal plain; N.J. to Fla. and Miss. July–Sept.

16. **Polygala cruciata** L. Drum-heads. Erect annual 1–3 dm, sometimes simple, but usually with a few decussate branches; lvs chiefly in whorls of 3 or 4, linear to oblanceolate or narrowly elliptic, 1–4 cm × 1–5 mm, obtuse or rounded, often apiculate; racemes cylindric, 1–6 × 1–1.5 cm, rounded at the summit, usually sessile or nearly so; fls pale rose-purple or greenish-purple; wings deltoid, 3–4 mm wide at base, 4–6 mm long, terminating in a slender cusp 0.5–1.5 mm; $2n=36, 40$. Damp or wet soil, marshes, pine-barrens, etc., chiefly but not wholly on the coastal plain; Me. to Fla. and Tex., also in Ky. and from O. and Mich. to Ill. and Minn. July–Sept. (*P. ramosior.*)

FAMILY **STAPHYLEACEAE**, The Bladder-nut Family

Fls regular, hypogynous, pentamerous, perfect or seldom unisexual; sep and pet distinct, imbricate; stamens alternate with the pet, seated on or outside the annular nectary-disk, in which the ovary may be basally embedded; ovary mostly 2–3-locular, laterally and sometimes also apically lobed, with distinct styles or a single style (carpels distinct in *Euscaphis*); ovules (1–)6–12 in 2 rows in each locule on axile or basal-axile placentas, apotropous, commonly ascending, often only 1 or 2 per locule maturing; embryo straight, with 2 large, flat cotyledons, embedded in the copious, oily endosperm; shrubs or trees with opposite, pinnately compound or trifoliolate lvs, caducous stipules, and small fls in terminal or axillary drooping panicles or racemes. 5/50, irregularly widespread. The description based on the 3 genera of Staphyleoideae.

1. **STAPHYLEA** L. Bladder-nut. Fls perfect; sep nearly as long as the erect pet; filaments slender, elongate; ovary and fr 3-lobed; styles erect, distinct below, connate above; fr a large, thin-walled, much inflated capsule, dehiscent at the top along the ventral sutures of the free tips of the carpels; shrubs or small trees with drooping terminal panicles of white or greenish-white fls. 10, N. Temp.

1. **Staphylea trifolia** L. Erect shrub to 5 m, with striped bark; lvs long-petioled, 3-foliolate, the lfls oblong to elliptic or ovate, 5–10 cm, serrate, the terminal one long-stalked, the lateral sessile or short-stalked and often oblique at base; infls 4–10 cm (peduncle included); fls long-pedicellate, greenish-white, campanulate, 8–10 mm; fr 3-lobed, 5 cm; $2n=78$. Moist woods and thickets; s. Que. to Minn., s. to Ga., w. Fla., and Okla. May.

FAMILY SAPINDACEAE, the Soapberry Family

Fls hypogynous, regular or more often slightly irregular, perfect or more often functionally unisexual; sep 4 or 5, usually imbricate; pet mostly 4 or 5 (seldom wanting), distinct, imbricate, often clawed, often with an internal, scale-like appendage near the base; stamens 4–10, often 8, apparently in a single cycle, seated acropetally to (or upon) the annular or unilateral nectary-disk; filaments often hairy; ovary (2)3(–6)-carpellate, generally plurilocular; ovules apotropous, 1 or less often 2 (seldom several) per locule, typically ascending; seeds often arillate; embryo curved, often with plicate or twisted cotyledons, the radicle often folded into a pocket in the seed-coat; endosperm wanting; shrubs or trees or less often herbs or vines, with alternate, pinnately or ternately compound, mostly exstipulate lvs and small fls in mostly cymose-paniculate infls. 140/1500, mostly trop. and subtrop.

1. **CARDIOSPERMUM** L. Balloon-vine. Sep 4, 2 large, 2 small; pet 4, somewhat unequal, each with an erect petaloid appendage at base; nectary-disk unilateral, biglandular; stamens 8; fr a bladdery-inflated, 3-locular and 3-lobed capsule; herbs, climbing by axillary tendrils, with ternate or biternate lvs and small, unisexual, white to red fls in small clusters, the long peduncle bearing tendrils near the fls. 14, mainly trop. Amer.

1. **Cardiospermum halicacabum** L. Much-branched, to 2 m; lfls ovate, incised, cuneate to the base; fls white, 5 mm wide; fr globose to ovoid, 3–4 cm thick; $2n=22$. Native of trop. Amer., escaped from cult. in se. U.S. and occasionally n. to N.J., O., Ill., and Mo. Summer.
Koelreuteria paniculata Laxm., Golden-rain tree, a tree with 7–15 toothed lfls 3–6 cm, fls in panicles 2–4 dm, and inflated, partly 3-locular capsules, occasionally escapes from cult.

FAMILY HIPPOCASTANACEAE, the Horse-chestnut Family

Fls hypogynous, evidently irregular, polygamo-monoecious; sep 5, imbricate, often connate below; pet 4 or 5, distinct, unequal, clawed, imbricate; nectary-disk small, extrastaminal, often unilateral; stamens (5)6–8, distinct, the inner whorl of 5 complete, the outer whorl incomplete; ovary (2)3(4)-locular, with a long terminal style and simple or obscurely lobed stigma; ovules 2 per locule, superposed on axile placentas; fr a loculicidal capsule, often unilocular and 1-seeded by abortion; seeds large, with a very large hilum; embryo massive, curved, one cotyledon generally much larger than the other, the radicle commonly folded into a deep pocket in the hard seed-coat; endosperm wanting; trees or shrubs with opposite, exstipulate, palmately 3–11-foliolate lvs and terminal mixed panicles of rather large and showy fls. 2/16, mainly N. Hemisphere.

1. **AESCULUS** L. Buckeye. Sepals connate at least half their length to form a lobed tube; lvs deciduous, 5–11-foliolate. 13, N. Hemisphere.

1 Pet usually 5, all but the lowermost with cordate-based blade above the claw .... 1. *A. hippocastanum.*
1 Pet 4, at least the 2 upper with tapering (or scarcely developed) blade.
  2 Stamens long-exserted; pet subequal; fr usually spiny .... 2. *A. glabra.*

2 Stamens included or barely exserted; pet very unequal; fr smooth.
  3 Pet villous at the margin.
    4 Cal and pedicels stipitate-glandular .... 3. *A. flava.*
    4 Cal and pedicels finely pubescent, not glandular .... 4. *A. sylvatica.*
  3 Pet stipitate-glandular on the margin .... 5. *A. pavia.*

  1. Aesculus hippocastanum L. Horse-chestnut. Tree to 25 m; winter-buds glutinous; lfls commonly 7(9), wedge-obovate, 1–2.5 dm, abruptly acute, irregularly serrate or biserrate, pubescent beneath when young, later glabrous; infls many-fld, ovoid-conic, 2–3 dm; cal 5–7 mm; upper and lateral pet with white rotund blade marked with red or yellow at the cordate base, on slender claws; fifth petal obovate, tapering to a broad claw, or wanting; fr echinate, 5 cm thick; $2n=40$. Native of se. Europe and adj. Asia, occasionally escaped from cult. in our range. May.

  2. Aesculus glabra Willd. Ohio-b. Small tree to 15 m; lfls 5(–7), bright or yellowish-green, oblanceolate to obovate, 6–16 × 3–6 cm, acute or acuminate, serrate, with conspicuous tufts of hair in the vein-axils beneath; infls 1–1.5 dm; cal 5–8 mm; pet 4, greenish-yellow, 1–2 cm, the 2 uppermost oblanceolate, slightly exceeding the 2 lateral, gradually tapering to the claw, the lateral with elliptic-oblong to elliptic-ovate blade and slender claw; stamens 7, often twice as long as the cor; fr generally echinate, 3–4 cm thick; $2n=40$. Moist but fairly well drained, chiefly alluvial, often calcareous soil; w. Pa. to s. Ont., s. Wis., Io., and Kans., s. to Tenn., n. Ala., Ark., and Tex. Apr., May. Our plants, as here described, are var. *glabra.*

  3. Aesculus flava Aiton. Sweet or yellow b. Canopy-tree to 30 m; lfls 5, dark bluish-green, ± drooping, oblanceolate to narrowly obovate, 10–15 cm, finely serrate, pubescent beneath when young; infls 1–1.5 dm, the pedicels and cal beset with small stipitate glands; cal tubular, 7–16 mm, shallowly lobed; pet 4, dull yellow to sometimes tawny or purplish or reddish, villous on the margin, the lateral ones 1.5–2.5 cm, with broadly elliptic blade and exsert claw, the upper 2–3 cm, with a much smaller blade on a long claw; stamens 7 or 8, equaling or shorter than the pet; fr smooth, 5–8 cm thick; $2n=40$. Rich moist woods; sw. Pa. and s. O. to s. Ill., s. to n. Ga. and n. Ala. May, June. (*A. octandra.*) Hybridizes with no. 2.

  4. Aesculus sylvatica Bartram. Painted or Georgia-b. Tree to 15 m, or blooming as a shrub; lfls 5, narrowly obovate, usually pubescent beneath until anthesis; fls much as in no. 3, the cal and pedicels finely pubescent, not glandular. S. Va. to Ga., e. Ala., and e. Tenn., especially on the piedmont. Apr. (*A. neglecta.*) Hybridizes with nos. 3 and 5.

  5. Aesculus pavia L. Red b. Shrub or small tree; lfls 5, rarely more, thin, oblanceolate to obovate, glabrous to tomentulose beneath; cal tubular, red or yellow, 10–18 mm, shallowly lobed, finely puberulent; pet red or rarely yellow, glandular-ciliate, the lateral 2–3 cm, with oblong-obovate blade and long claw, the upper 2.5–4 cm, with minute, oblong-obovate to suborbicular blade; stamens 6–8, sometimes slightly longer than the upper pet; fr smooth, 3–6 cm thick; $2n=40$. Rich damp woods; coastal plain from N.C. to Fla. and Tex, and inland to s. Mo., s. Ill., and Ky. Apr. (*A. discolor,* the phase with more tomentose lvs, commoner westward.)

FAMILY **ACERACEAE**, the Maple Family

  Fls regular, some or all of them functionally (or fully) unisexual, hypogynous or the staminate ones sometimes perigynous; sep (4)5(6), imbricate, distinct or sometimes connate below; pet (4)5(6), imbricate, distinct, often much like the sep, or sometimes wanting; stamens typically 8, but sometimes 4, 5, or 10–12, seated inside or outside or upon the nectary-disk, or the disk obsolete; ovary mostly bilocular, visibly compressed at right angles to the septum; styles distinct, or the style solitary and shortly to deeply lobed; ovules (1)2 per locule; fr generally a double samara, the usually 1-seeded, winged mericarps eventually separating from a persistent carpophore; embryo with elongate radicle and 2 flat or plicate, green cotyledons; endosperm wanting; trees or shrubs with opposite, usually exstipulate, simple or less often compound lvs and small fls in various sorts of infls. Only *Acer* and *Dipteronia,* the latter with 2 Chinese spp.

  1. ACER L. Maple. Lvs simple and usually palmately veined and lobed, or less often pinnately or palmately compound with few lfls; fr normally a double samara with winged mericarps, the seed-bearing base of the mericarp thickened and not winged. (*Argentacer, Negundo, Rufacer, Saccharodendron.*) 110, cosmop.

1 Lvs simple.
  2 Sinuses between the principal lf-lobes rounded.
    3 Juice milky; pet present; samara-wings very widely divergent .... 1. *A. platanoides.*
    3 Juice watery; pet none; samara-wings diverging at 120° or less.
      4 Lobes of the lvs acute or acuminate; bark ± roughened, not beech-like; widespread spp.

5 Lvs flat or nearly so; bark grayish, with loose-edged plates .... 2. *A. saccharum.*
5 Lvs with drooping sides; bark darker, thicker, ± deeply furrowed .... 3. *A. nigrum.*
    4 Lobes of the lvs blunt; bark pale, smooth and beech-like; s. fringe of our range .... 4. *A. barbatum.*
2 Sinuses sharp, forming a definite angle (least so in no. 9, marked by its deeply lobed lvs).
    6 Fls appearing during or after the unfolding of the lvs, in terminal infls, the fr maturing in
        midsummer or later; pet present.
        7 Lobes and teeth of the lvs rounded to subacute; infl a drooping panicle .... 5. *A. pseudoplatanus.*
        7 Lobes and teeth of the lvs with sharp divergent points; infl otherwise.
            8 Lvs coarsely and usually simply serrate, with 2–3 teeth per cm; infl an erect panicle .... 6. *A. spicatum.*
            8 Lvs finely double-serrate, with 7–12 points per cm; infl a drooping raceme .... 7. *A. pensylvanicum.*
    6 Fls appearing much before the opening of the lf-buds, in lateral leafless umbels; fr maturing in
        spring.
        9 Pet present; terminal lf-lobe comprising less to slightly more than half the length of the blade,
            broadest at base .... 8. *A. rubrum.*
        9 Pet none; terminal lf-lobe comprising well over half the length of the blade and much narrowed
            at base .... 9. *A. saccharinum.*
1 Lvs compound .... 10. *A. negundo.*

1. **Acer platanoides** L.   Norway-m. Tree with widely spreading crown; lvs much like those of no. 2, with 5–7 sharply acuminate lobes and few large teeth; juice milky (best seen at base of a detached petiole); fls yellow-green, in erect, pedunculate, rounded corymbs, pedicles glabrous; pet obovate, 5–6 mm, spreading; stamens seated on the disk; mericarps of the fr 3.5–5 cm, scarcely distended over the seed, the wings divergent at an angle of nearly 180°; $2n=26$. Native of Europe, often found as an escape from cult. in vacant lots, and now spreading into successional forests. Apr., May.

A. campestre L., the Hedge-m., with milky juice, smaller lvs with blunt or rounded lobes, petaliferous fls, and widely divergent mericarps, occasionally escapes from cult.

2. **Acer saccharum** Marshall.   Sugar-m.; hard m. Climax tree to 40 m; bark medium-gray, becoming roughened with loose-edged plates; lvs flat, about as wide as long, usually glabrous beneath except for a few tufts of hairs in the vein-axils, (3)5-lobed with rounded sinuses, the lobes usually bearing a few large sharp teeth, the central lobe usually with nearly parallel sides up to a pair of large teeth at about mid-length; fls in umbels from the terminal or uppermost lateral buds, appearing as the lf-buds open, drooping on slender, hairy pedicels to 8 cm; cal gamosepalous, 2.5–6 mm, ± hirsute; pet none; disk extrastaminal; ovary and fr glabrous; mericarps of the fr 2.5–4 cm, the seed-bearing basal parts diverging at right angles to the pedicel, the wings curved forward, divergent at an angle of 120° or less; $2n=26$. Rich to fairly dry woods, especially in calcareous soils; N.S. and N.B. to Minn. and e. S.D., s. to N.J., Del., w. Va., n. Ga., Tenn., and Mo. (*Saccharodendron s.; Acer saccharophorum.*) Plants intermediate toward *A. barbatum,* occurring along the s. boundary of our range, have been called var. *schneckii* Rehder, or var. *regelii* (Pax) Rehder, the latter name however based on a specimen of *A. barbatum.* Spp. 2–4 often treated as parts of a single sp.

3. **Acer nigrum** Michx. f.   Black m. Similar to no. 2, and not always sharply distinct; bark darker, thicker, more deeply furrowed; lvs darker and duller above, usually pubescent beneath, commonly with somewhat drooping sides and tip, 5-lobed as in no. 2 or often the lowermost lobes reduced or obsolete, the lobes generally with fewer and shorter, often obtuse or rounded teeth, the central lobe tending to taper from the base. Typically in somewhat moister sites than no. 2, often along streams; s. Que. to s. Minn., s. to L.I., N.J., Del., w. Va., w. N.C., Tenn., and n. Ark., commonest in the midwest. (*A. saccharum* var. *nigrum* and var. *viride; Saccharodendron n.*) Plants intermediate toward no. 4, occurring chiefly in the s. Appalachians, have been called *A. leucoderme* Small.

4. **Acer barbatum** Michx.   Southern sugar-m. Tree with smooth, pale gray, beech-like bark, becoming shallowly furrowed and scaly-ridged only on large trunks; lvs rarely over 10 cm wide, finely pubescent and pale or glaucous beneath, 5-lobed, with or without a few low teeth, both teeth and lobes obtuse or rounded; cal densely hirsute within; ovary and young fr pubescent; otherwise much like no. 2. Forests, especially on the piedmont and coastal plain; se. Va. to n. Fla., e. Tex., Ark., and e. Okla. (*A. floridanum.*)

5. **Acer pseudoplatanus** L.   Sycamore-m. Tall tree with rough-looking, coarsely crenate and 5-lobed lvs, the lobes and teeth never sharply acuminate; fls yellowish-green, numerous in slender drooping panicles 6–15 cm; sep and pet subequal, 4–5 mm; disk extrastaminal; style deeply cleft; mericarps 3–4 cm, somewhat distended over the seed but not reticulate, the wings diverging at an angle of 60–90°; $2n=52$. Native of Europe and w. Asia, occasionally escaped from cult. May.

6. **Acer spicatum** Lam.   Mt.-m. Tall shrub or clumped small tree to 10 m; winter-buds with 2–4 valvate scales; lvs 3-lobed or obscurely 5-lobed, softly hairy beneath, coarsely and irregularly serrate, the teeth 2–3 per cm, each tipped with a minute sharp gland; fls in fascicles of 2–4 along an erect axis, forming a slender, terminal, long-peduncled panicle 3–8 cm, long-pediceled, the terminal one of each fascicle usually perfect, the others sterile; pet greenish, very narrowly linear-oblanceolate, 3 mm, much exceeding the sep; disk extrastaminal; style barely notched; mericarps 1.8–2.5 cm, conspicuously reticulate over the seed, the wings diverging at about a right angle; $2n=26$. Moist woods; Nf. to Sask., s. to Conn., Pa., O. and ne. Io., and in the mts. to N.C. and Tenn. June.

7. **Acer pensylvanicum** L.   Moosewood; striped m. Ascending slender tree to 12 m; bark of middle-sized stems with slender white stripes on a greenish background; winter-buds with 2–4 valvate scales; lvs 3-lobed, glabrous at maturity, finely and sharply serrate; infl a slender, peduncled, drooping terminal raceme 3–10 cm, the fls slender-pedicellate, all in each infl commonly of one sex; pet bright yellow, narrowly obovate, 5–

8 mm, scarcely surpassing the oblong-oblanceolate sep; disk intrastaminal; mericarps 2.5–3 cm, scarcely veined over the seed, the wings diverging at 90–120°; $2n=26$. Moist woods; N.S. and s. Que. to n. Minn., s. to N.Y., Pa., Mich., and in the mts. to N.C., Tenn., and n. Ga. May, June.

8. Acer rubrum L. Red m. Tree to 35 m; winter buds with 6–10 imbricate scales; young twigs red; lvs sharply but shallowly lobed, coarsely double-serrate, or with a few minor lobes, acute or short-acuminate, ± pubescent beneath when young; functionally dioecious; fls generally red, unisexual, short-pedicellate in fascicles from a cluster of lateral buds, opening much before the lvs; sep oblong, distinct, 1 mm, the pet a little narrower and longer, oblong-linear; ovary glabrous; mericarps mostly 2–4 cm; $2n=78, 91, 104$, perhaps other nos. Swamps, alluvial soils and moist uplands, often as a successional tree; Nf. to se. Man., s. to Fla. and e. Tex., but missing from Io., most of Ill., and n. Mo. (*Rufacer r.; R. drummondii; Acer carolinianum; A. stenocarpum.*) Morphologically, cytologically and ecologically variable, but indivisible.

9. Acer saccharinum L. Silver-m; soft m. Tree to 30 m, with light gray bark separating in large plates; winter-buds with 6–10 imbricate scales; lvs deeply 5-lobed, silvery-white beneath, the lobes acuminate and ± sharply toothed or with minor lobes along the sides, the terminal lobe concavely narrowed to the base; fls from clusters of lateral buds, opening in earliest spring, greenish-yellow or reddish, each cluster unisexual; pedicels short, scarcely exceeding the strongly ciliate bud-scales; cal gamosepalous, shallowly lobed; stamens long-exsert; disk vestigial or wanting; ovary densely hairy; frs falling before the lvs are fully grown, the mericarps 3.5–5+ cm, sparsely hairy; $2n= 52$. Moist or wet soil, especially along riverbanks; N.B. and s. Que. to Minn., and e. S.D., s. to Ga., w. Fla., La., and Okla. (*Argentacer s.*)

10. Acer negundo L. Boxelder; ash-leaved m. Tree to 20 m, the trunk soon deliquescent; lfls 3–5 or on rapidly growing shoots up to 9, pinnate, lanceolate to ovate or oblong, coarsely and irregularly serrate or with a few shallow lobes; dioecious; fls apetalous, appearing with or before the lvs, the staminate in sessile, umbel-like fascicles, drooping on slender pedicels, the pistillate in drooping racemes; disk wanting; mericarps 3–4.5 cm; $2n=26$. Moist, especially alluvial soil; N.H. to Fla., w. to the Pacific, and irregularly in Mex. and Guat. Var. *negundo*, the common e. Amer. var., has glabrous, often glaucous twigs. (*Negundo n.; N. nuttallii.*) Var. *interius* (Britton) Sarg. (*Negundo i.*), with velutinous twigs and usually with tufts of hairs in the vein-axils beneath, is chiefly western, occasionally extending to Minn. and Mo. Var. *texanum* Pax, with similarly hairy twigs, rather densely puberulent frs, and short-acuminate lfls pubescent over the whole lower surface, occurs chiefly from the Ozarks to Tex., extending e. to s. Ind.

FAMILY **ANACARDIACEAE**, the Cashew Family

Fls small, regular, mostly hypogynous, perfect or more often unisexual, mostly 5-merous as to the sep, pet, and stamens, the sep generally connate below, the pet distinct or rarely wanting, the stamens alternate with (or twice as many as) the pet and generally seated outside the annular (often 5-lobed) nectary-disk; ovary mostly of 3 carpels, plurilocular or more often only one locule fully developed, the styles distinct or ± united; ovule solitary (or solitary in each locule), apotropous, in ours basal with a long funiculus; fr commonly drupaceous, with a ± resinous and sometimes waxy or oily mesocarp; seed with oily, curved or less often straight embryo and 2 expanded cotyledons, the endosperm scanty or none; woody plants with well developed resin-ducts (or latex-channels) at least in the bark; lvs mostly alternate and exstipulate, pinnately compound or trifoliolate, rarely simple. 60–80/600, mostly tropical.

1 Infls dense, terminal, or lateral on last year's twigs; frs red, glandular-pubescent .... 1. *Rhus.*
1 Infls loose, axillary; frs whitish or yellowish, smooth or inconspicuously hairy .... 2. *Toxicodendron.*

1. RHUS L. Sumac. Pet 5, often hairy, especially on the inner surface; stamens 5; ovary 3-carpellate but unilocular, with a short, terminal, 3-lobed style; ovule basal; frs red or reddish, glandular-hairy; innocuous, polygamo-dioecious shrubs or small trees with dense, crowded infls terminal or lateral on last-year's twigs. (*Schmaltzia.*) 100, cosmop.

1 Lfls several to many; sparingly branched (often tall) shrubs or small trees.
  2 Rachis of the lf not winged.
    3 Twigs and petioles glabrous; hairs of the fr short, obovoid .... 1. *R. glabra.*
    3 Twigs and petioles densely hairy; hairs of the fr elongate, tapering .... 2. *R. typhina.*
  2 Rachis of the lf winged .... 3. *R. copallinum.*
1 Lfls 3, all sessile or nearly so; bushy shrubs .... 4. *R. aromatica.*

1. Rhus glabra L. Smooth s. Usually a sparingly branched shrub, but sometimes to 6 m, vigorously colonial; petioles and younger branches glabrous and somewhat glaucous; lfls 11–31, lanceolate to narrowly oblong, 5–10 cm, acuminate, commonly serrate, much paler beneath; infl terminal, often 2 dm, frs 4–5 mm, somewhat flattened, bright red, densely beset with thick, minute red hairs 0.2 mm. Abundant in upland sites, old fields, roadsides, and margins of woods; N. Engl. and s. Que. to B.C., s. to n. Fla., Tex., and Mex. June, July. Forms with bipinnate lvs occur. A hybrid with no. 2 has been called *R.* × *borealis* (Britton) Greene.

2. Rhus typhina L.   Staghorn-s. Tall shrub or small tree to 10 m; younger branches, petioles, and lf-rachis densely and softly hirsute; lfls 9–29, lanceolate to narrowly oblong, 5–12 cm, acuminate, serrate, paler beneath; frs 4–5 mm, somewhat flattened, red, densely covered with slender, tapering hairs 1–2 mm. Dry, open places; N.S. and s. Que. to Minn., s. to W.Va. and O., and irregularly to n. Ga., n. Ala., and Io. June, July. (*R. hirta.*)

3. Rhus copallinum L.   Shining s. Shrub or small tree to 6 m; young twigs, petioles, and lf-rachis closely hairy; lfls 7–21, shining above, firm, oblong to lanceolate, 3–8 cm, inequilateral, entire or with a few teeth along the distal margin; lf-rachis winged, the wings 1–5 mm wide, interrupted at the base of each pair of lfls; infl to 15 cm; fr 4–5 mm, somewhat flattened, red, pilose with simple hairs and also beset with minute ovoid hairs. Open, dry places; s. Me. to Fla., w. to Ind., s. Ill., se. Neb., and Tex., and irregularly to Mich., Wis., and s. Io. June, July.

4. Rhus aromatica Aiton.   Squaw-bush. Malodorous, frequently branched, rounded shrubs to 2 m, often forming thickets; lfls 3, sessile or nearly so, at least the terminal one few-toothed or -cleft above the middle; fls pale yellow, in several short (1–2 cm), spike-like lateral clusters formed from axillary buds in late summer the previous year and expanding with (or just after) or before the lvs in the spring; frs 4–5 mm, bright red, densely hairy. Open woods, sand-dunes, and other open places; sw. Que. and w. Vt. to nw. Fla., w. to Alta., Oreg., Calif., and Mex. (*R. trilobata.*) Variable, with 4 vars. in our range:

a Lfls relatively large, the terminal one mostly 4–8 cm; fls with or before the lvs.
  b Copiously hairy, the twigs tomentulose, the lvs permanently appressed-hairy above, velvety beneath;
     Ill. and Mo. to Kans. and Okla. (*R. nortonii*) . . . . var. *illinoensis* (Greene) Rehder.
  b Much less hairy or glabrous.
    c Terminal lfl elliptic to rhombic-ovate, tapering about equally to the subcuneate base and acute
      or acutish tip; outer margin of lateral lfls distinctly convex toward the base; Vt. and sw.
      Que. to nw. Fla., w. to Mich., Io., Kans., and Tex. (*R. canadensis; Schmaltzia crenata.*) . . . . . . var. *aromatica.*
    c Terminal lfl more flabelliform-ovate, with cuneate base and rounded tip; margins of the lateral
      lfls nearly straight, meeting at an angle of about 90°; Ill. and Io. to Neb. and Tex. (*Schmaltzia*
      s.*) . . . . var. *serotina* (Greene) Rehder.
a Lfls smaller, the terminal one mostly 1.5–4 cm; fls expanding when the lvs are nearly fully grown;
  dunes mainly along the s. side of the Great Lakes; n. O., n. Ind., and ne. Ill. (*Schmaltzia a.*) . .
  . . . . var. *arenaria* (Greene) Fern.

2. TOXICODENDRON Miller.

Pet glabrous; fr a white or greenish-white to yellowish drupe, shining and glabrous or inconspicuously short-hairy, the hairs not glandular; allergenic shrubs or vines, with axillary, raceme-like, rather loose infls often drooping in fr; otherwise much like *Rhus,* and often included therein. Ca 10, New World and e. Asia.

1 Lfls 7–13, entire . . . . 1. *T. vernix.*
1 Lfs 3, entire to often toothed or lobed.
  2 Simple or sparingly branched, suberect shrubs, without aerial roots, not climbing.
    3 Petiole and young fr pubescent; lfls and their lobes or teeth mostly blunt; southern . . . . 2. *T. pubescens.*
    3 Petiole and fr glabrous; lfls and their teeth ± pointed; northern or western . . . . 3. *T. rydbergii.*
  2 Climbing or straggling vines with aerial roots; lfls pointed; widespread . . . . 4. *T. radicans.*

1. Toxicodendron vernix (L.) Kuntze.   Poison-sumac. Shrub to 5 m, often branched from the base; lfls 7–13, oblong to obovate or elliptic, 4–5 cm, acute to acuminate, entire, acute at base, glabrous or nearly so; infls to 2 dm; fr grayish-white, 4–5 mm. Swamps, usually in shade; s. Me. and sw. N.S. to Minn. and sw. Ont., s. to Md., O., and Ind.; Del. to Fla. and Tex. (*Rhus v.*)

2. Toxicodendron pubescens Miller.   Poison-oak. Simple or sparingly branched, suberect shrub, seldom over 1 m; young twigs pubescent, as also the petioles; lfls 3, the terminal one long-petiolulate, the lateral ones usually subsessile, pubescent on both sides or eventually glabrate above, ovate to obovate or subrotund, 4–8 cm, mostly blunt or rounded at the tip, commonly with 2–4 blunt or rounded deep teeth or shallow lobes on each side, or the margins merely wavy; fr yellowish-white, 5 mm, pubescent at first, becoming minutely papillose; 2n=30. Sandy woods; L.I. and N.J. to n. Fla., w. to W.Va., s. Mo., s. Kans., and Tex. May. (*T. toxicaria; Rhus quercifolia; R. toxicodendron.*)

3. Toxicodendron rydbergii (Small) Greene.   Western poison-ivy. Vigorously rhizomatous and colonial, suberect, simple or sparingly branched shrub to 1(–3) m; lvs tending to be approximate toward the top of the stem; petioles elongate, glabrous; lfls broadly ovate to suborbicular, tending to be openly folded along the midrib rather than flat, glabrous on both sides, or strigose beneath and often with a line of minute, curly hairs along the midrib above; infl unbranched (racemose) or sparingly branched, usually with fewer than 25 fls; frs 4–7 mm thick, smooth, sessile or subsessile and crowded in a ± erect infl; otherwise much like no.

4, with which it intergrades to some extent; $2n=30$. Open, sunny habitats; widespread in w. U.S., extending e. in n. U.S. (especially north of the 42nd parallel) and s. Can. to N.S. and N.Y., thence s. at upper elev. to Va. and W.Va.

4. **Toxicodendron radicans** (L.) Kuntze. Common poison-ivy. Climbing or straggling vine, producing abundant aerial roots; lfls 3, mostly flat, 5–15 cm, ovate to elliptic, acute or acuminate, entire or often with a few irregular, ± pointed teeth or shallow lobes; terminal lfl long-petiolulate, the others subsessile or short-petiolulate; infls up to 1 dm, loose, branched, with mostly more than 25 fls, often ± pendulous in fr; frs pedicellate, mostly 3–5 mm thick; $2n=30$. Commonly in open woods, often in disturbed habitats, sometimes in more open sites with nothing to climb on; s. N.S. to Fla., w. to Mich., se. Minn., e. Neb., Okla., and Tex., and s. to Guat.; also in China and Japan. May–July. (*Rhus r.*) Several confluent vars., 2 in our range.

a Frs puberulent, scabrous, or papillose; petioles glabrous or pubescent; lfls often entire; lower surface of the lfls with tufts of hairs in the vein-axils; s. N.S. and coastal Me. to Fla. and Bahama Isl., w. to Vt., Pa., W.Va., s. Ill., and Tex., commoner on the coastal plain or e. of the Appalachian divide .... var. *radicans.*
a Frs glabrous or with a few scattered hairs; petioles pubescent; lfls generally toothed; lower surface of the lfls glabrous or strigose or with the hairs confined to the main veins, but without axillary tufts; w. Vt. to w. Va., w. to Mich., se. Minn., Io., e. Neb., Kans., Okla., and Tex., mostly w. of the Appalachian divide .... var. *negundo* (Greene) Reveal

Family **SIMAROUBACEAE**, the Quassia Family

Fls small, regular, hypogynous, often unisexual; pet most commonly 5; stamens twice as many as the pet, or as many as and alternate with the pet, the filaments often with a ventral appendage near the base; nectary-disk intrastaminal, sometimes modified into a gynophore; ovary of 2–5 weakly to firmly united carpels, with distinct styles or a single common style, or seldom the carpels distinct; ovules 1(2) per carpel, apical and pendulous to basal and ascending-erect, usually epitropous; fr usually a capsule or samara, often schizocarpic; embryo oily, with 2 large, expanded cotyledons; endosperm wanting; trees or shrubs with very bitter bark, wood, and seeds, the lvs alternate, usually exstipulate and pinnately compound. 25/150, mostly of warm regions.

1. **AILANTHUS** Desf. Incompletely dioecious; fls 5-merous; stamens 10 (fewer or none on fls of female trees); carpels 2–5, united at the axis only, each with a single ovule, maturing into separate samaras with the seed near the center; trees with pinnately compound lvs and malodorous fls in large terminal panicles. 10, Old World.

1. **Ailanthus altissima** (Miller) Swingle. Tree of heaven. Rapidly growing tree with few branches and coarse twigs; lvs large, to 1 m; lfls 11–41 (the terminal one often imperfectly developed or wanting), narrowly oblong, entire except for one or more coarse teeth near the base, each tooth with a large gland beneath; infl pyramidal, 1–3 dm; fls greenish or greenish-yellow, 5 mm wide; frs 3–5 cm × 7–13 mm, twisted; $2n=80$. Native of e. Asia, often escaped from cult. in our range, especially in cities. (*A. glandulosa.*)

Family **MELIACEAE**, the Mahogany Family

Fls regular, hypogynous, perfect or unisexual; sep commonly connate at least below; pet usually as many as the sep, distinct or seldom connate below; stamens twice as many as the pet, or seldom as many as and alternate with the pet, usually monadelphous, the filament-tube often with appendages; nectary-disk intrastaminal, sometimes adnate to the ovary or modified into an androgynophore; ovary mostly 2–5-locular, with a single style; ovules typically 2 per locule, axile, generally pendulous and epitropous; fr a capsule, or less often a berry or drupe; embryo spatulate, with 2 cotyledons; endosperm oily; woody plants with bitter bark and alternate, exstipulate, pinnately compound or trifoliolate lvs. 51/550, mostly of warm reg.

1. **MELIA** L. Fls 5–6-merous; stamen-tube with 10–12 anthers and 20–24 narrow lobes; ovary on a short disk; fr a drupe with a 5-locular stone; trees with bipinnate lvs. 5, warm part of Old World.

1. **Melia azedarach** L.　Chinaberry. Tree to 15 m, with broadly rounded crown; lfls numerous, ovate or lance-ovate, 3–7 cm, serrate, acuminate; fls lilac, in axillary panicles, the slender, spreading pet 1 cm; fr yellow, 1.5 cm thick, persistent; $2n=28$. Native of Asia, escaped from cult. in se. U.S., as far n. as se. Va.

FAMILY RUTACEAE, the Rue Family

Fls perfect or unisexual, usually hypogynous and regular and 4–5-merous, with distinct pet; stamens usually as many (and alternate with) or twice as many as the pet, the filaments often connate below; nectary-disk intrastaminal, sometimes unilateral, sometimes modified into a gynophore; ovary plurilocular, often apically indented, or the carpels sometimes distinct except for the coherent styles; ovules ± epitropous, (1)2 (–several) in each carpel on axile (or marginal) placentas; embryo large, spatulate or sometimes linear, with 2 cotyledons; endosperm oily or wanting; plants commonly with secretory cavities containing aromatic ethereal oils scattered throughout the parenchymatous tissues, the lvs pellucid-punctate, exstipulate, usually pinnately compound or dissected or trifoliolate. 150/1500, mostly of warm reg.

1 Trees or shrubs; lvs trifoliolate or pinnately compound, with distinct lfls.
  2 Lfls 5 or more .... 1. *Zanthoxylum.*
  2 Lfls 3 .... 2. *Ptelea.*
1 Herb or half-shrub to 1 m; principal lvs twice pinnatifid .... 3. *Ruta.*

1. **ZANTHOXYLUM** L.　Prickly ash. Fls unisexual, our spp. dioecious, (3)4–5-merous, the stamens as many as and alternate with the pet; carpels 2–5, the ovaries distinct, the styles coherent above; ovules 2, collateral, pendulous; fr from each carpel a firm-walled or somewhat fleshy follicle, dehiscent across the top, the 1 or 2 seeds often hanging from a long funiculus; shrubs or trees with prickly stems, alternate, odd-pinnately compound lvs and clusters of small greenish or whitish fls. 200+, mainly of warm reg.

Phellodendron japonicum Maxim., the cork-tree, a small, unarmed tree with opposite, pinnately compound lvs, the fr a 5-stoned drupe, occasionally escapes from cult.

1 Fls in axillary clusters; follicles stipitate .... 1. *Z. americanum.*
1 Fls in terminal compound cymes; follicles subsessile .... 2. *Z. clava-herculis.*

1. **Zanthoxylum americanum** Miller.　Common p. a. Tall shrub, or rarely a small tree to 8 m, with prickly stems, paired pseudostipular prickles, and strongly aromatic foliage; lfls 5–11, oblong to elliptic or ovate, pubescent beneath at least when young; fls in sessile, umbel-like, precocious axillary clusters on branches of the previous year; sep none; pet yellow-green, fringed at the tip; ovaries 3–5; follicles stipitate, ellipsoid, 5 mm, the surface pitted; $2n=68, 136$. Moist woods and thickets; s. Que. to e. N.D., s. to S.C., Ga., and Okla., but only irregularly e. of O. and nw. N.Y.

2. **Zanthoxylum clava-herculis** L.　Hercules' club. Tall shrub or small tree; lfls 5–19, obliquely lanceolate to ovate, shallowly crenate, glabrous; fls numerous in large terminal cymes; sep present; ovaries 2 or 3; frs nearly sessile, subglobose, 5 mm, wrinkled or pitted; $2n=72$. Damp woods; Delmarva Penins., on outer coastal plain to Fla., thence w. to Tex., Ark., and se. Okla. Apr., May.

2. **PTELEA** L.　Hop-tree. Polygamous; sep, pet, and stamens 4 or 5, the latter imperfect or abortive in pistillate fls; ovary bilocular, with a short style and capitate, ± 2-lobed stigma; ovules 2 per locule, superposed on axile placentas; fr a thin, flat, orbicular samara, the broad wing surrounding the indehiscent bilocular body; unarmed shrubs or small trees with alternate (opposite), petiolate, trifoliolate lvs and terminal cymes of small, greenish-white or yellowish-white fls. 3, temp. N. Amer.

1. **Ptelea trifoliata** L.　Common h.-t. Shrub or small tree; lvs long-petioled; lfls sessile, ovate or elliptic to rhombic or ovate-oblong, the lateral ones oblique; cymes 4–8 cm wide; pet oblong, pubescent, 4–7 mm; filaments pubescent; fr 15–25 mm wide, reticulate-veiny, with the odor of hops; $2n=42$. Moist or rich woods and thickets; widespread in Mex. and e. to Kans., s. Wis., s. Mich., O., N.C., and Fla., and irregularly to s. Que., Vt., w. N.Y., N.J., and Va. May, June. Our plants all belong to ssp. *trifoliata,* most of them to the glabrous or inconspicuously pubescent var. *trifoliata.* The mainly more southern var. *mollis* T. & G., with the lvs and twigs evidently pubescent, the lvs densely so beneath, extends n. into our range as

far as D.C. and is disjunct on sand-dunes along the s. shore of Lake Michigan, where it has been called var. *deamiana*.

3. RUTA L.   Rue. Fls perfect; sep and pet each 4 or 5; stamens 8 or 10; ovary 4- or 5-locular and -lobed, with several axile ovules in each locule; style 1; fr a lobate loculicidal capsule; perennial herbs or half-shrubs, strongly glandular-punctate and heavily scented, with alternate, simple to bipinnatifid lvs and small fls in ample paniculiform cymes, terminal and from the upper axils. 60, Eurasia and N. Afr.

1. **Ruta graveolens** L.   Common rue. Erect, branched at base, 3–10 dm; lvs pale green or bluish, the principal ones twice pinnatifid into oblanceolate segments; fls greenish-yellow, 1 cm wide, the pet inflexed at the tip; 2n=72, 81. Native of Europe, cult. in old-fashioned gardens and sometimes found as an escape along roadsides and in waste places. June–Aug.

FAMILY **ZYGOPHYLLACEAE**, the Creosote-bush Family

Fls hypogynous, mostly perfect and regular, 4- or 5-merous, with distinct (or no) pet; stamens mostly 2 or 3 times as many as the pet; nectary-disk variously intra- or extrastaminal, or of separate glands; carpels mostly 4 or 5, united to form a compound, plurilocular ovary with axile placentas and a slender style with 1–several stigmas (the locules each divided into 2 locelli in *Kallstroemia*); ovules 1–several per locule, mostly epitropous; fr most commonly a capsule or schizocarp; embryo straight or slightly curved, with 2 cotyledons; endosperm hard and oily, or (in our genera) wanting; herbs (ours) or more often woody plants, the stem often swollen at the nodes; lvs evidently stipulate, bifoliolate or pinnately or often bipinnately compound, commonly without a terminal lfl. 30/250, mostly of warm dry reg.

1 Fr spiny, not beaked, splitting into 5 segments . . . . 1. *Tribulus.*
1 Fr tuberculate, long-beaked, splitting into 10 segments . . . . 2. *Kallstroemia.*

1. TRIBULUS L.   Sep caducous; stamens mostly 2 or 3 times as many as the pet; nectary-glands both intra- and extrastaminal; ovary 5-locular, with 5–10 ovules per locule; fr spiny, beakless, splitting at maturity into 5 segments, each transversely septate into 3–5 1-seeded compartments; germination hypogeal; diffusely branched, prostrate herbs with opposite, even-pinnate lvs and small yellow fls solitary at the axils. 30, Old World.

1. **Tribulus terrestris** L.   Puncture-vine. Annual; stems hirsute, branched from the base, forming mats to 1 m wide; lvs short-petioled, 2–6 cm, one of each pair distinctly the larger; lfls usually 6–8 pairs, oblong, 5–15 mm; peduncles 5–10 mm; fls 8–10 mm wide; intrastaminal glands distinct, not forming a ring; body of the fr 1 cm thick, each segment with 2 stout divergent spines and a longitudinal row of tubercles; 2n=12, 24, 26, 48. Native of the Mediterranean region, well established as a roadside weed in w. U.S., and occasionally found in our range.

2. KALLSTROEMIA Scop.   Caltrop. Sep persistent under the fr; ovary 10-locellar; style accrescent and forming a columnar beak at maturity; fr separating into 10 1-seeded, half-carpellary mericarps, these tuberculate on the back; germination epigeal; diffusely branched, prostrate to ascending herbs allied to *Tribulus*. 17, New World, mainly trop. and subtrop.

1. **Kallstroemia parviflora** Norton.   Small-fld c. Annual; stems 3–6 dm, finely pubescent with incurved ascending hairs and ± coarsely hirsute; lvs 3–6 cm, short-petioled; lfls usually 4 pairs, elliptic or oblong, pubescent; peduncles 1–3 cm; fls 1 cm wide; fr-body ovoid, strigose, 4 mm, the beak slightly longer. Disturbed dry places, often along railways; Ill. to Ariz., s. to Mex. and C. Amer. (*K. intermedia.*)

FAMILY **OXALIDACEAE**, the Wood sorrel Family

Fls perfect, hypogynous, regular, 5-merous; pet mostly convolute and distinct; stamens generally bicyclic, the filaments all connate at base; nectary-disk wanting but the outer filaments often glandular-thickened below or subtended by a gland; carpels 5, united to form a compound, plurilocular ovary with axile placentas and distinct, terminal styles; ovules 2–several per locule, epitropous and ± pendulous; fr generally a loculicidal capsule; seeds commonly with a large basal aril; embryo large, straight, spatulate, with 2 cotyledons; endosperm oily, usually copious; mostly herbs with alternate (or all basal), compound (often trifoliolate) lvs and axillary, often pedunculate, cymose to umbelliform infls that may be reduced to a single fl. 7–8/900, ± cosmop.

1. **OXALIS** L.   Wood-sorrel. Stamens 10, the outer (antepetalous) 5 shorter; fr capsular; our spp. herbs with digitately 3-foliolate lvs and obcordate infls. (*Ionoxalis, Xanthoxalis*.) 800, widespread. The nomenclature of our weedy spp. follows George Eiten.

1 Plants caulescent; lvs alternate; fls yellow.
  2 Pet 5–11 mm; plants weedy.
    3 Hairs of the stem and petioles (or many of them) septate, blunt; plants rhizomatous; stipules
      wanting .... 1. *O. stricta.*
    3 Hairs of the stem and petioles pointed, nonseptate; plants without rhizomes, though often with
      stolons; stipules present, often conspicuous.
      4 Stems erect or eventually decumbent, not creeping; seeds brown with transverse whitish ridges
      .... 2. *O. dillenii.*
      4 Stems creeping, rooting; seeds wholly brown .... 3. *O. corniculata.*
  2 Pet 12–20 mm; plants not weedy.
    5 Infl cymose; lfls 2–5 cm wide; fr very minutely hairy .... 4. *O. grandis.*
    5 Infl umbelliform; lfls to 1.5 cm wide; fr villous .... 5. *O. macrantha.*
1 Plants acaulescent; lvs basal; fls white or purple.
  6 Peduncles 1-fld .... 6. *O. acetosella.*
  6 Peduncles 2–several-fld .... 7. *O. violacea.*

1. **Oxalis stricta** L.   Common yellow w.-s. Perennial from long slender rhizomes, prostrate to more often erect and to 5 dm; stems pubescent with spreading, blunt-tipped, septate hairs, sometimes also strigose, seldom glabrous; stipules none; lfls 1–2 cm wide, usually glabrous to merely ciliate; peduncles usually exceeding the fls, 2–7-fld; pet 4–9 mm, yellow; fr 8–15 mm, provided with spreading, flexuous, septate hairs; 2*n*=24. Nearly cosmopolitan weed, often in natural habitats as well, probably originally native to N. Amer. Numerous infraspecific taxa are often defined on the distribution and orientation of the hairs. (*O. europaea; O. corniculata,* misapplied; *Xanthoxalis stricta; X. bushii; X. cymosa; X. rufa.*)

2. **Oxalis dillenii** Jacq.   Southern yellow w.-s. Cespitose perennial without rhizomes, erect to decumbent, but scarcely creeping, mostly antrorsely strigose with nonseptate, pointed hairs, the erect part of the stem light greenish, to 4 dm; stipules to 3 mm, or much smaller; lfls 1–2 cm wide; infl umbelliform, the pedicels abruptly divaricate or deflexed at maturity, but with erect frs; pet yellow, 4–10 mm; fr 1.5–2.5 cm, usually grayish-strigose; seeds brownish, conspicuously rugose transversely, the ridges whitish; 2*n*=18–24. Nearly cosmopolitan weed, often in natural habitats as well, probably originally native to N. Amer. Numerous infraspecific segregates are often defined on the habit and the amount and distribution of the pubescence. (*O. filipes; O. florida; O. stricta,* missapplied; *Xanthoxalis brittoniae; X. colorea.*)

3. **Oxalis corniculata** L.   Creeping yellow w.-s. Much like no. 2; stems trailing and freely rooting; pubescence variously oriented; herbage tending to have a purplish cast; infl cymose when well developed, the mature pedicels erect or ascending; pubescence of the fr mostly short and spreading, sometimes with intermingled longer hairs; seeds wholly brown; 2*n*=24, 42, 48. A widely distributed weed in the tropics and abundant in s. U.S., in our range often a weed in greenhouses and occasional in gardens. (*O. repens; Xanthoxalis c.*)

4. **Oxalis grandis** Small.   Big yellow w.-s. Rhizomatous; stems 3–10 dm, erect, simple or sparingly branched, sparsely villous below with hairs to 1 mm; lfls 2–5 cm wide, usually purplish at the margin; infl cymose, usually much surpassing the lvs; pedicels strigose or spreading-hairy or both; pet yellow, 12–18 mm; fr 8–10 mm, very minutely spreading-hairy or glabrous; seeds 2 mm. Usually in rich woods; Pa. to O. and s. Ill., s. to Ga. and Tenn. May, June. (*Xanthoxalis g.*)

5. **Oxalis macrantha** (Trel.) Small.   Showy yellow w.-s. Rhizomatous; stems 1–3 dm, eventually much branched, villous below with hairs 1–1.5 mm; lfls 6–15 mm wide; infl umbelliform, scarcely surpassing the lvs; pedicels sparsely villous with hairs 0.5–1 mm; pet yellow, 13–20 mm; fr to 20 mm, villous with hairs 1 mm; seeds 1.5 mm. Open woods; Ky. to Fla. and Miss. Apr., May. (*O. priceae; Xanthoxalis p.*)

6. **Oxalis acetosella** L.   Northern w.-s. Perennial from a slender scaly rhizome; lvs few, all basal, long-petioled; lfls sparsely pilose; peduncles 1-fld, 6–15 cm, slightly surpassing the lvs, 2-bracteate above the middle; pet mostly white, veined with pink, 10–15 mm, shallowly retuse at the oblique tip; cleistogamous fls produced late in the season on short recurved peduncles; 2*n*=22. Rich moist woods (especially spruce-fir); e. Que. to N.Y., Mich., Wis., and Sask., s. in the mts. to N.C. and Tenn.; widely distributed in Eurasia. (*O. montana,* the Amer. phase.)

7. **Oxalis violacea** L.   Violet w.-s. Perennial from a scaly bulbous base; lvs glabrous, or the petioles sometimes hairy; peduncles erect, 1–2 dm, much surpassing the lvs, with an umbelliform cluster of pedicellate fls subtended by small bracts; pet commonly rose-violet (white), 10–18 mm, not retuse; sep callous-tipped; $2n=28$. Dry upland woods and prairies; Mass. to N.D., s. to Fla. and Tex. Apr.–June. (*Ionoxalis v.*)

FAMILY **GERANIACEAE**, the Geranium Family

Fls mostly perfect, regular or (notably in *Pelargonium*) somewhat irregular, hypogynous, mostly 5-merous; pet distinct, mostly imbricate; extrastaminal nectary-glands usually alternate with the pet; stamens mostly 10 in 2 cycles, the outer (antepetalous) cycle often staminodial; carpels mostly 5, united to form a compound, plurilocular ovary with axile placentas and a single style with distinct stigmas; ovules mostly 2 per locule, superposed, epitropous, pendulous; in the tribe Geranieae (including our genera) the pistil with a prominent, elongating, gynobasic stylar column to which the fertile locular portion of the ovary appears to be attached in a lobed ring at the base, the fr with 5 1-seeded mericarps that usually separate elastically and acropetally from the persistent column, the mericarps often opening ventrally to discharge the seed; cotyledons 2; endosperm mostly scanty or none; mostly herbs with stipulate, lobed or compound or dissected lvs, the fls in cymose (often umbelliform) infls or solitary and axillary. 11/700, cosmop.

1 Lvs palmately cleft or palmately compound; fertile stamens in most spp. 10 .... 1. *Geranium.*
1 Lvs pinnatifid or pinnately compound; fertile stamens 5 .... 2. *Erodium.*

1. **GERANIUM** L.   Wild geranium; crane's bill. Pet obovate to obcordate; stamens 10, usually all with anthers (some staminodial in *G. pusillum*); filaments usually broadened below; stigmas 5, linear; each carpel-body (the basal, thickened portion of the carpel, filled by a seed) prolonged at maturity into a long, slender beak, these covering and adnate to the stylar column, which may project beyond the thickened portion as a slender stylar beak; carpel-bodies and their beaks typically separating acropetally from the stylar column at maturity, becoming outwardly coiled, the tip of the carpel-beak commonly remaining attached to the column at the summit, the carpel-body persistent at the basal (reflexed) end of its beak and opening ventrally to eject its seed, or sometimes the carpel-body disarticulating from its beak and separately deciduous; herbs with palmately lobed to divided or even compound lvs, the cauline ones chiefly opposite; fls pink or purple to white, commonly pedicellate in pairs at the end of axillary branches or peduncles. 300, cosmop.

1 True perennial from a stout, caudex-like rhizome; pet 12–20 mm; anthers 2 mm or more.
  2 Pedicels, cal, and fr not glandular; native .... 1. *G. maculatum.*
  2 Pedicels, cal, and fr-beak finely glandular; intr .... 2. *G. pratense.*
1 Annual to short-lived perennial, commonly taprooted; pet seldom over 12 mm; anthers to 1 mm.
  3 Lvs compound, at least the terminal segment distinctly petiolulate .... 3. *G. robertianum.*
  3 Lvs merely cleft, the segments confluent.
    4 Sep terminating in a subulate tip or awn 0.7–3 mm.
      5 Pedicels at maturity not more than twice as long as the cal.
        6 Carpel-bodies with long, antrorse hairs ca 1 mm; seeds obscurely reticulate .... 4. *G. carolinianum.*
        6 Carpel-bodies with spreading, often glandular hairs 0.5 mm; seeds strongly reticulate .... 5. *G. dissectum.*
      5 Pedicels at maturity more than twice as long as the cal.
        7 Pedicels spreading-hairy, eglandular; mature stylar beak ca 1 mm .... 6. *G. sibiricum.*
        7 Pedicels either strigose or glandular-villous; stylar beak 3–5 mm.
          8 Pedicels with closely appressed retrorse hairs .... 7. *G. columbinum.*
          8 Pedicels glandular-villous .... 8. *G. bicknellii.*
    4 Sep blunt to merely acute, or terminating in a minute callous point.
      9 Carpel-bodies glabrous, though generally wrinkled; stylar beak 2–5 mm .... 9. *G. molle.*
      9 Carpel-bodies hairy, not wrinkled; stylar beak (excluding stigmas) very short or none.
        10 Pet 2–4 mm; 3–5 outer stamens lacking anthers; complete fr 9–12 mm .... 10. *G. pusillum.*
        10 Pet 7–10 mm; all 10 stamens with anthers; complete fr 15–30 mm .... 11. *G. pyrenaicum.*

1. **Geranium maculatum** L.   Wild geranium. Perennial from a stout rhizome, with a few long-petioled basal lvs, these pedately cleft into 5–7 cuneately laciniate segments; stems erect, 3–7 cm, with a single pair of short-petioled lvs; fls few–several; cal and pedicels pubescent but not glandular; pet pink (white), 12–18

mm, entire; mature pedicels suberect; frs erect, 3–4 cm, including the 5–8 mm stylar beak. Common in woods; Me. to S.C. and n. Ga., w. to Man., Neb., and ne. Okla. Apr.–June.

G. sanguineum L., a Eurasian sp. with very leafy stems, retuse pet, and short stylar beak, is rarely found in our range as an escape from cult.

2. Geranium pratense L.   Meadow-g. Much like no. 1, the lvs generally more deeply cleft, with narrower and more sharply laciniate segments; pedicels, cal, and beak of the fr densely glandular-tomentose; peduncles usually 1-fld; pedicels deflexed after anthesis, becoming erect again at maturity; pet 15–20 mm, bright blue-violet, entire or slightly retuse; $2n=28$. Native of Eurasia, occasionally escaped or adventive with us, especially northeastward. June, July.

3. Geranium robertianum L.   Herb-Robert. Weak annual or biennial with branched, spreading, villous or villosulous stems to 6 dm; lvs 3–5-cleft to the base, at least the terminal segment evidently petiolulate, all segments pinnately lobed or cleft; peduncles from most of the upper nodes, mostly 2-fld; sep 7–9 mm, shortly aristate, erect at anthesis; pet 9–13 mm, bright pink or red-purple, entire, long-clawed; carpel-bodies (unlike our other spp.), disarticulating from the beak but tending to remain attached for some time to the stylar column by a pair of elongate, subapical filaments, the proper beak separating from the stylar column ± as in other spp.; $2n=32$, 64. Damp rich woods; widespread in the Old World (esp. Eurasia) and widely naturalized in our range. May–Sept. (*Robertiella r.*)

4. Geranium carolinianum L.   Carolina crane's-bill. Annual; stems several, freely branched, eventually to 6 dm, villous with spreading or somewhat retrorse hairs, and becoming glandular above, lvs rotund-reniform, 3–7 cm wide, deeply 5–9-cleft with oblong to obovate, deeply toothed or lobed segments; peduncles mostly 2-fld, the pedicels up to twice as long as the cal; sep to 1 cm, short-awned, equaling the retuse pet; fr 2–5 cm, the stylar beak 1–2 mm, the carpel-bodies hirsute with long antrorse hairs ca 1 mm; seeds very obscurely reticulate; $2n=52$. Dry, barren or sandy soil and waste places; Me. to B.C., s. to Fla., Tex., and Calif. May–Aug. Var. *carolinianum,* chiefly southern, but extending n. to Mass. and Mich., has a relatively loose and open infl; var. *confertiflorum* Fern., the common form in our range, has very short upper internodes, the fls thus in compact, many-fld, umbel-like terminal clusters. (*G. sphaerospermum* Fern., plants with seeds of maximum plumpness and sep of maximum width—an unusual combination.)

5. Geranium dissectum L.   Spreading to erect annual to 6 dm, retrorsely to spreading-hirsute; lvs round-cordate, 2–6 cm wide, the 5–7 divisions from rather shallowly lobed (on the lower lvs) to incised-dissected into linear segments; pedicels commonly 2, subequal to the fruiting cal and peduncle; sep 4–5 mm, shortly awned, about equaling the retuse pet, elongating to 6–8 mm in fr; carpel-bodies with short, spreading, often glandular hairs ca 0.5 mm; seeds strongly reticulate-pitted; $2n=22$. Native of Europe, intr. in our range from Mass. to Mich., s. to Ga. and Tex. Apr.–Aug.

6. Geranium sibiricum L.   Siberian crane's-bill. Annual or short-lived perennial; stems weak, spreading to erect, to 1 m, villous; cauline lvs reniform, 4–7 cm wide, palmately 3–5-cleft, the segments coarsely toothed or lobed, chiefly above the middle; peduncles with 1 or occasionally 2 fls, on villous pedicels; sep villous, 5–7 mm, shortly awned; pet upcurved, ± emarginate, scarcely surpassing the sep; mature fr, including the cal, 15–20 mm, the stylar beak ca 1 mm or less, the carpel-bodies finely pubescent on the sides, pilose on the back. Native of Asia and e. Russia, intr. here and there in our range. Aug., Sept.

G. nepalense Sweet, much like no. 6 but vigorously rhizomatous and with mostly 2-fld peduncles, is locally established in Mass.

7. Geranium columbinum L.   Longstalk crane's-bill. Annual or biennial; stems several, diffusely ascending, 2–4 dm; basal lvs reniform, 3–5 cm wide, 5–7-cleft nearly to the base into deeply and narrowly cleft segments; cauline lvs similar, on shorter petioles, even the uppermost ones opposite; peduncles elongate; pedicels 2, elongate, minutely pubescent with appressed retrorse hairs; sep 5–8 mm excluding the subulate tip, antrorsely strigillose; pet purple, retuse, 8–10 mm; mature fr 2.5 cm, the stylar beak 4–5 mm, the carpel-bodies glabrous or sparsely pilose on the back, not wrinkled; $2n=18$. Native of Europe, established as a weed from N.Y. to N.D., s to Ga. June–Aug.

8. Geranium bicknellii Britton.   Erect annual or biennial, usually with many ascending branches, eventually to 5 dm; lvs pentagonal, the principal ones cleft nearly to the base with usually 5 segments, these deeply incised into several narrowly oblong lobes; peduncles 2-fld, the elongate pedicels glandular-villous; sep at anthesis 7–9 mm including the conspicuous subulate tip; pet retuse, equaling the sep; mature fr, including the cal, 20–25 mm, the stylar beak 4–5 mm, the carpel-bodies 3 mm, sparsely hirsute, not wrinkled. Open woods and fields; Nf. and Que. to B.C., s. to N.Y., Pa., Ind., Mo., and S.D. May–Sept.

G. rotundifolium L., a native of Eurasia, with more rotund lvs cleft only to about the middle, the segments shallowly lobed about the rounded or truncate summit, and with smaller fls, the sep at anthesis 4–5.5 mm, blunt or shortly subulate-tipped, the fr 15–20 mm, is rarely adventive in our range. It might sometimes key to *G. molle,* but has the mature carpel-bodies hirsute and not wrinkled.

9. Geranium molle L.   Dove's-foot crane's-bill. Spreading or ascending annual, branched from the base, 2–5 dm; basal lvs rotund or reniform, 2–5 cm wide, 5–9 cleft, the segments usually 3-lobed at the summit; upper lvs progressively reduced, the uppermost alternate; fls numerous; cal and pedicels densely glandular-puberulent and also hirsute; sep acute, 3–4 mm at anthesis; pet 5–7 mm, deeply notched; fr (including the cal) 9–13 mm, the stylar beak 2–5 mm, the carpel-bodies glabrous but usually with oblique, curved wrinkles; $2n=26$. Native of Europe and w. Asia, now widespread as a weed and found here and there in our range. Apr.–Sept.

10. Geranium pusillum L. Small-fld crane's-bill. Diffusely branched annual or biennial, spreading or ascending, to 5 dm; basal lvs long-petioled, the rotund, 3–6 cm blade deeply 7–9-cleft with cuneate divisions palmately lobed at the summit; upper lvs progressively reduced; fls numerous on densely but minutely glandular pedicels; sep acute or shortly mucronate, 2.5–4 mm at anthesis, ± hirsute, especially at the margins, nearly as long as the retuse pet; 3–5 outer stamens lacking anthers; fr, including the cal, 9–12 mm, the stylar beak very short or none, the carpel-bodies strigillose, not wrinkled; seeds smooth; 2*n*=26. Native of Europe, established as a weed in fields and waste places from N. Engl. to B.C., s. to N.C., Ark., and Oreg. All summer.

11. Geranium pyrenaicum Burman f. Erect or ascending, short-lived perennial, 3–6 dm, little branched below the infl; sep at anthesis not more pubescent than the pedicels, 4–6 mm; pet 7–10 mm, deeply notched; fr 15–20 mm, the carpel-bodies strigose, not wrinkled. Native of Europe, rarely adventive in our range. May–Sept.

2. ERODIUM L. Stork's bill; filaree. Pet obovate, the 2 upper often differing in size from the 3 lower; outer series of stamens sterile or reduced to scales, the inner series with normal anthers; fr differing from *Geranium* in that the carpel-beaks separate completely from the stylar column (usually from the top downwards) each such beak typically remaining attached to the basally sharp-pointed carpel-body as a terminal awn, the lower part of which coils hygroscopically; seed smooth, retained within the carpel-body; small herbs with basal or alternate, serrate to more often pinnately divided or dissected lvs and umbel-like cymes of small, pink to purple fls. 75, mostly Mediterranean reg. and Middle East.

1 Pinnae sessile and sometimes confluent, mostly cleft to beyond the middle . . . . 1. *E. cicutarium.*
1 Pinnae usually short-petiolulate, less deeply cleft or merely toothed . . . . 2. *E. moschatum.*

1. Erodium cicutarium (L.) L'Her. Redstem-filaree. Winter-annual or biennial; stems at first anthesis very short, with the lvs mostly basal, later diffusely branched, to 4 dm; lvs elongate-oblanceolate, the principal ones pinnately compound with several sessile, ovate or oblong, deeply and irregularly cleft pinnae each 1–2.5 cm; infls long-peduncled, 2–8-fld, the pedicels 1–2 mm; sep 5–7 mm, mucronate or shortly awned; pet off-pink, 5–8 mm; anther-bearing filaments without teeth; fr 2–4 cm; 2*n* mostly = 40 in ours. Native of the Mediterranean region, now established as a common weed throughout most of the U.S. Apr.–Sept.

2. Erodium moschatum (L.) L'Hér. Whitestem-filaree. Much larger than no. 1; stems to 6 dm, the lvs to 3 dm; pinnae tending to be short-stalked, ovate-oblong, 2–4 cm, sharply and shallowly toothed or irregularly shallowly lobed; sep lacking a mucro; anther-bearing filaments with 2 teeth at the enlarged base; fr 3–4 cm; 2*n*=20. Native of the Mediterranean region, occasionally found as a weed in our range. Apr.–Sept.

Family **LIMNANTHACEAE**, the Meadow-foam Family

Fls perfect, regular, hypogynous, 3–5-merous; sep valvate; pet distinct, convolute; stamens as many as and alternate with the pet, or more often twice as many; filaments distinct, the antesepalous ones with an adnate nectary-gland internally at base; carpels 2–5, united by a gynobasic style, the ovary deeply lobed into globular segments; ovules solitary in each locule, basal, erect or ascending, apotropous; fr separating into indehiscent mericarps; seeds with straight embryo, 2 large cotyledons, small radicle, and no endosperm; small annual herbs, producing mustard-oil, with alternate, exstipulate, pinnately parted or compound to dissected lvs and solitary fls on long axillary pedicels. 2/11, temp. N. Amer.

1. FLOERKEA Willd. False mermaid. Fls 3-merous, autogamous; pet shorter than the sep; stamens 3–6; carpels 2 or 3; style shortly cleft at the top; delicate autogamous annuals with small white fls. Monotypic.

1. Floerkea proserpinacoides Willd. Glabrous, weak, decumbent to suberect, 0.5–3 dm; lvs compound or deeply divided into 3–7 linear, oblanceolate or elliptic segments 0.5–2 cm; pedicels axillary, at first about equaling the petiole, much longer in fr; sep lance-ovate, 2.5–3 mm at anthesis, to 6 mm at maturity; pet oblanceolate, 1–2 mm; mericarps 2.5 mm, ovoid-globose, warty. Damp woods in rich soil; N.S. and w. Que. to B.C., s. to Del., Tenn., and Calif. Apr., May.

FAMILY **BALSAMINACEAE**, the Touch-me-not Family

Fls perfect, hypogynous, resupinate, strongly irregular; sep 3(5), the lowest one (as seen) somewhat petaloid and often saccate, produced backward into a slender spur-nectary; upper petal (as seen) distinct, external in bud, concave and often partly sepaloid, the 4 others distinct or connate into 2 lateral pairs; nectary-disk wanting; stamens 5, the short, thick filaments connate at least above, and the short anthers connate or connivent, the androecium forming a deciduous calyptra over the ovary; carpels (4)5, united to form a compound, plurilocular ovary with axile placentas, a short style, and a single stigma or 5 stigmas; ovules pendulous, apotropous; fr a loculicidal, explosive capsule or a berry-like drupe; embryo straight, with a short hypocotyl and 2 expanded cotyledons; endosperm scanty or none; mostly herbs, commonly subsucculent, with simple, entire or merely toothed, pinnately veined lvs, the stipules represented only by a pair of small petiolar glands, or wanting. Only *Impatiens* and the monotypic *Hydrocera* of India and Java.

1. **IMPATIENS** L.   Touch-me-not. Sep in ours 3, the 2 upper (as the fl hangs on the pedicel) small, the lower one saccate, open in front, spurred at the closed end of the sac; pet apparently 3, the upper one often broader than long, each of the 2 lateral ones lobed in our spp. and regarded as 2 pet united; fr dehiscing elastically into 5 valves. 450, widespread but mainly trop. Asia and Afr.

1 Lvs alternate.
  2 Fls in axillary racemes; spurred sep and fr glabrous.
    3 Fls mostly orange-yellow, with brown spots, seldom pale; saccate sep conic, at its mouth half as
      wide as long; spur strongly curved forward .... 1. *I. capensis.*
    3 Fls mostly pale yellow with brown spots; saccate sep broadly obtuse at its summit, two-thirds
      to fully as wide as long; spur spreading at right angles .... 2. *I. pallida.*
  2 Fls solitary or paired in the axils; spurred sep and fr pubescent .... 3. *I. balsamina.*
1 Lvs all or mostly whorled or opposite .... 4. *I. glandulifera.*

1. **Impatiens capensis** Meerb.   Orange t.-m.-n. or jewel-weed. Glabrous annual, 5–15 dm, branched above; lvs soft, pale or glaucous beneath, ovate or elliptic, 3–10 cm, crenate-serrate with mucronate teeth, on long wingless petioles; racemes widely spreading, few-fld; fls drooping on slender pedicels, 2–3 cm, commonly orange-yellow, thickly spotted with reddish-brown, rarely lemon-yellow or white, or without spots; saccate sep conic, the slender spur 7–10 mm, curved forward close to the body; capsules 2 cm, dehiscing explosively when jarred; $2n=20$. Moist woods, brooksides, wet roadside ditches, and springy places; Nf. and Que. to Sask., s. to S.C., Ala., and Okla. June–Sept. (*I. biflora; I. nortonii.*)

2. **Impatiens pallida** Nutt.   Yellow t.-m.-n. or jewel-weed. Much like no. 1, avg larger; fls 2.5–4 cm, commonly pale yellow, ± dotted with reddish-brown, seldom ochroleucous or white; saccate sep obtuse, nearly or quite as wide as long, the short (4–6 mm) spur abruptly deflexed at about right angles; $2n=20$. Wet woods and meadows; Que. and N.S. to Sask., s. to N.C., Tenn., and Okla., mostly less abundant than no. 1, and often in shadier places. June–Sept.

I. parviflora DC., a Eurasian sp. with yellow (and red-dotted) fls 8–13 mm, the short spur prolonged straight back from the saccate sep, is sparingly intr. in N.S. and P.E.I.

3. **Impatiens balsamina** L.   Balsam. Annual, 3–8 dm, ± pubescent; lvs oblanceolate, 6–15 cm, sharply serrate, gradually tapering to the short petiole; pedicels 1 or 2 in an axil, 1–2 cm, 1-fld; fls purple or rose, varying to white, the pet 2–2.5 cm, the saccate sep wider than long and shorter than the spur; fr villous; $2n=14$. Native of s. Asia, widely naturalized in the tropics, and occasionally escaped from cult. in our range.

4. **Impatiens glandulifera** Royle.   Coarse, erect annual to 2 m; stems hexagonal; lvs mostly or all whorled or opposite, lanceolate to lance-ovate, acuminate, sharply serrate; racemes ascending on long peduncles, few-fld; fls purple or blue, varying to white, 3 cm; $2n=18, 20$. Native of the Himalyan region, becoming established as an escape from cult. in the ne. part of our range. Summer. (*I. roylei.*)

FAMILY **ARALIACEAE**, the Ginseng Family

Fls commonly small, mostly in umbels that may be arranged into secondary infls, usually perfect, regular (or the outermost ones irregular), epigynous, usually 5-merous; cal commonly of small teeth around the top of the ovary, or obsolete; pet distinct, mostly valvate; stamens alternate with the pet; carpels mostly 2–5, united into a compound, inferior, plurilocular ovary; style or styles generally swollen at base to form a ± definite stylopodium confluent with the epigynous nectary-disk; ovules solitary in each locule, apical-axile, pendulous, epitropous; fr a berry, or often a drupe with several pyrenes; seeds with small, dicotyledonous embryo and abundant, oily endosperm; woody plants or less often perennial herbs, with mostly alternate (rarely whorled), compound or lobed lvs and commonly broad or sheathing petiole that may have evident stipular appendages. 70/700, mostly trop. and subtrop.

1 Lvs compound.
  2 Lvs strictly alternate or basal, 2–3 times compound; umbels 2 or more .... 1. *Aralia*.
  2 Lvs in a single whorl, once palmately compound, umbel solitary .... 2. *Panax*.
1 Lvs simple.
  3 Stems strongly prickly .... 3. *Oplopanax*.
  3 Stems unarmed .... 4. *Hedera*.

1. **ARALIA** L. Pet and stamens each 5; carpels (4)5(6), the styles distinct or somewhat connate below; fr a berry-like drupe, tipped by the persistent styles, usually with as many pyrenes as carpels; prickly or unarmed herbs, shrubs, or trees, with pinnately or ternately compound or decompound lvs and 2–many umbels of white or greenish fls in each infl. 35, Indomal., e. Asia, N. Amer.

1 Herbs or half-shrubs, unarmed or in one sp. bristly toward the base.
  2 Plants scapose, the lvs and peduncle arising from the rhizome .... 1. *A. nudicaulis*.
  2 Plant leafy-stemmed.
    3 Umbels relatively few (2–25) in a loose cluster; stem bristly below .... 2. *A. hispida*.
    3 Umbels very numerous, in a larger, terminal, compound panicle; unarmed .... 3. *A. racemosa*.
1 Shrub or tree, armed with stout prickles; umbels numerous in a compound panicle .... 4. *A. spinosa*.

1. **Aralia nudicaulis** L. Wild sarsaparilla. Acaulescent perennial herb, the lvs and peduncle arising from a long rhizome; petiole erect, to 5 dm; lvs ternate, each division pinnately 3–5-foliolate; lfls lance-elliptic to obovate, to 15 × 8 cm, acuminate, finely serrate, the lateral ones asymmetric at base; peduncles usually much shorter than the petioles, bearing (2)3(–7) umbels; styles distinct; fr nearly black; 2*n*=24. Woods; Nf. to D.C., Ind., Neb., and Colo., and in the mts. to Ga. May, June.

2. **Aralia hispida** Vent. Bristly sarsaparilla. Perennial to 1.5 m from a stout rhizome, sometimes overwintering 1–5 dm above the ground, sharply bristly near the base; lvs few, on petioles usually shorter than the blade, bipinnate; lfls oblong to ovate or lanceolate, to 10 cm, acute or short-acuminate, sharply serrate; umbels several in a loose, open, terminal infl; styles connate half-length; fr globose, nearly black; 2*n*=24. Dry woods, especially in sandy or sterile soil; Nf. and Que. to Hudson Bay and Minn., s. to N.J., W.Va., and n. Ind. June, July.

3. **Aralia racemosa** L. Spikenard. Stout unarmed perennial herb to 2 m; lvs few, widely spreading, to 8 dm, the 3 primary divisions pinnately compound; lfls ovate, variable in size on the same lf, the larger to 15 cm, sharply and often doubly serrate, acuminate, obliquely cordate at base; infl a large panicle with numerous umbels; styles connate about half-length; fr dark purple; 2*n*=24. Rich woods; Que. and N.B. to Minn. and S.D., s. to N.C. and n. Mex. July.

4. **Aralia spinosa** L. Hercules' club. Shrub or small tree to 12 m, the stem, branches, and often the petioles and lf-rachis armed with stout prickles; lvs to 1 m, twice or partly thrice compound, the rather firm lfls ovate, 4–10 cm, acute or acuminate, serrate, acute to broadly rounded at base; umbels very numerous in a terminal compound panicle; fr black; 2*n*=24. Moist or wet woods; Del. to s. Ind. and Mo., s. to Fla. and Tex., and intr. n. to N.Y., s. N. Engl., and Mich. June, July.

2. **PANAX** L. Pet and stamens each 5; carpels (1)2 or 3, the styles distinct; fr a small, trigonous or flattened, berry-like drupe with 2 or 3 pyrenes; perennial herbs, the unbranched stems arising from a deep-seated, thickened or tuber-like root, bearing a single whorl of once palmately compound lvs (these usually 3), and one long-peduncled terminal umbel of white or greenish fls. 8, Asia and N. Amer.

1 Lfls long-petiolulate, acuminate; fls in early summer .... 1. *P. quinquefolium*.
1 Lfls sessile or subsessile, obtuse or subacute; fls in spring .... 2. *P. trifolium*.

1. **Panax quinquefolium** L. American ginseng. Root elongate, fusiform; stem solitary, 2–6 dm, with 1–4(5) lvs; lfls (3)5(7), oblong-obovate to obovate, 6–15 cm, acuminate, conspicuously serrate, on long petiolules; peduncle 1–12 cm; fls greenish-white, all or mostly perfect; styles usually 2; fr bright red, 1 cm thick; 2*n*=44,

48. Rich woods; Que. to Minn. and S.D., s. to Ga., La., and Okla., now rare. June, July. Some individuals in some populations have fls with only one style, locule, and seed.

2. Panax trifolium L.   Dwarf ginseng. Androdioecious; root globose; stem 1–2 dm; lfls 3–5, sessile or nearly so, lanceolate to elliptic or oblanceolate, 4–8 cm, obtuse or subacute, finely serrate; peduncle 2–8 cm; fls white or tinged with pink, often unisexual; styles usually 3; fr yellow, 5 mm thick; $2n=24$. rich woods and bottomlands; Que. and N.S. to Minn., s. to Pa., Ind., and Io., and along the mts. to Ga. Apr., May.

3. OPLOPANAX (T. & G.) Miq.   Pet and stamens each 5; carpels 2(3), the styles distinct; fr a berry-like drupe with 2(3) pyrenes; very prickly shrubs with large, simple, petiolate, lobed lvs and numerous crowded, head-like umbels in ample terminal racemes. Monotypic.

1. Oplopanax horridum (J. E. Smith) Miq.   Devil's club. Shrub 1–3 m, densely prickly on the stem, petioles, and lf-veins; lvs long-petioled, subrotund, 5–7-lobed, to 3.5 dm, cordate at base, the lobes acuminate or cuspidate, serrate; $2n=48$. Wet woods; Isle Royal, Mich. and adj. s. Ont.; Oreg. to Mont., n. to Alas., and in e. Asia. June. The Amer. plants are var. *horridum. (Fatsia h.; Echinopanax h.)*

4. HEDERA L.   Ivy; English ivy. Sep 5, very short; pet 5, fleshy, green; stamens 5; ovary usually 5-locular; style 1; fr a berry-like drupe with 2–5 pyrenes; unarmed, woody vine, climbing by numerous adventitious roots, with firm, evergreen, simple lvs and small fls in solitary or racemosely arranged umbels. 15, mainly Eurasia.

1. Hedera helix L.   Climbing to 30 m; young shoots and infl densely covered with small stellate to peltate hairs; lvs shining, dark green, dimorphic, those of flowering shoots narrowly elliptic to suborbicular-cordate, entire, the others broad, palmately 3- or 5-lobed; fr nearly black, with 2 or 3 pyrenes; $2n=48$. Native of Europe, cult. in various forms and occasionally escaped, especially southward.

FAMILY **APIACEAE**, the Carrot Family

Fls individually small, epigynous, 5-merous, all or many of them perfect and regular, typically borne in compound umbels, less often in heads or simple umbels, or the umbels irregular and proliferating; cal commonly of small teeth around the top of the ovary, or obsolete; pet distinct, valvate, typically inflexed at the tip; stamens alternate with the pet; carpels 2, united to form a compound, inferior, bilocular ovary with distinct (often short) styles, these often swollen at the base to form a stylopodium that is confluent with the epigynous nectary-disk; ovules solitary in each locule, apical-axile, pendulous, epitropous; fr a dry schizocarp, the 2 mericarps typically separating acropetally to reveal a slender (often bifid) ± persistent *carpophore,* or the carpophore less often wanting; embryo dicotyledonous; endosperm copious and oily; aromatic herbs with mostly alternate, often very large, simple to much more often compound or dissected leaves, the petiole commonly with a broad, sheathing base, exstipulate or with mere stipular flanges. (Umbelliferae, nom. altern.) 300/3000, cosmop., especially N. Temp.

The primary branches of the umbel are called *rays;* the members of the involucre (abbreviated invol) are *bracts,* and the members of the involucel (subtending an umbellet) are *bractlets.* The part of the mericarp that adjoins the other mericarp is the *commissure;* the ribs next to the commissure are the *lateral* ribs, and the others are *dorsal.* Laterally compressed frs have a narrow commissure, dorsally compressed frs a broad one. Generic delimitation is difficult, and the key here presented is strictly artificial.

1 Infl neither a true umbel nor a compound umbel . . . . Group I.
1 Infl a true umbel or a compound umbel.
  2 Lvs all simple, or reduced to petioles only, or represented by phyllodes . . . . Group II.
  2 Lvs, or many of them, compound, dissected, or deeply divided.
    3 Fr and ovary pubescent, tuberculate, bristly, or prickly . . . . Group III.
    3 Fr and ovary glabrous.
      4 Lvs organized into distinct and separate lfls of ± uniform shape, these often more than 2 cm
        wide . . . . Group IV.
      4 Lvs much dissected or decompound, the lfls not immediately obvious, being deeply lobed or
        divided into ovate, oblong, linear, or filiform segments less than 1 cm wide.
        5 Plants flowering . . . . Group V.
        5 Plants fruiting . . . . Group VI.

GROUP I

1 Peduncles arising from the nodes of a creeping or floating stem; lvs simple.
  2 Lvs orbicular to reniform; invol none or minute .... 1. *Hydrocotyle.*
  2 Lvs lanceolate to broadly oblong or ovate; invol evident.
    3 Invol of 2 ovate bracts; fls few, commonly 2–5 .... 2. *Centella.*
    3 Invol of 3 or more linear-oblong bracts; fls numerous, densely crowded .... 3. *Eryngium.*
1 Peduncles terminal or lateral, the stems not creeping or floating; lvs various.
  4 Plants with a rosette of pinnately dissected lvs on a pseudoscape; fr winged all around, but not
    tuberculate or bristly or scaly .... 48. *Cymopterus.*
  4 Plants ± leafy-stemmed; fr rough-tuberculate or bristly or scaly.
    5 Infl densely capitate, each fl subtended by a spine-tipped bractlet .... 3. *Eryngium.*
    5 Infl otherwise; bractlets none, or forming an involucel, unarmed.
      6 Cauline lvs simple and entire; fr smooth or granular .... 25. *Bupleurum.*
      6 Cauline lvs divided or dissected; fr rough-tuberculate or bristly.
        7 Cauline lvs deeply once palmately divided .... 4. *Sanicula.*
        7 Cauline lvs pinnately dissected .... 18. *Torilis.*

GROUP II

1 Umbels simple.
  2 Lvs oblong to reniform or orbicular, flat.
    3 Invol very small or none; lvs orbicular to reniform or broadly ovate .... 1. *Hydrocotyle.*
    3 Invol of 2 ovate bracts; lvs ovate to broadly oblong .... 2. *Centella.*
  2 Lvs reduced to hollow, septate, linear to spathulate phyllodes .... 40. *Lilaeopsis.*
1 Umbels compound.
  4 Bractlets ovate to lanceolate, conspicuous, surpassing the fls and frs .... 25. *Bupleurum.*
  4 Bractlets none, or linear and inconspicuous.
    5 Perennial; fr 4–6 mm, dorsally flattened, wing-margined .... 37. *Oxypolis.*
    5 Annual; fr 1.5–2 mm, subterete, merely ribbed .... 38. *Ptilimnium.*

GROUP III

1 Principal lvs once palmately or ternately or once pinnately compound or divided, the lfls sometimes
  again ± cleft.
  2 Fr merely pubescent; lfls mostly 1 dm wide or more (narrower only in a rare intr. sp.) .... 42. *Heracleum.*
  2 Fr bristly or spiny; lfls or segments less than 1 dm wide.
    3 Lvs palmately divided into 3–7 broad, toothed to incised segments .... 4. *Sanicula.*
    3 Lvs once pinnate, some of the primary pinnae again cleft .... 18. *Torilis.*
1 Principal lvs twice compound to decompound.
  4 Lvs with distinct, serrate lfls 1 cm wide or more.
    5 Rays 2–8; frs slightly flattened laterally, narrow and elongate, not winged .... 6. *Osmorhiza.*
    5 Rays 18–35; frs dorsally compressed, winged .... 43. *Angelica.*
  4 Lvs much dissected or decompound into segments less than 1 cm wide.
    6 Lvs all basal; low western perennial .... 47. *Lomatium.*
    6 Lvs cauline as well as basal, or all cauline.
      7 Invol of several conspicuous bracts; rays of the main umbel 20 or more.
        8 Bracts of the invol entire; fr merely pubescent .... 16. *Seseli.*
        8 Bracts of the invol pinnatifid; fr bristly or spiny .... 19. *Daucus.*
      7 Invol none, or of a few inconspicuous linear bracts; rays 15 or fewer.
        9 Bractlets of the involucel foliaceous, oblong to elliptic or obovate.
          10 Bractlets entire; fr short-beaked or beakless, merely pubescent .... 12. *Chaerophyllum.*
          10 Bractlets bifid; fr with a long, bristly beak .... 13. *Scandix.*
        9 Bractlets of the involucel narrow, inconspicuous, lanceolate to linear.
          11 Foliage glabrous; ultimate lf-segments filiform .... 17. *Spermolepis.*
          11 Foliage ± pubescent; ultimate lf-segments broader.
            12 Fr with hooked or glochidiate bristles or prickles; annual.
              13 Fr with a short unarmed beak; upper lvs with villous-ciliate sheath .... 8. *Anthriscus.*
              13 Fr beakless, covered with prickles; upper lvs with an entire sheath .... 18. *Torilis.*
            12 Fr merely puberulent; perennial .... 26. *Pimpinella.*

GROUP IV

1 Principal lvs once compound (or some of them simple in spp. of *Zizia* and *Thaspium*).
  2 Upper lf-sheaths much dilated, 1 cm or more wide when flattened; fls white; fr dorsally flattened,
    wing-margined.
    3 Outer fls of the infl irregular, radiant, the outer pet enlarged and often bifid; fr mostly 7–12 mm
      .... 42. *Heracleum.*

3 Outer fls regular like the others; fr mostly 3–5 mm .... 45. *Peucedanum*.
2 Upper if sheaths not dilated, less than 1 cm wide; fls and frs various.
4 Taprooted intr. weeds of waste places and disturbed habitats.
  5 Fls yellow; fr dorsally flattened, wing-margined .... 41. *Pastinaca*.
  5 Fls white; fr laterally flattened, scarcely winged.
    6 Lf-teeth sharply acute; lvs basally disposed, the cauline ones progressively reduced; infls few .... 26. *Pimpinella*.
    6 Lf-teeth blunt to rounded, except for the minute, mucronate tip; most spp. with leafy stems and numerous infls when well developed .... 20. *Apium*.
4 Native plants with a cluster of fibrous or tuberous-thickened roots, commonly occurring in woods or wet places.
  7 Lvs with 3 lfls (or the basal ones simple and merely toothed).
    8 Fls white or greenish-white.
      9 Lfls linear.
        10 Fr dorsally flattened, wing-margined; lfls not septate .... 37. *Oxypolis*.
        10 Fr scarcely compressed, with corky-thickened marginal ribs; lfls cross-septate ..... 39. *Cynosciadium*.
      9 Lfls much broader; fr slightly flattened laterally, with low ribs, not winged .... 5. *Cryptotaenia*.
    8 Fls yellow or cream-colored or purple.
      11 Central fl of each umbellet sessile; fr ribbed but not winged .... 28. *Zizia*.
      11 Fls all pedicellate; fr winged .... 29. *Thaspium*.
  7 Lvs with 5 or more lfls; fls white.
    12 Lfls entire, or with not more than 5 coarse teeth on each side.
      13 Lfls cross-septate; fr scarcely compressed, with corky-thickened marginal ribs ...... 39. *Cynosciadium*.
      13 Lfls not septate; fr dorsally flattened, wing-margined .... 37. *Oxypolis*.
    12 Lfls with numerous teeth, or irregularly incised; fr laterally flattened or subterete, not winged.
      14 Ribs of the fr obscure; lfls of the upper lvs irregularly incised .... 33. *Berula*.
      14 Ribs of the fr prominent; lfls of the upper lvs usually regularly serrate .... 34. *Sium*.
1 Principal lvs twice or thrice compound.
 15 Lfls entire; fls yellow .... 27. *Taenidia*.
 15 Lfls toothed or lobed; fls yellow or white.
  16 Plants following.
    17 Fls yellow or cream-colored.
      18 Invol none, or of 1 or 2 inconspicuous linear bracts.
        19 Central fl of each umbellet sessile .... 28. *Zizia*.
        19 Fls all pedicellate.
          20 Sep evident at 10×; native perennials with a cluster of fibrous or fleshy-fibrous roots, not weedy .... 29. *Thaspium*.
          20 Sep obsolete; taprooted intr. biennial weeds.
            21 Lvs 2–4 times dissected, with numerous small ultimate segments .... 21. *Petroselinum*.
            21 Lvs barely bipinnate, the principal pinnae with a few large lfls .... 41. *Pastinaca*.
      18 Invol of a few conspicuous, lanceolate, reflexed bracts .... 44. *Levisticum*.
    17 Fls white.
      22 Sheaths of the upper lvs dilated, 1 cm or more wide when flattened.
        23 Lfls commonly 2–3-lobed .... 45. *Peucedanum*.
        23 Lfls mostly merely toothed, only occasionally some of them lobed .... 43. *Angelica*.
      22 Sheaths of the upper lvs less than 1 cm wide.
        24 Rachis of the lf or pinna conspicuously foliaceous-winged below each triad of lfls .... 11. *Falcaria*.
        24 Rachis wingless.
          25 Primary lateral veins of the lfls tending to be directed to the sinuses between the teeth; some of the roots tuberous-thickened; base of stem thickened, hollow. cross-partitioned .... 35. *Cicuta*.
          25 Primary lateral veins mostly directed to the teeth; stem and roots not modified as in *Cicuta*.
            26 Sep none; bractlets none; plants with creeping rhizomes .... 7. *Aegopodium*.
            26 Sep evident at 10×; bractlets linear; plants with a stout taproot and/or caudex .... 30. *Ligusticum*.
  16 Plants fruiting.
    27 Fr evidently winged.
      28 Dorsal well as lateral ribs winged; frs not much compressed.
        29 Plants with a cluster of fibrous or fleshy-fibrous roots; fr rather broadly winged .... 29. *Thaspium*.
        29 Plants with a stout taproot and/or branching caudex; fr only narrowly winged ..... 30. *Ligusticum*.
      28 Dorsal ribs wingless, only the lateral ribs winged; fr dorsally compressed.
        30 Dorsal ribs strong, prominently elevated; plants usually or always taprooted.
          31 Invol none, or of a few inconspicuous, linear, deciduous bracts .... 43. *Angelica*.
          31 Invol of several conspicuous, lanceolate, persistent, reflexed bracts .... 44. *Levisticum*.
        30 Dorsal ribs filiform, scarcely elevated.
          32 Coarsely rhizomatous perennial; lower lvs ± biternate .... 45. *Peucedanum*.
          32 Taprooted biennial; lower lvs barely bipinnate .... 41. *Pastinaca*.
    27 Fr scarcely or not at all winged, flattened laterally or subterete.
      33 Rachis of the lf or pinna broadly winged below each triad of lfls; fr linear-oblong .... 11. *Falcaria*.

33 Rachis not winged; fr various, often broader.
  34 Primary lateral veins of the lfls tending to be directed to the sinuses between the
    teeth; base of the stem thickened, hollow and cross-partitioned .... 35. *Cicuta*.
  34 Primary lateral veins of the lfls tending to be directed to the teeth, or anastomosing
    and bearing no obvious relationship to the teeth or sinuses; stem-base not
    modified as in *Cicuta*.
    35 Stylopodium wanting; stem arising from a cluster of fibrous or fleshy-fibrous
      roots .... 28. *Zizia*.
    35 Stylopodium well developed.
      36 Reflexed styles extending back nearly to the middle of the fr; fr incon-
        spicuously ribbed; rhizomatous perennial .... 7. *Aegopodium*.
      36 Reflexed styles scarcely longer than the stylopodia; fr strongly ribbed;
        not rhizomatous.
        37 Taprooted biennial weed; ribs of the fr raised but not at all wing-
          like .... 21. *Petroselinum*.
        37 Strong perennials from a taproot and/or branching caudex, not weedy;
          ribs of the fr elevated, narrowly thin-winged .... 30. *Ligusticum*.

Group V

1 Fls yellow.
  2 Low, scapose, western perennial, mostly less than 3 dm .... 47. *Lomatium*.
  2 Taller, ± leafy-stemmed, 3 dm or usually more.
    3 Sep wanting; intr., weedy plants.
      4 Ultimate lf-segments filiform, mostly less than 1 mm wide.
        5 Plants perennial .... 22. *Foeniculum*.
        5 Plants annual .... 23. *Anethum*.
      4 Ultimate lf-segments broader, linear to cuneate-obovate, many or all of them more than 1
        mm wide; biennial .... 21. *Petroselinum*.
    3 Sep evident at 10×, ovate to obovate; native perennials, not weedy.
      6 Plants taprooted, western and midwestern .... 46. *Polytaenia*.
      6 Plants with a cluster of fibrous or fleshy-fibrous roots; Appalachian .... 29. *Thaspium*.
1 Fls white (or pink).
  7 Stem purple-spotted; coarse, freely branched biennial weed to 3 m .... 24. *Conium*.
  7 Stem not purple-spotted; habit various, but not as in *Conium*.
    8 Plants annual or biennial.
      9 Sep wanting.
        10 Bractlets ciliate or fimbriate .... 8. *Anthriscus*.
        10 Bractlets entire-margined, or wanting.
          11 Involucel scarcely developed, or wanting.
            12 Umbels terminating the stem and branches, not very numerous; lvs ± basally
              disposed, the cauline ones progressively reduced .... 10. *Carum*.
            12 Umbels in well developed plants very numerous and irregularly disposed, lateral
              as well as terminal; plants leafy-stemmed .... 20. *Apium*.
          11 Involucel well developed.
            13 Bractlets disposed all around the umbellet .... 12. *Chaerophyllum*.
            13 Bractlets all on the outer side of the umbellet .... 14. *Aethusa*.
      9 Sep evident at 10×.
        14 Outer fls of the umbel enlarged and radiant .... 15. *Coriandrum*.
        14 Outer fls like the others.
          15 Rays of the primary umbel mostly 2–4 .... 9. *Trepocarpus*.
          15 Rays of the primary umbel mostly 5–30 .... 38. *Ptilimnium*.
    8 Plants perennial.
      16 Plants dwarf, seldom over 2 dm at anthesis.
        17 Pet not inflexed at the tip .... 49. *Erigenia*.
        17 Pet inflexed at the tip.
          18 Herbage finely spreading-hairy; bractlets narrow, distinct .... 47. *Lomatium*.
          18 Herbage glabrous (and often glutinous); bractlets broad, connate below .... 48. *Cymopterus*.
      16 Plants taller, generally well over 2 dm at anthesis.
        19 Plants with bulbils in the axils of some of the upper lvs .... 35. *Cicuta*.
        19 Plants not bulbiliferous.
          20 Sep evident at 10×.
            21 Decumbent or floating aquatic with dimorphic lvs; adventive .... 32. *Oenanthe*.
            21 Erect terrestrial native western plants .... 31. *Perideridia*.
          20 Sep obsolete or nearly so.
            22 Involucel wanting .... 20. *Apium*.
            22 Involucel present.
              23 Bractlets lance-ovate, ciliate; rays few, mostly 3–10 .... 8. *Anthriscus*.
              23 Bractlets linear or filiform, not ciliate; rays in well developed plants more
                than 10 .... 36. *Conioselinum*.

GROUP VI

1 Fr dorsally flattened.
  2 Intr. annual weed with filiform lf-segments . . . . 23. *Anethum.*
  2 Native perennials, not weedy, with mostly broader lf-segments.
    3 Low western plants, less than 3 dm, scapose or with a pseudoscape.
      4 Fr with lateral wings only, the dorsal ribs wingless . . . . 47. *Lomatium.*
      4 Fr with one or more dorsal wings in addition to the 2 lateral ones . . . . 48. *Cymopterus.*
    3 Taller plants, at least 3 dm when well developed.
      5 Plants taprooted; stylopodium wanting; carpophore bifid to the base . . . . 46. *Polytaenia.*
      5 Plants with a cluster of fibrous or fleshy-fibrous roots.
        6 Carpophore and stylopodium wanting . . . . 29. *Thaspium.*
        6 Carpophore present, bifid nearly or quite to the base; stylopodium conic . . . . 36. *Conioselinum.*
1 Fr subterete or more or less compressed laterally.
  7 Stem purple-spotted; coarse, freely branching biennial weed to 3 m . . . . 24. *Conium.*
  7 Stem not purple-spotted; habit various, but not as in *Conium.*
    8 Fr lanceolate or linear and with an evident beak 1–3 mm, the ribs obsolete; bractlets ciliate or
       fimbriate . . . . 8. *Anthriscus.*
    8 Fr not beaked, often broader, its ribs evident; bractlets entire, or wanting.
      9 Plants distinctly perennial.
        10 Fr a little wider than long; rays of the primary umbel commonly only 2–4 . . . . 49. *Erigenia.*
        10 Fr as long as or longer than wide; rays several or many.
          11 Plants with bulbils in the axils of some of the upper lvs . . . . 35. *Cicuta.*
          11 Plants not bulbiliferous.
            12 Decumbent or floating aquatic with dimorphic lvs . . . . 32. *Oenanthe.*
            12 Erect terrestrial plants; lvs not notably dimorphic.
              13 Involucel wanting; intr., weedy spp.
                14 Ultimate segments of the lvs filiform, mostly well under 1 mm wide . . . . . 22. *Foeniculum.*
                14 Ultimate segments of the lvs evidently broader . . . . 20. *Apium.*
              13 Involucel present; native spp., not weedy.
                15 Some or all of the ribs of the fr evidently winged; carpophore wanting . . . . . . 29. *Thaspium.*
                15 Ribs of the fr filiform, not winged; carpophore present, bifid . . . . 31. *Perideridia.*
      9 Plants annual or biennial.
        16 Ultimate segments of the lvs linear or filiform.
          17 Fr mostly 8–10 mm, oblong-linear . . . . 9. *Trepocarpus.*
          17 Fr mostly 1.5–4.5 mm, often broader in shape.
             18 Involucel of well developed, slender bractlets all on one side of the umbellet; sep
               evident at 10×; plants of wet places, not weedy . . . . 38. *Ptilimnium.*
            18 Involucel wanting, or nearly so; sep obsolete; weedy plants of disturbed habitats.
              19 Fr ca 1.5 mm long, not more than half again as long as wide . . . . 20. *Apium.*
              19 Fr 3–4.5 mm, about twice as long as wide . . . . 10. *Carum.*
        16 Ultimate segments of the lvs mostly broader than linear.
          20 Fr relatively narrow, lanceolate to oblong or elliptic, 5–10 mm . . . . 12. *Chaerophyllum.*
          20 Fr broader, from broadly ovate to subglobose, up to ca 5 mm.
             21 Fr subglobose, scarcely separable into mericarps; sep persistent, those of the outer
              row of frs 1 mm or more . . . . 15. *Coriandrum.*
             21 Fr broadly ovate, readily separable into mericarps at maturity; sep obsolete.
               22 Ribs of the fr narrower than the intervals . . . . 21. *Petroselinum.*
               22 Ribs of the fr much wider than the intervals, or contiguous . . . . 14. *Aethusa.*

1. HYDROCOTYLE L.   Pennywort. Fr orbicular to ellipsoid, strongly flattened laterally, its ribs evident to obsolete, the secondary ones not developed; each mericarp with a layer of sclerenchyma surrounding the seed-cavity; carpophore wanting; peduncles axillary, elongate to nearly obsolete, bearing a simple umbel, or the umbel proliferous into an interrupted (sometimes forked) spike; invol small or none; sep minute or obsolete; pet white in our spp; small perennials, the slender stem prostrate or arched and rooting at the nodes, with simple, long-petiolate, broadly ovate to orbicular or reniform, sometimes peltate lvs. 75, mostly warm reg.

1 Lvs peltate.
  2 Infl usually a simple umbel; fr distinctly notched or cordate at base . . . . 1. *H. umbellata.*
  2 Infl usually proliferous, forming a simple or forked spike of 2 or more verticils; fr truncate to
    rounded at base . . . . 2. *H. verticillata.*
1 Lvs not peltate, the petiole attached at the base of the blade.
  3 Infl shorter than the subtending lf; fls and frs pedicellate.
    4 Umbels sessile or subsessile; lvs shallowly lobed . . . . 3. *H. americana.*
    4 Umbels long-peduncled; lvs lobed to about the middle . . . . 4. *H. ranunculoides.*
  3 Infl exceeding the subtending lf; fls and frs sessile . . . . 5. *H. sibthorpioides.*

1. Hydrocotyle umbellata L.   Water-p. Stems floating or creepng; lvs orbicular, peltate, to 7 cm, crenate or shallowly lobed; peduncles usually longer than the lvs; umbels usually simple, the fls rarely less than 10, the pedicels spreading or reflexed, variable, to 1 cm; fr 2–3 mm wide, distinctly notched or cordate at base, the margins acute. N.S. to Minn., s. to Fla. and Tex., Pacific coast and trop. Amer. July–Sept.

2. Hydrocotyle verticillata Thunb.   Whorled p. Stems filiform, creeping; lvs orbicular, peltate, to 6 cm, commonly with 8–14 shallow lobes; infl (except in depauperate plants) a simple or once to thrice forked spike to 15 cm or even more, bearing 2–several whorls of fls, these rarely more than 7 per whorl; fr 3–4 mm wide, truncate to rounded at base, the margins acute. Mass. to Fla. and Tex., chiefly near the coast; w. U.S. and trop. Amer. June–Aug. Typical *H. verticillata* has the fls and frs sessile or subsessile. Plants with distinctly pedicellate fls and frs, the pedicels to 1 cm, occur over the same range and have been called var. *triradiata* (A. Rich.) Fern. (*H. canbyi; H. australis.*)

3. Hydrocotyle americana L.   Marsh-p. Stems very slender, creeping; lvs orbicular to broadly ovate, 1–5 cm wide, shallowly 6–10-lobed; umbels sessile or subsessile, 2–7-fld, the pedicels under 1 mm; fr 1.5 mm wide, the dorsal surface acute, the dorsal and lateral ribs acute. Moist low ground; Nf. and Que. to Wis., s. to N.C., W.Va., and Ind. June–Sept.

4. Hydrocotyle ranunculoides L. f.   Buttercup-p. Stems floating or creeping, rooting at the nodes; lvs round-reniform with a basal sinus, to 7 cm wide, 5–6-lobed to about the middle; peduncles well developed but shorter than the lvs; fls 5–10 per umbel, on pedicels 1–3 mm; fr suborbicular, 2–3 mm wide, the dorsal surface rounded and virtually ribless, the lateral ribs low and obtuse; $2n=24$. Pa. and s. N.J. to Fla., w. through the s. states to the Pacific; S. Amer. May–Aug.

5. Hydrocotyle sibthorpioides Lam.   Lawn-p. Stems filiform, creeping; lvs subrotund, 1 cm wide or less, shallowly 7-lobed, on petioles 1–2 cm; peduncles surpassing the lvs; fls 3–10 per umbel; fr sessile, 1–1.5 mm wide, the ribs low and obscure; $2n=24$. Native of trop. Asia, established in lawns from N.J. and Pa. to S.C., w. to Ind. and Ky. Apr.–Sept. (*H. rotundifolia* Roxb., not Wallich.)

2. CENTELLA L.   Much like *Hydrocotyle*; fr orbicular to reniform, flattened laterally, constricted at the commissure, with conspicuous primary dorsal ribs and 2 well developed secondary ones; umbels 1 or few, arising from the base, simple, head-like, with few (2–5) subsessile to short-pedicellate white fls; bracts 2, ovate; sep obsolete; perennial herbs with simple lvs. 40, mainly S. Hemisphere.

1. Centella erecta (L. f.) Fern.   Stems prostrate and creeping, sending up at each node a cluster of petiolate lvs, the petioles often pubescent; blades ovate to broadly oblong, 1–6 cm, obtuse, entire or repand, truncate to shallowly cordate at base; peduncles 1–3, 1–10 cm; fr flat-ellipsoid, 3–5 mm wide. Marshes and wet soil along the coast; s. N.J. and Del. to Fla. and Tex.; trop. Amer. Summer. (*C. asiatica,* misapplied; *C. repanda.*)

3. ERYNGIUM L.   Eryngo. Infl composed of dense heads, each head subtended by bracts, each fl by a bractlet; fr globose to obovoid, not or scarcely compressed, variously covered with scales or tubercles, the ribs obsolete; stylopodium and carpophore wanting; sep usually conspicuous; pet white to purple; biennial or perennial, often spiny herbs. 200, widespread.

1 Bractlets spine-tipped, exceeding the fls; heads large, terminal or in a terminal cluster.
  2 Lvs evidently net-veined, not monocotyledonous in aspect.
    3 Plants taprooted; locally intr. European weed .... 1. *E. planum.*
    3 Plants fibrous-rooted; native coastal plain sp. .... 2. *E. aquaticum.*
  2 Lvs strictly parallel-veined, very monocotyledonous in aspect .... 3. *E. yuccifolium.*
1 Bractlets minute, subulate, hidden by the fls; heads small, axillary, solitary .... 4. *E. prostratum.*

1. Eryngium planum L.   Perennial to 1 m from a thickened, woody taproot; lvs net-veined, the basal and lower cauline with elliptic-oblong to elliptic-ovate, crenate to spinulose-serrate blade 4–10 cm rounded or cordate at base to the long petiole, the others progressively less petiolate (or sessile), more sharply toothed or cleft, and often broader; heads bluish, 1.5–2 cm, subtended by spreading, narrow, spinulose-toothed to entire bracts 1.5–3 cm; bractlets firm, 5–6 mm, the lower tricuspidate; $2n=16$. Native of Europe, sparingly intr. in fields and waste places, as in N.Y. June, July.

  Two other taprooted European spp. are also rarely intr. with us. *E. maritimum* L. has very firm, broad, palmately veined (and often lobed), very strongly spiny-margined lvs. *E. campestre* L., also very spiny, has more pinnate and more dissected lvs, the principal ones with a broad, green, spiny-toothed midstrip.

2. Eryngium aquaticum L.   Marsh-e. Erect, fibrous-rooted perennial to 1 m, branched above; lvs elongate, pinnately net-veined, linear to lance-oblong, the lower long-petioled, 1–3 dm, the upper smaller and less petiolate, varying from entire to spinulose-toothed or even pinnatifid; bracts spreading or deflexed, mostly exceeding the heads, entire to spinose-serrate or pinnatifid; heads shortly ovoid-cylindric, 1–1.5 cm; bractlets trifid above and spinulose-tipped; $2n=16$. Bogs and marshes near the coast; N.J. to ne. Fla. July–Sept. (*E. virginianum.*)

3. Eryngium yuccifolium Michx.   Rattlesnake-master.   Coarse, erect perennial to 1 m or more from fascicled roots, usually simple to the infl; lvs linear, much elongate, parallel-veined, entire or remotely spinulose, the lower to 8 dm × 2 cm, the upper gradually reduced; heads round-ovoid, 1–2 cm long and thick; bracts lance-ovate, entire or nearly so, cuspidate, rarely projecting from beneath the head; bractlets similar but smaller; $2n=96$. Moist or dry sandy soil, open woods, and prairies; s. U.S., n. to Va., Ind., Minn., and Kans., and occasionally adventive farther north. July, Aug. (*E. aquaticum,* misapplied.)

4. Eryngium prostratum Nutt.   Spreading e. Fibrous-rooted perennial, slender, prostrate and rooting at the nodes, or ascending to 3 or 4 dm; lvs petiolate, ovate to lanceolate, 1–4 cm, entire or toothed, innocuous; peduncles axillary, 1–3 cm; bracts linear-oblong, 2–10 mm; head cylindric, 4–8 mm; bractlets minute, subulate, not protruding beyond the fls; $2n=16$. Wet, sandy or muddy soil, shores, and bogs; Fla. to Tex., n. to se. Va., Ky., and Mo. July–Sept.

4. SANICULA L.   Sanicle. Fr ovoid or oblong to subglobose, slightly flattened laterally, not readily separating into mericarps, densely covered with hooked bristles sometimes arranged in rows; ribs obsolete; oil-tubes in our spp. (except no. 5) solitary in the intervals, 2–3 on the commissure; umbel irregular, with unequal spreading primary branches, the umbellets dense or subcapitate, commonly with 3 sessile or short-pedicellate perfect fls, their ovary bristly, and several staminate fls with smooth ovary, these all on much longer pedicels; sep relatively well developed, narrow, connate at base, persistent; pet greenish-white to greenish-yellow; stylopodium wanting; stems in our spp. arising from a cluster of fibrous or fleshy-fibrous roots; lvs palmately divided into 3–5(–7) broad segments, the basal long-petioled, the cauline progressively reduced and less petiolate; invol foliaceous. 40, nearly cosmop.

1 Styles distinctly longer than the bristles of the fr; staminate fls 12–25.
  2 Sep lance-subulate, 1–1.5 mm; staminate fls longer than the frs .... 1. *S. marilandica.*
  2 Sep lanceolate or ovate, 0.5 mm; staminate fls shorter than the frs .... 2. *S. gregaria.*
1 Styles shorter than the bristles of the fr; staminate fls 2–7 per umbellet.
  3 Cal-lobes at maturity of the fr inconspicuous, not exceeding the bristles.
    4 Roots slender, fibrous; plants biennial .... 3. *S. canadensis.*
    4 Roots thicker and cord-like; plants perennial .... 4. *S. smallii.*
  3 Cal-lobes at maturity connivent into a beak much exceeding the bristles .... 5. *S. trifoliata.*

1. Sanicula marilandica L.   Black snakeroot. Perennial; lvs 5(7)-parted (or even compound), doubly serrate, or toward the apex incised; bractlets like the bracts but smaller; fls greenish-white, the fertile ones sessile, the staminate 12–25 in each umbellet and surpassing the frs; sep lance-subulate, 1–1.5 mm, shortly connate below, equaling the pet, or nearly so; anthers greenish-white; fr subsessile, 4–6 mm, narrowed and with rudimentary bristles toward the base; styles recurved, exceeding the bristles. Woods; Que. and Nf. to B.C., s. to n. Fla., Mo., and N. M. June–Aug.

2. Sanicula gregaria E. Bickn.   Cluster-sanicle. Perennial; lvs 3–5-parted, the segments sharply serrate to incised; bractlets small, subscarious; fls greenish-yellow, the fertile on pedicels 0.5–1 mm, the staminate 12–25 per umbellet, surpassed by the frs; sep lanceolate to ovate, obtuse to subacute, ca 0.5 mm, connate to the middle; anthers bright yellow; fr subglobose, 3–5 mm, bristly to the base; styles as in no. 1; $2n=16$. Woods; N.S. and Que. to Minn. and e. S.D., s. to Fla. and Tex. June–Aug.

3. Sanicula canadensis L.   Canada-s. Fibrous-rooted biennial; lvs 3–5-parted; bractlets like the bracts but smaller; fls white, the fertile on pedicels 0.5–1 mm, the sterile few, mostly concealed by the fertile; sep subulate, surpassing the pet; anthers white; fr subglobose, on pedicels 1–1.5 mm, the bristles surpassing the inconspicuous styles and about equaling the sep; $2n=16$. Woods; Vt. and s. Ont. to Minn. and S.D., s. to Fla. and Tex. June–Aug.

4. Sanicula smallii E. Bickn.   Southern s. Perennial with somewhat thickened and cord-like roots; lvs 3-parted, the lateral segments usually deeply lobed; bractlets small, ovate; fls white, the fertile sessile, the staminate few, scarcely surpassing the fertile; sep subulate, about equaling the pet; fr sessile, subglobose to ovoid, 3–5 mm, the bristles exceeding the inconspicuous cal and styles; $2n=16$. Rich woods; n. Fla. to e. Tex., n. to se Mo., N.C., and e. Va. May, June.

5. Sanicula trifoliata E. Bickn.   Beaked s. Biennial; lvs 3-parted, the lfs coarsely and doubly serrate to incised, the lateral often deeply lobed; bractlets ovate, subscarious; fls white, the fertile sessile, the staminate few, on slender pedicels to 8 mm and much exceeding the fertile; sep lance-subulate, exceeding the pet; fr ovoid to oblong, 6–8 mm, including the comparatively few bristles, the sep lanceolate, connivent, exceeding the bristles, forming a conspicuous beak 2–2.5 mm; oil-tubes numerous. Woods; Vt. and adj. Que. to s. Wis. and ne. Io., s. to N.C. and Tenn. June–Aug.

5. CRYPTOTAENIA DC., nom. conserv. Honewort. Fr linear-oblong, slightly flattened laterally, tipped by the slender stylopodia, the ribs low and obtuse; oil-tubes 1–4 in the intervals, 2 on the commissure; carpophore bifid to the base; umbels compound, the rays ± unequal; invol none or of a single small bract; umbellets few-fld, the 0–3 bractlets minute, the pedicels very unequal; sep minute or obsolete; pet white; branching, glabrous herbs with trifoliolate lvs and numerous loose, irregular umbels arising terminally and from the upper axils. 4, the others Old World.

1. **Cryptotaenia canadensis** (L.) DC. Single-stemmed, fibrous-rooted perennial 3–10 dm; lower lvs long-petioled, the upper on short petioles dilated as far as the lfls; lfls lanceolate to rhombic or ovate, 4–15 cm, sharply and irregularly (often doubly) serrate or incised to sometimes lobed; rays 2–7, ascending, 1–5 cm; pedicels several, to 3 cm; fr dark, 5–8 mm; $2n=20$. Woods, Que. and N.B. to Man., s. to Ga., Ala., and Tex.; Japan. June, July. (*Deringa c.*)

6. OSMORHIZA Raf., nom. conserv. Sweet cicely. Fr linear to narrowly clavate, slightly flattened laterally, the base prolonged (in our spp.) into bristly tails, the ribs slender, acute, ± appressed-bristly; oil-tubes obscure or none; carpophore bifid less than half length; fls white or greenish-white; umbels peduncled, terminal and lateral, usually surpassing the lvs, few-rayed, with or without invol; umbellets few-fld; sep obsolete; pet white; erect perennials from thickened roots; lvs twice ternately compound, the lower petiolate, the upper subsessile, the ultimate segments ovate to lanceolate, serrate or lobed. 11, N. Amer., Andes, and e. Asia. Our spp. are habitally similar, 4–8 dm, pubescent to subglabrous, with mostly 3–6 widely ascending to divaricate rays, each ray with 3–6 fertile pedicels to 15 mm, usually mixed with some sterile ones.

1 Involucel present (but often ± deciduous in fr); mature styles nearly straight and parallel, mostly 1–
    4 mm.
  2 Styles, even in fr, not over 1.5 mm; plants inodorous .... 1. *O. claytonii.*
  2 Styles (including the stylopodium) ca 2 mm, in fr to 4 mm; plants anise-scented .... 2. *O. longistylis.*
1 Involucel none; mature styles strongly outcurved, to 1 mm (stylopodium included).
  3 Fr convexly narrowed to the summit .... 3. *O. depauperata.*
  3 Fr concavely narrowed to the summit .... 4. *O. chilensis.*

1. **Osmorhiza claytonii** (Michx.) C. B. Clarke. Bland s.c. Inodorous; styles at anthesis shorter than the pet, in fr nearly straight and parallel, 1.2–1.5 mm; mericarps 15–22 mm; bractlets several, linear-attenuate, reflexed, 3–8 mm; $2n=22$. Moist woods; Que. and N.S. to Sask., s. to N.C., Ala., and Ark., commoner n. and e. May, June. (*Washingtonia c.*)

2. **Osmorhiza longistylis** (Torr.) DC. Long-styled s.c. Much like no. 1, but the fls a little larger and more numerous in each umbellet, the styles at anthesis ca 2 mm, much exceeding the pet, in fr nearly straight and parallel, 3–4 mm; plants anise-scented; $2n=22$. Moist woods; Que. to Sask., s. to Ga., Tex. and Colo., commoner s. and w. May, June. (*Washingtonia l.*)

3. **Osmorhiza depauperata** Philippi. Blunt s.c. Much like no. 4; pedicels and rays more widely spreading and stiffly divaricate; fr mostly 10–15 mm, convexly narrowed to the rounded or obtuse to merely acutish summit, the tip not beak-like; stylopodium low-conic to depressed, as wide as or oftener wider than high; $2n=22$. Moist woods; Nf. and e. Que. to n. Vt.; Lake Superior reg. of n. Mich., ne. Minn., and s. Ont.; widespread in the w. cordillera; Chile and Argentina. (*O. obtusa.*)

4. **Osmorhiza chilensis** Hook. & Arn. Tapering s.c. Involucel none; fr 12–22 mm, concavely narrowed toward the summit, the terminal 1–2 mm distinctly set off as a broadly beak-like tip; styles at maturity outcurved, 0.3–1 mm (including the stylopodium, which is conic and commonly as high as or higher than wide); $2n=22$. Moist woods, N.S. and s. Que. to Me. and N.H.; Great Lakes reg. from Ont. to n. Wis. and ne. Minn.; widespread in the w. cordillera; Chile and Argentina. June. (*O. divaricata; Washingtonia d.*)

7. AEGOPODIUM L. Fr oblong-ovoid, flattened laterally, tipped by the 2 conspicuous stylopodia; ribs slender, inconspicuous, 5 per mericarp; styles conspicuous, reflexed, 1.5 mm; oil-tubes none; carpophore ± bifid; umbels long-peduncled, terminal and lateral; invol usually none; rays numerous, ascending; involucel none; pedicels many, short; sep obsolete; pet white; rhizomatous perennials with ternately once or twice compound lvs and large serrate lfls. 7, temp. Eurasia.

1. **Aegopodium podagraria** L. Goutweed. Erect, branched, 4–10 dm; lower lvs long-petiolate, mostly biternate with 9 lfls but often irregular; lfls oblong to ovate, 3–8 cm, sharply serrate; upper lvs reduced, short-

petiolate, chiefly once ternate; umbels dense, 6–12 cm wide, the 15–25 rays subequal; fr 3–4 mm; $2n$=22, 44. Native of Eurasia, cult. in old-fashioned gardens and often escaped in ne. U.S., especially in moist, partly shaded places. Most horticultural forms have white-margined lvs.

8. ANTHRISCUS Pers., nom. conserv. Fr ovoid to linear, laterally compressed, constricted at the commissure, distinctly beaked, the ribs obsolete, the oil-tubes obscure; carpophore entire or cleft at the tip; umbels compound, terminal (or lf-opposed) and from the upper axils, usually without invol; umbellets few-fld; bractlets ciliate or fimbriate, linear to ovate; sep obsolete; pet white; stylopodium conic; branching herbs with twice or thrice compound lvs and dentate to incised infls. 20, Eurasia.

1 Fr glabrous, linear to lanceolate; foliage glabrous to sparsely villosulous.
  2 Fr lanceolate, the beak a sixth as long as the body; bractlets lance-ovate .... 1. *A. sylvestris.*
  2 Fr linear, the beak a third as long as the body; bractlets lance-linear .... 2. *A. cerefolium.*
1 Fr hispid with hooked hairs, ovoid; foliage hispidulous .... 3. *A. caucalis.*

1. Anthriscus sylvestris (L.) Hoffm. Wild chervil. Freely branched biennial or short-lived perennial to 1 m; lf-segments 1.5–5 cm; umbels large, with 6–15 rays to 4 cm; bractlets lance-ovate, 3–6 mm; fr glabrous, lanceolate, 6 mm, the beak 1 mm; $2n$=16. Native of Europe, occasionally found as a weed in our range. May–July.

2. Anthriscus cerefolium (L.) Hoffm. Chervil. Freely branching annual to 7 dm; lf-segments 5–10 mm; umbels lax, with 3–6 rays to 3 cm; bractlets lance-linear, reflexed, 2–3 mm; fr glabrous, linear, the body 5–6 mm, the beak 2–3 mm; $2n$=18. Native of Europe, occasionally escaped from cult. in e. U.S. May–July. (*Cerefolium c.*)

3. Anthriscus caucalis M. Bieb. Bur-chervil. Branching annual to 1 m; lvs sparsely hispidulous, pinnately dissected, the ultimate segments 3–8 mm; umbels lf-opposed, lax, with 3–5 rays 1–2 cm; bractlets lance-ovate, aristate, 2–3 mm; pedicels elongating and becoming thicker than the rays in fr; fr ovoid, 3–4 mm, the body covered with short, thick, uncinate prickles, the short stout beak unarmed; $2n$=14. Native of Europe, intr. here and there in the U.S. June, July. (*A. scandicina.*)

9. TREPOCARPUS Nutt. Fr oblong-linear, slightly flattened laterally, glabrous; primary ribs obsolete, but 4 secondary ones and the face of the commissure prominently corky; oil-tubes small, solitary under the secondary ribs, 2 on the commissure, ± embedded in the seed; carpophore bifid to the base; umbels compound, lf-opposed; invol of a few foliaceous, entire or divided bracts; bractlets resembling the bracts; rays few, spreading; pedicels short; sep prominent, linear, unequal; pet white; styles very short, the stylopodium conic; slender, taprooted, leafy-stemmed annuals with pinnately decompound lvs and short, linear ultimate segments. Monotypic.

1. Trepocarpus aethusae Nutt. Plants 3–6 dm; petioles 0.5–2 cm; lf-blades 8–10 cm; peduncles 4–10 cm; bracts linear or lflike, 3–15 mm; bractlets 3–8 mm; rays 2–4, 5–15 mm, unequal; pedicels 2–8, to 1 mm; sep to 1 mm; fr 8–10 mm; $2n$=18. Moist low ground; S.C. to Miss., and up the Mississippi embayment to w. Ky.; rare. May, June.

10. CARUM L. Fr oblong to elliptic, flattened laterally, glabrous, the ribs narrow but prominent; oil-tubes solitary in the intervals, 2 on the commissure; carpophore bifid to the base; umbels compound, terminal and lateral, pedunculate; bracts few and subulate, or none; rays several to many; bractlets few and minute, or none; umbellets small, with very unequal pedicels; sep none; pet white (pink); stylopodium low-conic; taprooted biennials or perennials, the lvs pinnately dissected into linear or filiform segments. 30, widespread.

1. Carum carvi L. Caraway. Glabrous biennial to 1 m; lfls ovate, pinnately dissected into linear segments 5–15 mm; peduncles 5–13 cm; rays 7–14, ascending, 2–4 cm; pedicels ascending, 1–12 mm; involucel wanting or nearly so; fr 3–4.5 mm, half as wide, prominently ribbed; $2n$=20. Native of Eurasia, sometimes cult. and often established as a weed in waste places; Nf. to B.C., s. to Va., Ind., and Mo., commoner northward. June–Aug.

11. FALCARIA Fabr., nom. conserv. Fr linear-oblong, laterally compressed, slightly constricted on the commissure; ribs 5 per mericarp, broad and depressed, wider than the intervals; oil-tubes solitary in the intervals, 2 on the commissure; carpophore bifid to the base; umbels lateral and terminal, compound, long-pedunculate; bracts several, filiform; rays 10–25; bractlets filiform; umbellets many-fld; sep triangular; pet white; branching, glabrous herbs with once or twice ternately or ternately-pinnately compound lvs. 5, Medit. reg. and adj. Eurasia.

1. **Falcaria vulgaris** Bernh. Stem erect from a long taproot, to 1 m; lower lvs petiolate, the upper sessile at the top of the sheath, each of the 3 principal divisions either simple or again divided into 2–5 closely toothed, broadly linear segments 5–20 cm × 5–15 mm and often falcate; umbels 5–10 cm wide; bracts 5–20 mm; bractlets 2–10 mm; fr 2–5 × 1 mm; $2n=22$. Native of Eurasia, established as a weed in much of our range. July–Sept. (*F. sioides.*)

12. CHAEROPHYLLUM L.

Fr linear-oblong to elliptic or lanceolate, flattened laterally; ribs paler than the intervals; oil-tubes usually solitary in the intervals, 2 on the commissure; carpophore ± cleft; umbels lateral and terminal, compound or by reduction simple; invol none; bractlets conspicuous; sep obsolete; fls white, the marginal ones sometimes irregular; stylopodium conic; annuals or biennials with ternate-pinnately decompound lvs. 40, N. Temp.

1 Fr narrowly elliptic or oblong, broadest near the middle, with filiform ribs . . . . 1. *C. procumbens.*
1 Fr lanceolate, broadest below the middle, with stout ribs . . . . 2. *C. tainturieri.*

1. **Chaerophyllum procumbens** (L.) Crantz. Spreading chervil. Annual; stems spreading, often weak, usually branched from the base, 2–6 dm, glabrous or sparsely hairy; lf-segments oblong to ovate, glabrous beneath or with a few hairs; rays 1–3, the pedicels 2–6, very short at anthesis, later to 1 cm, filiform; fr narrowly elliptic or oblong, 5.5–10 mm, convexly narrowed to the summit, the ribs filiform, much narrower than the intervals. Moist woods and alluvial soil; N.Y. and s. Ont. to Mich., s. Wis., and e. Neb., s. to Ga., Ark., and Okla. Apr., May. Typical *C. procumbens* has glabrous fr. Occasional plants from Ind., O. and Ky. have pubescent fr and have been called var. *shortii* T. & G.

2. **Chaerophyllum tainturieri** Hook. Southern chervil. Annual; stems tending to be erect, usually branched, to 8 dm, mostly pubescent; lf-segments elliptic to narrowly oblong, hirsutulous beneath; rays 1–4, mostly 3, in fr stiffly erect or ascending; pedicels 3–10, becoming much thickened and often clavate; fr lanceolate, 5.5–8 mm, concavely narrowed to a short beak, the ribs stout, from a little narrower than to more often equaling or wider than the intervals; $2n=22$. Moist to dry sandy soil; se. Va., Ky., s. Ind., and e. Neb., s. to Fla., Tex., and Ariz. Mar.–May. (*C. floridanum.*)

Two European biennials have been reported as casuals along the Atlantic seaboard. *C. temulum* L. has 6–10 rays, lance-ovate bractlets, and oblong to ovate lf-segments. *C. bulbosum* L. has 10–20 rays, lance-linear bractlets, and lanceolate to linear lf-segments.

13. SCANDIX L.

Venus' comb. Fr linear to oblong, hispid or scabrous, the ± quadrate to subcylindric, evidently ribbed body tipped by a long, nearly ribless beak; oil-tubes solitary in the intervals, or obsolete; carpophore entire, or cleft at the top; infl of compound or simple umbels, the ultimate umbels aggregated into a usually few-rayed primary umbel or ± scattered, often paired; invol wanting, or of a small leafy bract; involucel of several lobed or dissected foliaceous bractlets; sep minute or obsolete; pet white, the outer ones of the outer fls enlarged; small annuals; lvs pinnately dissected, with ± linear, small and rather short ultimate segments. 10, Old World.

1. **Scandix pecten-veneris** L. Hispidulous to subglabrous, 1–4 dm, usually branched from the base; rays 1–3; involucel of 2-lobed bractlets; fr-body quadrangular, 6–15 mm, the flat, straight beak 2–7 cm; $2n=16$. Native of the Mediterranean region, widely established as a weed in the U.S., n. to N.J. and S.D. May–July.

14. AETHUSA L.

Fool's parsley. Fr broadly ovoid, subterete, the prominent corky ribs much wider than the intervals, or contiguous; oil-tubes solitary in the intervals, 2 on the commissure; carpophore bifid to the base; umbels compound, terminal and lateral; brcts few or none; bractlets 2–5, all on the outer side of the few-fld umbellet; sep none; pet white; stylopodium conic; poisonous annuals with pinnately decompound lvs. Monotypic.

1. **Aethusa cynapium** L. Freely branched, 2–7 dm; lvs shining, broadly deltoid, twice or thrice pinnately dissected into narrow acute segments; umbels 2–5 cm wide; rays 10–20; pedicels numerous, 1–5 mm; fr

rather acute, 3 × 2 mm; $2n=20$. Native of Eurasia, established as a weed here and there from N.S. and Me. to Pa. and O. June–Sept.

15. CORIANDRUM L. Fr subglobose, hard, scarcely separable into mericarps; primary and secondary ribs about equally devleoped, low, narrow, obtuse; oil-tubes none; carpophore bifid to the base; umbels terminal and lateral, compound, with 3–8 rays; invol none or of one narrow bract; umbellets small, few-fld; bractlets 3, small, linear, all at the outer side of the umbellet; sep unequal, those toward the periphery of the umbellet much the larger; pet white, pink, or purple, obcordate, the outer ones of the peripheral fls much the larger; stylopodium conic; annuals with pinnately decompound lvs. 2, Medit.

1. **Coriandrum sativum L.** Coriander. Erect, branched above, 3–7 dm; lowest lvs simple or variously divided into obovate incised segments; upper lvs pinnately dissected into linear segments; outer enlarged pet to 4 mm; fr 3–5 mm wide; sep persistent, those of the outer row of frs 1 mm or more; $2n=22$. Native of the Mediterranean region, established as a weed over much of the U.S., especially southward. June, July.

16. SESELI L. Fr ovoid, subterete, densely puberulent, with conspicuous narrow ribs; oil-tubes solitary in the intervals, 2–4 on the commissure; carpophore bifid to the base; umbels long-pedunculate, very dense; bracts several, linear; rays 20–40, about equal; umbellets many-fld; bractlets numerous, linear; sep linear; pet white; stylopodium depressed-conic; biennials or perennials with pinnate to decompound lvs and deeply incised lfls. 80, Eurasia.

1. **Seseli libanotis (L.) Koch.** Biennial or short-lived (usually monocarpic) perennial to 1 m, pubescent above; bracts and bractlets numerous, 5–8 mm; ovary and fr densely pubescent; umbels 5–10 cm wide; $2n=22$. Native of s. Europe, adventive in Md. July–Sept.

17. SPERMOLEPIS Raf. Fr ovoid, flattened laterally, constricted at the commissure, smooth to tuberculate or bristly, with narrow, inconspicuous, rounded ribs; carpophore cleft at the top; oil-tubes 1–3 in the intervals, 2 on the commissure; umbels peduncled, terminal and from the upper axils, loose and open, few-rayed; invol wanting, or of 1 or 2 slender bracts; involucel of a few small linear bractlets; umbellets 2–6-fld; sep obsolete; pet white; stylopodium low-conic; annuals with glabrous, ternately decompound lvs, the ultimate segments linear or filiform. 4, New World.

1 Fr glabrous or merely tuberculate.
  2 Primary rays of the umbel strongly ascending, unequal .... 1. *S. inermis.*
  2 Primary rays of the umbel ± widely spreading, subequal .... 2. *S. divaricata.*
1 Fr covered with uncinate bristles .... 3. *S. echinata.*

1. **Spermolepis inermis (Nutt.) Mathias & Constance.** Western s. Erect, 2–6 dm, commonly branched above the middle only; rays 5–11, strongly ascending, 2–15 mm, very unequal; pedicels 2–6, unequal, to 6 mm or those of the central umbellets sessile; fr 1.5–2 mm, tuberculate; $2n=22$. Dry or moist sandy soil; Tex. and n. Mex., e. to Neb. and s. Ill., and occasionally adventive farther e. May, June. (*S. patens.*)

2. **Spermolepis divaricata (Walter) Britton.** Southern s. Stems erect, 2–7 dm, widely branched above; rays 4–7, rarely fewer, 1–3 cm, ± widely spreading, subequal; pedicels 2–6, usually unequal, to 1.5 cm, or the central fl of each umbellet sessile; fr 1.5–2 mm, varying from strongly tuberculate to nearly smooth; $2n=16$. Dry sandy soil, chiefly on the coastal plain; Va. to Fla. and Tex., n. in the interior to Kans. Apr., May.

3. **Spermolepis echinata (Nutt.) A. A. Heller.** Hooked s. Low and often spreading, to 4 dm, branched from the base or only above; rays 5–14, ascending or suberect, unequal, to 1.5 cm; pedicels 2–6, to 7 mm, the central umbellets 1-fld, sessile; fr 1.5–2 mm, covered with short uncinate bristles; $2n=16$. Dry prairies and barrens, and a weed in waste places; Ariz. and n. Mex. to Okla., La., Mo., and w. Ky., and rarely intr. e. to N.Y. May, June.

18. TORILIS Adans. Hedge-parsely. Fr ovoid to narrowly oblong, flattened laterally, mostly ± beset with glochidiate and minutely barbellate prickles concealing the ribs; carpophore cleft up to half its length; umbels lateral or also terminal; invol of a few small bracts, or wanting; rays few–10; bractlets few, narrow; pedicels few and usually very short; pet white; stylopodium broadly conic; branching annuals with once to thrice pinnate, ternate-pinnate, or decompound lvs. 15, Eurasia, n. Afr.

1 Invol wanting, or of a single narrow bract .... 1. *T. arvensis.*
1 Invol of several bracts, commonly 1 to each ray .... 2. *T. japonica.*

1. Torilis arvensis (Hudson) Link.   Field h.-p. Freely branched, taprooted annual to 1 m, the stem retrorsely, the lvs antrorsely rough-strigose or partly glabrous; lvs ovate or lance-ovate in outline, 2–3 times pinnate, or the upper only once pinnate, at least the lower pinnae generally evidently petiolulate, the ultimate segments rather small; peduncles terminal and lf-opposed, 3–16 cm, invol wanting, or of a single slender bract; rays 2–10, 0.5–2.5 cm; pedicels 1–4 mm; fr ovoid-oblong, 3–5 mm, covered with widely spreading, nearly straight, cylindric-subconic, slightly upcurved prickles approaching 1 mm; $2n=12$. Native of Europe, now well established as a weed in s. U.S., n. to D.C., O., Ill., and Kans. June, July.

2. Torilis japonica (Houtt.) DC.   Japanese h.-p. Much like no. 1; lvs often a little less dissected; invol of several bracts, commonly 1 to each ray; fr with shorter, stouter, more obviously subconic, less widely spreading, more upcurved prickles ca 0.5 mm or a little less; $2n=16$. Native to the Old World, now widely established as a weed in our range, but less common than no. 1. (*T. anthriscus*, a preoccupied name.)

19. DAUCUS L.   Fr-body oblong or ovoid, flattened dorsally, the primary ribs low, with a row of short inconspicuous bristles, the 4 secondary ribs prominently winged, the wing divided into a row of flattened-subulate, hooked or straight spines; oil-tubes 1 under each secondary wing, 2 on the commissure; carpophore entire or bifid at the top; umbels compound, terminal and from the upper axils, long-pedunculate; bracts commonly large, pinnately dissected; bractlets linear or rarely pinnate; fls mostly white, the marginal ones often enlarged and irregular; stylopodium conic; annual or biennial herbs with pinnately decompound lvs. 60, widespread.

1 Biennial; invol bracts scarious-margined below, spreading or reflexed in fr .... 1. *D. carota.*
1 Annual; invol bracts not scarious-margined, appressed to the umbel in fr .... 2. *D. pusillus.*

1. Daucus carota L.   Wild carrot; Queen Anne's lace. Biennial with a stout taproot, 4–10 dm, hirsute to subglabrous; lvs oblong, pinnately decompound, the ultimate segments linear, lanceolate, or oblong; infl at anthesis showy, 4–12 cm wide, commonly narrower in fr, the outer rays longer than the others (to 7 cm) and arching inwards; bracts pinnatifid into firm, elongate, filiform-subulate segments, scarious-margined below the segments, spreading or reflexed in fr; umbellets with mostly (10–)20–numerous white or faintly yellowish fls (the central fl of the whole infl usually, the others rarely, purple or pink); fr 3–4 mm, broadest at the middle; $2n=18$. Native of Eurasia, now a weed throughout most of N. Amer. The cultivated carrot is a race of this sp.

2. Daucus pusillus Michx.   Slender annual, 0.5–6(–9) dm, ± hirsute; ultimate lf-segments linear, to 1 mm wide; invol bracts not scarious-margined, with segments under 1 cm, closely appressed to the infl in fr; fls all white (seldom purplish), the umbellets mostly 5–12-fld; fr 3–5 mm, usually broadest below the middle; $2n=22$. Dry, open places, somewhat weedy; widespread in s. U.S., n. to Mo. and N.C., and to be expected in s. Va. May–July.

20. APIUM L.   Fr oval to rotund, flattened laterally, somewhat constricted along the commissure; ribs 5 per mericarp, elevated; oil-tubes solitary in the intervals, 2 on the commissure; carpophore entire or not cleft beyond the middle; umbels terminal and lateral, compound, the lateral sometimes sessile; umbellets small; sep minute or none; pet white or greenish-white; stylopodium short-conic to depressed; lvs once pinnate to decompound. 10, widespread.

1. Apium graveolens L.   Celery. Taprooted perennial to 1 m, branched above; lvs once-pinnate; lfls 3–9, broadly triangular to suborbicular, usually deeply 3-lobed into cuneate-obovate, serrate or incised segments; umbels sessile or short-pedunculate, the several rays 1–2 cm; proper invol and involucel none, but some of the numerous umbels sometimes sessile and closely subtended by one or more reduced leaves; $2n=22$. Native of Eurasia, occasionally escaped in our range, but scarcely persistent. May–July. (*Celeri g.*)
    Three other spp. are merely casual with us. *A. leptophyllum* (Pers.) F. Muell., of trop. and subtrop. Amer., is an annual with filiform lf-segments. *A. repens* (Jacq.) Lagasca and *A. nodiflorum* (L.) Lagasca are Mediterranean rhizomatous perennials. The former has a well developed involucre, the latter does not.

21. PETROSELINUM Hill.   Fr broadly ovate, flattened laterally, constricted at the commissure; ribs 5 per mericarp, narrow, prominent; oil-tubes solitary in the intervals, 2 on the commissure; carpophore bifid at least to the middle; umbels long-peduncled, terminal and lateral; invol none or of 1–3 lance-subulate bracts; rays many; umbellets many-fld; bractlets 4–6, shorter than the pedicels; sep none; pet yellow or greenish-white;

stylopodium low-conic; taprooted biennials with pinnately or ternate-pinnately decompound lvs. 5, Eur., Medit.

1. **Petroselinum crispum** (Miller) Mansf. Parsley. Stem at anthesis to 1 m, much branched; lvs deltoid, the numerous ultimate segments linear to cuneate-obovate, variously toothed or lobed; umbels 5–8 cm wide, the rays about equal; pet yellow; fr 2–4 mm; $2n=22$. Native of the Mediterranean region, occasionally escaped from cult. in our range. (*P. hortense; Apium petroselinum.*)

22. FOENICULUM Miller. Fr oblong, slightly flattened laterally or subterete, the ribs slender but prominent; oil-tubes solitary in the intervals, 2 on the commissure; carpophore bifid to the base; umbels large, compound, on stout peduncles; invol and involucel none; umbellets numerous; sep none; pet yellow; stylopodium conic; coarse perennials with the lvs dissected into linear or filiform segments. 5, Europe and Medit. reg.

1. **Foeniculum vulgare** Miller. Fennel. Short-lived perennial, stout, 1–2 m, glabrous and glaucous; lf-segments mostly 1–4(–5) cm, well under 1 mm wide; rays 10–40, unequal, mostly 2–8 cm at maturity; fr 3.5–4 mm, the ribs acute; $2n=22$. Native of the Mediterranean region, now found throughout much of the U.S., especially southward, and elsewhere in warm regions.

23. ANETHUM L. Fr oblong or elliptic, dorsally flattened; ribs prominent, the dorsal narrow, the lateral narrowly to evidently winged; oil-tubes solitary in the intervals, 2–4 on the commissure; carpophore bifid to the base; umbels compound, terminal and lateral, overtopping the lvs; rays numerous; bracts and bractlets usually none; sep none; pet yellow; stylopodium conic; strongly scented annuals, the lvs pinnately dissected into numerous filiform segments. 4, Old World.

1. **Anethum graveolens** L. Dill. To 15 dm, branched above, glabrous and ± glaucous throughout; lvs ovate in outline, the lower long-petioled, the upper less so and smaller; ultimate lf-segments 5–20 mm; umbels to 15 cm wide; rays mostly 30–40, spreading, subequal; fr 3–5 mm, half as wide; $2n=22$. Native of s. Europe, escaped from cult. throughout most of the U.S. and elsewhere. July, Aug.

24. CONIUM L. Fr broadly ovoid, flattened laterally, barely constricted at the commissure; ribs 5 per mericarp, prominent, often undulate; oil-tubes very small and numerous; carpophore entire; umbels numerous, mostly terminal, often 2–4 together from the uppermost node of the stem; bracts few, ovate to lanceolate; rays numerous; bractlets like the bracts but smaller; sep none; pet white; stylopodium depressed-conic; biennials with large, pinnately decompound lvs. 4, temp. Eurasia, S. Afr.

1. **Conium maculatum** L. Poison hemlock. Stem freely branched, to 3 m, purple-spotted; lvs 2–4 dm, broadly triangular-ovate, 3–4 times pinnately compound, the ultimate segments ovate-oblong, 4–10 mm, toothed or incised; umbels 4–6 cm wide, the terminal one blooming first but soon overtopped by others; fr broadly ovoid, 3 mm, the pale brown ribs very prominent when dry; $2n=22$. Native of Eurasia, now widely intr. as a weed in waste places from Que. to Fla., w. to the Pacific. This is the hemlock of classical antiquity. All parts of the plant are very poisonous.

25. BUPLEURUM L. Fr orbicular to ellipsoid, slightly flattened laterally, somewhat constricted at the commissure, ribs narrow, scarcely prominent; oil-tubes obscure or none; carpophore bifid to the base; umbels compound, terminal and lateral; rays few; umbellets small, the involucel of conspicuously foliaceous bractlets exceeding the short pedicels; sep obsolete; pet yellow or greenish-yellow; stylopodium depressed-conic; plants with simple entire lvs. 150, N. Hemisphere.

1. **Bupleurum rotundifolium** L. Thoroughwax. Annual, 3–6 dm, glabrous and glaucous; lower lvs oblong to obovate, sessile or somewhat petiolate; upper lvs perfoliate, ovate, 2–5 cm; umbels peduncled, the 4–10 rays spreading or ascending, 0.5–2 cm; bracts none; bractlets broadly ovate to elliptic, acuminate, 8–12 mm, much surpassing the subsessile fls; fr dark brown or black, oblong-elliptic, 2.5–3 mm, obscurely ribbed; $2n=16$. Native of Europe and w. Asia, established as a weed in fields and waste places from s. N.Y. to s. Ind. and Kans., s. to N.C. and Ark., and occasionally adventive elsewhere. May, June.
   B. lancifolium Hornem. (*B. subovatum*), with rugose or tuberculate fr, and *B. fontanesii* Caruel, with linear sessile lvs and with a well developed involucre, are merely casual or waifs with us.

26. PIMPINELLA L. Fr ovate to orbicular, flattened laterally, the ribs 5 per mericarp, prominent and sometimes very narrowly winged; oil-tubes 2–4 in the intervals, 2 or more on the commissure; carpophore bifid at least to the middle; umbels peduncled, compound, terminal and lateral; invol usually none; rays 8–20; sep minute or none; fls white, pink, or purple, the marginal ones sometimes irregular; stylopodium ± conic; perennials with pinnate or decompound lvs. 150, mainly Old World.

1. **Pimpinella saxifraga L.** Burnet-saxifrage. Perennial, 3–6 dm, the stem pithy; lower lvs once pinnate, the lfls varying from ovate or subrotund and merely serrate to pinnately deeply dissected; upper lvs much reduced, sometimes consisting only of the sheath; fls white; fr glabrous, ovoid, 2–2.5 mm, obscurely ribbed, each mericarp nearly semicircular in cross-section; $2n=40$. Native of Eurasia, escaped or adventive in waste places; Nf. to DC., inland to Wis. and Ind. July–Sept.

P. major (L.) Hudson, with hollow stem and prominently ribbed, pentagonal mericarps, and *P. anisum* L., with puberulent frs, are merely casual or waifs with us.

27. TAENIDIA (T. & G.) Drude. Fr elliptic to broadly ovate-oblong, slightly compressed laterally or somewhat compressed dorsally, in the latter case with marginal wings; dorsal ribs low, oil-tubes mostly 3 in the intervals and 4 on the commissure; carpophore bifid to the base; umbels terminal, often some of them elongate, lf-opposed pedicels; invol and involucel generally wanting; umbels loose and open, with mostly 7–16 spreading or ascending rays, the outer rays the longer (especially in fr) and bearing umbellets with marginal fertile fls and inner staminate ones; the inner rays bearing staminate umbellets; sep minute or obsolete; pet yellow; stylopodium wanting; glabrous, somewhat glaucous perennials from tuberous-thickened roots, with basal and cauline, mostly 2 or 3 times ternate lvs and entire, mostly discrete lfls; $2n=22$. 2, e. U.S.

1 Fr laterally slightly compressed, not winged; widespread .... 1. *T. integerrima.*
1 Fr dorsally somewhat compressed, with marginal wings; mainly shale-barrens .... 2. *T. montana.*

1. **Taenidia integerrima** (L.) Drude. Yellow pimpernel. Plants with a light, pleasant, celery-like odor; stems branched, 4–8 dm; lower lvs long-petioled, commonly thrice compound, the upper once or twice compound and with wholly sheathing petioles; lfls ovate to oblong or elliptic, mostly 1–4 cm, sometimes some of the upper confluent; longer rays 4–9 cm in fr; fr mostly 4–5 × 3–4 mm, elliptic, slightly compressed laterally, not winged, the ribs all low; $2n=22$. Dry woods and open, rocky slopes, reputedly sometimes on shale-barrens with the next sp., s. Que. to c. Ga., w. to Minn., e. Kans., Okla., and Tex. Apr.–June.

2. **Taenidia montana** (Mackenzie) Cronq. Shale-barren p. Plants with a heavy, rather unpleasant, somewhat anise-like odor; fr somewhat compressed dorsally, mostly 4–7 × 3.5 mm, with lateral wings mostly less than 1 mm wide; otherwise very much like no. 1. Mainly on shale-barrens; s. Pa. to w. Va. and e. W.Va. (*Pseudotaenidia m.*)

28. ZIZIA Koch. Golden alexanders. Fr ovate to oblong, flattened laterally; ribs 5 per mericarp; oil-tubes solitary in the intervals, 2 on the commissure; carpophore bifid half its length; umbels compound, without invol; rays several to many, the inner often short; umbellets many-fld, with very unequal pedicels, the central fl commonly sessile; bractlets few, short, lance-linear; sep short, triangular; pet bright yellow; stylopodium wanting; branching perennials 3–8 dm, glabrous or nearly so, from a cluster of thickened roots, some or all of the lvs once to thrice ternately compound. 4, N. Amer.

1 Basal lvs once or twice ternately compound.
2 Lf-teeth fine, averaging 5–10 per cm of margin .... 1. *Z. aurea.*
2 Lf-teeth coarse, averaging 2–3 per cm of margin .... 2. *Z. trifoliata.*
1 Basal lvs simple, cordate at base .... 3. *Z. aptera.*

1. **Zizia aurea** (L.) Koch. Common g. a. Lower lvs twice ternate, the upper once ternate or irregularly compound; lfls ovate to lanceolate, finely serrate, the ascending teeth averaging 5–10 per cm of margin; rays 10–18, the outer ones of the terminal umbel becoming 3–5 cm and stiffly ascending; fr oblong-ovoid, 3–4 mm, half as wide; $2n=22$. Moist fields and meadows; Que. and Me. to Sask., s. to Fla. and Tex. May, June.

2. **Zizia trifoliata** (Michx) Fern. Mt. g.a. Lfls serrate with comparatively coarse teeth averaging 2–3 per cm of margin; lfls of upper lvs mostly lanceolate; rays 4–10, rarely more, at maturity slender, loosely spreading, the longest 6–12 cm; fr round-ovate, 3–4 mm, two-thirds as wide; otherwise much like no. 1; $2n=22$. Mt. woods; Va. to Ky. and Ark., s. to Fla. May. (*Z. bebbii.*)

3. **Zizia aptera** (A. Gray) Fern. Heart-lvd g. a. Basal lvs and occasionally the lower cauline ones simple, long-petioled, deltoid-ovate or round-ovate to oblong-ovate, cordate at base; cauline lvs once or twice ternate, the lfls lance-ovate to obovate-oblong; rays at anthesis 1–3 cm, at maturity ascending, to 5 cm; fr oblong-ovate, 3–4 mm; $2n=22$. Moist meadows and open woods; N.Y. to Man. and B.C., s. to Ga., Mo., and Colo. May, June. Our plants are var. *aptera*. (*Z. cordata.*)

29. **THASPIUM** Nutt. Meadow-parsnip. Fr oblong or ovoid, subterete or slightly flattened dorsally, some or all of the ribs prominently winged; oil-tubes solitary in the intervals, 2 on the commissure; carpophore wanting; umbels terminal and lateral, large, compound; bracts none; umbellets small; fls all pedicellate; involucel of small, narrow bractlets; sep ovate or obovate; pet yellow (or cream) or purple; stylopodium wanting; branching perennials with a rootcrown or short rhizome and a cluster of fibrous or fleshy-fibrous roots, and with variously compound lvs (or the lowest lvs simple), resembling *Zizia* in aspect. 3, N. Amer.

1 Basal lvs simple or once ternate; rays 6–10, unequal . . . . 1. *T. trifoliatum.*
1 Basal lvs twice or more ternate or pinnate; rays 8–16, subequal.
　2 Lfls ovate to lanceolate, serrate to incised; bractlets acute, 1–4 mm . . . . 2. *T. barbinode.*
　2 Lfls pinnately divided, with linear to oblong lobes; bractlets attenuate, 3–6 mm . . . . 3. *T. pinnatifidum.*

1. **Thaspium trifoliatum** (L.) A. Gray. Smooth m.-p. Sparingly branched, 3–8 dm, glabrous or nearly so; basal lvs simple, broadly ovate, usually cordate at base, or occasionally ternate like the cauline; cauline lvs ordinarily pinnate with 3(5) ovate or lanceolate, toothed lfls 4–8 cm, rarely the lateral lfls again 2–3-lobed; umbels long-peduncled, 3–8 cm wide, the 6–10 rays unequal; fr ellipsoid, 3.5–5 mm, three-fourths as wide, including the broad wings; $2n=22$. Woods, R.I. to Minn. and S.D., s. to Ga., La., and Okla. May, June. Typical *T. trifoliatum* (*T. atropurpureum*), with dark purple fls, is widespread, but more common eastward. Yellow-fld plants, usually distinguished as var. *flavum* S. F. Blake (*T. aureum,* misapplied), are also widespread, but more common westward.

2. **Thaspium barbinode** (Michx.) Nutt. Bearded m.-p. Stem to 1 m, branched above, pubescent around the upper nodes with minute stiffish hairs; basal and principal cauline lvs twice pinnate or ternate-pinnate, the lfls ovate to lanceolate, serrate to incised; umbels commonly 3–6 cm wide, at anthesis scarcely surpassing the lvs, the 8–16 rays subequal; fls pale yellow or cream-color; bractlets linear, acute, 1–4 mm; fr glabrous, ellipsoid, 4–6 mm, the lateral and some of the dorsal and intermediate ribs broadly winged; $2n=22$. Woods and prairies; N.Y. and Ont. to Minn., s. to Fla. and Okla. May, June.

3. **Thaspium pinnatifidum** (Buckley) A. Gray. Mt. m.-p. Stems to 8 dm, minutely puberulent at the nodes; basal and principal cauline lvs twice pinnate or ternate-pinnate; lfls 1–3 cm, deeply lobed or divided into linear or narrowly oblong segments; umbels 2–5 cm wide; bractlets narrowly linear, attenuate, 3–6 mm, mostly longer than the pedicels; fls yellow; fr ± puberulent, ellipsoid, 3–4 mm, narrowly winged; $2n=22$. Mt. woods; N.C., Tenn., s. Va., and Ky., and rep from sw. O. June.

30. **LIGUSTICUM** L. Lovage. Fr oblong to elliptic, somewhat flattened laterally to subterete, the ribs narrowly winged; oil-tubes in our spp. 2–4 in the intervals, 6 on the commissure; carpophore bifid to the base, its halves sometimes ± adnate to the commissural face of the mericarps; umbels compound, mostly terminal, sometimes also lateral; bracts narrow and deciduous, or none; rays few to many, ascending; umbellets small, few-fld; bractlets few–several, linear; sep triangular; pet white; stylopodium low-conic; perennials from a taproot and/or branching caudex, with ternately or pinnately compound lvs. 20, mostly N. Temp.

1 Fr half to two-thirds as wide as long; inland plants, from Pa. to Mo. . . . . 1. *L. canadense.*
1 Fr a third as wide as long; sea-coast plants from L.I. northward . . . . 2. *L. scothicum.*

1. **Ligusticum canadense** (L.) Britton. Nondo, angelico. Stout, freely branched, to 15 dm; lower lvs 3–4 times ternately compound, the upper less so and the uppermost sometimes simple; lfls thin, lanceolate to oblong or ovate, 3–8 cm, acute or acuminate, sharply serrate, broadly acute to rounded at base; terminal umbels often doubly compound, the primary rays to 2 dm, the secondary 2–5 cm, the pedicels 2–4 mm; fr elliptic, 5–7 mm, half to two-thirds as wide, the ribs narrowly winged; $2n=22$. Woods, chiefly in the mts.; Pa. and Ky. to Ga. and Ala.; s. Mo. and Ark. June, July.

2. **Ligusticum scothicum** L. Scotch lovage. Stout, simple or branched, 3–6 dm; lvs thick or fleshy, somewhat shining, chiefly twice ternate with 9 lfls, the upper smaller or once ternate; lfls rhombic to obovate, 3–10 cm, obtuse or acute, sharply serrate above the middle, mostly entire and cuneate at base; umbels compound; rays 10–20, 2–5 cm; pedicels 5–10 mm; fr oblong, 6–10 mm, a third as wide, the ribs prominent and narrowly winged; $2n=22$. Sandy or rocky seashores; Greenl. and Lab. to L.I.; n. Europe. July–Sept.

31. PERIDERIDIA Reichb.   Fr oblong to orbicular, laterally flattened, constricted at the commissure, the ribs filiform; oil-tubes in our sp. 3 in the intervals, 4 on the commissure; carpophore bifid to the base; umbels commonly 2 or 3 terminating the stem, with one or more lateral umbels below; bracts none or few and narrow; umbellets fewfld; bractlets few and narrow; sep triangular; pet white or pink; stylopodium ± conic; branching, glabrous perennials from fusiform or tuberously thickened roots; lvs pinnately decompound, or the uppermost once compound to simple. 15, mainly w. U.S., also one in e. Asia.

1. **Perideridia americana** (Nutt.) Reichb.   Stems 6–15 dm, sparingly branched; lvs ovate-oblong, the lower to 2 dm, thrice pinnately compound, with narrowly oblong ultimate segments; upper lvs once or twice compound with linear segments 1–3 cm, the uppermost lvs often simple; rays 7–14; pet white; fr broadly ovate to suborbicular, 3–5 mm; $2n=40$. Moist prairies and open woods; Mo., se. Kans., and n. Ark. to Ill., O., Ky., Tenn., and Ala. May, June. (*Eulophus a.*)

32. OENANTHE L.   Water-fennel. Fr subterete, oval to oblong, the ribs low, broader than the intervals; oil-tubes solitary in the intervals, 2 on the commissure; carpophore wanting; umbels terminal and lateral, compound, the rays spreading; bracts few and small, or none; umbellets small; with unequal pedicels and a few lance-linear bractlets; sep conspicuous, lanceolate, persistent; pet white; stylopodium conic; glabrous branching perennials with pinnate or decompound lvs. 40, Old World.

1. **Oenanthe aquatica** (L.) Lam.   Stem decumbent or floating; lvs decompound; lfls of submersed lvs filiform, of aerial lvs ovate, 1–2 cm, deeply pinnatifid into obtuse segments; fr ovate-oblong, 4–5 mm, tipped by the sep and ascending styles; $2n=22$. Aquatic; native of Eurasia, adventive at Washington, D.C., in O., and doubtless elsewhere in our range.

33. BERULA Besser ex Koch.   Fr broadly ovate to orbicular, laterally flattened, constricted at the commissure, the outer wall thickened and corky, obscuring the very narrow ribs; oil-tubes very numerous, carpophore wanting; umbels compound, terminal and lfopposed; involucre conspicuous, of narrow bracts; umbellets 1–2 cm wide; bractlets several, narrow, sep minute or none; pet white, stylopodium conic; fibrous-rooted aquatic perennials with once-pinnate lvs, the lfls ± incised. Only the following sp.

1. **Berula erecta** (Hudson) Coville.   Stem sparsely branched, to 8 dm; lvs oblong, the 9–23 lfls lanceolate to oblong or ovate, serrate or lobed, those of the upper lvs commonly lanceolate and deeply incised; rays 5–15, 1–3 cm; umbellets 1 cm wide; fr 1.5–2 mm, rarely maturing; $2n=12$. Springs, bogs, and shallow water; N.Y. and s. Ont. to Minn. and B.C., s. to Fla. and Mex.; Europe and the Mediterranean region. June–Sept. The Amer. plants are var. *incisa* (Torr.) Cronq. (*B. pusilla*.)

34. SIUM L.   Water-parsnip. Fr somewhat flattened laterally, ovate to orbicular, slightly constricted at the commissure, the ribs prominent and corky; oil-tubes 1–3 per interval, 2–6 on the commissure; umbels terminal and lateral, compound, with 6–many rays; bracts 2–several, lanceolate, reflexed; umbellets compact; bractlets slightly smaller than the bracts; sep minute or none; pet white, the outer sometimes slightly enlarged; stylopodium mostly depressed; usually tall, branched, glabrous perennials (sometimes annual?) with once pinnate lvs and 5–many elongate serrate lfls. 8, N. Temp. and S. Afr.

1. **Sium suave** Walter.   Fibrous-rooted from a very short erect crown; stem solitary, angular, stout, to 2 m; principal lvs with 7–17 linear to lance-ovate lfls 5–10 cm, sharply serrate, the primary lateral veins bearing no obvious relationship to the teeth or sinuses; submersed lvs, if present, bipinnately dissected; umbels 3–12 cm wide; bracts and bractlets usually 5 or 8; fr oval, 2–3 mm, nearly as wide; carpophore wanting; $2n=12$. Sunny wet meadows and swamps; Nf. to Alas. and Siberia, s. to Fla. and Calif. July–Sept. (*S. floridanum*, a lax shade-form; *S. carsonii*, a lax aquatic form with few lfls or even simple lvs.)

35. CICUTA L.   Water-hemlock. Fr ovoid to orbicular, flattened laterally, with thick corky ribs; oil-tubes solitary in the intervals, 2 on the commissure; umbels compound; bracts none or few, narrow, and inconspicuous; bractlets narrow, several, rarely none; sep well developed, triangular; pet white; stylopodium depressed or low-conic; carpophore bifid to the base, deciduous; glabrous, violently poisonous perennials, the tuberous-thickened base of the stem hollow, with well developed transverse partitions; roots clustered, generally some tuberous-thickened; lvs once to thrice pinnate or ternate-pinnate, mostly spp. with well defined lfls. 4, 1 circumboreal, the others N. Amer.

1 Axils not bulbiliferous; lfls often but not always well over 5 mm wide .... 1. *C. maculata.*
1 Axils of at least the upper lvs bulbiliferous; lfls to 5 mm wide .... 2. *C. bulbifera.*

    1. Cicuta maculata L.   Common w.-h. Stem stout, much branched, to 2 m; principal lvs twice or thrice pinnately compound, with wholly separate lfls, the uppermost less dissected or even simple; lfls linear to lanceolate, 3–10 cm, sharply and coarsely serrate to nearly entire, the primary lateral veins directed to the sinuses, whence one fork often extends into a tooth; umbels numerous, surpassing the leafy shoot, 5–12 cm wide; fr ovoid to orbicular, 2–4 mm, with prominent, rounded, pale brown ribs separated by darker intervals; $2n=22$. Swamps, marshes, and ditches; N.S. and Que. to Alas., s. to Fla., Calif., and Mex. June–Aug. (*C. mexicana.*) We have only the var. *maculata.*

    2. Cicuta bulbifera L.   Bulbiliferous w.-h. Stems slender, 3–10 dm, not much thickened at base and sometimes without thickened roots; lvs all cauline, with linear or very narrowly lanceolate, sparsely and saliently dentate to entire lfls or segments to 5 mm wide; umbels rarely to 5 mm wide, often none, surpassed by the leafy shoots; fr rarely matured, orbicular, 1.5–2 mm, the lateral ribs separated by a constriction at the commissure; $2n=22$. Swamps and marshes; Nf. to Alas., s. to Pa., W.Va., Ky., Kans., and Oreg., and at higher altitudes to Va. July–Sept.

    36. CONIOSELINUM Hoffm.   Hemlock-parsley. Fr elliptic or elliptic-oblong, flattened dorsally; dorsal ribs corky, prominent, acute or narrowly winged, the lateral ribs more evidently winged; oil-tubes 1–4 in the intervals, 2–8 on the commissure; carpophore bifid to the base, or nearly so; umbels large, compound; bracts foliaceous or narrow or none; rays numerous; umbellets small and densely fld; bractlets several, narrow, ± scarious; sep none; pet white; stylopodium conic; erect perennials from 1–several thickened roots, glabrous below, often scabrous-puberulent in the infl, with twice or thrice ternate-pinnately compound lvs and deeply lobed or incised lfls. 10, N. Amer. and Eurasia.

    1. Conioselinum chinense (L.) BSP.   Stout to very slender, 4–15 dm; lf-blades deltoid, the larger 1–2 dm, on elongate petioles, mostly 2–3 times compound, the upper once or twice compound, on short, broadly winged petioles; lfls lanceolate to ovate, 1.5–4 cm, pinnatifid; umbels few, 3–12 cm wide, long-peduncled; rays and pedicels many; fr elliptic or oblong, 2.5–4.5 mm; $2n=44$. Swamps, bogs, and wet meadows; Lab. and Nf. to Minn., s. to Pa., Ind., and Io., and in the mts. to N.C. Aug., Sept.

    37. OXYPOLIS Raf.   Water-dropwort. Fr elliptic or oblong to suborbicular, strongly flattened dorsally; dorsal ribs filiform; lateral ribs expanded into a thin or thick wing bearing a longitudinal nerve near the pericarp, the fr therefore apparently exhibiting 5 filiform ribs on each side; oil-tubes solitary in the intervals, 2–6 on the commissure; carpophore bifid to the base; umbels few, loose, open, compound; bracts and bractlets few, linear or filiform, or none; sep minute or none; pet white; stylopodium conic; erect glabrous herbs of marshes, from a cluster of fibrous or tuberous-thickened roots, sometimes also rhizomatous; lvs once pinnate or reduced to bladeless phyllodes. 6, N. Amer.

1 Lvs compound.
  2 Lfls 5–9 .... 1. *O. rigidior.*
  2 Lfls 1–3 .... 2. *O. ternata.*
1 Lvs reduced to bladeless phyllodes .... 3. *O. canbyi.*

    1. Oxypolis rigidior (L.) Raf.   Common w.-d. Stout or slender, to 1.5 m, with few lvs and branches; some of the roots often distally tuberous-thickened; lvs once pinnate, sessile or nearly so, oblong to deltoid, the 5–9 lfls narrowly linear to oblanceolate or elliptic, 6–14 cm × 5–40 mm, entire to strongly serrate; rays mostly 20+, spreading, forming a loose umbel to 15 cm wide; pedicels 5–20 mm; fr rounded at both ends, 4.5–6 mm, more than half as wide; $2n=32$. Swamps, marshes, ditch-banks, and wet prairies; L.I. to s. Ont. and Minn., s. to Fla. and Tex. July–Sept. (*O. turgida.*)

    2. Oxypolis ternata (Nutt.) Heller.   Three-lvd w.-d. Very slender, 4–8 dm; lvs 1–3, with elongate petioles from the summit of which arise 1–3 linear lfls; fr elliptic, 4–6 mm. In bogs on the coastal plain; se. Va. to Fla. Sept., Oct.

    3. Oxypolis canbyi (Coulter & Rose) Fern.   Corn-lvd or Canby's w.-d. Perennial from elongate, deep-seated rhizomes; stem very slender, to 1.5 m, with a few narrowly linear septate phyllodes 1–2 dm, these with basal sheaths 5–10 mm; rays 3–6; fr broadly elliptic, 4–6 mm, notched at the summit, the mericarps somewhat unequal; commissural oil-tubes 2, bowed away from the carpophore near the middle, lateral wings thick and corky with truncate margins; $2n=28$. Bogs on the coastal plain; Del. and Md.; S.C. and Ga. Aug.

38. PTILIMNIUM Raf. Mock bishop-weed. Fr ovoid to subglobose, slightly flattened laterally; ribs 5 per mericarp, about equal, corky, raised, pale brown, separated by dark brown or purple intervals, the adjacent lateral nerves of the 2 mericarps continuous to form a broad, pale brown corky strip; oil-tubes solitary in the intervals, 2 on the commissure, carpophore bifid up to half length; umbels compound, pedunculate, terminal and lateral; bracts well developed, filiform or cleft, umbellets small, bractlets filiform, larger on the peripheral side of the umbellet; sep deltoid to lanceolate; pet white, stylopodium conic; erect branching annuals, with the lvs dissected into filiform segments or reduced to the petiole. (*Harperella*.) 4, N. Amer.

1 Lvs with the blade dissected into many filiform segments.
  2 Bracts mostly 3-cleft; sep deltoid; styles 0.2–0.5 mm .... 1. *P. capillaceum*.
  2 Bracts rarely cleft; sep lance-triangular; styles 0.5--3 mm.
    3 Lf-segments crowded, appearing verticillate on the rachis .... 2. *P. costatum*.
    3 Lf-segments alternate or opposite, not crowded .... 3. *P. nuttallii*.
1 Lvs reduced to very slender petioles .... 4. *P. nodosum*.

    1. Ptilimnium capillaceum (Michx.) Raf. Atlantic m. b.-w. Erect, slender, 2–8 dm, freely branched above; lvs 4–10 cm, pinnately dissected into filiform segments, 5–25 mm, commonly 3 divisions at a node along the rachis; umbels 2–5 cm wide, overtopping the lvs; bracts few, 1–2 cm, simple or divided into filiform segments; rays 5–15 or more; sep minute, deltoid; fr broadly ovoid, subacute, 2–3 mm; styles 0.2–0.5 mm, not strongly recurved, scarcely as long as the stylopodium; $2n=14, 28$. Fresh or brackish swamps and marshes; Mass. to Fla. and Tex., chiefly near the coast in our range, n. in the interior to Ky., Mo., and Kans. June–Oct.

    2. Ptilimnium costatum (Elliott) Raf. Big m. b.-w. Robust, to 1.5 m, branched above; lvs 5–15 cm, pinnately dissected into innumerable filiform segments 5–10 mm, the primary divisions verticillate along the rachis; petiole well developed, the stipular wings none or barely developed at the base only; umbels long-pedunculate, 4–8 cm wide; bracts usually filiform and undivided; rays 12–30, 2–4 cm, fr broadly ovoid, 3–3.5 mm; styles 1.5–3 mm, spreading; $2n=22, 32$. Swamps and wet woods; irregularly from s. Ill. and Mo. to N.C., Ga., and Tex. July–Sept.

    3. Ptilimnium nuttallii (DC.) Britton. Ozark m. b.-w. Slender, 3–7 dm, branched above; lvs 3–10 cm, pinnately dissected into filiform segments 1–6 cm, the primary divisions alternate or opposite on the rachis; petiole to 1 cm, winged its whole length; umbels 2–6 cm wide; bracts filiform, usually entire; rays 15–30, slender, 1–3 cm; fr subglobose, 1.5 mm; styles 0.5–1.5 mm, strongly recurved at maturity; $2n=14$. Wet soil; Tex. and La. to Kans. and Mo., almost entirely s. of the Missouri R. June, July.

    4. Ptilimnium nodosum (Rose) Mathias. River-bank m. b.-w.; Harperella. Slender, sparsely branched, 1–10 dm; lvs consisting of petioles only, to 35 cm, with a scarious stipular margin toward the base, sometimes with axillary bulblets; rays 3–15, 5–20 mm; bracts minute or none; pedicels 3–15, 1–6 mm; bractlets lanceolate, 1 mm; fr ovoid, 1.5–2 mm; $2n=12$. Wet riverbanks and highly saturated soils; Potomac Valley of Md.; Va. and W.Va., s. to S.C., Ga., and n. Ala. July–Oct. (*P. fluviatile; P. viviparum; Harperella f.; Harperella n.*)

39. CYNOSCIADIUM DC. Fr broadly ovoid or globose-ovoid, evidently short-beaked, subterete or slightly flattened laterally; dorsal ribs narrow, prominent, the lateral ones low but broadly corky-thickened; oil-tubes solitary in the intervals, 2 on the commissure; carpophore merely notched at the top; glabrous annual from fascicled fibrous roots; basal lvs slender, petiolate, entire, cross-septate; cauline lvs petiolate, with 3–5 narrow, entire, flat but cross-septate lfls, invol and involucels wanting or of a few slender members; rays and pedicels rather few, unequal, spreading-ascending; cal-teeth evident, persistent; pet white; stylopodium conic. Monotypic.

    1. Cynosciadium digitatum DC. Plants 3–12 dm, freely (often subdichotomously) branched above; lfls to 10 cm × 5 mm, long-pointed; fr ca 2.5 mm, including the stout, 0.5+ mm beak, nearly or fully as wide. Roadside ditches and other wet, low places; s. Ill. and s. Mo. to Miss., La., and Tex.

40. LILAEOPSIS Greene. Fr broadly ovoid to suborbicular, slightly flattened laterally, the dorsal and intermediate ribs narrow and elevated, the lateral ones contiguous, much broader and usually rounded; oil-tubes solitary in the intervals, 2 on the commissure; carpophore wanting; umbels axillary, simple, peduncled; bracts few, minute; pedicels short, at maturity spreading or recurved; sep obsolete; pet white; stylopodium depressed or none; small aquatic perennials with creeping rhizomatous stems emitting from each node one or a few linear-spathulate, transversely septate phyllodes. 20, New World and Australia.

1 Phyllodes with 4–6 transverse septa, about as long as the peduncles .... 1. *L. chinensis.*
1 Phyllodes with 7–15 transverse septa, much longer than the peduncles .... 2. *L. attenuata.*

    1. Lilaeopsis chinensis (L.) Kuntze.   Phyllodes 2–6 cm, with 4–6 transverse septa; peduncle about as long, with 4–9 fls; fr broadly ovoid, 2 mm, nearly or fully as wide; $2n=22$. Tidal mud and brackish marshes; N.S. to Fla. and Miss. Summer. (*L. lineata.*)

    2. Lilaeopsis attenuata (Hook. & Arn.) Fern.   Phyllodes to 25 cm, with 7–15 transverse septa; peduncles 1–5 cm, with 5–15 fls; fr as in no. 1; $2n=22$. Tidal mud; se. Va. to Fla. and La.; S. Amer. Summer. (*L. carolinensis.*)

    41. PASTINACA L.   Fr glabrous, elliptic to obovate, strongly flattened dorsally, the lateral ribs narrowly winged, the others filiform; oil-tubes solitary in the intervals and visible to the naked eye through the pericarp, 2–4 on the commissure; carpophore bifid to the base; umbels large, compound, the terminal soon overtopped by the lateral ones; bracts and bractlets usually none; rays numerous, unequal; sep minute or none; pet yellow; stylopodium depressed-conic; tall stout herbs, the lower lvs long-petioled, the upper on shorter, wholly sheathing petioles, all typically once pinnate, the lfls variously serrate to lobed or divided, those of the larger lvs sometimes again few-pinnate. 15, temp. Eurasia.

    1. Pastinaca sativa L.   Parsnip. Taprooted biennial to 1.5 m; lfls 5–15, oblong to ovate, 5–10 cm, serrate or lobed, or, in robust pls the larger ones even divided into 2–several lfls; rays 15–25; umbel 1–2 dm wide; fr 5–7 mm; $2n=22$. Native of Eurasia, long cult. and thoroughly established as a weed in waste places, fields, and along roadsides throughout most of N. Amer. The common cult. parsnip and similar wild plants with smaller roots are considered conspecific, and some of the wild plants may actually be recent escapes.

    42. HERACLEUM L.   Fr elliptic to obovate, strongly flattened dorsally, usually pubescent, the lateral ribs broadly winged, the others filiform; oil-tubes 2–4 on the commissure, solitary in the intervals, extending only part-way from the stylopodium toward the base of the fr, readily visible; carpophore bifid to the base; umbels large and compound, with numerous unequal rays; bracts lanceolate, deciduous; umbellets many-fld; bractlets linear; sep minute or none; pet white; outer fls of at least the marginal umbellets irregular, the outer pet enlarged and often bifid; tall, stout biennials and perennials with large, ternate or once pinnate lvs and broad, toothed or cleft lfls, the petioles sheathing and usually conspicuously expanded. 60, circumboreal, only one native to N. Amer.

1 Rays of the principal umbel very numerous, mostly 50–150; intr. weed .... 3. *H. mantegazzianum.*
1 Rays fewer, mostly 15–30(–45).
  2 Lvs once ternate, with broad, petiolulate lfls; common native sp. .... 1. *H. lanatum.*
  2 Lvs pinnate, the pinnae sessile or the lowest short-petiolulate, rare weed .... 2. *H. sphondylium.*

    1. Heracleum lanatum Michx.   Cow-parsnip. Robust, single-stemmed perennial 1–3 m, generally ± pubescent or even tomentose; lvs once ternate, with broad, petiolulate, coarsely toothed and palmately lobed lfls mostly 1–3(–4) dm long and wide, the lateral ones mostly narrower than the central and often asymmetrical; terminal umbel commonly 1–2 dm wide at anthesis, with mostly 15–30 rays; bracts 5–10, narrow, deciduous; fr on pedicels 8–20 mm, obovate to obcordate, 7–12 mm, often pubescent; $2n=22$. Rich damp soil; Lab. to Alas. and Siberia, s. to Ga. and Ariz. June, July. (*H. sphondylium* ssp. *montanum*, as to Amer. plants; *H. maximum.*)

    2. Heracleum sphondylium L.   Lvs pinnate with mostly 5(–9) toothed (and often also cleft) segments; bracts few or none; otherwise much like no. 1, but the lf-segments often smaller and the umbel sometimes with as many as 45 rays; $2n=22$. Native of Eurasia, casually intr. in waste places from Nf. to N.Y.

    3. Heracleum mantegazzianum Sommier & Levier.   Giant hogweed. Very robust biennial or monocarpic perennial 2–5 m, stem usually purple-blotched; lvs to 3 m, ternate or ternate-pinnate, with pinnately lobed lateral segments to 13 dm; umbels to 5 dm wide, with 50–150 rays; pet ca 1 cm; fr 8–11 mm, on pedicels 1.5–4 cm; $2n=22$. Native of sw. Asia, recently established as a weed in c. and w. N.Y., and to be expected to spread.

43. ANGELICA L. Angelica. Fr oval, oblong, elliptic, or orbicular, usually flattened dorsally, the lateral and often also the dorsal ribs evidently winged, or the ribs sometimes all merely elevated and corky-thickened; carpophore bifid to the base; umbels very large, compound; bracts none or few and linear; rays many; umbellets densely many-fld; bractlets linear or filiform; sep minute or none; stylopodium broadly conic; pet in ours white or greenish-white; stout, taprooted, usually single-stemmed perennials with long-petioled, pinnately decompound basal lvs and progressively reduced upper lvs, the uppermost often bladeless, the lfls otherwise broad and toothed or cleft. 50, circumboreal.

1 Lateral ribs of the fr with thin flat wings; plants not maritime.
  2 Lfls acute or acuminate; ovary and fr glabrous or very minutely granular.
    3 Lfls merely acute, with a narrow, pale, mostly glabrous margin . . . . 1. *A. atropurpurea.*
    3 Lfls acuminate, without differentiated margin, roughly ciliolate . . . . 2. *A. triquinata.*
  2 Lfls obtuse; ovary and fr distinctly puberulent . . . . 3. *A. venenosa.*
1 Lateral ribs of the fr thick and corky; plants maritime . . . . 4. *A. lucida.*

   1. Angelica atropurpurea L. Purplestem-a. Stout, to 2 m, glabrous or nearly so to the infl; lower lvs 1–3 dm, long-petiolate, the upper progressively reduced, with broad, sheathing petioles; lfls ovate to lanceolate, 4–10 cm, acute, sharply serrate, thinly margined and only rarely ciliate, glabrous or minutely hairy beneath; umbels 1–2 dm wide, with 20–45 thinly hairy rays; fr oblong-elliptic, 4–6.5 mm, rounded at base, glabrous, the lateral wings thin, flat, broad; oil-tubes numerous; seed loose in the pericarp; $2n=22$. Swamps and wet woods; Lab. to Minn., s. to Del., W.Va., and Ind. June–Aug.

   2. Angelica triquinata Michx. Mt.-a. Stout, to 1.5 m, glabrous below and at base of the rays, usually scaberulous below the umbellets; lower lvs 1–3 dm, the upper much reduced, with broad sheathing petioles; lfls lanceolate to oblong, 3–8 cm, acuminate, coarsely toothed, the margin minutely ciliate and not differentiated; umbels 6–15 cm wide; rays 13–25, the outer spreading but not decurved; fr oblong-elliptic, 4–7 mm, cordate at base, glabrous, the lateral ribs prominently winged; oil-tubes few; seed adherent to the pericarp; $2n=22$. Mt. woods; Pa. to N.C. and Tenn. July–Sept. (*A. curtisii.*)

   3. Angelica venenosa (Greenway) Fern. Hairy a. Stout, to 2 m, becoming tomentulose-canescent in the infl; principal lvs long-petioled, 1–2 dm, the upper reduced, with sheathing petioles; lfls oblong to elliptic, varying to lanceolate, 2–4 cm, obtuse, finely serrate; umbels 5–15 cm wide, rays 18–35; pedicels and ovary pubescent; fr oblong-elliptic, 4–7 mm, cordate at base, sparsely pubescent along the wingless dorsal ribs; lateral ribs winged; oil-tubes few; seed adherent to the pericarp; $2n=22$. Dry woods and thickets; Mass. to Minn., s. to Fla., Miss., and Ark. July–Aug. (*A. villosa* Walter, not Lag.)

   4. Angelica lucida L. Seabeach-a. Stout, 5–10 dm; herbage glabrous; lower lvs long-petioled, 1–3 dm, the uppermost much reduced, with short, broad, sheathing petioles; lfls ovate, 4–7 cm, acute or obtuse, sharply and irregularly serrate, especially above the middle; umbels few, long-pedunculate; rays 20+, 3–10 cm, ascending, puberulent; fr elliptic or oblong, 5–9 mm, the ribs all very prominent, corky, acute but not winged, contiguous at base, the lateral ones somewhat the wider; oil-tubes numerous; seed loose in the pericarp. Beaches and rocks along the coast; L.I. to Lab.; margin of the Pacific Basin from Calif. to n. Asia. June–Aug. (*Coelopleurum actaeifolium.*)

   44. LEVISTICUM Hill, nom. conserv. Lovage. Fr oblong or elliptic, dorsally flattened; lateral ribs winged; dorsal and intermediate ribs merely acute; oil-tubes 1 or 2 in the intervals, 2–4 on the commissure; carpophore bifid to the base; umbels compound; bracts and bractlets several, lanceolate, reflexed; sep none; pet yellow or greenish-yellow; stylopodium conic; perennials with the aspect of *Angelica,* the lvs twice or thrice pinnately compound. 3, SW Asia.

   1. Levisticum officinale Koch. Stem 1–2 m; upper lvs progressively reduced and less compound, the uppermost sometimes simple; lfls narrowly to broadly cuneate and entire in the basal half, the distal half triangular, acute, sharply serrate or incised; umbels 3–10 cm wide; fr elliptic, 5–7 mm, half as wide; $2n=22$. Native of the Old World, escaped from cult. here and there in our range. July, Aug. (*Hipposelinum levisticum.*)

   45. PEUCEDANUM L. Fr strongly flattened dorsally, the lateral ribs evidently winged, the others wingless; oil-tubes 1–3 in the intervals, 2 or more on the commissure; carpophore bifid to the base; umbels compound; pet white or yellow; stylopodium conic; perennials, usually coarse and erect from a taproot or stout rhizome, with pinnately or ternately compound or dissected lvs, often with well defined lfls. 100+, Old World.

   1. Peucedanum ostruthium (L.) Koch. Masterwort. Aromatic, subglabrous stout perennial 5–15 dm from a coarse, branching rhizome; lvs large, the lower broadly deltoid and ± biternate, the lfls oblong or obovate, sharply and irregularly serrate (often doubly so), and commonly each 2–3-lobed or -parted; umbels large, long-peduncled, to 15 cm wide; bracts none; bractlets setaceous, deciduous; sep none; pet white; fr suborbicular, 4–5 mm; oil-tubes solitary in the intervals, 2 on the commissure; $2n=22$. Native of Europe, adventive here and there in the n. part of our range. June, July. (*Imperatoria o.*)

P. *palustre* (L.) Moench, a coarsely fibrous-rooted European biennial with more dissected carrot-like lvs has been collected as a casual introduction in Mass. It is not provided for in the key to genera.

46. POLYTAENIA DC.   Fr elliptic or oblong, strongly flattened dorsally, the lateral ribs with a thick corky wing, the others obscure; oil-tubes obscure; carpophore bifid to the base; umbels terminal and axillary, compound; invol none; umbellets small, with numerous short pedicels; bractlets linear; sep ovate; pet yellow; stylopodium wanting; stout, erect, taprooted, single-stemmed perennials with pinnately dissected lvs. 2, N. Amer.

1. **Polytaenia nuttallii** DC.   Prairie-parsley. Plants 5–10 dm, puberulent above; lower lvs long-petioled, ovate, the ultimate segments narrowly ovate to linear-oblong, each usually with a few sharp teeth; rays 10–20, scaberulous on the upper side; pedicels 3–5 mm, fr elliptic or oblong, 6–8 mm, depressed in the center over the dorsal ribs; $2n=22$. Prairies and plains; Mich. and Wis. to Neb., s. to Miss., Tex., and N.M. May, June. (*Pleiotaenia n.*)

47. LOMATIUM Raf.   Fr oblong to suborbicular, strongly flattened dorsally, the lateral ribs with conspicuous flat wings, the others slender; carpophore bifid to the base; umbels compound, with several to many usually unequal rays; invol minute or none; umbellets densely many-fld; bractlets in ours conspicuous, all on the outer side of the umbellet, often connate; sep minute or none; pet yellow or white (or pinkish); stylopodium scarcely developed; acaulescent perennials from a vertical thickened root, with several pinnately or ternate-pinnately decompound basal lvs, rarely also a lf near the base of the scape. (*Cogswellia.*) 70, w. and c. N. Amer.

1 Fls yellow; bractlets connate, commonly to the middle or beyond .... 1. *L. foeniculaceum.*
1 Fls white or pinkish, bractlets separate .... 2. *L. orientale.*

1. **Lomatium foeniculaceum** (Nutt.) Coulter & Rose.   Yellow wild parsley. Lvs numerous, to 2 dm (petiole included), cleft to the base into 3 divisions, each of these 3-pinnate, the ultimate segments linear, 2–5 mm; scape 1–2.5 dm; bractlets united to or beyond the middle; pet yellow; fr 6–9 mm; $2n=22$. Prairies and plains; Man., Mo., and Tex. to Mont., Ida., and Ariz. barely entering our range in Io. and Mo. Apr. Var. *foeniculaceum,* mainly of the n. Great Plains, has evidently pubescent herbage, ovaries, and frs, or the frs are sometimes glabrous at maturity. Var. *daucifolium* (T. & G.) Cronq., mainly of the s. Great Plains, has scantily pubescent or glabrate herbage, and glabrous ovaries and frs.

2. **Lomatium orientale** Coulter & Rose.   White wild parsley. Much like no. 1; ultimate lf-segments oblong-linear, 2–10 mm; bractlets glabrous or nearly so, scarious-margined, wholly separate; fls white or pinkish. Prairies and plains; w. Minn. and Io. to Mont. and Colo. Apr.

48. CYMOPTERUS Raf.   Fr somewhat flattened dorsally, the lateral and usually 1 or more of the dorsal ribs conspicuously winged; umbels compound, often head-like; involucre present and usually conspicuous, only on the outer margin of the umbellets; stylopodium wanting; low, acaulescent or short-stemmed perennials with compound lvs and white, yellow, or purple fls. 45, w. N. Amer.

1. **Cymopterus acaulis** (Pursh) Raf.   Wild parsley. Lvs clustered near the ground level on a pseudoscape arising from the subterranean crown of a long, stout taproot, pinnately dissected, the blade to 8 cm; scape-like peduncles usually several, 2–10 cm at maturity; infl dense, the 3–5 rays seldom 1 cm, the pedicels 1 mm; fls white; bractlets rather broad, connate below; fr ovoid to broadly oblong, 5–10 mm, without a carpophore. Dry plains and prairies; w. Minn. to Sask., Mont., Oreg., Ariz., and n. Mex. May, June. Ours is var. *acaulis.*

49. ERIGENIA Nutt.   Harbinger of spring. Fr transversely ellipsoid, cordate at base, notched at summit, laterally flattened, the ribs low and obtuse; oil-tubes 1–3 in the intervals, 9–11 on the commissure; carpophore wanting; umbel sessile, compound, subtended by one reduced foliage lf; umbellets 2–4; involucel of spatulate foliaceous bractlets; sep obsolete; pet white, not inflexed; stylopodium none; perennial from a globose tuber, bearing a few ternately decompound lvs and a simple terminal umbel of comparatively large, nearly sessile fls. Monotypic.

1. **Erigenia bulbosa** (Michx.) Nutt. Delicate herbs 5–15 cm at anthesis, taller at maturity; lvs broadly ovate in outline, 10–20 cm at maturity, much smaller at anthesis, their segments linear or spatulate; primary rays 1–2 cm; pet 3–4 mm; fr 5 mm wide, usually exceeded by the persistent bractlets. Rich woods; N.Y. to s. Mich., s. to D.C., Ala., and Ark. Mar., Apr.

Family **LOGANIACEAE**, the Logania Family

Fls perfect, regular, hypogynous, sympetalous, in ours 5-merous; cal ± deeply cleft, the lobes (or distinct sep) in most genera imbricate; cor salverform to tubular or campanulate; stamens on the cor-tube, alternate with the lobes; ovary superior, mostly bilocular, with few–many ovules on axile (or distally parietal) placentas; style terminal, simple or shortly 2-lobed; fr a capsule (in ours) or berry, seldom a drupe; embryo dicotyledonous, embedded in the endosperm; herbs or woody plants with opposite, simple, entire lvs; stipules usually interpetiolar, sometimes well developed and provided with colleters, sometimes (as in ours) inconspicuous or reduced to a mere line. 20/500, mostly of warm reg.

1 Twining woody vines; fls axillary .... 1. *Gelsemium.*
1 Herbs; fls in terminal cymes.
  2 Cor tubular, red and yellow, 3–6 cm .... 2. *Spigelia.*
  2 Cor funnelform, white, well under 1 cm .... 3. *Cynoctonum.*

1. **GELSEMIUM** A. L. Juss. Yellow jessamine. Fls 5-merous, heterostylic; sep elliptic, obtuse; cor funnelform, its ovate lobes imbricate; stamens inserted near the base of the cor-tube; ovary bilocular; stigmas 4, linear; fr a septicidal capsule, flattened contrary to the septum; seeds few, flat, narrowly winged; twining woody vines with narrow evergreen lvs and handsome, fragrant, yellow fls crowded in short axillary cymes or solitary; stipules reduced to an interpetiolar line. 2 se. U.S., n. Mex.; 1 sw. Pacific.

1. **Gelsemium sempervirens** (L.) Aiton f. Climbing to 5 m; lvs lanceolate, 4–7 × 1–2 cm, on petioles 3–8 mm; pedicels 3–5 mm, bearing 2 or more pairs of scale-like bracts; cor 2–3 cm; fr oblong or elliptic, 1.5–2 cm, short-beaked; 2n=16. Woods and thickets; se. Va. to Tenn. and Ark., s. to Fla. and Mex. Apr. All parts of the plant contain poisonous alkaloids.

2. **SPIGELIA** L. Pinkroot; wormgrass. Fls 5-merous; cal-lobes valvate; cor tubular or salverform, its 5 short lobes valvate in bud; stamens inserted below the middle of the cor-tube, the linear anthers somewhat exsert; ovary bilocular, the style eventually long-exsert; stigma 1; capsule strongly didymous, each half subglobose, loculicidal; seeds angularly subglobose; herbs with fls in terminal secund cymes. 50, warmer parts of the New World.

1. **Spigelia marilandica** L. Perennial, 3–7 dm; lvs sessile, lanceolate to lance-ovate or narrowly deltoid, 5–11 cm, acute to slenderly acuminate, the base obtuse to rounded or truncate; cyme usually solitary, racemiform, peduncled, few-fld, elongating during anthesis; sep linear-attenuate; cor erect, 3–6 cm, scarlet outside, yellow inside; fr 5–8 mm; 2n=48. Moist woods and thickets; N.C. to s. Ind., s. Mo., and Okla., s. to Fla. and Tex. May, June. Contains a poisonous alkaloid.

3. **CYNOCTONUM** J. F. Gmelin. Mitrewort. Fls 5-merous; sep ovate, spreading, united at base; cor funnelform, the short tube thicker than long, hairy in the throat; stamens included; anthers ovate; style very short; stigma capitate; capsule deeply 2-lobed, the lobes divergent or incurved, each longitudinally dehiscent along its inner face; seeds numerous; herbs with broad lvs connected at base by a stipular line or flap; fls small, sessile, white, in terminal, peduncled, branched, eventually secund and spike-like cymes. 6, mainly warm reg. (*Mitreola.*)

1 Lvs mostly 3–6 cm .... 1. *C. mitreola.*
1 Lvs mostly 1–2 cm .... 2. *C. sessilifolium.*

1. **Cynoctonum mitreola** (L.) Britton. Caribbean mitrewort. Annual 2–6 dm, branched above; lvs lanceolate to elliptic or ovate, 3–6 cm, the lower petioled, the upper sessile; cymes eventually 3–5 cm; bracteal

lvs small, resembling the sep; cor and fr each 3 mm; $2n=20$. Moist or wet soil, chiefly on the coastal plain; se. Va. to Fla., Ark., Mex., W.I., and n. S. Amer. (*Mitreola petiolata.*)

2. **Cynoctonum sessilifolium** (Walter) J. F. Gmelin. Small-lvd m. Much like no. 1, but with ovate or elliptic sessile lvs mostly 1–2 cm; infl a little more crowded, especially in fr. Moist low places, chiefly on the coastal plain; se. Va. to Fla. and Tex. July, Aug.

FAMILY **GENTIANACEAE**, the Gentian Family

Fls perfect, hypogynous, sympetalous, essentially regular except sometimes the cal, 4–5(–12)-merous; cal lobed or ± deeply cleft, or seldom the proper lobes much reduced or suppressed and the tube then sometimes 2-cleft; cor often with scales or nectary-pits within the tube or near the base of the lobes; cor-tube sometimes plicate above, the plaits even appearing as additional lobes alternate with the proper lobes; cor-lobes in most genera convolute; stamens on the cor-tube, alternate with the lobes; ovary superior, unilocular, with 2 parietal, sometimes ± deeply intruded placentas; style terminal, with entire or 2-lobed stigma, seldom more deeply cleft or wanting; ovules ± numerous, fr usually a septicidal (seldom irregularly dehiscent) capsule; seeds with a small, dicotyledonous embryo embedded in the oily endosperm; mostly (incl. all our spp.) herbs, usually glabrous or nearly so, with opposite or whorled, simple, entire, exstipulate lvs (in *Bartonia* the lvs reduced to opposite or alternate scales). 75/1000, cosmop., espec. temp. and subtrop.

1 Cor rotate or nearly so, the lobes much longer than the short tube.
  2 Style well developed.
    3 Cor-lobes without glands; lvs normally opposite .... 1. *Sabatia.*
    3 Cor-lobes with a large glandular spot surrounded by a fringe; lvs whorled .... 8. *Frasera.*
  2 Style none, the stigmas decurrent along the carpel-sutures .... 7. *Lomatogonium.*
1 Cor tubular to funnelform or campanulate or salverform, the lobes shorter than to only slightly longer than the tube.
  4 Cal-lobes 4 or 5(6); fl in summer and fall.
    5 Lvs normally developed, with linear to ovate blade.
      6 Cor not spurred at base.
        7 Style filiform; cor with abruptly spreading lobes and a slender tube 2 mm wide or less ...... 2. *Centaurium.*
        7 Style short and thick; cor-tube 3 mm wide or more, the lobes more gradually (or not at all) spreading.
          8 Cor-lobes alternating with folds or plaits in the sinuses, these often of different color or texture, sometimes as long as the proper lobes; perennials .... 3. *Gentiana.*
          8 Cor-lobes simple, without plaits, folds, or accessory lobes in the sinuses; annuals and biennials.
            9 Ovules covering most of the inner surface of the ovary; cor-lobes evidently fringed-fimbriate along the sides or around the top .... 4. *Gentianopsis.*
            9 Ovules in a row or band along each of the 2 sutures of the ovary; cor-lobes not fringed on the margins .... 5. *Gentianella.*
      6 Cor, at least in the larger fls, 4-spurred at base .... 6. *Halenia.*
    5 Lvs reduced to small, opposite or alternate scales .... 9. *Bartonia.*
  4 Cal-lobes apparently 2, foliaceous; fl spring .... 10. *Obolaria.*

1. **SABATIA** Adans. Marsh-pink; sea-pink. Fls (4)5–12-merous; cal-tube short, the lobes narrow and elongate; cor rotate, the tube very short, the spreading lobes oblanceolate to obovate or elliptic; stamens at the summit of the cor-tube, the linear anthers recurved or circinately coiled after anthesis; style elongate, 2-lobed or -cleft; capsule long-enclosed in the marcescent cor, eventually 2-valved; herbs with linear to ovate, often sessile lvs and commonly numerous handsome, rose or lilac to white fls in a loose terminal cyme. 17, N. Amer.

1 Fls (8)9–12-merous; perennial from long, thick rhizomes; fls pink (white).
  2 Cal-lobes narrowly oblanceolate to spatulate, with 3 or more distinct nerves .... 9. *S. dodecandra.*
  2 Cal-lobes linear, widest at base, with 1–3 obscure nerves .... 10. *S. kennedyana.*
1 Fls 5(4–7)-merous; duration and fls various.
  3 Branches from the main axis all or chiefly opposite, the whole cyme therefore apparently trichotomous.
    4 Actual pedicels, beyond the uppermost subtending bracts or branches, commonly only 1–2-(–5) mm; fls white.
      5 Perennial with the stems clustered on a short rhizome .... 1. *S. difformis.*
      5 Biennial, mostly single-stemmed .... 2. *S. quadrangula.*

4 Actual pedicels well developed beyond the uppermost subtending lvs or branches, some of them
   5 mm or more; fls pink (white); biennial.
   6 Lvs ovate or lance-ovate, 1.5–3 times as long as wide, broadest near the rounded or subcordate
     base . . . . 3. *S. angularis.*
   6 Lvs lanceolate to linear, over 3 times as long as wide, broadest near the middle, tapering to
     a narrow base . . . . 4. *S. brachiata.*
3 Branches of the infl alternate, the prolongation of the axis usually somewhat deflected to one side
   and the cyme therefore apparently dichotomous.
   7 Cal-tube narrowly 5-winged or strongly 5-ribbed, the ribs ending abruptly at the sinus; lateral
     veins of the cal-lobes much stronger than the midvein . . . . 5. *S. campestris.*
   7 Cal-tube weakly ribbed or ribless, the midnerve of the cal-lobes as prominent as or commonly
     much more prominent than the lateral nerves.
     8 Cal-lobes tapering from base to tip.
       9 Perennial from a short, branched caudex; cal-lobes about equaling the cor . . . . 6. *S. campanulata.*
       9 Annual; cal-lobes usually distinctly shorter than the cor . . . . 7. *S. stellaris.*
     8 Cal-lobes foliaceous, oblanceolate, 2–5 mm wide above the middle; perennial . . . . 8. *S. calycina.*

1. **Sabatia difformis** (L.) Druce. Stems to 1 m, angular, clustered on a stout rhizome; internodes progressively longer distally, the lower ones of the branches commonly much elongate; lvs lanceolate or lance-linear to (especially the lower) elliptic or oblong, sessile, 2–5 cm, acuminate, usually distinctly 5-nerved; pedicels 1–3(–5) mm; cal-lobes linear, long-tapering, 4–9 mm; cor-lobes white, drying creamy, oblanceolate, 8–18 mm; 2n=36. Wet pine-barrens on the coastal plain, often somewhat weedy; N.J. to Fla. June–Aug. (*S. lanceolata.*)

2. **Sabiata quadrangula** Wilbur. Much like no. 1, but biennial and mostly single-stemmed; lvs commonly 3-nerved, lanceolate to linear, broadest near the middle, abruptly narrowed to an obtuse or subacute base, or the lower more oblong or even ovate and often clasping; cal-lobes 3–8 mm, less than 0.5 mm wide; cor-lobes white, drying saffron, 6–13 mm; 2n=32, 34. Wet or dry pine-lands, fields, sand-hills, and flat-woods; se. Va. to Fla., more often on the piedmont than the coastal plain. June, July. (*S. paniculata,* misapplied.)

3. **Sabatia angularis** (L.) Pursh. Common marsh-pink. Stout biennial, 3–8 dm, commonly with many opposite branches at and beyond the middle, sharply wing-angled; lvs ovate or lance-ovate, 1–4 cm, broadly rounded or subcordate at base, 3–7-nerved; pedicels 5–30 mm, the conspicuous wings continuing upon the base of the cal; cal-lobes narrowly oblong to oblanceolate, 6–15 mm, usually at least 1.5 mm wide; cor pink (white) with a greenish center, the lobes 1–2 cm, oblanceolate to obovate; 2n=38. In a wide range of habitats, sometimes weedy; Conn. to s. Mich. and Kans., s. to Fla. and Tex. July, Aug.

4. **Sabatia brachiata** Elliott. Slender biennial with usually only a few opposite, erect branches beyond the middle, subterete or somewhat angled but scarcely winged; lvs sessile, lanceolate to narrowly oblong or linear, 1.5–4 cm, less than a third as wide, acute, narrowed at base, cyme usually few-fld, the pedicels 2–10 mm; cal-lobes linear, 2–12 mm, less than 1 mm wide; cor pink (white) with a yellowish or greenish eye, the lobes oblanceolate, 8–13 mm; 2n=32. In variously wet or dry, open or wooded habitats, sometimes weedy; se. Va. to Tenn. and se. Mo., s. to Ga. and La. June, July.

5. **Sabatia campestris** Nutt. Western marsh-pink. Annual, 2–4 dm, commonly branched from below the middle; lower lvs lance-ovate, 2–4 cm, broadly rounded to truncate or subcordate at the sessile base; rameal lvs lanceolate; fls all long-pedicellate, cal-tube 5-winged, the wings truncate or slightly projecting at the top; cal-lobes lanceolate, gradually tapering from base to tip, (10–)15–20(–25) mm, the strong lateral nerves much more prominent than the midnerve; cor-lobes lilac, mostly 15–22 mm; 2n=26. Variously wet or dry, usually ± open places; Ill. and Mo. to Miss. and Tex. June, July.

6. **Sabatia campanulata** (L.) Torr. Slender marsh-pink. Perennial from a short branched caudex, slender, 2–6 dm, simple to freely and divaricately branched; lvs of the main axis below the branches oblong or lance-oblong, 2–4 cm, obtuse, widest actually or apparently below the middle, rounded to a sessile base, 3-nerved; upper lvs much narrower; cal-tube inconspicuously 10-nerved, the lobes linear, 7–15 (avg 13) mm; cor pink (white) with a yellow eye, the lobes oblanceolate, 10–15 mm, averaging less than half as wide; style usually cleft to slightly below the middle. Moist or wet places on the coastal plain from Mass. to n. Fla. and w. to La. and Ark., locally inland to Ky., Ind., and Ill. July, Aug. (*S. gracilis.*)

7. **Sabatia stellaris** Pursh. Annual sea-pink. Much like no. 6, but annual; lvs of the main axis typically oblanceolate, 2–4 cm, acute, actually or apparently widest above the middle, concavely narrowed to a slender base; upper lvs linear; cal inconspicuously nerved, the linear lobes 5–11 (avg 9) mm; cor-lobes obovate, 9–18 mm, more than half as wide; style usually cleft to near the base; 2n=36. Salt or brackish marshes along the coast from Mass. to Fla. and La.; inland in Mex. July, Aug.

8. **Sabatia calycina** (Lam.) A. A. Heller. Perennial with solitary or sometimes clustered stems, 1–4 dm, commonly widely branched; lvs oblanceolate to oblong-elliptic, 3–6 cm, obtuse or subacute, cuneately narrowed to a petiole-like base; fls all long-pedicellate, 5–6(7)-merous; cal-lobes foliaceous, oblanceolate, 10–20 × 2–5 mm, nearly equaling to exceeding the pink or sometimes white cor-lobes; 2n=32. Wet soil on the coastal plain; se. Va. to Tex. and W.I. June, July.

9. **Sabatia dodecandra** (L.) BSP. Perennial sea-pink. Perennial with solitary or clustered stems 3–7 dm, with alternate branches or dichotomous above the middle, rarely with a few opposite branches; lvs lanceolate or narrowly elliptic, 2–5 cm; fls (8)9–12-merous, cal-lobes narrowly oblanceolate to spatulate, 1–3 mm wide,

conspicuously 3–5-nerved; cor-lobes pink (white), red and yellow at base, rather narrowly oblanceolate, 16–22 mm, usually ¼ or ⅓ as wide; $2n=38$. Salt or brackish marshes near the coast; Conn. to Fla. and La. July, Aug. Ours is var. *dodecandra*.

10. Sabatia kennedyana Fern.   Plymouth g. Much like no. 9; branches mostly opposite; lvs lanceolate to lance-linear; cal-lobes broadest at the base, the middle less than 1 mm wide, obscurely 1–3-nerved or apparently nerveless; cor-lobes 15–30 mm, at least a third as wide; $2n=40$. Fresh-water marshes and margins of streams or ponds on the coastal plain; interruptedly from N.S. to S.C. July–Sept.

2. CENTAURIUM Hill.   Centaury. Fls 4–5-merous; cal tubular, deeply cleft into carinate, linear or linear-attenuate segments; cor salverform, the slender tube slightly expanded above the cal, the lobes lanceolate to elliptic; stamens on the cor-throat, the slender filaments exsert, the oblong or linear anthers often spirally twisted after anthesis; ovary elongate, often 2-furrowed; style slender; stigma capitate, 2-lobed; capsule thin-walled, invested by the persistent cal and marcescent cor; annual or biennial herbs with numerous small, pink to white fls in terminal, corymbiform or spiciform cymes. (*Erythraea*.) 40–50, ± cosmop.

1 Fls sessile or subsessile (the terminal fls may appear pediceled, but their sessile nature is revealed by
    the bracteal lvs at the base of the cal).
  2 Fls in spike-like cymes .... 1. *C. spicatum.*
  2 Fls in corymbiform cymes .... 2. *C. erythraea.*
1 Fls on pedicels 3 mm or longer .... 3. *C. pulchellum.*

1. Centaurium spicatum (L.) Fritsch.   Spike-centaury. Slender, erect annual or biennial, 1–4 dm, simple or branched above; basal rosette caducous; cauline lvs sessile, lance-elliptic, 1–3 cm, narrowed at base; stem and closely ascending branches prolonged into interrupted, spike-like cymes 5–10 cm, each node normally with one sessile fl opposite an erect linear bract; fls pink to nearly white; cal 9–11 mm, distended over the ovary; cor-tube slightly (2 mm) surpassing the cal; cor-lobes 4–5 mm; $2n=22$. Native of s. Eur., intr. here and there from Mass. to Va. July, Aug.

2. Centaurium erythraea Raf.   Forking centaury. Erect biennial, 2–5 dm, branched above; cauline lvs lanceolate to oblong or narrowly elliptic, sessile, 2–4 cm; basal lvs smaller, petioled, forming a rosette; cymes terminating the stem and branches, trichotomously forked, corymbiform, each node bearing a pair of small linear bracts, a central fl, and 2 lateral branches; fls sessile, rose-purple; cal 3.5–5 mm; cor-tube 6–8 mm, the lobes elliptic, 5–6 mm; $2n$ mostly = 40. Native of Eur. and w. Asia, intr. in waste places here and there in the n. part of our range. June–Sept. (*C. centaurium; C. umbellatum.*)

3. Centaurium pulchellum (Swartz) Druce.   Branching centaury. Annual, much branched, often from the base, up to 2 dm, without a rosette; principal lvs sessile, lanceolate or lance-ovate, 1–2 cm; infl many-fld, comparatively dense, with short internodes; pedicels mostly 3–5 mm; cal ca 9 mm; cor-tube slightly exceeding the cal, its lobes pink, 4 mm; anthers oval, under 1 mm, not coiled; $2n=36$. Native of Eur., locally intr. in fields and waste places in our range. June–Sept.

3. GENTIANA L.   Gentian. Fls (4)5-merous; cal-tube campanulate, obconic, or tubular, with a membrane or rim around the inside at the summit, the lobes arising from just beneath the rim, well developed or sometimes ± reduced; cor tubular to funnelform or campanulate, sometimes remaining closed, convolute in bud, the lobes alternating with folds or plaits in the sinuses, these often of different color or texture, sometimes as long as or even longer than the proper lobes; a whorl of nectary-glands present around the base of the ovary, not on the cor; ovary ellipsoid to cylindric, often ± stipitate; style short and stout; stigma 2-lobed; seeds many, small; herbs, usually glabrous or only inconspicuously papillate-puberulent, most spp. perennial, ours all from a cluster of fleshy-fibrous to sometimes tuberous-thickened or merely fibrous roots; lvs sessile or subsessile; fls ± showy, blue or less often white, yellowish, or greenish. 300, ± cosmop., except Afr. (*Dasystephana*.)

1 Anthers separate; cor wide open; lvs and cal-lobes not ciliolate.
  2 Fls solitary and distinctly pediceled (unique among our spp.) .... 1. *G. autumnalis.*
  2 Fls sessile or short-pediceled, many or all of them in an involucrate or leafy-bracteate terminal
      cluster (as in all of the remaining spp.).
    3 Cor largely or wholly blue; lvs mostly widest below the middle; stem minutely scabro-puberulent
        .... 2. *G. puberulenta.*
    3 Cor greenish-white, striped with pale purple inside; lvs mostly widest above the middle and
        tapering to the base; stem glabrous .... 12. *G. villosa.*

1 Anthers connate, cor variously open or closed; lvs and cal-lobes ciliolate or not.
  4 Lvs and cal-lobes evidently ciliolate at 10×.
    5 Cor ± open at anthesis; plaits narrower and usually conspicuously shorter than the lobes.
      6 Cal-tube glabrous or merely puberulent in lines; cal-lobes oblanceolate.
        7 Lvs lanceolate to ovate, widest near the base; cal-lobes longer than the tube; cor-lobes
          somewhat spreading, usually 2–4 mm longer than the plaits .... 3. *G. catesbaei.*
        7 Lvs linear to elliptic, widest near or above the middle; cal-lobes shorter than to about as
          long as the tube; cor-lobes usually incurved above, seldom exceeding the plaits by more
          than 2 mm .... 4. *G. saponaria.*
      6 Cal-tube densely papillate-puberulent; cal-lobes linear-subulate, shorter than the tube .... 5. *G. decora.*
    5 Cor closed at anthesis; plaits about as long as or longer than the lobes.
      8 Cal-tube and stem densely papillate-puberulent .... 6. *G. austromontana.*
      8 Cal-tube (and usually also the stem) glabrous.
        9 Cor-lobes broadly rounded, about as wide and long as the 2–3-cleft plaits.
          10 Cal-lobes broad, commonly orbicular to obovate; eastern .... 7. *G. clausa.*
          10 Cal-lobes narrow, commonly lanceolate; western (var. *dakotica* of) .... 8. *G. andrewsii.*
        9 Cor-lobes truncate and often apiculate, narrower and much shorter than the fimbriate plaits;
          widespread .... 8. *G. andrewsii.*
  4 Lvs and cal-lobes smooth, not ciliolate at 10×.
    11 Cal-lobes with a decurrent keel; cor yellowish-white or greenish-white .... 11. *G. flavida.*
    11 Cal-lobes not keeled.
      12 Seeds winged; cal-lobes up to 12(–15) mm; cor usually blue or bluish upwards.
        13 Lvs dark green, the involucral ones linear to narrowly lanceolate, spreading .... 9. *G. linearis.*
        13 Lvs light green, the involucral ones lance-ovate, ascending, enveloping the calyces .... 10. *G. rubricaulis.*
      12 Seeds wingless (unique among our spp.); cal-lobes mostly 15–30 mm, seldom less; cor
        mostly yellowish-white or greenish-white and striped with pale purple inside .... 12. *G. villosa.*

1. Gentiana autumnalis L. One-fld g., pine-barren g. Slender, erect, 2–5 dm, rarely branched; lvs linear or narrowly oblanceolate, 2–7 cm × 1–5(–8) mm, narrowed to the base; fls solitary and terminal on the stem (and branches, if any), the peduncle to 3 cm; cal-lobes linear, 10–25 mm, not ciliate; cor blue, spotted with bronze-green inside the tube, the loosely spreading lobes ovate or elliptic, obtuse, their free portion (beyond the fimbriate plaits) 10–15 mm; anthers separate; $2n=26$. Moist pine-barrens on the coastal plain, locally from N.J. to S.C. Sept., Oct. (*G. porphyrio.*)

2. Gentiana puberulenta J. Pringle. Prairie-g. Stems 2–6 dm, mostly simple, minutely papillate-puberulent, often to some extent in lines; lvs firm, lanceolate or lance-oblong, mostly 2–7 × 0.5–2 cm, tending to be minutely scabro-ciliolate or papillate-ciliolate along the margins toward the base only, and along the midrib on the lower side toward the base only; cal glabrous or very nearly so, its lobes linear or lance-linear, 4–18 mm; cor bright blue, 3.5–6 cm, wide open, the free lobes, above the fimbriate and shallowly bifid plaits, 4–8 mm, broadly ovate or elliptic, acute or abruptly acuminate, ± spreading; anthers separate; $2n=26$. Prairies and dry upland woods; w. N.Y. to Man. and N.D., s. to W.Va., Ky., n. Ark., and Kans.; La. Aug.–Oct. (*G. puberula*, misapplied.)
    The closely related cordilleran sp. *G. affinis* A. Gray has been reported from n. c. Minn. and Huron Co., Ont. It differs in its smaller cor (mostly 2.5–4 cm) and unstable cal, the lobes variously large or small, commonly unequal, sometimes virtually suppressed, the tube often once or twice cleft especially when the lobes are reduced. The lvs are also a little more definitely scaberulous-margined ± throughout their length.

3. Gentiana catesbaei Walter. Coastal plain g. Stems 2–6 dm, mostly simple, usually papillate-puberulent; lvs lanceolate to ovate, 2–6 cm, thin but firm, finely scabro-ciliolate especially toward the base; fls few in a crowded terminal cluster; cal-tube glabrous or papillate-puberulent in lines, the lobes oblanceolate, often foliaceous, 10–25+ mm, finely scabro-ciliolate; cor blue, 3.5–5.5 cm, distinctly open to nearly closed, the lobes erect to often somewhat outcurved, ovate, obtuse to acute, usually 2–4 mm longer than the ± bifid plaits; anthers connate. Moist or wet woods and swamps on the coastal plain; N.J. to n. Fla. Sept., Oct. (*Dasystephana parvifolia.*)

4. Gentiana saponaria L. Soapwort-g. Stems 2–7 dm, glabrous or occasionally papillate-puberulent; lvs linear to elliptic or obovate, mostly 5–10 cm, scabro-ciliolate; fls few–several in a dense terminal cluster and 1–few in the upper axils or on short lateral branches; cal-tube glabrous, the lobes ciliolate, usually narrowly oblanceolate, shorter than to about as long as the tube; cor blue at least distally, 3–5 cm, ventricose, narrowly open, the lobes usually incurved above, rounded, seldom surpassing the bifid plaits by more than 2 mm; anthers connate; $2n=26$. Moist woods and thickets; s. N.Y. to Fla., w. irregularly to Ill., Ark., e. Okla., and e. Tex. Sept., Oct. (*G. cherokeensis; Dasystephana latifolia.*)

5. Gentiana decora Pollard. Appalachian g. Much like no. 4; stem and cal-tube papillate-puberulent; cal-lobes mostly linear-subulate and shorter than the tube, up to 8(–12) mm; cor 2.5–4.5 cm, more open, paler, the tube generally white and the lobes suffused with blue, the plaits asymmetrically cleft, one segment usually bent inward; $2n=26$. Moist woods at upper elev. in the mts. from w. Va. to Ga. Sept., Oct.

6. Gentiana austromontana J. Pringle & A. Sharp. Blue-Ridge g. Stems 1–5 dm, papillate-puberulent; lvs ovate or elliptic-ovate to lanceolate, 3–11 × 1–3 cm, scabro-ciliolate; fls in terminal clusters and often sessile in the upper axils or on short lateral branches; cal-tube papillate-puberulent, its lobes mostly deltoid-ovate to elliptic or lanceolate, to 12 mm, ciliolate-margined; cor 3–5 cm, closed, tapering to a narrow top, whitish below, suffused with blue above, the deltoid lobes only 1.5–3 mm, about equaling but only half as wide as the shortly bifid appendages; anthers connate. Upper elev., especially on grassy balds, in the Blue Ridge of Tenn. and N.C. and adj. Va. Sept., Oct.

7. **Gentiana clausa** Raf. Meadow closed or bottle-g. Much like no. 8; cal-lobes foliaceous, broadly lanceolate to obovate, two-fifths to five-sixths as wide as long; cor-lobes prolonged beyond their adnation to the plaits into firm, ovate or oblong, rounded lobes, the plaits about as long as the cor-lobes, shallowly 2–3-toothed or -lobed; $2n=26$. Moist meadows, woods, and thickets; Me. and adj. Que. to Md. and O., and s. in the mts. to N.C. and Tenn. Sept., Oct.

8. **Gentiana andrewsii** Griseb. Prairie closed or bottle-g. Stems 3–10 dm, glabrous, usually simple; lvs lanceolate or lance-ovate, up to 15 × 4 cm, scabro-ciliolate; fls in terminal clusters and often sessile in the upper axils or on short branches; cal-tube glabrous, the lobes mostly lanceolate, sometimes broader, 4–10(–15) mm, finely scabro-ciliolate; cor whitish below, generally blue upwards, 3–4.5 cm, closed, the thin lobes adnate to the plaits nearly to the narrow, truncate and usually apiculate summit, the free tip scarcely 1 mm, the plaits dilated above, fimbriate, evidently exceeding the cor-lobes and forming most of the summit of the cor; anthers connate; $2n=26$. Moist prairies, open woods, and swamps. July–Oct. The var. *andrewsii,* as principally described above, occurs from N.H. and s. Que. to Md., and W.Va., w. to s. Man., Minn., and Mo., and sporadically to Wyo. W. of the Mississippi R. it passes into the var. *dakotica* A. Nels., found mainly from Minn. to Sask., s. to Mo. and Kans. Var. *dakotica,* which may reflect introgression from no. 2, has better developed cor-lobes than var. *andrewsii,* these triangular to rounded, often mucronate, up to 3 mm, and it occasionally has papillate-puberulent stems.

9. **Gentiana linearis** Froelich. Narrow-lvd g. Stems 2–8 dm; herbage glabrous; lvs dark green, linear-oblong to narrowly lanceolate, 4–9 cm, seldom 1 cm wide, narrowed to the base; fls mostly few in a terminal cluster, often also some in the upper axils; involucral lvs spreading; cal-lobes linear or narrowly oblong, 4–12 mm; cor blue to sometimes white, 3–5 cm, narrowly open, the incurved lobes broadly ovate, obtuse or rounded, exceeding the obliquely triangular, entire or 1-toothed plaits by 3–5 mm; anthers connate; $2n=26$. Wet woods and meadows; Que. and sw. Lab. to N.B., N.J., Pa., and s. Ont., s. locally to Tenn., and w. on granitic or otherwise acid substrates to the Lake Superior region of Mich. and Ont. Aug., Sept.

10. **Gentiana rubricaulis** Schwein. Great Lakes g. Stems 3–7 dm; herbage glabrous; lvs pale green, lanceolate or lance-oblong, 4–8 cm, commonly over a third as wide, rounded or broadly obtuse at the sessile base, the involucral ones more lance-ovate, closely ascending, enveloping the calyces; fls usually all terminal; cal-lobes linear-oblong or oblanceolate, mostly 4–12 mm, acuminate, hyaline-scarious below; cor 2.5–5 cm, pale below, generally suffused with blue upwards, narrowly open, its lobes acute, surpassing the obliquely triangular, entire to 1-toothed plaits by 3–5 mm; anthers connate. Wet meadows, especially on calcareous or otherwise nongranitic substrates; s. and w. Ont. to s. Sask., Minn., Wis. and n. Mich.; isolated stations in Me. and N.B. Aug., Sept. (*G. linearis* var. *latifolia; Dasystephana grayi.*)

11. **Gentiana flavida** A. Gray. Pale g. Stout, 3–10 dm, almost always unbranched; herbage glabrous; lvs yellow-green, spreading, lanceolate or lance-ovate, 5–15 × 1.5–5 cm, long-acuminate, widest near the broadly rounded or subcordate sessile base, often 3-nerved; infl usually compact and many-fld; cal-lobes lance-ovate to deltoid-ovate, 4–15 mm, often spreading, with a prominent keel decurrent onto the tube; cor yellowish-white or greenish-white, 3.5–5 cm, narrowly open, its broadly ovate lobes commonly 4–6 mm longer than the oblique, ± erose plaits; anthers connate. Moist prairies and open woods; irregularly from s. Ont. to Pa. and N.C., w. to Minn., e. Neb., e. Kans., and n. Ark., more abundant westward. Sept., Oct. (*G. alba* Muhl., nomen subnudum.)

12. **Gentiana villosa** L. Striped g. Stems erect or ascending, 2–5 dm, mostly simple; herbage glabrous or nearly so; lvs dark green, oblanceolate to oblong-obovate, 4–10 cm, a fifth to two-fifths as wide, usually widest above the middle, tapering to a sessile base; fls subsessile, usually numerous and crowded in terminal (or also axillary) infls; cal-lobes irregular, linear, 10–30 × 1–3 mm; cor 3–5 cm, tubular with erect lobes, greenish-white, striped with pale purple inside, its lobes 3–6 mm longer than the oblique, entire or 1–2-toothed plaits; anthers connate or separate; seeds wingless; $2n=36$. Woods; s. N.J. to s. O. and s. Ind., s. to Ga., n. Fla., and La. Sept., Oct.

4. **GENTIANOPSIS** Ma. Gentian, fringed gentian. Fls 4-merous; cal with a small inner membrane across the base of each sinus, but without a continuous internal rim; cal-lobes with thin, hyaline margins, usually alternately dissimilar; cor funnelform to campanulate, without folds or plaits in the sinuses and without internal scales, but regularly with nectary glands alternating with the stamens toward the base within; ovary stipitate, the stigmas large, sessile or on a short style; ovules covering most of the inner surface of the ovary; seeds numerous, papillate (in ours) or reticulate and caudate; glabrous herbs, ours taprooted annuals or biennials, with showy, long-pedicellate fls solitary or in open, cymose infls. (*Anthopogon* Raf.) (Often included in *Gentiana* or *Gentianella.*) 20, temp. and boreal N. Amer. and Eurasia.

1 Upper lvs distinctly lanceolate or lance-ovate; cor-lobes fringed across the summit (as well as along the sides) with linear fringe-segments (1–)2–5 mm .... 1. *G. crinita.*
1 Upper lvs linear or very narrowly lanceolate; cor-lobes merely erose-toothed across the summit (often fringed along the sides) .... 2. *G. procera.*

1. **Gentianopsis crinita** (Froelich) Ma. Fringed gentian. Plants 3–8 dm, often branched above; lvs sessile, lance-ovate to lanceolate or rarely lance-linear, broadly rounded at base, typically 3–6 × 1–2 cm; fls bright blue, solitary on peduncles 5–20 cm terminating the axis and lateral branches; cal 2.5–4 cm, cleft to the middle into acuminate lobes of very unequal width, smooth or nearly so; cor (3)4–5.5 cm, the obovate lobes deeply fringed around the summit and part way down the sides; 2n=78. Low woods, wet meadows, and brook-banks; s. Me. to Md. and locally along the mts. to Ga., w. to Man., S.D., and Io. Aug.–Oct. (*Gentiana c.; Anthopogon c.*)

2. **Gentianopsis procera** (Holm) Ma. Lesser fringed gentian. Much like no. 1, and perhaps better subordinated to it; lvs typically lance-linear and less than 1 cm wide; cal 1.5–3 cm, its lobes strongly papillate-scabrous on the keels toward the base; cor-lobes more ascending than in no. 1; erose-dentate across the broad summit, often more deeply fringed along the sides; 2n=78. Bogs, meadows, and wet shores, especially in calcareous regions; chiefly from w. N.Y. and s. Ont. to s. Man. and Io., and in another var. to Alta. and on the Gaspé Peninsula. (*Gentiana p.*) The var. *victorinii* (Fern.) Iltis, a local ecotype in the intertidal zone along the St. Lawrence R., barely enters our range. It has small fls only 3–3.5 cm, with orbicular cor-lobes.

5. **GENTIANELLA** Moench. Gentian. Fls 4–5-merous; cal without an internal membrane; cal-lobes wholly green; cor marcescent, tubular to funnelform or campanulate, without folds or plaits in the sinuses but sometimes with fimbriate scales at the orifice, and regularly with nectary-glands alternating with the stamens toward the base within; ovary sessile or short-stipitate, the stigmas sessile or on a short, indistinct style; ovules in two rows or bands along the sutures of the ovary; seeds numerous, smooth; glabrous herbs, ours taprooted annuals or biennials with sessile or short-pedicellate fls in thyrsoid to umbelliform cymes. 100, widespread.

1 Cor-lobes with a transverse fringe across the base .... 1. *G. amarella.*
1 Cor-lobes not fringed .... 2. *G. quinquefolia.*

1. **Gentianella amarella** (L.) Boerner. Northern g., felwort. Plants 1.5–6 dm, seldom branched; lvs sessile, lanceolate or lance-ovate; fls solitary in the middle axils, several in a close cyme from the upper axils, the pedicels seldom 1 cm; fls 4–5-merous; cal-lobes linear-oblong, 3–5 mm; cor blue or bluish, 10–15 mm, tubular funnelform to subsalverform, the obtuse or acute lobes 3–5 mm, bearing at base a fringe of hairs 2 mm; 2n=36. Moist rocky or gravelly soil; interruptedly circumboreal, s. to Me., Vt., w. S.D., and Mex. July, Aug. (*G. acuta; Gentiana amarella.*) Highly variable, with some geographic correlation, but only doubtfully divisible into vars.

2. **Gentianella quinquefolia** (L.) Small. Stiff g., ague-weed. Plants 2–8 dm, simple or often freely branched; lvs sessile, lanceolate or lance-ovate, acute or subacuminate, 2–7 cm; fls blue or seldom white, in dense cymes terminating the stem or short axillary branches, on pedicels to 1 cm, or a few lower fls solitary; cal-lobes linear to lance-ovate; cor 15–23 mm, tubular-funnelform, blue to white, its lobes 4–7 mm, ovate or lance-ovate, acute or acuminate; 2n=38. Woods and moist or wet open places. Aug., Sept. Var. *quinquefolia*, with the cal-tube 1.5–3 mm, and the cal-lobes 1.5–5 mm, occurs in the Appalachian region from Me. to Ga., w. to w. N.Y. and adj. Ont. (*Gentiana q.*) Var. *occidentalis* (A. Gray) Small, with the cal-tube 2.5–5 mm and the lobes 4.5–10 mm (often foliaceous), is more western, occurring from O. and s. Ont. to Minn., s. to Ky., Ark., and se. Kansas. (*G. o.; Amarella o.*)

6. **HALENIA** Borkh. Fls 4-merous; cal-tube short, campanulate, with foliaceous lobes; cor (in ours) nearly tubular to somewhat narrowed distally, lobed to below the middle, prolonged at base (at least in some fls) into short, divergently retrorse spurs; stamens at the summit of the cor-tube, shorter than the lobes; ovary conic, gradually tapering to the short style; stigma 2-lobed; fr lanceolate, acute; herbs with opposite or basal lvs and small fls in cymose clusters terminating the stem and branches or forming a somewhat thyrsoid infl. 100, mainly New World.

1. **Halenia deflexa** (J. E. Smith) Griseb. Erect annual, simple or few branched; lower lvs spatulate or oblanceolate; principal lvs sessile or subsessile, lanceolate to ovate, 2–5 cm, acuminate; cymes loose, commonly 5–9-fld, the pedicels to 4 cm; fls purplish-green, 10–12 mm; cor-lobes lance-ovate, acute; spurs present in most fls, up to 5 mm. July, Aug. Var. *deflexa*, mostly 2–9 dm, with the principal internodes 5–10 cm, occurs in moist or wet woods and bogs from Lab. to B.C., s. to Mass., N.Y., n. Ill., Minn., and Mex. Var. *brentoniana* (Griseb.) A. Gray, 0.5–2 dm, with the principal internodes 1–3(–5) cm, occurs from N.S. to Lab. and Que., chiefly in exposed sites near the sea.

7. **LOMATOGONIUM** A. Braun. Marsh-felwort. Fls 4–5-merous; cal deeply parted into narrow lobes about equaling the cor; cor rotate, deeply parted, each lobe bearing near the base a pair of small nectariferous appendages; stamens on the short cor-tube; style

none, the 2 stigmas decurrent along the sides of the narrowly ovoid ovary; narrow-lvd annuals with conspicuous blue or white fls terminating the stem and pediceled from the upper axils, forming a racemiform infl. 15, boreal.

1. **Lomatogonium rotatum** (L.) Fries. Erect, simple or sparingly branched, 1–2 dm; lvs sessile, lanceolate to linear, 1–2.5 cm; cor 1.5–3 cm wide at anthesis, marcescent-persistent and becoming tubular; fr narrowly oblong; $2n=10$. Margins of salt-marshes and other wet places, often near the sea, but also well inland; circumboreal, s. to Me. and Sask. (*Pleurogyne r.*)

8. **FRASERA** L. Fls 4-merous; cal deeply cleft into narrow segments; cor rotate, the tube short, the lobes with 1 or 2 large glands below the middle, the gland ± completely surrounded by a usually fimbriate membrane; stamens on the cor-tube, the filaments dilated below and often connate at base; some fimbriate scales often present just internal to the stamens or between them; ovary with a ± elongate style and capitate stigma; perennial herbs with opposite or whorled lvs and fairly large fls in large terminal paniculiform cymes. 12, mostly w. U.S. (Often included in *Swertia*.)

1. **Frasera caroliniensis** Walter. American columbo. Short-lived perennial 1–2 m; lower lvs oblanceolate, to 4 dm, in whorls of (3)4(5), the upper progressively smaller; infl paniculiform, the lower branches 5–15 cm, the lower bracteal lvs to 10 cm; cor greenish-yellow, purple-dotted, the lance-ovate lobes acute, 10–18 mm, each bearing below the middle a large elliptic gland surrounded by a long fringe; fr elliptic, flat, 1.5–2 cm; $2n=78$. Rich woods; w. N.Y. to S.C. and n. Ga., w. to Mich., Ill., Mo., e. Okla., and La. May, June. (*Swertia c.*)

9. **BARTONIA** Muhl. Fls 4-merous; cal cleft nearly or quite to the base; cor campanulate, the lobes almost equaling the tube, imbricate in bud; stamens inserted at the sinuses of the cor; style very short and stout; stigma large, ± 2-lobed; slender, mycotrophic annuals or biennials, the lvs reduced to minute scales, the stem often spiral or even tending to twine, green to yellowish or purplish, prolonged into a slender terminal panicle (or in small plants a raceme) of small, white to purplish or yellowish fls. 3, e. N. Amer.

1 Lf-scales mostly opposite . . . . 1. *B. virginica.*
1 Lf-scales mostly alternate, or the lower opposite . . . . 2. *B. paniculata.*

1. **Bartonia virginica** (L.) BSP. Stem simple to the infl, erect, 1–4 dm; lf-scales usually opposite; infl racemose or paniculate, the branches and pedicels commonly opposite and erect; fls 3–4 mm; sep lance-subulate; pet oblong, usually denticulate, abruptly narrowed to a rounded or obtuse, often mucronate tip; anthers minutely apiculate; $2n=52$. Sphagnum bogs and wet meadows; Que. and N.S. to Wis., s. to Fla. and La. Late summer.

2. **Bartonia paniculata** (Michx.) Muhl. Screw-stem. Stems 2–4 dm, erect or lax, sometimes scrambling or almost twining; scale-lvs mostly alternate, or the lower opposite; infl usually paniculate, commonly 5–10(–25) cm, the slender pedicels arcuately ascending or divergent; fls 2.5–4 mm; sep lance-subulate, 2–2.5 mm, distinct; pet lanceolate, long-acuminate; anthers yellow, blunt; $2n=52$. Swamps, bogs, and wet meadows; Nf. to Fla. and Miss., along the coast, or on the coastal plain, n. in the interior to Ky., Mo., and Okla.; isolated in n. Mich. and s. Ont. Late summer. (*B. lanceolata.*) Our plants are mostly var. *paniculata,* as described above. Plants from N.S. and s. even to Mass. often diverge toward the otherwise more northern var. *iodandra* (Robinson) Fern., which has the sep connate below, the pet blunter, and the anthers often purplish. Such intermediate plants have been called var. *intermedia* Fern.

10. **OBOLARIA** L. Pennywort. Cal divided nearly to the base into 2 obovate or oblong-obovate, foliaceous lobes; cor funnelform, divided about to the middle into 4 oblong-obovate lobes; stamens on the cor-tube; pistil about as long as the stamens, the ellipsoid ovary tapering into the slender style; stigma 2-cleft; fr rupturing irregularly; delicate, fleshy, yellowish or purplish-green mycotrophic perennial from a cluster of thickened roots, the lower lvs scale-like, the bracteal ones with expanded blade; fls subsessile in clusters of 1–3 in the axils, and the terminal fl solitary. Monotypic.

1. **Obolaria virginica** L. Stems solitary or several, simple or sparingly branched, 8–15 cm, bracteal lvs 8–15 mm, sep shorter than to nearly equaling the cor; fls dull white, 10–12 mm; $2n=56$. Rich woods; N.J. and Pa. to O., s. Ind., and s. Ill., s. to Fla. and Tex. Apr., May.

FAMILY **APOCYNACEAE**, the Dogbane Family

Fls perfect, hypogynous, sympetalous, regular or nearly so, mostly 5-merous; cor commonly funnelform or salverform to urceolate or cylindric, with mostly convolute lobes; stamens on the cor-tube, alternate with the lobes; anthers distinct or ± closely connivent around the style-head; ovaries 2 in our genera, with a common style and large stigma; ovules few to many; fr in ours of 2 slender, many-seeded follicles; ours herbs or twining woody vines with milky juice and opposite or (*Amsonia*) alternate, simple lvs. 200/2000, mostly tropical.

1 Infl or fls axillary.
  2 Fls solitary in the axils; seeds naked .... 1. *Vinca.*
  2 Fls in axillary cymes; seeds comose .... 2. *Trachelospermum.*
1 Infl terminating the stem or its branches.
  3 Lvs alternate or irregularly scattered; cor salverform; seeds naked .... 3. *Amsonia.*
  3 Lvs strictly opposite; cor campanulate to tubular; seeds comose .... 4. *Apocynum.*

1. **VINCA** L. Cal deeply parted; cor large, salverform; anthers separate; ovaries 2, accompanied by 2 nectaries; ovules numerous; follicles linear or fusiform, few-seeded; seeds naked; perennial herbs or semishrubs with opposite, petiolate lvs and conspicuous, blue or rarely white fls solitary in one axil only of a pair of lvs. 12, Old World.

1 Lvs lance-elliptic, broadest near the middle, not ciliate .... 1. *V. minor.*
1 Lvs deltoid-ovate, finely ciliate .... 2. *V. major.*

1. **Vinca minor** L. Periwinkle. Stems somewhat woody, trailing or scrambling, to 1 m, forming mats, the flowering branches ± erect; lvs glabrous, coriaceous, lance-elliptic, 3–5 cm, entire; cor-tube 8–12 mm, the limb 2–3 cm wide; $2n=46$. Native of s. Europe, often escaped from cult. into roadsides and open woods in our range. Apr., May.
    V. herbacea Waldst. & Kit., more herbaceous, with scabrous lvs that have narrowly ascending (or obscure) primary lateral veins, the flowering branches not erect, is reported to be locally established in Mass.

2. **Vinca major** L. Greater periwinkle. More robust than no. 1, the flowering branches to 5 dm tall; lvs deltoid-ovate, 3–7 cm, ciliate; cor-tube 12–20 mm, the limb 3.5–5 cm wide; $2n=92$. Native of s. Europe, occasionally escaped in s. U.S., n. to Va. and Ill. Apr., May.

2. **TRACHELOSPERMUM** Lemaire. Cal deeply parted into narrow segments; cor salverform; anthers connivent and adherent to the stigma; ovaries 2, surrounded at base by 5 nectaries; ovules many; follicles elongate, slender; seeds comose; twining vines with opposite lvs and small, yellow or ochroleucous fls in peduncled axillary cymes. 30, most spp. in e. and s. Asia.

1. **Trachelospermum difforme** (Walter) A. Gray. Slender, woody, high-climbing; lvs variable, even on the same plant, lanceolate to broadly obovate, 5–12 cm, on petioles 2–15 mm; cymes long-peduncled, manyfld, produced in only one axil of a pair of lvs; cor-tube 6 mm, the lobes 3–4 mm; fr 15–25 cm. Moist or wet woods on the coast plain; Del. to Tex., n. in the interior to sw. Ind. and se. Mo. July.

3. **AMSONIA** Walter. Bluestar. Cal deeply 5-parted; cor salverform, villous in the throat, its lobes lanceolate; anthers separate; ovaries 2, without nectaries; ovules many; follicles elongate, linear, cylindric or somewhat moniliform; seeds naked; erect perennial herbs with alternate or irregularly scattered lvs and terminal cymes of small but handsome blue fls. 25, N. Amer. and Japan.

1 Cal glabrous; fr cylindric; lvs not shining .... *A. tabernaemontana.*
1 Cal sparsely hairy; fr tending to be moniliform; lvs shining above .... 2. *A. illustris.*

1. **Amsonia tabernaemontana** Walter. Common b. Erect, 4–10 dm; lvs thin, opaque, narrowly lanceolate to ovate or broadly elliptic, 8–15 cm, acuminate, finely hairy or glabrous beneath; cymes flat to pyramidal, many-fld; cor-tube 6–10 mm, villous, especially distally; cor-limb 1 cm wide; fr cylindric, erect, 8–12 cm; $2n=32$. Moist or wet woods; coastal plain from N.J. southward, more widespread in s. U.S., and n. in the interior to s. Ind., c. Ill., and Kans. May, June. (*A. amsonia.*)

2. **Amsonia illustris** Woodson. Ozark b. Much like no. 1; lvs firm, shining above, narrowly lanceolate, to 15 mm wide, glabrous beneath; cal-lobes sparsely pilose, especially distally; cor-tube sparsely villous near the summit externally; follicles tending to be slightly constricted between the seeds. Moist sandy or gravelly soil; c. Mo. to Kans. and Tex. May.

A. ciliata Walter, a southern sp. with narrower lvs and with the cor-tube strictly glabrous outside, probably does not reach our range.

4. **APOCYNUM** L. Dogbane. Cal deeply divided into triangular or lanceolate lobes; cor campanulate, short-cylindric, or urceolate, with short lobes, bearing within a tooth or scale near the base of the tube opposite each lobe; filaments very short; anthers lance-triangular, connate, adherent to the stigma and prolonged into a cone beyond it; ovaries 2, subtended by 5 nectaries; ovules numerous; seeds numerous; comose; colonial perennial herbs with tough fibrous bark, mostly with erect or ascending stems, opposite mucronate lvs, and small white or pink fls in terminal cymes. 3, N. Amer.

Hybridization is rampant in the group. Hybrids and hybrid segregates derived from *A. androsaemifolium* × either *A. cannabinum* or *A. sibiricum* are called *A.* × *floribundum* Greene (*A. medium*).

1 Cor 6–10, pinkish, with spreading or recurved lobes .... 1. *A. androsaemifolium*.
1 Cor 3–6 mm, white or greenish-white or yellowish, with erect or slightly divergent lobes.
  2 Lvs of the main axis on petioles commonly 5–10 mm, acute to rounded at base .... 2. *A. cannabinum*.
  2 Lvs of the main axis sessile or on obscure petioles rarely to 3 mm, obtuse to rounded or cordate
    at base .... 3. *A. sibiricum*.

1. **Apocynum androsaemifolium** L. Spreading d. Stems 2–8 dm, often ± inclined from the vertical, either simple or branched and then lacking a central axis; lvs petiolate, drooping, oblong-lanceolate to ovate, 3–8 cm, usually hairy beneath; principal cymes terminal; secondary smaller cymes in the upper axils; fls declined or nodding; cal-lobes triangular, to half as long as the cor-tube; cor campanulate, 6–10 mm, pinkish, marked with red inside, the lobes spreading or recurved; fr 5–15 cm; coma 1–2 cm; $2n=16$. Upland woods, occasionally in fields and roadsides; Nf. to B.C., s. to Ga., Tex., and Ariz. May–Aug.

2. **Apocynum cannabinum** L. Hemp-d. Erect, branched above, 0.5–1.5 m, with a well developed main axis; lvs on short but evident petioles 3–10 mm, erect or ascending, 5–11 cm, oblong-lanceolate to oval or broadly elliptic, acute to rounded at the base and mucronate apex, glabrous or hairy beneath; fls erect; cal-lobes usually reaching beyond the middle of the cor-tube; cor cylindric to urceolate, 3–6 mm, white or greenish-white, the lobes erect or slightly divergent; fr (5–)10–15(–20) cm; coma mostly 1–3 cm; $2n=16$. Open places; rare or local in e. Can. and N. Engl., abundant from N.Y. to N.D., s. to Fla. and Tex., and w. occasionally to the Pacific. May–Sept. Highly variable in lf-shape and pubescence.

3. **Apocynum sibiricum** Jacq. Clasping d. Erect to sometimes nearly prostrate; lvs usually glabrous beneath, sessile or on short petioles rarely to 3 mm, the base broadly obtuse to rounded or often subcordate and clasping; infl and cal usually glabrous; cor greenish-white or pale yellow; otherwise as in no. 2. Open places; Nf. to N.J. and W.Va., w. to Man. and Mo., and to the Pacific, more common westward. May–Sept. (*A. hypericifolium.*) Perhaps better treated as *A. cannabinum* var. *hypericifolium* A. Gray.

FAMILY **ASCLEPIADACEAE**, the Milkweed Family

Fls perfect, hypogynous, sympetalous, regular or nearly so, 5-merous; corolla with short to sometimes elongate tube and initially mostly convolute, often spreading or reflexed lobes, sometimes with a thickened ring of distinct or connate scales within the throat, and usually with a well developed *corona* arising from the external base of the filaments or from the region of union of the filaments and corolla; filaments on the corolla-tube, distinct (as in *Periploca*) or more often connate into a usually short sheath around (and adherent to) the style, the anthers in any case coherent or connate and adherent to the thickened style-head; the combined filaments, anthers and style forming a structure called the *gynostegium*; pollen of each pollen-sac typically united into a waxy mass known as a *pollinium*, the right-hand pollinium of one anther connected to the left-hand pollinium of the adjacent anther by a transverse, acellular, often ornate structure called a *translator,* the pollinium-pair characteristically extracted from the anthers by means of the translator, which becomes entangled in the legs of visiting insects; pollinia and translators of somewhat different structure from the above in a few genera, as described under *Periploca*; ovaries 2, distinct, each with its own style, these united only by the common style-head, which is thickened and has restricted lateral stigmatic surfaces between the anthers; ovules very numerous; fr in all our genera an ovoid to linear follicle, often only one of a pair of ovaries maturing into fr; seeds numerous, almost always with a coma; herbs, shrubs or vines, usually with milky juice; lvs simple, opposite or occasionally whorled or alternate. 250/2000, mostly of warm regions.

1 Stems erect or merely decumbent, not at all twining.
  2 Lobes of the cal and cor ± reflexed at anthesis, except in *A. viridis,* in which the cor-lobes are 1
    cm or more .... 2. *Asclepias.*
  2 Lobes of the cal and cor ascending or merely spreading, the cor-lobes mostly 2.5–4.5 mm (forms
    of *V. hirundinaria*) .... 4. *Vincetoxicum.*
1 Stems ± twining; lobes of the cal and cor spreading or ascending to erect, not reflexed.
  3 Upper surface of the cor-lobes conspicuously bearded with hairs ca 0.5–1+ mm.
    4 Corona with 5 slender, elongate, conspicuous lobes 4–8 mm and 5 very short, broadly rounded,
      deeply bifid lobes; style not prolonged beyond the anthers; pollinia 2 in each pollen-sac,
      open at the summit (unique among our genera) .... 1. *Periploca.*
    4 Corona with very short, broadly rounded lobes only; style with a slender tip prolonged well past
      the anthers; pollinia solitary in each pollen-sac, not open .... 3. *Metaplexis.*
  3 Upper surface of the cor-lobes glabrous or with some minute hairs less than 0.5 mm.
    5 Corona with 5 deeply bifid, slender and elongate lobes nearly equaling the cor-lobes .... 5. *Ampelamus.*
    5 Corona inconspicuous, shorter than to barely surpassing the gynostegium.
      6 Cor-lobes short and broad, only 1.5–4.5 mm long; anthers tipped with an evident scarious
      membrane; pollinia suspended .... 4. *Vincetoxicum.*
      6 Cor-lobes longer and mostly narrower, (5)6–18 mm; anthers with short or no scarious tip;
      pollinia suspended or often horizontal .... 6. *Matelea.*

1. **PERIPLOCA** L. Cor rotate, deeply lobed; corona rotate, its margin prolonged into 5 short, broadly rounded, deeply bifid lobes opposite the cor-lobes and 5 long, ascending, linear lobes alternating with them; filaments distinct; pollen-tetrads in each pollen-sac loosely coherent to form 2 distally open, cornucopia-shaped pollinia, each of which is entered by an expanded, sticky branch of the 2-cleft translator-arm; twining or erect shrubs with fls in terminal or apparently axillary, paniculiform cymes. 10, Old World.

    1. **Periploca graeca** L. Silk-vine. Deciduous, glabrous twiner to 10 m; lvs petiolate, lance-ovate to oblong, 5–10 cm; fls brown-purple, 2–2.5 cm wide, the narrowly oblong cor-lobes densely villous on the upper side especially toward the margins and base, and with a central, densely and shortly papillate-hairy strip toward the base; principal lobes of the corona slender, almost thread-like, 4–8 mm; anthers villous on the back; frs 10–15 cm; 2*n*=24. Native of s. Eur., rarely escaped from cult. in our range.

2. **ASCLEPIAS** L. Milkweed. Cor deeply divided, spreading or more often reflexed and concealing the cal; corona of 5 prominent *hoods* arising from near the top of the filament-column and subtending the anthers, alternate with the cor-lobes, the hoods diverse in structure and providing specific criteria, straight or curved, erect or ascending, or spreading at the base and upcurved at the tip, often bearing internally near the base a slender or flattened exsert *horn;* anthers tipped with a triangular, hyaline, inflexed appendage; pollinia solitary in each pollen-sac, pendulous, apically attached to the translator; follicles commonly erect, acuminate; seeds comose, except in *A. perennis;* perennial herbs mostly from a stout root or caudex, with usually simple stems (never twining), opposite (in some spp. alternate or whorled) lvs, and small or medium-sized fls in peduncled, terminal or extra-axillary cymose umbels. 150, mainly New World, esp. N. Amer. (*Acerates, Asclepiodora.*)

1 Horns present, subulate or flattened, exsert from the hoods of the corona.
  2 Horn recognizable only as a short free tip barely exserted between the lobes of the hood, which
    thus appears to be shortly 3-lobed at the summit .... 18. *A. stenophylla.*
  2 Horn distinctly otherwise, attached well below the top of the hood.
    3 Lvs below the infl alternate; juice not milky; cor bright yellow or orange .... 1. *A. tuberosa.*
    3 Lvs below the infl opposite or whorled; juice milky; color various.
      4 Hoods about equaling the gynostegium, the horns conspicuously exsert and surpassing the
      hoods.
        5 Lvs narrowly linear, all or mostly in whorls of 3–6 .... 2. *A. verticillata.*
        5 Lvs broader, opposite.
          6 Lateral margins of the hood divergent from near or below the middle, the hood therefore
          ± scoop-shaped.

    7 Lvs abruptly narrowed to an acutish to rounded or subcordate base; pet pink to red;
       seeds comose . . . . 3. *A. incarnata.*
    7 Lvs long-tapering to the base; pet white; coma none . . . . 4. *A. perennis.*
  6 Lateral margins of the hood approximate and essentially parallel, the hood truncate and
     therefore ± tubular.
    8 Margins of the hood terminating in a sharp erect tooth; lvs ± tapering to the shortly
       petiolar base; umbels usually several . . . . 5. *A. exaltata.*
    8 Margins of the hood not prolonged; lvs broadly rounded to cordate at base, sessile to
       subsessile; umbel mostly solitary and terminal . . . . 6. *A. amplexicaulis.*
 4 Hoods conspicuously surpassing the gynostegium; horns not surpassing the hoods.
  9 Lower lf-surface distinctly pubescent, sometimes even tomentose.
   10 Hoods 4–8 mm, the summit (beyond the horn) ovate or oblong.
     11 Hood with a sharp, triangular, ascending or inflexed lateral lobe at or near the
        middle of the margin; cor greenish-purple to nearly white.
       12 Tall, coarse plants, 1–2 m when well developed; umbels commonly several;
          widespread . . . . 7. *A. syriaca.*
       12 Smaller plants, 2–6 dm; umbels solitary, occasionally 2 or 3; prairies . . . . 8. *A. ovalifolia.*
     11 Hood without lateral lobes, or merely slightly widened at or near the middle; cor
        commonly purple . . . . 9. *A. purpurascens.*
   10 Hoods 10–15 mm, abruptly narrowed below the middle to a linear-oblong tip . . . . 10. *A. speciosa.*
  9 Lower lf-surface glabrous, or with a few hairs along the midvein.
   13 Lvs tapering or cuneate at base, the petiole distinct and usually well developed.
     14 Pet white to pink or purplish; horns flattened laterally.
       15 Lvs lanceolate to lance-ovate, acuminate; hoods scoop-shaped, with prominent
          lateral teeth . . . . 11. *A. quadrifolia.*
       15 Lvs broadly ovate or oblong, short-acuminate or apiculate; hoods saccate, with
          lateral teeth obscure or none . . . . 12. *A. variegata.*
     14 Pet red or red-purple; horns subulate . . . . 13. *A. lanceolata.*
   13 Lvs broadly rounded to subcordate at the sessile or subsessile base.
     16 Lvs and pedicels strictly glabrous . . . . 14. *A. sullivantii.*
     16 Lvs scabrous near the margin or scabrous-ciliolate.
       17 Cor normally red; coastal plain . . . . 15. *A. rubra.*
       17 Cor greenish-white; middle-West . . . . 16. *A. meadii.*
1 Horns none, or rudimentary, not exsert.
  18 Cor-lobes 10–15 mm, spreading; horn represented by a small, flat, median plate near the summit
     of the hood (*Asclepiodora*) . . . . 17. *A. viridis.*
  18 Cor-lobes 4–8 mm, reflexed; horn none.(*Acerates.*)
    19 Base of the hoods separated from the cor by a distinct column; summit of the hoods not
       reaching beyond the projecting center of the triangular anther-wings.
     20 Pedicels densely spreading-hairy . . . . 19. *A. hirtella.*
     20 Pedicels with incurved hairs . . . . 20. *A. longifolia.*
    19 Base of the hood adjacent to the cor, without a distinct column; summit of the hoods reaching
       distinctly beyond the projecting center of the anther-wings or even to their summit.
     21 Umbels lateral, sessile or nearly so . . . . 21. *A. viridiflora.*
     21 Umbel solitary, terminal, peduncled . . . . 22. *A. lanuginosa.*

   1. Asclepias tuberosa L. Butterfly-weed. Ascending or erect, 3–7 dm, rather coarsely hairy, simple to
much branched above; lvs numerous, alternate or on the branches opposite, linear to lanceolate or oblanceolate, 5–10 cm, cuneate to subcordate at base; juice not milky; umbels solitary and terminal to numerous;
fls bright yellow to orange-red; cor-lobes 7–10 mm, the hoods 5–7 mm, greatly exceeding the gynostegium,
nearly straight, erect, the lateral margins bearing an obscure tooth below the middle; fr 8–12 cm; erect;
$2n=22$. Prairies and upland woods, especially in sandy soil; s. N.H. to Fla., w. to Minn., S.D., Ariz., and
Mex. June–Aug. Var. *tuberosa*, with the lvs typically widest above the middle, hence abruptly pointed, is
the eastern phase of the sp., chiefly in and e. of the Appalachian Mts. in our range. Var. *interior* (Woodson)
Shinners, with the lvs typically widest below the middle, hence tapering gradually to the tip, is more western,
chiefly w. of the mts. in our range. Intermediates are abundant.

   2. Asclepias verticillata L. Whorled m. Stems slender, erect from a cluster of fibrous roots, 2–5 dm,
simple to the infl, pubescent in lines; lvs very numerous in whorls of 3–6, narrowly linear, 2–5 cm × 1–2
mm, revolute; umbels several from the upper nodes; peduncles 1–3 cm; fls white or greenish; cor-lobes 4–5
mm; hoods somewhat divergent, 1.5–2 mm, about equaling the gynostegium, their margins entire; horns
subulate, much surpassing the hoods, inflexed over the gynostegium; fr slender, erect on erect pedicels, 7–
10 cm; $2n=22$. Fields, roadsides, upland woods, and prairies; Mass. to Fla., w. to Sask., Kans., and Ariz.
June–Aug.

   3. Asclepias incarnata L. Swamp-m. Erect, stout, to 1.5 m, branched above; lvs lanceolate to oblong or
linear-oblong, usually acuminate, at base acutish to subcordate; umbels usually several or many; cor pink to
red (white), the lobes 4.5–6 mm; hoods 2.5–3 mm, about equaling the gynostegium, the lateral margins entire,
divergent from near the base; horn subulate, incurved, much surpassing the hood; $2n=22$. Open swamps,
ditches, and wet prairies; N.S. to Fla., w. to Sask., Utah, and N.M. June–Aug. Var. *incarnata*, with nearly
the range of the sp., but not common in the Atlantic seaboard states, is sparsely and inconspicuously hairy
to essentially glabrous, and the stems are usually repeatedly branched. Var. *pulchra* (Ehrh.) Pers. largely

replacing var. *incarnata* in N.S., N. Engl., and the coastal states southward, is evidently hairy and seldom much branched, with mostly broader lvs more abruptly contracted at base.

4. Asclepias perennis Walter. Smoothseed-m. Stems slender, often branched, 4–8 dm, pubescent in narrow lines; lvs thin, lanceolate to narrowly elliptic, 5–14 cm, gradually tapering at base into a conspicuous petiole; umbels solitary or few; cor white, the lobes 3–4 mm; hoods and horns as in no. 3, 2–3 mm; seeds without a coma. Swamps and wet woods; s. coastal plain from S.C. to Tex., n. in the Mississippi and Ohio valleys to s. Mo. and s. Ind. July, Aug.

5. Asclepias exaltata L. Tall m. Stems 8–15 dm, glabrous or puberulent in narrow lines; lvs thin, broadly elliptic, 1–2 dm, acuminate at both ends, glabrous, or puberulent beneath; petioles 1–2 cm; umbels loosely few-fld, the slender pedicels spreading or often drooping; cor white to pale dull purple, the lobes 7–10 mm; hoods white or pink, 4 mm, about equaling the gynostegium, the lateral margins adjacent, each terminating in an erect tooth 1–1.5 mm, the rest of the hood truncate; horns subulate, suberect, conspicuously exsert; fr erect on deflexed pedicels, puberulent, 12–15 cm. Moist upland woods; s. Me. to Va. and in the mts. to Ga., w. to Minn., e. Io., Ill., and Tenn. June, July. (*A. phytolaccoides.*)

6. Asclepias amplexicaulis J. E. Smith. Clasping m. Simple, erect or sometimes decumbent, 3–8 dm, with 2–5 pairs of lvs and a single (rarely 2) long-peduncled terminal umbel; lvs oval or broadly oblong, 7–15 cm, obtuse or rounded at summit, broadly rounded or commonly cordate at the sessile or subsessile base; peduncle 1–3 dm; umbel large, usually many-fld; cor greenish-purple, its lobes 8–11 mm; hoods pink, 5 mm, about equaling the gynostegium, the lateral margins adjacent, the broad summit truncate, entire or with a few low obtuse lobes; horns subulate, much exsert; fr 10–13 cm, erect on deflexed pedicels. Dry fields, prairies, and open woods, usually in sandy soil; N.H. to se. Minn., s. to Fla., Neb., and Tex. June–Aug. The hybrid with no. 7 is *A.* × *intermedia* Vail.

7. Asclepias syriaca L. Common m. Colonial by roots; stems stout, 1–2 m, mostly simple, hairy; lvs thick, narrowly or broadly elliptic to ovate or oblong, 10–15 cm, acute or apiculate, soft-hairy beneath; petioles 5–15 mm; umbels often numerous, terminal and lateral, compactly many-fld; peduncles 3–10 cm; cor nearly purple to nearly green, its lobes 7–10 mm; hoods pale purple, somewhat divergent, 4–5 mm, surpassing the gynostegium, the lateral margins with a prominent, sharp, triangular lobe near the middle; horns short, inflexed; fr erect on deflexed pedicels, 7–12 cm, tomentose and beset with soft, filiform to conic processes; $2n=22$. Fields, meadows, and roadsides; N.B. to Va. and n. Ga., w. to Man., Nebr., and Okla. June–Aug. (*A. kansana.*) Native, the specific epithet in error.

8. Asclepias ovalifolia Decne. Dwarf m. Slender, 2–6 dm; lvs firm, lance-ovate to oblong or elliptic, 3–6(–10) cm, cuneate to rounded at base, finely hairy beneath; petioles 2–8 mm; umbels solitary and terminal, or 1 or 2 lateral, loosely few-fld; cor greenish-white to greenish-purple, its lobes 5–7 mm; hoods as in no. 7, even in size; pods merely thinly hairy; $2n=22$. Dry prairies; n. Ill. to Wis., Minn., s. Sask., and w. S.D. June, July.

9. Asclepias purpurascens L. Purple m. Stems stout, erect, to 1 m, puberulent; lvs elliptic to ovate-oblong, 10–15 cm, hairy beneath, broadly cuneate to a petiole 8–25 mm; umbels few or one, terminal and subterminal, many-fld, on peduncles to 5 cm; cor normally purple, its lobes 7–10 mm; hoods pale purple, 5–7 mm, surpassing the gynostegium, without lateral teeth but often somewhat widened near the middle; horns short, flat, incurved; fr downy, without processes. Dry soil; s. N.H. to Va., w. to Wis., Io., Kans., and Okla. June, July.

10. Asclepias speciosa Torr. Showy m. Colonial by roots; stem stout, erect, to 1 m, hairy; lvs ovate to lance-ovate or ovate-oblong, 10–15 cm, broadly rounded or subcordate at base, hairy beneath; petioles 3–6 mm; umbels usually few, 5–7 cm wide; peduncles stout, 3–7 cm; cor greenish-purple, its lobes 10–15 mm; hoods 10–15 mm, abruptly narrowed below the middle to a linear-oblong tip; horns short, inflexed; fr densely tomentose and beset with soft filiform processes; $2n=22$. Moist prairies; Man. to Minn., Io., and Okla., w. to B.C. and Calif. July, Aug.

11. Asclepias quadrifolia Jacq. Four-lvd m. Slender, usually simple, 2–5 dm, with 1 or 2(3) terminal and subterminal umbels, normally with 3 lf-bearing nodes, the lower and upper each with a pair of small lvs, the middle with a whorl of 4 much larger ones, or the lvs rarely all opposite in 4 pairs; lvs thin, lanceolate or lance-ovate, the larger 6–12 cm, acuminate, cuneate to a conspicuous petiole; peduncles 1–4 cm; cor pink to white; its lobes 4.5–6 mm; hoods 4–5 mm, much exceeding the gynostegium, the lateral margins bearing a prominent, sharp, inflexed tooth near or below the middle; horn flattened, sword-shaped, falcately incurved; fr erect on erect pedicels, very slender, 8–12 cm. Dry upland woods; Vt. to Va. and in the mts. to n. Ga., w. to s. Ont., Ill., s. Io., and Okla., with a minor disjunction e. of the Mississippi R. May, June.

12. Asclepias variegata L. White m. Stout, usually simple, to 1 m; lvs several pairs, almost always opposite, broadly oblong to ovate, 8–12 cm, abruptly acuminate to obtuse and apiculate, cuneate to a petiole; umbels 1–4, compact, usually many-fld; cor white or pink-tinged, its lobes 7–9 mm; column very thick, expanded above; hoods divergent, strongly ventricose or subglobose, the body saccate, the lateral margins infolded around the horn, with or without an obscure lateral lobe near the middle; horns strongly flattened, obliquely obovate, abruptly contracted to a short inflexed beak; fr erect on deflexed pedicels. Upland woods and thickets; Conn. and s. N.Y. to n. Fla., w. to O., s. Ill., se. Mo., and Tex. June. (*Biventraria v.*)

13. Asclepias lanceolata Walter. Few-fld m. Stems slender, often 1 m, simple, with a few pairs of elongate, linear or narrowly lanceolate lvs separated by internodes 1–2 dm; umbels 1–4, terminal or subterminal, loosely few-fld; cor red to orange-red or reddish-purple, its lobes 9–12 mm; hoods 5–7 mm, much exceeding the gynostegium, the lateral margins with a small tooth below the middle; horns subulate, incurved; fr erect on deflexed pedicels. Swamps, bogs, and brackish marshes on the coastal plain; N.J. to Fla. and Tex. June, July.

14. **Asclepias sullivantii** Engelm. Smooth m. Glabrous throughout; stem rather stout, simple, to 1 m; lvs thick, 10–15 cm, narrowly or broadly elliptic to ovate or oblong, sessile or nearly so and somewhat cordate-clasping; umbels lateral at few to several upper nodes; cor purplish-rose, its lobes 9–12 mm; hoods 5–6 mm, much surpassing the gynostegium, the lateral margins entire, each hood flanked by a minute process at base; horns horizontally inflexed; fr glabrous or slightly verrucose, erect on deflexed pedicels. Moist prairies; O. and s. Ont. to Minn., N.D., and Okla. June, July.

15. **Asclepias rubra** L. Red m. Usually simple, 5–12 dm, with several pairs of lvs and 1–few small terminal umbels; lvs sessile or nearly so, lanceolate or lance-ovate, 8–12 cm, acuminate, rounded at base, minutely scabrellate near the margin above, otherwise glabrous; cor red, its lobes 8–10 mm; hoods slender, erect, red or orange-red, 6–8 mm, the lateral margins with an obscure lobe near the base; horn subulate, nearly erect, almost equaling the hood and at base scarcely adnate to it; fr 8–10 cm, erect on deflexed pedicels. Swamps, bogs, and wet woods on or near the coastal plain; L.I. and N.J. to Tex. June, July.

16. **Asclepias meadii** Torr. Simple, 3–6 dm; lvs 3–5 pairs, lance-ovate, 5–8 cm, acute or obtuse and mucronate, rounded to the sessile base, scabrous-ciliolate, otherwise glabrous; umbel solitary, terminal, few-fld; peduncle 5–15 cm; cor greenish-white, its lobes 8–10 mm; hoods purple, much distended at base; 5–7 mm, the lateral margins with an obtuse tooth near the middle; horns subulate, incurved, shorter than the hood. Prairies; originally from c. Ill. and c. Mo. to e. Kans., with outlying sites in ne. Ind., s. Wis., and se. Io.; now rare and restricted to Ill., Mo., and Kans. June, July.

17. **Asclepias viridis** Walter. Ozark-m. Decumbent or ascending, 2–6 dm; lvs oblong, lance-oblong, or elliptic, 6–12 cm, rounded or obtuse, abruptly narrowed at base to a 4–10 mm petiole; cor greenish, rotate (the lobes spreading or with ascending tip, not reflexed), 2–3.5 cm wide; hoods violet or purple, 4–6 mm, the horn rudimentary, represented by a small, flat, median plate near the summit of the hood. Prairies, barrens, and dry upland woods; O. to Mo. and Nebr., s. to Fla. and Tex. May–June, sometimes also in late summer. (*Asclepiodora v.*)

18. **Ascleplas stenophylla** A. Gray. Narrow-lvd m. Puberulent to subglabrous; stems 1 or 2 on a carrot-like or tuberous vertical root, 3–10 dm; lvs rather numerous, alternate or scattered or some opposite or subverticillate, linear, 6–15 cm, to 5(–8) mm wide; umbels subsessile in the upper axils, 10–25-fld; cor pale greenish to yellow, its lobes reflexed, 4.5–5.5 mm; hoods pale, 3–4 mm, ca equaling the gynostegium; horn adnate to the hood for most of its length (forming a ventral ridge), only the short tip free between the short terminal lobes of the hood; frs slender, erect, 9–12 cm. Dry prairies and glades; se. Minn.; Great Plains, from se. Mont. to n. Tex., e. to se. S.D., s. Mo., and w. Ark. June–Aug. (*Acerates s.*)

19. **Asclepias hirtella** (Pennell) Woodson. Prairie-m. Stout, 4–10 dm; lvs numerous, linear or lance-linear, 1–2 dm, scabrellate; umbels 2–10, densely fld (fls often 30–100) on peduncles 1–2(–3) cm; pedicels hirsutulous with spreading hairs; cor pale greenish or slightly purplish, its lobes reflexed, 4–6 mm; hoods 2–2.5 mm, the base separated from the cor by a distinct column, appressed to the gynostegium, obtuse at the summit, reaching just above the base of the anther-wings; horns none. Dry sandy soil and prairies; W.Va., O., and Mich. to Wis., n. Io., Kans., Okla., Ark., and sw. Ky. June–Aug. (*Acerates h.*)

20. **Asclepias longifolia** Michx. Slender, 3–6 dm; lvs numerous, linear, 8–15 cm, glabrous or nearly so; umbels 1–6, 10–30-fld, on peduncles 1–5 cm; pedicels with incurved hairs; cor pale greenish-white and tinted with purple, its lobes 5–6 mm; hoods as in no. 19, but reaching just to the projecting angle of the anther-wings. Moist or wet pine-lands on the coastal plain; Del. to Fla. and La. July, Aug. (*Acerates l.; A. floridanus,* misapplied.)

21. **Asclepias viridiflora** Raf. Green m. Erect to prostrate, 3–8 dm, thinly hairy; lvs lanceolate to linear or broadly oblong, elliptic, or ovate-oblong, scabrous on the margin, thinly hairy beneath; umbels lateral, densely fld, sessile or on peduncles to 2 cm; cor pale green, its lobes 6–8 mm; hoods narrowly elliptic-oblong, 4–5.5 mm long, appressed to the gynostegium, the base adjacent to the cor, the obtuse or subacute summit reaching well beyond the salient angle of the anther-wings or even to the top of the gynostegium; horns none. Dry upland woods, prairies, and barrens, especially in sandy soil; Conn. and s. N.Y. to Mich. and Man., s. to Ga., Ariz., and n. Mex., commoner westward. July, Aug. (*Acerates v.*)

22. **Asclepias lanuginosa** Nutt. Woolly m. Stems 1.5–3 dm, villous; lvs linear-oblong to more often lanceolate, 4–7 cm, obtuse, sparsely villous on both sides; umbel solitary, erect, terminal, short-peduncled; pedicels hirsutulous; cor pale greenish-yellow or somewhat purplish, its lobes 4–5 mm; hoods 2.5–3.5 mm, erect, appressed to the gynostegium, the base adjacent to the cor, the obtuse summit reaching distinctly beyond the projecting angle of the anther-wings or nearly to the top of the gynostegium; horns none. Dry upland woods and prairies; not common; Wis. and n. Ill. to Io., Kans., and N.D. (*A. nuttalliana; Acerates l.*)

3. **METAPLEXIS** R. Br. Cor campanulately subrotate, deeply cleft, the narrow lobes densely bearded on the upper side; corona very short, the 5 broadly rounded lobes alternating with the anther-bases; anthers terminating in a conspicuous, inflexed membrane; pollinia solitary in each pollen-sac, pendulous, apically attached to the translator; style

with a slender tip prolonged far past the anthers; twining shrubs or herbs with cordate lvs and pedunculate extra-axillary cymes solitary at the nodes. 6, e. Asia.

1. Metaplexis japonica (Thunb.) Makino. Twining perennial herb to 4 m, glabrous or nearly so; lvs cordate-ovate, 5–10 cm, glaucous beneath; peduncles 3–6 cm; fls in short, raceme-like or irregular cymes on pedicels 5–10 mm; cor lavender or purplish or whitish, the flaring tube only 2–3 mm, the lobes 7–8 mm; style-tip prolonged 6–8 mm, shortly bifurcate; fr lance-ovoid, 4–7 cm, tuberculate; $2n=22$. Native of e. Asia, intr. as a cornfield-weed in Io. June, July.

4. VINCETOXICUM N. Wolf. Swallow-wort. Cor small, campanulately subrotate, with short, broad lobes; corona a fleshy, lobed cup, from a little shorter to a little longer than the gynostegium; anthers terminating in a conspicuous, erect to inflexed membrane; pollinia solitary in each pollen-sac, pendulous, apically attached to the translator; perennial, usually ± strongly viny and twining (seldom erect) herbs with pedunculate extra-axillary cymes solitary at the nodes. 15, temp. Eurasia. (Often included in *Cynanchum*.)

1 Cor purple-black, its lobes deltoid, 1.5–3 mm, minutely hairy above .... 1. *V. nigrum.*
1 Cor paler, its lobes 2.5–4.5 mm, twice as long as wide, glabrous .... 2. *V. hirundinaria.*

1. Vincetoxicum nigrum (L.) Moench. Black s.-w. Climbing, 1–2 m; lvs short-petaloid, oblong to ovate, 5–10 cm, acuminate, rounded to subcordate at base; fls mostly 6–10 in umbelliform cymes on peduncles 0.5–1.5 cm; cor purple-black, its lobes rather fleshy, ± deltoid 1.5–3 mm, with tiny hairs 0.1–0.2 mm on the upper surface; corona inconspicuously 5-lobed; fr slender, 4–7 cm; $2n=44$. Native of s. Europe, occasionally escaped from cult. in our range, and locally established in woods and moist, sunny places. June–Sept. (*Cynanchum n.*)

2. Vincetoxicum hirundinaria Medikus. White s.-w. Slender, suberect or ± twining or scrambling; peduncles 1–3 cm; cor ochroleucous to maroon or pinkish, its lobes lanceolate, 2.5–4.5 mm, half as wide, glabrous, scarcely fleshy; corona deeply 5-lobed; $2n=22$; otherwise much like no. 1, for which it is often mistaken. Native of Europe, occasionally intr. into thickets and waste places in our range. June–Sept. (? *V. medium* and *Cynanchum medium,* the former name illegitimate.)

5. AMPELAMUS Raf. Sandvine. Cor deeply parted, its lobes erect at anthesis; corona of 5 erect, petaloid segments about equaling the cor-lobes, each divided to about the middle into 2 slender lobes; anthers with a prominent, membranous appendage; pollinia solitary in each pollen-sac, pendulous, apically attached to the translator; twining perennial herbaceous vines with white fls in pedunculate, extra-axillary infls solitary at the nodes. Monotypic.

1. Ampelamus albidus (Nutt.) Britton. Subglabrous, climbing to 10 m; lvs long-petioled, acute or acuminate, deeply cordate-ovate, the rounded basal lobes separated by a broad sinus; infl umbelliform or shortly racemiform; cor 5–6 mm; its lobes creamy or greenish-white, the corona bright white, fr slender, lanceolate, smooth, 10–15 cm, low moist woods or fields, often weedy; Pa. to Ind., Mo., and e. Neb., s. to Ga. and Tex. July, Aug. (*Cynanchum laeve; Gonolobus l.*)

6. MATELEA Aublet. Cor deeply divided, its lobes strongly ascending to widely spreading, linear to elliptic, always much longer than wide; corona a flat disk or shallow cup, the latter shorter than to slightly surpassing the gynostegium; margin of the corona distinctly or obscurely 10-lobed; anthers with a short, broad, membranous terminal appendage, or the appendage virtually obsolete; pollinia solitary in each pollen-sac, pendulous or ± horizontally placed, with a slender attachment to the translator; perennial twining herbs with broad lvs deeply cordate at base, the fls medium-sized, in pedunculate extra-axillary cymes solitary at the nodes; our spp. all with short-hairy herbage. 125, New World, mainly tropical. (*Gonolobus; Odontostephana; Vincetoxicum* Walter, not N. Wolf.)

1 Cal-lobes glabrous except often for a few hairs near the tip; corona nearly rotate, much shorter than
  the gynostegium; fr smooth.
  2 Cor-lobes glabrous above, about three times as long as the cal .... 1. *M. gonocarpa.*
  2 Cor-lobes finely hairy above, about twice as long as the cal .... 2. *M. suberosa.*
1 Cal-lobes hairy throughout; corona forming a cup around the gynostegium; fr muricate.
  3 Cor-lobes broader at or beyond the middle than at the sinus; alternate lobes of the corona prolonged
    into acute or acuminate, commonly bifid tips.

4 Fl-buds ovoid; cor-lobes 7–12 mm, two-fifths to half as wide .... 3. *M. carolinensis.*
4 Fl-buds conic; cor-lobes 11–18 mm, a fourth to a third as wide .... 4. *M. decipiens.*
3 Cor-lobes distinctly wider at the sinus than at the middle; lobes of the corona all truncate or emarginate .... 5. *M. obliqua.*

1. Matelea gonocarpa (Walter) Shinners. Common angle-pod. Lvs usually oblong or obovate-oblong, often twice as long as wide, cal-lobes hirsutulous at the tip, otherwise glabrous; cor conic in bud, gradually acute to acuminate, scarcely twisted, at anthesis brownish-purple to greenish-purple, its lobes narrowly lanceolate, 8–14 mm, thrice as long as the cal; corona nearly rotate, much shorter than the gynostegium; pollen-sacs inconspicuous, with narrow slits, the pollinia ± pendulous; fr glabrous, 8–13 mm, sharply angled but not winged. Moist woods and thickets; se. Va. to Fla. and Tex., and up the Mississippi embayment to s. Mo. and s. Ind. July. (*Gonolobus g.; Vincetoxicum g.*)

2. Matelea suberosa (L.) Shinners. Atlantic angle-pod. Cal glabrous or nearly so; cor short-conic in bud, abruptly acuminate, twisted, its lobes at anthesis finely hairy above, 5–8 mm, about twice as long as the cal; fr wing-angled; otherwise much like no. 1, and in similar habitats. Se. Va. to Fla. (*Gonolobus s.; Vincetoxicum s.*)

3. Matelea carolinensis (Jacq.) Woodson. Lvs broadly ovate to subrotund, abruptly acuminate, with rounded basal sinus; cor ovoid in bud, at anthesis brownish-purple, its lobe widely spreading, elliptic, 7–12 mm, two-fifths to half as wide, minutely hairy outside; corona cupulate, variable, alternate lobes short and broad or narrower, somewhat prolonged, and bifid; pollen-sacs as in spp. 4 and 5, obvious, with open slits, the pollinia ± horizontal; fr 10–16 mm, sharply muricate, not angled. Moist woods and thickets; s. U.S., n. to Del., Md., Ky., and s. Mo. June, July. (*Gonolobus c.; Odontostephana c.; Vincetoxicum c.; V. hirsutum.*)

4. Matelea decipiens (Alexander) Woodson. Much like no. 3; fl-buds conic; cor-lobes more ascending, 11–18 mm, a fourth or a third as wide, usually widest well above the middle. Open woods and thickets; s. U.S., n. in the interior to s. Ill. and Mo. May, June. (*Gonolobus d.; Odontostephana d.*)

5. Matelea obliqua (Jacq.) Woodson. Lvs ovate to rotund, the basal lobes sometimes separated by a wide, rounded sinus, sometimes overlapping; fl-buds conic; cor brownish-purple, its lobes ascending, 10–16 mm, 1.5–2.5 mm wide, gradually tapering from base to apex; corona fleshy, equaling or slightly surpassing the gynostegium, its lobes ± truncate or emarginate. Rocky woods and thickets; Pa. to O., Ind., and Mo., s. to N.C. and Tenn. June, July. (*Gonolobus o.; Odontostephana o.; Vincetoxicum o.; G. shortii; O. shortii.*)

FAMILY **SOLANACEAE**, the Nightshade Family

Fls perfect, hypogynous, sympetalous, typically 5-merous; cal usually gamosepalous, generally persistent; cor regular in most genera, rotate to tubular, the lobes commonly plicate (and sometimes also convolute) in bud, or seldom merely convolute, imbricate, or valvate; stamens borne on the cor-tube, as many as and alternate with the lobes, rarely fewer; ovary superior, mostly bicarpellate and bilocular, with the carpels obliquely oriented to the median plane of the fl, or seldom (as in *Nicandra*) the carpels and locules 3–5; style terminal, with a lobed stigma; ovules ± numerous on axile, often thickened placentas; fr a capsule or berry; seeds with a dicotyledonous, generally linear, often curved and subperipheral embryo, usually with an oily endosperm; herbs or less often shrubs or even trees, with alternate or (as in *Petunia*) falsely opposite lvs, the fls in various sorts of basically cymose infls, sometimes solitary at the nodes. 85/2800, nearly cosmop.

1 Stems woody; fr a red berry; cor lavender or purplish, 9–14 mm .... 6. *Lycium.*
1 Stems essentially herbaceous; fls and fr various.
  2 Plants flowering.
    3 Cor widely spreading, rotate or nearly so.
      4 Anthers opening by terminal pores or short slits .... 4. *Solanum.*
      4 Anthers dehiscing longitudinally.
        5 Lvs compound; fls bright yellow .... 5. *Lycopersicon.*
        5 Lvs simple; fls white or whitish .... 2. *Leucophysalis.*
    3 Cor campanulate to tubular or funnelform.
      6 Fls in branched, paniculiform or racemiform clusters .... 9. *Nicotiana.*
      6 Fls otherwise, solitary at the nodes.
        7 Fls sessile or subsessile; middle and upper lvs mostly sessile .... 7. *Hyoscyamus.*
        7 Fls pedicellate; lvs mostly petiolate.
          8 Cal tubular, 3.5–10 cm .... 8. *Datura.*
          8 Cal short-tubular or of separate sep, well under 3.5 cm.
            9 Fls ascending or erect .... 10. *Petunia.*
            9 Fls nodding.
              10 Cor white to yellow or greenish-yellow; carpels 2 .... 3. *Physalis.*
              10 Cor blue; carpels 3–5 .... 1. *Nicandra.*

2 Plants fruiting.
  11 Fr a berry (sometimes enclosed by the accrescent cal).
    12 Cal accrescent, wholly or partly enclosing the berry.
      13 Mature cal glabrous or merely pubescent.
        14 Cal parted nearly to the auricled base; carpels 3–5 .... 1. *Nicandra.*
        14 Cal lobed only at the summit, not auricled at base; carpels 2.
          15 Cal scarcely inflated, nearly filled by the berry, open at the end .... 2. *Leucophysalis.*
          15 Cal strongly inflated, usually not filled by the berry, virtually closed .... 3. *Physalis.*
      13 Mature cal spiny .... 4. *Solanum.*
    12 Cal not much if all accrescent, subtending but not enclosing the berry.
      16 Fr up to 2.5 cm thick, usually pulpy rather than juicy .... 4. *Solanum.*
      16 Fr over 3 cm thick, very juicy .... 5. *Lycopersicon.*
  11 Fr a capsule.
    17 Capsules sessile, secund, solitary at the nodes, circumscissile above .... 7. *Hyoscyamus.*
    17 Capsules pedicellate, longitudinally dehiscent.
      18 Capsules in paniculiform or racemiform terminal clusters .... 9. *Nicotiana.*
      18 Capsules solitary at the nodes, sometimes in the forks of the stem.
        19 Cal circumscissile at maturity, the base persisting as a collar below the usually
          spiny capsule .... 8. *Datura.*
        19 Cal persistent, the sep foliaceous; capsule not spiny .... 10. *Petunia.*

1. NICANDRA Adans. Apple of Peru. Cal deeply 5-parted, its lobes imbricate, accrescent and reticulate-veiny in fr, much surpassing the berry, with projecting margins and auriculate base; cor broadly campanulate, shallowly 5-lobed; stamens included; filaments hairy below; anthers short, longitudinally dehiscent; ovary 3–5-locular, with a slender style and 3–5-lobed stigma; berry enclosed by the cal, dry, many-seeded, eventually opening irregularly. Monotypic.

  1. **Nicandra physalodes** (L.) Gaertner. Glabrous annual to 1.5 m; lvs ovate or lance-ovate, 1–2 dm, coarsely and unevenly toothed, long-petiolate; pedicels ca 1 cm, arising from the stem at the side of a petiole, at maturity recurved at the tip; sep broadly cordate-ovate, 1 cm at anthesis, 2–3 cm in fr; cor blue, 2–2.5 cm long and wide; $2n=20$. Native of Peru; occasionally cult. for ornament and escaped along roadsides and in waste places throughout much of our range. July–Sept. (*Physalodes p.*)

2. LEUCOPHYSALIS Rydb. Much like *Physalis,* but the cal shallowly lobed, accrescent but not inflated, at maturity largely enclosing the berry but open at the end; cor wide open, nearly rotate, white or whitish; fls often more than 1 per node. 3, Asian, 2 N. Amer.

  1. **Leucophysalis grandiflora** (Hook.) Rydb. White-fld ground-cherry. Taprooted, short-lived perennial to 1 m, thinly villous and ± viscid; lvs ovate or lance-ovate, 5–12 cm, acute or short-acuminate, entire; fls commonly 2–4 from the upper nodes, on pedicels 10–15 mm; cal-lobes narrowly triangular, acuminate; cor white with a pale yellow center, 3–4 cm wide; filaments slender; anthers 3 mm; fruiting cal 1.5 cm, round-ovoid, open at the end, nearly filled by the berry. Dry sandy soil; Que. to Sask., s. to Vt., Mich., Wis., and Minn. June–Aug. (*Physalis g.; Chamaesaracha g.*)

3. PHYSALIS L. Ground-cherry. Cal at anthesis small, 5-lobed, the tube enlarging promptly thereafter, becoming membranous and strongly veiny, loosely but completely enclosing the fr, often 5-angled and often retuse at base, its lobes connivent in fr and scarcely enlarged; cor broadly funnelform or campanulate, shallowly lobed to entire, usually yellow or yellowish, often with 5 large dark spots within, seldom white; stamens inserted near the base of the cor, opening longitudinally; berry pulpy, many-seeded; mostly herbs, commonly widely branching, with alternate or seemingly opposite lvs; fls solitary or less often 2+ at the nodes on drooping pedicels. 90, ± cosmop., but esp. New World.

1 Rhizomatous, colonial perennials; cor commonly 12–20 mm.
  2 Cor white, distinctly 5-lobed; fruiting cal bright red-orange .... 1. *P. alkekengi.*
  2 Cor yellow or yellowish, often scarcely lobed; fruiting cal green or brown.
    3 Lvs pubescent beneath with simple hairs, or glabrous.
      4 Upper part of the stem with soft, slender, widely spreading hairs .... 2. *P. heterophylla.*
      4 Upper part of the stem with short, stiffish hairs, or glabrous.
        5 Cal-tube minutely strigose, chiefly in 10 narrow, longitudinal strips, with minute, appressed
          hairs rarely 0.5 mm; pubescence of the lvs minute or none, almost wholly restricted
          to the main veins, that of upper part of the stem sparse and erect-appressed .... 3. *P. longifolia.*

5 Cal-tube hirsute or hirsutulous over its whole surface with loose or spreading hairs 0.5–1.5
   mm; lvs usually sparsely hairy across the surface on both sides.
   6 Hairs of the upper part of the stem mostly or all ± recurved .... 4. *P. virginiana*.
   6 Hairs of the upper part of the stem ascending or ascending-spreading .... 5. *P. hispida*.
3 Lvs pubescent beneath with forked or stellate hairs.
   7 Cal hirsute; lf-blade decurrent on the petiole; inland .... 6. *P. pumila*.
   7 Cal finely cinereous-stellate; lf-blade not decurrent; maritime .... 7. *P. walteri*.
1 Taprooted annuals; cor (except in no. 11) seldom over about 1 cm.
   8 Upper part of the stem with rather long, soft, spreading hairs.
      9 Cor wholly yellow; cal-lobes deltoid; lvs not cordate at base .... 8. *P. missouriensis*.
      9 Cor with a dark center; cal-lobes more elongate and tapering; principal lvs usually cordate at
        base .... 9. *P. pubescens*.
   8 Upper part of the stem glabrous to merely strigillose.
      10 Cor wholly yellow; anthers 1–2.3 mm .... 10. *P. angulata*.
      10 Cor with a dark center; anthers mostly 2.5–3.5 mm .... 11. *P. philadelphica*.

1. **Physalis alkekengi** L. Chinese-lantern plant. Rhizomatous perennial 4–6 dm, usually erect and un-branched; lvs ovate, petioled; fls white, 1.5 cm wide; fruiting cal bright red, 5 cm; 2n=24. Native of Eurasia, occasionally escaped from cult. into waste places.

2. **Physalis heterophylla** Nees. Clammy g.-c. Erect or spreading, rhizomatous perennial 2–9 dm; pubes-cence of the upper parts, pedicels, and cal distinctly villous, of slender, spreading hairs; lvs chiefly ovate or rhombic, 3–8 cm, acute, shallowly and irregularly sinuate-dentate or sometimes entire, broadly rounded or subcordate at base, not decurrent, hairy on both sides; pedicels to 1 cm at anthesis, to 3 cm in fr; cal-lobes deltoid or ovate, the margins straight or slightly concave; cor 12–20 mm; filaments dilated at the summit, about as broad as the anthers, these 3–4.5 mm long; fruiting cal ovoid, 3–4 cm, retuse at base; fr green; 2n=24. Dry or sandy soil, upland woods, and prairies, probably our most abundant sp.; Que. and N.S. to Minn., Colo., and Utah, s. to Fla. and Tex. June–Sept. Typical *P. heterophylla* has the hairs of the stem and lower lf-surface very fine, viscid, and densely crowded, seldom over 0.5 mm, more copious on the lf-veins than across the surface. Plants with sparser, more uniformly distributed pubescence, of jointed hairs 1–2 mm, with a similar range but not extending quite so far n., have been called var. *ambigua* (A. Gray) Rydb. (*P. ambigua, P. nyctaginea*), but the difference is only doubtfully significant.

P. peruviana L. with slender filaments, acuminate lvs, and deep purple-blue (instead of yellow to light blue) anthers, occasionally escapes from cult. in our range.

3. **Physalis longifolia** Nutt. Longlf g.-c. Erect, rhizomatous perennial 4–8 dm, usually divergently branched, the younger parts glabrous to densely puberulent with appressed or ascending hairs rarely over 0.5 mm; lvs 4–10 cm, glabrous or very sparsely and minutely puberulent, the hairs of the lower surface restricted to the veins; pedicels 1–2 cm at anthesis, not much elongating in fr; cal minutely pubescent in 10 strips along the nerves, the hairs appressed, rarely 0.5 mm; cal-lobes triangular or ovate, 3–4 mm, densely ciliate; cor 11–17 mm, yellow, dark-blotched within; anthers 3–4 mm; filaments dilated; fruiting cal ovoid or short-cylindric, acuminate, 3–4 cm, scarcely retuse at base; 2n= 24. Fields, open woods, and prairies; Vt. and Ont. to Mont., s. to Va., Tenn., La., and Ariz. July–Aug. Plants of our range are mostly var. *subglabrata* (Mackenzie & Bush) Cronq., with rather thin, ± ovate, sinuate-toothed lvs abruptly narrowed to the long petiole. (*P. subglabrata; P. macrophysa*.) The more western var. *longifolia*, with firmer, narrower (lanceolate to lance-elliptic or narrowly rhombic), mostly entire lvs more tapering to the often shorter petiole, is occasionally intr. in our range. (*P. virginiana* var. *sonorae*.)

4. **Physalis virginiana** Miller. Virginia g.-c. Rhizomatous perennial 3–6 dm, the stems usually forked, with ascending branches; pubescence of the younger stems, the petioles, and often also the pedicels, consisting wholly or in considerable part of short, stiff, ± decurved hairs; lvs ovate to narrowly lanceolate, acute, sinuately toothed to entire, narrowed to the base and ± decurrent on the long petiole, sparsely to abundantly hirsutulous on both sides, minutely ciliolate; pedicels at anthesis 1–2 cm, not much elongating in fr; cal hirsute or hirsutulous with spreading hairs, its lobes at anthesis 2.5–5 mm, densely ciliolate; cor 12–18 mm, yellow, blotched within, filaments dilated below, gradually narrowed above; anthers 2–3 mm; fruiting calyx sunken at base, 5-angled, notably longer than thick; fr orange; 2n=24. Fields, upland woods, and prairies; Conn. to Mich., Minn., and Colo., s. to S.C., Ala., and Ariz. (*P. monticola*.)

5. **Physalis hispida** (Waterfall) Cronq. Plains-sandhill g.-c. Much like no. 6, and ± fertile with it exper-imentally, but differing notably in its sparser pubescence with the hairs all or nearly all simple, those of the upper part of the stem ascending-spreading to appressed-ascending, those of the lvs scattered over the surface; lvs thick, entire, ± elliptic, 2.5–7 cm; pedicels mostly 1–3 cm at anthesis, later often elongating to 3–5 cm; cal rough-hirsute with flattened, septate hairs much as in no. 4; cor 12–18 mm, with faint or no dark blotches inside; filaments dilated; anthers 2.5–3.5 mm; 2n=24. Sandy places on the Great Plains, from S.D. to Kans. to Wyo., Colo., and Tex., and occasionally intr. eastward at least as far as Ind., mainly along railroads. May–Aug. (*P. virginiana* var. *hispida; P. lanceolata*, misapplied.)

6. **Physalis pumila** Nutt. Prairie g.-c. Erect or spreading, rhizomatous perennial 2–5 dm; pubescence of the upper parts, including the petioles, pedicels, and lower lf-surfaces dense to less often sparse, rough to the touch, composed of loose or spreading, rather short hairs once or twice dichotomously forked; lvs lance-ovate to elliptic, 4–8 cm, entire or merely undulate, long-cuneate to the base and decurrent as a narrow wing along the petiole; pedicels at anthesis 1–2.5 cm, later sometimes to 3–5 cm; cal hirsute, its lobes triangular,

notably longer than wide; cor 12–17 mm, usually only faintly or scarcely blotched within; filaments dilated; anthers 2.5–3.5 mm; fruiting cal ovoid, 2–3 cm, retuse at base; $2n=24$. Prairies, and sometimes in disturbed habitats, as along railroads; Ill. and Mo. to Neb., Tex., and Ark. Summer.

7. **Physalis walteri** Nutt. Spreading or prostrate, rhizomatous perennial to 4 dm; pubescence of the upper parts, including the pedicels, cal and both sides of the lvs, cinereous, composed of minute, stellate or branching hairs; lvs ovate to elliptic, 3–10 cm, commonly obtuse, entire, rounded to obtuse or abruptly short-cuneate at base; petioles wingless; cal-lobes deltoid, little longer than wide; cor 11–17 mm; anthers 2–3 mm, equaling or shorter than the filaments; fruiting cal ovoid, 2–3 cm; $2n=24$. Sandy soil along the coast, from se. Va. to the Gulf of Mexico. May, June. (*P. viscosa* var. *maritime.*)

8. **Physalis missouriensis** Mackenzie & Bush. Ozark g.-c. Annual to 7 dm, widely branched from the base, the upper parts and petioles viscid-villosulous, the hairs ca 0.5 mm; lvs ovate, 3–6 cm, shallowly and unevenly repand-dentate, broadly acute to rounded at base, often inequilateral; pedicels at anthesis 3–5 mm, to 1 cm at maturity; fls 7–10 mm, yellow; cal-lobes deltoid, about as wide as long; anthers 0.7–1.2 mm; fruiting cal obscurely angled, 2–3 cm. Dry upland woods and barrens; Mo. to se. Neb., Kans., Okla., and Ark. May–Sept. (*P. pubescens* var. *m.*)

9. **Physalis pubescens** L. Downy g.-c. Annual to 6 dm, widely branched from near the base, villous or viscid-villous; lvs ovate, 3–10 cm; pedicels 5–10 at anthesis, 5–20 mm in fr; cor 6–10 mm, yellow with a dark center; cal-lobes narrow, 2–4 mm; anthers 1.5–2 mm; fruiting cal 2–4 cm, 5-angled; $2n=24$. Moist soil; pantropical, n. to Mass., Ont., and Wis. May–Sept. Three vars. in our range:

1 Lvs grayish with pubescence, often also with sessile glands, mostly toothed to near the base; chiefly
of ne. U.S., and essentially throughout our range . . . . . . var. *grisea* Waterfall (*P. pruinosa* of authors, perhaps not L.)
1 Lvs greenish, less densely hairy, and without sessile glands.
  2 Lvs usually toothed nearly to the base with 5–8 teeth on each side of the seldom translucent blade;
  pantropical, n. to our s. borders . . . . var. *pubescens* (*P. barbadensis.*)
  2 Lvs with few teeth, 3–4 on each side, or entire; blades mostly flaccid and translucent; Pa. to Io.
  and s. . . . . var. *integrifolia* (Dunal) Waterfall.

10. **Physalis angulata** L. Annual, 2–10 dm, glabrous, or with a few short appressed hairs especially on the younger parts; lvs mostly 4–10 cm, ovate to lance-ovate, irregularly and often coarsely toothed to entire; cor yellowish, not dark in the center, 4–10 mm; anthers 1–2.3 mm; $2n=24, 48$. Tropical Amer., n. to se. Va. and Mo., and rarely adventive northward. Summer. Var. *angulata*, with essentially the range of the sp., has pedicels 5–15 mm in fl, 20–30 mm in fr, the cal-teeth 2–2.5 mm in fl, the mature cal 2.5–3.5 cm. The well marked var. *pendula* (Rydb.) Waterfall, of Mo. to Kans. and Tex., and occasionally intr. eastward, has pedicels 15–40 mm in fl, 20–40 mm in fr, the cal-teeth 1 mm in fl, the mature cal 2–2.5 cm. (*P. pendula.*)

11. **Physalis philadelphica** Lam. Tomatillo. Branching annual 2–6 dm, nearly glabrous except on the younger parts, where it is strigillose, often in 2 strips, with decurved hairs; lvs ovate or rhombic, 2–6 cm, entire to sinuate-toothed; pedicels at anthesis 3–5 mm, scarcely longer in fr; cor 7–15 mm, with dark center; anthers blue-purple, mostly 2.5–3.5 mm, becoming twisted or contorted; fruiting cal rounded at base; berry edible, purplish, viscid, nearly filling the cal; $2n=24$. Native of Mexico, occasionally escaped from cult. in our range. (*P. ixocarpa.*)

4. **SOLANUM** L. Nightshade. Cor rotate or broadly campanulate; filaments very short; anthers oblong to lanceolate or linear, connivent or connate around the style, opening (in our spp.) by terminal pores or short terminal clefts; fr a many-seeded berry, the seeds mostly wingless; herbs, shrubs, or small trees, the infls in our spp. generally arising opposite or between the lvs. Perhaps 1500 spp., widespread, but best developed in trop. Amer. All our spp. bloom in summer, often continuing until the fall.

1 Pubescence of the lower lf-surface entirely of simple hairs, or none; anthers oblong or linear, not
  tapering to the tip; plants unarmed.
  2 Stems climbing or twining; fls mostly light blue or violet . . . . 1. *S. dulcamara.*
  2 Stems erect or diffuse; fls white.
    3 Lvs simple, entire to deeply lobed; plants annual.
      4 Lvs entire to merely toothed or wavy-margined.
        5 Cal scarcely accrescent, not cupping the fr; ours with black frs and mostly glabrous or
        somewhat appressed-hairy stems . . . . 2. *S. nigrum.*
        5 Cal accrescent, cupping the lower half of the greenish or yellowish fr; stems spreading-hairy
        . . . . 3. *S. sarrachoides.*
      4 Lvs evidently pinnatilobate . . . . 4. *S. triflorum.*
    3 Lvs pinnately compound; plants perennial . . . . 5. *S. jamesii.*
1 Pubescence of the lower lf-surface all or chiefly of stellate hairs; anthers tapering to the tip; stem
  generally prickly or spiny.
  6 Cal-tube spiny, even at anthesis; lvs deeply pinnately or bipinnately lobed; annual.
    7 Cor yellow; infl stellate-hairy, not glandular-villous . . . . 6. *S. rostratum.*
    7 Cor violet or blue; infl glandular-villous as well as stellate . . . . 7. *S. sisymbriifolium.*

6 Cal-tube merely stellate, or with 1 or 2 short spines; lvs entire to shallowly lobed; perennial.
8 Lvs ovate, half as wide as long, with a few large teeth or shallow lobes .... 8. *S. carolinense.*
8 Lvs linear to oblong or lance-oblong, a sixth to a third as wide as long, entire or sinuate-margined
.... 9. *S. elaeagnifolium.*

1. Solanum dulcamara L. Bittersweet. Rhizomatous perennial, shrubby below, climbing or scrambling to 1–3 m, moderately short-hairy to glabrous; lvs petiolate, some simple and with rather broadly ovate-subcordate blade 2.5–8 × 1.5–5 cm, others with a pair of smaller basal lobes or lfls; peduncles 1.5–4 cm, 10–25-fld, the infl 3–8 cm wide, jointed, bractless, often subdichotomously branched; cor light blue or violet, the lobes 5–9 mm (each with 2 shiny green basal spots), soon reflexed; anthers conspicuous, yellow; fr poisonous, bright red, 8–11 mm; $2n=24, 48, 72$. Thickets, clearings, and open woods, often in moist soil; native of Eurasia, naturalized throughout our range.

2. Solanum nigrum L. Black n. Branching annual, 1.5–6 dm, glabrous, or somewhat strigose or incurved-puberulent especially above; lvs petiolate; ovate to deltoid, irregularly blunt-toothed or subentire, 2–8 × 1–5.5 cm; peduncles numerous, ascending, up to 3 cm, the pedicels closely clustered (forming an umbelliform infl), mostly deflexed at least in fr; mature cal 2–3 mm, the lobes often unequal, sometimes reflexed; cor white or faintly bluish, 5–10 mm wide; frs globose, black, 8 mm, poisonous at least when young, many-seeded and often with 1–10 subglobose concretions up to half as long as the seeds; polyploid series based on $x=12$. A cosmopolitan weed of disturbed habitats, highly diversified, but not yet satisfactorily resolved into discrete taxa, in spite of many attempts. The native N. Amer. plants, as here described, are all diploid, so far as known; the oldest name for these at the specific level may be *S. ptychanthum* Dunal (*S. americanum* Miller, probably misapplied). At the varietal level the name would be *S. nigrum* var. *virginicum* L. Typical European *S. nigrum* is only casual with us, mainly about our large Atlantic ports. It is hexaploid and more pubescent (the hairs short, ± spreading and somewhat viscid), with a more nearly racemiform (but still compact) infl.

3. Solanum sarrachoides Sendtner. Hairy n. Much like no. 2; stem softly spreading-hairy, the hairs flattened, viscid, often gland-tipped; lvs evidently hairy along the main veins beneath, sometimes over the surfaces as well; cal viscid hairy, accrescent, at maturity 4–6(–9) mm and cupping the lower half of the greenish or yellowish fr, the lobes evidently connate below; $2n=24$. A weed in disturbed habitats, native to S. Amer., now widely intr. elsewhere in the world, and sometimes found in our range. Recent work suggests that the name *S. sarrachoides* should be restricted to a S. Amer. sp. with even more accrescent cal. Our plants would then take the name *S. physalifolium* Rusby.

S. villosum Miller, a European tetraploid, is casually intr. in our range. It is much like *S. sarrachoides,* but has yellow to orange-red frs and non-accrescent cal.

4. Solanum triflorum Nutt. Cut-lvd n. Foetid annual, ± prostrate, branched from the base, 1–6 dm, short-hairy or eventually glabrate; lvs short-petiolate, the blade 2–5 × 1–3 cm, evidently pinnatilobate, the rachis seldom wider than the length of the lobes; cal accrescent, the tube short, the rather narrow lobes to 6 mm in fr; fr globose, greenish, 9–14 mm; $2n=24$; otherwise like nos. 2 and 3. Native to w. U.S., occasionally intr. as a weed in our range.

5. Solanum jamesii Torr. Wild potato. Tuberiferous perennial, 1–4 dm; lvs pinnately compound, the lateral lfls lanceolate, sessile or short-stalked, not alternating with smaller ones; cor-lobes lance-triangular; $2n=24$. Native of w. U.S., reported to be established in Io.

S. tuberosum L., the potato, with stalked, ovate lfls irregularly alternating with much smaller sessile ones, and with broader cor-lobes, is occasionally found on rubbish-heaps and in waste places, but seldom survives the winter.

6. Solanum rostratum Dunal. Buffalo-bur. Coarse branching annual 3–10 dm, stellate-hairy throughout, the stems, cals, and to a lesser extent the lvs also beset with yellow spines 3–12 mm; lvs ovate or oblong in outline, deeply pinnately lobed, or the segments again lobed in larger lvs; racemes short-pedunculate, 3–15-fld, condensed at first, elongating to as much as 15 cm; fls light yellow, 2–3 cm wide; 4 anthers alike, yellow, the fifth much longer, curved, often purplish; cal accrescent and enclosing the berry; $2n=24$. Dry prairies and plains; native to the Great Plains, now occasionally intr. throughout much of our range, especially westward.

7. Solanum sisymbriifolium Lam. Sticky n. Much like no 6; stellate-hairy in the infl and also glandular-villous; lvs to 2 dm, stellate-hairy, deeply and irregularly pinnatifid, the segments coarsely toothed or shallowly lobed; cor pale violet or blue, 3 cm wide; anthers yellow, equal; $2n=24$. Native of S. Amer., established as a weed in s. U.S., and often adventive in waste places along the coast in our range.

S. citrullifolium A. Braun, a related sp. mainly of Mex. and Tex., has been reported, perhaps mistakenly, from our range. It has unequal anthers, one of them much the longer, outcurved, and violet.

8. Solanum carolinense L. Horse-nettle. Coarse, erect, branching, rhizomatous perennial to 1 m, spiny and loosely stellate-hairy; lvs ovate in outline, 7–12 cm, half as wide, with 2–5 large teeth or shallow lobes on each side, ± spiny along the main veins, stellate-hairy on both sides, the hairs sessile with 4–8 branches, the central branch often elongate; infl several-fld, elongating at maturity and forming a simple racemiform cluster; cor pale violet to white, 2 cm wide; anthers equal; fr deadly poisonous, yellow, 1–1.5 cm, subtended but not enclosed by the mostly unarmed cal; $2n=24$. Fields and waste places, especially in sandy soil, originally native to se. U.S., n. to Va. and Ky., now established as a weed n. to Vt., Ont., Mich., Minn., and westward.

S. dimidiatum Raf., of Kans. to Ark. and Tex., occurs in Mo. apparently as an introduction. It has the stellate hairs of the lower lf-surface distinctly stipitate, with 9–13 subequal branches, and the infl is commonly branched from the top of the peduncle, forming 2 or 3 raceme-like clusters. (*S. torreyi.*)

9. Solanum elaeagnifolium Cav.   Silverlf-n. Coarse, branching, rhizomatous perennial to 1 m, spineless or sparsely spiny, silvery-canescent with stellate hairs throughout; lvs linear to oblong or lance-oblong, 5–15 cm, a sixth to a third as wide, entire or merely sinuate; fls violet, 2 cm wide; $2n=24$, 72. Dry soil; Mo. and Kans. to Tex. and Ariz., and occasionally adventive eastward.

5. LYCOPERSICON Miller.   Tomato. Differing from *Solanum* chiefly in its anthers, which are connate into a slender cone, elongate, tapering to a long sterile tip, and opening by longitudinal slits; ovary in cultivated forms usually with more than 2 (commonly 4) locules; seeds winged by agglutinated hairs; unarmed herbs with pinnate compound lvs and yellow fls. 7, Pacific S. Amer.

1. Lycopersicon esculentum Miller.   Sprawling and freely branched, clammy-pubescent annual (potentially perennial but tender); lvs pinnately or bipinnately compound, to 3 dm, the lfls variable, the larger ovate or lanceolate, toothed or shallowly lobed; fls yellow, 1 cm wide; fr very juicy, several cm thick, in most cultivars scarlet; $2n=24$, 48. Native to the Andes, widely cult.; seedlings frequently appear about gardens, roadsides, and waste places, but do not persist. July–Sept.

6. LYCIUM L.   Matrimony-vine. Cal 4–5-lobed, campanulate to tubular, ruptured by the growing fr; cor tubular to funnelform, 4–5-lobed; anthers longitudinally dehiscent, much shorter than the slender filaments; fr a fleshy or dry berry; seeds 2–many, somewhat compressed, with strongly curved embryo; shrubs or small trees, usually thorny, with entire or minutely toothed lvs, often fascicled lvs, the fls mostly 1–4 in the axils. 100, widespread.

1. Lycium barbarum L.   Glabrous shrubs with long, weak, generally sparsely thorny, arched or climbing branches, 1–6 m; lvs short-petiolate, entire, elliptic to lanceolate, ovate, or oblanceolate, to 7 × 3.5 cm on vigorous young shoots, or only 1.5 × 0.3 cm on older ones; pedicels 0.7–2 cm; cor lavender or purplish, 9–14 mm, with (4)5 broad, spreading lobes shorter than to about equaling the tube; berry ellipsoid or ovoid, 1–2 cm red, 10–20-seeded; $2n=24$. Native of Eurasia, occasionally escaped from cult. in our range. June–Sept. (*L. chinense; L. halimifolium.*)

7. HYOSCYAMUS L.   Henbane. Cal campanulate or urceolate, 5-toothed, accrescent and enclosing the fr; cor funnelform, with an oblique, 5-lobed, slightly irregular limb; stamens mostly exserted, the anthers longitudinally dehiscent, much shorter than the filaments; capsule ± bilocular, circumscissile well above the middle; seeds numerous, flattened; embryo strongly curved; narcotic, poisonous herbs with ample, mostly toothed or incised-pinnatifid lvs; fls showy, borne in terminal, mostly secund and ± leafy-bracteate apparent racemes or spikes, the lower merely axillary (or adjacent to the axils). 15, mainly Mediterranean.

1. Hyoscyamus niger L.   Coarse, strong-scented biennial or annual to 1 m, conspicuously villous, especially the stem; lvs large, sessile, 5–20 × 2–14 cm, rather shallowly pinnatilobate, with unequal, triangular segments; mature cal 2.5 cm, urceolate, reticulate-veiny; cor 2.5–4.5 cm, nearly or quite as wide, prominently purple-reticulate on a pale, often greenish-yellow background, more distinctly purple in the throat; fr 1–1.5 cm, with a strongly thickened lid; $2n=34$. Native of Europe, now established as a weed of roadsides and waste places in n. U.S. and s. Can., more especially westward. May–Sept. A source of hyoscyamine and scopolamine.

8. DATURA L.   Cal cylindric or prismatic, generally circumscissile near the base, leaving a flaring collar under the fr; cor elongate-funnelform, very large, the lobes well developed or represented by slender projections; anthers longitudinally dehiscent, much shorter than the filaments; fr a 2-carpellate, 4-locellar, mostly spiny capsule, generally opening by 4 apical valves; seeds numerous, flattened, with curved embryo; narcotic, poisonous herbs or woody plants with large, entire to toothed or lobed lvs; fls terminal and solitary in origin, later appearing to be borne in the forks of the branches. 25, neotrop. and warm-temp., now widely intr.

1 Cal 3–5 cm, the tube strongly prismatic and narrowly 5-winged; cor 7–10 cm .... 1. *D. stramonium.*
1 Cal 6–15 cm, the tube terete or slightly angled, wingless; cor 12–20 cm .... 2. *D. wrightii.*

1. Datura stramonium L. Jimson-weed. Coarse, heavy-scented, inconspicuously puberulent annual to 1.5 m, often divaricately branched; lvs petiolate, with large, coarsely few-toothed or sublobate blade to 2 × 1.5 dm; cal 3–5 cm, strongly prismatic and narrowly 5-winged, unequally 5-toothed, the persistent base 4–6 mm and spreading to reflexed; cor white or anthocyanic, 7–10 cm, the limb 3–5 cm wide, shallowly 5-lobed, each lobe with a slender, projecting tooth to 1 cm; fr erect, ovoid, 3–5 cm, generally covered with short prickles, the lower prickles often shorter than the upper (or the fr smooth); 2n=24. Dry soil and waste places; widespread in temperate and warm regions, perhaps of American origin. June–Aug. (*D. tatula,* with anthocyanic fls.)

2. Datura wrightii Regel. Indian apple. Coarse, heavy-scented perennial to 1.5 m, often as broad as high, grayish-pubescent throughout with fine, ± crisp short hairs, the lvs velvety beneath; lvs to 2 dm, broadly ovate, rounded to cordate at base, entire or merely repand; cal 6–15 cm, terete or slightly angled, wingless, unequally 5-toothed, the persistent portion 1–2 cm, reflexed; cor 12–20 cm, white, the limb to 12 cm wide, slenderly 5-toothed; fr reflexed, subglobose, 3–4 cm, densely hairy and copiously short-spiny; 2n=24. Native of Mex. and sw. U.S., casually intr. here and there in our range. July–Oct. Confused in the past with *D. inoxia* Miller (*D. meteloides*), a Mexican sp. with narrower, 10-toothed cor-limb.

Datura metel L., a sp. of uncertain geographic origin, is rarely adventive in waste places. It has glabrous to subglabrous herbage, coarsely few-toothed, basally truncate to broadly cuneate lvs with a few large coarse teeth on the margins, and a 10-lobed cor-limb.

9. NICOTIANA L. Tobacco. Cal toothed or cleft; cor funnelform or salverform, with mostly spreading limb; stamens with long filaments and short, longitudinally dehiscent anthers; fr capsular, ordinarily 2-locular and dehiscent by 2–4 apical valves; seeds numerous, scarcely flattened; narcotic herbs or shrubs with entire or merely toothed lvs, the fls borne in terminal panicles or racemes. 60, mainly New World.

1. Nicotiana tabacum L. Annual to 2.5 m; lvs lanceolate to ovate, elliptic or nearly obovate, acute or acuminate, the lower often 5 dm, the upper gradually reduced; cal 1–2 cm, its lobes narrowly triangular, acute to long-attenuate; cor 5–6 cm, the tube dilated from the middle upward, the limb pink to red, varying to white; 2n mostly = 48. Native of trop Amer., occasionally escaped from cult. in tobacco-growing districts.
N. rustica L., a smaller sp. (to 1 m) with fls 1.5–2 cm, the wide tube dilated from below the top of the cal, the limb greenish-yellow, was cult. by Amerindians, but is now very rare or extinct in our range.

10. PETUNIA A. L. Juss. Petunia. Cal divided to below the middle, the lobes foliaceous, often unequal; cor funnelform to salverform, shallowly 5-lobed; stamens inserted at or below the middle of the cor-tube; anthers ovoid, longitudinally dehiscent; capsule 2-valved, many-seeded; herbs, usually clammy-pubescent; lvs alternate, or falsely opposite on the flowering branches, entire; fls small to large, solitary on short lateral peduncles, violet to red or white. 40, New World.

1 Cor 3–6+ cm; lvs ovate, oblong, or elliptic ....1. *P. × hybrida.*
1 Cor 0.6–0.8 cm; lvs narrowly spatulate .... 2. *P. parviflora.*

1. Petunia × hybrida Vilm. Common or garden p. Annual; lvs ovate to oblong or broadly elliptic, 3–8 cm; fls 3–6+ cm, violet or red to white or particolored; 2n=14, 28. Occasionally escaped from cult., but not persistent. Some forms of this cultigen approach one or the other of the original parents: *P. axillaris* (Lam.) BSP, with white fls, the cor-tube 4–6 cm, only slightly expanded upwards, the anthers inserted near the middle of the tube and extending nearly to its throat; and *P. integrifolia* (Hook.) Schinz & Thell. (*P. violacea*), with purple or violet fls, the cor-tube 3 cm, conspicuously expanded from base to summit, the stamens inserted near the base and extending barely past the middle of the tube.
2. Petunia parviflora A. L. Juss. Wild p. Prostrate, diffusely branched annual; lvs fleshy, spatulate or oblanceolate, 1–2 cm × 1–4 mm; cor blue, 6–8 mm, scarcely surpassing the oblanceolate foliaceous sep; 2n=18. Mainly subtropical in both N. and S. Amer., and occasionally n. into our range as a waif.

FAMILY **CONVOLVULACEAE**, the Morning-glory Family

Fls perfect, hypogynous, sympetalous, 5-merous; sep in most genera distinct or connate only at the base, often unequal; cor regular, commonly funnelform, scarcely to evidently lobed, commonly induplicate-valvate and often also convolute in bud, or merely convolute when more strongly lobed; stamens as many as and alternate with the lobes or connate members of the cor, attached toward the base of the cor-tube, the filaments often unequal; ovary superior, with 2(–5) carpels and locules and with a terminal, simple or often deeply cleft style or distinct styles, seldom (as in *Dichondra*) the ovary deeply bilobed with the segments united mainly by the base of the deeply cleft gynobasic style; ovules mostly 2 per carpel, basal or basal-axile, erect, apotropous, the micropyle directed downward and outward; fr usually capsular, seldom indehiscent; embryo large, with 2 plicate, often bifid cotyledons, embedded in the hard, often cartilaginous endosperm; mostly (incl. all our spp.) herbs, commonly twining and climbing or prostrate, seldom erect or even arborescent shrubs, with alternate, simple, exstipulate lvs; fls often showy, commonly subtended by a pair of bracts, these sometimes enlarged and forming an involucre. 50/1500, nearly cosmop.

1 Ovary divided to the base; styles 2, basal, virtually separate; cor minute .... 1. *Dichondra.*
1 Ovary 1, not divided; style or styles terminal; cor well developed.
  2 Styles 2, each 2-cleft, the stigmas therefore 4, linear .... 2. *Evolvulus.*
  2 Style 1, sometimes 2-cleft.
    3 Style shortly to deeply bifid, the 2 stigmas capitate; cor 1–2.5 cm .... 3. *Stylisma.*
    3 Style undivided, though sometimes with separate stigmas; cor often larger.
      4 Stigmas 2, linear to fusiform or ovoid.
        5 Bracts commonly small or minute, inserted well below the cal .... 4. *Convolvulus.*
        5 Bracts foliaceous, inserted just beneath the cal and ± concealing it .... 5. *Calystegia.*
      4 Stigma 1, capitate, sometimes lobed .... 6. *Ipomoea.*

1. **DICHONDRA** J. R. & J. G. Forster.   Sep 5, subequal, oblong or spatulate; cor campanulate, equaling or shorter than the cal, induplicate-valvate in bud, deeply lobed; stamens surpassed by the cor; ovary deeply divided into 2 essentially separate carpels, the 2 styles arising between them at the base; each carpel with 2 ovules; stigmas capitate; fr thin-walled, indehiscent or irregularly dehiscent, subtended by the persistent cal; creeping herbs with reniform to rotund lvs and small, axillary, greenish-white fls solitary on pedicels usually shorter than the lvs. 15+, warm reg.

  1. **Dichondra carolinensis** Michx.   Perennial; lvs round-reniform, 1–2 cm wide, often retuse, commonly hairy, on petioles usually exceeding the blades and bractless pedicels; sep spatulate, strongly veined, at maturity 3 mm, equaling the hirsutulous frs, each carpel developing into an indehiscent, usually 1-seeded utricle; $2n=30$. Moist low ground; se. Va. to Fla., Ark., and Tex.; Bermuda; Bahama Isl. (*D. repens,* misapplied.)

2. **EVOLVULUS** L.   Sep 5, ovate to subulate; cor mostly small, funnelform to rotate, entire to 5-lobed; stamens at the top of the short cor-tube; ovary with 2 locules and 4 ovules; styles 2, distinct or barely united at base, each 2-cleft, the fl with 4 linear stigmas; capsule typically 4-valved and 4-seeded, or with fewer seeds by abortion; plants never twining; ours small, hairy perennials with several stems from a thick root, and with small axillary fls usually solitary or occasionally few together at the end of the peduncle. 100, warm reg., mainly Amer.

1 Fls sessile or on pedicels much shorter than the lvs .... 1. *E. nuttallianus.*
1 Fls on filiform pedicels longer than the subtending lvs .... 2. *E. alsinoides.*

  1. **Evolvulus nuttalianus** Roemer & Schultes.   Erect or ascending, 1–2 dm, densely hairy, lvs crowded, ascending, lanceolate or narrowly oblong, 1–2 cm; fls solitary in the axils, sessile or on pedicels shorter than the narrowly lanceolate, villous sep; cor blue, broadly funnelform, 1 cm wide, its margin nearly entire; pedicels deflexed in fr; fr ovoid, 1–2-seeded. Dry prairies and barrens; N.D. to Mont., s. to Mo., Tenn., Tex., and Ariz. May, June. (*E. argenteus* Pursh, not R. Br.; *E. pilosus.*)

  2. **Evolvulus alsinoides** L.   Slender, prostrate or ascending, 2–4 dm, thinly appressed-hairy; lvs scattered, commonly shorter than the internodes, spreading, oblong, 1–2 cm; fls numerous, either solitary or few together at the end of filiform, spreading peduncles longer than the subtending lvs; cor pale blue to white, rotate, 6 mm wide; $2n=26$. A weed in the warmer parts of the world, intr. in Mo. Spring–summer.

3. **STYLISMA** Raf.   Sep 5, distinct, broad and herbaceous, persistent in fr; cor funnelform or shortly tubular-campanulate, resembling a small white morning-glory, the

margin subentire or inconspicuously lobed, often 5-angled, in bud plicate and only inconspicuously twisted; stamens included or partly exsert; ovary bilocular, the style ± deeply bifid, the stigmas small and depressed-capitate; fr ovoid or globose, 2–4-valved, with (1)2(–4) seeds; cotyledons deeply bifid into linear segments; slender herbaceous perennial vines with soft, rather narrow lvs; fls in axillary, rather long-pedunculate cymes of 2–7 (or solitary at the end of the peduncle), the pedicels generally with a pair of small to rather large bracts somewhat below the fl. (Often included in *Breweria* or *Bonamia*.) 6, Atlantic and c. U.S.

1 Lvs relatively broad, mostly well over 5 mm wide, broad-based; bracts 1–2 mm . . . . 1. *S. humistrata.*
1 Lvs linear, up to about 3(–5) mm wide, tapering to the base; bracts 5–25 mm . . . . 2. *S. pickeringii.*

1. **Stylisma humistrata** (Walter) Chapman. Stems elongate, mostly prostrate, commonly 1–2.5 m, villous to subglabrous; petioles mostly 2–5 mm; lf-blades oblong to elliptic, 3–6 × (0.5–)1–2(–3.5) cm, obtuse and mucronate, rounded below to an often shallowly cordate base; peduncles twice as long as the subtending lvs, bearing at the summit 2–several subulate bracts 1–2 mm and 1–several pedicellate fls; sep lance-ovate to elliptic, glabrous or nearly so outside, 6–9 mm; cor 15–33 mm; style cleft up to about half-length, the branches usually more than 3 mm; 2*n*=28. Dry pine-barrens, chiefly on the coastal plain; e. Va. to n. Fla., w. to Ark. and e. Tex. July, Aug. (*Breweria h.*)

2. **Stylisma pickeringii** (Torr.) A. Gray. Stems procumbent or trailing, freely branched, often 1–2 m, minutely pubescent; lvs linear, 3–6 × 1–3(–5) mm, tapering to a petiolar base 1–2 mm; peduncles as long as or longer than the subtending lvs, bearing at the summit 2–several linear bracts 5–25 mm and 1–5 pedicellate fls; sep broadly oval, 4–6 mm, densely hairy outside; cor 12–18 mm; style only shortly cleft; 2*n*=8. Dry pine-barrens on the coastal plain; s. N.J. to Ga. and Ala.; sand-prairies in w. Ill. and e. Io. and from Kans. to Tex. July–Sept. (*Breweria p.*) The coastal plain plants, with the style branches mostly 2–3 mm and subequal, and with mostly obtuse sep, are var. *pickeringii*. The western plants, with the style branches mostly 1–1.5 mm and unequal, and with mostly acute or acutish sep, are var. *pattersonii* (Fern. & Schubert) Myint.

4. **CONVOLVULUS** L. Sep 5, equal or nearly so, appressed; cor funnelform; stamens included, inserted near the base of the cor; pollen prolate to subspheroidal, with 3(4) elongate colpi; ovary 4-ovulate, bilocular or becoming 4-locellar by the development of secondary partitions between the ovules; style 1, slender, elongate; stigmas 2, linear, somewhat flattened, gradually tapering; erect, twining, or trailing perennial herbs with oblong to cordate-ovate or hastate lvs and pink and white fls borne singly or often several together on axillary peduncles; bracts 2, mostly small and well removed from the fl, not concealing the cal. 200+, cosmop., mainly Old World.

1. **Convolvulus arvensis** L. Field-bindweed. Aggressive perennial from deep-seated spreading roots; trailing or climbing to 1 m, often forming tangled mats; lvs variable, triangular to oblong in outline, 1–5 cm, the basal lobes spreading or descending; peduncles axillary, exceeding the subtending lvs, 1–2-fld; bracts subulate to ovate, 1–10 mm, inserted 5-20 mm below the fl; sep elliptic to subrotund, 3–5 mm; cor white or pink, 15–25 mm; 2*n*=48, 50. Native of Europe, now a common weed in fields, roadsides, and waste places in the U.S. and s. Can. May–Sept. (*Strophocaulos a.; C. ambigens,* a hairy form.)

5. **CALYSTEGIA** R. Br. Much like *Convolvulus* in aspect, but the bracts usually large, inserted just beneath the cal, and ± concealing it; pollen spheroidal, pantoporate with 20–40 apertures; ovary ± unilocular, the partition incomplete; stigmas oblong, ± cylindric, blunt; fls usually solitary; our spp. rhizomatous perennials. 30, widespread.

Jacquemontia tamnifolia (L.) Choisy, a widespread weed in tropical Amer. and s. U.S., has been collected as far n. as s. Ill. and se. Va. It will key to *Calystegia,* from which it obviously differs in its densely cymose-capitate clusters of blue fls ca 1 cm.

1 Cor 4–7 cm; fls single.
  2 Lvs subtending the fls with the petiole distinctly more than half as long as the midvein; stems
    including the floriferous part trailing or twining . . . . 1. *C. sepium.*
  2 Lvs subtending the fls with the petiole distinctly less than half as long as the midvein; stems up to
    and including the floriferous part erect . . . . 2. *C. spithamea.*
1 Cor 2–3.5 cm; fls mostly double . . . . 3. *C. hederacea.*

1. **Calystegia sepium** (L.) R. Br. Hedge-bindweed. Twining or occasionally trailing, to 3 m; lvs long-petioled, triangular to oblong in outline, hastate or sagittate, 5–10 cm; peduncles 5–15 cm; fls solitary; bracts

ovate or oblong, 1–2 cm, commonly cordate at base; cor pink or white, 4–7 cm; $2n=22$, 24. Thickets, shores, and disturbed sites; temperate regions of N. Amer. and Eurasia. We apparently have both native and intr. forms. Many confluent vars. have been described. (*Convolvulus s.*)

2. **Calystegia spithamea** (L.) Pursh. Low bindweed. Stems erect, at least to and including the flowering part, the remainder often elongating and eventually declined; lvs short-petioled, oblong to obovate-oblong, 3–8 cm, acute to rounded, at base rounded to truncate or cordate, always ± hairy; peduncles few, 2–8 cm, produced 5–20 cm above the base of the plant; bracts oblong or ovate, seldom cordulate at base; cor white or pink, 4–7 cm; $2n=22$. Dry, rocky or sandy soil, fields, and open woods; Que. and Me. to Minn., s. to Va. and Mo., and in the mts. to Ga.; May–July. (*Convolvulus catesbeianus, purshianus,* and *spithameus.*) The more strongly hairy forms, with usually cordate-based lvs, occur chiefly in the s. Appalachian region and may merit varietal recognition. (*Convolvulus spithameus* var. *pubescens.*)

3. **Calystegia hederacea** Wallich. Japanese bindweed. Stems climbing or trailing; lvs hastate, 4–8 cm, the terminal lobe oblong, the basal lobes acute, often 2–3-angled; fls pink, 2–3.5 cm; $2n=22$. Native of Japan, the double-fld form ("California rose") occasionally cult. and escaped along roadsides etc. in much of our range. The rarely cult. single-fld form is reported as an adventive weed in Pa. (*C. japonica,* a preoccupied name; *Convolvulus japonicus, pellitus,* and *wallichianus.*)

6. **IPOMOEA** L. Morning-glory. Sep 5, imbricate, often unequal; cor funnelform to campanulate or salverform, the margin entire, angled, or shallowly lobed; ovary 1–3-locular, with 2 ovules per locule, or sometimes 4-locellar by the development of partitions between the ovules; style 1, elongate; stigma entire or 2–3-lobed; capsule 2–4-valved; ours twining herbs with broad, cordate or lobed lvs and handsome fls produced singly or few together at the summit of axillary peduncles. 600+, cosmop. (*Pharbitis, Quamoclit.*)

1 Cor funnelform to nearly campanulate, variously blue, purple, or pink to white; stamens and style not exserted.
  2 Ovary 3-locular; stigma 3-lobed; sep conspicuously hirsute on the basal half (*Pharbitis*).
    3 Sep 10–15 mm, acute to acuminate; lvs almost always entire .... 1. *I. purpurea.*
    3 Sep 15–25 mm, with linear, recurved tip; lvs usually 3-lobed .... 2. *I. hederacea.*
  2 Ovary 2-locular; stigma 2-lobed or entire; sep not especially hairy below.
    4 Cor 5–8 cm; sep 13–20 mm, obtuse or rounded, glabrous .... 3. *I. pandurata.*
    4 Cor 1–2 cm; sep 10–15 mm, sharp-pointed, ciliate .... 4. *I. lacunosa.*
1 Cor ± salverform, usually scarlet; stamens and style exserted. (*Quamoclit.*)
  5 Lvs cordate, entire or shallowly lobed .... 5. *I. coccinea.*
  5 Lvs deeply pinnately divided into numerous lobes .... 6. *I. quamoclit.*

1. **Ipomoea purpurea** (L.) Roth. Common m.-g. Annual; stems pubescent, to 5 m; lvs round-cordate to ovate-cordate, entire (very rarely 3-lobed), abruptly short-acuminate or merely apiculate, glabrous to sparsely hairy; peduncles nearly or quite equaling the subtending petioles, 1–5-fld; sep lance-ovate to oblong, acute or acuminate, hirsute toward the base, commonly 10–15 mm; cor blue, purple, white, or variegated, 4–6 cm; ovary 3-locular; stigma 3-lobed; $2n=30$. Native of trop. Amer., escaped from cult. and now a weed in fields, roadsides, and waste places in much of our range. July–Sept. (*Pharbitis p.*)

2. **Ipomoea hederacea** Jacq. Ivy-lvd m.-g. Annual; stems pubescent, 1–2 m; lvs orbicular in outline, 5–12 cm, cordate at base with rounded basal lobes, usually deeply 3-lobed with rounded sinuses (occasionally 5-lobed or entire), the lobes acuminate; peduncles shorter than or equaling the petioles, with usually 1–3 fls; sep lanceolate, commonly 15–25 mm, narrowed from below the middle into a slender, linear, recurved tip, the basal part densely hirsute; cor purple or blue, varying to white, 3–5 cm; ovary 3-locular; stigma 3-lobed; $2n=30$. Native of warm-temp. Amer., now established as a weed in waste places, along roads, and in cornfields throughout much of our range, especially southward. July–Sept. (*Pharbitis h.*) Plants with entire lvs, rare with us but abundant southward, have been called var. *integriuscula* A. Gray. (*Pharbitis barbigera.*)

3. **Ipomoea pandurata** (L.) G. Meyer. Wild potato. Perennial from a large, deep-seated, tuber-like root; stems trailing or twining, to 5 m, glabrous or nearly so; lvs glabrous or more often hairy beneath, ovate, entire, acuminate, deeply cordate at base, 5–15 cm, rarely obscurely lobed or angled; peduncles stout, stiff, usually exceeding the subtending petioles, bearing 1–7 fls in a terminal cluster; sep ovate or oblong, 13–20 mm at anthesis, obtuse to broadly rounded, glabrous; cor white with red-purple center, 5–8 cm; ovary bilocular; $2n=30$. Dry woods and thickets; Conn. to s. Ont., Mich., and Kans., s. to Fla. and Tex. June–Sept.

4. **Ipomoea lacunosa** L. White m.-g. Annual; stems 1–3 m, glabrous or sparsely hairy; lvs ovate, 3–8 cm, short-acuminate, deeply cordate at base, entire to angular or 3-lobed; peduncles mostly shorter than the subtending petioles, rarely longer, with 1–5 fls; sep lanceolate or lance-ovate, 10–15 mm, sharply or abruptly acuminate, hirsute-ciliate, the outer surface toward the base glabrous or sparsely villous; cor white, occasionally pink or pale purple, 1–2 cm; ovary bilocular; $2n=30$. Moist fields and thickets, especially in alluvial soil; N.J. to O., Ill. and Kans., s. to Fla. and Tex. July–Sept.

5. **Ipomoea coccinea** L. Red m.-g.; scarlet creeper. Annual; stems twining, 1–3 m; lvs broadly ovate in outline, acuminate, deeply cordate at base, 5–10 cm, entire or sometimes coarsely toothed or angularly lobed;

fls few or several in a short, racemiform or cymose cluster at the end of long axillary peduncles; sep obtuse, with a long subulate appendage just below the tip; cor ± salverform, scarlet, often with a yellow throat, 2–3.5 cm; anthers and style exsert; fr 4-locellar; $2n=28$. Native of trop. Amer., escaped from cult. into moist soil and waste places as far n. as Mass. and Mich. July–Oct. (*Quamoclit c.*)

6. Ipomoea quamoclit L. Cypress-vine. Annual; stems slender, glabrous, twining, 1–5 m; lvs broadly ovate in outline, 4–8 cm, pinnately divided to the midvein into numerous narrowly linear lobes; petioles 1–3 cm; peduncles elongate, much longer than the subtending petioles, with 1–few fls at the top; sep obtuse or rounded, mucronate; cor ± salverform, scarlet, rarely varying to white, 2.5–3.5 cm; anthers and style exsert; fr 4-locellar; $2n=30$. Native of trop. Amer., occasionally escaped into fields and waste places, especially in the s. part of our range. Summer and fall. (*Quamoclit q.; Q. vulgaris.*)

FAMILY CUSCUTACEAE, the Dodder Family

Fls small, regular, perfect, sympetalous, hypogynous; cal persistent, deeply 4–5-lobed or the sep distinct; cor 4–5-lobed, the lobes imbricate in bud, the typically somewhat fleshy tube usually aligned with a set of small, crenulate or fringed scale-like appendages just beneath (and aligned with) the stamens; the stamens inserted just below the sinuses of the cor, with short filaments; ovary superior, with 2 carpels and locules (the partition sometimes incomplete) and 2 distinct styles; fr a small, membranous, globose or depressed-globose to ovoid capsule or utricle, circumscissile near the base or irregularly dehiscent or indehiscent; seeds 1–4; embryo mostly slender, filiform-cylindric, nearly or quite acotyledonous, peripheral and strongly curved or coiled around the starchy endosperm; slender, twining, achlorophyllous, parasitic annual or infrequently perennial herbs with pinkish-yellow to orange or white, glabrous stems attached to their host by intrusive haustoria, the small terrestrial root-system soon degenerating; lvs alternate, much reduced and scale-like; infls lateral, 1–many-fld, usually compact. Only one genus.

1. CUSCUTA L. Dodder. Characters of the family. 100, cosmop. (*Grammica.*) Our spp. bloom in late summer; most have a wide range of hosts.

1 Stigma capitate, not evidently longer than thick; fr indehiscent or rupturing irregularly. (*Grammica.*)
  2 Each fl subtended by 1–several bracts; sep distinct.
    3 Fls pediceled in loose panicles; bracts 1–few per fl, ovate to subrotund . . . . 1. *C. cuspidata.*
    3 Fls sessile or nearly so in dense, often rope-like masses; bracts usually several.
      4 Bracts imbricate, appressed, broadly rounded, mostly wider than long . . . . 2. *C. compacta.*
      4 Bracts all about the same length, loosely ascending with recurved tip . . . . 3. *C. glomerata.*
  2 Individual fls without bracts; sep united at base.
    5 Fls all or mostly 5-merous.
      6 Cor-lobes acute.
        7 Fr merely cupped at base by the withered cor and strongly protruding from it; cal-lobes very obtuse . . . . 4. *C. pentagona.*
        7 Fr ± enveloped by the withered, calyptrate cor, not protruding; cal-lobes acute to somewhat obtuse . . . . 5. *C. indecora.*
      6 Cor-lobes obtuse.
        8 Cal-lobes not reaching the sinuses of the cor; fr commonly higher than thick.
          9 Fls 2–4 mm; fr not beaked, or sometimes beaked in no. 7.
            10 Syles 1–1.5 mm; seeds ± 1.5 mm; widespread . . . . 6. *C. gronovii.*
            10 Styles 0.3–0.5(–0.7) mm; seeds ± 2.5 mm; western . . . . 7. *C. megalocarpa.*
          9 Fls 4–6 mm; fr with a ± conspicuous beak 1–1.5 mm; s. Appalachians . . . . 8. *C. rostrata.*
        8 Cal-lobes reaching the sinuses of the cor; fr commonly as thick as or thicker than high; southern . . . . 9. *C. obtusiflora.*
    5 Fls all or mostly 4-merous.
      11 Cor-lobes acute.
        12 Tip of the cor-lobes erect or ascending . . . . 10. *C. polygonorum.*
        12 Tip of the cor-lobes inflexed . . . . 11. *C. coryli.*
      11 Cor-lobes obtuse or rounded . . . . 12. *C. cephalanthi.*
1 Stigma appearing as a slender, tapering-cylindric continuation of its style, several times as long as thick; fr circumscissile near the base. (*Cuscuta.*)
  13 Cal-lobes acute; fls mostly 5-merous.
    14 Style (stigma included) elongate, much longer than the ovary, soon exsert from the short-cylindric cor; fr ca 1.5 mm . . . . 13. *C. epithymum.*
    14 Style about equaling the ovary, included in the urceolate cor; fr 2–2.5 mm . . . . 14. *C. epilinum.*
  13 Cal-lobes obtuse; fls mostly 4-merous . . . . 15. *C. europaea.*

1. Cuscuta cuspidata Engelm. Fls 5-merous, 3–4 mm, pediceled or subsessile in loose, open panicles; bracts 1–3, broadly round-ovate, appressed, much shorter than the cal; sep distinct, ovate to orbicular, very obtuse, sometimes minutely cuspidate; cor-lobes ovate, 1.5 mm; styles 1.5–3 mm; stigma capitate; fr globose or depressed-globose, 2–3 mm thick, indehiscent, 4-seeded, the withered cor calyptrate at the top. Ind. and Wis. to N.D., and Utah, and Tex. Occasional specimens of the widespread *C. gronovii* might be traced to this sp. in the key, but differ in their shorter styles and in having the fr protruding from the basally cupulate-persistent cor.

2. Cuscuta compacta A. L. Juss. Fls 5-merous, 4–5 mm, sessile in small dense clusters; bracts 3–5, closely appressed, broadly rounded and commonly wider than long, the inner progressively longer; cor-lobes ovate, obtuse; styles 0.7–2 mm; stigma capitate; fr ovoid to globose, 3–5 mm thick, not beaked, indehiscent, 1–2-seeded, the withered cor calyptrate at the top. Mass. to Pa., Ill., and Neb., s. to Fla. and Tex. (*C. paradoxa.*)

3. Cuscuta glomerata Choisy. Rope-d. Fls 5-merous, 4–5 mm, in dense, twisted, rope-like masses, sessile or nearly so; bracts oval or oblong, erect or loosely ascending, about equaling the cal, the tips squarrose or recurved; cor-lobes ovate, subacute, 2 mm; styles elongate, 1.5–3.5 mm; stigma capitate; fr concealed by the withered cor, 3–4 mm thick, subglobose with a short neck, mostly 2-seeded; $2n=30$. Ind. to Wis. and N.D., s. to Miss. and Tex.

4. Cuscuta pentagona Engelm. Field-d. Fls 5-merous, sessile or short-pedicellate in compact glomerules; cal about equaling the cor-tube, its lobes very broadly ovate or depressed, often broader than long; cor-lobes 1 mm, lanceolate to triangular, acute, about as long as the tube, spreading to reflexed; styles 0.5–1 mm; stigma capitate; fr depressed-globose, thin-walled, mostly 2-seeded, cupped at base by the withered cor and strongly protruding from it, not evidently thickened around the styles; $2n=56$. Throughout the U.S. and almost cosmop., living on a wide variety of hosts. (*C. arvensis; C. campestris.*)

5. Cuscuta indecora Choisy. Collared d. Fls 5-merous, short-pedicellate in loose to compact glomerules; cal-lobes triangular, acute or somewhat obtuse, shorter than to about equaling the broad cor-tube; cor-lobes broadly triangular, erect or ascending, usually shorter than the tube, acute with inflexed tip; styles 0.5–1.5 mm; stigma capitate; fr depressed-globose, enveloped by the withered, calyptrate cor, evidently thickened around the styles, which appear to be set in a shallow pit surrounded by a collar; seeds mostly 4; $2n=30$. On a wide variety of hosts; Ill. to N.D., w. to Calif., s. to Fla. and S. Amer.

C. suaveolens Ser., a S. Amer. sp., has been reported from our range on alfalfa. It differs from no. 5 most notably in that the ovoid-globose fr is not thickened around the styles.

6. Cuscuta gronovii Willd. Common d. Fls 5-merous, 2.5–4 mm, sessile or subsessile in dense clusters; cal short, its lobes broadly round-ovate or subrotund, overlapping, often scarcely reaching the middle of the cor-tube; scales commonly copiously fringed, reaching the sinuses of the cor; styles 1–1.5 mm, nearly or fully as long as the ovary; stigma capitate; fr cupped below by the withered cor, commonly globose-ovoid, 2.5–3.5 mm, with a thickened collar about the styles; seeds 2–4, ± 1.5 mm; $2n=60$. N.S. to Man., s. to Fla. and Ariz.; our commonest sp.

7. Cuscuta megalocarpa Rydb. Bigfr.-d. Fls 5-merous, 2.5–4 mm, short-pedicellate in few-fld glomerules; cal short, often less than half the length of the cor-tube, its lobes 1–1.5 mm, ovate, obtuse, overlapping basally; cor-lobes 0.7–0.9 mm, obtuse or rounded, shorter than the tube, spreading to reflexed; scales broad and truncate or sometimes narrower and fringed, shorter than the cor-tube, not reaching the filaments; ovary ovoid, much longer than the short styles, these 0.3–0.5(–0.7) mm; stigma capitate, fr ovoid and often beaked, only loosely cupped below by the withered cor and strongly protruding from it, 3–6 mm, with a thickened collar about the styles; seeds 2–4, 2–2.8 mm. Minn. and Man. to Sask. and Mont., s. to Kans., N.M., and Utah. (*C. curta; C. umbrosa,* misapplied.)

8. Cuscuta rostrata Shuttlew. Beaked d. Fls 5-merous, 4–6 mm, distinctly pediceled in small head-like umbels; cal much shorter than the relatively long cor-tube, its lobes triangular-ovate to broadly ovate, very obtuse, cor-lobes broadly ovate, erect or eventually reflexed in fr; styles 1.5–2.5 mm; stigma capitate; fr loosely cupped by the persistent cor and protruding from it, broadly ovoid, 4–5 mm, with a short stout beak ca 1–1.5 mm, but without a thickened collar; seeds (1)2–4, 2–3 mm. Mts. from W.Va. to S.C.

9. Cuscuta obtusiflora HBK. Southern d. Fls 5-merous, 2–2.5 mm, subsessile in dense clusters; cal often glandular, its lobes broadly rounded, reaching the sinuses of the cor; cor-lobes ovate, obtuse, spreading or eventually reflexed; styles short 0.5–1 mm; stigma capitate; fr depressed-globse to ovoid-globose, somewhat thickened around the styles, cupped at base by the withered cor; seeds ± 1.5 mm. Widely distributed in the s. states and s. to S. Amer. In our range known only in Ind. Our plants may be distinguished from the typical S. Amer. phase as var. *glandulosa* Engelm. (*C. glandulosa.*)

10. Cuscuta polygonorum Engelm. Smartweed-d. Fls mostly 4-merous, 2 mm, sessile or subsessile in small, dense clusters; cal as long as or longer than the cor-tube, its lobes very obtuse; cor-lobes triangular, acute, erect or ascending, commonly as long as or longer than the tube; styles very short, 0.5–0.8 mm; stigma capitate; fr cupped at base by the persistent cor, depressed-globose, commonly 2–2.5 mm high and 3 mm thick; seeds 2–4, 1–1.5 mm. N.Y. and Ont. to N.D., s. to Tenn. and Tex. On *Polygonum* and other hosts.

11. Cuscuta coryli Engelm. Hazel-d. Fls mostly 4-merous, 2.5 mm, in dense or loose clusters, some or all distinctly pediceled; cal half as long as the cor, its lobes acute; cor cylindric, its lobes narrowly triangular with acute, inflexed tip, about as long as the tube; styles 0.5–1 mm; stigma capitate; withered cor persistent around the base of the fr or calyptrate and pushed off; fr ± globose, 2–2.5 mm, slightly thickened at the top to form a collar around the styles; seeds 1 or 2, 1.5 mm; $2n=30$. N.Y. to N.D., s. to Va. and Okla.

12. Cuscuta cephalanthi Engelm. Buttonbush-d. Fls mostly 4-merous, 2–2.5 mm, sessile or subsessile in compact glomerules; cal shorter than the cor-tube, its lobes obtuse, strongly overlapping basally; cor cylindric-campanulate, the erect or spreading lobes obtuse or broadly rounded, shorter than the tube; styles 0.5–1.5 mm; stigma capitate; fr 3 mm, globose or depressed-globose, closely invested by the persistent, withered, eventually calyptrate cor; seeds 1–2, 1.5–2 mm; $2n=60$. Me. to Wash., s. to Fla. and N.M.

13. Cuscuta epithymum L. Legume-d. Fls 5-merous, subsessile in dense, globose, head-like clusters; cal-lobes triangular-ovate, acute, generally a little shorter than the short-cylindric cor; cor-lobes triangular, 1–1.5 mm, shorter than the tube, acute, spreading; stigma slender, appearing as a continuation of its style, the two together longer than the ovary; fr globose, 1.5 mm, capped by the withered cor, circumscissile near the base; seeds 4, scarcely 1 mm; $2n=14$. Native of Europe, intr. throughout much of the U.S., parasitic chiefly on legumes.

14. Cuscuta epilinum Weihe. Flax-d. Fls sessile in compact globose heads, all or mostly 5-merous; cal-lobes very broadly ovate, acute, about equaling the urceolate cor; stigma slender, as a continuation of its style, the two together up to about as long as the ovary; included in the eventually calyptrate cor; fr depressed-globose, 2–2.5 mm high and 2.5–3 mm thick, circumscissile near the base; seeds 4, ca 1.2 mm; $2n=42$. Native of Europe, parasitic on flax and to be expected in this country wherever flax is grown.

15. Cuscuta europaea L. Fls on short stout pedicels in dense head-like clusters, all or mostly 4-merous; cal-lobes broadly ovate, very obtuse, about equaling the tube of the short-cylindric, eventually calyptrate cor; styles and stigmas slender, together much shorter than the ovary; fr globose, ca 3 mm high and thick, circumscissile near the base; seeds mostly 4, ca 1.5 mm. Native of Europe, intr. here and there in our range.

FAMILY **MENYANTHACEAE**, the Buckbean Family

Fls perfect (or sometimes functionally unisexual and the plants dioecious), regular, sympetalous, 5-merous; cor-lobes valvate or induplicate-valvate or imbricate, the margins of the inner surface often fimbriate or crested; stamens attached to the cor-tube alternate with the lobes, sometimes alternating with a set of fringed scales; ovary bicarpellate, superior to half inferior, unilocular, with a terminal style and 2-lobed stigma; ovules numerous on the 2 parietal (often intruded) placentas; fr a capsule or berry; seeds with a linear, axile, dicotyledonous embryo and copious, oily endosperm; aquatic or semi-aquatic herbs, often with scattered vascular bundles in the stem; lvs alternate, simple or trifoliolate, exstipulate (but the petiole sometimes wing-margined toward the base). 5/30–35, cosmop.

1 Lvs trifoliolate, emergent; fls racemose .... 1. *Menyanthes.*
1 Lvs simple, floating; fls umbellate .... 2. *Nymphoides.*

1. MENYANTHES L. Buckbean. Cal deeply cleft; cor salverform, its lobes and throat densely and coarsely hairy within; ovary about ⅓ inferior; fr tardily dehiscent by 2 valves; glabrous, rather fleshy herbs from coarse rhizomes; lvs all basal, emergent, trifoliolate, the long petiole with a conspicuously expanded sheathing base; fls heterostylic, in a bracteate raceme terminating a shortly emergent, naked scape. Monotypic.

1. Menyanthes trifoliata L. Petioles 5–30 cm, lfls emergent, short-stalked or virtually sessile, entire or wavy-margined, elliptic ovate to elliptic-obovate, 3–6 cm at anthesis, larger later; infl crowded at anthesis; pedicels 5–20 mm, sometimes 2 from one axil; cor whitish, usually purple-tinged, the tube ca twice as long as the 3–5 mm cal, the lobes lance-ovate, 5–7 mm; fr thick-walled, 6–9 mm; $2n=54$, 108. In quiet shallow cold water; circumboreal, s. to N.J., Va., O., Ind., Mo., and Calif. May, June.

2. NYMPHOIDES Hill. Floating heart. Fls white or yellow, perfect or often functionally unisexual; cal parted nearly to the base into oblong lobes; cor broadly campanulate or subrotate, deeply lobed, often hairy or with a glandular appendage near the base of each lobe within; ovary tapering to a short style, or the broad, 2-lobed, persistent stigma virtually sessile; fr firm-walled, indehiscent or eventually rupturing irregularly; rhizomatous aquatic herbs with broad, cordate, floating lf-blades, some lvs long-petiolate and arising directly from the rhizome, others short-petiolate and arising from slender, petiole-like stems that also bear one or more bracteate umbels above, so that the infl may appear to arise a little below the top of a long petiole (*Limnanthemum.*) 20, widespread.

1 Fertile stem bearing a single lf beyond the umbel; fls white or creamy, 5–14 mm.
  2 Cal-lobes at anthesis 2–3 mm, green; capsule 4–5 mm .... 1. *N. cordata.*
  2 Cal-lobes at anthesis 4–5 mm, flecked with purple; capsule 10–14 mm .... 2. *N. aquatica.*
1 Fertile stem with a pair of opposite lvs subtending the umbel, often extending beyond the umbel and
  bearing one or more additional umbels; fls bright yellow, 20–25 mm .... 3. *N. peltata.*

1. Nymphoides cordata (Elliott) Fern. Little f.h. Stems often much elongate, green, very slender, near the summit less than 1 mm thick; lf broadly ovate, deeply cordate, 3–7 cm; functionally dioecious, some fls usually replaced by spur-like tuberous-thickened roots; cal-lobes at anthesis 2–3 mm, in fr extending well beyond the middle of the capsule; cor 5–8 mm, white or creamy, with a crest-like yellow gland at the base of each lobe; anthers ca 1 cm or less; fr 4–5 mm, not beaked; seeds smooth or sparsely papillate; 2*n*=36. In quiet water; Nf. and Que. to Ont., s. to Conn. and N.Y., and locally along the coastal plain to Fla. and La. July–Sept. (*N. lacunosa,* misapplied.)

2. Nymphoides aquatica (Walter) Kuntze. Big f.h. Much like no. 1, but larger and coarse, with thicker and firmer lvs 5–12 cm long and wide; stems 1.3–2.5 mm thick near the top; spur-roots seldom developed; pedicels and cal conspicuously flecked with purple; cal-lobes at anthesis 4–5 mm, at maturity reaching to or below the middle of the fr; cor 10–14 mm; fr 10–14 mm; seeds conspicuously papillate; 2*n*=36. Quiet water on the coastal plain; s. N.J. to Fla. and Tex. July–Sept. (*N. lacunosa.*)

3. Nymphoides peltata (S. G. Gmelin) Kuntze. Yellow f.h. Stems stout, commonly 2–3 mm thick, bearing a pair of opposite, usually unequal lvs subtending the umbel, often extended and bearing additional umbels; lvs subrotund, 5–15 cm long and wide; fls heterostylic; cor bright yellow, 2–2.5 cm, its lobes somewhat fringed below; anthers 4–5 mm; fr 12–25 mm, strongly beaked; seeds flat, the margin ciliate and narrowly winged; 2*n*=54. Native of Europe, intr. in quiet waters here and there in our range. June–Sept.

Family **POLEMONIACEAE**, the Phlox Family

Fls perfect, hypogynous, sympetalous, mostly (incl. all ours) 5-merous; cal usually gamo-sepalous, with equal or unequal lobes, the tube often with alternating green ribs and hyaline intervals (between the lobes); cor regular in most genera (incl. all ours), its lobes convolute in bud; stamens as many as and alternate with the cor-lobes, borne on the tube, sometimes at differing levels; ovary superior, mostly tricarpellate and trilocular, with a terminal style and usually 3 separate stigmas; ovules 1–many in each locule, on axile placentas; fr usually a loculicidal capsule; embryo straight or slightly curved, usually spatulate, with 2 cotyledons; endosperm usually oily, seldom scanty or none; mostly (incl. all ours) herbaceous or merely suffrutescent; lvs exstipulate. 18/300, best developed in w. U.S.

1 Lvs simple, entire.
  2 Lvs or most of them opposite; cor-limb at least 1 cm wide .... 1. *Phlox.*
  2 Lvs alternate; cor-limb in our sp. smaller .... 2. *Collomia.*
1 Lvs compound, dissected, or deeply lobed.
  3 Cor campanulate; lvs pinnately compound, with lanceolate to ovate lfls .... 3. *Polemonium.*
  3 Cor salverform or trumpet-shaped; lvs in our sp. dissected into narrow segments .... 4. *Gilia.*

1. PHLOX L. Phlox; Sweet William. Cal tubular, scarcely accrescent, with scarious or hyaline intervals between the 5 green and more herbaceous ribs, ruptured by the developing capsule; corolla salverform, with slender tube and abruptly spreading lobes; filaments short, unequally placed in the cor-tube, the anthers included or some of them partly exserted; capsule 3-valved, with 1(–4) seeds per locule, the seeds unchanged when moistened; perennial (seldom annual) herbs or suffrutescent plants with entire, chiefly opposite lvs and handsome fls in terminal or terminal and axillary cymes that may be reduced to a single fl. 50–60, N. Amer. and n. Asia. Allied spp. often hybridize.

P. drummondii Hook., a showy annual with the upper lvs alternate, occasionally escapes from cult.

1 Style short, scarcely if at all longer than the stigmas or the ovary.
  2 Lvs flat, linear to ovate; stems herbaceous, erect, ascending, or decumbent at base.
    3 Cymes loose and open, the 2 lowest branches usually 1 cm or more; widespread.
      4 Lvs abruptly narrowed or acuminate to a sharp, stiff point; cor-tube usually pubescent .... 1. *P. pilosa.*
      4 Lvs obtuse or acute, not sharp-pointed; cor glabrous .... 2. *P. divaricata.*

3 Cymes compact, closely subtended by bracteal lvs, the lowest branches rarely 5 mm; s. Ky. and
    s. .... 3. *P. amoena.*
2 Lvs subulate to lance-linear, usually with axillary fascicles; stems woody, diffusely branched, pros-
    trate or decumbent; s. Va. and s. .... 4. *P. nivalis.*
1 Style elongate, much exceeding the stigmas or the ovary.
    5 Cor-lobes conspicuously notched at the tip.
      6 Main lvs mostly 5–10 mm; cor-lobes notched up to ¼ length; eastern .... 5. *P. subulata.*
      6 Main lvs mostly 15–40 mm; cor-lobes notched ¼ to ½ length; western .... 6. *P. bifida.*
    5 Cor-lobes entire or merely erose across the tip.
      7 Lvs veiny, and with a submarginal connecting vein; lf-margin ciliolate.
        8 Infl puberulent, not glandular; cor-tube usually sparsely hairy .... 7. *P. paniculata.*
        8 Infl glandular-hairy; cor-tube glabrous .... 8. *P. amplifoliᾳ.*
      7 Lvs not veiny, and without a submarginal connecting vein, the margin not ciliolate.
        9 Sterile basal stolons present at anthesis, their lvs spatulate .... 9. *P. stolonifera.*
        9 Sterile basal stolons present or absent, but not with spatulate lvs.
          10 Cymes usually numerous, one terminal, the others axillary but on peduncles of ± uniform
            length, forming a subcylindric infl .... 10. *P. maculata.*
          10 Cymes solitary or few, the lower on long peduncles, forming a flattened or broadly rounded
            infl.
            11 Nodes of the flowering stems 7 or fewer below the infl.
              12 Infl minutely puberulent, not glandular; cor-tube glabrous .... 11. *P. ovata.*
              12 Infl and cor-tube glandular-hairy .... 12. *P. buckleyi.*
            11 Nodes of the flowering stems 8 or more below in infl (except in depauperate plants)
              .... 13. *P. glaberrima.*

1. **Phlox pilosa** L.   Prairie-p.   Erect, 3–6 dm, the sterile basal shoots also erect or ascending; lvs linear to lanceolate or lance-ovate, 3–8 cm, narrowed to a sharp, indurate tip; infl a loosely branched cyme, the branches conspicuously stalked, the fls all or mostly on distinct pedicels; cor commonly pale red-purple, 1.5–2 cm wide, the tube usually ± hairy; style short; 2n=14. Prairies and upland woods; Conn. to Fla., w. to Man., Nebr., and Tex. Apr.–June. The widespread var. *pilosa,* which however extends nw. only to s. Wis. and e. Io., is evidently glandular-hairy in the infl (var. *virens*). Var. *fulgida* Wherry, a nw. phase (Wis. and Ill. to Man. and Kans.), is eglandular, with the hairs of the cal very fine and lustrous, 0.5 mm. Var. *amplexicaulis* (Raf.) Wherry, chiefly of Tex. to La. and Ark., but extending n. to n. Mo. and s. Ind., is likewise eglandular, but the hairs of the cal are coarse and 1–1.5 mm. (*P. pulcherrima.*) An extensive but local population along the Sangamon R. in c. Ill. differs from var. *pilosa* in its glabrous cor-tube and has been described as ssp. *sangamonensis* Levin and D. M. Smith, but plants with glabrous cor-tube also occur elsewhere in the range of var. *pilosa.* Plants thought to form a stabilized derivative of *P. pilosa* × *P. amoena,* found in sw. Ind., w. Ky., and nw. Tenn., have been named *P. pilosa* sp. *deamii* Levin. These plants approach *P. amoena* in aspect, but have the pubescent cor-tube of *P. pilosa.* Other vars. occur southward.

2. **Phlox divaricata** L.   Forest-p.   Erect or decumbent at base, 3–5 dm, with decumbent basal stolons; lvs lance-ovate to oblong, 3–5 cm, obtuse or acute but not sharp-pointed; infl a loosely branched, glandular-hairy cyme, the branches on distinct peduncles, the pedicels often 5–10 mm; cor usually pale blue-purple, varying to red-purple or white, 2–3 cm wide, the glabrous tube 1–2 cm; style short; 2n=14. Rich moist woods; nw. Vt. and adj. Que. to Minn., s. to Ga. and Tex. Apr.–June. Var. *divaricata,* with the cor-lobes notched at the tip, is eastern, ranging w. to Mich. and e. Ill. Var. *laphamii* A. Wood, with entire cor-lobes, is more western, but occasional plants in the range of one var. have the cor-lobes of the other.

3. **Phlox amoena** Sims.   Hairy p.   Stems erect, ascending, or decumbent at base, 1.5–3 dm, finely hairy; lvs narrowly oblong-oblanceolate to lanceolate, tending to be erect or ascending, ± pubescent, the margins ciliate; infl very dense and compact; its branches very short, the pedicels 1–4 mm; cal pubescent, not glandular; cor 1.5–2 cm wide, usually red-purple, varying to blue-purple or white, the tube glabrous; style short; 2n=14. Dry woods, sandy pine-lands, and open slopes; s. Ky. to w. N.C., n. Fla., and e. Miss. Apr.–June.

4. **Phlox nivalis** Lodd.   Piney-woods p.   Plant with trailing-decumbent, evergreen sterile shoots and erect, deciduous flowering shoots, 1–3 dm with 4–8 nodes below the infl, the axillary fascicles of reduced lvs conspicuous; main lvs subulate to lance-linear, 1–2 cm; cal glandular-puberulent; cor light purple, varying to white, 2–2.5 cm wide, its lobes entire or erose; style short; 2n=14. Sandy slopes and flats, often in open pine woods; s. Va. to Fla. and Ala.; Tex. Apr.–May. (*P. hentzii.*)

5. **Phlox subulata** L.   Moss-pink.   Perennial by a prostrate suffruticose stem, freely branched and with numerous flowering branches 5–20 cm; lvs numerous and crowded, subulate, mostly 5–20 mm, sharp-pointed, usually ciliate, often with axillary fascicles; cymes few-fld; cor rose-purple to pink or white, 12–20 mm wide, its lobes notched an eighth to a fourth their length; style elongate; 2n=14. Sandy or gravelly soil and rock-ledges; s. N.Y. to s. Mich., s. to the mts. of N.C. and Tenn., and occasionally escaped from cult. as far n. as Me. Apr., May. Var. *subulata,* without glands in the infl, is northern. (*P. ciliata.*) Var. *setacea* (L.) A. Brand, with glandular-hairy infl, is southern, extending n. to s. O. and southernmost Pa. (*P. s.* var. *australis; P. brittonii.*)

6. **Phlox bifida** Beck.   Sand-p.   Stems suffruticose, prostrate or ascending, with several stiff, ascending or erect flowering branches 1–3 dm; internodes mostly 2–5 cm; lvs stiff, linear to narrowly lanceolate, 1.5–4 cm, 1-nerved, seldom with axillary fascicles; cymes few-fld; cor pale blue-violet, varying to white, 14–20

mm wide, its lobes notched a fourth to half their length; style elongate; $2n=14$. Dry sandy soil and rock-ledges; s. Mich. and s. Wis. to Tenn., n. Ark., and Kans. Apr., May. Typical plants have glandular-hairy infl; plants with eglandular or even glabrous infl occur more locally, especially in Ky. and Tenn., and have been called var. *cedraria* (Brand) Fern. (*P. stellaria.*)

7. **Phlox paniculata** L. Summer-p. Erect, to 2 m; lvs narrowly oblong to lanceolate or elliptic, 8–15 cm, to two-fifths as wide, acuminate, ciliolate, usually glabrous above, the conspicuous lateral veins confluent to form a submarginal connecting vein; infl often large, of several panicled cymes, densely short-hairy; cal glabrous, or the tube puberulent; cor red-purple, varying to white, 1.5–2 cm wide, the tube usually sparsely hairy; one or more anthers at least partly exsert from the cor-tube; style elongate; $2n=14$. Rich moist soil; s. N.Y. to n. Ga., w. to Ill., Mo. and Ark., and often escaped from cult. elsewhere in our range. July–Sept.

8. **Phlox amplifolia** Britton. Wide-lvd p. Much like no. 7; lvs wider and usually hirsutulous above; infl and cal conspicuously glandular-hairy; cor red-purple, 1.5–2 cm wide, its tube glabrous; anthers usually included; $2n=14$. Woods; s. Ind. to Mo., s. to w. N.C. and Ala. June–Sept.

9. **Phlox stolonifera** Sims. Crawling p. Sterile basal shoots conspicuous at anthesis, bearing petiolate spatulate lvs 1.5–4 cm; flowering stems diffusely spreading, 1–4 dm; lower lvs often spatulate and with petiole-like base, the upper ones lanceolate to oblong, sessile or nearly so; cymes loose and open, few-fld, glandular-hairy; cal glandular-hairy, its linear lobes usually outcurved; cor red-purple, 2.5–3 cm wide; style elongate; $2n=24$. Moist woods, chiefly in the mts., also on the piedmont; Pa. and s. O. to Ga. Apr., May.

10. **Phlox maculata** L. Meadow-p., spotted p. Stems erect, 3–8 dm, usually red-spotted; lvs numerous, lance-linear to lanceolate or narrowly oblong, 5–12 cm; infl of a terminal and several to many axillary cymes, the latter all short-peduncled, forming a subcylindric cluster 5–25 cm; infl densely but minutely hairy, not glandular; cal usually glabrous, 6–8 mm, cor red-purple, 12–22 mm wide; style elongate; $2n=14$. Low woods and wet meadows; s. Que. to Va. and in the mts. to Ga., w. to s. Minn., Io., and Mo. June, July. (*P. pyramidalis.*)

11. **Phlox ovata** L. Allegheny-p. Stems slender, 3–5 dm, often decumbent at base, usually glabrous; lvs few, seldom more than 4 pairs below the infl at anthesis, with a few lflss nodes near the base, the lower lvs narrowly lanceolate and long-petioled, the upper broader and becoming sessile, the larger ones 6–10 cm; sterile basal shoots often present, with lance-oblong, long-petioled lvs tapering to both ends; infl a few small cymes forming a rounded or flattened corymb; pedicels glabrous or minutely glandular-hairy; cal 10–13 mm; glabrous or sparingly villosulous; cor red-purple, 2.5–3 cm wide; style elongate; $2n=14$. Open woods and thickets, chiefly in the mts.; Pa. to n. Ga., also in O. and Ind. May, June. (*P. latifolia.*)

12. **Phlox buckleyi** Wherry. Shale-barren p. Flowering stems erect, 1.5–4 dm; lvs linear to lanceolate, to 12 cm, the upper much smaller, those of the sterile shoots evergreen; infl a compact cyme, glandular-hairy as also the cal and cor-tube; cor purple, 15–25 mm wide; $2n=28$. Shale-barrens; Va. and W.Va. May–June.

13. **Phlox glaberrima** L. Smooth p. Stems slender, erect, 5–12 dm, usually glabrous; lvs numerous, firm, spreading, linear to narrowly lanceolate or sometimes even ovate, gradually tapering to a sharp tip, 5–12(–15) cm × 3–15(–20) mm; infl usually of several cymes, terminal and from the upper 1–4 pairs of axils, the lower on long peduncles; cal 5–10 mm; cor red-purple, 1.5–2 cm wide; style elongate; $2n=14$. Wet woods and open prairies; Md. to O. and s. Wis., s. to S.C., Ga., Ala., and Ark. June, July. Variable, but the infraspecific taxa, if any, not yet clear. (*P. triflora; P. carolina,* at least as to plants from our range.)

2. **COLLOMIA** Nutt. Cal-tube chartaceous, of nearly uniform texture throughout, not ruptured by the developing fr, the lobes greenish and commonly more herbaceous; cor tubular-funnelform to nearly salverform; stamens included, often unequal or unequally inserted; seeds 1–3 per locule, those of the annual spp. becoming mucilaginous when wet; annual (seldom perennial) herbs with alternate, entire (ours) to variously dissected lvs, but without well defined lfls; fls mostly in terminal, head-like cymes. 13, temp. New World, mostly w. U.S.

1. **Collomia linearis** Nutt. Annual, 1–6 dm, hairy especially above, the main stem and each of the branches (when present) terminating in a dense, leafy-bracteate cluster of essentially sessile fls; lvs numerous, lanceolate to linear, sessile or nearly so, entire, 1–7 cm × 1–15 mm, the uppermost often broadest; cor pink or bluish to white, much longer than the cal, 8–15 mm, with slender tube and short (1.5–3 mm) lobes; filaments short, unequally inserted; seeds 1 per locule; $2n=16$. Mostly in dry soil; B.C. to Calif., e. to Wis., Neb., and N.M., and in the Gaspé region of Que.; occasionally intr. elsewhere. May–Aug. (*Gilia l.*)

3. **POLEMONIUM** L. Jacob's ladder. Cal herbaceous, somewhat accrescent and becoming chartaceous; cor tubular-funnelform or subsalverform to campanulate or nearly rotate; stamens about equally inserted; seeds 1–10 per locule, sometimes becoming mucilaginous when wet; perennial (1 sp. annual) herbs with alternate, pinnately compound or very deeply pinnatifid lvs, the fls in diverse sorts of basically cymose infls, mostly blue. 20, New World and Eurasia, esp. w. U.S.

1 Stamens well exserted from the cor .... 1. *P. van-bruntiae.*
1 Stamens equaling or shorter than the cor.
  2 Infl loose, paniculiform, the pedicels at anthesis nearly as long to longer than the cal; fl spr .... 2. *P. reptans.*
  2 Infl thyrsoid, the pedicels at anthesis shorter than the cal; fl summer .... 3. *P. occidentale.*

1. **Polemonium van-bruntiae** Britton. Appalachian J. l. Erect, 4–10 dm, the upper internodes elongate; lvs as in no. 2, the lfls sharply acute; infl compact, few-fld; cal at anthesis 8–10 mm, usually longer than the pedicel, the lobes conspicuously longer than wide; cor 14–20 mm, lobed to about the middle, the lobes ± erose; stamens surpassing the cor by 5–7 mm; 2*n*=18. Swamps and stream-banks; Me., Vt. and n. N.Y. to W.Va.; local. June, July.

2. **Polemonium reptans** L. Spreading J. l. Stems loosely clustered or solitary, slender, 2–5 dm, erect to diffuse, branched above; basal lvs long-petioled, cauline less so or the upper sessile; lfls oblanceolate to oblong or elliptic, 2–4(–7) cm, those of the principal lvs 7–17; infl few-fld, loose and open, the pedicels nearly as long to longer than the cal; cal 5–6 mm, the broadly triangular lobes nearly as long as the tube; cor campanulate, lobed to about the middle, the lobes entire; stamens shorter than the cor; 2*n*=18. Rich, moist woods; N.Y. to Minn., s. to Va., Ala., and e. Okla., most abundant w. of the mts. Apr.–June. (*P. longii,* a pathological form.) Var. *reptans,* with nearly the range of the sp., has the infl and cal minutely puberulent and the stem usually glabrous; the cor is mostly 12–16 mm. Var. *villosum* E. Braun, local in s. O. and adj. Ky., and largely replacing var. *reptans* there, has the infl densely glandular-villous, the stem usually glandular-villous as well, and the cor mostly 8–12(–13) mm.

3. **Polemonium occidentale** Green. Western J. l. Stems solitary from the upturned end of a mostly short and simple horizontal rhizome, erect, to 10 dm; lower lvs with 1–27 lfls 1–4 cm, the middle and upper smaller; fls crowded in a typically elongate and somewhat thyrsiform infl; cor 10–16 mm, fully as wide, the lobes longer than the tube; stamens about equaling or a little shorter than the cor; style conspicuously surpassing the stamens. Arbor-vitae bogs in n. Minn.; widespread in the w. cordillera. July.

P. caeruleum L., a related European sp. with descending rootstock or root, and at least frequently with clustered stems, occasionally escapes from cult.

4. **GILIA** Ruiz. & Pavón. Cal tubular to campanulate, with prominent scarious or hyaline intervals between the more herbaceous ribs, commonly ruptured by the fr; cor funnelform or salverform to tubular; filaments generally equal and equally inserted; seeds 1–many per locule, sometimes becoming mucilaginous when wet; taprooted herbs with alternate (or all basal, or the lower opposite), entire to pinnatifid or ternate lvs, but without definite lfls; infl basically cymose. 60, New World, mainly w. U.S.

1. **Gilia rubra** (L.) A. A. Heller. Standing cypress. Erect biennial often 1 m, usually simple to the infl; lvs numerous, pinnately parted into narrowly linear segments; infl elongate, simple or branched, forming a slender panicle 2–4 dm; cal-tube 3–5 mm, shorter than the linear lobes, cor scarlet outside, yellow inside, 2–4 cm, the lobes 7–10 mm; 2*n*=14. Native of s. U.S., occasionally escaped from cult. in our range. (*Ipomopsis r.*)

Several western annual spp. of *Gilia,* with smaller fls, these not scarlet, are occasionally intr. but do not persist. Spp. of *Navarretia,* with unequal, commonly spinulose-tipped cal-lobes, the rather small fls in dense cymose heads, are also sometimes temporarily intr.

FAMILY **HYDROPHYLLACEAE**, the Waterleaf Family

Fls perfect, hypogynous, sympetalous, mostly (incl. all ours) 5-merous; cal cleft to the middle or often to the base; cor regular or nearly so, with imbricate or less often convolute lobes; stamens as many as and alternate with the cor-lobes, attached toward the base or well up in the tube, very often flanked by a pair of small scales; ovary superior, bicarpellate but mostly unilocular, the style ± deeply bifid, or the styles 2 (as in *Hydrolea*); stigmas mostly capitate; placentas 2, parietal but often ± intruded, sometimes expanded, seldom (as in *Hydrolea*) joined in the middle so that the ovary is bilocular and the placentation axile; ovules 2–many on each placenta; fr usually a loculicidal capsule, but sometimes irregularly dehiscent; embryo straight, spatulate or sometimes linear, with 2 cotyledons; endosperm scanty to abundant and oily; mostly (incl. all ours) herbs; lvs exstipulate. 20/250, best developed in w. U.S.

1 Lvs deeply toothed to compound; style 1, bifid; ovary unilocular.
  2 Fls several to many in a cyme.
    3 Infl with 1 or more prominent, sympodial axes; plants taprooted .... 1. *Phacelia.*
    3 Infl subdichotomously branched; plants (except 1 sp.) fibrous-rooted .... 2. *Hydrophyllum.*

2 Fls solitary on short pedicels opposite the lvs.
    4 Cal exauriculate; cor 5–8 mm .... 3. *Ellisia.*
    4 Cal generally with auricles at the sinuses; cor in ours 3 mm .... 4. *Nemophila.*
1 Lvs entire; styles 2; ovary bilocular .... 5. *Hydrolea.*

1. PHACELIA A. L. Juss. Phacelia. Cal divided nearly to the base; cor tubular to rotate; filaments usually equal, attached to the cor-tube near its base; style shortly to deeply bifid; ovules 2–many on each of the 2 intruded, partition-like placentas; capsule 2-valved; taprooted herbs (ours annual or biennial), usually hairy, with entire (in extralimital spp.) to pinnately dissected lvs and mostly blue or white fls in 1 or more helicoid cymes that elongate and straighten with maturity. 150+, New World, esp. w. U.S.

1 Ovules and seeds numerous, 25 or more; Isle Royale and ne. Minn. with us .... 1. *P. franklinii.*
1 Ovules and seeds few, not more than 8; more southern spp.
  2 Cor-lobes fimbriate or coarsely denticulate.
    3 Cor-lobes conspicuously hairy outside; ovules mostly 8 and seeds 6–8 .... 2. *P. gilioides.*
    3 Cor-lobes glabrous or glabrate; ovules 4 and seeds 2–4.
      4 Pubescence of the stems and infl closely appressed; fls mostly blue-lavender .... 3. *P. purshii.*
      4 Pubescence of the stems and infl spreading; fls mostly white ....4. *P. fimbriata.*
  2 Cor-lobes minutely crenulate to entire.
    5 Filaments villous, about as long as the cor, or longer; style 4–15 mm at maturity.
      6 Cauline lvs all petiolate; seeds 3–4 mm; cor 10–15 mm wide .... 5. *P. bipinnatifida.*
      6 Upper cauline lvs sessile; seeds 1.5–1.75 mm; cor 5–10 mm wide .... 6. *P. dubia.*
    5 Filaments glabrous, shorter than the cor-tube; style (incl. branches) 1.5–2 mm .... 7. *P. ranunculacea.*

1. **Phacelia franklinii** (R. Br.) A. Gray. Canadian p. Annual or biennial, 1–10 dm, hirsute-puberulent and somewhat viscid; lvs basal and cauline, only gradually reduced upwards, petiolate (less so upwards) with pinnatifid or subbipinnatifid blade 1.5–10 cm; infls scattered or loosely aggregated; cor purplish, broadly campanulate, 6–10 mm, hairy outside, glabrous inside, the lobes entire; filaments hairy, equaling or slightly surpassing the cor; style elongate, hairy below; seeds numerous, 1.5 mm; $2n=22$. Open places in wooded regions; Isle Royale and the n. shore of Lake Superior, nw. to Mack. and Yukon, thence s. to B.C. and n. Utah. June, July.

2. **Phacelia gilioides** A. Brand. Ozark-p. Much like no. 3, but the denticulate to fimbriate cor-lobes pilose on the back, the ovules mostly 8, and the 6–8 seeds 1.5–2 mm; $2n=18$. Deciduous woods and limestone-barrens; c. Mo. to adj. Kans. and Okla. Apr., May.

3. **Phacelia purshii** Buckley. Miami-mist. Lax annual 2–4 dm, strigose throughout; lower lvs petioled, often pinnately compound, the upper sessile and clasping, coarsely pinnately lobed, the lobes 3–11, triangular or oblong, often falcate, sharply acute; cymes secund, 6–20-fld; cor commonly blue, broadly campanulate or subrotate, 5–13 mm wide, its lobes conspicuously fringed, glabrous; filaments villous below the middle; cor with very small appendages in the tube; ovules 4; seeds 2–4, 1.5–3 mm; $2n=18$. Rich moist woods; e. Pa. to S.C. and Ga., w. to se. Ont., n. Ind., e.c. Ill., e. Mo., and Ala. Apr.–June.

4. **Phacelia fimbriata** Michx. Blue-Ridge p. Much like no. 3; stem sparsely spreading-hirsute, becoming hirtellous in the infl and there often with scattered longer hairs as well; lvs avg smaller, the lobes broadly acute to obtuse, sparsely strigose on the upper surface, often glabrate beneath; cor white, seldom tinged with blue or lavender; seeds 3–3.5 mm; $2n=18$. Mt. woods; sw. Va. to N.C. and Tenn. May, June.

5. **Phacelia bipinnatifida** Michx. Forest-p. Biennial, 2–5 dm, spreading-hirsute and (especially in the infl) also glandular; lvs petiolate, broadly triangular to ovate in outline, the larger twice pinnatifid, the upper less divided, the segments ovate, acute, incised or coarsely toothed; cor blue, subrotate, 10–15 mm wide, the broad lobes entire or minutely erose, pilose on the back, the basal appendages in the tube 2–2.5 mm; filaments villous, mostly long-exsert; ovules 4; seeds 4, black, 3–4 mm; $2n=18$. Moist woods; Va. to s. O., n. Ill., and se. Mo., s. to Ga., Ala., and Ark. Apr.–June.

6. **Phacelia dubia** (L.) Trel. Appalachian phacelia. Lax annual 1–4 dm, strigose and stipitate-glandular throughout; basal lvs petiolate, 1.5–6 cm, pinnate or pinnatifid with 1–5 pairs of oval to orbicular, entire or toothed lfls or segments; cauline lvs short-petiolate to sessile, usually deeply lobed with 1–4 pairs of lanceolate to ovate acute segments; cor blue to white, rotate-campanulate, 5–10 mm wide, its lobes orbicular, entire, pilose on the back; stamens slightly exserted, the filaments densely villous on the lower ⅔; ovules 4–8; seeds 4–6, brown, 1.5–1.75 mm; $2n=10$. Woods, fields, and barrens; c. N.Y.; c. Pa. and n. Del. to Ga. and Ala., w. to W.Va. and Tenn. Apr., May.

7. **Phacelia ranunculacea** (Nutt.) Constance. Annual 0.5–2 dm, spreading-hairy and sparsely glandular in the infl, otherwise more strigose; lvs all petioled, deeply divided into 3–7 oblong or obovate, obtuse segments; cymes small and few-fld, eventually to 5 cm; cor blue, tubular-campanulate or almost funnelform, 4–5 mm, the tube exceeding the glabrous, entire lobes; stamens 1.5–2 mm at maturity; ovules 4; seeds 2–4, brown, 2–2.5 mm; $2n=12$ (Ozarkian); 28 (Appalachian). Moist alluvial woods; ne. Va. and adj. Md.; c. N.C.; s. Ind. and s. Ill., to w. Tenn., se. Mo., and n. Ark. Apr., May. (*P. covillei.*)

2. HYDROPHYLLUM L.   Waterleaf. Cal divided to below the middle or near the base; cor campanulate to tubular, white to purple; filaments exserted, each flanked at base by a pair of linear appendages of the cor; style shortly bifid; ovary unilocular, with 2 dilated placentas; ovules 4; seeds 1–3; capsule globose, 2-valved; herbs, mostly perennial and with fleshy-fibrous roots from a very short to well developed rhizome, the lvs variously cleft; fls in compact (often capitate) mostly subdichotomously branched cymes that lack a well developed main axis. 8, N. Amer.

1 Principal cauline lvs deeply pinnately divided.
  2 Infl spreading-hirsute, the hairs commonly 1–2 mm .... 1. *H. macrophyllum.*
  2 Infl strigose, the hairs seldom as much as 0.5 mm .... 2. *H. virginianum.*
1 Principal cauline lvs orbicular in general outline, palmately lobed.
  3 Fibrous-rooted perennial from long rhizomes; stamens exserted 3–6 mm .... 3. *H. canadense.*
  3 Taprooted biennial; stamens exserted only 1–3 mm .... 4. *H. appendiculatum.*

1. **Hydrophyllum macrophyllum** Nutt.   Hairy w. Stems 3–7 dm, the upper part, pedicels, and sep densely gray-hirsute with stout spreading hairs commonly 1–2 mm; cauline lvs commonly 1–2 dm, half to two-thirds as wide, pinnately divided to near the midvein, the segments usually 7 or 9, spreading at nearly right angles, the apex and the few coarse teeth broadly acute or obtuse; infl dense at anthesis; cor white or pinkish, 8–13 mm, lobed to about the middle; stamens and style long-exsert. Rich moist woods; O. to Ill., s. to w. Va., N.C., and Ala. May, June.

2. **Hydrophyllum virginianum** L.   Eastern w. Stems 3–8 dm, the upper part, pedicels, and sep with appressed or ascending hairs rarely to 0.5 mm; cauline lvs broadly ovate or broadly triangular in outline, 1–2 dm and usually somewhat wider, pinnately divided almost to the midvein, the segments usually 5(7–9), the terminal one and the basal pair often 2–3-lobed, all with sharply acute or acuminate apex and similar, strongly ascending teeth; infls dense at anthesis; sep sparsely hirsute-ciliate; cor 7–10 mm; stamens and style long-exsert; $2n=18$. Moist or wet woods, or open wet places; Que. to N.D., s. to N.C., Ala., and ne. Okla. May, June. The widespread var. *virginianum* has white to lavender or pale purple fls. Var. *atranthum* (Alexander) Constance, of the s. Appalachians from Va. and W. Va. to N.C., has deep violet fls and often has 7–9 lfls.

3. **Hydrophyllum canadense** L.   Maple-lvd w. Fibrous-rooted perennial from long rhizomes; stems 3–5 dm, the upper part, infl, and cal glabrous or with scattered stout spreading hairs to 1 mm; cauline lvs usually overtopping the cymes, 1–2 dm wide and nearly as long, cordate at base, palmately 5–9-lobed, seldom to beyond the middle, the lobes acuminate, coarsely toothed or incised; cal rarely with minute appendages at the sinuses; cor white to pink-purple, 8–12 mm, the lobes shorter than the tube; stamens exsert 3–6 mm; style exsert 4–5 mm; $2n=18$. Rich moist woods; s. Vt. and w. Mass. to Md. and in the mts. to Ga., w. to Wis., Ill., Mo. and n. Ark. May, June.

4. **Hydrophyllum appendiculatum** Michx.   Biennial w. Taprooted biennial; stems 3–6 dm, the upper part and infl densely pubescent with short slender hairs 0.3–0.5 mm and conspicuously hirsute with spreading hairs usually 2–3 mm; cauline lvs mostly overtopped by the cymes, orbicular in general outline, 6–15 cm wide at anthesis, truncate or with a broadly V-shaped base, shallowly 5–7-lobed, with obtuse or rounded sinuses; sep separate nearly to the base, densely hirsute, alternating with conspicuous small reflexed appendages; cor lavender or pink-purple, 9–13 mm, the lobes about equaling the tube; stamens and style exsert 1–3 mm; $2n=18$. Rich moist woods; s. Ont. to s. Minn., s. to Pa., Tenn., Mo., and e. Kans. May, June. (*Decemium a.*)

3. ELLISIA L., nom conserv.   Cal cleft to near the base, exappendiculate, accrescent and subrotate at maturity; cor campanulate, white to lavender, about equaling or slightly surpassed by the cor, cleft up to half its length; ovules 2 on each of the 2 expanded parietal placentas; fr unilocular, 2-valved; seeds finely reticulate, without a cucullus; small, branching annuals with pinnatifid lvs, at least the lower ones opposite, the fls solitary in or opposite the axils, the stem sometimes also terminating in a lax, few-fld infl. Monotypic.

1. **Ellisia nyctelea** (L.) L.   Water-pod. Plants 0.5–4 dm, usually freely branched, rather sparsely strigose or in part hirsute, the petioles coarsely ciliate toward the base; lf-blades to 6 × 5 cm, with wing-margined rachis and mostly 3–6 pairs of rather narrow, entire or few-toothed lateral segments; pedicels mostly under 1 cm at anthesis, sometimes to 5 cm in fruit; cal foliaceous-accrescent, the ovate or lance-ovate lobes veiny and commonly 1 cm at maturity; cor 5–8 mm; seeds mostly 4, 2.5–3 mm; $2n=20$. Moist, shaded bottoms and open fields; Mich. and Ind. to Sask. and the w. cordillera; also irregularly near the coast from s. N.Y. to Va. and occasionally adventive elsewhere in our range. May, June. (*Nyctelea n.*)

4. NEMOPHILA Nutt., nom. conserv.   Cal deeply cleft, commonly with small reflexed auricles in the sinuses; cor campanulate to rotate, white to blue or purple; stamens and style included; ovules 2–several on each of the 2 large, parietal placentas; fr unilocular, 2-valved; seeds with an obscure to evident, partial or complete covering (cucullus) external to the seed-coat; delicate, taprooted annuals with opposite or less often alternate, mostly pinnatifid or pinnatilobate lvs, the fls solitary in the axils or some in a loose, few-fld terminal cyme. 11, N. Amer.

1. **Nemophila triloba** (Raf.) Thieret. Stems weak, diffuse, branched, 1–4 dm; lvs alternate, long-petioled, the blade broadly triangular in outline, 1–3 cm, deeply divided into 3–5 obovate segments, these also lobed or deeply toothed; fls solitary on pedicels 3–10 mm opposite the lvs; cal 2 mm, the appendages irregularly developed, often missing from some sinuses; cor campanulate, white or bluish, 3 mm; fr much exceeding the persistent cal; cucullus small, persistent, at one end of the seed; $2n=18$. Damp rich woods; Md. and se. Va. to w. Ky. and se. Mo., s. to Fla. and Tex. Apr., May. (*N. microcalyx.*)

5. **HYDROLEA** L., nom. conserv. Sep narrow, distinct, unequal in width; cor broadly campanulate to subrotate, lobed to below the middle; filaments on the cor-tube, abruptly dilated at base; ovary bilocular; ovules very numerous on the broad axile placentas; styles 2; capsule globose, subtended by the persistent cal, mostly breaking irregularly or 2-valved; seeds minute; ours perennial herbs with alternate, entire lvs, usually axillary spines, and small or medium-sized blue fls in axillary or terminal cymes. 11, mainly tropical.

1 Fls in a terminal panicle of cymes; styles and stamens ca 10+ mm .... 1. *H. ovata.*
1 Fls in subsessile, few-fld axillary cymes; styles and stamens ca 5 mm or less.
  2 Cal glabrous or very minutely and inconspicuously puberulent .... 2. *H. uniflora.*
  2 Cal sparsely hirsute with spreading hairs 2–3 mm .... 3. *H. quadrivalvis.*

1. **Hydrolea ovata** Nutt. Rhizomatous; stems to 1 m, the upper part densely and softly villosulous with very slender hairs 0.2–0.3 mm, with or without a few scattered longer bristles; lvs ovate-lanceolate to ovate, 2–6 cm, minutely villosulous on both sides; lower axillary cymes on long ascending peduncles, the upper peduncles progressively shorter, forming a terminal pyramidal panicle; sep lanceolate or lanceolate-linear, 6–9 mm, ± glandular-villous; cor 11–17 mm; styles and stamens 10–15 mm; capsule ca 5 mm thick, hispidulous; $2n=20$. Swamps and wet woods; w. Ky. and s. Mo. to Tex. and Ga. July, Aug. (*Nama o.*) Hybridizes with no. 2.

2. **Hydrolea uniflora** Raf. Stems decumbent or creeping at base, 3–8 dm; upper part of the stem, infl, sep, and both sides of the lvs glabrous or very minutely puberulent, the hairs not over 0.1 mm; lvs lanceolate, acuminate at both ends, 4–10 cm; cymes 1–10-fld, dense, subsessile in the axils of foliage lvs; sep lanceolate to ovate, 5–8 × 1.5–4 mm; cor 7–11 mm; styles and stamens ca 3.5–5 mm; fr glabrous, 5 mm thick; $2n=20$. Wooded swamps; s. Ind. and se. Mo. to Ala. and Tex. June–Sept. (*H. affinis; Nama a.*)

3. **Hydrolea quadrivalvis** Walter. Much like no. 2; lvs glabrous beneath, sparsely bristly above, especially toward the base; upper part of the stem and cal sparsely hirsute with spreading hairs 2–3 mm, some of them glandular; sep lanceolate or lance-linear, 6–10 mm; $2n=20$. Wooded swamps, mainly on the coastal plain; se. Va. to Fla. and La., n. inland to Tenn. May–Aug. (*Nama q.*)

Family **BORAGINACEAE**, the Borage Family

Fls perfect, hypogynous, sympetalous, mostly (incl. all ours) 5-merous; sep distinct or connate at base or seldom to the middle or beyond; cor regular or in a few genera somewhat irregular, in most genera salverform, but sometimes tubular or funnelform or campanulate, the lobes usually imbricate or convolute; cor-tube in most genera with ± evident, often hairy appendages (the fornices) at the orifice opposite the lobes; stamens as many as and alternate with the cor-lobes, borne on the tube; ovary superior, basically 2-carpellate, each carpel ordinarily 2-ovulate and with a secondary partition; fr typically (incl. all our genera) of (1–)4 nutlets that are attached individually to the short or elongate gynobase; style simple or sometimes 2(4)-fid, typically attached directly to the gynobase and arising between the essentially distinct lobes of the ovary, less commonly borne on the summit of the entire or merely 4-lobed ovary which then separates only tardily into individual nutlets (or the stigma seldom sessile); embryo with 2 cotyledons; endosperm in most genera scanty or none; herbs (all ours) or less often woody plants, often rough-hairy, with simple, mostly entire, exstipulate, usually alternate lvs (the lower sometimes opposite). 100/2000, cosmop.

1 Ovary entire or shallowly 2–4-lobed; style (or sessile stigma) terminal .... 1. *Heliotropium.*
1 Ovary deeply 4-parted, the style arising between the essentially distinct lobes.
  2 Cor rotate; anthers large, conspicuous, with a prominent dorsal appendage .... 15. *Borago.*
  2 Cor with a well developed tube; anthers without a dorsal appendage, in most genera (notably
    excepting *Echium*) included.
    3 Cor irregular, the limb oblique, with unequal lobes, the tube sometimes bent.
      4 Stamens exsert; throat of the cor open .... 12. *Echium.*
      4 Stamens included; throat of the salverform cor closed by the fornices .... 13. *Anchusa.*
    3 Cor regular.
      5 Scrambling climbers with weak, retrorsely prickly-hispid stems; fruiting cal much enlarged
        and prominently veiny .... 3. *Asperugo.*
      5 Plants not climbing; fruiting calyx not greatly enlarged, seldom evidently veiny.
        6 Cor tubular or tubular-campanulate, the lobes erect or scarcely spreading.
          7 Cor-lobes distinctly acute or acuminate; style long-exsert .... 11. *Onosmodium.*
          7 Cor-lobes blunt to nearly obsolete; style included or barely exsert.
            8 Pedicels and cal glabrous or finely appressed-hairy; nutlets attached laterally at or
              below the middle .... 2. *Mertensia.*
            8 Pedicels and cal with spreading hairs or bristles; nutlets with a prominent, toothed,
              basal rim surrounding a stipe-like plug .... 14. *Symphytum.*
        6 Cor salverform or broadly funnelform.
          9 Nutlets bearing glochidiate prickles, these visible well before maturity.
            10 Fls regularly associated each with a bract; annual; nutlets narrowly attached along
              the median ventral keel .... 8. *Lappula.*
            10 Fls bractless, or only a few bracted, or the small bracts alternating with the fls; our
              spp. biennial.
              11 Nutlets attached on the middle third of the inner face .... 7. *Hackelia.*
              11 Nutlets attached on the apical third of the inner face .... 9. *Cynoglossum.*
          9 Nutlets smooth or variously roughened, but without prickles.
            12 Nutlets with a thickened basal rim surrounding a stipe-like plug; fls blue, ours with
              the limb 6–20 mm wide .... 13. *Anchusa.*
            12 Nutlets without a basal rim or stipe-like plug; fls various.
              13 Nutlets with a broad, basal attachment; fls white or greenish to yellow or orange
              .... 10. *Lithospermum.*
              13 Nutlets with a small, lateral or basal attachment.
                14 Cor mostly blue or white (sometimes yellow in *Myosotis*).
                  15 Nutlets smooth and completely encircled by an evident dorsomarginal
                    ridge .... 4. *Myosotis.*
                  15 Nutlets usually roughened on the back, and in any case not encircled by
                    a dorsomarginal ridge .... 5. *Plagiobothrys.*
              14 Cor yellow or orange; nutlets roughened, not margined .... 6. *Amsinckia.*

1. HELIOTROPIUM L.

1. HELIOTROPIUM L. Heliotrope. Cor salverform or funnelform, often with 5 small teeth alternating with the lobes; fornices wanting; anthers included, often connivent; ovary entire or merely shallowly lobed, the style terminal (or wanting and the stigma sessile); stigma with a broad, disk-like base commonly surmounted by a mostly short, entire or 2-cleft cone; fr separating at maturity into 4 nutlets, or the nutlets cohering in pairs; herbs (ours) or shrubs with blue or white fls mostly in terminal helicoid cymes, or sometimes solitary on the branches. 200+, mainly of warm regions.

1 Fls in second, helicoid cymes that elongate and straighten as they mature.
  2 Nutlets cohering in pairs; cal and axis of the infl ± hirsute.
    3 Fr longitudinally striate, 2-lobed and separable into 2 halves, each half separable into 2 nutlets,
      each nutlet with an abaxial fertile chamber and an adaxial empty one .... 1. *H. indicum.*
    3 Fr tuberculate, truncate, separating at maturity into 2 2-seeded halves .... 2. *H. amplexicaule.*
  2 Nutlets separating at maturity; fr 4-lobed before the nutlets separate.
    4 Cal and axis of the infl glabrous; plants perennial .... 3. *H. curassavicum.*
    4 Cal and axis of the infl densely white-hirsute; plants annual .... 4. *H. europaeum.*
1 Fls solitary at the end of short leafy branches .... 5. *H. tenellum.*

1. Heliotropium indicum L. Turnsole. Erect, sparsely hirsute, taprooted annual 3–8 dm; lvs ovate or deltoid-ovate, 4–10 cm, scabrellate, decurrent on the long petiole; spikes secund, mostly solitary at the end of the branches, eventually 8–15 cm; cor blue, 3–4 mm wide; fr 2.5–4 mm, divergently 2-lobed to about the middle, each lobe composed of 2 united (but separable) nutlets, each nutlet sharply ridged longitudinally on the back, tapering to the summit, with an abaxial fertile chamber and a swollen, adaxial empty one. A pantropic weed, supposed to be native to Brazil, extending n. to Va., s. Ind., and Mo., and rarely farther as a waif. May–Oct. (*Tiaridium i.*)

2. Heliotropium amplexicaule M. Vahl. Deep-rooted, diffusely branched, sparsely hirsute perennial 2–5 dm; lvs sessile, lanceolate to oblanceolate, the larger to 8 × 2 cm; spikes 2–5 on a naked common peduncle, each eventually 4–10 cm; cor blue or purple, shallowly lobed, 4–6 mm wide; fr depressed-ovoid, 2–3 mm, rough-tuberculate, separating into 2 halves, each half bilocular and 2-seeded, half-ovoid, truncate above, distended on the inner face over the seeds, the distal margin incurved and bearing 2 minute lobes; $2n=26$, 28. Native of S. Amer., intr. as a weed in se. U.S., n. occasionally to Va. and Mo., or farther n. as a waif. (*Cochranea anchusaefolia*.)

3. Heliotropium curassavicum L. Seaside-heliotrope. Glabrous perennial with scattered stems from stout creeping roots, somewhat fleshy, prostrate or decumbent, 2–5 dm; lvs linear or linear-oblanceolate, 1–4 cm × 2–5 mm; spikes seldom over 5 cm, the terminal ones usually paired on a peduncle, the lateral usually solitary and sessile or nearly so; cor 2–3.5 mm wide, white with a yellow eye; mature cal spreading; fr depressed-ovoid, 1.5–2.5 mm, soon splitting into 4 nutlets; $2n=26$, 28. Native of tropical Amer., established as a weed, especially in saline soil, in s. U.S. n. to Del. and occasionally as a weed farther n. May–Sept. Ours are var. *curassavicum*. The well marked var. *obovatum* A. DC. (*H. spathulatum*) of interior w. U.S., with broader, more oblanceolate or even oblate lvs 6–18 mm wide and with the cor 5–9 mm wide, often with a purple eye, may possibly extend to w. Minn.

4. Heliotropium europaeum L. European heliotrope. Erect or ascending annual 2–5 dm; stems closely hairy below, becoming hirsute above; lvs elliptic, 3–6 cm, long-petioled; spikes in peduncled groups of 2–5, eventually 4–10 cm, the axis and sep densely white-hirsute; cor white, 2–4 mm wide; fr depressed-ovoid or globose, 1.5–3 mm, soon splitting into 4 1-seeded nutlets; $2n=24$, 32. Native of s. Europe, established as a weed in s. U.S. and occasionally in our range n. to Mass.

5. Heliotropium tenellum (Nutt.) Torr. Pasture-heliotrope. Erect, freely branched annual 1–4 dm, white-strigose throughout; lvs linear, 1.5–3 cm × 1–3 mm; fls solitary at the end of numerous short branches; cal-lobes foliaceous, strigose, very unequal, the longest nearly equaling the cor; fr depressed-globose, 2 mm long, 3 mm thick, splitting into 4 1-seeded nutlets, finely hairy outside. Dry soil, upland woods, prairies and barrens; w. Io. and Kans. to Ala. and Tex. June–Aug. (*Lithococca t.*)

2. MERTENSIA Roth, nom. conserv. Bluebell. Cor tubular, funnelform, or campanulate, with a definite distinction between tube and limb, the tube short or elongate, the limb shallowly lobed; stamens inserted at the summit of the cor-tube (in ours), included or barely exsert; nutlets smooth or often wrinkled, attached laterally to the gynobase; leafy perennials with blue (pink or white) fls in modified, bractless, mostly small cymes ending the stem and branches. 35–40, N. Amer. and extratropical Eurasia.

1 Lvs and cal glabrous.
  2 Fls 18–25 mm, never leafy-bracted; woods .... 1. *M. virginica.*
  2 Fls 6–9 mm, often leafy-bracted; beaches .... 2. *M. maritima.*
1 Lvs and cal pubescent; fls 10–15 mm .... 3. *M. paniculata.*

1. Mertensia virginica (L.) Pers. Eastern b. Erect or ascending, 3–7 dm, glabrous throughout; cauline lvs elliptic to obovate or oblanceolate, 5–15 cm, rounded or obtuse at the apex, the lower long-tapering at base, the upper less so and the uppermost often rounded and sessile at base; sep ovate-oblong, blunt, 2.5–3.5 mm; cor 18–25 mm, the tube longer than the limb; anthers elliptic, 2 mm; $2n=24$. Moist or wet woods; N.Y. to Mich., Wis., Io., and e. Kans., s. to Ala. and Mo. Apr., May.

2. Mertensia maritima (L.) S. F. Gray. Seaside-b. Spreading or decumbent, to 1 m, glabrous throughout; principal cauline lvs ovate to spatulate-obovate, 2–6 cm, obtuse to acuminate, broadly petioled; cymes usually numerous, often divaricate, lax, and conspicuously leafy-bracted; cal-lobes broadly triangular-ovate, 2–3 mm, larger in fr; cor 6–9 mm, the tube longer than the limb; anthers elliptic, 1–1.5 mm; $2n=24$. Along the seacoast; Greenland to Mass.; Alas. to B.C.; n. Europe. June–Aug. (*Pneumaria m.*)

3. Mertensia paniculata (Aiton) G. Don. Northern b. Stems erect, 3–10 dm, glabrous or sparsely hairy; cauline lvs lanceolate to ovate, 5–14 cm, acuminate, tapering to a winged petiole, hairy on both sides; cal-lobes lanceolate, 3–6 mm, rather densely strigose; cor 10–15 mm, the limb slightly longer than the tube; anthers linear, 2.5–3.5 mm. Damp woods; Hudson Bay to n. Mich., n. Wis., and n. Minn., w. to Alas. and Oreg.; ne. Io. June, July.

3. ASPERUGO L. Madwort, catchweed. Cal 5-lobed to about the middle, each lobe with a smaller tooth on each side near the base, the whole strongly accrescent, becoming compressed, firm, strongly reticulate-veiny, and shortly prickly-hispid with curved or hooked hairs; cor small, blue, ± campanulate, with well developed fornices; anthers included; nutlets obliquely compressed, narrowly ovate, rough, attached to an elevated gynobase, the small scar situated just within the margin above the middle; annual weeds with weak, climbing-scrambling, retrorsely prickly-hispid stems and opposite to partly alternate or partly whorled lvs that are subentire and often remote; fls on short, stout, recurved pedicels in or near the axils of lvs or bracts, and in the forks of the branches. Monotypic.

1. Asperugo procumbens L.   Stems 3–12 dm, freely branched; lvs oblanceolate, 3–6(–10) cm, or the upper more elliptic; cor 2–3 mm long and wide; fruiting cal 1–2 cm wide; $2n$=48. Native of Eurasia, occasionally found in waste places in our range, especially northward. May–Aug.

4. MYOSOTIS L.   Forget-me-not; scorpion grass. Cor salverform to broadly funnelform, equaling to much exceeding the cal, the throat obstructed by the fornices; stamens included; nutlets smooth and shining, with an evident raised margin all the way around, and with a small, basilateral scar, the gynobase low and broad; glabrous or strigose (but not hispid or setose) herbs with blue or less often white (rarely yellow) fls in terminal, naked, helicoid cymes ("racemes"), or the lower fls scattered among the lvs; fruiting pedicels erect or spreading. 50, temp. and boreal.

1 Cal strigose, the hairs neither spreading nor uncinate; mostly perennial.
  2 Cor-limb 5–10 mm wide; nutlets not surpassing the style .... 1. *M. scorpioides.*
  2 Cor-limb 2–5 mm wide; nutlets distinctly surpassing the style .... 2. *M. laxa.*
1 Cal with some loose or spreading, uncinate hairs.
  3 Cor rather showy, the limb 5–8 mm wide; short-lived perennial .... 3. *M. sylvatica.*
  3 Cor not showy, the limb 1–4 mm wide; annual or biennial.
    4 Cal sub-bilabiate, 3 lobes shorter than the other 2; cor white.
      5 Fruiting pedicels ± erect; nutlets 1.2–1.5 mm .... 4. *M. verna.*
      5 Fruiting pedicels divergent; nutlets 1.4–2.2 mm .... 5. *M. macrosperma.*
    4 Cal symmetrical, with equal lobes; cor blue, less often white or yellow.
      6 Fruiting pedicels equaling or generally surpassing the cal .... 6. *M. arvensis.*
      6 Fruiting pedicels distinctly shorter than the cal.
        7 Plants floriferous nearly to the base, the lower fls scattered among the lvs; style distinctly
          surpassed by the nutlets .... 7. *M. micrantha.*
        7 Plants floriferous to not much if at all below the middle, the infl ordinarily essentially
          naked; style often surpassing the nutlets .... 8. *M. discolor.*

1. Myosotis scorpioides L.   Water scorpion-grass. Fibrous-rooted perennial 2–6 dm, often creeping at base and commonly stoloniferous, inconspicuously strigose; lvs 2.5–8 cm × 7–20 mm, the lower mostly oblanceolate, the others more oblong or elliptic to lance-elliptic; infl terminal, becoming open; mature pedicels spreading, about equaling or somwhat surpassing the 3–5 mm cal; cal closely strigose, the lobes equaling or shorter than the tube, sometimes unequal; cor blue, the limb 5–10 mm wide, flat; style equaling or more often surpassing the nutlets, rarely surpassed by them; $2n$=64. Shallow water and wet soil; native of Europe, naturalized throughout our range. May–Sept. (*M. palustris.*)

2. Myosotis laxa Lehm.   Smaller forget-me-not. Short-lived perennial or even annual, 1–4 dm, slender, often decumbent at base, but not creeping or stoloniferous, inconspicuously strigose; lvs much like no. 1, averaging a little smaller; infl terminal, becoming open; mature pedicels spreading, equaling or mostly longer than the 3–5 mm cal; cal closely strigose, the lobes shorter or a little longer than the tube, equal or unequal; cor blue, the limb 2–5 mm wide; style distinctly surpassed by the nutlets. Moist soil and shallow water; interruptedly circumboreal, and found throughout our range. May–Sept.

3. Myosotis sylvatica Hoffm.   Garden forget-me-not. Short-lived perennial to 5 mm; lvs oblong to lanceolate or spatulate, the larger 3–7 cm; racemes naked, seldom over 1 dm; fruiting pedicels 5–15 mm apart, ascending or spreading, to 9 mm; mature cal 4–5 mm, densely pubescent with hooked hairs, the lobes much longer than the tube; cor blue (white), salverform, the limb 5–8 mm wide; nutlets 1.5–2 mm; $2n$=14–48. Native of Eurasia, occasionally escaped from cult. in our range. Apr.–Sept.

4. Myosotis verna Nutt.   Early scorpion grass. Annual or winter annual 0.5–4 dm, hirsute-puberulent or hirsute-strigose throughout; lvs 1–5 cm × 2–10 mm, the lowermost mostly oblanceolate and petiolate, the others more oblong or narrowly elliptic and sessile; infl seldom comprising much more than the upper half of the plant, often irregularly leafy-bracteate below; fruiting pedicels erect or suberect, up to nearly half as long as the 4–7 mm cal; cal uncinate-hairy below, stiffly ascending-hirsute above, sub-bilabiate, 3 lobes shorter than the other 2; cor white, the limb 1–2 mm wide, not flat; nutlets 1.2–1.5 mm, surpassing the style. Upland woods and fields; Me. to Mich. and S.D., s. to Ga. and Tex.; also B.C. to Oreg. and Ida. Apr.–July. (*M. virginica,* misapplied.)

5. Myosotis macrosperma Engelm.   Big-seed scorpion-grass. Much like no. 4, but averaging larger, to 6 dm, the lvs 2–4 cm; pedicels at maturity 1–3 cm apart ascending or outcurved; mature cal 5–8 mm; nutlets 1.4–2.2 mm. Moist woods; Md. to s. Ind. and Mo., s. to Fla. and Tex.; disjunct (?.) in s. Ont.

6. Myosotis arvensis (L.) Hill.   Field scorpion-grass. Biennial or sometimes annual, 1–4 dm, often branched above, strigose to hirsute-puberulent throughout; lvs 1–6 cm × 3–16 mm, the lower mostly oblanceolate, the upper often more oblong or lanceolate; infl slightly if at all longer than the leafy part of the plant; mature pedicels ascending or spreading, equaling or generally surpassing the 3–5 mm cal; cal hirsute-puberulent and shortly uncinate-hispid; cor blue or occasionally white, the limb 2–4 mm wide; nutlets distinctly surpassing the style; $2n$=50–54. Native of Eurasia, established in fields and roadsides from Nf. to Minn., s. sometimes to N.H. and W.Va. Summer.

7. Myosotis micrantha Pallas. Blue scorpion-grass. Annual or winter-annual to 2 dm, often branched from the base, ± hirsute-puberulent throughout; lvs seldom over 2 cm × 7 mm; plants floriferous nearly to the base, the lower fls scattered among the lvs; fruiting pedicels ascending or somewhat spreading, under 2 mm, shorter than the 3–5 mm cal; cal strigose at least above, the tube also shortly uncinate-hispid; cor blue, the limb 1–2 mm wide, not flat; nutlets surpassing the style; $2n=36$–40. Native of Eurasia, locally intr. in dry waste places; Que. to Va., Mich., and Io. Apr.–July. (*M. stricta.*)

8. Myosotis discolor Pers. Yellow and blue scorpion-grass. Slender annual 1–5 dm, obscurely strigose to evidently hirsute-puberulent; lvs mostly 1–4 cm × 2–8 mm; "racemes" naked or with one or two leafy bracts near the base, not much if at all longer than the leafy part of the plant; fruiting pedicels 1–3 mm, evidently shorter than the 3–5 mm cal, ascending or somewhat spreading; cal-tube shortly uncinate-hispid; cor at first yellow or yellowish, ordinarily changing to blue, the limb 1–2 mm wide, not flat; nutlets shorter to occasionally longer than the style; $2n=64$. Native of Europe, locally intr. in fields and roadsides: Que. to Va. May–July. (*M. versicolor.*)

5. PLAGIOBOTHRYS Fischer & C. A. Meyer.

Cor rather small, ± salverform, white, the well developed fornices sometimes yellow; stamens included; nutlets tending to be keeled on the back, and with a well developed ventral keel; scar generally elevated and caruncle-like, mostly small, lateral to virtually basal, placed at the lower end of the ventral keel, or sometimes extending along the keel; gynobase short and broad to pyramidal; narrow-lvd annuals or sometimes perennials, with the fls in a series of sympodial, helicoid, naked or irregularly bracteate, elongating false racemes or spikes. 50, mostly w. N. Amer.

1. Plagiobothrys scouleri (Hook. & Arn.) I. M. Johnston. Meadow-plagiobothrys. Slender annual with several or many prostrate or ascending stems to 20 cm; lvs essentially all cauline, linear, to 6 cm × 5 mm, the lower 1–4 pairs opposite; stems terminating in an elongate, loosely fld false raceme or spike that may be irregularly bracteate below; cal 2–4 mm in fr; cor small, the limb mostly 1–2 mm wide; nutlets ovate or lance-ovate, 1.5–2.2 mm, usually roughened, with a small, basilateral scar; $2n=54$. Widespread in w. U.S. and Can., casually intr. here and there in our range. May–Aug. Ours is var. *penicillatus* (Greene) Cronq.

6. AMSINCKIA Lehm., nom., conserv.

Fiddleneck. Cor funnelform or campanulate, with or without well developed fornices; stamens included; nutlets in most spp. evidently roughened, keeled ventrally and often also dorsally, the scar typically placed at the lower end of the ventral keel, often elevated and caruncle-like; gynobase mostly short-pyramidal; cotyledons deeply bifid, thus apparently 4; taprooted, rough-hairy annuals with rather small, often narrow lvs, the orange or yellow fls in elongating, mostly naked, helicoid false spikes or racemes. 15, w. N. Amer. and w. and s. S. Amer.

1 Cor-throat obstructed by the well developed, hairy fornices .... 1. *A. lycopsoides.*
1 Cor-throat open, glabrous, the fornices scarcely developed.
  2 Sep 5, essentially distinct, not very unequal in width .... 2. *A. retrorsa.*
  2 Sep of many or all fls unequal in width and reduced in number by lateral fusion .... 3. *A. tessellata.*

1. Amsinckia lycopsoides Lehm. Rough-hairy annual, 1–6 dm; lvs linear to linear oblong or the upper more lanceolate, to 10 × 1.5 cm, often crowded at the base; sep 5, 6–10 mm at maturity; cor 7–9 mm, the tube somewhat exserted, the throat obstructed by the hairy fornices, the limb 3–6 mm wide; stamens inserted well down in the cor-tube, commonly below the middle; nutlets 2.5–3 mm; $2n=30$. Native of w. U.S., occasionally adventive in our range. Apr.–June. (*A. barbata.*)

2. Amsinckia retrorsa Suksd. Rough-hairy annual 1–6 dm; stem spreading-hispid and also evidently puberulent or strigose throughout with shorter and softer, ± retrorse hairs; lvs linear or linear-oblong, to 12 cm, seldom over 1 cm wide, the basal ones often crowded and wider than the others; sep 5, 5–12 mm at maturity; cor 5–8 mm, the tube 10-nerved below the insertion of the stamens, not much exserted from the cal; stamens inserted near the middle of the cor-tube; nutlets 2–3 mm; $2n=16$, 26. Widespread in w. U.S., occasional with us as a weed in disturbed sites. Apr.–July.

3. Amsinckia tessellata A. Gray. Devil's lettuce. Rough-hairy annual 1–6 dm; stem spreading-hispid, the upper part also with shorter and softer, somewhat retrorse hairs; lvs linear or generally broader, often lance-oblong or lance-ovate, up to 10 × 3 cm; sep 7–14 mm at maturity, commonly only (2–)4 by lateral fusion, the broader one(s) often apically bidentate; cor 7–12 mm, the tube 20-nerved below the insertion of the stamens; stamens inserted near the middle of the cor-tube; nutlets 2.5–3.5 mm, roughened, often tessellate-tuberculate; $2n=24$. Widespread in w. U.S., occasional with us as a weed in disturbed sites. Apr.–June.

7. HACKELIA Opiz.

Stickseed; beggar's lice. Cor salverform or broadly funnelform, the tube often shorter than the cal, the throat obstructed by the fornices; stamens and style included; fruiting pedicels short, recurved or reflexed; nutlets attached by a lanceolate to ovate area occupying the middle third only, the dorsal surface lance-ovate, bordered by a row of glochidiate prickles, and in some spp. with similar prickles on the surface; mostly taprooted perennials, or a few spp. biennial or even annual, with numerous, usually paired false racemes terminating axillary branches, these racemes naked above, leafy-bracteate at base; fls blue or white, small. 45, N. and S. Amer. and Eurasia.

1 Dorsal area of the nutlets with several (usually 10–25) erect glochidiate prickles about as long as the
    marginal ones .... 1. *H. virginiana.*
1 Dorsal area of the nutlets unarmed, or rarely with 1–3 short prickles .... 2. *H. deflexa.*

1. **Hackelia virginiana** (L.) I. M. Johnston.   Stout biennial to 1(–1.5) m, freely branched above; lower lvs to 3 dm, narrowed to a petiole; middle and upper lvs oblong-elliptic, 5–15 cm, sharply narrowed to both ends, sessile, the uppermost passing into the small, lanceolate to linear bracts; racemes spreading, eventually 5–15 cm; bracts often alternating with the fls, those beyond the middle minute or lacking; fruiting pedicels 2–10 mm; cor white or pale blue, 2–3 mm wide; dorsal area of the nutlets ovate, 2–3 mm, verrucose and bearing 10–25 erect prickles as long as the marginal ones; 2*n*=24. Upland woods; s. Que. to N.D., s. to Ga., La., and Tex. July–Sept. (*Lappula v.*)

2. **Hackelia deflexa** (Wahlenb.) Opiz.   Much like no. 1, but averaging more slender and with smaller lvs; dorsal area of the nutlets unarmed or with only a few short prickles. Moist woods, thickets, and hillsides; circumboreal, in Amer. from s. Que. to Mack. and B.C., s. to Vt., Io., and Colo. May–Aug. (*Lappula d.; L. americana; Hackelia a.*) The Amer. plants are var. *americana* (A. Gray) Fern. and I. M. Johnston.

8. **LAPPULA** Moench.   Stickseed; beggar's lice. Cor salverform or broadly funnel-form, the tube about as long as the cal, the throat closed by the fornices; stamens and style included; fruiting pedicels erect or spreading; nutlets erect and connivent, narrowly attached to the elongate gynobase along the median ventral keel (the lower part rounded and free), bearing one or more rows of glochidiate prickles along the continuous dorso-marginal ridge or cupulate border; rough-hairy annuals with erect stems, each branch ending in an elongate, bracteate false raceme (helicoid cyme) of small, blue or occasionally white fls. Ca. 12, mainly N. Hemisphere.

1 Marginal prickles of the nutlets in a single row .... 1. *L. redowskii.*
1 Marginal prickles of the nutlets in 2(3) rows .... 2. *L. squarrosa.*

1. **Lappula redowskii** (Hornem.) Greene.   Western stickseed. Much like no. 2; stem often more spreading hairy and lvs often more softly hairy; nutlets 2–3 mm, the ovate dorsal area surrounded by a single row of prickles. Native of Eurasia and w. N. Amer., intr. as a weed in waste places and along railways e. to Mich., Io., and Mo., and occasionally farther e. Most of our plants, with the marginal prickles distinct to the base, are var. *redowskii.* (*L. occidentalis.*) Var. *cupulata* (A. Gray) M. E. Jones, mainly of sw. U.S., also occasionally reaches the w. part of our range. It has the marginal prickles of the nutlets fused to form a cupulate (sometimes much swollen) border. (*L. texana.*)

2. **Lappula squarrosa** (Retz.) Dumort.   Two-row stickseed. Stems 2–8 dm, tending to be appressed-hairy, mostly freely-branched above; lvs linear to linear-oblong or linear-oblanceolate, 2–5 cm, usually ascending, sessile, rough-hairy; racemes numerous, eventually 5–10 cm; mature pedicels erect or ascending, 1–2 mm; bracts linear or lanceolate, 3–10 mm; cor blue, 2–3(–4) mm wide; nutlets 3–4 mm, the outer face lanceolate, roughened, surrounded by 2(3) rows of prickles, those of the inner row usually the longer; 2*n*=48. Native of Eurasia, established as a weed in waste places in much of the U.S. and s. Can., in our range more common northward. May–Sept. (*L. echinata; L. lappula.*)

9. **CYNOGLOSSUM** L.   Cor salverform or broadly funnelform, the short tube closed by the fornices; stamens included; nutlets bearing conspicuous, stout, glochidiate prickles, attached to the gynobase near their connivent summits, their rounded bases divergent; gynobase prolonged beyond the nutlets into a conspicuous persistent subulate organ; taprooted, ± robust, leafy herbs, the fls pedicellate in axillary and terminal bractless false racemes. 80, cosmop.

1 Perennial, not weedy; infl terminal, with a naked common peduncle; nutlets not margined .... 1. *C. virginianum.*
1 Biennial; weedy; infl of numerous false racemes axillary to lvs or terminating short axillary branches;
    nutlets with a raised margin .... 2. *C. officinale.*

1. Cynoglossum virginianum L. Wild comfrey. Var. *virginianum*: Erect, unbranched perennial 4–8 dm; basal lvs elliptic-oblong, the blade 1–2 dm, decurrent on the long petiole; cauline lvs sessile, progressively smaller, some broadly clasping at base, often some narrowed below and expanded at the very base; racemes (1–)3(4), 1–2 dm long at maturity, terminating a long, erect common peduncle; cal at anthesis 3–4 mm; cor light blue, 8–12(–16) mm wide, its broadly rounded lobes ± overlapping; mature pedicels 5–15 mm, recurved; nutlets 6–8 mm, uniformly bristly over the rounded outer surface, not margined. Upland woods; s. Conn. and s. N.Y. to O., Ill., Mo., and Okla., s. to Fla. and La. May, June. Var. *boreale* (Fern.) Cooperrider: Smaller; some cauline lvs usually distinctly petioled; cal at anthesis 2–2.5 mm; cor 5–8 mm wide, its lobes oblong, not overlapping; nutlets 3.5–5 mm. More northern, from Que. and N.B. to n. Conn., N.Y., Mich., Wis., and Minn., w. to B.C. May, June. (*C. b.*)

2. Cynoglossum officinale L. Hound's tongue. Coarse, single-stemmed biennial 3–12 dm, leafy to the top, villous or villous-hirsute; lowest lvs oblanceolate or narrowly elliptic, 1–3 dm overall, 2–5 cm wide, the others sessile and more oblong or lanceolate, numerous, gradually reduced; infl of numerous false racemes in the upper axils or on short axillary branches, the mature pedicels curved-spreading; cal at anthesis 3–5 mm; cor dull reddish-purple, nearly salverform, the limb 1 cm wide or a little less, the fornices exsert; nutlets 5–7 mm, ovate, descending-spreading, forming a broadly low-pyramidal fr, remaining attached to the style after splitting from the gynobase, the dorsal surface flattened, surrounded by a raised margin; 2n=24. Native of Eurasia, established as a weed in fields, meadows, and open woods throughout our range and westward.

Cynoglossum amabile Stapf & J. R. Drumm., Chinese forget-me-not, occasionally escapes from cult. It is less leafy than *C. officinale,* with a smaller cal (2–3 mm at anthesis) and differs most obviously in its bright blue (seldom pink or white) fls.

10. LITHOSPERMUM L. Stoneseed; gromwell; puccoon. Cor funnelform or salverform, with or without fornices; anthers included or partly exserted; gynobase low-pyramidal or flat or depressed; nutlets smooth to pitted or wrinkled, basally attached, the large scar often surrounded by a sharp rim, sometimes only one nutlet maturing; annual to more often perennial herbs, seldom pungently hairy, with mostly yellow or white or greenish-white fls in modified leafy-bracteate cymes, or solitary in or near the upper axils, often heterostylic; fruiting pedicels mostly erect or ascending. (*Buglossoides.*) 75, widespread, mostly temperate and mountainous regions.

1 Cor bright yellow or orange, the tube 7–30 mm, distinctly surpassing the cal.
  2 Cor-lobes entire; cor-tube 7–14 mm; nutlets smooth.
    3 Cor-lobes at anthesis 3–6 mm; foliage densely and softly canescent .... 1. *L. canescens.*
    3 Cal-lobes at anthesis 8–11 mm; foliage roughly hirsute .... 2. *L. caroliniense.*
  2 Cal-lobes erose-denticulate; cor-tube 13–30 mm; nutlets pitted .... 3. *L. incisum.*
1 Cor white to ochroleucous or greenish-white, 4–8 mm; cal (except in *L. tuberosum*) as long as or longer than the cor-tube.
  4 Perennial; lvs with some evident lateral veins.
    5 Basal lvs forming a rosette at anthesis; lower cauline lvs about as large as the upper .... 4. *L. tuberosum.*
    5 Basal lvs lacking at anthesis; lower cauline lvs reduced.
      6 Lvs on the upper half of the stem crowded, the internodes mostly under 2 cm; blades 6–15(–20) mm wide, gradually acute .... 5. *L. officinale.*
      6 Lvs fewer, the internodes commonly 3–6 cm; blades usually 2–4 cm wide, rarely less, distinctly acuminate .... 6. *L. latifolium.*
  4 Annual; lvs without evident lateral veins .... 7. *L. arvense.*

1. Lithospermum canescens (Michx.) Lehm. Hoary puccoon. Stems from a stout taproot, often several (up to ca 5), 1–4 dm at anthesis, usually simple; lvs lanceolate to narrowly oblong, 2–6 cm (or the lowest reduced), usually less than 25 below the infl, softly and densely canescent-strigose, obtuse; infl of 1–3 densely fld, terminal or subterminal, leafy-bracted cymes; fls heterostylic; cor-lobes linear, flat, 3–6 mm at anthesis, 6–8 mm at maturity, densely villous; cor orange to light golden-yellow, the tube 7–10 mm, the limb 1–1.5 cm wide; nutlets ochroleucous, smooth and shining; 2n=14. Prairies and dry, open woods; sw. Ont. to Sask., s. to Tenn. and Tex., and in the mts. from Pa. to Ga. and Ala. Apr., May. The Appalachian plants tend to have more spreading pubescence than those from farther west.

2. Lithospermum caroliniense (Walter) MacMillan. Plains puccoon. Stems erect, clustered (often a dozen or more) on a stout woody root, 3–6 dm at anthesis, very leafy, simple or branched above, villous or hirsute; lvs linear to lanceolate, 3–6 cm, rough-hirsute, the hairs papillate at base in var. *croceum*; cymes leafy-bracted, at first dense, becoming elongate and racemiform; fls heterostylic; cal-lobes linear, 8–11 mm at anthesis, 10–15 mm at maturity; cor bright orange-yellow, the tube 7–14 mm, hairy at base within, the limb 1.5–2.5 cm wide; nutlets ivory-white, smooth and shining. Upland woods, shores, and prairies, especially in sandy soil; sw. Ont. and n. O. to Minn. and Wyo., s. to Ill., Nebr., and Okla. (var. *croceum*), and from se. Va. to Fla. and Tex. (var. *caroliniense*). May–July. (*L. gmelini; L. croceum.*) Most of our plants are var. *croceum* (Fern.) Cronq., as described above, with keeled cal-lobes and usually more than 30 lvs below the infl. Var. *caroliniense,* on the se. coastal plain, has flat cal-lobes and usually less than 25 lvs below the infl.

3. **Lithospermum incisum** Lehm. Narrow-lvd puccoon; plains-stoneseed. Strigose perennial from a woody taproot, 0.5–4 dm; lvs wholly cauline, the lowest often reduced, the others linear-oblong to narrowly lanceolate or linear, rather numerous, 2–6 cm × 2–6 mm; well developed fls crowded in the uppermost axils, short-pedicellate; cor bright yellow, salverform, the tube 1.5–3.5(–4) cm, the limb 1–1.5(–2) cm wide, with erose or almost fimbriate lobes, these fls long-styled and mostly sterile; cleistogamous, highly fertile, short-styled fls commonly developed later farther down the stem, the plant becoming slenderly much-branched; nutlets ivory-white, shining, sparsely pitted; 2n=24, 36. Dry plains and barrens; s. Ont. to Man. and Wis., w. to B.C., Utah, s. Nev., and n. Mex. Apr., May. (*L. angustifolium; L. linearifolium.*)

4. **Lithospermum tuberosum** Rugel. Southern stoneseed. Stems 3–5 dm from tuberously thickened roots; basal lvs obovate-oblong, 5–10 cm, forming a rosette; cauline lvs much smaller, lanceolate to oblong or oblanceolate, obtuse or acute; fls ochroleucous, 6–8 mm, the cor-tube surpassing the cal; nutlets obovoid, 2–3 mm, white, sparsely pitted, otherwise smooth and shining. Woods; Ky. to Fla. and La. May, June.

5. **Lithospermum officinale** L. Gromwell. Erect, usually much branched perennial to 1 m, the principal internodes usually less than 2 cm; lvs subsessile, lanceolate to oblong or oblanceolate, 6–15(–20) mm wide, gradually acute, with 2 or 3 conspicuous lateral veins on each side; fls solitary in the crowded upper axils, 3–15 mm apart at maturity, white or nearly so, 4–5 mm; cal-lobes nearly as long as the cor; nutlets ovoid, 3–3.5 mm, white to pale brown, shining, smooth or sparsely pitted; 2n=28. Native of Eurasia, intr. as a weed of waste places from Que. to Minn., s. to N.J. and Ill. May–Aug.

6. **Lithospermum latifolium** Michx. American gromwell. Erect perennial, 4–8 dm, simple or branched above, the main internodes commonly 3–6 cm; lvs subsessile, lanceolate to lance-ovate, usually 2–4 cm wide, acuminate, with 2 or 3 prominent lateral veins on each side; fls solitary in the upper axils, becoming distant, ochroleucous, 5–7 mm, nearly as wide; cal-lobes nearly as long as the cor; nutlets ovoid, white, shining, 3.5–5 mm, smooth or sparsely pitted; 2n=28. Dry woods and thickets; N.Y. to Minn., s. to W.Va., Tenn., and Mo. May, June.

7. **Lithospermum arvense** L. Corn-gromwell. Strigose annual 1–8 dm, often branched at the base; lvs mostly linear to narrowly oblanceolate, 3–8(–15) mm wide, acute or obtuse, 1-nerved; fls solitary in the crowded upper axils, becoming remote, whitish or bluish-white, 5–8 mm, 2–4 mm wide; cal-lobes nearly as long as the cor; nutlets ovoid, pale dull brown, 3 mm, deeply wrinkled and pitted; 2n=28. Native of Eurasia, intr. as a weed of waste places over most of the U.S. Apr.–July. (*Buglossoides a.*)

11. **ONOSMODIUM** Michx. False gromwell. Fls precociously sexual, the style exserted and the anthers dehiscent well before the cor is mature, the style remaining exsert; cor hairy outside, nearly tubular, with rather narrow, erect, pointed lobes and thickened, basally inflexed sinuses; fornices wanting; anthers barely or only partly included; nutlets turgid, smooth or merely pitted, broadly attached at the base to the flattish or depressed gynobase, commonly only 1 or 2 maturing; coarse, often rough-hairy perennials with largely or wholly cauline lvs, the fls white to yellow, commonly with somewhat greenish lobes, borne in leafy-bracteate helicoid cymes. 5, N. Amer.

1 Cor-lobes yellow, acuminate, 2–3 times as long as wide .... 1. *O. virginianum.*
1 Cor-lobes dull white or greenish, acute, 1½–2 times as long as wide .... 2. *O. molle.*

1. **Onosmodium virginianum** (L.) A. DC. Eastern f. g. Stem 3–6 dm, strigose; lvs often oblanceolate and distinctly narrowed to the base; bracteal lvs and cal hispid-strigose with ascending hairs; cor yellow, 7–10 mm, its lobes 3 mm, narrowly triangular, acuminate; nutlets dull white, ± pitted, 2.5 mm. Dry woods and barrens; se. Mass. to e. N.Y., s. to Va. and Fla., thence w. to La. June, July.

2. **Onosmodium molle** Michx. Western f. g. Several-stemmed perennial from a woody root, 3–12 dm, hairy throughout; lower lvs reduced and deciduous, the others rather numerous and uniform, sessile, lanceolate or rather narrowly ovate; cor 8–16 mm, the dull white or greenish white lobes acute, 1½–2 times as long as wide. Open, moderately dry places; N.Y. and s. Ont. to Va., Tenn., and La., w. to Alta. and N.M. June, July. Three vars. in our range:

a Nutlets constricted just above the base to form a collar, mostly dull and smooth or nearly so, 2.5–3 mm; plants robust, often 1 m or more, more coarsely and conspicuously hairy than the other vars., the lvs to 15 × 4 cm; Appalachian region to the w. edge of the deciduous forest region . ... var. *hispidissimum* (Mackenzie) Cronq.
a Nutlets without a collar; plants smaller, commonly 3–7 dm, the lvs to 8 × 2 cm.
b Nutlets smooth and shining, 3.5–5 mm; pubescence relatively coarse and loose; plains and prairie region of c. U.S. and adj. Can., and reputedly e. to w. Ky. .... var. *occidentale* (Mackenzie) I. M. Johnston.
b Nutlets dull and pitted, 2.5–3 mm; barrens of Ky. and Tenn., extending w. to se. Mo. .... var. *molle.*

12. **ECHIUM** L. Cor funnelform, irregular, the upper side evidently the longer, the lobes unequal; fornices wanting, the throat open; filaments slender, ± unequal, some or

all strongly exsert; gynobase flat or nearly so; nutlets ± roughened, attached at base, the large scar sometimes surrounded by a low rim; style exsert; annual to perennial herbs or subshrubs with blue to purple or red (white) fls in a series of helicoid, bracteate cymes. 50, Old World.

1. **Echium vulgare** L. Blue-weed. Erect, taprooted biennial 3–8 dm, rough-hairy especially above; basal lvs ± oblanceolate, 6–25 cm (petiole included) × 0.5–3 cm, the cauline progressively smaller, becoming sessile; helicoid cymes numerous and often short, ± aggregated into a pyramidal or elongate, often virgate infl; cor bright blue (pink or white), 12–20 mm; 4 filaments long-exsert, the fifth barely or scarcely so; style hairy; $2n=16, 32$. Native of s. Europe, now a common weed in waste places, roadsides, and meadows in our range. June–Oct.

13. ANCHUSA L. Alkanet, bugloss. Cor funnelform or salverform, the throat often poorly defined, not much if at all longer than the ± spreading, equal or unequal, apically rounded lobes; nutlets with a stipe-like basal attachment that fits into a pit in the otherwise flattish receptacle, the basal margin of the nutlet forming a prominent thickened rim; leafy herbs, often pungently hairy, the blue fls borne in terminal, helicoid, bracteate false racemes with persistently erect or ascending pedicels. (*Lycopsis.*) 40, Old·World.

1 Cor regular; perennial.
  2 Nutlets erect, 5–9 mm high and 3–5 mm thick .... 1. *A. azurea.*
  2 Nutlets oblique, about 2 mm high and 3–4 mm long, the tip directed inward .... 2. *A. officinalis.*
1 Cor irregular, the tube bent, the lobes somewhat unequal; annual .... 3. *A. arvensis.*

1. **Anchusa azurea** Miller. Showy bugloss. Taprooted perennial, mostly single-stemmed, spreading-hispid throughout; basal lvs oblanceolate, petiolate, the others more lanceolate or lance-ovate to oblong, sessile and often clasping, sometimes to 30 × 8 cm; bracts narrow, resembling the cal-lobes; cal 8–10 mm at anthesis, to 15 mm in fr, cleft to well below the middle, the lobes slender; cor showy, the limb 12–20 mm wide; stamens inserted at the top of the tube; nutlets erect, 5–9 mm high and scarcely over half as thick, coarsely reticulate-ridged with mostly vertically elongate areolae; $2n=32$. Mediterranean sp. occasionally escaped from cult. in our region. May–July.

2. **Anchusa officinalis** L. Common bugloss. Taprooted, spreading-hirsute perennial 3–8 dm, often several-stemmed; lower lvs oblanceolate, petiolate, mostly 6–20 cm (petiole included) × 1–1.5 cm, the others gradually reduced, becoming sessile and more lanceolate; bracts lanceolate or lance-triangular; cal 5–7 mm at anthesis, scarcely longer in fr, the lanceolate or narrowly triangular lobes about equaling or a little longer than the tube; cor-tube 6–11 mm, the limb 6–11 mm wide; stamens inserted in the upper part of the tube; nutlets ± rugose and tuberculate, oblique, 2 mm high and 3–4 mm long, the tip directed inward; $2n=16$. Native of the Mediterranean region, now found as an occasional weed in much of our range.

3. **Anchusa arvensis** (L.) M. Bieb. Small bugloss. Erect or ascending, usually branched, hispid annual 2–6 dm; lvs narrowly oblong to oblanceolate, 4–8 cm, repand-dentate or entire; cor-tube 4–7 mm, distinctly bent, hardly surpassing the cal, the limb oblique, 4–6 mm, with unequal lobes; stamens inserted at or below the middle of the tube; nutlets obliquely ovoid, 1.5–2 × 3–4 mm; $2n=48$. Native of Europe, intr. as a weed in waste places in our range; $2n=48$. (*Lycopsis a.*)

14. SYMPHYTUM L. Comfrey. Cor tubular-campanulate, the well defined tube much longer than the short, erect or apically spreading lobes; fornices narrow and elongate, erect; filaments inserted at the level of the fornices; anthers included; style elongate, shortly exserted; nutlets incurved, ventrally keeled, with a stipe-like basal attachment that fits into the otherwise flattish receptacle, the basal margin of the nutlet forming a prominent, toothed rim; broad-lvd perennials with nodding, white or ochroleucous to pink or blue fls in several or many small, sympodial, naked modified cymes. 25, Europe and Medit. reg.

1 Lvs conspicuously decurrent on the stem; nutlets smooth .... 1. *S. officinale.*
1 Lvs not decurrent, or only very weakly so; nutlets evidently roughened .... 2. *S. asperum.*

1. **Symphytum officinale** L. Common c. Taprooted perennial 3–12 dm; stem and infl hispid-hirsute with spreading or recurved, subterete hairs; lvs large, the basal petiolate, with ovate or lance-ovate blade 15–30 × 7–12 cm, the cauline gradually reduced but still ample, the upper commonly sessile; stem evidently winged by the conspicuous decurrent lf-bases; cal 5–7 mm, cleft to below the middle; cor ochroleucous or dull blue, 12–18 mm; filaments as wide as the anthers; connective projecting beyond the thecae; nutlets 5–6 mm, black,

very smooth, shining; $2n=24-54$. Native of Eurasia, escaped or adventive in waste places here and there in our range. June–Aug.

2. **Symphytum asperum** Lepechin. Rough c. Much like no. 1; hairs of the stem and infl stouter, flattened, often recurved; lvs not decurrent, or only shortly (to 1 cm) and narrowly so; cor pink, turning blue; filaments narrower than the anthers; connective not projecting; nutlets roughened. Native of the Caucasus region, occasional in waste places across the n. part of our range, Que. to B.C. June–Aug. (*S. asperrimum.*)

S. tuberosum L., with a rhizome that is tuberous-thickened at intervals and with slender cauline hairs as in no.; 1, but nondecurrent lvs as in no. 2, has been reported as adventive in N. Engl.

15. BORAGO L. Borage. Cor rotate, with elongate, acute lobes and prominent fornices; filaments prolonged beyond the base of the anther into a prominent dorsal appendage; anthers elongate, conspicuous, connivent around the style; nutlets with a swollen, stipe-like basal attachment that fits into a pit in the otherwise flattish gynobase, the attachment surrounded by a prominent thickened rim (the basal margin of the nutlet) that fits closely to the gynobase; broad-leaved, coarsely hairy herbs with blue fls in loose, terminal modified sympodial cymes, these leafy-bracteate below, the long pedicels recurved in fr. 3, Old World.

1. **Borago officinalis** L. Taprooted annual 2–6 dm, hispid-setose and hispidulous; lower lvs petiolate, with broadly elliptic or ovate blade 3–11 × 2–6 cm, the others progressively reduced and the upper often sessile and clasping; pedicels 1–4 cm; sep densely bristly, 1–1.5 cm in fr; cor 2 cm wide; anthers dark, 5–9 mm, the linear appendages 3 mm; nutlets subcylindric, 4–5 mm, rough and wrinkled; $2n=16$. Native of Europe, sometimes cult., and established as a casual weed here and there in our range. May–Sept.

FAMILY **VERBENACEAE**, the Vervain Family

Fls sympetalous, hypogynous, mostly perfect; cal (2)4–5(–7)-lobed, sometimes irregular; cor salverform or funnelform, mostly 4–5-lobed, ± irregular in most genera, sometimes bilabiate; stamens (2)4(5), inserted on the cor-tube; ovary typically bicarpellate and with 4 uniovulate chambers, shallowly 4-lobed and separating in fr into 4 nutlets, but varying in diverse genera, sometimes with only 2 uniovulate chambers and ripening into 2 nutlets, or ripening into a drupe with 2 or 4 stones, and in *Phryma* unilocular and uniovulate, ripening into an achene; style terminal, or often arising between the 4 short distal lobes of the ovary, the stigma often unequally bilobed; embryo with 2 cotyledons; endosperm mostly wanting; herbs or woody plants, usually not aromatic, with mostly 4-angled stems and opposite, exstipulate lvs. (Phrymataceae.) 100/2600, mostly of warm regions.

1 Lvs simple.
  2 Infls of racemose type (i.e., indeterminate); ours herbs.
    3 Fr of 2 or 4 separating nutlets; ovary with 2 or 4 uniovulate chambers.
      4 Fr of 4 nutlets; cor 5-lobed, usually only slightly irregular .... 1. *Verbena.*
      4 Fr of 2 nutlets; cor 4-lobed, usually plainly 2-lipped .... 2. *Phyla.*
    3 Fr an achene; ovary unilocular and uniovulate .... 3. *Phryma.*
  2 Infls of cymose type (i.e., determinate); shrubs .... 4. *Callicarpa.*
1 Lvs palmately compound (rarely 1-foliolate); shrubs or trees .... 5. *Vitex.*

1. VERBENA L. Vervain. Fls in spikes (or heads) ending the stem and branches; cal tubular, 5-angled and unequally 5-toothed, persistent, scarcely changing in fr; cor salverform or funnelform, the tube straight or curved, the limb flat, weakly 2-lipped, with 5 obtuse or emarginate lobes; stamens 4, paired, included; anther-connective often with a glandular appendage; style terminal, shortly 2-lobed, only one lobe stigmatic; ovary with 4 uniovulate chambers, often rather shallowly 4-lobed, separating at maturity into 4 nutlets, these often ridged or reticulate; ours herbs. (*Glandularia.*) 250, mostly New World.

1 Sterile style-lobe not protruding beyond the stigmatic surface; anthers unappendaged; cor mostly small, often not showy; cal rarely over twice as long as the fr, usually not contorted beyond it.
  2 Spikes or heads crowded and short, at least during anthesis, never elongate and open, generally disposed in compound cymes .... 1. *V. brasiliensis.*

2 Spikes slender, open to compact at anthesis, greatly elongating in fr, solitary or in simple cymes or
    panicles.
  3 Spikes panicled at the ends of the stem and branches, mostly subtended by inconspicuous bracts;
     floral bracts inconspicuous.
    4 Spikes very slender and elongate, usually with remote (at least not contiguous) frs.
      5 Lvs once or twice pinnatifid, 3–5-cleft, or deeply incised.
        6 Schizocarp a little longer than broad; introduced .... 2. *V. officinalis.*
        6 Schizocarp about twice as long as broad; native .... 3. *V. riparia.*
      5 Lvs merely serrate, coarsely dentate, or shallowly incised.
        7 Lvs very scabrous above; fruiting cal spreading; cal-lobes connivent .... 4. *V. scabra.*
        7 Lvs not very scabrous above; fruiting calyx ascending; cal-lobes not connivent .... 5. *V. urticifolia.*
    4 Spikes thicker, densely fld, usually with contiguous and imbricate frs .... 6. *V. hastata.*
  3 Spikes solitary or in 3's ending the stems and branches, or, if panicled, subtended by leafy bracts;
     floral bracts often conspicuous.
    8 Lvs entire, serrate-dentate, or shallowly incised.
      9 Plants coarse, long-hairy; lvs mostly ovate-orbicular; spikes stout .... 7. *V. stricta.*
      9 Plants slender, short-hairy; lvs linear to elliptic-ovate; spikes slender .... 8. *V. simplex.*
    8 Lvs deeply incised, pinnatifid, or 3-cleft .... 9. *V. bracteata.*
1 Sterile style-lobe protruding well beyond the stigmatic surface; connective of the upper anthers mostly
  appendaged; cor mostly large and showy; cal usually more than twice as long as the fr and contorted
  beyond it. (*Glandularia.*)
  10 Cor 2–3 cm, the tube twice as long as the cal, the limb 10–15 mm wide .... 10. *V. canadensis.*
  10 Cor 1–1.5 cm, the tube up to 1.5 times as long as the cal, the limb 7–10 mm wide .... 11. *V. bipinnatifida.*

1. Verbena brasiliensis Vell. Conc. Brazilian v. Stout annual to 2.5 m, the stem 4-angled, somewhat scabrous-hairy above; lvs elliptic to linear-elliptic or lanceolate, tapering to a subsessile or petiolar base, sharply serrate or incised, at least above the middle, strigillose above, sparsely hairy beneath; spikes compact, mostly short, usually sessile in open cymes, not crowded; bracts scarcely equaling the cal, lance-subulate, ciliate; cal 3–4 mm; cor purple or lilac, its tube a little longer than the cal, hairy outside, the inconspicuous limb 2.5 mm wide; nutlets 1.5–2 mm; 2*n*=28. Native of S. Amer., intr. in dry sandy soil and waste places from Va. to Fla. and La. Mostly Apr.–Sept.

V. bonariensis L., a S. Amer. sp. with the lvs semi-amplexicaul and subcordate at base, is intr. in se. U.S. and rarely found in our range.

2. Verbena officinalis L. European v. Slender annual 2–6 dm, ascending or erect, the stem glabrous or nearly so; lvs 2–7 cm, strigillose, the lower ones ± ovate, narrowed to a petiole, 1–2-pinnatifid or 3–5-cleft with the parts again incised, the upper ones smaller and less dissected; spikes elongate, very slender, paniculately disposed or in 3's or even solitary; bracts usually half as long as the cal; cal 2–2.5 mm, glandular-hairy, subtruncate, its teeth minute; cor blue, purple, or lilac, the tube to 3 mm, the limb 4 mm wide, with ± rounded lobes; nutlets 2 mm; 2*n*=14, 28. Native of Europe, naturalized along roadsides and in fields and waste places in se. U.S., n. to Mass. and possibly Me. June–Sept.

3. Verbena riparia Raf. Bank-v. Erect annual 6–15 dm, widely branched, sparsely hairy or glabrate; lvs oblong to ovate, 4–14 cm, petiolate, 1–2-pinnatifid or nearly tripartite toward the base, sparsely strigillose, veiny beneath; spikes slender, elongate, paniculately disposed; bracts lance-ovate, equaling the cal at anthesis, acuminate; cal to 3 mm, minutely glandular hairy, subtruncate, the teeth minute; cor light blue, the tube slightly longer than the cal, puberulent, the limb 3.5 mm wide, with rounded lobes (one emarginate); nutlets 2–2.5 mm. Riverbanks; Va. and N.C. June, July. (*V. urticifolia* var. *r.*)

4. Verbena scabra M. Vahl. Rough v. Erect, single-stemmed annual or perennial to 1 m or more; lvs ovate to elongate-ovate, 3–13 × 2.5–5 cm, petiolate, acute or obtusish, toothed, very harshly scabrous and commonly strigillose above, less scabrous and paler beneath but also hispidulous along the veins; spikes paniculately disposed, slender, closely fld and frtd; bracts ovate-acuminate, half as long as the cal, hispidulous; cor white to pinkish or blue, the tube scarcely longer than the cal, the limb 2–3 mm wide, the lobes obtuse; stigmatic surface midway between the 2 subequal sterile lobes; mature cal 2.5–3 mm, hispidulous, divergent, the somewhat unequal lobes connivent; nutlets 1–1.5 mm, faintly striate, reticulate above. Various usually moist habitats; Va. and W.Va. to Fla. and trop. Amer. Mar.–Dec.

5. Verbena urticifolia L. White v. Erect, single-stemmed annual or perennial, 4–15 dm, often branching from near the base; lvs broadly lanceolate to oblong-ovate, petiolate, 5–12 cm, coarsely and somewhat doubly crenate-serrate; spikes paniculately disposed, slender; bracts ovate, acuminate, ciliate; cal-teeth short, sub-equal; cor white, the tube scarcely exserted, the limb 2(–4) mm wide, the lobes obtuse; fr exposed at the top; nutlets 1–2 mm; 2*n*=14. Thickets, moist fields, meadows, and waste places; N.B. (?) and Que. to N.D., s. to Fla. and Tex. June–Oct. Var. *urticifolia*, with the range of the sp., has the lvs hirtellous or glabrous on both sides, the hairs whitish, 1–1.3 mm, the cal strigose, 2–2.3 mm at maturity, and the nutlets 2 mm, corrugated on the back. Var. *leiocarpa* Perry & Fern., with more restricted range, from Conn. and Pa. to N.D., S.C., and Okla., has the lvs densely velutinous with hairs to 0.3 mm, the mature cal puberulent, to 2 mm, the nutlets 1.5 mm, smooth. A hybrid with *V. stricta* is *V. × illicita* Moldenke; one with *V. hastata* is *V. × engelmannii* Moldenke.

6. Verbena hastata L. Common v. Perennial; stems 4–15 dm, branched above, rough-hairy, with short, spreading or antrorse hairs; lvs lanceolate to lance-oblong or lance-ovate, 4–18 cm, gradually acuminate, petiolate, coarsely serrate or incised, often hastately 3-lobed at base, glabrous or strigillose on both sides; spikes strict, usually many in a terminal panicle, short and compact, thicker than in no. 5; bracts lance-subulate, commonly a little shorter than the cal; cal 2.5–3 mm, hairy, with ± connivent, shortly subulate-tipped lobes; cor usually blue, the tube somewhat longer than the cal, hairy, the limb 2.5–4.5 mm wide; nutlets linear, 1.5–2 mm, faintly striate or smooth; $2n=14$. Moist fields, meadows, prairies, and swamps; N.S. to B.C., s. to Fla. and Ariz. June–Oct. A hybrid with *V. simplex* is *V.* × *blanchardii* Moldenke; one with *V. stricta* is *V.* × *rydbergii* Moldenke.

7. **Verbena stricta** Vent. Hoary v. Perennial; stems 2–12 dm, subterete, rather densely pale-hairy; lvs ovate, elliptic, or suborbicular, 3–10 × 1.5–5 cm, sessile or nearly so, thick, sharply serrate to incised, hirsute and rugose above, densely hirsute-villous and veiny beneath; spikes solitary or several, long, thick, remaining compact; bracts lance-subulate, about equaling the cal, hirsute, ciliate; cal 4–5 mm, densely hirsute, its lobes acuminate; cor mostly deep blue or purple, the tube a little exserted, hairy, the limb 7–9 mm wide; nutlets ellipsoid, 2–3 mm, raised-reticulate above, strongly striate below; $2n=14$. Prairies, barrens, fields, and road-sides; Ont. and O. to N.D. and Wyo., s. to Tex. and N.M., and intr. into e. U.S. from Mass. to Del. and W.Va. June–Sept.

8. **Verbena simplex** Lehm. Narrow-lvd v. Perennial; stems chiefly erect, 1–7 dm, the few branches ascending, usually sparsely strigillose; lvs linear to narrowly oblong, lanceolate, or spatulate, 3–10 cm, to 1.5 cm wide, tapering to a short petiole or a subsessile base, obtuse or subacute, toothed, rugose above and veiny beneath, glabrate or sparsely strigillose; spikes slender, solitary on the stem and branches, usually rather dense; bracts lance-subulate, commonly shorter than the cal; mature cal 2(–5) mm, sparsely hairy, its lobes acuminate; cor deep lavender or purple, the tube scarcely longer than the cal, with scattered hairs at the mouth, the limb 4–6 mm wide; nutlets linear, 2–3 mm, raised-reticulate above, striate toward the base; $2n=14$. Dry soil of woods, fields, rocky places and roadsides; Ont., Vt. and Mass. to Minn., s. to Fla. and Tex. June–Aug. (*V. angustifolia* Michx., not Miller.) A hybrid with *V. stricta* is *V.* × *moechina* Moldenke.

9. **Verbena bracteata** Lagasca & Rodriguez. Prostrate v. Hirsute annual or perennial, usually with several diffusely branched, prostrate to ascending stems 1–5 dm; lvs 1–6.5 cm, pinnately incised or usually 3-lobed, narrowed to the short margined petiole, the lateral lobes smaller than the large, cuneate-obovate, incisely toothed central one; spikes terminal, sessile, commonly 10–15 mm thick, hispid-hirsute; bracts 8–15 mm, much longer than the cal, the lower often lf-like, the upper lance-linear and entire; cal 3–4 mm, its short lobes connivent over the fr; cor bluish to purple, the tube slightly exserted, the limb 2–3 mm wide; nutlets linear, 2–2.5 mm; $2n=14, 28$. Prairies, fields, roadsides, and waste places; Me. to B.C., s. to Fla. and Mex., probably not native in the ne. part of the range. Apr.–Oct. (*V. bracteosa.*) A hybrid with *V. stricta* is *V.* × *deamii* Moldenke; one with *V. urticifolia* is *V.* × *perriana* Moldenke.

V. lasiostachys Link, a western sp., often more erect, with shorter bracts not surpassing the cal, and with more conspicuous fls, the limb 3–5 mm wide, is intr. in Kings Co., N.Y.

10. **Verbena canadensis** (L.) Britton. Rose-v. Hirsute to glabrate perennial, decumbent or ascending, rooting at the lower nodes, 3–6 dm; lvs strigose or glabrate, variable, ovate or narrower, 3–9 × 1.5–4 cm, incised or incised-pinnatifid to trifid; spikes pedunculate, depressed-capitate at anthesis, elongating in fr; bracts up to as long as the cal, linear-attenuate, hirsute, usually ciliate; cal 10–13 mm, glandular-hirsute, with very slender, unequal lobes; cor blue to purple or white, the tube twice as long as the cal, the limb 11–15 mm wide, the lobes emarginate; anther appendages large (occasionally absent); fr constricted along the lines of cleavage, the nutlets 3–3.5 mm, reticulate-scrobiculate; $2n=30$. Various habitats, often in disturbed soil; Ill. to Tenn., s. Pa., Va., and La., w. to Colo. and Tex., and intr. in Mich. and Minn. Feb.–Oct. (*Glandularia c.; G. lambertii; G. drummondii; Verbena drummondii.*)

V. hybrida Voss, with more densely hairy lvs and often larger fls, occasionally escapes from cult. in our range.

11. **Verbena bipinnatifida** Nutt. Dakota v. Diffusely branched perennial 1–4 dm, the stems hispid-hirsute, the lvs appressed-hirsute, the cal hispid-hirsute along the nerves; lvs petiolate, 2–6 cm, pinnatifid, or ternate with ± bipinnatifid segments; spikes pedunculate, short at anthesis, later elongating; bracts linear-subulate, mostly longer than the cal; cal 8.5–10 mm, the slender lobes unequal; cor pink to purple, the tube half again as long as the cal, the limb 7–10 mm wide, the lobes emarginate; nutlets 3 mm; $2n=30$. Dry plains and other open places; S.D. to Ariz. and La., barely entering our range in nw. Mo. Feb.–Dec.

V. tenuisecta Briquet, a S. Amer. sp. with the bracts shorter than the strigose cal, is intr. in se. U.S., and n. as far as St. Louis, Mo., and may be expected in our range.

2. **PHYLA** Lour. Fogfr. Fls in dense, elongating, pedunculate axillary spikes; bracts small, cuneate-obovate to flabelliform; cal small, membranous, often compressed or even winged, with 2 or 4 teeth or segments; cor-tube very slender, straight or incurved, slightly exserted from the cal, the limb oblique, spreading, 4-parted and somewhat 2-lipped; stamens 4, paired, included or slightly exserted; anthers unappendaged; ovary with 2 uniovulate chambers; stigma thickened, oblique or recurved; fr included in the cal or even adnate to it, dividing into 2 nutlets, sometimes with separating thin exocarp; prostrate or creeping perennials with ± ascending or trailing branches, ± canescent with malpighian hairs. 10, originally N. Amer.

P. cuneifolia (Torr.) Greene, a western sp., is intr. at St. Louis and may be expected in our region. It differs from both of the following spp. in its narrow, linear-oblanceolate or narrowly cuneate lvs 4–10 times as long as wide, less than 1 cm wide, with 1–3 teeth on each side toward the summit.

1 Lvs broadest near or below the middle ± distinctly acute .... 1. *P. lanceolata.*
1 Lvs broadest above the middle, mostly obtuse or rounded at the tip .... 2. *P. nodiflora.*

    1. Phyla lanceolata (Michx.) Greene.   Trailing perennial herb with ascending or erect stems to 6 dm from a prostrate or distally assurgent, rooting aerial stem or superficial rhizome; lvs short-petiolate, with lance-elliptic or ovate, ± distinctly acute blade mostly 2–6 × 1–4.5 cm, 1.5–5 times as long as wide, broadest below or near the middle, saliently toothed to below the middle, with 5–10 teeth per side; spikes pedunculate from the axils, at first capitate, less than 1 cm, and fully as thick as long, but elongating to 1–2 cm during anthesis; bracts broad, 2.5–3 mm, equaling or usually surpassing the cal; cor white or pink, often with a yellow eye, the tube ca 2 mm, the lower lip 1.5–2 mm, the upper much shorter; $2n=32, 36$. Moist, low ground and muddy flats, both coastal and inland; Ont., to Minn. and S.D., s. to Fla. and Mex. May–Oct. (*Lippia l.*)

    2. Phyla nodiflora (L.) Greene.   Much like no. 1, but the lvs spatulate to obovate or elliptic-obovate, generally broadest a little above the middle, mostly rounded or obtuse at the tip and narrowed to a cuneiform, petiolar or subpetiolar base, often toothed only above the middle; $2n=36$. Moist soil in many habitats, especially in disturbed sites; pantropical, n. in Amer. to Va. and Calif., rarely further. May–Oct. (*Lippia n.*)

    3. PHRYMA L.   Lopseed. Fls in elongate, long-peduncled, interrupted spike-like racemes terminating the stem and also arising from a few upper axils; cal 2-lipped, the upper 3 lobes subulate, about equaling the tube, the lower 2 very short, broadly triangular; cor-tube scarcely widened upward, the upper lip straight, concave, emarginate, the lower much longer, spreading, 3-lobed; stamens 4, paired, included; ovary obliquely cylindric, with a single locule and a single erect, orthotropous ovule; style elongate, obliquely terminal; stigma 2-lobed; fr an achene, contained in the persistent cal, this closely reflexed against the axis, its upper lobes becoming hooked at the tip and indurate; simple-lvd perennial herbs. Monotypic.

    1. Phryma leptostachya L.   Erect, 5–10 dm, simple or with a few divergent branches; lower petioles to 5 cm, the upper shorter or the uppermost lvs sessile; lvs ovate, 6–15 cm; fls opposite and horizontal, pale purple to white, 6–8 mm, each subtended by 3 small subulate bracts; cal to 1 cm in fr; $2n=28$. Moist woods; Que. to Man., s. to Fla. and Tex.; e. Asia. June–Aug.

    4. CALLICARPA L.   Beautyberry. Fls 4(5)-merous; cal toothed, cleft, or entire; cor funnelform or salverform, the straight tube ampliate above, the narrow, spreading lobes all alike; stamens always present, attached to the base of the cor-tube, equal, usually exserted; pistil absent from staminate fls; ovary with 4 uniovulate chambers; stigma depressed-capitate or peltate; fr globose, drupaceous, with a fleshy exocarp and a hard endocarp, the latter separating into usually 4 nutlets; trees and shrubs with axillary or supra-axillary cymes of regular, usually polygamous fls. 140, warm reg.

    1. Callicarpa americana L.   American b. Shrub to 3 m; twigs densely stellate-scurfy or tomentose; lvs opposite or rarely ternate, the petiole 0.5–3.5 cm, the blade ovate to elliptic, 8–23 × 3.5–13 cm, coarsely toothed, stellate-scurfy at first, later glabrate above; cymes compact, many-fld, usually shorter than the petiole; cal obconic, 1.6–1.8 mm, very shortly 4-toothed; cor bluish to reddish (white), the tube 2.6-3 mm, the lobes 1.3–1.5 mm, blunt; frs rose-pink to purple or blue, 3–6 mm, densely clustered; $2n=36$. Moist, often wooded places; sw. Md. to N.C. and Ark., s. to Mex. and the W.I. June–Dec.

    C. dichotoma (Lour.) K. Koch, the Chinese b., with smaller lvs 2–6.5 × 1–2.3 cm, glabrous or subglabrous beneath, rarely escapes from cult. (*C. purpurea.*)

    Clerodendrum japonicum (Thunb.) Sweet, the Glory-bower, a shrub with large, cordate-ovate lvs and large terminal infl of bright-red, slightly irregular fls, the cor 1.5–2 cm, the stamens and style long-exsert, rarely escapes from cult. in our range.

    5. VITEX L.   Fls perfect; cal 5-toothed; cor salverform, irregular, white to blue or yellowish, with short, often slightly incurved tube and oblique, weakly bilabiate limb, the upper lip bifid; stamens 4, often exsert; stigma shortly bifid; ovary with 4 uniovulate chambers; fr drupaceous; trees or shrubs with palmately compound lvs and small cymose infls that may be aggregated into a terminal mixed panicle. 250, nearly cosmop.

1. **Vitex agnus-castus** L. Chaste tree. Aromatic shrub or low tree to 5 m, with densely short-hairy twigs; lvs (3)5–9-foliolate; lfls narrowly elliptic, long-pointed at both ends, the central one 4.5–11.5 × 1–2 cm, the others smaller, short-hairy to tomentose at least beneath; cymules small, sessile or subsessile, aggregated into a long terminal mixed panicle; cal 2–2.5 mm; cor lavender or lilac, the tube 6–7 mm, white-puberulent above the cal, the limb 5–13 mm wide; fr subglobose, 3 mm. Native to Eurasia, often escaped from cult. in s. U.S., n. to Md. Apr.–Oct.

V. negundo L., an oriental sp. with 3–5 deeply and irregularly cleft lfls, rarely escapes from cult. in our range.

FAMILY LAMIACEAE, the Mint Family

Fls sympetalous, hypogynous, perfect (seldom some unisexual), usually irregular, the cor typically bilabiate and 5-lobed, sometimes 4-lobed by fusion; cal gamosepalous, often oblique at the throat, regular or irregular; stamens 2 or 4, inserted on the cor-tube, the anthers often connivent, the pollen-sacs sometimes partly confluent or one sac suppressed; ovary of 4 essentially distinct, uniovulate segments united only by their gynobasic style, or the ovary less commonly merely 4-lobed for a third or more of its length, so that the style arises from between its lobes but is not gynobasic; style shortly bifid at the tip, one lobe often reduced; fr of (1–)4 1-seeded nutlets; embryo with 2 cotyledons; endosperm none or very scanty; aromatic plants with quadrangular stems and opposite, simple lvs, ours all herbs or low shrubs. (Labiatae, nom. altern.) 200/3200, cosmop.

SYNOPSIS OF THE SUBFAMILIES AND TRIBES

1 Ovary merely lobed; nutlets laterally attached; style terminal. (Genera 1–4.) .... Subfamily Ajugoideae.
1 Ovary deeply 4-parted; nutlets basally attached; style gynobasic.
  2 Cal with a distinct transverse protuberance on the upper side. (Genus 5.) .... Subfamily Scutellarioideae.
  2 Cal lacking a transverse protuberance.
    3 Stamens ascending under the upper lip of the cor, or spreading or directed forward. (Genera 6–39.)
      4 Style and stamens included within the cor-tube. (Genera 6–7.) .... Subfamily Lamioideae.
      4 Style and stamens usually exsert beyond the cor-tube. .... Tribe Marrubieae.
      5 Cor 2-lipped to nearly regular, the lips weakly differentiated, the upper one flat or nearly so.
        6 Anthers ovoid, with parallel or divergent distinct pollen-sacs. (Genera 8–21.) .... Tribe Saturejeae.
        6 Anthers spherical, the pollen-sacs confluent at the summit. (Genus 22.) .... Tribe Pogostemoneae.
      5 Cor strongly 2-lipped, the upper lip ± concave or hooded.
        7 Anthers ovoid; stamens 2 or 4.
          8 Upper stamens longer than the lower. (Genera 23–27.) .... Tribe Nepeteae.
          8 Upper stamens shorter than the lower. (Genera 28–36.) .... Tribe Lamieae.
        7 Anthers linear; stamens 2.
          9 Connective of the anthers very small. (Genera 37–38.) .... Tribe Monardeae.
          9 Connective of the anthers elongate, thread-like. (Genus 39.) .... Tribe Salvieae.
    3 Stamens descending along the lower lip. (Genus 40.) .... Subfamily Ocimoideae.

ARTIFICIAL KEY TO THE GENERA

1 Cal with a distinct cap or protuberance on the upper side of the tube.
  2 Cal 2-lipped, each lip entire, the tube with an erect protuberance .... 5. *Scutellaria*.
  2 Cal 5-lobed, the upper lobe extending over the others and decurrent on the tube, forming a cap .... 40. *Ocimum*.
1 Cal without a cap or protuberance on the tube.
  3 Upper lip of the cor very short and scarcely discernible, or its lobes borne laterally on the margins of the lower lip, the cor thus appearing one-lipped.
    4 Lower lip 5-lobed, the 2 lobes nearest its base representing the upper lip .... 2. *Teucrium*.
    4 Lower lip 3-lobed, or apparently 4-lobed if the median one is notched .... 1. *Ajuga*.
  3 Upper lip of the cor well developed and manifest, entire or 2-lobed, or the cor nearly regular.
    5 Stamens included in the cor-tube, not directly visible.
      6 Cal-lobes 10, subulate, hooked at the tip .... 7. *Marrubium*.
      6 Cal-lobes 5, broader, not hooked.
        7 Cal-lobes 1 + 4, the uppermost one much the largest; branching annual .... 6. *Sideritis*.
        7 Cal-lobes 2 + 3, the 2 upper somewhat larger than the 3 lower; creeping perennial .... 26. *Glechoma*.
    5 Stamens exsert beyond the cor-throat, sometimes even surpassing the lips.
      8 Stamens 2 .... Group I.
      8 Stamens 4.

9 Infl essentially axillary in appearance, the verticils several to many, subtended by foliage lvs and separated by normal internodes, or the uppermost subtending lvs smaller and internodes shorter. (Plants with axillary spikes or racemes are not included here.) .... Group II.
9 Infl essentially terminal in appearance, the verticils 1–many, all or mostly subtended by bracteal lvs conspicuously different from the foliage lvs, or separated by much shorter internodes. (Plants with lateral or axillary spikes or racemes are also included here.)
  10 Infls forming a panicle.
    11 Infls forming a loose, irregularly branched panicle; lvs lanceolate to linear; cal 2-lipped; fls blue .... 4. *Trichostema.*
    11 Infls individually small, collectively forming a many-fld, rounded or pyramidal panicle; cal regular; fls pink to purple .... 12. *Origanum.*
  10 Infls of one or more terminal racemes, spikes, or heads.
    12 Fls single in the axils of each bracteal lf, the verticils therefore with not more than 2 fls .... Group III.
    12 Fls 2–many in the axil of each bracteal lf, the verticils therefore with 4 or more fls .... Group IV.

GROUP I

1 Cal distinctly 2-lipped.
  2 Fls in loose, few-fld verticils in the axils of foliage lvs, blue, 3–4 mm .... 8. *Hedeoma.*
  2 Fls in terminal infls.
    3 Fls in terminal panicles, yellow, 12–15 mm; lower lip fringed .... 17. *Collinsonia.*
    3 Fls in terminal racemes or spikes or terminal or subterminal heads.
      4 Fls borne singly or in pairs at each node of the slender raceme.
        5 Upper lip of the cal entire .... 39. *Salvia.*
        5 Upper lip of the cal 3-lobed, the middle lobe the smallest .... 19. *Orthodon.*
      4 Fls 3 or more at each fl-bearing node (but not all in bloom at once).
        6 Verticils usually numerous and loosely fld, the fls at each node seldom more than 12 and then distinctly pediceled .... 39. *Salvia.*
        6 Verticils 1–5, dense and head-like, composed of numerous crowded fls .... 38. *Blephilia.*
1 Cal regular or nearly so, the lobes essentially all alike in size and shape.
  7 Fls in dense, head-like clusters or verticils, the pedicels obscure or none.
    8 Cor very irregular, 15–50 mm .... 37. *Monarda.*
    8 Cor nearly regular, 5 mm or less .... 21. *Lycopus.*
  7 Fls in loose panicles or paniculiform cymes, each one distinctly pediceled.
    9 Fls yellow, the lower lip fringed .... 17. *Collinsonia.*
    9 Fls pink-purple to white, the lower lip entire .... 20. *Cunila.*

GROUP II

1 Cal regular or nearly so, the lobes of the upper and lower lips similar in shape and equal or scarcely different in size.
  2 Cor almost equally 4-lobed or 5-lobed.
    3 Fls 1–3 in each axil, thus 2–6 in each verticil .... 3. *Isanthus.*
    3 Fls numerous in each axil .... 16. *Mentha.*
  2 Cor strongly 2-lipped, the upper lip concave and arched over the stamens.
    4 Fls distinctly pediceled, forming relatively loose cymules.
      5 Stems lax, creeping; fls usually 3 in each axil .... 26. *Glechoma.*
      5 Stems erect; fls usually 4–8 in each axil .... 33. *Ballota.*
    4 Fls sessile in the cymules.
      6 Cal-lobes prolonged into short but stiff spines.
        7 Lower cor-lip with 2 yellow or white protuberances at its base .... 35. *Galeopsis.*
        7 Lower cor-lip without protuberances .... 34. *Leonurus.*
      6 Cal-lobes slenderly acuminate but not spiny .... 32. *Lamium.*
1 Cal distinctly 2-lipped, the lobes of the upper and lower lips differing in size and shape.
  8 Stamens projecting beyond the cor .... 16. *Mentha.*
  8 Stamens ascending under the upper cor-lip and not surpassing it.
    9 Lvs linear, entire, revolute, finely tomentose beneath .... 10. *Conradina.*
    9 Lvs either broader, or toothed, or not revolute, or not tomentose beneath.
      10 Lobes of the upper cal-lip narrowly triangular, long-acuminate .... 11. *Satureja.*
      10 Lobes of the upper cal-lip wider than long, merely apiculate .... 9. *Melissa.*

GROUP III

1 Lvs cordate at base; bracts foliaceous, most of them longer than the cal.
  2 Fls blue .... 24. *Meehania.*
  2 Fls yellowish-white or greenish-white .... 30. *Synandra.*

1 Lvs not cordate at base; bracts small, shorter than to about equaling the cal.
  3 Principal lvs linear to narrowly oblong, sessile or subsessile; fls normally pink-purple .... 28. *Physostegia.*
  3 Principal lvs broadly ovate to oblong-ovate, long-petioled.
    4 Lower cor-lip yellow, fringed; racemes all terminal .... 17. *Collinsonia.*
    4 Lower cor-lip blue to white, not fringed; racemes terminal and lateral .... 18. *Perilla.*

Group IV

1 Stamens ascending under the upper cor-lip and not surpassing it.
  2 Cal distinctly 2-lipped and irregular.
    3 One cal-lobe (the upper median) longer and wider than the other 4 .... 27. *Dracocephalum.*
    3 Three cal-lobes (the upper lip) differing from the other 2.
      4 Bracts broadly rounded, apiculate .... 29. *Prunella.*
      4 Bracts setaceous, hirsute .... 11. *Satureja.*
  2 Cal nearly or quite regular, the lobes all alike or differing in size only.
    5 Bracts (not the subtending bracteal lvs) numerous, linear 1 cm, forming an enclosing involucre
      about each verticil .... 31. *Phlomis.*
    5 Bracts few, small, and inconspicuous.
      6 Lvs linear, entire, sessile.
        7 Stem finely pubescent .... 11. *Satureja.*
        7 Stem glabrous, or minutely pubescent on the angles only .... 36. *Stachys.*
      6 Lvs broader than linear, or serrate, or petioled.
        8 Cal 15-nerved; lower verticils often distinctly pedunculate .... 25. *Nepeta.*
        8 Cal 5–10-nerved; lower verticils sessile .... 36. *Stachys.*
1 Stamens, or some of them, protruding from the cor.
  9 Infl a dense or loose raceme in which the component verticils are plainly discernible; fls distinctly
    but shortly pediceled.
    10 Principal foliage lvs entire.
      11 Lower cor-lip about as long as the upper, its lobes nearly equal .... 13. *Thymus.*
      11 Lower cor-lip twice as long as the upper, its median lobe broadly dilated .... 14. *Hyssopus.*
    10 Principal foliage lvs toothed.
      12 Fls blue to lavender, varying to white .... 16. *Mentha.*
      12 Fls yellow .... 17. *Collinsonia.*
  9 Infl otherwise.
    13 Infl a number of terminal heads or crowded cymes, often with secondary heads or cymes in
      some of the upper axils, never a spike or raceme .... 15. *Pycnanthemum.*
    13 Infl a dense spike, or with one or 2 lower verticils sometimes separate; fls sessile or subsessile.
      14 Spike terete .... 23. *Agastache.*
      14 Spike secund .... 22. *Elsholtzia.*

1. AJUGA L. Bugle. Cal-lobes subequal, about as long as the tube; upper cor-lip very short, 2-lobed or entire, the lower lip elongate and dilated, its lateral lobes shorter and narrower than the broad, shallowly 2-lobed median segment; stamens 4, unequal, shorter than the lower lip; ovary rather shallowly 4-lobed, the style arising between the lobes; herbs, usually with toothed or lobed foliage lvs and medium-sized fls in whorls of 2–6+ axillary to bracteal lvs, forming a terminal leafy spike. 40, Old World.

1 Lvs sinuate-dentate to entire; fls blue (pink or white).
  2 Plants ± erect, not stoloniferous, not forming mats .... 1. *A. genevensis.*
  2 Plants stoloniferous and mat-forming as well as with ± erect flowering stems .... 2. *A. reptans.*
1 Lvs deeply 3-cleft into narrow segments; fls mostly yellow .... 3. *A. chamaepitys.*

1. Ajuga genevensis L. Standing b. Plants 1–4 dm, ± erect, rhizomatous but without stolons; stems ± villous all around, varying to subglabrous; lvs wavy-toothed or subentire, the basal ones commonly soon deciduous, the lower cauline mostly 3–7 × 1–3 cm, elliptic to obovate and tapering to a narrow or petiolar base, the upper generally broad-based and sessile, passing into the obovate bracts of the usually somewhat arachnoid-villous infl, this thyrsoid-spicate, its well developed bracts often suffused with blue; verticils 6–many-fld; cor bright blue (pink or white), 12–20 mm, the upper lip very short and minutely 2-toothed, the lower lip 5–8 mm; stamens exserted; filaments hairy; 2n=32. Native of Eurasia, escaped into lawns, gardens, and roadsides in our range. Apr.–June. Hybridizes with no. 2.

2. Ajuga reptans L. Carpet-b. Plants vigorously spreading by leafy stolons and forming loose mats; flowering stems ± erect, 1–3 dm, commonly pubescent in strips decurrent from the lf-bases; lvs often coppery or purplish; basal lvs often persistent, relatively large and long-petiolate; verticils usually 6-fld; upper cor-lip entire; otherwise much like no. 1; 2n=32. Native of Eurasia, sometimes escaped into lawns etc. in our range. May, June.

3. Ajuga chamaepitys (L.) Schreber. Ground-pine. Much-branched annual 1–2 dm; lvs cleft into 3(5) linear segments; lower lip of the yellow (purple) cor much prolonged, its base narrow, with 2 small lateral lobes, the median lobe dilated and merely retuse; $2n=28$. Native of the Mediterranean region, intr. in se. Va. May–Sept.

2. TEUCRIUM L. Germander; wood-sage.

Cal campanulate, 10-nerved, slightly oblique at the throat, with 5 scarcely to evidently unequal teeth; cor seemingly 1-lipped, the upper lip represented only by its 2 lobes, which are separated and displaced so as to arise from the lateral margins of the well developed, declined, otherwise 3-lobed lower lip, of which the central lobe is much the largest; stamens 4, exserted, the lower pair the longer; ovary merely 4-lobed, the style terminal, the nutlets laterally attached and almost completely united. 100, cosmop.

1 Lvs merely toothed.
  2 Fls in dense verticils, forming a crowded terete infl . . . . 1. *T. canadense.*
  2 Fls 1 or 2 at each node, forming a secund raceme . . . . 2. *T. scorodonia.*
1 Lvs deeply pinnatifid . . . . 3. *T. botrys.*

1. Teucrium canadense L. American g. Rhizomatous perennial with pubescent, solitary stems 3–10 dm; lvs lance-ovate to oblong, 5–12 cm, toothed, obtuse or rounded at base; petioles 5–15 mm; infl a crowded, spiciform raceme 5–20 cm with slender bracts to 1 cm, the pedicels 1–3 mm; cal 5–8 mm, 2-lipped, the 3 upper lobes wider and blunter than the 2 lower; cor pink-purple, 11–18 mm, cleft to the summit of the cal; stamens arching over the cor, diverging from the summit of the cal; fruiting cal not reticulate. Moist or wet soil throughout most of the U.S. and s. Can. June–Aug. Three vars.:

a Hairs all eglandular, those of the lower lf-surface curled or crooked.
  b Lvs firm, rarely over 2 cm wide, rugose above, densely hairy beneath; near the coast, Me. to Fla.
    (*T. littorale.*) . . . . var. *canadense.*
  b Lvs thin, broader, to 5 cm wide, not rugose, sparsely hairy beneath; throughout our range . . var. *virginicum* (L.) Eaton.
a Many of the hairs, especially in the infl, gland-tipped, those of the lower lf-surface straight and
  spreading; n. and w. part of our range, and w. (*T. boreale.*) . . . . var. *occidentale* (A. Gray) McClintock & Epling.

2. Teucrium scorodonia L. Wood-sage; wood-g. Rhizomatous perennial 3–5 dm; petioles 1 cm; lvs ovate-oblong or triangular-ovate, 3–6 cm, crenate, truncate to shallowly cordate at base; racemes few to several, terminal and axillary, secund, 1 dm; bracts ovate or lance-ovate; pedicels 1–3 mm; cal 5–6 mm, the upper lobe broadly depressed-ovate, apiculate, the others triangular, subulate-tipped; cal reticulate-veiny at maturity; cor greenish-yellow, 1 cm, the tube much exceeding the cal; filaments nearly straight; $2n=32$. Native of Europe, rarely intr. in our range, reported from Ont. and O.

3. Teucrium botrys L. Much branched annual 1–3 dm; stems glandular-villosulous; lvs 1–2 cm, deeply and ± ternately divided into short, obtuse, oblong segments; petioles 5–10 mm; bracts not much smaller than the lvs; fls usually 4–6 at each node, on pedicels 4–8 mm; cal strongly saccate at base, especially in age, eventually to 1 cm, its lobes shorter than the tube, broadly triangular, the 2 lower somewhat the smaller; cor 15–20 mm; $2n=32$. Native of Europe, occasionally intr. in waste places from N. Engl. to O. and W.Va. July–Sept.

3. ISANTHUS Michx. False pennyroyal.

Cal campanulate, nearly regular, the narrowly triangular lobes exceeding the tube; cor-tube included in the cal, the upper 4 lobes spreading or ascending, the lower one deflexed, usually slightly longer; stamens 4; filaments nearly straight, shortly exsert from the tube; fruiting cal accrescent, obscurely reticulate; nutlets prominently reticulate-areolate; annual with 1–3 fls from the axils of the scarcely reduced bracteal lvs, forming a large leafy panicle. Monotypic.

1. Isanthus brachiatus (L.) BSP. Stems much branched, 2–4 dm, finely puberulent, glandular in the infl; lvs short-petioled, mostly elliptic to lanceolate, usually 3-nerved, acute; pedicels to 1 cm, cal-lobes 2–3 mm, in fr 3–5 mm; cor blue, the oblong lobes 1.5–2 mm; nutlets 2.5–3 mm, puberulent at the top. Dry soil; Vt. and Ont. to Minn. and Nebr., s. to Fla. and Ariz. Aug., Sept. (*Trichostema b.*)

4. TRICHOSTEMA L. Blue curls.

Cal 10-nerved, campanulate, regular or (ours) 2-lipped and becoming inverted, the morphological upper lip with 3 partly connate teeth, the lower evidently shorter, with 2 distinct teeth; cor irregular, the lowest lobe the largest, declined, the other 4 subequal and more than half as long as the lowest one, the sinus between the 2 upper lobes scarcely more prominent than the lateral sinuses; stamens 4, exserted, the filaments often (as in our spp.) arcuate; pollen-sacs divaricate, often confluent in dehiscence; ovary rather deeply 4-lobed, the sculptured nutlets laterally attached, united above ⅓ their length; herbs or near-shrubs with mostly entire lvs; fls borne in cymose axillary clusters. 16, N. Amer.

1 Lvs oblong to ovate, mostly 5–25 mm wide, not more than 5 times as long as wide .... 1. *T. dichotomum*.
1 Lvs linear or nearly so, mostly 1–5 mm wide, at least 6 times as long as wide .... 2. *T. setaceum*.

1. **Trichostema dichotomum** L.   Weedy annual; stems much branched, to 7 dm, minutely glandular-hairy, especially upwards; lvs 1.5–6 × 0.5–2.5 cm, oblong to elliptic or ovate, with ± evident lateral veins; pedicels 3–5 mm; cal 3–5 mm, 5–9 mm in fr, becoming inverted at maturity so that the 3 long teeth are lowermost; cor-tube about equaling the cal, the lobes 3–6 mm; filaments finally to 15 mm; nutlets glabrous, 1.5–3 mm; $2n=38$. Dry soil, upland or sandy woods, and old fields; Me. to Mich. and Mo., s. to Fla. and Tex. Aug., Sept.

2. **Trichostema setaceum** Houtt.   Lvs linear, 1-nerved, up to 5 mm wide; otherwise much like no. 1, but apparently sharply distinct; $2n=38$. Chiefly in the coastal states from Conn. to Fla. and Tex., and irregularly inland to O. and Tenn. July–Oct. (*T. dichotomum* var. *lineare*.)

5. **SCUTELLARIA** L.   Skullcap. Cal obscurely nerved, strongly bilabiate, with entire lips, bearing a prominent transverse appendage on or proximal to the upper lip; cor 2-lipped, the upper lip generally ± galeate, the lateral lobes of the lower lip somewhat separated from the broad central lobe and marginally partly connate with the upper lip; stamens 4, ascending under the upper lip, the anthers of the upper pair with 2 pollen sacs, those of the lower pair with only one; style gynobasic, its upper branch reduced or wanting; ours all perennial, mostly rhizomatous; fls blue or violet, seldom pink or white, solitary in the axils or in terminal or axillary racemes. 300, cosmop.

1 Fls in axillary or terminal racemes, subtended individually by bracteal lvs obviously smaller than the foliage lvs.
  2 Fls chiefly in terminal racemes, or also in accessory lateral racemes arising from the upper axils but regularly subordinate to the terminal.
    3 Lvs cordate to broadly truncate at base; cor-tube without an internal ring of hairs.
      4 Stem pubescent with simple or glandular, spreading or decurved hairs.
        5 Racemes well differentiated, the bracts entire, quite unlike the crenate lvs .... 1. *S. ovata*.
        5 Racemes poorly differentiated, the lowest bracts much like the upper lvs, many or most of the bracts dentate .... 2. *S. arguta*.
      4 Stem (at the middle) glabrous or very sparsely and minutely pubescent with eglandular curved-ascending hairs .... 3. *S. saxatilis*.
    3 Lvs (median and upper) narrowed to rounded or rarely subcordate at base, but the actual summit of the petiole always ± V-shaped, with the lf-tissue extending down the petiole; cor-tube with an internal ring of hairs.
      6 Median and upper lvs crenate, the larger more than 15 mm wide.
        7 Cal densely or sparsely pubescent with appressed eglandular hairs.
          8 Racemes (3)5–many; lowest pair of fls subtended by bracts .... 4. *S. incana*.
          8 Racemes 1(3); lowest pair of fls commonly subtended by foliage lvs .... 5. *S. serrata*.
        7 Cal densely or sparsely pubescent with spreading glandular hairs .... 6. *S. elliptica*.
      6 Median and upper lvs entire, 4–15 mm wide .... 7. *S. integrifolia*.
  2 Fls chiefly or all in lateral racemes arising from the axils of foliage lvs.
    9 Cor nearly straight, 5–8 mm .... 8. *S. lateriflora*.
    9 Cor curved upward from the cal, 8–12 mm .... 9. *S.* × *churchilliana*.
1 Fls mostly solitary in the axils of ordinary foliage lvs.
  10 Cor 1.5–2 cm; lvs commonly 2–4 times as long as wide .... 10. *S. galericulata*.
  10 Cor 6–12 mm; lvs very often less than twice as long as wide.
    11 Petioles of typical median lvs 3–10 mm; northeastern .... 9. *S.* × *churchilliana*.
    11 Petioles of typical median lvs mostly under 3 mm; widespread.
      12 Uppermost sterile lvs 1–2 cm, entire or with a few low teeth, those subtending the fls similar in shape and not much smaller.
        13 Stem and cal eglandular; principal lvs commonly with 2 pairs of lateral veins .... 11. *S. leonardii*.
        13 Stems and cal glandular and hairy; main lvs with 3–5 pairs of lateral veins .... 12. *S. parvula*.
      12 Uppermost sterile lvs 2–4.5 cm, conspicuously serrate, those subtending the fls progressively smaller, narrower, and entire .... 13. *S. nervosa*.

1. **Scutellaria ovata** Hill.   Forest-s. Erect, usually stout, 3–7 dm; lvs long-petioled, ovate to round-ovate, crenate (usually more than 12 teeth on each side), cordate at base; racemes 1 or more, to 1 dm, even the lowest bracts very different from the adjacent foliage lvs; cor (10)15–25 mm. Woods; Md. to n. O. and Minn., s. to S.C., Ala., and Mex., most abundant w. of the Appalachians. June, July. (*S. cordifolia, S. versicolor.*) Three vars. with us:

a Stems with spreading glandular hairs.
  b Lvs firm, 5–8 cm when well developed; bracts regularly exceeding the cal, sometimes exceeding
    the cor, sometimes petiolate and dentate; Mo. and s. Ill. to Ala. and Tex. .... var. *ovata.*
  b Lvs thin, 8–12 cm when well developed; bracts regularly shorter than the cal, almost always sessile
    and entire; Va. and O. to Minn., s. to S.C. and Ark. .... var. *versicolor* (Nutt.) Fern.
a Stem pubescence very sparse, of curved eglandular hairs; otherwise as var. *ovata*; Ky., Tenn., and
  N.C. .... var. *calcarea* (Epling) Gleason.

2. **Scutellaria arguta** Buckley.   Appalachian s. Stems slender, freely branched, often ascending or decumbent, finely pubescent with spreading glandular hairs; lf-blades shorter than to twice as long as the petiole,
ovate, 1–3 cm, broadly rounded to subcordate at base, coarsely serrate, rarely as many as 10 teeth on a side;
racemes few-fld; lower bracts resembling foliage lvs and scarcely smaller, the upper progressively reduced;
cal glandular-villous; cor 15–20 mm. Mt. woods; Va., N.C. and Tenn. July, Aug.

3. **Scutellaria saxatilis** Riddell.   Rock-s. Stems slender, declined or decumbent, 2–5 dm, commonly
glabrous, sometimes obscurely pubescent with curved-ascending hairs; lvs ovate or deltoid-ovate, 2–4 cm,
with a few (less than 10) rounded teeth on each side, broadly rounded to cordate at base, long-petioled; bracts
gradually reduced, all or all but the lowest entire; pedicels and cal densely soft-hairy; cor 13–20 mm. Woods,
hillsides and moist cliffs, chiefly in the mts.; Del to O. and s. Ind., s. to S.C. and Tenn. July, Aug.

4. **Scutellaria incana** Biehler.   Stems erect, to 1 m, usually simple below the infl, minutely pubescent with
curved-ascending hairs, with (3) several to many racemes; lvs lance-ovate to ovate or rhombic-ovate, 5–10
cm, crenate except near the base, obtuse, broadly rounded, or subcordate at base, sparsely pilose above;
petioles 1–3 cm; bracts all much reduced, entire, about equaling the appressed-canescent cal; cor 18–25 mm.
Upland woods; w. N.Y. to s. Wis., s. to Ga., Ark., and Kans. June–Aug. Var. *incana,* widespread but missing
from most of the s. Appalachians, has the lvs soft-hairy across the surface beneath. Var. *punctata* (Chapman)
Mohr, of the mts. from W.Va. and Ky. to Ga. and Ala., has the lvs glabrous beneath except for the sparsely
hairy veins.

5. **Scutellaria serrata** Andrews.   Allegheny-s. Stems simple, erect, 3–6 dm, essentially glabrous; lvs seldom
more than 5 pairs, thin, ovate to rhombic-elliptic, 5–11 cm, crenate, broadly rounded to a cuneate base,
glabrous or nearly so on both sides; petioles 5–20 mm; raceme commonly solitary, 6–15 cm, or with a pair
of much shorter lateral ones as well; lowest pair of fls often subtended by foliage lvs; bracts entire, commonly
surpassing the cal; cor 2.5 cm. Woods in the piedmont and mts.; Pa., and O. to N.C. and Tenn. May, June.

6. **Scutellaria elliptica** Muhl.   Erect, 3–6 dm, usually simple, with racemes in the upper 1–3 pairs of axils;
median and upper lvs (petiole included) 4–7 cm, half as wide, much shorter than the internode above,
rhombic-ovate to rarely deltoid, crenate from the widest part to the tip, broadly acute at base and passing
into the short petiole; racemes 3–10 cm; bracts about equaling the glandular-hairy cal; cor 1.5–2 cm. Dry
upland woods and fields; s. N.Y. to s. Mich., s. to Fla. and Tex. May–Aug. (*S. pilosa; S. ovalifolia.*) Var.
elliptica, occurring from s. N.Y. to s. Mo. and southward, especially e. of the mts., has the middle and upper
parts of the stem finely pubescent with short, curved-ascending hairs. Var. *hirsuta* (Short) Fern., occurring
from w. Pa. to sw. Mich., and s. especially w. of the mts., has the stem pubescent with fine, spreading,
glandular hairs 1–2 mm.

7. **Scutellaria integrifolia** L.   Stems erect, 3–7 dm, finely hairy; lower lvs often petioled and dentate, but
usually soon deciduous; median and upper lvs lanceolate to oblanceolate, 2–6 cm × 4–15 mm, usually obtuse,
entire, tapering to a sessile or shortly petiolar base; lower bracts similar to the lvs, the upper progressively
smaller, the uppermost about equaling the cal; cor blue (pink), 18–28 mm, ascending or suberect; $2n=32$.
Fields and open woods; e. Mass. to Fla. and Tex., especially on the coastal plain; also inland to s. O., Ky.,
and Tenn. May–July.

8. **Scutellaria lateriflora** L.   Stems solitary from slender rhizomes, 3–7 dm, puberulent in lines with
ascending hairs, or sometimes glabrous; petioles 0.5–2.5 cm; lvs thin, pinnately veined, ovate or lance-ovate,
with broadly rounded or subcordate base, toothed, 3–8 × 1.5–5 cm; racemes mostly axillary or on axillary
branches, 3–10 cm, bracteate but scarcely leafy except sometimes near the base; cal 1.5–2.5 mm; cor blue
(pink or white), 5–8 mm, the short galea evidently surpassed by the lower lip; $2n=88$. Moist bottomlands;
Nf. to B.C., s. to Ga. and Calif. July–Sept.

9. **Scutellaria × churchilliana** Fern.   Intermediate between nos. 8 and 10, largely sterile, but propagating
vegetatively; stems slender, ascending, to 4 dm, with ascending pubescence as in no. 8; lvs lance-ovate to
lance-oblong, acuminate, coarsely serrate, rounded to the truncate or subcordate base; petioles 3–10 mm; fls
axillary or in elongate leafy racemes in which the bracts are progressively reduced; cor blue, somewhat
upcurved, 8–12 mm; $2n=60$. Moist woods; Me. to N.B., N.S., and Que., often with the parents.

10. **Scutellaria galericulata** L.   Marsh-s. Rhizomatous perennial; stems weak but mostly erect, 2–8 dm,
retrorsely strigose-puberulent especially along the angles, or less often glabrous or glandular; lvs barely or
scarcely petiolate, the blade lanceolate to narrowly ovate-oblong, pinnately veined, mostly blunt-toothed, 2–
6 cm × 6–20 mm, truncate-subcordate at base, glabrous above, usually puberulent beneath; fls solitary in
the axils of slightly reduced lvs; cal 3.5–4.5 mm; cor blue marked with white, 1.2–2 cm; $2n=32$. Wet soil;
circumboreal, s. to Del., Ind., Mo., and Calif. June–Aug. (*S. epilobiifolia.*)

11. **Scutellaria leonardii** Epling.   Stems erect, 1–2 dm, often several from the end of a moniliform rhizome,
minutely pubescent on the angles with eglandular curved-ascending hairs; main cauline lvs sessile, lance-

ovate or somewhat deltoid, 10–16 mm, 2–3 times as long as wide, entire, glabrous, or somewhat scabrous above, especially near the tip and margin, the lateral veins 1 or commonly 2 on each side of the midrib, not anastomosing; fls axillary, 7–9 mm, the short pedicels with curved-ascending hairs; cal not glandular, lower lip of the cor hirtellous in 2 lines. Dry upland woods and prairies; Mass. to Mich. and N.D., s. to Va., Tenn., Ark., and Okla., most abundant in the Middle West. May–July. (*S. ambigua*, misapplied.)

12. **Scutellaria parvula** Michx. Little s. Much like no. 11; stems with spreading glandular hairs, and often also minutely hairy on the angles; principal lvs sessile, ovate to rotund, 10–15 mm, half to three-fourths as wide, pubescent over the whole surface; lateral veins 3–5 on each side of the midvein; cal glandular-hairy. Prairies, upland woods, and rock-ledges; Que. to Minn., s. to Fla. and Tex. Two vars.:
  Var. *parvula*. Lateral veins of the lf not anastomosing, or very inconspicuously so; stem with short, retrorse, eglandular hairs as well as the glandular ones. Que. and Ont. to Minn., s. to w. Va., Ala., and Tex., more common northward.
  Var. *australis* Fassett. Lateral veins of the lf arched and anastomosing to form a continuous submarginal vein; stem with short, curved-ascending, eglandular hairs on the angles as well as the glandular ones. Chiefly southern, extending n. to L.I., W.Va., s. Ind., and Kans.

13. **Scutellaria nervosa** Pursh. Stems erect from slender rhizomes, with spreading glandular hairs below, glabrous above or sparsely puberulent on the angles with curved-ascending hairs; foliage lvs lance-ovate to round-ovate, the larger 2–4.5 cm, two-thirds as wide or wider, few-toothed, broadly rounded to subcordate at base, the lowest short-petioled, the upper sessile; lvs subtending the fls progressively reduced and narrower, often acute, the upper entire; pedicels 2–3.5 mm; cor blue, 8–10 mm; its lower lip hirtellous in 2 lines. Moist woods; May, June. Two vars.:
  Var. *nervosa*. Lvs strigose on the upper side. N.J. to Pa., O., and s. Ind., s. in the mts. to N.C. and Tenn.
  Var. *calvifolia* Fern. Lvs glabrous above. N.J. to Ont. and Io., s. to e. Va. and Tenn., excluding the highlands.

6. SIDERITIS L. Ironwort.

Cal broadly tubular, the sharply pointed lobes equal or unequal; cor mostly shorter than the cal, 2-lipped, the upper lip entire, narrowly oblong, the lower larger, 3-lobed; stamens 4, short, included in the cor-tube, the lower pair slightly the longer, in some spp. without anthers; nutlets rounded above; herbs with small, red, pink or yellow fls in dense axillary clusters; bracteoles wanting. 100, mainly Eurasia.

1. **Sideritis romana** L. Lanate-villous annual to 3 dm, branched from the base; lvs spatulate or oblanceolate, 1–2.5 cm, toothed; cal 6–10 mm, veiny, densely villous in the throat, the 4 lower lobes subequal, shorter than the tube, aristate, the upper lobe much larger, ovate, its 3 strong nerves confluent into an awn; cor 7–10 mm, yellow, white, or purple; 2*n*=28. Native of the Mediterranean region, naturalized as a weed in Pa. and W.Va. June–Aug.

7. MARRUBIUM L.

Cal nearly regular, 5–10-toothed, densely villous in the throat; cor strongly 2-lipped, the upper lip ascending, oblong, nearly straight, entire or shortly bifid, the lower lip spreading, 3-lobed, the central lobe the largest and often emarginate; stamens 4, included in the cor-tube, the lower pair slightly the longer; pollen-sacs divaricate, soon confluent; style gynobasic; nutlets truncate; perennials, generally woolly, with toothed or incised, rugose lvs, the fls in dense axillary clusters; bracteoles usually present. 35, Old World.

1. **Marrubium vulgare** L. Horehound. Taprooted perennial, the stems generally several, 3–10 dm, nearly prostrate to suberect, conspicuously white-woolly; lvs wholly cauline, canescent-woolly or partly subglabrate, not much reduced upward, petiolate, the blade broadly elliptic to rotund-ovate, evidently crenate, 2–5.5 cm; cal stellate and often ± long-hairy, the tube 4–5 mm, the 10 narrow, firm teeth somewhat shorter, eventually widely spreading, their spinulose tips recurved from the first; cor whitish, slightly exserted, with subequal lips; 2*n*=34, 36. Native of Eurasia, escaped and established in disturbed habitats nearly throughout the U.S. and s. Can. June–Aug.

8. HEDEOMA Pers.

Cal tubular, becoming flask-shaped, gibbous on the lower side at base, strongly 13-ribbed, 2-lipped, villous in the throat; cor weakly 2-lipped, the upper lip erect, emarginate, the lower spreading, 3-lobed; stamens 2, ascending under the upper cor-lip and about equaling it; nutlets ovoid, smooth; small annuals (our spp.) with numerous axillary, few-fld verticils of small, blue, pediceled fls. 38, New World.

1 Lvs lanceolate or broader, often serrate; upper cal-lobes triangular, entire .... 1. *H. pulegioides.*
1 Lvs linear, entire; cal-lobes subulate, ciliate .... 2. *H. hispidum.*

1. **Hedeoma pulegioides** (L.) Pers. American pennyroyal. Stems erect, usually branched, 1–4 dm, provided with fine, retrorsely curled hairs at least above; lvs lanceolate or elliptic to ovate or obovate, 1–3 cm, entire

or serrulate, the main ones distinctly petiolate; cal-tube 2–2.5 mm, sparsely hirsutulous, the lips 2–2.5 mm, the upper lobed a third or half its length into entire, triangular teeth, the lower cleft to the base into subulate, ciliate teeth; cor scarcely surpassing the cal; $2n=36$. Upland woods; N.B. and Que. to Wis. and Neb., s. to Ga. and Ark. July–Sept.

2. **Hedeoma hispidum** Pursh. Much like no. 1; stems simple or branched from the base, occasionally branched above; lvs linear or linear-oblong, 1–2 cm, sessile, entire; upper cal-lip cleft to the middle or below into subulate, ciliate teeth; $2n=34$. Dry soil, sand dunes, and barrens; Vt. to Mich. and Alta., s. to Conn., N.Y., O., Ala., Tex., and Colo. May–Aug.

9. **MELISSA** L. Balm. Cal 2-lipped, strongly 13-nerved, angular, sparsely villous, the upper lip flattened, oblong, with 3 conspicuous nerves prolonged into very short, broadly triangular lobes, the lower lip about as long, cleft nearly to the base into narrow, subulate-tipped lobes; cor-tube upcurved, the upper lip concave, 2-lobed, the lower lip with rounded lateral lobes and a longer, oblong or obovate median lobe; stamens 4, ascending under the upper cor-lip and about equaling it; nutlets obovoid, smooth. 3, Eurasia.

1. **Melissa officinalis** L. Common or lemon-b. Lemon-scented perennial, 4–8 dm; lvs long-petioled, ovate or deltoid-ovate, coarsely crenate, broadly obtuse to truncate at base, the main ones 4–7 cm, those of the branches much smaller and often cuneate at base; verticils few-fld, axillary to foliage lvs; pedicels 3–5 mm; cal 7–9 mm, the lips two thirds as long as the tube; cor pale yellow, becoming white or pinkish, 8–15 mm, the lips 3 mm; $2n=32$. Native of w. Asia, escaped from cult. and established in much of our range. June–Aug.

10. **CONRADINA** A. Gray. Cal 13-nerved, strongly 2-lipped, the upper lip broad, shallowly 3-lobed, the lower cleft to the base into triangular-subulate lobes; cor 2-lipped, the tube upcurved, widened to the throat, the upper lip arched, retuse, the lower lip 3-lobed; stamens 4, ascending under the upper lip, not exsert; nutlets subglobose, smooth; shrubs or half-shrubs with narrow, revolute lvs and medium-sized blue or lavender fls in small verticils from the upper axils. 4, se. U.S.

1. **Conradina verticillata** Jennison. Diffusely branched shrub to 5 dm, rooting below, bearing short, 2-lvd branches in most axils, the lvs therefore appearing whorled; lvs linear, 1–2 cm, revolute, green and glabrous above, finely gray-tomentose beneath; fls 1–3 in several upper axils, on pedicels 3 mm; cal glandular-hirsutulous, 7–8 mm, the lobes of the lower lip slightly surpassing the upper; cor lavender or pink, spotted within, 15–18 mm. Sandy riverbanks in the Cumberland Mts.; e. Ky. and e. Tenn. May, June. (*C. montana.*)

11. **SATUREJA** L. Cal conspicuously 10–13-nerved, often hairy in the throat, sometimes 2-lipped, the lobes subulate to triangular; cor-tube widened distally, the upper lip flat or slightly concave, straight or somewhat spreading, entire or emarginate, the lower lip deflexed, 3-lobed; stamens 4, ascending under the upper lip, the lower pair the longer; nutlets smooth; herbs of diverse habit. (*Acinos, Calamintha, Clinopodium.*) 100+, widespread.

1. **Satureja glabella** (Michx.) Briquet. Perennial to 6 dm; stems glabrous, except commonly for a minute pubescent area at each node; fls 2–8 at each node in the upper half of the plant, subtended by progressively reduced lvs and each fl by 2 linear bracts; pedicels 3–10 mm; cal glabrous, 4–6 mm, the lips half as long as the tube, the lobes acuminate; cor pale purple, 8–15 mm; style-branches short, slender, unequal. Calcareous soil; May–Aug. Two vars.:

Var. *glabella.* Stems weak, loosely branched, but not stoloniferous; lvs oblanceolate, obtuse, tapering to the base, the larger 2–4 cm, to 1 cm wide, with 2–4 conspicuous teeth on each side. Limestone barrens; Ky., Tenn., Mo., and Ark. (*Calamintha g.; Clinopodium g.*)

Var. *angustifolia* (Torr.) Svenson. Stems from short stolons; stolons for the next year usually developed while the plant is in bloom, bearing ovate petiolate lvs 3–10 mm; lvs of flowering stems linear, entire, 1–2 cm. Beaches of lakes Erie, Huron, and Michigan, where the plants are stiffly erect, simple or branched, 1–2 dm; also inland in O., W.Va. and Ill., and s. Wis., and from Mo. to Okla. and Tex., where the stems are diffusely branched and up to 4 dm. (*S. arkansana; Clinopodium glabrum; Calamintha nuttallii.*)

2. **Satureja hortensis** L. Summer-savory. Much branched annual; stems 1–3 dm, finely hairy; lvs linear to narrowly oblanceolate, often revolute, 1–2 cm; fls few in the uppermost axils, subsessile, subtended by reduced lvs and separated by short internodes, forming compact, spike-like, often secund infls; cal nearly regular, 5 mm, the lobes about equaling the tube; cor pale pink-purple to white, 5–7 mm; style-branches short, slender, subequal; 2*n*=45–48. Native of the Mediterranean region and sw. Asia, escaped from cult. here and there in our range. July–Sept.

3. **Satureja vulgaris** (L.) Fritsch. Wild basil. Erect perennial 2–5 dm from short stolons; lvs ovate or lance-ovate to ovate-oblong, 2–4 cm, entire or with a few low teeth, on petioles to 1 cm, or the upper subsessile; fls numerous in a dense, subglobose, terminal, head-like infl, or in vigorous plants with 1 or 2 smaller glomerules in the uppermost axils, mingled with numerous setaceous, hirsute bracts about as long as the cal; cal 9–10 mm, hirsute, the lips nearly as long as the tube, the upper lip cleft half way, the lower cleft to the base; cor rose-purple to pink or white, 12–15 mm; style-branches short, unequal, the upper subulate, the lower longer and wider; 2*n*=20. Upland woods; Nf. and Que. to Man., s. to N.C., Tenn., and Kans., and scattered in w. U.S.; also widespread in Europe. Plants from the s. part of our range may be intr. from Europe. (*Clinopodium v.*)

4. **Satureja calamintha** (L.) Scheele. Basil-thyme. Perennial to 8 dm; stems pubescent; lvs petioled, ovate-oblong to deltoid-ovate; cymules axillary, or by reduction of the upper lvs tending to form a panicle, each cymule peduncled and branched; cal 2-lipped, 4–6 mm; cor pale purple or nearly white, 9–12 mm; style-branches short, unequal, as in no. 3; 2*n* mostly = 24. Native of Europe, intr. in our range from Md. and Ky. s. June–Sept. Highly variable, and divided by European botanists into an indefinite number of infraspecific taxa or segregate spp. (*Clinopodium nepeta; Calamintha nepeta.*)

5. **Satureja acinos** (L.) Scheele. Mother-of-thyme. Erect annual 1–2 dm, the finely hairy stems usually branched at base; lvs obovate-oblong to elliptic, 6–12 mm, entire or with a few low teeth, scabrous, on petioles 1–2 mm; fls 1–3 in each upper axil; pedicels 2–4 mm; cal 5–6 mm, very gibbous beneath, constricted at the throat, hirsutulous on the ribs, the lips about equal and half as long as the tube, the lower lip bifid to the base, the upper widened to the summit and subtruncate, with 3 small teeth 0.5 mm; cor pale purple, 7–10 mm; 2*n*=18. Native of Europe, intr. as a weed of roadsides and waste places from N. Engl. to Wis. and N.J. June–Sept. (*Acinos arvensis.*)

12. ORIGANUM L.

Cal regular, about 13-nerved, very villous in the throat, the lobes ovate-triangular; cor-tube scarcely widened distally, the upper lip straight, flat or nearly so, 2-lobed, the lower lip somewhat longer, deflexed, deeply 3-lobed; stamens 4, the lower pair well exsert, divergent from each other, the upper pair shorter, about equaling the upper lip; nutlets ovoid, smooth; perennial herbs with large, repeatedly trichotomous terminal infl, each node and each fl subtended by a subsessile oval bract, the upper bracts progressively smaller. 15, Eur. and Medit. reg. to c. Asia.

1. **Origanum vulgare** L. Wild marjoram. Rhizomatous perennial 4–8 dm; lvs ovate, 1.5–3 cm, entire or nearly so, on petioles half as long; infl rounded or pyramidal, 5–15 cm wide, the ultimate cymules crowded; smaller bracts usually reddish-tinged; cal 2.5 mm, the lobes a fourth as long as the tube; cor purple-red to pale pink, 6–8 mm; 2*n*=30, 32. Native of Eurasia, intr. in disturbed sites, especially in calcareous soil, from N.S. to Ont., s. to N.C. July–Sept.

13. THYMUS L.

Thyme. Cal 10–13-nerved, 2-lipped, the tube densely villous in the throat, the lips about equaling the tube and each other, the lower lip bifid to the base, the upper much wider, 3-lobed to the middle with triangular upturned lobes; cor 2-lipped, the upper lip flat or nearly so, straight, broadly rounded, the lower lip spreading, slightly the longer, its 3 lobes subequal; stamens 4, straight, exsert, subequal; shrubs or half-shrubs with small lvs and small blue, purple, or white fls in axillary clusters often aggregated into a loose terminal spike-like infl. 50+, temp. Eurasia.

1. **Thymus serpyllum** L. Wild thyme. Diffusely branched, mat-forming perennial, rooting at the nodes; stems woody at base; fl stems ± erect, to 1 dm, hairy all around; lvs linear to subrotund, in our wild forms commonly elliptic to oblong-ovate or oblong-obovate, 5–10 mm, short-petioled; fls aggregated into a continuous or interrupted terminal "spike" 1–4(–15) cm; cal 3–4 mm; cor purple, 4–6 mm; 2*n*=24+. Native of Europe, commonly cult. and escaped into upland woods and fields; N.S. to w. N.Y. and Ont., s. to W.Va. and N.C. June–Sept.

14. HYSSOPUS L. Cal 15-nerved, regular, not hairy in the throat, the lobes triangular, cuspidate; cor strongly 2-lipped, the upper lip small, obovate, shallowly 2-lobed, the lower lip much larger, its lateral lobes bent upward, the terminal lobe broadly dilated; stamens 4, straight, somewhat divergent, exsert, the lower pair slightly the longer; nutlets ovoid, smooth. 15, s. Eur. and Medit. reg. to c. Asia.

1. **Hyssopus officinalis** L. Hyssop. Perennial from a woody rhizome; stems erect, 3–6 dm, finely puberulent; lvs subsessile, often with smaller axillary ones, lanceolate to oblanceolate, 1–3 cm, entire; fls 3–7 in each upper axil, the clusters sessile or subsessile, ± secund, subtended by reduced lvs, forming a spike-like terminal infl; cal 6.5–8 mm, the lobes up to half as long as the tube; cor blue (white) 7–12 mm, the median lobe of the lower lip 6 mm wide; $2n=12$. Native of Eurasia, found here and there as a weed along roadsides and in waste places in our range. July–Oct.

15. PYCNANTHEMUM Michx., nom. conserv. Mt.-mint. Cal 10–13-nerved, often 2-lipped, the lobes shorter than the tube; cor ± distinctly bilabiate, the upper lip entire or merely emarginate, the lower 3-lobed; stamens 4, commonly exsert, straight and divergent, the pollen-sacs parallel; rhizomatous herbs with linear to ovate lvs and small fls in crowded or head-like cymes terminating the stem and branches or also sessile in the axils of the upper lvs; cor purple to white, the lower lip commonly purple-spotted. Fl in mid- and late summer. (*Koellia*.) 20, N. Amer. Spp. 4–6 confluent.

1 Infls relatively loose, many of the branchlets within the fl-clusters evident in fr.
  2 Lvs relatively narrow, mostly more than 3 times as long as wide.
    3 Cal 3.5–5 mm, its teeth deltoid to acuminate, the lower ones 0.5–2 mm .... 1. *P. clinopodioides.*
    3 Cal 6–10 mm, its teeth attenuate to aristate, the lower 2.5–5 mm .... 2. *P. flexuosum.*
  2 Lvs wider, mostly not more than 3 times as long as wide.
    4 Cal-teeth, and often also some of the slender bracts immediately subtending the fls, tending to
      be tipped with 1–several slender bristles at anthesis.
      5 Cal strongly bilabiate; ± widespread in our range.
        6 Cal-teeth deltoid, obtuse or acute to acuminate, the lower less than half as long as the tube
          .... 4. *P. incanum.*
        6 Cal-teeth acuminate to attenuate, the lower usually more than half as long as the tube.
          7 Nutlets 0.5–1.3 mm, smooth or with a few short hairs at the tip .... 5. *P. loomisii.*
          7 Nutlets 1.2–1.5 mm, rugose or pitted, densely hairy at the tip .... 6. *P. pycnanthemoides.*
      5 Cal only slightly or scarcely bilabiate; sw. Va. to N.C. and S.C. .... 7. *P. beadlei.*
    4 Cal-teeth and subtending bracts not bristle-tipped; southern, barely reaching our range in s. Ill.
      .... 8. *P. albescens.*
1 Infls densely capitate, only the lower (or no) branchlets within the fl-clusters evident in fr.
  8 Lvs relatively broad, the principal ones either at least 2 cm wide, or less than 3 times as long as
    wide, or often both.
    9 Bracts long-ciliate, otherwise glabrous or nearly so .... 9. *P. montanum.*
    9 Bracts finely puberulent across the upper surface, not ciliate.
      10 Cal-teeth 0.5–1.5 mm, deltoid to narrowly acute .... 10. *P. muticum.*
      10 Cal-teeth 1.5–2.5 mm, attenuate to aristate .... 3. *P. setosum.*
  8 Lvs narrower, mostly at least 3 times as long as wide and not more than 1.5 cm wide.
    11 Lvs, or at least the bracts of the infl, evidently hairy above .... 11. *P. verticillatum.*
    11 Lvs and bracts glabrous or nearly so on the upper surface.
      12 Herbage hairy at least in part; bracts lf-like, not cuspidate.
        13 Pubescence of the stem chiefly confined to the angles .... 13. *P. virginianum.*
        13 Pubescence of the stem more uniformly distributed .... 12. *P. torrei.*
      12 Herbage glabrous; bracts scarcely lf-like, with a stiff cuspidate tip .... 14. *P. tenuifolium.*

1. **Pycnanthemum clinopodioides** T. & G. Stems to 1 m, with short, leafy axillary branches, pubescent with short, usually downcurved hairs or longer spreading hairs or both; lvs narrowly lanceolate, entire or shallowly serrate, the largest ones median, 4.5–9.5 × 1–2.5 cm, on petioles 3–6 mm; upper lf-surfaces with scattered hairs, the lower with longer spreading hairs especially on the main veins; glomerules relatively loose and open, each subtended by a pair of lf-like, evenly pubescent bracts; cal 3.5–5 mm, short-hairy, weakly bilabiate, its teeth narrowly deltoid, somewhat acuminate, ciliate with several jointed bristles at the tip, the upper 3 ca ⅔ as long as the lower pair, these 0.5–2 mm; nutlets glabrous, or hairy at the tip; $2n=76$. Locally near the coast from Mass. to N.C. Sometimes thought to reflect hybridization of nos. 4 and 11.

2. **Pycnanthemum flexuosum** (Walter) BSP. Stems to 1+ m, commonly with short, leafy axillary branches, minutely puberulent; lvs narrowly oblong or lance-oblong, entire or with a few low callous teeth, finely hairy on the veins beneath, the main ones 2–4 cm, ca a fourth as wide; lateral veins 3–4 pairs, obscure; heads as in no. 3, but the cal 6–10 mm, its lower teeth 2.5–5 mm; $2n=36$. Dry woods and barrens, chiefly in sandy soil on the coastal plain; se. Va. to Fla. (*P. hyssopifolium.*)

3. Pycnanthemum setosum Nutt. Stems to 1 + m, commonly with some sterile axillary branches, minutely puberulent; lvs ovate to lance-ovate or oblong, subsessile or on petioles to 3 mm, obtuse or acute, entire or with a few obscure teeth, broadly rounded to truncate at base, the main ones 3–6 cm, a third to half as wide, glabrous, the uppermost ones minutely canescent; lateral veins 5–7 pairs; heads few, dense, 15–25 mm wide; bracts lanceolate, long-acuminate, with prominent midvein, minutely puberulent; cal minutely puberulent, its nerves evident, its teeth 1.5–2.5 mm, attenuate to aristate, $2n=76$. Dry fields and upland woods on the coastal plain; N.J. to Ga. (*P. aristatum; P. umbratile.*) Putative hybrids with nos. 4 or 6 have been called *P.* × *monotrichum* Fern.

4. Pycnanthemum incanum (L.) Michx. Stems to 1 + m, branched above, finely puberulent or even tomentose above; lvs ovate, lance-ovate, or oblong, acute or short-acuminate, serrate, obtuse or rounded at base, whitened beneath with very fine hairs, or also hirsutulous on the veins, the main ones 5–10 × 1.5–3.5 cm, on petioles 5–15 mm; heads relatively loose, 1.5–3.5 cm wide; bracts and cal finely canescent or to-mentulose; cal-lobes triangular, acute or barely acuminate, tending to be bristle-tipped, the lower 1–1.5 mm, the upper 0.5–1 mm; cor white, mottled with purple, the lower lip 2–3 mm; $2n=76$. Upland woods; Vt. and N.Y. to s. O. and s. Ill., s. to N.C. and Tenn.

5. Pycnanthemum loomisii Nutt. Much like *P. pycnanthemoides*; stem often with only short hairs; mature cal 3–4.5 mm, its lower teeth 1–2 mm, usually longer than the upper; nutlets 0.5–1.3 mm, smooth or with a few short hairs at the summit; $2n=38$. Woods; Va. to s. Ill., s. to n. Fla.

6. Pycnanthemum pycnanthemoides (Leavenw.) Fern. Stems to 1 + m, loosely branched above; lvs broad-ly lanceolate to lance-ovate or lance-oblong, serrate, 4–10 × 1.5–4 cm, on petioles 3–12 mm, the upper and bracteal ones canescent on both surfaces with short appressed hairs and some longer jointed bristles, the others glabrous above, hairy beneath; heads relatively loose, 1.5–3.5 cm wide; mature cal 4–6 mm, its lobes distinctly acuminate and tending to be tipped with 1–several bristles, the 2 lower 1.5–3 mm, more than half as long as the tube, the upper 3 much shorter; cor 4–5 mm, pink to purple; nutlets 1.2–1.5 mm, rugose or pitted, with dense straight or curled hairs at the top; $2n=72$. Woods; Va. and W.Va. to Ga. and Ala., w. to w. Tenn. and s. Ill. (*P. tullia; P. viridifolium.*)

7. Pycnanthemum beadlei (Small) Fern. Stems to 8 dm, simple or sparingly branched, puberulent es-pecially on the angles with downcurved hairs; lvs lance-ovate, serrulate, the larger ones 5–8.5 × 2–3.5 cm, evidently glandular-punctate or atomiferous-glandular on both sides, otherwise subglabrous, the bracteal ones minutely canescent; infls a little loose, many of the branchlets visible in fr; cal 5–7 mm long, only slightly bilabiate, with narrowly deltoid, acuminate teeth, the lower 1–2 mm, the upper a little shorter; cor 5–6 mm; $2n=76$. Mts. of sw. Va. (Grayson Co.) to N.C. and S.C.

8. Pycnanthemum albescens T. & G. Much like no. 4, but the bracts and cal-teeth not bristle-tipped; cal-lobes deltoid, the lower ca 1 mm, the upper distinctly shorter; $2n=38$. Dry upland woods; se. Mo. and s. Ill. to Fla. and Tex.

9. Pycnanthemum montanum Michx. Much like no. 10; principal lvs broadly lanceolate or lance-elliptic to rather narrowly ovate, mostly 6.5–12 × 2–4.5 cm, (2–)2.5–3.5(–4) times as long as wide; bracts notably long-ciliate, otherwise glabrous or nearly so; cal with scattered long hairs, at least toward the summit, the teeth bristle-tipped; $2n=38$. Mt. woods; s. W.Va. to Va., N.C., and Ga.

10. Pycnanthemum muticum (Michx.) Pers. Stems to 1 m, short-hairy, branched mainly above; lvs oblong or lance-oblong to lance-ovate, glabrous, obscurely serrulate or with a few low teeth, acute or acuminate, rounded at the base, the main ones 4–7 cm, on petioles 1–3 mm; lvs subtending the heads smaller and proportionately wider, velvety above, glabrous beneath; heads dense, hemispheric, 8–15 mm wide; bracts narrowly lanceolate, acuminate, hairy; cal densely short-hairy; its triangular lobes 0.5–1.5 mm; $2n=$ca 108. Moist woods and meadows; Mass. and Vt. to Mich. and Mo., s. to Fla. and La.

11. Pycnanthemum verticillatum (Michx.) Pers. Stems to 1.5 m, branched above and usually with some short axillary branches, ± hairy; lvs narrowly lanceolate, acute to acuminate, mostly with a few low teeth, the main ones 3–5 cm × 8–12(–15) mm, with 4–7 pairs of lateral veins; heads numerous, dense, 8–15(–20) mm thick; outer foliaceous bracts velvety above; inner bracts lanceolate, commonly sur-passing the cals, conspicuously acuminate, thinly hairy on the back, densely ciliolate, with a conspicuous midvein; cal-lobes narrowly deltoid, 0.5–1 mm; $2n=76$–78. Upland woods and thickets; Vt. to N.C., w. to Mich., Io., w. Kans., and w. Okla. The e. var. *verticillatum,* extending w. to Mich., O. and e. Ky., is not very hairy, the lvs hairy chiefly along the veins beneath. The more w. var. *pilosum* (Nutt.) Cooperrider, extending e. to O. and s. Mich., is more hairy, the stem conspicuously spreading-pilose, the lower lf-surfaces evenly spreading-hairy, the cal ± shaggy. (*P. pilosum.*)

12. Pycnanthemum torrei Benth. Much like *P. verticillatum* var. *verticillatum,* but the lvs softer and less veiny, and the cal-teeth longer and sharper, 1–1.5 mm; invol bracts lanceolate, acuminate to aristate; $2n=78$–80. Dry upland woods; s. N.H. to Ga., w. to Ill. (*P. leptodon.*)

13. Pycnanthemum virginianum (L.) Durand & B. D. Jackson. Stems to 1 m, branched above, short-hairy mainly on the angles; lvs numerous, lance-linear, glabrous above, often puberulent on the midvein beneath, scaberulous on the margins, the main ones 3–6 cm × 3–10 mm, those subtending the heads much smaller; lateral veins 3–4 pairs, the uppermost arising near the middle of the lf; inner bracts thin, acute or short-acuminate, densely canescent or tomentulose; cal densely canescent distally, the deltoid lobes 0.5–1 mm; $2n=80$. Upland woods and moist prairies; Me. to N.D., s. to Ga. and Okla.

14. Pycnanthemum tenuifolium Schrader.   Nearly inodorous; stems 5–8 dm, very leafy, with many short axillary branches; herbage glabrous; lvs linear, entire, those of the main axis 2–5 cm × 2–4 mm; lateral veins 1–2(3) pairs, all arising in the basal fourth of the lf; heads very numerous and dense, hemispheric 3–8 mm wide, on peduncles 3–15 mm; outer bracts lanceolate, sometimes surpassing the heads; inner bracts numerous, closely appressed, about as long as the cals, firm, lanceolate, long-acuminate, the conspicuous midvein produced into a stiff, subulate point; cal-lobes narrowly triangular, acuminate, puberulent, usually 1–1.5 mm; cor 5–8 mm; 2n=80. Chiefly in dry soil of upland woods and prairies; Me. to Fla., w. to Minn., Kans., Okla., and Tex. (*P. flexuosum*, misapplied; *P. linifolium*.)

16. MENTHA L.   Mint. Cal 10–13-nerved, regular or weakly 2-lipped, the broadly triangular to subulate teeth equal or unequal; cor with short tube and nearly regularly 4-lobed limb, the upper lobe formed by fusion of 2, tending to be broader than the others and often apically emarginate, rarely and casually the cor equally 5-lobed; stamens 4, straight, somewhat divergent, exsert, about alike, the pollen-sacs parallel; rhizomatous perennial herbs with toothed, wholly cauline lvs and small, blue to lavender or white fls in the axils of lvs or in terminal "spikes" or heads. Many spp. produce small-fld pistillate plants as well as normal perfect-fld ones. Ca 25 spp., mainly of Eurasia and Australia, 1 circumboreal. The European spp., long in cultivation, have given rise through hybridization, polyploidy, and other kinds of chromosomal aberration to numerous ± stabilized additional populations that have become established in the wild as well as being retained in cult.

1 Fls in axillary verticils subtended by ordinary foliage lvs and separated by internodes of ordinary
  length.
  2 Cal pubescent throughout its length.
    3 Fertile; odor undistinguished; common and widespread .... 1. *M. arvensis*.
    3 Mostly sterile; odor sickly; rarely intr. .... 2. *M.* × *verticillata*.
  2 Cal glabrous throughout, or pubescent in the upper half only .... 3. *M.* × *gentilis*.
1 Fls in terminal spikes or heads, the subtending lvs of the component verticils, if present, distinctly
  smaller than the foliage lvs.
  4 Infl a terminal, globose to ovoid head of 1–3 verticils (the lower verticil sometimes separate;
    pedunculate axillary heads sometimes present).
    5 Pedicels, cal, and lvs pubescent .... 4. *M. aquatica*.
    5 Pedicels and cal glabrous; lvs glabrous or nearly so .... 5. *M.* × *citrata*.
  4 Infl a spike of several to many verticils.
    6 Bracteal lvs much longer than the fls, resembling foliage lvs but smaller or narrower.
      7 Cal pubescent throughout its length .... 2. *M.* × *verticillata*.
      7 Cal glabrous throughout, or pubescent in the upper half only .... 3. *M.* × *gentilis*.
    6 Bracteal lvs linear to lance-linear, little surpassing the fls.
      8 Cal-tube glabrous; lvs glabrous or with a few scattered hairs beneath.
        9 Petioles of the main lvs 4–15 mm; spikes stout; sterile .... 6. *M.* × *piperita*.
        9 Petioles 0–3 mm; spikes slender; fertile .... 7. *M. spicata*.
      8 Cal-tube pubescent; lvs distinctly and usually densely hairy beneath.
        10 Lvs 2–3 times as long as wide, broadly acute to rounded at base .... 8. *M. longifolia*.
        10 Lvs 1–2 times as long as wide, broadly rounded to subcordate at base .... 9. *M. suaveolens*.

1. Mentha arvensis L.   Field-m. Stems ascending or erect, 2–8 dm, pubescent with few to numerous, short and retrorse to longer and more spreading hairs, often glabrous between the angles; lvs short-petiolate, the blade 2–8 cm × 6–40 mm, glabrous or hairy, serrate, acuminate, with several pairs of lateral veins, rather narrowly ovate to more often somewhat rhombic-elliptic, those of the infl, at least, tending to be cuneately tapered to the petiole; verticils compact, axillary to the scarcely reduced (middle and) upper lvs, and separated by internodes of fairly ordinary length; cal pubescent, 2.5–3 mm; cor white to light purple or pink, 4–7 mm, rarely casually 5-lobed; 2n=24–96. Moist places, often along streams and shores; circumboreal, in Amer. s. to N.C., Mo., and Calif. Native Amer. and e. Asian plants, as here described, are var. *canadensis* (L.) Kuntze (var. *glabrata*; var. *villosa*). The European var. *arvensis*, with the lvs of the infl relatively broader, more ovate, and somewhat broadly rounded to the petiole, is intr. from Nf. to Que. and Pa., but extremes of one var. could pass for the other.

    M. pulegium L., pennyroyal, a usually densely velvety native of Europe, is rarely adventive with us. It has smaller lvs (commonly 1–2.5 cm) with only 2–3 pairs of lateral veins, those of the infl often somewhat reduced; it differs from all our other spp. in having the cal hairy in the throat and with distinctly unequal teeth.

    2. Mentha × **verticillata** L.   Largely sterile hybrid of nos. 1 and 4, approaching no. 1 in the form of the infl, but more robust, often hairier, with narrower cal-teeth, and with a sickly odor; 2n=42–132. European cultigen, rarely intr. with us.

    3. Mentha × **gentilis** L.   Red or Scotch m. Largely sterile hybrid of nos. 1 and 7, resembling the smoother forms of no. 1, but usually with smaller floral lvs; lvs lanceolate to ovate, glabrous or nearly so, sharply toothed in the distal half; cal-tube glabrous, or sparsely pubscent in the upper half, the lobes pilose; $2n=54$–120. European cultigen, locally escaped from cult. in our range. (*M. cardiaca.*)

4. Mentha aquatica L.   Water-m. Stems 3–8 dm, with recurved hairs; lvs broadly ovate, 3–7 cm, commonly more than half as wide, toothed, rounded or even subcordate at base; fls numerous in 1–3 verticils closely aggregated to form a globose or ovoid to rarely short-cylindric spike; accessory heads on peduncles 2–10 mm may develop from the upper axils; cal and pedicels hairy; $2n=96$. Native of Eurasia, found as an escape here and there in our range. A hybrid with no. 8 is *M. × dumetorum* Schultes.

5. Mentha × citrata Ehrh.   Lemon-m. Largely sterile hybrid derivative of nos. 4 and 7, much like no. 4, but glabrous; terminal heads averaging somewhat smaller, and accessory heads in the upper axils more often developed; odor lemon-like; $2n=84$, 120. European cultigen, occasionally escaped in our range.

6. Mentha × piperita L.   Peppermint. Sterile hybrid of nos. 4 and 7; rhizomatous and often also stoloniferous, 3–10 dm, glabrous or glandular, or the lvs often hirsute along the main veins beneath; main lvs ovate or lance-ovate to elliptic, serrate, acute to occasionally obtuse, 3–6 × 1.5–3 cm, on petioles 4–15 mm; fls crowded in dense terminal spikes (sometimes interrupted below) 2–7 cm long and 1–1.5 cm thick at anthesis; cal 3–4 mm, the lobes hispid-ciliate, the tube without hairs; cor 3.5–5 mm; $2n=66$, 72. European cultigen, now widely but rather sparingly established in moist low ground in our range and elsewhere in the U.S.

7. Mentha spicata L.   Spearmint. Rhizomatous perennial, 3–10 dm, glabrous or subglabrous (often glandular), the lvs often hirsute along the main veins beneath; lvs sessile or subsessile, lance-ovate or elliptic, 3–7 × 0.8–2.5 cm, 2–3.5 times as wide, serrate, ± acute; fls crowded into slender terminal spikes (sometimes interrupted below) 3–12 cm long and 0.5–1 cm thick at anthesis; cal 1–3 mm, the lobes generally hispid-ciliate, the tube without hairs; cor 2–4 mm; $2n=48$. Banks of streams and ditches, and other moist places; native of Europe, now widely established in N. Amer. and throughout our range.

8. Mentha longifolia (L.) L.   Horse-m. Stem stout and erect, to 1 m, finely hairy; lvs sessile or subsessile, lance-oblong, 2–7 cm, 2–3 times as long as wide, with sharp subappressed teeth, broadly acute to rounded at base, thinly hairy on both sides; spikes usually several, 3–10 × 0.5–1 cm, sometimes interrupted at base; cal 1.5–2 mm, hairy throughout, the narrow lobes about as long as wide; $2n=24$. Native of Eurasia, escaped from cult. into fields, roadsides, and waste places here and there in our range.

9. Mentha suaveolens Ehrh.   Apple-m. Stout and erect, 4–10 dm, with a sickly-sweet odor; stem sparsely hairy to densely white-tomentose; lvs sessile or nearly so, broad-based and often clasping, obtuse or rounded above, 2–5 × 1–4 cm, 1–2 times as long as wide, crenate-serrate, strongly rugose-reticulate, strongly villous-tomentose beneath, less so above; verticils crowded into elongate terminal spikes (sometimes interrupted below) 3–15 × 0.5–1 cm; cal 1–2 mm, short-hairy throughout; $2n=24$. Native of s. Europe, escaped into waste places and along roadsides throughout s. U.S. and n. occasionally to Me. and O.

M. × rotundifolia (L.) Hudson, $2n=24$, is a frequent and fertile hybrid of nos. 8 and 9.

M. × villosa Hudson (incl. *M. alopecuroides* Hull), $2n=36$, is a mostly sterile hybrid of nos. 7 and 9.

17. COLLINSONIA L.   Horse-balm; stone-root. Cal 15-nerved, regular or 2-lipped at anthesis, in fr enlarged, indurate, and evidently 2-lipped; cor very irregular, the tube long-exsert, widened from the middle to the summit, the 4 upper lobes about alike, the lower median lobe much longer, deeply fringed; stamens 2 or 4, divergent, long-exsert, the pollen-sacs widely divergent; nutlet usually solitary, globose, smooth; perennial herbs from a thick woody rhizome, with ample, petiolate, serrate lvs and terminal panicles of yellow or yellowish fls. 5, e. N. Amer.

1 Stamens 4; lvs only 2 or 3 pairs, usually closely approximate .... 1. *C. verticillata.*
1 Stamens 2; lvs several pairs, scattered on the stem .... 2. *C. canadensis.*

1. Collinsonia verticillata Baldwin.   Whorled h.-b. Stems 2–5 dm; lvs only 2 or 3 pairs, usually closely approximate and appearing whorled at the base of the infl (or the pairs seldom separated by as much as 4 cm), ovate to oblong-obovate, short-acuminate, hairy along the veins above and beneath, ciliate-margined, on petioles 1.5–4 cm; infl usually with a single elongate axis, the fls in the lower and middle part mostly 3–6 at a node, not bracteate; cal 4–6 mm at anthesis, nearly regular, the lanceolate lobes twice as long as the tube, in fr 2-lipped, 12–15 mm, hirsute within; cor 15–20 mm, the lower lobe about as wide as long; stamens 4. Low woods; s. Va. to Ga. and Ala.; disjunct in s. O. May–July. (*Micheliella v.*)

2. Collinsonia canadensis L.   Northern h.-b. Erect, branched above, to 12 dm; lvs several pairs, ovate or ovate-oblong, acuminate, glabrous, the lower 1–2 dm, on petioles to 1 cm, the upper progressively smaller and the uppermost commonly sessile; panicles 1–3 dm, divergently branched, often compound with branches from the uppermost lf-axils; fls all paired at the minutely bracteate nodes of the infl; cal 2.5–4 mm at anthesis, weakly bilabiate, 8 mm in fr; cor 12–15 mm, the lower lobe oblong, nearly as long as the tube; stamens 2; $2n=50$. Rich woods; Que. to Mich. and s. Wis., s. to Fla. and Ark. July–Sept.

18. PERILLA L.   Cal campanulate, 10-nerved, at anthesis weakly 2-lipped, the lips slightly longer than the villous tube, the lower cleft to the base into narrowly triangular segments, the upper cleft a third its length into broadly triangular lobes, the middle one the shortest; fruiting cal enlarged, distinctly 2-lipped, distended on the lower side; cortube shorter than the cal, the lips about equal, the 5 lobes broadly rounded; stamens 4, nearly equal, straight, not connivent, about as long as the cor; nutlets globose, areolate; annuals, often purple or purplish, with large, long-petioled lvs and small, purple or white fls borne singly in the axils of small bracteal lvs, forming a loose, elongate, spike-like infl. 5, India to Japan.

1. **Perilla frutescens** (L.) Britton.   Perilla-mint. Lvs ovate-oblong to broadly ovate, 8–15 cm, short-acuminate, coarsely serrate or incised; infls 5–15 cm, terminal and from the upper axils; bracts oval, folded; pedicels 1–3 mm; cal at anthesis 3 mm, in fr 9–12 mm and hairy within. Native of India, cult. for its ornamental foliage and sometimes used as a condiment, but in larger amounts causing severe pulmonary edema; escaped into waste places and roadsides, Mass. to Io. and Kans., s. to Fla. and Tex. Aug., Sept.

19. ORTHODON Bentham & Oliver.   Cal usually 10-nerved, 2-lipped, the lips longer than the broad tube, the lower cleft to its base into narrowly triangular lobes, the upper cleft half length into broadly triangular lobes, the middle lobe the shortest; fruiting cal enlarged, hairy within; cor-tube about equaling the cal, the lips short, shallowly lobed; fertile stamens 2, straight; nutlets globose, areolate; annuals with small fls borne 2 at each node in the axils of small bracts, forming a slender, interrupted terminal spike. (*Mosla.*) 15, e. Asia.

1. **Orthodon dianthera** (Buch.-Ham.) Hand.-Mazz.   Stems erect, 5–8 dm, finely hairy; lvs long-petioled, lance-ovate, 3–5 cm, acute at both ends, serrate; cor 3.5–4 mm. Native of e. Asia, established in moist places in se. Ky. Aug., Sept. (*Mosla d.*)

20. CUNILA L., nom. conserv.   Stone-mint; dittany. Cal obconic, regular, strongly 10-nerved, villous in the throat, the triangular lobes a fourth as long as the tube; cor nearly regular, the tube exsert from the cal, pubescent, the upper lip erect, shallowly obcordate, the lower lip spreading, 3-lobed; stamens 2, the filaments straight, elongate, surpassing the cor; nutlets ellipsoid, smooth; perennial herbs or low shrubs, with entire or dentate lvs and small, purple to white fls in peduncled, loosely branched, trichotomous clusters terminating the stem and in the upper axils. 15, New World.

1. **Cunila origanoides** (L.) Britton.   Stems woody at base, freely branched, 2–4 dm, glabrous; lvs subsessile, ovate or deltoid-ovate, 2–4 cm, acute, glabrous, commonly with a few teeth; axillary cymes 3–9-fld, usually shorter than the subtending lf, mingled with linear bracts 1 mm; terminal cymes larger, more loosely branched; cal glabrous, 3 mm; cor rose-purple to white, 6–8 mm. Dry or rocky woods; s. N.Y. and Pa. to Ind., Ill., and Mo., s. to S.C. and Okla. Aug.–Oct.

21. LYCOPUS L.   Water-horehound; bugle-weed. Cal evidently to obscurely 4–5-nerved, sometimes with additional lesser nerves, the 4–5 teeth equal or nearly so; cor small, the tube short, internally hairy at the throat, the limb nearly regularly 4–5-lobed, when 4-lobed the upper lobe tending to be broader than the others and emarginate; fertile stamens 2, slightly exserted, the upper pair obsolete or represented by small staminodes; pollen-sacs parallel, separately attached to the short connective; nutlets widened upward, bearing a corky ridge along the lateral angles and often across the top, the outer surface smooth and nearly plane, the inner convex and commonly glandular; rhizomatous and/ or stoloniferous perennials, scarcely aromatic, with serrate to pinnatifid, wholly cauline lvs and numerous dense axillary clusters of small white fls in the summer. Our spp. are habitally much alike, 1–10 dm, growing in wet places. 14, widespread but mainly N. Amer.

1 Cal-lobes broadly triangular, under 1 mm, more than half as wide at base, obtuse or barely acute, at
    maturity surpassed by the nutlets.
2 Cal and cor 4-lobed; no tubers; inner angle of the nutlet essentially as long as the lateral angles . . . . . 1. *L. virginicus.*
2 Cal and cor 5-lobed; tuberiferous; inner angle of the nutlet much shorter than the lateral ones, the
    summit of the set of 4 nutlets therefore concave . . . . 2. *L. uniflorus.*

1 Cal-lobes narrowly triangular, 1–2+ mm, much less than half as wide at base, tapering to a slenderly
   acuminate tip, at maturity surpassing the nutlets.
  3 Inner angle of each nutlet much shorter than the lateral angles, not ending in a tubercle, the set of
     4 nutlets therefore concave at the top; cor 4-lobed.
    4 Lvs sessile or nearly so; rhizome ending in a tuber .... 3. *L. asper.*
    4 Lvs tapering to a short petiole or petiolar base; no tubers.
      5 Hairs of the lower lf-surface short, 0.1–0.5 mm; anthers 0.25–0.5 mm; abundant and wide-
         spread native sp. .... 4. *L. americanus.*
      5 Hairs of the lower lf-surface longer 0.5–1.5 mm; anthers 0.5–0.7 mm; European sp., intr. in
         a limited part of our range .... 5. *L. europaeus.*
  3 Inner angle of each nutlet about as long as the lateral angles and ending in a tubercle, the set of
     nutlets thus essentially flat across the top; cor 5-lobed, or 4-lobed with the upper lobe deeply
     notched.
      6 Lvs petiolate, or concavely narrowed to a petiolar base .... 6. *L. rubellus.*
      6 Lvs sessile or nearly so.
        7 Lvs lanceolate or linear, gradually narrowed to the base, the upper evidently narrower than
           the lower and seldom as much as 1 cm wide .... 7. *L. angustifolius.*
        7 Lvs lance-ovate, usually rounded to the base, even the upper ones generally at least 1 cm wide
           .... 8. *L. amplectens.*

1. **Lycopus virginicus** L. Virginia w.-h. Plants stoloniferous, generally without tubers; stems ± densely appressed-hairy; lvs ± long hairy and on the lower surface usually also short-felty, lanceolate or lance-ovate to elliptic or narrowly rhomboid, 5–12 × 1.5–5 cm, acuminate, coarsely serrate, the lowest tooth not far below the middle of the blade, the margin below it concave and continued along the midvein nearly or quite to the stem; bracts minute; cal 1–1.2 mm, the 4 lobes ovate or triangular-ovate, half to three-fourths as long as the tube; cor generally 4-lobed, but the upper lobe often notched; stamens included or barely reaching the orifice of the cor; nutlets surpassing the cal, the set of 4 almost flat across the tuberculate summit, each 1.5– 2 × 0.7–1.3 mm, the inner angle nearly as long as the outer and ending in a tubercle. Mass. to n. Fla., w. to N.Y., Pa., s. Ind., Mo., e. Okla., and e. Tex. (*L. membranaceus.*) Hybridizes extensively with no. 2 where their ranges overlap, producing a hybrid swarm called *L.* × *sherardii* Steele.

2. **Lycopus uniflorus** Michx. Northern w.-h. Plants stoloniferous, each stolon ending in a shallow tuber from which the solitary stem arises the next year; herbage inconspicuously hairy or glabrous; lvs lanceolate or oblong, 2–6 cm, acute or short-acuminate, with a few low teeth, the proximal margins straight or slightly convex, rarely slightly concave; cal-lobes 5; cor 5-lobed, the lobes ± spreading; stamens exsert; nutlets surpassing the cal, each 1.1–1.8 mm, three-fourths as wide, the tubercles commonly 3 and restricted to the peripheral margin, the interior angle distinctly shorter than the outer ones and not ending in a tubercle, the set of 4 nutlets therefore with a depressed center; $2n=22$. Nf. and e. Que. to Alas., s. to N.C., Ark., and Calif. (*L. virginicus* var. *pauciflorus.*)

3. **Lycopus asper** Greene. Western w.-h. Rhizomes tuberous-thickened at the tip, the tuber producing a single stem the next year; stem spreading-hairy, at least on the angles; lvs numerous, narrowly oblong to oblanceolate, 4–10 × 0.6–3.5 cm, sessile or subsessile and usually broad-based, rather coarsely but fairly evenly serrate, glabrous or often scabrous to hirsute-puberulent; cal-lobes 5, narrow, firm, slenderly subulate-pointed, with evident midnerve, distinctly surpassing the mature nutlets; cor 4-lobed, barely surpassing the cal; nutlets mostly 1.6–2.1 × 1.4–1.8 mm, the outer apical margin truncate and often irregularly toothed, the inner angle much shorter and not tuberculate, the set of 4 nutlets therefore with a depressed center; $2n=22$. Minn., Io., and Sask. to B.C. and Calif., and intr. about the w. end of Lake Erie in Mich. and Ont.

4. **Lycopus americanus** Muhl. American w.-h. Rhizomes elongate, without tubers; lvs glandular-punctate and usually glabrous (or scabrous above) except along the veins beneath, where provided with short hairs only 0.1–1.5 mm, the blade lanceolate or narrowly oblong to nearly linear, 3–8 cm, tapering to a short petiole or narrow, subpetiolar base, coarsely and irregularly incised-toothed or subpinnatisect, the lower teeth the larger; bracts lanceolate and about equaling the cal, avg 3 mm; cal 2–3.3 mm, its 5 lobes narrow, firm, slenderly subulate-pointed, surpassing the nutlets; cor 4-lobed, the upper lobe broader and notched; anthers 0.25–0.5 mm; nutlets 1.0–1.4 × 0.7–1.0 mm, the outer apical margin smooth and broadly rounded, the corky lateral ridges confluent around the tip, the inner angle much shorter than the outer ones, so that the center of the set of nutlets is depressed; $2n=22$. Our most abundant sp.; Nf. and Que. to B.C., s. to Fla. and Calif.

5. **Lycopus europaeus** L. European w.-h. Much like no. 4, and hybridizing freely with it in the Great Lakes-St. Lawrence region, but often with slender stolons as well as rhizomes, and averaging more hairy, the hairs along the veins of the lower lf-surface mostly 0.5–1.5 mm; lvs averaging wider and more bluntly toothed, commonly strigose above; bracts avg longer (5.5 mm); cal a little longer (3–4.5 mm) its lobes with long, subulate tip; anthers longer (0.5–0.7 mm); nutlets a little larger (1.3–1.7 × 0.9–1.25 mm) and not so strongly tapering toward the base; $2n=22$. European sp., casually intr. along the coast from Mass. to N.C., and now well established along the St. Lawrence R. and about lakes Erie and Ontario and w. Lake Huron. Hybridizes with no. 4.

6. **Lycopus rubellus** Moench. Stalked w.-h. Plants glabrous to densely pubescent, stoloniferous, each stolon ending in a shallow, slender tuber from which a solitary stem arises the next year; lvs thin, lanceolate to elliptic, 5–10 × 1–3 cm, sharply acute to long-acuminate, subentire to sharply serrate, the margins below

the lowest tooth concave, forming a petiolar base, or the lf petioled; bracts minute; cal-lobes 5, narrowly triangular, often acuminate but not subulate-tipped; cor 5-lobed, twice as long as the cal; nutlets 1.2–1.5 mm, two-thirds as wide, very flat across the truncate-tuberculate summit, without a distinct dorsal area. Me. to Fla. and Tex., mainly on or near the coastal plain, w. irregularly to s. Mich., and common thence to Ill., Mo., and Ark.

7. **Lycopus angustifolius** Elliott.   Narrow-lvd w.-h. Plants with stolons and tubers as in no. 6; stem shortly and densely hairy; lvs felty on both sides with slightly longer hairs on the veins, lanceolate to narrowly linear, with ascending to spreading teeth, the lower 6–12 × 1.5–3 cm, the upper somewhat smaller and distinctly narrower, 4–7 × 0.3–1 cm; base of blade concavely narrowed to a sessile base; bracts 1–2 mm; cal 2.3–3 mm, the 5 attenuate (not subulate) teeth well surpassing the nutlets; cor 4-lobed, the upper lobe deeply notched; nutlets 1–1.2 mm, with low rounded teeth on the crest and upper surface. Coastal plain from se. Va. to Fla. and Miss., and inland irregularly to Tenn. and s. Mo.

8. **Lycopus amplectens** Raf.   Much like no. 6; lvs oblong to lance-oblong, 3–10 × 1–3.5 cm, with 4–6 salient teeth on each side, convexly narrowed from the lowest tooth to the sessile, obtuse or rounded base. Coastal plain; e. Mass. to Fla.; also in the mts. of N.C. and at the s. end of Lake Michigan.

22. ELSHOLTZIA Willd.

Cal regular, campanulate or broadly tubular, 10-nerved, its lobes triangular; cor almost regular, the tube exsert, gradually widened upwards, the upper lip shallowly 2-lobed, straight or upcurved, the lower lip spreading, its 3 lobes short and rounded; stamens 4, straight or somewhat divergent, exsert, the lower pair the longer; nutlets oblong or ovoid; herbs or half-shrubs with small fls subtended by overlapping bracts in dense terminal and axillary spikes. 35, mainly Eurasia.

1. **Elsholtzia ciliata** (Thunb.) Hylander.   Branching annual 3–10 dm; lvs petioled, ovate to lanceolate, 3–7 cm, acute or short-acuminate, crenate-serrate, tapering to the base; spikes 2–5 cm; bracts rotund, 3 mm, cuspidate, ciliate, closely imbricate, each subtending 2 or 3 pale blue fls 4 mm; cor villous around the lips. Native of Asia, established in moist soil from Que. to N.Y. and N.J., and sometimes inland to Wis. July–Sept.

23. AGASTACHE Clayton.

Giant hyssop. Cal nearly regular, the 15-nerved tube somewhat oblique at the throat, the 3-nerved lobes about alike; upper cor-lip directed forward, shallowly 2-lobed, the lower lip 3-lobed, its central lobe expanded and decurved, but its lateral lobes displaced onto the summit of the cor-tube; stamens 4, exsert, the 2 lower curved upward under the upper cor-lip, the 2 upper longer and downcurved; pollensacs parallel or nearly so; nutlets minutely hairy at the tip; perennial herbs with numerous rather small fls crowded in dense verticils subtended by inconspicuous bracts, forming terminal, continuous or interrupted spike-like infls. 20, N. Amer. and se. Asia.

1 Lvs green beneath; cal glabrous, green, pale green or pink-tinged.
  2 Cor greenish-yellow; cal-lobes at anthesis ovate, 1–1.5 mm .... 1. *A. nepetoides.*
  2 Cor purplish; cal-lobes at anthesis lance-triangular, 2–2.5 mm .... 2. *A. scrophulariaefolia.*
1 Lvs closely white-tomentulose beneath; cal pubescent, blue at least distally .... 3. *A. foeniculum.*

1. **Agastache nepetoides** (L.) Kuntze.   Catnip g.-h. Erect, 1–1.5 m, branched above; lvs thin, green, ovate or lance-ovate, to 15 cm, reduced upwards, coarsely serrate, rounded or subcordate at base, finely hairy beneath; petioles of the larger lvs to 6 cm, the upper reduced; spikes subcylindric, to 2 dm, 1–1.5 cm thick, occasionally interrupted at base; bracts broadly ovate, acuminate, not large; cal at anthesis glabrous, 6 mm, the lance-ovate lobes 1–1.5 mm, with convex margins, obtuse or subacute; cor greenish-yellow; 2n=18. Open woods; Vt. to s. Ont., Minn., and se. S.D., s. to Ga. and Okla. Aug., Sept.

2. **Agastache scrophulariaefolia** (Willd.) Kuntze.   Purple g.-h. Much like no. 1; lvs glabrous to villous beneath; spikes cylindric or tapering, to 15 cm, 1.5–2 cm thick (including the cors); bracts round-ovate, caudate-acuminate, often projecting; cal 7–9 mm at anthesis, the lobes 2–2.5 mm, lance-triangular, with straight or concave sides, sharply acute or acuminate; cor purplish. Upland woods; Vt. and N.Y. to Minn. and e. S.D., s. to N.C. and e. Kans. Aug., Sept.

3. **Agastache foeniculum** (Pursh) Kuntze.   Lavender g.-h. Erect, to 1 m, simple or branched above; lvs ovate to rhombic-ovate or deltoid-ovate, the larger to 9 cm, coarsely serrate, broadly obtuse to truncate at base, whitened beneath, with a very fine close pubescence, the hairs scarcely discernible at 10×; petioles rarely to 1.5 cm, even on the largest lvs; spikes cylindric, to 15 cm, 2–2.5 cm thick, commonly interrupted at base; bracts broadly ovate, acuminate; cal puberulent at anthesis, 5–7 mm, its lobes blue, lance-triangular, 1.5–2 mm; cor blue, nearly 1 cm; 2n=18. Dry upland woods and prairies; Wis. to n. Io. and Man., s. to Colo. and Alta., and occasionally e. (probably only intr.) to Que. July, Aug. (*A. anethiodora.*)

24. MEEHANIA Britton.   Cal campanulate, 15-nerved, slightly oblique at the throat, the lobes triangular, acute, the lower 2 slightly shorter than the upper 3; cor large, the tube very slender at base, distended distally, especially on the lower side, the upper lip 2-lobed, the lower slightly larger, more deeply 3-lobed, villous; stamens 4, ascending under the upper lip and about equaling it, the upper pair slightly the longer; pollen-sacs parallel or nearly so; trailing perennial herbs with handsome blue fls in a terminal, leafy, secund spike. 2, the other in e. Asia.

1. **Meehania cordata** (Nutt.) Britton.   Trailing stems to 6 dm, the erect or ascending flowering stems 1–2 dm; lvs long-petioled, broadly ovate, 3–6 cm, obtuse, shallowly crenate, cordate at base; spikes few-fld, 3–5 cm, the elliptic bracts little surpassing the cal; cal-tube 1 cm, the upper lobes 5–6 mm, the lower two 1 mm shorter; cor ascending, 2.5–3 cm. Rich mt. woods; sw. Pa. and s. O. to Tenn. and N.C.; rep. from Ill. May–July.

25. NEPETA L.   Cal 15-nerved, curved, oblique at the mouth, 5-toothed; cor bilabiate, the upper lip somewhat concave and often subgaleate, entire or bifid, the lower lip spreading, 3-lobed, the large central lobe sometimes notched; stamens 4, ascending under the upper lip and scarcely exserted, the upper pair slightly the longer; pollen-sacs widely divaricate, confluent in dehiscence; herbs, mostly erect, with toothed or cleft lvs and numerous small, white to blue fls in terminal, compound, often spike-like infls. 150, Eurasia.

1. **Nepeta cataria** L.   Catnip. Taprooted perennial; 3–10 dm, branched above, finely and densely canescent throughout; lvs all cauline, triangular-ovate, cordate or truncate at base, 3–8 cm, coarsely toothed, the petiole half as long as the blade; fl-clusters 2–6 cm, continuous or interrupted, rather loosely many-fld; cal at anthesis 5–7 mm, weakly bilabiate, the lobes half as long as the tube; cor 8–12 mm, dull white, dotted with pink or purple, the lower lip crenulate; $2n=32, 34, 36$. Native of Eurasia, now established in disturbed habitats throughout our range and w. July–Oct.

26. GLECHOMA L., nom. conserv.   Ground-ivy. Cal 15-nerved, slightly oblique at the throat, unequally 5-toothed, the midnerve of each tooth excurrent into a short awn; cor 2-lipped, the upper lip somewhat concave, 2-lobed, the lower lip spreading, with a large central lobe and smaller lateral ones; stamens 4, the upper pair the longer, normally ascending under the upper lip and scarcely exserted; pollen-sacs set end to end along one side of the filament just below its summit (which commonly surpasses the upper pollen-sac), divergent at a wide angle; opening by separate slits; creeping perennials with long-petioled, rotund or reniform lvs and small or medium-sized, short-pediceled, blue fls usually 3 per axil. 10, temp. Eurasia.

1. **Glechoma hederacea** L.   Gill-over-the-ground. Fibrous-rooted perennial from slender stolons or superficial rhizomes, the stems lax, 1–4 dm, retrorsely scabrous (or hirsute) to subglabrous, pilose at the nodes; lvs all about alike, petiolate, the blade glabrous or hirsute, rotund-cordate to cordate-reniform, strongly crenate, 1–3 cm; fls short-pedicellate; cal narrow, 5–6 mm, hirtellous-scabrous, the upper teeth the longer; cor blue-violet, purple-maculate, 13–23 mm, or in pistillate plants only 8–15 mm, the anthers then minute and included in the cor-tube; $2n=18, 36$. Moist woods and various disturbed habitats; native of Eurasia, now found throughout our range. Apr.–June. (*Nepeta h.*)

27. DRACOCEPHALUM L., nom. conserv.   Dragon-head. Cal tubular, 15-nerved, 2-lipped, the uppermost lobe much wider than the other 4; cor weakly 2-lipped (in ours), the elongate tube gradually widened distally, the limb much shorter; upper lip straight, emarginate, and subgaleately rounded, the lower lip 3-lobed, with the large central lobe sometimes 2-cleft; stamens 4, ascending under the upper lip, the upper pair slightly the longer; pollen-sacs divaricate at a wide angle, confluent in dehiscence; nutlets oblong, smooth; erect herbs with serrate lvs and verticils of small blue fls. (*Moldavica.*) (Name often misapplied to *Physostegia.*) 40, N. Hemisphere.

1 Fls in dense, terminal, head-like clusters; bracts pectinately serrate; native .... 1. *D. parviflorum.*
1 Fls in loose, elongate, interrupted racemes; bracts entire or ciliate; intr. .... 2. *D. thymiflorum.*

1. **Dracocephalum parviflorum** Nutt. American d.-h. Taprooted, short-lived, 2–8 dm, inconspicuously short-hairy; lvs petiolate, the blade mostly lance-elliptic to broadly lance-triangular, 2.5–8 × 1–2.5 cm, coarsely serrate; infl dense and spike-like, 1.5–3.5 cm thick, often interrupted below, the terminal segment 2–10 cm; bracts subfoliaceous, mostly 1–3 cm sessile or nearly so, aristately few-toothed; cal 1 cm, the tube about equaling the aristate-tipped lobes; cor purplish, barely surpassing the cal, with short lips; 2*n*=14. Dry soil; Que. to Alas., s. to n. N.Y., Ill., and Ariz. May–July.

2. **Dracocephalum thymiflorum** L. Thyme-lvd d.-h. Smaller-lvd and more slender than no. 1, with the bracts of the narrower and less compact infl entire or nearly so; cal 6–8 mm, the tube obviously longer than the teeth; 2*n*=20. Native of Eurasia, occasionally intr. in our range. June–Aug.

D. moldavica L., with lax, narrowly leafy-bracteate infl and large cor 2–2.5 cm much exceeding the cal, has been reported as casually adventive.

28. PHYSOSTEGIA Benth.

Cal 10-nerved, equally or unequally 5-toothed; cor elongate, bilabiate, the upper lip galeate-hooded and barely or scarcely notched, the lower lip about as long as the upper, spreading, 3-lobed; stamens 4, ascending under the upper lip, the lower pair the longer; pollen-sacs nearly parallel, opening separately; nutlets smooth; fibrous-rooted perennials with glabrous herbage and showy pink-purple (white) fls borne singly in the axils of small bracts, forming elongate terminal racemes. 12, N. Amer. (*Dracocephalum,* misapplied.)

In addition to the following spp., *P. ledinghamii* Cantino and *P. intermedia* (Nutt.) Engelm. & A. Gray may be sought at the margins of our range. The former is an apparent amphiploid of nos. 1 and 3, known from N.D. and Man. to s. Mack. It has fls mostly 14–23 mm and is in general intermediate between its presumed ancestors. *P. intermedia,* of the lower Mississippi Valley, extends n. to se. Missouri. It has a short cal (tube 2–4 mm) without glandular hairs and a relatively small (to 20 mm) cor lacking glandular hairs; the base of the stem is often strongly swollen, and the uppermost lvs are generally clasping and widest at the base of the blade.

1 Fls relatively large, the cor (14–)16–35 mm, not glandular.
  2 Generally none of the lvs clasping the stem; widespread . . . . 1. *P. virginiana.*
  2 Generally one or more pairs of the upper lvs clasping the stem; southeastern . . . . 2. *P. leptophylla.*
1 Fls smaller, the cor mostly 12–16 mm, ± stipitate-glandular; northwestern . . . . 3. *P. parviflora.*

1. **Physostegia virginiana** (L.) Benth. Obedience. Erect perennial to 15 dm, with 10–34 nodes below the infl, often branched at the top and bearing several racemes 5–15 cm; lvs all sessile, or the lowest 1–7 pairs petiolate but soon deciduous; sessile lvs 2–18 cm, mostly oblanceolate or elliptic, generally not clasping, sharply or less often bluntly toothed, rarely entire; upper lvs scarcely to evidently reduced; raceme-axis short-hairy and sometimes also stipitate-glandular; cal often glandular-punctate and sometimes also stipitate-glandular; cor (14–)16–35 mm, densely puberulent to glabrous, not glandular; 2*n*=38. In a wide range of mostly ± open habitats; Me. and Que. to Man. and N.D., s. to Fla. and n. Mex., seldom near the coast except as an escape from cult. July–Sept. Highly variable, here divided into 2 well marked geographic vars.: The relatively northern var. *virginiana,* seldom occurring as far s. as Va., Tenn., Mo., and Kans., is generally clonal, with the stems scattered on long, horizontal rhizomes, and lacks empty floral bracts below the infl. (*P. denticulata; P. formosior.*) The relatively southern var. *arenaria* Shimek, seldom occurring n. to n. O., n. Ill., and Io., generally has the stems clustered on a short caudex (without long rhizomes), and very often has some empty floral bracts below the infl; it typically grows in drier sites than var. *virginiana.* (*P. praemorsa; P. v.* ssp. *praemorsa; P. v.* var. *reducta; P. serotina.*) Many garden and escaped plants are ± intermediate, with the rhizomes of var. *virginiana* and the empty bracts of var. *arenaria,* but primary intergradation occurs mainly in c. and e. Tenn., c. and e. Ky., and adjoining territory.

2. **Physostegia leptophylla** Small. Rhizomatous, clonal, erect perennial to 15 dm, with 7–15 nodes below the infl; lvs crenate or repand to seldom entire or sharply toothed; lowest 4–9 pairs of lvs petiolate and often persistent, the petiole to 6 cm, the blade 3–10 × 1–3 cm and mostly tapering at base; middle and upper lvs sessile, mostly widest at or below the middle (but generally not at the base), at least some of them clasping; floral bracts abruptly smaller than the upper leaves; raceme-axis minutely puberulent; fls loosely spaced; cal-tube 3.5–6 mm at anthesis, often glandular-punctate but not stipitate-glandular; cor (14–)16–30 mm, densely puberulent to subglabrous; 2*n*=76. Wooded riverine swamps and fresh or brackish marshes on the coastal plain, often in deep shade; se. Va. to Fla. July–Sept. (*P. aboriginorum; P. denticulata,* misapplied.) Closely allied to the diploid (2*n*=38) *P. purpurea* (Walter) S.F. Blake, also of the coastal plain but extending n. only to N.C. and usually found in sunnier habitats, which has the upper lvs more reduced, and the principal lvs widest at or above the middle.

3. **Physostegia parviflora** Nutt. Western obedience. Stems to 10 dm, arising singly from a rhizome or vertical rhizome-like base, with 9–15 nodes below the infl; lower lvs reduced and deciduous, the others 3–

10 × 0.5–2.5 cm, serrate or subentire, sessile, linear-oblong to elliptic-oblong, those directly beneath the infl more lanceolate or lance-ovate, broadest near the base and somewhat clasping; racemes to 10 cm, closely fld; cal finely glandular-puberulent but not conspicuously punctate; cor 12–16 mm, usually stipitate-glandular; $2n=38$. Wet low places, as along the shores of rivers and lakes; w. Minn. and w. Io. to Man., B.C., Oreg., and n. Utah. July–Sept. (*Dracocephalum nuttallii*.)

29. PRUNELLA L.   Self-heal. Cal irregularly 10-nerved, bilabiate, the upper lip broad, shallowly 3-toothed, the lower deeply cleft into 2 narrow segments; cor bilabiate, the upper lip galeate-hooded and entire or nearly so, the lower shorter and 3-lobed; cor-tube with a ring of hairs inside; stamens 4, ascending under the galea, the lower pair the longer, scarcely exserted; filaments ± bidentate at the tip, the anther borne on the lower tooth; pollen-sacs set end to end on the expanded connective, separately dehiscent; nutlets smooth; perennial herbs with blue or purple to white fls crowded into a dense, evidently bracteate terminal spike, the bracts sharply differentiated from the lvs. 4, cosmop. (*Brunella*.)

1. **Prunella vulgaris** L.   Perennial, nearly prostrate to suberect, 1–5 dm; lvs few, petiolate, entire or obscurely toothed, the blade lanceolate or elliptic to broadly ovate, 2–9 × 0.7–4 cm, the lower mostly broader and with more rounded base than the upper; spikes 2–5 × 1.5–2 cm, the depressed-ovate, abruptly short-acuminate bracts 1 cm, strongly ciliate; cal 7–10 mm, green or purple, the lips longer than the tube, the teeth spinulose-tipped; cor blue-violet (pink or white), 1–2 cm (or smaller in plants with reduced anthers), the tube equaling or surpassing the cal, the lips short; $2n=28$, 32. Nearly cosmopolitan. The European var. *vulgaris*, intr. into disturbed sites in our range, tends to have relatively broad lvs, the middle cauline ones half as wide as long, with broadly rounded base. The native var. *lanceolata* (Barton) Fern., in both disturbed and natural sites, has narrower lvs, the middle cauline ones a third as wide as long, with tapering base. Some conspicuously white-hirsute plants along the s. border of our range may prove to represent the Asiatic var. *hispida* Benth., or may be extreme forms of var. *lanceolata*, which they otherwise resemble.

P. laciniata L., a Eurasian sp. with the upper (sometimes also the median) lvs pinnatifid, usually ochroleucous fls, and strongly hirsute herbage, has been found as a waif in Pa. and D.C.

30. SYNANDRA Nutt.   Cal-tube obconic, very thin-walled, loosely enclosing the cor; cal-lobes green, loosely spreading, lance-triangular, the 2 lateral exceeding the 2 lower, the median upper one much reduced or even wanting; cor strongly 2-lipped, the tube much dilated distally, the large upper lip subgaleately rounded, entire, the lower longer, 3-lobed; stamens 4, ascending under the upper lip, the lower pair slightly the longer; filaments hairy, incurved at the summit, bringing the anthers into contact; nutlets smooth, biconvex; perennial with solitary fls in the axils of bracteal lvs, forming a terminal spike. Monotypic.

1. **Synandra hispidula** (Michx.) Britton.   Stems 2–6 dm, villous; cauline lvs 2–3 pairs, broadly cordate-ovate, 4–8 cm, coarsely toothed, sparsely hairy, the petiole often longer than the blade; lowest pair of bracteal lvs resembling the foliage lvs and scarcely smaller, but sessile, the upper ones progressively reduced; cal 8–10 mm, the lobes slightly shorter than the tube; cor pale greenish-yellow or greenish-white, 3–3.5 cm, the lower lip with purple lines. Rich woods; w. Va., s. O., s. Ind., and s. Ill., s. to e. Tenn. and n. Ala. May–July.

31. PHLOMIS L.   Cal strongly 5-nerved, with subulate lobes, the upper 3 ascending or erect, the lower 2 slightly longer, spreading or deflexed; cor strongly 2-lipped, the upper lip galeate-hooded, emarginate, densely white-villous, the lower deflexed, 3-lobed; stamens 4, about equal, ascending under the upper lip; nutlets hairy at the tip; perennial herbs or shrubs with large fls in dense clusters axillary to bracteal lvs. 100+, temp. Old World.

1. **Phlomis tuberosa** L.   Jerusalem-sage. Erect, fibrous-rooted perennial 5–15 dm, the roots bearing small tubers; lower lvs long-petioled, cordate-ovate, 1–2 dm, coarsely crenate, the others progressively reduced and less petiolate, the bracteal ones subsessile; spikes 1–3 dm; each verticil subtended by numerous subulate bracts 1 mm; cal-tube 1 cm, hirsute on the nerves, the lobes 2–3.5 mm; cor purple, 2 cm; $2n=22$. Native of Eurasia, naturalized in w. N.Y. June–Sept.

32. LAMIUM L.   Dead nettle. Cal nearly regular, 5-nerved, the lobes equal or the upper the largest; cor bilabiate, the upper lip entire or 2-lobed, galeate-hooded, the lower spreading, constricted at the base of the enlarged, cleft or emarginate central lobe, the lateral lobes broad and low, seemingly borne on the cor-throat; stamens 4, ascending under the upper lip, the lower pair the longer; pollen-sacs conspicuously hairy, ± divergent or divaricate; nutlets trigonous, truncate above; herbs with toothed or pinnatifid, mostly cordate lvs and white to red or purple fls in verticils of 6–12 subtended by scarcely reduced lvs. 40, temp. Eurasia and n. Afr.

1 Upper lip of the cor 7–12 mm; perennial; lvs all petioled.
  2 Cor mostly pink-purple; terminal tooth of the upper lvs ± obtuse .... 1. *L. maculatum.*
  2 Cor white; terminal tooth of the upper lvs sharply acute .... 2. *L. album.*
1 Upper lip of the cor 2–5 mm; annual.
  3 Lvs all petiolate .... 3. *L. purpureum.*
  3 Lvs subtending the fl-clusters mostly or all sessile and clasping .... 4. *L. amplexicaule.*

    1. Lamium maculatum L. Spotted d.n. Rhizomatous perennial, lax and tending to creep at the base, 1.5–7 dm, somewhat hirsute; lvs all petiolate, with cordate-ovate or triangular-ovate, crenate-serrate blade 1.5–6 × 1–5 cm, generally with an irregular broad white stripe about the midvein; verticils several, less crowded than in no. 3, the lower subtended by full-sized lvs, the upper by smaller and less petiolate ones; cal 5–10 mm, with firm, unequal, somewhat divaricate slender lobes; cor pink-purple (white), 2–3 cm, the tube curved, hairy outside and with a ring of hairs inside, the upper lip 7–12 mm, the lateral lobes each with a single terminal tooth; $2n=18$. Native of Europe and adj. Asia, escaped from cult. on roadsides and waste places; e. Can. to Va. and W.Va. Apr.–Sept.

    2. Lamium album L. White d.n. Rhizomatous perennial, curved at base; stems pubescent, 2–5 dm; lvs all petioled, green, ovate or more often deltoid, 3–10 cm, coarsely crenate, narrowed with flat or concave sides to an acute tip; cal 10–13 mm, the lobes conspicuously longer than the tube, the terminal setaceous part longer than the triangular base; cor white, 2–3 cm, the tube curved, with an oblique ring of hairs internally near the base, the upper lip more than half as long as the tube, long-villous, the lateral lobes each with 2–3 terminal teeth; $2n=18$. Native of Eurasia, occasionally intr. in gardens, roadsides, and waste places; Ont. to Va. Apr.–Sept.

    3. Lamium purpureum L. Red d.n. Annual from a short taproot, often branched at base, decumbent or nearly prostrate, inconspicuously hairy, the stems 1–4 dm; lvs all petiolate, crenate or crenate-serrate, the lower with rotund-cordate or broader blade 0.7–3 cm and separated by 1–3 very elongate internodes from the several more ovate-cordate, closely crowded, shorter-petiolate, and eventually smaller upper pairs which subtend the verticils of fls; cal 5–6 mm, with firm, unequal, somewhat divaricate slender lobes about as long as the tube; cor pink-purple, mostly 1–1.5 cm, the tube straight, hairy outside and with a ring of hairs inside near the base; upper lip 2–4 mm; $2n=18$. Native of Eurasia, now established as an occasional weed in fields, gardens, and waste places throughout much of our range and elsewhere in Amer. Apr.–Oct.

    L. hybridum Vill., an allopolyploid ($2n=36$) with *L. purpureum* as one parent, is rarely adventive in the ne. part of our range. It differs from the latter in its irregularly incised-toothed lvs, the upper often ± decurrent onto the petiole.

    4. Lamium amplexicaule L. Henbit. Annual 1–4 dm from a short taproot, generally branched at the base, the several weak stems decumbent below; herbage inconspicuously hairy or subglabrous; proper lvs restricted to the lower part of the stem, petiolate, with broad, rounded, ± cordate, coarsely crenate or lobulate blade seldom 1.5(–2) cm; lvs subtending the fl-clusters sessile, broad-based, clasping, often 1.5(–2.5) cm, surpassing the cals but usually surpassed by the cors; verticils few and (except sometimes the upper) mostly well spaced, the lowest fully developed one often borne at or below the middle of the stem; lowest verticil sometimes few-fld and subtended by petiolate lvs; cal hirsute, 5–8 mm, the narrow, erect lobes about equaling the tube; cor purplish, 12–18 mm, glabrous inside, hairy outside, the tube straight; upper lip 3–5 mm, with purple hairs; occasional plants produce small, cleistogamous fls; $2n=18$. A weed in fields and waste places, especially in fertile soil; native to Eurasia and n. Africa, now well established in our range and elsewhere in Amer. Mar.–Nov.

    33. BALLOTA L. Black horehound. Cal regular, with 10 or more strong nerves, the lobes short and broadly triangular to linear; cor strongly 2-lipped, the tube about equaling the cal, the upper lip subgaleately rounded, hairy, the lower lip about as long as the upper, with prominent lateral lobes; stamens 4, ascending under the upper lip, the lower pair the longer; anthers glabrous; nutlets clavate, glabrous; perennial herbs or shrubs with pink or purple (white) fls in loose, short-peduncled cymes from the axils of the reduced upper lvs. 35, Eur., Medit., w. Asia.

    1. Ballota nigra L. Fibrous-rooted perennial herb 3–8 dm; main lvs short-petioled, ovate or round-ovate, 3–6 cm, coarsely crenate, the bracteal ones smaller and narrower; cymes 4–8-fld, 1–2 cm; peduncles 2–5 mm; fls subtended by setaceous bracts; cal 10-nerved, the tube 7–10 mm, the lobes nearly equal; cor 10–15 mm, pink to white, the upper lip entire, two-thirds as long as the tube, the lower median lobe transversely oblong, emarginate; $2n=22$. Native of the Mediterranean region, intr. in waste places in our range. June–Sept.

34. LEONURUS L. Cal turbinate, 5–10-nerved, the lobes prolonged into stiff slender spines, the 2 lower often slightly larger or deflexed; cor strongly bilabiate, the upper lip entire and galeately rounded, the lower spreading or deflexed, 3-lobed, with obcordate central lobe and oblong lateral lobes; stamens 4, ascending under the upper lip, the lower pair equaling or longer than the upper; pollen-sacs mostly parallel; nutlets triquetrous, truncate and hairy at the top; erect herbs with toothed or cleft, petiolate lvs and white to pink fls crowded in dense verticils subtended by bracteal lvs and by linear to subulate bracts; forming long, interrupted terminal spikes. 10, mostly temp. Eurasia.

1 Cal strongly 5-angled and 5-ribbed; upper cor-lip densely long-villous .... 1. *L. cardiaca.*
1 Cal mostly 10-nerved, scarcely angled; upper cor-lip short-hairy.
  2 Lvs coarsely dentate or serrate .... 2. *L. marrubiastrum.*
  2 Lvs palmately laciniate .... 3. *L. sibiricus.*

1. Leonurus cardiaca L. Motherwort; lion's tail. Fibrous-rooted perennial, the ± clustered stems 4–15 dm, retrorsely short-hairy on the angles (or subglabrous); lvs all cauline, the main ones with palmately cleft and again coarsely toothed blade 5–10 cm, the petiole about as long, the middle and upper gradually reduced and less cleft; fls subsessile; cal-tube firm, 5-angled, 3–4 mm, the lobes about equaling the tube, the 2 lower ones deflexed; cor 8–12 mm, white or pale pink, the upper lip conspicuously white-villous, the tube with a ring of hairs inside; $2n=18$. Native of Asia, formerly cult. as a home-remedy, now established as a casual weed over much of the U.S. and adj. Can. June–Aug.

2. Leonurus marrubiastrum L. Horehound-motherwort. Biennial; stems 10–15 dm, finely hairy; lvs lance-ovate to ovate, toothed, those subtending the verticils smaller, proportionately narrower, few-toothed to entire; bracts half as long to longer than the inconspicuously 10-nerved cal-tube; cor 5–7 mm, scarcely surpassing the cal, without an internal ring of hairs; $2n=24$. Native of Europe and n. Asia, intr. in waste places here and there in our range. June–Sept.

3. Leonurus sibiricus L. Annual or biennial; stems 10–15 dm, softly retrorse-hairy; lvs broadly ovate to rotund in outline, deeply 3-parted, each division laciniately toothed or cleft; lvs subtending the fls smaller, more deeply lobed into narrow segments; bracts half to fully as long as the weakly 10-nerved cal-tube; cor 10–14 mm, with a ring of hairs inside, the upper lip densely but finely hairy; $2n=20$. Native of Asia, widely intr. in tropical Amer. and occasionally found in waste places in our range. May–Sept.

35. GALEOPSIS L. Hemp-nettle. Cal mostly 10-ribbed, with equal, firm, spinescent teeth; cor bilabiate, the upper lip entire, subgaleately rounded, the lower lip 3-lobed, and with a pair of prominent projections ("nipples") on the upper side near the base; stamens 4, ascending under the upper lip, the lower pair the longer; pollen-sacs set end to end, with a transverse line of contact, at the expanded summit of the filament, the proximal one with its back to the filament, the distal one surpassing the filament; each pollen-sac opening by 2 valves, the larger valve smooth and next to the filament, the smaller valve minutely bristly-ciliate and thrown back from the tip, remaining attached at the line of contact of the pollen-sacs; annuals with toothed or entire lvs, the small fls crowded in dense verticils in the upper axils. 10, temp. Eurasia.

1 Stem hispid-setose with long, spreading or slightly reflexed hairs, often with shorter gland-tipped hairs
  as well .... 1. *G. tetrahit.*
1 Stem with fine, short, recurved hairs, and often with longer glandular hairs as well, but not hispid-
  setose .... 2. *G. ladanum.*

1. Galeopsis tetrahit L. Taprooted annual 2–7 dm, branched when well developed; stem swollen at the nodes, hispid-setose (notably on the angles) with long, spreading or slightly reflexed hairs, and often also with shorter but still prominent, gland-tipped hairs, especially just beneath the upper nodes; lvs petiolate, with rhombic-elliptic to ± ovate, coarsely blunt-serrate, basally rounded or cuneate, ± strigose or hirsute blade 3–10 × 1–5 cm; verticils dense, generally several, the lower often remote; cal ca 1 cm at anthesis, somewhat larger in fr, the erect spinescent lobes about equaling the tube, elongating to 5–10 mm; cor white or pink or variegated, commonly with 2 yellow or white spots; $2n=32$. Native of Eurasia, intr. and often appearing as a native in our range, especially northward. June–Sept. Var. *tetrahit* has the cor mostly 15–23 mm, the middle lobe of the lower lip nearly square, not emarginate. Var. *bifida* (Boenn.) Lej. & Courtois has smaller fls, seldom over 15 mm, the central lobe of the lower lip emarginate or cleft. (*G. bifida.*)

2. Galeopsis ladanum L. Stems freely branched, 1–4 dm, finely pubescent with minute, recurved hairs, often with spreading glandular hairs as well, but not hispid-setose; lvs 1.5–3 cm; verticils usually 2–4, few-fld; cal-lobes half as long as the canescent tube, only 3 mm even in fr; cor deep pink, 15–28 mm; $2n=16$. Native of Eurasia, intr. in waste places or along the seashore, chiefly in the coastal states, but sometimes

inland to Mich. June–Sept. Var. *ladanum* has lanceolate to ovate lvs to 15 mm wide and crenate-serrate with a few low teeth. (*Var. latifolia.*) Var. *angustifolia* (Ehrh.) Wallr., much more common with us, has mostly linear and entire lvs rarely over 5 mm wide. (*G. angustifolia.*)

36. STACHYS L.   Hedge-nettle. Cal nearly regular, 5–10-nerved, the lobes triangular to nearly lanceolate, acute or aristate; cor strongly 2-lipped, the tube about equaling the cal, the upper lip subgaleately rounded, entire or emarginate, the lower lip spreading or deflexed, 3-lobed; stamens 4, ascending under the upper lip, the lower pair slightly the longer; nutlets ovoid or oblong, usually hairy; herbs, usually hairy, with anthocyanic to white or yellow fls in small, dense axillary cymes subtended by reduced lvs, forming a terminal, loose or dense, usually interrupted spike. Perhaps 200, mainly N. Temp. Spp. 2–8 are intergradient, and further study might lead to a more conservative treatment.

1 Herbage glabrous or hairy, but not white-woolly.
  2 Rhizomatous perennials, mostly native (some forms of no. 1 intr.).
    3 Cal-lobes two-thirds to fully as long as the tube.
      4 Stem pubescent on the sides as well as on the angles . . . . 1. *S. palustris.*
      4 Stem glabrous, or pubescent on the angles only with retrorse bristles.
        5 Cal-lobes smooth; lvs commonly glabrous or nearly so above.
          6 Principal lvs sessile, or subsessile on petioles not over 8 mm.
            7 Lvs linear to narrowly oblong, entire or nearly so . . . . 2. *S. hyssopifolia.*
            7 Lvs oblong, distinctly serrate . . . . 3. *S. aspera.*
          6 Principal lvs with well developed petioles mostly 8–25 mm . . . . 4. *S. tenuifolia.*
        5 Cal-lobes hirsute or hirsute-ciliate; lvs usually evidently hairy above . . . . 5. *S. hispida.*
    3 Cal-lobes about half as long as the tube.
      8 Stem glabrous, or with reflexed bristles on the angles, not glandular; petioles glabrous on the upper side.
        9 Bristles of the stem none, or few and minute, to 0.5 mm, appressed.
          10 Petioles short, those of the principal lvs mostly 3–8 mm . . . . 6. *S. latidens.*
          10 Petioles longer, those of the principal lvs mostly 1–3 cm . . . . 7. *S. subcordata.*
        9 Bristles of the stem conspicuous, 1–2 mm, somewhat spreading . . . . 8. *S. clingmanii.*
      8 Stem, at least the younger parts, minutely but copiously stipitate-glandular beneath the bristles; petioles hairy on the upper side.
        11 Petioles of the principal lvs mostly 1–5 cm . . . . 9. *S. cordata.*
        11 Petioles of the principal lvs very short, mostly 2–5 mm . . . . 10. *S. eplingii.*
  2 Taprooted annuals, casually intr. from Europe.
    12 Cor 10–16 mm, pale yellow to white . . . . 11. *S. annua.*
    12 Cor 6–8 mm, anthocyanic to white . . . . 12. *S. arvensis.*
1 Herbage densely and softly white-woolly; intr.
  13 Cal-lobes projecting beyond the pubescence; lvs conspicuously toothed . . . . 13. *S. germanica.*
  13 Cal-lobes concealed by the tomentum; lvs entire or the teeth concealed by tomentum . . . . 14. *S. byzantina.*

1. Stachys palustris L.   Hedge-nettle. Rhizomatous perennial 3–10 dm, hairy throughout and often glandular; lvs sessile or some short-petioled, the main ones lance-triangular to lance-ovate or elliptic, broadly rounded to truncate-subcordate at base, 3.5–9 × 1–4 cm, crenate; infl a series of usually 6-fld verticils, the lower often subtended by foliage lvs, the upper by progressively reduced bracts; cal 6–9 mm, the narrow lobes nearly as long as the tube; cor purplish, white-maculate, 11–16 mm, the tube abruptly expanded on the lower side at the base of the oblique internal hairy ring, and often with a small, saccate gibbosity. Moist or wet places; circumboreal, in Amer. from Que. to Alas., s. to N.Y., Ill., Mo., and Ariz. June–Aug. (*S. ampla; S. arenicola; S. arguta; S. borealis; S. brevidens; S. homotricha; S. pustulosa; S. schweinitzii.*) Three vars. in our range:

1 Cal shortly glandular-hairy, the hairs about alike; cal-teeth with a deltoid base tapering abruptly to a long-attenuate tip; cor purple; hairs of the stem angles reflexed, conspicuously pustulate at base, longer and stouter than those of the sides; lvs usually oblong or linear, seldom over 2 cm wide; 2n=102; European, weedy and commonly intr. throughout our range . . . . var. *palustris.*
1 Cal with slender, gland-tipped hairs and long, stout, glandless ones; cal-teeth more lance-ovate, tapering gradually to the slender tip; cor pink or lavender; 2n=34, 68; native, not weedy; perhaps better treated as a distinct sp., *S. pilosa* Nutt.
  2 Hairs of the stem-angles reflexed, pustulate, distinctly longer but scarcely stouter than those of the sides; lvs lance-ovate or broadly oblong to ovate, mostly 2–4 cm wide; chiefly western, extending e. to Minn. and Ill., and adventive eastward . . . . var. *pilosa* (Nutt.) Fern.
  2 Hairs of the angles and sides of the stem copious, about alike, tending to be somewhat reflexed; lvs narrowly oblong or linear-oblong, rarely over 2 cm wide; Que. to Wis., s. to N.Y. and Ill. . . . . var. *homotricha* Fern.

2. Stachys hyssopifolia Michx.   Hyssop-h.-n. Stems 3–5 dm, often branched from the base, bearded at the nodes and sometimes sparsely hairy on the angles; lvs firm, sessile or subsessile, linear or linear-oblong, 2–7 cm × 3–7(–12) mm, tapering at base, entire or with a few low teeth; verticils usually remote, the bracts

entire, lance-linear; cal glabrous or with a few hairs on the upper side, the lobes glabrous, narrowly deltoid, nearly as long as the tube. Moist, usually sandy soil near the coast; Mass. to Ga.; also around the head of Lake Michigan. July–Sept. (*S. atlantica.*)

3. **Stachys aspera** Michx. Stem erect, 4–8 dm, bearded at the nodes and usually sparsely spreading-bristly on the angles; lvs subsessile, narrowly oblong, 3–7 cm, acute or acuminate, sharply serrate, usually acute at base, glabrous; bracts ovate or lance-ovate, denticulate and sparsely ciliate; cal-tube 3–4 mm, glabrous or commonly sparsely villous, the lobes glabrous, deltoid, 2.3–3.4 mm, more than half as wide. Moist soil; Pa. to Io., s. to Ga. and Mo. July–Sept. (*S. hyssopifolia* var. *ambigua; S. ambigua* (A. Gray) Britton, not J. E. Smith; *S. grayana.*)

4. **Stachys tenuifolia** Willd. Smooth h.-n. Stems to 1 m, usually simple, glabrous or with some short stout hairs on the angles; lvs thin, oblanceolate to oblong, mostly 6–15 cm, acuminate, crenate-serrate, usually broadly obtuse to subcordate at base, glabrous or rarely with a few hairs; petioles of the main lvs 8–25 mm; bracts usually with a few low teeth; cal-tube 3–4.5 mm, almost always glabrous, the lobes lanceolate, glabrous, 2.5–3.5 mm; $2n=32$. Moist soil, especially in shaded places; N.Y. to Mich. and Minn., s. to S.C. and Tex. June–Aug. Intermediates to no. 5 are not uncommon northward.

5. **Stachys hispida** Pursh. Hispid h.-n. Stems erect, 5–10 dm, conspicuously hirsute on the angles with deflexed bristles to 2 mm; petioles to 1 cm, or none; lvs narrowly oblong or lanceolate, 4–12 × 1–4 cm, acuminate, serrate, usually distinctly hairy above; bracts long-ciliate and denticulate; cal-tube hirsute, rarely also minutely hairy, 3.3–4.7 mm, the lobes narrowly lanceolate to almost linear, at least three-fourths as long, hirsute or ciliate. Wet soil; Mass. and Vt. to Ont. and Man., s. to Md., Ky., and Ark. July, Aug. (*S. tenuifolia* var. *h.*)

6. **Stachys latidens** Small. Stems 5–8 dm, glabrous or the angles with a few stiff reflexed hairs less than 0.5 mm; lvs lance-ovate to narrowly oblong, 7–12 cm, a fourth to a third as wide, acute or usually acuminate, crenate-serrate, obtuse or rounded at base, glabrous or very sparsely hirsutulous on both sides; petiole 3–8 mm, smooth on the upper side; cal-tube 3.5–4.5 mm, the lobes pubescent, half as long. Rocky slopes in the mts.; Va., W.Va., and N.C. June, July.

7. **Stachys subcordata** Rydb. Stems 6–10 dm, glabrous or the angles with a few short, stiff, reflexed hairs; lvs very thin, oblong-ovate or lance-ovate, 8–15 cm, a third to half as wide, acuminate, crenate-serrate, broadly rounded to cordate at base, sparsely and minutely hairy; petioles 1–3 cm, smooth above; cal-tube 4 mm, the lobes half as long (or a little more), deltoid-acuminate to an aristate tip. Mt. woods; sw. Va. to e. Tenn. June, July.

8. **Stachys clingmanii** Small. Stem 5–8 dm, hirsute on the angles only with pustulate hairs 1–2 mm deflexed at 45°; lvs lance-ovate to oblong, 6–12 cm, acuminate, sharply serrate, obtuse to subcordate at base, hirsutulous on both sides; petioles slender, 1–3 cm, smooth on the upper side; verticils usually 6-fld; cal-tube hairy, 3.3–5.3 mm, the lobes deltoid-acuminate, half as long, tapering to a subulate tip 1 mm. Mostly in mt. woods at high altitudes; Va. and W.Va. to N.C. and Tenn.; plants from Ill. and Ind., with the bristles scarcely deflexed, are also referred here.

9. **Stachys cordata** Riddell. Stems 4–10 dm, usually simple, ± hirsute, chiefly on the sides but also on the angles, with pustulate, spreading hairs overlying a minute, stipitate-glandular pubcrulcnce; lvs lance-oblong or oblong-obovate, 6–14 cm, a third to half as wide, broadest near the middle, acute to acuminate, crenate, cordate at base, sparsely hirsute on both sides; petioles 1–5 cm; cal-tube 3–4.5 mm, the lobes half as long. Woods, chiefly in the mts.; Md. and Va. to Tenn.; also s. O. to s. Ill. June–Aug. (*S. nuttallii; S. riddellii.*)

10. **Stachys eplingii** J. Nelson. Stem erect, simple, 5–10 dm, villous with spreading, pustulate hairs on the sides and angles and also minutely glandular-puberulent; petioles 2–5 mm; lvs lance-ovate to oblong, 6–12 cm, a third to half as wide, crenate-serrate, obtuse or rounded at base, hirsutulous above, villosulous on the veins beneath; verticils usually many-fld; cal-tube 3.5–5.5 mm, the lobes half as long, deltoid, scarcely apiculate. Mt. woods; w. Va. to Ky. and N.C., Ark. and Okla.; also near D.C. June, July. (*S. nuttallii,* misapplied.)

11. **Stachys annua** L. Taprooted annual 1–3 dm; stems glabrous or nearly so, usually with several branches; lvs lanceolate or ovate to oblong or elliptic, 1–6 cm; cal-lobes narrowly triangular, half to three-fourths as long as the tube, pubescent to the acuminate tip; cor pale yellow to white, 10–16 mm; $2n=34$. Native of Europe, rarely intr. in our range.

12. **Stachys arvensis** L. Field h.-n. Taprooted annual; stems diffuse or decumbent, 2–6 dm, sparsely hirsute; lvs broadly ovate, 2–4 cm, broadly obtuse or rounded at the summit, coarsely serrate, truncate or subcordate at base; petioles of the main lvs a third to half as long as the blade; cal-lobes nearly as long as the tube; cor anthocyanic to white, 6–8 mm, scarcely exceeding the cal; $2n=10, 20$. Native of Europe, now an occasional weed from Me. to Va. and casually inland. July–Oct.

13. **Stachys germanica** L. Coarse, densely white-woolly biennial 3–10 dm; lvs lanceolate to oblong, 4–10 cm, conspicuously serrate; verticils several, many-fld, the sharp cal-lobes distinctly protruding from the dense pubescence; cor red, 15–20 mm; $2n=30$. Native of Europe, rarely intr. in our range. June–Aug.

14. **Stachys byzantina** C. Koch. Perennial 4–8 dm, densely white-tomentose throughout with long matted hairs; lvs lanceolate to oblong, 5–10 cm, the inconspicuous teeth mostly covered by the tomentum; verticils

several, forming a nearly continuous spike, only the pink or purple cors (these 15–25 mm) projecting from the dense tomentum. Mediterranean sp., occasionally escaped from cult. in our range. (*S. olympica,* perhaps misapplied.)

37. MONARDA L.   Bee-Balm. Cal tubular, 13–15-nerved, essentially regular, the lobes in our spp. much shorter than the tube; cor strongly bilabiate, the upper lip narrow, entire, straight or curved and often galeate, the lower broader, spreading or deflexed, 3-lobed or with a central projecting tooth; stamens 2, ascending under the upper lip and in most spp. surpassing it; style exsert; erect herbs with lanceolate to ovate lvs and showy fls densely aggregated into head-like clusters terminating the branches or borne also in the upper axils, subtended by foliaceous bracts and mingled with narrow bractlets. 15, N. Amer.

1 Fl clusters (except on depauperate stems) both terminal and axillary; stamens not exsert beyond the
   strongly arched upper cor-lip . . . . 1. *M. punctata.*
1 Fl clusters solitary and terminal on the flowering branches; stamens surpassing the nearly straight or
   somewhat arcuate upper lip.
   2 Lvs distinctly petioled, the petioles over 5 mm, commonly over 1 cm.
     3 Cor 1.5–3(–3.5) cm, red-purple to pink or white.
       4 Upper cor-lip softly pubescent and at the apex ± densely villous; cor lavender . . . . 2. *M. fistulosa.*
       4 Upper cor-lip glabrous or merely puberulent.
         5 Cor white or ochroleucous; bracteal lvs largely green . . . . 3. *M. clinopodia.*
         5 Cor rose-purple; bracteal lvs strongly tinged with purple . . . . 4. *M. media.*
     3 Cor 3–4.5 cm, scarlet to crimson, seldom deep red-purple . . . . 5. *M. didyma.*
   2 Lvs sessile or nearly so, the petioles not over 5 mm.
     6 Upper cor-lip about as long as the tube . . . . 6. *M. russeliana.*
     6 Upper cor-lip half as long as the tube . . . . 7. *M. virgata.*

1. **Monarda punctata** L.   Horse-mint. Perennial; stems 3–10 dm, thinly canescent; lvs lanceolate or narrowly oblong, 2–8 cm, ± hairy; glomerules 2–5, or solitary on depauperate plants, the bracts lanceolate to ovate, much exceeding the cal, spreading or reflexed, often pale green to nearly white, or purple-tinged; cal 5–9 mm, densely villous in the throat, ± villous or hirsute externally at the summit, the lobes 1–1.5 mm; cor pale yellow, spotted with purple, 1.5–2.5 cm, the strongly arched upper lip about as long as the throat and slender tube together; $2n=22$. Dry, especially sandy soil; Vt. to se. Minn., s. to Fla. and Tex., but missing from most of the Ohio drainage. Var. *punctata,* with the stem pubescent with short recurved hairs and with the lvs glabrous to minutely puberulent beneath, the hairs not concealing the glands, occurs from N.J. to Fla. and Tex., chiefly on the coastal plain in our range. Var. *villicaulis* Pennell, with the hairs of the stem longer and more spreading, and with the lvs densely hairy beneath, the hairs concealing the glands, occurs from Mich. to e. Minn., s. to Ill. and Mo., and also near Lake Champlain in Vt. and N.Y.
   M. citriodora Cerv., lemon-b.-b. a southwestern annual with setaceous cal-lobes 4–8 mm and with white or pink, purple-dotted cor, extends n. to the Missouri R. and may be found in our range.

2. **Monarda fistulosa** L.   Wild bergamot. Pleasantly aromatic rhizomatous perennial 5–12 dm, often branched; stems usually hairy at least above; lvs deltoid-lanceolate to lanceolate or rarely ovate, acuminate, ± serrate, rounded or truncate to broadly acute at base, thinly hairy or glabrous to conspicuously hairy above or canescent beneath, the larger 6–10 cm on petioles 1–1.5 cm; heads 1–3.5 cm wide (excluding the cors); bracteal lvs lanceolate or ovate; cal 7–10 mm, puberulent, its throat densely hirsute internally with erect white hairs a third to half as long as the cal-lobes, these subulate, 1–2 mm; cor bright (but rather pale) lavender, 2–3.5 cm, the somewhat arcuate upper lip densely villous at the tip; $2n=32, 34, 36$. Upland woods, thickets, and prairies. Que. to Man. and B.C., s. to Ga., La., and Ariz. June–Sept. (*M. mollis,* the common form with the hairs of the stem decurved.) Most of our plants are var.*fistulosa,* as principally described above. The western var. *menthaefolia* (Graham) Fern., a shorter, less branched plant, often with a single head, the longest petioles seldom over 1 cm, enters our range in w. Minn. (*M. menthaefolia.*)

3. **Monarda clinopodia** L.   Basil-b.-b. Simple or branched perennial to 1 m, the stems glabrous or sparsely hairy; lvs ovate to lance-deltoid, 6–12 cm, acuminate, serrate, obtuse or more often rounded at base, the main ones usually ovate and twice as long as wide; petioles 1–3 cm; heads 1.3–3 cm thick (excluding cors); bracts lanceolate, largely green but often whitish or anthocyanic toward the base; fls fragrant, entomophilous; cal-tube glabrous or minutely puberulent, or villous either outside or inside at the summit, the hairs when present quite unlike those of no. 2; cal-lobes glandular; cor white or ochroleucous, 1.5–3 cm, the upper lip slender, nearly straight, glabrous or sparsely puberulent. Woods and thickets; Conn. to O., Ill., and Mo., s. to N.C. and Ala. June, July. Intergrades with no. 2.

4. **Monarda media** Willd.   Purple bergamot. Perennial; stems usually branched, to 1 m, glabrous or sparsely hairy; lvs ovate or broadly lance-deltoid, the larger ones often half as wide as long; petioles 1–2 cm; bracts partly or wholly purple; cal usually purple, glabrous to villous in the throat, the lobes minutely glandular-puberulent; cor red-purple, 2–3 cm, the upper lip slender, nearly straight, thinly hairy but not villous at the tip. Moist woods and thickets; Me. to s. Ont. and Va.; also in w. N.C. and reported from Ind. July, Aug.

5. Monarda didyma L.   Oswego-tea. Perennial 7–15 dm, often pilose at the nodes; lvs thin, ovate or deltoid-ovate to nearly lanceolate, 7–15 × 2.5–6 cm, acuminate, serrate, broadly acute to commonly rounded at base; petioles 1–4 cm; heads 2–4 cm thick (excluding the cors); bracteal lvs lanceolate, longer than the cals, usually red-tinted; fls inodorous, ornithophilous; cal 10–14 mm, glabrous to minutely puberulent, nearly or quite glabrous in the throat, the lobes 1–2 mm, subulate above a triangular base; cor scarlet to crimson, seldom deep red-purple, 3–4.5 cm, the nearly straight upper lip half as long as the tube; 2n=36. Moist woods and thickets; Me. to Mich., s. to N.J., W.Va., and O., and along the mts. to n. Ga. July–Sept. Hybridizes with no. 3.

6. Monarda russeliana Nutt.   Perennial; stems 3–6 dm usually simple, glabrous or very sparsely villous; petioles 0–5 mm; lvs ovate or deltoid-ovate to lance-deltoid, 5–10 × 2–5 cm, acuminate, serrate, rounded or subcordate at base, sparsely hairy above, villous on the veins beneath; heads 1.5–2.5 cm thick (excluding the cors); bracteal lvs commonly tinged with pink or purple, spreading or reflexed, lanceolate; cal hirsute in the throat, its lobes subulate, 1.5–4 mm, often stipitate-glandular; cor pale rose-purple to white, dotted with purple, 2.5–3.5 cm, the upper lip slender, nearly straight, about as long as the tube. Woods and thickets; s. Ind. to Io. and e. Kans., s. to Ala. and Tex. May, June. (*M. bradburiana.*)

7. Monarda virgata Raf.   Much like no. 6, but more slender and with longer internodes, the lvs lanceolate, inconspicuously serrate, glabrous or minutely pilosulous on the veins beneath; cal-lobes conspicuously stipitate-glandular, the upper cor-lip ca half as long as the tube. Upland woods; Ky.; Mo. to Ark., Okla., and Tex. May, June, (*M. russeliana,* misapplied.)

38. BLEPHILIA Raf.   Cal 13-nerved, naked in the throat, 2-lipped, the upper lip much the longer; cal-lobes narrowly triangular to subulate; cor 2-lipped, villous, the tube elongate, usually outcurved, the upper lip straight, subgaleately concave, entire, the lower about as long, spreading or deflexed, with 2 broad lateral lobes and an oblong median one; stamens 2, ascending under the upper lip and eventually exsert, their anthers connivent; rhizomatous (sometimes also stoloniferous) perennials with numerous pale purple fls densely crowded in the axils of the upper lvs. Only the following 2 spp.

1 Lobes of the lower cal-lip extending beyond the sinuses of the upper lip .... 1. *B. ciliata.*
1 Lobes of the lower cal-lip not reaching the sinuses of the upper lip .... 2. *B. hirsuta.*

1. Blephilia ciliata (L.) Benth.   Stems 4–8 dm, closely puberulent with recurved hairs 0.1–0.5 mm; main lvs lanceolate to ovate, 3–6 cm, entire or sometimes with a few teeth, sessile or subsessile; verticils usually 3–5, crowded into a continuous terminal spike 2–5 cm, or the lowest verticil sometimes separate, hemispheric and subtended by numerous ovate to obovate, closely appressed bracts about equaling the cals; cal 7–9 mm, the lower lobes 1.1–1.7 mm, extending past the sinuses of the upper lip; cor pale blue with purple spots, 11–14 mm. Woods. Mass. to s. Mich. and Wis., s. to Ga. and Ark. Plants of Ill., Ky., and Mo. grow in moister, more shaded places and have wider lvs with broader, often rounded base than the more n. and e. plants.

2. Blephilia hirsuta (Pursh) Benth.   Much like no. 1; stems hirsute, especially above, with spreading hairs commonly 1–2 mm; main lvs lance-ovate to broadly ovate, 4–8 cm, ± serrate, obtuse to rounded or subcordate at base, on petioles 1–2 cm; verticils well separated or the upper pair contiguous, the bracts narrower and shorter than in no. 1 and soon reflexed; lower cal-lobes 1 mm or less, not reaching the upper sinuses. Moist, shaded places; Que. to Minn., s. to N.C., Ala., and Ark. May–Aug.

39. SALVIA L.   Sage. Cal tubular to campanulate, 2-lipped, 10–15-nerved, the upper lip entire to 3-lobed, the lower deeply 2-lobed; cor 2-lipped, the upper lip straight or arched, often subgaleate, the lower 3-lobed; stamens 2, ascending under the upper lip and sometimes surpassing it; filaments short; connective much elongate, attached at or near its middle to the filament, bearing a single pollen-sac (half anther) at the upper end, the lower end either sterile or with an imperfect half-anther; ours herbs, the medium-sized to large fls in verticils subtended by usually much reduced bracteal lvs, forming a terminal, sometimes interrupted, spike-like infl. 500, cosmop.

1 Upper cal-lip entire; lvs narrow, up to a third as wide as long .... 1. *S. reflexa.*
1 Upper cal-lip 3-toothed, 3-lobed, or 3-awned, the teeth often small; lvs various.
  2 Lobes of the upper cal-lip short and close together.
    3 Principal lvs broadly rhombic or deltoid, decurrent on the petiole; native .... 2. *S. urticifolia.*
    3 Principal lvs lanceolate to oblong or elliptic, not decurrent; intr., weedy.
      4 Lvs chiefly cauline, several to many pairs .... 3. *S. nemorosa.*
      4 Lvs chiefly basal, the cauline much smaller, only 1–3 pairs .... 4. *S. pratensis.*
  2 Lobes of the upper cal-lip conspicuous, separated by distinct, usually flat or concave sinuses.
    5 Upper cal-lip more than half as long as the tube; fls 4–12 per node of the infl.
      6 Lvs chiefly basal, commonly lyrate, not weedy or somewhat weedy .... 5. *S. lyrata.*

6 Lvs chiefly cauline, merely toothed; intr., weedy .... 6. *S. sclarea.*
5 Upper cal-lip up to a third as long as the tube; fls 12–30 per node of the infl .... 7. *S. verticillata.*

1. Salvia reflexa Hornem.   Branching annual 3–6 dm, the stem with minute recurved hairs; principal petioles 8–20 mm; lvs lance-linear to lanceolate, 3–5 cm × 4–12 mm, entire or with a few low teeth, gradually narrowed to the base; racemes erect, 5–10 cm, the internodes 8–15 mm; bracteal lvs lance-linear, 1–3 mm; fls 2(4) per node; cal at anthesis 6–7 mm, minutely hairy on the nerves only, the upper lip entire, 5-nerved, half as long as the tube; cor blue, 8–12 mm, the tube no longer than the cal; $2n=20$. Dry sandy or gravelly soil of hillsides and prairies; O. to Io. and Mo., w. to N.C., Utah, and Mex., and sometimes adventive eastward. (*S. lanceifolia,* misapplied; *S. lanceolata.*)
    S. pitcheri Torr., an Ozarkian perennial with 6–20 fls per node, the cor 2 cm, the tube well exsert from the canescent cal, scarcely reaches our range as a native, but is occasionally found as an escape. (*S. azurea* var. *grandiflora.*)

2. Salvia urticifolia L.   Nettle-lvd s. Perennial 3–5 dm; stems one or few from a thick woody rhizome-caudex, mostly simple, finely glandular-villous above, sparsely hairy or glabrate below; lvs cauline, rhombic-ovate to deltoid, 5–8 cm, acute, coarsely crenate-serrate above, toward the base entire and abruptly narrowed to a cuneate base decurrent along the petiole almost to the stem; infl interrupted, long-peduncled above the scarcely reduced upper lvs, 1–2 dm; fls 6–10 per node, the bracteal lvs ovate, 2–3 mm, caducous; cal at anthesis 4–5 mm, glandular-punctate, the upper lip 3-toothed, half as long as the tube and about as long as the lower lip; cor blue, 10–13 mm, the upper lip much shorter than the lower. Woods and thickets; Pa. to S.C. and Ga., w. to Ala., Tenn., and w. Ky. May, June.

3. Salvia nemorosa L.   Fibrous-rooted perennial 3–7 dm; stem branched above, softly villosulous; lvs chiefly cauline, lance-ovate or lance-triangular, 5–10 cm, crenate, truncate to subcordate at base, rugose and glabrescent above, canescent beneath; fls very numerous in crowded whorls of 2–6 subtended by ovate anthocyanic bracteal lvs 4–10 mm, the dense spike-like infl 1–2 dm; cal villous, 6–8 mm, the upper lip minutely 3-toothed, shorter than the lanceolate lobes of the lower lip; cor blue, 9–12 mm, the tube no longer than the cal; $2n=12$, 14. Native of Europe and w. Asia, intr. in fields and waste places here and there in our range. June, July. (*S. sylvestris,* misapplied, the name properly applied to a hybrid of nos. 3 and 4.)

4. Salvia pratensis L.   Meadow-s. Fibrous-rooted perennial 3–6 dm; lvs mostly basal, long-petioled, ovate-oblong, 7–12 cm, irregularly serrate or crenate; cauline lvs few, smaller; racemes 1–2 dm, interrupted; fls 4–8 per node, the bracteal lvs broadly ovate, green, mostly under 1 cm; cal villosulous, the upper lip broadly ovate, minutely 3-toothed, two-thirds as long as the tube; and much shorter than the acute or aristate lobes of the lower lip; cor blue (1.5–)2–3 cm (smaller in pistillate fls), the upper lip arched into a half-circle and laterally flattened; $2n=18$. Native of s. and c. Europe, occasionally found as a weed in fields and pastures in our range. June–Aug.
    S. verbenacea L., another European weed, commonly with lobed or toothed lvs, the cor 1 cm, with the upper lip nearly straight, is rare with us and probably not persistent.

5. Salvia lyrata L.   Fibrous-rooted perennial 3–6 dm; principal lvs in a basal rosette, oblong or obovate-oblong in outline, 1–2 dm, deeply pinnately lobed into rounded segments (rarely unlobed), on petioles 3–10 cm; cauline lvs 1(–3) pair, much smaller, short-petioled or sessile; primary infl 1–3 dm, the few verticils widely separated, ca 6-fld; cal at anthesis 1 cm, the upper lip truncate, 3-aristate, about as long as the tube, the lower lip longer; cor blue or violet, 2–2.5 cm, the lips much shorter than the tube; $2n=36$. Upland woods and thickets, sometimes weedy; Conn. to Pa., s. O., s. Ill., and Mo., s. to Fla. and Tex. May, June.

6. Salvia sclarea L.   Clary. Coarse, spreading-hairy (and toward the top glandular) biennial 5–15 dm, freely branched above, the branches with scattered verticils of fls and conspicuous, round-ovate, caudate-acuminate, often dry and anthocyanic bracteal lvs 1–3 cm; lowest lvs long-petiolate, with rugose, ovate to ovate-oblong, basally subcordate, toothed or doubly toothed blade 7–20 cm, the cauline lvs progressively smaller and less petiolate, in small plants often few; cal glandular and coarsely hairy, the upper lip with aristate lateral teeth 1.5–3 mm well separated from the shorter central tooth; cor blue (white) or marked with yellow, 1.5–3 cm, the upper lip strongly arched, longer than the tube and surpassing the lower lip; stamens exsert; $2n=22$. Native of the Mediterranean region, occasionally escaped from cult. in the s. part of our range. June, July.
    S. officinalis L., the garden-sage, a smaller plant with narrow, canescent lvs and with the bracteal lvs lance-ovate, scarcely if at all surpassing the cal, occasionally escapes in our range.

7. Salvia verticillata L.   Perennial 4–8 dm; lvs broadly ovate-oblong to deltoid, 5–10 cm, coarsely and irregularly serrate, truncate or more often subcordate at base, sparsely hirsutulous, long-petioled; infls usually several, 1–2 dm, interrupted, with 12–30 fls per node; cal almost tubular, 6 mm at anthesis, the upper lip shortly 3-lobed, a fourth to a third as long as the tube and about equaling the lower lip; cor blue (white), 10–15 mm, its upper lip narrowed to the base; $2n=16$. Native of s. and e. Europe and w. Asia, intr. in fields and waste places here and there in our range. July–Sept.

40. OCIMUM L.   Cal very irregular, the 2 lowest lobes resembling the lateral but narrower, triangular, the uppermost lobe broad, rounded, its margins decurrent along the cal-tube, and the whole forming a round, nearly flat cap on the upper side of the cal; cal enlarged in fr; cor 2-lipped, the lower lip deflexed, entire, the upper lip 4-lobed; stamens

4, deflexed along the lower cor-lip; herbs or shrubs with small fls in verticils of 6, subtended by bracteal lvs, forming a slender, terminal, spike-like infl. 150, warm reg.

1. **Ocimum basilicum** L.   Basil. Annual 3–5 dm, with ovate or lance-ovate lvs tapering to the base and with long spikes of small white fls; fruiting cal 7–9 mm; $2n=48$. Native of trop. Asia and Afr., established as a weed in trop. Amer., and occasionally escaped in our range.

FAMILY HIPPURIDACEAE, the Mare's-tail Family

Fls inconspicuous, anemophilous, solitary in the axils of the upper lvs, epigynous, perfect or sometimes some or all of them unisexual; cal reduced to a rim around the top of the ovary; cor none; stamen solitary, atop the ovary; pistil apparently of a single carpel, the inferior, unilocular ovary with a terminal style that is stigmatic for its whole length and generally lies in the groove between the 2 pollen-sacs of the sagittate anther; ovule solitary, apical and pendulous; fr an achene or a drupelet with very thin, fleshy exocarp; emergent aquatic rhizomatous herbs, glabrous except for scattered minute, deciduous, peltate, glandular trichomes; lvs small, narrow, sessile, whorled in sets of (4–)6–12(–16), without stipules. A single genus and species.

1. **HIPPURIS** L.   Characters of the family.

1. **Hippuris vulgaris** L.   Mare's tail. Stems soft but stiffly erect, unbranched, mostly 2–6(–10) dm, the internodes of the submersed part often longer than the lax, thin lvs, which may soon degenerate; emersed lvs thicker and firmer, numerous and rather crowded, usually widely spreading, linear, 1–3 cm × 1–3 mm; fls sessile; frs. 1.7–2.5 mm; $2n=32$. In shallow, quiet water, or seldom on mud; circumboreal, in Amer. s. to Me., n. N.Y., n. Ind., Io., and N.M.; Australia and s. S. Amer. A salt-marsh form with elliptic obtuse lvs in 4's has been called *f. maritima* Hellenius. (*H. tetraphylla.*)

FAMILY CALLITRICHACEAE, the Water-starwort Family

Fls without perianth, unisexual, in most spp. subtended by a pair of horn-shaped bracteoles, sessile or short-pedicellate, borne singly in the lf-axils, or one fl of each sex in the same axil; staminate fl of 1(–3) stamens; anthers opening by lateral, apically confluent slits; pistillate fl of a single somewhat compressed, laterally 4-lobed (and sometimes winged), apically indented ovary, crowned by a pair of slender, often persistent styles or a single deeply cleft style; carpels 2, collateral, with a broad commissure; each primary locule divided into 2 locelli, each of these containing a single pendulous, axile, anatropous ovule with ventral raphe; fr dry, separating at maturity into 4 1-seeded half-carpellary nutlets; embryo with 2 short cotyledons, embedded in the thin, oily endosperm; low, slender, commonly much branched herbs, mostly annual, most spp. aquatic; lvs opposite, small, entire, without stipules. A single genus.

1. **CALLITRICHE** L.   Water-starwort. Characters of the family. 35, nearly cosmop.

1 Plants terrestrial; fr on a short pedicel . . . . 1. *C. terrestris.*
1 Plants mostly aquatic; fr sessile.
  2 Halves of the fr separated only by a shallow groove; shoot-tips usually floating or slightly emergent,
     the upper lvs then broadened and 3–several-nerved.
    3 Margins of the fr wingless, with inconspicuous (or no) commissural groove . . . . 2. *C. heterophylla.*
    3 Margins of the fr winged (at least distally), and with an evident commissural groove between
     the wings.
      4 Fr distinctly obovate, 1–1.4 mm, somewhat longer than wide . . . . 3. *C. palustris.*
      4 Fr suborbicular, 1.5–2 mm long and wide . . . . 4. *C. stagnalis.*
  2 Halves of the fr separated by a deep furrow extending almost to the commissure; plants wholly
    submerged, the lvs all slender and 1-nerved . . . . 5. *C. hermaphroditica.*

1. **Callitriche terrestris** Raf.   Stems tufted, 2–5 cm; lvs spatulate or oblanceolate, 2–5 mm, obscurely 3-nerved, not lepidote; pedicels ± erect, half to fully as long as the fr; fr 0.5–0.7 mm, 0.7–1 mm wide, deeply

notched at both ends, its lobes very narrowly winged and separated by a deep groove; style persistent, recurved, shorter than the fr; $2n=10$. Damp, usually shaded soil; Mass. to O., s. to Fla. and Tex. (*C. deflexa* var. *austinii.*)

2. **Callitriche heterophylla** Pursh. Much like no. 3, differing mainly in the frs, these ca 1 mm or a little less, nearly to fully as long as wide (not more than 0.1 mm longer than wide), often but not always a little wider above the middle than below, the margins rounded or obtuse or very shallowly and inconspicuously grooved between the carpels, not at all sharp-edged or winged; pit-like markings on the surface of the fr not obviously aligned in vertical rows. In quiet water throughout most of N. Amer., and also in S. Amer. (*C. anceps* Fern., the rare, mainly boreal, wholly submersed phase.)

3. **Callitriche palustris** L. Slender annuals, rooting at the lower nodes, with loosely clustered stems 1–2 dm, typically growing mainly submerged with only the shoot-tips floating, but sometimes stranded in mud and wholly emersed, rarely wholly submersed; submersed lvs linear, 0.5–1.5 cm; floating or emersed lvs spatulate to obovate, to 5 mm wide, triplinerved, appearing finely and sparsely lepidote at 10× or 20×; bracteoles whitish, 0.5–1.5 mm, deciduous; usually one male and one female fl per axil; frs mostly 1–1.4 mm long and ca 0.2 mm longer than wide, widest above the middle, compressed and with a broad, shallow groove down the middle of each side, and also with an evident commissural groove on each edge, narrowly winged (or at least sharp-edged) on each side of the commissural groove toward the summit, less (or scarcely) so below; pit-like markings on the fr tending to be aligned in vertical rows; $2n=20$. In quiet water; circumboreal, in Amer. s. to W.Va., Nebr., Tex., and Calif.; also in S. Amer. (*C. verna.*)

4. **Callitriche stagnalis** Scop. Fr suborbicular, 1.5–2 mm long and wide, the carpels distinctly wing-margined their whole length; $2n=10$; otherwise much like no. 3. In cold ponds and streams; native of Europe, intr. at scattered stations in our range and elsewhere in n. U.S. and s. Can.

5. **Callitriche hermaphroditica** L. Slender annuals, rooting at the lower nodes; stems loosely clustered, mostly 1–4 dm, typically wholly submersed; lvs all about alike, dark metallic green, wholly glabrous, mostly 0.5–1.5(–2) cm, up to about 1.5 mm wide, 1-nerved, broadest near the base and tapering very gradually to the bifid or broadly notched tip; fls solitary in the axils, without bracteoles; frs mostly 1–2 mm, flattened, suborbicular, the 4 segments more nearly distinct than in our other spp., the 2 major halves separated in life by a broad, deep median furrow extending almost to the commissure, each of these halves also with a deep, narrow commissural groove between the narrow but evident wing-margins; $2n=6$. In ponds and slow streams; circumboreal, s. in Amer. to N. Engl., N.Y., Mich., Minn., and Calif. (*C. autumnalis.*)

Family **PLANTAGINACEAE**, the Plantain Family

Fls regular (or the cal sometimes irregular), perfect or seldom unisexual, sympetalous, ordinarily 4-merous as to the cal, cor, and stamens, or the stamens seldom only 1–3; cor scarious, persistent, not large; stamens epipetalous, alternate with the cor-lobes; ovary superior, bilocular (or 4-locellar) with axile placentation, or unilocular and with a single basal ovule; style terminal, with a usually 2-lobed stigma; fr a circumscissile capsule, or an achene or small nut; embryo straight, dicotyledonous, ± spathulate; endosperm well developed; herbs; lvs simple, ± parallel-veined (the blade phyllodial), usually all basal in a close spiral, in a few spp. cauline and opposite. Only *Plantago, Littorella,* and *Bougeria* (1).

1 Fls numerous in spikes or heads; fr a capsule with (1)2–many seeds .... 1. *Plantago.*
1 Fls solitary or in groups of 3; fr indehiscent, 1-seeded .... 2. *Littorella.*

1. **PLANTAGO** L. Plantain. Fls sessile or subsessile in the axils of bracts, aggregated into spikes or heads; sep 4, the 2 next the bract often slightly different from the 2 next the axis (sometimes ± connate); cor long-persistent, its tube investing the summit of the fr; capsule circumscissile at or below the middle; ours annual or perennial herbs, most spp. acaulescent. 200+, cosmop.

1 Lvs cauline, opposite; heads axillary, long-peduncled .... 1. *P. psyllium.*
1 Lvs all basal, linear to broadly ovate or rotund; plants scapose.
  2 Bracts and sep glabrous or inconspicuously ciliate.
    3 Plants perennial (but the weedy spp. blooming the first year); stamens 4.
      4 Maritime plants with fleshy, linear lvs; cor-tube hairy outside .... 8. *P. maritima.*
      4 Plants not maritime; lvs broader than linear; cor-tube glabrous.
        5 Bracts and sep prominently raised-keeled; cor-lobes 1 mm or less; seeds 4–30.
          6 Fr dehiscent near the middle; bracts broadly ovate .... 2. *P. major.*
          6 Fr dehiscent far below the middle; bracts narrowly lance-triangular .... 3. *P. rugelii.*
        5 Bracts and sep scarcely if at all keeled; cor-lobes mostly over 1 mm; seeds 2–4.

7 Spike sparsely fld for most or all of its length, the axis exposed.
  8 Lvs with broadly rounded or cordate base; lateral nerves tending to arise from the
    midnerve within the blade of the lf .... 4. *P. cordata.*
  8 Lvs with tapering base; lateral nerves arising with the petiole .... 5. *P. eriopoda.*
7 Spike densely fld throughout, the axis completely hidden.
  9 Sep 4, separate; lvs (petiole included) at least a fifth as wide as long .... 6. *P. media.*
  9 Sep apparently 3, the 2 next the bract united into one with a double midvein; lvs
    rarely as much as a sixth as wide as long .... 7. *P. lanceolata.*
3 Plants annual; stamens 2.
  10 Seeds mostly 10–25; bracts usually distinctly longer than the cal .... 9. *P. heterophylla.*
  10 Seeds mostly 4; bracts barely if at all longer than the cal.
    11 Taproot well developed; cor-lobes spreading or reflexed in fr .... 10. *P. elongata.*
    11 Taproot very short and quickly deliquescent; cor-lobes ± erect in fr .... 11. *P. pusilla.*
2 Bracts or sep or both distinctly pubescent to long-villous.
  12 Lvs oblanceolate to obovate; cor-lobes typically erect after anthesis .... 12. *P. virginica.*
  12 Lvs linear or nearly so; cor-lobes spreading or reflexed after anthesis.
    13 Bracts inconspicuous, scarcely if at all exsert from the woolly spike .... 13. *P. patagonica.*
    13 Bracts conspicuously exserted from the not very woolly spike .... 14. *P. aristata.*

1. Plantago psyllium L. Leafy-stemmed p. Short-lived, mostly annual, 1–6 dm, evidently hirsute and tending to be a little viscid; lvs numerous, cauline, opposite, linear or nearly so, entire, 2–8 cm × 1–3 mm; peduncles axillary, 2–8 cm; spikes 0.5–1.5 cm, dense, nearly 1 cm thick; bracts broad, rounded, conspicuously scarious-margined; the lowermost abruptly and firmly foliaceous-caudate; cor-lobes 1.5–2 mm, soon reflexed; seeds 1 or 2, brown, 2–3 mm; $2n=12$. A weed in waste places, especially along railroad tracks; native to the e. Mediterranean region, now well established in e. U.S. (*P. indica; P. arenaria.*) The name has often been misapplied to a related sp. without caudate tips on the bracts.

2. Plantago major L. Common p. Fibrous-rooted perennial from a short, stout, erect caudex, glabrous or rather inconspicuously hairy, especially below; lvs with broadly elliptic to broadly ovate or cordate-ovate blade ± abruptly contracted to the well defined petiole, the blade entire or irregularly toothed, mostly 4–18 × 1.5–11 cm, 1.3–2(–3) times as long as wide, strongly 3–several-nerved, the nerves diverging at the base of the blade, ± parallel to the margin; petiole only seldom anthocyanic at base; scapes 5–25 cm; spikes dense but narrow, less than 1 cm thick, commonly 5–30 cm long, essentially glabrous; bracts broad, ovate-obtuse, mostly 2–4 mm, with a prominent, acute keel and thin margins; sep ovate, obtuse, the rounded keel about as wide as the scarious margins; cor-lobes ca 1 mm or less, reflexed after anthesis; stamens 4, exsert; fr rhombic-ovoid, 2.5–4 mm, circumscissile near or a little below the middle; seeds 6–30, 1 mm, strongly reticulate; $2n=12$, 24. Native of Eurasia and probably also parts of N. Amer., naturalized throughout the U.S. and s. Can. in lawns, roadsides, and waste places. (*P. asiatica.*)

3. Plantago rugelii Decne. American p. Much like no. 2; petiole generally glabrous and anthocyanic at base; bracts lance-attenuate; sep ovate or oblong, acute, the sharp keel much wider than the scarious margin; fr slenderly ellipsoid or narrowly ovoid, 4–6 mm, circumscissile far below the middle; seeds 4–10, 1.5–2 mm; $2n=24$. Lawns, gardens, roadsides, and waste places throughout our range, w. to Mont., s. to Fla. and Tex.

4. Plantago cordata Lam. King-root. Glabrous perennial with several long, fleshy roots 0.5–1.3 cm thick; principal lvs cordate-ovate, mostly 12–25 × 8–20 cm, the main lateral veins not parallel to the margin, tending to arise from the proximal part of the midvein within the blade, scapes to 3 dm, stout and hollow; spikes to 3 dm, interrupted, the axis exposed; bracts and sep about equal, broadly round-ovate, nearly flat, with very narrow keel, herbaceous sides, and narrow scarious margins, obtuse or rounded at the tip; fr ovoid, 5–10 mm, circumscissile at or just below the middle; seeds 2(–4), 3–4 mm, smooth, mucilaginous; $2n=24$. Semiaquatic, in marshes and along streams, especially on calcareous substrate; O. and s. Ont. to Wis. and Mo., and occasionally to N.Y., Va., N.C., Ga. and Ala.; now rare. Mainly spring.

5. Plantago eriopoda Torr. Woolly-crowned p. Perennial with 1–several thickened roots, conspicuously brown-woolly at the crown; lvs thick and fleshy, elliptic, 5–15 cm, half as wide, acute, tapering to the margined petiole; lateral veins all descending into the petiole; scapes 1–3 dm, glabrous or nearly so; spikes slender, 5–20 cm, the axis exposed; bracts and sep about equal, nearly flat, rounded or obtuse, with narrow scarious margins; cor-lobes 1–1.5 mm, spreading or reflexed; fr ovoid, rounded above, 3–4 mm, circumscissile just below the middle; seeds 2–4, 2–2.7 mm; $2n=24$. Salt marshes along the coast, Que. to N.S.; alkaline soil in the interior, Minn. and Io. to Mack., Yukon, and Wash., s. to Nebr., Colo., and Calif.

6. Plantago media L. Fibrous-rooted perennial; lvs spreading, elliptic to obovate or oblanceolate, acute or tapering at each end, 1–2 dm (petiole included); scapes 1–2 dm, strigose above; spikes dense, narrowly conic, becoming cylindric, 3–10 cm at maturity; bracts and sep subequal, broadly ovate, with broad scarious margins; sep ovate, acute; fr thickly ellipsoid, 3 mm, circumscissile at the middle; seeds 2–4, 2 mm; superficially like no. 7, except for the broader lvs; $2n=12$, 24. Native of Eurasia, occasionally found as a weed in waste places in our range.

7. Plantago lanceolata L. English p.; ribgrass. Fibrous-rooted perennial with a stout erect caudex, often flowering the first year, ± tan-woolly at the crown; lvs usually ascending, villous to glabrous, 3–several-nerved, narrowly elliptic or lance-elliptic, long-acute, 1–4 dm × 1–4 cm; scapes strongly 5-sulcate, strigose above, 1.5–6 dm; spikes very dense, ovoid-conic at first, at maturity cylindric, 1.5–8 cm × nearly or fully 1 cm; bracts thin, ovate, broadly scarious-margined, often caudate-tipped; sep often villous-ciliate toward the tip of the strong midnerve, the 2 outer connate with separate midveins, merely notched at the tip, or entire; cor-lobes 2–2.5 mm, spreading or reflexed; fr 3–4 mm, circumscissile near the base, the (1)2 seeds shining, blackish, 2 mm, deeply concave on the adaxial face; some fls or plants sometimes pistillate; $2n=12$. Native of Eurasia, now a cosmopolitan weed, particularly in moister parts of the temperate zone, common in lawns and along roadsides.

P. altissima L., a more robust European sp., 6–10 dm, with a stout rhizome, the scapes 6–12-sulcate, the seeds 3.5 mm, has been reported from a few localities in our range.

8. Plantago maritima L. Seaside-p. Perennial; lvs thick and fleshy, linear, to 15 cm, entire or nearly so; scapes 5–20 cm, hairy above, equaling or surpassing the lvs; spikes 2–10 cm, loosely or densely fld, usually interrupted below; bracts broadly ovate, acute, equaling or slightly shorter than the cal; sep broadly ovate, acute or subacute, often ciliolate, rounded on the back, the scarious margins as wide as the herbaceous center; cor-tube hairy; cor-lobes 1–1.5 mm, spreading after anthesis; fr ovoid, surpassing the cal, circumscissile near the middle; seeds 2–4, ellipsoid, 2–2.5 mm, flat on the inner face; $2n=12$. Salt marshes, beaches, and coastal rocks; circumboreal, s. along the Atlantic to N.J. The Amer. plants are ssp. *juncoides* (Lam.) Hultén. (*P. juncoides; P. oliganthos.*)

9. Plantago heterophylla Nutt. Many-seeded p. Annual, much like nos. 10 and 11, fibrous-rooted or seldom with an evident slender taproot; lvs linear, 3–10 cm, often with a few slender salient teeth; scapes several, to 10 cm; spikes slender, 2–6 cm, loosely fld; bracts ovate, broadly scarious-margined, commonly longer than the broadly rounded sep, only obscurely swollen at base; cor-lobes less than 1 mm, usually spreading in fr; stamens 2; fr slenderly ovoid, longer than the bracts and commonly nearly twice as long as the cal; seeds mostly 10–25, black, angular, coarsely pitted, 0.5–0.8 mm; $2n=12$. A weed of fields and waste places; se. U.S., n. to Va. and Mo., and rarely adventive n. to N.Y.

10. Plantago elongata Pursh. Much like no. 11, but a little larger in all respects, and with a well developed slender taproot; subdioecious; cal to 2 mm; cor-lobes 0.5–1 mm, in age usually spreading or reflexed; fr 2–3 mm; seeds 1–2 mm; $2n=12$. Moist, somewhat alkaline places; Man. and w. Minn. to B.C., s. to Okla. and Calif.

11. Plantago pusilla Nutt. Little p. Annual with a short, quickly deliquescent taproot; lvs narrowly linear, 3–8 cm; scapes several or many, 3–10 cm; spikes 2–6 cm, seldom longer, loosely fld; bracts ovate, seldom equaling the cal, the central herbaceous part about as wide as the scarious margins, distinctly saccate at base; anterior sep inequilateral, with narrow midvein and wide scarious margins; posterior sep similar but conduplicate and sharply keeled; cor-lobes 0.5 mm, in age ± erect and forming a beak above the fr; stamens 2; fr ovoid, convexly rounded at the tip, 1.5–2 mm, circumscissile shortly below the middle; seeds mostly 4, 0.8–1.3 mm; $2n=12$. Dry sandy soil; Mass. and s. N.Y. to Ill. and Kans., s. to Fla. and Tex., most common westward.

12. Plantago virginica L. Taprooted, hirsutulous annual or biennial; lvs oblanceolate to obovate, 5–10(–15) cm, obtuse, entire or inconspicuously toothed; scape to 2 dm; spikes 3–10(–20) cm × 6–8 mm; bracts lanceolate, mostly shorter than the cal, hirsutulous, a scarious margin scarcely developed; sep oblong-obovate, rounded above, the keel hirsutulous, the scarious margin broad and glabrous; cor-lobes 1–3 mm, erect and connivent even in fr; seeds 2, 1.3–2 mm, less than half as wide, convex on the back, strongly concave on the inner face, the hilum nearly as long as the seed; $2n=24$. Occasional plants have chasmogamous fls with spreading cor-lobes. A spring ephemeral of dry or sandy soil, often weedy; Mass. and N.Y. to Wis., Io., and S.D., s. to Fla. and Tex., and intr. westward.

P. rhodosperma Decne., a southwestern sp. with the keel of the sep excurrent, and with the seeds with nearly flat sides and short hilum, probably does not reach our range.

13. Plantago patagonica Jacq. Woolly p. Annual, much like no. 14, but smaller, more strongly woolly-villous, and with the bracts seldom much over 5 mm and scarcely or not at all exserted from the spike; $2n=20$. Dry prairies and plains; native in the cordilleran region and in s. S. Amer., intr. e. to Ill. and occasionally to the Atlantic states. Our plants are apparently identical with those from S. Amer., but the name var. *gnaphalioides* (Nutt.) A. Gray is available although a segregation prove possible. (*P. purshii.*) Occasional plants with the lower bracts conspicuously short-exserted from the spike have been called *P. patagonica* var. *spinulosa* (Decne.) A. Gray, and may indicate introgression from *P. aristata.* (*P. spinulosa.*) *P. coronopus* L., a Eurasian sp. with pinnatifid lvs and with short bracts broadly scarious-margined at base, has been repeatedly introduced in ballast without becoming established.

14. Plantago aristata Michx. Buckhorn. Taprooted, short-lived, usually annual, rather thinly hairy; lvs linear or nearly so, to 18 cm, the petioles dilated and papery at the sheathing, striate base; spikes cylindric, 3–6(–10) cm, the linear bracts conspicuously long-exsert, the lower ones exsert 5–25 mm, the upper ones often shorter but still conspicuous; sep narrowly oblong-obovate, rounded at the tip; cor-lobes 2 mm, spreading; stamens 4, typically only barely or scarcely exsert and the fls selfed, seldom longer and the fls open-pollinated; seeds 2, brown, elliptic, 2–3 mm, very convex on the outer side, concave on the inner; $2n=20$. Disturbed habitats; native from Ill. to La. and Tex., now naturalized over most of e. U.S. and adj. Can.

2. LITTORELLA P. Bergius.

Fls unisexual, basal in groups of 3, the central one staminate and long-pedicellate with a minute bract near the middle of the pedicel, the 2 lateral ones pistillate and sessile; staminate fls with 4 sep, 4 short, spreading cor-lobes, and 4 long-exsert stamens; pistillate fls with 3 or 4 sep and narrowly urceolate, 3–4-lobed cor; fr an achene, enclosed by the persistent cal; acaulescent herbs. 3, New World and Europe.

1. **Littorella uniflora** (L.) Asch.   Delicate, fibrous-rooted perennial; lvs linear, to 5 cm; staminate pedicels to 4 cm; sep 2.5–4 mm, lanceolate; achenes cylindric, 2 mm, half as wide; $2n=24$. Sandy shores or shallow water; Nf. and Que. to Ont. and n. N.Y.; also in Wis., Minn., and Europe. Summer. Only emersed plants flower. American plants are slightly smaller than the European, and may be distinguished as var. *americana* (Fern.) Gleason.

FAMILY **BUDDLEJACEAE**, the Butterfly-bush Family

Fls regular, sympetalous, mostly 4-merous, perfect or often functionally dioecious; cal lobed to very deeply cleft; stamens on the cor-tube, alternate with the imbricate lobes; ovary superior or (*Polypremum*) half-inferior, bilocular, with numerous ovules on thickened, axile placentas; style terminal, with capitate or 2-lobed stigma; fr usually capsular, often septicidal; seeds often winged, with small to large dicotyledonous embryo embedded in the endosperm; woody plants, or seldom (as in *Polypremum*) herbs, with simple, opposite or less often whorled (rarely alternate) lvs, often with lepidote or stellate or branching hairs; interpetiolar stipules commonly represented by a mere line. 10/150.

1 Shrub with lanceolate lvs . . . . 1. *Buddleja*.
1 Herb with diffusely branched stem and linear lvs . . . . 2. *Polypremum*.

1. **BUDDLEJA** L.   Butterfly-bush. Cal campanulate; cor broadly campanulate to salverform; stamens included; ovary superior; style short; capsule many-seeded; ornamental shrubs, usually with serrate lvs and numerous fls in heads or panicles. 100, mainly of warm reg.

1. **Buddleja davidii** Franchet.   Orange-eye b.-b.; summer-lilac. Shrubs 2–4 m; lvs lanceolate or lance-ovate, 1–2 dm, serrate, tomentose beneath; fls in slender panicles 1–2 dm; cor 1 cm, salverform, lilac with orange center; $2n=76$. Native of China, escaped from cult. here and there in our range. July–Sept.

2. **POLYPREMUM** L.   Sep lance-subulate, distinct to the base; cor funnelform, barely surpassing the cal, hairy in the throat; stamens included; anthers rotund; ovary half-inferior; style very short; stigma capitate; capsule somewhat flattened, loculicidal, the carpels eventually separating; seeds numerous, minute; diffusely branched annual with narrow lvs connected at base by a stipular membrane, the fls small, sessile, white, in leafy terminal cymes. Monotypic.

1. **Polypremum procumbens** L.   Branched from the base, 1–3 dm; lvs linear to narrowly oblanceolate, 1–2 cm; cymes dichotomous, often congested, bearing a fl at each fork; bracteal lvs resembling the cauline and scarcely smaller; cor 3 mm; fr oval, 2–3 mm; $2n=22$. Dry sandy soil and waste places; Del. to Fla., Tex., and Mex., n. in the interior to Ky. and Mo. and s. Ill., and occasionally on ballast near eastern cities. June–Sept.

FAMILY **OLEACEAE**, the Olive Family

Fls, regular, hypogynous, perfect or unisexual; cal 4-lobed and valvate, or $\pm$ reduced or even wanting; cor sympetalous and 4-lobed, or the pet small and distinct or even wanting; stamens 2(–4), borne on the cor-tube when the cor is sympetalous, aligned with the sinuses; ovary superior, bilocular, with terminal style and 2-lobed stigma, or the stigma sessile; ovules mostly 2 per locule, sometimes 1–4, seldom (as in *Forsythia*) $\pm$ numerous, the placentation axile; seeds with straight, spathulate, dicotyledonous embryo embedded in the oily endosperm, or without endosperm; woody plants with opposite, exstipulate lvs. 30/600.

Jasminum nudiflorum Lindley, winter-jasmine, occasionally escapes from cult. It is a lax, deciduous shrub with trifoliolate lvs (lfls 1–3 cm) and ± precocious yellow fls solitary or paired at the nodes of the green twigs.

1 Fls conspicuous, with well developed cor, perfect or sometimes unisexual.
  2 Cor salverform, the lobes no longer than the tube.
    3 Infls terminal; fls perfect.
      4 Fls generally lilac (white); anthers included; fr a capsule .... 1. *Syringa.*
      4 Fls white; anthers ± exsert from the cor-tube; fr a drupe .... 2. *Ligustrum.*
    3 Infls from old axils; fls mostly unisexual; fr a drupe .... 3. *Osmanthus.*
  2 Cor-lobes elongate, much exceeding the short tube; fls in lateral clusters.
    5 Fls bright yellow; fr a many-seeded capsule .... 4. *Forsythia.*
    5 Fls white; fr a drupe .... 5. *Chionanthus.*
1 Fls individually inconspicuous, generally unisexual; cor none.
  6 Shrubs with simple lvs; fr a drupe .... 6. *Forestiera.*
  6 Trees with pinnately compound lvs; fr a samara .... 7. *Fraxinus.*

1. SYRINGA L.

Lilac. Fls perfect; cal small, campanulate, truncate or 4-toothed; cor salverform, 4-lobed; stamens 2, included in our spp.; fr a 2-locular capsule with 2 seeds per locule; much branched deciduous shrubs with simple, entire lvs and fragrant fls in dense mixed panicles. 30, Eurasia.

  1. **Syringa vulgaris** L.   Common l. Shrub to 6 m; lvs ovate, 5–10 cm, short-acuminate, truncate to cordate at base; infls 1–2 dm; cor lilac (white), 1 cm wide; $2n=44$–48. Native of se. Europe, persisting indefinitely after cult. and often appearing spontaneously, especially in regions of abandoned farms. May.
    S. × *persica* L., Persian l., less frequently cult., has smaller, narrower lvs tapering at the base.

2. LIGUSTRUM L.

Privet. Fls perfect; cal small, obconic or campanulate, truncate or obscurely 4-lobed; cor salverform, the 4 lobes spreading or recurved; stamens 2; anthers large; ovules 2 per locule; stigma capitate; fr a small black drupe; shrubs with simple, short-petioled, entire lvs and small panicles of white fls terminating the main axis and short lateral branches. 40, Old World.

A number of spp. are cult. and occasionally escape, but only *L. vulgare* is commonly established. Some of the other spp. are keyed, but only *L. vulgare* is described.

1 Cor-tube 2.5–3 mm, about as long as the lobes; anthers 2 mm.
  2 Anthers not reaching the tip of the cor-lobes; twigs minutely puberulent .... 1. *L. vulgare.*
  2 Anthers exsert, the tip surpassing the cor-lobes; twigs densely pubescent .... *L. sinense.*
1 Cor tube 5–8 mm, twice as long as the lobes; anthers 3 mm.
  3 Filaments exsert from the cor-tube; anthers surpassing the cor-lobes; twigs glabrous .... *L. ovalifolium.*
  3 Filaments included in the cor-tube; twigs pubescent.
    4 Anthers not reaching the middle of the cor-lobes .... *L. amurense.*
    4 Anthers reaching nearly to the tip of the cor-lobes .... *L. obtusifolium.*

  1. **Ligustrum vulgare** L.   Common p. Much branched shrub to 5 m, with dark green, elliptic to ovate, long-persistent but eventually deciduous lvs commonly 2–4 cm; cor-tube wide, almost campanulate, 2.5–3 mm, about as long as the lobes; filaments mostly included or barely exsert; twigs minutely puberulent; $2n=46$. Native of Europe; our most commonly cult. and escaped sp. June.

3. OSMANTHUS Lour.

Dioecious or polygamo-dioecious; cal and cor rather small, 4-lobed; stamens 2(4), included; ovules 2 per locule; fr a 1-seeded bitter drupe; evergreen shrubs or small trees with simple lvs and small, scaly-bracted, axillary (in ours) or terminal panicles of greenish-white or white fls. 15, the others Asiatic.

  1. **Osmanthus americanus** (L.) A. Gray. Devil-wood. Shrub or tree to 15 m; lvs entire, firm and shining, lance-oblong to obovate, 5–15 cm, on petioles 1–2 cm; panicles axillary to and surpassed by older lvs; cor dull white, the tube 3–4 mm, the ovate spreading lobes nearly as long; fr dark purple, 1–1.5 cm; $2n=138$. Rich woods and swamps; se. Va. to Fla. and La.; Mex. Apr., May. Ours is var. *americanus* (*Amarolea a.*)

4. FORSYTHIA Vahl.

Golden bell. Fls perfect; cal deeply 4-lobed; cor deeply 4-parted, the short tube campanulate, the lobes oblong, widely spreading; stamens 2, very short, borne on the short cor-tube; fr a bilocular, many-seeded capsule; much branched shrubs with simple, entire or toothed lvs. 7, Eurasia.

Several spp. are commonly cult. and sometimes persist, but apparently do not become naturalized. The most common, with many cultivars, are *F. viridissima* Lind., suberect, with cross-chambered pith, merely toothed lvs, and rather light yellow or greenish-yellow fls; *F. suspensa* (Thunb.) Vahl, more arching, with hollow branches solid at the nodes, toothed or often 3-parted lvs, and golden yellow fls; and their hybrid, *F.* × *intermedia* Zabel.

5. **CHIONANTHUS** L.   Fringe-tree. Fls seemingly perfect; cal minute; cor divided nearly to the base, the 4 lobes narrowly linear; stamens 2, inserted on the short cor-tube, the anthers large, oblong, the filaments very short; ovules 2 per locule; style short, slender; fr an ovoid or ellipsoid drupe with very thin exocarp; shrubs or small trees with simple, entire lvs and conspicuous white fls in loose panicles from the axils of the previous year. 2, the other in e. Asia.

    1. **Chionanthus virginicus** L.   Tall shrub or small tree to 10 m; lvs lance-elliptic to oblong or somewhat obovate, 8–15 cm; panicles many-fld, drooping, 1–2 dm; cor white, its lobes 15–30 mm; fr dark blue, 10–15 mm; $2n=46$. Rich woods and along streams; N.J. to Fla., w. irregularly to s. O., s. Mo., e. Okla., and e. Tex. May. Functionally dioecious, the pistil of the fertile fls much larger than that of the sterile ones.

6. **FORESTIERA** Poiret.   Dioecious or polygamo-dioecious; cal none or minute; cor none; stamens 2–4; ovary 2-locular with 2 ovules per locule, that of the staminate fls none or vestigial; fr a small, black, usually 1-seeded drupe; shrubs or trees, often thorny, with simple lvs and lateral fascicles of nearly sessile fls or small lateral panicles; the fascicles of fls, subtended by bud-scales, are easily mistaken for single fls with numerous stamens and pistils. 15, N. Amer.

1 Fls precocious; lvs long-tapering at both ends; fr slenderly ellipsoid .... 1. *F. acuminata.*
1 Fls summer; lvs obtuse or subacute; fr thickly ellipsoid .... 2. *F. ligustrina.*

    1. **Forestiera acuminata** (Michx.) Poiret.   Swamp-privet. Shrub, or southward a small tree; lvs lanceolate or oblong, 4–8 cm, slenderly acuminate, entire or remotely serrulate, gradually tapering to the petiole; fls before the lvs, the staminate in dense, subsessile, lateral fascicles, the pistillate in panicles 1–2 cm; fr slenderly ellipsoid, often curved, 1–2 cm, not more than half as thick; $2n=46$. Swamps and wet woods; c. Ill. to s. Ind. and s. Mo., s. to La. and Miss., and irregularly to S.C., Fla., and Tex. (*Adelia a.*)

    2. **Forestiera ligustrina** (Michx.) Poiret.   Shrub 1–3 m; lvs elliptic to obovate, 1–4 cm, half as wide, obtuse or subacute, finely serrulate, abruptly narrowed to the short petiole; fls axillary, subtended by minute ovate bracts, both staminate and pistillate crowded and subsessile in dense fascicles; cal minute but present; fr broadly ellipsoid, 7–8 mm, more than half as thick. Usually in sandy or rocky soil; c. Ky. and Tenn. to Ala. and Fla. Aug.

7. **FRAXINUS** L.   Ash. Fls unisexual or seldom perfect; cal minute, 4-lobed or irregularly erose, or none; cor in our spp. none; stamens mostly 2; ovary bilocular, with 2 ovules per locule; fr a 1-seeded samara, the wing chiefly or wholly distal; trees with deciduous, pinnately compound lvs, the fls in dense fascicles, short racemes, or panicles, from the axils of the previous season. 65, N. Hemisphere.

1 Samaras differentiated into a flat wing and a terete or subterete body.
  2 Lvs papillose beneath; wing of the samara only shortly decurrent onto the upper third of the body
    .... 1. *F. americana.*
  2 Lvs not papillose; wing of the samara decurrent to the middle of the body, or beyond.
    3 Lateral lfls short-petiolulate or subsessile; samara-wing mostly less than 7 mm wide .... 2. *F. pennsylvanica.*
    3 Lateral lfls conspicuously slender-petiolulate; samara-wing more than 7 mm wide .... 3. *F. profunda.*
1 Samaras flat or nearly so, winged to the base.
  4 Twigs terete; fls all or partly unisexual.
    5 Samaras subtended by a minute cal; lfls petiolulate; southern .... 4. *F. caroliniana.*
    5 Samaras lacking a cal; lfls sessile; chiefly northern plants .... 5. *F. nigra.*
  4 Twigs sharply quadrangular; fls perfect .... 6. *F. quadrangulata.*

    1. **Fraxinus americana** L.   White ash. Tree to 40 m; bark becoming rather finely and closely reticulate-ridged; lfls 5–9, usually 7, oblong to ovate or obovate, usually abruptly acuminate, crenulate to sometimes

entire, paler and papillose beneath; petiolules mostly wingless; twigs and lvs mostly glabrous; terminal bud generally blunt, wider than high; old lf-scars commonly with a concave upper margin; anthers linear to linear-oblong, apiculate; fr linear to oblanceolate, 3–5 cm, the wing extending a third of the length of the terete body, the free part, above the body, longer than the body itself; subtending cal 1–1.5(–2) mm, seldom cleft on one side only; 2n=46, 92, 138. A valuable timber tree of rich, moist (not wet) woods; N.S. to Minn., s. to Fla. and Tex. (*F. biltmoreana,* a ± pubescent, chiefly southern form, perhaps reflecting hybridization with no. 2.)

2. **Fraxinus pennsylvanica** Marshall. Tree to 25 m; bark flaky; lfls 5–9, usually 7, lanceolate or lance-ovate to oblong or elliptic, acuminate to a blunt or acute tip, serrate to crenulate or subentire, not papillose beneath, acute or broadly cuneate and often inequilateral at base, usually ± decurrent onto the short petiolule or subsessile; twigs and lvs densely pubescent to often glabrous; terminal bud acute, higher than wide; lf-scars with truncate or barely concave upper margin; anthers of no. 1; fr linear to spatulate, the wing extending to about the middle of the body, its free part, above the terete or subterete body (this mostly less than 2 mm wide) commonly shorter than the body itself and 3–8, usually 5–6 mm wide; cal 1–1.5 mm, usually cleft on one side; 2n=46. Moist or often wet woods; N.S. and Que. to Alta., s. to Fla. and Tex. (*F. campestris.*) Typical *F. pennsylvanica,* with the twigs, infl, and lower surfaces of the lvs densely hairy, is called red ash. Glabrous plants are called green ash; these have been segregated as var. *subintegerrima* (Vahl) Fern. (*F. lanceolata*), but are probably not taxonomically significant.

3. **Fraxinus profunda** (Bush) Bush. Pumpkin-ash. Tree to 40 m; bark ridged-reticulate as in no. 1; twigs hairy; lfls 5–9, usually 7, lanceolate to oblong or elliptic, long-acuminate, entire or nearly so, hairy beneath, broadly acute to rounded and often inequilateral, on conspicuous wingless petiolules 8–15 mm; fr linear-oblong to spatulate, 4–7.5 (avg 5.5) cm × 7–12 (avg. 9) mm, the wing extending at least to the middle of the terete body (this mostly more than 2 mm wide); cal subtending the fr 3–4 mm, or short as in no. 2; 2n=138. Swamps and wet woods; irregularly from s. N.J. to Fla. and w. to O., Ind., Ill., Mo., and La. (*F. michauxii; F. tomentosa.*)

4. **Fraxinus caroliniana** Miller. Water-ash; Carolina-ash. Tree to 15 m; twigs terete or nearly so, glabrous or thinly hairy; lfls 5–7, petiolulate, lanceolate to elliptic, abruptly acuminate to an obtuse tip; fr flat, lanceolate to elliptic, 3.5–5 cm × 8–20 mm, subtended by a minute cal usually less than 1 mm; wing extending nearly or quite to the base. Swamps and bottom-lands, mainly on the coastal plain; Va. to Fla. and Tex.

5. **Fraxinus nigra** Marshall. Black ash. Tree to 25 m, twigs glabrous, terete or nearly so; lfls 7–11, sessile, lanceolate to oblong, long-acuminate, often conspicuously serrate, broadly acute at base; fr flat, lanceolate to oblanceolate, 2.5–4 cm × 6–10 mm, the wing rounded to emarginate, extending nearly or quite to the base; no cal beneath the fr; 2n=46. Wet woods and swamps; Nf. and Que. to Man., s. to Del., W.Va., Ind., and Io.

6. **Fraxinus quadrangulata** Michx. Blue ash. Tree to 30 m, with sharply 4-angled or narrowly 4-winged twigs; lfls 7–11, commonly lanceolate, usually long-acuminate, broadly cuneate to subrotund and usually inequilateral at the petiolulate base; fls perfect; anthers blunt or cleft at the tip; fr flat, elliptic to narrowly oblong-obovate, 2.5–4 cm × 6–10(–12) mm, usually emarginate, the wing extending nearly or quite to the base; cal minute, soon deciduous. Moist woods; s. Ont. to s. Mich., s. Wis., and e. Kans., s. to W.Va., Ga., Ala., and Okla.

Family **SCROPHULARIACEAE**, the Figwort Family

Fls perfect, hypogynous, sympetalous (rarely apetalous); cal 4–5-lobed or cleft; cor 5-lobed or less often 4-lobed, slightly to usually evidently irregular, often evidently bilabiate, sometimes spurred or saccate at base; stamens borne on the cor-tube alterante with the lobes, typically 4, or only 4 functional and the fifth (uppermost) one staminodial, sometimes only 2 (the upper lateral pair), seldom (*Verbascum*) 5 and all polleniferous; ovary superior, bilocular, with a terminal style and simple or 2-lobed stigma; placentation axile; fr mostly capsular; seeds (2–) ± numerous, with a straight or slightly curved dicotyledonous embryo and oily endosperm; herbs (all ours) or seldom woody plants, not infrequently hemiparasitic on the roots of other plants, but usually green even so; lvs exstipulate, opposite or alternate, simple to sometimes pinnately dissected, the fls in various sorts of infls, often large and showy. 190/4000.

Synopsis of the Subfamilies and Tribes

1 Upper lip of the cor overlapping the lower lip in bud. (Scrophularioideae.)
  2 Stigmas mostly 2 and distinct (solitary in *Limosella*). (Genera 1–10.) .... Tribe Gratioleae.
  2 Stigma 1.
    3 Anther-bearing stamens 5. (Genus 11.) .... Tribe Verbasceae.
    3 Anther-bearing stamens 4.
      4 Filaments 5, the upper median one sterile. (Genera 12–14.) .... Tribe Scrophularieae.

    4 Filaments 4.
       5 Lower median lobe of the cor folded downward, forming a pouch enclosing the anthers.
         (Genus 15.) .... Tribe Collinsieae.
       5 Lower median lobe of the cor flat, or arched upward to form a palate. (Genera 16–20.) .. Tribe Antirrhineae.
1 Lower lip of the cor overlapping the upper lip in bud. (Rhinanthoideae.)
    6 Upper lip of the cor flattened or merely arched, not enclosing the anthers.
       7 Pollen-sacs divergent; stigmas 2. (Genus 21.) .... Tribe Digitalieae.
       7 Pollen-sacs parallel; stigma 1 (2 in *Buchnera* and *Dasistoma*).
         8 Stamens 4; mostly hemiparasitic. (Genera 22–26.) .... Tribe Buchnereae.
         8 Stamens 2; not parasitic. (Genera 27–29.) ....Tribe Veroniceae.
    6 Upper lip of the cor very concave, galeate, enclosing and commonly concealing the anthers; mostly
       hemiparasitic. (Genera 30–37.) .... Tribe Euphrasieae.

Artificial Key to the Genera

1 Cor with a distinct spur or saccate gibbosity at the base protruding between the 2 lower cal-lobes.
    2 Cor with a broad, saccate gibbosity at base .... 19. *Antirrhinum.*
    2 Cor with a slender spur at base.
       3 Fls in terminal racemes .... 16. *Linaria.*
       3 Fls solitary in the axils of the lvs.
         4 Stems prostrate or trailing (or climbing by adventitious roots).
            5 Lvs pinnately veined, entire; stems prostrate but not rooting .... 17. *Kickxia.*
            5 Lvs palmately veined and lobed; stems trailing, rooting .... 18. *Cymbalaria.*
         4 Stems erect .... 20. *Chaenorrhinum.*
1 Cor without a spur or saccate gibbosity at base.
    6 Foliage lvs (not bracteal lvs subtending the fls) alternate, or the lvs all basal.
       7 Stems prostrate, the lvs and 1-fld peduncles clustered at the nodes; fls tiny, not more than about
         5 mm .... 10. *Limosella.*
       7 Stems ± erect; fls larger, mostly 1–5 cm long or wide.
         8 Stamens 5; cor nearly regular .... 11. *Verbascum.*
         8 Stamens 4 or 2; cor ± strongly irregular.
            9 Stamens 2 .... 29. *Besseya.*
            9 Stamens 4.
              10 Upper lip of the cor galeate, enfolding the stamens.
                11 Fls subtended by a pair of bractlets at the base of the cal; cal-lobes 5 .... 30. *Schwalbea.*
                11 Fls not bracteolate; cal, except in *Pedicularis furbishiae*, with 2 or 4 lobes.
                  12 Cor at least twice as long as the cal.
                    13 Foliage lvs deeply toothed to pinnatifid; pollen-sacs similar in size and
                      position .... 34. *Pedicularis.*
                    13 Foliage lvs entire or rarely trifid; pollen-sacs dissimilar (see descr.) .... 36. *Orthocarpus.*
                  12 Cor much less than twice as long as the cal; pollen-sacs dissimilar ....37. *Castilleja.*
               10 Upper lip of the cor not at all galeate, not enfolding the stamens .... 23. *Digitalis.*
    6 Foliage-lvs opposite or whorled.
       14 Stamens 2.
         15 Cal 4-lobed, or sep 4.
            16 Fls in the axils of opposite lvs.
              17 Cal-lobes all separate to the base .... 8. *Micranthemum.*
              17 Cal-lobes separate to the middle, except on the lower side, where they are separate
                 to the base .... 9. *Hemianthus.*
            16 Fls in the axils of alternate lvs, or in lateral racemes, or in dense terminal spikes or
             spike-like racemes.
              18 Cor-lobes longer than the tube; lvs only rarely whorled .... 28. *Veronica.*
              18 Cor-lobes much shorter than the tube; lvs whorled .... 27. *Veronicastrum.*
         15 Cal 5-lobed, or sep 5.
            19 Lvs whorled; fls in dense, spike-like racemes .... 27. *Veronicastrum.*
            19 Lvs opposite; fls solitary in the axils of foliage-lvs, pedicellate.
              20 Cor nearly regular, its lobes about as long as the tube; outer sep ovate, much wider
                 than the inner lateral ones .... 1. *Bacopa.*
              20 Cor distinctly irregular and bilabiate; sep of equal or subequal width.
                21 Sterile stamens minute or lacking; fls in some spp. subtended by a pair of
                  bracteoles just beneath the cal .... 3. *Gratiola.*
                21 Sterile stamens (the lower pair) represented by a pair or slender filaments at
                  the cor-throat; bractlets none .... 7. *Lindernia.*
      14 Stamens 4 (an additional staminode sometimes present).
         22 Cor galeate, i.e., the upper lip forming a hood or beak (the galea) that enfolds the anthers,
          the teeth of the upper lip short or obsolete (galeate condition least marked in *Euphrasia*).
            23 Lvs palmately veined, seldom over twice as long as wide .... 32. *Euphrasia.*
            23 Lvs elongate and pinnately veined or pinnatifid.
              24 Foliage-lvs (not bracteal lvs) entire or obscurely few-toothed.
                25 Cor pubescent, light red .... 31. *Odontites.*
                25 Cor glabrous, white with a yellow palate .... 35. *Melampyrum.*
              24 Foliage-lvs with numerous prominent teeth, or pinnatifid.

26 Cal somewhat inflated at anthesis, accrescent and very conspicuously inflated
    in fr; cal-lobes 4; lvs merely serrate .... 33 *Rhinanthus*.
26 Cal not much, if at all, inflated; cal-lobes 2, or if more than 2 then the lvs
    pinnatifid or bipinnatifid .... 34 *Pedicularis*.
22 Cor not galeate, though often bilabiate, the upper lip, if differentiated, not forming a hood
  or beak and not enfolding the anthers.
  27 Sep distinct nearly or quite to the base.
    28 Bracteal lvs abruptly much reduced in contrast to the foliage-lvs below them, the
      fls thus in a terminal infl.
      29 Fls closely subtended by 2 or 3 large, sep-like bractlets .... 12 *Chelone*.
      29 Fls not bracteolate.
        30 Staminode reduced to a mere projecting knob or scale on the upper cor-
          lip; cor greenish or brownish, not very showy .... 14 *Scrophularia*.
        30 Staminode well developed, attached well down in the cor-tube; cor showy,
          blue or purple to white .... 13 *Penstemon*.
    28 Bracteal lvs gradually reduced, the fls therefore generally appearing axillary or in
      lateral racemes.
      31 Cor nearly regular, its lobes subequal, about as long as the tube .... 1 *Bacopa*.
      31 Cor distinctly bilabiate, its lobes shorter than the tube.
        32 Lvs deeply pinnatifid .... 4 *Leucospora*.
        32 Lvs entire or merely toothed.
          33 Annual; pedicels without bractlets at base .... 7 *Lindernia*.
          33 Perennial; pedicels with 2 bractlets at base, in addition to the sub-
            tending bracteal lf .... 2 *Mecardonia*.
  27 Sep evidently connate to form a lobed tube.
    34 Cor strongly bilabiate.
      35 Lower lip of the cor with the median lobe folded downard between the lateral
        lobes, forming a pouch enclosing the stamens .... 15 *Collinsia*.
      35 Lower lip of the cor typically arched upwards to form a palate, in any case not
        downfolded to enclose the stamens.
        36 Cal cleft to below the middle; bracteal lvs alternate; intr. weeds .... 6 *Mazus*.
        36 Cal not cleft as far as the middle; bracteal lvs opposite; native, not weedy ...... 5 *Mimulus*.
    34 Cor only slightly irregular, scarcely or not at all bilabiate.
      37 Fls wholly or predominantly yellow.
        38 Lvs entire to pinnatifid, the segments broad.
          39 Anthers glabrous .... 22 *Dasistoma*.
          39 Anthers pubescent .... 24 *Aureolaria*.
        38 Lvs pinnatifid or bipinnatifid into filiform segments .... 23 *Seymeria*.
      37 Fls some other color than yellow.
        40 Cor ± campanulate; anthers dithecal, villous .... 25 *Agalinis*.
        40 Cor salverform; anthers monothecal, glabrous .... 26 *Buchnera*.

1. BACOPA Aublet, nom. conserv.

Water-hyssop. Sep 5 (in our spp.), distinct, imbricate, the upper median one about twice as wide as the 2 lower and 4 times as wide as the lateral ones; cor tubular or campanulate to salverform, in our spp. virtually regular or with the 2 upper lobes connate half-length; stamens (2–)4, inserted below the middle of the cor-tube; pollen-sacs parallel; stigmas 2 and distinct or one and bilobed; fr globose, septicidal and/or loculicidal; small perennial herbs of wet shores or shallow water, with sessile, opposite, entire or toothed lvs and small, white or blue fls borne singly or in pairs in the lf-axils. 100, warm reg.

1 Fls closely subtended by a pair of minute bractlets.
  2 Lvs oblanceolate to cuneate-obovate, 1-veined, with a few obscure lateral veins .... 1 *B. monnieri*.
  2 Lvs ovate, palmately 5–9-veined .... 2 *B. caroliniana*.
1 Fls not subtended by bractlets; lvs palmately veined.
  3 Pedicels mostly 2–4 times as long as the sep; lvs 5–13-veined .... 3 *B. rotundifolia*.
  3 Pedicels little if at all longer than the sep; lvs obscurely 3–5-veined .... 4 *B. innominata*.

  1. Bacopa monnieri (L.) Wettst. Stems creeping and forming mats, glabrous; lvs oblanceolate to cuneate-obovate, entire, 0.5–2.5 cm, with a single midvein and a few obscure pinnately disposed lateral veins; fls solitary at the nodes on pedicels 1–2.5 cm; cal subtended by a pair of linear bractlets 2–3 mm; upper sep ovate to lanceolate, ca 6 × 2.5 mm; cor white to sometimes pink or blue, campanulate, 7–10 mm, glabrous within, the 5 lobes subequal, retuse; stamens included; stigma one, shallowly bilobed; $2n\ 2n=64$. Wet sandy shores on the coastal plain; se. Va. to Fla. and Tex., and widespread in the tropics. All summer. (*Bramia m.*)

  2. Bacopa caroliniana (Walter) Robinson. Plants lemon-scented; stems creeping or floating, smooth below, villous distally; lvs ovate, obtuse, broadly rounded to subcordate at base, 1–2.5 cm, crenate, palmately 5–9-veined; fls solitary from some upper nodes on pedicels 5–15 mm; cal subtended by a pair of minute bractlets

1-2 mm; sep glandular-punctate, the upper one broadly ovate-cordate, ca 8 mm; cor blue, campanulate, 1 cm, pubescent within, the 5 lobes a little shorter than the tube, the 2 upper united half-length; ovary surrounded by a circle of short bristles; stigma one, shallowly bilobed. Wet shores and shallow water; se. Va. to Fla. and Tex.; disjunct in Ky. All summer. (*Hydrotida c.*)

3. **Bacopa rotundifolia** (Michx.) Wettst.   Stems typically submersed with the tips floating, less often emersed, 2-6 dm, the younger parts usually hairy but soon glabrate; lvs very thin, obovate to subrotund, 1-3.5 cm, entire, palmately 5-13-veined; fls usually 2-4 from the uppermost nodes, on pedicels (0.5)1-1.5 cm, without bractlets; sep 3-5 mm, obtuse, the upper one rotund-elliptic, obscurely to evidently reticulate-veiny; cor white witih a yellow throat, narrowly campanulate, mostly 4-10 mm, its lobes equaling or a little shorter than the tube, the 2 upper united half-length; stigmas 2, distinct; 2*n*=36. Shallow water and wet soil; Ind. to Io., N.D., and Mont., s. to Miss. and Tex., and occasionally intr. elsewhere, as in tidal Va. and Md. June–Sept. (*B. simulans; Bramia r.; Macuillamia r.*)

4. **Bacopa innominata** (M. Gomez) Alain.   Much like no. 3; stems glabrous, more often emersed; lvs 0.5-1 cm, fairly thick, obscurely 3-5-veined; pedicels 3-6 mm; upper sep broadly ovate-cordate, 4-6 mm; cor 3-5 mm, the lobes half as long as the tube; stamens 2-4. Wet places and shallow water; W.I. and C. Amer., n. to N.C., and in the intertidal zone to Va. and Md. (*B. cyclophylla; B. stragula; Macuillamia obovata.*)

2. MECARDONIA Ruiz & Pavon.

Sep 5, distinct or nearly so, elongate, unequal in width; cor tubular-campanulate, distinctly irregular, the lobes of the upper lip united to beyond the middle, the tube villous within on the upper side; stamens 4; stigmas 2; fr narrowly ovoid, septicidal; perennial herbs with opposite, serrate lvs and small fls, the long pedicels solitary from the axils, each 2-bracteolate at base. 15, New World.

1. **Mecardonia acuminata** (Walter) Small.   Stem erect or ascending, 2-5 dm, not much branched, glabrous; lvs oblanceolate, 2-4 cm, serrate above the middle, cuneate at base; pedicels 1-3 cm, ascending, the basal bractlets shorter than the subtending lvs; cal 6-9 mm; sep narrowly lanceolate, attenuate, the outer twice as wide as the inner; cor 1 cm, white with purple lines on the lower lip; 2*n*=42. Moist woods; Del. and Md. to Ky. and Mo., s. to Fla. and Tex. All summer. (*Bacopa a.; Pagesia a.*)

3. GRATIOLA L.

Hedge-hyssop. Sep 5, distinct or nearly so, elongate, usually unequal; cor tubular or narrowly campanulate, bilabiate, the lobes of the upper lip united nearly to the tip, hairy at the base within, those of the lower lip rounded; stamens 2, inserted near the middle of the cor-tube; second pair of stamens represented by minute sterile filaments, or none; connective of anthers in most spp. expanded and saucer-shaped, longer and wider than the transverse pollen-sacs, or in *G. pilosa* not expanded, shorter and narrower than the longitudinal pollen-sacs; stigmas 2, plate-like; fr septicidal or also loculicidal; small annual or perennial herbs with opposite lvs and small yellow or white fls on solitary axillary pedicels, usually subtended by a pair of sep-like bractlets just beneath the cal. 20, widespread.

1 Perennial; lvs linear to ovate; lf linear, maintaining the full width to the base, if wider, broadest near
   the rounded, truncate, or even subcordate base.
  2 Plants villous-hirsute with joined hairs, not glandular .... 1. *G. pilosa.*
  2 Plants glabrous or glandular or glandular-hairy.
    3 Lvs entire, lanceolate to ovate; cor (with rare exceptions) bright yellow .... 2. *G. aurea.*
    3 Lvs (or most of them) sharply toothed; cor-lobes white or tinged with purple.
      4 Lf-teeth several per side; lf-blades lanceolate to ovate, thin, conspicuously 3-5-nerved from
        the base .... 3. *G. viscidula.*
      4 Lf-teeth 1-3 per side, above the middle; lf-blades linear to narrowly lanceolate, thick, obscurely
        1-nerved .... 4. *G. ramosa.*
1 Annual; lvs commonly lanceolate to oblanceolate, broadest near or above the middle and gradually
   tapering to the base.
  5 Pedicels slender, divergent, 1 cm or more .... 5. *G. neglecta.*
  5 Pedicels stout, erect, usually 1-4 mm .... 6. *G. virginiana.*

1. **Gratiola pilosa** Michx.   Hirsute perennial; stem 2-4 dm, not much branched; lvs ovate or lance-ovate, sessile by a broad base, 1-2 cm, entire or with a few low teeth; fls sessile or on pedicels to 5 mm, or to 9 mm in fr; bractlets lance-linear, 9 mm; sep lance-linear, the outer 7 mm, the inner 4 mm; cor white, 1 cm; fr conic, 4-5 mm; 2*n* =22. Pine-barrens, chiefly on the coastal plain; Va. to Fla. and Tex.; also inland from N.C. to Ky. and Ark. Summer and fall. (*Sophronanthe p.; Tragiola p.*) Sometimes excluded from *Gratiola* because of its simple anther-connective and longitudinal pollen-sacs.

2. **Gratiola aurea** Pursh.   Yellow h.-h. Perennial, glabrous or glandular-pubescent above; stems ascending, or creeping at base, 1-3 dm; lvs glandular-punctate, sessile by a broad base, lanceolate to ovate, 1-2.5 cm,

entire or barely denticulate; pedicels filiform, 5–15 mm; bractlets 2, sep-like; sep lanceolate, 4–5 mm; cor bright yellow (white), 12–16 mm; fr globose, 2–3 mm, slightly shorter than the cal; $2n=28$. Muddy or sandy shores, especially in acid soils; Nf. and Que., s. along the coast to Fla., inland irregularly to Ill. and N.D. Summer and fall. (?*G. lutea* Raf.)

3. **Gratiola viscidula** Pennell. Perennial; stems puberulent or glandular above, to 6 dm; lvs sessile and somewhat clasping, lanceolate to ovate, 1–2.5 cm, with several small sharp teeth, 3–5-nerved; pedicels 1–1.5 cm; bractlets usually present and resembling the sep; sep 4–7 mm; cor 8–13 mm, white or tinged with purple, marked with purple lines internally; fr 2 mm, much shorter than the sep; $2n=14$. June–Aug. Moist or wet places; Del. to S.C., n. Ga., and e. Tenn., and irregularly to s. O., W.Va., Ky., Mo., and n. Fla. Summer and fall. (*G. viscosa* Schwein., not Hornem.)

4. **Gratiola ramosa** Walter. Perennial; stems glabrous, 1–3 dm, not much branched; lvs sessile by a broad base, linear to narrowly lanceolate, thick, 0.8–1.5 cm, obscurely 1-nerved, bearing 1–3 thick glandular-callous teeth on each side above the middle; pedicels 8–12 mm, minutely glandular; bractlets usually none; sep thick, narrowly linear, 3–5 mm; cor 10–12 mm, yellow with white lobes; fr much shorter than the sep; $2n=14$. Moist or wet places in pine-barrens on the coastal plain; Md. to Fla. and La. Summer.

5. **Gratiola neglecta** Torr. Annual; stems 1–3 dm, glandular-pubescent; lvs lanceolate to oblanceolate, 2–5(–6) cm, narrowed from near or above the middle to the base; pedicels very slender, widely divergent, 1–2 (+) cm; bractlets equaling or surpassing the cal; sep lanceolate, 4–6 mm; cor 8–10 mm, the lobes white, the tube yellowish; fr 3–5 mm; $2n=16$. Wet soil; Que. to B.C., s. to Ga. and Ariz.; our most abundant sp. May–July, also again in autumn, and often then with cleistogamous fls. (*G. virginica,* misapplied.)

6. **Gratiola virginiana** L. Annual; stem 1–4 dm, glabrous or nearly so; lvs sessile, lanceolate to oblong, elliptic, or oblanceolate, 2–4 cm, narrowed to the base; pedicels stout, erect, 1–4(–10) mm; bractlets closely subtending and about equaling the cal; sep narrowly lanceolate, 4–6 mm; cor white, marked with purple lines internally, 8–12 mm; fr 5–9 mm; $2n=16$. Wet woods; N.J. to O., Ill., and Io., s. to Fla. and Tex. May–Aug., again in autumn from seedlings of the current year; late fls often cleistogamous. (*G. sphaerocarpa.*)

4. **LEUCOSPORA** Nutt.
Cal deeply 5-parted, the sep linear; cor tubular, bilabiate, the lips slightly spreading, much shorter than the tube, the upper shorter than the lower; stamens 4, inserted near the middle of the cor-tube and included in it; style shortly 2-lobed, the stigmas wedge-shaped; fr ovoid, septicidal; annual with opposite, pinnatifid lvs and small fls on axillary, solitary or paired, bractless pedicels. Monotypic.

1. **Leucospora multifida** (Michx.) Nutt. Villosulous throughout, erect or decumbent at base, diffusely branched, 1–2 dm; lvs triangular-ovate in outline, 2–3 cm, deeply pinnatifid into 3–7 linear or narrowly oblong segments, the larger of these also few-toothed or occasionally pinnatifid, at base somewhat decurrent on the long petiole; pedicels 5–10 mm; cal 3–4 mm at anthesis; cor pale lavender, 3–5 mm. Wet sandy or muddy shores; s. Ont. and O. to Io. and Nebr., s. to Ga. and Tex. July–Oct. (*Conobea m.*)

5. **MIMULUS** L.
Monkey-fl. Cal tubular, angled or prismatic, the lobes usually shorter than the tube; cor bilabiate, the lower lip arched in the throat, or with 2 elevated ridges partly or wholly closing the orifice; stamens 4, inserted near the middle of the cor-tube; pollen-sacs divergent; style elongate; stigmas 2, plate-like; fr cylindric, loculicidal; ours perennial herbs with rhizomes or stolons; lvs opposite; fls showy, yellow to red or blue, on solitary axillary pedicels. 120, widespread but mainly w. N. Amer.

1 Fls blue.
  2 Lvs sessile; pedicels 20–45 mm, longer than the cal .... 1. *M. ringens.*
  2 Lvs petiolate; pedicels at anthesis 2–10 mm, shorter than the cal .... 2. *M. alatus.*
1 Fls yellow.
  3 Cal-lobes narrowly triangular, all about equal and much longer than wide .... 3. *M. moschatus.*
  3 Cal-lobes broadly triangular, or depressed-triangular, the uppermost the largest, the other 4 much
     wider than long or scarcely developed.
    4 Stem weak, diffuse or creeping; cal at anthesis 4.5–12 mm; cor 8–25 mm .... 4. *M. glabratus.*
    4 Stem ± erect, stout; cal at anthesis 11–17 mm; cor 25–45 mm .... 5. *M. guttatus.*

1. **Mimulus ringens** L. Allegheny m.-fl. Glabrous; stems 2–13 dm, 4-angled or very narrowly winged; lvs sessile, lanceolate, narrowly oblong or oblanceolate, acute or acuminate, obscurely crenate, either rounded or narrowed at base, the larger 5–10 cm, the upper progressively reduced and the uppermost sometimes bract-like; pedicels solitary in the upper axils, divergent, 2–4.5 cm, at maturity erect or upcurved; bractlets none; cal 10–16 mm, the short lobes broadly ovate to semicircular at base, abruptly prolonged into a short, nearly linear tip; cor blue, 2–3 cm; $2n=24$. Wet woods and swamps, and open marshy places; Que. and N.S. to Sask., s. to Ga., La., and Okla. Plants of tidal mud from Que. to Me., with considerably shorter internodes, pedicels, and cal have been distinguished as var. *colpophilus* Fern.

2. Mimulus alatus Aiton.   Sharpwing m.-fl. Much like no. 1, and hybridizing with it, but the lvs tapering to a narrowly winged petiole 1–2 cm, the blade lanceolate to ovate, more sharply and coarsely toothed; pedicels at anthesis mostly 2–10 mm, to 15 mm in fr; $2n=22$. Wet woods and shady streambanks; Conn. to s. Ont., s. Mich., Io., and Kans., s. to Fla. and Tex. July, Aug. Hybridizes with no. 1.

3. Mimulus moschatus Douglas.   Musky m.-fl. Villous and viscid, musk-scented; stems creeping at base, with ascending tips, 2–4 dm; lvs thin, petioled, ovate or lance-ovate, 3–6 cm, entire or remotely dentate, rounded or subcordate at base, pinnately veined; pedicels filiform, 1–2 cm; cal 8–11 mm, the throat oblique, the lobes about equal, narrowly triangular, 3–4.5 mm; cor yellow, open in the throat, 17–22 mm; $2n=32$. Cool wet soil, especially along brooks and springs; Nf. and Que. to Mich., s. through N.Y. to W.Va., and widespread in the w. cordillera, sometimes escaped from cult. elsewhere in our range. July, Aug.

4. Mimulus glabratus HBK.   Round-lf m.-fl. Stems weak, prostrate to ascending, glabrous or minutely villosulous above; lvs palmately veined, the lower petioled, the upper sessile or nearly so; fls from the upper axils, on pedicels 1–4 cm; cal accrescent in fr, very irregular, the throat oblique, the upper median lobe broadly ovate and blunt (sometimes mucronate), the other 4 lobes very short or obsolete; cor bright yellow, sometimes with reddish-brown dots, the throat open, the lower lip bearded; $2n=28, 30, 60, 62, 90, 92$. In shallow water or very wet places, especially in calcareous situations; Mich. to Man. and Mont., s. to S. Amer. June–Sept. Most of our plants belong to var. *fremontii* (Benth.) A. L. Grant, with entire or repand to irregularly denticulate, subrotund to ± reniform or occasionally rotund-ovate lvs, the cor mostly 8–15 mm; c. and n. Great Plains, e. to Ill., Mich., and Ont. Var. *michiganensis* (Pennell) Fassett, local near the Straits of Mackinac, Mich., has more ovate or oval, more obviously toothed lvs and mostly larger fls, the cor 15–25 mm; it may reflect ancient introgression from no. 5.

5. Mimulus guttatus DC.   Common yellow m.-fl. More or less erect, branched, 3–6 dm, glabrous or glandular-puberulent above; lvs variable, mostly obovate to subrotund, sinuate-dentate, the principal veins arising near the base; lower lvs short-petioled, the upper sessile; fls bright yellow, 2.5–4.5 cm, the cor-throat constricted by the well developed, bearded palate; cal accrescent in fr, the lobes all broadly triangular, the upper median one the largest, the lateral ones tending to fold inward in fr; $2n=16, 28, 30, 32, 48, 56$. Wet places; native mainly in the w. cordillera, occasionally escaped from cult. in our range. July, Aug. (*M. langsdorfii*). The cordilleran population is more variable than the cult. forms here described, often with smaller fls.

6. MAZUS Lour.

Cal 5-parted to about the middle or a little beyond, the narrow segments equal, with ± evident midrib; cor bilabiate, the short upper lip subacutely 2-toothed, the lower lip much longer and 3-lobed; stamens 4; stigmas 2, plate-like; fr loculicidal; low herbs with basal and opposite lvs, or the uppermost lvs alternate, as are the minute or obsolescent bracts of the terminal racemes; fls anthocyanic. 20, Australasia.

1 Lax, short-hairy annual; cor 7–10 mm . . . . 1. *M. pumilus.*
1 Creeping, glabrous perennial; cor ca 1.5 cm . . . . 2. *M. miquelii.*

1. Mazus pumilus (Burm. f.) van Steenis.   Lax, short-hairy, apparently annual, to 15 cm, often several-stemmed; basal lvs spatulate or broader, to 4 × 1.5 cm, irregularly toothed; cauline lvs often smaller; raceme longer at maturity than the proper stem, openly 4–10-fld; cal 4–5 mm at anthesis, to 9 mm in fr; cor 7–10 mm, blue-violet, marked with yellow and white; $2n=16–52$. Lawns and wet bottomlands; native of e. Asia, widely but sparingly intr. in our range. May–Nov. (*M. japonicus.*)

2. Mazus miquelii Makino.   Glabrous, creeping perennial, rooting at the nodes; cor ca 1.5 cm. Native of e. Asia, rarely intr. in lawns in our range. (*M. reptans.*)

7. LINDERNIA All.

False pimpernel. Cal regular, the 5 sep separate nearly or quite to the base; cor bilabiate, the upper lip much narrower than the lower, erect, shallowly 2-lobed, the lower somewhat deflexed, 3-lobed; stamens 2 (the upper pair) or 4, the upper pair fertile, inserted at the middle of the cor-tube and included in it; lower pair of stamens sterile or fertile, inserted at the throat of the cor-tube, the basal part of the filament bent back on itself, the terminal part free, the whole resembling a forked filament; style elongate; stigmas 2, plate-like; fr fusiform to ellipsoid, commonly asymmetrical, septicidal; small annual (ours) or biennial herbs with opposite, 3–5-nerved, entire or denticulate lvs and small, white to violet fls on solitary, axillary, bractless peduncles. (*Ilysanthes.*) 70, cosmop.

1 Anthers 2; lower pair of stamens sterile; widespread native sp. . . . . 1. *L. dubia.*
1 Anthers 4; lower pair of stamens fertile; rare intr. sp. . . . . 2. *L. procumbens.*

1. Lindernia dubia (L.) Pennell.   Low, branching, glabrous or glandular annual 0.5–2(–3) dm; lvs entire or obscurely few-toothed; sep narrow, 3-nerved, 3–4.5 mm; cor 5–10 mm; later fls often cleistogamous;

$2n=18$, 32. Wet sandy or muddy shores; Que. and N.H. to N.D., s. to Fla. and Tex., and in the Pacific states. July–Sept. Two rather poorly defined vars., both widespread in our range. Var. *dubia*: Lvs up to 3 cm, at least the lower ones narrowed to the base; pedicels relatively short, mostly 0.5–1.5 cm, surpassed by or only slightly surpassing their subtending lvs; seeds mostly 2–3 times as long as wide. (*Ilysanthes d.; I. attenuata.*) Var. *anagallidea* (Michx.) Cooperrider: Lvs mostly 6–15 × 3–10 mm, all broadly rounded at base, or the lowermost ones a little narrowed; pedicels mostly 1–2.5 cm, all except sometimes the lowermost ones conspicuously surpassing the subtending lvs; seeds 1.5–2 times as long as wide. (*L. anagallidea; Ilysanthes a.; I. inaequalis.*)

2. **Lindernia procumbens** (Krocker) Philcox.   Resembling no. 1, but fls smaller, commonly 5–6 mm, the sep about as long as the cor-tube; stamens 4, all with anthers. Casual with us, intr. from Eurasia. (*L. pyxidaria.*)

8. **MICRANTHEMUM** Michx., nom. conserv.   Cal regular, 4-parted nearly to the base; cor obliquely salverform, the 3 upper lobes about equal, the lower one longer; stamens 2 (the lower pair), inserted at the cor-throat; filaments exsert, thickened at base; pollen-sacs divergent; stigmas 2, divergent; fr globose, enclosed in the cal, rupturing irregularly; small creeping herbs with opposite lvs and minute white fls solitary in the axils. 2–4, New World.

1. **Micranthemum umbrosum** (Walter) S. F. Blake.   Stems creeping, freely branched, glabrous, to 3 dm; lvs sessile, broadly elliptic to rotund, 4–10 mm, obscurely 3–5-nerved from the base; pedicels 1 mm or less; cal at anthesis 1.5–2 mm; cor shorter than the cal, its lower lobe 1 mm; fr 1 mm. Wet soil or shallow water on the coastal plain; tropical Amer., n. along the coast to se. Va. (*Globifera u.*)

9. **HEMIANTHUS** Nutt.   Cal narrowly obconic, 4-lobed, the lobes separate to the middle except on the lower side, where the cleft extends to the base; upper lip of the cor lacking, the lower lip 3-lobed, its median lobe much longer than the other 2 and incurved; stamens 2 (the lower pair), inserted near the cor-throat, the filaments like those of *Lindernia*; style deeply bifid; fr rupturing irregularly; small glabrous annuals with opposite or whorled, sessile lvs and minute axillary fls. 10, the others in tropical Amer.

1. **Hemianthus micranthemoides** Nutt.   Creeping, 5–20 cm, with many ascending branches; lvs numerous, elliptic, 3–5 mm, entire, obtuse, obscurely 3-nerved; pedicels solitary at many nodes, 1 mm, becoming 2–3 mm and recurved in fr; cor 2 mm, falling without opening; fr 1.5 mm, surpassed by the cal. Muddy shores of major rivers, between high and low tides, beyond the direct influence of salt water; Hudson R. to s. Va.; not recently collected. (*Micranthemum m.*)

10. **LIMOSELLA** L.   Mudwort. Cal campanulate, equally 5-lobed; cor campanulate, regular, the 5 lobes shorter than the tube; stamens 4; pollen-sacs confluent into 1; stigma solitary and ± capitate; fr septicidal; seeds many; small annuals, the stems prostrate, at each node emitting roots, a cluster of basal lvs, and few–several basal, 1-fld peduncles; cor minute, white or suffused with pink. 15, nearly cosmop.

1 Lf with a lanceolate to elliptic blade . . . . 1. *L. aquatica.*
1 Lf with no distinction of petiole and blade . . . . 2.   *L. subulata.*

1. **Limosella aquatica** L.   Northern m. Lvs with an elongate, slender petiole 1.5–10+ cm and a lanceolate to elliptic blade 1–3 cm × 2.5–12 mm; cor 2.5–3.5 mm; otherwise much like no. 2; $2n=40$. Muddy shores; interruptedly circumboreal, reaching our range in w. Minn. June–Sept.

2. **Limosella subulata** Ives.   Atlantic m. Lvs usually 5–10 in each tuft, linear, obtuse, 2–5 cm × 1–2 mm, without distinction of petiole and blade; peduncles shorter than the lvs; cal 2 mm; cor 3.5–4 mm; fr globose; $2n=20$. Muddy or sandy shores between tide-levels, in fresh or brackish sites along the coast from Nf. and Que. to Md.; also in S. Amer. Aug., Sept. (*L. aquatica* var. *tenuifolia.*)

11. **VERBASCUM** L.   Mullein. Cal regular, deeply 5-parted; cor rotate, nearly regular, the 3 lower lobes slightly longer than the 2 upper; stamens 5, all with anthers, ± dimorphic, the 2 lower filaments commonly differing from the 3 upper in length and pubescence, the anthers also often dimorphic; stigma capitate; fr septicidal, the 2 valves ± cleft above; seeds longitudinally ridged; ours mostly biennials (no. 2 often perennial), producing a rosette the first year and a tall flowering stem the next; lvs alternate, entire or toothed; fls yellow, white, or blue, in 1–many elongate racemes or panicles or congested spike-like infls. 300, Eurasia.

1 Cal and other parts of the plant glandular-hairy, without branched hairs.
  2 Cor yellow or white .... 1. *V. blattaria.*
  2 Cor purple .... 2. *V. phoeniceum.*
1 Cal and other parts of the plant glandless, covered with branched hairs.
  3 Infl freely branched .... 3. *V. lychnitis.*
  3 Infl densely crowded and usually simple (plants may have more than one infl).
    4 Lvs not decurrent on the stem, or only very shortly so .... 4. *V. phlomoides.*
    4 Lvs decurrent on the stem, usually as far as the next lf below.
      5 Fls 1–2.5 cm wide; widespread weed .... 5. *V. thapsus.*
      5 Fls 2.5–4.5 cm wide; locally intr. in Wis. .... 6. *V. densiflorum.*

1. Verbascum blattaria L.   Moth-m. Stems slender, to 1.5 m, simple or branched, glandular-hairy above, without branched hairs; lvs variable, narrowly triangular to oblong or lanceolate, sessile, not decurrent, coarsely toothed to subentire, glabrous, the basal larger, oblanceolate; racemes elongate, loose, with a single fl at each node on a pedicel 8–15 mm; cor 2–3 cm wide, yellow or white, usually with an anthocyanic center, the filaments all about equally beset with purple-knobbed hairs; 2n=18, 30, 32. Native of Eurasia, established as a weed in disturbed sites throughout our range. June–Oct.

V. virgatum Stokes, glandular-hairy throughout, and with the short (commonly 3–5 mm) pedicels usually several at the lower nodes of the infl, has been found several times in our range.

2. Verbascum phoeniceum L.   Purple m. Much like no. 1, but often perennial; lvs commonly sparsely hairy beneath; pedicels 1–2.5 cm; cor purple, 2–3 cm wide; 2n=32, 36. Native of Eurasia, established as a weed in waste places on L.I. May–Aug.

3. Verbascum lychnitis L.   White m. Stems erect, to 1.5 m, often branched above, white-tomentose; lvs oblong-oblanceolate, entire or obscurely toothed, sessile but not decurrent, softly tomentose beneath, green and glabrous or nearly so above; infl commonly freely branched, forming a large pyramidal panicle of loose interrupted racemes, the fls distinctly pedicellate, several at each node; cor yellow (white), 10–20 mm wide; filament-hairs yellow or white; 2n=32, 34. Native of Eurasia, found as an occasional weed in fields and waste places in our range, especially eastward. June–Oct.

4. Verbascum phlomoides L.   Clasping m. Stems 5–12+ dm; lvs oblong to lance-ovate or ovate-oblan-ceolate, toothed, tomentose on both sides but only thinly so above, sessile, not decurrent or only very shortly so; infl simple or sometimes branched, at first dense and spike-like, at full maturity usually elongating and exposing the axis between the fl-clusters; cor 2.5–3.5 cm wide; stigma spatulate, decurrent on the style; otherwise much like no. 5; 2n=32, 34. Native of Europe, found here and there in our range as a weed.

5. Verbascum thapsus L.   Common m. Stout, erect, 1–2 m, usually densely gray-tomentose throughout; lvs entire or shallowly crenate, the lower oblong or oblanceolate, to 3 dm, petioled, the upper progressively reduced, sessile, decurrent along the stem to the next lf below; spike-like infl very dense, 2–5 dm × 3 cm, usually solitary; cor yellow, 1–2.5 cm wide; upper 3 filaments short, densely white-villous, with short anthers; lower 2 filaments much longer, glabrous or nearly so, with linear anthers; stigma capitate; 2n=32, 36. Native of Europe, now abundant throughout most of temperate N. Amer., especially in disturbed sites. June–Sept. A hybrid with no. 4 is *V.* × *kerneri* Fritsch.

6. Verbascum densiflorum Bertol.   Much like no. 5, but with larger fls (cor 2.5–4.5 cm wide) and spatulate stigma; often more branching, with several longer, less dense infls; lvs often dentate and more acuminate. Native of Europe, locally intr. and abundant in se. Wis. (*V. thapsiforme.*)

12. CHELONE L.   Turtlehead. Cal deeply 5 parted, regular, its lobes broadly elliptic, obtuse or rounded above; cor bilabiate, the upper lip shallowly 2-lobed or emarginate, arched, the lower shortly 3-lobed, the middle lower lobe elevated into a villous palate nearly or quite closing the throat; fertile stamens 4, inserted at the base of the cor, the filaments flat, villous, the anthers densely villous-tomentose; sterile stamen much shorter and narrower, glabrous; fr septicidal; seeds flat, suborbicular, with a broad wing; perennial herbs, simple or sparingly branched, with glabrous stem and large, opposite, serrate, usually glabrous lvs; fls large, white to pink or purple, in dense spikes terminating the stem and branches, each fl closely subtended by 2 or 3 large, sep-like bracts. 4, e. N. Amer.

1 Lvs sessile, broadly rounded at base .... 1. *C. cuthbertii.*
1 Lvs either distinctly petioled or long-cuneate to the base.
  2 Cor red-purple throughout .... 2. *C. obliqua.*
  2 Cor largely or wholly white or pale .... 3. *C. glabra.*

1. Chelone cuthbertii Small.   Stems 4–8 dm; lvs lanceolate or lance-oblong to lance-linear, 5–10 × 1–4 cm, long-acuminate, sharply serrate with ascending teeth, rounded to the sessile base; spike 3–6 cm, often short-peduncled, the fls conspicuously 4-ranked; cor violet-purple, 2–3 cm, its beard yellow; staminode purple. Swampy woods and bogs; se. Va. and w. N.C. Aug.–Sept.

2. Chelone obliqua L.   Purple t. Stems 3–7 dm; lvs lanceolate or lance-oblong, 6–12 × 2–4 cm, acuminate, serrate, obtuse or broadly cuneate at base, on distinct petioles 5–15 mm; cor 2.5–3.5 cm, purple or red-purple throughout, with a white or yellowish beard; staminode white or nearly so. Wet woods on the coastal plain; Md. to Ala.; also Ind. to s. Minn. and Ark. Aug.–Oct.

C. lyoni Pursh, a s. Appalachian sp. with broad, ovate lvs rounded or truncate at base, on well developed petioles 1.5–3 cm, the cor with a yellow beard, occasionally escapes from cult. in N. Engl.

3. Chelone glabra L.   White t. Erect, 5–8 dm, simple or branched above; lvs linear to lance-ovate, to 15 cm, acuminate, serrate with salient or ascending teeth, on a short indistinct petiole, or tapering to a sessile base; spikes 3–8 cm, subtended by scarcely reduced foliage-lvs; cor 2.5–3.5 cm, white, or partly greenish-yellow, or tinged with pink, or even purple toward the summit, the palate white-bearded; staminode green; $2n=28$. Wet woods; Nf. to Minn., s. to Ga. and Ala. July–Sept. Highly variable, with ± evident regional trends of variation.

13. PENSTEMON Schmidel.   Beard-tongue. Cal herbaceous, deeply 5-parted; cor tubular or trumpet-shaped, ± distinctly bilabiate, the tube much longer than the lobes; fertile stamens 4, about as long as the tube, the pollen-sacs divaricate; sterile stamen about as long as the fertile ones, usually yellow-bearded toward the tip; style elongate; stigma capitate; fr septicidal; ours (biennial or) perennial herbs, the erect stems arising from a rosette of petioled basal lvs, the cauline lvs oblong to lanceolate or ovate, sessile and often clasping, the upper progressively reduced; fls white to blue-violet or red-violet (red in some extralimital spp.), in terminal paniculiform or racemiform infls. Our spp. all with highly variable lvs, much alike from sp. to sp. 280, mainly N. Amer.

1 Fls large, the cor 3.5–5 cm, the cal at anthesis 7–15 mm . . . . 1. *P. grandiflorus.*
1 Fls smaller, the cor 1.5–3.5 cm, the cal at anthesis 2–10 mm.
  2 Cor only weakly bilabiate, its throat covered internally with minute glandular hairs.
    3 Stems glabrous below the infl, or sometimes puberulent in 2 lines . . . . 2. *P. tubaeflorus.*
    3 Stems evidently pubescent all around . . . . 3. *P. albidus.*
  2 Cor evidently bilabiate, its throat with few to many short bristles within.
    4 Base of the lower lip of the cor arched upward, forming a palate ± closing the throat.
      5 Cor violet; widespread sp . . . . 4. *P. hirsutus.*
      5 Cor white; local from Ky. to Ala. . . . . 5. *P. tenuiflorus.*
    4 Base of the cor not arched, the throat open.
      6 Cor differentiated (above the cal) into a basal tube and a distal dilated throat, the latter much inflated and only slightly ridged within; lvs glabrous beneath, or minutely puberulent along the midvein.
        7 Cor white . . . . 6. *P. digitalis.*
        7 Cor pale violet or pale purple outside . . . . 7. *P. laevigatus.*
      6 Cor only gradually and slightly dilated upwards, the throat flattened and strongly ridged within; lvs (except *P. gracilis*) always pubescent on the midvein beneath and usually on the surface also.
        8 Cor white, the lower lip commonly marked with purple lines . . . . 8. *P. pallidus.*
        8 Cor pale violet or blue-purple or reddish-purple.
          9 Cor reddish-purple; se. coastal plain (in our area) . . . . 9. *P. australis.*
          9 Cor pale violet or blue-purple; inland.
            10 Lateral branches of the infl divergent or ascending; Pa. to Ind. and s. . . . . 10. *P. canescens.*
            10 Lateral branches of the infl short and erect; Wis. to Io. and w. . . . . 11. *P. gracilis.*

1. Penstemon grandiflorus Nutt.   Large b.-t. Biennial or short-lived perennial, glabrous and glaucous throughout, to 2 m; lvs entire, the lower obovate-oblong, 2–4 cm wide, the upper progressively shorter but scarcely narrower, becoming broadly ovate to rotund, subcordate and somewhat clasping; bracteal lvs similar but smaller; fls 2–4 per axil, short-pediceled in a racemiform infl 1.5–3 dm; cal at anthesis 7–11 mm; cor 3.5–5 cm, pale purple outside, glabrous inside and out, widely dilated; fr 1.5–2 cm; seeds 2.5–4 mm; $2n=16$. Dry prairies and barrens; Wis. to N.D. and Wyo., s. to Ill., nw. Mo., and Tex. May, June. (*P. bradburyi.*)

P. cobaea Nutt., foxglove b.-t., a similarly large-fld sp. of the southern plains and Ozarkian region, with glandular-hairy infl and cor, narrower bracteal lvs, and often toothed foliage lvs, scarcely reaches our range as a native, but occasionally escapes from cult.

2. Penstemon tubaeflorus Nutt.   Tube b.-t. Stem 3–10 dm, glabrous below the infl (or puberulent in 2 lines); lvs glabrous, the principal cauline ones elliptic to oblong or lance-oblong, 8–12 × 2–4 cm, rounded at base, entire, the upper cauline ones evidently reduced and distant; infl 1–3 dm, slender, cylindric, often interrupted, with short branches; bracteal lvs small, lance-linear, callous-denticulate; cal sparsely glandular or subglabrous, 2.5–5 mm at anthesis; cor 1.5–2.5 cm, white, finely and densely glandular-puberulent all around both inside and out, the tube gradually dilated from base to top, the limb nearly regular, the throat not pleated; seeds 0.9–1.5 mm; $2n=32$. Prairies and moist woods; Ind. to Wis. and Nebr., s. to Miss. and Tex.; intr. eastward, as from Me. to Pa. May, June.

3. Penstemon albidus Nutt. Prairie white b.-t. Stems usually 2–5 together, 1.5–5 dm, retrorsely puberulent below, glandular-hairy in the infl; cauline lvs lance-oblong, 4–8 × 0.8–2 cm, entire or with a few low teeth, rounded at base, glabrous to puberulent or scabrous; infl strict, 0.6–2.5 dm; 100 pedicels, cal and cor glandular-pubescent; cal 4–7 mm at anthesis; cor white or faintly violet-tinged, 1.5–2 cm, weakly bilabiate, the tube gradually dilated upward, finely glandular-puberulent within; seeds 2–3 mm; 2n=16. Dry prairies; w. Minn. and w. Io., w. to Alta., Wyo., and N.M. May, June.

4. Penstemon hirsutus (L.) Willd. Northeastern b.-t. Stems erect, usually several, 4–8 dm, glandular-puberulent in the infl, glabrous or often villous below; cauline lvs lanceolate to oblong, 5–12 cm, toothed or subentire, rounded or truncate at base; cal at anthesis 4–7 mm, at maturity more than half as long as the fr; cor 2–3 cm, the tube very slender, not widened distally, pale violet outside, pale-hairy inside but lacking purple lines, the mouth ± closed by the arched base of the lower lip, the lobes white. Dry woods and fields; Que. and Me. to Mich. and Wis., s. to Va. and Ky. May–July.

5. Penstemon tenuiflorus Pennell. Kentucky b.-t. Much like no. 4; stems usually solitary, glabrous below, finely puberulent in the infl; lvs entire to obscurely denticulate; cal 2–4 mm at anthesis, less than half as long as the fr; cor white throughout, evidently flaring in age. Dry calcareous soil; Ky. to Ala.

6. Penstemon digitalis Nutt. Tall white b.-t. Stems to 1.5 m, glabrous and shining, often glaucous or purplish, varying to finely puberulent; cauline lvs narrowly oblong or lance-oblong to narrowly triangular, the larger 10–15 cm, glabrous or finely puberulent (especially along the midvein); infl 1–3 dm, with erect or strongly ascending branches, often glandular; cal 3–7 mm at anthesis; cor 1.5–3 cm, white or very faintly suffused with violet, usually marked with purple lines inside, the tube abruptly and strongly dilated into a wide throat; anthers pubescent or sometimes glabrous; 2n=96. Moist open woods and prairies; N.S. and Me. to Minn. and S.D., s. to Va., Ala., and Tex. May–July. (*P. alluviorum* and *P. deamii,* with relatively small fls and often puberulent stems.)

7. Penstemon laevigatus Aiton. Eastern b.-t. Stem 6–12 dm, glabrous to hirsutulous or minutely puberulent, often in narrow strips, the hairs up to 0.5 mm, often ± reflexed; lvs lanceolate or lance-ovate to narrowly oblong, 2–5 cm wide, glabrous, or rarely minutely hairy along the midvein beneath; cal 3–8 mm at anthesis; cor violet-purple (sometimes rather pale) outside, pale or white inside, 1.5–3.5 cm, the throat abruptly dilated above the middle; anthers glabrous. Meadows and moist or dry woods; Me. to Mich. and Ill., s. to Ga., Miss., and Ark. (*P. calycosus.*)

8. Penstemon pallidus Small. Eastern white b.-t. Stems 3–7 dm, hairy throughout with straight, subulate, ± reflexed hairs 0.1 mm, the same pubescence ± developed on both sides of the narrowly lanceolate or lance-oblong lvs; infl glandular, 1–2.5 dm; cal 2.5–5 mm at anthesis; cor 1.6–2.2 cm, white, marked with fine purple lines in the throat; seeds 0.5–0.7 mm; 2n=16. Dry woods and fields; Me. to Minn., s. to Va. and Ark. Apr.–June. (*P. arkansanus.*)

9. Penstemon australis Small. Southeastern b.-t. Stems 4–8 dm, finely pubescent with minute, somewhat reflexed hairs, or also hirsutulous; lvs minutely hairy on both sides; infl slender, the lateral branches shorter than the internodes; cal at anthesis 5–8 mm; cor 2–2.5 cm, reddish-purple outside, the upper lobes also reddish-purple inside, the lower white inside with purple lines. Sandy soil on the coastal plain (in our area); se. Va. to Fla. and Ala. May, June.

10. Penstemon canescens (Britton) Britton. Appalachian b.-t. Stems erect, often clustered, 4–8 dm, finely pubescent or hirsutulous with spreading hairs ca 0.5–1 mm, or occasionally only with minute, ± reflexed hairs; basal lvs broadly elliptic or ovate; cauline lvs mostly lance-ovate or ovate, sometimes narrower, usually sharply serrate, ± pubescent on both sides; cal 2–6.5 mm at anthesis; cor 2–3 cm, pale purple outside, white with fine violet lines inside, the throat moderately inflated. Cliffs and woods; Pa. to e. Tenn. and n. Ala., chiefly in the mts., but also w. to s. Ind. May, June. (*P. brevisepalus.*)

11. Penstemon gracilis Nutt. Slender b.-t. Stems 2–5 dm, finely puberulent below with minute reflexed hairs, often in 2 strips, or glabrous; cauline lvs lanceolate to lance-linear, glabrous or less often finely hairy; infl glandular-pubescent, slender, the short lateral branches erect; cal 4–6 mm at anthesis; cor 1.5–2.2 cm, pale violet; seeds 0.6–0.8 mm; 2n=16. Prairies and open woods; Wis. to Man. and Alta., s. to Io., Nebr., and N.M. June, July. (*P. wisconsinensis,* the phase with puberulent lvs.)

14. SCROPHULARIA L. Figwort. Cal regular, deeply saucer-shaped, divided to the middle into 5 broad obtuse lobes; cor bilabiate, the tube wide, the upper lip directed forward, 2-lobed, the lateral lobes of the lower lip also directed forward, the median lobe deflexed; stamens 4; sterile filament dilated at the summit, lying under the upper cor-lip; stigma capitate; fr septicidal; seeds angular, rugose; perennial herbs with opposite, petiolate, serrate or incised lvs and small, dull reddish-brown to greenish fls in large, terminal, paniculiform infls; cal not subtended by conspicuous bracts. 150, mainly temperate Eurasia.

1 Sterile filament yellowish-green, often wider than long .... 1. *S. lanceolata.*
1 Sterile filament dark purple or brown, commonly longer than wide .... 2. *S. marilandica.*

1. **Scrophularia lanceolata** Pursh.   American f. Stems erect, to 2 m, glabrous, or minutely glandular in the infl, the sides flat or shallowly grooved; petioles 1.5–3 cm, rarely a third as long as the blade, narrowly margined to the base; blades ovate or lance-ovate, 8–20 cm, sharply serrate or incised or doubly serrate, acuminate, truncate to broadly rounded at base, glabrous beneath; infl 1–3 dm, tending to be cylindric, rarely over 8 cm wide; cor 7–11 mm, dull reddish-brown except the yellowish-green lower lobe, sterile filament yellowish-green, roundly dilated at the summit, often wider than long; fr dull brown, 6–10 mm; 2*n*=92–96. Open woods, roadsides, and fence-rows; Que. and N.S. to B.C., s. to Va., Mo., and N.M. Late May–July. (*S. leporella; S. occidentalis.*)

2. **Scrophularia marilandica** L.   Eastern f. Stems erect, to 3 m, glabrous below, sparsely glandular in the infl, the sides, especially in older parts, roundly angled and prominently grooved; petioles slender, not margined, 1.5–5 cm, the main ones mostly a third to half as long as the blade; blades ovate to lance-ovate, to 25 cm, acuminate, serrate, broadly rounded to shallowly cordate at base; infl loosely and irregularly branched, tending to be pyramidal, often 10–15 cm wide; cor 5–8 mm, reddish-brown; sterile stamen dark purple or brown, narrowly dilated, commonly longer than wide; fr 4–7 mm, somewhat shining; 2*n*=ca 86. Open woods; Que. to Minn., s. to S.C. and La. July, Aug., the period rarely overlapping that of no. 1. (*S. neglecta.*)

S. nodosa L., a European sp. with the main petioles less than a third as long as the blade, narrowly wing-margined throughout, with mostly smaller and blunter lvs, otherwise like no. 2, is locally adventive in N. Engl.

15. COLLINSIA Nutt.   Blue-eyed Mary.

Cal campanulate, somewhat irregular, the lobes longer than the tube, the 2 lower shorter and broader than the 3 upper; cor bilabiate, the tube conspicuously gibbous on the upper side near the base, the lower lip deflexed, its median lobe shorter than the lateral ones, folded longitudinally into a pouch enclosing the 4 stamens, the upper lip erect, 2-lobed, bearing a double ridge at its base or at the sinus; pollen-sacs divaricate; fifth stamen vestigial; style elongate; stigma capitate or slightly 2-lobed; fr septicidal, the valves cleft at the tip; seeds few, convex on one side; annuals or winter-annuals, the cauline lvs opposite, the uppermost often whorled; fls often bicolored, in whorls of 2–8 from a few upper nodes. 17, temp. N. Amer., espec. Calif.

1 Cor 9–16 mm, its lobes (except the folded medial one) emarginate or bifid.
  2 Principal lvs broadest just above the base; cor-lobes merely emarginate .... 1. *C. verna.*
  2 Principal lvs broadest well above the base; cor-lobes cleft to 2–3 mm .... 2. *C. violacea.*
1 Cor 4–7 mm, the lobes (except the folded medial one) entire or minutely erose .... 3. *C. parviflora.*

1. **Collinsia verna** Nutt.   Eastern b.-e. M. Stems weak, 2–4 dm, often decumbent below, glabrous below, finely glandular-puberulent above; lower lvs petioled; main lvs sessile, triangular-ovate or oblong-ovate, 2–5 cm; fls mostly in 1–3 whorls of 4–6 fls each, the pedicels 5–20 mm; cal-lobes narrowly triangular, 4–5 mm; cor-tube 3–4 mm; upper lip normally white, varying to pale blue, 6–10 mm; lower lip bright blue, 8–12 mm; cor-lobes emarginate; upper stamens with pubescent filaments; sterile filament 1 mm, fleshy, projecting at right angles from near the base of the cor-tube; seeds usually 4, 2.5–3 mm, the ventral concavity large and conspicuous; 2*n*=14. Rich moist woods, especially in alluvial soil; N.Y. to Mich. and s. Wis., s. to w. Va., Ky. and Ark. Apr., May.

2. **Collinsia violacea** Nutt.   Ozarkian b.-e. M. Much like no. 1; principal lvs oblong, lanceolate, or elliptic, 1.5–4 cm, widest near the middle; cal-lobes lance-ovate, 3–4 mm; cor slightly smaller, each lobe cleft 2–3 mm, the lower lip violet, the upper pale violet to sometimes white, much shorter than the lower; seeds 6–12, 1.5 mm, the ventral concavity small and inconspicuous; 2*n*=14. Rich woods; c. Ill. to e. Kans., s. to Ark. and Tex. Apr., May.

3. **Collinsia parviflora** Dougl.   Cordilleran b.-e. M. Erect, 1–4 dm, often branched from the base; cauline lvs linear to narrowly oblong, 1–4 cm, entire or nearly so, narrowed below to a petiolar base; lowest fls solitary, upper in whorls of 2–6 on slender pedicels 5–15 mm; cal-lobes lance-linear, acuminate, 2.5–4 mm; cor barely surpassing the cal-lobes; upper lip usually white or whitish, the lower blue; cor-lobes oblong-spatulate, not retuse; filaments all glabrous; seeds 2; 2*n*=14. Sterile rocky soil; widespread in the w. cordillera, e. to Man. and w. Neb.; n. Mich.; also s. Ont., Vt., and N.S., where probably only intr. May–July.

16. LINARIA Miller.   Toadflax.

Cal deeply 5-parted; cor very irregular, spurred at base, strongly bilabiate, the upper lip erect, 2-lobed, the lower 3-lobed, the throat open (in our native sp.) or closed by a conspicuous elevated palate on the lower lip; stamens 4; stigma capitate; fr ovoid to globose, each locule rupturing irregularly at the top; annual to perennial herbs, generally glabrous, wtih erect stems, numerous lvs, and several to many fls in terminal racemes. 100+, mostly Eurasian. Several spp., in addition to those listed below, occasionally escape from cult., but do not generally persist.

1 Fls mainly or wholly yellow.
  2 Lvs linear, narrowed to a petiole-like base .... 1 *L. vulgaris.*
  2 Lvs lance-linear to ovate, broadest near the sessile, obtuse to rounded base.
    3 Cor (spur included) 2.5–4 cm; lvs ovate or lance-ovate .... 2. *L. dalmatica.*
    3 Cor (spur included) 1.5–2.3 cm; lvs lanceolate .... 3. *L. genistifolia.*
1 Fls mainly or wholly blue or white.
  4 Native annual with a short taproot .... 4. *L. canadensis.*
  4 Intr. colonial perennial with slender rhizomes .... 5. *L. repens.*

1. Linaria vulgaris Miller.   Butter-and-eggs. Perennial, 3–8 dm, colonial by creeping roots; lvs very numerous, pale green, 2–5 cm × 2–4 mm, narrowed below to a petiole-like base; fls numerous in a compact spike, yellow with orange palate, 2–3.5 cm, including the spur; fr round-ovoid, 8–12 mm; seeds winged; $2n=12$. Native of Europe, established in fields, roadsides, and waste places throughout temperate N. Amer. May–Sept. (*L. linaria.*)

2. Linaria dalmatica (L.) Miller.   Dalmatian t. Stout, glaucous perennial, branched above, 4–12 dm, colonial by creeping roots; lvs numerous, ovate or lance-ovate, sessile and clasping, palmately veined, 2–5 × 1–2+ cm; fls short-pedicellate or subsessile in elongate racemes, bright yellow, with well developed, orange-bearded palate, 2.5–4 cm, the spur about as long as the rest of the cor; fr broadly ovoid-cylindric, 6–8 mm; seeds irregularly wing-angled; $2n=12$. Roadsides and other disturbed sites; e. Mediterranean sp., now well established in arid w. U.S., and occasional with us. July, Aug.

3. Linaria genistifolia (L.) Miller.   Colonial perennial, much like no. 2, but the lvs narrower, lanceolate, tapering gradually to the pungent tip, and the fls smaller, 1.5–2.3 cm overall; $2n=12$. Native of Europe, occasionally intr. in waste places in our range.

4. Linaria canadensis (L.) Dum.-Cours.   Annual t. Slender annual or winter-annual from a short taproot, 1–5 dm, essentially glabrous throughout, producing a rosette of short, prostrate stems with assurgent tips, the lvs of which are mostly opposite or ternate and broader than those of the 1–several erect main stems, these with linear lvs 1–3.5 cm × 1–2.5 mm, all but sometimes the lowermost alternate; racemes nearly naked; cor rather light blue, the lower lip bearing 2 short white ridges that form a weakly developed palate barely closing the throat; fr 2–4 mm; seeds prismatic-angled; $2n=12$. Mostly in sandy soil; Mass. and even Que. to Minn., s. to Fla. and Mex., and on the Pacific coast. Apr.–June. The common form in our range is var. *canadensis,* with the cor mostly 8–10 mm (exclusive of the 2–6 mm spur) and the seeds smooth or nearly so. The well marked var. *texana* (Scheele) Pennell, with larger fls (10–12 mm exclusive of the 5–9 mm spur) and densely tuberculate seeds, is more southern, barely reaching our range.

5. Linaria repens (L.) Miller.   Perennial to 8 dm, colonial from slender creepers; lvs linear-oblanceolate, gradually tapering from above the middle to the base, the lower ones commonly whorled; flowering part of the raceme compact, the axis soon elongating and the globose frs remote; fls blue, on pedicels 2–4 mm; cor 8–10 mm, excluding the blunt, broadly conic, strongly curved 2–5 mm spur; palate well developed, closing the cor-throat; seeds sharply 3-angled, transversely rugose; $2n=12$. Native of Europe, becoming established as a weed in Nf. and N.B., s. locally to Mass. Summer.

17. KICKXIA Dumort.   Cancerwort. Cal regular, deeply 5-parted; cor strongly irregular, spurred at base, bilabiate, the lobes much longer than the tube, the throat closed by the well developed palate; stamens 4; stigma capitate; fr subglobose, each cell circumscissile on the side near its summit, nearly half the valve being separated; our spp. pubescent annuals, prostrate but not creeping, with short-petioled, alternate, broadly ovate lvs and small, yellow or purple, long-pediceled fls solitary in the axils. 30, Old World.

1 Lvs round-ovate, broadly rounded to cordate at base; pedicels villous throughout .... 1. *K. spuria.*
1 Lvs broadly triangular-ovate or hastate, truncate at base; pedicels glabrous over most of their length
  .... 2. *K. elatine.*

1. Kickxia spuria (L.) Dumort.   Villous throughout; stems slender, to 6 dm, with many branches from the base and few distally; petioles 2–6 mm; lvs broadly round-ovate, the larger 2–3 cm, entire, rounded or cordate at base; pedicels 5–20 mm, villous; cor 6–8 mm, yellow, the upper lip purple within; spur decurved, 5 mm. Native of the Mediterranean reg., established as a weed in moist sandy soil in most of our range, s. to Fla. June–Sept. (*Linaria s.*)

2. Kickxia elatine (L.) Dumort.   Stems freely branched, villous, to 5 dm; petioles 1–5 mm; lvs broadly ovate to triangular-ovate, 1–3 cm, truncate at base, ± hastate by the development of 1–2 low teeth at the lateral angles; pedicels very slender, 1–3 cm, glabrous throughout or villosulous near the base and summit only; fls as in no. 1; $2n=18, 36$. Native of Eurasia, established as a weed in moist sandy soil in most of our range, s. to Fla. and La. June–Sept. (*Linaria e.*)

18. CYMBALARIA Hill.   Kenilworth-ivy. Cal regular, deeply 5-parted; cor bilabiate, distinctly spurred at base, the throat closed by the prominent palate, the 2 upper lobes erect, the 3 lower spreading; stamens 4; fr globose, rupturing into 2 terminal pores that later extend to the base; trailing herbs with alternate, palmately lobed lvs. 10, mostly Mediterranean.

1. **Cymbalaria muralis** P. Gaertner, Meyer, & Scherb. Glabrous annual; stems trailing, rooting at the nodes, 1–4 dm; lvs long-petioled, suborbicular in outline, with 3–7 shallow lobes; fls solitary in the axils on long slender pedicels; cor blue with yellow palate, 7–10 mm, the spur obtuse, 2–3 mm; fr 3–4 mm thick; $2n=14$. Native of Eurasia, often escaped from cult. in our range. Summer. (*C. cymbalaria; Linaria cymbalaria.*)

19. ANTIRRHINUM L.

Snapdragon. Cal regular, deeply 5-parted; cor tubular or campanulate, very irregular, saccate-gibbous at base between the 2 lower sep, 2-lipped, the lower lip with a prominent palate closing the throat; stamens 4; stigma capitate; fr asymmetrical, opening by subterminal pores; herbs with handsome, often bicolored fls and narrow, entire lvs, the lower opposite, the upper alternate. 35, w. Eurasia and w. N. Amer.

1. **Antirrhinum majus** L. Common s. Perennial, blooming the first year, 4–8 dm, glandular-hairy above; lvs lanceolate, to 15 mm wide; fls in terminal racemes; cor variously colored, 2.5–4 cm; cal-lobes broadly ovate, obtuse, 3–5 mm; fr glandular, 1.5 cm; $2n=16, 32$. Native of the Mediterranean region, occasionally escaped from cult. in our range. Summer.
    A. orontium L., lesser s., a Eurasian annual with narrower lvs (to 5 mm wide), elongate, linear cal-lobes (8–20 mm) and smaller (10–13 mm) pink-purple fls, rarely escapes.

20. CHAENORRHINUM Reichb.

Cal deeply 5-parted; cor very irregular, spurred at base, strongly bilabiate, the lower lip with a well developed palate that does not close the throat; stamens 4; stigma capitate; fr subglobose, oblique, each locule opening by a single large, irregular, terminal pore; annual or perennial herbs with alternate, narrow, entire lvs and small, pediceled fls solitary in the axils. 20, Mediterranean region.

1. **Chaenorrhinum minus** (L.) Lange. Lesser toadflax. Erect, branching, glandular-hairy annual 1–3 dm; lvs linear, 1–2 cm, obtuse, narrowed below but scarcely petiolate; pedicels 10–15 mm, arising from many or most axils; sep linear-spatulate, unequal, 3 mm; cor 5–6 mm, blue-purple, with yellow on the palate, the spur 1.5–2 mm; fr subglobose, 5 mm; $2n=14, 28$. Native of Europe, now widely established in waste places, especially on cinder railway-ballast, throughout our range. June–Sept. (*Linaria m.*)

21. DIGITALIS L.

Foxglove. Cal 5-parted nearly to the base, the segments unequal; cor irregular, tubular to campanulate, 5-lobed; stamens 4, the pollen-sacs divergent; style elongate; stigmas 2, distinct; fr ovoid to globose, primarily septicidal; tall biennial or perennial herbs with alternate lvs and medium-sized to large fls in long, terminal, usually secund racemes. 20, Europe and Mediterranean region.

1 Cor large, mostly 4–5.5 cm, the lower median lobe only slightly exceeding the others .... 1. *D. purpurea.*
1 Cor smaller, mostly 2–3 cm, the lower median lobe much the longest .... 2. *D. lanata.*

1. **Digitalis purpurea** L. Common f. Biennial or sometimes perennial, 6–18 dm; cauline lvs lanceolate to lance-ovate, gradually tapering to the base, crenate, hairy beneath; sep 10–15 mm, overlapping, the largest broadly elliptic to obovate; cor 4–5.5 cm, narrowly campanulate, purple or white, spotted with dark purple inside; $2n=56$. Native of Europe, occasionally escaped from cult. in our range. Summer. Classical source of the drug digitalis.

2. **Digitalis lanata** Ehrh. Grecian f. Biennial or more often perennial, 3–10 dm, villous in the infl; cauline lvs narrowly oblong, sessile, entire or nearly so, glabrous beneath; sep lanceolate, densely villous, 7–10 mm; cor white to pale yellow 2–3 cm, the tube strongly inflated, veined with brown or violet, the lower median lobe prolonged, 8–13 mm, much surpassing the lateral ones; $2n=56$. Native of s. Europe, rarely escaped from cult. in our range. Summer. Produces a variant form of digitalis.
    D. lutea L., another European sp., essentially glabrous except for the glandular-ciliolate sep, and with yellow fls 1.5 cm, the 5 cor-lobes subequal, rarely escapes.

22. DASISTOMA Raf.

Mullein-foxglove. Cal-tube cupulate, the 4 lower lobes oblong, obtuse, about equaling the tube, the upper median lobe shorter and narrower; cor-tube narrowly campanulate, longer than the lobes, densely hairy within, the lobes all about alike and widely spreading; stamens 4, paired; filaments villous, inserted near the middle of the cor-tube, scarcely exsert; anthers glabrous, oblong, dehiscent throughout; style short; stigmas 2; fr round-ovoid, loculicidal, each valve with a short, flat, triangular beak; seeds angular, with papery, reticulate outer coat. Monotypic.

1. **Dasistoma macrophylla** (Nutt.) Raf.   Robust hemiparasitic perennial 1–2 m, somewhat hairy; lvs opposite, the lower broadly ovate, 2–4 dm, deeply pinnatifid or bipinnatifid, the upper progressively reduced to lanceolate and entire; fls sessile or subsessile in the upper axils, forming elongate, interrupted, leafy spikes; cal 6–10 mm; cor ephemeral, yellow, 1.5 cm, the tube ampliate, longer than the spreading lobes, densely villous within; fr 6–11 mm; seeds 2–2.5 mm. Moist rich woods; O. and c. Pa. to Io. and Ians., s. to Ga. and Tex. June–Sept. (*Seymeria m.; Afzelia m.*)

23. SEYMERIA Pursh, nom. conserv.

Cal regular, the cupulate tube shorter than the 5 narrow lobes; cor slightly irregular, the tube short and broad, the lips widely spreading or the upper arched, the lobes of the upper lip partly united, those of the lower free to the tube; stamens 4, equal; filaments short, inserted in the cor-tube, hairy at base; anthers stout, linear-oblong, opening by short (in our sp.) terminal clefts; style about equaling the stamens; stigma minute; fr ovoid, loculicidal; herbs with opposite lvs and yellow fls, often marked with purple. 20, Mexico and se. U.S.

1. **Seymeria cassioides** (J. F. Gmelin) S. F. Blake.   Much branched annual, 5–10 dm, finely hairy; lvs ovate in outline, 1–2 cm, pinnatifid or bipinnatifid into filiform segments; fls solitary in the upper axils, on pedicels 3–8 mm, forming racemes terminating all the branches; cal-lobes linear-subulate, 2–3 mm; cor pale yellow, ± marked with purple internally, 1 cm wide; anthers 3 mm; fr acuminate, 6–7 mm; $2n=26$. Pine-barrens on the coastal plain; se. Va. to Fla. and La., and farther inland in the s. states. Sept., Oct. (*Afzelia c.*)

24. AUREOLARIA Raf.

False foxglove. Cal-tube campanulate to cupulate, the lobes triangular to linear, often unequal; cor irregular, the tube somewhat oblique, the 5 lobes shorter than the tube, all equally distinct, broadly rounded; stamens 4, paired, the lower pair the longer; filaments slender, villous; anthers pubescent, each sac tipped with a short awn at base; style slender; stigma 1, ovoid-capitate; fr ovoid to ellipsoid, loculicidal; seeds winged except in *A. pedicularia*; large, branching hemiparasitic herbs, the principal lvs opposite, entire to deeply lobed, the upper reduced and often scattered, subtending large, yellow, solitary, pediceled fls. 9, e. N. Amer., and 1 Mex.

1 Axis of the infl, pedicels, and cal glabrous; perennial.
  2 Pedicels at anthesis straight, stout, 1–4 mm; stem not glaucous; lower lvs rarely lobed . . . . 1. *A. laevigata.*
  2 Pedicels at anthesis upcurved, 4 mm or more; stems glaucous; lower lvs deeply pinnately lobed . . . . 2. *A. flava.*
1 Axis of the infl, pedicels, cal, or at least some of these parts, pubescent.
  3 Perennial; pubescence wholly of simple, eglandular hairs.
    4 Pedicels at anthesis 4–15 mm; fr glabrous.
      5 Pedicels slender, 0.5 mm thick, straight or flexuous; infl very sparsely puberulent; cal-lobes
        usually longer than the tube . . . . 3. *A. patula.*
      5 Pedicels stout, 1 mm thick, abruptly upcurved; infl finely and densely pubescent; cal-lobes
        almost always shorter than the tube . . . . 4. *A. grandiflora.*
    4 Pedicels at anthesis 1–3 mm, straight and stout, fr hairy . . . . 5. *A. virginica.*
  3 Annual; pubescence, at least in part (especially the cal-tube and pedicels) of stipitate-glandular hairs
    . . . . 6. *A. pedicularia.*

1. **Aureolaria laevigata** (Raf.) Raf.   Appalachian f.f. Perennial 0.5–1.5 m; stems usually green, glabrous but not glaucous; lowest lvs lance-ovate, entire or sometimes with a few lobes or teeth; middle and upper lvs lanceolate, entire or sometimes serrate; pedicels stout, straight, suberect, 1–4 mm at anthesis, not much more in fr; cal glabrous, its lobes lance-acuminate, entire; cor 3–3.5 cm, its lobes finely ciliate; fr glabrous, narrowly ovoid, 10–15 mm. Upland woods, chiefly in the mts.; Pa. and w. Md. and s. O. to Ga. July–Sept. (*Gerardia l.; Dasistoma l.*)

2. **Aureolaria flava** (L.) Farw.   Smooth f.f. Perennial 1–2 m; stem glabrous and glaucous, branched above; lower lvs deeply pinnately lobed, with entire or toothed, widely divergent lobes and broad sinuses; upper lvs less deeply lobed, the bracteal ones narrowly lanceolate, entire or serrulate; pedicels stout, abruptly upcurved, 4–10 mm at anthesis; cal glabrous; cor 3.5–5 cm; fr glabrous, ovoid, 12–20 mm; $2n=24$. Dry upland woods; Me. to Mich. and Wis., s. to Fla. and La. July–Sept. (*Gerardia virginica,* misapplied.) Midwestern plants tend to have much longer cal-lobes (5–14 mm, as opposed to 2–5 mm) than plants of the e. and n. states, and have been segregated as var. *macrantha* Pennell.

3. **Aureolaria patula** (Chapman) Pennell.   Cumberland f.f. Differing from no. 4 in its much sparser pubescence, entire or merely serrulate bracts, glabrous or subglabrous cal, long, entire cal-lobes considerably exceeding the tube, and slender pedicels commonly 1–2.5 cm at anthesis. Woods; c. Ky. to n. Ga. Aug.–Oct.

4. **Aureolaria grandiflora** (Benth.) Pennell.   Western f.f. Widely branched, densely cinereous-hairy perennial 0.5–1.5 m; lower lvs ovate in outline, ± pinnatifid, the upper progressively reduced, the bracteal ones laciniate to entire; pedicels stout, 4–10 mm at anthesis, abruptly upcurved; cal-tube hemispheric, longer than the lanceolate or triangular, entire or toothed lobes; cor 4–5 cm; fr glabrous, ovoid, 15–20 mm. Upland woods; nw. Ind. to s. Wis. and se. Minn., s. to La. and Tex. July–Sept. (*Gerardia g.; Dasistoma g.*) Our plants are mostly var. *pulchra* Pennell, with the upper and bracteal lvs sharply serrate to deeply laciniate. The southern and Ozarkian var. *serrata* (Torr.) Pennell, with the upper and bracteal lvs entire to shallowly serrate, barely enters our range in Mo. (var. *cinerea.*)

5. **Aureolaria virginica** (L.) Pennell.   Downy f.f. Perennial 0.5–1.5 m, with ascending branches, finely downy throughout; lower lvs lance-ovate, usually with one or 2 pairs of large obtuse lobes below the middle, the upper progressively reduced and less lobed, the bracteal ones lanceolate and commonly entire; pedicels very stout, straight, ascending, 1–3 mm; fr ovoid, 10–15 mm, densely pubescent; $2n=26$. Dry woods; Mass. to Ont. and Mich., s. to Fla. and Ala. June–Sept. (*Gerardia v.; G. flava,* misapplied.)

6. **Aureolaria pedicularia** (L.) Raf.   Annual f.f. Much branched annual to 1 m, puberulent to glandular-hairy; principal lvs 3–6 cm, sessile or subsessile, pinnatifid, with 5–8 pairs of irregularly serrate or deeply cleft pinnae; pedicels upcurved, 1–2 cm at anthesis, stipitate-glandular; cal and cor stipitate-glandular or glandular-hairy, the cal-lobes spreading, foliaceous, 7–10 mm, crenate or pinnatifid to entire; cor yellow, tinged or marked with brown (the other spp. merely yellow), 2.5–4 cm; fr ellipsoid or ovoid, glandular-hairy, 10–15 mm; seeds 0.8–1 mm, wingless; $2n=28$. Dry upland woods; s. Me. to e. Minn., s. to Fla. and La. Aug., Sept. (*Gerardia p.; Dasistoma p.*) Four vars. in our range:

a Upper part of the stem with scattered or no glands; Me. to N.C., w. to N.Y., Pa., w. Va., and occasionally
   to e. Minn. and n. Ill. . . . . var. *pedicularia.*
a Upper part of the stem conspicuously glandular.
  b Cal-tube turbinate; fr ellipsoid; lf-teeth mostly rounded.
    c Cal-lobes 5–10 mm, lanceolate, relatively less deeply lobed; lvs tending to be ovate, less deeply
      and sharply cut; midwestern, from Mich. and n. O., to Wis. and Minn. . . . . .var. *ambigens* (Fern.) Farw.
    c Cal-lobes 8–16 mm, lance-linear to lanceolate, relatively deeply lobed; lvs more lance-ovate and
      more deeply and sharply cut; sw. Va. and adj. Ky. to Ga. . . . . var. *austromontana* Pennell.
  b Cal-tube hemispheric; fr ovoid; lf-teeth acute or acutish; plants mostly more densely and loosely
    hairy than the other vars.; Ky. and s. Mo. to N.C., Fla., and La. . . . . var. *pectinata* (Nutt.) Gleason.

25. **AGALINIS** Raf., nom. conserv.   Cal regular, the hemispheric or campanulate tube usually longer than the lobes; cor slightly irregular, the campanulate tube often bulged beneath, the 5 lobes all equally distinct; stamens 4, the lower pair the longer; filaments hairy at least toward the base; pollen-sacs villous, dithecal, obtuse to cuspidate at base; stigma 1, flattened and ± elongate; fr globose to subglobose (except in *A. aspera*), loculicidal; small hemiparasitic herbs with opposite lvs, tending to become alternate on the branches; fls pink to purple (white), in most spp. marked with 2 yellow lines and some darker purple spots in the throat, medium-sized to large, axillary to the ± reduced upper lvs, forming a raceme (or spike), or by reduction terminal. Except for *A. auriculata,* our spp. all slender, erect, branching, with linear, entire, 1-nerved lvs rarely over 5 cm and almost always scabrous or scaberulous on the upper side, the fls produced in late summer, chiefly Aug. and Sept., each lasting only a day. 60, New World. (*Gerardia,* misapplied, the name now typified by a member of the Acanthaceae.)

1 Stem retrorsely hairy; usually some of the lvs lobed at base . . . . 12. *A. auriculata.*
1 Stem glabrous or with ascending hairs; lvs entire.
  2 Perennial from a creeping rhizome; se. coastal plain only . . . . 1. *A. linifolia.*
  2 Annual, fibrous-rooted; widespread.
    3 Cal-lobes oblong to semi-orbicular, obtuse or rounded; salt marshes . . . . 2. *A. maritima.*
    3 Cal-lobes triangular to lanceolate or subulate, acute or acuminate.
      4 Plants relatively dark green and often purple-tinged, tending to blacken in drying; cal-tube
        scarcely or not at all veiny; cor mostly purple, drying dark purplish.
        5 Fr ellipsoid-cylindric, 7–11 mm; sinuses between the cal-lobes mostly V- to narrowly
          U-shaped, densely puberulent within and along the margins . . . . 3. *A. aspera.*
        5 Fr subglobose, 4–7 mm; sinuses between the cal-lobes mostly broader, not or only slightly
          pubescent within.
          6 Pedicels 2–5 mm, up to about as long as the cal.
            7 Stem glabrous or nearly so; axillary fascicles in most vars. wanting or poorly developed
              . . . . 4. *A. purpurea.*
            7 Stem very scabrous with short, antrorse hairs; axillary fascicles of lvs well developed
              . . . . 5. *A. fasciculata.*

6 Pedicesl 1–3 cm, evidently longer than the cal.
  8 Upper lip of the cor erect or somewhat reflexed, its lobes pubescent across the base
     within; coastal plain and piedmont only .... 6. *A setacea.*
  8 Upper lip of the cor directed forward and arched over the stamens, its lobes glabrous
     within; common and widespread .... 7. *A. tenuifolia.*
4 Plants persistently yellowish-green; cal-tube distinctly reticulate-veined; cor pink, drying pale.
  9 Cal-teeth only 0.1–0.4 mm .... 8. *A. obtusifolia.*
  9 Cal-teeth 0.5 or more.
    10 Plants very abundantly branched, most branches with a single terminal fl; Middle West
      .... 11. *A. gattingeri.*
    10 Plants more sparsely branched, the main axis terminating in a normal raceme and the
      floriferous lateral branches with a raceme of 2 or more fls.
      11 Cor-lobes retuse or emarginate; e. Mass. to Md. .... 9. *A. acuta.*
      11 Cor-lobes rounded; Middle West .... 10. *A. skinneriana.*

1. Agalinis linifolia (Nutt.) Britton. Perennial agalinis. Rhizomatous perennial 8–15 dm; lvs often broadly linear, 3–5 cm, to 3 mm wide, gradually tapering; pedicels at anthesis 5–20 mm, very erect; cal-tube 3.5–5 mm, the lobes broadly subulate, 0.3–0.8 mm; cor 3–4 cm, without yellow marks; pollen-sacs 4 mm; fr subglobose, 6–7 mm; seeds brown-reticulate; $2n=28$. Wet sandy soil on the coastal plain; Del. to Fla. and La. (*Gerardia l.*)

2. Agalinis maritima (Raf.) Raf. Salt-marsh a. Seldom over 3.5 dm; lvs rather broadly linear, to 2 mm wide; racemes often short, with 2–5 pairs of fls; pedicels 2–10 mm; cal-tube 2–3 mm, the lobes semi-orbicular to oblong, 0.5–1 mm, broadly rounded or obtuse; cor 1.2–2 cm; pollen-sacs 1.3–1.8 mm; fr subglobose, 5–6 mm; seeds brown-reticulate; $2n=28$. Salt marshes; N.S. and Me. to Fla., Mex., and W.I., more robust southward. (*Gerardia m.*)

3. Agalinis aspera (Douglas) Britton. Rough a. Stems 2–6(–8) dm, with ascending branches; lvs narrowly linear, very scabrous above, commonly with small axillary fascicles; pedicels suberect, 5–11 mm (to 18 mm in fr); cal-tube 3–5 mm, the lobes triangular-acute, 1.5–3 mm, the sinuses mostly V-shaped or narrowly U-shaped, densely puberulent within and along the margin; cor 1.8–2.5 cm, purple, the upper lobes only slightly spreading; pollen-sacs 1.8–3 mm; fr ellipsoid-cylindric, 7–11 mm; seeds dark. Dry prairies; Wis. and Ill. to Man., Neb., and Okla. (*Gerardia a.*)

4. Agalinis purpurea (L.) Pennell. Smooth a. Stems 1–12 dm, simple to much branched, quadrangular, smooth to the touch; largest lvs to 4 mm wide; axillary fascicles of lvs wanting or only weakly developed, except in var. "*racemulosa*"; fls few-many, on pedicels 1–4 mm; cal-tube 2–4 mm, the lobes triangular, usually acuminate, 1 mm or more, the sinuses generally broadly rounded; cor purple (white), its upper lobes reflexed-spreading, pubescent within; fr subglobose, 4–6 mm; seeds dark; $2n=28$. Moist, especially sandy soil of bogs, shores, and barrens, usually in sun; N.S. to Minn. and S.D., s. to Fla., Mex., and W.I. (*Gerardia p.*) Three or four poorly marked vars.:

a Plants small, rarely over 3 dm, with broadly linear lvs and few, small fls, the cor mostly 1–1.5 cm,
  the pollen-sacs ca 1 mm, the cal-lobes relatively long, mostly 1.8–5 mm and three-fourths to
  twice as long as the tube; N.S. and Sable Island. (*Gerardia neoscotica.*) .... var. *neoscotica* (Greene) B. Boivin.
a Plants larger and more floriferous when well developed; cor and pollen-sacs mostly larger, but cal-
  lobes averaging smaller, rarely as much as 2.8 mm.
  b Cor mostly 2–4 cm; cal-lobes up to half as long as the tube; pollen-sacs mostly 2.5–3.5 mm; common
    and widespread (*A. virgata* and *Gerardia racemulosa*, both names applying to a slender, narrow-
    lvd coastal plain phase with more evident axillary fascicles, sometimes called var. *racemulosa*
    (Pennell) B. Boivin) .... var. *purpurea.*
  b Cor mostly (1–)1.5–2 cm; cal-lobes four-tenths to fully as long as the tube; pollen-sacs 1.4–2.0 mm;
    northern, from Me. and Pa. to Minn. and Io., always n. of 40°. (*Gerardia paupercula.*) .....
    .... var. *parviflora* (Benth.) B. Boivin.

5. Agalinis fasciculata (Elliott) Raf. Fascicled a. Stems tall, often 1 m, rather stout, scabrous to the touch; lvs commonly 1–2 mm wide, very often with axillary fascicles of smaller ones; pedicels 2–5 mm at anthesis; cal-tube 2.8–4 mm, the lobes 0.5–2 mm, averaging a third as long as the tube; cor 2.5–3.5 cm, purple; pollen-sacs 2–3.5 mm; fr subglobose, 5–6 mm; seeds dark; $2n=28$. Dry sandy soil, often weedy; s. Md. to Fla. and Tex., n. in the interior to s. Mo., and sometimes adventive farther n. (*Gerardia f.*)

6. Agalinis setacea (J. F. Gmelin) Raf. Thread-a. Stems 1.5–7 dm, slender, with many ascending branches; lvs glabrous or nearly so (as also the stem), narrowly linear or setaceous, usually under 1 mm wide; pedicels widely divergent, 1–3 cm; cal-tube 2–3 mm, the lobes minute, 0.2–0.5 mm; cor 1.5–2.5 mm, its upper lip spreading, hairy across the base of the lobes within; pollen-sacs 2–2.5 mm; fr globose, 3–4 mm; seeds dark; $2n=28$. Dry sandy soil on the coastal plain and adj. piedmont; L.I. to Ala. (*Gerardia s.*)

7. Agalinis tenuifolia (M. Vahl) Raf. Common a. Stems 2–6 dm, glabrous or nearly so, usually much branched; lvs linear, 1–6 mm wide; pedicels filiform, widely divergent, 1–2 cm at anthesis; cal-tube 2–4 mm, its lobes broadly triangular to subulate, 0.3–2.2 mm; cor 1–1.5 cm, its upper lip arching over the stamens, glabrous within; pollen-sacs 1.5–2.2 mm; $2n=28$. Que. to Minn., s. to Fla. and Tex. (*A. besseyana.*)

8. Agalinis obtusifolia Raf. Blunt-lvd a. Herbage persistently yellow-green; stems 2–6 dm, glabrous or nearly so, with stiffly ascending branches; lvs linear or widened distally, obtuse to acute or acuminate; pedicels

5–15 mm at anthesis, ascending at 30–60°; cal-tube 1.8–3 mm, reticulate-veiny, the lobes thickened, narrow, minutely hairy, projecting 0.1–0.4 mm; cor 1–1.5 cm, pink, with or without yellow lines and red-purple spots in the throat, the lobes rounded, sometimes emarginate so that the cor appears 10-lobed; pollen-sacs 1.2–2.1 mm; fr ellipsoid-globose, 3–5 mm; seeds yellow-brown; 2*n*=26. Dry, open places, usually in sandy soil; Pa. and Del. to Fla. and La., and inland to Ky. and Tenn. (*Gerardia o.; G. decemloba; G. parvifolia; ? G. erecta.*)

9. **Agalinis acuta** Pennell. Sand-plain a. Stems 1–4 dm; herbage persistently yellow-green; lvs linear-filiform, 1–2.5 cm; racemes developed on the lateral branches, but always short; pedicels mostly 1–2 times as long as the bracts, diverging at an angle of about 45°; cal-tube 2.2–3.2 mm, reticulate-veiny, the lobes broadly triangular, 0.7–1.3 mm; cor 1–1.5 cm, pink, with 2 yellow lines and some faint purple spots in the throat, the lobes emarginate or retuse; pollen-sacs 1.5–1.7 mm; fr subglobose, 4 mm; seeds yellow-brown; 2*n*=26. Sandy soil; e. Mass. to Md. Now rarely seen. (*Gerardia a.*)

10. **Agalinis skinneriana** (A. Wood.) Britton. Midwestern a. Stems 2–5 dm, with a few stiffly ascending branches, strongly striate-angled or narrowly 4-winged, usually somewhat scabrous on the angles; herbage persistently yellow-green; pedicels 1–2 times as long as the bracts, ascending at usually 20–30°; cal-tube 2.5–3 mm, reticulate-veiny, the lobes broadly triangular, 0.7–1 mm; cor 1–1.5 cm, pink, with 2 yellow lines and some purple spots in the throat, the lobes ± truncate; pollen-sacs 0.5–1.3 mm; fr subglobose, 4–5 mm; seeds yellow-brown; 2*n*=26. Dry prairies, open woods, and barrens, especially in sandy soil; sw. Ont. to O., Wis., Mo., and Kans., rare. (*Gerardia s.*)

11. **Agalinis gattingeri** (Small) Small. Very much branched, the very slender lateral branches with a few small lvs and rarely more than a single apparently terminal fl; fls also very few on the central axis, but a terminal raceme scarcely developed; herbage persistently yellow-green; pedicels at anthesis 5–25 mm; cal-tube 2.5–3.5 mm, reticulate-veiny, the lobes broadly triangular, 0.6–0.8 mm; cor 1.2–1.8 cm, pink; pollen-sacs 1.3–2 mm; fr subglobose, 4–5 mm; seeds yellow-brown; 2*n*=26. Woods, hills, and barrens; sw. Ont. to Ala., w. to Minn., Okla., and Tex. (*Gerardia g.*)

12. **Agalinis auriculata** (Michx.) S. F. Blake. Stems simple or sparingly branched, 3–8 dm, rough-hairy with retrorse hairs and some longer spreading ones; lvs sessile, lanceolate, 2–5 cm, rounded at base, entire or some, especially the upper, bearing near the base 1 or 2 small, divergent, lanceolate lateral lobes; fls sessile in the upper axils, forming a leafy spike; cal 15–20 mm, the tube densely retrorse-hairy, the lobes lanceolate, longer than the tube, usually very unequal in width; cor 2–2.5 cm, purple; fr round-ovoid, 1–1.5 cm; 2*n*=26. Prairies or open upland woods; O. to Wis. and Minn., s. to Mo., Kans., and Okla., also locally in s. U.S. and apparently intr. from N.J. to Pa. and Va. Aug., Sept. (*Otophylla a.; Tomanthera a.*)

26. **BUCHNERA** L. Blue-hearts. Cal regular, tubular, shortly 5-lobed; cor salverform, nearly regular, the slender tube villous within, the 5 lobes widely spreading, subequal, oblong, rounded or emarginate; stamens 4, half as long as the cor-tube; anthers glabrous, monothecal, the pollen-sacs confluent; stigma clavate; fr ovoid-oblong, usually gibbous at base, loculicidal; hemiparasitic perennial herbs with sessile opposite lvs, the upper progressively reduced, the uppermost subtending solitary sessile fls, forming a terminal spike. 100, mainly Old World tropics.

1. **Buchnera americana** L. Stem mostly simple, 4–9 dm, roughly short-hairy; lower lvs lanceolate, 5–10 cm, acuminate, coarsely few-toothed, very scabrous, the upper progressively smaller and narrower; spikes peduncled, elongating in fr to 1–2 dm; cal hairy, 6–9 mm, much surpassing the subulate bracts; cor deep purple, the tube 1–1.5 cm, the lobes 5–8 mm; fr 6–8 mm, barely surpassing the persistent cal; 2*n*=42. Sandy or gravelly soil of upland woods or prairies; N.Y. and s. Ont. to Mich., Ill., and se. Kans., s. to Ga. and Tex. July–Sept.

27. **VERONICASTRUM** Fabr. Culver's root. Cal deeply 4–5-parted, the 2 lobes of the lower lip longer than the 2 or 3 of the upper; cor tubular, nearly regular, the lobes much shorter than the tube, the upper lobe about equaling the lateral ones and wider than the lower one; stamens 2, inserted below the middle of the cor-tube, long-exsert; pollen-sacs parallel; style slender, about equaling the stamens; stigma 1, minute; fr narrowly ovoid or ellipsoid, not flattened, opening by 4 short terminal slits; tall perennials with narrow whorled lvs and slender terminal spikes of pink or white fls. 2, the other closely related, in e. Asia.

1. **Veronicastrum virginicum** (L.) Farw. Erect, 8–20 dm, usually with a few erect branches; lvs in whorls of 3–6, on petioles 3–10 mm, lanceolate to narrowly oblong or oblanceolate, acuminate, finely and sharply serrate, glabrous to villous beneath; spikes erect, 5–15 cm, with numerous, crowded, divergent fls; cor 7–9 mm; fr narrowly ovoid, 4–5 mm; 2*n*=34. Moist or dry upland woods and prairies; Vt. to Ont. and Man., s. to Ga. and La. June–Aug. (*Veronica v.; Leptandra v.*)

28. VERONICA L.

28. VERONICA L.   Speedwell. Cal deeply 4-parted in our spp. (5-parted in no. 12), the lobes equal or the upper shorter; cor-tube much shorter than the lobes (except in no. 1), the limb rotate, obscurely 2-lipped, 4-lobed by fusion of the upper 2; stamens 2 (the upper pair), inserted at the orifice of the cor; pollen-sacs parallel; style exsert; stigma 1; fr short, often flattened, loculicidal and sometimes also septicidal, subtended by the sep and tipped by the persistent style; herbs with opposite (rarely whorled) foliage-lvs, the bracteal lvs mostly alternate or scattered; fls small, white to blue or purple. 300, N. Temp., especially Old World.

1 Main stem terminating in an infl, its fls densely crowded or more remote and axillary, the upper bract-lvs commonly alternate. (*Veronicella.*)
  2 Bracts abruptly smaller than the foliage-lvs, the fls thus in a terminal raceme or spike; perennial.
    3 Lvs sharply serrate, the larger 4 cm or more; fls crowded in a spike-like infl . . . . 1. *V. longifolia.*
    3 Lvs entire or obscurely toothed, not over 4 cm; infl looser, more raceme-like.
      4 Fr higher than wide; stem loosely spreading-hairy, not creeping . . . . 2. *V. wormskjoldii.*
      4 Fr wider than high; stem closely puberulent, often creeping at base . . . . 3. *V. serpyllifolia.*
  2 Bracts only gradually reduced, the fls thus axillary; annual, except *V. filiformis.*
    5 Pedicels very short, only 0.5–2 mm even in fr.
      6 Fls white or whitish; lvs 3–10 times as long as wide, not pinnatifid . . . . 4. *V. peregrina.*
      6 Fls blue; lvs either broader or some of them pinnatifid.
        7 Lvs merely toothed; fr ciliate, the sides subglabrous . . . . 5. *V. arvensis.*
        7 Some of the lvs pinnatifid; fr glandular-hairy . . . . 6. *V. verna.*
    5 Pedicels longer, 5–40 mm at least in fr.
      8 Plants annual; lf-blades very often over 1 cm.
        9 Lvs, or many of them, palmately 3–5-lobulate; fr slightly if at all notched . . . . 7. *V. hederaefolia.*
        9 Lvs mostly ± toothed, not lobulate; fr evidently notched.
          10 Mature pedicels 6–15 mm; cor 4–8 mm wide
            11 Fr 4–6 × 3–4 mm, sparsely glandular-hairy . . . . 8. *V. agrestis.*
            11 Fr 2.5–4 × 3.5–6 mm, densely pubescent with long glandular and short eglandular hairs . . . . 9. *V. polita.*
          10 Mature pedicels 15–40 mm; cor 8–11 mm wide . . . . 10. *V. persica.*
      8 Plants perennial, rhizomatous and trailing; lf-blades under 1 cm . . . . 11. *V. filiformis.*
1 Main stems never terminating in an infl, the lvs opposite throughout and the fls all in axillary racemes.
  12 Plants evidently pubescent; lvs 1–3 times as long as wide; mesophytes.
    13 Lvs oval or elliptic, narrowed to the base; pedicels much shorter than the subtending bracts
      . . . . 12. *V. officinalis.*
    13 Lvs ovate or triangular-ovate, broadly rounded or truncate at the sessile or subsessile base; pedicels equaling or longer than the subtending bracts.
      14 Stems prostrate or ascending; cal-lobes about equal; style 3–5 mm . . . . 13. *V. chamaedrys.*
      14 Stems erect; upper sep shorter than the lower; style 6–8 mm . . . . 14. *V. teucrium.*
  12 Plants essentially glabrous, or merely finely glandular in the infl (occasionally evidently hairy in no. 19, which has the lvs 3–20 times as long as wide); plants growing in water or wet places.
    15 Lvs all short-petiolate.
      16 Lvs mostly broadest near the base; style 2.5–3.5 mm . . . . 15. *V. americana.*
      16 Lvs broadest near or above the middle; style 1.8–2.2 mm . . . . 16. *V. beccabunga.*
    15 Lvs (at least the middle and upper ones of the flowering shoots) sessile.
      17 Fr turgid, slightly or scarcely notched; seeds numerous, 0.5 mm or less.
        18 Lvs 1.5–3 times as long as wide; mature pedicels strongly ascending or upcurved . . . . 17. *V. anagallis-aquatica.*
        18 Lvs (2.5–)3–5(–8) times as long as wide; mature pedicels divaricately spreading . . . . 18. *V. catenata.*
      17 Fr flattened, conspicuously notched; seeds 5–9 per locule, 1.2–1.8 mm; lvs (3)4–20 times as long as wide . . . . 19. *V. scutellata.*

1. Veronica longifolia L.   Long-lvd s. Erect perennial to 1 m; lvs opposite or ternate, short-petioled, lanceolate or lance-ovate, 4–10 cm, usually acute or acuminate, very sharply serrate; racemes 1 or few, erect, spike-like, the axis hairy but not glandular; cor blue, hairy in the throat, its lobes 4–5 mm; fr little-flattened, 3 mm, smooth or puberulent, half as long as the persistent style; 2n=34, 68. Native of Europe; intr. in fields, roadsides and waste places from Nf. to N.D., Md., and Ill. June–Aug. (*V. maritima; V. spicata,* misapplied to plants with obtuse lvs.)

2. Veronica wormskjoldii Roemer & Schultes.   Northern s. Rhizomatous perennial 0.7–3 dm; stems simple, erect, or curved at base, loosely villous-hirsute, the infl more densely so and somewhat viscid or glandular; lvs all cauline, opposite (or some reduced upper ones alternate), elliptic to lanceolate or ovate, 1–4 × 0.5–2 cm, hairy like the stem, or often glabrous, slightly toothed or entire; fls pedicellate in well defined, elongating terminal racemes, at least the upper bracts usually alternate; cor deep blue-violet, 6–10 mm wide; filaments 1–1.5 mm; style 1–3 mm; fr glandular-hairy, broadly notched, 4–7 mm, nearly as wide; seeds numerous, 1 mm; 2n=18, 36. Moist meadows, streambanks, bogs and moist open slopes; Alas. to Greenl. s. to the mts. of Me., N.H., and N.M. July, Aug. (*V. alpina,* misapplied.)

3. Veronica serpyllifolia L. Thyme-lvd s. Rhizomatous perennial; stems 1–3 dm, finely and closely puberulent, tending to creep at base or to produce prostrate lower branches, otherwise simple; lvs mostly opposite, elliptic to broadly ovate, 1–2.5 × 0.5–1.5 cm, rounded to acutish, glabrous to nearly so, slightly toothed or entire; fls pedicellate in elongating terminal racemes, at least the upper bracts generally alternate; cor 4–8 mm wide; style 2–3.5 mm; fr finely and sometimes sparsely glandular-hairy, evidently notched, 3–4 mm, broader than high; seeds numerous, under 1 mm; $2n=14$. Nearly cosmopolitan. May–Aug. Var. *serpyllifolia,* native of Europe, and established in fields, meadows, and lawns throughout our range, or even appearing native in open woods, has pale fls with blue lines, finely and closely puberulent pedicels, and filaments 1–2.5 mm. Var. *humifusa* (Dickson) M. Vahl, irregularly distributed in moist woods and alpine meadows, and native in our range from N.S. and n. N.Y. to n. Mich. and n. Minn., has bright blue fls, the pedicels with some spreading, viscid or glandular hairs, the filaments 2–4 mm. (*V. humifusa; V. tenella.*)

4. Veronica peregrina L. Purslane-s. Annual 0.5–3 dm, simple or branched especially below; main lvs oblong or linear-oblong to oblanceolate, 0.5–3 cm × 1–8 mm, irregularly toothed or entire; infl terminal, elongate, lax; bracts alternate, gradually reduced upwards, often not much different from the lvs, each subtending a single subsessile fl; cor white or whitish, only 2 mm wide; mature pedicels 1–2 mm; fr 3–4 mm, ± obcordate, the notch varying in depth; style very short, 0.1–0.3 mm; seeds numerous, 0.4–0.8 mm; $2n=52$. Moist places; temperate N. and S. Amer., and intr. in Europe. Apr.–Sept. Our common form, the var. *peregrina,* is glabrous. The chiefly western var. *xalapensis* (HBK.) St. John & Warren, has the stem and commonly also the sep and frs ± pubescent with short, gland-tipped hairs; it reaches the w. edge of our range, and extends e. to the Atlantic along our n. border.

5. Veronica arvensis L. Corn-s. Erect or ascending annual 0.5–3 dm, somewhat villous-hirsute below, usually more puberulent above; main lvs ovate or broadly elliptic, palmately veined, 0.5–1.5 cm, 1–2 times as long as wide, crenate-serrate, the lower generally short-petiolate; infl terminal, the bracts alternate; fls subsessile; cor blue-violet, 2–2.5 mm wide; mature pedicels only 1–2 mm; fr 3 mm, obcordate, ciliate-margined, with subglabrous sides; style 0.4–1 mm; seeds (5–)8–11 per locule, 0.8–1.2 mm; $2n=14, 16$. Native of Eurasia, established as an inconspicuous weed in gardens, lawns, and fields in much of N. Amer. Apr.–Sept.

6. Veronica verna L. Much like no. 5, but some of the lvs pinnatifid; fr glandular-hairy, slightly broader than long; style ca 0.5–0.6 mm, barely or scarcely reaching the top of the fr; $2n=16$. Native of Europe, now established in campgrounds, along roadsides, and in other ± open, usually disturbed sites in parts of our range, as in Mich. and s. Ont. Apr.–June.

V. dillenii Crantz, an closely related European sp. with longer style, ca 1.5 mm, has also been reported from our range.

7. Veronica hederaefolia L. Ivy-lvd s. Annual, spreading-hirsute, branched at base, with prostrate or weakly ascending stems 0.5–4 dm; lvs 0.5–2 cm wide, mostly wider than long, palmately veined and most of them palmately 3–5-lobulate, the petiole up to as long as the blade; lowest lvs opposite; floriferous part of the stem elongate, with alternate, well developed lvs, each with a single axillary fl, the short pedicel elongating to 5–15 mm in fr; sep strongly ciliate, ovate, truncate or subcordate at base, accrescent to 5–7 mm; cor pale bluish, 3–5 mm wide; style 0.6–1.0 mm; fr glabrous, 2.5–3.5 mm, slightly or scarcely notched, globose or broader than high; seeds 1 or 2 per locule, transversely rugose; $2n=56$. Native of Europe, established in moist fields and waste ground from N.Y. to O., Ky., and S.C. Apr., May.

8. Veronica agrestis L. Field-s. Prostrate or ascending annual 1–3 dm; lvs oblong to commonly ovate or even rotund, 1–2 cm, obtuse, crenately serrate, broadly obtuse to rounded or truncate at base, short-petiolate; pedicels 6–10 mm, elongating to 15 mm in fr; sep ovate or lance-ovate, 3.5–5 mm at anthesis, minutely ciliate, in fr somewhat longer and 3-nerved; fls usually whitish, 4–8 mm wide; style ca 1 mm, not exceeding the sinus of the fr; fr evidently reticulate-veiny, sparsely glandular-pubescent, 4–6 × 3–4 mm, its lobes parallel, the sinus narrow, extending ca one-third of the way to the base; $2n=28$. Native of Eurasia, now established in fields, lawns, and roadsides from Nf. and N. Engl. to N.Y. and Pa., w. to Mich.

9. Veronica polita Fries. Much like no. 8; fls blue; style up to 1.5 mm, surpassing the sinus of the fr; fr densely pubescent with long glandular and short eglandular hairs, 2.5–4 × 3.5–6 mm, with rounded lobes, the sinus broader and shallower than in no. 8; $2n=14$. Native of Eurasia, now widespread in our range and southward. (*V. didyma,* misapplied.)

10. Veronica persica Poir. Birdseye-s. Somewhat villous-hirsute annual, the stems 1–4 dm, simple or branched especially below, loosely ascending and often rooting below; main lvs opposite, short-petiolate, the blade broadly elliptic or ovate, crenate-serrate, 1–2 cm, 1–1.5 times as long as wide; floriferous part of the stem elongate, with alternate bracts like the lvs, each with a single long-pedicellate axillary fl, the pedicel becoming 1.5–4 cm; cal-lobes lance-ovate, accrescent to 6–8 mm, becoming strongly 3-nerved; cor blue, (5–)8–11 mm wide; fr 5–9 mm wide, 3–5 mm high, divergently lobed, reticulately veined; style 1.8–3 mm; seeds 5–10 per locule, 1.2–1.8 mm, transversely rugose; $2n=28$. Native of sw. Asia, intr. in gardens, lawns, roadsides and waste places throughout much of n. Amer. Apr.–Aug. (*V. tournefortii.*)

11. Veronica filiformis J. E. Smith. Creeping s. Slender, trailing perennial from slender rhizomes, also rooting at the nodes; lf-blades cordate-orbicular to reniform, slightly toothed, mostly under 1 cm, short-petiolate; rarely or never fruiting with us; otherwise much like *V. persica;* $2n=14$. Native of Eurasia, sometimes grown in rock-gardens, and now a frequent lawn-weed in ne. U.S.

12. **Veronica officinalis** L.   Common s. Spreading-hairy perennial, the stems with creeping base and assurgent tips and branches, 0.5–2.5 dm; lvs all opposite, elliptic or elliptic-obovate, narrowed to a ± petiolar base, 1.5–5 × 0.6–3 cm, 1.5–3 times as long as wide, rather finely serrate, the larger ones with mostly 12–20 teeth per side; racemes axillary, pedunculate, spiciform, the pedicels 1–2 mm, surpassed by the small, alternate bracts; cor light blue, sometimes with darker lines, 4–8 mm wide; fr 4 mm, broadly triangular-subcordate, shallowly retuse; style 2.5–4.5 mm; seeds 6–12 per locule, 1 mm; $2n=18$, 32–36. Dry fields and upland woods; native of Europe, now well established throughout our range. Apr.–July.

13. **Veronica chamaedrys** L.   Germander-s. Rhizomatous perennial, 1–3 dm, prostrate or loosely ascending, spreading-hairy, at least on the stem; lvs all opposite, ovate, broad-based and sessile or nearly so, 1.5–3 × 0.8–2 cm, 1–2 times as long as wide, coarsely crenate-serrate with mostly 5–11 teeth per side; racemes axillary, pedunculate, loose, the 10–20 fls evidently pedicellate; pedicels 5–9 mm at maturity, surpassing the subtending, alternate bracts; cor blue, 8–12 mm wide; style 3–5 mm; fr (rarely produced) 3–3.5 mm, broadly triangular-obcordate, with apparently about 6 seeds per locule; $2n=16$, 32. A weed in lawns and open, mostly disturbed sites; native of Europe, intr. in much of our range. May–June.

14. **Veronica teucrium** L.   Erect, pubescent perennial 3–8 dm; lvs all opposite, sessile or nearly so, lanceolate to ovate, broad-based, 2–4 cm, 2–4 times as long as wide, coarsely serrate; racemes commonly 2–4 from the upper axils, 5–10 cm, pedunculate, becoming loose, the bracts shorter than the pedicels; sep (4)5, the lower 2.5–4 mm, the upper usually two-thirds as long; cor blue, 10–13 mm wide; style 6–8 mm; fr obovate, slightly surpassing the cal; $2n=64$. Native of Eurasia, occasionally escaped from cult. in the n. part of our range, s. to Md. and O. June, July. (*V. latifolia*, perhaps misapplied.)

15. **Veronica americana** (Raf.) Schwein.   American s. Rhizomatous perennial, glabrous throughout, with erect or ascending simple stems 1–10 dm; lvs opposite, all short-petiolate, serrate to subentire, lanceolate or lance-ovate to narrowly subtriangular, or the lower more elliptic, 1.5–8 × 0.6–3 cm, generally 2–4 times as long as wide; racemes axillary, pedunculate, open, mostly 10–15-fld; cor 5–10 mm wide, blue; style 2.5–3.5 mm; mature pedicels divaricate, 5–15 mm; fr turgid, 3 mm long and about as wide, scarcely notched; seeds numerous, 0.5 mm or less; $2n=36$. Swamps and streambanks; Nf. to Alas., s. to N.C., Mo. and Calif.; also ne. Asia. May–Aug.

16. **Veronica beccabunga** L.   Brooklime. Much like no. 15; lvs more elliptic or obovate, 2–6 cm, broadest near or above the middle, usually broadly rounded above, obtuse to rounded at base, crenate; fls 5–7 mm wide; fr turgid, 3–4 mm wide, not notched; style 1.8–2.2 mm; $2n=18$, 36. Native of Eurasia, sparingly established on muddy shores and streambanks from Que. to Mich., s. to N.J. and W.Va. Summer.

17. **Veronica anagallis-aquatica** L.   Water-s. Fibrous-rooted, probably biennial or short-lived perennial, ± erect, 2–10 dm, glabrous, or slightly glandular in the infl; lvs all opposite, mostly elliptic or elliptic-ovate to elliptic-oblong, sessile and mostly clasping (or the lower with subpetiolar base), 2–10 × 0.7–5 cm, 1.5–3 times as long as wide, sharply serrate to entire (sterile autumnal shoots with more rounded and petiolate lvs); racemes axillary, pedunculate, many-fld; sep highly variable in form and size; cor rather light blue, 5–8 mm wide; mature pedicels generally strongly ascending or upcurved, 3–8 mm; fr turgid, 2.5–4 mm, scarcely notched, nearly or fully as wide; style 1.5–2.5 mm; seeds numerous, 0.5 mm or less; $2n=36$. Along ditches and slowly moving streams, or in other wet places, frequently in shallow water, but largely emersed; native of Eurasia, now established throughout our range. May–Sept. (*V. glandifera*.) Hybridizes with no. 18.

18. **Veronica catenata** Pennell.   Much like no. 17, often finely stipitate-glandular upwards; lvs entire or subentire, (2.5–)3–5(–8) times as long as wide; racemes often with fewer fls; fls white to pink or pale bluish, 3–5 mm wide; sep less variable, mostly broader and blunter; fruiting pedicels divaricately spreading; fr mostly a little wider than high, sometimes more evidently notched; petiolate-lvd autumnal shoots not formed; $2n=36$. Slow-flowing streams and ditches; widespread in the U.S. and s. Can., and in Eurasia. May–Sept. (*V. aquatica* Bernh., not S. F. Gray; *V. comosa, V. connata,* and *V. salina,* all probably misapplied.)

19. **Veronica scutellata** L.   Narrow-lvd s. Rhizomatous perennial, glabrous throughout, or occasionally hairy; stems erect or ascending, 1–4 dm; lvs all opposite, sessile, linear to lanceolate, 2–8 cm × 2–15 mm, (3–)4–20 times as long as wide, entire or with a few remote, divergent, slender small teeth; racemes axillary, pedunculate, mostly 5–20-fld, the pedicels becoming 6–17 mm; cor bluish, 6–10 mm wide; fr flattened, 2.5–4 mm, evidently wider than high, conspicuously and rather broadly notched; style 2–4 mm; seeds 5–9 per locule, 1.2–1.8 mm; $2n=18$. Swamps and bogs; Eurasia and the n. two-thirds of the temperate zone in N. Amer., blanketing our range. May–Sept.

29. **BESSEYA** Rydb.   Cal deeply 4-parted, nearly regular; cor strongly bilabiate, the upper lip oblong, acute, the lower lip decurved, nearly flat, shallowly or irregularly 3-lobed; stamens 2 (the upper pair), inserted at the sinus between the cor-lips, the stout filaments straight, exsert; style slender, stigma minute, capitellate; fr loculicidal, suborbicular, slightly flattened; perennial with a rosette of basal lvs from which arises a simple stem with several much reduced lvs and a dense, terminal spike-like raceme of small yellow fls. 7, N. Amer.

1. **Besseya bullii** (Eaton) Rydb.   Kitten-tails. Basal lvs ovate, 6–12 cm, crenate or obtusely serrate, rounded to cordate at base, palmately veined, long-petiolate; stems 2–4 dm, hairy; cauline lvs alternate, sessile or

nearly so, ovate or oblong, 1–2 cm; infl 5–15 cm; pedicels 1–2 mm; upper cor lip surpassing the villous cal, lower lip about equaling the cal, its middle lobe the smallest; fr 4–7 mm, hirsutulous. Sandy prairies and barrens; Mich. and O. to Minn. and Io., rare and local. May, June. (*Synthyris b.; Wulfenia b.*)

30. SCHWALBEA L. Chaffseed. Cal tubular, very irregular, strongly 12-nerved, 2-lipped, the tube much longer on the lower side, the 2 lobes of the long lower lip united nearly to the tip, the 3 lobes of the upper lip narrowly lance-triangular, the uppermost the shortest; bractlets 2, narrow, closely subtending the cal; cor tubular, strongly irregular, hairy, the upper lip nearly straight, oblong, concave, entire or emarginate, the lower decurved, shallowly 3-lobed, with 2 hairy folds extending inward from the sinuses; stamens 4, ascending under the upper cor-lip and included in it; style elongate, with a solitary, cylindric stigma; fr narrowly ellipsoid, included in the cal, septicidal; seeds nearly linear. Monotypic.

1. **Schwalbea americana** L.   Finely hairy perennial, 3–6 dm, simple; lvs alternate, sessile, lanceolate, 2–5 cm, acute, entire, 3-nerved, the upper progressively smaller; fls solitary in the uppermost axils on short pedicels, forming a leafy spike-like raceme; bractlets 1 cm; cal 2 cm; cor 3–3.5 cm, yellowish on the tube, purplish distally; fr 10–12 mm, the fruiting cal with 12 strongly projecting ribs; $2n=32, 36$. Moist sandy soil; s. N. Engl. to Fla. and Tex., and locally in Tenn. and e. Ky.; very rare. June, July. (*S. australis.*)

31. ODONTITES Ludwig.   Eyebright. Cal regular, the 4 lobes triangular, shorter than the tube; cor tubular, irregular, the upper lip nearly straight, concave, entire, the lower spreading, shallowly 3-lobed; stamens 4, slightly unequal, ascending under the upper lip and included in it; pollen-sacs mucronate at base; stigma capitate; fr included in the cal, laterally compressed, loculicidal, few-seeded; annuals; mostly hemiparasitic, with small opposite lvs and small, red, yellow, or purple fls on short pedicels from the upper axils, forming a terminal, often secund infl. 30, mainly European and Mediterranean.

1. **Odontites serotina** (Lam.) Dumort.   Stems 1–4 dm, often widely branched, finely retrorse-hairy; lvs lanceolate, broad-based, sessile, 1–3 cm, rough-hairy, with 2–3 blunt teeth on each side; bracts 8–20 mm, fls subsessile; cal campanulate, 5 mm; cor 1 cm, hairy, light red; fr elliptic, hairy, 7 mm; $2n=20$. Native of Europe, established as a weed in fields and waste places; Nf., e. Que. and N.S. to Me. and n. N.Y.; Wis., and perhaps elsewhere in our range. July–Sept. (*O. odontites.*)

32. EUPHRASIA L.   Eyebright. Cal 4-lobed (the upper median lobe lacking), more deeply cleft above and below than on the sides; cor bilabiate, the upper lip concave, shallowly 2-lobed or merely notched, its sides often reflexed or revolute, the lower lip spreading, 3-lobed, the lobes often emarginate; stamens 4, included, ascending under the upper lip; pollen-sacs acuminate or mucronate at base; ovary laterally compressed; style elongate; stigma capitellate; fr loculicidal, truncate or retuse above; seeds several, fusiform, with ca 10 narrow wings; low, hemiparasitic herbs, ours all annuals with small, thinly retrorse-hairy stem and opposite, sessile, ovate to rotund, palmately veined, coarsely toothed lvs, the upper bracteal lvs similar but tending to be alternate; fls small, sessile or subsessile. Hundreds of described spp., only a few reasonably distinct; irregularly wide-spread.

1 Bracteal lvs with subulate or bristle-tipped teeth; cor relatively large, mostly 5–10 mm; cal and fr and
    usually also the lvs and bracts glabrous or nearly so; intr. .... 1. *E. officinalis.*
1 Bracteal lvs with acute or obtuse teeth; cor smaller, mostly 2.5–4.5 mm (or up to 6 mm in no. 2,
    which has the lvs, bracts, cal and fr pubescent); native, northern.
  2 At least some of the internodes more than 3 times as long as the lvs; infl ± elongate .... 2. *E. disjuncta.*
  2 None of the internodes more than 3 times as long as the lvs, infl ± condensed, at least at first.
    3 Foliage-lvs several pairs, gradually decreasing in size upwards; infl in well developed plants of
      several verticils, becoming elongate at maturity; stems simple or often branched; northern
      but scarcely alpine .... 3. *E. randii.*
    3 Foliage-lvs 2–5 pairs, gradually increasing in size upwards and commonly smaller than the bracts;
      infl densely crowded; stems regularly simple; alpine .... 4. *E. oakesii.*

1. **Euphrasia officinalis** L.   European e. Stems 1–4 dm, simple or freely branched, floriferous in the upper ⅓ to ⅔; lvs ovate to suborbicular, 5–15(–20) mm, sharply 3–5-toothed on each margin, glabrous or sometimes hairy beneath; bracteal lvs similar, but the teeth more distinctly subulate or bristle-tipped; cal glabrous or

nearly so, the lobes 2.5–3.5 mm at maturity; cor 5–10 mm, typically pale lavender with deeper purple guide-lines, the lower lip also marked with yellow, or seldom the cor deeper violet; upper cor-lip evidently bilobed; fr glabrous or nearly so; seeds 1.3–2 mm; $2n=22, 44$. Dry or moist fields, roadsides, and waste places; native of Europe, early intr. into N. Engl., and now well established from Nf. and Que. to Mass., N.Y., and Pa., and inland to Ont. June–Oct. Divided by European botanists into a myriad of scarcely distinguishable microspecies, of which the following have also been attributed to our range: *E. borealis, E. curta, E. micrantha, E. nemorosa, E. rigidula, E. stricta, E. tatarica, E. tetraquetra.* The names *E. americana* and *E. canadensis* were based on intr. forms of this same complex.

2. **Euphrasia disjuncta** Fern. & Wieg.   Arctic e. Stems 0.5–3 dm, in well developed plants at least some of the internodes more than 3 times as long as the lvs; lvs several pairs, ovate, 5–10(–15) mm, with 3–5 acute or obtuse teeth on each side, ± hairy on both surfaces, the bracts basically similar; cal crisp-hairy, its broadly triangular lobes ultimately 2.5–3 mm; cor mostly 4–6 mm, white with purple lines, essentially intermediate in size and form between that of no. 1 and those of nos. 3 and 4; fr ± hairy; $2n=44$. Streambanks, bogs, and other wet places; Lab. & Nf. to Alas., s. to Me., n. Mich., ne. Minn., and Mont. July–Sept. (*E. arctica* var. *d.,* assuming a circumboreal distribution for *E. arctica,* but that name may not properly apply, the nomenclature confused.)

3. **Euphrasia randii** Robinson.   Nova Scotian e. Simple or sparingly branched from the base, 0.5–3 dm; lvs ovate to orbicular, 5–15 mm, with 3–5 obtuse or rounded teeth on each side, glabrous to evidently pubescent on both surfaces; bracteal lvs similar but smaller; infl soon elongating, the lower internodes 1–2 cm and only the uppermost verticils remaining approximate; cal-lobes eventually 1.5–2 mm, broadly tri-angular; cor 2.5–4.5 mm, pale with darker lines, or sometimes deep violet, the upper lip only very shallowly bilobed; seeds 1–1.3 mm. Moist peaty soil or sod; Nf. and Que. to N.S. and Me. (*E. bottnica,* misapplied.)

4. **Euphrasia oakesii** Wettst.   White-Mt. e. Stem almost always simple, 3–10 cm; foliage-lvs 2–5 pairs, ovate to suborbicular, 2–7 mm, increasing in size toward the summit, with 2–3 rounded teeth on each side; infl densely crowded and head-like; otherwise as no. 3. Alpine situations from Lab. and Nf. to the higher peaks of Me. and N.H. (*E. williamsii.*)

33. RHINANTHUS L.   Yellow rattle. Cal membranous, reticulate-veiny, shallowly 4-lobed, more deeply cleft above and below than on the sides, in fr strongly accrescent and becoming suborbicular; cor very irregular, the upper lip very concave, with a lateral lobe-appendage on each side near the summit, the lower lip 3-lobed, the throat with 2 longitudinal ridges; stamens 4, paired, ascending under the upper lip; stigma capitate; fr flat, suborbicular, loculicidal; seeds large, flat, narrowly winged; hemiparasitic annuals with opposite, sessile or subsessile, serrate, pinnately veined lvs and rather small yellow fls sessile or subsessile in the axils of the uppermost or bracteal lvs, forming a secund leafy spike. 2–several, temperate Eurasia and N. Amer.

1. **Rhinanthus crista-galli** L.   Stem 1.5–8 dm, simple or few-branched, thinly villous-puberulent on 2 of the 4 sides; lvs lance-triangular to oblong, firm, serrate or crenate-serrate, 2–6 cm × 4–15 mm, scabrous above, scabrous-hispidulous beneath; bracts gradually reduced upwards, mostly shorter and broader than the lvs, with longer and sharper teeth; cor 9–14 mm; mature cal 12–17 mm; $2n=14$, + 0–8 B. Various habitats; circumboreal, s. in Amer. to N.Y., Colo., and Oreg., our lowland plants intr. from Europe, the alpine ones native. (*R. borealis; R. kyrollae; R. minor; R. oblongifolius; R. stenophyllus.*)

34. PEDICULARIS L.   Lousewort. Cal campanulate to tubular, entire or lobed, usu-ally oblique and longer on the upper side; cor very irregular, the tube gradually enlarged distally, the upper lip (galea) as long as or longer than the lower, curved forward and very concave or arched, often laterally compressed, entire or with 2 lateral teeth near the tip, the lower lip ± expanded, with 2 longitudinal folds below the sinuses; stamens 4, paired, ascending under the upper lip, included; stigma capitate; fr much compressed, ovate or oblong, usually falcate or inequilateral, loculicidal, but usually opening only or most deeply on the upper side; hemiparasitic herbs with opposite or alternate, sharply toothed to bipinnatifid lvs and medium-sized yellow to red or purple fls in terminal, leafy-bracteate spikes or racemes. 500, mainly North Temp. and boreal, also in the Andes.

1 Cal-lobes 5, subequal; Me. and N.B. . . . . 1. *P. furbishiae.*
1 Cal-lobes on each side of the vertical median plane completely united, each lateral half entire or with
small appendages; widespread.
  2 Galea with a small marginal tooth on each side near the tip.
    3 Each lateral half of the cal entire; galea with 2 subterminal teeth only . . . . 2. *P. canadensis.*
    3 Each half of the cal with several small, rounded appendages; galea with a pair of small lateral
    teeth near the middle, as well as the subapical teeth . . . . 3. *P. palustris.*
  2 Galea entire; each half of the cal with a foliaceous appendage . . . . 4. *P. lanceolata.*

1. Pedicularis furbishiae S. Wats.   Furbish's l.   Stems 4–6 dm; lvs lanceolate, 2–5 cm wide, deeply pinnatifid, the lance-oblong segments again shallowly pinnatifid; cal subequally 5-lobed, the lobes oblong, foliaceous, 4–6 mm, erose; cor yellow, 16–20 mm, the galea with 4 very short teeth near the tip. Banks of rivers, typically in fairly wet but well drained soil, along with *Alnus viridis*; St. John R. valley, Me. and N.B. July, Aug.

2. Pedicularis canadensis L.   Forest-l.; woody-betony.   Perennial by short rhizomes, forming large clumps; stems 1.5–4 dm, sparsely villous; lvs chiefly basal, lanceolate to narrowly oblong or oblanceolate, pinnately lobed into several oblong or ovate, crenate segments; lower lvs on petioles often longer than the blade, the cauline progressively reduced and the upper subsessile; spikes commonly solitary, 3–5 cm; in fr to 20 cm, the bracts oblanceolate, usually toothed only at the tip; cal very oblique, the 2 lateral halves entire, separated by a short cleft above and a deeper one below; cor yellow to purple, 18–25 mm, the galea with 2 slender teeth just below the rounded tip, the lower lip shorter than the galea; $2n=16$. Upland woods and prairies; Que. and Me. to Man., s. to Fla. and Tex.; also Colo. to n. Mex. Apr.–June.

3. Pedicularis palustris L.   Glabrous or villosulous biennial or annual; stems 3–5 dm, commonly branched; lvs cauline, short-petioled, lanceolate, deeply pinnatifid into several oblong or ovate, crenate segments; spikes 3–8 cm, in fr to 20 cm, the bracts resembling the lvs but smaller; cal 6–9 mm, somewhat oblique, each lateral half ornamented along the margin with half a dozen short, fleshy, rounded lobes; cor 18–25 mm, pink-purple, or with purple galea, the galea with 2 short subapical teeth and another reflexed tooth near the middle of each margin; $2n=16$. Wet soil; Lab. to Que. and N.S.: Eurasia. July, Aug.

4. Pedicularis lanceolata Michx.   Swamp-l.   Short-lived; stems 3–8 dm, mostly solitary, glabrous or nearly so, simple or few-branched; lvs mostly opposite, sessile or short-petioled, 5–10 cm, pinnately lobed less than half way to the midvein, each lobe crenate; spikes terminal and from the upper axils; cal 1 cm, very oblique, each lateral half bearing near its lower side a foliaceous, oblong or ovate, entire or crenate appendage; cor 15–25 mm, yellow, the galea about as long as the lower lip, truncate or very shallowly notched at the tip, otherwise entire. Swamps and wet soil; Mass. to Mich., Minn., and N.D., s. to N.C., Mo., and Neb. Aug., Sept.

35. MELAMPYRUM L.   Cow-wheat.

Cal-tube cupulate, with 4(5) lobes, the upper slightly the longer; cor-tube longer than the lips, straight, slightly widened distally, the upper lip straight, very concave, truncate across the narrow summit between the 2 short teeth, the lower lip about equaling the upper, 3-lobed, with 2 low ridges below the sinuses; stamens 4, paired, included, ascending under the upper lip; pollen-sacs acute at base; stigma entire; fr flattened, obliquely ovate, pointed, loculicidal; seeds normally 4, ellipsoid or oblong, hemiparasitic annuals with opposite lvs, the small fls borne in the upper axils. 25, N. Hemisphere.

1. Melampyrum lineare Desr.   Stems simple or commonly branched, 1–4 dm; cauline lvs below the branches (usually lacking at anthesis) oblanceolate to spatulate; lvs subtending the branches sessile or short-petiolate, 2–6 cm, narrowly linear to lance-ovate; bracteal lvs often laciniately toothed near the base; cal-lobes subulate, longer than the tube; cor 6–12 mm, white with yellow palate; fr 8–10 mm; $2n=18$. Woods, or toward the n. also in bogs or wet soil; Lab. to B.C., s. to Va., O., n. Ind., and Minn., and along the mts to Ga. Four vars.:

a Bracteal lvs entire or nearly so, or the uppermost rarely with a few short basal teeth.
  b Lvs mostly linear, rarely over 5 mm wide; Lab. to B.C., s. only to n. N. Engl., n. Mich., and n.
    Minn. . . . . var. *lineare*.
  b Lvs lanceolate to ovate, the larger 10–30 mm wide; Mass. and N.Y. to Ga., avoiding the coastal
    plain . . . . var. *latifolium* Barton.
a Bracteal lvs generally with a few prominent, sharp teeth or segments.
  c Plants few-branched or simple; teeth of the middle and upper bracts shorter than the width of the
    undivided portion of the bracts; lvs lanceolate, (2–)5–10 mm wide; the common form in most
    of our range; Que. to Minn., s. to Va. and Tenn. . . . . var. *americanum* (Michx.) Beauverd.
  c Plants much branched; teeth of the middle and upper bracts commonly about as long as the width
    of the undivided portion of the bracts; lvs linear to lanceolate, 2–6(–8) mm wide; sandy soil
    and barrens on the coastal plain; e. Mass. to Va.; also nw. Ind. . . . . var. *pectinatum* (Pennell) Fern.

36. ORTHOCARPUS Nutt.   Owl-clover.

Cal tubular, 4-cleft, most deeply on the lower side, the lobes shorter than the tube; cor very irregular, the tube dilated above, the upper lip laterally compressed, broad-based, narrowed to the summit, the lower lip about as long as the upper but notably wider, shortly 3-lobed; stamens 4, paired, ascending under the upper lip, included; pollen-sacs separate, one attached by its middle, the other suspended from its summit; stigma capitate; fr shorter than the cal, symmetrical, ellipsoid, loculicidal; hemiparasitic annuals with alternate lvs and small fls in dense terminal spikes. 25, mostly w. U.S., especially Calif.

1. Orthocarpus luteus Nutt. Finely glandular-hairy; stems 1–3 dm, simple or occasionally branched; lvs numerous, linear or rarely trifid, 1–2 cm; bracteal lvs trifid or rarely linear; fls bright yellow, 9–12 mm; 2n=28. Dry prairies and plains; Minn. to Atla. and B.C., s. to Nebr., N.M., and Calif. July–Sept.

37. CASTILLEJA Mutis. Indian paintbrush. Cal tubular, membranous, deeply divided into 2 lateral halves; cor-tube slender, elongate, scarcely dilated distally, the upper lip (galea) triangular-acuminate, laterally compressed, the lower lip much shorter than the upper, appressed or deflexed; stamens 4, paired, ascending under the galea, included; one pollen-sac attached at its center, the other suspended from its summit; stigma capitate or shallowly 2-lobed; fr ovoid or oblong, loculicidal; hemiparasitic herbs with alternate, often pinnatifid lvs and large fls in dense terminal spikes, each fl subtended by a large, entire or pinnatifid, sometimes brightly colored bracteal lf. 150+, mostly w. N. Amer.

1 Foliage and bracteal lvs (or most of them) lobed or cleft.
  2 Each lateral half of the cal widened distally, entire or barely emarginate at its rounded tip; bracteal
    lvs usually scarlet .... 1. *C. coccinea.*
  2 Each lateral half of the cal distinctly bifid; bracteal lvs usually green .... 2. *C. sessiliflora.*
1 Foliage and bracteal lvs entire, or the latter with 1–3 small teeth only .... 3. *C. septentrionalis.*

1. Castilleja coccinea (L.) Sprengel. Painted cup. Annual or biennial, ± hairy, usually simple, 2–6 dm; principal cauline lvs very diverse, entire to commonly 3–5-cleft, with linear or oblong segments, the terminal one usually the largest; spike dense, 4–6 cm, elongating to 1–2 dm in fr; bracts wholly or mostly scarlet (rarely yellowish), commonly deeply 3-lobed, occasionally 5-lobed, cal 17–25 mm, thin, often ± scarlet, deeply divided into lateral halves, each half widened distally and the summit broadly rounded to truncate or barely emarginate; cor greenish-yellow, 21–27 mm, little longer than the cal, the minute lower lip less than a third as long as the galea; 2n=48. Meadows, moist prairies, and damp sandy soil; Mass. to Ont. and Man., s. to S.C., Miss., and Okla. May–Aug.

2. Castilleja sessiliflora Pursh. Downy paintbrush. Soft-hairy perennial; stems usually clustered, simple, 1–4 dm, very leafy; principal cauline lvs 3–5 cm, cleft to below the middle into 3 linear-linear divergent lobes, the middle lobe sometimes again cleft; spikes dense, 3–5 cm; bracts like the lvs but somewhat smaller, green or sometimes pink-tipped; cal 25–40 mm, cleft to about the middle, each half also cleft into 2 linear-attenuate segments 8–14 mm; cor curved, exserted from the bracts, purplish to yellow or white, 35–55 mm, the galea 9–12 mm, the lobes of the lower lip 5–6 mm; 2n=24. Dry prairies and plains; Wis. and n. Ill. to Sask., s. to Mo., Tex., and Ariz. May–July.

3. Castilleja septentrionalis Lindl. Northeastern p. Perennial, villous in the infl, otherwise glabrous; stems 2–6 dm, simple or branched; lvs linear or lance-linear, 3–10 cm, entire; spike compact or interrupted at the base, later elongating; bracts pale yellow to nearly white or tinged with purple, oblong to narrowly elliptic, about equaling the fls, entire or with a few teeth at the rounded tip; cal 15–20 mm, the lateral halves separate to near the middle, each half ending in 2 blunt lobes 2–3 mm; galea barely surpassing the cal, the very short, appressed lower cor-lip included; 2n=24. Damp rocky soil; Lab. and Nf. to Vt.; Keweenaw Point, Mich. July, Aug. (*C. acuminata; C. pallida* var. *s.*)

Family **OROBANCHACEAE**, the Broom-rape Family

Root-parasitic herbs without chlorophyll; fls hypogynous, sympetalous, irregular, perfect (except *Epifagus*); cal 2–5-lobed or toothed, persistent; cor bilabiate, withering persistent, (4–)5-lobed; stamens 4, paired, borne on the cor-tube alternate with the lobes, the fifth (upper) one staminodial or wanting; ovary superior, unilocular, with (2)4(6) intruded parietal placentas; style terminal, with a capitate or crateriform or lobed stigma; capsule 2-valved, many-seeded; embryo minute, undifferentiated, embedded in the oily endosperm; small, often fleshy, yellowish or brownish plants, the lvs reduced to scales. 17/150.

1 Stems simple or few-branched; infl various; fls all alike.
  2 Cal nearly regular, or deeply cleft both above and beneath .... 1. *Orobanche.*
  2 Cal deeply cleft on the lower side only .... 2. *Conopholis.*
1 Stems abundantly and paniculately branched (except when depauperate), the fls subsessile in long,
    spike-like racemes, the upper with well developed cor, the lower with minute, calyptriform cor
    .... 3. *Epifagus.*

1. OROBANCHE L. Broom-rape. Cal campanulate, variously lobed; cor tubular, the tube usually downcurved, much longer than the lobes; stamens about equaling the cor-tube; fls in racemes or spikes or solitary. 100+, cosmop.

1 Fls solitary or few (rarely up to 10), the pedicel always much longer than the cal; cal subequally 5-
   lobed, not subtended by bractlets.
  2 Pedicels mostly 1–3, much longer than the mostly very short stem .... 1. *O. uniflora.*
  2 Pedicels mostly 4–10, slightly if at all longer than the ± elongate stem .... 2. *O. fasciculata.*
1 Fls several to many, sessile or subsessile in a dense spike or spike-like raceme; cal commonly subtended
   by 1–3 bractlets.
  3 Cal 5-lobed, the lobes subequal, linear, well developed .... 3. *O. ludoviciana.*
  3 Cal 4-lobed (rarely 2-lobed, or with a minute 5th lobe).
    4 Cal divided to the base into 2 lateral segments, these generally bifid .... 4. *O. minor.*
    4 Cal-tube not much shorter than the 4 subequal lobes .... 5. *O. ramosa.*

    1. Orobanche uniflora L. Cancer-root. Parasitic on many kinds of plants; proper stem 1–3(–5) cm, all or mostly underground, with a few overlapping, oblong-obovate, glabrous, blunt or apiculate scales; pedicels 1–3, 6–20 cm, erect, finely glandular-hairy, without bractlets, each with a single white to violet fl 2 cm; cal-lobes 5, about equal, triangular-acuminate, slightly longer than the tube; 2*n*=36, 48, 72. Moist woods and streambanks; Nf. and Que. to Yukon, s. to Fla. and Calif. Our plants are var. *uniflora.* (*Thalesia u.; Anoplanthus u.*)

    2. Orobanche fasciculata Nutt. Clustered b.-r. Parasitic on many spp., but especially on *Artemisia*; proper stem 5–15 cm; scale-lvs ovate, pubescent, all or at least the upper acuminate; pedicels 4–10, 2–15 cm, not much if at all longer than the stem, forming a loose, flat-topped corymb much surpassing the stem; bractlets none; cor 1.5–3 cm; cal-lobes about equaling the tube; 2*n*=24. Dry soil, prairies, and plains, in our range especially on sand-dunes; Mich. and nw. Ind. to B.C., Calif., and n. Mex. June. (*Thalesia f.; Anoplanthus f.*)

    3. Orobanche ludoviciana Nutt. Prairie b.-r. Parasitic on many kinds of plants, especially *Artemisia* and other Asteraceae; plants 1–3 dm, simple or rarely branched, with numerous appressed scale-lvs; spikes many-fld, dense, forming ⅓ to ⅔ of the shoot; fls mostly sessile or the lower on erect pedicels to 15 mm; cal subtended by (1)2 bractlets, as well as by a bract, bilabiate, the upper lip with a single lobe, the lower with 4; cal-lobes linear, 8–12 mm; cor often purplish, 1.5–2.5 cm; 2*n*=24, 48, 72, 96. Chiefly in sandy soil; Ill. and Ind. to Sask. and Tex., w. to Calif. and n. Mex. June–Aug. Our plants are var. *ludoviciana.* (*Myzorrhiza l.*)

    4. Orobanche minor J. E. Smith. Lesser b.-r. Parasitic on many hosts, especially *Trifolium*; plants 1.5–5 dm, strict and simple from a ± bulbous-thickened base, ± glandular-villous; infl spicate, ± elongate, not very dense, especially below, the axis exposed; fls sessile, subtended by a bract but without bractlets; cal split to the base above and below, the 2 lateral segments slender and elongate, mostly bifurcate, commonly unequally so, the lower fork the smaller; cor white or yellowish, marked with violet, 10–15 mm; 2*n*=38. Native of Europe and w. Asia, intr. from N.J. to N.C.

    5. Orobanche ramosa L. Branching b.-r. Parasitic on many hosts, especially *Lycopersicon, Cannabis,* and *Nicotiana*; stems slender, 1–3 dm, usually with 1 or 2 branches; spikes slender, rather loose and open, the axis often exposed; cal-tube broad, equaling or a little shorter than the 4 equal, caudate-acuminate lobes (a minute fifth lobe rarely developed); cor 12 mm, the tube pale yellow, the limb blue; 2*n*=24. Native of Europe and w. Asia, rarely intr. in our range, as in Ky.

    O. purpurea Jacq., with larger (2–2.5 cm) fls and 5-lobed cal, parasitic chiefly on *Achillea,* has been reported from Ont.

2. CONOPHOLIS Wallr. Squaw-root. Cal subtended by 1 or 2 minute bractlets, tubular, split down the lower side, irregularly toothed; cor tubular, curved downward, very irregular, the upper lip straight, concave, entire or nearly so, the lower lip decurved, 3-lobed; stamens and style about equaling the cor; fr ovoid, tipped by the persistent style and capitate stigma; unbranched herbs, the stout stem mostly or wholly concealed by the numerous, fleshy, overlapping lf-scales; fls numerous; subsessile, crowded in a dense spike, each subtended by a bract smaller than the lf-scales. 2, N. Amer.

    1. Conopholis americana (L.) Wallr. Parasitic on several kinds of oaks, pale brown or yellowish through-out; stems stout, erect, 5–20 cm; lf-scales ovate or broadly lanceolate, to 2 cm; spike usually constituting half or more of the shoot, 1.3–2.8 cm thick, the bracts about equaling the 8–13 mm cal; cor 8–14 mm; 2*n*=40. Rich woods; N.S. to Fla., w. to n. Mich., Wis., Ill., and Ala. May, June.

3. EPIFAGUS Nutt. Beech-drops. Fls dimorphic, the lower small, pistillate, fertile, the upper seemingly perfect but functionally staminate; lower fls with cupulate, strongly

5-ribbed cal, the 5 short lobes triangular-subulate, the cor calyptriform, not opening, promptly forced off by the developing ovary, the stamens none and the style short; upper fls with somewhat larger cal and tubular, colored, distinctly zygomorphic, shortly 4-lobed cor about equaled by the stamens and style; freely branched herb with small, scattered, alternate lf-scales and numerous subsessile, solitary, axillary fls, forming a large panicle. Monotypic.

1. **Epifagus virginiana** (L.) Barton.   Parasitic on the roots of beech; stems generally pale brown, usually with fine brown-purple lines, 1–5 dm, with numerous long, ascending branches; lf-scales triangular-ovate, 2–4 mm; lower fls 5 mm; cor of upper fls 1 cm, white, commonly with 2 brown-purple stripes; fr somewhat oblique, 5 mm, dehiscent across the top. Under beech trees; Que. and N.S. to Wis., s. to Fla. and La. Aug.–Oct. (*Leptamnium v.; Epiphegus v.*)

FAMILY **ACANTHACEAE**, the Acanthus Family

Fls perfect, hypogynous, sympetalous; cal ± deeply (4)5-lobed, or the lobes sometimes suppressed; cor from essentially regular to more often irregular, commonly bilabiate and (4)5-lobed; stamens on the cor-tube, alternate with the lobes, commonly 4 or 2 and paired, the missing (upper) member(s) sometimes represented by staminodes; ovary superior, bilocular, with a terminal style; placentation axile; ovules most commonly 2 in each locule, superposed or sometimes collateral, less often (*Ruellia*) several, each with a modified funiculus that is typically developed into a hook-shaped jaculator and functions in flinging out the seeds; fr an explosively dehiscent loculicidal capsule; seeds with a very thin testa (often becoming mucilaginous when wet) and large, dicotyledonous embryo, usually without endosperm; herbs or less often woody plants with simple, opposite, exstipulate lvs and often with showy fls. 250/2500.

1 Stamens 4; cor subequally 5-lobed . . . . 1. *Ruellia.*
1 Stamens 2; cor strongly 2-lipped.
  2 Peduncles elongate; each fl subtended by 3 minute triangular bracts . . . . 2. *Justicia.*
  2 Peduncles shorter than the subtending petioles; bracts foliaceous . . . . 3. *Dicliptera.*

1. RUELLIA L.   Cal regular, the narrow lobes much longer than the tube; cor large, funnelform, the tube equaling or much longer than the campanulate or obconic throat, the limb slightly oblique, spreading, subequally 5-lobed; stamens 4, weakly united in pairs toward the base, the lower pair somewhat the longer; a staminodium commonly present; fr clavate or obovate; seeds 3–8 per locule, compressed, suborbicular; ours perennial herbs with large blue-lavender fls in axillary few-fld cymes or terminating short axillary branches. 300+, mostly trop. and warm-temp.

1 Cal-lobes distinctly lanceolate, 2 mm wide or more . . . . 1. *R. strepens.*
1 Cal-lobes linear, subulate, or setaceous, 1 mm wide or less.
  2 Lvs sessile or subsessile, the petiole, if any, less than 3 mm; fr glabrous . . . . 2. *R. humilis.*
  2 Lvs narrowed or rounded below to a petiole or petiole-like base 3 mm or more.
    3 Stem equally hairy on all 4 sides with minute curved hairs only; fr hairy . . . . 3. *R. pedunculata.*
    3 Stem more densely hairy on 2 opposite sides than on the other 2, and commonly with long
       bristles as well as short recurved hairs, or seldom glabrous; fr usually glabrous . . . . 4. *R. caroliniensis.*

1. **Ruellia strepens** L.   Stems 3–10 dm, minutely puberulent in narrow strips on 2 opposite sides; petioles 5–20 mm; main lvs thin, ovate or lance-ovate, 5–15 cm, acute or short-acuminate, abruptly narrowed or cuneate at base, glabrous or sparsely strigose; fls from a few nodes near the middle of the stem, 1–3, subtended by 2 ovate leafy bracts, on an axillary common peduncle 5–15 mm or terminal on a leafy axillary branch; cal-lobes lanceolate, 1–2 cm × 2–4 mm; cor 3.5–5 cm, the slender tube about equaling the throat; fr glabrous; 2*n*=34. Moist woods; N.J. to O. and Io., s. to S.C., Ala., and Tex. May–July. Cleistogamous forms with smaller fls in cymose clusters from several axils occur occasionally.

2. **Ruellia humilis** Nutt.   Fringelf-ruellia. Stems 2–6 dm; lvs sessile or subsessile, 3–8 cm, acute to very obtuse; fls in sessile or subsessile, crowded cymose clusters from the axils of several upper lvs, subtended by lanceolate bracts; cal-lobes narrowly linear to subsetaceous, less than 1 mm wide; cor 3–7 cm; fr glabrous. Prairies and dry upland woods; Pa. to n. Ind., se. Minn., and Nebr., s. to w. N.C., Ala., and Tex. (*R. ciliosa,*

misapplied.) Highly variable in lf-form, pubescence, and fl-size. The common and widespread var. *humilis* is copiously hairy on the young internodes, the pubescence commonly extending also to the cal-lobes and the veins and margins of the lvs. Var. *calvescens* Fern., of the Appalachian region from s. O. and s. Ind. to Ky., Tenn., Ga., and Ala., is glabrous or only sparsely hairy.

3. Ruellia pedunculata Torr.   Stems 3–7 dm, equally pubescent on all 4 sides with minute recurved hairs; lvs lanceolate to lance-ovate, the main ones 5–10 cm, on petioles 3–10 mm; peduncles axillary from a few median nodes, 2–10 cm, bearing at the summit a pair of foliaceous bracts and 1–3 short-pediceled fls; cal-lobes very narrowly linear; cor 3–5 cm, the tube equaling or slightly longer than the throat; fr finely hairy; $2n=34$. Dry or rocky upland woods; s. Ill. and Mo., s. to La. and Tex.; also in the Appalachian region from Md. to Ga. and Ala. June–Aug. (*R. purshiana* Fern., the Appalachian plants.) There are cleistogamous forms, as in no. 1.

4. Ruellia caroliniensis (Walter) Steudel.   Stems 3–8 dm, villous with spreading hairs, or minutely re-curved-hairy, or both; lvs lanceolate to ovate, the upper rather crowded, the main ones 4–12 cm, distinctly petioled, villous to subglabrous; cor 2.5–5 cm, the tube somewhat longer than to twice as long as the throat; fr glabrous or occasionally hairy; $2n=34$. Moist or dry woods; N.J. to s. O. and s. Ind., s. to Fla. and Tex. (*R. parviflora.*) Highly variable.

2. JUSTICIA L.   Water-willow. Cal nearly regular, cleft to near the base into linear or lanceolate lobes; cor 2-lipped, the upper lip concave, entire or emarginate, the lower spreading or deflexed, 3-lobed; stamens 2; pollen-sacs separate, one terminal, transverse or oblique, the other lateral; fr clavate or cylindric, the fertile part elevated on a long stipe-like base; seeds normally 2 per locule, orbicular; ours perennial herbs with narrow, sessile or subsessile, entire lvs and long-peduncled axillary spikes, the fls small, white to violet, each subtended by 3 minute triangular bracts at the base of the cal. 300, warm regions.

1 Fls in a head-like spike; seeds rough, not margined; widespread .... 1. *J. americana.*
1 Fls in a loose spike; seeds smooth or nearly so, margined; se. coastal plain .... 2. *J. ovata.*

1. Justicia americana (L.) M. Vahl.   American w.-w. Colonial by rhizomes; stems 5–10 dm, often rooting below, usually simple, glabrous; lvs linear to lanceolate or narrowly oblong, 8–16 cm × 8–25 mm, long-cuneate at base; peduncles from several upper axils, stiffly ascending, 5–15 cm; spikes 1–3 cm, crowded with opposite fls, the cal longer than the internode above it; cal-lobes narrowly lanceolate, 4–8 mm; cor 8–13 mm, the lobes about equaling or a little longer than the tube, pale violet to nearly white, marked with purple on the lower lip; terminal pollen-sac transverse; seeds densely verrucose, 3 mm wide; $2n=26$. Mud and shallow water; w. Que. to Mich. and Wis., s. and sw. to Ga., Kans., and Tex. June–Aug. (*J. mortuifluminis; Dianthera a.*)

2. Justicia ovata (Walter) Lindau.   Coastal plain w.-w. Much like no. 1; stems 2–6 dm; lvs lanceolate to elliptic or oblanceolate 3–10 × 1–2.5 cm, gradually or abruptly cuneate at base; peduncles 3–10 cm at anthesis; spike soon becoming elongate, the internodes longer than the cal of the fl just beneath; cal-lobes 7–10 mm, linear; terminal pollen-sac oblique; seeds smooth or nearly so, with a distinctly differentiated margin. In mud and shallow water; coastal plain from se. Va. to Fla. and Tex., n. in the Mississippi Valley to se. Mo. and sw. Ky. Ours is var. *ovata.* (*J. humilis; Dianthera ovata.*)

3. DICLIPTERA Juss., nom. conserv.   Cal regular, deeply 5-cleft; cor strongly bila-biate, the tube scarcely enlarged upward, the lips entire or inconspicuously toothed at the top; stamens 2, inserted near the top of the cor-tube; pollen-sacs distinct, one apparently lateral, the other terminal; fr elliptic, flat, 2–4-seeded, the thin sides torn loose by the recurving of the indurate margins; herbs with long-petioled lvs and axillary, cymose, sessile or subsessile, few-fld clusters of small, pink to purple or red fls subtended by foliaceous bracts. 150, warm regions.

1. Dicliptera brachiata (Pursh) Sprengel.   Freely branched annual 3–7 dm; petioles 1–3 cm; lvs ovate or lance-ovate, 5–10 cm, acuminate; infls developed from most upper axils and often accompanied by floriferous branches; bracts oblong to obovate, 6–10 mm, cuspidate; cal-lobes 3 mm; cor pink or purple, 12–15 mm, or much smaller and cleistogamous, the lips about as long as the tube; $2n=ca$ 80. Moist or wet woods; s. Ind. to s. Mo., s. to Fla. and Tex., and n. along the coast to se. Va. (*Diapedium b.*)

FAMILY **PEDALIACEAE**, The Sesame Family

Fls perfect, mostly hypogynous, sympetalous; cor irregular, sometimes spurred or saccate at base, the limb often oblique or ± bilabiate, the 5 lobes imbricate; stamens on the cor-tube, alternate with the lobes, 4 and paired, the fifth (upper) one commonly represented by a small staminode, or sometimes only the 2 upper-lateral stamens fully developed; ovary mostly superior, bicarpellate, unilocular with 2 intruded, forked parietal placentas, or bilocular with axile placentas, or 4-locellar; fr variously capsular or indehiscent, very often armed with horns or hooks or prickles; seeds with scanty or no endosperm and a straight, spatulate, dicotyledonous embryo; mostly herbs, beset (at least when young) with characteristic trichomes that have a short stalk and a broad head composed of 4 or more mucilage-filled cells, often with other kinds of hairs as well; lvs simple, exstipulate, opposite or the upper alternate. (Martyniaceae, Trapellaceae.) 20/80, mostly tropical.

1 Fls solitary in the axils of the lvs; fr 2–3 cm, unarmed .... 1. *Sesamum.*
1 Fls in racemes; fr 1–2 dm, ending in a pair of long, arcuate horns .... 2. *Proboscidea.*

1. SESAMUM L.   Fr a columnar, unarmed, 4-locellar capsule, dehiscent only at the summit; cal 5-cleft, nearly regular; cor campanulate, the limb oblique, the lower lobes the larger; stamens 4, straight, with parallel pollen-sacs; herbs with solitary, axillary fls, the upper lvs generally alternate. 20, Africa and India.

1. **Sesamum indicum** L.   Sesame. Erect annual to 1.5 m, evidently to inconspicuously hairy; lvs lanceolate to ovate, entire or in part coarsely toothed; fls pink or yellowish to white, 2–3 cm, on short, ascending-erect pedicels; fr 2–3 cm, roundly 4-angled; seeds oily, aromatic; $2n=26$. Native probably of India, but unknown in the wild; long in cult., and occasionally found in our range as a casual escape.

2. PROBOSCIDEA Schmidel.   Unicorn-plant. Fr a 4-locellar drupaceous capsule, the exocarp fleshy, separating from the woody, reticulately sculptured, dehiscent endocarp, the thickened, seed-bearing portion crested on the upper (and sometimes also the lower) side, tapering into a long, stout, arcuately incurved beak that splits into 2 halves; cal irregular, cleft nearly to the base on the lower side; cor ventricose, weakly bilabiate; stamens 4; pollen-sacs widely divaricate, joined only at the summit; ovary initially unilocular, with 2 deeply intruded parietal placentas, each of these forked into 2 vertical lamellae, the ovules uniseriate along the edge of each lamella; herbs with a heavy odor; upper lvs often alternate; fls in terminal racemes that may appear lateral because of the continuing sympodial growth of the stem. 10, mostly N. Amer.

1. **Proboscidea louisiana** (Miller) Thell.   Devil's Claw. Densely glandular-pubescent annual, beginning to bloom when only 1–2 dm, but becoming coarse, freely branched, and sprawling-ascending, to ca 6 dm tall and 2 m wide; lvs long-petioled, subrotund to reniform-cordate, irregularly sinuate to entire, the later ones to 25 cm; racemes mostly 8–20-fld; cal 1 cm or more, somewhat accrescent, but deciduous; cor 3.5–5.5 cm long and wide, dull whitish or yellowish, mottled and spotted with purple, or golden-yellow and marked with vermilion; fr 1–2 dm. A common weed in much of s. and c. U.S., frequently in cattle feed-lots, native perhaps as far n. as Ind., cult. for its curious frs (used as pickles) and occasionally escaped as far n. as Minn. and Me. (*Martynia l.*)

FAMILY **BIGNONIACEAE**, the Trumpet-Creeper Family

Fls perfect, hypogynous, sympetalous, usually ± strongly irregular, the cor often bilabiate, cal with mostly 5 teeth or lobes, sometimes bilabiate, or the lobes sometimes suppressed; stamens on the cor-tube, alternate with the 5 lobes, commonly 4 and paired, the fifth (upper) one staminodial or wanting, or sometimes only 2 stamens polliniferous and the other 3 staminodial; ovary superior, bilocular with 2 axile placentas in each locule, or unilocular with 2 or 4 ± intruded parietal placentas; ovules ± numerous on each placenta; style terminal, with a 2-lobed stigma; fr usually capsular, often with a replum; seeds mostly flat, very often winged, mostly without endosperm (*Paulownia* excepted); embryo straight, the 2 cotyledons usually enlarged and ± foliaceous; mostly woody plants, very often vines, with opposite or whorled, exstipulate, simple or more often compound lvs and mostly large and showy fls in various sorts of infls. 100+/800.

1 Lvs simple, cordate to rotund; trees.
  2 Fr short, ovoid; stamens 4 .... 1. *Paulownia*.
  2 Fr elongate, cylindric; stamens mostly 2 .... 2. *Catalpa*.
1 Lvs compound; climbing or trailing vines.
  3 Lfls 5–13, sharply serrate; tendrils none .... 3. *Campsis*.
  3 Lfls 2, entire; lvs with a terminal tendril .... 4. *Bignonia*.

1. **PAULOWNIA** Siebert & Zucc.   Empress-tree. Cal nearly regular, coriaceous, the 5 broad lobes separate to the middle; cor evidently bilabiate, its lobes shorter than the tube, the upper reflexed, the lower spreading; stamens 4, included; pollen-sac divergent; style elongate, very shortly bilobed; fr woody, ovoid 2-locular, loculicidal; seeds flattened, winged, with rather scanty, oily endosperm; trees with opposite or whorled simple lvs and showy, blue or violet fls in terminal panicles. 17, e. Asia.

  1. **Paulownia tomentosa** (Thunb.) Steudel.   Tree to 15 m; lvs long-petioled, broadly cordate-ovate, entire or 3–5-angled, 1.5–2.5(–8) dm, hairy or even tomentose beneath with stellate or branched hairs; fls 5–7 cm; fr pointed, 3–4 cm; 2*n*=40. Native of China, escaped from cult. in s. U.S. and n. to se. Conn., Pa., O., Ind., and Mo. Apr., May.

2. **CATALPA** Scop.   Catalpa. Cal splitting irregularly or more commonly into 2 unequal lobes; cor large, evidently bilabiate, with an oblique, campanulate tube and 5 spreading lobes erose or crisped on the margin; stamens commonly 2, about as long as the cor-tube, accompanied by 3 minute staminodia, or occasionally 4 with 1 staminodium; fr elongate, cylindric, bilocular, loculicidal; seeds flat, narrowly oblong, the 2 wings terminating in hairs; trees with opposite or whorled, simple, cordate-ovate to rotund, entire or shallowly lobed lvs and terminal panicles of large, white or yellowish fls ± spotted or mottled with yellow and purple. 10+, e. N. Amer. and e. Asia.

1 Cor limb 5–6 cm wide, sparsely spotted with purple .... 1. *C. speciosa*.
1 Cor-limb 3–4 cm wide, densely spotted with purple .... 2. *C. bignonioides*.

  1. **Catalpa speciosa** Warder.   Northern c. Tree to 30 m, with well developed trunk; lvs broadly cordate-ovate or rotund, to 3 dm, distinctly acuminate, hairy beneath; infls to 2 dm; cor white, marked with 2 yellow stripes and faintly purple-spotted, the limb 5–6 cm wide; fr 2–5 dm × 1–1.5 cm, with a stout quadrangular placenta; seeds 2.5 cm, the wings rounded at the end and with a flat fringe of hairs; 2*n*=40. Alluvial forests; s. Ind. to Ark. and Tex., often cult. farther n. and e., and sometimes escaped. May, June.

  2. **Catalpa bignonioides** Walter.   Southern c. Tree to 15 m, with short trunk and widely spreading branches; lvs barely short-acuminate; infls to 3 dm; cor more conspicuously purple-spotted, the limb 3–4 cm wide; fr more slender, 6–10 mm thick, with a thin flat partition; seed-wings gradually narrowed to the end and with a narrow tuft of short hairs; otherwise as in no. 1; 2*n*=40. Native of s. U.S., occasionally escaped from cult. in our range.

    C. ovata G. Don, Chinese c., with smaller, mainly yellow fls, and with the lvs glabrous beneath and often lobed, rarely escapes in our range.

3. **CAMPSIS** Lour., nom. conserv.   Trumpet-creeper. Cal-lobes 5, triangular, acute or acuminate, shorter than the coriaceous obconic tube; cor-tube coriaceous, narrowly campanulate or almost tubular, the limb spreading, 5-lobed, only slightly bilabiate; stamens 4, paired, about equaling the cor-tube; fr elongate, narrowly clavate, tapering to both ends, scarcely flattened; seeds in several rows on each side of the broad, thin partition, flat, oblong, broadly winged on each side; woody vines, trailing or climbing by rootlets along the stem, with odd-pinnately compound lvs and large orange-red fls on very short pedicels. 2, the other in e. Asia.

  1. **Campsis radicans** (L.) Seemann.   Stems to 10 m or more; petioles 2–8 cm; lfls 5–13, lanceolate to ovate, 4–8 cm, acuminate, sharply and coarsely serrate, rounded at base; infls terminal, crowded; cal-tube 10–15 mm, the lobes 5 mm; cor 6–8 cm; fr 10–15 cm, with 2 longitudinal ridges; seeds 1.5 cm; 2*n*=40. Moist woods, fence-rows, and roadsides; N.J. to O. and Io., s. to Fla. and Tex., and often escaped from cult. farther n. July, Aug. (*Tecoma r.; Bignonia r.*)

4. BIGNONIA L.   Cross-vine. Cal-tube cupulate, the lobes blunt, short, irregular or nearly obsolete; cor-tube widely tubular above a narrow base, the limb only slightly bilabiate; stamens 4, paired, shorter than the cor-tube; fr elongate, linear, flattened parallel to the partition, septicidal; seeds flat, broadly winged at each side; woody vine with 2-foliolate lvs terminated by a branched tendril that clings by terminal adhesive disks; vascular bundles forming a cross in cross-section of the stem; fls yellow, long-pediceled in sessile axillary cymes. Monotypic.

1. **Bignonia capreolata** L.   Climbing to 10–15 m; petioles 1–2 cm; lfls 2, oblong or elliptic, subacuminate, entire, narrowed to a cordate base; cal 5–8 mm; cor 5 cm; fr 15–20 cm; seeds 3 cm; $2n=40$. Moist woods; s. Md. to s. O. and s. Mo., s. to Fla. and La. May, June. (*Anisostichus c.; A. crucigera.*)

FAMILY **LENTIBULARIACEAE**, the Bladderwort Family

Fls perfect, hypogynous, sympetalous; cal variously lobed or divided, often bilabiate; cor evidently bilabiate, the lower lip in ours prolonged into a basal spur; stamens 2, inserted on the cor-tube near its base, with ± confluent pollen-sacs; ovary superior, bicarpellate, unilocular; style wanting or very short; ovules (2–) numerous and somewhat sunken into the free-central (or basal) placenta; fr mostly capsular; seeds small, without endosperm, the embryo undifferentiated or nearly so; herbs, mostly insectivorous, aquatic or of wet places. 5/200.

1 Cal 5-lobed; fls solitary on bractless peduncles . . . . 1. *Pinguicula.*
1 Cal 2-lobed; fls solitary or more commonly racemose, each subtended by a bract . . . . 2. *Utricularia.*

1. PINGUICULA L.   Butterwort. Cal 5-lobed, somewhat 2-lipped; cor 2-lipped, 5-lobed, the lower lip lacking a palate, longer than the upper, and prolonged into a conspicuous basal spur; acaulescent perennials with a rosette of entire basal lvs and one or more, 1-fld bractless scapes. 35, mainly temp. and boreal N. Hemisphere, and Andes.

1. **Pinguicula vulgaris** L.   Violet b. Lvs commonly 3–6, ovate or elliptic, 2–5 cm, very obtuse, narrowed to the base, the upper surface viscid; scapes 1–3, 5–15 cm, glabrous or minutely glandular; cor violet, 1.5–2 cm (spur included); $2n=64$. Wet rocks and shores, and farther n. also in bogs and wet meadows; Lab. to Alas., s. to n. N. Engl., N.Y., Mich., Minn., and Oreg. June, July. Small insects, caught on the slimy lf-surface, are digested.

2. UTRICULARIA L.   Bladderwort. Cal parted to the base into upper and lower segments, the upper often wider; cor bilabiate, the upper lip usually subentire or shallowly 2-lobed, the lower entire or 3-lobed, usually elevated at the base into a prominent palate, the tube prolonged at base into a spur or sac; herbs of water or wet soil, the lvs (lf-like branches?) linear and entire or variously dissected, in most spp. provided with small bladders that catch tiny aquatic animals; fls yellow or violet, in short racemes (or apparently solitary and terminal), each subtended by a small bract. 150, cosmop.

1 Lvs minute and linear or none; bladders minute or none.
  2 Bracts accompanied each by a pair of bractlets; fls yellow.
    3 Bractlets oblong, acute; spur (7–)9–12(–14) mm; lower cor-lip 9–16 mm . . . . 1. *U. cornuta.*
    3 Bractlets subulate, very narrow; spur to 8 mm; lower lip to 12 mm . . . . 2. *U. juncea.*
  2 Bracts not accompanied by bractlets.
    4 Fls violet, solitary and terminal, subtended by a tabular bract . . . . 3. *U. resupinata.*
    4 Fls yellow, few or solitary, the bract entirely lateral and peltate . . . . 4. *U. subulata.*
1 Lvs dichotomously branched or dissected, usually conspicuous on submersed stems; bladders usually
    well developed.
  5 Lvs (at least the upper ones) in whorls of 4–9.
    6 Fls rose-purple; all parts of the lvs slender . . . . 5. *U. purpurea.*
    6 Fls yellow; uppermost lvs with the rachis inflated into a whorl of floats.
      7 Peduncles 1–5(–7)-fld, 3–5 cm; floats 1–4 cm . . . . 6. *U. radiata.*
      7 Peduncles usually 6–14(–17)-fld, 10–25 cm; floats 3–8 cm . . . . 7. *U. inflata.*
  5 Lvs all alternate, none of them inflated; fls yellow.

8 Lower cor-lip about twice as long as the upper; ultimate lf-segments flat, nearly or fully as wide as the primary ones.
  9 Bladders on specialized branches distinct from the dissected lvs; lower cor-lip 8–12 mm, with well developed palate .... 8. *U. intermedia.*
  9 Bladders on ordinary lvs; lower lip 4–8 mm, with small or no palate .... 9. *U. minor.*
8 Lower cor-lip little if any longer than the upper; lf-segments progressively narrowed in successive dichotomies, the ultimate ones filiform.
  10 Leafy stems loosely floating beneath the surface of the water; fls mostly 6–20 per peduncle.
    11 Peduncles, below the lowest bract, with 1–few widely scattered bract-like scales; lower lip 1–2 cm .... 10. *U. vulgaris.*
    11 Peduncles without scales below the bracts; lower lip mostly 6–8 mm .... 11. *U. geminiscapa.*
  10 Leafy stems forming tangled mats below the surface of the water; fls 1–6 per peduncle.
    12 Lower cor-lip 5–6 mm, well exceeding the thick, very blunt spur .... 12. *U. gibba.*
    12 Lower lip 8–10 mm, about equaling or slightly shorter than the conic spur.
      13 Lvs dimorphic, some twice dichotomous and with bladders, others thrice dichotomous and without bladders .... 13. *U. fibrosa.*
      13 Lvs all alike, bladder-bearing, usually twice forked .... 14. *U. biflora.*

1. **Utricularia cornuta** Michx. Naked b. Lvs small, mostly subterranean, seldom seen; nodes mostly 6–14; roots finely branched, with minute bladders; flowering branches green to yellow-green, erect, straight, 10–25 cm; fls 1–6, all chasmogamous; bracts ovate, 1–2 mm, facing the axis; bractlets 2, edgewise to the axis, facing each other, oblong, acute; pedicels 1–2 mm; fls yellow, the spur 7–14 mm, directed downward, the lower lip 9–16 (avg. 13) mm, with greatly elevated palate surrounded by a spreading margin; $2n=18$. Wet shores; Nf. and Que. to n. Mich. and Minn., s. to Fla. and Tex.; W.I. July, Aug. (*Stomoisia c.*)

2. **Utricularia juncea** M. Vahl. Rushlike b. Much like no. 1; flowering branches greenish-purple to purple; nodes mostly 4–32; bractlets slenderly subulate; fls smaller, some of them cleistogamous; palate without spreading margin, or with only a very narrow one. Wet shores of bogs and ponds on or near the coastal plain; s. N.Y. to Fla. and Miss. Plants blooming in summer have the spur 6–8 mm and the lower lip 8–12 (avg. 9) mm. (*Stomoisia j.*) Plants blooming in fall are smaller, with much smaller fls, the spur only 2–3 mm. (*U. virgulata; Stomoisia v.*)

3. **Utricularia resupinata** B. D. Greene. Resupinate b. Stems delicate, horizontal just below the surface of the soil or on the surface in shallow water, emitting linear lvs to 3 cm and from the base of each 1 or 2 delicate roots with or without bladders; flowering branches erect, 2–10 cm, uniflorous; bract tubular, surrounding the stem, its margin notched; fls violet, 1 cm, tipped backward on the pedicel and facing upward, the spur almost horizontal; lower lip 3-lobed, the palate scarcely developed; $2n=44$. Muddy soil or very shallow water at the edge of ponds; N.S. and n. N.Y. to n. Mich., n. Minn., and sw. Ont., s. to L.I., Del. and Ind.; Ga. and Fla. July, Aug. (*Lecticula r.*)

4. **Utricularia subulata** L. Slender b. Underground parts very delicate, rarely collected, with bladders 0.5 mm; aerial lvs, if present, linear, to 1 cm; flowering branches very slender, erect, 3–20 cm, 1–10-fld; bracts ovate or elliptic, 1–2 mm, peltate, attached at or below their middle; pedicels 4–15 mm; cor commonly yellow, the lower lip usually 4–7 mm, with prominent palate and widely spreading limb; spur about equaling the lip and appressed to its lower surface. Wet soil and very shallow water, often among mosses; N.S. to Fla. along the coastal plain, thence w. to Ark. and Tex. May–Sept. (*Setiscapella s.*) A depauperate form 2–5 cm with 1–3 whitish fls 1–2 mm sometimes occurs with the normal form. (*U. cleistogama; Setiscapella c.*)

5. **Utricularia purpurea** Walter. Spotted b. Stems submersed, to 1 m; lvs in whorls of 5–7, verticillately branched into filiform segments, many of which bear a terminal bladder; flowering branches 3–15 cm, 1–4-fld; cor 1 cm, violet or red-violet, the lower lip with a basal yellow spot, 3-lobed, the lateral lobes strongly and separately elevated at base into a palate; spur shorter than and appressed to the lower lip. Quiet water; Que. and N.S. to Minn. and n. Ind., s. along the coastal plain to Fla. and La.; W.I. July–Sept. (*Vesiculina p.*)

6. **Utricularia radiata** Small. Floating b. Stems elongate, submersed; lower lvs alternate, 3 cm, repeatedly dichotomous into filiform segments, each bearing bladders; uppermost lvs subtending the peduncle in whorls of 4–7, the rachis of each 1–4 cm, inflated, usually thickest at or below the middle, bearing at the tip a series of finely dissected branches; peduncles 3–5 cm; bracts 2–3 mm, at least as wide as long, sometimes lobed; lower pedicels 1–2 cm; fls (1–)3–5(–7), yellow; sep 3–4 mm; upper cor-lip subrotund; lower lip subrotund, 8–10 mm, shallowly 3-lobed, with a prominent small palate; spur shorter than and appressed to the lower lip; $2n=28$. Ponds along or near the coastal plain; Me. to Fla. and Tex.; nw. Ind. May–Aug. (*U. inflata* var. *minor.*)

7. **Utricularia inflata** Walter. Inflated b. Much like no. 6, but much larger; floats 3–8 (commonly 5) cm, usually thickest well beyond the middle, with finely dissected branches in the distal half; peduncles 10–25 cm; bracts 3–4 mm, longer than wide, never lobed; lower pedicels often 3 cm; fls mostly 6–14(–17); sep 4–7 mm; lower lip 10–15 mm, not lobed; $2n=18, 36$. Ponds along the coastal plain; Del. to Fla. and Tex. May–July.

8. **Utricularia intermedia** Hayne. Northern b. Stems free-floating or creeping on the soil under shallow water; lvs numerous, mostly 0.5–2 cm, commonly 3-parted at base and then 1–3 times dichotomous, the segments slender, flat, not much narrower in successive dichotomies, the ultimate ones rather blunt; bladders on specialized branches distinct from the lvs; fls mostly 2–4 in a lax raceme on an emergent peduncle 6–20

cm; pedicels to 15 mm, remaining erect; cor yellow, the lower lip 8–12 mm, with well developed palate, nearly twice as long as the upper lip; spur nearly as long as the lower lip; $2n=44$. Shallow water; circumboreal, s. to Del., Ind., and Calif. Summer.

9. Utricularia minor L.   Lesser b. Stems mostly creeping on the soil and forming mats under shallow water; lvs numerous, mostly 0.3–1 cm, commonly 3-parted at base and then dichotomous or irregularly 1–3 times divided, the segment slender, flat, the ultimate ones strongly acuminate; fls mostly 2–9(–15) in a lax raceme on an emergent peduncle 4–15 cm; pedicels soon arcuate-recurved; cor yellow, the lower lip 4–8 mm, twice as long as the upper, the palate scarcely developed; the spur small, up to half as long as the lower lip; $2n=44$. Shallow water; circumboreal, s. to N.J., Ind., and Calif.

10. Utricularia vulgaris L.   Common b. Submersed, free-floating; lvs numerous, 1–5 cm, mostly 2-parted at base and then repeatedly and unequally (quasipinnately) dichotomous, the segments terete, progressively narrower, the ultimate ones filiform and strongly acuminate; bladders numerous, on the lvs; peduncles emergent, 6–20 cm, with 1–several scattered bracts below the infl; fls mostly 6–20 in a lax raceme, yellow, the lower lip mostly about equaling the upper, slightly lobed, 1–2 cm, with well developed palate; spur falcate, directed forward, two-thirds as long as the lower lip; pedicels arcuate-recurved in fr; $2n=44$. Quiet water; circumboreal, s. to Fla. and Calif. June–Aug. (*U. macrorhiza.*)

11. Utricularia geminiscapa Benj.   Mixed b. Resembling no. 10, but much smaller; lvs 1–2 cm, usually with little trace of a central rachis; flowering branches 5–15 cm, without scales below the bracts; bracts 2–3 mm; pedicels to 8 mm, arched-recurving in fr; petaliferous fls with the lower lip 6–8 mm, somewhat surpassing the upper; spur not or scarcely falcate; cleistogamous fls without cor, solitary on pedicels 5–15 mm, often along the leafy submersed stems; peduncle naked. Quiet water; Nf. and Que. to n. Mich. and n. Wis., s. to Pa. and e. Va. July, Aug. (*U. clandestina.*)

12. Utricularia gibba L.   Creeping b. Creeping stems delicate, seldom over 10 cm, forming mats near the bottom in shallow water; lvs scattered, rarely over 5 mm, commonly once (twice) forked, with a few bladders; flowering branches 5–10 cm, with 1–3 yellow fls 5–6 mm; palate prominent; spur very thick and obtuse, only ca half as long as the lower lip. Shallow water; Que. to Wis., s. to Fla. and La.; also Pacific states, W.I., and c. Amer.

13. Utricularia fibrosa Walter.   Stems creeping and forming tangled mats near the bottom in shallow water, radiating from the base of the scape, dimorphic, some with rather crowded, thrice-forked, bladderless lvs to 15 mm, the others with smaller, twice-forked lvs bearing numerous bladders; flowering branches 1.5–4 dm, with 2–6 long-pediceled fls; cor yellow, the lower lip 8–10 mm, with prominent palate; spur equaling or slightly exceeding the lip. Shallow ponds on the coastal plain; e. Mass. to Fla. and Tex., n. in the interior to Ark. and Okla. June–Aug. (*U. pumila.*)

14. Utricularia biflora Lam.   Much like no. 13; vegetative stems all alike; lvs very delicate, rarely over 5 mm, usually twice-forked; flowering branches 5–12 cm; pedicels to 15 mm; fls 1–4. Shallow water on the coastal plain; e. Mass. to Fla. and Tex., n. to Okla. July, Aug.

Family CAMPANULACEAE, the Bellflower Family

Fls sympetalous, epigynous, generally perfect; cor regular to highly irregular, with valvate lobes; stamens attached to the annular, epigynous nectary-disk or to the base of the cor, as many as and alternate with the cor-lobes; pollen presentation mechanism complex, the anthers connivent or connate to form a tube into which the pollen is shed, the style provided with a fringe of collecting hairs below the initially appressed stigmatic branches, growing up through the anther-tube and pushing out the pollen; ovary inferior, of 2–3(–5) carpels, typically plurilocular with axile placentation, seldom unilocular with parietal placentation; fr commonly capsular, often poricidal or opening by longitudinal slits that may be more numerous than the carpels; seeds numerous, small, with a straight, dicotyledonous embryo embedded in the oily endosperm; herbs (all ours) or less often ± woody plants, with simple, mostly alternate lvs and variously cymose to racemose infls. 70/2000. (Lobeliaceae.)

There are 2 well marked subfamilies, connected by some transitional genera in the Old World. The Campanuloideae (including our first 3 genera) have a regular, often campanulate cor and usually merely connivent anthers; the carpels usually number 3, sometimes 4 or 5. The Lobelioideae (including only *Lobelia,* among our genera) have a highly irregular cor, often with a fenestrate tube, and connate anthers; the carpels number 2; the fls are morphologically upside down to those of the Campanuloideae, but resupinate by twisting of the pedicel so that one sepal is adaxial (as in the nonresupinate Campanuloideae) and the 3-lobed morphological upper lip of the cor is visually the lower lip, and the visual upper lip is 2-lobed or 2-cleft, sometimes so deeply so as to lose its identity, the cor then appearing unilabiate.

1 Cor regular; carpels 3–5.
  2 Anthers distinct; cor campanulate to rotate; capsule opening by lateral pores.
    3 Cor campanulate to funnelform, or in one sp. rotate, the fls then in long terminal spikes; biennial
      or perennial . . . . 1. *Campanula.*
    3 Cor widely spreading; fls sessile or subsessile in the lf-axils or in loose peduncled cymes; annual
     . . . . .2. *Triodanis.*
  2 Anthers connate at base; cor slenderly tubular; capsule opening by terminal valves . . . . 3. *Jasione.*
1 Cor highly irregular, bilabiate; carpels 2 . . . . 4. *Lobelia.*

1. CAMPANULA L. Bellflower; harebell. Sep 5; cor regular, campanulate or funnelform to rotate, valvate in bud, 5-lobed, in our sp. blue to violet or white; stamens attached to the very base of the cor; filaments widened at base; anthers distinct; ovary 3- or 5-locular with numerous ovules and elongate style; fr short, usually strongly ribbed, opening by 3 or 5 lateral pores. 300, chiefly N. Temp. and Arctic.

1 Cor rotate; fls in elongate terminal spikes . . . . 1. *C. americana.*
1 Cor campanulate or funnelform; infl various.
  2 Stems erect or ascending, terete or obscurely angled.
    3 Fls solitary, or in loose open clusters on slender pedicels.
      4 Hypanthium bristly with long pale hairs . . . . 2. *C. trachelium.*
      4 Hypanthium glabrous or minutely puberulent.
        5 Cauline lvs linear to narrowly lanceolate, entire or very minutely toothed; style not exsert
         . . . . 3. *C. rotundifolia.*
        5 Cauline lvs lanceolate to lance-ovate, coarsely toothed; style exsert . . . . 4. *C. divaricata.*
    3 Fls otherwise.
      6 Fls short-pediceled in an erect, slender, secund raceme . . . . 5. *C. rapunculoides.*
      6 Fls sessile in an involucrate terminal glomerule . . . . 6. *C. glomerata.*
  2 Stems weak, very slender, 3-angled, reclining on other plants . . . . 7. *C. aparinoides.*

1. Campanula americana L. Tall b. Coarse, erect, often freely branched winter-annual or biennial, mostly 0.5–2 m; lvs thin, lanceolate to ovate-oblong, 7–15 cm including the margined petiole, acuminate, serrate; spike(s) 1–4+ dm, the lower bracts lf-like, the upper reduced; fls blue, solitary or in small clusters, the rotate cor 2.5 cm wide; sep linear, spreading, 5–10 mm; pollen pantoporate (unique among our spp.); style downcurved, with upcurved tip; fr 7–12 mm, opening by round pores near the top; $2n=58$. Moist borders and open woods; Ont. to Minn., s. to Fla. and Okla. July–Sept. (*Campanulastrum a.*)

2. Campanula trachelium L. Throatwort. Erect perennial 3–10 dm, the angular stem often bristly above; lvs short-bristly, coarsely and irregularly serrate, the lower on long petioles, cordate or triangular, the upper narrower but scarcely shorter, lanceolate, narrowed to a very short petiole; fls in loose terminal or axillary clusters; hypanthium and lanceolate sep (10–13 mm) bristly with pale hairs; cor blue or violet, campanulate, 2.5–4 cm; $2n=34$. Native of Eurasia and n. Afr., intr. in waste places and along roadsides from Que. to Mass. and O. Aug.

3. Campanula rotundifolia L. Harebell. Perennial, 1–8 dm, glabrous or inconspicuously hispidulous; basal lvs petiolate, often with broadly ovate to subrotund or cordate-rotund, angular-toothed blade to 2 cm, sometimes merely oblanceolate, often deciduous; cauline lvs ± numerous, commonly linear or nearly so, 1.5–8 cm, seldom 1 cm wide; fls typically several or rather many in a lax, racemiform or elongate-paniculiform infl, solitary in depauperate or subalpine specimens; cal-lobes 4–12 mm; cor blue, 1.5–3 cm, campanulate, the lobes much shorter than the tube; style not exceeding the cor; fr nodding, opening near the base; $2n=34$, 56, 68, 102, etc. Dry woods, meadows, cliffs, and beaches; circumboreal, s. in Amer. to N.J., Ind., Io. and Mex. June–Sept.

C. patula L., a biennial European sp. with minutely roughened stem and with the fr opening near the top, is escaped (but perhaps not established) in Conn.

4. Campanula divaricata Michx. Appalachian b. Perennial; stems 3–7 dm; lvs thin, lance-linear to lance-ovate, 2–8 cm, sharply and coarsely toothed, narrowed at base to a short petiole or nearly sessile; infl widely branched, many-fld; bracts linear or subulate; fls drooping on slender pedicels; sep lance-triangular, 1.5–3 mm; cor pale blue, 6–8 mm; style exsert 3–5 mm; fr opening near the base; $2n=34$. Rocky woods in the mts., chiefly at low elev.; w. Md. and e. Ky. to Ga. and Ala. July–Sept. (*C. flexuosa.*)

5. Campanula rapunculoides L. Rover-b. Perennial 4–10 dm from a rhizome with many fleshy vertical roots, usually unbranched; stem smooth or sparsely hairy; lvs coarse, irregularly serrate, usually sparsely hairy beneath, the lower long-petioled, ovate, rounded to subcordate at base, the upper progressively narrowed to lanceolate and often subsessile; infl strict, unbranched, secund, with much reduced bracts; sep and hypanthium glabrous to finely hairy; cor blue, somewhat nodding, 2–3 cm; $2n=102$. A persistent weed in lawns, roadsides and waste ground; native of Eurasia, intr. from N.S. to Minn., s. to Del., Ky., and Mo. July, Aug.

6. Campanula glomerata L. Clustered b. Perennial 3–7 dm; lvs denticulate, hirtellous, the lower petioled, lance-oblong, the upper ovate to narrowly triangular, sessile and ± clasping; fls sessile in terminal glomerules subtended by leafy bracts, or a few sessile in the axils of the uppermost lvs; sep narrowly lanceolate, 6–10 mm; cor blue-violet, 2–3 cm; $2n=34$, 68. Native of Eurasia, escaped from cult. into roadsides and waste places in N. Engl. and adj. Can., w. to Minn. June, July.

7. Campanula aparinoides Pursh.   Marsh-b. Perennial from filiform rhizomes, the stems weak, slender, usually reclining on other plants, somewhat 3-angled, scabrellate on the angles to nearly smooth; lvs linear or narrowly lanceolate, the lower to 9 cm × 8 mm, the upper shorter and proportionately narrower, often roughened on the margins and midvein beneath; fls solitary on long slender pedicels; sep triangular to lanceolate, 1.5–5 mm; cor funnelform, pale blue to white, 4–13 mm; fr opening at base; $2n=34$, 68, 170. Wet sunny meadows; N.S. and Que. to Sask., s. to Ga., Ky., Mo., and Neb. June–Aug. Two vars.: Var. *aparinoides* with lanceolate lvs (the longer ones 1–5 cm, avg 6 times as long as wide), the cor 4–10 mm, the divergent pedicels often upcurved at the tip and with naked summit 0.4–4 cm, is the slender s. phase, extending n. to the Canadian boundary. Var. *grandiflora* Holzinger (*C. uliginosa*), with linear lvs (the longer ones 2–9 cm, avg 12 times as long as wide), the cor 5–13 mm, the ascending pedicels with naked summit 1–8 cm, is the more robust n. phase, extending s. to Pa. and Io.

2. TRIODANIS Raf.   Fls sessile or subsessile in the middle and upper axils, 1–several per axil, forming a dense or interrupted false spike, the lower fls cleistogamous, with reduced or abortive cor, the upper fls with normal, regular, wide open cor, the 5 lobes longer than the tube; cal-lobes 5, or only 3 or 4 in cleistogamous fls; stamens free, the short filaments expanded and ciliate at base; ovary mostly trilocular with axile placentas, varying to unilocular and with parietal placentas; stigma (2)3-lobed; capsules linear to ellipsoid or clavate, opening by pores at or usually above the middle; seeds numerous; annuals with toothed lvs. 8, mostly N. Amer.

Specularia speculum-veneris (L.) Tanfani, the Venus' looking-glass, with the fls all normally developed, borne in dichasia at the ends of the stem and branches, the filaments glabrous, is occasionally intr. (from s. and w. Europe) in our range, but apparently does not persist.

1 Middle and upper lvs (floral bracts) seldom over 2(–2.5) times as long as wide.
  2 Distal end of the pores mostly 1–1.5 mm below the top of the fr; bracts about as broad as long, or
    broader; open cors usually several; widespread .... 1. *T. perfoliata.*
  2 Distal end of the pores rarely 0.5 mm below the top of the fr; bracts usually longer than broad;
    open cor usually 1 per main branch; southern .... 2. *T. biflora.*
1 Middle and upper lvs narrower, mostly 5–10 times as long as wide .... 3. *T. leptocarpa.*

1. Triodanis perfoliata (L.) Nieuwl.   Round-lvd triodanis. Erect, 1–10 dm, simple or with few long branches, often floriferous to near the base, ± scabrous or hairy at least below; lvs (and floral bracts) sessile and cordate-clasping, rotund-ovate or broader, usually toothed, palmately veined, 0.5–3 cm, nearly as wide, or wider, or some of the lower ones narrower; cal cleft to the hypanthium, the lobes narrowly triangular-acuminate, 5–8 mm in open fls; upper fls with deep purple to pale lavender cor 8–13 mm, the tube 2–4 mm; frs oblong or narrowly obovoid, those of cleistogamous fls 4–7 mm, those of open fls to 1 cm, all bilocular or trilocular; pores usually 1–1.5 mm below the tip of the fr; seeds 0.5–0.6 mm; $2n=60$. Various habitats, often in disturbed sites; Me. to B.C., s. to trop. Amer. May, June. (*Specularia p.*)

2. Triodanis biflora (Ruiz & Pavon) Greene.   Much like no. 1 and hybridizing freely with it; lvs and floral bracts a little narrower and less veiny, the middle bracts 1.3–1.8(–2.5) times as long as wide; each stem or main branch usually producing only 1 open fl (the terminal one), the other fls cleistogamous; pores reaching the top of the fr, or rarely as much as 0.5 mm below it. Dry soil: s. Va. to Ky. and Mo., s. to Mex., and in S. Amer. May, June. (*Specularia b.*)

3. Triodanis leptocarpa (Nutt.) Nieuwl.   Narrow-lvd t. Simple or basally branched, 1–8 dm, often floriferous to near the base, scabrous or shortly spreading-hairy at least below; lvs sessile, lanceolate or oblanceolate to lance-elliptic, or the upper (floral) ones linear, inconspicuously crenate or subentire, mostly 1.5–3.5 cm × 2–7 mm and 5–10 times as long as wide, the lowest ones short-petiolate and a little broader; cal divided to the hypanthium, the narrow lobes 6–15 mm, or smaller in the cleistogamous fls; upper fls with blue-violet cor 7–10 mm, the tube scarcely 2 mm; fr linear, those of the cleistogamous fls 8–15 mm and unilocular, those of the open fls 15–25 mm and often bilocular; seeds 0.7–1.0 mm. Dry open places; s. Minn. to Mont., s. to Ark. and Tex.; adventive in Ind. May, June. (*Specularia l.*)

3. JASIONE L.   Sep 5; cor tubular, soon becoming deeply cleft; anthers united at base into a ring around the style; ovary 2-locular; style clavate, undivided; fr obovoid, opening by terminal valves; fls in dense head-like umbels subtended by a leafy involucre. 20, Europe and Medit. Reg.

1. Jasione montana L.   Sheep's bit. Stems several from a long stout root, 2–4 dm, branched above; lvs sessile, nearly linear, 1–3 cm, hairy; umbels long-peduncled, 1–2 cm wide; bracts ovate, toothed; pedicels short at anthesis, elongating to 5 mm in fr; sep linear, eventually 2 mm; cor blue, soon split into linear lobes, surpassed by the style; $2n=12$. Native of Europe, intr. in waste places and fields, usually in sandy soil; Mass. to N.J. June–Sept.

4. LOBELIA L. Lobelia.

Cor irregular, inverted so that the 3-lobed morphological upper lip is on the lower (abaxial) side; cor-tube split (between the 2 lobes of the visually upper lip) to below the middle, commonly nearly to the base, often fenestrate on the sides near the base as well; filaments and anthers connate, the lower 2 anthers shorter than the others and bearded at the tip; ovary 2-locular, with axile placentation; style growing up through the anther-tube and pushing out the pollen; stigma 2-lobed; capsule opening at or near the top; ours fibrous-rooted, mostly single-stemmed herbs with the fls in terminal, bracteate racemes, rarely in the uppermost axils. Most of our spp. perennate by short offsets, a few by rhizomes; *L. inflata* appears to be annual. 300, widespread.

1 Cor fenestrate (i.e., with a slit or "window" on each side near the base); fls 15–45 mm, including the
   hypanthium.
   2 Fls ordinarily scarlet, mostly 30–45 mm .... 13. *L. cardinalis.*
   2 Fls blue to white, mostly 15–33 mm.
      3 Cal with prominent foliaceous auricles; pedicels bracteolate at or above the middle .... 12. *L. siphilitica.*
      3 Auricles small and inconspicuous, or none; pedicels bracteolate near the base (or sometimes at
         the middle in no. 11).
         4 Herbage evidently short-hairy throughout, or at least in the infl .... 11. *L. puberula.*
         4 Herbage glabrous or nearly so, or the plant sparsely hairy at the base.
            5 Lower lip of the cor glabrous within.
               6 Sep entire or nearly so .... 10. *L. elongata.*
               6 Sep evidently callous-denticulate .... 9. *L. georgiana.*
            5 Lower lip of the cor bearded at the base within .... 8. *L. glandulosa.*
1 Cor not fenestrate; fls smaller, mostly 7–18 mm, including the hypanthium.
   7 Lvs hollow, subterete, submersed, all in a basal rosette .... 7. *L. dortmanna.*
   7 Lvs flat or thread-like, rarely submersed, all or mostly cauline (except in one sp.).
      8 Lvs linear, filiform, or narrowly lanceolate.
         9 Lower lip of the cor glabrous within.
            10 Pedicels bracteolate at or above the middle; northern .... 6. *L. kalmii.*
            10 Pedicels bracteolate at the base; southern .... 5. *L. nuttallii.*
         9 Lower lip of the cor bearded toward the base within.
            11 Pedicels glabrous, ebracteolate; rhizomatous .... 4. *L. boykinii.*
            11 Pedicels scabrous, bracteolate near the base; not rhizomatous .... 3. *L. canbyi.*
      8 Lvs broader, lance-oblong to obovate, some or all of them more than 1 cm wide; pedicels bracteate
         at or near the base; lower lip of the cor bearded within.
         12 Hypanthium obovoid, nearly as long as the cor, much inflated in fr .... 2. *L. inflata.*
         12 Hypanthium obconic, shorter than the cor, not much inflated in fr .... 1. *L. spicata.*

1. Lobelia spicata Lam. Spiked l. Perennial 3–10 dm, simple, often hairy, especially below; lvs ascending, obovate to oblanceolate or lanceolate, 5–10 cm, hairy, the upper gradually reduced to bracts; raceme slender, crowded; bracts lanceolate to linear; pedicels 2–5 mm, bibracteolate at base; hypanthium short, obconic, this and the narrowly lanceolate sep smooth or hirtellous, the sep often with conspicuous reflexed basal auricles; cor blue to white, 7–11 mm, the lower lip bearded at base within; $2n=14$. Various habitats, often weedy; Que. and N.S. to Minn., s. to Ga. and Ark. Four vars.:

a Lvs chiefly basal, the cauline 1–5, much smaller than the ovate to obovate or suborbicular basal ones;
      s. Pa. to S.C. and La., chiefly on the piedmont .... var. *scaposa* McVaugh.
a Lvs chiefly cauline, the basal ones, if present, no larger than the cauline.
   b Plants rough-hairy, including the stem, bracts, and cal-lobes; Great Plains, e. irregularly to nw.
         Ind., Mich., n. Ont., and the Gaspé region .... var. *hirtella* A. Gray.
   b Plants glabrous or sparingly hairy, the cal-lobes glabrous or occasionally ciliate.
      c Cal-lobes with long, slender, deflexed auricles, often as long as the hypanthium; southern, n. to
            W.Va. and Kans. .... var. *leptostachys* (A. DC.) Mackenzie & Bush.
      c Cal-lobes with very short or no auricles; Me. and N.B. to N.D., s. to Pa. and Ark., and in the
            mts. to Ga.; the common phase in most of our range. (Var. *campanulata*.) .... var. *spicata.*

2. Lobelia inflata L. Indian tobacco. Annual to 1 m, usually branched, loosely hairy throughout, or the upper part of the stem and upper surfaces of the lvs subglabrous or merely strigose; lvs sessile or subsessile, ovate-oblong to oblong-obovate, 5–8 × 1.5–3.5 cm, ± serrate; racemes terminating the branches, 1–2 dm; lower bracts foliaceous, the upper gradually reduced; pedicels 3–9 mm, bibracteolate at base; sep linear, 3–5 mm; auricles wanting; cor blue or white, 6–8 mm, the lower lip bearded at base within; fruiting hypanthium inflated, 7–11 × 3.5–8 mm; $2n=14$. Open woods, or sometimes a garden-weed; P.E.I. to Minn., s. to Ga. and Miss. July–Oct. Poisonous.

3. Lobelia canbyi A. Gray.   Perennial by short offsets; stem 4–10 dm, erect, simple or branched above, often somewhat hairy below; lvs linear, to 5 cm, 2–5 mm wide; racemes slender, few-fld, 1–3 dm; pedicels ascending or erect, scabrous, 3–10 mm, about equaling the narrowly linear bracts, bibracteolate at base; sep narrowly lance-linear, erect, 3–4 mm; auricles wanting; cor blue, 1 cm, the lower lip bearded; hypanthium ± elliptic or oblong in fr, 4–7 × 2.5–4 mm; $2n=14$. Swamps on the coastal plain; N.J. to Ga.; also in Tenn. and w. N.C. Aug., Sept.

4. Lobelia boykinii T. & G.   Glabrous perennial 4–8 dm from a rather short rhizome, simple or sparingly branched, often immersed at base; lvs filiform, 5–20 mm, often deciduous before anthesis; raceme open, 1–2 dm; pedicels slender, spreading, 1–2 cm, exceeding the filiform bracts; bracteoles none; sep loose or spreading, filiform, 4–5 mm; cor 1 cm, blue with a white center, the lower lip bearded at base; fruiting hypanthium subglobose, 3 mm; $2n=14$. Ponds and swamps on the coastal plain; N.J. to Fla. Rare. June.

5. Lobelia nuttallii Roemer & Schultes.   Perennial by short offsets, 2–6 dm, simple or with a few erect branches, ± pubescent below, glabrous or nearly so above; lower lvs narrowly oblanceolate, to 4 cm, the upper smaller, linear, often closely ascending; racemes slender, loose; bracts linear or subulate, 2–5 mm, often with 1–4 callous teeth; pedicels 3–8 mm, bibracteolate at base; sep lance-linear, 2–3 mm; auricles none; cor 1 cm, pale blue with a white center and 2 greenish spots; lower lip glabrous, but the tube sometimes hairy within; fruiting hypanthium subglobose, 3 mm; $2n=14$, 28. Sandy swamps on the coastal plain; L.I. to Fla. and Tex., inland in the Appalachian region from s. Ky. southward. July–Sept.

6. Lobelia kalmii L.   Glabrous or subglabrous perennial 1–4 dm; basal lvs spatulate, 1–3 cm, sometimes immersed, often deciduous; cauline lvs mostly linear, 1–5 cm × 1–5 mm; racemes loose, sometimes secund; bracts linear or filiform, the lower to 15 mm; pedicels bibracteolate at or above the middle; sep 2–3 mm; cor 7–13 mm, blue to partly or wholly white, the lower lip glabrous; $2n=14$. Calcareous shores and swamps; Nf. to B.C., s. to Pa., O., Ill., and Minn. July, Aug. (*L. strictiflora.*)

7. Lobelia dortmanna L.   Water-l. Glabrous aquatic perennial to 1 m, the hollow upright stem usually partly submersed; lvs in a basal rosette, linear, fleshy, hollow, 2–9 cm, outcurved near the tip; cauline lvs minute, filiform; raceme emersed, few-fld; bracts ovate, 2–3 mm; bracteoles none; sep 2 mm; cor pale blue or white, 1–2 cm, the lower lip bearded at base within; $2n=14$. Borders of ponds, usually in sand; Nf. to N.J., w. to Minn.; also on the Pacific coast and in Europe. July–Sept.

8. Lobelia glandulosa Walter.   Glabrous; stems weak, simple, 3–14 dm; lvs all cauline, decurrent, narrowly linear to broadly lanceolate, callous-denticulate to subentire, 3–15 cm × 2–14 mm; raceme open, secund, to 1.5 dm; pedicels stout, straight, erect, rough-hairy, bibracteolate at base, 5–13 mm in fr; sep 3–15 mm, strongly callous-dentate to less often entire; auricles small or none; fls 2–3.3 cm, blue with a white eye, fenestrate, the lower lip bearded at base within; $2n=28$. Wet places, mainly on the coastal plain; se. Va. to Fla. July–Oct.

9. Lobelia georgiana McVaugh.   Glabrous, 3–10 dm, simple; lower lvs elliptic-oblong to rarely ovate, to 10 × 3 cm, glaucous-green beneath, denticulate or repand, narrowed to a petiole, the upper smaller, sessile or nearly so; racemes elongate, the lowest fl well above the uppermost lf; bracts lance-linear, 5–15 mm, glandular-denticulate; pedicels 2–4 mm, bibracteolate at base; fls 2–2.5 cm, blue, fenestrate, the lower lip glabrous; sep linear, 5–8 mm, glandular-denticulate; auricles minute or none; $2n=14$. Wet soil; s. Va. to w. N.C., s. to Fla. and Miss. July–Nov. (*L. amoena* var. *glandulifera.*)

10. Lobelia elongata Small.   Glabrous, 5–12 dm, mostly simple; lvs linear to lance-linear, to 16 × 2 cm, sharply toothed or the narrower ones subentire, the upper gradually reduced; raceme secund; bracts lance-linear, 1–2 cm, callous-toothed; pedicels 3–6 mm, minutely scabrellate, bibracteolate at base; fls 2–2.5 cm, blue, fenestrate, the lower lip glabrous within; sep linear, 10–14 mm; auricles minute or none; $2n=28$. Wet ground near the coast; s. Del. to Ga. Aug.–Oct.

11. Lobelia puberula Michx.   Perennial, 4–12 dm, usually simple, short-hairy ± throughout, or rarely largely glabrate; lvs oblong to oblong-ovate or lanceolate, 5–10 cm, dentate, the lower obtuse or rounded, the upper smaller and often acute; raceme 1–3 dm, usually secund, crowded; bracts lanceolate to ovate, 1–2 cm; pedicels 2–5 mm, bibracteolate near the base or sometimes at the middle; fls 1.5–2 cm, blue, fenestrate, the lower lip glabrous within; sep lance-linear, 6–10 mm; auricles small; $2n=14$. Wet soil; s. N.J. to s. O. and s. Ill., s. to Fla. and Tex. Aug.–Oct.

12. Lobelia siphilitica L.   Stout, erect, 5–15 dm; lvs thin, narrowly oblong or elliptic to lanceolate or oblanceolate, mostly 8–12 cm, narrowed to a sessile base; racemes crowded, 1–3 dm, the lower bracts lanceolate, foliaceous, to 5 cm, the upper reduced; pedicels ascending, 4–10 mm, bibracteolate at or above the middle; fls 2–3 cm, blue (white), fenestrate, glabrous; sep lanceolate to ovate, 8–12 mm, with foliaceous auricles; $2n=14$. Swamps and wet ground; Me. to Man. and Colo., s. to N.C. and Tex. Aug., Sept. The common phase in our range is var. *siphilitica,* with the herbage, hypanthium, and cal ± hirsute, the lvs mostly 2–6 cm wide, and the fls usually more than 20. The var. *ludoviciana* D.C., glabrous or nearly so, with the lvs seldom over 1.5 cm wide, and the fls 6–20, occurs principally on the Great Plains, extending e. about to the Mississippi R.

13. Lobelia cardinalis L.   Cardinal-fl. Erect, usually unbranched, 5–15 dm, ± hairy or largely glabrous; lvs thin, lanceolate to oblong, to 16 × 5 cm, acute or acuminate, dentate, mostly 3–5 times as long as wide, the lower short-petioled, the upper smaller, sessile or nearly so; racemes 1–4 dm; bracts linear or narrowly lanceolate; pedicels hairy, 5–15 mm, bibracteolate at base; fls 3–4.5 cm, scarlet (rarely pink to white), the lower lip glabrous within; sep linear, 8–12 mm; auricles none or minute; $2n=14$. Wet soil; N.B. to Minn., s. to the Gulf of Mexico. July–Sept. Our plants, as described above, are var. *cardinalis.*

FAMILY **RUBIACEAE**, the Madder Family

Fls perfect, mostly 4(–8)-merous, epigynous, sympetalous, regular; cal mostly small, with open aestivation, or obsolete; stamens on the cor-tube, as many as and alternate with the lobes; ovary inferior, mostly 2(–5)-carpellate and plurilocular, with axile (or axile-basal or axile-apical) placentation; style slender, terminal, with lobed or capitate stigma, or (as in *Galium*) the styles distinct; ovules and seeds 1–many per locule; seeds with a dicotyledonous, usually straight embryo embedded in the oily endosperm; herbs or more often woody plants with simple, mostly entire, commonly decussately opposite lvs and interpetiolar (usually connate) stipules, or with whorled lvs and no stipules, the stipules typically bearing colleters within, or seldom the stipules reduced to mere interpetiolar lines; infls basically cymose. 450/6500, cosmop. but mostly trop. and subtrop.

1 Shrubs with globose, peduncled heads of fls . . . . 3. *Cephalanthus.*
1 Herbs; infl various, seldom head-like.
  2 Principal lvs opposite.
    3 Ovules and seeds several to many per locule; fr capsular . . . . 1. *Hedyotis.*
    3 Ovules solitary in each locule; fr various, often a schizocarp or berry.
      4 Fls paired, united by their hypanthia; fr a berry; creeper . . . . 2. *Mitchella.*
      4 Fls otherwise; fr dry, eventually splitting lengthwise.
        5 Ovary bilocular; fls in axillary glomerules or solitary in the axils.
          6 Both halves of the fr remaining closed after splitting; lvs not glabrous . . . . 4. *Diodia.*
          6 One half of the fr closed, the other open; lvs glabrous . . . . 5. *Spermacoce.*
        5 Ovary mostly trilocular; fls in terminal involucrate glomerules . . . . 6. *Richardia.*
  2 Principal lvs whorled.
    7 Sep triangular; fls in involucrate heads . . . . 7. *Sherardia.*
    7 Sep obsolete; infl various.
      8 Cor funnelform, with slender tube evidently longer than the lobes . . . . 8. *Asperula.*
      8 Cor rotate to cupulate or short-funnelform, the tube no longer than the lobes . . . . 9. *Galium.*

1. **HEDYOTIS** L.   Fls 4-merous; sep lance-linear to ovate; cor salverform or funnel-form to subrotate; stigmas 2; capsule partly to wholly inferior, with several to many seeds in each of the 2 locules, loculicidally dehiscent across the top; mostly herbs with small, opposite lvs and rather small to minute fls. (*Houstonia; Oldenlandia.*) 300, cosmop., especially in warm regions. Spp. 1–4 apparently hybridize.

1 Cor minute (2 mm wide), white, ± rotate; capsule wholly inferior.
  2 Annual; lvs lanceolate to oval; fr white-hairy . . . . 9. *H. uniflora.*
  2 Perennial; lvs linear or nearly so; fr smooth or verrucose . . . . 10. *H. boscii.*
1 Cor larger, salverform or funnelform, often colored; capsule only partly inferior.
  3 Cor funnelform, fls cymose; fr not flattened.
    4 Plants fibrous-rooted; fr globose or depressed-globose.
      5 Basal lvs numerous at anthesis, distinctly ciliate . . . . 1. *H. canadensis.*
      5 Basal lvs usually none at anthesis; if present, then not ciliate.
        6 Cauline lvs rounded to subcordate at base; sep at anthesis 1.7–6.5 mm . . . . 2. *H. purpurea.*
        6 Cauline lvs tapering or narrowed at base; sep at anthesis 1–2 mm.
          7 Lvs broadly linear to narrowly oblong, 2–5 mm wide; pedicels 1.5–4 mm . . . . 3. *H. longifolia.*
          7 Lvs narrowly linear, 1–3 mm wide; pedicels 5–15 mm . . . . 4. *H. nuttalliana.*
    4 Plants taprooted; fr obovoid-cylindric, longer than thick . . . . 5. *H. nigricans.*
  3 Cor salverform; peduncles 1-fld; fr didymous-flattened.
    8 Perennial with rhizomes or creeping stems; cor with a yellow eye.
      9 Stems prostrate and creeping; corolla-tube pubescent within . . . . 6. *H. michauxii.*
      9 Stems erect; corolla-tube glabrous within . . . . 7. *H. caerulea.*
    8 Annual, not creeping; cor with a reddish eye . . . . 8. *H. crassifolia.*

1. **Hedyotis canadensis** (Willd.) Fosb.   Stems few or solitary from a fibrous-rooted perennial base, mostly simple to the infl, 8–30 cm, glabrous or short-hairy; basal lvs ± numerous, persistent, petiolate, lanceolate to obovate, conspicuously ciliate; cauline lvs sessile, narrowly oblong to lanceolate or oblanceolate, 1–3 cm, glabrous or sparsely (rarely more densely) hairy, usually ciliate; infls terminal, small, few-fld; fls heterostylic; sep lance-linear, 1.8–3 mm; cor purple, hairy within, 5–9 mm, funnelform, the lobes more than half as long as the tube; stamens included or barely exserted; fr globose or depressed-globose, half-inferior, 2.5–3.5 mm; seeds meniscoid, with a ridge across the hollowed inner face; $2n=12, 14$. Rocky woods and hillsides; Me. and s. Ont. to Minn., s. to sw. Va., Tenn. and Ark. May, June. (*Houstonia c.; Houstonia ciliolata; H. setiscaphia.*)

2. **Hedyotis purpurea** (L.) T. & G.   Stems usually several from a fibrous-rooted perennial base, simple or branched above, 1.5–5 dm, villous, especially below, to glabrous; lvs sessile, ovate to lance-oblong, 20–50

× 5–30 mm, rounded to subcordate at base, 3–5-nerved; fls numerous in terminal cymes, short-pediceled, heterostylic; sep lanceolate, 1.7–6.5 mm at anthesis, sometimes ciliate; cor purple to nearly white, hairy within, funnelform, 5.5–9.5 mm, the lobes half as long as the tube; stamens, frs and seeds of no. 1; $2n=12$, 24. Dry woods, pine-barrens, prairies, and bottomlands; N.J. to Io., s. to Ga., Miss., and Tex. May–July. (*Houstonia p.; Houstonia lanceolata.*)

3. **Hedyotis longifolia** (Gaertner) Hook. Stems numerous from a fibrous-rooted perennial base, simple or branched above, 1–2.5 dm, glabrous or finely hairy, especially at the nodes; lvs sessile, broadly linear to narrowly oblong, 10–30 × 2–5 mm, narrowed to the base, glabrous or nearly so, 1-nerved; fls short-pediceled, numerous in loose or crowded terminal cymes, heterostylic; sep lance-linear, 1–2 mm, in fr equaling or exceeding the capsule; cor purplish to white, hairy within, funnelform, 5.5–9 mm, the lobes half as long as the tube; stamens, frs and seeds of no. 1; $2n=12$, 24. Dry, commonly gravely or sterile soil; Me. to Ont. and Sask., s. to S.C., Miss., and Ark. June–Aug. (*Houstonia l.*)

4. **Hedyotis nuttalliana** Fosb. Fibrous-rooted perennial 1.5–5 dm; stems diffusely branched, very slender, glabrous or minutely hairy on the angles; lvs linear, 1.5–4 cm × 1–3 mm, tapering to the base, 1-nerved, glabrous or nearly so; pedicels capillary, 5–15 mm; fls heterostylic; sep lance-linear, 1–2 mm at anthesis; cor purple, hairy within, funnelform, 5.5–7 mm, the lobes half as long as the tube; frs 1.8–2.8 mm, otherwise as in no. 1; stamens and seeds of no. 1. Dry soil; Pa. to Mo. and Okla., s. to Va., Ga., and Tex. May–July. (*Houstonia tenuifolia.*)

5. **Hedyotis nigricans** (Lam.) Fosb. Taprooted perennial; stems few or several, 2–6 dm, ± erect, branched from the base, often somewhat hairy below; lvs linear, sessile, mostly 20–30 × 1–3 mm, 1-nerved, slightly roughened, often with axillary fascicles; middle and upper stipules setiform or bristle-tipped; fls very numerous, on pedicels to 5 mm (some sessile), forming a crowded cymose panicle, heterostylic; sep narrowly triangular, often ciliate, 1–2 mm; cor purple to nearly white, hairy within, funnelform, 5–8 mm, the lobes two-thirds as long as the tube; fr obovoid-cylindric, narrowed to the base, two-thirds inferior, 2.5–3.5 mm; stamens and seeds of no. 1; $2n=18$, 20. Dry soil and barrens; O. and Ind. to s. Mich., Io., and Nebr., s. to Fla. and Mex. June–Aug. (*Houstonia n.; Houstonia angustifolia.*)

6. **Hedyotis michauxii** Fosb. Mt.-bluets. Perennial; stems prostrate, diffusely branched, elongate, rooting at the nodes, 1–2 dm, glabrous; lvs ovate to orbicular, 3–7 mm, abruptly rounded or broadly cuneate at base, short-petioled; fls heterostylic, solitary on peduncles 1–4 cm, terminal or from the upper axils; sep oblong, obtuse, 1–2 mm; cor blue-violet with a light yellow eye, salverform, the tube hairy within, the limb 10–14 mm wide, the lobes usually longer than the tube; stamens included; fr didymous-flattened, 4–5 mm wide; seeds of no. 7; $2n=16$. Moist soil in the mts.; Pa. to Ga. May, June. (*Houstonia serpyllifolia.*)

7. **Hedyotis caerulea** (L.) Hook. Bluets. Delicate perennial with very slender, fragile rhizomes, eventually forming clumps up to 1 dm wide, but apparently blooming the first year and thus potentially annual; lower lvs oblanceolate to spatulate, 5–12 mm, narrowed to a petiole often as long; upper lvs subsessile, oblong-spatulate to linear; peduncles erect, slender, terminal and from the upper axils, 2–7 cm, 1-fld; fls heterostylic; sep narrowly oblong, 1–2 mm, acute; cor typically rather light blue-lavender with a pale yellow eye, salverform, the tube 5–10 mm, glabrous within, the limb 10–14 mm wide; stamens included; fr didymous-flattened, 3–4 mm wide, much broader than long; seeds globular, with a deep round cavity occupying the inner face; $2n=16$, 32, 48. Moist soil and meadows; N.S. and Que. to Wis., s. to Ga. and Ark. Apr.–June. (*Houstonia c.; Houstonia faxonorum.*)

8. **Hedyotis crassifolia** Raf. Small bluets. Much branched annual 5–15 cm, the stems glabrous; basal lvs ovate to elliptic, 5–10 mm, tapering to a short petiole, scaberulous on the margin; upper lvs narrower, subsessile; fls short-styled, solitary, terminal or also from the upper axils, on peduncles 1–3 cm; sep narrowly oblong to lance-ovate, 1.2–3 mm, half to fully as long as the cor-tube; cor blue or purple with a reddish eye, salverform, the tube 2.5–5 mm, glabrous within, the limb 4–8 mm wide; fr and seeds of no. 7; $2n=16$. Dry soil; coastal plain from se. Va. to Tex.; inland from Ill. to Nebr., s. to Tenn. and Tex. Apr. (*Houstonia minima; Houstonia patens; Houstonia pusilla.*)

9. **Hedyotis uniflora** (L.) Lam. Erect or spreading annual 1–5 dm, the stem simple to diffusely branched, usually ± hirsute-villous, especially above; lvs sessile or subsessile, lanceolate to oval, 1–2.5 cm, 1–3-nerved, white-hairy on the margins and midvein; fls homostylous, subsessile in terminal or axillary glomerules; sep lanceolate to ovate, ciliolate, acute, 1 mm; cor white, ± rotate, 2 mm wide, its lobes shorter than the sep; fr wholly inferior, densely white-hairy; seeds very numerous, angular; $2n=36$, 72. Wet soil, mostly on the coastal plain; L.I. to Fla. and Tex., n. to se. Mo. Summer. (*Oldenlandia u.*)

10. **Hedyotis boscii** DC. Glabrous perennial 1–3 dm with numerous prostrate or spreading, diffusely branched stems; lvs linear to very narrowly elliptic, 1–3 cm × 1–3 mm, narrowed to a sessile or very shortly petiolar base; fls subsessile in small axillary or terminal glomerules (or solitary); sep narrowly lanceolate, long-acuminate, 1 mm; cor, fr, and seeds as in no. 9, but the fr mostly glabrous or merely verrucose; $2n=36$. Wet soil of the coastal plain; se. Va. to Fla. and Tex. (*Oldenlandia b.*)

2. MITCHELLA L.

Fls heterostylic, mostly 4-merous, paired, their hypanthia united; cor funnelform, with elongate tube and short, spreading or recurved lobes villous on the inner face; ovary 4-locular, with a single erect ovule in each locule; stigmas 4, slender; fr a twin berry, with 8 seeds; creeping evergreen herbs with axillary or terminal white fls and scarlet (white) berries. 2, the other in ne. Asia.

1. **Mitchella repens** L.   Partridge-berry. Stems rooting at the nodes, 1–3 dm, forming mats; lvs petioled, round-ovate, 1–2 cm; fls mostly terminal, the common peduncle shorter than the subtending lvs; cor 10–14 mm, occasionally with 3, 5, or 6 lobes; fr insipid, 5–8 mm thick, crowned with the short sep; $2n=22$. Woods; N.S. to Ont. and Minn., s. to Fla. and Tex. May–July.

3. **CEPHALANTHUS** L.   Fls 4-merous; hypanthium obovoid; cal short; cor slenderly funnelform, the short lobes imbricate in bud; filaments short; ovary bilocular, with 1–3 pendulous ovules per locule; style elongate, long-exsert; stigma capitate; fr eventually splitting from the base upward into indehiscent nutlets; shrubs or small trees with fls densely crowded into globose pedunculate heads. 6, widespread.

1. **Cephalanthus occidentalis** L.   Buttonbush. Shrub 1–3 m, or even a small tree; lvs opposite or occasionally in whorls of 3, ovate-oblong or lance-oblong, 8–15 cm; stipules deltoid, 2–3 mm, glandular-dentate; heads 3 cm thick, on elongate peduncles terminating the stems or from the upper axils, the fls intermingled with filiform-clavate bractlets; cor white, 5–8 mm; fr obconic, angular, 5 mm; $2n=44$. Swamps and streamsides; N.S., N.B. and Que. to Minn., s. to Mex. and W.I. June–Aug. Most of our plants are glabrous or inconspicuously hairy, and represent var. *occidentalis*. The chiefly southern var. *pubescens* Raf., with the twigs and lower lf-surfaces evidently soft-hairy, extends n. to Va., Ind. and Mo.

4. **DIODIA** L.   Fls mostly 4-merous; sep 2–4, sometimes very unequal; cor salverform or funnelform, with slender tube and 4 short lobes; ovary bilocular, with a single axile ovule in each locule; fr dry, splitting into 2 nutlets; herbs, ours low and branched, with narrow lvs and small fls sessile in the axils. 50, mainly warm regions.

1 Sep 4; cor funnelform; style undivided, with capitate stigma . . . . 1. *D. teres.*
1 Sep 2; cor salverform; style cleft into 2 linear stigmas . . . . 2. *D. virginiana.*

1. **Diodia teres** Walter.   Poorjoe, buttonweed. Annual with prostrate to spreading or ascending stems 2–8 dm; lvs stiff, sessile, linear to narrowly lanceolate, 2–4 cm, scabrous, aristate; stipules 4–10 mm, with a short sheath and 5–8 ascending or erect setaceous bristles; fls in many of the upper axils; sep 4, lanceolate, 2 mm; cor funnelform, white to pale purple, 5–6 mm; style capitate; fr obovoid, hairy, 2.5–6 mm, crowned by the persistent sep; $2n=28$. Dry or sandy soil, often weedy; s. N. Engl. to s. Wis., Io. and Kans., s. to trop Amer. June–Sept. (*Diodella t.*)

2. **Diodia virginiana** L.   Virginia-buttonweed. Annual 2–8 dm, branched and spreading, the stems ± hirtellous on the angles; lvs thin, narrowly elliptic to lance-oblong, 3–6(–10) cm, acute, gradually narrowed to an acute or obtuse base; stipules linear, 3–5 mm; sep 2, linear or lance-linear, 4–6 mm; cor salverform, with very slender tube, white or pink, 8–11 mm; stigmas 2; fr ellipsoid, obtusely 8-ribbed, 6–9 mm; $2n=28$. Wet ground; s. N.J. to s. O., s. Ill., and Mo., s. to Fla. and Tex. June–Aug.

5. **SPERMACOCE** L.   Fls 4-merous; sep ± connate at base, persistent; cor funnelform; stamens and style short, deeply included; ovary bilocular, with a single axile ovule in each locule; fr capsular, septicidal, one half retaining the septum and closed, the other half separating from the septum; herbs with dissected stipular sheaths and small white fls sessile or nearly so in axillary glomerules. 100, New World, mainly warm.

1. **Spermacoce glabra** Michx.   Buttonweed. Perennial 2–6 dm, often decumbent; stipules dissected into several filiform segments 2–3 mm; lvs narrowly elliptic to lanceolate or oblanceolate, 3–8 cm, acute, tapering to a very short petiole; fls glomerate in the upper axils; sep lance-oblong, eventually 2–3 mm; cor white, 2–3 mm, densely bearded within; fr obovoid, 3 mm; $2n=28$. Swamps and wet ground; se. O. to e. Kans., s. to Fla. and Tex. July–Sept.

6. **RICHARDIA** L.   Fls 4–8-merous; sep foliaceous, ± connate at base; cor funnelform, with short lobes; stamens included; ovary (2)3–4(–6)-locular, with a single axile ovule in each locule; fr separating into (2)3–4(–6) nutlets; herbs, mostly diffusely branched, with fls in terminal involucrate glomerules. 15, trop. Amer.

1 Lvs glabrous or nearly so except for the scabrous margins and the midrib beneath . . . . 1. *R. scabra.*
1 Lvs appressed-hispid on both sides . . . . 2. *R. brasiliensis.*

1. **Richardia scabra** L.   Florida-pusley. Much branched, hairy annual 2–8 dm; lvs elliptic or ovate to lance-ovate, 2–6 cm, glabrous or nearly so except for the scabrous margins and the midrib on the lower

surface; fls mostly 6-merous, in dense terminal glomerules surrounded by 2 or 4 involucral lvs with broad, sessile base; cor-tube 2–8 mm, lobes 0.5–2.5 mm; fr 2–3.5 mm, muricate-tuberculate, its 3(–6) cocci nearly cylindric, the adaxial scar nearly closed; $2n=28, 56$. Native of tropical Amer., established as a weed in se. U.S., and occasionally adventive in our range. Summer.

2. **Richardia brasiliensis** Gomes. Much like no. 1, but becoming perennial, and more hairy, the lvs appressed-hispid on both sides; fr 2.5–3(–4) mm, its obovoid cocci with broad, open, medially keeled adaxial scar; $2n=28, 84$. Disturbed, often sandy sites; native to tropical Amer., and extending ne. to se. Va. as a weed.

7. SHERARDIA L.

Fls 4(–6)-merous; sep erect; cor funnelform with slender tube; stamens about equaling the cor-lobes; ovary bilocular, with a single axile ovule in each locule; style unequally bifid; fr dry, longitudinally 2-ribbed, crowned by the persistent sep; herbs with whorled lvs and small fls in terminal heads, surrounded by an involucre of lanceolate, basally connate lvs. Monotypic.

1. **Sherardia arvensis** L. Field-madder. Diffusely branched annual to 4 dm, the square stems procumbent at base, ± rough-hairy; lvs mostly in whorls of 6, linear to narrowly elliptic, 0.5–2 cm, sharp-pointed, hairy; involucral lvs usually 8, lance-linear; sep triangular; cor pink or blue, 4–5 mm; fr obovoid, scabrous, 2–7 mm; $2n=22$. Waste places, or a weed in cultivated ground; native of w. Eurasia and n. Afr., now found here and there in our range.

8. ASPERULA L.

Differing from *Galium* in its funnelform cor, the tube usually distinctly longer than the lobes; pedicels often bracteolate. 200, Old World.

1. **Asperula arvensis** L. Erect annual 1–5 dm, the angled stems retrorsely scabrous at least below; lvs in whorls of 4–8, 1–3 cm, the upper linear to lanceolate, blunt; fls subsessile in a terminal head subtended by an involucre of separate hairy lvs; cor blue-purple, 5–6.5 mm, salverform, the slender tube much longer than the lobes; fr smooth, 2–3 mm thick; $2n=22$. Fields and waste places; native of Europe, adventive in s. N.Y. May, June.

9. GALIUM L.

Bedstraw; cleavers. Cal-lobes none; cor rotate, with (3)4 short lobes, varying in a few spp. to cupulate or funnelform with the lobes longer than or about equaling the tube; stamens mostly shorter than the cor; ovary bilocular, with a single axile ovule in each cell; styles 2, short; stigmas capitate; fr dry or seldom fleshy, of 2 globose or subglobose carpels, or one carpel abortive; each carpel indehiscent, 1-seeded, sometimes bristly; herbs with slender, 4-angled stems and whorled lvs, the small cymose fls ebracteolate, terminating the branches or axillary. 300, cosmop.

1 Plants annual from a short taproot.
  2 Fls greenish-white, or somewhat reddish outside.
    3 Frs relatively large, (1.5–)2–4(–5) mm; plants generally not much branched .... 22. *G. aparine.*
    3 Frs tiny, ca 1 mm or less; plants tending to be freely branched ....23. *G. parisiense.*
  2 Fls bright yellow .... 24. *G. pedemontanum.*
1 Plants perennial from creeping rhizomes.
  4 Fr a colored berry 3–3.5 mm; lvs in 4's.
    5 Lvs oblong or elliptic, 2–3 times as long as wide .... 18. *G. hispidulum.*
    5 Lvs linear-oblong, 5–7 times as long as wide .... 19. *G. uniflorum.*
  4 Fr dry, usually smaller, in some spp. bristly or muricate.
    6 Fr smooth to granular, verrucose, or muricate, but without hairs or bristles.
      7 Stems erect or nearly so.
        8 Principal lvs 3-nerved, in whorls of 4; native.
          9 Fls white .... 1. *G. boreale.*
          9 Fls purple .... 2. *G. latifolium.*
        8 Principal lvs 1-nerved, in whorls of 5 or more; intr.
          10 Fls white or greenish.
            11 Cor broadly cupulate, 4–6 mm wide; lvs linear, 0.5–2 mm wide .... 3. *G. glaucum.*
            11 Cor ± rotate, 2–5 mm wide; lvs often more than 2 mm wide.
              12 Lvs concolorous, mostly 1–2.5 cm and broadest above the middle .... 4. *G. mollugo.*
              12 Lvs pale beneath, mostly 2.5–5 cm and broadest at or below the middle .... 5. *G. sylvaticum.*
          10 Fls bright yellow .... 6. *G. verum.*
      7 Stems weak and ± matted, ascending or reclining or scrambling.
        13 Lvs sharply acute or cuspidate, the main ones in whorls of 5–8.
          14 Lvs elliptic or oblanceolate, retrorsely scabrous on the margins .... 7. *G. asprellum.*
          14 Lvs linear, antrorsely scabrous (or smooth) on the margins .... 8. *G. concinnum.*
        13 Lvs blunt, obtuse or rounded to merely acutish, mostly in whorls of 4–6.
          15 Cors mostly 4-lobed, the lobes longer than wide.

16 Cymes repeatedly branched, bearing 5–many fls; nodes not bearded .... 9. *G. palustre.*
16 Cymes once or twice branched, bearing 2–4 fls; nodes short-bearded.
    17 Lvs ascending or loosely spreading, often over 2.5 mm wide; widespread ...... 10. *G. obtusum.*
    17 Lvs soon ± recurved or reflexed, mostly 1–2.5 mm wide; northern ...... 11. *G. labradoricum.*
15 Cors mostly 3-lobed, the lobes about as wide as or wider than long.
    18 Lvs mostly 4 in a whorl; fls solitary (2) at the ends of flexuous, ± elongate peduncles
        often over 1 cm (these commonly borne 3 together at the node), or the
        peduncles shorter and firmer in depauperate plants .... 12. *G. trifidum.*
    18 Lvs 4–6 in a whorl, usually some of the whorls with at least 5 lvs; fls mostly 2–
        3 on each peduncle, the pedicels and peduncles relatively short and straight,
        the pedicels less than 1 cm .... 13. *G. tinctorium.*
6 Fr bristly or hairy.
  19 Main lvs in 4's; stems erect or ascending, not retrorsely scabrous.
    20 Fls, or some of them, sessile or subsessile along the side of the infl.
      21 Lvs oval or elliptic, broadest near the middle, obtuse .... 14. *G. circaezans.*
      21 Lvs lanceolate, broadest below the middle, acute or acuminate .... 15. *G. lanceolatum.*
    20 Fls all pediceled, terminating the branches of the infl.
      22 Main lvs oval to round or oblong; fr uncinate-bristly.
        23 Lvs firm, almost uniform in size, in numerous whorls .... 16. *G. pilosum.*
        23 Lvs very thin, in 2–4(5) whorls, the uppermost the largest ....17. *G. kamtschaticum.*
      22 Main lvs lance-linear; hairs of the frs not hooked .... 1. *G. boreale.*
  19 Main lvs in 6's or 8's.
    24 Cor ± rotate, 2–3 mm wide; stems reclining or scrambling, often retrorsely scabrous,
      not hairy at the nodes; native .... 20. *G. triflorum.*
    24 Cor funnelform, 4–7 mm wide; stems ± erect, hairy at the nodes, otherwise glabrous;
      intr. .... 21. *G. odoratum.*

1. Galium boreale L.   Northern b. Erect perennial 2–8 dm, the numerous stems commonly short-bearded just beneath the nodes, otherwise glabrous or scaberulous; lvs in 4's, glabrous or scabrous, lance-linear, 1.5–4.5 cm, 3-nerved, the tip minutely rounded, not mucronate; sterile axillary branches with smaller lvs often developed; fls numerous in terminal, rather showy cymose panicles, with white or slightly creamy cor 3.5–7 mm wide; fr 2 mm, glabrous or with short, straight or curled (not hooked) hairs; $2n=44$, 66. In a wide variety of not too dry habitats; circumboreal, s. in Amer. to Del., Ky., Mo. and Calif. June, July.

2. Galium latifolium Michx.   Wide-lvd b. Perennial; stems 1.5–6 dm, smooth, or hairy at and below the nodes; lvs in 4's, lanceolate or lance-ovate, 3–6 cm, 3–4 times as long as wide, acute or acuminate, 3-nerved, hairy on the veins and margins; infls 1–4 from each of the upper 1–3 nodes, divaricately branched; cor purple; fr smooth or granular, 3–4 mm; often only one carpel fully developed, the other maturing into a small elaiosome. Dry woods in the mts.; Pa. to Va. and Ky., s. to Ga. and Ala. June, July. Hairy-stemmed plants have been called var. *hispidum* Small, in contrast to the smooth-stemmed var. *latifolium.*

3. Galium glaucum L.   Perennial 3–8 dm; stem roundly 4-angled; lvs mostly in whorls of 8–10, linear, 1-nerved, 2–4 cm × 0.5–2 mm, minutely roughened on the margins, revolute to the midrib; peduncles terminal and from the upper axils, trichotomously branched, each forming an umbelliform cluster and together a cymose panicle; cor white, broadly cupulate, 4–6 mm wide, the lobes somewhat longer than the tube; fr smooth, 1.5–2 mm; $2n=22$, 44. Native of Eurasia, intr. in our range from N. Engl. and s. Que. to N.J. May–July. (*Asperula g.; A. galioides.*)

4. Galium mollugo L.   Perennial 3–12 dm, the numerous weak stems usually ± erect from a decumbent base, glabrous or shortly spreading-hairy below; short, slender, leafy perennial offshoots produced in summer or fall; lvs in 6's or 8's, antrorsely scabro-ciliate, otherwise glabrous, linear-oblong to oblanceolate, mostly 1–2.5 cm and broadest above the middle, 1-nerved, cuspidate, concolorous; fls numerous in terminal, often divaricately branched, rather showy cymose panicles, the cor white or nearly so, 2–5 mm wide; fr glabrous, 1–1.5 mm; $2n=22$, 44, 66, 88. A Eurasian weed, now well established in our range. The typical phase has the cor 2–3 mm wide, the pedicels usually longer than the width of the fls, and lax infls with the branches strongly divaricate after anthesis. The var. *erectum* (Hudson) Domin (*G. album* Miller) has the cor 3–5 mm wide, the pedicels usually shorter than the width of the fls, and rather dense infl with less divaricate branches. Both are reported in our range.

5. Galium sylvaticum L.   Scotch mist. Perennial 4–8 dm, erect or nearly so, glabrous except for the antrorsely scabro-ciliate lf-margins; lvs in 6's or 8's, thin, narrowly lanceolate, 2.5–5 cm × 5–8 mm, broadest near or below the middle, acute, pale beneath; infls several, divaricately spreading from the upper axils, forming a large diffuse panicle; pedicels very slender; cor white, 2–3 mm wide, its lobes acuminate; fr smooth; $2n=22$. Native of Europe, occasionally found along roadsides and in thickets in N. Engl. and N.Y. Aug.

6. Galium verum L.   Yellow b. Perennial; stems clustered, 4–10 dm, roundly 4-angled, usually finely puberulent; lvs in whorls of 8–12, ± linear, 1-nerved, 1.5–4 cm × 0.5–3 mm, acute, shining and usually hairy above, densely puberulent beneath, the margins usually ± inrolled; fls bright yellow, forming a dense, showy cymose panicle; $2n=22$, 44. Native of Eurasia, now a weed of roadsides and fields in our range, especially northward. June–Sept. (A strictly diploid phase with odorless, lemon-yellow fls and open ± interrupted infl, has been called *G. wirtgenii,* or *G. verum* ssp. *wirtgenii,* in contrast to typical *G. verum,* diploid or tetraploid, with fragrant golden-yellow fls and denser infl. We have both.)

7. **Galium asprellum** Michx. Rough b. Perennial; stems much branched, spreading or scrambling to 2 m, retrorse-scabrous on the angles; lvs in 6's or those of the branches in 4's or 5's, narrowly elliptic to oblanceolate, usually widest above the middle, to 20 × 6 mm, sharply acute or cuspidate, retrorsely scabrous on the margins and midvein beneath; infls terminal and from the upper axils, usually ascending, rarely over 2 cm, 1–3 times forked, few-fld, forming a small loose panicle; cor 4-lobed, white, 3 mm wide; fr smooth, 2 mm. Wet woods and thickets; Nf. to Minn., s. to N.C. and Mo., most abundant northeastward. May–Aug.

8. **Galium concinnum** T. & G. Shining b. Perennial, the slender, spreading or ascending stems 2–5 dm, much branched, often sparsely retrorse-scabrous on the angles; lvs in 6's or those of the branches in 4's, linear or linear-elliptic, 1–2 cm, sharply acute or cuspidate, antrorsely scabrous on the margins; infls terminal and divaricately spreading from the upper axils, 2–3 times branched, the branches and short pedicels very slender; cor 4-lobed, white, 2.5–3 mm wide; fr smooth, 2 mm. Dry woods; N.J. to Minn. and Nebr., s. to Tenn. and Ark., most abundant westward. June–Aug.

9. **Galium palustre** L. Marsh-b. Perennial; stems slender, 2–6 dm, minutely and sparsely retrorse-scabrous on the angles, not bearded; lvs in whorls of 2–6, linear to narrowly oblanceolate, 5–15 mm, blunt, antrorsely ± scabrous on the margins; infls many-fld, repeatedly forked, the short slender pedicels mostly ascending at anthesis, widely spreading or somewhat reflexed in fr; cor white, acutely 4-lobed, 4 mm wide; fr smooth, 2 mm; $2n=24, 48$. Wet soil; Nf. and Que. to Conn. and N.J., and also in Europe. June–Aug.

10. **Galium obtusum** Bigelow. Bluntlf-b. Matted perennial, strict and branched from the base or diffusely branched throughout, 2–8 dm; stems smooth on the angles but densely short-bearded at the nodes; principal lvs in 4's, linear to lanceolate or oblanceolate, 10–30 × 1–6 mm, obtuse, the margins and the midvein beneath usually ± scabrous or hispid-ciliolate; infls short, almost all terminal, with 3–5 fls on short ascending pedicels, the pedicels in fr 5–10 mm, divaricate, stiff, smooth; fls white, 4-merous, 2–3.5 mm wide, the lobes longer than wide; frs 4–5 mm, ± tuberculate, often only one mericarp developed; $2n=48$. Swampy thickets and moist meadows; N.S. and s. Que. to Minn. and S. D., s. to Fla. and Tex. May–July. (*G. tinctorium,* misapplied.) The widespread var. *obtusum* has the lvs mostly (2)3–5 mm wide, and has consistently bearded nodes. The var. *filifolium* (Wiegand) Fern., from s. N.J. to Ga., often near the coast, has lvs mostly 1–2 mm wide, and the nodes are not always bearded.

11. **Galium labradoricum** (Wieg.) Wieg. Labrador-b. Perennial 1–4 dm, simple or branched above; stems slender, densely short-bearded at the nodes, otherwise glabrous or nearly so; lvs in 4's, soon recurved or deflexed, linear-oblanceolate, 8–15 × 1–2.5(–3) mm, hispidulous on the margins, usually smooth on the midrib; infls few, mostly terminal, 3-fld, 3–10 mm; cor white, 2–3 mm wide, 4-lobed, the lobes longer than wide; fr smooth, 2–3 mm, usually both mericarps developed; $2n=24$. Cold bogs, swamps, and wet thickets; Nf. and Lab. to the Mackenzie R., s. to Mass., N.Y., N.J., Ind. and Io. June, July.

12. **Galium trifidum** L. Northern three-lobed b. Perennial; stems numerous, often much branched, slender and weak, tending to scramble on other plants, retrorse-scabrous on the angles or sometimes essentially glabrous, not evidently bearded at the nodes, mostly in 4's, rarely some in 5's or 6's, linear to narrowly elliptic, blunt, 1-nerved, 5–20 mm, often spreading-scabrous on the margins and sometimes also on the midrib beneath, otherwise glabrous; peduncles numerous, axillary or terminal, commonly 1–3 together on short axillary branches, ± elongate and flexuous, often over 1 cm, each 1(2)-fld; cor whitish, 1–1.5(–1.8) mm wide, 3-parted, the lobes about as wide as or wider than long; fr glabrous, the segments divergent and nearly distinct at maturity, each 1–1.75 mm thick; $2n=24$. Moist places at various altitudes; circumpolar, in Amer. s. to N.Y., Pa., Ill., Nebr., and Calif. June–Aug. (*G. brevipes; G. brandegei,* misapplied.) Ours is the circumboreal (but scarcely cordilleran) var. *trifidum.*

13. **Galium tinctorium** L. Southern three-lobed b. Very much like *G. trifidum,* of which it is often considered to be a var. or ssp., but said to be distinct in the field; differing as indicated in the key; pedicels consistently glabrous. Moist or swampy places, usually in circumneutral or somewhat alkaline soils; Nf. to Fla. and Hispaniola, w. to Minn., Mo., Tex., and Mex. Most plants of our range belong to the var. *tinctorium,* with the mericarps 2–3 mm and the pedicels rarely over 5 mm. The chiefly more southern var. *floridanum* Wieg. extends n. to the s. fringe of our range and along the coast to Mass.; it is more robust, with the mericarps 3–4 mm, and pedicels commonly 5–8 mm in fr.

14. **Galium circaezans** Michx. Forest-b.; wild licorice. Erect or ascending perennial 2–6 dm, simple or branched from the base, the stems in 4's, oval, elliptic, or ovate-oblong, broadest near the middle, 2–5 × 1–2.5 cm, obtuse, 3–5-nerved; infls terminal and from the upper axils, simple or with 1 or 2 divaricate forks, the fls remote, sessile or subsessile; cor greenish-purple, pilose, the lobes acute; fr reflexed, uncinate-hispid, 3 mm; often only one carpel fully developed, the other maturing into a small elaiosome; $2n=22$. Dry woods and thickets; Me. to n. Mich. and Minn., s. to Fla., Kans., and Tex. June, July. Northern plants, with ± densely hairy lvs, have been segregated as var. *hypomalacum* Fern.; they pass freely into the chiefly more southern var. *circaezans,* with more thinly hairy or glabrous lvs.

15. **Galium lanceolatum** Torr. Wild licorice. Erect or ascending slender perennial 3–7 dm, branched from the base; stem glabrous or nearly so; lvs in 4's, thin, the lower elliptic, the upper lanceolate, 3–8 × 1–2.5 cm, long-tapering to an acute or acuminate tip, 3–5-nerved, minutely ciliate, finely hairy on the midvein (and sometimes other veins) beneath; infl widely divaricate, 1–3-forked; cor glabrous, turning purple in age, its lobes acuminate; fr deflexed, uncinate-hispid, 3 mm. Dry woods and thickets; Que. to Minn., s. to N.C. and Tenn. June, July.

16. Galium pilosum Aiton. Erect or ascending perennial, 2–10 dm, often with many basal branches, otherwise simple to the infl; lvs numerous, in 4's, elliptic to oval, 1–2.5 cm, half as wide, firm, usually 3-nerved; infls terminal and from the upper axils, each divaricately 2–3-forked, the fls terminating the branchlets; fr uncinate-bristly; $2n=22$. Dry woods; N.H. to Mich. and Kans., s. to Fla. and Tex. June–Aug. Most of our plants are the northern phase, var. *pilosum,* with the stem and lvs ± pubescent with straight hairs. The southern phase, var. *puncticulosum* (Michx.) T. & G., with the stem and lvs ± pubescent with short, upwardly incurved hairs, extends n. to Mo. and on the coastal plain to s. N.J.

17. Galium kamtschaticum Steller. Stems scattered, erect, 1–2 dm, glabrous; lvs in 2–4(5) whorls of 4, antrorsely scabro-ciliate and commonly with scattered similar hairs at least on the upper surface, thin, evidently trinerved, broadly obovate to broadly ovate-elliptic, 1–3 cm, half or two-thirds as wide, mucronate at the blunt apex, mostly concavely narrowed to the cuneate base; peduncles 1–2 from the uppermost whorl of lvs, rather elongate, 2–3(–6)-fld; fr 1.5 mm, uncinate-bristly; $2n=22$, 44. Mossy woods; n. N. Engl. and n. N.Y. to se. Can.; ne. side of Lake Superior; near the Pacific coast from Korea to Wash. July.

18. Galium hispidulum Michx. Berry-b. Diffusely branched, rough-hairy perennial, 1.5–6 dm, evergreen toward the s.; lvs in 4's, thick, elliptic or oblong, 5–12 mm, half as wide, apiculate, 1-nerved; fls 2–5 together on hairy peduncles from the upper axils; fr a purple berry 3.5 mm. Sandy soil, especially pine-land on the coastal plain; s. N.J. to La. Summer. (*G. bermudense,* misapplied.)

19. Galium uniflorum Michx. Berry-b. Stiffish evergreen perennial, diffusely branched, 1–4 dm, the slender stems glabrous; lvs in 4's, thick, narrowly linear-oblong, 12–30 × 2–4 mm, acute, 1-nerved, minutely roughened along the margin; fls 1–3 at a node, terminating a lateral branch 1–5 mm; fr a purple-black berry 3 mm. Dry or sandy soil, especially pine-land; coastal plain from se. Va. to Tex. Summer.

20. Galium triflorum Michx. Sweet-scented b. Perennial; stems 2–8 dm, prostrate or scrambling, usually retrorsely hooked-scabrous on the angles at least below; lvs vanilla-scented, mostly in 6's, or only 4 on the smaller branches, narrowly elliptic to oblanceolate, cuspidate, 1.5–5 cm, 1-nerved, generally antrorsely scabro-ciliate on the margins and retrorsely hooked-scabrous on the midrib beneath, otherwise mostly glabrous; infls axillary or also terminal, the peduncles simple and 3-fld to repeatedly branched; fls greenish-white, 2–3 mm wide; fr. 1.5–2 mm, uncinate-bristly; $2n=44$, 66. Woods; circumboreal, s. in Amer. to Fla. and Mex. June–Aug. Southeastern plants, occasionally extending n. to our range, have reduced upper lvs and branching peduncles, forming a diffuse infl, and have been segregated as var. *asprelliforme* Fern., but are not well set off from the more northern phase, common in our range (var. *triflorum*) with merely 3-fld axillary peduncles.

21. Galium odoratum (L.) Scop. Sweet woodruff. Perennial; stems ± erect, 1.5–5 dm, retrorsely hispid at the nodes, otherwise glabrous or nearly so; lvs vanilla-scented, mostly in whorls of (6)8(10), oblanceolate or narrowly elliptic, cuspidate, 1.5–5 cm × 4–12 mm, antrorsely scabro-ciliate on the margins and sometimes on the midrib beneath, othrwise glabrous or nearly so; infl terminal, branched, nearly naked; cor white, funnelform, 3–5 mm long and 4–7 mm wide, the lobes somewhat longer than the tube; fr 3 mm, covered with hooked bristles; $2n=22$. Woods, native of Europe, sparingly intr. in our range. May, June. (*Asperula o.*)

22. Galium aparine L. Cleavers. Annual 1–10 dm, the stems weak, retrorsely hooked-scabrous on the angles, ± scrambling, seldom much branched; lvs in whorls of (6)8, narrow, 1-nerved, cuspidate, 1–8 cm, retrorsely scabro-ciliate on the margins and often also on the midvein beneath, uncinate-hispid above; infls mostly 3–5-fld, on axillary peduncles or in 3's on short branches; cor greenish-white, 4-parted, 1–2 mm wide; fr uncinate-hispid, or rarely smooth. Damp ground, usually in shade; circumpolar, and found over most of temperate N. Amer. Plants with relatively large frs (mostly 2.8–4 mm, exclusive of the bristles), and with lvs to 8 cm, said to be mainly tetraploid to octoploid (on $x=11$) are var. *aparine.* Plants with smaller frs (mostly 2–3 mm), and the lvs not over 4 cm, said to have usually $2n=20$, but sometimes up to $2n=66$, are var. *echinospermum* (Wallr.) Farw. (*G. spurium; G. vaillantii.*)

G. *virgatum* Nutt., Ozark b., with the lvs in 4's, the fls sessile in the axils, barely extends n. of the Missouri R. in c. Mo.

23. Galium parisiense L. Slender annual 1–4 dm, the stem often retrorse-scabrous on the angles; lvs in whorls of 5–8, often 6, linear or linear-oblanceolate, mucronate, 1-nerved, 4–10 mm, antrorsely scabrous on the margins; fls in small cymes ending the mostly numerous branches; cor minute; fr 1 mm, the segments approximate; $2n=22$, 44, 55, 66. Native of Europe, intr. as a weed in fields and roadsides in the s. part of our range. June–Aug. Our plants, with the slightly granular-roughened, otherwise glabrous frs, are var. *leiocarpum* Tausch. (*G. divaricatum.*) Var. *parisiense* has hooked hairs on the frs.

24. Galium pedemontanum All. Annual 1–3.5 dm, the stems with spreading hairs and small recurved prickles; lvs 3–11 × 2–4 mm, ovate or elliptic, acute, 1-veined, with revolute margins, slightly hairy; peduncles much shorter than the subtending lvs, 2–3-fld; fls yellow, minute, 0.5–1 mm; styles united below; frs 1 mm, glabrous; $2n=18$. Native of Europe, intr. in disturbed sites from Va. and W.Va. to N.C., w. to O., Ill., Kans., and Okla. (*Cruciata p.*)

Family **CAPRIFOLIACEAE**, the Honeysuckle Family

Fls perfect, mostly (3–)5-merous, with a constriction just beneath the cal-limb, sympetalous, regular or irregular; stamens on the cor-tube, alternate with the lobes (only 4 in

Linnaea); ovary (half to) wholly inferior, mostly (2)3–5-carpellate and plurilocular, with axile placentation, but only one carpel well developed in *Viburnum*; style terminal, or the stigma(s) sessile; ovules and seeds 1–many per locule; seeds with a small, straight, dicotyledonous, basal to axile embryo and oily endosperm; shrubs (or small trees) or woody vines, seldom (as in *Triosteum*) herbs, with opposite, simple (pinnately compound in *Sambucus*) lvs; stipules usually wanting or vestigial, in any case without colleters; infls mostly cymose or partly so. 15/400, mostly N. Temp. and boreal.

1 Style very short or none; cor small, rotate to broadly campanulate, mostly white and regular.
  2 Lvs simple; fr a 1-seeded drupe .... 6. *Viburnum.*
  2 Lvs pinnately compound; fr a 3–5-seeded berry-like drupe .... 7. *Sambucus.*
1 Style elongate; cor various.
  3 Stamens as many as the cor-lobes, usually 5.
    4 Shrubs or climbing vines.
      5 Fr a capsule; lvs serrate .... 1. *Diervilla.*
      5 Fr fleshy, a berry or berry-like; lvs mostly entire.
        6 Ovary 2–3-locular, each locule with several ovules; cor seldom less than 10 mm .... 2. *Lonicera.*
        6 Ovary 4-locular, 2 locules with several abortive ovules, 2 with solitary fertile ovules; cor
          3–8 mm .... 3. *Symphoricarpos.*
    4 Erect herbs .... 4. *Triosteum.*
  3 Stamens 4; cor-lobes 5; stems trailing; fls paired .... 5. *Linnaea.*

1. DIERVILLA Miller.

Bush-honeysuckle. Cor funnelform, nearly regular, 5-lobed; stamens 5, barely exsert; hypanthium slender, prolonged above the bilocular, multiovulate ovary into a persistent beak; style elongate, exsert; stigma capitate; fr a conic or slender septicidal capsule; low shrubs with opposite, serrate lvs and yellow fls in terminal or subterminal cymes. 3, N. Amer.

1. Diervilla lonicera Miller. Shrub to 12 dm; twigs terete; petioles 3–10 mm; lvs oblong-lanceolate to oblong-ovate, 8–15 cm, acuminate, serrulate, ciliate-margined, sometimes also hairy beneath; peduncles 3–7-fld; cal-lobes 4–5 mm; cor 12–20 mm, yellow, turning reddish, hairy within; fr slender, beaked, 8–15 mm, tardily dehiscent; $2n=36$. Dry or rocky soil; Nf. to Sask., s. to N.C., Tenn., Ind. and Io. June, July. (*D. diervilla.*)

2. LONICERA L.

Honeysuckle. Cor tubular or funnelform, mostly conspicuous, deeply or shallowly 5-lobed, often bilabiate with a 4-lobed and a 1-lobed lip, the tube often gibbous or spurred at base; cal short, 5-toothed; stamens 5; ovary 2–3-locular with several ovules per locule; style elongate; stigma capitate; fr a few-seeded berry; erect shrubs or trailing or twining vines, with simple, mostly entire lvs; infl terminal, of 1–few 6-fld whorls, or axillary, the peduncles then bearing a pair of terminal sessile fls, their ovaries sometimes united. 180, N. Hemisphere. (*Nintooa; Phaenanthus; Xylosteon.*)

1 Fls in pairs terminating axillary peduncles.
  2 Plants erect, not twining or climbing; cor 0.7–2.5 cm.
    3 Style glabrous; cor not strongly bilabiate; native spp.
      4 Bracts at the top of the peduncle subulate to spatulate.
        5 Ovaries evidently separate, divergent; fr red .... 1. *L. canadensis.*
        5 Ovaries apparently wholly united; fr blue .... 2. *L. caerulea.*
      4 Bracts broadly oval, foliaceous, often anthocyanic; fr black .... 3. *L. involucrata.*
    3 Style hirsute; cor bilabiate or not; fr red; intr. spp., except no. 8.
      6 Peduncles very short, less than 5 mm, shorter than the petioles .... 7. *L. maackii.*
      6 Peduncles longer, 5+ mm.
        7 Cor scarcely bilabiate, the upper lip 4-lobed nearly or quite to its base.
          8 Lvs glabrous beneath; peduncles 1.5–2.5 cm .... 4. *L. tatarica.*
          8 Lvs pubescent beneath (thinly so in no. 5); peduncles 0.5–1.5 cm.
            9 Cor glabrous outside, barely gibbous at base .... 5. *L.* × *bella.*
            9 Cor pubescent outside, ± saccate at base .... 6. *L. morrowii.*
        7 Cor strongly bilabiate, the upper lip lobed up to half its length.
          10 Ovaries glabrous, partly or wholly connate .... 8. *L. oblongifolia.*
          10 Ovaries glandular, separate .... 9. *L. xylosteum.*
  2 Plants trailing or twining-climbing; cor 3–5 cm; fr black .... 10. *L. japonica.*
1 Fls in opposite, sessile, 3-fld cymules, producing 6-fld whorls; woody climbers.
  11 None of the lvs connate .... 11. *L. periclymenum.*
  11 Uppermost lvs below the fls connate.
    12 Cor scarcely bilabiate, the 5 lobes nearly equal .... 12. *L. sempervirens.*

12 Cor strongly bilabiate.
  13 Lvs glabrous above.
    14 Fls whitish to pale yellow or greenish-yellow, or purplish.
      15 Uppermost connate lvs glaucous above, rounded or emarginate at the end .... 13. *L. prolifera.*
      15 Uppermost connate lvs green above, pointed or mucronate .... 14. *L. dioica.*
    14 Fls golden to orange .... 15. *L. flava.*
  13 Lvs pubescent above; fls yellow to orange .... 16. *L. hirsuta.*

    1. Lonicera canadensis Marshall. Fly-honeysuckle. Shrub to 2 m, with straggling branches; lvs triangular-ovate to oblong, 3–12 cm, acute or obtuse, broadly cuneate to rounded at base, ciliate, glabrous to sparsely hairy beneath; peduncles axillary, 2–3 cm; bracts linear or subulate, from much shorter to slightly longer than the divergent separate ovaries; bractlets orbicular to elliptic, to 0.5 mm, or obsolete; cor yellowish, 12–22 mm, spurred at base, glabrous, its lobes subequal, a third to half as long as the tube; style glabrous; fr red; $2n=18$. Dry or moist woods, seldom swamps; N.S. and e. Que. to Sask., s. to Pa., O., Ind. and Minn., and in the mts. to N.C. May, June. (*Xylosteon canadense; X. ciliatum.*)

    2. Lonicera caerulea L. Waterberry. Shrub to 1 m, with upright branches, variable in pubescence; lvs oval to oblong, 2–8 cm, usually widest above the middle, obtuse or rounded at base, veiny, hairy beneath; peduncles axillary, 3–10 mm; bracts narrow, often much exceeding the ovaries; bractlets wholly concrescent into a cupule closely surrounding and ripening with the ovaries, the cal barely emerging; cor 10–15 mm, yellowish, gibbous at base, the 5 subequal lobes about equaling the tube; style glabrous; fr blue; $2n=18$. Swamps and wet woods; ± circumboreal, in Amer. from Lab. to Mack., s. to Pa., Mich., Minn., and Calif. May–July. Ours may be called var. *villosa* (Michx.) T. & G. (*L. villosa.*)

    3. Lonicera involucrata (Richardson) Banks. Bearberry-h. Shrub 1–3 m; lvs ovate to obovate, 6–12 cm, short-acuminate, tapering to a short petiole, villosulous beneath; peduncles 2–4 cm; bracts oval, foliaceous, 1–2 cm, persistent, often anthocyanic; cor yellow, 10–15 mm, pubescent, saccate at base, the lobes subequal, scarcely half as long as the tube; ovaries 3-locular, distinct; style glabrous; fr lustrous black at full maturity; $2n=18$. Cold, moist woods and mt. meadows; Que. to n. Mich.; s. Alas. to Calif., N.M., and n. Mex. (*Distegia i.*) Ours is var. *involucrata.*

    4. Lonicera tatarica L. Tartarian h. Glabrous shrub to 3 m, the vigorous shoots hollow; lvs ovate to oblong, 3–6 cm, rounded to subcordate at base; peduncles axillary, 15–25 mm; bracts subulate, shorter or longer than the separate ovaries; bractlets broadly ovate, about a third as long as the ovary; cor white to pink, 1.5–2 cm, glabrous, nearly regular, the lobes equaling or longer than the tube; style hirsute; fr red, rarely yellow; $2n=18$. Native of Eurasia, escaped from cult. throughout our range. May, June. (*Xylosteum t.*)

    5. Lonicera × bella Zabel. Hybrid of nos. 4 and 6, often taller, to 6 m; lvs slightly hairy beneath; peduncles 5–15 mm, sparsely hairy; cor and margins of the sep glabrous; cor pink, fading yellow. Escaped from cult. and well established in parts of our range, as in Mich., Wis., and Minn.

    6. Lonicera morrowii A. Gray. Lvs oblong to narrowly elliptic, softly pubescent beneath; peduncles 5–15 mm, densely hairy; bractlets pubescent; sep ciliate; cor pubescent, white, turning yellow; otherwise much like no. 4. May, June. Native of Japan, occasionally escaped in our range.

    7. Lonicera maackii (Rupr.) Maxim. Shrub to 5 m; lvs ovate to lance-ovate, 3.5–8.5 cm, acuminate, short-petioled, pubescent at least on the veins; fls paired on peduncles shorter than the petioles, the cor white, turning yellow, 15–20 mm, ± distinctly bilabiate, the tube short and not gibbous, usually glabrous outside; style hairy; fr dark red; $2n=18$. Native of Asia, escaped and becoming established in our range, as in N.Y., Md., Ky., and O.

    8. Lonicera oblongifolia (Goldie) Hook. Swamp fly-h. Shrub to 2 m; twigs solid; lvs oblong to oblanceolate, 3–7 cm, obtuse or acute at both ends, downy beneath, puberulent above when young; peduncles axillary, 2–4 cm; bracts and bractlets minute or none; ovaries glabrous, partly or wholly connate; cor yellow, 10–15 mm, hairy, strongly bilabiate, the tube 2.5–5 mm, the upper lip 4-lobed one fourth its length; style (and often filaments) hirsute; fr red. Swamps and bogs; N.B. to Man. s. to Pa., O., Mich., and Minn. May, June. (*Xylosteon o.*)

    9. Lonicera xylosteum L. European fly-h. Much like no. 8, but with hollow twigs, shorter (12–20 mm) peduncles, well developed bracts and bractlets, separate, glandular ovaries, and mostly shorter (7–12 mm) cor; $2n=18$. Native of Europe, occasionally escaped and ± established in our range. May, June.

    10. Lonicera japonica Thunb. Japanese h. Trailing or climbing vine, the young stems hairy; lvs ovate to oblong, 4–8 cm, rounded or broadly cuneate at base, sometimes toothed or lobed, glabrescent or hairy; peduncles axillary, 5–10 mm; bracts ovate, foliaceous; bractlets rotund, ciliate; cor 3–5 cm, strongly bilabiate, white or cream, the tube about equaling the lips; fr black; $2n=18$. Native of e. Asia, now well established in woods and fields in our range. (*Nintooa j.*)

    11. Lonicera periclymenum L. Woodbine. Woody climber, the younger branches usually pubescent and often glandular; lvs all separate, glabrous, elliptic to obovate, acute or obtuse, or the uppermost broadly ovate and acuminate; whorls of fls aggregated into a capitate cluster, the bractlets glandular; cor pale yellow, 3–5 cm, the tube glandular, longer than the lips; $2n=18, 36, 54$. Native of Europe, n. Afr., and w. Asia, sometimes escaped from cult., especially eastward.

    12. Lonicera sempervirens L. Trumpet-h. Woody twiner; lvs narrowly oblong to broadly oval, the uppermost pair (seldom 2 pairs) connate into a rhombic-elliptic disk, glaucous and glabrous, or villous especially beneath; fls in 1–4 whorls; cor 3–5 cm, narrow, red or yellow outside, yellow inside, the 5 lobes about equal, much shorter than the tube; stamens and style barely exserted; $2n=18$, 36. Woods and thickets, native at least from Conn. to Fla. and w. to Okla., widely escaped from cult. elsewhere. May–Fall. The widespread var. *sempervirens* has the stems, upper side of lvs, hypanthium, and outside of the cor glabrous. Var. *hirsutula* Rehder, occurring from Ala. to N.C., and reported from Va., has ciliate lvs hairy above, sometimes glandular stems, hairy cor, and glandular hypanthium. (*Phaenanthus s.*)

13. Lonicera prolifera (Kirchner) Rehder.   Grape-h. Woody climber with glabrous stems; lvs glaucous beneath, the lower broadly oval, sessile or nearly so, 4–8 cm, obtuse to emarginate, the upper 2–4 pairs ± connate, the uppermost forming a suborbicular disk rounded or retuse at the ends and glaucous above; spike short-peduncled; cor pale yellow, 2–3 cm, gibbous at the base, hairy inside, the lips scarcely as long as the tube; style hairy; $2n=18$. Moist woods and thickets. May, June. Var. *prolifera,* occurring from c. N.Y. to Wis. and Ill., has the lvs thinly villous beneath. Var. *glabra* Gleason, of s. Mo. and Ark., has the lvs glabrous. (*L. sullivantii.*)

The European *L. caprifolium* L., with the cor 3–5 cm, white or yellowish, the tube sometimes purplish, not gibbous, glabrous within, occasionally escapes from cult.

14. Lonicera dioica L.   Wild h. Climbing shrub with glabrous branches; lvs 5–12 cm, variable in shape, glaucous beneath, the uppermost one or two pairs united into a rhombic or doubly ovate disk, narrowed to an obtuse tip or rounded and mucronate; spike short-peduncled; cor 1.5–2.5 cm, pale yellowish to purplish, gibbous near the base, hairy inside, the lips about equaling the tube. Moist woods and thickets, occasionally on dunes or in swamps. May, June. Four vars.:

a Hypanthium glabrous.
  b Lvs glabrous beneath; cor-tube externally and style glabrous or sparsely hairy; Que. and Mass. to
    Wis., s. to N.J., N.C., and Ind. . . . . var. *dioica.*
  b Lvs sparsely to densely villous beneath; cor-tube glandular and villous; style hirsute; n. Ont. to
    Mack. and B.C., s. to Mich., Ill., Io., and Okla.; $2n=18$. (*L. glaucescens.*) . . . . var. *glaucescens* (Rydb.) Butters.
a Hypanthium densely glandular and sometimes also hairy; lvs and style of var. *glaucescens.*
  c Hypanthium merely glandular; s. Ont. to s. Mich., s. in the mts. to Va. and N.C. . . . . var. *orientalis* Gleason.
  c Hypanthium densely long-hirsute above the glands; O. and Ind. . . . . var. *dasygyna* (Rehder) Gleason.

15. Lonicera flava Sims.   Yellow h. Woody climber with glabrous branches; lvs not glaucous, the lower oblanceolate to broadly oval, acute or obtuse, the uppermost pair connate into an oval or subrotund disk with obtuse ends; fls golden to orange, 3–3.5 cm; cor-tube glabrous outside, glabrous or sparsely hairy inside, gradually expanded upwards, about as long as the lips; style glabrous; $2n=18$. Rocky woods and thickets; s. Mo. and Ky. to N.C., Ga., and Ark. Apr., May. Plants with the cor merely yellow, mostly shorter, and more hairy within have been distinguished as var. *flavescens* (Small) Gleason. They may be hybrids with *L. prolifera.* (*L. flavescens* Small, not Dippel; *L. flavida.*)

16. Lonicera hirsuta Eaton.   Hairy h. Twining shrub, the young stems hairy and glandular; lvs dull green, sessile or short-petioled, broadly oval to rhombic-ovate, 6–12 cm, hairy on both sides, the upper 1 or 2 pairs acuminate, connate into a rhombic-elliptic to subrotund disk; spikes with 1–4 crowded whorls of fls; cor yellow to orange, the pubescent tube slightly gibbous. Moist woods. June, July. Var. *hirsuta,* of w. N. Engl. and e. N.Y., has the cor-tube 15–22 mm, and the hypanthium densely glandular. Var. *interior* Gleason, of c. N.Y. and Pa. to s. Ont., Minn., and Man., has the cor-tube 10–18 mm, and the hypanthium glabrous or sparsely glandular.

3. SYMPHORICARPOS Duhamel.   Fls regular or the cor somewhat ventricose, (4)5-merous as to the cal, cor, and stamens, the cor short-campanulate to elongate-campanulate or subsalverform; ovary 4-locular, 2 locules with several abortive ovules, 2 with each a single functional ovule; style well developed; stigma ± capitate; fr a 2-stoned white or red berry-like drupe; low bushy shrubs with short-petioled, ovate-oblong to rotund, entire or sometimes toothed or lobed lvs, and clusters of small white or pink fls terminating the stem or also in the upper axils. 8 N. Amer., 1 China.

1 Cor 2–4 mm; fr red . . . . 1. *S. orbiculatus.*
1 Cor 5–8 mm; fr white.
  2 Style mostly 4–7 mm, usually long-hairy near the middle . . . . 2. *S. occidentalis.*
  2 Style 2–3 mm, glabrous . . . . 3. *S. albus.*

1. Symphoricarpos orbiculatus Moench.   Coralberry. Branching shrub to 1.5 m, the slender purplish stems hairy above; lvs oval or ovate, 2–4 cm, obtuse or rounded at both ends, hairy beneath; fls in dense clusters from the uppermost axils, sessile or nearly so; cor 2–4 mm, the lobes half as long as the tube; anthers and villous style included; fr red, persistent, 5–7 mm; $2n=18$. Dry or rocky soil and margins of woods; Conn. to N.C. and La., w. to Mich. and Colo. June–Aug. (*S. symphoricarpos.*)

2. Symphoricarpos occidentalis Hook.   Wolfberry. Shrub to 1 m, colonial by roots, the younger parts finely hairy; lvs ovate to ovate-oblong. 2.5–8 cm, often coarsely crenate; fls sessile in short dense spikes terminal and in the upper axils; cor 5–8 mm, the lobes somewhat arcuate-spreading, about equaling the tube; anthers 1.5–2 mm, shorter than the filaments; style 4–7 mm, long-hairy near the middle, seldom glabrous; fr 6–9 mm, white; $2n=36$. Dry prairies and moist low ground; Mich. to nw. Mo., w. to B.C. and Colo. June–Aug.

3. **Symphoricarpos albus** (L.) S. F. Blake. Snowberry. Shrub to 1 m, the younger parts finely hairy or glabrous; lvs ovate or oval, usually hairy beneath, mostly 2–3 cm; fls in pairs on short pedicels or in few-fld, interrupted spikes; cor 5–8 mm, ventricose, the lobes equaling or merely half as long as the tube; anthers 1–1.5 mm, nearly or fully as long as the filaments; style glabrous, 2–3 mm; fr 6–10 mm, white; $2n=36, 54$. Our native plants, as here described, are var. *albus*. The var. *laevigatus* (Fern.) S. F. Blake, of the Pacific slope, mostly 1–2 m, with the fr mostly 1–1.5 cm, and with the lvs usually glabrous beneath, often escapes from cult. Dry or rocky soil; Que. to s. Alas., s. to Va., Mich., Minn., and Calif. May–July. (*S. racemosus.*)

4. **TRIOSTEUM** L. Horse-gentian. Cor tubular-campanulate, gibbous at the base, unequally 5-lobed; sep 5, linear, elongate; stamens 5, the linear anthers included, on very short filaments; ovary 4-locular, one locule empty, the others with each a single ovule; style slender; stigma ± capitate; fr a yellow, red or greenish dry drupe with 3 oblong stones, crowned by the persistent sep; coarse, erect, hairy perennial herbs with large lvs connate or united by a ridge around the stem, and greenish-yellow to dull red fls solitary or in small clusters in the axils. 10, N. Amer. and e. Asia.

1 Sep uniformly pubescent on back and margin.
  2 Main lvs broadly connate-perfoliate; hairs of the stem mostly less than 0.5 mm .... 1. *T. perfoliatum.*
  2 Lvs not connate; hairs of the stem mostly more than 0.5 mm .... 2. *T. aurantiacum.*
1 Sep hispid-ciliate, pubescent to glabrous on the back .... 3. *T. angustifolium.*

1. **Triosteum perfoliatum** L. Perfoliate h.-g. Coarse herb to 13 dm; stems crisp-hairy, the abundant hairs mostly less than 0.5 mm; lvs obovate, ovate-oblong, or subrhombic, the main ones 10–30 × 4–15 cm, narrowed below the middle but broadly connate-perfoliate at base (3–9 cm wide where united), sparsely setose above, usually densely and softly hairy beneath; fls (1–)3–4 per axil; sep 10–18 mm, ca 1.5 mm wide, finely and uniformly hairy on the back and margin, often also glandular; cor 8–17 mm, crisp-hairy, purplish to dull greenish-yellow, the style exsert ca 2 mm; frs subglobose, dull orange-yellow; $2n=18$. Woods and thickets, often in thin or rocky soils; Mass. to s. Ont. and Minn., s. to S.C. and Okla. May–July.

2. **Triosteum aurantiacum** E. Bickn. Much like no. 1, and not always sharply separable, but the lvs distinct, tapering to a narrow base (seldom 1–3 pairs with undilated connate base 1–2 cm wide), the hairs of the stem mostly over 0.5 mm, the sep. ca 2 mm wide, the cor purplish-red, the style about equaling the cor or shortly included, and the fr bright orange-red. Rich woods and thickets; Que. and N.B. to Minn., s. to Ga., Ky., and Okla. May–July, a little after no. 1. Three vars.:

a Stems sparsely hirsute-pubescent with hairs 0.5–1.5 mm overlying shorter, partly glandular hairs;
  upper lf-surface glabrous or inconspicuously hairy.
  b Lvs evidently hairy beneath; widespread, but giving way in the Ozarkian region to var. *illinoense*
    .... var. *aurantiacum.*
  b Lvs glabrous or nearly so; c. N.Y., w. to Pa. and Ind. .... var. *glaucescens* Wieg.
a Stems shaggy-pubescent with hairs up to 3 mm, the short glandular hairs few or none; lvs often with
  appressed setae to 1 mm on the upper surface; O. to Kans. and Okla. (*T. illinoense.*) ....
  .... var. *illinoense* (Wieg.) Palmer & Steyermark.

3. **Triosteum angustifolium** L. Lesser h.-g. More slender than the previous 2 spp., the stems 3–8 dm, sparsely and retrorsely setose-hispid with hairs 1.5–3 mm and also glandular with shorter hairs; lvs oblanceolate to obovate, the main ones 8–18 × 2–6 cm, narrowed below the middle to a sessile base, setose-strigose above with hairs 1–2 mm; fls usually solitary in the axils; sep 9–12 mm, hispid-ciliate, the back glabrous or pubescent; cor greenish-yellow, loosely villous; fr orange-red. Moist woods and low ground; May, June. Two vars.: Var. *angustifolium*, pubescence of the lower lf-surface none or restricted to the veins; lvs mostly about a fourth as wide as long; Conn. to s. Ont., O. and Mo., s. to N.C., Ala., and La. Var. *eamesii* Wieg., lvs softly pubescent over the surface beneath, often obovate, ca half as wide as long; Conn. to N.J. and N.C.

5. **LINNAEA** L. Cor funnelform to campanulate, regular or nearly so, 5-lobed; cal 5-lobed; stamens 4, paired, included, inserted near the base of the cor-tube; ovary 3-locular, two locules each with several abortive ovules, the third with a single functional ovule; style slender, exserted; stigma capitate; fr small, dry, indehiscent, unequally 3-locular, 1-seeded; trailing, barely woody evergreen with erect, few-lvd branches bearing a terminal 2-fld peduncle. Monotypic.

1. Linnaea borealis L.   Twinflower. Stems trailing and creeping, emitting numerous short, suberect, leafy branches to 10 cm, these with a slender terminal peduncle to 10 cm; lvs firm, short-petioled, broadly oval to obovate, 1–2 cm; fls nodding; cor pink to white, 10–15 mm, shallowly 5-lobed, flaring from near the top of the cal, hairy inside; 2n=32. Moist or dry woods and cold bogs; circumpolar, s. in Amer. to N.J., W.Va., n. Ind., Minn., and Calif. June–Aug. Our plants are var. *longiflora* Torr. (*L. americana.*)

6. VIBURNUM L.   Cor regular (or the marginal ones sometime irregular), rotate to broadly campanulate, 5-lobed; stamens 5, exsert; stigma 3-lobed, sessile on a short stylopodium at the top of the ovary; ovary with 2 abortive locules and one fertile locule with a single ovule; fr a 1-stoned drupe; shrubs or trees with entire to lobed, simple lvs and many small, mostly white fls in umbelliform (or more ordinary) cymes. 250, widespread. Spp. 1–7 have 2 pairs of winter bud scales; spp. 8–11 have one pair; and spp. 12 and 13 have naked buds.

V. dilatatum Thunb., of e. Asia, rarely escapes from cult., as in N.J. and Pa. It has broad, toothed exstipulate lvs, hairy on both sides, veined as in spp. 1–4, pedunculate umbelliform cymes, a glabrous stylopodium, and bright red frs.

1 Marginal fls neutral, with greatly enlarged, somewhat irregular cor.
  2 Lvs pinnately veined, not lobed .... 13. *V. alnifolium.*
  2 Lvs palmately veined and lobed .... 7. *V. opulus.*
1 Marginal fls like the central ones.
  3 Lvs (or many of them) generally lobed.
    4 Lvs pilose on the veins beneath; fr red; filaments up to 1 mm .... 6. *V. edule.*
    4 Lvs stellate-pubescent beneath; fr blue-black; filaments 3–4 mm .... 5. *V. acerifolium.*
  3 Lvs not lobed.
    5 Lateral veins simple or 1-2-forked, each extending to a tooth.
      6 Infl oppositely branched even at the base; rare escape .... 1. *V. sieboldii.*
      6 Infl umbelliform at the base; common native spp.
        7 Stylopodium glabrous; petioles mostly with linear stipules; stone flattened.
          8 Petioles 3–15(–20) mm, up to a fifth as long as the basally obtuse to shallowly subcordate
            blade .... 2. *V. rafinesquianum.*
          8 Petioles (12–)20–50 mm, about a third as long as the basally cordate blade .... 3. *V. molle.*
        7 Stylopodium pubescent; stipules only seldom present; stone ellipsoid .... 4. *V. dentatum.*
    5 Lateral veins branching and anastomosing before reaching the lf-margin.
      9 Cymes on peduncles 5–50 mm.
        10 Lvs stellate-hairy beneath, at least when young; fr red, turning dark; intr. .... 12. *V. lantana.*
        10 Lvs not stellate; fr blue-back; native .... 8. *V. nudum.*
      9 Cymes sessile or subsessile on peduncles rarely as much as 5 mm.
        11 Lvs, or some of them, sharply acuminate .... 9. *V. lentago.*
        11 Lvs rounded to acute or barely acuminate.
          12 Petioles winged, densely red-tomentose at least when young; lvs shining .... 10. *V. rufidulum.*
          12 Petioles narrowly winged or wingless, glabrous or minutely scurfy; lvs dull .... 11. *V. prunifolium.*

1. Viburnum sieboldii Miq.   Coarse shrub or small tree; lvs exstipulate, elliptic-obovate, 5–15 cm, tapering to a short petiole, veined as in nos. 2–4, toothed; infl rounded, oppositely branched, the branches becoming bright red in autumn; frs at first pink, becoming blue-black; 2n=16, 32. Spring. Native of Japan, rarely escaped in our range.

2. Viburnum rafinesquianum Schultes.   Downy arrow-wood. Shrub to 1.5 m, the young stems glabrous or very sparsely stellate; petiole short, pubescent, mostly with a pair of linear stipules; lvs ovate-lanceolate to subrotund, 3–7 cm, glabrous or nearly so above, acuminate or acute, basally obtuse to subcordate, coarsely serrate, the teeth usually 6–10 on each side; cymes 4–7-rayed; sessile or on peduncles to 6 cm; hypanthium glandular; stylopodium glabrous; fr blue-black, flat-ellipsoid, 6–8 mm; stone flattened, shallowly grooved on both sides; 2n=36. May, June. Two vars.:

Var. *rafinesquianum*.   Petioles 2–15 (avg 6) mm, often exceeded by the stipules, avg ca a tenth as long as the blade; lvs softly stellate beneath. Dry, especially calcareous woods; Vt. to Man., s. to N.J., O. and Mo. (*V. pubescens.*)

Var. *affine* (Bush) House.   Petioles 4–20 (avg 10) mm, usually exceeding the stipules, a fifth to a tenth as long as the blade; lvs pilose on the veins beneath, at least near the base, otherwise glabrous. Rocky woods; s. Ont. to N.D., sw. to Ark. and Okla. (*V. affine.*)

3. Viburnum molle Michx.   Missouri arrow-wood. Shrub to 4 m, the young stems glabrous to sparsely pilose, the older ones with gray exfoliating bark; petioles glabrous, 1–5 cm, avg a third as long as the blade, mostly with linear stipules; lvs round-ovate, 5–10 cm, acute, coarsely serrate with 10–20 teeth on a side, all or most deeply cordate, glabrous or sparsely hairy above, thinly hairy beneath at least on the veins; cyme 5–7-rayed, the peduncle 3–5 cm; hypanthium glandular; stylopodium glabrous; fr blue-black, flat-ellipsoid, 8–10 mm; stone flattened, slightly grooved on both sides; 2n=18, 36. Rocky woods and hills; e. Pa.; Ind. to Io., s. to Ky. and Ark. May.

4. Viburnum dentatum L.   Arrow-wood. Shrub 1–5 m, with close gray-brown or reddish bark; petioles 8–25 mm, glabrous or stellate, usually exstipulate; lvs lance-ovate to rotund, 4–10 cm, sharply acute or short-acuminate to broadly rounded, sharply toothed; cyme mostly 5–7-rayed, the peduncle 3–6 cm; hypanthium glabrous, sparsely glandular, or setose; stylopodium pubescent; fr blue-black, subglobose to ovoid, 5–10 mm; stone ellipsoid, plump, deeply grooved on one side; $2n=36, 54, 72$. Me. to Ill., s. to Fla. and Tex., and elsewhere as an escape from cult. May–July. Variable in morphology and habitat; divisible into several vars.:

a Petioles glabrous; lower lf-surface glabrous, or pubescent in the axils of the veins; hypanthium glabrous;
    moist woods and swamps; Me. to N.Y. and O., s. in the mts. to Ga.; the common northern form.
    (*V. recognitum.*) .... var. *lucidum* Aiton.
a Petioles stellate; lvs usually stellate beneath, at least on the veins.
  b Stipules and stipitate-glandular hairs none.
    c Lvs thin, mostly ovate, the lower surface thinly stellate to glabrate; hypanthium glabrous or
      glandular; mostly on the coastal plain, N.J. to Fla. and Tex. (*V. semitomentosum.*) .... var. *dentatum.*
    c Lvs thick, mostly rotund, the lower surface thickly stellate and with strongly raised veins.
      d Hypanthium glabrous or glandular; sandy soil of the coastal plain, e. Mass. to N.J. (*V. ve-*
        nosum.) .... var. *venosum* (Britton) Gleason.
      d Hypanthium and cor densely setose or hirsute; coastal plain, Fla. to La., and in s. O. ... var. *scabrellum* T. & G.
  b Stipules present on many petioles; stipitate-glandular hairs often present.
    e Lvs ± stellate beneath; petioles stellate on top, sometimes also stipitate-glandular; s. O. to s.
      Mo. .... var. *deamii* (Rehder) Fern.
    e Lvs nearly or quite glabrous beneath; petioles stellate only in the furrow, stipitate-glandular on
      top; s. Ind. and Ill. .... var. *indianense* (Rehder) Fern.

5. Viburnum acerifolium L.   Dockmackie; flowering maple. Shrub 1–2 m; young stems, petioles, lower lf-surfaces and infl finely stellate; petioles 1–2 cm; lvs maple-like, palmately veined, shallowly 3-lobed (rarely lobeless), 6–12 cm, coarsely toothed, basally rounded to subcordate; cymes 3–5 cm wide, mostly 7-rayed, on peduncles 3–5 cm; cor 4–5 mm wide; filaments 3–4 mm; fr purple-black, ellipsoid or subglobose, 8 mm; stone lenticular, with 3 shallow grooves on one side, 2 on the other; $2n=18$. Moist or dry woods; Que. and N.B. to Minn., s. to Fla. and La. May, June.

6. Viburnum edule (Michx.) Raf.   Squashberry. Straggling shrub 0.5–2 m, with glabrous or finely glandular stems and petioles; petioles 1–3 cm; lvs palmately veined, shallowly 3-lobed, 5–10 cm, sharply serrate, basally rounded to subcordate, sometimes some lvs lobeless and pinnately veined; cymes 1–2.5 cm wide, with less than 50 fls, mostly 5-rayed, short-peduncled; cor 4–7 mm wide; filaments up to 1 mm; fr red, 1–1.5 cm; stone flattened, not grooved; $2n=18$. Moist woods; Lab. to Alas., s. to Pa., Mich., Minn., Colo., and Ore. June, July. (*V. pauciflorum; V. eradiatum.*)

7. Viburnum opulus L.   Shrub 1–5 m; young stems smooth; petioles 1–3 cm, with 1–6 large glands near the apex; stipules slender; lvs maple-like, palmately veined, 3-lobed, coarsely toothed, basally obtuse to truncate, hairy beneath, especially on the veins; cymes 5–10 cm wide, on peduncles 2–5 cm; marginal fls neutral, with enlarged, slightly irregular cor 1.5–2.5 cm wide; fr red, 10–15 mm; stone flat, suborbicular, not grooved; $2n=18$. Var. *americanum* Aiton, the high-bush cranberry, with the petiolar glands stalked, round-topped, and mostly higher than wide, occurs in moist woods from Nf. to B.C., s. to Pa., n. O., Io., and Wash. June. (*V. trilobum.*) The Eurasian var. *opulus,* the Guelder rose, with the petiolar glands mostly sessile, concave-topped, and wider than high, occasionally escapes from cult. June, July. The cult. Snowball is a form of var. *opulus* with all fls neutral and enlarged.

8. Viburnum nudum L.   Shrub to 4 m; young stems smooth or brown-scurfy; petioles 5–15 mm; lvs ovate to lanceolate, oblong, or oblanceolate, to 12 cm, narrowly revolute or narrowly cartilaginous-thickened, the lower side with scattered minute appressed red-brown scales, or glabrous; fr ovoid to subglobose, 6–12 mm, blue-black, glaucous; stone flattened, the sides not grooved; $2n=18$. Mostly in wet woods and swamps; Nf. to Man., s. to Fla. and Tex. May–July. Two geographic vars., often treated as distinct spp., but apparently overlapping in all characters:
Var. *nudum.*   Possum-haw. Lvs shiny above, evidently veiny beneath, mostly oblong or oblong-lanceolate, varying to ovate, acute or tapering at base, fairly uniformly tapering or rounded to the tip, mostly entire or merely undulate; winter-buds reddish-brown to fuscous; peduncles 2–5 cm, avg 3.5 cm at anthesis; drupe with round-obovate stone and usually bitter pulp; swamps of the coastal plain, Conn. to Fla. and Tex., and inland from Va. and Ky. southward.
Var. *cassinoides* (L.) T. & G.   Withe-rod. Lvs dull, firmer than in var. *nudum,* indistinctly veiny, ovate to lanceolate or oblanceolate, tending to be bluntly short-acuminate, rounded to tapering at base, crenulate, varying to seldom entire; winter-buds yellow-brown or golden; peduncles 5–25 mm, avg 13 mm at anthesis; drupe with elliptic or oblong-elliptic stone and sweet pulp; more northern, extending s. mainly in the mts. to N.C., Ga., and Ala. (*V. cassinoides.*)

9. Viburnum lentago L.   Sheepberry; nannyberry. Tall shrub or small tree to 10 m, glabrous throughout or minutely scurfy on the infl or petioles; petioles 1–2 cm, some or all irregularly wing-margined; lvs ovate to oblong or orbicular, 5–8 cm, some or all sharply acuminate, sharply serrate, the teeth 6–10 per cm of margin, often incurved and callous-tipped; cymes sessile, 5–10 cm wide, with 3–5(7) rays; cor 4–8 mm wide, the lobes acute; fr blue-black with a whitish bloom, ellipsoid to subglobose, 8–15 mm, with sweet pulp; stone flat, nearly smooth; $2n=18$. Woods and roadsides; Que. to se. Sask. and sc. Mont., s. to N.J., Va., Ill., Mo., Nebr., Wyo. and Colo. May, June.

10. Viburnum rufidulum Raf.   Southern black haw. Shrub or small tree to 10 m; petioles and to some extent the lvs and cymes red-brown-tomentose, especially when young; petioles ± wing-margined, especially toward the base; lvs firm, shining, elliptic to obovate, 5–8 cm, rounded to acute at apex, serrulate, broadly acute to less often rounded at base; cymes sessile, (3)4-rayed, 5–10 cm wide; cor ca 8 mm wide; fr sweetish to tart, mostly blue to purple, glaucous, ellipsoid, 10–15 mm; cal-tube not produced at the end of the fr; $2n=18$. Dry or rocky woods and banks; s. Va. to s. O., Mo., and Kans., s. to Fla. and Tex. Apr., May.

11. Viburnum prunifolium L.   Black haw. Shrub or small tree to 8 m; stem and foliage glabrous, or sparsely brown-scurfy when young; petioles 5–12 mm, wingless or nearly so; lvs oblong or elliptic to obovate, rounded to apiculate or acute at the tip, serrulate, obtuse to rounded at the base, the floral ones mostly 3–5 cm at anthesis, those on sterile stems eventually 6–8 cm; cyme sessile (3)4-rayed, 5–10 cm wide; cor 4–7 mm wide; fr sweetish, blue-black, 9–15 mm, ellipsoid to subglobose; cal-tube forming a ring at the tip of the fr; stone flat, scarcely grooved; $2n=18$. Woods, thickets, and roadsides; Conn. to s. Wis., s. Io., and Kans., s. to Ga. and Tex. Apr., May. (*V. bushii.*)

12. Viburnum lantana L.   Wayfaring tree. Shrub to 5 m; young stems, petioles, and lower lf-surface finely and loosely gray-stellate; petioles 1–3 cm; lvs oblong to ovate, 5–12 cm, acute or obtuse, finely serrate, basally rounded or cordate; cyme short-pedunculate, about 7-rayed; cor 5–8 mm wide; fr red, turning dark, 8–10 mm; stone furrowed on both sides; $2n=18$. Native of Eurasia, sometimes escaped from cult. June.

13. Viburnum alnifolium Marshall.   Hobble-bush. Shrub 1–2 m, the lower branches often rooting; young stems, buds, infl, petioles and lf-veins beneath tomentosely rusty-stellate; petioles 1.5–3 cm; lvs pinnately veined, broadly ovate, 10–18 cm (or smaller at anthesis), short-acuminate, basally cordate, serrulate; summer lvs sometimes narrower and coarsely toothed; cyme sessile; rays mostly 5, 2–5 cm; marginal fls neutral, with enlarged, slightly irregular, sometimes pink cor 2–3.5 cm wide; fr red, turning dark, 8–10 mm, the stone ovoid, flattened on one side, several-grooved; $2n=18$. Moist woods; N.S. to Mich., s. in the mts. to N.C., and Tenn. May, June. (*V. lantanoides.*)

V. plicatum Thunb., an east-Asian sp. with cymes mostly 7-rayed and pedunculate on short lateral branches, and with mostly smaller (6–10 cm) lvs and blue-black fr, sometimes escapes from cult. (*V. tomentosum.*) The Japanese snowball is a form with all fls sterile.

7. SAMBUCUS L.   Elder. Cor regular, rotate to saucer-shaped, 5-lobed; stamens 5; style very short, 3–5-lobed; ovary 3–5-locular, with one ovule per locule; fr berry-like, juicy, with 3–5 small seed-like stones; coarse herbs, shrubs, or small trees, with large, pinnately compound lvs and serrate (or cleft) lfls, the small white fls numerous in large terminal cymes. 20, widespread.

1 Infl 5-rayed from the base, lacking a central axis beyond its lowermost branches .... 1. *S. canadensis.*
1 Infl panicle-like, with the main axis extending well above the lowermost branches .... 2. *S. racemosa.*

1. Sambucus canadensis L.   Common e. Shrub to 3 m, spreading underground and eventually forming thickets; younger stems scarcely woody, with long internodes and large white pith; lfls 5–11, usually 7, lanceolate to ovate, variable in size, acuminate, sharply serrate (rarely laciniate), the lower occasionally divided into 3, glabrous or more often hairy beneath; infl 5-rayed from the base, flat or convex, 5–15 cm wide; fls white, 3–5 mm wide; fr purple-black (seldom red, green, or yellow), edible, 5 mm; $2n=36$. Moist woods, fields, and roadsides; N.S. and Que. to Man. and S.D., s. to Mex. and the W. Ind. July, Aug. The common phase in most of our range, with the lf-pubescence mostly setulose and largely restricted to the veins, is var. *canadensis.* Ozarkian plants, from Ill. to Tex., with fine short pubescence on both the veins and the lf-surface, may be distinguished as var. *submollis* Rehder.

A cut-lvd form of *S. nigra* L., a more arborescent European sp. with jet black frs and typically with only 5 lfls, is reported to be rarely escaped in our range.

2. Sambucus racemosa L.   Red-berried e. Shrub to 3 m, the younger parts usually finely hairy; pith brown; stipules minute or none; lfls 5–7, lance-ovate to narrowly oblong, acuminate, finely toothed, usually soft-hairy beneath; infl pyramidal or convex, panicle-like, with an evident main axis extending beyond the usually paired lowermost branches; fls white, 3–4 mm wide; fr red (seldom yellow or white), 5 mm; $2n=36 + 0$–2 B. Rich woods; circumboreal, in Amer. from Nf. to B.C., s. in our range to Pa., Ind., Ill., and in the mts. to N.C., May, June. The N. American ssp. *pubens* (Michx.) House is represented in our range by var. *pubens* (Michx.) Koehne. (*S. pubens.*) Other vars. are cordilleran.

S. ebulus L., the dwarf elder, a coarse herb to 2 m, with well developed ovate stipules, purple anthers, and black fr, occasionally escapes from cult.

FAMILY **ADOXACEAE**, the Moschatel Family

Fls perfect, sympetalous; each cor-lobe with a nectary at the base, the nectaries composed of cushion-like groups of multicellular, sessile or short-stalked, clavate glands; stamens attached at the sinuses of the cor, either paired and each with a single pollen-sac, or solitary at the sinuses, but with the filament divided above so that the pollen-sacs are well separated; ovary plurilocular, each locule with a solitary, apical-axile, pendulous ovule; fr a small, dry drupe with distinct stones; seeds with a small, dicotyledonous embryo and copious oily endosperm; small, perennial herbs with alternate basal lvs and a single pair of opposite cauline lvs; stipules wanting; 3/3, N. Hemisphere.

1. **ADOXA** L.   Moschatel. Fls half-epigynous, with a shortly prolonged hypanthium, mostly 5 in a compact, head-like terminal cyme; sep of the central fl commonly 2, of the lateral ones 3; cor rotate, that of the central fl regular and commonly 4-lobed, that of the lateral ones slightly irregular and commonly 5-lobed; stamens paired at the sinuses of the cor, each bearing a single pollen-sac; carpels several, most often 4 in the central fl and 5 in the lateral ones, united to form a compound ovary with distinct styles and capitate stigmas. Monotypic.

    1. **Adoxa moschatellina** L.   Delicate herb with a musky odor, 5–20 cm; basal lvs long-petiolate, ternate, the primary divisions discrete, again 1–2-ternate or parted, the ultimate segments rather broad, thin, round-toothed and mucronate; cauline lvs similar but smaller and less dissected, commonly borne a little above the middle of the shoot; cor yellowish-green, 5–8 mm wide; $2n=36, 54$. Moist, often mossy places in forested regions; circumboreal, extending s. to s. Minn., Wis., n. Ill., n. Io., and Delaware Co., N.Y. May–July.

Family **VALERIANACEAE**, the Valerian Family

    Fls sympetalous, epigynous, regular or irregular, perfect or unisexual; cal-segments either inrolled at anthesis and later expanded and pappus-like, or much reduced or obsolete; cor mostly 5-lobed, often bilabiate, the tube often spurred or gibbous; stamens borne on the cor-tube, alternating with but fewer than the lobes, typically 3; pistil basically tricarpellate, one carpel fertile, the other 2 sterile and sometimes obsolete; style with a simple, entire or bilobed or more often trilobed stigma; ovule solitary, pendulous; fr dry, indehiscent; seeds with a large, straight, oily dicotyledonous embryo, without endosperm; opposite-lvd, exstipulate herbs with the fls in various sorts of basically determinate infls (but not in involucrate heads). 13/300.

1 Cauline lvs pinnately divided; cal-segments 9–20, inrolled at anthesis, expanded on the mature fr;
    ours perennial . . . . 1. *Valeriana.*
1 Cauline lvs simple cal-segments minute or obsolete; annual or biennial . . . . 2. *Valerianella.*

    1. **VALERIANA** L.   Valerian. Fls perfect or unisexual; cal initially involute and in-conspicuous, later enlarged and spreading, usually with several or many long, setaceous, plumose, pappus-like segments; cor-tube sometimes gibbous at base, the 5 lobes equal or subequal; stamens 3; ovary basically tricarpellate, but 2 of the carpels vestigial; stigma 3-lobed; fr a nerved achene; annual or (ours) perennial herbs with entire to bipinnatifid lvs, the fls in corymbiform to paniculiform or thyrsoid infls of basically determinate nature. 200, nearly cosmop.

1 Cor-tube 12–16 mm . . . . 1. *V. pauciflora.*
1 Cor-tube 0.5–4 mm.
  2 Plants with a stout rhizome or caudex and numerous fibrous roots.
    3 Basal lvs undivided, or with a single pair of small basal lobes . . . . 2. *V. uliginosa.*
    3 Basal lvs divided into numerous segments . . . . 3. *V. officinalis.*
  2 Plants with a stout taproot and short branched caudex . . . . 4. *V. edulis.*

    1. **Valeriana pauciflora** Michx.   Long-tube v. Stems 3–8 dm from a slender rhizome; basal lvs slender-petioled, broadly cordate, repand, 5–8 cm; cauline lvs short-petioled, pinnately divided into 3–7 segments, the terminal lobe broadly ovate or deltoid, much larger than the lateral ones; infl small with short, approximate branches, becoming diffusely pyramidal in age; cor-tube slender, 12–16 mm, its lobes 2–3 mm; fr lance-oblong, 5 mm, sparsely villous at least when young. Moist rich soil; Pa. and Md. to Va., Ky., and s. Ill. May, June.

2. **Valeriana uliginosa** (T. & G.) Rydb.   Fibrous-rooted perennial from a stout branched rhizome or caudex, 3–10 dm, glabrous or nearly so; basal lvs long-petiolate, the blade ovate-elliptic to obovate, 6–14 cm, toothed or entire, or with a single pair of small basilateral lobes; cauline lvs 3–6 pairs, well developed, pinnatifid, ciliate; fls all perfect, the cor 5–7 mm; achenes lanceolate to ovate-oblong, 3–4 mm, glabrous. Marshy meadows, swamps, and bogs; Me. and s. Que. to N.Y., n. O., and w. Mich. May–July. (*V. septentrionalis* var. *u.*)

3. **Valeriana officinalis** L.   Garden-heliotrope. Fibrous-rooted perennial 5–15 dm from a short rhizome; stems hairy at least at the nodes; basal and cauline lvs similar, pinnately divided into 11–21 lanceolate, dentate segments; petioles of the upper lvs progressively shorter; infl large and open, the lower branches often remote from the upper; cor obconic, the tube 4 mm, the lobes 1 mm; fr lance-oblong, 3–5 mm, glabrous; $2n=14, 28, 56$. Native of Eurasia, escaped from cult. here and there in our range. May–Aug.

4. **Valeriana edulis** Nutt.   Taprooted v. Perennial from a long stout taproot and short branched caudex, 3–12 dm; lvs thick, nearly parallel-veined, densely ciliate, the basal linear-oblanceolate, 1–3 dm, entire or rarely with 1–2 basal divisions; cauline lvs pinnately parted into a few ascending or falcately divergent segments from a broad flat rachis; infl elongate, with numerous lateral branches, becoming diffuse in age; plants polygamo-dioecious; cor of perfect and staminate fls 2.5–3.5 mm, of pistillate fls scarcely 1 mm; fr ovate or ovate-oblong, 3–4 mm. Swamps and wet open soil; c. O. and s. Ont. to Minn. and Io.; cordilleran region from s. B.C. to Mex. May, June. Our plants are usually short-hairy, and may be distinguished from the usually glabrous cordilleran plants as *V. edulis* var. *ciliata* (T. & G.) Cronq. (*V. ciliata.*)

3. **VALERIANELLA** Miller.   Corn-salad. Cal minute or obsolete; cor funnelform, narrowly campanulate, or tubular, the 5 lobes subequal; stamens 3; stamens and style exsert; fr dry, 3-locular, the one locule fertile and 1-seeded, the other 2 empty; dichotomously forked, annual or biennial herbs with simple, mostly sessile lvs, the lower often connate; fls in cymose glomerules ending the branches. Our spp. much alike in aspect and habitat. 60, N. Hemisphere.

1 Fertile locule of the fr greatly thickened by a corky mass on the back .... 1. *V. locusta.*
1 Fertile locule without a corky mass.
  2 Fertile locule much wider than the 2 sterile locules together; sterile locules closely approximate,
    with no groove between, but separated in the center by a narrow pit; fr sharply triangular in
    cross-section .... 2. *V. chenopodifolia.*
  2 Fertile locule much narrower to slightly wider than the 2 sterile ones together; fr not sharply
    triangular in cross-section.
    3 Cor tiny, 1.5–2 mm, its lobes 0.4–0.8 mm .... 3. *V. radiata.*
    3 Cor a little larger, 3–5 mm, its lobes 1–2 mm .... 4. *V. umbilicata.*

1. **Valerianella locusta** (L.) Betcke.   European c.-s. Annual 1–4 dm; lvs ciliolate and sometimes short-hairy, the lower broadly oblanceolate or broader, and ± petiolate, the others sessile and more oblong, entire or the upper with a few teeth near the base, 1–7 cm × 3–18 mm; cor white or pale bluish, 1.5–2 mm; fr 2(–4) mm, the fertile locule bearing a thick corky mass on the back; groove between the sterile locules narrow, shallow, and inconspicuous; $2n=14$ (–18). Moist, open places, often in disturbed soil; native of Europe, now widely established in the U.S. Apr.–June. (*V. olitoria.*)

2. **Valerianella chenopodifolia** (Pursh) DC.   Great Lakes c.-s. Bracts and uppermost lvs lanceolate, glabrous, acute; cor white, 3–4 mm; fr sharply triangular in cross-section, the fertile locule flat on the back, much wider than the 2 sterile ones, which are barely separated by a short pit near the center. S. Ont. to Wis., s. to Pa. and Ind. May, June.

3. **Valerianella radiata** (L.) Dufr.   Bracts lanceolate, glabrous or sparsely ciliate; cor white, 1.5–2 mm, its lobes 0.4–0.8 mm; fr ovoid, 2–2.5 mm, the fertile locule a little wider to a little narrower than the 2 sterile ones together, these separated by a ventral groove, as in one form of no. 4. Va. to O., s. Ill. and Kans., s. to Fla. and Tex. Apr., May.

4. **Valerianella umbilicata** (Sulliv.) A. Wood.   Infl loosely fld, with elongate branches; cor white, 3–5 mm; its lobes 1–2 mm; fr trimorphic, 3–5 mm, 2 forms near orbicular in dorsal (abaxial) view, the fertile locule ovoid, the sterile ones much dilated, their lateral walls relatively firm, ± incurved on the adaxial side, the ventral walls very thin and membranous, soon ruptured to expose a deep and narrow or shallow and broad adaxial pit; in the third type the sterile locules not expanded, together forming an ovoid, medially grooved protuberance hidden from abaxial view by the fertile locule. Local from s. N.Y. and s. Ont. to Ill., s. to N.C. and Tenn. May, June. (*V. intermedia; V. patellaria.*)

FAMILY **DIPSACACEAE**, the Teasel Family

Fls sympetalous, mostly perfect, epigynous, ± irregular; cal small, cupulate or deeply cut into 4 or 5 segments or into more numerous teeth or bristles; cor 4–5-lobed, often ±

2-lipped; stamens (2–)4, borne near the top of the cor-tube, alternate with the lobes, exsert, not connate; style with a simple or 2-lobed stigma; ovary inferior, unilocular, with a solitary, apical, pendulous ovule; fr dry, indehiscent, enclosed (except at the tip) by a gamophyllous, apically cupulate-toothed or subentire involucel that may be adnate to the ovary below; seeds with large, spatulate, dicotyledonous embryo and rather scanty, oily endosperm; herbs with opposite or whorled, exstipulate lvs, the fls ordinarily in dense, involucrate, cymose heads. 10/270, Old World.

1 Stem (and usually also the lvs and involucre) prickly .... 1. *Dipsacus.*
1 Stem not prickly.
  2 Cal-setae or -teeth 4 or 5; receptacle bracteate, not hairy.
    3 Marginal and central fls subequal; cor 4-lobed; none of the lvs pinnatifid.
      4 Cal 4-lobed, without awns; involucel 8-ribbed, smooth .... 2. *Succisella.*
      4 Cal with 4 or 5 short awns; involucel 4-angled, pubescent .... 3. *Succisa.*
    3 Marginal fls ± enlarged, subradiate; cor 5-lobed; some of the lvs pinnatifid .... 4. *Scabiosa.*
  2 Cal-setae or -teeth 8 or more; receptacle hairy; not bracteate .... 5. *Knautia.*

1. DIPSACUS L. Teasel. Cal short, 4-angled or 4-lobed, hairy; cor 4-lobed, the marginal ones not enlarged; involucel 4-angled, truncate or 4-toothed at the top; involucral bracts slender, often elongate; receptacular bracts ovate or lanceolate, acuminate into an awn surpassing the fls; tall biennial or perennial herbs, little branched, with prickly stems, large, sessile or even connate lvs, and small fls in dense, ovoid to cylindric heads. 10+, Europe, w. Asia, n. Afr.

1. **Dipsacus sylvestris** Hudson. Common or wild t. Stout, taprooted biennial (or short-lived, monocarpic perennial) 0.5–2 m, the stem increasingly prickly on the angles above; lvs ± prickly, especially on the midrib beneath, otherwise smooth or nearly so, the basal ones oblanceolate, crenate, generally dying early in the second season, the cauline lanceolate, to 3 dm, becoming entire upwards, commonly connate at base; heads erect, ovoid or subcylindric, 3–10 cm, on long naked peduncles; involucral bracts upcurved, unequal, the longer ones surpassing the heads; receptacular bracts ending in a conspicuous stout straight awn; cal silky, 1 mm; cor slender, hairy, 10–15 mm, the tube whitish, the short (1 mm) lobes generally pale purple; fr 5 mm; $2n=16$, 18. Roadsides and waste ground, especially in moist low places; native of Europe, now a widespread weed in N. Amer. July–Sept.

    D. fullonum L. long cult. as the fullers' teasel, is rarely adventive with us. It differs from the closely allied wild *D. sylvestris* in the stouter, strongly recurved spine-tips of its receptacular bracts. Linnaeus included both in his *D. fullonum,* and the name has subsequently been restricted by different authors to each of them.

2. **Dipsacus laciniatus** L. Cut-lf t. Much like *D. sylvestris,* and hybridizing with it, but the principal cauline lvs irregularly laciniate-pinnatifid; $2n=16$, 18. Native of Europe, sparsely and irregularly intr. in our range.

2. SUCCISELLA G. Beck. Cal 4-lobed, without awns; involucel 8-ribbed, urceolate, smooth; otherwise like *Succisa.* Probably monotypic.

1. **Succisella inflexa** (Kluk) G. Beck. Devil's bit. Much branched, fibrous-rooted perennial 4–10 dm; lvs lance-linear, only those of the branches much reduced; heads 1–1.5 cm thick; cor pale blue; fr 5 mm, glabrous, strongly 8-ribbed; cal short, awnless; $2n=20$. Native of Europe, occasionally found in fields and waste places from N. Engl. to Pa. June–Sept. (*Scabiosa australis; Succisa inflexa.*)

3. SUCCISA Haller. Cal shallowly cupuliform, with (4)5 persistent setae; involucel 4-angled; marginal and central fls subequal; cor-lobes 4, subequal; perennial herbs with hemispheric, long-pedunculate heads. 3, Eurasia, Afr.

1. **Succisa pratensis** Moench. Devil's bit. Slender, fibrous-rooted perennial 3–10 dm; lvs mostly in the lower half of the stem, entire or nearly so, oblanceolate, slender-petiolate, the upper ones much reduced; gynodioecious, the perfect heads up to 3 cm thick, terminal, the female smaller; cor 4–7 mm, blue to white; fr 5 mm, hairy; involucel toothed; cal short, with 5 short awns; $2n=20$. Native of Eurasia and n. Afr., occasionally found as a weed in the ne. part of our range. Aug. (*Scabiosa succisa.*)

4. SCABIOSA L. Cal cupuliform, the upper part usually prolonged into 5 setae; involucel 8-ribbed and cylindric below, expanded above into an orbicular or funnelform, scarious corona with many (sometimes excurrent) veins; cor with short tube and 5 unequal lobes, the marginal ones usually enlarged and subradiate; herbs with long-pedunculate heads and 1–3 rows of herbaceous involucral bracts. 80, Eurasia, Afr.

1. **Scabiosa columbaria** L.   Scabious. Biennial or perennial 3–8 dm; basal lvs oblanceolate or obovate to lyrate; cauline lvs once or twice pinnately divided, the divisions often laciniate; heads 3 cm thick, blue, pink, or white; achenes 4 mm, deeply ribbed-sulcate; corona 1.5 mm, ca 24-veined; $2n=16$. Native of Europe, n. Afr., and w. Asia, rarely adventive in our range. July–Sept.
   S. atropurpurea L., sweet scabious, pincushions; annual or biennial, with dark purple to rose or white fls and 8-veined corona, is rarely adventive with us and probably never persistent.

5. **KNAUTIA** L.   Cal short, with 8–12 setaceous teeth; cor 4-lobed, ± irregular, the marginal ones distinctly so; involucral bracts about equaling the head; receptacle densely hairy, without bracts; involucel 2-toothed at the top, 4-ribbed; herbs with fls in dense hemispheric heads on long peduncles. 60, Europe, w. Asia, n. Afr.

1. **Knautia arvensis** (L.) Duby.   Blue buttons. Taprooted, hairy perennial 3–10 dm, lowest lvs usually merely coarsely toothed, the others ± deeply pinnatifid with narrow lateral and broader terminal segments, to 2.5 cm, reduced upwards; heads 1.5–4 cm thick; involucral bracts 8–12 mm; cal 3–4 mm; the 8–12 teeth bristle-like; cor lilac-purple, the marginal ones the largest; fr 5–6 mm, densely hairy; $2n=20, 40, 43, 46$. A European weed, naturalized from Nf. and Que. to N.J. and W.Va. June–Sept. (*Scabiosa a.*)

FAMILY **ASTERACEAE**, the Aster Family

Individual fls epigynous, perfect or unisexual, sympetalous, regular or irregular, commonly 5-merous, without definite cal; stamens alternate with the cor-lobes, epipetalous, usually with elongate anthers united into a tube; ovary bicarpellate but unilocular, with a single erect ovule; style usually 2-cleft; fr an achene, unappendaged, or more commonly crowned with a pappus consisting of hairs or scales; fls sessile in a close head on a common receptacle, sometimes individually subtended by a small bract (chaff), and almost always collectively subtended by an invol of few to many bracts; our spp. annual to perennial herbs or rarely shrubs, the heads arranged in various sorts of basically determinate infls. (Compositae, nom. altern.) 1100/20,000.

The *invol bracts* are usually herbaceous or subherbaceous in texture, varying to scarious, hyaline, coriaceous, or cartilaginous. They may be few and in a single row, or numerous and imbricate, or modified into spines, or even (as in *Xanthium*) concrescent into a bur.

The *receptacle* may be *chaffy*, with a bract behind each fl (as in many Heliantheae), or may be covered with long, stout bristles (as in most Cynareae), or may be *naked*, without chaff or bristles. When naked it may sometimes be minutely pitted, with slender, chaffy partitions separating the pits, and is then said to be *alveolate*. It may even be softly hairy, as in some spp. of *Artemisia*.

The fls are of several general types. In one type they are perfect (or functionally staminate) and the cor is tubular or trumpet-shaped or goblet-shaped, with typically 5 short lobes or teeth. This type of fl is called a *disk-fl*. A head composed wholly of disk-fls is said to be *discoid*.

In another type the fl is pistillate or neutral (without a style), and the cor is tubular only at the very base, above which it is flat and usually bent backward so as to spread away from the center of the head. The flattened part of a cor of this type is called a *ray* or *ligule*, and the fl bearing it is called a *ray-fl* or *ligulate fl*. Often the ligules exhibit traces of 2 or 3 cor-lobes as small terminal teeth. Except for the pistillate heads of a few dioecious groups, the head is never composed wholly of fls of this type. Instead these pistillate or neutral ray-fls are found at the margin of the head, the center being occupied by the disk-fls. Such a head, with both ray-fls and disk-fls, is said to be *radiate*.

In some spp. the ray or ligule of the marginal, pistillate fls does not develop, so that the cor is tubular. In addition to not bearing stamens, a cor of this type differs from the cor of an ordinary disk-fl in the absence of the regular terminal teeth, and often also in being more slender. A head in which the pistillate fls lack rays is said to be *disciform,* although the term discoid is sometimes loosely extended to cover this type.

Another type of fl superficially resembles the ray-fl of a radiate head, but differs in being perfect and in usually having 5 terminal teeth on the ligule. The heads of the tribe Lactuceae consist wholly of fls of this type and are called *ligulate* heads. Ligulate perfect fls are rare in other tribes, and almost never make up the whole head.

In some spp. of *Centaurea* the marginal fls are neutral and have an enlarged, irregular, ray-like cor. These fls are considered to be modified disk-fls. Members of the tribe Mutisieae (not represented in our region) usually have some fls with a bilabiate cor.

The pollen-presentation mechanism is a characteristic feature of the family. The anthers are coherent by their lateral margins, or rarely merely connivent. The base of the anther varies from obtuse or subtruncate to broadly rounded, sagittate, or distinctly caudate (tailed). The anthers dehisce introrsely, and the pollen is pushed out through the anther-tube by growth of the style. The style-branches commonly diverge above the anther-tube, have various distinctive forms and texture, and tend to be stigmatic only on limited parts of their surface. The characteristic style-branches of the various tribes are to be sought only in the fertile disk-fls. The styles of ray-fls are mostly very similar in all groups, and those of sterile disk-fls are often reduced and undivided. The sterile disk-fls, when present, are said to be *functionally staminate.* Strictly staminate fls, with no pistillate parts, do not normally occur in the Asteraceae, because the style is necessary as a piston or plunger to eject the pollen.

Synopsis of the Tribes

1 Fls all ligulate and perfect; juice usually milky. (Genera 87–107) .... Tribe 9. Lactuceae.
1 Fls, or some of them, tubular and eligulate; juice usually watery.
  2 Heads mostly radiate, or not infrequently disciform, or in scattered genera, species, and individuals discoid; cors predominantly yellow, although various anthocyanic shades and white are not uncommon.
    3 Lower (or all) lvs in most genera opposite, although a number of genera have wholly alternate lvs; pappus usually chaffy, or of a few firm awns, or none (of capillary bristles in *Arnica*); receptacle chaffy or less often naked; invol bracts tending to be herbaceous and several-seriate, but varying in groups of genera to uniseriate, or subchartaceous, or even concrescent into a bur; style-branches often ± hispidulous and with the stigmatic lines poorly developed, varying to as seen in the Astereae or Senecioneae; anthers not tailed; rays tending to be relatively broad, often much larger than is usual for the other tribes. (Genera 1–33) .... Tribe 1. Heliantheae.
    3 Lvs all alternate; receptacle naked in most genera, but sometimes chaffy; style-branches mostly flattened and with well developed ventromarginal stigmatic lines; invol, pappus, anthers, and rays various.
      4 Anthers obtuse at base to merely sagittate, not tailed; styles diverse, but not as in the Inuleae; heads radiate or less often disciform or discoid; pappus and receptacle various.
        5 Invol bracts rather dry and scarcely herbaceous, becoming hyaline-scarious toward the margins and tip; pappus none, or minute and chaffy or coroniform (of definite small scales in *Hymenopappus*); style-branches mostly truncate and penicillate; receptacle chaffy or naked; lvs generally ± dissected, varying to sometimes entire and then mostly small; plants mostly with a characteristic odor. (Genera 34–42) .... Tribe 2. Anthemideae.
        5 Invol bracts herbaceous to chartaceous, rarely evidently hyaline-scarious toward the margins and tip; pappus capillary, chaffy, or of awns, or occasionally none; style branches of diverse sorts; receptacle naked; lvs usually entire or merely toothed, sometimes lobed or pinnatifid; plants mostly inodorous, or at least not with the odor of the Anthemideae.
          6 Invol bracts mostly equal and subuniseriate, a calyculus of much reduced outer bracts often also present; style-branches usually truncate and penicillate, without a definite terminal appendage; pappus capillary, or seldom wanting. (Genera 43–48) ...... Tribe 3. Senecioneae.
          6 Invol bracts mostly in several series, imbricate or subequal (or the outer larger); calyculus wanting; style-branches usually with a terminal appendage; pappus capillary or chaffy or of awns, or seldom none. (Genera 49–63) .... Tribe 4. Astereae.
      4 Anthers ± strongly tailed at the base; style-branches generally glabrous or merely papillate, subtruncate to broadly rounded at the tip, the stigmatic lines often confluent around the end; heads very often disciform or discoid, the rays when present yellow like the disk; pappus mostly capillary, rarely chaffy or none; receptacle chaffy or more often naked. (Genera 64–69) .... Tribe 5. Inuleae.
  2 Heads mostly discoid (marginal fls enlarged, subradiate, and neutral in a few spp. of *Centaurea*); cors mostly anthocyanic or white, yellow only in a few spp. of the Cynareae.
    7 Anthers obtuse to sagittate at the base; plants not spiny; receptacle naked or rarely chaffy, not densely bristly.

8 Style-branches cylindric to often clavate, obtuse, minutely papillate, not hairy, with inconspicuous ventromarginal stigmatic lines near the base; lvs mostly opposite, less often whorled, only occasionally alternate. (Genera 70–75) ....Tribe 6. Eupatorieae.
8 Style-branches slender and gradually attenuate, minutely hispidulous outside, smooth and stigmatic inside; lvs alternate. (Genera 76–77) .... Tribe 7. Vernonieae.
7 Anthers ± distinctly tailed at the base; plants very often with the lvs or invol ± spiny, but sometimes wholly unarmed; style with a thickened, often hairy ring below the branches, changing abruptly in texture at that level and papillate thence to the tip, the branches commonly connate at least below; receptacle in most genera densely bristly; lvs alternate. (Genera 78–86) .... Tribe 8. Cynareae.

Artificial Keys to the Genera

1 Fls all ligulate and perfect; juice milky .... Key 6.
1 Fls not all ligulate; ray (ligulate) fls when present marginal, either pistillate or neutral; juice ordinarily watery.
  2 Heads radiate; rays restricted to the margin of the heads, never perfect.
    3 Rays yellow or orange (sometimes marked with purple or brown at the base).
      4 Receptacle chaffy, at least toward the margin; pappus chaffy, or of awns, or none .... Key 1.
      4 Receptacle naked; pappus various .... Key 2.
    3 Rays some other color than yellow or orange .... Key 3.
  2 Heads discoid or disciform, without rays. (Some plants with very small and inconspicuous rays are keyed here as well as with the radiate group.)
    5 Receptacle bristly or chaffy, at least toward the margin; plants often spiny .... Key 4.
    5 Receptacle naked (deeply honeycombed in *Onopordum*); plants not spiny (except *Onopordum*) .... Key 5.

Key 1

(Heads radiate; rays yellow or orange; receptacle chaffy)

1 Disk fls sterile, with undivided style, their ovaries much smaller than those of the fertile ray-fls.
  2 Inner bracts of the invol (or outer bracts of the receptacle) conspicuously prickly; annual ...... 27. *Acanthospermum.*
  2 Invol and receptacular bracts without prickles; perennial.
    3 Ray-achenes in 2–3 series, even at flowering time .... 28. *Silphium.*
    3 Ray-achenes uniseriate.
      4 Coarse plants seldom under 1 m, with large, ± lobed lvs commonly well over 1 dm .... 26. *Polymnia.*
      4 More slender plants, mostly 0.5–5 dm, with merely crenate lvs less than 1 dm .... 29. *Chrysogonum.*
1 Disk-fls fertile, with divided style, their ovaries as large as or larger than those of the ray-fls, which may be either fertile or sterile.
  5 Lvs, or some of them, opposite (the middle and upper lvs in some spp. regularly alternate).
    6 Rays persistent on the achenes and becoming papery .... 12. *Heliopsis.*
    6 Rays deciduous from the achenes at or before maturity.
      7 Disk-cors densely woolly below the middle .... 15. *Guizotia.*
      7 Disk-cors not woolly, usually glabrous.
        8 Achenes of the disk not obviously flattened, or if so, then flattened at right angles to the invol bracts; receptacular bracts mostly concave and clasping; invol bracts not obviously dimorphic and biseriate except in *Sigesbeckia* and *Tetragonotheca.*
          9 Invol bracts 4(5), large, broadly ovate, leafy .... 2. *Tetragonotheca.*
          9 Invol bracts distinctly otherwise.
            10 Invol bracts biseriate, dimorphic, the outer linear-oblanceolate and spreading, the inner shorter and broader, subtending the achenes, all conspicuously stipitate-glandular .... 16. *Sigesbeckia.*
            10 Invol bracts in one or more subequal or ± imbricate series, not evidently biseriate and dimorphic, not at all stipitate-glandular.
              11 Receptacle columnar .... 14. *Spilanthes.*
              11 Receptacle flat or convex to occasionally low-conic.
                12 Low fleshy seacoast shrub .... 13. *Borrichia.*
                12 Herbs, not fleshy.
                  13 Receptacle chaffy only near the margin; invol bracts uniseriate, equal, each enfolding a ray-achene; heavy-scented annuals .... 22. *Madia.*
                  13 Receptacle chaffy throughout; invol bracts in more than one series, not enfolding achenes; annual or perennial.
                    14 Achenes of the disk-fls ± strongly flattened, usually also winged; pappus-awns not paleaceous-dilated at base .... 10. *Verbesina.*
                    14 Achenes only moderately compressed, not winged; pappus-awns paleaceous-dilated at base .... 1. *Helianthus.*
        8 Achenes flattened parallel to the invol bracts (terete in *Bidens beckii,* an aquatic plant with

filiform-dissected lvs); receptacular bracts flat or only slightly concave; invol double, the bracts dimorphic.

    15 Pappus of 2–6 awns or teeth, these barbed or hispid, usually retrorsely so (rarely smooth or even wanting); achenes not wing-margined.

        16 Achenes beaked .... 19. *Cosmos.*

        16 Achenes not beaked .... 18. *Bidens.*

    15 Pappus of 2 short teeth or awns, barbed upwardly or not at all, or a mere border, or none; achenes wing-margined, except in *C. tinctoria* .... 17. *Coreopsis.*

5 Lvs alternate.

    17 Receptacle chaffy only near the margin; invol bracts uniseriate, equal, each enfolding a ray-achene .... 22. *Madia.*

    17 Receptacle chaffy or bristly throughout; invol bracts not enfolding achenes.

        18 Invol bracts dry, scarious-margined .... 35. *Anthemis.*

        18 Invol bracts ± herbaceous, without scarious margins.

            19 Receptacle merely bristly; pappus of ca 6–10 awned scales .... 6. *Gaillardia.*

            19 Receptacle definitely chaffy; pappus otherwise.

                20 Receptacle strongly conic or columnar.

                    21 Rays subtended by receptacular bracts .... 5. *Ratibida.*

                    21 Rays not subtended by receptacular bracts.

                        22 Receptacular bracts spinescent, surpassing the disk-fls .... 4. *Echinacea.*

                        22 Receptacular bracts not spinescent, but sometimes shortly awn-pointed .... 3. *Rudbeckia.*

                20 Receptacle merely convex or low-conic.

                    23 Invol bracts subequal or slightly imbricate, all about alike .... 10. *Verbesina.*

                    23 Invol bracts biseriate, dimorphic, the outer much narrower and shorter than the inner .... 17. *Coreopsis.*

KEY 2

(Heads radiate; rays yellow or orange; receptacle naked)

1 Pappus of scales, or of awns, or a crown, or none.

  2 Pappus none, or a mere tiny border or crown.

    3 Lvs opposite .... 23. *Flaveria.*

    3 Lvs alternate.

      4 Lvs pinnately dissected .... 39. *Tanacetum.*

      4 Lvs merely toothed or pinnatifid, scarcely dissected .... 38. *Chrysanthemum.*

  2 Pappus of distinct scales or awns.

    5 Lvs (at least the lower) opposite.

      6 Pappus of ca 3–6 dissimilar scales .... 25. *Tagetes.*

      6 Pappus of ca 10–20 scales, each deeply cleft into 5–10 bristle-tips .... 24. *Dyssodia.*

    5 Lvs alternate, or all basal.

      7 Lvs all basal .... 8. *Hymenoxys.*

      7 Lvs not all basal.

        8 Pappus of 2–several firm awns, not scaly .... 52. *Grindelia.*

        8 Pappus of several scales, sometimes awn-tipped.

          9 Heads small, the disk 5 mm wide or less, the rays not over ca 5 mm long.

            10 Subshrubby perennial; disk-fls perfect and fertile .... 55. *Gutierrezia.*

            10 Annual; disk-fls functionally staminate, without an ovary .... 56. *Amphiachyris.*

          9 Heads evidently larger.

            11 Style-branches with a subulate appendage .... 6. *Gaillardia.*

            11 Style-branches truncate, exappendiculate .... 7. *Helenium.*

1 Pappus (at least of the disk-fls) of capillary bristles, sometimes also with some short outer scales.

  12 Lvs opposite (or the uppermost ones alternate) .... 9. *Arnica.*

  12 Lvs alternate or all basal.

    13 Invol bracts uniseriate, equal, commonly with a few very much shorter ones at the base.

      14 Disk-fls sterile; stem merely bracteate .... 46. *Tussilago.*

      14 Disk-fls fertile; stem usually leafy .... 43. *Senecio.*

    13 Invol bracts ± imbricate in several series.

      15 Pappus simple, not divided into an outer and inner series.

        16 Anthers tailed; inuleous style .... 64. *Inula.*

        16 Anthers not tailed; astereous style.

          17 Plants with a well developed taproot; s. and w. edges of our range .... 49. *Haplopappus.*

          17 Plants fibrous-rooted, without a taproot; widespread in our range.

            18 Lvs ± resinous-punctate, narrow, entire, sessile or subsessile, the basal and lower cauline ones soon deciduous; heads tending to be sessile or subsessile in small glomerules, arranged in a terminal, corymbiform, flat-topped infl .... 54. *Euthamia.*

            18 Lvs variously shaped and disposed, but not resinous-punctate (translucent-punctate in *S. odora*, which has a paniculiform infl with recurved-secund branches) .... 53. *Solidago.*

      15 Pappus double, the outer of short bristles, the inner of much longer bristles.

KEY 3

(Heads radiate; rays not yellow)

KEY 4

(Heads discoid or disciform; receptacle bristly or chaffy)

1 Pappus of scales, or awns, or none.
  2 Pappus present.
    3 Lvs, or most of them, opposite .... 18. *Bidens.*
    3 Lvs alternate or all basal.
      4 Lvs prickly or spiny on the margins, the plants thistles.
        5 Heads many-fld, not aggregated into secondary heads; fls yellow .... 86. *Cnicus.*
        5 Heads 1-fld, aggregated into globose secondary heads; fls white or bluish .... 84. *Echinops.*
      4 Lvs not prickly or spiny.
        6 Invol bracts distinctly hooked at the tip .... 78. *Arctium.*
        6 Invol bracts not hooked.
          7 Invol bracts (or some of them) fimbrillate, dentate, or
            spiny, evidently imbricate .... 85. *Centaurea.*
          7 Invol bracts entire, not much imbricate.
            8 Lvs entire; heads strictly discoid, the fls all tubular and perfect .... 21. *Marshallia.*
            8 Lvs toothed or pinnatifid; heads minutely radiate, only the ray fls fertile; disk fls with
              undivided style .... 30. *Parthenium.*
  2 Pappus none.
    9 Fls either all perfect and fertile with divided style, or the central ones thus and the marginal
      ones pistillate or neutral.
      10 Plants annual, slender, with linear entire lvs 1–4 mm wide .... 22. *Madia.*
      10 Plants of various habit, but mostly more robust and with the lvs either pinnatifid or well
        over 4 mm wide.
        11 Lvs alternate.
          12 Suffruticose, with pinnatifid lvs not over ca 3 cm; receptacle chaffy; anthemideous
            style .... 37. *Santolina.*
          12 Herbs with entire to occasionally pinnatifid lvs well over 3 cm; receptacle densely
            bristly; cynareous style .... 85. *Centaurea.*
        11 Lvs opposite.
          13 Invol stipitate-glandular, the outer bracts spreading or reflexed .... 16. *Sigesbeckia.*
          13 Invol not stipitate-glandular, the bracts all appressed .... 11. *Eclipta.*
    9 Fls of 2 kinds, some pistillate and fertile, others hermaphrodite and sterile with undivided style.
      14 Procumbent or creeping annual .... 27. *Acanthospermum.*
      14 Erect plants, annual to perennial.
        15 Staminate (sterile hermaphrodite) and pistillate fls in the same head; invol not tuberculate
          or prickly.
          16 Coarse perennials with medium-sized heads (disk 6–13 mm wide) borne in congested
            or open and leafy cymes .... 26. *Polymnia.*
          16 Plants either perennial with heads solitary in the lf-axils, or annual; heads small,
            mostly under 6 mm wide .... 31. *Iva.*
        15 Staminate and pistillate fls in separate heads, the staminate generally uppermost; invol
          of pistillate heads nut-like or bur-like.
          17 Invol of the pistillate heads a bur with hooked prickles .... 33. *Xanthium.*
          17 Invol of the pistillate heads with one or 2–several series of tubercles or straight
            spines .... 32. *Ambrosia.*
1 Pappus of capillary (sometimes plumose) bristles.
  18 Lvs not prickly or spiny; invol bracts entire; style not cynareous.
    19 Outer fls pistillate; lvs mostly 1–5 mm wide .... 65. *Filago.*
    19 Fls all perfect; basal lvs over 5 mm wide .... 74. *Carphephorus.*
  18 Plants either with somewhat prickly-margined or spiny lvs, or with the invol bracts fimbrillate,
    dentate, or spiny; cynareous style.
    20 Achenes distinctly obliquely attached to the receptacle; lvs unarmed .... 85. *Centaurea.*
    20 Achenes basally attached; lvs prickly or spiny, the plants thistles.
      21 Filaments united into a tube below; lvs ± white-mottled .... 81. *Silybum.*
      21 Filaments separate; lvs mostly not mottled.
        22 Pappus-bristles merely barbellate, not plumose .... 79. *Carduus.*
        22 Pappus-bristles plumose.
          23 Inner invol bracts with elongate, narrow, eventually spreading, stramineous, ray-
            like tip; achenes densely hairy .... 83. *Carlina.*
          23 Inner invol bracts not modified as above; achenes glabrous .... 80. *Cirsium.*

KEY 5

(Heads discoid or disciform; receptacle naked)

1 Pappus chaffy, or of awns, or none.
  2 Lvs, at least the lower, opposite or whorled.

   3 Lvs whorled .... 72. *Sclerolepis.*
   3 Lvs opposite.
      4 Lvs entire or remotely serrulate; invol and lvs not dotted with oil-glands .... 23. *Flaveria.*
      4 Lvs pinnatifid or bipinnatifid; invol and lvs dotted with embedded oil-glands.
         5 Pappus-scales fewer than 10, not divided into bristles .... 25. *Tagetes.*
         5 Pappus-scales ca 10–20, each parted into 5–10 bristles .... 24. *Dyssodia.*
  2 Lvs alternate or all basal.
      6 Pappus of 2–several long, firm, deciduous awns .... 52. *Grindelia.*
      6 Pappus of a short cup or crown, or short rounded scales, or none.
         7 Invol bracts nearly equal, colored, petal-like at least distally .... 34. *Hymenopappus.*
         7 Invol bracts not at all petal-like, evidently imbricate except in *Adenocaulon.*
            8 Invol bracts only 4 or 5; achenes with stalked purple glands .... 48. *Adenocaulon.*
            8 Invol bracts more numerous; achenes without such glands.
               9 Infl spiciform, racemiform, or paniculiform .... 42. *Artemisia.*
               9 Infl corymbiform, or the heads solitary.
                  10 Achenes, especially the marginal ones, conspicuously stipitate .... 41. *Cotula.*
                  10 Achenes sessile.
                     11 Lvs merely toothed .... 38. *Chrysanthemum.*
                     11 Lvs pinnately dissected.
                        12 Receptacle flat or merely convex .... 39. *Tanacetum.*
                        12 Receptacle strongly convex, pointed .... 40. *Matricaria.*
1 Pappus (at least of most of the fls) of capillary bristles.
 13 Outer fls, or all the fls of some heads, pistillate.
    14 Shrub .... 60. *Baccharis.*
    14 Herbs.
      15 Plants ± white-woolly, at least on the lower lf-surface.
        16 Basal lvs cordate, hastate, or reniform, often very large .... 47. *Petasites.*
        16 Basal lvs rounded or narrowed to the base, 5 cm wide or less.
           17 Plants dioecious or nearly so (female plant commonly with a few central staminate
              fls in *Anaphalis*).
              18 Basal lvs conspicuous, tufted, persistent, the cauline ones few and reduced;
                  fl spring .... 69. *Antennaria.*
              18 Basal lvs soon deciduous, scarcely if at all larger than the numerous cauline
                  ones; fl late summer .... 68. *Anaphalis.*
           17 Plants not dioecious; fls all fertile, the outer numerous and pistillate, the inner
              few and perfect .... 67. *Gnaphalium.*
      15 Plants not at all white-woolly, though sometimes otherwise pubescent.
        19 Outer (pistillate) fls essentially epappose .... 58. *Erigeron.*
        19 Fls all with well developed capillary pappus.
           20 Principal invol bracts in one series, sometimes with some short outer ones at the
              base .... 44. *Erechtites.*
           20 Bracts in more than one series, ± imbricate.
              21 Lvs ample, chiefly lanceolate to ovate; disk-fls functionally staminate; anthers
                tailed .... 66. *Pluchea.*
              21 Lvs narrow, chiefly linear or linear-oblanceolate; disk-fls perfect; anthers not
                tailed.
                  22 Invol bracts herbaceous or herbaceous-tipped .... 57. *Aster.*
                  22 Invol bracts sometimes greenish, but scarcely herbaceous .... 59. *Conyza.*
 13 Fls all perfect.
    23 Plants prickly or spiny; thistles .... 82. *Onopordum.*
    23 Plants unarmed.
      24 Fls bright yellow or orange .... 43. *Senecio.*
      24 Fls white, pink, purple, or pale creamy.
        25 Heads few-fld, aggregated into secondary heads .... 77. *Elephantopus.*
        25 Heads separate.
          26 Lvs alternate, entire to broadly lobed.
            27 Invol of equal, essentially uniseriate bracts, often with a few much smaller
              bracteoles at base; senecioneous style .... 45. *Cacalia.*
            27 Invol bracts evidently imbricate in several series (or the outer with prolonged
              setaceous tip in forms of *Kuhnia*).
              28 Pappus double, the inner of long bristles, the outer of very short ones;
                vernonieous style .... 76. *Vernonia.*
               28 Pappus simple, of similar bristles; eupatorieous style.
                  29 Infl simple, spiciform or racemiform to subcorymbiform, the heads
                    sessile or pedunculate on the main axis of the plant .... 75. *Liatris.*
                  29 Infl more branching .... 73. *Kuhnia.*
          26 Lvs mostly or all opposite or whorled, or if mainly alternate, then dissected into
           filiform segments; eupatorieous style.
            30 Twining herbaceous vines; principal invol bracts 4; fls 4 .... 71. *Mikania.*
            30 Erect herbs; bracts and fls in most spp. more than 4 .... 70. *Eupatorium.*

Key 6

(Heads ligulate; juice milky)

1 Pappus of simple capillary bristles only.
  2 Achenes terete or prismatic, scarcely flattened.
    3 Achenes smooth or nearly so, not evidently spinulose or muricate.
      4 Fls pink or purple to white or ochroleucous or chloroleucous; heads several or many; cauline
        lvs present or reduced to mere scales; achenes beakless.
        5 Cauline lvs well developed, mostly well over 1 cm wide; widespread .... 87. *Prenanthes.*
        5 Cauline lvs less than 1 cm wide, often reduced to scales; western .... 88. *Lygodesmia.*
      4 Fls bright yellow to orange or orange-red (sometimes drying pink or purple in *Agoseris,* which
        is strictly scapose, with solitary terminal heads).
        6 Perennial from an elongate or very short rhizome or from a short caudex or crown, strictly
          fibrous-rooted .... 92. *Hieracium.*
        6 Annual, biennial, or perennial from a taproot or several strong roots, without rhizomes.
          7 Plants strictly scapose, with solitary heads .... 95. *Agoseris.*
          7 Plants not at once scapose and with solitary heads, usually with several or many heads
            and at least a few cauline lvs.
            8 Achenes with a long, slender beak that bears a ring of soft, white, reflexed hairs at the
              summit, just beneath the pappus .... 97. *Pyrrhopappus.*
            8 Achenes beaked or beakless, without any ring of hairs just beneath the pappus.
              9 Heads numerous and small, the invol 3.5–5 mm, with ca 10–20 fls, the achenes
                1.5–2.5 mm .... 94. *Youngia.*
              9 Heads few to numerous, larger, the invol 5–12 mm, often with more than 20 fls,
                the achenes often more than 2.5 mm .... 93. *Crepis.*
    3 Achenes spinulose, or with some short processes near the summit of the body, tipped by a slender
      beak; fls yellow.
      10 Plants strictly scapose, with solitary terminal heads; heads many-fld .... 99. *Taraxacum.*
      10 Plants branching, with narrow cauline lvs and several or many heads, these 7–15-fld .... 100. *Chondrilla.*
  2 Achenes ± strongly flattened.
    11 Achenes beakless, without any enlarged pappiferous disk at the summit; heads many-fld (fls
      ca 80–250 in our spp.) .... 91. *Sonchus.*
    11 Achenes beaked or beakless, in either case somewhat enlarged at the summit where the pappus
      is attached; heads relatively few-fld (ca 5–56 fls in our spp.).
      12 Erect, ± leafy-stemmed plants; invol imbricate (except *L. muralis*) .... 89. *Lactuca.*
      12 Delicate, creeping plants with erect scapes or peduncles up to ca 1 dm bearing 1 or 2 heads
        each; invol calyculate .... 90. *Ixeris.*
1 Pappus of plumose bristles, or bristles and scales, or none, the scales very slender and bristle-like in
  Microseris.
  13 Pappus of scales, or bristles and scales, or none, not plumose.
    14 Fls blue (pink or white); pappus of minute narrow scales only .... 101. *Cichorium.*
    14 Fls yellow or orange; pappus various, but not as above.
      15 Pappus well developed; annual or perennial plants.
        16 Pappus of mixed capillary bristles and long, very slender, gradually attenuate scales;
          invol 17–25 mm; taprooted perennial .... 96. *Microseris.*
        16 Pappus of 5 or more prominent or sometimes very inconspicuous scales and generally
          5–40 longer capillary bristles; invol 4–14 mm; annual or perennial, tending to
          be fibrous-rooted (definitely so when perennial) .... 98. *Krigia.*
      15 Pappus none or vestigial; annual.
        17 Invol bracts membranous or herbaceous; peduncles not inflated; stem leafy.
          18 Invol calyculate; achenes 3–5 mm .... 102. *Lapsana.*
          18 Invol not calyculate, the bracts all equal; achenes mostly under 3 mm .... 98. *Krigia.*
        17 Invol bracts conspicuously thickened and keeled after flowering; peduncles conspic-
          uously inflated; cauline lvs reduced to inconspicuous bracts, or none .... 103. *Arnoseris.*
  13 Pappus of plumose bristles, at least in part.
    19 Plume-branches of the pappus interwebbed; invol uniseriate; leafy-stemmed plants with grass-
      like lvs .... 107. *Tragopogon.*
    19 Plume-branches of the pappus not interwebbed; invol various but not strictly uniseriate; lvs
      not grass-like.
      20 Plants leafy-stemmed .... 104. *Picris.*
      20 Plants scapose, the scape naked or minutely scaly-bracted.
        21 Receptacle chaffy-bracted .... 105. *Hypochoeris.*
        21 Receptacle naked .... 106. *Leontodon.*

1. HELIANTHUS L.   Sunflower. Heads radiate, the rays large, yellow, neutral; invol
bracts subequal to imbricate, usually green; receptacle flat to low-conic, chaffy, its bracts

clasping the achenes; disk-fls perfect and fertile; style-branches flattened, hispidulous on both sides (at least distally), the stigmatic lines poorly developed; achenes thick, moderately compressed at right angles to the invol bracts, glabrous or sometimes hairy; pappus of 2 readily deciduous awns with enlarged, thin, paleaceous base, rarely with some additional short scales; coarse annual or more often perennial herbs with simple lvs, at least the lowermost ones opposite; $x=17$. 50, N. Amer. Hybrids abound.

1 Plants annual; receptacle flat or nearly so; disk nearly always red-purple.
  2 Central receptacular bracts inconspicuously short-hairy; invol bracts chiefly ovate or ovate-oblong, abruptly contracted above .... 1. *H. annuus.*
  2 Central receptacular bracts white-bearded at the tip; invol bracts narrower, tapering .... 2. *H. petiolaris.*
1 Plants perennial; receptacle generally convex or low-conic; disk various.
  3 Lvs very narrow, not over 1/10 as wide as long, rarely over 1 cm wide (except sometimes the lowest ones); disk usually red-purple.
    4 Stem glabrous, sometimes glaucous; rhizome elongate .... 20. *H. salicifolius.*
    4 Stem ± hairy; rhizome scarcely developed .... 21. *H. angustifolius.*
  3 Lvs wider, at least some of them over 1 cm wide and less than 10 times as long.
    5 Invol bracts evidently imbricate, broad, firm, appressed, rounded to sharply acute.
      6 Cauline lvs well developed, (2–)2.5–8 times as long as wide; rhizomatous.
        7 Disk red-purple .... 14. *H. pauciflorus.*
        7 Disk yellow .... 15. *H.* × *laetiflorus.*
      6 Cauline lvs either reduced or not more than twice as long as wide; rhizome scarcely developed; disk red-purple.
        8 Stem relatively leafy, often to above the middle, the blades mostly 1–1.7(–2) times as long as wide; western .... 13. *H. silphioides.*
        8 Stem less leafy, the lvs basally disposed, mostly (1.3–)1.7–2.5(–3) times as long as wide; eastern .... 12. *H. atrorubens.*
    5 Invol bracts narrow, seldom much imbricate, some or all with loose, acuminate or attenuate tip; disk yellow.
      9 Lvs basally disposed (see descr.), 3–8 pairs below the infl .... 11. *H. occidentalis.*
      9 Lvs ± well distributed along the stem, generally more than 8 pairs below the infl.
        10 Stem evidently hairy below as well as in the infl.
          11 Herbage densely and softly short-hairy; lvs sessile, broad-based .... 10. *H. mollis.*
          11 Herbage more coarsely or sparsely hairy or scabrous, or in part glabrous; lvs otherwise.
            12 Lvs large, the larger 4–12 cm wide on a petiole (1.5–)2–8 cm .... 16. *H. tuberosus.*
            12 Lvs smaller, or with shorter petioles, or both.
              13 Lvs tapering to the base; rhizomes short; roots ± thickened.
                14 Stem spreading-hirsute; lvs flat, triplinerved at base .... 17. *H. giganteus.*
                14 Stem white-strigose; lvs not triplinerved, usually some folded .... 19. *H. maximilianii.*
              13 Lvs broad-based; rhizome elongate; roots fibrous .... 4. *H. hirsutus.*
        10 Stem essentially glabrous below the infl, often also glaucous.
          15 Heads small, the disk mostly 0.5–1.5 cm wide, the rays 5–10(–13) in number; rhizomes short or scarcely developed.
            16 Lvs abruptly narrowed to the 1–3 cm petiole, resin-dotted beneath and usually also evidently short-hairy .... 8. *H. microcephalus.*
            16 Lvs narrowed to a subsessile base or short petiole under 1 cm, glabrous or nearly so beneath, not resin-dotted .... 9. *H. laevigatus.*
          15 Heads larger, the disk 1.5–3.5 cm wide, the rays (8–)10–25; rhizomes short to elongate.
            17 Petiole short (seldom as much as 1 cm) or none.
              18 Lvs glaucous beneath, somewhat tapering toward the base .... 7. *H. eggertii.*
              18 Lvs somewhat hairy to subglabrous beneath, but not glaucous.
                19 Lvs widest near the truncate or broadly rounded base .... 3. *H. divaricatus.*
                19 Lvs tapering to a narrow base (occasional forms of) .... 17. *H. giganteus.*
            17 Petiole longer, commonly 1–4 cm on well developed lvs.
              20 Lvs relatively narrow, usually at least 3 times as long as wide, the middle and upper ones seldom over 4 cm wide.
                21 Lower lf-surface pale, ± glaucous, only sparsely or not at all hairy (occasional forms of) .... 5. *H. strumosus.*
                21 Lower lf-surface green and generally evidently hairy (sometimes sparsely so), not glaucous.
                  22 Upper surface of the lvs distinctly scabrous; lower surface rather coarsely (and often sparsely) long-hairy, many or all of the hairs on the order of 1 mm long or more (occasional forms of) .... 17. *H. giganteus.*
                  22 Upper surface of the lvs usually scarcely scabrous, the short hairs tending to be ± appressed; lower surface more finely and shortly hairy than in no. 17, most or all of the hairs well under 1 mm .... 18. *H. grosseserratus.*
            20 Lvs wider, seldom over 3 times as long as wide, often over 4 cm wide.
              23 Invol bracts very loose, evidently surpassing the disk; lvs thin, generally strongly serrate .... 6. *H. decapetalus.*

23 Invol bracts only moderately loose, slightly or not at all surpassing the
disk; lvs fairly firm, inconspicuously serrulate or subentire .... 5. *H. strumosus.*

1. **Helianthus annuus** L.   Common s. Coarse, rough-hairy annual (0.5–)1–3 m; lvs chiefly alternate (except
the lowermost), mostly toothed, long-petiolate, ovate or broader, at least the lower cordate in well developed
plants; heads large, the red-purple (yellow) disk seldom under 3 cm wide; invol bracts ovate or ovate-oblong
and abruptly narrowed above the middle to the acuminate tip, ciliate and with some rather long coarse hairs
on the back; receptacle flat or nearly so, its bracts inconspicuously hairy at the tip; $2n=34$. A weed in disturbed
sites, especially in moist, low ground, throughout the U.S. and adj. Can. and Mex. July–Sept. Typical wild
plants are branched, with several or many heads. Cult. forms, which readily escape, have solitary (or few),
often much larger heads. (*H. lenticularis; H. aridus.*)
   H. debilis Nutt. var. *cucumerifolius* (T. & G.) A. Gray, with a natural range in Tex., is occasionally found
as a roadside weed in our range, especially southward. It is a much smaller and more delicate plant than *H.
annuus,* seldom as much as 1 m tall, and the stem is strongly mottled. (*H. cucumerifolius.*)

2. **Helianthus petiolaris** Nutt.   Plains-s. Similar to *H. annuus,* but smaller, seldom over 1 m, with narrower, more often entire, rarely cordate lvs, which may be more densely hairy beneath, and with smaller
heads, the disk 1–2.5 cm wide; invol bracts lanceolate or lance-ovate, tapering gradually to the tip, shortly
scabrous-hispid, seldom at all ciliate or with any long hairs; central receptacular bracts conspicuously white-
bearded at the tip; $2n=34$. Prairies, plains, and waste places, especially in sandy soil; native mainly on the
Great Plains and in sw. U.S., but occasionally eastward as a weed. June–Sept. Ours is var. *petiolaris.*

3. **Helianthus divaricatus** L.   Divaricate s. Fibrous-rooted perennial from long rhizomes; stems 0.5–1.5
m, glabrous below the infl, often glaucous; lvs all opposite, sessile or rarely on a short petiole to 5 mm,
scabrous above, loosely hirsute or hispidulous (at least on the main veins) beneath, narrowly lanceolate to
broadly lance-ovate, 5–18 × 1–5(–8) cm, broadest near the truncate or broadly rounded base, tapering to
the slender, acuminate tip, shallowly toothed or subentire, trinerved near the base; heads 1–several at the
tips of stiff cymose branches; disk yellow, 1–1.5 cm wide; invol bracts lance-acuminate or -attenuate, ciliolate,
rather loose, often with reflexed tip; rays 8–15, 1.5–3 cm; $2n=34$. Abundant in dry woods and other open
places; widespread in e. U.S. and adj. Can., from Mass. and N.H. to Wis., s. to Fla., La., and e. Okla. July–
Sept.

4. **Helianthus hirsutus** Raf.   Hairy s. Fibrous-rooted perennial from long rhizomes; stems 0.6–2 m,
spreading-hairy; lvs all or mostly opposite, generally ascending on a short petiole 5–15(–20) mm, hirsute on
both sides or scabrous above, narrowly lanceolate to ovate, 7–16 × (1.5–)2–6 cm, serrate to entire, triplinerved
at the abruptly contracted to often broadly rounded or subcordate base; heads 1–several on short, stout
peduncles; disk yellow, (1.2–)1.5–2(–2.5) cm wide; invol bracts conspicuously ciliate and often also hairy on
the back, slender, long-pointed, often with loose or reflexed tip; rays 10–15, 1.5–3.5 cm; $2n=68$. Dry, wooded
or open places; Pa. to Minn., s. to n. Fla. and Tex. July–Oct. (*H. chartaceus.*)

5. **Helianthus strumosus** L.   Rough-lvd s. Rhizomatous perennial; stems 1–2 m, glabrous below the infl
or with a few long hairs, often glaucous; lvs opposite or the uppermost alternate, relatively thick and firm,
scabrous-hispid above, green and moderately short-hairy to more often glaucous and subglabrous beneath,
mostly broadly lanceolate to ovate and 8–20 × 2.5–10 cm (sometimes narrower), long-acuminate, shallowly
toothed or subentire, commonly decurrent onto the 0.5–3 cm petiole; disk yellow, 1.2–2.5 cm wide; invol
bracts subequal, lanceolate, somewhat loose, especially the long-acuminate tips, which commonly equal or
slightly surpass the disk; rays 8–15, 1.5–4 cm; $2n=68$, 102. Woods and open places; Me. to Fla., w. to Minn.,
e. Kans., and e. Tex. July–Sept.

6. **Helianthus decapetalus** L.   Forest-s. Perennial from slender rhizomes; stem 0.5–1.5(–2) m, glabrous
below the short-hairy infl; lvs thin, pale beneath, moderately scabrous to subglabrous, broadly lanceolate to
ovate, 8–20 × 3–8 cm, long-acuminate, serrate (generally sharply so), ± abruptly contracted near the base
and decurrent onto the 1.5–6 cm petiole; upper lvs usually alternate; disk yellow, 1–2 cm wide; invol bracts
very loose, thin, green, conspicuously ciliate, occasionally hispidulous on the back, attenuate-acuminate, at
least some of them usually conspicuously surpassing the disk; rays 8–15, 1.5–3.5 cm; $2n=34$, 68. Woods and
along streams; Me. and Que. to Wis. and Io., s. to Ga. and Mo. Aug.–Oct. (*H. trachelifolius.*)

7. **Helianthus eggertii** Small.   Tennessee-s. Fibrous-rooted perennial from long rhizomes; stems 1–2 m,
glabrous and glaucous; lvs mostly or all opposite, glabrous or nearly so above, glabrous and strongly glaucous
beneath, lanceolate, mostly 9–15 × 1.5–3.5 cm, entire or serrulate, tapering to a sessile or subsessile base;
heads rather few, on long peduncles often 1–1.5 dm; disk yellow, 1.2–2 cm wide; invol bracts firm, lance-
acuminate, ciliolate, about equaling the disk; rays 10–14, up to 2 cm; $2n=102$. Rocky hills and barrens; ec.
Ky. to c. Tenn. Aug., Sept.

8. **Helianthus microcephalus** T. & G.   Small-headed s. Fibrous-rooted perennial with crown-buds and a
short (seldom over 5 cm) rhizome; stems (0.7–)1–2 m, glabrous and generally glaucous; lvs scabrous above,
resinous-dotted and usually also loosely short-hairy beneath, sometimes also glaucous, lanceolate or lance-
ovate, 7–15(–20) × (1–)2–5(–6) cm, toothed or entire, gradually tapering distally, ± abruptly narrowed to
the 1–3 cm petiole; upper lvs alternate; heads on long, slender peduncles, small, the yellow disk 0.5–1 cm
wide; invol bracts few, lanceolate, acuminate or attenuate, ciliolate, otherwise glabrous or nearly so; rays 5–
8, 1–1.5 cm; $2n=34$, 68. Woods and brushlands; N.J. to nw. Fla., w. to s. Minn., e. Ark., and se. La. Aug.,
Sept. (*H. glaucus; H. smithii.*)

9. **Helianthus laevigatus** T. & G.   Shale-barren s. Fibrous-rooted perennial from a crown or short rhizome; stems 1–2 m, glabrous, often glaucous; lvs firm, essentially glabrous, serrulate to entire, lanceolate to sometimes lance-elliptic or lance-ovate, 6–18 × 1–4 cm, narrowed to a subsessile base or short petiole less than 1 cm; upper lvs alternate; heads rather small, the yellow disk 1–1.5 cm wide; invol bracts subequal or slightly imbricate, lanceolate, acuminate-attenuate, often ciliolate, otherwise glabrous; rays 5–10(–13), 1–2 cm; $2n=68$. Mainly on shale-barrens in our range; mts. of Va. and W.Va. to N.C. and S.C. Aug.–Oct. (*H. reindutus.*)

10. **Helianthus mollis** Lam.   Ashy s. Perennial from stout rhizomes, usually colonial, 0.5–1(–1.2) m, densely and softly hairy throughout; lvs sessile, subcordate, ascending, broadly lanceolate to broadly ovate or oblong, 6–15 × 2–8 cm, 1.4–3(–4) times as long as wide, serrulate or entire; uppermost lvs sometimes alternate; heads few or solitary; disk yellow, 2–3 cm, wide; invol bracts slightly imbricate, lanceolate, acuminate, often finally glandular as well as densely white-hairy, the upper part loose or spreading; rays mostly 16–35, 1.5–3.5 cm; $2n=34$. Prairies and other dry places; chiefly Ozarkian and midwestern, from O. to Wis., s. Io., and e. Kans. s. to Ga. and Tex., and occasionally intr. e. to the Atlantic. July–Sept. A probable hybrid with *H. giganteus* has been named *H. doronicoides* Lam.

11. **Helianthus occidentalis** Riddell.   Naked-stemmed s. Rhizomatous and often stoloniferous perennial 0.5–1.5 m; lvs basally disposed, the lower ones much the largest, with ovate to lance-elliptic or lanceolate, entire or subentire blade 6–15 × 2–8 cm, 1.5–5 times as long as wide, sharply set off from the long petiole, the others ± reduced and distant; disk yellow, 1–1.5 cm wide; invol bracts ciliolate, ± imbricate, lanceolate or lance-ovate, (1.5–)2–3 mm wide, at least the inner with loose, slender tip; rays 10–15, (1–)1.5–3 cm; $2n=34$. Dry, often sandy soil; Md. and DC. to Minn., s. to Ga., w. Fla., and Tex. Aug.–Oct. Appalachian plants, from DC., Md., and W.Va. to n. Ga., form the var. *dowellianus* T. & G. (*H. dowellianus*), with glabrous or closely strigose lvs that are often not as quickly reduced upwards as in var. *occidentalis*; the basal and lowermost cauline lvs are often deciduous, and some of the lower cauline ones (but well above the base) are often fairly well developed. More western plants in our range, with scabrous to hirsute, more consistently basally disposed lvs, belong to var. *occidentalis*.

12. **Helianthus atrorubens** L.   Appalachian s. Fibrous-rooted perennial from a very short, stout rhizome or crown; stems mostly solitary, 0.5–2 m, usually conspicuously spreading-hairy, at least below; lvs nearly all opposite, 3–8 pairs below the infl, commonly hairy on both sides, especially on the main veins beneath, lance-ovate to sometimes broadly ovate, mostly (1.3–)1.7–2.5(–3) times as long as wide, abruptly contracted to the petiole, the largest ones near the base, these commonly 6–20(–25) cm, toothed; petioles tending to be conspicuously wing-flared upward, often conspicuously spreading-hairy, the lower generally ⅓ to fully as long as the blade; heads several on long, naked peduncles in a corymbiform infl; disk red-purple, 1–1.5(–2) cm wide; invol bracts evidently imbricate, broad, firm, appressed, mostly oblong or elliptic, rounded to acutish, sometimes with an abrupt, very short acumination, 2.5–4(–5) mm wide, ciliolate, otherwise glabrous or nearly so; rays 10–15, 1–3 cm; pappus without accessory scales; $2n=34$. Dry, open woods; basically Appalachian and Atlantic, from Va. and e. Ky. to c. Ga. and Ala., w. to w. Tenn. and se. La. July–Oct.

13. **Helianthus silphioides** Nutt.   Ozark s. Similar to *H. atrorubens*, and passing into it where their ranges overlap; stems commonly several, up to 3 m, relatively more leafy, often to above the middle; blades relatively broad, generally only 1–1.7(–2) times as long as wide, on short, generally inconspicuously hairy petioles that are seldom more than ⅓ as long as the blade and seldom conspicuously wing-flared upward; invol bracts not always ciliolate; $2n=34$. Basically Ozarkian, from Ark. and s. Mo. to c. La., e. to s. Ill., w. Ky., and Tenn. July–Oct. Perhaps better called *H. atrorubens* var. *pubescens* Kuntze.

14. **Helianthus pauciflorus** Nutt.   Stiff s. Perennial with long rhizomes; stems scabrous or hispid to subglabrous, ± naked above, the heads long-peduncolate; lvs nearly all opposite, up to 15 pairs below the infl, scabrous to shortly hispid on both sides, mostly (2–)2.5–8 times as long as wide, mostly trinerved, toothed or entire, tapering to a short petiole or petiolar base, the middle ones seldom much smaller than the lowest ones present at anthesis; disk red-purple, 1.5–2.5(–3) cm wide; invol bracts evidently imbricate, broad, firm, appressed, mostly ovate or broadly lanceolate, sharply acute to obtuse, conspicuously ciliolate, otherwise generally glabrous; rays 10–21, 1.5–3(–3.5) cm; pappus nearly always with some short scales in addition to the 2 longer awns; $2n=102$. Prairies and roadsides; Mich. and Ind. to Alta., N.M., and Tex., and e. occasionally to Mass. and N.B. Aug., Sept. Two vars.:

Var. *pauciflorus*.   Robust, mostly 1–2 m; lvs 9–15 pairs below the infl, oblong-lanceolate to lance-ovate, acuminate, 8–27 cm; chiefly midwestern, in Mich., Wis., e. Minn., Io., e. Neb., e. Kans., Mo., and Okla., and intr. eastward. (*H. rigidus.*)

Var. *subrhomboideus* (Rydb.) Cronq.   Smaller, 0.3–1.2 m; lvs 5–10 pairs below the infl, rhombic-ovate to lance-linear, acute or obtuse, 5–12 cm; chiefly on the Great Plains and in Minn. (*H. subrhomboideus.*)

15. **Helianthus × laetiflorus** Pers.   Much like *H. rigidus* var. *rigidus*, but with yellow disk; lvs often larger and with much longer (to 5 cm) petiole; invol bracts avg a little narrower, generally (2–)2.5–4 mm wide, less imbricate, more pointed, and occasionally sparsely short-hairy on the back; $2n=102$. Roadsides and other disturbed sites, mainly or wholly as an escape from cult., often ± sterile; scattered in e. and midwestern U.S. Aug., Sept. Thought to consist of hybrids and hybrid progeny of *H. pauciflorus* var. *subrhomboideus* and *H. tuberosus*, perhaps mainly a cultigen.

16. **Helianthus tuberosus** L.   Jerusalem-artichoke. Perennial with well developed, commonly tuber-bearing rhizomes; stems stout, 1–3 m, ± spreading-hairy; lvs numerous, those of the upper half or ⅔ of the stem alternate in well developed plants, scabrous above, short-hairy beneath, broadly lanceolate to broadly ovate, the better developed ones mostly 10–25 × 4–12 cm, serrate, abruptly contracted or somewhat tapering to

the winged petiole, this 2–8 cm; heads often many in a corymbiform infl, the disk yellow, 1.5–2.5 cm wide; invol bracts rather dark, especially near the base, narrowly lanceolate, acuminate or subattenuate, loose especially above the middle, often hispidulous; rays 10–20, 2–4 cm; $2n=102$. Moist soil and waste places; nearly throughout e. U.S. and adj. Can., and w. across the Great Plains. (*H. subcanescens,* small plants with chiefly opposite lvs.)

17. Helianthus giganteus L.   Swamp-s. Perennial with crown-buds, short rhizomes, and thickened, often fleshy roots; stems 1–3 m, coarsely spreading-hairy to occasionally subglabrous; lvs flat, strongly scabrous above, usually hirsute beneath, ± triplinerved at the base, strongly toothed to subentire, lanceolate, acuminate, 8–20 × 1–3.5 cm, tapering to the short petiole or petiolar base, the upper generally alternate; heads several or many in an open infl; disk yellow, 1.5–2.5 cm wide; invol bracts narrow, thin, green (or dark below), acuminate or attenuate, loose, often conspicuously surpassing the disk, strongly hirsute-ciliate and often hairy on the back; rays 10–20, 1.5–3 cm; $2n=34$. Swamps and other moist places; Me. and N.B. to n. S.C. and n. Ga., w. to s. Alta. and Neb. Aug.–Oct. (*H. subtuberosus; H. rydbergii,* a nw. phase approaching the cordilleran *H. nuttallii* T. & G.)

18. Helianthus grosseserratus Martens.   Sawtooth s. Coarsely fibrous-rooted, rhizomatous perennial; stems 1–4 m, strigose in the infl, otherwise glabrous and often glaucous; lvs lanceolate, 10–20 × 1.5–4 cm (or the lower larger), acuminate, tapering to an often winged petiole 1–4 cm, sharply toothed to sometimes subentire, strigose on both sides, or more hirtellous or puberulent especially on the paler lower surface, usually slightly trinerved at base, the middle and upper commonly alternate; disk yellow, 1.5–2.5 cm wide; invol bracts lance-linear, loose, attenuate, surpassing the disk, ± ciliate, at least near the base, and sometimes short-hairy on the back; rays 10–20, 2–4.5 cm; $2n=34$. Bottomlands, damp prairies, and other moist places; Me. to Minn., s. to Ga. and Tex., perhaps only intr. e. of O. July–Oct. A hybrid with *H. salicifolius* has been called *H. × kellermani* Britton.

19. Helianthus maximilianii Schrader.   Maximilian-s. Perennial with crown-buds, short rhizomes, and thickened, often fleshy-roots; stems usually several, 0.5–3 m, conspicuously pubescent, especially upwards, with mostly short, white, ± appressed hairs; lvs strongly scabrous on both sides, lanceolate, acuminate, 7–20 × 1–3 cm, entire or slightly toothed, firm, pinnately veined, not triplinerved, gradually narrowed to the short, winged petiole or subpetiolar base, generally at least some of them openly folded along the midrib and often falcate, the upper often mainly alternate; heads generally several, the infl tending to be elongate and racemiform; invol bracts narrow, firm, attenuate or subcaudate, loose, often much exceeding the disk, canescent with short white hairs and sometimes basally ciliate; disk yellow, 1.5–2.5 cm wide; rays mostly 10–25, 1.5–4 cm; $2n=34$. Prairies and waste ground, often in sandy soil; Man. and Minn. to Mo., w. to B.C. and Tex., and sparingly intr. eastward; perhaps native as far e. as Mich. July–Oct. (*H. dalyi,* a depauperate form.)

20. Helianthus salicifolius A. Dietr.   Willow-lvd s. Perennial with coarse roots and long stout rhizomes; stems 1.5–3 m, glabrous, sometimes glaucous; lvs numerous, commonly alternate except near the base, subglabrous or sparsely hairy, linear or nearly so, long-acuminate, to 20+ cm, seldom over 5(–10) mm wide, about the same color on both sides; disk red-purple (yellow), 1–2 cm wide; invol bracts loose, all subulate or lance-subulate, caudate, usually some of the inner exceeding the disk, glabrous or slightly hairy, often ciliolate; rays 10–20, 1.5–3 cm; $2n=34$. Prairies and dry places, especially on calcareous substrate; Mo. to Kans. and Tex., and intr. into O. and n. Ill. Aug.–Oct. (*H. orgyalis.*)

21. Helianthus angustifolius L.   Narrow-lvd s. Fibrous-rooted perennial with crown-buds, nearly or quite without rhizomes; stem solitary, 0.5–1.5(–2) m, ± hairy, especially below; lvs numerous, sessile or nearly so, commonly alternate except near the base, linear or nearly so and revolute-margined, mostly 5–15(–20) cm × 2–10(–15) mm and 10–30 times as long as wide, dark green and scabrous above, pale beneath with fine, loose, sometimes deciduous hairs, and often also atomiferous-glandular; petiolate basal lvs with broader blade sometimes present; disk red-purple (yellow), (1–)1.5–2 cm wide; invol bracts narrow, at least the inner with loose, narrow tip seldom surpassing the disk; rays (8–)10–15(–21), 1.5–3 cm; $2n=34$. Swamps and moist places; L.I. to Fla. and Tex., chiefly near the coast in our range, but inland to s. O., s. Ind., and s. Mo. Aug.–Oct.

2. TETRAGONOTHECA L.   Squarehead. Heads radiate, the rays pistillate and fertile, light yellow; invol of 4(5) relatively large and leafy bracts, united at base; receptacle conic, especially in fr, its bracts subtending rays as well as disk-fls, thin, partly clasping the achenes; disk-fls perfect and fertile, the cor-tube expanded at base and covering the top of the achene; style-branches flattened, with rather long, hairy appendage; achenes quadrangular or subterete, scarcely compressed; pappus of several short scales, or none; coarse, taprooted perennial herbs with clustered stems, opposite, ± toothed, often clasping lvs and few or solitary heads. 4, U.S., Mex.

1. Tetragonotheca helianthoides L.   Plants 3–8 dm, viscid-villous, especially the stem; lvs large and thin, to 20 × 10 cm, ovate to elliptic or rhombic, narrowed to a sometimes petioliform, often clasping base, coarsely toothed; invol bracts broadly ovate, 2–3 cm, ciliate, otherwise subglabrous; rays 6–10, 1.5–3 cm, often 1 cm wide; achenes subterete; pappus none; $2n=34$. Dry, open woods, often in sandy soil; s. Va. to Fla. and Ala., and inland to e. Tenn. May–June, sometimes again in autumn.

3. RUDBECKIA L.

3. RUDBECKIA L.   Conefl. Heads radiate, the rays mostly 5–21, neutral, yellow or orange to sometimes anthocyanic, or rarely absent and the heads discoid; invol of subequal or irregularly unequal, green and ± herbaceous, mostly spreading or reflexed bracts in 2–3 series; receptacle enlarged, conic or columnar, chaffy, its bracts partly enfolding the achenes, not spinescent, but sometimes awn-pointed; disk-fls perfect and fertile, narrowed to a ± distinct tube at base; style-branches flattened, with very short and blunt or elongate and acuminate, externally hairy appendage, without well marked stigmatic lines; achenes ± quadrangular, glabrous; pappus a short crown, or none; herbs with alternate, entire to pinnatifid lvs. 16, mainly U.S.

1 Pappus none; style-appendages elongate, subulate; herbage coarsely hirsute . . . . 1. *R. hirta.*
1 Pappus present, sometimes minute; style-appendages short and blunt.
  2 Receptacular bracts glabrous, or ciliolate on the margins, rarely with a few appressed hairs on the
    back, not at all canescent.
    3 Receptacular bracts obtuse or acute, not awn-pointed; lvs merely toothed or subentire.
      4 Lvs narrow, the basal ones up to 2 cm wide; herbage strongly hirsute . . . . 2. *R. missouriensis.*
      4 Lvs wider, some of them more than 2 cm wide; herbage seldom strongly hirsute . . . . 3. *R. fulgida.*
    3 Receptacular bracts conspicuously awn-pointed; some of the lvs generally deeply trilobed or
      even pinnatifid . . . . 4. *R. triloba.*
  2 Receptacular bracts canescent near the tip with short, viscidulous hairs.
    5 Lvs entire or merely toothed . . . . 5. *R. heliopsidis.*
    5 Lvs, or some of them, pinnatifid or deeply trilobed.
      6 Stem ± densely short-hairy, at least above the middle . . . . 6. *R. subtomentosa.*
      6 Stem essentially glabrous . . . . 7. *R. laciniata.*

1. Rudbeckia hirta L.   Black-eyed Susan. Biennial or short-lived perennial 3–10 dm, ± hispid or hirsute throughout; lvs variable, the lower mostly oblanceolate to elliptic and long-petiolate, the others lance-linear to oblong or ovate, mostly sessile; heads mostly long-pedunculate, the hemispheric or ovoid disk 12–20 mm wide, dark purple or brown (yellow); invol bracts copiously hirsute or hispid, sometimes much reduced; rays 8–21, orange or orange-yellow, sometimes darker or marked with purple at base, 2–4 cm; receptacular bracts acute, ± hispid or hispidulous near the tip; style-appendages elongate, subulate; pappus none; $2n=38$. Various habitats; Nf. to Fla., w. to B.C. and Mex. June–Oct. Var. *hirta* occurs mostly in relatively undisturbed habitats in the mts. from Pa. to Ga. and Ala., and sparingly n. to Me. and w. to Ill.; it has coarsely toothed lvs, the basal ones with the blade 2.5–7 cm wide and about twice as long, the cauline ones lance-ovate to ovate or pandurate. Var. *pulcherrima* Farw. (*R. serotina*) is widespread especially in disturbed habitats; it has entire or finely toothed lvs, the basal ones with lanceolate or oblanceolate blade 1–2.5(–5) cm wide and (3–)4–5 times as long, the cauline ones spatulate or oblanceolate to broadly linear.

2. Rudbeckia missouriensis Engelm.   Missouri-conefl. Densely spreading-hirsute throughout, including the invol bracts; plants not stoloniferous; branches closely ascending; lvs narrow, the basal ones broadly linear to lance-spatulate, to 2 cm wide, the cauline ones linear-spatulate, entire; otherwise much like no. 3; $2n=38$. Mostly in dry, open places; Mo. and s. Ill. to Ark. and Tex., barely entering our range. June–Oct. (*R. fulgida* var. *m.*)

3. Rudbeckia fulgida Aiton.   Eastern c. Perennial 3–10 dm, commonly stoloniferous, sparsely to moderately hairy, seldom densely hirsute; lower lvs lanceolate to cordate, long-petiolate, the others similar or gradually reduced, short-petiolate or sessile; heads commonly long-pedunculate, the hemispheric or ovoid disk dark purple or brown, 10–18 mm wide; rays 8–21, yellow to orange; receptacular bracts obtuse or acute, smooth or ± ciliolate-margined, rarely with a few appressed hairs on the back; pappus an inconspicuous low crown; $2n=38, 76$. Chiefly in woods or moist places; Pa. to Mich., Ill., and s. Mo., s. to Fla. and Tex., and occasionally adventive elsewhere, as in Conn. July–Oct. Three morphologically and geographically overlapping regional vars. that seem sharply distinct at some points of contact:

a Rays mostly 2.5–4 cm; lvs usually sharply toothed, varying in shape from nearly as in var. *umbrosa*
  to as in the wider-lvd forms of var. *fulgida*; streambanks, swamps, and other wet, open or shady
  places; midwestern phase, mainly from Mich. and Ill. to W.Va. and s. Mo. (*R. speciosa; R. deamii;*
  R. sullivantii) . . . . var. *speciosa* (Wender.) Perdue.
a Rays mostly 1–2.5(–3) cm.
  b Cauline lvs mostly ovate or subcordate, sharply toothed to sometimes entire, abruptly contracted
    to a narrowly or scarcely winged petiole; woods; s. Appalachian phase, mainly from s. O. to
    Ga. and Ala. (*R. umbrosa*) . . . . var. *umbrosa* (Boynton & Beadle) Cronq.
  b Cauline lvs mostly narrower than ovate and sessile or merely narrowed to a winged petiole or
    petioliform base, usually entire or merely denticulate; moist, low ground in shaded or open
    places; southeastern phase, widespread from Fla. to La. and n. at lower elev. to s. N.Y., Pa.
    and Ill. (*R. spathulata; R. tenax; R. truncata*) . . . . var. *fulgida.*

4. Rudbeckia triloba L.   Three-lobed c. Biennial or short-lived perennial 5–15 dm, moderately hirsute or strigose to subglabrous; lvs thin, sharply toothed to subentire, the basal ones broadly ovate or subcordate and long-petiolate, the cauline mostly narrower and shorter-petiolate or sessile, usually some of the larger ones deeply trilobed or seldom even pinnatifid; rays 6–13, yellow or orange, 1–2(–2.5) cm; disk dark purple or brown, hemispheric or ovoid, 8–15 mm wide; receptacular bracts glabrous, abruptly narrowed to a short but distinct awn-tip often shortly surpassing the disk-cors; pappus a minute crown; $2n=38$, 57, the triploid more northeastern. Woods and moist soil; Conn. to Mich., Io., and Nebr., s. to Fla. and Tex. June–Oct. Most of our plants belong to the widespread var. *triloba,* with the lvs often (not always) well over 5 cm, some of them generally trilobed with the lobes generally acute, rarely any of them pinnately lobed. The var. *pinnatiloba* T. & G. occurs on the Gulf coast of w. Fla. and disjunct in the Blue Ridge of N.C. and sw. Va. (where it has been called *R. beadlei* Small). It is a smaller and more delicate plant, with the lvs up to 5(–8) cm, some of them generally pinnately few-lobed, the lobes often obtuse.

5. **Rudbeckia heliopsidis** T. & G.   Piney-woods c. Perennial from a woody rhizome; stem 6–10 dm, spreading-villous to subglabrous, sparsely leafy; lvs thin, subglabrous or sparsely hairy, serrate or subentire, the basal very long-petiolate, the cauline progressively less so, the principal ones with ovate to elliptic-ovate or lance-ovate blade mostly 6–10 × 3–5 cm; heads several; disk dark, hemispheric or ovoid, 10–15 mm wide; invol bracts short, with loose or spreading tip; rays ca 13, yellow, 1.5–2.5 cm; receptacular bracts broadly rounded, distally viscidulous-canescent; pappus a minute crown. Pine and oak-hickory woods; rare and local from se. Va. to S.C., Ga., and ne. Ala. July–Sept.

R. *grandiflora* (Sweet) DC., an Ozarkian sp. with larger heads (rays 3–5 cm) and with the lvs hirsute on both sides, barely enters our range in c. Mo. as an introduction.

R. *maxima* Nutt., a glabrous Ozarkian sp. with conspicuously elongating disk, probably does not enter our range.

6. **Rudbeckia subtomentosa** Pursh.   Sweet c. Perennial from a stout rhizome; stem 6–20 dm, glabrous below, ± densely short-hairy above; lvs firm, densely short-hairy, especially beneath, ovate to sometimes lance-elliptic, petiolate, serrate, generally some of the larger ones deeply trilobed; disk dark purple or brown, 8–16 mm wide, not elongating; rays 12–21, yellow, 2–4 cm; receptacular bracts obtuse or acutish, distally viscidulous-canescent; pappus a minute crown; $2n=38$. Prairies and low ground; Mich. to Wis., s. to w. Tenn., La., and Okla. July–Sept.

7. **Rudbeckia laciniata** L.   Cutlf-c.; golden glow (a cult. double form). Perennial from a woody base, 5–30 cm; stem glabrous, often glaucous; lvs large, petiolate, coarsely toothed or laciniate, some or most of them pinnatifid or sometimes merely trilobed, subglabrous, or hairy beneath; disk yellow or grayish, 1–2 cm wide, hemispheric at first; rays drooping, lemon-yellow, 6–16, 3–6 cm; receptacular bracts blunt, distally viscidulous-canescent; pappus a short, usually toothed crown; $2n=36$, 54, 72, 102+. Moist places; Que. to Fla., w. to Mont. and Ariz. July–Sept. Three vars. with us, a fourth cordilleran.

a Heads relatively small and several or numerous on slender, flexuous peduncles, the disk mostly 1–
    1.5 cm wide, not elongating, the rays mostly ca 5 or 8; lvs usually pinnatifid as in var. *laciniata,*
    varying to merely trifid as in var. *humilis*; plants mostly 1–2 m, more often sexual than apomictic;
    Va. to Fla. and La. .... var. *digitata* (Miller) Fiori.
a Heads larger, the disk mostly 1.5–2 cm wide, the rays mostly ca 8 or 13.
    b Tall, 1.5–3 m when well developed, with several or many heads, the disk generally elongating to
        2–3 cm in age; lvs evidently pinnatifid or even subbipinnatifid; more often apomictic than
        sexual; widespread, but not at highest elev. .... var. *laciniata.*
    b Smaller, 0.5–1.5 m, with only 1–6 heads, the disk not elongating; lvs mostly merely 3-cleft or
        undivided, occasionally some of them pinnatifid; more often sexual than apomictic; high elev.
        in s. Appalachian Mts. of Va. and Ky. to N.C. .... var. *humilis* A. Gray.

4. ECHINACEA Moench.   Conefl. Heads radiate, the rays mostly ca 13(21), neutral or with a vestigial style, often drooping, our spp. purple to white; invol bracts in 2–4 subequal or slightly imbricate series, with firm base and spreading or reflexed green tip, the inner sometimes passing into those of the receptacle, but not sharply divisible into 2 types; receptacle conic, its bracts partly enfolding the achenes, firm, with stout, spinescent tip conspicuously exceeding the disk-cor; disk-cors slightly bulbous-thickened at base, not narrowed to a slender tube; style-branches flattened, without well marked stigmatic lines, and with slender, acuminate, hairy appendage; achenes quadrangular, glabrous or sparsely hairy on the angles; pappus a short, toothed crown; perennial herbs with simple, alternate, entire or toothed, mostly trinerved lvs and solitary or few, mostly long-pedunculate heads. 5, U.S. and adj. Can. (*Brauneria.*)

1 Lf-blades mostly 1.5–5 times as long as wide; plants tending to be fibrous-rooted.
    2 Lvs ± pubescent on both sides .... 1. *E. purpurea.*
    2 Lvs glabrous on both sides, or somewhat scabrous or short-hairy above .... 2. *E. laevigata.*
1 Lf-blades mostly 5–20 times as long as wide; plants strongly taprooted .... 3. *E. pallida.*

1. **Echinacea purpurea** (L.) Moench.   Purple c. Stems 1–few from a coarsely fibrous-rooted crown, caudex, or short, stout rhizome, hirsute to glabrous, 6–18 dm, simple or often few-branched; lvs ± hairy on both

sides, toothed or less commonly entire, the main ones with broadly lanceolate to elliptic or broadly ovate blade up to 15 × 10 cm, mostly 1.5–5 times as long as wide, rather abruptly contracted (or even rounded or subcordate) to the petiole; disk 1.5–3.5 cm wide; rays reddish-purple to occasionally pale pink, 3–8 cm, drooping; $2n=22$. Woods and prairies, generally in moister sites than no. 3; chiefly Ozarkian and midwestern, from Ill. and s. Io. to e. Okla., extreme ne. Tex., and c. La., e. irregularly to s. Mich., Ky., Tenn., and Ga., and less commonly to Va. and N.C. June–Oct.

2. **Echinacea laevigata** (Boynton & Beadle) S. F. Blake.   Appalachian c. Much like no. 1, and perhaps only a geographic var. of it, but glabrous and ± glaucous, or the lvs sometimes scabrous to short-hairy above; stems consistently simple, 6–12 dm; rays up to 8 cm; plants tending to have a short, quickly deliquescent taproot or vertical caudex; $2n=22$. Woods and fields; locally from se. Pa. to n. Ga., from the Piedmont to the Folded Appalachians. (*E. purpurea* var. *l.*)

3. **Echinacea pallida** Nutt.   Prairie c. Stems clustered on a strong taproot, usually simple; herbage coarsely spreading-hirsute; lvs entire or nearly so, basally disposed, elongate and narrow, the blade to 20 × 4 cm, mostly 5–20 times as long as wide (or the basal a little wider), tapering to the petiole; disk 1.5–3 cm wide; rays pink, varying to purple or white; $2n=22, 44$. Dry, open places, especially on the prairies and plains; e. Mont. to Tex., e. to Wis., Ill., Ark., and La., and irregularly, mainly as an intr., to Mich., N.Y., N.C., and Ga. Var. *pallida,* the more eastern segment of the sp., the principal phase in our range, is mostly tetraploid and robust, 4–10 dm, with drooping rays mostly 4–8 cm, and typically with white pollen. (*E. simulata* McGregor, of se. Mo., ne. Ark., s. Ill., and w. Ky., diploid with yellow pollen, otherwise as var. *pallida,* should perhaps be recognized as another var.) Var. *angustifolia* (DC.) Cronq., the more western segment of the sp., is diploid and smaller, mostly 1–5 dm, with spreading to drooping rays 2–4 cm, and yellow pollen. (*E. angustifolia.*) An eastern outlier of var. *angustifolia,* in the cedar glades of c. Tenn., has been called *E. tennesseensis.*

5. **RATIBIDA** Raf.   Conefl. Heads radiate, the rays mostly 3–13, neutral, relatively broad, yellow or partly or wholly brown-purple; invol a single series of green, subherbaceous, linear or lance-linear bracts; receptacle columnar, chaffy, its bracts subtending rays as well as disk-fls, ± clasping the achenes, the tip densely velutinous and incurved; disk-cors numerous, short, cylindric, scarcely narrowed at base; style-branches flattened, with ovate to subulate, hairy appendage; achenes compressed at right angles to the invol bracts, often also evidently quadrangular, glabrous except for the sometimes ciliate margins; pappus coroniform, with 1 or 2 prolonged, awn-like teeth, or of teeth only, or absent; perennial herbs with alternate, pinnatifid lvs and naked-pedunculate heads. 6, N. Amer. (*Lepachys.*)

1 Disk columnar, 2–4.5 times as long as thick; plants taprooted .... 2. *R. columnifera.*
1 Disk shorter, 1–1.6 times as long as thick; plants fibrous-rooted .... 1. *R. pinnata.*

1. **Ratibida pinnata** (Vent.) Barnhart.   Globular c. Fibrous-rooted from a stout, woody rhizome or sometimes a short caudex, 4–12 dm, ± hirsute, or the stem strigose above; lower lvs long-petioled, the upper short-petioled or sessile; lf-segments lanceolate, acute, coarsely toothed or entire; heads usually several, naked-pedunculate; disk ellipsoid-globular, 1–2 cm, 1–1.6 times as long as thick, much shorter than the rays, these pale yellow, (2.5–)3–6 cm, spreading or often reflexed; style-appendages elongate, acuminate; achenes smooth; pappus none; $2n=28$. Prairies, old fields, and dry woods, often on limestone; s. Ont. to Minn. and S.D., s. to Tenn., Ga., w. Fla., La., and Okla., and adventive e. to Vt. and Mass. June–Aug.

2. **Ratibida columnifera** (Nutt.) Wooton & Standley.   Columnar c. Taprooted, with clustered stems 3–12 dm, strigose or partly hirsute, generally leafy to above the middle; lvs pinnatifid or partly bipinnatifid, the ultimate segments linear or lanceolate, entire or nearly so, relatively few and often very unequal, generally many or all of them 1.5 cm or more; heads (1–)several or many, naked-pedunculate, the disk columnar, 1.5–4.5 cm, 2–4.5 times as long as thick; rays yellow or (f. *pulcherrima* (DC.) Fern.) partly or wholly brown-purple, 1–3.5 cm, spreading or reflexed; style-appendages very short and blunt; achenes with the inner margin fringed-ciliate to nearly smooth, usually slightly winged; pappus an evident awn-tooth on the inner angle of the achene, and often also a shorter one on the other angle; $2n=26, 27, 28, 34–38$. Prairies and other dry, open places; Minn. to Mo. and La., w. to Mont., Tex., and n. Mex., and occasionally intr. e. June–Aug.

6. **GAILLARDIA** Foug.   Blanket-fl. Heads radiate or occasionally discoid, the rays yellow to purple, broad, 3-cleft, usually neutral, sometimes pistillate and fertile; invol bracts in 2–3 series, herbaceous above the chartaceous base, ± spreading, becoming reflexed in fr; receptacle convex to subglobose, provided with numerous soft or more often chaffy or spine-like setae that do not individually subtend the disk-fls, or these rarely obsolete; disk-fls perfect and fertile, the cor-lobes woolly-villous; style-branches flattened, with introrsely marginal stigmatic lines and a usually ± elongate and externally hairy

appendage; acheness broadly obpyramidal, partly or wholly covered by a basal tuft of long, ascending hairs; pappus of 6–10 awned scales; taprooted herbs with alternate (or all basal), entire to pinnatifid lvs and rather large, mostly long-pedunculate heads. 15, New World.

1 Rays yellow, sometimes with purplish base; true perennial .... 1. *G. aristata.*
1 Rays red-purple, often yellow at the tip; annual or short-lived perennial .... 2. *G. pulchella.*

1. Gaillardia aristata Pursh.   Common b.-f. Hairy perennial 2–7 dm from a slender taproot, also spreading by slender creeping roots; lvs narrow, linear-oblong to lance-ovate, or the lower oblanceolate, to 15 × 2.5 cm, entire to somewhat pinnatifid; heads solitary or few, long-pedunculate, the disk 1.5–3 cm wide, purple or brownish-purple (yellow); rays 6–16, yellow, usually purplish at base, 1–3.5 cm; setae of the receptacle evidently exceeding the achenes; style-appendages elongate; 2*n*=36, 72. Plains, meadows, and other open places; B.C. to Sask. and Minn., s. to Ariz. and N.M., and occasionally intr. or escaped eastward. May–Sept.
      G. aestivalis (Walter) H. Rock var. *flavovirens* (C. Mohr) Cronq., with the setae of the receptacle obsolete, and with yellow disk, barely enters our range in s. Ill. (*G. lanceolata* var. *f.; G. lutea.*)

2. Gaillardia pulchella Foug.   Rosering-b.-f. Glandular-villous annual or sometimes short-lived perennial, 1–6 dm, simple to more often freely branched and with several or many heads, often decumbent at base, the littoral forms commonly somewhat succulent; disk 1–2.5 cm wide; rays purple or the tip yellow, mostly 1–2 cm; setae of the receptacle about equaling or only slightly exceeding the achenes; otherwise much like no. 1; 2*n*=34, 68. Dry, often sandy places, sometimes on sea-beaches; Mo. and s. Neb. to Colo., N.M., and n. Mex., and along the coast from Tex. to Fla. and se. Va. May–Sept. (*G. drummondii; G. picta* Sweet, the littoral phase, perhaps properly to be treated as a distinct var.)

7. HELENIUM L.   Sneezeweed. Heads mostly radiate, the rays pistillate or neutral, mostly yellow, cuneate, 3-lobed, not very numerous; invol bracts in 2–3 series, subequal or the inner shorter, ± herbaceous, generally soon deflexed, the outer sometimes joined at the base; receptacle convex to ovoid or conic, naked; disk-fls very numerous, perfect, the cor-lobes glandular-hairy; style branches flattened, with ventromarginal stigmatic lines and dilated, subtruncate, penicillate tip; achenes 4–5-angled, with as many intermediate ribs, pubescent on the angles and ribs, or glabrous; pappus of 5–10 scarious or hyaline, often awn-tipped scales; herbs with alternate, glandular-punctate, usually decurrent lvs and solitary to numerous heads. 40, New World.

1 Fibrous-rooted perennials; at least the larger lvs generally 5 mm wide or more.
  2 Rays pistillate; disk yellow.
    3 Lvs chiefly or all cauline, numerous, the lower generally deciduous; widespread .... 1. *H. autumnale.*
    3 Lvs ± basally disposed; local in Augusta and Rockingham cos., Va. .... 2. *H. virginicum.*
  2 Rays neutral (rarely wanting); disk red-brown or purple-brown, rarely yellow.
    4 Heads several or many in a corymbiform, leafy-bracteate infl; stem and generally also the lvs
       puberulent or villous-puberulent; widespread .... 3. *H. flexuosum.*
    4 Heads 1–3(–6) on long, naked peduncles; middle and lower part of stem and middle and lower
       lvs not hairy; Va. and s. .... 4. *H. brevifolium.*
1 Taprooted annual with numerous, narrow lvs rarely over 2 mm wide .... 5. *H. amarum.*

1. Helenium autumnale L.   Common s. Fibrous-rooted perennial 5–15 dm, glabrous to finely strigose or puberulent; lvs numerous, all or mainly cauline, lance-linear to elliptic or narrowly ovate, narrowed to a sessile or subpetiolar base, decurrent as wings on the stem, not much reduced upwards, 4–15 cm × 5–40 mm, the lower ones deciduous; heads (1–) several or many in a leafy infl, hemispheric or subglobose, the disk yellow, 8–20 mm wide; rays ca 13 to ca 21, pistillate, mostly 1.5–2.5 cm; pappus-scales generally brownish, ovate or lanceolate, tapering to a short awn, up to ca 1 mm overall; 2*n*=32, 34, 36. Moist low ground; Que. to Fla., w. to B.C. and Ariz. Aug.–Oct. The var. *autumnale,* with elliptic to oblong or lanceolate, generally toothed lvs 3–6 times as long as wide, occurs from Conn. to Minn., s. to Fla. and Tex. (*H. altissimum; H. latifolium; H. parviflorum.*) The var. *canaliculatum* (Lam.) T. & G., a usually smaller, sometimes single-headed plant with narrower, mostly subentire lvs 7–12 times as long as wide, occurs from Que. to N.Y., w. to Wis. Other vars. occur westward.

2. Helenium virginicum S. F. Blake.   Virginia s. Lvs tending to be basally disposed, the elongate, often persistent basal ones larger than the progressively reduced and relatively few cauline ones; pappus-scales white-hyaline, 1–2 mm. Very local around sink-hole ponds in Augusta and Rockingham cos., Va. July–Sept.

3. Helenium flexuosum Raf.   Southern s. Fibrous-rooted perennial 2–10 dm, ± puberulent or villous-puberulent at least on the stem, the lower part of the stem generally spreading-villous; lvs decurrent but smaller, fewer, and more erect than in no. 1, entire or subentire, the lowermost ones oblanceolate, commonly deciduous, the others oblong or lanceolate to lance-linear, sessile, generally not much reduced upwards, 3–

12 × 0.5–2 cm; heads in well developed plants numerous in an open, corymbiform, leafy-bracteate infl, the disk subglobose or ovoid-globose, 6–15 mm wide, red-brown or purplish; rays ca 8 to ca 13, neutral, sometimes purplish at base, (0.5–)1–2 cm (rarely wanting); disk-fls predominantly 4-merous; pappus-scales 5(–8), ovate or lanceolate, shortly awn-tipped, ca (0.5–)1 mm overall; 2n=28. Moist ground and waste places; Mass. and N.H. to Fla., w. to Wis., Ill., Mo. and Tex., apparently only intr. northward. June–Oct. (*H. nudiflorum; H. polyphyllum.*)

4. **Helenium brevifolium** (Nutt.) A. Wood.   Few-headed s. Fibrous-rooted perennial 3–8 dm, glabrous or merely atomiferous-glandular below, the peduncles and involucres loosely puberulent; lvs relatively few, the basal tufted and generally persistent, oblanceolate or elliptic, usually petioled, 2–17 cm overall and 0.6–2(–3) cm wide; cauline lvs few, reduced, seldom exceeding the internodes, narrowly decurrent; heads 1–3(–6), naked-pedunculate, the disk hemispheric to subglobose, 1–2 cm wide, usually purple-brown or red-brown; rays ca 8 to ca 13, neutral, 1–2.5 cm, trilobed; pappus of 5–10 awnless, often distally rounded scales 1–1.7 mm; 2n=26, 28. Swampy or boggy places and moist pine woods; se. Va. to Fla., La., and Tenn. May, June. (*H. curtisii,* the rare form with yellow disk.)

5. **Helenium amarum** (Raf.) H. Rock.   Narrow-lvd s. Taprooted, glabrous or atomiferous-glandular annual 2–5 dm; lvs very numerous, linear or linear-filiform, 1.5–8 cm, seldom over 2 mm wide; heads on short, naked peduncles extending above the leafy part of the plant; rays 5–10, pistillate, 5–12 mm; disk subglobose, yellow, 6–12 mm wide; pappus-scales commonly 6–8, ca 1.5 mm, the awn-tip about equaling the hyaline body; 2n=30. Prairies, open woods, fields, and waste places, especially in sandy soil; mainly Texan and Ozarkian, n. to Mo., but irregularly intr. e. to Conn., Md., Va., Ga., and Fla., and in Mex. and W.I. June–Oct. (*H. tenuifolium.*) We have only the var. *amarum,* with yellow disk and entire lvs.

8. **HYMENOXYS** Cass.   Heads mostly radiate, the rays pistillate and fertile, broad, yellow, mostly 5–35; invol bracts permanently erect, partly or wholly herbaceous, in 2–3 similar or sharply differentiated series; receptacle naked, hemispheric to merely convex; disk-fls yellow, perfect and fertile; style-branches flattened, with introrsely marginal stigmatic lines, truncate, minutely penicillate; achenes turbinate, hairy, mostly 5-angled; pappus of a few (typically 5) hyaline, often awned scales; aromatic herbs with alternate or all basal, entire to ternate or pinnatifid, commonly punctate lvs. 20, New World.

1. **Hymenoxys herbacea** (Greene) Cronq.   Lakeside daisy. Perennial from a stout, ± fibrous-rooted caudex; lvs tufted, all basal, oblanceolate, 1–8 cm × 1.5–10 mm, ± densely villous or sericeous when young, soon glabrate, strongly punctate; scape 5–25 cm; head solitary, the disk 8–20 mm wide; invol bracts broadly rounded at the tip; rays 0.5–2 cm. Rare and local in prairie remnants and in open, rocky places about Lake Erie; Ill. to O. and s. Ont.; now perhaps restricted to Marblehead Peninsula in O. May, June. (*Tetraneuris h.; Hymenoxys acaulis* var. *glabra,* misapplied.)

9. **ARNICA** L.   Arnica. Heads radiate (in our spp.) or sometimes discoid, the rays pistillate, yellow or orange, relatively few and large; invol bracts herbaceous, subequal but ± evidently biseriate; receptacle convex, naked; disk-fls perfect and fertile, yellow; style-branches flattened, truncate, penicillate; achenes cylindric, 5–10-nerved; pappus of numerous white to tawny, barbellate to subplumose, capillary bristles; perennial herbs with opposite lvs (uppermost lvs rarely alternate) and solitary to numerous, rather large, turbinate to hemispheric heads. 27, temperate and boreal parts of N. Hemisphere.

1 Perennial from a rhizome or caudex well over 2 cm; basal lvs petiolate.
  2 None of the lvs cordate; plants boreal, but not of Mich.
    3 Pappus white or whitish, merely barbellate; lower lvs persistent, conspicuously larger than the
      progressively reduced middle and upper ones .... 1. *A. lonchophylla.*
    3 Pappus tawny, subplumose; lower lvs not conspicuously larger than those above, often deciduous,
      the stem appearing relatively leafy .... 2. *A. lanceolata.*
  2 Basal lvs strongly cordate; Keweenaw peninsula, Mich. .... 3. *A. cordifolia.*
1 Perennial from a short simple erect caudex 5–20 mm; basal lvs sessile or subsessile, rosulate; Del.
  and se. Pa. to Fla. .... 4. *A. acaulis.*

1. **Arnica lonchophylla** Greene.   Stems 1.5–4 dm, solitary or sometimes loosely clustered, on a densely scaly rhizome; herbage finely stipitate-glandular and often also spreading-hairy or scabrous-puberulent, or the lvs subglabrous; lvs dentate, the lowest mostly persistent, 5–15 × 1.5–3.5 cm; middle and upper lvs few and reduced, sessile, the stem appearing sparsely leafy or subnaked; heads (1–)3–7(–9), turbinate or campanulate, the disk 1–2 cm wide; invol 8–12 mm; rays (6–)8(–10), 1–2 cm; pappus-bristles white, merely barbellate; ours apomictic triploids and tetraploids, 2n=57, 76. Rocky places, in our range chiefly or wholly in calcareous situations; cordilleran region of Can. and nw. U.S.; Ont. and ne. Minn.; N.S. to Que. and Nf. Ours are var. *lonchophylla.* (*A. chionopappa.*) Var. *arnoglossa* (Greene) B. Boivin is restricted to Wyo. and w. S.D. (*A. arnoglossa.*)

2. **Arnica lanceolata** Nutt. New England a. Stems mostly solitary, 2–5 dm from a long, nearly naked rhizome, spreading-villous and often glandular above, mostly glabrous below; lvs sparsely long-hairy or subglabrous, dentate or denticulate, the lower ones evidently petiolate, elliptic or oblanceolate, to 20 cm (petiole included) × 4 cm, the upper sessile and often more lance-ovate or lance-oblong, the stem appearing leafy; heads 1–3, campanulate, the disk 1.5–2 cm wide; invol 10–13 mm, spreading villous and stipitate-glandular, the bracts more strongly acuminate than in no. 1; rays 10–15, 1.5–2 cm; pappus-bristles tawny, subplumose; $2n=76$ (apomictic tetraploid). Streambanks and other moist, protected places especially in acid soils; mts. of n. N.Y., N.H., and Me. to N.B. and the Gaspé penins. of Que. July, Aug. (*A. mollis,* misapplied.)

3. **Arnica cordifolia** Hook. Heart-lvd a. Rhizomatous perennial 1.5–4.5 dm, hairy or glandular or both; basal and lowest cauline lvs conspicuously petiolate, with strongly cordate base, to 11 × 8 cm, generally dentate; cauline lvs reduced upwards, becoming sessile; heads 1–3, the disk 1.5–2.5 cm wide; invol 12–20 mm; rays 10–15, 2–3 cm; pappus white, strongly barbellate; $2n=(38)$, 57, 76 (95), the polyploids apomictic. Dry woods; Keweenaw penins., Mich.; widespread in the w. cordillera. June, July. (*A. whitneyi.*)

4. **Arnica acaulis** (Walter) BSP. Southeastern a. Perennial 2–8 dm from a short, simple, erect caudex 5–20 mm, with numerous fibrous roots; herbage glandular and hirsute; basal lvs rosulate, sessile or nearly so, broadly elliptic to ovate or even rhombic, obscurely toothed or subentire, 4–15 × 1.5–8 cm; cauline lvs few and reduced, the uppermost often alternate; heads 3–20, campanulate, the disk 13–20 mm wide; invol 10–12 mm; rays 10–15, 1.5–2.5 cm; pappus white, merely barbellate; $2n=38$. Sandy pine woods; Del. and se. Pa., s. mainly along the coastal plain to Fla. May–July.

10. VERBESINA L.

Heads radiate, the rays yellow or white, pistillate or neutral, or rarely wanting and the heads discoid; invol of subequal or slightly imbricate, often somewhat herbaceous bracts; receptacle chaffy, shortly conic to sometimes merely polsterform or nearly flat; disk-fls perfect and fertile; style-branches flattened, with papillate or hairy, acute appendage; achenes ± strongly flattened at right angles to the invol bracts, very often winged; pappus mostly of 2 short to well developed awns, occasionally with a few shorter outer scales or minute awns; leaves simple; our spp. herbs. (*Actinomeris, Phaethusa, Pterophyton, Ridan, Ximenesia.*) 60, New World.

1 Plants evidently perennial; roots fibrous or fleshy-fibrous.
  2 Invol bracts well developed, appressed or somewhat loose, not deflexed.
    3 Heads 1–10; disk 9–16 mm wide in fl; rays 8–15, yellow; lvs alternate (*Pterophyton*) .... 1. *V. helianthoides.*
    3 Heads mostly 20–100+, the disk 3–7 mm wide in fl; rays 1–5 (*Phaethusa*).
      4 Rays yellow; lvs opposite .... 2. *V. occidentalis.*
      4 Rays white; lvs alternate .... 3. *V. virginica.*
  2 Invol bracts few, narrow, soon deflexed; rays yellow; lvs alternate (*Actinomeris*) .... 4. *V. alternifolia.*
1 Plants annual, taprooted (*Ximenesia*) .... 5. *V. encelioides.*

1. **Verbesina helianthoides** Michx. Ozark flatseed-sunflower. Hirsute perennial, 5–12 dm, the stem winged, leafy throughout; lvs alternate, lanceolate to narrowly ovate, serrate, 6–15 × 2–6 cm, sessile or broadly short-petiolate; heads few, mostly 1–10, the disk 9–16 mm wide in fl; invol bracts erect but rather loose; receptacle shortly conic; rays 8–15, pistillate or neutral, yellow, 1–3 cm; achenes winged or wingless, often a little spreading, but not reflexed; $2n=34$. Prairies and dry woods; O. to Ga., w. to Io., Kans., and Tex. June–Oct. (*Phaethusa h.; Pterophyton h.*)

2. **Verbesina occidentalis** (L.) Walter. Southern flatseed-sunflower. Coarse perennial 1–2 m, the stem leafy, glabrous below, puberulent above; lvs opposite, ovate, serrate, 7–17 × 4–11 cm, strigillose to scabrous, the petioles ± winged, decurrent on the stem; heads small, numerous, 20–100+ in a congested or rather open infl, the disk 3–7 mm wide in fl; invol bracts erect, the tips loose, often imbricate, commonly sparsely puberulent; receptacle small, convex; rays 2–5, yellow, usually pistillate and fertile, 0.5–2 cm; achenes wingless; $2n=34$. Bottomlands, thickets, woods, and waste places; Md. to O. and Mo., s. to Fla. and Tex. Aug.–Oct. (*Phaethusa o.*)

3. **Verbesina virginica** L. Frostweed. Coarse perennial to 2 m, the stem leafy, puberulent; lvs alternate, ovate to lance-elliptic, serrate to entire, 9–20 × 3–10 cm, scabrous to subglabrous above, velutinous or sometimes appressed-puberulent beneath, the petioles ± winged, often decurrent on the stem; heads small, numerous, crowded, 20–100+ in a dense, generally ± flat-topped infl, the disk 3–7 mm wide in fl; invol bracts appressed, evidently imbricate, puberulent; rays 1–5, pistillate and fertile, white, under 1 cm; achenes erect, flattened, winged or wingless; $2n=32$, 34. Bottomlands, thickets, woods and waste places; Va. to e. Kans., s. to Fla. and Tex. Aug.–Oct. (*Phaethusa v.*) Ours is the widespread var. *virginica.*

4. **Verbesina alternifolia** (L.) Britton. Wingstem. Perennial 1–3 m, the stem leafy, spreading-hirsute to subglabrous, usually ± winged; lvs alternate, lanceolate or lance-elliptic to occasionally ovate, usually gradually narrowed to a petiolar base, sharply serrate or subentire, 10–25 × 2–8 cm, scabrous-hirsute, especially above; heads 10–100+ in an open infl; disk 1–1.5 cm wide in fl; invol bracts few, glabrous or subglabrous, rather small, narrow, soon deflexed, the disk-fls loosely spreading even before anthesis; rays 2–10, neutral, yellow, 1–3 cm; achenes broadly winged or sometimes wingless, spreading in all directions, forming a globose

head 8–15 mm thick; $2n=68$. Thickets, woods, and bottomlands; N.Y. and s. Ont. to e. Neb., s. to Fla., La., and Okla., rarely adventive elsewhere. Aug.–Oct. (*Actinomeris a.; Ridan a.*)

5. **Verbesina encelioides** (Cav.) Benth. & Hook. Golden crownbeard. Erect annual 2–10 dm, the stem and lower lf-surfaces strigose-canescent, the upper lf-surfaces sometimes greener but still strigose; lvs alternate (except the lower), ovate or deltoid, sometimes rather narrowly so, rather coarsely toothed, especially near the base, 4–13 × 2–10 cm; petiole well developed, commonly auriculate-dilated at base; heads long-pedunculate in an open infl, the disk 13–20 mm wide in fl; invol bracts loose or a little spreading, scarcely imbricate, canescently strigose or hirsute; rays 10–15, pistillate, yellow, evidently trilobed, 1–1.5 cm; achenes winged, a little spreading but not reflexed, the fruiting head hemispheric; $2n=34$. Open, often waste places; native of Mex. and sw. U.S., casually intr. with us, especially westward. May–Oct. (*Ximenesia e.; X. exauriculata.*) Ours is var. *encelioides*.

11. **ECLIPTA** L., nom. conserv. Heads radiate, the rays pistillate, usually white, very short; invol bracts 1–2-seriate, herbaceous or herbaceous-tipped, subequal, or the inner narrower and shorter; receptacle flat or slightly convex, its bracts slender and fragile, or the inner ones wanting; disk-fls perfect and fertile, 4(5)-toothed; style-branches flattened, with short, obtuse, hairy appendage; achenes thick, commonly transversely rugose, 3–4-angled, those of the disk somewhat compressed at right angles to the invol bracts; pappus none or an obscure crown or 2 short awns; branching herbs with opposite simple lvs. 4, mostly tropical.

1. **Eclipta prostrata** (L.) L. Yerba-de-tajo. Weak or spreading, strigose annual, often rooting at the nodes; lvs lanceolate or lance-elliptic to lance-linear, acute, narrowed to a sessile or shortly petiolar base, remotely serrulate, 2–10 cm × 4–25 mm; heads 1–3 in terminal or axillary clusters on the many branches, the disk 4–6 mm wide in fl; invol bracts, or some of them, usually a little exceeding the disk; rays whitish, minute; achenes 2–2.5 mm, rugose or warty, slightly hairy above; pappus a nearly obsolete crown well removed from the margins of the truncate-topped achene; $2n=22$. A weed in bottomlands and muddy places, native to the New World, now pantropical, and n. in our range to Mass., s. Ont., and Wis. Aug.–Oct. (*E. alba; E. erecta; Verbesina a.*)

12. **HELIOPSIS** Pers. Sunfl-everlasting. Heads radiate, the rays yellow, pistillate and fertile, persistent and becoming papery; invol bracts herbaceous at least distally, in 1–3 subequal series, or the outer ones enlarged and leafy; receptacle convex to conic, chaffy throughout, its bracts concave and clasping, subtending the rays as well as the disk-fls; disk-fls perfect and fertile; style-branches flattened, with a short, hairy appendage; achenes nearly equably quadrangular (or those of the rays triquetrous); pappus none, or a short irregular crown or a few teeth; herbs with opposite, petiolate lvs. (*Kallias.*) 13, New World.

1. **Heliopsis helianthoides** (L.) Sweet. Robust, fibrous-rooted perennial 5–15 dm; lvs ovate or lance-ovate, often subtruncate at base, serrate, 5–15 × 2.5–8 cm, on petioles 5–35 mm; heads naked-pedunculate, the disk 1–2.5 cm wide; rays 8–16, rather pale yellow, 1.5–4 cm; achenes essentially glabrous; $2n=28$. Rich to dry woods, prairies, and waste places; Que. to B.C., s. to Ga. and N.M. July–Oct. Var. *helianthoides*, with glabrous, often glaucous stem and rather thin lvs smooth on both sides or merely slightly scabrous above, occurs from N.Y. to Mich. and Ill., and s. to Ga. and Miss. Var. *scabra* (Dunal) Fern., with firmer lvs scabrous on both sides, the stem often scabrous as well, occurs in the Great Plains and prairie region, and e. across the n. part of our range to the Atlantic. (*H. scabra.*)
Zinnia elegans Jacq. the garden zinnia, is native to Mex. and occasionally escapes from cult. in our range; it has clasping, entire, ovate lvs and large, mostly red or orange heads with long-persistent rays.

13. **BORRICHIA** Adans. Sea oxeye. Heads radiate, the rays yellow, pistillate and fertile; invol bracts in 2 or 3 slightly unequal series, the outer subherbaceous, the inner more chartaceous and sometimes transitional to those of the receptacle; receptacle slightly convex, its bracts rigid, clasping the achenes; disk-fls perfect and fertile; style-branches flattened, with elongate, hairy appendage; achenes glabrous, quadrangular, or those of the rays triquetrous; pappus a short dentate crown; ± fleshy seacoast shrubs with opposite, usually canescent lvs. 7, warm Amer.

1. **Borrichia frutescens** (L.) DC. Rhizomatous, colonial, sparingly branched shrubs 2–12 dm, strongly canescent, or the stems often subglabrous; lvs oblanceolate or spatulate, entire or spinose-denticulate, mucronate, 2–7 cm × 3–25 mm; heads solitary or several, the disk 1–1.5 cm wide; invol bracts canescent, the outer ovate and somewhat spreading, the inner less hairy, more appressed, becoming shortly spine-tipped

like those of the receptacle; rays ca 13, to 1 cm; $2n=28$. Seacoast, especially in salt marshes; Va. to Fla., Tex., and Mex.; Bermuda. Mar.–Oct.

14. SPILANTHES Jacq. Heads radiate, the rays pistillate, yellow or white, or rarely wanting and the heads discoid; invol bracts in 1 or 2 series, subherbaceous and subequal; receptacle elongate, conic or subcylindric, its bracts clasping the disk-achenes and eventually deciduous with them; disk-fls perfect and fertile style-branches flattened, truncate, exappendiculate; achenes of the rays 3-angled or tangentially flattened, those of the disk radially flattened, commonly ciliate-margined; pappus of 1–3 slender setiform awns, or none; annual or rarely perennial herbs with opposite, commonly toothed lvs. 40, mainly warm reg.

1. **Spilanthes americana** (Mutis) Hieron. Subglabrous to hirsute perennial, the stems weak, often rooting at the lower nodes, 2–6 dm; lvs ovate or lanceolate, 2.5–8 × 0.5–3 cm, on petioles 0.5–2 cm; heads few on naked peduncles, the disk 5–9 mm wide, elongating conspicuously in fr; rays several, yellow, 3-toothed, 3–10 mm; receptacle narrowly conic; pappus of 1 or 2 very short awns, or none. $2n=52$. Woods and moist or swampy places; tropical Amer., n. to S.C., s. Ill., and Mo. June–Oct. (*S. repens.*)

15. GUIZOTIA Cass. Heads radiate, the rays pistillate and fertile, yellow; invol bracts few, subherbaceous, uniseriate; receptacle convex or conic, its bracts flat or nearly so, membranous or subscarious, striate, subtending rays as well as disk-fls; disk-fls perfect and fertile, the cor densely woolly below the middle; style-branches flattened, with subulate, hairy appendage; achenes glabrous, ± compressed parallel to the invol bracts, often also quadrangular; pappus none; annuals with opposite (or the upper alternate) simple lvs and campanulate or subhemispheric heads. 12, trop. Afr.

1. **Guizotia abyssinica** (L. f.) Cass. Niger-seed. Scabrous or subglabrous annual to 1 m; lvs sessile and clasping, lance-oblong, acuminate, serrate, to 10 × 3 cm; heads several or many, the disk 1.5 cm wide; outer invol bracts ca 5, ovate or obovate; rays 8–13, broad, 1.5–2 cm; $2n=30$. Native of trop. Afr., occasionally adventive in waste places in our range. Sept., Oct.

16. SIGESBECKIA L. Outer fls of the head pistillate, with short and inconspicuous, 2–3-toothed, radiate or subcampanulate, whitish or yellowish cor, the inner fls perfect, tubular, regularly 3–5-toothed; invol biseriate and dimorphic, the outer series of ca 5 linear-oblanceolate, spreading or reflexed green bracts, the inner shorter and broader, erect, each subtending a pistillate fl; receptacle very small, its membranous bracts concave and clasping the achenes; style-branches flattened, with short, hairy appendage; achenes quadrangular, incurved, glabrous; pappus none; branching, mostly annual, ± glandular herbs with opposite, toothed lvs and small heads. 6, mainly warm reg.

1. **Sigesbeckia orientalis** L. Annual 3–10 dm, strigose to hirsutulous; lvs resinous-dotted beneath, ovate or deltoid, coarsely toothed, 4–12 × 2–7 cm, the petiole up to half as long, with slender base and upwardly expanding wing-margins; heads rather numerous in small cymose clusters ending the branches; disk under 1 cm wide; invol very conspicuously stipitate-glandular, the loose outer bracts 5–15 mm; rays short and inconspicuous, yellowish or dirty white; achenes 3–3.5 mm; $2n=30$. Waste places; native of e. Asia, sparingly intr. in our range. Summer.

17. COREOPSIS L. Tickseed. Heads radiate, the rays few, often 8, conspicuous, usually neutral, yellow or rarely pink or white, sometimes marked with reddish-brown at base; invol bracts biseriate and dimorphic, all joined at the base, the outer narrower and usually shorter, commonly more herbaceous than the inner; receptacle flat or slightly convex, its bracts thin and flat; disk-fls perfect and fertile; style-branches flattened, with short or elongate, subtruncate to caudate, hairy appendage; achenes flattened parallel to the invol bracts, usually winged, not beaked; pappus of 2 smooth or upwardly barbed, short awns or teeth, or a minute crown, or obsolete; herbs (ours) with opposite or rarely alternate, entire to pinnatifid or ternate or dissected lvs. 100, mainly New World.

1 Style-appendages evidently acute, often cuspidately so; disk cors mostly 5-toothed.
2 Receptacular bracts chaffy-flattened near the base, caudate-attenuate above; lvs simple or pinnatifid.

3 Wings of the achenes thin, not strongly incurved; not stoloniferous.
  4 Leafy to the middle or usually beyond; peduncles not greatly elongate.
    5 Lvs broad, simple or with 1 or 2 pairs of broad lateral lobes .... 1. *C. pubescens*.
    5 Lvs pinnately parted, the lateral segments linear or lance-linear .... 2. *C. grandiflora*.
  4 Leafy only near the base, with long, subnaked, pedunculoid branches .... 3. *C. lanceolata*.
3 Wings strongly incurved, finally callous-thickened; stoloniferous .... 4. *C. auriculata*.
2 Receptacular bracts linear or linear-clavate, rounded to merely acute; lvs mostly ternately or palmately lobed or compound.
  6 Lvs divided to the base, essentially trifoliolate, or rarely simple and entire.
    7 Lvs evidently petiolate, with 3 lfls, or the upper (rarely all) entire .... 5. *C. tripteris*.
    7 Lvs essentially sessile.
      8 Lf-segments lanceolate or broader, entire, mostly 5–30 mm wide .... 6. *C. major*.
      8 Lf-segments, or some of them, parted, the ultimate ones 0.3–1 mm wide .... 7. *C. verticillata*.
  6 Lvs trifid distinctly above the base, the segments mostly 2–7 mm wide .... 8. *C. palmata*.
1 Style appendages short and blunt, broader than long, or nearly obsolete; disk-cors mostly 4 toothed.
  9 Rays largely or wholly yellow; achenes winged except often in no. 10.
    10 Wings of the achenes deeply lacerate; rays wholly yellow; perennial .... 9. *C. gladiata*.
    10 Wings entire or wanting; rays usually red-brown at base; annual or biennial .... 10. *C. tinctoria*.
  9 Rays pink (white); achenes wingless; rhizomatous perennial .... 11. *C. rosea*.

1. **Coreopsis pubescens** Elliott. Hairy t. Clustered perennial 6–12 dm, leafy throughout, evidently pubescent with short, loosely spreading hairs, rarely glabrous, lvs relatively broad, mostly elliptic or ovate,
short-petiolate, 4–10 × (1–)1.5–4 cm, entire or very often with 1(2) pair of much smaller but similarly shaped
pinnae at base; pappus commonly of 2 short, chaffy teeth; otherwise much like no. 2; $2n=26$, 28. Chiefly in
woods, especially in sandy soil; Va. to s. Ill. and Okla., s. to Fla. and La.; occasionally intr. elsewhere. June–
Sept.

2. **Coreopsis grandiflora** Hogg. Bigfl-t. Similar to no. 3, occasionally annual, often taller (to 1 m), leafy
nearly to the summit, the 1–many slender peduncles 0.5–2 dm, seldom more than half as long as the leafy
part of the stem; lvs mostly pinnatifid into linear-filiform to narrowly lanceolate segments, the lateral lobes
rarely more than 5 mm wide, the terminal one sometimes to 1 cm; outer bracts lance-subulate; rays averaging
a little smaller; achenes often with a large callous ventral excrescence at top and bottom; $2n=26 + 0–4$ B.
Native to se. U.S., escaped from cult. and ± established in much of our range, especially southward. May,
June. Ours is the widespread var. *grandiflora*.

3. **Coreopsis lanceolata** L. Longstalk-t. Clustered perennial 2–6 dm from a short caudex, glabrous or
spreading-villous; stems leafy below, elongate and naked above; lvs spatulate to linear or lance-linear, simple
or with 1 or 2 pairs of small lateral lobes, the lower long-petiolate, mostly 5–20 cm overall and 0.5–2 cm
wide; heads few or solitary on long, naked peduncles, the disk yellow, 1–2 cm wide; outer bracts 8–10, lanceolate to oblong-ovate, ± scarious-margined, 5–10 mm; inner
bracts longer and broader; rays 1.5–3 cm; receptacular bracts flat and chaffy below, caudate-attenuate above;
disk-fls 5-toothed; style appendages cuspidately acute; achenes 2–3(–4) mm, orbicular or broader, with thin,
flat wings; pappus of 2 short, chaffy teeth; $2n=24$, 26, 48. Dry, often sandy places; Mich. and the n. shore
of Lake Superior to Fla. and N.M.; cult. and often escaped. Apr.–June. (*C. crassifolia; C. heterogyna.*)

4. **Coreopsis auriculata** L. Lobed t. Perennial with slender naked stolons or stoloniform rhizomes; stems
erect or ascending, 1–6 dm, leafy below; lvs petiolate, the blade ovate to broadly elliptic or suborbicular, to
8 × 3.5 cm, often with a pair of much smaller lateral lobes at base; heads solitary or few, on long, naked
peduncles; disk yellow, 7–15 mm wide; outer bracts ca 8, lance-oblong or ovate-oblong, often narrowly thin-
margined; inner bracts longer and more acute, ovate; rays yellow, 1.5–2.5 cm; receptacular bracts narrowly
linear, attenuate upwards; disk-cors 5-toothed; style appendages cuspidately acute; achenes 2–3 mm, the
rather narrow wings finally cartilaginous and involute; pappus of 2 minute, deciduous scales; $2n=24$. Woods;
Va. and Ky. to Fla. and La. Apr.–June.

5. **Coreopsis tripteris** L. Tall t. Single-stemmed perennial 1–3 m from a short, stout rhizome, usually
glabrous and somewhat glaucous; lvs mainly cauline, numerous, on evident petioles 0.5–3 cm, mostly
trifoliolate, or the upper entire, the lfls lanceolate or narrowly elliptic, 5–10(–13) cm × 6–25(–30) mm, the
terminal one often again divided; heads several or many, the disk up to 1 cm wide, yellow, becoming purplish
or deep red, its cors 5-toothed; outer bracts ca 8, linear-oblong, obtuse, 2–3 mm, the inner much broader
and 2–3 times as long, more acute; rays yellow, 1–2.5 cm; receptacular bracts narrowly linear or linear-
clavate; style-appendages cuspidately acute; achenes obovate, 4–7 mm; pappus of a few minute erect bristles,
sometimes also with 2 short, upwardly barbed awns; $2n=26$. Mostly in moist or wet low places and in woods;
Mass. and s. Ont. to Wis., s. to Fla. and Tex. July–Sept.

6. **Coreopsis major** Walter. Forest-t. Much like no. 5, but smaller, 5–10 dm, vigorously rhizomatous but
the stems commonly tufted, the herbage often short-hairy; lvs sessile, the lfls averaging a little shorter, 3–8
cm; outer invol bracts longer and often leafier, sometimes equaling the inner; $2n=26$, 78, 104. Mostly in
dry, open woods; s. Pa. and O. s. to Fla. and Tex. July–Sept. Most of our plants belong to the widespread
var. *major*, with lfls 1–3 cm wide. The var. *rigida* (Nutt.) Boynton, with firmer lfls mostly 5–10 mm wide,
often in more exposed or less favorable sites, occurs from Va. to S.C. and Ga., but not to the exclusion of
var. *major*; it has $2n=104$ and may reflect alloploid introgression from *C. delphinifolia* Lam., which is itself
thought to be an alloploid of nos. 6 and 7.

7. **Coreopsis verticillata** L. Threadlf-t. Rhizomatous perennial 3–9 dm, often hispidulous at the nodes, otherwise glabrous; lvs sessile, 3–6 cm, ternately parted to the base, at least the central segment pinnatifid or again ternate, the ultimate segments linear-filiform, 0.3–1 mm wide; heads on very slender and rather short peduncles, the disk 5–10 mm wide, dark reddish-purple; outer invol bracts oblong or linear-oblong, 3–7 mm; rays yellow, ca 8, 1–2.5 cm; receptacular bracts linear-clavate, acutish; disk-cors 5-toothed; style-appendages cuspidately acute; achenes 3–5 mm, narrowly winged; pappus obsolete; $2n=26, 52, 78$. Dry, open woods; Md. and DC. to S.C., mainly on the coastal plain, but also inland to the Folded Appalachians; escaped elsewhere. June–Aug.

8. **Coreopsis palmata** Nutt. Finger-t. Rhizomatous perennial 4–9 dm, glabrous except for the scabro-ciliate lf-margins and sometimes hairy nodes; lvs rather numerous, firm, narrow, 3–8 cm, essentially sessile, deeply trilobed at or somewhat below the middle, the lobes linear-oblong, 2–7 mm wide, the central one sometimes again lobed; heads few or solitary, short-pedunculate, the disk yellow, 8–15 mm wide; outer invol bracts 8–12, linear-oblong, nearly equaling the much wider and thinner, broadly rounded or obtuse inner ones; rays 1.5–3 cm, yellow; receptacular bracts linear-clavate, acutish; disk-cors 5-toothed; style appendages cuspidately acute; achenes 5–6.5 mm, narrowly winged; pappus obsolete or of 2 callous teeth; $2n=26$. Prairies and open woods; Mich. to Man. s. to La. and Tex. June, July.

9. **Coreopsis gladiata** Walter. Swamp-t. Short-lived, fibrous-rooted, glabrous perennial 5–12 dm; lvs mostly entire, the lower oblanceolate or spatulate, conspicuously long-petiolate, to 30 × 3.5 cm, or much smaller, middle cauline lvs reduced, linear, the upper mere upper bracts; heads several, the disk 6–18 mm wide, dark purplish; outer invol bracts much shorter than the inner; rays yellow, 1–2.5 cm; receptacular bracts linear, acute; disk-fls 4-toothed; style-appendages short and blunt; achenes oblong, 2–4.5 mm, the wings deeply and subpectinately lacerate; pappus of 2 short, upward barbed awns; $2n=26, 52, 156$. Wet, often acid places; coastal plain from s. Va. to Fla. and Tex. Sept., Oct. Ours is var. *linifolia* (Nutt.) Cronq., with the lvs minutely dark-dotted in transmitted light in life, many or all of the lvs opposite, the lower up to ca 1 cm wide, not greatly elongate. (*C. angustifolia; C. oniscicarpa.*) The var. *gladiata,* with the lvs not dark-dotted, all or nearly all alternate, the lower much elongate and mostly 1–3.5 cm wide, probably does not reach our range. (*C. falcata; C. longifolia.*)

10. **Coreopsis tinctoria** Nutt. Plains-t. Glabrous, leafy and branching, taprooted annual 4–12 dm; lvs subsessile or short-petiolate, 5–10 cm, once or twice pinnatifid, the ultimate segments linear or lance-linear, 1–4 mm wide; heads numerous, the disk red-purple, 5–12 mm wide; outer invol bracts 2 mm, the inner 5–8 mm, deltoid-oblong or ovate; rays 1–2 cm, yellow, usually red-brown at base; receptacular bracts linear-subulate, attenuate-acuminate, few-striate; disk cors 4-toothed; style-appendages short and blunt; achenes 1–4 mm, wingless or narrowly to broadly winged; pappus obsolete or nearly so; $2n=24$. Moist, low places and disturbed sites; primarily c. and s. Great Plains and Tex., e. to Ark. and La., but widely cult., escaped, and irregularly established elsewhere; June–Aug. Ours is var. *tinctoria.*

C. basalis (Otto & Dietr.) S. F. Blake, a usually smaller sp. with 5-toothed disk-cors and with the outer invol bracts half to fully as long as the inner, is occasionally found in our range as an escape from cult.

11. **Coreopsis rosea** Nutt. Pink t. Glabrous, rhizomatous perennial 2–6 dm; lvs 2–5 cm, linear, ± entire or occasionally some irregularly lobed; heads rather short-pedunculate, the yellow disk 5–10 mm wide; outer invol bracts short, eventually spreading; rays pink (white), ca 1 cm; receptacular bracts linear, acute to obtuse; disk-cors 4-toothed; style appendages short and blunt; achenes narrow, wingless, ca 2 mm; pappus a minute cup, or nearly obsolete; $2n=26$. Wet, often sandy or acid soil, or in shallow water; N.S. and Mass. to Ga., mainly on the coastal plain. Aug., Sept.

18. BIDENS L. Beggar-ticks.

Heads radiate or discoid, the rays when present few, often 8, sterile, generally neutral, yellow or less often white or pink; invol bracts biseriate and dimorphic, the outer ± herbaceous, often very large, the inner membranous, often striate; receptacle flat or slightly convex, its bracts narrow, flat or nearly so; disk-fls perfect and fertile; style-branches flattened, with externally hairy, usually short appendage; achenes flattened parallel to the invol bracts, but not (or scarcely) winged, or sometimes almost regularly tetragonal, rarely subterete; pappus of (1)2–4(6) awns or teeth, commonly retrorsely barbed, but sometimes antrorsely barbed or even barbless, in a few spp. obsolete; herbs (ours) with opposite, simple to dissected lvs. 200, cosmop.

Thelesperma megapotamicum (Sprengel) Kuntze, of the Great Plains, has been reported as adventive in Ill. and Mo. It is a glabrous perennial with the lvs once or twice pinnatifid into slender segments; the discoid (or inconspicuously radiate) heads terminate long, naked peduncles; the outer invol bracts are very short, and the inner ones are connate to the middle or above; the wingless achenes have a pappus of 2 retrorsely barbed awn-scales. (*T. gracile.*)

1 Aquatic with the submersed lvs filiform-dissected .... 16. *B. beckii.*
1 Semi-aquatic or terrestrial, without filiform-dissected lvs.
  2 Lvs simple, though sometimes rather deeply 3(–7)-cleft.

3 Lvs (except sometimes the lower) sessile, occasionally narrowed to a short, somewhat petioliform base. .
    4 Outer invol bracts loosely spreading or reflexed; heads mostly nodding in age; anthers shortly exserted; disk-corollas mostly 5-lobed.
        5 Rays 1.5–3 cm; receptacular bracts reddish-tipped; chiefly coastal .... 1. *B. laevis.*
        5 Rays up to 1.5 cm, or none; receptacular bracts yellowish-tipped; widespread .... 2. *B. cernua.*
    4 Outer invol bracts ascending or suberect; heads mostly erect; anthers included; disk-corollas mostly 4-lobed; rays wanting or to 1 cm.
        6 Estuarine; heads campanulate or narrower; achenes coarsely striate .... 3. *B. hyperborea.*
        6 Not (or only casually) estuarine; heads campanulate or broader; achenes not striate .... 4. *B. comosa.*
3 Lvs with a distinct (sometimes winged) petiole mostly 1–4 cm.
    7 Heads with well developed rays mostly 1–2 cm .... 13. *B. mitis.*
    7 Heads discoid or nearly so, the rays, if present, not over ca 5 mm.
        8 Estuarine; heads narrow, with mostly 7–30 fls.
            9 Achenes with only a few long hairs and strong median ribs; N.Y. to Que. .... 6. *B. eatonii.*
            9 Achenes copiously hairy; median ribs obscure or none; N.Y. to Md. .... 7. *B. bidentoides.*
        8 Not (or only casually) estuarine; heads broader, the terminal one mostly with 30–150 fls.
            10 Achenes consistently 3-awned, very flat and smooth, the midribs scarcely evident; disk-cors 4-lobed, light yellow; lvs merely toothed .... 4. *B. comosa.*
            10 Achenes, or at least the central ones, more often 4-awned, compressed-quadrangular, with the midrib evident (at least distally) on each side, commonly sparsely hispid-strigose and/or tuberculate; disk-cors 5-lobed, orange-yellow; lvs often deeply cleft as well as toothed .... 5. *B. connata.*
2 Lvs once to thrice pinnately divided or compound (or trifoliolate), the terminal segment or lfl often stalked.
  11 Heads discoid, or with very short rays not over 5 mm.
    12 Achenes ± cuneate, broadened upwards; lfls 3(5), merely toothed.
      13 Outer invol bracts 3–5, not ciliate .... 8. *B. discoidea.*
      13 Outer invol bracts 5–16+, ciliate at least near the base.
        14 Outer invol bracts mostly 5–10, typically 8 .... 9. *B. frondosa.*
        14 Outer invol bracts mostly 10–16(–21), typically ca 13 .... 10. *B. vulgata.*
    12 Achenes linear, narrowed above; lvs mostly 2–3 times pinnate .... 11. *B. bipinnata.*
  11 Heads with well developed rays ca 1 cm or more.
    15 Achenes 2.5–5 mm, scarcely ciliate; coastal states from Md. s. .... 13. *B. mitis.*
    15 Achenes 5–10 mm, ciliate on the margins; widespread.
      16 Achenes narrowly cuneate-oblong, 2.5–4 times as long as wide .... 12. *B. coronata.*
      16 Achenes broader, mostly 1.5–2(–2.5) times as long as wide.
        17 Outer invol bracts mostly ca 8(–10), not longer than the inner .... 14. *B. aristosa.*
        17 Outer invol bracts mostly 12–25, commonly longer than the inner .... 15. *B. polylepis.*

1. **Bidens laevis** (L.) BSP.   Showy bur-marigold. Much like no. 2, wholly glabrous; sometimes perennial; heads not so consistently nodding; outer invol bracts scarcely leafy, seldom surpassing the disk; rays 1.5–3 cm; receptacular bracts reddish at the tip; achenes 3–4-angled, or often flat; pappus of 2–4 retrorsely barbed awns; $2n=22, 24$. Low, wet places, sometimes in shallow water, chiefly but not wholly coastal; N.H. and Mass. to Fla. and Calif., s. to S. Amer. Aug.–Nov. Passes into no. 2.

2. **Bidens cernua** L.   Bur-marigold. Annual 1–10(–40) dm, glabrous or the stem scabrous-hispid; lvs sessile, often basally connate, simple, lance-linear to lance-ovate, coarsely serrate to subentire, 4–20 × 0.5–4.5 cm; heads hemispheric, many-fld, the disk 12–25 mm wide, commonly nodding at least in age; outer invol bracts 5–10, lance-linear, unequal, usually rather leafy and spreading, commonly surpassing the disk; rays (6–)8, yellow, to 1.5 cm, or none; receptacular bracts thin and soft, mostly yellowish at the blunt tip; disk-corollas mostly 5-lobed; anthers shortly exserted; achenes 5–8 mm, the margins tending to be thickened, cartilaginous, and pale at maturity, as also the finally convex (broadly dome-shaped) summit; pappus of (2–)4 retrorsely barbed awns; $2n=24$. Low, wet places; widespread in temp. parts of the N. Hemisphere, and throughout our range. Aug.–Oct. (*B. elliptica; B. filamentosa; B. glaucescens; B. gracilenta; B. leptopoda; B. prionophylla.*)

3. **Bidens hyperborea** Greene.   Northern estuarine b.-t. Annual 1–7 dm, glabrous or the stem scabrous-hispid; lvs oblanceolate to narrowly lance-elliptic or narrowly oblong, not ciliate, sessile except sometimes the lower, acute to more often acuminate, 2–10 cm × 3–10 mm, with a few small teeth or subentire; heads erect or occasionally nodding, discoid or with rays up to nearly 1 cm, narrow, campanulate or turbinate, the disk 5–15 mm wide; outer invol bracts 4–8, leafy, ascending-erect, lance-linear, 1.5–4 cm; disk-corollas (3)4(5)-lobed; anthers included; achenes narrowly cuneate, flat or nearly so, blackish, coarsely striate, 4–7 mm, the summit finally cartilaginous and somewhat convex; pappus of 2–4 retrorsely barbed awns; $2n=24, 36$. Estuaries; Mass. (and reputedly N.Y.) to N.S. and Que.; James Bay. Aug., Sept.

4. **Bidens comosa** (A. Gray) Wieg.   Strawstem b.-t. Much like no. 5, but differing in a fairly well correlated set of characters; stem stramineous; lvs sessile or petiolate, merely toothed; outer invol bracts relatively large and leafy; disk-cors 4(5)-lobed, pale yellow or greenish-yellow; anthers included; achenes 5–10 mm, smooth except for the retrorse marginal barbels, very flat, the median nerves inconspicuous (or the outer one sometimes a little raised distally); pappus of (2)3(4) awns, the median awn(s) shorter than the lateral ones; $2n=24$. Disturbed habitats, often somewhat drier than those of no. 5; widespread in c. and e. U.S. and adj. Can., s. to N.C. and Mo., and w. sporadically into the w. cordillera. Aug.–Oct. (*B. acuta.*)

5. **Bidens connata** Muhl. Purplestem b.-t. Annual 1–20 dm, glabrous or nearly so; stem green to purplish; lvs simple, serrate, often some of them deeply 3(–7)-cleft, 3–15 cm, to 4 cm wide (exclusive of the lobes), on evident, sometimes winged petioles to 3 cm; heads erect, discoid or with rays to ca 4 mm, rather broadly campanulate to hemispheric, at least the terminal one generally with more than 30 fls, the disk 8–20 mm wide; outer invol bracts 4–9, herbaceous, but not so large as in no. 4; disk-cors (4)5-lobed, orange-yellow; anthers shortly exserted; achenes 3–7 mm, cuneate or cuneate-obovate, compressed-quadrangular, with the midribs evident and raised at least distally (or the marginal ones flatter), the surface mostly tuberculate and rather sparsely bristly-strigose; pappus of (2–)4(–6) retrorsely (rarely antrorsely) barbed awns, the median ones generally shorter than the lateral; $2n=24, 48, 60$. Wet waste places, widespread in c. and e. U.S. and adj. Can., w. to N.D., and Kans., s. to N.C. and Mo., and sporadically elsewhere. Aug.–Oct. Only with difficulty to be distinguished from the Eurasian *B. tripartita* L., with more consistently cleft lvs and mostly smoother, often flatter achenes.

B. heterodoxa (Fern.) Fern. & St. John consists of a series of rare and local coastal and subcoastal populations from Conn. to Que., to some extent estuarine, approaching *B. eatonii* in achenes and pappus (variously antrorsely or retrorsely barbed or barbless) but with broader, more numerously fld heads as in *B. connata*. The proper taxonomic status of these plants is uncertain.

6. **Bidens eatonii** Fern. New England estuarine b.-t. Glabrous annual 2–7 dm; lvs lanceolate, long-acuminate, serrate, often with 1 or 2 prominent lateral lobes, the blade 4–12 cm × 7–20 mm, attenuate into a slender or wing-margined petiole 1–3.5 cm; heads erect, discoid or shortly radiate, subcylindric to narrowly campanulate, the disk 5–12 mm wide, the fls 7–30; outer invol bracts mostly 3–5, foliaceous, narrow, erect and surpassing the disk; achenes flat or compressed-quadrangular, ± striate and with a strong median rib on each face, smooth except for a few scattered appressed hairs, 6–11 × 1–2 mm; pappus of 2–4 retrorsely or antrorsely barbed awns, mostly less than half as long as the achene, sometimes much reduced; $2n=48$. Estuaries; Que. to N.Y. Aug., Sept.

7. **Bidens bidentoides** (Nutt.) Britton. Southern estuarine b.-t. Much like no. 6; achenes copiously appressed-hispidulous, narrowly linear-cuneate, 6–13 × 1–1.5 mm, the median ribs obscure or none; awns 2(–4), antrorsely barbed, from a little less than half to usually evidently more than half as long as the achene; $2n=48$. Estuaries; N.Y. to Md. July–Oct. (*B. mariana*.)

8. **Bidens discoidea** (T. & G.) Britton. Few-bracted b.-t. Glabrous annual 3–8 dm; lvs thin, trifoliolate, on petioles 1–6 cm, the lfls (or at least the terminal one) petiolulate, lanceolate to lance-ovate, serrate, acuminate, the terminal one the largest, to 10 × 4 cm; heads numerous, small, discoid, the disk 3–10 mm wide; outer invol bracts 3–5, linear-spatulate, leafy, much surpassing the disk, scarcely or not at all ciliate-margined; achenes narrowly cuneate, short-hairy, 3–6 mm; pappus of 2 antrorsely setose awns to 2 mm, or nearly obsolete; $2n=24$. Wet places; n. N.S., s. Que. and Minn., s. to D.C., Ala., Okla., and Tex. Aug., Sept.

9. **Bidens frondosa** L. Devil's b.-t. Annual, generally glabrous or nearly so, 2–12 dm; lvs on petioles 1–6 cm, pinnately compound, with 3–5 lanceolate, acuminate, serrate lfls to 10 × 3 cm, sometimes sparsely short-hairy beneath, at least the terminal one slender-petiolulate; heads campanulate to hemispheric, or narrower in depauperate plants, discoid or nearly so, the disk to 1 cm wide in fl; outer invol bracts 5–10, typically 8, green and ± leafy, usually conspicuously surpassing the orange disk, evidently ciliate on the margins, at least toward the base; anthers exserted; achenes flat, narrowly cuneate, strongly 1-nerved on each face, commonly dark brown or blackish, subglabrous or appressed-hairy, 5–10 mm; pappus of 2 retrorsely (seldom antrorsely) barbed awns; $2n=24, 36, 48, 72$. Waste places, especially in wet soil; Nf. and N.S. to Wash., s. to Ga., La., and Calif. June–Oct.

10. **Bidens vulgata** Greene. Tall b.-t. Much like no. 9, averaging a little more robust and larger-headed, the herbage glabrous to densely villous-puberulent; outer invol bracts in well developed heads 10–16(–21), typically 13, averaging a little more leafy; disk yellow; anthers mostly included; achenes up to 12 mm, dark as in no. 9, or more commonly olivaceous or somewhat yellowish; $2n=24, 48$. Wet to dryish waste places; widespread in n. U.S. and adj. Can., s. to N.C., Mo., and Calif. Aug.–Oct. (*B. puberula*.)

11. **Bidens bipinnata** L. Spanish needles. Glabrous or minutely hairy annual 3–17 dm; lvs 4–20 cm including the 2–5 cm petiole, mostly 2–3 times pinnately dissected, the ultimate segments tending to be rounded; heads narrow, disciform, the disk only 4–6 mm wide at anthesis, the short, ochroleucous rays not surpassing the disk; outer invol bracts 7–10, linear, ± acute, not evidently expanded upwards, shorter than the inner; achenes linear, tetragonal, narrowed above, often sparsely hairy, 10–13 mm, or some of the outer shorter; pappus of (2)3 or 4 yellowish awns; $2n=24, 72$. Moist to fairly dry, disturbed habitats; Mass. to Fla., w. to Neb., Calif., and Mex., e. Asia. July–Oct.

12. **Bidens coronata** (L.) Britton. Northern tickseed-sunfl. Glabrous annual or biennial, 3–15 dm; lvs to 15 cm, pinnately parted, with mostly 3–7 lance-linear to linear, incised-dentate to entire, acute or acuminate segments; petioles 0.3–1.5 cm; heads radiate, with ca 8 rays 1–2.5 cm, the disk 8–15 mm wide; invol glabrous or nearly so, the 6–10 outer bracts linear or linear-spatulate, sometimes ciliate-margined, very rarely exceeding the disk; achenes flat or nearly so, narrowly cuneate-oblong or the inner cuneate-linear, mostly 5–9 mm and 2.5–4 times as long as wide, dark, smooth or hairy, the margins antrorsely ciliate; pappus of 2 short, erectly setose strong awns or awn-scales mostly 1–2 mm (seldom obsolete); $2n=24$. Wet places; Mass., s. Ont., n. Wis., and e. Minn., s. to N.C., Ky., and Neb. Aug.–Oct. (*B. trichosperma*.)

13. **Bidens mitis** (Michx.) Sherff. Coastal plain tickseed-sunfl. Annual 3–10 dm, glabrous or nearly so; lvs 4–12 cm including the 0.3–3 cm petiole, highly variable in form, pinnately compound to more often

pinnately 3–7-parted with the terminal segment the largest, varying to simple, ovate, and merely toothed, or even narrow and subentire; heads rather small, the disk ca 1 cm wide or less; outer invol bracts 7–10, linear or linear-spatulate, 5–10 mm, mostly short-ciliate, glabrous on the back, or the invol sparsely hispid at base; rays ca 8, yellow, 1–2 cm,; achenes rather broadly cuneate, 2.5–4.5(–5) mm, glabrous or nearly so, scarcely ciliate; pappus of 2 sharp, antrorsely barbed teeth to 1 mm, or obsolete; $2n=24$. Swamps and other wet places; coastal plain from Md. to Fla. and Tex. July–Oct. (*B. coronata,* misapplied.)

14. Bidens aristosa (Michx.) Britton.   Midwestern tickseed-sunfl. Annual or biennial, 3–15 dm, glabrous or slightly hairy; lvs pinnately or bipinnately divided or compound, 5–15 cm including the 1–3 cm petiole, the segments lanceolate or lance-linear, acuminate, incised-serrate or pinnatifid; rays ca 8, 1–2.5 cm; disk 8–15 mm wide; outer invol bracts mostly ca 8(–10), linear, 5–12 mm, mostly shorter than the inner, straight, the back glabrous or finely short-hairy, the margins smooth or moderately ciliate; achenes flat, 5–7 mm, mostly 1.5–2(–2.5) times as long as wide, strigose, the margins antrorsely ciliate and commonly narrow and interruptedly thickish-winged; pappus of 2(–4) antrorsely or retrorsely barbed awns, or none. Wet places, most often in shade; Me. to Minn., s. to Va., Okla., and Tex., probably only intr. eastward. Aug.–Nov.

15. Bidens polylepis S. F. Blake.   Ozark tickseed-sunfl. Much like no. 14, but the outer invol bracts mostly 12–25(–30), 10–25 mm, mostly surpassing the inner, somewhat curled and twisted, conspicuously hispid-ciliate and often also coarsely short-hairy on the back; pappus more commonly short or obsolete; $2n=24$. Wet places, mostly in sun; Ozark, plains, and prairie sp., n. to Neb., Io., Ill., and Ind., and occasionally adventive eastward. Aug.–Oct. (*B. involucrata,* a preoccupied name.)

16. Bidens beckii Torr.   Water b.-t. Aquatic perennial; lvs 2–4 cm, the submersed ones filiform-dissected, the emersed ones simple, lanceolate to ovate, sessile, serrate; heads terminal and solitary, the disk 1 cm wide; rays 1–1.5 cm; outer invol bracts more herbaceous and less striate than the rather similar inner ones; achenes subterete, 10–14 mm; pappus of 3–6 awns, longer than the achenes, retrorsely barbed above; $2n=26$. In ponds and streams; Que. to N.J., w. to Man. and Mo.; disjunct in Oreg., Wash., and B.C. July–Oct. (*Megalodonta b.*)

19. COSMOS Cav.   Cosmos. Heads radiate, the rays neutral, white to pink, red, or yellow; invol bracts biseriate and dimorphic, the outer subherbaceous, the inner membranous or almost hyaline; receptacle flat, chaffy, its bracts plane; disk-fls perfect; style-branches slender, flattened, with short, hairy appendage; achenes quadrangular, not much compressed, linear, beaked; pappus of 2–8 retrorsely or rarely antrorsely barbed awns; herbs with opposite, simple to pinnately dissected lvs and subhemispheric heads. 20, warm New World.

1 Rays anthocyanic or white; ultimate lf-segments ca 1 mm wide or less . . . . 1. *C. bipinnatus.*
1 Rays intensely orange-yellow or orange-red; lf-segments wider . . . . 2. *C. sulphureus.*

1. Cosmos bipinnatus Cav.   Common c. Annual 6–20 dm, glabrous or minutely scabrous; lvs sessile or short-petiolate, 6–11 cm, pinnately dissected, the ultimate segments linear or linear-filiform, ca 1 mm wide or less; heads rather numerous, the disk 1–1.5 cm wide; rays ca 8, rose or lilac (white), 1.5–4 cm, often half as wide; achenes 7–16 mm, the body longer than the beak; pappus of 2 or 3 short awns, or none; $2n=24$. Native of Mex. and adj. U.S., commonly cult. and casually escaped. Aug.–Oct.

2. Cosmos sulphureus Cav.   Orange c. Much like no. 1; ultimate lf-segments broadly linear to lanceolate, mostly (1.5–)2 mm wide or more; rays intensely orange-yellow or orange-red, 2–3 cm; achenes 15–30 mm; $2n=24$. Native of trop. Amer., cult. and casually escaped toward the s. part of our range, and n. to Pa. Aug.–Oct.

20. GALINSOGA Ruíz & Pavón.   Quickweed. Heads small, radiate, the rays few, short, broad, only slightly surpassing the disk, white (pink), pistillate and fertile; invol bracts few, relatively broad, greenish at least in part, several-nerved, each subtending a ray, and sometimes joined at base with the 2 adjacent receptacular bracts; a few shorter and narrower outer bracts often present; receptacle conic, chaffy, its bracts membranous, rather narrow, nearly flat; disk-fls perfect; style-branches flattened, with short, minutely hairy appendage; achenes 4-angled, scarcely compressed, or, especially the outer, somewhat flattened parallel to the invol bracts; pappus of several or many scales, often fimbriate or awn-tipped, that of the rays ± reduced or none; annuals with opposite lvs. 14, originally New World.

1 Rays with well developed pappus about equaling the tube; pappus-scales of the disk-fls tapering to a
      short awn-tip . . . . 1. *G. quadriradiata.*
1 Rays without pappus, or nearly so; pappus-scales of the disk not awn-tipped . . . . 2. *G. parviflora.*

1. **Galinsoga quadriradiata** Ruiz & Pavon.   Common q. Freely branching, somewhat hairy, 2–7 dm, the hairs of the stem generally fairly coarse and spreading; lvs petiolate, with ovate (often broadly so), rather coarsely toothed blade 2.5–7 × 1.5–5 cm; heads mostly rather numerous (in well developed plants) in open, leafy cymes, slender-pedunculate, the peduncles and often also the invols spreading-villous, with gland-tipped hairs; outer invol bracts 1–2, herbaceous, deciduous, the inner 3–4 mm, deciduous with their attached receptacular bracts; rays (3–)5)(6), white (pink), strongly 3-toothed, 2–3 mm, nearly as wide; inner receptacular bracts mostly entire; disk 3–6 mm wide; achenes black, hispidulous with appressed or ascending hairs; pappus of the disk of slender, fimbriate scales tapering to a short but definite awn-tip, often shorter than the cor, that of the rays of short but well developed fimbriate scales, about equaling the tube; $2n=32, 48, 64$. Native of C. and S. Amer., now a cosmopolitan weed. June–Nov. (*G. bicolorata* and *G. caracasana,* with pink rays; *G. ciliata.*)

2. **Galinsoga parviflora** Cav.   Lesser q. Much like no. 1, less hairy, the stem glabrous or sparsely pubescent with appressed or sometimes spreading hairs, the peduncles appressed-hairy, or finely villous with spreading, gland-tipped hairs; lvs ovate or lance-ovate, mostly less coarsely toothed; outer invol bracts 2–4, scarious-margined, persistent, the inner ones and their attached pales also persistent; inner receptacular bracts deeply trifid; achenes sparsely appressed-hairy or glabrous; pappus scales of the disk-fls conspicuously fimbriate, generally blunt, nearly or quite as long as the cor; rays to 1.5(–2) mm, nearly or quite epappose; $2n=16$. Native from sw. U.S. to S. Amer., but now a cosmopolitan weed.

21. MARSHALLIA Schreber.   Barbara's buttons.

Heads discoid, the fls all tubular and perfect, with slender, elongate, hairy, conspicuously lobed, white or more often anthocyanic cor; invol of 1 or 2 series of narrow, subequal, ± herbaceous bracts; receptacle conic or convex, chaffy, its bracts linear to linear-spatulate, commonly herbaceous distally; style-branches flattened, elongate, with short, minutely hairy appendage; achenes 5-angled, 10-ribbed, the ribs commonly hairy; pappus of 5(6) short, scarious or hyaline scales; perennial herbs with alternate (or all basal) entire lvs and long-pedunculate, broadly campanulate heads. 8, mainly se. U.S.

1. **Marshallia trinervia** (Walter) Trel.   Colonial B.b. Glabrous, ± colonial from short-creeping rhizomes, the stems 4–8 dm; lvs thin, strongly trinerved, mostly 5–10 × 1–3.5 cm, equably distributed, the lower oblanceolate or spatulate, petiolate, obtuse or rounded, often deciduous before anthesis, those above fully as large, lance-ovate or lance-elliptic to ovate, acuminate or acute, sessile or nearly so; internodes 10–25; heads 2–3 cm wide in fl, solitary (few) on a terminal peduncle 1–2 dm; receptacular and invol bracts acute; $2n=18$. Woods, streambanks, and cliffs, often on calcareous clay; Va. to S.C., Tenn., and La. May, June.

2. **Marshallia grandiflora** Beadle & Boynton.   Appalachian B.b. Single-stemmed or clustered from a short caudex, 2–9 dm, puberulent under the heads, otherwise glabrous; lvs 3-nerved, somewhat basally disposed, the lower evidently petiolate, with broadly oblanceolate or spatulate to elliptic blade 5–15 × 1–3 cm, the others sessile and ± reduced upwards; internodes 5–12; heads 2–3.5 cm wide in fl, solitary on a terminal peduncle 1–3 dm; receptacular and invol bracts acute; $2n=18$. Moist to wet places in woods, meadows, and along streams; Appalachian region from sw. Pa. to N.C. and Tenn. June–Aug.

3. **Marshallia obovata** (Walter) Beadle & Boynton.   Somewhat resembling no. 2, but the lvs a little smaller, seldom over 2 cm wide, and not strongly reduced upwards, the leafy part of the stem giving way abruptly to the long terminal peduncle; invol and receptacular bracts obtuse (sometimes minutely mucronulate), the latter dilated upwards; $2n=18$. Meadows and open woods; s. Va. to Ga., Ala., and w. Fla. Apr.–June. Ours is var. *obovata.*

22. MADIA Molina.   Tarweed.

Heads radiate, the rays pistillate and fertile, yellow, often short, or rarely wanting and the heads discoid; invol bracts subherbaceous, uniseriate, equal, enfolding the achenes, the invol appearing deeply sulcate; receptacle convex, chaffy only near the margin; disk-fls perfect but sometimes sterile; style-branches flattened, with an acute, hairy appendage; ray-achenes generally ± compressed radially, finely striate, commonly incurved, those of the disk similar or empty; pappus none, or a short crown or a few short scales; ± glandular and heavy-scented, annual to perennial herbs with chiefly alternate, entire or merely toothed lvs and small or middle-sized heads. 17, w. N. Amer. and Chile.

1. **Madia sativa** Molina. Chile-t. Coarse annual 2–10 dm, often simple, conspicuously spreading-hirsute and stipitate-glandular; lvs mostly lance-linear or linear-oblong, the lower linear-oblanceolate, 4–18 cm × 4–12 mm, entire or nearly so, sessile, crowded, erect; heads clustered along the upper part of the main stem, or at the ends of the branches, the invol ovoid or broadly urn-shaped, 6–12 mm high and wide; rays ca (8)13, 2–7 mm, disk-fls fertile; pappus none, $2n=32$. Native from Wash. to Calif., and in Chile; becoming widespread as a roadside weed in e. U.S. June–Aug. Ours is var. *congesta* T. & G. (*M. capitata.*)

2. **Madia glomerata** Hook. Mt.-t. More slender than no. 1, 1–5 dm, with shorter, generally more appressed pubescence, scarcely or not at all glandular below the infl; lvs linear or lance-linear, 2–7 cm × 1–4 mm; heads smaller and relatively narrower, 2–5 mm wide; rays mostly 1–2, or wanting from some heads, 2 mm; $2n=28$. Open places; cordilleran region, occasionally intr. along our w. border. July–Sept.

23. FLAVERIA A. L. Juss.

Rays pistillate and fertile, yellow, inconspicuous, commonly solitary, or sometimes wanting; invol of 2–8 subequal greenish bracts, occasionally with 1 or 2 smaller outer ones; receptacle small, naked or sometimes bristly; disk-fls 1–15, perfect and fertile, yellow; style-branches flattened, truncate, minutely penicillate; achenes glabrous, ca 10-ribbed; pappus wanting, or of 2–4 irregular scales; herbs with narrow, opposite, sessile and often connate lvs, and numerous small, sessile or subsessile glomerate heads. 20, New World.

1. **Flaveria campestris** J. R. Johnston. Rather coarse annual 3–9 dm, glabrous, or sometimes slightly villous at the nodes; lvs lanceolate, sessile, remotely serrulate, 3–7 cm × 6–20 mm; heads cylindric, in glomerules subtended by leafy bracts; invol bracts 3, 5–7 mm, with 2 much smaller calyculate ones at base; rays solitary, 2 mm; disk-fls 3–4; receptacle naked; pappus none. Low ground; Colo. and Kans. to Mex., and intr. in Mo. Aug.–Oct.

24. DYSSODIA Cav.

Heads radiate, the rays pistillate and fertile, yellow or orange; invol bracts biseriate, the inner distinct or weakly connate at the base; receptacle naked, flat to convex; disk-fls numerous, tubular and perfect; style-branches flattened, subtruncate, shortly penicillate around the tip, and with a minute terminal cone; achenes slender, ± striate; pappus of 12–22 scales, each scale cleft into 5–10 bristles; aromatic herbs with pinnatifid to pinnately dissected, mainly opposite lvs, these and the invol beset with scattered embedded oil-glands. 7, New World.

1. **Dyssodia papposa** (Vent.) A. Hitchc. Stinking marigold. Much branched, malodorous annual 1–4 dm, puberulent or subglabrous; lvs 2–4 cm, pinnatifid or bipinnatifid, the ultimate segments linear or filiform; heads numerous, campanulate, the disk 4–10 mm wide; invol 6–8 mm, the outer bracts linear, subherbaceous, ⅔ as long as the broader and more chartaceous inner, which are weakly joined at base; rays few, inconspicuous, erect, orange-yellow, 1–2 mm; receptacle pitted, slightly convex; style-branches truncate; pappus scales ca 20, deeply divided into 5–10 bristles; $2n=26$. Dry open places, often weedy; w. U.S. to s. Mex. and intr. e. to N.H. and Tenn. July–Oct. (*Boebera p.*)

25. TAGETES L.

Marigold. Heads mostly radiate, the rays few (except in double forms), pistillate and fertile, mostly yellow to red-orange; invol bracts uniseriate, 3–10, united to near the apex and each with a row of evident embedded oil-glands on each side; receptacle flat, generally small, naked; disk-fls tubular and perfect; style-branches ± elongate, flattened, with introrsely marginal stigmatic lines and a short, often expanded, hirsutulous appendage; achenes slender and elongate; pappus of several very unequal, often ± connate scales, generally 1 or 2 elongate and acute or awn-tipped; glabrous, aromatic herbs with conspicuously gland-dotted, usually pinnatifid lvs, at least the lower opposite. 30, warm New World.

1. **Tagetes minuta** L. Strong-smelling annual 3–10 dm; lvs pinnately compound; lfls 9–17, lance-linear, 1.5–6 cm, sharply serrate; heads numerous, tending to form flat-topped clusters; invol narrowly cylindric, 8–12 mm, with 3–5 short teeth, few-fld, the rays commonly 3, only 1–2 mm; achenes 5–6 mm; longer pappus-scales 2–3 mm, the others less than 1 mm; $2n=48$. Native of S. Amer., now ± established as a weed in disturbed sites n. to Va. and rarely beyond. Aug.–Oct.

T. erecta L., the common marigold, and *T. patula* L., the "French" marigold, both native to Mex., are widely cult. in double forms and occasionally escape. *T. erecta* has yellow or orange heads mostly 6–10 cm wide on fistulous peduncles, the invol ca 2 cm. *T. patula* has smaller, usually partly or wholly deep red-orange heads 3.5–5 cm wide on scarcely fistulous peduncles, the invol ca. 1.5 cm.

26. POLYMNIA L. Lf-cup. Heads radiate or disciform; invol a single series of green bracts; outer fls pistillate and fertile, the cor with or without an expanded, yellow to white ray; receptacle flat or nearly so, chaffy throughout, the bracts subtending the pistillate fls larger and more herbaceous than those of the disk; disk-fls sterile, with undivided style; ray-achenes thick, not much compressed, glabrous or shortly hairy distally; pappus none; perennial herbs (ours) or shrubs with large, opposite (or the upper alternate) lvs and several to many medium-sized subhemispheric heads. 20, New World.

1 Achenes trigonous, not striate; rays none, or whitish and to 1(–1.5) cm .... 1. *P. canadensis.*
1 Achenes impressed-striate; rays yellow, 1–2(–3) cm, rarely wanting .... 2. *P. uvedalia.*

1. Polymnia canadensis L. Pale-fld lf-c. Coarse perennial 0.6–2 m, glabrate below, viscid-villous or stipitate-glandular above; lvs large, thin, to 3 dm, broadly oblong to ovate, pinnately few-lobed, also toothed, the petiole wingless or winged only near the blade; heads in congested cymes ending the branches, the pale yellow disk 6–13 mm wide; invol bracts lanceolate to lance-linear, narrower and often shorter than the bracts that subtend the ray-achenes; cor of pistillate fls minute and tubular, or expanded into a short, whitish ray up to 10(–15) mm; achenes 3–4 mm, unequally 3-ribbed and -angled, not striate; $2n=30$. Moist woods, especially in calcareous regions; Vt. and Ont. to Minn., s. to Ga. and Ark. June–Oct. (*P. radiata,* the radiate form.)

2. Polymnia uvedalia L. Yellow-fld lf-c. Coarse perennial 1–3 m, the stem glandular or spreading-hairy beneath the heads, otherwise generally glabrous; lvs large, sometimes over 3 dm, deltoid-ovate, subpalmately lobed and veined, scabrous-hispid to subglabrous above, more finely hairy and often glandular beneath, with broadly winged, sometimes runcinate petiole; heads in moderately open, leafy cymes, the bright yellow disk ca 1.5 cm wide; invol bracts lance-ovate to ovate or elliptic, leafy, 1–2 cm, much broader and generally longer than the outer receptacular bracts; rays ca 8–11, bright yellow, 1–2(–3) cm, rarely reduced and inconspicuous; achenes ca 6 mm, impressed-striate (or shallowly ribbed and grooved), with many nerves; $2n=32$. Woods and meadows; N.Y. to Ill. and Mo., s. to Fla. and Tex. July–Sept. (*Smallanthus u.*)

27. ACANTHOSPERMUM Schrank. Heads radiate, the rays mostly 5–10, minute, yellow, pistillate and fertile; invol biseriate and dimorphic, the outer bracts 4–6, herbaceous elliptic to ovate, the inner larger, ± prickly or spiny, as many as and individually enclosing the ray-achenes, ripening with them into burs; receptacle small, convex, chaffy throughout, its bracts soft, loosely folded or convex and embracing the 5–15 disk-fls, these sterile, with undivided style; ray-achenes thick, radially somewhat compressed, glabrous; pappus none; diffuse, often dichotomous annuals with opposite, toothed or entire lvs and small heads solitary in the axils or in the forks of the stem. 10, trop. Amer.

1. Acanthospermum australe (Loefl.) Kuntze. Paraguay-bur. Stems prostrate and often rooting, to 6 dm or more, short-hairy, especially distally; lvs broadly rhombic-ovate or triangular, irregularly toothed above the cuneate, entire base, 1.5–3.5 × 1–3 cm, glandular-punctate, on petioles 3–15 mm; heads 4–6 mm wide at anthesis; burs 7–9 mm, ellipsoid-fusiform to nearly oblong, slightly compressed, densely glandular, 5–7-ribbed, the ribs bearing 1 or 2 rows of hooked prickles 1–2 mm, the top abruptly invaginated and with an evident rim; $2n=22$. A weed in sandy soil and waste places, native to trop. Amer., and intr. n. occasionally as far as Mass. and Oreg.

28. SILPHIUM L. Rosin-weed. Heads radiate, the rays yellow, pistillate and fertile, their ovaries imbricate in 2–3 series; invol bracts subequal or imbricate in 2–several series, firm, herbaceous or partly membranous-chartaceous; receptacle flat, chaffy, its bracts (or sometimes in part the inner bracts of the invol) subtending rays as well as disk-fls; disk-fls sterile, with undivided style; ray-achenes glabrous, strongly flattened parallel to the invol bracts, wing-margined, pappus none, or of 2 awns confluent with the wings of the achene; coarse perennial herbs with medium-sized or large subhemispheric heads. 15, U.S.

1 Stem ± leafy.
  2 Lvs, or their petiolar bases, connate-perfoliate; stem square .... 1. *S. perfoliatum.*
  2 Lvs sessile or petiolate, but not connate-perfoliate; stem not square.
    3 Lvs entire or merely toothed, seldom over 2.5(–3) dm; fibrous-rooted.
      4 Heads relatively large, with 16–35 (commonly ca 21 or 34) rays; lvs sessile or nearly so, broad-based and often clasping, commonly 2–4 times as long as wide .... 2. *S. integrifolium.*

4 Heads smaller, with mostly 6–14(–21) rays; lvs various, but mostly not as above.
  5 Stem essentially glabrous, often glaucous.
    6 Lvs truncate or broadly rounded at base, or the lower cordate, on petioles 1.5–4.5 cm;
      Cumberland sp. . . . . 3. *S. brachiatum.*
    6 Lvs otherwise, many of them either narrowed to the base, or sessile, or both; widespread
      sp. . . . . 4. *S. trifoliatum.*
  5 Stem evidently spreading-hairy, not glaucous.
    7 Stem conspicuously shaggy-hispid, many or most of the hairs more than 2 (to 5) mm;
      Cumberland sp. . . . . 6. *S. mohrii.*
    7 Stem not shaggy, seldom any of the hairs over 2 mm; widespread southern sp. . . . . 5. *S. asteriscus.*
  3 Lvs deeply pinnatifid or bipinnatifid, the larger commonly 2.5–5 dm; taprooted . . . . 7. *S. laciniatum.*
1 Stem nearly naked, the principal lvs all at or near the base; taprooted.
  8 Heads larger; rays ca 13 to ca 21; invol 13–25 mm; disk 1.5–2.5 cm wide . . . . 8. *S. terebinthinaceum.*
  8 Heads smaller; rays ca 5 or 8(–10); invol 6–11 mm; disk 0.8–1.5 cm wide . . . . 9. *S. compositum.*

1. Silphium perfoliatum L.   Cup-plant. Stems 1–2.5 m, square; lvs or their petiolar bases connate-perfoliate, the blade deltoid to ovate, coarsely toothed, to 3 × 1.5 dm; heads in an open infl, the disk 1.5–2.5 cm wide; invol 12–25 mm, its bracts elliptic or ovate, subequal or the outer longer, rays 1.5–4 cm; $2n=14$. Woods and low ground; s. Ont. to N.D., s. to N.C., Miss., La., and Okla., and intr. into N. Engl. July–Sept. Two vars:

Var. *perfoliatum.*   Stem generally glabrous or nearly so, sometimes spreading-hairy toward the base; most of the lvs, except for a few upper ones, generally ± evidently wing-petiolate, the lower surface glabrous or short-hairy, seldom any of the hairs as much as 1 mm; head with mostly 16–35 (commonly ca 21 or 34) rays; invol bracts glabrous except for the ciliate margins; widespread, mainly to the w. or n. of var. *connatum.*

Var. *connatum* (L.) Cronq.   Stem usually spreading-hispid, varying to largely glabrous; many or most of the lvs sessile, only the lower or lowermost ones evidently petiolate; lower lf-surface, at least along the midrib, relatively long-hairy; heads with fewer (mostly ca 8 or ca 13) rays; invol usually finely hairy, varying to glabrous; scattered stations in the mts. and adj. piedmont of Va., W.Va., and N.C. (*S. connatum.*)

2. Silphium integrifolium Michx.   Prairie r.-w. Fibrous-rooted from a short rhizome or caudex; stems 5–15 dm; lvs firm, mostly opposite, seldom ternate or alternate, entire or toothed, ovate or lance-ovate to elliptic, mostly 7–15 × 2–6.5 cm, 2–4 times as long as wide, scabrous above, sessile and often clasping; heads relatively large and many-fld, with mostly 16–35 (commonly ca 21 or 34) rays 2–5 cm, the disk mostly 1.5–2.5 cm wide; $2n=14$. Prairies and roadsides, less often in open woods; s. Mich. to Neb., s. to Ala., Miss., and Tex. July–Sept. Two vars.:

Var. *integrifolium.*   The more hairy phase; peduncles, backs of the invol bracts, and usually also the lower surfaces of the lvs and at least the upper part of the stem velvety or scabrous (most or all of the hairs ca 0.5 mm or less); typically in the tallgrass prairie.

Var. *laeve* T. & G.   The less hairy phase; peduncles, invol bracts, stem, and lower surfaces of the lvs glabrous or nearly so, often somewhat glaucous; typically in the mixed-grass prairie, to the w. of var. *integrifolium*, and barely entering our range; an outlier in c. Tenn. (*S. speciosum.*)

3. Silphium brachiatum Gattinger.   Cumberland r.-w. Fibrous-rooted from a short rhizome or caudex; stem 10–15 dm, glabrous and glaucous; lvs opposite, narrowly ovate to triangular, coarsely toothed, mostly 13–25 × 5–10 cm, scabrous above, scabrous or hispid beneath, truncate or broadly rounded at base, or the lower cordate, on petioles 1.5–4.5 cm; heads rather numerous on slender peduncles in an open infl, relatively small, the disk ca 1 cm wide, with 4–6(–8) rays 1–1.5 cm; invol bracts glabrous except the ciliate margins; $2n=14$. Woods in the Cumberland Plateau region; Ky. and Tenn. to Ala. and Ga. July–Sept.

4. Silphium trifoliatum L.   Whorled r.-w. Fibrous-rooted from a short rhizome or caudex; stem robust, (0.7–)1–2 m, glabrous, often glaucous; lvs variously whorled or opposite or seldom alternate, entire or merely toothed, petiolate or tapering to a subpetiolar base; heads several to commonly ± numerous in an open infl, rather small, with ca 8 or ca 13 rays 1.5–3 cm, the disk ca 1–1.5(–2) cm wide; invol bracts glabrous except the ciliate margins; $2n=14$. Open woods, prairies, and disturbed open places; se. Pa. to O. and Ind., s. to N.C., Ga., and Miss. Two vars.:

Var. *trifoliatum.*   Lvs mostly ternate or quaternate, occasionally opposite, seldom partly alternate, scabrous above, hirsute at least on the midrib beneath, the blade lanceolate or lance-elliptic to rather narrowly lance-ovate, (2.5–)3–5 times as long as wide, 7–20 × 1.5–5 cm, tapering to a hirsute-ciliate petiole up to 1.5 cm, or to a shortly subpetiolar base; n. var., from se. Pa. to O. and Ind., s. to Va. and in the mt. region to Ga.

Var. *latifolium* A. Gray.   Lvs prevailingly opposite, often wider than in var. *trifoliatum*, up to 7.5 cm wide and sometimes only twice as long as wide, tending to be less pubescent, often essentially glabrous, often more petiolate, the petiole sometimes to 5 cm; s. var., broadly overlapping the range of var. *trifoliatum* in Tenn., Ky., and sw. Va., and extending s. to Ga., Ala., and Miss. (*S. glabrum; S. laevigatum.*)

5. Silphium asteriscus L.   Southern r.-w. Fibrous-rooted from a short rhizome or caudex, 5–12 dm; herbage sparsely to densely spreading-hispid, or the lvs merely hispid-scabrous, many of the hairs of the stem ca 1 mm or more; stem leafy, the lvs opposite or alternate, coarsely toothed or entire, mostly 6–15 × 1.5–5 cm, (2–)2.5–5 times as long as wide, the lower often rather large and evidently long-petiolate, the others variously sessile or short-petiolate, rounded or tapering at base; heads several or many in a ± leafy-bracteate infl, with mostly ca 8 or ca 13(–21) rays 1.5–3 cm, the disk mostly 1–2 cm wide; $2n=14$. Open woods, glades, and clearings; Va. to n. Fla., w. to Mo. and Tex. June–Sept. Our plants, with the stem lacking under-pubescence and with the receptacular bracts blunt and neither densely white-hairy nor glandular-hairy toward the tip, are var. *asteriscus*.

6. Silphium mohrii Small. Shaggy r.-w. Fibrous-rooted from a short rhizome or crown, 6–15 dm, copiously shaggy-hispid, many or most of the hairs of the stem over 2 (to 5) mm; lvs alternate, opposite, or in part ternate, entire or toothed, the basal and lower cauline ones well developed and persistent, with broadly ovate to merely lanceolate, basally rounded or abruptly contracted blade 12–30 × 3–20 cm on a petiole 5– 20 cm; cauline lvs progressively reduced upwards, becoming sessile, often ovate; heads several or rather numerous, with ca 13 rays 1.5–2 cm, the disk 1.5–2.5 cm wide. Glades and open woods; Cumberland Plateau region of Ky., Tenn., Ala., and Ga. July–Oct. The Ky. plants, with broad, basally truncate-cordate lower lvs, glabrous invol, and not quite so long and shaggy pubescence on the stem (hairs 2–3 mm), have been described as *S. wasiotensis* Medley and may be distinct.

7. Silphium laciniatum L. Compass-plant. Coarse, taprooted, rough-hairy, 1–3 m; lvs alternate, deeply pinnatifid or bipinnatifid, the lower very large, to 5 dm, progressively reduced upwards, the uppermost entire and well under 1 dm; heads in a narrow, sometimes racemiform infl, large, the disk 2–3 cm wide; invol 2– 4 cm, exceeding the disk, its bracts ovate, acuminate, squarrose, not much imbricate; rays (13–)17–25(–34), 2–5 cm; 2*n*=14. Prairies; O. to Minn. and S.D., s. to Ala. and Tex.; locally intr. e. along railroads to N.Y. July–Sept. The basal lvs tend to align themselves in a north-south direction.

8. Silphium terebinthinaceum Jacq. Basal-lvd r.-w. Taprooted; stem 0.7–3 m, essentially glabrous, nearly naked, its lvs, except those near the base, few and reduced to mere large bracts; principal lvs large, long-petiolate, the blade narrowly to broadly ovate, oblong, or elliptic, usually cordate (seldom only subtruncate) at the base, 1–5 × 0.7–3 dm, usually sharply toothed, glabrous or scabrous, sometimes hirsute on the midrib beneath; heads several or rather numerous in an open-corymbiform infl, relatively large, with ca 13 to ca 21 rays 2–3 cm, the disk 1.5–3.5 cm wide; invol 13–25 mm, its bracts essentially glabrous, strongly imbricate, the outer broadly elliptic, the inner elongate, more oblong, loose and broad-tipped; 2*n*=14. Prairies; s. Ont. and O. to Minn., s. to Ga. and Miss. July–Sept. (*S. rumicifolium.*)

9. Silphium compositum Michx. Lesser basal-lvd r.-w. Much like no. 8, but with smaller and often more numerous heads, the disk 0.8–1.5 cm wide, the invol 6–11 mm, the rays 5–10, 1–2 cm; lvs often wider than long, variously merely toothed to deeply pinnatifid or palmatifid and again lobed or cleft; 2*n*=14. Dry, often sandy places, esp. in pine woods; Va. to Fla., w. to Tenn. and Ala. June–Sept. (*S. lapsuum; S. orae; S. reniforme; S. venosum.*)

29. CHRYSOGONUM L. Gold-star. Heads radiate, the rays few, uniseriate, yellow, pistillate and fertile; invol bracts in 2 series of ca 5, the outer herbaceous and surpassing the disk, the inner more chartaceous and subtending the rays; receptacle flat, chaffy; disk-fls sterile, with undivided style; achenes ± compressed parallel to the invol bracts, wingless; pappus a short, half-cup-shaped crown; each inner invol bract grown to 2 or 3 adjacent receptacular bracts at base, thus partly enclosing the achenes, all falling as a unit; perennial herbs with opposite lvs and rather small heads. Monotypic.

1. Chrysogonum virginianum L. Fibrous-rooted perennial to 5 dm, often beginning to fl when very small; stems glandular and spreading-villous; lvs ovate to suborbicular, hairy, long-petiolate, crenate, 2.5–10 × 1.5–6 cm, the middle and upper generally deltoid or cordate at base; heads few or solitary on slender terminal and axillary peduncles, the disk 7–10 mm wide; outer invol bracts villous like the stem; rays ca 5, broad, 7– 15 mm; 2*n*=16, 32. Woods, s. Pa. and se. O. to Fla. and Miss. Mar.–July. Ours is var. *virginianum*.

30. PARTHENIUM L. Heads radiate, the rays few, short and rather inconspicuous, white or yellowish, pistillate and fertile, the expanded part rarely wanting and the heads thus disciform; invol of 2–4 series of broad, ± imbricate, dry and scarcely herbaceous bracts, the inner subtending the rays; receptacle small, convex or conic, chaffy; disk-fls staminate, with undivided style; ray-achenes flattened parallel to the invol bracts, their subtending bracts joined at base to the 2 or 3 adjacent receptacular bracts, thus partly enclosing the achene, the whole commonly falling as a unit; pappus of 2 or 3 short or elongate awns or scales, or nearly obsolete; bitter, aromatic herbs (ours) or shrubs with alternate, entire to pinnatifid lfs and rather small hemispheric heads. 16, New World.

1 Perennials; lvs merely toothed, or some of them lyrate at the base.
  2 Plants with a tuberous-thickened, usually short root; heads relatively small, the disk mostly 4–7
    mm wide.
    3 Herbage rather shortly and closely hairy, or partly glabrous .... 1. *P. integrifolium.*
    3 Herbage loosely spreading-hairy, the stem often somewhat shaggy .... 2. *P. auriculatum.*
  2 Plants with a creeping rhizome; heads larger, the disk mostly 7–10 mm wide .... 3. *P. hispidum.*
1 Annual; lvs pinnatifid or bipinnatifid .... 4. *P. hysterophorus.*

1. **Parthenium integrifolium** L.   Eastern parthenium. Perennial 3–10 dm from a tuberous-thickened, usually short root; stem simple or branched above, strigose-puberulent above, often glabrous below; lvs large and sometimes few, crenate-serrate, or sublyrate at base, scabrous to subglabrous, the basal long-petiolate, with lance-elliptic to broadly ovate blade 7–20 × 4–10 cm, the cauline progressively shorter-petiolate and generally reduced, the upper often sessile and clasping, but less conspicuously so than in the next 2 spp.; heads numerous in a flat-topped infl, the disk 4–7 mm wide; rays scarcely 2 mm; achenes obovate, black, 3 mm; $2n=72$. Prairies and dry woods; Va. to se. Minn., s. to Ga. and Ark., and adventive in Mass. June–Sept.

2. **Parthenium auriculatum** Britton.   Appalachian parthenium. Similar to no. 1, but the stem and to a lesser extent the lvs coarsely spreading-hairy; cauline lvs ± auriculate-clasping. Dry woods and old fields; mts. and piedmont from n. Va. and adj. W.Va. to N.C. and Tenn. May–July.

3. **Parthenium hispidum** Raf.   Ozark parthenium. Perennial from a creeping rhizome, this sometimes coarse and woody, but not tuberous-thickened; herbage usually loosely spreading-hairy as in no. 2, or the pubescence seldom approaching that of no. 1; middle and upper lvs tending to be broad-based and ± strongly auriculate-clasping; heads larger than in no. 1 (the disk mostly 7–10 mm wide) and averaging fewer; achenes 3–5 mm; otherwise as in no. 1; $2n=72$. Prairies and limestone glades; Mo. and adj. Kans. to n. La., Okla., and Tex., and adventive e. as far as s. Mich. May–July.

4. **Parthenium hysterophorus** L.   Santa Maria. Annual to 1 m, usually much branched, ± hairy, and often glandular above; lvs pinnatifid or usually bipinnatifid, to 2 × 1 dm; heads small, numerous in an often leafy infl, the disk 3–5 mm wide; achenes obovate, black, 2–2.5 mm; $2n=34$. A weed of waste places; trop. Amer., n. to se. U.S. and occasionally to Mass., Mich., and Kans. July–Oct.

31. IVA L.   Marsh-elder. Heads disciform, the pistillate fls few, with tubular or obsolete cor; invol of a few equal or imbricate, ± herbaceous bracts in 1–3 series; receptacle small, chaffy, its bracts commonly linear or spatulate, often subtending the pistillate as well as the staminate fls, these outer pales sometimes larger and more like the bracts of the invol; disk fls 6–20 in our spp., staminate, with undivided style, connate filaments, scarcely united anthers, and abortive or no ovary; achenes obovate, thick but somewhat compressed parallel to the invol bracts; pappus none; herbs or shrubs with opposite (or the upper alternate) lvs and small heads of greenish-white fls sessile or short pedunculate in the axils of the upper lvs or bracts (infl bractless in one sp.). 15, N. Amer.

1 Maritime perennials.
  2 Lvs 4–10 cm; invol 2–4 mm, with 4–5(6) equal bracts .... 1. *I. frutescens.*
  2 Lvs 1.5–4.5(–6) cm; invol 4–7 mm, with 6–9 imbricate bracts .... 2. *I. imbricata.*
1 Inland annual weeds.
  3 Heads axillary to evident bracts .... 3. *I. annua.*
  3 Heads not subtended by bracts .... 4. *I. xanthifolia.*

1. **Iva frutescens** L.   Maritime m.-e. Perennial, somewhat fleshy shrub or coarse herb 0.5–3.5 m, puberulent or strigose at least above; lvs generally all opposite except the reduced upper ones, the main ones 4–10 cm, evidently short-petiolate; invol 2–4 mm, of 4–5(6) equal bracts; pistillate fls 4–5(6), with tubular cor 1 mm tending to persist on the copiously resin-dotted achene; $2n=34$. Marshes and other moist places along the seashore; N.S. and Mass. to Fla. and Tex. Our plants are mostly var. *oraria* (Bartlett) Fern. & Griscom, with lvs less than 4 times as long as wide, up to 10 × 4 cm, and with 8–17 teeth to a side. In Va. this passes into the more southern var. *frutescens,* with smaller, narrower lvs (4–7 × 0.7–1.5 cm, 4–8 times as long as wide) with up to ca 8 teeth to a side (or subentire).

2. **Iva imbricata** Walter.   Beach-dune m.-e. Perennial, somewhat fleshy, glabrous herb or subshrub 3–6(–10) dm, commonly decumbent and branching at base; lvs oblanceolate to linear oblong or elliptic, mostly 1.5–4(–6) cm, entire or with a few spreading teeth, scarcely or obscurely petiolate, only the lower opposite; invol 4–7 mm, accrescent, with 6–9 broad, rounded, imbricate bracts; pistillate fls 2–4, with tubular cor 1–1.5 mm tending to persist on the copiously resin-dotted achene; $2n=34$. Dunes along the seashore; Va. to Fla., and La.; Bahamas; Cuba. Aug., Sept.

3. **Iva annua** L.   Rough m.-e. Annual weed 0.5–2 m; stem glabrous below, strigose near the middle, commonly spreading-hirsute above; lvs chiefly opposite (only some of the reduced upper ones alternate), petiolate, lanceolate to broadly ovate, ± serrate, acuminate, 5–15 × 2–7 cm, scaberulous-strigose; infl of several or many spiciform branches, the heads practically sessile in the axils of reduced, ovate to lance-linear, conspicuously ciliate-margined lvs 5–20 mm; invol 2–3.5 mm, its bracts 3–5, subtending the achenes, sparsely long-hirsute, broad and ciliate distally; pistillate fls with tubular cor 1–1.5 mm tending to persist on the resinous-dotted achene, sometimes subtended by very slender and inconspicuous receptacular bracts; $2n=34$. Waste ground, especially in moist soil; Ind. to N.D., s. to Miss. and N.M., and occasionally intr. eastward. Sept., Oct. (*I. caudata; I. ciliata.*)

4. Iva xanthifolia Nutt.   Big m.-e. Coarse annual weed 0.5–2 m; stem simple and glabrous below, viscid-villous in the infl; lvs chiefly opposite (only some of the upper alternate), long-petiolate, ovate, often very broadly so, coarsely and often doubly serrate, trinerved or triplinerved, 5–20 × 2.5–15 cm, strigose-scaber-ulous above, paler and often finely sericeous beneath; infl large, paniculiform, the numerous heads subsessile, not subtended by bracts; invol 1.5–3 mm, its 5 subherbaceous bracts somewhat larger than the 5 more membranous receptacular bracts (like an inner invol) that partly enfold the glabrous or distally hispidulous achenes; cor of the pistillate fls less than 0.5 mm, or obsolete; $2n=36$. Bottomlands and moist waste places; Wis. to B.C., s. to Tex. and N.M., and occasionally intr. eastward. Aug.–Oct. (*Cyclachaena x.*)

32. AMBROSIA L.   Ragweed. Heads numerous, small, unisexual, rayless; staminate heads in a racemiform or spiciform, bractless infl, with subherbaceous, 5–12-lobed invol, flat receptacle with slender, filiform-setose pales, monadelphous filaments, scarcely united anthers, and undivided style; pistillate heads borne below the staminate, in the axils of lvs or bracts, with closed, nutlike or burlike invol bearing 1–several rows of tubercles or spines and containing 1 (–several) pistillate fls without cor; pappus none; herbs (ours) or shrubs; lvs often lobed or dissected, in most spp. opposite at least below. 40, New World. (*Franseria.*)

1 Pistillate invols with 2 fls and 2 sharp beaks; casual intr. westward .... 1. *A. tomentosa*.
1 Pistillate invols with 1 fl and 1 beak; includes some common and widespread spp.
  2 Staminate heads sessile, and with the invol produced above into a conspicuous lobe .... 6. *A. bidentata*.
  2 Staminate heads pedunculate; invol not conspicuously oblique.
    3 Lvs palmately 3–5-lobed, or undivided; annual to 2(–5) m .... 5. *A. trifida*.
    3 Lvs once or twice pinnatifid; plants seldom over 1 m.
      4 Perennial from creeping roots; invol with a single cycle of tubercles .... 3. *A. psilostachya*.
      4 Taprooted annuals.
        5 Pistillate invol with several series of spines .... 2. *A. acanthicarpa*.
        5 Pistillate invol with a single cycle of short spines .... 4. *A. artemisiifolia*.

1. Ambrosia tomentosa Nutt.   Skeletonlf-bursage. Colonial perennial from creeping roots; lvs all or nearly all alternate, densely and conspicuously short-white-hairy beneath, greener and less hairy above, openly subbipinnatifid to subtripinnatifid with short and broad ultimate segments, the slender midstrip beset with small segments between the major ones; fruiting invols 4–6 mm, 2-seeded, with 2 sharp beaks and up to 10 short spines in one or more cycles. Plains sp., casually intr. as far e. as Wis. and Ill. May–Aug. (*Franseria discolor; Gaertneria d.* Not *Franseria tomentosa,* properly *A. grayi* (A. Nels.) Shinners, another Great Plains sp., with less dissected lvs that have a large terminal segment.)

2. Ambrosia acanthicarpa Hook.   Annual bursage. Branching annual 1–8 dm, coarsely strigose to sca-brous-hispid or both; lvs petioled, opposite below, alternate above, bipinnatifid (or the upper sessile and once pinnatifid), the blade 2–8 cm; heads numerous in terminal-racemiform clusters, sometimes chiefly staminate or chiefly pistillate; fruiting invols solitary or clustered in the upper axils, 5–10 mm, 1-fld and 1-beaked, with several series of flattened, divergent spines, these sometimes reduced in the chiefly staminate plants; $2n=36$. Open places, especially in sandy soil; Wash. to Calif., e. to Sask. and Tex., and occasionally intr. eastward, as in Minn. and Mo. July–Oct. (*Franseria a.; Gaertneria a.*)

3. Ambrosia psilostachya DC.   Western ragweed. Resembling no. 4, but a colonial perennial from creeping roots; lvs thicker, short-petiolate or subsessile, usually only once pinnatifid, averaging narrower in outline; fruiting invol merely tuberculate above, sometimes obscurely so; $2n=36, 72, 108, 144.$ Waste places, usually in dry or sandy soil; Mich. and Ill. to La., w. to the Rocky Mts., and casually intr. eastward. (July) Aug.–Oct. (*A. coronopifolia.*)

4. Ambrosia artemisiifolia L.   Common ragweed. Annual weed 3–10 dm, branching at least above, var-iously hairy or in part subglabrous; lvs opposite below, alternate above, once or more often twice pinnatifid, the blade narrowly to broadly ovate or elliptic in outline, 4–10 cm, the middle and lower ones, at least, evidently petiolate; staminate invols short-pedunculate, symmetrical or slightly oblique, inconspicuously nerved; fruiting invols 3–5 mm, 1-fld and 1 beaked, with a single series of short, sharp, erect spines near or above the middle; $2n=34, 36.$ Waste places; throughout our range, w. to the Pacific. (July) Aug.–Oct. (*A. elatior; A. media.*)

5. Ambrosia trifida L.   Giant ragweed. Coarse annual weed to 2(–5) m; stem spreading-hairy above, often glabrous below; lvs opposite, petiolate, ± scabrous, broadly elliptic to more often ovate or suborbicular, often 2 dm, serrate, palmately 3–5-lobed, or, especially in depauperate individuals, lobeless; staminate invols ca 1.5 mm, slightly oblique, unilaterally evidently 3-nerved; fruiting invols 5–10 mm, several-ribbed, each rib ending in a tubercle or spine; $2n=24.$ Moist soil and waste places throughout our range, s. to Fla., w. to B.C. and n. Mex. (July) Aug.–Oct. The common form in most of our range is var. *trifida,* with the petioles of at least the upper lvs ± wing-margined, and the ribs of the fr ending in short spines. (*A. striata.*) From s. Ill. and Mo. to Miss., Ariz., and n. Mex. is found var. *texana* Scheele, with wingless petioles, and with the ribs of the mostly smaller fr ending in blunt or almost obsolete tubercles. (*A. aptera.*)

6. Ambrosia bidentata Michx.   Lancelf ragweed. Branching annual weed 3–10 dm, the stem conspicuously spreading-hirsute, especially above; lvs numerous, sessile, opposite below, alternate above, lanceolate, acuminate, 2.5–7 cm × 4–10 mm, usually with a single pair of sharp large teeth below the middle, hirsute to scabrous, or the upper side glabrous; staminate heads sessile, the invol its upper side produced into a conspicuous, retrorsely spreading, hispid-hirsute, lanceolate to triangular-ovate lobe, the bractless infl thus appearing retrorse-bracteate; fruiting invol 5–8 mm, villous-hirsute, several-ribbed, the ribs produced into short, stiff spines; 2n=34. Prairies and waste places; O. and Ky. to Nebr., La., and Tex., and occasionally intr. elsewhere. July–Oct.

33. XANTHIUM L.   Cocklebur.

Heads small, unisexual; staminate heads uppermost, many-fld, with poorly developed or no invol, the receptacle short, cylindric, chaffy, the filaments monadelphous, the pistil composed mainly of the undivided style; invol of the pistillate heads completely enclosing the 2 fls, forming a conspicuous 2-chambered bur with hooked prickles; pistillate fls lacking a cor, and with the styles exsert from the invol; achenes thick, solitary in the chambers of the bur; pappus none; coarse, taprooted, annual weeds with alternate lvs and solitary or clustered axillary and terminal heads; 3, originally New World.

1 Lvs broad, generally cordate or deltoid at base, spineless . . . . 1. *X. strumarium.*
1 Lvs lanceolate, tapering to the base, bearing a 3-forked axillary spine . . . . 2. *X. spinosum.*

1. Xanthium strumarium L.   Common c. Plants 2–20 dm, appressed-hairy or subglabrous; lvs long-petiolate, broadly ovate to suborbicular or reniform, generally cordate or subcordate at base, sometimes shallowly 3–5-lobed, often 15 cm; staminate heads in a terminal cluster, the pistillate ones in several or many short axillary infls; bur broadly cylindric to ovoid or subglobose, 1–3.5 cm, covered with stout hooked prickles, terminated by 2 straight or in ours ± incurved beaks; 2n=36. Fields, waste places, flood-plains, and lake- and sea-beaches; now a cosmopolitan weed, probably originally native only to the New World. July–Sept. Our most common phase is var. *canadense* (Miller) T. & G., with the burs brownish or yellowish-brownish, 2–3.5 cm, the lower part of the prickles conspicuously spreading-hairy as well as ± stipitate-glandular. (*X. echinatum; X. italicum; X. pensylvanicum; X. speciosum.*) Also found throughout our range is var. *glabratum* (DC.) Cronq., with merely atomiferous-glandular or glandular-puberulent to subglabrous, commonly paler burs seldom over ca 2 cm. (*X. americanum; X. chinense; X. cylindraceum; X. echinellum; X. globosum; X. inflexum.*) Var. *strumarium*, with straight-beaked, yellow-green, merely puberulent burs to 2 cm, occurs in tropical Amer. and in s. Europe, and rarely in our range as a waif.

2. Xanthium spinosum L.   Spiny c. Plants 3–12 dm, the stems strigose or puberulent; lvs lanceolate, entire or with a few coarse teeth or pinnate lobes, tapering to each end, short-petiolate, 2.5–6 × 0.5–2.5 cm, sparsely strigose or glabrate above, except for the usually more hairy main veins, densely silvery-sericeous beneath, bearing a tripartite yellow spine 1–2 cm in the axil; burs mostly solitary or few in the axils, cylindric, ca 1 cm, beakless or with a single short beak, finely puberulent and provided with slender, hooked prickles; 2n=36. Waste places; now a cosmopolitan weed in the warmer parts of the world, and occasionally found in our range. July–Oct. (*Acanthoxanthium s.*)

34. HYMENOPAPPUS L'Hér.

Heads mostly discoid; invol bracts appressed, in 2–3 equal series, at least the inner with broad, obtuse, scarious, somewhat petaloid, yellowish or whitish tip, ± herbaceous below; receptacle small, naked; disk-fls perfect, the cor yellow or whitish, with campanulate 4–5-lobed throat, the lobes mostly ovate; style-branches flattened, with very short, minutely hairy appendage; achenes ca 15–20-nerved and 4–5-angled; pappus of 12–20 membranous or hyaline small scales, or rarely wanting; herbs with alternate, pinnatifid lvs and several or many small or middle-sized, hemispheric or campanulate heads. 10, N. Amer.

1. Hymenopappus scabiosaeus L'Hér.   Old plainsman. Taprooted biennial 3–7 dm, floccose-tomentose and partly glabrate, becoming more villous-puberulent above; lvs pinnatifid or the larger bipinnatifid, the lower 8–25 cm including the short petiole, 3–12 cm wide, or occasionally narrower and entire, the middle and upper generally smaller, often scattered and progressively reduced, at least some always pinnatifid; heads in an open-corymbiform infl, the disk 7–12 mm wide at anthesis; invol 7–10 mm; lobes of the disk-cor ca as long as the throat; pappus-scales hyaline, 1 mm or less; 2n=34. Prairies and dry, open woods; Tex. to s. Neb., e. irregularly to n. Ind., S.C., and Fla. May. (*H. carolinensis.*) Ours is var. *scabiosaeus.*

35. ANTHEMIS L.   Chamomile.

Heads radiate or rarely discoid, the rays elongate, white or yellow, pistillate or neutral; invol bracts subequal or more often imbricate in several series, dry, the margins ± scarious or hyaline; receptacle convex or conic or hemispheric, chaffy at least toward the middle; disk-fls numerous, perfect, the tube usually cylindric at base, occasionally flattened; style-branches flattened, truncate, penicillate; achenes terete or 4–5-angled, or occasionally ± compressed but not callous-margined; pappus a short crown, or more often none; herbs, usually aromatic, with alternate, incised-dentate to pinnately dissected lvs and medium-sized to rather large heads terminating the branches. 60, Old World.

Besides the spp. treated below, *A. mixta* L. (annual; rays pistillate, white with yellow base) and *A. nobilis* L. (perennial; rays pistillate, white, receptacular bracts blunt) have been found as waifs in our range.

1 Rays yellow .... 1. *A. tinctoria.*
1 Rays white.
  2 Rays pistillate and fertile; receptacle chaffy throughout .... 2. *A. arvensis.*
  2 Rays sterile and usually neutral; receptacle chaffy only toward the middle .... 3. *A. cotula.*

1. Anthemis tinctoria L. Yellow c. Short-lived perennial 3–7 dm; stem sparingly branched above or simple, villous-puberulent at least above; lvs pinnatifid, 2–6 cm, with winged rachis and deeply toothed or pinnatifid segments, villous or almost floccose beneath; heads solitary and long-pedunculate at the ends of the branches; disk 12–18 mm wide; invol thinly villous-tomentose; rays 20–30, pistillate and fertile, yellow, 7–15 mm; receptacle chaffy throughout, its bracts narrow, with firm yellow awn-tip equalling the disk-fls; achenes compressed-quadrangular, ± striate-nerved; $2n=18$. Fields and waste places; native of Europe, sparingly naturalized in much of our range. June–Sept. (*Cota t.*)

2. Anthemis arvensis L. Corn-c. Similar to no. 3, commonly a little more hairy, and not ill-scented; lvs appearing a little less finely dissected; heads averaging a little larger; rays pistillate and fertile; receptacle chaffy throughout, its bracts softer, paleaceous, with short, cuspidate awn-tip; achenes not tuberculate; $2n=18$. Fields and waste places; native of Europe, naturalized over most of the U.S., and widespread in our range. May–Aug.

A. secundiramea Biv., a Mediterranean sp., has been collected on railroad ballast in Va., but may not be fully naturalized. The heads avg smaller than in no. 2, the lvs are glandular-punctate beneath, and the achenes are tuberculate.

3. Anthemis cotula L. Dogfennel. Taprooted, usually subglabrous, malodorous annual 1–9 dm; lvs 2–6 cm, 2–3 times pinnatifid, with narrow segments; heads short-pedunculate, the disk 5–10 mm wide, becoming ovoid or short-cylindric at maturity; invol sparsely villous; rays mostly 10–16, sterile and usually neutral, white, 5–11 mm; receptacle chaffy only toward the middle, its firm, narrow, subulate bracts tapering to the apex; achenes subterete or nearly quadrangular, obscurely ca 10-ribbed, strongly glandular-tuberculate; $2n=18$. Fields and waste places; native of Europe, now a cosmopolitan weed. (*Maruta c.*)

36. ACHILLEA L. Yarrow. Heads radiate, the rays 3–13, pistillate and fertile (rarely neutral), white, sometimes pink or yellow, short and broad; invol bracts imbricate in several series, dry, with scarious or hyaline margins and often greenish midrib; receptacle conic or convex, chaffy; disk-fls mostly 10–75, perfect and fertile; style-branches flattened, truncate, penicillate; achenes compressed parallel to the invol bracts, callous-margined, glabrous; pappus none; perennial herbs with alternate, subentire to pinnately dissected lvs and several or many relatively small heads in a ± corymbiform infl. 75, N. Hemisphere, mainly Old World.

In addition to the following spp., *A. tomentosa* L., 2–4 dm, with bright yellow fls, casually escapes from cult.

1 Lvs subentire to incised, not dissected.
  2 Rays 3–5 mm; lvs subentire or shallowly serrate; intr. .... 1. *A. ptarmica.*
  2 Rays 1–2 mm; lvs incised; native, boreal .... 2. *A. sibirica.*
1 Lvs pinnately dissected .... 3. *A. millefolium.*

1. Achillea ptarmica L. Sneezeweed. Rhizomatous perennial 3–6 dm; stem villous-tomentose at least above; lvs glabrous or nearly so, linear or lance-linear, 3–10 cm × 2–6 mm, sessile, acuminate, closely and rather shallowly serrate to subentire; heads several or numerous in an open-corymbiform infl, the disk 4–8 mm wide; invol 4–5 mm; rays commonly 8–10, white, 3–5 mm; disk-fls 50–75; $2n=18$. Beaches, roadsides, and waste places; native to n. Eurasia, naturalized in ne. Amer. from Lab. to N.Y., w. to Minn. and Mo. July–Sept. Forms escaped from cult. often have numerous rays.

2. Achillea sibirica Ledeb. Resembling no. 1; stems 5–8 dm; lvs 5–10 cm × 4–8 mm, incised; heads in a short, crowded, corymbiform infl; rays 6–13, 1–2 mm; disk-fls 25–30; $2n=36$. Woods; ne. Asia to Man. and Minn.; Gaspé peninsula, Que. July, Aug.

3. **Achillea millefolium** L.   Common y. Aromatic, rhizomatous perennial 2–10 dm, very sparsely to rather densely villous or woolly-villous; lvs pinnately dissected, the blade 3–15 cm, to 2.5 cm wide, the basal petiolate, all but the lowest cauline sessile; heads numerous in a flat or round-topped, short and broad, paniculate-corymbiform infl, the disk 2–4 mm wide; invol in ours 4–5 mm; rays ca 5, white or occasionally pink, 2–3 mm; disk-fls 10–30; $x=9$. Various habitats, but especially in disturbed sites, circumboreal, and throughout our range. June–Nov. A highly variable circumboreal polyploid complex, not yet satisfactorily sorted out into infraspecific taxa. Typical European *A. millefolium,* a hexaploid with flat lvs, the relatively broad ultimate segments all in nearly the same plane, is rarely intr. in our range. Our native forms are mostly tetraploid, with narrower lf-segments disposed in various planes so that the lf is airily 3-dimensional. These represent the transcontinental ssp. *lanulosa* (Nutt.) Piper, consisting of several vars. as yet not well understood. Another native tetraploid, called var. *nigrescens* E. Meyer, extends into our range from the ne., reaching Mass. and n. Mich. It is a relatively small plant with flat lvs like typical *millefolium,* distinguished by the dark brown or nearly black margins of the invol bracts.

37. **SANTOLINA** L.   Heads discoid, the fls all tubular and perfect, or the outer rarely with abortive anthers; invol bracts imbricate in several series, dry, scarious on the margins or throughout; receptacle convex, chaffy throughout; style-branches flattened, truncate, penicillate; achenes evidently 3–5-angled, glabrous; pappus none; aromatic subshrubs, branched at the base, with alternate, pectinate or pinnatifid lvs and small or middle-sized, long-pedunculate, hemispheric yellow heads. 10, Mediterranean.

1. **Santolina chamaecyparissus** L.   Lavender cotton. Tomentose-puberulent subshrub 1–3 dm; lvs 3–10 mm, pinnatifid, with crowded narrow rounded segments, sometimes more elongate, to 3 cm, with more distant segments; heads solitary or few on naked peduncles, the disk 5–12 mm wide; cors conspicuously exceeding the invol; $2n=$ca 45. Native of s. Europe, cult. for ornament, and established in waste places near Boston. Oct.

38. **CHRYSANTHEMUM** L.   Chrysanthemum. Heads radiate or seldom discoid, the rays pistillate and fertile, white or sometimes yellow or anthocyanic; invol bracts ± imbricate in 2–4 series, dry, at least the margins and tips scarious or hyaline, the midrib sometimes greenish; receptacle flat or convex, naked; disk-fls tubular and perfect, the cor with (4)5 lobes; style-branches flattened, truncate, penicillate; achenes subterete or angular, 5–10-ribbed, or those of the rays with 2–3 wing-angles; pappus a short crown, or none; herbs or seldom shrubs with alternate, entire or toothed to occasionally pinnatifid lvs and hemispheric or depressed-hemispheric heads. 100, N. Hemisphere, espec. Old World.

1 Heads solitary or few, large, the disk 1–2.5 cm wide.
   2 Rays white; perennial; achenes all ca 10-ribbed.
     3 Herbs.
       4 Lvs somewhat basally disposed, the middle and upper ones progressively reduced; some of
         the lvs often somewhat lobed or cleft as well as toothed .... 1. *C. leucanthemum.*
       4 Lvs well distributed along the stem, the largest ones near or somewhat below the middle, all
         merely toothed .... 2. *C. lacustre.*
     3 Soft shrub with the lvs crowded toward the branch-tips .... 3. *C. nipponicum.*
   2 Rays yellow; annual; achenes of the rays with 2 wing-angles .... 4. *C. segetum.*
1 Heads several or numerous, small, the disk 4–9 mm wide.
   5 Lvs pinnatifid, the segments incised or again pinnate; rays white .... 5. *C. parthenium.*
   5 Lvs crenate, or with a few small basal pinnae; rays commonly wanting .... 6. *C. balsamita.*

1. **Chrysanthemum leucanthemum** L.   Ox-eye daisy. Rhizomatous perennial 2–8 dm, simple or nearly so, glabrous or inconspicuously hairy; basal lvs oblanceolate or spatulate, 4–15 cm (petiole included), crenate and often also ± lobed or cleft, the cauline reduced and becoming sessile, deeply and distantly blunt-toothed or sometimes more closely toothed or subentire; heads solitary at the ends of the branches, naked-pedunculate, the disk 1–2 cm wide; rays 15–35, white, 1–2 cm; achenes terete, ca 10-ribbed; $x=9$; typically diploid, sometimes polyploid. Fields, roadsides, and waste places; native of Eurasia, naturalized throughout most of temp. N. Amer. May–Oct. (*Leucanthemum l.; L. vulgare.*)

2. **Chrysanthemum lacustre** Brot.   Portuguese daisy. More robust and leafy stemmed than no. 1, with closely and sharply serrate, not at all pinnatifid lvs, the stem in well developed plants few-branched above, the heads averaging a bit larger, the disk to 2.5 cm wide; basal lvs wanting or inconspicuous; lower cauline lvs narrowed to a petiolar base but mostly soon deciduous; middle cauline lvs well developed, elliptic or elliptic-oblanceolate, commonly 6–11 × 1.5–3 cm, tending to clasp at base; $2n=18$. Native to sw. Europe, occasionally escaped from cult. into fields and moist, low ground in our range. June–Aug. (*Leucanthemum l.*)

3. Chrysanthemum nipponicum (Franchet) Sprenger. Nippon-daisy, Montauk d. Soft shrub to 1 m; lvs crowded toward the ends of the stout branches, firm, broadly oblanceolate to oblong-spatulate, sessile or nearly so, 3–10 × 1–3 cm, toothed above the middle; heads naked-pedunculate, solitary at the branch-tips, much like those of nos. 1 and 2; $2n=18$. Native of Japan, escaped from cult. onto beaches in N.Y. and N.J. Sept., Oct. (*Nipponanthemum n.*)

4. Chrysanthemum segetum L. Corn-c. Glabrous annual 2–6 dm; lvs oblong to ovate or the lower spatulate, coarsely and sharply toothed or irregularly pinnatifid, the lower subpetiolate, the others sessile and clasping, to 6 × 2 cm; heads few, terminating the branches, the disk 1–2 cm wide; rays 10–20, yellow, 8–15 mm; disk-achenes subterete and ca 10-ribbed; ray-achenes flattened and with 2 wing-angles; $2n=18$. Fields and waste places; native of the Old World, occasionally intr. in our range. June–Oct.

5. Chrysanthemum parthenium (L.) Bernh. Feverfew. Perennial from a taproot or stout caudex; stem 3–8 dm, puberulent above; lvs mainly cauline, finely hairy at least beneath, pinnatifid, with incised or again pinnate, rounded segments, evidently petiolate, the blade to 8 × 6 cm; heads several or many in a corymbiform infl, relatively small, the disk 5–9 mm wide; rays 10–21, or more numerous in double forms, 4–8 mm, white; achenes subterete, ca 7–10-ribbed; $2n=18$. Waste places; native of Europe, occasionally intr. or escaped in our range. June–Sept. Generic position uncertain. (*Matricaria p.; Tanacetum p.*)

6. Chrysanthemum balsamita L. Costmary. Coarse, fragrant perennial 5–15 dm; stem strigose above; lvs silvery-strigose or subsericeous when young, later ± glabrate, crenate, sometimes with a few reduced basal pinnae, the basal with elliptic or broadly oblanceolate blade 10–25 × 2.5–8 cm on a petiole of ca equal length, the cauline smaller, numerous, sessile or nearly so, seldom over 10 cm; heads numerous in a corymbiform infl, the disk 4–7 mm wide; rays wanting, or occasionally present, white, and well under 1 cm; $2n=18, 54$. Roadsides and other waste places; native of s. Europe and the Orient, occasionally intr. or escaped in our range. Aug.–Oct. Generic position uncertain. (*Tanacetum b.; Balsamita major.*)

39. TANACETUM L. Tansy. Much like *Chrysanthemum,* but the heads disciform or nearly so, the outer fls pistillate, with short tubular cor which in some spp. is expanded into a short yellow ray, or the pistillate fls rarely wanting; disk-fls 5-toothed, yellow; achenes mostly 5-ribbed or -angled, commonly glandular; lvs pinnately dissected to rarely entire. 50, N. Hemisphere.

1 Lvs glabrous or nearly so; heads relatively many; disk 5–10 mm wide .... 1. *T. vulgare.*
1 Lvs ± villous; heads relatively few and large, the disk 10–18 mm wide .... 2. *T. huronense.*

1. Tanacetum vulgare L. Common t. Coarse, aromatic perennial 4–15 dm from a stout rhizome, glabrous or nearly so; lvs numerous, 1–2 dm, nearly half as wide, sessile or short-petiolate, punctate; pinnatifid with evidently winged rachis, the pinnae again pinnatifid or deeply lobed, with broadly winged rachis, the pinnules often again toothed; heads disciform, numerous, commonly 20–200, the disk 5–10 mm wide; $2n=18$. Roadsides, fields, and waste places; native of the Old World, established throughout most of the U.S. and adj. Can. Aug.–Oct.

2. Tanacetum huronense Nutt. Eastern t. Rhizomatous perennial 1–8 dm, ± villous throughout; lvs scarcely punctate, the cauline 5–20 × 1.5–8 cm, the basal often longer and persistent, all bi- or tripinnatifid, with narrow ultimate segments, the main rachis scarcely or not at all winged, that of each pinna only narrowly so; heads 1–15, rarely more, the disk 10–18 mm wide; rays inconspicuous or more commonly evident, to 4 mm; $2n=54$. Beaches and riverbanks; Me. and Nf. to n. Mich., n. Wis., and Alas. July–Aug. Closely allied to the Siberian and Alaskan *T. bipinnatum* (L.) Schultz-Bip.

40. MATRICARIA L. Heads radiate or discoid, the rays white, pistillate and usually fertile; invol bracts dry, 2–3-seriate, not much imbricate, with ± scarious or hyaline margins; receptacle naked, hemispheric to more often conic or elongate; disk-cors yellow, 4–5-toothed; style-branches flattened, truncate, penicillate; achenes glabrous or roughened, generally nerved on the margins and ventrally, nerveless dorsally; pappus a short crown or none; herbs with alternate, pinnatifid or pinnately dissected lvs and small or middle-sized heads terminating the branches or in a corymbiform infl. 40, N. Hemisphere and S. Afr.

1 Heads with evident white rays.
  2 Achenes with 2 marginal and 1 ventral, strongly callous-thickened, almost wing-like ribs, minutely
      roughened on the back and between the ribs .... 1. *M. maritima.*
  2 Achenes with 2 nearly marginal and 3 ventral, raised but not at all wing-like ribs, otherwise smooth
      .... 2. *M. recutita.*
1 Heads discoid .... 3. *M. matricarioides.*

1. Matricaria maritima L.   Scentless chamomile. Short-lived, often annual, nearly inodorous herb 1–7 dm, glabrous or nearly so; lvs 2–8 cm, bipinnatifid, the ultimate segments mostly elongate, linear or linear-filiform, sometimes shorter and broader; heads several or numerous, the disk 8–15 mm wide; outer invol bracts narrow and scarcely or barely scarious-margined; rays 12–25, 6–13 mm; disk-cors 5-toothed; receptacle hemispheric, rounded; achenes as described in the key; $2n=18$, 36. Native of Europe, established along roadsides and in waste places in our range. July–Sept. (*M. perforata; M. inodora; Chamomilla i.; C. maritima.*)

2. Matricaria recutita L.   Chamomile. Glabrous branching aromatic annual 2–8 dm; lvs 2–6 cm, bipinnatifid, the ultimate segments linear or filiform; heads ± numerous, the disk 6–10 mm wide; invol bracts all evidently scarious-margined; rays 10–20, 4–10 mm; disk-cors 5-toothed; receptacle conic, acute; achenes as described in the key; $2n=18$. Native of Eurasia, now widely intr. along roadsides and in waste places in N. Amer., but less common than no. 1. (*M. chamomilla,* misapplied.)

3. Matricaria matricarioides (Less.) Porter.   Pineapple-weed. Branching, leafy, pineapple-scented, glabrous annual 5–40 cm; lvs 1–5 cm, 1–3 times pinnatifid, the ultimate segments short, linear or filiform; heads several or many, rayless, the disk 5–9 mm wide; invol bracts with broad hyaline margins; disk-cors 4-toothed; receptacle conic, pointed; achenes with 2 marginal and one or sometimes more rather weak ventral nerves; pappus a short crown; $2n=18$. Roadsides and waste places; cordilleran region, intr. e. to the Atlantic. May–Sept. (*M. discoidea; M. suaveolens* (Pursh) Buchenau, not L.; *Chamomilla s.*)

41. COTULA L.   Heads disciform, the outer 1–several series of fls pistillate, with short tubular cor, or rarely all the fls perfect; invol bracts slightly unequal, ca 2-seriate, membranous or subherbaceous, commonly with narrow scarious margins; receptacle naked, flat to conic; disk-fls usually fertile, with 4-toothed yellow cor; style-branches flattened, truncate, penicillate; achenes, especially of the outer fls, ± stipitate, compressed parallel to the invol bracts, commonly minutely striate, otherwise 2–4-nerved or nerveless; pappus a short crown or none; annual or occasionally perennial herbs with alternate, entire to more often pinnatifid or pinnately dissected lvs and small or medium-sized heads. 50, mainly S. Hemisphere.

1. Cotula coronopifolia L.   Essentially glabrous, somewhat succulent perennial 5–30 cm tall, often trailing; lvs sessile, with subscarious sheathing base, entire or with a few coarse teeth or deep narrow lobes, narrowly oblong or lanceolate to linear, 1–6 cm × 1–10 mm, or broader when lobed; heads solitary at the ends of the branches; disk 5–11 mm wide; pistillate fls in a single series, without cor; achenes of the outer fls broadly winged, shorter than the stipe, the others merely 2-nerved, short-stipitate, the stipes persistent on the receptacle; $2n=20$. Tidal flats, wet meadows, and other moist places; native of S. Afr., occasionally found from Que. to Mass. in our range. July–Aug.

42. ARTEMISIA L.   Wormwood; sage; mugwort. Heads disciform or discoid, the outer fls in our spp. pistillate and fertile, the central ones perfect or sometimes sterile; invol bracts dry, imbricate, at least the inner scarious or with scarious margins; receptacle flat to convex or hemispheric, naked or beset with long hairs; style-branches flattened, truncate, penicillate; achenes ellipsoid or obovoid to nearly prismatic, scarcely compressed, usually glabrous; pappus none; herbs or shrubs, usually aromatic, with alternate, entire to dissected lvs and few to numerous small, ovoid to campanulate or hemispheric heads in a spiciform, racemiform, or paniculiform infl. 100+, N. Hemisphere and S. Amer.

1 Disk-fls fertile, with normal ovary.
  2 Receptacle not hairy.
    3 Perennials from a rhizome or woody caudex, sometimes shrubby; lvs ± tomentose or tomentulose beneath.
      4 Principal lvs 2–3 times pinnatifid, the ultimate segments up to 1.5 mm wide.
        5 Lvs green above, 3–6 cm, with slender ascending lobes .... 1. *A. abrotanum.*
        5 Lvs generally white-tomentose on both sides, 1–3 cm, with short spreading lobes .... 2. *A. pontica.*
      4 Principal lvs entire or once pinnatifid, or if bipinnatifid then the ultimate segments well over 1.5 mm wide.
        6 Heads large, the invol 6–7.5 mm, the disk-cors 3.2–4 mm .... 3. *A. stelleriana.*
        6 Heads smaller, the invol 2.5–5 mm, the disk-cors 1–3 mm.
          7 Divisions of the principal lvs again toothed, cleft, or lobed; lvs ordinarily with 1 or 2 pairs of stipule-like lobes at base; widespread intr. weed .... 4. *A. vulgaris.*
          7 Divisions of the lvs entire or nearly so, or the lvs entire; lvs usually without stipule-like lobes; native, chiefly western.
            8 Lvs evenly serrate, glabrous above; plants commonly 1–3 m .... 5. *A. serrata.*
            8 Lvs entire or variously toothed or lobed, but not evenly serrate, glabrous or more commonly hairy above; seldom over 1 m .... 6. *A. ludoviciana.*
    3 Annuals or biennials with a taproot; lvs green, essentially glabrous.

9 Infl dense, spike-like or of spiciform branches, the heads crowded, obscurely pedunculate;
    invol 2–3 mm .... 7. *A. biennis.*
9 Infl loose, paniculiform, the heads on evident slender peduncles; invol 1–2 mm .... 3. *A. annua.*
2 Receptacle beset with numerous long hairs between the disk-fls.
  10 Plants robust, 4–10 dm; lf-segments 1.5–4 mm wide; intr. .... 9. *A. absinthium.*
  10 Plants smaller, 1–4 dm; lf-segments ca 1 mm wide or less; native, mainly western .... 10. *A. frigida.*
1 Disk-fls sterile, with abortive ovary; receptacle naked, without hairs.
  11 Lvs mostly entire, or the lower 3–5-cleft; perennial, but not taprooted .... 11. *A. dracunculus.*
  11 Lvs pinnatifid to dissected; taprooted biennial or perennial .... 12. *A. campestris.*

1. **Artemisia abrotanum** L.   Southernwood. Perennial and ± shrubby, 0.5–2 m, much branched; lvs 3–6 cm, thinly tomentose beneath, green and glabrous or nearly so above, 2–3 times pinnatifid with elongate, linear or filiform, ascending segments 0.5–1.5 mm wide; infl ample; invol 2–3.5 mm; achenes 4–5-angled, broadest at the truncate summit; $2n=18$. Roadsides and waste places; native of the Old World, cult. and sparingly established throughout much of the U.S., especially toward the ne. Aug., Sept.

2. **Artemisia pontica** L.   Rhizomatous perennial 4–10 dm, generally suffrutescent, simple or nearly so; lvs 1–3 cm, white-tomentose on both sides, more thinly so and sometimes eventually glabrate above, twice or thrice pinnatifid with short divergent segments scarcely 1 mm wide, ordinarily with a pair of stipule-like lobes or auricles at base; infl relatively narrow, elongate; invol 2–3 mm; achenes as in no. 1; $2n=18$. Dry open places; native of Europe, escaped and sparingly established in ne. U.S. and adj. Can. Aug., Sept.

3. **Artemisia stelleriana** Besser.   Beach-wormwood; dusty miller. Inodorous, rhizomatous perennial 3–7 dm, simple to the infl, densely white-tomentose or floccose; lvs 3–10 cm (petiole included) × 1–5 cm, with a few relatively broad, rounded lobes, which may be again slightly lobed; infl narrow and often dense; heads relatively large, the invol 6–7.5 mm; disk-cors 3.2–4.0 mm; achenes subterete, but narrowed at base and rounded above; $2n=18$. Sandy beaches; native from Kamtchatka to Japan, escaped from cult. and established from Que. to Va., and occasionally inland to Minn. May–Sept.

4. **Artemisia vulgaris** L.   Mugwort. Rhizomatous perennial 0.5–2 m, simple or branched above, the stem glabrous or nearly so below the infl; lvs green and glabrous or nearly so above, densely white-tomentose beneath, chiefly obovate or ovate in outline, 5–10 × 3–7 cm, the principal ones cleft nearly to the midrib into ascending, acute, unequal segments that are again toothed or cleft, and ordinarily with 1 or 2 pairs of stipule-like lobes at base; infl generally ample and leafy; invol 3.5–4.5 mm; disk-cors 2.0–2.8 mm; achenes ellipsoid, not nerved or angled; $2n=16$. Fields, roadsides, and waste places; native of Eurasia, now established throughout most of e. U.S. and adj. Can. July–Oct.

5. **Artemisia serrata** Nutt.   Toothed sage. Rhizomatous perennial 1–3 m; stem glabrous or nearly so below the infl; lvs numerous, lanceolate or lance-linear, acuminate, sharply and regularly serrate, 8–15 cm × 8–25 mm, green and essentially glabrous above, densely white-tomentose beneath, sometimes with a pair of small, stipule-like lobes at base; infl ample, generally leafy, invol 2.5–3.5 mm; disk-cors 2 mm; achenes ellipsoid, not nerved or ridged; $2n=36$. Prairies and low ground; Minn. and Io. to Ill., and intr. in N.Y. Aug.–Oct.

6. **Artemisia ludoviciana** Nutt.   White sage. Rhizomatous perennial 3–10 dm, simple up to the infl, the stem ± white-tomentose at least above; lvs lanceolate or lance-elliptic, 3–10 cm, entire or irregularly toothed to coarsely few-lobed or deeply parted, the undivided portion up to 1(–1.5) cm wide, persistently white-tomentose on both sides or becoming glabrous above; invol 2.5–3.5 mm; disk-cors 1.9–2.8 mm; $2n=18, 36$. July–Oct. Prairies, dry ground, and waste places. A widespread, variable sp. of w. U.S. and n. Mex., native e. as far as Ill., and occasionally intr. eastward. Most of our plants are var. *ludoviciana*, with entire to coarsely few-lobed lvs and a mostly compact and elongate infl. (*A. gnaphalodes,* the form with the lvs persistently tomentose above; *A. herriotii; A. pabularis.*) Var. *mexicana* (Willd.) Fern., with many of the lvs deeply parted and with a strong tendency toward a more diffuse, often leafy infl, is chiefly southwestern, occasionally reaching our range as an introduction. (*A. mexicana.*) Other vars. occur westward.
    Artemisia carruthii A. Wood, a southwestern sp. with the lvs divided nearly or quite to the midrib into elongate slender lobes seldom over 1(–2) mm wide, is occasionally intr. in our range.

7. **Artemisia biennis** Willd.   Biennial wormwood. Coarse, nearly inodorous, glabrous annual or biennial 3–30 dm; lvs 5–15 cm, pinnatisect nearly to the midrib into several narrow lobes that are usually again sharply toothed, or the lower bipinnatifid; infl dense, spike-like or of spiciform branches, the heads usually numerous, crowded, scarcely pedunculate; invol glabrous, 2–3 mm; achenes ellipsoid, 4–5-nerved; $2n=18$. Waste places and stream-banks, especially in sandy soil; native to nw. U.S. but widely distributed in our range and elsewhere as a weed. Aug.–Oct.

8. **Artemisia annua** L.   Annual w. Sweet-scented glabrous annual 3–30 dm, usually bushy-branched; lvs 2–10 cm, mostly twice or thrice pinnatifid, the ultimate segments linear or lanceolate, sharply toothed; infl broad and open, panicle-like, the heads loose, often nodding, on evident short peduncles; invol glabrous, 1–2 mm; achenes narrowly turbinate, obscurely or scarcely nerved; $2n=18$. Fields and waste places; native to Asia and e. Europe, now naturalized in N. Amer., especially in e. and c. U.S. and adj. Can. Aug.–Nov. Source of a potentially important antimalarial drug, artemisinin.

9. **Artemisia absinthium** L.   Common w.; absinthium. Perennial herb or near-shrub 4–10 dm, the stem finely sericeous or eventually glabrate; lvs silvery-sericeous, sometimes eventually subglabrate above, the lower long-petiolate and 2–3 times pinnatifid, with mostly oblong obtuse segments 1.5–4 mm wide, the blade

rounded-ovate in outline; upper lvs progressively less divided and shorter-petiolate, the segments often more acute; infl ample, leafy; involucre 2–3 mm; receptacle beset with numerous long white hairs between the fls; achenes nearly cylindric, but narrowed to the base and rounded at the top; $2n=18$. Fields and waste places; native of Europe, now established across n. U.S. and adj. Can., throughout our range. July–Sept.

10. **Artemisia frigida** Willd.   Prairie-sagewort. Mat-forming perennial 1–4 dm, from a stout caudex or woody crown, somewhat shrubby at base; lvs small and numerous, clustered at base and well distributed along the stem, sericeous-tomentose, the blade 5–12 mm, twice or thrice ternately divided into linear or linear-filiform divisions 1 mm wide or less, commonly with a pair of simple or trifid stipule-like divisions at the base; infl paniculiform, or racemiform in depauperate plants; invol 2–3 mm; receptacle with numerous long hairs; achenes narrowed to the base, truncate above, obscurely if at all nerved; $2n=18$. Prairies and dry open places; sw. Wis., Io. and Kans. to Ariz., Alaska, and Siberia, and occasionally intr. eastward. July–Sept.

11. **Artemisia dracunculus** L.   Fibrous-rooted perennial 5–15 dm, the stems clustered on short, coarse rhizomes or a branching caudex; herbage glabrous or occasionally villous-puberulent; lvs narrowly linear to lance-linear, 3–8 cm × 1–6 mm, mostly entire, but occasionally some 2–5-cleft, the lower generally deciduous before flowering; infl paniculiform, usually open and ample; invol glabrous or nearly so, 2–3 mm; disk-fls sterile; achenes ellipsoid, nerveless; $2n=18$. Dry open places; Ill. to Man. and Tex., s. to Yukon, B.C., and n. Mex.; also in Eurasia. July–Sept. (*A. cernua; A. dracunculoides; A. glauca.*)

12. **Artemisia campestris** L.   Scarcely odorous biennial or perennial 1–10 dm from a taproot; basal lvs crowded, 2–10 cm (petiole included) × 0.7–4 cm, twice or thrice pinnatifid or ternate, with mostly linear or linear-filiform segments seldom over 2 mm wide, glabrous to sericeous or villous, persistent or deciduous; cauline lvs similar but smaller and less divided, the uppermost often ternate or simple; infl small and spike-like to diffuse and panicle-like; invol glabrous to densely villous-tomentose, 2–4.5 mm, disk-fls sterile, with abortive ovary; achenes subcylindric; $2n=18$, 36. Open places, often in sandy soil; circumboreal, extending s. to Fla. and Ariz. July–Sept. A complex sp., composed of many races, the taxonomy still confused. Ssp. *caudata* (Michx.) Hall & Clem., a robust, mostly single-stemmed biennial (occasionally short-lived perennial) to 1 m or more tall, occurring on dunes and other very sandy places along the coast and irregularly inland throughout our range, is well marked. Typically it is glabrous or subglabrous. A more hairy, scarcely definable phase of ssp. *caudata,* which occurs on the n. Great Plains and enters our range along the Great Lakes, has been called *A. forwoodii* S. Wats., but has no name under *A. campestris.* All or nearly all of the rest of our material of *A. campestris* belongs to a phase that has been called *A. canadensis* Michx., or *A. campestris* var. *canadensis* (Michx.) Welsh. It is a ± multicipital perennial, seldom 6 dm, with fewer and often larger heads than ssp. *caudata,* occurring from Vt. and Minn. northward, often in the mts. In the broad sense the var. *canadensis* may be considered a part of *A. campestris* ssp. *borealis* (Pallas) Hall & Clem.

43. SENECIO L.   Groundsel, ragwort. Heads radiate or sometimes discoid, the rays pistillate and fertile, yellow to orange or occasionally reddish, invol bracts ± herbaceous, essentially equal, uniseriate or subbiseriate, often with some bracteoles at base; receptacle flat or convex, naked; disk-fls perfect and fertile, yellow to orange or reddish; style-branches flattened, truncate, penicillate; achenes subterete, 5–10-nerved; pappus of numerous, usually white, entire or rarely barbellulate capillary bristles; herbs (ours) or sometimes woody plants, with alternate (or all basal), entire to variously dissected lvs and mostly small to medium-sized heads. 1000+, cosmop. Spp. 4–14 ill-defined, their chromosomes also subject to miscount because of B-chromosomes.

S. millefolium T. & G., Blue Ridge ragwort, grows in the mts. of N.C., S.C., and Ga., and has recently been recorded from sw. Va. It would key to *S. jacobaea,* but it is a true perennial from a rhizome-caudex, with basally disposed leaves that are dissected into small and mostly narrow ultimate segments.

1 Perennial (or only biennial in no. 7).
  2 Heads very large, the disk 2–4 cm wide; invol 10–16 mm; maritime .... 1. *S. pseudo-arnica.*
  2 Heads smaller, the disk 5–18 mm wide; invol 4–10 mm; not maritime.
    3 Principal lvs mostly 2–3 times pinnatifid; intr. .... 16. *S. jacobaea.*
    3 Principal lvs entire or toothed to once pinnatifid; native.
      4 Plants ± persistently tomentose, at least on the stem, lower surface of the lvs, and invols (least
        so in no. 7).
        5 Tomentum fine and close; basal lf-blades to 5 cm.
          6 Achenes glabrous; western .... 4. *S. canus.*
          6 Achenes hirtellous along the angles; shale-barrens .... 5. *S. antennariifolius.*
        5 Tomentum loose; basal lf-blades often (esp. in no. 6) over 5 cm.
          7 Cauline lvs entire or ± toothed; mainly coastal plain .... 6. *S. tomentosus.*
          7 Cauline lvs (at least the larger ones) evidently pinnatifid; inland .... 7. *S. plattensis.*
      4 Plants not tomentose, or not persistently so, except sometimes at the base and in the lf-axils.
        8 Pubescence, if any, crisp and spreading; lvs all entire or merely slightly toothed; stem solitary
          from a very short, erect caudex or crown; western .... 2. *S. integerrimus.*

8 Pubescence, if any, floccose-tomentose; cauline lvs generally ± pinnatifid; habit mostly
   otherwise.
9 Heads discoid (rare radiate forms occur, chiefly with the discoid ones); n. Mich. .... 11. *S. indecorus.*
9 Heads radiate (rare discoid forms occur, chiefly with the radiate ones).
   10 Plants with long, slender stolons or superficial rhizomes; basal lvs narrowly obovate
      to orbicular .... 10. *S. obovatus.*
   10 Plants otherwise.
     11 Basal lvs chiefly oblanceolate or elliptic and tapering to the petiole.
       12 Heads few, rarely over 20; stem often glabrous at base; northern .... 8. *S. pauperculus.*
       12 Heads many, mostly 20–100+; stem densely woolly at base; southern ...... 9. *S. anonymus.*
     11 Basal lvs broadly lanceolate to reniform-cordate, mostly subtruncate to cordate
       at base.
       13 Basal lvs only slightly or scarcely cordate at base, often (in no. 12 consistently)
         more than 1.5 times as long as wide.
         14 Stems arising from ± elongate, branching rhizomes; mainly Appalachian
          .... 12. *S. schweinitzianus.*
         14 Stems arising from a short, often simple and erect or ascending rhizome-
          caudex; western .... 13. *S. pseudaureus.*
       13 Basal lvs ± strongly cordate, mostly 0.8–1.5 times as long as wide; wide-
        spread .... 14. *S. aureus.*
1 Annual, or sometimes biennial, ± leafy throughout.
  15 Rays well developed and conspicuous.
    16 Lvs entire or merely toothed; native, northwestern .... 3. *S. congestus.*
    16 Lvs, at least the cauline ones, evidently pinnatifid or dissected.
      17 Plants glabrous or somewhat tomentose, but not evidently glandular.
       18 Native, distinctly fibrous-rooted, southern annual or winter-annual with once or in
        part twice pinnatifid lvs .... 15. *S. glabellus.*
       18 Intr., taprooted to ± fibrous-rooted, northern biennial or short-lived perennial with
        2–3 times pinnatifid lvs .... 16. *S. jacobaea.*
      17 Plants conspicuously glandular-hairy .... 18. *S. viscosus.*
  15 Rays very short and inconspicuous, or none.
    19 Rays present, minute; bracteoles not black-tipped.
      20 Plants sparsely hairy or subglabrate, scarcely or obscurely glandular .... 17. *S. sylvaticus.*
      20 Plants densely and conspicuously glandular-hairy .... 18. *S. viscosus.*
    19 Rays none; bracteoles black-tipped .... 19. *S. vulgaris.*

1. Senecio pseudo-arnica Less. Seaside g. Stout perennial 1–7 dm, thinly tomentose or partly glabrate; lvs early deciduous near the base of the stem, otherwise persistent, crowded, and only gradually if at all reduced upwards, oblanceolate to obovate, 7–20 × 1.5–7 cm, coarsely toothed to subentire, the lower commonly tapering to the petiole or petiolar base; heads 1–5, very large, the disk 2–4 cm wide; invol 10–16 mm; rays ca 21, 1.5–2.5 cm (rarely wanting); achenes glabrous; pappus very copious; $2n=40$. Sea-beaches; Me., N.B., and the lower St. Lawrence R., n. to the Arctic; Alaska and Siberia. July, Aug.

2. Senecio integerrimus Nutt. Single-stemmed g. Stout, single-stemmed, fibrous-rooted perennial 2.5–7 dm from a very short, erect caudex or crown, villous or hirsute with crisp loose hairs when young, commonly nearly glabrous by flowering time; lvs entire, denticulate, or sometimes irregularly dentate, the basal mostly oblanceolate or elliptic and short-petiolate, 8–25 cm (petiole included) × 1–5 cm; cauline lvs progressively reduced, becoming sessile; heads 5–25 in a rather congested infl, the terminal peduncle often shorter and thicker than the others; disk 8–17 mm wide; invol 5–10 mm, its bracts often minutely and irregularly black-tipped; rays 6–10 mm; achenes glabrous; $2n=40$. Prairies and other open places; Sask., Minn., and Io., w. to B.C. and Calif. Apr.–June. Our plants are var. *integerrimus.*

3. Senecio congestus (R. Br.) DC. Northern swamp-g. Coarse, single-stemmed, fibrous-rooted annual or biennial 1.5–15 dm; pubescence spreading, crisp-villous, commonly persisting in large part until flowering time or beyond, especially in the infl; lvs entire or coarsely toothed, scarcely pinnatifid, rather equably distributed, 3–20 × 0.5–4.5 cm, the lower petiolate and often soon deciduous, the upper becoming sessile and ± clasping; heads several or numerous in an often congested infl, the disk 7–14 mm wide, or larger in fr; invol 7–10 mm, its bracts ca 21, very thin and generally pale, commonly with darker base; rays pale yellow, 4–9 mm; pappus accrescent, very fine and copious; achenes glabrous; $2n=48$. Swamps and edges of ponds; circumpolar, s. in our range to Minn. and extreme n. Io. June–Aug. (*S. palustris,* a preoccupied name.)

4. Senecio canus Hook. Gray g. Several-stemmed perennial 1–4 dm from a ± branched caudex, often with an evident short taproot; herbage ± strongly white-tomentose, often less so in age, the upper surfaces of the lvs sometimes glabrate; basal and lowermost cauline lvs ± tufted, from narrowly oblanceolate to broadly elliptic or subovate, the blade mostly 1–4 or 5 cm × 4–30(–45) mm, entire to irregularly subpinnately lobed, borne on a short or elongate petiole; middle and upper lvs few, strongly and progressively reduced, becoming bract-like; heads several; invol 4–8 mm; rays mostly 6–13 mm; achenes glabrous; $2n=46, 92, 138.$ Dry, open often rocky places; widespread in the w. cordillera, and e. to Minn. May–July.

5. Senecio antennariifolius Britton. Shale-barren g. Much like no. 4, even more closely tomentose, but the upper lf-surfaces commonly soon glabrate; basal lf-blades mostly elliptic or obovate, 1–4 cm × 6–20 mm, shallowly and rather remotely dentate to occasionally entire; cauline lvs few and much reduced, linear and entire or pinnatifid with linear segments; achenes hirtellous on the angles; $2n=46$. Shale-barrens; s. Pa. to Va. and W.Va. Apr.–June.

6. Senecio tomentosus Michx. Southern woolly g. Perennial 2–7 dm from a short caudex, sometimes also stoloniferous, persistently floccose-tomentose until flowering time or later, generally very densely so at the base, the upper lf-surfaces sometimes soon glabrate; basal lf-blades chiefly lance-ovate to elliptic or ovate and abruptly contracted to the petiole, to 20 × 5 cm, crenate or subentire; cauline lvs conspicuously reduced upwards, becoming sessile, entire or crenate, scarcely pinnatifid; heads several or rather many, the disk 7–12 mm wide; invol 4–6 mm; achenes hispidulous; $2n=46$. Dry open places and pine woods, especially in sandy soil; s. N.J. to Fla. and Tex., chiefly on or near the coastal plain. Apr.–June.

7. Senecio plattensis Nutt. Platte g. Biennial or short-lived perennial 2–7 dm from a short caudex, sometimes also stoloniferous, ± persistently floccose-tomentose until flowering time or later, at least as to the stem, lower lf-surfaces, and invols; basal lvs narrowly or broadly elliptic or ovate to suborbicular or broadly oblanceolate, crenate-serrate or some deeply lobed, the blade and petiole to 10 × 3 cm; cauline lvs conspicuously reduced upwards, becoming sessile, ± pinnatifid; heads several or rather many, the disk 6–12 mm wide; invol 4–6 mm; achenes usually hirtellous; $2n=44, 46$. Prairies and other dry, open places; Great Plains, and irregularly e. to s. O., Tenn., N.C., and Va., mostly inland from the coastal plain. May–July. Passes into nos. 6 and 8.

8. Senecio pauperculus Michx. Northern meadow-g. Perennial 1–5 dm from a rather short caudex, occasionally also with very short slender stolons or superficial rhizomes, lightly floccose-tomentose when young, generally soon glabrate, except often at the very base and in the axils; basal lvs mostly oblanceolate to elliptic, generally tapering to the petiolar base, crenate or serrate to subentire, seldom over 12 × 2 cm; cauline lvs ± pinnatifid, the lower sometimes much larger than the basal, the others conspicuously reduced and becoming sessile; heads seldom more than 20; disk 5–12 mm wide; invol 4–7 mm, its bracts often purple-tipped; rays 5–10 mm (rarely wanting); achenes glabrous or hispidulous; $2n=40–92$, often 44 or 46. Meadows, moist prairies, streambanks, beaches and cliffs; Lab. to Alaska, s. to Ga. and Oreg., more common northward. May–July. (*S. balsamitae*.)

9. Senecio anonymus A. Wood. Appalachian g. Perennial 3–8 dm from a rather short caudex, densely and persistently woolly at base, otherwise soon glabrate except sometimes in the axils; basal lvs mostly elliptic-oblanceolate, tapering to the petiole, crenate or serrate, to 30 cm (petiole included) × 3.5 cm; cauline lvs deeply pinnatifid, reduced and becoming sessile upwards; heads numerous, commonly 20–100+, small, the disk 5–9 mm wide; invol 5–7.5 mm; rays 4–8 mm (rarely none); achenes usually hispidulous; $2n=44$. Meadows, pastures, roadsides, and dry woods; s. Pa. and s. O. to Fla., chiefly in the mts. May, June. (*s. smallii*.)

10. Senecio obovatus Muhl. Running g. Perennial 2–7 dm with well developed slender stolons or superficial rhizomes and generally a short caudex, lightly floccose-tomentose when young, generally soon glabrate; basal lvs narrowly obovate to orbicular, rounded at the tip, tapering or abruptly contracted to the petiolar base, crenate-serrate or sometimes (especially toward the base) more deeply cut, to 20 cm (petiole included) × 6 cm; cauline lvs conspicuously reduced, becoming sessile and generally pinnatifid; heads several or rather many, the disk 7–14 mm wide; invol 4–6 mm, conspicuously shorter than the disk, the bracts often purple-tipped; rays 5–10 mm (rarely wanting); achenes usually glabrous; $2n=44$, ca 90. Rich woods and rocky outcrops, especially in calcareous situations; Vt. to Fla., w. to Kans. and Tex., commoner southward. Apr.–June. (*S. rotundus*.)

11. Senecio indecorus Greene. Taller discoid g. Perennial 3–8 dm from a short, branching caudex, glabrous or soon glabrate except sometimes in the axils; lvs thin, the basal elliptic or broadly ovate, tapering or subtruncate at base, serrate, evidently petiolate, the blade to 6 × 4 cm; cauline lvs sharply incised-pinnatifid, the lobes irregularly again few-toothed, reduced and becoming sessile upwards; heads mostly 6–40, yellow, discoid or rarely with short rays; invol 7–10 mm, its bracts often purple-tipped; achenes glabrous; $2n=46$. Moist woods, streambanks, swales, and bogs; Que. to Alas., s. to n. Mich., Wyo., and Wash.; n. Calif. July, Aug. (*s. discoideus*, misapplied.)

12. Senecio schweinitzianus Nutt. New England g. Blades of basal lvs lanceolate or lance-ovate, acute or acutish, subtruncate or shallowly cordate at base, sharply and generally rather finely toothed, up to 8 × 5 cm, commonly 1.75–3.5 times as long as wide; otherwise much like no. 14; $2n=46$. Moist meadows and swampy woods; N.S. and Que. to n. N.Y.; disjunct on and near Roan Mt., N.C.–Tenn. May–Aug. (*S. robbinsii*.)

13. Senecio pseudaureus Rydb. Western heart-lvd g. Stems solitary or few from a simple or weakly branched, short, mostly erect or ascending rhizome-caudex with many fibrous roots; basal lf-blades mostly 2–4 × 1–2 cm, seldom larger, ovate to broadly lanceolate, subcordate to subtruncate at base, rather bluntly toothed, normally held erect, at right angles to the ground; otherwise much like no. 14; $2n=46$. Moist prairies and other open places; widespread in the w. cordillera, e. to s. Man., w. Minn., and nw. Mo. May–July. Ours is var. *semicordatus* (Mackenzie & Bush) T. M. Barkley.

14. Senecio aureus L. Heart-lvd g. Perennial 3–8 dm from a branched, rhizomatous caudex or creeping rhizome, often with stoloniform but coarse, leafy basal offshoots, lightly floccose-tomentose when young, soon essentially glabrous; basal lvs long-petiolate, some or all with ± strongly cordate blade, these generally broadly rounded distally, crenate, up to 11 × 11 cm, mostly 0.75–1.5(–1.75) times as long as wide, often held at an angle to the petiole, parallel to the ground; cauline lvs conspicuously reduced and generally ±

pinnatifid, becoming sessile upward; heads several or rather many, the disk 5–12 mm wide; invol 5–8 mm, its bracts often strongly purple-tipped; rays 6–13 mm; achenes glabrous; $2n$=44, 66, ca 132. Moist woods and swampy places; Lab. to Minn., s. to N.C., n. Ga., and c. Ark.; Fla. Apr.–Aug. (*S. gracilis.*)

S. clivorum Maxim., a showy oriental sp. with large, reniform to cordate basal lvs and large heads with rays 3–4 cm, is locally established as an escape from cult. in Md. (*Ligularia c.*)

15. **Senecio glabellus** Poir. Yellowtop. Fibrous-rooted, mostly single-stemmed annual or winter-annual 1.5–8 dm, glabrous or obscurely floccose-tomentose in the axils; lvs mostly once pinnatifid (or in part twice pinnatifid) with generally rounded teeth and lobes, or the basal sometimes merely round-toothed, the largest ones at or near the base of the stem, up to 20 × 7 cm, progressively reduced upward; heads ± numerous, the disk 5–10 mm wide; invol 4–6 mm, its bracts ca 21 or 13; rays 5–12 mm; achenes minutely hairy or glabrous; $2n$=46. Moist, open or shaded places, often a weed in low fields; N.C. and Ky. to Fla., w. to S.D. and Tex., and adventive n. to s. Ont. May–July.

16. **Senecio jacobaea** L. Tansy-ragwort. Biennial or short-lived perennial 2–10 dm ± fibrous-rooted to evidently taprooted, simple to the infl, thinly floccose-tomentose at first, generally glabrous or nearly so by anthesis, except often in the infl; lvs equably distributed, mostly 2–3 times pinnatifid with ± pointed segments or teeth, 4–20 × 2–6 cm, the lower petiolate and often deciduous, sometimes less dissected than the others, the upper becoming sessile; heads ± numerous in a short broad infl; disk 7–10 mm wide; invol 4 mm, its bracts ca 13, over 1 mm wide, generally dark-tipped; rays 4–8 mm; achenes minutely hairy; $2n$=40, 80. Native of Europe, now established as a weed in dry soil and waste places in parts of n. U.S. and adj. Can., in our range chiefly from N.J. to se. Can. July–Oct.

17. **Senecio sylvaticus** L. Woodland-g. Annual 1.5–8 dm from a ± evident taproot, leafy throughout, generally simple to the infl, sparsely or moderately pubescent with crisp loose hairs, or subglabrous; lvs all ± pinnatifid and irregularly toothed, 2–12 cm × 4–40 mm; heads several or many, the disk 3–7 mm wide; invol 5–7 mm, its principal bracts ca 13; bracteoles inconspicuous or wanting, not black-tipped; rays reduced and inconspicuous, less than 2 mm; pappus very copious, equaling or surpassing the disk-fls; achenes strigil-lose-canescent; $2n$=40. A weed in disturbed soil and waste places; native of Europe, established here and there in the n. part of our range, especially in n. N. Engl. July–Sept.

18. **Senecio viscosus** L. Sticky g. Strong-scented annual 1–6 dm from a taproot, copiously and conspic-uously glandular-hairy throughout; lvs equably distributed, pinnatifid and toothed, 3–12 × 1–5 cm, the lower petiolate; heads less crowded than in nos. 17 and 19, the disk 7–15 mm wide; invol 6–8 mm, with ca 21 or seldom only 13 principal bracts and generally a few well developed but not black-tipped bracteoles; rays short but evident and loosely spreading in life, in herbarium specimens becoming rather circinate and inconspicuous, scarcely exceeding the pappus, which generally surpasses the disk-fls; achenes glabrous or nearly so; $2n$=40. Native of Europe, now ± established as a weed in waste places from Que. and N.S. to Pa. and N.J., near the coast. July–Sept.

19. **Senecio vulgaris** L. Common g. Annual 1–4 dm from a ± evident taproot, leafy throughout, sparsely crisp-hairy or subglabrous; lvs coarsely and irregularly toothed to more often pinnatifid, 2–10 × 0.5–4.5 cm, the lower tapering to the petiole or petiolar base, the upper sessile and clasping; heads several or many, discoid, the fls all tubular and perfect; disk 5–10 mm wide; invol 5–8 mm, with ca 21 principal bracts and some short but well developed, black-tipped bracteoles; pappus very copious, from a little shorter than to equaling or generally surpassing the cors; achenes short-hairy chiefly along the angles; $2n$=40. A weed in disturbed soil and waste places; native of the Old World, now widely distributed in n. temp. regions, and throughout our range. May–Oct.

44. ERECHTITES Raf.

Heads disciform, whitish (in ours) or dull yellow; invol a single series of narrow, equal, ± herbaceous bracts, sometimes with a few bracteoles at base; receptacle flat, naked; outer fls pistillate, filiform-tubular, eligulate, in 2–several series; inner fls perfect but sometimes sterile, the cor narrowly tubular, 4–5-toothed; style-branches flattened, truncate or with a very short hairy appendage; achenes 5-angled or 10–20-nerved; pappus of numerous slender bristles; erect herbs with alternate, entire to pinnately dissected lvs and cylindric to ovoid heads. 12, N. and S. Amer. and Australia.

1. **Erechtites hieracifolia** (L.) Raf. Fireweed. Fibrous-rooted annual weed 0.1–2.5 m, glabrous or some-times ± spreading-hairy throughout; lvs to 20 × 8 cm, sharply serrate with callous-tipped teeth, sometimes also irregularly lobed, the lower oblanceolate to obovate, tapering to a short petiolar base, the middle and upper more elliptic, lanceolate, or oblong, and, especially in robust specimens, often auriculate-clasping; heads in a flat-topped or elongate infl, or solitary in depauperate plants, turbinate-cylindric; invol 1–1.5 cm; pappus copious, bright white, eventually deciduous; $2n$=40. Various habitats, including dry woods, marshes, and waste places; Nf. to Fla., w. to Nebr. and Tex., and intr. elsewhere. Aug., Sept. Most of our plants are var. *hieracifolia*, with the denuded receptacle of old heads 5–8 mm wide and the achenes 2–3 mm, with mostly 10–12 ribs. The var. *megalocarpa* (Fern.) Cronq., a well marked ecotype of saline coastal marshes from Mass. to N.J., is somewhat fleshy, with slightly larger heads that are more conspicuously swollen at base (the denuded receptacle 9–12 mm wide), and with larger achenes, these 4–5 mm, 16–20-ribbed.

45. CACALIA L. Indian plantain. Heads discoid, cylindric or nearly so, white or flesh-colored, not yellow; invol a single series of herbaceous or subherbaceous bracts, often with scarious margins, sometimes also with some reduced basal bracteoles; receptacle naked, flat, or sometimes with a conic central projection; fls all perfect, the cor deeply 5-lobed; style-branches flattened, truncate or with a very short appendage, penicillate, the ventromarginal stigmatic lines expanded and confluent on the inner surface; achenes cylindric, several-nerved; pappus of numerous capillary bristles; perennial herbs with alternate, petiolate lvs and medium-sized heads, sometimes with ± milky juice. (*Mesadenia, Synosma.*) 50, N. Amer., 1 Eurasia.

1 Heads many-fld; principal invol bracts ca 13; larger lvs hastate .... 1. *C. suaveolens.*
1 Heads 5-fld; invol bractsx 5; lvs not hastate.
  2 Larger lvs palmately veined, the principal nerves divergent.
    3 Lvs green on both sides; stem conspicuously angled and sulcate, not glaucous .... 2. *C. muhlenbergii.*
    3 Lvs pale and glaucous beneath; stem glaucous, terete or slightly striate .... 3. *C. atriplicifolia.*
  2 Lvs strongly several-nerved, the nerves longitudinal, converging distally .... 4. *C. plantaginea.*

    1. Cacalia suaveolens L. Hastate I.-p. Perennial 1–2.5 m with long, fleshy-fibrous roots, glabrous or nearly so; stem striate or grooved, simple to the infl, leafy; middle and lower lvs triangular-hastate, 5–20 cm, nearly or quite as wide, sharply toothed, conspicuously petioled; upper lvs progressively less hastate and with shorter, more winged petioles; heads in a flat-topped infl, commonly 20–40-fld; the disk 7–11 mm wide; invol 1 cm, its principal bracts 10–15, commonly with some reduced but conspicuous, loose, subulate outer ones; receptacle flat, deeply pitted; 2*n*=40. Riverbanks and moist low ground; R.I. and Conn., w. to se. Minn., s. to Md., Ky., Tenn., Ill., and in the mts. to Ga. July–Oct. (*Synosma s.*)

    2. Cacalia muhlenbergii (Schultz-Bip.) Fern. Great I.-p. Stout, fibrous-rooted, glabrous perennial 1–3 m; stem conspicuously 6–8-angled and sulcate; lvs green on both sides, irregularly dentate and sometimes shallowly lobed, commonly ± ciliolate in the sinuses, palmately veined, the lower very large and long-petioled, to 8 dm wide, often reniform, the upper reduced and more flabellate or ovate; heads numerous in a short and broad, flat-topped infl, 5-fld, narrowly cylindric, the disk 3–7 mm wide as pressed; invol 7–12 mm, its principal bracts 5, sometimes with a few minute basal bracteoles; receptacle with a short conic projection in the center; 2*n*=50. Open woods; N.J. and Pa. to Minn. and Mo., s. to Ga. and Miss. June–Sept. (*C. reniformis; Mesadenia r.; M. muhlenbergii.*)

    3. Cacalia atriplicifolia L. Pale I.-p. Much like no. 2, and hybridizing with it; stem glaucous, terete or slightly striate; lvs pale and glaucous beneath, averaging smaller, proportionately longer, and more pointed (the lower triangular-ovate), with fewer and larger teeth, or merely shallowly lobed, the sinuses smooth; 2*n*=50–56. Woods and moist or rather dry open places; N.J. and Pa. to s. Minn., s. to Ga., w. Fla., and Okla. (*Mesadenia a.*)

    4. Cacalia plantaginea (Raf.) Shinners. Tuberous I.-p. Stout, glabrous perennial 6–18 dm from a short, tuberous-thickened base and fleshy-fibrous roots; stem striate-angled; lvs thick and firm, entire or nearly so, with several prominent longitudinal nerves converging toward the summit; basal and lowermost cauline lvs conspicuously long-petioled, the blade 6–20 × 2–10 cm, commonly elliptic and tapering to the base, or deltoid-ovate and subtruncate in robust specimens; cauline lvs few, conspicuously reduced upwards, becoming sessile or subsessile; otherwise as in no. 2; 2*n*=54. Wet prairies and marshy or boggy places; O. and s. Ont. to Minn. and e. S.D., s. to Ala. and Tex. June–Aug. (*C. tuberosa; Mesadenia t.*)

46. TUSSILAGO L. Coltsfoot. Heads radiate, the rays yellow, pistillate and fertile, numerous in several series; invol a single series of equal, ± herbaceous bracts, sometimes with a few basal bracteoles; receptacle flat, naked; central fls sterile, with undivided, merely lobed style; anthers entire or minutely sagittate; achenes linear, 5–10-ribbed; pappus of numerous capillary bristles, that of the sterile fls ± reduced; perennial herb with large basal lvs, scaly-bracted stems (the bracts alternate), and medium-sized solitary heads, blooming before the basal lvs develop. Monotypic.

    1. Tussilgao farfara L. Rhizomatous perennial 0.5–5 dm, the cauline bracts ca 1 cm; lvs long-petioled, cordate to suborbicular, with deep narrow sinus, callous-denticulate and shallowly lobed, 5–20 cm long and wide, glabrous above, persistently white-tomentose beneath; heads at first cylindric, expanding with maturity, the disk up to 3 cm wide; invol 8–15 mm; rays very numerous, narrow, not much exceeding the invol and pappus; 2*n*=60. Native of Eurasia, naturalized in disturbed and waste places in ne. U.S. and adj. Can., w. as far as Minn., and s. to Md., Va., and Ky. Apr.–June.

47. PETASITES Miller. Subdioecious; heads of male plants with the fls all tubular and functionally staminate, or sometimes with a few outer pistillate fls like those of the female heads; heads of female plants with the fls all pistillate and fertile, or sometimes with a few central functionally staminate fls; pistillate fls with tubular-filiform cor, the outer ones sometimes provided also with a ligule; invol a single series of equal, ± herbaceous bracts, sometimes with a few basal bracteoles; receptacle flat, naked; achenes linear, 5–10-ribbed; pappus of numerous capillary bristles, elongating in fr, that of the sterile fls ± reduced; ± white-tomentose or woolly perennial herbs with large basal lvs, merely bracteate stems (the bracts alternate), and several or many medium-sized, purple, white, or rarely yellowish heads, flowering before the lvs expand. 15, N. Hemisphere.

1 Fls whitish, the marginal ones of the pistillate heads radiate; native.
  2 Lvs evidently lobed . . . . 1. *P. frigidus.*
  2 Lvs merely dentate or even subentire . . . . 2. *P. sagittatus.*
1 Fls usually purple, all rayless; intr. . . . . 3. *P. hybridus.*

    1. Petasites frigidus (L.) Fries. Northern sweet coltsfoot. Rhizomatous perennial; basal lvs long-petioled, glabrous or sparsely hirsute above, loosely white-tomentulose beneath, large, 5–40 cm wide, palmately or pinnipalmately veined and lobed and also ± toothed; stem 1–5 dm, with parallel-veined bracts 2.5–6 cm, the lower sometimes with an abortive blade at the end; heads campanulate in a corymbiform or racemiform infl; invol 5–9 mm; fls whitish, the outer 10–25 in pistillate heads with short rays 2–7 mm; $2n=60(-62)$. Meadows, swampy places, and moist woods; circumboreal, s. in Amer. to Mass., Mich., Minn., and Calif. Apr.–July. Nearly all of our plants belong to var. *palmatus* (Aiton) Cronq., with the lvs palmately cleft well over half way to the base. (*P. palmatus.*) A few plants from n. Minn. and n. Mich. (and northward) with the lvs palmately or pinnipalmately cleft not more than about half way to the base, may reflect hybridization with no. 2. The name *P. × vitifolius* Greene is available for these.

    2. Petasites sagittatus (Banks) A. Gray. Arrowhead sweet-coltsfoot. Much like no. 1, but the basal lvs triangular-ovate, cordate or more commonly sagittate at base, to 3 dm, pinnipalmately veined, merely dentate, with 20–45 teeth on each side, varying to merely a little wavy and callous-denticulate; bracts of the stem averaging longer and narrower, more often with abortive blade; rays mostly a little longer, commonly 8–9 mm; $2n=58-60$. Wet places; Lab. to Alas., s. to Wis., Minn., S.D., and Wash. May, June.

    3. Petasites hybridus (L.) Gaertner, Meyer & Scherb. Butterfly-dock. Perennial from a tuberous-thickened base; basal lvs large, long-petioled, reniform or cordate, to 3+ dm wide, callous-dentate or -denticulate, glabrous or nearly so above, arachnoid-tomentose and eventually glabrate beneath; stem 1–4 dm, or to 1 m in fr; heads numerous in a narrow, racemiform infl, the fls purple (rarely whitish), all rayless; invol of male plants 6–8 mm, of female ones 3–4 mm; $2n=60$. Native of Europe, escaped from cult. in moist waste places; Mass. to Del. and Pa. Apr., May.

    48. ADENOCAULON Hook. Invol small, of less than 10 nearly equal green bracts; heads disciform; fls whitish, tubular, the outer 3–7 pistillate, the inner as many or a few more, staminate, with undivided style; receptacle naked; anthers sagittate; pappus none; herbs with large alternate lvs and ample subnaked infl, the branches and achenes ± stipitate glandular. 4, the others in Guatemala, Chile, and e. Asia.

    1. Adenocaulon bicolor Hook. Trail-plant. Fibrous-rooted slender perennial to 1 m; lvs mostly near the base, long-petiolate, large and thin, deltoid-ovate to cordate or subreniform, 3–15 cm wide, essentially glabrous above, closely white-woolly beneath, entire to more often coarsely toothed or shallowly lobed; invol bracts 2 mm or less, reflexed in fr and eventually deciduous; achenes clavate, 5–8 mm; $2n=46$. Moist, shady woods; n. Minn. and n. Mich.; s. B.C. to Calif. and Mont. May–Aug.

    49. HAPLOPAPPUS Cass., nom. conserv. Heads radiate (in ours), the rays yellow, pistillate (in ours); invol bracts ± numerous, subequal to imbricate, generally greenish at least distally; receptacle flat or somewhat convex, naked; disk-fls perfect but sometimes sterile; style-branches flattened, with ovate to subulate, sometimes very long, externally short-hairy appendage; achenes 4–5-angled or striate; pappus of numerous capillary bristles, these unequal but not divided into distinct long inner and short outer series; taprooted herbs (ours) or shrubs with alternate, simple to pinnatifid lvs and medium-sized to rather large heads. (*Aplopappus, Isopappus, Prionopsis, Sideranthus.*) 150, New World.

1 Heads large, the disk 1.5–3 cm wide; achenes glabrous . . . . 1. *H. ciliatus.*
1 Heads smaller, the disk up to 1.5 cm wide; achenes strongly hairy.
  2 Perennial; lvs pinnatifid . . . . 2. *H. spinulosus.*
  2 Annual; lvs merely spine-toothed or even entire . . . . 3. *H. divaricatus.*

1. **Haplopappus ciliatus** (Nutt.) DC.   Wax-goldenweed. Glabrous annual or biennial 4–15 dm, commonly simple to near the top, leafy throughout, the lvs sessile, clasping, oblong to ovate or elliptic-obovate, rounded or obtuse at the tip, mostly 3–8 × 1–4 cm, spinose-dentate; heads few, large, the disk 1.5–3 cm wide; invol bracts slightly or moderately imbricate, the outer ones (at least) with long, loose or spreading green tip, at least the larger bracts generally well over 1 mm wide; rays ca 25–50, 1–2 cm; achenes glabrous, the central ones sterile; pappus rather scanty, eventually deciduous; 2n=12. Open or waste places; w. Mo. to La., Tex., and N.M., and intr. e. to Ill. Aug., Sept. (*Prionopsis c.*)

2. **Haplopappus spinulosus** (Pursh) DC.   Cutlf-ironplant. Several-stemmed perennial 2–6 dm, thinly to-mentose at least when young, sometimes later subglabrate; lvs 1.5–6 cm × 2–10 mm, pinnatifid, with narrow lobes and narrow rachis less than 3 mm wide; achenes hairy, the nerves obscure; pappus persistent; 2n=16 + 2 B. Dry plains and foothills; Minn. to Alta. and Mex. July–Sept. Our plants are var. *spinulosus*. (*Sideranthus s.*)

3. **Haplopappus divaricatus** (Nutt.) A. Gray.   Annual weed 2–10 dm, generally much branched above, subglabrous to more often coarsely scabrous or spreading-hairy, sometimes also glandular; lvs oblanceolate or linear-oblanceolate, with a few spiny teeth or occasionally entire, 3–7 cm × 3–10 mm, or the lower sometimes larger; heads numerous in a diffuse infl, the disk less than 1 cm wide; invol 5–7 mm, its bracts well imbricate, 1 mm wide or less; rays 7–18, 4–6 mm; achenes sericeous-canescent; pappus persistent; 2n=8, 10, 12, 14. Waste places, especially in sandy soil; Va. to Fla., w. to Tex., and inland to Okla. and Kans. Aug.–Nov. (*Isopappus d.*)

50. CHRYSOPSIS (Nutt.) Elliott, nom. conserv.   Golden aster. Heads radiate (in ours), the rays yellow, pistillate and fertile; invol bracts imbricate, greenish at least in part; receptacle flat or a little convex, naked; disk-fls numerous, perfect and fertile; style-branches flattened, with externally short-hairy, mostly elongate appendage; achenes ± compressed, several- to many-nerved; pappus double, the inner of elongate capillary bristles, the outer of short, coarse bristles or scales; herbs with alternate, entire or incon-spicuously toothed lvs and medium-sized or rather large heads. 20, N. Amer. (*Pityopsis.*)

1 Lvs elongate, linear, parallel-veined; achenes ± linear; heads campanulate.
  2 Plants 3–9 dm; middle lvs evidently smaller than the lower; Del. and s. . . . . 5. *C. graminifolia.*
  2 Plants 1–3.5 dm; middle lvs as large as the lower; Mass. to N.J. . . . . 6. *C. falcata.*
1 Lvs mostly broader than linear, in any case pinnately veined; achenes ± obovate; heads hemispheric or nearly so.
  3 Herbage evidently woolly or floccose with long, slender, flexuous hairs at least when young, later sometimes glabrous; eastern spp., not weedy.
    4 Peduncles and invols glandular, not hairy; herbage ± glabrate in age . . . . 3. *C. mariana.*
    4 Peduncles and usually also the invols ± woolly; herbage permanently woolly . . . . 4. *C. gossypina.*
  3 Herbage variously hairy but not woolly, the pubescence persistent; western or weedy.
    5 Taprooted, not rhizomatous, mostly 2–4 dm; lvs mostly entire . . . . 1. *C. villosa.*
    5 Rhizomatous as well as taprooted, mostly 4–10 dm; lvs mostly few-toothed . . . . 2. *C. camporum.*

1. **Chrysopsis villosa** (Pursh) Nutt.   Taprooted, several-stemmed perennial 2–4(–5) dm, ± pubescent with long or short, appressed or spreading hairs; lvs numerous, nearly alike, oblong-elliptic to linear-oblanceolate, acute to rounded, seldom over 5 × 1 cm, mostly entire, the lower tending to be short-petiolate, the lowermost deciduous; heads several, the invol 5–10 mm, ± strigose or hirsute and sometimes also glandular, its bracts regularly imbricate, gradually tapering to a point, often purple-tipped; rays mostly 10–35, 6–10 mm, achenes narrowly obovate, 3–5-nerved; 2n=18, 36. Dry, open, often sandy places; Minn. to Mo., w. to B.C., Calif., and Tex. July–Oct. (*Heterotheca v.*) Two or more vars. with us, others farther w.

a Heads small, the invol 6–7 mm, the disk less than 1.5 cm wide; lvs narrow, 5 mm wide or less; southwestern, barely or scarcely reaching our range.
  b Lvs silvery-silky with fine, loosely appressed hairs; invol copiously silky-strigose. (*C. berlandieri.*)
    . . . . var. *canescens* A. Gray.
  b Lvs more coarsely and sparsely hairy, the hairs thickened toward the base; invol sparsely to densely strigose. (*C. angustifolia.*) . . . . var. *angustifolia* (Rydb.) Cronq.
a Heads larger, the invol 8–11 mm, the disk often over 1.5 cm wide; lvs broader, commonly over 5 mm wide.
  c Lvs relatively broad, mostly oblong to elliptic or obovate and scarcely petioled, generally some rather closely subtending the heads; Minn. to Mont. and Ariz. (*C. foliosa.*) . . . . var. *foliosa* (Nutt.) D. C. Eat.
  c Lvs narrower, more oblanceolate or spatulate, and mostly subpetiolate; heads generally more evi-dently pedunculate; nearly the range of the sp. (*C. bakeri; C. ballardii.*) . . . . var. *villosa.*

2. **Chrysopsis camporum** Green.   Prairie g. a. Like a robust form of no. 1, 4–10 dm, but with slender creeping rhizomes as well as a taproot; rather coarsely and often only thinly hairy; lvs 3–7 cm × 8–20 mm, generally with a few sharp, small teeth; invol 8–11 mm; disk mostly 1.2–2.5 cm wide; rays ca 21 to ca 34, ca 1 cm; achenes 6–12-nerved; 2n=36. (*C. villosa* var. *c.*) Typically a prairie sp. of Mo. to Ind. and Ill., but

recently intr. and becoming abundant as a weed along roadsides and in fields in se. U.S., well established in Va., and to be expected elsewhere along the s. margin of our range. The weedy form is more glandular and has been described as var. *glandulissima* (Semple) Cronq. July–Sept.

3. Chrysopsis mariana (L.) Elliott.   Shaggy g. a. Fibrous-rooted perennial from a short, woody rhizome or branching caudex, 2–8 dm, loosely shaggy-villous with long, flexuous hairs when young, later often ± glabrate; lvs basally disposed, the lower oblanceolate to ovate, petiolate, to 18 × 3.5 cm (petiole included), generally denticulate; middle and upper lvs smaller, lanceloate to oblong or elliptic, sessile; heads in an often congested infl, nearly hemispheric, the disk 1–2 cm wide; invol and peduncles stipitate-glandular, the invol 7–10 mm; rays ca 13 to ca 21, ca 1 cm; achenes obovate, 3–5-nerved; 2*n*=8, 16, 24, 32, but only 24 with us. Woods and sandy places; s. N.Y. to s. O. and e. Ky., s. to Fla. and La. Aug.–Oct. (*Heterotheca m.*)

4. Chrysopsis gossypina Nutt.   Woolly-headed g. a. Weakly taprooted to fibrous-rooted biennial or more often short-lived perennial 3–8 dm, often decumbent or ascending, floccose-woolly throughout; lvs numerous, oblanceolate or oblong-elliptic, obtuse or rounded, ± entire, the lower up to ca 6 cm × 18 mm, gradually reduced upward; heads few in an open infl, hemispheric, the disk 1.2–2.5 cm wide; invol 7–11 mm; rays mostly ca 34, and ca 1 cm; achenes narrowly obovate; 2*n*=18. Sandy places, often with pine or scrub oak; coastal plain from Va. to Fla. and Ala. Sept., Oct. (*C. longii.*)

5. Chrysopsis graminifolia (Michx.) Elliott.   Grass-lvd g. a. Fibrous-rooted perennial 3–9 dm, often with slender, stoloniform rhizomes, ± silvery-silky with appressed hairs, at least below; lvs basally disposed, the larger ones elongate, parallel-veined, grass-like, to 35 × 2(–3.5) cm, the others numerous and progressively reduced; heads several or many, terminating the branches, turbinate or campanulate; achenes linear. Sandy, usually dry places; Del. to s. O., s. to Fla., Bahama Isl., Tex., and Guat. Aug., Sept. Two well marked vars. with us:

Var. *latifolia* Fern.   Peduncles and invol silky, not glandular, or the invol glandular above; invol 8–12 mm; rays mostly ca 13, 7–12 mm; disk-fls 30–50, the dry cors 6.5–9 mm; 2*n*=36. Nearly the range of the sp. (*Heterotheca g.; Pityopsis g.*)

Var. *aspera* (Shuttlew.) A. Gray.   Peduncles, as well as the invol, stipitate-glandular, scarcely silky; heads avg smaller, invol 5–8(–10) mm; rays more often 8 than 13, mostly 4–8 mm; disk-fls mostly 15–30, the dry cors 4–6.5 mm; 2*n*=18, 36. Va. to n. Fla. and s. Miss., commonest on the piedmont. (*Heterotheca a.; Pityopsis a.*)

The var. *graminifolia,* mostly diploid, with the heads of var. *aspera* and the pubescence of var. *latifolia,* occurs mainly on the coastal plain from N.C. to Fla. and Tex. (var. *microcephala.*)

6. Chrysopsis falcata (Pursh) Elliott.   Falcate g. a. Cespitose, fibrous-rooted perennial 1–3.5 dm, branched above; stem white-woolly with usually loose hairs, or glabrate in age; lvs crowded, linear, ± arcuate, glabrous or glabrate, parallel-veined, often folded, to 10 cm × 5 mm, or the lowermost oblanceolate and a little wider, but no longer than the others; heads several, terminating the branches, campanulate; invol 5–8 mm, slightly white-hairy; achenes linear-fusiform; 2*n*=18. Dry sandy soil, mostly near the coast; Mass. to N.J. July–Sept.

51. HETEROTHECA Cass.   Golden aster. Much like *Chrysopsis,* but the ray-achenes thick, commonly 3-angled, glabrous or nearly so, without pappus; annual or biennial, hairy herbs. 3–5, N. Amer.

1. Heterotheca subaxillaris (Lam.) Britton & Rusby.   Camphor-weed. Taprooted, weedy annual or biennial 2–25 dm, erect or in some habitats low and spreading, glandular, especially above, and spreading-hairy; lvs ovate or oblong, dentate or subentire, the lower ones petiolate (some with a stipule-like basal expansion of the petiole), deciduous, the middle and upper sessile and clasping; invol 6–8 mm, its bracts finely glandular, and usually slightly hairy at the tip; 2*n*=18. Dry, often sandy places; Del. to Fla., w. to Kans. and Mex., and adventive n. to L.I. July–Sept.

52. GRINDELIA Willd.   Gum-weed. Heads radiate or occasionally discoid, the rays mostly 12–45, yellow, pistillate and fertile; invol ± resinous, its bracts herbaceous-tipped, imbricate or subequal; receptacle flat or convex, naked; disk-fls yellow, the inner and often also the outer sterile; style-branches flattened, with externally hairy, lance-linear or occasionally very short appendages; achenes compressed to subquadrangular, scarcely nerved; pappus of 2–several firm, deciduous, often minutely serrulate awns; taprooted herbs, sometimes woody at the base, with alternate, ± resinous-punctate lvs and relatively large, hemispheric heads, the disk 1–3 cm wide. 50, New World.

1 Invol bracts loose but not squarrose, not markedly imbricate . . . . 1. *G. lanceolata.*
1 Invol bracts distinctly squarrose-reflexed, markedly imbricate . . . . 2. *G. squarrosa.*

1. Grindelia lanceolata Nutt.   Spiny-tooth g.-w. Short-lived, monocarpic perennial sometimes sparsely hairy below, 3–15(–20) dm; lvs scarcely punctate, sharply serrate or serrulate with bristle-tipped teeth (seldom entire), acute to acuminate, the middle ones linear or lance-oblong, 4–11 cm × (2–)4–28 mm; disk 1–2 cm wide; invol bracts only slightly resinous, loose, not markedly imbricate; rays 15–30, 10–16 mm; achenes 4–

6 mm; pappus-awns mostly 2, entire; $2n=12$. Dry, open places, often on limestone; c. Tenn., s. Ind., and s. Ill. to Kans., Tex., and N.M. June–Sept. Ours is var. *lanceolata*.

2. **Grindelia squarrosa** (Pursh) Dunal. Curly-top g.-w. Glabrous biennial or short-lived perennial 1–10 dm, branched at least above; lvs abundantly punctate; heads several or many, the disk 1–2 cm wide; invol bracts strongly resinous, imbricate in several series, the green tips (especially of the outer) squarrose-reflexed; rays 25–40, 7–15 mm, rarely absent; achenes 2–3 mm; pappus-awns 2–8, finely serrulate to subentire; $2n=12$, 24. Open or waste places; Man. and Minn. to B.C., Nev., and Tex., and widely intr. elsewhere. July–Sept. The southwestern var. *nuda* (A. Wood) A. Gray, with discoid heads, is not a regular member of our flora. Three other vars., all radiate, occur with us:

a Lvs closely and evenly serrulate or crenate-serrulate.
  b Upper and middle lvs 2–4 times as long as wide, most ovate or oblong .... var. *squarrosa*.
  b Upper and middle lvs 5–8 times as long as wide, mostly linear-oblong or oblong to oblanceolate
  .... var. *serrulata* (Rydb.) Steyerm.
a Lvs entire or remotely serrulate, or, especially the lower, coarsely and irregularly toothed or pinnatifid.
  (*G. perennis*.) .... var. *quasiperennis* Lunell.

53. SOLIDAGO L.   Goldenrod. Heads radiate, the rays pistillate and fertile, yellow or in 2 spp. white; invol bracts usually ± imbricate in several series, chartaceous at base, commonly with ± herbaceous green tip; receptacle small, flat or a little convex, alveolate, naked or seldom with a few phyllary-like bracts near the margin; disk-fls perfect and fertile, yellow, seldom more than 25(–60); style-branches flattened, with lanceolate, externally hairy appendage; achenes several-nerved, subterete or angled; pappus of numerous equal or sometimes unequal, capillary (in one sp. short and firm) bristles, usually white; fibrous-rooted perennial herbs with simple, alternate, entire or variously toothed lvs and few to many, mostly rather small, campanulate to subcylindric heads; $x=9$. Nearly 100, mainly N. Amer. (*Oligoneuron, Unamia*.) Hybrids are often found. All our spp. bloom in mid- or late summer and fall, *S. juncea* being one of the earliest, and *S. speciosa* one of the latest. Several spp. may have one or two B-chromosomes in addition to the number given.

Goldenrods can be divided into several groups on the basis of three sets of characters that are independently distributed with respect to each other: the nature of the underground parts, the nature and distribution of the lvs, and the nature of the infl. Several terms explained below are used without further comment in the key and descriptions. In many spp. the rhizome is very short, stout, and densely rooting, sometimes being more nearly a caudex than a proper rhizome. Such spp. often have several stems clustered together, although the stems may also be solitary. In other spp. the rhizome is elongate and generally more slender, and the stems are usually scattered. Some few spp. have both a short, stout rhizome or caudex and more elongate, slender, sometimes stoloniform rhizomes.

In many spp. the lvs are *basally disposed*. The radical and lowermost cauline lvs are relatively large and usually ± persistent, with the blade either gradually or abruptly contracted to a definite petiole. The cauline lvs (either numerous or often few) are progressively reduced and less petiolate upward, those near and above the middle of the stem usually being sessile or nearly so and of different shape from those below, often relatively as well as actually narrower. In other spp. the lvs are *chiefly cauline*. The radical lvs are small and relatively inconspicuous, or more often wanting. The lowermost cauline lvs are reduced and generally soon deciduous, so that the stem appears to be naked toward the base at flowering time. The largest leaves are somewhat above the base but evidently below the middle of the stem. The middle and upper lvs are gradually reduced but essentially similar in shape to the larger ones. Most spp. of this group have numerous, sessile or subsessile lvs. A few spp. fall between the two habit types, or range from one to the other. Measurements of lf-length in the descriptions include the blade and petiole, unless only the blade is specified.

Infls are mostly of 3 general types. In one group the heads are in *axillary* clusters or in a terminal, ± elongate *thyrse* that is straight, cylindrical, and not at all secund, or the infl consists of several such thyrsoid branches. In another group the infl is *paniculiform*, with at least the lower branches recurved-secund, or is slender and elongate (sometimes *racemiform*) and ± one-sided, or at least nodding at the tip. In a third group the infl is short and broad, flat or round-topped, but not at all secund, and is said to be *corymbiform*.

Key to the Groups of Species

Group I

Group II

Group III

2 Invol bracts appressed and obtuse or broadly rounded; widespread sp. .... 11. *S. speciosa.*
2 Outer invol bracts acute and tending to be squarrose-tipped; southwestern .... 22. *S. petiolaris.*
1 Achenes persistently short-hairy.
  3 Lvs ± strongly triple-nerved; rays 6–18.
    4 Herbage pubescent throughout with short spreading hairs; rays 6–9, mostly 3–5 mm; Great
      Plains, e. to Io. and w. Minn. .... 44. *S. mollis.*
    4 Herbage subglabrous or ± hairy, but not as above; rays 10–17; more eastern or northern.
      5 Perennial from a short stout caudex or crown, without creeping rhizomes .... 38. *S. calcicola.*
      5 Perennial from a creeping rhizome, without any well developed caudex .... 40. *S. canadensis.*
  3 Lvs not triple-nerved; rays 3–5(–8). (Extreme forms of *S. elliottii* might be sought here, except for
    the more numerous rays.)
    6 Lvs relatively broad, the blade mostly 1–2.2(–2.5) times as long as wide and abruptly contracted
      to the definite though usually winged petiole.
      7 Stem glabrous below the infl; widespread .... 16. *S. flexicaulis.*
      7 Stem conspicuously spreading-hairy; Ky. .... 17. *S. albopilosa.*
    6 Lvs narrower, mostly (2.2–)2.5–10 times as long as wide and tapering to the sessile or only
      obscurely short-petiolate base.
      8 Lvs mostly (2.2–)2.5–3(–3.5) times as long as wide; s. Appalachian .... 18. *S. flaccidifolia.*
      8 Lvs mostly 3–10 times as long as wide.
        9 Stem striate-angled, grooved, not glaucous; s. Appalachian region.
          10 Heads relatively small and delicate (see descr.); infl generally appearing more axillary
            than terminal ....19. *S. curtisii.*
          10 Heads relatively large and coarse (see descr.); infl generally appearing more terminal
            than axillary .... 20. *S. lancifolia.*
        9 Stem terete, glaucous; widespread .... 21. *S. caesia.*

GROUP IV

A single species .... 32. *S. sphacelata.*

GROUP V

1 Herbage minutely spreading-hirtellous throughout .... 23. *S. nemoralis.*
1 Stem (and often also the lvs) usually glabrous below the infl, or the stem occasionally with long
  scattered hairs.
  2 Lvs strongly scabrous above; stem ± strongly angled, at least below; bog-sp. .... 24. *S. patula.*
  2 Lvs glabrous, faintly scabrous, or occasionally somewhat hirsute above; stem terete or striate, not
    much angled; various habitats.
    3 Basal and lowermost cauline lvs gradually tapering to the petiole.
      4 Plants with long, slender, stoloniform rhizomes; coastal plain from N.J. s. .... 14. *S. stricta.*
      4 Plants without long rhizomes.
        5 Lvs somewhat succulent; plants mostly maritime, but sometimes inland .... 15. *S. sempervirens.*
        5 Lvs not succulent; plants not maritime.
          6 Infl mostly much longer than broad; plants of wet or poorly drained sites.
            7 Plants usually puberulent in the infl; achenes usually glabrous; middle and upper lvs
              not especially numerous; widespread .... 12. *S. uliginosa.*
            7 Plants glabrous even in the infl; achenes rather thinly hairy; middle and upper lvs ±
              numerous; Va. and W.Va. and s. .... 13. *S. gracillima.*
          6 Infl about as broad as long; upland plants.
            8 Heads (or many of them) with some slender, phyllary-like receptacular bracts internal
              to the ray-fls; rays mostly 7–13; disk-fls mostly 8–14.
              9 Lvs scarcely or not at all triple-nerved; basal lvs 2–7.5 cm wide; achenes persistently
                short-hairy .... 25. *S. juncea.*
              9 Lvs ± strongly triple-nerved; basal lvs often less than 2 cm wide (or wanting);
                achenes glabrous or sparsely hairy .... 26. *S. missouriensis.*
            8 Heads without receptacular bracts; rays 3–7; disk-fls 5–9 .... 27. *S. pinetorum.*
    3 Basal and lowermost cauline lvs rather abruptly contracted to the petiole.
      10 Lvs glabrous or occasionally strigose; disk-fls mostly 8–14.
        11 Plants with slender, stoloniform rhizomes in addition to the more deep-seated main
          rhizome or caudex .... 29. *S. tarda.*
        11 Plants without slender, stoloniform rhizomes .... 28. *S. arguta.*
      10 Lvs loosely hirsute at least on the midrib and main veins beneath; disk-fls mostly 4–7 ...... 30. *S. ulmifolia.*

GROUP VI

1 Lvs chiefly petiolate, with broadly ovate or elliptic blade 1.3–2 times as long as wide .... 31. *S. drummondii.*
1 Lvs sessile or nearly so, generally more than twice as long as wide.
  2 Lvs not triple-nerved.

3 Lvs minutely translucent-punctate, ordinarily anise-scented when bruised, all entire .... 33. *S. odora.*
3 Lvs not translucent-punctate, not anise-scented; usually at least some ± toothed.
  4 Perennial from a branched caudex, without creeping rhizomes .... 30. *S. ulmifolia.*
  4 Perennial from creeping rhizomes, without any well developed caudex.
    5 Stem ± hairy, at least above the middle; mostly not of swamps.
      6 Lvs narrow, the larger ones 2–7(–10) mm wide; stem uniformly puberulent at least above
        the middle; rays mostly 2–6; Va. and s. .... 34. *S. tortifolia.*
      6 Lvs broader, the larger ones 1–4 cm wide; stem spreading-hirsute, at least above the
        middle; rays mostly 6–12.
        7 Lvs sessile and ± clasping, obscurely serrulate or subentire, not at all rugose; coastal
          states from N.J. s. .... 35. *S. fistulosa.*
        7 Lvs merely subsessile, not clasping, ± strongly toothed and rugose-veiny; wide-
          spread .... 36. *S. rugosa.*
    5 Stem glabrous below the infl; swamp-plants.
      8 Invol 2.5–4 mm, its bracts narrow, not over 0.6 mm wide; coastal or inland; s. Me. and
        Mass. to w. N.Y. .... 36. *S. rugosa.*
      8 Invol 4–6 mm, its bracts broad, the larger ones 0.7–1.2 mm wide; near the coast; N.S.
        and Mass. to Fla. .... 37. *S. elliottii.*
2 Lvs ± strongly triple-nerved (least so in *S. radula*).
  9 Plants glabrous even in the infl; heads, or many of them, with some phyllary-like receptacular
    bracts near the margin (internal to the rays) .... 26. *S. missouriensis.*
  9 Plants puberulent at least in the infl; receptacular bracts generally wanting.
    10 Stem glaucous, glabrous below the infl .... 39. *S. gigantea.*
    10 Stem not glaucous, and commonly puberulent down to about the middle or below.
      11 Lvs relatively narrow, seldom less than 5 times as long as wide, seldom over 2 cm wide.
        12 Rays 5–8; invol 4–5 mm; local in Ky. .... 42. *S. shortii.*
        12 Rays either more than 8, or the invol less than 4 mm, or both; widespread.
          13 Rays 7–10(11); lvs glabrous, or occasionally puberulent on the midrib and main
            veins beneath; stem glabrous below the middle; local from Pa., Md., and
            D.C. to Tenn., Ky., and s. Ind. .... 41. *S. rupestris.*
          13 Rays 10–17, or if fewer (as sometimes in var. *hargeri*), then the lvs evidently
            pubescent across the surface beneath and the stem pubescent to the base;
            lvs always pubescent at least on the midrib and main veins beneath; wide-
            spread .... 40. *S. canadensis.*
      11 Lvs wider, mostly 2–5 times as long as wide, not infrequently more than 2 cm wide;
        rays 4–10.
        14 Lvs green, scarcely canescent, their pubescent coarse, mostly ± scabrous; Ozark-
          ian .... 43. *S. radula.*
        14 Lvs canescent, the pubescence finer and denser than in *S. radula*; Great Plains .... 44. *S. mollis.*

Group VII

1 Rays yellow; pappus bristles not clavellate-thickened (except to some degree in no. 48).
  2 Infl ± puberulent.
    3 Invol bracts not striate; few-headed alpine plants udner 4 dm .... 3. *S. cutleri.*
    3 Invol bracts longitudinally few-striate; not alpine, often larger and with more numerous heads.
      (*Oligoneuron.*)
      4 Lvs relatively broad, the middle cauline 2–6 times as long as wide; achenes conspicuously
        10–15-ribbed; mostly in dry places .... 45. *S. rigida.*
      4 Lvs narrower, the middle cauline 6–15 times as long as wide; achenes not more than ca
        7-nerved; mostly in moist places.
        5 Lvs flat, not at all triple-nerved .... 46. *S. ohioensis.*
        5 Lvs mostly conduplicate, tending to be triple-nerved .... 47. *S. riddellii.*
  2 Infl glabrous; slender plants 2–5 dm, with mostly less than 50 heads .... 48. *S. houghtonii.*
1 Rays white; many of the pappus-bristles clavellate-thickened toward the tip. (*Unamia.*) .... 49. *S. ptarmicoides.*

1. Solidago macrophylla Pursh. Big-lvd g. Stems 1–10 dm from a branched caudex, glabrous or nearly so below the infl; lvs thin, ± hirsute on the midrib and main veins beneath and sometimes more shortly so above, ± basally disposed, but often only gradually reduced upwards, the larger ones with elliptic or ovate to subcordate, acuminate, sharply serrate blade 2–15 × 1–7 cm, abruptly contracted to the long petiole; infl narrow and elongate, conspicuously leafy-bracteate at least below; invol 8–11 mm, its bracts imbricate, acuminate or attenuate, thin and loose, often some with squarrose tip; rays 7–12, 5–8 mm; achenes glabrous, 4–5 mm; 2*n*=18, 54. Moist, cool, often shaded places; Catskill Mts., N.Y. and Mt. Greylock, Mass., to Nf., Lab., and Hudson Bay.

2. Solidago squarrosa Muhl. Squarrose g. Stems 3–15 dm from a branched caudex, glabrous or nearly so below the rough-puberulent infl; lvs basally disposed, glabrous, or somewhat scabrous above, the larger ones with broadly oblanceolate to obovate, elliptic, or elliptic-ovate, sharply serrate blade mostly 5–20 × 2–10 cm, tapering or sometimes abruptly contracted to the long petiole; infl narrow and elongate, generally leafy-bracteate at least below; invol 5–9 mm, its bracts firm, at least the outer with squarrose, commonly herbaceous tip; rays 10–17, 3.5–5 mm; disk fls mostly 13–24; achenes glabrous, 3–3.5 mm; $2n=18$. Rocky woods; N.B. to s. Ont., s. to O., s. Ind., and in the mts. to N.C.

3. **Solidago cutleri** Fern.   New England g. Stems 5–35 cm from a branched caudex, hirtellous-puberulent in the infl; lvs basally disposed, the larger ones mostly oblanceolate or spatulate to elliptic, rounded to acute, toothed, 2–15 cm × 5–40 mm; infl few-headed and corymbiform, or more elongate and leafy-bracteate in larger plants; invol 5–8(–9) mm, glabrous or finely atomiferous-glandular, its bracts firm, strongly imbricate, obtuse or merely acutish; fls 30–50 per head, the rays 10–13; achenes hairy. Alpine situations; Me. to N.Y.

4. **Solidago puberula** Nutt.   Dusty g. Plants 2–10 dm from a branched caudex, covered with minute, stiffly spreading viscidulous hairs, or glabrate below; lvs basally disposed, the larger ones broadly oblanceolate to elliptic or obovate, serrate, obtuse or acute, mostly 5–15 × 1–3.5 cm, the others more lance-elliptic to lance-linear and entire; infl thyrsoid, dense, often leafy-bracteate, with stiffly ascending, not at all secund branches, or unbranched in small plants; invol 3–5 mm, its bracts narrow, ca 0.5 mm wide or less at midlength, with slender, acuminate, ± subulate tip; rays 9–16; disk-fls (8–)10–18; achenes glabrous or occasionally sparsely hairy; $2n=18$. Mostly in open places, generally in sandy or acid soil or rocks; N.S. and s. Que., s. through the coastal and Appalachian states to Fla. and La. Most of our plants are var. *puberula,* as principally described above. The well marked var. *pulverulenta* (Nutt.) Chapman, on the coastal plain from s. Va. to Fla. and Ala., has more numerous and smaller lvs (the middle cauline ones commonly 1–4 cm) that are less hairy on the upper surface than on the lower, and has more evenly tapering, scarcely subulate invol bracts up to 0.75 mm wide at midlength. (*S. pulverulenta.*)

5. **Solidago roanensis** T. C. Porter.   Roan Mt. g. Stems 2–10 dm from a branched, sometimes elongate caudex, hirsute-puberulent in the infl and sometimes irregularly or decurrently so below, lvs basally disposed, thin, glabrous, or scaberulous above, tending to be acuminate, the larger mostly 6–15 × 2–5 cm, with elliptic to elliptic-obovate or subrhombic, serrate blade, the others mostly rhombic to lance-elliptic; infl long and narrow, leafy-bracteate below, not secund; invol 4–5 mm, its bracts thin and slender (less than 0.75 mm wide at midlength), tapering to a narrowly acute or minutely obtuse tip; rays 6–9, 2–3 mm; disk fls 8–12; achenes glabrous or sometimes sparsely hairy; $2n=18$. Woods and clearing in the mts.; Va., Md., sw. Pa., and W.Va. to Tenn., S.C., and Ga. (*S. monticola,* a preoccupied name.)

6. **Solidago simplex** HBK.   Plants 1–9 dm from a branched caudex, essentially glabrous except for some puberulence in the infl (or more hairy in var. *gillmanii*); lvs basally disposed, the larger ones narrowly oblanceolate to narrowly obovate, 2–30 cm × 4–40 mm, toothed or subentire, generally acute or acutish, often irregularly ciliolate-margined; cauline lvs progressively reduced, mostly lance-elliptic to oblanceolate or linear; infl racemiform to dense and thyrsoid, or even ample and branched, but not secund, its lvs mostly reduced and inconspicuous; invol bracts imbricate, acutish or obtuse, often glutinous; rays 7–10; achenes short-hairy; $2n=18$, 36, 54, in ours mostly 36. Rock-crevices and sand-dunes; N.S. to n. and sw. Va. and reputedly Ky., w. to Mich. and n. Ind., and across s. Can. to the cordillera and the Pacific. Our plants are ssp. *randii* (Porter) Ringius, with 3 vars. Robust plants, 3–9 dm, often more hairy than the other vars., growing on sand dunes along Lake Michigan and w. Lake Huron, with large heads (invol 6–9 mm) in a long, often branched infl, are var. *gillmanii* (A. Gray) Ringius. (*S. deamii; S. gillmanii.*) Of the remaining, more widely distributed plants, those with the basal lvs mostly 3–8 times as long as wide, tending to be sharply toothed, the infl tending to be compact and thyrsoid, are var. *monticola* (Porter) Ringius. (*S. randii; S. maxonii*), while those with the basal lvs 7–10 times as long as wide, tending to be subentire, and the infl tending to be loose and subracemiform, are var. *racemosa* (Green) Ringius. (*S. racemosa.*)

7. **Solidago sciaphila** E. Steele.   Driftless area g. Plants 4–11 dm from a branched caudex, glabrous except for a little puberulence in the infl; lvs basally disposed, the larger ones with broadly oblanceolate to obovate or elliptic blade to 15 × 8 cm, crenate, serrate, or rarely subentire, tapering to the petiole; middle and cauline lvs progressively reduced, becoming sessile or subsessile and ± rhombic, toothed or sometimes entire; infl an elongate, bracteate, terminal thyrse, rarely branched; invol 5–7 mm, its bracts strongly imbricate, glabrous, obtuse; rays 7–8, 2.5–4 mm; achenes short-hairy. Calcareous or sandy cliffs; sw. Wis. and adj. Ill., Minn., and Io.

8. **Solidago hispida** Muhl.   Hairy g. Plants 1–10 dm from a branched caudex, generally spreading-hirsute throughout, sometimes sparsely or scarcely so in northern plants (these called var. *arnoglossa* Fern.) or the hairs occasionally mostly appressed; lvs basally disposed, the larger ones broadly oblanceolate to obovate or elliptic, toothed or entire, 8–20 × 1.5–6 cm; infl elongate and narrow, generally ± leafy-bracteate below, not at all secund, some of the lower branches sometimes elongate and stiffly ascending; invol 4–6 mm, its bracts obtuse or rounded, appearing narrower than in *S. erecta,* tending to be rather yellowish, the green tip often ill-defined; rays 7–14, usually deep yellow; achenes glabrous; $2n=18$. Dry woods and open, often rocky places; Nf. to Minn. and S.D., s. to Ga. and Ark.

9. **Solidago bicolor** L.   Silver-rod. Much like no. 8, but with white or whitish rays; invol 3–5 mm, its bracts whitish or light straw-colored except for the generally well defined light green tip; $2n=18$. Dry woods and open, often rocky places; N.S. and Que. to Wis., s. to Ga. and La. Hybridizes with nos. 8 and 10.

10. **Solidago erecta** Pursh.   Plants 3–12 dm from a branched caudex, essentially glabrous below the puberulent infl; lvs basally disposed, the larger ones broadly oblanceolate to obovate or elliptic, 7–30 × 1.5–5 cm, the middle cauline ones commonly 0.5–2 cm wide; infl elongate and narrow, often interrupted below, not at all secund, sometimes with a few long, straight or arching, cylindrical branches like the main axis; invol 3.5–6.5 mm; rays 5–9, averaging less deeply yellow than in *S. hispida;* disk-fls 6–10; achenes glabrous,

seldom less than 2.5 mm long; $2n=18$. Dry woods; coastal Mass.; N.J. to s. O. and Ind., s. to Ga. and Miss. Very distinct from *S. hispida* where their ranges overlap.

11. Solidago speciosa Nutt. Showy g. Plants 3–15 dm from a stout, woody caudex, coarsely puberulent in the infl, otherwise glabrous or slightly scabrous; lvs thick and firm, entire or the lower slightly toothed, sometimes gradually increasing in size toward the base, the persistent lower ones then often broad and abruptly petiolate, to 30 × 10 cm, sometimes all smaller and nearly uniform, the lower then generally deciduous; infl not at all secund, usually dense, simple or more commonly with rather crowded, stiffly ascending branches, sometimes loose and more open, with conspicuously pedicellate heads; invol 3–5 mm, its bracts obtuse or rounded, glutinous, yellowish; rays 6–8, 3–5 mm; disk-fls 7–9; achenes glabrous, seldom over 2 mm; $2n=18, 36$. Open woods, fields, prairies, and plains; Mass. and s. N.H. to Minn. and Wyo., s. to Ga., Ark., Tex., and N.M. Three vars. in our range.

a Lvs numerous; infl dense.
  b Robust and, when well developed, very broad-lvd; lower lvs mostly persistent; throughout that part
    of the range of the sp. that was originally forested. (*S. conferta*.) . . . . var. *speciosa*.
  b Smaller, more rigid, and often more scabrous, with smaller lvs seldom over 2 cm wide, the lower
    generally deciduous; replacing var. *speciosa* on the plains and occasionally found as far e. as
    O. and Tenn. (*S. rigidiuscula; S. speciosa* var. *angusta*, misapplied.) . . . . var. *rigidiuscula* T. & G.
a Lvs relatively few; infl tending to become open-paniculate; lower lvs tending to be persistent and long-
  petiolate, as in var. *speciosa*, but narrow, as in var. *rigidiuscula*; fl earlier than other vars.; sandy
  places; n. Ind. to Mich., Wis., Ill. and Minn. (*S. j.*) . . . . var. *jejunifolia*. (E. Steele) Cronq.

12. Solidago uliginosa Nutt. Northern bog-g. Plants to 15 dm from a rather long, branched caudex, glabrous except for the usually puberulent infl; lvs basally disposed, the larger ones elongate, oblanceolate to narrowly elliptic, subentire to evidently serrate, tapering to a long petiole with ± sheathing base, the blade and petiole 6–35 cm × 6–60 mm, mostly 6–15 times as long as wide; cauline lvs progressively reduced, not especially numerous; infl much longer than broad, the branches straight and not secund to evidently recurved-secund, but then rather short; invol 3–5 mm, its inner bracts mostly obtuse or rounded, the outer often acute; rays 1–8; achenes glabrous or occasionally hirtellous; $2n=-18, 36$, the tetraploids more southern or in more calcareous habitats. Bogs; Nf. and Que. to Minn., s. to Md., O., Ind., and in the mts. to N.C. and Tenn. (*S. linoides; S. neglecta; S. purshii; S. uniligulata*.)

13. Solidago gracillima T. & G. Southern bog-g. Similar to no. 12; glabrous throughout; middle and upper lvs mostly more numerous and more conspicuously reduced; infl elongate and narrow, varying from straight and nonsecund to more often secund or with short, secund branches and recurved tip, or with several long, slender, ascending, evidently secund branches; invol (3.5–)4–7 mm; achenes rather thinly strigose; $2n=18$. Swamps and other moist places; Va. and W.Va. to Ga., Ala., and n. Fla. (*S. austrina; S. perlonga*.)

14. Solidago stricta Aiton. Wand-g. Glabrous perennial 3–20 dm with a short, simple caudex and long stoloniform rhizomes; lvs basally disposed, thick and firm, the lowest ones oblanceolate or elliptic-oblan-ceolate, sometimes very narrowly so, 6–30 × 0.3–2(–5) cm, entire or obscurely serrate; cauline lvs abruptly reduced and sessile, entire, the middle and upper ones numerous, erect, often scarcely more than mere bracts; infl narrow, elongate, naked, sometimes nodding at the tip, the short branches occasionally recurved-secund; heads on slender, flexuous, minutely bracteolate peduncles; invol 4–6 mm; rays 3–7; disk-fls 8–12; achenes hairy, sometimes sparsely so; $2n=18, 36, 54$. Sandy, usually moist places, especially among pines, or some-times in coastal marshes, where it hybridizes with no. 15; coastal plain from N.J. to Fla., Tex., W.I., and s. Mex. (*S. petiolata*, misapplied.)

15. Solidago sempervirens L. Seaside-g. Plants somewhat succulent, 4–20 dm, usually with a very short and compact caudex, essentially glabrous, or scabrous-puberulent in the infl; lvs ± basally disposed, entire, the largest ones oblanceolate, 10–40 × 1–6 cm, the cauline ones generally rather numerous; infl dense, paniculiform, sometimes leafy at base, at least the lower branches ± recurved-secund; invol 3–7 mm, its bracts acute or acuminate; rays 3–5 mm; achenes hairy; $2n=18, 36$. Saline places along the coast from the Gulf of St. Lawrence to trop. Amer., and recently spreading inland locally, especially along highways that are salted in winter, even as far w. as Mich. Var. *sempervirens*, with relatively large heads (invol 4–7 mm, rays 12–17, disk-fls 17–22) is northern, s. to N.J. and locally to Va. Var. *mexicana* (L.) Fern., with smaller heads (invol 3–4 mm or seldom 5 mm, rays 7–11, disk-fls 10–16), and commonly also with narrower lvs, is southern, rarely extending as far n. as Mass. Hybridizes with no. 36 as well as with no. 14.

16. Solidago flexicaulis L. Zigzag g. Plants 3–12 dm from long rhizomes; stem glabrous below the infl, striate-angled and grooved, tending to be somewhat zigzag; lvs chiefly cauline, sharply toothed, acuminate, usually hirsute beneath at least on the midrib and main veins, mostly with ovate or elliptic blade 7–15 × 3–10 cm, 1–2.2(–2.5) times as long as wide, abruptly contracted to the broadly winged petiole; infl a series of mostly short clusters, the lower axillary to foliage-lvs, but these progressively reduced upwards, the terminal part of the infl often appearing as a naked thyrse; invol 4–6 mm, the outer bracts obtuse, the inner broadly rounded; rays mostly 3–4; disk-fls 5–9; achenes short-hairy; $2n=18, 36$, the tetraploids more western. Woods; N.S. and N.B. to N.D., s. to Va., Ky., and Ark., and in the mts. to Ga.

17. Solidago albopilosa E. L. Braun. Cave-g. Much like no. 16, but shorter and weaker (mostly 3–5 dm), the stem conspicuously spreading-hirsute, the lvs more hairy, inclined to be subcordate at base, and avg smaller (blade 4–9 × 2.5–5 cm); rays 4–5; $2n=36$. Under overhanging cliffs in Powell and Menifee counties, Ky.

18. Solidago flaccidifolia Small. Much like no. 19, but the lvs wider, mostly (2.2–)2.5–3(–3.5) times as long as wide, and the infl usually in large part terminal, slender, and thyrsoid (sometimes branched), with only the lower clusters of heads evidently surpassed by their reduced subtending lvs, but varying to fully axillary and leafy; invol puberulent; plants often with elongate rhizomes; $2n=36$, 54. Moist woods and edges of clearings; sw. Va. and se. Ky. to Ga., Ala., and Miss. (*S. latissimifolia,* misapplied.)

19. Solidago curtisii T. & G. Plants 3–15 dm from a short and caudex-like to occasionally more elongate rhizome, the stem striate-angled and grooved, glabrous or hairy; lvs chiefly cauline, ± serrate, acuminate, glabrous or evidently hairy, numerous, lanceolate to narrowly elliptic, mostly 10–18 × 1–4(–4.5) cm, 3–10 times as long as wide, tapering to the sessile or obscurely short-petiolate base; infl largely or wholly of axillary clusters shorter than their subtending lvs (or in robust forms many of the clusters elongate and leafy-bracteate), the terminal segment sometimes thyrsoid and nearly naked; invol (2.5–)3–5(–6) mm, glabrous or short-hairy but not glandular, its bracts 1-nerved or obscurely nerved to ± evidently 3-nerved, seldom any of them as much as 1 mm wide at or above midlength; rays (2)3–5(6); disk-fls 4–9; achenes hairy; $2n=18$. Mt. woods, but not at the highest elev.; Va., W.Va., Tenn., and Ga. (*S. pubens,* the hairy phase.)

20. Solidago lancifolia (T. & G.) Chapman. Much like no. 19; rhizome commonly elongate; infl in considerable or large part terminal and thyrsoid, with much reduced lvs, only the lower part consisting of axillary clusters; heads relatively large and coarse, the invol 4.5–7 mm, tending to be finely granular-glandular or glandular-puberulent, its bracts strongly 3-nerved, the larger ones ca 1 mm wide or more at or above midlength; rays 5–8; disk-fls 6–12. Upper elev. (mostly above 1500 m) in the mts. of Va., N.C., and Tenn.

21. Solidago caesia L. Axillary g. Plants 3–10 dm from a short, stout, caudex-like rhizome, sometimes with long, creeping rhizomes as well; stem terete, glaucous; lvs chiefly cauline, ± serrate, acuminate, glabrous or slightly hairy, lanceolate or lance-elliptic, 6–12 × 1–3 cm, 3–10 times as long as wide, tapering to the sessile or obscurely short-petiolate base; infl chiefly axillary (sometimes branched) as in no. 19; invol glabrous, 3–4.5 mm, its bracts narrow, obtuse or rounded, tending to be obscurely several-nerved; rays (1–)3–4(–5); disk-fls 5–8; achenes hairy; $2n=18$. Woods; N.S. and s. Que. to Wis., s. to Fla. and Tex. An apparent hybrid with no. 30 is *S. × ulmicaesia* Friesner.

22. Solidago petiolaris Aiton. Plants 4–15 dm from a stout caudex, sometimes with long, slender rhizomes as well; stems finely puberulent or scabrous-puberulent above, generally glabrous below; lvs firm, lance-linear to more often lance-elliptic, elliptic, or ovate, 3–12 cm × 5–30 mm, entire or occasionally few-toothed, chiefly cauline, thick and strongly glutinous, glabrous, or scabrous along the midrib and main veins on both sides with short, loose, commonly subconic hairs 0.1–0.4 mm, or these rarely thinly scattered across the surface beneath; infl narrow and generally elongate, usually ± leafy-bracteate, the lower clusters sometimes elongate and stiffly ascending but not secund; peduncles 3–15 mm; invol 4.5–7.5 mm, atomiferous-glandular to occasionally glabrous, its bracts strongly imbricate, the outer with acute, sometimes squarrose tip, the inner erect and often blunter; rays (5–)7–9, 3–7 mm; disk-fls (8–)10–16; achenes glabrous or nearly so. $2n=18$, 36. Rocky woods and open places, especially in sandy soil; N.C. to Neb., s. to Fla., La., and N.M., reaching our range in c. Mo. All or nearly all of our plants belong to the Ozarkian var. *angusta* (T. & G.) A. Gray, as described above. (*S. angusta.*) The widespread but chiefly more southern var. *petiolaris,* more puberulent (especially on the lower surface of the lvs) and less glutinous, probably does not reach our range.

23. Solidago nemoralis Aiton. Gray g. Plants 1–10 dm from a branched caudex; herbage densely and finely puberulent with minute, loosely spreading hairs; lvs basally disposed, weakly or scarcely triple-nerved, the larger ones oblanceolate or a bit wider, 5–25 × 0.8–4 cm, ± toothed; infl paniculiform, sometimes long, narrow, and merely nodding at the tip, varying to more ample and sometimes with long divergent, recurved-secund branches; invol bracts imbricate, glabrous except for the ciliolate margins; rays 5–9, short; disk-fls 3–6; achenes short-hairy; $2n=18$, 36. Dry woods and open places, especially in sandy soil; N.S. to Fla., w. to Alta. and Tex. Three vars.:

a Heads smaller, the invol 3–4.5 mm; pubescence faintly viscidulous; achenes hirtellous or strigose.
  b Infl generally rather compact, or narrow and elongate; upper lvs usually rather gradually reduced;
    plants not very robust; the common phase in our range, passing into var. *haleana* along our s.
    border, and into var. *longipetiolata* at the w . . . . var. *nemoralis.*
  b Infl ample and open, with long, divergent branches; upper lvs numerous and conspicuously reduced;
    plants relatively robust; southeastern phase . . . . var. *haleana* Fern.
a Heads larger, the invol 4.5–6 mm; pubescence not at all viscidulous; achenes subsericeous; Great
  Plains phase, reaching our range in Minn., Io., and Mo. (*S. n.* var. *decemflora; S. pulcherrima.*)
  . . . . var. *longipetiolata* (Mackenzie & Bush) Palmer & Steyerm.

24. Solidago patula Muhl. Rough-lvd g. Stems 5–20 dm from a short caudex, glabrous below the rough-puberulent infl, angular at least below; lvs basally disposed, glabrous beneath, strongly scabrous on the upper surface, the lower with somewhat sheathing petiole and elliptic, elliptic-ovate, or elliptic-obovate, sharply toothed blade 8–30 × 4–10 cm, the middle and upper gradually reduced but still generally toothed; infl paniculiform, generally with widely spreading, recurved-secund branches, in smaller plants sometimes narrower, denser, and elongate, but still secund; invol 3–4.5 mm, its bracts acute (especially the outer) to obtuse (especially the inner); rays 5–12; disk-fls 8–23; achenes sparsely hairy; $2n=18$. Swamps and wet meadows;

Vt. to Wis., s. to Ga., Miss., and Tex. Most of our plants, as described above, belong to var. *patula*. Southward, from Va. to Ga. and Tex., this gives way to var. *strictula* T. & G., a smaller plant with smaller, narrower (to 5 or 6 cm wide), less strongly toothed lvs, the upper notably numerous, much reduced, and commonly entire. (*S. salicina.*)

25. Solidago juncea Aiton.   Early g. Plants 3–12 dm from a stout, branched caudex or short rhizome, commonly with ± deep-seated creeping rhizomes as well, essentially glabrous, or sometimes ± short-hirsute on the lvs or in the infl; lvs basally disposed, the larger ones 15–40 × 2–7.5 cm, with rather narrowly elliptic-acuminate, ± serrate blade tapering to the long petiole; infl dense, mostly about as broad as long, with recurved-secund branches; invol 3–5 mm; rays 7–12, minute; disk fls 9–14; receptacle with some slender, chaffy, phyllary-like bracts near the margin internal to the rays; achenes short-hairy; $2n=18$. Dry, open places and open woods, especially in sandy soil; N.S. and N.B. to Minn., s. to Va., Tenn., ne. Miss., Mo., and in the mts. to n. Ga. and n. Ala.

26. Solidago missouriensis Nutt.   Missouri g. Plants (3–)5–10 dm from creeping rhizomes, sometimes with a caudex as well, glabrous throughout; lvs firm, strongly triple-nerved (at least the middle and lower), entire or some (especially the lower) serrate, the lowest ones oblanceolate and conspicuously serrate but mostly soon deciduous, the others slightly to strongly reduced upward, lance-elliptic to broadly linear, tapering to a sessile or obscurely petiolar base, often with axillary fascicles of much reduced lvs; infl paniculiform, with ± strongly recurved-secund branches, mostly short and broad; invol 3–5 mm, its bracts firm, broadly rounded to occasionally acutish; rays 7–13; disk-fls 8–13; receptacle commonly with some bracts near the margin as in no. 25; achenes glabrous or sparsely hairy; $2n=18$, 36. Prairies and other dry, open or sparsely wooded places; widespread in w. U.S., entering our range from Minn. to Mo., and locally e. to nw. Ind. and s. Ont. Our plants, as here described, are var. *fasciculata* Holzinger. (*S. glaberrima; S. moritura.*)

27. Solidago pinetorum Small.   Fascicled piney-woods g. Plants 4–11 dm from a branched caudex, slender, glabrous throughout, or some of the lvs ± ciliate-margined; lvs basally disposed, the lower linear-oblanceolate, serrate or subentire, ± strongly triple-nerved; upper lvs spreading or reflexed, and bearing axillary fascicles of small lvs; infl paniculiform, with recurved-secund branches, commonly as broad as long; invol 3–4 mm, its bracts obtuse or rounded, with evident midrib; rays 3–7; disk-fls 5–9; achenes glabrous, or slightly hairy distally; $2n=18$. Open places and dry woods, especially in sandy soil; coastal plain and piedmont from s. Va. to S.C.

28. Solidago arguta Aiton.   Forest-g. Stems 5–15 dm from a stout, branched caudex, glabrous except for the somewhat puberulent infl; lvs basally disposed, glabrous, or slightly scaberulous above, toothed to subentire, the larger ones 10–30 × 3–12 cm, the broadly elliptic or ovate blade rather abruptly contracted to the long petiole; infl paniculiform, with recurved-secund branches, sometimes elongate and narrow, more often broad and open, wtih long, divergent branches; invol 3–7 mm, its bracts acute or obtuse; rays 2–8; disk-fls 8–20; $2n=18$, 36. Woods, forest-openings, and dry meadows; Me. to Fla., w. to Ky., Mo., and La. Three vars. in our range:

a Achenes glabrous; northern, from s. Me. to Va. and occasionally to N.C., Ky., Tenn., and s. Mo.;
  $2n=18$ .... var. *arguta*.
a Achenes hairy, at least distally; more southern, from Md., Va., W.Va., and Ky. s.
  b Shale-barren ecotype of w. Md., w. Va., e. W.V.a, and e. Ky., with relatively very firm lvs, the basal
    ones commonly ± truncate at base; $2n=18$. (*S. harrisii.*) .... var. *harrisii* (E. S. Steele) Cronq.
  b Series of non-shale-barren ecotypes with less firm, basally somewhat tapering lvs; $2n=18$, 36, the
    tetraploids with somewhat larger, more numerously fld heads than the diploids; Va. to n. Fla.,
    w. to W.Va., Ky., and occasionally Mo., Ark., and La. (*S. boottii* and *S. yadkinensis,* both
    misapplied.) .... var. *caroliniana* A. Gray.

29. Solidago tarda Mackenzie.   Much like *S. arguta* var. *caroliniana,* but with well developed, slender, stoloniform rhizomes in addition to the more deep-seated main rhizome or short caudex; lvs glabrous or occasionally strigose, the lower mostly ovate, acute or acuminate, with truncate to obtuse base; upper cauline lvs gradually reduced; $2n=54$. In sandy soil in more xeric places than no. 28; s. N.J. and se. Pa. to n. Fla. and Ala., mainly on the coastal plain. (*S. ludoviciana,* in part.)

30. Solidago ulmifolia Muhl.   Elm-lvd g. Stems 4–12 dm, glabrous or nearly so below the infl, mostly solitary or paired from a caudex or short rhizome; lvs numerous, thin, sharply serrate, loosely hirsute at least on the midrib and main veins beneath; basal lvs wanting or with well developed, elliptic to elliptic-obovate blade abruptly contracted to the petiole, but generally soon deciduous; lowermost cauline lvs tending to be soon deciduous and smaller than the persistent ones just above, which are ovate or rhombic-ovate to elliptic or lance-elliptic, acute or acuminate, broadly short-petiolate or tapering and subsessile, mostly 6–12 × 1.2–5.5 cm, the lvs thence ± reduced upward, those at the base of the infl generally small, relatively broad, and often numerous; infl paniculiform, with recurved-secund branches, these generally few, elongate, and divergent, the heads crowded; invol 2.5–4.5 mm; rays 3–5, minute; disk-fls 4–7; achenes short-hairy; $2n=18$. Woods; N.S. to Ga. and Fla., w. to Minn., Kans., and Tex. Ours is var. *ulmifolia*.

31. Solidago drummondii T. & G.   Ozark-g. Stems 3–10 dm from a stout, branched caudex, uniformly pubescent with short, spreading hairs, occasionally glabrate near the base; lvs chiefly cauline, broadly ovate or elliptic-ovate, all except sometimes the uppermost evidently short-petiolate, ± triple-nerved, but also pinnately veined, finely and usually densely spreading-hairy at least on the lower side, generally only those near the infl reduced, the others 1.3–2 times as long as wide, the larger ones 3.5–9 × 2.5–7 cm; infl paniculiform, with recurved-secund branches, or the heads apparently sometimes drooping; invol 3–4.5 mm, its bracts obtuse or rounded; rays 3–7, well developed; disk-fls 4–7; achenes short-hairy; $2n=18$. Cliff-crevices and rocky woods, especially in calcareous soil; Mo., Ill., and Ark.

32. Solidago sphacelata Raf.   Short-pappus g. Stems 5–12 dm from a short and caudex-like to somewhat elongate rhizome, densely spreading-puberulent to occasionally subglabrous; lvs basally disposed, ± densely spreading-puberulent beneath, sparsely so or more often glabrous above, the basal ones tufted and persistent, conspicuously petiolate, with serrate, cordate blade 4–12 × 4–11 cm, the cauline ones progressively reduced, less pteiolate, and less cordate; infl paniculiform, with a few widely spreading, secund branches, the narrow heads densely crowded and often subglomerate; invol 3–4.5 mm, its bracts firm, somewhat keeled, shortly green-tipped; rays 3–6, short; disk-fls 3–6; pappus-bristles firm, reduced, much shorter than the hairy achenes; $2n=18$. Open woods and rocky places, especially in calcareous soil; Va. to Ga. and Ala., w. to Ill. and w. Ky. and Tenn. (*Brachychaeta s.*)

33. Solidago odora Aiton.   Licorice-g. Stems 6–16 dm from a short, stout caudex, rough-puberulent in the infl and in lines decurrent from at least the upper lf-bases; lvs chiefly cauline, sessile, entire, glabrous except for the scabrous margins, finely translucent-punctate, anise-scented when bruised (rarely inodorous), not prominently veined, the main ones 4–11 × 0.5–1.5(–2) cm, 5–15 times as long as wide; infl paniculiform, with recurved-secund branches; invol 3.5–5 mm, its bracts slender, acute, yellowish; rays 3–5(6), fairly showy; disk-fls 3–5; achenes short-hairy or subglabrous; $2n=18$. Dry, open woods, especially in sandy soil; Mass., N.H., and Vt. to s. O. and s. Mo., s. to Fla. and Tex. Ours is var. *odora*.

34. Solidago tortifolia Elliott.   Leafy piney-woods g. Stems 3–13 dm from long creeping rhizomes, uniformly puberulent above the middle; lvs chiefly cauline, very numerous, sessile, linear to lance-linear or narrowly oblong, the larger ones 2.5–7 cm × 2–7(–10) mm, glabrous or ± scabrous on one or both surfaces, not prominently veined (but sometimes obscurely triple-nerved), usually at least some of the lower ones remotely serrulate; infl paniculiform, with recurved-secund branches; invol 2.5–3.5 mm, yellowish; rays 2–6, small; disk-fls 2–6; achenes short-hairy; $2n=18$. Dry, usually sandy soil, often in pinelands; coastal plain from Va. to Fla. and Tex.

35. Solidago fistulosa Miller.   Hairy piney-woods g. Stems 7–15 dm from long creeping rhizomes, stout, conspicuously spreading-hirsute, at least above the middle; lvs chiefly cauline, numerous, crowded, strongly hirsute on the midrib beneath, and often less densely so across the surface, less hairy or more often glabrous above, sessile, broad-based and somewhat clasping, lance-ovate to elliptic-oblong, obscurely serrulate or subentire, the larger ones 3.5–12 × 1–3.5 cm; infl paniculiform, usually dense, with recurved-secund branches; invol glabrous, 3.5–5 mm, its bracts thin and slender; rays 7–12, small; disk fls 4–7; achenes short-hairy; $2n=18$. Wet or dry places, often in pinelands; N.J. to Fla. and La., mainly on the coastal plain.

36. Solidago rugosa Miller.   Wrinkle-lvd g. Stems 3–15(–25) dm from long creeping rhizomes, evidently spreading-hirsute (glabrous in var. *sphagnophila*); lvs chiefly cauline, numerous, crowded, slightly to very strongly rugose-veiny, not triple-nerved, glabrous or scabrous above, hirsute at least on the midrib and main veins beneath (except var. *sphagnophila*), lance-elliptic to lance-ovate or rhombic-elliptic, serrate, subsessile, not clasping, the larger ones 3.5–13 × 1.3–5 cm; infl paniculiform, with recurved-secund branches; invol 2.5–4(–5) mm, its bracts slender, not more than ca 0.6 mm wide; rays 6–11, small; disk-fls (3)4–8; achenes short-hairy; $2n=18, 36, 54$. Various habitats; Nf. to Fla., w. to Mich., Mo., and Tex. (*S. altissima,* misapplied.) Two ssp. may be distinguished, and the typical one may be divided into 3 vars., as follows:

a Lvs relatively thin, not very strongly rugose, tending to be sharply toothed and acuminate; pubescence
   tending to be long and relatively soft, or wanting; invol bracts mostly acute or acutish; rays 8–
   11; mostly in relatively moist, often wooded places; northern, the common phase in most of our
   range . . . . ssp. *rugosa.*
   b Stem evidently spreading-hairy, rarely only in the infl; lvs hairy at least on the midrib and main
      veins beneath.
      c Infl relatively ample, with divaricate, often very long branches all evidently exceeding their
         strongly reduced subtending lvs; approximately the range of the ssp. . . . . var. *rugosa.*
      c Infl narrow and leafy, the lower branches scarcely or not at all exceeding their well developed
         subtending lvs. Nf. to Ont., N.Y., and Mass. . . . . var. *villosa* (Pursh) Fern.
   b Stem and lvs essentially glabrous; otherwise much like var. *rugosa*; swampy places, blooming earlier
      than the other vars.; s. Me. to N.J., w. N.Y. and irregularly to N.C., S.C., and perhaps Mich.;
      possibly better called a distinct sp., *S. aestivalis* E. Bicknell . . . . var. *sphagnophila* Graves.
a Lvs relatively thick and firm, strongly rugose-veiny, tending to be blunt-toothed or even subentire,
   and often merely acutish at the tip, the pubescence tending to be relatively short and harsh; invol
   bracts generally obtuse or rounded; rays 6–8; mostly in rather dry places; mainly southern, but
   occasionally to Mass., Ont., and Mich. Perhaps to be divided into several vars. (*S. aspera; S.
   celtidifolia*) . . . . ssp. *aspera* (Aiton) Cronq.

37. Solidago elliottii T. & G.   Coastal swamp-g. Plants stout, (4–)10–30(–40) dm from long creeping rhizomes, wholly glabrous, or puberulent in the infl; lvs chiefly cauline, numerous, sessile or nearly so (the base of the blade often subauriculate rounded to a very short petiole), elliptic or lance-elliptic, ± serrate or the upper entire, not triple-nerved, not rugose, the larger ones 6–15 × 1.5–3.5 cm; infl paniculiform, sometimes conspicuously leafy-bracteate, with short or elongate, slightly to strongly recurved-secund branches; invol 4–6 mm, its bracts obtuse or rounded, the larger ones 0.7–1.2 mm wide; rays 6–10(–12); disk-fls 4–7; achenes short-hairy; $2n=18$, 54. Fresh or brackish swamps near the coast; N.S.; Mass. to Fla.

38. Solidago calcicola Fern. Stems 5–12 dm from a stout caudex, puberulent above, glabrous below; lvs chiefly cauline, glabrous or scabrous above, hirsute on the midrib and main veins beneath, ± triple-nerved, sharply serrate, commonly lance-elliptic, acuminate, sessile or nearly so, 6–18 × 1.5–3.5 cm; infl a series of short or elongate, thyrsoid, not at all secund branches, the lower subtended by scarcely or slightly reduced lvs; invol 4–6 mm, its bracts slender, imbricate, acute or acutish, yellowish, the green tip often obscure; rays 10–14, well developed. Cliffs and moist woods; n. Vt. and N.H. to the Gaspé penins. of Que. and to Nf.

39. Solidago gigantea Aiton. Smooth g. Similar to no. 40, but puberulent only in the infl, the stem glaucous; lvs glabrous or with a line of hairs along the 3 main veins beneath, the larger ones 6–17 × 1–4.5 cm; infl paniculiform, with recurved-secund branches; invol 2.5–4 mm, its bracts mostly firmer, blunter, and greener than in no. 40; rays (8–)10–17; disk-fls 6–12; $2n=18$, 36, 54. Moist open places; N.S. and Que. to Ga., w. to B.C. and N.M. (*S. serotina.*)

40. Solidago canadensis L. Common g. Perennial, mostly with long, creeping rhizomes (rhizome short in var. *gilvocanescens*), 2.5–20 dm, the stem ± puberulent at least above the middle; lvs chiefly cauline, numerous, triple-nerved, lance-linear to lance-elliptic or narrowly elliptic, sessile, tapering to both ends, the larger ones 3–15 cm × 5–22 mm, puberulent at least on the midrib and main veins beneath; infl mostly paniculiform, with strongly recurved-secund branches (varying to thyrsoid and not secund in var. *salebrosa*); invol 2–4(–5) mm, its bracts imbricate, thin and slender, acute or acuminate, yellowish, without a well defined green tip; rays 10–17 (or fewer in var. *hargeri*), 1–3 mm; disk-fls 2–8 (to 13 in var. *salebrosa*); achenes short-hairy; $2n=18$, 36, 54. Moist or dry open places and thin woods; throughout most of the U.S. and s. Can. Five vars. in our range:

a Pubescence relatively sparse and restricted, the lower half (or nearly half) of the stem generally glabrous or nearly so, the lvs puberulent on the midrib and main veins beneath, occasionally sparsely so across the surface, glabrous to somewhat scabrous above; lvs thin, sharply serrate to sometimes entire.

  b Heads relatively small, the invol 2–3 mm, the rays 1.0–1.5 mm, the disk-fls 3–7 in number; infl paniculiform, with recurved-secund branches; $2n=18$; Nf. to Mass. and n. NIY., w. to n. Mich. and n. Minn. . . . . var. *canadensis*.

  b Heads larger, the invol 3–4(–5) mm, the rays 1.5–2.5 mm, the disk-fls 8–13 in number; infl paniculiform or thyrsoid, often somewhat leafy, the branches varying from not at all recurved-secund to evidently so; $2n=18$, 54; cordilleran region of w. U.S., e. across s. Can. to n. Minn., Me., and Lab. (*S. elongata.*) . . . . var. *salebrosa* (Piper) M. E. Jones.

a Pubescence relatively dense and covering more of the plant, the stem ± pubescent nearly or quite to the base, the lvs densely spreading-puberulent across the surface beneath, similarly pubescent to merely scabrous above; lvs, except in var. *hargeri*, firm and shallowly few-toothed to entire.

  c Upper surface of the lvs merely scabrous or even subglabrous, the hairs much fewer, shorter, and stiffer than those on the lower surface; plants 1–2 m when well developed; larger lvs usually over 7 cm; rhizome generally elongate; Que. to Fla., w. to N.D., Kans., and Ariz., chiefly in the deciduous forest region of e. U.S.

    d Heads very small and very numerous, the invol up to ca 2.5(3) mm, the rays up to ca 2.5 mm, the disk-fls 2–4(5) in number, the rays avg ca 9; lvs thin, sharply toothed to sometimes entire; $2n=18$ . . . . var. *hargeri* Fern.

    d Heads somewhat larger and not so numerous, the invol 2.5+ mm, the rays 2.5+ mm, the disk-fls 3–7 in number, the rays avg 13; lvs firm and usually shallowly few-toothed or entire; $2n=54$. (*S. altissima.*) . . . . var. *scabra* T. & G.

  c Upper surface of the lvs densely spreading-puberulent, the hairs only slightly if at all fewer, shorter, and stiffer than those on the lower surface; plants 2.5–8(–12) dm; lvs mostly 3.5–7 cm; rhizome typically short; $2n=18$, 36; Great Plains region, e. to Minn. and Mo. (*S. pruinosa.*) . . . . . var. *gilvocanescens* Rydb.

41. Solidago rupestris Raf. Riverbank-g. Stems 5–15 dm from long creeping rhizomes, glabrous below, slightly puberulent above the middle, more definitely so in the infl; lvs chiefly cauline, numerous, crowded, thin, triple-nerved, glabrous, or occasionally puberulent along the midrib and main veins beneath, remotely serrulate or subentire, sessile or subsessile, slender, tapering to both ends, mostly 5–12 × 0.6–1.2 cm; infl paniculiform, with recurved-secund branches; invol 2–3 mm, the invol bracts blunter and more oblong than in no. 40; rays 7–10(11), 1–2 mm; disk-fls 2–7; achenes short-hairy; $2n=18$. Riverbanks from Pa., Md. and D.C. to Tenn., Ky., s. O., and s. Ind.

42. Solidago shortii T. & G. Kentucky-g. Stems 6–13 dm from a short, stout rhizome, scabrous-puberulent at least above the middle; lvs chiefly cauline, numerous, firm, glabrous, triple-nerved, remotely serrulate, narrowly elliptic or lanceolate, acuminate or sharply acute, tapering to the subsessile or obscurely petiolate base, the larger ones 7–10 × 1–1.5 cm; infl paniculiform, with recurved-secund branches; invol 4–5 mm, its bracts firm, acute or obtusish; rays 5–8, 2–3 mm; disk-fls 5–9; achenes short-hairy; $2n=36$. Dry, open places at Blue Licks, in Fleming, Nicholas and Robertson counties, Ky., and formerly on Rock Island, at the falls of the Ohio, near Louisville.

43. Solidago radula Nutt. Rough g. Plants 4–12 dm from a caudex, at least sometimes with creeping rhizomes as well; stem scabrous to shortly and loosely hirsute; lvs chiefly cauline, numerous, firm, elliptic or lance-elliptic to rather narrowly elliptic-obovate, subsessile, obscurely to evidently toothed or the upper entire, ± evidently trinerved, subglabrous, or more commonly scabrous-hirsute (seldom more softly spread-ing-hairy), mostly 2–5 times as long as wide, the larger ones 3–8 × 1–3 cm; infl paniculiform, with densely fld, ± recurved-secund branches, or occasionally simple and nodding; invol glabrous, 3.5–5.5 mm, its bracts relatively broad and firm, acutish to more often obtuse or broadly rounded; rays 4–7, 2–3.5 mm; disk-fls 4–6; achenes short-hairy; $2n=18$. Open, rocky places and dry woods, especially in calcareous soil; Mo. and s. Ill. and w. Ky. to Okla., La., and Tex.; N.C.; Ga.

44. Solidago mollis Bartling.   Velvety g. Plants rhizomatous, 1–6 dm; herbage densely spreading-hirtellous throughout, or the stem glabrous below; lvs chiefly cauline, numerous and crowded, triple-nerved, elliptic and sessile (or the lower cauline oblanceolate and obscurely petiolate), rounded to acute at the tip, the larger ones 3–10 cm × 8–40 mm, the middle ones 2.5–4 times as long as wide; infl dense, paniculiform or occasionally compactly thyrsoid, the lower branches often ± recurved-secund; invol 3.5–6 mm, glabrous, its bracts imbricate, rounded to acutish, firm, the broader ones 0.8–1.3 mm wide; rays 6–9, 3–5 mm, relatively broad; disk-fls 3–8; achenes short-hairy; $2n=36, 54$. Plains, dry hills, and dry, open woods; w. Minn. and Io. to Man. and Okla., w. to Mont. and Colo.

45. Solidago rigida L.   Stiff g. Plants 2.5–15 dm from a stout branched caudex; herbage densely pubescent with short spreading hairs, varying to subglabrous; lvs basally disposed, firm, slightly toothed or entire, the larger ones with elliptic, elliptic-oblong, or broadly lanceolate to broadly ovate, rounded to acutish blade 6–25 × 2–10 cm, often exceeded by the long petiole; middle cauline lvs sessile or nearly so, 2–6 times as long as wide; infl dense, corymbiform, 5–25 cm wide; heads relatively large, the disk 5–10 mm wide; invol 5–9 mm, glabrous or puberulent, its bracts firm, broadly rounded, conspicuously striate; rays 7–14, 3–5 mm; disk-fls 17–35; achenes turgid or angular, 10–20-nerved; $2n=18, 36$. Prairies and other dry, open places, especially in sandy soil; R.I., Conn., w. Mass., and N.Y. to Ga., w. to Alta. and N.M., more common westward. (*Oligoneuron r.*) Three vars.:

a Lvs and stem ± densely pubescent.
  b Plants relatively large and robust, 6–15 dm; achenes glabrous; $2n$ mostly = 36; our common form,
    w. to Minn. and Mo. .... var. *rigida*.
  b Plants smaller, 2.5–6(–8) dm; achenes ordinarily with some short, loose hairs near the tip; $2n=18$;
    Great Plains region, from Minn. and Io. westward. (*S. parvirigida; Oligoneuron canescens.*)
    .... var. *humilis* T. C. Porter.
a Lvs glabrous or nearly so except for the margins and often the midrib beneath; stem glabrous or hairy;
  otherwise as var. *rigida*; $2n=18$; se. phase, from s. O. to N.C., Tenn., Ga., Ala., and Tex. (*S. jacksonii; Oligoneuron j.*) .... var. *glabrata* E. L. Braun.

46. Solidago ohioensis Riddell.   Ohio g. Plants 4–9 dm from a stout branched caudex, glabrous except for the scabro-ciliate lf-margins; lvs basally disposed, flat, not triple-nerved, the larger ones with lance-elliptic or narrowly elliptic, slightly toothed to entire blade 8–22 cm × 8–50 mm, often exceeded by the long petiole; cauline lvs few or rather numerous, the middle ones sessile or short-petiolate, 6–12 times as long as wide; infl corymbiform, the heads numerous, rarely less than 50, often several hundred; invol glabrous, 4–5 mm, its bracts firm, obtuse or broadly rounded, ± striate; rays 6–8, short; disk-fls ca 20; achenes glabrous, 3–5-angled, the angles scarcely or obscurely nerved; $2n=18$. Swamps, beaches, and other moist places; c. N.Y. and s. Ont. to Ill., Mich., and Wis.

47. Solidago riddellii Frank.   Plants 4–10 dm from a ± well developed caudex, sometimes with creeping rhizomes as well; stem generally stout, glabrous except for some puberulence in the infl; lvs ± basally disposed, glabrous except for the scabrous margins, entire, firm, mostly conduplicate, tending to be triple-nerved, the lowest ones with elongate, narrow blade much exceeded by the petiole, but often deciduous; cauline lvs numerous, becoming sessile but otherwise not strongly reduced upwards, clasping or sheathing at base, often arcuate; infl corymbiform, the heads numerous and crowded, rarely less than 50, often several-hundred; invol 5–6 mm, glabrous, its bracts firm, obtuse or broadly rounded, ± striate; rays 7–9; achenes glabrous or nearly so, evidently 5–7-nerved; $2n=18$. Swamps, wet meadows, and moist prairies; s. Ont. to O., w. to Minn. and Mo.

48. Solidago houghtonii T. & G.   Great Lakes g. Slender perennial from a caudex, 2–5 dm, rough-puberulent in the infl, otherwise glabrous except for the scabro-ciliolate lf-margins; lvs basally disposed, tending to be weakly triple-nerved and often conduplicate, the larger ones linear-oblanceolate to linear, to 20 × 2 cm, the cauline few, 7–15 below the infl; infl corymbiform, its heads crowded, relatively few, mostly less than 50; invol 5.5–8 mm, its bracts firm, blunt, scarcely striate; rays ca 6 and disk-fls ca 13; achenes glabrous, 3–5-angled, the angles weakly nerved; many of the pappus-bristles slightly clavate and subplumose; $2n=54$. Swamps and moist beaches, often in moist sandy swales behind the dunes; Genesee Co., N.Y., and along the shores of the n. parts of Lake Michigan and Huron, rarely inland in ne. Mich.

49. Solidago ptarmicoides (Nees) B. Boivin.   Stems 1–7 dm from a branched caudex, scabrous at least above; lvs firm, glabrous or scabrous, entire or with a few remote salient teeth, tending to be trinerved, 3–20 cm × 1.5–10 mm, the lower linear-oblanceolate and petiolate, sometimes tufted, persistent, and larger than those above, sometimes smaller and deciduous, the others becoming sessile upward and linear or nearly so; heads mostly 3–60 in an open, minutely bracteate, corymbiform infl; invol (4)5–7 mm, glabrous, its bracts imbricate, firm, greenish above but scarcely herbaceous, often with strongly thickened midrib; rays 10–25, white, 5–9 mm; disk-fls numerous, white; achenes glabrous; pappus copious, many of the bristles clavellate-thickened and subplumose toward the tip; $2n=18$. Prairies and other open, usually dry, commonly calcareous places; Vt., N.Y., and w. Que. to Ga., w. to Sask., Colo., and Ark. (*Aster p.; Unamia p.*) Hybrids with *S. rigida* var. *humilis* have been called *Aster* or *Unamia* or *S. × lutescens*; those with *S. ohioensis* have been called *S. × krotkovii*; and those with *S. riddellii* have been called *S. × bernardii*; European garden-hybrids, presumably with *S. canadensis*, have been called *Solidaster luteus*.

54. EUTHAMIA Nutt. Flat-topped goldenrod. Heads radiate, the rays 7–30(–35), small, pistillate and fertile; invol ± glutinous, its small, chartaceous, yellowish or green-tipped bracts imbricate in several series; receptacle small, usually ± fimbrillate; disk-fls perfect and fertile, 2–12(–20), in most spp. fewer than the rays; style-branches flattened, with lanceolate, externally short-hairy appendage; achenes several-nerved, short-hairy; pappus of numerous white capillary bristles; rhizomatous perennial herbs with numerous, alternate, resinous-punctate, narrow, entire, sessile or subsessile, mainly or wholly cauline lvs and ± numerous small yellow heads pedunculate or often sessile in small glomerules, commonly forming a terminal, corymbiform, flat-topped infl. 8, N. Amer. All spp. bloom in mid- or late summer and fall. (Often included in *Solidago*.)

1 Disk-fls 12–20, usually as many as or more numerous than the rays; N.S. .... 1. *E. galetorum.*
1 Disk-fls 2–12, mostly fewer than the rays; widespread.
  2 Lvs evidently 3-nerved, the larger ones ordinarily with one or 2 additional pairs of fainter lateral
    nerves; heads relatively broad, mostly 20–35(–45)-fld .... 2. *E. graminifolia.*
  2 Lvs 1-nerved, or obscurely to sometimes evidently 3-nerved, without any additional lateral nerves;
    heads narrower, mostly 10–21-fld.
    3 Lvs relatively thin and lax, and very often with well developed axillary fascicles; invol 3–4.5(–
      5) mm; coastal states .... 3. *E. tenuifolia.*
    3 Lvs relatively thick and firm, ordinarily without axillary fascicles (or these present to some extent
      in *E. remota*); inland.
      4 Invol 3.5–4.5(–5) mm; near the s. shores of lakes Michigan and Erie .... 4. *E. remota.*
      4 Invol (4.5–)5–6.5 mm; w. Ill. to s. Minn. to Colo., s. to Tex., w. Tenn., and w. Fla.
        5 Plants copiously resinous; relatively northern .... 5. *E. gymnospermoides.*
        5 Plants scantily resinous; relatively southern .... 6. *E. leptocephala.*

1. Euthamia galetorum Greene. Nova Scotia f.-t. g. Glabrous, 3–8 dm; lvs evidently to obscurely 3-nerved, 2–6 cm × 1.5–4 mm; infl rather small, the heads sessile in small glomerules; invol 3–4 mm, its bracts atomiferous-glandular, evidently to rather obscurely green-tipped; disk-fls 12–20, as many as or a few more than the small rays; receptacle scarcely fimbrillate; 2*n*=18. Moist lake-shores; s. N.S. (*Solidago g.*)

2. Euthamia graminifolia (L.) Nutt. Common f.-t. g. Glabrous to densely spreading-hirtellous, 3–15 dm; lvs 4–13 cm × 3–12 mm, evidently 3-nerved, the larger ones ordinarily with 1 or 2 additional pairs of fainter lateral nerves; heads mostly sessile or subsessile in small glomerules, mostly 20–35(–45)-fld, with 15–25(–35) rays and (4)5–10(–13) disk-fls; invol turbinate, 3–5 mm; rays minute and often scarcely spreading, ca 1 mm; 2*n*=18. Open, usually moist ground; Nf. and Que. to B.C., s. to N.C., Mo., and N.M. (*Solidago g.*) The common form in our range is var. *graminifolia*, with relatively narrow lvs, the middle cauline ones 11–20 times as long as wide, acuminate. Typical var. *graminifolia* is essentially glabrous. Similar plants with evidently spreading-hirtellous herbage have been called var. *nuttallii* (Greene) W. Stone, but do not form an ecogeographically distinctive population. The var. *major* (Michx.) Moldenke is northern, entering our range only in n. Minn.; it has broader lvs, the middle cauline ones mostly 7–11 times as long as wide, merely acute, and is often smaller, with less ample infl.

    E. × hirtipes (Fern.) Sieren, from se. Va., has the aspect of a hybrid between *S. tenuifolia* var. *tenuifolia* and the *nuttallii* phase of *E. graminifolia*, and was tentatively so considered by Fernald. Some ± similar plants from N.C. are reported to have 2*n*=54 chromosomes. Further study is in order.

3. Euthamia tenuifolia (Pursh) Nutt. Coastal plain f.-t. g. Glabrous to somewhat scabro-hirtellous, 3–10 dm, often sweetly aromatic; lvs relatively thin and lax, linear or nearly filiform, 1–3(–4) mm wide, (15–)20–50 times as long as wide, 1-nerved or also with a pair of weak lateral nerves, very often with well developed axillary fascicles; many or most of the heads commonly pedunculate; invol 3–4.5(–5) mm, evidently resinous, mostly 10–21-fld; rays ca 2 mm, spreading; 2*n*=18. Open, sandy places, especially near the coast; N.S. to Fla. and La. Two well marked vars., often treated as spp.:

    Var. *tenuifolia*. Larger lvs mostly 2–3(–4) mm wide, often with a pair of weak lateral nerves, with or without axillary fascicles; rays mostly (8–)10–16 and disk-fls 5–7(–9), the fls 17–21 in all. Blooming later than var. *microcephala*, where their ranges overlap. Northern, s. to Va. and irregularly to Ga. and n. Fla. (*Solidago t.*)
    Var. *microcephala*. Lvs linear or nearly filiform, 1-nerved or seldom also with a faint pair of lateral nerves, the larger ones 3–6 cm × 1–2(–3) mm, very often with axillary fascicles; rays mostly 7–11(–13) and disk-fls 3–4(5); the fls 11–15(–17) in all. Southern, ranging n. to Md. (*Solidago m.; E. minor.*)

4. Euthamia remota Greene. Lakes f.-t. g. Much like no. 5, but with smaller heads, and with some tendency to develop axillary fascicles of lvs, the lvs often narrower and the heads often more evidently pedunculate; invol 3.5–4.5(–5) mm; 2*n*=18. Moist, open, sandy or mucky places, commonly about lake-shores; near the s. end of lakes Erie and Mich., from N. O. to s. Wis. (*Solidago r.*)

5. **Euthamia gymnospermoides** Greene. Great Plains f.-t. g. Glabrous except for the slightly scabrous lf-margins, 4–10 dm; lvs densely and strongly glandular punctate, obscurely to sometimes evidently 3-nerved, without any additional lateral nerves, linear, 4–10 cm × 1.5–4(–6) mm; heads sessile in small glomerules or somewhat pedunculate, mostly 14–20-fld, the short rays 10–14, the disk-fls 4–6; invol (4.5–)5–6.5 mm, narrow, ± strongly glutinous, its bracts obtuse or the inner acute; $2n=36$. Open, often sandy, moist to rather dry places; s. Minn. and w. Ill. to Mo., Colo., and Tex. (*Solidago g.*)

6. **Euthamia leptocephala** (T. & G.) Greene. Mississippi Valley f.-t. g. Similar to no. 5, but much less resinous, the lvs only sparsely or obscurely punctate, or sometimes more closely and evidently so, but the punctae then with scanty resin and appearing somewhat pustulate; lvs 4–8 cm × 3–6 mm, 10–20 times as long as wide; invol only slightly or scarcely glutinous; lvs 3–7 mm wide; $2n=18, 54$. Open, often moist and sandy places and thin woods; s. Mo. to s. Ill. and c. Tenn., s. to Tex., La., Miss., and w. Fla. (*Solidago l.*)

55. GUTIERREZIA Lagasca.

Heads radiate, the rays pistillate and fertile, often few; invol bracts strongly imbricate, glutinous, herbaceous-tipped; receptacle flat to conic or hemispheric, alveolate and evidently pubescent; disk-fls perfect and fertile; style-branches flattened, with long, slender, externally short-hairy appendage; achenes ± clavate, pubescent, usually several-nerved; pappus of the disk-fls of 1–2 series of white, erose scales evidently shortly than the cor, often united at base (rarely none), that of the rays ± reduced or even obsolete; taprooted herbs or shrubs, generally glabrous and often glutinous, with alternate, narrow, entire, commonly punctate lvs and usually numerous small yellow heads in a terminal, often flat-topped infl. 15, N. Amer.

1. **Gutierrezia sarothrae** (Pursh) Britton & Rusby. Matchbrush. Subshrub 1–6 dm, branched below; lvs ± linear, to 3 cm × 2 mm; heads numerous, narrow, obconic; invol 3–4 mm; rays 3–8, 2–3 mm; pappus of the disk-fls of several short but evident scales, that of the rays similar but the scales shorter and broader; $2n=8$. Dry soil in open places, especially where overgrazed; Man. and w. Minn. to Kans. and Okla., w. to Wash. and Mex. July–Sept.

56. AMPHIACHYRIS (A. DC.) Nutt.

Heads radiate, the rays 7–12, pistillate and fertile; invol bracts strongly imbricate, glutinous, shining, with stramineous base and abrupt green tip; receptacle flattish or convex, alveolate and sometimes also pubescent; disk-fls 10–50, functionally staminate, without an ovary, the style-branches short, slender, short-hairy, not stigmatic; achenes clavate, hairy; pappus of the disk-fls of 5(–8) long, white, spatulate scales united up to half-length, that of the rays a minute, toothed crown or several short, ± concrescent scales; taprooted, glabrous annuals with alternate, narrow, entire, punctate lvs and numerous small yellow heads in a ± flat-topped infl. 2, N. Amer.

1. **Amphiachyris dracunculoides** (DC.) Nutt. Broomweed. Annual, 3–8 dm, simple below, bushy-branched above; lvs numerous, linear, finely glandular-punctate, to 6 cm × 3 mm; heads numerous, campanulate or hemispheric; invol 3–6 mm; rays 6–10, 3–5 mm; achenes multinerved, villous-puberulent; $2n=8, 10$. Dry soil on plains and prairies; Mo. and Kans. to Tex. and N.M.; Tenn.; Ala.; intr. in Ky. July–Oct. (*Gutierrezia d.*)

57. ASTER L.

Wild aster; Michaelmas-daisy. Heads generally radiate, the rays pistillate and fertile, anthocyanic or white, in a few spp. reduced and inconspicuous or even wanting, the cor of the pistillate fls then a mere slender tube; invol bracts in 2 or more series, equal or more often imbricate, usually ± herbaceous at the tip and chartaceous below, sometimes herbaceous or chartaceous throughout; receptacle naked, flat or a little convex; disk-fls perfect and fertile, red or purple to yellow; style-branches flattened, with mostly narrow and acute or acuminate, externally short-hairy appendage; achenes several-nerved; pappus of numerous capillary bristles, sometimes with an additional short outer series; perennial (seldom annual) herbs, most spp. fibrous-rooted, with simple, alternate, entire or variously toothed lvs and solitary to more often several or numerous, hemispheric to subcylindric heads. (*Brachyactis, Doellingeria, Ionactis, Sericocarpus.*) 175+, mainly N. Amer. Our spp. bloom in mid- or late summer and fall. Hybrids abound; some of the more notable ones are here formally treated. Some of the species with chromosome-numbers based on $x=8$ (especially in the *Heterophylli*) tend to have additional B-chromosomes, and have often been reported to have numbers based on 9. The disks-cors characteristically consist of a slender (often short) basal tube and a more swollen limb;

the lobes are part of the limb. The terms *lvs basally disposed* and *lvs chiefly cauline* are here used as explained under *Solidago*.

The plant traditionally called *Aster* (or *Unamia*) *ptarmicoides* is here referred to *Solidago*, in spite of its white rays. It hybridizes with several spp. of *Solidago* (especially of the Sect. *Oligoneuron*), but not with *Aster*. It is marked by its 10–25 white rays, white disk-cors and copious (but not double) pappus with many of the bristles clavellate-thickened toward the tip.

KEYS TO THE GROUPS OF SPECIES

1 At least the basal or lower cauline lvs cordate or subcordate at base and also evidently petioled.
  2 Infl corymbiform, commonly flat or round-topped (occasionally more elongate), with few and often
    leafy bracts; invol bracts relatively broad and firm, the outer often over 1 mm wide and seldom
    more than 2.5 times as long as wide; plants ordinarily becoming colonial by creeping rhizomes,
    and usually either glandular or with white rays (*Biotia*) .... Group VIII.
  2 Infl more paniculiform, often elongate, its bracts either very narrow or very small or both, often
    numerous; invol bracts relatively narrow, the outer not over 1 mm wide and usually at least
    3 times as long as wide; plants rarely becoming colonial (except *A. ciliolatus*), eglandular,
    usually with anthocyanic rays (except *A. sagittifolius*) (*Heterophylli*) .... Group V.
1 None of the lvs at once cordate and evidently petioled.
  3 Lvs sessile and conspicuously cordate-clasping; invol bracts well imbricate, glandular or hairy .... Group VI.
  3 Lvs sessile or petiolate, sometimes strongly auriculate-clasping (chiefly in spp. with either glabrous
    or scarcely imbricate invol bracts), but not cordate.
    4 Plants with either the invol or the achenes distinctly glandular, the herbage sometimes glandular
      as well.
      5 Achenes glandular; invol only obscurely or not at all glandular .... Group X.
      5 Achenes glabrous or hairy, not glandular; invol glandular .... Group VII.
    4 Plants not glandular.
      6 Lvs (except sometimes the lower) ± strongly auriculate-clasping .... Group I.
      6 Lvs scarcely or not at all auriculate-clasping, though sometimes with broad or slightly sheathing
        base.
        7 Lvs silvery-silky on both sides (sometimes glabrate in age), entire (*Concolores*) .... Group IV.
        7 Lvs glabrous or hairy, but not all all silvery-silky, sometimes toothed.
          8 Taprooted annuals with slender and short rays scarcely exceeding the mature pappus,
            or the rays wanting .... Group XIV.
          8 Fibrous-rooted perennials (from a rhizome, caudex, or crown) with relatively well de-
            veloped and conspicuous rays.
            9 Pappus double, the inner of long capillary bristles, the outer of short bristles ca 1 mm
              or less .... Group XI.
            9 Pappus simple, not divided into distinct inner and outer series.
              10 Rays few, mostly 3–8; achenes densely sericeous (*Sericocarpus*) .... Group XII.
              10 Rays more numerous, mostly 8–50; achenes glabrous or ± hairy, but not densely
                sericeous.
                11 Plants with fleshy, very narrow lvs seldom as much as 1 cm wide, strictly
                  glabrous; invol bracts scarcely herbaceous; coastal salt-marshes .... Group XIII.
                11 Plants otherwise, the lvs not fleshy, though sometimes very narrow; invol
                  bracts mostly with ± herbaceous tip; glabrous or more often ± hairy;
                  widespread.
                  12 Invol bracts, or some of them, with subulate, marginally inrolled green
                    tip; perennial from a caudex or very short and stout rhizome .... Group III.
                  12 Invol bracts flat, not with subulate, marginally inrolled tip (but the tip
                    spinulose-mucronate in *A. ericoides* and *A. falcatus*); plants often but
                    not always with creeping rhizomes.
                    13 Invols 7–12 mm, or occasionally only 6 mm in some spp. with broad
                      invol bracts; larger invol bracts generally 1 mm wide or more
                      (except in *A. elliottii*); rays commonly violet or blue, occasionally
                      pink (regularly so in *A. elliottii*) or white .... Group IX.
                    13 Invols 2.5–6 mm; ordinarily none of the bracts over 1 mm wide
                      (occasionally a little more in *A. falcatus*); rays most often white,
                      sometimes pink or blue .... Group II.

GROUP I

1 Invol bracts, at least the inner ones, long-acuminate to attenuate.
  2 Perennial from a short, stout rhizome or caudex, occasionally with short stolons as well; stem
    uniformly hairy at least under the heads, densely spreading-hispid to occasionally glabrous
    below the infl; lvs of the infl not conspicuously crowded .... 5. *A. puniceus*.

  2 Perennial from long, creeping rhizomes; stem and branches pubescent in lines above, glabrous or
      sparingly hispid below the infl; lvs, at least those of the infl, conspicuously crowded . . . . 6. *A. firmus.*
1 Invol bracts obtuse or merely acute, occasionally acuminate.
  3 Outer invol bracts, or some of them, enlarged and leafy throughout; Me. and n. N.H. to Lab. . . . . . . 1. *A. crenifolius.*
  3 Outer invol bracts equaling or shorter than the inner, generally chartaceous at base.
    4 Perennial from a short, stout rhizome or branching caudex, seldom with shortly creeping rhizomes
      as well; invol bracts appressed, strongly imbricate; herbage glabrous, or with some lines of
      hairs in the infl, generally glaucous . . . . 9. *A. laevis.*
    4 Perennial from long creeping rhizomes, without any well developed caudex; invol bracts equal
      or imbricate, often loose or spreading; herbage glabrous or hairy, not glaucous.
      5 Lvs with ovate or lanceolate, serrate blade abruptly contracted or occasionally more tapering
        into a relatively long, generally entire, strongly auriculate-clasping petiolar base . . . . 8. *A. prenanthoides.*
      5 Lvs otherwise, commonly sessile or nearly so.
        6 Stem coarsely spreading-hairy, much as in *A. puniceus* . . . . 4. *A.* × *longulus.*
        6 Stem glabrous or inconspicuously hairy (often short-hairy in lines).
          7 Invol bracts imbricate or equal; plants slender to more often stout,
          the stem often over 2.5 mm thick and the rhizome often over 2 mm thick, the lvs
          mostly over 5 mm wide . . . . 2. *A. novi-belgii.*
          7 Invol bracts appressed, imbricate; very slender plants, the stem rarely over 2.5 mm thick,
          the rhizome rarely over 2 mm thick, the lvs 1.5–5(–9) mm wide . . . . 3. *A. borealis.*

Group II

1 Invol bracts, at least the outer, with loose or squarrose, minutely spinulose-mucronate tip; lvs entire.
  2 Heads numerous, often secund on the branches, small, the invol 3–5 mm; rays 8–20; widespread
    . . . . 20. *A. ericoides.*
  2 Heads fewer, solitary or clustered at the ends of the branches, or the stem scarcely branched and
    the infl subracemiform; invol 5–7 mm; rays 20–35; western . . . . 19. *A. falcatus.*
1 Invol bracts appressed or a little loose, not spinulose-mucronate; lvs various.
  3 Lobes of the disk-cors comprising 45–75% of the limb; lvs usually hairy beneath, at least along the
    midrib.
    4 Plants with creeping rhizomes; lvs generally hairy across the surface beneath . . . . 15. *A. ontarionis.*
    4 Plants without creeping rhizomes; lvs hairy only along the midrib beneath, or seldom glabrous
    . . . . 16. *A. lateriflorus.*
  3 Lobes of the disk-cors comprising 15–45% of the limb; lvs glabrous beneath except in forms of *A.*
    praealtus.
    5 Heads very small (invol 2.5–3.5 or 4 mm; rays 3–6 mm), numerous, and often unilaterally
      racemiform on the branches of the infl.
      6 Rameal lvs much reduced, seldom 1.5 cm . . . . 17. *A. racemosus.*
      6 Rameal lvs much larger, usually many of them 1.5 cm or more . . . . 12. *A. lanceolatus.*
    5 Heads either larger or few, seldom unilaterally racemiform.
      7 Veinlets of the lf forming a conspicuous reticulum beneath, the areolae mostly nearly isodi-
        ametric; lobes of the disk-cors comprising 17–25% of the limb; chiefly western . . . . 13. *A. praealtus.*
      7 Veinlets of the lf forming an obscure reticulum, or, if the reticulum is evident, then the areolae
        clearly longer than broad (except sometimes in the Appalachian, nonrhizomatous sp. *A.*
        concinnus).
        8 Invol bracts mostly not strongly imbricate, the outer usually at least two-thirds as long as
          the inner; lobes of the disk-cors comprising 19–36% of the limb; western . . . . 11. *A. hesperius.*
        8 Invol bracts ± strongly imbricate, the outer seldom as much as two-thirds as long as the
          inner.
          9 Peduncles conspicuously elongate, copiously bracteate, usually at least some of them 2
            cm or more; green tips of the invol bracts mostly short and broad.
            10 Lvs less than 1 cm wide . . . . 18. *A. dumosus.*
            10 Lvs usually well over 1 cm wide . . . . 10. *A. concinnus.*
          9 Peduncles either short, or sparsely bracteate, or the bracts lf-like; green tips of the invol
            bracts mostly more elongate.
            11 Infl mostly short and broad; invol 5–7 mm; rhizome slender, seldom over 2 mm
              thick; lobes of the disk-cors comprising 15–30% of the limb; very slender plants
              of cold bogs . . . . 3. *A. borealis.*
            11 Infl mostly elongate; invol 3.5–6 mm high; rhizome often over 2 mm thick; lobes
              of the disk-cors comprising 30–45% of the limb; stouter plants, not of cold bogs.
              12 Low plants, mostly 1.5–6 dm; northeastern . . . . 14. *A. tradescantii.*
              12 Taller plants, mostly 6–15 dm; widespread . . . . 12. *A. lanceolatus.*

Group III

1 Invol broadly urn-shaped, with mostly 40–100 fls . . . . 21. *A. pilosus.*
1 Invol narrower, mostly narrowly obconic when pressed, with 16–32 fls.
  2 Ray-fls 4–10 more than those of the disk; plants usually hairy; western . . . . 22. *A. parviceps.*
  2 Ray-fls 0–3 more than those of the disk; plants glabrous; se. Pa. and Md. . . . . 23. *A. depauperatus.*

Group IV

1 Achenes densely sericeous; infl mostly elongate-racemiform .... 24. *A. concolor.*
1 Achenes glabrous; infl loosely corymbiform or paniculiform .... 25. *A. sericeus.*

Group V

1 Invol bracts with reflexed tip; rays mostly 20–45 .... 35. *A. anomalus.*
1 Invol bracts with erect, often appressed tip; rays mostly 10–25.
  2 Cauline lvs, or some of them, either sessile and cordate-clasping or with conspicuously auriculate-
     clasping petiole .... 34. *A. undulatus.*
  2 Cauline lvs not at all clasping.
    3 Lvs ordinarily entire or subentire; invol bracts with short, broad, diamond-shaped green tip.
      4 Plant with nearly all the lvs below the infl cordate or subcordate; invol bracts minutely hairy
        on the back .... 28. *A. shortii.*
      4 Plants with only the lower lvs cordate or subcordate; invol bracts glabrous except for the
        sometimes ciliolate margins .... 27. *A. oolentangiensis.*
    3 Lvs evidently toothed (at least the lower ones); bracts various.
      5 Peduncles and branches of the infl generally sparsely or scarcely bracteate, the peduncles very
        unequal, some usually over 1 cm; heads relatively few, often less than 50, rarely over 100
        .... 26. *A. ciliolatus.*
      5 Peduncles and branches of the infl ordinarily copiously bracteate, especially when the peduncles
        approach or exceed 1 cm; infl often with more than 100 heads.
        6 Plants essentially glabrous, or sometimes slightly puberulent in the infl.
          7 Green tips of the invol bracts very narrow and elongate, the bracts acuminate or very
            acute; infl narrow, with mostly strongly ascending branches and short, crowded
            peduncles; widespread .... 31. *A. sagittifolius.*
          7 Green tips of the invol bracts relatively short and broad, ± diamond-shaped, the bracts
            obtuse or barely acute; infl broader and more open, with longer peduncles; Appa-
            lachian region from N.Y. s. .... 30. *A. lowrieanus.*
        6 Plants with one or both sides of the lvs, or the stem, or both, ± hairy or scabrous.
          8 Invol bracts obtuse or merely acute, the tip short and broad, its green color usually partly
            or even wholly replaced or obscured by anthocyanin; lvs usually sharply and deeply
            toothed, and with scarcely or obscurely winged petiole .... 29. *A. cordifolius.*
          8 Invol bracts strongly acute to acuminate, the green tip ± elongate and often narrow,
            ordinarily not at all purple to the naked eye; lvs usually shallowly and often bluntly
            toothed, the cauline ones often with strongly winged petiole.
            9 Branches of the infl with long, copiously and minutely bracteate, nonfloriferous base,
              or the peduncles mostly well over 1 cm and copiously bracteate, or both; Ozarkian
              .... 33. *A. texanus.*
            9 Branches of the infl shorter and more densely floriferous, either more leafy or floriferous
              toward the base, the peduncles seldom 1 cm; widespread.
              10 Plants relatively densely hairy (see text); green tips of the invol bracts tending to
                be elongate-rhombic; rays commonly bright blue .... 32. *A. drummondii.*
              10 Plants relatively thinly hairy (see text); green tips of the invol bracts very narrow;
                rays pale bluish or lilac to white .... 31. *A. sagittifolius.*

Group VI

A single species .... 36. *A. patens.*

Group VII

1 Lvs basally disposed.
  2 Invol strongly glandular; Mass. to S.C., chiefly near the coast.
    3 Lowest lvs abruptly contracted to the petiole; invol bracts appressed, with greenish but scarcely
      herbaceous tip; Mass. to L.I. .... 49. *A. × herveyi.*
    3 Lowest lvs tapering to the petiole; invol bracts with loose or spreading herbaceous tip; Mass. to
      S.C. .... 50. *A. spectabilis.*
  2 Invol only slightly glandular; s. Appalachian reg. .... 51. *A. surculosus.*
1 Lvs chiefly cauline.
  4 Lvs strongly auriculate-clasping.
    5 Stems clustered on a caudex or short rhizome; rays 45–100; widespread .... 39. *A. novae-angliae.*
    5 Stems scattered, from creeping rhizomes; rays 20–40; northwestern .... 40. *A. modestus.*
  4 Lvs only slightly or scarcely auriculate-clasping.
    6 Lvs and invol bracts thin; outer invol bracts herbaceous to the base; northwestern .... 40. *A. modestus.*

6 Lvs and invol bracts thick and firm; invol bracts ordinarily distinctly chartaceous at base; widespread.
  7 Heads small, the invol 5–8 mm; widespread, but not in e. Va. .... 37. *A. oblongifolius.*
  7 Heads large, the invol 10–15 mm; e. Va. and N.C. .... 38. *A. grandiflorus.*

Group VIII

1 Plants glandular in the infl; rays somewhat purplish in life .... 41. *A. macrophyllus.*
1 Plants not glandular; rays ordinarily white.
  2 Plants with well developed tufts of basal lvs on separate short shoots .... 42. *A. schreberi.*
  2 Plants ordinarily without well developed tufts of basal lvs.
    3 Lvs glabrous or sparsely hairy; eastern .... 43. *A. divaricatus.*
    3 Lvs scabrous above, ± densely spreading-hairy beneath; western .... 44. *A. furcatus.*

Group IX

1 Pappus very coarse and firm; lvs linear, 14–40 times as long as wide (*Heleastrum*) .... 53. *A. hemisphericus.*
1 Pappus relatively soft and fine; lvs usually wider.
  2 Outer invol bracts much narrower than the inner and very small, passing into the minute peduncular bracts; western. (Occasional large-headed plants of the Appalachian sp. *A. concinnus* might be sought here) .... 45. *A. turbinellus.*
  2 Outer invol bracts not much if at all narrower than the inner, generally well differentiated from the usually more lf-like peduncular ones.
    3 Invol bracts long-attenuate, not much imbricate; rays commonly pink .... 7. *A. elliottii.*
    3 Invol bracts rounded or obtuse to strongly acute, but not attenuate, often strongly imbricate; rays mostly blue or purple.
      4 Invol bracts, or some of them, with rather loose or spreading tip; herbage glabrous or inconspicuously hairy; mainly near the coast .... 2. *A. novi-belgii.*
      4 Invol bracts appressed; herbage variously glabrous or hairy.
        5 Infl sparsely leafy, corymbiform, ± flat-topped, with usually few heads; lvs ± basally disposed, except in *A. radula.*
          6 Lvs very obscurely veined, usually glabrous beneath and entire or nearly so; N.J. and s.
            7 Invol narrowly obconic or turbinate; rays mostly 9–14 .... 52. *A. gracilis.*
            7 Invol broader, more campanulate; rays mostly 15–30 .... 51. *A. surculosus.*
          6 Lvs evidently veiny beneath, usually toothed and ± hairy beneath.
            8 Lower lvs the largest, 3–10 cm wide; casual escape .... 46. *A. tataricus.*
            8 Lower lvs reduced and deciduous; native.
              9 Larger lvs mostly 4–6 cm wide; local in Ky .... 47. *A. saxicastellii.*
              9 Larger lvs up to 2.5(–3) cm wide; Nf. to Va. and Ky. .... 48. *A. radula.*
        5 Infl densely leafy, usually elongate, the heads ± numerous.
          10 Veinlets of the lf forming a conspicuous reticulum beneath, with nearly isodiametric areolae .... 13. *A. praealtus.*
          10 Veinlets of the lf not forming a conspicuous reticulum beneath, the areolae, if evident, clearly longer than broad .... 11. *A. hesperius.*

Group X

1 Lvs (including the reduced lower ones) mostly 10–22 below the infl, the largest ones 2–6 cm wide .. 54. *A. acuminatus.*
1 Lvs mostly 25–75 below the infl, not over 2.5 cm wide.
  2 Lvs mostly 25–40 below the infl, the largest ones 7–24 mm wide .... 55. *A. × blakei.*
  2 Lvs mostly 40–75 below the infl, the largest ones 5–12 mm wide .... 56. *A. nemoralis.*

Group XI

1 Inner pappus-bristles clavate; lvs veiny, scarcely rigid; rays mostly white (*Doellingeria*).
  2 Achenes short-hairy; invol 3–5 mm; creeping rhizomes present.
    3 Heads 23–54-fld, with 7–14 rays and 16–40 disk-fls; widespread .... 57. *A. umbellatus.*
    3 Heads 12–22-fld, with 4–7 rays and 8–15 disk-fls; western .... 58. *A. pubentior.*
  2 Achenes glabrous; invol 4.5–7 mm; no creeping rhizomes .... 59. *A. infirmus.*
1 Inner pappus-bristles not clavate; lvs rigid, 1-nerved, otherwise veinless; rays mostly violet (*Ianthe, Ionactis*) .... 60. *A. linariifolius.*

Group XII

1 Lvs, or some of them, toothed, the largest ones 1–4.5 cm wide; disk-fls 9–20 .... 61. *A. paternus.*
1 Lvs entire, 2–12 mm wide; disk-fls 5–10 .... 62. *A. solidagineus.*

Group XIII

A single species (*Oxytripolium,* in part) .... 63. *A. tenuifolius.*

Group XIV

1 Pistillate cor surpassing the style, the ligule over 1 mm, becoming circinately rolled outwards; invol
    bracts only slightly or scarcely herbaceous (*Oxytripolium,* in part) .... 64. *A. subulatus.*
1 Pistillate cor shorter than the style, the ligule wanting or a vestige less than 1 mm; outer invol bracts
    distinctly herbaceous (*Brachyactis, Conyzopsis*).
    2 Lvs mostly 5–10 times as long as wide; plants under 2.5 dm; northeastern .... 65. *A. laurentianus.*
    2 Lvs mostly 13–30 times as long as wide; plants often over 2.5 dm; more western .... 66. *A. brachyactis.*

    1. Aster crenifolius (Fern.) Cronq.   Newfoundland-a. Rhizomatous; stem 1–7 dm, puberulent in lines decurrent from the lf-bases, at least above; lvs glabrous, or hairy along the midrib beneath, lanceolate to oblong or elliptic, 5–15 cm × 9–40 mm, rather distantly serrulate to occasionally entire, sessile and auriculate-clasping, or the lower tapering and subpetiolate but not enlarged and mostly deciduous; heads few in a corymbiform infl, or solitary; invol glabrous, its bracts scarcely or not at all imbricate, at least some of the outer leafy throughout and surpassing the inner; rays 15–40(–60), blue, 8–17 mm; 2*n*=32, 48, 64? Moist, exposed or protected places; Nf. to N.B., Me., n. N.H., and apparently n. Vt. (*A. tardiflorus,* misapplied.)

    2. Aster novi-belgii L.   New York a. Large and stout to sometimes small and slender, from long creeping rhizomes; stem 2–14 dm, sometimes puberulent in lines, sometimes glabrous except just under the heads; lvs chiefly cauline, lanceolate to elliptic or lance-linear, sessile and usually ± auriculate-clasping, though often narrowed toward the base, sharply serrate to entire, glabrous except for the scabrous-ciliate margins, often thick and firm, 4–17 cm × 4–25 mm; heads several or many in an open or more often leafy-bracteate infl; invol glabrous, 5–10 mm, its obtuse to sometimes sharply acute bracts ± imbricate or subequal, generally at least some of them with noticeably loose or spreading tip; green portion of the invol bracts evidently tapering and cuneate at the base, or the outermost bracts wholly green; rays 20–50, blue (rose or white), 6–14 mm; 2*n*=48, 64. Moist places, often in salt-marshes; Nf. to S.C., chiefly near the coast. (*A. elodes; A. longifolius; A. tardiflorus.*)

    A. retroflexus Lindley (*A. curtisii* T. & G.), known from the Blue Ridge Province of N.C., S.C., Ga., and Tenn., may be expected in s. Va. It will key to *A. novi-belgii,* from which it differs most notably in that the green tip of the invol bracts is subtruncate at the base.

    A. anticostensis Fern. (*A. gaspensis* Victorin), 2*n*=80, is thought to be an alloploid derivative of nos. 2 and 3. Previously known from the Gaspé Peninsula and Anticosti I., it has recently been reported from ne. Me. and adj. N.B.; it grows on unstable, gravelly, usually calcareous river-banks that are subject to periodic flooding. It suggests a few-headed, sparsely leafy-branched form of no. 2 with narrow, only inconspicuously clasping lvs, but the invol bracts are ± appressed.

    3. Aster borealis Prov.   Northern bog-a. Plants slender, from long rhizomes seldom over 2 mm thick; stem 1.5–10 dm, glabrous below, puberulent above in decurrent lines, seldom over 2.5 mm thick; lvs chiefly cauline, linear to lance-linear or oblong-linear, 4–13 cm × 2–5(–9) mm, sessile and usually slightly auriculate-clasping, entire or subentire, generally scabrous on the margins and sometimes puberulent along the midrib beneath, ascending, mostly thin and lax; heads few to sometimes rather many in an open, usually short and broad infl, or occasionally solitary in small plants; invol 5–7 mm, glabrous, its slender, mostly acute bracts ± strongly imbricate, often with purple tip or margins; rays 20–50, white to pale blue or lavender, 7–15 mm; lobes of the disk-cors comprising 15–30% of the limb; 2*n*=16, 32, 48, 64. Cold bogs; n. N.J. to Que., w. to Minn., Colo., and Alas. (*A. junceus,* misapplied; *A. junciformis.*)

    4. Aster × longulus E. Sheldon.   Thought to be a set of hybrids between nos. 3 and 5; rhizomatous; stem coarsely hairy as in no. 5; lvs 10–16 times as long as wide, only slightly clasping, long-hairy beneath at least along the midrib and toward the base. N. Wis. and n. Minn. to Sask.; apparently also in n. N.J. in a slightly less hairy form. (? *A. lanceolatus* var. *hirsuticaulis.*)

    5. Aster puniceus L.   Bristly a. Plants with a short stout rhizome or caudex, sometimes with short thick stolons as well; stem stout, 0.5–2.5 m, simple or much branched above, uniformly spreading-hairy at least under the heads, conspicuously spreading-hispid (or occasionally glabrous) below; lvs chiefly cauline, sessile, auriculate-clasping, rather distantly serrate to occasionally entire, scabrous to subglabrous above, glabrous or spreading-hairy along the midrib beneath, lanceolate to oblong or elliptic-oblong, 7–16 cm × 12–40 mm;

heads few to many in a leafy infl; invol 6–12 mm, its bracts slender and loose, scarcely or not at all imbricate, at least the inner long-acuminate to attenuate, often some of the outer enlarged and leafy, but still narrow; rays 30–60, blue (rose or white), 7–18 mm; achenes glabrous or nearly so; $2n=16$. Swamps and other moist places; Nf. to Sask., s. to Va., Ill., and Neb., and in the mts. to Ga. and Ala.

6. Aster firmus Nees. Shining a. Similar to no. 5, but colonial by long creeping rhizomes; stem and branches puberulent in lines above, glabrous or sparingly hispid below the infl; lvs more crowded, especially upwards, firmer, often shining, entire or nearly so; invol bracts often less attenuate; rays blue or lavender (mostly rather pale) to occasionally white; $2n=16, 32$. Moist places; W.Va. and w. N.Y. to Minn., S.D., and s. Mo. (*A. lucidulus.*)

7. Aster elliottii T. & G. Southern swamp-a. Colonial from well developed creeping rhizomes, 6–15 dm, tending to be somewhat succulent; stem puberulent in lines, at least above; lvs glabrous beneath, scaberulous above, serrate, the lowermost ones enlarged, long-petiolate, with elliptic blade to 25 × 5 cm, but sometimes early deciduous, the cauline ones ± reduced upwards, becoming sessile or nearly so with ± sheathing but not auriculate base; infl corymbiform or paniculiform; invol 8–11 mm, its bracts not much if at all imbricate, glabrous except for the sometimes ciliolate margins, narrow with long-attenuate, loose or somewhat squarrose green tip (this well under 1 mm wide), or the inner purplish instead of green; rays 25–45, pink or sometimes lavender, 7–12 mm; achenes glabrous or sparsely hairy; $2n=16$. Swamps and other moist low places; se. Va. to Fla. and Ala.

8. Aster prenanthoides Muhl. Zigzag a. Colonial from well developed creeping rhizomes; stems 2–10 dm, often zigzag, pubescent in lines, becoming uniformly so under the heads and glabrate toward the base; lvs chiefly cauline, scabrous to glabrous above, glabrous or loosely hairy along the midrib beneath, the main ones mostly 6–20 × 1–5 cm, with ovate or lanceolate, serrate, acuminate blade abruptly contracted into a relatively long, generally entire, broadly winged, strongly auriculate-clasping petiolar base; heads several or many in an open, sparsely leafy-bracteate infl; invol glabrous, 5–7 mm, its bracts acute or obtusish, seldom much imbricate, rather loose, often with spreading tip, the outer sometimes wholly herbaceous; rays 20–35, blue or pale purple (white), 7–15 mm; achenes strigose; pappus yellowish; $2n=32, 48, 64$. Streambanks, meadows, and moist woods; N.Y. to Minn., s. to D.C., Ky., Io., and in the mts. to N.C. and Tenn. (*A. oticus.*)

9. Aster laevis L. Smooth a. Plants 3–10 dm from a short, stout rhizome or branched caudex, sometimes with short, slender, creeping red rhizomes as well; herbage glabrous except occasionally for some puberulent lines in the infl, commonly ± glaucous; lvs mainly cauline, thick and firm, highly variable in size and shape, but the larger ones more than 1 cm (often more than 2.5 cm) wide and often less than 5 times as long as wide, entire or sometimes toothed, sessile and ± strongly auriculate-clasping, the lower tapering to a winged petiole and scarcely clasping, those of the infl reduced and bract-like, broadest at the clasping or subclasping base; heads several or many in an open infl; invol 5–9 mm, its firm, appressed, acute bracts conspicuously imbricate in several series, with short, diamond-shaped green tip mostly 1–2 mm; rays 15–30, blue or purple, 8–15 mm; achenes glabrous or nearly so; pappus usually reddish; $2n=48$. Open, usually dry places; Me. to B.C., s. to Ga., Ark., and N.M. Most of our plants are var. *laevis,* as described above. The western var. *geyeri* A. Gray, with the invol bracts fewer-seriate, narrower, and more sharply pointed, their green tips more elongate, extends into our range from Minn. to Mo. (*A. geyeri; A. laeviformis.*)

10. Aster concinnus Willd. Narrow-lvd smooth a. Much like no. 9 and perhaps only varietally distinct, but with narrower, often less evidently glaucous lvs, the principal ones only slightly or not at all clasping, often petiolate, seldom more than 2.5 cm wide and seldom less than 5 times as long as wide (petiole included in the length); infl avg more diffusely branched, with more numerous and smaller bracts, the heads mostly long-pedunculate; $2n=48$. Dry woods and open places; s. Appalachian region from s. Pa. to Ga. and Ala.

11. Aster hesperius A. Gray. Western lined a. Similar to no. 12, and passing into it; lvs usually entire; invol avg larger, to 7.5 mm; its bracts seldom much imbricate, avg broader; rays blue or white; lobes of the disk-cors comprising 19–36% of the limb; $2n=48, 64$. Moist low places; Wis. and Mo. to Alta. and s. Calif.

12. Aster lanceolatus Willd. Eastern lined a. Colonial by long rhizomes; stems 6–15 dm, pubescent in lines above; lvs all cauline, lanceolate or lance-elliptic to linear, serrate or occasionally entire, glabrous or somewhat scabrous above, sessile or tapering to a petiole-like base, sometimes a little clasping, but scarcely auriculate, the main ones 8–15 cm × 3–35 mm, mostly not strongly reticulate, the areolae, if visible, generally irregular and longer than wide; heads ± numerous in an elongate leafy infl, the invol 3–6 mm, its bracts narrow, sharply acute to acutish, glabrous except for the sometimes ciliolate margins, ± strongly imbricate, with elongate, usually appressed green tip; rays 20–40, white or occasionally lavender or blue, 4.5–12 mm; lobes of the disk-cors comprising 30–45% of the limb; $2n=32, 48, 64$. Moist low places; N.S. to N.C., w. to N.D. and Tex. (*A. paniculatus,* a preoccupied name; *A. lamarckianus; A. simplex* misapplied, the type apparently a garden hybrid of *A. laevis* and *A. lanceolatus* var.) Three vars.:

a Heads very small, the invol 3–4 mm; w. N.Y. to Wis. and Mo. (*A. interior.*) . . . .
. . . . var. *interior* (Wieg.) Semple & Chmielewski.
a Heads larger, the invol mostly 4–6 mm.
  b Lvs mostly 1–3.5 cm wide, seldom over 11 times as long as wide; widespread, but mostly giving
     way in N. Engl. and Can. to the next. (*A. acutidens; A. simplex.*) . . . . var. *simplex* (Willd.) A. G. Jones.
  b Lvs 3–12 mm wide, at least 12 times as long as wide; invol bracts avg narrower than in var. *simplex;*
    N.S. and Que. to Pa. and n. N.J., w. to Minn. (*A. simplex* var. *ramosissimus.*) . . . . var. *lanceolatus.*

13. Aster praealtus Poiret.   Veiny lined a. Much like no. 12; lvs thick and firm, conspicuously reticulate beneath with nearly isodiametric areolae, entire or nearly so, scabrous to subglabrous above, glabrous or scabrous-puberulent beneath, the main ones 7–13 cm × 8–18 mm; invol 5–7(–8) mm; rays 6–15 mm, bluish-purple or rarely white; lobes of the disk-cors comprising 17–25% of the limb; $2n=32$, 64. Moist low ground; Mich. to Ga., w. to Neb., s. Ariz., and n. Mex.; irregularly and rarely e. to Mass., N.Y., N.J., and Md. (*A. salicifolius,* a preoccupied name.) The otherwise more western var. *nebraskensis* (Britton) Wieg., with the stem and often also the lower lf-surfaces more uniformly and conspicuously spreading-puberulent, has been reported from Io. (*A. nebraskensis; ? A. woldeni.*)

14. Aster tradescantii L.   Shore-a. Rhizomatous, 1.5–6 dm; stem glabrous below, usually hairy in lines above; lvs glabrous, entire or shallowly toothed, linear or lance-linear to lance-elliptic, the basal petiolate, the others sessile or nearly so, 3–10 cm × 3–10 mm, the upper reduced and mostly rather few; heads relatively few in an elongate, open infl; invol 3.5–5 mm high, glabrous, its narrow bracts well imbricate, with ± elongate green tip; rays 15–30, 3–8 mm, white; lobes of the disk-cors comprising 30–45% of the limb; $2n=16$, 32. Seasonally inundated streambanks and shores, often among rocks; s. Que., s. N.S. and c. Me. to N.Y. and the Delaware R. in N.J.; n. Mich.? (*A. saxatilis.*)

15. Aster ontarionis Wieg.   Bottomland-a. Similar to no. 16, but with long, creeping rhizomes, and often verging toward no. 12 in habit; stem uniformly spreading-puberulent, at least above; lvs uniformly (sometimes sparsely) puberulent beneath, or sometimes more densely so and almost villous along the midrib, more scabrous or subglabrous above; invol glabrous or sometimes puberulent; lobes of the disk-cors comprising 45–65% of the limb; $2n=32$, 48. River bottom-lands; n. N.Y. to Ky., Tenn., and n. Miss., w. to Minn., S.D., and Okla. (*A. missouriensis,* a preoccupied name.)

16. Aster lateriflorus (L.) Britton.   Goblet-a. Stems 3–12 dm from a branched caudex or short stout rhizome, ± curly-villous to glabrous; lvs scabrous or glabrous above, apparently glabrous beneath except for the usually villous or puberulent midrib; basal and lower cauline lvs soon deciduous, or the basal sometimes persistent, petiolate, and with obovate to elliptic or subrotund blade to 8 × 4 cm, those above sessile or nearly so, broadly linear to more often lanceolate, lance-elliptic, or subrhombic, tending to taper from the middle to both ends, entire or serrate, the main ones 5–15 × 0.5–3 cm, those of the branches often abruptly reduced; heads ± numerous in a widely branched or occasionally more simple infl, commonly subracemiform on the branches; invol glabrous, 4–5.5 mm, its bracts imbricate in few series, obtuse or acute, with revolute, fairly broad green tip, often suffused with purple upward; rays 9–14, white or slightly purplish, 4–6.5 mm; disk-cors goblet-shaped, the recurved lobes comprising 50–75% of the limb; $2n=16$, 32, 48, 64. Various habitats, most commonly in open woods, dry open places, and on beaches; Magdalen I. to Fla., w. to Minn., e. S.D., e. Kans., and Tex. (*A. agrostifolius; A. hirsuticaulis.*)

A. acadiensis Shinners, a slender lax plant from N.B., N.S., and P.E.I., differs from no. 16 in its delicately open-paniculate infl, with not at all secund branches, and often larger heads, the invol to 6.5 mm. It is perhaps better subordinated to *A. lateriflorus* as var. *tenuipes* Wieg.

17. Aster racemosus Elliott.   Small-headed a. Colonial by long rhizomes; stems 4–15 dm, glabrous or ± puberulent in lines; lvs all cauline, glabrous, or slightly scabrous above, linear to narrowly lanceolate, acute, tapering to the sessile base, entire or slightly toothed, to 11 × 1 cm, those of the branches becoming much reduced; heads mostly numerous in an open, ample infl with long, divaricate, divergently bracteate, often recurved branches that tend to be secund, the shortly bracteate peduncules short or up to 1.5 cm; invol 2.5–3.5(–4) mm, glabrous, its bracts imbricate, with mostly elongate green tip; rays 15–30, white or seldom purplish, 3–6 mm; lobes of the disk-cors comprising ca 40% of the limb. Mostly in moist, open places, and in floodplain-forests; Me. to Fla. and La., chiefly near the coast, and upstream along river-bottoms in the Mississippi R. drainage to w. Tenn., s. O., and w. Mo. (*A. brachypholis; A. fragilis,* misapplied, the type at B! probably a garden hybrid of *A. lanceolatus* and some other sp. such as *A. dumosus; A. vimineus* Lam., misapplied, the type at P! better referred to *A. lateriflorus* or a hybrid derivative thereof.)

18. Aster dumosus L.   Long-stalked a. Plants with creeping rhizomes or sometimes a short rhizome or caudex, the stems 3–10 dm, glabrous or puberulent upward; lvs linear to lance-linear or linear-elliptic, sessile, entire or nearly so, ± scabrous above, glabrous beneath, all or nearly all cauline, the main ones 3–11 cm × 3–10 mm, those of the branches much reduced; infl open, usually ample, often diffuse, the branches with numerous spreading or ascending, oblong or spatulate bracts; heads ± numerous, conspicuously long-pedunculate, the peduncles bracteate like the branches of the infl, generally at lest some of them 2 cm or more (even to 15 cm); invol 4–6(–7) mm, glabrous, its bracts strongly imbricate, with short broad green tip, tending to be obtuse, often some of them dilated upward; rays 13–30, pale lavender or bluish to sometimes white, 5–9 mm; lobes of the disk-cors comprising 20–35% of the limb; $2n=16$, 32. Dry or moist, often sandy places; s. Me. to Fla., w. to Mich., Ark., and La., most commonly on or near the coastal plain. (*A. coridifolius.*)

19. Aster falcatus Lindley.   White prairie-a. Similar to no. 20, but smaller, seldom over 6 dm, with fewer and larger heads; heads solitary or clustered at the ends of the not at all secund branches, or the stem scarcely branched and the infl subracemiform; invol 5–7 mm, its bracts seldom much imbricate, the outer often acute as well as the inner; rays 20–35, 4–8 mm; disk-fls (14–)18–30; $2n=10$, 20, 30. Prairies and other open places; Mack., Yukon, and e. Alas. s. to w. Minn., Io., Tex. and Ariz. Typical *A. falcatus* is northern, tufted, and not rhizomatous. More or less rhizomatous plants, commoner with us, have been segregated as var. *commutatus* (T. & G.) A. G. Jones. (*A. commutatus.*)

20. Aster ericoides L.   Squarrose white a. Plants 3–10 dm, hairy (the hairs appressed or spreading), colonial by long creeping rhizomes, the stems arising singly; lvs numerous, linear, sessile, to 6 cm × 7 mm, the lower and often also the middle ones soon deciduous, those of the branches reduced and divaricate; heads numerous, small, commonly somewhat secund on the divergent or recurved branches; invol 3–5 mm, its bracts ± strongly imbricate, the outer obtuse or acutish, spinulose-mucronate, and ± squarrose, some or all of the bracts coarsely ciliolate-margined, and usually also short-hairy on the back; rays 8–20, white (blue or pink), 3–6 mm; disk-fls 4–15(–20); achenes sericeous-strigose; $2n$=10, 20, 30. Dry, open places; s. Me. to se. Man., s. to Del., n. Va., Tenn., s. Ill., Ark., Tex., n. Mex., and se. Ariz. The closely allied, more northwestern sp. *A. pansus* (S. F. Blake) Cronq., with the stems clustered on a caudex or very short rhizome, may be sought in nw. Minn.

21. Aster pilosus Willd.   Awl-a. Stems 1–15 dm from a stout caudex; basal and lower cauline lvs soon deciduous, or the basal persistent, oblanceolate, and petiolate, those above sessile or nearly so, entire or slightly toothed, linear to lance-elliptic, seldom over 10 × 1(–2) cm, the upper and rameal ones numerous and reduced, often subulate; heads in an often diffuse infl, sometimes secund on the branches, their copiously subulate-bracteate to nearly naked peduncles 3–40 mm; invol broadly urn-shaped, constricted above the middle, then flaring, glabrous, 3.5–8 mm, its bracts imbricate or subequal, with loose, subulate, marginally inrolled green tip; heads mostly 40–100-fld, the rays 16–35, 5–10 mm; $2n$ mostly = 32, 48. Open, rather dry places, often in sandy soil; common and widespread from Me. and N.S. to Ga. and nw. Fla., w. to se. Minn., Neb., Kans., and La. Three vars.:

a Rays white, rarely pink or purple, 5–10 mm; invol 3.5–5 mm, rarely more; heads 40–70-fld.
  b Stem and often also the lvs sparsely to more often densely spreading-hirsute; plants generally robust
    and many-headed; $2n$ mainly = 32 or 48; nearly the range of the sp., but ne. only as far as
    Mass. (var. *platyphyllus.*) .... var. *pilosus.*
  b Stem and lvs nearly or quite glabrous; $2n$ mainly = 48; chiefly northern, and especially northeastern,
    but sometimes s. to Ga. and Tenn. (var. *demotus; A. faxonii; A. glabellus; A. polyphyllus; A.*
    pringlei; A. ramosissimus.) ....var. *pringlei* (A. Gray) S. F. Blake.
a Rays purple to pink to lavender, 8–15 mm; invol 5–8 mm; fls up to 100+; herbage glabrous, or the
  stem sparsely spreading-hirsute; invol seldom much imbricate; Ky. to Ga. and Ala. (*A. priceae.*)
  .... var. *priceae* (Britton) Cronq.

22. Aster parviceps (Burgess) Mackenzie & Bush.   Small-headed a. Plants with a short simple caudex, the stems rather slender, 3–8 dm, ± spreading-hirsute to occasionally glabrous; lvs glabrous, or ± hairy like the stem, entire or slightly toothed, the basal and lower cauline soon deciduous, or the basal persistent, oblanceolate, and petiolate, those above persistent, sessile or occasionally narrowed to a petiole-like base, linear or nearly so, to 10 cm × 6 mm, the upper and rameal ones numerous and much reduced; heads many in an often dense infl, on short or elongate, often copiously subulate-bracteate peduncles, narrow, the glabrous invol cylindric in life, rather narrowly obconic when pressed, 4 mm, its green-tipped, evidently imbricate bracts (or some of them) with subulate, marginally inrolled tip; heads 18–32-fld, the rays 12–18, white, 2–5 mm, 4–10 more than the disk-fls; $2n$=16, 32. Open woods, fields, and pastures; Io., Ill., Mo., and e. Kans.

23. Aster depauperatus Fern.   Eastern serpentine-a. Similar to no. 22, but glabrous throughout, and more slender, wiry, and diffuse, only 1–4 dm, all the cauline lvs (except the numerous, much reduced, subulate ones of the infl) generally deciduous by anthesis; heads 16–32-fld, the rays 9–16, from 2 fewer to 3 more than the disk-fls; $2n$=16. On serpentine in se. Pa. and adj. Md.; reputedly disjunct on diabase in Granville Co., N.C.

24. Aster concolor L.   Eastern silvery a. Plants with a short caudex or crown, often with creeping rhizomes as well; stems slender, 3–10 dm, simple or sparingly branched, thinly sericeous or sometimes merely strigose, rarely spreading-villous, glabrate below; lvs sericeous, sometimes glabrate in age, entire, the lower soon deciduous, the others sessile and broad-based but not strongly clasping, lanceolate or oblong to broadly elliptic, to 5 × 1.5 cm; infl narrow and racemiform, or occasionally with racemiform branches, the peduncles minutely bracteate; invol 5–9 mm, densely and finely sericeous, its bracts narrower and more imbricate than in no. 25; rays 8–16, blue (pink), 7–12 mm; achenes densely sericeous, the pubescence obscuring the nerves; $2n$=8, 16. Dry sandy places, often among pines; coastal states from Mass. to Fla. and La., and up the Mississippi embayment to sw. Tenn.; less commonly inland in the mts. of Ky. and Tenn.

25. Aster sericeus Vent.   Western silvery a. Stems clustered on a short branched caudex, brittle, wiry, 3–7 dm, ± branched upwards, thinly sericeous, or glabrate below; lvs sericeous, entire, the basal oblanceolate and petiolate, but these and the ones on the lower half of the stem soon deciduous, the others sessile but only slightly or not at all clasping, lanceolate or lance-ovate to oblong or elliptic, to 4 × 1 cm; heads several or many in a widely branched corymbiform or paniculiform infl, often clustered toward the branch-tips; invol sericeous, 6–10 mm, its broad, acute bracts several-seriate but seldom much imbricate, their leafy tips often loose or spreading, the larger ones mostly 1.5–3 mm wide; rays 15–25, deep violet to rose-purple (white), 8–15 mm; achenes glabrous, closely 8–12-nerved; $2n$=10. Dry prairies and other open places; Mich. to S.D., s. to Mo. and Tex., and irregularly e. in Tenn.

26. Aster ciliolatus Lindley.   Northern heart-lvd a. Rhizomatous, sometimes with a short branched caudex as well, 2–12 dm, ± hirsute or hirsute-puberulent especially in the infl and on the lower surfaces of the lvs, varying to essentially glabrous; lvs acute or acuminate, the basal and commonly some of the lower cauline ones petiolate and cordate or often merely subcordate, 4–12 × 2–6 cm, sharply serrate, often deciduous, those above becoming merely abruptly narrowed to the broadly winged petiole, and often less toothed, or the upper sessile and entire; infl open, with relatively few heads (often less than 50, rarely over 100), the branches and peduncles sparsely or scarcely bracteate, the bracts narrow, the peduncles very unequal, generally some over 1 cm; invol 5–8 mm, its slender bracts slightly or moderately imbricate, glabrous except the sometimes ciliolate margins, sharply acute or acuminate, the green tip narrow and elongate; rays 12–25, blue, 8–15 mm; achenes glabrous or nearly so; $2n$=48, 72. Woods and clearings; Me. and Que. to N.Y., w. to Minn., S.D., Wyo., Mont., B.C., and Mack. (*A. lindleyanus; A. saundersii; A. wilsonii.*)

27. **Aster oolentangiensis** Riddell. Prairie heart-lvd a. Plants with a branched caudex or short rhizome; stems 2–15 dm, scabrous-puberulent to occasionally subglabrous; lvs basally disposed, thick and firm, entire or occasionally shallowly serrate, scabrous-hispid above, the hairs on the lower surface softer, and usually longer and looser; basal and usually also some of the lowest cauline lvs long-petiolate, cordate (usually shallowly so) or subcordate, lanceolate or ovate in outline, 4–13 × 1.2–6 cm, some of the basal ones sometimes smaller and with more tapering base, those above ± abruptly smaller, narrower, and less petiolate, and generally not at all cordate, the upper sessile and lanceolate or linear; infl open-paniculiform, copiously and narrowly bracteate; invol 4.5–8 mm, its bracts well imbricate, obtusish to sharply acute, with a diamond-shaped green tip shorter than the chartaceous base, glabrous except for the often ciliolate margins; rays 10–25, blue (pink), 5–12 mm; achenes glabrous or nearly so; $2n$=32. Prairies and dry open woods; w. N.Y. and s. Ont. to Minn. and S.D., s. to w. Tenn., nw. Miss., La., and e. Tex. (*A. azureus; A. poaceus.*)

28. **Aster shortii** Lindley. Midwestern blue heart-lvd a. Plants with a short, branched caudex; stems 3–12 dm, glabrous or nearly so below, spreading-hirtellous above; lvs entire or occasionally few-toothed, glabrous or scaberulous above, spreading-hirtellous beneath, lanceolate or rather narrowly ovate, acute or acuminate, petiolate, nearly all of those below the infl cordate or subcordate, the lower cauline ones usually deeply so, 6–15 × 2–6 cm, the basal ones, if persistent, often shorter, broader, more rounded, and somewhat toothed, the middle and upper cauline ones only gradually reduced; infl open, paniculiform, with numerous usually narrow bracts, the bracteate peduncles often very long; invol 4–6 mm, its bracts narrow, acute or the outer obtusish, well imbricate, usually minutely pubescent, with small, ± diamond-shaped green tip shorter than the chartaceous base; rays 10–20, blue (rose-red or white), 5–14 mm; achenes glabrous; $2n$=16, ?18, ?36. Woods; sw. Pa. and adj. W.Va. and nw. Md. to Ky., Tenn., w. Ga., and ec. Ala., w. to se. Minn., ne. Io., and ne. Miss.; Ark. (*A. camptosorus.*)

29. **Aster cordifolius** L. Common blue heart-lvd a. Plants with a branched caudex or short rhizome, occasionally with creeping rhizomes as well; stems 2–12 dm, glabrous below the infl or occasionally loosely hairy (chiefly in lines); lvs rather thin, sharply toothed, acuminate, ± scabrous or scaberulous above, at least toward the margins, and ± hirsute beneath with mostly long, flattened hairs, narrowly to broadly ovate, the larger ones 3.5–15 × 2.5–7.5 cm, all but the reduced ones of the infl cordate (usually deeply so) and petiolate, the petioles shorter upward but only slightly if at all winged; infl paniculiform, with loosely ascending to widely spreading branches and often very numerous heads, the peduncles usually well developed and ± bracteolate, sometimes to 1.5 cm; invol 3–6 mm, glabrous, its narrow bracts imbricate, obtuse or merely acute, the green tip short and broad (or in the innermost more elongate), the green usually partly (or wholly) replaced or obscured by anthocyanin, visibly so to the naked eye in some of them; rays 8–20, blue or purple (white), 5–10 mm; achenes glabrous; $2n$=16, ?18, 32, ?36. Woods and clearings; N.S. to Minn., s. to Va., n. Ga., nw. Ala., and Mo. An apparent hybrid with no. 28 has been named *A.* × *finkii* Rydb.

30. **Aster lowrieanus** T. C. Porter. Smooth heart-lvd a. Resembling no. 29, but glabrous except for the sometimes puberulent peduncles, the lvs very smooth to the touch, avg narrower and more elongate, the middle and upper ones generally only shallowly cordate or merely subcordate or abruptly narrowed to the commonly conspicuously winged petiole, or even sessile and lanceolate; invol bracts often lacking anthocyanin; $2n$=?36. Woods; s. Ont. and s. Que., s. through N.Y., N.J., and Pa. to the mts. of Md., W.Va., Va., N.C., Tenn., and Ga. Perhaps originating by hybridization of nos. 9 and 29, but now self-perpetuating. Some plants of the shale-barrens of Va. have slender creeping rhizomes and have been distinguished as *A. schistosus* Steele.

31. **Aster sagittifolius** Willd. Arrow-lvd a. Plants with a branched caudex or short rhizome; stems 4–12 dm, ± glabrous below the infl, or the upper part occasionally villous-puberulent in lines; lvs rather thick, shallowly toothed, glabrous or scabrous above, glabrous or ± villous-hirsute with flattened hairs beneath, the lowest ones ovate or more often lance-ovate, acuminate, cordate, 6–15 × 2–6 cm, long-petiolate, those above progressively less cordate or merely abruptly narrowed to the shorter, often broadly winged petiole, or the upper tapering and sessile; infl paniculiform, elongate, with ± strongly ascending, narrowly bracteate branches, the heads often very numerous, on narrowly bracteate peduncles rarely over 1 cm and usually under 5 mm, thus appearing crowded; invol 4–6 mm, its imbricate bracts glabrous except for the sometimes ciliolate margins, very slender and long-pointed, with elongate, narrow green tip, often minutely purple-tipped under a lens; rays 8–20, usually pale blue or lilac, sometimes white, 4–8 mm; achenes glabrous; $2n$=16, ?18, ?36, 48. Streambanks, woods, and less often in open places, often in more mesic habitats than no. 32, with which it hybridizes extensively; sw. Vt. to Minn., s. to Ga., w. Fla., Miss., and Mo. (*A. hirtellus; A. urophyllus.*) My interpretation of the name *A. sagittifolius* is based on examination of the type material at Berlin.

32. **Aster drummondii** Lindley. Hairy heart-lvd a. Stouter than no. 31; stems usually densely hairy at least above the middle with minute, stiffly spreading hairs; lvs relatively firm, shallowly toothed, scabrous above, densely pubescent with short spreading hairs beneath, the lowest ones ovate or lance-ovate, acuminate,

cordate, 6–14 × 2.5–6.5 cm, long-petiolate, those above progressively less cordate (or the upper merely broadly rounded at base) and with shorter, usually broadly winged petiole; infl paniculiform, with spreading or ascending, bracteate branches, the heads often numerous, on bracteate peduncles usually well under 1 cm; glabrous or puberulent, 4.5–7 mm, its bracts firm, imbricate, sharply acute or acuminate but with broader and proportionately longer chartaceous base than in no. 31, the green tip tending to be elongate-rhombic; rays 10–20, bright blue, 5–10 mm; achenes sparsely short-hairy or glabrous; 2n=16, 32, ?36. Typically in clearings and open woods; s. O. to Minn., s. to w. Ky., Miss., La., Kans., and Tex.

33. Aster texanus Burgess. Ozark heart-lvd a. Resembling no. 32 but with more open infl, the branches commonly very long, with copiously and minutely bracteate nonfloriferous base, or in more compact plants merely the peduncles conspicuously elongate (mostly well over 1 cm) and copiously bracteate; plants fully as hairy as no. 32, or less so; 2n=16, ?18. Bottomlands and open woods; w. Ky. to e. Kans., s. to s. Miss. and e. Tex.

34. Aster undulatus L. Clasping heart-lvd a. Plants with a branched caudex or short rhizomes; stem 3–12 dm, densely pubescent with short, spreading hairs, varying to subglabrous below the infl; lvs entire or toothed, scabrous to glabrous above, usually shortly and rather loosely hairy beneath, at least the lower ones cordate or subcordate at base and petioled, lance-ovate to ovate, 3.5–14 × 1.5–7 cm, those above extremely variable in size and shape, but always at least some of them either sessile and cordate-clasping or with the petiole enlarged and auriculate-clasping at base; infl open, paniculiform, the branches and mostly well developed peduncles ± spreading and bracteate; invol 4–7 mm, usually minutely puberulent, its bracts imbricate, sharply acute or acuminate, sometimes very slender, more often a little wider as in no. 32, often minutely purple-tipped; rays 10–20, blue or lilac; achenes minutely hairy, at least above; 2n=?18, 32, ?36. Mostly in dry, open woods and clearings; Me. to Fla., w. to O., se. Ind., e. Tenn. and c. Miss., and se. La. (*A. claviger; A. corrigiatus; A. gracilescens; A. loriformis; A. sylvestris; A. triangularis; A. truellius.*)

35. Aster anomalus Engelm. Limestone heart-lvd a. Plants with a short, stout rhizome or branched caudex; stem 2–10 dm, pubescent with short, spreading, often coarse hairs, or merely scabrous; lvs thick and firm, scabrous-hirsute above, more loosely and softly hairy beneath, entire or nearly so, petiolate, the basal and lower cauline ovate or lanceolate, deeply cordate, 4–9 × 2–5.5 cm, the middle and upper gradually or abruptly reduced; infl open, paniculiform, with numerous narrow bracts, the bracteate peduncles often very long, invol 6–10 mm, short-hairy (subglabrous), the elongate, narrow, well imbricate bracts with reflexed green tip; rays 20–45, bright blue, 7–15 mm; achenes glabrous; 2n=16, ?18. Dry woods, usually associated with limestone; Ozark region from s. Ill. to e. Kans., Okla., and Ark.

36. Aster patens Aiton. Clasping a. Plants with a short caudex, sometimes with creeping rhizomes as well; stems rather slender and brittle, 2–15 dm, shortly and loosely hairy; lvs ± hairy or scabrous, at least beneath, sessile and conspicuously cordate-clasping, broadly ovate to oblong, entire, 2.5–15 × 0.8–4.5 cm, the lower soon deciduous; heads few to rather numerous in an open, divaricately branched infl; invol 5–9 mm, its bracts well imbricate, mostly acute, ± glandular or short-hairy or both; rays 15–25(–30), blue (pink), 8–15 mm; achenes shortly sericeous; 2n=10, 20. Woods and dry, open places; Mass. and s. N.H. to s. Mich. and Kans., s. to Fla. and Tex. Four vars.:

a Invol ± strongly glandular, often also hairy; chiefly e. of the Miss. R.
  b Lvs relatively thin, often constricted above the large clasping base, the middle and upper ones 3.5–
    6 times as long as wide; pubescence mostly a little softer than in the 2 following vars.; woods;
    se. N.Y. to O., s. to Ga., Ala., and n. Miss. (*A. phlogifolius.*) . . . . var. *phlogifolius* (Muhl.) Nees.
  b Lvs relatively thick and firm, rarely constricted above the base, the middle and upper ones 1.5–
    3.5 times as long as wide; in drier or more open places than var. *phlogifolius.*
    c Branches very long, conspicuously and minutely bracteate; plants avg smaller and more slender
      than in var. *patens;* N.J. to Fla. and Tex., and up the Mississippi embayment to w. Tenn.
      (*A. tenuicaulis.*) . . . . var. *gracilis* Hook.
    c Branches shorter, with fewer and larger bracts; Mass. and s. N.H., to s. Mich., s. to Ga. and Ala.
      . . . . var. *patens.*
a Invol only slightly or scarcely glandular, ± strongly sericeous; invol bracts broader and blunter than
    in the other vars.; habit varying from that of var. *patens* to var. *gracilis;* Mo. and adj. Ky. to
    Kans., Miss., and Tex. (*A. patentissimus.*) . . . . var. *patentissimus* (Lindley) T. & G.

37. Aster oblongifolius Nutt. Aromatic a. Plants rhizomatous, sometiems also with a short caudex; stem 1–10 dm, rigid, brittle, usually ± branched, glandular upward, commonly also ± hairy or scabrous; lvs firm, entire, sessile and obscurely to evidently auriculate-clasping, narrowly to broadly oblong or lance-oblong, to 8 × 2 cm, scabrous or short-hirsute, or sometimes glabrous except the margins, the lower soon deciduous, those of the branches numerous and reduced, becoming mere spreading bracts; heads several or many, terminating the branches; invol densely glandular, 5–8 mm, its bracts in several series but not much imbricate, firm, with chartaceous base and long, green, loose or spreading, acute or acuminate tip; rays 15–40, blue or purple (rose), 1–1.5 cm; achenes strigose or finely sericeous; 2n=10, 20. Dry, usually open places; Pa. and DC., s. in the mts. to N.C. and n. Ala., w. to N.D., Wyo., and N.M. (*A. kumlienii.*)

38. Aster grandiflorus L. Big-headed a. Similar to the larger forms of no. 37, but more openly branched, with smaller lvs, and with conspicuously larger heads, the invol 10–15 mm, its bracts wider and often more imbricate, the outer sometimes blunter; rays 12–25 mm; stem rather sparsely or sometimes scarcely pubescent, with usually spreading hairs, as well as glandular upward; 2n=60. Sandy places, often in pine woods or old fields; e. of the mts. in Va. and N.C.

39. Aster novae-angliae L.   New England a. Plants with a stout caudex and short thick rhizome, occasionally with creeping rhizomes as well; stems clustered, 3–20 dm, usually strongly spreading-hirsute at least above, and becoming glandular upward as well; lvs chiefly cauline, lanceolate, entire, 3–12 cm × 6–20 mm, sessile and conspicuously auriculate-clasping, scabrous or stiffly appressed-hairy above, more softly hairy beneath, or the upper lvs becoming glandular; heads several or many in a leafy, usually short infl, the invol and peduncles densely glandular; invol 6–10 mm, its numerous slender bracts about equal, often purplish, with chartaceous base and loose or spreading, attenuate tip, the outer sometimes a little broader and more leafy; rays 45–100, usually bright reddish-purple or rosy, 1–2 cm; achenes densely sericeous or appressed-hirsute, their nerves obscure; $2n=10$. Moist, open or sometimes wooded places; Mass. and Vt. to N.D. and Wyo., s. generally to DC., Tenn., Ark., and N.M., and in the mts. to Va., N.C., and Ala.; prairies in Ala. and Miss. A hybrid with *A. ericoides* has been named *A.* × *amethystinus* Nutt.

40. Aster modestus Lindley.   Northwestern sticky a. Plants with long creeping rhizomes, the stems mostly arising singly, 3–10 dm, densely stipitate-glandular, at least upwards, and often ± spreading-hairy as well; lvs chiefly cauline, thin, lanceolate, remotely and rather saliently dentate to entire, 4.5–13 cm × 8–40 mm, sessile and obscurely to usually ± evidently auriculate-clasping, glabrous or scaberulous above, softly hairy to glabrous beneath; heads several or occasionally rather many in a short infl, the invol 7–11 mm, its thin, slender, bracts acuminate or attenuate, often deep purplish, about equal, loose, the outer herbaceous to the base; rays 20-40, dark purple, 1–1.5 cm; achenes evidently nerved, sparsely short-hairy; $2n=18$. Moist, open or wooded places; Isle Royale, n. Minn., and w. Ont. to Alas. and Oreg. (*A. major.*)

41. Aster macrophyllus L.   Big-lvd a. Rhizomatous and colonial, 2–12 dm, with abundant clutsers of basal lvs on short sterile shoots, glandular at least in the infl, otherwise glabrous or rough-hairy; lvs thick and firm, crenate or serrate with usually mucronulate teeth, the basal and lower cauline ones cordate, 4–30 × 3–20 cm, commonly short-acuminate to obtuse, long-petiolate, the middle and upper gradually or abruptly reduced, becoming sessile and ovate to lanceolate or elliptic; infl corymbiform, its bracts few and broad; invol 7–11 mm, generally glandular, sometimes also short-hairy, its bracts firm, well imbricate, appressed, rounded to sharply acute, the green tip sometimes obscure, the outer 1–2.5 mm wide and not over 2.5 times as long as wide; rays 9–20, 7–15 mm, ± tinged with lilac or purple; $2n=72$. Woods; N.S., N.B. and Que. to Wis. and Minn. and se. Man., s. to Pa. and Ky., and in the mts. to Ga. (*A. ianthinus; A. multiformis; A. nobilis; A. roscidus; A. violaris.*)

42. Aster schreberi Nees.   Similar to no. 41 but eglandular, with merely puberulent infl and normally with white rays; basal lvs tending to have a rectangular sinus; invol 5–10 mm, often very narrow, its bracts avg narrower than in no. 41 and sometimes a little loose, those of the inner row much the longest; rays 6–14; $2n=54$. Woods; N.H. and se. Me. to Del. and Va., w. to O. and W.Va., and irregularly to Tenn., Ala., Ill., and e. Wis. (*A. chasei; A. curvescens; A. glomeratus.*) The midwestern plants (*A. chasei* G. N. Jones) may be alloploids of nos. 41 and 44, whereas the more eastern ones (*A. schreberi* proper) may be alloploids of nos. 41 and 43.

43. Aster divaricatus L.   Common white heart-lvd a. Rhizomatous and colonial, 2–10 dm, usually without tufts of radical lvs; stem puberulent at least in the infl, not glandular; lvs rather thin, sharply serrate with usually mucronulate teeth, glabrous or with some long, mostly appressed hairs, especially along the main veins beneath; lower lvs ovate (often narrowly so), with cordate base, conspicuously acuminate, 4–20 × 2–10 cm, petiolate, the lowest often smaller than those above and commonly deciduous, the middle and upper ones progressively less cordate, less petiolate, and ± reduced; infl corymbiform, occasionally becoming elongate, its bracts few and often broad; invol 5–10 mm, its bracts firm, well imbricate, rounded to acute, very shortly green-tipped, otherwise mostly whitish, the outer 0.7–1.5 mm wide (or a little wider in var. *chlorolepis*) and seldom more than 2.5 times as long; rays 5–16(–20), white, or in var. *chlorolepis* sometimes lilac-tinged; $2n=18$, 36. Woods; n. N.H. and sw. Que. (just n. of Lake Champlain), w. to the Niagara peninsula of s. Ont., s. to D.C. and s. O., and in the mt. region to n. Ga. and e. Ala. (*A. carmesinus; A. castaneus; A. excavatus; A. stilletiformis; A. tenebrosus.*) The widespread var. *divaricatus*, diploid, with relatively small heads, the invol 5–8 mm, the rays 5–10(–12), 5–15 mm, gives way at upper elev. in the Blue Ridge of sw. Va., N.C., Tenn., and n. Ga. to the var. *chlorolepis* (Burgess) Ahles, tetraploid, with relatively large heads, the invol 7–10 mm, its bracts avg broader, and with not so pale chartaceous part, the inner ones often purplish distally, the rays (10–)12–16(–20), 10–20 mm. (*A. chlorolepis.*)

44. Aster furcatus Burgess.   Midwestern white heart-lvd a. Rhizomatous, 3–8 dm, usually without tufts of basal lvs; stem puberulent above; lvs thick and firm, scabrous above, ± densely spreading-hairy across the surface beneath, serrate with mucronulate teeth, the lower petiolate, ovate (often narrowly so), acuminate, with cordate base, 6–15 × 3–8 cm, often deciduous, the middle and upper ones rather gradually reduced, often with winged laciniate petiole, eventually sessile and lanceolate or lance-elliptic; infl corymbiform, flat or round-topped, with few and rather leafy bracts; invol 7–10 mm, its bracts firm, well imbricate, often puberulent, mostly obtuse or rounded, very shortly green-tipped, the outer 1–2 mm wide and not over 2.5 times as long; rays 9–18, white, or becoming lilac or rosy in age, rarely so when young, 8–15 mm. $2n=18$. Woods; s. Mich. and s. Wis. to s. Ind. and s. Mo. and nw. Ark.

45. Aster turbinellus Lindl.   Prairie a. Stems 4–12 dm from a branched caudex, glabrous or ± spreading-hirsute; lvs chiefly cauline, entire, firm, glabrous except for the scabrous-ciliolate margins and sometimes some coarse hairs on the midrib beneath, broadly linear to oblong or lance-elliptic, sessile or subpetiolate, not clasping, 6–10 cm × 8–20 mm, the upper more reduced, those of the often long and wiry branches becoming bract-like; heads usually ± numerous, turbinate, conspicuously long-peduncalate, the peduncles beset with minute, mostly appressed bracts; invol glabrous, 7–12 mm, its bracts obtuse or rounded, multiseriate, the outer narrow and small, passing into those of the peduncles, the inner much broader, shortly green-tipped; rays 15–20, violet, 8–12 mm; achenes strigillose and minutely punctate; pappus tawny-rufescent; $2n$=96–100. Mostly in dry, open places and open woods; c. Ill. to La. and e. Kans.

46. Aster tataricus L. f.   Tatarian a.   Coarse, rough-hairy plants, 5–20 dm from a stout caudex; lower lvs long-petiolate, with large, elliptic, conspicuously toothed blade 8–40 × 2.5–15 cm, the middle and upper smaller, sessile or nearly so, and mostly entire; infl corymbiform, flat-topped; invol 7–10 mm, strigosepuberulent or subglabrous, its bracts not much imbricate, the larger ones mostly 1–2 mm wide; rays 15–20, purple or blue, 1–2 cm; $2n$=54. Native of s. Siberia, casually escaped from cult. in e. U.S.

47. Aster saxicastellii J. Campbell & Medley.   Much like no. 48, but with larger, more coarsely toothed lvs, those mostly (7–)9–14 × (3–)4–6 cm, on a winged petiole 1–3 cm, or the upper tapering to a narrow sessile base; rays 10–30, white or pale blue; achenes pubescent; $2n$=54. Along the banks of rivers from Pulaski and Laurel cos., Ky. to Scott Co., Tenn.

48. Aster radula Aiton.   Low rough a.   Rhizomatous, 1–12 dm, the stems glabrous except for some puberulence beneath the heads; lvs chiefly cauline, veiny, with several pairs of evident divergent primary lateral veins, ± serrate, sessile or nearly so, scabrous above, hairy or occasionally glabrous beneath, elliptic to oblong or lance-oblong, 3–10 × 0.6–2.5(–3) cm; heads several in a short and broad, sparsely leafy-bracteate, corymbiform infl, or seldom solitary; invol 6–11 mm, its relatively broad, firm bracts ± imbricate, rounded to acute, puberulent or more often glabrous except for the ciliate or fimbriate margins, often squarrose-upper, the green tip often ill-defined or wanting; rays 15–40, violet, 8–15 mm; achenes glabrous; $2n$=18. Bogs, streambanks, and other moist places; Lab. and Nf. to Md., Va., W.Va., and Ky.

49. Aster × herveyi A. Gray.   Hybrid of nos. 41 and 50; rhizomatous; stem 2–10 dm, glabrous below, becoming glandular upwards; lvs firm, glabrous or slightly glandular, toothed or subentire, the basal ones tufted on sterile shoots, broadly ovate to elliptic, 6–13 × 3–8 cm, abruptly contracted to the long petiole, the lower cauline ones similar, becoming reduced, narrower, and less petiolate upwards; infl corymbiform, sparsely leafy-bracteate; invol 18–12 mm, its bracts well imbricate, glandular or glandular-puberulent, relatively broad, appressed, with short, obtuse or acutish, green or greenish but scarcely herbaceous tip; rays 10–20, purple, 10–15 mm; achenes short-hairy; $2n$=72. Woods; local in Mass., Conn., and e. L.I.

50. Aster spectabilis Aiton.   Low showy a.   Rhizomatous; stems 1–9 dm, usually densely glandular, at least above, or occasionally merely spreading-villous; lvs basally disposed, firm, scabrous (especially above) or glabrous, entire or remotely and shallowly toothed, the lower ones with elliptic blade 2–15 cm × 8–40 mm tapering to the well developed petiole, those above more oblong, becoming sessile and somewhat reduced, well spaced; heads few to rather many in an open, corymbiform, sparsely leafy-bracteate infl; invol 8–16 mm, its bracts broad and firm, ± imbricate, with chartaceous base and loose or spreading green tip, densely glandular (at least the inner), the outer sometimes viscid-villous as well; rays 15–35, rich violet-purple, 1–2.5 cm; achenes short-hairy; $2n$=72. Dry, sandy soil, often among pines; Mass. to S.C. and Ala., in our range near the coast.

51. Aster surculosus Michx.   Similar to no. 50, but not glandular, or only slightly and inconspicuously so; rhizome sometimes with nodular-thickened, woody portions; stem spreading-hirtellous (or merely strigose below), often slightly viscidulous, but scarcely glandular, or sometimes slightly so on the peduncles; heads avg smaller, the invol 7–12 mm, its bracts avg narrower and blunter, only obscurely or not at all glandular, otherwise glabrous except for the often ciliolate margins; rays paler, more bluish. Various habitats, especially in sandy soil; Appalachian region from Ky. and Va. to Tenn., N.C., S.C., and Ga.

52. Aster gracilis Nutt.   Slender a. Plants with a thickened, hard, corm-like base, sometimes rhizomatous as well, the several puberulent or hirtellous or occasionally subglabrous stems 1.5–5 dm, lvs thick and firm, obscurely veined except for the evident midrib and sometimes a single pair of narrowly divergent laterals, entire or nearly so, not evidently sheathing below, the basal with elliptic blade 2–6 cm × 8–20 mm, shorter than the petiole, often deciduous, the cauline narrower and most of them sessile, 1.5–9 cm × 2.5–14 mm, 4–12(–15) times as long as wide; heads several or rather many in a short and broad, usually sparsely bracteate, corymbiform infl, narrow, the invol narrowly obconic or turbinate, 7–12 mm, its relatively broad, firm bracts glabrous or obscurely puberulent, imbricate, the outer shortly green-tipped, the inner scarcely so or merely purple margined, commonly some or all of them shortly squarrose; rays sparsely bracteate, corymbiform infl, narrow, the invol narrowly obconic or turbinate, 7–12 mm, its relatively broad, firm bracts glabrous or obscurely puberulent, imbricate, the outer shortly green-tipped, the inner scarcely so or merely purple-margined, commonly some or all of them shortly squarrose; rays 8–14, blue-violet to rose-purple, often rather pale, 5–8 mm; achenes thinly strigillose or glabrate; $2n$=18. Dry, sandy places, often among pines; coastal plain from N.J. to Ga.

53. Aster hemisphericus Alexander.   Perennial from scaly, often nodular-woody creeping rhizomes, 2–8 dm; stem commonly glabrous or nearly so, or short-hairy only just beneath the heads; lvs ± basally disposed, firm, glabrous or nearly so, grass-like, linear or nearly so, not veiny, only the midrib evident, entire or occasionally some of them remotely spinulose-toothed, to 20 cm (overall) and ca 1 cm wide, the lowermost sometimes deciduous, the lower tapering to a ± petiolar, sheathing base, the others progressively reduced and sessile; heads several in an elongate-racemiform infl (or the infl with spiciform or racemiform branches); invol campanulate or campanulate-hemispheric, 9–12 mm, glabrous, its bracts seldom much imbricate, linear-oblong, loose and often distally spreading, firm with chartaceous base and green tip, or the outer nearly wholly green, the outer more acute than the inner, the larger ones 1.5–2.5 mm wide; rays 15–35, deep lavender or purple, 1–2.5 cm; achenes glabrous or slightly strigose; pappus coarse and firm, the larger bristles flattened and often slightly clavellate; $2n=18 + 0$–4 B. Prairies and open woods, less commonly in moist, low ground; Mo. to Tex., La., Tenn., and w. Fla. Closely allied to *A. paludosus* Aiton, of the coastal plain from N.C. to Ga., with broader, more corymbiform infl.

54. Aster acuminatus Michx. Whorled a. Plants with slender creeping rhizomes, these becoming apically enlarged and scaly; stem 2–8 dm, loosely villous or puberulent and commonly somewhat viscidulous; lvs all cauline, mostly 10–22 below the infl, thin, glabrous or slightly scabrous above, viscidulous-puberulent beneath, at least along the larger veins, the lower reduced and soon deciduous, the others elliptic or obovate, acuminate, ± sharply and saliently few-toothed, tapering to a sessile or shortly petiolar base, often forming a pseudowhorl beneath the infl, the largest ones 6–17 × (1.5–)2–6 cm; heads several or many in an open, slenderly branched, sparsely subulate-bracteate, mostly corymbiform infl; invol 6–9 mm, its slender, sharply pointed bracts well imbricate, thin, pale stramineous to faintly greenish and somewhat purple-tinted, scarcely herbaceous, the outer ciliate-margined, otherwise glabrous; rays 10–21, white or faintly pinkish, 9–15 mm; achenes copiously glandular; $2n=18$. Woods; Nf., Que., and s. Ont. to N.J., N.Y., and Pa., and s. in the mts. to n. Ga.

55. Aster × blakei (T. C. Porter) House. Hybrid of nos. 54 and 56, but sometimes at least briefly self-perpetuating; lvs thin or a little thickish, elliptic or lance-oblong, slightly toothed, acute or acuminate, tapering to a sessile or subsessile base, 25–40 below the infl, the largest ones 7–24 mm wide; $2n=18$. Bogs and moist woods; Nf., Que., and N.S. to N.Y. and N.J.

56. Aster nemoralis Aiton. Leafy bog-a. Rhizomatous; stem 1–6 dm, harshly puberulent with faintly viscidulous hairs; lvs all cauline, 40–75+ below the infl, firm, scabrous above, ± puberulent beneath at least along the main veins (and sometimes also glandular), linear or lance-linear to elliptic or oblong, acute or obtuse, entire or nearly so, the margins commonly revolute, sessile, 12–50 × 2–12 mm; heads conspicuously slender-pedunculate, solitary or several in a minutely bracteate corymbiform infl; invol 5–7.5 mm, its usually thin bracts well imbricate, sharply pointed, purplish-tinged or faintly greenish, but scarcely herbaceous, viscidulous-puberulent or subglabrous; rays 13–27, broad, pink or liliac-purple, 9–15 mm; achenes glandular; $2n=18$. Sphagnum-bogs; Nf. and s. Lab. to James Bay, s. to N.J., N.Y., and n. Mich.

57. Aster umbellatus Miller. Tall flat-topped white a. Plants (4–)10–20 dm from well developed creeping rhizomes; stem usually glabrous or nearly so below the infl; lvs all cauline, entire, finely reticulate beneath with well developed primary lateral veins, mostly 4–16 cm × 7–35 mm, sessile or nearly so and tapering to both ends, mostly rather narrowly elliptic or lance-elliptic and 4–6 times as long as wide, seldom broader; heads ± numerous (30–300+) in a corymbiform, generally dense infl that tends to be flat-topped; invol 3–5 mm, glabrous or somewhat puberulent, its bracts well imbricate, greenish but scarcely herbaceous, relatively thin and slender, seldom any of them as much as 1 mm wide, acute or obtuse; fls 23–54 per head, the rays (6)7–14, white, 5–8 mm, the disk-fls 16–40, ochroleucous; achenes sparsely to evidently strigose or puberulent; pappus double, the inner of firm, rather sordid bristles, the larger ones ± clavellate-thickened toward the tip, the outer less than 1 mm; $2n=18$. Moist, low places; Nf. to Minn., s. generally to Va., Ky., and Ill., and in the mts. to n. Ga. and ne. Ala. (*Doellingeria u.*) Low and few-headed plants toward the ne. end of the range, with the stem and lower lf-surfaces often puberulent, may be varietally separable from the widespread, normally robust and many-headed typical phase, which has the stem generally glabrous or subglabrous below the infl, the lvs glabrous beneath (and often glaucous) or merely puberulent along the midrib and main veins, rarely sparsely so over the surface.

58. Aster pubentior Cronq. Northwestern flat-topped white a. Resembling no. 57; stem puberulent; lvs sparsely to usually rather densely puberulent across the surface beneath; invol puberulent, its bracts mostly long-triangular and acute; heads 17–22-fld, the rays 4–7; $2n=18$. Moist low places and moist woods; n. Mich. to Neb. and Alta. (*A. umbellatus* var. *pubens; Doellingeria pubens.*)

59. Aster infirmus Michx. Appalachian flat-topped white a. Plants 4–11 dm; stems glabrous, mostly solitary from a short, fibrous-rooted crown; lvs chiefly or all cauline, entire, reticulate-veiny, glabrous or scabrous above, glabrous or more often short-hirsute along the midrib and main veins beneath, seldom sparsely short-hairy across the whole lower surface, mostly elliptic or elliptic-ovate, (3–)6–13 × 1.5–5 cm, 2–4(–5) times as long as wide, acute or acuminate, with shortly petiolar or subsessile base, the reduced lower ones obovate or broadly oblanceolate and rounded or obtuse; heads (2–)5–35(–75) in an open, sparsely or scarcely leafy, corymbiform infl; invol 4.5–7 mm, glabrous or puberulent, its bracts well imbricate, relatively firm and broad, the larger ones commonly 1 mm wide or a bit more, often longitudinally striate; heads 25–45-fld, the rays 5–9, white, 6–10 mm, broad and showy, the disk-fls ca 18–36, ochroleucous; achenes glabrous; pappus double, the inner of rather firm, sordid bristles, most of which are clavellate-thickened above, the outer less than 1 mm; $2n=18$. Woods; Mass. to Va., W.Va., and Ky., s. to Tenn., Ga., and Ala. (*Doellingeria i.*)

60. Aster linariifolius L. Stiff a. Plants with a short caudex, rarely with creeping rhizomes as well; stems several, wiry, 1–5(–7) dm, finely puberulent, becoming tomentose-puberulent upward; lvs numerous and similar, firm, linear or nearly so, entire, minutely scaberulous, scabrous-ciliate on the margins, nerveless except for the prominent midrib, 1.2–4 cm × 1.2–4 mm, the lowest ones soon deciduous; heads solitary, or more often several in a mostly corymbiform infl; invol 6–9 mm, its bracts strongly imbricate, firm, keeled, greenish upward but scarcely herbaceous, finely scaberulous like the lvs, acute to broadly rounded, the inner ± fringed-ciliate upward and usually purple-margined; rays 10–20, violet (white), broad and showy, 7–12 mm; disk yellow or anthocyanic; achenes copiously long-hairy; pappus double, the inner bristles elongate, firm, and tawny, the outer ca 1 mm or less; $2n=18$. Dry ground and open woods, especially in sandy soil; Me. and Que. to n. Fla., w. to Wis., Mo., Ark., and Tex. (*Ionactis l.*)

61. Aster paternus Cronq. Toothed white-topped a. Stems 1.5–6 dm from a branched caudex, generally scabrous-puberulent in the infl; lvs ciliate-margined and sometimes hairy over the surface as well, at least some of them evidently toothed, the basal and lower cauline ones generally enlarged and persistent, broadly oblanceolate to obovate, elliptic, or even subrotund, petiolate, the blade 1.5–10 × 1–4.5 cm; cauline lvs becoming sessile upward, otherwise scarcely to strongly reduced; infl corymbiform, flat-topped, the heads commonly in small glomerules; invol glabrous, narrow, 5–9 mm, its bracts well imbricate, broad, with short spreading green tip, or the inner wholly chartaceous; rays 4–8, white (pink), 4–8 mm; disk-fls 9–20, 4–5.5 mm (dry), white or ochroleucous, or seldom lavender; achenes densely sericeous, the upper hairs simulating an outer pappus; pappus-bristles usually reddish, obscurely clavellate above; $2n=18$. Dry woods; Me. and Vt. to Ga., w. to s. O., W.Va., e. Ky., e. Tenn., and e. Ala. (*Sericocarpus asteroides.*)

62. Aster solidagineus Michx. Narrow-lvd white-topped a. Essentially glabrous, 2–6 dm from a short stout caudex; lvs entire or narrowly oblong (or the lower oblong-oblanceolate), 2–8 cm × 2–12 mm, more than 5 times as long as wide, sessile, or the lower petiolate (but only slightly if at all larger than the others); infl flat-topped, the heads tending to be in small glomerules; invol cylindric, 4–7 mm, its bracts well imbricate, broad and firm, the inner chartaceous and erose-margined; rays 3–6, white, 5–10 mm; disk-fls 5–10, 4–5.5 mm (dry), white or ochroleucous; achenes ± densely sericeous; pappus white, some of the bristles obscurely clavellate above; $2n=18$. Dry woods and open ground; Mass. to Ga., w. to W.Va., s. Ind., w. Ky., and La. (*Sericocarpus linifolius.*)

63. Aster tenuifolius L. Perennial salt-marsh a. Glabrous perennial 2–7 dm, the stems often zigzag, arising singly from slender creeping rhizomes; lvs few, fleshy, linear or nearly so, 4–15 cm × 1–5(–12) mm, the lower soon deciduous, the upper reduced and often bract-like; heads (solitary) several or many in an open infl, the peduncles subulate-bracteate; invol 6–9 mm, its bracts ± imbricate, firm, chartaceous, sometimes greenish upward but scarcely herbaceous, acuminate to sharply acute, the inner often purple-margined; rays mostly 15–25, blue to pink (whitish), 4–7 mm, sometimes circinately outrolled; $2n=10$, 20. Salt-marshes along the coast; Mass. to Fla. and La.

64. Aster subulatus Michx. Annual salt-marsh a. Glabrous, somewhat fleshy annual 1–10 dm from a short taproot; lvs linear, entire or nearly so, to 20 × 1 cm, entire; heads remotely solitary to usually several or many in an open infl; invol rather narrow, 5–8 mm, its usually well imbricate bracts acuminate, often somewhat greenish, but scarcely herbaceous, frequently purplish toward the margins and tips, all except sometimes the outer with scarious or hyaline margins; rays inconspicuous, bluish, more numerous than the 5–15 disk-fls, becoming circinately rolled outwards, longer than the style, but scarcely or not at all exceeding the copious fine white pappus; $2n=10$, 20. Coastal, mostly saline marshes; N.B. and se. Me. and N.H. to trop. Amer., and rarely and irregularly inland to e. N.Y. and s. Mich.; also in a nonmaritime, often weedy var. in the Ozark reg., n. to Nebr. Most of our plants are var. *subulatus*, as here described; plants of N.B. have blunter lvs, and a less imbricate, slightly more herbaceous invol, and have been distinguished as var. *obtusifolius* Fern.

65. Aster laurentianus Fern. St. Lawrence annual a. Similar to no. 66, but smaller, and with proportionately broader lvs and invol bracts; stem 3–25 cm; lvs narrowly oblong to spatulate, 5–10 times as long as wide, not ciliate; invol bracts subequal, or the outer longer than the inner, not ciliate or only obscurely so, fewer and thinner than in no. 66, avg broader and sometimes blunter; pistillate fls sometimes with a vestigial erect ray under 1 mm; $2n=14$. Damp brackish sand or mud; P.E.I., Magdalen I. and ne. N.B. (*A. frondosus,* misapplied.)

66. Aster brachyactis S. F. Blake. Western annual a. Taprooted annual 1–7 dm, glabrous except for the coarsely and remotely ciliolate margins of the lvs and sometimes also the invol bracts; lvs linear or nearly so, 3–12 cm × 1–9 mm, 13–30 times as long as wide, the lower soon deciduous; heads several or many in an open-paniculiform to spiciform infl; invol 5–11 mm, its bracts loose, linear, acute or acuminate, herbaceous (especially the outer), equal or slightly imbricate, or the outer occasionally longer than the inner; pistillate fls more numerous than the disk-fls, with tubular eligulate cor much shorter than the style; achenes appressed-hairy; pappus soft and copious; $2n=14$. Waste places, especially where moist and saline; sparingly intr. across the n. part of our range; native in w. U.S. and Can., and apparently Siberia. (*Brachyactis angusta.*)

58. ERIGERON L. Daisy; fleabane.

Heads usually radiate, the rays pistillate and fertile, white or anthocyanic; invol bracts narrow, herbaceous and equal to scarcely herbaceous and evidently imbricate, but without *Aster*-like green tips; receptacle flat or nearly so, naked; disk-fls numerous, perfect and fertile, yellow; some spp. with rayless pistillate fls between the disk-fls and ray-fls; style-branches flattened, the appendage short (not over 0.3 mm in ours), acute to more often obtuse, or even obsolete; achenes in our spp. 2–4-nerved; pappus of capillary bristles, with or without a short outer series of minute bristles or scales; herbs with alternate lvs and solitary to numerous, hemispheric to turbinate heads. Typically blooming in spring and early summer (unlike *Aster*), but some weedy spp. continuing until fall. 200, mainly N. Temp.

1 Pappus of the pistillate (ray) fls reduced, under 1 mm; weedy annuals.
  2 Foliage ample; plants 6–15 dm; middle part of stem long-spreading-hairy . . . . 7. *E. annuus.*
  2 Foliage sparse; plants mostly 3–7 dm; pubescence often shorter and appressed . . . . 6. *E. strigosus.*
1 Pappus of all fls of evident long bristles, sometimes with a few short outer bristles or scales as well;
    seldom annual, most spp. not weedy.
  3 Rays short and inconspicuous, only slightly if at all exceeding the disk . . . . 9. *E. acris.*
  3 Rays well developed and conspicuous, evidently exceeding the disk.
    4 Lvs numerous and narrow, all nearly alike, not over 3 cm × 5 mm . . . . 8. *E. hyssopifolius.*
    4 Lvs not especially numerous nor all alike, some generally over 3 cm × 5 mm.
      5 Disk-cors (except sometimes the outermost) 4–6 mm; rays ca 1 mm wide.
        6 Plants with superficial rhizomes or stolons; rays 50–100; widespread . . . . 2. *E. pulchellus.*
        6 Plants without rhizomes or stolons; rays 125–175; western . . . . 1. *E. glabellus.*
      5 Disk-cors less than 4 mm; rays narrower, except in no. 5.
        7 Rays less than 50, 0.5 mm wide or more; achenes 4-nerved . . . . 5. *E. vernus.*
        7 Rays 100 or more, 0.5 mm wide or less; achenes 2-nerved.
          8 Invol 4–6 mm; rays 150–400, white to deep pink, 5–10 mm . . . . 3. *E. philadelphicus.*
          8 Invol 2.5–4 mm; rays 100–250, mostly blue, to 5 mm . . . . 4. *E. quercifolius.*

1. Erigeron glabellus Nutt. Hairy biennial or perennial 1–5 dm; lower lvs oblanceolate, to 15 cm × 15 mm or rarely more, the middle and upper ones evidently reduced, linear or lanceolate; heads solitary or several; disk 10–20 mm wide; invol 5–9 mm; rays 125–175, 8–15 mm, ca 1 mm wide, blue, pink or white; disk-cors 4.0–5.5 mm; pappus double; 2*n*=18, 36. Meadows, prairies, and open ground; Wis. to Alas. and Utah. Our plants, with spreading hairs, are the mostly northern var. *pubescens* Hook. (*E. asper; E. abruptorum; E. drummondii; E. oligodontus; E. oxyodontus.*)

2. Erigeron pulchellus Michx. Robin's plantain. Plants 1.5–6 dm from a short, fibrous-rooted caudex, perennating by slender, stoloniform rhizomes; basal lvs oblanceolate to suborbicular, 2–13 cm × 6–50 mm, commonly toothed; cauline lvs ovate to lanceolate or oblong, reduced upwards; heads solitary or few; invol 5–7 mm; disk 10–20 mm wide; rays 50–100, 6–10 mm, ca 1 mm wide or a little more, blue, or sometimes pink or white; disk-cors 4.0–6.0 mm; pappus simple; 2*n*=18. Woods and streambanks; Me. to Ont. and e.c. Minn., s. to Ga., Miss. and e. Tex. Three vars.:

a Rays white; achenes hairy; disk-cors 4–4.5 mm; se. Minn. . . . . var. *tolsteadii* Cronq.
a Rays blue or occasionally pink, very rarely white; achenes glabrous or nearly so; disk-cors 4.5–6 mm.
  b Plant spreading-hairy; range of the sp. . . . . var. *pulchellus.*
  b Plant essentially glabrous except for the ciliate lf-margins and glandular invols; local in ne. Ky. and
    adj. O. . . . . var. *brauniae* Fern.

3. Erigeron philadelphicus L. Philadelphia daisy. Biennial to short-lived perennial 2–7 dm, pubescent with long, spreading hairs, varying to subglabrous; basal lvs narrowly oblanceolate to obovate, coarsely crenate-toothed or lobed, rounded above, seldom over 15 × 3 cm; cauline lvs clasping, mostly oblong or ovate, ± reduced; heads 1–many; invol 4–6 mm, ± hirsute with flattened hairs, or subglabrous; disk 6–15 mm wide; rays mostly 150–400, 5–10 mm, 0.5 mm wide or less; deep pink or rose-purple to white; disk-cors 2.5–3.2 mm; achenes 2-nerved; pappus of ca 20–30 bristles; 2*n*=18. Various habitats, widespread in N. Amer., and throughout our range, somewhat weedy, varying in size with the habitat. (*E. purpureus; E. provancheri.*)

4. Erigeron quercifolius Lam. Oak-lvd d. Resembling no. 3, but ± villous-hirsute, the hairs usually shorter, those of the upper part (up to ½ or even more) of the stem sometimes appressed; stems slender, 1–4(–6) dm; basal lvs oblanceolate to obovate, sinuately lobed to subentire, 1–14 × 0.5–4 cm; cauline lvs ± reduced, sometimes clasping; heads 1–many, the infl more nearly naked than in no. 3; invol 2.5–4 mm, viscid-villous, the bracts darker than in no. 3; disk 5–10 mm wide; rays 100–250, up to 5 × 0.5 mm, blue or light blue-lavender, varying to sometimes white or seldom pinkish; disk-cors 1.5–2.5 mm; pappus-bristles ca 10–15; 2*n*=18, 36. Moist, sandy places and pine woods; coastal plain from Va. to Tex.

5. Erigeron vernus (L.) T. & G. Southern white d. Biennial or short-lived perennial 1.5–5 dm with a short subsimple caudex and rosulate offsets, glabrous or the stem sparsely appressed-hairy; basal lvs thick, oblanceolate to suborbicular, obtuse or rounded, 2–15 × 0.4–2 cm, denticulate; cauline lvs very few and small; heads 1–many; invol 3–4 mm, glutinous or sometimes sparsely hairy; disk 5–11 mm wide; rays 25–40, white, 4–8 × 0.5–1.3 mm; disk-cors 2.5–3.8 mm; achenes 4-nerved; pappus simple; 2*n*=18, ca 27. Sandy or peaty places, sphagnum-bogs, and pine-barrens; coastal plain from Va. to Fla. and La.

6. Erigeron strigosus Muhl. Rough fleabane. Annual or rarely biennial, 3–7(–12) dm, sparsely leafy, ± hairy, the hairs spreading or usually mostly appressed; basal lvs mostly oblanceolate to elliptic, entire or toothed, not over 15 cm (petiole included) × 2.5 cm; cauline lvs linear to lanceolate, mostly entire; heads several to very many; invol 2–5 mm, obscurely glandular and ± hairy; disk 5–12 mm wide; rays 50–100, white, or occasionally pinkish or bluish, to 6 mm, 0.4–1.0 mm wide, or much reduced; disk-cors 1.5–2.5 mm; pappus as in no. 7; 2*n*=18, 27, 36, 54. A weed in disturbed sites over most of the U.S. and s. Can. (*E. ramosus.*) Four vars.:

a Rays very short, about equaling the disk; s. N. Engl. and N.Y. . . . . var. *discoideus* Robbins.
a Rays well developed, evidently exceeding the disk.

b Hairs of the invol flattened, over 1 mm; hairs of the stem mostly long and spreading; morphologically
  transitional to no. 7, found chiefly in N. Engl. and adj. Can. . . . . var. *septentrionalis* (Fern. & Wieg.) Fern.
b Hairs of the invol not obviously flattened, less than 1 mm; hairs of the stem mostly short and
  appressed, except near the base.
  c Infl diffuse and subnaked, the pedicels often flexuous; heads tiny, the invol only 2–3 mm; coastal
    states from s. N.J. to Tex. . . . . var. *beyrichii* (Fischer & C. Meyer) A. Gray.
  c Infl not diffuse, or if so, then somewhat leafy; heads avg larger, the invol (2.5–)3–4 mm; widespread
    . . . . var. *strigosus*.

7. Erigeron annuus (L.) Pers. Annual fleabane. Annual or rarely biennial, 6–15 dm, amply leafy; stem hirsute, the hairs spreading except near the top; basal lf-blades elliptic to suborbicular, coarsely toothed, to 10 × 7 cm, ± abruptly long-petiolate; cauline lvs numerous, broadly lanceolate or broader, all except sometimes the uppermost usually sharply toothed; heads several to very numerous; invol 3–5 mm, finely glandular and sparsely beset with long, flattened, transparent hairs; disk 6–10 mm wide; rays 80–125, white or rarely anthocyanic, 4–10 × 0.5–1.0 mm; disk-cors 2.0–2.8 mm; achenes 2-nerved; pappus of the disk-fls double, with 10–15 bristles and several very short, slender scales (visible at 20×), that of the rays-fls of short scales only; $2n=27, 54$. A weed in disturbed sites over most of the U.S. and adj. Can., more common northward.

8. Erigeron hyssopifolius Michx. Hyssop-d. Perennial; stems numerous, 15–35 cm, subglabrous or sparsely strigose to occasionally densely spreading-villous; lvs thin, lax, subglabrous, all about alike except for the reduced and scale-like lowest ones, linear or nearly so, to 3 cm × 5 mm, often with short leafy axillary shoots; heads 1–5, on conspicuous, long, subnaked peduncles to which the cauline lvs are abruptly reduced; disk 6–12 mm wide; invol 4–6 mm; rays 20–50, white or sometimes pink or rose-purple, 4–8 mm; pappus obscurely double; $2n=18$. Mostly on rocky shores and banks; n. Vt. to Me., se. Can., and n. Mack. Summer. Ours is var. *hyssopifolius*.

9. Erigeron acris L. Trimorphic d. Biennial or perennial 3–8 dm, ± hairy, or the lvs subglabrous; lowest lvs mostly oblanceolate, to 10 × 1.5 cm, the cauline ones smaller, narrowly lanceolate to ovate; heads several or many on arcuate or obliquely ascending peduncles; invol 5–12 mm, finely glandular and often ± hirsute, the bracts very narrow and long-attenuate; pistillate fls numerous, in several series, the outer with narrow and inconspicuous, pinkish to white, erect rays equaling or slightly exceeding the pappus, not over 5 × 0.5 mm; inner pistillate fls merely filiform-tubular, without rays; disk-fls usually fewer than the pistillate fls; pappus copious, simple or obscurely double. Circumboreal, s. to n. Me., n. Mich. and n. Minn. Our plants, as here described, are the boreal N. Amer. and Siberian var. *kamtschaticus* (DC.) Herder (*E. asteroides* and *E. droebachensis*, both misapplied.)

59. CONYZA Less., nom. conserv. Invol bracts ± imbricate, scarcely herbaceous; pistillate fls numerous and slender, without rays, or in some spp. (all ours) with very short, narrow, and inconspicuous white or purplish rays barely or scarcely exceeding the pappus; disk-fls few, in ours not more than ca 20; style-appendages short, as in *Erigeron*; receptacle flat or nearly so, naked; achenes 1–2-nerved, or nerveless; pappus of capillary bristles, sometimes with a short outer series; herbs, often weedy, with alternate lvs and several to many, mostly rather small heads. 50+, mainly trop. and subtrop.

1 Plant simple or nearly so up to the infl, with a well defined central axis, often over 3 dm.
  2 Invol 3–4 mm, glabrous or nearly so; pistillate fls mostly 25–40 . . . . 1. *C. canadensis*.
  2 Invol 4–6 mm, copiously short-hairy; pistillate fls mostly (50–)70–200+ . . . . 2. *C. bonariensis*.
1 Plant diffusely branched from near the base, without a central axis, mostly 1–3 dm . . . . 3. *C. ramosissima*.

1. Conyza canadensis (L.) Cronq. Horseweed. Annual or winter-annual, 1–15 dm, simple or nearly so to the infl; lvs numerous, ± pubescent, or at least coarsely ciliate at the base, oblanceolate to linear, acute, toothed (especially the lower) or entire, gradually reduced upwards; the cauline ones to 8 cm × 8 mm, the basal ones larger and broader but generally deciduous; heads, except in depauperate plants, numerous in a long and open infl; invol 3–4 mm, glabrous or with a few small scattered hairs, the bracts strongly imbricate, brown or with a brown or pale midvein and greener sides; rays mostly 25–40, commonly ca 34, white or sometimes pinkish, 0.5–1.0 mm, equaling or shortly surpassing the style and pappus; $2n=18$. A weed in waste places and old fields, throughout the U.S. and s. Can. and to trop. Amer. Late summer and autumn. (*Leptilon c.; Erigeron c.*) Var. *canadensis*, our common phase, has the stem coarsely spreading-hirsute, and lacks purple tips on the bracts. Var. *pusilla* (Nutt.) Cronq., chiefly along the coast or on the coastal plain, from Conn. to trop. Amer., has the stem glabrous or nearly so, and some or all of the invol bracts are minutely purple-tipped. (*Leptilon p.*)

2. Conyza bonariensis (L.) Cronq. Annual, 1–10+ dm, copiously and loosely hairy, habitally like no. 1, or often with some of the lateral branches elongate and overtopping the central axis; lower lvs in robust plants sometimes to 15 × 2 cm; heads larger than in no. 1, the disk often over 1 cm wide; invol 4–6 mm, copiously short-hairy; pistillate fls (50–)70–200+, with a very short or scarcely developed ray up to 0.5 mm, this generally surpassed by the style and equaling to more often surpassed by the often tawny or reddish pappus; $2n=54$. A weed in waste places, widespread in trop. Amer. and occasionally intr. in se. U.S., n. to Va. Summer. (*C. floribunda; Erigeron b.; Leptilon b.*)

3. **Conyza ramosissima** Cronq. Dwarf fleabane. Diffusely branched, slender, ± hairy annual 1–3 dm, with no well defined central axis; lvs narrowly linear, to 4 cm × 2 mm, the uppermost reduced to mere bracts; heads numerous, much like those of no. 1, often covering the broad, much branched summit of the plant; invol 3–4 mm, the outer bracts short-hairy, the inner glabrous; rays minute, purplish, about equaling or slightly exceeding the pappus; $2n=18$. A weed in waste places, especially in sandy soil or along streams; O. and s. Ont. to Minn., s. to Ala. and Tex. Summer and fall. (*Erigeron divaricatus; Leptilon d.*)

60. BACCHARIS L.

Dioecious; fertile heads with ± numerous tubular-filiform pistillate fls; sterile heads with ± numerous seemingly perfect but functionally staminate fls, the ovary abortive, the style-branches sometimes connate; fls white to yellowish or greenish; receptacle flat or merely convex, naked; invol bracts subequal to strongly imbricate, chartaceous or subherbaceous; pappus of numerous capillary bristles, those of the sterile heads fewer and shorter than those of the fertile ones; achenes usually somewhat compressed and ribbed; shrubs or subshrubs with alternate lvs. 400, New World, mainly trop. and warm-temp.

1. **Baccharis halimifolia** L. Groundsel-tree; sea-myrtle. Freely branched shrubs 1–3(–4) m, glabrous and somewhat glutinous, the branches and invols sometimes minutely scurfy; lvs short-petiolate, the blade thick and firm, puncticulate, elliptic to broadly obovate, coarsely few-toothed, especially distally, up to ca 6 × 4 cm, those of the infl smaller, narrower, and mostly entire; heads numerous in small, pedunculate clusters, forming terminal, leafy-bracteate infls on the major branches; pistillate invols 4–6.5 mm, the bracts strongly imbricate in several series, the inner ones narrow and ± acute; pappus white, that of the pistillate fls much exceeding the cors and invols in fr, that of the sterile heads shorter than the fls and with the bristles somewhat fringed distally; $2n=18$. Marshes and beaches, especially near the seashore; Mass. to Fla., Ark., and Tex.; W.I. Aug.–Nov.

61. BOLTONIA L.'Hér.

Heads radiate, the rays white to pink or blue, pistillate; invol bracts subequal to strongly imbricate, scarious-margined, with green or greenish midrib or tip; receptacle hemispheric or conic, naked; disk-fls yellow, fertile; style-branches flattened, with short, lanceolate, externally hairy appendage; achenes obovate, strongly flattened, ± wing-margined; pappus of several minute bristles and 2(4) longer awns, these commonly reduced or wanting in the ray-achenes, sometimes also in the disk; glabrous, often stoloniferous, short-lived perennials with relatively narrow, chiefly entire lvs and several to very many small to middle-sized hemispheric heads. 5, U.S.A.

1 Infl narrow and few-headed to broader and corymbiform, but not diffuse, always ± leafy-bracteate,
    at least some of the lvs over 1 cm; heads larger, the rays (5)8–15 mm, the disk 6–10 mm wide.
  2 Cauline lvs not at all decurrent; widespread .... 1. *B. asteroides.*
  2 Cauline lvs conspicuously decurrent; Ill. and Mo. .... 2. *B. decurrens.*
1 Infl diffusely branched, leafy-bracteate or not; heads mostly smaller, the rays 5–8 mm, the disk 3–
    6(–8) mm wide.
  3 Infl ± leafy-bracteate, at least some of its lvs more than 1 cm; some of the lvs present at anthesis
    1 cm wide or more .... 3. *B. caroliniana.*
  3 Infl merely subulate-bracteate, its bracts commonly less than 1 cm; lvs present at anthesis seldom
    over 0.5(–1) cm wide .... 4. *B. diffusa.*

1. **Boltonia asteroides** (L.) L'Hér. Plants 3–15(–20) dm, with or without rhizomes; lvs broadly linear to lanceolate or narrowly lance-elliptic, reduced upward, the larger ones 5–15 cm × 5–20 mm; infl commonly corymbiform, somewhat leafy, the lvs narrow, at least some of them over 1 cm; invol bracts ± imbricate; rays white to pink, purple or blue, (5–)8–15 mm; disk mostly 6–10 mm wide, with fewer than 200 fls; achenes evidently wing-margined; pappus-awns usually well developed, (0.2–)0.6–2 mm; $2n=18, 36$. Moist or wet places; N.J. to Fla., w. to N.D., Okla., and Tex.; escaped and locally established elsewhere. July–Oct. Three vars.:

a Heads few, seldom more than 25; plants avg smaller and less leafy than the other vars.; coastal states
    from N.J. to Fla. and La. .... var. *asteroides.*
a Heads more numerous, more than 25 except in occasional depauperates.
  b Invol bracts mostly linear or nearly so and acute or acutish; chiefly Upper Mississippi Valley and
    ne. to the w. end of Lake Erie .... var. *recognita* (Fern. & Griscom) Cronq.
  b Invol bracts, except the inner, spatulate and rounded-obtuse; chiefly Ozarkian. (*B. latisquama.*) .
    .... var. *latisquama* (A. Gray) Cronq.

2. **Boltonia decurrens** (T. & G.) A. Wood. Similar to *B. asteroides* var. *recognita,* but with broader and more numerous lvs, those of the stem conspicuously decurrent; rhizomes wanting; invol bracts subequal or only slightly imbricate; disk-fls numerous, ca 250–400; pappus-awns of the disk apparently always well developed. Flood plains along major rivers; Mo. and Ill.; now rare. Sept., Oct.

3. **Boltonia caroliniana** (Walter) Fern. Mostly 8–20 dm; lvs lanceolate to linear-oblanceolate, or the lower broader, tapering to both ends, the larger ones mostly 8–15 × 1–2.5 cm, some of those present at full anthesis generally at least 1 cm wide; infl diffusely branched, somewhat leafy, the lvs mostly narrower and over 1 cm; invol bracts narrow, acute, subequal or slightly imbricate, rays white (lilac), mostly 5–8 mm; disk mostly 3–6(–8) mm wide; achenes scarcely or only very narrowly wing-margined; pappus-awns minute, less than 0.5 mm; 2*n*=18. Moist lowlands; coastal plain and piedmont of s. Va. to S.C. (*B. ravenelii,* with unusually large heads.)

4. **Boltonia diffusa** Elliott. Doll's daisy. Mostly 5–15 dm; lvs mostly linear or nearly so, the larger ones mostly 3–11 cm and up to 0.5 or 1(–2) cm wide, but the lower lvs commonly deciduous, so that the lvs in well grown flowering plants are rarely more than 0.5(–1) cm wide; infl diffusely branched, merely subulate-bracteate, the bracts commonly less than 1 cm; invol bracts narrow, acute or acuminate, evidently imbricate, rays white or lilac, mostly 5–8 mm; disk mostly 3–6 mm wide; achenes evidently wing-margined; pappus-awns less than 1 mm, or sometimes obsolete; 2*n*=18, 36. Moist or wet to sometimes rather dry places; se. U.S., n. to N.C., Ky., s. Ill., and Mo., and w. to Okla. and e. Tex. July–Oct.

62. ASTRANTHIUM Nutt.

Daisy. Heads solitary or several, radiate, mostly long-pedunculate; invol bracts slightly imbricate, green, with membranous or scarious margins; receptacle strongly convex, cushion-like, naked; rays pistillate and fertile, white or anthocyanic; disk-fls perfect, yellow; style-appendages ovate to lance-ovate; achenes ± obovate, compressed, mostly 2-nerved; pappus obsolete; caulescent herbs with alternate lvs. 10, Mex. and s. U.S.

1. **Astranthium integrifolium** (Michx.) Nutt. Annual 5–45 cm, sparingly to freely branched (often from near the base) when well developed, sparsely hairy, more densely so at base; lvs small, to 8 × 2 cm, the lower oblanceolate or spatulate, the others mostly linear to elliptic; disk 5–10 mm wide; rays 8–30, mostly blue or purple, 6–12 mm; achenes minutely glandular-glochidiate; 2*n*=8. Wooded, often sandy places, or on barrens or along roadsides; Ky. to nw. Ga., w. to se. Kans., Okla., Tex., and n. Mex. Apr.–July. (*Bellis i.*) Ours is var. *integrifolium.*

63. BELLIS L.

Heads solitary; invol bracts herbaceous, equal; receptacle conic, naked; rays pistillate and fertile, white or anthocyanic; disk-fls perfect, yellow; style-appendages deltoid or ovate, scarcely longer than broad; achenes compressed, mostly 2-nerved; pappus obsolete; herbs, scapose or nearly so.

1. **Bellis perennis** L. English daisy. Fibrous-rooted, scapose perennial, ± spreading-hairy; lf-blades elliptic or ovate to orbicular, dentate or denticulate, to 4 × 2 cm, narrowed to a margined petiole of equal or greater length; scape 5–15 cm; disk 5–10 mm wide; rays numerous, white to pink, up to 1 cm; 2*n*=18. A weed in lawns and waste places, intr. from n. Europe and ± established throughout n. U.S. Apr.–Nov.

64. INULA L.

Heads radiate, yellow, the rays pistillate; invol hemispheric or campanulate, its bracts imbricate in several series, the inner commonly narrow and scarious, the outer more herbaceous and often broader; receptacle flat or convex, naked; disk-fls tubular and perfect; anthers sagittate-tailed; style-branches flattened, externally slightly papillate distally, with well developed ventromarginal stigmatic lines extending all the way around the rounded tip; achenes subterete or ± prominently 4–5 ribbed or angled; pappus a single series of few to numerous capillary bristles, sometimes connate at base; ± hairy or glandular herbs with alternate or basal, simple lvs. 200, Old World.

1 Broad-lvd, coarse, velvety-woolly perennial with large heads and showy rays . . . . 1. *I. helenium.*
1 Narrow-lvd, glandular annual with small heads and short rays . . . . 2. *I. graveolens.*

1. **Inula helenium** L. Elecampane. Coarse perennial herb to 2 m, the stem finely spreading-hairy; lvs irregularly and shallowly dentate, densely velvety-woolly beneath, sparsely spreading-hairy or subglabrous above, the lower long-petiolate and elliptic, with blade to 5 × 2 dm, the upper becoming ovate, sessile and cordate-clasping; heads few, pedunculate, large, the disk 3–5 cm wide, the invol 2–2.5 cm; outer bracts broad, herbaceous, densely short-hairy; inner bracts narrow, subscarious, glabrous; rays numerous, slender, 1.5–2.5 cm; achenes glabrous, columnar and ± distinctly quadrangular, multistriate; pappus-bristles united at base,

often in groups; 2n=20. Intr. from Europe; cult., escaped, and sparingly naturalized in moist or wet disturbed sites. May–Aug.

2. **Inula graveolens** (L.) Desf. Taprooted, freely branching (when well developed), aromatic and copiously glandular annual to 1 m; lvs all cauline, broadly linear to linear-elliptic, entire or few-toothed, up to ca 10 × 1 cm; heads numerous in a diffuse, leafy-bracteate infl, small, the disk up to ca 1.5 cm wide, the invol 6–8 mm, its bracts all narrow, the outer greener; rays few and inconspicuous, hardly exceeding the disk; achenes loosely strigose, constricted at the summit beneath the expanded, pappiferous disk; pappus-bristles distinct; 2n=18, 20. Native of Europe, casually intr. in disturbed sites in our range, as in N.Y. and N.J. Aug.–Oct. (*Cupularia g.; Dittrichia g.; Pulicaria g.*)

65. FILAGO L. Heads ± numerous, disciform, white or whitish; invol scanty or none, its bracts (when present) imbricate, scarious, resembling the bracts of the cylindric or obconic to merely convex receptacle; outer fls pistillate, fertile, filiform-tubular, in several series, the outermost of these epappose and each subtended and partly enclosed by a concave or boat-shaped bract, the others bractless and with a pappus of capillary bristles; central fls 2–11, appearing perfect but often sterile, with capillary pappus; anthers shortly tailed; achenes small, scarcely nerved; low white-woolly annuals with alternate, entire lvs. 10+, Eurasia, n. Afr.

1 Heads glomerate at the ends of the stem and branches .... 1. *F. germanica.*
1 Heads in a racemiform or paniculiform infl .... 2. *F. arvensis.*

1. **Filago germanica** (L.) Hudson. Herba impia. Stems 1–several from the base, 0.5–4 dm; lvs linear or lance-linear, 1–2 cm × 1–3 mm, erect or nearly so; stem terminating by a sessile dense cluster of heads, this usually subtended by several leafy branches terminating in similar clusters, and these often again proliferous; heads 3.5–4.5 mm; woolly only toward the base, proper involucral bracts few or none, when present resembling the bracts of the receptacle, these largely hyaline-scarious and shining, with a firmer, darker central portion, each tipped by a conspicuous, short, often somewhat spreading bristle; receptacle shortly cylindric-obconic; inner pistillate flowers ca 20–25; disk (perfect) fls 2–3; 2n=28. Native of Europe, intr. in disturbed sites from N.Y. and O. s. May–Sept. (*F. vulgaris; Gifola g.*) Wagenitz' argument that the species should be called *F. vulgaris* is here rejected.

2. **Filago arvensis** L. Simple or branching annual 0.3–5 dm; lvs erect, linear or nearly so, up to ca 4 cm × 5 mm; heads mostly 3–5 mm high, in a racemiform or paniculiform, leafy-bracteate infl; proper invol very scanty; receptacular bracts woolly, simulating an invol, somewhat concave-clasping, obscurely hyaline-margined and -tipped; receptacle nearly flat; 2n=28. European weed, now well established in nw. U.S. and casually in our range, as in Mich. July, Aug. (*Logfia a.*)

66. PLUCHEA Cass. Marsh-fleabane. Heads disciform, white to yellow or pink-purple; invol ovoid to broadly campanulate or hemispheric, its bracts firm, imbricate; receptacle flat, naked; outer fls in several series, pistillate, the filiform cor shorter than the style; central fls tubular, perfect, mainly sterile, often with undivided style; anthers filiform-tailed; achenes tiny, 4–6-angled; pappus a single series of capillary bristles; herbs (all ours) or shrubs with alternate, usually toothed lvs. 50, mostly warm reg.

1 Lvs sessile, broad-based and ± clasping .... 1. *P. foetida.*
1 Lvs petioled, or sometimes merely tapering to the base.
  2 Invol evidently pubescent with short, several-celled, glandular hairs .... 2. *P. odorata.*
  2 Invol merely atomiferous-glandular, sometimes very sparsely so .... 3. *P. camphorata.*

1. **Pluchea foetida** (L.) DC. Fibrous-rooted perennial 3–10 dm, finely glandular and often somewhat cobwebby-puberulent; lvs oblong to elliptic, lance-ovate, or ovate, 4–10 × 1–4 cm, sharply callous-denticulate, reticulate-veiny, rounded to acute at the tip, closely sessile and ± cordate-clasping; heads several or many in a short and broad, leafy infl; invol 5–8 mm, finely glandular or viscid-puberulent; disk 6–12 mm wide; cors ochroleucous; 2n=20. Permanently wet soil, often in meadows or swampy woods; s. N.J. to Fla. and Tex., chiefly near the coast, but sometimes inland, as in Mo.; W.I. Aug., Sept., or in the south all year.

2. **Pluchea odorata** (L.) Cass. Annual, ± fibrous-rooted, to 10(–15) dm, glandular-puberulent to occasionally subglabrate; lvs ample, lanceolate to elliptic or ovate, 4–15 × 1–7 cm, acute or acuminate, ± serrate, short-petiolate, or sometimes, especially the upper, merely tapering to the base; heads pink-purple, numerous in a generally ± flat-topped or layered infl; invol evidently pubescent with short, multicellular, glandular-viscid hairs, the bracts imbricate in several series, commonly ± pink or purple, at least distally; 2n=20. Salt or brackish (seldom freshwater) marshes, chiefly along the coast; Mass. to Fla., Tex., and trop. Amer., and occasionally inland, as in s. Ill. and se. Mich. Aug., Sept. Most of our plants are var. *succulenta* (Fern.) Cronq.,

seldom over 6 dm, with large heads, the invol 5.5–7 mm high, the disk 5–9 mm wide. In Md. and Va. this passes into the more southern var. *odorata*, taller, with smaller heads, the invol 4–5.5 mm high and the disk 4–7 mm wide. (*P. purpurascens.*)

3. **Pluchea camphorata** (L.) DC.   Similar to no. 2 in aspect, avg taller, to 2 m, sometimes perennial, and more glabrate; lvs thinner, on the avg more serrate, and more evidently petiolate, avg a little narrower and more acuminate; infl generally round-topped, often ± elongate; invol 4–6 mm, sometimes purplish, but more often not, merely granular-glandular, or nearly glabrous; disk 3–6 mm wide; $2n=20$. Wet or moist, nonsaline places; Del. and Md. to n. Fla., w. to s. O., s. Ill., and e. Okla. and Tex. Aug.–Oct. (*P. petiolata.*)

67. GNAPHALIUM L.   Cudweed; everlasting.

Heads disciform, yellow or whitish, the numerous outer fls pistillate, with tubular-filiform cor about equaling the pappus, the few inner ones coarser and perfect; invol ovoid or campanulate, its bracts ± imbricate, scarious at least at the tip; receptacle flat to subconic, naked; anthers caudate; style-branches slender, minutely papillate outside, slightly flattened, with ventromarginal stigmatic lines extending to the truncate or slightly expanded tip; achenes small, terete or slightly compressed, nerveless; pappus of capillary bristles, these sometimes thickened at the summit, sometimes connate at the base; woolly herbs with alternate, entire lvs. (*Gamochaeta.*) 100+, cosmop.

1 Perennial with narrow, spiciform or subcapitate infl; achenes sparsely strigose; boreal.
  2 Pappus-bristles distinct, falling separately; plants 2–10 cm .... 1. *G. supinum.*
  2 Pappus-bristles united at base into a ring; plants mostly 10-60 cm .... 2. *G. sylvaticum.*
1 Annual or biennial; infl various; achenes smooth or papillate; widespread.
  3 Pappus-bristles united at base; infl narrow, usually spiciform-thyrsoid .... 3. *G. purpureum.*
  3 Pappus-bristles distinct, falling separately; infl otherwise.
    4 Invol 2–3 mm; infl of numerous small axillary and terminal clusters, overtopped by their subtending lvs; plants to 25 cm ....4. *G. uliginosum.*
    4 Invol 5–7 mm; infl ample, panicle-like; plants mostly over 25 cm.
      5 Lvs not decurrent at base.
        6 Stem woolly, scarcely glandular except sometimes near the base .... 5. *G. obtusifolium.*
        6 Stem glandular-hairy, scarcely woolly except in the infl .... 6. *G. helleri.*
      5 Lvs decurrent at base; stem glandular-hairy, sometimes also woolly .... 7. *G. macounii.*

1. **Gnaphalium supinum** L.   Alpine c. Dwarf, thinly woolly perennial 2–10 cm; basal lvs tufted, linear or linear-oblanceolate, acute, to 2.5 cm × 3 mm; cauline lvs similar, few; heads 1–8 in a terminal, spiciform or subcapitate infl; invol 5–6 mm, scarcely woolly, its bracts light greenish or tan, with dark brown margins and tips, imbricate, the outer obtuse, the inner mostly acute; pappus-bristles separate; achenes sparsely strigose; $2n=28$. Alpine places; circumboreal, s. to n. Me. and n. N.H. July–Sept.

2. **Gnaphalium sylvaticum** L.   Woodland-c. Erect, simple, thinly woolly perennial 1–6 dm; lvs commonly subglabrate above, linear or narrowly oblanceolate, the larger basal and lower cauline ones 3–8 mm wide, those at the base of the infl 2–3 mm wide; infl narrow, spiciform-thyrsoid, somewhat leafy-bracteate, with 10–many heads; invol scarcely woolly, or woolly only at the base, 5–7 mm, its bract rounded or obtuse, light stramineous or greenish-stramineous toward the base, some or all with a conspicuous dark brown, commonly reverse-V-shaped spot above the middle, the tip paler; pappus-bristles united at base, falling in a ring; achenes sparsely strigose; $2n=56$. Open woods and waste places; circumboreal, s. to n. Me., n. N.H., n. N.Y., and n. Wis. July–Sept.

3. **Gnaphalium purpureum** L.   Purple c. Thinly woolly annual or biennial 1–4 dm; lowest lvs spatulate or oblanceolate, rounded at the tip, generally mucronate, to 10 × 2 cm, often forming a persistent basal cluster or rosette; heads numerous in a terminal, somewhat leafy-bracteate, spiciform-thyrsoid, seldom branched, sometimes interrupted infl; invol 3–5 mm, woolly below, its bracts imbricate, mostly acute to acuminate, light brown, often tinged with anthocyanin; pappus-bristles united at base, deciduous in a ring; achenes papillate; $2n=14$, 28. Sandy soil and waste places; Me. to B.C., s. to Mex. and S. Amer., more common southwards. May–Oct. The widespread, polymorphic var. *purpureum* has the lvs, except the uppermost, oblanceolate to spatulate, tending to be obviously greener and less hairy on the upper surface than on the lower. (*G. pensylvanicum; G. peregrinum; G. spathulatum.*) The more stable var. *falcatum* (Lam.) T. & G., occurring from Va. s., has the lvs narrowly oblanceolate to linear or merely linear-oblanceolate, tending to be about equally hairy on both sides. (*G. falcatum; G. calviceps.*)

4. **Gnaphalium uliginosum** L.   Low c. Branching annual, 0.5–2.5 dm, generally diffuse; stem densely and often rather loosely white-woolly, the lvs sparsely so; lvs numerous, mainly or wholly cauline, linear or oblanceolate, to 4 cm × 5 mm; heads glomerate in numerous small clusters in the axils and at the ends of the branches, overtopped by their subtending lvs; invol 2–3 mm, woolly at base, its bracts greenish or brown, often paler at the tip, not much imbricate, acute, or the outer obtuse; achenes papillate or smooth; $2n=14$. Streambanks and waste places, wet or dry; a European weed, now established nearly throughout our range, n. to Nf. and w. to B.C. July–Oct.

5. Gnaphalium obtusifolium L.   Fragrant c. Erect, fragrant annual or winter annual, (1–)3–10 dm; stem thinly white-woolly, commonly becoming subglabrous (or even a little glandular) near the base; lvs numerous, essentially all cauline, lance-linear, up to 10 × 1 cm, sessile but not decurrent, white-woolly beneath, green and from glabrous to slightly glandular or slightly woolly above; infl ample, branched, and many-headed in well developed plants, flat or round-topped and often elongate; invol ochroleucous or dingy, campanulate, woolly only near the base, 5–7 mm, its bracts acutish to obtuse or somewhat rounded; fls mostly ca 75–125, 3–10 of them perfect; achenes glabrous; pappus-bristles distinct, falling separately, or sometimes temporarily coherent in small groups by means of tiny, interlocking hairs near the base; 2n=28. A common native N. Amer. weed, found throughout our range, s. to Fla. and w. to Neb. and Tex. The var. *saxicola* (Fassett) Cronq., found along cliffs and ravines in s.c. Wis., is a lax, slender from less than 2.5 dm, with broader lvs less hairy beneath than in var. *obtusifolium*. (*G. saxicola*.)

6. Gnaphalium helleri Britton.   Similar to no. 5, less common, avg a little smaller; stem glandular-hairy, becoming woolly in the infl; lvs woolly beneath, ± glandular-hairy above, the hairs shorter and sparser than those of the stem. Dry, commonly sandy soil, often in woods; Me. to Ga., w. to Ind., Ark., and Tex. Aug.–Oct.

7. Gnaphalium macounii Greene.   Clammy c. Similar to no. 5; stem ± glandular-hairy, becoming woolly in the infl, rarely somewhat woolly to near the base, as well as glandular; lvs distinctly decurrent at base, the upper surface ± glandular-hairy, the lower woolly or sometimes glandular-hairy; fls mostly 60–120, 6–20 of them perfect; 2n=28. Open places; Que. to B.C., s. to W.Va., Tenn., and Mex. July–Sept. (*G. viscosum*, misapplied.)

The mainly western sp. *G. stramineum* HBK. is casually intr. in our range, as on the coast of Va. It has basally adnate-auriculate, not at all glandular lvs and very numerous (commonly 150–200+) fls per head, 8–20(–26) of them perfect. (*G. chilense*.)

68. ANAPHALIS DC.   Pearly everlasting. Dioecious or polygamo-dioecious, the pistillate heads sometimes with a few central staminate fls; heads disciform or discoid; invol bracts imbricate, almost wholly dry and scarious; receptacle naked, flat or convex; pistillate fls with tubular-filiform cor and bifid style; staminate fls with coarser, tubular cor and generally undivided style; anthers caudate; pappus of distinct capillary bristles, neither clavellate nor conspicuously barbellate; white-woolly perennials with alternate entire lvs. 25+, N. Temp., mostly e. Asia.

1. Anaphalis margaritacea (L.) Benth. & Hook.   Rhizomatous, 3–9 dm, leafy, loosely white-woolly, or the pubescence rusty in age; lvs numerous, all cauline, lanceolate or linear, to 12 × 2 cm, sessile, commonly less pubescent above than beneath, or glabrous above, the margins often revolute; heads 1 cm wide or less, numerous and crowded in a short, broad infl; invol 5–7 mm, its bracts pearly-white; achenes papillate; 2n=26. Various habitats, chiefly dry and open; n. N. Amer. and e. Asia, s. to Va., W.Va., Neb., and Calif. July, Aug. Highly variable.

69. ANTENNARIA Gaertner.   Pussytoes. Dioecious; invol bracts imbricate in several series, scarious at least at the tip, white or often colored; receptacle flat or convex, naked; staminate heads discoid, with tubular corollas, usually undivided styles, and scanty pappus, the bristles commonly barbellate or clavellate; anthers tailed; pistillate heads disciform, with filiform-tubular corollas, bifid styles, and copious pappus of capillary, naked bristles weakly united at base; achenes nerveless, terete or nearly so, glabrous or papillate; woolly perennial herbs, ours fibrous-rooted, stoloniferous, and mat-forming or colonial; lvs alternate, simple, entire, in our spp. the largest ones basal and at the ends of the stolons; heads rather small, (1–) several in terminal infls; x=14. 30+, mainly N. Amer., especially the w. cordillera. Our spp. 3–6 form a polyploid-apomictic complex. Absence of staminate plants from a local population indicates apomixis.

1 Upper surface of the lvs nearly or quite as densely hairy as the lower, becoming glabrate, if at all, only in extreme age; n. Mich., w. Minn., and westward.
  2 Invols 4–7 mm; dry pistillate cors 2.5–4.5 mm .... 1. *A. microphylla*.
  2 Invols 7–11 mm; dry pistillate cors 5–8 mm .... 2. *A. parvifolia*.
1 Upper surface of the principal lvs distinctly less pubescent than the lower, sooner or later glabrate (or glabrous from the first); widespread.
  3 Principal lvs relatively small, less than 1.5 cm wide, 1-nerved or obscurely 3-nerved.
    4 Heads small, the pistillate invols 5–7 mm; regularly sexual; mainly shale-barrens .... 3. *A. virginica*.
    4 Heads larger, the pistillate invols 7–10 mm; sexual or often apomictic; widespread .... 4. *A. neglecta*.
  3 Principal lvs larger, 1.5 cm wide or more, prominently 3- or 5-nerved.
    5 Heads several .... 5. *A. plantaginifolia*.
    5 Heads solitary .... 6. *A. solitaria*.

1. **Antennaria microphylla** Rydb.   Rosy p.   Mat-forming, stoloniferous, 5–40 cm; herbage closely and persistently gray-tomentose, or some of the lvs rarely finally glabrate above; basal lvs oblanceolate or spatulate, 8–30 × 2–7 mm; heads several in a subcapitate or rather loose cyme; pistillate invols 4–7 mm, the scarious part of the bracts deep pink to bright white or dull whitish, generally appearing striate at 20×; dry pistillate cors 2.5–4.5 mm; sexual or apomictic; 2*n*=28, 56. Dry, open places, meadows, and open woods; chiefly in the w. cordillera, e. to w. Minn., w. Ont., and n. Mich. May–Aug. (*A. rosea*.) The name *A. microphylla* may eventually have to give way to the older *A. chilensis* Remy, from Chilean Patagonia.

2. **Antennaria parvifolia** Nutt.   Plains p.   Similar to no. 1, but commonly appearing stouter, with larger, mostly closely aggregated heads, and on the avg with larger, relatively broader lvs; plants seldom over 15 cm; basal lvs spatulate or oblanceolate, 1–3.5 cm × 2.5–10 mm; pistillate invols 7–11 mm, the bracts usually blunt, the scarious part bright white or less often dull white, seldom pink; dry pistillate cors 5–8 mm; sexual or apomictic; 2*n*=56, 84, 112, ca 140. Dry, open places; w. Minn. to Man., B.C., and Ariz. May–July. (*A. aprica*.)

3. **Antennaria virginica** Stebbins.   Shale-barren p.   Heads small, the pistillate invols 5–7 mm; consistently sexual; diploid or tetraploid; otherwise like a small and delicate form of *A. neglecta* var. *neodioica*. Mainly shale-barrens; w. Va. and e. W.Va. to Pa., but also to Ky. and s. and e.c. O. Apr., May. (*A. neglecta* var. *argillicola*.)

4. **Antennaria neglecta** Greene.   Field-p.   Plants 1–4 dm, with short and leafy or longer and merely bracteate stolons; basal lvs and those at the ends of the stolons densely and persistently tomentose beneath, only sparsely so (and eventually glabrate) above, or the upper side glabrous from the first, relatively small, mostly under 1.5 cm wide, 1-nerved or obscurely 3-nerved; pistillate invols 7–10 mm; variously sexual or apomictic, diploid or polyploid. Woods and open places; Nf. and Que. to Yukon, s. to D.C., Va., W.Va., s. Ind., Mo., and Calif. Apr.–June. Four vars. in our range:

a Basal lvs and those at the ends of the stolons glabrous above nearly or quite from the first; habit
   mainly of var. *neglecta*, but sometimes varying toward that of var. *neodioica*; mainly polyploid
   and apomictic; mostly northern, s. to N.Y. and N.J. (*A. canadensis; A. neglecta* var. *randii*.) ..
   ....var. *canadensis* (Greene) Cronq.
a Basal lvs and those at the ends of the stolons tardily glabrate.
  b Stolons long, procumbent, with small and often few lvs; basal lf-blades tending to taper gradually
    to the base, scarcely petiolate.
    c Upper cauline lvs tipped by a scarious, flat or curled, flag-like appendage; mainly or wholly
     sexual and diploid; widespread. (*A. campestris; A. longifolia*.) .... var. *neglecta*.
    c Upper cauline lvs with a slender, subulate tip, or only those about the infl scarious-tipped; mainly
     polyploid and apomictic; mostly northern, seldom s. of N. Engl., N.Y., and Mich. (*A.
     petaloidea*.) .... var. *petaloidea* (Fern.) Cronq.
  b Stolons shorter, generally merely decumbent at the base, and more leafy; basal leaf-blades tending
    to be more abruptly contracted and somewhat petiolate; mainly apomictic and polyploid;
    widespread. (*A. brainerdii; A. neglecta* var. *attenuata; A. neodioica; A. rupicola*.) ... var. *neodioica* (Greene) Cronq.

5. **Antennaria plantaginifolia** (L.) Richardson.   Plantain-p.   Plants 1–4 dm, stoloniferous, the stolons sparsely leafy or merely bracteate; basal lvs and those at the ends of the stolons densely and persistently tomentose beneath, only sparsely so (and eventually glabrate) above, or the upper side glabrous (or somewhat glandular) from the first, tending to persist throughout the winter, large, 3- or 5-nerved, evidently petiolate, the blade ovate to elliptic or obovate, mucronate, the larger ones 2–6 × 1.5–5 cm; cauline lvs reduced, mostly linear or oblanceolate; heads several in a generally subcapitate cyme; pistillate invols 5–10 mm, the bracts white-tipped (often pinkish toward the base), striate; staminate invols avg smaller, with broader and more conspicuous white tips to the bracts; styles often crimson; variously sexual or apomictic, diploid or polyploid. Open woods and dry ground; N.S. and N.B. to N.D. and e. Mont., s. to Ga., La., and Tex. Apr.–June. Three vars. all widespread.

a Heads small, the pistillate invols 5–7 mm; bracts of the pistillate invols often narrower and duller
  than in the other vars.; regularly sexual and diploid .... var. *plantaginifolia*.
a Heads larger, the pistillate invols 7–10 mm; polyploid, (2*n*=56, 70, 84, 112); sexual or more often
  apomictic.
  b Basal lvs and those at the ends of the stolons tardily glabrate (as in var. *plantaginifolia*). (*A. fallax;
    A. farwellii; A. munda; A. obovata; A. occidentalis*.) .... var. *ambigens* (Greene) Cronq.
  b Basal lvs and those at the ends of the stolons glabrous above (or somewhat glandular) nearly or
    quite from the first. (*A. p.; A. brainerdii*.) .... var. *parlinii* (Fern.) Cronq.

6. **Antennaria solitaria** Rydb.   Southern single-head p.   Much like no. 5, but differing sharply in its solitary heads; stems 1–2.5 dm, nearly naked; basal lvs avg longer (to 7.5 cm) and somewhat narrower in shape, often obovate; elongate stolons nearly naked; invol 8–10 mm; diploid and sexual. Woods; sw. Pa. and s. O. to s. Ind., s. to Va., Ga., and La. Apr., May.

70. EUPATORIUM L.   Heads discoid; the fls all tubular and perfect; invol bracts variously imbricate or subequal; receptacle naked; fls pink or purple to blue or white; style-branches elongate, linear or linear-clavate, obtuse, papillate, with inconspicuously

ventromarginal stigmatic lines near the base (or sometimes for much of their length); achenes prismatic, mostly 5(–8)-angled and -nerved, glabrous, or inconspicuously hairy along the veins, or in most of our spp. atomiferous-glandular; pappus a single series of capillary bristles; perennial herbs (all ours) or shrubs, with entire or toothed to occasionally dissected, often glandular-punctate, mostly opposite, sometimes whorled or alternate lvs, all our spp. with the lowest lvs reduced and sometimes deciduous; heads small to large, in a mostly corymbiform infl, rarely solitary or forming a true panicle. Nearly 1000, mainly New World. Divided by some authors into numerous much smaller genera, these perhaps better regarded as sections or subgenera. Some of the segregates are indicated in parentheses in the key. In the system of King and Robinson our sp. 6 is referred to *Conoclinium*, 7 to *Fleischmannia*, and 8–10 to *Ageratina*; the others remain in *Eupatorium*.

Spp. 1–5 form a closely related, hybridizing, mainly diploid group, with $x=10$. Spp. 8–10 form a closely related group of diploids with $x=17$. Spp. 11–24 form an intricately reticulate complex of diploids, autopolyploids, established or temporary allopolyploids, and temporary hybrids, based on $x=10$. The diploids are mainly sexual and outcrossing, the polyploids mainly apomictic.

Some spp. here recognized are wholly diploid; others consist of both diploid and polyploid elements, the polyploids being taxonomically associated with the diploids they most resemble. Only *E. godfreyanum*, among our spp., appears to be wholly polyploid. In addition to the taxa recognized here, there are scattered small populations representing diverse sorts of allopolyploids that may not be permanently established. It does not seem useful to provide these with formal names.

1 Lvs all or mostly in whorls of 3–7, generally at least 2 cm wide. (*Eupatoriadelphus*.)
  2 Lvs ± strongly triplinerved, rather abruptly contracted to the petiole, thick and firm, up to 12(–15) cm; fls (4–)5–8(–10) per head .... 2. *E. dubium*.
  2 Lvs otherwise, either more gradually narrowed to the petiole, or not at all triplinerved, or commonly with both of these differences, also often over 15 cm.
    3 Fls mostly (8–)10–16(–22) per head; infl or its segments flat-topped .... 1. *E. maculatum*.
    3 Fls mostly 4–7 per head; infl convex.
      4 Stem persistently spreading-glandular hairy; local in sw. Va. and adj. Ky. and N.C. .... 3. *E. steelei*.
      4 Stem glabrous or nearly so below the infl; widespread.
        5 Stem purplish only or chiefly at the nodes, solid or sometimes eventually developing a slender central cavity .... 4. *E. purpureum*.
        5 Stem ± purplish essentially throughout, hollow with a large central cavity .... 5. *E. fistulosum*.
1 Lvs mostly opposite, sometimes some of them alternate, ternate in rare individuals of some spp., or, if regularly whorled, then well under 2 cm wide.
  6 Fls ± numerous, usually at least 9 per head.
    7 Fls anthocyanic.
      8 Receptacle conic; fls mostly 35–70, commonly blue. (*Conoclinium*.) .... 6. *E. coelestinum*.
      8 Receptacle flat; fls mostly (13–)18–24, pink-purple (seldom pale blue). (*Fleischmannia*.) ..... 7. *E. incarnatum*.
    7 Fls white.
      9 Invol bracts not strongly imbricate, the principal ones subequal and subbiseriate, a few shorter outer ones sometimes also present. (*Ageratina*.)
        10 Lf-blade obviously longer than the petiole, not extremely thin and delicate; widespread spp.
          11 Lvs relatively thick and firm, mostly crenate or crenate-serrate, acute or obtuse, the larger ones 3–7(–10) × 2–5 cm .... 8. *E. aromaticum*.
          11 Lvs thinner, mostly sharply serrate, acuminate, the larger ones mostly 6–18 × 3–12 cm .... 9. *E. rugosum*.
        10 Lf-blade about as long as the petiole, extremely thin and delicate; local in Ky. and Tenn. ....10. *E. luciae-brauniae*.
      9 Invol bracts evidently imbricate in 3 or more series, the outer less than half as long as the inner. (*Uncasia*.)
        12 Lvs evidently petiolate .... 11. *E. serotinum*.
        12 Lvs sessile.
          13 Lvs neither especially broad at base nor at all connate, seldom as much as 1.5 cm wide .... 12. *E. resinosum*.
          13 Lvs broad-based and nearly always connate-perfoliate, the larger ones seldom less than 1.5 cm wide .... 13. *E. perfoliatum*.
  6 Fls mostly 5 per head, rarely 3–7.
    14 Principal lvs pinnatifid or pinnately dissected, with filiform segments. (*Traganthes*.) .... 14. *E. capillifolium*.
    14 Lvs all entire to deeply toothed, not at all pinnatifid. (*Uncasia*.)
      15 Invol bracts acuminate to attenuate (or the inner rounded and strongly mucronate in forms of *E. album*.)

16 Larger lvs mostly 1.5–3 cm wide; invol 8–11 mm .... 15. *E. album.*
16 Larger lvs up to 1(–1.3) cm wide; invol 5–7 mm .... 16. *E. leucolepis.*
15 Invol bracts rounded to acute, but scarcely acuminate.
  17 Lvs broadly cuneate or subtruncate to rounded or subcordate at base.
    18 Plants glabrous below the infl; lvs acuminate or gradually and narrowly acute,
      pinnately veined, not trinerved or triplinerved .... 17. *E. sessilifolium.*
    18 Plants evidently hairy below as well as in the infl; lvs various.
      19 Lvs strictly pinnately veined, as in no. 17 .... 18. *E. godfreyanum.*
      19 Lvs ± evidently trinerved or triplinerved from near the base
        20 Lvs ovate (mostly broadly so) to subrotund, mostly evenly toothed except
          in var. *cordigerum*; upper lvs and main branches of the infl usually
          all opposite, seldom alternate .... 19. *E. rotundifolium.*
        20 Lvs narrower, mostly lanceolate or lance-ovate to elliptic-ovate, mostly
          coarsely and unevenly toothed; upper lvs and main branches of the
          infl tending to be alternate .... 20. *E. pilosum.*
  17 Lvs tapering to a narrow base.
    21 Plants with conspicuously tuberous-thickened rhizomes; lvs tending to be deflexed;
      se. Va. and southward .... 21. *E. mohrii.*
    21 Plants without tuberous-thickened rhizomes, mostly from a short crown or caudex
      or caudex-like rhizome instead; lvs seldom deflexed.
      22 Plants generally branched at or near the ground-level; slender sterile axillary
        shoots of the middle and lower lvs commonly elongating .... 22. *E. cuneifolium.*
      22 Plants generally simple below the middle, the stems generally distinct down
        to the caudex or rhizomes, the middle and lower axillary shoots generally
        not elongating.
        23 Lvs mostly 2.5–7 times as long as wide, the larger ones seldom less than
          1 cm wide, all opposite, or the upper alternate.
          24 Invol 2.5–4 mm; lvs obtuse to acute, but tending to carry their width
            well above the middle, triplinerved, the principal pair of lateral
            veins emerging as evident branches from the midvein .... 23. *E. semiserratum.*
          24 Invol 4.5–7 mm; lvs gradually and very narrowly acute or acuminate,
            trinerved, the principal pair of lateral veins distinct from the
            midvein all the way to the base of the lf .... 24. *E. altissimum.*
        23 Lvs mostly 6–40 times as long as wide, seldom over 1 cm wide, mostly
          ternate or quaternate, but sometimes merely opposite, or even alter-
          nate above .... 25. *E. hyssopifolium.*

1. **Eupatorium maculatum** L.   Spotted joe-pye weed. Stems 6–20 dm, speckled or sometimes more evenly purplish, seldom glaucous; lvs mostly in 4's or 5's, lance-elliptic to lanceolate or lance-ovate, relatively gradually narrowed to the short petiole (sometimes more abruptly contracted in ssp. *bruneri*), 6–20 × 2–9 cm, seldom evidently triplinerved, sharply serrate; infl or its divisions ± flat-topped in life; invol 6.5–9 mm, often purplish, its bracts well imbricate in several series, often 3–5-nerved, obtuse, essentially glabrous, or the outer often ± inconspicuously short-hairy; fls purple to rather pale lavender, (8)9–22 per head; $2n=20$. Moist places, especially in calcareous soils; Nf. to B.C., s. to Md., W.Va., Ill., Neb., N.M., and Utah, and along the higher mts. to N.C. and Tenn. July–Sept. Two sspp., one with two vars.:

a Lvs relatively thin and not very hairy, only seldom strongly puberulent beneath; eastern, from the
  Atlantic to Mich., Ill., and e. Minn. .... ssp. *maculatum.*
  b Lvs of the uppermost whorl relatively large and elongate, surpassing the infl; high northeastern,
    entering our range in Me. and n. N.H. and n. Vt., occurring at higher elev. than var. *maculatum*
    where their ranges overlap .... var. *foliosum* (Fern.) Wieg.
  b Lvs of the uppermost whorl smaller, not surpassing the infl; widespread in our range, but giving
    way to var. *foliosum* at the ne. .... var. *maculatum.*
a Lvs relatively firm and more strongly hairy, densely covered with short spreading hairs beneath;
  western, extending e. to Minn. and Io. (*E. bruneri.*) .... ssp. *bruneri* (A. Gray) G. W. Douglas.

2. **Eupatorium dubium** Willd.   Three-nerved joe-pye weed. Stems 4–10(–15) dm, generally purple-speck-led, viscid-puberulent at least near the top, scarcely or not at all glaucous; lvs mostly in 3's or 4's, thick and firm, often somewhat rugose, ovate or lance-ovate, relatively abruptly contracted to the short petiole, 5–12(–15) × 2–7 cm, coarsely serrate, ± strongly triplinerved; infl dense, slightly to strongly convex, not large; invol 6.5–9 mm, often purplish, its bracts well imbricate, obtuse, often 3–5-nerved; fls purple, (4–)5–8(–10) per head; $2n=20$. Moist places, especially in sandy or gravelly, acid soil; near the coast from N.S. and s. N.H. to S.C. July–Sept.

3. **Eupatorium steelei** E. Lamont.   Steele's joe-pye weed. Much like no. 4, but the stems persistently spreading-glandular-hairy nearly or quite to the base, the lvs firmer, glandular-hairy beneath (like the stems), and avg broader in shape and more abruptly contracted to the petiole. Mt. woods in the Folded Appalachians Province of sw. Va. and adj. Ky. and N.C., and Tenn. July–Sept.

4. **Eupatorium purpureum** L.   Purple-node joe-pye weed. Stems 6–20 dm, glabrous or nearly so below the infl, slightly glaucous, usually purple only or chiefly at the nodes, otherwise greenish, usually solid, the pith remaining intact or ultimately developing a slender central cavity; lvs mostly in 3's or 4's, lanceolate to

ovate or elliptic, 8–30 × 2.5–15 cm, gradually or sometimes rather abruptly narrowed to the short petiole, pinnately veined, usually sharply toothed, the lower surface glabrous or sparsely and inconspicuously short-hairy; infl convex; invol imbricate, 6.5–9 mm, its bracts well imbricate, commonly 3-nerved, obtuse or acutish; fls 4–7 per head, generally very pale pinkish or purplish, but variable; 2n=20, 40. Thickets and open woods, often in drier habitats than related spp.; s. N.H. to Va. and in the mts. to Ga., w. to Wis., Io., Okla., and w. Fla. July–Sept. (*E. trifoliatum.*) Plants from our western border, as in Minn., Io., and Mo., have the lvs more evidently short-hairy across the surface beneath, and have been distinguished as var. *holzingeri* (Rydb.) Lamont.

5. **Eupatorium fistulosum** Barratt. Hollow-stemmed joe-pye weed. Similar to no. 4, often more robust, to 3 m; stem strongly glaucous, usually purplish throughout, hollow with a large central cavity; lvs mostly in 4's to 7's, more elliptic, generally narrowly so, with mostly finer, more rounded, blunter teeth; cor generally bright pink-purple; 2n=20. Bottomlands and moist woods; s. Me. to Io., s. to Fla. and Tex. July–Sept.

6. **Eupatorium coelestinum** L. Mist-fl. Rhizomatous; stems 3–9 dm, puberulent; lvs opposite, petiolate, crenate or crenate-serrate, deltoid-ovate, sometimes narrowly so, 3–10 × 2–5 cm, trinerved or triplinerved, sparsely appressed-hairy or subglabrous, often atomiferous-glandular; invol 3–5 mm, its bracts narrow, firm, long-pointed, ± imbricate; receptacle conic; fls 35–70 per head, bright blue or violet (rarely reddish-purple), often purplish when dry; pappus scanty; 2n=20. Woods, streambanks, meadows, and fields; N.Y. to Ill. and Kans., s. to Fla., Tex., and W.I. July–Oct. (*Conoclinium c.*)

7. **Eupatorium incarnatum** Walter. Pink eupatorium. Stems weak, 3–12 dm, freely branching, each branch ending in a small infl; lvs opposite, petiolate, deltoid or somewhat cordate at base, tapering to the acuminate or acute tip, corsely crenate-serrate, to 7 × 5.5 cm, trinerved; invol 3–5 mm, glabrous or nearly so, its bracts sharply acute to blunt, the main ones subequal, but some irregularly shorter outer ones also present; receptacle flat; fls (13–)18–24 per head, mostly pink-purple, seldom pale blue; achenes 5–8-nerved, glabrous or nearly so; 2n=20. Woods, ditch-banks, and swamps; se. Va. to Fla., w. to W.Va., s. Ind., s. Mo., Okla., Tex., and Mex. Sept., Oct.

8. **Eupatorium aromaticum** L. Small-lvd white snakeroot. Resembling no. 9, and not always sharply distinct, avg smaller, seldom over 8 dm, and with shorter, firmer, more densely distributed pubescence; lvs relatively thick and firm, obtuse or acute, crenate or crenate-serrate, 3–7(–10) × 2–5 cm, more than 4 times as long as the short petiole; invol bracts obtuse or rounded to acute; fls mostly 10–19 per head, the cor-lobes commonly villous externally, at least near the tip; 2n=34. Dry woods, especially in sandy soil; Mass. to Fla., w. to s. O., Ky., Tenn., and e. La. Aug.–Oct.

9. **Eupatorium rugosum** Houttuyn. White snakeroot. Stems 1–3, 3–15 dm, glabrous below the infl and sometimes glaucous, or shortly and loosely hairy; lvs opposite, narrowly to broadly ovate or even subcordate, rather thin, glabrous or hairy especially on the main veins beneath, serrate, usually sharply and coarsely so, mostly acuminate, the larger ones mostly 6–18 × 3–12 cm (or smaller in depauperate plants), 1.5–5 times as long as the well developed petiole; infl flat-topped or more rounded, invol 3–5 mm, glabrous or short-hairy, its principal bracts subequal and subbiseriate, acuminate to obtuse; fls mostly (9–)12–25 per head, cor bright white, mostly 3–4 mm, its lobes often short-hairy; achenes generally glabrous; 2n=34. Woods; N.S. to Sask., s. to Ga. and Tex. July–Oct. (*E. urticaefolium.*) Poisonous. Our plants are var. *rugosum.* Var. *roanense* (Small) Fern., with larger heads (invol to 7 mm, fls to 34), occurs on the highest mts. of Tenn. and N.C.

10. **Eupatorium luciae-brauniae** Fern. Rock-house white snakeroot. Resembling no. 9, but more delicate and slender, 3–6 dm, often glabrous throughout; lvs very thin, deltoid or subcordate, about as broad as long, coarsely and irregularly dentate, the petiole nearly or fully as long as the blade; invol bracts attenuate-acuminate. Under overhanging sandstone cliffs; Cumberland Plateau region of s. Ky. and n. Tenn. Sept.

11. **Eupatorium serotinum** Michx. Late eupatorium. Stems 4–20 dm, puberulent especially above; lvs opposite, with an evident petiole 1 cm or more, the blade lanceolate to ovate, serrate, mostly acuminate, 5–20 × 1.5–10 cm, 3–5-nerved or -plinerved, commonly less hairy than the stem, the upper surface often subglabrous; heads numerous; invol 3–4 mm, densely villous-puberulent, its bracts broadly rounded to merely obtuse, evidently imbricate in ca 3 series; fls 9–15 per head, white; 2n=20. Mostly in bottomlands and moist woods, sometimes in drier or more open places; s. N.Y. to Ill. and reputedly Minn., s. to Fla. and Tex.; casually intr. elsewhere, as in Conn. Aug.–Oct. *E. serotinum* var. *polyneuron* F. J. Hermann is a hybrid with *E. perfoliatum.*

12. **Eupatorium resinosum** Torr. PIne-barren eupatorium. Plants 4–10 dm, viscidulous-puberulent throughout, the lvs commonly subtomentosely so beneath; lvs opposite, sessile, slender, 5–13 × 0.5–1.5 cm, evenly serrate, long-acuminate; heads ± numerous in a flat-topped infl; invol 3.5–5 mm, its bracts broadly rounded to acutish, imbricate in ca 3 series, the inner tending to be somewhat whitish distally, the outer less than half as long as the inner; fls white, 9–14 per head; 2n=20. Pocosins, bogs, and other wet places, often in pine-barrens; L.I. (where now probably extinct) and N.J. to Del. and rarely N.C. Aug.–Oct.

13. **Eupatorium perfoliatum** L. Boneset. Stems 4–15 dm, conspicuously crisp-villous with long spreading hairs; lvs opposite (very rarely ternate), broad-based and strongly connate-perfoliate, tapering gradually to the acuminate tip, crenate-serrate to the base, 7–20 × 1.5–4.5 cm, sparsely pubescent or subglabrous above, more evidently hairy beneath; infl flat-topped; invol 4–6 mm, its bracts imbricate in ca 3 series, the outer less than half as long as the inner, mostly obtuse, the inner more acuminate and often pale distally; fls 9–23 per head, dull white; 2n=20. Moist or wet low grounds; N.S. and Que. to n. Fla., w. to N.D., Neb., Okla.,

and Tex. July–Oct. Estuarine plants from Me. and Que. avg narrower-lvd than var. *perfoliatum* (as described above) and have a sparser, shorter, and stiffer pubescence; these are var. *colpophilum* Fern. & Griscom. Forms with purple fls, form with ternate lvs, and forms with the lvs not connate have been observed.

14. **Eupatorium capillifolium** (Lam.) Small.   Dog-fennel.   Coarse, the stems 5–20 dm, clustered on a thick, woody caudex, puberulent, or glabrate below, freely branched upward; lvs very numerous and narrow, delicate, glandular-punctate, glabrous, the lowest ones opposite, the others alternate, mostly 2–10 cm, often with axillary fascicles or with short, sterile, leafy axillary branches, the main ones pinnately divided into a few filiform segments mostly ca 0.5(–1.0) mm wide, those of the infl mostly simple; heads very numerous in an elongate true panicle; invol 2–3.5 mm, the inner bracts much longer than the outer, usually mucronate or abruptly acuminate; fls 3–6, white or chloroleucous; $2n=20$. Open places, often in old fields and pastures; coastal states from N.J. to Fla., w. to Tex. and s. Ark., in our range wholly on the coastal plain. Sept., Oct.

15. **Eupatorium album** L.   White-bracted eupatorium. Stems mostly solitary from a crown or very short, stout rhizome, 4–10 dm, conspicuously spreading-villous at least below (except often in var. *vaseyi*), often merely villous-puberulent above; lvs opposite, elliptic to elliptic-ovate, lance-elliptic, or elliptic-oblanceolate, sessile or nearly so, 4–13 × 1–4 cm, the larger ones seldom less than 1.5 cm wide except sometimes in var. *subvenosum*, glandular-punctate, evidently hairy to sometimes subglabrous; infl dense, corymbiform, ± flat-topped; invol 8–11 mm, often with dark sessile glands, otherwise generally glabrous or only slightly hairy, its bracts imbricate, conspicuously white-scarious upward (especially the inner), all narrow and long-acuminate, or the inner with broader, more rounded, mucronate tip; fls 5, the cor white, 4–5.5 mm; $2n=20$, 30, 40. Dry, open woods, especially in sandy pinelands; coastal states from s. Conn. to c. Fla. and w. to Miss., also inland in the mt. regions to s. O. and e. Ky. and Tenn., and in Ark. The sp. consists of a widespread, common diploid phase and two more sporadic or local polyploid phases that reflect hybridization with other spp.:

a Lvs usually evidently pubescent, mostly coarsely serrate, tending to be obtuse or rounded at the tip, the larger ones seldom less than 1.5 cm wide; diploid; common and widespread, with the range of the sp. (*E. petaloideum; E. album* var. *glandulosum*.) .... var. *album*.
a Lvs sparsely pubescent or subglabrous, serrate to subentire, often acute, somtimes less than 1.5 cm wide; apomictic polyploids, local, not abundant.
  b Lvs small, mostly 4–7 × 1–2 cm, acute to obtuse, ± trinerved or triplinerved, few-toothed, with up to 10 teeth per side, or even entire; s. N.Y. to N.J. and Del. .... var. *subvenosum* A. Gray.
  b Lvs larger, mostly 5–11 × 2–4 cm, acute or somewhat acuminate, more pinnately veined, rather closely toothed, with mostly 10–20 teeth per side; very probably originating by hybridization with no. 17; D.C. and Md., s. to the mts. of n. Ga. and n. Ala. (*E. album* var. *monardifolium; E. fernaldii.*) .... var. *vaseyi* (T. C. Porter) Cronq.

16. **Eupatorium leucolepis** (DC.) T. & G.   Stem 4–10 dm from a crown or short, stout rhizome, hirtellous or puberulent to sometimes merely strigose; lvs opposite, lance-oblong to linear-oblong or oblanceolate, 3–8 cm × 3–10(–13) mm, toothed or entire, sessile, glandular-punctate; invol 5–7 mm, its bracts imbricate, narrow, tapering to a long-acuminate or mucronately subattenuate point, conspicuously villous-puberulent and often also atomiferous-glandular, their scarious margins mostly inconspicuous; fls 5, the cor white, 3–4 mm; $2n=20$, 30, 40. Pine-barrens, wet meadows, and margins of ponds, especially in sandy soil; Mass. to Fla. and La., on the coastal plain or near the coast. July–Oct. Var. *leucolepis*, from N.Y. s., has bluntly few-toothed to entire lvs, often folded along the midrib. Var. *novae-angliae* Fern., local in Mass. and R.I., has sharply toothed, flat lvs, often with slightly longer pubescence than in var. *leucolepis*.

17. **Eupatorium sessilifolium** L.   Plants 6–15 dm, single-stemmed from a crown or short caudex, puberulent in the infl, the herbage otherwise glabrous except that the lvs are gland-dotted; lvs opposite, sessile or subsessile, lanceolate, usually broadly rounded at the base, serrate, acuminate or long-acute, mostly 7–18 × 1.5–5 cm, 2.5–7 times as long as wide, the venation strictly pinnate; invol 4.5–6.5 mm, its bract imbricate, broadly rounded to merely obtuse, villous-puberulent and usually also atomiferous-glandular; fls 5(6), white; $2n=20$, 30. Woods, especially in sandy, acid soils; s. N.H. to se. Minn., s. to Ga. and Ark. Aug., Sept.

18. **Eupatorium godfreyanum** Cronq.   Much like no. 17, but more hairy, the stem loosely puberulent down to the middle or even to the base, the leaves also somewhat puberulent, especially beneath; lvs mostly 5–11 × 1.5–4.5 cm, 2–3(–3.5) times as long as wide, the base often more cuneate and not so broadly rounded as in no. 17, the venation strictly pinnate; $2n=30$, 40. Woods and disturbed, open sites, usually in less acid soils than no. 17; N.J., Md., Va., and W.Va. to N.C. and Tenn. July–Sept. (*E. vaseyi*, misapplied.) Possibly originating through hybridization of nos. 17 and 19, but now widespread and stable.

19. **Eupatorium rotundifolium** L.   Round-lvd eupatorium. Stems 3–15 dm, mostly 1–2 from a short rhizome or crown; herbage pubescent throughout with spreading, usually short and soft hairs, commonly atomiferous-glandular as well; lvs opposite, or the upper seldom alternate, sessile or subsessile, ovate (mostly broadly so) to subrotund, evenly toothed except in var. *cordigerum*, broad-based and distinctly trinerved or triplinerved, 2–12 × 1–6 cm; invol 4.5–6.5 mm, its bracts imbricate, sharply acute to obtuse, evidently villous-puberulent and often also atomiferous-glandular; fls 5 (to 7 in var. *ovatum*), white; $2n=20$, 30, 40. Woods, in dry or seldom wet soil; Me. to Fla., w., to s. O., e. Ky., Tenn., Okla., and Tex. July–Sept. Sp. here considered to consist of 3 vars.:

a Lvs with broadly rounded, cordate-clasping base, rough-textured and strongly pubescent, regularly to often irregularly toothed or even lobulate; apomictic tetraploid, probably originating by hybridization of *E. perfoliatum* with *E. rotundifolium* var. *rotundifolium* and possibly not long-persistent, but frequently created when propinquity permits; Va. to S.C., w. to Ark. and c. Miss. (*E. cordigerum.*) .... var. *cordigerum* Fern.

a Lvs with broadly cuneate to broadly rounded or subtruncate, but not cordate-clasping base.

  b Principal pair of lateral veins of the lf diverging from the base of the midrib; lvs mostly 1–1.5(–1.7) times as long as wide, blunt, the margins commonly crenate; diploids and chiefly autoploid triploids and tetraploids, common and widespread, nearly throughout the range of the sp., except at the north .... var. *rotundifolium*.

  b Principal pair of lateral veins diverging distinctly above the base of the midrib; lvs mostly 1.5–2 times as long as wide, mostly acute and serrate; triploids and tetraploids reflecting hybridization of var. *rotundifolium* with other spp., perhaps mainly *E. sessilifolium*; well established and persistent from Me. to Va. and e. Ky., and s. in and near the mts. to n. Ga. and n. Ala., seldom elsewhere. (*E. pubescens.*) .... var. *ovatum* (Bigel.) Torr.

20. Eupatorium pilosum Walter. Ragged eupatorium. Much like no. 19, but the lvs narrower, mostly lanceolate or lance-ovate to elliptic-ovate, mostly coarsely and unevenly toothed; upper lvs and main branches of the infl tending to be alternate; diploids and polyploids. Chiefly in wet soil; Mass. to Va. and Fla., w. to e. Ky., s. Tenn., and Miss. (*E. verbenaefolium; E. rotundifolium* var. *saundersii.*)

21. Eupatorium mohrii Greene. Tuberous eupatorium. Plants with short, conspicuously tuberous-thickened rhizomes, the erect branches therefrom often branching near ground-level into two or more aerial stems, these 3–10(–12) dm, puberulent or strigose-puberulent; lvs opposite or the upper alternate, rarely ternate, mostly lance-elliptic, tapering to the sessile or subsessile base, bluntly few-toothed, tending to be recurved or deflexed, small, mostly 1.5–6 cm × 3–10(–12) mm, glandular-punctate, strigose or subglabrous; invol 3–5 mm, its bracts imbricate, broadly obtuse to acutish, atomiferous-glandular and slightly to evidently villous-puberulent, the inner inconspicuously scarious-margined and often somewhat pale distally; fls 5(6), white; $2n=20$, 30, 40. Pond-margins, ditches, shores, and moist low ground, often in sandy or peaty soil; coastal plain from se. Va. to Fla. and La. July–Sept. (*E. recurvans,* the sexual diploid phase.)

22. Eupatorium cuneifolium Willd. Bushy-branched eupatorium. Stems 3–10 dm, ± clustered on a crown or caudex, commonly branched at or near the ground-level, copiously provided with short, mostly loosely spreading hairs; lvs opposite, or the upper alternate, tending to be broadest above the middle, oblanceolate to obovate or nearly elliptic, tapering to the base, few-toothed or entire, triplinerved, 2–5 cm × 5–18 mm, glandular-punctate, often less hairy than the stem, the axillary shoots of the middle and lower ones commonly elongating into slender, mostly sterile branches with reduced lvs; invol 4–7 mm, coarsely villous-puberulent and atomiferous-glandular, its bracts imbricate, broadly rounded to acute, the inner inconspicuously scarious-margined; fls 5; cor white, 3–5 mm; $2n=20$, 30, 40. Pine and oak woods, mostly in dry, sandy soil or in sand-hills; se. Va. to c. Fla. and w. to Miss. Aug., Sept. Triploids with oblanceolate lvs ca 5 mm wide may reflect hybridization with *E. hyssopifolium. E. tortifolium* Chapm. and *E. linearifolium* Walter (as used by Fernald) may be such hybrids.

23. Eupatorium semiserratum DC. Stems 5–12 dm, mostly solitary from a very short, stout rhizome or crown, densely villous-puberulent, sometimes also atomiferous-glandular, often with loose axillary fascicles of a few reduced lvs; lvs opposite or the uppermost scattered, firm, twisted at the base to bring the blade into a vertical plane, elliptic or elliptic-oblanceolate, mostly 4–8 cm × 8–30 mm, 2.5–6 times as long as wide, gradually narrowed to the sessile or shortly petiolar base, usually serrate or crenate-serrate, especially above the middle, finely and densely puberulent (and often also atomiferous-glandular), sometimes more shortly so above than beneath, triplinerved, the principal pair of lateral veins arising as branches from the midrib; invol 2.5–4 mm, softly short-hairy like the peduncles, its bracts imbricate, broadly rounded or obtuse to sometimes submucronately acute, the inner obscurely scarious-margined; fls 5; cor white, 2.5–3.5 mm; $2n=20$. Low woods, clearings, and swampy places; Va. to s. Fla., w. to Tex., and n. into Ark., se. Mo., and s. Tenn. (*E. cuneifolium* var. *s.*)

24. Eupatorium altissimum L. Tall eupatorium. Plants stout, 8–20 dm; stems arising singly or in pairs from a system of stout, branching rhizomes, softly villous-puberulent, or glabrate below; lvs numerous, opposite, hairy like the stem, strongly trinerved to the base, lance-elliptic, gradually and narrowly acute or acuminate, and gradually narrowed to a sessile or shortly petioliform base, serrate above the middle and subentire, mostly 5–12 cm × 8–30 mm; invol 4.5–7 mm, short-hairy like the herbage, its bracts imbricate, broadly rounded or obtuse; fls 5; cor white, 4–5 mm; $2n=20$, 30, 40. Woods, thickets, savannas, glades, and clearings; N.J. and Pa. to s. Minn. and e. Neb., s. to S.C., Miss., and Tex.; casually intr. elsewhere, as in Conn. *E. saltuense* Fern., from se. Va., with triplinerved lvs, may reflect hybridization with some other sp. such as *E. album.*

25. Eupatorium hyssopifolium L. Stems solitary (seldom several) from a short crown, strigose or scabrous-puberulent, especially above; lvs verticillate in 3's or 4's, or sometimes merely opposite, or even alternate above, narrow, mostly 6–40 times as long as wide, sessile or tapering to a slender base, spreading or ascending, glandular-punctate, otherwise glabrous or slightly hairy chiefly on the main veins beneath, the principal ones subtending conspicuous axillary fascicles of reduced lvs; invol 4–7 mm, its bracts imbricate, broadly rounded to acute, the inner narrowly scarious-margined; fls 5, the cor white, 3.5–4 mm; $2n=20$, 30, 40. Fields and other open places, especially in dry, sandy soil; coastal states from Mass. to n. Fla. and w. to La., and sometimes inland to O., Ky., and Tenn. Aug.–Oct. Two well marked vars.:

a Plants to 1 m; lvs usually quaternate but sometimes ternate, opposite, or even alternate above, entire or inconspicuously and irregularly few-toothed, linear or nearly so, the main ones 2–7 cm × 1–5 mm and 10–40 times as long as wide; diploids and polyploids; Mass. to Ga. and w. to La., and sometimes inland as far as Tenn. (*E. h.* var. *calcaratum.*) .... var. *hyssopifolium.*
a Plants more robust, to 1.5 m; lvs sometimes quaternate, but more often ternate or merely opposite (or alternate above), conspicuously and divergently toothed, to 10 cm, the main ones mostly 5–10(–12) mm wide and 6–15 times as long as wide; all polyploids, mainly tetraploid; s. N.Y. to Ga. and n. Fla. and w. to La., inland occasionally as far as s. O., c. Ky., and e. Tenn., probably originating through hybridization with some other sp., but now a stable entity. (*E. torreyanum.*) .... var. *laciniatum* A. Gray.

71. MIKANIA Willd., nom. conserv.   Climbing hempweed. Heads discoid, the fls 4, all tubular and perfect; invol narrow, of 4 principal bracts and occasionally a few short outer ones; receptacle small, naked; cor white to pink or ochroleucous; style branches papillate, elongate, linear, acutish, with short ventromarginal stigmatic lines near the base; achenes 5-angled; pappus a single series of numerous capillary bristles, usually connate at base; mostly perennial twining vines, with opposite, simple, usually petiolate lvs and numerous small heads. 200+, mainly trop. Amer.

1. **Mikania scandens** (L.) Willd.   Twining herbaceous vine to 5 m, puberulent and somewhat atomiferous to subglabrous; roots fleshy, fascicled; leaves petiolate, deeply cordate, acuminate, entire or sinuately few-toothed, palmately veined, 2.5–14 × 1.5–8.5 cm; infls small and numerous, corymbiform, on axillary peduncles; invol 4–5.5 mm; fls white or pinkish; achenes 1.5–2.5 mm; $2n=38$. Climbing on bushes in moist places; on the coastal plain (or near the coast) from Me. to Fla., s. Ill., and Tex., s. to trop. Amer.; locally inland to Mich., Ind., and Ky.; pantropical. July–Oct.

72. SCLEROLEPIS Cass.   Heads discoid, the rather numerous fls all tubular and perfect; invol campanulate or subhemispheric, its bracts subequal; receptacle hemispheric or conic, naked; cors purplish; style-branches with short, ventromarginal stigmatic lines and an elongate, papillate, linear-clavate appendage; achenes 5-angled, atomiferous-glandular; pappus of 5 short, broad, obtuse scales; slender, rhizomatous, aquatic herbs with narrow, entire lvs verticillate in 4's to 6's, the heads rather small, naked-pedunculate, mostly solitary. Monotypic.

1. **Sclerolepis uniflora** (Walter) BSP.   Stems commonly emergent for 1–3 dm, simple or nearly so, obscurely atomiferous; lvs 7–23 mm, to 2 mm wide, punctate, glabrous; invol 3–4 mm; disk 4–12 mm wide; pappus 1 mm. In still, shallow water on the coastal plain (or near the coast) from Ala. and Fla. to N.J., and disjunct in Mass. and s. N.H. July–Oct.

73. KUHNIA L.   Heads discoid, the fls all tubular and perfect; invol bracts striate, obviously imbricate, the outer several series generally much shorter or differently shaped (or both) than the only slightly unequal inner ones, thus somewhat calyculate; receptacle flat or nearly so, naked; cor creamy white to dull yellow or red; anthers tending to separate by the time of anthesis; style-branches papillate, elongate, linear-clavate, rounded-obtuse, with inconspicuous, short, ventromarginal stigmatic lines near the base; achenes 10-ribbed; pappus of 10, 15, or 20 uniseriate, equal, plumose bristles; perennial herbs from a stout taproot, with opposite or more often subopposite or unevenly alternate lvs and solitary to numerous, subcylindric to campanulate, medium-sized heads. 3, Mex., U.S. (Sometimes included in *Brickellia,* which has merely barbellate pappus-bristles.)

1. **Kuhnia eupatorioides** L.   False boneset. Plants 3–13 dm, densely puberulent to subglabrous; stems numerous; lvs narrowly lanceolate to broadly rhombic-lanceolate, the main ones 2.5–10 cm × 0.5–4 cm, gland-dotted beneath, entire or toothed, sessile or the lower short-petiolate; heads mostly in small corymbiform clusters terminating the branches; invol 7–14 mm, the inner bracts mostly linear or oblong, the outer subulate-deltoid to lanceolate, with slender attenuate tip; fls creamy-white; pappus-bristles 20; $2n=18$. Dry open places, especially in sandy soil; N.J. to O., N.D., and Mont., s. to Fla. and Ariz. Aug.–Oct. (*Brickellia e.*) Three vars. in our range:

a Slender tips of the outer bracts not greatly prolonged.
  b Fls 7–14 per head; plants only slightly hairy, with relatively thin, seldom much toothed lvs; N.J.
    to s. Ind., Fla., and se. Tex. .... var. *eupatorioides.*

b Fls 14–23 per head; plants more hairy, stouter, with firmer, more strongly toothed lvs; w. O. to
  Mont. and N.M. (*K. glutinosa; K. suaveolens.*) .... var. *corymbulosa* T. & G.
a Slender tips of the outer bracts much prolonged, only slightly if at all shorter than the inner bracts;
  otherwise much like var. *eupatorioides*; s. Ill., Mo., and Ark. (var. *ozarkana.*) .... var. *texana* Shinners.

Brickellia grandiflora (Hook.) Nutt., a western sp. with deltoid or cordate lvs and merely barbellate pappus, extends e. to s. and c. Mo., and may prove to occur in our range. It would key to *Kuhnia.*

74. CARPHEPHORUS Cass.   Heads discoid, the fls all tubular and perfect; invol bracts from few and subequal to numerous and imbricate; receptacle flat or nearly so, chaffy toward the margin in our spp.; cor pink-purple; style-branches papillate, elongate, linear or linear-clavate, with short ventromarginal stigmatic lines near the base; achenes 10-ribbed; pappus of 20–55 coarse, unequal, barbellate bristles in a single series; fibrous-rooted perennial herbs with alternate, entire, basally disposed lvs and several or many heads in a corymbiform to thyrsoid-paniculate infl. 7, se. U.S.

1 Invol bracts acute or acutish; plant evidently hairy .... 1. *C. tomentosus.*
1 Invol bracts broadly rounded; plant ± glabrous below the puberulent infl .... 2. *C. bellidifolius.*

1. Carphephorus tomentosus (Michx.) T. & G.   Plants 2–8 dm, the stem and generally also the lvs ± spreading-hirsute below, the hairs commonly shorter and more appressed upwards; tufted lower lvs oblanceolate or a little broader, 3.5–15 × 0.5–2 cm (petiole included); cauline lvs quickly reduced and becoming sessile, erect; heads 2–40 in an open-corymbiform, often flat-topped infl; invol campanulate, 7–11 mm, conspicuously viscid-hairy, its 20–40 bracts imbricate, firm, with rather lax or spreading, thick, often callous-pointed, acute or acutish tip; fls 15–30; achenes 3–4 mm; $2n=20$. Sandy soil, especially in pine-barrens; se. Va. to S.C. and Ga. Aug.–Oct.

2. Carphephorus bellidifolius (Michx.) T. & G.   Resembling no. 1, but glabrous or subglabrous below the puberulent and often more ample infl; invol bracts ca 15–30, merely ciliate-margined, not especially thick, broadly rounded above; achenes 4–6 mm; $2n=20$. Sandy soil, especially in pine-barrens; se. Va. to Ga. Aug.–Oct.

75. LIATRIS Schreb., nom. conserv.   Blazing star, gay feather. Heads discoid, the fls all tubular and perfect; invol bracts imbricate, often with petaloid scarious margins; receptacle naked; fls pink-purple (white); style-branches elongate-clavate, obtuse, papillate, with inconspicuous ventromarginal stigmatic lines near the base; achenes ca 10-ribbed; pappus of 1 or 2 series of stout, barbellate or plumose bristles; perennial herbs from a thickened, usually corm-like rootstock; lvs alternate, entire, ± punctate, narrow and sessile or with the blade tapering to the petiole, the basal or lower cauline usually the largest; heads smallish to fairly large, in a mostly spiciform or racemiform (seldom corymbiform or evidently branched) infl, seldom solitary. (*Lacinaria.*) 30, N. Amer.

1 Pappus barbellate, the barbels mostly 0.1–0.3(–0.4) mm long.
  2 Heads relatively broad, with mostly 14–many fls; larger lvs mostly 1–4 cm wide.
    3 Middle invol bracts herbaceous (or somewhat coriaceous) throughout, or with narrow, entire or
      slightly erose but scarcely lacerate scarious margins.
      4 Fls mostly 25–80 per head; widespread .... 1. *L. scariosa.*
      4 Fls mostly 14–24 per head; southern.
        5 Middle and outer invol bracts generally loose or squarrose above the middle; heads tending
          to be turned away from the axis .... 2. *L. squarrulosa.*
        5 Middle and outer invol bracts, like the inner, appressed; heads closely ascending, not turned
          away from the axis .... 5. *L. turgida.*
    3 Middle invol bracts with wide, uneven, irregularly lacerate scarious margins.
      6 Heads with 30–100 fls, the terminal head distinctly larger than the others; cor glabrous within
        .... 3. *L. ligulistylis.*
      6 Heads with 14–35 fls, uniform in size; cor hairy toward the base within .... 4. *L. aspera.*
  2 Heads narrower, with mostly 3–14 fls; lvs very often all less than 1 cm wide, though broader in
    some spp.
    7 Invol bracts with broadly rounded to merely acute, erect or sometimes loose and bluntly mu-
      cronate tip; infl not densely spicate except in no. 9.
      8 Cor hairy toward the base within.
        9 Larger lvs mostly (7–)10–20 mm wide; lvs and invol bracts only weakly or scarcely punctate;
          montane .... 5. *L. turgida.*
        9 Larger lvs rarely more than 7(–10) mm wide; lvs and invol bracts evidently to scarcely
          punctate; only seldom montane.

10 Invol bracts firm, coriaceous, strongly glandular-punctate, tending to be keeled or cupped
   above and often acutish, the midrib generally excurrent as a short, broad, blunt
   mucro .... 6. *L. regimontis.*
10 Invol bracts thinner and more chartaceous, punctate or not, but only slightly or not at
   all keeled, cupped, or mucronate, instead with mostly flat, broadly rounded tip ... 7. *L. graminifolia.*
8 Cor glabrous (or very nearly so) within.
   11 Heads evidently pedunculate; fls 4–5(6) per head .... 8. *L. microcephala.*
   11 Heads strictly sessile; fls (5)6–10(–14) per head .... 9. *L. spicata.*
7 Invol bracts with acuminate, spreading tip; infl densely spicate, the heads numerous and sessile;
   mainly western .... 10. *L. pycnostachya.*
1 Pappus evidently plumose, the barbels mostly 0.5–1 mm long.
   12 Fls 10–60 per head; cor-lobes coarsely hairy within.
      13 Invol bracts appressed or merely a little loose, the tip mostly broadly rounded and shortly
         mucronate .... 11. *L. cylindracea.*
      13 Invol bracts with loosely spreading or squarrose, acuminate tip .... 12. *L. squarrosa.*
   12 Fls 4–6 per head; cor-lobes glabrous .... 13. *L. punctata.*

1. **Liatris scariosa** (L.) Willd.   Northern b.s. Glabrous or hairy, 3–8(–10) dm; lowermost lvs mostly 10–35 cm, long-petiolate, with elliptic to linear-elliptic or broadly oblanceolate blade; heads seldom more than 20(–35), subsessile or more often ascending on arcuate or sometimes spreading peduncles to 5 cm; invol ± hemispheric, 9–17 mm, its bracts appressed to more often loose or sometimes distally squarrose, broadly rounded, often anthocyanic distally, the middle ones only narrowly or not at all scarious-margined, often ciliolate but not erose-lacerate, the innermost ones sometimes more obviously scarious and erose; fls (21–)25–80 per head; cor-tube ± hairy toward the base (or near the middle) within; pappus strongly barbellate; $2n=20$. Prairies, open woods, and other dry, open places; Me. to Mich., s. to Pa., Mo., Ark., and in the mts. to n. Ga. Aug., Sept. Three vars.:

a Cauline lvs relatively few, mostly 8–20(–25) below the infl; larger lvs 2–5 cm wide, tending to be lax
  and spreading; fls seldom more than 40 per head; mts. from s. Pa. to n. Ga., especially on shale-
  barrens .... var. *scariosa.*
a Cauline lvs more numerous, mostly 20–60 below the infl, the larger ones, like the others, tending to
  be more stiffly ascending; fls up to 80 per head.
   b Lower lvs narrow, mostly 1–2(–2.5) cm wide; N. Engl. and e. N.Y., especially near the coast. (*L.
     novae-angliae; L. borealis,* misapplied.) .... var. *novae-angliae* Lunell.
   b Lower lvs wider, mostly 2–5 cm wide; inland ne. and c. U.S., from N.Y., Pa., and n. W.Va. to n.
     Mich., Mo., and Ark. .... var. *nieuwlandii* Lunell.

2. **Liatris squarrulosa** Michx.   Southern b.s. Similar to no. 1 and evidently hybridizing with it, but avg more robust, to 15 dm, and often more evidently hairy; heads 6–60, often more than 20, commonly in a spiciform infl with the heads turned away from the axis, or the peduncles sometimes more elongate and ascending-spreading, to 5 cm; invol glabrous or often evidently short-hairy, the middle and outer bracts commonly notably loose or squarrose above the middle; fls (11–)14–24(–28) per head. Dry woods and open places, especially in rocky or sandy soil; s. W.Va. and s. O. to s. Mo., s. to Fla. and La. July–Sept. (*L. earlei; L. ruthii; L. scabra*—the more hairy phase; *L. tracyi.*)

3. **Liatris ligulistylis** (A. Nels.) K. Schum.   Northern plains b.s. Plants 2–11 dm, usually glabrous below the infl; lvs 8–100 below the infl, the lowest ones 8–27 × 0.5–4 cm, the others rather abruptly reduced and becoming sessile; heads (1–)3–10(–35), evidently pedunculate to occasionally subsessile, the terminal one obviously the largest; invol 13–20 mm, broadly campanulate or hemispheric, glabrous, its bracts seldom much squarrose, with conspicuous, lacerate, scarious margins, seldom at all crisped, often purplish upwards; fls 30–100 per head; cor glabrous within; pappus barbellate; $2n=20$. Mostly in damp, low places, occasionally in drier soil; Wis. to Alta., Colo., and N.M. Aug., Sept.

4. **Liatris aspera** Michx.   Lacerate b.s. Short-hairy, or glabrous throughout, 4–15 dm; lowest lvs 5–40 × 1–4.5 cm, long-petiolate; cauline lvs 25–90 below the infl; heads (10–) ± numerous in an elongate-spiciform infl, or the peduncles occasionally more elongate and to 5 cm; invol 8–15 mm, campanulate or subhemispheric, glabrous, its bracts loosely spreading or squarrose, tending to be bullate, often purplish upward, with conspicuous, lacerate, often crisped, scarious margins that are sometimes folded under; fls 14–35 per head; cor hairy toward the base within; pappus barbellate; $2n=20$. Dry, open places and thin woods, especially in sandy soil; N.D. to Okla. and Tex., e. to Mich. and Miss., and occasionally to s. Ont., O., w. Va. and n. Fla. Aug.–Oct.

5. **Liatris turgida** Gaiser.   Appalachian b.s. Relatively small, mostly 2–8 dm, glabrous or somewhat hirsute; lvs relatively few, weakly or scarcely punctate, the lowest ones oblanceolate, 7–30 × 0.7–2 cm, usually irregularly ciliate-margined toward the base; heads mostly 5–40, closely ascending in a ± spiciform infl, subsessile or on peduncles to 1(–3) cm; invol turbinate, 7–12 mm, its bracts appressed, broadly rounded, scarious-margined, ciliate, not strongly punctate, generally purplish above; fls mostly 8–15(–20), often ca 13; cor hairy toward the base within; pappus barbellate; $2n=20$. Dry, rocky woods; mts. of Va., W.Va. and N.C. to n. Ga. and ne. Ala. Aug., Sept.

6. **Liatris regimontis** (Small) K. Schum.   Piedmont b.s. Stem glabrous, 4–10 dm; lvs markedly punctate, linear or nearly so, 6–30 cm × 2–7(–10) mm, the margins irregularly ciliate below, the surfaces sometimes

also sparsely hairy; heads sessile or on stout, often bracteate peduncles mostly under 1 cm, closely ascending or somewhat divergent, seldom fewer than 20, forming an elongate, spike-like (sometimes branched), sometimes secund infl; invol cylindric or obconic, 7–11 mm, its bracts firm, coriaceous-herbaceous, strongly punctate, obscurely scarious-margined and slightly or scarcely ciliate, with keeled or cupped, slightly spreading, often acutish (or distinctly acute) tip, the midvein commonly projecting as a short, broad, blunt mucro; fls 5–12 per head; cor 6–8 mm, hairy toward the base within; pappus barbellate; $2n=20$. Pine or oak woods, especially in sandy soil; s. Va. to Ga. and S.C., mainly on the piedmont and coastal plain, seldom in the mts. Sept., Oct. Hybridizes with or passes into nos. 5 and 7. (*Lacinaria smallii.*)

7. Liatris graminifolia Willd. Grass-lvd b.s. Plants 2–12 dm, hairy or subglabrous; lvs numerous, strongly or scarcely punctate, all linear or nearly so, the lowest ones 6–30 cm × 2–7 mm (or a little wider in very robust plants), the margins irregularly ciliate toward the base; heds 10–many in an elongate, spiciform (seldom branched) infl, suberect on short, closely ascending peduncles, or subsessile, only seldom turned away from the axis; invol turbinate or subcylindric, 7–12 mm, its bracts relatively firm, scarcely to sometimes evidently punctate, ± ciliolate on the scarious margins, mostly broadly rounded and often purplish upward; fls mostly (6)7–14 per head, the cor hairy toward the base within; pappus barbellate; $2n=20$. Dry, open woods, especially in sandy soil among pines; coastal plain and piedmont from N.J. and s. Pa. to Ga., nw. Fla:, and Ala., and occasionally extending into the mts. Aug.–Oct.

8. Liatris microcephala (Small) K. Schum. Small-headed b.s. Slender, 3–8 dm, essentially glabrous; lvs numerous, linear, the lower ones 3–16 cm × 1–5 mm, the others fairly gradually reduced upward; infl racemiform and sometimes secund, the ± numerous heads on slender, erect peduncles 3–20 mm (or a few of the uppermost subsessile), often turned away from the axis; invol narrow, subcylindric, 6–8 mm, its green or greenish bracts narrowly scarious-margined, blunt, commonly broadly rounded above; fls 4–5(6) per head; cor 6–8 mm, including the 1.5–2 mm lobes, glabrous within; pappus barbellate, short, seldom reaching the sinuses of the cor; $2n=20$. Exposed, rocky places, glades, and sandy shores; mts. of Ky. to S.C., Ga., and Ala. Aug., Sept.

9. Liatris spicata (L.) Willd. Sessile b.s. Glabrous or seldom somewhat hirsute, 6–20 dm; lvs numerous, linear or nearly so, the lowest ones 10–40 cm × (2–)5–20 mm, the others reduced upward; heads sessile, crowded into an elongate, densely spiciform infl; axis of the infl glabrous; invol subcylindric or narrowly turbinate-campanulate, 7–11 mm, its bracts narrowly scarious-margined, often anthocyanic, at lest the inner ones mostly rounded or broadly obtuse, the outer sometimes acutish; fls (5)6–10(–14) per head, the cor glabrous within, 7–11 mm including the 2–3 mm lobes; pappus barbellate; $2n=20$. Wet meadows and other moist, open places; N.Y. to Mich. and se. Wis., s. to Fla. and La., and occasionally w. to Wyo. and N.M. July–Sept.

10. Liatris pycnostachya Michx. Thick-spike b.s. Strict, 5–15 dm, ± hirsute in the infl (including the axis) or throughout, seldom glabrous; lvs numerous, linear or nearly so, the lower ones 10–50 cm × 3–13 mm, the others reduced upward; heads sessile, crowded in an elongate, densely spiciform infl; invol subcylindric or narrowly turbinate, 8–11 mm, its bracts tapering to an acuminate, conspicuously squarrose tip, or the inner ones sometimes blunter and loosely erect; fls 5–7(–12) per head, the cor 7–9 mm including the 2–2.5 mm lobes, glabrous or very nearly so within; pappus barbellate; $2n=20$, 40. Moist or dry prairies and open woods; Ind. and Ky. to Minn., N.D., Tex., and Miss., and intr. e. to N.J. and w. N.Y. July–Sept.

11. Liatris cylindracea Michx. Few-headed b.s. Glabrous or rarely short-hairy, 2–6 dm; lvs ± numerous, firm, linear or nearly so, the lowest small and subsheathing, the next longer, 10–25 cm × 2–12 mm, the rest reduced upward; heads few or even solitary, stiffly pedunculate or sessile; invol 11–20 mm, broadly cylindric or cylindric-campanulate, its bracts firm, appressed or a little loose, generally broadly rounded and shortly mucronate, occasionally more tapering or without the mucro; fls 10–35 per head, the cor-lobes coarsely hairy within; pappus evidently plumose; $2n=20$. Dry, open places; w. N.Y. and s. Ont. to s. O., n. Ind., Mich., and Minn., s. to Mo. and irregularly to Tenn. and Ala. July–Sept.

12. Liatris squarrosa (L.) Michx. Plains b.s. Glabrous or hairy, 3–8 dm; lvs linear or a little broader, firm, those near the base 6–25 cm × 4–13 mm, often partly sheathing, or the very lowest often smaller and deciduous; heads mostly few or even solitary, on stiff, erect peduncles or sessile; invol 12–25(–30) mm, its bracts firm, with loose or squarrose, acuminate tip; fls 20–45 per head, or up to 60 in the terminal head; inner surface of the cor-lobes coarsely hairy; pappus evidently plumose; $2n=20$. Dry, open places; Del. to S.D., s. to Fla. and Tex. July–Aug. Three vars. in our range:

a Plants hairy, at least in the infl.
  b Invol bracts mostly tapering to a long, loosely spreading tip; pubescence mostly short and curly or
    appressed; widespread eastward and southward, w. to Ill. and Mo. . . . . var. *squarrosa.*
  b Invol bracts abruptly contracted to the short, squarrose tip; pubescence tending to be longer and
    more spreading; chiefly western, from s. Io. to n. La., w. to e. Neb., e. Kans., and se. Tex. (*L.*
    hirsuta.) . . . . var. *hirsuta* (Rydb.) Gaiser.
a Plants essentially glabrous (except the cor-lobes); invol similar to that of var. *hirsuta*, avg a little
  smaller, Io. and Mo. to S.D. and Tex. (*L. glabrata.*) . . . . var. *glabrata* (Rydb.) Gaiser.

13. Liatris punctata Hook. Rootstock (unlike all our other spp.) ± elongate and pointed at the base, like a short, fleshy-thickened taproot, or seldom horizontal and resembling a thickened rhizome, producing scattered aerial stems; lvs numerous, punctate, glabrous except for the often coarsely ciliate margins, linear or linear-oblanceolate, the lowest smaller than those just above and often deciduous; heads several or many in a spiciform infl, sessile or nearly so; invol subcylindric, 10–18 mm, its bracts punctate, mostly sharply-mucronate-acuminate, commonly some of them ± ciliate-margined; fls mostly 4–6 per head, the cor-lobes glabrous, the tube hairy toward the base within; pappus evidently plumose; $2n=20$. Dry, open places, often in sandy soil; sw. Mich. to Man., Alta., Ark., N.M., and n. Mex. Aug.–Oct. Highly variable, and perhaps taxonomically divisible. (*L. angustifolia; L. densispicata.*) Our common phase is var. *nebraskana* Gaiser, to 8 dm, with scarcely ciliate lvs mostly 2–3 mm wide. The chiefly more western var. *punctata* is 1–4 dm, with more ciliate lvs to 7 mm wide.

76. **VERNONIA** Schreber.   Ironweed. Heads discoid, the fls all tubular and perfect, purple (white); invol of numerous ± appressed, imbricate bracts, in our spp. constricted at the summit; receptacle flat or convex, naked; style-branches slender, gradually tapering distally, minutely hispidulous outside, smooth inside, inconspicuously stigmatic near the base within; achenes (in ours) ribbed, commonly hairy; pappus double, the inner of numerous long, slender bristles, the outer of short bristles or short narrow scales; our spp. perennial herbs, usually simple to the infl; lvs alternate, in our spp. sessile or subsessile and serrate or entire; heads in ours ± numerous, medium-sized, in corymbiform cymes. 1000, N. Amer., S. Amer., Afr., Madagascar, and se. Asia. Our spp. bloom in late summer and fall. Hybrids abound. A hybrid swarm in the midwest, derived chiefly from *V. missurica* and *V. gigantea,* but also involving *V. baldwinii* and *V. fasciculata,* has been called *V.* × *illinoensis* Gleason.

1 Prinicpal invol bracts prolonged into a filiform tip.
  2 Fls numerous, mostly 55–100+ per head; lvs pitted-punctate beneath; Ozarkian .... 1. *V. arkansana.*
  2 Fls fewer, mostly 29–55; lvs plane beneath; Appalachian and Atlantic.
    3 Pappus brownish-purple, or sometimes merely dark-tawny; lowlands .... 2. *V. noveboracensis.*
    3 Pappus bright stramineous or pale tawny to nearly white; upland woods .... 3. *V. glauca.*
1 Principal invol bracts obtuse to acuminate.
  4 Lower surface of the lvs hairy, the pubescence commonly obscuring any punctae.
    5 Inner invol bracts rounded or obtuse, sometimes minutely apiculate, only seldom conspicuously resinous.
      6 Fls mostly 32–58; hairs of the lower lf-surface long, crooked .... 4. *V. missurica.*
      6 Fls fewer, mostly 13–30; hairs of the lower lf-surface short, straight .... 5. *V. gigantea.*
    5 Inner invol bracts acuminate, densely resinous on both sides of the prominent, often keeled midvein .... 6. *V. baldwinii.*
  4 Lower surface of the lvs glabrous and strongly pitted-punctate .... 7. *V. fasciculata.*

1. **Vernonia arkansana** DC.   Ozark-i. Stems 0.7–1.5(–2) m, glabrous or nearly so (or more puberulent above) and somewhat glaucous; lvs linear or lance-linear, 10–20 × 0.5–2.5 cm, acuminate, denticulate or sometimes entire, glabrous or rarely thinly hairy, pitted-punctate beneath; infl very irregular, the pedicels 1–5 cm, thickened at the top; fls mostly 55–100+; invol 9–15 mm, its bracts arachnoid-ciliate, linear or narrowly lanceolate at the appressed base, with loosely flexuous filiform tip 2–4 mm; pappus brownish-purple; $2n=34$. Low woods and streambanks; Ozark region of Mo., s. Ill., Ark., Okla., and Kans., occasionally intr. n. and e. to Wis. (*V. crinita.*)

2. **Vernonia noveboracensis** (L.) Michx.   New York i. Stems 1–2 m, thinly hairy or glabrous; lvs lanceolate or lance-linear, 10–28 × 1.5–6 cm, long-acuminate, sharply serrate to subentire, gradually narrowed to the base, scabrellate above, puberulent to thinly tomentose near the base, the lower side not much paler than the upper; infl loose and open, flat-topped or concave; fls 30–55(–65); invol 7–17 mm, its principal bracts ovate or oblong at base, abruptly narrowed to a filiform tip 2–8 mm, arachnoid-ciliate, glabrous on the back; pappus purple or brownish-purple, or sometimes merely dark-tawny; $2n=34$. Low wet woods and marshes, especially near the coast; Mass. to Fla., inland to w. Pa., W.Va., e. Ky., Tenn., and Ala. Mostly Aug., Sept.

3. **Vernonia glauca** (L.) Willd.   Appalachian i. Stems 0.7–1.5 m, puberulent above, otherwise glabrous or nearly so; lvs lance-ovate to ovate, 1–2.5 dm × 2–10 cm, shortly acuminate, saliently serrate, glabrous or barely scabrellate above, distinctly paler and puberulent beneath, rather abruptly narrowed to a petiole-like base; infl loose, very irregular, some branches commonly elongate; fls 32–48; invol 5–9 mm, its principal bracts arachnoid-ciliate, lanceolate to ovate at the appressed base, narrowed to a filiform tip 2–5 mm; pappus bright-stramineous or pale tawny to nearly white; $2n=34$. Upland woods; se. Pa., s. N.J., Md., and e. W.Va., s. to Va., Ga. and Ala., avoiding the coastal plain southward. Mostly July.

4. **Vernonia missurica** Raf.   Missouri-i. Stem stout, 1–2 m, pubescent or thinly tomentose; lvs lanceolate, 6–20 × 1.5–6 cm, coarsely serrate to subentire, scabrellate above, thinly to densely tomentose beneath, at least along the veins, with long, crooked hairs; infl widely branched, flattened or concave; fls 32–58; invol bracts appressed, purple, or greenish along the inconspicuous midvein, rounded to acute, often apiculate, arachnoid-ciliate, glabrous or puberulent on the back; pappus tawny to brown; $2n=34$. Moist prairies, bottomland pastures, and low roadsides; sw. Ont. to Io. and Nebr., s. to Ala. and Tex. Our plants mostly contaminated by introgression from no. 5.

5. **Vernonia gigantea** (Walter) Trel. Tall i. Stems 1–3 m, glabrous or nearly so; lvs thin, lanceolate to lance-oblong or lance-ovate, 10–30 × 3–7 cm, long-acuminate, gradually narrowed to the base, finely serrate to subentire, glabrous or nearly so above, thinly pubescent with minute (rarely 0.3 mm) straight hairs beneath; infl loose, flattened to concave or irregular, 2–5 dm wide; fls 13–30; invol 4–7 mm, its principal bracts appressed, regularly imbricate, obtuse, rounded, or rarely apiculate, usually purple, arachnoid-ciliate or entire, commonly glabrous and not resinous on the back; pappus purple to brown or tan; $2n=34$. Moist or wet woods, and becoming weedy in pastures; w. N.Y. to s. Mich. and e. Nebr., s. to S.C., Fla., and Tex. Ours is the widespread var. *gigantea.* (*V. altissima.*)

6. **Vernonia baldwinii** Torr. Western i. Stems 6–15 dm, thinly pubescent to commonly tomentose; lvs lance-ovate to ovate, 7–17 × 2–6 cm, acute or shortly acuminate, sharply serrate, puberulent above, tomentose beneath, with long, crooked hairs; infl 1–2 dm wide, flattened or irregular; heads mostly very numerous; fls 17–34; invol 5–8 mm, its principal bracts lanceolate or lance-ovate, acuminate, arachnoid-ciliate, densely resinous toward the tip on both sides of the prominent (often keeled) midvein; pappus rather light brown or brownish-purple; $2n=34$. Dry prairies, plains, and upland woods. The Ozarkian var. *baldwinii,* occurring from w. Ill. to Mo., Ark., and e. Okla., has mostly 23–34 fls; the tips of the invol bracts are recurved, those of the inner pubescent on the inner surface. The var. *interior* (Small) Schubert, mainly on the Great Plains from s. Io. and Nebr. to e. Colo. and s. Tex., often weedy, has mostly 17–27 fls; the tips of the invol bracts are merely loosely spreading, not pubescent within. (*V. interior.*)

7. **Vernonia fasciculata** Michx. Smooth i. Stems 3–14 dm, often red or purple, puberulent above, otherwise glabrous or nearly so; lvs denticulate to sharply serrate, glabrous, conspicuously pitted-punctate beneath; infl usually flat and dense, 4–10 cm wide; fls 10–26; invol 5–9 mm, its principal bracts rounded to subacute, entire or sparsely ciliate, commonly glabrous on the back; pappus tawny to purple; $2n=34$. Wet prairies and marshes. The midwestern var. *fasciculata,* occurring from O., Ind., and Ky. to Minn., e. Nebr., e. Kans., and e. Okla., is 5–14 dm tall, with the middle cauline lvs 8–18 cm, narrowly to broadly lanceolate, glabrous above or minutely puberulent near the margin. The more western var. *nebraskensis* Gleason (*V. corymbosa*), occurring from s. Man. to Nebr. and ne. Colo., barely entering our range in w. Minn. and w. Io., is 3–6 dm tall, with the middle cauline lvs 4–10 cm, lance-ovate, scabrous above near the margins.

77. **ELEPHANTOPUS** L. Elephant's foot. Heads discoid, 1–5-fld, in glomerules subtended by 1–3 sessile, foliaceous bracts and resembling a single head; invol of 4 pairs of decussate bracts, the 2 outer pairs shorter, the alternate pairs conduplicate; fls all perfect and fertile, anthocyanic or sometimes white; receptacle flat or nearly so; cor unequally 5-cleft; style of *Vernonia;* pappus of 5–8 short, rigid, flattened scales prolonged at the summit (in our spp.) into a straight terminal bristle; perennial herbs with small glomerules of heads terminating the branches; our spp. fibrous-rooted and mostly single-stemmed. 30, mainly trop.

1 Lvs cauline, or both cauline and basal .... 1. *E. carolinianus.*
1 Lvs all or chiefly basal, those of the stem reduced to mere bracts.
  2 Midvein of the lvs strigose beneath .... 2. *E. nudatus.*
  2 Midvein of the lvs villous beneath, with spreading or reflexed hairs .... 3. *E. tomentosus.*

1. **Elephantopus carolinianus** Willd. Leafy e.f. Stems 3–10 dm, densely hirsute below, usually divaricately branched above and with numerous glomerules; lvs or most of them cauline, broadly elliptic to obovate, 9–25 × 3–10 cm, crenate, pilose beneath; glomerules 10–15 mm; bracts triangular-ovate, often longer than the glomerules; invol 8–10 mm, its bracts acuminate, resinous, thinly strigose; pappus-bristles gradually dilated into a narrowly triangular base; $2n=22$. Dry woods; N.J. to O., Ill., and Kans., s. to Fla., Tex., and W.I. Aug., Sept.

2. **Elephantopus nudatus** A. Gray. Stems 2–8 dm, glabrate to softly pilose, branched above but seldom with more than a dozen glomerules; lvs basal, prostrate in a rosette, oblanceolate or oblong-oblanceolate, 8–25 × 1.5–6 cm, acute or obtuse, crenate, long-tapering to the base, sparsely pilose on both sides, densely strigose along the midvein beneath; glomerules 6–12 mm, scarcely surpassed by the bracts; invol 6–8 mm, its bracts thinly strigose, densely resinous; pappus-bristles abruptly dilated into a broadly triangular base; $2n=22$. Sandy soil on the coastal plain; Del. to Fla., La., Ark., and Tex. Aug., Sept.

3. **Elephantopus tomentosus** L. Stems stout, 3–8 dm, hirsute or villous; lvs basal, prostrate in a rosette, broadly elliptic to obovate-oblong, 10–25 × 3–10 cm, obtuse or rounded, crenate, gradually narrowed to the sessile base, softly villous beneath with spreading or somewhat retrorse hairs; infl divaricately branched, usually with 3–10 glomerules; bracts triangular-ovate, equaling or barely surpassing the glomerules; invol 9–13 mm, its bracts thinly strigose and sparsely resinous; pappus-bristles gradually dilated into a narrowly triangular base; $2n=22$. Open, sandy woods; coastal plain from se. Va. to Fla. and Tex., n. to the interior of Ky., and s. to Chiapas. Aug., Sept.

78. **ARCTIUM** L. Burdock. Fls all tubular and perfect, the cor pink or purplish, with a slender tube, short throat, and long, narrow lobes; invol subglobose, its bracts multi-

seriate, narrow, appressed at base, with a spreading, subulate, inwardly hooked tip; receptacle flat, densely bristly; anthers very shortly awn-pointed at the tip, evidently tailed at the base; style with an abrupt change of texture below the minutely papillate branches; achenes oblong, slightly compressed, few-angled, many nerved, truncate, glabrous; pappus of numerous short, subpaleaceous, separately deciduous bristles; coarse biennial herbs with large, alternate, entire or toothed to rarely laciniate, mostly cordate-based lvs and several or many heads. 5, temp. Eurasia.

1 Infl mostly of racemiform or thyrsoid branches, with mostly short-peduncled or subsessile heads, not
    corymbiform . . . . 1. *A. minus.*
1 Infl ± corymbiform, the heads mostly long-pedunculate.
  2 Invol evidently tomentose, mostly 1.5–2.5 cm thick . . . . 2. *A. tomentosum.*
  2 Invol not tomentose, mostly 3–4 cm thick . . . . 3. *A. lappa.*

1. Arctium minus Schk. Common b. To 1.5 m, rarely more; lvs petiolate, the lower petioles mostly hollow, the blade narrowly to very broadly ovate, cordate at base, to 5 × 4 dm, thinly tomentose or eventually glabrate beneath, subglabrous above; branches of the infl ascending to widely spreading, racemiform or subthyrsoid, the heads mostly short-pedunculate or subsessile; invol 1.5–2.5 cm thick, glabrous or slightly glandular to sometimes evidently arachnoid-tomentose, usually a little shorter than the fls, greenish-stramineous or a little purplish, the inner bracts often more flattened than the others and scarcely hooked; achenes 4–5.5 mm; 2*n*=32, 36. Native of Eurasia, now established on roadsides and in waste places throughout most of the U.S. and s. Can.
    A. nemorosum Lej., with larger heads mostly 3–3.5 cm thick, is only casually intr. with us. It may reflect past hybridization of nos. 1 and 3.

2. Arctium tomentosum Miller. Cotton-b. Resembling no. 3 but smaller, seldom over 13 dm; lower petioles mostly hollow; invol mostly 1.5–2.5 cm thick, ± strongly arachnoid-tomentose and the bracts only weakly or scarcely hooked; corolla-limb glandular outside; 2*n*=36. Native of Eurasia, sparingly established in our range. June–Oct.

3. Arctium lappa L. Great b. To 1.5 or even 3 m; lvs petiolate, the petioles mostly solid, progressively shorter upwards, the blade ovate or broader, cordate, to 5 × 3 dm, thinly tomentose beneath, subglabrous above; infl corymbiform, with long, glandular or glandular-hairy peduncles commonly 3–10 cm; heads large, the invol (2.5–)3–4 cm wide, generally equaling or surpassing the fls, glabrous or slightly glandular, and often with a few long cobwebby hairs; achenes 6–7 mm; 2*n*=32, 36. Native of Eurasia, sparingly established as a weed along roadsides and in waste places over most of the n. U.S. and adj. Can. Aug.–Oct.

79. CARDUUS L.   Thistle. Much like *Cirsium,* differing chiefly in its pappus of merely capillary, not at all plumose bristles; stem commonly spiny-winged; achenes quadrangular or somewhat flattened, with 5–10 or more nerves, or nerveless. 100, Old World.

1 Invol bracts mostly 2 mm wide or more; heads mostly nodding . . . . 1. *C. nutans.*
1 Invol bracts rarely as much as 2 mm wide; heads not nodding.
  2 Plants very strongly spiny; stem tough . . . . 2. *C. acanthoides.*
  2 Plants weakly spiny; stem brittle . . . . 3. *C. crispus.*

1. Carduus nutans L.   Musk-t. Biennial, 3–20 dm; lvs glabrous, or long-villous chiefly along the main veins beneath, deeply lobed, to 25 × 10 cm; heads mostly solitary and nodding at the ends of the branches, usually large, the disk (1.5–)4–8 cm wide (pressed); peduncles naked for some distance below the head; middle and outer invol bracts conspicuously broad (2–8 mm), with long, flat, spreading or reflexed, spine-pointed tip; inner bracts narrower and softer; 2*n*=16. Roadsides and waste places; native of Europe, now widely established in the U.S. and adj. Can., commoner westward. June–Oct. (*C. thoermeri.*) Hybridizes with no. 2, the hybrids partly fertile.

2. Carduus acanthoides L.   Plumeless t. Biennial, 3–15 dm, very strongly spiny, the stem tough; lvs deeply lobed or pinnatifid, to 25 × 8 cm, loosely villous beneath, chiefly along the main veins, with long multicellular hairs, or glabrous, the upper surface glabrous or similarly hairy; heads clustered or solitary at the ends of the branches, erect, small, the invol 1.4–2 cm high, the disk 1.5–2.5 cm wide (pressed); peduncles, if any, winged all the way to the heads; invol bracts narrow, rarely as much as 2 mm wide, erect or loosely spreading, the middle and outer spine-tipped, the inner softer and flatter; 2*n*=22. Roadsides, pastures, and waste places; native of Europe, now widely but sparingly established in our range. July–Oct.

3. Carduus crispus L.   Welted t. Biennial, 6–20 dm, rather weakly spiny, the stem brittle; lvs broader and less deeply cleft than in no. 2, cottony-tomentose beneath at least when young; heads clustered at the ends of the often short branches, similar to those of no. 2, but less pungent and avg a little smaller, the invol 1.2–1.7 cm; peduncles often wingless in the distal several mm; 2*n*=16. Roadsides and waste places; native of Eurasia, sparingly intr. in our range, chiefly about the larger ports. June–Sept.

80. CIRSIUM Miller. Thistle. Heads discoid, the fls all tubular and perfect, or the plants sometimes dioecious by abortion; invol bracts imbricate, usually some or all of them spine-tipped, and in many spp. with a thickened, glutinous, dorsal ridge; receptacle flat to subconic, densely bristly; cors purple to yellowish or white, with slender tube and long, narrow lobes; filaments usually papillose-hairy; anthers with a firm, narrow, apical appendage, evidently tailed at the base; style with a thickened, hairy ring and an abrupt change of texture below the ± connate, papillate branches; achenes basifixed or nearly so, glabrous, firm, thick-compressed, often curved, nerveless, commonly with a cartilaginous, pale yellow collar; pappus of numerous plumose bristles, deciduous in a ring; spiny herbs with alternate, toothed to more often pinnatifid lvs and medium-sized to large heads. 200, N. Hemisphere. (Spp. all sometimes included in *Carduus*.)

1 Polygamo-dioecious, the heads nearly unisexual; colonial by creeping roots .... 16. *C. arvense.*
1 Heads all alike, with perfect fls; most spp. not colonial.
  2 None of the invol bracts spine-tipped.
    3 Invol 2–3.5 cm; fruiting pappus 1.5–2.5 cm; stem not winged .... 13. *C. muticum.*
    3 Invol 1–2 cm; fruiting pappus ca 1 cm or less; stem spiny-winged .... 14. *C. palustre.*
  2 At least the outer bracts evidently (though often shortly) spine-tipped.
    4 Stem conspicuously winged by the decurrent lf-bases .... 15. *C. vulgare.*
    4 Stem scarcely or not at all winged.
      5 Heads conspicuously and closely invested by a series of narrow, strongly spiny-toothed leaves; inner bracts with attenuate narrow tip. (Occasional forms of *C. pumilum* that might otherwise be sought here have the tips of the inner bracts expanded, erose, and crisped.)
      .... 1. *C. horridulum.*
      5 Heads otherwise, though sometimes subtended by 1 or 2 reduced leaves.
        6 Stem merely crisp-hirsute to subglabrate, or sometimes ± arachnoid when young; upper surfaces of the lvs glabrous or hirsute, not tomentose.
          7 Lvs merely arachnoid-tomentose beneath, often eventually glabrate.
            8 Heads relatively small, the invol 1.5–2.5 cm .... 12. *C. nuttallii.*
            8 Heads larger, the invol 2.5–5 cm.
              9 Plants without well developed basal lvs at flowering time; cauline lvs crowded (internodes commonly under 1 cm), rarely over 2.5 cm wide .... 2. *C. repandum.*
              9 Plants generally with well developed and persistent basal lvs, these, and the larger lower cauline ones, seldom less than 3 cm wide; lvs not especially crowded except sometimes at the base.
                10 True perennial with thickened, hollow roots .... 3. *C. hillii.*
                10 Biennial (sometimes short-lived perennial?) with solid roots .... 4. *C. pumilum.*
          7 Lvs closely and persistently white-tomentose beneath.
            11 Invol mostly 2.5–3.5 cm; achenes 4–6 mm; plants robust, 1–3 m.
              12 Lvs deeply pinnatifid .... 5. *C. discolor.*
              12 Lvs toothed or shallowly lobed .... 6. *C. altissimum.*
            11 Invol 1.5–2.5 cm; achenes 3–4 mm; more slender, 5–15 dm.
              13 Cauline lvs numerous, mostly 30–70; coastal plain; fl Aug.–Sept. .... 7. *C. virginianum.*
              13 Cauline lvs relatively few, mostly 10–25; inland; fl May, June .... 8. *C. carolinianum.*
         6 Stem densely, closely, and persistently white-tomentose; upper surfaces of the lvs thinly white-tomentose or eventually glabrate.
          14 Lvs toothed to pinnatifid, the lobes lanceolate or oblong to triangular-ovate, often crowded; fls mostly purple; western.
            15 Relatively slender plants of poorly drained sites, with a well developed tendency to spread by slender, short-lived creeping roots .... 9. *C. flodmanii.*
            15 Stouter plants of well drained sites, creeping roots poorly or scarcely developed ... 10. *C. undulatum.*
          14 Lvs deeply pinnatifid, with long, linear, remote lobes; fls white .... 11. *C. pitcheri.*

1. Cirsium horridulum Michx. Yellow t. Stout biennial 2–15 dm, simple or with short, stout, ascending, peduncle-like branches, some of the roots commonly fleshy-thickened; herbage thinly tomentose and eventually generally glabrate, or the lvs merely arachnoid and glabrate; lvs strongly spiny, broad and pinnatifid to seldom narrow and merely spiny-toothed, the largest ones at or near the base; heads light yellow or white to lavender or purple, several or solitary, each subtended by a number of narrow, erect, strongly spiny, reduced lvs; invol (2.5–)3–5 cm, the outer bracts with erect spine-tip, the inner merely attenuate, all with modified, shortly scabrous or scabrous-ciliate margin; achenes 4–6 mm; 2*n*=32, 34. Open places, especially in sandy soil or along salt or fresh marshes; coastal states (and Pa.) from Me. to Fla. and Tex.; Mex. May–Aug. (*C. spinosissimum.*) Ours are the widespread var. *horridulum.*

2. Cirsium repandum Michx. Sand-hill t. Deep-rooted perennial, 2–6 dm, the herbage loosely and rather copiously arachnoid when young, subglabrate or merely hirsute in age; lvs crowded (internodes commonly less than 1 cm), narrow, sessile, 6–15 × 1–2.5 cm, densely spiny-margined, coarsely toothed or shallowly lobed, and often finely toothed as well; heads (solitary) several, terminating short branches from near the summit, only shortly or scarcely pedunculate, purple; invol 2.5–4 cm, the middle and outer bracts with short, mostly erect spine-tip, the glutinous dorsal ridge poorly developed or wanting; inner bracts looser and merely attenuate; achenes 3.5–4 mm; $2n=30$. Dry, sandy soil, especially in sand-hills and pine-barrens; se. Va. to Ga. June–Sept.

3. **Cirsium hillii** (Canby) Fern. Hollow-rooted t. Much like no. 4, but perennial with long, thickened roots, these traversed longitudinally by several cavities; glutinous dorsal ridge of the invol bracts avg larger and better developed; achenes mostly 4–5 mm; $2n=30$. Prairies and other open places, often in sandy soil; s. Ont. to Pa., w. to Minn. and S.D. June–Aug.

4. **Cirsium pumilum** (Nutt.) Sprengel. Pasture-t. Stout biennial (sometimes short-lived perennial?) 3–8 dm, with 1–several coarse, slightly thickened roots; herbage green and coarsely arachnoid-villous or crisp-hirsute; lvs lobed or pinnatifid, beset with numerous short marginal spines in addition to the scattered longer ones; basal lvs well developed and generally persistent, these and the lower cauline ones 12–25 × 2–7 cm; cauline lvs not especially crowded except sometimes near the base; heads few or solitary, large, the invol 3.5–5 cm, usually some of its bracts with a narrow, glutinous dorsal ridge; middle and outer invol bracts tipped by a short, ± erect spine; inner bracts elongate, with expanded, chartaceous, crisped and erose tip; fls sweet-scented, purple (white); achenes 3–4 mm; $2n=30$. Pastures, old fields, and open woods; s. Me. to w. N.Y., s. to Del., Pa., and in the mts. to Va., e. W.Va., and N.C. (*C. odoratum.*)

5. **Cirsium discolor** (Muhl.) Sprengel. Field-t. Much like no. 6, and not always sharply distinct; lvs deeply pinnatifid, generally firmer and spinier; plants avg smaller (1–2 m), the peduncles tending to be a little leafier, and the heads often a little broader-based; $2n=20$. Fields, open woods, river-bottoms, and waste places; s. Que. and Ont. to Minn. and adj. Man., s. to N.C., Miss., La., and Kans. July–Oct.

6. **Cirsium altissimum** (L.) Sprengel. Tall t. Robust, fibrous-rooted perennial, 1–3(–4) m, openly branched when well developed; stem crisply spreading-hirsute to subglabrate, sometimes slightly tomentose in the infl; lvs large, the lower to 5 × 2 dm, broadly oblanceolate to obovate or elliptic, densely white-tomentose beneath, scabrous-hirsute to subglabrous above, merely spiny-toothed or coarsely toothed to sometimes lobed (seldom more than halfway to the midrib), the reduced ones of the infl sometimes more evidently lobed than the others; heads several or numerous on the ± leafy peduncles; invol (2–)2.5–3.5(–4) cm; middle and outer bracts tipped with a spine 2–5 mm; inner bracts merely attenuate or often with a scarious, slightly dilated and erose tip; fls mostly pink-purple; achenes 4.5–6 mm; $2n=18(20)$. Fields, waste places, river-bottoms, and open woods; Mass. to N.D., s. to Fla. and Tex. July–Oct. (*C. iowense.*)

7. **Cirsium virginianum** (L.) Michx. Virginia-t. Slender biennial 5–10(–15) dm from a cluster of fleshy-fibrous roots; stem glabrous or arachnoid, sometimes more evidently tomentose when young; lvs closely white-tomentose beneath, glabrous or hirsute on the upper surface, the basal ones soon deciduous, the others numerous, mostly 30–70, 5–20 cm, sometimes merely spinose-ciliate and up to 1 cm wide, sometimes evidently pinnatifid and to 5 cm wide; peduncles with a few reduced lvs or bracts; heads as in no. 8, but avg larger, the invol to 2.5 cm; $2n=28$. Savannas, bogs, and wet pinelands; coastal plain from s. N.J. to Fla. Aug., Sept. (*C. revolutum.*)

8. **Cirsium carolinianum** (Walter) Fern. & Schubert. Spring-t. Slender, fibrous-rooted biennial 5–15(–18) dm; stem glabrous or arachnoid, sometimes more evidently tomentose when young; lvs closely white-tomentose beneath, glabrous or hirsute on the upper surface, from merely spinose-ciliate and up to ca 2.5 cm wide to evidently pinnatifid and to 5 cm wide, the basal ones to 3 dm, the cauline relatively few, mostly 10–25, 8–15 cm and (except when lobed) seldom more than 1.5 cm wide, narrow-based, reduced upward; heads (1–) several, on long, naked peduncles terminating the branches; invol 1.5–2 cm, its middle and outer bracts with a glutinous dorsal ridge and a slender, suberect or spreading spine 1.5–4 mm, the inner merely attenuate and commonly crisped; fls pink-purple; achenes 3–4 mm; $2n=20$. Open woods and dry, sandy soil; s. O. to the mts. of N.C., Ga., and Ala., w. to Mo. and Tex. May, June. (*C. flaccidum.*)

9. **Cirsium flodmanii** (Rydb.) Arthur. Prairie-t. Resembling no. 10, but smaller and more delicate, 3–8 dm, and vigorously spreading by short-lived creeping roots, the individual plants thus established generally becoming taprooted; lvs usually rather deeply pinnatifid, the lobes commonly lance-triangular and seldom over 7 mm wide, or sometimes fully as wide as those of no. 10, varying to subentire and merely spiny-margined (especially the basal ones); heads merely rounded (not invaginated) at the base, and avg smaller, the invol 2–2.5(–3) cm, its bracts appearing darker, narrower, more numerous, and more closely imbricate, their spine-tips sometimes under 3 mm; fls rich purple (white); achenes 3–5 mm, with a conspicuous apical yellow band; $2n=22$. Swales and other moderately moist, poorly drained sites on the Great Plains, extending e. to Man., Minn., and Io., and occasionally intr. eastward. June–Sept. (*C. canescens,* misapplied.)

10. **Cirsium undulatum** (Nutt.) Sprengel. Wavy-lvd t. Stout, short-lived perennial from a taproot, this sometimes branching below-ground and producing more than one stem, but without the well developed creeping roots of no. 9; stem 3–12 dm, usually branched above, densely, closely and persistently white-tomentose; lower lf-surfaces similarly tomentose, the upper surfaces more thinly so and sometimes eventually glabrate; lvs coarsely toothed to pinnatifid, the lobes ovate, deltoid, or occasionally oblong, seldom under 7 mm wide; heads mostly several or many, tending to be broadly and shallowly invaginated at the base; invol 2.5–4(–5) cm, its bracts with a glandular-glutinous dorsal ridge and commonly a little marginal tomentum, well imbricate, the inner with attenuate and often crisped tips, the others with spine-tip 3–5 mm; fls pink-purple, sometimes rather pale; achenes 4–7 mm, inconspicuously or not at all banded across the top, becoming mucilaginous when wet (unlike all related spp.); $2n=26$. Hillsides, prairies, railroad-tracks, and other open places, in well drained soil; B.C. to Ariz., e. to s. Man., Minn., and Mo., and occasionally intr. eastward. June–Sept. (*C. megacephalum.*)

11. **Cirsium pitcheri** (Torr.) T. & G. Dune-t. Taprooted biennial or monocarpic perennial, 5–10 dm; stem and lower lf-surfaces densely and persistently white-tomentose, the upper lf-surfaces more thinly so; lvs deeply pinnatifid, with narrow rachis and long, remote, linear, entire or few-toothed, weakly spine-tipped lobes to 8 cm × 7 mm; heads several, white or whitish; invol 2.5–3 cm, its bracts slightly tomentose especially marginally, well imbricate, the inner long-acuminate, the others with weak spine-tip 1–2 mm; achenes 5–7.5 mm; $2n=34$. Sand dunes around lakes Michigan and Huron and e. Lake Superior, in Ont., Mich., Wis., Ind., and formerly Ill. May–Sept.

12. **Cirsium nuttallii** DC. Coastal tall t. Coarse biennial (1–)1.5–3.5 m, branching and many-headed when well developed, the heads commonly solitary at the ends of long, slender, subnaked branches; stem glabrous or with crisp spreading hairs; lvs arachnoid-tomentose beneath when young, generally eventually glabrate, smooth or somewhat crisp-hairy on the upper surface, thin, deeply pinnatifid, the lobes generally again toothed or cleft; larger (lower) lvs to 6 × 1.5 dm, but often soon deciduous; invol 1.5–2.5 cm, the middle and outer bracts with a glutinous dorsal ridge and tipped with a weak, abruptly spreading spine mostly 1–2(–3) mm, the inner bracts innocuous and merely attenuate, often crisped but not expanded; fls pink or lavender (often very pale) to white; achenes 3–4 mm; $2n=24$, 28. Wet or dry, usually sandy soil, often in thickets; se. Va.; S.C. to Fla. and La. June–Aug.

13. **Cirsium muticum** Michx. Swamp-t. Coarse biennial with several stout roots, 5–20 dm, branching and many-headed when well developed, villous or arachnoid when young, later glabrescent; lvs deeply pinnatifid into laceolate or oblong, entire, lobed, or dentate segments, only weakly spiny, often very large, to 5.5 × 2 dm, sometimes thinly tomentose beneath, otherwise arachnoid-hirsute to subglabrous; invol 2–3.5 cm, multiseriate, its bracts ± tomentose, especially marginally, generally with a glutinous dorsal ridge, innocuous or with a vestigial spinule up to ca 0.5 mm, the inner commonly with loose, crisped tips, but not expanded; fls purple, pink, or deep lavender, rarely white; achenes 5 mm; $2n=20$. Swamps and bogs, wet meadows, and moist woods and thickets; Nf. to Sask., s. to Del. and Mo. and in the mts. to N.C. and Tenn., and irregularly to Fla. and Tex. July–Oct.

14. **Cirsium palustre** (L.) Scop. Marsh-t. Biennial 3–20 dm, the stem spiny-winged by the decurrent lf-bases; herbage ± arachnoid; heads purple, several or many, clustered, seldom any of the peduncles over 1 cm; invol 1–2 cm, its bracts not at all spine-tipped; mature pappus 1 cm or less; achenes 3 mm; $2n=34$. Native of Eurasia, now widely intr. in ne. U.S. and se. Can., often invading woods and seemingly native. June, July.

15. **Cirsium vulgare** (Savi) Tenore. Bull-t. Biennial weed 5–15 dm; stem conspicuously spiny-winged by the decurrent lf-bases, copiously spreading-hirsute to sometimes arachnoid; lvs strongly spiny, pinnatifid, the larger ones with the lobes again toothed or lobed, scabrous-hispid above, thinly white-tomentulose to sometimes green and merely hirsute beneath; heads several, purple; invol 2.5–4 cm, its bracts all spine-tipped, without any well developed glutinous dorsal ridge; achenes 3–4 mm; $2n=68$. Pastures, fields, roadsides, and waste places; native of Eurasia, now widely established in N. Amer. June–Oct. (*C. lanceolatum,* misapplied.)

16. **Cirsium arvense** (L.) Scop. Canada-t. Colonial perennial 3–15(–20) dm from deep-seated creeping roots, subglabrous, or the lvs ± white-tomentose beneath; heads ± numerous in an often flat-topped infl, unisexual or nearly so, the plants polygamo-dioecious; invol 1–2 cm high, its bracts all innocuous, or the outer with a weak spine-tip ca 1 mm; fls pink-purple (white); achenes 2.5–4 mm; pappus of the pistillate heads surpassing the corollas, that of the staminate heads surpassed by the corollas; $2n=34$. A noxious weed of fields and waste places; native of Eurasia, now widely intr. in n. U.S. and s. Can. July, Aug. The var. *arvense* has merely toothed or shallowly lobed, weakly spiny lvs. (*C. setosum; C. a.* var. *mite.*) The more common var. *horridum* Wimmer & Graebner has more spiny, deeply pinnatifid lvs.

81. SILYBUM Adans., nom. conserv. Milk-thistle.

Heads discoid, the fls all tubular and perfect; invol bracts imbricate, broad and firm, most of them spiny-margined and strongly spine-tipped; receptacle flat, densely setose; cors purple, with slender tube and long narrow lobes; filaments glabrous, connate at least below; anthers with a firm, slender, terminal appendage, shortly tailed at the base; style with an abrupt change of texture below the connate, papillate branches; achenes basifixed, somewhat compressed, glabrous; pappus of numerous slender, unequal, subpaleaceous bristles, deciduous in a ring; spiny winter-annual or biennial with alternate lvs and large, globose heads terminating the branches. 2, Mediterranean.

1. **Silybum marianum** (L.) Gaertner. Glabrous or slightly tomentose, 6–15 dm; lvs pinnately lobed, less so upwards, to 4 dm and nearly half as wide, petiolate below, becoming sessile and strongly auriculate-clasping above, spiny-margined, ± marked with white along the main veins; coarse, spreading tips of the invol bracts basally expanded; disk 3–6 cm wide; achenes 6–7 mm; $2n=34$. A weed in waste places and about ballast-dumps; native of the Mediterranean region, rarely seen here and there in our range. May–July. (*Mariana m.*)

82. ONOPORDUM L. Similar to *Cirsium,* differing chiefly in the receptacle, which is flat, fleshy, and honey-combed, often with short bristle-tips on the partitions, but not densely bristly; pappus-bristles naked or plumose; spiny biennials, mostly tomentose or woolly. 40, sw. and c. Asia.

1. **Onopordum acanthium** L. Scotch thistle. Coarse, branching biennial to 2 m, very strongly spiny, with broadly winged stem; herbage ± tomentose; lvs toothed or slightly lobed, sessile and decurrent, or the lower petiolate, the blade to 6 × 3 dm; heads 2.5–5 cm wide, the invol bracts all spine-tipped; achenes 4–5 mm, slightly compressed, transversely rugulose; pappus merely barbellate; 2*n*=34. A Eurasian weed, sparingly naturalized over much of the U.S. and s. Can. July–Oct.

83. CARLINA L. Heads discoid, the fls all tubular and perfect; invol bracts several-seriate, the outer spiny, the inner with innocuous, elongate, eventually spreading, stramineous tip; receptacle flat, chaffy, its bracts connate below, each one split above into bristles; cor hardly divided into tube and limb, with short lobes; filaments glabrous; anthers with a firm, narrow, apical appendage, evidently tailed at the base; style with an abrupt change of texture below the very short, erect, papillate branches; achenes basifixed, cylindric, truncate, nerveless, densely hairy; pappus a single series of coarse, plumose bristles, these chaffy-flattened toward the base and tending to be irregularly connate into groups; spiny herbs with alternate, toothed to pinnatifid lvs and medium-sized to rather large heads. 28, Eurasia and n. Afr.

1. **Carlina vulgaris** L. Taprooted biennial to 4(–7) dm, arachnoid-tomentose to subglabrous, simple or more often freely branched, the heads terminating the branches; lvs to 10(–15) cm, the lower tapering to a petiolar base, the others broadly sessile; stem not winged; invol 2–2.5 cm, the outer bracts like reduced lvs, the inner with long, slender, acuminate, brightly stramineous, eventually spreading tip; fls purple; achenes 3–4 mm; pappus 1 cm; 2*n*=20. Native of Europe, now well established as a weed in pastures and old fields near Ithaca, N.Y. and in nw. N.J., and to be expected elsewhere. Aug.

84. ECHINOPS L. Globe-thistle. Heads 1-fld, numerous, closely aggregated into globose secondary heads that often have a common invol of a few narrow, reflexed bracts; proper invol subtended by a tuft of capillary bristles, narrow, its bracts well imbricate, firm, often awn-tipped; fls tubular and perfect, the cor blue to purple or white, with slender tube and long, narrow lobes; anthers with a firm, narrow terminal appendage, shortly tailed at the base, the tails fringed-hairy; style with a ring of hairs and an abrupt change of texture below the rather short, papillate branches; achenes fusiform, hairy; pappus of numerous short setae or narrow scales, these distinct or ± united, forming a crown; spiny perennial thistles, generally ± white-tomentose, with alternate, toothed or pinnatifid lvs. 100, Old World.

1. **Echinops sphaerocephalus** L. Coarse, branching, to 2.5 m; stem spreading-hairy (some of the hairs gland-tipped), also tomentose above; lvs white-tomentose beneath, green and merely scabrous or hirsute above (many of the hairs gland-tipped), sessile and clasping (at least the middle and upper) not decurrent, pinnatifid, to 3.5 × 2 dm; secondary heads naked-pedunculate, 3.5–6 cm thick, pale bluish; proper invol 1.5–2 cm, its bracts puberulent on the back; subtending bristles seldom half as long as the invol; 2*n*=30, 32. Waste places; native of Eurasia, sometimes cult., and casually established as a weed here and there in e. U.S. and adj. Can. July–Sept.

E. exaltatus Schrader is also casual with us. It is much like *E. sphaerocephalus,* but lacks the glandular hairs, and the invol bracts are glabrous on the back.

85. CENTAUREA L. Star-thistle; knapweed. Heads discoid, the fls sometimes all tubular and perfect, or more often the marginal ones neutral, with enlarged, irregular, falsely radiate cor; invol bracts imbricate in several series, dry, sometimes spine-tipped, more often some of them with enlarged, scarious or hyaline, erose or lacerate or pectinate terminal appendage, seldom all merely rounded above and entire; receptacle nearly flat, densely bristly; cors anthocyanic to yellow or white, with slender tube, short throat, and long narrow lobes; anthers with a narrow, firm, apical appendage, evidently (though often shortly) tailed at base; styles with a thickened, often hairy ring and an abrupt change of texture at the base of the papillate branches; achenes obliquely or laterally attached to the receptacle, seldom evidently nerved; pappus of several series of graduated bristles or narrow scales, often much reduced, or wanting; herbs with alternate or all basal, entire to pinnatifid lvs. 500+, mostly Old World.

Carthamus tinctorius L., the safflower, readily distinguished by its foliaceous outer invol bracts and bright orange fls, occasionally escapes from cult.

1 Invol bracts not at all spiny, commonly ± lacerate or fringed.
  2 Perennial with a caudex and strong taproot; summit of the caudex conspicuously clothed with long,
    chaffy-fibrous, old lf-bases; pappus 3–6 mm .... 1. *C. scabiosa.*
  2 Annual to perennial, the perennials, except sometimes in no. 3, chiefly fibrous-rooted; old lf-bases,
    if persistent, very short.
    3 Pappus well developed, mostly 6–11 mm .... 2. *C. repens.*
    3 Pappus much reduced, ca 3 mm or less, or none.
      4 Lvs pinnatifid, with narrow lobes .... 3. *C. maculosa.*
      4 Lvs entire or toothed, or some of the larger ones few-lobed.
        5 Annual or winter-annual; lvs linear or nearly so, under 1 cm wide .... 4. *C. cyanus.*
        5 Perennial; lvs broader, some of the lower ones, at least, over 1 cm wide.
          6 Scarious tips of the invol bracts conspicuously blackish at least in part, those of the
          middle and outer ones regularly pectinate, seldom any of them obviously bifid.
            7 Scarious tips of the invol bracts small, 1–3 mm; heads relatively narrow, the pressed
            invol generally higher than broad; marginal fls enlarged and ray-like .... 5. *C. dubia.*
            7 Scarious tips of the invol bracts large, the larger ones (3–)4–6 mm; heads relatively
            broad, the invol broader than high; marginal fls not enlarged, except in a frequent
            hybrid with *C. jacea* .... 6. *C. nigra.*
          6 Scarious tips of the invol bracts tan to dark brown, those of the middle and outer ones
          irregularly lacerate, those of the inner ones expanded and often strongly bifid; mar-
          ginal fls generally enlarged and ray-like .... 7. *C. jacea.*
1 Invol bracts, or some of them, evidently spine-tipped; annual or biennial.
  8 Stem merely angled, not winged; pappus none.
    9 Central spine of each invol bract slender, mostly 1.5–4 mm .... 8. *C. diffusa.*
    9 Central spine stout, the larger ones mostly 1–3 cm .... 9. *C. calcitrapa.*
  8 Stem evidently winged by the decurrent lf-bases; pappus present, at least in the central fls.
    10 Larger spines of the invol bracts 11–22 mm; marginal fls epappose .... 10. *C. solstitialis.*
    10 Larger spines 5–9 mm; marginal fls, like the others, with pappus .... 11. *C. melitensis.*

1. **Centaurea scabiosa** L. Hard-heads. Taprooted perennial 3–15 dm, the caudex clothed with the long, chaffy-fibrous, persistent lf-bases of former years; herbage hirsute or subglabrous, and slightly arachnoid when young; lvs deeply pinnatifid, or the basal ones sometimes merely toothed; heads few or solitary, naked pedunculate, hemispheric or ovoid; invol 16–25 mm, its bracts with a conspicuous, blackish, scarious, pectinate-fringed tip; fls generally purple, the marginal ones usually enlarged; pappus 3–6 mm; $2n=20, 30, 40$. Fields and waste places; native of Europe, escaped from cult. and found here and there in the n. part of our range. June–Sept.

2. **Centaurea repens** L. Russian knapweed. Coarse, bushy-branched, colonial perennial from deep-seated creeping roots, 4–8 dm, finely arachnoid-tomentose, becoming glabrate; lvs rather small, the lower cauline ones to 15 × 4 cm and often few-toothed, the others numerous, smaller, entire or few-toothed; heads numerous, terminating the branches; invol pale, 9–15 mm, the middle and outer bracts broad, striate, glabrous, with large, broadly rounded, subentire hyaline tip, the inner bracts narrower, tapering to a plumose-hairy tip; fls purple, the marginal ones not enlarged; larger pappus-bristles subplumose above, 6–11 mm; achenes basilaterally attached; $2n=26$. Fields, roadsides, and waste places; native of Asia, now widely established in w. U.S. and occasional in our range. June–Sept. (*C. picris; Acroptilon r.*)

C. americana Nutt., basket-fl, an Ozarkian annual with large heads, long pappus, and conspicuously enlarged marginal fls, occasionally extends into our range as a casual introduction.

3. **Centaurea maculosa** Lam. Spotted knapweed. Biennial or short-lived perennial 3–15 dm, sparsely scabrous-puberulent, and with a thin and loose evanescent tomentum; lvs pinnatifid with narrow lobes, or the reduced ones of the infl entire; heads terminating the numerous branches, constricted upward in life; invol 10–13 mm, its bracts striate, the middle and outer ones with short, dark, pectinate tip; fls pink-purple, the marginal ones enlarged; pappus to 3 mm, or rarely none; $2n=18, 36$. Fields, roadsides, and waste places; native of Europe, now commonly established in our range and westward. June–Oct. (*C. rhenana; ? C. stoebe* L., an older name.)

C. moschata L., sweet sultan, rarely escapes from cult. It is a glabrous or obscurely tomentulose annual or biennial with toothed to pinnatifid lvs, long-pedunculate, white, yellow, or purple heads with enlarged marginal fls, and entire invol bracts with broadly rounded, narrowly scarious tip.

4. **Centaurea cyanus** L. Cornfl.; bachelor's button. Annual or winter-annual 2–12 dm, usually loosely white-tomentose when young, the lower lf-surfaces often persistently so; lvs narrow, often linear, entire, or the lower ones a little toothed or with a few narrow lobes, to 13 × 1 cm (excluding the lobes); heads terminating the branches; invol 11–16 mm, its bracts ± striate, with a relatively narrow, often darkened, pectinate or lacerate fringe near the tip; fls mostly blue, sometimes pink, purple, or white, the marginal ones enlarged;

pappus 2–3 mm; $2n=24$. Fields, roadsides, and waste places; native to the Mediterranean region, widely cult. and now a cosmopolitan weed.

5. **Centaurea dubia** Suter. Short-fringed knapweed. Habitally like the next 2 spp.; invol 11–18 mm high, generally higher than broad, even as pressed; appendages of the invol bracts conspicuously blackish at least in part, 1–3 mm, the middle and outer ones deeply and rather regularly pectinate; marginal fls enlarged and ray-like; pappus mostly wanting; $2n=22$, 24. Fields, roadsides, and waste places, now widely established in se. Can. and ne. U.S., s. to Va. and W.Va. July–Oct. (*C. nigrescens; C. vochinensis.*)

6. **Centaurea nigra** L. Black knapweed. Perennial, 2–8 dm, rough-puberulent and sometimes arachnoid when young; lvs entire or toothed, the basal ones broadly oblanceolate or elliptic, entire or toothed to sometimes few-lobed, mostly (1–)1.5–4(–6) cm wide, petiolate, the cauline ones reduced upward and becoming sessile; heads terminating the often numerous branches; invol 12–19 mm, broader than high; appendages of the invol bracts well developed; conspicuously blackish at least in part, the middle and outer deeply and fairly regularly pectinate, the larger ones mostly (3–)4–6 mm long, seldom any of them markedly bifid; fls pink-purple, the marginal ones typically not enlarged; pappus ca 1 mm or less; $2n=22$, 44. Fields, roadsides, and waste places; native of Europe, now widely established in s. Can. and n. U.S., s. to Va. and W.Va. July–Oct. Hybridizes with no. 7, producing segregating or stabilized intermediates called *C. × pratensis* Thuill., these often approaching *C. nigra* as to invol, but subradiate as in *C. jacea*. (*C. nigra* var. *radiata*.)

7. **Centaurea jacea** L. Brown knapweed. Habitally similar to *C. nigra*; invol 12–18 mm high, from a little narrower to a little broader than high; appendages of the invol bracts well developed, broad, tan to dark brown, the middle and outer ones rather irregularly lacerate, the inner ones less so and often deeply bifid; marginal fls almost always enlarged; pappus none; $2n=22$, 44. Fields, roadsides, and waste places; native of Europe, now widely established in s. Can. and n. U.S., s. to W.Va. June–Sept.

8. **Centaurea diffusa** Lam. Tumble-knapweed. Diffusely branched annual or biennial 1–6 dm; herbage sparsely scabrous-puberulent under the thin and deciduous arachnoid tomentum; lvs small, ± pinnatifid, the lower ones deciduous, the reduced ones of the infl mostly entire; heads numerous, narrow; invol 8–10 mm, the middle and outer bracts coarsely pectinate or weakly spinose-ciliate, and tipped with a defininte slender spine 1.5–4 mm; fls few, white, seldom yellowish or pinkish or red, the marginal ones not enlarged; pappus none; $2n=18$. A weed in waste places, native to se. Europe, and now found here and there in our range. July–Sept.

9. **Centaurea calcitrapa** L. Purple star-thistle. Branching biennial 1–8 dm, arachnoid-villous or glabrate; lvs small, pinnatifid with narrow lobes, or the upper entire; heads numerous, narrow; invol 10–18 mm, its bracts weakly spinose-ciliate and stoutly spine-tipped, the larger spines 1–3 cm; fls few, purple; pappus none; $2n=18$. Native of Europe, now widespread in the U.S. as a roadside weed. June–Sept.

10. **Centaurea solstitialis** L. Yellow star-thistle. Annual or biennial, 2–8 dm, thinly but persistently tomentose, the stem evidently winged by the decurrent lf-bases; basal lvs lyrate or pinnatifid, to 20 × 5 cm, the middle and upper smaller, becoming linear and entire; invol 10–15 mm, broad-based; middle and outer bracts spine-tipped, the larger central spines 11–22 mm; inner bracts with a small hyaline appendage; fls yellow, the marginal ones not enlarged; pappus of the marginal fls wanting, that of the others 3–5 mm; $2n=16$. A weed in fields and waste places, native to the Mediterranean region, occasionally found in our range.

11. **Centaurea melitensis** L. Maltese star-thistle. Annual 1.5–7 dm, lightly arachnoid when young and with some more persistent coarser crisp hairs; stem narrowly winged by the decurrent lf-bases; basal and lower cauline lvs oblanceolate, toothed to lyrate-pinnatifid, 3–15 cm × 5–35 mm, usually deciduous; middle and upper lvs smaller, becoming linear-oblong and entire; invol 8–15 mm, broad-based, its middle and outer bracts slenderly spine-tipped, the central spine 5–9 mm; inner bracts weakly spinose or merely tapering, not at all enlarged apically, generally purple-tinged; fls yellow; pappus 1.5–3 mm; $2n=24$, 36. A weed in waste places; native of the Mediterranean region, occasionally found in our range. June–Sept.

86. CNICUS L.

Blessed thistle. Heads essentially discoid, the fls yellow, all tubular and perfect except for a few inconspicuous marginal neutral ones with very slender, 2–3-lobed cor; invol of several series of broad, firm, spine-tipped bracts, the spines of the inner ones pinnatisect; receptacle flat, densely bristly; anthers with a firm, narrow, terminal appendage, shortly sagittate-tailed at the base; style with a ring of hairs and an abrupt change of texture at the base of the very short, scarcely divergent, papillate branches; achenes obliquely attached to the receptacle, subterete, strongly 20-ribbed, glabrous or sparsely hairy, with a firm, 10-toothed crown; pappus biseriate, the outer of 10 firm, smooth awns about as long as the achene, alternating with as many much shorter, minutely hairy and sparsely pectinate inner ones; prickly annuals with alternate, toothed or pinnatifid lvs, the heads terminating the branches, closely subtended by foliage lvs crowded beneath the proper invol. Monotypic.

1. **Cnicus benedictus** L. Branching annual 1.5–8 dm; step spreading-villous; lvs to 20 × 5 cm, the lower petiolate, the middle and upper sessile, scarcely or not at all decurrent; invol 3–4 cm, generally surpassed by the closely subtending ovate or lance-ovate lvs; achenes 6–8 mm; $2n=22$. Waste places; native to the

Mediterranean region, now sparingly established here and there in the U.S. and s. Can. May–Aug. (*Centaurea b.*)

87. PRENANTHES L. White lettuce; rattlesnake-root. Fls all ligulate and perfect, 5–35, pink or purple to white or pale yellow, the cor-tube more than half as long as the ligule; invol cylindric or seldom campanulate, of 4–15 principal bracts and some ± reduced outer ones; receptacle small, naked; achenes elongate, ± cylindric, glabrous, mostly reddish-brown, in our spp. ± ribbed striate; pappus of numerous deciduous capillary bristles; perennial herbs with milky juice, slightly to strongly tuberous-thickened roots, well developed alternate lvs (the larger ones at least 1 cm wide) and corymbiform or paniculiform to thyrsoid or subracemiform infl, the heads nodding or less often ± erect. 27, mainly N. Amer. and Eurasia. (*Nabalus.*)

1 Invol evidently pubescent with long coarse hairs.
  2 Infl cylindric, ± thyrsoid, the branches all very short.
    3 Heads nodding; principal invol bracts 5–6(–9); fls 5–8(–13) .... 10. *P. roanensis.*
    3 Heads loosely ascending or suberect, rarely nodding; fls and principal invol bracts mostly more
      numerous.
      4 Stem and lvs glabrous; fls usually purplish .... 1. *P. racemosa.*
      4 Stem and lower surfaces of the lvs rough-hairy; fls cream-color .... 2. *P. aspera.*
  2 Infl broader, corymbiform or paniculiform, at least some of its branches elongate.
    5 Fls (15–)19–27(–38); principal invol bracts 12–13 .... 3 *P. crepidinea.*
    5 Fls (8–)10–11(–14); principal invol bracts ca 8 .... 4. *P. serpentaria.*
1 Invol glabrous or apparently so.
  6 Fls mostly 8–18; principal invol bracts mostly 7–11, often 8.
    7 Plants 1–4 dm, alpine; invol black or blackish; n. N.Y. and northward.
      8 Lower lvs generally deeply lobed; peduncles and stem glabrous .... 6. *P. nana.*
      8 Lvs entire or merely toothed; infl villous-puberulent .... 7. *P. boottii.*
    7 Plants 4–15 dm, not alpine; invol generally not blackish; widespread.
      9 Infl (or its main branches) very slender and elongate; fls mostly pink or pale lavender .... 9. *P. autumnalis.*
      9 Infl broader and more corymbiform or paniculiform; fls only seldom pink.
        10 Pappus stramineous or light brown .... 5. *P. trifoliolata.*
        10 Pappus cinnamon-brown .... 8. *P. alba.*
  6 Fls mostly 5 or 6; principal invol bracts (4)5(6) .... 11. *P. altissima.*

    1. Prenanthes racemosa Michx. Glaucous w.l. Plants 3–17 dm, long-hairy in the infl, otherwise glabrous and glaucous; lowest lvs petiolate, broadly oblanceolate to obovate or elliptic, the blade and petiole 7–40 × 1.5–10 cm, generally persistent; cauline lvs ± reduced upwards, becoming sessile and clasping; infl narrow and elongate, thyrsoid, the heads rather loosely ascending or occasionally nodding, with pink or purplish (white) fls; invol 9–14 mm, purplish or blackish, sparsely to usually rather densely long-hirsute; $2n=16$. Streambanks, moist meadows, and prairies; Que. to N.J., w. to Alta. and Colo. Aug., Sep. Var. *racemosa,* with ca 8(7–10) invol bracts and ca 13(9–16) fls per head, occurs from Que. to N.J., w. to Minn. and Io. Var. *multiflora* (Cronq.) Cronq., with ca 13(10–14) invol bracts and ca 21(17–26) fls per head, occurs from Alta. and Colo. to Minn. and Io., and occasionally through the n. part of our range to Que. and n. Me.

    2. Prenanthes aspera Michx. Rough w.l. Stem strict, 5–17 dm, rough-hairy or scabrous at least above; lvs scabrous or coarsely hirsute on the lower and often also the upper side, toothed or entire, the lower ones well developed and somewhat obovate, tapering to the petiole, but soon deciduous, the others sessile or nearly so and often clasping, oblong to elliptic or lanceolate, gradually reduced upward, the larger ones 4–11 × 1–5 cm; infl narrow and elongate, thyrsoid-racemiform, the heads crowded, loosely ascending to suberect; invol 12–17 mm, coarsely and usually densely long-hairy, rather pale, with (6–)8(–10) principal bracts, fls (8–)11–14(–19), ochroleucous; pappus stramineous. Dry prairies; O. to Minn., S.D., Okla., and La. Aug., Sept.

    3. Prenanthes crepidinea Michx. Midwestern w.l. Stem 1–2.5 m, glabrous below the somewhat glandular-puberulent or villous-puberulent infl; lvs glabrous or scabrous above, generally short-hirsute at least on the midrib and main veins, most of them petiolate, elliptic to deltoid, cordate, or hastate, coarsely dentate to occasionally shallowly lobed or entire, only gradually reduced upward, the blade to 25 × 20 cm; infl corymbose-paniculiform; heads nodding; invol 12–16 mm, corasely hirsute-setose, with (10–)12–13(–15) pale green or usually ± black-tinged principal bracts; fls (15–)19–27(–38), ochroleucous; pappus tan or sordid. Moist woods, thickets, and prairies; w. N.Y., w. Pa., and W.Va. to s. Wis., n. Ark., and Ky. Aug., Sept.

    4. Prenanthes serpentaria Pursh. Lion's foot. Stem mostly 5–15 dm, glabrous, or often rough-hairy in the infl; lvs glabrous or inconspicuously hairy, well distributed along the stem, mostly wing-petiolate and with a broad-based, usually pinnately few-lobed blade to 17 × 10 cm, often with large, ± rounded lobes expanded above the base, varying to occasionally merely toothed and tapering to a shortly wing-petiolar base, the basal ones sometimes trifoliate and again cleft; infl mostly rather open-paniculiform, with elongate, ascending branches; heads nodding; invol 10–13(–15) mm, its principal bracts ca 8, with a few (sometimes very few) long, coarse hairs, often speckled with fine black dots, but without the larger, waxy-looking papillae of no. 5; reduced outer bracts avg narrower and a little longer than in no. 5, often more than 2 mm, and often more than 2.5 times as long as wide; fls (8–)10–11(–14), generally ochroleucous or chloroleucous; pappus stramineous; $2n=16$. Woods, especially in sandy soil; Mass. to n. Fla., w. to Ky., Tenn., and Miss.; rare n. of Md. Aug.–Oct. (*Nabalus integrifolius,* with merely toothed lvs.)

5. Prenanthes trifoliolata (Cass.) Fern.   Gall-of-the-earth. Stem stout, 4.5–12+ dm, glabrous; lvs glabrous above, paler and sometimes a little hairy beneath, highly variable in size and shape, the lower long-petioled, pinnately or palmately few-lobed, usually rather deeply so., or trifoliolate and again cleft, varying to occasionally hastate or merely toothed, the middle and upper ones progressively smaller, less petiolate, and less cut; infl elongate-paniculiform; heads nodding; invol 10–14 mm, with (7)8(9) principal bracts, these glabrous, often purplish, generally provided with minute, white, waxy-appearing papillae (often barely visible at 10×) and sometimes also black-spotted (as in no. 4); reduced outer bracts relatively short and broad, seldom over 2(2.5) mm or more than 2(2.5) times as long as wide; fls (9)10–12(13), white to ochroleucous or chloroleucous or grayish, seldom pale lavender; pappus stramineous or pale brown; $2n=16$. Woods, especially in sandy soil; Nf. to Md. and Pa., and s. along the coast to N.C., and in the mts. to N.C., Tenn., and n. Ga. Aug., Sept. An apparent hybrid with no. 1, found in Me., N.S., and N.B., is *P. × mainensis* A. Gray.

6. Prenanthes nana (Bigelow) Torr.   Dwarf cut-lf w.l. Plant 1–4 dm, glabrous, or the lvs occasionally hairy on the midrib and main veins beneath; lvs pale beneath, the lower long-petiolate, deeply few-lobed, the others becoming short-petiolate or subsessile and merely toothed; infl narrow and usually elongate, somewhat thyrsoid, the heads nodding; invol 8–13 mm, blackish-green or black, the pigment spots minute and usually crowded, generally resolvable into scarcely elongate papillae at 50×; principal bracts (7)8(–10), glabrous, the reduced outer ones varying from calyculate as in no. 5 to longer as in no. 7; achenes ribbed-striate; fls 9–12, ochroleucous; pappus stramineous; $2n=16$. Alpine sites; n. N.Y. to Nf. July, Aug. (*P. trifoliolata* var. *n.*)

7. Prenanthes boottii (DC.) A. Gray.   Dwarf simple w.l. Stem 1–3(–4) dm, glabrous below, villous-puberulent above, at least in the infl; lvs glabrous, pale beneath, entire or merely toothed, the lower with deltoid or cordate blade 1.5–5 × 1–3 cm, exceeded by the petiole; infl narrow, somewhat thyrsoid; heads mostly nodding; invol 8–12 mm, black or blackish-green, the pigment-spots minute, crowded, generally resolvable into short parallel lines at 50×; principal bracts 8–11(–13), minutely puberulent at the tip, otherwise glabrous; some of the reduced outer bracts generally at least half as long as the main ones; fls 9–18, whitish; achenes evidently several-ribbed, but only obscurely striate; $2n=32$. Alpine sites; n. N.Y. to Me. July, Aug.

8. Prenanthes alba L.   Rattlesnake-root. Stem stout, 4–17 dm; herbage ± glaucous; lvs glabrous above, often hairy beneath,, highly variable in size and shape, the lower long-petioled, palmately or pinnately few-lobed to sagittate or hastate-reniform and merely coarsely toothed, becoming smaller, less cut, and less petiolate upward; infl elongate-paniculiform; heads nodding; invol 11–14 mm, with mostly 8 principal bracts, these glabrous, generally somewhat purplish, ± densely papillate with white, waxy-appearing cells that are usually distinguishable at 10×; reduced outer bracts, as in no. 4, mostly a little longer and narrower than in no. 5; fls 9–11(–15), mostly white to pinkish or lavender; pappus cinnamon-brown; $2n=32$. Woods; Me. to Man., s. to N.C., W.Va., and Mo. Aug., Sept.

9. Prenanthes autumnalis Walter.   Slender rattlesnake-root. Plants 4–14 dm, glabrous throughout; basal and lowermost cauline lvs elongate, 7–35 × 1–12 cm, few-toothed to generally pinnately lobed, the lobes often rather narrow and distant; cauline lvs conspicuously reduced upward, often becoming linear and entire; infl very narrow and elongate, either subracemiform, or in more robust plants with closely ascending, subracemiform branches; heads nodding; invol 10–13 mm, generally purplish, without long hairs, with mostly 8 principal bracts, the reduced outer ones often passing into those of the slender peduncle; fls (8–)10–11(–13), pink or pale lavender; pappus stramineous. Sandy, usually moist places, often among pines; coastal plain from N.J. to n. Fla. Aug.–Oct. (*P. virgata.*)

10. Prenanthes roanensis (Chick.) Chick.   Appalachian w.l. Stem 1.5–6(–10) dm, glabrous below the infl, or sometimes conspicuously long-hairy; lvs mostly glabrous or nearly so, seldom evidently hairy, the lower and middle ones petiolate, with deltoid to mostly sagittate or hastate, toothed and sometimes deeply lobed or cleft blade commonly 3–10 × 2–8 cm; infl elongate and narrow, thyrsoid-racemiform, leafy-bracteate at least below, the branches all very short; heads nodding; invol 10–12 mm, coarsely hirsute-setose, with 5–6(–9) principal bracts, these pale green with a darker median line and tip; fls 5–8(–13), ochroleucous; pappus stramineous; $2n=16$. Rich woods and moist, open slopes at high elev. in the s. Appalachian Mts., mainly or wholly in the Blue Ridge province; N.C. and Tenn. to sw. Va. and reputedly Ky. Aug., Sept. (*P. cylindrica.*) Intergrades with no. 11.

11. Prenanthes altissima L.   Tall w.l. Stem 4–20 dm, glabrous or often spreading-hirsute toward the base; lvs thin, glabrous above, often hirsute on the midrib and main veins beneath, the lower long-petioled, highly variable in size and shape, commonly with deltoid to sagittate or cordate, merely mucronate-toothed blade 4–15 × 2.5–16 cm, or sometimes deeply few-lobed, gradually reduced upward, becoming short-petiolate or subsessile and elliptic; infl elongate-paniculiform; heads nodding; invol glabrous, 9–14 mm, with (4)5(6) principal bracts, generally dark-tipped; fls 5–6, ochroleucous or chloroleucous; $2n=16$. Woods; Nf. to Ga., w. to Mich., Mo., Ark., and La. Aug., Sept. Most of our plants are var. *altissima,* with whitish to pale brown pappus. Southwestward this passes into var. *cinnamomea* Fern., chiefly of Mo., Ark., and La., with bright yellow-brown or almost orange pappus.

88. LYGODESMIA D. Don. Rush-pink. Fls all ligulate and perfect, 5–12, pink or purple, rarely white; invol cylindric, with 4–10 principal bracts and a few much reduced outer ones; achenes glabrous or scabrous, slender, columnar to fusiform; pappus of numerous capillary bristles; ± rush-like herbs with milky juice and alternate (or the lower opposite), narrow, mostly linear or subulate lvs; heads erect, terminating the branches, or sometimes in a racemiform infl. 10, N. Amer.

1 Colonial perennial from deep-seated creeping roots; ligules showy, 10–12 × 4 mm .... 1. *L. juncea.*
1 Taprooted annual; ligules inconspicuous, 5–7 × 1 mm .... 2. *L. rostrata.*

1. Lygodesmia juncea (Pursh) D. Don. Skeleton-weed. Glabrous colonial perennial from deep-seated creeping roots; stem 1.5–6 dm, grooved-striate, much branched; lvs all alternate, linear, to 4 cm × 3 mm, the upper reduced to subulate scales; heads terminating the branches; invol 10–16 mm, with 4–8 (chiefly 5) principal bracts; fls (4)5(–10), the ligule pink (white), spreading, mostly 10–12 × 4 mm; achenes columnar, several-nerved; 2*n*=18. Prairies and other dry, open places, often in sandy soil; Minn. to Ark., w. to Man., B.C., and Ariz. June–Sept.

2. Lygodesmia rostrata (A. Gray). A. Gray. Annual skeleton-weed. Taprooted, glabrous annual to 8 dm, freely branched at least above; lvs linear, to 15 cm × 4 mm, the lower several pairs opposite, the others alternate, the uppermost reduced to linear bracts; heads terminating the branches; invol 12–17 mm, with 7 or 8 principal bracts; fls 8–11, the ligule lavender with white tip, 5–7 × 1 mm, ± erect; achenes columnar-fusiform, 8–10-ribbed, scabrellous near the tip; 2*n*=12. Open, sandy places; w. Io. to w. Minn., N.D., n. Tex., Wyo., and Utah. July–Sept. (*Shinnersoseris r.*)

89. LACTUCA L. Lettuce. Fls all ligulate and perfect, yellow, blue, or white, the cor-tube generally more than half as long as the ligule; invol cylindric, often broadened at the base in fr, generally imbricate, occasionally merely calyculate; achenes compressed, winged or strongly nerved marginally, with 1 or several lesser nerves in each face, beaked or sometimes beakless, but in any case expanded at the summit where the pappus is attached; pappus of 2 rows of usually equal capillary bristles, none markedly coarser than the others; lactiferous herbs with alternate, entire to pinnatifid lvs and usually numerous heads with relatively few fls (5–56 per head in our spp.), the infl mostly paniculiform. (*Mulgedium, Mycelis.*) 50, Eurasia, Afr., and N. Amer.

1 Achenes with only a median nerve on each face, occasionally with an additional pair of very obscure
  ones.
  2 Heads relatively large, the invol 15–22 mm; achenes 7–10 mm (beak included).
    3 Lvs ± prickly-toothed; fls 20–56 per head .... 1. *L. ludoviciana.*
    3 Lvs toothed but scarcely prickly; fls 13–25 per head .... 2. *L. hirsuta.*
  2 Heads smaller, the invol 10–15 mm in fr; achenes 4.5–6 mm .... 3. *L. canadensis.*
1 Achenes evidently to very prominently several-nerved on each face.
  4 Perennial; heads large, the invol 15–20 mm in fr; fls blue; lvs not sagittate .... 4. *L. pulchella.*
  4 Annual or biennial; heads small, the invol 9–15 mm in fr (to 18 mm in no. 8, which has distinctly
    sagittate lvs); fls yellow, blue, or whitish.
    5 Achenes with a short stout beak less than half as long as the body, or beakless.
      6 Fls mostly 11–34(–55) per head.
        7 Pappus white; infl ample, open-paniculiform .... 5. *L. floridana.*
        7 Pappus light brown; infl elongate, narrowly paniculiform .... 6. *L. biennis.*
      6 Fls 5 per head (unique among our spp.) .... 9. *L. muralis.*
    5 Achenes with a filiform beak from nearly as long to twice as long as the body.
      8 Lvs prickly-margined; achenes spinulose above, at least at maturity .... 7. *L. serriola.*
      8 Lvs not prickly-margined; achenes not spinulose .... 8. *L. saligna.*

1. Lactuca ludoviciana (Nutt.) DC. Prairie-l. Leafy-stemmed biennial or possibly short-lived perennial, the stem glabrous, 3–15 dm; lvs prickly or coarsely hairy on the midrib and sometimes also on the main veins beneath, ± prickly-toothed on the margins and usually also pinnately lobed, mostly 7–30 × 3–20 cm; heads relatively large, with 20–56 yellow or occasionally blue or pale lilac fls; invol 15–22 mm in fr; achenes strongly flattened, blackish, with a median nerve on each face, transversely rugulose, 7–10 mm overall, the slender beak equaling or a little shorter than the body; pappus 7–10 mm; 2*n*=34. Prairies and other open places; w. Ind. to Man., Mont., Colo., and Tex., casually intr. eastw. July–Sept. (*L. campestris,* the blue-fid form.)

2. Lactuca hirsuta Muhl. Hairy tall l. Similar to no. 3, often more hairy, and with some tendency for the lvs to be basally disposed; lvs pinnatisect, with broad, toothed lobes that are commonly more than 1 cm wide and often somewhat narrowed at the base; heads larger, the invol 15–22 mm in fr, the achenes 7–9 mm, the mature pappus 8–12 mm; $2n=34$. Dry open woods and clearings; P.E.I. to n. Fla., w. to Mich., Mo., and Tex. June–Sept. (Var. *sanguinea,* the less hairy phase.)

3. **Lactuca canadensis** L.   Tall l. Leafy-stemmed annual or usually biennial 3–25 dm, glabrous or occasionally coarsely hirsute, often ± glaucous; lvs highly variable, entire or toothed to pinnately lobed or cleft, sagittate or sometimes narrowed to the base, 10–35 × 1.5–12 cm; heads numerous, small, with 13–22 yellow fls; invol 10–15 mm in fr; achenes blackish, very flat, with a median nerve on each face, transversely rugulose, 4.5–6.5 mm overall, the slender beak from a little more than half to about as long as the body; mature pappus 5–7 mm; $2n=34$. Fields, waste places, and woods; Que. and P.E.I. to Sask., s. to Fla. and Tex. July–Sept. (*L. sagittifolia; L. steelei.*)

4. **Lactuca pulchella** (Pursh) DC.   Blue l. Perennial 2–10 dm from deep-seated creeping roots, glabrous or glabrate; lvs elongate, 5–18 cm × 6–35 mm, entire, or the lower ones ± lobed, rarely sharply denticulate, often glaucous beneath; heads several or numerous, very showy, with 18–50 blue fls; invol 15–20 mm in fr; achenes 4–7 mm overall, the slender body moderately compressed, prominently several-nerved on each face, the beak stout, often whitish, up to as long as the body; $2n=34$. Mostly in meadows, thickets, prairies, and other moist low places; Alas. to Calif., e. to Minn. and Mo., and occasionally intr. eastward. June–Sept. (*L. tatarica* ssp. *p.; L. oblongifolia.*)

5. **Lactuca floridana** (L.) Gaertner.   Woodland l. Robust, leafy-stemmed annual or biennial 5–20 dm; lvs mostly petiolate and not sagittate at the base, often hairy along the main veins beneath, otherwise glabrous or nearly so, the blade 8–30 × 2.5–20 cm, from elliptic (or cordate) and merely toothed to evidently pinnatifid; heads numerous in an ample, paniculiform infl, with 11–17(–27) blue (white) fls; invol 9–14 mm in fr; achenes 4–6 mm, narrowed upward, beakless or with a stout beak sometimes a third as long as the body, several-nerved on each face, tending to be thickened on the margins; pappus white; $2n=34$. Thickets, woods, and moist, open places; N.Y. to Fla., w. to Minn., Kans., and Tex. June–Sept. (*L. villosa,* the form with merely toothed lvs.)

6. **Lactuca biennis** (Moench) Fern.   Tall blue l. Robust, leafy-stemmed annual or biennial, 6–20 dm, glabrous, or the lvs hairy on the main veins beneath; lvs pinnatifid or occasionally merely toothed, 10–40 × 4–20 cm, sometimes sagittate at the base; heads numerous in an elongate, rather narrow, paniculiform infl, often crowded, with 15–34(–55) bluish to white or occasionally yellow fls; invol 10–14 mm in fr; achenes 4–5.5 mm, thin-edged, prominently several-nerved on each face, tapering to a beakless or shortly stout-beaked tip; pappus light brown; $2n=34$. Moist places; Nf. to B.C., s. to N.C. (chiefly in the mts.), Ill., Colo., and Calif. (*L. spicata,* misapplied.) An apparent hybrid with no. 3 has been called *L.* × *morssii* Robinson.

7. **Lactuca serriola** L.   Prickly l. Leafy-stemmed annual or biennial, 3–15 dm, the stem often prickly below, otherwise glabrous; lvs prickly on the midrib beneath, and more finely prickly-toothed on the margins, otherwise generally glabrous, pinnately lobed or lobeless, commonly twisted at base to stand vertically, sagittate-clasping, oblong or oblong-lanceolate in outline, the upper much reduced; heads numerous in a long, often diffuse infl, with (13–)18–24(–27) light yellow fls, drying blue; invol 10–15 mm in fr; achenes gray or yellowish-gray, the body compressed, 3–4 mm, a third as wide, prominently several-nerved on each face, spinulose or hispidulous above at least marginally, the slender beak about equaling the body, rarely twice as long; $2n=18$. A weed in fields and waste places; native of Europe, now naturalized throughout most of the U.S. July–Sept. The common form with lobeless lvs, called var. *integrata* Gren. & Gordon, may reflect introgression from *L. sativa* L., cultivated lettuce, with which *L. serriola* hybridizes freely. (*L. scariola.*)

8. **Lactuca saligna** L.   Willowlf.-l. Leafy-stemmed annual or biennial, the stem glabrous, 3–10 dm, more slender than in no. 7; lvs conspicuously sagittate, linear and entire or with scattered, narrow, sometimes slightly toothed lobes, 6–15 cm, the rachis (or the lf, when entire) 3–8 mm wide, glabrous or with a few prickles on the midrib beneath and sometimes minutely stellate marginally; heads numerous in a long infl; fls 8–16, light yellow, drying blue; invol 12–18 mm in fr; achenes as in no. 7, but avg slightly smaller and narrower, merely scabrous distally, the beak community twice as long as the body; $2n=18$. A weed in waste places; native of Europe, now found here and there in our range. Aug., Sept.

9. **Lactuca muralis** (L.) Fresen.   Wall-l. Slender, glabrous annual or biennial 3–10 dm, the lower surfaces of the lvs and often also the stem ± glaucous; basal and lower cauline lvs 6–18 × 3–8 cm, pinnatifid, with a broad, somewhat ivy-like or maple-like terminal segment, tending to be auriculate at base; middle and upper lvs few and much reduced; heads several or usually numerous in a corymbiform or more often paniculiform infl, narrow, with 5 yellow fls; invol merely calyculate, 9–11 mm in fr; achenes 4–5 mm, including the scarcely 1 mm beak, several-nerved on each face; outer row of pappus-bristles shorter than the inner; $2n=18$. Moist places; native of n. Europe, now known from Que. and Me. to N.Y., w. to Mich. (*Mycelis m.*)

90. **IXERIS** (Cass.) Cass.   Fls all ligulate and perfect, yellow, blue, or white, the tube one-fourth to one-half as long as the ligule; invol cylindric, mostly calyculate rather than imbricate; achenes somewhat compressed, long-beaked, the 10 nerves ± wing-elevated; pappus of capillary bristles, multicellular in cross-section at least near the base; slender, lactiferous herbs, mostly scapose or subscapose, with solitary or few heads in a corymbiform infl. 50, Asia and E. Indies.

1. **Ixeris stolonifera** A. Gray. Creeping lettuce. Delicate, creeping, glabrous perennial; lvs petiolate, with broadly elliptic to orbicular, entire or obscurely toothed blade 0.5–3 cm; heads solitary or paired on peduncles ca 1 dm or less; invol 7–12 mm; fls yellow; achenes reddish-brown, moderately compressed, 4–6 mm, the slender beak about equal to the body; $2n=24$. A weed in lawns; native of Japan, occasionally intr. in e. U.S., as in Del., Pa., and N.Y. June–Aug. (*Lactuca s.*)

91. SONCHUS L. Sow-thistle.

Fls all ligulate and perfect, yellow, few to more often numerous (80+ in all our spp.); invol ovoid or campanulate, rarely narrower, its bracts generally imbricate, occasionally merely calyculate, often basally thickened in age; achenes flattened, 6–20-ribbed, merely narrowed at the tip, beakless, glabrous, often transversely rugulose; pappus of numerous white, capillary, often somewhat crisped bristles that tend to fall connected, and some stouter outer bristles that fall separately; lactiferous herbs with alternate or all basal, entire to dissected, mostly auriculate, often prickly-margined lvs, and solitary to usually several or many, medium-sized to rather large heads in an irregularly corymbose-paniculiform to subumbelliform infl. 50, Old World.

1 Perennial, deep-rooted and creeping below-ground; heads relatively large, mostly 3–5 cm wide in fl,
   the fruiting invol 14–22 mm high . . . . 1. *S. arvensis.*
1 Annual; heads smaller, 1.5–2.5 cm wide in fl, the fruiting invol 9–13 mm.
   2 Mature achenes transversely tuberculate-rugose as well as several-nerved . . . . 2 *S. oleraceus.*
   2 Mature achenes merely several-nerved, not rugulose . . . . 3. *S. asper.*

1. **Sonchus arvensis** L. Perennial s.-t. Perennial, 4–20 dm, with long vertical roots and spreading by creeping roots, glabrous at least below the infl and often somewhat glaucous; lvs prickly-margined, the lower and middle ones usually pinnately lobed or pinnatifid, 6–40 × 2–15 cm, becoming less lobed and more auriculate upward, the upper reduced and distant; heads several in an open-corymbiform infl, relatively large, 3–5 cm wide in fl; fls ca 150–235 per head; fruiting invol 14–22 mm; achenes 2.5–3.5 mm, with 5+ prominent ribs on each face, strongly rugulose; $2n=36, 54$. A European weed, now widely intr. in temperate N. Am. July–Oct. Var. *arvensis* is ± copiously provided with coarse, spreading, gland-tipped hairs on the invol and peduncles. Var. *glabrescens* (Guenther) Grab. & Wimmer lacks these hairs. (*S. uliginosus.*)

2. **Sonchus oleraceus** L. Common s.-t. Annual 1–20 dm from a short taproot, glabrous except sometimes for a few spreading gland-tipped hairs on the invols and peduncles; lvs pinnatifid to occasionally merely toothed, soft, the margins only weakly or scarcely prickly, 6–30 × 1–15 cm, all but the lowermost ones prominently auriculate, the auricles well rounded but eventually sharply acute; lvs progressively less divided upward and ± reduced; heads several in a corymbiform infl, relatively small, 1.5–2.5 cm wide in fl; invol 9–13 mm in fr; fls 80–250 per head; cor-tube about equaling the ligule; achenes 2.5–3.5 mm, transversely rugulose and evidently to rather obscurely 3–5-ribbed on each face; $2n=32, 36$. A cosmopolitan weed, native to Europe. July–Oct.

3. **Sonchus asper** (L.) Hill. Prickly s.-t. Much like no. 2, commonly more prickly; lvs pinnatifid, or frequently obovate and lobeless, with rounded and conspicuously prickly-toothed, not acute auricles; cor-tube somewhat longer than the ligule; achenes flatter, with 3(4–5) evident ribs on each face, not rugulose, although there may be minute projections from the marginal ribs; $2n=18$. A cosmopolitan weed, native to Europe. July–Oct.

92. HIERACIUM L. Hawkweed.

Fls all ligulate and perfect, yellow to sometimes orange or red; invol cylindric to hemispheric, its bracts ± imbricate; achenes terete or prismatic, mostly narrowed below, truncate or occasionally narrowed toward the summit, ± strongly ribbed-sulcate; pappus of numerous whitish to more often sordid or brownish capillary bristles; fibrous-rooted perennial herbs from an elongate or very short rhizome (or mere crown), with milky juice, and alternate or all basal, entire or ± toothed lvs; heads in various sorts of infls, or solitary; most spp. with at least a few stellate hairs on the herbage or invol. (*Pilosella.*) Thousands of names, perhaps hundreds of spp., mostly in temperate (or mountainous tropical) regions. Spp. 1–6 (all intr.) belong to a huge polyploid-apomictic complex without clear specific boundaries, and spp. 7–12 (some native, some intr.) to another. Spp. 13–18, all native, are diploid so far as known; they produce occasional hybrids, some of which have been named.

1 Plants with a well developed cluster of basal lvs at fl time, these generally as large as or larger than
   any of the upper ones; infl corymbiform (or the heads solitary), not elongate.
2 Plants stoloniferous; rhizome generally elongate.
   3 Heads solitary (rarely 2 or 3 on some individuals) . . . . 1. *H. pilosella.*
   3 Heads 2–many (rarely solitary on some individuals of no. 2).

4 Heads mostly 2–6, long-pedunculate; lvs glabrous or nearly so above; fls yellow .... 2. *H. flagellare.*
4 Heads (3–)5–50, short-pedunculate in a compact, corymbiform infl.
  5 Fls yellow.
    6 Herbage glaucous; lvs glabrous or nearly so above .... 3. *H. floribundum.*
    6 Herbage not glaucous; lvs hairy on both sides .... 4. *H. caespitosum.*
  5 Fls red-orange, becoming deeper red in drying .... 5. *H. aurantiacum.*
2 Plants not stoloniferous; rhizome generally short and praemorse.
  7 Achenes 1.5–2.0 mm; basal lvs 5–12 times as long as wide (petiole included), rarely if ever over
    3 cm wide.
    8 Herbage glaucous, sparsely hairy to subglabrous .... 6. *H. piloselloides.*
    8 Herbage not glaucous; leaves hairy on both sides .... 4. *H. caespitosum.*
  7 Achenes 2.2–4 mm; basal lvs broader, 1.7–5 times as long as wide, often over 3 cm wide.
    9 Fls 40–80 per head; at least some of the lvs generally sharply toothed near the base; intr.
      weeds, except no. 9.
      10 Basal lvs broadly rounded or truncate to cordate at base .... 7. *H. murorum.*
      10 Basal lvs (as well as the cauline) tapering to the petiole.
        11 Spreading hairs of the peduncle mostly gland-tipped; basal lvs 1.5–5 cm wide; plants
          1.5–10 dm; widespread .... 8. *H. lachenalii.*
        11 Spreading hairs of the peduncle glandless; basal lvs 2 cm wide or less; plants 1–3.5
          dm; n. N.H. and northward .... 9. *H. robinsonii.*
    9 Fls 15–40 per head; lvs entire or merely callous-denticulate.
      12 Basal lvs evidently purple-veined above in life; invol glabrous to sometimes evidently
        stipitate-glandular, but without long, glandless setae .... 14. *H. venosum.*
      12 Basal lvs not purple-veined; invol stipitate-glandular and with long, glandless setae .... 13. *H. traillii.*
1 Plants either without a well developed cluster of basal lvs at fl time, or with an elongate, cylindric
  infl, or both.
  13 Achenes distinctly narrowed toward the summit.
    14 Hairs of the lower part of the stem mostly 1 cm or more; fls 40–90 per head ....17. *H. longipilum.*
    14 Hairs seldom approaching 1 cm; fls mostly 20–40 per head .... 16. *H. gronovii.*
  13 Achenes truncate, only very obscurely if at all narrowed upwards.
    15 Fls mostly 8–30 per head .... 15. *H. paniculatum.*
    15 Fls mostly 40–110 per head.
      16 Peduncles densely and conspicuously long-stipitate-glandular; lower lvs the largest, the
        others progressively reduced upwards .... 18. *H. scabrum.*
      16 Peduncles merely stellate, with few or usually no long gland-tipped hairs; middle and
        lower lvs rather similar in size and shape.
        17 Lower part of the stem and lower lf-surfaces provided with long, firmly bulbous-
          based hairs, not at all stellate; intr. weed .... 10. *H. sabaudum.*
        17 Plant without long, bulbous-based hairs; lvs usually but not always finely stellate;
          widespread native, occasionally weedy spp.
          18 Lvs copiously provided, at least marginally, with short, stout, blunt, subconic
            hairs, as well as usually ± stellate; boreal .... 11. *H. umbellatum.*
          18 Lvs without subconic hairs, often stellate or long-hairy or both.
            19 Invol bracts gradually tapering to a narrow but obtuse or rounded tip; plants
              1.5–15 dm; widespread .... 12. *H. kalmii.*
            19 Invol bracts gradually tapering to a sharp point; plants 1–3.5 dm; northeast-
              ern .... 9. *H. robinsonii.*

1. **Hieracium pilosella** L.   Mouse-ear h. Abundantly stoloniferous, and with a long, slender rhizome; stem 3–25(–40) cm, lfless or with a single generally much reduced lf, viscid-puberulent or subtomentose, sparsely or moderately spreading-hispid with gland-tipped, usually blackish hairs, often long-setose as well; basal lvs oblanceolate or a little broader, 2–13 × 0.6–2 cm, tawny-tomentose with stellate hairs beneath, at least when young, and with some long glandless setae as well, green and glabrous above except for the long setae; lvs of the stolons similar but smaller; heads solitary, or rarely 2–3 and long-pedunculate on some individuals; invol 7–11 mm, stellate, short-hispid with black, sometimes gland-tipped hairs, and occasionally long-setose as well; achenes 1.5–2 mm, truncate; $2n=18, 36, 45, 54, 63$. A weed in pastures and fields; native of Europe, now widespread in our range. May–Sept.

2. **Hieracium flagellare** Willd.   Whiplash-h. Vigorously stoloniferous and with a slender, elongate or short and praemorse rhizome, 1.5–4 dm tall, somewhat glaucous; stem nearly naked, pale, finely stellate and with scattered spreading bristles, generally also with some shorter, gland-tipped setae, especially upwards; basal lvs oblanceolate, 3–13 × 0.5–2.5 cm, green and sparsely setose or virtually glabrous above, pale, thinly stellate, and with scattered long setae (especially along the midrib) beneath; heads 2–6, on ± elongate peduncles in an open infl; invol 8–12 mm, inconspicuously stellate, copiously beset with rather short, blackish, gland-tipped setae and often also with some longer glandless ones; $2n=26, 45, 54$. A weed in fields and along roadsides, etc.; native of Europe, now sparingly intr. from N.Y. to Que. May, June.

H. lactucella Wallr., another European sp., has also been collected in the ne. part of our range. It differs from no. 2 in its smaller, more congested heads (invol 5–8 mm) and in the absence of stellate hairs from the lvs. (*H. auricula,* misapplied.)

3. **Hieracium floribundum** Wimmer & Grab.   Glaucous h. Stoloniferous and with a long rhizome, the stem 2–8 dm, naked or with one or 2 reduced lvs, sparsely long-setose, the peduncles commonly also hispid

with blackish, gland-tipped hairs and somewhat stellate; herbage ± glaucous; basal lvs oblanceolate or narrowly elliptic, 3–14 cm (petiole included) × 6–20 mm, generally long-setose below (at least along the midrib) and on the margins toward the base, glabrous or nearly so above; lvs of the stolons much smaller and sometimes more hairy; heads 3–50 in a corymbiform, generally compact infl; invol 5–9 mm, hispid with blackish, generally gland-tipped hairs, slightly stellate-tomentose and sometimes also long-setose; achenes to 2 mm, truncate; $2n=27$. Meadows, roadsides, and fields; native of Europe, established from N.B. and Me. to N.Y. and reputedly O. June–Aug. Thought to have originated by hybridization of no. 4 with *H. lactucella*.

4. Hieracium caespitosum Dumort.   Yellow king-devil. Plants with a short or more often elongate rhizome and commonly with short, stout stolons; stems 1–several, 2.5–9 dm, long-setose, becoming stellate-tomentose and blackish-glandular-hispid above, naked or with only 1 or 2(3) reduced lvs; basal lvs oblanceolate or narrowly elliptic, 5–25 cm (petiole included) × 1–3 cm, mostly 5–12 times as long as wide, long-setose on both sides, sometimes sparsely so above, commonly slightly stellate beneath; heads mostly 5–30 in a compact, corymbiform infl; invol 6–8 mm, hispid with blackish, gland-tipped hairs, commonly also sparsely long-setose and slightly stellate; fls bright yellow; achenes 1.5–2 mm, truncate; $2n=18, 27, 36, 45$. A weed in fields, pastures, and along roadsides, occasionally in dry woods; native of Europe, now widespread in se. Can. and ne. U.S., s. to N.C., n. Ga., and Tenn. May–Sept. (*H. pratense.*)

5. Hieracium aurantiacum L.   Orange-red king-devil. Much like no. 4, but the fls red-orange (unique among our spp.), becoming deeper red in drying; plants 1–6 dm, with slender stolons and ordinarily with a slender, elongate rhizome; lvs sometimes a little wider, up to 3.5 cm; invol long-setose, hispid with blackish, gland-tipped hairs, and slightly tomentose; $2n=18, 27, 36, 45, 54, 63, 72$. A weed of fields, roadsides, and meadows; native of Europe, now widespread in se. Can. and ne. U.S., s. to N.C. and W.Va., and w. to Minn. and Io. June–Sept.

6. Hieracium piloselloides Villars.   Glaucous king-devil. Plants 2–10 dm from a usually rather short, praemorse rhizome; stem naked or with 1 or 2(–5) small lvs; herbage glaucous, sparsely long-setose or subglabrous, the peduncles becoming stipitate-glandular and somewhat stellate; basal lvs oblanceolate, 3–18 cm (petiole included) × 0.5–2 cm, 5–12 times as long as wide; infl corymbiform, compact to fairly open; invol 6–8 mm, hispid with blackish, mostly gland-tipped hairs and somewhat stellate; achenes 1.5–2 mm, truncate; $2n=36, 45$. A weed of fields, meadows, pastures, and roadsides; native of Europe, now well established nearly throughout our range. June–Sept. (*H. florentinum; H. praealtum.*)

7. Hieracium murorum L.   Wall-h. Similar to no. 8, and passing into it; basal lvs with well defined blade, broadly rounded or subtruncate to cordate at base; stem commonly naked or with only 1 or 2 lvs; $2n=27, 36$. Native of Europe, sparingly intr. along roadsides and in other waste places from N.S. and Que. to N.Y. and N.J., and reportedly w. to Mich. June, July.

8. Hieracium lachenalii C. Gmelin.   European h. Plant with a short stout rhizome or caudex; stem 1.5–10 dm, sparsely setose and stellate to subglabrous; lvs long-hirsute, with narrowly to broadly elliptic or even lance-ovate blade 2–13 × 1.5–5.5 cm, tapering to the base, generally with at least a few sharp coarse teeth toward the base, mostly exceeding the petiole; cauline lvs 2-several, reduced and scattered upwards; infl. open, corymbiform, the stout peduncles long-stipitate-glandular and ± stellate; invol 8–12 mm, hispid with blackish, gland-tipped hairs and often also stellate; fls 40–80; achenes 2.5–3.5 mm; $2n=27$. Fields, roadsides, and other waste places; native to Europe, intr. from Nf. and Que. to N.J. and Minn. May–Aug. (*H. maculatum,* with spotted lvs; *H. vulgatum.*)

9. Hieracium robinsonii (Zahn) Fern.   Northeastern h. Similar to no. 8, but smaller, 1–3.5 dm; lf-blades to 8 × 2 cm; basal lvs persistent or deciduous; spreading hairs of the peduncles glandless, often few; heads solitary or few (to 10); invol bracts stellate and shortly hispid with blackish, glandless or gland-tipped hairs, gradually tapering to a very sharply pointed tip; achenes 3–5 mm. Ledges and rock-crevices along streams; n. N.H. to Nf. and Que. June–Sept.

10. Hieracium sabaudum L.   Similar to no. 12, the lvs avg a little larger; lower part of the stem and lower surfaces of the lvs sparsely to densely provided with long, firm, bulbous-based hairs, not at all stellate, the upper surface of the lvs similarly hairy or more often glabrous; infl generally narrow, the lower branches and peduncles elongate but closely ascending; $2n=18, 27$. Native of Europe; established as a weed from Mass. to Pa. Aug.–Oct.

11. Hieracium umbellatum L.   Northern h. Similar to no. 12 and passing into it; lvs mostly 4–12 times as long as wide, scarcely or not at all clasping, often narrowed to the base, ± copiously provided, at least marginally, with short, stiff, blunt, subconic hairs, as well as generally ± stellate; peduncles commonly also with short, stout, subconic hairs scarcely if at all surpassing the stellate puberulence; $2n=18, 27$. Thickets, open woods, and beaches; circumboreal, in Amer. s. to P.E.I., N.B., Mich., Minn., Colo., and Oreg.; disjunct to Spruce Knob, W.Va. July–Sept. (*H. scabriusculum*; Amer. plants possibly to be recognized as a distinct var., but the nomenclature recondite.)

12. Hieracium kalmii L.   Canada h. Perennial from a short caudex; stem 1.5–15 dm, often spreading-hairy below, sometimes becoming stellate-puberulent above; lvs stellate-puberulent to subglabrous, and often long-hairy beneath, the basal and lowermost cauline ones small and soon deciduous; the others (except for the strongly reduced upper ones), mostly rather numerous, nearly alike in size and shape, sessile and tending to be broadly rounded and somewhat clasping at base, elliptic to ovate, lanceolate, or oblong, 3–12 cm × 7–40 mm, 2–5 times as long as wide, usually with a few irregularly spaced sharp teeth; infl loosely corymbiform to often umbelliform, at least in part, or the heads occasionally solitary; peduncles stellate-puberulent, occasionally with some longer hairs as well; invol 6–13 mm, its bracts imbricate in several series, glabrous or obscurely puberulent, occasionally with a few longer hairs, gradually tapering to a narrow but obtuse or rounded tip; achenes truncate, 2.5–3.5 mm; $2n=27$. Woods, beaches, and fields, especially in sandy soil; Nf. and Lab. to N.J., w. through Man. and Io. to B.C. and Oreg. July–Sept. Var. *fasciculatum* (Pursh) LePage, ranging from Que. to Minn., s. to N.J., O., Ill., and Mo., is robust, 5–15 dm, with 25–50 rather firm lvs that seldom have any long hairs; the infl tends to be umbelliform, with stiff peduncles 2–4 cm. Var. *kalmii*, ranging from Me. and N.H. to Nf., Mich., Io., Oreg., and B.C., is more slender, 1.5–10 dm, with 5–30 rather thinner lvs that often have some long hairs beneath, the infl more open, with laxer peduncles to 10 cm (or the heads even solitary). (*H. canadense.*)

13. Hieracium traillii Greene. Shale-barren h. Similar to no. 14, but more hairy; lvs long-setose over the surface beneath and often sparsely so above, not purple-veined; peduncles avg stouter, densely stellate-tomentose as well as copiously provided with spreading, blackish, gland-tipped hairs; the invol to 12 mm, conspicuously hairy, with long, eglandular setae, shorter, blackish, gland-tipped hairs, and also some small, stellate hairs. Dry, open woods, often on shale-barrens; mts. of s. Pa. to Va. and W.Va. May–Aug. (*H. greenii* Porter & Britton, a preoccupied name.)

14. Hieracium venosum L. Veiny h. Stems 1–few from a short, praemorse rhizome, 2–8 dm, glabrous or very nearly so, naked or with 1–3(–6) reduced lvs; basal lvs elliptic to ovate or broadly oblanceolate, 3–16 cm (short petiole included) × 0.8–5 cm, 1.7–5 times as long as wide, often densely long-setose along the margins and toward the base, otherwise sparsely so or subglabrous, the midrib and main veins generally reddish-purple above in life, the whole undersurface sometimes reddish-purple; infl open, corymbiform, the peduncles elongate, usually slender, and often rather flexuous; invol 7–10 mm, glabrous or sometimes evidently stipitate-glandular, obscurely or not at all stellate, the peduncles likewise; fls 15–40 per head; achenes 2.2–4 mm, truncate or more often distinctly narrowed near the summit; $2n=18$. Mostly in dry, open woods; N.Y. to Va. and n. Ga., w. to Mich., Ky., Tenn., and Ala.; c. Fla. May–July. *H.* × *scribneri* Small is a hybrid with no. 15, and *H.* × *marianum* Willd. is thought to be a hybrid with no. 16 or 18.

15. Hieracium paniculatum L. Panicled h. Stems solitary (2) from a short caudex or crown, 3–15 dm, long-hairy below, otherwise glabrous or nearly so, leafy up to the infl; lvs thin, glabrous or with a few long hairs on the persistently glaucous lower surface, irregularly callous-toothed or subentire, the lowest ones petiolate, only slightly if at all enlarged, mostly soon deciduous, the others elliptic, often narrowly so, narrowed to a sessile or subsessile base, only gradually reduced upward, 4–12 × 1–2 cm; infl open-paniculiform, with long, flexuous, very slender peduncles, these, like the invol, glabrous or occasionally with some gland-tipped hairs; invol narrow, 5–9 mm; fls 8–30; achenes truncate, 1.8–2.5 mm; $2n=18$. Woods; N.S. and Que. to Minn., s. to Va., O., and in the mts. to n. Ga. July–Sept. *H.* × *alleghaniense* Britton is a hybrid with no. 16.

16. Hieracium gronovii L. Beaked h. Stems mostly solitary on a crown or short, praemorse rhizome, 3–15 dm, conspicuously spreading-hairy toward the base, the hairs rarely approaching 1 cm, becoming merely strigose-puberulent or obscurely stellate to subglabrous upward, sometimes also finely glandular; lvs finely stellate and usually also ± long-hairy, the basal or lowest cauline ones broadly oblanceolate to obovate or elliptic, 4–20 cm (including the usually short petiole) × 1–2.5 cm, deciduous or persistent, the cauline ones progressively reduced upward, soon becoming sessile and commonly somewhat clasping, the upper part of the stem naked or nearly so; infl ± elongate and openly cylindric, at least in well developed plants; the peduncles puberulent and sparsely to fairly copiously long-stipitate-glandular; invol 6–9 mm; fls 20–40; achenes 2.5–4 mm, distinctly narrowed toward the top; $2n=18$. Dry, open woods, especially in sandy soil, sometimes fields and pastures; Mass. to Fla., w. to s. Ont., Kans., and Tex. July–Oct.

17. Hieracium longipilum Torr. Long-haired h. Plant with a short stout caudex or crown; stem 6–20 dm, densely long-hairy below, the hairs mostly 1 cm or more, sometimes 2 cm, becoming glabrous or nearly so above; lvs pubescent like the stem, or the hairs shorter; basal and lower cauline lvs rather numerous, oblanceolate or narrowly elliptic, 9–30 × 1.5–4.5 cm, crowded, the lowest ones often deciduous, the others progressively reduced upwards, the upper half of the stem commonly naked or merely bracteate; infl elongate, cylindric, the branches and peduncles stellate, long-stipitate-glandular and sometimes sparsely setose; invol 7–10 mm, stellate-puberulent and hispid with blackish, mostly gland-tipped hairs; fls 40–90; achenes 3–4.5 mm, narrowed above. Dry prairies, open woods, and fields, especially in sandy soil; Mich. and Ind. to Minn., Kans., and Okla. July, Aug.

18. Hieracium scabrum Michx. Sticky h. Stems mostly solitary from a short, simple caudex or crown, 2–15 dm, setose at least near the base with spreading hairs seldom as much as 5 mm, becoming stellate and long-stipitate-glandular, densely so in the infl; lvs setose on both sides, more densely so on the petiole and along the midrib beneath; basal and often also the lowermost cauline lvs ordinarily deciduous, the lower lvs broadly oblanceolate to elliptic, 5–20 cm (including the usually short petiole) × 1–4.5 cm, the others progressively reduced upward, soon becoming sessile, so that the upper part of the stem does not appear to be very leafy; infl open-corymbiform (especially in smaller plants) to more often elongate and cylindric; invol 6–9 mm, hispid with blackish, mostly gland-tipped hairs, especially toward the also stellate-hairy base; fls 40–100; achenes 2–3 mm, truncate, only very obscurely if at all narrowed upward; $2n=18$. Open ground and dry woods; especially in sandy soil; N.S. and Que. to Minn., s. to Va., Ky., Mo., and in the mts. to n. Ga. July–Sept.

93. CREPIS L. Hawk's beard. Fls all ligulate and perfect, yellow; invol cylindric or campanulate, the principal bracts in 1 or 2 equal or subequal series, the reduced outer ones few or many; receptacle naked; achenes terete or subterete, fusiform or nearly columnar, often beaked, 10–20-ribbed; pappus of numerous whitish capillary bristles; herbs with milky juice, mostly (incl. all our spp.) taprooted or with several strong roots; lvs alternate, entire to bipinnatifid, ± basally disposed, the cauline progressively reduced, the uppermost bract-like. Nearly 200, N. Hemisphere.

1 Native western perennial, not weedy; invol 9–16 mm .... 1. *C. runcinata.*
1 Introduced weeds, chiefly annual or biennial; invol 5–12 mm.
  2 Achenes narrowed toward the top, but scarcely beaked.
    3 Mature achenes dark purplish-brown; inner invol bracts hairy within .... 2. *C. tectorum.*
    3 Mature achenes pale; inner invol bracts glabrous within.
      4 Invol 8–12 mm, glabrous; achenes 4–6 mm .... 3. *C. pulchra.*
      4 Invol 5–8 mm, pubescent; achenes 1.5–2.5 mm .... 4. *C. capillaris.*
  2 Achenes distinctly slender-beaked .... 5. *C. vesicaria.*

    1. Crepis runcinata (James) T. & G. Perennial (1–)2.5–5(–8) dm with one or several strong roots; herbage glabrous or occasionally glandular-hispidulous; basal lf-blades narrowly obovate, elliptic, lanceolate, or spatulate, 0.5–3.5 cm wide, 4–8 times as long, sessile or longer than the petiole, entire to remotely dentate or runcinate-pinnatifid; cauline lvs few and much reduced; heads (1–)3–7(–12); invol 9–16 mm, ± glandular-hairy or hispid, sometimes also tomentose, the outer bracts seldom over half as long as the 10–15 inner ones; fls 20–50; achenes (3.5–)4–5.5(–7.5) cm, contracted at the top, but not beaked, brown, 10–13-ribbed; 2*n*=22. Moist open places; w. Minn. to Alta., Wash., Calif., and n. Mex. June, July. Our plants, as here described, are var. *runcinata.* (*C. glaucella.*)

    2. Crepis tectorum L. Taprooted annual 1–10 dm; herbage glabrous or puberulent; basal lvs petiolate, with lanceolate to oblanceolate, acute, denticulate to pinnately parted blade to 15 × 4 cm; cauline lvs sessile, auriculate, mostly linear or nearly so; heads several; invol 6–9 mm; inner bracts 12–15, tomentose-puberulent and sometimes also glandular-hispid outside, strigose or puberulent within; outer bracts subulate, a third as long as the inner; fls 30–70; receptacle very finely ciliate; achenes 2.5–4.5 mm, dark purplish-brown at maturity, fusiform, rather strongly attenuate but scarcely beaked, 10-ribbed, the ribs roughened; 2*n*=8. Native of Eurasia, sparingly established here and there in our range. June, July.

    3. Crepis pulchra L. Taprooted annual 2–10 dm, the stem bearing spreading glandular hairs below, glabrous above; lvs spreading-hairy like the lower part of the stem, the lowermost ones to 20 × 4 cm, mostly oblanceolate or spatulate, short-petiolate, and coarsely toothed or runcinate-pinnatifid; heads mostly numerous; invol 8–12 mm, glabrous; inner bracts ca 13, generally with raised and thickened, basally expanded midrib; outer bracts very short; receptacle glabrous; fls ca 15–30; achenes 4–6 mm, contracted above but not beaked, stramineous or pale greenish, 10–12-ribbed, the outer ones scabrous-hirtellous; 2*n*=8. Native of Eurasia, locally established in waste places from Va. to O. and Ind. and southward. May, June.

    C. nicaeensis Balbis, a s. European sp. with golden-brown achenes 2.5–3.8 mm, and with finely ciliate receptacle, is occasonally found in our range, but may not be persistent.

    4. Crepis capillaris (L.) Wallr. Taprooted annual or biennial 2–9 dm, often much branched; stem hispidulous at least near the base; lvs glabrous or hispidulous, the basal to 30 × 4.5 cm, petiolate, lanceolate or oblanceolate, denticulate to runcinate-pinnatifid or even bipinnatifid; cauline lvs progressively reduced, clasping and acutely auriculate; heads several to usually numerous; invol 5–8 mm, tomentose and often glandular-bristly with black hairs as well; inner bracts 8–16, becoming spongy-thickened on the back; outer bracts linear, up to half as long as the inner; receptacle glabrous; fls 20–60; achenes mostly tawny or pale brown, 1.5–2.5 mm, narrowed at both ends, ca 10-ribbed; 2*n*=6. Native of Europe, sparingly intr. in meadows, pastures, lawns, and waste places in our range. July–Oct.

    5. Crepis vesicaria L. Annual, biennial, or occasionally short-lived perennial; stem 1–8 dm, purple near the base, ± hispid or tomentulose or both; lvs finely short-hairy on both sides, the basal to 20 × 4 cm, narrowly oblanceolate to spatulate or obovate, dentate to runcinate-pinnatifid or pectinately parted; cauline lvs progressively reduced, auriculate-clasping; heads several to ± numerous; invol 8–12 mm, its inner bracts 9–13, tomentose and often glandular, sometimes with some short black setae distally, becoming carinate-thickened in fr; outer bracts up to ca half as long as the inner; fls numerous; achenes pale brown, 4–5.9 mm, 10-ribbed, gradually attenuate into a slender beak equaling or a little longer than the body; 2*n*=8, 16. Fields and waste places; native of the Medit. region and w. Europe, sparingly intr. into e. U.S. s. to N.C. May–July. Amer. plants are ssp. *taraxacifolia* (Thuill.) Thellung.

    C. setosa Haller f., with the invol and stems strongly setose with stiff yellow bristles, and with achenes 3–5 mm, is probably not persistent with us.

    94. YOUNGIA Cass. Fls all ligulate and perfect, yellow; invol calyculate at base; achenes slightly compressed, fusiform, tapering above but scarcely beaked, slightly expanded to the pappiferous disk at the summit, strongly and somewhat unequally 11–13-ribbed; pappus of numerous fine white capillary bristles; herbs with milky juices and alternate, often mainly basal lvs. 35, Asia.

1. Youngia japonica (L.) DC. Asiatic hawk's beard. Polymorphic, subscapose annual 1–9 dm, scabrous or ± hairy toward the base; lvs mainly or all basal, mostly lyrate-pinnatifid, or subentire in small plants, up to ca 20 × 6 cm; heads small, numerous in a corymbiform or paniculiform infl; fls ca 10–20, the tube ca ¼ as long as the ligule; invol 3.5–5 mm, glabrous, with 4 short outer and ca 8 longer inner bracts; achenes brownish, 1.5–2.5 mm; pappus 2.5–3.5 mm; $2n=16$. Native to se. Asia, now a pantropical weed, and becoming common on the coastal plain in se. U.S., n. to Pa. (*Crepis j.*)

95. AGOSERIS Raf. False dandelion. Fls all ligulate and perfect, ± numerous, yellow or occasionally orange-red, often turning pinkish or purple in age or in drying; invol campanulate, its bracts imbricate or subequal; achenes oblong or linear, terete or angular, ca 10-nerved, commonly narrowed to an indurate whitish base, distinctly beaked at the tip when mature; pappus of numerous capillary bristles; taprooted lactiferous scapose herbs with entire to pinnatifid basal lvs and medium-sized to large solitary heads. 7, w. U.S., 1 S. Amer.

1. Agoseris glauca (Pursh) D. Dietr. Perennial 1–7 dm, glabrous or nearly so and somewhat glaucous; lvs linear or oblanceolate, 5–35 cm × 1–30 mm, entire or occasionally with a few scattered teeth or shallow lobes; invol bracts imbricate or subequal, mostly sharply pointed, sometimes purple-spotted; fls yellow, often drying pinkish; achenes 5–12 mm, often hirtellous-puberulent, the body tapering gradually to a stout, evidently striate beak up to half as long; $2n=18, 36$. Prairies and meadows; w. Minn. and Man. to B.C. and Calif. June, July. We have only the var. *glauca,* as described above.

96. MICROSERIS D. Don. Fls all ligulate and perfect, yellow; invol campanulate or narrower; achenes columnar to fusiform, but scarcely beaked, with a whitish basal callosity, ca 10-ribbed; pappus of 5–many members, these commonly with paleaceous base and slender, bristle-like tip, varying (in spp. approaching *Agoseris*) to merely obscurely flattened capillary bristles; lactiferous herbs with a taproot, entire to pinnatifid lvs, and 1–many heads on long naked peduncles. 25, chiefly w. U.S.

1. Microseris cuspidata (Pursh) Schultz-Bip. False dandelion. Scapose perennial 5–35 cm; lvs crowded, narrow, 7–30 cm × 3–20 mm, long-acuminate, entire, the margins villous-ciliate and often crisped; invol 17–25 mm, its bracts subequal or slightly imbricate; achenes 8 mm, strongly striate throughout, tapering slightly to the truncate beakless apex; pappus of numerous mixed capillary bristles and very slender, gradually attenuate scales; $2n=18$. Prairies and other dry open places, often in gravelly soil; Ill. and Wis. to Mont., Colo., and Okla. Apr.–June. (*Agoseris c.; Nothocalais c.*)

97. PYRRHOPAPPUS DC., nom. conserv. Fls all ligulate and perfect, yellow, numerous; invol campanulate or narrower, calyculate, the inner bracts corniculate-appendaged near the tip; achenes oblong or linear, subterete or fusiform, with 5 broad, cross-rugulose ridges and as many narrow grooves, contracted above into a long, slender beak, which bears a ring of minute, soft, white, reflexed hairs just beneath the pappus of numerous sordid or rufescent capillary bristles. 3, N. Amer.

1. Pyrrhopappus carolinianus (Walter) DC. Caulescent annual or biennial 2–10 dm, commonly minutely hirtellous-puberulent under the heads, otherwise generally glabrous or nearly so; lvs entire to pinnatisect, the basal sometimes much the largest and persistent, to 25 × 6 cm, sometimes deciduous and scarcely larger than the well developed cauline ones; heads several or occasionally solitary; invol 1–2 cm at anthesis; body of the achene 4–5 mm, the filiform, subapically very fragile beak longer, often twice as long; $2n=12$. Fields, dry woods, bottomlands, and waste places; Del. and Md. to Ill. and Kans., s. to Fla. and Tex. June–Sept. (*Sitilias c.*)

98. KRIGIA Schreb. Dwarf dandelion. Fls all ligulate and perfect, yellow or orange; invol campanulate, with 5–18 essentially equal bracts, not calyculate; achenes oblong or turbinate, 10–20-nerved or -ribbed, tending to be transversely rugulose; pappus of 5+ prominent to sometimes very inconspicuous scales and usually 5–40 longer bristles, or obsolete; lactiferous herbs, commonly with some spreading, gland-tipped hairs at least under the heads; lvs alternate, subopposite, or all basal, entire to pinnatifid; heads 1–several, mostly long-pedunculate. 7, N. Amer.

1 Perennial; pappus-bristles 20–40; invol 7–14 mm.
2 Caulescent, branched above and with several heads .... 1. *K. biflora.*
2 Scapose; heads solitary .... 2. *K. dandelion.*
1 Annual; pappus bristles 5–10 or none; invol 3–7 mm.
3 Pappus evident; plants scapose, or sometimes leafy near the base .... 3. *K. virginica.*
3 Pappus wanting, or a minute scaly vestige; lvs cauline and basal .... 4. *K. oppositifolia.*

1. Krigia biflora (Walter) S. F. Blake.   Orange d.d. Fibrous-rooted perennial 2–8 dm from a short caudex, glabrous except generally under the heads, somewhat glaucous; basal lvs oblanceolate to broadly elliptic, mostly 4–25 cm (petiole included) × 1–5 cm, entire or toothed, or sometimes pinnatifid below but with a large, broad terminal segment; cauline lvs few, sessile and clasping, often much reduced, the uppermost often subopposite and with several long peduncles in their common axil; heads several; invol 7–14 mm, much surpassed by the orange fls; bracts 9–18, narrow, reflexed in age; pappus of 20–35 very fragile, unequal bristles and ca 10 inconspicuous hyaline scales less than 0.5 mm; $2n=10$. Woods, roadsides, and fields; Mass. to Ga., w. to Man., Colo., and Ariz. May–Oct. (*K. amplexicaulis.*)

2. Krigia dandelion (L.) Nutt.   Colonial d.d. Colonial perennial with slender, tuberiferous rhizomes, generally glabrous except generally under the heads; scape 1–5 dm, usually solitary; lvs all basal (sometimes a single pair present just above the base), linear to oblanceolate, entire or pinnately few-lobed, 3–20 cm × 2–25 mm; head solitary, 3–4.5 cm wide in fl; invol 9–14 mm, much surpassed by the golden-yellow fls; bracts 9–18, reflexed in age; pappus of 25–40 unequal capillary bristles and ca 10 inconspicuous scales 0.6–1 mm; polyploids based on $x=5$. Low prairies, fields, and other moist, chiefly open places, often in sandy soil; N.J. to Fla., w. to Kans. and Tex. Apr.–June. (*Cynthia d.*)

3. Krigia virginica (L.) Willd.   Virginia d.d. Slender annual 3–40 dm, scapose or leafy only near the base, the several scapes or scapiform peduncles generally spreading-glandular-hairy above, sometimes throughout; lvs linear to oblanceolate or obovate, 1.5–12 cm × 1–12 mm, entire to pinnatifid, loosely villous-hirsute to glabrous, the hairs sometimes glandular; invol 4–7 mm, its (5–)9–18 bracts lanceolate or narrower, reflexed in age, not keeled or ribbed; pappus of 5 short, thin scales alternating with as many scabrous bristles several times as long; $2n=10, 20$. Sandy places; Me. and Vt. to Wis., s. to Fla. and Tex. Mar.–July.

4. Krigia oppositifolia Raf.   Opposite-lvd d.d. Slender, branching, usually several-stemmed annual 5–45 cm; lvs basal and cauline, linear-oblong to oblanceolate or a little broader, entire to pinnatifid, 1.5–15 cm × 2–20 mm; some of the upper internodes commonly shortened so that the lvs appear opposite, with several long, sometimes similarly branched and bracteate peduncles in their common axil; heads 8–15 mm wide in fl, yellow with a golden center; invol 3–5 mm high, its bracts 6–10, lanceolate or broader, persistently erect, becoming keeled in fr; pappus wanting, or a mere scaly vestige; $2n=8$. Generally in moist, low places; s. Va. to Fla., s. to s. Ill., se. Neb., Kans. and Tex. Apr.–July. (*Serinia o.*)

99. TARAXACUM Wiggers, nom. conserv.

Dandelion. Fls all ligulate and perfect, yellow, mostly numerous; invol bracts biseriate, the outer usually shorter than the inner and often reflexed; achenes columnar or thickly fusiform, terete or 4–5-angled, longitudinally sulcate or ribbed, ordinarily muricate or tuberculate at least above, commonly topped by a smooth, conic or pyramidal cusp that tapers to a slender beak, or rarely beakless; pappus of numerous white capillary bristles; taprooted, lactiferous, perennial, scapose herbs, the lvs all basal and rosulate, entire to pinnatifid or subbipinnatifid. 60, mostly N. Temp. Spp. confluent through polyploidy and apomixis. Taxonomy and nomenclature in utter confusion. We here define spp. broadly and follow traditional nomenclature.

1 Inner invol bracts generally ± corniculate (i.e., with a hooded appendage near the tip).
2 Achenes red or purplish-red or brownish red at maturity; lvs tending to be very deeply cleft for
their whole length, the lobes narrow; intr., not very common weed .... 1. *T. laevigatum.*
2 Achenes brown or olivaceous or stramineous (as in all the following spp.); lvs generally less cleft;
native, northern, scarcely weedy .... 2. *T. ceratophorum.*
1 Inner invol bracts generally not corniculate; all intr. weeds.
3 Outer invol bracts soon becoming reflexed.
4 Outer bracts narrow, scarcely wider than the inner; common .... 3. *T. officinale.*
4 Outer bracts broader, ovate to lanceolate; rare .... 4. *T. spectabile.*
3 Outer invol bracts persistently appressed; rare .... 5. *T. palustre.*

1. Taraxacum laevigatum (Willd.) DC.   Red-seeded d. Similar to no. 3, often more slender; lvs generally very deeply cut for their whole length, the lobes narrow, the terminal lobe seldom much larger than the others; heads a little smaller; invol 1–2 cm, its inner bracts 11–13, often somewhat corniculate, the outer bracts appressed to reflexed, a third to a little over half as long as the inner; body of the achene becoming bright red or purplish-red or brownish-red at maturity, commonly somewhat rugulose below as well as muricate above, abruptly contracted into a cylindrical cusp just below the beak, the beak usually stramineous, ½–3 times as long as the body; 2n=16, 24, 32. Fields, pastures, lawns, and other disturbed sites; native of Eurasia, now established throughout most of n. U.S. and adj. Can., but much less common than no. 3. Mar.–Dec. (*T. erythrospermum; Leontodon e.*)

2. **Taraxacum ceratophorum** (Ledeb.) DC. Northern d. Similar to no. 3, avg a little less robust; scape arising laterally, not terminal (unique among our spp.) lvs with a broadly winged petiolar base, often less lobed, commonly glabrous or nearly so; heads sometimes a little smaller; invol bracts, at least the inner, commonly corniculate, the outer appressed to loosely spreading, from much shorter than to nearly as long as the inner, and often wider; 2n=16, 32. Meadows and other moist places in the mts.; circumboreal, s. at high altitudes to N.H., Mass., and N.M. July, Aug. (*T. lapponicum; T. latilobum,* misapplied.)

3. **Taraxacum officinale** Weber. Common d. Lvs commonly sparsely hairy beneath and on the midrib, otherwise generally glabrous, or sometimes completely so, oblanceolate, 6–40 × 0.7–15 cm, ± runcinate-pinnatifid or lobed, the terminal lobe tending to be larger than the others, tapering to a narrow, scarcely or obscurely winged petiolar base; scapes 5–50 cm, glabrous or ± villous, especially upward; heads usually large, the invol mostly 1.5–2.5 cm, inner bracts mostly 13–20, usually not or scarcely corniculate, at first erect, finally reflexed, the mature achenes and pappus then forming a conspicuous ball; outer bracts reflexed, a little shorter and scarcely wider than the inner; body of the achene 3–4 mm, pale gray-brown or stramineous, muricate above or sometimes to near the base, the beak 2.5–4 times as long as the body; 2n=16–48, often 24. Lawns and disturbed sites; native of Eurasia, now a cosmopolitan weed of temperate climates. Apr.–Dec. (*Leontodon taraxacum.*)

4. **Taraxacum spectabile** Dahlst. Lvs hairy, often dark-spotted, often purple-veined beneath; outer invol bracts ovate to lanceolate; 2n=40, and probably other nos.; otherwise much like no 3. Native of n. and w. Europe, occasionally intr. in moist places in Ont. and N.Y.

5. **Taraxacum palustre** (Lyons) Simons. Marsh-d. Lvs suberect, rather narrow, remotely toothed; outer invol bracts very dark, broad, strongly and permanently appressed; beak twice as long as the body of the achene; otherwise much like no. 3; 2n=24, 32, 40. Seasonally flooded highway-ditches and wet or flooded sites in calcareous gravel or clay over limestone or marble bedrock; native of Europe, rarely intr. in se. Ont. and c. N.Y. Apr.–June. (*T. turfosum.*)

100. CHONDRILLA L.   Fls all ligulate and perfect, yellow, 7–15 per head; invol cylindric, calyculate; achenes multinerved, glabrous, muricate above, and with a beak that is expanded at the top into a pappiferous disk; pappus of numerous capillary bristles, generally white; lactiferous, branching, often rush-like herbs with well-developed, usually pinnatifid basal lvs and reduced, mostly scattered and entire cauline lvs. 25, temp. Eurasia.

1. **Chondrilla juncea** L.   Skeleton-weed. Branching, rush-like, taprooted perennial with runcinate-pinnatifid, often deciduous basal lvs 5–13 × 1.5–3.5 cm and reduced linear cauline lvs 2–10 cm × 1–8 mm, the stem 3–15 dm, strongly spreading-hispid near the base, the invols white-tomentose, the herbage otherwise glabrous; heads scattered along the branches, commonly with 9–12 fls; invol 9–12 mm, with ca 8 principal bracts; body of the achene 3 mm, muricate above and bearing a circle of scales at the base of the long, slender beak; 2n=15, 30. Roadsides, fields, and waste places; native of Eurasia, intr. in our range from N.Y. to Va., W.Va. and Mich. July–Sept.

101. CICHORIUM L.   Fls all ligulate and perfect, blue (pink or white); invol bracts biseriate, the outer shorter; achenes glabrous, striate-nerved; sub-5-angled, or the outer slightly compressed; pappus of 1–2 series of scales, sometimes minute; branching, leafy-stemmed, lactiferous herbs with alternate, entire to pinnatifid lvs and several showy heads. 9, mainly Europe and Medit. reg.

1. **Cichorium intybus** L.   Chicory. Perennial 3–17 dm from a long taproot; lower lvs oblanceolate, petiolate, toothed or more often pinnatifid, 8–25 × 1–7 cm, becoming reduced, sessile, and entire or merely toothed upwards; heads to 4 cm wide in fl, mainly matutinal, sessile or short-pedunculate, borne 1–3 together in the axils of the much reduced upper lvs, the long branches thus racemiform; invol 9–15 mm, its outer bracts loose, fewer than the inner and at least half as long, becoming callous-thickened at base; achenes 2–3 mm, pappus-scales numerous, minute, narrow; 2n=18. Roadsides, fields, and waste places; native of Europe, now a cosmopolitan weed. June–Oct.

102. LAPSANA L.   Fls all ligulate and perfect, yellow; invol cylindric-campanulate, minutely calyculate, the inner bracts subequal, uniseriate, keeled; achenes narrow, sub-terete or slightly compressed, often curved, narrowed to both ends, usually more gradually so downwards, 18–30-nerved, 5 or 6 of the nerves generally stronger; pappus none; branching, annual, lactiferous herbs, with alternate, entire to pinnatifid lvs and several to many small, rather few-fld heads, the peduncles tending to be apically pale and indurate. 9, temp. Eurasia.

1. Lapsana communis L.   Nipplewort. Hirsute to subglabrous weed 1.5–15 dm; lvs thin, petiolate, with ovate to subrotund, obtuse or rounded, roothed or occasionally basally lyrate blade 2.5–10 × 2–7 cm, progressively less petiolate and eventually narrowed upwards; heads naked-pedunculate, several or many in a corymbiform or paniculiform infl; invol 5–8 mm, with 8 inner bracts; fls 8–15; achenes 3–5 mm, glabrous, curved; $2n=12, 14, 16$. Woods, fields, and waste ground; native of Eurasia, now found throughout our range. June–Sept.

103. ARNOSERIS Gaertner.

Fls all ligulate and perfect, yellow; invol campanulate, becoming broader in fr, its bracts equal and uniseriate, becoming keeled with an enlarged and indurate midrib in fr, rarely minutely and scantily calyculate; achenes obovoid, 8–10-ribbed, some of the ribs commonly stronger than the others and making the achenes subangular; pappus none; branching, annual, lactiferous herbs with well developed basal lvs, the stem sparsely and minutely bracteate or essentially naked, becoming simple, monocephalous, and strictly scapose in depauperate plants. Monotypic.

1. Arnoseris minima (L.) Schweigger & Koerte.   Dwarf nipplewort. Plant 5–30 dm, glabrous or minutely granular-hairy; basal lvs oblanceolate or broader, mostly toothed, 0.5–7 cm × 2–15 mm; peduncles conspicuously inflated for some distance below the heads; invol 4–7 mm, its 10–22 bracts narrow; achenes 1.5–2 mm, often slightly rugose as well as strongly ribbed, otherwise glabrous, shining; $2n=18$. Fields and waste places; native of Europe, but found occasionally from N.S., N.B., and Me. to Pa., O., and Mich., perhaps not becoming fully established. June–Sept.

104. PICRIS L.

Ox-tongue. Fls all ligulate and perfect, yellow; invol campanulate or ovoid-urceolate, the inner bracts uniseriate, subequal, after anthesis becoming thickened and often clasping the achenes; outer bracts narrow and successively shorter, or sometimes uniseriate, broad and foliaceous; achenes subterete, angled, or somewhat compressed, often incurved, commonly 5–10-ribbed (sometimes obscurely so), tending to be transversely rugulose; pappus of plumose bristles, commonly with some outer, shorter, merely barbellate members, the outer achenes sometimes with the pappus reduced or obsolete; lactiferous herbs with alternate or all basal, entire to pinnatifid lvs. 40, temp. Eurasia, n. Afr.

1 Invol imbricate, the bracts all narrow, less than 3 mm wide .... 1. *P. hieracioides.*
1 Invol biseriate, the outer bracts foliaceous, 3.5–8 mm wide .... 2. *P. echioides.*

1. Picris hieracioides L.   Hawkweed-oxtongue. Biennial or short-lived perennial 2–10 dm, with the aspect of *Hieracium*, spreading-hispid to subglabrous; lowest lvs oblanceolate, 7–30 cm (petiole included) × 0.5–5 cm, often deciduous, the others lanceolate or oblong, sessile and often clasping, reduced; heads several in a corymbiform infl; invol 8–15 mm, its bracts imbricate, less than 3 mm wide; achenes evidently rugulose, 3.5–6 mm, narrowed above but only very shortly and stoutly if at all beaked; pappus readily deciduous as a unit; $2n=10$. Waste places; native of Eurasia, occasionally found in our range. July–Sept.

2. Picris echioides L.   Bristly o.-t. Coarse, somewhat spiny-hispid annual 3–8 dm; lvs toothed or entire, the lower oblanceolate and petioled, the others more lanceolate or oblong, sessile and somewhat clasping, the blade to 25 × 7 cm; invol 1–2 cm, biseriate, the inner bracts narrow, commonly with a subterminal spiny awn-tip, the outer 3–5 foliaceous, ovate or lance-ovate, 3.5–8 mm wide, not much if at all shorter than the inner; achenes distinctly slender-beaked, the body 2–4 mm, the beak nearly as long, or longer, very fragile near its base; outer achenes half-moon-shaped, pale, woolly, partly enfolded by the bracts, the others narrower, shining, brown, rugulose, obscurely nerved; $2n=10$. Fields and waste places; native of Europe, occasionally found in our range. July–Sept.

105. HYPOCHOERIS L.

Cat's ear. Similar to *Leontodon,* from which it differs primarily by having long, chaffy bracts on the receptacle; stem sometimes leafy below. 100, cosmop.

1 Essentially glabrous annual .... 1. *H. glabra.*
1 Perennial; lvs hispid .... 2. *H. radicata.*

1. **Hypochoeris glabra** L.   Smooth c.-e. Taprooted annual or winter-annual, essentially glabrous; stem 1–4 dm, simple or sparingly branched, naked or only sparsely and minutely bracteate; basal lvs oblanceolate, toothed or pinnatifid, 2.5–15 × 0.7–3.5 cm; heads several or solitary, terminating the branches, not very showy, opening only in full sun, the ligules ca equaling the invol and only ca twice as long as wide; invol mostly 8–10 mm at anthesis, up to 17 mm in fr, its bracts imbricate; body of the achenes mostly 4–5 mm, multinerved, the nerves muricate upward; outermost achenes usually beakless, the others with a well developed slender beak; shorter outer pappus bristles commonly merely barbellate; $2n=8, 10, 12$. Disturbed and waste places, especially in sandy soil; European weed, intr. in se. U.S., n. to s. Ill.

2. **Hypochoeris radicata** L.   Spotted c.-e. Perennial from a caudex, fibrous-rooted, or more often several of the roots enlarged; stem 1.5–6 dm, branched above or in small plants simple, naked or only sparsely and minutely bracteate, often spreading-hispid below; basal lvs hispid, oblanceolate, toothed or pinnatifid, 3–35 × 0.5–7 cm; heads usually several, terminating the branches, rather showy, the ligules surpassing the invol and ca 4 times as long as wide; invol 10–15 mm at anthesis, to ca 25 mm in fr, its bracts imbricate, glabrous or hispid; body of the achenes mostly 4–7 mm, multinerved, the nerves muricate; achenes all with a slender beak from a little shorter to more often much longer than the body, or the outer ones only short-beaked; some of the outer pappus-bristles often shorter than the inner and merely barbellate; $2n=8$. Roadsides, pastures, fields, and waste places; native of Eurasia, now widely established in the U.S. and s. Can. May–Sept.

106. LEONTODON L.   Fls all ligulate and perfect, yellow; invol ovoid or oblong, imbricate or calyculate; receptacle alveolate or fimbriately villous, not chaffy-bracted; achenes narrow, subterete, several- or many-nerved, long-beaked or merely narrowed upwards; pappus of plumose bristles, sometimes with some shorter outer nonplumose bristles or scales, or that of the marginal achenes wholly of the latter type; lactiferous herbs with well developed basal lvs and a naked or merely scaly-bracted scape, monocephalous, or the heads terminating the several branches. 50, temp. Eurasia.

1 Pappus wholly of plumose bristles; heads usually several .... 1. *L. autumnalis.*
1 Pappus not wholly of plumose bristles; heads solitary.
　2 Pappus alike in all the fls, of long, plumose bristles and commonly some short outer barbellate
　　bristles or scales; invol imbricate, 10–18 mm .... 2. *L. hispidus.*
　2 Pappus of outer fls reduced to a crown; invol calyculate, 6–11 mm .... 3. *L. taraxacoides.*

1. **Leontodon autumnalis** L.   Fall-dandelion. Fibrous-rooted perennial from a short caudex or crown; scapes 1–8 dm, commonly decumbent at the base, scaly-bracted at least above, ordinarily branched, commonly tomentose-puberulent at the summit, otherwise glabrous; basal lvs oblanceolate, 4–35 × 0.5–4 cm, glabrous or moderately hirsute, deeply, narrowly, and rather distantly lobed to occasionally entire; heads terminating the branches; invol 7–13 mm, with narrow, imbricate bracts, scarcely elongating in fr; achenes fusiform-columnar, not beaked, weakly nerved, transversely rugulose, 4–7.5 mm; pappus wholly of plumose bristles, these chaffy-flattened at the base; $2n=12, 24$. Roadsides, pastures, fields, meadows, and waste places; native of Eurasia, now established from Del. to Greenl. and inland sometimes to Wis. June–Oct. (*Apargia a.*) Var. *autumnalis* has the invol merely tomentose-puberulent to glabrous. Var. *pratensis* (Less.) Koch, more boreal, has the invol also spreading-hirsute.

2. **Leontodon hispidus** L.   Big hawkbit. Fibrous-rooted perennial 1–6 dm, hairy or glabrous, the scape simple and ordinarily naked; basal lvs oblanceolate, 2–25 cm × 6–30 mm, pinnately lobed or subentire; invol imbricate, 10–15 cm, or to 2 cm in fr; achenes 6–12 mm, fusiform, sometimes ± beaked, scabrous; pappus of plumose bristles and some shorter, outer, slender scales or barbellate bristles; $2n=14$. Fields, meadows, and waste places; native of Europe, occasionally found in our range from Conn. and R.I. to Pa., Ont., and O. May–Sept. (*Apargia h.; Leontodon hastilis.*)

3. **Leontodon taraxacoides** (Villars) Merat.   Little hawkbit. Fibrous-rooted, chiefly perennial, 1–3.5 dm, the scapes simple, curved-ascending, ordinarily naked; basal lvs oblanceolate, hispid-hirsute, 4–15 × 0.6–2.5 cm, usually shallowly lobed; head solitary; invol shortly calyculate, 6–11 mm, scarcely larger in fr, glabrous or hairy; achenes fusiform, scarcely or shortly beaked, 3–6 mm, scabrous; pappus of inner fls of plumose bristles and some shorter outer scales that may be tipped with a scabrous bristle; pappus of outer fls reduced to a short, laciniate crown; $2n=8$. A weed in lawns and waste places; native of Europe, intr. at scattered localities in e. U.S. June–Sept. (*L. leysseri; L. nudicaulis,* probably misapplied.)

107. TRAGOPOGON L.   Goat's beard. Fls all ligulate and perfect, yellow or purple; invol cylindric or campanulate, its bracts uniseriate and equal; receptacle naked; achenes linear, terete or angled, 5–10-nerved, narrowed at base, slender-beaked, or the outer occasionally beakless; pappus a single series of plumose bristles, united at base, the plume-branches interwebbed, several of the bristles commonly longer than the others and naked at the tip; taprooted lactiferous herbs with alternate, linear, entire, clasping, commonly somewhat grass-like lvs, the heads solitary at the ends of the branches. 50, mainly Eurasia and n. Afr.

1 Fls purple .... 1. *T. porrifolius.*
1 Fls yellow.
  2 Peduncles enlarged and fistulous above in fl and fr; bracts surpassing the rays .... 2. *T. dubius.*
  2 Peduncles not enlarged in fl, scarcely so in fr; bracts not surpassing the rays .... 3. *T. pratensis.*

1. Tragopogon porrifolius L. Salsify; vegetable-oyster. Glabrous biennial 4–10 dm; lvs to 30 cm, and nearly 2 cm wide, tapering rather gradually from the base, not recurved at the tip; peduncles evidently enlarged and fistulous under the heads in fl and fr; invol bracts mostly ca 8, 2.5–4 cm in fl, slightly to strongly surpassing the purple rays, elongating to 4–7 cm in fr; achenes 25–40 mm, the body thicker than in no. 2 and usually only 10–16 mm, abruptly contracted to the long, slender beak; pappus brownish; $2n=12$. Roadsides and waste places, mostly in rather moist soil; European cultigen, established as a weed here and there over much of the U.S. Apr.–Aug.

2. Tragopogon dubius Scop. Fistulous g.-b. Mostly biennial, 3–10 dm; lvs elongate, generally tapering fairly uniformly from base to apex, not recurved, evidently floccose when young, later ± glabrate except commonly in the axils; peduncles evidently enlarged and fistulous under the heads in fl and fr; invol bracts typically 13, down to 8 on later heads or small plants, 2.5–4 cm in fl, distinctly surpassing the rather pale lemon-yellow rays, elongating to 4–7 cm in fr; achenes slender, 25–36 mm, gradually narrowed to the stout beak; pappus whitish; $2n=12$. Roadsides and other open, relatively dry places; native of Europe, now widely established over most of the U.S., especially westward. May–July. (*T. major.*)

3. Tragopogon pratensis L. Showy g.-b. Mostly biennial, 1.5–8 dm; lvs elongate, to 30 × 2 cm, rather abruptly narrowed a little above the base, and tending to have the margins somewhat crisped, cirrhose-recurved at the tip, slightly floccose when young, soon glabrous; peduncles not enlarging in fl and scarcely so in fr; invol bracts typically 8, 12–24 mm in fl, equaling or shorter than the chrome-yellow rays, elongating to 18–38 mm in fr; achenes 15–25 mm, rather abruptly contracted to the slender, relatively short beak, the body not much shorter than in our other spp.; pappus whitish; $2n=12$. Roadsides, fields, and waste places, commonly in slightly moister habitats than no. 2; native of Europe, now widely established in our range and westward. May–Aug.

CLASS *LILIOPSIDA* Monocotyledons

Embryo mostly with a single cotyledon; floral parts, when of definite number, typically borne in sets of 3, seldom 4, almost never 5 (carpels often fewer); pollen of uniaperturate or uniaperturate-derived type; ours all herbs; plants without intrafascicular cambium, the vascular bundles often scattered; vessels often confined to the roots, or even wanting; leaves in most species parallel-veined. 5 subclasses, 19 orders, 66 families.

FAMILY **BUTOMACEAE**, the Flowering Rush Family

Fls axillary to bracts, hypogynous, regular, perfect, trimerous; sep 3; pet 3, pink; stamens 9; pollen monosulcate; pistils 6, connate at the very base into a ring, otherwise distinct, distally unsealed, each with a short, terminal style and shortly bilobed, shortly decurrent stigma; ovules numerous, scattered over the inner surface of the carpel; fr of separate follicles; endosperm wanting; embryo straight, with a terminal cotyledon and lateral plumule; glabrous, perennial, emergent aquatic herbs from a stout, creeping, dorsiventral, edible rhizome; lvs distichous in origin at the rhizome-tip, parallel-veined, linear, erect and ± triquetrous, not differentiated into blade and petiole, but the base somewhat expanded and sheathing; scape axillary, erect, terminating in a cymose umbel subtended by 3 bracts. A single genus and species.

1. BUTOMUS L. Flowering rush. Sep and pet persistent, the sep nearly as long as the pet but more greenish; stamens 9, the outer cycle of 3 pairs of obliquely antesepalous members, the inner of 3 directly antepetalous members, so that all spread at equal angles from the receptacle.

1. Butomus umbellatus L. Lvs basal, erect, floating, or submersed in water up to several m deep, linear, to 1 m, 5–10 mm wide; scape 1–1.5+ m; fls numerous, 2–2.5 cm wide, on pedicels 5–10 cm; $2n$=20, 26, 39. Native of Eurasia, thoroughly established on shores and riverbanks in the St. Lawrence R. valley and in Lake Champlain, and more recently spread inland, even to N.D., S.D., Mont., and Ida. June–Aug.

FAMILY **ALISMATACEAE**, the Water-Plantain Family

Flx axillary to bracts, hypogynous, regular, perfect or unisexual; sep 3, green; pet 3, white (pink), deciduous; stamens (3)6–many; anthers extrorse; pollen mostly pantoporate; pistils 3–28 in a single cycle, or ± numerous in an apparent spiral, each with a terminal or basilateral style and often with a decurrent stigma; ovule solitary, ventral-basal; fruit mostly of achenes; endosperm wanting; embryo horseshoe-shaped, with a terminal cotyledon and lateral plumule; rhizomatous perennial herbs, aquatic or of marshes; lvs all basal, with an open petiolar sheath and usually a fairly broad blade with an acrodromous or campylodromous variant of parallel venation, or the blade suppressed and the petiole flattened and parallel-veined or terete; scape ending in a terminal infl with the pedicels or primary branches commonly ternate. 12/75.

1 Pistils in a single whorl, on a small, flat receptacle .... 1. *Alisma.*
1 Pistils in several series, seemingly spiraled, forming a globose head; receptacle large.
  2 Achenes flattened and distinctly winged; verticils 3-bracted, without bracteoles .... 2. *Sagittaria.*
  2 Achenes turgid, ribbed or ridged, not winged; verticils 3-bracted with many bracteoles .... 3. *Echinodorus.*

1. ALISMA L. Water-plantain. Fls perfect; receptacle small, flat; sep 3, ovate to ovate-oblong, persistent; pet 3; stamens 6, the filaments equaling or longer than the anthers; pistils 10–28, in a single whorl; style lateral; achenes flattened but not winged, grooved on the curved back; infl ample, paniculate, the axis and larger branches bearing a whorl of pedicels or branches at each node; perennial; plants erect, or lax in deep-water forms; lvs never sagittate. 9, mainly N. Hemisphere.

1 Back of the achene ordinarily with a single median dorsal groove; lvs emersed or terrestrial, the blade
   mostly 1.3–2.5(–2.7) times as long as wide, usually rounded to truncate or somewhat cordate at
   base.
  2 Fls larger, the pet 3.8–4.5 × 3.0–3.9 mm; more northern .... 1. *A. triviale.*
  2 Fls smaller, the pet 1.8–2.5 × 1.4–2.0 mm; more southern .... 2. *A. subcordatum.*
1 Back of the achene with a median dorsal ridge flanked by a groove on each side; lvs typically submersed
   and elongate, less often terrestrial or emersed, the blade then 2.7–5 times as long as wide, tapering
   at the base .... 3. *A. gramineum.*

1. Alisma triviale Pursh. Northern w.-p. Rhizome 1–2 cm thick; lvs usually long-petioled, the blade elliptic to broadly ovate, rounded to subcordate at base, 3–18 × 2–12 cm, 1.5–2.5(–2.7) times as long as wide; scape 1–10 dm; pedicels in whorls of 3–10; sep obtuse, 2–3 mm; pet white, 3.8–4.5 × 3.0–3.9 mm, half again as long as the sep; style short, 0.4–0.6 mm, the stigmatic part ± curved; achenes 15–23, 1.8–3.0 mm, ordinarily with a median dorsal groove, rarely with 2 grooves as in no. 3; $2n$=28. Marshes, ponds, and streams; N.S. and e. Que. to Pa., w. to s. B.C., Mo., N.M., and Calif. June–Sept. (*A. brevipes; A. plantago-aquatica* var. *americana.*)

2. Alisma subcordatum Raf. Southern w.-p. Much like no. 1, but with smaller fls; sep 1.5–2.3 mm; pet 1.8–2.5 × 1.4–2.0 mm, barely or scarcely longer than the sep; anthers 0.4–0.6 mm; style very short, 0.2–0.4 mm; achenes 1.5–2.2 mm; $2n$=14. Usually in shallow water; Vt. and Mass. to N.D., s. to Fla. and Tex. (*A. plantago-aquatica* var. *parviflorum.*) Variously reported to intergrade with no. 1 or to be genetically isolated from it.

3. Alisma gramineum Lej. Grass-lvd w.-p. Rhizome 0.3–1.5 cm thick; lvs usually submerged, very thin, linear, 15–100 cm × 2–13 mm, but sometimes terrestrial or emersed, the blade then rather narrowly lance-elliptic, tapering to the base, 2–6 cm × 4–15 mm, 2.7–5 times as long as wide; pedicels stout; sep 1.3–2.2 mm; pet faintly anthocyanic, 2.3–3.7 × 2.1–3.7 mm; anthers 0.3–0.6 mm; style short, 0.3–0.6 mm, uncinate-curved; achenes 11–20, 1.8–2.6 mm, with a median dorsal ridge flanked by a groove on each side; $2n$=14. Circumboreal, in Amer. s. to N.Y., Wis., Mo., Nebr., Colo., and Calif. June–Aug. (*A. geyeri.*)

2. SAGITTARIA L. Arrow-head. Fls unisexual (the plants monoecious or rarely dioecious) or some of them perfect; receptacle large, convex; sep 3, reflexed in fr or persistently erect or spreading; pet 3, deciduous, white; stamens 7–many; pistils numerous, aggregated into a subglobose head on a large receptacle, appearing to be spirally arranged; achenes flattened, winged at least on the margins; infl racemose, with 1–12 mostly 3-fld whorls, each whorl subtended by 3 bracts, occasionally with floriferous branches substituting for fls at the lowest whorl(s), or the fl solitary and terminal in depauperate plants; rooted annual or perennial aquatics, the infl and usually also the lvs emersed; rhizomes sometimes with apical tubers. The lvs are all basal, typically with an aerial, expanded, often sagittate blade on a petiole about as long as the water is deep. In deep or swiftly moving water, wholly submersed lvs are produced by some spp.; these often take the form of ribbon-like bladeless phyllodia (the flattened petioles). Similar short, stiff phyllodia are produced in tidal sites. 25+, cosmop., mainly New World. (*Lophotocarpus*.)

1 Sep appressed in fr; lower fls perfect; fruiting pedicels recurved.
  2 Lvs usually with basal lobes, seldom linear and bladeless; fls in 3–12 whorls; inland .... 1. *S. calycina.*
  2 Lvs mostly without basal lobes, often lanceolate or spatulate; fls in 1–2(3) whorls; tidal ....2. *S. spatulata.*
1 Sep reflexed (widely spreading in no. 3) in fr; lower fls pistillate; fruiting pedicels in most spp. ± ascending.
  3 Filaments smooth, without scales; lvs often sagittate.
    4 Stamens mostly 7–15; lvs ordinarily without basal lobes; fruiting pedicels recurved; chiefly tidal .... 3. *S. subulata.*
    4 Stamens 15–many; lvs usually with basal lobes, but sometimes (especially when submersed) without; fruiting pedicels ascending; not tidal.
      5 Beak of the achene spreading at right angles to the body; bracts ± boat-shaped .... 8. *S. latifolia.*
      5 Beak of the achene ascending to erect (or distally incurved); bracts not boat-shaped.
        6 Beak of the achene erect, short, to 0.5 mm; achene nearly or quite without facial wings .... 7. *S. cuneata.*
        6 Beak of the achene ascending, more than 0.5 mm (except sometimes in no. 6); achenes usually with 1 or more facial wings.
          7 Fls in 2–4 whorls; bracts relatively thick and herbaceous; near the coast .... 4. *S. engelmanniana.*
          7 Fls in 5–12 whorls; braacts stramineous, firm and papery; mostly inland.
            8 Petiole sharply 5-winged in section; head of achenes globose, to 1.5 cm thick; infl not branched .... 5. *S. australis.*
            8 Petiole corrugated in cross-section, at least in its lower half; head of achenes tending to be depressed-globose, (1.2–)1.7–2.2 cm thick; infl often branched at base ...... 6. *S. brevirostra.*
  3 Filaments roughened with minute scales; lvs only rarely sagittate.
    9 Bracts of the infl evidently papillose-roughened or corrugate-ridged; filaments linear .... 9. *S. lancifolia.*
    9 Bracts smooth; filaments subulate or broad.
      10 Pistillate fls sessile or nearly so; scape often bent at the lowest fl-whorl .... 10. *S. rigida.*
      10 Pistillate fls evidently pedicellate; scape not bent.
        11 Phyllodia flat, not nodose; widespread .... 11. *S. graminea.*
        11 Phyllodia terete, nodose; Mass. to N.Y. .... 12. *S. teres.*

1. **Sagittaria calycina** Engelm.   Mississippi a.-h. Annual, erect, or lax in deep-water forms; petioles elongate, terete, very spongy, typically with emersed or floating, hastate or sagittate blade 4–60 cm, the terminal lobe linear to deltoid-orbicular, the basal lobes large, pointed, divergent; small or deep-water forms with only submersed, linear lvs to 0.5 cm wide; scape stout, terete, 1–15 dm; fls in 3–12 whorls, the upper staminate, on long, slender pedicels, the middle and lower perfect, on somewhat shorter, stout, recurved pedicels; bracts mostly ovate, connate below, very thin and soon withering; sep 4–12 mm, broad, blunt, appressed to the fruiting head; pet 7–15 mm; stamens 9–12 in perfect fls, 12–many in staminate ones; filaments roughened with minute scales; achenes 2–3 mm, narrowly winged on the dorsal and ventral margins only, the sides wingless, the oblique or horizontal beak often more than half as long as the width of the body; $2n=22$. Marshes, ponds, and streams, in circumneutral or alkaline waters; drainage of the Mississippi R. (sens. lat.), n. to lakes Michigan and Erie, s. to Tex., N.M., and Mex., and apparently e. sporadically to Del.; Calif. June–Oct. (*S. montevidensis* ssp. *c.; Lophotocarpus c.; L. depauperatus.*)

2. **Sagittaria spatulata** (J. G. Smith) Buchenau.   Tidal sagittaria. Annual(?), mostly submersed at high tide, emersed or stranded at low tide, the rare, permanently emersed lvs sagittate with short basal lobes, the more characteristic lvs modified into narrow phyllodia 4–18 × 0.5–1 cm, the free end often spatulate and to 1.5 cm wide; petioles and scapes terete; scape thick but weak, to 1 dm, with 1–2(3) whorls of fls or only a solitary fl; bracts ovate, obtuse, scarious, 2–5 mm; lower (or all) fls perfect; sep 3–5 mm, oblong-orbicular, appressed to the fruiting head; pet ovate, 4–5 mm; stamens ca 12, with linear, glabrous filament; fruiting pedicels recurved; achenes 1.5–2 mm, narrowly winged on the margins, the oblique to horizontal beak ca half as long as the width of the body; $2n=22$. Brackish to nearly fresh tidal waters and salt marshes; N.B. and adj. Que. to N.C. July–Sept. (*S. montevidensis* var. *spongiosa; Lophotocarpus spatulatus; L. spongiosus.*)

3. **Sagittaria subulata** (L.) Buchenau.   Hudson sagittaria. Plants of tidal waters, or submersed in inland waters mainly near the coast; lvs typically phyllodial and bladeless, 5–30 cm × 1–6 mm, or the floating end not infrequently dilated into an ovate or elliptic (rarely cordate or sagittate) blade 2–6 × 1–2 cm; scape erect or floating, to 40(–90) cm; fls in 1–10 whorls (or solitary and terminal), the upper generally staminate and slender-pedicellate, the lower pistillate, on stout, recurved pedicels; bracts thin, mostly connate at base, the

lance-ovate body 2–4 mm, with a caudate tip, soon evanescent; sep 2–5 mm, widely spreading in fr; pet 5–10 mm; stamens 7–15, the glabrous or subglabrous filaments subulate, equaling or somewhat longer than the anthers; achenes 1.5–2.5 mm, crenulate-winged on the margins and with 1–3 crenulate wings on each face, the beak subulate, oblique, 0.2–0.5 mm; $2n=22$, 44. Coastal states from Mass. and N.Y. (to the upper limits of tidal effect on the Hudson R.) to Fla. and Ala. July–Sept. (*S. natans; S. subulata* var. *gracillima* = *S. lorata* = *S. stagnorum*, a sporadic deep-water riverine non-tidal form with phyllodial lvs 30–90 cm, the floating ends rarely expanded.)

4. Sagittaria engelmanniana J. G. Smith.   Acid-water a.-h.   Erect; lvs long-petioled, the blade sagittate (rarely merely lanceolate), its main portion mostly narrow, 4–10 × 0.2–2(–5) cm, (1–)2–many times as long as wide, with narrow, linear to lance-attenuate basal lobes; petioles rounded or with 2 blunt ridges on the abaxial side; scape 2–8 dm, with 2–4 whorls of fls, the upper staminate, the lower pistillate; pedicels 0.5–3.5 cm, ascending in fr; bracts herbaceous and relatively thick, 5–25 mm, mostly shorter than the pedicels; sep short, 4–7 mm, reflexed in fr; pet 8–12 mm; stamens 15–25, the slender, glabrous filaments as long as or longer than the anthers; mature receptacle markedly echinate; achenes 2.5–4 mm, with 1 or 2(3) wings on each face, these extending onto the obliquely ascending, 1–2 mm beak; achene-faces with resin-ducts; $2n=22$. Acid waters of bogs, ponds, and streams near the coast from Mass. to Fla. and Miss., mainly on the coastal plain, but inland to w.c. N.Y. Aug., Sept.

5. Sagittaria australis (J. G. Smith) Small.   Appalachian a.-h.   Main portion of the lf-blade deltoid-ovate, relatively blunt, 3–13 × 2.5–10 cm, 0.7–2.2 times as long as wide, the basal lobes broad; petiole sharply 5-winged in section; fls in 5–12 whorls; bracts 7–30 mm, equaling or longer than the 7–23 mm pedicels, stramineous, papery-firm, broad-based and long-acuminate; receptacle not markedly echinate; achenes 2.3–3.2 mm, without resin-ducts, usually with 1 wing on each face, these not extending into the beak; beak mostly 0.7–1.7 mm, often ± incurved above the obliquely ascending base; $2n=22$; otherwise much like no. 4. Mostly in circumneutral water of lakes, ponds, or swamps; N.Y., Pa., and N.J. to S.C. and Ga., w. to c. O., s. Ind., se. Mo., and Miss., mainly inland. July–Sept. (*S. longirostra,* misapplied.)

6. Sagittaria brevirostra Mackenzie & Bush.   Midwestern a.-h.   Main portion of the lf-blade mostly lanceolate, sometimes broader, mostly 5–20 × 2–8 cm, 0.8–8 times as long as wide; petioled corrugated in x-section, at least in the lower half; infl often branched at the lowest whorl; fls in 5–12 whorls; bracts 1.5–4 cm, equaling or longer than the 1.3–2.5 cm pedicels, stramineous, papery-firm, broad-based and long-attenuate; fruiting head usually oblate-spheroidal; receptacle not markedly echinate; achenes 2–3 mm, lacking resin-ducts, with a short wing (not extending into the beak) on one or both faces, or the faces wingless; dorsal wing entire to crenulate, usually truncate at the top and not confluent with the beak; ventral wing narrower and confluent with the beak; beak 0.4–1.7 mm; $2n=22$; otherwise much like no. 4. In circumneutral to somewhat alkaline shallow water of ponds and swamps; O. and Mich. to Minn. and S.D., s. to Tenn., Ala., and Tex. July–Oct. (*S. engelmanniana* var. *b.*)

7. Sagittaria cuneata Sheldon.   Northern a.-h.; wapato.   Emersed or submersed, and vegetatively plastic, with large, edible tubers; lvs long-petioled, sagittate to lanceolate or phyllodial, the blade (when emersed) with lanceolate to deltoid-ovate central lobe 10–18 × 1–10 cm and triangular, pointed, usually much shorter basal lobes, commonly (and unlike all our other spp.) cordate when floating; submersed phyllodia, when present, commonly ribbon-shaped, to 5 dm; scape 1–5+ dm, with 2–10 whorls of fls, occasionally branching at the lowest whorl, the upper fls staminate, the lower pistillate, on ascending pedicels 0.5–2 cm; bracts 1–4 cm, lanceolate, acute or acuminate, whitish and delicate; sep ovate, acuminate, 4–8 mm, reflexed in fr; pet 7–10 mm; stamens 15–25, the filaments subulate, glabrous, about as long as the anthers; achenes 1.8–2.6 mm, with a prominent, rounded wing on the back, the ventral wing nearly or quite as large as the dorsal and confluent with the beak, nearly or quite without facial wings, but with resin-ducts; beak very short, erect, 0.1–0.5 mm; $2n=22$. Ponds and marshes, in water or mud (often sandy silt); N.S. and Que. to Mack. and Alas., s. to N.Y., n. O., Ill., Okla., Tex., and Calif. July–Sept. (*S. arifolia.*)

8. Sagittaria latifolia Willd.   Common a.-h.; wapato.   Vegetatively very plastic, ± erect and with large, edible tubers; lvs sagittate or sometimes narrow and phyllodial, the blade 5–40 × 0.5–25 cm, with narrow or broad lobes; scape 1–12 dm, with 2–15 whorls of fls, sometimes branching at the lowest whorl, the upper fls staminate on short pedicels, the lower pistillate on longer, ± ascending pedicels; bracts 4–15 mm, boat-shaped, broad-based, rounded above or broadly acute, 4–15 mm, papery but friable at the tip; sep ovate, obtuse, 5–11 mm, reflexed in fr; pet 1–2 cm; stamens 20–40, on slender, glabrous filaments; achenes 2.5–4 mm, winged on the margins only, or with a poorly developed wing on each face, with (var. *latifolia*) or without (var. *pubescens*) resin-ducts; beak 0.6–1.8 mm, set at essentially right angles to the body; $2n=22$. Abundant in swamps, ponds, and streams; N.S. and Que. to s. B.C., s. to trop. Amer. July–Sept. The widespread, glabrous var. *latifolia* is uncommon in the principal range of the stellate-hairy var. *pubescens* (Muhl.) J.G. Smith, which occurs chiefly from the mts. to the coast, from s. Pa. to W.Va. and Ga., w. occasionally to O., Tenn., and Tex.

9. Sagittaria lancifolia L.   Lance-lvd sagittaria.   Erect emersed aquatic; lvs lanceolate or ovate, 12–35 × 1–10 cm, without basal lobes; scape 5–20 dm, with 4–10 whorls of fls, sometimes branching at the lowest whorl, the upper fls staminate on filiform pedicels 2–3 cm, the lower pistillate on stouter, ascending pedicels 1–1.5 cm; bracts oblong-ovate, obtuse or acute, somewhat boat-shaped, 0.5–1.5 cm, evidently, copiously, and irregularly papillate-roughened or somewhat corrugate-ridged on the back; sep reflexed, oblong-ovate, 3–6 mm, roughened like the bracts; pet 8–15 mm; stamens 12–30, the filaments slender, linear, longer than the anthers, roughened with minute scales; achenes 1.5–2.3 mm; narrowly obovate to often falcate, with marginal wings only, or with 1 or 2 poorly developed facial wings, sometimes with a resin-duct on each face; beak broad-based, obliquely ascending, 0.3–0.7 mm; $2n=22$. In shallow water and swamps; near the coast from Del. to Fla., Tex., Mex., W.I., and n. S. Amer. Our plants, as here described, are var. *media* Micheli. (*S. falcata*.) Typical *S. lancifolia* is mainly tropical and subtropical.

10. Sagittaria rigida Pursh.   Sessile-fruited a.-h. Erect or lax, emersed or submersed, highly variable according to the habitat; lvs lanceolate and entire, or shortly sagittate with narrow basal lobes, or phyllodial; scape 1–8 dm, with 2–8 whorls of fls, typically geniculate at the lowest whorl, the upper fls staminate on filiform pedicels 1.5–3 cm, the lower pistillate, sessile or nearly so (the pedicel sometimes to 5 mm, rarely to 1 cm); bracts ovate, obtuse, 4–6 mm, connate below, smooth; sep ovate, 4–7 mm, reflexed at maturity; pet 1–3 cm; stamens 15–many; filaments subulate, roughened with minute scales; achenes obovate to oblong, 2–3 mm, narrowly winged on the margins only, usually with a resin-duct on each face; beak 0.8–1.4 mm; $2n=22$. In shallow water of swamps and ponds; Me. and Que. to Minn., s. to Va., Tenn., Mo., and Neb. July–Sept. (*S. heterophylla*.)

11. Sagittaria graminea Michx.   Grass-lvd sagittaria. Erect and emersed or lax and submersed; rhizome short; lvs bladeless, flat, phyllodial, or with lance-linear to broadly elliptic-ovate blade, very rarely hastate or sagittate, 4–30 × 0.5–10 cm; scape simple or sometimes branched at the base of the infl, 0.3–5 dm; fls in 2–12 whorls, the upper usually staminate on slender, erect pedicels, the lower pistillate on thicker, ascending pedicels; bracts 3–15 mm, connate up to half-length; sep ovate, 3–6 mm, reflexed in fr; pet white or pink, 1–2 cm, often as wide; stamens 12–many; filaments dilated, roughened with minute scales; achenes 1.2–3 mm, winged on the margins and with one or more facial wings, the beak less than 0.75 mm; $2n=22$. Swamps, mud, and shallow water; Nf. and s. Lab. to Minn. and S.D., s. to Cuba and Tex. July–Sept. Three vars. with us:

a Phyllodia 1–2.5 cm wide, obtuse or rounded at the tip; fruiting pedicels 4–6.5 cm; beak 0.1–0.3 mm;
  coastal plain from s. Va. to Fla. ....var. *weatherbiana* (Fern.) Bogin.
a Phyllodia to 1.5 cm wide, acute; fruiting pedicels (0.5–)1–4 cm.
  b Beak of the achene 0.4–0.7 mm; dorsal wing crenulate; filaments 1.2–1.7 mm, longer than the
    anthers; Minn. to n. Io., e. to n. Mich. and the n. shore of Lake Huron. (*S. cristata*.) ....
    .... var. *cristata* (Englem.) Bogin.
  b Beak 0.1–0.3 mm, or obsolete; dorsal wing entire or occasionally erose; filaments up to ca 1 mm,
    about equaling or shorter than the anthers; widespread, but uncommon in the range of var.
    cristata. (*S. eatonii*.) .... var. *graminea*.

12. Sagittaria teres S. Watson.   Quill-lvd sagittaria. Erect aquatic with long, slender rhizomes; lvs phyllodial, nodose, terete, attenuate, 3–20 cm, occasionally with a small blade; scape 1–8 dm; fls in 2–4 whorls, the upper staminate, the lower pistillate, both on slender, ascending-spreading pedicels 1–3 cm, usually only 1 or 2 fruiting; bracts small, ovate, 2–3 mm, connate nearly or fully throughout; sep ovate, obtuse, 3–5 mm, reflexed in fr; pet 3–6 mm; stamens 12–15; filaments dilated, roughened with minute scales; achenes 2–3 mm, with 1–3 prominent wings on each face, these and the dorsal wing crenulate; beak erect to horizontal, 0.3–0.4 mm; $2n=22$. On sandy soil in shallow, acid water from Mass. to N.J., often with no. 4. Aug., Sept.

3. ECHINODORUS Rich.   Bur-head. Fls perfect; receptacle elevated, convex or globose; sep 3, persistent, sometimes accrescent; pet 3, white, deciduous; stamens 6–30; filaments elongate; pistils borne in several series and aggregated into a globose head; achenes turgid, evenly ribbed or ridged; scape erect or prostrate, bearing an infl of 1–many verticils of fls, sometimes branched from the lower verticils; verticils 3-bracted and commonly with additional bracteoles. (*Helianthium*.) 20+, widespread, mainly tropical.

1 Achenes 10–20 per head, stamens 6 or 9; pet 1–3 mm .... 1. *E. tenellus*.
1 Achenes 40 or more per head; stamens (9)12 or more; pet 5–12 mm.
  2 Scape erect; style longer than the ovary .... 2. *E. berteroi*.
  2 Scape soon becoming ± prostrate; style shorter than the ovary .... 3. *E. cordifolius*.

1. Echinodorus tenellus (Martius) Buchenau.   Little b.-h. Delicate stoloniferous perennial, simple, erect, to 1 dm; lvs with long-petiolate, linear to more often elliptic or lanceolate blade 1–3 cm, or phyllodial and bladeless when submsersed; fls in 1 (or 2) whorls of 3–6, the slender pedicels usually recurved at the tip; bracts minute, 1–3 mm; sep orbicular-ovate, 1–2 mm, appressed in fr; pet white or pink, suborbicular, 1–3 mm; stamens 6 or 9; anthers basifixed, 0.2 mm; achenes 10–20 in a loose head, dark red to nearly black, 1 mm, 8-ribbed, virtually beakless. In mud or wet sand or shallow water; trop. Amer., n. to se. U.S. and irregularly in the Atlantic states (mainly on the coastal plain) to Mass. (where now wanting), and up the Mississippi R. to Ill. and Mo.; irregularly to Kans., sw. Ky., and formerly Mich. Our plants, as here described, are var. *parvulus* (Engelm.) Fassett. (*E. parvulus*.)

2. Echinodorus berteroi (Sprengel) Fassett.   Tall b.-h. Erect annual, slender or often robust, 1–6 dm; lvs erect, long-petiolate, with ± ovate, basally ± cordate blade mostly 4–20 × 2–12 cm, or in small forms the blade smaller, more lanceolate, and tapering to the base, in submersed forms the blade very thin and crisped;

pellucid lines of the lvs (visible in transmitted light) mostly less than 1 mm apart and often several mm long; scape simple or branched, with usually several or many whorls of 3–8 fls on stiffly ascending pedicels; bracts lanceolate or linear, 3–6 mm; sep ovate, acute, 4–5 mm, ± reflexed in fr, the veins smooth, scarcely raised; pet white, broadly ovate, 5–10 mm; stamens (9)12 or more; anthers 0.5–0.8 mm; versatile; achenes 40+, brown, 2.5–3.5 mm, the erect to oblique beak 0.5–1 mm, up to half as long as the body, the fruiting head 3–7 mm thick, echinate in profile; each side of the achene with 5 arching ribs, 2 of them wing-like and alternating with the others, the keel entire; each side of the nutlet with a lanceolate, acuminate gland toward the top, entering the base of the beak. In swamps and ditches; often on sandy soil; C. Amer. and W.I. to Calif., Tex., and n. in the Mississippi R. drainage to O., Ill., and S.D. June–Oct. (*E. rostratus; E. cordifolius,* misapplied.) Our plants, as here described, are var. *lanceolatus* (Engelm.) Fassett.

3. **Echinodorus cordifolius** (L.) Griseb.   Creeping b.-h. Perennial, the fibrous roots often prominently tuberculate; scape at first erect, but soon prostrate and even rooting, often producing small lvs and bulbils at the nodes of the infl; lvs erect, broadly ovate, obtuse, ± cordate at base, 5–20 × 3–15 cm; pellucid lines of the lvs mostly 1 mm or more apart and rarely over 1 mm long; fls in whorls of 5–15 on long, lax, often ± recurved pedicels; bracts lance-linear, long-acuminate, 1–2.5 cm; sep ovate to orbicular, obtuse, 5–7 mm, ± reflexed in fr, the veins evidently raised-roughened on the back; pet white, obovate, 6–12 mm; stamens ca 21; anthers 1–1.5 mm, versatile; achenes 40+, brown, 1.8–2.5 mm, with an ascending beak only 0.2–0.5 mm, the fruiting head ca 1 cm thick, nearly smooth in profile; each side of the achene with 3–4 abruptly curved and sometimes joining ribs of which the 1 or 2 toward the dorsal edge are wing-like distally; summit of the keel often crested; glands of the achene rounded at both ends and not closely approaching the beak; $2n=22$. Swamps; D.C. to Fla., along the coastal plain, w. to Tex., and n. in the interior to s. Ind., Ill., Mo., and Kans., s. to tropical Amer. July–Oct. (*E. radicans.*) Ours is nomenclaturally typical.

FAMILY **HYDROCHARITACEAE,** the Frog's-bit Family

Fls regular or (*Vallisneria*) slightly irregular, perfect or more often unisexual; perianth of 3 distinct, green sep and 3 distinct pet, sometimes elevated on a long, slender hypanthium, or the pet or the whole perianth wanting; stamens 3–many, sometimes some of them staminodial; pistil of 3–6+ carpels weakly united to form a compound, inferior, unilocular ovary, often with ± deeply intruded partial partitions; styles as many as the carpels, often lobed or bifid, sometimes shortly connate below into a common style; placentation laminar, the ovules scattered over the surface of the partial partitions, or merely parietal (or nearly basal) when the partitions are not intruded; fr submerged, usually opening irregularly; seeds several to many, usually without endosperm; embryo straight, with an obliquely terminal cotyledon; submersed or partly emergent, rooted or free-floating aquatic herbs; infl a compact, usually few-fld cyme (especially the staminate fls) or a single fl (especially the pistillate fls), collectively subtended by (1)2 distinct or more often ± connate bracts that form a sessile to long-pedunculate spathe. 15/100.

1 Lvs cauline.
  2 Pet showy, much larger than the sep; fls nectariferous, pollinated by insects; principal lvs in whorls
       of 4–8 .... 1. *Egeria.*
  2 Pet only slightly if at all larger than the sep, or reduced or obsolete; fls with explosive anthers,
       pollinated by water.
    3 Stamens 9; principal lvs in whorls of 3, or merely opposite .... 2. *Elodea.*
    3 Stamens 3; principal lvs in whorls of 3–8 .... 3. *Hydrilla.*
1 Lvs basal or essentially so.
  4 Lvs ribbon-shaped, much elongate, not differentiated into blade and petiole .... 4. *Vallisneria.*
  4 Lvs differentiated into blade and petiole.
    5 Filaments connate for most of their length into a central column; pet less than 1.5 times as long
       as the sep; stipule 1, axillary, adnate below to the petiole .... 5. *Limnobium.*
    5 Filaments united below into radial pairs, pet more than 1.5 times as long as the sep; stipules 2,
       lateral, free from the petiole .... 6. *Hydrocharis.*

1. **EGERIA** Planchon.   Much like *Elodea,* but the fls more showy, nectariferous, and insect-pollinated; pet much longer than the sep; dioecious; fls of both sexes elevated to the surface by the slender, pedicel-like hypanthium; staminate spathes 2–4-fld; staminate fls with 9 distinct stamens and a small central nectary; filaments elongate; anthers short, each pollen-sac opening longitudinally, not explosive; pistillate spathes split half way down one side; style deeply trifid, each segment deeply 2–3-cleft and with a small nectary externally at the base of each lobe; lower lvs opposite or ternate, the others in whorls of 4–8. 2, S. Amer.

1. **Egeria densa** Planchon. Brazilian water-weed. Principal lvs in whorls of 4–6, oblong or broadly linear, 12–40 × 1.5–5 mm, spreading, the internodes short; staminate spathes 11–13 mm, deeply cleft down one side; hypanthium becoming 3–6 cm, the sep 3–4 mm, the pet 9–11 mm; pistillate plants with smaller pet, rare, unknown in our cult. and escaped plants; $2n=48$. Native from se. Brazil to n. Arg., commonly cult. in aquaria and occasionally established in ponds in our range. (*Anacharis d.; Elodea d.*)

2. **ELODEA** Michx. Water-weed. Fls unisexual (and the plants dioecious) or seldom perfect, borne singly in submersed, sessile or subsessile, axillary, bidentate or bifid spathes, elevated to the surface of the water on a long, slender, pedicel-like hypanthium (or the staminate fls sessile, breaking loose, and floating); fls ephemeral and delicate; pet white or purple, not much (if at all) longer than the sep, or wanting; no nectaries; staminate fls with typically 9 stamens, an outer cycle of 6 distinct, an inner cycle of 3 with the filaments evidently connate below, the filaments shorter than the explosively dehiscent, eventually explanate anthers; pollen floating; pistillate fls with 3 staminodia, and an elongate style with 3 slender, simple or often bifid stigmas; ovules few, parietal or nearly basal; fr capsular; submersed aquatic perennials, rooted to the bottom or free-drifting when broken loose; stems simple or sparsely branched; lvs all cauline, sessile, slender, minutely and sharply serrulate, 1-nerved, at least the middle and upper opposite or whorled; stipules minute, evanescent. (*Anacharis, Philotria.*) 5, New World.

E. schweinitzii (Planchon) Caspary, with perfect fls and 3 stamens, originally native apparently to s. N.Y. and e. Pa., has not been recollected since 1832 and is apparently extinct. It may reflect hybridization between *E. canadensis* and *E. nuttallii*.

1 Staminate spathe 7–10+ mm, its fl at anthesis elevated on a long, capillary hypanthium; lvs avg 2
    mm wide, firm, the upper closely imbricate . . . . 1. *E. canadensis*.
1 Staminate spathe 2–4 mm, its fl sessile, separating from the plant and floating at anthesis; lvs avg 1.3
    mm wide, flaccid, the upper not closely imbricate . . . . 2. *E. nuttallii*.

1. **Elodea canadensis** Michx. Common w.-w. Lower lvs small, opposite, ovate or lance-ovate, the others ternate, 6–17 × 1–5 (avg 2) mm, linear to lance-oblong, bright green, ± firm, the upper closely imbricate; fls elevated on a very slender, pedicel-like hypanthium 2–30 cm; staminate spathes 7–10+ mm (including the slender base), expanded and 2-lobed above; pistillate spathes 8–20 mm, cylindric; sep of staminate fls 3–4 mm, of pistillate fls 2–3.5 mm; pet white, clawed, 2–4.5 mm; fr turbinate, 5–6.5 mm with a 5–6 mm beak; $2n=48$. In quiet water, Que. to B.C., s., to N.C., Ala., Ark., and Calif. July–Sept. (*Anacharis c.; Philotria c.*)

2. **Elodea nuttallii** (Planchon) St. John. Free-fld w.-w. Habitally like no. 1, but the lvs narrower, paler, and softer, mostly 0.3–1.75 (avg 1.3) mm wide, not imbricate at the stem-tip; staminate spathes sessile, broadly ovoid 2–4 mm, the sessile staminate fl breaking loose and floating to the surface at anthesis, its sep 2 mm, the pet wanting or minute (to 1.5 mm); pistillate fls with sep 1–2 mm and pet 1–2(–2.5) mm; $2n=48$. In quiet, usually noncalcareous water; Me. and Que. to Minn. and Nebr., s. to N.C., Miss., and Mo., and irregularly in w. U.S. July–Sept. (*Anacharis n.; A. occidentalis; Philotria angustifolia; P. minor.*)

3. **HYDRILLA** Rich. Much like *Elodea*; monoecious or dioecious; stamens 3, principal lvs in whorls of 3–8; stipules fringed. 1, originally Old World.

1. **Hydrilla verticillata** (L. f.) Royle. Plants producing small tubers; lvs up to 20(–40) × 2(–5) mm, dentriculate; stipules fringed with orange-brown hairs; staminate fls breaking loose as in *Elodea nuttallii*; sep 1.5–3 mm; pet narrower, transparent, with a few red streaks; $2n=16$ (diploid), 24 (triploid). Rivers, lakes, and ponds; intr. in se. U.S. and spreading n. near the coast to Va., Md., D.C., and Del., locally abundant. June–Aug.

4. **VALLISNERIA** L. Tape-grass. Dioecious; staminate fls minute, numerous in a head subtended by a short-pedunculate, ovoid, bivalved spathe arising from the base, with 3 sep (one smaller), one pet-vestige, and 2 stamens, the fls at anthesis separating individually and floating to the surface; pistillate fls mostly solitary and sessile in a tubular, apically bifid spathe on a long, slender scape (reaching the surface), with 3 sep, 3 pet-vestiges, 3 staminodia, and 3 broad, bifid stigmas; pistillate scape coiling after anthesis and retracting the fr; ovules scattered over the ovary-wall; fr elongate, cylindric, indehiscent, many-seeded; vigorously stoloniferous, aquatic perennials with long, ribbon-like basal submersed lvs from a very short, erect crown. 2, cosmop.

1. **Vallisneria americana** L.   Water-celery. Lvs very thin, to 2 m, 3–10 mm wide, seldom more; staminate scape stout, 3–15 cm, the fls only 1–1.5 mm wide; stamens erect, with short, ± connate filaments, the androecium surrounded by a ring of short hairs; pistillate fls white, the sep 3.5–6.5 mm; staminodia adnate to the short style; fr 5–12 cm. In quiet water; N.S. and Que. to Minn. and S.D., s. to trop. Amer., and irregularly widespread elsewhere. July–Oct. Ours are var. *americana*.

5. **LIMNOBIUM** Rich.   Frog's bit. Fls unisexual, the plants monoecious; spathes basal and sessile or short-pedunculate, the staminate 1-lvd and subtending 2–9 fls, the pistillate 2-lvd and usually subtending a solitary fl; fls long-pedicellate; sep 3; pet 3, narrow, not more than ca 1.5 times as long as the sep, or wanting; filaments united into a central column, from which the 9–12(18) anthers diverge in the upper half; ovules scattered over the surfaces of the 6–9 deeply intruded (or axially joined?) parietal placentas; styles 6–9 bifid nearly to the base, each cleft into 2 linear, hairy lobes; fr fleshy, many-seeded; aquatic or marsh herbs, rooting and producing tufts of lvs at the nodes of the long stolons, or in deeper water occasionally free-floating; lvs with long petiole and floating or emergent, dilated blade; stipule 1, axillary, adnate below to the petiole; roots usually with numerous simple branches. Monotypic.

1. **Limnobium spongia** (Bosc) Steudel.   American f.b. Lf-blades broadly ovate (especially when emergent) to deeply cordate-orbicular but usually acute, 3–7 cm long and wide, 5–7-veined, the lateral veins arcuate-ascending; floating lvs aerenchymatous and spongy toward the base beneath; pedicels 3–10 cm; pet white, linear or linear-oblong, ca 1 cm, not much longer than the slender sep; anthers elongate, ca 3.5 mm; stigmas conspicuous, 10–15 mm; fr 4–12 mm thick. In ponds and bayous; trop. Amer., n. along the Atlantic coastal plain to N.J. and up the Mississippi embayment to s. Ill.; also in n. Ind. and w. N.Y. June–Sept.

6. **HYDROCHARIS** L.   Frog's bit. Pet broadly obovate, more than 1.5 times as long as the sep; stamens in 4 alternating trimerous whorls (those of the innermost antepetalous whorl staminodial), their filaments connate in pairs below, forming 6 radial pairs; styles usually 6, flat, bifid up to half-length; stipules (1 or) 2, mostly lateral and free from the petiole; roots unbranched; otherwise much like *Limnobium*. 3, Old World.

1. **Hydrocharis morsus-ranae** L.   European f.b. Much like *Limnobium spongia* in aspect, but the lvs more consistently cordate-orbicular, less obviously spongy beneath, and with more broadly arching (even initially descending) lateral veins; dioecious; stolons with seasonally dimorphic terminal buds; pet ca 1 cm, 2–3 times as long as the sep, white with a yellow basal spot; outer (antesepalous) stamens basally connate with those of the third cycle; anthers ca 1 mm long and wide; styles 6, ca 4 mm overall; 2*n*=28. Slow-moving water, ditches, and pools; native of Europe, intr. and spreading, from Que. to s. Ont. and n. N.Y. along the St. Lawrence and Ottawa rivers and along lakes Erie and Ontario.

FAMILY **SCHEUCHZERIACEAE**, the Scheuchzeria Family

Fls in terminal, bracteate racemes, perfect, anemophilous, hypogynous, trimerous; perianth of 2 similar cycles of 3 distinct, yellow-green tep; stamens 6, free and distinct; filaments short; anthers elongate, extrorse; pollen in dyads, inaperturate; carpels 3(–6), slightly connate at base, otherwise distinct, each with an extrorse, sessile stigma; ovules 2 (– several), basal-marginal; fr of recurved-spreading follicles; seed without endosperm; embryo with a single rounded cotyledon; rhizomatous perennial herbs; lvs alternate, cauline and basal, with an open sheath, prominent ligule, and elongate, semiterete blade with an evident pore on the inner side at the tip; stem with numerous long, intravaginal hairs at the nodes. Monotypic.

1. **SCHEUCHZERIA** L.   Characters of the family. Monotypic.

1. **Scheuchzeria palustris** L.   Pod-grass. Stems 2–4 dm; lvs with conspicuously dilated sheath, the basal ones clustered, the cauline 1–3, well separated; lf-blades erect, 5–30 cm × 1–3 mm; racemes 3–10 cm, the lowest bract foliaceous, the second with small or no blade, the others reduced to small sheaths; tep lance-ovate, 2–3 mm; fr 6–8 mm; 2*n*=22. Cold sphagnum-bogs; circumboreal, in Amer. s. to N.J., Pa., Ind., Io., and Calif. Late spring.

FAMILY JUNCAGINACEAE, the Arrow-grass Family

Fls small and inconspicuous, perfect or seldom unisexual, anemophilous, borne in terminal, bractless spikes or racemes; perianth of (4)6 similar and distinct tep in 2 cycles, or sometimes 3 in a single cycle, or of a single tep, or wanting; stamens (1–)6(8), the very short filament generally adnate to the base of the tep; anthers extrorse; pollen in monads, inaperturate; gynoecium of 3, 4, or most commonly 6 carpels, these generally adnate to a central axis from which they later separate, each with its own stigma sessile or on a short style; ovules solitary in each carpel or locule, mostly ventral-basal; individual frs usually basically follicular; endosperm wanting; embryo straight, with a single cotyledon; herbs, mostly semiaquatic and emergent, commonly cyanogenic, most spp. perennial and rhizomatous; lvs mainly or wholly basal, with an evident ligule at the juncture of the slender blade and open sheath, or the elongate lf not divided into blade and sheath. 5 genera, 20 spp.

1. **TRIGLOCHIN** L. Arrow-grass. Fls perfect; perianth of 1 or 2 whorls of 3 tep; anthers broad; ovaries 3 or 6, closely approximate, attached along their inner margin to an erect, elongate axis, each unilocular; stigmas minutely plumose; follicles eventually separating from each other and from the persistent axis; seeds cylindric; scape erect; lvs with an evident sheath and long, narrow blade. 15, cosmop.

1 Ovaries and stigmas 6; axis of the fr-cluster slender not winged .... 1. *T. maritimum.*
1 Ovaries and stigmas 3; axis of the fr-cluster triquetrous-winged.
  2 Tep 6; mature carpels elongate, basally attenuate .... 2. *T. palustre.*
  2 Tep 3; mature carpels short, rounded .... 3. *T. striatum.*

1. **Triglochin maritimum** L. Common a.-g. Plants 2–8 dm; lvs erect, to 5 dm, 1–3 mm wide; raceme 1–4 dm, the curved erect pedicels 5 mm, decurrent as narrow ridges along the rachis; tep 6, 1–2 mm; stamens 6; ovaries and fr ovoid-oblong, 5 mm at maturity; carpels rounded at the base, each one concave along the back and sharply angled on the margins, the fr-cluster thus appearing narrowly 12-winged; seldom the alternate carpels small and sterile; stigmas radiating, persistent; axis between the carpels very slender; $2n=12, 24, 36, 48, 76, 96, 120, 144$. Brackish or fresh marshes and bogs; circumboreal, s. in Amer. to N.J., O., Io., Nebr., and Mex.; Patagonia. May–Aug. (*T. elatum, T. gaspense,* below high tide from Nf. to ne. Me. and sw. N.S.)

2. **Triglochin palustre** L. Slender a.-g. Very slender, 2–4 dm; lvs to 3 dm, 1–2 mm wide; raceme 1–2 dm, the very slender pedicels strictly erect; tep 6, 1–1.5 mm; stamens 6; ovaries 3, ovoid, in fr-linear-clavate, 6–9 mm, at maturity parting from the axis from the base upward, remaining attached at the summit, the base very sharply pointed; axis broadly 3-winged, the wings extending between the carpels; $2n=24$. Brackish marshes along the coast, and in bogs inland; circumboreal, in Amer. s. to Pa., Ind., Io., Nebr., and N.M. May–July.

3. **Triglochin striatum** Ruiz & Pavon. Southern a.-g. Habitally like no. 2; lvs often as long as the scapes; tep and stamens each 3; ovaries 3; fr-cluster subglobose, 3 mm, the separated carpels rounded at both ends, the axis broadly 3-winged. Salt-marshes and wet coastal sands; tropical Amer., n. locally to Del. and Md.; also Old World. May–Sept.

FAMILY POTAMOGETONACEAE, the Pondweed Family

Fls small, perfect, regular, tetramerous throughout, anemophilous or less often hydrophilous, borne in axillary or terminal, bractless, often somewhat fleshy spikes that in most spp. are elevated a little above the water; perianth a single whorl of 4 distinct, fleshy-firm, valvate, shortly clawed tep; stamens 4, opposite the tep and basally adnate to the claw; anthers virtually sessile, extrorse; pollen inaperturate; ovaries (1–)4, distinct, alternating with the stamens, each with a short terminal style or a sessile stigma; ovule solitary, marginal or basal-marginal; fr of distinct achenes or drupelets, generally buoyant in the water, the pericarp aerenchymatous; endosperm wanting; embryo with a large hypocotyl and an obliquely terminal cotyledon that encloses the plumule; glabrous perennial herbs of fresh (seldom brackish) water, rooted in the substrate, most spp. with creeping sympodial rhizomes; lvs alternate or sometimes (especially the uppermost) subopposite, all submersed or some floating, parallel-veined (or with a single midvein), with a well developed basal open or closed sheath, the blade or petiole sometimes attached near the top of the sheath, which projects beyond as a ligule, more often attached farther down the sheath or directly to the node, so that the sheath appears to form an intrapetiolar stipule. A single genus.

1. **POTAMOGETON** L.   Pondweed. Characters of the family. 100, cosmop. The plants normally produce lvs (or phyllodia) on the submersed part of the stem. If the stem reaches the surface, some spp. also produce floating lvs of different shape and texture. In some spp. the midrib is paralleled by one or more rows of colorless, translucent *lacunar cells,* forming lacunar strips or bands that are usually easily observed by transmitted light. Lf-measurements are taken from the principal lvs of the main axis, not from those of short, lateral branches. In some spp. the sides of the achenes are depressed at the center, and the achenes are said to be *pitted.*

1 Submersed lvs, or many of them, 4 mm wide or wider.
  2 Achenes with a beak 2–2.5 mm, and with a prominent basal appendage; lf-margins undulate-crisped
    and sharply serrulate .... 5. *P. crispus.*
  2 Achenes with a beak 1.5 mm or less, and lacking a basal appendage; lvs entire or minutely toothed,
    not undulate-margined, or only weakly so.
    3 Blade of the submersed lvs cordate or auriculate at the sessile base; lvs all submersed.
      4 Base of the lf adnate to the stipular sheath; lf usually minutely serrulate at least toward the
        tip .... 4. *P. robbinsii.*
      4 Base of the lf free from the stipular sheath; lf entire.
        5 Frs 4–5.7 mm, with a sharp, narrow dorsal keel; lf-tip cucullate .... 27. *P. praelongus.*
        5 Frs 1.6–4.2 mm, only obscurely or scarcely keeled; lf tip flat.
          6 Stipules disintegrating into persistent fibers; lvs lance-ovate or narrower, mostly (3–)5–
            12 cm, coarsely 13–21-veined; achenes 2.2–4.2 mm .... 28. *P. richardsonii.*
          6 Stipules evanescent; lvs ovate to suborbicular, mostly 1–5(–8) cm, delicately 7–15-veined;
            achenes 1.6–3 mm .... 29. *P. perfoliatus.*
    3 Blade of the submersed lvs sessile or petiolate, not cordate or auriculate at base.
      7 Stem strongly flattened, two-thirds to three-fourths as wide as the submersed lvs; lvs all
        submersed, up to 5 mm wide .... 7. *P. zosteriformis.*
      7 Stem terete or only slightly flattened, less than half as wide as the submersed lvs, these often
        more than 5 mm wide; floating lvs usually but not always present.
        8 Petioles and stem evidently black-spotted .... 21. *P. pulcher.*
        8 Petioles and stem not black-spotted.
          9 Stipular sheath 4–10 cm; submersed lvs with mostly 7–37 veins.
            10 Principal submersed lvs falcate-folded, mostly 23–37-veined; achenes 4–5.5 mm
              (beak included) .... 20. *P. amplifolius.*
            10 Submersed lvs not falcate-folded, mostly 7–19-veined; achenes 3–4 mm.
              11 Larger submersed lvs with the blade 1–2.5(–3) cm wide, on a petiole (2–)5–13
                cm; dorsal keel of the fr often tuberculate .... 22. *P. nodosus.*
              11 Larger submersed lvs with the blade 2–5 cm wide, sessile or on a petiole to 2
                cm; keel smooth .... 26. *P. illinoensis.*
          9 Stipular sheath 1–3 cm; submersed lvs with 3–9 veins (or up to 13 in no. 18).
            12 Lacunar strips (one on each side of the midvein) in the submersed lvs 1–2+ mm
              wide, consisting of 9–18 rows of cells .... 18. *P. epihydrus.*
            12 Lacunar strips in the submersed lvs to 0.5 mm wide, consisting of 1–4 rows of cells,
             or wanting.
             13 Submersed lvs 5–20 cm; achenes pitted; plants often red-tinged .... 19. *P. alpinus.*
             13 Submersed lvs 3–8 cm; achenes not pitted; plants not red-tinged .... 25. *P. gramineus.*
1 Submersed lvs less than 4 mm wide.
  14 Stipular sheath of submersed lvs adnate to the base of the lf, the tip projecting as a ligule.
    15 Achenes strongly flattened, and with a conspicuous dorsal keel; floating lvs commonly (not
      always) produced.
      16 Adnate portion of the stipular sheath mostly longer than the free ligule; frs with a single
        dorsal keel and smoothly rounded sides .... 16. *P. spirillus.*
      16 Adnate portion mostly shorter than the free ligule; frs usually with a lateral, entire or
        toothed ridge on each side of the dorsal keel .... 17. *P. diversifolius.*
    15 Achenes turgid, not flattened; lvs all submerged.
      17 Lvs finely 20–60-veined, roundly auriculate at base, usually minutely spinulose-serrulate
        distally .... 4. *P. robbinsii.*
      17 Lvs 1–5-veined, entire, not auriculate.
        18 Achenes mostly 3–4.5 mm, with a tiny beak (persistent short style); lvs ± sharp-
          pointed .... 3. *P. pectinatus.*
        18 Achenes mostly 2–3 mm, beakless (the broad stigma sessile and wart-like); lvs mi-
          nutely blunt or rounded or retuse.
          19 Stipular sheaths tightly clasping, only slightly thicker than the stem, 0.5–2 cm;
            spikes with mostly unequally spaced whorls of fls .... 1. *P. filiformis.*

19 Stipular sheaths (at least of the principal lvs) subinflated, much thicker than the
stem, 2–5 cm; spikes with ± equidistant whorls of fls .... 2. *P. vaginatus.*
14 Stipule or stipular sheath free, not adnate to the lf-base.
20 Floating, expanded lvs generally developed on at least some plants in each colony.
21 Floating lvs 0.8–1.5 cm; spikes 3–8 mm; submersed lvs thin and flat, 1(3)-veined, up to
1 mm wide .... 15. *P. vaseyi.*
21 Floating lvs 1.5–12 cm; spikes 10–50 mm; submersed lvs otherwise.
22 Submersed lvs flat, 2–4 mm wide, evidently 5–several-veined and with well developed
lacunar bands; fr with 1 sharp and 2 low keels .... 18. *P. epihydrus.*
22 Submersed lvs petiole-like, up to 2 mm wide, with 3–5 obscure veins, lacking lacunar
bands; fr with a single (or no) keel.
23 Frs 3.5–5 mm, deeply wrinkled on each side; plants coarser, with larger lvs (see
descr.) .... 23. *P. natans.*
23 Frs 2–3.5 mm, the sides smoothish; plants more slender (see descr.) .... 24. *P. oakesianus.*
20 Lvs all submersed.
24 Peduncles elongate, mostly (5–)10–25 cm; lvs very slender, up to 0.5 mm wide, 1-veined
.... 6. *P. confervoides.*
24 Peduncles shorter, or the lvs broader and with more than 1 vein.
25 Stem evidently flattened, 2–3 times as wide as thick; lvs 2–5 mm wide, 15–25-veined;
spikes with mostly 7–11 verticils of fls .... 7. *P. zosteriformis.*
25 Stem terete or only slightly flattened; lvs often narrower, 1–9(–13)-veined; spikes with
1–5 verticils of fls.
26 Fr keeled; spikes capitate, 4–7 mm, with only 1 or 2 whorls of fls, on peduncles
5–15 mm.
27 Fr rounded on the sides and with 3 low, ridge-like keels; rhizome undeveloped
.... 12. *P. hillii.*
27 Fr with concave sides and a single wing-like keel; rhizome elongate .... 14. *P. foliosus.*
26 Fr not keeled (except sometimes in no. 13); peduncles and spikes often longer
and with more than 2 whorls of fls.
28 Stipular sheaths ± strongly fibrous, tending to become shreddy in age.
29 Lvs rounded or merely apiculate at the tip; winter-buds with the inner
lvs modified into a fan-shaped structure .... 8. *P. friesii.*
29 Lvs narrowly pointed; winter-buds scarcely differentiated, or with the
inner lvs modified into a fusiform structure.
30 Bands of lacunar cells wanting or poorly developed; stipular sheaths
white .... 9. *P. strictifolius.*
30 A band of 1 or 2 rows of lacunar cells usually present on each side
of the midrib; stipular sheaths usually brown .... 10. *P. ogdenii.*
28 Stipular sheaths membranous or scarious, delicate, not becoming shreddy.
31 Body of the fr 1.5–2.2 mm; lvs 0.2–2.5 mm wide, acute to sometimes
obtuse .... 11. *P. pusillus.*
31 Body of the fr 2.5–3.5 mm; lvs 1–3.5 mm wide, obtuse or rounded,
sometimes minutely cuspidate .... 13. *P. obtusifolius.*

1. Potamogeton filiformis Pers. Threadlf-p. Stems 1–4 dm, usually erect, branched from the base; rhizome elongate, ending in a slender white tuber 1–2 cm; lvs all submersed, narrowly linear, 5–12 cm × 0.2–1.5 mm, obtuse to merely acute, 1-nerved, with remote cross-veins; lacunar cells none; stipular sheaths 0.5–2 cm, tightly clasping the stem, adnate to the blade for 2–10 mm, the margins connate below; peduncles slender, 2–15 cm; spikes submersed and hydrophilous, 1–5 cm, with mostly 2–5 unequally spaced whorls of fls, at least the lower whorls generally well separated; frs obovoid, 2–3 mm, the beak very low and truncate, the dorsal keel low and rounded, the lateral keels obscure or none; $2n=78$. Shallow calcareous waters; circumboreal, in Amer. s. to Me., N.H., Pa., Mich., Minn., and Ariz. Our plants are mostly var. *borealis* (Raf.) St. John, with the uppermost whorls of fls in the spike adjacent, and the lower ones at most 7 mm apart. (*P. interior.*) The chiefly northern var. *filiformis,* seldom found in our range, has laxer spikes, even the upper whorls separated.

2. Potamogeton vaginatus Turcz. Bigsheath-p. Much like no. 1, avg coarser; tubers 3–5 cm; lvs 1–2 mm wide; stipular sheaths (at least of the primary lvs) subinflated, 2–3 times as thick as the stem, 2–5 cm, open to the base; spike with 5–12 nearly equidistant whorls of fls, 3–8 cm at maturity; fr 3–3.5 mm; $2n=78$. Deep, alkaline or brackish waters; circumboreal, in Amer. s. to N.J., Mich., N.D., and Utah.

3. Potamogeton pectinatus L. Sago-p. Stems 3–8 dm, simple or sparingly branched below with elongate internodes, above usually flexuous, freely dichotomously branched, with internodes 1–3 cm; rhizome slender, ending in a white tuber 1–1.5 cm; lvs all submersed, narrowly linear, 3–10 cm × 0.5–1.5 mm, 1-nerved, tapering to an acute tip; stipular sheaths adnate to the blade and rather closely clasping the stem for 1–3 cm, the margins connate at least below, the free tip 2–10 mm; spikes submersed and hydrophilous, 1–6 cm, on peduncles 3–10 cm, with several whorls of fls, the lower separated; frs obliquely obovoid, 3–4.5 mm, distinclty short-beaked, rounded on the back, with a low dorsal keel only, or with 2 lateral keels only, or with 3 keels; $2n=78$. Calcareous or alkaline, usually shallow water; nearly cosmopolitan, and widespread in our range. A major food for ducks.

4. Potamogeton robbinsii Oakes. Fern-p. Stems to 1 m, sparingly branched below, repeatedly branched above when flowering; rhizome not tuberous; lvs all submersed, crowded on sterile stems (the internodes often only 1 cm), rather firm, distichous, linear, 3–10 cm × 3–8 mm, abruptly contracted and with rounded auricles at the juncture with the stipule, with prominent midvein and numerous (20–60) fine lateral veins, the margin pale and somewhat cartilaginous, usually minutely spinulose-serrulate at least toward the acute tip; stipular sheath open, adnate to the blade for 5–15 mm, the free part as long or longer, soon disintegrating into coarse fibers; peduncles numerous, almost paniculate, 2–5 cm, subtended by stipular sheaths with the blades much reduced or none and often accompanied by winter-buds; spikes slender, 7–15 mm, with 3–5 separated whorls of fls; achenes rarely produced, obovoid, 3.5–4.5 mm, with a narrow, sharp dorsal keel and 2 rounded lateral ones; $2n=52$. Quiet water; Que. to B.C., s. to Del., Ala., Minn., and Oreg.

5. **Potamogeton crispus** L. Curly p. Stems compressed, 3–8 dm, sparingly branched; rhizomes elongate and slender; lvs all submersed, linear-oblong, 3–8 cm × 5–12 mm, rounded or obtuse to minutely cuspidate, finely and irregularly toothed and rather crisply undulate-margined, narrowed or rounded to a sessile base, 3–5-veined; stipular sheaths 3–8 mm, adnate at base to the lf, scarious, soon disintegrating; hard winter-buds commonly produced; peduncles 2–5 cm, often recurved in fr; spike dense, 1–2 cm; body of the achene ovoid, 3 mm, shallowly pitted, with 3 round dorsal keels, the central one prolonged at base into a projecting appendage; achene-beak erect, conic, 2–2.5 mm; $2n=52$. Native of Europe, locally intr. in alkaline or high-nutrient waters nearly throughout our range.

6. **Potamogeton confervoides** Reichb. Alga-p. Stems 1–8 dm, slender and delicate, repeatedly branched; rhizome elongate; lvs numerous, all submersed, very soft and delicate, flat, 2–5 cm × 0.25 mm, tapering to a hair-like tip, 1-nerved, the space from the midrib to the margin entirely lacunar; stipular sheaths axillary, obtuse, 1–5 cm, seldom longer, evanescent; winter-buds fusiform, 1–2 cm; peduncles usually solitary and terminal below, (3–)5–25 cm, slender below, somewhat clavate; spike capitate, few-fld, 5–12 mm; achenes turgid, broadly obovoid, 2–3 mm, with a sharp dorsal keel flanked by a pair of more obscure ones. Shallow, acid waters; Nf. to Pa. and N.J.

7. **Potamogeton zosteriformis** Fern. Flatstem-p. Stems elongate, freely branched, strongly compressed, 1–3 mm wide, mostly without nodal glands; rhizome scarcely developed; lvs all submersed, linear, 1–2 dm × 2–5 mm, acute or cuspidate, scarcely narrowed at base, with 1 or 2 main veins and numerous (to 35) very fine ones; stipular sheaths axillary, free, rather firm, 1–2 cm, obtuse to acuminate; winter-buds commonly produced, 4–7.5 cm, their lvs not strongly differentiated; peduncles stout, 2–5 cm, often curved; spike cylindric, 1.5–2.5 cm, with 7–11 whorls of fls; achenes broadly elliptic-ovoid, 4–4.5 mm, with a sharp but narrow, slightly dentate dorsal keel; $2n=52$. Ponds and slow streams; Que. and N.B. to B.C., s. to Va., Ind., Io., and Calif. (*P. compressus* and *P. zosterifolius,* misapplied.) A putative hybrid with no. 9 is *P.* × *haynesii* Hellquist. (*P. longiligulatus,* misapplied.)

8. **Potamogeton friesii** Rupr. Fries' p. Stem to 1 m or more, sparingly branched below, often more freely so above, slightly compressed, to 1 mm wide, commonly with paired glands at the nodes; rhizome scarcely developed; lvs all submersed, linear, 3–7 cm × 1.5–3 mm, minutely cuspidate at the obtuse or rounded tip, 5–7(9)-veined, the lateral veins very delicate, the midvein flanked by a row of lacunar cells on each side extending two-thirds its length, or these wanting; stipular sheaths axillary, free, white, 0.5–2 cm, closed at base when young, becoming fibrous in age; winter-buds commonly produced, 1.5–5 cm, the inner lvs reduced and arranged into a fan-shaped structure, set at right angles to the outer lvs; peduncles 1.5–5 cm, flattened and clavate; spike slender, 8–18 mm, with 2–5 remote whorls of fls; frs ovoid to obovoid, 2–3 mm, with rounded sides, not or scarcely keeled; $2n=26$. Chiefly in calcareous waters; Que. and Nf. to Mack. and Alas., s. to N.J., Pa., and Ind., n. Io., Nebr., and Utah.

9. **Potamogeton strictifolius** Ar. Bennett. Straight-lvd p. Stems very slender; to 1 m, branched above, commonly with paired glands at the nodes; rhizome scarcely developed; lvs all submersed, linear, rather stiff, 2–6 cm × 0.5–2 mm, acute or attenuate or often bristle-tipped, with bold, vein-like margins, 3–5(7)-veined, the lateral veins very delicate, the midvein not bordered by lacunae or with 1 row on each side in the basal third only; stipular sheaths axillary, free, white, 7–15 mm, closed at base when young, becoming conspicuously fibrous in age; winter-buds commonly produced, flattened, the outer lvs divergent, arcuate, inrolled; peduncles mostly terminal, very slender, usually 3–4 cm, not thickened distally; spike slender, 10–15 mm, with 3 or 4 remote whorls of fls; frs obovoid, 2–3 mm, not keeled; $2n=52$. Alkaline ponds and streams; Que. to Mack., s. to Conn., N.Y., O., n. Ind., Minn., and Utah. (*P. rutilus,* misapplied.)

10. **Potamogeton ogdenii** Hellquist & Hilton. Ogden's p. Stems freely branching, slender but rigid, commonly with paired glands at the nodes; rhizome scarcely developed; lvs all submersed, linear, cuspidate to aristate, 5–10 cm × 1–3 mm, 3–9(–13)-veined, commonly with 1 or 2 rows of lacunar cells on each side of the midvein; stipular sheaths axillary, free, brown (white), 1–2 cm, becoming fibrous at the tip; winter-buds uncommon, sometimes with the inner lvs rolled into a hard, fusiform structure; peduncles mostly terminal and erect, 1–2 cm, slightly clavate; spike cylindric, 5–11 mm, with 2–4 whorls of fls; fr rarely produced, suborbicular, plump, 2.5–3 mm excluding the prominent 0.5 mm beak, obscurely or not at all keeled. Alkaline waters; Mass., Vt., and N.Y.

11. **Potamogeton pusillus** L. Slender p. Stems very slender, to 1.5 m, with numerous branches distally, with or without nodal glands; rhizome scarcely developed; lvs all submersed, narrowly linear, 1–7 cm × 0.2–2.5 mm, 1- or 3(5)-veined, with 0–5 rows of lacunar cells on each side of the midvein; stipular sheaths axillary, free, to 1 cm, delicate, the margins connate or merely overlapping; winter-buds commonly produced, 1–3 cm, the inner lvs rolled into a hard, fusiform structure; peduncles filiform to slightly clavate, axillary or terminal, mostly erect, 0.5–6 cm; spikes capitate to cylindric, 2–10 mm, with 1–4 whorls of fls; body of the fr ovoid to obovoid, usually plump, not keeled, 1.5–2.2 mm, the beak 0.1–0.6 mm; $2n=26$. Common and

widespread in both acid and alkaline waters from Nf. to Alas., s. to Fla. and Mex. Highly variable. (*P. berchtoldii; P. panormitanus; P. tenuissimus.*) The most narrow-lvd phase, with subulate, 1-nerved lvs 0.2–0.6 mm wide, considered to be restricted to N. Engl. and s. Que., has been called var. *gemmiparus* Robbins. (*P. gemmiparus.*)

12. Potamogeton hillii Morong. Hill's p. Stems very slender, usually freely branched, to 1 m, with or more often without nodal glands; rhizome scarcely developed; lvs all submersed, narrowly linear, 3–7 cm × 1–2 mm, usually sharply acute or cuspidate, 3(5)-veined, the lateral veins very delicate, the midvein flanked by one or 2 rows of lacunar cells on each side; stipular sheaths axillary, free, white or light brown, 1–1.5 cm, becoming fibrous; winter-buds ± sharply differentiated; peduncles mostly axillary, 5–15 mm, clavate, commonly recurved; spike capitate, 4–7 mm, with 1(2) whorls of fls; body of the fr obovate, flattened, 2.5–4 mm, with a low, sharp dorsal keel and a pair of lateral keels, the beak ca 0.5 mm; $2n=26$. Chiefly in clear, cold, calcareous waters; irregularly from Mass. and Vt. to Pa. and w. to Ont., O., and Mich. (*P. porteri.*)

13. Potamogeton obtusifolius Mert. & Koch. Bluntlf-p. Stems very slender, to 1 m, much branched, commonly with paired glands at the nodes; rhizome scarcely developed; lvs all submersed, often suffused with red, linear, 3–10 cm × 1–3.5 mm, obtuse or rounded (sometimes minutely cuspidate) at the tip, the lateral veins 2(4), very delicate, the midvein flanked by 1–4 rows of lacunar cells on each side; stipular sheaths axillary, free, 1–2 cm, very thin and scarious, the margins merely overlapping; winter-buds commonly produced, 3.5–8 cm, but the lvs not strongly differentiated; peduncles axillary, 1–2(–4) cm; spikes 8–15 mm, thick-cylindric, with mostly 3 closely set whorls of fls; body of the fr obovoid, slightly compressed, with or without a low dorsal ridge-keel, the beak 0.6–0.7 mm; $2n=26$. In ponds and streams; Que. and N.S. to Mack. and B.C., s. to n. N.J., Minn., and Wyo.

14. Potamogeton foliosus Raf. Leafy p. Stems to 7 dm, usually freely branched, slightly compressed, to 1 mm wide, usually without nodal glands; rhizome slender and elongate; lvs all submersed, linear, 1.5–8 × 0.5–2 mm, mostly acute or apiculate, 1–3(5)-veined, with 0–2 rows of lacunar cells flanking the midvein; stipular sheaths axillary, free, greenish to brown, 0.5–2 cm, tubular, with connate margins, delicate to slightly fibrous; winter-buds uncommon, 1–2 cm, the inner lvs rolled into hard, fusiform structure; peduncles axillary, usually clavate, recurved, 5–15 mm, seldom longer; spikes mostly capitate, 2–7 mm, with 1 or 2 whorls of fls; body of the fr 1.5–2.7 mm, with rounded to often concave sides and an undulate, wing-like dorsal keel, the beak 0.2–0.6 mm; $2n=28$. Streams and ponds; N.S. and Que. to Alas., s. to W.I. and C. Amer. Ours is the widespread var. *foliosus.*

15. Potamogeton vaseyi Robbins. Vasey's p. Annual; stems filiform, freely branched, 2–5 dm, often with paired glands at the nodes; submersed lvs thin and transparent, narrowly linear, 2–6 cm × 0.2–1 mm, sharply acute, 1-nerved or rarely with 2 weak lateral nerves, without lacunar cells; stipular sheaths free, linear, scarcely encircling the stem, 4–10 mm, scarious, weakly fibrous in age; floating lvs rather sparingly produced on at least some plants of a colony, the blade spatulate to obovate, 8–15 mm, 5–9-veined, the spaces between the veins filled by minute, elliptic lacunar cells, the petiole ca as long as the blade; winter-buds 1–3 cm, the inner lvs rolled into a hard, fusiform structure; peduncles slender, 0.5–3 cm; spikes cyclindric, 3–8 mm, with 1–4 ± contiguous whorls of fls; fr obovoid, 1.5–2.5 mm, with flat or shallowly pitted sides, with or without a low, rounded dorsal keel; $2n=28$. In quiet water; N.B. and s. Que. to Minn., s. to Pa., O., and Ill. (*P. lateralis,* the name based on a mixture of *P. pusillus* and *P. vaseyi.*)

16. Potamogeton spirillus Tuckerman. Northern snailseed-p. Much like no. 17; submersed lvs 1–8 cm × 0.5–2 mm, mostly 15–80 times as long as wide, rounded or obtuse to sometimes acutish at the tip, 1- or 3-veined, usually with 2 or 3 rows of lacunar cells flanking the midvein; adnate part of the stipular sheath longer than the free part; floating lvs (not always produced) 7–35 × 2–13 mm, with 5–13 veins; submersed spikes globular, with 1–8 frs, on recurved peduncles to 3 mm, the aerial ones more cylindric, to 13 mm, with 4–35 frs; fr 1.3–2.4 mm, with a sharp, entire or sinuate dorsal keel, the smooth sides conforming to the embryo. Shallow water; Nf. to Que. and sw. Ont., s. to Va., n. O., and n. Io. (*P. dimorphus,* misapplied.)

17. Potamogeton diversifolius Raf. Common snailseed-p. Stems slender, to 15 dm, clustered, sparsely branched above the base; submersed lvs linear to filiform, 1–10 cm × 0.1–0.5 mm, obtuse to more often acute or even setaceous-tipped, 1(3)-veined, the larger ones with narrow lacunar bands flanking the midvein; stipular sheaths membranous, 2–18 mm, the adnate part mostly shorter than the free tip; floating lvs (not always produced) lance-elliptic to suborbicular, acute to rounded at each end, 5–40 × 2–20 mm, the 3–17 veins strongly impressed beneath, their stipular sheath free from the 5–40 mm petiole; fruiting spikes dimorphic, those axillary to submersed lvs globular to ellipsoid, with 1–15 frs, on ± recurved, slightly clavate peduncles 1–10 mm, those axillary to floating lvs more cylindric, sometimes to 3 cm, with 5–120 frs, on peduncles 3–32 mm; frs compressed, ± orbicular, 1–2 mm, with a minute beak, the entire or toothed dorsal keel usually flanked by a pair of lateral keels that may be represented only by a row of teeth; embryo snailcoiled. Shallow water; Me. to Mont. and Oreg., s. to Fla. and Mex. Two fairly well marked vars.:

Var. *diversifolius.* Middle cauline lvs 20–180 times as long as wide, mostly 0.3–1.5 (avg 0.5+) mm wide; floating lvs 3–17-veined, 2–20 mm wide; widespread, but in our range n. only to Mass., s. N.Y., n. O., and c. Minn. (*P. capillaceus; P. hybridus,* misapplied.)

Var. *trichophyllus* Morong. Lvs more slender, the middle cauline ones 180–500 times as long as wide, mostly 0.1–0.35 (avg 0.2) mm wide; floating lvs 3–7-veined, to 11 mm wide; northeastern, mainly from Me. to N.J. and Pa., w. irregularly to s. Ont., Mich., n. Ind., Wis., and c. Minn. (*P. bicupulatus; P. capillaceus,* misapplied.)

18. Potamogeton epihydrus Raf. Ribbonlf-p. Stem slender, somewhat flattened, often branched, to 2 m; rhizomes elongate; lvs sessile, linear, to 2 dm, 2–10 mm wide, 5–13-nerved, the midrib flanked by a pair of conspicuous bands (each 1–2+ mm wide) of pale green or translucent lacunar cells; stipules 1–3 cm, free; floating lvs usually numerous, linear-elliptic to somewhat spatulate or narrowly obovate, 3–7 cm × 8–20 mm, 11–27-veined, mostly obtuse, narrowed below to a slender petiole about as long as the blade; peduncles about as thick as the stem, 2–5 cm, seldom longer; spikes numerous, cylindric, dense, 1–3 cm; frs obliquely obovoid, 2.5–4 mm, shallowly pitted, the sharp dorsal keel flanked by a pair of lower ones; $2n=26$. Ponds and slow streams; Nf. to Que. to s. Alas., s. to Ga., La., and Calif.

19. Potamogeton alpinus Balbis. Red p. Plants reddish-tinted; stems elongate, simple or branching, the upper internodes often 1–2 dm; submersed lvs lance-linear to narrowly oblong, (5–)8–14(–20) cm × 5–20 (usually ca 10) mm, mostly 7–9-veined, narrowed to a sessile base, the midrib flanked by narrow bands of lacunar cells; floating lvs (often lacking) mostly opposite, oblanceolate to narrowly obovate, 5–7 cm × 10–25 mm, obtuse, many-nerved, narrowed to a slender petiole, their stipules free, 2–3 cm, very broad (to 16 mm), obtuse or subacute; peduncles 8–15 cm, ca as thick as the stem; spikes dense, cylindric, 1.5–3 cm; frs minutely pedicellate, flattened-ellipsoid, 3 mm, with 1 sharp and narrow and 2 obscure rounded dorsal keels, pitted when young; $2n=26, 52$. Ponds and slow streams; circumboreal, s. in Amer. to Mass., N.Y., Wis., Minn., and Calif. The plants of Amer. and e. Asia have been segregated as var. *tenuifolius* (Raf.) Ogden.

20. Potamogeton amplifolius Tuckerman. Biglf-p. Stem elongate, usually simple; principal submersed lvs falcately folded, 25–37-veined, 8–20 cm, on petioles 1–6 cm, acute at both ends, the lower lanceolate, 2 cm wide, the upper gradually wider; floating lvs elliptic, 5–10 cm, half as wide, obtuse, rounded at base, many-nerved, on petioles 8–20 cm; stipules axillary, free, open, 2-keeled, 3.5–12 cm, tapering to a sharp slender point; peduncles stout, 8–20 cm, somewhat clavate, 3–4 mm thick at the top; spikes dense, 2–5 cm; frs obliquely obovoid, 4–5.5 mm (incl. the 1 mm beak), with 3 low, rounded keels; $2n=52$. Ponds and lakes; Que. to B.C., s. to Ala. and Calif.

21. Potamogeton pulcher Tuckerman. Spotted p. Stems rarely over 5 dm, simple, black-spotted; submersed lvs narrowly lanceolate, undulate-margined, 8–15 × 1–3 cm, 9–19-veined, long-acuminate, tapering at base to a poorly defined petiole; floating lvs broadly elliptic or oval, 4–8 × 2–5 cm, many-veined, obtuse, the base broadly obtuse to subcordate; petioles stout, black-spotted, 2–8 cm; stipules axillary, free, to 6 cm, cuspidate to long-acuminate; peduncles 5–12 cm, somewhat thicker than the stem; spike dense, cylindric, 2–4 cm; frs obliquely obovoid, 3.5–4.5 mm (beak included), with flat sides and 3 low, sharp dorsal keels. Shallow, acid waters and muddy shores; s. Me. to Fla. and Ala., chiefly near the coast; s. Ind. to Mo. and Ark. Shore-forms may lack lvs of the submersed type.

22. Potamogeton nodosus Poiret. Longlf-p. Stem branched, to 2 m; submersed lvs thin, narrowly lanceolate to linear, to 3 dm, 1–2.5(–3) cm wide, 7–15-veined, acute, usually with a pair of conspicuous lacunar bands along the midvein, gradually tapering into a long (2–13 cm) petiole; floating lvs elliptic, 5–13 × 1–4 cm, acute or acutish, on petioles 5–20 cm; stipules axillary, free, 4–10 cm, attenuate to obtuse; peduncles 5–12 cm, stout, often thicker than the stem; spikes dense, cylindric, 3–5 cm; frs broadly semi-obovoid, 3–4 mm (including the short beak), with a sharp, narrow, often tuberculate dorsal keel sometimes flanked by 2 low ridges. Variously slow or fast, deep or shallow, often alkaline water; nearly cosmopolitan, and throughout our range. (*P. americanus; ? P. fluitans.*)

23. Potamogeton natans L. Floating p. Stems 1–2 m, simple or sparingly branched; submersed lvs petiole-like, narrowly linear, 1–4 dm × 1–2 mm, with 3–5-obscure veins; floating lvs elliptic or elliptic-ovate, 5–10 × 2–4.5 cm, obtuse or bluntly short-cuspidate, rounded or subcordate at base, many-nerved; petiole usually much exceeding the blade and flexibly attached to it; stipules axillary, free, 4–10 cm, acute; peduncles 8–15 cm, usually thicker than the stem; spikes dense, cylindric, 2–5 cm; frs obovoid, turgid, with a rather loose, shiny pericarp, 3.5–5 mm, deeply wrinkled on each side, not keeled, or with an obscure dorsal keel; $2n=52$. Ponds and slow streams; circumboreal, s. in Amer. to Pa., Ind., Io., and Calif.

24. Potamogeton oakesianus Robbins. Oakes' p. Much like no. 23; stem to 1 m, often freely branched; submersed lvs petiole-like but very delicate, 0.25–1 mm wide, with 3 veins; floating lvs lanceolate to elliptic or ovate, 2.5–6 × 1–2.5 cm, the broader ones obtuse; stipules 2.5–4 cm; peduncles 3–8 cm, conspicuously thicker than the stem; spike 1.5–3 cm; frs obovoid, 2–3.5 mm, with a tight, dull pericarp, not pitted or wrinkled. In quiet, acid waters; Nf. and Que. to N.J., Pa., and Wis.; rep. from W.Va.

25. Potamogeton gramineus L. Variable p. Stems slender, somewhat compressed, much branched, 3–7 dm; submersed lvs sessile, linear, lance-linear to oblanceolate, 3–8 cm × 3–10 mm, acute to cuspidate, 3–7-veined; floating lvs narrowly to broadly elliptic, 2–5 cm, a fourth to half as wide, 11–19-nerved, obtuse to rounded or subcordate at base, the slender petiole often exceeding the blade; stipules axillary, free, 1–3 cm, obtuse; peduncles 2–4(–15) cm, thicker than the stem; spikes dense, cylindric, 1.5–3 cm; frs obovoid, 2–2.5 mm, the sharp dorsal keel and the obscure lateral ones developed chiefly beyond the middle, the sides shallowly pitted; $2n=52$. A highly variable circumboreal sp., in Amer. extending s. to N.Y., Io., and Calif. (*P. heterophyllus*, misapplied.) Putative hybrids with various other spp. have been called *P.* × *hagstroemii*, *P.* × *spathulaeformis*, *P.* × *subnitens*, and *P.* × *varians*.)

26. Potamogeton illinoensis Morong. Illinois-p. Stem stout, usually much branched, to 2 m; submersed lvs lanceolate to oblanceolate or narrowly obovate, 8–20 × 2–5 cm, 9–19-nerved, acute, tapering to a sessile or short-petioled base; floating lvs not always developed, lanceolate to elliptic, 7–13 × 2–6 cm, the stout petiole 4–11 cm; stipules axillary, free, obtuse, 4–10 cm; peduncles 4–12 cm, usually thicker than the stem, but not clavate; spikes dense, cylindric, 2.5–6 cm; frs broadly semi-obovoid, 3–4 mm (including the 0.5 mm beak), with 3 conspicuous but low dorsal keels; $2n=104$. In quiet water; N.S. to Ont., Minn., and Calif., s. to Fla. and Mex. (*P. angustifolius; P. lucens*, misapplied.)

27. **Potamogeton praelongus** Wulfen. Whitestem-p. Stems to 3 m, freely branched, usually flexuous or zigzag when the internodes are short (3–5 cm); lvs all submersed, sessile, lanceolate to lance-linear, 8–30 × 1–4.5 cm, obtuse or subacute and usually hooded at the tip, rounded and slightly clasping at base, with 11–35 veins, 3 or 5 of them stronger; stipules axillary, white, scarious, 3–8 cm, free but closely appressed to the stem, persistent, shredding at the tip; peduncles about as thick as the stem, usually elongate, to 5 dm; spikes cylindric, 3–7.5 cm, dense or interrupted; frs obovoid, turgid, 4–5.7 mm, short-beaked, with a sharp, narrow dorsal keel and often 2 obscure or rounded lateral ones; $2n=52$. Usually in deep water; circumboreal, s. in Amer. to Conn., Md., Ind., Minn., Colo., and Calif.

28. **Potamogeton richardsonii** (Ar. Bennett) Rydb. Much like no. 29; lvs narrower, lance-ovate to nearly linear, (3–)5–12 cm × 5–20 mm, acute, sessile and cordate-clasping, (3–)13–21(–35)-veined, 3(5) veins stronger; stipules 1–2 cm, soon disintegrating into ± persistent white fibers; peduncles 1.5–25 cm; spikes 1.5–3 cm; frs 2.2–4.2 mm; $2n=52$. Ponds and streams; Que. to Alas., s. to Pa., Ind., Io., Colo., and Calif.

29. **Potamogeton perfoliatus** L. Redhead-grass; perfoliate p. Stem slender, to 2.5 mm, often much branched; internodes mostly 1–3 cm; lvs all submersed, ovate to suborbicular, 1–5(–9) cm × 5–25(–40) mm, mostly broadly rounded or obtuse, sessile, with cordate-clasping, often perfoliate base, delicately 3–25-veined, 3(5) veins stronger; stipules axillary, free, evanescent; peduncles 1–7 cm, about as thick as the stem, not clavate; spikes dense, cylindric, 1–4.5 cm; frs obliquely obovoid, 1.6–3 mm (including the short beak), very shallowly pitted, with 3 obscure dorsal keels, or keelless; $2n=52$. Ponds and slow streams; irregularly cosmopolitan, in Amer. from Nf. and Ont. to N.C. and O., disjunct along the Gulf Coast.

FAMILY **RUPPIACEAE**, the Ditch-grass Family

Fls perfect, very small and inconspicuous, borne in very short, mostly 2-fld, terminal spikes, each spike initially concealed in the sheath of the uppermost vegetative lf; peduncle eventually much elongate and commonly becoming spirally twisted; tep wanting; stamens 2, opposite; anthers subsessile, extrorse, the pollen-sacs well separated on the expanded connective, which has a tiny abaxial appendage near the tip; pollen isobilateral; pistils (2–)4(–8), each with an expanded, ventro-apical stigma, becoming elevated on a slender stipe in fr, so that the several pistils of a fl form an umbelliform cluster; ovule solitary, pendulous, ventral-apical; fr an ovoid, commonly asymmetrical drupelet; endosperm wanting; embryo with an obliquely terminal cotyledon and a lateral plumule; glabrous, submersed aquatic herbs, rooted in the substrate; shoot-tip very compact and congested, but branched just beneath the terminal infl, the stem elongating and again terminating in an infl, etc.; typically 4 lvs between successive infls; lvs alternate or opposite, linear or setaceous, with a single midvein, expanded below into a well developed, distally open sheath. A single genus and sp.

1. **RUPPIA** L. Ditch-grass, widgeon-grass. Characters of the family.

1. **Ruppia maritima** L. Stem to 8 dm; lvs 3–10 cm × 0.5 mm; fr 2–3 mm; $2n=14, 16, 20, 24, 28, 40$. Interruptedly cosmopolitan; along both coasts of the U.S. in saline or brackish (rarely truly marine) water, and scattered inland in saline or brackish (rarely fresh) water. Highly variable, and by some authors divided into several apparently confluent spp. or vars.

FAMILY **NAJADACEAE**, the Water-nymph Family

Fls very small and inconspicuous, hydrophilous, solitary (–few) in the axils, unisexual, without a proper perianth; staminate fl subtended by an outer invol of scales (these often ± connate into a cup or tube) and by a very thin and membranous, flask-shaped inner invol with a narrow, ± bilobed mouth; anther solitary, virtually sessile (but the stalk beneath the inner invol lengthening in anthesis); pollen spheroidal, without an exine; pistillate fl without an invol, or with an inconspicuous, membranous invol ± adherent to the ovary; pistil solitary, the unilocular ovary surmounted by 2–4 elongate stigmas; ovule solitary, basal, erect; fr indehiscent, with a very thin pericarp closely surrounding the seed; endosperm wanting; embryo straight, with an obliquely terminal cotyledon and lateral plumule; slender, usually freely branched, glabrous, submersed aquatic herbs, rooted in the substrate but not rhizomatous, most spp. annual; lvs opposite or apparently whorled, linear or subulate, minutely serrulate or evidently serrate, 1-nerved, somewhat expanded and sheathing at the base, not ligulate. Only the genus *Najas*.

1. **NAJAS** L. Water-nymph. Characters of the family; vegetatively plastic. 35, cosmop. Counts of lf-teeth in the descriptions do not include those of the sheath.

1 Dioecious; lower side of the midvein of the lvs (and often also the internodes) prickly . . . . 1. *N. marina.*
1 Monoecious; lf-surface and internodes smooth.
  2 Lf-teeth multicellular, evident at 10×, 7–15 per side; lvs becoming recurved in late season; seed-
    coat pitted, the areolae in ca 12–18 ladder-like rows, distinctly wider than long . . . . 2. *N. minor.*
  2 Lf-teeth unicellular, minute, 20 or more per side (except in no. 5); lvs spreading or ascending; seed-
    coat smooth or pitted with areolae in ca 20 or more rows, the areolae about as long as or longer
    than wide.
    3 Seeds pitted, dull, fusiform or nearly cylindric; anthers 1- or 4-locular.
      4 Style apical.
        5 Anthers 4-locular; seeds mostly 1.2–2.5 mm, with areolae in 20–45 rows; widespread . . . 3. *N. guadelupensis.*
        5 Anthers unilocular; seeds mostly 3.3–3.8 mm, with areolae in 50–60 rows; Hudson R. . . . . 4. *N. muenscheri.*
      4 Style offset from the apex of the fr and seed; anthers unilocuar . . . . 5. *N. gracillima.*
    3 Seeds smooth, glossy, broadest above the middle; anthers unilocular . . . . 6. *N. flexilis.*

1. Najas marina L. Alkaline w.-n. Dioecious; stems 0.5–4.5 dm, 0.5–4 mm thick, often prickly; lvs 0.5–4 cm × 0.5–4.5 mm, spreading or ascending, prickly along the midvein beneath, coarsely serrate with 8–13 multicellular teeth projecting 0.5–1 mm on each side; anthers dithecal, with 4 microsporangia; seeds 2.2–4.5 mm, reddish-brown, ovoid, pitted, with irregular aerolae; $2n=12$. Brackish or highly alkaline water of ponds and lakes; irregularly cosmop.; in our range from N.Y. and Pa. to Wis., Ill., and Minn.

2. Najas minor Allioni. Eutrophic w.-n. Monoecious; dark green; stems 1–2 dm, 0.2–1 mm thick; lvs 0.5–3.5 cm × 0.1–0.2 mm, becoming recurved in age, serrate with 7–15 multicellular teeth on each side, the base abruptly expanded and minutely fringe-toothed across the subtruncate summit, the teeth extending part-way down the sides of the expanded base; anthers monothecal and with a single microsporangium; seeds 1.5–3 mm, purplish, rather slender and somewhat curved distally, pitted, the areolae quadrangular, ca 0.03 mm high and 0.1 mm wide, in ca 12–18 ladder-like longitudinal rows; $2n=12$. Ponds, lakes, and slow-moving streams, often in eutrophic or alkaline waters; native of the Old World, intr. in the U.S. from N.Y. and w. Mass. to Ga. and w. Fla., w. to Ill. and Ark.

3. Najas guadelupensis (Sprengel) Magnus. Southern w.-n. Monoecious; stems profusely branched, 1–8 dm; lvs 0.5–3 cm × 0.2–2 mm, spreading, not so sharply pointed as in our other spp., with sloping shoulders at the base, minutely serrulate with 20–100 unicellular teeth per side; anthers dithecal, with 4 microsporangia; style apical; seeds 1.2–2.5 mm, purple-tinged, fusiform, pitted, the areolae 4–6-angled, about as long (0.08–0.1 mm) as wide, in 20–40 rows; $2n=12$, 36, 48, 54, 60. Widespread in N. and S. Amer., n. to Me., s. Que., Man., Atla., and Wash. The widespread var. *guadelupensis*, slender, with stems 0.2–1 mm thick, and with 50–100 teeth on each side of the lf, tends to give way at the n. to var. *olivacea* (C. Rosend. & Butters) Haynes, which occurs from N.Y. and s. Que. to n. Ind., n. Io., and s. Man., and is stouter, with stems mostly 1–2 mm thick, and with 20–40 teeth per side of the lf. (*N. olivacea.*)

4. Najas muenscheri R. T. Clausen. Hudson R. w.-n. Monoecious; stems 3–9 dm, ca 1 mm thick; lvs 1–1.5 mm, spreading, minutely serrulate with 50–100 unicellular teeth per side; anthers monothecal and with a single microsporangium; seeds 3.3–3.8 mm, slender, fusiform-cylindric, with 50–60 rows of minute, rectangular areolae. Abundant on tidal mudflats along the Hudson R. (*N. guadelupensis* var. *m.*)

5. Najas gracillima (A. Braun) Magnus. Slender w.-n. Monoecious; light green; stems 0.5–5 dm, only 0.2–0.7 mm thick, sparingly branched; lvs lax, 0.5–3 cm × 0.1–0.5 mm, spreading to ascending, minutely serrulate with 13–17 unicellular teeth per side, the base abruptly expanded and minutely fringe-toothed across the subtruncate summit (but not down its sides); anthers monothecal and with a single microsporangium; style offset from the apex of the fr; seeds 2.0–3.2 mm, light brown, fusiform-cylindric, pitted with 20–45 longitudinal rows of minute areolae a little higher than wide. Oligotrophic, soft-water lakes; N.S. to Ala., w. to Minn. and Io., intolerant of pollution and now become rare.

6. Najas flexilis (Willd.) Rostkov and Schmidt. Northern w.-n. Monoecious; stems 0.5–5 dm, 0.2–0.6 mm thick, often profusely branched above; lvs 1–4 cm × 0.2–0.6 mm, spreading or ascending, with sloping shoulders at the base and tapering above to a long, slender point, minutely serrulate with 35–80 unicellular teeth per side; anthers monothecal and with a single microsporangium; seeds (1.2–)2.5–3.7 mm, smooth and glossy, yellow to deep brown, ± obovate, widest above the middle, very lightly areolate in ca 30–50 rows. In lakes and rivers; Nf. to B.C., s. to Va., W.Va., O., Ind., Io., Nebr., and Utah.

FAMILY **ZANNICHELLIACEAE**, the Horned Pondweed Family

Fls small and inconspicuous, hydrophilous, in small, axillary, usually complex and sympodial infls, unisexual, the plants monoecious; perianth wanting, or of 3 tiny, scale-like, unvasculated, basally connate tep; stamen actually or apparently solitary, with 1–6 thecae and 2–12 microsporangia; pollen spheroidal, monosulcate or more often lacking an exine; pistils (1–)3–4(–9), distinct; ovule solitary, ventral-apical, pendulous; fr of achenes or drupelets; endosperm wanting; cotyledon circinately coiled above the swollen hypocotyl; slender, much branched, glabrous, submersed aquatic herbs, ± rhizomatous and rooted in the substrate; lvs alternate or opposite, or the distal ones seemingly in whorls of 3–4, with a linear or filiform, 1-nerved or partly 3-nerved blade and a sometimes stipule-like sheath. 4/7–8.

1. **ZANNICHELLIA** L. Horned pondweed. Fls paired, one of each sex on an axillary, bifurcated common stalk; perianth none; filament elongate; anther with (1)2(4) thecae and (2)4(8) microsporangia; pistillate fl subtended by a minute, membranous, initially closed cupule; ovaries (1–)4 or 5(–9), distinct, each with a slender style and peltate-funnelform stigma; fr a short-stipitate achene or drupelet with a warty dorsal ridge, terminated by the persistent, beak-like style; most lvs opposite, but the distal ones in seeming whorls of 3 or 4, the prominent, stipule-like sheath completely wrapped around the stem but free from the slender blade. Monotypic.

1. **Zannichellia palustris** L. Rhizome scarcely differentiated from the very slender and fragile leafy stem; lvs 3–10 cm × 0.5 mm, 1-nerved, or partly 3-nerved in robust plants; achenes keeled on both the outer and inner sides, the body 2–3 mm, the style 1 mm; $2n=24, 48, 32, 36$. In fresh (often eutrophic) or brackish water almost throughout N. Amer., and widespread in the Old World.

Family **ZOSTERACEAE**, The Eel-grass Family

Fls small and inconspicuous, hydrophilous, unisexual, arranged in 2 rows on one side of a flattened spadix subtended (and often largely enclosed by) a spathe, the axis of the spadix often with a series of ± well developed marginal lobes or appendages (retinacula) that fold over and ± cover the fls, the fls otherwise bractless and without perianth; stamen solitary; anther sessile, tetrasporangiate and dithecal; pollen filamentous, up to 2 mm, without exine; ovary solitary, bicarpellate, unilocular, with a single apical, pendulous ovule and 2 basally connate styles; fr small, drupaceous or firm and eventually opening irregularly; endosperm wanting; embryo with a basally closed and sheathing cotyledon lying in the groove of an enlarged hypocotyl; glabrous, rhizomatous, perennial marine herbs; lvs alternate, commonly distichous, parallel-veined or with only a midrib, generally ligulate at the juncture of the sheath and narrow blade. 3/15–20.

1. **ZOSTERA** L. Eel-grass. Monoecious, the staminate and pistillate fls typically alternating in each row on the spadix, which is largely hidden at anthesis in the slender, open, sheathing spathe; anthers deciduous, discharged from the spathe at anthesis; stigmas elongate, bristle-like, exserted from the spathe, deciduous; fr oblong-ovoid, the thin wall eventually rupturing irregularly; lvs elongate, ribbon-like; spathe with a deciduous blade. 10+, cosmop.

1. **Zostera marina** L. Common e.-g. Stems freely branched, to 2.5 m, lvs to 1 m, with 3–7 strong nerves and many minute ones, on sterile stems to 1 cm wide, on fertile ones 2–5 mm wide; lf-sheaths closed, splitting in age; spadix 2–8 cm, lacking retinacula; frs ± exserted, the body 3–4 mm, the beak 1–2 mm; seed strongly ca 20-ribbed, the ribs visible through the pericarp; $2n=12$. Shallow water (to 10 m) in sheltered bays and coves, usually wholly submersed; circumboreal, on the Atlantic coast from Greenl. to Fla. (*Z. stenophylla.*)

Family **ACORACEAE**, the Sweet Flag Family

Fls perfect, very numerous and tiny, covering the elongate, linear-cylindric spadix; perianth of 6 short, broad tep; stamens 6, with linear filament and broadly reniform, introrse anther; ovary superior, 2–3-locular, with a single apical-axile ovule per locule; style very short and broad, undivided, subpersistent on the developing fr; fr narrowly obpyramidal, hard and dry, indehiscent, 2–3-seeded; embryo monocotyledonous, axially embedded in the hard endosperm; perennial from stout rhizomes; lvs erect, elongate, linear-ensiform, unifacial, the midvein usually off-center; shoot with ethereal oil cells, but lacking raphides; spadix diverging laterally from a tall, 3-angled scape that is surmounted by an erect, long green spathe resembling the foliage lvs. 1/2. A single genus.

1. **ACORUS** L.   Sweet flag. Characters of the family. 2, the other Asian.

1. **Acorus calamus** L.   Sweetly aromatic; lvs crowded at the base, erect, to 2 m, 8–25 mm wide; scape resembling the lvs, prolonged into an erect green spathe 2–6 dm; spadix 5–10 cm, 1 cm thick at anthesis, 2 cm at maturity, covered with yellowish-brown fls; fr 4–5 × 2 mm; 2n=18, 24, 36, 48. Swamps and shallow water; irregularly circumboreal, in Amer. from N.S. and Que. to Minn., Alta., and e. Wash., s. to Fla., Tex., and Colo. Spring and early summer. (*A. americanus.*)

FAMILY **ARACEAE**, the Arum Family

Fls very numerous and tiny, unisexual or seldom perfect, closely aggregated over all or part of a fleshy axis, forming a cylindric to capitate spadix usually subtended by a colored or foliaceous spathe; plants commonly monoecious with the staminate fls uppermost in the spadix, rarely dioecious; perianth very small, the mostly 4 or 6 similar sep distinct or connate, ± in 2 cycles, or often wanting; stamens mostly 2–6; filaments very short; anthers extrorse; ovary superior but ± embedded in the spadix, 1–3(+)-locular, with 1–several ovules per locule; placentation in plurilocular ovaries axile or basal-axile; style very short or none; fr usually a berry, seldom dry or leathery and opening irregularly, or the whole spadix ripening as a multiple fr; embryo rather large, monocotyledonous, axially embedded in the usually copious, oily endosperm, or the endosperm sometimes wanting; mostly herbs, with raphides but without ethereal oil cells, very diverse in habit and foliage, the lvs in most genera with a distinct, sheathing petiole and an expanded, bifacial, often ± net-veined blade. 110/1800.

1 Lvs simple; fls nearly or quite covering the spadix.
  2 Spathe consisting of a tubular lf surrounding the base of the spadix .... 1. *Orontium.*
  2 Spathe well developed, with a ± expanded blade.
    3 Spathe green or brown, convolute, wholly or nearly surrounding the spadix at base.
      4 Spadix subglobose; perianth present; lvs merely cordate or rounded .... 2. *Symplocarpus.*
      4 Spadix elongate; perianth none; lvs usually sagittate .... 3. *Peltandra.*
    3 Spathe white, open, flat; perianth none; lvs cordate at base .... 4. *Calla.*
1 Lvs compound, fls covering only the basal part of the spadix.
  5 Spadix free from the spathe at base, with fls all around; no bulblets .... 5. *Arisaema.*
  5 Spadix adnate to the spathe on one side at base, bearing pistillate fls on the free side; lvs producing
    bulblets at the base of the lfls .... 6. *Pinellia.*

1. **ORONTIUM** L.   Golden club. Fls perfect, bright yellow, covering the linear spadix; spathe consisting of a tubular lf surrounding the base of the scape; stamens 4 or 6; filaments wide and thin; pollen-sacs divergent; ovary partly embedded, unilocular, with 1 ovule; fr a green or brown utricle; stoutly rhizomatous; lvs basal, the stout petiole broadly dilated at base, the blade floating or submerged, with wide midvein. Monotypic.

1. **Orontium aquaticum** L.   Petioles 1–2 dm; blades elliptic, 6–20 cm, a third as wide, acute, narrowed below to an obtuse or acute base, the veins parallel from base to apex, with numerous transverse connecting veinlets; scape 2–4 dm, flattened above; spadix 2–5 cm; 2n=26. Swamps and shallow water, especially on the coastal plain; Mass. to Fla., w. to c. N.Y., sw. Pa., e. Ky., w. Tenn., and La.

2. **SYMPLOCARPUS** Salisb.   Skunk-cabbage. Fls perfect, covering the subglobose spadix, this subtended and mostly enclosed by a fleshy, ovate, pointed spathe; perianth of 4 erect, connivent tep; stamens 4; ovaries buried in the spadix, unilocular, uniovulate; style stout, 4-angled, subulate; seeds embedded in the enlarged, spongy spadix, covered by the persistent perianth and style; herb from a thick rhizome; spathe partly underground, with very short peduncle, the lvs appearing later and becoming very large. Monotypic.

1. Symplocarpus foetidus (L.) Nutt.   Malodorous; spathe ovoid with incurved summit, 8–15 cm, green, purplish, or spotted or striped with both; fr globose, 8–12 cm thick; seeds 1 cm thick; lvs basal, short-petioled, ovate, cordate, to 6 dm, the veinlets conspicuously anastomosing; $2n=30, 60$. Swamps and moist low ground; Que. and N.S. to N.C., w. to Minn. and Io.; e. Asia. Feb.–Apr. (*Spathyema f.*)

3. PELTANDRA Raf.   Arrow-arum. Fls covering all or most of the slender spadix; spathe elongate, oblong, convolute below and covering the pistillate fls, opening distally and exposing the staminate ones; perianth none; ovary unilocular, subtended by a few white staminodia; ovules 1–few; stamens 4–5, embedded in a sterile ovary; perennial from thick fibrous roots, with long-petioled, basal, 3-nerved, pinnately veined lvs and a long scape bearing a slender, green or greenish spathe. 2, Atlantic N. Amer.

1. Peltandra virginica (L.) Schott & Endl.   Tuckahoe. Lvs oblong to broadly triangular, 1–3 dm at anthesis, larger later, a strong nerve descending into each of the divergent or parallel, acute to rounded basal lobes; scape 2–4 dm, recurved in fr; spathe green with pale margin, 1–2 dm, the part above the pistillate fls deliquescent in age, the lower part enclosing the fr; spadix nearly as long as the spathe, white to orange, the lower fifth pistillate; fr a head of brown berries, the 1–3 seeds surrounded by gelatinous material; $2n=88$, 112. Swamps and shallow water; s. Me. to Fla., w. to Mich., Mo., and Tex. Late spring. (*P. luteospadix.*)

4. CALLA L.   Water-arum. Fls all perfect, or the upper staminate, covering the spadix; spache expanded, not enclosing the spadix, ovate to elliptic; perianth none; stamens 6; filaments narrow, flat; anthers short and broad; ovary unilocular, with several erect anatropus ovules; fr a red berry, the few seeds surrounded by gelatinous material; perennial herbs with petioled basal lvs and solitary spathes. Monotypic.

1. Calla palustris L.   Rhizome elongate; petioles stout, 1–2 dm; lvs ovate to subrotund, 5–10 cm, cordate at base, short-acuminate, the veins and veinlets curved-ascending, parallel, not anastomosing; scape 1–3 dm; spathe white, 3–6 cm, prolonged into a linear involute tip 5–10 mm; spadix short-cylindric, 1.5–2.5 cm, on a thick short stipe; berries 8–12 mm; $2n=36, 72$. Swamps and shallow water; circumboreal, in our range s. to Md., Ind. and Io. Early summer.

5. ARISAEMA Martius.   Monoecious or dioecious; fls covering the basal part of a fleshy spadix that is subtended by a green or purple-brown spathe; perianth none; staminate fls above the pistillate, composed of 2–5 subsessile anthers opening apically; pistillate fls consisting of a unilocular ovary with 1–several basal orthotropous ovules and a broad stigma; fr a cluster of globose red berries, each with 1–3 seeds; cormose perennials with long-petioled compound lvs. 150, irregularly cosmop.

1 Lfs 3; spathe expanded above, arching over the blunt spadix .... 1. *A. triphyllum.*
1 Lfls 7–13; spadix protruding, long-acuminate .... 2. *A. dracontium.*

1. Arisaema triphyllum (L.) Schott.   Jack-in-the-pulpit; Indian turnip. Corm very acrid; lvs mostly 2, the petiole 3–6 dm at anthesis, sometimes later to 15 dm; lfls 3, acuminate, the terminal elliptic to rhombic-ovate, the lateral often asymmetrical; veins of the lfls parallel from the midrib to the marginal vein, connected by numerous anastomosing veinlets; peduncle 3–20 cm; spathe convolute below, expanded above and arched over the spadix, abruptly acuminate; spadix cylindric or barely clavate, yellow, blunt; fr 1 cm; $2n=28, 56$. N.S. to N.D., s. to Fla. and Tex. Spring. Four vars., 3 in our range, the fourth (*quinatum*) southern:

a Plants mostly of rich moist woods; lateral lfs distinctly asymmetrical, the proximal part broadly
 rounded on the lower side, acute or acuminate on the upper, in robust plants sometimes 2-parted;
 deflexed flange at the top of the spathe-tube not obviously ridged or fluted; mostly tetraploid;
 widespread. (*A. atrorubens.*) .... var. *triphyllum.*
a Plants mostly of boggy places or wet woods; lateral lfls not strongly asymmetrical, the lower side
 tending to be acute or acuminate like the upper; deflexed flange at the top of the spathe-tube
 narrow, mostly 1–3 mm wide; diploid, mostly blooming later than var. *triphyllum.*
  b Spathe-tube strongly ridged and fluted, the ridges white; spathe 2–6 cm wide on the expanded limb,
   commonly green with purple stripes in the throat; northeastern, s. to N.J., Pa., and in the mts.
   to N.C., w. rarely to Ind. .... var. *stewardsonii* (Britton) O. A. Stevens.
  b Spathe-tube obscurely if at all ridged or fluted; spathe 1.5–3 cm wide on the expanded limb, wholly
   purple or wholly green within; southern, n. to Ind., N.Y., and se. Conn. .... var. *pusillum* Peck.

2. Arisaema dracontium (L.) Schott.   Green dragon; dragon-root. Lf usually solitary, the long petiole 2–5 dm at anthesis, later to 1 m, pedately divided into 7–15 lfls, these elliptic to oblanceolate, acuminate, sometimes confluent below, the central ones 1–2 dm, the outer progressively smaller; peduncle arising from a basal sheath, shorter than the petiole; spathe green, slender, convolute, 3–6 cm, pointed; spadix tapering to a long slender point exserted 5–10 cm beyond the spathe; fr orange-red; $2n=28, 56$. Damp woods; Que. to Minn., s. to Fla. and Tex., commoner westward. Spring. (*Muricanda d.*)

6. PINELLIA Tenore, nom. conserv. Spadix adherent to the spathe on one side of the base, bearing pistillate fls on the free side; staminate fls separated from the pistillate by a swollen ring; upper part of the spadix naked; ovary with a single ovule; small stoloniferous herbs, perennial from corms; lvs compound. 7, China and Japan.

1. Pinellia ternata (Thunb.) Breitenb. Lf solitary, long-petioled, producing bulblets at the base of the lfls; lfls 3, lanceolate or lance-ovate, 3–6 cm, acute, the terminal sessile, the lateral nearly so; scape exceeding the lf; spathe 5 cm, green, obtuse; spadix prolonged into a slender naked tip surpassing the spathe. Native of Japan, adventive under Rhododendrons in s. N.Y. and N.J. Summer.

FAMILY **LEMNACEAE**, the Duckweed Family

Small, free-floating (sometimes barely submersed) aquatics, consisting of a flat or thick-ened thallus (often called a frond), with or without 1–several short unbranched roots; frond with a reproductive pouch on each margin near the base, or a single such pouch on the upper surface; reproduction mainly by vegetative budding in the pouch(es), the new plants sooner or later breaking away; infl (seldom produced) in one of the 2 marginal pouches or in the single pouch on the upper surface, subtended by a tiny, membranous spathe when in a marginal pouch; perianth none; staminate fls 1 or 2, each consisting of a single anther on a short filament; pistillate fl solitary, naked, composed of a unilocular ovary with a short terminal style and 1–8 basal ovules, ripening into a utricle; embryo straight, monocotyledonous, sometimes without a radicle; endosperm scanty and sheath-ing the embryo, or wanting. 6/30.

1 Roots present, developing from the lower side of the thallus.
  2 Each thallus normally with 2 or more roots . . . . 1. *Spirodela.*
  2 Each thallus with a single root . . . . 2. *Lemna.*
1 Roots none.
  3 Thallus thin, flat, linear or strap-shaped, often curved, well over 2 mm . . . . 3. *Wolffiella.*
  3 Thallus thick, globose to ellipsoid or ovoid, or flattened on top, less than 2 mm . . . . 4. *Wolffia.*

1. SPIRODELA Schleiden. Duckweed. Thallus flat or slightly convex on both sides, usually somewhat asymmetrical, marked with 3–15 obscure to evident nerves diverging from a nodal point near the base; roots 2–12+, descending from the nodal point, each with a single vascular strand; reproductive pouches 2, lateral, opposite the nodal point; infl (seldom produced) in one of the lateral pouches, with 2 staminate fls and one pistillate, collectively subtended by a tiny spathe; anther bilocular; ovules 1–4. 4, cosmop.

1 Roots mostly 5–12; fronds (3–)5–8(–10) mm, ⅔ to fully as wide; common . . . . 1. *S. polyrhiza.*
1 Roots 2–5; fronds 2.5–5 mm, half to ¾ as wide; rare . . . . 2. *S. punctata.*

1. Spirodela polyrhiza (L.) Schleiden. Greater d. Thallus broadly oval to obovate, anthocyanic beneath, (3–)5–8(–10) mm, ± evidently 5–15-nerved, forming turions especially in the fall; roots mostly 5–12; spathe open only at the top; seeds 1 or 2, smooth or minutely reticulate; $2n=30, 40, 50$. Quiet water; nearly cosmopolitan, and throughout our range.

2. Spirodela punctata (G. Meyer) C. Thompson. Like a small version of no. 1; thallus 2.5–5 mm, half to ¾ as wide, obscurely 3–5-nerved; roots 2–5; seed solitary; $2n=40, 50$. Quiet water; widely scattered in warm regions, rare and irregular with us. (*S. oligorhiza.*)

2. LEMNA L. Duckweed. Thallus flat to strongly convex beneath, variously rotund, ovate, obovate, oblong, or stipitate, with 1–5 obscure nerves from a nodal point near the base, or nerveless; root 1, opposite the nodal point, without vascular tissue, surrounded by an inconspicuous short sheath at base; reproduction as in *Spirodela*; ovules 1–6; seed longitudinally ribbed and transversely striate, or smooth. 10, cosmop.

1 Thallus tapering to a long (4–16 mm) stipe persistently attached to the parent .... 1. *L. trisulca.*
1 Thallus without a stipe, or the stipe very short and inconspicuous.
  2 Thallus symmetrical or nearly so.
    3 Fronds minute, mostly 1–2.5 mm, without anthocyanin, 1-nerved or nerveless; spathe open,
       reduced .... 4. *L. minuta.*
    3 Fronds larger, (1.5–)2.5–6 mm, very often anthocyanic, typically 3(5)-nerved; spathe sac-like,
       open at the top.
      4 Fronds usually gibbous, the upper surface nearly flat, the lower strongly bulged; fr narrowly
        winged; seeds (1)2–3(–6), distinctly ribbed .... 2. *L. gibba.*
      4 Fronds usually flat or nearly so on both sides; fr wingless; seed solitary, smooth or only very
        lightly ribbed .... 3. *L. minor.*
  2 Thallus evidently oblique or falcate; no anthocyanin; spathe open, reduced.
    5 Thallus obliquely obovate, with rounded sides, 1–2.5 mm, commonly 1–2 times as long as wide
      .... 5. *L. perpusilla.*
    5 Thallus more oblong, with nearly parallel sides, 2.5–5 mm, commonly 1.5–3 times as long as
      wide .... 6. *L. valdiviana.*

1. Lemna trisulca L.   Star-d. Thallus thin, oval or oblong, denticulate or erose toward the tip, obscurely 3-nerved, tapering to a 4–16 mm stipe that remains attached to the parent plant, forming tangled colonies, mostly floating just beneath the water-surface, emersed in fl and fr; spathe open; seed solitary; $2n=20, 40, 60, 80$. Widespread in both the Old and New Worlds, and throughout our range.

2. Lemna gibba L.   Thallus orbicular to obovate, 3–6 mm, nearly symmetrical, solitary or often in groups of 2–4, very often anthocyanic beneath, mottled yellow-green above, or anthocyanic on both sides, ± evidently 3–5-nerved, usually gibbous, the upper surface flat or nearly so and with ± prominent papillae, the lower usually strongly bulged, the tissue with larger and more prominent air-spaces than in no. 3; spathe sac-like, open only at the top; fr narrowly winged; seeds (1)2–3(–6), distinclty many-ribbed; $2n=40, 50, 70, 80$. Nearly cosmop., but rare and irregular with us.

3. Lemna minor L.   Lesser d. Thallus rotund to elliptic or obovate, (1.5–)2.5–6 mm, nearly symmetrical, commonly in groups of 2–5, very often anthocyanic beneath, dark green above, or anthocyanic on both sides, ± evidently 3-nerved, usually flat or only slightly convex on both sides, the upper surface with ± prominent papillae; spathe sac-like, open only at the top; fr not winged; seed solitary, smooth or only inconspicuously ribbed; $2n=20, 30, 40, 50$. Nearly cosmop., and common throughout our range. (*L. obscura,* a small form; *L. turionifera,* overwintering by small, rootless turions at the bottom of the water.)

4. Lemna minuta HBK.   Thallus flat or nearly so on both sides, thin-margined, smooth or with a few low papillae along the midline above, ovate to elliptic, symmetrical or nearly so, solitary or sometimes 2(+) together, not anthocyanic, 1–2.5 mm, mostly ⅗ to fully as wide, the single obscure nerve reaching at most ⅔ of the way from node to apex; spathe open; fr exserted, elongate, tapering to the persistent terminal style; seed solitary, elongate, evidently ribbed; $2n=36, 40$. Mainly w. U.S. and temp. S. Amer., perhaps not in our range. (*L. minima; L. minuscula.*)

5. Lemna perpusilla Torr.   Thallus ± convex on both sides, with a prominent papilla at node and apex above, and usually with smaller papillae along the midline, obliquely obovate, asymmetrical and ± falcate, 1–2.5 mm, commonly 1–2 times as long as wide, solitary or up to 5 together, not anthocyanic, with 1–3 obscure nerves or nerveless; root-sheath winged at base (unique in the genus); spathe reduced, open; fr asymmetrical, with a persistent lateral style; seed solitary, ovoid, strongly ribbed; $2n=40$. Nearly cosmop., and found in most of our range. (*L. trinervis.*)

6. Lemna valdiviana Philippi.   Thallus flat or nearly so on both sides, usually smooth above, rather narrowly oblong, mostly 2.5–5 mm, commonly 1.5–3 times as long as wide, mostly 2–10 together, not anthocyanic, nerveless or indistinctly 1-nerved, the nerve often reaching more than ¾ the distance from node to tip; spathe reduced, open; fr exserted, elongate-ovate, the persistent style obliquely terminal; seed solitary, oblong-ovoid, evidently ribbed; $2n=40$. Widespread in the W. Hemisphere, and scattered in our range. (*L. cyclostasa.*)

3. WOLFFIELLA Hegelm.   Thallus thin and flat, linear or strap-shaped, straight or curved, without veins or roots; epidermis punctate (at least post mortem) with brown pigment-cells; reproductive pouch solitary, on the upper surface near the base, to one side of the midline; reproduction almost entirely by budding, 2 or more plants often remaining attached; fls (rarely produced) without a spathe, paired, one of each sex, the pistillate one nearer the base, with a single orthotropous ovule; anther unilocular, opening across the top. 6(+), cosmop.

1. Wolffiella floridana (Donnell-Smith) C. Thompson.   Thallus falcate or often doubly so, tapering at least distally to a long, slender tip, 4–14 × 0.4–0.7 mm; colonies of numerous thalli commonly coherent at base, submersed or floating with only the base above water, often several together in a tangled mass; $2n=40$. Stagnant water; Mass. to n. Ill., s. to Fla. and Tex., more common southward.

4. WOLFFIA Horkel, nom. conserv.   Water-meal. Thallus minute, not over ca 1.5 mm, thickened, globose to ellipsoid or ovoid, or flattened on the upper side only, without veins or roots; reproductive pouch solitary, on the upper surface near the basal end, to one side of the midline; reproduction almost entirely by budding, the young plants soon detached; fls (rarely produced) without a spathe, paired, one of each sex, the pistillate one nearer the base, with a single orthotropous ovule; anther unilocular, opening across the top. 8, cosmop. The smallest angiosperm.

1   Thallus flattened above, the upper surface finely brown-punctate, with one papilla.
   2   Upper side of the thallus with a prominent papilla near the middle .... 1. *W. papulifera.*
   2   Upper side of the thallus elevated distally into an evident terminal papilla .... 2. *W. punctata.*
1   Thallus rounded above, not punctate, the upper surface without an obvious papilla .... 3. *W. columbiana.*

     1. Wolffia papulifera C. Thompson.   Thallus broadly ovoid, slightly asymmetrical, mostly 0.5–1.5 × 0.3–1.0 mm, 1–1.5 times as long as wide, rounded at the tip, the upper side floating just above the water, punctate (post mortem) with brown pigment-cells, elevated near the middle into a conical papilla; $2n=20$, 40, 50, 60, 80. Quiet water; N.J. and Md. to O., Ind., Mich., and Kans., s. to Fla., Tex., and S. Amer. *W. brasiliensis,* perhaps misapplied.

     2. Wolffia punctata Griseb.   Thallus bright green and shiny above, pale beneath, ellipsoid to oblong-ovoid, symmetrical, mostly 0.1–1.2 × 0.2–0.5 mm, 1.3–2+ times as long as wide, ± pointed at both ends, the upper surface floating just above the water, punctate (post mortem) with brown pigment-cells, elevated distally into a terminal papilla; stomates 25–50; $2n=20$, 30, 40. Quiet water; widespread in the U.S. and W.I., n. to Conn., s. N.H., s. Ont., and Minn., often with no. 3, but more consistently forming a single layer with the upper surface exposed. (*W. borealis.*)

     3. Wolffia columbiana Karsten.   Thallus light green and translucent, broadly ellipsoid to globose, asymmetrical, mostly 0.8–1.4 × 0.6–1.2 mm, 1–1.3 times as long as wide, not punctate, floating low in the water, only a small circular portion of the upper surface in contact with the air; upper surface slightly roughened, sometimes to form ca 3 obscure papillae; stomates 3–5; $2n=30$, 40, 50, 70. Quiet water; Mass. and N.H. to s. Ont. and Minn., s. to S. Amer.; often forming more than one layer, some wholly submersed.

FAMILY **XYRIDACEAE**, the Yellow-eyed Grass Family

     Fls hypogynous, perfect, usually entomophilous but without nectaries or nectar; sep typically 3, the upper (anterior, adaxial) one thin and membranous, ± enclosing the pet and abscising as the fl opens, or sometimes reduced or obsolete; lateral sep 2, boat-shaped, keeled, chaffy-scarious, persistent and clasping the capsular fr; pet 3, ephemeral, nearly or quite alike, yellow or sometimes white or blue, long-clawed and distinct, or connate below into a slender tube; stamens mostly 3, opposite and basally adnate to the pet, often alternating with staminodes; ovary tricarpellate, typically (incl. our spp.) unilocular with 3 parietal (often ± intruded) placentas each bearing several or many ovules; embryo small, scarcely differentiated into parts, lying alongside the base of the copious, mealy endosperm; herbs; lvs mostly or all basal, with an open sheath and narrow, parallel-veined (or uni-veined), very often equitant and unifacial blade; scapes or peduncles each usually ter-minating in a head or dense spike with the fls sessile in the axils of firm, spirally arranged, closely imbricate bracts. 4/nearly 300 spp.

     1. XYRIS L.   Yellow-eyed grass. Fls trimerous; anterior sep well developed, deciduous; pet distinct, with slender claw (hidden by the bract) and broad, yellow (white), toothed blade; staminodes present, bifid; placentas in most spp. (incl. ours) parietal, ± intruded distally; style-branches 3, alternate with the staminodes; seeds longitudinally lined. 250+, mainly warm reg.

1   Keel of the lateral sep merely lacerate, or entire.
   2   Mature spikes mostly 1.5–3.5 cm.
     3   Lateral sep ± exserted beyond the tip of the subtending bract .... 1. *X. smalliana.*
     3   Lateral sep shortly included, hidden by the subtending bract.
       4   Lf-bases thickened, rounded; lvs 5–10 mm wide, evidently twisted .... 2. *X. platylepis.*
       4   Lf-bases thin, keeled; lvs 10–25 mm wide, flat or only slightly twisted .... 3. *X. iridifolia.*
   2   Mature spikes mostly 0.5–1.5 cm.

5 Keel of the lateral sep narrow, rather thick, its margin entire or nearly so; lvs 4–15 cm × 1–2.5
   mm; seeds nearly 1 mm; northern .... 6. *X. montana*.
5 Keel broader, thin, usually ± lacerate at least distally; lvs often (not always) evidently larger;
   seeds ca. 0.5 mm.
  6 Plant-base stramineous or yellow-green; lvs ascending; N.J. and s. .... 4. *X. jupicai*.
  6 Plant-base generally anthocyanic; lvs spreading; widespread .... 5. *X. difformis*.
1 Keel of the lateral sep evidently ciliate or fimbriate.
  7 Tips of the lateral sep not exserted beyond the subtending bract, not fimbriate.
    8 Mature spikes under 1 cm; lvs 4–15 cm × 1–2.5 mm, rough; northern .... 6. *X. montana*.
    8 Mature spikes mostly 1–3 cm; lvs mostly 10–50 cm × 2–20 mm, smooth.
      9 Bracts and lateral sep with a small apical tuft of short, usually reddish-brown hairs; lvs
        ascending, twisted; plants tending to be bulbous-based; widespread .... 7. *X. torta*.
      9 No such hairs; lvs spreading, scarcely twisted; plants tending to be invested at the hard base
        by the stubble of old fibrous lf bases; Va. and s. .... 8. *X. ambigua*.
  7 Tips of the lateral sep evidently exserted beyond the subtending bracts, broad, fimbriate, usually
    crisped; N.J. and s.
    10 Lf-bases firm, castaneous, long-persistent as scales; scape smooth to the touch .... 9. *X. caroliniana*.
    10 Lf-bases soft, stramineous to pale greenish or pinkish; scape-ridges harsh .... 10. *X. fimbriata*.

1. **Xyris smalliana** Nash. Lvs linear, mostly (20–)30–50(–60) cm × 5–15 mm, deep lustrous green, flat or nearly so, usually anthocyanic at the dilated, thin, soft, keeled base; scapes 5–15 dm, smooth, terete and with very low ridges below, 1–2-ridged and somewhat flattened distally; spikes 1–2(–2.5) cm at maturity, ellipsoid or narrowly ovoid; lateral sep ± exserted distally, the keel broad, thin, lacerate toward the tip; pet-blades yellow, obovate, 5–6 mm, unfolding in the afternoon; 2*n*=18. Wet low ground, especially near the coast; Me. to Fla. and Miss. Early summer–fall. (*X. congdoni*.)

2. **Xyris platylepis** Chapman. Lvs shallowly set, linear, mostly 20–40(–50) cm × 5–10 mm, deep lustrous green, twisted, ascending, flexuous, usually anthocyanic at the abruptly expanded and thickened, fleshy, scale-like base; scapes 5–11 dm, twisted, flexuous, terete and smooth or with very low ribs below (the ribs obscurely papillate), slightly compressed and often 1-ridged above; spikes 1.5–3(–4) cm at maturity, ellipsoid to ovoid or oblong; lateral sep shortly included, the keel narrow except toward the expanded, lacerate tip; pet-blades yellow or white, obovate, 5 mm, unfolding in the afternoon; 2*n*=18. Moist but not really wet, often sandy places, somewhat weedy; coastal plain from Va. to Fla. and La. Midsummer–fall.

3. **Xyris iridifolia** Chapman. Lvs shallowly set, linear or broader, iris-like, mostly 40–70 cm × 10–25 mm, deep lustrous green, flat or slightly twisted, anthocyanic at the somewhat dilated, soft, thin, keeled base; scapes 6–8 dm, smooth, straight or slightly twisted, terete and 2-ridged below, conspicuously broadened and flattened above; spikes 2–3.5 cm at maturity, broadly ellipsoid or oblong, blunt; lateral sep shortly included, almost straight, with a broad, lacerate keel; pet-blades cuneate, 3 mm, unfolding in the morning; seeds 0.8–1 mm, farinose; 2*n*=18. Wet, low places, especially in heavy soil, the base commonly submersed; coastal plain from Va. to Fla. and Tex.

4. **Xyris jupicai** Rich. Short-lived, often annual; lvs linear, mostly 10–60 cm × 2–5 mm, ascending, smooth, lustrous yellow-green, pale (not anthocyanic) at the dilated, soft, thin, keeled base; scapes 2–8 dm, terete and finely many-ridged below, becoming flattened (but narrow) and usually 1- or 2-edged above; spikes 0.5–1.5 cm at maturity, narrowly ovoid to oblong, blunt; lateral sep shortly included, the broad keel lacerate for ½–⅔ its length; pet-blades cuneate, 3 mm, yellow, unfolding in the morning; seeds 0.5 mm, faintly ribbed; 2*n*=18. Wet, sunny, disturbed sites, often weedy; coastal plain from N.J. to Fla., Tex., and trop. Amer., perhaps only intr. in U.S. Early summer–fall.

5. **Xyris difformis** Chapman. Lvs broadly linear or elliptic-linear, mostly 10–50 cm × 5–15 mm, spreading, lustrous dark green, anthocyanic at the dilated, soft, thin, keeled base; scapes 1.5–7 dm, terete and twisted below, straighter above and somewhat compressed, with 2 broad thin ridges that may be collectively as wide as the body; spikes 0.5–1.5 cm at maturity, ovoid to subglobose, blunt or acute; lateral sep shortly included, the broad keel lacerate in its upper half; pet-blades cuneate, 4 mm, yellow, unfolding in the morning; seeds 0.5–0.6 mm, broadly ellipsoid, very finely ribbed; 2*n*=18. Wet, low, often sandy places, often in river-swamps; coastal region from s. Me. to Fla. and Tex., and less commonly inland to Mich., Ind., Wis., and Okla. Midsummer–fall. Two vars. with us, both widespread. Var. *difformis* is ± robust, with smooth lvs. Var. *curtissii* (Malme) Kral is smaller and more tufted, seldom over 2 dm, the lf-surfaces evidently papillate. (*X. bayardii*; *X. curtissii*; *X. papillosa*.)

6. **Xyris montana** H. Ries. Densely tufted, lvs narrowly linear, mostly 4–15 cm × 1–2.5 mm, ascending, straight or slightly twisted, with pistillate or scabrous-tuberculate surface, dark green except at the dilated, soft, thin, keeled, anthocyanic base; scapes 0.5–3 dm, terete and slightly twisted below, usually 2–4-keeled just below the spike, the ridges papillate; spikes few-fld, less than 1 cm long; lateral sep shortly included, slightly curved, narrow, the thickened, narrow keel entire or slightly ragged or ciliate apically; seeds 0.8–1 mm, somewhat caudate, finely papillate-ridged, often some (or most) of them abortive. Sphagnum-bogs, tamarack-swamps, etc.; N.S. and s. Me. to Pa., w. to Mich., Wis., Minn., and s. Ont. Midsummer–fall.

7. **Xyris torta** J. E. Smith. Lvs rather shallowly set, linear, ascending, mostly 20–50 cm × 2–5 mm, twisted, longitudinally grooved, lustrous dark green, fleshy-thickened at the pale to often purplish or casta-neous base, some of the outer lvs short and with notably dilated and dark base, the plants thus appearing

bulbous; scapes 1.5–8(–10) dm, ± twisted, flexuous, and many-ridged below, 2–4(–6)-ridged above and somewhat flattened under the spike; spikes mostly 1–2.5 cm at maturity, many-fld, ovoid; lateral sep shortly included, lunate, the broad, thickened keel ciliate-scabrous from near the base to the tip, where it bears a small tuft of reddish-brown (seldom pale) hairs; a similar tuft of hairs at the tip of each bract; pet-blades yellow, obovate, 4 mm, unfolding in the morning; seeds 0.5 mm, plump, finely ridged; $2n=18$. Wet, acid soil; N.H. to Wis., s. to Va., Ga., Okla., and Tex. Spring–summer.

8. Xyris ambigua Beyr. Lvs broadly linear, spreading, scarcely twisted, mostly 10–40 cm × 3–20 mm, lustrous dark green, the firm, thickened base brownish or stramineous, only rarely anthocyanic, the plant tending to be invested with the stubble of old fibrous lf-bases; inner surface of inner lvs with prominent dark longitudinal veins; scapes (2–)5–10 dm, twisted, many-ribbed below, becoming flattened and 2-edged above; spikes 1–3 cm at maturity, many-fld, lance-ovoid or ellipsoid; lateral sep shortly included, curved, the broad, thickened keel ciliate-scabrid; pet-blades yellow, obovate, 8 mm, unfolding in the morning; seeds plump, 0.5–0. 6 mm, finely ridged; $2n=18$. Moist or wet low ground, often in grass-sedge bogs or savannas or along the upper edges of wet places; coastal plain from Va. to Fla. and Tex. Early summer–fall.

9. Xyris caroliniana Walter. Lvs deeply set, the outer ones short, scale-like, castaneous, the principal ones linear, 20–50 cm × 2–5 mm, twisted and flexuous, the abruptly dilated, fleshy-thickened, dark brown, shiny bases long-persistent as scales; scapes 5–11 dm, twisted, flexuous, smooth to the touch, terete and minutely ridged below, somewhat flattened (and often 1-edged) above; spikes 1.5–3 cm at maturity, few- to many-fld, narrowly ellipsoid or lance-ovoid; lateral sep almost linear, nearly straight, the broad keel ciliate below, long-fimbriate (and often crisped) at the prominently exserted tip; pet-blades yellow or white, obovate, 8–9 mm, usually opening in afternoon; seeds 0.7–1 mm, narrow, with ca 20 strip-like longitudinal lines; $2n=18$. Moist to well drained, sandy soil on the coastal plain, from Va. to Fla. and Tex. Early summer–fall. (*X. flexuosa.*)

10. Xyris fimbriata Elliott. Lvs rather deeply set, ascending, linear, mostly 5–70 cm × 0.5–2.5 mm, flat or slightly twisted, with soft, thin, keeled, pale to anthocyanic base; scapes 8–15 dm, twisted and evidently multiribbed below, flattened and 2–3(4)-ridged above, the ridges tuberculate-scabrid, harsh to the touch; spikes 1–2.5 cm at maturity, ovoid or broadly ellipsoid, many-fld; keel of the lateral sep broad, slightly thickened, long-fimbriate (and usually crisped) especially toward the prominently exserted tip, making the spike appear dull and fuzzy; pet-blades yellow, 5–6 mm, obovate, unfolding in the morning; seeds 0.8–1 mm, slender, with ca 12 broad, pale, flattened, longitudinal wavy lines nearly as broad as the intervals; $2n=18$. Wet low places, the base often submersed; coastal plain from N.J. to Fla. and Miss. Midsummer–fall.

FAMILY **COMMELINACEAE**, the Spiderwort Family

Fls hypogynous, commonly perfect, usually entomophilous but without nectaries or nectar, some genera self-pollinated; sep 3, usually herbaceous; pet 3, ephemeral, mostly blue (or pink) or white, distinct and sometimes clawed, or connate below to form a tube, often one differently colored and/or ± reduced; stamens mostly 3 + 3, but sometimes only (1–)3 functional and the others staminodial or suppressed; ovary trilocular (often unilocular above), or one or 2 locules reduced or even suppressed; ovules 1–several in each functional locule; placentation axile; fr commonly a loculicidal capsule, seldom indehiscent; embryo small, capping one end of the abundant, mealy endosperm, with a single ± terminal cotyledon and a lateral plumule; seed-coat with a disk-like or conical opercular swelling (embryostega) over the embryo; herbs with alt lvs, a closed sheath, and well defined (narrow or broad), parallel-veined, commonly somewhat succulent blade; infls often ± cymose and subtended by a folded, spathaceous leafy bract, or the fls sometimes solitary in the axils. 50/700.

1 Cor regular; fls or infls axillary to normal or reduced foliage lvs.
  2 Fls in umbel-like cymes; fertile stamens 6; principal lvs 10–40 cm .... 1. *Tradescantia.*
  2 Fls solitary, or some in short racemes; fertile stamens 2 or 3; principal lvs 3–6 cm .... 2. *Murdannia.*
1 Cor ± irregular, one pet smaller; fertile stamens 3; infls subtended by a folded spathe .... 3. *Commelina.*

1. TRADESCANTIA L. Spiderwort. Sep 3, herbaceous, green or anthocyanic; pet 3, all alike, obovate to elliptic, blue or pink (white); stamens 6; filaments usually villous; ovary 3-locular, with 1–2 ovules per locule; style slender; stigma capitate; fr loculicidal; perennial herbs, usually somewhat succulent, with elongate, linear to lanceolate lvs dilated into conspicuous basal sheaths, the umbel-like cymes of several to many handsome fls subtended by elongate foliaceous bracts (except for the last sp.). Except for *T. virginiana*, our spp. bloom in summer and early fall. 50, New World.

1 Bracts elongate, resembling the lvs in shape and scarcely differing in size.
  2 Principal lf-blades wider than the circumference of the sheath, often over 2 cm wide, usually less
      than 10 times as long as wide .... 1. *T. subaspera.*
  2 Principal lf-blades up to about as wide as the circumference of the sheath, usually under 2 cm wide,
      usually more than 10 times as long as wide.
    3 Sep glabrous or with only nonglandular hairs.
      4 Pedicels and sep conspicuously pubescent .... 2. *T. virginiana.*
      4 Pedicels glabrous; sep glabrous, or hairy only at the tip .... 3. *T. ohiensis.*
    3 Sep ± pubescent with glandular hairs (sometimes with eglandular hairs as well).
      5 Sep and pedicels densely and softly villous with both glandular and eglandular hairs commonly
        1–1.5 mm; sep 10–15 mm .... 4. *T. bracteata.*
      5 Sep and pedicels sparsely pubescent with glandular hairs only, these ça 0.5 mm; sep 6–10 mm
        .... 5. *T. occidentalis.*
1 Bracts inconspicuous, 1–10 mm .... 6. *T. rosea.*

    1. Tradescantia subaspera Ker Gawler. Wide-lvd s. Stem rather stout, 4–10 dm, with 4–10 nodes, glabrous or sparsely pilose; lf-blades firm, dark green, lanceolate, glabrous or sparsely pilose, ciliate-margined, the larger 10–20 × 2–4 cm, tapering to a petiole-like part much narrower than the sheath; cymes terminal and usually also lateral from the upper nodes, sessile or short-peduncled; pedicels 1–2 cm, thinly hairy or glabrate; sep 5–10 mm, thinly to densely hairy with glandular or eglandular hairs or both; pet 10–15 mm; $2n=12, 24$. Rich moist woods. Var. *subaspera,* with the stem conspicuously flexuous above, and with the uppermost lateral cymes sessile or nearly so, occurs chiefly w. of the mts., from W.Va. to Tenn., w. to Ill. and Mo. (*T. pilosa.*) Var. *montana* (Shuttlew.) E. S. Anderson & Woodson, with the stem straight or nearly so, the uppermost lateral cymes pedunculate, occurs chiefly in the mts., from Va. to W.Va. to Ala. and Fla.

    2. Tradescantia virginiana L. Virginia s.; widow's tears. Stems straight, erect, 0.5–4 dm at anthesis, glabrous or obscurely puberulent; lvs usually 2–4, elongate, glabrous or nearly so, 5–15 mm wide, rarely wider; bracts often wider and longer than the lvs; cyme terminal and usually solitary; pedicels 15–35 mm, sparsely pilose with eglandular hairs 1 mm; sep 10–15 mm, rather densely villous throughout with eglandular hairs 1–2 mm; pet blue or purple, 12–18 mm; $2n=12, 24$. Moist woods and prairies; Me. to Pa., Mich., and Minn., s. to Ga. and Mo. Spring.

    3. Tradescantia ohiensis Raf. Smooth s. Stem slender, straight or nearly so, often branched, 4–10 dm, glabrous and glaucous; lvs narrowly linear, flat, firm, glabrous, glaucous, usually under 1 cm wide, conspicuously dilated into a sheath; cymes solitary, terminating the stem and branches; pedicels 7–25 cm, glabrous; sep glaucous, 8–12 mm, often red-margined, glabrous throughout or occasionally minutely pilose with eglandular hairs at the tip; pet blue or rose (white), 1–2 cm; $2n=12, 24$. Meadows, thickets, and prairies; Mass. to Minn., s. to Fla. and Tex., commonest in the Middle West. (*T. barbata; T. canaliculata; T. reflexa.*)

    4. Tradescantia bracteata Small. Sticky s. Stem rather stout, straight, rarely branched, (1–)2–4 dm at anthesis, glabrous or minutely puberulent; lvs glabrous or sparsely pilose at base, the larger 8–15 mm wide; bracts often longer and wider than the lvs; cyme usually solitary and terminal, an additional lateral one rarely present; pedicels and sep densely and softly villous with mingled glandular and eglandular hairs 1–1.5 mm; sep obtuse or subacute, 10–14 mm; pet rose or blue, 15–20 mm; $2n=12, 24$. Prairies; s. Ind. to Minn., w. to s. Mont., e. Colo., and n. Okla.

    5. Tradescantia occidentalis (Britton) Smyth. Prairie-s. Stem slender, straight, often branched, 2–6 dm at anthesis, glabrous and glaucous; lvs firm, glabrous, involute, usually narrowly linear and under 1 cm wide; bracts like the lvs; cymes solitary and terminal, or with another one peduncled from an upper node; pedicels and sep sparsely pubescent with glandular hairs 0.5 mm; sep acute to acuminate, 6–10 mm; pet rose to blue, 12–16 mm; $2n=12, 24$. Dry prairies and plains; w. Wis. and Minn. to La., w. to Mont., Utah, and Tex.

    6. Tradescantia rosea Vent. Pink s. Glabrous or nearly so; stems cespitose, very slender, usually branched, 1–4 dm; lvs erect, narrowly linear, 1–5 mm wide, equaling or exceeding the stems; bracts scarious or green, linear, rarely 1 cm; pedicels 10–15 mm; sep 5 mm; pet bright pink, 1 cm; capsule subglobose, 4 mm thick; $2n=12, 24, 36$. Sandy soil; se. Va. to Fla. Our plants are the well marked var. *graminea* (Small) E. S. Anderson & Woodson. (*Cuthbertia g.*)

    2. MURDANNIA Royle. Fls virtually regular, produced singly or in clusters from the axils of normal or somewhat reduced foliage lvs; stamens 3 + 3, regularly disposed, but only the antesepalous ones (or 2 of them) fertile, the others staminodial; ovary and fr trilocular; herbs with leafy stems, often freely branched. 50, trop.

    1. Murdannia keisak (Hassk.) Hand.-Mazz. Stems weak, decumbent or prostrate at base, 3–8 dm; lvs lance-linear, 3–6 cm; fls solitary, the pedicels arising from one to several of the upper sheaths, or some of them in 2–4-fld axillary racemes; sep 5–6 mm; pet pink, ca 8 mm. Fresh tidal marshes and margins of lakes and ponds; native to e. Asia, intr. from Md. to Fla., and inland to Ky., Tenn., and Ark. Late summer and fall. (*Aneilema k.*)

3. COMMELINA L.   Day-fl.   Sep 3, herbaceous, somewhat unequal, 2 sometimes connate at base; pet 3, the upper 2 usually blue, ovate to reniform, clawed at base, the lower one smaller, in most spp. white; fertile stamens 3, 2 with oblong anthers, one incurved with a longer anther; sterile stamens (2)3, smaller, with imperfect, cruciform anthers; filaments not hairy; ovary sessile, 3-locular, the lower 2 locules fertile, with 1 or 2 ovules, the upper median locule smaller, 1-ovuled or empty, or abortive; fr a loculicidal, 2–3-valved capsule; herbs with succulent stem and linear to lance-ovate lvs; infl of small cymes, each closely subtended by a folded cordate spathe from which the pedicels protrude (or sometimes 2 cymes to a spathe); usually self-pollinated. 200+, cosmop., mostly warm reg.

1 Spathe open to the base along one side; annual.
  2 Lower median pet white or nearly so; fr bilocular; lf-sheaths mostly 1–2 cm .... 1. *C. communis.*
  2 Lower median pet blue like the others; fr trilocular; lf-sheaths mostly 0.5–1 cm.
    3 Spathes distinctly falcate; seeds deeply reticulate .... 2. *C. diffusa.*
    3 Spathes scarcely or not at all falcate; seeds smoothish .... 3. *C. caroliniana.*
1 Spathe-margins connate in the lower third; perennial.
  4 Sheaths red-brown-ciliate; lower median pet blue, nearly as large as the others .... 4. *C. virginica.*
  4 Sheaths white-ciliate; lower median pet white, evidently reduced in size .... 5. *C. erecta.*

1. Commelina communis L.   Common d. Fibrous-rooted annual; stems at first erect, later diffuse and rooting from the lower nodes, to 8 dm; lf-blades lance-ovate, the larger 5–12 × 1.5–4 cm; sheaths 1–2 cm; spathe (folded) broadly semicordate, 1.5–3 cm, half as wide, acute or short-acuminate, glabrous or minutely hairy, basally with dark green veins on a paler background, its margins free, its stalk 1–7 cm; blade of upper pet 8–15 mm; lower median pet white or nearly so; anthers 6 (3 sterile); ovary 3-locular; fr 2-locular and 4-seeded, the upper locule of the ovary abortive; $2n=36$–90. Moist or shaded ground, often a garden-weed; native of e. Asia, intr. from N.H. to N.D. and Tex. Several forms, often locally constant, differing in details of pubescence, size and intensity of color of pet, and color of sterile stamens.

2. Commelina diffusa Burman f.   Creeping d. Fibrous-rooted annual, diffusely branched, decumbent and rooting from the lower nodes, to 1 m; lf-blades lanceolate, the larger 3–8(–11) × 1–1.5(–2) cm; sheaths 0.5–1 cm; spathe (folded) semicordate and ± falcate, 1.5–2.5 cm, nearly as wide, acute or short-acuminate, glabrous or finely ciliate, its margins free, its stalk 1–2 cm; larger spathes usually with a 1–few-fld upper cyme in addition to the well developed lower one; blade of upper pet 6–8 mm; lower median pet blue; anthers 5 or 6 (2 or 3 sterile); fr 3-locular, the lower locules each 2-seeded, the upper one 1-seeded; seeds of the lower locules 2–2.8(–3.2) mm, deeply reticulate; $2n=28$–60. Wet woods and river-banks; native to the Old World, only intr. in N. Amer., where mainly in se. U.S., n. sometimes to Del., O., Ill., Minn., and Kans.

3. Commelina caroliniana Walter.   Spathes not at all or only slightly falcate; upper cyme generally vestigial (rarely well developed and 1-fld); fr (5)6–8 mm; seeds of the lower locules 2.4–4.3(–4.6) mm, smooth or faintly alveolate; otherwise much like no. 2. Native of India, long intr. and weedy in se. U.S., n. to Md.

4. Commelina virginica L.   Virginia-d. Rhizomatous perennial; stem stout, erect or nearly so, to 12 dm, often widely branched; lf-blades lanceolate, long-acuminate, scabrous above, glabrous or finely hairy beneath, the main ones 10–20 × 2.5–5 cm; sheaths pilose-ciliate with red-brown hairs 2–5 mm, not prolonged into distinct auricles; spathes usually clustered toward the summit, sessile or short-peduncled, 2–3(–3.5) cm, nearly as wide when folded, the nerves connected by numerous cross-veins, the margins connate in the lower third; lower pet blue, scarcely reduced; fr 3-locular, the lower locules each 2-seeded, the upper one 1-seeded; $2n=60$. Moist or wet woods; N.J. to Fla., w. to Ill., Kans., Okla., and Tex. (*C. hirtella.*)

5. Commelina erecta L.   Erect d. Perennial from a cluster of thickened fibrous roots; stems erect or ascending, to 1 m, usually branched; principal lvs linear to lanceolate, 4–15 × 0.5–4 cm, the sheaths white-ciliate, somewhat prolonged at the summit into rounded, often flaring auricles; spathes arising near the summit of the culm, solitary or in small clusters, short-peduncled, broadly semi-deltoid, often with con-spicuous radiating cross-veins, the margins connate in the lower third, upper pet 10–25 mm, the lower one much smaller and white; $2n=56$–120. Dry, usually sandy soil; s. N.Y. to Pa., O., Mich., se. Minn., Io., and Wyo., s. to S. Amer.; also Old World. Highly variable, but only with difficulty divisible into vars. Var. *erecta* (*C. virginica,* misapplied; *C. elegans*), with lanceolate or lance-ovate lvs 10–15 × 1.5–4 cm, and with the spathes glabrous or nearly so, is mainly Ozarkian (with us), but extends ± throughout the range of the sp. Var. *angustifolia* (Mich.) Fern. (incl. *C. crispa*), with linear lvs 4–10 cm and small spathes 1–2 cm, centers in Tex., but extends irregularly into our range. Var. *deamiana* Fern., with linear lvs 10–15 cm, the spathes 2.5–3.5 cm and often pilose at base, occurs mainly on sand dunes from Ind. and s. Mich. to Neb., occasionally s. and w. to Tex. and Ariz.

FAMILY **ERIOCAULACEAE**, the Pipewort Family

Fls individually very small, hypogynous, unisexual, regular or irregular, closely crowded into a dense, commonly white or grayish or leaden, terminal, usually bisexual head subtended by an invol of chaffy bracts, the common receptacle beset with chaffy bracts subtending the fls, or merely hairy, or naked; sep 2 or 3, distinct or variously connate; pet 2 or 3, distinct or variously connate, or wanting; staminate fls with the filaments adnate to the cor-tube, or often with a slender, stipe-like androphore (above the cal) from the top of which the filaments and pet (when these are present) diverge; stamens bicyclic, or the antesepalous cycle wanting; ovary superior, often stipitate, 2- or 3-locular; style once or twice bifid or trifid, often with prominent appendages below the primary branches; ovules solitary in each locule, apical-axile, pendulous; fr a loculicidal capsule; embryo small, lenticular, capping the copious, mealy endosperm; scapose herbs of wet places or shallow water; lvs all basal, parallel-veined, narrow and grass-like, but without a well differentiated sheath, only the scapes with a single nearly or quite bladeless sheath. 13/1200.

1 Pet 2; sep 2; stamens 4; anthers with 2 pollen-sacs; scapes glabrous .... 1. *Eriocaulon*.
1 Pet none; sep 3; stamens 3; anthers with one pollen-sac; scapes villous above .... 2. *Lachnocaulon*.

1. ERIOCAULON L. Pipewort. Fls each subtended by a small bract, dimerous in our spp., the sep 2, distinct or connate, the pet 2, each with a nectariferous gland just within the tip, the ovary bilocular and the style bifid; staminate fls with an androphore; stamens bicyclic (4 in our spp.), exsert at anthesis; anthers with 2 pollen-sacs, black at maturity in our spp.; ovary on a gynophore which also bears the pet; lvs with evident lacunar tissue especially toward the base; roots fleshy-fibrous, septate, pale, unbranched. 400, mainly warm reg. The hairs of the receptacle, mentioned in the key below, are very slender and much elongate, in contrast to the thick short hairs of the perianth and receptacular bracts.

1 Receptacle glabrous; delicate, low plants to 2(–3) dm, with heads 3–6 mm thick.
  2 Perianth and receptacular bracts densely white-hairy .... 1. *E. aquaticum*.
  2 Perianth and receptacular bracts with few or no hairs .... 2. *E. parkeri*.
1 Receptacle villous between the fls; coarser plants, 2–10 dm, with heads 7–15+ mm thick.
  3 Receptacular bracts with pale, sharp-pointed, protruding tip; heads hard .... 3. *E. decangulare*.
  3 Receptacular bracts with dark, obtuse to subacute tip, not protruding; heads softer .... 4. *E. compressum*.

1. Eriocaulon aquaticum (Hill) Druce. Lvs very thin, often pellucid, 2–10 cm × 2–5 mm, 3–9-nerved with conspicuous cross-veinlets, tapering evenly from a pale, spongy-aerenchymatous base; scapes usually solitary for each rosette, 3–20 cm (or much longer in deep water), slender, slightly twisted, 5–7-ridged; mature heads subglobose, 4–6 mm thick, gray except for the exserted white-hairy tips of the tep and receptacular bracts; invol bracts ovate to obovate, obtuse, lustrous dark gray, glabrous, reflexed and concealed at maturity; receptacle glabrous; $2n=32$. Usually in shallow water, occasionally in deeper water or on miry shores; Nf. to Minn., s. to Del. and Ind., and in the mts. to N.C.; also in Ireland and Scotland, where $2n=64$. July–Sept. (*E. articulatum*; *E. pellucidum*; *E. septangulare*.)

2. Eriocaulon parkeri Robinson. Much like no. 1; scapes 4–5-ridged, tending to be straight, often 2–4 from the same rosette; heads a bit smaller, mostly 3–4 mm thick, hemispheric, the outer invol bracts very pale, dull gray or stramineous, tending to remain ascending even at maturity; perianth and receptacular bracts with few or no hairs; $2n=48$. Tidal flats and muddy shores, often submerged, in fresh to slightly brackish water; Me. and Que. to Mass. and c. N.Y., and s. near the coast to N.C. July–Oct.

3. Eriocaulon decangulare L. Densely tufted; lvs firm, mostly 10–40 cm, tapering evenly from the broad (1–4 cm), spongy, translucent base to the narrow but blunt tip, many-nerved, the cross-veinlets inconspicuous or wanting; scapes 1–3 per rosette, rigid, 3–10 dm, twisted, 8–12-ridged, 1–3 mm wide just below the head, scape-sheath surpassed by the lvs; heads bisexual, at maturity subglobose, 7–15+ mm thick, hard, dull white, the outermost fls reflexed and obscuring the invol bracts, these stramineous, lanceolate to narrowly ovate, 3–4 mm, acute, white-hairy distally; receptacle copiously long-villous, its bracts narrowly acute to acuminate, with pale, shortly protruding, often hairy tip; perianth pubescent distally with cylindric-clavate hairs 0.2–0.5 mm. Moist but seldom permanently wet places; N.J. to Fla. and Tex., mainly on the coastal plain, but also in the mts. of N.C. July–Sept.

4. Eriocaulon compressum Lam. Much like no. 3; foliage softer and more spongy; scape-sheath usually surpassing the lvs; heads softer, unisexual or nearly so, the plants subdioecious; outer invol bracts grayish-translucent, 2–3 mm, broadly ovate to oblong or elliptic, blunt, often squarrose; receptacle less copiously hairy, its dark-tipped bracts blunt or subacute, not protruding. Wet places or shallow water on the coastal plain from N.J. to Fla. and Tex. and in the mts. of N.C. May–July.

2. LACHNOCAULON Kunth.   Fls each subtended and partly enfolded by a small, scarious bract; common receptacle long-villous; sep 3; pet none or vestigial; staminate fls with an androphore bearing (2)3 stamens and as many fringed or entire staminodes; anthers pale, with a single pollen-sac; pistillate fls with a (2)3-locular ovary and (bifid) trifid style, each primary style-branch again bifid; lvs without evident lacunar tissue; roots very slender, fibrous, branched. 7, se. U.S., W.I.

1. **Lachnocaulon anceps** (Walter) Morong.   Tufted, beset at the base with old fibrous lf-bases; lvs 3–6 cm, 2–3.5 mm wide at base, tapering gradually to the slender tip, smooth or with scattered multicellular hairs; scapes 1–several, 1–4 dm, twisted, usually sparsely villous with ascending hairs 1–2 mm; heads globose to short-cylindric, 4–7 mm thick, whitish or pale gray; invol bracts mostly 1–1.5 mm, obtuse, fringed above with clavate white hairs; receptacular bracts 1.5–2 mm, white-hairy distally, as also the sep; staminodes short, fringed. Moist or wet to fairly dry places on the coastal plain from Va. to Fla., Tex., and Cuba, seldom in the mts. of N.C., Ga., and Ala. June–Aug.

Family **JUNCACEAE**, the Rush Family

Fls perfect or seldom unisexual, ordinarily (including all our spp.) trimerous; perianth regular, small, mostly green or brown, ± chaffy or scale-like, the sep and pet essentially alike but in 2 separate whorls, commonly persistent into fr; stamens 6 or 3; ovary superior, with 3 axile to parietal or basal placentae; stigmas 3; fr a loculicidal, 3-valved capsule; embryo small, straight, monocotyledonous, embedded in the starchy endosperm; herbs of sedge-like aspect, lfless or with narrow, terete or grass-like lvs. 8/300.

1 Ovules and seeds several to usually many; plants glabrous (rarely scabrous) .... 1. *Juncus.*
1 Ovules and seeds 3; lvs long-hairy along the margins, at least when young .... 2. *Luzula.*

1. JUNCUS L.   Rush. Tep narrow, lance-subulate to lance-ovate, dry, often firm and even sharp; stamens 6 or 3, opposite the sep when only 3; ovary and fr trilocular, subtrilocular (with incomplete partitions) or unilocular; seeds several to usually numerous (in any case more than 3), commonly ellipsoid or fusiform and minutely apiculate, sometimes with each end prolonged into a slender tail that may be longer than the body; smooth (1 of our spp. scabrous) herbs with usually simple stems and a few flat or terete, basal or cauline lvs, sometimes with bladeless sheaths only, and with a terminal, compact to loosely branched cyme of few–many fls, these solitary, paired, or often in glomerules; lf-sheaths open. 200, cosmop.

1 Individual fls prophyllate, i.e., subtended by 2 small opposite bracteoles at the base of the perianth;
   lvs, when present, not septate.
  2 Basal sheaths, or some of them, with blades; infl (except in no. 7) terminal. (*Poiophylli.*) .... Group I.
  2 Basal sheaths bladeless; involucral lf erect, resembling a continuation of the stem, the infl therefore
     apparently lateral. (*Genuini.*) .... Group II.
1 Individual fls eprophyllate, without subtending bracteoles; lvs often septate.
  3 Stems prostrate and creeping, or floating .... Group III.
  3 Stems mostly erect or nearly so.
    4 Lvs flat or terete, or channeled on the upper side, solid, not (except in no. 18) cross-septate .... Group IV.
    4 Lvs aerenchymatously hollow, nodulose at regular intervals with transverse septa. (*Septati.*)
      5 Seeds 0.7–2.3 mm, fusiform, with a ± definite slender appendage at each end .... Group V.
      5 Seeds smaller, obovoid or thickly ellipsoid, not caudate-appendaged.
        6 Capsule subulate, thickest near the base and tapering to a slender beak, about 5 times as
          long as thick; placentae fertile only below the middle .... Group VI.
        6 Capsule of various shapes, but never subulate; placentae various .... Group VII.

Group I

1 Annual; infl a third or more of the total height of the plant .... 10. *J. bufonius.*
1 Perennial, variously tufted or rhizomatous; infl proportionately much smaller.
  2 Cauline lvs present, arising in the upper three-fourths of the stem.
    3 Infl branched, many-fld; cauline blades divergent near or below the middle of the stem; often
     in salt-marshes.

    4 Fr barely if at all surpassing the sep; stamens nearly equaling the tep; anthers 1 mm or more
    .... 1. *J. gerardii.*
    4 Fr strongly surpassing the sep; stamens half as long as the tep; anthers 1 mm or less .... 2. *J. compressus.*
  3 Infl simple, 1–4-fld; cauline blades divergent near the top of the stem; montane .... 3. *J. trifidus.*
2 Cauline lvs none (except the invol), the foliage lvs all basal or nearly so.
  5 Lvs flat (though narrow), or becoming involute when dry.
    6 Fruit definitely 3-locular.
      7 Sep 4–5.5 mm; infl generally overtopped by the involucral lf; western .... 4. *J. brachyphyllus.*
      7 Sep 3–4 mm; infl generally surpassing the involucral lf; widespread .... 5. *J. secundus.*
    6 Fr unilocular or only imperfectly 3-locular, the intruded partitions not meeting in the center
    .... 6. *J. tenuis.*
  5 Lvs terete, or narrowly channeled along the upper side.
    8 Fr twice as long as thick, the valves trough-shaped at the summit; tep erect or appressed even
      at maturity.
      9 Seeds evidently caudate-appendaged at both ends, mostly 1–1.5 mm overall .... 9. *J. vaseyi.*
      9 Seeds merely apiculate, ca 0.5 mm .... 8. *J. greenei.*
    8 Fr less than twice as long as thick, the valves scarcely or not at all trough-shaped; tep becoming
      divergent by growth of the fr.
      10 Infl appearing lateral, much surpassed by the erect involucral lf .... 7. *J. coriaceus.*
      10 Infl appearing terminal, not much surpassed by the inconspicuous involucral lf .... 6. *J. tenuis.*

Group II

1 Longer (inner) prophyll broadly round-obovate; stem without regular longitudinal ribs, merely irreg-
    ularly longitudinally ridged when dry .... 11. *J. arcticus.*
1 Longer prophyll deltoid-ovate, acute to obtuse; stem often with regular longitudinal ribs (best observed
    when dry) traceable for several cm without interruption.
  2 Stamens 6; fr acute or mucronate or beaked.
    3 Sep 2.6–4.3 mm, nearly or fully equaling the fr.
      4 Infl 3–6 cm, much branched; involucral lf up to a third as long as the stem .... 12. *J. inflexus.*
      4 Infl 1–2 cm, not much branched; involucral lf at least half as long as the stem .... 13. *J. filiformis.*
    3 Sep 1.5–2.2 mm, reaching scarcely beyond the middle of the fr .... 14. *J. gymnocarpus.*
  2 Stamens 3; fr usually broadly rounded or retuse except in no. 15.
    5 Stem and lowest bract rather coarsely ridged-sulcate with 10–25(–30) ridges.
      6 Infl very dense and head-like, mostly 1–1.5 cm wide; uncommon introduction .... 15. *J. conglomeratus.*
      6 Infl loose to fairly compact, but not head-like, mostly 1.5–6+ cm wide; common native .... 16. *J. pylaei.*
    5 Stem and lowest bract smooth to finely striate with 30–60 very low ridges .... 17. *J. effusus.*

Group III

1 Pet 5–10 mm, much surpassing the sep; southern sp .... 24. *J. repens.*
1 Pet less than 4 mm, barely surpassing the sep; northern spp.
  2 Fls solitary, or in pairs; stamens 6 .... 36. *J. subtilis.*
  2 Fls in glomerules of 3–10; stamens 3(6) .... 37. *J. bulbosus.*

Group IV

1 Lvs equitant, and with irregular partial septa (*Ensifolii*) .... 18. *J. ensifolius.*
1 Lvs neither equitant nor septate.
  2 Gynodioecious plants of coastal salt-marshes and beaches. (*Thalassi.*) .... 19. *J. roemerianus.*
  2 Plants with perfect fls, not maritime.
    3 Seeds 2–3 mm, caudate-appendaged at both ends; stamens 6; lvs very slender. (*Stygii.*) .... 20. *J. stygius.*
    3 Seeds less than 1 mm, not caudate.
      4 Fls borne singly or in pairs; stamens 6; lvs slender and terete .... 35. *J. pelocarpus.*
      4 Fls borne in head-like clusters; lvs flat and grass-like. (*Graminifolii.*)
        5 Stamens 6 .... 21. *J. longistylis.*
        5 Stamens 3.
          6 Plants mostly 2–5(–7) dm, with 5–20 heads; main lf-blades 1–3 mm wide .... 22. *J. marginatus.*
          6 Plants mostly 6–12 dm, with 20–100 heads; main lf-blades 4–6 mm wide .... 23. *J. biflorus.*

Group V

1 Lvs scabrous; seeds 2–2.3 mm .... 25. *J. caesariensis.*
1 Lvs glabrous; seeds 0.7–1.9 mm.
  2 Pet lance-subulate, their margins not scarious or only very narrowly so.

3 Seeds 1.2–1.9 mm, the body constituting less than half the total length .... 26. *J. canadensis.*
3 Seeds 0.7–1.2 mm, the body constituting more than half the total length.
  4 Infl open, not over twice as long as wide; glomerules 5–20-fld .... 27. *J. subcaudatus.*
  4 Infl strict, 3–6 times as long as wide; glomerules 2–7-fld .... 28. *J. brevicaudatus.*
2 Pet obtuse or barely acute, with conspicuous scarious margins .... 29. *J. brachycephalus.*

GROUP VI

1 Lvs terete or nearly so.
  2 Uppermost cauline lf with a well developed, septate blade longer than its sheath.
    3 Stamens 6.
      4 Auricles 2–5 mm; sep mostly 3.5–5.5 mm; plants robust (2–)4–10 dm .... 30. *J. torreyi.*
      4 Auricles 0.5–1 mm; sep mostly 2.5–3.5 mm; plants slender, 1–4 dm .... 31. *J. nodosus.*
    3 Stamens 3; auricles 1–2.5 mm; sep 2.2–3.2 mm .... 32. *J. scirpoides.*
  2 Uppermost cauline if bladeless or with a vestigial blade shorter than its sheath .... 33. *J. megacephalus.*
1 Lvs evidently flattened, somewhat equitant .... 34. *J. validus.*

GROUP VII

1 Stamens 6.
  2 Fls solitary or in pairs .... 35. *J. pelocarpus.*
  2 Fls glomerulate.
    3 Fully developed cauline lvs 2 or more, 0.5–1.5 mm thick, not overtopping the infl.
      4 Branches of the infl divaricate, or ascending at an angle of more than 30°; infl usually less
        than twice as high as wide; tep ± acute (or the inner obtuse), the inner equaling or longer
        than the outer .... 38. *J. articulatus.*
      4 Branches of the infl erect or ascending at an angle of usually less than 30°, the infl usually
        more than twice as high as wide; tep obtuse or rounded, the inner shorter .... 39. *J. alpinoarticulatus.*
    3 Fully developed cauline lf 1(2), 2–4 mm thick, overtopping the infl .... 40. *J. militaris.*
1 Stamens 3 (rarely 6 in some fls).
  5 Glomerules globose, densely many-fld; fr much shorter than the sep .... 41. *J. brachycarpus.*
  5 Glomerules obpyramidal to hemispheric, 2–50-fld; fr scarcely shorter than the sep.
    6 Perianth about as long as the mature fr.
      7 Infl over twice as long as wide, with 3–100 glomerules; stem slender.
        8 Fr stramineous to light brown; heads 5–50-fld .... 42. *J. acuminatus.*
        8 Fr dark purple-brown; heads 2–7-fld .... 43. *J. elliottii.*
      7 Infl under twice as long as wide, with 200 or more glomerules; stem stout .... 44. *J. nodatus.*
    6 Perianth only half to three-fourths as long as the fr.
      9 Fr 4–6 mm, linear-trigonous, a fifth as thick .... 5. *J. diffusissimus.*
      9 Fr 2.8–4 mm, a third as thick .... 46. *J. debilis.*

1. Juncus gerardii Loisel. Black grass. Stems 2–6 dm, from long, slender rhizomes; lvs elongate, to 2 dm; cauline lvs 1 or 2, the uppermost divergent near the middle of the stem; lf-sheaths entire at the summit; infl 2–8 cm, many-fld, with ascending or erect branches, seldom surpassed by the invol lf; fls prophyllate, tep oblong, purple-brown with green midstrip, the sep 2.4–3.2 mm, narrowed to an obtuse incurved tip, the pet slightly shorter, scarious at the broad obtuse summit; stamens 6, reaching nearly to the tip of the tep, the anthers ca 1.5 mm, 3–4 times as long as the filaments; fr trilocular, ovoid-ellipsoid, 2.4–3.3 mm, rounded above and short-beaked, equaling or barely surpassing the sep; seeds ca 0.5 mm, ellipsoid or obovoid-ellipsoid, longitudinally ribbed; $2n=80, 84$. Salt-marshes along the coast; interruptedly circumboreal, s. to Va.; now widely intr. inland in our range, in both saline and disturbed nonsaline habitats.

2. Juncus compressus Jacq. Much like no. 1; primary invol bract usually surpassing the infl; tep oblong to obovate, obtuse, light brown with greenish midstrip and membranous margin, the sep 2–2.5 mm; stamens 6, extending to about the middle of the tep or a bit beyond, the anthers up to ca 1 mm, 1–2 times as long as the filaments; fr obovoid-globose, 2.1–2.6 mm, short-beaked, obviously surpassing the sep; $2n=40, 44$. Wet meadows and salt-marshes; native to Europe, now irregularly established from Nf. to Me. and N.H., w. to Wis., Minn., Mont., and Utah.

3. Juncus trifidus L. Stems densely cespitose from a short rhizome, 5–30 cm up to the infl, surrounded at base by crowded sheaths ca 3 cm, naked for most of their length, and bearing toward the summit 2 or 3 filiform, minutely serrulate lvs 4–10 cm and 1 or sometimes 2 simple infls of 1–4 fls, the fls prophyllate, with lacerate bracteoles; basal sheaths with fimbriate-lacerate auricles and a short bristle-tip; tep lanceolate or lance-ovate, acuminate or aristate, mostly 3–4 mm, the pet acute or acuminate, slightly shorter; anthers 6, ca twice as long as the filaments; fr trilocular, 2.5–3.5 mm, broadly obovoid, with a slender beak ca 1 mm; seeds few, irregularly polyhedral, 0.6–1.3 mm; $2n=30$. Rock crevices and alpine meadows; Europe and ne. Amer., s. in the mts. to Va. and N.C. Plants from N.Y. southward mostly have relatively large fls for the sp., with only 1 or 2 fls per infl, and have been segregated as ssp. *carolinianus* Hamet-Ahti (var. *monanthos,* misapplied). More northern plants in e. N. Amer. mostly have 2–4 fls per infl and pass as typical *J. trifidus.*

4. Juncus brachyphyllus Wieg.   Stems cespitose from a short rhizome, 3–5 dm; lvs on the basal part of the stem, the sheaths projecting into prominent, firmly membranous auricles, the blades thick but ± flattened, ca 1–1.5 mm wide; infl (2–)3–6(–7) cm, loose to compact, usually somewhat surpassed by the invol bract; fls numerous, prophyllate; tep pale greenish to stramineous, lance-subulate, long-acuminate and pungent, the pet 3.5–4 mm and evidently scarious-margined, the sep less so, 4–5.5 mm, ± closely investing the fr; stamens 6; fr trilocular, oblong-ovoid, 3.7–4.6 mm, truncate, ca equaling the pet but shorter than the sep; seeds 0.5 mm, merely apiculate, becoming mucilaginous when wetted. Sandy prairies and sandstone glades; Mo. and w. Ark. to Neb., Kans., Okla., and ne. Tex.; disjunct in Oreg., Wash., and n. Idaho. (*J. kansanus.*)

5. Juncus secundus P. Beauv.   Stems loosely cespitose, 3–6 dm; basal lvs flat or involute, narrowly linear, 8–20 cm, usually less than a third as long as the stems; auricles pale, membranous, gradually rounded at the summit; invol lf inconspicuous, commonly shorter than the infl; branches of the infl ascending, commonly incurved above, each with 3–8 secund, prophyllate fls; tep greenish, subequal, 3–4 mm, the sep lance-ovate, the pet lance-ovate; anthers 6, longer than the filaments; fr trilocular, oblong-ovoid or short-cylindric, 2.6–3.3 mm, blunt or truncate; seeds 0.5 mm, merely apiculate. Usually in clay soil; Me. to s. Ont., s. to Ga.; also s. O. to Mo., Kans., Tenn., and Ark.

6. Juncus tenuis Willd.   Path-r. Stems ± cespitose, 1–8 dm; lvs basal, 1–3 dm, a third to half the height of the stems, flat and 1–1.5 mm wide, or becoming involute, varying to subterete and merely narrowly channeled on the upper side; sheaths scarious-margined, auriculate; invol lf 1–10 cm, often surpassing the infl, but not appearing as a continuation of the stem; infl loosely branched, not secund, the fls prophyllate; tep lance-subulate, the sep 3–5.5 mm, the pet 2.8–5 mm; anthers 6, shorter than the filaments; fr oblong-ovoid or slightly obovoid, obtuse to truncate, 2.6–4.2 mm, shorter than the sep, only imperfectly trilocular, the partitions not meeting in the center; seeds 0.3–0.5 mm, merely apiculate, mucilaginous when wetted; 2n=40, 42, 80, 84. Dry or moist (often compacted) soil, abundant along forest paths; almost throughout N. Amer., and naturalized elsewhere. Highly variable, and divisible with some difficulty into 3 vars.:

a Lvs subterete; auricles short, firm, rounded; fr shining mahogany-brown; mainly on the coastal plain.
   (*J. dichotomus; J. platyphyllus.*) . . . . var. *dichotomus* (Elliott) A. Wood.
a Lvs mostly flat or involute; fr paler, dull.
   b Auricles thin, slender, pointed; widespread. (*J. interior.*) . . . . var. *tenuis.*
   b Auricles rounded, cartilaginous, often becoming yellowish; mainly inland. (*J. dudleyi.*) . . . .
   . . . . var. *dudleyi* (Wieg.) F. J. Herm.

7. Juncus coriaceus Mackenzie.   Stems densely cespitose, 4–10 dm, erect or decurved above, often surrounded by old basal sheaths; lvs all basal except the invol one; sheaths firm, often darker upwards, with firm rounded auricles; blades elongate, subterete, narrowly channeled on the upper side; invol lf erect, resembling the stem but channeled, 1–2 dm; infl 2–4 cm; fls prophyllate; tep lance-ovate, acuminate, widely spread by the fr, 3.3–4.5 mm; fr 3–4 mm, globose-ovoid, only obscurely angled, shining brown, unilocular, with narrow partial partitions, scarcely dehiscent, occasionally with a slender beak to 1 mm; seeds 0.5–0.8 mm, substipitate, angularly obconic or turbinate, evidently ribbed vertically and with finer cross-ridges. Wet ground, swamps, or brackish marshes; N.J. to Fla., and Tex., n. in the interior to Ky. and Okla. (*J. setaceus,* misapplied.)

8. Juncus greenei Oakes & Tuckerman.   Stems cespitose from a short rhizome, 2–8 dm; basal lvs filiform, subterete, 5–20 cm; invol lvs similar, 2–15(–20) cm; infl small and compact, obpyramidal, 2–5 cm; fls prophyllate; tep lance-oblong, permanently appressed, acute or aristulate, the sep 2.3–4 mm, the pet 1.9–3.4 mm; fr trilocular, ovoid-cylindric, 3–4+ mm, half as thick, truncate, conspicuously surpassing the tep; seeds oblong to obovoid, 0.3–0.6 mm, merely apiculate; 2n=80. Moist to dry, clay or sandy soil, sometimes on dunes; N.B. to N.H., mostly near the coast; also inland from sw. Ont. to Ind., Io., and Minn.

9. Juncus vaseyi Engelm.   Stems cespitose from a short rhizome, 3–8 dm; basal lvs very slender, terete or nearly so, to 3 cm; invol lf shorter than the infl or up to 5(–7) cm; infl obpyramidal, compact, 1–4 cm; fls prophyllate; tep lanceolate, acute or acuminate, appressed even at maturity, the sep 3.5–4.5 mm, the pet 3.2–4.2 mm; anthers 6, about equaling the filaments; fr trilocular, oblong-cylindric, 4.1–5.4 mm, truncate; seeds fusiform, becoming mucilaginous when wetted, the body 0.5–0.8 mm, each end with a slender pale appendage usually 0.2–0.4 mm; 2n= 42. Moist, often sandy soil; N.S. to Sask. and B.C., s. to N.Y., Ind., Minn., and Colo. An apparent hybrid with no. 6, in Me., has been called *J.* × *oronensis* Fern.

10. Juncus bufonius L.   Toad-r. Annual to 3 dm, the stems terete, sometimes branched; lvs basal and usually also cauline, capillary, 0.5–10 cm × 0.2–1.1 mm, convex on the lower surface, flat or channeled on the upper; fls often cleistogamous, scattered singly along the upper part (often more than half) of the stem, short-pedicellate or subsessile in the axils of hyaline-scarious or partly leafy bracts, each one also closely subtended by a pair of broad, hyaline-scarious bracteoles 1–2.5 mm; tep slender, with greenish midstrip and hyaline margins, mostly 3–6 mm, the outer a little the longer; anthers 6, 0.3–1 mm, up to as long as the filaments; style very short, fr imperfectly trilocular; seeds 0.3–0.6 mm, apiculate; 2n=26–120, a small polyploid pillar-complex. Moist or vernally wet or moist places; nearly cosmopolitan, and throughout our range. Most of our plants are polyploid and have acuminate pet slightly longer than the fr. A ± diploid halophytic phase, found in salt-marshes along the coast from Lab. to Mass., and widespread in Europe, differs morphologically from typical, widespread *J. bufonius* in its consistently small size (rarely to 15 cm) and obtuse or mucronate pet equaling or slightly shorter than the fr. This phase has been segregated as var. *halophilus* Fern. or *J. ambiguus* Guss.

11. Juncus arcticus Willd. Wire-r. Stems arising in rows from long rhizomes, simple, 4–8 dm, finely and irregularly striate, without foliage lvs; basal sheaths to 12 or 15 cm, mucronate, bladeless; invol lf to 18 cm, a fifth to a third as long as the stem and appearing as a continuation of it; infl apparently lateral, capitate to diffuse, (1–)3–5(–12) cm; inner prophyll broadly round-obovate; tep lanceolate, 3.3–4.7 mm, acuminate, with a conspicuous dark stripe on each side of the midrib; anthers 6, 2–4 times as long as the filaments; fr trilocular, acute, short-beaked, exceeding the perianth by 0.5–1 mm; $2n$=40, 80, 84. Calcareous or brackish shores and dunes, and also inland; circumboreal, in our range s. to Pa. and Mo., especially abundant along the shores of the Great Lakes; S. Amer. (Incl. *J. balticus.*) Polymorphic; our plants may be referred to var. *littoralis* (Engelm.) B. Boivin.

12. Juncus inflexus L. Stems cespitose from a short rhizome, 3–8 dm, glaucous, finely striate, without foliage lvs; basal sheaths to 10 cm, purple-brown at base, bristle-tipped; invol lf 1–2 dm, a fourth to a third as long as the stem; infl appearing lateral, 3–6 cm, freely branched and many-fld; fls prophyllate; tep lance-acuminate, 2.6–4 mm, slightly shorter to slightly longer than the dark purple-brown, shining, trilocular, beaked fr; anthers 6, about as long as the filaments; $2n$=40, 42. Native of Eurasia and n. Afr., intr. in wet meadows and along roads in our range, as in N.Y., Ont., and Mich.

13. Juncus filiformis L. Stems cespitose or in rows on a short or long rhizome, finely striate, without foliage lvs, 1–4 dm to the infl, surmounted by an erect invol lf half to fully as long; basal sheaths pale brown, seldom over 6 cm; infl 1–2 cm, sparingly branched, each primary branch with 1–4 prophyllate fls; sep lanceolate, 2.7–4.3 mm; pet 2.3–4.1 mm, acutish or obtuse; anthers 6, much shorter than the filaments; fr trilocular, 2.4–3.7 mm, obovoid, short-beaked; $2n$=40, 70, 80, 84. Sandy shores, bogs, and alpine meadows; circumboreal, s. in our range to the mts. of N. Engl., N.Y., Pa., and W.Va., and to n. Mich. and n. Minn.

14. Juncus gymnocarpus Cov. Stems arising in rows from long rhizomes, 4–8 dm, finely striate, without foliage lvs; basal sheaths to 8 cm, reddish-purple at base, bristle-tipped; infl 2–4 cm, few-fld, very lax and open; fls prophyllate; sep lanceolate, 1.5–2.5 mm, acute; pet lance-ovate, somewhat shorter, obtuse; stamens 6; fr trilocular, broadly ovoid-trigonous, 2.4–3 mm, chestnut-brown, shining, well exserted, shortly beaked. Rare and local in sphagnum-bogs; e. Pa. to the mts. of Tenn., N.C., and S.C.; w. Fla.

15. Juncus conglomeratus L. Much like no. 16, but with very dense and head-like infl mostly 1.5–2.5 × 1–1.5 cm; invol lf 5–15 cm; tep a little looser, the outer as long as the inner; fr usually mucronate; basal sheaths usually rather pale brown; stem avg with more numerous (up to 25 or even 30) ridges; fr truncate, apiculate; $2n$=42. Low, marshy or boggy or peaty places; native of Europe, locally intr. from Nf. and N.S. to N.Y. and Que. (*J. effusus* var. *c.*)

16. Juncus pylaei Laharpe. Densely cespitose; stems rather soft, to 1 m, rather coarsely ridged-sulcate with 10–20 ridges, the epidermal cells over the ridges enlarged; basal sheaths to 2 dm, bladeless, mostly dark reddish-brown or blackish; invol lf 10–20 cm, appearing like a continuation of the stem; infl apparently lateral, many-fld, open to fairly compact, mostly 1.5–6+ cm wide; fls prophyllate; tep 2.5–3.5 mm, pale brown, rigid, lanceolate, long-acuminate, ± closely investing the fr, the outer a little longer than the inner and often loose or slightly spreading at the tip; anthers 3; fr trilocular, shorter than to about equaling the tep, broadly obtuse or rounded to truncate or retuse at the tip, usually not mucronate; $2n$=40. Wet, boggy or sandy, rather low-nutrient soils; Nf. to Minn. and sw. Mont., s. to N.C. and Ind. (*J. effusus* var. *p.*)

17. Juncus effusus L. Soft r. Densely cespitose; stems to 1+ m, smooth or with 30–60 very low, inconspicuous ridges, the epidermal cells over the ridges not enlarged; basal sheaths to 2 dm, bladeless, mostly reddish-brown; invol lf (10–)15–25(–35) cm, appearing like a continuation of the stem; infl apparently lateral, many-fld, usually open, the longest branches (3–)4–10 cm; fls prophyllate; tep 2–2.5(–3) mm, stramineous, broadly lanceolate, acute, the outer commonly a bit longer than the inner; anthers 3; fr trilocular, slightly shorter to slightly longer than the tep, mostly obtuse to truncate, seldom (*J. griscomi*) mucronate; $2n$=40, 42. Open marshes and wet meadows; nearly cosmop., and throughout our range. Our common native plant, throughout e. U.S. and se. Can., is var. *solutus* Fern. & Wieg., with rigid tep closely investing the fr. The European var. *effusus*, with soft tep ± spreading from the base, is locally intr. from Nf. to Me. and inland to Mich. and Minn. (Var. *compactus*, a form with compact infl.)

18. Juncus ensifolius Wikström. Rhizomatous; stem 2–6 dm, flattened; lvs mostly crowded below, the larger ones 10–40 cm × 1.5–6 mm, all equitant; blade with scattered, irregular, partial or complete partitions; heads few, many-fld, 1+ cm thick, mostly dark; fls axillary to evident, hyaline-scarious bracts but not prophyllate; outer tep 2.5–4 mm, acuminate, inner a little shorter and often a little wider and less pointed; stamens 3 (in ours); fr incompletely 3-locular; seeds 0.4–0.6 mm, apiculate at both ends; $2n$=20. Wet places; widespread in the w. cordillera, disjunct in Ashland Co., Wis. and along the Delaware R. in N.Y. Ours is var. *ensifolius.*

19. Juncus roemerianus Scheele. Needlerush. Stout, rigid, erect, gyno-dioecious, colonial perennial from long rhizomes, mostly (4–)10–15(–20) dm; basal sheaths inflated, mostly bladeless, only the inner ones bearing rigid, erect, terete blades about as long as or longer than the stem, the plants otherwise lfless except for the erect invol lf, this up to 3(–9) dm and appearing like a continuation of the stem; infl apparently lateral, 2–45 cm, usually with numerous repeatedly forked, spreading branches, each branchlet terminated by 2–6 sessile, 2–8-fld glomerules that are subtended by short, ovate bracts; fls not prophyllate, the perfect ones with pale brown, glossy, indurate tep ca 4 mm, the 6 stamens nearly as long, with anthers ca 5 times as long as the filaments, the fls producing relatively few good seeds; pistillate fls with shorter, red-brown tep ca 3 mm, short staminodes, and long red stigmas, the abruptly short-acuminate frs trilocular, ca 5.5 mm, much surpassing the tep, the seeds obovoid, 0.4–0.6 mm; $2n=18$. Tidal marshes from s. N.J. to Fla. and Tex.

20. Juncus stygius L. Stems 1–3.5 dm, solitary or few together, from slender rhizomes; lvs 1–3, basal, or one cauline above the middle; blades very narrow; sheaths firm, usually with slightly prolonged auricles; invol lf erect, 1–2 cm; infl 1(2), compact, 1–4-fld, 1 cm; tep with broad scarious margins, subequal, 4.3–6.2 mm; sep lance-subulate, the pet wider and obtusish; stamens 6, nearly as long as the tep, the anthers much shorter than the filaments; fr ellipsoid, 6–8.5 mm, conspicuously mucronate, much exceeding the tep; seeds 2–3 mm, with a thick pale appendage at each end about as long as the body. Bogs, marshes, and shallow pools; interruptedly circumboreal, in our range extending s. to the mts. of N. Engl. and N.Y. and n. Minn. Our plants are var. *americanus* Buchenau.

21. Juncus longistylis Torr. Stems arising singly from a slender rhizome, 2–6 dm, with 1–5 adjacent hemispheric heads, these 2–8-fld, 8–15 mm thick; lvs mostly basal; 1 or 2 short cauline lvs often developed; sep carinate, lanceolate, 4.4–5.7 mm, acuminate, narrowly scarious-margined; pet oblong with broad scarious margins, 4.5–6 mm, acute; stamens 6, considerably shorter than the tep; anthers light yellow, 2–2.5 mm, longer than the filaments; fr trilocular, ellipsoid, 4–5 mm, half as thick, with a beak 0.5–1 mm; seeds ca 0.5 mm, apiculate at both ends; $2n=40$. Damp meadows and shores; widespread in the w. cordillera, e. to Neb. and nw. Minn.; w. Ont.; Nf.

22. Juncus marginatus Rostk. Stems cespitose, bulbous-thickened at base, 2–5 dm, seldom taller; lf-sheaths with rounded scarious auricles; principal blades 1–3 mm wide, with 3 prominent veins; invol lf shorter to somewhat longer than the infl; heads 5–20, 4–6 mm thick, subtended by lance-attenuate bracts; fls eprophyllate; sep lanceolate, 2.1–3.1 mm, acuminate or short-aristate; pet 2.3–3.3 mm, oblong with broadly scarious margins, obtuse or rounded; stamens 3, nearly as long as the tep, the anthers reddish-brown, much shorter than the filaments, usually soon shriveling; fr incompletely 3-locular (the partitions not meeting in the center), somewhat turgid-inflated, broadly obovoid, 1.8–2.9 mm, nearly or quite as thick, broadly rounded or somewhat retuse at the tip; $2n=38$, 40. Wet meadows and swales; N.S. to Minn. and S.D., s. to Fla. and Tex.

23. Juncus biflorus Elliott. Stems stout, 6–12 dm, bulbous-thickened at base, arising singly but close together from a stout rhizome; lf-sheaths with rounded auricles; principal blades 4–6 mm wide, with 5 main veins; infl obpyramidal, 4–10 cm, often much overtopped by the invol lf; heads 20–100, 4–6 mm, subtended by ovate acuminate bracts; fls eprophyllate; sep lanceolate, 1.8–2.5 mm, acuminate or short-aristate; pet 2–2.8 mm, oblong with broad scarious margins, stamens 3, nearly as long as the tep, the anthers much shorter than the filaments, often reddish and exsert; fr as in no. 22; $2n=40$. Moist or wet meadows and shores; Mass. to Mich. and Mo., s. to Fla. and Tex. (*J. aristulatus,* misapplied; *J. longii,* with long rhizomes.) Perhaps better treated as *J. marginatus* var. *biflorus* (Elliott) Torr.

24. Juncus repens Michx. Creeping r. Stems at first erect or ascending, 5–20 cm, with numerous soft basal lvs, a few cauline lvs, and a terminal infl of 2–8 sessile or peduncled heads, each 8–15 mm thick and composed of 5–15 green, often falcately curved fls; stem later becoming elongate, prostrate or floating, and producing additional flowering branches from the nodes; fls eprophyllate; tep rigid, subulate, the sep 4–5 mm, the pet 5–10 mm; stamens 3; fr trilocular, slender, obtuse, about equaling or a little shorter than the sep. Wet shores, marshes, and shallow water, on the coastal plain; Del. to Fla., Tex., and Okla.

25. Juncus caesariensis Cov. Scabrous r. Cespitose, 4–7 dm; lvs scabrous (visibly so at 10×), conspicuously septate, the lower to 3 dm, the upper much shorter; invol lf inconspicuous, to 2 cm; infl divaricately branched, 6–12 cm, two-thirds as wide, the glomerules terminal and lateral, 2–6-fld, 1 cm thick; fls eprophyllate; tep lance-subulate, rigid, striate, the sep 3.7–4.3 mm, the pet 3.9–5.2 mm; fr 4.5–5.7 mm, strongly trigonous, ovoid, tapering to a short beak, subtrilocular, the partitions incomplete distally; seeds slenderly fusiform, caudate at both ends, 2–2.3 mm, the body two-fifths as long. Sphagnum-bogs in the pine-barrens; N.S.; s. N.J. to se. Va. (*J. asper,* a preoccupied name.)

26. Juncus canadensis J. Gay. Stems cespitose, stout, rigid, 4–10 dm; lvs erect, terete, septate, 1.5–2.5 mm thick; infl compact to open, 2–20 cm; heads few–many, obpyramidal and 5–10-fld to globose and 50-fld, fls eprophyllate; tep lance-subulate, 3-nerved, the sep 2.7–3.8 mm, the pet very slightly longer; stamens 3 or sometimes 6; fr trigonously prismatic or subtriquetrous, 3.3–4.5 mm, abruptly narrowed to a short beak, subtrilocular, the partitions incomplete distally; seeds slenderly fusiform, 1.2–1.9 mm, the slender whitish appendages constituting more than half the length; $2n=80$. Swamps, marshes, and wet shores; Que. and N.S. to Minn., s. to S.C., Ind., Neb., and reputedly La.

27. Juncus subcaudatus (Engelm.) Cov. & S. F. Blake. Slender, 3–8 dm; lvs terete, septate, 1 mm thick; infl open, divaricately branched, the lower branches nearly horizontal; heads few–many, obpyramidal and 5-fld to hemispheric and 20-fld; fls eprophyllate; perianth as in no. 26, but smaller, the sep 1.9–3 mm, the pet 2.1–3.2 mm, the fr 3–3.6 mm; seeds fusiform, 0.7–1.2 mm, with a short appendage at each end. Swamps and wet shores; N.S. to Va. and in the mts. to n. Ga.; inland to O., W.Va., Tenn., and Mo.

28. Juncus brevicaudatus (Engelm.) Fern. Stems slender, densely cespitose, 1–5 dm; lvs erect, terete, septate, 1–2 mm thick; infl strict, 3–12 cm, 3–6 times as long as wide, with few–many erect or ascending branches and few–many 2–7-fld heads; fls eprophyllate; tep lance-subulate, 3-nerved, the sep 2.3–3.9 mm, the pet 2.6–3.2 mm; stamens 3 or sometimes 6; fr trigonously prismatic, much exceeding the perianth, 3.5–4.8 mm, abruptly tapering to the very short beak or merely acute, unilocular, with narrow partial partitions;

seeds fusiform, 0.8–1.2 mm, the body occupying three-fifths of the total length; $2n=80$. Marshes, wet meadows, and shores; Que. and N.S. to w. Ont. and Minn., s. to Mass. and N.Y., and in the mts. to N.C. and Tenn.

29. **Juncus brachycephalus** (Engelm.) Buchenau. Stems slender, densely cespitose, 3–7 dm; lvs terete, septate, often spreading, 1–2 mm thick; infl ovoid or pyramidal, 3–15 cm, half to two-thirds as wide, usually with numerous spreading-ascending or somewhat outcurved branches and numerous turbinate, 2–5-fld heads, in small plants the branches few and shorter and the heads few; tep lance-oblong, 3-nerved, conspicuously scarious on the margins, obtuse or subacute, the sep 1.8–2.5 mm, the pet a bit longer; stamens 3 or sometimes 6; fr trigonously prismatic or subtriquetrous, evidently exceeding the perianth, 2.4–2.8 mm, abruptly narrowed to the short beak, subtrilocular, the partitions complete only near the base; seeds fusiform, 0.8–1.2 mm, the body occupying three-fifths of the length; $2n=80$. Wet meadows and sandy shores; Me. to n. Ont. and Wis., s. to Pa., O., and Ind.; reported from n. Ga.

30. **Juncus torreyi** Cov. Stems stout, rigid, erect, 4–10 dm, arising singly from the nodes of cord-like rhizomes that are evidently tuberous-thickened at intervals; cauline lvs 2–5, the blades terete, septate, usually divaricate, 1–3 dm × 1–3 mm; margins of the sheaths pale and scarious, prolonged into scarious auricles 2–5 mm; infl usually dense, 2–4 cm, occasionally open and to 8 cm, with 2–10(–20) globose, 25–100-fld heads 10–15 mm thick; fls eprophyllate; tep linear-subulate, tapering to very slender, rigid points, the sep 3.5–5.5 mm, the pet evidently shorter; stamens 6; fr 4.5–5.7 mm, otherwise as in no. 31; $2n=40$. Wet sunny places or shallow standing water, especially on prairies; N.Y. and s. Ont. to Ky. and Tex., w. to Sask., Wash., Calif., and n. Mex.; also at scattered stations e. to Me. and N.J. and in the Gulf states.

31. **Juncus nodosus** L. Stems slender, erect, 1.5–4 dm, arising singly from the nodes of slender rhizomes that are evidently tuberous-thickened at intervals; cauline lvs 2 or 3, the blades terete, septate, 0.5–2 dm × 0.7–1.5 mm, sheath-membrane yellowish, prolonged into membranous auricles 0.5–1 mm; heads mostly 2–10(–15) in a loose or congested, proliferating infl, globose, 5–25-fld, mostly 6–9 mm thick; fls eprophyllate; tep acuminate, but broader than in no. 30, 2.5–3.5 mm, the pet about as long as the sep; stamens 6; fr slender, sharply triquetrous, 3.5–4.5 mm, equaling or commonly surpassing the tep, tapering into a slender, only tardily dehiscent stylar beak, unilocular, the placental partitions only slightly intruded; seeds apiculate, ca 0.5 mm; $2n=40$. Bogs, marshes, and wet shores; Nf. to Mack. and B.C., s. to Va., Ind., Mo., Tex., and Calif.

32. **Juncus scirpoides** Lam. Stems slender, 3–8 dm, erect from stout rhizomes; lvs terete, conspicuously septate, 1–2 mm thick, the uppermost with a normal blade; auricles ovate-oblong, 1–2.5 mm; infl compact to divaricately branched, 3–12 cm, with 4–15 globose, many-fld heads 8–12 mm thick; fls eprophyllate; tep rigid, usually lance-subulate, 2.2–3.2 mm, green or turning brown in age; stamens 3; anthers shortly exsert; fr equaling or usually slightly surpassing the tep, trigonous-subulate, tapering to a prominent beak 0.5–1 mm, unilocular as in no. 31, dehiscent in the basal two-thirds. Wet sandy soil, meadows, and shores; s. N.Y. to Fla. and Tex., mostly on the coastal plain, n. in the interior to w. Ky., s. Mo., and Okla.; also in n. Ind. and sw. Mich.

33. **Juncus megacephalus** M. A. Curtis. Much like no. 32, but stouter, with thicker lvs; blade of the uppermost cauline lf shorter than its sheath, usually nonseptate, seldom over 2 cm, or sometimes lacking; heads 10–15 mm thick; tep soft, deep reddish-brown, 3.5–4.5 mm, the pet slightly shorter than the sep; anthers included. Wet, often sandy soil on the coastal plain; se. Va. to Fla. and Miss.

34. **Juncus validus** Cov. Stems stout and tough, clustered, 3–10 dm; lvs aerenchymatously hollow and transversely evidently septate, but also flattened and ± equitant, basal and cauline, the basal ones 1–5 dm × 3–6(–8) mm; cauline lvs 3–several, well spaced, the lower like the basal, the upper smaller; infl terminal, well surpassing the invol bracts, these mostly 1–8 cm; heads mostly 15–75, often individually sessile on the ± elongate branches, hemispheric to subglobose, many-fld, 7–15 mm thick; fls eprophyllate; sep green to brownish-stramineous, rigid, slender, gradually tapering, 3.5–4.5 mm, the pet similar but a bit shorter; stamens 3, included; fr slender, trigonous-subulate, unilocular, surpassing the sep by 1–2 mm; seeds rather few, plump, 0.3–0.4 mm. Wet places, as along ditches, streams, and ponds; se. Va. to Fla. and Tex., and inland to w. Tenn., Mo., and Okla.

35. **Juncus pelocarpus** E. Meyer. Rhizomatous, colonial, ± erect, mostly 1–5 dm, with very slender, obscurely septate lvs; infl much branched, 5–15 cm, broadly ovoid or obpyramidal with ascending branches to flattened with divaricate branches, bearing numerous solitary or paired, distinctly secund fls, some or all of the fls usually replaced by subulate bulbils; fls eprophyllate; tep oblong, scarious on the margins and at the obtuse tip, the sep 1.6–2.3 mm, the pet 1.8–2.8 mm; anthers 6, longer than the filaments; fr narrowly ovoid-ellipsoid, 2.4–3.1 mm, gradually acuminate into a slender beak, unilocular, the placentae fertile only below the middle; seeds few, plump, 0.4 mm; $2n=40$. Moist boggy or sandy soil and shores, in soft-water habitats with a seasonally variable water-level. Var. *pelocarpus*, ranging from Lab., Nf. and Que. to Minn., s. to Del., Md., and n. Ind., is dwarf, mostly 1–3 dm, erect or prostrate (but not repent), the rhizome 1 mm thick or less. The ill-defined clinal var. *crassicaudex* Engelm. (*J. abortivus*), occurring irregularly from se. Va. to Fla., is stouter, 3–8 dm, erect, the rhizome 2–5 mm thick.

36. **Juncus subtilis** E. Meyer. Stems very slender, 0.5–5 dm, floating or prostrate and repent, mat-forming and scarcely arising above the substrate, bearing at many nodes small fascicles of short capillary lvs; infl 1–4 dm, simple or nearly so, usually with fls at only 1 or 2 nodes, often many of them replaced by subulate bulbils; tep 2.2–2.8 mm; pet 2.5–4.4 mm; anthers 6, usually shorter than the filaments; fr 3–5 mm; otherwise much like no. 35. Muddy shores; Nf. to Me. and w. Que.; Greenl.

37. Juncus bulbosus L. Stems usually elongate, creeping or floating, producing several capillary lvs and a short flowering branch from the nodes; infl small, scarcely branched, with 1–6 obpyramidal glomerules of 3–10 fls each; fls eprophyllate; tep linear-oblong, acute, subequal, 3–3.5 mm; stamens 3(6); fr unilocular, trigonously oblong-prismatic, blunt or mucronulate, 3.2–4 mm; 2n=40. Shallow water and wet shores; N.S. and Nf.; Pacific coast; Europe. (*J. supinus.*)

38. Juncus articulatus L. Stems usually erect, rather closely set on a coarse rhizome, 1–6 dm, with chiefly cauline lvs, these 2–4, the blade mostly 2–15 cm × 0.7–1.5 mm, terete, aerenchymatously hollow, septate-nodulose; infl broadly ovoid to depressed, with divergent to widely ascending branches, up to nearly twice as long as wide, with few to many obpyramidal to subhemispheric, 3–10-fld glomerules; fls eprophyllate; tep lanceolate to lance-subulate, (2–)2.5–3 mm, subequal (or the pet a bit longer than the sep), the sep acute to subacuminate, the pet more often obtuse; stamens 6; fr unilocular, chestnut-brown to purple-brown, exsert, sharply ovoid-trigonous, tapering in the distal half, 2.5–4 mm, acute or seldom obtuse below the mucronate tip; seeds 0.5 mm; 2n=80. Bogs, wet meadows, and shores; circumboreal, s. in our range to R.I., W.Va., s. O., n. Ind., and Minn. A form with creeping stems, rooting at the nodes is rarely found.

39. Juncus alpinoarticulatus Chaix. Much like no. 38, avg a little smaller, mostly 0.5–3 dm; infl slender, 5–15 cm, less than half as wide, its branches ± closely ascending or erect; tep mostly 1.5–2.5 mm, the sep obtuse or rounded but sometimes minutely apiculate, the pet a little shorter, generally broadly rounded at the tip; fr obtuse or rounded to the stylar apiculation; 2n=40, 80. Wet meadows and sandy or gravelly shores; circumboreal, s. in our range to Pa., Ind., and Mo. (*J. alpinus.*) The name *J. alpinoarticulatus* ssp. *americanus* (Farw.) Hämet-Ahti has been proposed to cover most of the American plants, together with those from Kamtchatka and easternmost Siberia, but the taxonomy and nomenclature are complex. The hybrid with no. 38 is *J.* × *alpiniformis* Fern.; that with no. 30. is *J.* × *stuckeyi* Reinking.

40. Juncus militaris Bigelow. Stems stout, erect from a rhizome, (3–)5–10 dm, with a few (or no) bladeless sheaths near the base and a single (seldom two) long foliage lf near the middle, its stout stiff blade overtopping the infl; rhizome, when submersed, often producing many long capillary lvs; infl obpyramidal, 4–15 cm, freely branched (the branches ascending), the obpyramidal to subhemispheric glomerules 5–13(–25)-fld; fls eprophyllate; tep lance-subulate to lanceolate, subequal, 2.3–3.5(–4) mm, the sep often aristulate; stamens 6; fr unilocular, trigonously ovoid-prismatic, 2.4–3.3 mm, acuminate into a conspicuous beak. Shallow water and wet shores; N.S. to Del.; inland in n. N.Y., s. Ont., and n. Mich.

J. subnodulosus Schrank, a European species, was collected as a waif in Mass. many years ago and described as *J. pervetus* Fern. It has obtuse tep and widely spreading primary branches of the infl.

41. Juncus brachycarpus Engelm. Stems erect, 3–8 dm, aligned on a stout rhizome; lvs terete, septate; infl 2–10 cm, open or congested, with 3–10(–20) bristly, globose, densely many-fld heads 1 cm thick; fls eprophyllate; tep lance-subulate, the sep 3.2–3.8 mm, the pet 2.2–3.2 mm; stamens 3, shorter than the tep; fr unilocular, trigonously oblong-prismatic, abruptly acute and mucronate, 1.5–2.7 mm, nearly concealed within the perianth; 2n=44. Damp or wet soil; Mass. to Ga.; s. Ont. to O., Minn., and Kans., s. to Miss. and Tex.

42. Juncus acuminatus Michx. Stems cespitose, erect, slender, 2–8 dm; lvs terete, septate, 1–3 mm thick; infl rarely congested, usually narrowly ovoid or pyramidal, 5–12 cm, less than half as wide, with (3–)5–20(–50) hemispheric or broadly obpyramidal heads 6–10 mm thick, each with (5–)10–50 eprophyllate fls; tep green or stramineous, lance-subulate, subequal, 2.6–3.9 mm, at least the sep broadly scarious-margined, in the lower half; stamens 3(6), shorter than the tep; fr stramineous or light brown, unilocular, about equaling the tep, trigonously ovoid-prismatic, 2.8–4 mm, acute or rarely obtuse, mucronate; 2n=40. Wet soil, meadows, shores, and low woods; Me. and N.S. to Wis., s. to Fla. and Mex.; B.C. to Oreg.

43. Juncus elliottii Chapman. Stems slender, cespitose, erect, 2–7 dm; some of the roots with subterminal tuberous thickenings; lvs terete, septate, 1–3 mm thick; infl narrowly pyramidal or ovoid, 1–12 cm, less than half as thick, congested to loosely branched, with 3–100 hemispheric or obpyramidal, 2–7-fld heads 3–5 mm thick; fls eprophyllate; tep subequal, lanceolate, 2.2–3 mm, aristulate, very narrowly scarious-margined; stamens 3; fr unilocular, dark purple-brown, prismatically oblong-trigonous, about equaling the tep, very abruptly short-acuminate or merely apiculate. Damp or wet, sandy or peaty soil, especially in pine-barrens; s. Del. to Fla. and Tex., chiefly on the coastal plain. Spring.

44. Juncus nodatus Cov. Stems clustered, stout, erect, 6–12 dm, 3–10 mm thick at base; lvs coarse, elongate, terete and septate, the upper often reaching to the infl, the larger ones 3–5 mm thick; infl broadly pyramidal with numerous divaricate branches, decompound, 1–2 dm, with 200 or more obpyramidal, 2–7-fld heads 3–5 mm thick; fls eprophyllate; tep subequal, linear-subulate, much narrower than in the 2 preceding spp., scarious-margined, 1.9–2.6 mm; stamens 3; fr unilocular, pale brown, prismatically ovoid or long-trigonous, 2.1–2.8 mm, obtuse or subacute. Swamps and shallow water; n. Ind. to Kans., s. to Miss. and Tex. (*J. robustus,* a preoccupied name.)

45. Juncus diffusissimus Buckley. Slimpod-r. Stems clustered, slender, erect, 2–5 dm, with usually 2–4 terete, septate lvs; infl decompound, very diffusely branched, widely spreading, 1–2 dm, usually constituting a third of the height of the plant, bearing very many obpyramidal, 3–10-fld heads; fls eprophyllate; tep subequal, linear-subulate, 2.3–2.8 mm; stamens 3; fr unilocular, golden-brown, prismatically linear-trigonous or subtriquetrous, 4–6 mm, a fifth as thick, gradually attenuate, very acute, much surpassing the tep. Wet soil and muddy shores; s. Ind. to Mo. and Okla., s. to Ala. and Tex.; coastal plain from se. Va. to S.C.; reported from W.Va.

46. Juncus debilis A. Gray. Stems cespitose, very slender, erect, 1–4 dm; lvs terete, septate, 0.5–1 mm thick; infl loosely and divaricately branched, a third to half the height of the plant, with 5–50 obpyramidal, 2–5 (–10)-fld heads 3–6 mm thick; fls eprophyllate; tep linear-subulate, subequal, 1.2–2.8 mm, very narrowly scarious-margined; stamens 3; fr unilocular, pale cinnamon-brown, ellipsoid-trigonous or narrowly ovoid-trigonous, acute, 2.8–4.2 mm, distinctly surpassing the tep. Moist sandy soil; R.I. to Mo., s. to Fla. and Miss.

2. LUZULA DC., nom. conserv. Wood-rush. Fls generally prophyllate, as in spp. of *Juncus*; perianth as in *Juncus*; ovary and loculicidal capsule unilocular; ovules and seeds 3, basal; infl varying from open, lax, and subumbellately paniculiform to composed of several compact spikes or heads, these sometimes clustered to form a secondary spike or head; lvs with closed sheath and flat, grass-like blade provided (at least when young) with some long, loose hairs along the margins, especially at the throat. 75, mainly in temp. and cool parts of the N. Hemisphere.

Spp. 3–7 belong to a circumboreal pillar-complex, perhaps better treated with fewer binomials and more trinomials, but the taxonomy and nomenclature not yet clear.

1 Fls solitary (rarely paired) at the ends of the branches of the infl.
  2 Infl simple, the primary branches 1(–4)-fld; perianth (2.5–)3–4.5 mm .... 1. *L. acuminata*.
  2 Infl loosely and repeatedly branched; perianth 1.7–2.5 mm .... 2. *L. parviflora*.
1 Fls few to many in glomerules.
  3 Fls (2)3–6(–8) in each glomerule; perianth white or rose-tinged .... 10. *L. luzuloides*.
  3 Fls several to many in each glomerule; perianth sordid to deep brown.
    4 Bracts subtending the glomerules obtuse to acuminate, scarcely if at all projecting beyond the
      fls; glomerules (or some of them) peduncled, or all sessile in an erect spike.
     5 Bracts subtending the fls entire or lacerate; seed with a basal appendage.
      6 Plants only loosely cespitose, with short but evident rhizomes and/or stolons; casual intr.
        from Europe .... 3. *L. campestris*.
      6 Plants more densely cespitose, without evident rhizomes or stolons.
       7 Plants producing basal bulblets .... 7. *L. bulbosa*.
       7 Plants not producing bulblets.
        8 Branches of the infl strongly ascending.
         9 Tep 1.5–2.5 mm, the inner evidently shorter than the outer; style very short, 0.2–
           0.3 mm; intr. in the ne. part of our range .... 4. *L. pallidula*.
         9 Tep 2.5–3.5 mm, subequal, style 0.4–0.7 mm; widespread .... 5. *L. multiflora*.
        8 Some of the shorter branches of the infl widely divergent; chiefly southern .... 6. *L. echinata*.
     5 Bracts subtending the fls conspicuously ciliate; seed exappendiculate .... 8. *L. confusa*.
    4 Bracts subtending the glomerules terminating in a long, very slender point projecting beyond the
      fls; glomerules sessile on a nodding spike .... 9. *L. spicata*.

1. Luzula acuminata Raf. Clustered perennial 1–4 dm, with short stolons and rhizomes; lvs ± persistently hairy along the margins, all with a blunt callous tip, the basal ones elongate, to 3 dm × 1 cm, the cauline 2–4, shorter and somewhat narrower; infl (1–)3–6 cm, the loosely spreading, almost filiform primary branches with a single terminal fl, or few-branched and with 2–4 fls; tep lance-ovate, usually chestnut-brown in the center with scarious margins, (2.6–)3–4.5 mm, subequal or the pet a bit longer than the sep; anthers ca twice as long as the filaments; fr ovoid, 3.2–4.5 mm, mucronate; seeds purple-brown, the body subglobose, 1–1.5 mm, with a pale terminal appendage nearly as long; 2n=18. Moist woods, less often along roadsides or in other open places; Nf. to Sask., s. to Ga., Ala., and Io. Spring. Two vars.:

Var. *acuminata*. Primary branches of the infl almost entirely simple and with a single terminal fl, a few rarely with an additional lateral fl; invol lf none or short and inconspicuous. Northern, s. to Md., Va., W.Va., n. Ky., Ind., Ill. and Io. (*L. saltuensis*.)

Var. *carolinae* (S. Wats.) Fern. Some or many primary branches of the infl branched and bearing (2)3–4 fls; invol lf conspicuous, up to 4 cm. Southern, from se. Pa. to W.Va. and s. O., s. to Ga. and Ala.

2. Luzula parviflora (Ehrh.) Desv. Stems ± clustered on short rhizomes, 3–9 dm; lvs glabrous at maturity, basal and cauline (the cauline often 4 or more), the blade 4–20 cm × 5–13 mm, gradually tapering to a firm, slender tip; infl 5–12 cm, decompound, with loosely spreading filiform branches, the fls terminal, solitary or seldom paired; bracts entire to slightly lacerate; tep pale brown, almost translucent and virtually nerveless, 1.7–2.5 mm, lance-ovate, acute or mucronulate; anthers from a little shorter to a little longer than the filaments; fr dark-brown to purple-brown or blackish-brown, 2–2.7 mm, mucronulate, surpassing the tep; seeds ellipsoid, 1.1–1.4 mm, remaining attached to the placenta by a tuft of long hairs; 2n=24. Moist or wet, wooded or open, often rocky places; circumboreal, s. to the mts. of N. Engl. and N.Y., Isle Royale in Mich., n. Minn., and the w. cordillera. Our plants are var. *melanocarpa* (Michx.) Buchenau.

3. Luzula campestris L. Plants only loosely cespitose, with short but evident rhizomes or stolons; anthers mostly 1–1.5 mm long and 2–4 times as long as the filament; seed-body globular; 2n=12; otherwise much like no. 5. Native of Europe, casually intr. in disturbed habitats at least in Mass. and perhaps elsewhere.

4. Luzula pallidula Kirschner.   Herbage pale; branches of the infl densely papillose; tep notably pale, 1.5–2.5 mm, the inner shorter than the outer; anthers 0.5–0.6 mm, ca equaling the filament; seed-body under 1 mm; $2n=12$; otherwise much like no. 5. Native of Eurasia, intr. in our range from N.S. and N.B. to n. N.Y. (*L. pallescens* and *L. campestris* var. *pallescens,* misapplied.)

5. Luzula multiflora (Retz.) Lej.   Densely cespitose, 2–4(–5) dm; basal lvs several; cauline lvs 2 or 3, flat except toward the slender, callous-pointed tip, 2–6 mm wide; infl of 1(–3) subsessile and (3–)5–10 (–16) pedunculate, ovoid to short-cylindric glomerules, the peduncles strongly ascending, ± smooth, sometimes branched; fls 7–15 per cluster; tep lanceolate, pale or brownish to dark brown, the outer 2.5–3.5 mm, equaling or a little longer than the inner; anthers 0.8–1.5 mm, 1–2.5 times as long as the filaments; style 0.4–0.7 mm, a little shorter than the ovary; fr shorter than the tep; seed-body ovoid to broadly elliptic, 0.9–1.2 × 0.6–0.9 mm, with an evident caruncle 0.3–0.4 mm; $2n=(24)$ 36. In wooded or open (often disturbed) places; widespread and apparently native in most of our range, s. to Del., Ind., and Mo.; also in Eurasia. (*L. campestris* var. *m.*)

6. Luzula echinata (Small) F. J. Herm.   Some of the shorter branches of the infl widely divergent; glomerules capitate to broadly ovoid; anthers mostly 0.7–1.2 + mm long and (1.5–)2.4 times as long as the filament; $2n=12$; otherwise much like no. 5. Woods, thickets, and clearings; Ga. to Miss. and Tex., n. to n. N.J., Pa., s. O., and Io. (*L. campestris* var. *e.*)

7. Luzula bulbosa (A. Wood.) Rydb.   Plants producing basal bulblets; infl varying from as in no. 5 to as in no. 6; glomerules capitate to often short-cylindric; anthers mostly 0.5–1 mm and 1–2 times as long as the filament; $2n=12$; otherwise much like no. 5. Woods, thickets, and clearings; Fla. to Tex., n. to Mass., Conn., N.J., Pa., Ind., Mo., and Kans. (*L. campestris* var. *b.*)

8. Luzula confusa Lindeberg.   Densely cespitose, 1–3 dm; basal lvs numerous; cauline lvs 2 or 3, narrowly linear, 1–2(3) mm wide, involute toward the tip; glomerules 2–5, short and headlike, some usually elevated on slender peduncles, or occasionally all sessile and forming a short erect spike; bracts subtending the glomerules short, acute or acuminate; bracts subtending the fls shorter than the fls, conspicuously ciliate; tep mostly brown, 2.5–3 mm, acuminate, with narrow scarious margins, equaling or somewhat exceeding the red-brown fr; seeds ellipsoid, 1–1.2 mm, not appendaged, but with a tuft of minute hairs at base; $2n=36$, 48. Alpine and arctic meadows and hillsides; circumboreal, s. to the high mts. of Me. and N.H.

9. Luzula spicata (L.) DC.   Densely cespitose, 1–3 dm; basal lvs numerous; cauline lvs 2 or 3, narrowly linear; fls in small, dense, sessile glomerules, forming a nodding spike 1–3 cm; bracts subtending the fls acuminate, red-brown below, scarious above, often ciliate; tep 2–2.5 mm, terminating in a slender fragile bristle; fr purple-brown, about equaling the tep; seeds ellipsoid, 1–1.2 mm, with a narrow raphe and a very short, rounded, basal appendage; $2n=12$, 14, 18, 24. Arctic and alpine meadows, ledges, and slopes; circumboreal, s. to the high mts. of N.Y. and N. Engl.

10. Luzula luzuloides (Lam.) Dandy & Wilmott.   Loosely tufted, slender, 3–7 dm; lvs narrowly linear, 15–30 cm, tapering to a very long slender point, the larger ones 3–5 mm wide; infl compound, loosely branched, the fls in terminal clusters of (2)3–6(–8); tep white or rose-tinged, lance-acuminate, the pet 3–3.8 mm, the sep 0.5 mm shorter; fr red-brown, round-ovoid, about equaling the sep; seed ellipsoid, 1–1.3 mm, with a pale, ridge-like raphe extending along one side and slightly prolonged at the tip; $2n=12$, 24. Native of Europe, intr. in Vt., Ont., N.Y., and Minn. (*L. nemorosa,* a preoccupied name.)

FAMILY CYPERACEAE, the Sedge Family

Fls perfect or often unisexual, spirally or less often distichously arranged on the axis of a spike or spikelet, axillary (or at least seemingly so) to small bracts called scales, usually without an evident bract between the fl and the axis; perianth of 1–many (often 6) short to much elongate bristles, or often wanting; stamens (1–)3, exserted at anthesis, the plants wind-pollinated; ovary superior, tricarpellate or less often bicarpellate, with an accordingly trifid or bifid style, unilocular, with a single basal anatropous ovule, ripening into an achene; embryo monocotyledonous, embedded in the well developed endosperm; herbs, often grass-like in aspect, with solid or seldom hollow, triangular to less often terete stems; lvs mostly 3-ranked, with closed (rarely open) sheath and parallel-veined, typically elongate and grass-like blade, or some or all the lvs with reduced or no blade. 100/4500.

1 Achene enclosed in a sac, the perigynium, the style at anthesis protruding through the terminal orifice.
  2 Lf-blades with a midrib, never as much as 2 cm wide .... 14. *Carex.*
  2 Lf-blades without a differentiated midrib, very broad, mostly 2–4 cm wide .... 15. *Cymophyllus.*
1 Achene not enclosed in a perigynium.
  3 Scales distichous; spikelets grouped into spikes or heads.
    4 Infl terminal; achene not subtended by bristles .... 8. *Cyperus.*
    4 Infl axillary; achene subtended by bristles .... 9. *Dulichium.*
  3 Scales spirally imbricate, except in a few spp. that do not have the spikelets in spikes or heads.

5 Spikelets dimorphic, the pistillate ones with a single fl and eventually a conspicuous white achene, the staminate ones several-fld, with narrower scales .... 13. *Scleria.*
5 Spikelets all alike, except for minor differences in size; achenes only rarely white.
  6 Achenes crowned with a tubercle (the persistent, evidently differentiated style-base).
    7 Spikelet solitary and terminal, without obvious subtending bracts; lvs bladeless .... 3. *Eleocharis.*
    7 Spikelets few to numerous, subtended by foliaceous or setaceous bracts, rarely solitary and then appearing lateral; lf-blades present.
      8 Tubercle prominent, as broad as the summit of the achene .... 6. *Rhynchospora.*
      8 Tubercle minute .... 5. *Bulbostylis.*
  6 Achenes blunt or merely apiculate, the style-base, if persistent, short, slender, merging with the achene and not evidently differentiated as a unitary structure.
    9 Achene subtended by bristles or scales or both (in addition to the scales of the spikelet).
      10 Achene subtended by bristles only.
        11 Bristles 1–6, short to sometimes rather elongate .... 1. *Scirpus.*
        11 Bristles numerous (more than 10), much elongate .... 2. *Eriophorum.*
      10 Achene subtended by scales, or by bristles and scales.
        12 Achene subtended by 3 broad, stipitate scales and 3 bristles .... 12. *Fuirena.*
        12 Achene subtended by a scale or scales only.
          13 Subtending scale 1, adaxial, minute; bract 1 .... 10. *Hemicarpha.*
          13 Subtending scales 2, abaxial and adaxial; bracts 2 or 3 .... 11. *Lipocarpha.*
    9 Achene not subtended by bristles or scales (aside from the spikelet-scales).
      14 Bract 1, erect or nearly so, appearing like a continuation of the culm.
        15 Achenes tiny, 0.5–0.7 mm; spikelets 2–6 mm .... 10. *Hemicarpha.*
        15 Achenes somewhat larger, 1.3 mm or more; spikelets mostly 5–10 mm .... 1. *Scirpus.*
      14 Bracts 2 or more, setaceous or foliaceous, the infl terminal.
        16 Fertile fls or achenes 1 per spikelet, terminal .... 7. *Cladium.*
        16 Fertile fls several or many per spikelet.
          17 Principal lf-blades 5 mm wide or wider .... 1. *Scirpus.*
          17 Principal lf-blades 3 mm wide or narrower.
            18 Lf-sheaths long-ciliate or fimbriate; achenes trigonous .... 5. *Bulbostylis.*
            18 Lf-sheaths short-ciliate or entire; achenes trigonous or lenticular .... 4. *Fimbristylis.*

1. SCIRPUS L.

1. SCIRPUS L. Bulrush. Scales spirally arranged, ± scarious; fls perfect, each in the axil of a scale; perianth of (1–)3–6 short to elongate bristles, or seldom obsolete; stamens 3, or sometimes fewer; style 2–3-cleft, either completely deciduous or more often deciduous above the base and leaving a slender tip on the achene; herbs, mostly of wet places, with variously arranged spikelets. 200+, cosmop. The diffuse centromere may contribute to the frequent interfertility of plants with different numbers of chromosomes.

1 Invol consisting merely of the 1–3 lowest, slightly modified, empty scales of the solitary, terminal spikelet. (*Baeothryon.*)
  2 Culms subterete, smooth .... 1. *S. cespitosus.*
  2 Culms triangular, scabrous on the angles.
    3 Bristles 6, flat, white, elongate, 1–3 cm at maturity .... 2. *S. hudsonianus.*
    3 Bristles 3–6, terete, not white, short, only about 2 mm.
      4 Midvein of the floriferous scales prolonged into a mucro 0.5–1 mm .... 3. *S. verecundus.*
      4 Midvein not prolonged into a mucro .... 4. *S. clintonii.*
1 Invol very different from the scales in form, size, and texture.
  5 Principal invol bract erect or nearly so, like a continuation of the stem, the infl thus appearing lateral.
    6 Spikelets 1–15, closely sessile at one point on the culm.
      7 Rhizomatous perennials. (*Schoenoplectus.*)
        8 Spikelets (1)2–several; plants sometimes growing in shallow water, but the culms stiff and emergent for most of their length.
          9 Lvs relatively short, less than half as long as the culms; style bifid or less often trifid.
            10 Bract solitary; culms sharply triquetrous .... 5. *S. americanus.*
            10 Bracts 2 or 3, the second and third ones resembling enlarged scales of the spikelet, but empty; culms merely trigonous .... 6. *S. pungens.*
          9 Lvs elongate, more than half as long as the culms; style trifid .... 7. *S. torreyi.*
        8 Spikelet solitary; plants usually aquatic, with flaccid, only shortly emergent or distally floating stems and lvs .... 8. *S. subterminalis.*
      7 Tufted, annual or apparently so.
        11 Scales few, mostly 5–14, notably keeled and boat-shaped (*Isolepis.*) .... 9. *S. koilolepis.*
        11 Scales more numerous, more than 15 (at least in the larger spikelets), neither boat-shaped nor strongly keeled. (*Actaeogeton.*)
          12 Achenes smooth or obscurely pitted .... 10. *S. smithii.*
          12 Achenes finely but conspicuously cross-ridged .... 11. *S. supinus.*
  6 Spikelets several to many in a spicate or branched cluster.
    13 Slender, maritime plants mostly 1–4 dm; spikelets crowded, sessile in a short spike. (*Blysmus.*) .... 12. *S. rufus.*

13 Coarser, nonmaritime plants mostly 1–3 m; spikelets not in a spike.
  14 Stems terete; sheaths bladeless or with the blade much reduced. (*Pterolepis.*)
    15 Mainly tricarpellate; spikelets mostly or all individually pedicellate .... 13. *S. heterochaetus.*
    15 Mainly bicarpellate; spikelets mostly or all sessile in small clusters.
      16 Spikelets appearing dull gray-brown, the individual scales at 10× with prominent
        red-brown striolae on a pale, gray-white background; scales, at least the
        middle and lower ones, mostly (3–)3.5–4 mm .... 14. *S. acutus.*
      16 Spikelets appearing more reddish-brown, the scales with a brown or tawny
        ground-color, so that the red-brown striolae are not prominent; scales mostly
        (2–)2.5–3(–3.5) mm .... 15. *S. validus.*
  14 Stems sharply triangular; at least the upper sheath with normal blade. (*Malacogeton.*)   16. *S. etuberculatus.*
5 Principal invol bracts spreading, foliaceous, 2 or more, the infl thus clearly terminal.
  17 Spikelets 3–50, 10–40 × 6–12 mm; scales puberulent; achenes 2.5–5 mm. (*Bulboschoenus.*)
    18 Lf-sheaths convex at the mouth, the ventral nerves abruptly divergent above.
      19 Bristles mostly equaling or surpassing the distinctly trigonous achene .... 17. *S. fluviatilis.*
      19 Bristles shorter than or up to about as long as the usually lenticular or merely plano-
        convex achene.
        20 Infl relatively open, with (10–)15–50 spikelets; bristles persistent .... 18. *S. cylindricus.*
        20 Infl more congested, with mostly 5–20 spikelets; bristles ± caducous .... 19. *S. robustus.*
    18 Lf-sheaths truncate or concave, the ventral nerves gradually divergent; infl congested and
      bristles short or obsolete .... 20. *S. maritimus.*
  17 Spikelets very numerous, 2.5–10 × 2–4 mm; scales glabrous; achenes less than 2 mm.
    21 Mature bristles evidently surpassing the scales and giving the infl a woolly look. (*Tricho-*
      phorum.)
      22 In dense tussocks; achenes pale yellowish-gray to whitish .... 28. *S. cyperinus.*
      22 Colonial from long rhizomes; achenes reddish-brown .... 29. *S. longii.*
    21 Mature bristles ± contained within the scales, the infl not woolly.
      23 Bristles smooth, contorted, well surpassing the achenes when extended. (*Androcoma.*)
        24 Mature stems ± erect; rays lacking bulblets .... 26. *S. pendulus.*
        24 Mature stems lax, with the infls drooping toward the ground; rays with axillary
        bulblets .... 27. *S. lineatus.*
      23 Bristles barbellate at least distally, straight or contorted, shorter or somewhat longer
        than the achenes, or obsolete. (*Taphrogeton.*)
        25 Spikelets all pedicellate; lvs 10–20 .... 25. *S. divaricatus.*
        25 Some or all spikelets glomerate; lvs often fewer than 10.
          26 All or most of the fls bicarpellate .... 21. *S. microcarpus.*
          26 All or most of the fls tricarpellate.
            27 Lvs 10–20; scales reddish-brown, the body about as wide as long .... 22. *S. polyphyllus.*
            27 Lvs 2–10; scales brown or black, the body mostly longer than wide.
              28 With long rhizomes; lower sheaths anthocyanic near the base .... 23. *S. expansus.*
              28 Densely cespitose; sheaths not anthocyanic .... 24. *S. atrovirens.*

    1. Scirpus cespitosus L. Stems very densely tufted on a short, freely rooting rhizome, 1–4 dm, smooth, subterete, clothed with several conspicuous, light brown scale-lvs at the base, and commonly with a single more normal lf a little higher, this with a typical sheath but the blade slender and only 4–6 mm; spikelet 1, terminal, brown, 4–6 mm, several-fld, the invol represented only by 2 or 3 empty scales at the base (these often deciduous as the spikelet approaches full maturity), the lowest scale with a prominent, broad, blunt awn 1–3 mm that may shortly surpass the spikelet; scales ovate, acute to shortly mucronate; bristles 6, very fragile, usually shortly surpassing the achene; anthers 1.1–2.5 mm; achene trigonous, brown, 1.5–1.7 mm, minutely apiculate; $2n=104$. Tundra, alpine mats, and acid bogs; circumboreal, s. in Amer. to the mts. of N. Engl. and N.Y. and to n. Ill., Minn., Utah, and Oreg.; disjunct in the mts. of N.C. and Tenn. Fr June–Aug.

    2. Scirpus hudsonianus (Michx.) Fern. Stems clustered on a rather short, freely rooting rhizome, 1–4 dm, sharply triquetrous, antrorsely scaberulous on the angles, bearing several ± reduced scale-lvs at the base and 1 or 2 more normal lvs a little higher, these with narrow blade 5–12 mm; spikelet 1, terminal, brown, 5–7 mm, mostly 10–20-fld, the invol represented by 2 or 3 empty scales at the base (these often deciduous as the spikelet approaches full maturity), the lowest scale with a strong midrib prolonged into a blunt mucro 0.5–2 mm; scales lance-ovate, blunt; bristles 6, white, flattened, crisped, elongate and much surpassing the scales, at maturity forming a silky-white tuft extending 1–2 cm beyond the end of the spikelet; anthers 0.6–1.1 mm; achene trigonous, brown, narrowly obovoid, 1–5 mm; apiculate; $2n=58$. Bogs; circumboreal, extending s. to Conn., N.Y., Mich., Minn., and B.C. Fr June–Aug. (*Eriophorum alpinum.*) Transitional to *Eriophorum.*

    3. Scirpus verecundus Fern. Cespitose perennial from short rhizomes; stems slender, erect, scabrous on the 3 angles; lvs several, the lower bladeless, the upper elongate, often surpassing the stem, to 1.5 mm wide; spikelet 1, terminal, ovoid, 5 mm, 4–8-fld; bract erect, ovate, prolonged into a mucro equaling or surpassing the spikelet; scales ovate, the sides brown, the broad green midrib prolonged into a mucro 0.5–1 mm; bristles 3–6, about equaling the achene; achene brown, trigonous, oblong, 2 mm, obtuse. Dry fields and open woods; Me. to Va., w. to Ont., O., Ky., and Mo. Fr May, June.

4. Scirpus clintonii A. Gray. Cespitose perennial from short rhizomes; stems slender, erect, scabrous on the 3 angles; lower sheaths bladeless or nearly so, the uppermost usually with a blade shorter than the culm and to 1 mm wide; spikelet 1, terminal, ovoid, 4–5 mm, 4–7-fld; bract erect, ovate, prolonged into a stout mucro much shorter than the spike; scales ovate, acute or obtuse, the green midvein often not reaching the tip; anthers ca 1 mm; bristles 3–6, equaling or shortly surpassing the achene; achene pale brown, trigonous, obovoid, 1.4–2 mm, obtuse. Dry woods; Que. and N.B. to N.Y. and Minn. Fr May–July.

5. Scirpus americanus Pers. Olney-threesquare. Rhizomatous, colonial perennial 5–20 dm; stems avg stouter than in no. 6, often 1+ cm thick below, sharply triquetrous, with conspicuously concave sides, easily flattened in pressing; lvs few, all on the lower part of the stem, with normal sheath and short blade seldom over 1 dm long, sometimes 1 cm wide; spikelets 2–15, essentially sessile in a compact cluster, mostly 6–15 mm, subtended by a prominent, rather blunt, green bract 1–3.5(–5.5) cm that appears like a continuation of the culm, without additional bracts; scales thin, largely hyaline-scarious, the brown midrib firmer and commonly exserted as a short mucro about equaling the distal notch; bristles 4(–6), often unequal, retrorsely barbellate, not much if at all surpassing the achene; style bifid; achene planoconvex, 1.8–2.5 mm (including the 0.3 mm apiculus); $2n=78$. Marshes, wet meadows, and other wet, low places, tolerant of alkali; N.S. to Wash., s. to S. Amer. Fr June–Sept. (*S. olneyi.*) The sterile hybrid with no. 6 is *S.* × *contortus* (A. Eames) T. Koyama.

6. Scirpus pungens Vahl. Common threesquare. Rhizomatous, colonial perennial 1.5–15 dm; stems trigonous, with flat to slightly concave or slightly convex sides; lvs usually several, all borne near the base, the blade elongate (but scarcely surpassing the middle of the stem) or reduced, firm, flat and 2–4 mm wide, or more often channeled above or folded; spikelets 1–6, essentially sessile in a compact cluster, mostly 7–20 mm, subtended by a prominent, sharply pointed green bract (1–)2–15 cm that resembles a continuation of the stem; 1 or 2 smaller bracts nearly always present, resembling enlarged scales but not subtending fls; scales thin, largely hyaline-scarious, the brown midrib firmer, exserted beyond the broad apical notch as a prominent mucro or short awn; bristles 4–6, often unequal, retrorsely barbellate, not much if at all surpassing the achene, this 2.2–3.3 mm (including the 0.5 mm stylar apiculus); $2n=78$. Marshes and wet, low ground, tolerant of alkali; interruptedly cosmopolitan, and throughout our range. Fr May–Aug. (*S. americanus,* misapplied.) Two vars. in our range. Var. *pungens,* with strictly bicarpellate fls, occurs mainly in and e. of the Appalachian region. Var. *longispicatus* (Britton) Cronq., with many or all the fls tricarpellate, extends e. from w. U.S. into the Great Lakes region and the Ohio Valley. (*S. longispicatus.*)

7. Scirpus torreyi Olney. Torrey-threesquare. Colonial perennial from soft rhizomes; stems erect, 5–10 dm, sharply 3-angled; lvs several, elongate, often surpassing the culm; bract erect, like a continuation of the culm, 6–15 cm, blunt; spikes 1–4 in a head, light brown, ovoid, 8–15 mm; scales ovate, the greenish midrib excurrent as a short (to 0.5 mm) mucro beyond the tapering or minutely notched body; bristles longer than the achene; style trifid; achene compressed-trigonous, obovoid, 3.2–4 mm including the prominent (0.5+ mm) apiculus; $2n=42, 70$. Marshes, muddy shores, and quiet water; N.B. to Man., s. to N.J., Pa., Va. (mts.), Ind., Mo., and Nebr. Fr July–Sept.

8. Scirpus subterminalis Torr. Water-b. Rhizomatous, aquatic perennial with slender, subterete, flaccid stems to 1+ m, shortly emergent or merely floating distally; lvs ± numerous, capillary, arising from near the base and usually trailing just below the surface; seldom the plant ± terrestrial and with ± erect stem and lvs; bract (1–)1.5–6 cm, like a continuation of the culm; spikelet 1, light brown, ovoid to cylindric, 7–12 mm; scales 4–6 mm, very thin and almost hyaline, acute, with a firmer midrib that may be minutely exserted; bristles shorter than to occasionally shortly surpassing the achene; anthers (2–)2.5–3.5 mm; achene brown, trigonous, 2.5–3.8 mm including the prominent (0.5 mm) slender beak; $2n=74$. Quiet or flowing water; Nf. to Alas., s. to S.C., Ga., Mo., Utah, and Calif. Fr July–Sept.

9. Scirpus koilolepis (Steudel) Gleason. Tufted annual; stems very slender, 5–20 cm; lvs basal, 2–5 cm, arcuate-setaceous; spikelet 1(2), overtopped by a filiform bract 1–3 cm; spikelet 3–7 mm, with few (5–14) fls; scales boat-shaped, compressed-keeled, scarious on the margins, the midline strongly curved-convex; bristles none; achene brown, trigonous, obovoid, ca 1.3 mm, dull, minutely roughened and pitted in vertical rows. Moist ground; Tex. to Ala., n. to Okla., Kans., Mo., and Ky.; N.C., Calif. Fr Apr.–June. (*S. carinatus,* a preoccupied name.)

10. Scirpus smithii A. Gray. Bluntscale-b. Tufted annual; stems to 6 dm, terete or obtusely trigonous; lf-sheaths variously bladeless or with a short, slender blade; bract 2–10 cm, ± erect, like a continuation of the culm; spikelets 1–several, capitately aggregated, ovoid, 5–10 mm, at least the larger ones with mostly 15–30 fls; scales obovate, broadly acute to obtuse and mucronate; style bifid; achene glossy, brown to nearly black, broadly obovate, 1.3–1.9 mm, lenticular or planoconvex; bristles barbed or smooth, well developed and surpassing the achene, or ± reduced or obsolete; $2n=38, 40$. Wet shores, often in the intertidal zone of estuaries, but also along inland lakes; Que. to Minn., s. to Ga. and w. Tenn. Fr July–Sept. Fluctuating water-levels may favor plants with smooth or no bristles.

Some authors segregate *S. purshianus* Fern. (*S. debilis,* a preoccupied name) on the basis of its unequally biconvex, rounded-obovate, more or less pitted achenes 1.75–2 mm, stouter bristles, scales with a distinct midvein, lacking a green central strip, and $2n=38$ chromosomes, as opposed to *S. smithii* proper, with planoconvex, somewhat cuneate-obovate and subtruncate, smooth achenes 1.5–1.8 mm, more delicate and slender bristles, scales with a broad green midstrip and very obscure midvein, and $2n=40$ chromosomes. The correlations are imperfect, and identification is often difficult.

11. Scirpus supinus L. Sharpscale-b. Tufted annual (or seemingly annual); stems slender and wiry, to 3 dm; lvs few, borne mostly below the middle, the sheath bladeless and merely mucronate-tipped, or the upper ones with a slender blade to 10 cm; bract 2–10 cm, ± erect, like a continuation of the culm; spikelets 2–several, closely clustered, 3–15 mm, at least the larger ones mostly with more than 15 fls; scales hyaline-scarious, with firmer, greenish, shortly excurrent midrib; bristles present or absent; achene blackish, 1.5 mm, finely densely, and conspicuously cross-ridged as seen at $10\times$, tricarpellate and trigonous, or bicarpellate and planoconvex. Swales and shores, tolerant of alkali; interruptedly circumboreal, and nearly cosmopolitan, but local and irregular. Fr June–Sept. Typical *S. supinus* is tricarpellate and Eurasian; $2n=28$. Tricarpellate Amer. plants with greenish brown scales, $2n=50$, mainly in c. and w. U.S., but e. as far as O., are var. *saximontanus* (Fern.) T. Koyama. (*S. saximontanus.*) Bicarpellate plants, with brown scales, $2n=22$, mainly in the coastal states from Mass. to Tex., but also inland to Mich. and even Kans., are var. *hallii* (A. Gray). (*S. hallii.*)

12. Scirpus rufus (Hudson) Schrader. Seaside-b. Perennial from slender rhizomes; stems very slender, 1–4 dm; blades 1–2 mm wide, shorter than the stem; bract erect or ascending, 1–2 cm; spike 1, 1–2 cm; spikelets distichously arranged, spreading or ascending, crowded, 5–8 mm, 2–5-fld; scales purple-brown, lance-ovate, acute; achene flattened-fusiform, narrowed to the base, long-tapering into the slender beak, 5–6 mm including the beak; bristles 0–6, soon deciduous, shorter than the achene; $2n=40$. Salt or brackish shores and marshes; Nf., N.S., N.B. and Que.; shore of Hudson Bay; n. Europe. Fr July–Sept. The Amer. plants have been segregated as var. *neogaeus* Fern. (*Blysmus r.*)

13. Scirpus heterochaetus Chase. Slender b. Much like *S. acutus*; stems hard but more slender; infl notably looser and laxer, most or all of the spikelets individually pedicellate; spikes light to medium brown, the scales with short, red-brown striolae on a usually tawny background; most or all of the fls tricarpellate, the style trifid, the achene unequally trigonous; $2n=38$. Margins of freshwater lakes and streams; Mass. and Que., w. to Wash. and Oreg., and s. in the plains region to Okla. and n. Tex. Not common. Fr June–Aug.

14. Scirpus acutus Muhl. Hardstem-b. Stout, rhizomatous, colonial perennial 1–3 m; stems terete, stout, firm, not easily crushed between the fingers; lvs few, mainly or all toward the base, commonly with prominent, well developed sheath and short, poorly developed blade (or bladeless); principal bract 2–10 cm, ± erect, like a continuation of the stem; subsidiary bracts small and inconspicuous; spikelets ± numerous in a subumbellately branched infl, all or nearly all sessile in small clusters at the ends of the rather stiff and ascending or horizontal branches of the infl, mostly 8–15 mm, dull gray-brown; scales mostly (3–)3.5–4 mm, thin and largely hyaline-scarious, with numerous short, linear, reddish-brown striolae on a pale, gray-white background, the midrib firm, commonly shortly exserted as a contorted fragile awn ca 1 mm; equaling or a bit longer than the achene; achenes 2.2–2.5 mm, ± completely hidden by the scales, bicarpellate and planoconvex, or occasionally some tricarpellate and unequally trigonous; $2n=38, 40, 42$. Marshes and muddy shores of lakes and streams, tolerant of alkali; widespread in temp. N. Amer., and throughout our range. Fr June–Aug.

15. Scirpus validus Vahl. Softstem-b. Similar to *S. acutus,* avg smaller and more slender, the stems soft, easily crushed between the fingers; infl tending to be looser and more open, with longer, laxer, sometimes drooping rays, some of the spikelets often borne singly; spikelets avg smaller, seldom over 1 cm, more shining and reddish-brown; scales mostly (2–)2.5–3(–3.5) mm, with a brown or tawny ground-color, so that the striolae are not usually prominent at $10\times$, the awn-tip ± straight, under 0.5 mm; achenes 1.8–2.3 mm, not wholly concealed by the scales; $2n=42$. Marshes and muddy shores of lakes and streams, tolerant of alkali; widespread in temperate N. Amer. and s. into tropical Amer. Fr June–Aug. (*S. steinmetzii*). Perhaps properly to be included in the Eurasian sp. *S. tabernaemontani* or a broadly defined *S. lacustris* L.

16. Scirpus etuberculatus (Steudel) Kuntze. Swamp-b. Perennial from stout rhizomes; stems erect, sharply triangular, 1–2 m, nearly or quite equaled by the 2–3 elongate, channeled and keeled aerial lvs; submerged, ribbonlike lvs usually present; primary bract erect, 1–2 dm; the primary peduncles either elongate with solitary spikelets or short and producing a secondary bract up to 1 dm with other peduncled spikelets; spikelets narrowly obovoid, 1–2 cm; scales lance-ovate, 6 mm, acuminate or acute and mucronulate; achene obovoid, trigonous, nearly black, the body 2.5–3 mm, conspicuously (ca 1 mm) apiculate; bristles 6, 4 mm. Swamps and quiet or flowing shallow water on the coastal plain; Del. to Fla. and Tex., n. to s. Mo. Fr June–Aug.

17. Scirpus fluviatilis (Torr.) A. Gray River-b. Stout perennial with tuber-bearing rhizomes; stems mostly 6–15 dm, sharply triquetrous; lvs several, well distributed, elongate, ± flat, mostly 6–16 mm wide; lf-sheath convex, the ventral veins continuing nearly to the summit and abruptly divergent; invol bracts several, very unequal, some elongate and lflike; spikelets mostly 10–40, 12–25 × 6–10 mm, some sessile in a central cluster, others generally single or clustered on several slender, flexuous rays to 7 cm; scales tan or light brown, thin and scarious, hirtellous on the back, short-awned from a tapering to more often lacerate or notched tip, the awn flexuous or ± recurved, commonly 2–4 mm; bristles 6, persistent, retrorsely barbellate, variably developed, commonly equaling or surpassing the achene; style trifid; achene trigonous, 3.5–5 mm, obpyriform to obovate, tapering to the prominent beak-apiculus; $2n=104, 110$. Marshes, standing water, and fresh-tidal or freshwater shores, tolerant of alkali; widespread in n. U.S. and s. Can., s. to Va., Mo., Kans. and Calif. Fr June–Aug.

18. Scirpus cylindricus (Torr.) Britton. Morphologically and ecologically intermediate between nos. 17 and 19, but apparently forming a persistent set of populations; infl usually relatively open, as in no. 17; bristles persistent, half to fully as long as the achene; style mostly bifid and achene planoconvex, intermediate in size and shape. Brackish transitional zones of tidal river-systems, above no. 19 and below no. 17; Me. to Ga. (*S. novae-angliae,* with distinctly trigonous achenes.)

19. Scirpus robustus Pursh.   Saltmarsh-b. Stout perennial with stout, tuber-bearing rhizomes; stems most-ly 6–15 dm, sharply triquetrous; lf-sheath as in no. 17; blade ca 1 cm wide; bracts 2–4, elongate; spikes mostly (3)5–20, ovoid to cylindric, all sessile or some clustered or solitary on rays to 4 cm, the infl appearing more compact than in no. 17; scales brown with thin margins, puberulent, the apex often fissured in age, the midvein prolonged into a recurved awn 2 mm; style bifid or trifid; achene glossy, dark brown to nearly black, 3–3.5 mm, broadly obovate, abruptly rounded-truncate to the minute beak-apiculus; $2n=106$–110. Brackish or saline marshes along the coast; N.S. to Tex.; Calif. Fr July–Oct.

20. Scirpus maritimus L.   Alkali-b. Stout perennial with tuber-bearing rhizomes; stems mostly 6–15 dm, sharply triquetrous; lvs usually several, elongate, $\pm$ flat, to ca 1 cm wide; orifice of the lf-sheaths truncate or concave, the ventral veins gradually diverging well below the summit and the intervening thin triangle easily torn; bracts 2–5, the outer elongate, the inner dilated at base and contracted to a narrow blade; spikelets 3–20+, some or all sessile in a central cluster, some often single or clustered on rays to 5 cm, ovate to cylindric, 10–40 mm; scales puberulent, thin-margined, notched at the tip, the midvein prolonged into a curved awn 1–3 mm; bristles 2–6, shorter than the achene, or obsolete; style bifid or trifid; achene brown to black, glossy, lenticular or planoconvex to trigonous, 2.5–4 mm, obovate, broadly rounded to the minute beak-apiculus; $2n=80$–114. Fr June–Sept. Fresh, saline, or alkaline swamps and marshes; interruptedly circumboreal, with 2 vars. in our range. Atlantic coastal plants represent the var. *maritimus*, found also in Europe, bicarpellate or often tricarpellate, with $\pm$ acute spikelets and with the inner invol bracts often basally anthocyanic. Western plants, extending e. to Minn., Mich., and Mo., represent the dubious var. *paludosus* (A. Nels.) Kuk., bicarpellate, with blunt spikelets, the inner invol bracts only rarely anthocyanic at base. (*S. paludosus*.)

21. Scirpus microcarpus C. Presl.   Coarse, leafy-stemmed perennial from stout creeping rhizomes; stems arising singly or few together, 6–15 dm, obscurely trigonous; lvs several, with flat, grass-like blade (6–)8–15 mm wide and up to several dm long, the sheaths $\pm$ anthocyanic; spikelets 4–6(–8) mm, very numerous, sessile in small, pedunculate clusters in a compound, umbelliform, terminal cyme that is subtended by several conspicuous, lf-like, unequal, sheathless bracts, the longest of these mostly 1–3 dm; scales numerous, 1–2 mm, largely hyaline and cellular-reticulate, blackish or greenish-black with greener, ill-defined midrib that may be very shortly excurrent; bristles 4–6, slender, minutely and retrorsely barbellate, slightly surpassing the achene; style bifid (or a few trifid); achenes pale, 1.0–1.2 mm, including the minute stylar apiculus, lenticular or a few trigonous; $2n=62$, 66. Wet low ground; Nf. to Alas., s. to W.Va., Io., N.M., and s. Calif.; e. Siberia. Fr June, July. (*S. rubrotinctus.*)

22. Scirpus polyphyllus Vahl.   Cespitose perennial from short rhizomes; stems to 1.5 m, sometimes with nodal bulblets; lvs 10–20, the main blades 3–8 mm wide; infl widely twice or thrice branched, usually with bulblets at base; spikelets broadly ovate, 3–4 mm, sessile in numerous small glomerules; scales rotund, mucronate, with conspicuous green midstrip and red-brown or red-purple sides; bristles 6, brown, surpassing the achene, straight and smooth below, contorted and retrorsely barbellate above; style trifid; achenes 1.1–1.3 mm, obovate or nearly obtriangular, compressed-trigonous; $2n=58$. Swamps and marshes; Mass. and Vt. to Ill. and s. Mo., s. to Ga. and Ala. Fr July, Aug.

23. Scirpus expansus Fern.   Perennial from long reddish rhizomes; stems arising singly, stout, to 2 m; lower sheaths anthocyanic at base and usually also at the orifice; blades 1–2 cm wide; infl twice or thrice branched with numerous long rays mostly at acute angles; spikelets 2.5–5 mm, ovoid to seldom cylindric, sessile in small glomerules; scales 1.3–2.2 mm, broadly oval, obtuse or minutely mucronulate, with broad green midstrip, the thin sides greenish, becoming black; bristles 6, pale or light brown, shorter to somewhat longer than the achene, nearly straight, retrorsely barbellate almost to the base, deciduous; style trifid; achene pale brown or purplish-brown, 1–1.6 mm, compressed-trigonous, broadly obovoid, minutely beaked; $2n=64$. Swamps and streamsides; Me. to Ga., w. irregularly to O. and Mich. Fr July, Aug. Closely allied to the Eurasian *S. sylvaticus* L. (*S. sylvaticus* var. *bissellii,* a hybrid of nos. 21 and 23.)

24. Scirpus atrovirens Willd.   Black b. Cespitose perennial to 1.5 m from short tough rhizomes; main lvs to 18 mm wide, mostly on the lower half of the stem, the lower sheaths and blades usually septate-nodulose; infl open or sometimes very compact, once or usually twice widely branched, often with axillary bulblets; spikelets ovoid or short-cylindric, 2–8 mm, all densely crowded in subglobose heads; scales 1.4–2.1 mm, broadly elliptic or obovate, brownish or blackish, the pale midvein prolonged into a mucro less than 0.4 mm; bristles mostly 6, shorter to slightly longer than the achene, $\pm$ straight, the upper $\frac{2}{3}$ retrorsely barbellate; style trifid; achene very pale to white, compressed-trigonous, 0.8–1.2 mm; $2n=50$, 52, 54, 56. Swamps and wet meadows; Nf. and Que. to Wash., s. to Ga., Tex., and Ariz. Fr June, July. Most of our plants belong to the highly variable var. *atrovirens,* as described above. (*S. ancistrochaetus,* with the bristles barbellate nearly to the base; *S. flaccidifolius,* local in se. Va. and adj. N.C., with lax stems, the infls lopping over toward the ground; *S. georgianus,* with 0–3 short bristles; *S. hattorianus,* with lower blades and sheaths nearly smooth.) Westward, mainly w. of Mississippi R., var. *atrovirens* passes into var. *pallidus* Britton, with fewer and larger glomerules and more consistently blackish scales 1.8–2.8 mm, the midrib excurrent into an awn-point 0.4–0.7 mm. (*S. pallidus.*)

25. Scirpus divaricatus Elliott.   Cespitose perennial; stems slender and weak, to 1.5 m, often lopping over to the ground, often with bulblets at the nodes; lvs 10–20, mostly 5–10 mm wide; infl large and lax, often bearing tufts of lvs; spikelets cylindric, to 1 cm, individually pedicellate; scales numerous, in few rows on the rachilla, broadly ovate, with green midrib and pale or hyaline sides; bristles 6, ca equal to achene, contorted, retrosely barbellate above the middle; achenes 1 mm, elliptic or obovate, trigonous with protruding angles and concave sides; $2n=28$. Swamps and wet woods; n. Fla. to La., n. to se. Va., s. Tenn., and se. Mo. Fr May, June.

26. Scirpus pendulus Muhl. Cespitose perennial to 1.5 m from short tough rhizomes; stems rigid, upright, with only a terminal infl or sometimes 1 or rarely 2 lateral infls; blades 3–10 mm wide; infl decompound; bracts usually shorter than the rays; spikelets elongating to 6–13 mm, usually 1 in each terminal cluster sessile and the others pedicellate, often drooping; scales obovate-elliptic, 1.7–2.2 mm, acuminate or conspicuously mucronate, with green midstrip and chestnut sides; bristles 6, brown, contorted, smooth, twice as long as the achene; style trifid; achene pale purple-brown to brown, compressed-trigonous, 1–1.3 mm, sharply short-beaked; $2n=40$. Marshes and wet meadows; Me. to Minn., S.D., and Colo., s. to Fla., N.M., and n. Mex. Fr June, July. (*S. lineatus,* misapplied.)

27. Scirpus lineatus Michx. Much like no. 26, but the mature stems lax and with the infls lopping over toward the ground; infls in the axils of 2 or 3 upper lvs as well as terminal; rays more divergent, and with axillary bulblets; spikelets individually pedicellate; $2n=36$. Wet (often calcareous) woods; se. Va. to Fla. and La. Fr. May, June. (*S. frontinalis.*)

28. Scirpus cyperinus (L.) Kunth. Wool-grass. Cespitose perennial, forming dense tussocks on short tough rhizomes; stems to 2 m; principal blades 3–10 mm wide; bracts lf-like, unequal, spreading, usually drooping at the tip, pigmented at base but not glutinous; rays of the infl several, ascending below, ± divaricate above, branched and involucellate toward the tip, the ultimate branches with 1–several sessile spikelets and often 1 or 2 pedicellate spikelets; spikelets ovoid to cylindric, 3–5(–10) mm; scales elliptic or oval, 1–2 mm, obtuse and mucronulate to broadly acute, marked (at high magnification) with numerous fine red lines; bristles 6, smooth, contorted, evidently surpassing the scales and giving the mature spikelets a woolly look; style trifid; achene pale yellowish-gray to nearly white, obovate, compressed-trigonous, 0.7–1 mm, sharply short-beaked; $2n=60, 64, 66, 68, 70$. Bogs, marshes, and wet meadows; Nf. to B.C., s. to Fla. and Tex. Fr June–Sept. Highly variable, consisting of several geographically and ecologically widely overlapping phases that intergrade freely without wholly losing their identity. (*S. atrocinctus,* mainly northern, and extending w. to B.C., in meadows, with pedicellate spikelets, blackish scales, and early fr; *S. pedicellatus,* northern but not western, in alluvial marshes, with pedicellate spikelets, pale brown scales, and late fr. (*S. eriophorum; S. rubricosus.*) Putative hybrids with no. 24 have been called *S.* × *peckii* Britton.

29. Scirpus longii Fern. Much like no. 28, but colonial from long, stout rhizomes, and rarely flowering; base of invol blackish and glutinous; scales 2–3 mm, blackish, rounded and not mucronulate at the tip; achenes reddish-brown; $2n=66, 68$. Marshes near the coast from N.S. and Me. to s. N.J. Fr June. Hybridizes with no. 28.

2. ERIOPHORUM L.   Cotton-grass. Scales spirally arranged, scarious, not awned; fls perfect, each in the axil of a scale; perianth of numerous (more than 10) persistent bristles, these much elongate at maturity, so that the mature spiklet forms a dense, cottony tuft commonly 2–4 cm; stamens 1–3; style trifid, deciduous; achene unequally trigonous, often with a short, slender stylar apiculus; perennial herbs of wet places, with grass-like lvs, the upper sheaths often bladeless; spikelets many-fld, solitary and terminal or few–many in an umbelliform cyme or head-like cluster; foliaceous bracts present in spp. with more than 1 spikelet. 20, N. Hemisphere.

1 Spikelet solitary; lower scales sterile; foliaceous bracts none.
  2 Densely cespitose, without creeping rhizomes; sterile scales 10–15.
    3 Achene twice as long as wide; lower scales whitish-margined .... 1. *E. vaginatum.*
    3 Achene thrice as long as wide; scales not whitish-margined .... 2. *E. brachyantherum.*
  2 Colonial from creeping rhizomes; sterile scales 7 or fewer .... 3. *E. chamissonis.*
1 Spikelets 2 or more; lower scales fertile; foliaceous bracts present.
  4 Foliaceous bract 1, erect, the infl appearing lateral; lvs 1–2 mm wide.
    5 Blade of the uppermost stem-lf at least as long as its sheath .... 4. *E. tenellum.*
    5 Blade of the uppermost stem-lf much shorter than its sheath .... 5. *E. gracile.*
  4 Foliaceous bracts 2 or 3, the infl appearing terminal; lvs wider.
    6 Scales prominently 3–7-nerved, the sides brown or coppery .... 6. *E. virginicum.*
    6 Scales 1-nerved, the sides usually olive-green to drab or blackish.
      7 Midvein prominent, usually widening distally, reaching the tip of the scale .... 7. *E. viridicarinatum.*
      7 Midvein very slender, not reaching the very thin tip of the scale .... 8. *E. polystachion.*

1. Eriophorum vaginatum. L.   Tussock c.-g. Very densely cespitose, forming large tussocks; stems 2–7 dm; lvs clustered at base, with filiform blade 1 mm wide; 1 or 2 dilated, bladeless sheaths mostly below the middle of the stem; spike solitary; basal sterile scales 10–15, lance-ovate, blackish, long-acuminate, with white or pale margins, spreading or reflexed at maturity; bristles white or seldom brownish or reddish; anthers 2–3 mm; achene distinctly obovate, 2.5–3.5 mm, slightly over half as wide, minutely (0.1–0.2 mm) apiculate; $2n=58, 60$. Bogs and open conifer-swamps; circumboreal, s. to N.J., Pa., and n. Ind., and Alta. Fr Apr.–July. (*E. callithrix,* misapplied.) Our plants are var. *spissum* (Fern.) B. Boivin. (*E. spissum.*)

2. **Eriophorum brachyantherum** Trautv. & C. A. Meyer. Short-anthered c.-g. Much like no. 1, but more loosely cespitose, with fewer stems; bladeless sheaths 1–3, the uppermost one above the middle; basal sterile scales of the spikelet blackish, scarcely paler at the margins, ascending; 0.5–1.5 mm; bristles pale brown or yellowish-white; anthers ca 0.5–1.5 mm; achene narrowly obovate, 2–2.7 mm, a third as wide, minutely (to 0.1 mm) apiculate; $2n=58$. Bogs and muskeg; circumboreal, reaching our range only in Ont. Fr June, July. (*E. opacum.*)

3. **Eriophorum chamissonis** C. A. Meyer. Rusty c.-g. Extensively colonial from creeping rhizomes, stems rather stout, 3–7 dm; lvs few, mostly near the base, ± with narrow, channeled or triangular blade 1–2 mm wide; bladeless sheaths 1 or 2, the upper near or below the middle; sterile scales few, obovate to triangular-ovate, blackish, pale-margined, ascending; anthers 1–2.5 mm; bristles reddish-brown to nearly white; achene dark, oblong-obovate, 2–2.7 mm, a third as wide, distinctly (0.3–0.5 mm) apiculate; $2n=58$. Bogs; circumboreal, s. to N.B., Minn., and Oreg. Fr July, Aug. (*E. russeolum.*)

4. **Eriophorum tenellum** Nutt. Conifer c.-g. Colonial from slender rhizomes; stems arising singly, slender, erect, 3–8 dm, obtusely trigonous, scaberulous above; blades 1–2 mm wide, channeled, the uppermost to 15 cm, equaling or longer than its sheath; bract erect, usually shorter than the umbel; spikelets 3–6, short-pedicellate in a head-like infl, or 1 or 2 on scabrous pedicels to 5 cm; scales stramineous to reddish-brown, ovate, obtuse or rounded; bristles sordid white; achene brown, narrowly obovate-oblong, 2.5–3 mm. Bogs and swamps, often under conifers; Nf. to Minn., s. to N.J., Pa., and Ill. Fr June–Sept.

5. **Eriophorum gracile** Koch. Slender c.-g. Colonial from slender rhizomes; stems arising singly, 2–6 dm, weak, often spreading or reclining, smooth, subterete below, somewhat trigonous above; blades 1–2 mm wide, channeled, the uppermost blunt, 1–5 cm, shorter than its sheath; bract erect, shorter than its umbel; spikelets 2–5, the slender, subterete, softly and minutely hairy pedicels to 3 cm; scales ovate, obtuse, drab or blackish, at least toward the margins and tip; bristles white or sordid; achene elliptic-oblong to oblanceolate, 2.5–3.5 mm, 3–4.5 times as long as wide; $2n=60, 76$. Bogs and swamps; circumboreal, s. in Amer. to Pa., Ind., Io., Colo., and Calif. Fr Apr.–July.

6. **Eriophorum virginicum** L. Tawny c.-g. Stems stiff, erect, to 1 m, smooth or scabrous only at the summit, solitary or few together from a freely rooting base, the plant also with more slender, spreading rhizomes; blades flat, elongate, 2–4 mm wide; foliaceous bracts 2(3), unequal, much exceeding the infl; spikelets several, on short subequal pedicels, forming a crowded cluster; scales relatively thick and firm, coppery or brown, obtuse or acute, prominently 3–7-nerved; bristles tawny or coppery, seldom white; anthers 1–1.5 mm; achene 3–3.5 mm, a third as wide; $2n=58$. Swamps and bogs; abundant; Nf. and Que. to Man. and Minn., s. to Fla. and Ky. Fr Aug.–Oct.

7. **Eriophorum viridicarinatum** (Engelm.) Fern. Dark-scale c.-g. Very much like no. 8, but the scales consistently blackish-green, with well developed, notably paler midrib that tends to be expanded distally and reaches the tip of the scale; upper lf-sheath without a dark rim; anthers 0.8–1.5 mm; achenes 2.5–3.5 mm; $2n=58$. Swamps, bogs, and wet meadows; Nf. to B.C., s. to Conn., n. O., n. Ind., Minn., Colo., and Wash. Fr May–Aug.

8. **Eriophorum polystachion** L. Thin-scale c.-g. Extensively colonial from creeping rhizomes; stems subterete, 2–6(–9) dm; sheaths with a dark border at the top; blades ± flat for most of their length, 2–8 mm wide, the uppermost one mostly as long or longer than its sheath; invol bracts unequal, 2 or 3 foliaceous, the longest one equaling or generally surpassing the infl; spikelets 3–several, on compressed, smooth or minutely scabrous-hirtellous pedicels to 5 cm; scales tawny to drab or blackish-green, lance-ovate, with a slender midvein not extending to the very thin, hyaline-scarious tip; bristles white; anthers 2–4.5 mm; achene blackish, rather narrowly obovate, 2–3 mm, 2–3 times as long as wide; $2n=58, 60$. Bogs and marshes; circumboreal, s. to Me., N.Y., Mich., Io., Colo., N.M., and Oreg. Fr June–Aug. (*E. angustifolium.*)

3. **ELEOCHARIS** R. Br. Spike-rush. Spikelet solitary, terminal, ovoid to linear, terete or somewhat flattened, few–many-fld; scales spirally imbricate (rarely distichous), often deciduous at maturity, the lower 1–3 empty in most spp., the others floriferous; fls perfect; stamens 1–3; style 2–3-cleft, its expanded base (tubercle) usually enlarged at maturity and persisting on the lenticular or trigonous achene; bristles typically 6, occasionally more, often reduced in number or size, or lacking; herbs of water or wet soil; stems simple, erect to prostrate, the lvs basal, reduced to bladeless sheaths. 150, cosmop. (*Heleocharis.*) The diffuse centromere may contribute to the frequent infraspecific variation in chromosome number.

1 Tubercle confluent with the achene as a stylar beak, not forming a distinct apical cap.
  2 Scales 2.5–5 mm; achenes 1.9–2.8 mm; plants seldom under 1 dm.
    3 Lowest scale empty; stems (2–)4–10+ dm; spikelet with (5–)10–20+ fls .... 1. *E. rostellata.*
    3 Lowest scale usually floriferous; stems (0.5–)1–3(–4) dm; fls 3–9 .... 2. *E. pauciflora.*
  2 Scales 1.5–2(–2.5) mm; achenes 0.9–1.3 mm; plants mostly under 1 dm .... 3. *E. parvula.*
1 Tubercle obviously differentiated from the achene, forming a distinct apical cap.

4 Spikelet scarcely thicker than the stem; scales persistent.
  5 Stems angular, not septate-nodose.
    6 Stems triangular; spikelet 1–2(–2.5) cm .... 5. *E. robbinsii*.
    6 Stems quadrangular; spikelet 2–5 cm .... 6. *E. quadrangulata*.
  5 Stems terete, septate-nodose at intervals .... 7. *E. equisetoides*.
4 Spikelet evidently thicker than the stem just beneath it; scales deciduous.
  7 Achene trigonous to nearly terete; style trifid.
    8 Achene obpyramidal, truncate at the top; tubercle very low and flat, closely covering the entire
      summit of the achene .... 4. *E. melanocarpa*.
    8 Achene obovoid, rounded at the top; tubercle either conspicuously narrower than the achene
      or separated from it by a distinct constriction.
      9 Achene twice as long as wide, nearly circular in section, gray, with several longitudinal
        ridges and numerous fine cross-lines; basal scale often floriferous.
        10 Stems flattened, 1–1.5 mm wide when fresh; scales ca 3 mm .... 8. *E. wolfii*.
        10 Stems scarcely flattened, 0.5 mm wide or less; scales ca 2 mm.
          11 Stems spongy, 0.25–0.5 mm thick, rugose-striate when dry .... 9. *E. radicans*.
          11 Stems not spongy, up to 0.25 mm thick, not wrinkled when dry .... 10. *E. acicularis*.
      9 Achene two-thirds to fully as wide as long, trigonous; basal scale usually empty.
        12 Surface of the achene smooth under a hand-lens.
          13 Achene brown to olive or yellowish; perennial.
            14 Scales spirally arranged.
              15 Angles of the achene not winged; bristles usually present.
                16 Tubercle about as wide as high; achene brown; coastal .... 11. *E. albida*.
                16 Tubercle conic-subulate; achene pale olive or yellowish; widespread 12. *E. intermedia*.
              15 Angles of the achene narowly but distinctly winged; bristles none .... 13. *E. costata*.
            14 Scales distichously arranged (unique among our spp.) .... 16. *E. baldwinii*.
          13 Achene pale greenish-gray to nearly white; annual .... 17. *E. microcarpa*.
        12 Surface of the achene distinctly roughened or pitted under a hand-lens.
          17 Sheaths truncate at the orifice.
            18 Stems 4–8-angled, with a vascular bundle under each angle .... 14. *E. tenuis*.
            18 Stems flattened, with 9–14 vascular bundles .... 15. *E. compressa*.
          17 Sheaths oblique at the orifice, the projecting lobe as long as its diameter.
            19 Achene and tubercle together scarcely over 1 mm; se. Va. and s. .... 18. *E. vivipara*.
            19 Achene and tubercle together 1.6–3 mm; more widespread.
              20 Tubercle nearly or quite as long as the achene; annual .... 19. *E. tuberculosa*.
              20 Tubercle very much shorter and narrower than the achene; perennial.
                21 Stems sharply triangular, usually twisted; cespitose .... 20. *E. tortilis*.
                21 Stems subterete, not twisted; rhizomatous .... 21. *E. fallax*.
  7 Achene lenticular or biconvex; style bifid (sometimes trifid in nos. 22 and 25).
    22 Evidently perennial, rhizomatous.
      23 Lf-sheaths obliquely truncate and firm at the summit; anthers 1.3–2.5 mm.
        24 Achene prominently rugose-punctate; coastal .... 21. *E. fallax*.
        24 Achene only very finely cellular-roughened; widespread .... 22. *E. palustris*.
      23 Lf-sheaths prolonged into a loose, white, scarious tip; anthers 0.7–1 mm .... 23. *E. flavescens*.
    22 Annual, or apparently so; anthers 0.3–0.8 mm.
      25 Achene black or very dark red at maturity; tubercle minute.
        26 Achene 0.7–1 mm, with brown bristles .... 24. *E. caribaea*.
        26 Achene ca 0.5 mm, with white bristles .... 25. *E. atropurpurea*.
      25 Achenes stramineous to dark brown, 1–1.5 mm; tubercle broad, flat, appressed .... 26. *E. ovata*.

1. Eleocharis rostellata (Torr.) Torr. Cespitose perennial; stems (2–)4–10+ dm, ± flattened at least distally and 1–2 mm wide, generally some much elongate, recurved to the ground, and rooting; spikelet fusiform, (5–)8–13 mm, with (5–)10–20(–25) fls; scales equaling or surpassing the achene, firm, ovate, obtuse or the upper acute, scarious-margined, the lowest one empty; bristles ca equal to achene, retrorse-serrate; style trifid; achene light greenish to brown, rounded-trigonous to planoconvex, smooth or slightly cellular-roughened, 1.9–2.8 mm including the prominent, pale tubercle, this up to 0.75 mm and confluent with the body. Saline or calcareous swamps and marshes along the coast from N.S. to Fla. and locally inland to Ill., Mich., Wis., and Minn.; widespread in w. U.S. and trop. Amer.

2. Eleocharis pauciflora (Lightf.) Link. Perennial with clustered stems on a rather short, stout rhizome, also producing long, slender rhizomes with a thickened terminal bud; stems (0.5–)1–3(–4) dm, slender, seldom 1 mm thick distally, not flattened, not proliferous; spikelet ovoid, 4–8 mm, with 3–9 fls; scales ovate, acute to acuminate, 2.5–5.5 mm, the 2 lowest the largest, usually all floriferous; bristles variously longer or shorter than the achene, or obsolete; anthers ca 1.5–2.5 mm; style trifid; achene gray-brown, rounded-trigonous to planocovex, obovoid, cellular-roughened, 1.9–2.6 mm including the short (to 0.6 mm), thickened, confluent tubercle; $2n=20$. Wet calcareous shores and boggy places; circumboreal, s. in Amer. to N.J., Pa., Ind., Io., N.M., and Calif., and in S. Amer.

3. Eleocharis parvula (Roemer & Schultes) Link. Diminutive, very slender perennial from inconspicuous slender rhizomes, forming dense mats; stems filiform, 2–6(–10) cm; spikelet ovoid, 2.5–4(–6) mm, with 2–9(–20) fls; scales ovate, obtuse or acute, green to stramineous or brown, 1.5–2(–2.5) mm, the lowest one empty; style trifid, achene stramineous, ± trigonous, 0.9–1.3 mm including the short (0.15–0.3 mm), thick-

CYPERACEAE

ened, confluent tubercle; $2n=8$, 10. Wet, saline or alkaline sites; irregularly cosmop. The var. *parvula,* with bristles equaling or surpassing the smooth achene, occurs in salt-marshes on both coasts of N. Amer. and at scattered inland stations chiefly e. of the Mississippi R., as well as in the Old World. The var. *anachaeta* (Torr.) Svenson, with the bristles ± reduced or obsolete, and often with cellular-roughened achenes, occurs at scattered inland stations w. of the Mississippi, s. to Mex. and the Caribbean. (*E. coloradoensis.*)

4. Eleocharis melanocarpa Torr.   Densely cespitose perennial; stems 2–6 dm, wiry, flattened, sometimes proliferous at the tip; sheaths truncate and prominently mucronulate; spikelet narrowly ovoid, 6–15 mm, obtuse, many-fld; scales firm, obtuse; bristles short or vestigial; anthers ca. 1.3–2 mm; achene obpyramidal, trigonous with rounded angles, dark brown, 1 mm, truncate above; tubercle very flat, covering the summit of the achene and somewhat projecting at the margin, slightly elevated in the middle. Wet sand and pine-barrens, mostly near the coast; Mass. to Fla. and Tex., and inland in Mich. and Ind.

5. Eleocharis robbinsii Oakes.   Cespitose and rhizomatous perennial, in water often producing floating tufts of capillary stems; fertile stems slender, triangular, 2–7 dm, 1–2 mm thick; sheaths brown, obliquely truncate; spikelet lanceolate, 1–2(–2.5) cm × 2–3 mm, acute to almost subulate, 4–8-fld; scales few, lance-ovate, tapering to an obtuse tip, scarious-margined; bristles conspicuously toothed, surpassing the achene; style trifid; achene biconvex, obovoid, brown, 2–2.5 mm, constricted above to an urn-shaped neck, marked with vertical rows of transversely elongate cells; tubercle flattened, acuminate, a quarter to half as long as the achene. In mud or shallow water; along the coast from N.B. and N.S. to Fla., and inland from N.Y. to Ind., Mich., and Wis.

6. Eleocharis quadrangulata (Michx.) Roemer & Schultes.   Cespitose and rhizomatous perennial; stems stout, to 1 m, 2–6 mm thick, sharply quadrangular; spikelet cylindric, (1.5–)2–5 cm, about as thick as the stem; scales in 4 rows, elliptic to obovate, 5–6 mm, broadly rounded, scarious-margined; bristles equaling or surpassing the achene; style bifid or trifid; achenes biconvex, obovoid, brown, 2–3 mm, constricted to an urn-shaped neck, marked with vertical rows of transversely elongate cells; tubercle dark, flattened, a third to half as long as the achene. Shallow water; Mass. to Wis., s. to Fla. and Mex., commoner southward.

7. Eleocharis equisetoides (Elliott) Torr.   Rhizomatous perennial; stems to 1 m, 5 mm thick, cross-septate at intervals of 2–5 cm; spikelet cylindric, 1.5–3 cm, about as thick as the stem; scales oval, obtuse to broadly rounded, scarious-margined, purple-dotted inside; bristles few, crooked, weak, up to as long as the achene; style trifid or bifid; achene biconvex, obovoid, golden-brown, 2 mm; tubercle dark, flattened, triangular, 1 mm. Shallow water; near the coast from Mass. to Fla., w. to Wis., Mo., and Tex. (*E. interstincta,* misapplied.)

8. Eleocharis wolfii A. Gray.   Rhizomatous perennial; stems in small tufts, erect, 1–4 dm, flattened, 1–1.5 mm wide; spikelet narrowly ovoid, 4–9 mm; scales 3 mm, imbricate in several rows, lance-ovate, acute or acuminate, somewhat marked with purple-brown, with broad scarious margins and tip; basal scale empty or floriferous; bristles none; style trifid; achene obpyramidal, 0.7–1 mm, pale or gray, with several longitudinal ridges and numerous transverse lines; tubercle conic, a fourth as long. Marshes; Ind. to Alta., s. to La. and Colo.; N.Y.

9. Eleocharis radicans (Poiret) Kunth.   Stems 2–8 dm, 0.25–0.5 mm thick, spongy, becoming rugose-striate when dry; spikelet ovate, usually flattened, 2–4 mm; anthers not over 0.6 mm; bristles not overtopping the tubercle; otherwise much like no. 10. Wet sand; trop. Amer., n. to se. Va. and Okla.; Mich.

10. Eleocharis acicularis (L.) Roemer & Schultes.   Diminutive, very slender perennial, commonly forming dense tufts on slender rhizomes; stems 3–12 cm, filiform, up to 0.25 mm thick; spikelet 2.5–7(–9) mm, 3–15-fld; scales (1.3–)1.5–2.2 mm, with a greenish midrib and ± hyaline, paler or partly anthocyanic margins, the basal one floriferous; bristles mostly 3 or 4, equaling or surpassing the achene, or reduced or obsolete; anthers 0.7–1.3 mm (dry); style trifid; achene white to pale gray or faintly yellowish, 0.7–1 mm, rounded-trigonous, longitudinally 8–18-ribbed and with very numerous fine, straight cross-ridges; tubercle short, ± triangular-conic; $2n=20–58$. Marshes, muddy shores, and other wet places; circumboreal, s. to Fla. and Mex.

11. Eleocharis albida Torr.   Cespitose perennial, sometimes also rhizomatous; stems slender, erect, 1–3 dm; sheaths very oblique, the projecting lobe twice as long as its diameter; spikelet ovoid, 2–10 mm; scales broadly rounded above, cartilaginous, pale brown, closely appressed; bristles dark brown, longer than the achene; style trifid; achene brown, trigonous, 1 mm; tubercle depressed-pyramidal, a third as wide and a fourth as long as the achene. Brackish marshes along the coast; Md. to Mex.; Bermuda.

12. Eleocharis intermedia (Muhl.) Schultes.   Densely cespitose; stems reclining or ascending, capillary, very unequal, 5–25 cm; sheaths truncate or somewhat oblique, the projecting lobe acute, no longer than its diameter; scales obtuse or the upper acute, with brown sides, the lowest one empty; bristles pale brown, usually slightly longer than the achene; anthers less than 0.5 mm; style trifid; achene trigonous, light olive or yellowish, 1 mm, the narrow tubercle conic-subulate, a third to half as long; $2n=22$. Wet soil; Que. to Minn., s. to Pa., w. Va., Tenn., and Io. (*E. macounii.*)

13. Eleocharis tricostata Torr.   Perennial from a short stout rhizome; stems 2–6 dm, compressed or subterete; sheaths truncate and mucronate; spikelet cylindric, 6–16 mm; scales closely imbricate in many rows, broadly rounded above; bristles none; style trifid; achene yellow or yellowish-olive, 0.8–1 mm, obovoid, trigonous, its angles narrowly winged; tubercle minute, depressed-pyramidal to nearly flat. Pine-barren ponds; se. Mass. to Fla.; Mich.

14. Eleocharis tenuis (Willd.) Schultes.   Stems slender, 4–8-angled, with a vascular bundle under each angle, scattered or loosely clustered on creeping rhizomes; sheaths truncate, conspicuously anthocyanic below;

spikelet ellipsoid to ovoid, 3–10 mm, 10–30-fld; scales ovate, obtuse to acute, purplish-brown to black, scarious-margined, mostly 2–3 mm, the lowest one larger; bristles ± reduced or wanting; style trifid; achene unequally trigonous, ovoid, distinctly roughened, 0.6–1.3 mm, often persistent after the scales have fallen. Wet places; Nf. to B.C., s. to S.C. and Tex. Variable, and ± divisible into 4 geographic vars.; the chromosome-numbers are based on few counts.

a Achenes yellow to dull orange; stems 6–8-angled, less often (*E. nitida*) 4-angled; $2n=38$; northern, s.
    to Pa. and Ill. (*E. elliptica*.) .... var. *borealis* (Svenson) Gleason.
a Achenes olivaceous; stems 4-angled or sometimes 5-angled; more southern.
  b Tubercle prominent, acute; stems capillary, merely angled; $2n=24$, Atlantic drainage .... var. *tenuis*.
  b Tubercle depressed, with an apiculate center.
    c Stems sharply 4-winged; $2n=38$; N.Y. to N.C. and Tenn., chiefly in the mts. ....
      .... var. *pseudoptera* (Weatherby) Svenson.
    c Stems capillary, merely angled; $2n=20$; chiefly Mississippi drainage, but locally in Pa. and N.J.
      (*E. verrucosa*.) .... var. *verrucosa* (Svenson) Svenson.

15. Eleocharis compressa Sullivant. Stems slender, compressed and often twisted, with 9–14 vascular bundles; scales lance-ovate, prolonged into a scarious acuminate tip, often bifid or lacerate in age; achene golden to brown, 1–1.5 mm; tubercle depressed to rounded-conic; $2n=18, 20, 24, 36$; otherwise much like no. 14. Marshes and shores; Ont. to Sask., s. to Ga. and Tex. (*E. acuminata*.) Passes into *E. tenuis* var. *borealis*.

16. Eleocharis baldwinii (Torr.) Chapman. Low, tufted or mat-forming perennial with slender, wiry, often proliferous stems 3–20 cm, lacking rhizomes; spikelets flattened, linear to ovate, 3–6 mm, 3–8-fld, the scales distichous, linear, keeled, acute, red-brown; achene sharply trigonous, 1 mm, shining dark olive-brown, sometimes obscurely striolate, the tubercle pale, pyramidal and subulate-tipped; bristles shorter than the achene. Wet sandy soil in pine-barrens on the coastal plain; se. Va. to Fla. and La.

17. Eleocharis microcarpa Torr. Cespitose annual with slender stems to 3 dm (or more elongate in water); sheaths oblique at the orifice; spikelet ovoid or oblong, 2–7 mm, several-fld; scales rounded above, pale at the margin or throughout; bristles half to nearly as long as the achene; style trifid; achene obovoid, trigonous, pale greenish-gray to nearly white, 0.6–0.8 mm; tubercle minute, depressed-pyramidal. Acid swamps near the coast; Conn. to Fla. and Tex.; Ind. and Tenn. (*E. torreyana; E. brittonii* Svenson, coarser, to 5 dm, with flatter tubercle, perhaps to be held as distinct.)

18. Eleocharis vivipara Link. Cespitose perennial; stems 1–3 dm, filiform, erect to reclining and commonly proliferous at the tip; sheaths oblique at the orifice; spikelet linear-cylindric to narrowly lanceolate, 3–8 mm, or often proliferous and sterile; scales oblong, obtuse, 2 mm; bristles nearly equaling the achene; style trifid; achene trigonous, obovate, honeycomb-reticulate, scarcely over 1 mm; tubercle small, pyramidal. Swamps on the coastal plain; se. Va. to Fla.

19. Eleocharis tuberculosa (Michx.) Roemer & Schultes. Cespitose annual; stems flattened, 2–8 dm; spikelet ovoid, 5–15 mm, acute; scales firm, appressed, broadly rounded above, stramineous or dark; bristles usually exceeding the achene but shorter than the tubercle; style trifid; achene trigonous, broadly obovoid, 1.2–1.5 mm, tubercle obscurely trigonous to dome-shaped, obtuse or acute, attached by a narrow base, about as long and thick as the achene; $2n=30$. Wet, especially sandy soil, chiefly near the coast; N.S. to Fla., Tenn., Tex., and Ark.

20. Eleocharis tortilis (Link) Schultes. Cespitose perennial; stems very slender, sharply trigonous, 2–5 dm, usually conspicuously twisted; spikelet ovoid, 2–8 mm, acute; scales firm, appressed, usually with brown sides, round or obtuse; bristles about equaling the tubercle; style trifid; achene trigonous, broadly obovoid, 1.1–1.5 mm; tubercle trigonous-subulate, half as long and wide as the achene. Swamps and bogs, chiefly near the coast; L.I. to Fla. and Tex. (*E. simplex*, misapplied.)

21. Eleocharis fallax Weatherby. Much like the uniglumate forms of *E. palustris*, but the achenes prominently punctate-reticulate, often some or most of those on a given plant tricarpellate and obtusely 3-angled; culms maroon at the base; $2n=42$. Fresh and brackish swamps along the coast; s. Mass. to La. and Tex.; Cuba. (*E. ambigens*.)

22. Eleocharis palustris L. Rhizomatous perennial; stems scattered or in small clusters, 1–10 dm, slender to very stout; spikelet 5–40 mm, lanceolate to lance-ovate in outline, light to dark brown or chestnut; empty scales (glumes) at the base of the spikelet 1 or 2(3), when solitary commonly ± encircling the stem at base, when 2 or 3 generally extending only ca half-way around, but the lowest one even then sometimes encircling; fertile scales mostly 2–4.5 mm; bristles 4(–8), retrorsely barbed, shortly surpassing the achene, or sometimes reduced or obsolete; anthers 1.3–2.5 mm (dry); style bifid; achene lenticular, yellow to medium-brown, 1–2 mm, very finely cellular-roughened; tubercle 0.4–0.7, constricted at base, narrow and conic to depressed-deltoid; $2n=10$–92; a polyploid complex with a degree of morphologic-cytologic-geographic differentiation, but the subgroups apparently inextricably linked, and not yet satisfactorily sorted out into proper infraspecific taxa. Wet places, tolerant of salt and alkali; circumboreal, s. in e. U.S. to Va., Tenn., and Mo. (*E. smallii*, widespread with us, biglumate, much like European *E. palustris* proper; *E. macrostachya*, western, biglumate; *E. uniglumis*, European, with a single encircling glume; *E. halophila*, mainly of coastal salt-marshes, much like *E. uniglumis*, uniglumate, the scales ± tapering to an acute or subobtuse tip; *E. erythropoda* (*E. calva*), widespread with us, uniglumate, the scales all broadly rounded or obtuse.)

23. **Eleocharis flavescens** (Poiret) Urban.   Tufted perennial from a compact system of slender rhizomes; stems divaricate, 3–15 cm; sheaths tenuous, pale, prolonged into a loose, white, scarious tip; spikelet ovoid or oblong-ovoid, 2–7 mm, with fewer than 20 fls; scales ovate or elliptic, obtuse or subacute, with brown sides, the basal one empty; bristles white or pale green; anthers 0.7–1 mm; style bifid or occasionally some trifid; achene lenticular, 1 mm; tubercle pale, short-conic above a swollen base. Var. *flavescens,* with the largest scales ca 2 mm, the tubercle a fifth as long as the black or purplish-brown achene, and the bristles usually shorter than the achene, is mainly tropical, extending n. chiefly on the coastal plain to Va. and reputedly N.J. (*E. flaccida; E. ochreata.*) Var. *olivacea* (Torr.) Gleason, with the largest scales ca 2.5 mm, the tubercle a fourth as long as the olivaceous achene, the bristles usually exceeding the achene, occurs from N.S. to Minn., s. chiefly in the e. states to Fla. (*E. olivacea.*)

24. **Eleocharis caribaea** (Rottb.) S. F. Blake.   Tufted annual; stems erect or divaricate, 3–20 cm, rarely more; sheaths firm, very oblique at the top; spikelet ovoid, 2–6 mm; scales numerous, round-ovate, obtuse; bristles brown, usually exceeding the achene, seldom shorter; anthers ca 0.5 mm; achene lenticular, black, shining, obovoid, 0.7–1 mm; tubercle pale, minute, very short. Pantropical, n. to S.C. and irregularly to Ont., Mich., and Ind. (*E. capitata; E. geniculata,* misapplied.)

25. **Eleocharis atropurpurea** (Retz.) Kunth.   Tufted annual; stems divaricate or ascending, 3–15 cm; sheaths firm and very oblique at the top; spikelet ovoid, 2–8 mm, many-fld; scales closely imbricate, 1–1.5 mm, ovate, obtuse, with broad green or stramineous midstrip and deep brown or purplish sides; bristles whitish, scarcely barbellate, shorter than the achene, or obsolete; anthers less than 0.5 mm; style bifid; achene lenticular, obovoid, smooth and shining, black to dark cherry-red, 0.5 mm tubercle pale, minute; $2n=20$. Pantropical, n. to Ga., Io., and B.C.

26. **Eleocharis ovata** (Roth) Roemer & Schultes.   Blunt s.-r. Tufted and apparently annual, but sometimes with slender, short, inconspicuous rhizomes within the tuft; stems 0.5–5 dm, several- to many-ribbed, 0.5–2 mm thick; spikelet 5–13 mm, ovoid, with many (seldom less than 40) fls; scales 1.7–2.5 mm, purplish or brownish, with greenish midstrip and paler, hyaline margins, the lowest one empty or floriferous; bristles mostly 6 or 7, brownish, retrorsely barbellate, equaling or surpassing the achene, or reduced or wanting; anthers 0.3–0.8 mm; style bifid or sometimes trifid; achene lenticular (even when the style is trifid), 1–1.5 mm, stramineous to olive or dark brown, smooth and shining; tubercle short, flattened, broad-based, appressed to the broad top of the achene; $2n=10$. Marshes and other wet places, sometimes intertidal, widespread in the N. Hemisphere, and found throughout our range. (*E. diandra,* an intertidal ecotype; *E. engelmannii; E. obtusa.*)

4. **FIMBRISTYLIS** Vahl, nom. conserv.   Spikelets several to many in simple or compound, often umbelliform cymes or glomerules, forming a terminal infl subtended by a cluster of sheathish, leafy or scarious invol bracts; scales spirally imbricate; fls perfect; perianth none; stamens 1–3; style 2–3-cleft, wholly deciduous, usually enlarged at base, the unbranched part often flattened and fimbriate especially distally; radicle lateral; herbs with a few long, grass-like lvs near the base, the sheath entire or short-ciliate. 200, warm reg. Spp. 3–5 are closely related and might with some reason be treated as well marked vars. of the tropical *F. spadicea* (L.) Vahl.

1 Style trifid; achene trigonous or merely plump; annual.
  2 Spikelets slender, linear-oblong to lanceolate, 3–7 mm .... 1. *F. autumnalis.*
  2 Spikelets stout, subglobose to ovoid or short-cylindric, 2–4 mm .... 2. *F. miliacea.*
1 Style bifid; achene lenticular (often plump) or terete; annual or perennial.
  3 Perennial; stamens (2)3; anthers ca 2–3 mm.
    4 Rhizomatous.
      5 Rhizomes long and slender; scapes tending to be flattened and scabrous-edged above; mostly
        submaritime .... 3. *F. caroliniana.*
      5 Rhizomes short, stout, and knotty; scapes terete or broadly oval in section, smooth; wide-
        spread .... 4. *F. puberula.*
    4 Cespitose, not rhizomatous; brackish coastal marshes .... 5. *F. castanea.*
  3 Annual; stamens 1 or 2; anthers ca 1 mm or less.
    6 Spikelets (or many of them) on ± elongate rays.
      7 Achene plumply obovate; lvs 1–2(–4) mm wide; ligule a line of short hairs .... 6. *F. annua.*
      7 Achene banana-shaped; lvs less than 1 mm wide; ligule wanting .... 7. *F. perpusilla.*
    6 Spikelets 3–8 in a terminal glomerule .... 8. *F. vahlii.*

1. **Fimbristylis autumnalis** (L.) Roemer & Schultes.   Tufted annual to 2 dm; stems flattened, often harsh-edged; lvs subdistichous, to 4 mm wide; ligule a line of short hairs; longest invol bract suberect, sometimes surpassing the infl; spikelets 3–7 mm, slender, in a dense or open system of often subumbelliform cymes; scales lance-ovate, usually keeled, excurrently mucronate; stamens (1)2; anthers minute; style trifid, much longer than the achene, smooth; achene 1 mm, obovoid-trigonous, smooth to verrucose; $2n=10$. Moist or wet, often disturbed sites; pantropical, n. to Me. and Minn.

2. **Fimbristylis miliacea** (L.) Vahl.   Tufted annual to 5(–10) dm; lvs distichous and equitant, smooth, rigid; ligule wanting; longest invol bract shorter than the infl; spikelets 2–4 mm, stout, in a loose to congested system of cymes; scales ovate, usually blunt or emarginate; stamens 1 or 2; anthers less than 1 mm; style trifid, not much longer than the achene, ± fimbriate below the branches; achene 1 mm, obovoid, plump but scarcely trigonous, with several smooth or often warty vertical ribs connected by numerous fine raised ridges; $2n=10$. Wet places, often a rice-field weed; pantropical, n. to Ky.

3. **Fimbristylis caroliniana** (Lam.) Fern.   Perennial from long, slender rhizomes; stems solitary or in small tufts, to 1.5(–2) m, tending to be flattened and scabrous-edged above; lvs subdistichous, mostly 2–5 mm wide, scabrous on the margins distally, usually otherwise glabrous; ligule a line of short hairs; longest invol bract from much shorter to slightly longer than the infl; spikelets 5–15 mm, ellipsoid or lance-ovoid, few–many in a compound system of subumbellate cymes, the edges of the peduncles scabrous; scales ovate, glabrous or often puberulent distally, sometimes shortly excurrent-mucronate; stamens 3; anthers ca 3 mm; style bifid, fimbriate from near the base to just above the branch-point; achene lenticular, obovate, 1 mm, finely cellular-reticulate in vertical rows; $2n=20$, 40, 60. Coastal marshes and dune-swales, or sometimes farther inland on the coastal plain, landward from no. 5, seaward from no. 4; N.J. to Fla. and Mex.

4. **Fimbristylis puberula** (Michx.) Vahl.   Perennial from short, stout, knotty rhizomes; stems solitary or in small tufts, to 1 m, smooth, slender, terete or broadly oval in section; lvs smooth or often hairy, usually involute, ca 1 mm wide; ligule inconspicuous or wanting; longest invol bract usually much shorter than the infl; spikelets 5–10 mm, ellipsoid to lance-ovoid or ovoid, not very numerous, in a usually compound system of subumbellate cymes; scales ovate to obovate or reniform, usually smooth, minutely excurrent-mucronate; stamens 3; anthers 2–2.5 mm; style bifid, fimbriate from near the middle to the branch-point; achene lenticular, obovate, 1 mm, finely cellular-reticulate in vertical rows; $2n=20$, 40. Meadows, prairie swales, savannas, and upper edges of marshes and bogs, not or scarcely maritime; s. U.S., n. mainly on the coastal plain irregularly to L.I., Pa., Mich., and Ill., more commonly to Mo. and Neb., and irregularly to Utah. Ours is var. *puberula*. West of our range this gives way to var. *interior* (Britton) Kral, with numerous more slender rhizomes and with the longest bract usually surpassing the infl. (*F. interior*).

5. **Fimbristylis castanea** (Michx.) Vahl.   Densely cespitose perennial to 1.5(–2) m; stems terete to elliptic in section; lvs slender, narrowly linear, rarely as much as 2 or 3 mm wide, involute or often semicircular in section; ligule wanting or incomplete; longest invol bract up to as long as the infl, seldom more; spikelets 5–10+ mm, mostly ovoid or lance-ovoid, in a dense to open compound system of subumbellate cymes; scales broadly ovate, smooth, rounded above; stamens 2 or 3; anthers ca 2 mm; style bifid, fimbriate from the base to the branch-point; achene lenticular, obovate, 1.5–2 mm, finely and densely cellular-reticulate; $2n=20$, 40. Brackish coastal marshes, seldom in alkaline sites inland; coastal plain from L.I. to Fla., Tex., Mex., and W.I.

6. **Fimbristylis annua** (All.) Roemer & Schultes.   Clustered annual to 3(–5) dm; lvs narrowly linear, 1–2(–4) mm wide, glabrous to tomentose; ligule a line of short hairs; spikelets 3–8 mm, lance-ovoid or oblong, few–many in a simple or compound system of umbelliform cymes; scales broadly oblong to ovate, smooth, only seldom shortly excurrent-mucronate; stamens 1 or 2; anthers 1 mm; style bifid, fimbriate from near the base to the branch-point; achene lenticular, plump, obovate, 1 mm, rather coarsely cellular-reticulate in vertical rows, sometimes warty; $2n=30$. Moist, sunny, sometimes disturbed sites; widespread in trop. and warm-temp. regions, n. in U.S. to Del., Pa., and Mo. Often included in the mainly tropical, coarser perennial *F. dichotoma* (L.) Vahl.

7. **Fimbristylis perpusilla** Harper.   Much like no. 8, glabrous, even smaller, to 8 cm, with solitary or clustered stems; spikelets 2–4 mm, ovoid or subglobose, in a more open, umbelliform infl; style equaling or only slightly longer than the pale, terete, slightly curved, banana-shaped, finely reticulate achene; $2n=10$. Alluvium of borders of pineland ponds; rare and irregular from Ga. to Md.

8. **Fimbristylis vahlii** (Lam.) Link.   Diminutive, tufted annual to 1.5 dm; lvs slender, spreading-recurved, less than 1 mm wide, often stiffly short-hairy; ligule wanting; scapes slender, wiry, subterete; spikelets 5–10 mm, linear-elliptic to lance-ovoid, 3–8 in a dense terminal cluster subtended by several long, slender invol bracts; scales lance-ovate or lance-oblong, glabrous, with an often slightly spreading mucro; stamen 1; anther less than 0.5 mm; style bifid, much longer than the achene, smooth or merely papillate; achene lenticular, thick, obovate, 0.5–0.7 mm, rather coarsely cellular-reticulate in vertical rows; $2n=20$. Weedy in fine soil in moist places; trop. Amer., n. irregularly to N.J., Ky., Ill., and Kans.

5. **BULBOSTYLIS** Kunth, nom. conserv.   Spikelets mostly in simple or compound, often umbelliform cymes (or virtually a head) forming a terminal infl subtended by 2–several slender, ± lf-like invol bracts; scales spirally imbricate; fls perfect; perianth none; stamens 1–3; style trifid, the unbranched portion smooth, the base enlarged and usually persistent on the trigonous achene as a tubercle (minute in our spp.); radicle basal; herbs with mostly basal, filiform to narrowly linear lvs, the sheath long-ciliate or fimbriate at the top. 80+, mainly trop.

B. stenophylla (Elliott) C. B. Clarke, a coastal plain sp., closely approaches our range in N.C. and may be sought in se. Va. The stems have a single dense terminal cluster of virtually sessile spikelets, subtended by elongate, setaceous invol bracts.

1 Achenes stramineous, finely cross-rugulose; widespread .... 1. *B. capillaris.*
1 Achenes grayish or grayish-blue, finely papillate; southeastern .... 2. *B. ciliatifolia.*

    1. Bulbostylis capillaris (L.) C. B. Clarke.   Tufted annual to 3 dm; scapes capillary; lvs short, slender, to 0.5 mm wide, involute, smooth or with scabrous margins; spikelets lance-ovoid or ovate, acute, 3–5 mm, few-fld, 2–7 in open, umbelliform cymes; scales ovate, curved-keeled, with a prominent, often shortly excurrent midrib; anthers 2, 0.5 mm; achene obovoid-trigonous, 1 mm, stramineous, finely cross-rugulose; tubercle minute, depressed-globose. Dry, open, often rocky or sandy places; Me. to Minn., s. to Fla., Tex., Ariz., C. Amer., and W.I.

    2. Bulbostylis ciliatifolia (Elliott) Fern.   Tufted, to 4 dm; lvs up to half as long as the culms, slender, to 0.5 mm wide, often involute; spikelets lance-ovoid, 2–6 mm, few-fld, in umbelliform cymes; scales ovate or broadly ovate, curved-keeled; anthers 2 or 3, 1 mm; achene obovoid-trigonous, 1 mm, grayish or grayish-blue, finely papillate; tubercle minute, depressed-globose; $2n=60$. Sandy places; se. Va. to Fla. and Tex. and n. to Tenn.; Cuba. Two ecologically differentiated vars. of nearly coextensive range. Var. *ciliatifolia,* of less extreme habitats, often weedy, is annual, with few spikelets in a simple or subsimple infl that is seldom surpassed by the longest bracts; the veins and edges of the lvs are usually hispidulous, and the midrib of the scale is only slightly or not at all excurrent. Var. *coarctata* (Elliott) Kral, usually in ant-hills with longlf pine, is ± perennial, with more numerous spikelets in a more compound infl that is commonly surpassed by the longest bract; the lvs are mostly smooth except for the scabrous-tuberculate to hispidulous edges, and the midrib of the scale is usually shortly mucronate-excurrent. (*B. coarctata.*)

6. RHYNCHOSPORA Vahl, nom. conserv.   Beak-rush.

Spikelets several to many, cymosely (or in part subumbellately) arranged in leafy-bracteate, open to congested infls, often forming 1–several dense glomerules; scales spirally imbricate, sometimes each subtending a fl, but more often the lower empty and only 1–few of the upper ones subtending fls; fls perfect or some of them staminate; perianth of (1–)6(–20) bristles, or sometimes wanting; stamens (1–)3; style bifid, its expanded base (tubercle) enlarged at maturity and persistent on the lenticular achene; our spp. mostly perennial (2 spp. annual), with ± leafy (seldom lfless), often trigonous stems and narrow, grass-like lvs with closed sheath. (*Dichromena, Psilocarya.*) 200+, chiefly of warm regions.

1 Each of the numerous scales of the spikelet subtending a perfect fl; annual. (*Psilocarya.*)
  2 Tubercle triangular-subulate, nearly or quite as long as the achene .... 1. *R. scirpoides.*
  2 Tubercle much depressed .... 2. *R. nitens.*
1 All but 1–5 (to 10 in *R. miliacea*) of the scales of the spikelet empty or subtending staminate fls; mostly perennial.
  3 Scales keeled; spikelets borne in a dense terminal head subtended by several spreading, leafy bracts with a conspicuous white base. (*Dichromena.*) .... 3. *R. colorata.*
  3 Scales not keeled; infl various, only seldom a single head, in any case not subtended by conspicuously white-based invol bracts. (*Rhynchospora.*)
    4 Achenes 3.5–5.5 mm, with a long, subulate tubercle 13–22 mm. (Horned rushes.)
      5 Bristles normally 6, about equal, 8–13 mm, conspicuously longer than the achene.
        6 Spikelets in dense glomerules of mostly 10–30; no rhizomes .... 4. *R. macrostachya.*
        6 Spikelets solitary or in glomerules of 2–6; rhizomatous .... 5. *R. inundata.*
      5 Bristles normally 5, unequal, shorter than or merely equaling the achene .... 6. *R. corniculata.*
    4 Achenes less than 3 mm, the tubercle also less than 3 mm.
      7 Bristles strongly plumose below with long ascending silky hairs .... 7. *R. oligantha.*
      7 Bristles merely barbellate or minutely hairy or smooth, or wanting.
        8 Bristles retrorsely barbellate (seldom antrorsely barbellate or smooth); spikelets in 1–several dense glomerules; achene concavely narrowed to a stipe-like base.
          9 Spikelets with a solitary, terminal, perfect fl.
            10 Glomerules globose or subglobose, the lowest spikelets deflexed .... 8. *R. cephalantha.*
            10 Glomerules turbinate to hemispheric, the lowest spikelets ascending or spreading
              .... 9. *R. chalarocephala.*
          9 Spikelets with 2–5 fls (often 1 or 2 of them staminate) and 1–5 frs.
            11 Bristles 8–14, usually minutely hairy at base .... 10. *R. alba.*
            11 Bristles 6, generally not hairy at base.
              12 Glomerules commonly with ± numerous spikelets; achene fully (or more than) half as wide as long.
                13 Bristles nearly or fully equaling the tubercle, which is more than half as long as the achene.
                  14 Achene uniformly brown, scarcely umbonate; bristles 1.8–2.8 mm ..... 11. *R. capitellata.*
                  14 Achene with a prominent pale umbo on each side; bristles 3.2–3.7 mm .... 12. *R. glomerata.*
              13 Bristles equaling or shorter than the achene; tubercle up to half as long as the achene .... 13. *R. knieskernii.*
              12 Glomerules with 1–10 spikelets; achene scarcely (or less than) half as wide as long .... 14. *R. capillacea.*

8 Bristles antrorsely barbellate or lacking; infl various; achene-base not obviously stipe-like.
  15 Achenes appearing smooth at 10×.
    16 Achene obovate, broadest distinctly above the middle.
      17 Bristles 0–3, to 0.3 mm; tubercle a fifth as long as the achene .... 15. *R. pallida.*
      17 Bristles 5 or 6, 1–3 mm; tubercle half to fully as long as the achene.
        18 Achenes uniformly pale brown; colonial by slender rhizomes .... 16. *R. fusca.*
        18 Achenes with a lustrous white central spot; densely cespitose .... 17. *R. filifolia.*
    16 Achene elliptic to rotund, broadest at the middle.
      19 Tubercle flat-subulate, elongate, 1–2 mm .... 18. *R. gracilenta.*
      19 Tubercle deltoid, 0.3–0.8 mm.
        20 Stems 1.5–3 mm thick at base, much surpassing the lvs .... 19. *R. fascicularis.*
        20 Stems up to 1 mm thick at base, scarcely longer than the lvs .... 20. *R. debilis.*
  15 Achenes evidently sculptured at 10×.
    21 Achenes marked with numerous minute pits but not transversely ridged; infl of 1–
      several glomerules.
      22 Spikelets 4–6 mm; achenes 2.1–2.6 mm .... 21. *R. grayi.*
      22 Spikelets 2.5–3 mm; achenes 1.5–1.8 mm .... 22. *R. harveyi.*
    21 Achenes transversely ridged; infl in most spp. more open and branching.
      23 Bristles conspicuous, persistent, surpassing the tubercle.
        24 Achenes narrowly elliptic-obovate, half as wide as long; lvs 1.5–4(–5) mm
          wide .... 23. *R. inexpansa.*
        24 Achenes broadly obovate to subrotund, nearly or quite as wide as long; lvs
          mostly 4–10 mm wide.
          25 Pedicels and branches of the cyme widely divaricate; tubercle de-
            pressed, 0.2–0.3 mm .... 24. *R. miliacea.*
          25 Pedicels and branches of the cyme ascending; tubercle elevated, 0.4–
            0.7 mm .... 25. *R. caduca.*
      23 Bristles inconspicuous, often caducous or none, much shorter than the achene.
        26 Spikelets in very loose, open clusters, all or most of them much shorter
          than their capillary pedicels .... 26. *R. rariflora.*
        26 Spikelets in cymes or glomerules, all or most of them longer than their
          pedicels.
          27 Achene thickened distally; tubercle almost conic .... 27. *R. globularis.*
          27 Achene strongly flattened throughout; tubercle flat.
            28 Spikelets 3–4 mm; achenes 1.3–1.6 mm, with ca 12 transverse
              ridges .... 28. *R. torreyana.*
            28 Spikelets 2–3 mm; achenes 1–1.3 mm, with fewer than 8 ridges ..... 29. *R. perplexa.*

1. Rhynchospora scirpoides (Vahl) Griseb.   Tufted annual 1–6 dm; blades 1–3 mm wide; spikes ovoid to cylindric, 3–7 mm, loosely clustered at the ends of the stem and branches; scales numerous, thin, 1-nerved, ovate, acute, 3 mm, each subtending a perfect fl, perianth wanting; stamens 1 or 2; achenes 0.7–1 mm, rotund or a little wider than long, contracted to a short broad stipe, longitudinally finely striate and sometimes obscurely cross-rugulose, pale brown, becoming nearly black, with raised pale margins; tubercle flat, triangular-subulate, nearly or quite as long as the achene. Wet sandy soil; e. Mass. and R.I.; nw. Ind. and sw. Mich.; se. Va. to e. N.C. (*Psilocarya s.*)

2. Rhynchospora nitens (Vahl) A. Gray.   Much like no. 1; achene scarcely stipitate, transversely conspicuously rugose, only inconspicuously margined; tubercle very short and closely appressed. Wet sandy soil and bogs; se. Mass. to Tex.; nw. Ind. (*Psilocarya n.*)

3. Rhynchospora colorata (L.) Pfeiffer.   White-topped sedge. Stems slender, 2–6 dm, solitary or few together on slender stoloniform rhizomes; blades 2–3 mm wide, usually shorter than the stem; spikelets whitish, compressed at anthesis, more fusiform at maturity, 4–7 mm, numerous in a dense terminal head subtended by 4–7 widely spreading, linear, foliaceous, very unequal, conspicuously bicolored bracts with a white base; scales keeled, ca 8–12, not much imbricate, only 1–3 subtending perfect fls, the others with staminate fls or some empty; perianth wanting; achenes obovate, 1 mm, transversely rugose, capped by the paler, depressed tubercle. Damp, often sandy soil; se. Va. to Mex. and W.I. (*Dichromena c.*)

4. Rhynchospora macrostachya Torr.   Coarse, erect, annual or short-lived perennial (0.5)1–2 m, with few or solitary, triquetrous stems, lacking rhizomes; main lvs 1 cm wide; infl of a terminal glomerule, a few erect secondary peduncles terminated by glomerules, and often also tertiary peduncles and glomerules; spikelets mostly 10–30 in each glomerule, widely spreading, 15–23 mm; lower scales empty; a few of the upper larger and each subtending a fl; bristles normally 6, 3 on each side, 10–13 mm, subequal; anthers 3.5–4 mm; achenes flat, brown, obovate, 4.5–5.5 mm, the tubercle forming a beak 16–22 mm. Swamps near the coast from e. Mass. to Fla. and Tex., and n. in the interior to Mo. and s. Mich.

5. Rhynchospora inundata (Oakes) Fern.   Much like no. 4, often smaller and more slender, and producing long, coarse, stoloniform rhizomes; main lvs mostly 4–7 mm wide; infl diffusely branched; spikelets usually divergent or even reflexed, solitary or in glomerules of 2–6; bristles 8–12 mm; achenes 3.5–4.5 mm, the beak 14–19 mm. Swamps and pond-margins; e. Mass. to Del.

6. Rhynchospora corniculata (Lam.) A. Gray.   Stout and leafy perennial to 2 m, with or without rhizomes; lvs commonly 8–20 mm wide; infl diffusely branched, to 2 dm wide, the glomerules short-spiciform, usually with 4–10 spreading or ascending spikelets 15–23 mm; bristles (3–)5(6), normally 2 equal ones on one side and a central one flanked by 2 shorter ones on the other side, all usually shorter than the achene, this flat, brown, obovate, 4–5.5 mm, the tubercle forming a beak 13–18 mm. Swamps and marshes; near to coast from Del. to Fla. and Tex., n. in the interior to s. Ind., s. Mo., and Okla.; W.I. Var. *corniculata*, of the e. seaboard, has the base of the beak half as wide as the top of the achene. Var. *interior* Fern., of the Mississippi Valley, has the base of the beak nearly as wide as the top of the achene.

7. **Rhynchospora oligantha** A. Gray. Densely cespitose perennial 1.5–4 dm; stems very slender, trique-trous, lfless or 1-lvd near the base; lvs setaceous, about equaling the stem; infl of (1)2–5 spikelets on 1–2(3) short branches, subtended by an erect bract resembling a continuation of the stem, one branch diverging at a right angle, the other 1 or 2 erect or arcuate; spikelets light brown, 4–7 mm, with 2–4 fls and 1–3 frs; bristles 6, a little shorter to a little longer than the achene, strongly plumose toward the base; achenes brown, plump, elliptic to obovate, 2–2.7 mm, constricted just below the top, transversely rugulose; tubercle conic, 0.3–0.6 mm. Pine-barren bogs; N.J. to Fla., Tex., W.I., and C. Amer.

8. **Rhynchospora cephalantha** A. Gray. Erect perennial to 1 m, with clustered stems; lvs 1.5–4.5 mm wide; glomerules 1–7, globose or subglobose (the lower spikelets ± reflexed), the terminal one 8–20 mm thick, the lateral subsessile or on peduncles to 2 cm; spikelets dark brown, 3–5.5 mm, the solitary fl terminal and perfect; bristles 6, about equaling the tubercle, usually retrorsely barbellate; achenes obovate-pyriform, 1.3–2.5 mm, two-thirds as wide, smooth, dark brown and somewhat concave near the raised margin, elevated toward the center into a prominent pale umbo; tubercle flattened, triangular-subulate, two-thirds to fully as long as the achene. Wet acid soil on the coastal plain; N.J. to Fla., La., and Cuba.

Many authors maintain that *R. microcephala* Britton, with small, round-topped achenes 1.3–1.6 mm, small spikelets 3–4 mm, and smaller glomerules, not over 1 cm wide, should be distinguished from *R. cephalantha* proper, with larger, often truncate achenes 1.8–2.5 mm, larger spikelets 4–5.5 mm, and larger glomerules, some 1.5–2 cm wide.

9. **Rhynchospora chalarocephala** Fern. & Gale. Much like the more slender and narrow-lvd forms of no. 8, but the glomerules hemispheric to broadly turbinate (the lower spikelets ascending or merely spreading), mostly 8–15 mm thick; achenes 1.4–1.8 mm, half to three-fifths as wide, the basal stipe-like part longer than in no. 8. Acid swamps and bogs on the coastal plain; N.J. to Fla.

10. **Rhynchospora alba** (L.) Vahl. Erect perennial to 7 dm, the clustered, very slender stems usually overtopping the lvs, these 0.5–2.5 mm wide; glomerules 1–3, broadly turbinate, 6–20 mm thick, the uppermost barely or scarcely surpassed by its bracts, the lateral ones usually remote and long-pedunculate; spikelets 4–5 mm, whitish, becoming pale brown, with 2(3) fls and 1 or 2 frs; bristles 8–14, biseriate, stout, flattened, about equaling the tubercle, usually retrorsely barbellate, often minutely antrorse-hairy at base; achenes flattened-pyriform, 1.5–2 mm, contracted at base, brownish-green with very faintly transverse brown lines; tubercle subulate, half to two-thirds as long as the achene; $2n=26, 42$. Sphagnum-bogs and open conifer-swamps; circumboreal, s. to N.C., O., n. Ind., Minn., and Calif.

11. **Rhynchospora capitellata** (Michx.) Vahl. Erect, cespitose perennial 3–8 dm; lvs flat, 1.5–3.5 mm wide; glomerules 2–several, loosely turbinate or hemispheric, the terminal to 1.5 cm thick, the lateral some-what smaller, often on paired or branched peduncles; spikelets 3.5–5 mm, castaneous, with mostly 2–5 fls and frs; bristles 6, 1.8–2.8 mm, usually retrorsely barbellate, about equaling or a little shorter than the tubercle; achenes plump, pyriform-obovate, 1.2–1.8 mm, two-thirds as wide, with narrowly raised margins and without a prominent umbo, smooth, ± uniformly dark brown; tubercle 0.8–1.6 mm, much widened at base. Bogs and wet sands; N.S. to Ont., Mich., and Mo., s. to Fla. and Tex.; Pacific coast.

12. **Rhynchospora glomerata** (L.) Vahl. Much like no. 11; lvs to 5 mm wide; frs (1)2–3; bristles 3.2–3.7 mm, equalizing or surpassing the tubercle; achenes with an evident pale umbo and heavy, wire-like raised margin; tubercle 1.3–1.8 mm. Bogs and wet sand; Fla. to Tex., n. on the coastal plain to N.J. and inland to Ky. and Ark.

13. **Rhynchospora knieskernii** Carey. Cespitose perennial to 5 dm; stems slender, flexuous; lvs 1–2 mm wide, soon becoming involute; terminal glomerule turbinate, 5–10 mm thick; lateral glomerules 2 or 3, remote, somewhat smaller, subsessile; spikelets 2–3 mm, brown, with 2–3 frs and a terminal staminate fl; bristles 6, retrorsely barbellate, half to fully as long as the achene; achenes plump, elliptic or obovate, 1.1–1.3 mm, slightly or scarcely more than half as wide, obscurely roughened, shining yellow-brown centrally, darker toward the margins; tubercle triangular, up to half as long as the achene. Pine-barren bogs; N.J. to Del.

14. **Rhynchospora capillacea** Torr. Cespitose; lvs involute, filiform, 0.2–0.4 mm wide; glomerules 1 or commonly 2, the terminal ovoid or oblong, with 2–10 erect or ascending spikelets, the lateral remote, subsessile, with 1–4 spikelets; bristles 6, retrorsely barbellate or seldom smooth, surpassing the achene; achenes elliptic-oblong to narrowly elliptic-obovate, 1.7–2.1 mm, scarcely (or less than) half as wide, conspicuously narrowed toward the base, pale (and often obscurely darker-lined) centrally, darker toward the margins; tubercle subulate-attenuate, 0.8–1.6 mm; $2n=26$. Calcareous swamps, bogs, and shores; Nf. to Sask., s. to Va., Tenn., and Mo.

15. **Rhynchospora pallida** M. A. Curtis. Stems 3–8 dm, usually surpassing the lvs, loosely clustered or solitary, bulbous-thickened at base and producing short, bulbous-tipped rhizomes; infl a dense hemispheric fascicle 1–2.5 cm thick, subtended by 2 unequal divaricate bracts; spikelets silvery to reddish, ovoid-attenuate, 4.5–5 mm, with a solitary perfect terminal fl; bristles 0–3, up to a fourth as long as the achene; stamens 2;

achenes obovate, 1.2–1.6 mm, nearly as wide, smooth, brown with a pale central spot; tubercle broadly conic, 0.2–0.3 mm. Acid bogs along the coast; L.I. to N.C.

16. **Rhynchospora fusca** (L.) Aiton f.   Colonial by slender rhizomes; stems 1.5–4 dm; lvs very slender, involute, mostly shorter than the stems; infl of 1–3 turbinate or ovoid glomerules 10–15 × 5–12 mm, the lower on exsert peduncles, each subtended and much surpassed by an erect foliaceous bract; spikelets 4–6 mm, dark brown, with 2–3 fls and frs; bristles antrorsely barbellate, 3 about equaling the tubercle, 2 or 3 about equaling the achene; achenes obovate or triangular-obovate to pyriform, 1.1–1.4 mm, three-fourths as wide, light brown, smooth at 10×, tubercle flat-subulate, 0.7–1.5 mm; $2n=26, 32$. Bogs and marshes; Nf. and Que. to Sask., s. to Del., Md., N.Y., Mich., and Minn.; also in Europe.

17. **Rhynchospora filifolia** A. Gray.   Densely cespitose, to 7 dm; lvs filiform or rarely to 2 mm wide; glomerules 2–4, broadly turbinate to hemispheric, the terminal one 8–15 mm thick, the lateral remote, subsessile to short-peduncled, somewhat surpassed by their bracts; spikelets 3–4 mm, dark brown, with 3–6 fls and 1–4 frs; bristles 6, 1.2–1.7 mm, antrorsely barbellate; achenes obovate, 0.9–1.3 mm, two-thirds as wide, pale brown toward the margin, lustrous and nearly white at the center; tubercle deltoid, half as long as the achene. Pine-barren bogs on the coastal plain; N.J. to Fla. and Tex.; Cuba and C. Amer.

18. **Rhynchospora gracilenta** A. Gray.   Cespitose, to 1 m; lvs filiform or to 2 mm wide; terminal glomerule ovoid to subglobose, 6–12 mm wide; lateral glomerules 1–2, short-peduncled; spikelets 3–3.5 mm, castaneous, with 2–3 fls and 1–2 frs; bristles 6, 1.8–3.4 mm, antrorsely barbellate; achenes ellipsoid to rotund, dull brown, often with a pale central spot, 1.3–1.7 mm, two-thirds to fully as wide; tubercle flat-subulate above a triangular base, 1–2 mm, bogs and wet soil on the coastal plain; N.J. to Fla. and Tex.; rarely inland from w. Va. s. and w.; Mex. and C. Amer.

19. **Rhynchospora fascicularis** (Michx.) Vahl.   Cespitose, erect to 1 m, relatively coarse, the stems 1.5–3 mm thick near the base; lvs filiform or to 4 mm wide, no more than half as long as the stems; terminal glomerules broadly turbinate to corymbiform, 5–25 mm thick, or compound and to 5 cm wide; lateral glomerules 0.3, usually remote, conspicuously pedunculate; spikelets 2.5–4 mm, with 2–4 fls and 1–3 frs; bristles (0–)6, antrorsely barbellate, variable, up to as long as the achene; achenes elliptic to subrotund, brown, sometimes pale at the center, 1.3–1.7 mm, three-fourths to fully as wide; tubercle deltoid, 0.3–0.8 mm, up to half as long or sometimes fully as long as the achene. Wet acid soil on the coastal plain; trop. Amer., n. to Tex. and se. Va.

20. **Rhynchospora debilis** Gale.   Much like no. 19, but the stems slender and lax, no more than 1 mm thick; longer lvs equaling the stems; bristles never more than half as long as the achene. Coastal plain from se. Va. to n. Fla. and Ala.

21. **Rhynchospora grayi** Kunth.   Cespitose, 3–8 dm; lvs flat, 2–3 mm wide; infl of 1–3 terminal and subterminal, hemispheric to subglobose glomerules, with 1–3 remote, slenderly peduncled lateral glomerules; spikelets lanceolate to narrowly ovoid, 4–6 mm, with 2–3 fls and 1 fr; bristles 6, antrorsely barbellate, shorter to longer than the achene; achenes olivaceous to nearly black, broadly elliptic to rotund, much thickened distally, 2.1–2.6 mm, nearly or quite as wide, finely marked with numerous pits, tubercle conic, not flattened, 0.3–0.8 mm. Wet sandy soil on the coastal plain; se. Va. to Fla., Tex., and Cuba.

22. **Rhynchospora harveyi** W. Boott.   Much like no. 21; spikelets globose-ovoid, 2.5–3 mm; bristles delicate, often caducous, half to nearly as long as the achene; achenes brown, 1.5–1.8 mm; tubercle 0.3–0.5 mm. Acid soils, chiefly on the coastal plain; se. Va. to Fla. and Tex., n. inland to Mo.

23. **Rhynchospora inexpansa** (Michx.) Vahl.   Cespitose, to 1 m; lvs flat, 2–5 mm wide; infl of 2–6 loosely branched elongate cymes on filiform peduncles; spikelets lanceolate, 4–6 mm, castaneous, with 2–5 fls and 1–4 frs; bristles 6, stiffly erect, antrorsely barbellate, surpassing the tubercle; achenes flat, brown, narrowly elliptic-obovate, 1.7–2.1 mm, half as wide; transversely strongly ridged; tubercle flat, 0.7–1 mm. Moist acid soil on the coastal plain or occasionally inland; se. Va. to Ga. and Tex.

24. **Rhynchospora miliacea** (Lam.) A. Gray.   Stout, erect, commonly 1 m or more, loosely rhizomatous, with 1–few stems together; lvs numerous, flat, to 1 cm wide; infl of 4–8 often overlapping cymes, each decompound, with much elongate, widely divaricate or even reflexed, capillary branches, each terminating in a single ovoid spikelet 3–4 mm, the scales soon falling and exposing the 3–10 persistent achenes; bristles 6, fragile, antrorsely barbellate, surpassing the tubercle; achenes pale brown, obovate, 0.9–1.2 mm, nearly or quite as wide, transversely ridged; tubercle depressed-conic, 0.2–0.3 mm. Swamps on the coastal plain; se. Va. to Fla., La., and the W.I.

25. **Rhynchospora caduca** Elliott.   Stout, erect, 7–11 dm, the stems loosely clustered on short rhizomes; lvs flat, to 7 mm wide; infl elongate, of several peduncled, loosely branched, broadly turbinate or corymbiform cymes; spikelets ovoid, 4–6 mm, dark brown, with 3–6 fls and 2–5 frs; bristles 6, brittle, antrorsely barbellate, surpassing the tubercle; achenes light brown, broadly obovate to subrotund, 1.3–1.6 mm, nearly or quite as wide, transversely sharply ridged; tubercle acute, 0.4–0.7 mm. Marshes and wet soil, chiefly on the coastal plain; se. Va. to Fla. and Tex.

26. **Rhynchospora rariflora** (Michx.) Elliott.   Cespitose, 2–5 dm, the stems and lvs filiform; cymes 1–3, small and loosely branched, each of the 3–6 spikelets on an elongate, capillary pedicel; spikelets ovoid, 3–4 mm, castaneous, 2–4-fld, with 1–3 frs; bristles 6, antrorsely barbellate, unequal, the longest to 0.5 mm; achenes brown, plump, broadly obovoid or subrotund, 1.2–1.5 mm, transversely ridged; tubercle flat, deltoid, 0.3–0.5 mm. Bogs and wet soil on the coastal plain; N.J. to Fla., Tex., and trop. Amer.

27. Rhynchospora globularis (Chapman) Small. Cespitose, 3–9 dm; lvs coarse, flat, 2–4 mm wide; cymes 1–4, compact, the lateral ones pedunceld, the branchlets ending in a small glomerule of a few ovoid spikelets 2–4 mm with 1–3 frs; bristles 5–6, antrorsely barbellate, half to three-fourths as long as the achene; achenes brown, broadly obovoid or subrotund, much thickened distally, 1.1–1.5 mm, nearly as wide or slightly wider, transversely ridged; tubercle conic-deltoid, 0.3–0.6 mm. Swamps, bogs, and wet soil. Var. *globularis,* with few spikelets, seldom over 6 per glomerule, occurs on the coastal plain from Del. to Fla. and Tex., and in Calif. Var. *recognita* Gale (*R. cymosa,* misapplied), with dense glomerules, the spikelets numerous and crowded, occurs from N.J. to S.C. and w. through the s. states, also inland from O. to n. Ill. and Mo.; trop. Amer.

28. Rhynchospora torreyana A. Gray. Cespitose, very slender, 5–10 dm; lvs 1–2 mm wide; cymes 2–4, freely branched, corymbiform, the terminal one 1–3 cm wide, the lateral smaller, on exsert peduncles; spikelets mostly pediceled, narrowly ovoid, 3–4 mm, castaneous, with 3–6 fls and 1–5 frs; bristles 6, antrorsely barbellate, to 1 mm; achenes castaneous, very flat, obovate, 1.3–1.6 × 1–1.2 mm, with ca 12 closely spaced transverse ridges; tubercle depressed-triangular, 0.3–0.4 mm. Wet sandy or peaty soil on the coastal plain; e. Mass. to Ga.

29. Rhynchospora perplexa Britton. Much like no. 28; spikelets ovoid, 2–3 mm; bristles lacking, or caducous, or 1–3 and up to a third as long as the achene; achenes broadly obovate, 1–1.3 × 0.9–1.1 mm, with a few (less than 8) transverse ridges; tubercle 0.2–0.3 mm. Moist or wet soil or shallow water on the coastal plain; se. Va. to Fla., Tex., and the W.I.

7. CLADIUM P. Browne. Scales spirally imbricate, deep brown, forming a lanceolate to ovoid spikelet; perianth none; uppermost fl of each spikelet perfect, with 2 stamens, a terete ovary, and 2–3 stigmas; middle fls staminate or abortive; lowest scale of each spikelet empty; achene ovoid, terete, dull, pointed but not tuberculate; infl compound, cymosely branched, bearing spikelets in small terminal capitate clusters. 50+, widespread.

1 Lf-blades 1–3 mm wide, smooth or nearly so .... 1. *C. mariscoides.*
1 Lf-blades 5–10 mm wide, very rough on the margins and midrib beneath .... 2. *C. jamaicense.*

1. Cladium mariscoides (Muhl.) Torr. Twig-rush. Culms stiff, slender, to 1 m, solitary or few together from rhizomes, often colonial; blades 1–3 mm wide, channeled toward the base, flat or nearly so in the middle, becoming terete distally, smooth or nearly so; infl slender, the terminal cyme 5–10 cm, the lower ones remote; spikelets lanceolate, becoming ovoid, 3–5 mm; lower scales short, subrotund, the upper ovate; achene dull brown, 2.5–3.5 mm, conspicuously pointed, the base either contracted or broadly truncate. Swamps and marshes, usually in calcareous or saline places; N.S. to Minn., s. to w. Fla. and Ky.; e. Tex. (*Mariscus m.*) Some plants have the infl much congested.

2. Cladium jamaicense Crantz. Saw-grass. Culms stout, 5–10 mm thick, to 3 m; lf-blades 5–10 mm wide, very rough on the margins and along the midrib beneath, flat or folded in the middle, becoming trigonous distally; infl decompound, 3–8 × 1–3 dm; achene contracted at base. Swamps and shallow water; near the coast from se. Va. to Fla., Tex., and W.I.

8. CYPERUS L. Flatsedge. Scales distichous, the lowermost one empty and ± modified; fls perfect, each in the axil of a scale; perianth wanting; stamens 3, less often 1 or 2; style 2–3-cleft, the beakless (or nearly beakless) achene accordingly lenticular or ± trigonous; spikelets few–many in dense or loose spikes or heads, each subtended by 2 small bracts; spikes or heads commonly in a simple or compound terminal umbel that is subtended by sheathless, leafy invol bracts; each ray of the umbel surrounded at base by a tubular prophyll; herbs with solid, ± triangular stems, the lvs with closed sheath and usually an elongate, grasslike blade. 600+, cosmop. Our spp. belong to 5 subgenera, as in the following key. Spp. 14–22 and 23–28 appear to be intergradient, presumably reflecting extensive hybridization.

1 Achenes trigonous; stigmas 3.
  2 Rachilla articulate at the base of each scale, at maturity separating or readily separable into joints, its internodes outwardly arcuate, firm or cartilaginous, conspicuously winged, the wings enfolding the achene. (Sp. 29.) .... Subg. 3. *Torulinium.*
  2 Rachilla not separating into joints, its internodes straight or nearly so, winged or wingless.
    3 Spikelets at maturity deciduous from the rachis of the spike just above an elevated, scarlike base; scales persistent, or sometimes deciduous from the rachilla before maturity. (Spp. 14–28.) .... Subg. 2. *Mariscus.*
    3 Spikelets not deciduous from the rachis; scales at maturity deciduous from the persistent rachilla. (Spp. 1–13.) .... Subg. 1. *Cyperus.*
1 Achenes lenticular; stigmas 2.

4 Spikelets 1-fld, deciduous intact from the rachis of the spike, the rachilla articulate at the base. (Spp. 36–37.) .... Subg. 5. *Kyllinga*.
4 Spikelets with several or many fls, the scales individually deciduous; rachilla not articulate at the base. (Spp. 30–35.) .... Subg. 4. *Pycreus*.

ARTIFICAL KEY TO THE SPECIES

1 Achenes trigonous; stigmas 3.
  2 Plants perennial .... Group I.
  2 Plants annual .... Group II.
1 Achenes lenticular; stigmas 2 .... Group III.

GROUP I

1 Spikelets (except often the uppermost) reflexed, the dense spikes obovoid or obconic.
  2 Plants rough or hairy at least on the stems and lower side of the bracts.
    3 Peduncles, prophylls, and upper side of the bracts rough or hairy .... 17. *C. plukenetii*.
    3 Peduncles, prophylls, and upper side of the bracts smooth .... 18. *C. retrofractus*.
  2 Plants glabrous, except the lf-margins .... 19. *C. hystricinus*.
1 Spikelets (except often the lowermost) not reflexed.
  4 Spikelets closely aggregated into dense heads, mainly the tips exposed.
    5 Spikelets with (6–)10–22 fls .... 11. *C. pseudovegetus*.
    5 Spikelets with 1–3(–5) fls.
      6 Lowest flowering scale 2–2.5 mm; heads ellipsoid-cylindric .... 20. *C. retrorsus*.
      6 Lowest flowering scale 3.5–4.3 mm; heads ± globose .... 21. *C. echinatus*.
  4 Spikelets more loosely arranged in subcapitate to cylindric spikes.
    7 Spikelets serially disposed on ± elongate axes, forming short-cylindric spikes.
      8 Spikelets subulate, round or nearly so in section; scales 4–6 mm.
        9 Scale evidently overlapping the next one above on the same side .... 15. *C. lancastriensis*.
        9 Scale just reaching the base of the next one above on the same side .... 16. *C. refractus*.
      8 Spikelets ± compressed, or the scales less than 4 mm, or both.
        10 Scales golden-yellow at maturity; widespread spp.
          11 Scales (3–)3.5–4.5(–5) mm, keeled .... 14. *C. strigosus*.
          11 Scales (2–)2.5–3(–3.5) mm, scarcely keeled .... 2. *C. esculentus*.
        10 Scales purple-brown or red-brown, 2.5–3.5 mm, evidently keeled; southern .... 1. *C. rotundus*.
    7 Spikelets radiating from the summits of very short axes, forming loose or irregular subcapitate spikes.
      12 Scales with 3–5 nerves close together in the center, the sides nerveless.
        13 Scales 2–3 mm; lower cauline lvs bearing blades; bracts 3–5 .... 3. *C. dentatus*.
        13 Scales 1–1.5 mm; lower cauline lvs reduced to bladeless sheaths; bracts 2 .... 4. *C. haspan*.
      12 Scales with 7–13 well distributed nerves.
        14 Scales sharply acute; rachilla-joints winged .... 22. *C. croceus*.
        14 Scales obtuse or rounded, or obtuse and mucronate; rachilla wingless except in no. 23.
          15 Achene stout, ca ⅔ as thick as long, or thicker, with markedly concave sides ...... 26. *C. houghtonii*.
          15 Achene more slender, only slightly (if at all) more than half as thick as long, with flat or slightly concave sides.
            16 Invol bracts ascending or divergent-ascending, sometimes with more arcuate tip; infl usually with 1–several evident rays in addition to the central cluster.
              17 Stem smooth; scales blunt or with a minute mucro up to 0.2 mm.
                18 Rachilla winged; scales approximate; achene 1.5–2 mm; coastal plain
                .... 23. *C. grayi*.
                18 Rachilla wingless; scales subremote, barely overlapping the next one above; achenes 2–2.4 mm; Ill. to Tex. and La. .... 24. *C. grayioides*.
              17 Stem scabrous near the top; scales with an evident mucro 0.5–1 mm ..... 25. *C. schweinitzii*.
            16 Invol bracts spreading horizontally or more often reflexed; infl a dense head, only rarely with 1–4 rays.
              19 Heads drab-green to light reddish-brown; widespread .... 27. *C. lupulinus*.
              19 Heads stramineous to yellowish-brown; se. coastal plain .... 28. *C. filiculmis*.

GROUP II

1 Scales outcurved at the tip; stamen 1.
  2 Scales (5)7–9-nerved, with a short but distinct awn-tip .... 13. *C. squarrosus*.
  2 Scales 3-nerved, somewhat recurved-acuminate, but not awn-tipped .... 12. *C. acuminatus*.
1 Scales not outcurved at the tip; stamens 3, or less often 2.
  3 Scales (1.5–)2–5 mm, with 7–13 well distributed nerves.
    4 Spikelets serially disposed on ± elongate axes, forming short-cylindric spikes.

5 Scales (3–)3.5–4.5(–5) mm; spikelets deciduous intact at maturity. (First-year forms of) .... 14. *C. strigosus.*
5 Scales (1.5–)2–3 mm; spikelets disarticulating into joints at maturity .... 29. *C. odoratus.*
4 Spikelets radiating from the summits of very short axes, forming loose or irregular subcapitate
    spikes.. 5. *C. compressus.*
3 Scales 0.5–1.7 mm, with 3–5 nerves close together in the center, the sides nerveless.
    6 Spikelets radiating from the summits of very short axes, forming dense heads (no. 7) or loosely
        subcapitate clusters (no. 6).
        7 Scales ovate, 0.9–1.5 mm, keeled, shortly mucronate .... 6. *C. fuscus.*
        7 Scales broadly obovate-rotund, 0.5–0.8 mm, neither keeled nor mucronate .... 7. *C. difformis.*
    6 Spikelets serially disposed on ± elongate axes, forming short-cylindric spikes.
        8 Scales ovate to elliptic; achenes pearly-white .... 8. *C. erythrorhizos.*
        8 Scales broadly obovate; achenes blackish.
            9 Rachilla wingless; scales scarcely mucronate .... 9. *C. iria.*
            9 Rachilla narrowly winged; scales evidently mucronate .... 10. *C. microiria.*

Group III

1 Spikelets 1-fld, deciduous intact from the rachis of the dense, capitate spike; the rachilla articulate at
    the base, not grooved as in the next group.
    2 Perennial from long rhizomes .... 36. *C. brevifolioides.*
    2 Tufted annual .... 37. *C. tenuifolius.*
1 Spikelets several-fld, the scales individually deciduous; one edge of the achene fitting into a groove
    in the persistent rachilla; all annual.
    3 Spikelets serially disposed on a ± elongate axis, forming short-cylindric spikes; southern .... 30. *C. flavicomus.*
    3 Spikelets radiating from the summits of very short axes, forming loose or irregular subcapitate
        spikes; widespread.
        4 Scales ± stramineous.
            5 Scales broadly ovate, very blunt, in side-view hardly over twice as long as wide .... 31. *C. flavescens.*
            5 Scales narrower, in side-view ca 3 times as long as wide.
                6 Achene oblong; scales 1.5–2.3 mm .... 32. *C. polystachyos.*
                6 Achene obovoid; scales 2.5–3.5 mm ....33. *C. filicinus.*
        4 Scales at maturity pigmented with red-purple or deep brown.
            7 Style cleft almost to the base; scale more pigmented distally .... 34. *C. diandrus.*
            7 Style undivided in the basal third; scale more pigmented proximally .... 35. *C. bipartitus.*

1. Cyperus rotundus L. Nutsedge. Perennial with numerous slender rhizomes ending in small tubers; stems 2–5 dm, smooth; lvs basally disposed, mostly shorter than the stem, 3–6 mm wide; bracts few, all or all but one shorter than the infl; rays 3–7, simple or sometimes branched at the top, to 10 cm; spikelets 3–10 in short-cylindric spikes, 1–4 cm; scales closely imbricate, ovate, 2.5–3.5 mm, obtuse, keeled, 7-nerved (3 in the midvein, 2 nearby on each side), the sides purple-brown or red-brown; rachilla persistent, winged; achenes nearly black, narrowly trigonous-obovoid, 1.5 mm; $2n=108$. A weed of sandy soil; pantrop., n. to Va. and Ky., and occasionally adventive farther n.
    C. setigerus Torr. & Hook., a cespitose, non-tuberous sw. sp. with the bracts mostly surpassing the infl, has been collected just to the s. of our range in w. Mo., and may prove to occur in our range as well.

2. Cyperus esculentus L. Yellow nutsedge. Sweet-scented perennial with numerous slender rhizomes ending in small tubers; stems stout, triquetrous, 1–7 dm, smooth; lvs basally disposed, elongate, 3–8 mm wide; at least the lower invol bracts surpassing the infl; rays usually several, to 7 cm but more often not surpassing the sessile spikes, often again branched at the top; prophyll obliquely truncate or prolonged 1–3 mm beyond the orifice; spikes numerous in short-cylindric spikes, slender, 0.5–5 cm, only 1–2 mm wide, with 8–30+ fls; scales yellow-brown, (2–)2.5–3(–3.5) mm, thin, scarcely keeled, ovate, acute, conspicuously 7–9-nerved; rachilla persistent, narrowly hyaline-winged; achenes oblong, unequally trigonous, 1.3–2 mm, tan to golden-brown; $2n=108$. Damp or wet soil, sometimes a weed; widespread in trop. and temp. (but not cold-temp.) regions, and nearly throughout our range.

3. Cyperus dentatus Torr. Perennial from short rhizomes ending in tubers; stems solitary, 1–5 dm; lvs numerous, 2–5 mm wide, about equaling the stem; bracts 3–5, the longer surpassing the infl; rays numerous, 1–5(–10) cm, often branched above; spikes short, loosely hemisperic, 1–3 cm wide; spikelets few to many, flattened, 3–15 mm, with 8–25 fls; some spikelets usually proliferated, bearing much elongate scales; scales normally ovate, 2–3 mm, the midvein prolonged as a short stout point; rachilla wingless; achenes obovoid-trigonous, 1 mm or less, half as thick; $2n=34$. Sandy shores; Que. and N.S. to n. N.Y., s. to W.Va. and N.C.; nw. Ind.

4. Cyperus haspan L. Somewhat cespitose perennial 2–7 dm; lowest cauline lvs ordinarily reduced to bladeless sheaths, the uppermost either bladeless or rarely with well developed blade; bracts 2, erect or ascending, usually shorter than the infl; rays none or to 10 cm, often branched at the top; spikes short, loosely subglobose, 1–2 cm thick; spikelets to 1 cm, flattened with 5–25 fls; rachilla persistent; scales ovate or elliptic, 1–1.5 mm, 3-nerved in the center, obtuse, often red-purple; rachilla wingless; achenes obovoid-trigonous, pearly white, 0.4–0.7 mm; $2n=26$. Swamps and shallow water, mostly near the coast; pantrop., n. to se. Va.

5. Cyperus compressus L. Tufted annual 1–3 dm; lvs ± basally disposed, flat or folded, 1–2 mm wide; bracts 3–5, surpassing the infl, 1–2 mm wide; spikes few and sessile, or 1–4 rays present and up to 10 cm; spikelets 3–10, digitately spreading, 1–2.5 cm × 2–4 mm, flattened, 10–40-fld; rachilla persistent; scales 2.5–3.5 mm, broadly ovate, keeled, acuminate and with shortly excurrent midrib, multinerved, greenish, with broad, thin, coppery or pale (often also red-lineolate) margins; achenes shining brownish, broadly obovoid-trigonous, 1–1.4 mm. Sandy woods and fields; pantrop., n. near the coast to Del. and inland to Mo.

6. Cyperus fuscus L. Tufted annual 1–4 dm; lvs 1–4 mm wide; bracts 2–5, surpassing the infl; rays few, 1–3 cm; spikelets 3–6 mm, 10–20(–40)-fld; rachilla not winged; scales ovate, 0.9–1.5 mm, with reddish-brown sides, keeled, shortly mucronate; stamens 2; achenes whitish, trigonous, 1 mm; $2n=72$. Native of Eurasia, intr. at scattered stations in the e. part of our range.

7. Cyperus difformis L. Tufted annual 2–6 dm, bearing 2–4 bracts and several sessile or short-peduncled, globose, densely fld heads 8–12 mm thick; spikelets 4–8 mm, scarcely flattened, 10–40-fld; rachilla not winged; scales obovate-rotund, often broader than long, 0.5–0.8 mm, with reddish or brownish sides; stamen 1(–3); achenes light yellowish, trigonous, about equaling the scales. Wet soil; native of the Old World tropics, intr. in se. Va.

8. Cyperus erythrorhizos Muhl. Redroot flatsedge. Stout, tufted annual 1–7 dm with red roots; lvs crowded toward the base, with elongate blade 2–9 mm wide; invol bracts several, elongate, unequal; infl of clustered, cylindric spikes, the terminal cluster sessile, the others on rays to 7 cm, each spike 1–4 cm, with numerous serially arranged spikelets on an elongate rachis; spikelets 3–12 × 1–1.5 mm, 6–many-fld; rachilla persistent, with narrow, deciduous, hyaline wings; scales 1.2–1.5 mm, ovate to elliptic, with prominent, green, minutely excurrent keel, 1 or 2 adjacent pairs of much finer lateral nerves, and narrow, golden-brown margins; achenes pearly-white, unequally trigonous, 0.7–1 mm; $2n=96$. Streambanks and other wet places; Mass., N.H., and N.Y. to Ont. and N.D., s. to Ark., Tex., and Ariz., and irregularly to the Pacific.

9. Cyperus iria L. Tufted annual to 6 dm, with rather short, basally disposed lvs; rays usually well developed and commonly branched at the top; spikes loose and open, 1–2 cm, bearing several rather distant, ascending spikelets; spikelets 10–20-fld, or to 1 cm; rachilla persistent; scales broadly obovate, rounded-emarginate, scarcely mucronate, 1.2–1.7 mm, with green midrib and golden-brown sides; rachilla wingless; stamens 2 or 3; achenes trigonous, 1–1.5 mm, blackish. Swamps and muddy ground; native of Eurasia, intr. in se. U.S. as far n. as s. Ill., Ky. and se. Va.

10. Cyperus microiria Steudel. Much like no. 9; spikes 2–4 cm; rachilla narrowly winged; scales truncate, with an evident green mucro to 0.4 mm. Native to e. Asia, intro. on L.I. and in e. Pa. (*C. amuricus,* at least as applied to our plants.)

11. Cyperus pseudovegetus Steudel. Tufted perennial 3–8 dm; lvs elongate, often surpassing the stem, mostly 4–6 mm wide but conduplicate; bracts 4–8, much surpassing the infl; rays usually well developed, each bearing a terminal glomerule of spikes, or occasionally suppressed, producing a compact infl; spikes ovoid or subglobose, very dense, 7–15 mm; spikelets flat, ovate, 2–5 × 1.5–3 mm, closely imbricate with their flat sides together, (6–)10–22-fld; scales 1.5–2.5 mm, falcately incurved-ascending with divergent tip, in side-view with parallel sides, distinctly bicarinate basally; rachilla persistent, wingless; stamen 1; achenes linear-trigonous, 1–1.5 mm. Swamps and marshes; N.J. to Fla., w. to Ill., Mo., Okla., and Tex. Sometimes confused with *C. virens,* a more s. sp. not reaching our range.

12. Cyperus acuminatus Torr. & Hook. Tufted annual 0.5–4 dm; lvs few, all near the base, slender and ± elongate, 1–2(–3) mm wide; invol bracts 3–6, unequal, elongate, some or all much surpassing the infl; spikelets 3–7 mm, strongly flattened, 8–20(–40)-fld, borne in very dense, globose clusters on very short axes; scales 1.3–2 mm, ovate, bicarinate basally, strongly 3-nerved, otherwise subhyaline and cellular-reticulate, shortly recurved-acuminate at the tip; rachilla persistent, wingless; stamen 1; anthers 0.5–0.7 mm; achenes rather broadly trigonous, 0.6–1 mm. Streambanks and other wet, low places, tolerant of alkali; Ill. to N.D., Mo., Tex., and n. Mex., e. occasionally to O., Ky. and w. N.C., and w. irregularly to the Pacific.

13. Cyperus squarrosus L. Slender, tufted, sweet-scented annual 0.5–2 dm; lvs few, all borne near the base, mostly 0.5–2(–2.5) mm wide, up to as long as or a little longer than the stems; invol bracts elongate, some or all of them much surpassing the infl; spikelets in dense, capitate clusters, the terminal cluster sessile, the others (if present) on slender rays to 3 cm; spikelets 4–10 mm, flattened, 6–20-fld; scales evidently (5)7–9-nerved, 1.0–1.7 mm exclusive of the recurved, slender awn-tip, this 0.3–1 mm; rachilla wingless, ± persistent, the scales falling from it at maturity; stamen 1; anthers 0.2–0.3 mm; achenes trigonous, 0.6–1 mm; $2n=48, 56, 64$. Moist or wet, often sandy soil; nearly cosmop, and throughout our range. (*C. aristatus; C. inflexus.*)

14. Cyperus strigosus L. False nutsedge. Short-lived perennial without well developed rhizomes, sometime blooming the first year; stems few or solitary, smooth, sharply triangular, 1–6(–10) dm; lvs crowded toward the base, mostly 2–10(–15) mm wide; spikes usually branched at base; spikelets numerous in short-cylindric spikes, spreading ± at right angles from the ± elongate rachis, 6–25 × 1–2 mm, strongly compressed, deciduous at maturity, the rachilla disarticulating just above the reduced lowermost pair of scales; scales (3–)3.5–4.5(–5) mm, keeled and multinerved, with greenish midrib and ± golden sides, tending to be narrowly hyaline-margined distally, persistent, each one overlapping the next one above on the same side; anthers less than 0.5 mm; achene 1.5–2 mm, unequally linear-trigonous, held between the hyaline wings of the readily deciduous rachilla. Moist fields, swamps, and shores; Que. to Minn. and S.D., s. to Fla. and Tex., and w. irregularly to the Pacific.

15. Cyperus lancastriensis Porter. Stout perennial 3–9 dm with short, knotty rhizomes; lvs 4–10 mm wide, usually shorter than the stem; at least the lowest bract surpassing the infl; rays several, usually well developed, to 12 cm; spikes unbranched, short-cylindric, rounded at both ends, 1–3 cm; spikelets 50–100+, horizontally radiating except the uppermost and lowermost, subulate, 3–8-fld, acuminate, deciduous at maturity; scales oblong, 4–5 mm, finely multiribbed, the tip of one reaching well beyond the base of the next one on the same side; anthers 0.7–1 mm; achenes trigonously obovoid-oblong, 2–2.6 mm, a third as thick, held between the hyaline wings of the rachilla. Woods and fields, N.J. to W.Va., O., and Mo., s. to Ga. and Ark.

16. Cyperus refractus Engelm. Stout perennial 3–8 dm, with short, knotty rhizomes; lvs firm, to 10 mm wide, often equaling the stem; longer bracts usually surpassing the infl; rays 4–10, to 15 cm; spikes unbranched, mostly rather loose, short-cylindric, rounded at both ends, 2–5 cm; spikelets 15–50, the upper ascending or erect, the middle spreading, the lower deflexed, subulate, 3–10-fld, 1 mm thick, acuminate, deciduous at maturity; scales 4–6 mm, appressed, finely multiribbed, obtuse, the tip of one reaching just to the base of the next one above on the same side, the lower empty and often more persistent than the others; anthers 1–1.5 mm; achenes trigonously linear-oblong, 2.5–3 mm, held between the broad hyaline wings of the rachilla. Woods and fields; N.J. to O. and Mo., s. to S.C., Ga., Ala., and Ark.

17. Cyperus plukenetii Fern. Much like no. 18, but the rays, prophylls, and upper side of the bracts shortly rough-puberulent or scabrid; stem obtusely trigonous to rounded, rough-hairy to the base; spikes tight, bur-like, only slightly or scarcely longer than thick, tapering downward from the broad top, the axis 7–13 mm; spikelets 6–8 mm, with 1(2) fertile scale; scales 4–4.5 mm; bracts up to as long as the rays; rachilla-wings clasping the achene for half its length. Dry, sandy places; N.J. to Fla., w. to Ky., s. Mo., se. Okla., and e. Tex. (*C. retrofractus* var. *p.*)

18. Cyperus retrofractus (L.) Torr. Tufted perennial, the cormose stem-bases joined by short rhizomes; stems rigid, 3–9 dm, acutely trigonous, antrorsely scabrous above, often smooth near the base; lvs 4–8, 3.5–8 mm wide, minutely rough-puberulent across the surface beneath and on the midrib above; bracts 3–12, the longest one at least as long as the longest ray; rays 4–13, glabrous, the longest 1–2 dm; prophylls smooth to ciliate; spikes obovoid or shortly obconic, 2–3 cm, less than twice as long as thick, the axis 10–22 mm; spikelets stramineous or dull greenish, often suffused with anthocyanin, 1–2(–4)-fld, subulate-acuminate, 8–10(–15) mm, crowded, the uppermost widely spreading or recurved, the others reflexed; scales closely appressed, lance-ovate, acute, (4–)4.5–5(–5.5) mm; achenes trigonously oblong, held between the hyaline wings of the rachilla for ¾ their length. Pastures, roadsides, and other disturbed places; N.J. to Ga., w. to s. O., Mo., and Ark. (*C. dipsaciformis.*)

19. Cyperus hystricinus Fern. Much like no. 18, but smooth except for a few remote marginal teeth on the lvs; spikes golden-brown at maturity, twice as long as thick; scales lanceolate; rachilla-wings clasping the achene for its whole length. Dry, sandy places; N.J. to n. Fla. and e. Tex., especially on the coastal plain. (*C. retrofractus* var. *h.*)

20. Cyperus retrorsus Chapm. Perennial; stems very slender, smooth, 3–10 dm; lvs shorter than the stems, 3–5 mm wide; bracts 4–8, usually surpassing the infl; rays 5–9, to 10 cm; spikes short-cylindric, 6–25 × 6–12 mm; spikelets very numerous, crowded, radiating horizontally, 3–6 mm, 1–3-fld; scales thin, ovate, 2–2.5 mm, multinerved, obtuse or minutely mucronulate; achenes oblong-trigonous, 1–1.5 mm, a third to half as thick. Sandy barrens and coasts; s. N.Y. to Fla. and Tex., mostly on the coastal plain, n. in the interior to Ky. and Okla. (*C. cylindricus,* a preoccupied name; *C. ovularis* var. *c.; C. torreyi.*)

C. flavus (Vahl) Nees, of trop. Amer., with the spikes ovoid and all sessile, has been collected as a waif in N.J. and Pa. (*C. cayennensis.*)

21. Cyperus echinatus (L.) Wood. Globe-flatsedge. Perennial; stems 3–10 dm, smooth; lvs flat, 3–8 mm wide; bracts 4–7, the lowest much elongate and surpassing the infl; infl with 1 or 2 sessile terminal spikes and usually 2–10 ascending rays to 10 cm; spikes globose, 8–15 mm thick; spikelets very numerous, closely packed and radiating in all directions, 3–7.5 mm, 1–3-fld; scales oblong-elliptic, obtuse, multinerved, the lower 3.5–4.2 × 1.5–2 mm; achenes trigonous, 1.8–2.2 mm, half as thick. Dry woods and fields; sc. Conn. and s. N.Y. to s. O., Ill., and e. Kan., s. to n. Fla. and ne. Mex. (*C. ovularis.*)

22. Cyperus croceus Vahl. Tufted perennial 2–6 dm; stems smooth, scarcely thickened at base; lvs flat, soft, 3–6 mm wide; bracts 4–10, the longest surpassing the infl; rays 3–8, to 10 cm; spikes subglobose, 8–15 mm; spikelets 3–6-fld, to 10 mm, radiating in all directions from the short axis, deciduous at maturity; scales ovate, acute, 2.5–3 × 1–1.5 mm, multinerved; rachilla broadly winged; achenes trigonous, 1.3–1.7 mm, half as thick. Dry soil; trop. Amer., n. mostly on the coastal plain to N.J. and in the interior to Mo. (*C. echinatus* and *C. globulosus,* both misapplied.)

23. Cyperus grayi Torr. Slender perennial 1–4 dm, the basal corms prominent; lvs rigid, folded, ± recurved, shorter than the glabrous stem; bracts 3–7, divergently ascending, the longer usually surpassing the infl; rays (2–)4–10(–16), to 8 cm; spikes loosely fld, subglobose, 6–25 mm thick; spikelets radiating in all directions from a very short axis, to 12 mm, 2–10-fld; scales approximate, clearly overlapping vertically, ovate, 1.8–2.8 mm, blunt, scarcely mucronulate, multinerved; rachilla-joints with conspicuous hyaline wings; achenes trigonous with flat sides, 1.5–2 mm, a third to half as thick; 2n = ca 166. Dunes and dry, sandy soil, chiefly on the coastal plain; see. N.H. to La.

24. Cyperus grayioides Mohlenbrock. Much like no. 23; rachilla wingless; scales subremote, barely reaching the base of the next scale above; achenes 2–2.4 mm; 2n = ca 166. Rare and local; sandy prairie blowouts in c. and nw. Ill.; disjunct to sandy places in Tex. and La.

25. Cyperus schweinitzii Torr.   Perennial from short, knotty rhizomes; stems 1–10 dm, scabrous above on the sides and sharp angles, 1–2.5 mm thick below the infl; lvs basally disposed, 2–8 mm wide, scabrous on the margins; bracts 3–8, strongly ascending, usually much longer than the infl; sessile spike obconic to oblong; rays 1–8, rarely over 10 cm; spikelets 5–15 in the sessile spike, fewer in the reduced ones, all crowded, ascending, flattened, 5–25 × 3–4.5 mm, 5–18-fld; scales broadly ovate-elliptic to rotund, the body 2.5–3.5 mm, in half-view nearly or fully as wide as long, multinerved, with a conspicuous mucro 0.5–1 mm; rachilla wingless or nearly so; anthers 0.6–1 mm; achenes dark, trigonously oblong with plane or only slightly concave faces, 2–3 mm, half as thick; $2n =$ ca 166. Sandy soil; Mass. and Que. to Minn., s. and w. to N.J., O., Mo., Utah, N.M., and adj. Mex.

26. Cyperus houghtonii Torr.   Much like no. 25; stems smooth, rather obtusely angled, mostly 0.5–1.5 mm thick below the infl; lvs smooth or scabrous on the margins; invol bracts divergently ascending or ± spreading, scales rotund, 2–2.5 mm, in side-view over half as wide as long, multinerved, obtuse or the uppermost minutely mucronulate; achenes 1.5–2 mm, two-thirds as thick (or a little thicker), trigonous with evidently concave faces; $2n =$ ca 168–172. Dry, often sandy soil, commonly with *Pinus banksiana;* Mass., N.H., Vt., and s. Que. to Minn. and nw. Ind.; isolated (as a waif?) in n. Va. and e. W.Va. Thought to be derived by stabilizing selection from hybrids between nos. 25 and 27 but the achenes different from both.

27. Cyperus lupulinus (Sprengel) Marcks.   Perennial from short, tuberous-knotty rhizomes; stems 1–5 dm, smooth, rather obtusely angled, 0.4–1.2 mm thick; lvs light green, 1–3.5 mm wide, flat or folded, shorter than the stem; invol bracts 2–4, widely spreading to more often deflexed, the margins of these and the cauline lvs scabrous; infl usually a single subglobose or hemispheric sessile spike, or seldom with a few rays to 7 cm, each bearing a similar but smaller spike; spikelets drab-green or light reddish-brown, very crowded, radiating from the axis, flattened, 2.5–4 mm wide, (3–)6–22-fld; scales oblong-elliptic, 2–3.5 mm, in half-view from the side about a third as wide as long, multinerved, obtuse or with a minute mucro 0.1–0.4 mm, readily deciduous; rachilla-joints sharp-edged but not winged; anthers 0.3–0.6 mm; achene trigonous, with flat or only slightly concave sides, 1.4–2.2 mm, about half as wide, or a bit wider; $2n =$ ca 166. Dry woods and fields; abundant; Me. and s. Que. to Minn., S.D., and Colo., s. to Fla. and Tex. (*C. bushii; C. macilentus; C. filiculmis* var. *m.*)

28. Cyperus filiculmis Vahl.   Much like no. 27; lvs gray-green, 1–2.5 mm wide; heads stramineous to yellowish-brown; spikelets 2–3 mm wide; anthers 0.6–1 mm; achene a little narrower, a third to nearly half as wide as long; $2n =$ ca 166. Dunes, pine-barrens, and disturbed sandy places, mainly on the coastal plain; se. Md. and se. Va. to Fla. and Tex. (*C. martindalei.*)

29. Cyperus odoratus L.   Stout annual 1–8 dm; lvs 2–10 mm wide; bracts several, mostly surpassing the infl and some much elongate, rays (0–)3–10, 2–10 cm, branched at the top, producing a congested infl; prophylls prolonged 5–15 mm beyond the orifice into a bilobed tip; spikelets linear-cylindric, brownish at maturity, 1–2 cm, 5–20-fld, divaricately spreading; scales ovate, (1.5–)2–3 mm, finely several-nerved, obtuse or acute; rachilla articulate at the base of each scale, at maturity readily separating into joints, its internodes outwardly arcuate, conspicuously winged, the wings enfolding the achene; achenes brown, trigonously outwardly arcuate, conspicuously winged, the wings enfolding the achene; achenes brown, trigonously oblong or obovoid-oblong, 1–2 mm. Moist or wet soil; pantrop, n. to Mass., se. Me., Ont., Minn., and Kans. (*C. engelmannii; C. ferax; C. ferruginescens; C. speciosus.*)

30. Cyperus flavicomus Michx.   Annual to 8 dm, stouter than the next 5 spp.; lvs elongate, often surpassing the stem; bracts 3–7, to 3 dm × 8 mm; some spikes sessile, others on 1–6 rays to 15 cm; spikelets very numerous, 1–2 cm × 2–3 mm, scales very broadly oblong-ovate, 1.4–1.7 mm, as wide or slightly wider when flattened, 5-nerved near the center, broadly hyaline-margined, rounded and ± erose at the tip; stamens 2 or 3; achenes lenticular, black, obovate, 1.2–1.6 mm wide, one edge fitting into a groove in the persistent rachilla (as in the next 5 spp.). Wet soil; trop. Amer., extending n. along the coast to se. Va. and inland to Ky. (*C. albomarginatus; C. sabulosus,* misapplied.)

31. Cyperus flavescens L.   Tufted annual 0.5--3 dm; cauline lvs short and usually erect; bracts 2–4, much longer than the infl; rays 1–5 on peduncles to 2(–5) cm; spikelets 5–15, 10–15 × 2–2.5 mm; scales yellowish or stramineous with narrow hyaline margins, broadly ovate, 1.7–2.5 mm, obtuse; stamens mostly 3; achenes lenticular, black, obovate, 1 mm, apiculate, at maturity marked (at high magnification) with elongate superficial cells and numerous pale, irregular, transverse lines. Wet soil; pantrop., n. to Mass., Mich., and Mo.

32. Cyperus polystachyos Roth.   Tufted annual 2–6 dm; lvs 1–3 mm wide, shorter than the stem; bracts 3–5, the largest 5–18 cm; umbel with 1–6 rays to 6 cm, or the spikes all sessile; spikelets 10–15, 8–15 × 1–1.8 mm; scales stramineous, oblong, 1.5–2.3 mm, subacute; stamens 2; achenes lenticular, black, oblong, 1 mm, half as wide. Wet soil, mostly near the coast; pantrop., extending n. to Mass. and in the interior to Mo. Our plants are var. *texensis* (Torr.) Fern. (*C. gatesii; C. microdontus; C. paniculatus.*)

33. Cyperus filicinus Vahl.   Tufted annual 1–4 dm; lvs 1–3 mm wide, shorter than the stem; bracts several, to 2.5 dm, 1–3 mm wide; spikes all sessile, or the 1–6 rays to 10 cm; spikelets 5–10, 15–20 × 2.5–3 mm, narrowly lanceolate in outline and acute; scales stramineous, narrowly oblong, 2.5–3.5 mm, acute or obtuse and minutely apiculate; stamens 2; achenes lenticular, obovate-elliptic, 1.2–1.7 mm. Brackish marshes and sandy beaches along the coast; Me. to Fla., La., and the W.I. (*C. nuttallii; C. polystachyos,* misapplied.)

34. Cyperus diandrus Torr.   Tufted annual (0.5)1–2(–4) dm; cauline lvs often equaling the stem, to 3 mm wide; bracts usually 3, the lowest to 15 cm; rays 1–3, to 6 cm; spikelets 6–10, 8–15 × 2–3.5 mm; scales ovate, 2.5–3 mm, as seen from the side in half-view over 3 times as long as wide and obscurely acuminate, at maturity red-purple at the tip and in a tapering marginal band almost or quite to the base that is separated from the midrib by a colorless furrowed strip; stamens 2 (or 3 in upper fls); achenes lenticular, elliptic-obovate, 1–1.3 mm; style cleft almost to the base, persistent. Wet ground, especially along shores; Me. and Que. to N.D., s. to Va. and Io.

35. Cyperus bipartitus Torr. Tufted annual 1–2(–4) dm; cauline lvs 1–3 mm wide, usually shorter than the stem; spikes all sessile, or the 1–5 rays to 10 cm; spikelets 3–10, 8–15 × 2–3.5 mm; scales ovate, 2–2.5 mm, as seen from the side in half-view usually less than 3 times as long as wide, obtuse, and with the prominent midvein incurved distally to a somewhat cucullate tip, at maturity strongly red-brown from midrib to margin (less so distally), not sulcate; stamens 2(3); achenes lenticular, elliptic-obovate, 1–1.5 mm; style undivided in the basal third, deciduous. Wet ground, especially along shores; Me. and Que. to N.D., s. to Ga. and Tex. (*C. niger* var. *castaneus; C. rivularis.*)

36. Cyperus brevifolioides Thieret & Delahoussaye. Perennial 1–3 dm with long rhizomes; lvs ± basally disposed, the lower sheaths bladeless, the upper with blades 2–4 mm wide; bracts 3, one erect, 2 spreading; spike solitary, sessile, round-ovoid, 5–10 mm; spikelets very numerous, 3–4.5 mm, 1-fld; scales pale, acuminate, with a smooth keel; stamens 2–3; anthers 0.6–0.8 mm; achenes lenticular, obovate, 1.5–1.8 mm. Wet ground; native of e. Asia, casually intr. from Conn. to Ga., Ala. and Tenn. (*C. brevifolius* misapplied; *Kyllinga brevifolioides.*)

37. Cyperus tenuifolius (Steudel) Dandy. Tufted slender annual 0.5–3 dm; lower sheaths bladeless, the upper with soft blades 1–3 mm wide; bracts 2–4, widely spreading, 3–10 cm; spikes 1 or sometimes 2 or 3, ovoid to subglobose, strictly sessile, 5–10 mm; spikelets very numerous, flattened, 2–3 mm, 1-fld; scales very thin, with a ± denticulate keel and hyaline margins; stamens 2; anthers 0.2–0.4 mm; achenes tan, lenticular, elliptic, 1.1–1.4 mm, half as wide. Muddy or sandy shores; trop. Amer. and Afr., n. to Del., Pa., O., Mo., and Kan. (*C. densicaespitosus; Kyllinga pumila.*)

C. sesquiflorus (Torr.) Mattf. & Kük., another widespread tropical sp. of the subg. *Kyllinga,* extends on the coastal plain to n. N.C. and may be sought in se. Va. It is a mat-forming, aromatic perennial with conspicuously white spikelets, and with the achenes more than half as wide as long, purple-black except for the ochroleucous apiculus and short stipe.

9. DULICHIUM Pers. Spikelets crowded in rather short, axillary spikes of mostly 7–10, distichous on the rachis; scales distichous, each subtending a perfect fl, their margins hyaline and decurrent on the eventually disarticulating rachilla as a hyaline wing; bristles 6–9, retrorsely barbed, exceeding the achene; stamens 3; style bifid; achene flattened, linear-oblong; tall, rhizomatous perennial sedges with subterete, hollow, jointed, leafy stems, numerous short blades and numerous short-peduncled spikes solitary in the upper axils. Monotypic.

1. Dulichium arundinaceum (L.) Britton. Erect, 3–10 dm; lvs numerous, the lower bladeless, the upper sheaths often overlapping; blades mostly 5–15 cm × 2.5–8 mm; peduncles scarcely exserted; spikes 1–3 cm; spikelets 10–25 mm; scales 5–8 mm, narrow, acuminate, several-nerved; achene short-stipitate, 2.5–3 mm, beaked with the long, slender, persistent style; $2n=32$. Swamps and marshes; Nf. to Minn., s. to Fla. and Tex.; B.C. to Calif. and Mont.

10. HEMICARPHA Nees. Scales spirally imbricate, ± scarious; fls perfect, each in the axil of a scale and subtended also by a very thin adaxial inner scale, or the inner scale often reduced or obsolete; stamen 1(2); style bifid; achene lenticular or subterete, minutely apiculate; delicate, glabrous annuals with few and slender lvs; spikes solitary to several, subsessile in a terminal head subtended by 2–3 unequal foliaceous bracts, the lowest of which appears like a continuation of the stem. ± 5, mainly trop. Genus perhaps better submerged in *Lipocarpha.*

1. Hemicarpha micrantha (Vahl) Britton. Tufted annual, 2–10(–20) cm; lvs to 10 cm × 0.5 mm, resembling the stems; spikes mostly 1–3 per stem, ovoid, 2–6 mm; scales numerous, 1–2 mm, firm, obovate, mucronate or (especially the lower) with a short awn-tip; achene 0.5–0.7 mm. Moist sandy soil; trop. Amer., n. to Me., Minn., and Wash. Most of our plants (as well as the tropical ones) belong to the var. *micrantha,* with the inner scale ± strongly reduced (sometimes bifid) or obsolete. Var. *aristulata* Coville (*H. drummondii*), with a relatively well developed inner scale ± equaling the achene, occurs chiefly in w. U.S., extending e. sporadically to Ohio.

11. LIPOCARPHA R. Br. nom. conserv. Scales spirally imbricate, ± scarious; fls perfect, each in the axil of a scale and also subtended by an additional pair of hyaline scales, one adaxial and persistent on the rachilla, the other abaxial and deciduous with the achene; stamens 1 or 2; style bifid or trifid; achene subtrigonous or lenticular; infl a terminal head of 3–several spikelets subtended by 2 or 3 foliaceous bracts. ± 12, mainly trop.

1. Lipocarpha maculata (Michx.) Torr. Tufted slender annual 5–30 cm; lvs near the base, with dilated sheath and narrow (1–2 mm) blade often equaling the stem, bracts spreading, unequal, the longest to 12 cm; spikelets 3–7 mm; scales numerous, green, becoming brown in age, dark-spotted, narrowly obovate, 1.5–2 mm, acute; achene oblong, 1 mm. Damp or wet, chiefly sandy soil; se. Va. to Fla. and Tex.; rep. from Cass Co., Ill.

12. FUIRENA Rottb. **Umbrella-grass.** Scales numerous, spirally arranged, usually hairy, with an often spreading or recurved awn or mucro, the lowest 1–3 empty; fls perfect, with 1–3 stamens and trifid style; ovary subtended by 3 bristles alternating with 3 perianth-scales ("petals"), the latter long-stipitate, with expanded blade, often awned; achenes trigonous, long-stipitate, beaked-apiculate; leafy-stemmed sedges with strongly veined sheaths, ours with a few spikelets in a terminal head subtended by 2 or 3 ± foliaceous bracts, often with some additional pedunculate axillary clusters. 30, warm reg.

1 Sheaths bladeless or nearly so; herbage essentially glabrous .... 1. *F. scirpoidea.*
1 Sheaths with well developed blade; plants ± hairy, at least below.
  2 Pet blunt but with the midnerve excurrent as a subapical-dorsal bristle .... 2. *F. simplex.*
  2 Pet without a subapical-dorsal bristle, though sometimes long-tapering at the tip.
    3 Rhizomatous perennials; anthers 0.8–2+ mm.
      4 Upper sheaths glabrous; pet obtuse to acutish .... 3. *F. breviseta.*
      4 All sheaths hairy; pet narrowly acute to acuminate .... 4. *F. squarrosa.*
    3 Tufted annuals; anthers 0.5–0.7 mm .... 5. *F. pumila.*

1. Fuirena scirpoidea Michx. Rhizomatous perennial 2–6 dm; herbage essentially glabrous; sheaths loose, oblique at the orifice but virtually bladeless; spikelets 7–10(–15) mm, in a single terminal cluster of (1)2–5 subtended by a sheathing short-bladed bract; scales 2.5–3.5 mm, with at least 5 strong median nerves convergent to form a stiff, erect, scabrid mucro less than half as long as the scale; bristles retrorsely barbellate; pet-blades ovate, acutish-apiculate; anthers 3, ca 2 mm; 2n=46. Sandy or sandy-peaty marshes, swales, and seeps, sometimes in slightly brackish sites; along the coast from N.C. to Fla., Tex., and Cuba; disjunct in s. Ill.

2. Fuirena simplex Vahl. Rhizomatous perennial 2–5(–10) dm; mid-stem sheaths usually smooth, the others hispid or hirsute; larger blades 5–20 cm × 3–7 mm, hispid-ciliate and often also hairy on the surfaces; spikelets 8–15(–20) mm, in 1–3(–5) clusters; scales 2.5–3.5 mm, 3 of the 5–7 strong raised nerves convergent and exserted to form a spreading awn usually at least ⅔ the length of the scale; bristles retrorsely barbellate, reaching or often surpassing the base of the pet-blades, these ovate, blunt or retuse, with the midrib excurrent as a retrorsely barbellate dorsal-subapical bristle; anthers 3, 0.9–1.2 mm; 2n=(30), 46. Wet sunny places; C. Amer. and W.I., n. to Ill., Mo., and Nebr. Most of our plants are var. *simplex*, as described above. The var. *aristulata* (Torr.) Kral, mostly on the s. Great Plains and entering our range in w. Mo., lacks rhizomes and is usually annual, seldom over 3 dm, with 1–3 anthers 0.5–0.6 mm.

3. Fuirena breviseta (Coville) Coville. Rhizomatous perennial to 5(–10) dm, the stems arising from axillary, cormose offshoots; lower sheaths hispid, the upper (and usually also the middle) ones glabrous; larger blades 5–15 cm × 3–10 mm, usually hispid-ciliate at least proximally, often also hairy on the surfaces, but the foliage smoother and stiffer than in no. 4; spikelets 10–15(–20) mm, in 1–4 clusters; scales 3–3.5 mm, with mostly 3 strong median nerves convergent and exserted to form a spreading-recurved awn as long as the scale; bristles smoothish, short, incurved, rarely reaching the base of the pet-blades, these ovate or oblong, broadly acute to obtuse, sometimes apiculate; anthers 3, 1–1.3 mm; 2n=46. Bogs and wet sandy places; coastal plain from se. Va. to Fla. and Tex.; Cuba.

4. Fuirena squarrosa Michx. Rhizomatous perennial to 10 dm, the stems arising from axillary, cormose offshoots; sheaths all hispid-hirsute; larger blades 8–20 cm × 4–10 mm; hispid-ciliate and scabrous or hairy on the surfaces; spikelets to 20 mm, in 1–3 clusters; scales 2.5–3.5 mm, with 3(5) strong median nerves convergent and exserted to form a spreading-recurved awn more than half as long as the scale; bristles antrorsely barbellate, reaching the base or even the tip of the pet-blades, these ovate, usually incurved-acuminate; anthers 3, ca 1 mm; 2n= 46. Moist or wet, often sandy places; L.I. to n. Fla., w. to Tenn., Ark., and Tex., mainly on the coastal plain with us.

5. Fuirena pumila (Torr.) Sprengel. Tufted annual 1–6 dm; sheaths hispid-hirsute to subglabrous; larger blades 5–12 cm × 3–5 mm, hairy or scabrous at least along the margins; spikelets 7–15 mm, in 1–3 clusters (or solitary); scales 2.5–3.5 mm, the usually 3 strong median nerves convergent and exserted to form a recurved-spreading awn nearly as long as the scale; bristles retrorsely barbellate, reaching at least to the base of the pet-blades, these ovate, slenderly incurved into an apical bristle; anther 1, 0.5–0.7 mm; 2n=46. Moist to wet, usually sandy places, sometimes weedy; Mass. to Fla. and Tex., chiefly on and near the coastal plain, and disjunct to s. Mich. and n. Ind.

13. SCLERIA Bergius. **Stone-rush.** Fls unisexual, the plant monoecious; spikelets small, the staminate few-fld, the pistillate with the lower scales empty and only the uppermost one fertile; perianth none; stamens 1–3; style trifid; achene bony or crustaceous, usually white or whitish, globose to ovoid or obscurely trigonous, obtuse or apiculate, in our spp. subtended by a simple or variously ornamented disk (hypogynium); sedges with trigonous culms and solitary or few, small, compact cymes. 200, warm reg.

1 Achenes smooth and shining.
  2 Hypogynium white, papillose-crustaceous, achene without tubercles.
    3 Main blades 4–8 mm wide; achenes (hypogynium included) 2.5–3.5 mm .... 1. *S. triglomerata.*
    3 Main blades 1–3 mm wide; achenes 1.5–2 mm .... 2. *S. minor.*
  2 Hypogynium brown, smooth, shortly truncate-connate, with ca 9 rounded tubercles between it and
    the body of the achene .... 3. *S. oligantha.*
1 Achenes rough or reticulate.
  4 Hypogynium with 3 or more conspicuous rounded tubercles; achene-body rough-papillate.
    5 Tubercles 3 (each of them may be 2-lobed) .... 4. *S. ciliata.*
    5 Tubercles 6 .... 5. *S. pauciflora.*
  4 Hypogynium without rounded tubercles; achene-body pitted or ridged.
    6 Hypogynium with 3 oblong lobes appressed to the achene and resembling a calyx .... 6. *S. reticularis.*
    6 Hypogynium flat, very low, unlobed .... 7. *S. verticillata.*

1. Scleria triglomerata Michx. Whipgrass. Culms to 1 m, from hard, knotty rhizomes; main blades 4–8 mm wide, often hairy, abruptly attenuate; cymes 1–3, the lower pedunculate; bracts foliaceous, the lowest erect, 5–15 cm; staminate scales lanceolate, acuminate; anthers 2.5–4 mm; pistillate scales ovate, the midrib prolonged into a short awn; achenes white, rarely drab or gray, 2.5–3.5 mm, apiculate, the body as long as or a little longer than thick, blunt or apiculate; hypogynium white, papillose-crustaceous. Moist or dry sandy soil and pine-barrens; Mass. to Wis., s. to Fla., Tex., Puerto Rico, and Mex. (*S. flaccida; S. nitida.*)

2. Scleria minor W. Stone. Very slender, to 8 dm, from long rhizomes; main blades 1–3 mm wide, long-attenuate, usually wholly glabrous; cymes 1, occasionally 2 or 3, the lower short-peduncled; bracts foliaceous, the lowest erect, 1–2 mm wide; scales as in no. 1; achenes white, apiculate, 1.5–2 mm, the body about as thick as long. Damp sandy or peaty soil on the coastal plain; L.I. to S.C. Perhaps better treated as *S. triglomerata* var. *minor* Britton. (*S. triglomerata* var. *gracilis.*)

3. Scleria oligantha Michx. Slender, 3–8 dm, cespitose from knotty rhizomes; main blades 3–5 mm wide, attenuate, often sparsely villous; cymes often solitary, the lower ones when present long-peduncled, the upper subtended and partly sheathed by foliaceous bracts to 10 cm × 5 mm; pistillate scales ovate, acuminate or short-aristate; achene bright white, ovoid, 3–3.5 mm, apiculate, the short, broad, smooth, brown hypogynium separated from the body by 8 or 9 rounded white tubercles. Wet meadows and woods; C. Amer., Mex., and Puerto Rico, n. to Tex., Mo., Ind., Va., and reputedly N.J.

4. Scleria ciliata Michx. Erect from knotty rhizomes, 3–8 dm; lvs glabrous or hairy, long-attenuate, the larger 2–7 mm wide; cymes 2–3 and approximate or solitary, short-peduncled or sessile, 1–3 cm; pistillate scales ovate, acuminate or short-awned, glabrous or hairy, the prominent midrib projecting as a wing-like keel toward the base; achene globose, white, roughly papillose, (1.5–)2–3 mm thick, minutely apiculate; hypogynium with 3 prominent, transversely ellipsoid tubercles, each often ± 2-lobed. Damp sandy soil and pine-barrens; se. Va. to Fla., Tex., and Cuba, n. in the interior to Mo. (*S. elliottii.*)

5. Scleria pauciflora Muhl. Carolina-whipgrass. Erect from knotty rhizomes, 2–5 dm; lvs attenuate, 1–3 mm wide, glabrous to strongly hairy; cyme usually solitary and sessile, subtended by a foliaceous bract 2–5 cm; a lower, short-peduncled cyme sometimes also present; anthers 2–2.5 mm; scales ovate, acuminate to aristate; achene white, subglobose, 1–2(–2.5) mm, roughly verrucose, minutely apiculate; hypogynium with 6 rounded tubercles. Damp or dry, sandy or sterile soil; N.H. to Mich., and Mo., s. to Fla., Tex., W.I., and Mex. Perhaps better subordinated to *S. ciliata.*

6. Scleria reticularis Michx. Annual, or perennial from short, slender rhizomes, very slender, to 8 dm; main blades 2–4 mm wide, sometimes hairy; cymes (1)2–4, the lateral remote; pistillate scales lanceolate to ovate, acute or acuminate; achene subglobose, white or sordid-gray, 1.5–2 mm thick, minutely apiculate, marked with narrow ridges enclosing shallow, irregularly polygonal pits; hypogynium double, the inner 3-lobed, resembling a calyx, each lobe oblong and closely appressed to the glabrous or minutely hairy achene. Damp sandy soil and pine-barrens; tropical Amer. and s. U.S., n. mainly along the coast to Mass., and inland to sw. Mich., e. Wis., and Mo. (*S. r.* var. *pubescens; S. muhlenbergii; S. setacea,* misapplied.)

7. Scleria verticillata Muhl. Annual; very slender, 2–6 dm, often tufted; sheaths usually long-villous; main blades 1 mm wide; infl of 2–8 short, capitate cymes each 2–4 mm, sessile, subtended by inconspicuous bracts 4–7 mm, forming an erect interrupted spike; anthers 1 mm; scales glabrous, lanceolate, keeled; achene white, subglobose, 1 mm thick, conspicuously apiculate, verrucose or transversely ridged, constricted below to a short smooth hypogynium. Wet sandy soil; widespread in trop. Amer., n. to Conn. and Minn.

14. CAREX L. Sedge. Fls unisexual, without perianth, borne in spikes, each fl solitary in the axil of a scale; spikes unisexual or bisexual, when bisexual the staminate fls usually either terminal (spikes *androgynous*) or basal (spikes *gynaecandrous*); stamens (2)3; pistillate fls individually enclosed by a sac-like scale (the *perigynium*), from the mouth of which the style or stigmas protrude, as well as being subtended by the open pistillate scale; stigmas 2 or 3, the achene accordingly lenticular or trigonous; grass-like perennial herbs with 3-ranked lvs, closed sheaths, and triangular or terete, mostly solid stems, the lower sheaths bladeless (plants *aphyllopodic*) or with ± well developed blade (plants *phyllopodic*); spikes solitary and terminal, or much more often racemosely arranged in a terminal infl that is rarely again branched, sometimes some of them well removed from the others and axillary to lvs near the base. 1500+, cosmop, especially in moist North Temperate and Arctic regions.

Some spp. of subg. *Primocarex* have a definite *rachilla* alongside the achene within the perigynium, showing that each pistillate fl represents a branch of an infl that has been reduced to a single fl. The perigynium is a highly modified bract on the adaxial side of this short, uniflorous branch. The bract wraps around the pistillate fl, and its margins are connate so that the fl is enclosed in a sac with a minute apical opening. In two of the four subgenera (*Primocarex* and *Vignea*) the line of fusion of the bract-margins can be seen as a suture or imperfection toward the tip of the dorsal side of perigynium. The side of the perigynium next to the pistillate scale is called the dorsal side, and the side next to the axis of the spike is called the ventral side. Many of the characteristically tristigmatic spp. of *Carex* occasionally have a few distigmatic fls intermingled; such specimens should be keyed as tristigmatic.

Our 3 subgenera may be characterized as follows:

1 Spike solitary, terminating the stem; perigynium often containing a vestigial rachilla as well as the
    achene. (Spp. 1–9.) .... *Primocarex.*
1 Spikes usually more than one (but solitary in a few species); no rachilla.
  2 Stigmas 2 and achenes lenticular, in nearly all spp; spikes sessile and relatively short (seldom over
    1.5 cm), not elongate and cylindric; perigynium usually with a ± evident dorsal suture, at least
    distally. (Spp. 10–87.) .... *Vignea.*
  2 Stigmas 2 or 3, the achene accordingly lenticular or trigonous; if stigmas 2, then the spikes elongate
    and cylindric (at least the larger ones seldom under 1.5 cm), or at least some of them evidently
    pedunculate, or both; perigynium without a dorsal suture. (Spp. 88–230.) .... *Carex.*

Artificial Key to the Species

1 Spike 1 on each stem .... Group I.
1 Spikes 2 or more (depauperate plants may bear only a single spike).
  2 Achenes lenticular or planoconvex; stigmas 2 .... Group II.
  2 Achenes trigonous or nearly round in section; stigmas 3.
    3 Style persistent on the achene at maturity, becoming bony or cartilaginous .... Group III.
    3 Style withering and usually deciduous as the achene matures.
      4 Perigynium beaked, its teeth at the orifice well developed, stiff, sharp, slender or stout. (For
        convenience some spp. with conspicuous teeth, though scarcely bony or sharp, are included
        here as well as in the following leg of the key.) .... Group IV.
      4 Perigynium beaked or beakless, its orifice entire or minutely toothed or with soft and blunt
        teeth. (For convenience some spp. with stiff and sharp but inconspicuous teeth are included
        here as well as in the preceding leg of the key.)
        5 Bracts of the pistillate spikes (excluding the pistillate spikes arising from the base) bladeless,
          consisting of sheath only, or with small scale-like blade .... Group V.
        5 Bracts of the pistillate spikes (or at least the lowest nonbasal pistillate spike) with well
          developed blade.
          6 Terminal spike bearing some perigynia .... Group VI.
          6 Terminal spike entirely staminate.
            7 Perigynia pubescent, at least around the base of the beak .... Group VII.
            7 Perigynia glabrous.
              8 Bract of the lowest pistillate spike with a well developed sheath .... Group VIII.
              8 Bract of the lowest pistillate spike sheathless or very nearly so .... Group IX.

Group I

1 Achenes lenticular or planoconvex; stigmas 2.
  2 Plants loosely rhizomatous .... 2. *Dioicae.*
  2 Plants ± densely cespitose, without long rhizomes.

3 Spike bisexual, androgynous; rachilla present . . . . 4. *Capitatae.*
3 Spike unisexual, or bisexual and gynaecandrous; rachilla wanting. (*C. exilis.*) . . . . 20. *Stellulatae.*
1 Achenes trigonous, or round in cross-section; stigmas 3.
  4 Spikes bisexual.
    5 Spike gynaecandrous . . . . 54. *Squarrosae.*
    5 Spike androgynous.
      6 Perigynia linear or subulate, spreading or reflexed . . . . 3. *Orthocerates.*
      6 Perigynia elliptic or oblong to globose, ascending.
        7 Lowest pistillate scale foliaceous, much surpassing the strongly beaked perigynium . . . . . . 25. *Phyllostachyae.*
        7 Lowest pistillate scales smaller and not foliaceous, varying from shorter to a little longer
          than the beaked or beakless perigynium.
          8 Perigynia white; culm-lf solitary, undeveloped at anthesis, becoming 2–5 cm wide
            . . . . See *Cymophyllus fraseri.*
          8 Perigynia green to brown; lvs numerous, up to 3 mm wide.
            9 Perigynia 5–6 mm, scarcely beaked, filled by the 4–5 mm achene . . . . 6. *Firmiculmes.*
            9 Perigynia 3–4.5 (–5) mm, beaked or beakless.
              10 Achene filling the evidently short-beaked perigynium; rachilla well developed.
                11 Plants loosely rhizomatous . . . . 1. *Obtusatae.*
                11 Plants densely tufted, not rhizomatous . . . . 5. *Filifoliae.*
              10 Achene far from filling the beakless perigynium; rachilla wanting . . . . 7. *Polytrichoidae.*
  4 Spikes unisexual.
    12 Perigynia planoconvex, with lacerate winged margins . . . . 10. *Macrocephalae.*
    12 Perigynia not wing-margined.
      13 Spikes 20–40 mm.
        14 Perigynium 4–5 mm, multinerved, surpassed by the acute or aristate scale . . . . 8. *Pictae.*
        14 Perigynium 2.5–3 mm, virtually nerveless, about equaling the obtuse scale . . . . 9. *Scirpinae.*
      13 Spikes 4–12 mm. (*C. umbellata.*) . . . . 26. *Montanae.*

Group II

1 Spikes all much alike, sessile, relatively short, seldom over 1.5 cm.
  2 Stems arising singly or few together from ± elongate rhizomes, stolons, or decumbent stems.
    3 Stems becoming decumbent and sending up new stems from old axils; plants of sphagnum-
      bogs . . . . 14. *Chordorrhizae.*
    3 Stems arising from elongate rhizomes; mostly not of sphagnum-bogs.
      4 Spikes closely crowded into a dense infl.
        5 Plants phyllopodic to somewhat aphyllopodic, the lvs all clustered at or near the base;
          perigynia usually nerveless or nearly so ventrally, often sharp-edged but not thin-
          margined . . . . 11. *Divisae.*
        5 Plants strongly aphyllopodic, the lvs not forming a basal cluster; perigynia ± evidently
          nerved ventrally, the body tending to be thin-margined distally.
          6 Sheaths firm, ventrally partly greenish or greenish-striate; perigynia 2.3–4 mm. . . . . 12. *Intermediae.*
          6 Sheaths distinctly thin-hyaline ventrally, not greenish; perigynia 4–6.2 mm . . . . 13. *Arenariae.*
      4 Spikes ± remote. (*C. disperma* and *C. socialis.*) . . . . 15. *Bracteosae.*
  2 Stems closely clustered, the rhizomes none or very short.
    7 Plants androgynous, with staminate fls at the top of some or all spikes.
      8 Infl simple, the spikes single at each node, seldom more than 10 in the infl.
        9 Lateral spikes with a few staminate fls at the tip; perigynia usually green or greenish even
          at maturity . . . . 15. *Bracteosae.*
        9 Lateral spikes wholly staminate or pistillate; perigynia usually brown at maturity . . . . 20. *Stellulatae.*
      8 Infl compound, at least the lowest node bearing 2 or more spikes.
        10 Perigynia very firm and thick-walled, not spongy, ± abruptly beaked, dark in color, ranging
          from medium or deep brown to deep olive-green or nearly black . . . . 18. *Paniculatae.*
        10 Perigynia thin-walled, or often ± spongy-thickened toward the base, gradually or some-
          times more abruptly beaked, pale in color, ranging from greenish or stramineous to
          light brown or golden-brown.
          11 Pistillate scales with a definite awn 1–5 mm; perigynia rounded toward the only slightly
            or not at all spongy-thickened base . . . . 16. *Multiflorae.*
          11 Pistillate scales awnless, or if with a definite awn then the perigynium with broad,
            subtruncate, notably spongy-thickened base . . . . 17. *Vulpinae.*
    7 Plants gynaecandrous, the staminate fls produced at the base of some or all spikes.
      12 Perigynia plump, at most only sharp-edged, never thin-winged, distended to the margins by
        the achene.
        13 Perigynium-body nearly or quite filled by the achene; perigynia ± ascending, or merely
          ascending-spreading . . . . 19. *Heleonastes.*
        13 Perigynium-body spongy at base, the achene occupying only the upper half or two-thirds.
          14 Perigynia, or at least the lower ones in the spike, ± widely spreading (or even
           reflexed), often smaller and/or broader and/or shorter-beaked than in the next
           group . . . . 20. *Stellulatae.*
          14 Perigynia appressed-ascending, relatively long (4–5.5 mm), narrow (3–5 times as
           long as wide) and long-beaked (1–2 mm) . . . . 21. *Deweyanae.*

12 Perigynia with thin-winged margins, at least along the upper part of the body and the lower
 part of the beak, and often extending to the base and apex .... 22. *Ovales.*
1 Spikes differentiated, the lowest and uppermost distinctly unlike; lateral spikes sessile or peduncled,
 if sessile then ± elongate.
 15 Lower pistillate spikes spreading or drooping on slender peduncles .... 48. *Cryptocarpae.*
 15 Lower pistillate spikes erect or strongly ascending, sessile or short-peduncled.
  16 Perigynia turgid, obovoid, elliptic in cross-section .... 27. *Bicolores.*
  16 Perigynia flat to planoconvex or biconvex.
   17 Achene jointed with the style; perigynia not lustrous .... 49. *Acutae.*
   17 Achene continuous with the persistent style; perigynia lustrous. (*C. saxatilis.*) .... 55. *Vesicariae.*

GROUP III

1 Perigynia slenderly subulate, 8–15 × 1.5–3 mm, gradually tapering into the beak.
 2 Perigynium-teeth reflexed .... 50. *Collinsiae.*
 2 Perigynium-teeth erect to merely spreading .... 51. *Folliculatae.*
1 Perigynia broader in shape and usually also in measurement, the body lance-ovoid to broadly ovoid
 or ellipsoid or broadly obconic.
 3 Pistillate spikes (at least the lower ones) nodding or drooping on long, slender peduncles .... 52. *Pseudocypereae.*
 3 Pistillate spikes ascending to erect, sessile or pedunculate.
  4 Perigynia firm-walled, generally scarcely or not at all inflated .... 53. *Paludosae.*
  4 Perigynia thin-walled, ± inflated.
   5 Body of the perigynium obconic or obconic-obovoid, very broadly rounded or ± truncate (or
    even indented) above and very abruptly beaked .... 54. *Squarrosae.*
   5 Body of the perigynium ovoid to ellipsoid or globose, tapering or rounded above, not so
    abruptly beaked.
    6 Bract of the lowest pistillate spike not or scarcely sheathing, or if definitely sheathing then
     the perigynia less than 10 mm .... 55. *Vesicariae.*
    6 Bract of the lowest pistillate spike definitely sheathing at base; perigynia 10–20 mm .... 56. *Lupulinae.*

GROUP IV

1 Perigynium planoconvex, with lacerate-winged margins .... 10. *Macrocephalae.*
1 Perigynium not wing-margined.
 2 Perigynia 10–20 mm .... 56. *Lupulinae.*
 2 Perigynia not more than about 8 mm.
  3 Some of the pistillate spikes on nearly basal peduncles. (*C. rossii.*) .... 26. *Montanae.*
  3 None of the spikes on nearly basal peduncles.
   4 Pistillate spikes stout, erect, densely fld, 6–15 mm thick, sessile or short-peduncled .... 41. *Hirtae.*
   4 Pistillate spikes either loosely fld, or slender and elongate, or spreading or drooping (or all of
    these).
    5 Pistillate scales with an evident awn-tip; terminal spike pistillate for about the upper third,
     staminate below. (*C. davisii.*) .... 35. *Gracillimae.*
    5 Pistillate scales no more than cuspidate; terminal spike wholly staminate, or occasionally
     with a few distal or basal perigynia.
     6 Plants brownish at base; pistillate scales narrow, acute or acuminate, pale, largely hyaline-
      scarious, surpassing the body of the perigynium; beak 1.7–4 mm .... 38. *Longirostres.*
     6 Plants purplish at base; pistillate scales otherwise; beak often shorter .... 36. *Sylvaticae.*

GROUP V

1 Staminate spike long-peduncled; lf-blades 2 mm wide or more.
 2 Lf-blades 2–4 mm wide .... 28. *Digitatae.*
 2 Lf-blades 10–25 mm wide. (*C. plantaginea.*) .... 32. *Laxiflorae.*
1 Staminate spike sessile or nearly so; lf-blades various, often under 2 mm wide.
 3 Perigynia 2.5–4.4 mm, prominently beaked.
  4 Plants densely tufted, without creeping rhizomes. (*C. umbellata.*) .... 26. *Montanae.*
  4 Plants with long creeping rhizomes, not densely tufted .... 23. *Lamprochlaenae.*
 3 Perigynia 1.5–2 mm, beakless .... 30. *Albae.*

GROUP VI

1 Perigynia of the terminal spike borne at its base.
 2 Perigynia subtended by long lf-like scales .... 25. *Phyllostachyae.*
 2 Perigynia little or not exceeded by the scales. (*C. assiniboinensis.*) .... 36. *Sylvaticae.*
1 Perigynia of the terminal spike borne at its summit or middle.

3 Perigynium distinctly beaked.
  4 Lateral spikes elongate-cylindric, spreading or drooping on slender peduncles .... 36. *Sylvaticae.*
  4 Lateral spikes short-cylindric, erect, most or all of them sessile or nearly so .... 39. *Extensae.*
3 Perigynium beakless, or with a short, obscure, or poorly differentiated beak.
  5 Pistillate spikes slender, spreading or drooping, 2–6 cm, loosely fld with perigynia in few rows
      and often exposing the axis; perigynia often over 4 mm .... 35. *Gracillimae.*
  5 Pistillate spikes short-cylindric, 1–2(–4) cm; perigynia densely and compactly arranged, 1.8–
    4 mm.
    6 Sheaths or lower lf-surfaces or both pubescent .... 40. *Virescentes.*
    6 Sheaths and blades glabrous.
      7 Perigynia nearly or quite as wide as long, firm, diverging at nearly right angles .... 43. *Shortianae.*
      7 Perigynia obviously longer than wide, thin, ascending.
        8 Scales conspicuously surpassing the perigynia; spikes spreading or drooping on slender
          peduncles. (*C. paupercula.*) .... 46. *Limosae.*
        8 Scales shorter than the perigynia, or the spikes sessile.
          9 Terminal spike about half pistillate .... 47. *Atratae.*
          9 Terminal spike with 1 or very few perigynia. (*C. livida.*) .... 31. *Paniceae.*

GROUP VII

1 Perigynia with 20+ sharp fine nerves extending nearly or quite to the tip .... 32. *Laxiflorae.*
1 Perigynia with fewer nerves or nerveless.
  2 Perigynium beakless or minutely apiculate .... 45. *Scitae.*
  2 Perigynium beaked, the body obovoid in general outline.
    3 Perigynium 2-ribbed only, the intermediate nerves none, or very obscure, or developed only at
      the base.
      4 Stems and lvs glabrous.
        5 Beak of the perigynium 0.5 mm or more; style-base not expanded .... 26. *Montanae.*
        5 Beak of the perigynium shorter, conic; style-base expanded into a ring 0.5 mm in diam-
          eter .... 24. *Praecoces.*
      4 Stem and lvs pubescent .... 29. *Triquetrae.*
    3 Perigynium 2-ribbed and conspicuously several-nerved.
      6 Pistillate spikes slender, peduncled, spreading or drooping.
        7 Perigynium lanceolate or narrowly ovoid, 4.7–8 mm; plants tufted .... 36. *Sylvaticae.*
        7 Perigynium obovoid, 3.2–4.4 mm; plants rhizomatous .... 42. *Anomalae.*
      6 Pistillate spikes short-cylindric, sessile or nearly so, erect. (*C. vestita.*) .... 41. *Hirtae.*

GROUP VIII

1 Perigynium 2-ribbed, otherwise nerveless or with usually less than 10 faint or obscure nerves.
  2 Perigynium beakless, the orifice entire .... 31. *Paniceae.*
  2 Perigynium beaked.
    3 Pistillate spikes erect or nearly so; perigynium abruptly narrowed to the beak.
      4 Rhizomatous; perigynia 4.3–6.7 mm, including the often well developed beak .... 31. *Paniceae.*
      4 Tufted; perigynia 2.4–4.1 mm, the beak very short .... 32. *Laxiflorae.*
    3 Pistillate spikes spreading or drooping, mostly on slender peduncles; perigynia various.
      5 Pistillate spikes 1.5–6 cm, or if only 1 cm, then the perigynia 3.5–5.8 mm.
        6 Teeth of the perigynium-beak minute, not soft and scarious .... 36. *Sylvaticae.*
        6 Teeth of the perigynium-beak distinct under a hand-lens, soft and scarious .... 38. *Longirostres.*
      5 Pistillate spikes 0.5–1.5 cm; perigynia 2.4–3.2 mm .... 37. *Capillares.*
1 Perigynium 2-ribbed and also conspicuously nerved, the nerves numerous, normally more than 10,
  extending the length of the perigynium.
  7 Pistillate spikes sessile or nearly so, short-cylindric to subglobose, very densely fld; perigynia
    spreading or squarrose, conspicuously beaked; bracts of the pistillate spikes with short or no
    sheath .... 39. *Extensae.*
  7 Pistillate spikes short-cylindric to elongate, at least the lower peduncled, densely or loosely fld, the
    perigynia mostly ascending, never squarrose or reflexed; bracts of the pistillate spikes with well
    developed sheath.
    8 Perigynia obovoid to fusiform or lanceolate, distinctly narrowed to the base.
      9 Pistillate spikes short-cylindric, usually erect, in a few spp. drooping and then with sharply
        trigonous perigynia.
        10 Perigynia with fine elevated nerves; awns of the pistillate scales smooth or none .... 32. *Laxiflorae.*
        10 Perigynia with very numerous impressed nerves, with the appearance of a longitudinally
          wrinkled surface; awns of the pistillate scales rough .... 34. *Oligocarpae.*
      9 Pistillate spikes elongate, at least the lower widely spreading or drooping; perigynia not sharply
        trigonous .... 36. *Sylvaticae.*
    8 Perigynia ovoid or ellipsoid or rarely obovoid, rounded at base.
      11 Perigynium beakless.
        12 Pistillate spikes short-cylindric, 1–2(–4) cm, erect or nearly so on short peduncles .... 34. *Oligocarpae.*

GROUP IX

1. OBTUSATAE

2. DIOICAE

3. ORTHOCERATES

4. CAPITATAE

5. FILIFOLIAE

6. FIRMICULMES

7. POLYTRICHOIDAE

One sp. in our range . . . . 7. *C. leptalea.*

8. PICTAE

One sp. in our range . . . . 8. *C. picta.*

9. SCIRPINAE

One sp. in our range . . . . 9. *C. scirpoidea.*

10. MACROCEPHALAE

One sp. in our range . . . . .10. *C. kobomugi.*

11. DIVISAE

1 Plants normally monoecious, some or all of the spikes androgynous.
  2 Rhizome coarse, with blackish scales; plants relatively coarse, (1–)3–7 dm, the lvs 1–3 mm wide;
    stems scabrous or scaberulous on the angles above.
    3 Beak of the perigynium short, a fifth to a third as long as the body; seacoast . . . . 11. *C. divisa.*
    3 Beak prominent, a third to half as long as the body; inland, mainly western . . . . 12. *C. praegracilis.*
  2 Rhizome slender, with brownish scales; plants slender, 0.5–2 dm, the lvs only 0.3–1.5 mm wide;
    stems smooth . . . . 13. *C. stenophylla.*
1 Plants normally dioecious.
  4 Stems scaberulous on the angles above; rhizomes with blackish scales . . . . 12. *C. praegracilis.*
  4 Stems smooth; rhizomes with brownish scales . . . . 14. *C. douglasii.*

12. INTERMEDIAE

One sp. in our range . . . . 15. *C. sartwellii.*

13. ARENARIAE

1 Spikes 4–12; scales shorter than the perigynia; inland . . . . 16. *C. siccata.*
1 Spikes numerous; scales equaling or mostly surpassing the perigynia; seacoast . . . . 17. *C. arenaria.*

14. CHORDORRHIZAE

One sp. in our range . . . . 18. *C. chordorrhiza.*

15. BRACTEOSAE (PHAESTOGLOCHIN)

1 Sheaths loose, septate-nodulose and usually mottled or striped green and white on the dorsal side.
  2 Ripe perigynium yellowish-brown; scales with a slender tip surpassing the base of the beak of the
    subtended perigynium . . . . 30. *C. gravida.*
  2 Ripe perigynium green; scales barely if at all reaching the base of the beak of the subtended
    perigynium . . . . 31. *C. sparganioides.*
1 Sheaths tight, not septate-nodulose on the dorsal surface, not mottled or striped.
  3 Infl capitately ovoid, the densely aggregated spikes hardly distinguishable except by the projecting
    setaceous bracts.
    4 Perigynia two-fifths to three-fifths as wide as long, widest just below the middle, rounded or
      broadly cuneate at base . . . . 19. *C. cephalophora.*
    4 Perigynia three-fifths to three-fourths as wide as long, widest near the broadly rounded or truncate
      base . . . . 20. *C. leavenworthii.*

3 Infl spicate, the spikes plainly distinguishable and often separated by an exposed internode of the axis.
  5 Pistillate scales strongly tinged with brownish or reddish-purple.
    6 Main lvs 3–6 mm wide; beak of the perigynium two-thirds to fully as long as the body; widespread native sp. .... 38. *C. alopecoidea.*
    6 Main lvs 2–3 mm wide; beak of the perigynium up to ca half as long as the body; locally intr., weedy European spp.
      7 Roots and often the basal sheaths and base of stems purplish-tinged; ligule longer than wide, acute; perigynia corky-thickened at base .... 21. *C. spicata.*
      7 Roots and basal sheaths brown to blackish, not purplish-tinged; ligule ca as wide as or wider than long, blunt; perigynia not corky-thickened.
        8 Perigynia widely spreading, even before maturity, over half as wide as long, rounded at base, rather abruptly beaked .... 22. *C. muricata.*
        8 Perigynia appressed-ascending or the lowermost somewhat spreading at maturity, less than half as wide as long, almost equally tapering at both ends .... 23. *C. divulsa.*
  5 Pistillate scales greenish or greenish-hyaline or white-hyaline, or becoming stramineous or pale brown when dried; native spp.
    9 Perigynia densely white-punctate; plants loosely rhizomatous .... 32. *C. disperma.*
    9 Perigynia not white-punctate; plants tufted, except in no. 25.
      10 Perigynium conspicuously spongy-thickened at base, nerveless or finely striate or nerved on the spongy base only.
        11 Beak of the perigynium smooth-margined; scales pointed, early deciduous .... 24. *C. retroflexa.*
        11 Beak of the perigynium serrulate-margined; scales obtuse, persistent.
          12 Plants colonial by well developed creeping rhizomes; perigynia mostly 4–5 times as long as wide; local in s. Ill. and southward .... 25. *C. socialis.*
          12 Plants densely tufted, not rhizomatous; perigynia 2–3 times as long as wide.
            13 Stigmas straight or only slightly twisted; achene occupying only the upper half of the basally subtruncate perigynium; perigynia reflexed when ripe .... 26. *C. radiata.*
            13 Stigmas mostly once or twice coiled, or if occasionally not so, then the plants otherwise not as above.
              14 Broadest lvs 0.9–1.5 mm wide; base of fertile stems 0.7–1.3 mm thick; Appalachian .... 27. *C. appalachica.*
              14 Broadest lvs 1.8–2.6 mm wide; base of fertile stems 1.5–2.2 mm thick, widespread .... 28. *C. rosea.*
      10 Perigynium not spongy-thickened, prominently nerved dorsally .... 29. *C. muhlenbergii.*

16. MULTIFLORAE

One sp. in our range .... 33. *C. vulpinoidea.*

17. VULPINAE

1 Perigynia lanceolate to lance-ovate, broadest near the truncate or retuse base, the beak often equaling or longer than the body.
  2 Perigynium with the spongy base continuous with the body.
    3 Sheaths prolonged beyond the base of the blade, ventrally thin, fragile, and cross-corrugated ...... 34. *C. stipata.*
    3 Sheaths with thick-margined, concave mouth, not prolonged, not corrugated .... 35. *C. laevivaginata.*
  2 Perigynium dilated at base into a disk-shaped spongy structure notably wider than the short, subconic body .... 36. *C. crus-corvi.*
1 Perigynia distinctly ovate, rounded at base, the beak no longer than the body.
  4 Sheaths ventrally cross-corrugated; perigynia green or becoming ± stramineous, strongly few-nerved dorsally .... 37. *C. conjuncta.*
  4 Sheaths not (or scarcely) corrugated; perigynia becoming largely or wholly golden-brown, inconspicuously few-nerved dorsally .... 38. *C. alopecoidea.*

18. PANICULATAE

1 Perigynia obovoid; sheaths concave at the mouth, not prolonged .... 39. *C. decomposita.*
1 Perigynia lance-ovoid to ovoid; sheaths distinctly prolonged beyond the base of the blade.
  2 Dorsal surface of the perigynium with a median thin, often depressed and pale strip bordered by a pair of veins and extending to the base; sheaths ventrally pale and red-dotted .... 40. *C. diandra.*
  2 Dorsal surface of the perigynium without such a strip, the principal veins divergent well above the base; sheaths strongly coppery toward the mouth ventrally .... 41. *C. prairea.*

19. Heleonastes

1 Spikes 2–4, closely aggregated and overlapping in a short, ovoid or oblong head.
  2 Pistillate scales white-hyaline, except the greenish midrib; perigynia beakless .... 42. *C. tenuiflora.*
  2 Pistillate scales strongly brown-tinged; perigynia distinctly short-beaked.
    3 Lateral spikes gynaecandrous; inland .... 43. *C. heleonastes.*
    3 Lateral spikes pistillate; maritime .... 44. *C. glareosa.*
1 Spikes (1–)2–9, at least the lower ones generally separate and not overlapping.
  4 Scales reddish-brown, equaling and ± concealing the perigynia; maritime .... 45. *C. mackenziei.*
  4 Scales paler, mostly shorter than the distally exposed perigynia; not maritime.
    5 Spikes 1–3, usually 2; perigynia 1–5, 2.5–4 mm .... 46. *C. trisperma.*
    5 Spikes 4–9; perigynia 5–30, 1.7–2.5 mm.
      6 Perigynia mostly 5–10(–15), ventrally nerveless or obscurely nerved, loosely somewhat spread-
        ing, the beak-apiculations interrupting the outline of the spike .... 47. *C. brunnescens.*
      6 Perigynia (10–)15–30, evidently nerved ventrally, more appressed, the beak-apiculations not
        interrupting the outline of the spike .... 48. *C. canescens.*

20. Stellulatae

1 Spike usually solitary; lvs involute; anthers large, 2–3.6 mm .... 49. *C. exilis.*
1 Spikes 2–many; lvs flat or plicate; anthers smaller, 0.6–2.2 mm.
  2 Spikes 7–15, crowded into an elongate head, many-fld, with (20–)25–40 perigynia .... 50. *C. arcta.*
  2 Spikes 2–8, often less crowded and/or with fewer fls.
    3 Plants subdioecious; spikes unisexual or nearly so .... 51. *C. sterilis.*
    3 Plants monoecious, at least the terminal spike gynaecandrous.
      4 Beak of the perigynium smooth-margined .... 57. *C. seorsa.*
      4 Beak sparsely to densely serrulate on the margins. (Spp. 51–55 ± confluent.)
        5 Perigynia mostly 2–3 mm wide and 1–1.7 times as long as wide .... 55. *C. atlantica.*
        5 Perigynia mostly 1–2 mm wide, often more than 1.7 times as long as wide.
          6 Lvs relatively broad, the widest ones 2.8–5 mm.
            7 Infls 1.5–3 cm; lowest 2 spikes only 1–10 mm distant .... 52. *C. wiegandii.*
            7 Infls 3–8.5 cm; lower spikes 1–4 cm distant .... 53. *C. ruthii.*
          6 Lvs narrower, the widest ones not over 2.7 mm wide.
            8 Perigynia relatively long, narrow, and long-beaked, mostly 2.8–3.5(–4) mm, 1.8–3.2
              times as long as wide, the beak half as long to nearly as long as the body .... 56. *C. echinata.*
            8 Perigynia shorter, or relatively wider, and with a shorter beak up to half as long as
              the body.
              9 Perigynia nerveless (seldom few-nerved) ventrally, tending to be somewhat convexly
                tapered from widest point to beak, forming a shoulder; beak conspicuously
                setulose-serrulate .... 54. *C. interior.*
              9 Perigynia mostly several-nerved ventrally, ± cuneate or even concavely tapered
                from widest point to beak; beak more sparsely serrulate .... 55. *C. atlantica.*

21. Deweyanae

1 Perigynia ca 0.8–1.2 mm wide and 4–5 times as long as wide, strongly nerved on both sides, or only
  weakly so ventrally .... 58. *C. bromoides.*
1 Perigynia ca 1.3–1.6 mm wide and 3–4 times as long as wide, faintly nerved or nerveless on both
  sides .... 59. *C. deweyana.*

22. Ovales

(Narrowly defined spp., some ± confluent)

1 Bracts leaflike and many times longer than the infl .... 60. *C. synchnocephala.*
1 Bracts much shorter and ± setaceous, seldom as much as twice as long as the infl.
  2 Scales nearly or fully as wide and long as the perigynia and ± concealing them.
    3 Perigynia evidently slender-beaked, at least the distal 0.5+ mm of the beak generally subterete,
      marginless, and smooth or nearly so.
      4 Perigynia slender, mostly 2.6–5 times as long as wide; native spp.
        5 Infl loose or moniliform, ± nodding; widespread northern sp. .... 61. *C. praticola.*
        5 Infl congested and erect; local in Me. .... 62. *C. oronensis.*
      4 Perigynia broader, mostly 2–2.5 times as long as wide; infl congested; intr., weedy sp. .... 63. *C. leporina.*
    3 Perigynia with a flattened, broadly margined and serrulate, often ill-defined beak, this sometimes
      with a minute subterete tip less than 0.5 mm.
      6 Infl ± elongate and flexuous or moniliform, not stiff.

7 Ventral side of the perigynium nerveless or obscurely few-nerved .... 64. *C. foenea.*
7 Ventral side of the perigynium strongly 5–8-nerved .... 65. *C. argyrantha.*
6 Infl stiff and congested, the spikes ± aggregated or approximate.
  8 Lateral spikes ovoid-elliptic; perigynia closely appressed; bracts not prominent, the lowest
    one never as long as the head; western, barely reaching our range .... 66. *C. xerantica.*
  8 Lateral spikes suborbicular; perigynia looser; lowest bract prominent, often equaling or
    surpassing the head; widespread sp. .... 67. *C. adusta.*
2 Scales evidently shorter and/or narrower than the perigynia, largely exposing at least the distal
  margins as well as the beaks of the perigynia.
9 Achenes narrow, mostly only 0.5–0.8 mm wide; perigynia often but not always more than 2.5
  times as long as wide; body of the perigynium never obovate.
10 Perigynium 6.5–10 mm .... 68. *C. muskingumensis.*
10 Perigynium 2.4–5.5 mm.
  11 Perigynium 2.4–3.9 × 1.1–1.5 mm, up to 3 times as long as wide.
    12 Perigynium with stiffly spreading to recurved beak, the body often not winged to
      the base .... 73. *C. cristatella.*
    12 Perigynium with stiffly ascending beak, the body winged to the base .... 74. *C. bebbii.*
  11 Perigynium either at least 4 mm long, or more than 3 times as long as wide, or both.
    13 Principal lvs 1–3 mm wide.
      14 Perigynia 1.5–2.5 times as long as wide; sheaths ventrally green-veined almost
        to the summit, with only a very short hyaline area .... 82. *C. straminea.*
      14 Perigynia 2.5–5 times as long as wide; sheaths ventrally hyaline.
        15 Perigynia 4–5.5 × 1.5–2 mm, 2.5–3 times as long as wide, strongly flattened
          and much wider than the achene .... 69. *C. scoparia.*
        15 Perigynia 3.3–5 × 0.8–1 mm, 3.5–5 times as long as wide, planoconvex and
          not much wider than the achene .... 70. *C. crawfordii.*
    13 Principal lvs 3–7 mm wide.
      16 Spikes 8–12 mm, overlapping and crowded; perigynia more than 30 .... 71. *C. tribuloides.*
      16 Spikes 5–8 mm, separate (at least the lower) in an elongate infl; perigynia 15–30 ... 72. *C. projecta.*
9 Achenes broader, 0.9–1.5+ mm wide (if a little narrower, then the perigynium body obovate);
  perigynia up to ca 2.5 times as long as wide.
17 Perigynia less than 4 mm long and also less than 2 mm wide.
  18 Perigynium-body obovate; achene not over 1 mm wide.
    19 Perigynium-beak broad, its winged margins extending to the tip; pistillate scales
      distally boat-shaped and usually obtuse, the midvein not reaching the tip .... 84. *C. longii.*
    19 Perigynium-beak slender, its winged margins not reaching the tip; pistillate scales
      distally flat and acute, the midvein reaching the tip .... 85. *C. albolutescens.*
  18 Perigynium-body ovate to oblong, elliptic, or orbicular; achene often over 1 mm wide.
    20 Perigynium-body orbicular to broadly elliptic, abruptly rounded to the slender
      beak .... 79. *C. festucacea.*
    20 Perigynium-body ovate to elliptic or oblong, more gradually tapering into the often
      broader beak.
      21 Scales about as long as but distinctly narrower than the perigynia.
        22 Ventral side of the perigynium nerveless or obscurely few-nerved .... 64. *C. foenea.*
        22 Ventral side of the perigynium strongly 5–8-nerved .... 65. *C. argyrantha.*
      21 Scales distinctly shorter (as well as usually narrower) than the perigynia.
        23 Infl mostly lax and elongate, the lower spikelets well separated; main lvs
          mostly 1.5–2.5 mm wide .... 75. *C. tenera.*
        23 Infl more condensed; main lvs mostly 2.5–6 mm wide .... 78. *C. normalis.*
17 Perigynia more than 4 mm long, or more than 2 mm wide, or both.
  24 Sheaths with an elongate hyaline ventral area; perigynia in most spp. ± ovate or broadly
    ovate in outline, widest at a fourth to two-fifths of their total length.
    25 Perigynia planoconvex, not over 4.4 mm.
      26 Perigynia mostly 2–2.5 times as long as wide, widest at a fourth to a third of
        its length .... 78. *C. normalis.*
      26 Perigynia mostly 1.3–2 times as long as wide, widest at a third to half its
        length .... 79. *C. festucacea.*
    25 Perigynia flat, often but not always over 4.4 mm.
      27 Perigynium very thin, ± translucent, evidently nerved on both sides, 4.2–7 mm;
        pistillate scales acute to acuminate .... 80. *C. bicknellii.*
      27 Perigynium thicker and firmer, opaque, often nerveless ventrally; pistillate scales
        merely acute.
        28 Perigynium 3.2–4.8 × 2.2–3.4 mm, the body no wider than long; widespread
          sp. .... 76. *C. brevior.*
        28 Perigynium 4.5–5.7 × 3.3–4.5 mm, the body wider than long; se. coastal
          plain .... 77. *C. reniformis.*
  24 Sheaths ventrally green-veined almost to the summit, with only a short hyaline area;
    perigynium in most spp. widest at two-fifths to half its length from the base and at
    or beyond the middle of the orbicular or obovate body.
    29 Pistillate scales acuminate into a subulate or awn-like tip.
      30 Spikes separated in an elongate infl, the lateral ones conspicuously tapering to
        the base .... 82. *C. straminea.*
      30 Spikes crowded and overlapping, the lateral ones obtuse to short-clavate at
        base .... 83. *C. alata.*

29 Pistillate scales inconspicuous, obtuse or merely acute.
  31 Perigynium 3.9–5.1 × 2.1–2.6 mm, 1.8–2.4 times as long as wide, broadest at
    a third to two-fifths of its length .... 81. *C. suberecta.*
  31 Perigynium otherwise, broadest at two-fifths to half its length, and often less
    than 1.8 times as long as wide.
    32 Lateral spikes obtuse or rounded at base; perigynia 3–4.5 mm, widest at or
      above the summit of the achene.
      33 Perigynia evidently nerved on both faces.
        34 Perigynium-beak broad, its winged margins extending to the tip;
          pistillate scales distally boat-shaped and mostly obtuse, the
          midvein not reaching the tip .... 84. *C. longii.*
        34 Perigynium-beak slender, its winged margins not reaching the tip;
          pistillate scales distally flat and acute, the midvein reaching the
          tip .... 85. *C. albolutescens.*
      33 Perigynia ventrally nerveless .... 86. *C. cumulata.*
    32 Lateral spikes clavate at base; perigynia 4–5.2 mm, widest at about the
      middle of the achene .... 87. *C. silicea.*

23. Lamprochlaenae

One sp. in our range .... 88. *C. supina.*

24. Praecoces

One sp. in our range .... 89. *C. caryophyllea.*

25. Phyllostachyae

1 Pistillate scales with hyaline margins.
  2 Perigynium-body obovoid-oblong, tapering into the stoutly pyramidal beak .... 90. *C. willdenovii.*
  2 Perigynium-body subglobose above the stipe-like base, abruptly prolonged into the slender
    beak .... 91. *C. jamesii.*
1 Pistillate scales wholly green .... 92. *C. backii.*

26. Montanae

1 Spikes usually all borne ± close together above the middle of the stem, the terminal one staminate,
  the lateral ones pistillate, only rarely any of the spikes near-basal.
  2 Perigynium-body (beak and contracted base excluded) ellipsoid to obovoid, distinctly longer than
    wide, usually wider than thick, obscurely trigonous at maturity.
    3 Lowest pistillate spike short-pedunculate, ± remote from and not overlapping the spike next
      above it, though still borne above the middle of the stem .... 93. *C. novae-angliae.*
    3 Lowest pistillate spike sessile (like the others) and usually overlapping the spike next above.
      4 Pistillate scales suborbicular, obtusish to short-mucronate, ca half as long as the body of the
        perigynium .... 94. *C. peckii.*
      4 Pistillate scales ovate to lance-ovate, acuminate or cuspidate, nearly equaling to surpassing
        the body of the perigynium.
        5 Perigynia not concealed by the scales, these green, or with a purple strip on each side of
          the broad green midstrip .... 95. *C. albicans.*
        5 Perigynia largely concealed by the scales, these dark purple except for the narrow midnerve
          .... 96. *C. nigromarginata.*
  2 Perigynium-body subglobose, nearly or quite as wide and thick as long.
    6 Plants cespitose, without long rhizomes; lvs relatively short and broad, at least the larger ones
      3–7 mm wide .... 97. *C. communis.*
    6 Plants producing long rhizomes; lvs longer and more slender, mostly 1–3 mm wide.
      7 Beak 0.2–1 mm, one-eighth to half as long as the rest of the perigynium .... 98. *C. pensylvanica.*
      7 Beak 1–2 mm, half to fully as long as the rest of the perigynium .... 99. *C. lucorum.*
1 Some of the pistillate spikes borne singly on short to ± elongate peduncles that originate near the
  base of the stem, well removed from the other spikes.
  8 Terminal staminate spike closely associated with one or more pistillate spikes, the lowest of these
    subtended by a foliaceous bract mostly surpassing the staminate spike.
    9 Perigynium-beak 1–1.7 mm, half to fully as long as the body; mainly cordilleran .... 100. *C. rossii.*
    9 Perigynium-beak 0.4–0.7 mm, less than half as long as the body; widespread in the n.
      with us .... 101. *C. deflexa.*
  8 Terminal staminate spike alone or associated with a pistillate spike, in the latter case the subtending
    bract-scale not surpassing the staminate spike .... 102. *C. umbellata.*

27. BICOLORES

28. DIGITATAE

29. TRIQUETRAE

30. ALBAE

31. PANICEAE

32. LAXIFLORAE

12 Lvs of the fertile and sterile shoots similar, seldom any of them over 10 mm wide.
  13 Basal scales of the pistillate spikelets, like the others, subtending pistillate fls.
    14 Staminate spike sessile or nearly so, often hidden by the bracts and pistillate spikes
      .... 128. *C. abscondita.*
    14 Staminate spike on an evident peduncle 0.5–8.5 cm .... 129. *C. digitalis.*
  13 Basal 1–3 scales of the pistillate spikes empty or subtending staminate fls .... 130. *C. laxiculmis.*

33. Granulares

1 Staminate peduncle none or shorter than the uppermost pistillate spike .... 131. *C. granularis.*
1 Staminate peduncle elongate, overtopping the uppermost pistillate spike ....132. *C. crawei.*

34. Oligocarpae

1 Lf-sheaths hispidulous; perigynia broadest well above the middle .... 133. *C. hitchcockiana.*
1 Lf-sheaths glabrous; perigynia broadest near the middle.
  2 Pistillate peduncles smooth.
    3 Lower pistillate scales with awns about as long as the body or longer, the whole scale distinctly
      longer than the perigynium.
      4 Basal sheaths conspicuously purplish; main lvs 2–4 mm wide .... 134. *C. oligocarpa.*
      4 Basal sheaths green or brown-tinged; main lvs 4–8 mm wide .... 135. *C. amphibola.*
    3 Lower pistillate scales awnless or with an inconspicuous awn to 1 mm, the whole scale much
      shorter than the perigynium .... 136. *C. flaccosperma.*
  2 Pistillate peduncles rough .... 137. *C. conoidea.*

35. Gracillimae

1 Perigynia strongly angled at the lateral ribs, otherwise nerveless; plants brownish or greenish at
  base .... 138. *C. prasina.*
1 Perigynia obscurely trigonous, the lateral ribs not at the angles; plants purplish at base.
  2 Perigynium beakless, 2.4–3.7 mm.
    3 Sheaths glabrous on the hyaline ventral strip; main lvs 3–8 mm wide .... 139. *C. gracillima.*
    3 Sheaths hairy on the ventral strip; main lvs 2–3 mm wide .... 140. *C. aestivalis.*
  2 Perigynium acuminate to a short beak; sheaths hairy.
    4 Perigynia 3–3.5 mm .... 141. *C. aestivaliformis.*
    4 Perigynia 3.5–6 mm, avg more than 4 mm.
      5 Pistillate scales ovate, acute or obtuse or minutely cuspidate; lateral spikes with a few staminate
      fls at base .... 142. *C. formosa.*
      5 Pistillate scales lanceolate to oblong, long-acuminate or awned; lateral spikes wholly pistillate.
        6 Upper pistillate scales with a definite awn; widespread forest sp. .... 143. *C. davisii.*
        6 Upper pistillate scales merely acuminate; southern sp. .... 144. *C. oxylepis.*

36. Sylvaticae

1 Achene sessile in the base of the perigynium.
  2 Perigynia pubescent, coriaceous ....145. *C. assiniboinensis.*
  2 Perigynia glabrous.
    3 Spikes short-cylindric, 1–2(–2.5) cm, closely fld; lvs soft-hairy .... 146. *C. castanea.*
    3 Spikes linear, slender, (2)2.5–7 cm, loosely fld; lvs glabrous or nearly so.
      4 Perigynium stipitate, its beak much shorter than the body .... 147. *C. arctata.*
      4 Perigynium sessile, its beaks about as long as the body .... 148. *C. sylvatica.*
1 Achene elevated above the perigynium-base on a slender stipe 0.5–1.5 mm.
  5 Spikes loose, the perigynia separated by internodes 2–4(–6) mm .... 149. *C. debilis.*
  5 Spikes denser, the perigynia separated by internodes 1–1.5 mm .... 150. *C. venusta.*

37. Capillares

One sp. in our range .... 151. *C. capillaris.*

38. Longirostres

One sp. in our range .... 152. *C. sprengelii.*

39. EXTENSAE

1 Perigynia mostly spreading or reflexed; sheaths white ventrally; widespread, not maritime.
  2 Perigynia 2.2–3.3 mm, straight or nearly so; achene nearly filling the perigynium .... 153. *C. viridula.*
  2 Perigynia 3.5–6.2 mm (or as little as 3.2 mm in *C. cryptolepis*), many or most of them strongly
    recurved-falcate; achene in the lower half of the perigynium.
    3 Pistillate scales conspicuous, coppery-brown; perigynium-beak rough-margined.
      4 Staminate spike short-peduncled or subsessile; perigynium gradually tapering to the beak;
        widespread .... 154. *C. flava.*
      4 Staminate spike on a peduncle surpassing the uppermost pistillate spike; perigynium abruptly
        contracted to the beak; local ne. .... 155. *C. lepidocarpa.*
    3 Pistillate scales inconspicuous, about the same color as the perigynia; beak smooth .... 156. *C. cryptolepis.*
1 Perigynia ascending; sheaths red-dotted ventrally; coastal salt-marshes .... 157. *C. extensa.*

40. VIRESCENTES

1 Terminal spike wholly staminate.
  2 Perigynia faintly and finely nerved, ellipsoid, beakless .... 158. *C. pallescens.*
  2 Perigynia conspicuously nerved, obovoid, with a beak 0.2–0.5 mm .... 159. *C. torreyi.*
1 Terminal spike pistillate at the top, staminate below.
  3 Perigynia densely pubescent.
    4 Pistillate spikes 2–4 cm, loosely fld at base; anthers 1.5–2.5 mm .... 160. *C. virescens.*
    4 Pistillate spikes 1–2 cm, closely fld throughout; anthers 0.7–1.5 mm .... 161. *C. swanii.*
  3 Perigynia glabrous.
    5 Perigynia appressed-ascending, much wider than thick, blunt, obscurely nerved .... 162. *C. complanata.*
    5 Perigynia spreading, nearly or quite as thick as wide, short-pointed, conspicuously nerved.
      6 Sheaths glabrous or nearly so ventrally; pistillate scales obtuse to short-cuspidate .... 163. *C. caroliniana.*
      6 Sheaths hairy all around; pistillate scales long-acuminate .... 164. *C. bushii.*

41. HIRTAE (Carex proper)

1 Lf-sheaths and pistillate scales hairy; teeth of the perigynium 1–1.5 mm; intr., weedy .... 165. *C. hirta.*
1 Lf-sheaths and pistillate scales glabrous; teeth mostly under 1 mm; native, not weedy.
  2 Perigynia glabrous, or only inconspicuously hairy, impressed-nerved .... 166. *C. striata.*
  2 Perigynia pubescent, ribbed (but the ribs often obscured by the pubescence).
    3 Beak of the perigynium distinctly bidentate with stiff teeth.
      4 Perigynia 4–7 mm, conspicuously ribbed, short-hirtellous; dry habitats .... 167. *C. houghtoniana.*
      4 Perigynia 2.8–5 mm, the ribs hidden by the dense pubescence; mostly in wet habitats.
        5 Lvs flat or nearly so, mostly 2–5 mm wide .... 168. *C. pellita.*
        5 Lvs folded along the midrib, only 1–1.5(–2) mm wide as folded .... 169. *C. lasiocarpa.*
    3 Beak of the perigynium soft and hyaline at the only obscurely bidentate tip .... 170. *C. vestita.*

42. ANOMALAE

One sp. in our range .... 171. *C. scabrata.*

43. SHORTIANAE

One sp. in our range .... 172. *C. shortiana.*

44. PENDULINAE

1 Pistillate scales acute or acuminate, sometimes also with an awn or cusp; perigynium strongly
  nerved .... 173. *C. joorii.*
1 Pistillate scales broadly rounded or usually retuse, the midvein prolonged into an awn; perigynium
  inconspicuously or scarcely nerved .... 174. *C. glaucescens.*

45. SCITAE

1 Perigynia minutely hispidulous distally; pistillate scales with brown sides and broad, much paler
  midvein .... 175. *C. flacca.*
1 Perigynia glabrous; pistillate scales wholly dark brown-purple .... 176. *C. barrattii.*

46. LIMOSAE

1 Pistillate scales ovate or elliptic, mostly about as wide and long as the perigynia.
  2 Stems obtusely 3-angled, very smooth; scales enclosing the perigynium-base .... 177. *C. rariflora.*
  2 Stems sharply 3-angled, usually rough above; scales not enclosing the perigynium-base .... 178. *C. limosa.*
1 Pistillate scales lanceolate, narrower and longer than the perigynia .... 179. *C. paupercula.*

47. ATRATAE

1 Pistillate spikes mostly sessile or nearly so.
  2 Stems of the season strongly aphyllopodic, not surrounded by dried sheaths of previous years;
    perigynia densely and conspicuously papillate .... 180. *C. buxbaumii.*
  2 Stems of the season phyllopodic or somewhat aphyllopodic, surrounded by the dried sheaths of
    previous years; perigynia not notably papillate.
    3 Spikes short, the terminal one 6–14 mm; pistillate scales atropurpureous; boreal .... 181. *C. norvegica.*
    3 Spikes longer, the terminal one (10–)15–30 mm; pistillate scales stramineous or brown; Great
      Plains .... 182. *C. parryana.*
1 Pistillate spikes on slender peduncles 1–4 cm .... 183. *C. atratiformis.*

48. CRYPTOCARPAE

1 Pistillate scales awnless, the sides blackish or deep purple-brown .... 184. *C. torta.*
1 Scales awned, the sides merely brown.
  2 Densely tufted plants, without long rhizomes, not of salt-marshes.
    3 Sheaths glabrous, perigynia somewhat inflated, obovoid, rounded above to an abrupt beak .... 185. *C. crinita.*
    3 Sheaths scabrous-hispidulous; perigynia flattened, elliptic, tapering from near or below the middle
      to a minute beak.
      4 Perigynia smooth to slightly papillate distally .... 186. *C. gynandra.*
      4 Perigynia densely granular-papillate throughout .... 187. *C. mitchelliana.*
  2 Salt-marsh plants with elongate rhizomes .... 188. *C. paleacea.*

49. ACUTAE

1 Longest bract well overtopping the spikes.
  2 Salt-marsh plants with long rhizomes .... 189. *C. salina.*
  2 Densely tufted plants, without long rhizomes, mostly not of salt-marshes.
    3 Perigynia essentially nerveless on both faces .... 190. *C. aquatilis.*
    3 Perigynia with a few sharp, elevated nerves on each face .... 191. *C. lenticularis.*
1 Longest bracts shorter than or merely equaling the spikes.
  4 Lower sheaths scabrous, red-brown, splitting to form a pinnate network .... 196. *C. stricta.*
  4 Lower sheaths glabrous, not filamentose.
    5 Scales black; perigynia distally purple-brown.
      6 Perigynia essentially nerveless on both faces .... 192. *C. bigelowii.*
      6 Perigynia with a few sharp, elevated nerves on each face .... 193. *C. nigra.*
    5 Scales brown or red-brown; perigynia not purplish-brown.
      7 Perigynia red-dotted, each face nerveless or faintly 1–3-nerved .... 194. *C. haydenii.*
      7 Perigynia not dotted, each face evidently several-nerved .... 195. *C. emoryi.*

50. COLLINSIAE

One sp. in our range .... 197. *C. collinsii.*

51. FOLLICULATAE

1 Main lvs 1.5–4 mm wide; bract-sheaths concave at the mouth .... 198. *C. michauxiana.*
1 Main lvs 4–16 mm wide; bract-sheaths prolonged at the mouth .... 199. *C. folliculata.*

52. PSEUDOCYPEREAE

1 Mature perigynia spreading or ascending, rather thin-textured, nearly round in cross-section, rather
  abruptly narrowed to the beak, scarcely stipitate .... 200. *C. hystericina.*

1 Mature perigynia ± reflexed, rigidly coriaceous, obtusely trigonous, very gradually tapering to the beak, slender-stipitate.
  2 Teeth of the perigynium-beak curved-divergent, 1.2–2.3 mm .... 201. *C. comosa.*
  2 Teeth of the perigynium-beak nearly straight and parallel, 0.5–1 mm .... 202. *C. pseudocyperus.*

53. PALUDOSAE

1 Perigynia glabrous.
  2 Teeth of the perigynium-beak 1 mm or less.
    3 Lvs narrow, mostly 2–4 mm wide; pistillate spikes 1.5–3(–4) cm; perigynium 3.5–5.5 mm, with impressed nerves .... 203. *C. melanostachya.*
    3 Lvs wider, mostly 5–15 mm wide; pistillate spikes 3–11 cm; perigynium with ± raised nerves, except often in *C. hyalinolepis.*
      4 Perigynium 2.5–4.5 mm; lvs mostly 5–8 mm wide .... 204. *C. acutiformis.*
      4 Perigynium 4.7–8 mm; lvs mostly 8–15 mm wide.
        5 Stems aphyllopodic, lateral; ligules much longer than wide .... 205. *C. lacustris.*
        5 Stems phyllopodic, central; ligules scarcely if at all longer than wide .... 206. *C. hyalinolepis.*
  2 Teeth 1–3 mm; perigynia 5.5–10 mm.
    6 Sheaths and blades glabrous .... 207. *C. laeviconica.*
    6 Sheaths pubescent; blades usually pubescent beneath near the base .... 208. *C. atherodes.*
1 Perigynia pubescent .... 209. *C. trichocarpa.*

54. SQUARROSAE

1 Terminal spike generally staminate; pistillate scales with an awn equaling or surpassing the perigynium .... 210. *C. frankii.*
1 Terminal spike gynaecandrous, mainly pistillate; pistillate scales awnless or with an awn surpassed by the perigynium-beak.
  2 Achene a little less than half as wide as long; style much curved near its base .... 211. *C. squarrosa.*
  2 Achene half to three-fifths as wide as long; style straight or nearly so .... 212. *C. typhina.*

55. VESICARIAE

1 Perigynium-beak merely emarginate; lvs narrow, mostly 1–4 mm wide.
  2 Achene trigonous; stigmas 3; faces of the perigynia evidently nerved .... 213. *C. oligosperma.*
  2 Achene lenticular; stigmas 2; faces of the perigynia nearly or quite nerveless .... 214. *C. saxatilis.*
1 Perigynium-beak with 2 slender sharp teeth; lvs often but not always wider.
  3 Pistillate scales acute to acuminate or cuspidate.
    4 Perigynia ascending at maturity, ± arranged into 6(8) vertical rows.
      5 Achene symmetrical; perigynia up to ca 3 mm thick.
        6 Perigynium-beak 0.5–1 mm, its teeth ca 0.25 mm; lvs mostly 2–3 mm wide .... 215. *C. × mainensis.*
        6 Perigynium-beak 1–2 mm, its teeth (0.3–)0.5–1.2 mm; lvs mostly 3–8 mm wide .... 216. *C. vesicaria.*
      5 Achene deeply indented in the middle of one angle; perigynia (4–)4.5–7 mm thick .... 217. *C. tuckermanii.*
    4 Perigynia spreading or reflexed at maturity, ± arranged into (6)8–12 vertical rows.
      7 Bract of the lowest pistillate spike 1–2 times as long as the infl; plants with long rhizomes.
        8 Perigynium 4–7 mm, with smooth beak.
          9 Lvs flat, yellow-green, not glaucous, the larger ones 5–12 mm wide; widespread .... 218. *C. utriculata.*
          9 Lvs ± involute, papillate-glaucous above, 1.5–4 mm wide; boreal .... 219. *C. rostrata.*
        8 Perigynium 6–9 mm, its beak scabrous on the margins; coastal states .... 220. *C. bullata.*
      7 Bract of the lowest pistillate spike 2–several times as long as the infl; plants tufted, without long rhizomes .... 221. *C. retrorsa.*
  3 Pistillate scales with a slender awn equaling or longer than the body.
    10 Staminate scales with an evident rough awn; no long rhizomes.
      11 Pistillate spikes 14–20 mm thick; perigynium-beak shorter than the body .... 222. *C. lurida.*
      11 Pistillate spikes 8–13 mm thick; perigynium-beak longer than the body .... 223. *C. baileyi.*
    10 Staminate scales merely acuminate; plants with long rhizomes .... 224. *C. schweinitzii.*

56. LUPULINAE

1 Sheath of the uppermost nonbracteal lf wanting or less than 1.5 cm; perigynium-beak 1.5–4.2 mm; no rhizomes.
  2 Perigynia dull, cuneate to the base, mostly 8–35 per spike, radiating in all directions .... 225. *C. grayi.*
  2 Perigynia lustrous, convexly rounded to the base, 1–12 per spike, ascending or spreading ...... 226. *C. intumescens.*
1 Sheath of the uppermost nonbracteal lf usually more than 1.5 cm; perigynium-beak 4.5–10 mm; plants usually with long rhizomes.

3 Achene longer than wide, or about as wide as long, widest near the middle; perigynia usually
   ascending.
   4 Staminate peduncle much surpassing the uppermost pistillate spike .... 227. *C. louisianica.*
   4 Staminate peduncle surpassed by or merely equaling the uppermost pistillate spike.
      5 Angles of the achene smoothly curved; faces flat to slightly concave .... 228. *C. lupulina.*
      5 Angles of the achene pointed; faces strongly concave .... 229. *C. lupuliformis.*
3 Achene distinctly wider than long, widest above the middle; perigynia usually spreading .... 230. *C. gigantea.*

1. **Carex obtusata** Lilj. Stems (0.5–)1–1.5(–2) dm, scattered on long rhizomes; lvs crowded toward the base, slender but ± flat, (0.5–)1–1.5 mm wide; spike 1, androgynous, bractless, mostly 8–15 mm, the upper half or three-quarters staminate; pistillate scales ovate or lance-ovate, sharply acute or cuspidate, from a little shorter to a little longer than the perigynia; perigynia few, mostly 1–6, ascending, plump, ellipsoid, thick-walled, obscurely to conspicuously ribbed and sulcate, usually becoming rich, shining brown, 3–4 mm including the evident, hyaline beak; achene trigonous, filling the perigynium; rachilla more than half as long as the achene; 2n=52. Dry plains and hills; w. Ont., Minn. and Nebr. to Utah, Alas., and Eurasia. If the rachilla is wanting, see sp. 88.

2. **Carex dioica** L. Loosely rhizomatous; stems 0.5–2 dm; lvs basally disposed, 3–15 cm, very slender, to 1 mm wide; spike 1, bractless, 10–15 mm, androgynous, sometimes nearly unisexual; pistillate scales short, broad; perigynia crowded, plump, 3–3.5 mm, multistriate, thick-walled, glabrous and shining, abruptly contracted to a short beak, soon widely spreading and often distally falcate-recurved, filled by the lenticular achene; rachilla obsolete; 2n=48, 52. Sphagnum-bogs; circumboreal, s. to Pa., Mich., Minn., and Utah. Amer. and Siberian plants are var. *gynocrates* (Wormsk.) Ostenf. (*C. gynocrates.*)

3. **Carex pauciflora** Lightf. Stems solitary or few together from long slender rhizomes, 1–4 dm, aphyllopodic; main lvs 1–2 mm wide, shorter than the culm; spike 1, bractless, to 1 cm; staminate scales closely infolded into a slender terminal cone; pistillate scales lanceolate, 4–6 mm, pale brown, soon deciduous; perigynia few, mostly 1–6, soon deflexed, deciduous at maturity, light green, becoming stramineous or pale brown, finely several-nerved, 6–7.5 mm, slender and long-tapering, spongy at base, nearly round in section; achene trigonous, not filling the perigynium, continuous with the exserted, persistent style; rachilla obsolete; 2n=46, 76. Sphagnum-bogs; circumboreal, s. to Conn., Pa., W.Va., Mich., Minn., and Wash.

4. **Carex capitata** L. Cespitose, aphyllopodic, with setaceous lvs 5–20 cm and slender, subnaked stems 1–3 dm; spike 1, bractless, ovoid, 5–15 mm, androgynous; pistillate scales subrotund, with broad hyaline margins; perigynia 6–25, often surpassing the scales, ascending-spreading, ovate to suborbicular, 2–3.5 mm, flattened, sharply margined, smooth, nerveless, basally rounded, tapering into a beak 0.3–0.7 mm, not filled by the lenticular achene; rachilla more than half as long as the achene; 2n=50. Mostly acidic rocky or gravelly soil; circumboreal, s. to the White Mts. of N.H.; s. S. Amer.

5. **Carex filifolia** Nutt. Very densely tufted, (0.5–)1–3 dm; lvs wiry; involute, acicular, resembling the stems and nearly or fully as long, to 0.7 mm wide; spike 1, the upper half or two-thirds staminate; scales broadly obovate, apically rounded or broadly obtuse, or the lower sometimes cuspidate, the brown, scarious central part fading into the broad white-hyaline margins; pistillate scales equaling or a little longer than the ± concealed perigynia; perigynia 5–15, plump, obscurely trigonous and obscurely several-nerved, puberulent at least above, 3–4.5 mm including the 0.2–0.5 mm beak, filled by the trigonous achene; rachilla half to fully as long as the achene; 2n=50. Dry plains and hills; w. Minn. to Yukon, Wash., Tex., and N.M.

6. **Carex geyeri** F. Boott. Elk-sedge. Loosely cespitose, the aphyllopodic stems 1.5–5 dm tall from short rhizomes; lvs elongate, flat, 1.5–3 mm wide; spike 1, the terminal staminate part slender, 1–2.5 cm; pistillate scales oblong-obovate, longer and wider than the perigynia, acute or the lower short-cuspidate, with hyaline margins; perigynia 1–3, ellipsoid or obovoid, obscurely trigonous, 5–6 mm, 2-ribbed, otherwise nerveless, tapering to a spongy base, rather abruptly contracted above to the scarcely beaked tip; achene trigonous, 4–5 mm, filling with perigynium; rachilla up to half as long as the achene, or obsolete. Dry woods in calcareous soil; Centre Co., Pa.; Alta. to Colo., w. to B.C. and Calif.

7. **Carex leptalea** Wahlenb. Stems densely clustered on slender, freely branched rhizomes, very slender, 1.5–6 dm; lvs shorter than the stems, slender but ± flat, 0.7–1.2 mm wide; spike 1, 0.5–1.5 cm, the terminal staminate part often short; pistillate scales obtuse to acute or short-awned, mostly shorter than the perigynia, or the lowest prolonged into a slender tip surpassing the perigynia; perigynia 1–10, 2.5–4.5 mm, appressed-ascending, often rather remote, membranaceous, elliptic or lance-elliptic, with a narrow, often substipitate, spongy base 0.5–1 mm, beakless, with 2 marginal nerves and many finer nerves on each face; achene trigonous, 1.3–1.8 mm, not filling the perigynium; rachilla wanting; 2n=52. Bogs and wet soil; Lab. to Alas., s. to Fla., Tex., and Calif. (*C. harperi*, a form with the perigynia up to 5 mm.)

8. **Carex picta** Steudel. Dioecious; stems 1–3 dm, cespitose from stout woody rhizomes, much shorter than the lvs; main lvs 3–5 mm wide; spike 1, linear, 2–4 cm, the basal scales usually sterile and somewhat elongate, the flowering scales usually purple with paler midnerve and narrow hyaline margins, the uppermost often short-cuspidate; perigynia narrowly oblong-obovoid, 3-angled, many-nerved, 4–5 mm, a third as wide, yellowish-green or suffused with purple, acute, hairy distally; achene 3–4 mm, filling the perigynium; no rachilla. Dry woods; s. Ind., Tenn., Ala., and La.; fl Mar., Apr.

9. **Carex scirpoidea** Michx. Dioecious; stems 1–4 dm, surpassing the lvs, 1–few together from a rhizome; main lvs 1–3 mm wide; peduncle often with a minute bract ca 1 cm from the top; spike 1, erect, slender, 1–

3 cm; pistillate scales deep brown, a little shorter and much narrower than the perigynia, thinly hairy; perigynia appressed-ascending, obovate-oblong, 2.5–3 mm, densely short-hairy, 2-keeled, otherwise virtually nerveless, subtriangular in cross-section, the minute beak 0.2 mm; achene trigonous; ± filling the perigynium; no rachilla; $2n$=62–68. Dry soil, especially in calcareous regions; irregularly circumboreal, widespread in n. N. Amer. and s. in our range to the n. parts of N. Engl., N.Y., and Mich. Our plants are var. *scirpoidea*.

10. Carex kobomugi Ohwi.   Usually dioecious; stems very stout, 1–3 dm; lvs yellow-green, stiff, 3–6 mm wide, rough-margined, surpassing the stems; spikes numerous, sessile, densely aggregated into a head that appears almost as a single spike, the staminate heads oblong-cylindric, 3–4 × 1–2 cm, the pistillate ovoid, 3–6 × 2–4 cm; scales acuminate into a stout, often elongate cusp; perigynia ± erect, dark, lance-ovate to elliptic, thickly planoconvex, spongy-thickened below, 10–14 mm, evidently multinerved, the infolded margins lacerately toothed, the beak smooth, sharply bidentate, nearly as long as the body; achene 4–7 mm, unequally trigonous; stigmas 3. Native of e. Asia, intr. in coastal sands from Mass. to Va.

11. Carex divisa Hudson.   Very much like no. 12; spikes mostly 3–8(–12), all or mostly androgynous; perigynium ovate or broadly ovate, evidently nerved dorsally, the body 3–5 times as long as the short beak. Native of Europe, locally intr. in brackish marshes along the coast in Md. and N.C.

12. Carex praegracilis W. Boott.   Freeway-sedge. Stems (1–)3–7 dm, scaberulous on the angles above, scattered on long, stout, black or brownish-black, monopodial rhizomes; lvs basally disposed, ± elongate, ± flat, mostly 1–3 mm wide; spikes mostly 6–25, androgynous or some of them wholly pistillate (often some plants largely pistillate and others largely staminate), sessile, inconspicuously bracteate, less than 1 cm, aggregated into a thick-cylindric to ± ellipsoid or ovoid head 1.5–3.5(–5) cm; scales generally ± concealing the perigynia, these ± ovate or lance-ovate to elliptic, 2.9–3.9 mm, planoconvex, sharp-edged and ± serrulate distally, inconspicuously several-nerved dorsally, nerveless ventrally, with a prominent beak 0.6–1.3 mm; achene lenticular; $2n$=60. Open, moist or wet, often alkaline places, chiefly on prairies; B.C. to Calif., e. to Io. and Minn., and intr. e. along salted highways to s. Que. and s. N.Y.

13. Carex stenophylla Wahlenb.   Stems 0.5–2 dm, smooth, scattered on long, slender, brownish, sympodial rhizomes; lvs basally disposed, 0.3–1.5 mm wide, often involute or canaliculate; spikes sessile, normally androgynous, inconspicuously bracteate, 4–9 mm, closely aggregated into an ovoid or oblong-cylindric head 8–17 mm; perigynia fusiform-elliptic to broadly elliptic-ovate, 2.6–3.3(–3.5) mm, planoconvex, sharp-edged and ± serrulate distally, ± evidently several-nerved dorsally and sometimes also ventrally, or virtually nerveless, with an evident beak 0.5–1 mm; achene lenticular. Open, often grassy places; irregularly circumboreal, in Amer. from Minn., Io., and Kans. w. across the plains and irregularly in the cordillera n. to Alas.; intr. along railways in Mo. and Ill. (*C. eleocharis*, the Amer. pls.)

14. Carex douglasii F. Boott.   Stems 1–3(–4) dm, smooth, scattered on long, slender, sympodial rhizomes; lvs basally disposed, 1–2.5 mm wide, flat or involute; dioecious; spikes sessile, inconspicuously bracteate, ovoid-fusiform, 1–1.5 cm, closely aggregated into a rhomboidal or ovoid to subcylindric head 1.5–4.5 cm; perigynia mostly ± hidden by the scales, rather narrowly elliptic to elliptic-obovate, planoconvex, becoming sharp-edged and serrulate distally, obscurely multistriate and with evident marginal nerves, 3.5–4.5 mm, including the prominent, 1–1.5 mm beak; achene lenticular; $2n$=60. Dry soil; Man., Io., and Nebr. to B.C., Calif., and N.M.; intr. along railways in Mo.

15. Carex sartwellii Dewey.   Stems 3–8 dm, scattered on long, coarse, dark rhizomes, strongly aphyllopodic, the lvs well distributed along the lower half of the stem, flat, 2–5 mm wide; sheaths elongate and firm, the ventral part tending to be partly greenish or greenish-striate; spikes numerous, 20+, sessile, inconspicuously bracteate (or the lowest one more evidently so), androgynous or some wholly staminate or wholly pistillate, the middle ones, especially, often staminate, ovoid, ca 1 cm or less, closely crowded into a ± cylindric to sometimes ovoid infl 2–5 cm (only the lowest ones somewhat separate); pistillate scales usually pale brown to stramineous; perigynia 2.3–4 mm, broadly lanceolate to round-ovate or ± elliptic, planoconvex, sharp-edged, tending to be serrulate distally, evidently several-nerved dorsally, usually less so ventrally, spongy at the base, ± abruptly contracted into the short (0.4–1 mm), minutely bidentate beak; achene lenticular; $2n$=62. Wet meadows, open swamps, and shallow water; N.Y. and Ont. to B.C., s. to Ind., Mo., and Colo.

C. disticha Hudson, a closely related Eurasian sp., is rarely adventive along our n. boundary, and apparently persistent at the s. end of Georgian Bay in Ont. It differs in its more prominent basal spikes, reddish-brown pistillate scales, and larger perigynia, the larger ones 4–5.5 mm, with a beak 1–2 mm.

16. Carex siccata Dewey.   Stems 1.5–4(–6) dm, scattered on tough brown rhizomes, distinctly aphyllopodic; lvs ± flat, (1–)1.5–3 mm wide, generally all on the lower fourth of the stem; sheaths thin-hyaline ventrally; spikes mostly 4–12, sessile, ca 1 cm or less, inconspicuously bracteate, aggregated into a cylindric or clavate-cylindric infl 1.5–3.5 cm, mostly androgynous, but the lower often largely or wholly staminate, and the uppermost one sometimes wholly pistillate and closely subtended by one or more very short staminate spikes so as to appear gynaecandrous; perigynia generally conspicuously surpassing the scales, lance-ovate to ± elliptic, planoconvex, sharp-edged proximally, becoming thin-margined distally, evidently about 10-nerved dorsally, evidently to obscurely few-nerved ventrally, 4.5–6.2 mm including the prominent beak that is more than half as long as the body; achene lenticular; $2n$=70. Dry, sterile or sandy soil, in open places in light shade; Me. to Mack., s. to N.J., O., Ill., Minn., and Ariz. (*C. foenea*, misapplied.)

17. Carex arenaria L.   Stems 1.5–5 dm, scattered on long tough rhizomes, aphyllopodic; lvs thick, 1–3 mm wide; spikes numerous, 5–12 mm, closely aggregated into an oblong head 2–6 cm, usually the lowest bract, and sometimes the others, long-cuspidate; perigynia surpassed by or merely equaling the broadly ovate scales, planoconvex, lance-ovate, multinerved on both sides, 4–5.5 mm, broadly winged above, the conspicuously bidentate beak a third as long as the body; achene lenticular. Native of Europe, intr. in our range on coastal sands from Md. to N.C.

18. Carex chordorrhiza L. f.   Old stems much elongate, prostrate, clothed with parts of the old lvs; fertile stems 1–3 dm, with 1–3 lvs near the base, their blades 1–5 cm; sterile stems with several much longer lvs; lf-blades 1–2 mm wide; head 5–12 cm, crowded; spikes 3–8, androgynous; scales broadly ovate, about equaling the perigynia, these 1–5, plump, oblong-ovoid, 2.5–3.5 mm, nearly as thick as wide, strongly many-nerved, obscurely margined, the beak emarginate, a fourth as long as the body; achene thick-lenticular; $2n$=62. Sphagnum-bogs; circumboreal, s. to Mass., N.Y., Ind., Io., and Mont.

19. Carex cephalophora Muhl.   Densely cespitose, the stems 3–6 dm, slightly shorter to more than twice as long as the lvs; lvs 2–5 mm wide; spikes in a dense ovoid head 1–2 cm, scarcely distinguishable except by the setaceous projecting bracts, subglobose, 5 mm, androgynous; perigynia greenish or somewhat stramineous, spreading, ovate, planoconvex, 2.5–3.5 mm, two-fifths to three-fifths as wide shortly below the middle of the body, rounded or cuneate at base, conspicuously serrulate distally, the sharply bidentate beak a third as long as the body. Var. *cephalophora,* abundant in dry or moist woods, occasionally in open places, from Me. and sw. Que. to Man., s. to Fla. and Tex., has the pistillate scales broadly ovate, hyaline at the margin, green along the center, with green midnerve and 2 obscure lateral nerves, acuminate to short-awned, and the body much shorter than and mostly concealed by the perigynia. Var. *mesochorea* (Mackenzie) Gleason, of dry soil and open woods from Mass. to Va., Ind., Mo., Tenn., and Tex., has the stems usually much longer than the lvs, and has much larger pistillate scales, the body nearly or quite as long as the body of the perigynium, distinctly 3-nerved near the center. (*C. mesochorea.*)

20. Carex leavenworthii Dewey.   Densely cespitose, the stems 2–5 dm, conspicuously exceeding the lvs, or the lvs more elongate in moist ground or shade; lvs 1–3 mm wide; infl, spikes, and pistillate scales as in the typical var. of no. 19; perigynia greenish-stramineous, spreading, broadly ovate, planoconvex, 2.5–3.3 mm, three-fifths to three-fourths as wide above the broadly rounded, truncate, or even subcordate base, inconspicuously rough-margined or smooth distally, the sharply bidentate beak a fourth as long as the body. Dry open ground or dry woods; N.Y. and N.J. to Fla. and Tex., thence n. to Io., s. Mich., and sw. Ont.

21. Carex spicata Hudson.   Densely cespitose, the stems slender, 2–6(–10) dm, usually conspicuously exceeding the lvs; roots and often the basal sheaths and base of stems purplish-tinged; lvs thin, 2–3 mm wide, with a lanceolate, acute ligule longer than wide; infl cylindric, 2–3.5 cm; spikes 3–10, androgynous, contiguous, subglobose, 5–8 mm; bracts setaceous, to 2 cm, or none; scales triangular-ovate, nearly as long as the perigynia, acute to short-aristate, suffused with orange-brown or purplish-red, the midrib green; perigynia green to stramineous, lance-ovate, planoconvex, corky-thickened at base, 4–5.5 mm, a third to half as wide, finely several-nerved dorsally, nerveless ventrally, gradually narrowed into a rough-margined beak ca half as long as the body; $2n$=58, 60. Native of Europe, naturalized from N.S. to Va., w. to O. and Mich. (*C. muricata,* misapplied.)

22. Carex muricata L.   Much like no. 21: roots and basal sheaths blackish-brown, not anthocyanic; ligule short and blunt, about as wide as long; infl a little looser; scales brown to blackish-brown, much narrower and shorter than the body of the perigynium; perigynia widely spreading, 3.5–4.5 mm, over half as wide as long, rounded at base, rather abruptly short-beaked, not corky-thickened; $2n$=56, 58. Native of Europe, naturalized in disturbed sites from N.B. to Pa. and Ont. (*C. pairaei; C. echinata,* misapplied.)

23. Carex divulsa Stokes.   Much like no. 22; roots brown or blackish-brown; basal sheaths at first pale brown, becoming dark to blackish-brown; ligule short and blunt, usually wider than long; infl looser and more elongate, 3–10+ cm, the lowest spikes often 1–3 cm apart; perigynia appressed-ascending (or the lowest ones spreading), 3.5–4.5 mm, usually less than half as wide, ± equally narrowed to both ends, not corky; $2n$=56, 58. Native of Europe, established in disturbed sites from Mass. to D.C. and Ont. (*C. virens.*)

24. Carex retroflexa Muhl.   Densely cespitose; stems 2–4 dm, rather stiff; lvs elongate, flat, 1–3 mm wide; spikes 4–8, ovoid or subglobose, 5–8 mm, androgynous, the lower separate and subtended by setaceous bracts 1–5 cm, the upper approximate, with much shorter or no bracts; pistillate scales ovate, acute to short-acuminate or cuspidate, nearly as long as the perigynia and usually deciduous before maturity; perigynia green to brownish-green, spongy-thickened at base, 2.5–3 mm, the beak smooth-margined, sharply but shortly bidentate. Dry woods, often somewhat weedy. Var. *retroflexa,* of Vt. to Ont., O., and Mo., s. to Fla. and Tex., has the spongy perigynium-base distinctly nerved, tumid on the ventral face, the perigynium somewhat biconvex, two-fifths to half as wide as long, acuminately narrowed into the beak. Var. *texensis* (Torr.) Fern., of O. and nw. Ind. to se. Mo., s. to S.C., Miss., and Tex., has the spongy perigynium-base nerveless or nearly so, scarcely tumid on the ventral face, the perigynium planoconvex, a third as wide as long, the beak with sides straight or nearly so.

25. Carex socialis Mohlenbrock & Schwegman.   Much like no. 26, but loosely colonial, with well developed creeping rhizomes; spikes well spaced; perigynia more ascending, 3.6–4 mm, narrow, mostly 3–5 times as long as wide; achene set ca 0.5 mm above the base of the perigynium; stigmas flexuous or recurved, not coiled. Low woods; coastal plain and interior low plateaus from S.C. and Ga. to Tex., n. to s. Ind., s. Ill., and se. Mo.

26. Carex radiata (Wahlenb.) Small.   Cespitose; stems 2–8 dm, erect, slender, the fertile ones 0.8–1.5 mm thick at base; lvs elongate, the widest ones 1.3–1.9 mm wide; spikes 4–7, androgynous, sessile, closely aggregated or the lower often distinctly separate; bracts setaceous, the lowest often surpassing the infl; pistillate scales persistent, shorter than the perygynia, rounded to acute, seldom short-awned; perigynia 3–20, light green, widely spreading or reflexed at maturity, planoconvex, lance-ovate, 2.6–3.8 mm as wide, spongy and somewhat swollen at the subtruncate or broadly rounded base, nerveless, or obscurely nerved near the base, gradually tapering or slightly acuminate into a serrulate, very shortly bidentate beak scarcely a third as long as the body; achene lenticular, ovate to obovate, 1.5–2 mm, occupying the upper half of the body of the perigynium, its base 0.5–0.9 mm above the base of the perigynium; stigmas very slender, straight to slightly twisted, only 0.03–0.06 mm wide; $2n=58$. Moist woods and edges of ponds, requiring more moisture than the next 2 spp.; N.S. and s. Que. to Man. and N.D., s. to N.C., Ala., and Mo. (*C. rosea,* misapplied.) The 2 following spp. are closely allied to *C. radiata.* The characters are confluent, but the populations apparently not.

27. Carex appalachica J. M. Webber & P. Ball. Much like nos. 26 and 28; fertile stems arching in fr, 0.7–1.3 mm thick at base; widest lvs 0.9–1.5 mm wide; lower spikes distinctly separate; perigynia ascending or suberect, 2–3.4 mm, cuneate to somewhat rounded at base; achene set low in the perigynium, its base 0.1–0.5 mm above that of the perigynium; stigmas 0.05–0.08 mm wide, coiled up to 3 times, seldom merely twisted; $2n=52$. Well drained sites in open forests and at forest-margins; Me. to s. Que. and s. Ont., s. to N.J. and Pa., and in the mts. to N.C. and Tenn. (*C. radiata,* misapplied.)

28. Carex rosea Schk. Much like nos. 26 and 27, but more robust; fertile stems erect, 1.5–2.2 mm thick at base; widest lvs 1.8–2.6 mm wide; perigynia radiating in all directions, 2.6–4.2 mm, the base cuneate to rounded; achene set low in the perigynium, as in no. 27; stigmas a little stouter, 0.07–0.1 mm wide, once or twice coiled; $2n=52$. Mostly in habitats a little drier than those of no. 26, a littler moister than those of no. 27; N.S. and s. Que. to Minn. and e. Nebr., s. to Ga., Tenn., and Ark. (*C. convoluta.*)

29. Carex muhlenbergii Schk. Cespitose; stems slender but stiff and wiry, 2–10 dm, usually much surpassing the lvs; lvs flat but thick, 2–5 mm wide; infl ovoid to cylindric, 2–4 cm; spikes androgynous, ovoid or subglobose, 5–8 mm, densely aggregated but still easily distinguishable, the lower often separated but still overlapping; scales greenish, ovate, acute to acuminate or cuspidate; perigynia 8–20, greenish, 3–4 mm, broadly ovate, planoconvex, half to three-fourths as wide as long and the body often suborbicular, strongly 5–11-nerved dorsally, abruptly narrowed into a rough-margined beak a fourth to a third as long as the body. Dry, sterile or sandy soil, usually in full sun; Me. to Minn., s. to Fla., Nebr., and Tex. Two vars.:

a Perigynia spreading, nerved on both faces, or (var. *enervis; C. plana*) nerveless ventrally; scales, excluding the cusp, about as long as the body of the perigynium; bracts short and delicate, usually 5–20 mm, the triangular base about half as long as the lowest perigynium, the tip setaceous ..... var. *muhlenbergii.*
a Perigynia ascending, nerveless ventrally; scales acuminate into a cusp, the body equaling or often surpassing the whole perigynium; lowest bract elongate, the dilated base strongly nerved, equaling or surpassing the lowest perigynium, the spreading or ascending linear tip 1–5 cm; Mo. and Nebr. to Kans., Miss. and Tex. (*C. austrina.*) .... var. *australis* Olney.

30. Carex gravida L. Bailey. Densely cespitose; stems 3–6 dm, rough above; lvs 4–8 mm wide, the sheath loose, ventrally thin and pale, dorsally septate-nodulose and white with green veins or mottled; infl dense, ovoid or oblong, 1–3 cm, at least the lower spikes distinguishable; bracts setaceous, shorter than the head; scales with a slender tip surpassing the base of the beak of the subtended perigynium; perigynia yellowish-brown, shining, ovate, abruptly contracted into a rough-margined, bidentate beak about a third as long as the body; achene suborbicular, the style-base expanded into an ovoid body 0.8×0.3–0.5 mm; $2n=58$. Dry open soil and prairies; O. and sw. Ont. to Minn., S.D., and Wyo., s. to Ky. and Tex. Two weak vars.:

a Perigynia 4–5 mm, half as wide, slightly acuminate or gradually tapering into the beak, dorsally nerveless or very obscurely nerved; northern, s. to Ky. and Mo. .... var. *gravida.*
a Perigynia 3–4.5 mm, two-thirds to three-fourths as wide, abruptly narrowed into the beak, sharply few-nerved dorsally; southern, n. to sw. Mich., s. Io., and Kans. (*C. lunelliana.*) ....
.... var. *lunelliana* (Mackenzie) Hermann.

31. Carex sparganioides Muhl. Cespitose; stems 4–8 dm, usually rough above, equaling or surpassing the lvs; main lvs 4–10 mm wide, the sheath loose, ventrally pale and often thin, dorsally septate-nodulose and white with green veins or mottled; spikes androgynous, subglobose or ovoid; bracts none or setaceous, shorter than the heads; scales ovate, with broad hyaline margins, acute to cuspidate, barely if at all reaching the base of the beak of the subtended perigynium; perigynia planoconvex, green, ovate or lance-ovate, 3–4.5 mm, a third to two-thirds as wide, acuminate into a rough-margined beak a third to half as long as the body; achene lenticular, broadly ovate, the style-base very slightly thickened; $2n=46$, 48. Dry woods and thickets, especially in calcareous regions; N.B. and Que. to Minn., S.D., and Okla. Three vars.:

a Lower spikes well separated, sometimes by as much as 2–3 cm, the infl elongate, 4–10 cm; scales acuminate and usually distinctly cuspidate, about equaling the body of the perigynium; N.H. and Que. to Minn., s. to Va. and Mo. .... var. *sparganioides.*
a Spikes all aggregated into an ovoid or narrowly ovoid head 2–4 cm.
  b Scales as in var. *sparganioides*; culms sometimes smooth on the angles; N.Y. to Va. and Tenn., w. to s. Ont., Mich. and Okla. (*C. aggregata.*) .... var. *aggregata* (Mackenzie) Gleason.
  b Scales acute or with a very short cusp, scarcely more than half as long as the body of the perigynium;

stems strongly roughened on the wing-like angles; N.B. to Mich. and Minn., s. to N.J., O., and
Ill. (*C. cephaloidea.*) ....var. *cephaloidea* (Dewey) Carey.

32. Carex disperma Dewey.   Soft, slender, the stems 1–4 dm, scattered on slender, branching rhizomes;
lvs flat, 1–2 mm wide; spikes 2–5, sessile, separate or the upper approximate, 3–6 mm, with 1–4(–6) perigynia
and 1–3 terminal staminate fls; bract obsolete, or filiform and up to 2 cm; scales triangular-ovate, stramineous
to white-hyaline except the green midrib, equaling or more often shorter than the perigynia; perigynia ellipsoid,
2–3 mm, densely white-punctate, nearly round in cross-section, the margins appearing merely as 2 stronger
nerves, the minute beak 0.2 mm; achene thick-lenticular, filling the perigynium, its style-base semipersistent
as a slender apiculus; $2n=70$. Bogs and wet woods, usually in shade; circumboreal, s. to Pa., Ind., Minn.,
Utah, and Calif.

33. Carex vulpinoidea Michx.   Stems stout, clustered, 3–10 dm, aphyllopodic; lvs scattered along the
lower half or two-thirds of the stem, flat or nearly so, 2–5 mm wide; ventral side of the sheaths sparsely red-
dotted and usually conspicuously cross-rugulose; spikes androgynous, numerous, sessile, few-fld, densely aggregated into an often irregular or interrupted infl that is 5–10 cm long, up to 1.5 cm wide, and generally
compound at least toward the base; upper spikes hardly distinguishable; bracts small, setaceous, only the
lower ones sometimes elongate to 5 cm; pistillate scales slender, the firm midrib excurrent as an often greenish
awn 1–5 mm; perigynia flattened or planoconvex, ± stramineous to light brown or partly greenish, 2–3.5
mm, the body narrowly ovate to rotund-ovate, nerveless or inconspicuously few-nerved, serrulate-margined
distally; achene lenticular; $2n=52$. In marshes and other wet low places; Nf. to Fla., w. to B.C., Wash., and
Ariz. The widespread var. *vulpinoidea* usually has the lvs surpassing the stems; the perigynium tapers into
a prominent beak sometimes as long as the body. (*C. setacea.*) The var. *ambigua* F. Boott, mainly in our
coastal states, but also inland nearly throughout our range, has the lvs mostly surpassed by the stems; the
relatively broad perigynia are more abruptly contracted to the short beak up to half as long as the body. (*C.
annectens.*)

34. Carex stipata Muhl.   Stems stout, triangular, densely clustered, 3–10 dm; lvs coarse and often elongate;
sheaths prolonged beyond the base of the blade, ventrally thin, fragile, and ± conspicuously cross-corrugate;
spikes numerous, androgynous, small, few-fld, sessile, closely aggregated into a dense, compound infl 3–10
× 1–3 cm; bracts short and inconspicuous, or some of them setaceous and surpassing the spikes; scales mostly
shorter than the perigynia, these greenish or greenish-stramineous, widely spreading (the infl appearing prickly
because of the prominent beaks), lance-triangular or lance-ovate, gradually narrowed from the base to the
tip, or a little more abruptly so near the middle, planoconvex, spongy-thickened at the base, prominently
nerved-striate on both sides or seldom nerveless ventrally; achene lenticular; $2n=48$, 52. Wet low ground;
Nf. to Alas., s. to Fla., N.M., and Calif. Var. *stipata,* the common phase in our range, extending s. to N.C.
and Tenn., has the main blades mostly 5–10 mm wide, the scales acuminate or merely cuspidate, the perigynia
4–5.2 (avg 4.7) mm and a third as wide. Var. *maxima* Chapman, the Gulf states phase, extending n. on the
coastal plain to s. N.J. and inland to Mo. and Ind., is stouter, with larger and thicker heads and stems, the
main blades 8–15 mm wide, the scales cuspidate or short-awned, the perigynia 5–6 (avg 5.4) mm and less
than a third as wide, due to the relatively longer beak. (*C. uberior.*)

35. Carex laevivaginata (Kük.) Mackenzie.   Much like no. 34; sheaths not cross-corrugated, not prolonged,
the mouth distinctly concave and thickened; spike shorter and less compound, 2–5 × 1–1.5 cm, green or
somewhat stramineous at maturity; perigynia 4.9–6.2 (avg 5.2) mm, usually less than a third as wide. Wet
woods; Mass. to Mich. and Minn., s. to n. Fla., Ala., and Mo.

36. Carex crus-corvi Shuttlew.   Stems very stout, densely clustered, 4–8 dm, sharply triangular and narrowly winged, shorter than the lvs; main lvs 5–10 mm wide; sheaths thin and truncate at the mouth, not
corrugated; infl ovoid to cylindric, 8–18 cm, the lower branches ± separate, to 5 cm, the upper shorter and
contiguous; scales triangular-ovate, equaling or shorter than the body of the perigynium; perigynia divaricate,
5.6–8.2 mm, dilated at base into a suborbicular spongy disk 1.5–2.5 mm wide, the body ovate, planoconvex,
narrower than the disk, sharply nerved dorsally, obscurely nerved or nerveless ventrally, the beak much
longer than the body, deeply bidentate. Swampy woods and meadows throughout the Gulf states and n. in
the interior to Ind., s. Ont., s. Mich., and s. Minn. (*C. bayardii* Fern., a form with the perigynium-base
scarcely nerved.)

37. Carex conjuncta F. Boott.   Stems clustered, 4–8 dm, stout but soft, 2–4 mm wide when pressed, about
as long as the lvs; main lvs 5–10 mm wide; sheaths somewhat prolonged beyond the base of the blade, cross-
corrugate ventrally; infl slender, 2–5 cm, compound, the lower branches distinct, the upper spikes scarcely
distinguishable; scales ovate-triangular, about equaling the perigynia; perigynia green or becoming ± stramineous, flatly planoconvex, ovate, 3.5–4 mm, half as wide, rounded at the spongy base, sharply few-nerved
dorsally, nerveless ventrally or with 1–3 short nerves at the base, tapering to a rough beak half as long as
the body. Damp woods; N.Y. to Minn. and S.D., s. to Va. and e. Kans.

38. Carex alopecoidea Tuckerman.   Stems clustered, 4–8 dm, stout but soft, 2–3 mm wide when pressed,
usually shorter than the lvs; main lvs 3–6 mm wide; sheaths somewhat prolonged beyond the base of the
blade, not cross-corrugated; infl 2–4 cm, slender, simple or compound, the lower spikes sometimes slightly
separated; scales triangular-ovate, usually shorter than the perigynia; perigynia becoming largely or wholly
golden-brown, flatly planoconvex, ovate, 3.2–3.8 mm, half as wide, rounded at base, inconspicuously 2–3-
nerved dorsally, nerveless ventrally, tapering into a rough beak two-thirds to nearly as long as the body. Wet
meadows; Que. and Me. to Mich., Minn. and Sask., s. to N.J., Ind., and Mo.

39. Carex decomposita Muhl. Stems stout, clustered, 5–10 dm, aphyllopodic; lvs ± elongate, mostly 5–8 mm wide; sheaths ventrally pale, red-dotted, and concave at the mouth; spikes androgynous, small, sessile, aggregated into a compound, evidently spicate-paniculate infl 7–15 cm, the lower branches to 4 cm and with numerous spikes; scales triangular-ovate, mucronate, about equaling the body of the perigynium; perigynia deep olive-green, coriaceous, biconvex, obovoid, 2–2.8 mm, three-fifths or two-thirds as wide, obscurely few-nerved ventrally, strongly several-nerved in the basal half dorsally, abruptly rounded above to a rough-margined beak a third as long as the body. Swamps; N.Y. to Mich., s. to Fla. and La.; very rare in the north.

40. Carex diandra Schrank. Stems densely clustered, 3–10 dm, aphyllopodic, rather slender above; lvs ± elongate, ± flat, mostly 1–2.5 mm wide; sheaths ventrally pale and red-dotted, prolonged 2–3 mm beyond the base of the blade; spikes androgynous, ± numerous, small, sessile, aggregated into a compact but branched infl 2–3.5(–5) cm long and seldom over 1 cm thick; bracts small, setaceous or almost like the pistillate scales; stamens 3 (in spite of the name); scales largely hyaline-scarious and stramineous or brownish, with firmer, sometimes shortly excurrent midrib; perigynia ovate or lance-ovate, 2.4–3 mm, half as wide, usually dark brown, shining, ± spreading, very firm and thick-walled, the nerveless ventral surface slightly convex and often exposed in the spike, the dorsal surface strongly convex and with a median thin, often depressed and pale strip (an extension of the dorsal suture) bordered by a pair of veins and reaching to the base; beak ± abruptly differentiated from the body of the perigynium, coarse, serrulate-margined, often pale or greenish; achene lenticular; $2n=60$. Swamps, wet meadows, and sphagnum-bogs; circumboreal, s. to N.J., Pa., Ind., Mo., Colo., and Calif.

41. Carex prairea Dewey. Stems clustered, aphyllopodic, 5–10 dm, somewhat exceeding the lvs; main lvs 2–3 mm wide; sheaths prolonged 2–3 mm beyond the base of the blade, strongly coppery toward the mouth ventrally; spikes androgynous, numerous, small, sessile, aggregated into a thick-spiciform infl, this 3–8 cm, usually interrupted below, compound, at least the lowest node with 2 or 3 spikes; scales reddish-brown, lance-ovate, as long as and largely concealing the ± appressed perigynia, these brown, dull, lance-ovate, 2.5–3 mm, half as wide, tapering to a serrulate, pale beak nearly as long as the body, flat to low-convex and nearly nerveless ventrally, broadly rounded and strongly few-nerved dorsally, the main nerves divergent well above the base; $2n=66$. Swamps, wet meadows, and wet prairies; Que. to Sask., s. to N.J., O., and Io. (*C. diandra* var. *ramosa*.)

42. Carex tenuiflora Wahlenb. Tufted and also with long, slender rhizomes; stems very slender, 2–6 dm; main lvs 1–2 mm wide; spikes 2–4, gynaecandrous, sessile, closely approximate into an ovoid head 8–15 mm; scales white-hyaline with green center, as long and wide as the perigynia; perigynia 3–15, densely white-punctate, ellipsoid or somewhat obovoid, planoconvex 3–3.5 mm, half as wide, sharp-edged, obscurely nerved, tapering, beakless; achene lenticular-ellipsoid, almost filling the perigynium; $2n=58$. Wet woods and bogs; circumboreal, s. to Me., N.Y., Mich., and Minn. An apparent hybrid with no. 46 has been called *C. × trichina* Fern.

Carex loliacea L., a circumboreal sp. with 2–5 ± remote spikes, the pistillate scales white-hyaline except the midrib, about thrice the 3–8 ± widely spreading, evidently multinerved, tapering, beakless perigynia 2.5–3 mm, approaches our range along the n. shore of Lake Superior and may be sought in n. Minn.

43. Carex heleonastes L. f. In small tufts and also with long, slender rhizomes; stems slender but stiff, usually surpassing the lvs, sharply triangular, very rough on the angles above; lvs 4–8 on the lower fourth of the stem, 6–12 cm × 1–2 mm, scabrous distally; spikes 2–4, gynaecandrous, aggregated into a head 1–2 cm × 5–10 mm; scales strongly brown-tinged with pale margins, nearly as long as the perigynia; perigynia 5–10, elliptic-obovate, planoconvex, 2.5–3 mm, abruptly sharp-margined, finely several-nerved on both sides, tapering to a distinct (0.5 mm) beak; achene lenticular, filling the perigynium. Wet open places, especially in calcareous regions; circumboreal, and reported from n. Mich.

44. Carex glareosa Wahlenb. Tufted on short rhizomes; stems very slender, 2–4 dm, surpassing the lvs; main lvs 1–2 mm wide; spikes 2–4, closely approximate in an ovoid or oblong head 1–2 cm, the lateral pistillate, the terminal with slender staminate base; scales ovate, slightly shorter than the perigynia, brown with narrow hyaline margins; perigynia obovate-fusiform, multinerved, 1.5–3 mm, less than 2.6 times as long as wide, the beak a fifth as long as the body; achene lenticular; $2n=66$. Salt-marshes; interruptedly circumboreal, s. to N.B. and Que. Our plants represent the irregularly distributed var. *amphigena* Fern. (*C. marina*, misapplied.)

45. Carex mackenziei Krecz. Tufted and with long, slender rhizomes; stems erect, 1.5–4 dm; main lvs 1–3 mm wide; spikes 3–6, ovoid or short-cylindric, sessile, separate with the 2 uppermost approximate, 1 cm, the terminal much longer and half-staminate; scales reddish-brown, broadly ovate, about as long and wide as the perigynia; perigynia planoconvex, oval, stipitate, 2.5–3.3 mm, half as wide, lightly nerved on both faces, the smooth short beak a sixth as long as the body; achene lenticular. Salt or brackish marshes; circumboreal, s. to Me. and N.S. (*C. norvegica*, a preoccupied name.)

46. Carex trisperma Dewey. Loosely tufted on short, slender rhizomes; stems very slender and weak, 2–7 dm; spikes (1)2(3), sessile, 1–4 cm apart in a slender, often flexuous infl, each with 1–5 perigynia and a few basal staminate fls; lowest spike subtended by a setaceous bract 2–4 cm; scales ovate, acute, hyaline with a green center, shorter than or about equaling the perigynia, these thickly planoconvex, oval, 2.6–4 mm, finely many-nerved, the slender, smooth, emarginate beak 0.5 mm; achene lenticular, oval-oblong, filling the perigynium; $2n=60$. Typical plants have lvs 1–2 mm wide and usually 2–5 perigynia per spike. Plants from Nf. to Vt. and Pa., mostly near the coast, with setaceous lvs 0.5 mm wide, and with only 1 or 2 somewhat smaller perigynia, have been called var. *billingsii* Knight.

47. Carex brunnescens (Pers.) Poiret. Much like no. 48; lvs green, 1–2.5 mm wide; spikes avg smaller, 4–8 mm, brownish, or sometimes pale as in no. 48; perigynia mostly 5–10(–15), a little more spreading, so that the beak-apiculations interrupt the outline of the spike, evidently but rather finely several-nerved dorsally, nerveless or obscurely nerved ventrally, smoother-textured, the wall very thin distally and easily ruptured; dorsal suture well developed, 0.4–0.8 mm, extending onto the distal end of the body, often with a narrow, white-hyaline, overlapping longitudinal flap; $2n=56$. Bogs and wet woods; circumboreal, s. to N.J., O., Mich., Minn., and in the mts. to N.C.

48. Carex canescens L. Tufted, 2–6 dm; lvs mostly somewhat glaucous or grayish-green in life, clustered near the base, often shorter than the stems, ± flat, 1.5–3.5 mm wide; bracts inconspicuous, or the lowest ones setaceous-tipped and surpassing the spikes; spikes 4–8, gynaecandrous, 5–10 mm, sessile, the upper approximate, the lower usually somewhat remote, silvery-greenish to pale grayish or pale-stramineous; pistillate scales mostly shorter than the perigynia, these mostly (10–)15–30, ascending (the short beaks not interrupting the outline of the spike), mostly 1.8–2.5 mm, planoconvex, not thin-edged, minutely rough-textured, somewhat corky-thickened below, not especially thin-walled distally, elliptic or elliptic-ovate, with a beak-apiculation up to 0.5 mm, evidently but rather finely nerved on both faces, the ventral nerves fewer than the dorsal; dorsal suture short and inconspicuous, up to ca 0.4(–0.7) mm, or practically obsolete; achene lenticular, filling the perigynium; $2n=54$, 56. Swamps and bogs; circumboreal, s. to Va., O., Minn., Ariz., and Calif.

49. Carex exilis Dewey. Densely tufted, the stiff, smooth stems 2–7 dm, usually surpassing the slender, involute lvs; spike usually solitary, 0.8–2.5(–4) cm, gynaecandrous or unisexual; lateral spikes 1 or 2 and much smaller, or usually none; lower 2 scales empty and erect; anthers 2–3.5 mm; pistillate scales ovate, about equaling the body of the perigynium, usually acute, reddish-brown with hyaline margins; perigynia up to ca 25, at least the lower ones spreading or reflexed, castaneous, planoconvex, ovate, 2.6–4.7 mm, spongy-thickened at base, with up to 15 faint nerves dorsally and up to 7 ventrally, the serrulate-margined beak a fifth to two-thirds as long as the body; achene lenticular; $2n=$ca 62. Sphagnum-bogs and other open, wet, low places; Lab. and Nf. to Del. and Md., w. to Ont., n. Mich., and n. Minn.; disjunct in N.C. and Miss.

50. Carex arcta F. Boott. Stems densely tufted, (1–)3–6 dm, sharply triangular, scabrous on the angles above; lvs all on the lower part of the stem but not closely tufted, ± flat, 1.5–4 mm wide, elongate, often surpassing the stems; spikes 7–15, gynaecandrous, sessile, 5–10 mm, all ± closely aggregated into a pale greenish to brownish-green, narrowly ovoid to cylindric head 1.5–4 cm; bracts short and inconspicuous, or the lowest ones setaceous and to 5 cm; perigynia (20–)25–40, ± strongly spreading, planoconvex, somewhat spongy at base, ovate or lance-ovate, 2.2–3.4 mm, several-nerved dorsally, few-nerved or nerveless ventrally, with a prominent, serrulate-margined beak 0.6–1.2 mm; achene lenticular; $2n=60$. Streambanks, meadows, and other wet places; Que. to s. Yukon, s. to N.Y., Mich., Minn., and Calif.

51. Carex sterilis Willd. Stems tufted, aphyllopodic, 1–7 dm, stiff, scabrous on the angles above; lvs 3–5 per stem, all in the basal third, shorter than the stems, plicate, 1–2.5 mm wide, scabrous above; spikes 3–8, 3–13 mm, sessile, all crowded or the lower remote; subdioecious, occasionally with a few intermingled or basal fls of opposite sex, rarely with staminate and pistillate spikes in the same infl; anthers 1.2–2.2 mm; pistillate scales castaneous with green midrib and hyaline margins; perigynia 5–25, at least the lower ones spreading to reflexed, castaneous, planoconvex, ovate to deltoid, spongy-thickened at base, 2.1–3.8 mm, two-fifths to two-thirds as wide, 5–12-nerved on both sides or nerveless ventrally, serrulate distally and on the prominent, softly bidentate beak 0.6–1.6 mm; achene lenticular. Calciphile, in wet places; Nf. to Sask., s. to Pa., W.Va., Tenn., Ill., and Mo. (*C. muricata* var. *s.; C. elachycarpa,* teratological.)

52. Carex wiegandii Mackenzie. Stems tufted, aphyllopodic, 1–10 dm, smooth to slightly scabrous above; lvs 3–8 per stem, all in the basal third, usually shorter than the stems, plicate, 2–5 mm wide, scabrous above; spikes 4–6, 4.5–8 mm (or the terminal one to 15 mm), gynaecandrous, sessile, aggregated to an infl 1–3 cm, the 2 lowest sometimes set as much as 10 mm apart; anthers 0.7–1.3 mm; pistillate scales castaneous with green midrib and hyaline margins; perigynia 5–25, at least the lower ones spreading to reflexed, green to castaneous, planoconvex, broadly ovate, spongy-thickened at base, 2.5–3.7 mm, two-fifths to two-thirds as wide, 5–18-nerved dorsally, nerveless or with up to 10 faint nerves ventrally, serrulate distally and on the bidentate beak 0.6–1.2 mm; achene lenticular. Sphagnum-bogs and other wet, open places; rare; Nf. and Lab. to Mass., nw. Pa., Ont., and n. Mich.

53. Carex ruthii Mackenzie. Much like no. 52, but with a longer, more open infl mostly 3–8.5 cm, the lowest spikes set 1–4 cm apart; lvs ± equaling the stems, plicate to flat; perigynia 3–4.5 mm, olive-green, and only the prominent, 0.9–1.8 mm beak serrulate. Moist woods and boggy meadows and balds; Blue Ridge, from sw. Va. to nw. S.C. and ne. Ga. (*C. muricata* var. *r.*)

54. Carex interior L. Bailey. Stems tufted, aphyllopodic, 2–9 dm, smooth; lvs 3–5 per stem, all in the basal third, shorter than to about equaling the stems, plicate, 1–2.5 mm wide; spikes (2)3–6, sessile, contiguous or ± remote, small and few-fld, the terminal one with a conspicuous, slender, staminate base, some of the lateral ones often wholly pistillate; bracts small and inconspicuous; anthers 0.6–1.4 mm; perigynia mostly 5–15, crowded, widely spreading or the lower reflexed, green or tan, planoconvex, coriaceous and shining, spongy-thickened at base, 4–12-nerved dorsally, nerveless or occasionally few-nerved ventrally, ovate or triangular-ovate, 2.2–3.2 × 1.1–1.8 mm, 1.4–2 times as long as wide, tending to be ± convexly tapered from the widest point to the beak, conspicuously serrulate-margined distally and on the beak, this short, broad, very shallowly bidentate, 0.5–1 mm, a fourth to half as long as the body; achene lenticular; $2n=54$. Swamps, bogs, and other wet places; Nf. and Lab. to s. Yukon and Alas., s. to Va., Mo., and Mex.

55. Carex atlantica L. Bailey. Very much like no. 54; stems scabrous on the angles above; lvs 0.5–4 mm wide; perigynia green, seldom castaneous, usually evidently several-nerved ventrally, 1.9–3.8 × 1.3–3 mm, 1.1–1.7 times as long as wide, broadly ovate to suborbicular, cuneately or somewhat concavely narrowed from the widest point to the sparsely and less conspicuously serrulate, sharply bidentate beak 0.5–1.2 mm. Swamps and bogs; N.S. to Fla., especially on the coastal plain, but also w. to Mich., Ill., Mo., Ark., and Tex. Var. *atlantica*, with nearly the range of the species, is relatively robust, with the widest lvs 1.5–4 mm, the infl mostly 1.8–4.5 cm, with 3–8 spikes, and the perigynia 2.3–3.8 × 1.5–3 mm. (*C. incomperta.*) Var. *capillacea* (L. Bailey) Cronq., mainly on the coastal plain, but also occasionally inland to O., s. Ont., s. Mich., and n. Ind., is smaller and more slender, with lvs 0.5–1.5 mm wide, the infl 0.8–2 cm with 2–5 spikes, the perigynia 1.9–3 × 1.3–2 mm. (*C. howei.*)

56. Carex echinata Murray. Star-sedge, Stems densely tufted, aphyllopodic, 1–6 dm, scabrous on the angles above; lvs 3–6 per stem, all in the basal third, up to about as long as the stems, plicate, mostly 1–2.5 mm wide; spikes 3–7, sessile, small and few-fld, the terminal one with a conspicuous, slender, staminate base, some of the lateral ones often wholly pistillate; bracts small and inconspicuous; anthers 0.8–1.6 mm; perigynia 5–15, crowded, widely spreading or the lower reflexed, green or tan, planoconvex, spongy-thickened at base, lightly several- to many-nerved dorsally, few- to several-nerved or virtually nerveless ventrally, narrowly lance-triangular to move commonly lance-ovate or even ovate, 2.8–3.5(–4) mm, 1.8–3.2 times as long as wide, often serrulate-margined distally, with a prominent, slender, serrulate-margined, sharply bidentate beak 1–1.6 mm, half as long to almost as long as the body; achene lenticular; $2n$=50–58. Swamps, bogs, and other wet places; circumboreal, s. to Va. (and in the mts. to N.C.), Ind., Io., Utah, and Calif. Ours is the widespread var. *echinata*. (*C. angustior; C. cephalantha; C. josselynii,* teratological; *C. laricina; C. muricata,* misapplied; *C. stellulata.*)

57. Carex seorsa Howe. Stems densely tufted, aphyllopodic, 2–7 dm, smooth; lvs 2–4 per stem, all in the basal third, usually shorter than the stems, plicate to flat, mostly 1–4 mm wide; spikes 4–8, sessile, small and few-fld, gynaecandrous or the lateral ones often wholly pistillate; anthers 1–2 mm; perigynia 5–25, crowded, widely spreading or the lower reflexed, green, planoconvex, spongy-thickened at base, 6–14-nerved dorsally, 0–6-nerved ventrally, elliptic-ovate, 2–3 × 1–2 mm, 1.2–2.1 times as long as wide, smooth-margined, with a smooth, truncate or obscurely bidentate beak 0.2–0.6 mm, up to a third as long as the body; achene lenticular; $2n$=48. Wet woods; s. N.H. to Ga., especially on the coastal plain, and irregularly inland to n. O., s. Ont., Mich., nw. Ind., and Tenn.

58. Carex bromoides Willd. Densely cespitose, very slender, erect or ascending, 3–8 dm; main lvs 1–2 mm wide; spikes 3–7, sessile, to 2 cm, usually approximate or overlapping, at least the terminal one gynaecandrous; bracts short; pistillate scales ovate-oblong, shorter than or about equaling the perigynium-body, pale brown or orange-tinged, with hyaline margins, acutish to short-cuspidate; perigynia 6–15, light green, loosely ascending or appressed, planoconvex, lanceolate, 4–5.5 mm, a sixth to a fourth as wide, serrulate distally, sharply nerved on both sides, or only weakly so ventrally, tapering at the spongy-thickened base, the rough-margined, sharply bidentate beak half to two-thirds as long as the body; achene narrow, lenticular, 2 mm. Wet woods, swamps, and bogs; Que. and N.S. to Ont. and Minn., s. to Fla. and Mex.

59. Carex deweyana Schwein. Much like no. 58; main lvs 2–3+ mm wide; spikes pale green or silvery, the lower one often remote and conspicuously surpassed by a subtending bract; pistillate scales largely hyaline or scarious and whitish or light brown, with firmer, greenish, sometimes shortly excurrent midrib, often wholly covering the body of the perigynium, but shorter than the beak; perigynia mostly (5–)10–25, lance-elliptic to rather narrowly elliptic, 4–5.5 mm, a fourth to a third as wide, faintly nerved or nerveless on both sides; achene broad, ± orbicular-obovate, 2–2.5 mm; $2n$=54. Woods; Lab. and Nf. to Mack. and B.C., s. to Pa., Mich., Io., Ariz., Calif., and Mex. Our plants are var. *deweyana;* two other vars. are more western.

60. Carex synchnocephala Carey. Densely cespitose, erect, 1–5 dm; main lvs 2–4 mm wide; bracteal lvs 2–4, dilated at base, the larger usually 1–2 dm; spikes ca 10, ovoid, 1 cm, closely crowded and almost indistinguishable in an ovoid, lobed head 2–3 cm; pistillate scales lanceolate, two-thirds as long as the perigynia, hyaline except the midnerve, acuminate; perigynia very thin, lanceolate, 5–6 mm, one sixth as wide, lightly nerved on both sides, gradually tapering to the long, slenderly bidentate beak; achene lenticular, elliptic, 1–1.5 mm. Wet meadows and shores; N.Y. to Mo., Io., Mont., and Sask.

61. Carex praticola Rydb. Densely tufted, aphyllopodic, 3–8 dm; main lvs 2–4 mm wide, shorter than the stems; spikes 3–7, gynaecandrous, 7–18 mm, sessile in a slender, rather loose and flexuous, ± nodding spike; pistillate scales largely concealing the mature perigynia, except often beak-tip; perigynia loosely ascending-appressed, pale green to brownish (4–)4.3–5.7(–6) × (1.2–)1.3–2(2.1) mm, 2.6–4 times as long as wide, planoconvex, wing-margined and serrulate, evidently nerved dorsally, nerveless or nearly so (seldom evidently nerved) ventrally, the beak with a subterete, ± entire tip 0.5+ mm; achene lenticular, 1.7–2.2 mm. Meadows; Greenl. to Alas., s. to Que., n. Mich., N.D., and Calif. (*C. pratensis,* a preoccupied name.)

62. Carex oronensis Fern. Loosely tufted, 5–10 dm; main lvs 3–5 mm wide, shorter than the stems; spikes 3–9, gynaecandrous, 5–10 mm, short-clavate at base, dark brown, sessile in an erect oblong head 2–3 cm; pistillate scales ± concealing the perigynia; perigynia lance-subulate, 4–4.5 mm, 3–4 times as long as wide, planoconvex, tapering to a slender beak, scarcely winged below the middle. Penobscot R. valley in Me.

63. Carex leporina L. Tufted, erect, 3–8 dm; main lvs 2–4 mm wide, shorter than the stems; spikes 3–7, gynaecandrous, ovoid, 1 cm, sessile and closely crowded; pistillate scales ovate, reddish-brown with broad hyaline margins, largely concealing the perigynia; perigynia 4–5.5 mm, 2–2.5 times as long as wide, planoconvex, narrowly scarious-margined, finely but sharply nerved on both sides, the beak distally terete, reddish-brown at the tip; achene lenticular, 1.5–2 mm; $2n=64, 66, 68$. Native of Eurasia, intr. in wet meadows and pastures from Nf. to N.Y. and reputedly to N.C. (*C. ovalis.*)

64. Carex foenea Willd.   Densely tufted, aphyllopodic, 3–10 dm; main lvs 3–5 mm wide, mostly shorter than the stems; bracts inconspicuous; spikes 3–15, gynaecandrous, 6–25 mm, often basally clavate, sessile in a loose, commonly flexuous spike that may be interrupted below; pistillate scales as long as but often evidently narrower than the perigynia; perigynia planoconvex, usually turning pale to dark brown (especially over the achene ventrally) at maturity, smooth, 3.4–5 × 1.2–2.2 mm, ventrally nerveless or obscurely few-nerved, more strongly (1–)3–8-nerved dorsally, usually widest at ⅓ to ⅔ of their total length, tapering to a flattened, often ill-defined, sometimes minutely slender-tipped beak, wing-margined, the wings finely and evenly serrulate-ciliolate, not expanded above the middle of the body; achene lenticular, 1.5–2 mm. Variously in wet to dry places; Lab., and Nf. to Yukon and B.C., s. to Conn., N.Y., Pa., Mich., and Ida. (*C. aenea.*)

65. Carex argyrantha Tuckerman.   Much like no. 64, but the perigynia uniformly silvery-green, finely granular-papillose, 3.1–4.2 × 1.2–2.2 mm, strongly and evenly (4)5–8(–10)-nerved on both faces, usually widest at ⅔ to ½ of their total length, the wing-margins expanded above the middle of the body, more coarsely and irregularly serrulate or erose-ciliolate. Woods and clearings, often in dry, sandy or rocky soil; N.S. and N.B. to Va., w. to s. Que., s. Ont., O., and Mich. (*C. foenea,* misapplied.)

66. Carex xerantica L. Bailey.   Densely tufted, aphyllopodic, 3–7 dm; main lvs generally much shorter than the stem, 2–4 mm wide; bracts inconspicuous; spikes 3–6, gynaecandrous, 8–17 mm, ovoid-elliptic, sessile in a short stiff spike 1–2 cm; pistillate scales brown with paler margins, or largely white-hyaline, ± concealing the perigynia, these ± strongly appressed, elliptic or ovate, slightly to strongly planoconvex, 4.5–7 × 1.9–2.8 mm, 2.1–2.9 times as long as wide, lightly to evidently multinerved on both sides, or the ventral side nearly nerveless, the body wing-margined and distally serrulate, tapering to the mostly ill-defined, flattened, strongly margined and serrulate beak; achene lenticular, 2–3 × 1.2–1.6 mm; $2n=68$. Bluffs, dry hillsides, and grassy plains; Alta. to Ariz., e. to Man., Minn. and Neb.

67. Carex adusta F. Boott.   Densely tufted, aphyllopodic, 2–8 dm; main lvs surpassed by the stem, 3–4 mm wide; lowest bract dilated at base and often twice as long as its spike; spikes 4–15, gynaecandrous, 6–12 mm, subglobose, short-clavate at base, sessile and densely crowded in an ovoid or oblong cluster 2–3 cm; pistillate scales brown with narrow white-hyaline margins, largely concealing the perigynia; perigynia appressed-ascending or in age looser, oblong-ovate, strongly planoconvex, 4.2–5.2 mm, 2–2.5 times as long as wide, finely nerved dorsally, obscurely or scarcely so ventrally, narrowly serrulate-winged above the middle, merely sharp-edged below, rather abruptly narrowed to the flat, serrulate beak; achene lenticular, 2–2.5 × 1.5–2 mm; $2n=64$. Dry, open places; Nf. to Mack., s. to N.Y., Mich., Minn., and B.C.

68. Carex muskingumensis Schwein.   Cespitose, with numerous very leafy sterile stems; fertile stems stout, 5–10 dm; main lvs 3–5 mm wide; sheaths ventrally green-veined almost to the summit; spikes 5–10, gynaecandrous, fusiform, pointed at both ends, 15–25 × 4–6 mm, closely aggregated into a dense cluster 4–8 cm; pistillate scales lanceolate, half as long as the perigynia, pale brown with hyaline margins; perigynia appressed, lanceolate, thin, 6.5–10 mm, 3–4 times as long as wide, finely nerved on both sides, gradually tapering to the flat, serrulate, deeply bidentate beak half as long as the body; achene lenticular, 2–2.5 mm, less than 1 mm wide. Low woods, wet meadows, and river-bottomlands; O. and Ky. to Mich., Minn., Man., Kans., and Okla.

69. Carex scoparia Schk.   Tufted, 3–10 dm, aphyllopodic; main lvs 1.5–3 mm wide, shorter than the stems; sheaths ventrally hyaline; spikes 3–8, gynaecandrous, 8–14 mm, ovoid to fusiform or subglobose, pale greenish to dull-stramineous or tan, sessile in an open to more often condensed spike 2–4 cm; bracts inconspicuous, or the lowest one up to about as long as the infl; pistillate scales lance-ovate, acuminate or shortly awn-pointed, narrower and shorter than the perigynia; perigynia lanceolate, very flat, 4–5.5 × 1.5–2 mm, 2.5–3 times as long as wide, much wider than the achene, several-nerved on both sides, wing-margined and serrulate, tapering gradually to the ill-defined flat beak; achene lenticular, 1.3–1.8 × 0.7–0.8 mm; $2n=60, 64, 68$. Open swamps, wet meadows, and shores; Nf. to Fla., w. to B.C., Oreg., and N.M.

70. Carex crawfordii Fern.   Tufted, 2–7 dm, aphyllopodic; main lvs 1–3 mm wide, shorter than the stems; sheaths ventrally hyaline; spikes 5–12, gynaecandrous, 5–10 mm, subglobose to oblong, pale greenish to dull-stramineous or tan, sessile in a compact spike or narrow head 1.5–3 cm; bracts inconspicuous, or the lowest one up to about as long as the infl; pistillate scales shorter and narrower than the perigynia, sometimes mucronulate; perigynia appressed-ascending, planoconvex and not much wider than the achene, 3.3–4 × 0.8–1 mm, 3.5–5 times as long as wide, lightly few-nerved on both sides or nearly nerveless especially ventrally, narrowly wing-margined and serrulate, especially distally, tapering gradually into the slender, distally terete beak; achene lenticular, 1–1.3 × 0.6–0.7 mm; $2n=68, 70$. Wet soil, meadows, swamps, and shores; Nf. to B.C., s. to N.J., Mich., Minn., and Wash.

71. Carex tribuloides Wahlenb.   Tufted, aphyllopodic, usually stout, 5–10 dm; main lvs 3–7 mm wide, surpassed by the stems; sheaths ventrally green-veined almost to the summit; spikes 5–15, gynaecandrous, 6–12 mm, ovoid to subglobose or obovoid, usually blunt, rounded to acute at base, sessile, densely to loosely aggregated into an ovoid or oblong cluster 2–5 cm; pistillate scales lanceolate, half to two-thirds as long as the perigynia, merely acute, pale brown with hyaline margins and tip; perigynia numerous (more than 30), appressed-ascending, green or becoming stramineous, 4–5(–5.5) × 1–1.3(–1.6) mm, 3–4 times as long as wide, lanceolate, flat or barely distended over the achene, the wing rather abruptly narrowed about the middle and often obsolete in the lower fourth, obscurely nerved on both sides, gradually tapering into the beak;

achene lenticular, 1.5 × 0.5–0.75 mm. Wet woods and meadows; Que. and Me. to Minn., and Neb., s. to Fla., La., and Okla., more common s.

72. **Carex projecta** Mackenzie.   Much like no. 71; spikes subglobose to obovoid, 5–8 mm, about as thick, distinct and ± separated (at least the lower) in a rather lax and flexuous infl 3–5 cm; perigynia mostly 15–30, loosely ascending-spreading at maturity, dull brown, 3–5 mm, the wing usually gradually narrowed from the middle to the base, sharply nerved dorsally, obscurely so ventrally; 2*n*=64. Open swamps and wet meadows; Nf. to Man., s. to W.Va., Ill., and Mo. (*C. tribuloides* var. *reducta*.)

73. **Carex cristatella** Britton.   Stout, tufted, aphyllopodic, 4–9 dm; main lvs 3–7 mm wide, shorter than or equaling the stems; sheaths ventrally green-veined almost to the summit; spikes 6–12, gynaecandrous, 5–8 mm long and thick, subglobose or short-ovoid, sessile, densely crowded in an ovoid cluster 2–4 cm; lowest bract setaceous-prolonged but shorter than the infl; pistillate scales lanceolate, much shorter than the perigynia, hyaline with a green midnerve, merely acute; perigynia greenish-stramineous to pale brown, spreading or slightly recurved, oblong, 2.4–3.9 mm, 2–3 times as long as wide, distended over the achene, the wing usually broadest above the achene and tapering gradually to the base, finely nerved on both sides, gradually tapering to the beak; achene lenticular, 1.5 × 0.5 mm; 2*n*=70. Open swamps, wet meadows, and shores; N.H. and w. Que. to Va., w. to N.D., Neb., and e. Kans. (*C. cristata*, a preoccupied name.)

74. **Carex bebbii** (L. H. Bailey) Fern.   Densely tufted, 2–9 dm, aphyllopodic; lvs elongate, mostly 2–4 mm wide; sheaths ventrally hyaline; spikes 4–12, gynaecandrous, 5–9 mm, pale greenish to stramineous or light brown, sessile in a compact, crowded spike or head 1.5–3 cm; bracts inconspicuous, even the lowest one shorter than the infl, pistillate scales shorter and narrower than the perigynia, largely hyaline-scarious except for the firmer, often greenish midrib; perigynia crowded, stiffly ascending (the beaks often standing out from the body of the spike), ovate, 2.7–3.7 × 1.1–1.5 mm, 2–2.7(–3) times as long as wide, planoconvex, evidently nerved on both sides or nerveless ventrally, wing-margined and serrulate, tapering to an ill-defined, flattened, serrulate beak; achene lenticular, 1.1–1.5 × 0.6–0.8 mm; 2*n*=68. Wet meadows and shores, especially in calcareous soil; Nf. to B.C., s. to N.J., Ill., Nebr., and Colo.

75. **Carex tenera** Dewey.   Much like no. 74; lvs 1.5–2.5 mm wide; spikes mostly 4–8(–12), in a usually moniliform or interrupted spike 1.5–5 cm, the individual spikes a little stouter and more ragged-looking than in *C. bebbii*; perigynia 2.8–4(–4.3) × 1.4–1.9 mm, (1.5–)1.7–2.5(–3) times as long as wide, planoconvex or flattened, achene 1.3–2.1 × 0.9–1.3 mm; 2*n*=56. Moist or wet soil, meadows, and thickets; Que. and Me. to N.C., w. to Mo., S.D., and Mont.

76. **Carex brevior** (Dewey) Mackenzie.   Densely tufted, 2–10 dm, aphyllopodic; lvs coarse and firm, 2–4 mm wide, much shorter than the stems; spikes mostly 3–6, gynaecandrous, stout, 6–10(–15) mm, sessile in a moniliform or interrupted to occasionally more congested spike 1.5–5 cm, greenish-stramineous to light-brown; bracts inconspicuous, shorter than the infl; pistillate scales largely hyaline-scarious, with firmer, often greenish midrib, distinctly narrower and tending to be shorter than the perigynia; perigynia crowded, stiffly ascending (the beaks tending to stand out from the body of the spike), 3.2–4.8 × 2.2–3.4 mm, 1.3–1.7 times as long as wide, the body ± strongly flattened, suborbicular, nerved on both sides (*C. molesta*) or more commonly nearly nerveless ventrally, wing-margined all around and serrulate distally, ± abruptly contracted to the flattened, serrulate beak 0.8–1.5 mm; achene lenticular, 1.6–2.2 × 1.3–1.9 mm; 2*n*=68. Widespread and abundant, usually in dry soil; Que. to Va., w. to the Pacific, (*C. merritt-fernaldii*; *C. festucacea*, misapplied.)

77. **Carex reniformis** (L. Bailey) Small.   Much like no. 76; lvs 2–3 mm wide; pistillate scales much narrower and shorter than the perigynia; perigynia 4.5–5.7 × 3.3–4.5 mm, the body generally wider than long, nerveless or nearly so ventrally. Moist soil on the coastal plain; se. Va. to Ga., Ark., and Tex., and inland to Tenn. and Mo.

78. **Carex normalis** Mackenzie.   Tufted, aphyllopodic, leafy-stemmed, 3–8 dm; main lvs 2.5–6 mm wide, shorter than the stem; sheaths ventrally hyaline; spikes 5–10, gynaecandrous, subglobose, 6–9 mm, sessile, loosely aggregated in an infl 3–5 cm, or the lowermost occasionally separate; pistillate scales ovate, much (ca 1.5 mm) shorter than the perigynia, hyaline, slightly brown-tinged, with green midnerve, acute or obtuse; perigynia ascending, planoconvex, ovate, green or pale greenish-brown, 3.3–4.4 mm, 2–2.5 times as long as wide, broadest at a fourth to a third their length, finely nerved dorsally, lightly or obscurely nerved ventrally, gradually tapering to the flat, serrulate beak; achene lenticular, 1.5–2 × 1 mm. Open woods and meadows; Me. to Mich. and S.D., s. to N.C., O., Mo., and Okla. (*C. mirabilis*, a preoccupied name.) *C. tincta* Fern., with brown scales pale at the margin, may be a hybrid with *C. foenea*.

79. **Carex festucacea** Schk.   Tufted, aphyllopodic, the slender stems 5–10 dm, surpassing the lvs; main lvs 2–5 mm wide; sheaths ventrally hyaline; spikes 4–10, gynaecandrous, ovoid to subglobose, 6–16 mm, often distinctly clavate at base, sessile, distinct but crowded in a compact cluster or moniliform infl 2.5–6 cm; pistillate scales ovate, much shorter and narrower than the perigynia, hyaline and lightly brown-tinged, acute or acuminate; perigynia planoconvex and conspicuously winged, 2.5–3.5 × 1.5–2.2 mm, less than twice as long as wide, broadest at a third to half their length, the body suborbicular to broadly elliptic, finely nerved on both sides (the ventral nerves 5 or fewer and often indistinct), abruptly narrowed to the flat, serrulate beak, the wing-margins not reaching the tip; achene lenticular, 0.95–1.2 mm; style straight or bent at the base; 2*n*=ca 68. Marshes, woods and low ground; Mass. to Ont. and Ill., s. to Ga., Miss., and Tex. (*C. straminea*, misapplied.)

80. **Carex bicknellii** Britton.   In small tufts from a short, stout rhizome, aphyllopodic, 3–12 dm; main lvs 2–4 mm wide, shorter than the stem; sheaths ventrally hyaline; spikes 3–7, gynaecandrous, the pistillate part globose to ovoid, 8–12 mm, often clavate at base and up to 18 mm including the staminate part, sessile,

separate or somewhat aggregated into an oblong to linear cluster 3–7 cm; anthers 3–4 mm; pistillate scales lance-ovate, acute to acuminate, shorter (by 1–2 mm) and much narrower than the perigynia, pale brown with green midnerve and narrow hyaline margins; perigynia stramineous, broadly ovate, 4.2–7 (avg 5.7) mm, 1.3–2 times as long as wide, very flat, thin and almost translucent, broadly winged, sharply several-nerved on both sides, abruptly contracted to the flat, serrulate beak; achene lenticular, 1.6–2 × 1.5 mm; 2*n*=76. Open woods, fields, and meadows in moist or dry soil; Me. to Sask., s. to Del., O., Mo., Okla., and N.M.

81. Carex suberecta (Olney) Britton.   Tufted, aphyllopodic, 3–7 dm; main lvs 2–3 mm wide, shorter or longer than the stem, their sheaths ventrally green-veined almost to the summit, with only a short hyaline area; spikes 2–5, gynaecandrous, ovoid, 7–12 mm, sessile, distinct but closely aggregated into an ovoid or short-oblong infl; pistillate scales shorter and narrower than the perigynia, yellowish-brown with pale midnerve and narrow hyaline margins, acute to cuspidate; perigynia numerous, appressed, prominently distended over the achene, 3.9–5.1 × 2.1–2.6 mm, 1.8–2.4 times as long as wide, broadest at a third to two-fifths their length, the broad body straight-tapered to the base, faintly nerved dorsally, nerveless or nearly so ventrally, abruptly contracted to the flat, serrulate beak; achene lenticular, 1.5 × 1 mm. Swamps and moist or wet meadows and shores, calciphile; s. Ont. to w. Va., s. Minn., and Mo.

82. Carex straminea Willd.   Densely tufted, aphyllopodic, 4–10 dm; main lvs 2–3 mm wide, shorter than the stem, their sheaths ventrally green-veined almost to the summit, with only a short hyaline area; spikes 4–8, gynaecandrous, sessile, well separated in an infl 3–6 cm, or the uppermost aggregated, the pistillate part short-ovoid or subglobose, 7–10 mm, often with an elongate staminate base; pistillate scales lanceolate, much narrower and somewhat shorter than the perigynia, hyaline or nearly so, brown-tinged with a paler midnerve, acuminate to shortly aristate; perigynia very flat and thin, lightly but sharply nerved on both faces, the flattened, serrulate beak half as long as the body; achene lenticular, 1.5 × 0.75 mm. Var. *straminea,* of nonsaline swamps and wet meadows, from Mass. to D.C., w. to Mich. and Ind., has the perigynia 4–5.2 mm, half to two-thirds as wide, broadest at a third to two-fifths of their length, the body ovate to orbicular. (*C. richii.*) Var. *invisa* W. Boott, of salt marshes from Nf. to Va., has the perigynia 4.2–5.5 mm, two-fifths or three-fifths as wide, broadest at two-fifths to half of their length, the body orbicular to obovate. (*C. hormathodes.*)

83. Carex alata T. & G.   Tufted, aphyllopodic, 3–10 dm; main lvs 2–4 mm wide, shorter than the stems, their sheaths ventrally green-veined almost to the summit, with only a short hyaline area; spikes 4–8, gynaecandrous, silvery-green or silvery-brown, subglobose to ovoid, 8–13 mm, obtuse at base with few staminate fls, sessile and closely aggregated in an erect cluster 2–4 cm; pistillate scales shorter than the perigynia, ovate or lance-ovate, nearly hyaline with a narrow green center, short-aristate; perigynia flat, 3.7–4.9 × 2.8–3.5 mm, 1.25–1.7 times as long as wide, several-nerved on both faces, broadest near or even above the middle, broadly rounded to the narrow beak less than half as long as the obovate body; achene lenticular, 1.5–2 × 1 mm. Wet soil, mostly near the coast, from Mass. to Fla. and Tex., also inland in N.Y., O., s. Ont., Mich., Ind., and Mo.

84. Carex longii Mackenzie.   Densely tufted, aphyllopodic, 3–12 dm; main lvs 2–3 mm wide, shorter than the stiff stems, their sheaths ventrally green-veined almost to the summit; spikes 3–10, gynaecandrous, sessile, crowded or the lower separate, ovoid, 6–13 mm, densely fld, the lateral with few staminate fls, hence obtuse or rounded at base; pistillate scales ovate, narrower and shorter than to nearly as long as the perigynia, the midrib not reaching the boat-shaped, mostly obtuse tip; perigynia flatly planoconvex, appressed to somewhat spreading, 3–4.5 × 1.6–2.6 mm, many-nerved on both sides, the body obovate, broadest above the summit of the achene and tapering to the broad beak with the wings reaching the tip; achene lenticular, 1.3–1.7 × 0.75–1 mm; style straight. Swamps and bogs, mostly near the coast, from N.S. to Fla. and Tex.; less commonly inland to Ind., Mich., and Sask., and s. to trop. Amer. (*C. albolutescens,* misapplied.)

85. Carex albolutescens Schwein.   Much like no. 84, but the perigynium-beak more slender, spreading, its winged margins not reaching the tip; pistillate scales distally flat and acute, the midrib reaching the tip; style laterally flexuous; 2*n*=66. Swamps and low woods and thickets; Mass. to Ga., w. to s. Mich., s. Ill., s. Mo., La., and Tex. (*C. straminea,* misapplied.)

86. Carex cumulata (L. Bailey) Mackenzie.   Tufted, aphyllopodic, 4–8 dm; main lvs 3–5 mm wide, shorter than the stiff stems, their sheaths ventrally green-veined almost to the summit; spikes 5–10, gynaecandrous, ovoid with rounded base, 7–10 mm, sessile, densely aggregated in an ovoid cluster 2–3 cm; pistillate scales much (usually 1–1.5 mm) shorter than the perigynia, hyaline, pale green or ± stramineous, obtuse; perigynia appressed, flatly planoconvex, 3.3–4 × 2.4–3.8 mm, nerveless on the ventral face of the obovate-orbicular body, abruptly rounded to the 0.75 mm beak; achene lenticular, 2 × 1.25 mm. Dry, rocky or sandy soil; P.E.I. to Wis., s. to N.J. and Ill.

87. Carex silicea Olney.   Densely tufted, aphyllopodic, 3–8 dm; main lvs 2–4 mm, equaling or shorter than the stiff stems, their sheaths ventrally green-veined almost to the summit; spikes 3–10, gynaecandrous, stoutly fusiform, 6–15 mm, densely fld, usually well separated in an erect infl 4–8 cm, the lateral ones tapering to the base, with numerous staminate fls; pistillate scales lance-ovate, nearly or quite as long as the perigynia, but distinctly narrower, pale brown or silvery, acute; perigynia mostly appressed, flatly planoconvex, ovate, 4–5.2 mm, three-fifths as wide, sharply nerved dorsally, obscurely so ventrally, broadest about the middle of the achene, tapering to the short, flat, serrulate beak; achene lenticular, 1.5 × 1 mm. Sand or sandy soil; near the coast from Nf. to Del.

88. Carex supina Willd. Stems 0.5–3 dm, in small tufts on long rhizomes; lvs crowded toward the base, slender but flat above, 1–1.5 mm wide; spikes mostly 2 or 3, the terminal one staminate, 6–25 mm, the lateral one(s) pistillate, sessile, approximate, suborbicular or short-oblong, 4–12 mm, with mostly 4–15 ascending or eventually spreading perigynia; bracts short, sheathless; pistillate scales broadly ovate, wider than and nearly as long as the perigynia, reddish-brown, with stramineous 3-nerved center and conspicuously white-hyaline margins; perigynia coriaceous, brown and shining, obovoid, 2-ribbed, nerveless or obscurely few-nerved, 2.5–3.5 mm including the smooth, cylindric beak; achene trigonous, filling the perigynium; $2n$=38. Dry, often rocky slopes; circumboreal, s. irregularly to ne. Minn. and adj. Ont. The infl. is occasionally reduced to a solitary staminate spike with 1 or 2 basal perigynia; such plants look much like *C. obtusata,* but lack a rachilla in the perigynium.

89. Carex caryophyllea Latour. Laxly cespitose, with short creeping rhizomes; stems 1–4 dm, usually much surpassing the short lvs, these all near the base, 1.5–3.5 mm wide; spikes 3 or 4, approximate, the terminal one staminate, 1–2 cm, the lateral ones pistillate, 0.5–1.5 cm, sessile or short-peduncled; pistillate scales brown, ovate-oblong, as wide and long as the perigynia; perigynia narrowly obovoid, 2.5–3 mm, finely hairy, obtusely trigonous, tapering toward the base, short-beaked; achene rounded-trigonous; $2n$=62–68. Native of Eurasia, intr. in dry soil from N.B. and Me. to D.C., especially abundant in e. Mass.

90. Carex willdenovii Schk. Tufted, the stems weak and slender, 0.5–3 dm, much shorter than the lvs, distally winged triquetrous; main lvs 2–4 mm wide; spikes 1–3, androgynous, the lower widely separated, nearly basal, on long, capillary peduncles; staminate scales obtuse or acute; pistillate scales dilated at base, with hyaline margins, the lowest ± foliaceous, surpassing the perigynium and sometimes the whole spike, the others narrower than the perigynium; perigynia 3–8, green or greenish, 4.5–6 mm, the body obovoid-oblong above the short stipe-like base, nerveless, sharply 2-edged above, tapering into the very rough, triangular-pyramidal beak, which constitutes a third to half of the entire length; achene rounded-trigonous, with a short stylar apiculus. Moist woods, especially in acid soils; chiefly in the coastal states from Mass. to Ga. and Tex., but also inland to O., Tenn., and s. Ind.

91. Carex jamesii Schwein. Much like no. 90; main lvs 2–3 mm wide; staminate scales truncate, erose, with a dark transverse apical band; lowest pistillate scales dilated-saccate at base; perigynia 2–4, obpyriform, the body subglobose above a stipe-like base, distally rounded and abruptly prolonged into the slender 3-angled beak. Rich woods, especially in calcareous soil; chiefly midwestern, from e. Nebr., e. Kans., and e. Okla. to Minn., Mich. and Tenn., occasionally s. to La. and e. to Md. and N.Y.

92. Carex backii F. Boott. Much like no. 90; main lvs 3–6 mm wide; staminate scales ca 3, the margins connate to above the middle; pistillate scales wholly green, wider than and nearly concealing the perigynia, the lower elongate and leaf-like, (1.5–)2–7 cm, the upper progressively reduced but still surpassing the perigynia; perigynia 2–5, turgid, rounded-trigonous, inconspicuously nerved, 4–5.4 mm, rhombic-ovoid and obscurely beaked to subglobose or obovoid and with a conspicuous, smooth or obscurely serrulate beak to 2 m; achene without a stylar apiculus; $2n$=66. Woods and thickets; Que. to N.J., w. across the n. states to Minn., Neb., B.C., Oreg., and Utah. (*C. durifolia; C. saximontana.*)

93. Carex novae-angliae Schwein. Loosely cespitose, reputedly sometimes rhizomatous; stems very slender, 1–4 dm, shorter than the lvs; main lvs 1–2 mm wide; terminal spike staminate, slender, 4–16 mm; pistillate spikes usually 2 or 3, 3–6 mm, 1–3 cm apart, the 2 lowest not overlapping, the lowest one distinctly peduncled; lowest bract foliaceous, often equaling or surpassing the infl; pistillate scales stramineous or reddish-brown, with hyaline margins, acute or short-cuspidate, usually wider but somewhat shorter than the 2–10 perigynia, these light green to brownish-stramineous, minutely pubescent, 2.2–2.7 mm, the body slightly flattened, obovoid above a stipe-like base, 2-keeled, abruptly prolonged into a beak a fourth to two-fifths as long as the body; achene rounded-trigonous. Moist woods; Nf. to s. Ont. and reputedly Wis., s. in the mts. to Conn. and Pa.

94. Carex peckii Howe. Loosely cespitose; stems slender but erect, 2–6 dm, ± strongly surpassing the lvs; main lvs 1–1.5 mm wide; infl longer or shorter than the lowest bract, 1–2 cm, with a small terminal staminate spike and 2 or 3 ± closely aggregated pistillate ones; pistillate scales reddish-brown with broad hyaline margins and pale center, wider but evidently shorter than the perigynia, blunt or rounded, sometimes shortly mucronate; perigynia 3–12, dull green or yellowish-green, short-hairy, 2.7–4.1 mm, slightly flattened, narrowly obovoid above a stipe-like base, 2-keeled, abruptly contracted into a slender beak 0.5 mm; achene rounded-trigonous; $2n$=36. Open woods, ± calciphile; Que. to Yukon, s. to N.J., Mich., Minn., Nebr., and B.C. (*C. nigromarginata* var. *elliptica; C. albicans,* misapplied.)

95. Carex albicans Willd. Densely cespitose; lvs 0.5–2.5 mm wide; terminal spike staminate, 4–14 mm; pistillate spikes (1)2–3(4), sessile, 2–7 mm, ± closely aggregated and overlapping; lowest bract usually shorter than the infl; pistillate scales green, or with a purple strip on each side of the broad green center, ovate or lance-ovate, acuminate or cuspidate, up to as long as the perigynia, but distinctly narrower above; perigynia 4–12, dull green or yellowish-green, puberulent, 2–3.5 mm, slightly flattened, narrowly obovoid above a stipe-like base, 2-keeled, abruptly contracted into a slender beak 0.5–1 mm; achene rounded-trigonous; $2n$=40. N.S. to Fla., w. to s. Ont., Mich., Wis., Nebr., and Okla. Two vars. with us:

Var. *albicans.* Stems firm, erect, 2–5 dm, evidently surpassing the lvs; staminate scales ± blunt, the midrib distally weak or fading out; ± calciphile; Mass. to S.C., w. to Mich., Io., Nebr., and Okla. (*C. artitecta; C. pensylvanica* var. *muhlenbergii; C. nigromarginata* var. *m.*)

Var. *emmonsii* (Dewey) Rettig.   Stems lax, 0.5–2 dm, typically shorter than the lvs; staminate scales acute or acuminate, the strong midnerve extending to the tip or prolonged as a short cusp; mostly in acid soils on the coastal plain from N.S. to Fla., but also inland to s. Ont., Mich., Wis., Ind., and Ill. (*C. emmonsii; C. nigromarginata* var. *minor.*)

96. Carex nigromarginata Schwein.   Cespitose and mat-forming, fibrillose at base; stems all short but of varying length on the same plant, up to ca 2 dm, mostly well surpassed by the lvs; lvs up to 4 dm at maturity, mostly 2–4 mm wide; terminal spike staminate, 5–8 mm; pistillate spikes 2 or 3, 4–7 mm, sessile, all ± closely associated with the staminate spike, none basal; lowest bract often surpassing the infl; staminate scales short-cuspidate to obtusish; pistillate scales mostly dark purple except the narrow midrib, short-cuspidate to merely acute, wider and from a littler shorter to a little longer than the largely concealed perigynia; perigynia 6–15, dull green or yellowish-green, sparsely short-hairy, 3–4 mm, slightly flattened, oblong-obovoid above a stipe-like base, 2-keeled, abruptly contracted into a slender beak 1 mm; achene rounded-trigonous. Dry woods, chiefly in acid soils on the coastal plain; Conn. to Fla. and La., n. in the Mississippi Valley to s. Ind. and s. Mo.; disjunct in s. Ont.

97. Carex communis L. Bailey.   Densely cespitose, smooth and purplish at base; stems 2–5 dm, equaling or surpassing the lvs; main lvs 3–6 mm wide; terminal spike staminate, 4–18 mm; pistillate spikes 2 or 3, sessile or nearly so, 5–10 mm, the lowest one a little removed from the others, but none basal; middle and lower bracts largely green, but tending to be scarious-lobed at base; pistillate scales lance-ovate, about as long as the perigynia, obtuse to acuminate, reddish-purple with 3-nerved green center and narrow hyaline margins; perigynia 3–10, 2.5–4 mm, the body subglobose above a stipe-like base, 2-keeled, finely hairy, abruptly prolonged into a slender beak 0.5–1 mm; achene rounded-trigonous. Deciduous woods; N.S. to Minn., s. to S.C., Ky., and Ark.

98. Carex pensylvanica Lam.   Stems (1–)2–5 dm, tufted in small to large, basally fibrillose clumps and also with long rhizomes; lvs 1–3 mm wide, shorter than the stems in tall plants; staminate spike terminal, 1–2.5 cm; pistillate spikes 1–3, sessile or short-pedunculate and loosely ascending, borne fairly close to each other and to the staminate spike, but not closely crowded, typically short-oblong, up to ca 1.5 cm; lowest bract 1–3 cm, ± leafy, surpassed by the staminate spike; pistillate scales castaneous to stramineous, longer or shorter than the perigynia; perigynia 2.6–4.5 mm, short-hairy, 2-keeled and with several evident to obscure facial nerves, the body subglobose above the contracted base, abruptly prolonged into the sharply bidentate beak 0.2–1 mm; achene rounded-trigonous. N.Y and s. Me. to Va. (and in the mts. to N.C. and Tenn.), w. to s. Sask., B.C., Wash., Calif., Oreg., and N.M. Two vars. with us, a third on the Pacific coast:

Var. *pensylvanica*.   Perigynium 1.2–1.5(–1.7) mm wide; stems usually smooth; ventral face of sheath of upper cauline lf shallowly to more often deeply concave, its cleft often extending more than 1 mm (to 2.3 mm) below the junction with the blade; $2n=36$. Upland woods in e. U.S., widespread with us.

Var. *digyna* Boeckeler.   Perigynium (1.5–)1.7–2.2 mm wide; stems usually scabrous below the infl; ventral face of sheath of upper cauline lf only shallowly concave, its cleft not extending more than 1(–1.5) mm below the junction with the blade; $2n=40$. Mainly on the prairies and plains to the w. of our range, but extending e. to Minn., Io., n. Mo., and occasionally Ind., Mich., and s. Ont. (*C. heliophila; C. inops* ssp. *h.*)

99. Carex lucorum Willd.   Much like no. 98; stem usually strongly scabrous beneath the infl; sheath of upper cauline lf usually deeply concave, its cleft extending 1–6 mm below the junction with the blade; perigynia 1.2–1.7 mm wide, abruptly contracted into a prominent beak 1–2 mm; $2n=40$. Dry, open places or open woods of oak or pine; N.S. and N.B. to N.J., Pa., and s. Ont., w. less commonly to Mich., Wis., and Minn., and s. irregularly to N.C. and Tenn. (*C. pensylvanica* var. *distans.*)

100. Carex rossii F. Boott.   Much like no. 101, but more densely cespitose, without creeping rhizomes; stems 1–3 dm; staminate spike longer, 5–12(–15) mm; perigynium 3–4.5 mm, its beak 1–1.7 mm, half to fully as long as the body, often strongly bidentate with teeth up to 0.6 mm; $2n=36$. Dry, rocky, exposed places; Yukon to Calif. and Colo., e. irregularly to Minn. and to Keweenaw Point, Mich.

101. Carex deflexa Hornem.   Loosely cespitose and with rather short, slender rhizomes; stems slender, 1–2 dm, purple-tinged at base, usually surpassed by the lvs; lvs soft and thin, 1–3 mm wide; slender-pedunculate, near-basal pistillate spikes usually abundant; staminate spike terminal, 2–5 mm, often over-topped by one or more of the 2–4 associated pistillate spikes, the lowest one of these subtended by a green bract 5–20 mm; pistillate scales lance-ovate to oblong, as wide as but shorter than the perigynia; perigynia green, stipitate, 2–3 mm, the body obovoid or pyriform, 2-keeled, finely short-hairy, abruptly contracted to the minutely bidentate beak 0.4–0.7 mm; $2n=36$. Moist open places, and margins of woods and swamps; Greenl. to Alas., s. in our range to Mass., n. N.Y., n. Mich., and n. Minn.

102. Carex umbellata Schk.   Densely cespitose and basally fibrillose; stems 5–10 (–20) cm, much surpassed and often concealed by the lvs; lvs 1–5 mm wide; staminate spike terminal, 5–10 mm, sometimes with a short pistillate spike just beneath it, the subtending bract of this pistillate spike scale-like and shorter than the staminate spike; 1–3 near-basal, short-pedunculate pistillate spikes 4–10 mm generally associated with each flowering stem, ± removed from the terminal staminate spike; pistillate scales acute to short-cuspidate, about as long and wide as the greenish perigynia; perigynia 2.5–4 mm, finely hairy or glabrous, obovoid, 2-keeled above, tipped with a prominent, 2-edged, often curved, bidentate beak 0.5–1.5 mm; achene rounded-trigonous; $2n=32$. Dry to moist soil, in shade or sun; Nf. to B.C., s. to Va., Ga., Mo., and Tex. (*C. abdita; C. rugosperma*, with a relatively long perigynium-beak; *C. tonsa.*)

103. Carex aurea Nutt.   Stems slender, 0.3–4 dm, solitary or in small tufts on long creeping rhizomes; lvs mostly 1–4 mm wide, often surpassing the stems; spikes 2–several, 5–20 mm, approximate or ± remote, erect or loose on slender, sometimes elongate peduncles, the terminal one staminate, less often gynaecandrous, rarely androgynous, the others pistillate, often some of them near-basal; pistillate scales half to fully as long as the perigynia; perigynia ellipsoid to more often obovoid or obovoid-globose, ± rounded and beakless distally, only slightly or not at all compressed, 1.7–3 mm, evidently 12–20-ribbed, varying to nearly nerveless, strongly whitish-papillate on a light green background when young, tending to turn golden or yellow-brown with increasing maturity, becoming somewhat fleshy and obscurely papillate, or sometimes remaining whitish-papillate and not becoming fleshy; stigmas 2 (or 3 in some fls), the achene accordingly lenticular (often plump) or trigonous; 2n=52. Moist or wet places; Nf. to Alas., s. to Pa., n. Ind., Minn., Nebr., N.M. and Calif. (*C. garberi* and *C. hassei,* with the perigynia only obscurely nerved and remaining dry and white-papillate at maturity, but the characters confluent.)

104. **Carex pedunculata** Muhl. Tufted and often shortly rhizomatous; most lvs basal, thickish and stiff, 1–3 dm × 2–3 mm; fertile stems 1–3 dm, with 4–5 spikes 6–15 mm; uppermost spike staminate or commonly with a few basal perigynia; anthers 2–3 mm; pistillate spikes often with a few terminal staminate fls, the upper 1 or 2 not far from the summit, the lower 1 or 2 on long slender peduncles from near the base, often concealed by the lvs; pistillate scales obovate-oblong, the body broadly obtuse to truncate, extending to or beyond the broadest part of the perigynium, the thin sides purple-brown with hyaline margins, the green center prolonged into a cusp equaling or surpassing the perigynium; perigynia with a basal elaiosome, ant-dispersed, obovoid, 3.5–5 mm, roundly angled on the back, keeled on the lateral angles, thinly hairy to glabrate, minutely beaked; achene sharply trigonous; 2n=26. Rich woods, usually in calcareous soil; Nf. to Sask. and S.D., s. to N.C., Ala., and Io., abundant in the Great Lakes region. Fl early, fr May.

105. **Carex concinna** R. Br. Rhizomatous and sometimes also tufted, 5–15 cm; lvs mostly basal, often recurved-spreading, 5–10 cm × 1–3 mm; staminate spike terminal, solitary, only 3–7 mm; anthers 1–1.5 mm; pistillate spikes 2 or 3, approximate, sessile or short-peduncled, 4–8 mm; bract subtending the lowest spike 1 cm or less, with an expanded, usually sheathing base 2–5 mm and an awn-tip of similar or lesser length; pistillate scales blunt, brown with hyaline margins, distinctly shorter than the perigynia, often only half as long; perigynia 5–12, 2–3(–3.5) mm, puberulent, plump, 2-ribbed, abruptly contracted to the very short beak less than 0.5 mm; basal spines none; achene rounded-trigonous; 2n=54. Coniferous woods; Nf. to Yukon, s. to n. Mich., n. Wis., S.D., Colo., and Oreg.

106. **Carex richardsonii** R. Br. Loosely rhizomatous; fertile stems 1–2.5 dm, scabrous-puberulent above; lvs mostly basal, thick and stiff, 1–2 dm × 1.5–2.5 mm; terminal spike staminate, linear, 12–20 mm, its scales rounded or blunt, with narrow purple center and broad hyaline margins; anthers 2–3.5 mm; pistillate spikes 2(3), erect, approximate or slightly separate, short-peduncled, 1–2 cm, their scales ± acute, wider than and equaling or more often surpassing the perigynia, brown-purple with hyaline margins, perigynia 10–25, obovoid, 2.5–3.5 mm, roundly angled on the back, keeled on the lateral angles, thinly pubescent; basal spines none; achene sharply trigonous; 2n=52. Dry or rocky upland woods; Vt. to Alta., s. to D.C., O., Ind., Io. and S.D.; rare and local in the east. Fl early, fr May.

107. **Carex hirtifolia** Mackenzie. Loosely tufted, 3–6 dm; lvs soft and flaccid, softly hairy, 3–7 mm wide, the 2 mid-lateral veins conspicuous on the upper side; spikes 2–4, approximate or the lower somewhat separate; lowest bract sheathless, with blade surpassing the spike; upper bracts smaller; terminal spike staminate, slender, 1–2 cm; lateral spikes pistillate, sessile or short-peduncled, 8–15 mm; pistillate scales obovate, with broad hyaline margins and green midvein excurrent into a short awn nearly or quite equaling the perigynium; perigynia trigonous, hairy, nerveless, the obovoid body 2.6–3.5 mm, abruptly rounded into a slender erect beak 0.8–1.3 mm; achene trigonous, filling the perigynium. Woods; N.B. and Que. to Minn., s. to Md., W.Va., Ky., Mo., and e. Kans.

108. **Carex eburnea** F. Boott. Densely tufted and with long, slender rhizomes; stems very slender, 1–3 dm; lvs shorter than the stems, involute, 0.5 mm wide or less; spikes subtended by short bladeless sheaths, the pistillate 2 or 3, 3–6 mm, 2–6-fld, on long erect peduncles overtopping the terminal linear staminate spike; pistillate scales broadly ovate, much shorter than the perigynia, whitish with narrow green midvein, obtuse or acutish; perigynium-body light green, some becoming brown, 1.5–2 mm, obovoid, trigonous, ribbed on the lateral angles, finely few-nerved, glabrous, filled by the trigonous achene, abruptly narrowed to a conic beak 0.2 mm. Calcareous soil; Nf. to Mack. and B.C., s. to Va., Ala., Nebr., and Tex.

109. **Carex livida** (Wahlenb.) Willd. Glaucous; stems 2–5 dm, in small tufts from long rhizomes; lvs 1–3 mm wide, channeled or involute, often equaling or surpassing the stems; terminal spike staminate, rarely pistillate at the tip, 1–2 cm; lateral spikes 1–3, pistillate, 1–2 cm, sessile or short-peduncled, contiguous to shortly overlapping or only shortly separate near the summit of the stem, or the lowest one rarely on an elongate basal peduncle; bracts with short sheath and narrow blade usually surpassing the spike, pistillate scales ovate, shorter and distally narrower than the perigynia, with purplish sides and green center, obtuse to acute or mucronate; perigynia glaucous-green, fusiform, 3–5 mm, ribbed on the lateral angles, finely and obscurely nerved, beakless; achene broadly ovoid-trigonous, filling the perigynium; 2n=32. Wet soil; circumboreal, s. to N.J. and Minn. (*C. grayana.*)

110. **Carex panicea** L. Much like no. 109; somewhat bluish-green, but scarcely glaucous; main lvs 2–5 mm wide, flat, shorter than the stems; sheaths conspicuous, the lowest 1–2 cm; terminal spike 1.5–3 cm, long-peduncled; lateral spikes generally well separated, the lower long-peduncled (but not basal), the upper less so or subsessile; pistillate scales ovate, shorter than but about as wide as the perigynia, these stoutly fusiform to obovoid, obscurely trigonous, ribbed on the lateral angles, beakless or minutely beaked; achene nearly filling the perigynium; 2n=32. Native of Europe, intr. in fields, lawns, and meadows from N.S. and N.B. to N.J.

111. Carex meadii Dewey. Much like no. 112 and perhaps not sharply distinct; lvs 3–7 mm wide; pistillate spikes 10–30 × 5–7 mm, the crowded perigynia in ca 6 rows, obovoid, 3.3–4.2 mm, broadest above the middle and abruptly rounded to the tip, sometimes with a very short, outcurved beak; $2n=52$. Dry open prairies and meadows; N.J. to Mich. and Sask., s. to N.C., Ga., Ark., and Tex.

112. Carex tetanica Schk. Stems 2–6 dm, in small clumps on long, slender, deep-seated, pale rhizomes, phyllopodic (at least the fertile stems) and often fibrillose at base, only seldom purplish below; lvs flat, 2–4.5 mm wide, surpassed by the stems; terminal spike staminate, strongly rough-pedunculate, 1.5–4 cm; pistillate spikes 1–3, mostly widely separated, erect on slender peduncles, 7–40 × 3–5 mm; bracts long-sheathing, the blade shorter than the stem; pistillate scales ovate, obtuse to acuminate or short-awned, brownish-purple-tinged with 3-nerved green center and hyaline margins; perigynia overlapping in mostly 3 rows, obovoid-fusiform, 2.5–3.5 mm, obscurely trigonous, with numerous fine nerves and 2 more prominent lateral ribs, broadest (1.5–2 mm) above the middle and evenly and broadly tapered or rounded to the beakless tip; achene broadly obovoid-trigonous, with concave sides, filling the perigynium; $2n=56$. Meadows and low woods; Mass. to N.J. and Va., w. to Minn., Alta., and Nebr.

113. Carex woodii Dewey. Much like no. 112; rhizomes shallow, stout, reddish; strongly aphyllopodic, with bladeless red basal lf-sheaths; perigynia ± 2-ranked, tapering above, at least the lower not overlapping; $2n=44$. Lf-mold in rich woods; N.Y. to D.C., w. to Man. and Mo. (*C. colorata; C. tetanica* var. *w.*)

114. Carex vaginata Tausch. Stems 2–6 dm, few together, from long rhizomes, phyllopodic, the main lvs 2–5 mm wide; terminal spike staminate, 1–2 cm; pistillate spikes 1–3, often staminate at the tip, rather loosely spreading, widely separated, the lower peduncles elongate, the upper shorter; bracts with rather loose sheaths 1–2 cm and short blades shorter than the spikes; pistillate scales shorter and narrower than the perigynia, purplish-brown, with or without a narrow green center, usually acute; perigynia usually in 2 rows, the lower separated by internodes 2–5 mm, the upper ± overlapping, glabrous, 2-keeled, otherwise rather obscurely nerved, 3.5–5 mm, fusiform-obovoid, with a somewhat outcurved beak ca 1 mm; achene obovoid-trigonous with concave sides, nearly filling the perigynium; $2n=32$. Wet woods and bogs, chiefly in calcareous districts; circumboreal, s. in Amer. to Me., n. N.Y., n. Mich., n. Minn., Sask., and B.C. (*C. saltuensis,* the Amer. plants.)

115. Carex polymorpha Muhl. Stems stout, 3–6 dm, in dense colonies from stout rhizomes, aphyllopodic, the basal lvs reduced to bladeless sheaths; stem-lvs firm, flat, 3–5 mm wide; terminal spike peduncled, staminate, 1–3 cm, often with 1 or 2 smaller ones at its base; pistillate spikes 1 or 2, erect, widely separated, the lower peduncle longer than the upper; bracts with conspicuous loose sheaths and well developed blades usually exceeding the spikes; pistillate scales ovate, much shorter than the perigynia, with purplish-brown sides and conspicuous green center; perigynia numerous, crowded in several rows, glabrous, 4–5.5 mm, stoutly ellipsoid, obscurely nerved except the 2 prominent lateral keels, the prominent beak straight or somewhat outcurved, 1.3–2.2 mm, with a very oblique orifice; achene broadly obovoid-trigonous, loosely enveloped. Dry open woods, mostly in acid soils; Me. to Md. and W.Va.

116. Carex leptonervia (Fern.) Fern. Tufted, 2–6 dm; basal sheaths brown; largest lvs to 10 mm wide; upper bract-sheaths minutely serrulate on the angles; terminal spike staminate, 0.7–1.6 cm, sessile or on a short peduncle to 2 mm, surpassed by or only slightly surpassing the uppermost pistillate spike; pistillate spikes 2–4(5), 1–2 cm, the upper 1 or 2 approximate to the staminate one, sessile or short-peduncled, the others more remote, evidently pedunculate, none basal; second pistillate bract often overtopping all the spikes; pistillate scales usually short-awned or apiculate; perigynia 5–14, 2.2–4.1 mm, 2-ribbed, otherwise nerveless or only obscurely few-nerved, obovoid and obscurely trigonous, abruptly contracted to the short, entire beak; achene trigonous; $2n=36$. Moist woods; Nf. to ne. Minn., s. to N.J., Pa., Ind., and Wis., and in the mts. to N.C. and Tenn.

117. Carex crebriflora Wiegand. Plants 3–7 dm, forming large tufts; stems sharply triangular and slightly winged, retrorse-serrate; basal sheaths ± brown; lvs 3–8 mm wide; terminal spike staminate, 1–2 cm, sessile or nearly so, little if at all surpassing the 2 contiguous upper pistillate spikes, overtopped by the bracts; 1 or 2 pistillate spikes also borne well below the summit, but not basal, on short or elongate peduncles; pistillate scales ± acute to often cuspidate; perigynia 4–20, ochroleucous or yellowish, 3.5–5 mm, finely many-nerved as well as 2-ribbed, obtusely trigonous, fusiform, tapering to the ill-defined beak with oblique, entire orifice; achene trigonous. $2n=42$. Wet woods, especially in sandy or silty soil; e. Va. to Fla. and Tex.

118. Carex styloflexa Buckley. Plants 3–8 dm, forming large tufts; fertile stems lax or ascending, sharply triangular but not winged, minutely roughened above; basal sheaths ± brown; largest lvs to 7 mm wide, seldom more; terminal spike staminate, 1–3.5 cm, its peduncle usually and its summit always surpassing the short (ca 5 mm), sessile, uppermost pistillate spike; pistillate spikes 2–5, 0.5–2 cm, widely separated but none basal, the lowest on a long, drooping peduncle, the upper on short, scarcely exserted peduncles; pistillate scales acute to acuminate; perigynia 4–20, closely overlapping, greenish-stramineous, 3.5–5.5 mm, finely many-nerved as well as 2-ribbed, obtusely trigonous, fusiform, long-tapering to both ends, the elongate but ill-defined beak somewhat outcurved, with an oblique, entire orifice; achene trigonous; $2n=48$. Wet woods and bogs, often in sandy or silty soil; Conn. to Fla., w. to s. O., Ky., Tenn., La., and se. Tex.

119. Carex striatula Michx. Tufted, 2–6 dm; fertile stems ascending, sharply triangular but not winged, smooth or nearly so; basal sheaths white or light brown; lvs of the sterile shoots 7–14 mm wide, of the fertile shoots somewhat narrower; terminal spike staminate, 2.5–3.5 cm, on a peduncle 0.5–12 cm; pistillate spikes 2 or usually 3, 2–6 cm, slender, on short to elongate ascending peduncles, scattered but none basal; pistillate scales acute to short-awned; perigynia 6–18, only slightly or scarcely overlapping, 3.5–5 mm, finely many-nerved as well as 2-ribbed, obtusely trigonous, fusiform, tapering to an ill-defined, often somewhat outcurved beak with an oblique, entire orifice; achene trigonous. Dry to mesic woods; L.I. and Pa. to Fla. w. to Tenn. and Tex. (*C. laxiflora* var. *angustifolia*.) Perhaps properly to be included in *C. laxiflora*.

120. Carex laxiflora Lam. Tufted, 3–7 dm; fertile stems erect or ascending, lateral or sometimes central, sharply triangular and narrowly winged, smooth or nearly so; basal sheaths light brown; lvs of the short sterile shoots 5–25 mm wide, of the fertile shoots narrower, 2–8 mm; bract-sheaths smooth or sometimes serrulate on the angles; terminal spike staminate, 1–2.5 cm, on a peduncle 1–2 cm; pistillate spikes 2–4, 1–3 cm, slender, scattered on short or elongate, erect peduncles, none basal; at least the lower pistillate scales cuspidate or short-awned; perigynia 5–15, only slightly or scarcely overlapping, 2.6–3.8 mm, finely many-nerved as well as 2-ribbed, obtusely trigonous, obovoid, tapering to a rather short, straight or slightly outcurved beak with an oblique, entire orifice; achene trigonous; $2n=40$. Dry to mesic woods; Me. and s. Que. to Del. and Md., and in the mts. to N.C., Tenn., and Ala., w. to Wis. and Ind.; s. Mex.

121. Carex purpurifera Mackenzie. Tufted, 3–7 dm; fertile stems lateral, ascending or decumbent, papillate on the angles; basal sheaths purple; lvs glaucous-green, those of the elongate sterile shoots 5–10 mm wide, of the fertile ones somewhat smaller; terminal spike staminate, 1–3 cm, purplish, on a peduncle 0.5–5 cm; pistillate spikes mostly 3, slender, widely separated but none basal, 1.5–4.5 cm; pistillate scales acute to short-awned; perigynia 4–16, loosely alternating, scarcely overlapping, 3.5–4.5 mm, finely many-nerved as well as 2-ribbed, obtusely trigonous, fusiform, often obliquely so, the short, often ill-defined beak with an oblique, entire orifice; achene trigonous; $2n=34$–38. Dry to mesic woods, especially on limestone escarpments; Ky. and w. Va. to n. Ga. and n. Ala. (*C. laxiflora* var. *p.*)

122. Carex gracilescens Steudel. Tufted, 2–8 dm, fertile stems mostly roughened on the angles; basal sheaths purple; lvs not glaucous, those of both fertile and sterile shoots 1–5 mm wide, or the latter larger, to 10 mm wide; angles of the bract-sheaths usually minutely ciliate-serrulate; terminal spike staminate, 0.6–2.5 cm, subsessile to evidently pedunculate; pistillate spikes 2–4, scattered but none basal, slender, 1–3 cm; pistillate scales cuspidate to short-awned; perigynia 4–18, only slightly overlapping (or the lowest scarcely so), 2.2–3 mm, finely many-nerved as well as 2-ribbed, obtusely trigonous, ellipsoid to obovoid, abruptly contracted to a short, abruptly bent beak with an entire orifice; achene trigonous; $2n=40$. Sandy or rocky woods and open, sometimes disturbed sites, often calciphile; Que. and Vt. to Ga. and nw. Fla., w. to Wis., e. Nebr., and e. Okla. (*C. laxiflora* var. *gracillima; C. ormostachya,* largely northern, with the angles of the bract-sheaths merely granular.)

123. Carex blanda Dewey. Tufted, 2–6 dm; fertile stems triangular and slightly winged, minutely serrulate above; basal sheaths ± brown; lvs of the sterile shoots up to ca 10 mm wide, those of the fertile ones a little narrower; angles of the bract-sheaths minutely ciliate-serrulate; terminal spike staminate 1–2 cm, on a short to elongate peduncle, often overtopped by the uppermost bract; pistillate spikes 3–4, 1.5–2 cm, scattered or the upper 2 approximate, none basal; pistillate scales prominently cuspidate or short-awned; perigynia 4–18, crowded and overlapping, greenish or yellowish-green, 2.5–4 mm, finely many-nerved as well as 2-ribbed, obtusely trigonous, obovoid, abruptly contracted to a short, abruptly bent beak with entire orifice; achene obtusely trigonous; $2n=36, 38, 40$. Dry to mesic woods, sometimes weedy; Me. and s. Que. to N.D., s. to Ga., nw. Fla., and Tex. (*C. amphibola; C. laxiflora* var. *b.*)

124. Carex albursina Sheldon. Tufted, 2–6 dm; fertile stems triangular and slightly winged, roughened on the angles; basal sheaths mostly brown, but the immediately suprabasal ones often purple; lvs of the sterile shoots 10–50 mm wide, of the fertile ones 4–16 mm; angles of the bract-sheaths minutely ciliate-serrulate; terminal spike staminate, 0.5–2 cm, sessile or nearly so, often hidden by the uppermost pistillate spike and its bract; pistillate spikes 3 or 4, 1–3 cm, the uppermost one or 2 near the staminate peduncle, the others remote, on short to elongate peduncles, none basal; pistillate scales broadly obtuse or subtruncate, only minutely apiculate; perigynia 3–20, ± overlapping, 3–4.2 mm, greenish-stramineous, finely many-nerved as well as 2-ribbed, obtusely trigonous, obovoid, with a short, abruptly bent beak; achene convexly trigonous; $2n=44$. Rich woods, especially in calcareous regions; Vt. and s. Que. to Minn., s. to S.C. and Ark. (*C. laxiflora* var. *latifolia*.)

125. Carex plantaginea Lam. Tufted, 3–6 dm; fertile stems triangular, roughened on the angles, purple at base; basal sheaths purple; lvs of the sterile shoots elongate, often surpassing the fertile stems, 10–30 mm wide, roughened on the margins and toward the tip on the main veins; lvs of the fertile stem reduced to bladeless or nearly bladeless purple sheaths; terminal spike staminate, 1–2 cm, purplish, long-peduncled; pistillate spikes 2–4, 1–3 cm, scattered, the lowest on a basal peduncle to 2 cm; pistillate scales acuminate to cuspidate; perigynia 4–15, crowded and overlapping, 3.7–5 mm, finely many-nerved as well as 2-ribbed, sharply trigonous, elliptic in outline, constricted into a short, oblique beak with entire orifice; achene sharply trigonous; $2n=50, 52$. Rich moist woods; N.B. and s. Que. to Minn., s. Ind., Ky., N.J., Md., and in the mts. to n. Ga.

126. Carex careyana Torr. Tufted, 3–6 dm; fertile stems triangular, roughened on the angles, purple at base; basal sheaths purple; lvs smooth, those of the sterile shoots 10–25 mm wide, of the fertile ones only 2–6 mm wide; terminal spike staminate, 1–2 cm, purplish, pedunculate; pistillate spikes 2 or 3, 0.7–2 cm, scattered, the lowest sometimes basal; pistillate scales acute to cuspidate; perigynia 4–7, overlapping, 5–6.5 mm, finely many-nerved as well as 2-ribbed, sharply trigonous, elliptic-ovate in outline, tapering to a slightly oblique beak with entire orifice; achene sharply trigonous; $2n=68$. Rich woods, often calciphile; N.Y. to s. Ont., Mich. and Io., s. to Va., Ala., and Mo.

127. Carex platyphylla Carey.   Tufted, 2–4 dm, the fertile stems lateral, roughened on the angles; basal sheaths white or light brown; lvs smooth, somewhat glaucous, those of the sterile shoots 10–25 mm wide, of the fertile shoots 2–6 mm wide; terminal spike staminate, 0.6–1.5 cm, on a peduncle 1–8 mm; pistillate spikes 2–4, 0.6–1.5 cm, scattered, the lowest one basal; pistillate scales apiculate or with an awn to 2 mm; perigynia 2–9, overlapping, 3–4.5 mm, finely many-nerved as well as 2-ribbed, sharply trigonous, elliptic in outline, rather abruptly contracted to the short, erect or outcurved beak with entire orifice; achene sharply trigonous; 2n=68, 70. Woods; Me. and s. Que. to Va. and in the mts. to N.C., w. to Wis. and Mo.

128. Carex abscondita Mackenzie.   Loosely tufted, seldom also with long rhizomes; fertile stems 0.5–2.5 dm, minutely hispidulous, much surpassed by the lvs; basal sheaths white to light brown; lvs roughened on the margins and toward the apex on the veins, those of the sterile shoots 3–9 mm wide, of the fertile ones 1.5–4 mm wide; terminal spike staminate, 0.5–1.2 cm, sessile or nearly so, often ± hidden by the bracts and pistillate spikes; pistillate spikes 2 or 3, 0.6–1.2 cm, often hidden in the foliage, the upper sessile or nearly so, the lowermost one usually basal, on a peduncle to 1.5 cm; pistillate scales acute; perigynia 3–9, overlapping, 2.8–4.2 mm, finely many-nerved as well as 2-ribbed, evidently trigonous, short-ellipsoid, rather abruptly narrowed to a short, slightly oblique beak with entire orifice; achene trigonous. Moist to wet woods and swamps; n. Fla. to La., n. to s. Ark., s. Ind., Va., and (mainly near the coast) Mass. (*C. ptychocarpa,* a preoccupied name.)

129. Carex digitalis Willd.   Densely tufted, 1–5 dm; fertile stems weak, hispidulous, triangular but not winged; basal sheaths white or light brown; lvs roughened on the margins and hispidulous on the veins toward the tip, those of the sterile shoots 1–5 mm wide, of the fertile ones 1–3 mm; terminal spike staminate, 1.2–5 cm, evidently pedunculate; pistillate spikes 1–3, scattered, 0.6–2 cm, erect to drooping, on short to elongate peduncles, the lowest usually basal; pistillate scales acute; perigynia 3–9, finely many-nerved as well as 2-ribbed, 2–4 mm, obovoid and sharply trigonous, scarcely beaked; achene trigonous; 2n=48. Dry woods; Me. to Fla., w. to Wis., Ill., Mo., and e. Tex.

130. Carex laxiculmis Schwein.   Tufted; stems very weak and slender, 2–5 dm, ascending to reclining; basal sheaths white or light brown; lvs roughened on the margins and hispidulous on the veins toward the tip, those of the sterile shoots 3–10(–12) mm wide, of the fertile ones a little narrower; terminal spike staminate, 1–2 cm, on a very short to elongate (10 cm) peduncle; pistillate spikes 2–4, 0.6–2 cm, scattered, on short and erect to long (9 cm) and drooping peduncles, the lowest one basal; pistillate scales cuspidate to acute, the lowest 1 or 2(3) empty or subtending staminate fls; perigynia 4–9, 2.5–4 mm, finely many-nerved as well as 2-ribbed, elliptic-ovoid and sharply trigonous, abruptly contracted to a short, slightly oblique beak with entire orifice; achene sharply trigonous; 2n=44, 46. Moist or wet woods, especially in heavy clay soils, chiefly in calcareous districts; s. Me. to Mich., s. Wis., and s. Io., s. to N.C., n. Ga., n. Ala., and Mo.

131. Carex granularis Muhl.   Tufted on very short rhizomes, 3–8 dm; larger lvs 4–10(–13) mm wide, often surpassing the stem; terminal spike staminate, sessile or nearly so, scarcely surpassing the uppermost pistillate spike and much overtopped by the bracts; pistillate spikes short-cylindric, 1–3 cm, the lower on long-exserted (but not basal) peduncles, the upper short-peduncled or subsessile, the upper 1 or 2 contiguous to the staminate spike; pistillate scales triangular-ovate, half to fully as long as the perigynium, acute or cuspidate; perigynia crowded in several rows, ellipsoid to obovoid, 2.2–4 mm, sharply several-nerved, abruptly contracted to a short, straight or abruptly outcurved beak; achene concavely trigonous, loosely enveloped; 2n=42. Wet meadows and swales, chiefly in calcareous districts; Que. and Me. to Sask., s. to Fla., Okla., and ne. Tex. (*C. haleana; C. rectior; C. shriveri.*)

132. Carex crawei Dewey.   Stems 1–3 dm, stiff, solitary to loosely clustered on long rhizomes; lvs thick and stiff, pale green, 1–4 mm wide, usually curved or recurved; terminal spike staminate, 1–2 cm, its peduncle elongate, overtopping the pistillate spike and usually the bracts; pistillate spikes 2–4, the lowest almost basal, on short-exsert peduncles, the upper shorter-pedunculate or subsessile, short-cylindric, 1–1.5 cm; pistillate scales triangular-ovate, much shorter than the perigynia, acute, acuminate or short-cuspidate; perigynia crowded in several rows, ovoid to ellipsoid, 2.1–3.5 mm, sharply or obscurely nerved, rather abruptly tapering into a very short straight beak; 2n=38. Wet meadows, shores, and rock-ledges in calcareous districts; Que. to B.C., s. to N.J., Tenn., Ala., and Ark.

Carex microdonta Torr. & Hook., a closely allied, more sw. sp. may be sought in our range in Mo. It has lvs 3–6 mm wide, and evidently ribbed, shortly but conspicuously beaked perigynia 3–4.5 mm.

133. Carex hitchcockiana Dewey.   Much like no. 134, brownish at base; lvs wider, to 7 mm wide; sheaths hispidulous; bracts distinctly roughened with short stiff hairs; pistillate scales 4.7–9 mm; perigynia obovoid-fusiform, tapering gradually to the base and abruptly to the straight or minutely outcurved beak, 4.3–5.9 mm, finely many-nerved; achene with a minute, sharply bent beak. Rich moist woods; Mass., Vt., and s. Que. to Minn., s. to Va. and Ark.

134. Carex oligocarpa Schk.   Densely tufted, 2–8 dm, purplish at base; lvs 2–4 mm wide, usually surpassing the stems; sheaths glabrous; staminate spike 1–3 cm, on a peduncle ca as long as the uppermost pistillate spike; bracts glabrous or rarely minutely granular-puberulent; pistillate spikes 2–4, 0.5–2 cm, very loosely fld; pistillate scales ovate, 3–6 mm, all or at least the lower with an awn longer than the body and usually longer than the perigynia; perigynia 7–12, alternately arranged, fusiform or slenderly ellipsoid, 3.5–5.4 mm, nearly straight, tapering almost equally to both ends, finely many-nerved; achene trigonous, filling the perigynium, with a minute straight beak. Moist rich woods; Mass., Vt., s. Que., and Ont. to Minn., s. to Fla. and Tex.

135. Carex amphibola Steudel. Tufted, 3–8 dm, leafy-stemmed; main lvs 4–8 mm wide; basal sheaths becoming green or brown; peduncles and axis of infl ± smooth; staminate spike sessile or nearly so, usually overtopping the uppermost pistillate one; pistillate spikes 2–5, 1–2 cm, loosely few-fld, mostly widely separate, the lowest near the middle of the stem on exsert peduncles, the upper shorter-peduncled; pistillate scales ovate, the body half as long the the perigynia, the midvein of the lower prolonged into an awn equaling or surpassing the perigynium; perigynia ellipsoid, 4–5.3 mm, two-fifths to half as wide, finely many-nerved, beakless; achene sharply trigonous with concave sides. Woods and fields, abundant; N.B. and Que. to Minn., s. to Ga. and Tex. (*C. a.* var. *turgida; C. bulbostylis; C. corrugata.*) The older name *C. grisea* Wahlenb. may prove to apply to this sp.

136. Carex flaccosperma Dewey. Tufted, 1.5–6 dm; lvs green or glaucous, the basal ones to 15 mm wide, the bracts to 10 mm wide; staminate spike 1–2 cm, usually about equaling or somewhat overtopping the uppermost pistillate one; pistillate spikes 2–4, 1.5–4 cm, densely many-fld, widely separate, the lowest often nearly basal on exsert peduncles, the upper on progressively shorter peduncles or sessile; bracts foliaceous, surpassing the stems; pistillate scales ovate, a third to half as long as the perigynia, acute or with an awn to 1 mm; perigynia ellipsoid, 3–5.3 mm, two-fifths to half as thick, finely many-nerved, beakless; achene ± concavely trigonous. Wet woods or swamps or moist fields; Mass. and Ont. to s. Ind. and Mo., s. to N.C. and Ark. (*C. glaucodea.*)

137. Carex conoidea Schk. Tufted, 1–7 dm; the stems usually much surpassing the lvs; main lvs 2–4 mm wide; peduncles and axis of the infl scabrous; staminate spike 1–2 cm, linear, long-peduncled, usually much overtopping the uppermost pistillate spike; pistillate spikes 2–4, widely separate or the upper 2 contiguous, short-cylindric, 1–2 cm, on short rough peduncles; bracts foliaceous, the margins of their sheaths rough; pistillate scales ovate, much shorter than the perigynia, the green midvein usually excurrent into a short awn; perigynia ellipsoid, 2.5–3.8 mm, a third to two-fifths as thick; achene concavely trigonous. Moist open places; Nf. and Que. to Minn., s. to N.C. and Mo.

Carex katahdinensis Fern., occurring on gravelly shores from Nf. to Me. and Minn., may be only an ecotype of no. 137. It is smaller, to 2 dm, the stems much overtopped by the lvs and elongate bracts; staminate spike sessile or nearly so, little if at all overtopping the pistillate; pistillate spikes 0.5–1.5 cm, nearly or quite contiguous.

138. Carex prasina Wahlenb. Tufted, 3–8 dm, brown or greenish at base; main lvs 3–5 mm wide; sheaths glabrous; terminal spike staminate or with a few apical perigynia; pistillate spikes 2–4, widely separated, cylindric, 2–5 cm, 5 mm thick, curved or nodding, the lower on long peduncles, the upper on much shorter ones; upper bract sheathless or nearly so; pistillate scales ovate to obovate-oblong, shorter than the perigynia, mucronate or short-cuspidate; perigynia 2.9–4.2 mm, rhombic-ovoid, trigonous, sharply angled at the prominent lateral ribs, obtusely angled on the back, otherwise essentially nerveless, tapering to an acute, at first nearly flat, later trigonous beak; achene concavely trigonous. Moist or wet woods and streambanks; Que. and Me. to Mich., s. to S.C., Ga., and Ala.

139. Carex gracillima Schwein. Tufted, 4–9 dm, strongly purple-tinged at base; main lvs 3–8 mm wide; sheaths glabrous, at least ventrally; terminal spike pistillate in the distal third, or rarely wholly staminate; pistillate spikes 3–5, mostly widely separate, cylindric, 3–6 cm × 3 mm, spreading or nodding on slender peduncles nearly as long; bracts all sheathing; pistillate scale ovate to obovate, half to nearly as long as the perigynia, usually obtuse, sometimes short-cuspidate; perigynia ellipsoid or somewhat obovoid, 2.4–3.7 mm, obscurely trigonous, sharply several-nerved, obtuse or rounded at the beakless summit; achene concavely trigonous; $2n$=50, 52, 54. Woods, abundant; Nf. and Que. to Man., s. to Va. and Mo.

140. Carex aestivalis M. A. Curtis. Tufted, 3–7 dm, ± purplish at base; main lvs 2–3 mm wide, usually hairy; sheaths hairy, at least on the ventral strip; terminal spike pistillate in the distal half; lateral spikes 2–4, pistillate, usually approximate and overlapping, cylindric, 2–4 cm × 2–3 mm, erect or nearly so, the lower on peduncles to 2 cm, the upper subsessile; upper bracts sheathless or nearly so; pistillate scales oval to obovate, half as long the perigynium, obtuse or short-cuspidate; perigynia narrowly ellipsoid, 2.6–3.5 mm, a third as wide, obscurely several-nerved, tapering to an obtuse beakless tip; achene concavely trigonous. Woods, especially in the mts.; N.H. and Vt. to N.C., Ky., Ga., and Ala.

141. Carex aestivaliformis Mackenzie. Tufted, 4–9 dm, purplish at base; main lvs 2–4 mm wide, the blade glabrous or hairy, the lower (or all) sheaths hairy at least on the back; terminal spike pistillate toward the tip; lateral spikes 2 or 3, usually approximate, ascending or spreading, cylindric, 2–4 cm × 3 mm, long-peduncled; uppermost bracts sheathless or nearly so; pistillate scales ovate-oblong, slightly shorter than the perigynia, acute to acuminate or short-cuspidate; perigynia narrowly ovoid-ellipsoid, 3–3.5 mm, obscurely several-nerved, acuminately tapering to a very short, barely bidentate beak; achene concavely trigonous. Wet woods and meadows; N.H. to Va. and Ga. Rare, perhaps a hybrid of *C. aestivalis* and *C. gracillima*.

142. Carex formosa Dewey. Tufted, 3–7 dm, strongly purplish at base; main lvs 3–7 mm wide, hairy beneath; sheaths hairy at least on the back; terminal spike with a few distal perigynia; lateral spikes pistillate, with a few basal staminate fls, well separated, cylindric, 1–3 cm × 5 mm, densely many-fld, erect or spreading, on peduncles about as long; bracts all sheathing, the uppermost much reduced; pistillate scales ovate to obovate, slightly shorter than the perigynia, obtuse to acuminate or cuspidate; perigynia ovoid, 4–5 mm, half as wide, obscurely trigonous, sharply several-nerved, acuminately narrowed to a distinct short beak; achene sharply and concavely trigonous. Moist soil in woods and thickets; Mass., Conn. and s. Que. to Mich., Wis., and N.D.

143. Carex davisii Schwein. & Torr. Tufted, 3–9 dm, purplish at base; main lvs 4–8 mm wide, hairy beneath; sheaths hairy; terminal spike peduncled, with distal perigynia; pistillate spikes 2 or 3, the upper 2 often approximate, cylindric, 2–4 cm × 6 mm, erect or spreading or even nodding, short-peduncled; bracts all sheathing; pistillate scales white or hyaline with green center, oblong or ovate-oblong, shorter to somewhat longer than the perigynia, narrowed into a prominent awn; perigynia dull orange at maturity (unlike related spp.), ovoid, 4.5–6 mm, half as wide, obscurely trigonous, sharply several-nerved, acuminately tapering into a sharply bidentate beak 0.7 mm; achene concavely trigonous. Woods; Mass., Vt. and s. Ont. to Mich. and Minn., s. to Md., Nebr., and Tex., abundant in the Middle West.

144. Carex oxylepis Torr. & Hook. Much like no. 143; main lvs 3–5 mm wide; pistillate scales ovate, much narrower and usually shorter than the perigynia, acuminate or the lower short-awned; perigynia ovoid, 3.5–4.6 mm, sharply several-nerved, tapering into a distinct short-beak. Rich, moist or wet woods; se. Va. to Fla. and Tex., n. in the interior to s. Ill. and s. Mo.

145. Carex assiniboinensis W. Boott. Plants tufted, with 3 kinds of shoots; rosettes, long, arching stolons rooting at the tip, and weak, slender fertile culms 2–6 dm with lvs at base; main lvs 1–3 mm wide; terminal spike 2–3 cm, long-pedunculate, staminate but usually with a single basal perigynium subtended by a prolonged, attenuate scale; pistillate spikes ca 3, widely separate, very loosely fld, 2–4 cm, on slender, spreading to drooping peduncles; pistillate scales ca equaling the perigynia; perigynia narrowly lanceolate, coriaceous, 5.5–8 mm, densely hispidulous; the slender beak as long as the body; achene concavely trigonous; 2*n*=32. Moist open woods; n. Mich. and n. Wis. to nw. Ont., Sask., the Dakotas, and n. Io.

146. Carex castanea Wahl. Tufted, 3–9 dm, purplish at base, the stems overtopping the lvs; main lvs 3–6 mm wide, soft-hairy; spikes approximate at the top of the stem, the terminal one staminate, erect and long-pedunculate, the lateral ones (usually 3) pistillate, on slender, spreading or drooping peduncles, short-cylindric, 1–2(–2.5) cm × 6–8 mm, closely 10–40-fld in several rows; lower bracts setaceous, villous-ciliate, with very short sheath, the upper usually bladeless; pistillate scales ovate or ovate-oblong, chestnut-tinged, about equaling the body of the perigynium, acute, acuminate, or minutely cuspidate; perigynia lanceolate, 3.5–5.8 mm, glabrous, obscurely trigonous, strongly 2-ribbed, finely several-nerved, tapering to a slender bidentate beak half as long as the body; achene concavely trigonous; 2*n*=64. Swamps, bogs, wet meadows, and margins of coniferous woods; Nf. to w. Ont., s. to Conn., Mich., and Minn. A hybrid with *C. arctata* has been called *C.* × *knieskernii* Dewey.

147. Carex arctata W. Boott. Densely tufted, 3–8 dm, purplish at base; main stem-lvs 3–5 mm wide, those of the sterile shoots 3–8 mm wide; staminate spike 1–3 cm, rarely pistillate distally; pistillate spikes 3–5, very slender, 2–6 cm, the upper approximate, the lower well separated, drooping or spreading, very loosely fld; bracts all sheathing, the upper with much reduced setaceous blade; pistillate scales ovate or oblong, the body much shorter than the perigynium, many or all tipped with a short cusp, the whole usually three-fourths as long as the perigynium, but sometimes distinctly surpassing it; perigynia glabrous, narrowly ovoid, trigonous, 3.2–4.8 mm, abruptly narrowed to a short stipe, conspicuously 2-ribbed, finely and usually obscurely several-nerved, narrowed above to a short beak; achene concavely trigonous. Moist rich woods; Nf. to Minn., s. to Pa. and O.

148. Carex sylvatica Hudson. Much like no. 147; perigynia 5–6 mm, the body sessile, obovoid, conspicuously 2-ribbed, otherwise nerveless, the beak slender and nearly as long as the body. Native of Europe, naturalized from L.I. to s. Ont.

149. Carex debilis Michx. Tufted, to 1 m, purplish at base; main lvs 2–4 mm wide; staminate spike 2–4 cm, very slender, sometimes with a few distal perigynia; pistillate spikes 2–4, well separated, very slender, spreading or nodding, 3–6 cm × 4 mm, very loosely fld, the perigynia separated by internodes 2–4(–6) mm; pistillate scales oblong to obovate, half as long as the perigynia, with hyaline or brownish margins and narrow green midvein; perigynia lanceolate to narrowly ovoid, obscurely trigonous, 4.7–8.3 mm, conspicuously 2-ribbed, obscurely several-nerved, narrowed to a beak; achene trigonous, the angles thickened, the sides concave below, borne on a slender stipe 0.5–1.5 mm; 2*n*=60. Nf. to n. Mich. and Wis., s. to Fla. and Tex. Three vars.:

a Perigynia pubescent; pistillate scales with the midvein usually excurrent as a cusp or short awn; mt.
    woods, Pa. to Ga. (*C. allegheniensis*.) .... var. *pubera* A. Gray.
a Perigynia glabrous; pistillate scales various.
  b Perigynia avg nearly 7 mm, broadest below the middle, tapering with straight or slightly convex
    sides to a conspicuous beak with white-hyaline tip; chiefly in moist or wet ground near the
    coast, Mass. to Fla. and Tex., n. in the interior to s. Ind. .... var. *debilis.*
  b Perigynia avg 5.6 mm, broadest and somewhat ventricose at or near the middle, tapering with
    concave sides to a short tip; woods; Nf. to n. Mich. and Minn., s. to Va. and Mo., and in the
    mts. to N.C. and Tenn. (*C. flexuosa*.) .... var. *rudgei* L. Bailey.

150. Carex venusta Dewey. Densely tufted, 3–9 dm, purplish at base; main stem lvs 2–3 mm wide, those of the sterile shoots somewhat wider; staminate spike 2–3 cm, often with a few distal perigynia; pistillate spikes 2–4, spreading or nodding, slender-peduncled, at least the lower well separated, cylindric, 2–5 cm × 5 mm, densely fld, the perigynia separated by internodes 1–1.5 mm; pistillate scales oblong to ovate, half as long as the perigynia, ± red-tinged, obtuse to acute or mucronate; perigynia lanceolate, conspicuously nerved, tapering into a beak 0.5 mm; achene concavely trigonous, on a stipe 0.6–1.5 mm. Swamps and wet woods near the coast; N.Y. to Fla. and La. Var. *venusta,* with pubescent perigynia 6.4–8.1 (avg 7.2) mm, occurs from se. Va. to Fla. Var. *minor* Boeckeler, with glabrous perigynia 5.2–7.7 (avg 6.2) mm, occurs throughout the range of the sp. (*C. oblita.*)

151. Carex capillaris L.   Densely tufted; stems slender and lax, 1–6 dm; lvs mainly in a basal cluster, to ca 15 cm, 1–3 mm wide; terminal spike staminate or rarely gynaecandrous, 4–10 mm; pistillate spikes 1–4, usually ± remote, on lax, capillary, nodding or loosely spreading peduncles 0.5–1.5 cm, 5–25-fld, sometimes compound, all subtended by sheathing bracts, of which the upper may have reduced or no blade; pistillate scales mostly shorter but often wider than the perigynia, white-hyaline distally and usually also marginally, otherwise light brown or greenish; perigynia ± elliptic or lance-ovate, 2.4–3.3 mm, tapering to a short, poorly defined beak, obliquely 2-nerved, not much if at all compressed, loosely enclosing the trigonous achene and empty distally, glabrous, or faintly scabrous-serrulate toward the beak, shining brown to olive-green; $2n=36$, 54, 56. Streambanks, wet meadows, and wet ledges; circumboreal, s. in Amer. to n. N.Y., Mich., Wis., Minn., and N.M.

152. Carex sprengelii Dewey.   Stems 4–8 dm, clustered on a stout, short or somewhat elongate rhizome that is densely covered with old fibrous lf-remains; lvs cauline as well as basal, elongate, mostly 1.5–4 mm wide; terminal 1 or 2 spikes staminate or with a few basal perigynia, 1–2 cm, the others 2–4, pistillate or the upper with a few distal staminate fls, 1.5–3.5 cm, 10–30-fld, not crowded, at least the lower ones loose to spreading or nodding on long, capillary peduncles; lowest bract with sheathing base 1–10 mm and long, slender blade, the others more reduced and often sheathless; pistillate scales narrow, acute or acuminate, pale, largely scarious or hyaline, often scabrous along the midrib, surpassing the body of the perigynium; perigynia 4.1–7.7 mm, the body broadly ellipsoid or ellipsoid-obovoid, prominently 2-nerved, 2.2–3.8 mm, abruptly contracted to the slender softly bidentate beak 1.7–4 mm; achene trigonous, the persistent style-base commonly contorted; $2n=42$. Open woods, meadows, and rocky ledges; N.B. to B.C., s. to N.J., Io., and Colo. (*C. longirostris,* a preoccupied name.)

153. Carex viridula Michx.   Tufted, 1–4 dm; lvs flat to canaliculate, often equaling or surpassing the stems, mostly 1–3 mm wide; terminal spike sessile or short-peduncled, slender, staminate, 7–21 mm; lateral spikes 2–4, pistillate (or the upper androgynous), short and stout, 5–15 mm, sometimes compound, all sessile or short-pedunculate and crowded at the summit, or 1 or 2 of the lower ± remote and more evidently pedunculate; bracts sheathless or nearly so (except those subtending any remote spikes), one or more of them with an elongate blade much surpassing the infl; pistillate scales shorter than the perigynia, thin and pale except the green midrib, varying to medium brown; perigynia 2.2–3.3 mm, straight or nearly so, mostly spreading or the lower reflexed, pale green to stramineous or yellowish, obovoid, distended by the trigonous, 1.1–1.4 mm achene but distally empty and slightly inflated, irregularly several-nerved, or with 2 ± evidently marginal nerves and 2–6 often less prominent ones on each face, the slender, ± smooth beak 0.8–1.2 mm; $2n=66$–72. Abundant on boggy or marly shores in calcareous districts; circumboreal, s. in Amer. to N.J., Ind., S.D., and N.M. (*C. chlorophila; C. oederi,* misapplied.) Hybridizes with no. 154.

154. Carex flava L.   Tufted, 1–8 dm; lvs flat, mostly 3–5.5 mm wide, the plants appearing leafier than no. 153; terminal spike sessile or short-pedunculate, slender, staminate or with some distal perigynia, 6–24 mm; lateral spikes 2–5, pistillate, short and stout, 6–17 mm, all sessile or short-pedunculate and crowded at the summit, or 1 or 2 of the lower ± remote and more evidently pedunculate; bracts sheathless or nearly so (except those subtending any remote spikes), one or more of them with an elongate blade much surpassing the infl; pistillate scales strongly tinged with coppery brown, thus conspicuous in the spike; perigynia 3.7–6.2 mm, most of them spreading and evidently falcate-recurved, relatively slender and gradually tapering to the poorly defined, distally rough-margined beak, this 1.4–2.3 mm, set at a ± divergent angle to the body; perigynium-body strongly yellowish toward the base, usually more greenish (or eventually brownish) distally, prominently several-nerved on the upper surface, more obscurely so on the lower, achene hardly larger than in no. 153, a larger part of the perigynium thus empty; $2n=60, 64, 68, 70$. Bogs and wet meadows in calcareous districts; circumboreal, s. in Amer. to N.J., Ind., Ida., and B.C. (*C. laxior.*)

155. Carex lepidocarpa Tausch.   Much like no. 154; staminate spike on a long peduncle surpassing the uppermost pistillate spike; most pistillate spikes remote; perigynia 3.5–5 mm, conspicuously swollen, the body nearly as wide as long, abruptly contracted to the beak, the central ones spreading, only the lower deflexed; $2n=58, 68$. Wet places in calcareous districts; Europe, and locally from N.S. to Nf. and e. Que. (*C. flava* var. *l.*)

156. Carex cryptolepis Mackenzie.   Much like no. 154; stems 2–6 dm, surpassing the lvs, these 1.5–3.5 mm wide; pistillate scales lance-ovate, about equaling the perigynium-body, about the same color as the perigynium and thus inconspicuous; perigynia greenish to yellowish or golden-brown, 3.2–4.8 mm, abruptly contracted to the smooth beak 1.2–1.5 mm. Wet meadows and shores in calcareous districts; Nf. to Minn., s. to N.J., O., and Ind. (*C. flava* var. *fertilis.*)

157. Carex extensa Gooden.   Stems tufted, stiff and wiry, 2–5 dm; lvs involute, 1–3 mm wide, glaucous or grayish-green, the sheaths red-dotted ventrally; staminate spike sessile or short-peduncled, 1–2.5 cm; pistillate spikes 2–4, slightly or widely separated, 0.5–2 cm, erect, sessile or nearly so, cylindric; lowest bract elongate; pistillate scales ovate, two-thirds as long as the perigynia, reddish-brown; perigynia ascending, ovoid, brown, 2.5–4 mm, firm, strongly ribbed and nerved, tapering into a beak a fourth as long as the body; $2n=60$. Native of Europe, intr. at the edge of salt-marshes from L.I. to Va.

158. Carex pallescens L.   Tufted, 2–5 dm; main lvs 2–4 mm wide, soft-hairy, especially on the lower side and on the sheaths, varying to glabrous; terminal spike peduncled, staminate, 1–2 cm; pistillate spikes

2–4, densely fld, short-cylindric, 8–15 mm, the lower short-peduncled, the upper subsessile; lower bracts exceeding the infl, the upper much smaller; anthers 1.7–2.8 mm; pistillate scales ovate, about as long as the perigynia, pale green, faintly brown-tinged, acute to cuspidate; perigynia ascending, ellipsoid, 2.1–2.8 mm, nearly round in cross-section, obscurely and finely nerved, beakless, obtuse; achene concavely trigonous. Moist woods and meadows; Nf. to Wis., s. to N.J. and O.

159. Carex torreyi Tuckerman. Much like no. 158; pistillate spikes 5–12 mm; pistillate scales subrotund, much shorter than the perigynia, minutely cuspidate, conspicuously hyaline-margined; perigynia obovoid, 2.5–3.5 mm, narrowed to the base, the body conspicuously 15–25-nerved, rounded to a cylindric beak 0.2–0.5 mm; $2n=58$. Thickets and moist meadows, mainly on the n. Great Plains; Wis. and Minn. to Alta., Mont., and Colo.

160. Carex virescens Muhl. Tufted, 5–10 dm, reddish-purple at base, the stems usually overtopping the lvs; main lvs 2–4 mm wide, hairy; terminal spike 2–4 cm, pistillate above; pistillate spikes 2–4, erect, 2–4 cm, linear-cylindric, loosely fld at base, sessile or short-peduncled; bracts sheathless or nearly so, the lowest foliaceous, the upper much reduced; anthers 1.5–2.5 mm; pistillate scales ovate, shorter than to surpassing the perigynia, hyaline-margined, with green midvein, acute to cuspidate; perigynia obscurely trigonous, ± ellipsoid, 2–3 mm, densely hairy, conspicuously few-ribbed, acute, beakless; achene concavely trigonous. Dry woods; Me. to Mich., s. to N.C., Ga., and Mo.

161. Carex swanii (Fern.) Mackenzie. Much like no. 160; lvs often overtopping the stems; terminal spike 1–2 cm; pistillate spikes 1–2 cm, ellipsoid or thick-cylindric, densely fld throughout; anthers 0.7–1.5 mm; body of the pistillate scales much shorter than the perigynia, often ending in a cusp equaling the perigynium; perigynia obovoid, obscurely trigonous, 1.8–2.5 mm, blunt. Dry woods and fields; N.S. to Wis., s. to N.C. and Ark. (*C. virescens* var. *minima*.)

162. Carex complanata Torr. & Hook. Tufted, red-purple at base; main lvs erect, equaling the stems, 2–4 mm wide; spikes 2–5, usually 3, the terminal one pistillate above; pistillate spikes approximate, short-cylindric or narrowly ovoid, 8–15 mm, sessile or nearly so; lowest bract elongate, much surpassing the spikes, the others small; anthers 1.3–2 mm; pistillate scales usually much shorter than the perigynia, the body with brownish hyaline margins, acute to short-cuspidate; perigynia flattened-obovoid, 2–2.8 × 1.3–2 mm, appressed-ascending, 2-ribbed, rounded at the beakless tip. Dry woods and fields; Me. to Fla., w. to Mich., Io., Tex., and Mex. Var. *complanata*, with glabrous blades and glabrous to hairy sheaths, the perigynia usually nerveless ventrally, occurs on the coastal plain from N.J. s. Var. *hirsuta* (L. Bailey) Gleason, with soft-hairy blades and sheaths, the perigynia with fine but distinct nerves on both sides, occurs nearly throughout the range of the sp. (*C. hirsutella*.)

163. Carex caroliniana Schwein. Tufted, 3–8 dm, tinged with red-purple at base; lf-sheaths glabrous or hairy on the back, glabrous or nearly so ventrally; blades elongate, 2–4 mm wide, glabrous except at base; spikes usually 3, ± approximate, short-cylindric or narrowly ovoid, 1–2 cm, densely fld, the terminal one pistillate above, the others wholly pistillate; pistillate scales ovate, usually much shorter than the perigynia, brown-tinged, acute or short-cuspidate; perigynia obovoid, 2.2–2.8 mm, two-thirds as wide and nearly as thick as wide, turgid, spreading, strongly nerved, minutely beaked; achene concavely trigonous, bent-apiculate. Dry woods and meadows; N.J. and Pa. to O., Mo., and Okla., s. to N.C. and Tex.

164. Carex bushii Mackenzie. Much like no. 163; lf-sheaths densely short-hairy all around; blades hairy to sometimes nearly glabrous above the base; pistillate scales lanceolate, often minutely pilose, long-acuminate into a cusp or awn usually protruding from the spike and often surpassing the perigynium; perigynia 2.4–3.3 mm. Moist woods. prairies, and meadows; Mass. and N.Y. to Mich. and Kans., s. to Va., Miss., and Tex.

165. Carex hirta L. Vigorously colonial by creeping rhizomes; aphyllopodic, 3–10 dm; sheaths hairy at the mouth and usually on the back; main lvs 3–6 mm wide, often sparsely hairy; staminate spikes 1–3, 2–3 cm, only the upper one peduncled; pistillate spikes 2–3, widely separated, cylindric, 2–5 cm, densely fld, erect, short-peduncled; bracts lf-like, short-spreading at base, often surpassing the stem; pistillate scales lance-ovate, half to fully as long as the perigynia, usually sparsely villous, acute to acuminate or awned; perigynia narrowly ovoid, 5–8 mm, hairy, conspicuously nerved, acuminately tapering into a beak over half as long as the body, its divergent teeth 1–1.5 mm; achene trigonous. Native of Europe, intr. in waste places and dry fields from P.E.I. to Wis. and D.C.

166. Carex striata Michx. Vigorously colonial by creeping rhizomes, strongly aphyllopodic, 4–12 dm; main lvs 2–5 mm wide; staminate spikes 1 or 2, the terminal one 3–5 cm; pistillate spikes 1 or 2, densely fld, cylindric, erect, 2–4 cm, sessile or nearly so; lowest bract elongate, surpassing the stem; pistillate scales ovate, half to nearly as long as the perigynia, with red-purple sides and hyaline margins, acute to acuminate; perigynia coriaceous, ovoid, (4–)4.5–6.5 mm, conspicuously many-nerved (the nerves impressed), acuminately tapering into a bidentate beak a fourth as long as the body; achene concavely trigonous. Sandy swamps on the coastal plain; se. Mass. to Fla. Plants of our range have glabrous perigynia and represent the var. *brevis* L. Bailey. The rather ill-defined var. *striata*, with minutely hairy perigynia, is more southern. (*C. walteriana*.)

167. Carex houghtoniana Torrey. Vigorously colonial by creeping rhizomes, strongly aphyllopodic, 3–6 dm; main lvs 3–6 mm wide; staminate spike 2–4 cm, long-peduncled, sometimes with 1 or 2 much smaller ones at its base; pistillate spikes 1–3, widely separated, cylindric, 1–4 cm, densely fld, erect, short-peduncled or subsessile; lowest bract lf-like, usually surpassing the stem; pistillate scales shorter than the perigynia, lance-ovate, with red-purple sides and hyaline margins, acute or acuminate, or the midvein prolonged into a cusp; perigynia ovoid, (4–)4.5–7 mm, finely hairy, conspicuously many-nerved, acuminately tapering into a bidentate beak two-fifths as long as the body; achene concavely trigonous; $2n=56$. Sandy or rocky soil; Nf. to Sask., s. to N. Engl., n. N.Y., Mich., and Minn. (*C. houghtonii.*)

168. Carex pellita Muhl. Vigorously colonial by creeping rhizomes, 3–10 dm, strongly aphyllopodic; lvs flat or nearly so, 2–5 mm wide; terminal spike staminate, 2–5 cm, often closely subtended by one or 2 shorter, sessile staminate spikes; pistillate spikes 2 or 3, 1–4 cm, rather remote, sessile or the lowest one on an erect slender peduncle, leafy-bracteate, often surpassing the staminate spikes; pistillate scales partly or wholly brownish or purplish, usually narrower than the perigynia and acute or shortly awn-tipped; perigynia 3–3.5 mm, densely velutinous or velutinous-sericeous, multiribbed, but the ribs ± obscured by the pubescence, the body ovoid to broadly ellipsoid or subglobose, turgid, not much compressed, firm, abruptly contracted to the beak, this 0.8–1.5 mm, including the 0.3–0.9 mm teeth; achene concavely trigonous. Wet meadows and other wet places, sometimes in shallow water, or seldom on stabilized dunes or dry ditchbanks; N.B. and Que. to B.C., s. to Va., Tenn., Ark., and Calif. (*C. lanuginosa,* misapplied.)

169. Carex lasiocarpa Ehrh. Much like no. 168, differing most prominently in lvs, these permanently folded along the midrib and tending to appear subterete, only 1–1.5(–2) mm wide as folded; perigynia 2.8–4.3 mm; $2n=56$. Bogs, marginal sedge-mats, and shallow water; circumboreal, s. in Amer. to N.J., W.Va., Io., and Wash. (*C. lanuginosa.*) The Amer. plants have been called var. *americana* Fern.

170. Carex vestita Willd. Vigorously colonial by creeping rhizomes, the slender stems 4–8 dm, strongly aphyllopodic; main lvs 3–5 mm wide; staminate spikes 1–3, the terminal one usually 2–5 cm, the lateral much shorter; pistillate spike 1 and subterminal or 2–3 and ± remote, erect, short-cylindric, 1.5–2.5 cm, densely fld, sessile or nearly so; lowest bract shorter than the stem, 2–5 cm; pistillate scales lance-ovate, up to as long as the perigynia, acute or acuminate, with purple sides and hyaline margins; perigynia ovoid, 2.8–4.4 mm, densely hairy, abruptly tapering into the beak; beak a third as long as the body, marked with purple on the back, hyaline-tipped, eventually split; achene concavely trigonous. Woods; s. Me. to se. Va.

171. Carex scabrata Schwein. Vigorously colonial by creeping rhizomes, 4–9 dm, phyllopodic; main stem-lvs 4–8 mm wide, those of the large sterile shoots to 14 mm wide; staminate spike 2–4 cm, longer than its peduncle; pistillate spikes 3–several, cylindric, 2–4 cm, erect, the lower long-peduncled, the upper progressively less so or subsessile; bracts lf-like, sheathless or nearly so; pistillate scales lanceolate, nearly equaling the perigynia, acuminate or acute; perigynia obovoid, obtusely trigonous, 3.2–4.4 mm, minutely scabrous-puberulent, conspicuously 2-ribbed, few-nerved, abruptly narrowed into a somewhat outcurved beak half as long as the body and minutely bidentate; achene concavely trigonous. Moist shaded ground and swamps; N.S. and Que. to Mich., s. to N.J., O., and Mo., and in the mts. to Ga.

172. Carex shortiana Dewey. Stout, tufted, 4–8 dm; main lvs 4–8 mm wide; spikes 4–6, cylindric, erect, the terminal gynaecandrous, the lateral pistillate, the lowest on long slender peduncles, the others progressively less so or subsessile; bracts lf-like, sheathless or nearly so; pistillate scales ovate, nearly or quite as long as the perigynia, reddish-brown, acute or rounded and mucronate; perigynia spreading, flattened-trigonous, broadly obovate, 1.8–2.6 mm, nearly as wide, cuneate to the base, conspicuously 2-ribbed at the lateral angles, otherwise nerveless, minutely apiculate; achene concavely trigonous. Moist woods and meadows; Pa. and sw. Ont. to Ind., Io., Mo., Kans., and Tenn.

173. Carex joorii L. Bailey. Vigorously colonial from stout creeping rhizomes, 4–13 dm, stout; main lvs elongate, glaucous, 5–10 mm wide; staminate spike 3–5 cm; pistillate spikes 3–5, separate but overlapping, 2–6 cm, staminate at the tip, usually ± erect, the lower long-peduncled, the upper less so or subsessile; lowest bract lf-like, the upper much reduced and setaceous; pistillate scales oblong, the body shorter than the perigynium, brown-tinged, the midvein prolonged into a rough awn; perigynia dull green and glaucous when young, becoming brown, broadly obovoid, 3.2–5 mm, tapering to the base, conspicuously several-nerved, broadly rounded above into a prominent, short beak with an entire orifice; achene concavely trigonous. Wet woods and swamps, mainly on the coastal plain; se. Va. (and reputedly Md.) to Fla. and Tex., n. in the interior to se. Mo. and McCreary Co., Ky.

174. Carex glaucescens Elliott. Much like no. 173; pistillate spikes 3–6, mostly nodding, remote, often staminate at the tip; pistillate scales oblong to obovate, retuse, the thin sides purple-brown or reddish, the strong midvein prolonged into a cusp often as long as the body; perigynia broadly obovoid or rhomboid, very glaucous when young, 3.5–5 mm, 2-ribbed, otherwise only obscurely nerved or nerveless. Swamps and wet woods, chiefly on the coastal plain; se. Va. to Fla. and La. (*C. verrucosa,* misapplied.)

175. Carex flacca Schreber. Vigorously colonial by creeping rhizomes, 2–6 dm; main lvs 2–5 mm wide; terminal spike staminate, 1–3 cm, long-peduncled; pistillate spikes 1–3 cm; pistillate scales ovate, with broad, pale midvein and brown sides, acute to short-acuminate or mucronate; perigynia elliptic-obovoid, minutely hispidulous distally, nerveless or nearly so, even the lateral ribs inconspicuous, abruptly narrowed to a minute beak; achene rounded-trigonous, bent-apiculate. Native of Eurasia and n. Afr., intr. in fields and meadows from N.S. and Que. to Ont. and N.Y. (*C. glauca.*)

176. Carex barrattii Schwein. & Torr. Loosely tufted on long rhizomes, 3–8 dm; main lvs 2–4 mm wide; terminal spike staminate, 3–5 cm, long-peduncled, often with a smaller one at its base; pistillate spikes 2–5, crowded to well separated, linear-cylindric, 2–4 cm, spreading or drooping, staminate at the top, the lowest with peduncles somewhat shorter than the spike, the upper with shorter peduncles or subsessile; lowest bract foliaceous to setaceous, 2–5(–10) cm; upper bracts scale-like; pistillate scales ovate, about as long as but usually narrower than the perigynia, dark brown-purple, with slender, concolorous midvein; perigynia stramineous, or darker at the summit, ovoid, 2.4–3.7 mm, obscurely trigonous, 2-ribbed and obscurely few-nerved, glabrous, very minutely beaked; achene concavely trigonous, straight-apiculate. Wet ground, especially in pine-barren swamps near the coast; Conn. to N.C.

177. Carex rariflora (Wahlenb.) J. E. Smith. Much like no. 178; stems rarely over 3 dm, smooth, obtusely triangular; staminate spike 6–15 mm, short-peduncled; pistillate spikes 6–16 mm, with up to 10 perigynia, on peduncles of their own length, separated by internodes up to 15 mm; lowest bract to 2 cm; pistillate scales dark purple, usually concealing the perigynia and partly enclosing them at base; perigynia 2.8–3.9 mm, more than half as thick as wide, acutish but beakless; $2n=52$. Circumboreal, s. in our range to cold bogs in N.B. and on Mt. Katahdin, Me.

178. Carex limosa L. Stems arising singly or few together from long, creeping rhizomes, 2–6 dm, strongly aphyllopodic; roots covered with a yellowish-brown, felty tomentum; lvs few, 1–2 mm wide, tending to be canaliculate; terminal spike staminate, 1–3 cm; pistillate spikes 1–3, nodding on slender peduncles, occasionally with a few staminate fls at the tip, 1–2.5 cm, the lowest usually subtended by a nearly or quite sheathless leafy bract 2–10 cm; pistillate scales light to dark brown, commonly about as long and wide as the perigynia, obtuse to acute or acuminate; perigynia pale, commonly greenish or stramineous, densely papillate, elliptic to ovate, somewhat compressed, 2-ribbed and with 4–7 ± evident nerves on each face, 2.3–4.2 mm, beakless or with a minute beak to 0.2 mm; achene trigonous, rather loosely enveloped by the distally empty perigynium; $2n=56, 62, 64$. Sphagnum bogs; circumboreal, s. to N.J. (and reputedly Del.), O., Io., and Calif.

179. Carex paupercula Michx. Stems loosely clustered in small tufts on short or long rhizomes, 2–7 dm, phyllopodic, the remains of old lvs commonly persistent at base; roots covered with a yellowish-brown, felty tomentum; lvs flat, 1–3 mm wide; terminal spike staminate, 0.7–1.5 cm; pistillate spikes 1–4, nodding on slender peduncles, often with a few staminate fls at the base, 0.7–1.5 cm; pistillate spikes 1–4, the lowest subtended by a nearly or quite sheathless leafy bract 2–10 cm; pistillate scales light to dark brown, often with green midstripe, generally longer and narrower than the perigynia and tapering to a long, narrow point or short awn; perigynia much as in no. 178, $2n=58$, ca 60. Acid swamps and sphagnum-bogs; circumboreal, s. in our range to N.J., Pa., Mich., and Minn.

180. Carex buxbaumii Wahlenb. Stems 3–10 dm, arising singly or few together from long creeping rhizomes, strongly aphyllopodic, not surrounded by old sheaths from previous years (but these often persistent separately from the new stems); lvs elongate, 2–4 mm wide; spikes mostly 2–5, approximate or somewhat remote, erect or closely ascending, sessile or (especially the lower) with ± well developed peduncle, the terminal one gynaecandrous, 1–3 cm, the lateral ones pistillate, about as long or somewhat shorter; bract subtending the lowest spike sheathless or nearly so, shorter to longer than the infl; pistillate scales lanceolate to lance-ovate, brown to purplish-black with a usually paler midrib, surpassing the perigynia, tapering to an awn-tip 0.5–3 mm; perigynia 2.7–4.3 mm, beakless or very shortly beaked, rather narrowly elliptic to elliptic-obovate or elliptic-ovate, firm-walled, not strongly papillate, light gray-green, densely papillate, 2-ribbed and with 6–8 inconspicuous or obscure nerves on each face; achene trigonous, somewhat narrower and much shorter than the perigynial cavity. $2n=74$, ca 100, 106. Peat-bogs, marshes, wet meadows, and other wet places; circumboreal, s. to N.C., Ky., Ark., and Calif.

181. Carex norvegica Retz. Stems slender and lax, 2–7 dm, loosely to densely tufted on a compact system of rather slender rhizomes, somewhat aphyllopodic; lvs ± flat, 1.5–3 mm wide; spikes mostly 2–5, approximate, erect or closely ascending, sessile or short-pedunculate, relatively small, the terminal one gynaecandrous, 6–14 mm, the lateral ones pistillate, shorter, seldom over 10 mm; bract subtending the lowest spike sheathless or nearly so, shorter to longer than the infl; pistillate scales purplish-black or brownish-black, prominently white-hyaline-margined, up to as long as the perigynia and usually fully as wide; perigynia coppery-yellowish or light green to dark purple, elliptic, commonly rather narrowly so, to ± obovate, 2.1–3 mm, distended and nearly filled by the trigonous achene, empty only just beneath the short (0.3–0.4 mm) generally blackish beak, 2-ribbed, otherwise only inconspicuously or scarcely nerved, minutely cellular-reticulate; $2n=54, 56$. Streambanks, seepage areas, and moist meadows; circumboreal, s. in Amer. to Que., Wis., Minn., and Utah. (*C. halleri; C. media; C. vahlii.*)

182. Carex parryana Dewey. Stems 2–6 dm, loosely tufted on short, scaly creeping rhizomes, phyllopodic, the principal lvs crowded near the base, seldom over 15(–25) cm, flat to channeled and revolute-margined, up to ca 4 mm wide; spikes 1–5, the terminal one (1–)1.5–3 cm, often longer than the others, these sessile or stiffly short-pedunculate and ± erect; spikes often all wholly pistillate, or the terminal one frequently staminate, or gynaecandrous, or with intermingled male and female fls; bract subtending the lowest spike scarcely sheathing, usually short and inconspicuous, rarely as long as the infl; pistillate scales stramineous to more often brown with pale, hyaline margins, broadly obtuse and shorter than the perigynia to narrower, acute, and surpassing the perigynia; perigynia ± distinctly obovate, 1.9–3 mm including the short (0.2–0.6 mm) or obsolete beak, often short-hairy distally, 2-ribbed, otherwise nerveless or inconspicuously several-nerved; achene trigonous, rather loosely filling the perigynium. Meadows, swales, and moist low ground on the prairies and high plains; widespread in interior w. U.S. and s. Can., e. to w. Minn. (*C. hallii.*)

183. Carex atratiformis Britton. Tufted, phyllopodic, 3–9 dm; main lvs 2–5 mm, much shorter than the slender stems; terminal spike mostly pistillate; lateral spikes 2–5, short-cylindric, 1–2 cm, spreading or drooping, often with a few basal staminate fls; peduncles slender, the lower 1–4 cm, the upper shorter; lowest bract lf-like, about equaling the infl, the upper very small; pistillate scales ovate to oblong, as wide and long or a little longer than the perigynia, dark brown throughout, acute to minutely cuspidate; perigynia elliptic, much flattened, 2.6–3.9 mm, half to two-thirds as wide, sharp-edged at the 2 ribs, otherwise nerveless or nearly so, rounded above to a very short, shallowly bidentate beak. Open meadows and exposed ledges; Lab. to Yukon, s. to the mts. of New England and the vicinity of Lake Superior.

184. Carex torta F. Boott.   Plants densely tufted, the stout stems 2–7 dm, arising laterally; lowest lvs reduced to bladeless sheaths; main lvs 3–5 mm wide, inversely W-shaped in x-section; terminal spike staminate, 2–4 cm, pedunculate; pistillate spikes 3–6, usually approximate and much overlapping, linear-cylindric, 3–8 cm, the lower short-peduncled and curved-spreading or drooping, the upper sessile or nearly so; bracts sheathless, the lowest lf-like, usually shorter than the stem, the others much reduced; pistillate scales elliptic to oblong, obtuse, about as long as but narrower than the perigynia, with broad, greenish or stramineous midstrip and blackish or deep brown-purple sides; perigynia ovate, 2.5–4.2 mm, half as wide, 2-ribbed, otherwise nerveless, gradually tapering to a minute, bent or twisted beak; achenes lenticular, filling the lower three-fourths of the perigynium; $2n=66$. Streambanks, sandbars, and shallow water; Que. and N.S. to N.C., w. to c. Ont., c. O., Tenn., and Ala.

185. Carex crinita Lam.   Densely tufted, 4–16 dm, the strongly aphyllopodic stems surpassing the lvs; main lvs 7–13 mm wide, the sheaths glabrous, bracts lf-like, sheathless or nearly so, the lowest one 2–5.5 dm; spikes loosely spreading to drooping on slender peduncles, often curved, the staminate ones 1–3, slender, 4–9 cm; pistillate spikes 2–5, below the staminate, linear-cylindric, 4–11 cm; pistillate scales (at least the lower) with truncate or often retuse body, the sides thin and coppery-brown, the conspicuous pale midvein prolonged into a rough flat awn to 10 mm; upper scales usually shorter than the lower, perigynia silky-green, 2-ribbed, otherwise nerveless or faintly nerved, smooth to slightly papillate distally, somewhat inflated, nearly circular in x-section, rounded to an abrupt, minute beak; achene lenticular; $2n=66$. Wet woods and swales; Nf. and Que. to Minn., s. to Ga. and Tex. The widespread var. *crinita* has ellipsoid to obovoid perigynia 2–3(–3.5) × 1–2 mm, achene constricted on one side or edge, with a bent style, varying to straight and symmetrical. The var. *brevicrinis* Fern., mainly along the coastal plain, n. locally to Mass., Ky., and Mo., has strongly obovoid perigynia 3–4.5 × 2–3 mm, the achene symmetrical.

186. Carex gynandra Schwein.   Much like no. 185; lf-sheaths hispidulous; lowest bract 1–3.5 dm; lower pistillate scales with acute or acuminate body; perigynia elliptic to ovate, acute, somewhat compressed, scarcely or not at all inflated, 2.4–4.2 mm; $2n=68$. Woods and swales; Nf. to ne. Minn., s. to Wis. and Va., and in the mts. to n. Ga. (*C. crinita* var. *g.*)

187. Carex mitchelliana M. A. Curtis.   Much like no. 186; stems 5–8 dm; main lvs 3–8 mm wide; sheaths hispidulous; lowest bract 1–2.5 dm; staminate spikes 1–2, to 5 cm; pistillate spikes 2–4, 3–8 cm, often with some distal staminate fls; pistillate scales with truncate or retuse body, the awn to 4 mm; perigynia 2.5–4 mm, ovate to broadly ovate, acute but scarcely beaked, closely enveloping the symmetrical achene, stramineous to light brown, 1–4-nerved on each side; $2n=66$. Swamp forests and wet thickets and meadows; Mass. to n. Fla., w. to Pa., Ky., and Ala.; rare. (*C. crinita* var. *m.*)

188. Carex paleacea Wahlenb.   Stems stout, 3–8 dm, aphyllopodic, borne singly or few together on long stout rhizomes; main lvs 3–8 mm wide; bracts lf-like, sheathless; spikes widely spreading or drooping on slender peduncles, the staminate 1–3, mostly 2–4 cm, the pistillate below the staminate, 2–4, stout-cylindric, 2–5 cm, 5–8 mm thick excluding the scales; body of the pistillate scales ovate and obtuse to retuse, somewhat shorter than the perigynia, with brown sides, the conspicuous pale midvein a third as wide as the scale and prolonged into a rough flat awn to 1 cm; perigynia biconvex to planoconvex, glaucous-green, firm, elliptic, 2.2–3.5 mm, three-fifths as wide, 2-ribbed, otherwise nerveless or dorsally several-nerved, abruptly apiculate to a beak 0.2–0.5 mm; achene lenticular, loosely enveloped, strongly constricted on one side near the middle; $2n=71–73$. Coastal salt-marshes; Greenl. and w. Can. to Mass.; n. Europe.

189. Carex salina Wahlenb.   Much like no. 188, more tufted, but still producing long rhizomes; main lvs 2–5 mm wide, involute in age; spikes all ± erect, the lower short-peduncled, the upper sessile or nearly so; pistillate scales ovate to lanceolate, the purple-brown body usually longer than the perigynia, retuse to acute, the green midvein often prolonged into a short awn; perigynia 1.9–3.3 mm, strongly 2-ribbed, otherwise nerveless or obscurely nerved; $2n=78–83$. Saline or brackish marshes along the coast; circumboreal, s. in Amer. to Mass. and James Bay. (*C. recta; C. subspathacea.*) Said to consist at least in considerable part of a series of variably fertile or sterile hybrids between no. 188 and members of the *Acutae*, but now ± stable and forming large populations.

190. Carex aquatilis Wahlenb.   Vigorously and densely rhizomatous, commonly forming large clumps or a turf, the flowering stems 3–10(–15) dm, surrounded by the dried-up lf-bases of the previous year; lower sheaths red-tinged, lvs mostly 2–7 mm wide; spikes mostly 3–7, cylindric, approximate to ± remote, mostly (1–)1.5–5 cm, the terminal one staminate (sometimes with a few basal perigynia), sometimes with a smaller staminate spike at its base, the others pistillate or (especially the upper) sometimes androgynous, erect, all sessile or nearly so, or the lower ones often evidently pedunculate; bracts sheathless or nearly so, the lowest one 7–25 cm and mostly surpassing the infl; pistillate scales reddish-brown or tawny to often purplish-black, usually with a paler midrib, generally narrower and often shorter than the perigynia; perigynia spotted or suffused with reddish-brown or purple, elliptic to obovate, ± strongly flattened, 2–3.3 mm including the 0.1–0.3 mm entire beak, 2-ribbed, otherwise nerveless; achene lenticular, distinctly shorter and often narrower than the perigynial cavity; $2n=74–84$. In shallow water or wet soil, sometimes in salt-marshes; circumboreal, s. to N.J., Ind., Io., and N.M. Our plants have been called var. *substricta* Kük. (*C. substricta.*)

191. Carex lenticularis Michx.   Densely tufted, with very short or no rhizomes, the flowering stems 2–8 dm, surrounded by lf-bases of the previous year; lvs mostly 2–4 mm wide, V-shaped in x-section, with papillae and stomates on both sides, lacking palisade tissue; spikes 3–6, approximate or ± remote, erect, sessile or the lower ± evidently pedunculate, all slender (more so than in no. 188) and ± elongate, mostly (1–)1.5–5 cm, seldom appreciably over 5 mm thick, the terminal one staminate, less often gynaecandrous or with a few basal perigynia, the others generally pistillate; bract subtending the lowest spike elongate, commonly surpassing the infl, sheathless or with a short white sheath, the others more reduced; pistillate scales reddish-brown to sometimes blackish or dark brown, with a conspicuous broad pale or green midstrip, commonly shorter than and exposing the perigynia, these 1.7–3.2 mm, somewhat compressed, lanceolate to ovate or broadly elliptic, 2-ribbed, 3–7-nerved on each face, short-stipitate and with a slender, terete, entire, often black-tipped beak 0.1–0.3 mm, the body commonly greenish distally, papillate only on the upper third; achene lenticular, whitish-iridescent, distinctly shorter than the perigynial cavity, adnate below to the swollen perigynium-base; $2n=86–92$. Wet shores and beaches; Lab. to Alas., s. to Mass., N.Y., Wis., Minn., and Calif.; e. Asia. Ours is var. *lenticularis*.

192. Carex bigelowii Torr.   Stems erect, 1–4 dm, solitary or few together from stout rhizomes, surrounded by the dried-up lf-bases of the previous year, phyllopodic; main lvs 2–6 mm wide; terminal spike staminate, 0.5–2 cm, usually long-pedunculate; pistillate spikes 2 or 3, cylindric, 0.5–3.5 cm, erect, the lower short-peduncled, the upper sessile or nearly so; bracts sheathless, the lowest usually shorter than the infl, the upper much smaller; pistillate scales elliptic to obovate, obtuse, about as long and wide as the perigynia and concealing them, dark brown throughout or with a very narrow pale midvein; perigynia pale green, usually strongly suffused with purple or even wholly purple, ovate to obovate, biconvex, 2.1–3.8 mm, rounded or obtuse at the summit, minutely apiculate, 2-ribbed, otherwise nerveless; achene lenticular, nearly filling the perigynium; $2n=76–80$. Rocky ledges and mt. meadows; circumboreal, s. in our range to the higher mts. of N.Engl. and N.Y. (*C. concolor*, perhaps misapplied.)

193. Carex nigra (L.) Reichard.   Loosely tufted, with both long and short rhizomes, the stems 1–8 dm, surpassing the lvs; main lvs lacking palisade-tissue, 2–3 mm wide, V-shaped in section, with stomates and papillae only on the upper side; staminate spikes peduncled, 1–3 dm, often with 1 or 2 smaller ones at its base; pistillate spikes 2–5, contiguous or separate, cylindric, 1–4 cm × 5 mm, often acute, erect, sessile or the lowest short-pedunculate, commonly with a few terminal staminate fls; bracts sheathless or with a very short purple-brown sheath, the lowest lf-like but shorter than or barely equaling the infl, the upper much reduced; pistillate scales ovate or oblong, usually slightly narrower and considerably shorter than the perigynia, with dark brown or atropurpureous sides and very narrow pale midvein scarcely reaching the obtuse tip; perigynia thin, planoconvex, ovate to obovate, 2.0–3.7 mm, half to three-fifths as wide, rounded to obtuse at the tip, finely few-nerved (as well as 2-ribbed) minutely beaked, green, becoming tawny, distally dark-blotched, papillate throughout; achene lenticular, loosely enveloped in the lower half of the perigynium; $2n=$ca 84. Wet meadows and salt-marshes, mostly near the coast; Lab. to Conn. and Vt.; disjunct in n. Mich.; widespread in Europe. (*C. acuta*, misapplied.)

194. Carex haydenii Dewey.   Plants tufted, the slender stems arising laterally, scabrous on the angles, 5–12 dm, usually much overtopping the lvs; lowest lvs reduced to dark reddish-brown, bladeless sheaths; main lvs 2–5 mm wide, inversely W-shaped in x-section, the sheaths glabrous, dorsally green, with a concave, red-dotted mouth; lowest bract lf-like, usually a little shorter than the infl, the others much reduced; staminate spikes 1–3, the terminal one 2–5 cm, the others much smaller; pistillate spikes 2 or 3, sessile or nearly so, erect, separate or somewhat overlapping, 1–5 cm, linear-cylindric, the uppermost with some distal staminate fls; pistillate scales with green midvein and dark brown sides, narrower than the perigynia, long-acuminate and projecting 0.5–1.5 mm beyond them; perigynia 1.5–2.6 mm, olive-green with red dots, biconvex, rounded-obovoid and slightly inflated, rounded above to a minute beak, with 1 or 2 faint nerves on each side; achene lenticular, loosely enveloped in the lower half of the perigynium; $2n=54$. Marshes, wet meadows, and wet, open woods; N.B. to Minn. and S.D., s. to N.J., Pa., Ind., and Mo.

195. Carex emoryi Dewey.   Rhizomatous, with both short and long rhizomes, the slender stems arising laterally, scabrous on the angles, 4–12 dm; lowest lvs reduced to reddish-brown, bladeless sheaths; main lvs 3.5–6(–8) mm wide, the sheaths glabrous, reddish dorsally, with a tawny, generally convex mouth; lowest bract lf-like, about equaling the infl, the others reduced; staminate spikes (1)2(3), 1.5–5 cm; pistillate spikes 3–5, short-pedunculate or sessile, erect, 2–10 cm, slender, tapering at the base, the uppermost with some distal staminate fls; pistillate scales with a broad green or tawny midstrip and a submarginal tawny to reddish-brown band, generally narrower and shorter than the perigynia; perigynia 1.7–3.2 mm, green becoming stramineous, not dotted, flattened and planoconvex, usually several-nerved on each side, short-beaked; achene lenticular, closely enveloped by the perigynium; $2n=72$. Streambanks and roadside ditches; s. N.Y. to N.D. and Man., s. to Va., n. Ark., and Tex. (*C. stricta* var. *e.*)

196. Carex stricta Lam.   Plants forming large clumps, with both long and short rhizomes, the stems arising laterally, 4–14 dm, scabrous on the angles, rather lax, usually surpassing the lvs; lowest lvs reduced to bladeless sheaths, these splitting ventrally and becoming pinnately fibrillose; foliage lvs 3–6 mm wide, inversely W-shaped in x-section, the sheaths shallowly concave (and often with a central mucro) at the thickened mouth; ligule acute, triangular, much longer than wide; bracts sheathless, the lowest lf-like, shorter to slightly longer than the infl, the others much smaller, pistillate spikes 2–4, up to 6 cm, often approximate and usually overlapping, erect, linear-cylindric, sessile or nearly so, sometimes apically staminate; pistillate scales lanceolate to oblong, from narrower and shorter to less often as wide as and longer than the perigynia, reddish-brown or purple-brown, usually with a conspicuous pale midrib, blunt to acute; perigynia planoconvex to nearly flat, ovate, 1.6–3.4 mm, three-fifths as wide, tapering to a minute, straight beak, 2-ribbed, otherwise nerveless or obscurely nerved on both faces; achene lenticular, filling the lower two-thirds of the perigynium; $2n=68$. Abundant in swales and marshes, especially where seasonally flooded; Que. and N.S. to Minn. and Man., s. to Va. and Tex. (*C. strictior.*) A putative hybrid with no. 190 has been called *C.* × *abitibiana* Lepage.

197. Carex collinsii Nutt. Densely tufted, 2–6 dm, the stems surpassing the lvs; main lvs 2–4 mm wide; terminal spike staminate, 5–10 mm, sessile or short-peduncled; pistillate spikes 2–4, separate, erect, pedunculate, usually staminate at the tip, the axis up to 1 cm; bracts lf-like, overtopping the spikes, strongly sheathing at base; pistillate scales ovate, 3–5 mm, hyaline with green midvein, acute to short-awned, persistent; perigynia slenderly subulate, 8–12 mm, nearly circular in cross-section, long-attenuate above, the beak with a long fissure and 2 conspicuous, abruptly reflexed teeth 1 mm; achene rounded-trigonous, continuous with the persistent slender style. Bogs, especially *Chamaecyparis*-swamps, mostly on the coastal plain; R.I. to Pa. and Ga.

198. Carex michauxiana Boeckeler. Tufted, 2–6 dm; main lvs 1.5–4 mm wide; terminal spike staminate, 0.6–1.5 cm, sessile or nearly so, scarcely projecting beyond the upper pistillate ones; pistillate spikes 2–4, broadly ovoid, 1.5–2.5 cm, erect, the lower distinctly peduncled, the upper less so; bracts lf-like, 1–3 mm wide, surpassing the stems, their sheaths concave at the mouth; pistillate scales ovate, a third to half as long as the perigynia, hyaline or brown-tinged, with green midstrip, acute or acuminate; perigynia slenderly subulate, 8–13 × 1.5–2 mm, nearly circular in cross-section, sharply many-nerved, long-attenuate into a slender beak with erect teeth 1 mm; achene loosely enveloped, rounded-trigonous, continuous with the persistent slender style. Bogs and wet meadows; Nf. to Ont., n. Mich., n. Wis., ne. Minn., and Sask., s. to N.Y.; e. Asia.

199. Carex folliculata L. Stems to 1 m, forming large clumps; terminal spike staminate, 1.2–2.5 cm, peduncled and often exserted beyond the perigynia; pistillate spikes 2–5, usually strongly separate, erect, subglobose, 1.5–3 cm long and wide, the lowest peduncles elongate, the upper shorter; bracts lf-like, 4–8 mm wide, surpassing the stems, their sheaths prolonged at the mouth into a rounded or truncate lobe; pistillate scales ovate, much shorter than the perigynia, hyaline or brown-tinged with green midstrip, the body acute or acuminate; perigynia slenderly subulate, 10–15 × 2–3 mm, nearly circular in cross-section, sharply many-nerved, long-attenuate into a stout beak with erect teeth; achene loosely enveloped, concavely trigonous, continuous with the persistent slender style. Wet or swampy woods; Nf. and Que. to Wis., s. to Fla. and La. Var. *folliculata*, a northern form occurring s. to s. Del., W.Va., and Ind., has the main lvs 6–16 mm wide, the pistillate scales usually rough-awned, the whole scale from two thirds as long to slightly longer than the perigynium, the pistillate spikes occasionally with a few distal staminate fls. Var. *australis* L. Bailey, of the coastal plain from s. Md. to Fla. and La., has the main lvs 4–10 mm wide, the pistillate scales usually long-acuminate, seldom awned, about half as long as the perigynia, the pistillate spikes frequently staminate at the tip; it is perhaps better treated as a distinct species, *C. lonchocarpa* Mackenzie.

200. Carex hystericina Muhl. Stems 3–10 dm, rather slender, clustered on a short, stout rhizome; lvs tending to be septate-nodulose (especially the sheaths), the long flat blade 3–9 mm wide; bracts subtending the lowest pistillate spike sheathless or with only a short sheath, and with a long blade somewhat surpassing the infl, the other bracts much reduced or wanting; terminal spike staminate, 2–4 cm; pistillate spikes several, approximate or the lowest one remote, 1.5–4 × 1–1.5 cm, the lower ones loosely nodding on slender,flexuous peduncles, the uppermost often subsessile and loosely ascending-spreading; pistillate scales with a short, largely scarious or hyaline body only 1–2 mm and a prominent rough awn-tip 2–6 mm; perigynia very numerous, densely crowded, mainly spreading or ascending-spreading, usually pale greenish, 5–7 mm, prominently (12–)15–20-nerved, thin-textured, slightly inflated and nearly round in cross-section, scarcely stipitate, lanceolate or lance-ovate, rather abruptly tapering to the conspicuous, slender beak with short straight teeth 0.3–0.9 mm, achene trigonous, loose in the lower half of the perigynium, continuous with the persistent, bony style, which becomes flexuous or contorted as the achene matures; $2n=58$. Swamps, wet meadows, and shores; N.B. and Que. to Wash., s. to Va., Ky., Tex., and Calif.

201. Carex comosa F. Boott. Much like no. 200; stems stout; lowest bract surpassing and sometimes several times as long as the infl; terminal spike to 6 cm, staminate or sometimes androgynous or gynaecandrous or with the perigynia in the middle; pistillate spikes 2–7, loose and ± nodding on slender peduncles, tending to be rather closely grouped together; perigynia ± reflexed, firm-textured and only slightly or scarcely inflated, obtusely trigonous, tapering to a short-stipitate base, and very gradually tapering above to the long beak with slender, firm, arcuate or divergent teeth 1.2–2.3 mm; mature style straight or seldom flexuous or contorted. Swamps and wet meadows; Que. to Minn., s. to Fla. and La.; Wash. to Calif. and n. Ida.

202. Carex pseudocyperus L. Much like nos. 200 and 201; stems stout; main lvs 5–15 mm wide; lowest bract commonly much surpassing the infl; pistillate spikes approximate or the lowest separate, 3–7 × 1 cm, spreading or drooping, the lowest long-peduncled, the upper peduncles shorter; pistillate scales with very small, reddish-brown body, the pale midvein excurrent into an awn nearly as long to longer than the perigynium; perigynia ± reflexed, 4.2–6.2 mm, firm-textured and only slightly or scarcely inflated, obtusely trigonous, stipitate, slenderly ovoid, gradually tapering to the long beak with straight, parallel teeth 0.6–1 mm; $2n=66$. Swamps and bogs; Nf. to Sask., s. to Pa., Ind., and Minn.

203. Carex melanostachya M. Bieb. Stems slender, mostly 3–6 dm, arising singly or few together from long rhizomes; lvs long and narrow, mostly 2–4 mm wide; bracts lf-like but scarcely or not at all sheathing,

± erect, the lowest one about equaling or somewhat surpassing the infl; terminal spike staminate, 2–3 cm, evidently pedunculate, or often with 1 or 2 shorter sessile ones at its base, the other 2 or 3 spikes a little more remote, pistillate, 1.5–3(–4) cm; pistillate scales castaneous, with a paler firm midrib prolonged as a short awn or cusp; perigynia plump, ellipsoid-trigonous, 3.5–5(–5.5) mm, firm and impressed-nerved at maturity, rather abruptly contracted to the short, evidently bidentate beak; achene trigonous. Wet grasslands and disturbed wet places; native of Europe, irregularly intr. in N. Amer., as in e. Kans. and near Montreal. (*C. nutans* Host, not J. F. Gmelin.)

204. Carex acutiformis Ehrh.   Stems stout, 5–15 dm, arising singly or few together from long, stout, scaly rhizomes, aphyllopodic, but clothed at the base with the dried-up lf-bases of the previous year; lower sheaths soon breaking and becoming fibrillose; main lvs 5–8 mm wide; ligule about as long as wide; staminate spikes 2–4; pistillate spikes 2–5, widely separate, cylindric, 3–8 cm × 7 mm, densely fld, erect, sessile or short-peduncled; bracts sheathless or nearly so, the lowest one elongate and lf-like, the upper much reduced; pistillate scales lanceolate, up to about as long as the perigynia, acute to acuminate or short-awned; perigynia broadly and obtusely ovoid-trigonous, glabrous but dull, with numerous ± evidently raised nerves, 2.5–4.5 mm, narrowed to a short beak with obscure triangular teeth wider than long; achene trigonous, continuous with the short, straight, slender style; $2n=38, 78$. Native of Eurasia and Afr., intr. in wet meadows here and there in our range, as in Mass. and Ind.

205. Carex lacustris Willd.   Stems stout, 5–15 dm, arising singly or few together from long, stout, scaly rhizomes, aphyllopodic, lateral, not surrounded by old lf-bases, the lower sheaths strongly reddened, soon breaking and becoming very fibrillose; main lvs 8–15 mm wide, glabrous except for the scabrous margins, usually conspicuously cross-septate; ligule much longer than wide; staminate spikes 2–4, distal, the lateral ones sessile; pistillate spikes 2–4, usually separate, cylindric, 3–10 × 1–1.5 cm, erect, densely fld, sessile or short-peduncled; bracts lf-like and scarcely or not at all sheathing, some or all surpassing the infl; body of the pistillate scales much shorter than the perigynia, ovate to oblong, the sides hyaline or pale brown, the green midrib often prolonged into an awn to 3 mm; perigynia rather slenderly ovoid or ellipsoid-ovoid, glabrous, with ± evidently raised nerves, 4.7–7.3 (avg 6) mm, gradually narrowed into the beak, this with divergent triangular teeth 0.4–1 mm; achene trigonous, continuous with the persistent, slender, abruptly bent or flexuous style; $2n=74$. Swamps and marshes; Que. to Va., w. to Sask. and Nebr.

206. Carex hyalinolepis Steudel.   Much like no. 205; stems phyllopodic, surrounded by old lf-bases, the lower sheaths scarcely if at all pinkish, seldom filamentous; ligule shorter to slightly longer than wide; mature perigynia 5–8 (avg 6.9) mm, the numerous nerves scarcely elevated, often somewhat impressed, or indistinguishable under a hand-lens. Swamps and marshes; coastal plain from N.J. to Fla. and Tex., n. in the interior to Ill., Mich., and s. Ont. (*C. lacustris* var. *laxiflora*.) An apparent hybrid with *C. pellita* has been called *C. × subimpressa* Clokey.

207. Carex laeviconica Dewey.   Stems stout, 5–12 dm, arising singly or few together from long rhizomes, aphyllopodic, the lower sheaths soon breaking and becoming strongly fibrillose; lvs glabrous, the main ones 3–6 mm wide, usually evidently cross-septate; staminate spikes ca 3, distal; pistillate spikes 2–4, very remote, cylindric, 3–7 cm, densely fld, erect, sessile or short-peduncled, sometimes staminate at the top; bracts lf-like but scarcely or not at all sheathing, equaling or surpassing the infl; pistillate scales thin, hyaline or pale brown, the body usually much shorter than the perigynium, either acuminate or obtuse with the green midvein prolonged into an awn nearly or quite equaling the perigynium; perigynia 5.5–9 mm, broadly ovoid, conspicuously many-nerved, tapering into a slender, minutely scabrous beak with nearly straight, moderately divergent, scabrous teeth 1–2.2 mm; achene trigonous, loosely enveloped, with a persistent, straight, slender style; $2n=110$. Swales and borders of ponds; Ill. and Mo. to Sask.

208. Carex atherodes Sprengel.   Stems stout, 3–15 dm, arising singly or few together from long, coarse, deep-seated rhizomes, aphyllopodic, the lower sheaths soon breaking and becoming fibrillose; sheaths villous-hirsute all around, or sometimes only toward the ventral summit; blades elongate, 4–10 mm, usually hirsute beneath toward the base, tending to be septate-nodulose; bracts subtending the pistillate spikes sheathless or nearly so, but with elongate blade; spikes several, sessile or inconspicuously short-pedunculate, the lower pistillate and the upper staminate, or the middle one(s) androgynous; staminate spikes 2–6 cm, pale, commonly stramineous, the scales short-awned; pistillate spikes cylindric, 2–10 × 1 cm, the scales stramineous or pale greenish and often largely scarious, narrow, tipped by a rough awn 1–5(–9) mm; perigynia crowded, ascending, pale greenish to stramineous, conspicuously 12–20-ribbed, glabrous, firm, 7–10 mm, lanceolate or lance-ovate, somewhat turgid-inflated below, more flattened toward the well developed beak, this with prominent, often divergent teeth 1.5–2.5(–3) mm; achene trigonous, loosely filling the lower part of the perigynium, continuous with the straight bony style; $2n=74$. Marshes and shallow water; circumboreal, extending s. in Amer. to N.Y., W.Va., Mo., Colo., Utah, and Oreg.

209. Carex trichocarpa Muhl.   Much like no. 208; avg more slender; sheaths globose, the mouth deeply concave and strongly purple-tinged; blades glabrous or merely scabrous; perigynia loosely short-white-hairy. Marshes and wet meadows; Que. to Minn., s. to Del., N.C., W.Va., Ind., and Mo. *C. caesariensis* MacKenzie, of s. N.J., appears to be a hybrid with *C. pellita*.

210. Carex frankii Kunth.   Stems 2–8 dm, in small tufts on rather short rhizomes, aphyllopodic; lvs to 1 cm wide, the ligule shorter to somewhat longer than wide; terminal spike 0.5–3 cm, usually wholly staminate, rather short-peduncled, often concealed among the pistillate ones, or short-peduncled and exsert; pistillate spikes 3–6, usually approximate or crowded, sometimes separate or the lower remote, cylindric or elliptic, 1.5–3 × 1 cm, very dense, rounded at both ends, sessile or nearly so; bracts lf-like, 2–4 times as long as the infl, at least the lower strongly sheathing; pistillate scales with concealed body and an exsert slender awn equaling or surpassing the perigynium-beaks; perigynia obconic, inflated, the body 2.2–3.6 mm, with 2 strong ribs and ca 10–18 fine sharp nerves, depressed-truncate above and very abruptly beaked, the beak 1.3–2.3 mm, its teeth 0.5 mm; achene loosely enveloped, obovoid-trigonous, 1.5–2 mm, with a persistent straight style. Swamps and wet woods; Va. to w. N.Y., s. Ont., Mich., and se. Nebr., s. to Ga. and Tex.

211. Carex squarrosa L. Stems 3–9 dm, rather stout, densely tufted, aphyllopodic; main lvs elongate, not clustered at base, 3–6 mm wide; ligule much longer than wide; spikes 1(–3), the lower third staminate, the pistillate part elliptic, 1–3 .× (1–)1–5.2 cm, rounded at both ends, very dense; lateral spikes, if present, pistillate, smaller, on erect short peduncles; bract of the terminal spike short and narrow, of the lateral ones foliaceous and elongate but not sheathing; staminate scales acute or acuminate; pistillate scales largely concealed, acuminate or short-awned; perigynia very numerous and crowded, squarrose-spreading, obconic or obconic-obovoid, inflated, the body 3.5–6 mm, its summit with 2 strong ribs and a few obscure nerves, the beak 2–3.5 mm, with teeth 0.2 mm; achene loosely enveloped, rather slenderly ellipsoid-trigonous, 2.2–3 mm, ca two-fifths as wide, the persistent style very strongly sinuous or abruptly bent below. Swampy woods and thickets; Conn. to N.C., w. to w. Que., se. Mich., Nebr., and Ark.

212. Carex typhina Michx. Much like no. 211; stems usually surpassed by the upper lvs; main lvs 5–10 mm wide; spikes 1–6, the uppermost one mostly pistillate and with a short staminate part at base, the pistillate part cylindric or ovoid-cylindric, 2–4 × 1–1.5 cm, subtended by a short narrow bract; lateral spikes pistillate, smaller, erect or spreading on short peduncles; staminate scales obtuse; pistillate scales usually concealed, obtuse to acute, not awned; achene wider, half to three-fifths as wide as long, the style straight or slightly curved near the top of the perigynium. Moist or wet woods and marshes; Me. and Que. to Wis. and se. Minn., s. to Ga. and La.

213. Carex oligosperma Michx. Colonial by creeping rhizomes, forming a turf; stems slender, 4–10 dm, purplish at base; lvs pale green, stiff, elongate, involute, 1–3 mm wide; ligule much wider than long; staminate spike usually solitary; pistillate 1, or 2 or 3 and widely separated, sessile or nearly so, ovoid to short-cylindric, 1–2 cm; lowest bract lf-like but not sheathing; perigynia 3–15(–18), ovoid and somewhat inflated but also compressed, 4–7 mm, half as thick, strongly several-nerved, abruptly narrowed into an emarginate beak 1–2 mm; achene trigonous, obovoid, 2–3 mm; style persistent, usually contorted at base; $2n=76$. Grassy bogs and wet meadows; Nf. to Mack., s. to Conn., Pa., Ind., and Wis.

214. Carex saxatilis L. Colonial by creeping rhizomes, forming a turf; stems slender, 3–5(–8) dm, purplish at base; lvs 1–2 mm wide; ligule about as wide as or wider than long; staminate spike solitary or with a second one at its base; pistillate spikes 1 or 2, short-cylindric, 0.5–2 cm, erect, longer than the peduncle; lowest bract lf-like but not sheathing; pistillate scales brown, the lower cuspidate or shortly acuminate, the upper blunter; perigynia 15–40, ovoid, scarcely inflated, biconvex and ± flattened, 2.5–3.5 mm, nerveless or nearly so except for the marginal ribs, the beak 0.5 mm, emarginate; achene lenticular, obovoid, 2–2.5 mm, the persistent style strongly bent or contorted below; $2n=80$. Moist or wet, sandy or gravelly soil; circumboreal, s. to Me., Hudson's Bay, and the w. cordillera. Our plants are var. *miliaris* (Michx.) L. Bailey, which ranges n. to Lab. (*C. miliaris.*)

215. Carex × mainensis Porter. Stems slender, 3–8 dm, tufted, often on a short stout rhizome; lvs 2–3 mm wide, long-attenuate and conspicuously involute; staminate spike solitary, or often with a much smaller second one at its base; pistillate spikes 1–3, well separated, cylindric, 1–3 cm × 6–8 mm; pistillate scales brownish, narrower and considerably shorter than the perigynia; perigynia ovoid or oblong-ovoid, 3.5–5 mm, scarcely inflated, only faintly nerved (except the marginal ribs), the beak 0.5–1 mm, emarginate, its broadly triangular teeth 0.25 mm. Beaches and streambanks; Me. to Nf. and Lab. Thought to be a hybrid of no. 214 and some other sp. of this group.

216. Carex vesicaria L. Stems 3–10 dm, sharply trigonous, strongly scabrous above, loosely to densely clustered on a system of rather short, stout, branching rhizomes; lvs tending to be septate-nodulose, the blade elongate, flat, 3–8 mm wide; those subtending the pistillate spikes sheathless or nearly so but with elongate blade; spikes several, sessile or inconspicuously short-pedunculate, ± erect, remote, the lower pistillate, the upper staminate, or one of them androgynous; staminate spikes 2–7 cm; pistillate spikes 2–7 × 1–1.5(–2) cm at maturity; pistillate scales somewhat shorter and narrower than the perigynia, thin, acutish to long-acuminate but scarcely awned; perigynia crowded, ascending, ± in 6(–8) vertical rows, 5–8 × 2–3 mm, strongly 10–20-ribbed, lanceolate or lance-ovate in outline, bladdery-inflated below, gradually tapering to the flatter, often poorly defined beak with evident short (0.3–1.2 mm) teeth; achene yellowish, trigonous, 1.7–2.4 mm, loose in the lower part of the perigynium, continuous with the bony, eventually flexuous to strongly contorted style; $2n=74, 82$. Wet soil or shallow water in bogs or swamps and the margins of ponds or streams; circumboreal, s. in Amer. to Del., Ky., Ind., Mo., and Calif. Our plants belong to the widespread var. *vesicaria*.

217. Carex tuckermanii F. Boott. Much like no. 216; stems 4–8 dm, tufted on short rhizomes; lvs soft and flat, the main ones 2–4 dm × 3–6 mm; staminate spikes usually 2, widely separated, much elevated above the pistillate, these 2–4 well separated, cylindric, 2–5 cm; perigynia numerous, imbricate and ascending in ca 6 rows, 7–10 × (4–)4.5–7 mm, inflated and very thin-walled, broadly ovoid, the conic summit tapering into a slender beak 2 mm, its teeth 0.5–1 mm; achene or trigonous, obovoid, 3–4 mm, deeply indented in the middle of one angle. Swampy woods, pond-margins, and wet meadows; N.B. to Minn., s. to N.J., O., and Io.

218. Carex utriculata F. Boott. Colonial by long, stout, deep-seated creeping rhizomes, sometimes forming a dense sod; stems 5–12 dm, bluntly trigonous, smoother than in no. 220; lvs septate-nodulose (especially the sheaths), flat, yellow-green, smooth above, the larger ones 5–12 mm wide; lvs subtending the pistillate spikes nearly or quite sheathless, but at least the lowest one with an elongate blade 1–2 times as long as the infl; spikes several, remote, forming an elongate infl, ± erect, the upper ones staminate and 2–7 cm, the lower pistillate, 2–10 cm, ca 1(–1.5) cm thick at maturity, only the lowest one sometimes evidently pedunculate; pistillate scales narrower and usually shorter than the perigynia, acutish to more often acuminate or short-awned; perigynia densely crowded, mostly in 8+ vertical rows, widely spreading at maturity, glabrous, shining, ± strongly 8–16-nerved, 4–7 mm, the inflated, broadly ellipsoid to ovoid or subglobose body rather abruptly contracted to an evident smooth beak 1–2 mm with short (0.2–0.8 mm) teeth; achene yellowish, trigonous, 1.3–2 mm, loose in the lower half of the perigynium, its persistent, bony style becoming strongly flexuous or contorted with maturity; $2n=82$. Wet soil or shallow water; boreal Amer., s. to Del., Ind., Nebr., N.M., and Calif. (*C. rostrata,* misapplied.)

219. Carex rostrata Stokes. Similar to no. 218, and often confused with it, but the lvs papillate-glaucous on the upper surface, mostly 1.5–4 mm wide, tending to be involute; perigynia avg smaller; $2n=60, 72, 74, 76$. Wet soil or shallow water; circumboreal, s. to n. Mich. and n. Minn.

220. Carex bullata Schk. Much like no. 218, the stems more slender, 3–9 dm, sharply trigonous and more evidently roughened above; lvs 2–5 mm wide; staminate spikes 1–3 on a long common peduncle surpassing the uppermost pistillate spike; pistillate spikes 1 or 2, well separated, thick-cylindric, 1.5–4 × 1.5–2 cm; perigynia numerous, closely imbricate, widely divergent, broadly ovoid, 6–9 × 2.5–3 mm, abruptly contracted into a stout, straight, rough-margined beak half to two-thirds as long as the body, the slender teeth 0.5–1 mm. Swamps and bogs, chiefly on the coastal plain; N.S. to Ga.

221. Carex retrorsa Schweinitz. Stems 3–10 dm, densely clustered on a very short rhizome; lvs septate-nodulose (especially the sheaths), mostly 4–10 mm wide; lvs subtending the pistillate spikes sheathless or only shortly sheathing, at least the lowest one generally several times as long as the infl; spikes several, crowded, sessile or nearly so, or the lowest one more remote and slender-pedunculate; lower spikes pistillate, the upper staminate or androgynous; pistillate spikes 1.5–5 × 1.5–2 cm; pistillate scales conspicuous, shorter and narrower than the perigynia; perigynia numerous, densely crowded in 8+ rows, widely spreading or the lowest ones retrorse, glabrous, shining, evidently 6–13-nerved, 7–10 mm, firm-walled but somewhat inflated, ellipsoid to subglobose and often somewhat oblique, narrowed to a prominent, slender, smooth beak 2–3(–4) mm with short (0.3–0.9 mm) teeth; achene dark brownish, narrowly trigonous, 2 mm, loose in the lower part of the perigynium, the persistent, bony style becoming contorted with maturity; $2n=70$. Swampy woods and wet meadows; Que. to B.C., s. to Del., Md., Ind., Io., and Oreg.

222. Carex lurida Wahlenb. Tufted, without long rhizomes; stems aphyllopodic, 2–10 dm, usually surpassed by the lvs, obtusely trigonous and smooth or nearly so, purplish at base; lvs septate-nodulose, with flat blade (2–)4–7 mm wide; ligule triangular, distinctly longer than wide; terminal spike staminate, 1–7 cm, its scales with the midrib prolonged into an evident awn; pistillate spikes 1–4, approximate, or the lower remote, sessile and erect or (especially the lower) evidently pedunculate and even drooping, 1–7.5 × 1.4–2 cm, densely fld; bracts leafy, surpassing the infl, with or without a sheathing base; pistillate scales rough-awned, or the upper merely acuminate; perigynia numerous, in many rows, 6–9 mm, ovoid or ovoid-globose, somewhat inflated, pale, smooth and shining, strongly ca 10-nerved, with a slender, bidentate beak half to almost as long as the body, the slender teeth 0.3–1.8 mm; achene concavely trigonous, densely granular, loosely enveloped in the lower part of the perigynium, the persistent style twisted or abruptly bent. Swamps and wet meadows and woods; N.S. to Minn., s. to Fla. and Mex.

223. Carex baileyi Britton. Much like no. 222, but seldom over 6 dm; main lvs 1–2 dm × 2–4 mm; ligule rounded, little if at all longer than wide; pistillate spikes 8–13 mm thick; perigynia 5–7 mm, the slender beak equaling or longer than the body; achene 1.5–2 mm. Swamps, woods, and wet meadows; N.H. to Va., Ky., and Tenn., chiefly in the mts.

224. Carex schweinitzii Dewey. Stems 3–7 dm, aphyllopodic, sharply triangular, arising singly or few together from slender rhizomes; lvs septate-nodulose, mostly 4–10 mm wide, very rough distally; terminal spike staminate, slender-pedunculate, usually with a conspicuous bract or smaller sessile spike some distance below, the scales acute or cuspidate; pistillate spikes 2–5, approximate or the lowest widely separate, cylindric, 3–8 × 1–1.5 cm, the lowest spreading or ascending and distinctly peduncled, the upper ascending to erect, short-peduncled to subsessile; lower bracts lf-like, ± strongly sheathing at base, much surpassing the infl; pistillate scales with small, entire, hyaline or brown-tinged body, the midvein excurrent into a rough awn often equaling or surpassing the perigynium; perigynia spreading or ascending, 5.5–7.2 mm, substipitate, ovoid, inflated, with 7–9 slender nerves, abruptly narrowed to a slender beak with erect or spreading teeth 0.4–1 mm; achene loosely enveloped, concavely trigonous, with a persistent flexuous style. Cool, shady streambanks; rare; Vt. to n. Mich., s. to N.J., N.C., Tenn., and Mo.

225. Carex grayi Carey. Stems 3–9 dm, solitary or in small clusters, scabrous on the angles above; rhizomes wanting; basal sheaths persistent, purplish-red; lvs 4–11 mm wide, the uppermost nonbracteal one (as also the bracts) sheathing or with a sheath seldom over 1 cm; terminal spike staminate, 0.5–5.5 cm, on a peduncle of 0.5–6 cm; pistillate spikes 1 or 2, densely fld, globular, ± approximate, on peduncles 0.7–3.5 cm, their bracts leafy, 8–26 cm; pistillate scales 4–11 mm, lance-ovate to orbicular-ovate, often tipped with a rough awn to 7 mm; perigynia 8–35, radiating in all directions from the short axis, dull, sometimes hispidulous, 12.5–20 × 4–8 mm, strongly 16–25-nerved, rhombic-ovoid, cuneate to the base, tapering from the widest point to a poorly defined, bidentate beak 1.5–3 mm, the teeth hispidulous internally; achene 3.3–4.8 × 2.6–3.7 mm, convexly trigonous, not thickened on the angles; style persistent but withering, sometimes contorted; $2n$=52, 54. Moist woods; Mass., Vt., s. Que., and s. Ont. to Wis. and se. Minn., s. to Ga. and Mo.

226. **Carex intumescens** Rudge. Much like no. 225; pistillate spikes 1–4, 1–2.7 cm long and wide, ovoid to obovoid, loosely fld, on peduncles to 1.5 cm, often closely aggregated; perigynia 1–12, mostly spreading or ascending, with a satiny lustre, 10–16.5 × 2.5–6.5 mm, convexly rounded to the base, the poorly defined beak 2–4.2 mm, achene 3.5–5.7 × 2.5–4 mm, with flat or convex faces; $2n$=48. Moist or wet woods; Nf. to se. Man., s. to Fla. and Tex.

227. **Carex louisianica** L. Bailey. Stems 2–8 dm, slender, smooth, solitary or few together from long, dark, scaly rhizomes; basal sheaths persistent, reddish to brownish; lvs 2–6 mm wide, the uppermost nonbracteal one with a sheath 2–10 cm; terminal spike staminate, 0.5–7 cm, on a peduncle 3–10 cm; pistillate spikes 1–4, short-cylindric, 1.5–4.5 × 1.5–3 cm, pedunculate, not crowded, their subtending bracts leafy, 10–30 cm, basally sheathing; pistillate scales 4.5–6.5 mm, lance-ovate, awnless; perigynia 10–30, smooth and shiny, strongly multinerved, 10–14 × 3.5–6 mm, rounded at base, with a conic beak 4.5–7 mm, the teeth smooth; achene 2.5–3.5 × 1.7–2 mm, widest near the middle, broadly stipitate, trigonous with ± flat faces and somewhat thickened angles; style persistent and becoming bony, strongly contorted near the base. Wet woods and swamps, chiefly on the coastal plain; N.J. to Fla. and Tex., n. in the Mississippi Valley to s. Ind.

228. **Carex lupulina** Muhl. Stems 2–13 dm, smooth, solitary or few together from long, dark, scaly, sympodial rhizomes; basal sheaths persistent, reddish to brownish; lvs evidently septate-nodulose, 4–15 mm wide, the uppermost nonbracteal one with a sheath 1.5–25 cm; terminal spike staminate, 1.5–8.5 cm, its peduncle 0.5–6 cm and surpassed by to about equaling the uppermost pistillate spike; pistillate spikes (1)2–5, ascending, ovoid to cylindric, 1.5–6.5 × 1.3–3 cm, pedunculate but the upper generally crowded, their subtending bracts leafy, 13–55 cm, basally sheathing; pistillate scales 6–15 mm, slender, acute or with a rough awn to 6 mm; perigynia 8–80, ascending, smooth and shiny, strongly 13–22-nerved, 11–19 × 3–6 mm, including the conic beak, this 6–10 mm, bidentate, its teeth smooth or nearly so; achene loosely enveloped, 3–4(–4.5) × 1.7–2.8 mm, ± stipitate, somewhat rhomboid, trigonous with flat to concave faces and smoothly curved, somewhat thickened lateral angles; style persistent and becoming bony, strongly contorted at or near the base; $2n$=56. Moist to wet woods, meadows, and marshes; N.S. and N.B. to Minn. and Nebr., s. to Fla. and Tex.

229. **Carex lupuliformis** Sartwell. Very much like no. 228; perigynia ascending to spreading; achene 3–4.5 × 2.4–3.4 mm, its faces strongly concave, diamond-shaped in outline, each lateral angle straight-tapering to each end from a hard, nipple-like knob near the middle; $2n$=60. Wetter places than no. 228, in open marshes or along shores, sometimes in shallow water, or in very wet floodplain forests; Vt. and adj. Que. to Minn., s. to Va., Ky., and Tex.

230. **Carex gigantea** Rudge. Stems 5–12 dm, smooth, solitary or few together from long, dark, scaly, sympodial rhizomes; basal sheaths persistent, brownish; lvs evidently septate-nodulose, 5–16 mm wide, the uppermost nonbracteal one with a sheath 5–20 cm; staminate spikes 1–5, 2–8 cm, on a collective peduncle 3–8 cm that is surpassed by or only slightly surpasses the uppermost pistillate spike; pistillate spikes 2–5, ascending, not crowded, their subtending bracts leafy, with a sheath 0.5–5 cm; pistillate scales 4.5–10.5 mm, slender, not awned; perigynia 20–75, ± spreading, smooth and shiny, strongly multinerved, 11–18 × 4–6 mm, with a short fat body and long conic beak 6–9 mm, this bidentate with smooth teeth; achene loosely enveloped, 2.2–2.6 × 2.7–3 mm, broadly stipitate, obpyramidal with rounded or truncate summit, concavely trigonous, with thickened angles; style persistent and becoming bony, straight or weakly contorted below. Swamps and wet woods, chiefly on the coastal plain; Del. to Fla. and Tex., n. in the Mississippi Valley to s. Ind.

15. **CYMOPHYLLUS** Mackenzie. Fls unisexual, each in the axil of a scale, in a solitary terminal spike, the pistillate below the staminate; bract none; stamens 3; ovary surrounded by a perigynium as in *Carex*, the style and 3 elongate stigmas persistent on the mature trigonous achene; shortly rhizomatous, the flowering culms bearing at anthesis a few imbricate basal sheaths without blades, the uppermost sheath later splitting and elongating to form a large solitary near-basal lf without midvein or sheath. Monotypic.

1. **Cymophyllus fraseri** (Andrews) Mackenzie. Stems at anthesis 1–4 dm, obscurely trigonous, covered at base by the overlapping striate sheaths; blade solitary, pale green, thick or coriaceous, oblong-linear, 2–6 dm × 2–5 cm, obtuse or acute, longitudinally striate; pistillate part of spike globose-ovoid, 1 cm thick at maturity; scales round-ovate, half as long as the perigynia; perigynia white, spreading, elliptic-ovoid, 5–6 mm; staminate part of spike short-cylindric, the scales white. Rich mt. woods; s. Pa., Va., and W.Va. to S.C. and Tenn. (*Carex f.*)

FAMILY **POACEAE**, the Grass Family

Fls perfect or sometimes unisexual, 1–many in distichously organized spikelets, each spikelet typically with a pair of subopposite small bracts (*glumes*) at the base and one to several or many florets alternating on opposite sides of an often zigzag axis (the *rachilla*) above the glumes; each floret typically consisting of a pair of subopposite enclosing or subtending scales (the *lemma* and *palea*), two or three much smaller scales (the *lodicules*) above these, and the stamens and pistil; midvein of the lemma often excurrent as a dorsal or terminal awn; palea placed with its back to the rachilla, typically with 2 main veins, generally enfolded by the lemma, rarely suppressed; lowest floret sometimes represented by an empty or staminate, glume-like sterile lemma, the spikelet then usually with a single perfect terminal fl; stamens most often 3, sometimes 6, rarely more, seldom only 1 or 2, typically exserted at anthesis and the plants wind-pollinated, but sometimes included and the plants selfed; anthers elongate, basifixed but so deeply sagittate as to appear versatile; ovary superior, unilocular, bicarpellate or less often tricarpellate, accordingly with 2 or 3 often large and feathery stigmas; ovule solitary, subapical to basilateral, orthotropous to hemitropous or almost anatropous; fr, called a *caryopsis* or *grain,* usually tightly enclosed by the persistent lemma and palea, indehiscent, usually dry, the seed-coat usually adnate to the pericarp; embryo basilateral, peripheral to the endosperm, complex in structure, with an enlarged lateral cotyledon (the scutellum); herbs or seldom somewhat woody plants, without secondary thickening, the culms (flowering stems) terete (seldom flattened) and usually with hollow internodes; lvs distichously or rarely spirally arranged, but not 3-ranked, with a usually open sheath and a parallel-veined, typically narrow and elongate blade, often with a pair of small *auricles* at the base of the blade, and commonly with a membranous *ligule* adaxially at the juncture of sheath and blade, or the ligule sometimes composed of hairs, or wanting; spikelets arranged in a determinate or mixed secondary infl that most commonly has the form of a panicle but is sometimes spike-like or raceme-like. 600/10,000. The term *rame* is here used for an unbranched infl that bears both sessile and pedicellate spikelets.

The arrangement of genera of grasses into subfamilies and tribes is in a state of flux. Many of the characters now being used are microscopic. The key to genera here presented (largely written by Richard Pohl) is artificial. The sequence of genera in the text is based mainly on that of Clayton and Renvoise, Genera Graminum, 1986.

Sclerochloa dura (L.) P. Beauv., a native of s. Europe, is intr. at scattered stations in w. U.S. and in O., and may be expected elsewhere in our range. Not very closely related to any of our genera, it falls between the first two alternatives of Group IV in the key, having only one obvious spike, as in the second alternative, but having the spikelets secundly arranged along two sides of the trigonous rachis, as in the first alternative. It is a tufted low annual, mostly under 2 dm, with broad, rounded, prominently 5-nerved lemmas. The persistent style-base forms a short beak on the grain.

Key to the Groups of Genera

1 Grasses with woody perennial culms; sheaths and blades deciduous; blades on short pseudopetioles; blooming very rare .... 1. *Arundinaria.*
1 Grasses with herbaceous, annual culms; sheaths usually persistent; blades sessile on the sheaths; annual bloomers.
  2 Tall, stout grasses, the culms commonly 2 m or more .... Group I.
  2 Grasses of small to moderate size, usually less than 2 m.
    3 Spikelets concealed by spines, hooks, or bony involucres.
      4 Spikelets concealed in hard, bony involucres, lacking spines; dwarf stoloniferous plants .... 82. *Buchloe.*
      4 Spikelets concealed by spines or hooks; plants not stoloniferous .... 95. *Cenchrus.*
    3 Spikelets never covered by spines or concealed in bony involucres.
      5 Spikelets disarticulating above the glumes, which remain on the inflorescence.
        6 Fertile floret 1, sometimes with reduced or sterile one(s) above or below it .... Group II.
        6 Fertile florets 2–several.
          7 Infl an open or dense panicle .... Group III.
          7 Infl of one or more spikes or rames .... Group IV.
      5 Spikelets disarticulating below the glumes, falling singly or in clusters .... Group V.

Group I

(Giant grasses, the culms 2 m or more.)

1 Staminate and pistillate spikelets separate, in the same infl.
  2 Infl of 1–several unbranched spikes, each pistillate near the base, staminate above; terrestrial ...... 105. *Tripsacum.*

2 Infl an open panicle; aquatic.
  3 Pistillate spikelets long-awned, erect, terminal; staminate spikelets drooping, awnless, on lower
       branches of panicle .... 4. *Zizania.*
  3 Pistillate and staminate spikelets intermingled on the same branches, short-awned .... 5. *Zizaniopsis.*
1 Some or all of the spikelets with perfect fls.
  4 Spikelets several-fld, disarticulating above the glumes and between the florets .... 61. *Phragmites.*
  4 Spikelets 1-fld, disarticulating below the glumes or falling in pairs.
    5 Infls conspicuously silky with grayish or silvery hairs that conceal the spikelets .... 97. *Miscanthus.*
    5 Infls glabrous or pubescent, but not silky.
      6 Infl a panicle of short spikelet-clusters.
        7 Each sessile, perfect-fld, awned spikelet accompanied by an awnless staminate one .... 99. *Sorghum.*
        7 Each sessile spikelet accompanied by a hairy sterile pedicel but no spikelet .... 100. *Sorghastrum.*
      6 Infl of several to many elongated racemes or rames of many spikelet-pairs.
        8 Spikelets of each pair equal and fertile .... 96. *Erianthus.*
        8 Spikelets dimorphic, the sessile one perfect-fld, awned, the pistillate one reduced, awnless,
          usually sterile .... 101. *Andropogon.*

GROUP II

(Spikelets with 1 fertile floret, this rarely accompanied by sterile or reduced ones;
disarticulation above the glumes.)

1 Infl of balanced or 1-sided spikes or rames.
  2 Spike solitary, balanced, the spikelets borne in triads on opposite sides of the rachis .... 53. *Hordeum.*
  2 Spikes 1-several, unilateral, the spikelets all on one side of the rachis.
    3 Low, stoloniferous grasses, dioecious, the staminate plants with 1-several racemose spikes, the
      pistillate ones with small basal hard burs .... 82. *Buchloe.*
    3 Plants perfect-fld, usually erect, not extensively stoloniferous.
      4 Spikes digitate; lemmas awnless; cult. and escaped .... 79. *Cynodon.*
      4 Spikes solitary, racemose, or paniculate; lemmas usually awned; wild plants.
        5 Spike solitary; second glume bearing a protruding spine on the keel .... 78. *Ctenium.*
        5 Spikes 1-many; glumes not spiny.
          6 Spikes borne in 1-several whorls .... 75. *Chloris.*
          6 Spikes racemosely arranged along the rachis.
            7 Spikes slender, the spikelets appressed to the rachis.
              8 Lemmas awned.
                9 Glumes minute or obsolete .... 8. *Nardus.*
                9 Glumes longer than the lemma .... 77. *Gymnopogon.*
              8 Lemmas awnless .... 76. *Schedonnardus.*
            7 Spikes short, dense, the spikelets placed at an angle to the rachis .... 81. *Bouteloua.*
1 Infl an open or dense panicle.
  10 Floret rigid, shining.
    11 Floret awnless, compressed.
      12 Fertile floret laterally compressed, with 2 little scales (sterile lemmas) at the base .... 39. *Phalaris.*
      12 Fertile floret dorsally compressed, solitary, without basal scales.
        13 Infl a lax, open panicle .... 12. *Milium.*
        13 Infl a dense, cylindrical panicle, the spikelets intermingled with sterile bristles .... 91. *Setaria.*
    11 Floret awned, cylindrical.
      14 Awns 3 on each floret .... 62. *Aristida.*
      14 Awns single.
        15 Glumes much reduced; rachilla prolonged behind the palea as a naked bristle ...... 7. *Brachyelytrum.*
        15 Glumes as long as the floret; rachilla not prolonged.
          16 Awn readily deciduous, not twisted .... 9. *Oryzopsis.*
          16 Awn twisted near the base, firmly attached.
            17 Edges of the lemma overlapping, flat .... 10. *Stipa.*
            17 Edges of the lemma infolded, clasping the keels of the palea .... 11. *Piptochaetium.*
  10 Floret soft, papery, not shining.
    18 Glumes minute or much shorter than the floret.
      19 Lemma 1-veined or 3-veined.
        20 Lemma 1-veined; caryopsis gelatinizing when wet .... 70. *Sporobolus.*
        20 Lemma 3-veined; caryopsis not gelatinizing when wet.
          21 Rachilla extending behind the floret; lemma with straight nerves, awnless .... 22. *Catabrosa.*
          21 Rachilla not so extended; lemma with converging nerves, usually awned .... 73. *Muhlenbergia.*
      19 Lemma 5-veined; floret strongly flattened laterally; aquatic or marsh grasses.
        22 Lemma awnless; glumes reduced to a minute cupule; wild plants .... 3. *Leersia.*
        22 Lemma awned or awnless; "glumes" (actually sterile lemmas) ca ⅓ as long as the
          lemma; crop plant .... 2. *Oryza.*
    18 At least one glume nearly as long as the spikelet.
      23 One or 2 neutral or staminate florets present below the fertile one, all falling as a unit.
        24 Lower florets neutral, awned .... 38. *Anthoxanthum.*
        24 Lower florets awnless.

Group III

(Fertile florets several to many; disarticulation above the glumes; infl a panicle.)

GROUP IV

(Infl of 1-sided or symmetrical spikes or rames.)

13 Spikelets not so crowded, not spreading away from the rachis; middle internodes of the
    spike mostly 4 mm or more.
    14 Self-pollinators with small anthers mostly 1–2 mm; bunch-grasses .... 50. *Elymus.*
    14 Outcrossers with larger anthers (2–)3–7 mm; most spp. with creeping rhizomes .... 51. *Elytrigia.*

GROUP V

(Spikelets falling entire, either solitary or in clusters.)

1 Spikelets falling separately, without attached stalks or rachis-internodes.
  2 Spikelets interspersed with stiff elongate bristles that partly conceal them; panicle dense, spike-
    like... 91. *Setaria.*
  2 Spikelets not interspersed with bristles; panicles various.
    3 Tall aquatic grasses; all spikelets unisexual.
      4 Tufted annual grass; pistillate spikelets all borne in an erect terminal cluster; staminate spikelets
        drooping, on lower panicle-branches .... 4. *Zizania.*
      4 Rhizomatous perennial; pistillate and staminate spikelets intermixed on the same branches
        .... 5. *Zizaniopsis.*
    3 Land- or marsh-grasses; some or all spikelets with perfect fls.
      5 Spikelets 2- to many-fld.
        6 Spikelets large, strongly flattened, keeled, many-fld; rhizomatous coastal dune-grass .... 63. *Uniola.*
        6 Spikelets various, but not as above; not rhizomatous dune-grasses.
          7 Sheaths with united edges .... 25. *Melica.*
          7 Sheaths with overlapping edges.
            8 Spikelets 3–8-fld; infl unbranched, spicate .... 51. *Elytrigia.*
            8 Spikelets mostly 2-fld; infl branched, paniculate (though sometimes dense).
              9 Both florets with perfect fls, awnless; rachilla prolonged .... 32. *Sphenopholis.*
              9 Upper floret staminate, awned; rachilla not prolonged .... 34. *Holcus.*
      5 Spikelets with only a single perfect floret, sometimes with neutral or staminate ones above or
        below the perfect one.
        10 Infl an open or dense panicle.
          11 Spikelets laterally compressed.
            12 Outer 2 bracts ("glumes") much shorter than the spikelets; cult. aquatic cereal .... 2. *Oryza.*
            12 Outer 2 bracts as long as the spikelet; wild plants.
              13 Glumes long-awned.
                  .... 44. *Polypogon.*
              13 Glumes awnless or short-awned.
                14 Bracts of the spikelet 2 .... 3. *Leersia.*
                14 Bracts of the spikelet 3 or 4.
                  15 Panicle open, with visible branches .... 45. *Cinna.*
                  15 Panicle dense, cylindrical, spike-like.
                    16 Glumes united near the base; lemma awned .... 46. *Alopecurus.*
                    16 Glumes separate; lemma awnless .... 47. *Phleum.*
           11 Spikelets dorsally compressed.
            17 Spikelets awned or awn-tipped, densely clustered; ligules absent .... 87. *Echinochloa.*
            17 Spikelets awnless; ligules present.
              18 Plants bearing terminal panicles and solitary large subterranean spikelets on
                rhizomes; rare on sandy coastal plains .... 85. *Amphicarpum.*
              18 Plants with aerial panicles only.
                19 Second glume inflated; fertile floret shorter than the outer bracts; panicle
                  dense, cylindrical; paludose plants .... 86. *Sacciolepis.*
                19 Second glume not inflated; fertile floret as long as the bracts; panicles
                  various; plants mostly of drier places.
                  20 Fertile floret hard; edges of lemma inrolled .... 84. *Panicum.*
                  20 Fertile floret leathery, edges of lemma flat, exposed .... 93. *Leptoloma.*
        10 Infl or 1–many spikes or racemes.
          21 Spikelets laterally compressed, lacking sterile florets.
            22 Infl a single erect balanced raceme; first glume absent; rhizomatous cult. lawn-
            grass .... 83. *Zoysia.*
            22 Infl of 1–many 1-sided spikes; both glumes present; wild plants.
              23 Spikelets orbicular; glumes equal, concealing the floret .... 48. *Beckmannia.*
               23 Spikelets lanceolate, acute; glumes dimorphic; floret visible .... 80. *Spartina.*
          21 Spikelets dorsally compressed, often with a sterile lemma below the fertile one.
            24 Infl silky; callus bearing long gray or silvery hairs .... 97. *Miscanthus.*
            24 Infl not silky-hairy.
              25 Spikelets with a protruding knob (first glume) at base .... 88. *Eriochloa.*
             25 Spikelets without a basal knob.
                26 Fertile lemma soft, flexible, its edges thin, exposed .... 92. *Digitaria.*
                26 Fertile lemma rigid, its edges inrolled at maturity.
                  27 Back of the fertile lemma turned toward the rachis .... 89. *Paspalum.*
                  27 Back of the fertile lemma turned away from the center of the rachis .... 90. *Axonopus.*
1 Spikelet falling with attached stalks, rachis-internodes, or bristles, or in clusters of 2–many.

28 Spikelets, or some of them, borne in thickened spikes and partly or entirely concealing its internodes.
  29 Each spike with pistillate solitary spikelets in the bony lower part and paired staminate ones
    in the flattened upper part .... 105. *Tripsacum.*
  29 All nodes of the rachis with perfect-fld spikelets.
    30 Fertile florets several; lemmas awned .... 57. *Aegilops.*
    30 Fertile floret 1; lemmas awnless.
      31 Spikelet solitary at each rachis-node; both glumes visible before the lemma .... 23. *Parapholis.*
      31 Spikelets paired at each rachis-node, one abortive; only the first glume visible ....104. *Coelorachis.*
28 Spikelets exposed, borne on a thin rachis.
  32 Infl a cylindrical plumose spike of detachable fascicles, each containing 1–several spikelets ..... 94. *Pennisetum.*
  32 Infl not of plumose detachable fascicles.
    33 Infl a raceme of numerous short, drooping, 1-sided spikes, these falling whole from the
      intact rachis .... 81. *Bouteloua.*
    33 Infl otherwise; rachis disarticulating into individual segments, each bearing 1–3 spikelets.
      34 Spikelets solitary; low creeping annual with cordate-ovate lf-blades .... 103. *Arthraxon.*
      34 Spikelets paired or in triads, or rarely solitary but accompanied by sterile pedicels;
        blades linear.
        35 Infl a single terminal spike or rame.
          36 Infl appearing as a mass of long, scabrous awns, concealing the spikelets (see
            also hybrids of *Elymus* and *Hordeum*) .... 53. *Hordeum.*
          36 Infl otherwise.
            37 Rachis conspicuously ciliate; each internode bearing a sessile, awned,
              perfect-fld spikelet and a pedicellate, awnless vestigial one .... 102. *Schizachyrium.*
            37 Rachis not ciliate; spikelets similar, the lower ones paired, the upper ones
              solitary .... Hybrids of *Elymus* and *Hordeum.*
        35 Infl not a single spike, instead variously a panicle or of spikes or rames.
          38 Infl a panicle of racemes or rames.
            39 Spikelets of each pair equal and fertile .... 96. *Erianthus.*
            39 One spikelet of each pair fertile, the other reduced or absent.
              40 Wild plants; each spikelet-pair composed of a sessile, awned, perfect-
                fld spikelet and a ciliate sterile pedicel .... 100. *Sorghastrum.*
              40 Weeds and cultigens; each spikelet-pair composed of a sessile, awned,
                perfect-fld spikelet and a pedicellate, awnless, staminate one .... 99. *Sorghum.*
          38 Infl a digitate cluster of several rames.
            41 Spikelets of each pair alike, laterally compressed; blades lanceolate; de-
              cumbent annual .... 98. *Microstegium.*
            41 Sessile spikelet awned, perfect-fld; pedicellate spikelet reduced and sta-
              minate or abortive; blades linear; erect perennials .... 101. *Andropogon.*

1. ARUNDINARIA Michx.

1. ARUNDINARIA Michx.   Cane. Spikelets many-fld, elongate, the rachilla disarticulating between the lemmas; glumes unequal, distant, shorter than the lemmas, these
lanceolate, 11–17-veined, about equaling the 2-keeled paleas; lodicules 3, relatively well
developed; stamens 3; stigmas 3; woody perennials with slender, hollow culms, freely
branched after the first year; lvs linear or lanceolate, with evident cross-veins, tapering
to a petiole and articulated with the sheath; spikelets large, in loose racemes or panicles.
150 (depending on generic circumscription), warm reg.

    1. Arundinaria gigantea (Walter) Chapman.   Giant cane. Culms to 3+ m from a system of rhizomes,
forming dense brakes; sheaths glabrous or hairy, ciliate, truncate at the summit and bearing a few long
deciduous bristles on each angle; blades 10–30 cm, usually hairy beneath, sometimes also above; aerial stems
monocarpic, sometimes some of them flowering the first year, but others only after several years, then forming
fastigiately branched panicles on the old wood; first glume to 6 mm, or wanting; second glume to 12 mm;
lemmas acuminate, the lower 15–20 mm; 2n=48. Damp woods, wet ground, and swamps; Md. and Va. to
s. O. and s. Ill., s. to Fla. and Tex. (*A. macrosperma; A. tecta.*)

2. ORYZA L.

2. ORYZA L.   Spikelets 1-fld, articulated below the glumes, laterally compressed;
glumes subequal, triangular-subulate, 1-veined, much shorter than the lemma; lemma
narrowly oblong at anthesis, boat-shaped at maturity, acute, awned or awnless, 5-nerved,
the lateral nerves near the involute margins; palea about equaling the lemma; plants of
wet soil, with wide flat lvs and large open panicle. 25, trop.

    1. Oryza sativa L.   Rice. Annual 1–2 m; blades elongate, 8–15 mm wide; panicle 1–3 dm; lemma 7–10
mm, reticulate rugose, short-pilose, its awn, when present, to 5 cm; 2n=24. Asiatic cultigen, cult. in s. U.S.
and occasionally adventive as far n. as se. Va., but not persistent.

3. LEERSIA Swartz, nom. conserv.   Spikelets 1-fld, laterally compressed, articulated at base in a small cupule atop the pedicel; glumes wanting; lemma awned or awnless, broad, chartaceous, oblong, semi-elliptic, or oval, keeled, 5-veined, the lateral nerves near the margin, the intermediate veins usually inconspicuous; palea similar to the lemma in texture and about as long, narrow, keeled, 3-veined, its margins clasping the inrolled margins of the lemma; stamens 1, 2, 3, or 6; stigmas 2; perennials (our spp. with long rhizomes) with membranous ligules, flat blades, and spikelets on short pedicels in small, spike-like racemes borne in a terminal panicle. 7, cosmop.

1 Stamens 6; panicle exserted, with erect or ascending branches .... 1. *L. hexandra.*
1 Stamens 2 or 3; panicle either with ± spreading branches or ± enclosed in the sheath.
  2 Spikelets elliptic to linear, at least twice as long as wide.
    3 Stamens 3; panicle-branches 2 or more at some nodes .... 2. *L. oryzoides.*
    3 Stamens 2; panicle-branches 1 per node .... 3. *L. virginica.*
  2 Spikelets suborbicular, less than twice as long as wide; stamens 2 .... 4. *L. lenticularis.*

1. Leersia hexandra Swartz.   Usually decumbent and rooting from the lower nodes, 3–15 dm; sheaths retrorse-scabrous, prolonged into 2 erect narrow lobes (1–)2–6 mm; ligules as long as the sheath-lobes and adnate to them; main lvs 3–15 mm wide, panicle 5–15 cm, its branches erect or ascending, some of them floriferous to the base, spikelets erect or appressed, little imbricate, 3–5 mm, a fourth to a third as wide; lemma ciliate on the keel and margins, glabrous or minutely hispidulous on the sides; stamens 6; seed seldom set; 2*n*=48. Swamps, marshes, and wet fields; pantropical, extending n. to Tenn. and s. Va. (*Homalocenchrus h.*)

2. Leersia oryzoides (L.) Swartz. Rice cut-grass. Culms decumbent and sprawling, rooting at the nodes, distally erect, to 15 dm, sheaths roughly retrorse-scabrous, ligule truncate, to 1 mm; blades very rough-margined with stiff colorless spinules, the main blades 15–30 cm × 6–15 mm; panicles pale green or whitish, the terminal one to 30 cm, exserted or partly included, its slender branches ± spreading when exserted; lower part of the panicle with 2 or 3(4) branches per node; spikelets ascending, 4–7.5 mm, a fourth to a third as wide, imbricate in spike-like clusters of 3–8 ending the branches; lemma pilose, stiffly ciliate on the keel; stamens 3; late-season axillary panicles reduced and ± included in the sheaths; 2*n*=48. Swamps, wet meadows, and muddy soil; Que. and N.S. to B.C., s. to Calif. and Fla.; Europe; e. Asia. Estuarine plants vary to a flaccid, virtually glabrous phase with included panicles.

3. Leersia virginica Willd.   White grass. Slender and weak, erect or decumbent and rooting at base, 5–15 dm; sheaths smooth or scaberulous, rarely retrorse-scabrous; ligule short, truncate; blades scaberulous, especially marginally, rarely hairy, 5–20 cm × 3–15 mm; panicle 5–20 cm, its branches solitary, distant, mostly straight, spreading, floriferous from about the middle; spikelets scarcely imbricate, the tip of one seldom extending to the middle of the next, appressed to ascending; lemma 2.9–4.1 mm, a fourth to a third as wide, sparsely ciliate on the keel and margins, smooth to pilose on the sides, stamens 2; 2*n*=48. Moist or wet woods; Que. to Minn. and S.D., s. to Fla. and Tex. (*Homalocenchrus v.*)

4. Leersia lenticularis Michx.   Catchfly-grass. Culms 5–15 dm; sheaths smooth or nearly so; ligule truncate, 1 mm; main blades 1–2 cm wide, glabrous or sometimes soft-hairy; panicle large, 1–2(–3) dm, freely branched, the slender branchlets spreading, each bearing 1–4 spike-like racemes 1–2 cm; spikelets short-pediceled, closely imbricate and appressed to each other, broadly elliptic to orbicular; lemmas 3.8–5.6 mm, more than half as wide, hispid-ciliate on the veins and keel; stamens 2, 2*n*=48. Swamps and streambanks on the coastal plain from se. Va. to Fla., and Tex., n. in the Mississippi Valley to Minn. and Wis.

4. ZIZANIA L.   Wild rice. Spikelets 1-fld, articulated at the base, readily deciduous, subterete, unisexual, the staminate on the lower, the pistillate on the upper branches of the large panicle; glumes none; lemma of the staminate spikelet thin, herbaceous, linear, acuminate or short-awned, 5-veined; pistillate spikelets inserted in a cup-shaped excavation at the summit of the clavate pedicels, the lemma awned, prominently 3-ribbed, closely clasping the palea to produce the illusion of a single sheath around the rod-like grain; grasses of marshes and shallow water, with tall stems and wide flat lvs. 4, incl. 1 in Tex. and 1 in e. Asia.

1 Pistillate lemmas minutely scabrous all over, the abortive ones less than 1.5 mm wide .... 1. *Z. aquatica.*
1 Pistillate lemmas glabrous or nearly so between the scabrous-hispid nerves, the abortive ones 1.5–2
  mm wide .... 2. *Z. palustris.*

1. Zizania aquatica L.   Robust annual; ligule membranous; blades large and soft; panicle erect, to 6 dm, the staminate branches widely spreading, the pistillate at first erect, at anthesis ascending; staminate spikelets pendulous, 6–11 mm, stramineous to purplish, glabrous or nearly so, the lemma awnless or with an awn to

3 mm; pistillate spikelets linear, with an awn to 7 cm, the lemma thin and membranous, minutely scabrous all over, the exposed surface of the palea likewise roughened; some spikelets in the pistillate part of the infl sterile and abortive, the lemma subulate, less than 1.5 mm wide, often twisted, tapering insensibly into the awn; $2n$=30. S. Que. and coastal states from Me. to Fla. and La., irregularly inland in n. N.Y. and from w. Lake Erie to Wis. and s. Ill. The widespread var. *aquatica*, southern w. r., is robust, (1)2–3(–5) m, with lvs mostly 2.5–5(–8) cm wide, the infl 2.5–6 dm, its pistillate lemmas with an awn mostly 2–7 cm. The var. *brevis* Fassett, estuarine w. r., occurring along the St. Lawrence estuary in Que., is smaller, to 1 m, with lvs to 1.5 cm wide and small, few-branched infls 1–2.5 dm, the pistillate lemmas with an awn under 1.5 cm.

   2. **Zizania palustris** L.   Much like no. 1, but the pistillate lemmas firm and tough, glabrous or nearly so between the scabrous-hispid nerves, the abortive ones 1.5–2 mm wide, linear, tapering abruptly at the pubescent tip into the awn; pistillate palea glabrous. Two vars.:

   Var. *palustris*.   Northern w. r. Stems arising 0.7–1.5 m above the water (this sometimes over 1 m deep); lvs seldom over 1 cm wide; ligule 3–5(–10) mm; panicle narrow, the lower branches with usually 5–15 staminate spikelets; awns of the pistillate lemmas 1.5–4 cm. Que. and N.B. to Pa., w. to Io. and Minn. (*Z. aquatica* var. *angustifolia*.)
   Var. *interior* (Fassett) Dore.   Interior w. r. Stems 1–3 m (in water seldom over 6 dm deep); lvs 1–3 cm wide; ligule 10–15 mm; panicle wider and more ample, the lower branches with usually 26–50 staminate spikelets; awns of the pistillate lemmas 4–7 cm. Ind. to Mo. and Nebr., n. to N.D. and se. Man. (*Z. interior; Z. aquatica* var. *i.*)

5. ZIZANIOPSIS Doell & Aschers.   Spikelets 1-fld, articulated at the base, readily deciduous, subterete, linear-subulate, unisexual, the staminate below the pistillate on the same branches of the panicle; glumes none; lemma lance-linear, acuminate, 7-veined, the pistillate one awned; palea 3-veined, somewhat exceeding the lemma; grain obovoid, tipped with the stiff persistent style; tall marsh-grasses, perennial from rhizomes, with large, flat lvs, long ligules, and large, much branched panicles. 4, New World.

   1. **Zizaniopsis miliacea** (Michx.) Doell & Aschers.   Southern wild rice. Culms 1–4 m, blades to 1 m × 3 cm, with wide midvein, scabrous on the margins; panicle nodding, 2–6 dm, with numerous ascending, whorled branches; lemmas scabrous on the nerves, the body 6–9 mm; awn of pistillate spikelets 1–5 mm; $2n$=24. Marshes and streambanks on the coastal plain; Md. to Fla. and Tex., n. in the interior to se. Mo.

6. DIARRHENA P. Beauv.   Spikelets 3–5-fld, disarticulating above the glumes and between the lemmas; florets at first appressed, at maturity spreading, the upper much reduced; first glume 1-nerved; second glume 3–5-nerved; lemmas smooth, subcoriaceous, the 3 conspicuous nerves convergent at the summit and excurrent into a short, stout cusp; palea shorter than the lemma, truncate, strongly veined; stamens 2; grain obovoid, abruptly contracted into a stout 2-lobed beak, soon spreading the lemma and palea and conspicuously projecting; seed loose within the pericarp, attached only at the base; perennial from a thick, scaly rhizome; stems slender, erect, with lvs mostly below the middle; panicle long and slender, few-fld. Monotypic.

   1. **Diarrhena americana** P. Beauv.   Culms 5–12 dm; sheaths pubescent toward the top; blades 2–4 dm × 10–18 mm, with excentric midvein; infl long-exsert, 1–3 dm, drooping; spikelets at first subcylindric, soon flattened by the spreading of the lemmas; first glume triangular; second glume oblong, cuspidate; $2n$=60. Moist woods. Two vars.:

   Var. *americana*.   Panicle hirsutulous; first glume 2.2–4 mm, the second 4–5.4 mm; lemmas ovate, 7–10 mm, acuminate; callus (except of the first lemma) ± evidently hairy. Ind. to W.Va. and Tenn.; Mo.
   Var. *obovata* Gleason.   Panicle scabrous; first glume 1.9–2.8 mm, the second 2.4–4.3 mm; lemmas 5.2–6.8 mm, abruptly rounded into a short cusp; callus glabrous. More western, Ind. to S.D., Kans., and Tex. Perhaps better treated as *D. obovata* (Gleason) Brandenburg.

7. BRACHYELYTRUM P. Beauv.   Spikelets 1-fld, articulated above the glumes; glumes minute, subulate to triangular, 1-veined; lemma linear-subulate, herbaceous, rounded on the back, gradually tapering into an elongate awn; palea 2-keeled; rachilla prolonged into an elongate bristle appressed to the furrow of the palea; perennial from knotty rhizomes, with solid culms, broad, flat lvs, short, membranous ligule, and narrow, few-fld panicles of readily deciduous spikelets. Monotypic.

   1. **Brachyelytrum erectum** (Schreber) P. Beauv.   Culms 5–10 dm, glabrous or hairy; sheaths retrorse-hairy; blades scabrous to hairy, ciliate-margined, 8–18 cm × 8–16 mm, panicle erect or declined, 5–15 cm, the few spikelets appressed; first glume none or up to 0.8 mm; second glume subulate, 1–4 mm; lemma 6–10 mm, scabrous to hispidulous on the nerves, its awn 12–25 mm; prolongation of the rachilla two-thirds

as long as the lemma; $2n=22$. Moist woods; Que. and Nf. to Ont., Minn., and Kans., s. to Ga., La., and Okla.; e. Asia. Two well marked vars. in N. Amer.:

Var. *erectum*. Lvs with less than 10 cilia per 0.5 cm of margin; larger florets 10–12 mm; lemmas (5)7- or 9-nerved, strongly hispid with hairs 0.5+ mm; anthers 5 mm or more. Relatively southern, n. to Mass., Mich., and Wis.

Var. *glabratum* (Vasey) T. Koyama & Kawano. Lvs with more than 15 cilia per 0.5 cm of margin; florets 8–10 mm; lemmas 3- or 5-nerved, glabrous to scabrous or hispidulous with hairs up to 0.2 mm; anthers up to 4 mm. Relatively northern, s. to Pa., W.Va., O., and Mich. (Var. *septentrionale*.)

8. NARDUS L. Spikelets 1-fld, solitary and sessile, arising from shallow cavities and alternating on 2 sides of a 3-sided rachis, with a single minute glume; lemma lance-subulate, obscurely 3-veined, prolonged into a short awn, sharply infolded near each margin, the edges meeting over the linear obtuse palea; anthers filiform, elongate; stigma 1, filiform; densely cespitose perennial, the lower sheaths bladeless or with much reduced blade, the upper with narrow, linear, involute blade, the bluish spikelets in a slender spike, ascending, all turned in one direction. Monotypic.

1. **Nardus stricta** L. Moor-matgrass. Culms rough, 1–4 dm; spikes 3–8 cm; lemmas 8–12 mm, of which the terminal awn constitutes a third; $2n=16$. Native of Europe, intr. here and there from Nf. and N.S. to Mich.

9. ORYZOPSIS Michx. Ricegrass. Spikelets 1-fld, articulated above the glumes; glumes equal or nearly so, herbaceous to membranous, broad, acute to acuminate, distinctly to obscurely 3–7-veined; lemma about equaling or a little shorter than the glumes, fusiform, nearly terete, obscurely veined, becoming indurate at maturity, with a short, oblique, blunt callus at base; margins of the lemma folded over a palea of similar or more membranous texture; awn terminal, articulated with the lemma, readily detached; tufted perennials with flat or involute lvs, membranous or no ligule, and open or contracted panicles or racemes of often large spikelets. 50, N. Temp. and subtrop.

1 Lemma densely and conspicuously long-silky; western .... 1. *O. hymenoides*.
1 Lemma with short, appressed hairs; widespread spp.
  2 Lvs involute, at least when dry, 1–2 mm wide; glumes 3.5–4.8 mm.
    3 Awn 6–11 mm .... 2. *O. canadensis*.
    3 Awn 1–2(–3) mm, or frequently lacking .... 3. *O. pungens*.
  2 Lvs flat or nearly so, the main ones 5–15 mm wide; glumes larger, mostly 6–9 mm.
    4 Blade of the uppermost lf large, usually 10–20 cm; infl evidently paniculate .... 4. *O. racemosa*.
    4 Blade of the uppermost lf reduced, 0–3 cm; infl racemiform .... 5. *O. asperifolia*.

1. **Oryzopsis hymenoides** (Roemer & Schultes) Ricker. Indian r. Densely tufted, 3–7 dm; ligules 2.5–7.5 mm, acuminate, becoming lacerate; blades numerous, smooth, strongly involute, ca 1 mm wide, nearly as long as the stems; panicle (5–)7–15(–19) cm, open, diffusely branched, with long-pedicellate spikelets; glumes ovate-acuminate or tapering to an awn to 2 mm, hyaline-margined, finely puberulent to subglabrous, the 3(5) nerves prominent toward the greenish glume-base, obscure in the anthocyanic distal part; first glume (4.5–)5–7.5(–8) mm, the second a bit shorter; lemma 2.5–4(–5), fusiform, turgid, maturing dark and shiny, densely white-pilose with long soft hairs ca = glumes; awn 3–5.5 mm, straight, readily deciduous; $2n=28$, 48, 65, 130. Dry, open often very sandy places; Man. and nw. Minn. to B.C., s. to Calif. and Tex. Hybridizes with several spp. of *Stipa*, producing sterile plants called × *Stiporyzopsis bloomeri* (Bolander) B. L. Johnson.

2. **Oryzopsis canadensis** (Poiret) Trin. Loosely tufted, 3–8 dm; ligules of upper lvs mostly 1.5–3 mm; blades 1–2 mm wide, involute at least when dry, those of the flowering stems to 1 dm, the basal ones longer; panicle lax and open, ovoid, 8–15 cm, with flexuous, capillary, widely spreading branches; glumes elliptic-obovate, 3.5–4.8 mm, perfectly smooth, very thin, the lateral veins inconspicuous; mature lemma 2.5–4 mm, dull brown, rather sparsely appressed-pilose; awn persistent, 6–11 mm, crooked, twisted and often somewhat coiled in the basal half. Dry, sandy or rocky woods; Nf. to Sask., s. to n. N.Y., n. Mich., n. Wis. and n. Minn. (*Stipa c.*)

3. **Oryzopsis pungens** (Torr.) A. Hitchc. Densely tufted, 1–5 dm; lvs much as in no. 2; panicle 3–8 cm, usually slender with appressed or strongly ascending branches, or ovoid and open at anthesis; glumes elliptic-obovate, 3.5–4 mm, minutely scabrous distally at 20×, very thin, the lateral veins inconspicuous; lemma 2.5–4 mm, gray or pale green, its awn 1–2(3) mm, straight or slightly bent, or obsolete; $2n=22$, 24. Dry, rocky or sandy woods; Nf. to B.C., s. to nw. N.J., Pa., Ind., S.D., and Colo.

4. **Oryzopsis racemosa** (Smith) Ricker. Blackseed-r. Loosely tufted from a knotty rhizome, 4–10 dm; principal lvs cauline, scaberulous above, pubescent beneath, the blades mostly 10–20 cm × 8–15 mm, with many fine veins; lower lvs ± reduced, the basal sometimes to mere sheaths; ligule minute or none; panicle sparsely branched, 1–2 dm, the few straight branches spreading or ascending, bearing the few appressed

spikelets toward the end; glumes herbaceous, narrowly elliptic, acute or short-acuminate, 7–9 mm, distinctly 7-veined; lemma 5.5–7 mm, dark brown and shining; awn 12–22 mm; $2n=46, 48$. Dry to moist woods and wooded dunes; Que. to Ont., Minn., and N.D., s. to Va., Ky., Mo., and Nebr.

5. **Oryzopsis asperifolia** Michx. Rough-lvd r. Loosely tufted, often widely radiating or prostrate, 3–8 dm; principal lvs basal or nearly so, evergreen, with long, stiffly erect blade to 4 dm, 4–10 mm wide, with strong, closely spaced veins, densely and finely rough-puberulent (or merely scabrous) beneath, glabrous and usually glaucous above, long-tapering at base; cauline lvs progressively reduced, the uppermost bladeless or with a blade to 3 cm; ligules 0.2–0.7 mm, truncate, ciliolate; panicle slender, racemiform, 3–7 cm, the paired branchlets each with a single spikelet; glumes elliptic or broadly ovate, herbaceous, distinctly 5- or 7-veined, 6–8.5 mm, obtuse to abruptly acute or mucronate; lemma 5–7 mm, pale green or yellowish, finely appressed-hairy, its flexuous awn 6–14 mm; $2n=46, 48$. Upland woods; Nf. and Que. to B.C., s. to Pa., n. Ind., S.D., N.M., and Utah.

10. **STIPA** L.   Needle-grass; porcupine-g. Spikelets 1-fld, articulated above the glumes; glumes about equal, lance-linear, papery, with broad scarious margins, the first 3-veined, the second 5-veined, the nerves parallel, only the middle one extending to the often slenderly prolonged tip; lemma shorter than or equaling the glumes, indurate, terete, linear to narrowly fusiform, obscurely nerved or nerveless, villous at least at the base, the margins meeting or overlapping around the membranous palea, and with a bearded, elongate, sharp-pointed callus; awn elongate, terminal, articulated with the lemma but persistent, usually twice geniculate, the lower one or two segments hygroscopic and twisted; perennials with narrow, elongate, often involute lvs and open or contracted panicles of large spikelets. 300, nearly cosmop.

1 Awn 2–3.5 cm, the body of the lemma 4.5–6.5 mm .... 1. *S. viridula.*
1 Awn 9–20 cm, the body of the lemma 9–23 mm.
  2 Lemma 9–14 mm, its awn very slender and flexuous, indistinctly once geniculate .... 2. *S. comata.*
  2 Lemma (15–)17–23 mm, its awn stiff, sharply twice geniculate .... 3. *S. spartea.*

1. **Stipa viridula** Trin.   Green n.-g. Perennial, forming large tufts, 5–11 dm tall; sheaths ± villous-ciliate at the margins and top; ligule rounded or truncate, 0.5–1.5 mm, or the upper to 3 mm; lvs 1–4(–6) mm wide, usually involute, ± scaberulous, the lower much elongate; panicle contracted, 1–3 dm, the ascending branches each bearing 2–several spikelets; glumes 8.5–12 mm, tapering to a very slender point; mature lemma fusiform, pale brown, 4.5–6.5 mm, appressed-hairy throughout, its awn 2–3.5 cm, weakly twice geniculate below the middle; some plants functionally pistillate; $2n=82$. Dry plains and prairies; Wis., Minn., and Io., w. to the Rocky Mts.; also in w. N.Y.

2. **Stipa comata** Trin. & Rupr.   Needle-and-thread grass. Tufted, 4–10 dm; sheaths glabrous or nearly so, the upper often inflated over the base of the panicle; ligule 2–5 mm; blades smooth or scaberulous, 1–3 mm wide, usually involute; panicle narrow, 2–3.5 dm, the ascending branches each with 1–few spikelets; glumes 15–35 mm, tapering to a long filiform point; mature lemma 9–14 mm, pale brown, villous at base, villosulous to glabrate above, its awn 9–16 cm, very slender, loosely flexuous or coiled, obscurely once geniculate; $2n=44$, 46. Dry plains and prairies, often in sandy soil; widespread in the w. cordillera, extending e. across the Great Plains to Minn. and Io., and irregularly to Mich. and n. Ind.

3. **Stipa spartea** Trin.   Porcupine-grass. Culms 6–12 dm, in small tufts; sheaths glabrous; ligules of the upper lvs 4–6 mm, the lower much shorter; blades 2–5 mm wide, glabrous beneath, scabrous and usually also hairy above, the lower elongate, tapering to a fine point; panicle narrow, ± nodding, 1–2 dm, usually wholly exserted, the few branches each with 1–few spikelets; glumes 28–45 mm, tapering to a very slender point; mature lemma (15–)17–23 mm, brown, pubescent at base, decreasingly so above; awn 12–20 cm, stout, stiff, twice geniculate near the middle, the central segment usually 1.5–3 cm; $2n=44$, 46. Dry prairies and open woods; Mich. and Ind. to N.D., Wyo., Mo., and N.M.; also in scattered dry habitats in O., Ont., and Pa.

11. **PIPTOCHAETIUM** J. S. Presl, nom. conserv.   Much like *Stipa,* but the margins of the lemma merely curled over the 2 keels of the firm palea, often leaving its intercostal sulcus exposed. 25, New World.

1. **Piptochaetium avenaceum** (L.) Parodi.   Loosely tufted, 5–10 dm; sheaths glabrous; ligules ovate, 2–3 mm; blades 2–3 dm × 1–2 mm, glabrous or nearly so; panicle 1–2 dm, loose, open, few-fld, the slender divergent branches each with 1 or 2 spikelets; glumes 9–13 mm, abruptly short-acuminate; mature lemma dark brown, 7.5–11 mm, villous in the basal third, glabrous in the middle, scabrous distally, its margins inrolled to form a deep ventral groove; awn scabrous, 4.5–7 cm, twice geniculate in the middle, the central segment ca 1 cm; $2n=22$, 28. Dry woods; Mass. and N.Y. to Fla., w. to Ky., Ark., and Tex.; also in Mich. and n. Ind., mainly near the w. side of Lake Michigan. (*Stipa a.*)

12. MILIUM L. Spikelets 1-fld, articulated above the glumes; glumes equal, herbaceous, ovate or elliptic, rounded on the back, obtuse or acute, 3-veined; lemma about as long as the glumes, elliptic, awnless, veinless, obtuse, rounded on the back, smooth, at first thin, at maturity firm, white, and shining, its margins partly covering a palea of similar texture, the whole resembling a fr of *Panicum*; grasses with broad flat lvs, membranous ligule, and widely branched panicle. 3–4, n. temp.

1. **Milium effusum** L. Glabrous and glaucous perennial 6–12 dm, erect from a bent base; ligules prominent and pale; main lvs 10–18 mm wide; panicle 1–2 dm; ovoid or pyramidal, the branches in fascicles of 2 or 3, widely spreading and bearing drooping spikelets beyond their middle; glumes scaberulous, ca 3 mm; $2n=14, 28$. Rich, moist or dry woods; Que. and N.S. to Ont. and Minn., s. to n. N.J., W.Va., and Ill.; also in Eurasia.

13. FESTUCA L. Fescue. Spikelets 2–15-fld, the rachilla disarticulating above the glumes and between the lemmas; glumes narrow, unequal, 1–3(–5)-veined, usually shorter than the lemmas; lemmas membranous or often indurate, rounded on the back, obscurely 5-veined, acute or acuminate, usually awned from the tip, sometimes awnless; upper lemmas considerably reduced; palea about equaling the lemma; stamens 3; grain obovoid-oblong; open-pollinated, tufted perennials; infl an open to contracted panicle. 200+, widespread in temp. and cool regions.

1 Lvs 3–12 mm wide, flat.
  2 Larger lemmas 5.5–9(–10) mm; lf-blades auriculate at base; anthers mostly 2–4 mm.
    3 Auricles glabrous; internodes of the rachilla smooth or very nearly so .... 1. *F. pratensis.*
    3 Auricles ciliate; internodes of the rachilla antrorsely scabrous .... 2. *F. elatior.*
  2 Larger lemmas 3.3–5.2 mm; lf-blades not auriculate; anthers mostly 0.8–1.5 mm.
    4 Principal lowermost panicle-branches usually bearing 2–7 spikelets scattered along the outer half; spikelets narrowly ovate, 2–4 mm wide .... 3. *F. subverticillata.*
    4 Principal lowermost panicle-branches bearing 8–20 spikelets clustered at the ends; spikelets broadly ovate, 4–6 mm wide .... 4. *F. paradoxa.*
1 Lf-blades 0.2–3 mm wide, often involute or conduplicate.
  5 Second glume three-fourths to fully as long as the spikelet; lowest lemma 6.5–10 mm .... 5. *F. altaica.*
  5 Second glume up to half as long as the spikelet; lowest lemma 2.5–6 mm.
    6 Awns well developed, mostly 4–12 mm .... 6. *F. occidentalis.*
    6 Awns short, up to ca 3 mm, or none.
      7 Basal sheaths persistent, remaining firm and entire; rhizomes wanting.
        8 Lemmas awnless, small, mostly 2.3–4 mm .... 7. *F. filiformis.*
        8 Lemmas short-awned, larger, the lowest one mostly (3)3.5–6 mm.
          9 Anthers short, up to 1.7(–2) mm; infl contracted, often spiciform; native, not weedy ... 8. *F. brachyphylla.*
          9 Anthers longer, 2–3 mm; infl more open; intr., weedy.
            10 Spikelets mostly 5–7 mm .... 9. *F. ovina.*
            10 Spikelets larger, mostly 7–10 mm .... 10. *F. trachyphylla.*
      7 Basal sheaths soon disintegrating into fibers; often rhizomatous .... 11. *F. rubra.*

1. **Festuca pratensis** Hudson. Meadow-f. Tufted, but often with basally decumbent culms; old sheaths brown, decaying to fibers, culms erect above a geniculate base, 3–12 dm, glabrous; sheaths smooth; blades lax, not evidently ridge-veined, glabrous or scabrous, 3–5(–7) mm wide, dilated at base into conspicuous smooth auricles; infl 1–2.5 dm, erect or nodding at the tip, contracted at least after anthesis, the internodes of the branches less than twice as long as the spikelets; spikelets 10–15 mm, 4–10-fld; first glume subulate, 2.5–4 mm, 1-veined, the second lanceolate, 3.5–5 mm, 3–5-veined, with hyaline margins; larger lemmas 5.5–7 mm, usually glabrous, 5-veined, the tip hyaline, acute, rarely with a short awn to 2 mm; rachilla-joints glossy, smooth or very nearly so; anthers 2–4 mm; $2n$ mostly = 14. Native of Europe, cult. for forage and established in fields, meadows, and moist soil throughout most of the U.S. and adj. Can. (*F. elatior,* misapplied.)

    F. gigantea (L.) Vill., a European sp. much like *F. pratensis* but with lvs 6–18 mm wide and awns 10–18 mm, is adventive in N.Y. and s. Que. and perhaps elsewhere in our range.

2. **Festuca elatior** L. Tall or alta-f. Much like no. 1, often taller, to 2 m; old sheaths pale stramineous, only tardily disintegrating; blades coarse and thick, prominently ridge-veined above, 4–10 mm wide, with ciliate auricles; first node of the panicle with 2–3 branches, each with 5–15 spikelets, these mostly 3–6-fld; first glume (3–)4–6 mm, the second (4–)5–9 mm; lemmas 7–8.5+ mm, usually scabrous distally, the awn 0.3–2(–4) mm; $2n$ most often = 42, sometimes 28, 56, 63, 70. Native of Europe, widely planted in our range and elsewhere in the U.S., and readily escaping. (*F. arundinacea.*)

3. **Festuca subverticillata** (Pers.) E. Alexeev. Nodding f. Culms few in a tuft, 6–12 dm, smooth; sheaths glabrous or pilose; blades flat, soft, minutely scaberulous on the veins, sometimes pilose, 3–8 mm wide, without auricles; infl long-exsert, 1.5–3 dm, the elongate, slender branches racemiform, eventually widely

spreading or deflexed, bearing spikelets only above the middle; spikelets relatively remote, the summit of one scarcely reaching the base of the next, 2–4(5)-fld, 4–8 × 2–4 mm; first glume subulate, carinate, 2.2–3.8 mm, the second ovate, 2.7–4.4 mm; lemmas broadly rounded on the back, obtuse to acute, 2.5–4.7 mm, appressed till maturity; palea acute; anthers 0.8–1.5 mm; ovary hairy at the tip; $2n=42$. Moist woods; Que. and N.Engl. to Fla., w. to Man. and Tex. (*F. nutans; F. obtusa.*)

4. **Festuca paradoxa** Desv.   Much like no. 3, the culms usually stouter and more tufted, and the infl more freely branched; spikelets separated by short internodes, approximate and overlapping, (3)4–5(6)-fld, 5–8 × 4–6 mm; first glume 2.4–4.2 mm, the second elliptic-oblong, 3.1–5.2 mm; lemmas turgid, obtuse, 3.6–5.2 mm, soon diverging and exposing the rachilla; palea obtuse; anthers ca 1–1.5 mm. Moist or wet open woods and prairies; Pa. and Va. to S.C., w. to Io., Okla., and Tex. (*F. shortii.*)

5. **Festuca altaica** Trin.   Rough f. Stout, tufted, glabrous perennial (3–)5–10 dm; lower sheaths long-persistent, 3–8 cm, 3–7 mm wide when folded, many-veined; blades 1.5–3 mm wide, with sclerenchymatous girders connecting the veins to the epidermis on both sides; panicle open or ± contracted, with straight, ± ascending branches; spikelets 4–6-fld, 8–11 mm, the second or both glumes three-fourths to fully as long; first glume 5–8.3 mm, the second 6.5–10.3 mm; lemmas thin with broad scarious margins, the lowest one 6.5–10 mm, the upper ones much reduced; awn 1 mm or less; anthers 3.5–5 mm; ovary-top with at least a few hairs; $2n=28$. Prairies, hillsides, and open woods; Asia and nw. Amer., s. to Colo., e. onto the Great Plains, and at scattered stations in n. Mich., Ont., Que., Nf., and Lab. Some infraspecific segregation may be desirable, but the taxonomy is not yet clear. (*F. hallii; F. scabrella.*)

6. **Festuca occidentalis** Hook.   Western f. Culms 4–11 dm, slender, glabrous, shining; sheaths and blades smooth, the blades filiform; infl open, flexuous, 10–20 cm; spikelets on slender pedicels, 6–12 mm, 3–5(–7)-fld; first glume 2.7–3.6 mm, the second 3.5–4.5 mm; lemmas green or purplish, soft and membranous, the body 4–6.5(8) mm, the awn two-thirds to almost twice as long; rachilla-joints 1–1.5 mm; anthers 1–2 mm; top of the ovary bristly-pubescent; $2n=14, 28, 42, 56, 64, 70$. Dry woods; n. Mich. and adj. Ont.; Wis.; Wyo. to B.C. and Calif.

7. **Festuca filiformis** Pourret.   Lvs mostly more than half as long as the stems, scabrous, delicately wiry-capillary, 0.3–0.6 mm wide, with 3 large and 0–4 small veins but only one distinct rib, the sclerenchyma forming a ± continuous ring; infl very slender; spikelets small, the first glume 1.1–2.1 mm, the second 1.8–3 mm, the lemmas mostly 2.3–4 mm and awnless; anthers 1.5–2.5 mm; $2n=14, 28$; otherwise much like spp. 8–10. A European weed intr. into lawns and waste places from Nf. to N.C., Ill., and Minn. (*F. tenuifolia; F. capillata; F. ovina* var. *c.*)

8. **Festuca brachyphylla** Schultes.   Short-lf f. Culms very slender, densely tufted, not stoloniferous, glabrous, 1–5 young shoots developing within the sheaths parallel to the main axis; basal sheaths closed up to a third or half length, open above with overlapping margins, mostly pale or drab brown, long-persistent, not becoming fibrous; blades setaceous, 0.3–0.7 mm wide, with 3 large and 0–5 small veins, the sclerenchyma in 3–5+ strands (var. *brachyphylla*), or forming a ± continuous ring (var. *rydbergii*); infl 1–10(–13) cm, narrow and spiciform, or somewhat open at anthesis, the first pedicel of the lowermost branches usually no more than 5(–7) mm from the base; spikelets 2–4(5)-fld, the first glume 2–3(–3.5) mm, 1-nerved, the second 2.5–4.5 mm, 3-nerved; lemmas mostly 3.5–5.5(–6) mm, with a short awn 1–3 mm; anthers up to ca 2 mm; $2n=28, 42$. Widespread in n. N. Amer. and the w. cordillera, s. to Que. and n. Mich., the var. *brachyphylla* also s. to alpine summits and subalpine cliffs of N.Engl. and N.Y. Two ecotypic, genetically differentiated vars.: Var. *brachyphylla*, ± alpine, dwarf, 0.5–2 dm, the culms usually less than twice as long as the relatively soft, glabrous basal lvs, which tend to become angular in drying as the soft tissue between the sclerenchymatous ribs collapses; infl 1–4 cm; anthers up to 1 mm. Var. *rydbergii* (St.-Yves) Cronq., of lower elev., taller, 2–5 dm, the culms mostly 2–3 times as long as the stiff, mostly 1-ribbed, often scabrous basal lvs, which tend to remain rounded in drying; infl 2–10 cm; anthers (1.0–)1.2–1.7(–2) mm. (*F. saximontana.*)

9. **Festuca ovina** L.   Sheep-f. Much like no. 8; sheaths open to the base; blades ca 0.5 mm wide, firm, the sclerenchyma forming a continuous, ± even band; infl more open and spreading; spikelets 5–7 mm, 3–8-fld; lemmas 3–3.5 mm; anthers 2–2.5 mm; $2n=14$. Native of Europe, weedy and occasionally intr. in the ne. part of our range. In the past often interpreted broadly to include our spp. 7, 8, and 10.

10. **Festuca trachyphylla** (Hackel) Krajina.   Hard f. Much like no. 9; blades scabrid, 0.6–1.2 mm wide, (3)5(7)-ribbed, the sclerenchyma forming a continuous or interrupted band of uneven thickness; spikelets 7–10 mm, 4–8-fld, the lemmas 3.8–5.5 mm; anthers 2.5–3 mm; $2n=28, 42$. Native of Europe, weedy and widely intr. in our range. (*F. duriuscula* and *F. longifolia*, both misapplied; *F. brevipila.*)

11. **Festuca rubra** L.   Red f. Culms glabrous, 3–10 dm, usually loosely tufted, often decumbent at base, frequently rhizomatous; young shoots diverging from the axils and breaking through the base of the sheath; lower sheaths closed and reddish-purple when young, soon disintegrating into loose reddish-brown fibers; blades glabrous, 0.7–2 mm wide, with (3)5 ribs, the sclerenchyma in 7–9(11) discrete strands opposite the veins; infl 5–20 cm, narrow, with ascending branches or in some forms loosely spreading, the lowermost branches (or at least one of them) naked for mostly (4)5 mm below the first pedicel; spikelets 4–7-fld; first glume subulate, 2.6–4.5 mm, the second broader, 3.5–5.5 mm; lemmas 4.8–6.1 mm, the longest awns 1–3 mm; $2n=14, 28, 41, 49, 50, 56, 70$. In a wide range of open habitats, widely distributed in n. Europe and in N. Amer., in our range from Que. to Wis., s. to Ga. and Mo.; plants of the coastal region thought to be native, but some or all of the inland forms may be intr. (*F. prolifera*, a viviparous form with some of the lemmas elongate and foliaceous, occurring from Nf. to the mts. of N.Engl.)

4. VULPIA C. Gmelin. Similar to *Festuca* and only marginally distinct, but now customarily held at generic rank; most spp. annual and cleistogamous or nearly so, with most commonly only one, less often 2, rarely 3 anthers, these small, up to ca 1.5 mm; grain linear-cylindric, tapering to both ends. 25, widespread, but especially Mediterranean.

1 First glume more than half as long as the second.
  2 Lemmas hairy, relatively small, the lowest one 2.5–3.5 mm; grains 1.5–2 mm .... 1. *V. elliotea.*
  2 Lemmas glabrous or merely scabrous, often larger, 2.7–7 mm; grains over 2 mm.
    3 Spikelets with 5–11+ closely imbricate florets, the rachilla-internodes typically 0.5–0.7 mm;
      awns of the lemmas up to 6(–9) mm .... 2. *V. octoflora.*
    3 Spikelets with 4–7 more loosely set florets, the rachilla-internodes ca 1 mm; awns of the lemmas
      3–12 mm .... 3. *V. bromoides.*
1 First glume up to half as long as the second .... 4. *V. myuros.*

1. Vulpia elliotea (Raf.) Fern. Squirrel-tail fescue. Slender, erect annual 2–6 dm with glabrous herbage; blades flat or involute, setaceous, 0.5–1 mm wide; infl slender, 5–20 cm, its branches scaberulous; spikelets 3–6-fld, 3–5 mm (excl. awns); first glume subulate, 1.3–2.7 mm, the second broader, 2.3–4 mm; lemmas narrow, ± involute, hairy, especially distally, gradually narrowed into a rough awn 4.5–9 mm; lowest lemma 2.5–3.5 mm; anther 1, ca 0.5 mm; grains 1.5–2 mm. Dry sandy soil; s. N.J. to Tex., chiefly on the coastal plain, n. in the interior to Mo. (*V. sciurea; Festuca s.*)

2. Vulpia octoflora (Walter) Rydb. Six-weeks fescue. Slender, decumbent to erect annual 1–6 dm, glabrous or hairy; blades flat or involute, setaceous, 0.5–1 mm wide; infl slender, 3–20 cm, with a few ascending or rarely spreading branches; spikelets flattened, with 5–11+ fls; first glume subulate, 1-veined, 1.7–4.5 mm, the second lanceolate, 3-veined, a fourth longer; florets closely imbricate, the internodes of the rachilla typically 0.5–0.7 mm; lemmas involute, straight, obscurely veined, glabrous or merely scabrous in our vars., soon diverging and exposing the rachilla, the lowest one 2.7–6.5 mm, with an awn 0.3–6(–9) mm; anther 1(–3), 0.3–1.5 mm; grains mostly 1.7–3.3 mm; 2n=14. Dry or sterile soil; Que. and N.Engl. to B.C., s. to Fla. and Calif. (*Festuca o.*) Two ill-defined vars. in our range: In the mostly southern var. *octoflora* the spikelets are 5.5–10 mm, and the awn of the lowest lemma is 0.3–3 mm. In the mostly more northern var. *glauca* (Nutt.) Fern. the spikelets are 4–5.5 mm, and the awn of the lowest lemma is 2.5–6+ mm.

3. Vulpia bromoides (L.) S. F. Gray. Much like *V. myuros* var. *myuros,* but with a longer first glume, this 3.5–5 mm, more than half as long as the second one; awn of the first lemma 3–12 mm; infl fully exserted from the sheath, its branches tightly erect-appressed; 2n=14. Native of Europe, now a cosmopolitan weed, only occasional with us. (*Festuca b.; F. dertonensis; V. d.*)

4. Vulpia myuros (L.) C. Gmelin. Rat-tail fescue. Erect or ascending annual 1–7 dm; blades slender, commonly involute, 0.5–3 mm wide, usually thinly hairy above; infl a contracted, often rather dense panicle or spike-like raceme 3–25 cm, often not fully exserted from the sheath, the branches variously erect to drooping; spikelets 3–7-fld, 5.5–12 mm (excl. awns); florets only loosely imbricate, the rachilla-joints ca 1 mm; glumes thin, glabrous, subulate, the first 1-veined and mostly 0.5–2.5 mm, the second 3-veined, at least twice as long, 2.3–5.5 mm; lowest lemma 4.5–7 mm, usually scabrous above, with an awn 7.5–22 mm; anther usually 1, under 1 mm; grains 3–4.5 mm; 2n=14, 42. Native of Europe, now a widespread weed in temp. and subtrop. regions. (*Festuca m.; Vulpia megalura,* with ciliate-margined lemmas.)

15. LOLIUM L. Spikelets solitary at each node of the spike, several-fld, placed edgewise to the rachis, the edge fitting into a concavity; disarticulation above the glumes and between the lemmas; first glume none except in the terminal spikelet; second glume (on the side away from the rachis) narrowly lance-oblong, strongly 3–11-veined; lemmas rounded on the back, herbaceous or subcoriaceous, obtuse to acute, awned from the tip or awnless, 3–7-veined, the intermediate veins usually obscure; palea thin; lvs flat, spikes elongate, the summit of each spikelet seldom reaching the base of the next one above it on the same side. 7, temp. Eurasia.

1 Plants perennial .... 1. *L. perenne.*
1 Plants annual.
  2 Florets mostly 11–22; glume herbaceous, distinctly shorter than the spikelet. (Forms of) .... 1. *L. perenne.*
  2 Florets 4–9; glume subcoriaceous, often as long as the spikelet.
    3 Lower lemmas 5–8.5 mm; grain plump, 2–3 times as long as thick; palea mostly shorter than
      the lemma .... 2. *L. temulentum.*
    3 Lower lemmas 8–12 mm; grain slender, 3.5–5 times as long as thick; palea mostly equaling or
      longer than the lemma .... 3. *L. persicum.*

1. Lolium perenne L. Ryegrass. Mostly short-lived perennial; culms slender, 3–9(–12) dm; spike slender, 1–2 dm, the rachis scabrous on the margins and sometimes also on the back; spikelets (2–)5–22-fld; glume

4–12 mm, herbaceous, distinctly shorter than the spikelet; lemmas awned or awnless, the lowest 5–8 mm, the upper progressively reduced; $2n=14, 28$. Native of Europe, cult. in meadows and lawns and often escaped onto roadsides and in waste places throughout our range and w. to the Pacific. Two well marked but wholly intergradient vars., apparently originating as agronomic strains:

Var. *perenne*. English r. Perennial; lvs 2–5 mm wide; rachis of the spike smooth except for the margins; florets mostly (2–)5–10; tip of the glume extending nearly or quite to the top of the lemma just above it; lemmas awnless.

Var. *aristatum* Willd. Italian r. Annual or perennial; stems rough below the spike; lvs to 8 mm wide; rachis rough on the side opposite the spikelet; florets mostly 11–22; tip of the glume seldom extending to the top of the lemma just above it; lemmas, at least the upper, with awns to 15 mm. (*L. p.* var. *italicum; L. multiflorum.*)

2. Lolium temulentum L. Darnel. Annual; culms solitary or few together, 4–10 dm, scaberulous; lvs usually glabrous beneath, scabrous above, 3–10 mm wide; spike 1–2(–4) dm with scabrous margins; spikelets 4–8-fld; glume firm, straight, strongly veined, equaling or surpassing the uppermost lemma, (12–)15–25 mm; lemmas obtuse, with an awn up to 17 mm; or sometimes awnless, the lower lemmas in the spikelet 5–8.5 mm; paleas shorter than the lemmas, grain 2–3 times as long as thick; $2n=14$. Probably native to the Mediterranean region, now widespread as a weed and occasionally found throughout most of our range.

3. Lolium persicum Boiss. & Hohen. Persian darnel. Much like no. 2; spike 5–12(–20) cm; spikelets 4–9-fld; glume 7.5–20 mm, usually surpassed by the uppermost lemma; lower lemmas 8–12 mm; paleas shorter than or merely equaling the lemmas; grain slender, 3.5–5 times as long as thick; $2n=14$. Native of the Middle East, intr. as a weed of wheat-fields and disturbed sites in s. Can. and adj. U.S., notably on the n. Great Plains, but also in Ont. and N.Y. (*L. dorei.*)

16. SCOLOCHLOA Link, nom. conserv. Spikelets large, 3–4-fld, disarticulating above the glumes and between the lemmas; glumes unequal, thin, the first 3-veined, the second 5-veined, about equaling the lemma above it; lemmas lanceolate, acute, 7-veined, very thin at the tip, densely villous on 2 sides at the callus; palea thin, 2-toothed, finely ciliate, equaling the lemma; tall grasses, perennial from a long thick rhizome, with flat elongate blades and widely spreading panicle. 2, the other in Siberia.

1. Scolochloa festucacea (Willd.) Link. Sprangletop, marsh-grass. Stout, erect, 10–18 dm; lvs 5–10 mm wide, elongate to a slender tip; ligule firm, 3–5 mm; panicle 1–2 dm, the branches ascending, bearing spikelets mostly above the middle, the lowest ones much elongate; spikelets 7–11 mm; glumes lanceolate, appressed, the first 5–7.5 mm, the second 6.5–10 mm; lemmas usually lacerate at the tip and with minutely projecting veins, the lowest 6–7 mm, the upper considerably smaller; $2n=28$. Marshes and shallow water; n. Eurasia and nw. N. Amer., s. and e. to Oreg., Mont., Kans., and Io. (*Fluminea f.*)

17. CYNOSURUS L. Dog's-tail grass. Spikelets dimorphic, the sterile and fertile paired; sterile spikelets consisting of 2 glumes and several narrow, scabrous, acuminate lemmas on a continuous rachilla, borne in front of the fertile spikelets and nearly covering them; fertile spikelets 2–4-fld, the rachilla disarticulating; glumes narrow, unequal; lemmas broader, rounded on the back, scabrous, awn-tipped; lvs narrow; spikelets densely crowded in a spike-like or capitate panicle. 4, Europe and Medit. reg.

1. Cynosurus cristatus L. Densely tufted perennial 3–8 dm; lvs few, 1–3 mm wide; infl slender, long-exsert, 3–10 cm; spikelets subsessile in short-peduncled pairs; lemmas 3–3.5 mm, short-awned; $2n=14$. Native of Europe, intr. in fields, roadsides, and waste places nearly throughout our range.

C. echinatus L., an annual with lvs 3–9 mm wide and with a subcapitate panicle 1–4 cm, the lemmas with awns 5–10 mm, is adventive in N.Y. and Md. and may be expected elsewhere.

18. PUCCINELLIA Parl., nom. conserv. Spikelets several-fld, terete or subterete, disarticulating at maturity above the glumes and between the lemmas, the rachilla-joints often elongate; first glume obtuse to acute, 1(3)-veined; second glume 3-veined, but the veins inconspicuous; lemmas exceeding the glumes, ovate to oblong, mostly rounded on the back, awnless, obscurely to evidently 5-veined, the summit often obtuse or truncate, thin, often hyaline and finely erose-ciliate; style wanting; lf-sheaths open for most of their length; perennial or less often annual grasses with glabrous herbage, most spp. occurring in alkaline or saline soils. (*Torreyochloa.*) 30, N. Temp. and Arctic.

1 Lemmas prominently 5-veined; plants not halophytic. (*Torreyochloa*.) .... 1. *P. pallida.*
1 Lemmas inconspicuously or obscurely veined; plants halophytic. (*Puccinellia* proper.)
   2 Anthers linear, 1.5–2.5 mm; lemmas mostly 3–4.5 mm .... 2. *P. maritima.*
   2 Anthers oblong, 0.3–1.2 mm; lemmas often under 3 mm.

3 Infl contracted, rather dense, the lowest branches bearing spikelets nearly or quite to the base; midvein of the lemma reaching the tip .... 3. *P. fasciculata.*
3 Infl open or contracted, the lowest branches all or chiefly sterile below the middle; midvein of the lemma not reaching the tip.
   4 Distal margin of the lemma prominently and regularly scabrous; panicle-branches prominently scabrous.
      5 First lemma 2.1–3.5 mm, acute to narrowly (seldom broadly) obtuse; lower branches of the infl erect or ascending to less often spreading or descending at maturity .... 4. *P. nuttalliana.*
      5 First lemma 1.7–2.5 mm, broadly obtuse to truncate; lower branches of the infl horizontal to descending at maturity .... 5. *P. distans.*
   4 Distal margin of the lemma entire, repand-erose, or with a few scattered scabrules, acute to obtuse; panicle-branches glabrous or only sparsely (rarely prominently) scabrous .... 6. *P. pumila.*

1. **Puccinellia pallida** (Torr.) R. T. Clausen.   Culms 3–10 dm, slender and weak, usually ± decumbent and creeping at base; lvs soft, 2–10 mm wide; ligules 2.5–9 mm; infl 5–15 cm, with relatively few branches, eventually diffuse; spikelets narrowly ovate, 4–7 mm, 4–6-fld; glumes broadly rounded at the scarious tip, the first 0.9–2.1 mm, the second 1.1–2.4 mm; lemmas 2–3.5 mm, ovate, sharply 5-veined, finely hairy or scaberulous, erose at the rounded tip; palea 4–5 times as long as wide; anthers 0.3–1.5 mm; 2*n*=14. Swamps and shallow water; Nf. to Minn., s. to N.C. and Mo. (*Torreyochloa p.; Glyceria p.; P. fernaldii,* plants at the small end of the range in all dimensions.)

2. **Puccinellia maritima** (Hudson) Parl.   Seaside alkali-grass. Coarse perennial, tufted and sometimes stoloniferous, with erect culms 3–8 dm; lvs 1–3 mm wide, often involute; ligules 1–2.5 mm; infl 5–25 cm, usually narrow, with ascending, smooth to scabrous branches, the lowest all or chiefly naked at base; spikelets 5–12 mm, 5–11-fld; glumes ovate, narrowed to an obtuse tip, the first 2–3.5 mm, the second 2.8–4 mm; lemmas firm, obscurely veined, 3–4.5 mm, tapering to a thin rounded summit, usually hairy toward the base; anthers linear, 1.5–2.5 mm; 2*n*=14–77, most often 56. Salt-marshes and shores; N.S. to R.I., and occasionally adventive s. to Phila.; Europe.

3. **Puccinellia fasciculata** (Torr.) E. Bickn.   Coarse, tufted perennial, or sometimes annual or biennial, 2–6 dm; lvs 2–5 mm wide, often involute in age; ligules 1–2 mm; infl narrowly ovoid, 4–15 cm, becoming long-exsert, its lowest branches floriferous to the base; spikelets 3–4 mm, 2–5-fld; first glume ovate, acute, 1 mm, the second broadly ovate, subacute, 1.2–1.8 mm; lemmas 1.9–2.4 mm, ovate-oblong, minutely hairy at the base, obscurely veined, but the midvein reaching the broadly obtuse tip; anthers 0.4–0.8 mm; 2*n*=28. Salt-marshes and sandy shores; Mass. to Va.; Europe. (*P. borreri.*)

4. **Puccinellia nuttalliana** (Schultes) A. Hitchc.   Nuttall's alkali-grass. Tufted, erect perennial 2–8 dm; lvs 1–3 mm wide, often involute; ligules 1–3 mm; infl 5–25 cm, narrowly contracted to broadly ovoid, the scabrous branches erect or ascending to less often spreading or descending at maturity, the lowest bearing spikelets chiefly above the middle; pedicels very scabrous; spikelets slender, 4–8 mm, 3–9-fld; glumes narrowly ovate, the first 1–2.4 mm, the second 1.6–2.9 mm; lemmas broadly oblong, obscurely veined, minutely hairy at base, scabrous on the margins toward the acute to narrowly (seldom broadly) obtuse tip; first lemma 2.1–3.5 mm; anthers 0.7–1.2 mm; 2*n*=28, 42, 56. Alkaline places; Wis. to B.C., s. to Kans., N.M., and Calif., and occasionally intr. e. to Me. and N.Y. (*P. airoides.*)

5. **Puccinellia distans** (Jacq.) Parl.   European alkali-grass. Tufted perennial 1–5 dm, decumbent at base; lvs 1–4 mm wide, flat, often becoming involute; ligules mostly under 2 mm; infl ovoid, 6–20 cm, loosely branched, its branches and pedicels scabrous, the lower branches spreading or reflexed, usually bearing spikelets only beyond the middle; spikelets 3–5 mm, 3–7-fld; glumes ovate, the first 0.7–1.8 mm, the second 1.2–2 mm; lemmas elliptic to obovate, obscurely veined, glabrous or short-pilose at base, scabrous on the margins toward the broadly obtuse to truncate tip; first lemma 1.7–2.5 mm; anthers 0.7–1.1 mm; 2*n*=14, 28, 42. Native of Europe, naturalized along the coast from N.B. to Del., and becoming common inland especially along highways in our range, w. to the Pacific.

6. **Puccinellia pumila** (Vasey) A. Hitchc.   Tufted perennial 1–5(–7) dm, sometimes stoloniferous in late season; lvs flat, 1–2 mm wide; ligules 1–3 mm; infl 3–20 cm, its branches usually smooth or nearly so (as also the pedicels), the upper erect, the lower often spreading or descending, usually bearing spikelets mainly or only beyond the middle; spikelets 3–6-fld; glumes ovate, the first 1–2.7 mm, the second 1.5–4 mm; lemmas ovate to obovate, obscurely veined, glabrous or short-pilose at base, acute to obtuse, the distal margins entire or erose or with a few scattered scabrules; first lemma 2.4–4.2 mm; anthers 0.7–1.5 mm; 2*n*=42, 56. Salt-marshes and gravelly beaches along the coast; Conn. to Lab. and Hudson's Bay; Calif. to Aleut. Isl. (*P. paupercula.*) The measurements above apply to the usual form of the sp. A taller, perhaps not persistent form from P.E.I., with diffuse infl and much elongate glumes and lemmas (up to 10 mm) has been called *P. paupercula* var. *longiglumis* Fern. & Weatherby. It may be teratological.

19. **BRIZA** L.   Quaking grass. Spikelets 4–20-fld, disarticulating above the glumes and between the lemmas, broad and laterally compressed, orbicular to cordate or triangular; glumes very broad, 3–9-veined, chartaceous, scarious-margined; lemmas closely imbricate, cordate, broader than long, 7–9-veined, chartaceous, the scarious margins flaring over the lemmas opposite; uppermost florets reduced; palea shorter than the lemma; panicle with

very slender branches and pedicels bearing usually nodding, often purplish or brownish spikelets. 20, N. Temp., S. Amer.

    1. Briza media L. Perennial q. g. Loosely tufted perennial 3–6 dm; lvs 2–4 mm wide, scabrous-margined; ligule truncate, ca 1 mm; infl 5–15 cm, lax, freely branched; spikelets numerous, round-ovate, 4–7 mm, 4–12-fld; glumes 2.5–3.5 mm, 3–5-veined; lemmas ca 4 mm; $2n=14, 28$. Native of Europe, sparsely intr. in meadows and waste places in the n. part of our range.

    2. Briza minor L. Little q. g. Annual 1–5 dm, the culms solitary or clustered; lvs 3–8 mm wide, rough on the margins and often also on the surface; ligule lanceolate, 3–6+ mm; infl 5–15 cm, lax, freely branched; spikelets numerous, triangular-ovate, 3–5 mm, 4–8-fld; glumes 2–3.5 mm, 3–5-veined; lemmas 1.5–2 mm; $2n=10$. Native of Europe, occasionally found in waste places in our range from se. N.Y. southward.

    3. Briza maxima L. Big q. g. Annual 2–6 dm; lvs 3–8 mm wide; ligule 2–5 mm; infl to 10 cm, secund, nodding above, with few (up to 12) spikelets, these ovate, 12–20 mm, 7–20-fld; glumes 5–7 mm, 5–9-veined; lemmas 6–8 mm; $2n=14$. Native of the Mediterranean region, cult. and rarely escaped in our range.

    20. POA L. Bluegrass. Spikelets 2–several-fld, ± compressed laterally, disarticulating above the glumes and between the lemmas; glumes lanceolate to ovate, acute or subacute, 3-veined or the first 1-veined; lemma in most spp. keeled, often scarious at the margins and tip, generally 5-veined (but the intermediate veins obscure or obsolete in some spp.), often with a tuft of long, cobwebby hairs (called a web) at the base; uppermost florets reduced, unisexual or vestigial; rachilla in a few spp. finely puberulent or scaberulous, otherwise glabrous; lvs ending in a boat-shaped tip; panicles open or contracted, the branches generally in fascicles of 2–5, sometimes more. 150+, widespread, mostly temp. and boreal. Many of the spp. confluent through polyploid (often apomictic) forms.

17 Lemmas apparently 3-veined, the intermediate veins obscure or obsolete.
  18 Ligule of cauline lvs (2-)3-5 mm, usually ovate or triangular; panicle open;
     rachilla glabrous .... 14. *P. palustris.*
  18 Ligule of cauline lvs truncate, 0.2-1(-1.5) mm.
    19 First glume nearly as wide as but obviously shorter than the first lemma;
      rachilla merely scaberulous at 30×; panicle often narrow and with
      ascending branches, sometimes more open .... 15. *P. interior.*
    19 First glume much narrower than but nearly as long as the first lemma;
      rachilla finely puberulent at 30×; panicle at maturity open, with
      spreading branches .... 16. *P. nemoralis.*
8 Lemmas not webbed.
  20 Infl slender, elongate, mostly 3-5 times as long as wide.
    21 Lemmas sharply keeled.
      22 Relatively small, slender plants, 2-5 dm, with lvs 1-3 mm wide; keel and marginal
        veins of the lemma villous .... 17. *P. glauca.*
      22 Coarse, tall plants, 5-12 dm, with lvs 6-10 mm wide; keel and marginal veins of
        the lemma scabrous to subglabrous .... 18. *P. chaixii.*
    21 Lemmas scarcely keeled, crisp-puberulent on the lower half, sometimes more strongly
      so on the main veins .... 19. *P. canbyi.*
  20 Infl at anthesis broad and open, not over twice as long as wide.
    23 Lemmas distinctly 5-veined; lower glume 1-veined .... 20. *P. autumnalis.*
    23 Lemmas apparently 3-veined, the intermediate veins obscure or obsolete; lower glume
      3-veined.
      24 First glume lanceolate, less than half as wide as long .... 21. *P. fernaldiana.*
      24 First glume ovate, more than half as wide as long .... 22. *P. alpina.*

1. Poa pratensis L.   Kentucky-b. Rhizomatous, forming a dense sod, or in tufts on long rhizomes in open ground; culms 3-10 dm; sheaths usually glabrous; blades soft, 2-5 mm wide, seldom narrower and involute; ligule shorter than wide; infl ovoid, fairly dense, with spreading or ascending branches, the lower mostly in sets of 5 or 4; spikelets 3-5-fld; with very short rachilla-joints; first glume 1.8-2.9 mm, the second 2.3-3 mm; lemmas distinctly 5-veined, thinly to densely hairy on the veins below but glabrous between them, webbed at base, the lowest 2.5-3.5 mm; anthers 1-1.4 mm; $2n=21-147$. Moist or dry soil, avoiding acid soils and heavy shade, throughout the U.S. and far n., often cult. in lawns and meadows; in most of our range intr. from Europe, but probably native along our n. boundary and in Can. (*P. angustifolia; P. subcaerulea.*) Some nematode-infested plants in the ne. part of our range have notably larger, 7-veined lemmas and often a reduced and stiff infl.

2. Poa arctica R. Br.   Arctic b. Rhizomatous, 1-4 dm, the culms in small tufts erect from a decumbent base; lvs flat or folded, erect, 1-3 mm wide, much shorter than the stems; ligule ovate, 1-3 mm; infl broadly ovoid, loosely branched, the branches paired, slender, spreading; spikelets ovate to oblong, 5-8 mm, 3-6-fld; first glume lanceolate, acute, 2.6-5.2 mm, the second ovate, 3-5.5 mm; lemmas 3.4-5.5 mm, usually purple, strongly suffused with orange below the scarious tip, pubescent on and between the rather obscure veins, especially toward the base, with or without a small web; anthers 1.5-2.3 mm; $2n=38, 56, 72, 76$. Circumboreal in cold meadows and on cold, stony slopes, s. to the Rocky Mts. and reputedly to N.S. and the mts. of Me. and N.H. (*P. alpigena,* at least as to our plants.)

3. Poa cuspidata Nutt.   Rhizomatous, 3-6 dm, the culms in large loose tufts; lower sheaths hairy at least at the top; lvs 2-4 mm wide; ligule shorter than wide, broadly rounded or truncate; infl loose and open, ovoid, its slender, widely divergent branches usually paired, bearing a few spikelets above the middle; spikelets 5-8 mm, 3-4-fld, the rachilla-joints 0.8-1.3 mm; first glume 2.3-4.3 mm, the second 3.1-4.8 mm; lemmas distinctly 3-veined (the intermediate veins obscure), ± hairy on the veins and glabrous between them, webbed at base; lowest lemma 3.6-5.2 mm; anthers 2.1-2.8 mm; $2n=28$. Moist woods; N.J. to s. Ind., s. to Ga. Early spring.

4. Poa compressa L.   Canada-b. Culms erect, 2-7 dm from long running rhizomes, not tufted, strongly flattened, 2-edged; lvs 2-4 mm wide; the cauline ones seldom over 1 dm; ligule short, mostly 1-2 mm; infl usually compact and narrow, bluish- or grayish-green, 2-8 cm, its branches usually paired, bearing spikelets to near the base; pedicels of the lateral spikelets 0.5 mm; spikelets 4-6 mm, 3-6-fld, first glume 1.7-2.4 mm, the second 1.8-2.8 mm; lemmas firm, 2-2.8 mm, with or without a small basal web, slightly pubescent on the keel and marginal veins below, the intermediate veins obscure; anthers 0.9-1.7 mm; $2n=14-56$. Native of Europe, established in open, usually dry, variously acid or (along highways) alkaline soil throughout our range, to Alas., Ga., and the Pacific.

5. Poa arida Vasey.   Plains-b. Tufted with short rhizomes, erect, 2-6 dm; lvs stiff, erect, usually folded, 1-2 mm wide; ligule 1.5-4 mm; infl compact, 3-12 cm, 3-4 times as long as wide; spikelets 5-9 mm, 4-7-fld; first glume 2-4 mm, the second 2.6-4.4 mm; lemmas hairy, often densely so, on and between the veins toward the base, but not webbed; lowest lemma 2.4-4.3 mm; anthers 1.2-2 mm; $2n=63-103$. Dry to moist prairies; w. Minn. and nw. Io. to Mont., Utah, and Ariz. (*P. pratensiformis.*)

6. Poa bulbosa L.   Bulbous b. Culms erect from a bulbous-thickened base, 2-5 dm, purplish below; panicles compact and crowded, ovoid, 4-8 cm; florets mostly converted into turgid purplish bulblets, some of the scales prolonged into linear tips 5-15 mm; $2n=14, 28, 35, 56$. Native of Europe, widely intr. in w.

U.S. and in our range occasionally found in fields, lawns, and meadows from N.Y. to N.C., w. to Ill. and s. Ont.

7. Poa trivialis L. Rough b. Culms slender to stout, erect from a decumbent (even shortly stoloneous) base, 5–10 dm, without rhizomes, smooth or often scaberulous below the infl; sheaths finely scabrous; blades soft, 2–6 mm wide, the cauline ones often 15 cm; ligule (2.5–)3–7+ mm, panicle soon long-exsert, ovoid or oblong, the ascending branches in sets of 5–8, with numerous crowded spikelets; pedicels scabrous; spikelets ovate or elliptic, 2-fld and 2.7–3.2 mm or 3-fld and 3–4 mm; glumes lanceolate, incurved, the first 1.7–2.9 mm, the second 2–3.3 mm; lemmas thin, narrowly ovate, sharply 5-veined, acute or acuminate, glabrous except the scaberulous keel and webbed base; anthers 1–2 mm; $2n=14, 28$. Native of Europe, intr. in meadows, moist woods, and along roadsides from Nf. to Minn. and S.D., s. to N.C. and Ill., and in the Pacific states.

8. Poa alsodes A. Gray. Culms loosely tufted, slender, 3–10 dm, without rhizomes; sheaths glabrous; lvs thin, soft, 2–4 mm wide, the cauline ones rarely over 10 cm; ligule 0.7–2.2(–3) mm; infl very lax, its branches mostly in 4's and 5's, the larger ones elongate, with 1–few spikelets at the tip; base of the infl tardily exsert from the sheath, the branches therefore persistently ascending; pedicels smooth or scaberulous; spikelets ovate, 2–3-fld, the 2-fld ones 3–4.5 mm, the 3-fld to 5.6 mm; first glume lanceolate, 2.1–3 mm, the second broader, 2.4–3.7 mm; lemmas thin, lance-ovate, acute, faintly 5-veined, 2.5–4.1 mm, glabrous except the webbed base and pubescent lower part of the keel. Moist woods; N.S. and Me. to Minn., s. to Del., Ind., and in the mts. to N.C. and Tenn.

9. Poa languida A. Hitchc. Culms slender, usually weak, 3–10 dm, without rhizomes; sheaths glabrous or nearly so; blades soft, 2–5 mm wide; ligule (2–)2.5–4 mm; infl loose, ± nodding, 5–10 cm, the slender branches bearing a few spikelets beyond the middle, the lower branches usually paired, seldom solitary or in 3's; spikelets ovate, 3–4 mm, 2–4-fld; glumes acute, the first lanceolate to ovate, 1.7–2.6 mm, the second broadly lanceolate to ovate or elliptic, 2.1–2.9 mm; lemmas firm, obscurely veined, oblong, 2.4–3.2 mm, glabrous except the webbed base, obtuse and somewhat cucullate at the tip; anthers 0.7–1 mm. Dry or rocky woods; Mass. to Minn., s. to Pa., Ky., and Io.

10. Poa saltuensis Fern. & Wieg. Much like no. 9; ligule 0.6–1.5(–3) mm; spikelets 3.5–5.5 mm, 3–5-fld; lemmas green, membranous, distinctly 5-veined, lance-ovate, 2.4–3.9 mm, acute or acuminate at the scarious tip; anthers 0.9–1.5 mm; $2n=28$. Woods and clearings; Nf. to n. Mich. and Minn., s. to Pa. and in the mts. to Va.

11. Poa wolfii Scribn. Culms slender, erect, 4–8 dm, without rhizomes; lvs 1–2 mm wide, the longer ones crowded toward the base of the stem, the upper only 3–8 cm; ligule round-ovate, 1–2 mm; infl loose and open, often nodding, the slender branches paired, bearing spikelets above the middle; spikelets 4–6 mm, 2–4-fld; glumes narrowly ovate, obtuse, 3-veined, the first 2.5–3.5 mm; the second 3–3.8 mm; lemmas 2.5–4.5 mm, distinctly 5-veined, villous on the keel and marginal veins and webbed at the base; anthers 1.1–1.4 mm; $2n=28$. Moist woods; O. to Minn., Mo., and e. Nebr.

12. Poa paludigena Fern. & Wieg. Marsh-b. Culms slender and weak, 2–6 dm, without rhizomes; sheaths scaberulous; lvs to 10 cm, 1–2 mm wide; ligule truncate, to 1(–1.5) mm; infl lax, 5–15 cm, the slender branches paired, bearing a few spikelets above the middle; spikelets 3.5–4.5 mm, 2–5-fld; glumes lanceolate, scarious-margined, the first 1.7–2.2 mm, the second 2–2.8 mm; lemmas 2.5–3.3 mm, short-villous on the keel and marginal veins, glabrous on the obscure intermediate veins, webbed at base; anthers 0.5–0.7 mm. Bogs and wet woods; N.Y. and Pa. to Wis., Ill., and e. Minn.; rarely collected.

13. Poa sylvestris A. Gray. Forest-b. Culms erect, 4–8 dm, without rhizomes; lvs soft, mostly cauline, the main ones 3–5 mm wide; ligule 1 mm; infl rather narrow but open, oblong, 1–2 dm, its slender flexuous branches in sets of 4–8, soon divaricate or reflexed, bearing a few spikelets well beyond the middle; spikelets 2–5-fld; glumes scarious-margined, acute, the first lanceolate, 1.5–2.7 mm, the second oblong, 1.9–3.4 mm; lemmas distinctly 5-veined, 2.1–3.5 mm, villous-puberulent on the marginal veins, at least toward the base, and nearly or quite to the tip of the keel, often also between the veins and on the intermediate pair, webbed at base; anthers 1.1–1.4 mm; $2n=28$. Moist woods; N.Y. to Wis. and Io., s. to Fla. and Tex.

14. Poa palustris L. Fowl meadow-grass. Culms stout, loosely tufted from a purplish, often somewhat decumbent base, 5–15 dm, without rhizomes; lvs 1–2 mm wide; ligule ovate or triangular, 2.5–5 mm; infl declined or nodding, 1–3 dm, fairly open, its branches in sets of ca 5, bearing numerous spikelets at and beyond the middle; spikelets 2–4-fld; glumes lance-ovate, the first 1.9–2.7 mm, the second 2–3.1 mm; lemmas 2.1–3 mm, 3-veined or with an obscure additional pair of veins, webbed at base, hairy on the lower half or two-thirds of the veins, glabrous or scabrous toward the golden tip; rachilla glabrous; anthers 0.8–1.2 mm; $2n=28, 42$. Wet meadows and damp soil; circumboreal, s. to Va., Mo., and N.M.

15. Poa interior Rydb. Inland-b. Tufted, stiffly erect, 3–8 dm, without rhizomes; lvs mostly crowded toward the base, erect, 1–2 mm wide; ligule truncate, 1 mm; infl 5–10 cm, narrow, its branches ascending; spikelets resembling those of no. 14; first glume nearly as wide as but distinctly shorter than the first lemma; rachilla merely scaberulous at $30\times$; $2n=28, 42, 56$. Dry woods; widespread in the cordilleran region and on the n. Great Plains, e. to Minn., Mich., Vt. and Que.

16. Poa nemoralis L. Wood-b. Culms slender, tufted, 4–8 dm, without rhizomes; lvs lax, 1–2 mm wide, often divergent, often some of them above the middle of the shoot, the upper scarcely shorter than the lower; ligule truncate, 0.5 mm; infl narrowly ovoid, 1–2 dm, eventually loose and open, the slender branches in sets of ca 5, bearing spikelets in the distal half; spikelets 2–4-fld; glumes narrowly lanceolate, long-acuminate, the first 2.2–3 mm, conspicuously narrower than but nearly equaling the first lemma, the second 2.3–3.3

mm; lemmas apparently 3-veined (the intermediate veins obscure), 2.1–3.1 mm, webbed at base, the straight keel and marginal veins sericeous or villosulous below; glabrous or scabrous in the distal third; rachilla evidently puberulent at 30×; anthers 1.2–1.6 mm; $2n=28$–70. Native of Europe, intr. across the continent in s. Can., and occasionally found in moist meadows and along roadsides throughout most of our range.

17. Poa glauca Vahl. Culms clustered, 2–5 dm, erect, without rhizomes; lvs glaucous, lax, 1–3 mm wide, the uppermost one at or below the middle of the shoot; ligule 1–2 mm; infl long-exsert, narrow, rather dense, 3–12 cm, the branches in sets of 2–5, at first ascending, later spreading, each with a few 2–4-fld spikelets; glumes lanceolate, subequal, 2.3–3.8 mm, less than half as wide; lemmas 2.5–3.5 mm; densely sericeous or villosulous on the lower half of the keel and marginal veins, not webbed, the intermediate veins obscure; rachilla puberulent at 30×; anthers 1–2.2 mm; $2n=42$, 49, 56, 63, 70. Gravelly or rocky soil and cliffs; circumboreal, s. to Vt., n. Mich., and ne. Minn. (*P. glaucantha; P. scopulorum.*)

18. Poa chaixii Villars. Tall b. Culms 5–12 dm, compressed below, without rhizomes; sheaths compressed, rough, the uppermost much elongate; ligule truncate, 1 mm; blades 6–10 mm wide, abruptly narrowed at the tip; infl elongate, the lower branches in 5's; spikelets 4–5-fld; first glume 2–3 mm, 1-veined, the second 3–4 mm, 3-veined; lemmas sharply 5-veined, scabrous on the veins and sometimes also between them, not webbed; rachilla glabrous or nearly so; $2n=14$. N. Europe; locally intr. with us, as in N.Y., about Ottawa, and about Duluth, Minn.

19. Poa canbyi (Scribn.) Howell. Canby-b. Culms tufted, 4–10 dm, often purplish below, without rhizomes; lvs forming large basal rosettes 15–30 cm high, usually involute and 1–2.5(–3) mm wide, the few cauline ones 5–7 cm; ligule 3–5 mm; infl narrow and ± dense, (7–)9–16(–20) cm, with short, erect branches, at least some of which bear spikelets to near the base; spikelets 2–5-fld; glumes acuminate, often scabrous, the first lanceolate, 2.4–3.7 mm, the second oblong, 2.9–4.5 mm; lemmas 4–5.5 mm, obscurely 5-veined, scarcely keeled, crisp-puberulent on the lower half, sometimes more strongly so on the central and marginal veins, not webbed; rachilla usually finely puberulent at 30×; $2n=72$–106. Dry soil; e. Que.; n. Mich. and Minn. to Man. and Neb., w. to Alas., Calif., and Tex.

20. Poa autumnalis Vahl. Culms in large loose tufts, 3–6 dm, without rhizomes, the soft blades crowded near the base, 2–3 mm wide, the cauline ones narrower and short; infl 6–15 cm, very loose and open, the elongate, capillary branches single or paired, widely spreading, bearing a few distal spikelets; spikelets 4–6-fld, the rachilla joints 1–1.3 mm, glabrous; first glume lanceolate, 1.8–2.8 mm, 1-veined, the second ovate or elliptic, 2.2–3.7 mm, 3-veined; lemmas oblong, 3.2–4.4 mm, usually distinctly 5-veined, short-hairy on the keel and marginal veins at least below the middle, usually also on the intermediate veins and interveinal areas near the base, scarious at the broadly rounded or subtruncate tip, not webbed; anthers ca 1 mm; $2n=28$. Moist woods; N.J. to Mich. and Ill., s. to Fla. and Tex.

21. Poa fernaldiana Nannf. Tufted, 1–2.5 dm, without rhizomes, leafy chiefly below the middle; lvs soft, 1–2 mm wide; ligule 2.5–3.5 mm; infl loose and few-fld, the branches usually paired, spreading or ascending, bearing usually a single spikelet; spikelets 4–6 mm, 2–4-fld; glumes lanceolate, tapering to the tip, 3-veined, the first 2.6–4.2 mm, the second 2.8–3.4 mm; lemmas 2.8–4.3 mm, obtuse, obscurely 5-veined, the keel and marginal veins stronger than the intermediate ones and densely sericeous below the middle, not webbed, or only scantily so; anthers 0.7–0.9 mm; $2n=42$. Summits of higher mts.; Nf. and e. Que. to Mt. Marcy, N.Y. (*P. laxa,* misapplied.)

22. Poa alpina L. Alpine b. Culms 1–4 dm, scarcely tufted, without rhizomes; lvs mostly on short basal shoots, 3–5 cm × 1.5–4 mm; cauline lvs 1 or 2, erect, 1–5 cm × 2–3 mm; ligule (1–)2–4 mm; panicle long-pedunculate, pyramidal, 3–6 cm, nearly as wide, its branches mostly paired, soon divaricately spreading; spikelets 4.5–6(–7) mm, (3)4–6-fld; glumes ovate, over half as wide as long, herbaceous-margined, the first 2.4–3.9 mm, the second 2.6–4.5 mm; lemmas 2.9–4.5 mm, spreading-hairy on the lower half of the keel and marginal veins, and to some extent in the interveinal areas below, not webbed, the intermediate veins obscure; anthers 1.5–2.2 mm; $2n=14$–74. Calcareous shores and ledges; circumboreal, s. to N.S., Que., the Bruce Peninsula of Ont., and Keweenaw Point, Mich. (*P. gaspensis.*)

23. Poa annua L. Speargrass. Tufted annual with few–several decumbent to ascending or seldom erect culms to 3 dm, sometimes rooting at the lower nodes; sheaths loose; blades soft, 2–3 mm wide; infl ovoid, 2–8 cm, with few ascending branches bearing rather crowded spikelets above the middle; spikelets green, 3–5 mm, 3–6-fld; glumes broadly lanceolate, acute, scarious-margined, indistinctly veined, the first 1.5–2.4 mm, the second 1.8–2.8 mm; lemmas thin, elliptic, 5-veined, obtuse, ± hairy on the veins only, not webbed, the lowest one 2.4–3.4 mm; anthers 0.8–1 mm; $2n=14$, 28. Native of Eurasia; abundant in moist soil nearly throughout the U.S. and n. to Lab. and Alas.; often a lawn-weed.

24. Poa chapmaniana Scribn. Much like no. 23, but the culms usually erect or strongly ascending and the sheaths close; spikelets avg a little smaller, the first glume 1.5–2.1 mm, the second 1.7–2.4 mm; lemmas hairy on the keel and marginal veins, and also webbed at base, the intermediate veins obscure; lowest lemma 1.6–2.5 mm; anthers 0.1–0.2 mm. Open ground and fields; Del. to Io., s. to the Gulf, and occasionally introduced northward. May, June.

21. DACTYLIS L. Orchard-grass. Spikelets few-fld, flat, disarticulating above the glumes and between the lemmas; glumes unequal, nearly as long as the lemmas, 1–3-veined, keeled; lemmas obtuse or acute, usually mucronate or short-awned, keeled, 5-veined, the intermediate veins often inconspicuous; uppermost lemmas reduced; infl open, the lower branches naked at base, erect or divergent, the branches and main axis bearing numerous short-pediceled spikelets in dense one-sided clusters near the ends; perennial tufted grasses with compressed sheaths, long ligules and flat blades. Monotypic.

1. **Dactylis glomerata** L.   Culms 5–12 dm; sheaths scaberulous; ligules 5–7 mm; blades elongate, 3–8 mm wide; infl long-exsert, 1–2 dm; spikelets 3–6-fld; glumes lance-acuminate, usually ciliate on the keel; lemmas 5–8 mm, usually ciliate on the keel, awnless or with a terminal awn to 1 mm; $2n=14, 28$. Native of Europe, intr. in moist fields, meadows, lawns, and roadsides throughout our range and most of N. Amer.

22. **CATABROSA** P. Beauv.   Brook-grass. Spikelets mostly 2-fld, disarticulating above the glumes and between the lemmas; rachilla-joint elongate; glumes unequal, obscurely nerved, the second erose at the top; lemmas strongly 3-veined, broadly truncate and ± erose at the scarious summit; palea resembling the lemma, strongly 2-veined; perennial aquatic grasses, prostrate and rooting at the lower nodes, with overlapping sheaths, flat soft blades, and open panicles. 7, mostly in n. Eurasia.

1. **Catabrosa aquatica** (L.) P. Beauv.   Culms 1–7 dm; lvs 3–8(–10) mm wide; infl 8–15 cm, the finely hairy branches divergent, in fascicles of ca 5; first glume ovate, 1–1.7 mm; second glume oblong, truncate and usually erose, 1.6–2 mm; lemmas broadly oblong, 2.2–3 mm, glabrous; anthers 1–1.3 mm; $2n=20$. Shallow water or wet soil; circumboreal, s. to N.B., Wis., Nebr., and Ariz.

23. **PARAPHOLIS** C. E. Hubbard.   Spikelets 1-fld, solitary, alternate, sessile and sunken in concavities of the rachis, forming a terete slender spike; glumes 2, borne side by side and covering the cavity except at anthesis, strongly 5-veined, coriaceous, the outer margins indurate, the overlapping inner margins scarious; lemma lanceolate, scarious, concealed by and much shorter than the glumes, with its back to the rachis, 3-veined, the lateral veins very short; palea scarious, barely shorter than the lemma; rachis disarticulating at maturity below each spikelet; annuals with overlapping sheaths and elongate spikes. 5, Eurasia.

1. **Parapholis incurva** (L.) C. E. Hubbard.   Densely tufted, the culms 5–25 cm, upcurved from a decumbent base; blades 1–5 cm; ligule to 1 mm; spike elongate, curved, to 10 cm, 2 mm thick, with 10–20 spikelets, the base included in the upper sheath; glumes 4.5–7 mm, in the uppermost spikelet opposite; anthers 0.5–0.9 mm; cleistogamous; $2n=28$. Muddy seashores and marshes; native of Europe, intr. in our range along the coast from N.J. to N.C. (*Pholiurus i.*)

24. **GLYCERIA** R. Br., nom. conserv.   Mannagrass. Spikelets 3-fld, ovate to oblong or cylindric, subterete or moderately flattened, disarticulating above the glumes and between the lemmas; glumes unequal, shorter than the lemmas, 1-veined, often scarious at the margins and tip; lemmas awnless, lance-ovate to elliptic or obovate, usually obtuse, rounded on the back, usually scarious at the tip, the (5)7(9) veins parallel, often raised; paleas elliptic to obovate, becoming indurate; stigmas elevated on naked slender styles; perennials of marshes, shallow water, or wet ground, erect or decumbent at base and often rooting from the lower nodes, with flat or folded blades, closed sheaths, and freely branched, often ample panicles; $x=10$. (*Panicularia.*) 35, mainly N. Temp. For *G. pallida*, see *Puccinellia pallida*.

1 Spikelets linear-cylindric, or becoming flattened at maturity, 10–40 mm, (5–)8–16-fld; joints of the
  rachilla 1–4 mm; stamens 3.
  2 Lemma blunt or rounded, the palea not projecting more than 0.5(–1) mm beyond it.
    3 Lemmas glabrous and shining between the veins, 3.1–3.9 mm; anthers 0.6–1 mm .... 1. *G. borealis.*
    3 Lemmas dull and scabrous between the veins; anthers 1–3 mm.
      4 Rachilla-joints 1.1–1.8 mm; lemmas 3.7–5.3 mm; anthers 1–1.7 mm .... 2. *G. septentrionalis.*
      4 Rachilla-joints 1.9–2.5 mm; lemmas 5.4–7 mm; anthers 2–3 mm .... 3. *G. fluitans.*
  2 Lemma acute, the palea projecting 1.5–3 mm beyond it .... 4. *G. acutiflora.*
1 Spikelet ovate or oblong, somewhat flattened, 3–10 mm, 3–10-fld; joints of the rachilla under 1 mm;
  stamens 2 except in no. 10.
  5 Panicle strict; ligule under 1 mm.
    6 Infl ovoid or oblong, dense, the 2 lowest internodes each less than 2.5 cm and usually less than
      2 cm .... 5. *G. obtusa.*
    6 Infl very slender, the 2 lowest internodes each 2–7 cm .... 6. *G. melicaria.*

5 Panicle lax and diffuse; ligule 2–6 mm, or only 1 mm in no. 9.
  7 Spikelets 3–4 mm wide; veins of the lemma visible but not raised; palea less than twice as long
    as wide.
    8 Lemma conspicuously projecting beyond the palea; palea round-obovate, its sides not com-
      pletely covered by the lemma .... 7. *G. canadensis.*
    8 Lemma equaling the palea in length and width, or barely longer .... 8. *G.* × *laxa.*
  7 Spikelets 2–2.5 mm wide; veins of the lemma prominently raised; palea 3–5 times as long as
    wide.
    9 Spikelets mostly 2.5–4 mm; glumes obtuse, minute, the first 0.5–1 mm, the second 0.8–1.3
      mm; stamens 2 .... 9. *G. striata.*
    9 Spikelets mostly 4–6.5 mm; glumes acute, the first 1.2–1.9 mm, the second 1.5–2.4 mm;
      stamens (2)3 .... 10. *G. grandis.*

1. Glyceria borealis (Nash) Batchelder.   Northern m. Culms 8–12 dm, often decumbent and rooting from the lower nodes; lvs 2–5 mm wide; ligule 4–12 mm; infl slender, 2–5 dm, the branches 8–12 cm, each with several appressed spikelets; spikelets 1–2 cm, 8–12-fld, at first cylindric, the lemmas tending to spread at maturity; rachilla-joints 1–1.3 mm; glumes elliptic, rounded at the tip, soon scarious, obscurely veined, the first 1.1–2.1 mm, the second 1.9–3.3 mm; lemmas obtuse, scarious at the margins and tip, 3.1–3.9 mm, green and shining between the minutely scaberulous veins; palea not projecting; anthers 3, 0.6–1 mm; grain 1–1.2 mm; $2n=20$. In shallow water and wet soil; Nf. to Alas., s. to Conn., N.J., Ind., Io., Nebr., and Calif.

2. Glyceria septentrionalis A. Hitchc.   Eastern m. Culms 10–15 dm, rather soft, often decumbent and rooting from the lower nodes; main lvs 6–10 mm wide; ligule prominent; infl narrow, 2–4 dm, often somewhat spreading at anthesis; spikelets subsessile or short-pedicelled, 1–2 cm, 8–16-fld, the lemmas spreading at maturity; rachilla-joints 1.1–1.8 mm; glumes scarious, obscurely veined, elliptic to obovate, the first 1.9–3.9 mm, the second 2.8–5.1 mm; lemmas elliptic, 3.7–5.3 mm, dull and scaberulous between the veins, blunt or rounded and usually erose at the scarious tip; palea sometimes projecting as much as 1 mm; anthers 3, 1–1.7 mm; grain 1.5–2 mm; $2n=40$. Shallow water or very wet soil; Mass. to Minn., s. to S.C. and Tex. (*G. arkansana,* a rare form with the lemmas minutely short-hairy instead of scabrous.)

3. Glyceria fluitans (L.) R. Br.   Differing from no. 2 essentially only in dimensions; lemmas 5.4–7 mm; rachilla-joints 1.9–2.5 mm; palea very rarely exceeding the lemma; anthers 2–3 mm; $2n=40$. Nf., N.B., N.S., and e. Que., where possibly native; intr. at New York City; Europe.

4. Glyceria acutiflora Torr.   Culms to 1 m, usually decumbent and often rooting from the lower nodes; lvs 3–6 mm wide; ligule 5–7+ mm, infl erect or declined, narrow, 2–3 dm, the branches erect, strongly flattened or trough-shaped; spikelets 2–4 cm, 5–12-fld; glumes narrowly ovate, the first 2–3.3 mm, the second 3.9–5.6 mm; lemmas lanceolate, 7–8.5 mm, acute, scabrous; palea acuminate, 2-toothed, projecting 1.5–3 mm; rachilla-joints 3–4 mm; stamens 3; $2n=20, 40$. Shallow water or very wet soil; Me. to Del., Va., and Tenn., w. to Mich. and Mo.; also in e. Asia.

5. Glyceria obtusa (Muhl.) Trin.   Coastal m. Culms 6–10 dm, often decumbent at base; main lvs 3–8 mm wide; ligule under 1 mm; panicle dense and compact, ovoid or oblong, 0.6–1.2 dm, usually with 3 strictly ascending branches from each node, the 2 lowest internodes each less than 2(–2.5) cm; spikelets 4–7 mm, 4–7-fld; glumes broad and scarious, the first 1.5–2 mm, the second 2–2.7 mm; lemmas broadly elliptic to obovate, 3–3.7 mm, scarious at the margins and the blunt summit, about as long as the palea, the veins visible but not raised; stamens 2; $2n=40$. Swamps, wet woods, and shallow water; N.S. to N.C., chiefly near the coast, and extending inland to the Catskills and e. Pa.

6. Glyceria melicaria (Michx.) C. E. Hubbard.   Northeastern m. Culms slender, erect, 5–10 dm; lvs 2–5 mm wide; ligule under 1 mm; panicle very narrow, its 2 lowest internodes each 2–7 cm; branches 1–3 from each node of the infl, strictly erect, the lower not much surpassing the next node; spikelets 4 mm, 3–4-fld; first glume narrowly lanceolate, 1.3–2.4 mm; second glume broader, 1.7–3 mm; lemmas broadly elliptic, abruptly acute, 1.9–2.8 mm, smooth, somewhat surpassing the indurate palea, the veins visible but not raised; stamens 2; $2n=40$. Swamps and wet ground; Que., N.S., and N.B. to Ky. and N.C. (*G. torreyana.*)

7. Glyceria canadensis (Michx.) Trin.   Rattlesnake-m. Culms solitary or few in a tuft, erect, to 1 m; main lvs 3–8 mm wide; ligule 2–6 mm; infl 1–3 dm, diffuse, with drooping branches bearing spikelets mostly toward the tip; spikelets broadly ovate, 4–8 mm, 5–10-fld; glumes scarious-margined, the first lanceolate, 1.6–2.4 mm, the second broadly ovate, 2.1–2.3 mm; lemmas broadly ovate, with visible but not raised veins, 2.9–4 mm, the thin or scarious margins not covering the sides of the round-obovate palea, the acute tip projecting 0.5 mm beyond the palea; stamens 2; $2n=60$. Swamps, bogs, and wet woods; Nf. to Minn., s. to N.J. and Ill.

8. Glyceria × **laxa** (Scribn.) Scribn.   Much like no. 7, but often taller, with larger (to 4 dm) and very diffuse panicles with smaller spikelets, glumes and lemmas; spikelets ovate, 4–5 mm; 4–6-fld; glumes scarious-margined, the first lanceolate, 1.2–2 mm, the second ovate, 1.8–2.4 mm; lemmas ovate, with visible but not raised veins, 2.3–2.9 mm, the scarious margins almost or quite covering the obovate palea, the obtuse tip scarcely or not at all projecting; male-sterile; $2n=42, 46$. Swamps and wet woods; P.E.I. to W.Va. (*G. canadensis* var. *laxa.*) Thought to be a hybrid of no. 7 with no. 10 or no. 9.

9. Glyceria striata (Lam.) A. Hitchc.   Fowl-m. Culms slender, tufted, erect, 5–12 dm; main lvs 2–5(–8) mm wide; ligule 1–3 mm; infl 1–2 dm, the numerous branches ascending, drooping at the tip, bearing spikelets mostly beyond the middle; spikelets green or purplish, ovate, 2.5–4 mm; 3–6-fld; glumes ovate to obovate, obtuse, thin, the first 0.5–1 mm, the second 0.8–1.3 mm; lemmas elliptic to obovate, with prominently raised veins, 1.4–2.1 mm, obtuse and ± scarious at the tip, palea narrowly obovate, 3–4 times as long as wide; anthers 2, 0.3–0.6 mm, $2n=20$. Open swamps and marshes, often gregarious; Nf. and Lab. to B.C., s. to Fla. and Calif. A smaller, stricter form of the northern part of the range, the lemmas usually purplish and with distinctly scarious tip, has been segregated as var. *stricta* (Scribn.) Fern.

10. Glyceria grandis S. Wats.   American m. Culms usually clustered, stout and erect, to 15 dm; lvs 8–12 mm wide; ligule 2–4+ mm; infl 2–4 dm, much branched, usually nodding at the tip; spikelets 4–6.5 mm, 5–9-fld, slightly flattened; rachilla-joints 0.6–1 mm; glumes membranous or scarious, pale or white, acute, the first 1.2–1.9 mm, the second 1.5–2.4 mm; lemmas usually purple, narrowly ovate, with prominently raised veins, gradually tapering from below the middle to the obtuse firm tip; anthers (2)3, 0.5–1 mm; $2n=20$. Swamps, marshes, brooksides, and shallow water; Que. and N.S. to Va., w. to the Pacific.

25. MELICA L.   Melic grass. Spikelets 2–5-fld, disarticulating above or below the glumes; glumes large, membranous, 3–5-veined; lemmas rounded on the back, 5–7-veined, or with additional veins near the base, obtuse or acute, awned or awnless; uppermost 1–4 lemmas much smaller, sterile and often distinctly unlike the fertile ones; slender, tufted to rhizomatous grasses with closed sheaths, flat lvs, conspicuous ligules, and sparsely branched, few-fld panicles. 50+, mostly temp. reigons.

1 Fertile lemmas acute, awned .... 1. *M. smithii.*
1 Fertile lemmas obtuse, awnless.
  2 First glume oblong, 2–4 times as long as wide; glumes subequal .... 2. *M. mutica.*
  2 First glume ovate, less than twice as long as wide; glumes distinctly unequal .... 3. *M. nitens.*

1. Melica smithii (Porter) Vasey.   Awned m. Culms 5–15 dm; lvs soft, scabrous or short-hairy at least above, 5–15 mm wide; sheaths retrorsely scabrous; infl few-fld, with a few divergent (eventually deflexed) branches bearing a few spikelets toward the summit; pedicels densely antrorse-hispid; spikelets green or often purplish, 12–20 mm; 3–6-fld; lemmas and glumes lanceolate, acute, sharply but finely veined; lemmas bifid at the tip, bearing an erect awn 2–5 mm; sterile lemma(s) similar but smaller. Moist hardwood forests; n. Mich. and adj. Ont.; Mont. and Wash. to Wash. and Oreg.

2. Melica mutica Walter.   Two-fl melic. Culms 5–10 dm; sheaths scabrous or pilose; lf-blades 3 or 4, flat, the upper 10–20 cm × 2–5 mm; infl 1–2 dm, with 1–few spreading lower branches to 8 cm; spikelets short-pedicellate, pendulous, 7–11 mm; glumes nearly equal, oblong, 6.5–9 mm, nearly or quite equaling the lemmas; fertile lemmas 2, strongly nerved, about even with each other at the blunt tips; sterile lemmas ca 2 mm, obconic or hooded, surpassed by the fertile ones; $2n=18$. Woods and banks; Md. to Ind. and Ill., s. to Fla. and Tex.

3. Melica nitens Nutt.   Three-fl m. Culms 5–15 dm; sheaths glabrous; blades 5–8, flat, 10–20 cm × 5–12 mm; infl 1–2 dm, more freely branched than in no. 2, with short lateral branches from most of the nodes; spikes short-pedicellate, pendulous, 9–12 mm; first glume broadly ovate or ovate-elliptic, 5–8 × 3.5–5 mm, its margins meeting around the spikelet; fertile lemmas (2)3, the second one usually distinctly projecting beyond the first; sterile lemmas 2, obconic, surpassed by the fertile ones; $2n=18$. Rocky or dry upland woods; Pa. to s. Minn. and Nebr., s. to Ga. and Tex.

26. SCHIZACHNE Hackel.   Spikelets 3–5-fld, disarticulating above the glumes and between the lemmas; glumes membranous, 1–5-veined; lemmas narrow, rounded on the back, 5–7-veined, bifid at the tip, with an awn between the teeth, densely short-bearded on the callus; uppermost lemma similar but much smaller and sterile; slender, loosely tufted perennial grasses with flat lvs and ample panicles. Monotypic.

1. Schizachne purpurascens (Torr.) Swallen.   Culms erect from a short-decumbent base, 4–10 dm; lvs mostly erect, elongate, 1–5 mm wide, usually glabrous; infl with a few drooping branches each bearing 1–3 slender spikelets 2 cm; glumes purple at base, unequal, 5–8 mm; fertile lemmas 8–10 mm, strongly veined, bifid a fourth or a fifth of their length; awns 8–15 mm, eventually divergent; $2n=20$. Dry rocky or sandy woods; s. Can. to Alas. and n. Eurasia, s. to Pa., Ky., Nebr., and Mex. Plants with the lvs white-hairy on the veins above, occurring mainly in and near s. Ont. and adj. N.Y., Vt., and N.H., have been described as var. *pubescens* Dore.

27. ARRHENATHERUM P. Beauv.   Oatgrass. Spikelets 2-fld, disarticulating above the glumes and between the lemmas; glumes unequal, chartaceous, hyaline in age, the first shorter, 1-veined, the second equaling the lemmas, 3-veined; lemmas thin, rounded on the back, short-bearded at base, 5–7-veined, the lower enclosing a staminate fl and bearing below the middle a long geniculate awn, the upper one enclosing a perfect fl and awnless or bearing just below the tip a much shorter straight awn; tall perennials with flat lvs and narrow panicles. 6, temp. Eurasia.

1. **Arrhenatherum elatius** (L.) J. & C. Presl. Tall o. Tufted, short-lived perennial to 2 m; culms smooth, or minutely hairy at the nodes; sheaths smooth; blades scabrous, 4–8 mm wide; infl shining, slender, 1–3 dm, the short branches fascicled; glumes lance-ovate, acute or acuminate, the first 4.5–8 mm, the second 6.5–10 mm; lemmas glabrous or sparsely pilose; awn of the lower lemma 10–20 mm, geniculate near the middle; awn of the upper lemma 0–6 mm; anthers 4 mm; $2n=14, 28, 42$. Meadows, roadsides, and waste ground, usually in moist soil; native of Europe, now widespread in the U.S. and found throughout our range. Plants with the lower internodes short, thickened into a series of subglobose corms 1 cm thick, are sporadic with us and have been distinguished as var. *bulbosum* (Willd.) Spenner.

28. HELICTOTRICHON Besser.

Perennial; spikelets erect or spreading; first glume 1–3-veined, the second 3(5)-veined; otherwise like *Avena*. (*Avena* in part; *Avenula*.) 90, Old World.

1. **Helictotrichon pubescens** (Hudson) Pilger. Loosely tufted perennial 3–10 dm; sheaths ± hairy; lvs with 2 lines of bulliform cells flanking the midvein above; ligule to 1 mm and truncate in basal lvs, 5–8 mm and acute in cauline ones; spikelets erect, 2–4-fld; glumes 9–18 mm, the first 1(3)-veined, the second 3-veined; rachilla and callus beset with copious long white hairs; lemmas 9–13 mm, scabrous, hyaline at the erose tip; awn 1.5–2 cm; $2n=14$. Native of Europe, rarely intr. or adventive in waste places from Que. to Del., and in w. Minn. (*Avena p.; Avenula p.*)

29. AVENA L.

Oats. Spikelets 2–5-fld, articulated above the glumes and usually between the lemmas; rachilla often hirsute, at least at the base of the lemmas; glumes nearly equal, membranous or chartaceous, exceeding the lemmas, mostly 5–11-veined; lemmas indurate, tapering to an entire or often 2-toothed tip, firm below, often scarious distally, rounded on the back, obscurely 5–9-veined or prominently so distally; awn arising from the back of the lemma, often geniculate near the middle, or wanting; annuals with broad flat lvs and ample panicles of large, often nodding spikelets. As now limited, ca 15 spp. of Eurasia.

1 Lemmas usually with brown hairs, the first two both with well developed awns .... 1. *A. fatua*.
1 Lemmas glabrous or scabrous, the second one awnless .... 2. *A. sativa*.

1. **Avena fatua** L. Wild oats. Stout annual to 10+ dm; ligules 2–6 mm; blades scabrous to somewhat pilose, 3–10(–15) mm wide; panicle 10–25+ cm, open, its branches horizontally spreading; spikelets ± nodding, (2)3-fld, the hirsute rachilla readily disarticulating; glumes prominently (7)9- or 11-veined, the first one 19–26+ mm, lanceolate, acuminate; lemmas 14–20+ mm, 5- or 7-veined, usually with brown hairs on the back, the 2 lower ones awned; awns (18–)28–45(–52) mm, stout, geniculate, twisted below, thin and tapering above, attached above the middle of the lemma; third lemma usually awnless; $2n=42$. Native of Europe, frequently adventive or established as a weed of cult. ground or waste places, but not common in most of our range except at the nw.

A. sterilis L., animated oats, is rarely adventive (from Europe) in our range. It has large spikelets, 25–45 mm, with awns 3–9 cm arising from below the middle of the lemmas; the rachilla readily disarticulates above the glumes but not between the lemmas, which therefore all fall together.

2. **Avena sativa** L. Oats. Much like no. 1, and hybridizing with it; herbage glabrous to scabrous; spikelets 2(3)-fld, the rachilla glabrous or only sparsely hirsute, not readily disarticulating; lemmas glabrous to sometimes scabrous, 3–7-veined, the callus naked or only sparsely bearded; awns, when present; (15–)22–35 mm, then only on the first lemma, not geniculate; $2n=42$. European cultigen derived from no. 1, cult. throughout our range and often adventive in disturbed sites, but probably not persistent.

30. TRISETUM Pers.

Spikelets 2(3–5)-fld, disarticulating above the glumes and between the lemmas; rachilla prolonged behind the second palea, often hairy; glumes thin or membranous, with broad hyaline margins, about as long as the spikelet, the first 1–3-veined, the second 3–5-veined; lemmas thin, elongate, rounded on the back or distally keeled, obscurely veined, short-bearded at base, entire and awnless, or with a straight or bent awn arising from above the middle of the back; seeds with minute embryo and liquid endosperm; ligules membranous, erose; tufted perennials (rarely annual) with spiciform or subcapitate to loose and elongate panicles. 50+, cosmopolitan.

1 Lemmas entire, awnless .... 1. *T. melicoides.*
1 Lemmas 2-toothed, at least the upper one awned.
  2 Panicle dense and contracted, spike-like, or interrupted near the base .... 2. *T. spicatum.*
  2 Panicle relatively lax and open, at least not spike-like .... 3. *T. flavescens.*

   1. Trisetum melicoides (Michx.) Scribn. Culms 4–8 dm, glabrous or scaberulous; sheaths usually glabrous; blades 3–6 mm wide, scaberulous or sparsely pilose; infl slender, 8–20 cm, the lower branches sometimes 5 cm; spikelets 2-fld or rarely with a much reduced third fl; glumes widest near or below the middle, scabrous on the keel only, the first 1-veined, 4–5.6 mm, the second 3-veined, 5–7 mm; rachilla-joints 1.3–2 mm, thickly beset with white hairs of about the same length; lemmas acute, undivided, awnless. Moist gravelly or rocky soil, usually in woods; Nf. and Que. to Ont., Mich., and Wis.

   2. Trisetum spicatum (L.) K. Richter. Culms tufted, 1–5 dm, antrorsely hairy above, retrorsely so (or glabrous) below; sheaths retrorsely hairy; blades flat or involute, glabrous or hairy, 1–3 mm wide; infl 3–10 cm, dense, or interrupted at base, spikelets mostly 2-fld; first glume linear-oblong, 3–5 mm, 1-veined; second glume lanceolate to oblanceolate, 4–6 mm, 3-veined; hairs of the rachilla 0.5 mm; lemmas slightly surpassing the glumes, the terminal teeth 1.3–2 mm, narrowly triangular; awn flexuous, 3.5–5 mm; anthers 0.6–1.4 mm; $2n=14, 28, 42$. Arctic and alpine meadows and shores; circumboreal, s. to the higher mts. of U.S., Hispaniola, Mex., and s. S. Amer.; in our range s. to N.Y., Pa., Mich., and Minn., and in N.C. The numerous proposed vars. are ill-defined and confluent.

   3. Trisetum flavescens (L.) Beauv. Culms loosely tufted, 3–8 dm, glabrous, or hairy at the nodes; basal sheaths ± hairy, the others glabrous, main blades 2–6 mm wide; infl shining, 8–15 cm, narrow, rather dense but hardly spike-like; spikelets 5–7.5 mm, 2–4-fld; first glume 3–4 mm, subulate or lance-subulate, tapering from near the base to the tip, 1-veined; second glume much wider, 3-veined, 5–6 mm; rachilla and callus with hairs 1 mm; lemmas bifid; awn 4.5–6.5 mm, inserted 2 mm below the lemma-tip; anthers 1.5–2.5 mm; $2n=28$. Native of Europe, intr. in waste places and along roads from Vt. to Mass. and N.Y., and occasionally elsewhere in our range.

   31. KOELERIA Pers. Spikelets normally 2-fld, disarticulating above the glumes and between the lemmas; rachilla prolonged behind the second palea and occasionally bearing a third, rudimentary or fertile, or a fourth rudimentary fl; glumes unequal, obscurely keeled, scarious-margined, the first 1-veined, the second 3–5-veined; lemmas about as long as the glumes, rounded on the back, acute, scarious at margin and tip, obscurely 5-veined; palea hyaline, nearly as long as the lemma; seeds with minute embryo and liquid endosperm; lvs narrow and shining; panicles contracted, silvery-green. 20, N. Temp. and Arctic.

   1. Koeleria pyramidata (Lam.) P. Beauv. Junegrass. Tufted perennial 3–6 dm; culms hairy below the infl; ligule short, erose; blades 1–3 mm wide, flat, or involute when dry; spike-like panicle 5–12 × 1–2 cm; spikelets subsessile, overlapping, ± scabrous; first glume lance-oblong, 2.6–4+ mm; second glume broadest above the middle, 3.2–5+ mm; lemma awnless; anthers 1–1.8 mm; $2n=14, 28, 42, 56, 70, 84$. Dry soil, prairies, sand-hills, and open woods; circumboreal, s. in Amer. to Del., Mo., and Tex.; s. S. Amer. (*K. cristata,* an illegitimate name; *K. gracilis; K. macrantha; K. nitida.*)

   32. SPHENOPHOLIS Scribn. Spikelets strongly compressed, mostly 2-fld, disarticulating below the glumes and between the lemmas; rachilla prolonged behind the upper palea, seldom bearing a third floret; glumes keeled, dissimilar, the first 1(3)-veined, linear or rather narrowly oblong, usually distinctly shorter than the firmer, much wider, 3(5)-veined, scarious-margined second one, which nearly equals the first lemma; lemmas firm, slender, rounded on the back or keeled distally, obscurely veined, awnless or (especially the upper one) with a straight or bent, subterminal awn; palea hyaline, nearly equalling the lemma; seeds with minute embryo and liquid endosperm; ligule membranous; annual or more often perennial, with tufted or solitary stems, not rhizomatous, and with shining, often slender and spike-like panicles. 4, N. Amer.

1 Spikelets small, 1.5–5 mm; second lemma awnless or with an awn less than 3.5 mm.
  2 First glume less than a third as wide as the second; second lemma smooth to scaberulous.
    3 Lower blades usually under 10 cm, flat, 2–8 mm wide; widespread.
      4 Spikelets awnless .... 1. *S. obtusata.*
      4 Spikelets awned .... 2. *S. × pallens.*
    3 Lower blades longer, mostly (10–)15–45 cm, filiform, less than 2 mm wide, strongly involute;
      southeastern .... 3. *S. filiformis.*
  2 First glume a third to two-thirds as wide as the second; second lemma strongly scabrous .... 4. *S. nitida.*
1 Spikelets larger, 5–9.5 mm; second lemma with an awn 3.5–7 mm .... 5. *S. pensylvanica.*

1. **Sphenopholis obtusata** (Michx.) Scribn. Wedge-grass. Annual or short-lived perennial 2–12 dm with tufted or solitary culms, the herbage glabrous to scabrous or pubescent; lvs flat, mostly 2–8 mm wide, mostly under 10 cm; ligules 1.5–3.5 mm, finely erose-ciliate; infl 0.5–2 dm, erect or nodding; glumes dissimilar, the first 1–3 mm, narrow, 0.1–0.3(–0.4) mm wide in side-view, with scabrous keel, the second 1.2–4.2 mm, broader, 0.5–1 mm wide in side-view; lemmas awnless, scabrous, smooth to scaberulous, the first 1.4–4.4 mm, the second a bit shorter; anthers small, mostly 0.3–0.7 mm; 2*n*=14. Moist meadows, streambanks, and shores of ponds or lakes; Nf. to Alas., s. to Fla., W.I., and Mex. Two vars. with similar ranges:

Var. *obtusata*. Panicle usually densely cylindric; second glume very broadly rounded or truncate at the tip, the length 2–3 times the folded width; lower rachilla-joint 0.5–0.7 mm.

Var. *major* (Torr.) K. S. Erdman. Panicle rather loose; second glume acute, the length 3–6 times the folded width; lower rachilla-joint 0.8–1 mm. (*S. intermedia*.)

2. **Sphenopholis × pallens** (Biehler) Scribn. A series of largely sterile hybrids between spp. 1 and 4; spikelets in the size range of no. 1, but the first lemma with an awn up to 3.5 mm.

3. **Sphenopholis filiformis** (Chapman) Scribn. Slender, tufted, usually perennial, 2–10 dm, often finely pubescent below; lvs very slender, usually less than 2 mm wide, with strongly involute margins, the lower with short sheath and long blade (10–)15–45 cm, the upper with longer sheath and shorter blade; panicle slender, with few ascending branches; spikelets 3–4 mm; glumes spreading at maturity, the first narrow, usually less than 0.2 mm wide in side-view, the second much wider, 1.5–3.5 × 0.5–0.8 mm, truncate; first lemma oval, 1.8–3.2 mm, scaberulous distally, acute or short-awned, the second shorter, scaberulous, occasionally short-awned; anthers 1–2 mm; 2*n*=14. Sandy soil in pine woods; se. Va. to Fla., Tenn., and e. Tex.

4. **Sphenopholis nitida** (Biehler) Scribn. Tufted, slender, 3–8 dm; lvs short, usually hairy (at least the lower), the upper sometimes merely scaberulous, the main ones 2–5 mm wide; infl slender to diffuse, never spike-like, often with spreading branches to 10 cm; spikelets 2.5–4 mm; glumes subequal, 1.5–3.5 mm, the first one blunt, wider than in the other spp., 0.2–0.5 mm wide in side-view and ⅓–⅔ the width of the second glume; second glume 0.5–1 wide in side-view, widest near the blunt tip; lemmas oval, blunt to acute, rarely short-awned, the first 2–3.5 mm, smooth to scabrous, the second smaller, very scabrous; anthers 1.2–1.6 mm; 2*n*=14. Dry or moist woods and hillsides; Mass. to Mich. and Ill., s. to n. Fla., Mo., and Tex.

5. **Sphenopholis pensylvanica** (L.) A. Hitchc. Culms slender to robust, 3–12 dm, glabrous; sheaths glabrous; blades scaberulous, to 10(–20) cm, 2–5 mm wide; infl lax, slender, 1–3 dm; spikelets 5–9.5 mm; glumes scabrous on the keel, the first narrowly oblong, often 3-veined, 3.5–5.5 mm, the second obovate-oblong, 3–5-veined, 0.7–1.1 mm wide in side-view, acute to mucronate; first lemma 3.9–7.2 mm, glabrous except for the scaberulous tip; awnless or with an awn to 2.5 mm; second lemma 3.6–6.5 mm, scabrous, with a bent awn 3.5–7 mm; third floret, when present, smaller, awned; anthers ca 1 mm; 2*n*=14. Swamps and wet woods; Mass. to n. Fla., w. to O., s. Mo., and La. (*Trisetum p.; S. palustris*.)

33. DESCHAMPSIA P. Beauv. Hairgrass.

Spikelets 2-fld, disarticulating above the glumes and between the lemmas; rachilla hairy, prolonged beyond the base of the upper lemma; glumes membranous, usually shining, equaling or longer than the lemmas; lemmas membranous, obtuse or often truncate and erose-toothed, rounded on the back, obscurely 5-veined, the midvein diverging at or below the middle into a short awn, the callus bearded; annuals or more often perennials with narrow lvs and conspicuous panicles of purplish to silvery or yellowish spikelets. 30+, temp. and cold regions.

D. danthonioides (Trin.) Munro, an annual with lemmas 2–3 mm and awns 5–7.5 mm, has been collected as an adventive in Ohio.

1 Glumes little if at all longer than the lemmas; anthers over 1 mm.
  2 Lemmas scabrous; awn geniculate below the middle; blades filiform .... 1. *D. flexuosa*.
  2 Lemmas glabrous; awn straight or nearly so; blades flat or folded .... 2. *D. cespitosa*.
1 Glumes about twice as long as the lemmas; anthers under 1 mm .... 3. *D. atropurpurea*.

1. **Deschampsia flexuosa** (L.) Trin. Densely tufted perennial 3–10 dm; lvs mostly at or near the base; ligules 1–2.5 mm; blades usually involute, 1–2 mm wide; infl loose and open, somewhat nodding, to 15 cm, the lowest branches in fascicles of 2–5; spikelets purplish or silvery, 4.3–6 mm, the first glume 3–4.5 mm, the second acuminate, 3.6–5.3 mm; rachilla prolonged less than one-fourth the length of the upper floret; lemmas scabrous, truncate and minutely toothed; awn arising near the middle, geniculate below the middle, the distal half somewhat divergent, surpassing the lemma by 1–3 mm; palea not bifid; rachilla prolonged less than one-fourth the length of the upper floret; anthers linear, 2–3 mm; 2*n*=26, 28, 56. Dry woods, fields, and sandhills; circumboreal, s. to Ga., O., Wis., and Minn.; Ark. and Okla.; Mex.

2. **Deschampsia cespitosa** (L.) P. Beauv. Tufted h. Densely tufted perennial 3–12 dm; lvs mostly below the middle of the stem, the blades flat or folded, 1–5 mm wide, ligules often elongate to 3–12 mm; infl loose and open to narrow and contracted, the lower branches in fascicles of 2–5; spikelets purplish or silvery, 2.3–5.7 mm, the first glume 2–5.1 mm, the second 2.7–5.3 mm, acute or obtuse and often erose-tipped; rachilla

prolonged one-fourth to one-half the length of the upper floret; lemmas glabrous, usually truncate and 4-toothed, the awn attached below the middle, straight or nearly so, from shorter than to barely exceeding the lemma; palea bifid, anthers 1.3–2.3 mm; $2n$=24–28, 52, 56. Wet or boggy ground, circumboreal, s. to N.J., W.Va., N.C., Ill., Minn., and Ariz. A complex sp., the vars. not yet well understood.

3. **Deschampsia atropurpurea** (Wahlenb.) Scheele.   Perennial; culms solitary or in small tufts, 1–5 dm; lvs not densely crowded at base, the short blades erect, flat, 2–5 mm wide; ligule 1–3 mm; infl loose, 4–10 cm, its lowest branches usually paired; glumes purplish with brownish-scarious margins, equal, acuminate, 4.3–6.1 mm, twice as long as and generally concealing the lemmas, these obtuse or erose-truncate, minutely short-hairy above, long-hairy near the base; awn inserted near the middle, usually slightly bent, about equaling the glumes; anthers oval or oblong, 0.4–0.8 mm; $2n$=14. Cold bogs and meadows; circumboreal, s. to the higher mts. of N.H. and N.Y., and to Colo. (*Vahlodea a.*)

34. HOLCUS L.   Velvet-grass.
Spikelets mostly 2-fld, disarticulating below the glumes; glumes thin, keeled, much exceeding and concealing the lemmas, the first 1-veined, the second broader, 3-veined; lemmas rounded on the back, obscurely veined, the lower awnless (in most spp., incl. ours), enclosing a perfect fl, the upper awned, with a staminate fl; awn arising below the lemma-tip; perennials (ours) with flat, soft, relatively wide lvs and dense, contracted, ovoid or subcylindric panicles of small spikelets. 8, Old World.

1. **Holcus lanatus** L.   Common v.-g. Culms 4–10 dm, without rhizomes, soft-villous below, the upper internode usually glabrous; sheaths soft-hairy; blades pale green, soft-hairy, 4–10 mm wide; infl narrowly ovoid, dense, 5–15 cm; glumes 4–5 mm, acute, villous on the surface, short-hirsute on the veins; lemmas smooth and shining, 2 mm, obtuse, the awn of the upper one hooked, 1–2 mm, scarcely exserted from the glumes; $2n$=14. Meadows and roadsides in moist soil; native of Europe, now well established throughout our range.

H. mollis L., another European sp. with slender rhizomes, glabrous culms (except the bearded nodes), and exsert awns 3–5 mm, is locally adventive with us.

35. CORYNEPHORUS P. Beauv., nom. conserv.
Spikelets 2-fld, articulated above the glumes and between the lemmas, also readily detachable from the pedicel below the glumes; glumes subequal, membranous, acuminate, somewhat keeled, obscurely 1–3-veined; lemmas half as long as the glumes, thin or hyaline, hairy at base, rounded on the back, the midvein divergent into an awn just above the base, otherwise veinless; awn twice as long as the lemma, jointed in the middle, the basal half brown, straight, stout, the distal half pale, gradually thickened to the tip; panicle narrow; blade involute-filiform. 6, Europe, Medit.

1. **Corynephorus canescens** (L.) P. Beauv.   Gray hairgrass. Densely tufted, glaucous perennial 2–6 dm; lvs rough, erect, mostly crowded near the base; sheaths often purplish; infl silvery, 5–12 cm, dense, with short rough branches; glumes 2.8–3.6 mm; $2n$=14. Native of Europe; intr. in dry sterile soil along the coast from Mass. to N.H. and Pa.; rarely inland, as in Mich.

36. AIRA L.   Hairgrass.
Spikelets 2-fld, disarticulating above the glumes and between the lemmas; rachilla very short, not prolonged beyond the upper palea; glumes subequal, exceeding the lemmas, ovate, obscurely 1–3-veined, acute or acuminate; lemmas firm, elongate, rounded on the back, obscurely 5-veined below, tapering into 2 pale setaceous teeth, one or both lemmas awned; awn arising from below the middle, bent, the straight stout lower part about equaling the body of the lemma, the slender terminal part longer, exsert; palea enclosed by the involute lemma; small annual grasses with short, very narrow blades and diffuse or spike-like panicles. 12, irregularly cosmop. (*Aspris.*)

1 Panicles contracted, almost spike-like . . . . 1. *A. praecox.*
1 Panicles lax and open, diffusely branched.
  2 Pedicels mostly 1–2 times as long as the spikelets; both lemmas awned . . . . 2. *A. caryophyllea.*
  2 Pedicels mostly 2–5 times as long as the spikelets; lower lemma often awnless . . . . 3. *A. elegantissima.*

1. **Aira praecox** L.   Culms tufted, 5–20 cm, scabrous below the nodes; sheaths and the very short, usually involute and twisted blades scaberulous; infl 1–5 cm, dense and almost spike-like, with short appressed branches, the pedicels mostly shorter than the spikelets; glumes 2.7–3.3 mm; lemmas 2.5–3 mm, both awned, the awn 3.3–3.7 mm; $2n$=14. Native of Europe, intr. in dry sandy soil near the coast, from s. N.Y. to Va.

2. Aira caryophyllea L.   Silver-h. Culms very slender, smooth, 1–3 dm; sheaths and filiform blades scaberulous; infl lax, pyramidal, nearly as broad as long, with ascending branches; pedicels mostly 1–2 times as long as the spikelets; glumes 2.3–3 mm; lemmas 1.8–2.3 mm, both awned, the awn 2.4–3.3 mm; $2n=28$. Native of Europe, intr. in dry soil in open places from Vt. and Mass. to Fla. and Tex., mostly near the coast.

3. Aira elegantissima Schur.   Culms smooth, 1–4 dm; sheaths and filiform blades scaberulous; infl ovoid or cylindric, diffusely branched, 1–2 dm; spikelets on long filiform pedicels; glumes 1.8–2.2 mm; lemmas 1.7–1.9 mm, the lower often awnless, the upper with an awn 1.7–2.4 mm; $2n=14$. Native of Europe, intr. along roadsides and in waste ground from Md. to Fla. (*A. capillaris; A. elegans.*)

37. HIEROCHLOE R. Br., nom. conserv.   Spikelets articulated above the glumes, with 2 lower staminate lemmas and one upper fertile lemma of different form; glumes broad, very thin, obscurely 3-veined, glabrous, subequal, the second slightly wider; staminate lemmas firm, brown, rounded on the back, densely ciliate, acute; fertile lemma indurate at maturity, awnless; palea 1–3-veined; fragrant perennials, the sterile shoots with elongate basal lvs, the fertile usually with small blades and a terminal panicle of broad spikelets. 10+, temp. and cool regions, mostly N. Hemisphere.

1 Lemmas all awnless . . . . 1. *H. odorata.*
1 Staminate lemmas awned, the awn of the upper one 5–9 mm . . . . 2. *H. alpina.*

1. Hierochloe odorata (L.) P. Beauv.   Sweet grass. Vigorously rhizomatous; culms usually 3–6 dm, their sheaths few, elongate, bladeless or with lanceolate blades rarely over 3 cm; panicle pyramidal, 5–10 cm, with widely spreading or somewhat drooping branches; glumes shiny, ovate, 4–6 mm; staminate lemmas equaling the glumes or slightly shorter, often hairy on the back, awnless, sharply acute; fertile lemma shorter, hairy at the tip; $2n=28, 42, 56$. Often apomictic or infertile. Moist soil, meadows, or bog-margins; circumboreal, s. to N.J., Md., O., Io., and Ariz. (*H. nashii.*)

2. Hierochloe alpina (Swartz) Roemer & Schultes.   Tufted on short rhizomes, 2–5 dm; panicle compact, 3–5 cm; glumes 6–8 mm; sterile lemmas awned below the tip, the awn of the lower lemma short and straight, that of the upper lemma 5–9 mm, geniculate; $2n=56-78$. Often apomictic. Circumboreal in arctic and alpine meadows, s. to alpine summits of N.Engl. and N.Y. (*Savastana a.*)

38. ANTHOXANTHUM L.   Sweet vernal grass. Spikelets 1-fld, articulated above the glumes; glumes unequal, much exceeding the lemmas, the first ovate, 1-veined, the second lanceolate, 3-veined; sterile lemmas 2, beneath the fertile one, subequal, 2-lobed, the first awned on the back near the tip, the second near the base, its awn twisted below, geniculate at about the tip of the lemma; fertile lemma broadly rounded, often suborbicular, smaller than and enclosed by the sterile ones, at first hyaline, at maturity brown and shining, palea 1-veined; stamens 2; grasses with sweet-scented foliage, flat lvs, and short, subcylindric, ± spike-like panicles. 20, N. Temp., and trop. mts.

1 Perennial; awn of the lower sterile lemma straight . . . . 1. *A. odoratum.*
1 Annual; awn of both sterile lemmas geniculate . . . . 2. *A. aristatum.*

1. Anthoxanthum odoratum L.   Tufted perennial 3–7 dm; lvs mostly near the base, 2–5 mm wide, the upper much shorter; ligules (1–)2–3 mm; panicle spike-like, long-exsert, 3–9 cm; glumes scabrous on the keel to villous throughout, the first 4 mm, the second 7–9 mm; sterile lemmas 3–3.5 mm; golden-silky, the awn of the second about equaling the second glume; fertile lemma suborbicular, 2 mm; $2n=10, 20$. Native of Europe, intr. in lawns, meadows, roadsides, and waste places from Nf. to Ga., w. to Mich., Ill., and La.

2. Anthoxanthum aristatum Boiss.   Annual 1–6 dm; culms geniculate and decumbent, branched above; lvs seldom as much as 2 mm wide; ligules 0.5–2 mm; infl 1.5–4 cm; glumes glabrous, the first one with an awn 1 mm; awns of the sterile lemmas geniculate and protruding from the glumes; fertile lemma 1–1.5 mm; $2n=10$. A European sp. intr. in waste places here and there nearly througout our range. (*A. puelii.*)

39. PHALARIS L.   Spikelets articulated above the glumes, with 1 perfect terminal fl and 1 or (in our spp.) 2 much smaller linear sterile lemmas below it; glumes subequal, compressed and keeled, often winged along the keel, the lateral veins usually stronger than the midvein; lemmas awnless, shorter than the glumes, the fertile one firm or coriaceous and often shining; ligules membranous, usually large; panicle dense or spike-like, with medium-sized to large spikelets. 15, cosmop., mainly temp.

1 Rhizomatous perennial; panicle 7–25+ cm, branched or lobed; keel not winged .... 1. *P. arundinacea.*
1 Annual; panicle 1.5–6 cm, very dense and compact; keel ± winged.
  2 Keel narrowly winged, the wing up to ca 0.5 mm wide; sterile lemmas 1.5–2.5 mm ....2. *P. caroliniana.*
  2 Keel broadly winged distally, the wing ca 1 mm wide; sterile lemmas 2.5–4.5 mm .... 3. *P. canariensis.*

1. Phalaris arundinacea L. Reed canary-grass. Colonial by rhizomes; culms stout, 7–15 dm; main blades usually 1–2 dm × 10–15(–20) mm; panicle 7–25+ cm, dense but branched or lobed, often lightly suffused with purple; glumes 4–6.5 mm, wingless, glabrous to scaberulous; sterile lemmas linear-subulate, evidently villous, 1–2 mm; fertile lemma 3–4.5 mm, appressed-hairy distally; $2n=14, 28, 35, 42$. Streambanks, lakeshores, marshes, and moist ground; circumboreal, in Amer. from Nf. to Alas., s. to N.C., Kans., and Calif., more abundant westward. Ribbon-grass, with green- and white-striped lvs, is a horticultural variant.

2. Phalaris caroliniana Walter. Erect annual 3–9 dm; main cauline lvs mostly 5–12 cm × 4–8 mm; ligules 1–3 mm; infl 2–6 cm, cylindric to ellipsoid, spike-like and scarcely or not at all lobed; glumes 4.5–6 mm, glabrous on both sides, scabrous on the narrowly winged keel; sterile lemmas subulate, 1.5–2.5 mm, pilose, fertile lemma 3–4.5 mm, appressed-pilose throughout; $2n=14$. Waste places, roadsides, and wet fields; widespread in Mex. and s. U.S., n. in our range to Va., Md., Ky., and Mo.

3. Phalaris canariensis L. Canary-grass. Erect annual 3–10 dm; main cauline lvs mostly 4–10 mm wide; ligules 4–8 mm; infl 1.5–4 cm, dense, ellipsoid; glumes 7–10 mm, strongly winged along the keel distally, the midvein marked by a broad green stripe; sterile lemmas 2.5–4.5 mm, at least 1.5 mm wide, rather sparsely and finely hairy or subglabrous; fertile lemma 5–7 mm, densely and finely appressed-hairy; $2n=12$. Native of Europe, intr. throughout most of N. Amer., but seldom abundant. Used commercially for canary-seed.

40. CALAMAGROSTIS Adans. Reed-grass. Spikelets 1-fld, articulated above the glumes; glumes subequal, acute or acuminate, the first usually a little the larger, lemma equaling or shorter than the glumes, inconspicuously 5-veined, tapering to an obtuse or truncate, erose tip, usually thin and membranous, subtended at base by white hairs arising from the callus, and bearing a slender dorsal awn; palea membranous, half to fully as long as the lemma; rachilla prolonged (in our native spp.) behind the palea as a slender, usually hairy, often inconspicuous bristle, perennials, often rhizomatous, with long narrow lvs and a loose and open to contracted and spike-like panicle. 150+, temp. and cool reg. Many of the plants at higher ploidy-levels are apomictic.

1 Callus-hairs well exceeding the lemma; rachilla not prolonged; intr., weedy .... 1. *C. epigejos.*
1 Callus-hairs up to about as long as the lemma; rachilla prolonged behind the palea; native.
  2 Infl dense, erect, its branches ± appressed except at anthesis.
    3 Awn attached well above the middle of the lemma .... 2. *C. cinnoides.*
    3 Awn attached at or below the middle of the lemma.
      4 Awn 4.5–9 mm, evidently longer than the lemma, twisted and geniculate .... 3. *C. purpurascens.*
      4 Awn shorter, variously straight or twisted and geniculate or not.
        5 Callus-hairs scanty, about a fourth as long as the lemma .... 4. *C. pickeringii.*
        5 Callus-hairs copious, most of them half to three-fourths as long as the lemma.
          6 Dryland plains sp, 2–4 dm, with basally disposed lvs only 1–2 mm wide .... 5. *C. montanensis.*
          6 More eastern or boreal spp. of moist habitats, 4–10 dm, with well distributed, mostly broader lvs.
            7 Awn straight or only slightly twisted and bent; collar glabrous .... 6. *C. stricta.*
            7 Awn evidently twisted and geniculate; collar usually bearded .... 7. *C. lacustris.*
  2 Infl looser and laxer, its branches somewhat spreading.
    8 Callus-hairs half as long as the lemma; awn distinctly geniculate.
      9 Collar bearded; Appalachian .... 8. *C. porteri.*
      9 Collar glabrous; midwestern .... 9. *C. insperata.*
    8 Callus-hairs nearly as long as the lemma; awn straight or nearly so .... 10. *C. canadensis.*

1. Calamagrostis epigejos (L.) Roth. Feathertop. Densely colonial by rhizomes, 6–20 dm; lvs 4–20 mm wide; infl 4–12 mm; infl dense, elongate, to 35 cm, at maturity silky from the protruding callus-hairs; glumes lance-subulate, 4–5(–8) mm; lemma membranous, 2–2.5(–3.5) mm, the slender awn inserted variously from near the base to above the middle; callus-hairs copious, much exceeding the lemma and nearly equaling the glumes; rachilla not prolonged; $2n=28, 42, 56, 70$. Native of Eurasia, intr. in sandy woods, salt-marshes, fields, and waste places from Mass. to N.J. and Pa., and casually w. to s. Ont., Mich., N.D., and Kans.

2. Calamagrostis cinnoides (Muhl.) Barton. Rhizomatous; culms 6–12 dm, rather stout, smooth or nearly so; sheaths smooth to scaberulous; main blades 5–10 mm wide; infl erect, dense, 1–2 dm × 2–5 cm, with scabrous branches; glumes lance-acuminate, usually outcurved, the first 4.5–8 mm, the second usually slightly shorter; lemma scabrous, acuminate, 3.7–6 mm; callus-hairs half as long as the lemma, mostly in 2 tufts, one on each side; awn straight, inserted well above the middle of the lemma and barely surpassing it; rachilla prolonged 1 mm, glabrous except for the terminal tuft of hairs that extend almost to the lemma-tip. Open swamps and wet woods; N.S.; coastal states from Me. to Ga., less often inland to W.Va., O., Ky., and Tenn.

3. **Calamagrostis purpurascens** R. Br. Purple r. Densely tufted perennial 3–7 dm, often developing short, thick rhizomes; sheaths scaberulous; ligules 2–5 mm; blades thick, scabrous, 2–4 mm wide, usually involute; infl contracted, usually pinkish or purplish, 4–10 cm; spikelets rarely 2-fld; glumes acuminate, 5–9 mm, the lemma three-fourths as long; callus-hairs a fourth as long as the lemma; awn inserted near the base, 4.5–9 mm, longer than the lemma, geniculate near the middle, the terminal part 3–4 mm, the basal part twisted; rachilla-vestige densely villous with hairs 1 mm; $2n=28-84$. Rocky slopes and well drained, sandy soil; Greenl. to Alas. and e. Asia, s. to Que., Colo., and Calif., entering our range only in se. Ont. and Cook Co., Minn.

4. **Calamagrostis pickeringii** A. Gray. Culms 2–7 dm, scattered on slender rhizomes; sheaths glabrous or nearly so; ligules 2–5 mm; blades flat, 3–7 mm wide, tapering to the base, scabrous beneath, nearly glabrous above; infl 4–15 cm, contracted, lanceolate to subcylindric; glumes acute, glabrous to scaberulous, the first 2.9–4.7 mm, the second 2.8–4.4 mm; lemma scabrous, 2.6–3.7 mm; callus-hairs scanty, a fourth as long as the lemma; awn inserted 1 mm above the base, twisted below, geniculate near the middle, about equaling the lemma; sexual, but seldom fruiting; $2n=28$. Bogs and wet shores, mainly in the mts.; Nf. to the mts. of Mass. and N.Y.; disjunct on L.I. and in N.J.

5. **Calamagrostis montanensis** Scribn. Plains-r. Strongly rhizomatous, stiffly erect, 2–4 dm; culms scabrous below the infl; lvs basally disposed, erect, mostly under 2 mm wide, ± involute, scabrous, sharp-pointed; ligules 3–5 mm; infl dense, erect, ± interrupted, 5–10 cm, usually pale; glumes 4–5 mm, acuminate, scabrous; lemma nearly as long as the glumes, the awn attached 1 mm above the base and about equaling it; palea nearly as long as the lemma; callus-hairs half as long as the lemma; rachilla-vestige 1.5 mm, its beard nearly equaling the lemma; $2n=28$. Plains and dry open ground; Minn. and Man. to se. B.C., s. to S.D. and Colo.

6. **Calamagrostis stricta** (Timm) Koeler. Rhizomatous, to 1 m, the culms scabrous below the infl, otherwise glabrous; sheaths glabrous or scaberulous; ligules 1–7+ mm, often lacerate; blades glabrous, or scabrous above, involute, 2–6 mm wide when unrolled; infl 5–15 mm, narrow and congested, sometimes lobed; glumes 2–6 mm, acute or abruptly acuminate, glabrous or scabrous; lemma scabrous, in most plants about equaling the second glume; callus-hairs copious, half to nearly as long as the lemma; awn inserted near the middle, straight or nearly so; rachilla-vestige to 1 mm, hairy throughout; $2n=28-126$, apomictic at the higher ploidy-levels. Wet meadows and marshes; circumboreal, s. to N.Engl. and N.Y., Pa., W.Va., O., Ind., Mo., Ariz., and Calif. (*C. inexpansa; C. neglecta.*)

7. **Calamagrostis lacustris** (Kearney) Nash. Culms slender, 4–10 dm, from creeping rhizomes; sheaths and blades scabrous, the collar usually shortly bearded; blades firm, 2–4 mm wide; infl 8–15 cm, tawny to purplish, rather dense; glumes firm, rather broad, 3.5–4.5 mm, scabrous; lemma 3.5 mm, scabrous, the awn inserted near the base, about equaling the lemma, evidently geniculate near the middle, twisted below; callus-hairs in 2 tufts, half to three-fourths as long as the lemma; palea nearly equaling the lemma; $2n=84-120$. Mossy rocky and marshy meadows and sandy shores; Nf. and Que. to Vt. and N.Y., w. to Mich. and Minn. (*C. fernaldii.*)

8. **Calamagrostis porteri** A. Gray. Culms slender, to 1 m, from creeping rhizomes; sheaths glabrous except at the densely hairy collar; ligules 2–5 mm, blades 4–7 mm wide, scabrous and glaucous above, darker green and less scabrous or smooth beneath; infl 1–2 dm, some of the loosely ascending, fascicled branches spikelet-bearing to the base; glumes long-acuminate, scabrous on the keel, otherwise subglabrous, the outer 4–6 mm; lemma 3.5–4.5 mm; callus-hairs in 2 tufts, half as long as the palea, which is nearly as long as the lemma; awn inserted 1 mm above the base, geniculate near the middle, twisted below, equaling the lemma; rachilla to 1 mm, long-bearded; $2n=84-104$. Dry rocky ridge-top forests s. of the glacial boundary in the Appalachian Mts.; s.c. N.Y. to Va., W.Va., Ky., and Tenn.

C. perplexa Scribn., known only from a few hundred plants in a single colony in Tompkins Co., N.Y., may reflect past hybridization of nos. 8 and 10. It has the foliage, the stout, geniculate and twisted awn, and the habitat of no. 8, combined with the laxer panicle, longer callus-hairs, and occasionally branching culms of no. 10. $2n=70$, but the plants sterile, without viable pollen.

9. **Calamagrostis insperata** Swallen. Much like no. 8; culms stouter; blades weakly scabrous, to 12 mm wide, light green and glaucous on both sides, the collar glabrous; $2n=56$. Dry rocky woods; local in s. O. (Jackson Co.) and apparently also in s. Mo.

10. **Calamagrostis canadensis** (Michx.) P. Beauv. Bluejoint. Culms 5–15 dm, often branched above, smooth, clustered on creeping rhizomes; sheaths glabrous or nearly so; ligules 3–8 mm; blades scabrous on both sides, flat, 4–8 mm wide; infl somewhat nodding, 8–25 cm, open or fairly dense, but not contracted, the longer branches 3–8 cm; spikelets 2–6 mm; glumes subequal or the second slightly longer, acute to acuminate, rounded or keeled, glabrous to scabrous-puberulent on the sides; lemma nearly or quite smooth, translucent at the dentate or erose tip, three-fourths to fully as long as the glumes; awn delicate, erect, straight or nearly so, inserted near the middle of the lemma; callus-hairs abundant, nearly as long as the lemma; rachilla-vestige 0.1–0.3 mm, hairy throughout; $2n=42-66$. Open swamps, wet meadows, and prairies, or moist or wet soil in the mts.; Greenl. to Alas., s. to N.C., Mo., and Ariz. Highly variable, but not readily divisible into vars. (*C. macouniana,* the small-fld extreme; *C. langsdorfii,* the large-fld extreme; *C. nubila.*)

41. AMMOPHILA Host. Beach-grass. Spikelets 1-fld, strongly flattened, articulated above the glumes; glumes chartaceous, subequal, lance-linear, keeled, the first 1-veined, the second 3-veined; lemma chartaceous, keeled, shorter than the glumes, obscurely 3–5-veined, awnless, subtended by a short tuft of callus-hairs; palea about equaling the lemma; coarse, stiff, erect perennials from long rhizomes; ligules membranous; lf-blades flat at base, involute above; panicle contracted and spike-like. 2, the other on the coast of Europe and N. Afr.

1. **Ammophila breviligulata** Fern. Culms glabrous, 5–10 dm; sheaths glabrous; ligules 1–3 mm; ovate or truncate; blades 4–8 mm wide when unrolled, scabrous above; infl dense, 1–4 dm × 1–2.5 cm, its base often enclosed in the upper sheath; glumes subequal, 10–15 mm, scabrous on the keel or throughout; lemma shorter than the glumes, scaberulous; callus-hairs 1–3 mm; prolongation of the rachilla 2 mm, hairy; $2n=28$. Dunes and dry sandy shores, along all the Great Lakes, along Lake Champlain, and along the Atlantic from Nf. to N.C.

A. arenaria (L.) Link., marram-grass, of coastal Europe, is said to be locally intr. along our Atlantic coast. It differs from *A. breviligulata* in its much longer ligule, 1–3 cm.

42. APERA Adans. Like *Agrostis,* but the palea strongly 2-nerved, nearly as long as the relatively firm lemma; lemma with a well developed subapical awn; rachilla shortly prolonged behind the palea; annual. 4, Eurasia.

1. **Apera spica-venti** (L.) P. Beauv. Annual 3–10 dm, branched from the base; sheaths smooth; ligules membranous, erose-lacerate, 2–5 mm; lvs scaberulous, elongate, 3–5 mm wide; infl a freely branching, fairly dense panicle 1–2 dm, the branches ascending or spreading at anthesis, naked at base, forking at or above the middle; spikelets 2.4–3.2 mm; glumes lance-acuminate, the lower 1-veined, the upper 3-veined; lemma about as long as the second glume, shortly bearded at base, scabrous distally, and bearing just below the tip a straight rough awn 5–12 mm; rachilla prolonged 0.5 mm; $2n=14$. Native of Europe, sparingly intr. in waste places from N.Engl. to Md., O., s. Ont., and Mich. (*Agrostis s.*)

A. interrupta (L.) P. Beauv., with narrower lvs 1–3 mm wide, and with more contracted infl, many of the branches spikelet-bearing to the base, is casually intr. in w. U.S. and rarely in our range, as in O. and Mass. (*Agrostis i.*)

43. AGROSTIS L. Bent-grass; bent. Spikelets 1-fld, articulated above the glumes; glumes about equal, narrow, acute or acuminate, 1-veined, somewhat keeled; lemma acute to obtuse, evidently shorter than to nearly equaling the glumes; rounded on the back, obscurely nerved, awnless or short-awned from the back; rarely the awn to 1 cm and borne just below the tip; callus often minutely bearded; palea delicate, ± reduced or obsolete, seldom as much as three-fourths as long as the lemma; rachilla in most spp. not prolonged behind the palea; annuals and perennials with expanded or contracted panicles of small spikelets. 100+, mainly temp. and subarctic.

1 Palea half to two-thirds as long as the lemma.
  2 Panicle-branches all naked toward the base; ligules up to ca 2 mm .... 1. *A. capillaris.*
  2 Some of the panicle-branches bearing spikelets to the base; larger ligules 2.5–6 mm.
    3 Rhizomatous; lvs 3–8 mm wide; infl triangular-ovoid, with widely spreading branches, usually
      red-purple .... 2. *A. gigantea.*
    3 Not rhizomatous, though sometimes stoloniferous; infl narrower, with more ascending branches,
      usually stramineous .... 3. *A. stolonifera.*
1 Palea obsolete or much reduced, distinctly less than half as long as the lemma.
  4 Perennial; lemma awned or awnless.
    5 Lemma awnless or with a short, nearly straight awn near the tip.
      6 Panicle-branches forking near or below the middle; lvs 2–6 mm wide .... 4. *A. perennans.*
      6 Panicle-branches mostly forking well beyond the middle; lvs 1–2(–3) mm wide .... 5. *A. hyemalis.*
    5 Lemma with a geniculate awn inserted at or near the middle, once to twice as long as the lemma.
      7 Panicle-branches scabrous; anthers 1–1.8 mm; lvs cauline and basal .... 6. *A. canina.*
      7 Panicle-branches glabrous or nearly so; anthers 0.5–0.7 mm; lvs mostly basal .... 7. *A. mertensii.*
  4 Annual; lemma usually with a long, flexuous awn (to 10 mm) from just below the tip .... 8. *A. elliottiana.*

1. **Agrostis capillaris** L. Rhode Island b. Loosely tufted, rhizomatous and sod-forming, sometimes with stolons as well, 2–6 dm; culms and sheaths smooth; ligule mostly 0.5–2 mm, wider than high, ± truncate; lvs usually spreading or ascending, 2–4 mm wide; panicle 4–15(–20) cm, permanently open, with smooth, divaricate branches and branchlets, even the pedicel-tips smooth; spikelets 2–3 mm, not crowded; glumes glabrous, or scabrous distally on the keel; lemma 1.5–2.5 mm, acute or acutish, rarely short-awned from the back; callus minutely bearded; palea half to two-thirds as long as the lemma; anthers 0.8–1.5 mm; $2n=28$–35. Native of Europe and perhaps e. Can., widely cult. in cooler Amer. for lawns and pastures, and escaped from Nf. to Mich., s. to N.C. and O., and in w. U.S. (*A. tenuis.*)

2. **Agrostis gigantea** Roth. Redtop. Rhizomatous and sod-forming, to 10 or even 15 dm, not stoloniferous; larger (upper) ligules mostly 2.5–6 mm, higher than wide; lvs 3–8 mm wide; panicle 10–20 cm, notably suffused with purplish-red, at anthesis triangular-ovoid, with widely spreading, unequal branches, sometimes later more contracted; at least some of the panicle-branches floriferous to the base; panicle-branches and often even the pedicels scabrous; spikelets rather crowded, 2–3.5 mm; glumes scabrous along the keel; lemma two-thirds as long as the glumes, distally scabrous, awnless or seldom with a short dorsal awn; callus minutely bearded; palea half to two-thirds as long as the lemma; anthers 0.8–1.5 mm; 2n=42. Native of Europe, cult. and escaped into moist meadows, shores, coastal marshes, and other moist places throughout most of the U.S. and s. Can. (*A. alba*, misapplied; *A. stolonifera* var. *major*.)

3. **Agrostis stolonifera** L. Creeping b. Much like no. 2, and perhaps not sharply distinct, but not rhizomatous, though sometimes with stolons; panicle greenish or stramineous, seldom over 10 cm, narrower, especially after anthesis, with ascending or erect branches; 2n=28–46. Native of Europe, cult. and intr. as in no. 2. Two confluent vars.:

Var. *stolonifera*. More or less tufted, with poorly developed or no stolons; lvs mostly less than 5 cm, infl mostly a little more open than in the next var.

Var. *palustris* (Hudson) Farw. Creeping bent. Evidently stoloniferous; lvs to 10+ cm; especially common in our coastal states, often appearing native. (*A. p.; A. s.* var. *compacta*.)

A. viridis Gouan, an Old-World sp. with broadly truncate lemma, dense, lobed infl, and wholly scabrous glumes, is adventive at some ports on the Atlantic coast. (*A. verticillata; A. semiverticillata; Polypogon viridis*.)

4. **Agrostis perennans** (Walter) Tuckerman. Autumn-b. Tufted perennial 5–10 dm; lvs flat, 2–6 mm wide, elongate, the uppermost one more than 5 cm; infl mostly pale greenish, 10–25 cm, notably longer than thick, the smooth or sparsely scabrous branches forking near or below the middle, soon divaricate, the longest ones often less than 6(–12) cm; spikelets 1.8–2.8 mm; glumes subequal, scabrous on the midvein; lemma 1.3–2 mm, awnless or rarely with a very short, slender awn near the tip; palea obsolete; anthers 0.3–0.6 mm; 2n=42. Various habitats, usually in dry soil; Que. and N.S. to Minn., s. to Fla. and Tex. The foregoing description is based primarily on the widespread and highly variable var. *perennans* (*A. oreophila; A. schweinitzii*). Plants of bogs on the coastal plain from N.J. and Md. to Miss. have somewhat larger spikelets (2.7–3.5 mm) on short pedicels, aggregated in spike-like clusters toward the ends of the panicle-branches; these are var. *elata* (Pursh) A. Hitchc. (*A. elata; A. altissima*.)

5. **Agrostis hyemalis** (Walter) BSP. Ticklegrass. Slender, tufted perennial 3–9 dm; lvs mostly basal or below the middle of the stem, usually erect, the blade flat or more often involute, 1–2(–3) mm wide; ligule short, mostly 1–2.5 mm; infl ± flushed with reddish, ovoid or pyramidal, very diffuse, 1–3+ dm, sometimes half as long as the whole plant, the scabrous filiform branches finally divaricate, mostly forking well above the middle; spikelets 1.2–3.2 mm; glumes scabrous on the midvein; lemma awnless or with a short straight awn; callus short-bearded; palea obsolete or to 0.3 mm; anthers 0.3–0.6 mm; 2n=28, 42. Abundant, widely distributed in many habitats, and highly variable; Lab. to Alas., s. to Fla. and Mex. Two vars.:

Var. *hyemalis*. Blooming early, March to July, occasionally persisting until autumn; spikelets short-pediceled, therefore aggregated into small terminal clusters, the pedicels 0.3–2 mm, avg under 1 mm; glumes 1.2–2.4 (avg 1.5) mm; Mass. to Fla. along the coastal plain, n. in the interior to Io. and Ind.

Var. *scabra* (Willd.) Blomq. Blooming later, June to November; spikelets more loosely arranged, the pedicels 0.5–5 (avg 2) mm; glumes 2–3 mm; Lab. to Alas., s. to Ga., Tex., and Mex. (*A. scabra; A. h.* var. *tenuis*.)

6. **Agrostis canina** L. Velvet b. Loosely tufted perennial 2–7 dm, sometimes with short leafy stolons; lvs not conspicuously crowded toward the base; blades of the cauline lvs flat, 4–6 cm × 1–3 mm; infl 5–20 cm, distinctly longer than thick even when fully expanded, the scabrous branches spreading at anthesis, later more erect; spikelets 1.9–2.5 mm; lemma 1.3–1.9 mm, bearing a bent awn to 4.5 mm near the middle; callus very shortly bearded; palea obsolete; anthers 1–1.8 mm; 2n=14, 28, 35, 42, 56. Native of Europe, intr. in fields and meadows at low altitudes from Nf. and Que. to Del., W.Va., Tenn., and Mich.

7. **Agrostis mertensii** Trin. Densely tufted perennial 1–5 dm; lvs mostly crowded at the base, the blades flat or more often involute, 1–3 mm wide; ligules mostly truncate and 1–2 mm; infl pyramidal, 5–15 cm, nearly as thick at base when fully expanded, the branches glabrous or nearly so, forking near the middle; spikelets purplish, 2–4 mm; first glume often scabrous along the midvein, the second usually glabrous; lemma 1.6–2.8 mm, bearing at or near the middle a bent awn 3–6 mm; callus very shortly bearded; palea short, up to 0.5 mm; anthers 0.5–0.7 mm; 2n=56. Peaty or rocky soil; circumboreal, s. to Que. and on the higher mts. of Me., N.H., Vt., N.Y., W.Va., and in the w. cordillera. (*A. borealis*.)

8. **Agrostis elliottiana** Schultes. Tufted annual 1–6 dm; lvs flat, 1–2 mm wide; infl diffuse, often half as long as the entire plant, the capillary branches eventually widely spreading and themselves branched near the middle; glumes lanceolate, becoming purple, subequal, 1.6–2 mm, nearly glabrous on the keel; lemma very thin, often minutely papillose-scabrous in age, 1.2–1.6 mm, sharply veined, the truncate tip bearing 2 minute setae; awn usually present, inserted just below the lemma-tip, pale, very delicate, flexuous, to 10 mm; callus short-bearded; palea obsolete; 2n=28. Open woods, barrens, fields, and roadsides; Md. to O., Mo., and Kans., s. to Fla. and Tex., and reported as intr. in Mass. and Me.

44. POLYPOGON Desf.   Spikelets 1-fld, articulated a little below the glumes; glumes equal, membranous, entire or 2-lobed, 1-veined, the vein excurrent as an awn from the tip or between the lobes; lemma much shorter than the glumes, very delicate in texture, often with a short, fragile, terminal awn; palea about equaling the lemma and of similar texture; tufted grasses with flat lvs, elongate, membranous ligule, and dense, ovoid to cylindric, spike-like panicles. 10, mainly temp. Eurasia.

1. Polypogon monspeliensis (L.) Desf.   Rabbitfoot-grass. Erect or ascending annual 1–4 dm; main lvs 3–8 mm wide, rather shoirt, scaberulous; infl ovoid-cylindric to cylindric, 2–5 × 1 cm, excluding the awns, tawny at maturity; glumes 2 mm, silky, shortly 2-lobed; awn rough, 3–6 mm; lemma 1 mm, its delicate, fragile awn 0.4–0.7 mm; $2n=28$. Native of Europe, established in waste places throughout most of our range, especially southward.

45. CINNA L.   Woodreed. Spikelets 1-fld, articulated below the glumes; glumes nearly equal, herbaceous or scarious-margined, linear to narrowly lanceolate, acute or acuminate, 1–3-veined; lemma herbaceous, about equaling the second glume, 3-veined, bearing a short straight awn just below the tip, distinctly stipitate above the glumes; palea 1-keeled; rachilla prolonged as a minute bristle behind the palea; anthers 2 or (our spp.) 1; tall perennials with wide, flat lvs, elongate, membranous ligule, and ample panicles of small spikelets. 5, N. and S. Amer., temp. Eurasia.

1 Second glume 3-veined; panicle-branches ascending, rather crowded .... 1. *C. arundinacea.*
2 Second glume 1-veined; panicle-branches divaricate or drooping, loosely fld .... 2. *C. latifolia.*

1. Cinna arundinacea L.   Common w. Culms erect, 10–15 dm, with usually 5–10 nodes; lvs 6–12 mm wide, scabrous at least on the margins; ligules tinged with reddish-brown; infl narrow, 1.5–3 dm, often somewhat drooping, with crowded, ascending branches, dull grayish-green; first glume 1-veined, 3.5–5.6 (avg 4) mm; second glume sharply 3-veined, usually distinctly scabrous, 4.1–6.6 (avg 5) mm; awn usually under 0.5 mm; anthers 0.8–1.8 mm; $2n=28$. Moist woods; Me. to Ont., Minn., and N.D., s. to Ga. and Tex.

2. Cinna latifolia (Trevir.) Griseb.   Drooping w. Much like no. 1; nodes usually 3–7; lvs to 15 mm wide; ligules colorless or nearly so; infl lax and open, with slender, divaricate or recurved branches, pale green and somewhat shining; first glume 2.3–3.8 (avg 3) mm; second glume 1-veined, 2.6–4.1 (avg 3.3) mm, nearly glabrous except on the keel, the scarious margins constituting half the total width; awn to 1.5 mm; $2n=28$. Moist woods; circumboreal, s. to Pa., n. Ill., Minn., and Calif., and in the mts. to N.C.

46. ALOPECURUS L.   Foxtail. Spikelets 1-fld, articulated below the glumes; pedicels conspicuously swollen at the joint; glumes equal, 3-veined, strongly compressed and keeled, ± connate at base, especially on the adaxial side; lemma thin and membranous, about as long as the glumes, 5-veined, the margins often ± connate, the midvein excurrent into an awn; palea in ours minute or none; low annuals and perennials with flat lvs, membranous ligule, and dense, cylindric or subcylindric, spike-like panicles., 25, N. Temp.

1 Glumes acute or acuminate, firm almost or quite to the tip, 4–6 mm.
  2 Glumes conspicuously ciliate on the keel, especially above the middle, with hairs 1–1.5 mm .... 1. *A. pratensis.*
  2 Glumes short-ciliate on the keel below the middle, merely scabrous beyond it .... 2. *A. myosuroides.*
1 Glumes obtuse or truncate and ± erose, 2–3.2 mm.
  3 Awn equaling the glumes or exsert by up to 0.5(–1) mm .... 3. *A. aequalis.*
  3 Awn surpassing the glumes by 1.5–3.5 mm.
    4 Perennial; anthers 1.3–2 mm .... 4. *A. geniculatus.*
    4 Annual; anthers 0.4–0.7 mm .... 5. *A. carolinianus.*

1. Alopecurus pratensis L.   Meadow-f. Perennial 4–8 dm, erect or decumbent at base; infl 2–8 cm × 5–10 mm, scarcely tapering; glumes acute or subacuminate, connate one-fourth their length, 4–5.5 mm, the keel narrowly winged, conspicuously ciliate, especially above the middle, with hairs 1–1.5 mm; awn attached below midlength of the lemma, geniculate near the top of the lemma and exserted 2.3–6.5 mm; anthers 2.2–3.5 mm; $2n=28$, 42. Native of Eurasia, naturalized in moist meadows, fields, and waste places nearly throughout our range, and westward.

2. Alopecurus myosuroides Hudson.   Tufted annual 3–6 dm, often decumbent at base; infl 5–10 cm × 3–5 mm, tapering to the tip; glumes acute or subacuminate, their margins connate (especially on the inner side) for one-third to half their length, 4.5–6 mm, the keel winged, ciliate below the middle, scabrous above; awn attached below midlength of the lemma, geniculate and exserted 5–7 mm; anthers ca 3 mm; $2n=14$. Native of Europe, occasionally found in fields and waste places at scattered stations nearly throughout our range.

3. **Alopecurus aequalis** Sobol. Short-awn f. Annual or short-lived perennial 2–5 dm; culms very slender, erect or decumbent at base, solitary or in small tufts; infl slender, 2–8 cm × 3–5 mm, scarcely tapering; glumes 2–2.7 mm, connate near the base, blunt and scarious at the tip, villous-ciliate on the keel; awn attached just below midlength of the lemma, straight, equaling the glumes or exsert up to 0.5(–1) mm; anthers 0.5–1 mm; $2n=14$. In mud or shallow water; circumboreal, s. to N.J., Ind., Mo., and Calif.

4. **Alopecurus geniculatus** L. Water- or marsh-f. Perennial; culms 2–8 dm, usually decumbent, sometimes rooting at the nodes; infl 3–5 cm × 4–6 mm, often purple-tinged; glumes as in no. 5; awn attached halfway between the base and the middle of the lemma, geniculate, exserted 1.5–3.5 mm; anthers 1.3–2 mm; $2n=28$. In mud and shallow water; a Eurasian or perhaps circumboreal sp., found throughout most of our range as an apparent introduction.

5. **Alopecurus carolinianus** Walter. Carolina-f. Densely tufted annual 1–6 dm, erect or ascending to rarely decumbent at base; infl cylindric, 2–5 cm × 4–5 mm; spikelets 2–2.4 mm; glumes connate only at the base, scarious and erose at the blunt tip, villous-ciliate on the keel, ± villous on the sides and base; awn attached halfway between the base and the middle of the lemma, ± geniculate, exserted 1.5–3 mm; anthers 0.3–0.7 mm; $2n=14$. Wet soil; Mass. to Minn., s. to Fla. and Tex., and w. to the Pacific.

47. PHLEUM L. Spikelets 1-fld, strongly flattened, articulated above the glumes; glumes equal, compressed and keeled, stoutly 3-veined, truncate or rounded to acuminate, with thin or scarious margins; lemma much shorter than the glumes, thin and delicate, broad, 3–5-veined, obtuse or truncate, awnless; palea narrow, up to nearly as long as the lemma; mostly tufted grasses with membranous, often elongate ligule, flat lvs, and very dense, cylindric to ellipsoid, spike-like panicles; our spp. perennials with awned glumes. 10, temp. and cool regions.

1 Infl cylindric, normally 5–10+ cm, several times as long as wide .... 1. *P. pratense.*
1 Infl ellipsoid or short-cylindric, mostly 1–5 cm, 2–3.5 times as long as wide .... 2. *P. alpinum.*

1. **Phleum pratense** L. Timothy. Tufted, usually 5–10 dm, the culms somewhat bulbous-thickened at base, minutely scabrous at the summit; upper sheaths terete, not dilated; blades ordinarily 5–8 mm wide, rough-margined; infl cylindric, (3–)5–10(–15) cm × 5–8 mm; glumes 2–3.2 mm, abruptly rounded-truncate to an awn 0.7–1.5 mm, hispidulous-ciliate on the keel, with thin, pale or green margins; $2n=14, 21, 28–84$. Native of Europe, cult. for hay and pasture, escaped and naturalized throughout most of the U.S. and s. Can.

2. **Phleum alpinum** L. Mountain-timothy. Tufted, 2–5 dm, the culms smooth at the summit, the base not bulbous, but tending to be decumbent or shortly creeping; upper sheaths distended, especially near their middle; blades usually 3–6 mm wide, rough-margined; infl ellipsoid to thick-cylindric, 1–5 cm × 8–12 mm; glumes 2.5–3.5 mm, hispid-ciliate on the keel, tipped with an awn 1.5–3 mm, the thin margins usually purple; $2n=14, 28$. Wet meadows and bogs; circumboreal at high latitudes, s. to the mts. of New England, and in n. Mich. and throughout the western cordillera; also in the colder parts of the s. hemisphere.

48. BECKMANNIA Host. Spikelets 1–2-fld, articulated below the glumes, strongly flattened, nearly circular in outline; glumes subequal, with approximate margins, 3-veined, the lateral veins obscure; lemma firm, ca as long as the glumes but much narrower, lanceolate, obscurely veined; palea lanceolate, partly enclosed by the lemma; stout grasses with overlapping sheaths, large ligule, usually flat blades, and elongate compound panicles of short, crowded, overlapping spikes. 2, the other in Europe.

1. **Beckmannia syzigachne** (Steud.) Fern. American sloughgrass. Annual 5–10 dm; lvs scaberulous, the larger 5–10 mm wide; ligule 5–8.5 mm; panicle 10–30 cm, erect, slender, with numerous overlapping branches 1–5 cm; spikes up to 2 cm, contiguous, those of a single branch resembling a 4-rowed spike; spikelets 2.5–3.5 mm long and wide; glumes cross-wrinkled, somewhat inflated toward the elevated midvein; lemma acuminate, its point barely protruding from the glumes; a second, sterile floret sometimes present; lodicules linear, nearly 1 mm; $2n=14$. Marshes and wet soil; Alas. to Calif. and N.M., e. to Ill.; O.; w. N.Y., where probably intr.; n. Asia.

49. BROMUS L. Brome, brome-grass (applied mostly to the perennial spp.); chess, cheat (applied mostly to the annual spp., from *B. secalinus*). Spikelets with several to many fls, eventually disarticulating between the lemmas and above the glumes, oval to narrowly oblong, subterete or laterally flattened; glumes somewhat unequal, shorter than the lemmas; lemmas 3–9-veined, 2-toothed at the tip, awnless or more often awned between the teeth; spikelets large, often in lax or drooping panicles; sheaths usually closed nearly to the top. 100, widespread in temp. reg.

1 Lemmas compressed and strongly keeled; first glume 3- or 5-veined, the second 5–9-veined; coarse, broad-lvd grasses. (*Ceratochloa*.)
  2 Awn mostly over 3 mm; palea nearly as long as the lemma .... 1. *B. carinatus*.
  2 Awn less than 3 mm, or none; palea ca half as long as the lemma .... 2. *B. wildenowii*.
1 Lemmas rounded or only weakly keeled on the back; glumes and habit various.
  3 Perennial; awns shorter than the lemmas or none; teeth of the lemma short, up to 0.5 mm.
    (*Bromopsis, Pnigma, Zerna*.)
    4 Plants with well developed creeping rhizomes, becoming colonial.
      5 Lemmas glabrous or scabrous; nodes and blades most glabrous .... 3. *B. inermis*.
      5 Lemmas hairy, at least marginally; nodes and upper lf-surfaces hairy .... 4. *B. pumpellianus*.
    4 Plants not rhizomatous, often tufted.
      6 Spikelets all or mostly shorter than their slender, often drooping pedicels; lvs broad, mostly 4–15 mm wide; native plants of natural habitats.
        7 Lemmas hairy mainly along the margins, or rarely glabrous throughout; anthers 0.9–1.7 mm .... 5. *B. ciliatus*.
        7 Lemmas hairy throughout.
          8 First glume 3-veined, the second 5-veined; anthers 1–2 mm .... 6. *B. kalmii*.
          8 First glume 1-veined (rarely with a pair of faint lateral veins), the second 3-veined (rarely with a faint additional pair).
            9 Lvs mostly 8–20 per culm; the sheaths overlapping, auriculate and densely hairy at the top; anthers 1.5–2.2 mm .... 7. *B. altissimus*.
            9 Lvs mostly 4–6 per culm, the sheaths shorter than at least the upper internodes, without an especially hairy apical zone; anthers 2.5–5 mm .... 8. *B. pubescens*.
      6 Spikelets mostly longer than their stiff, erect or ascending pedicels; lvs narrow, mostly 2–3 mm wide; weed of disturbed habitats .... 9. *B. erectus*.
  3 Annual, winter-annual, or sometimes biennial; awns often as long as or longer than the lemmas (but sometimes short or wanting); teeth of the lemma often more than 0.5 mm.
    10 First glume 3- or 5-veined, the second 5- or 7-veined; lemmas relatively short and broad, 6–12 mm, (1.2–)1.5–4 mm wide in side view, the apical teeth seldom over ca 2 mm; awns up to ca 12 mm. (*Bromus, Zeobromus*.)
      11 Lemmas very broad, at least the larger ones 2.5–4 mm wide in side view.
        12 Lemmas awnless or with an awn up to ca 1 mm .... 10. *B. brizaeformis*.
        12 Lemmas with a well developed awn mostly 5–12 mm .... 11. *B. squarrosus*.
      11 Lemmas narrower, less than 2.5 mm wide in side view.
        13 Panicle relatively compact, the lateral branches mostly closely ascending or erect and shorter than the spikelet that terminates them .... 12. *B. hordeaceus*.
        13 Panicle relatively open, the lateral branches mostly longer than the spikelet that terminates them.
          14 Margins of the lemmas soon involute, wrapping around the grain and exposing the rachilla .... 13. *B. secalinus*.
          14 Margins of the lemma gaping and overlapping in fr.
            15 Panicle-branches erect or ascending, straight or nearly so.
              16 Anthers (3–)4–5 mm; palea about equaling the lemma; panicle to 30 cm ...... 14. *B. arvensis*.
              16 Anthers 1.5–3 mm; palea shorter than the lemma; panicle seldom over 15 cm .... 15. *B. racemosus*.
            15 Panicle-branches (at least the lower) widely spreading, often flexuous.
              17 Awns straight or nearly so and erect-ascending, the longest ones of a spikelet not more than twice as long as that of the lowest lemma; branches of the infl stiffly spreading or ascending .... 16. *B. commutatus*.
              17 Awns flexuous or recurved-divergent, the longest ones of a spikelet more than twice as long as that of the lowest lemma; branches of the infl lax and flexuous .... 17. *B. japonicus*.
    10 First glume 1(–3)-veined, the second 3-veined; lemmas relatively long and narrow, mostly 0.5–1.5 mm wide in side-view, the apical teeth (1–)2–6 mm; longer awns (10–)12–60 mm. (*Genea, Stenobromus*.)
      18 Panicle dense, its branches to 1 cm .... 18. *B. rubens*.
      18 Panicle open, with long slender branches.
        19 Spikelets smaller, the first glume 5–14 mm, the second 8–17 mm, the lemmas 10–20 mm; awns 10–30 mm.
          20 First glume 5–7 mm, the second 8–11 mm; lemmas 10–12(–15) mm; awns (7–)10–17 mm .... 19. *B. tectorum*.
          20 First glume 7–14 mm, the second 9–17 mm; lemmas 13–18(–20) mm; awns 18–30 mm .... 20. *B. sterilis*.
        19 Spikelets larger, the first glume 13–20 mm, the second 20–30 mm, the lemmas 22–30 mm; awns (30–)35–60 mm .... 21. *B. rigidus*.

1. Bromus carinatus Hook. & Arnott. California b. Coarse, tufted, rather short-lived perennial 3–12 dm, often fl the first season, glabrous or hairy; blades flat and broad, 3–9 mm wide; ligule 1–3(–4) mm; panicles 10–20+ cm, loosely contracted, often nodding; spikelets 20–30(–40) mm, 5–10-fld, strongly compressed; glumes lanceolate, acute, strongly keeled, the first 7–11 mm, 3- or 5-veined, the second 9–13 mm, 5- or 7-veined; lemmas 11–15(–17) mm, keeled, ca 7-veined, soon spreading and exposing the elongate (2.5–5 mm) joints of the rachilla; awns (3–)4–6(–8) mm; palea nearly as long as the lemma; $2n=28$, 42, 56, 70. Moist open places; native to the w. cordillera, and casually intr. e. to Io. and Ill. (*B. marginatus.*)

2. Bromus wildenowii Kunth.   Rescue-grass. Stout annual or short-lived perennial to 1 m; blades flat and broad, 3–12 mm wide; ligule 2.5–5 mm; panicles 10–30 cm, loosely contracted; spikelets (16–)20–30(–35) mm, 4–12-fld, strongly compressed; glumes lanceolate, acuminate, strongly keeled, the first 8–11.5 mm, 3-veined, the second 9–13 mm, 5-veined; lemmas 12–20 mm, keeled, closely overlapping, glabrous (at least below) between the 11–13 long-ciliate veins; awns up to 2.5 mm, or none; palea half as long as the lemma; $2n=28$, 42, 56. Native of S. Amer., widely cult. for forage in warm regions, and occasionally intr. or escaped in the s. part of our range, n. to N.Y. and Mo. (*B. catharticus,* misapplied.) Perhaps properly to be submerged under *B. unioloides* HBK.

3. Bromus inermis Leysser.   Smooth b. Culms 5–10 dm from long rhizomes with dark brown scales; stems, sheaths, and blades glabrous, or the stems finely pubescent at the nodes; blades 8–15 mm wide; auricles vestigial or none; ligule 0.5–1 mm; infl 1–2 dm, open at anthesis, later contracted, often with 4–10 branches from a node; spikelets 15–30 mm, ca 3 mm wide, 7–11-fld; first glume 4–8 mm, 1-veined, the second 7–11 mm, 3-veined; lemmas 10–12 mm, 3- or 5-veined, the outer pair of veins often inconspicuous, the body obtuse or retuse, glabrous or scabrous, awnless or with an awn to 2.5 mm; $2n=28$, 42, 56, 70. Native of Europe, cult. for forage and often escaped; Que. to Minn., Alta., and the Pacific, s. to Md., O., Mo., and N.M.

4. Bromus pumpellianus Scribn.   Much like no. 3; nodes long-hairy; sheaths glabrous or hairy; lvs 5–8 mm wide, hairy above, commonly glabrous beneath; auricles sometimes well developed; ligule 1–2.5 mm; infl 6–15 cm, narrow; with 1–4 ascending branches from a node; glumes glabrous; lemmas 5- or 7-veined, evidently hairy at least at the base and margins, the awn 1.5–4(–5) mm; $2n=42$, 56. Native to w. U.S. and Can., disjunct in n. Mich. and s. Ont. Hybridizes readily with no. 3, and perhaps better called *B. inermis* ssp. *pumpellianus* (Scribn.) Wagnon, or var. *purpurascens* (Hook.) Wagnon.

5. Bromus ciliatus L.   Fringed b. Perennial 6–12 dm, the culms few or even solitary, glabrous, or hairy at the nodes; sheaths pilose or seldom glabrous, often overlapping; blades 4–10 mm wide, glabrous to sparsely villous on one or both sides; ligule 0.3–1(–1.5) mm; infl 1–2 dm, loose and open with slender, often flexuous, drooping or spreading branches to 15 cm; spikelets drooping, 15–25 mm, 4–10-fld; glumes glabrous or at most scabrous to minutely hispid, the first lance-subulate, 5–8 mm, 1-veined (rarely with a faint pair of lateral veins), the second lanceolate, 7–10 mm, 3-veined, often short-awned; lemmas 10–13 mm, rather long-hairy near the margins, especially toward the base, glabrous or nearly so on the back: awns 3–5 mm; anthers 0.9–1.7 mm; $2n=14$, 28, 56. Moist woods and other wet places; Lab. and Nf. to Alas., s. to Pa. (and in the mts. to Tenn.), Mo., and Tex. (*B. dudleyi.*)

6. B. kalmii A. Gray.   Perennial 5–10 dm, the culms loosely clustered or solitary, mostly glabrous except for the often hairy nodes; lvs mostly 3–6 per stem, the nodes mostly exposed; sheaths villous to occasionally glabrous; blades 1–2 dm × 5–10 mm, glabrous or hairy; ligule under 1 mm; infl nodding, 5–10 cm, the relatively few spikelets drooping on slender flexuous pedicels; spikelets 15–25 mm, 6–11-fld, softly villous; first glume 6–7 mm; 3-veined, the second 7–9 mm; 5-veined; lemmas 8–10 mm, 7-veined, obtuse, the awn 2–3 mm, the short lemma-teeth a third as wide as long; $2n=14$. Dry woods, rocky banks, and sandy soil; Me. to Minn. and S.D., s. to Md. and Io.

B. nottowayanus, Fern., of se. Va. and occasionally reported from elsewhere, differs in its longer awns (5–8 mm) and in having mostly 6–8 cauline lvs with the nodes covered by the sheaths. It may reflect hybridization with no. 7.

7. Bromus altissimus Pursh.   Much like no. 8; lvs more numerous, commonly 8–20 per culm, the sheaths longer than the internodes, densely villous externally in a ring at the top; blades with 2 well developed basal flanges usually prolonged into auricles or short divergent spurs; anthers 1.5–2.2 mm. Moist woods; Me. to N.C., w. to Mont. and Okla. (*B. latiglumis.*)

8. Bromus pubescens Muhl.   Perennial 6–15 dm, the culms solitary or few together, glabrous or nearly so, at least the upper nodes exposed; lvs 4–6 per stem; sheaths glabrous to densely canescent with spreading or somewhat retrorse hairs; blades 8–15 mm wide, narrowed at base, not auriculate, glabrous to sparsely villous; infl 1–2 dm, its elongate branches loosely spreading or drooping; spikelets 2–3 cm; first glume 5–8 mm, subulate, 1-veined, rarely with an additional pair of faint lateral veins; second glume broader, 6–10 mm, 3-veined, rarely with a faint additional pair; lemmas 7-veined, 9–12 mm, thinly to densely hairy throughout, the awn 2–8 mm; anthers 2.5–5 mm. Rich moist woods; Que. and N.Engl. to Fla., w. to Alta. and Ariz. (*B. purgans,* misapplied, a nomen rejiciendum.) Fl and fr before no. 7.

9. Bromus erectus Hudson.   Tufted perennial 3–12 dm; sheaths glabrous or minutely puberulent on the back, usually sparsely ciliate; blades 2–3 mm wide, sparsely pilose above; infl narrow, 5–15 cm, with short erect branches, spikelets 15–30 mm, usually longer than their pedicels, 7–10-fld; glumes subulate, the first 6–8 mm, 1-veined, the second 8–10 mm, 3-veined, lemmas 10–12 mm, conspicuously veined, glabrous or short-hairy, acuminate into an awn 4–7 mm; $2n=28$–112. Native of Europe, intr. along roadsides and in waste places from N.Engl. to Mich., Wis., and D.C.

10. Bromus brizaeformis Fischer & C. A. Meyer.   Rattlesnake-chess. Annual 2–6 dm; culms glabrous; sheaths and blades softly villous, the latter 3–6 mm wide; ligule 0.5–1.5(–2) mm; infl 6–15 cm, loose and open, tending to be nodding and secund, the slender branches often bearing only a single spikelet nearly or fully as long; spikelets somewhat flattened, lanceolate to ovate, 8–15-fld, 15–30 mm, to 13 mm wide; glumes broad, obtuse, several-veined, the first 5–8 mm, the second 6–9 mm; lemmas 9–12 mm, 7- or 9-veined, glabrous or minutely puberulent; very obtuse, somewhat inflated at base, 2.5–4 mm wide on a side; awn to 1 mm, or none; palea evidently shorter than the lemma; anthers scarcely 1 mm; $2n=14$. Native of sw. Asia, sparingly intr. throughout most of the U.S. except in the south.

11. Bromus squarrosus L.   Annual 2–6 dm; culms glabrous; sheaths and blades softly villous, the latter 3–6 mm wide; ligule 1–2 mm; infl to 20 cm, very lax, tending to be secund, subracemose and often with few spikelets, these broadly oblong or broadly lance-ovate, 22–34 × 6–10 mm, somewhat compressed, 8–30-fld, glabrous or occasionally hairy, often longer than their pedicels; glumes broad, obtuse or acutish, several-veined, the first 4.5–7 mm, the second 6–9 mm; lemmas 8–11 mm, several-veined, at least the larger ones 2.5–3.5 mm wide in side view and with a prominently angled shoulder near the middle of the broadly hyaline-scarious margin, the terminal teeth mostly 2–3 mm; awns of the upper lemmas 5–12 mm, usually divergent, those of the lower lemmas often shorter; palea evidently shorter than the lemma; anthers 1–1.2 mm; $2n=14$. Native of Europe, occasionally intr. in waste places in our range.

12. Bromus hordeaceus L.   Soft chess. Annual or sometimes biennial, 3–10 dm; culms glabrous or often puberulent; sheaths and blades hairy, the latter 2–6 mm wide; ligule 0.5–2 mm; infl rather dense, erect, 3–10 cm, its branches usually shorter than the spikelets; spikelets 10–25 mm, 6–9-fld, soft-hairy or rarely glabrous; first glume 5–7 mm, 3(5)-veined; second glume broader, 5–9-veined, 6–9 mm, often awned; lemmas 7–11 mm, ca 7-veined, 1.5–2.2 mm wide on a side, the terminal teeth 1–1.5 mm; awns straight, erect, 3–9 mm; palea shorter than the lemma; anthers 0.2–2 mm; $2n=28$. Native of Europe, sparingly intr. in our range in disturbed sites. (*B. mollis.*)

13. Bromus secalinus L.   Cheat; chess. Annual 3–12 dm; culms glabrous or sometimes retrorse-hairy below the nodes; middle and upper sheaths generally glabrous, the lower glabrous or hairy; blades 3–8 mm wide, glabrous or hairy; ligule 0.5–3 mm; infl loose and open, 7–15 cm, its branches several from a node, simple or again branched; spikelets 1–2.5 cm, usually drooping, 6–12-fld; first glume oblong, 4–6 mm, 3- or 5-veined, the second similar, 5–8 mm, 5- or 7-veined; lemmas elliptic, obtuse, 6–9 mm, obscurely 7-veined, glabrous or minutely scaberulous, rarely evidently hairy, awnless or with an awn to 6 mm between the two broad teeth; margins of the lemma soon involute, causing the florets to diverge and expose the flexuous rachilla; palea about as long as the lemma; anthers 1–2 mm; $2n=14, 28$. Native of Europe, where often a weed in wheat-fields; intr. in grain-fields, roadsides, and waste places throughout the U.S.

14. Bromus arvensis L.   Annual 3–10 dm, the culms glabrous; sheaths with scattered hairs; blades 10–20 cm × 2–6 mm, flat, sparsely hairy; panicle 15–30 cm, erect or eventually nodding, with long, slender, ascending to suberect branches; spikelets 10–25 mm, 4–10-fld; first glume 4–6 mm, 3(5)-veined, the second 6–8 mm, 5- or 7-veined; lemmas 7–9 mm, with a straight awn 6–10 mm; palea about equaling the lemma; anthers (3–)4–5 mm; $2n=14$. Native of s. Europe, sparingly intr. in our range and elsewhere in the U.S.

15. Bromus racemosus L.   Annual 3–10 dm, the culms usually glabrous; lower sheaths evidently hairy, the upper less so or glabrous; panicle 5–15 cm, narrow, erect or eventually nodding, with erect or strongly ascending branches; spikelets 10–15 mm, 4–10-fld; first glume 4–6 mm, the second 4–7 mm; lemmas 6.5–8 mm, elliptic or narrowly oblanceolate, rounded to a hooded tip; awn 5–9 mm, straight; palea slightly shorter than the lemma; rachilla-joints mostly 1–1.5 mm; anthers 1.5–3 mm; $2n=28$. Native of Europe, occasionally intr. in disturbed sites in our range, w. to the Pacific.

16. Bromus commutatus Schrader.   Hairy chess. Annual 3–12 dm, the culms glabrous, or puberulent below the nodes; lower sheaths softly retrorse-pilose, the upper glabrous or nearly so; blades flat, 1–7 mm wide, pilose; ligule 0.5–2.5 mm; panicle relatively short, 4–15(–20) cm, broad and open, the branches rather stiffly spreading or ascending, not flexuous or drooping, the pedicels mostly longer than the spikelets; spikelets 10–18(–30) mm, 4–10(12)-fld, slightly compressed; glumes scabrous or puberulent, the first 5–6(–7) mm, 3- or 5-veined, the second 6–7(–8.5) mm, 5–9-veined; lemmas 7–10 mm, 1.7–2.5 mm wide in side view, the veins faint, the terminal teeth not over 2 mm; awns 5–12 mm, straight or nearly so; palea less than 1.5 mm shorter than the lemma; anthers 0.7–1.7 mm; rachilla-joints 1.5–2 mm; $2n=14, 28, 56$. Native of Europe, commonly intr. in disturbed sites in our range, w. to the Pacific.

17. Bromus japonicus Thunb.   Japanese chess. Much like no. 16, and apparently passing into it, but the panicle avg a little larger, and with slender, flexuous, spreading to drooping branches; spikelets (13–)15–30 mm; lemmas (6–)7.5–9 mm, 1.2–2.2 mm wide in side view; awns 7–12 mm, flexuous or recurved-divergent, the longest ones of a spikelet more than twice as long as that of the lowest lemma; palea 1–2.5 mm shorter than the lemma; $2n=14$. Native of the Old World, intr. as a weed in waste places from N.H. and Vt. to Ga., w. to the Pacific.

18. Bromus rubens L.   Foxtail-chess. Annual 1–5(–8) dm, with fine retrorse pubescence; blades narrow, 1.5–4(–5) mm wide, sometimes involute; ligule 1–2.5(–3) mm; infl contracted, ovoid, short, to 10 cm, often purplish, the short branches and pedicels stout, ascending; spikelets 18–25(–30) mm, 3–8(–10)-fld, scaberulous to finely hirsute; glumes subulate, the first (5–)6.5–10 mm, 1-veined, the second 9–13 mm, 3-veined; lemmas subulate, 12–17 mm, 0.5–1.2 mm wide in side-view, 3- or 5-veined, the slender apical teeth 4–5 mm; awns 12–20(–24) mm, straight or slightly arcuate-spreading; palea shorter than the lemma; anthers 0.4–1 mm; $2n=28$. Native of the Mediterranean region, widely intr. as a weed in arid w. U.S., and casually adventive with us.

19. Bromus tectorum L.   Junegrass, downy chess. Annual 2–7 dm; sheaths and blades soft-hairy, the latter 2–4 mm wide; ligule 1–2.5(–4) mm; infl 1–2 dm, often nodding, repeatedly but rather openly branched, bearing somewhat crowded, mostly drooping, 3–8-fld spikelets to 2(–3.5) cm on slender pedicels; glumes subulate, the first 5–7 mm, 1-veined, the second 8–11 mm, 3-veined; lemmas narrowly lanceolate, 8–12(–15) mm, evidently hairy or seldom glabrous, 1–1.5 mm wide in side-view, 5–7-veined, acuminate into slender scarious teeth 1–3 mm; awns (7–)10–17 mm; palea shorter than the lemma; anthers 0.5–0.7 mm; 2n=14. Native of Europe, now a common weed nearly throughout the U.S., esp. westward.

20. Bromus sterilis L.   Barren brome. Annual 4–10 dm, the culm usually glabrous, but the sheaths and blades usually soft-hairy; blades 2–4 mm wide; ligule 1.5–2.5(–3.5) mm; infl relatively long, 1–2 dm, open, the elongate, capillary, often drooping branches mostly bearing only one or 2 spikelets, these mostly 25–31(–36) mm (to 5 cm incl. the awns), 4–8-fld, glabrous to strigose-puberulent; glumes subulate, the first 7–14 mm, 1-veined, the second 9–17 mm, 3-veined; lemmas very narrow, 13–18(–20) mm, 0.8–1.2 mm wide in side-view, 7-veined, the pair of veins nearest the midrib developed only distally, the outermost pair only proximally; awns 18–30 mm; palea almost equaling the lemma; anthers 0.9–1.6 mm; 2n=14, 28. Native of s. Europe, intr. in fields and waste places in much of the U.S., including most of our range.

21. Bromus rigidus Roth.   Ripgut-grass. Annual 3–8 dm; culms smooth; sheaths and blades pilose, the blades 3–6 mm wide; ligule 2–4(–5.5) mm; infl relatively long, 1–2 dm, with erect or spreading branches (or the lowest reflexed) each bearing 1 or 2 4–7-fld spikelets 30–40+ mm (to 9 cm incl. the awns); glumes subulate, the first 13–20 mm, 1-veined, the second 20–30 mm, 3-veined; lemmas subulate, 22–30 mm, 5-veined, scabrously short-hairy, 1–1.5 mm wide in side-view, the scarious teeth 3–6 mm; awns (30–)35–60 mm; palea shorter than the lemma; anthers 0.7–1.5 mm; 2n=28, 42, 56, 70. Native of the Mediterranean region, now widely intr. in w. U.S., and casually adventive in our range.

50. ELYMUS L.   Wild rye. Spikelets (1)2(–4) at each node of the spike, (1)2(–6)-fld, eventually disarticulating above (seldom below) the glumes and between the florets; glumes often narrow and awn-like, often displaced to form a false invol beneath the spikelets, seldom much reduced or even wanting; rachilla commonly twisted at base, bringing the florets into ± dorsiventral alignment with the axis; lemma long-awned or seldom awnless; self-pollinating perennial bunch-grasses with mostly flat lvs, short ligules, and small anthers, these mostly 1–3(–3.5) mm. (*Hystrix, Sitanion.*) 75, N. Hemisphere.

1 Spikelets all or nearly all solitary at the nodes of the rachis .... 1. *E. trachycaulus.*
1 Spikelets all or nearly all 2(–4) at the nodes of the rachis.
  2 Glumes well developed, those of a pair essentially similar.
    3 Spikes abruptly pendulous from near the base; lvs mostly 9–15, the larger ones mostly 15–25+
      mm wide .... 4. *E. wiegandii.*
    3 Spikes erect or merely arcuate; lvs usually narrower, often fewer.
      4 Glume-bases thin and striate, or indurated for only 1 mm; body of the glume not thickened
        adaxially.
        5 Awns of the lemmas mostly straight; larger paleas 5.5–8.5(–9) mm .... 2. *E. glaucus.*
        5 Awns of the lemmas mostly outcurved; larger paleas 8.5–12.5 mm .... 3. *E. canadensis.*
      4 Glume-bases clearly indurated and ± terete.
        6 Glumes 0.8–2 mm wide, ± expanded and flattened above the bowed-out base, swollen on
          half or all the adaxial surface as well as the abaxial base .... 5. *E. virginicus.*
        6 Glumes 0.2–1 mm wide, indurated and ± terete for 1–3 mm above the base, not or scarcely
          widened above the base.
          7 Lvs usually soft-hairy above; spike usually villous; rachis-internodes 1.5–3 mm .... 6. *E. villosus.*
          7 Lvs glabrous or scabrous above; spikes scabrous to hispidulous; rachis-internodes
            (3–)4.5(–7) mm .... 7. *E. riparius.*
  2 Glumes, or some of them, ± reduced and usually unequal, or wanting.
    8 Spikelets ± appressed to the rachis; glumes setaceous, tapering from an indurated base .... 8. *E. diversiglumis.*
    8 Spikelets widely spreading at maturity; glumes wanting, or occasionally represented by filiform
      bristles .... 9. *E. hystrix.*

1. Elymus trachycaulus (Link) Gould.   Slender wheatgrass. Tufted, 3–10(–15) dm; lvs usually flat, 2–6 (–8) mm wide; scabrous to pilose at least on the upper side; auricles short or none; spikes 4–15(–25) cm, compact, the mostly solitary spikelets 9–16 mm, overlapping, 3–5(–7)-fld, readily disintegrating at maturity; rachilla generally villosulous; glumes lanceolate to oblong-elliptic, acute to short-awned, broad and 5–7-veined, with membranous margins, the first one 6–10 mm, the second 7–12 mm; lemmas 7.5–10(–13) mm, glabrous or scabrous distally and sometimes short-hairy on the margins, awnless or with a straight awn up to as much as 3 cm; anthers 1–2 mm; 2n=28. Variable and often divided into several confluent vars. of dubious significance. Various habitats; Nf. to Alas., s. to W.Va., Ky., Ill., Mo., and Ariz. (*Agropyron pauciflorum; A. richardsonii; A. subsecundum; A. tenerum; A. trachycaulum; A. caninum,* misapplied.) A series of hybrids with various spp. of *Hordeum* has been called *E.* × *macounii* Vasey.

2. Elymus glaucus Buckley.   Blue w.r. Green to glaucous, tufted, 6–14 dm; lvs ± flat, 4–12 mm wide, glabrous to scabrous; auricles mostly well developed, ca 2 mm and clasping the stem; spikes 6–16 cm, erect, compact or loose below, spikelets mostly paired, 10–16 mm, 2–3(4)-fld; glumes subequal, (6.5–)9–14+ mm, narrowly lanceolate, broadest above the flat, sometimes slightly indurate base, 3–5-veined, almost parallel and concealing the base of the subtended florets, glabrous to scabrous, tapering into a short awn; lemmas 8.5–12+ mm, 5-nerved above, glabrous to scabrous on the nerves, tapering into a slender, usually straight awn (10–)14–22(–30) mm; larger paleas 5.5–8.5 mm; anthers 1.7–3 mm; $2n=28$. Open woods, moist meadows, and dry hillsides; widespread in the w. cordillera, e. less commonly to Ark., Io., n. Mich., Ont., and w. N.Y.

3. Elymus canadensis L. Canada w.r. Coarse, 8–15 dm, green or glaucous, in small clumps; lvs mostly 5–9 per stem, firm, flat, 3–12(–15) mm wide, glabrous or scabrous above; auricles 1–2 mm, often clasping the stem; spikes (–7)10–20(–30) cm, usually broadly arcuate or distally nodding, compact above, often loose or interrupted below, the internodes mostly 4–7 mm; spikelets mostly paired, or some of the lower in 3's or even 4's, 12–15 mm, 3–4-fld; glumes subequal, 15–30 mm overall, mostly 0.8–1.5 mm wide, 3–5-veined, broadest above the mostly flat and striate base, tapering into a straight awn usually equaling or longer than the body; lemmas 8.5–14 mm, usually strongly scabrous-hirtellous, 5–7-veined above, gradually tapering into a long, flexuous, spreading, scabrous awn 15–30(–35) mm; larger paleas 8.5–12.5 mm, the tip very acute and usually bidentate; anthers 2.5–4 mm; $2n=28$, 42. Streambanks and dry to moist fields and meadows; N.B. and Que. to Alas., s. to N.C., Tex., and Calif. A hybrid with *Hordeum jubatum* has been called × *Elyhordeum dakotense* (Bowden) Bowden.

4. Elymus wiegandii Fern. Broad-lvd w.r. Tufted, coarse, 10–20 dm, ± glaucous; lvs mostly 9–15, dark green, thin and flat, usually villosulous above, the larger ones 15–25+ mm wide, the sheaths often overlapping; spikes 10–35 cm, lax, abruptly pendulous from near the base, the middle internodes mostly 5–9 mm; spikelets mostly paired, 3–7-fld; glumes 15–30 mm overall, mostly 0.4–0.7(–1.2) mm wide, with abruptly narrowed, indurate but not terete base; lemmas villous-hirsute to glabrous or merely scabrous, the long awn outcurved at maturity; paleas 9–15 mm, the tip bidentate and densely hairy; $2n=28$. Rich alluvial soil in the shade; N.B. and Que. to Pa., w. less commonly to Mich., Minn., Io., Man., and Wyo.

5. Elymus virginicus L. Virginia w.r. Tufted perennial 5–12 dm; lvs mostly 6–10 per stem, flat, 4–10 mm wide, scabrous on both sides; auricles up to 1 mm, or wanting; spikes ridigly erect, 4–12(–16) cm, the base often included in the summit of the ± inflated uppermost sheath; spikelets mostly paired, 2–4(–5)-fld, disarticulating below the glumes; glumes subequal, 10–30 mm overall, firm, 0.8–2 mm wide, with a yellowish, cartilaginous, bowed-out base 1+ mm exposing the subtended florets, expanded above the base and tapering to the tip, flat and with 2–3 ciliolate veins abaxially, but swollen and uninerved for at least the basal half adaxially, usually ± long-awned, up to 4 cm overall; lemmas small, 6–9 mm, glabrous to crisp-puberulent, scabrous and becoming 5-veined above the middle, usually with a long straight awn up to 3.5 cm, but sometimes virtually awnless; larger paleas 6.5–8.5 mm, obtuse to truncate or slightly emarginate; anthers 1.5–3 mm; $2n=28$. Moist woods, meadows, and prairies; Nf. to Alta., s. to Fla. and Ariz. Highly variable, and often divided into a number of ± sympatric, intergrading vars. or forms. A hybrid with *Hordeum jubatum* has been called × *Elyhordeum montanense* (Scribn.) Bowden.

6. Elymus villosus Muhl. Downy w.r. Tufted, slender, 3–10 dm; lvs 5–7, usually softly villous on the upper side, the main ones 5–10 mm wide; sheaths glabrous to pilose; spikes 5–12 cm, arcuate, copiously villous-hirsute or sometimes glabrous to scabrous, closely fld, the internodes mostly 1.5–3 mm; spikelets 1(2)-fld, paired on the rachis; glumes setaceous, 15–30 mm overall, 0.4–1 mm wide, indurated and terete for 1–3 mm at the straight or somewhat bowed-out base, not widened above, strongly 1–3-nerved; lemmas 2–4 cm, including the straight, ascending awn; paleas mostly 5.5–6.5(–7) mm, obtuse, $2n=28$. Dry woods and banks; Vt. to N.D., s. to N.C. and Tex. A hybrid with *Hordeum jubatum* has been called × *Elyhordeum iowense* Pohl.

7. Elymus riparius Wieg. Streambank w.r. Much like no. 6; lvs 8–10, glabrous; spikes 6–15 cm, not so dense, the internodes (3–)4–5(–6) mm; spikelets scabrous to hispidulous, 2–4-fld; paleas mostly 7–8 mm, bidentate; $2n=28$. Moist woods and streambanks; Me. to N.C., w. to Wis., Io., Nebr., and Ark.

8. Elymus diversiglumis Scribn. & C. Ball. Minnesota w.r. Tufted, 7–15 dm; lvs 5–15 mm wide, usually hairy above, the sheaths commonly overlapping; spikes flexuous and ± strongly arcuate, 10–20 cm, the rachis flattened, ciliate-margined, with internodes (3–)5–7(–10) mm; spikelets usually paired, closely ascending, 2–3(4)-fld; glumes setaceous from an indurated base, very unequal, 2–15(–20) × 0.1–0.5 mm, or sometimes some of them obsolete; lemmas (7–)8–10(–12) mm, silvery-hirsute or rarely merely hirtellous, with a widely divergent awn 2–3 times as long; paleas 7.5–9 mm, obtuse, usually with an expanded, more densely hairy tip; anthers 2–3(–3.5) mm; $2n=28$. Borders of woods, clearings, river-terraces, and disturbed sites; Wis., Minn., and Io. to sw. Ont., Sask., and e. Wyo. (*E. interruptus*, misapplied.)

9. Elymus hystrix L. Bottlebrush-grass. Culms in small tufts or solitary, 6–15 dm; sheaths smooth to scabrous or hairy; lvs 8–15 mm wide; spikes ± erect, 5–12 cm, open, the internodes of the flexuous, 2-edged rachis 4–10 mm; spikelets (1)2(–4) per node, soon becoming horizontally spreading, disarticulating above the glumes (if any) and between the florets; glumes wanting, or unequal and filiform-setaceous, up to 16 mm; lemmas glabrous or hairy, linear, involute, rounded on the back, 5-veined distally, 8–11 mm, gradually tapering into a rough, straight awn 1–4 cm; paleas 8–10 mm, obtuse or truncate; $2n=28$. Woods; N.S. and Que. to N.D., s. to Va. and Okla. (*Hystrix patula*.)

51. ELYTRIGIA Desv. Wheatgrass. Spikelets solitary (seldom some of them paired) at each node of the bilateral spike, borne flatwise to the rachis, several-fld, eventually disarticulating above the glumes (below in *E. repens*) and between the florets; glumes broad or narrow, rounded on the back or keeled, usually several-veined, awnless or shortly awn-tipped; lemmas with a straight or divergent awn, or awnless; anthers elongate, (2–)3–8 mm; outcrossing perennial grasses, tufted or more often strongly rhizomatous. (*Lophopyrum, Pascopyrum, Pseudoroegneria.*) 80, mainly N. Hemisphere.

1 Plants with long creeping rhizomes, not tufted.
  2 Lemmas usually densely villous; glumes acute to acuminate, but not awned .... 2. *E. dasystachya.*
  2 Lemmas glabrous or merely scabrous or minutely short-hairy; glumes awnless or sometimes awn-tipped.
    3 Glumes slender, firm, gradually tapering from below the middle into an awn-tip .... 1. *E. smithii.*
    3 Glumes broader, oblong or lanceolate, with convexly curved margins.
      4 Culms hollow at anthesis; blades lax, flat, finely veined; widespread .... 3. *E. repens.*
      4 Culms pithy at anthesis; lvs stiff, coarsely veined, often involute; coastal .... 4. *E. pungens.*
1 Plants tufted, without rhizomes.
  5 Lemma evidently awned; glumes rounded to acute or shortly awn-tipped, but not thick or indurate .... 5. *E. spicata.*
  5 Lemma awnless; glumes thick, indurate, obtuse to truncate .... 6. *E. elongata.*

1. **Elytrigia smithii** (Rydb.) A. Löve. Western w. Strongly rhizomatous, usually glaucous, 3–8 dm; lvs flat, becoming involute in drying, 2–4.5 mm wide, the upper surface scabrous and sometimes pilose, strongly ridged and furrowed, with ca 7–14 furrows across the width; spikes stiff, erect, 5–15 cm, the middle internodes 5–10+ mm; spikelets 12–24 mm, mostly 4–8-fld, sometimes some of them in pairs; glumes slender and rigid, gradually tapering from below the middle and passing imperceptibly into a short awn-tip; lemmas 8–11+ mm, glabrous to sometimes densely short-hairy, awn-pointed or with an awn to 5 mm; glumes and lemmas only faintly veined; anthers 2–4.5 mm; 2*n*=56. Dry, open places, often in alkaline bottomlands; widespread in w. U.S., e. to Minn., Io., and Mo., and casually to N.Y. and Ky. Thought to be an alloploid of *E. dasystachya* and *Leymus triticoides.*

2. **Elytrigia dasystachya** (Hook.) A. & D. Löve. Thickspike w. Strongly rhizomatous, usually glaucous, 3–9 dm; lvs involute or sometimes flat, 1–3.5(–5) mm wide, firm, glabrous to scaberulous or sometimes pilose; spikes stiff, erect, 6–25 cm, the middle internodes 11–20+ mm, mostly 3–7-fld, sometimes some of them in pairs; glumes in var. *dasystachya* lance-oblong, acute to acuminate, broadest at or above midlength, usually more than half as long as the lemmas; in var. *psammophila* more slender and attenuate, approaching the form of *E. smithii,* up to half as long as the lemmas; lemmas 7–10 mm, acute, sometimes shortly awn-tipped, densely villous, or in var. *dasystachya* sometimes merely scabrous; anthers (3.5–)4–5 mm; 2*n*=28. (*Elymus lanceolatus.*) Var. *dasystachya* is widespread in the w. cordilleran region and on the n. Great Plains, and casually intr. eastward; var. *psammophila* (J. M. Gillett & Senn) Cronq. occurs on sand dunes along lakes Michigan and Huron in Ont., Mich., Ill., and Wis., disjunct from the natural range of var. *dasystachya.*

3. **Elytrigia repens** (L.) Nevski. Quack-grass. Strongly rhizomatous, green or occasionally glaucous, 5–10 dm, the culms hollow at anthesis; lvs mostly flat, 3–10+ mm wide, usually with scattered hairs above, the numerous slender veins ca 0.2 mm apart; spikes erect, 8–17 cm, the middle internodes 4–7 mm; rachis-joints usually flat on one side and rounded on the other; spikelets 10–18 mm, 3–8-fld, disarticulating below the glumes; glumes about half as long as the spikelet, 5–7-veined, lanceolate, acute and usually with an awn 0.5–4 mm; lemmas 7–10 mm, 5-veined, glabrous to apically scaberulous, tapering to a point or a short, straight awn to 5(–10) mm; anthers 4–5.5 mm; 2*n*=42. Native of Eurasia and possibly our Atlantic coast, now widespread as a weed in moderately moist, disturbed sites throughout most of the U.S. and Can. (*Agropyron r.*) A probable hybrid with *Elymus trachycaulus* has been called *Agropyron pseudorepens*; the concept of *Agropyron* is now restricted to the crested wheatgrasses.

4. **Elytrigia pungens** (Pers.) Tutin. Salt-marsh w. Strongly rhizomatous, very glaucous, 5–10 dm, the culms pithy at anthesis; lvs stiff, flat or often involute, scabrous, the nerves stout, convex, about as wide as the intervals; spikes compact, 8–12 cm; rachis-joints ± 4-angled; spikelets 13–20 mm, 7–11-fld; glumes with broader, stouter veins than no. 3; lemmas acute, sometimes short-awned; 2*n*=42. Borders of salt-marshes; N.S. to Mass.; also in Europe.

5. **Elytrigia spicata** (Pursh) D. Dewey. Bluebunch w. Green or glaucous, 4–10 dm, in dense tufts sometimes 1.5 dm wide at base; lvs numerous, mostly cauline, flat to loosely involute, 2–4 mm wide, usually pilose above; auricles clasping the stem; spikes 7–18 cm, slender, the middle internodes 9–17 mm, the remote spikelets shorter to slightly longer than the internodes, 12–16+ mm, 4–6(–9)-fld; glumes narrowly oblong to obovate, rounded to acute, rarely awn-tipped, scabrous-margined, glabrous to scabrous on the 4–5 veins; lemmas 8–10 mm, with divergent awns 9–15 mm, varying to sometimes (but not in our plants) awnless; anthers 4–6 mm; 2*n*=14, 28. Open, moderately dry places; widespread in w. N. Amer., e. to Sask., N.D., and Nebr.; disjunct on the Keweenaw Peninsula of Mich. (*Agropyron s.; Pseudoroegneria s.*)

6. **Elytrigia elongata** (Host) Nevski. Tall w. Tufted, glaucous perennial 7–15(–20) dm; lvs flat to loosely involute, 2.5–5(–8) mm wide, stiff, thick-veined, the auricles standing erect; spikes elongate, 10–30 cm, loose and open, the middle internodes (9–)15–20(–25) mm, the lower often longer; spikelets (13–)16–22 mm, 6–18-fld, becoming arcuate in age; glumes indurate, oblong 5–7-veined, obtuse to truncate; lemmas 8.5–13

mm, broadly lanceolate, obtuse or rounded, awnless, the midvein thickened; anthers 4–7 mm; 2n=14, 28, 42, 56, 70. Dry, open places, especially in alkaline soils; Mediterranean sp., escaped at scattered stations in w. U.S., and locally established in s. Ont. (*Agropyron e.; Elymus e.; Lophopyrum e.*)

52. LEYMUS Hochst.   Similar to *Elymus,* and like it with the rachilla twisted and the glumes displaced to stand side by side subtending the spikelets, but outcrossing, with anthers mostly over 3 mm, and with awnless or only shortly awned lemmas; plants tufted or more often on ± elongate rhizomes. 30, N. Hemisphere.

1 Densely tufted, scarcely rhizomatous; glumes setaceous, ca 0.5 mm wide; western .... 1. *L. cinereus.*
1 Vigorously rhizomatous; glumes 2–4 mm wide; beaches and dunes.
  2 Stems finely and densely hairy at the summit; glumes evidently hairy .... 2. *L. mollis.*
  2 Stems essentially glabrous at the summit; glumes glabrous or inconspicuously hairy .... 3. *L. arenarius.*

1. Leymus cinereus (Scribn. & Merr.) A. Löve.   Basin wild rye. Robust, densely tufted, 7–20 dm, with short or usually no rhizomes; lvs flat, 5–15 mm wide; ligule 2–7 mm; spikes straight and stout, 10–20 cm, sometimes slightly branched; spikelets (2)3–6 per node, 10–20 mm, (2)3–5-fld; glumes subequal, setaceous, 7–15 mm, only 0.5 mm wide; lemmas 8–11 mm, hirsutulous (glabrous), awnless or often with a short awn to 5 mm; anthers (3–)4.5–6 mm; 2n=28, 56. Moderately dry, open places; widespread in w. U.S. and adj. Can., e. to Sask., w. S.D., and w. Neb., and reported from Minn. (*Elymus c.; E. condensatus,* misapplied.)

2. Leymus mollis (Trin.) Pilger.   Amer. dunegrass. Stout, erect, green or somewhat glaucous, 5–15 dm, from long, stout rhizomes; culms finely hairy under the spike, otherwise glabrous; sheaths crowded and overlapping at base; lvs 6–15 mm wide; ligule scarcely 1 mm; spike stout, erect, dense, 1–3 dm × 1–2 cm; spikelets coarse, 4–6-fld, mostly paired at the nodes; glumes lanceolate, 13–30 × 2–4 mm, strongly 3–5-nerved, about as long as the spikelet, evidently villous-hirsute; lemmas hairy like the glumes, or more so, tapering to a slender, awnless tip; anthers 5–9 mm; 2n=28. Sandy beaches and dunes; both coasts of n. N. Amer., s. to Mass. and Calif., and on the shores of Lakes Huron, Michigan, and Superior, and on the Arctic shore and the coast of ne. Asia. (*Elymus m.*) Ours is the relatively southern var. *mollis.*

3. Leymus arenarius (L.) Hochst.   Lyme-grass. Much like no. 2 but more glaucous and less hairy, the culms glabrous or nearly so even at the summit, the glumes stiffer, glabrous or inconspicuously hairy; 2n=56. Native of n. and w. Europe, locally intr. about the Great Lakes. (*Elymus a.*)

53. HORDEUM L.   Barley. Spikelets 1(2)-fld, not disarticulating, borne in triads on opposite sides of the rachis, the lateral spikelets often pedicellate and sterile, the central one sessile and fertile; glumes elongate, awned or awn-like, setaceous throughout or widened at base, the 6 in each triad of spikelets forming a false invol to the florets; lemma of the lateral spikelets often reduced or abortive, that of the central spikelet indurate, obscurely veined, its rounded back turned away from the rachis, usually long-awned; rachilla prolonged behind the palea as a short bristle; grasses with mostly flat lvs, scarious, truncate ligules, and dense, bristly spikes which, except in *H. vulgare,* disarticulate at each joint. (*Critesion.*) 35, widespread.

1 Perennials with slender awns.
  2 Glumes 25–150 mm; awns of lemmas 10–60 mm .... 1. *H. jubatum.*
  2 Glumes 7–20 mm; awns of lemmas 5–10(–20) mm .... 2. *H. brachyantherum.*
1 Annuals with stout awns.
  3 Rachis not disarticulating; lvs 5–12+ mm wide, awns of lemmas 6–16 cm (rarely wanting) .... 7. *H. vulgare.*
  3 Rachis disarticulating; lvs 1–5 mm wide; awns of lemmas 0.5–3(–4.5) cm.
    4 Glumes of central spikelet at most scabrous-margined, not ciliate.
      5 Glumes unlike, those of the central spikelet and the inner glumes of the lateral spikelets
        broadened above the base, whereas the outermost glumes are awn-like from the base ...... 3. *H. pusillum.*
      5 Glumes all awn-like .... 4. *H. geniculatum.*
    4 Glumes of the central spikelet with ciliate margins.
      6 Floret of the central spikelet sessile or subsessile, the rachilla-joint not nearly as long as the
        pedicels of the lateral spikelets .... 5. *H. murinum.*
      6 Floret of the central spikelet with a rachilla-joint as long as or longer than the pedicels of the
        lateral spikelets .... 6. *H. leporinum.*

1. Hordeum jubatum L.   Foxtail-b. Tufted perennial (sometimes fl the first year) 3–7 dm, glabrous to densely soft-hairy; lvs flat to involute, 1.5–4(–5) mm wide; auricles wanting or less than 0.5 mm; spikes 4–10 cm (excl. awns), nodding, commonly purpurescent, often appearing as a mass of upwardly scabrous awns; central spikelet of a triad sessile, the lateral on curved pedicels 0.7–1.2 mm, their florets reduced, borne on rachilla-joints ca 0.7 mm; glumes long and awn-like, 2.5–6(–15) cm; fertile lemma 5.5–8 mm, faintly 5-veined, tapering into an awn 1–6 cm; anthers 1–1.5 mm; 2*n*=14, 28, 42. A weed along roadsides, and often becoming abundant in fields and moist or wet meadows; Nf. to Alas., s. to Del., Ill., Mo., Tex., and Calif. Hybridizes with several spp. of *Elymus*, q.v.

2. **Hordeum brachyantherum** Nevski.   Meadow-b. Tufted perennial 3–7 dm, glabrous to scabrous or hairy; lvs flat, 1–5(–7) mm wide, lacking auricles; spikes 2–5(–8) cm (excl. awns), erect, often becoming brownish-purple; central spikelet sessile, the lateral on curved pedicels 0.7–1 mm, their florets much reduced, rarely staminate; glumes awn-like 0.7–2 cm; fertile lemma 6.5–8 mm; faintly 5-veined, tapering into an awn 0.5–1(–2) cm; anthers 1–1.8 mm; 2*n*=28. Meadows, bottomlands, and salt-marshes; Alas. to Calif., e. to Mont. and N.M., and intr. here and there in our range. Often confused with the European *H. nodosum* L., with longer awns, longer anthers, and well developed auricles.

3. **Hordeum pusillum** L.   Little b. Annual to 4(–6) dm; lvs flat, 2–4 mm wide, erect, without auricles; spikes 2–7 cm (excl. awns), erect; central spikelet sessile, the lateral on curved pedicels 0.3–0.7(–1) mm, their florets reduced, awn-tipped, and borne on curved rachilla-joints ca 0.5 mm; glumes 7–12(–15) mm, those of the central spikelet and the inner glumes of the lateral spikelets slightly to conspicuously broadened above the base before narrowing into slender awns, the outermost glumes wholly awn-like and somewhat shorter, scabrous to puberulent; fertile lemma 5–7 mm, scabrous, faintly 5-veined, tapering into a short awn (4–)5–8 mm; anthers 0.5–0.8 mm; 2*n*=14. Native weed of pastures, roadsides, and moderately moist waste places; Del. to Fla., w. to Wash., Calif., and n. Mex., and casually intr. n. to Me., Mich., and Minn.; s. S. Amer.

4. **Hordeum geniculatum** All.   Mediterranean b. Much like no. 3; lvs narrow, 1–2.5(–4) mm wide; spikelets small, 1.5–4 cm (excl. awns); lateral spikelets on straight or slightly curved pedicels 0.7–1 mm, their florets much reduced, short-awned and borne on rachilla-joints 0.1–0.5 mm; glumes (6–)9–16 mm, all awn-like; fertile lemma glabrous, tapering into a short awn 8–16 mm, surpassing and stouter than the awns of the glumes; anthers 0.6–1 mm; 2*n*=14, 28. Moderately moist waste places; Mediterranean sp., commonly intr. in w. U.S. and sporadic with us. (*H. gussoneanum; H. hystrix.*)

5. **Hordeum murinum** L.   Much like no. 6; floret of the central spikelet sessile or subsessile, the rachilla-joint not nearly as long as the pedicels of the lateral spikelets; lemmas of the lateral spikelets larger, 10–18 mm; 2*n*=14, 28, 42. Roadsides, ditch-banks, and waste places; native of Europe, intr. and well established in w. U.S., sporadic with us.

6. **Hordeum leporinum** Link.   Annual to 4(–6) dm; lvs flat, 2–5 mm wide, glabrous or hairy; auricles well developed, to 3.5 mm; spikes 4–7(–10) cm (excl. awns), erect, often partly enclosed in the inflated upper sheath; all 3 spikelets of the triad well developed, the central one sessile, its floret on an elongate rachilla-joint 1–2 mm and equaling or even surpassing the pedicels of the lateral spikelets; lateral spikelets usually staminate; glumes 15–22(–28) mm, those of the central spikelet and the inner glumes of the lateral spikelets ± lanceolate, 3-nerved and ciliate on the margins, the nerves scabrous, the outer glumes of the lateral spikelets awn-like and somewhat longer, scabrous; fertile lemma (6–)7–10 mm, glabrous, 5-veined, those of the lateral spikelets somewhat larger, 7–20 mm, sterile, all 3 lemmas tapering into awns 18–30 mm; anthers 0.8–1.2 mm; 2*n*=14, 28, 42. Moderately moist waste places; native of s. Europe, intr. and well established in w. U.S., sporadic with us. Perhaps better subordinated to no. 5.

7. **Hordeum vulgare** L.   Barley. Glabrous annual 6–12 dm; lvs 5–12+ mm wide, with well developed auricles to 6 mm; spikes stout, 6–9(–12) cm (excl. awns), the rachis remaining intact; all 3 spikelets of a triad sessile and fertile (6-rowed barley), or the lateral pair sterile (2-rowed barley); glumes subequal, 6.5–20 mm, linear, broadened below, 3-veined, ascending-pilose, tapering into slender, scabrous awns; fertile lemmas 6.5–12 mm, 5-nerved, tapering into long, stout, flattened, scabrous-margined awns 6–16 cm, or (pearl-barley) awnless and 3-lobed at the tip; anthers 2–2.5 mm; 2*n*=14, 28. A European cultigen, occasionally found as a waif along roads and railways in our range. (*H. distichon* L., two-rowed barley.)

54. AGROPYRON Gaertn.   Crested wheatgrass. Spikelets solitary at each node of a bilateral spike, borne flatwise to the flexuous or zigzag rachis, several-fld, eventually disarticulating above the glumes and between the florets; internodes of the rachis very short, mostly 0.5–2(–3) mm; glumes strongly and somewhat excentrically keeled, the keel commonly produced into a short awn; lemmas firm, distally keeled and usually short-awned, with 1 or 2 pairs of less prominent lateral veins; anthers ± elongate, (2–)3–5 mm; outcrossing perennial bunchgrasses. One broadly defined (as here) or several closely allied spp., native to Eurasia.

1. **Agropyron cristatum** (L.) Gaertn.   Plants 3–8 dm; lvs 1.5–6(–10) mm wide; ligules very short; spikes short and stout, 3–7(–10) cm; spikelets 7–10 mm, variously glabrous or scabrous to evidently rough-hairy; glumes and lemmas short-awned or nearly awnless; 2*n*=14, 28, 42. Eurasian sp., widely planted in w. U.S. as a range grass, and locally established here and there in our range. (*A. pectiniforme*, with subglabrous spikelets; *A. sibiricum* and *A. desertorum*, with relatively small spikelets not quite so closely crowded and not so strongly divergent from the rachis. These segregates perhaps to be recognized at some level, but the taxonomy still not clear.)

55. SECALE L.   Rye. Infl a bilateral spike with sessile, solitary spikelets borne flatwise to the rachis; spikelets 2-fld, the rachilla prolonged beyond the upper floret as a bristle-tip that may bear a vestigial third floret; glumes narrow, rigid, keeled, lemmas lance-subulate, excentrically keeled, 5-veined, shortly bilobed, the keel and marginal nerves strongly ciliate, tapering above into a straight rough awn; caryopsis deeply furrowed in back, hairy at the top; annual or perennial grasses. 5, temperate Eurasia.

1. **Secale cereale** L.   Rye. Robust annual (biennial) 6–12 dm, branched from the base; lvs flat, 3–7(–10) mm wide, spikes stout, 6–15 cm, arcuate-nodding; spikelets disarticulating above the glumes and between the florets; glumes equal, 7.5–15 mm, the strong keel scabrociliate; lemmas 12–16 mm, pectinate-ciliate on the keel and margins, with awn 2–7 cm; anthers ca 8 mm; $2n=14$, 16, 27–29. European cultigen, often spontaneous in disturbed sites with us, but probably never persistent.

56. **TRITICUM** L.   Wheat. Infl a thick, bilateral spike with sessile, solitary spikelets borne flatwise to the rachis; spikelets 2–5-fld, turgid but laterally somewhat flattened, disarticulating above the glumes and between the florets; glumes thick and firm, 3–several-veined, excentrically keeled, mucronate or awned from the bidentate summit; lemmas broad, excentrically keeled, firm, acute to awned, with 5–7 nonconvergent veins; caryopsis deeply furrowed in back; annual or winter-annual grasses. 20, Eurasia.

1. **Triticum aestivum** L., nom. conserv.   Common w. Annual or winter-annual, 5–12 dm, usually branched below; lvs flat, 2–10(–20) mm wide; spikes mostly 5–10 cm, glumes subequal, 6–10 mm, lemmas 7–12 mm, similar to the glumes, indurate but with scarious margins, the tip broadly lobed and mucronate, or with an awn to 7 cm; anthers 2.5 mm; $2n=42$. Eurasian cultigen, often spontaneous in fields and along roadsides with us, probably never truly persistent, but repeatedly re-established.

57. **AEGILOPS** L.   Goat-grass. Spikelets 2–5-fld, turgid, borne flatwise singly and fitting closely into alternate hollows on 2 sides of the rachis, the joints thickened and exposed at the top, the whole spike compact and cylindrical, at maturity breaking off at the base or disarticulating into joints; glumes indurate, many-veined, not or scarcely keeled, 2–4-toothed at the truncate tip, 1 or more of the teeth sometimes produced into an awn; lemmas rounded on the back, indurate above, the veins not convergent, toothed at the truncate summit, one or more of the teeth often prolonged into an awn; caryopsis furrowed on the back, hairy at the top; annual grasses with flat lvs. 15, Medit. reg. and w. Asia.

1. **Aegilops cylindrica** Host.   Jointed g.-g. Tufted winter annual, 3–7 dm, decumbent at base; lvs 1–5 mm wide; ligule short; spikes 7–18 cm, 5 mm thick, disarticulating at the nodes when mature; glumes 7–10 mm, asymmetrical, unequally 9–13-veined, mostly bearing one long awn and one short awn or tooth, lemmas 8–11 mm, asymmetrical, 5-veined, those of the upper (usually sterile) spikelets with an awn to 10 cm, those of the lower with progressively shorter awns or none, $2n=28$. Native of s. Europe and w. Asia, established and becoming a noxious weed in wheat-fields in w. U.S., e. less commonly along roadsides and in other disturbed habitats to Mich., Ind., and even N.Y. Hybridizes with wheat. (*Cylindropyrum c.*)

A. ovata L., an e. Mediterranean sp. with short, ovoid spikes bearing 2–5 spikelets, the glumes each with 4 long awns, is reported to be established as a weed in Va.

58. **CHASMANTHIUM** Link.   Spikelets 2–many-fld, flat, disarticulating above the glumes and between the lemmas; glumes shorter than the lemmas, acute or acuminate, 3–7-veined, compressed-keeled, the keel serrulate; lower 1–4 lemmas empty; lemmas acuminate, entire or bifid, 5–15-veined, compressed-keeled, the keel serrate or ciliate; palea firm, sharply 2-keeled, the keels arching and exposed; tall perennials from short or long rhizomes, with flat, serrate lvs and large, branched or spike-like panicles; ligule a short, hyaline, ciliate membrane. 5, e. and c. U.S. (*Uniola*, in part.)

1 Spikelets (15–)20–40+ mm, 6–17+-fld, broadest below the middle, in an open infl .... 1. *C. latifolium.*
1 Spikelets 5–10 mm, 3–5(–7)-fld, broadest above the middle, in a narrow infl.
  2 Sheaths glabrous or nearly so except for the ciliate margins .... 2. *C. laxum.*
  2 Sheaths evidently pubescent at least at the summit .... 3. *C. sessiliflorum.*

1. **Chasmanthium latifolium** (Michx.) Yates.   Wild oats. Loosely colonial from short, stout rhizomes, 10–15 dm; sheaths glabrous; lf-blades 1–2 dm × 1–2 cm; infl open, nodding or drooping; spikelets green, nodding on slender pedicels, ovate or lance-ovate, 6–17-fld, (15–)20–40 mm; glumes and the 1–2 sterile lemmas lance-

linear, subequal, two-thirds as long as the lance-ovate fertile lemmas, these 11-15-veined, with pilose callus; palea much shorter than the lemma; grain 3–5 × 2–2.5 mm, mostly enclosed between the lemma and palea; $2n=48$. Moist woods and streambanks; N.J. to Ga. and nw. Fla., w. to s. O., Ind., Ill., Kans., and Tex. (*Uniola l.*)

2. **Chasmanthium laxum** (L.) Yates. Tufted on short rhizomes, 5–15 dm; sheaths essentially glabrous except for the ciliate margins; blades elongate, 3–7 mm wide, attenuate into a long, almost filiform tip; infl virgate, 1.5–4 dm; spikelets sessile or short-pedicellate, 3–5(–7)-fld, 5–10 mm, reverse-triangular or V-shaped; glumes 1.5–3 mm, equal or subequal; lemmas (the lower 1 or 2 empty) acuminate, 5–7(–9)-veined, with glabrous callus; palea shorter than the lemma; grain 2–3 × 1–2 mm, exposed, widely spreading the lemmas and paleas at maturity; $2n=24$. Woods, meadows, and swamps; Fla. to Tex., n. to L.I., N.J., Ky., and s. Mo. (*Uniola l.*)

3. **Chasmanthium sessiliflorum** (Poiret) Yates. Much like no. 2, but the sheaths villous at the summit or throughout; lvs mostly 6–12 mm wide; $2n=24$. Moist open woods and streambanks; Fla. to Tex., n. to se. Va., Tenn., and e.c. Mo. (*Uniola s.*)

59. DANTHONIA Lam. & DC., nom. conserv. Wild oatgrass. Spikelets several-fld, articulated above the glumes and between the lemmas; rachilla not prolonged beyond the upper palea; glumes extending beyond the lemmas, thin, lanceolate, acuminate, 3- or 5-veined, obscurely keeled (or more evidently so distally); lemmas closely imbricate, rounded on the back, obscurely several-veined, ± pilose at least on the margins, ending in 2 short to more often prominent teeth flanking a terminal awn (or mucro), the flat brown base of the awn slightly twisted when dry, the upper part straight, usually divergent from the spikelet; palea elliptic to obovate, extending about to the base of the awn; tufted perennials with narrow, often involute lvs, a ligule of hairs, and small, lax or racemiform panicles with large spikelets. 100+, widespread in temp. regions. (*Sieglingia.*)

1 Lemma awnless, minutely 3-toothed at the tip .... 1. *D. decumbens.*
1 Lemma with a terminal awn flanked by a pair of prominent teeth.
  2 Teeth of the lemmas 0.8–1.8 mm, triangular-acuminate .... 2. *D. spicata.*
  2 Teeth of the lemmas 2–4.5 mm, aristate or setaceous.
    3 Awn 4.5–10 mm.
      4 Lemmas pilose on the back as well as on the margins; infl pale green .... 3. *D. compressa.*
      4 Lemmas glabrous on the back, pilose only on the margins; infl purplish .... 4. *D. intermedia.*
    3 Awn 11–18 mm .... 5. *D. sericea.*

1. **Danthonia decumbens** (L.) DC. Heather-grass. Culms 2–5 dm; lvs 1–3 mm wide; raceme 2–4 cm, with 3–8 pediceled spikelets; glumes rounded below, keeled distally, 7–11 mm, the first slightly the longer; lemmas 4–6 mm, minutely 3-toothed; $2n=18, 24, 36, 124$. Boggy or peaty soil; Europe; also in Nf. and N.S., where possibly native. (*Sieglingia d.*)

2. **Danthonia spicata** (L.) F. Beauv. Poverty-o. Densely tufted, 2–6 dm; lvs borne mostly at or near the base, with a prominent tuft of white hairs at each side of the summit of the sheath, the blades curly, usually involute, seldom over 12 cm, 1–2 mm wide; lower sheaths often containing cleistogamous spikelets; infl contracted, racemiform, 2–5 cm, the short branches rarely with more than one spikelet; glumes 8.5–13 mm; lemmas broadly ovate, sparsely pilose on the back, 3.4–5.2 mm, including the triangular-acuminate teeth 0.8–1.8 mm; awn 4.5–7 mm; often apomictic, with abortive anthers; $2n=36$. Dry woods in sandy or stony soil; Nf. to Fla., w. to B.C. and N.M.

3. **Danthonia compressa** Austin. Much like no. 2 and perhaps not sharply distinct; stem very slender, flattened, 4–8 dm; sheaths glabrous; blades usually flat, often to 20 cm, 2–4 mm wide, scaberulous; infl lax, 5–10 cm, the filiform branches often reflexed at anthesis, at least the lower branches with 2 or 3 spikelets; glumes 9–13 mm; lemmas 4.5–8 mm, of which the teeth constitute half, sparsely pilose on the back; awn 5.5–8 mm; $2n=36$. Dry woods; N.S. and Que. to O., N.C., and Tenn.

4. **Danthonia intermedia** Vasey. Timber-o. Densely tufted, 2–5 dm, the culms slender, smooth; sheaths distally pilose, the lower often containing cleistogamous spikelets; blades flat or involute, 5–10 cm × 1–3 mm, glabrous or sparsely pilose; infl contracted, 2–6 cm, the branches appressed, mostly shorter than their single terminal spikelet, or the lowest bearing 2 spikelets; glumes 10–16 × 3–4 mm, suffused with purple; lemmas 6.5–9 mm, of which a third is the setaceous teeth, glabrous on the back, sparsely pilose on the margins, chiefly below the middle; awn 6–10 mm; often apomictic, with abortive anthers; $2n=18, 36$, ca 96. Meadows, bogs, and barrens; Lab. and Nf. to Alas., s. to Que., n. Mich., and Ariz.

5. **Danthonia sericea** Nutt. Downy o. Densely tufted, 5–12 dm; blades flat or involute, 2–5 mm wide, the cauline ones relatively broad and usually flat, those of the innovations narrower and often involute; infl lax, 5–10 cm; glumes narrowly lanceolate, 11–18 mm; lemmas 6.5–10 mm, including the setaceous teeth (these 2.8–4.5 mm) villous at least on the margin with copious white hairs 2–3 mm; awn 11–18 mm; $2n=36$. Var. *sericea*, with the sheaths villous and the lemmas villous on the back, occurs in dry sandy soil and pine

woods, chiefly on the coastal plain, in e. Mass. and from s. N.J. to Ky., Fla., and La. Var. *epilis* (Scribn.) Blomq., with glabrous sheaths and very thin lemmas that are villous on the margins only (or also at base), occurs in bogs, stream-borders, and seeps from s. N.J. to Va. and Ga. (*D. epilis.*)

60. MOLINIA Schrank.   Spikelets 2–4-fld, disarticulating above the glumes and between the lemmas; uppermost floret reduced or abortive; glumes subequal, shorter than the lemmas, the lower 1-veined or veinless, the upper 1–3-veined; lemmas chartaceous, rounded on the back, not scarious, obscurely 3-veined; palea about equaling the lemma; rachilla-joints elongate; densely tufted grasses with long flat blades and elongate upper internodes, bearing a slender panicle; ligule a row of hairs. 2–3, temp. Eurasia.

1. **Molinia caerulea** (L.) Moench.   Moorgrass. Perennial 5–12 dm, thickened at the nodes, the lower internodes very short, the lowest node often disarticulating; sheaths pilose at the throat; blades 2–4 dm × 3–6 mm; infl narrow, purplish, 1–3 dm, the ascending branches fascicled; glumes lanceolate, mostly 1-veined, the first 2.1–2.6 mm, the second 2.4–3 mm; lemmas glabrous, acute or acutish, 3.4--4.5 mm; rachilla-joints 1.3–2 mm, glabrous or hairy, usually short-pilose at the top; $2n=36$. Native of Europe, intr. in fields, meadows, and roadsides from Me. to Pa., w. to Ont. and Wis.

61. PHRAGMITES Adans.   Spikelets few-fld, disarticulating above the glumes and at the base of the rachilla-joints between the lemmas; rachilla (except below the first lemma) densely villous; glumes 3-veined, very unequal; lemmas narrow, elongate, acuminate, 3-veined, much longer than the palea; tall perennials from stout rhizomes, with short truncate ligules, broad blades, and large, dense, copiously branched panicles. 3, cosmop.

1. **Phragmites australis** (Cav.) Trin.   Common reed. Stout, 2–4 m, extensively colonial; ligule 1 mm; lvs flat, long-acuminate, usually 2–3 cm wide; infl subtended by a tuft of silky hairs, tawny, 2–4 dm, usually declined; spikelets 3–7-fld, narrowly linear before anthesis; first glume narrowly elliptic, rather blunt, the second linear, nearly twice as long; lemmas very narrow, 8–12 mm, the upper usually the shorter; hairs of the rachilla white, as long as the lemmas, exposed after anthesis as the lemmas diverge; $2n=36, 48, 54, 72, 84, $ ca 96. Swamps and wet shores, tolerant of salt, seldom producing seed; nearly cosmop, and throughout most of our range. (*P. communis.*)

62. ARISTIDA L.   Three-awn. Spikelets 1-fld, articulated above the glumes; glumes membranous, linear to lanceolate, 1(2–5)-nerved, acute, acuminate, or short-awned, the first sometimes deciduous; lemma indurate, linear, obscurely veined, closely convolute about the palea and grain, often scabrous on the obscure keel or toward the tapering summit, and with a sharp, usually bearded callus; awns of the lemma normally 3, elongate, similar or dissimilar; annual or perennial, often weedy, often branched from some or all the nodes, with narrow, often involute lvs, minute ligule, and terminal, usually slender or spike-like, dense or lax panicles or racemes. The awns, erect and parallel when immature or moist, diverge when dry and mature into characteristic positions. Lengths of the glumes and lemmas are here measured from the base of the spikelet. 200, mainly warm reg.

1 Plants perennial; awns as in spp. 6–7 below, not as in spp. 8–12.
  2 First glume ca half as long as the second; central awn 4–7 cm . . . . 1. *A. purpurea.*
  2 First glume nearly or fully as long as (or even longer than) the second; central awn mostly 1.5–
      3.5 cm.
    3 Lower sheaths glabrous, or merely pilose with straight (often appressed) hairs.
      4 Central and lateral awns, when dry, ascending and about equally divergent; glumes 1-nerved
        . . . . 2. *A. purpurascens.*
      4 Central awn, when dry, bent at base into about a quarter-circle, hence strongly deflexed in
        comparison with the erect or ascending lateral awns; first glume usually 2-nerved.
        5 Lemma 4–6 mm; first glume 5.5–9.5 mm, the second 6.5–10.5 mm . . . . 3. *A. virgata.*
        5 Lemma 7–10 mm; first glume 8–14 mm, the second 10–13 mm . . . . 4. *A. palustris.*
    3 Lower sheaths woolly with soft, tangled hairs . . . . 5. *A. lanosa.*
1 Plants annual.
  6 Awns persistent, not articulated with the lemma, not more than one of them coiled at base.
    7 Central awn, when dry, not coiled at base, merely outcurved and differentiated from the lateral
      ones only in length or position or not at all.
      8 First glume 3- or 5-veined, 12–29 mm; lemma 10–18 mm, its central awn (3.5–)4–7 cm . . . . 6. *A. oligantha.*
      8 First glume 1-veined, under 10 mm; lemma 3–10 mm, its central awn up to 2.5 cm . . . . 7. *A. longespica.*
    7 Central awn, when dry, coiled at base into a half to 3 full turns and therefore sharply differentiated
      from the straight or nearly straight, much shorter lateral awns.

9 First glume 10–21 mm, 3- or 5-veined; second glume 15–28 mm; lemma 11–21 mm; central
   awn 1.7–3.7 cm .... 8. *A. ramosissima.*
9 First glume 5–12 mm, 1-veined; second glume 6–15 mm; lemma 4.5–10.5 mm; central awn
   0.3–1.9 cm.
10 First and second glumes subequal, both surpassing the point of insertion of the awns of
   the lemma .... 9. *A. dichotoma.*
10 First glume one-half to three-fourths as long as the second, often not reaching the level
   of insertion of the awns of the lemma .... 10. *A. basiramea.*
6 Awns deciduous, articulate with the summit of the lemma, all similar, divergent when dry to a
  nearly horizontal position.
11 Awns separate almost to the articulation .... 11. *A. desmantha.*
11 Awns united at base into a twisted column ca 1 cm .... 12. *A. tuberculosa.*

1. **Aristida purpurea** Nutt.   Purple t.-a. Perennial; culms 2–5 dm, very numerous in large tufts; lvs mainly basal, very narrow, usually involute; panicle (excluding the awns) 3–10 cm, narrow or spike-like, with short ascending or appressed branches; first glume 8–13 mm; second glume 15–23 mm, much exceeding the lemma; awns essentially similar and equally divergent, 4–7 cm; $2n=22$. Dry prairies and plains; Minn. and Io. to Wash., Calif., and n. Mex. Ours is the widespread var. *longiseta* (Steudel) Vasey. (*A. longiseta.*)

2. **Aristida purpurascens** Poiret.   Arrowfeather t.-a. Perennial 4–10 dm, tufted on a knotty base; lvs to 2 dm, mostly flat, 1–4 mm wide; lower sheaths covering the nodes, usually ± pilose; infl 1–2 dm, slender, loosely or densely spike-like, with short, ascending or appressed branches; glumes 1-veined, the first 8–14 mm, scabrous on the keel and sometimes on the sides, the second 6.5–11.5 mm, almost always exceeding the lemma and exceeded by the first glume; lemma 5.5–9.5 mm; awns about equally divergent, the central one 2–3.5 cm, the lateral 1.5–2.5 cm. Dry sandy soil and prairies; Mass. to s. Ont., Wis., and Kans., s. to Fla. and Tex.

3. **Aristida virgata** Trin.   Perennial 5–10 dm, tufted on a knotty base and strictly erect; nodes mostly exposed; blades flat or drying involute, 1–2 mm wide, often 2–3 dm long, tapering to a very fine point; panicle elongate, 3–4 dm, strict, with short erect branches; glumes subequal, the first commonly with a lateral vein on one side only, 5.5–9.5 mm, the second 1-veined, 6–10.5 mm; lemma 4–6 mm; central awn 1.5–2.2 cm, strongly curved-divergent from the base; lateral awns 0.8–1.5 cm, erect and scarcely protruding from the panicle. Dry sandy soil of the coastal plain; s. N.J. to Fla. and Tex.

4. **Aristida palustris** (Chapman) Vasey.   Much like no. 3, often taller, to 15 dm; sheaths glabrous to pilose (the hairs often appressed); lvs 2–4 mm wide; first glume with a lateral vein on one side only, scabrous on the keel, 8–14 mm; second glume 1-veined, subglabrous, 10–13 mm, usually slightly exceeding the first; lemma 7–10 mm, its central awn 1.5–4 cm, when dry reflexed to a nearly horizontal position; lateral awns 1–3 cm, ascending. Cumberland Plateau in Ky.; damp woods of the coastal plain from N.C. to Fla. and Tex. (*A. affinis,* misapplied.)

5. **Aristida lanosa** Muhl.   Woolly t.-a. Perennial; culms stout, solitary or few together, 5–15 dm, the nodes mostly covered by the sheaths, these woolly with soft, tangled hairs; lvs flat, 3–6 dm × 2–5 mm; infl 2–5 dm, with woolly nodes and loose, slender, ascending branches; glumes 1-veined, the first slightly falcate-outcurved, scabrous on the keel and usually puberulent on the sides, 9.5–19 mm, the second glabrous or nearly so, somewhat shorter, 8–15 mm; lemma 7–12 mm, scabrous distally; central awn 1.5–3 cm, conspicuously deflexed to one side; lateral awns half to two-thirds as long. Dry sandy soil on the coastal plain; N.J. to Fla. and Tex., thence ne. to Mo. and Okla.

6. **Aristida oligantha** Michx.   Prairie t.-a. Annual 2–4 dm, branched from the base and usually from all the nodes; lvs 1 mm wide, flat or involute, tapering to a filiform point; terminal panicle 1–2 dm, very lax, few-fld, the lower spikelets usually paired, the upper solitary; lateral panicles few-fld, dense, with short internodes; first glume 3- or 5-veined, scabrous on the keel, 12–29 mm, the second 1-nerved, glabrous, slightly longer; lemma 10–18 mm; awns about equally divergent, 4–7 cm; $2n=22$. Dry open ground; Mass. and N.J., where probably intr., to Fla. and Tex.; abundant in the interior, from Ky. and O. to Mich., S.D., and Tex., and adventive in s. Ont.

7. **Aristida longespica** Poiret.   Slimspike t.-a. Annual, 2–5 dm, loosely tufted in small bunches, often branching from the lower nodes; blades flat or involute, 1–2 mm wide, the larger often only 4–8 cm long; panicle ± elongate, up to half the length of the plant, very slender, often raceme-like, with appressed or ascending spikelets; axillary panicles much reduced; glumes subequal, 2–11 mm, 1-veined or the first one sometimes obscurely 3-veined; lemmas 3–10 mm; awns terete, up to ca 2.5 cm, about equally divergent, or the cental one more deflexed. Moist or dry, often sterile or sandy soil; N.H. to s. Ont., Io., and Nebr., s. to the Gulf. Two vars.:

Var. *longespica.*   Central awn 1–10(–14) mm, the lateral awns 0–5(–8) mm. Range of the species, but more common eastward.
Var. *geniculata* (Raf.) Fern.   Central awn (8–)12–27 mm, the lateral ones (1–)6–18 mm; spikelets avg a little larger. More common westward, seldom e. to Ind. and s. Ont. (*A. intermedia; A. neocopina.*)

A. adscensionis L., a chiefly tropical sp. with the awns flat at the base, the central one 8–15 mm, the first glume half to two-thirds as long as the second, is rarely adventive in our range.

8. **Aristida ramosissima** Engelm.   Annual 2–5 dm; culms very slender, erect, branched from the base and all nodes; lvs very narrow, flat or involute; panicle lax, raceme-like, 5–10 cm, the spikelets solitary or rarely

paired; racemes of the branches usually smaller; glumes unequal, the first 3–5-veined, 10–21 (avg 14) mm, the second 15–28 (avg 20) mm; lemma 11–21 mm, tapering gradually into the awn; central awn stout, 1.7–3.7 cm, bent at the base into a quarter to a full turn; lateral awns erect, 0.5–6 mm, rarely absent. Dry soil; s. Ind. to Io., s. to Tenn. and Tex.

9. Aristida dichotoma Michx. Churchmouse-t.-a. Tufted annual 2–4 dm, erect or ascending, branched from the base; lvs filiform, mostly involute, or the lower flat; terminal panicle 3–8 cm, very slender, often reduced to a raceme; lateral infls much shorter, mostly enclosed in the subtending sheaths; glumes subequal, mucronate, 1-veined, 5.2–10.5 mm; lemma 4.4–9 mm; central awn 3–10 mm, its base nearly horizontally divergent and loosely coiled, usually in a half to one full turn; lateral awns straight, erect, 0.7–3.3 mm. Dry sandy or sterile soil; Me. to Mich., Io., and Kans., s. to Fla. and Tex.

10. Aristida basiramea Engelm. Forktip-t.-a. Erect, tufted annual 3–6 dm; lvs very narrow, ca 1 mm wide, often involute; panicles slender, the terminal one 5–10 cm, rather loose, the lateral shorter and more slender, scarcely surpassing the subtending sheaths; glumes 1-veined, distinctly unequal; lemma usually about equaling the first glume; central awn divergent, coiled at base when dry into one-half to 3 complete turns; lateral awns erect to curved-divergent but not coiled. Dry, often sandy soil; Mich. to Minn. and Wyo., s. to Ill., Ark., Okla., and Colo., and casually e. to Me., Pa., and Va., and Fla. Two vars. with nearly coextensive range:

Var. *basiramea.* First glume 6–12 mm, the second 9.5–15 mm; central awn 11–19 mm; lateral awns erect to curved-divergent, 7.5–13 mm.

Var. *curtissii* (A. Gray) Shinners. First glume 5–8 mm, the second 7–13 mm; central awn 7–12.5 mm, the lateral ones 2–4 mm, straight, erect. (*A. c.; A. dichotoma* var. *c.*)

11. Aristida desmantha Trin. & Rupr. Annual 3–8 dm, the culms erect, branched from the base and to some extent from the lower nodes; lvs 1–4 mm wide toward the base, tapering to a fine point, folded or involute, the sheaths often villous-margined; panicle loosely branched, 1–2 dm, with ascending branches; glumes subequal, 13–20 mm, including the short awn, the first one scabrous on the keel; lemma 9–12 mm; awns articulate with the lemma, scarcely united at base, loosely coiled and widely divergent when dry, subequal, 2.4–3.3 cm. Dry sand; Ill. to Nebr. and Tex.

12. Aristida tuberculosa Nutt. Annual 3–8 dm, the culms branched at base and lower nodes; lower sheaths villous, the upper glabrous or nearly so; blades 1–3 mm wide, involute at least when dry; panicle very lax and open, sparsely branched, 1–3 dm; glumes subequal, 20–28 mm, including the short awn, the first scabrous on the keel; lemma 11–14 mm, its awns about equal, united at base into a twisted column 7–13 mm, the free parts loosely coiled at base, about horizontal, 3–5 cm. Dry sterile soil, especially on dunes, along the coast from Mass. to Ga., and inland from Ind. and s. Mich. to Minn. and Io.

63. UNIOLA L. Spikelets 5–many-fld, flat, disarticulating below the glumes; glumes shorter than the lemmas, subequal, acute or slightly mucronate, 3-veined, compressed-keeled, the keel serrulate; lower 2–6 lemmas empty; lemmas acute and obtusish or slightly mucronate, 3–9-veined, compressed-keeled, the keel serrulate; palea shorter to longer than the lemma, 2-keeled; erect, rhizomatous or stoloniferous perennials, with numerous large, stramineous spikelets in open to narrow and strict infls; ligule a fringe of hairs. 2, the other from Baja Calif. to Ecuador.

1. Uniola paniculata L. Sea-oats. Stout, 1–2.5 m, from extensively creeping rhizomes; lf-blades 2–4 dm, mostly under 1 cm wide, broad-based, ± involute, attenuate into a long slender tip; infl 2–6 dm, rather crowded, or more open in age; spikelets ascending on stout pedicels, ovate or lance-ovate, 15–40 mm, 10–25-fld; glumes and the 3–5 sterile lemmas lance-linear, scarcely shorter than the lower fertile lemmas; lemmas lance-ovate, smooth or scaberulous on the keel, about equaling the palea; $2n=40$. Coastal sands; se. Va. to Fla., Tex., Mex., and W.I.

64. DISTICHLIS Raf. Salt-grass. Dioecious; spikelets flattened, 4–19-fld, the pistillate ones disarticulating above the glumes and between the lemmas; glumes unequal, obscurely veined, shorter than the lemmas; lemmas firm or coriaceous, smooth, obscurely many-veined, closely overlapping; low colonial perennial grasses from creeping scaly rhizomes with stiff, erect stems, numerous narrow, usually involute blades with overlapping sheaths, and small panicles of large, crowded spikelets. 4, New World.

1. Distichlis spicata (L.) Greene. Colonial from hard, white rhizomes with stiff scales; stems 1.5–4 dm, with numerous rigid, involute lf-blades 5–10 cm, the lvs often pilose at the collar; infl congested, short, to 8 cm; lemmas ovate or lance-ovate, about equaling the palea, this with 2 winged keels, the wings ciliolate and/or irregularly serrate; $2n=40$. Two vars.:

Var. *spicata.* Seashore s.-g. Infl very dense, the short pedicels mainly concealed; principal spikelets 8–12 mm, the pistillate with 4–9 (typically 5) fls, the staminate with 8–12 fls; first glume 1.7–3 mm, the second 2.4–3.8 mm; lemmas 3.3–4.5 mm. Salt marshes along the coast from N.S. and P.E.I. to tropical Amer., and on the Pacific coast.

Var. *stricta* (Torr.) Rydb. Infl less dense, the pedicels more exposed; principal spikelets 12–20 mm, avg more numerously fld, the pistillate with typically 7 fls, the staminate with 10–18 fls; first glume 2.8–5.1 mm, the second 3.6–6.6 mm; lemmas 4.4–7.5 mm; keels of the pistillate paleas more broadly winged; collar more conspicuously pilose. Saline soil or alkaline situations in the interior; w. U.S. and adj. Can., e. to Sask., Minn., and Mo. (*D. stricta.*)

65. TRIDENS Roemer & Schultes.   Spikelets 3–9-fld, disarticulating above the glumes and between the lemmas; first glume 1- or 3-veined, the second 1–5-veined; lemmas broad, rounded on the back, obtuse or retuse, densely hairy at base and on the 3 veins, at least in the lower half, the midrib usually shortly excurrent, and the 2 lateral veins produced as minute teeth; palea broad, its 2 veins near the margin, smooth or short-hairy; stigmas dark purple; grain strongly concave on the ventral side; perennial grasses with flat lvs and ample panicles and without long rhizomes. (*Triodia,* in part.) 15, New World.

1. **Tridens flavus** (L.) A. Hitchc.   Purpletop. Tufted and sometimes with very short rhizomes, 8–15 dm; sheaths bearded at the top, otherwise glabrous, as are the blades, these 3–8 mm wide, elongate to a slender tip; infl viscid, with loosely spreading-ascending branches villous at base above; spikelets purple (yellow), narrowly ovoid, scarcely compressed, 5–10 mm, 4–9-fld, the lateral ones appressed along the branchlets or on looser pedicels up to ca 3 mm; glumes unequal, firm, oblong or ovate, 2.5–3.5 mm, 1-veined, obtuse or often mucronate; lemmas regularly imbricate, densely villous on the basal half; $2n=40$. Fields, roadsides, and open woods; Mass. to s. Ont., s. Mich., and Nebr., s. to Fla. and Tex. (*Triodia f.*) Most of our plants belong to the widespread var. *flavus,* as principally described above. The more strictly southeastern var. *chapmanii* (Small) Shinners reaches our range in se. Va. and s. N.J. It has more divergent panicle-branches, villous all around at the more strongly buttressed base, and the spikelets are on longer, more divergent pedicels mostly 3–20 mm. (*T. c.*)

66. TRIPLASIS P. Beauv.   Sand-grass. Spikelets 2–6-fld, with long rachilla-joints, disarticulating above the glumes and between the lemmas; glumes chiefly equal, 1-veined, acute, smooth; lemmas oblong, rounded on the back, hairy on the 3 parallel veins, 2-lobed with an awn between the lobes; palea shorter than and soon diverging from the lemma, densely villous on the 2 keels in the upper half; tufted grasses with many short internodes, loose sheaths, a ligule of short hairs, and short panicles at first included in the upper sheath, eventually exsert with spreading branches. 2, N. Amer.

1. **Triplasis purpurea** (Walter) Chapman.   Purple s.-g. Slender annual 2–8 dm; blades 1–2 mm wide, shorter than their sheaths, the upper much reduced; terminal infl 2–8 cm, with a few branches each bearing a few 2–5-fld purple spikelets; glumes narrowly lanceolate, 2–4 mm; lemmas 3–4 mm, the short lobes rounded or erose, the hairy awn 1 cm or less; rachilla-joints half as long as the lemma; included panicles and solitary spikelets produced within the lower sheaths. Dry sand along the coast from N.H. to Tex.; also along the shores of the Great Lakes, and in the interior from Ind. to Minn., Colo., and Tex.

67. LEPTOCHLOA P. Beauv.   Sprangletop. Spikelets 2–12-fld, articulated above the glumes; glumes 1-veined; lemmas 3-veined, rounded on the back, obtuse or truncate to acute or shortly bifid at the tip, the veins sometimes excurrent into a short central awn or minute lateral teeth; palea broad, nearly as long as the lemma; annual (all ours) or perennial, usually weedy grasses, often branched from the base, with scarious ligule and numerous long spike-like racemes disposed along a central axis to form a large panicle. (*Diplachne.*) 40, warm reg.

1 Sheaths sparsely papillose-pilose; spikelets 1–2.5 mm, 2–4-fld .... 1. *L. filiformis.*
1 Sheaths glabrous to scabrous; spikelets 4–10 mm, 6–12-fld.
  2 Lemmas 2–3 mm, obtuse to truncate, minutely mucronate .... 2. *L. uninervia.*
  2 Lemmas 3–5 mm, acute to acuminate, short-awned .... 3. *L. fascicularis.*

1. **Leptochloa filiformis** P. (Lam.) Beauv.   Red s. Culms simple or branched from the base, 3–10 dm; sheaths usually overlapping, sparsely papillose-pilose; ligule 1–2 mm, fimbriate-ciliate; blades soft, flat, scabrous on the margins, the larger ones 6–10 mm wide; panicle 1–5 dm, with numerous spreading racemes 5–15 cm; spikelets 1–2.5 mm, 2–4-fld; glumes subulate or narrowly lanceolate, subequal or the second a little longer and broader, sometimes equaling or surpassing the uppermost lemma; lemmas subacute to obtuse or broadly rounded, 0.7–1.5 mm, puberulent on the veins toward the base; anthers 0.1–0.3 mm; grain 0.6–1 mm; $2n=20$. A weed of gardens and fields, especially in moist soil; trop Amer., n. to Va., s. Ind., s. Ill., and S.D. (*L. attenuata.*)

2. Leptochloa uninervia (C. Presl) A. Hitchc. & Chase. Culms 3–8 dm; main lvs 1–3 mm wide, involute, elongate, scabrous; panicle narrowly ovoid to subcylindric, 1–2 dm, with numerous ascending branches; spikelets 4–9 mm, 6–10-fld; first glume triangular-subulate, 1.1–1.7 mm, the second oblong-elliptic, 1.7–2.6 mm, rounded to the tip; lemmas 2–3 mm, broadly obtuse to subtruncate, minutely mucronate, the lateral veins usually excurrent as short points; $2n=20$. Moist ground and along ditches; trop. Amer., and sparingly intr. along our e. seaboard and in Ill. Hardly more than another var. of no. 3.

L. panicoides (C. Presl) A. Hitchc., another chiefly tropical sp., is reported to extend n. to s. Ind. and s. Ill. It differs in its wider, nearly smooth lvs, the main ones 5–10 mm wide, and in its 5–7-fld spikelets 3.5–5 mm. (*Diplachne p.*)

3. Leptochloa fascicularis (Lam.) A. Gray. Tufted, densely branched from the base, 1–10 dm, procumbent to erect; sheaths smooth or nearly so; blades scabrous, 1–3 mm wide; ligule 2.5–6 mm; panicle subcylindric, 1.5–3 dm × 2–5 cm, with numerous ascending branches; spikelets 5–10 mm, 6–12-fld; first glume narrowly triangular or lanceolate, the second lanceolate or oblong; lemmas 3–5 mm, usually acuminate, short-awned between the 2 minute teeth, pubescent at base and often on the lower part of the veins, the lateral veins often shortly excurrent; grain 1.5–2 mm. Widespread in the warmer parts of N. and S. Amer., especially in brackish or alkaline sites. Three well marked vars. with us, sometimes treated as distinct spp.:

a Low plants, seldom over 5 dm; first glume 2.5–3.5 mm, the second 4–7 mm, lemmas 4–5 mm, with
    an evident awn (1–)2.5–5 mm; near the coast from Mass. southward. (*L. m.; Diplachne m.*) ..
    .... var. *maritima* (Bicknell) Gleason.
a Taller plants, mostly 5–10 dm; awn up to 2.5 mm.
    b First glume 2.3–3.4 mm, the second 3.4–5 mm; lemmas 4–5 mm, with an awn 0.5–2.5 mm, c.
        N.Y.; Vt.; s. Ont.; s. O.; Minn., Io., and Mo. to Wash., Calif., and Tex. (*L. a.; Diplachne a.*)
        .... var. *acuminata* (Nash) Gleason.
    b First glume 1.3–2 mm, the second 2.2–3.5 mm; lemmas 3–4 mm, with an awn 0.5–1 mm; Mo.
        and Ill. to Fla. and Tex., s. to S. Amer. .... var. *fascicularis*.

68. ERAGROSTIS Wolf. Lovegrass. Spikelets 2–many-fld, slightly to strongly flattened, disarticulating above the glumes and between the lemmas, or eventually fragmenting irregularly, or the lemmas individually deciduous from the persistent rachilla (the grain falling free); glumes 1-veined or the second one 3-veined, usually unequal; lemma 3-veined (the lateral veins sometimes obscure), awnless, blunt to acute or short-acuminate, ± keeled but varying from V-shaped in cross-section to rounded on the back; palea somewhat shorter than the lemma, often ciliate on the 2 keels; grain short-cylindric to subglobose, ovoid, or pyriform; ligule a band of short hairs; sheaths almost always pilose or hirsute at the top; panicle open and ± diffusely branched, varying to dense and subglobose to cylindric or spike-like; only 2 of our spp. have anthers more than ca 0.5 mm. 250, cosmop.

In addition to the following spp., several spp. have been reported as waifs in our range. Among these are *E. barrelieri* Daveau, *E. curtipedicellata* Buckley, and *E. elliottii* S. Wats.

1 Mat-forming annuals, creeping by stolons.
    2 Fls perfect; lemmas 1.5–2 mm; pedicels 1–3 mm .... 12. *E. hypnoides.*
    2 Fls unisexual, the plants dioecious; lemmas 2.1–4 mm; pedicels under 1 mm .... 13. *E. reptans.*
1 Plants with decumbent or ascending to erect stems, not mat-forming.
    3 Perennial.
        4 Lemmas rounded on the back, their lateral veins inconspicuous or nearly obsolete .... 1. *E. hirsuta.*
        4 Lemmas compressed and with conspicuous lateral veins.
            5 Panicle-branches glabrous, the axils glabrous or nearly so; anthers 1.5 mm .... 2. *E. trichodes.*
            5 Panicle-branches scabrous, conspicuously pilose in the axils; anthers under 0.5 mm.
                6 Most of the lateral spikelets shorter than their loosely ascending to spreading pedicels;
                    spikelets 5–7 mm, 7–11-fld, purplish .... 3. *E. spectabilis.*
                6 Lateral spikelets appressed to the rachis and shorter than their pedicels.
                    7 Spikelets 5–9 mm, with 5–9(–11) fls; lemmas 2–3 mm .... 4. *E. curvula.*
                    7 Spikelets 5–16 mm, with 11–28 fls; lemmas 1.6–2 mm .... 5. *E. refracta.*
    3 Annual.
        8 Grain with a longitudinal furrow extending its whole length .... 11. *E. capillaris.*
        8 Grain without a furrow.
            9 Lvs with scattered warty glands along the margins; lemmas keeled but rounded on the back;
                branches of the infl relatively short and stiff, not much again branched.
                10 Lemma 2–2.8 mm, with glandular pits in the keel .... 6. *E. cilianensis.*
                10 Lemma 1.5–2 mm, without glandular pits .... 7. *E. minor.*
            9 Lvs without marginal glands; lemmas keeled and V-shaped in cross-section; infl more lax and
                branching.
                11 Paleas, or many of them, deciduous with the lemma; first glume less than half as long as the
                    lowest lemma; panicle-branches at one of the two lowest nodes usually whorled .... 8. *E. pilosa.*
                11 Paleas persistent on the rachilla after the lemmas have fallen; first glume more than half
                    as long as the lowest lemma; panicle-branches usually solitary or paired.

1. **Eragrostis hirsuta** (Michx.) Nees. Stout, erect, tufted perennial 4–10 dm; sheaths longer than the internodes, ciliate-margined and sometimes villous on the back; blades 4–10 mm wide, elongate and tapering to a fine point, involute when dry; infl diffuse, half the length of the entire shoot, its scabrous branches pilose in the axils; pedicels spreading, 8–18 mm; spikelets 2–4 mm, 2–6-fld; first glume 1.4–2 mm, the second 1.6–2.2 mm; lemmas 1.7–2.4 mm, rounded on the back, the lateral veins inconspicuous or nearly obsolete, at maturity falling individually from the intact rachilla, on which the paleas may persist; grain 0.7–1 mm, barrel-shaped; $2n=100$. Dry sandy soil near the coast; Md. and Va. to Fla. and Tex., n. in the interior to Mo., and rarely intr. northward, as in Mass. and Me.

E. intermedia A. Hitchc., a southwestern sp. with the lf-sheaths (including the margins) glabrous below the summit, and with a longitudinal furrow in the grain, extends n. to about the edge of our range in Mo., and is rarely intr. elsewhere with us.

2. **Eragrostis trichodes** (Nutt.) A. Wood. Sand-l. Slender, erect, tufted perennial 5–12 dm; lvs 2–6 mm wide, elongate, tapering to a fine point; infl diffusely branched, 2–5 dm, the branches spreading to ascending, glabrous, scarcely or not at all pilose in the axils; spikelets 4–10 mm, 4–18-fld; first glume 1.6–3.5 mm, the second 1.9–3.3 mm; lemmas 2.1–3.3 mm, compressed-keeled and with prominent lateral veins, at maturity falling individually, together with their paleas, from the intact rachis; anthers ca 1.5 mm. Dry sand, prairies, and open woods; Ill. to Nebr., Ark., and Tex. (*E. pilifera.*)

3. **Eragrostis spectabilis** (Pursh) Steudel. Purple l. Tufted, erect or ascending perennial 3–6 dm; lvs firm or stiff, 3–8 mm wide, tapering to a fine point; infl diffuse, two-thirds as long as the whole shoot, its base usually included in the upper sheath, its scabrous branches rigid, divaricate, pilose in the axils; spikelets purple, 5–7 mm, 7–11-fld, most of the lateral ones shorter than their loosely ascending to spreading pedicels; first glume 1–2 mm; lemmas persistent on the eventually fragmenting rachilla, 1.6–2.1 mm, compressed, scabrous on the keel, the lateral veins evident; $2n=20, 40$. Dry soil, fields, and open woods; Me. to N.D., s. to Fla. and Tex. (*E. pectinacea,* misapplied.)

4. **Eragrostis curvula** (Schrader) Nees. Weeping l. Erect, tufted perennial 6–18 dm; sheaths keeled and ± hispidulous basally; lvs firm, involute, narrow, arcuate-ascending, tapering to a fine point; infl 2–3.5 dm, pilose in the main axils, elongate and open but scarcely diffuse, the pedicels short and ± appressed to the branchlets; spikelets leaden-gray, 5–9 mm, 5–9(–11)-fld, the lemmas individually deciduous or more persistent and the rachilla fragmenting; first glume 1.5–2(2.5) mm, the second 2–2.8 mm; lemmas 2–3 mm, evidently 3-veined; $2n=20, 40, 60$. Native of S. Afr., cult. for forage or ornament and escaped at scattered stations as far n. as N.J. and Pa. (where becoming common).

5. **Eragrostis refracta** (Muhl.) Scribn. Erect, slender, tufted perennial 4–8 dm; lvs mostly crowded near the base, the blades elongate, 2–5 mm wide, tapering to a fine point; infl ovoid, two-thirds as long as the whole shoot, the scabrous branches pilose in the axils, spreading or eventually reflexed, with relatively few, scattered spikelets in the distal half; spikelets 5–16 mm, 10–28-fld, often purplish, short-pediceled, strongly ascending or appressed to the branches; first glume 0.9–1.7 mm, the second 1.4–1.8 mm; lemmas 1.6–1.9 mm, compressed-keeled and evidently 3-veined, persistent, the rachilla eventually fragmenting; grain 0.5–0.9 mm, half as thick. Moist or wet, sandy or muddy soil along the coast; Del. to Fla. and Tex.

6. **Eragrostis cilianensis** (All.) Janchen. Stink-grass. Densely tufted annual 1–4 dm, the culms spreading or ascending from a decumbent base, rarely erect, plants beset with elevated glands or glandular pits along the margins of the lvs and lower side of the midrib, along the midvein of the lemmas, and often also on the panicle-branches and sheaths, the latter otherwise glabrous below the summit; lf-blades 3–12 cm × 2–6 mm; infl ovoid or cylindric, 5–15 cm, its branches short and stiff, spreading, not much rebranched, glabrous or pilose in the axils; spikelets not strongly compressed, 4–25 mm, broadly linear, 2–3 mm wide, 10–40-fld; first glume 1.3–1.9 mm, the second 1.5–2 mm; lemmas 2–2.8 mm, broadly elliptic-ovate, keeled but otherwise rounded on the back, closely imbricate, falling individually from the intact rachilla, on which the paleas tend to persist; grain 0.7 mm, dull brown; $2n=20, 40$. Native of Europe, now a common weed throughout our range and most of the U.S. (*E. major; E. megastachya.*)

7. **Eragrostis minor** Host. Much like no. 6, glandular-warty on the lf-margins and to some extent elsewhere, but the lemmas without glands; sheaths usually pilose on the margins, sometimes also on the back, as well as at the summit; axils of the panicle-branches glabrous; spikelets more slender, 4–11 × 1.5–2 mm, 5–12(–20)-fld, the lemmas 1.5–2 mm; $2n=40, 80$. Native of Europe, intr. in moist soil, waste land, gardens, and roadsides from N.S. and Vt. to Wis. and Io., s. to Ga., Tex., and Calif. (*E. poaeoides.*)

8. **Eragrostis pilosa** (L.) P. Beauv. India-l. Annual 1–6 dm, erect or ascending from a decumbent base; lvs 1–3 mm wide; sheaths shorter than the internodes; infl 5–20 cm, ellipsoid to ovoid, diffusely branched, the branches at one of the two lowest nodes commonly whorled, seldom merely fascicled or paired; pedicels appressed to more often spreading; spikelets 5–17-fld, 1–2 mm wide; first glume 0.3–0.8 mm, less than half as long as the lowest lemma; lemmas 1.2–1.8 mm, individually deciduous from the persistent rachilla; paleas nearly always deciduous, sometimes only tardily so; grain pyriform to slightly compressed, 0.5–0.9 mm; $2n=40$. Pantropical weed, extending n. in our range to Me., Mich., Wis., and Mo. Ours is var. *pilosa.* (*E. multicaulis; E. peregrina.*)

9. **Eragrostis pectinacea** (Michx.) Nees. Carolina l. Annual 1–6 dm, the culms erect or ascending, often repeatedly branched; lvs 1–4 mm wide; middle sheaths mostly shorter than their internodes; infl 5–25 cm,

ovoid or pyramidal, diffusely branched with one or two branches at a node, or the lower branches rarely fascicled or whorled; pedicels appressed, rarely diverging as much as 20°; spikelets mostly 6–15-fld, 1.2–2.5 mm wide; first glume 0.5–1.1 mm, half to three-fourths as long as the lowest lemma; lemmas 1–2.2 mm, individually deciduous from the rachilla, on which the paleas persist; grain pyriform, slightly compressed, 0.5–1 mm; 2n=60. Moist ground, especially as a weed in gardens and waste places; widespread in our range, s. to C. Amer. and W.I.

E. tephrosanthes Schultes, a weed mainly of trop. N. Amer., occurs rarely and sporadically with us. It has spreading pedicels like *E. frankii,* but is otherwise more like *E. pectinacea.*

10. **Eragrostis frankii** C. A. Meyer. Annual 1–5 dm, the culms densely cespitose, repeatedly branched; lvs 1–4 mm wide; middle sheaths mostly longer than their internodes; infl 5–20 cm, ellipsoid, rather compactly branched, the middle branches the longest, the lower ones solitary or rarely paired; pedicels spreading; spikelets 3–6(–9)-fld, 1–2.5 mm wide; first glume 1–1.5 mm, three-fourths to one and one-fourth times as long as the lowest lemma; lemmas 1.1–1.6 mm, individually deciduous from the rachilla, on which the paleas persist; grain 0.5–0.7 mm, broadly ellipsoid to subglobose, somewhat compressed, sometimes shallowly indented on one side; 2n=40, 80. Riverbanks, sand-bars, and moist ground; Mass., Vt. and s. Que. to Minn., s. to Fla. and Ark.

11. **Eragrostis capillaris** (L.) Nees. Lace-grass. Tufted, erect or ascending annual 2–5 dm; lvs 2–4 mm wide; infl large and diffusely branched, often two-thirds the height of the whole shoot; with slender branches and elongate, capillary pedicels; spikelets 2–3 mm, 2–5-fld; first glume 0.9–1.5 mm, the second 1.1–1.4 mm; lemmas 1.2–1.6 mm, individually deciduous from the rachilla, on which the paleas persist; grain 0.6–0.7 mm, with a groove its entire length on one side; 2n=100. Open woods and dry soil; Me. to Wis., s. to Ga. and Tex.

Eragrostis mexicana (Hornem.) Link. and its close relative *E. neomexicana* Vasey, both with the grain furrowed as in *E. capillaris,* are rarely and sporadically adventive with us. Both differ from *E. capillaris* in their proportionately smaller and more congested infl and 7–15-fld spikelets with larger lemmas (1.7+ mm). *E. neomexicana* is robust under favorable conditions, 5–12 dm, with lvs 3–10 mm wide, and commonly has a ring of yellow glands below each node of the stem and glandular pits on the sheath (at least along the keel). *E. mexicana* averages smaller, seldom over 5 dm, with lvs 3–6 mm wide, and lacks the glands.

12. **Eragrostis hypnoides** (Lam.) BSP. Annual; stems creeping, rooting at the nodes, forming mats, sending up short culms 5–15 cm; lf-blades 1–4 cm × 1–3 mm; infl ovoid to subcylindric, 2–8 cm; peduncle glabrous; pedicels 1–3 mm; spikelets linear, 10–35-fld; fls perfect; lemmas 1.5–2 mm, glabrous, shining, almost hyaline, falling individually from the intact rachilla, on which the paleas persist; anthers 0.3 mm. Mud-flats and sandy shores; throughout most of the U.S. except the arid or mountainous regions, s. to Arg.

13. **Eragrostis reptans** (Michx.) Nees. Habitally like no. 12, forming larger mats; dioecious; infls short and crowded, ovoid to subglobose, 1–3 cm, the staminate mostly more open and fewer-fld than the pistillate; peduncle villous; pedicels very short, 0.2–0.4 mm; lemmas herbaceous, 2.1–4 mm, usually sparsely villous along the veins, not falling individually; anthers of the staminate plants 1.5–2 mm. Wet soil and sandy or mud shores; Ky. and Ill. to S.D., Tex., and n. Mex.

69. **ELEUSINE** Gaertn. Spikelets several-fld, articulated above the glumes, the terminal ones in each spike well developed; glumes unequal, the first shorter, 1-veined, the second strongly 3- or 5-veined, shorter than the lemmas but resembling them; lemmas compressed, ovate to lanceolate, acute to acuminate, strongly 3- or 5-veined, the veins approximate and parallel near the keel; grain ellipsoid to ovoid, rugose; annuals with compressed culms and keeled sheaths; ligule membranous, ciliate or erose; blades flat, soft; spikes digitate and terminal, or with 1 or 2 lateral ones, with flat rachis and densely crowded, divergent spikelets. 9, warm reg.

1. **Eleusine indica** (L.) Gaertn. Yard-grass. Culms 3–6 dm, branched from the base, spreading or ascending; sheaths papillose-ciliate distally; blades usually smooth; spikes usually 3–8, 4–10 cm × 5 mm; spreading or ascending; spikelets crowded, 3–6-fld; second glume 2–3 mm; lemmas 2.5–4 mm; 2n=18, 36. A common weed in lawns, gardens, and waste places; native of the Old World, now pantropical, extending n. to Mass., S.D., and Utah.

70. **SPOROBOLUS** R. Br. Dropseed. Spikelets 1-fld, articulated above the glumes; glumes lanceolate to ovate, 1-veined, from much shorter to somewhat longer than the lemma; lemma membranous or herbaceous, rounded on the back, veinless or 1(3)-veined, obtuse to acuminate, awnless; palea about as long as the lemma or longer, of similar texture, usually conspicuous and in some spp. wider than the lemma; grain in most spp. soon separating from the lemma and palea, falling free; pericarp free from the seed; annual or perennial grasses with narrow, often involute lvs, minute or no ligule, and open or contracted panicles of rather small spikelets. 100, mostly of warm reg., esp. in the New World.

1 Panicle open, ovoid or pyramidal to cylindric, with spreading or ascending branches.
  2 First glume subulate above an expanded base; second glume 3–6 mm .... 1. *S. heterolepis.*
  2 First glume merely acute or obtuse; second glume 1–3(–3.7) mm.
    3 Panicle-branches, at least the lower, definitely whorled.
      4 Panicle pyramidal; spikelets 1.2–2 mm; lvs flat, 2–4 mm wide .... 2. *S. pyramidatus.*
      4 Panicle narrower, cylindric; spikelets 2.5–3.7 mm; lvs involute, setaceous .... 3. *S. junceus.*
    3 Panicle branches alternate, or some of them irregularly clustered.
      5 Sheaths with a conspicuous tuft of long white hairs at the summit; panicle-branches usually
        spikelet-bearing to near the base .... 4. *S. cryptandrus.*
      5 Sheaths with only a few (or no) long hairs at the summit; panicle-branches with spikelets borne
        chiefly near the tips .... 5. *S. airoides.*
1 Panicle contracated and ± spike-like, the branches erect or appressed.
  6 Glumes and lemma of 3 distinctly different lengths, the first glume a third to two-thirds as long as
    the lemma; perennials, often tufted.
    7 Lemma 1.4–2.5 mm; first glume 0.4–1.2 mm .... 6. *S. indicus.*
    7 Lemma (2.5–)3–7 mm; first glume 1.5–6 mm.
      8 Lemma glabrous; pericarp gelatinous when wet .... 7. *S. asper.*
      8 Lemma pubescent; pericarp loose when wet, not gelatinous .... 8. *S. clandestinus.*
  6 Glumes and lemma nearly equal, the first glume three-fourths to fully as long as the lemma.
    9 Rhizomatous perennial; lvs ± spreading, 3–6 mm wide, conspicuously distichous .... 9. *S. virginicus.*
    9 Tufted annuals; lvs mostly erect, filiform-involute.
      10 Glumes a little longer than the 3-nerved lemma; palea obtuse; lower sheaths sparsely pus-
        tulate-pilose .... 10. *S. ozarkanus.*
      10 Glumes a little shorter than the 1(–3)-nerved lemma; palea acute or acuminate; lower sheaths
        glabrous or with a few long soft hairs near the collar.
        11 Lemma pubescent, mostly 3–5 mm .... 11. *S. vaginiflorus.*
        11 Lemma glabrous, 1.5–3 mm .... 12. *S. neglectus.*

1. Sporobolus heterolepis A. Gray. Prairie-d. Erect, tighty clustered perennial 4–10 dm; lvs very long, narrow, usually involute, crowded toward the base; panicle 1–2 dm × 1.5–3 cm, cylindric to narrowly ovoid, with whorled, ascending, mostly racemiform and basally naked branches, these and the pedicels irregularly interrupted by paler, slightly dilated segments; spikelets 3–6 mm; first glume half as long as the second, subulate above a broader base; second glume acuminate into a carinate or involute tip, usually slightly exceeding the lemma; palea usually slightly exceeding the lemma; fr globose, 1+ mm thick; 2*n*=72. Dry open ground, especially on prairies, often in slight depressions; O. to Sask. and Tex., e. occasionally to Pa., Conn., and Que.

2. Sporobolus pyramidatus (Lam.) A. Hitchc. Tufted perennial (often blooming the first year) 1–7 dm, usually ascending or prostrate; lvs crowded at base; sheaths pilose at the throat; blades flat, 5–20 cm × 2–4 mm, sparsely long-ciliate toward the base; panicle pale, pyramidal, 3–10(–20) × 1.5–6+ cm, the lower branches whorled, widely spreading, somewhat viscid, closely fld, naked toward the base; spikelets pale, 1.2–2 mm; first glume 0.3–1 mm, the second equaling the lemma; 2*n*=24, 36. Sandy or gravelly soil, along the seashore and in alkaline soil in the interior, sometimes a street-weed; trop. Amer., n. to Kans. and Mo., and as an adventive to N.Y. (*S. argutus.*)

3. Sporobolus junceus (Michx.) Kunth. Tufted perennial 3–10 dm, with erect slender culms; lvs 1–3 dm, very narrow, involute, crowded toward the base; panicle cylindric, 8–25 × 1–4 cm, its branches regularly whorled, spreading or ascending, 1–3 cm, racemiform, bearing scattered, ± secund, bronze-brown spikelets 2.5–3.7 mm on short (mostly under 1 mm) pedicels in the distal half; first glume lanceolate, acute or obtuse, 1.2–3 mm; second glume acute, about equaling the lemma. Pine-barrens of the coastal plain; se. Va. to Fla. and Tex. (*S. gracilis.*)

4. Sporobolus cryptandrus (Torr.) A. Gray. Sand-d. Tufted perennial (sometimes blooming the first year) 3–10 dm; lvs flat or drying involute, 2–6 mm wide, tapering to a long point, the lower sheaths overlapping; sheaths with a conspicuous tuft of long white hairs on each side at the top; panicle ovoid or pyramidal, mostly 10–20 × 2.5–5 cm, its branches mostly alternate or irregularly disposed, spikelet-bearing to near the base; spikelets cleistogamous, pale or olivaceous; glumes acute, the first up to about half as long as the broader second one, which about equals the lemma; lemma about equaling the palea; anthers ca 0.5 mm; 2*n*=18, 36, 72. Dry, especially sandy soil; widespread in w. and c. U.S. and nearby Can. and Mex., e. irregularly to Que., Me. N.J., and O. Doubtfully divisible into vars. Most of our plants have spikelets 2–3 mm in an exserted panicle with spreading branches; these have been called var. *fusicolor* (Hook.) Pohl, a name based on plants from Wash. Toward the w. part of our range, and w. to the Pacific, the common (and nomenclaturally typical) phase has spikelets 1.4–2 mm in a panicle with the lower part included in the enlarged and inflated upper sheath, and the panicle-branches are more ascending.

5. Sporobolus airoides (Torr.) Torr. Alkali-sacaton. Stout perennial 4–10 dm, the base of the large clumps densley clothed with the shiny, cream-colored old sheaths; sheaths glabrous or slightly pilose at the upper corners; blades narrow and mostly involute, finely scaberulous above and ± hirsute near the base; panicle open, pyramidal, 12–40 × (5–)10–20 cm, sometimes partly enclosed at base in the uppermost sheath, the branches spreading, alternate or irregularly disposed, bearing spikelets chiefly toward the tips; first glume 0.7–1.2+ mm, the second 1.5–2+ mm; lemma 1.8–2.5(–3) mm; palea about equaling the lemma; anthers 1.2–1.5 mm; 2*n*=ca 80, ca 90, 108, 126. Low-lying, alkaline meadows; widespread in interior w. U.S. and n. Mex., e. to the Dakotas, Mo., and Okla. Ours is the widespread var. *airoides.*

6. Sporobolus indicus (L.) R. Br.   Smut-grass. Erect, tufted, unbranched perennial 3–10 dm; lvs crowded toward the base, 15–25+ cm × 3–5 mm, tapering to a filiform point; infl 1–4 dm × 0.5–1 cm, dense and spike-like; spikelets numerous, small, crowded, the first glume 0.4–0.9 mm, the second 0.8–1.3 mm; lemma 1.4–2 mm, about equaling the palea; 2n=18, 24, 36. A weed of wet open soil, roadsides and ditchbanks; native of trop. Asia, naturalized in trop Amer., and extending n. to Ark., Ky., Va., and sometimes N.J. (*S. poiretii*.)

S. contractus A. Hitchc., a western sp. with the second glume 1.5–2.5 mm and the lemma 2–2.5 mm, is reportedly adventive in Me. and N.Y. Like *S. cryptandrus*, it has the sheaths conspicuously white-hairy at the top, but the infl is more contracted and like that of *S. indicus*.

7. Sporobolus asper (Michx.) Kunth.   Tall d. Stout, tufted, erect, 4–15 dm, the glabrous culms ± solid, mostly 2–5 mm thick; sheaths glabrous, overlapping, rather sparsely long-pilose at the throat and around the collar; blades elongate, flat, becoming involute, 1–4 mm wide, tapering to a scabrous point, pilose near the base at least on the upper side; panicles 2–6, pale or purplish, dense, cylindric, 10–50 cm, up to 1 cm thick, partly or mostly enclosed in the inflated sheaths; first glume 1.7–4.7 mm, the second 2.5–6 mm; lemma glabrous, (3.5–)4–6.5 mm, longer than the second glume, blunt, with a boat-shaped tip; palea about as long as the lemma; anthers (1.5–)2–2.5 mm; pericarp gelatinous when wet; 2n=54, 88, 108. Dry or sandy soil, especially on prairies, sometimes on beaches; Me. and Vt. to N.C., Ky., Tenn., and Ala., w. to N.D., Wash., Utah, and N.M. Nearly all of our plants are var. *asper*, as principally described above. The chiefly southwestern var. *drummondi* (Trin.) Vasey reaches its nw. limits in Mo. and Io. It has more slender culms, 1–2 mm thick, and a somewhat laxer, looser infl with 16–36 spikelets per square cm of surface, instead of 30–75+ as in var. *asper*.

8. Sporobolus clandestinus (Biehler) A. Hitchc.   Rough d. Much like no. 7; perennial, occasionally rhizomatous, with solitary or tufted culms 4–17 dm; lower sheaths frequently pubescent, the collar usually pubescent but only rarely pilose; lvs often hairy, lemma (2.5–)3–7 mm, pubescent; palea pubescent, equaling to sometimes much surpassing the lemma; pericarp becoming loose when wet, but not gelatinous; 2n=46, 48, 52, 54, 56. Dry sandy soil and prairies; Tex. to Fla., n. to Kans., Io., s. Wis., Ind., Va., and near the coast to Mass. (*S. asper* var. *c.; S. canovirens*.)

9. Sporobolus virginicus (L.) Kunth.   Perennial from long creeping rhizomes; culms erect, 1.5–6 dm; lvs conspicuously distichous; sheaths overlapping; blades spreading or ascending, 5–8 cm, broadest (3–6 mm) near the base and tapering to a fine point; panicle exsert, 3–8 cm, stramineous or purplish; glumes sharply acute, the first 1.6–2.9 mm, the second 1.9–2.9 mm; lemma 1.9–2.8 mm, ca equaling the palea; 2n=18, 30. Sandy or muddy shores and marshes, usually in saline soil; trop. Amer., n. along the coast to se. Va.

10. Sporobolus ozarkanus Fern.   Lower sheaths rather sparsely pustulate-pilose; glumes a little longer than the smooth, 3-nerved lemma; palea obtuse, shorter than the lemma; otherwise much like no. 11. Dry sandy or sterile soil; Mo. to Ky., w. and s. to Kans., Ark., and Tex. (*S. vaginiflorus* var. *o.; S. neglectus* var. *o.*)

11. Sporobolus vaginiflorus (Torr.) A. Wood.   Poverty-grass. Rather delicate, tufted, erect to spreading annual 2–5 dm, the lvs and culms glabrous or nearly so, the culms very thin and wiry, seldom over 1 mm thick; blades 1–2 mm wide, the lower elongate, the upper progressively shorter to only 1–2 cm; panicle 1–5 cm, slender, eventually exsert; axillary panicles also developed and mostly included in the lower sheaths; spikelets crowded; glumes and lemmas narrowly lance-triangular, straight, acuminate; first glume 2.8–4.1 mm, the second 2.9–4.6 mm; lemmas mostly 3–5 mm, minutely villous, 1(–3)-nerved; palea acute or acuminate, about equaling or somewhat exceeding the lemma; grain permanently enclosed by the lemma and palea; pericarp becoming gelatinous when wet; 2n=54. Dry sandy or sterile soil; N.S. and Me. to Minn., s. to Ga., Tex., and Ariz.

12. Sporobolus neglectus Nash.   Much like no. 11; lower sheaths generally glabrous; panicle rarely exsert, usually surpassed by the uppermost blade; spikelets smaller, the glumes and lemmas less acuminate and proportionally wider; first glume 1.5–2.4 mm, the second 1.7–2.7 mm; lemma 1-nerved, 1.5–3 mm, glabrous, about equaling the wide, acute, split palea; grain falling free at maturity, and freeing the seed when moistened; 2n=36. Dry sterile or sandy soil; Me. and s. Que. to N.D., s. to N.J., Tenn., La., and Tex.; also Wash. and Ariz.

71. CRYPSIS Aiton, nom. conserv.   Spikelets 1-fld, articulated above the glumes; glumes compressed and keeled, subequal, 1-veined, acute, scarious-margined; lemma membranous, somewhat longer than the glumes, 1-veined, awnless; palea about as long as the lemma; low annuals with erect to decumbent or prostrate stems, short blades, and dense, short, spike-like panicles of small spikelets; ligule a zone of hairs. (*Heleochloa*.) 10, Old World.

1. Crypsis schoenoides (L.) Lam.   Superficially resembling an awnless *Alopecurus*; culms tufted, erect to prostrate, 1–4 dm; spike-like panicles several, terminal and lateral, ovoid-cylindric, 1–4 cm × 6–10 mm, pale green, the base of each partly included in the distended sheaths of the 1 or 2 subtending lvs; spikelets 3 mm; 2n=36. Native of s. Europe, intr. in waste places here and there from Mass. to Del. and Pa., w. to Wis., Io., and Mo.

72. CALAMOVILFA (A. Gray) Hackel.   Spikelets 1-fld, articulated above the glumes; glumes chartaceous, 1-veined, the outer ovate, the inner much longer, lance-ovate; lemma chartaceous, obscurely 1-veined, acute, awnless, the callus bearded; palea about as long as the lemma; rachilla not prolonged; rhizomatous perennials with overlapping basal sheaths and long, narrow, usually involute blades; ligule a dense fringe of short hairs; infl an open or contracted, ovoid to oblong panicle. 4, N. Amer.

1 Lemma and palea glabrous on the back; rhizome elongate; inland .... 1. *C. longifolia.*
1 Lemma and palea hairy on the back; rhizome short; coastal plain .... 2. *C. brevipilis.*

1. Calamovilfa longifolia (Hook.) Scribn.   Sand-reed. Stout, to 2 m, from long rhizomes; sheaths to 15 cm; ligule 1–1.5 mm; blades to 60 cm × 12 mm; not articulated to the sheath; infl 1.5–7 dm; first glume 3.5–6.5 mm, the second 5–8 mm; lemma 4.5–7 mm, glabrous except for the callus-hairs half to seven-eighths as long; palea glabrous; 2*n*=40, 60. Two well marked vars.:

Var. *longifolia*.   Panicle slender, 6.5–14 times as long as wide, with crowded spikelets and ± ascending or erect branches; sheaths usually glabrous, seldom lightly pubescent, rarely densely so. Dry sandy soil and prairies from Wis. and Ill. to sw. Ont., se. B.C., Wyo., and Colo.

Var. *magna* Scribn. & Merrill.   Infl broader, 1.3–6 times as long as wide, with more spreading branches and less crowded spikelets; sheaths usually pubescent, often densely so, only rarely glabrous. Mainly on dunes along the shores of lakes Michigan and Huron, with scattered outlying stations w. to w. Wis. and e. Io.

2. Calamovilfa brevipilis (Torr.) Scribn.   Culms slender, 5–15 dm, from a short stout rhizome; sheaths glabrous, to 30 cm, the basal ones keeled, firm, persistent on the rhizome; blades to 50 cm, 2–5 mm wide, articulated to the sheath; ligule under 0.5 mm; infl ± loose and open, 1–4 dm × 4–20 cm, with ascending to spreading branches; pedicels often with an apical tuft of fine hairs; first glume 1.7–4 mm, the second 3.3–5 mm; lemma 4–5.4 mm, the callus-hairs a fourth to half as long; lemma and palea hairy on the back. Swamps and bogs in the pine-barrens of N.J. and from se. Va. to S.C.

73. MUHLENBERGIA Schreber.   Muhly. Spikelets 1-fld (casually 2-fld in a few spp.), articulated above the glumes; glumes usually equal or subequal, seldom very unequal, subulate to ovate, keeled, 1-veined or rarely 2–3-veined or veinless, membranous or scarious toward the margins, acute to subulate or rarely truncate, often awned, the body shorter to a little longer than that of the lemma; first glume rarely obsolete; lemma membranous, laterally compressed but rounded on the back, 3-veined, acute to subulate, sometimes awned from the tip; palea nearly as long as the lemma; callus often bearded; style-branches naked; rachilla not prolonged; fr permanently enclosed by the lemma; our spp. perennial, often freely branched, bearing open or contracted panicles of small spikelets. 100, mainly New World.

1 Panicle loose and open, 4 cm thick or more; spikelets on capillary pedicels of equal or greater length.
  2 Plants rhizomatous.
    3 Infl, when expanded at anthesis, nearly or quite as long as thick .... 16. *M. asperifolia.*
    3 Infl more elongate, mostly 2–4 times as long as thick .... 15. *M. torreyana.*
  2 Plants tufted or matted, not rhizomatous.
    4 Spikelets ca 1.5–2 mm .... 14. *M. uniflora.*
    4 Spikelets (excl. awns) 2.5–5 mm .... 13. *M. capillaris.*
1 Panicle slender, contracted, often dense, less than 2.5 cm thick.
  5 Callus glabrous; body of the lemma glabrous or minutely pubescent.
    6 Creeping rhizomes wanting; glumes a third to two-thirds as long as the lemma.
      7 Ligule up to 0.5 mm; lemma minutely pubescent on the back .... 12. *M. cuspidata.*
      7 Ligule 1.5–3 mm; lemma glabrous .... 11. *M. richardsonis.*
    6 Creeping rhizomes well developed; glumes nearly or fully as long as the lemmas .... 5. *M. glabriflora.*
  5 Callus slightly to strongly bearded, the lemma otherwise glabrous or hairy.
    8 Glumes with stiff awn-tips much exceeding the awnless lemma.
      9 Internodes dull, puberulent; lemma pilose along the margins .... 10. *M. glomerata.*
      9 Internodes smooth and polished except at the top; body of the lemma glabrous .... 9. *M. racemosa.*
    8 Glumes without protracted awn-tips, usually shorter than or merely equaling the body of the awned or awnless lemma.
      10 Glumes minute, veinless; no rhizomes .... 8. *M. schreberi.*
      10 Glumes at least a third as long as the lemma, 1(–3)-veined; rhizomatous.
        11 Infls terminal and axillary, the axillary ones usually basally included in the sheath; internodes smooth and shining.
          12 Glumes 1.4–2 mm, evidently shorter than the lemma; ligule 0.2–0.5 mm .... 7. *M. bushii.*
          12 Glumes 2–4(–5) mm, usually at least equaling the lemma; ligule 0.8–1.5 mm .... 6. *M. frondosa.*
        ·11 All infls terminal or on elongate, exserted lateral peduncles; internodes glabrous or puberulent.

13 Glumes broad near the base and overlapping, with an abrupt sigmoid curvature to
an acute tip, both much shorter than the lemma.
  14 Internodes and sheaths glabrous; lemma awnless or with an awn-tip to 4 mm   4. *M. sobolifera.*
  14 Internodes and usually the bases of the sheaths pubescent; lemma with an awn
  4–10 mm ....   3. *M. tenuiflora.*
13 Margins of the glumes straight, tapering from base to apex; glumes nearly as long
as the lemma.
  15 Ligule 1–2.5 mm; infl loose, the spikelets on pedicels up to their own lengths .....   2. *M. sylvatica.*
  15 Ligule 0.5–1 mm; infl dense, often lobulate, the spikelets densely clustered and
  usually subsessile ....   1. *M. mexicana.*

1. **Muhlenbergia mexicana** (L.) Trin.   Wirestem-m. Vigorously rhizomatous; culms 3–9 dm; internodes dull and puberulent, especially toward the top; sheaths glabrous; blades 5–15 cm × 2–6 mm, glabrous; ligule membranous, 0.5–1 mm; infls terminal on the main stem or also on long leafy branches, the peduncles exserted 2–12 cm; panicles very slender to dense and lobulate, the terminal one 7–21 cm × 2–10 mm; spikelets closely clustered, mostly subsessile, 1.7–4.4 mm, the glumes narrow, shorter than or equaling the lemma, often cuspidate or with an awn-tip to 1.5 mm; lemmas 1.5–3.4 mm, awnless or with an awn-tip to 9 mm, the callus bearded; anthers 0.3–0.5 mm; 2n=40. In a wide range of moist or wet, open or wooded habitats; Que. and N.S. to B.C., s. to N.C., Mo., Tex., and Calif. (*M. foliosa; M. ambigua,* the awned form.)

2. **Muhlenbergia sylvatica** (Torr.) Torr.   Forest-m. Vigorously rhizomatous, 4–8(–11) dm, erect, or sprawling when old; culms puberulent, especially on the upper half of an internode; sheaths glabrous; blades 5–14 cm × 3–5(–7) mm; ligule prominent, membranous, ciliate, often lacerate, 1–2.5 mm; infls slender, with appressed branches, the peduncle included or exserted to 8 cm; panicle 6–20 cm × 2–5(–10) mm, the spikelets on pedicels up to their own length, 2.2–3.7 mm; glumes subequal, tapering evenly to an acuminate tip, equaling or somewhat shorter than the lemma; lemma 2.2–3.7 mm, short-pilose on the callus, with a slender awn 5–18 mm; anthers 0.3–0.5 mm; 2n=40. Upland woods, rocky ledges, and streambanks; Me. and Que. to Minn., s. to N.C., Ark., and Tex. (*M. umbrosa.*)

3. **Muhlenbergia tenuiflora** (Willd.) BSP.   Vigorously rhizomatous; culms solitary or few together, erect, 5–12 dm; ligule 0.5–1 mm; blades 6–18 cm × (4–)6–10(–13) mm, tapering conspicuously to the base; infl arcuate, slender, 15–30 cm × 2–5 mm, with few appressed branches; spikelets 3–5 mm; glumes subequal, shorter than the lemma, broad and overlapping at base, abruptly curved to an acute or awned tip, the second often 2–3-veined; lemma sparsely bearded on the callus; anthers 1–2 mm; 2n=40. Rich upland woods; Mass. and N.H. to Wis. and Io., s. to Tenn., Ga. (in the mts.) and Mo. The widespread var. *tenuiflora* is retrorsely hirsute on the sheaths and stems especially at and just below the nodes, and the lemma is tipped with an awn 4–11 mm. Var. *variabilis* (Scribn.) Pohl, at middle elev. in the S. Appal. from W.Va. to e. Ky., w. N.C., and n. Ga., is virtually glabrous and has awnless or short-awned (to 4 mm) lemmas.

4. **Muhlenbergia sobolifera** (Muhl.) Trin.   Vigorously rhizomatous; culms arising singly or few together, erect, becoming much branched, the internodes glabrous; sheaths glabrous, frequently overlapping; blades stiff, 4–16 cm × 2–7 mm; ligule membranous, 0.5–1 mm, infls terminal or some on erect leafy lateral branches, included at base or more often exsert on slender erect peduncles to 11 cm; panicles very slender, arcuate, 5–15 cm × 2–4 mm, with short erect branches; spikelets whitish, glumes subequal, 1.3–2(–2.5) mm, shorter than the lemma, the margins usually sigmoid-curved; lemma 1.8–2.5 mm, awnless or with an awn-tip to 4 mm, the callus scantily bearded; anthers 0.4–0.8 mm; 2n=40. Dry upland woods or rocky outcrops; Mass. to Va., w. to Wis., Nebr. and Tex. (*M. setigera,* the awned, mainly western form.)

5. **Muhlenbergia glabriflora** Scribn.   Vigorously rhizomatous, 3–10 dm, becoming much branched and very leafy; internodes slightly puberulent near the top, otherwise glabrous; blades and sheaths glabrous; blades stiffish, 3–5(–8) cm × 1.5–4 mm; ligule 0.5–1.5 mm; infls numerous, terminal and axillary, dense, lobulate, cylindric, 2–5 cm × 3–5 mm, included at base or on peduncles exsert 1–4 cm; spikelets densely crowded, often subsessile; glumes slightly shorter to slightly longer than the lemma, this 2.2–3 mm, glabrous, awnless; anthers 0.3–0.5 mm; 2n=40. Mostly in shade on low ground in heavy clay; irregularly from Va. and N.C. to Io., Mo., Ala., and Tex.; rare.

6. **Muhlenbergia frondosa** (Poiret) Fern.   Vigorously rhizomatous, 5–10 dm, the culms at first simple and erect, later bushy-branched and sprawling; internodes smooth and shining; blades and sheaths glabrous; blades lax, the larger ones 4–12 cm × 2–7 mm; ligule membranous, lacerate-ciliate, 0.8–1.5 mm; panicles terminal and axillary, the latter borne at many nodes and only partly exserted, all narrow, up to 10 × 1 cm, the branches erect except during anthesis; terminal peduncle sometimes exsert to 5 cm; glumes 2–4(–5) mm, subequal, narrowly lanceolate to linear, tapering from the base to an acuminate or short-awned tip, usually surpassing the lemma or occasionally a fourth shorter; lemma 2.9–3.6 mm, glabrous except the slightly bearded callus, awnless or with an awn to 11 mm; anthers 0.3–0.6 mm; 2n=40. Forest-margins, thickets, alluvial plains, and moist prairies, sometimes a weed in fields and gardens; N.B. to N.D., s. to Ga. and Tex. (*M. mexicana,* misapplied; *M. commutata,* the awned form.)

7. **Muhlenbergia bushii** Pohl.   Much like no. 6; internodes mostly concealed by the overlapping sheaths; blades grayish-green, stiffish, the larger ones 5–10(–15) cm × 2–5 mm; ligule short, 0.2–0.6 mm; glumes 1.4–2(–2.5) mm, half to two-thirds as long as the lemma; lemma 2.6–3.3 mm, usually awnless or with a minute awn-tip, rarely with an awn to 7 mm; 2n=40. Prairies, plains, and dry woods, adapted to drier habitats than no. 6; Ill. and s. Ind. to Io., Nebr., and Tex. (*M. brachyphylla,* a preoccupied name.)

8. Muhlenbergia schreberi J. F. Gmelin.   Nimblewill. Perennial 2–6(–9) dm, at first erect and clustered, but the weak, slender culms later sprawling and rooting at the lower nodes; internodes glabrous; lvs glabrous except around the orifice of the sheath and base of the blade; sheaths usually shorter than the internodes; ligule minute, ciliolate, only 0.1–0.2 mm; blades 2–8 cm × 1–4 mm; infls very slender, usually on peduncles exsert 1–10 cm, 2–10 cm × 1.5–3(–6) mm, the branches appressed; glumes minute and veinless, membranous, the second 0.1–0.3 mm, the first even shorter or virtually obsolete; lemma 1.5–2 mm, slightly bearded on the callus; awn 1.5–4 mm; anthers 0.1–0.4 mm; 2n=40. Disturbed, moist or wet places, often a weed in lawns and gardens; N.H. and Mass. to Minn. and Nebr., s. to Fla. and Mex. (*M. palustris.*) Hybridizes with several other spp., possibly nos. 3, 6, and 7, producing plants with slightly larger glumes (these sometimes with 1 or 2 veins), with or without short rhizomes. These are sometimes collectively called *M.* × *curtisetosa* Scribn.

9. Muhlenbergia racemosa (Michx.) BSP.   Much like no. 10; culms becoming much branched from near the middle; internodes shining, glabrous or slightly puberulent near the top; ligule 0.6–1.5 mm, lacerate; infls 4–17.5 cm, terminal or on long leafy branches, the peduncle included or exsert up to 10 cm; florets pilose only on the callus and sometimes the keels of the palea, but the margins and back of the lemma glabrous; anthers 0.6–0.8 mm; grain 1.4–2.3 mm; 2n=40. Drier habitats than no. 10; prairies, rock-outcrops, open upland woods, or sometimes along roads or in other disturbed sites; Wis. and Ill. to Alta. and e. Wash., s. to Okla. and Ariz. Hybridizes with no. 10.

10. Muhlenbergia glomerata (Willd.) Trin.   Marsh-m. Vigorously rhizomatous, 3–12 dm, stiffly erect, unbranched or sparingly branched; internodes finely puberulent, especially near the top; blades and sheaths glabrous, the blades 6–15 cm × 2–6 mm; ligule 0.2–0.6 mm; somewhat ciliate, infl a dense, cylindric, somewhat lobed panicle 2–11 cm × 3–15 mm, on a peduncle 5–15 cm, spikelets green to often purplish, densely crowded, 3.2–5(–8) mm, the glumes subequal, much exceeding the lemma, tapering to a rigid awn nearly as long as the body; lemma 2–3.5 mm, awnless or rarely with a minute awn-tip, bearded on the callus and pubescent along the veins and margins, sometimes nearly to the tip; anthers 0.8–1.5 mm; grain 1–1.6 mm; 2n=20. Marshes, gravelly shores, wet meadows, seepage areas, and bogs; Nf. to B.C., s. to N.C., Ky., Io., Nebr., Utah, and Nev.

11. Muhlenbergia richardsonis (Trin.) Rydb.   Mat-m. Culms clustered, 3–6 dm, very slender and wiry, minutely roughened at the nodes, erect or often decumbent and rooting near the base, forming mats, but the plant without rhizomes; ligule 1.5–3 mm; blades usually involute, 1–2 mm wide, erect or ascending, the main ones 3–6 cm; panicles 1–4(–7) cm, slender, with short, ascending, scarcely overlapping branches; glumes lance-ovate, glabrous, the first 0.9–1.3 mm, the second 1–1.5 mm; lemma glabrous, obscurely veined, 2–3.3 mm, tapering to a subulate point; 2n=40. Wet gravelly soil; Anticosti I. to Me.; s. Mich.; Minn. to Wash. and N.M.

12. Muhlenbergia cuspidata (Torr.) Rydb.   Culms tufted, 2–7 dm, slender, stiff and strictly erect, somewhat bulbous-thickened at base or with bulb-like offsets; ligule 0.5 mm or less; blades erect or nearly so, flat or involute, 1–2 mm wide; panicle very slender and spike-like, the appressed lateral branches 5–15 mm; glumes lance-subulate, subequal, 1.7–2.8 mm; lemma 2.8–4.1 mm, slender, acuminate, awnless, minutely pubescent on the back, otherwise glabrous. Prairies and open hillsides, in dry or gravelly soil; Alta. to N.M., e. to Mich., O., Ky., and Mo.

13. Muhlenbergia capillaris (Lam.) Trin.   Hairgrass. Tufted, 5–10 dm; sheaths glabrous to scaberulous; ligule triangular, 2–5+ mm; larger blades 2–3 dm × 2–4 mm; panicle ample and open, divergently branched, usually a third to half as long as the entire shoot; spikelets purple; glumes lance-ovate, tapering to a slender tip or short awn, subequal or somewhat unequal; lemma slender, fusiform, 2.5–5 mm. W.I., e. Mex., and se. U.S., n. to Okla., e. Kans., and s. Ind., and in the Atlantic coastal states to Mass. Three vars., 2 in our range:

Var. *capillaris.*   Glumes a fourth to half as long as the spikelet, awnless or with an awn to 3 mm (first glume) or 5 mm (second glume); lemma with an awn 2–13(–18) mm, or seldom awnless; widespread, mostly in dry, sandy or rocky soil.

Var. *trichopodes* (Elliott) Vasey.   Glumes more than half as long as the lemma, awnless or the second one occasionally mucronate; lemma awnless or with an awn to 4 mm. Moist or wet, sandy pine woods and bogs; se. Va. to Fla. and Tex. (*M. t.; M. expansa.*)

14. Muhlenbergia uniflora (Muhl.) Fern.   Culms tufted, 2–4 dm, very slender, often decumbent and rooting at the base; sheaths and blades glabrous or nearly so, the latter flat or concave, rarely over 1 mm wide; ligule membranous, usually erose, 0.5–1.5 mm; infl slender but open, a fourth to half as long as the entire shoot; spikelets drab, ellipsoid, occasionally 2-fld, on pedicels mostly more than twice as long; glumes ovate, acute or obtuse, the second barely longer than the first and ca half as long as the spikelet; lemma acute, 1.2–2 mm, awnless; 2n=ca 42. Moist or wet, sandy or peaty soil, open meadows, and bogs; Nf. to w. Ont. and n. Mich., s. to N.J. (*Sporobolus u.*)

15. Muhlenbergia torreyana (Schultes) A. Hitchc.   Rhizomatous; culms and sheaths strongly compressed at base; stems erect, 3–7 dm; sheaths glabrous; ligule reduced to a zone of short hairs; blades usually conduplicate, 1–3 mm wide, glabrous; infl loose and open, 2–4 times as long as thick; spikelets greenish-drab to purple, 1.5–2 mm; glumes acute, the first scarcely shorter than the second, which nearly or quite equals the acute, minutely puberulent, awnless lemma. Moist pine-barrens; N.J. and Del.; Ga. (*Sporobolus t.; S. compressus.*)

16. Muhlenbergia asperifolia (Nees & Meyen) L. Parodi. Scratchgrass, alkali m. Rhizomatous; culms tufted, glaucous, 1–5 dm, spreading or decumbent at base, rarely erect; sheaths glabrous, usually overlapping and keeled; ligule truncate, erose, 0.2–1 mm; main blades 3–7 cm × 1–2.5 mm, abruptly expanded above the collar, scabrous, especially on the cartilaginous margin; base of the panicle at first enclosed in the sheath, later exserted and the branches widely spreading, the infl open, diffuse, about as thick as long; spikelets often purple or blackish, sometimes 2-fld; glumes acute, subequal, 1–1.4 mm; lemma obtuse or subacute, 1.2–1.7 mm, awnless; $2n=20$. Moist soil and streambanks, often in alkaline situations, sometimes a lawn-weed; widespread in w. U.S., e. to Minn., Io., and Mo., and intr. as far as Mich., O., s. Ont., and e. N.Y.; s. S. Amer. (*Sporobolus a.*)

74. DACTYLOCTENIUM Willd.

Spikelets crowded, divergent, flattened, few-fld, articulated between the glumes; glumes very broad, compressed and keeled, 1-veined, the first awnless, the second with a stout awn just below the rounded tip; lemmas broadly ovate, compressed and keeled, 3-veined, abruptly acuminate to a short awn-like point; grain subglobose, rugose, furrowed on one side; annuals with usually overlapping sheaths, flat blades, and sessile digitate spikes, the uppermost spikelets abortive and the rachis projecting 1–5 mm as a sharp point. 10, warm reg.

1. Dactyloctenium aegyptium (L.) P. Beauv. Crowfoot-grass. Culms widely spreading, flattened, rooting at the nodes; sheaths glabrous; blades to 8 mm wide, papillose-ciliate, especially at base; spikes 2–5, each 2–5 cm; second glume 2 mm, its awn 2–3 mm, often flexuous, usually divergent forward and over the adjacent spikelet; lemmas 2–5, 3 mm; $2n=20, 24, 36, 40, 48$. Open ground, waste places, and fields; native of the Old World tropics, now well established as a weed in trop. Amer. and abundant in se. U.S., occasionally adventive in our range, n. to Me. and Ill.

75. CHLORIS Swartz.

Finger-grass. Spikelets with 1 perfect fl and 1 or more empty lemmas, articulated above the glumes, sessile in 2 rows on 2 sides of a slender triangular rachis; glumes small, unequal, the first 1-veined, the second 1- or 3-veined; lemmas unequal, the lowest fertile, compressed and keeled, usually awned just below the tip, 3-veined, the lateral veins marginal; second lemma smaller, sterile, usually truncate or inflated, awned or awnless; other lemmas, if present, still smaller and mostly concealed within the second; usually tufted, bearing several elongate spikes in a terminal digitate cluster. 40, warm reg.

1 Perennial; fertile lemma inconspicuously appressed-pilose, spikelets not imbricate .... 1. *C. verticillata.*
1 Annual; fertile lemma conspicuously long-ciliate distally; spikelets imbricate .... 2. *C. virgata.*

1. Chloris verticillata Nutt. Windmill f.-g. Densely tufted perennial 1–4 dm, usually erect; lvs chiefly near the base; sheaths compressed; blades obtuse, 2–4 mm wide, usually only 5–15 cm long; spikes numerous, 6–15 cm, eventually widely spreading; glumes lance-subulate, awn-pointed, the first 2–3 mm, the second 3–4 mm; fertile lemma 2.4–2.8 mm, acute, appressed-pilose on the margins and keel, its awn 5–9 mm; sterile lemma wedge-obovate, truncate, 1.5 mm, its awn 3.5–5 mm. Dry prairies and plains; Io. and w. Mo. to Colo., s. to La. and N.M., and locally adventive elsewhere in our range.

2. Chloris virgata Swartz. Feather f.-g. Spreading to erect annual 2–5 dm; uppermost sheaths inflated; blades 3–6 mm wide, tapering to a sharp point; spikes crowded, erect or ascending, 3–8 cm; glumes acuminate, fertile lemma long-white-ciliate on the margins distally and on the keel toward the base, its awn 4–9 mm; sterile lemma narrowly wedge-obovate, its awn 3–7 mm. Native to trop Amer., widespread as a weed in sw. U.S., and occasionally adventive in our range as far n. as Me.

76. SCHEDONNARDUS Steud.

Spikelets 1-fld, articulated above the glumes, sessile and appressed along 2 sides of a slender triangular rachis; glumes slightly unequal, lance-subulate, 1-veined, acuminate; lemma narrowly lanceolate, acuminate, 3-veined; palea narrow, much shorter than the lemma; densely tufted perennial with short narrow blades, membranous ligule, and a large infl of a terminal and several lateral spikes. Monotypic.

1. Schedonnardus paniculatus (Nutt.) Trel. Tumble-grass. Culms 2–5 dm at anthesis; lvs usually clustered at base, the blade 3–5 cm × 1–2 mm; panicle constituting half to three-fourths the length of the shoot, rigid, with several very slender, widely divergent spikes 2–5 cm, the axis elongating after flowering, the whole panicle forming a loose spiral, breaking off and rolling before the wind; first glume 1.9–3.2 mm; second glume 2.6–4.3 mm; grain 2 mm; $2n=20, 30$. Dry prairies; Ill. to Sask. and Ariz.; Argentina.

77. GYMNOPOGON P. Beauv.   Beard-grass.   Spikelets 1-fld (in ours), articulated above the glumes, sessile, appressed and remote along two sides of a slender triangular rachis; glumes subequal, subulate, 1-veined; lemma shorter than the glumes, narrow, rounded on the back, obscurely 3-veined, tipped with a short straight awn; rachilla often prolonged behind the palea, bearing at its summit a greatly reduced lemma or short awn; ours perennial, with short erect culms from a knotty rhizome, minute ligule, stiffly spreading lvs, and a large panicle of very slender, elongate, divaricate spikes. 10, New World.

1 Awn of the lemma 4.5–9 mm; spikes floriferous nearly to the base .... 1. *G. ambiguus.*
1 Awn of the lemma 0.8–1.6 mm; spikes floriferous only above the middle .... 2. *G. brevifolius.*

   1. Gymnopogon ambiguus (Michx.) BSP.   Culms mostly solitary, 3–6 dm; lvs crowded toward the base; sheaths overlapping; blades divaricate, the larger 5–10 cm × 6–12 mm, rounded or subcordate at base, glabrous; panicle constituting a third to half of the entire shoot; spikes numerous, 10–15 cm, at first erect, later widely and stiffly divaricate; spikelets remote, the summit of the glumes barely or not reaching the base of the next above; glumes 4–6 mm; lemma 3.5–4.3 mm, sparsely pilose-ciliate; awn straight, 4.5–9 mm; rachilla prolonged nearly to the summit of the lemma, there bearing an awn 1–5 mm. Dry sandy woods and barrens; coastal plain from s. N.J. to Fla. and Tex., and in the interior from s. O. and Ky. to s. Mo. and southward.

   2. Gymnopogon brevifolius Trin.   Much like no. 1; lvs 3–6 mm wide, spikes floriferous only above the middle; glumes 2.8–3.7 mm; lemma 2.3–3 mm, conspicuously pilose on the back and margins; awn 0.8–1.6 mm. Dry pine-barrens on the coastal plain from s. N.J. to Fla., Ark., and La.; Pulaski Co., Ky.

78. CTENIUM Panzer, nom. conserv.   Spikelets sessile, pectinately arranged in 2 rows on one side of a flat rachis, articulated above the glumes, with several lemmas, of which only one is fertile; first glume small, nearly hyaline, 1-veined; second glume nearly as long as the spikelet, firm on the round exposed back, thin on the sides, conspicuously 3-veined, the midvein prolonged into a stout divergent awn from near the middle of the glume; lemmas thin, obscurely 3-veined, the two lower empty, without paleas, the third with a palea and a fl, the upper 1–3 sterile and much reduced; midvein of the lower 3 lemmas prolonged into an awn from below the tip; perennials with narrow lvs, membranous ligule, and a solitary, usually curved spike. 20, New World and Afr.

   1. Ctenium aromaticum (Walter) A. Wood.   Toothache-grass. Culms densely tufted, 8–15 dm, hairy at the top; lf-blades smooth, 1–3 mm wide; spike usually curved, 5–12 dm; first glume (on the inner side of the spike!) 1.5–2.5 mm; second glume 4–7 mm, pubescent, the veins glandular below the middle, the awn 4–6 mm; lemmas conspicuously ciliate, especially at the middle, 4–6 mm, the fertile one and the two lower sterile ones with awns 2–6 mm, the upper sterile lemma single and awnless. Wet pine-barrens; se. Va. to Fla. and La. (*Campulosus a.*)

79. CYNODON Rich., nom. conserv.   Spikelets 1-fld, articulated above the glumes; glumes narrow, subequal, 1-veined; lemma strongly flattened, 3-veined, acute, awnless, the lateral veins near the margin; palea narrow, nearly as long as the lemma; rachilla prolonged behind the palea as a slender bristle half as long as the lemma; perennials with short blades, the ligule a ring of hairs, and slender digitate spikes of small spikelets. 10, warm reg.

   1. Cynodon dactylon (L.) Pers.   Bermuda-grass. Widely spreading by stolons and rhizomes and forming mats, the culms erect or ascending, 1–3 dm; main lvs 2–4 mm wide; spikes 4–6, divergent, 3–5(–8) cm; glumes 1.5–2.2 mm, the first curved, the second nearly straight; lemma 2–2.5 mm, minutely pubescent to long-villous on the keel; $2n=36, 40$. Native of the Old World, abundant (and often cult.) in lawns and in disturbed sites in s. U.S. (n. to Del.) and adventive n. to N.Engl., Mich., and Nebr. (*Capriola d.*)

80. SPARTINA Schreber.   Cord-grass. Spikelets 1-fld, strongly flattened, articulated below the glumes, closely or loosely imbricate in 2 rows on 2 sides of a triangular rachis; glumes unequal, narrow, 1-veined, acute, acuminate, or awned, the second (exterior in the spike) exceeding the lemma; lemma firm, keeled, awnless, obscurely veined; palea equalling or usually exceeding the lemma; perennial, usually colonial from stout, scaly rhizomes, with a few to many dense, one-sided spikes appressed to spreading along a common axis; ligule a band of hairs; good seed only sparingly produced. 15, New World and w. Europe and Afr.

1 Main lf-blades mostly 5–20 mm wide, flat, or involute when dry.
  2 Spikelets appressed, loosely or scarcely imbricate, those on one side of the spike 3–8 mm apart .... 1. *S. alterniflora.*
  2 Spikelets ascending or divergent, densely imbricate, those on one side of the spike 1–3 mm apart.
    3 Second glume pointed but awnless; first glume two-fifths to two-thirds as long as the lemma .. 2. *S. cynosuroides.*
    3 Second glume awned; first glume three-fourths to fully as long as the lemma .... 3. *S. pectinatus.*
1 Main lf-blades 1–4 mm wide, involute .... 4. *S. patens.*

1. Spartina alterniflora Loisel. Smooth c.-g. Culms coarse, soft and fleshy, to 3 m, or in the north as low as 3 dm, sulfurously malodorous when bruised, rhizomes elongate, thick and soft, the closely imbricate scales inflated; lvs elongate, 5–15 mm wide, glabrous or nearly so; panicle narrow, 1–3 dm; spikes slender, appressed to the axis, 5–10 cm; spikelets scarcely imbricate, erect or nearly so, those on one side of the spike usually 3–8 mm apart; rachis prolonged beyond the uppermost spikelet and usually conspicuously exceeding it; spikelets glabrous or somewhat hairy; first glume linear, half to nearly as long as the lemma; second glume 10–14 mm, exceeding the lemma, its midvein scarcely or not at all prolonged beyond the thin margins; $2n$=42, 56, 70. Coastal salt-marshes, mainly intertidal, smaller at the upper tidal margins; Que. and Nf. to Fla. and Tex.; also on the Atlantic coast of S. Amer. and in n. Europe.

2. Spartina cynosuroides (L.) Roth. Big c.-g. Culms coarse and hard, 1–3 m, rhizomes stout, firm, elongate; main lvs 8–20 mm wide, cuttingly scabrous on the margins; panicle dense, 1–3 dm; spikes numerous, crowded, ascending, usually 3–7 cm, the projecting rachis not surpassing the uppermost spikelet; spikelets firm and harsh, closely imbricate, those on one side of the spike rarely over 3 mm apart; first glume linear, two-fifths to two-thirds as long as the lemma; second glume 10–15 mm, sharply acute to acuminate but not awned, scabrous on the keel, exceeding the obtuse lemma; $2n$=28, 42, ca 80. Salt or brackish marshes along the coast; Mass. to Fla. and Tex.

3. Spartina pectinata Link. Prairie cord-g. Stout, erect, 1–2 m, with long coarse rhizomes; main lvs 3–8 dm, 5–10+ mm wide, involute when dry, cuttingly scabrous on the margins, tapering to a long slender point; panicle 2–4 dm, the spikes 7–27, ascending or appressed, 5–12 cm, short-peduncled, the rachis prolonged but not surpassing the uppermost spikelets; spikelets crowded, closely imbricate, those on one side of the rachis less than 3 mm (usually ca 1.5 mm) apart; first glume 5–13 mm, three-fourths to fully as long as the lemma; second glume lanceolate, the body 10–13 mm, the rough midnerve prolonged into an awn 3–10 mm; $2n$=28, 42, 84. Marshes, shores, and wet prairies; Nf. and Que. to Alta. and Wash., s. to N.C. and Tex., especially abundant in the prairie states. (*S. michauxiana.*)

    S. caespitosa A. A. Eaton (*S. patens* var. *caespitosa*), occurring sporadically from Me. to Md., is thought to be a hybrid between nos. 3 and 4, but differs from both in being densely tufted, with poorly developed or no rhizomes.

4. Spartina patens (Aiton) Muhl. Salt-meadow c.-g. Culms slender, stiff and wiry, 3–10 dm, on long slender rhizomes; lvs 4–30 cm, involute (or flat at base), mostly 1–4 mm wide (when flattened); spikes mostly 2–7, 1.5–5 cm, ascending, not appressed; spikelets closely imbricate, scabro-ciliate on the keels; first glume linear, mucronate, 2–6 mm; second glume narrowly lanceolate, long-acuminate or almost aristate, 7.5–13 mm; lemma obtuse, 5.5–8 mm; $2n$=28, 42, 56. Salt-marshes and wet beaches; Que. to Fla. and Tex.; also in saline situations inland in c. N.Y., s. Ont., and e. Mich. (*S. juncea.*) The principal source of eastern salt-hay.

    S. gracilis Trin., alkali c.-g., is widespread in interior w. U.S. and barely enters our range in Clay Co., Minn. It differs from *S. patens* in its short, broader spikes closely appressed to the often flexuous stem, and in the more copiously and conspicuously long-hispid keels of the glumes and lemmas.

81. BOUTELOUA Lagasca. Gramma-grass. Spikelets with 1 perfect fl and 1 or more sterile vestiges, articulated above the glumes, inserted in 2 rows on one side of a narrow flat rachis; glumes unequal, narrow, 1-veined, acuminate to awn-pointed; fertile lemma rounded on the back, 3-veined, the lateral veins marginal or submarginal, usually excurrent below the tip into a short awn; vestige stipitate, reduced to an empty 3-awned lemma; usually tufted, the relatively short spikes 1–many, racemose on a common axis. 40, New World.

1 Spikes numerous (10–)40–70, forming a long, erect infl ....1. *B. curtipendula.*
1 Spikes 1–3.
  2 Rachis of the spike prolonged as a stiff point beyond the tip of the uppermost spikelets .... 2. *B. hirsuta.*
  2 Rachis of the spike not exceeding the uppermost spikelet .... 3. *B. gracilis.*

1. Bouteloua curtipendula (Michx.) Torr. Side-oats g. Erect perennial 3–10 dm from short slender rhizomes; sheaths smooth or nearly so; ligule a band of short hairs; blades elongate, 2–7 mm wide, scabrous on the margins; spikes (10–)40–70, spreading or nodding, 8–15 mm, secund along an axis 1–3 dm, falling entire; spikelets usually 3–7; first glume linear-subulate, 3–4 mm; second glume lanceolate, 4–7 mm; fertile lemma usually somewhat exceeding the glumes, acuminate, its lateral veins prolonged into awns 1 mm; vestige with a long central awn and 2 shorter lateral ones arising below the middle, or much reduced or obsolete; $2n$=20–103. Dry woods in the e. states, and dry prairies and sand-hills in the w.; Me. to Mont., s. to Ala., Calif., and C. and S. Amer. (*Atheropogon c.*) Ours is var. *curtipendula*.

2. **Bouteloua hirsuta** Lagasca.   Hairy g.   Densely tufted perennial 1.5–6 dm; lvs mostly crowded toward the base; sheaths pilose at the throat; blades flat, 1–3 mm wide; spikes 1–3, usually 2, straight or recurved, 2–4 cm; rachis projecting 2–5 mm beyond the uppermost spikelet as a straight stiff point; spikelets numerous, closely imbricate, divergent; first glume subulate; second glume lanceolate, 3–4 mm, densely papillose-hirsute on the prominent midvein; fertile lemma about equaling the second glume, sparsely villous throughout, its awn short and flattened; vestige long-stipitate, glabrous at base, obconic, often reaching the summit of the second glume, its 3 awns equal, exceeding the tips of the fertile lemma and second glume; $2n$=12–50. Dry prairies and sand-hills; Wis. and Ill. to N.D., s. and w. to Tex., Calif., and Mex.; Fla.

3. **Bouteloua gracilis** (HBK.) Lagasca.   Blue g.   Much like no. 2; lvs curled or flexuous, involute, 1–2 mm wide; rachis not surpassing the uppermost spikelet; second glume 3.5–5 mm, scabrous and sparsely villous on the midvein; fertile lemma densely villous at base and along the midvein, sparsely villous-ciliate, the terminal awn short, the lateral ones 1.5 mm, arising above the middle; vestige stipitate, densely long-villous around the base, bearing 3 rough awns which equal or surpass the second glume; $2n$=20–84. Dry prairies and sand-hills; Ill. to Alta., s. to Tex. and s. Calif., and rarely adventive eastward.

82. BUCHLOE Engelm.   Buffalo-grass. Dioecious; staminate spikes 1–3, short, terminating slender culms, the spikelets 2-fld, with thin glumes nearly as long as the lemmas, the first glume lance-subulate, 1-veined, the second ovate, 1- or 3-veined, the lemmas narrow, acute, 3-veined, hardly longer than the palea; pistillate spikes short and headlike, in a pair subtended by the dilated, many-veined bases of the 2 uppermost lvs and ± concealed by the foliage, each consisting of 3–5 closely imbricate, 1-fld spikelets of which only the modified second glumes are visible from the outside, the first glume (concealed) thin, 1-veined, sometimes obsolete, the second glume thick and indurate, convex on the back, contracted above and ending in 3 stiff, erect, narrowly triangular lobes nearly as long as the body, its margins folded over and enclosing the lemma, which is lance-ovate, indurate, and inconspicuously 3-lobed at the summit; stoloniferous, sod-forming perennial. Monotypic.

1. **Buchloe dactyloides** (Nutt.) Engelm.   Staminate culms 1–2 dm, the pistillate 2–5 cm; lf-blades sparsely pilose, 3–10 cm, ca 2 mm wide; staminate spikes 6–15 mm; pistillate spikelets (second glume) 6 mm; $2n$=40. Dry prairies and plains; w. Minn. and w. Io., n. to Alta., w. to the mts., s. to Ariz. and N.M.; the most important constitutent of the short-grass prairies. (*Bulbilis d.*)

83. ZOYSIA Willd.   Spikelets 1-fld, laterally somewhat compressed, shortly pedicellate, appressed flatwise to the slender rachis of the spike-like raceme, glabrous, disarticulating below the solitary (second) glume, this coriaceous, mucronate or short-awned, completely enfolding the thin lemma and palea, or the palea obsolete; low, rhizomatous or stoloniferous perennials. 10, Old World.

1. **Zoysia japonica** Steudel.   Korean or Japanese lawngrass. Plants with shallow rhizomes, forming a dense turf, growing vigorously in hot sun, but early becoming brown and dormant; lvs flat, 3–10 cm × 2–5 mm; infls 2–4 cm, arising only (or scarcely) ca 1 dm above the ground; spikelets 3–3.5 mm, pale purplish-brown; $2n$=40. Native of e. Asia, cult. in warm temperate regions as a lawngrass, casually escaped mainly in the s. part of our range.

84. PANICUM L.   Panic-grass. Spikelets with one perfect terminal fl, lanceolate or fusiform to ovoid, obovoid, or globose, usually somewhat compressed dorsiventrally and biconvex; first glume very short or minute and usually scarious or membranous, or in some spp. nearly as long as the second; second glume about as long as the spikelet, herbaceous, green or colored, often prominently nerved; sterile lemma similar to the second glume; enclosing a palea and sometimes a staminate fl; fertile lemma indurate, usually white or pale and smooth or shining, veinless (or sometimes with 5–7 evident veins at full maturity), the margins inrolled over the edges of the indurate or membranous palea, the apex typically obtuse; ligule in most spp. a band of hairs, or obsolete; spikelets usually small, in panicles or spike-like racemes. 500, cosmop. (*Dichanthelium.*) The spp. of subg.

Dichanthelium, and especially spp. 21–34, form an intergrading group in which the taxonomic limits are difficult to discern and are still debatable. The flag-lf referred to in the descriptions of some spp. of subg. *Dichanthelium* is the uppermost lf of the culm.

1 Blades of the basal and cauline lvs similar and elongate; winter rosette not formed; primary and
   secondary panicles similar or nearly so, their primary branches simple at base; spikelets all fertile;
   annual or perennial. (Subg. *Panicum.*) .... Group I.
1 Blades of the basal and cauline lvs usually different in shape, the former closely crowded and forming
   a winter-rosette (2 spp. excepted); primary panicles borne on simple culms in early summer, their
   branches normally forked at the very base, their spikelets tending to be open-pollinated but often
   or usually sterile; culms later usually freely branched, bearing numerous reduced panicles (often
   concealed by the lvs) with fertile cleistogamous spikelets; perennial. (Subg. *Dichanthelium.*)
   2 Lvs crowded toward the base, the basal and cauline similar, erect to narrowly ascending, bright
      green or bluish-green, long, narrow, linear, mostly 8–20 cm long and seldom over 5 mm wide,
      the longest ones ca 20 times as long as wide .... Group II.
   2 Lvs otherwise, either relatively or actually wider, or the cauline ones evidently different from the
      basal ones, or differing in more than one of these ways.
      3 Lvs at base of plant numerous, relatively long and soft, only slightly shorter than the soft,
         yellowish-green lower cauline lvs; plants branching only at the base, the slender stems
         remaining simple above; lvs mostly 4–10 mm wide .... Group III.
      3 Lvs at base of plant typically forming a rosette of short, relatively broad blades, usually con-
         spicuously shorter than the larger cauline ones; lvs 3–30 mm wide.
         4 Pubescence (at least of some of the sheaths and internodes) of 2 intermingled types: short,
            soft, crisp hairs and much longer, straight or merely flexuous, variously appressed to
            spreading hairs .... Group IV.
         4 Pubescence not as above, the hairs in any particular area of ± uniform type.
            5 Ligule a band of hairs 2–5 mm, conspicuously protruding from the sheath .... Group V.
            5 Ligule none or minute, or a band of hairs less than 2 mm.
               6 Main cauline lvs distinctly cordate at base, (7–)10–30(–40) mm wide; spikelets hairy .... Group VI.
               6 Main lvs rounded or narrowed at base, variously wide or narrow; spikelets hairy or
                  glabrous.
                  7 Second glume and sterile lemma short-acuminate or acute, conspicuously longer than
                     the fertile lemma .... Group VII.
                  7 Second glume shorter than, equaling, or barely exceeding the fr.
                     8 Sheaths viscid, at least on the back near the top .... Group VIII.
                     8 Sheaths not viscid, though in some spp. inconspicuously glandular-spotted.
                        9 Lf-blades long and slender, typically ca 15 times as long as wide .... Group IX.
                        9 Lf-blades broader, usually ca 10 times as long as wide, or less.
                           10 Spikelets small, mostly 0.9–2.9 mm .... Group X.
                           10 Spikelets larger, mostly 3.0–4.0 mm .... Group XI.

GROUP I

1 Spikelets subsessile, closely aggregated into spike-like secund racemes.
   2 Plants evidently perennial, with rhizomes and/or stolons; spikelets 2–4 mm.
      3 Second glume and sterile lemma obtuse or rounded; first glume three-fourths to nearly as long .... 1. *P. obtusum.*
      3 Second glume and sterile lemma acute, the first glume half as long .... 2. *P. hemitomon.*
   2 Plants annual; spikelets 5–6 mm .... 3. *P. texanum.*
1 Spikelets with long or short pedicels, in an open or contracted panicle.
   4 Spikelets distinctly verrucose .... 4. *P. verrucosum.*
   4 Spikelets glabrous to scabrous.
      5 Sterile palea at maturity enlarging and dilating the spikelet, becoming indurate and involute .... 5. *P. hians.*
      5 Sterile palea none, or membranous, or inconspicuous, not becoming large and indurate.
         6 Plants perennial from a rhizome or hard knotty crown.
            7 Spikelets secund along the smaller branches of the panicle, not over 4 mm.
               8 Rhizome developed; sheaths scarcely keeled; spikelets deflexed at an angle from the
                  pedicel .... 6. *P. anceps.*
               8 Rhizome not developed; sheaths strongly keeled; spikelets not deflexed.
                  9 Ligule long-ciliate .... 7. *P. longifolium.*
                  9 Ligule merely erose .... 8. *P. rigidulum.*
            7 Spikelets not at all secund, usually over 4 mm.
               10 Panicle narrow, with appressed branches.
                  11 First glume weakly 3- or 5-veined; rhizomes generally short, the stems ± densely
                     clustered .... 9. *P. amarulum.*
                  11 First glume strongly 7- or 9-veined; rhizomes elongate, the stems more scattered .... 10. *P. amarum.*
               10 Panicle diffuse or spreading, two or three times as long as thick .... 11. *P. virgatum.*
         6 Plants annual from a cluster of fibrous roots.
            12 Lf-sheaths glabrous .... 12. *P. dichotomiflorum.*
            12 Lf-sheaths papillose-pubescent.
               13 Spikelets 4.5–6 mm .... 13. *P. miliaceum.*
               13 Spikelets 1.6–3.6 mm.

14 Panicle relatively slender, 2–3 times as long as thick .... 14. *P. flexile.*
14 Panicle more diffuse, less than twice as long as thick.
   15 Fr nigrescent, half as wide as long, the margins of the lemma barely folded
      over the palea .... 15. *P. philadelphicum.*
   15 Fr stramineous, a fourth or a third as wide as long, the infolded margins of
      the lemma each as wide as the space between them .... 16. *P. capillare.*

Group II

1 Second glume and sterile lemma pointed, extending 0.5–1.5 mm beyond the fertile lemma ..... 17. *P. depauperatum.*
1 Second glume and sterile lemma blunt or merely acute, above equaling the fertile lemma .... 18. *P. linearifolium.*

Group III

1 Spikelets shortly spreading-hairy; sheaths with widely spreading or retrorse hairs .... 19. *P. laxiflorum.*
1 Spikelets glabrous; sheaths with ascending hairs .... 20. *P. strigosum.*

Group IV

1 Ligule "double," i.e., with a dense band of short hairs in front of a thin line of long hairs; nodes
   usually conspicuously bearded .... 21. *P. commonsianum.*
1 Ligule a single band of hairs; nodes bearded or not.
   2 Lvs glabrous above, or with only a few widely scattered hairs; ligule 0.5–1.5 mm.
      3 Spikelets 1.9–2.5 mm; blades ciliate, at least at base .... 22. *P. lancearium.*
      3 Spikelets 1.3–1.9 mm; blades not ciliate .... 23. *P. columbianum.*
   2 Lvs hairy above as well as beneath; ligule 0.5–4 mm; spikelets 1.3–1.6 mm .... 24. *P. leucothrix.*

Group V

1 Spikelets 3.5–4.2 mm; larger lvs 10–18 mm wide .... 44. *P. ravenelii.*
1 Spikelets 0.8–2.5 mm; lvs 3–13 mm wide.
   2 Panicle-branches narrowly ascending, the panicle ca 3–4 times as long as wide .... 25. *P. spretum.*
   2 Panicle branches widely ascending or spreading, the panicle ovoid, up to ca twice as long as wide.
      3 Spikelets minute, only 0.8–1.0 mm; lvs 3–5 mm wide; stems and sheaths only finely short-hairy;
         coastal plain .... 26. *P. wrightianum.*
      3 Spikelets larger, 1.1–2.5 mm; lvs 3–13 mm wide; stems and sheaths ± long-hairy, varying to
         glabrous.
         4 Sheaths glabrous or with ascending or appressed hairs usually under 2 mm .... 27. *P. lanuginosum.*
         4 Sheaths with soft, spreading or reflexed hairs 2–5 mm .... 28. *P. villosissimum.*

Group VI

1 Spikelets ellipsoid to ovoid or obovoid, 1.9–5.2 mm.
   2 Primary veins of the lf-blades, excluding the midvein, scarcely differentiated from the intervening
      secondary veins; spikelets 1.9–3.4 mm.
      3 Larger cauline lvs erect or ascending, broadest at or just below the middle .... 29. *P. boreale.*
      3 Larger cauline lvs spreading, broadest near the base .... 38. *P. commutatum.*
   2 Primary veins, including the midvein, raised above the lower surface of the blade and sharply
      differentiated from the much finer intervening secondary ones; spikelets 2.5–5.2 mm.
      4 Spikelets 2.5–3.7 mm; nodes not more densely pubescent than the sheaths and internodes.
         5 Sheaths, or some of them, strongly papillose-hispid .... 39. *P. clandestinum.*
         5 Sheaths glabrous or softly villous .... 40. *P. latifolium.*
      4 Spikelets 3.8–5.2 mm; nodes ± densely retrorse-bearded .... 41. *P. boscii.*
1 Spikelets nearly spherical, 1.3–1.9 mm.
   6 Nodes appressed-hairy; panicle nearly as wide as long .... 42. *P. sphaerocarpon.*
   6 Nodes glabrous; panicle ca twice as long as wide .... 43. *P. polyanthes.*

Group VII

1 First glume ± acute, ca two-fifths as long as the spikelet; stems glabrous, not glandular .... 33. *P. yadkinense.*
1 First glume blunt, mostly less than a third as long as the spikelet; stems often with a glandular band
   below each node .... 34. *P. scabriusculum.*

Group VIII

1 Only one species .... 35. *P. scoparium.*

Group IX

1 Lf-blades broadest near the middle, tapering to both ends, mostly erect or strongly ascending; northern,
   s. to N.J. .... 29. *P. boreale.*
1 Lf-blades broadest near the base, tapering regularly to the narrow or involute tip, variously ascending
   or suberect to widely spreading; coastal plain, n. to N.J.
  2 Plants densely villous on the nodes, internodes, sheaths, and both sides of the lf-blades ..... 37. *P. consanguinenum.*
  2 Plants much less hairy (see description for details) .... 36. *P. aciculare.*

Group X

1 Flag-lf ascending or erect; spikelets short-hairy.
  2 Lvs mostly 6–14 mm wide, glabrous on both sides or rarely hairy beneath; plants only sparsely
    branched in age; northern .... 29. *P. boreale.*
  2 Lvs mostly 2–6 mm wide, papillose-hirsute on both sides, varying to nearly glabrous above; plants
    freely branched in age, forming dense masses 1–2 dm; midwestern .... 48. *P. wilcoxianum.*
1 Flag-lf usually ± spreading; spikelets variously glabrous or short-hairy.
  3 Blades of main cauline lvs mostly 5–12 mm wide; spikelets 1.5–2.9 mm; widespread.
    4 Spikelets 2.7–4 mm; second glume and sterile lemma about equaling the fr; autumnal phase
      sparsely branched .... 47. *P. oligosanthes.*
    4 Spikelets 1.5–2.9 mm; second glume and sterile lemma a little shorter than the fr; autumnal
      phase much branched .... 30. *P. dichotomum.*
  3 Blades of main cauline lvs mostly 2–5 mm wide; spikelets 0.9–1.6 mm; coastal plain.
    5 Lvs with thin green margins .... 31. *P. ensifolium.*
    5 Lvs with thick, usually whitish-cartilaginous margins .... 32. *P. tenue.*

Group XI

1 Spikelets papillose-hirsute with hairs 0.5–1+ mm .... 45. *P. leibergii.*
1 Spikelets glabrous or only minutely hairy, the hairs all well under 0.5 mm.
  2 Panicle narrow, its branches strictly erect .... 46. *P. xanthophysum.*
  2 Panicle ovoid, its branches spreading or ascending .... 47. *P. oligosanthes.*

1. Panicum obtusum HBK.   Vine-mesquite. Slender, erect, 3–7 dm; basal stolons vine-like, to 2 m; lf-blades 1–2 dm × 3–6 mm; panicle slender, 5–15 cm, bearing several erect secund racemes 1–3 cm; spikelets ellipsoid to obovoid, 2.5–4 mm, brownish, obtuse; first glume three-fourths to nearly as long as the spikelet; second glume and sterile lemma subequal, obtuse or rounded, strongly veined; $2n=20, 36, 40$. Moist soil; Utah to Kans., nw. Mo. (where probably only intr.) and Ark., s. to Mex.

2. Panicum hemitomon Schultes.   Maiden-cane. Erect from a stout rhizome, 5–15 dm, often emitting roots from submersed nodes; lower sheaths loose, shorter than the internodes, the upper tighter, usually exceeding the internodes, ciliate, sometimes hirsute; blades 1–3 dm × 5–15 mm, abruptly contracted at base; panicle 6–20 cm, slender, with appressed branches, the lower ones often remote; spikelets ± secund, short-pediceled, flattened-ovoid, 2.2–2.9 mm, the first glume half as long, somewhat gibbous at base; second glume and sterile lemma sharply veined, acute, the latter with a conspicuous palea; $2n=36$. Swamps and ponds on the coastal plain; N.J. to Fla. and Tex.

3. Panicum texanum Buckley.   Texas millet. Annual 5–15+ dm, often decumbent and rooting below, softly pubescent ± throughout; blades 8–20 cm, × 7–15 mm; panicle 8–20 cm, with short, appressed branches; spikelets secund, subsessile, fusiform, 5–6 mm; fertile lemma rugose; $2n=36, 54$. Prairies and open ground, especially in low land along streams, often a weed in fields; Tex., Okla., and n. Mex., casually intr. e. to Ky., N.C. and Fla.

4. Panicum verrucosum Muhl.   Annual 2–9 dm; culms very slender, solitary or tufted, erect or spreading; blades thin, 5–15 cm × 5–10 mm, narrowed to the base, glabrous; panicle very lax and diffuse, 5–20 cm, its very slender branches widely spreading; spikelets narrowly obovoid, acute, 1.7–2.1 mm; first glume triangular, 0.5–0.8 mm; second glume and sterile lemma obscurely veined, distinctly verrucose; $2n=36$. Wet woods and shores; e. Mass. to Fla. and Tex., mostly on the coastal plain, and also inland in Tenn., Ky., O., ne. Ind., and sw. Mich.

5. Panicum hians Elliott.   Tufted perennial 1–6 dm, erect or decumbent at base; blades 6–15 cm × 2–4 mm, no wider than the sheaths, pilose near the base; panicle 4–12 cm, with very slender ascending branches; pedicels 1 mm or less; spikelets stramineous or purple-tinged, somewhat secund, 1.9–2.3 mm, at first lance-ovoid, soon becoming much dilated distally by the enlarged sterile palea; first glume broadly triangular-ovate, a third to half as long as the spikelet; second glume and sterile lemma sharply veined, subequal; sterile palea becoming larger, involute, and indurate in fruit, eventually exceeding the lemma; $2n=18$. Wet shores of ponds and streams, mostly on the coastal plain; se. Va. to Fla., Tex., and Mex., n. in the Mississippi Valley to se. Mo. (*Steinchisma h.*)

6. **Panicum anceps** Michx. Culms 5–10 dm, erect, several together from long, scaly rhizomes; sheaths glabrous to densely pilose; blades 2–4 dm × 6–12 mm, glabrous to pilose on both sides; panicle pyramidal, 1–4 dm, its branches spreading or ascending; branchlets scabrous; spikelets usually crowded and subsecund, ovoid to lanceolate, 2.2–3.9 mm, often slightly falcate, usually set at an angle to the pedicel; first glume appearing erect, triangular-ovate, a third to half as long as the spikelet; second glume and sterile lemma sharply veined, acute to acuminate; $2n=18$, 36. Moist, especially sandy soil; N.J. to s. O. and w. Mo., s. to the Gulf. (*P. rhizomatum.*)

7. **Panicum longifolium** Torr. Tufted perennial to 1 m; culms much compressed; sheaths compressed, glabrous or villous; ligule a row of fine hairs 1–3 mm, blades to 3 dm, no wider than the sheaths; panicle usually 1–2 dm, its branches spreading to subappressed; spikelets lance-ovoid, green or purple-tinged, 2–3.5 mm; first glume half to three-fourths as long as the spikelet; second glume and sterile lemma subequal, acuminate; $2n=18$. Moist or wet soil, open woods, bogs, and pond-margins, seldom in dry soil, mainly on the coastal plain; N.S.; e. Mass. to Fla., Ark., and Tex.; Ky. (*P. combsii.*)

8. **Panicum rigidulum** Nees. Coarse, densely tufted perennial to 1.5 m; lvs mostly crowded toward the base; sheaths glabrous, strongly compressed; ligule short, membranous, erose; blades narrow, 2–4 dm, sometimes exceeding the panicle, scaberulous on the margin; panicle 1–3 dm, pyramidal or narrow, its branches scabrous; spikelets lance-ovoid, green or purple, 1.8–3 mm, on pedicels half to twice as long; glumes and sterile lemma shortly to long-acuminate, sharply veined and somewhat keeled; first glume half to two-thirds as long as the spikelet; second glume and sterile lemma subequal, conspicuously longer than the often shortly stipitate fr; $2n=18$. Wet soil; Me. to Mich., s. to Fla. and Tex. (*P. agrostoides; P. condensum; P. stipitatum.*)

9. **Panicum amarulum** A. Hitchc. & Chase. Culms stout, 1–2(–3) m, glabrous and glaucous, ± densely clustered on a hard, knotty base, only seldom somewhat scattered on more elongate rhizomes; blades firm, much elongate, to 50 cm × 12 mm; panicles slender, 1.5–4 dm, densely fld, with long subappressed primary branches, the secondary branches mostly paired or whorled, usually with tertiary and quaternary branches; pedicels capillary, appressed; spikelets glabrous, ovate, acuminate, 4–6 mm; glumes acuminate, the first one half to two-thirds the length of the spikelet, rather weakly 3- or 5-veined, the second one 5–9-veined; sterile lemma 7- or 9-veined; fr smooth, shiny, 2.4–3.6 × 1–1.5 mm, with a well developed caryopsis inside; $2n=36$. Coastal beaches, usually in swale areas behind the foredunes; N.J. to Mex., seldom inland in N.C. and W.Va. Hybridizes with no. 11.

10. **Panicum amarum** Elliott. Much like no. 9, but with well developed creeping rhizomes and solitary or slightly clustered culms; panicles sparsely fld, the primary branches usually 1 or 2 per node, tertiary branches common; spikelets 4.7–7.7 mm; first glume prominently 7- or 9-veined; largely sterile, the caryopsis abortive; $2n=54$. Coastal foredune system; Conn. to Tex. Morphologically intergradient with no. 9.

11. **Panicum virgatum** L. Switchgrass. Stout, erect, to 2 m, from hard, scaly rhizomes, often forming large tufts; ligule a dense zone of silky hairs; blades 2–5 dm, to 15 mm wide; glabrous or pilose near the base; infl open, freely branched, pyramidal, 2–4 dm; spikelets ovoid, soon widened distally by spreading of the glumes and sterile lemma, 2.2–5.6 mm; first glume half to nearly as long; second glume and sterile lemma subequal, conspicuously veined, acute or long-acuminate; $2n=18$–108. Open woods, prairies, dunes, shores, and brackish marshes; N.S. and Que. to Man. and Mont., s. to Ariz., Mex., and W.I. Highly variable in size of spikelets, length and shape of glumes, and number of chromosomes.

12. **Panicum dichotomiflorum** Michx. Glabrous annual to 1 m or more, erect to decumbent or diffuse; panicle in large plants to 4 dm, widely branched; lf-blades 4–20 mm wide; spikelets 1.6–3.3 mm, green or purple-tinged, ellipsoid to oblong; first glume broad, obtuse or rounded, a third as long; second glume and sterile lemma acute, 7-veined, the latter with or without a palea; $2n=36$, 54. Moist soil and shores, often a weed in cult. land; N.S. and Que. to Minn. and S.D., s. to Fla. and Tex.

13. **Panicum miliaceum** L. Proso millet, broomcorn millet. Stout annual 2–6(–10) dm; sheaths overlapping, densely hirsute; blades elongate, rounded at base, 10–20 mm wide; panicle included at base, pyramidal to cylindric, dense, 8–20 cm, often nodding at maturity; spikelets turgid, acute, 4.5–6 mm; first glume half as long, acute or acuminate, 5-veined; second glume and sterile lemma equal, distinctly 7- or 9-veined; fr stramineous to brown, 3–3.5 mm; $2n=36$, 54, 72. Native of the Old World, occasionally cult. for forage and adventive along roadsides and in waste places. A wild-adapted type, widespread as a field-weed in the midwest (Wis., Minn., Ill., Io., N.D., S.D., Nebr., Kans., Colo.), has been named ssp. *spontaneum* (Kit.) Tzvelev. It is larger, 7–20 dm, with open panicles 10–50 cm, and deciduous spikelets.

14. **Panicum flexile** (Gattinger) Scribn. Wiry witch-grass. Slender annual 2–6 dm, erect or nearly so, branched from the base; blades erect, 2–6 mm wide; peduncle exsert, often elongate, the panicle 6–20 cm, half to a third as thick, with numerous ascending or spreading capillary branches, the axillary pulvini glabrous; spikelets narrowly lanceolate, 2.6–3.6 mm; first glume usually acute, nearly half as long as the acuminate second glume and sterile lemma; fr stramineous, narrowly ellipsoid, 1.8–2 mm, a fourth to a third as wide; $2n=18$. Moist or dry soil, often in open woods; N.Y. and s. Ont. to N.D., s. to Fla. and Tex.

15. **Panicum philadelphicum** Bernh. Slender annual 1–5 dm, erect or rarely decumbent, branched from the base; blades usually yellowish-green, 3–8 mm, wide; pedicel well exsert; panicle ovoid, ca a third as long

as the entire plant; spikelets tending to be paired at the ends of capillary branches; axillary pulvini short-villous to glabrous; spikelets ovoid to ellipsoid, 1.6–2.4 mm, the first glume obtuse or acute, two-fifths as long as the abruptly acuminate second glume and sterile lemma; fr plump, becoming blackish; margins of the lemma barely inrolled, the exposed part of the palea half to three-quarters as wide as the fr; $2n=18$. Dry soil and sand fields; N.S. and N.Y. to Minn., s. to Ga. and Tex. (*P. tuckermanii*, the form, chiefly northern, with the axillary pulvini glabrous.)

16. Panicum capillare L.   Witch-grass. Coarse, rough-hairy annual to 7 dm, branched from the base, erect or ascending or in depauperate forms spreading; lvs 6–17 mm wide; panicles diffusely branched, sometimes two-thirds as long as the whole plant; axillary pulvini glabrous to evidently hairy; spikelets all or mostly on long pedicels, acute to acuminate, 1.8–3.5 mm, the first glume a third to half as long as the sterile lemma; fr stramineous, 1.4–2 mm, a third as wide; margins of the lemma distinctly inrolled, the visible part of the palea a third to half as wide as the fr; $2n=18$. Dry or moist soil, often a weed in fields and gardens, widespread throughout most of the U.S. and s. Can. (*P. barbipulvinatum; P. gattingeri.*)

17. Panicum depauperatum Muhl.   Culms clustered, erect or nearly so, 1–4 dm, very slender, glabrous to puberulent; lvs bright green or blue-green, crowded toward the base, the basal and cauline similar, erect to narrowly ascending, glabrous to long-pilose (as also the sheaths), mostly 8–20 cm × 2–5 mm, the longest ones ca 20 times as long as wide; primary panicle narrow and relatively few-fld, with crooked, often short and appressed pedicels, 3–8 cm, eventually exsert, not much exceeding the lvs; spikelets glabrous or minutely hairy, ellipsoid, 2.7–4.1 mm, the first glume membranous, ovate or triangular, a third as long; second glume and sterile lemma sharply veined, pointed, projecting 0.5–1.5 mm beyond the fertile lemma; autumnal phase similar, the panicles small and usually concealed among the lvs near the base; $2n=18$. Dry or sandy soil, commonly in open woods; N.S. and Que. to Minn., s. to Ga. and Tex. (*Dichanthelium d.*)

18. Panicum linearifolium Scribn.   Vegetatively much like no. 17, avg somewhat taller and eventually more branched; primary panicle more numerously fld, usually much surpassing the lvs, at maturity open, the longest pedicels 8–18 mm; spikelets ellipsoid, 1.7–3.1 mm, glabrous to pilose; second glume and sterile lemma blunt, about equaling the fr; $2n=18$. Dry or stony soil, open woods, and banks; N.S. and Que. to Minn., s. to Ga. and Tex. (*P. werneri; Dichanthelium l.*) Occasional plants intermediate toward no. 17, called *P. perlongum* Nash, may be of hybrid origin.

19. Panicum laxiflorum Lam.   Culms clustered, 2–5 dm, simple or branched at the base only, the nodes bearded with soft, spreading hairs; sheaths pilose with spreading or more often retrorse hairs to 4 mm; ligule an entire or minutely fringed rim less than 0.5 mm, or obsolete; basal lvs numerous, relatively long and soft, similar to and only slightly if at all shorter than the soft, yellowish-green lower cauline lvs; blades 8–15 cm × 5–10 mm, sparsely pilose to glabrous on both sides, finely ciliate to glabrous on the margins, the uppermost one generally at least three-fourths as long as the basal ones; primary panicle 6–10 cm, its branches spreading, sparsely pilose to glabrous; spikelets soft-hairy, oblong-obovoid, 1.7–2.3 mm, obtuse; first glume broadly deltoid, a fourth to a third as long; second glume and sterile lemma soon separated and surpassed by the white or pale stramineous fr; autumnal phase basally branched, the blades scarcely reduced, much exceeding the small infls; $2n=18$. Woods; Md. to s. Ind. and s. Mo., s. to Fla. and Tex. (*P. xalapense; Dichanthelium l.*)

20. Panicum strigosum Muhl.   Vegetatively much like no. 19; culms sparsely to densely pilose; sheaths and blades densely ascending-pilose, the blade-margins coarsely papillose-ciliate; cauline lvs few, often solitary, the uppermost one 1.5–6 cm, less than three-fourths as long as the basal; primary panicle 4–8 cm, its widely ascending branches pilose; spikelets glabrous, oblong-obovoid, 1.1–1.5 mm; first glume a third to half as long, triangular; second glume and sterile lemma equalling the fr; autumnal phase densely matted, with much reduced panicles; $2n=18$. Sandy woods and pine-barrens; se. Va. to Fla., Tex., C. Amer., W.I., and n. S. Amer. (*P. laxiflorum* var. *pubescens; P. longepedunculatum; Dichanthelium leucoblepharis* var. *pubescens.*)

21. Panicum commonsianum Ashe.   Culms clustered, stiff, erect or ascending, 1–6 dm, usually at least some of the sheaths and internodes with distinctly bistratal pubescence consisting of intermingled short, soft, crisp hairs and much longer (1–2+ mm) coarser hairs; sheaths conspicuously striate; ligule a dense band of short (ca 1 mm or less) hairs in front of a thin line of longer (even to 5 mm) hairs; blades ascending, 4–9 cm × 3–7 mm, involute at the tip, glabrous or with a few scattered long hairs above, glabrous to more often hairy (with unequal hairs) beneath; primary panicle ovoid, 3–8 cm; spikelets finely hairy, ellipsoid, 1.7–2.6 mm; autumnal phase becoming widely spreading to prostrate, branched chiefly from the middle nodes, the blades scarcely reduced, the panicles much reduced and surpassed by the lvs; $2n=18$. Sandy soil. Var. *commonsianum*, of the coastal plain from Mass. to Fla., is rather thinly hairy. (*P. addisonii; P. mundum.*) The more copiously hairy var. *euchlamydeum* (Shinners) Pohl is more inland, occurring from nw. Pa. to n. Ill., Wis., and e. Minn. *P. commonsianum* might perhaps properly be subordinated to *P. ovale* Elliott, of the southern coastal plain, but the proper nomenclatural innovations have not been made.

22. Panicum lancearium Trin.   Much like no. 23, and likewise with bistratal pubescence, but often only sparsely hairy or in part glabrous; blades ciliate at least toward the base, those of the midstem often over 6 cm; primary panicles avg larger, often over 6 cm; spikelets finely hairy to subglabrous, 1.9–2.5 mm, the first glume nearly or quite half as long, obtuse to truncate; autumnal phase copiously branched from the lower and middle nodes, forming dense mats, the blades crowded, much reduced, involute, the panicles reduced and concealed among the lvs; $2n=18$. Pine woods; coastal plain from se. Va. to Fla., Miss., and Cuba. (*P. nashianum; P. patulum,* the form with hairy lvs; *Dichanthelium sabulorum* var. *patulum.*) Perhaps properly to be subordinated to *P. sabulorum* along with no. 23.

23. Panicum columbianum Scribn. Culms clustered, erect or ascending, often purplish, densely short-pubescent with minute hairs 0.1–0.4 mm, and generally with some intermingled longer hairs (often 1 mm) toward the summit of the lower internodes; sheaths likewise with bistratal pubescence; ligule a band of hairs 0.5–1.5 mm; blades 1–7 cm × 3–7 mm (those of the midstem commonly 2–5 cm), glabrous above or with a few widely scattered hairs, minutely puberulent beneath; primary panicle ovoid, mostly 2.5–4 cm, its axis puberulent; spikelets finely hairy, oblong-ovoid, obtuse, 1.3–1.9 mm, the first glume triangular-ovate, two-fifths as long; autumnal phase spreading or decumbent, branched early from most of the nodes, the blades scarcely reduced, the panicles smaller, surpassed by the lvs; 2*n*=18. Moist or dry, especially sandy soil; Me. to Minn., s. to Va., Tenn., and Ill. (*P. oricola; P. tsugetorum; Dichanthelium c.; D. sabulorum* var. *thinium.*) Perhaps properly to be subordinated to *P. portoricense* Desv. or to the still older *P. sabulorum* Lam. (the latter from Uruguay), but the proper nomenclatural innovations not yet made.

24. Panicum leucothrix Nash. Culms densely tufted, 1–4 dm, ± erect to spreading or ascending from a geniculate base and becoming prostrate; pubescence on at least some of the sheaths and internodes bistratal as in nos. 21–23, with intermingled short, crisp hairs and much longer, coarser, straighter ones, the lower internodes and sheaths evidently hairy, the upper often progressively less so; ligule a band of hairs 0.5–4 mm; blades erect or ascending, 2–6 cm × 2–5 mm, short-hairy on both sides; primary panicle ovoid, 2–5 cm, its axis puberulent to glabrate, with widely spreading branches; spikelets finely hairy, oblong-ovoid, 1.3–1.6 mm; first glume subrotund to triangular-ovate, a fourth to two-fifths as long; autumnal plants prostrate to erect, with numerous branches from all nodes, tending to form large mats, the blades slightly to evidently reduced, the infls varying from clusters of spikelets hidden among the lvs to exsert panicles 1–2 cm; 2*n*=18. (*P. albemarlense; P. auburne; P. meridionale.*) A group of uncertain status, possibly reflecting hybridization between the *P. sabulorum* group (represented with us by spp. 22–23) and the *P. acuminatum* group (represented with us by spp. 25–28).

25. Panicum spretum Schultes. Culms loosely clustered, erect or nearly so, 3–8 dm, glabrous; sheaths glabrous throughout or sparsely ciliate distally; ligule a band of hairs 2–3 mm; blades strongly ascending or subappressed, 6–10 cm × 3–6 mm, glabrous, varying to minutely puberulent beneath and sparsely ciliate at base; primary panicle 6–10 cm, narrow, ca 3–4 times as long as wide, with strongly ascending branches; glabrous or nearly so; spikelets pubescent, obovoid to ellipsoid, 1.3–1.8 mm, the first glume a third as long, broadly rounded, the second glume and sterile lemma equaling the fr; autumnal phase reclining, bearing numerous short ascending branches, these in turn fastigiately branched, the blades much smaller, often puberulent, the panicles small and the later ones progressively reduced; 2*n*=18. Moist to wet, sandy or peaty shores and pine-barrens near the coast from N.S. to Fla. and Tex.; also in nw. Ind. and in Mich. and s. Ont. (*P. eatonii; Dichanthelium acuminatum* var. *densiflorum.*)

26. Panicum wrightianum Scribn. Culms loosely clustered, slender and weak, 1.5–4 dm, somewhat geniculate at base, ascending or spreading, finely short-hairy; sheaths finely hairy; ligule a band of hairs 2–3 mm; blades thin, spreading, 2–4 cm × 2–5 mm, finely hairy beneath, short-pilose or rarely glabrous above; primary panicle ovoid, 3–5 cm, its axis puberulent; spikelets finely hairy, ovoid, 0.8–1 mm; the first glume inconspicuous, a fourth or a third as long; autumnal phase more decumbent, with numerous short, ascending branches, the crowded lvs 1.5–2.5 cm, the short-exsert panicles 1–3 cm; 2*n*=18. Wet ground, beaches, and borders of ponds; coastal plain from Mass. to Fla., Tex., and Cuba. (*Dichanthelium acuminatum* var. *w.*)

27. Panicum lanuginosum Elliott. Polymorphic; culms erect to prostrate, tufted, usually straight and radiating from the base, glabrous to papillose-pilose or villous with hairs of fairly uniform nature, the sheaths similarly pubescent (or glabrous); ligule a band of hairs 2–5 mm; blades 2–12 cm × 3–12 mm, glabrous or rather uniformly appressed-pilose on both sides; primary panicle ovoid, with divergent, often flexuous branches, the axis glabrate to villous; spikelets finely hairy, plump-ellipsoid to obovoid, 1.1–2.1 mm; first glume broadly angular-rotund, less than two-fifths (usually under one-third) as long; autumnal phase spreading or prostrate, copiously branched chiefly from the middle nodes, the blades half as large as the vernal ones, the panicles few-fld, mostly surpassed by the lvs; 2*n*=18. Widely distributed in moist or dry sites, open woods, dunes, shores, and prairies, throughout our range, s. to the Gulf, and w. to the Pacific. Six vars. in our range:

a Stems and sheaths glabrous, or minutely puberulent in the internerves; blades glabrous, not papillose, except the basal cilia; culms with elongate internodes, and eventually with strict fascicles of secondary branches; Mass. and Que. to Minn., s. to Fla. and La. (*P. lindheimeri; P. longiligulatum; Dichanthelium acuminatum* var. *lindheimeri.*) .... var. *lindheimeri* (Nash) Fern.
a Stems and sheaths more or less hairy.
  b Stems and sheaths softly villous; scarcely papillose; blades soft-hairy on both sides, not papillose; moist, especially sandy woods, chiefly on the coastal plain; N.J. to Ind., Fla., and Tex. (*P. chrysopsidifolium; Dichanthelium lanuginosum.*) .... var. *lanuginosum.*
  b Stems and sheaths distinctly papillose-pilose.
    c Blades glabrous on the upper surface; usually in moist soil; Me. to Minn., s. to Ga. and Ariz. (*P. tennesseense.*) .... var. *tennesseense* (Ashe) Gleason.
    c Blades distinctly papillose-pilose on the upper surface, like the stems and sheaths.
      d Axis of the panicle glabrate or sparsely pilose in the axils only; spikelets 1.6–2 mm; chiefly in moist soil; N.S. and N.B. to N.Y., Minn., and Mo. .... var. *septentrionale* Fern.
      d Axis of the panicle pilose.
        e Spikelets 1.6–2 mm; blades short-pilose above; the commonest var., chiefly in dry soil throughout the range of the sp. (*P. glutinoscabrum; P. huachucae; P. languidum.*) ...
        .... var. *fasciculatum* (Torr.) Fern.

e Spikelets 1.3–1.6 mm; hairs of the upper lf-surface 3–6 mm; wet woods, salt marshes, swamps, and bogs; N.S. to Del. and Io. (*P. implicatum; Dichanthelium acuminatum* var. *implicatum.*) . . . . var. *implicatum* (Scribn.) Fern.

28. Panicum villosissimum Nash.   Culms ± clustered, erect or ascending, 1–3 dm at first anthesis, soon elongating to as much as 6 dm, evidently papillose-pilose with spreading or retrorse hairs 1–5 mm, as also the sheaths; ligule a band of hairs 3–5 mm; blades 3–10 cm × 3–13 mm, ± papillose-pilose on both sides; primary panicle on a papillose-hairy to glabrate peduncle, 2–6 cm, ovoid, with widely divergent branches, its axis pilose to glabrate; spikelets finely hairy, ellipsoid or oblong-obovoid, 1.5–2.5 mm; first glume triangular-ovate, acute, 0.7–1.4 mm, a third to three-fifths as long as the spikelet; fr 1.5–2 mm, half or two-thirds as wide; autumnal phase developing early, the lateral branches and often the secondary panicles visible before the primary panicle has completed anthesis; branches several from the middle and lower nodes, the stems widely spreading or prostrate and often geniculate at the lower nodes, the scarcely reduced lvs equaling or surpassing the small panicles; 2*n*=18. Dry, especially sandy soil, open woods, and prairies; Mass. to Minn. and Kans., s. to Fla. and Tex. (*P. benneri; P. praecocius; P. pseudopubescens; P. scoparioides*, a less hairy or subglabrate phase, perhaps of hybrid origin; *P. subvillosum; Dichanthelium acuminatum* var. *villosum.*)

29. Panicum boreale Nash.   Culms 2–7 dm, usually erect or nearly so, in tufts of 2–10, glabrous, or villosulous at the nodes; lowest sheaths usually papillose-pilose, the upper merely ciliate, often longer than the internodes; ligule obsolete, or a band of short hairs not over 1 mm; blades erect or strongly ascending, firm, rounded at the ciliate base, tapering from below the middle, glabrous or rarely hairy beneath, mostly 5–15 cm × (4–)6–14(–16) mm, the upper scarcely reduced; primary panicle long-exsert, relatively few-fld, 5–10 cm, the lower branches ascending; spikelets long-pedicellate, minutely hairy, ellipsoid, 1.9–2.3 mm; first glume a fourth to two-fifths as long; second glume and sterile lemma usually purplish, subequal, equaling or barely shorter than the fr; autumnal phase similar, with a few lax branches at the upper nodes bearing somewhat smaller lvs and shorter, few-fld panicles; 2*n*=18. Open woods, fields, and shores; Que. and Nf. to Minn., s. to N.J. and n. Ind. (*P. bicknellii; P. bushii; P. calliphyllum; Dichanthelium b.*) It is possible that some of the plants usually referred here are hybrids between broad-lvd and narrow-lvd spp.

30. Panicum dichotomum L.   Delicate, slender plants of forests, or more robust, freely branching plants of disturbed habitats, 3–10 dm, the nodes glabrous to densely bearded, the herbage otherwise glabrous or sometimes sparsely hairy; sheaths usually shorter than the internodes; ligule a band of short hairs ca 1 mm or less; blades of main cauline lvs thin; 5–12(–15) cm × 3–10(–13) mm, spreading, the flag-lf nearly or fully as large as the next 2 below; panicles well exsert, open, with ± spreading branches, branchlets, and pedicels, the spikelets never secund; spikelets glabrous or occasionally short-hairy, ellipsoid, 1.5–2.9 mm; first glume a third as long, usually acute; second glume rounded above, shorter than the sterile lemma, both shorter than the fr; autumnal phase much branched above the middle, commonly with fascicles of branchlets, erect or reclining from its own weight, the blades 2–5 cm × 2–5 mm, the panicles much reduced; 2*n*=18. Woods and open, disturbed habitats; N.B. and s. Can. to Mich., s. to Fla. and Tex. (*P. annulum; P. barbulatum; P. clutei; P. lucidum; P. mattamuskeetense; P. microcarpon; P. nitidum; P. roanokense; Dichanthelium d.*)

31. Panicum ensifolium Baldwin.   Culms tufted, very slender, erect or reclining, 1.5–4 dm; sheaths glabrous or minutely ciliate; ligule a band of hairs to 0.5 mm, cauline lvs few and distant, spreading or reflexed, 1–5 cm × 2–5 mm, with thin green margins, much constricted at base, glabrous above or puberulent at base, glabrous to puberulent beneath; primary panicle soon long-exsert, 2–6 cm, ovoid with short, spreading branches and few spikelets; spikelets glabrous or minutely hairy, ellipsoid or obovoid, somewhat pointed, 0.9–1.5 mm; first glume a third to two-fifths as long, ovate, obtusely pointed; second glume and sterile lemma shorter than the fr; autumnal phase copiously and often fastigiately branched from the lower nodes, sometimes forming dense mats, the blades 1–2 cm, the panicles 1 cm, equaling or barely exceeding the lvs; 2*n*=18. Bogs and wet woods, or rarely in dry pine-land; coastal plain from s. N.J. to La. (*P. chamaelonche; Dichanthelium dichotomum* var. *e.*)

32. Panicum tenue Muhl.   Culms tufted, 1.5–4 dm, typically slender and delicate, mostly glabrous, sometimes appressed-hairy; lvs sometimes mainly crowded near the base, sometimes more distributed along the stem, the sheaths glabrous or sometimes sparsely ciliate; ligule a band of hairs less than 0.5 mm; blades ascending or spreading, 2–8 cm × 2–4 mm, usually rather thick, with thickened, usually whitish-cartilaginous margins; primary panicle soon long-exsert, 3–6 cm, ovoid with spreading branches; spikelets finely hairy, ellipsoid, 0.9–1.6 mm; first glume broadly ovate to subrotund, a fourth or a third as long; second glume and sterile lemma blunt, distinctly shorter than the fr; autumnal phase densely tufted, 1 dm, the blades scarcely reduced, the panicles half as large, equaling or somewhat surpassing the lvs; 2*n*=18. Damp sandy soil of the coastal plain; se. Va. to Fla., La., the W.I., and C. Amer. (*P. albomarginatum; P. trifolium; Dichanthelium dichotomum* var. *t.*) Apparently confluent with no. 31.

33. Panicum yadkinense Ashe.   Culms few, erect, or geniculate at base, 6–10 dm, glabrous; sheaths glabrous or sparsely ciliate, especially distally, those of the primary stems usually with pale, glandular spots on the back; blades thin, bright green, 6–12 mm wide, glabrous, or sparsely ciliate at base; primary panicle 6–10 cm, about as wide, its lower branches ascending; spikelets ellipsoid, pointed, glabrous, 2–2.5 mm; first glume subacute, two-fifths as long; second glume and sterile lemma distinctly surpassing the fr; autumnal phase erect or nearly so, sparsely and loosely branched from the middle nodes, the branches 1–2 dm, the blades 2–5 cm × 2–5 mm, the panicles much reduced, about equaling the uppermost lvs; 2*n*=18. Moist or wet woods; N.J. to s. Ill., s. to Ga. and La.Vegetatively resembling luxuriant plants of no. 30.

34. Panicum scabriusculum Elliott. Culms few, erect, mostly 7–15 dm, glabrous to puberulent, often with a light-colored or mottled band of glandular tissue below each node; sheaths shorter than the internodes, glabrous or hispid, often mottled or white-spotted; ligule a minute membrane, usually with a band of short hairs just above it; blades elongate, mostly 10–25 cm × 6–15 mm, rather stiff, long-tapering from near the base to an involute tip, scabrous on the margins, glabrous on both sides or the lower puberulent beneath, at base glabrous or sparsely papillose-ciliate; primary panicle ovoid, with spreading or ascending branches, 8–15(–20) cm, half as wide, or wider; spikelets glabrous or minutely villosulous, ellipsoid to ovoid or lance-ovoid, pointed, 2.2–3.4 mm; first glume short and blunt, mostly less than one-third as long; second glume and sterile lemma pointed and surpassing the fr; autumnal phase loosely branched from the middle and upper nodes, the branches eventually forming dense tufts with much reduced blades and small panicles partly or wholly included in the sheaths; 2n=18. Wet, often sandy soil on the coastal plain; Conn. and N.J. to Fla. and Tex. (*P. aculeatum; P. cryptanthum; Dichanthelium s.*)

35. Panicum scoparium Lam. Velvet p.-g. Culms solitary or few, erect or geniculate at base, tall and stout, 7–15 dm, with a broad viscid-glabrous band just below the nodes, otherwise softly long-spreading-hairy and with bearded nodes; sheaths loose, often constricted at the summit, densely reflexed-villous at base, softly hairy except along the middle of the back, where glabrous and viscid; blades lanceolate, the larger 10–25 cm × 10–20 mm, slightly narrowed to the rounded base, softly hairy on both sides; primary panicle ovoid, many-fld, 8–15 cm, softly hairy, glandular-spotted; spikelets soft-hairy, obovoid, abruptly apiculate, 2–2.7 mm; first glume two-fifths as long, triangular-ovate, acute; second glume slightly shorter than the fr and sterile lemma; autumnal phase copiously branched and rebranched from the middle and upper nodes, the blades progressively reduced and the uppermost often with the sheaths hairy throughout, the panicles small, few-fld, usually shorter than the blades; 2n=18. Wet soil; Cuba and Fla. to Tex., n. to Mo., Ky., N.C., and along the coastal plain to Mass. (*Dichanthelium s.*)

36. Panicum aciculare Desv. Culms clustered, spreading to often stiffly erect, 2–6 dm, smooth or sparsely villous below; lower sheaths villous to sometimes glabrous, the upper often ciliate; ligule a band of hairs to 1 mm; blades stiffly spreading to suberect, 3–12 cm × 2–7 mm, long-acuminate, glabrous to somewhat villous especially on the lower side; primary panicle 3–10 cm, with flexuous spreading or ascending branches; spikelets finely hairy (seldom glabrous), obovoid to fusiform, blunt, 1.8–3.6 mm; first glume a fourth to half as long, triangular or ovate, often acute; second glume and sterile lemma equal, equaling or slightly exceeding the fr; autumnal phase densely branched, often forming cushions 1–3 dm thick, with closely involute blades, the panicles very few-fld, hidden among the lvs; 2n=18. Sandy pine-woods and barrens of the coastal plain; N.J. to Fla. and Tex., and inland in Tenn. (*P. angustifolium; P. fusiforme; P. hirstii; Dichanthelium a.*)

37. Panicum consanguineum Kunth. Culms clustered, spreading or ascending, 2–6 dm, with bearded nodes and densely ascending-villous internodes; sheaths and blades villous; ligule a band of hairs not over 1 mm, or obsolete; blades most 6–10 cm × 4–8 mm; primary panicle 4–8 cm, with ascending branches; appearing ± compact; spikelets short-villous; obovoid, blunt, 2.3–2.8 mm; first glume a third as long, broadly triangular, blunt; second glume and sterile lemma slightly surpassed by the mature fr; autumnal phase with numerous short crowded branches, flat blades 3–5 cm, and very small panicles much surpassed by the upper lvs; 2n=18. Sandy woods on the coastal plain; se. Va. to Fla. and Tex., n. in the Mississippi Valley to s. Mo. (*Dichanthelium c.*)

38. Panicum commutatum Schultes. Culms few–several, erect, ascending, or rarely decumbent at base, 3–8 dm, glabrous to sparsely villous or crisp-puberulent; herbage often purple-tinged or ± glaucous; sheaths ciliate to puberulent on the back; ligule none or very short and erose; blades lanceolate, the larger ones 5–12 cm × 7–18 mm, many-nerved, glabrous on both sides, usually sparsely ciliate at the cordate-clasping base; flag-lf nearly or fully as large as those next below; primary panicle eventually exsert, ovoid with spreading or ascending branches, 5–12 cm; spikelets obscurely short-hairy or glabrous, ellipsoid or oblong-ellipsoid, 2.1–3.4 mm, obscurely pointed; first glume two-fifths or half as long (0.7–1.7 mm), triangular-ovate, subacute; second glume and sterile lemma about equaling or often slightly exceeding the fr; autumnal phase branched from the middle and upper nodes, the uppermost primary internode deciduous, the blades not much reduced, the panicles ± reduced and commonly partly included at base; 2n=18. Woods and thickets; Mass. to Mich. and Mo., s. to Fla. and Tex. (*P. ashei; P. joorii; P. mutabile; Dichanthelium c.*)

39. Panicum clandestinum L. Culms clustered, often in large colonies, stout, erect, 6–15 dm, glabrous to sparsely papillose-hirsute; sheaths varying from roughly papillose-hirsute on the back to glabrous, ciliate, pubescent at the top; blades lanceolate, spreading, glabrous, usually papillose-ciliate at the cordate base, the larger ones 10–25 × 2–3 cm, up to ca 6+ times as long as wide, with strongly elevated midvein and sharply differentiated primary veins, the intervening secondary nerves evidently smaller and finer; primary panicle tardily exsert, broadly ovoid, with widely spreading or ascending branches, 7–14 cm; spikelets usually more than 50, sparsely-hairy, oblong-obovoid, 2.5–3.4 mm; first glume a third to half as long, ovate, subacute; second glume slightly shorter than the fr and sterile lemma; autumnal phase early and sparsely branched, the lvs not much reduced, the bristly sheaths strongly overlapping, the panicles much reduced, ± included in the sheaths; 2n=36. Moist woods and thickets; abundant; Que. and N.S. to Mich., Mo., and Okla., s. to Fla. and Tex. (*Dichanthelium c.*)

40. Panicum latifolium L. Culms clustered, slender, erect, 4–10 dm, glabrous or rarely sparsely puberulent; sheaths ciliate hairy at the top, otherwise glabrous to softly villous; blades lanceolate, spreading, glabrous or nearly so, ciliate at the cordate base, the larger ones 10–16 cm × 15–40 mm, the lateral veins differentiated into 2 types as in no. 39; primary panicle tardily exsert, ovoid with ascending branches, 6–12 cm; spikelets glabrous to softly villosulous, oblong-obovoid, 2.9–3.7 mm, first glume half as long, acute; second glume and sterile lemma shorter than the fr; autumnal phase sparsely branched from the middle nodes, the blades not much reduced or crowded, the panicles small, included at base; $2n=18$. Woods and thickets; Que. and Me. to Minn., s. to N.C., Tenn., and Mo. (*Dichanthelium l.*)

41. Panicum boscii Poir. Culms clustered, erect or ascending, often somewhat geniculate, 3–7 dm, glabrous to puberulent, sparsely to usually densely retrorse-bearded at the nodes; sheaths ciliate, pubescent at the top, glabrous to softly villous on the back; blades lanceolate, spreading, ciliate toward the cordate base, glabrous to soft-hairy on both sides, the larger ones 6–12 cm × 15–30 mm, the lateral veins differentiated into 2 types as in no. 39; primary panicle usually sessile or partly included, or eventually exsert, 5–10 cm, ovoid with a few ascending or spreading branches, relatively few-fld; spikelets oblong-obovoid, sparsely hairy, usually papillose, 3.8–5.2 mm; first glume narrowly obovoid, acute, ca half as long; second glume slightly shorter than the sterile lemma; autumnal phase sparsely branched from the middle nodes, the blades half as large, the panicles much reduced, partly included; $2n=18, 36$. Woods; Mass. to Ill. and Mo., s. to Fla. and Tex. (*Dichanthelium b.*)

42. Panicum sphaerocarpon Elliott. Culms few-several, spreading, 1.5–5 dm, glabrous except the appressed-hairy nodes; sheaths glabrous (and sometimes viscous-spotted) on the back, villous-ciliate with hairs usually 1–2 mm, some of the sheaths shorter than the internodes; ligule typically none, but sometimes a band of hairs to 1 mm; blades cordate and papillose-ciliate at base, the largest ones 6–12 cm × 7–15 mm, the primary veins scarcely stronger than the sets of 3–5 intermediate veins; flag-lf mostly 3–9 cm, typically borne ± midway between the stem-base and the panicle-tip; primary panicle 5–10 × 4–8 cm, long-exsert at anthesis, with spreading branches; spikelets (excluding the first glume) almost spherical, minutely puberulent, 1.3–1.9 mm, the first glume a third as long, broadly ovate; autumnal phase more widely spreading or prostrate, the few branches mostly from the base and lower nodes, the lvs and panicles scarcely reduced; $2n=18$. Moist or dry, preferable shady places; Mass. and Vt. to O. and Kans., s. to Fla. and Tex. (*Dichanthelium s.*)

43. Panicum polyanthes Schultes. Culms few in a tuft, stout and erect, 4–10 dm, glabrous; sheaths usually all longer than the internodes, glabrous on the back, ciliate with hairs usually under 1 mm; ligule none; blades cordate and papillose-ciliate at base, the largest 12–20 cm × 15–30 mm, the secondary veins in sets of usually 6–9 between the well differentiated primary veins; flag-lf usually 10–15+ cm; panicle 8–18 cm, seldom over half as wide, with ascending branches; spikelets (excluding the first glume) almost spherical, minutely puberulent, 1.3–1.9 mm; first glume a third to two-fifths as long, broadly rounded; autumnal phase scarcely different, usually producing a few flowering branches from the lower nodes; $2n=18$. Dry or damp soil, usually in open woods; N.Y. to Ind., Mo., and Okla., s. to Ga. and Tex. (*Dichanthelium sphaerocarpon* var. *isophyllum.*)

44. Panicum ravenelii Scribn. & Merr. Culms few, 4–6 dm, usually purplish, coarsely pubescent with spreading or ascending hairs, the nodes densely bearded; sheaths densely papillose-pilose with ascending hairs; ligule a band of hairs 2.5–4 mm; blades spreading or ascending, the larger ones 8–12 cm × 10–18 mm, soft-hairy beneath, glabrous above or sparsely ciliate toward the abruptly rounded base; primary panicle ovoid, 6–10 cm, included at base or exsert as much as 5 cm; spikelets thinly to densely villosulous, ellipsoid-obovoid, 3.5–4.2 mm; first glume up to half as long, triangular-ovate, acute; second glume and sterile lemma subequal, shorter than the fr; autumnal phase sparsely branched from the middle and upper nodes, the blades two-thirds as wide as those of the primary stems, the panicles much reduced, surpassed by the upper blades; $2n=18$. Dry woods; Md. and Del. to Fla. and Tex., n. to s. Mo. (*Dichanthelium r.*)

45. Panicum leibergii (Vasey) Scribn. Culms erect or geniculate below, 3–6 dm, forming large tufts, minutely puberulent; sheaths shorter than the internodes, papillose-hirsute with spreading hairs; ligule a short, ciliate membrane, or nearly obsolete; blades erect or ascending, the larger ones 7–11 cm × 7–12 mm, slightly tapering to a broadly rounded base, papillose-hirsute on both sides, varying to subglabrous above, ± papillose-ciliate; primary panicle at first included, tardily becoming long-exsert, 5–10 cm, ovoid or oblong with few-fld ascending branches; spikelets papillose-hirsute with hairs 0.5–1+ mm, oblong-obovoid, 3–4 mm; first glume three-fifths as long, triangular-ovate, acute; second glume and sterile lemma subequal, slightly longer than the fr; sterile lemma normally staminate; autumnal phase bearing a few simple branches from the middle and lower nodes, the blades scarcely reduced, the uppermost barely exceeding the small but evidently chasmogamous panicles; $2n=18$. Dry prairies and open places; O. and Mich. to Man., S.D., and Kans.; c. N.Y. and Pa. (*Dichanthelium l.*)

46. Panicum xanthophysum A. Gray. Culms few-several in loose tufts, erect or ascending, 2–5 dm, glabrous; sheaths loose, often exceeding the internodes, glabrous to pilose or papillose-pilose; blades yellowish-green, erect or nearly so, the larger 10–15 cm × 10–20 mm, often 10+ times as long as wide, glabrous on both sides, slightly narrowed to the rounded, papillose-ciliate base; primary panicle 5–10 cm, very narrow, with erect or narrowly ascending branches; spikelets few, seldom more than 40, minutely puberulent, obovoid, 3.3–3.8 mm; first glume half as long (1.6–2.1 mm) or a little more, triangular-ovate, acute; second glume slightly shorter than the sterile lemma and fr; autumnal phase with 1 or 2 erect branches, bearing scarcely reduced blades equaling or exceeding the reduced panicles; $2n=36$. Dry sandy soil; Que. and Me. to Man., s. to N.J., Pa., W.Va., Mich., and Minn. (*Dichanthelium x.*)

47. Panicum oligosanthes Schultes. Culms loosely clustered, few–several, erect or ascending, 2–7 dm, often purplish, densely pubescent to glabrous; sheaths occasionally glabrous, sometimes papillose, more often papillose-pilose with ascending hairs; ligule a band of hairs 1–2 mm; blades spreading, lanceolate, glabrous or rarely sparsely papillose-pilose above, glabrous to softly hairy or sparsely papillose-pilose beneath, usually papillose-ciliate and densely long-hairy at base, the larger 6–12 cm × 7–12 mm; primary panicle short-exsert (to 5 cm), becoming long-exsert in age, ovoid, 5–10 cm; spikelets glabrous or minutely villosulous, with broad, heavy veins, turgid, broadly ellipsoid to obovoid, 2.7–4 mm; first glume two-fifths as long, broadly ovate; second glume and sterile lemma subequal, barely equaling the fr; autumnal phase sparsely branched, chiefly from the middle and upper nodes, forming loose bunches, the blades not much reduced, surpassing the few-fld panicles; 2n=18. Dry or moist, often sandy soil, open woods, and prairies; Me. to Fla., w. to Wyo. and Tex., and in the Pacific states. Highly variable, but not clearly divisible into vars. (*P. helleri; P. scribnerianum; Dichanthelium o.*)

   48. Panicum wilcoxianum Vasey.   Culms densely clustered, erect, 1–3 dm, papillose-hirsute with ascending hairs to sparsley villous; sheaths loose, usually longer than the internodes, papillose-hirsute with hairs usually 2–5 mm; ligule a band of short hairs 0.4–1.6 mm; blades erect, 4–8 cm × 2–6 mm, scarcely wider than the sheaths, involute distally, papillose-hirsute on both sides, varying to nearly glabrous above; primary panicle tardily exsert, ovoid, 2–4 cm, the flexuous axis and branches glabrate except in the axils; spikelets softly villosulous, ellipsoid-obovoid, 2.4–2.9 mm; first glume a third to nearly half as long, broadly ovate, obtuse to acute, second glume slightly shorter than the fr and sterile lemma; autumnal phase branching early, before maturity of the primary panicle, from all nodes, forming dense masses 1–2 dm, the erect blades scarcely reduced, much surpassing the small panicles; 2n=18. Dry prairies; Ill. and n. Ind. to Man., Kans., and N.M. (*P. deamii; Dichanthelium oligosanthes* var. *w.*)

85. AMPHICARPUM Kunth.
Spikelets dimorphic; aerial spikelets in narrow terminal panicles, with a single perfect but often sterile terminal fl and a lateral floret represented only by a glume-like sterile lemma; first glume short or obsolete; second glume and sterile lemma subequal, about as long as the cartilaginous second lemma; fertile lemma with scattered short hairs; subterranean spikelets plump, fusiform, fertile, solitary at the end of a slender elongate branch, the first glume lacking, the second glume and sterile lemma several-veined, equaling the indurate fertile lemma; annual or perennial grasses with relatively short lvs and ciliate ligule. 2, the other in se. U.S.

   1. Amphicarpum purshii Kunth.   Erect slender annual 3–7 dm; sheaths striate, hirsute; blades 5–15 cm, hirsute, the uppermost much reduced; panicle exsert, 5–15 cm, simple or with a few main branches, the branchlets appressed; spikelets elliptic, glabrous, 4 mm; subterranean spikelets 6–7 mm, sparsely hairy; 2n=18. Wet pine-barrens; N.J. to Ga.

86. SACCIOLEPIS Nash.
Spikelets with one perfect terminal fl, lanceolate, awnless, pedicellate, aggregated into a slender spiciform panicle; first glume short, triangular, 3-veined; second glume strongly saccate at base, sharply 11–13-veined; sterile lemma equaling the second glume, flat, 5-veined, the outer veins approximate in pairs near the margin, its palea nearly as long and often subtending a staminate fl; fertile lemma much shorter, short-stipitate, chartaceous, with involute margins clasping the margins (but not the tip) of the similar-textured palea.

   1. Sacciolepis striata (L.) Nash.   Perennial; culms glabrous, to 1 m or more, rooting from the lower nodes and often decumbent at base; sheaths ciliate distally, the lower often densely papillose-hirsute; ligule a band of hairs 2–3 mm; blades elongate, 6–10 mm wide, glabrous or the lower papillose-hirsute; panicles solitary, 8–15 cm; spikelets glabrous, 4–4.7 mm; first glume 1–1.4 mm; 2n=36. Swamps, ditches, and muddy ground, chiefly on the coastal plain, occasionally inland; s. N.J. to Fla., Tenn., Okla., Tex., and the W.I.

87. ECHINOCHLOA P. Beauv., nom. conserv.
Spikelets planoconvex, with one perfect terminal fl and one neuter (ours) or sometimes staminate floret below, sessile or nearly so, crowded in 4–several rows in thick racemes or spikes, these aggregated into a terminal panicle; first glume 3-veined, awnless or minutely awn-tipped, usually less than half as long as the second; second glume and sterile lemma prominently 5-veined, surpassing the fertile lemma, one or both often awned from the tip; sterile lemma with a well developed palea; fertile lemma with a firm, cartilaginous, rounded, smooth, shiny body narrowed to a ± differentiated tip, its palea of similar texture, with the margins (but not the tip) clasped by the lemma; ours annuals with compressed sheaths, no ligule, and soft, elongate lvs, these incised at the base. 40, mostly in warm reg.

1 Racemes few, distant, appressed or ascending, simple, up to 2(–3) cm; spikelets awnless; lvs 3–6(–9) mm wide .... 1. *E. colonum.*

1 Racemes numerous, spreading or ascending and overlapping; larger branches of the infl commonly
    over 2 cm and usually again branched; spikelets awned or awnless; lvs 5–30 mm wide.
  2 Lower sheaths usually papillate-hairy; fertile lemma relatively narrow, nearly or fully 3 times as
    long as wide .... 2. *E. walteri.*
  2 Sheaths glabrous; fertile lemma broader, ca twice as long as wide.
    3 Second glume and sterile lemma usually with stout, papillose-based hairs on the veins; fertile
      lemma abruptly narrowed to a firm, acuminate, persistent tip .... 3. *E. muricata.*
    3 Second glume and sterile lemma hairy or scabrous to subglabrous, but the hairs usually not
      papillose-based; fertile lemma obtuse or broadly acute, with a thin, membranous, withering
      tip set off from the body by a line of minute hairs .... 4. *E. crusgalli.*

    1. Echinochloa colonum (L.) Link. Jungle-rice. Culms weak, freely branched, 1–7 dm; lvs glabrous, the
blades 2–6(–9) mm wide; panicle slender, 5–12 cm, with several alternate racemes often 1 cm apart; racemes
erect or ascending, blunt, 1–2 cm; spikelets crowded and subsessile in 4 longitudinal rows, 2.5–3 mm, often
obovoid, awnless; second glume and sterile lemma uniformly pubescent, the hairs under 0.5 mm; fertile
lemma 2.5–2.9 mm, obtuse or acute; 2*n*=54. Native to the Old World tropics, now a pantrop. weed, extending
n. to Va., Mo., and Calif.

    2. Echinochloa walteri (Pursh) Heller. Tall and usually erect, mostly 1–2 m; lower sheaths beset with
coarse, papillose-based hairs, rarely glabrous; blades up to 25 mm wide; infl dense, often nodding, 1–3.5 dm,
the spikelets almost concealed in a mass of awns 1–3 cm on the sterile lemmas; second glume with an awn
2–10 mm; fertile lemma elliptic, nearly or fully 3 times as long as wide, with a minute, withering tip, but
this not set off by a line of hairs as in no. 4; 2*n*=36. Marshes and wet soil, especially along the coast; Mass.
to Fla. and Tex., and irregularly inland to Mich., Wis., Io. and Mo.

    3. Echinochloa muricata (P. Beauv.) Fern. Barnyard-grass. Erect or decumbent, branched from the base,
often 1 m or more; sheaths glabrous; blades 8–30 mm wide; infl 1–3 dm, erect, typically with spreading,
somewhat distant branches 2–8 cm, the longer ones rebranched, the main axis and branches glabrous or
variously hairy but without conspicuous long setae, the hairs rarely over 3 mm; spikelets commonly purplish;
second glume and sterile lemma about equal, usually beset with stout, papillose-based hairs at least on the
veins, the glume awnless or nearly so; fertile lemma ca twice as long as wide, abruptly narrowed to an
acuminate, persistent tip not sharply differentiated from the body; 2*n*=36. Damp or muddy ground, or a
weed in waste places; Me. and Que. to Alta. and Wash., s. to Fla., Tex., Calif., and n. Mex. Two vars., both
widespread in our range:

  Var. *muricata.* Spikelets 3.5 mm or more to the base of the awn or mucronate tip of the sterile lemma; sterile lemma with
an awn 6–25 mm, or seldom awnless. (*E. pungens.*)
  Var. *microstachya* Wieg. Spikelets less than 3.5 mm; sterile lemma awnless or with an awn to 6(–10) mm. (*E. microstachya;*
E. wiegandii, with few or no pustulate-based hairs on the spikelets.)

    4. Echinochloa crusgalli (L.) P. Beauv. Barnyard-grass. Stout, often branched from the decumbent base;
sheaths glabrous; blades 5–30 mm wide; infl 1–2.5 dm, erect, with usually 15–25 appressed or spreading
branches 2–4 cm (to 6 cm in var. *frumentacea*), the longer ones rebranched; main axis and branches of the
infl beset with stout, often papillate-based setae typically as long as or longer than the spikelets; spikelets
2.8–4 mm; second glume and sterile lemma variously scabrous, hispid, or hirsute to subglabrous, but the
hairs usually not papillate-based, the glume awnless or nearly so; fertile lemma 2.5–3.5 mm, ca twice as long
as wide, the short, soft beak withering, somewhat inflexed, sharply differentiated from the obtuse body,
marked by a line of minute bristles at base; 2*n*=54. Native of Eurasia, widespread in the U.S., Can., and
Mex. Two vars. with us:
  Var. *crusgalli.* Panicle-branches erect-appressed or spreading, with greenish or purple-tinged spikelets, these not so plump
and closely crowded as in the next var.; sterile lemmas awnless or often some of them with ± well-developed awns, sometimes
to 25 mm, plants mostly 3–7 dm. (*E. occidentalis.*)
  Var. *frumentacea* (Roxb.) W. F. Wight. Billion-dollar grass. Panicle-branches appressed or only slightly spreading, densely
fld with closely crowded, plump, awnless or only shortly awn-tipped, usually grayish-purple spikelets; plants relatively large
and coarse, 7–15 dm, the lvs mostly 15–30 mm wide. (*E. frumentacea.*) Cultigen from se. Asia, occasionally cult. with us for
forage, sometimes escaped, but not persistent.

    88. ERIOCHLOA HBK. Cup-grass. Spikelets 1-fld, racemose, very shortly pediceled
in 2 rows along one side of a rachis, lance-ovoid or ellipsoid; rachilla-joint thickened,
forming a ring-like callus below the second glume, the virtually obsolete first glume adnate
to it, the spikelet appearing to be set in a thickened, shallow cup atop the pedicel; second
glume and sterile lemma chartaceous, 5–7-veined, usually pilose, similar and subequal or
the glume a little the longer; fertile lemma surpassed by the second glume, cartilaginous,
finely rugulose or pappilate-roughened, usually mucronate or awned, the inrolled margins
clasping a palea of similar texture and about equal length; tufted grasses with terminal
panicles of several, usually appressed racemes. 25, warm reg.

1 Spikelets slenderly acuminate; lemma with an awn 0.3–1 mm .... 1. *E. contracta.*
1 Spikelets merely acute; lemma muticous or minutely mucronate .... 2. *E. villosa.*

1. Eriochloa contracta A. Hitchc.   Prairie c.-g. Tufted annual 3–8 dm, the culms hirsutulous; lvs flat, 4–7 mm wide, hairy; panicle 10–15 cm, contracted, its base included in the upper sheath; racemes several, overlapping, 1–2 cm, the rachis and pedicels (1 mm) hairy, the thickened pedicel-tip glabrous, purple-margined; spikelets borne singly at the nodes of the infl-branches, or paired only at the base, lance-ovoid, 3.5–4 mm, the second glume slenderly acuminate, the sterile lemma similar but a little shorter; fertile lemma with a slender awn 0.3–1 mm; $2n=36$. Damp soil; Nebr. to Colo., La., and Ariz., and intr. in Minn., Mo., s. Ill., and Va.

E. acuminata (Presl) Kunth, of Mex. and s. U.S., is casually intr. n. to Ky. and Ill. It has acuminate spikelets like no. 1, but the fertile lemma has a mucro only 0.1–0.3 mm. It differs from both of our other spp. in that many of the spikelets are paired on the branches of the infl.

2. Eriochloa villosa (Thunb.) Kunth.   Chinese c.-g. Rather coarse, leafy-stemmed annual to 1 m, erect from a decumbent base; lvs flat, 6–15 mm wide, inconspicuously puberulent; racemes several, appressed-erect, 2–5 cm, overlapping, the rachis densely spreading-villous, the short pedicels beset with longer hairs 1.5–3 mm; spikelets borne singly at the nodes of the infl-branches, or paired only at the base, 4.5–5 mm, the second glume and sterile lemma about equal, pointed but scarcely acuminate; fertile lemma obscurely mucronulate or muticous; $2n=54$. Native of e. Asia, intr. and becoming established in Ill., Io., Nebr., and Kans.

89. PASPALUM L.   Bead-grass. Spikelets ovate, elliptic, obovate, or orbicular, planoconvex, short-pediceled or subsessile, solitary or in pairs, forming a 2-rowed or 4-rowed, usually spike-like raceme on one side of a flattened or triquetrous rachis; first glume usually none, very short when present; second glume and sterile lemma of about equal length, 2–7-veined, acute to rounded; fertile lemma indurate, smooth or minutely papillose, its margins inrolled over the indurate palea; ours perennial (except *P. boscianum*), usually tufted or branched from the base, with usually soft lvs, membranous ligule, and a terminal infl of 1–many racemes along a common axis, often with additional racemes on axillary peduncles from the upper sheaths. 250, mainly trop. and warm-temp.

1 Rachis of the racemes broad, foliaceous or winged, wider than the spikelets and folded over them.
  2 Racemes numerous, usually 20–50; spikelets 1.1–1.5 mm .... 1. *P. fluitans.*
  2 Racemes few, normally 2–5; spikelets 1.7–2.1 mm .... 2. *P. dissectum.*
1 Rachis of the racemes narrow or wingless, never as wide as the rows of spikelets.
  3 Racemes regularly 2, approximate or less than 1 cm apart at the summit of the peduncle .... 12. *P. distichum.*
  3 Racemes 1 and terminal, or 2–several and alternate.
    4 Racemes loosely fld, the spikelets in rather distant (3–10 mm) clusters, the tip of the upper fl
      barely reaching the base of the next pedicel .... 11. *P. bifidum.*
    4 Racemes spike-like, with numerous, crowded, overlapping spikelets.
      5 Spikelets long-villous, especially at the margin.
        6 Racemes 3–6; spikelets 2.9–3.8 mm; sterile lemma ovate. .... 3. *P. dilatatum.*
        6 Racemes 8–30; spikelets 2.2–2.8 mm; sterile lemma elliptic .... 4. *P. urvillei.*
      5 Spikelets glabrous or minutely hairy.
        7 Fertile lemma dark brown at maturity; plant annual .... 10. *P. boscianum.*
        7 Fertile lemma stramineous or pale brown at maturity; plants perennial.
          8 Sterile lemma 5–7-nerved, the outer pair of nerves of each side approximate and sub-
            marginal, or the lemma 4-nerved by suppression of the midnerve.
            9 Spikelets solitary .... 5. *P. laeve.*
            9 Spikelets mostly or all paired.
              10 Spikelets 3.8–4.3 mm; racemes 2–5 .... 6. *P. floridanum.*
              10 Spikelets 2.8–3.1 mm; racemes 5–10 .... 7. *P. pubiflorum.*
          8 Sterile lemma 3-nerved, the outer nerves submarginal, or 2-nerved by suppression of
           the midnerve.
          11 Spikelets planoconvex, at least a third as thick as wide.
            12 Racemes 5–10; spikelets 2.8–3.1 mm .... 7. *P. pubiflorum.*
            12 Racemes 1–4; spikelets 1.4–2.3 mm .... 8. *P. setaceum.*
          11 Spikelets flattened on both sides, about a fourth as thick as wide .... 9. *P. praecox.*

1. Paspalum fluitans (Elliott) Kunth.   Culms glabrous, when submersed elongate and little branched, when terrestrial tufted from the base; sheaths loose, papillose-hispid to glabrous; lf-blades tapering at both ends, thin, (5–)8–12(–20) cm, often 1 cm wide or more; racemes numerous (usually 20–50), crowded, spreading or ascending, 2–4 cm; rachis 0.8–2 mm wide, its acuminate tip surpassing the uppermost spikelet by 1–3 mm; spikelets solitary (not paired), elliptic, acute, 1.1–1.5 mm; glume and sterile lemma 2-veined near the margin, the glume finely glandular-villosulous; $2n=20$. Shallow water, swamps, and muddy shores; widespread in s. U.S., n. on the coastal plain to se. Va. and in the Mississippi Valley to w. Ky., s. Ind., c. Ill., and c. Mo. (*P. mucronatum; P. repens,* misapplied.)

2. Paspalum dissectum (L.) L.   Stems creeping and forming mats, or ascending to erect and branched from the base, 2–5 dm, glabrous; sheaths loose, glabrous, often purplish; blades linear, glabrous, 3–10 cm ×

2–4 mm; panicle often overtopped by the uppermost lvs; racemes 2–5, 2–3 cm; rachis to 4 mm wide, ± folded over the spikelets, ending in a spikelet; spikelets solitary, crowded, elliptic to obovate, glabrous, 1.7– 2.1 mm, two-thirds as wide; glume and sterile lemma 3–5-veined; $2n{=}40$. Shallow water and muddy shores on the coastal plain from s. N.J. to Fla. and Tex., n. in the Mississippi Valley to s. Ill. and s. Mo.; Cuba.

3. Paspalum dilatatum Poir.   Dallis-grass. Stout, erect, 8–15 dm; foliage glabrous, or the lowest sheaths sparsely villous; blades to 3 dm, 8–12 mm wide; racemes 3–6, spreading or ascending, 8–12 cm, the panicle much surpassing the reduced upper lf; spikelets ovate, acute, 2.9–3.8 mm; glume pubescent and near the margin long-villous; sterile lemma nearly glabrous; fertile lemma shorter and blunt; $2n{=}40, 50, 60{+}$. Pastures, waste ground, and occasionally in moist fields or woods, native of S. Amer., cult. and widely escaped in s. U.S., extending n. to Ky. and in the coastal states to N.J.

4. Paspalum urvillei Steudel.   Densely cespitose, stout, erect, to 2 m; upper blades and sheaths glabrous, lower sheaths and base of the lower blades hirsute; panicle much surpassing the upper lf, dense, the numerous (8–30) racemes erect or appressed, 5–10 cm; spikelets elliptic, acute, 2.2–2.8 mm; glume white-villous; sterile lemma nearly glabrous; fertile lemma shorter and blunt; $2n{=}40, 60$. Waste ground; native of trop. Amer., now widespread from s. U.S. (n. to Va.) s. to Argentina.

5. Paspalum laeve Michx.   Culms erect or ascending, 3–8 dm, usually several from one base; sheaths keeled, usually loose, villous to pilose or glabrous; blades 5–25 cm × 3–10 mm, glabrous or pilose; racemes 2–6, spreading, 4–12 cm; spikelets borne singly, glabrous, planoconvex, obovate to elliptic, orbicular, or obovate, 2.4–3.4 mm, two-thirds to fully as wide; glume and sterile lemma 5-veined, the outer veins approximate near the margin; $2n{=}40, 80$. Various habitats; se. Mass. to Pa., O., s. Ind., Mo. and Kans., s. to the Gulf. Abundant and variable. (*P. angustifolium; P. circulare; P. longipilum; P. plenipilum*.)

6. Paspalum floridanum Michx.   Stout, erect, 1–2 m; lvs elongate, 5–10 mm wide, pubescent at least at base; uppermost sheath bladeless or nearly so; peduncle 2–3 dm; racemes 2–5, 6–10 cm, sparsely silky at base, the narrow axis scaberulous or rarely long-pilose, somewhat flexuous; spikelets in pairs or apparently solitary, one sometimes represented by a sterile pedicel, round-obovate, varying to elliptic or rarely ovate, blunt, glabrous, 3.8–4.3 mm, two-thirds to almost as wide; glume and sterile lemma 5-veined, the outer veins approximate near the margin; fertile lemma pale brown, shining, minutely rugulose; $2n{=}120, 160$. Moist, usually sandy soil on the coastal plain from s. N.J. to Fla. and Tex., n. in the interior to s. Mo. and e. Kans. (*P. glabratum,* a form with glabrous lower sheaths.)

7. Paspalum pubiflorum Rupr.   Culms stout, strongly compressed, to 1 m, usually decumbent, rooting at the nodes; sheaths loose, glabrous or sparsely villous; blades 8–17 mm wide, usually pilose at base; panicle often equaled or surpassed by the lvs; racemes 5–10, thick, 5–10 cm; rachis 1–2 mm wide, spikelets mostly in pairs, oblong-obovate, 2.8–3.1 mm; glume and sterile lemma 3–7-veined; $2n{=}60$. Moist or wet soil; s. O. to Kans., so to N.C., Fla., and Tex. Our plants, with glabrous spikelets, are var. *glabrum* Vasey. (*P. laeviglume*.)

8. Paspalum setaceum Michx.   Culms tufted, prostrate to erect, 3–10 dm from short, knotty rhizomes; herbage variously glabrous or puberulent or long-hairy; lvs 3–20 mm wide, racemes 1–4; spikelets usually paired, 1.4–2.5 mm, three-fourths to fully as wide, variously elliptic-obovate to obovate or suborbicular, glabrous or finely villosulous or with minute, capitate hairs, often brown-dotted; glume 2-veined or obscurely 3-veined; sterile lemma 2- or 3-veined, the midvein developed or not; $2n{=}20, 40, 50$. Dry or moist, open or lightly wooded places, often in sandy soil; N.H. to Fla. and the W.I., w. to Minn., Nebr., Colo., and Ariz. Several ill-defined vars., 5 in our range:

a Lf-surfaces essentially glabrous; eastern and southern, N.J. to Fla., w. to W.Va., s. Ky., e. Okla., and
  e. Tex. (*P. c.; P. debile; P. kentuckiense; P. longepedunculatum.*) .... var. *ciliatifolium* (Michx.) Vasey.
a Lf-surfaces ± hairy.
  b Lvs densely puberulent; plants prostrate; sterile lemma without a midvein; Mass. to D.C. (*P. p.*)
  .... var. *psammophilum* (Nash) D. Banks.
  b Plants otherwise, differing in one or more respects from the above.
    c Spikelets small, mostly 1.4–1.9 mm; lvs mostly erect or ascending, the surfaces villous, not
      puberulent; coastal states from Mass. to Fla. and e. Tex., and inland to s. W.Va., e. Ky.,
      and E. Tenn. .... var. *setaceum.*
    c Spikelets larger, mostly (1.6–)1.8–2.5 mm; lvs variously spreading or ascending to erect.
      d Upper surface of the lvs with only long hairs, not puberulent; sterile lemma generally with an
        evident midvein; widespread from N.H. to Fla., w. to Mich., Kans., and Tex. (*P. m.; P.
        pubescens.*) .... var. *muhlenbergii* (Nash) D. Banks.
      d Upper surface of the lvs finely puberlent, with or without some longer hairs; sterile lemma
        generally without a midvein; mainly midwestern and southern, e. to Mich., nw. Ind., and
        Mo. (*P. s.; P. bushii.*) .... var. *stramineum* (Nash) D. Banks.

9. Paspalum praecox Walter.   Culms erect, 5–15 dm, usually solitary; sheaths glabrous, villous, or sericeous; blades 3–5 mm wide, often pilose; panicles usually surpassing the lvs; racemes 3–5, spreading, 2–6 cm; spikelets paired or solitary even in the same raceme, flattened, less than a third as thick as wide, broadly oval to suborbicular, overlapping, 2.3–3.2 mm, glabrous; glume and sterile lemma 3-veined, the lateral veins near the margin; $2n{=}20, 40$. Swamps, wet woods, and pine-barrens on the coastal plain; se. Va. to Tex. The typical form has the lower sheaths glabrous or nearly so; our plants, with hairy sheaths, are var. *curtisianum* (Steudel) Vasey. (*P. lentiferum*.)

10. Paspalum boscianum Fluegge.   Annual; culms solitary or tufted, 6–10 dm, sometimes branched above; sheaths loose, glabrous; blades 6–17 mm wide, glabrous or sparsely pilose toward the base; panicle often equaled or exceeded by the upper lvs; racemes 5–10, spreading or ascending, 4–7 cm; rachis 1–2.5 mm wide; spikelets paired, crowded, obovate-oblong, acutish, glabrous, 2–2.8 mm, three-fourths as wide; glume 5-veined, the outer veins approximate near the margin; sterile lemma 3-veined; fertile lemma and palea dark brown; $2n=40$. Moist or wet soil, sometimes a weed; tropical Amer., n. to Va.

11. Paspalum bifidum (Bertol.) Nash.   Culms solitary, erect from short rhizomes, 6–10 dm; lower sheaths and base of the elongate blades villous; panicle on a peduncle 1–4 dm, much surpassing the short uppermost lf; racemes usually 2–4, erect or ascending, 5–10 cm, very loose or interrupted, the rachis very slender; spikelets elliptic, acutish, planoconvex, 3.3–4 mm; first glume often present, deltoid, to 1 mm; second glume and sterile lemma distinctly 5–7-veined. Dry pine woods; coastal plain from se. Va. to Fla. and Tex., and inland to Tenn. and Okla.

12. Paspalum distichum L.   Knotgrass. Stoloniferous; culms 2–5 dm, ascending from a creeping base with conspicuous sheaths; lf-blades relatively short, 2–8 cm × 2–5 mm, ciliate at base; racemes 2, one sessile, the other short-stalked, ascending and often incurved, 2–5 cm, the rachis 1–1.5 mm wide; spikelets solitary, 2.6–3 mm, half as wide, plump, planoconvex, ovate; first glume often developed, triangular, to 1 mm; second glume 5-veined, minutely pubescent, this and the sterile lemma somewhat leathery; $2n=40, 48, 60$. Swamps, wet ground, and waste places, often in coastal salt-marshes; mainly in subtropical and warm-temperate regions throughout the world, n. in our range to Ky. and se. Va., and adventive to N.Y. There is some controversy as to whether the name *P. distichum* should apply to our sp. or to the closely related, more tropical plant usually called *P. vaginatum* Swartz; in the latter case, our plants take the name *P. paspalodes* (Michx.) Scribn.

90. AXONOPUS P. Beauv.   Spikelets 1-fld, sessile in 2 rows on a 3-angled or 3-winged rachis, forming a slender, elongate raceme; first glume none; second glume equaling the sterile lemma; fertile lemma and palea indurate, the former with narrowly inrolled margins, its back turned from the axis of the raceme; lvs obtusish; peduncles long, terminal (or also axillary), bearing 2 or 3 terminal digitate racemes, or with 1 or 2 other racemes below the summit. 100+, warm Amer.

1. Axonopus furcatus (Fluegge) A. Hitchc.   Stoloniferous; culms erect or decumbent at base, compressed, 4–8 dm; lvs to 10 mm wide; racemes 2, divaricate, 5–10 cm, rarely with a third just below, often with another from the uppermost sheath; spikelets 4–5 mm, glabrous, the midvein of the sterile lemma obscure, the other veins approximate in 2 submarginal pairs; fertile lemma 2.5–3 mm; $2n=40$. Damp or wet soil of the coastal plain; se. Va. to Fla., Tex., and Ark.; reported from Md. (*Anastrophus f.*)

91. SETARIA P. Beauv., nom. conserv.   Foxtail-grass. Spikelets turgid, with one perfect terminal fl and a neuter or sometimes staminate floret below, sessile or subsessile, articulated and eventually deciduous above the persistent subtending bristles (or even above the glumes and sterile lemma), crowded into a dense and spike-like panicle with many nodes and short branches, each branch-system with numerous reduced sterile branches forming the prominent bristles; first glume triangular to ovate, 3- or 5-veined, up to nearly half as long as the spikelet; second glume several-veined, longer, up to as long as the spikelet; sterile lemma glume-like, about equaling the fertile one, usually with a well developed palea; fertile lemma indurate, smooth or more often transversely rugulose, its margins revolute and clasping a palea of similar texture. 140, cosmop., mostly in warm regions.

1 Bristles mostly 4–12 below each spikelet; infls erect.
  2 Perennial from short rhizomes; spikelets 2–2.8 mm .... 1. *S. geniculata.*
  2 Tufted annual; spikelets 3–3.5 mm .... 2. *S. glauca.*
1 Bristles mostly 1–3 below each spikelet (up to 6 in *S. faberi*, which has nodding infls.)
  3 Fertile lemma ± rugulose; disarticulation below the glumes.
    4 Panicle with verticillate branches and scabrous-hispid (not villous) axis .... 3. *S. verticillata.*
    4 Panicle-branches not verticillate; axis of the panicle densely villous as well as often more coarsely hairy.
      5 Lvs evidently long-hairy (as well as scabrous) above; spikelets 2.5–3 mm; infls at maturity nodding from near the base .... 4. *S. faberi.*
      5 Lvs merely scabrous above; spikelets 1.6–2.5 mm; infls erect or nodding from near the tip .... 5. *S. viridis.*
  3 Fertile lemma smooth, shiny, disarticulating above the glumes and sterile lemma.
    6 Sterile palea narrow, half as long as the lemma; spikelets ca 3 mm; cult. and weedy .... 6. *S. italica.*
    6 Sterile palea broad, equaling its lemma; spikelet ca 1 mm; native se. sp. .... 7. *S. magna.*

1. Setaria geniculata (Lam.) P. Beauv.   Knotroot-f. Perennial 3–12 dm from short, knotty rhizomes; sheaths glabrous; blades flat, to 25 cm, 2–8 mm wide, scabrous above and sometimes long-hairy at the throat;

infl dense, cylindric, 3–10 cm, green or tawny, its axis scabrous-hispid; bristles 4–12 below each spikelet, 2–12 mm, antrorsely barbed; spikelets 2–2.8 mm; first glume 3-veined, a third as long as the spikelet, the second 5-veined, half to two-thirds as long as the spikelet; sterile lemma equaling the fertile one, clasping a broad palea of equal length, often staminate; fertile lemma transversely rugulose; $2n=36, 72$. Moist ground, gardens, salt-marshes, and waste places; tropical N. Amer., n. to Calif., Tex., Kans., Io., Ill., W.Va., and Mass.

2. **Setaria glauca** (L.) P. Beauv.   Yellow f. Annual 3–13 dm, usually erect, often in large tufts; sheaths glabrous; blades loosely twisted, to 30 cm, 4–10 mm wide, scabrous above and with long papillose hairs near the throat, glabrous beneath; infl stiffly erect, cylindric, (3–)5–10(–15) cm, yellowish, the axis hispid; bristles 4–12 below each spikelet, 3–8 mm, antrorsely scabrous; spikelets 3–3.5 mm; first glume 3-veined, a third as long as the spikelet, the second 5-veined, half as long as the spikelet; sterile lemma about equaling the fertile one, clasping a broad hyaline palea of equal length, sometimes staminate; fertile lemma transversely strongly rugulose; $2n=36, 72$. Cult. soil and waste places; native of Europe, now a cosmopolitan weed, and abundant throughout our range. (*S. lutescens; S. pumila.*)

3. **Setaria verticillata** (L.) P. Beauv.   Bur-f. Annual 3–10 dm, scabrous just beneath the infl; sheaths ciliate-margined above; blades flat, 10–25 cm × 5–15 mm, scabrous on both sides, sometimes also with scattered hairs above; infl erect, yellow-green or purplish, 5–15 cm; cylindric but lobulate, the branches verticillate, the axis scabrous-hispid on the angles, not villous; bristle one below each spikelet, 4–7 mm; spikelets 2–2.2 mm; first glume 1(–3)-veined, a third as long as the spikelet, the second 5-veined, nearly as long as the spikelet; sterile lemma nearly as long as the spikelet, clasping a rather broad palea half as long; fertile lemma transversely finely rugulose; $2n=36$. Native of the Old World, now widespread in N. Amer. as a weed in gardens and greenhouses, and along the banks of streams and irrigation-ditches. Var. *verticillata* has the bristles retrorsely scabrous, and the axis of the panicle is retrorsely scabrous-hispid. The much less common var. *ambigua* (Guss.) Parl. has the bristles and panicle-axis antrorsely instead of retrorsely scabrous. (*S. viridis* var. *ambigua.*)

4. **Setaria faberi** R. Herrm.   Nodding or giant f. Much like *S. viridis* var. *viridis,* but more robust, 5–20 dm; blades 15–30 × 1–2 cm, scabrous on both sides and bearing long, soft, papillose-based hairs above; infls 6–20 cm, drooping from near the base; spikelets 2.5–3 mm, subtended by 1–6 (usually 3) bristles, the second glume two-thirds to three-fourths as long as the more strongly rugose fertile lemma; sterile lemma with a palea two-thirds as long; $2n=36$. Native of e. Asia, now widespread in our range as a weed of fields and waste places.

5. **Setaria viridis** (L.) P. Beauv.   Green f. Annual 2–25 dm; sheaths ciliate along the upper margins; blades flat, up to 40 × 2.5 cm, scabrous above, glabrous or scaberulous beneath; infl erect or slightly nodding near the tip, cylindric, the axis hispid and densely villous, the branches not verticillate; spikelets 1.6–2.5 mm; first glume a third the length of the spikelet, 3-veined, the second nearly equaling the spikelet, 5–6-veined; sterile lemma slightly surpassing the fertile one, with a narrow, hyaline palea a third as long; fertile lemma very pale green and transversely very finely rugulose; $2n=36$. A cosmop. weed, mostly of temp. regions, found in fields, gardens, and waste places throughout our range. Most of our plants belong to var. *viridis,* 2–10 dm, with lvs up to 20 cm × 12 mm, the infl 3–10(–15) cm, not lobed. The less common var. *major* (Gaudin) Posp. is vegetatively larger, 15–25 dm, with lvs to 40 × 2.5 cm, the infl to 20 cm and somewhat lobulate. It may reflect introgression from no. 6.

6. **Setaria italica** (L.) P. Beauv.   Foxtail-millet. Vegetatively like the larger forms of no. 5, with lobulate infls, but with larger spikelets ca 3 mm, the fertile lemma smooth and shiny, disarticulating above the more persistent glumes and sterile lemma; sterile palea half as long as its lemma, which is usually shorter than the fertile lemma; $2n=18$. Old-World cultigen, thought to be derived from no. 5, cult. for fodder or grain, and occasionally found as a weed nearly throughout our range.

7. **Setaria magna** Griseb.   Salt-marsh f. Coarse annual 1–4(–6) m, often branched, above, scabrous below the infl and below the smooth nodes; sheaths villous along the margins above; blades flat, scaberulous (especially above and along the cartilaginous margins), up to 60 × 3.5 cm; infl cylindric, to 4.5 dm, curved or nodding, often interrupted near the base, its axis scabrous and densely villous; spikelets 1.8–2.5 mm, each subtended by 1 or 2 antrorsely scabrous bristles 1–2 cm; first glume 3-veined, a third as long as the spikelet; second glume 7-veined, about equaling the fertile lemma; sterile lemma slightly exceeding the fertile one, 5- or 7-veined, with an ovate, hyaline palea the same length, sometimes staminate; fertile lemma concealed by the second glume and sterile lemma, smooth, shiny, brownish, disarticulating above the more persistent glumes and sterile lemma; $2n=36$. Coastal salt-marshes from N.J. to Fla. and Tex., and intr. in W.I. and C. Amer.

92. DIGITARIA Haller, nom. conserv.   Crab-grass. Spikelets with one perfect terminal fl, single or in clusters of 2 or 3 on unequal pedicels along one side of an elongate rachis, forming 2–several slender, spike-like racemes; first glume minute or lacking; second glume a third to fully as long as the spikelet, conspicuously 5- or 7-veined; sterile lemma (appearing as another glume when the first glume is obsolete) about as long as the fertile one; fertile lemma leathery with hyaline margins, acute, often shining, usually faintly marked with longitudinal rows of minute pits; palea flat, similar in texture to the fertile lemma, which clasps its margins; ours weedy annuals, branched from the base, with few–

several terminal, digitate or approximate, spike-like, one-sided racemes. (*Syntherisma.*) 200+, cosmop.

1 Rachis of the raceme slender, sharply triquetrous but scarcely winged; stems erect or ascending, not
   rooting .... 1. *D. filiformis.*
1 Rachis 0.5–1 mm wide, evidently winged, the wings as wide as or wider than the central axis; stems,
   at least in spp. 4–6, decumbent or prostrate and rooting at the lower nodes.
  2 Fertile lemma becoming dark brown or purple-black; second glume three-fourths to fully as long
    as the fertile lemma.
    3 Second glume about as long as the spikelet; hairs of the spikelet capitellate .... 2. *D. ischaemum.*
    3 Second glume three-fourths as long as the spikelet; hairs not capitellate .... 3. *D. violascens.*
  2 Fertile lemma paler, white to stramineous or grayish-brown; second glume a third to three-fifths
    as long as the fertile lemma (to four-fifths as long in no. 5).
    4 Spikelets 2.4–3.6 mm, glabrous or merely scabrous; pedicels triquetrous, scabrous.
     5 Blades papillose-pilose, the upper surface often densely so .... 4. *D. sanguinalis.*
     5 Blades glabrous or only sparsely papillose-pilose on the upper side near the throat .... 5. *D. ciliaris.*
    4 Spikelets 1.5–1.7 mm, minutely crinkled-villous; pedicels terete, glabrous .... 6. *D. serotina.*

    1. Digitaria filiformis (L.) Koeler.   Branched from the base, erect or ascending; blades 2–6 mm wide, flat, at least the lower commonly pilose; racemes 2–6, erect or ascending, often distinctly separated at base; rachis triquetrous, scarcely winged; spikelets in pairs or threes, well separated and scarcely overlapping; first glume lacking; the second three-fifths to three-fourths as long as the spikelet, usually pubescent and erose-ciliate with capitellate hairs; fertile lemma dark brown or purple; $2n=36$. Fields and open ground, often a troublesome weed southward. Forms with glabrous spikelets occur in both vars. (*D. laeviglumis.*)

    Var. *filiformis.* Stems 3–10 dm; upper sheaths glabrous, the lower glabrous to sparsely pilose; racemes seldom over 10 cm; spikelets 1.7–2.2 mm. N.H. to Mich., Io., and Kans., s. to Fla. and Tex.
    Var. *villosa* (Walter) Fern. Stems 8–15 dm; upper sheaths glabrous or pilose, the lower densely pilose; racemes up to 15 or even 25 cm; spikelets 2–2.8 mm. Coastal plain from Va. to Fla. and Tex., and n. in the interior to Ill. (*D. villosa.*)

    2. Digitaris ischaemum (Schreber) Muhl.   Smooth c.-g. Plants forming loose tufts, ascending or suberect, only seldom rooting at the lower nodes; herbage glabrous (or the upper margins of the sheaths ciliate); racemes 2–5(–8), 4–10(–15) cm; spikelets elliptic or somewhat obovate, 1.7–2.3 mm; becoming dark brown or purple-black; first glume lacking or minute and hyaline; second glume and sterile lemma equal and as long as the spikelet, both ± pubescent or subtomentose with capitellate hairs, especially in strips between the veins; $2n=36, 45$; otherwise much like no. 4. Native of Eurasia, now established as a weed over much of the U.S., and throughout our range, but less abundant than no. 4. (*D. humifusa.*)

    3. Digitaria violascens Link.   Distinguished from no. 2 by the somewhat smaller (1.5 mm), less pubescent spikelets, the hairs not capitellate, the second glume slightly shorter; $2n=36$. Pantropical, extending into our range in Ky. and s. Ind.

    4. Digitaria sanguinalis (L.) Scop.   Northern c.-g. Decumbent or prostrate, much branched, rooting at the nodes, usually 3–6 dm; blades and blades papillose-pilose, the blades 4–10 cm × 5–10 mm; racemes 3–6 in each of 1–3 whorls, 5–15 cm; rachis 1 mm wide, broadly winged, scabrous on the margins; pedicels triquetrous, scabrous; spikelets 2.4–3.2 mm; first glume minute, often deciduous, the second 0.8–1.8 mm, a third to three-fifths the length of the spikelet; sterile lemma usually scabrous on the 5 strong veins; fertile lemma grayish-brown; $2n=18$–76, mostly 36. Native of Europe, now cosmop. and established as a weed of lawns, fields, gardens, and waste places throughout our range and w. to the Pacific, giving way southward to no. 5.

    5. Digitaria ciliaris (Retz.) Koeler.   Southern c.-g. Much like no. 4, and intergrading with it; blades glabrous or only sparsely papillose-pilose on the upper surface near the throat; spikelets 2.7–3.6 mm; second glume (1–)1.5–2.5 mm, half to four-fifths the length of the spikelet; lateral veins of the sterile lemma smooth; $2n$ mostly = 54. A weed as no. 4, but more southern; trop. Amer., n. to Va., s. Ind., Mo., and s. Nebr. (*D. adscendens.*)

    6. Digitaria serotina (Walter) Michx.   Dwarf c.-g. Much like no. 4; lvs densely pilose; racemes 2–6, slender, often curved, 4–8 cm; pedicels terete, glabrous; spikelets 1.5–1.7 mm,m villous with minute crooked hairs; first glume lacking, the second a third to half as long as the spikelet; sterile lemma 5-veined; fertile lemma stramineous or brown-tinged to nearly white. Waste ground; coastal plain from Fla. and La. (and Cuba) to se. Va., and rarely adventive northward.

    93. LEPTOLOMA Chase.   Fall witch-grass. Spikelets with one perfect terminal fl, fusiform, solitary on long slender pedicels; first glume minute or obsolete; second glume nearly as long as the spikelet, 3-veined; sterile lemma like another glume, strongly 5- or 7-veined; fertile lemma minutely rugulose, leathery, with thin margins clasping the edges of the palea; perennials with flat blades and membranous ligule, the panicle very diffuse, eventually breaking away and rolling before the wind. Monotypic.

1. **Leptoloma cognatum** (Schultes) Chase.   Culms tufted and sometimes shortly rhizomatous, 4–7 dm; lower sheaths villous to papillose-hirsute; blades 5–10 cm × 2–6 mm; panicle diffusely branched, often a third to half the height of the plant, purplish, villous in the axils; pedicels 1–4 cm, 3-angled, scabrous; spikelets narrowly elliptic, acute, 2.5–3.5 mm, the second glume and sterile lemma commonly pubescent between the veins, the hairs spreading at maturity; 2n=36, 72. Dry, especially sandy soil; se. U.S. and n. Mex., n. to Nebr., Minn., s. Mich., s. Ont., Va., and irregularly to N.H. (*Digitaria c.*)

94. PENNISETUM Rich.

Spikelets with a single perfect terminal fl and a lateral fl represented only by a thin, sterile, glume-like lemma; first glume short or wanting; second glume about as long as the sterile lemma; fertile lemma chartaceous, narrowly lanceolate, awnless; panicles dense, spike-like, with short-pediceled, involucrate groups of 1–few spikelets, the group and its invol falling together, the invol of several–many distinct or basally connate, often plumose bristles (short sterile branches). 100+, warm Old World.

1. **Pennisetum villosum** R. Br.   Feathertop. Tufted perennial 3–7 dm; culms becoming villous distally; panicle tawny, ca 1 dm, the numerous slender bristles very plumose, spreading, 3–5 cm; spikelets ca 1 cm; 2n=18, 27, 36, 45, 54. Native of Afr. cult. for ornament and occasionally escaped along the s. border of our range.

95. CENCHRUS L.

Sandbur. Spikelets lance-oblong, usually dorsally compressed, sessile, with 1 perfect terminal fl above a neuter or staminate floret; glumes membranous or hyaline, the first 1- or 3-veined and equaling to much shorter than the 1–7-veined second one, or the first glume obsolete; sterile lemma 3–7-veined, about as long as the fertile one, with a well devloped palea; fertile lemma about as long as the spikelet, membranous, 5- or 7-veined, equaling and partly enclosing its palea; 1–several spikelets collectively enclosed (or the tips exserted) in a bur of concrescent, barbed or scabrous spines or bristles, the burs sessile or nearly so on a slender rachis, forming a solitary, spike-like panicle with the axis usually prolonged into a short point; grasses with solid culms, compressed-keeled sheaths, and mostly flat lvs, the ligule reduced to a ciliate rim. 20, widespread, mostly in warm reg.

Two spp. of *Tragus* are rarely found in our range, especially on ballast and about wool mills. The bur, instead of being composed of concrescent branchlets, as in *Cenchrus,* is composed of the enlarged second glumes, which are uncinate-spiny. *T. racemosus* (L.) All. has the second glume 3.6–4.5 mm, whereas *T. berteroanus* Schultes has it 2.3–3 mm.

1 Bur bearing one whorl of basally united, flattened spines, subtended by 1–several whorls of smaller and finer bristles . . . . 1. *C. echinatus.*
1 Bur bearing several whorls of spines, the spines emerging at irregular intervals.
  2 Spines slender, numerous, 45–75; spikelets 6–8 mm, 2–3 per bur; widespread . . . . 2. *C. longispinus.*
  2 Spines broader at base, fewer, not more than about 40.
    3 Bur densely long-villous, its usually solitary spikelet 6–9 mm; coastal . . . . 3. *C. tribuloides.*
    3 Bur short-hairy (or glabrous), its 2–4 spikelets 3.5–6 mm; southern . . . . 4. *C. incertus.*

1. **Cenchrus echinatus** L.   Hedgehog-grass. Annual, ascending from a geniculate base, 2–8 dm; lvs 4–25 cm × 4–10 mm; burs well spaced on a flexuous rachis, short-hairy, 5–10 × 3.5–6 mm, truncate at the base, the single row of coarse upper spines 2–5 mm, retrorsely scabrid, mostly erect, sometimes interlocking, those of the outer rows finer, half as long, more divergent or some of them reflexed; spikelets 2–3 per bur, 5–7.5 mm; 2n=34, 68. Sandy waste places and forest-margins; trop. Amer., n. to N.C. and even D.C.

2. **Cenchrus longispinus** (Hackel) Fern.   Common s. Spreading or ascending annual 2–8 dm, usually branched; sheaths villous-ciliate distally; blades 6–18 cm × 3–7 mm; burs hairy, 8–12 × 3.5–6 mm (spines excluded), with numerous (mostly 45–75) retrorsely barbed, slender spines 3.5–7 mm, the lower spines relatively short and pointed downward; spikelets 2–3(4) per bur, 6–8 mm, exserted at the tip, visible down to the middle through the lateral cleft in the bur; 2n=34. A weed in sandy soil or disturbed habitats; Me. to Fla., w. to N.D., Oreg., Calif. and Tex.

3. **Cenchrus tribuloides** L.   Dune-s. Stout, much branched, decumbent or trailing annual 2–10 dm; sheaths villous at the summit and ± villous-ciliate; blades 2–14 cm × 3–14 mm; burs densely villous, 9–16 × 4–8 mm (spines excluded), with mostly 15–40 basally flattened, distally retrorse-barbed spines 4–8 mm; spikelets 6–9 mm, 1(2) per bur and concealed in it; 2n=34. Coastal sands, especially on dunes; s. N.Y. to Fla. and Tex., and rarely in trop. Amer.

4. **Cenchrus incertus** M. A. Curtis.   Annual to short-lived perennial, slender, 3–10 dm; sheaths glabrous or sparsely villous at the summit; blades 2–18 cm × 2–6 mm; bur 5.5–10 × 2.5–5 mm (spines excluded),

usually rather shortly hairy, cleft on 2 sides, with mostly 8–40 basally flattened, retrorsely barbed spines 2–5 mm; spikelets 3.5–6 mm, 2–4 per bur, exserted at the tip; $2n=34$. Sandy soil; trop Amer., n. to se. Va., Ark., and Kans.

96. **ERIANTHUS** Michx. Plumegrass. Spikelets all alike, in pairs, one sessile, the other pediceled, subtended usually by a ring of hairs; rachis of the raceme eventually disarticulating; glumes equal or nearly so, glabrous to villous, awnless; lemmas 2, hyaline, shorter than or equaling the glume, the lower one sterile, the fertile one ending in a long awn; tall, coarse, tufted perennials with long lvs and long terminal panicles in later summer, these usually conspicuously silky (except in *E. strictus*) from the hairs subtending the spikelets. 25, warm Amer. and s. Eurasia.

1 Stem appressed-villous below the panicle and bearded at the nodes; subtending hairs ca equaling or
　　more often surpassing the spikelets.
　2 Awn flattened, spirally twisted below; panicle silvery to tawny . . . . 1. *E. alopecuroides.*
　2 Awn subterete, straight or slightly flexuous; panicle tawny to purple . . . . 2. *E. giganteus.*
1 Stem wholly glabrous below the panicle, or hirsute at the nodes only in *E. strictus.*
　3 Panicle-branches loosely ascending to somewhat spreading; subtending hairs copious and conspic-
　　uous, from half as long to fully as long as the spikelet.
　　4 Subtending hairs nearly or fully as long the spikelet; awn flattened, spirally twisted below . . . . 3. *E. contortus.*
　　4 Subtending hairs ca half as long as the spikelet; awn subterete, straight or slightly flexuous . . . . . . 4. *E. brevibarbis.*
　3 Panicle-branches closely appressed to the axis; subtending hairs few and short, or none . . . . 5. *E. strictus.*

1. **Erianthus alopecuroides** (L.) Elliott. Silver-p. Culms stout, 1–3 m, sericeous below the panicle and bearded at the nodes; blades 10–25 mm wide, densely pilose near the base; panicle 2–3 dm, silvery to tawny, its axis villous; spikelets 5–6 mm, cream-color, villous, shorter than and almost concealed by the numerous subtending hairs; awn 10–15 mm, flattened, spirally twisted below; $2n=60$. Moist soil, fence-rows, and old fields; N.J. to s. Ind., s. to Fla. and Tex. (*E. divaricatus,* misapplied.)

2. **Erianthus giganteus** (Walter) Muhl. Sugar-cane p. Culms 1–3 m, silky under the panicle and bearded at the nodes, smooth below; blades 6–15 mm wide, smooth or pilose; sheaths densely hairy at the summit; panicle narrow, 1.5–4 dm, tawny or purple, its axis and branches pilose; spikelets ca 6 mm, sparsely long-villous, equaling or usually considerably shorter than the subtending hairs; awn 15–25 mm, terete, straight or slightly flexuous. Moist ground and old fields; N.J. to Ky. and Ark., s. to Fla. and Tex. Late summer. (*E. saccharoides; E. compactus.*)

3. **Erianthus contortus** Elliott. Bent-awn p. Culms 1–2 m, glabrous below the panicle; blades 10–15 mm wide, pilose only at base; panicle narrow, 2–3 dm, brown or purple, its axis and branches sparsely long-villous; spikelets ca 8 mm, sparsely long-villous, about equaling the pale or white subtending hairs; awn 15–20 mm, flattened, spirally twisted below. Moist soil and waste ground, mostly on the coastal plain; Md. to Fla. and Tex., thence n. to Tenn. and Ark.

4. **Erianthus brevibarbis** Michx. Brown p. Culms 1–2 m, wholly glabrous, even the axis and branches of the panicle; blades to 15 mm wide, scaberulous, pilose toward the base; panicle 2–4 dm, narrow, brown or purple; spikelets 5–8 mm, scabrously puberulent toward the tip, the copious, pale or white subtending hairs 2–4 mm; awn 2 cm, subterete, straight or slightly flexuous. Damp soil of the coastal plain from Del. and Md. to La., n. to Ark. and s. Ill. (*E. coarctatus.*)

5. **Erianthus strictus** Baldwin. Narrow p. Culms 1–2 m, glabrous below the panicle except the often hirsute nodes; blades 6–10 mm wide, glabrous or sometimes pilose at the very base; panicle very strict and narrow, with closely appressed branches; spikelets brown or green, 8–10 mm, scabrous-puberulent, naked at base or with a few inconspicuous subtending hairs 1–2 mm; awn 15–20 mm, subterete, straight or loosely flexuous; $2n=30$. Moist or wet soil; se. Va. to s. Mo., Fla., and Tex.

97. **MISCANTHUS** Andersson. Spikelets all alike, one of each pair short-pediceled, the other long-pediceled, both eventually deciduous from the summit of the pedicels; rachis of the racemes continuous; glumes subequal; lemmas 2, hyaline, ciliate, shorter than the glumes, the fertile (upper) one often awned, the sterile one a little larger but awnless; tall perennials with numerous long racemes aggregated to form a silky terminal panicle. 20, warmer parts of the Old World.

1 Fertile lemma awned; plants tufted . . . . 1. *M. sinensis.*
1 Fertile lemma awnless; plants rhizomatous . . . . 2. *M. sacchariflorus.*

1. **Miscanthus sinensis** Andersson. Eulalia. Robust, 2–3 m, in large tufts; lvs elongate, to 1 m, ca 1 cm wide, scabrous-margined; racemes simple, rather closely approximate, forming a fan-shaped panicle; glumes

narrow, 3–4 mm, subtended and equaled or shortly exceeded by a ring of long silky hairs; lemmas 2–3 mm, the fertile one with an awn 6 mm; $2n=35$–57. Native of China, cult. for ornament and occasionally escaped. Autumn.

2. **Miscanthus sacchariflorus** (Maxim.) Hackel.   Amur silver-grass.   Colonial, 1.5–2 m from long stout rhizomes; lvs 1–2 cm wide; infl pure silky-white, the basal hairs ca twice as long as the glumes; lemmas awnless; $2n=38$–95. Native of e. Asia, cult. for ornament and occasionally escaped.

98. MICROSTEGIUM Nees.

Spikelets all alike, paired, one sessile, the other pedi-celed, both eventually deciduous; rachis of the racemes articulate; glumes equal, the lower one flat, 2–3-veined, the upper one keeled and 3-veined; lemmas 2, hyaline, the lower one sterile, the fertile one often awned. (*Eulalia,* in part.) 30, warmer parts of the Old World.

1. **Microstegium vimineum** (Trin.) A. Camus.   Straggling annual 6–10 dm; blades lanceolate, 5–8 cm, racemes 2–5 cm, approximate, few in the panicle; pedicel flattened, ciliate; glumes 5 mm, awnless; lemmas shorter than the glumes, the fertile one awnless or often with a slender awn 4–8 mm; $2n=40$. Native of trop. Asia, sparingly intr. from N.J., Pa., and O. southward. (*Eulalia v.*)

99. SORGHUM Moench, nom. conserv.

Spikelets in pairs, or at the ends of the branches in threes, numerous, dorsally compressed, forming a large, branching panicle that in wild-adapted plants disarticulates at maturity into joints bearing a single pair (or trio) of spikelets; one spikelet of each pair sessile and perfect, the other pedicellate and staminate or neuter; glumes indurate, about equal; lemmas hyaline, the palea ± reduced, both sorts of spikelets typically with a lower empty lemma and an upper floriferous one; upper lemma of the fertile spikelet usually with a geniculate and twisted, readily deciduous dorsal awn. 25, mostly African.

1 Rhizomatous perennial; lvs mostly 1–2 cm wide .... 1. *S. halepense.*
1 Annual; lvs mostly well over 2 cm wide .... 2. *S. bicolor.*

1. **Sorghum halepense** (L.) Pers.   Johnson-grass.   Robust perennial to 1.5 m, colonial by long rhizomes; lvs elongate, 1–2 cm wide; panicle 1.5–4 dm, open, its main axis glabrous or scabrous; sessile spikelet silky, 4–6 mm, the pediceled one staminate (or neuter), glabrous, slightly longer; awn 1–1.5 cm; $2n=40$. Native to the Mediterranean region, cult. for forage, escaped and well established as a weed especially in the s. part of our range, but also n. to Mass. and Mich. (*Holcus h.; S. miliaceum.*) Under some conditions the plant becomes poisonous through the production of prussic acid. Hybridizes with cult. sorghum.

2. **Sorghum bicolor** (L.) Moench.   Sorghum, milo.   Coarse, maize-like annual with broad lvs mostly (2–)3–5 cm wide; panicle 1.5–5 dm, its main axis usually ± hairy; pedicellate spikelet mostly shorter than the sessile one, which is spread open by the very turgid grain at maturity; $2n=20$. Originally African, now widely cult. in numerous and diverse cultivars. The cult. plants, with a mostly very dense and compact, nonshattering infl, may be called var. *bicolor.* Shattercane, a weed of sorghum-fields, is taller and matures earlier, with a more open, soon shattering infl. This may be called var. *drummondii* (Steudel) Mohlenb. It is wholly interfertile with var. *bicolor.* The native wild African plants, ancestral to the others, form a third var.

100. SORGHASTRUM Nash.

Spikelets in pairs, one perfect, sessile, and subterete, the other represented only by its pedicel (2 such pedicels flanking the terminal spikelet of a raceme); glumes coriaceous, the first one hirsute, clasping or enclosing the glabrous or ciliate second one; lemmas 2, small, thin and hyaline, the lower one empty (sometimes nearly obsolete), the upper fertile and ending in a bent, twisted awn; erect perennials with elongate blades tapering to the base, the sheaths auricled at the summit; panicle slender, its ultimate branches bearing short, disarticulating racemes of 1–5 pairs of spikelets. 12, warm Amer. and Afr.

1 Awn 9–15 mm, bent once .... 1. *S. nutans.*
1 Awn 20–35 mm, bent twice .... 2. *S. elliottii.*

1. **Sorghastrum nutans** (L.) Nash.   Indian grass.   Culms 1–2.5 m, in loose tufts from short rhizomes, smooth except the sericeous nodes; sheaths glabrous to hirsute; ligule well developed, firm, continuous with the auricles; blades 5–10 mm wide; panicle 1–3 dm, narrow, freely branched, golden, the nodes and smaller

branches ± villous; spikelets lanceolate, 6–8 mm, the first glume pale brown, villous; awn 9–15 mm, twisted below, bent at about a third of its length; sterile pedicels densely villous, 4–5.5 mm; $2n=20, 40, 80$. Moist or dry prairies, open woods, and fields; throughout our range, s. to the Gulf, w. to Utah and Ariz. (*S. avenaceum.*) An important constitutent of the tall-grass prairies.

2. **Sorghastrum elliottii** (C. Mohr) Nash. Much like no. 1, but without rhizomes and with narrower lvs and more open, less hairy (not golden) panicle; stems slender, often glabrous throughout; sheaths usually glabrous; spikelets 5–7 mm, chestnut-brown to nearly black, sparsely white-villous; awn 20–35 mm, bent twice and twisted below the second bend, its margins white-ciliate toward the base; $2n=20$. Dry or sandy soil, chiefly on the coastal plain; Md. to Fla. and Tex., inland to Tenn. and Ark.

101. ANDROPOGON L. Bluestem.

Spikelets of 2 kinds, in pairs (or trios) at joints of the rachis, one sessile and perfect, the other(s) pediceled and either staminate, neuter, abortive, or completely suppressed; glumes of the fertile spikelet equal or subequal, coriaceous, flat to concave on the back, lacking a midvein; fertile spikelet with 2 narrow, hyaline lemmas shorter than the glumes, the lower one empty and awnless, the upper one fertile and usually with an evident terminal awn; palea reduced and hyaline, or wanting; perennial, usually tufted, often glaucous, with elongate lvs; spikelets in racemes or spikes, these solitary, paired, digitate, or panicled, in our spp. mostly long-villous, the common peduncle usually subtended and often partly enclosed by a spathe-like lf. 100, widespread, mostly in warm reg.

1 Pediceled spikelet staminate, resembling the sessile one in size and shape, but awnless.
  2 Awn of the fertile spikelet 8–20 mm, twisted below .... 1. *A. gerardii.*
  2 Awn of the spikelet none, or to 5 mm, straight .... 2. *A. hallii.*
1 Pediceled spikelet much reduced or obsolete, often represented only by its pedicel.
  3 Hairs of the rachis of the raceme white or silvery.
    4 Sessile spikelet shorter than the adjacent sterile pedicel.
      5 Common peduncle 1 cm or less; upper nodes glabrous or sparsely villous; awn straight .... 3. *A. virginicus.*
      5 Common peduncle 1–several cm; upper internodes densely villous on the terminal 3–10 mm;
        awn loosely twisted below .... 4. *A. gyrans.*
    4 Sessile spikelet longer than the sterile pedicel .... 5. *A. ternarius.*
  3 Hairs of the rachis tawny .... 6. *A. mohrii.*

1. **Andropogon gerardii** Vitman. Big b. Robust, tufted, sometimes shortly rhizomatous perennial 1–3 m, sometimes sod-forming, often glaucous; blades usually 5–10 mm wide, the lower ones and the sheaths sometimes villous; racemes (2)3–4(–6), 5–10 cm, subdigitate, on a long-exserted peduncle; joints of the rachis and pedicels equal, subterete, sparsely or usually densely ciliate, densely bearded at the top; spikelets appressed, 7–10 mm, the glumes minutely scaberulous, often ciliate; awn of the fertile lemma 8–20 mm, twisted below and ± bent; pedicellate spikelet staminate, about as large as the sessile one; $2n=20, 40, 60, 80–86$. Moist or dry, open places, a major constitutent of the tall-grass prairie; Que. to Sask., s. to Fla. and Ariz. (*A. furcatus; A. provincialis.*)

2. **Andropogon hallii** Hackel. Sand-b.; turkeyfoot. Culms 1–2 m, from long rhizomes; pedicels and joints of the rachis densely bearded with long (usually 4–6 mm) pale to golden hairs; spikelets 9–12 mm; awn of the fertile lemma 0–5 mm, straight; $2n=60, 70, 100$. Sandy prairies; Io. to Mont., Ariz., Tex., and n. Mex.; disjunct in Bureau Co., Ill. Said to intergrade with no. 1.

3. **Andropogon virginicus** L. Broom-sedge. Culms tufted, 5–15 dm, branched above, mostly glabrous and often glaucous, the uppermost nodes sometimes sparsely villous; lvs often pilose on the sheath and ligule, the blade 3–8 mm wide; uppermost lvs spathe-like, enclosing the short (2–10 mm) peduncle and base of the few–many paired racemes, these 2–3 cm, with slender, flexuous, long-villous rachis; fertile spikelet 3–5 mm, shorter than the long-villous sterile pedicel, but longer than the internode just above it, the straight awn 1–2 cm; stamen mostly solitary; sterile spikelet none, or a delicate narrow glume to 2 mm; $2n=20$. Our two principal vars. are very distinct northward, but pass freely into each other along the coastal plain from Md. southward through a series of intermediate forms often called var. *glaucopsis* (Elliott) A. Hitchc. Typical material differs as follows:

Var. *virginicus.* Lvs seldom over 3 dm; stem-sheaths glabrous; infl slender, elongate, simple or racemosely branched, the branches when present rarely surpassing the third node above their base; uppermost nodes mostly glabrous; fertile spikelets avg 3.6 mm; sterile spikelet very seldom developed; dry soil of fields and open woods; Mass. to s. Ont., O., Mo., and Kans., s. to Fla. and Tex.

Var. *abbreviatus* (Hackel) Fern. & Griscom. Lvs usually 3.5 dm or more; stem-sheaths usually scabrous; infl compact, obconic or corymbiform, densely glomerately branched, the internodes greatly shortened; uppermost nodes mostly villous; fertile spikelets avg 4.2 mm; sterile spikelet commonly present; wet soil; W.I. and C. Amer., n. to Calif., Ark., Ky., Va., and along the coastal plain to Mass. (*A. glomeratus.*)

4. **Andropogon gyrans** Ashe. Culms tufted, 3–10 dm, often branched above, mostly glabrous below, the short uppermost internodes densely villous at the summit; lvs flat, 3–4 mm wide, the upper usually crowded,

broader (even to 15 mm) and spathe-like, brownish or coppery, 8–15 cm; racemes numerous, paired, 3–5 cm, white-villous, usually flexuous; sessile spikelet three-fourths as long as the sterile pedicel and shorter than the internode next above it; awn 1–2 cm, loosely twisted below; $2n=20$. Fields and open woods; N.J. to O. and Mo., s. to Fla. and Tex. Variable, but the vars. not well defined. (*A. elliottii,* misapplied.)

5. **Andropogon ternarius** Michx. Splitbeard-b. Culms tufted, 5–12 dm, branched above, the 2 uppermost nodes sparsely villous to glabrous; blades 2–3 mm wide, often purplish-glaucous; spathes varying from narrowly linear to boat-shaped, 3–6 cm, to 6 mm wide; racemes in 3–6 pairs on a long-exserted peduncle, 3–6 cm, with mostly fewer than 12 joints; rachis densely white-villous; sessile spikelet 4.5–7.5 mm, once and a half to twice as long as the internode next above it and longer than the long-villous sterile pedicel; awn 15–20 mm, twisted below; $2n=40, 60$. Open woods and dry fields; Del. to Ky. and s. Mo., s. to Fla. and Tex.

6. **Andropogon mohrii** Hackel. Culms stout, tufted, 6–12 dm, compressed; lower blades 3–5 mm wide, their sheaths densely villous, especially toward the summit; spathes narrowly boat-shaped, inflated, 3–6(– 10) cm; peduncles short (2–3 cm), mostly included, or the terminal ones longer and exsert; racemes 2–4 on the peduncle, 3–4 cm, the rachis densely tawny-villous; sessile spikelet 4.5–5.5 mm, about equaling the sterile pedicel and much longer than the internode next above it; awn 12–20 mm, straight or nearly so. Wet sandy soil on the coastal plain; se. Va. to La.

102. SCHIZACHYRIUM Nees. Much like *Andropogon* and only weakly separated from it, but now usually held at generic rank; rames solitary at the top of each peduncle; joints of the rachis flat, gradually widened distally, obliquely cup-shaped at the tip. 50, mostly warm reg.

1. **Schizachyrium scoparium** (Michx.) Nash. Little bluestem. Culms loosely or densely tufted, 5–12 dm, often freely branched above; herbage often glaucous; blades 3–7 mm wide; rames long-exsert, bearing 5–20 sets of spikelets on a straight or flexuous, white-ciliate rachis; glumes 5–10 mm; awn 7–14 mm, twisted near the base; sterile pedicels usually in pairs, densely ciliate, often spreading, each bearing a vestigial spikelet 3– 10 mm (awn included); $2n=40$. Dry soil, old fields, prairies, and open woods; N.B. and Que. to Alta., s. to Fla. and Mex. Abundant and highly variable, an important constituent of tall- and mixed-grass prairies. (*Andropogon* s.; *A. praematurus,* a smutted form of this or of some sp. of *Andropogon*.) Three vars. with us:

a Lf-sheaths not keeled, or only weakly so; cilia of the rachis 1–3(–4) mm; the common form, with the
    range of the sp. . . . . var. *scoparium*.
a Lf-sheaths broad and prominently keeled; cilia longer, ca 5 mm.
  b Sheaths and blades glabrous or nearly so; coastal sands from e. Mass. to N.C., and reported from
    the shores of the Great Lakes. (*S. littoralis.*) . . . . var. *littoralis* (Nash) Gould.
  b Sheaths and blades villous; Ky. to Tex. (*A. divergens.*) . . . . var. *divergens* (Hackel) Gould.

103. ARTHRAXON P. Beauv. Spikelets usually paired on the rachis of the spike, one member sessile and fully developed, the other pedicellate and usually ± reduced (or even obsolete); normal spikelets alternate and laterally appressed to the articulate rachis, the rachis-joint and spikelet commonly falling as a unit; normal spikelet with firm glumes, 2 very thin, membranous lemmas, and no palea, the lower lemma empty and awnless, the upper subtending a perfect fl and usually awned from near the base; low, decumbent or creeping grasses with lanceolate or lance-ovate lvs cordate-clasping at base, and with 2–many simple or often branched, usually subdigitately arranged spikes. 7, warm Old World.

1. **Arthraxon hispidus** (Thunb.) Makino. Slender, lax, decumbent annual to 5(–10) dm; blades 2–7 cm × 5–15 mm; spikes few–several, digitate, 2–8 cm; normal spikelet (2–)3–5(–8) mm, the lower glume 4–15-veined and with the veins usually spiculate-scabrous at least distally, the upper inconspicuously 3- or 5-veined; sterile lemma virtually veinless, the fertile one 1-veined, often with a supra-basal awn to 10 mm; anthers usually 2; sterile spikelet reduced to the 1–4 mm pedicel, or obsolete; $2n=10$. A weed of moist or wet soil, native to se. Asia, now widely intr. in warm regions, and found here and there in our range, n. as far as Mo. and s. N.Y.

104. COELORACHIS Brongn. Racemes spike-like, cartilaginous, cylindric, the rachis readily disarticulating, each internode thickened distally and concave at its summit; spikelets paired, the fertile one sessile, closely appressed to the hollow of the adjacent internode, its first glume coriaceous and pitted or wrinkled; sterile spikelet smaller, its stout pedicel appressed to the rachis; smooth perennials; growing in small tufts, the slender solitary racemes resembling the pistillate portion of *Tripsacum*. 12, mainly trop.

1. **Coelorachis rugosa** (Nutt.) Nash. Culms 6–10 dm, usually branched above; sheaths overlapping, narrowly but sharply keeled; blades usually folded, 4–8 mm wide; racemes slender; 3–8 cm, exserted or partly included in the subtending sheath; sessile spikelet 4–5 mm, usually surpassing the adjacent internode; its first glume coarsely transversely wrinkled on the back, narrowly 2-winged toward the summit, often notched; sterile spikelet 3–4 mm, often divaricate when dry, together with its pedicel (2–3 mm) conspicuously exceeding the fertile one. Wet pine-barrens; s. N.J. to Fla. and Tex. (*Manisurus r.; Rottboellia r.*)

    C. cylindrica (Michx.) Nash, a more strictly southern sp. with narrower lvs and with the first glume of the sessile spikelet pitted instead of transversely wrinkled, probably does not reach our range.

105. TRIPSACUM L.   Gama-grass. Spikelets unisexual, in spikes; staminate spikelets above the pistillate, in pairs at the nodes, 2-fld, with 2 glumes and 2 flowering lemmas; pistillate spikelets solitary at the nodes, sunken in hollows on the thickened articulate rachis, with 2 glumes, 1 sterile lemma, and 1 fertile lemma; stout perennials. 7, warm Amer.

    1. **Tripsacum dactyloides** (L.) L.   Culms 1–3 m, in large clumps from thick rhizomes; blades elongate, often to 2 cm wide; spikes 1–4, 10–25 cm, the lower fourth or third pistillate; spikelets 7–10 mm; first glume of the pistillate spikelets toward the outside of the pit, indurate, ovate, acute; both staminate spikelets of a pair sessile or subsessile, the outer glumes oblong; $2n$=18, 36, 45, 54, 72, 90,108. Swamps and wet soil; trop. Amer., n. in e. and c. U.S. to Mass., s. Mich. (only casual), Io., and Nebr.

FAMILY **SPARGANIACEAE**, the Bur-reed Family

Fls small, grouped into dense, unisexual, complex heads along the stem and branches, the staminate heads uppermost; perianth of (1–)3–6 oblong to spatulate, often scale-like, hypogynous tep in 1 or 2 series; stamens 1–8, often opposite the tep, the anthers on slender filaments; staminate heads through asymmetric growth sometimes coming to appear as a mass of irregularly disposed stamens and scales; ovary 1, superior, unilocular or seldom bilocular or even trilocular, with a single pendulous, apotropous ovule in each locule; fr sessile or short-stipitate, small, dry, indehiscent, with spongy exocarp and stony endocarp, usually subtended by the persistent tep as well as beaked by the persistent style; embryo straight, linear, with a terminal cotyledon, surrounded by copious endosperm and thin perisperm; rhizomatous, ± aquatic perennials; lvs emergent or floating, distichous, with sheathing base and long, linear, parallel-veined, often spongy blade. A single genus.

1. SPARGANIUM L.   Bur-reed. Characters of the family. 15, mainly N. Temp.

1 Stigmas and locules 2; fr broadly obpyramidal . . . . ´ 1. *S. eurycarpum.*
1 Stigma and locule 1; fr fusiform or ellipsoid.
  2 Achene with beak 0–1.5 mm and stipe 0–1 mm; staminate head solitary.
    3 Achene short-beaked; pistillate heads all axillary . . . . 9. *S. minimum.*
    3 Achene beakless; one or more of the pistillate heads supra-axillary . . . . 10. *S. hyperboreum.*
  2 Achene with beak 1.5–6 mm and stipe 1–5 mm; most spp. with 2 or more male heads.
    4 Anthers and stigma oblong or ovate, 0.4–0.8 mm; tep inserted from base to middle of the stipe;
      lvs scarcely or not at all keeled.
      5 Staminate heads several, separated from the pistillate; achene dull . . . . 7. *S. fluctuans.*
      5 Staminate head 1(2), next to the upper pistillate one; achene shining . . . . 8. *S. glomeratum.*
    4 Anthers and stigma linear, 0.8–4 mm; tep inserted chiefly above the middle of the stipe, the 3
      inner ones at its very summit; lvs often keeled.
      6 Pistillate heads all axillary and subtended by bracts.
        7 Achene lustrous, pale brown, its body 5.5–7 mm; stigma 1.5–3 mm . . . . 2. *S. androcladum.*
        7 Achene dull, sordid brown, its body 3–5 mm; stigma 1–1.5 mm . . . . 3. *S. americanum.*
      6 Pistillate heads partly supra-axillary.
        8 Achene-beak about as long as the body; plants normally erect . . . . 4. *S. chlorocarpum.*
        8 Achene-beak evidently shorter than the body; plants ordinarily with floating lvs.
          9 Lvs rounded on the back, 2–6 mm wide, the upper dilated at base . . . . 5. *S. angustifolium.*
          9 Lvs flat (or keeled toward the base), 6–12 mm wide, not dilated at base . . . . 6. *S. emersum.*

    1. **Sparganium eurycarpum** Engelm.   Giant b.-r. Stout, erect, 5–12 dm; lvs to 8 dm, 6–12 mm wide, scarcely dilated at base, the bracts similar but shorter; infl branched, the branches from the lower axils with

retuse, or abruptly rounded to the stout 3 mm beak; stigmas 2, linear; ovules, seeds, and locules 2; $2n=30$. In mud or shallow water; Que. and N.S. to B.C., s. to Va., Ind., Okla. and Calif.

2. Sparganium androcladum (Engelm.) Morong.   Stout, erect, 4–10 dm; lvs to 8 dm, 5–12 mm wide, some or all carinate or triangular in cross-section; bracteal lvs ascending; infl simple or sometimes branched; pistillate heads on the central axis 2–4, on the branches none or seldom 1 or 2, all sessile, 3 cm thick when ripe; tep spatulate, two-thirds as long as the achene; stipe 2–4 mm; achene-body shining, pale brown, ellipsoid-fusiform, 5.5–7 mm, slightly constricted, the beak straight, 4–6 mm; stigma 1.5–3 mm; staminate heads 3 or more on the branches, 5–8 on the central axis; anthers linear, 1–1.5 mm. Muddy shores and shallow water; Que. to Va.; O. to Minn. and Mo. (*S. lucidum.*)

3. Sparganium americanum Nutt.   Stout, erect, 3–10 dm; lvs to 1 m, 4–12 mm wide, usually thin and flat, occasionally weakly carinate; bracts often dilated at base; infl simple or branched; pistillate heads on the central axis 2–4, on the branches 1–3, all sessile, 2 cm thick when ripe; stipe 2–3 mm; tep two-thirds as long as the fr, dilated at the tip; achene-body fusiform, dull, sordid brown, 3–5 mm, slightly constricted, the beak straight, 1.5–4.5 mm; staminate heads 1–5 on the branches, 3–10 on the main axis. Mud or shallow water of swamps and ponds; Nf. and Que. to Minn., s. to Fla. and La.

4. Sparganium chlorocarpum Rydb.   Ordinarily erect, 0.5–6 dm; lvs usually much exceeding the stem, the lower distichously imbricate, erect or nearly so, flat or somewhat keeled, 2–7 mm wide, little or not dilated at base; bracts shorter, usually conspicuously erect or strongly ascending; infl unbranched; pistillate heads 1–4; usually crowded, sessile or the lowest short-peduncled, the lowest usually supra-axillary; tep spatulate, two-thirds as long as the achene; achene greenish-brown to brown, shining, fusiform, the body 4–6 mm, tapering to a conspicuous beak about as long, staminate part of the infl usually 4–10 cm, the heads few–several, not crowded; $2n=30$. Swamps, muddy shores, and shallow water; Nf. and Que. to Ida., s. to W.Va., Ind., and Io. (*S. acaule,* a depauperate form.)

5. Sparganium angustifolium Michx.   Stem usually floating and elongate; lvs floating, rounded on the back, 2–6 mm wide; upper lvs and bracts conspicuously dilated at base; infl unbranched; pistillate heads 1–3, the lowest supra-axillary and usually peduncled, the others axillary, usually sessile, 1–2 cm thick when ripe; tep spatulate, erose, two-thirds as long as the frs; achenes brown, dull, the body 3–5 mm, abruptly contracted to a 1 mm beak; staminate part of the infl 1–4 cm, the 1–6 heads close together; $2n=30$. In deep or shallow water, occasionally on mud; Nf. to Alas., s. to N.J., Pa., Mich., and Calif.

6. Sparganium emersum Rehmann.   Much like no. 5, but with broader (6–12 mm), flat, more coarsely veined lvs, the basal ones ± V-shaped in section below; bracts and upper lvs not dilated at base except by the scarious margins; pistillate heads 1.5–2(–2.5) cm thick when ripe; $2n=30$. In quiet water; circumboreal at high latitudes, s. to N. Engl., N.Y., Man., Colo., and Calif. Our plants, and most others from the conterminous U.S. and s. Can., belong to var. *multipedunculatum* (Morong) Reveal. (*S. multipedunculatum.*) The more boreal and Eurasian var. *emersum* is more emergent, with larger heads and more keeled basal lvs. (*S. simplex.*)

7. Sparganium fluctuans (Morong) Robinson.   Stem slender and elongate, to 15 dm; lvs floating, flat, thin and translucent, 3–10 mm wide, cross-reticulate beneath; infl often branched; pistillate heads 2–4, chiefly on the branches, sessile or short-peduncled, 1.5–2 cm thick when ripe; tep linear-oblong, reaching the middle of the achene, attached at or below the middle of the short (2–4 mm) stipe; achene dark reddish-brown, the body obovoid-oblong, 3–4 mm, obscurely constricted just below the middle, rounded above into a stout curved beak 2–3 mm; staminate heads several on the central axis, 1 or 2 on the branches. In quiet water; Nf. and Que. to Minn., s. to Conn., Pa., and Mich.

8. Sparganium glomeratum Laest.   Stems stout, floating to erect, 2–6 dm; lvs flat or weakly keeled, 3–8 mm wide, the bracteal ones much dilated at base, infl usually simple, occasionally with a basal branch; pistillate heads several, closely glomerate, sessile, 1.5–2 cm thick when ripe; tep linear-oblong, reaching just beyond the middle of the fr, attached at or near the base of the short stipe; achene brown, shining, the body fusiform, 3–4 mm, slightly constricted below the middle, tapering to a straight or slightly curved beak 1.5–2 mm; staminate head 1(2), contiguous with the upper pistillate head; $2n=30$. Bogs and shallow water; interruptedly circumboreal, reaching our range in N. Minn.

9. Sparganium minimum (Hartman) Fries.   Stems usually long and floating, varying to erect and 1–3 dm; lvs thin, flat, 2–7 mm wide, much elongate when floating; pistillate heads 1–3, axillary, sessile or the lowest rarely short-peduncled, 1 cm thick when ripe; tep narrowly spatulate, erose at the tip, reaching beyond the middle of the fr; achene sessile or nearly so, dull greenish or brownish, the body elliptic-obovoid, 3–4 mm, scarcely constricted, acute, the slender beak 0.5–1.5 mm; staminate head normally 1, separate; anthers oblong, 0.3–0.8 mm; $2n=30$. In shallow water; circumboreal, s. to N.J., Pa., n. Ind., and N.M.

10. Sparganium hyperboreum Laest.   Stems floating, or decumbent and ascending, very slender; lvs thick, flat, 1–4 mm wide, much elongate when floating; pistillate heads 2–4, the lower 1 or 2 supra-axillary, sometimes short-peduncled, 1 cm thick when ripe; tep narrowly spatulate, scarcely reaching the middle of the fr; achenes sessile or nearly so, dark yellow, the body ellipsoid, 4 mm, scarcely constricted, acute at both ends, tipped with the persistent, dark, papilliform stigma less than 0.5 mm; staminate head 1, closely adjacent to the pistillate; anthers oblong, 0.4–0.8 mm; $2n=30$. Cold quiet water; circumboreal, s. to N.S.

Family **TYPHACEAE**, the Cat-tail Family

Fls unisexual, very numerous, in a very dense, elongate-cylindric, complex spike, many or all of them axillary to short, bristle-like bracts, the staminate (upper) part of the spike sharply differentiated from the pistillate part; staminate fls with modified perianth of 0–3(–8) capillary bristles or slender scales; stamens (1–)3(–8), the short filaments distinct or connate; anthers with a broad, prolonged connective; pistillate fls with a hypogynous modified perianth of ± numerous capillary bristles or narrow scales in 1–4 irregular cycles, these commonly appearing as long hairs from the lower part of the gynophore; pistil a single carpel elevated on a gynophore that elongates in fr to form a slender stipe; style persistent and elongating in fr; ovule solitary, pendulous, apotropous; sterile pistillate fls intermingled with the fertile ones; fr small, dry, 1-seeded, eventually opening, distributed by wind; embryo straight, cylindric, with a terminal cotyledon, surrounded by endosperm and perisperm; glabrous, rhizomatous, often densely colonial perennials, emergent from shallow water or growing in very wet soil; stems erect, simple, terminating in an infl; lvs distichous, most of them basal or near-basal, with a strongly sheathing base and an elongate, flattened, parallel-veined, firm but spongy blade. A single genus.

1. **TYPHA** L.   Cat-tail. Characters of the family. 10, cosmop.

1 Stigma broad, spatulate; staminate and pistillate parts of the spike normally contiguous .... 1. *T. latifolia.*
1 Stigma linear; staminate and pistillate parts of the spike normally separated.
  2 Spikes deep brown; pistillate bracteoles distally rounded; widespread .... 2. *T. angustifolia.*
  2 Spikes pale brown; pistillate bracteoles acuminate; coastal from Del. southward .... 3. *T. domingensis.*

1. **Typha latifolia** L.   Common c.-t. Stems 1–3 m; lvs flat, (8–)10–23 mm wide; pistillate and staminate portions of the spike contiguous, or rarely separated by as much as 4 mm, the pistillate portion brown, 10–15 cm, 2–3 cm thick at maturity; compound pedicels long and slender; pistillate bracteoles wanting; stigma broad and thick, spatulate; fr 1 cm, with copious white hairs arising near the base (these linear, not expanded upwards), the achene 1 mm long, above the middle of the whole fr; sterile pistillate fls about as long as the fertile and similarly hairy, expanded into a spatulate tip; staminate bracteoles white, capillary; pollen in tetrads; 2n=30. Clean marshes; nearly cosmop., and throughout our range, the common sp. inland. Hybrids with the next 2 spp. have been called *T.* × *glauca* Godr.

2. **Typha angustifolia** L.   Narrow-lvd c.-t. Stems 1–1.5 mm; lvs 5–11 mm wide, auriculate at the juncture of sheath and blade; pistillate and staminate portions of the spike separated by (1–)2–12 cm, the pistillate portion deep brown, 10–20 cm, 1–2 cm thick at maturity; compound pedicels short and stout; pistillate fls each accompanied by a hair-like bracteole with an expanded, flat, spatulate tip; stigma linear; fr 5–8 mm, subtended by copious hairs with a slightly expanded brown tip, the achene usually distinctly above the middle; sterile pistillate fls about as long as the fertile ones, dilated into a cuneate truncate tip; staminate bracteoles brown and scale-like; pollen in monads; 2n=30. Marshes, more tolerant of salt and alkali than no. 1; nearly cosmop, and throughout our range, the common sp. along the coast, now becoming common inland as well.

3. **Typha domingensis** Pers.   Southern c.-t. Much like no. 2, but taller (often 2.5–4 m) and with more numerous lvs, these thicker, up to 15 mm wide, often some or all of them exauriculate; pistillate portion of the spike light brown, 1.3–2.5 cm thick at maturity, separated from the staminate portion by up to 8 cm, rarely contiguous with it; pistillate bracteoles acuminate; sterile pistillate fls narrowly cuneate at the tip; hairs subtending the fr white with usually a single large brown cell near the tip; pollen in monads; 2n=30. Pantrop., extending n. in coastal marshes to Md. and Del., and n. inland to Nebr. and Utah.

Family **BROMELIACEAE**, the Bromeliad Family

Fls mostly perfect, regular or nearly so, hypogynous to often epigynous; sep and pet each 3, unlike, the sep green and herbaceous to frequently ± petaloid, distinct or connate below; pet distinct or shortly connate below, commonly provided along the basal margins with a pair of scale-like appendages that sometimes function as nectaries, and the ovary regularly with septal nectaries; stamens 3 + 3, often connate, or adnate to the tep; ovary trilocular, with a trifid style and axile placentation, fr a berry, or less often a usually septicidal capsule, or seldom multiple and fleshy; seeds often plumose or winged; embryo monocotyledonous, usually peripheral at the base of the copious, mealy endosperm; epiphytes and terrestrial xerophytes, often acaulescent, with narrow, parallel-veined, often firm and spiny-margined lvs, in most genera provided with stalked, peltate, water-absorbing scales at least when young; infl often provided with showy bracts. 45/2000, mainly warm New World.

1. TILLANDSIA L.   Lvs entire; pet separate; ovary superior; fr capsular; seeds plumose; epiphytes. 500, warm Amer.

1. **Tillandsia usneoides** (L.) L.   Spanish moss. Plants rootless, pendent from trees, consisting of slender, wiry, branching stems to 5 m, bearing numerous scattered, densely cinereous-lepidote, filiform lvs 2–5 cm; infl reduced to a single subsessile fl terminating very short axillary branches; sep narrowly ovate, to 7 mm; pet narrow, recurved, green fading yellowish, 1 cm; fr cylindric, 2.5 cm; $2n=32$. Trop. Amer., n. on the coastal plain to Va. and historically to s. Md.

FAMILY **PONTEDERIACEAE**, the Water-hyacinth Family

Fls perfect, regular or irregular; perianth of 6 petaloid tep in 2 series, these usually connate at base to form a perianth-tube; stamens 3–6; filaments adnate to the perianth-tube; ovary superior, trilocular with axile placentas, or unilocular with intruded parietal placentas, or with 2 reduced, empty locules and a single fertile locule, containing a solitary, terminal, pendulous ovule; fr a loculicidal capsule, or dry, 1-seeded, and indehiscent; embryo cylindric, with a terminal cotyledon, surrounded by the copious, starchy, mealy endosperm; glabrous, aquatic or semi-aquatic herbs, free-floating or rooted in the substrate, the lvs mostly with a sheath, a distinct (sometimes inflated) petiole, and an expanded, floating or emersed blade with parallel, curved-convergent veins, or sometimes linear and virtually sheathless; infl terminal and subtended by a generally bladeless sheath, the fls usually in racemes or spikes or panicles, seldom solitary. 9/30, mainly trop. and subtrop.

1 Stamens 6, 3 exserted and 3 included; anthers versatile.
  2 Perianth-lobes 1 cm or less; ovary unilocular; plants rooted in the soil . . . . 1. *Pontederia*.
  2 Perianth-lobes 3–4 cm; ovary trilocular; plants usually free-floating . . . . 2. *Eichhornia*.
1 Stamens 3, exserted; anthers basifixed.
  3 Lvs narrowly linear; stamens all alike; fls yellow, solitary . . . . 3. *Zosterella*.
  3 Principal lvs broader, oblong to reniform; stamens dimorphic; fls white to blue, 1–several . . . . 4. *Heteranthera*.

1. PONTEDERIA L.   Pickerel-weed. Fls trimorphically heterostylic; perianth blue to white, funnelform with bilabiate limb, the 3 lower lobes nearly or quite distinct, the 3 upper united about half their length, the middle upper lobe broader than the lateral and marked with yellow; stamens 5, inserted on the perianth-tube, the 3 opposite the lower lip exsert, those opposite the upper lip included and sometimes reduced or sterile; ovary 3-locular, one locule with a single pendulous ovule, the other 2 smaller and empty; nectaries septal; fr achene-like, enclosed within the accrescent perianth-tube, and beaked with the persistent style-base; marsh or aquatic herbs, often colonial, with broad or narrow lvs; infl a contracted spike-like panicle, the long stalk arising from the rhizome and bearing one lf with an expanded blade and above it a bladeless sheath. 3, New World.

1. **Pontederia cordata** L.   Erect or nearly so, to 10 dm; basal lvs and lower stem-lf similar, long-petioled, the blade firm, broadly cordate to lanceolate, to 18 cm; spathe loosely sheathing, 3–6 cm; infl crowded, 5–15 cm; perianth funnelform, villous or shortly glandular-pubescent in bud, later often glabrate, violet-blue, rarely varying to white, the tube 5–7 mm, the lobes 7–10 mm; filaments pubescent; fr 5–10 mm, longer to sometimes shorter than wide; $2n=16$. Marshes and shallow water; N.S. to Ont. and Minn., s. to S. Amer. (*P. lanceolata.*)

2. EICHHORNIA Kunth, nom. conserv.   Perianth tubular at base, the limb spreading, somewhat bilabiate, the upper lobe marked with yellow; stamens 6, inserted on the perianth-tube, the 3 opposite the lower lip included; ovary 3-locular, ripening into a 3-valved, many-seeded capsule; nectaries septal. 6 or 7, New World.

1. **Eichhornia crassipes** (Martius) Solms-Laub.   Water-hyacinth. Plants normally floating, with long pendent roots and a cluster of radiating lvs; petioles usually greatly inflated; lf-blades orbicular to reniform, 4–

12 cm wide; infl a contracted panicle 4–15 cm; perianth 5–7 cm wide, lilac or rarely white, the upper lobe bearing a violet blotch with a yellow center; $2n=32$. Trop. Amer., n. (as an introduction) to Va. and Mo.

3. **ZOSTERELLA** Small.   Water star-grass. Infl of a single subsessile fl completely embraced by the convolute 1-lvd spathe; fls regular; perianth salverform, the tube very slender, elongate, the limb spreading, pale yellow, its 3 outer segments slightly narrower than the inner; stamens 3, all alike, with inflated filaments and sagittate anthers that coil downward after dehiscence; ovary incompletely 3-locular; nectaries wanting; freely branching plants with linear, pellucid, grass-like lvs, the spathes sessile in the axils, the small fls expanded on the water-surface. Monotypic.

1. **Zosterella dubia** (Jacq.) Small.   Perennial, sometimes fl the first year; stems usually submersed, elongate, branched, slender; lvs without petiole or midrib, linear, obtuse, to 15 cm, 2–6 mm wide, or smaller and thicker when emersed; spathe 2–5 cm, slender, abruptly short-caudate at the spreading tip; perianth-tube 1.5–11 cm, the limb spreading, 1–2 cm wide, pale yellow, the outer segments linear, the inner lance-linear; filaments inflated near the middle, tapering to the ends; fr narrowly ovoid, 1 cm, indehiscent, with 7–30 seeds; $2n=30$. In quiet water, or sometimes on mud-flats; s. Que. to N.D. and Wash., s. to Cuba and C. Amer. (*Heteranthera d.*)

4. **HETERANTHERA** Ruíz & Pavón.   Mud-plantain. Fls 1–several in a sessile or peduncled infl subtended by a sheathing bladeless spathe, below which is another blade-bearing lf; fls regular or with the perianth-segments unequally disposed; perianth-tube very slender, with 6 spreading, linear to lanceolate segments, the 3 outer narrower than the 3 inner; stamens 3, dimorphic, the 2 posterior (lower) ones short, with usually ovate anthers, the anterior (upper) one larger, with longer filament and oblong to sagittate anther; anthers not coiling after dehiscence; ovary incompletely 3-locular, the 3 parietal placentas intruded, with many ovules; nectaries wanting; fr capsular; plants submersed to partly emersed, or creeping in mud; lvs and stems dimorphic, the erect fertile stems differentiated from the very short to elongate, branched and creeping sterile ones; some lvs bladeless, narrow, sessile and often forming a rosette, others petiolate, with expanded, floating or emersed blade; fls small, white to blue or violet. 11, warm Amer. and Afr.

1 Fls 2–several in each spathe; perianth-tube 3–12 mm.
  2 Filament-hairs purple; spike elongating well out of the spathe .... 1. *H. multiflora.*
  2 Filament-hairs white; spike scarcely elongating out of the spathe .... 2. *H. reniformis.*
1 Fl solitary in each spathe; perianth-tube 11–45 mm.
  3 Emersed lvs with acutish blade ca half as wide as long; perianth-segments subequally disposed .... 3. *H. limosa.*
  3 Emersed lvs with obtuse or rounded blade nearly or fully as wide as long; perianth-segments
     unequally disposed, 3 above, 2 lateral and spreading, one below .... 4. *H. rotundifolia.*

1. **Heteranthera multiflora** (Griseb.) Horn.   Much like no. 2; blades usually a little longer than wide; fls 3–16, all opening in one or 2 days, the spike elongating well out of the spathe; perianth pale purple or white, scarcely glandular-pubescent, the tube 3–12 mm, the central upper lobe with 2 yellow spots in the dark purple basal area; hairs of the filaments copious, purple; $2n=32$. In shallow water or emersed on mud; c. U.S., from Nebr., Kans., and Okla. to Ill. and Miss.; disjunct on the coastal plain from N.J. to N.C., and in S. Amer.

2. **Heteranthera reniformis** Ruiz & Pavón.   Aerial lvs with floating or emersed, reniform blade 1–4 × 1–5 cm, as wide as or somewhat wider than long; petioles to 1.3 dm; spathe folded, 1–5 cm, loosely sheathing below; fls 2–8, all opening the same day, only the terminal one sometimes extending past the tip of the spathe; perianth largely white, densely glandular-pubescent outside, the tube 5–10 mm, the lobes 3–6.5 mm, 5 above and one below, the central upper one with 2 yellow (or green) spots at the base; filaments sparsely pubescent with spreading multicellular white hairs at least toward the tip; style pubescent with multicellular hairs; $2n=48$. In shallow water or emersed on mud; Conn. and N.Y. to Mo., s. to trop. Amer.

3. **Heteranthera limosa** (Swartz) Willd.   Aerial lvs emersed, the blade 1–5 cm long and about half as wide, oblong to ovate, acutish, with cuneate to truncate base; petioles to 1.5 dm; spathe folded, abruptly caudate-acuminate, 1–4.5 cm, enclosing 1 fl; fls purple to white, the perianth-tube 1.5–4.5 cm, the lobes 5–15 mm, about equally spreading, the 3 upper ones basally yellow; stamens subequal, with straight filaments, the lateral ones yellow, the central one purple or white; style glabrous; $2n=14$. In shallow water at the edges of ponds and ditches, or wholly emersed on mud; Mississippi R. drainage from Ky. to Io. and S.D., s. to S. Amer.; also Calif. to Ariz. and w. Mex.

4. **Heteranthera rotundifolia** (Kunth) Griseb.   Much like no. 3; aerial lvs with floating or emergent, relatively broad and blunt blade, commonly nearly or fully as broad as long, obtuse or rounded above,

rounded or subcordate at base; spathe closely clasping the fl; perianth-segments 3 above, 2 lateral and spreading, and one below, the central upper one with lateral flanges in the proximal one-fourth; lateral stamens with distally incurved filaments; $2n=14$. Exposed mud-flats or in shallow water of ephemeral pools, roadside ditches, and small ponds; Mississippi R. drainage from Ky. to S.D., s. to S. Amer., and in s. Ariz. and w. Mex.

Family **HAEMODORACEAE**, the Bloodwort Family

Fls perfect, regular or nearly so, the perianth of 6 similar, persistent, distinct or ± connate tep in 1 or 2 cycles; stamens 6 or 3, opposite the pet when 3; ovary superior to inferior, trilocular with axile placentas; style solitary, with a capitate stigma; fr a loculicidal capsule with (1–) several to many seeds; embryo small, with a terminal cotyledon, embedded in the copious, fleshy, starchy endosperm, which also contains stored protein, oil, and hemicellulose; perennial geophytic herbs of lilioid habit, often with a characteristic polyphenolic red pigment in the roots and rhizomes; foliage-lvs all basal, with sheathing base and parallel-veined, equitant, unifacial blade; infl usually conspicuously long-hairy, in our genera repeatedly branched. 16/less than 100, mostly S. Hemisphere.

1 Stamens 6; ovary only half inferior .... 1. *Lophiola.*
1 Stamens 3; ovary wholly inferior .... 2. *Lachnanthes.*

1. **LOPHIOLA** Ker Gawler.   Perianth deeply 6-parted, its segments spreading at anthesis, erect and persistent in fr; stamens 6, erect, shorter than the perianth; ovary half-inferior, with numerous ovules; fr subglobose, many-seeded, tipped by the persistent style which eventually splits longitudinally; seeds elliptic to fusiform; no red pigments. Monotypic.

1. **Lophiola aurea** Ker Gawler.   Gold-crest. Stems slender, 3–6 dm, thinly hairy below, becoming tomentose above; lvs erect, 2–5 mm wide, the basal to 3 dm, the upper smaller or bract-like; infl 5–10 cm wide, freely branched, rounded or low-pyramidal, densely white-wholly; fls golden yellow, 1 cm wide, densely woolly outside and at base within; seeds 1 mm, stramineous. Acid swamps and pine-barren bogs; N.S.; N.J. and Del.; N.C. to Fla. and Miss. July, Aug. (*L. americana*; *L. septentrionalis*, the Nova Scotia plants.)

2. **LACHNANTHES** Elliott.   Perianth 6-parted to the top of the ovary, narrowly campanulate, its segments nearly erect, persistent in fr, the sep shorter and narrower than the pet; stamens 3, exsert, erect, attached to the base of the pet; anthers contorted when dry; ovary wholly inferior; fr subglobose; seeds few in each locule, rounded or quadrate, flat, attached at the middle; sap red. Monotypic.

1. **Lachnanthes caroliniana** (Lam.) Dandy.   Red-root. Stems stout, erect, 2–8 dm, nearly glabrous below, tomentose above; lvs erect, 3–10 mm wide, the lower to 4 dm, the cauline smaller or bract-like; infl dense, flat or rounded, 3–8 cm wide, densely woolly; fls 10–12 mm, dingy yellow, densely tomentose; sep lance-subulate, two-thirds as long as the linear-oblong pet; seeds 2–3 mm wide; reddish-brown. Swamps and pine-barren bogs; N.S. to Fla., La., and Cuba; disjunct in w. Va. and c. Tenn. July, Aug. (*L. tinctoria*; *Gyrotheca t.*)

Family **LILIACEAE**, the Lily Family

Fls mostly perfect and regular, (2)3(4)-merous; tep in 2 usually similar and petaloid cycles, often very showy, occasionally (as in *Trillium*) the outer set narrower, greener, and more sepaloid; all tep distinct, or less often all joined below to form a perianth-tube that may also bear a corona; stamens free or adnate to the perianth, usually as many as the tep; ovary superior to inferior, often with septal nectaries, with mostly several or many ovules in each locule, the placentation axile; fr usually capsular, less often a berry or dry and indehiscent; embryo ± linear, with a terminal cotyledon, centrally embedded in the usually very hard, nonstarchy endosperm; perennial, mostly geophytic herbs from a rhizome, bulb, or corm; lvs simple, typically narrow and parallel-veined, but varying to broad and net-veined (as in *Trillium*), often all basal, only rarely (as in *Xerophyllum*) persistent throughout the year. 280/nearly 4000, cosmop. (Alliaceae, Amaryllidaceae, Asparagaceae, Convallariaceae, Funkiaceae, Hemerocallidaceae, Hyacinthaceae, Hypoxidaceae, Melanthiaceae, Trilliaceae, Uvulariaceae). The key is largely artificial and is intended only for our spp.

1 Ovary inferior.
  2 Fls with a conspicuous corona above the perianth.
    3 Corona between the pet and the stamens .... 35. *Narcissus.*
    3 Corona united to the filaments and connecting them .... 37. *Hymenocallis.*
  2 Fls without a corona.
    4 Perianth-segments (usually yellow) pilose outside; lvs usually hairy .... 33. *Hypoxis.*
    4 Perianth-segments (never yellow) and lvs glabrous.
      5 Fls erect, 7–10 cm .... 36. *Zephyranthes.*
      5 Fls nodding, ca 2 cm .... 34. *Leucojum.*
1 Ovary superior.
  6 Fls or infls produced laterally on the stem or its branches.
    7 Lvs reduced to minute scales; ultimate branches filiform .... 32. *Asparagus.*
    7 Lvs broad, flat, green.
      8 Tep united to form a tubular, 6-toothed perianth .... 31. *Polygonatum.*
      8 Tep distinct.
        9 Fl-stalk jointed well above the base; fr a berry .... 27. *Streptopus.*
        9 Fl-stalk not jointed; fr capsular .... 25. *Uvularia.*
  6 Fls or infls terminal, sometimes on a scape.
    10 Lvs all in one or 2 whorls.
      11 Lvs 3, in a single whorl at the top of the stem; fl solitary .... 22. *Trillium.*
      11 Lvs in 2 whorls; fls in a small cluster .... 23. *Medeola.*
    10 Lvs in several whorls, or alternate, or basal, or none at flowering time.
      12 Principal lvs cauline, alternate or whorled, never crowded toward the base.
        13 Stem branched, each branch bearing normal foliage lvs.
          14 Anthers much longer than the filaments; fr capsular .... 25. *Uvularia.*
          14 Anthers shorter than the filaments; fr a berry .... 26. *Disporum.*
        13 Stem not branched below the infl.
          15 Tep less than 2.5 cm.
            16 Fls in a simple raceme; fr a berry.
              17 Tep 4 .... 29. *Maianthemum.*
              17 Tep 6 .... 28. *Smilacina.*
            16 Fls in a panicle.
              18 Tep ca 2 mm; panicle rarely over 1 dm; fr a berry; style very short; stigma
                  obscurely 3-lobed .... 28. *Smilacina.*
              18 Tep 4 mm or more; panicle 2–6 dm; fr capsular; ovary 3-lobed, each lobe
                terminated by a short style.
                19 Stem pubescent; plants perennial from stout rhizomes.
                  20 Tep clawed and glandular at base .... 6. *Melanthium.*
                  20 Tep neither clawed nor glandular .... 7. *Veratrum.*
                19 Stem glabrous; plants perennial from a bulb .... 4. *Stenanthium.*
          15 Tep more than 3 cm.
             21 Lvs numerous on each stem .... 19. *Lilium.*
             21 Lvs only 2 or 3 on each stem .... 20. *Tulipa.*
      12 Principal lvs at or near the base of the plant (sometimes wanting at anthesis), the fl or infl
        borne on a scape or scape-like, merely bracteate stem.
        22 Fl solitary; lvs mostly 2 or 3.
          23 Fl erect; tep ascending-erect .... 20. *Tulipa.*
          23 Fl nodding; tep ± recurved .... 21. *Erythronium.*
        22 Fls several to many; lvs various.
          24 Fls in an apparent umbel or a short, irregularly branched cluster.
            25 Tep less than 3 cm.
              26 Plants with the taste and odor of onion or garlic .... 18. *Allium.*
              26 Plant without the taste or odor of onion or garlic.
                27 Lvs narrowly linear; fr capsular .... 17. *Nothoscordum.*
                27 Lvs oblong or obovate, 2–6 cm wide; fr a berry .... 24. *Clintonia.*
            25 Tep more than 5 cm .... 16. *Hemerocallis.*
          24 Fls in an apparent raceme or panicle or spike-like raceme.
            28 Tep united to about the middle or beyond.
              29 Perianth 1 cm or less.
                30 Perianth smooth outside.
                  31 Lvs elliptic; fls white, campanulate .... 30. *Convallaria.*
                  31 Lvs linear; fls blue or brownish, tubular or urceolate .... 14. *Muscari.*
                30 Perianth rough outside; fls white or yellow, tubular .... 11. *Aletris.*
              29 Perianth 2 cm or more, funnelform .... 15. *Hosta.*
            28 Tep separate nearly or quite to the base.
              32 Lvs elongate-spatulate to obovate; pedicels not bracteate.

33 Fls white, unisexual .... 10. *Chamaelirium.*
33 Fls pink, perfect .... 9. *Helonias.*
32 Lvs broadly or narrowly linear or filiform; pedicels bracteate.
  34 Style 1, short or elongate, or the stigma nearly sessile.
    35 Stems arising from stout rhizomes; fls yellow .... 8. *Narthecium.*
    35 Stems arising from a bulb; fls anthocyanic to white, or partly green.
      36 Tep with a broad green stripe outside .... 13. *Ornithogalum.*
      36 Tep not green-striped .... 12. *Camassia.*
  34 Styles 3, one terminating each lobe of the ovary.
    37 Pedicels bracteate well above the base; lvs evergreen .... 1. *Xerophyllum.*
    37 Pedicels bracteate at base; lvs seasonal.
      38 Tep with 1 or 2 dark or yellowish spots or glands at or near
        the base.
        39 Pet abruptly narrowed into a basal claw .... 6. *Melanthium.*
        39 Pet not clawed .... 5. *Zigadenus.*
      38 Tep not glandular or spotted at base.
        40 Lvs 2-ranked; stems arising from a rhizome .... 2. *Tofieldia.*
        40 Lvs several-ranked; stems arising from a thick bulbous
          base.
          41 Pedicels longer than the fls .... 3. *Amianthium.*
          41 Pedicels much shorter than the fls .... 4. *Stenanthium.*

1. XEROPHYLLUM Michx. Fls perfect; tep 6, spreading, oblong-elliptic, glandless and clawless, withering persistent around the fr; stamens 6, with long filaments and short, basifixed, latrorse anthers; ovary superior, with 3 distinct, elongate, slender, persistent styles stigmatic down the inner side; capsule loculicidal, 3-lobed, with 2(3) oblong, un-appendaged seeds per locule; perennial herbs from a stout, caudex-like rhizome, producing a dense cluster of long, slender, persistent, firm and sharp-tipped lvs, a tall stem beset with similar but smaller lvs, and a dense terminal white raceme. 2, the other in w. N. Amer.

    1. Xerophyllum asphodeloides (L.) Nutt. Turkey-beard. Basal lvs to 4 dm, ca 2 mm wide; stem erect, 8–15 dm; cauline lvs filiform, progressively reduced, the lower 10–15 cm; raceme 5–6 cm thick, compact at first, elongating to as much as 3 dm; pedicels 3–4 cm, at maturity suberect; fls 1 cm wide; fr 5 mm; $2n=30$. Pine-barrens from N.J. s. occasionally to N.C.; mt. woods from Va. to Tenn. and Ga. June.

2. TOFIELDIA Hudson. False asphodel. Fls perfect; tep separate, spreading or ascending, sessile, glandless; stamens 6, hypogynous, with subulate, flattened filaments and ovate or rotund, introrse, basifixed anthers; ovary superior, with 3 short, subulate styles; ovules numerous; capsule septicidal, subtended by the persistent perianth and tipped by the enlarged divergent styles; perennial herbs with several 2-ranked, linear lvs at or near the base, and an erect scape usually bearing a small lf near the middle and terminating in a dense raceme of white or greenish fls. 15, N. Hemisphere and Andes of S. Amer.

1 Pedicels in clusters of 2 or 3, each with 3 small bractlets at the top.
  2 Fr twice as long as the perianth; scape sticky-hairy; n. and interior states .... 1. *T. glutinosa.*
  2 Fr about equaling the perianth; scape short-hairy, not sticky; s. coastal plain ....2. *T. racemosa.*
1 Pedicels solitary, without bractlets at the top; scape glabrous; boreal .... 3. *T. pusilla.*

    1. Tofieldia glutinosa (Michx.) Pers. Basal lvs several, 8–20 cm, to 8 mm wide; cauline lf usually single and bract-like, near the middle of the stem, or absent; scape 2–5 dm, very sticky-hairy upwards, the raceme 2–5 cm; fls white, 2 or 3 together at each node, on sticky-hairy pedicels 3–6 mm, each subtended just below the perianth by small ovate bractlets; tep oblanceolate, 4 mm; fr ovoid, thin-walled, 5–6 mm; seeds fusiform, the body 1–1.3 mm, with a filiform contorted appendage at each end. Moist or wet places; Nf. to Alas., s. to N.Y., Ind., and Calif., and in the mts. to W.Va. and N.C. July. Our plants are var. *glutinosa.*

    2. Tofieldia racemosa (Walter) BSP. Basal lvs erect, 2–4 dm × 3–5 mm; cauline lf usually single and bract-like, below the middle of the stem; scape 3–7 dm, beset with short, stout, spreading hairs, especially upwards, but not sticky, the raceme 5–15 cm; fls ochroleucous, clustered and bracteate as in no. 1, the terminal ones opening first; tep oblong, obtuse, 4 mm; fr narrowly obovoid, 3 mm, firm-walled; seeds narrowly ellipsoid, 2 mm, appendaged at both ends, the body only 1 mm. Wet pine-lands and bogs on the coastal plain; N.J. to Fla. and Tex. July. (*Triantha r.*)

    3. Tofieldia pusilla (Michx.) Pers. Smaller than no. 1; basal lvs numerous, 1–6 cm; cauline lvs 0–2, below the middle of the scape, much reduced; scape glabrous, 5–20 cm, the dense terminal raceme 5–20 mm; tep greenish-white, narrowly oblong-obovate, 1.5–2.5 mm; fr 3 mm; seeds ellipsoid, angular, 0.6 mm. Wet rocks and alpine or arctic meadows; circumboreal, in Amer. s. to Que., Isle Royale, n. Minn., and Mont. July, Aug. (*T. palustris,* an illegitimate name.)

3. AMIANTHIUM A. Gray, nom. conserv.   Fly-poison.   Fls perfect; tep 6, glandless, oblong-obovate, several-nerved; stamens 6, with long, flattened filaments; anthers dorsifixed, introrse, cordate-ovate with confluent thecae; ovary superior, deeply 3-lobed, appearing like separate carpels, each lobe acuminate into a stout conic style with a minute stigma; capsule 3-lobed, septicidal, each lobe tipped by the persistent style; seeds 1 or 2 per locule, oblong, purple-brown; poisonous perennial herb from a thick bulb, with a cluster of long basal lvs, an erect stem bearing a few greatly reduced lvs, and a dense, terminal, bracteate, white (pink) raceme, the persistent tep commonly turning light green. Monotypic.

1. **Amianthium muscaetoxicum** (Walter) A. Gray.   Basal lvs linear, to 4 dm × 2 cm, obtuse or abruptly acute; cauline lvs much reduced or bract-like; stem 3–10 dm; raceme at first conic, becoming cylindric, 3–12 × 2–3 cm; fls 1 cm wide; $2n=32$. Open woods; s. N.Y., Pa., W.Va., Ky., Mo., and Okla., s. to Fla. and Ark. June, July. (*Chrosperma m.*)

4. STENANTHIUM (A. Gray) Kunth, nom. conserv.   Fls perfect or unisexual; tep 6, narrowly lanceolate, long-acuminate, adnate to the ovary-base; stamens 6, much shorter than the perianth, the filaments slender, the anthers obcordate, extrorse, with confluent thecae; ovary superior, ovoid, deeply 3-lobed, each lobe prolonged into a short, stout, outcurved style; capsule 3-lobed, short-cylindric, septicidal; perennial herb, the slender stem bulbous at base; lvs chiefly cauline, linear; infl a terminal white or greenish panicle. Ca 5, N. Amer. and Asia.

1. **Stenanthium gramineum** (Ker Gawler) Morong.   Stem erect, leafy, to 1.5 m; lvs numerous, the larger 2–4 dm × 5–15 mm; panicle 2–5 dm, freely branched, the central axis continued into a wand-like terminal raceme; pedicels and bracts 2 mm; fls varible in size, the tep 4–10 mm; fr 10–15 mm; seeds 3–5 mm; $2n=20$. Moist woods and meadows; Pa., O., Ind., and Mo., s. to Fla. and Ark.; locally intr. in Mich. July–Sept. (*S. robustum,* the form with erect instead of deflexed frs.).

5. ZIGADENUS Michx.   Death-camas. Fls perfect or polygamous; tep 6, all ± alike, with 1 or 2 small flat glands below the middle; stamens 6; filaments slender at least above; anthers extrorse, obcordate, the thecae confluent; ovary 3-lobed, the base sometimes adnate to the base of the perianth, each lobe prolonged into a short divergent style; ovules few; capsule 3-lobed, ovoid to conic, surrounded at base by the persistent perianth; poisonous perennial herbs, the erect stems arising from a bulb or thick rhizome, bearing long linear lvs near the base, reduced or bract-like ones above, and a simple raceme or several racemes or a panicle; fls white or suffused with brown on purple on the lower side. (*Anticlea; Oceanoros; Tracyanthus; Zygadenus,* a later spelling.) 15, N. Amer. and e. Asia.

1 Tep 7–17 mm, bearing a bilobed gland or 2 glands well above the base.
  2 Tep acuminate, bearing 2 glands; plants with a short, stout rhizome (*Zigadenus*) .... 1. *Z. glaberrimus.*
  2 Tep rounded-obtuse; gland bilobed; plants with a bulb (*Anticlea*) .... 2. *Z. elegans.*
1 Tep 3–6 mm, with a single basal gland (*Oceanoros*).
  3 Infl normally paniculate; fls often polygamous .... 3. *Z. leimanthoides.*
  3 Infl a simple raceme; fls perfect .... 4. *Z. densus.*

1. **Zigadenus glaberrimus** Michx.   Stems slender, erect, 3–12 dm from a short, stout rhizome; lvs mostly near the base, 2–5 dm × 6–20 mm, narrowed to the base, strongly nerved, the upper much reduced; panicle 1–6 dm, each branch ascending, bearing a raceme of 3–10 fls; bracts ovate, acuminate, 5 mm; pedicels 5–10 mm; tep 12–17 mm, acuminate, 3 lanceolate, 3 lance-ovate, each with 2 round glands just above the claw-like base; fr conic, acuminate, 12–18 mm; $2n=52$. Bogs and marshes on the coastal plain; Va. to Fla. and La. July, Aug.

2. **Zigadenus elegans** Pursh.   Stems erect, 2–6 dm from tunicated bulbs; herbage glaucous especially when young; lvs mostly crowded toward the base, linear, 2–4 dm, up to 12 mm wide; infl 1–3 dm, paniculate (seldom merely racemose), its branches subtended by large, lance-ovate bracts, these usually suffused with purple and marcescent at anthesis; tep 7–12 mm, oblong-obovate, rounded-obtuse, adnate at base to the ovary-base, white or greenish-yellow, usually strongly suffused with purple or brown toward the base beneath, and with a dark obcordate gland just below the middle on the upper side; fr ovoid, 10–15 mm; seeds angular,

3 mm; $2n=32$. Beaches and bogs and other wet, often calcareous places; Que. to N.Y., w. across n. U.S. and adj. Can. to the cordilleran region from Alas. to n. Mex.; disjunct in the s. Appalachian region (Va. to N.C.) and in the Ozarks (Mo. and Ark.). July, Aug. Our plants, as here described, mostly belong to the e. var. *glaucus* (Nutt.) Preece, called white camas. Along our w. border, in Minn. and Io. var. *glaucus* passes into the more western, mainly cordilleran var. *elegans,* with yellower fls, less consistently glaucous herbage, and often merely racemose infl. (*Anticlea e.*)

3. **Zigadenus leimanthoides** A. Gray. Stems erect, 5–25 dm, from tunicated bulbs with conspicuous fibrous coats; leaves mostly crowded near the base, linear, 2–5 dm × 4–12 mm, the upper reduced or bract-like; panicle pyramidal, 1–4 dm, with rather few spreading or ascending branches naked toward the base and bearing a dense raceme above; bracts ovate, 2 mm; pedicels 5–12 mm; fls white or ochroleucous or chloroleucous, the lower in each raceme fertile, the upper usually at least functionally staminate; tep ovate-elliptic, 3–5 mm, with an inconspicuous small yellow gland at base, sometimes basally adnate to the ovary-base; fr ovoid, 3-lobed, 1 cm. Bogs and wet woods on the coastal plain; s. N.Y. to Ga. and La.; also in the mts. from Va. to Ky., N.C., and Tenn. July, Aug. (*Oceanoros l.*)

4. **Zigadenus densus** (Desr.) Fern. Crow-poison. Much like no. 3, but smaller and more delicate, 4–10 dm, the infl a raceme; bulbs usually smooth; pedicels 10–20 mm; fls perfect. Moist or wet pine-barrens and open swamps on the coastal plain; se. Va. to Fla. and La. May, June. (*Tracyanthus angustifolius.*)

6. **MELANTHIUM** L.   Bunch-fl. Fls polygamous; tep 6, spreading, clawed at base, bearing a pair of glands at the base of the blade; filaments adnate to the claws; anthers extrorse, obcordate, with confluent thecae; ovary superior, 3-lobed, each lobe terminated by a short style; capsule ovoid, septicidal, subtended by the withered perianth; seeds several, elliptic, thick about the center, bordered by a firm wing; poisonous perennial herbs from stout rhizomes, with elongate sheathing lvs and a larger, terminal, pubescent panicle of green or greenish fls. 5, N. Amer.

1 Claw of the tep ca half as long as the flat, obtuse or distally rounded blade .... 1. *M. virginicum.*
1 Claw of the tep two-thirds to fully as long as the undulate-ruffled, ± acute blade .... 2. *M. hybridum.*

1. **Melanthium virginicum** L.   Stem stout, erect from a thick rhizome, 7–15 dm, hairy above; lvs linear, acuminate, the lower much elongate, often 3 dm, 1–2(–3) cm wide; panicle ovoid, 2–3 dm, hairy, its lower branches bearing a raceme of slender-pedicellate fls, the lower fls perfect, the upper staminate; tep spreading, pale green, 6–13 mm, the blade ca twice as long as the claw, flat, oblong or oval or obovate, obtuse, glandular at base; fr ellipsoid, 3-lobed, 10–15 mm; seeds 5–8 mm. Wet woods and meadows; N.Y. to Ind. and Minn., s. to Fla. and Tex. June, July. (*Veratrum v.*)

2. **Melanthium hybridum** Walter.   Much like no. 1; lvs oblanceolate, 3–6 cm wide; tep green or whitish, 4–7 mm, the blade equaling or slightly exceeding the claw, ruffled-undulate, deltoid-ovate, ± acute and often inflexed at the tip, the basal gland prominent; seeds 7–10 mm. Woods; Conn. to S.C. and W.Va. July, Aug. (*M. latifolium.*)

7. **VERATRUM** L.   False hellebore. Fls polygamous; tep 6, free from the ovary, narrowed at base, not glandular; stamens 6; filaments slender, free or adnate to the perianth at the very base; anthers extrorse, obcordate, with confluent thecae; ovary superior, 3-lobed, each lobe terminated by a short style; capsule ovoid, septicidal, subtended by the withered perianth; seeds large, flat, the small embryo surrounded by a broad wing; tall, coarse, poisonous herbs, perennial from a stout rhizome, with large broad lvs and a terminal panicle of green or purple fls. 30, N. Hemisphere.

1 Upper lvs broad; fls hairy; filaments free from the perianth .... 1. *V. viride.*
1 Upper lvs linear; fls glabrous; filaments adnate to the base of the perianth.
  2 Fls dark maroon, all alike; staminate tep obtuse .... 2. *V. woodii.*
  2 Fls greenish, the staminate and fertile different; staminate tep pointed .... 3. *V. parviflorum.*

1. **Veratrum viride** Aiton.   Stem stout, erect, leafy to the top, to 2 m; lvs sessile or nearly so, often somewhat clasping, plicate, oval or elliptic, to 3 dm, half as wide; panicle freely branched, 2–5 dm, hairy; pedicels 2–4 mm; tep yellowish-green, elliptic or broadly lanceolate, 8–13 mm, narrowed to the base, ciliate; filaments erect, shorter than and free from the perianth; fr ovoid, 18–25 mm; $2n=32$. Swamps and wet woods; Que. to Ont., s. to N.C.; Alas. to Oreg. June, July.

2. **Veratrum woodii** Robbins.   Stem 7–15 dm; lower lvs narrowly elliptic to broadly oblanceolate; upper lvs nearly linear, the uppermost much reduced or bract-like; panicle 3–6 dm, slender, the short lower branches bearing mostly staminate fls, the elongate end of the central axis chiefly pistillate; pedicels 3–6 mm; tep dark maroon, 6–9 mm, glabrous, oblanceolate, narrowed to the base, obtuse or subacute; filaments erect, about

equaling the perianth and adnate to it for 0.5 mm at base; fr as in no. 1. Rich woods; O. to Mo. and Okla. July, Aug.

3. **Veratrum parviflorum** Michx.   Stem 6–15 dm; lvs and infl as in no. 2; pedicels to 12 mm; tep greenish or yellowish-green, 4–7 mm, glabrous, conspicuously narrowed to the base, not glandular, those of the staminate fls oblanceolate, acuminate or sharply acute, those of the pistillate fls obovate, obtuse or subacute; filaments curved, a third as long as the perianth-segments and adnate to their base; fr as in no. 1; $2n=16$. Moist wooded slopes in the mts; Va., W.Va., Ky., N.C., and Tenn. Aug. (*Melanthium p.*)

8. **NARTHECIUM** Hudson, nom. conserv.   Bog-asphodel.   Tep 6, linear-oblong, 3-nerved, withering and persistent around the fr; stamens 6, the filaments elongate, filiform, densely woolly, the anthers short, introrse, dorsifix; ovary superior, gradually tapering into a poorly differentiated style; stigma minutely 3-lobed; ovules numerous; capsule elongate, loculicidal; seeds fusiform, prolonged at each end into a long filiform appendage; perennial herbs with stout rhizomes, with linear, basal, equitant lvs and an erect scape bearing a terminal raceme of yellow fls. 4, N. Hemisphere.

1. **Narthecium americanum** Ker Gawler.   Basal lvs erect, 1–2 dm × 1–2 mm; cauline lvs few, bract-like; raceme crowded, 3–6 cm; pedicels bracteate; tep 4–6 mm, the pet scarious-margined; fr fusiform, long-pointed, 10–14 mm; seeds ca 8 mm. Pine-barren bogs; N.J. to S.C.; rare. June, July. (*Abama a.*)

9. **HELONIAS** L.   Swamp-pink. Tep 6, persistent around the fr; stamens 6, the long filaments surpassing the perianth, the blue anthers short-oblong, extrorse; ovary superior, 3-lobed, with many ovules on the axile placentae; styles 3, separate to the base; fr an obcordate, 3-lobed, loculicidal capsule; seeds appendaged at both ends; perennial herb from a very short rhizome, with a cluster of long basal lvs and a hollow scape beset with reduced bract-like lvs and bearing a dense, terminal, spike-like raceme of pink fls. Monotypic.

1. **Helonias bullata** L.   Basal lvs evergreen, elongate-spatulate, becoming 3 dm; scape elongating to 1 m, covered with short bract-like lvs at base, these becoming remote and scale-like above; raceme ovoid, 3–10 cm, ca 3 cm thick; fls fragrant, 1 cm wide; seeds linear, 5 mm; $2n=14$. Swamps and bogs; s. N.Y. and (formerly) N.J. to se. Va. on the coastal plain; Blue Ridge Mts. of Va., N.C., and Ga. Apr., May.

10. **CHAMAELIRIUM** Willd.   Devil's bit. Dioecious; tep 6, narrowly linear-spatulate, 1-nerved; stamens 6, the slender, flattened filaments about equaling the perianth, the short white anthers extrorse; ovary ellipsoid, with many ovules, capped by 3 distinct, slender, style-like stigmas; capsule loculicidal; seeds the fusiform body set obliquely within the white, coarsely cellular wing; perennial from a short, stout, praemorse rhizome, with numerous basal lvs, an erect stem with several smaller lvs, and a terminal raceme. Monotypic.

1. **Chamaelirium luteum** (L.) A. Gray.   Staminate plant 3–7 dm, the pistillate taller, to 12 dm; basal lvs elongate-spatulate to obovate, 8–15 cm; the cauline progressively smaller and narrower, the upper linear; fls white, drying yellowish, the tep 3 mm; staminate spike 4–12 cm × 10–15 mm, with spreading pedicels; pistillate spike very slender, to 3 dm, the pedicels erect or ascending; fr ellipsoid or somewhat obovoid, 7–14 mm; seeds 3–5 mm. Moist woods and bogs; Mass. to Fla., w. to s. Ont., O., s. Ind., s. Ill., Ark., and La. June. (*C. obovale.*)

11. **ALETRIS** L.   Colic-root. Fls perfect; perianth tubular to campanulate, 6-lobed, adnate to the lower part of the ovary; stamens 6; filaments short, inserted at the top of the perianth-tube; anthers linear, introrse; style straight, barely 3-lobed, with minute stigmas; fr a loculicidal capsule enclosed by the persistent, withered perianth; seeds numerous, minute, fusiform; perennial herbs from a short, stout, praemorse rhizome or a mere crown, with a basal rosette of narrow lvs, an erect stem bearing bract-like lvs, and a terminal spike-like raceme; perianth roughened with numerous short, scale-like points, giving it a mealy appearance. 25, N. Amer. and e. Asia.

1 Perianth white, tubular, 8–12 mm .... 1. *A. farinosa.*
1 Perianth yellow, campanulate, 5–7 mm .... 2. *A. aurea.*

1. Aletris farinosa L.   Stem 5–10 dm, beset with scattered linear bracts 3–20 mm; lvs narrowly lanceolate or oblanceolate, 8–20 cm, acuminate; raceme 1–2 dm, its bracts linear or clavate; fls white, tubular, 8–10 mm, the perianth-lobes narrowly triangular, 2–2.5 mm; ovary ca a third inferior, the lower part becoming obscurely 3-lobed and slightly gibbous in fr, the upper part becoming round-ovoid; seeds 0.7 mm; $2n=26$. Sandy soil, open woods, and barrens; s. Me. to Fla., w. to Minn. and Tex. June, July.

2. Aletris aurea Walter.   Much like no. 1; raceme usually looser; fls yellow, campanulate, 5–7 mm, the perianth-lobes triangular, 1.5 mm, nearly as wide; $2n=26$. Pine-barrens, chiefly on the coastal plain; s. Md. to Fla. and Tex. June, July.

12. CAMASSIA Lindl., nom. conserv.

Camas. Fls regular or slightly irregular; tep essentially alike, spreading, separate, sessile or short-clawed, 3–7-nerved; stamens 6, hypogynous; filaments filiform, elongate; anthers linear-oblong, introrse, versatile; ovules numerous; style solitary, slender, the stigma 3-lobed; fr a loculicidal capsule with few to several seeds; perennial from coated bulbs, with several linear basal lvs, an erect scape, and a loose, terminal, bracteate raceme of showy white to blue or violet fls. 5, N. Amer.

1. Camassia scilloides (Raf.) Cory.   Wild hyacinth. Bulb 1–3 cm thick; lvs 2–4 dm × 5–10 mm; scape 3–6 dm, rather stout; raceme many-fld; pedicels 1–2 cm, spreading, about equaling the filiform bracts; tep blue or pale violet to white, 7–12 mm, withering and persistent at the base of the subglobose, transversely veined fr. Prairies and moist open woods; w. Pa. and s. Ont. to s. Wis. and e. Kans., s. to Ga. and Tex.; sometimes casually escaped elsewhere. Apr., May. (*C. angustata; Quamasia hyacinthina.*)

Scilla non-scripta (L.) Hoffsgg. & Link, the English bluebell, has an erect raceme of 6–12 nodding, blue to white fls, the perianth at first tubular, soon becoming campanulate. *S. sibirica* Andrews has ca 3 deep blue fls in a short raceme, the perianth rotate. Both spp. have 1-nerved tep (unlike *Camassia*) and are occasionally found in our range as escapes from cult.

13. ORNITHOGALUM L.

Fls perfect; tep 6, distinct; stamens 6, free from the perianth; filaments broad and flat, at least at base, the anthers oblong, versatile; ovary superior; style and stigma trigonous; capsule loculicidal, obtusely trigonous, with ± numerous seeds in each locule; perennial herbs from a poisonous, coated bulb, with linear basal lvs, an erect or ascending scape, and a bracteate raceme of white fls. 150, temp. Old World.

1 Lower pedicels 2–6 cm; filaments tapering to the tip .... 1. *O. umbellatum.*
1 Lower pedicels up to 1 cm; filaments flat, ending in 2 broad flat teeth .... 2. *O. nutans.*

1. Ornithogalum umbellatum L.   Star of Bethlehem. Bulb renewed each year; lvs elongate, 2–5+ mm wide, with a white stripe on the upper surface; scape 1–3 dm; raceme 3–10-fld, the ascending pedicels longer than the internodes; bracts up to as long as the pedicels; fls erect; tep spreading, lance-oblong, 1.5–2 cm, white above, with a broad green midstripe beneath; ovary longer than the style; $2n=18–108$. Native of Europe, escaped from cult. into roadsides and even fields and woods in our range. May, June.

2. Ornithogalum nutans L.   Bulb progressively renewed over 3–4 years; lvs elongate, 8–15 mm wide, with a white stripe on the upper surface; scape 2–5 dm; raceme elongate, 3–12-fld; bracts surpassing the 5–10 mm pedicels; fls at first erect or ascending, soon nodding; perianth broadly campanulate, the tep oblong, 1.5–3 cm, white above, with a broad green midstripe beneath; ovary shorter than the style; $2n=45$. Native of w. Asia, occasionally escaped from cult. in our range.

14. MUSCARI Miller.

Grape-hyacinth. Perianth tubular or urceolate, shortly 6-toothed; stamens 6, inserted on the perianth-tube, included; stigma 1; fr a loculicidal capsule, distinctly 3-angled or almost winged; seeds angular; perennial herbs from a coated bulb, the linear lvs basal, the short erect scape bearing a dense raceme in the spring. 60, Europe, Medit. reg., w. Asia.

1 Raceme elongate, loose, the terminal fls sterile, on long pedicels .... 1. *M. comosum.*
1 Raceme short and compact at anthesis; pedicels no longer than the fls.
  2 Lvs nearly terete, channeled on one side, 1–3 mm wide, recurved at the tip .... 2. *M. racemosum.*
  2 Lvs flat or channeled, 3–8 mm wide, erect .... 3. *M. botryoides.*

1. Muscari comosum (L.) Miller.   Lvs flat, 2–4 dm × 8–20 mm, long-attenuate; scape 3–7 dm; raceme 1–2 dm, loose; lower fls fertile, brown, 7–9 mm, the pedicels divaricate; terminal fls bright blue, sterile, on erect or ascending pedicels 4–6 times as long as the narrowly tubular perianth; $2n=18$. Native of Europe, rarely intr. in waste places. May. (*Leopoldia c.*)

2. **Muscari racemosum** (L.) Miller.   Blue-bottle. Lvs narrowly linear, nearly terete, channeled along one side, 1–3 mm wide, soon recurved at the tip; scale 1–3 dm; raceme 3–5 cm at anthesis, longer in fr; fls blue, most of them fertile, declined or nodding, exceeding their slender pedicels; uppermost few fls sterile, erect or ascending, somewhat smaller, on similarly short pedicels; perianth obovoid to oblong-urceolate; $2n=18$–72. Native of Europe, now a weed in fields and waste places in our range from Mass. to Mich. and southward. Apr, May. (*M. atlanticum; M. neglectum.*)

3. **Muscari botryoides** (L.) Miller.   Grape-hyacinth. Lvs flat, narrowly linear-oblanceolate, to 2.5 dm, 3–8(–10) mm wide, scapes 1–2 dm at anthesis, to 4 dm in fr; raceme ovoid-cylindric and 2–4 cm at anthesis, elongating in fr; fls all blue, all fertile except a few at the tip, nodding, exceeding their slender pedicels; perianth globular-urceolate, 4–5 mm; $2n=18$, 36. Native of Europe, escaped from cult. into waste places nearly throughout our range. Apr., May.

15. HOSTA Tratt., nom. conserv.   Plantain-lily. Perianth funnelform, the tep united at base into a long tube, erect or spreading above; stamens free from the perianth in most spp.; anthers versatile; ovules numerous; style 1; stigma undivided; perennial herbs from a cluster of thick roots, producing several large, strongly ribbed, plantain-like basal lvs and an erect scape or scape-like stem bearing a terminal bracteate raceme of showy, white to blue or purple fls in summer. (*Funkia.*) 10, China, Japan.

1. **Hosta ventricosa** (Salisb.) Stern.   Scape to 1 m; lvs ovate or cordate, ca 10 cm wide; fls blue or purple, 5 cm, the slender tube abruptly expanded into a campanulate limb; $2n=60$. Native of e. Asia, occasionally escaped from cult. in our range.
Hosta plantaginea (Lam.) Aschers., with cordate or ovate-subcordate lvs and fragrant white fls 10–13 cm, and *H. lancifolia* (Thunb.) Engl., with lanceolate lvs 3–5 cm wide and pale lilac to nearly white fls 4 cm, the tube gradually expanded into the obconic limb, also occasionally escape.

16. HEMEROCALLIS L.   Day-lily. Perianth funnelform, the tep spreading or recurved, connate below into a short tube; stamens inserted at the summit of the tube; filaments elongate, declined; anthers linear, introrse, versatile; ovules numerous; style slender, declined; stigma capitate; tall perennial herbs with numerous long, linear basal lvs and lfless scapes bearing a terminal cluster of large fls, each lasting a single day. 20, temp. Eurasia.

1. **Hemerocallis fulva** (L.) L. Scapes commonly 1 m tall; fls tawny-orange, not fragrant, ca 12 cm wide; sterile triploid, $2n=33$, at least as to the commonly cult. and escaped plants. Long in cult., and freely escaped in our range. June, July.
H. lilioasphodelus L. (*H. flava*), similar, but with fragrant lemon-yellow fls, also occasionally escapes, as do various hybrid cultigens derived from these and other spp.

17. NOTHOSCORDUM Kunth, nom. conserv.   Much like *Allium,* but inodorous; fls perfect; tep 6, all alike; stamens 6; filaments slender, adnate to the base of the tep and half as long; anthers linear, introrse, versatile; ovary shallowly 3-lobed, with 6–10 ovules per locule; style straight, slender; stigmas 3, minute; fr loculicidal; perennial herbs from a coated bulb, with basal linear lvs, an erect scape, and a terminal determinate umbel subtended by 2 membranous bracts. 35, New World.

1. **Nothoscordum bivalve** (L.) Britton.   Bulb ovoid, 2 cm, with membranous coats; lvs few, 1–2 dm × 2–3 mm; scape 1–3 dm; fls 5–12; pedicels 2–5 cm, often unequal; tep white or greenish-white, oblong, 9–13 mm; fr subglobose, 5 mm; seeds black, angularly ellipsoid, 3 mm; $2n=18$. Open woods, prairies, and barrens, usually in moist soil; se. Va. to s. O., c. Ill. and Kans., s. to S. Amer. Mar.–Sept.

18. ALLIUM L.   Onion. Fls perfect; tep 6, uniform in color but often somewhat unlike in shape or size, generally withering and persistent below the fr; stamens 6, often adnate to the base of the tep, the filaments of the epipetalous series often wider, or strongly flattened, or variously toothed; anthers short, introrse; ovules 1 or 2 per locule; capsule short, ovoid to globose or obovoid, 3-lobed, loculicidal; seeds black; herbs from a coated bulb, with a strong odor of onion or garlic, the lvs usually narrow, basal or on the lower part of the stem, the scape-like stem erect, terminated by a determinate umbel subtended by 1–3 bracts; fls white to pink or purple, sometimes replaced by sessile bulblets. 500, N. Hemisphere.

1 Lvs absent at anthesis, when present flat, lance-elliptic, mostly 2–8 cm wide .... 6. *A. tricoccum.*
1 Lvs present at anthesis, linear, rarely as much as 2 cm wide.
  2 Lvs terete or nearly so, hollow at least toward the base.
    3 Inner and outer filaments essentially alike, barely widened below; pedicels shorter than the fls;
      fls not replaced by bulblets .... 7. *A. schoenoprasum.*
    3 Inner filaments broad, flat, terminating in 2 exsert hair-like appendages; pedicels longer than the
      fls; often many fls replaced by bulblets .... 8. *A. vineale.*
  2 Lvs flat or keeled, not hollow.
    4 Bulb-coats persisting as a conspicuous reticulum of anastomosing fibers.
      5 Ovary and fr unappendaged.
        6 Tep withering in fr; fls often replaced by sessile bulblets; widespread .... 1. *A. canadense.*
        6 Tep becoming callous-keeled and permanently investing the fr; fls normal; western .... 2. *A. perdulce.*
      5 Ovary and fr with 3 prominent crests at the summit; widespread .... 3. *A. textile.*
    4 Bulb-coats otherwise, either nonfibrous or with ± parallel, nonreticulate fibers.
      7 Lvs arising near together at the soil-surface; native spp. of woods or prairies.
        8 Scape declined or decurved just below the tip, the umbel nodding .... 4. *A. cernuum.*
        8 Scape not declined, the umbel erect .... 5. *A. stellatum.*
      7 Lvs alternate on the lower half of the stem; intr. spp. of disturbed soil.
        9 Umbels without bulblets .... 9. *A. ampeloprasum.*
        9 Umbels producing bulblets, with or without fls also.
          10 Lvs 2–4 mm wide; bulbs simple; invol bracts 2 .... 10. *A. oleraceum.*
          10 Lvs 6–12 mm wide; bulbs compound; invol bract 1 .... 11. *A. sativum.*

  1. Allium canadense L. Bulb ovoid-conic, 1–3 cm, its coats fibrous-reticulate; stem erect, stout, 2–6 dm, leafy in the lower third, lvs elongate, flat, 2–4(–7) mm wide; bracts 2 or 3, broadly ovate, acuminate; fls when present pink or white, on pedicels 1–3 cm, often many or all of them replaced by sessile bulblets to 1 cm; tep withering in fr; fr not crested; seeds shining, finely alveolate, each alveolus with a minute central pustule; 2*n*=14, 21, 28. Open woods and prairies; N.B. to N.D., s. to Fla. and Tex. (*A. mutabile.*) The common phase in our range is var. *canadense*, with many or all of the fls replaced by sessile bulblets, the fls, when developed, rarely producing fr. The strictly floriferous var. *lavendulare* (Bates) Ownbey & Aase occurs from Mo. to S.D., Okla., and n. Ark. Other floriferous vars. occur to the s. and w.

  2. Allium perdulce S. V. Fraser. Strictly floriferous, without bulblets in the infl; tep persistent, becoming callous-keeled and permanently investing the fr; alveoli of the seed-coat without pustules; otherwise much like no. 1. Sandy soil on the plains; S.D. and w. Io. to Okla., Tex., and N.M. Our plants belong to the relatively northern var. *perdulce*, with fragrant fls and deep rose tep, fading purple.

  3. Allium textile A. Nels. & J. F. Macbr. Bulb ovoid-conic, 1–2 cm with fibrous-reticulate coats; stem 1–2.5 dm; lvs mostly 1 or 2, arising below the middle, 1–2 mm wide, bracts ovate, short-acuminate; bulblets none; pedicels 5–10 mm, or in fr to 20 mm; tep pink or white, lanceolate or lance-oblong, 6–8 mm, acute; filaments shorter than the perianth; fr retuse, 3–4 mm, its valves (and each lobe of the ovary) bearing 2 short erect projections on the back just below the top; 2*n*=14, 28. Dry plains and prairies; s. and w. Minn. and w. Io. to Alta., Wash., and N.M. May, June. (*A. reticulatum.*)

  4. Allium cernuum Roth. Nodding o. Bulb slenderly conic, very gradually tapering into the stem; lvs several, arising near together at the soil-surface, shorter than the stem, 2–4(–8) mm wide; scape 3–6 dm, abruptly declined near the top; umbel nodding (at least in bud), many-fld, without bulblets; pedicels 12–25 mm, becoming rigid; tep white to rose, ovate or elliptic, 4–6 mm, obtuse or subacute; stamens exsert; filaments barely widened at base; fr obovoid, 3-lobed, 4 mm, each valve (and each lobe of the ovary) bearing 2 erect triangular processes near the top; 2*n*=14. Dry woods, rocky banks, and prairies; N.Y. to Mich., Minn., and B.C., s. to Va., Ky., and Mo., and in the mts. to Ga., Ala., and Ariz. July, Aug. (*A. allegheniense; A. oxyphilum* Wherry, of shale-barrens, with few, whitish fls, may merit some recognition.)

  5. Allium stellatum Ker Gawler. Bulb ovoid or ovoid-conic, 1.5–3 cm; stem 3–7 dm; lvs 3–6, arising together at the soil-surface, 1–2 mm wide, keeled; bracts 2, ovate or lanceolate, acuminate; pedicels slender, 1–2 cm; tep pink, ovate to oblong, 4–7 mm, obtuse or subacute; stamens equaling or surpassing the tep; filaments slightly widened below; fr 3-lobed, subglobose, shorter than the perianth, crested as in no. 4; 2*n*=14. Prairies, barrens, and rocky hills; w. Ont. to Ill., w. to Sask., Wyo., and Okla. July–Sept.

  6. Allium tricoccum Aiton. Ramps, wild leek. Bulb ovoid-conic, its coats finely fibrous-reticulate; lvs flat, lance-elliptic, (1–)1.5–3+ dm, withering before anthesis; umbel erect, subtended by 2 ovate deciduous bracts; tep ovate to oblong-obovate, white, 4–7 mm, obtuse, about equaling the stamens; epipetalous filaments widened below; fr depressed, deeply 3-lobed, each valve often gibbous on the back below the middle; 2*n*=16. Rich woods; N.S. and s. Que. to Md. and in the mts. to n. Ga. and n. Ala., w. to e. N.D., s. S.D., and Mo. Two well marked vars.:

  Var. *tricoccum*. Plants larger, the bulb mostly (1.5–)2–3 cm thick, the lvs (3–)5–8+ cm wide, evidently petiolate, the scape arising (12–)25–35 cm above the ground, the bracts subtending the infl 2–3 cm, the fls (15–)30–50+, the fruiting pedicels (1–)1.5–2.5+ cm; sheaths and petioles and often also the scape and bracts commonly anthocyanic; range of the sp.; fl late, commonly July.

  Var. *burdickii* Hanes. Smaller, the bulb mostly 1–1.5(–2) cm thick, the lvs (1.5–)2–4(–4.5) cm wide, less strongly or scarcely petiolate, the scape arising (10–)13–16(–18) cm, the bracts 1–2 cm, the fls (6–)10–18(–25), the fruiting pedicels 1–2 cm; not anthocyanic; widespread, but chiefly midwestern, extending e. to w. Me., N.J., and the mts. of N.C. and Tenn.; fl early, commonly June. (*A. burdickii.*)

7. **Allium schoenoprasum** L.   Bulb slender, often scarcely thicker than the stem; lvs erect, terete, hollow, mostly cauline, 2–4 mm thick, the longest nearly equaling the stout, 2–5 dm stem; umbel compact, hemispheric, subtended by 2 ovate bracts; pedicels 3–7 mm; fls numerous; tep bright pink, ovate to lanceolate, 10–14 mm, acuminate, prominently 1-nerved; filaments all about alike, barely widened at base; fr ovoid, 3-lobed, half as long as the perianth; $2n=16, 24, 32$. Circumboreal, in Amer. s. to Me., n. N.Y., n. Mich., Minn., Colo., and Wash. June, July. Our native plants, as described above, are var. *sibiricum* (L.) Hartman. Var. *schoenoprasum*, a much smaller European plant only 1–2 dm, with the lvs basal and only 1–2 mm thick, is the cult. chives; it rarely escapes in our range.

8. **Allium vineale** L.   Field-garlic; scallions. Bulb round-ovoid, 1–2 cm; stem slender, 3–10 dm; lvs cauline, few–several, nearly terete except at the base, very slender, mostly 1–2 dm, hollow at least below; umbel erect, subtended by a short invol, often producing only bulblets, these ovoid, usually tipped by a long, fragile, rudimentary lf; normal fls on pedicels 1–2 cm, red-purple or pink, sometimes white or greenish; tep acute or acutish; inner stamens with broad flat filaments, ending in 2 exsert hair-like appendages exceeding the anther; $2n=16, 32, 40$. Native of Europe, established as a noxious weed of lawns, fields, and meadows in our range. June.
Allium cepa L., the cult. onion, occasionally escapes. It is a much coarser plant, with the stem inflated below, the lvs basal and 5–15 mm thick; the filaments are flattened only at the base, and lack the hair-like appendages.

9. **Allium ampeloprasum** L.   Wild leek. Bulbs round-ovoid; stem stout, 5–10 dm, bearing a few lvs in the lower half; lvs linear, 5–20 mm wide, sharply keeled; umbel globose, many-fld; bulblets none; pedicels 2–3 cm; tep lavender to red-purple, lance-ovate, acute, 5 mm; inner stamens with broad flat filaments, terminating in 2 hair-like appendages surpassing the anther; $2n=16-80$. Native of Europe, established as a weed at scattered stations in our range.

10. **Allium oleraceum** L.   Field-garlic. Bulbs ovoid-conic, 1–2 cm; stem slender, 3–6 dm; lvs linear, flat, 8–30 cm × 2–4 mm; umbel erect, subtended by 2 elongate linear bracts, bearing a few sessile bulblets and numerous nodding, red-purple or violet fls; stamens usually exsert, the filaments all slender; $2n=32, 40$. Native of Europe, rarely adventive in our range.

11. **Allium sativum** L.   Garlic. Bulbs usually compound; stem 3–8 dm, leafy to or beyond the middle; lf-blades flat, keeled on the back, 6–12 mm wide; umbel erect, producing ovoid bulblets and a few fls, the invol deciduous in one piece, terminating in a beak 3–10 cm; $2n=16$. Native of w. Asia, rarely escaped from cult. in our range.

19. **LILIUM** L.   Lily. Perianth campanulate or funnelform, its 6 tep clawed or sessile, erect to spreading or recurved, often connivent at base, in our spp. spotted with purple toward the base; stamens 6; filaments elongate, often divergent from the style; anthers linear, versatile; style elongate, with 3-lobed stigma; fr a loculicidal capsule with numerous closely packed, flat seeds; tall perennial herbs from a scaly bulb, our spp. with an erect stem bearing numerous narrow lvs (alternate or whorled) and at the summit 1–many erect or nodding, yellow to red fls. 80, N. Temp.
Lilium philippinense Baker has been recorded as casually escaped and established in Ky. It may be recognized by its alternate, grass-like lvs and 1–few horizontal, narrowly funnelform, white or greenish-white fls 17–25 cm.

1 Fls erect; tep distinctly clawed.
  2 Uppermost lvs, as well as the lower, alternate.
    3 Tep strongly papillose toward the base within .... 1. *L. bulbiferum.*
    3 Tep not papillose within .... 2. *L. catesbaei.*
  2 Uppermost lvs whorled .... 3. *L. philadelphicum.*
1 Fls nodding; tep not obviously clawed.
  4 Lvs all or chiefly whorled; no axillary bulblets.
    5 Anthers held close together, barely or not at all exserted from the only slightly or moderately recurved tep.
      6 Tep deep red outside, only slightly spreading at the tip, the anthers not readily visible in side-view; s. Appalachian Mts. .... 4. *L. grayi.*
      6 Tep usually yellow or orange-yellow, seldom nearly red, more evidently (though not strongly) recurved, the anthers barely exserted; widespread .... 5. *L. canadense.*
    5 Anthers well separated on strongly exserted, divergent filaments; tep strongly recurved-reflexed.
      7 Lvs lanceolate to nearly linear, broadest at or below the middle.
        8 Lvs spiculate-scabrous along the margins and the veins beneath; anthers relatively short, mostly 5–15 mm; midwestern .... 6. *L. michiganense.*
        8 Lvs smooth or slightly papillose along the margins; anthers longer, mostly 15–25 mm; Atlantic states .... 7. *L. superbum.*
      7 Lvs oblanceolate to obovate, broadest distinctly above the middle .... 8. *L. michauxii.*
  4 Lvs all alternate, the upper with axillary bulblets .... 9. *L. lancifolium.*

1. Lilium bulbiferum L.   Orange lily. Tall, with numerous linear alternate lvs and usually one erect fl; tep widely spreading, clawed, conspicuously papillose toward the base within; $2n=24$. Native of Europe, occasionally escaped from cult. in the ne. part of our range. June, July.

2. Lilium catesbaei Walter.   Pine-lily. Bulb 1–3 cm, composed of narrow scales; cauline lvs all alternate, erect or appressed, linear or linear-oblong, the lower to 8 cm, the upper progressively smaller; fl solitary, erect, tep conspicuously clawed, the blade red-purple, becoming yellow (or white) and purple-spotted toward the base; fr 3 cm; $2n=24$. Wet pinelands and bogs; se. Va. to Fla. and La. Our plants are var. *longii* Fern.: bulb-scales without prolonged lf-like tips; cauline lvs linear-oblong, acute; tep 7–10 × 2–3.5 cm, the blade twice as long as the claw.

3. Lilium philadelphicum L.   Wood-lily. Bulb 2–3 cm thick; stems erect, 3–8 dm; lvs 5–10 cm, acuminate, at least the uppermost in a whorl of 4–7; fls 1–5, erect; tep red-orange, varying to nearly red or nearly yellow, the blade ovate to lance-ovate, 1–2 times as long as the claw, purple-spotted toward the base, acuminate or bluntly apiculate; $2n=24$. June–Aug. Two well marked vars.: Var. *philadelphicum*, of dry open woods and thickets from N.H. to N.C. and Ky., has the lvs mostly whorled, with or without a few alternate lvs between the whorls, lanceolate to oblanceolate, 10–15(–25) mm wide, the fr 2.5–3.5 cm. Var. *andinum* (Nutt.) Ker Gawler (*L. umbellatum*), in meadows and along shores from O. to Minn., B.C., and N.M. has the lvs mostly alternate, except in the uppermost whorl, narrowly lanceolate to linear, 3–10 mm wide, the fr 4–7 cm.

4. Lilium grayi S. Wats.   Bell-lily. Stem stout, 6–12 dm, smooth; lvs in 3–8 whorls of 4–8, lanceolate, lance-oblong, or narrowly elliptic, the larger to 11 cm, 1–2.5 cm wide, tapering to an acute or obtuse tip, several-nerved; fls 1–8, nodding, narrowly bell-shaped; tep oblong-spatulate, slightly out-curved at the tip, 4–6 cm, deep red outside, lighter red within, becoming yellow in the throat, marked with numerous purple spots; anthers not readily visible in side-view; fr 3–5 cm; $2n=24$. Rich moist woods in the higher mts. of Va., N.C., and Tenn. June, July.

5. Lilium canadense L.   Wild yellow lily. Stem slender, erect, 6–15 dm, smooth; lvs mostly in 6–11 whorls of 4–12, the lowest regularly and a few of the uppermost occasionally alternate; lf-blades lanceolate to linear-elliptic, widest at or below the middle, tapering to both ends, often spiculate-scabrous along the margins and veins beneath, the largest 8–15 cm × 8–20 mm; fls 1–5, nodding from long pedicels; tep narrowly oblanceolate, acuminate, only slightly or moderately recurved, 5–8 cm, yellow or orange-yellow, varying to sometimes nearly red, marked with purple spots within; filaments straight or nearly so, only the juxtaposed anthers evidently exserted; fl-bud subterete; $2n=24$. Moist or wet meadows; Que. and Me. to Md. and in the mts. to Va., w. to O., Ky., s. Ind., and Ala. June–Aug.

6. Lilium michiganense Farw.   Michigan-lily. Much like no. 7, but usually spiculate-scabrous along the margins of the lvs and on the veins beneath; anthers mostly 5–15 mm; $2n=24$. Bogs, meadows, low woods, and wet prairies; w. N.Y. and s. Ont. to Man., s. to Tenn. and Ark. June, July.

7. Lilium superbum L.   Turk's-cap lily. Stout and erect, to 25 dm; principal lvs whorled, the upper alternate; lf-blades lanceolate, tapering to both ends, smooth or slightly rounded-papillose along the margins; fls 1–many, nodding from long, erect or ascending pedicels; tep strongly recurved, lanceolate, 6–9 cm, orange or orange-red, spotted with purple; filaments long-exserted, outcurved distally, the anthers widely separated, mostly 15–25 mm; fl-buds 3-angled; $2n=24$. Wet meadows and low ground; Mass. and se. N.H. to Ga. and Ala. July–Sept.

8. Lilium michauxii Poiret.   Carolina-lily. Stout, 5–12 dm; lvs in whorls of 3–7, or the uppermost 2 or 3 alternate, rather fleshy and glaucous, smooth, oblanceolate to obovate, obtuse or acute, the main ones 6–12 × 1.5–2.5 cm, the upper progressively smaller; fls 1 or 2, rarely more, nodding; tep orange-red, purple-spotted within and becoming yellow in the throat, recurved from about the middle, narrowly lanceolate, acuminate, 7–10 cm; stamens long-exserted. Dry sandy woods of the coastal plain from se. Va. to La.; also in the mts. from Va. and W.Va. southward. Aug. (*L. carolinianum*.)

9. Lilium lancifolium Thunb.   Tiger-lily. Stem stout, 6–12 cm, cobwebby especially above; lvs very numerous, alternate, linear to narrow lanceolate, sessile, the upper rounded and clasping at the base and with axillary bulblets, the lower 10–15 cm, fls several in a raceme, the stout, widely divergent pedicels subtended by small lvs; tep recurved, orange-red with many purple-brown dots, 7–10 cm, bearing a pubescent strip basally within; $2n=24$, 36. Native of e. Asia, often escaped from cult. about dwellings and on roadsides in our range. July, Aug. (*L. tigrinum*.)

20. TULIPA L.   Tulip. Tep 6, essentially alike, separate, erect or ascending at anthesis to form a campanulate or stoutly cylindric fl; stamens 6, filaments shorter than the perianth; anthers oblong-linear, introrse, basifixed; ovules numerous; style short or virtually wanting; capsule loculicidal; perennial herbs from coated bulbs, bearing an erect stem with a few oblong to linear lvs and a large erect, usually solitary fl. 100, temp. Eurasia.

1. Tulipa sylvestris L.   Lvs 2 or 3, linear-oblong, to 2 cm wide; stems 3–5 dm; tep lanceolate, acute, 5 cm, yellow within, the outer set green or greenish outside; $2n=24$, 48. Native of s. Europe, established in Pa. and Md. Apr., May.
   Numerous garden hybrids as well as various species-tulips are cult. and may persist locally, but do not appear to be established.

21. ERYTHRONIUM L.   Trout-lily, fawn-lily. Tep (4–)6, lanceolate, separate to the base but connivent, at anthesis spreading and usually eventually recurved; stamens 6; filaments elongate, flattened below; anthers linear-oblong, basifixed; ovules several to many; style slender below, usually thickened above to the 3 short stigmas; fr an obovoid to oblong loculicidal capsule; perennial from a deep solid corm, the slender stem about half subterranean; lvs borne near the middle of the stem and therefore appearing basal, usually mottled with brown, lanceolate to oblanceolate or elliptic. 15, all N. Amer. except *E. dens-canis* L., the European dog-tooth violet, with narrow, tapering, violet tep.

Colchicum autumnale L., The Autumn-crocus, rarely escapes from cult. Its poisonous bulbs produce in autumn 1–3 rose-purple fls the bases of which are below-ground; the perianth is funnelform with a long (to 2 dm) pedicel-like tube and expands above into a limb 5 cm wide; the lanceolate lvs appear the following spring.

1 Tep 25–50 mm; stoloniform offshoots, if present, produced from the corm.
  2 Perianth normally yellow; stigmas not separate.
    3 Ovary and fr convex or truncate or apiculate at the summit; widespread .... 1. *E. americanum.*
    3 Ovary and fr depressed at the summit; southern .... 2. *E. umbilicatum.*
  2 Perianth normally bluish-white; stigmas separate and outcurved.
    4 Lvs mottled in life, ± flat; tep strongly reflexed at anthesis; frs ± erect .... 3. *E. albidum.*
    4 Lvs not mottled, evidently folded; tep spreading or only slightly reflexed; frs resting on the ground
        .... 4. *E. mesochoreum.*
1 Tep 10–15 mm, pink, stem producing a stoloniform offshoot just below the lvs .... 5. *E. propullans.*

1. Erythronium americanum Ker Gawler.   Extensively colonial; sterile corms numerous, producing a single lf and 2–several stolon-like offshoots; fertile corms few, 2-lvd; lvs mottled in life; scape stout, 1–2 dm; tep 1.5–5 cm, recurved-reflexed at anthesis, yellow, often spotted toward the base within or darker-colored outside, the pet with a small rounded auricle on each side 3–6 mm above the base; ovary and fr convex or truncate or apiculate, the fr commonly held above the ground ± horizontally; stigmas short, scarcely separate, terminating the clavate style; $2n=48$. Moist woods; N.S. and w. Ont. to Minn., s. to Fla. and Ala., commoner eastward. Apr., May.

2. Erythronium umbilicatum Parks & Hardin.   Much like no. 1, but scarcely colonial, the sterile corms without offshoots, or producing only one; pet not auriculate; style slender throughout; ovary and fr umbilicate, the fr lying on or near the ground; $2n=24$. Moist woods and rocky slopes; Va., W.Va. and w. Md. to Ala. and n. Fla. Mar.–May.

3. Erythronium albidum Nutt.   Extensively colonial; sterile corms numerous, producing a single lf and usually 1–3 stolon-like offshoots; fertile corms few, 2-lvd; lvs mottled in life, flat or nearly so; scape stout, 1–2 dm; tep 2.5–5 cm, strongly recurved-reflexed at anthesis, normally bluish-white, varying to light pink, often suffused with green or blue outside, yellow at base within, not auriculate; stigmas stout, separate, divergent from the linear-clavate style; fr ± erect, held off the ground, rounded to slightly apiculate or slightly depressed at the summit; $2n=44$. Moist woods, especially on south slopes; s. Ont. to Minn., s. to Md., DC., Ga., Ky., Mo., and Okla. Mar.–May.

4. Erythronium mesochoreum Knerr.   Much like no. 3, but scarcely colonial, the sterile corms without offshoots, producing a new corm at the base of the old; lvs conduplicate, not mottled; tep spreading or slightly reflexed at anthesis; mature peduncles recurved, the fr resting on the ground, depressed at the summit; $2n=22$. Prairies, pastures, and dry open woods; Io., w. Ill., and e. Nebr. to Ark., Okla., and Tex. Mar., Apr.

5. Erythronium propullans A. Gray.   Extensively colonial; stoloniform offshoots fleshy, appearing laterally on the stem just below the lvs; scape filiform, 4–10 cm; tep 4–6, pink, 10–15 mm; largely sterile; $2n=44$. Rich woods, often on north slopes; se. Minn. Apr., May.

22. TRILLIUM L.   Trillium; wake-robin. Tep distinct, the sep herbaceous and generally green, the pet white or colored; stamens hypogynous; filaments mostly short; anthers linear, basifixed, sometimes with a prolonged connective; ovary trilocular, 3-lobed or 3- or 6-angled or -winged; style short or none; stigmas 3, ± elongate, usually divergent; fr a many-seeded berry; herbs from a usually short and stout rhizome, the erect stem bearing a single whorl of usually ample, ± net-veined lvs and a single large, terminal fl. Most spp. with white fls produce pink or rarely maroon forms, and spp. with maroon fls likewise have occasional forms with yellow, green, or more rarely white fls. 50, N. Amer. and e. Asia. Spp. 5–8 and 11–13 ± confluent.

1 Ovary ± evidently 3-angled or 3-lobed, not winged.
  2 Lvs sessile; fl sessile or on a peduncle up to 3 cm .... 1. *T. pusillum.*

2 Lvs rounded to a petiole 5–10 mm; peduncle mostly 1–5 cm.
   3 Lvs and pet obtuse to subacute .... 2. *T. nivale.*
   3 Lvs acuminate; pet acute or acuminate .... 3. *T. undulatum.*
1 Ovary evidently 6-angled or 6-ridged or 6-winged.
  4 Fl pedunculate.
    5 Stigmas straight or nearly so, slender and of uniform diameter throughout; filaments slender;
      pet white, turning pink in age .... 4. *T. grandiflorum.*
    5 Stigmas strongly recurved, stout, tapering distally; filaments stout.
      6 Filament mostly more than a fourth as long as the anther, often over 2 mm long.
        7 Fl held above the lvs on an erect or declined peduncle; pet normally maroon; anthers 5–
          12 mm.
          8 Pet ascending below, recurved spreading above; southern Appalachian region .... 5. *T. sulcatum.*
          8 Pet spreading from the base; more widespread .... 6. *T. erectum.*
        7 Fl held below the lvs on a short, reflexed or decurved peduncle; pet normally white; anthers
          3–7 mm .... 7. *T. cernuum.*
      6 Filament scarcely a fourth as long as the anther, seldom as much as 2 mm; pet normally
        white .... 8. *T. flexipes.*
  4 Fl sessile.
    9 Anther-connective strongly incurved; filament half to fully as long as the anther; lvs petio-
      late .... 9. *T. recurvatum.*
    9 Anther-connective erect or slightly incurved; filament less than half as long as the anther; lvs
      sessile.
      10 Prolongation of the anther-connective commonly 2–4 mm and nearly as long as the filament;
        stigmas 1.5–2 times as long as the ovary .... 10. *T. sessile.*
      10 Prolongation of the anther-connective up to ca 1 mm, much shorter than the filament;
        stigmas up to about as long as the ovary.
        11 Pet mostly less than 4 times as long as wide, normally of uniform color throughout; lvs
          mostly less than 1.5 times as long as wide.
          12 Fls musk- or spice-scented or foetid, usually containing some purple pigment .... 11. *T. cuneatum.*
          12 Fls strongly lemon-scented, yellow, without purple pigment .... 12. *T. luteum.*
        11 Pet mostly more than 5.5 times as long as wide, usually with the claw purplish and the
          blade greenish; lvs mostly more than 1.5 times as long as wide ....13. *T. viride.*

1. Trillium pusillum Michx. Least t. Slender, 1–3 dm; rhizomes up to 1 cm thick; lvs lance-elliptic or lance-ovate, sessile, at anthesis 2.5–6 cm, acute or subobtuse, about equaling or surpassing the pet; pet white, aging to pink or rose-purple, undulate, 1.5–3 cm, ovate or lance-ovate; filaments broadened at base, up to about as long as the linear anthers; style straight, erect, 2 mm, or nearly obsolete; stigmas recurved; ovary obtusely trigonous, flask-shaped, as also the fr; $2n=10$. Several slightly differing, ± disjunct populations have been segregated as vars. on doubtfully sufficient grounds. Plants from coastal N.C. and S.C. are var. *pusillum*; those of coastal Va. and Md. are var. *virginianum* Fern.; those of montane Va. and adj. W.Va. are var. *monticola* Bodkin & Reveal; those of c. Ky. to n. Ala. and sw. Tenn., and of the Ozark region of s. Mo. and Ark., avg larger than the others, are var. *ozarkanum* (E. J. Palmer & Steyerm.) Steyerm. (*T. o.*)

2. Trillium nivale Riddell. Snow-t. Diminutive; stem 8–15 cm at anthesis; lvs elliptic to elliptic- or ovate, at anthesis 3–5 cm, acute or usually obtuse, rounded at base to a petiole 5–10 mm; peduncle ± erect, 1–3 cm, recurved in fr; sep lanceolate, much shorter than the pet; pet white (sometimes pinkish at base), elliptic or elliptic-obovate, 2.5–4 cm, obtuse; anthers 7–10 mm, somewhat exceeding the slender filaments; ovary subglobose, roundly 3-lobed. Rich moist woods; w. Pa. and W.Va. to Minn., s. S.D., w. Nebr., and Mo. Mar., Apr.

3. Trillium undulatum Willd. Painted t. Stem 2–4 dm at anthesis; lvs ovate, thin, 5–10 cm at anthesis, much larger at maturity, sharply acuminate, broadly rounded to a 5–10 mm petiole; peduncle 2–5 cm, erect or declined; sep lanceolate, shorter than the pet; pet white with a streaky purple crescent near the base, lance-ovate to obovate-oblong, undulate, 2–4 cm, acute or acuminate; filaments very slender, longer than the extrorse anthers; ovary pale, 3-lobed; fr erect, 1.5–2 cm; $2n=10$. Moist or wet woods and streambanks; Que. and Ont. to N.J. and Pa., s. in the mts. to Tenn. and Ga., and w. to Mich. and Wis. May, June.

4. Trillium grandiflorum (Michx.) Salisb. Big white t. Stem 2–4 dm at anthesis; lvs ovate or rhombic or subrotund, at anthesis 8–12 cm, short-acuminate, narrowed from below the middle to an acute base (rarely short-petioled); peduncle erect or declined, 5–8 cm, the fl held above the lvs; sep lanceolate, ascending-spreading, 3–5 cm; pet normally white, turning pink in age, ascending from the base, spreading above, obovate or elliptic, evidently longer than the sep, 4–8 cm, acute; filaments more than half as long as the introrse anthers and scarcely (if at all) wider, the whole stamen 15–25 mm; stigmas slender, erect, slightly shorter than the white, 6-winged and -sulcate ovary; $2n=10$. Rich moist woods; Que. and Me. to Minn., s. to Pa., O., and Ind., and in the mts. to n. Ga. and ne. Ala. Apr., May.

5. Trillium sulcatum Patrick. Barksdale's t. Much like no. 6; fls scarcely foetid; peduncle longer, mostly 5–9+ cm and more than twice as long as the distally sulcate sep; pet ascending below, concealing the ovary from side-view, recurved-spreading above; filaments longer, ca half or two-thirds as long as the anthers, which usually surpass the ovary; $2n=10$. Moist rich woods; s. Appalachian region, from w. Va., s. W.Va., and e. Ky. to nw. Ga. and ne. Ala. Apr., May.

6. Trillium erectum L.   Purple t.; stinking Benjamin. Stem 2–4 dm; lvs usually broadly rhombic, about as wide as long, narrowed with nearly straight margins from near the middle to the acute base; peduncle erect to lateral, mostly 3–8 cm, up to ca twice as long as the sep, the ± foetid fls held ± above the lvs; sep lanceolate to lance-ovate or seldom ovate, flat or weakly sulcate-tipped, ca equaling or somewhat shorter than the pet; pet normally maroon (white, yellow, or green), lanceolate to lance-ovate or seldom ovate, 2.5–6 cm, acute, widely spreading from the base, exposing the dark ovary; stamens shorter than to somewhat longer than the ovary, the filaments mostly a fourth to half as long as the anthers; fr dark red, ovoid-globose, with 6 wing-ribs; $2n=10$. Moist woods; Que. and Ont. to Md. and O. and reputedly ne. Ill., s. to the mts. of N.C., Ga., and Tenn. Apr., May.

7. Trillium cernuum L.   Nodding t. Stem 2–4 dm; lvs broadly rhombic-ovate, 6–10 cm at anthesis, acuminate, narrowed from near the middle to an acute base and obscurely petioled; peduncle 1–5 cm, reflexed or recurved below the lvs; sep lance-acuminate, about equaling the pet; pet normally white, 1.5–2.5 cm; anthers 3–7 mm, a fourth or a third longer than the filaments; ovary white or pinkish; $2n=10$. Moist or wet woods in May and June. Var. *cernuum,* with lance-oblong, usually acute pet 5–10 mm wide, and with anthers 3–5 mm long, occurs from Nf. and Que. to Md., and Del., w. occasionally to Mich. Var. *macranthum* A. J. Eames & Wieg., with oblong-oval to obovate, often obtuse pet 10–18 mm wide, and with the anthers 4–7 mm, occurs from Vt. to Minn. and Sask., s. to Md., mainly allopatric with var. *cernuum.*

8. Trillium flexipes Raf.   Bent t. Stem 2–4 dm at anthesis; lvs broadly rhombic, at anthesis 8–15 cm, at least as wide as long, abruptly acuminate, narrowed from near the middle to a sessile base; peduncle 4–12 cm, nearly horizontal to somewhat declined; sep lanceolate, about equaling the pet; pet normally white, spreading, lance-ovate to ovate, usually obtuse, 2–5 cm; filaments short, seldom over 2(–4) mm, mostly less than a fourth as long as the whitish, 6–15 mm anthers; ovary white or pinkish, flask-shaped, sharply 6-angled; $2n=10$. Moist or wet woods, often in calcareous soil; Mich. to Minn., s. to n. Ark., n. Ala., and nw. Ga., and rep. from c. N.Y. Apr.–June. (*T. declinatum,* a preoccupied name; *T. gleasoni.*)

9. Trillium recurvatum Beck.   Rhizome horizontal, relatively long and slender; lvs with a distinct petiole 1–2 cm, the blade elliptic to ovate or subrotund, acute or short-acuminate, usually mottled; fl sessile; sep lance-triangular, commonly reflexed at anthesis, 2–2.5 cm; pet erect or arching inward, 2–3 cm, normally maroon, distinctly and slenderly clawed, the blade lanceolate to ovate, acute or short-acuminate; filaments half to fully as long as the distally incurved anther; stigmas slender, widely divergent, about as long as the 6-winged ovary; $2n=10$. Moist woods; w. O. to s. Mich., s. Wis., and e. Io., s. to Ala., La., and e. Tex. Apr., May.

10. Trillium sessile L.   Toadshade. Rhizome very thick and stout; stem 1–3 dm at anthesis; lvs broadly ovate to rotund, sessile, rounded, broadly obtuse or obscurely cuspidate, usually only inconspicuously or not at all mottled, seldom over 9 cm at anthesis; fl sessile, foetid; sep spreading or ascending, lance-ovate, 2–3 cm; pet normally maroon (yellow or green), ascending, narrowly to broadly elliptic, narrowed to a sessile base, equaling or slightly longer than the sep and twice as long as the stamens; filaments (2–)3–4(–5) mm, less than half as long as the introrse anthers, the flattened connective with a prolongation nearly as long as the filament; stigmas 1.5–2 times as long as the subglobose, sharply 6-angled or -winged ovary; $2n=10$. Rich moist woods; w. Pa. and sw. N.Y. to s. Mich., Ill., Mo., and e. Kans., s. to Md., Va., n. N.C., n. Ala., and n. Ark. Apr., May.

11. Trillium cuneatum Raf.   Rhizome very thick and stout; stem 2–4 dm at anthesis; lvs sessile, broadly to very broadly ovate-elliptic, mottled, 7–14 cm; fl sessile, musk- or spice-scented or foetid; sep spreading, lanceolate, mostly 3–3.5 cm; pet ± erect, maroon (varying to yellow), oblanceolate to elliptic-obovate, (3.5–)4–7(–9) cm, acutish to blunt at the tip; usually cuneate at base; filaments short, mostly 2–4 mm, basally dilated; connective shortly (to 1 mm) or scarcely prolonged; stigmas equaling or shorter than the 6-ridged, ovoid ovary; $2n=10$. Rich woods; c. N.C. and s. Ky. to Ga. and Miss., more widespread than the next. Apr., May (*T. hugeri.*)

12. Trillium luteum (Muhl.) Harbison.   Yellow t. Much like no. 11, but with lemon-scented, bright yellow (or initially greenish-yellow) fls; $2n=10$. Rich woods; s. Ky. and e. Tenn. to w. N.C., n. Ga., and n. Ala. Apr., May. Note that occasional yellow-fld forms occur in other spp.

13. Trillium viride Beck.   Green t. Rhizome thick and stout; stem 2–4 dm at anthesis; lvs sessile, narrowly to broadly elliptic, blunt to acute, 8–12 cm, not or obscurely mottled, minutely white-speckled on the upper side by the thickly scattered stomates (unique among our spp.); fl sessile, with the odor of rotten apples; sep lanceolate, spreading or recurved, 3.5–5 cm; pet erect, narrowly spatulate, 4–6 cm, usually greenish or yellowish beyond the more purplish claw; filaments 3.5–5+ mm, basally dilated; connective shortly (to 1 mm) or scarcely prolonged; stigmas equaling or a little shorter than the distally sharply 6-ridged or -winged, ovoid-ellipsoid ovary; $2n=10$. Rich woods and moist rocky slopes; w. Ill. and e. Mo. Apr., May.

23. MEDEOLA L.   Indian cucumber-root.

Tep 6, essentially similar, separate, recurved; stamens hypogynous; filaments slender; anthers oblong, extrorse, versatile; ovary ovoid, with 3 ovules per locule; styles 3, separate almost to the base, recurved-spreading, stigmatic along the inner side; fr a globose few-seeded berry; herb from a thick, tuber-like horizontal rhizome, the slender stem with 2 whorls of lvs and a sessile, few-fld, terminal umbel of greenish-yellow fls. Monotypic.

1. Medeola virginiana L.   Rhizome 3–8 cm, succulent, cucumber-flavored; stem erect, 3–7 dm, ± invested when young with flocculent wool which persists about the lf-bases; lower lvs 5–11 per whorl, oblong-oblanceolate, 6–12 cm, a fourth or a third as wide, acuminate at both ends; upper lvs normally 3 per whorl, ovate, 3–6 cm, half to two-thirds as wide, acuminate, becoming bright red at the more rounded base; umbel 3–9-fld; pedicels spreading or deflexed, 15–25 mm, becoming red; tep 7 mm; berry dark purple or black; $2n=14$. Rich woods; N.S. and Que. to Mich. and se. Wis., s. to Va. and n. Mo., and in the mts. to Ga. and Ala.; also in Fla. and reputedly se. La. May–July.

24. CLINTONIA Raf.   Bead-Lily.

Fls perfect; tep 6, distinct, narrow; stamens 6, inserted at the base of the perianth; filaments slender; anthers oblong, laterally dehiscent; ovary superior, 2–3-locular, with 2 or more ovules per locule; style long and slender; stigma obscurely lobed; fr a berry with few–several seeds; rhizomatous herbs with 2–5 ample basal lvs, the bases of which sheathe a lfless erect scape bearing a determinate umbel of showy fls. 6, N. Amer. and e. Asia.

1 Fls yellow, mostly nodding; berries normally blue .... 1. *C. borealis.*
1 Fls greenish-white, ± erect; berries normally black .... 2. *C. umbellulata.*

1. Clintonia borealis (Aiton) Raf.   Lvs 2–5, dark glossy green, oblong to elliptic or obovate, eventually to 3 dm, abruptly acuminate, ciliolate; scape 1.5–4 dm, usually hairy above at least when young; pedicels 1–3 cm, erect in fr, softly hairy; fls 3–8, nodding, the tep greenish-yellow, narrowly oblong, 15–18 mm; ovules 10 or more per locule; fr blue (white), 8 mm thick; $2n=28, 32$. Rich moist woods and wooded bogs; Lab. and Nf. to Man. and Minn., s. to N.J., Pa., and n. Ind., and in the mts. to N.C. and Tenn. May, June.

2. Clintonia umbellulata (Michx.) Morong.   Much like no. 1; fls mostly 10–24, ± erect, the tep 8–10 mm, greenish-white, spotted with purplish-brown; ovules 2 per locule; fr black, 6–8 mm thick; $2n=28$. Mt. woods.; c. N.Y. and e. O. to N.C., Tenn., and n. Ga. May, June. (*Xeniatrum u.*)

25. UVULARIA L.   Bellwort, merrybells.

Fls perfect; tep 6, elongate, distinct; stamens 6; filaments short; anthers elongate-linear, extrorse, longitudinally dehiscent; ovary shallowly 3-lobed, with several ovules per locule; styles separate or united to beyond the middle; capsule loculicidal, obovoid and 3-lobed or 3-winged; seeds subglobose; herbs from a slender rhizome, the erect stem forked above the middle, the lower part bearing a few bladeless sheaths and 0–4 lvs; lvs sessile or perfoliate, reaching full size after anthesis; fls stramineous to yellow or greenish-yellow, nodding, terminal but appearing axillary by prolongation of the branches. 5, N. Amer.

1 Lvs perfoliate.
  2 Tep glabrous .... 3. *U. grandiflora.*
  2 Tep glandular-papillose on the inner surface .... 4. *U. perfoliata.*
1 Lvs merely sessile.
  3 Styles separate to well below the middle .... 1. *U. puberula.*
  3 Styles separate only in the upper third or fourth .... 2. *U. sessilifolia.*

1. Uvularia puberula Michx.   Much like no. 2 at anthesis; stem puberulent in lines below the lf-bases; lvs at first rough-margined, green beneath, at maturity minutely serrulate, rather rigid, smooth and shining, rounded at base, often persistent through the winter; tep greenish-yellow, 15–27 mm; styles separate to well below the middle, about equaling to much exceeding the anthers; fr 3-angled, sessile, 20–30 mm; $2n=14$. Mt. woods; Md., sw. Pa., Va. and W.Va. to Ga.; also on the coastal plain from L.I. to S.C. in a form with usually smooth stem and thinner lvs, sometimes segregated as var. *nitida* (Britton) Fern. May. (*Oakesia puberula; Oakesiella puberula; U. pudica.*)

2. Uvularia sessilifolia L.   Stem 1–3 dm at anthesis, bearing 0–2 lvs below the fork and only 1 or 2 fls; lvs at anthesis lance-oblong, acute at both ends, glaucous beneath, at maturity elliptic, to 8 × 3 cm, obtuse or rounded at base, nearly smooth on the margins, tep pale stramineous, 12–25 mm; styles connate for three-fourths their length; about equaling the perianth, much exceeding the anthers; fr 3-angled, distinctly stipitate, 15–20 mm; $2n=14$. Woods; N.S. and N.B. to Minn. and S.D., s. especially in the mts., to S.C., w. Fla. and La. Apr., May. (*Oakesia s.; Oakesiella s.*)

3. Uvularia grandiflora J. E. Smith   Stem 2–5 dm at anthesis, at maturity to 1 m, forking above, bearing (0)1(2) lvs below the fork, 4–8 on the sterile branch, and several lvs and 1–4 fls on the fertile branch; lvs perfoliate, broadly oval to oblong, to 12 cm, usually minutely hairy beneath; fls yellow, nodding; tep 2.5–5 cm, acute or acuminate, smooth within; $2n=14$. Rich woods, preferring calcareous soil; Me. and s. Que. to Minn. and N.D., s. to Conn., Va., n. Ga., Ala., and Okla. Apr., May.

4. Uvularia perfoliata L.   Stem 2–4 dm at anthesis, later taller, forking above, bearing 2–4 lvs below the fork, 1–4 on the sterile branch, and several lvs and 1–3 fls on the fertile branch; lvs perfoliate, oval to lance-oblong, to 9 cm, glabrous beneath; fls yellow; tep 17–25 mm, acute or acuminate, conspicuously glandular-papillose within; $2n=14$. Moist woods, preferring acid soils; s. N.H. to s. Ont., and O., s. to S.C., Ga., w. Fla. and e. Tex. May, June.

26. DISPORUM Salisb.   Fairy-bells.   Fls perfect; tep 6, elongate, similar, distinct; stamens 6, inserted on the base of the perianth; filaments elongate; anthers linear-oblong, extrorse; ovules 2–6 per locule; style with 3 short, spreading stigmas, or undivided; fr a trilocular, ellipsoid to obpyriform, few-seeded berry; rhizomatous herbs, the stem forked above and bearing several sessile, reticulate-veiny lvs; fls rotate or campanulate, drooping or spreading, solitary or in small umbel-like clusters of 2 or 3 terminating the branches. 15, Asia and N. Amer. (*Prosartes,* the 5 N. Amer. spp., perhaps to be held as distinct.)

1 Tep spotted with purple; fr stramineous, obconic or obpyriform .... 1. *D. maculatum.*
1 Tep unspotted; fr red, ellipsoid.
  2 Lower lf-surface lanulose, the veins densely so; e. U.S. .... 2. *D. lanuginosum.*
  2 Lower lf-surface ± uniformly scabridulous or puberulent; w. U.S. and n. Mich. .... 3. *D. hookeri.*

1. Disporum maculatum (Buckley) Britton.   Stem 2–8 dm, forked above, hairy when young; lvs sessile, thin, oblong or ovate-oblong, 4–10 cm, thinly hairy beneath, commonly abruptly acuminate; fls in subsessile clusters of 1–3 ending the branches, the pedicels 6-15 mm; tep white, spotted with purple, lance-ovate, 15–25 × 4.5–7 mm; stamens nearly as long as the tep, the filaments 13–18 mm; ovary obpyriform or obconic, stramineous, covered with stellate-glandular hairs; ovules 2–4 per locule; fr 3-lobed, yellow, densely hairy; $2n=12$. Rich mt. woods; irregularly from nw. Ga. and ne. Ala. to w. N.C., e. Tenn., w. Md., e. Ky., and s. O.; disjunct in s. Mich. May (*D. cahnae; D. schaffneri.*)

2. Disporum lanuginosum (Mich.) Nicholson.   Stem 4–9 dm, dichotomously forked above, hairy when young; lvs sessile, thin, ovate or lance-ovate, 5–12 × 2–5 cm, acuminate, lanulose beneath, the veins densely so; fls in subsessile clusters of 1–3 ending the branches, the pedicels 15–30 mm, tep yellowish-green, unspotted, narrowly lanceolate, 15–25 × 2.5–4 mm; stamens half to two-thirds as long as the perianth, the filaments 6–10 mm; ovary narrowly ellipsoid, glabrous or slightly hairy; ovules pendulous, 2 per locule; stigma trifid; fr red, glabrous; $2n=18$. Rich woods; N.Y. to s. Ont., s. to Ga. and Ala., chiefly in the mts. May, early June.

3. Disporum hookeri (Torrey) Nicholson.   Much like no. 2; lvs long-acuminate, puberulent on the lower surface; tep creamy-white, short-acuminate, widest at or above the middle; stamens usually well exserted; ovary and lower half of the style usually hairy; stigma subentire; $2n=18$. Woods; B.C. to Calif. and w. Mont.; disjunct in n. Mich. May, June. Ours is var. *oreganum* (S. Wats.) Q. Jones.

27. STREPTOPUS Michx.   Twisted stalk. Perianth campanulate to rotate, the tep separate, essentially alike but the outer whorl usually slightly the wider; stamens 6, adnate to the base of the tep; filaments widened at base; anthers oblong to linear, apiculate or aristate; ovary 3-locular with many ovules; style slender (in our spp.), 3 cleft to entire; fr a red, ellipsoid to globose, many-seeded berry; rhizomatous herbs, often branched, with alternate, sessile or clasping lvs and small, greenish-white to pink, purple, or dark red, solitary or paired axillary fls, the axillary peduncle adnate to the stem over the next internode above its origin, with the free part jointed or geniculate well above its base. 10, Eurasia and N. Amer.

1 Lvs cordate-clasping; nodes glabrous .... 1. *S. amplexifolius.*
1 Lvs merely sessile; nodes pubescent .... 2. *S. roseus.*

1. Streptopus amplexifolius (L.) DC.   Stem 4–10 dm, glabrous except near the base; lvs ovate-oblong or lance-ovate, acuminate, cordate-clasping, entire or very minutely toothed, the main ones 6–12 × 2–5.5 cm; free part of the peduncle and pedicel together 3–5 cm, jointed at about two-thirds of its length, above the joint 1-fld (or sometimes 2-fld) and abruptly deflexed or twisted; tep greenish-white, 1 cm, spreading from near the middle; anthers 1-pointed, those of the outer series much longer than the filaments; stigma entire or barely 3-lobed; fr red, mostly ellipsoid, 1.5 cm; $2n=32$. Rich moist woods; circumboreal, in Amer. s. to Mass., N.Y., Mich., Wis., and Minn., and in the mts. to N.C. and Ariz. Our plants are var. *americanus* J. A. Schultes.

    S. × *oreopolus* Fern. is a sterile triploid hybrid between our 2 spp., differing from both in its dark wine-red fls.

2. Streptopus roseus Michx.   Stem simple or in larger plants branched, 3–8 dm, sparsely and finely hairy, especially at the nodes; lvs lanceolate or lance-ovate, acuminate, broadly rounded at the sessile base, finely ciliate, the main ones 5–9 × 2–3.5 cm; peduncle and pedicel combined 1–3 cm, jointed at or below the middle, always 1-fld; tep pink with darker red streaks, 1 cm, spreading only near the tip; anthers each double-pointed; lobes of the style nearly 1 mm; fr red, subglobose, 1 cm; $2n=16$. Rich woods; Lab. and Nf. to n. Minn., s. to N.J., Pa., and s. Mich., and in the mts. to N.C.; also in the Pacific states and w. Can. May–July. Three ill-defined vars. in our range: Var. *roseus,* of the s. Appalachians, from Pa. to N.C., has wholly glabrous pedicels; the other 2 vars. have ± ciliate pedicels. Var. *perspectus* Fassett, widespread in our range, but more common eastward, has short, matted rhizomes. Var. *longipes* (Fern.) Fassett, chiefly midwestern, has elongate rhizomes.

28. SMILACINA Desf., nom. conserv.   False Solomon's seal. Perianth regular, the 6 tep spreading, equal, distinct; stamens 6, hypogynous; filaments slender; anthers ovate, introrse; ovary globose, with 2 ovules per locule; style very short; stigma obscurely 3-lobed; fr a globose berry with usually only 1 or 2 seeds; herbs from a long creeping rhizome, the erect or ascending stems bearing alternate, sessile or subsessile lvs and a terminal raceme or panicle of small white fls. 25, N. Amer. and Asia. (*Vagnera.*)

1 Fls in a panicle .... 1. *S. racemosa.*
1 Fls in a raceme.
  2 Lvs 6 or more, hairy beneath; raceme nearly sessile .... 2. *S. stellata.*
  2 Lvs 1–4, glabrous; raceme long-peduncled .... 3. *S. trifolia.*

1. Smilacina racemosa (L.) Desf.   Stem usually curved-ascending, 4–8 dm, finely hairy; lvs spreading horizontally in 2 ranks, elliptic, 7–15 × 2–7 cm, obtuse or rounded at base, short-acuminate, finely hairy beneath; panicle pedunculate or rarely sessile, 3–15 cm; fls very numerous, short-pediceled, 2–5 mm wide; fr red, dotted with purple; $2n=36, 72, 144$. Rich woods; N.S. to B.C., s. to Ga. and Ariz. May, June. (*Vagnera r.*)

2. Smilacina stellata (L.) Desf.   Stem ascending or usually erect, 2–6 dm, finely hairy or glabrous; lvs spreading or more often strongly ascending, usually folded, sessile and somewhat clasping, lanceolate or lance-oblong, 6–15 × 2–5 cm, gradually tapering to the acute tip, finely hairy beneath; raceme short-peduncled or subsessile, 2–5 cm, with few–several fls 8–10 mm wide; fr 6–10 mm, at first green with blackish stripes, becoming uniformly dark red; $2n=36$. Moist, especially sandy soil of woods, shores, and prairies; Nf. to B.C., s. to N.J., Va., Ind., Mo., and Calif. May, June. (*Vagnera s.*) Coastal plants of se. Can., s. reputedly to Conn., tend to have fleshy, crowded, imbricate, obtuse lvs, and have been segregated as var. *crassa* Victorin.

3. Smilacina trifolia (L.) Desf.   Slender, erect, 1–4 dm at anthesis; lvs 1–4, commonly 3, sessile, oval to oblong or lanceolate, 6–12 × 1–4 cm, acute or acuminate, glabrous; raceme long-peduncled, surpassing the lvs; fls 3–8, 8 mm wide; fr dark red; $2n=36$. Wet woods and sphagnum-bogs; Nf. and Lab. to Mack. and B.C., s. to N.J., O., Mich., and Minn.; also in n. Asia. May–June. (*Vagnera t.*)

29. MAIANTHEMUM Wiggers, nom. conserv.   Tep 4, distinct, spreading; stamens 4, hypogynous; filaments slender; anthers introrse; ovary bilocular, with 2-lobed style and 2 ovules per locule; berry globose, 1–2-seeded; low herbs from a slender rhizome, the erect stem with a few lvs and a short terminal raceme of small white fls. 3, N. Temp.

1. Maianthemum canadense Desf.   Canada mayfl. Stem 5–20 cm; lvs (1)2(3), short-petioled to sessile, 3–10 cm, ovate to ovate-oblong, usually cordate at base; raceme erect, 2–5 cm; fls 4–6 mm wide; fr pale red, 3–4 mm thick; $2n=36$. Moist woods; Lab. and Nf. to Mack., s. to Md., S.D., and in the mts. to Ky. and N.C. May, June. The common form in most of our range, extending w. to Minn., has the lvs glabrous beneath, entire or very minutely crenulate, the transverse veins usually developed; this is var. *canadense.* The chiefly more w. var. *interius* Fern., which occasionally extends e. to O., w. N.Y., and even Mass., has the lvs hairy beneath and distinctly ciliate, with the transverse veins more obscure.

30. CONVALLARIA L.   Lily of the valley. Fls perfect; tep united into a globose-campanulate perianth with 6 short recurved lobes; stamens 6; filaments short, inserted on the perianth near its base; ovary with several ovules in each of the 3 locules; style straight, included; stigma scarcely lobed; fr a many-seeded berry; rhizomatous herb, the short stem bearing a few lfless sheaths and 2 or 3 broad lvs, the scape terminating in a bracteate raceme. Monotypic.

1. Convallaria majalis L.   Lvs narrowly elliptic, to 2 dm, acuminate; scape 1–2 dm; raceme loose, secund; bracts small, lanceolate; pedicels drooping; fls white, fragrant, 6–9 mm; fr (seldom produced) red, 1 cm; $2n=38$. Widespread in n. Eurasia, commonly cult. and escaped near gardens; also in rich mt. woods of Va., W.Va., N.C., and Tenn. May. The S. Appalachian plants, seemingly native, are only loosely colonial and are said to differ in other minor ways from the more densely colonial cult. and casually escaped plants.

Sometimes treated as a distinct sp. under the name *C. montana* Raf. or *C. majuscula* Greene, they may reflect an early escape of a different phase of the sp. from cult.

31. POLYGONATUM Miller.　Solomon's seal.　Perianth regular, tubular, shortly 6-lobed; stamens included, inserted on the perianth-tube; anthers introrse, oblong-linear or sagittate; ovary ovoid to globose, 3-locular; ovules numerous in each locule; style slender, shorter than the perianth; stigma capitate, obscurely lobed; fr a dark blue or black, several-seeded berry; herbs from a knotty rhizome with conspicuous large scars marking the position of previous stems; stem erect or arching, bearing in the upper part numerous alternate lvs in 2 ranks and short, axillary, 1–15-fld peduncles with pendulous, white to greenish or yellow fls. 50, N. Temp.

1 Lvs hairy on the smaller veins beneath .... 1. *P. pubescens.*
1 Lvs glabrous .... 2. *P. biflorum.*

　　1. Polygonatum pubescens (Willd.) Pursh.　Stem slender, 5–9 dm, mostly erect; cauline bract papery, caducous; lvs narrowly elliptic to broadly oval, 4–12 × 1–6 cm, narrowed below to a short petiole, glabrous above, glaucous and hairy on the smaller veins beneath, with 3–9 prominent nerves, peduncles slender, sharply deflexed, 1–2(–4)-fld; pedicels usually shorter than the peduncle; fls 10–13 mm, yellowish-green; $2n=20$. Moist woods and thickets; N.S. to se. Man., s. to Md., Ind., and Minn., and in the mts. to n. Ga. May–July. (*P. biflorum,* misapplied; *P. boreale.*)
　　2. Polygonatum biflorum (Walter) Elliott.　Stem slender or stout, 4–12(–25) dm, arching or erect; cauline bract foliaceous, persistent; lvs sessile and often clasping, lance-elliptic to broadly oval, 5–15 × 1–8 cm, glabrous, paler and glaucous beneath, with 7–19 prominent veins; peduncles slender or flattened, arcuate, not strongly deflexed, 1–15-fld; pedicels usually shorter than the peduncle; fls 14–22 mm, mostly greenish-white or yellowish-green; filaments mostly glabrous or minutely roughened; $2n=20$, 40. Moist woods, thickets, and roadsides; Mass. and s. N.H. to Minn., Man., and N.D., s. to Fla. and n. Mex. May–July. Tetraploids are usually larger and coarser than diploids, with more numerous fls on each peduncle and more numerous, stronger veins in the lvs, and have been called *P. commutatum* (*P. canaliculatum,* misapplied), but some diploids are as large as any tetraploid. A few plants from se. Mich. and adj. Ont. with honey-yellow fls and ± hairy filaments have been called var. *melleum* (Farw.) R. Ownbey.

32. ASPARAGUS L.　Asparagus.　Fls perfect or unisexual; perianth tubular to campanulate, the tep separate or united; stamens 6, inserted on the base of the perianth; filaments slender; anthers oblong, introrse; ovules 2 in each of the 3 locules; style slender; stigmas 3, short; fr a berry with a few large rounded seeds; herbs with the lvs reduced to inconspicuous scales and replaced functionally by branches incapable of further growth and sometimes flattened and lf-like in appearance; fls small. 300, Old World.

　　1. Asparagus officinalis L.　Perennial from a short rhizome, freely branched, to 2 m; ultimate branchlets filiform, 8–15 mm; pedicels solitary or paired; lateral, 5–10 mm, jointed at the middle; fls greenish-white, campanulate, 3–5 mm; berry red, 8 mm; $2n=20$, 40. Native of Europe, escaped from cult. into waste places or along salt marshes. May, June.

33. HYPOXIS L.　Star-grass.　Fls regular, the sep and pet distinct to the summit of the inferior ovary, similar or differing slightly in size, the sep usually pilose outside; anthers basifixed, but sometimes deeply sagittate; filaments short; ovules numerous in each of the 3 locules; fr indehiscent, or a loculicidal capsule; small perennial herbs from corms or rhizomes, with grass-like, usually hairy, basal lvs and an irregular umbel of 2–6 yellow fls or a single terminal fl. 100, irregularly cosmop.

1 Fr longitudinally dehiscent; seeds reticulate; anthers evidently basifixed .... 1. *H. sessilis.*
1 Fr indehiscent; seeds muricate; anthers so deeply sagittate as to appear versatile.
2 Seeds brown, with blunt projections; principal lvs 3-nerved .... 2. *H. micrantha.*
2 Seeds black, beset with sharp projections; principal lvs 5–9-nerved .... 3. *H. hirsuta.*

　　1. Hypoxis sessilis L.　Habitally like no. 2; peduncle only 1–2-fld; fr pyriform, dehiscent; seeds golden-brown, iridescent, coarsely reticulate. Sandy soil on the coastal plain; se. Va. to Fla. and Tex. June, July.
　　H. longii Fern., with green or whitish cleistogamous fls, the tep only ca 3 mm, may be only a form of *H. sessilis.* Originally described from se. Va., it has more recently been found also in a limited, coherent area of Ark., Okla., La., and Tex., associated with *Pinus taeda.*

2. **Hypoxis micrantha** Pollard. Much like no. 3, but with shorter, 3-nerved lvs only 1–3 mm wide; fr 4–8 mm, containing several brown seeds covered with low, blunt projections. Sandy soil in woods, chiefly on the coastal plain; se. Va. to Tex. and Cuba. May–July.

3. **Hypoxis hirsuta** (L.) Cov. Common s.-g. Lvs linear, pilose, 2–10 mm wide, to 6 dm at anthesis, the main ones 5–9-nerved; scape shorter than the lvs at anthesis, to 4 dm at maturity, bearing an irregular umbel of 2–6 fls on long pedicels; fls yellow, irregular in size, 1–2.5 cm wide; anthers deeply sagittate, the pollen-sacs divergent below; fr ellipsoid, 3–6 mm, indehiscent, containing several shiny, black, sharply muricate seeds 1–1.5 mm. Dry open woods; Me. to Man., s. to Ga. and Tex. Apr.–July. (*H. leptocarpa.*)

34. LEUCOJUM L.

Snowflake. Fls regular, the 6 equal, distinct tep forming a funnelform or campanulate perianth; anthers opening at the apex, on short filaments; ovules numerous in each of the 3 locules of the inferior ovary; seeds rounded; perennial herbs from a coated bulb, with elongate, linear, basal lvs and an erect hollow scape; fls solitary or few in an umbel, peduncled, subtended by one or 2 bracts. 12, mainly Mediterranean.

1. **Leucojum aestivum** L. Snowflake. Scape 2–4 dm; lvs 3–4 dm, ca 1 cm wide, obtuse; bract foliaceous, 3–5 cm; fls 2–8, on long drooping peduncles; perianth campanulate, ca 2 cm, the tep white, edged with green; $2n=22, 24$. Native of Europe, locally established in our range as an escape from cult.

35. NARCISSUS L.

Perianth slenderly tubular at base, with broad, widely spreading segments; corona saucer-shaped, of the same or different color, entire to erose or lobed; stamens inserted on the perianth-tube within the corona; ovary inferior, with numerous ovules in each of the 3 locules; perennial scapose herbs from coated bulbs, with narrow, usually erect, basal lvs and erect scapes bearing 1–few conspicuous fls. 60, Europe, Medit., w. Asia. Many of the plants in cult. reflect hybridization among diverse spp.

1 Fls wholly yellow . . . . 1. *N. pseudonarcissus.*
1 Fls mainly white . . . . 2. *N. poeticus.*

1. **Narcissus pseudonarcissus** L. Daffodil. Scape 2–4 dm, about equaling the lvs; fls yellow, solitary, horizontal, 4–6 (in cult. to 10) cm wide; corona about as long as the tep, usually frilled at the margin; $2n$ mostly = 14, 28. Native of Europe, occasionally escaped from cult. into moist meadows in our range. Spring.
$N. \times incomparabilis$ Miller (the hybrid of nos. 1 and 2) with the corona only half as long as the tep, and *N. jonquilla* L., the jonquil, with 2–6 yellow fls 2–4 cm wide, the corona less than a fourth as long as the tep, rarely escape.

2. **Narcissus poeticus** L. Poet's narcissus. Habitally like no. 1; fls usually several; perianth white, 5 cm wide; corona saucer-shaped, less than a fourth as long as the tep, usually yellow with a red margin; $2n=14$. Native of Europe, occasionally escaped from cult. into moist meadows in our range. Apr., May.

36. ZEPHYRANTHES Herbert.

Perianth tubular at base, with 6 broad, curved-ascending or spreading segments; filaments elongate, inserted on the throat of the perianth; corona none; ovary inferior, with numerous ovules in each of the 3 locules, seeds flat; perennial herbs from a coated bulb, with several flat, linear, basal lvs and a hollow scape bearing a single large erect fl on a short peduncle subtended by scarious bracts. 35, warm Amer.

1. **Zephyranthes atamasco** (L.) Herbert. Atamasco-lily. Lvs 3–5 cm wide, to 4 dm; scape about equaling the lvs; peduncle 5–15 mm; fls funnelform, white, tinged with purple, or purplish, 7–10 cm, the tep acute; $2n=24$. Moist woods and meadows, often in large colonies; se. Va. to Ala. May. (*Atamasco a.*)

37. HYMENOCALLIS Salisb.

Spider-lily. Perianth tubular at base, with narrow elongate segments; corona conspicuous, membranous, connecting the filaments and apparently bearing the stamens on its margin; ovary inferior, with 2 ovules in each of the 3 locules; style slender, elongate; perennial scapose herbs from a coated bulb, with long, sessile basal lvs and a terminal umbel of large, white, sessile fls subtended by scarious bracts. 50, warm Amer.

1. **Hymenocallis caroliniana** (L.) Herbert. Lvs linear-oblanceolate, obtuse, glaucous, to 6 dm × 5 cm; scape 3–7 dm, rather fleshy, with 2–6 fls; perianth-tube very slender, 5–10 cm, the segments linear, about as

long as the tube, widely curved-spreading; corona broadly funnelform, 3–4 cm, erose-margined. Wooded swamps; s. Ind. to s. Mo., s. to the Gulf. Aug. (*H. occidentalis.*)

FAMILY **AGAVACEAE**, the Agave Family

Much like the Liliaceae, but coarse and evergreen, variously acaulescent to shrubby or even arborescent, the stout stem commonly with secondary growth of monocotyledonous type; lvs perennial and ± strongly thickened, parallel-veined, usually leathery or firm-succulent, often spine-tipped and/or with prickly or loosely fibrous margins; vascular bundles with a well developed fibrous cap at the phloem pole; fls borne in rather massive, often dense racemes or panicles or heads terminating the stem(s), or axillary and subterminal; tep in 2 cycles of 3, all petaloid, often thick and fleshy, distinct or connate below to form a tube; stamens 6; filaments adnate to the base of the tep or to the perianth-tube; ovary mostly trilocular with axile placentation and several to many anatropous or sometimes campylotropous ovules per locule, sometimes each locule vertically divided into two locelli by an ingrowth from the carpellary midrib. 18/600.

1 Ovary superior .... 1. *Yucca.*
1 Ovary inferior .... 2. *Agave.*

1. YUCCA L.   Spanish bayonet. Tep distinct; stamens hypogynous, much shorter than the tep, the filaments outcurved above, the anthers sagittate; ovary superior, trilocular, with numerous ovules in 2 rows per locule, or (as in our spp.) 6-locellar, with a single row of ovules in each locellus; style short and stout, with a subglobose or 3-lobed stigma; fr in our spp. capsular; seeds numerous, flat, black; stem stout, from very short (a mere caudex) to several m tall, simple or sparingly branched, with numerous spine-tipped linear lvs crowded near its summit; infl a large, dense panicle or raceme, sessile or on a ± elongate, bracteate peduncle; fls large, white or ochroleucous or chloroleucous, moth-pollinated; $2n=60$, 10 large, 50 small. 35, N. Amer., mainly western.

1 Principal lvs well over 1 cm wide; infl elevated on a tall scape .... 1. *Y. filamentosa.*
1 Lvs narrow, rarely over 1 cm wide; base of the infl not elevated above the lvs .... 2. *Y. glauca.*

1. Yucca filamentosa L.   Adam's needle. Caudex short, to 3 or 4 dm; lvs numerous, stiff, linear-elliptic to linear-spatulate, to 8 dm, 2–7 cm wide, fibrous along the margins, abruptly prolonged into a short stout spine; ± scabrous; infl paniculate, many-fld, rising to 1–3 m, its base elevated well above the lvs; tep 5–7 × 2–3 cm, rounded above, short-acuminate; style nearly 1 cm; stigmas slender; fr thick-cylindric, 2–4 cm; seeds 6 mm. Sand-dunes and dry sandy soil, especially near the coast; Md. to Fla. and La.; often escaped from cult. farther n. June–Sept. (*Y. concava; Y. smalliana.*)

2. Yucca glauca Nutt.   Soap-plant. Caudex very short, sometimes prostrate; lvs widely radiating, stiff, linear, to 8 dm, rarely over 1 cm wide, fibrous along the white margin, gradually tapering to a short slender spine; infl racemose or with a few short branches, rising to 1–2 m, its base not or scarcely elevated above the lvs; tep ovate to lanceolate, 4–5 cm, usually acute; fr prismatic, 5–7 cm; seeds 8–10 mm. Dry prairies and plains; w. Io. and Mo. to Mont., Wyo., Ariz., and Tex. June (*Y. arkansana.*)

2. AGAVE L.   Agave (the larger, longer lived spp. called century-plants). Tep united below, withering persistent on the capsular, trilocular fr; filaments slender, exsert, with versatile linear anthers; ovary inferior; style subulate, with a 3-lobed stigma; seeds numerous, flat, black; monocarpic acaulescent perennials, ultimately producing an erect scape with an elongate, terminal infl of ochroleucous or chloroleucous to yellow or orange fls; $2n=60$, 10 large, 50 small. (*Manfreda.*) 300, warm New World.

1. Agave virginica L.   False aloe. Lvs in a basal rosette, lanceolate or oblanceolate, to 4 dm × 5 cm, acuminate, entire or denticulate; scape 1–2 m, bracteate, bearing a loose slender spike 2–5 dm; perianth erect or nearly so, greenish-white, tubular, 2–3 cm, the erect, narrow, triangular lobes half as long as the tube; fr subglobose. Dry sterile soil; N.C. and s. O. and s. Mo., s. to Fla. and Tex. June, July. (*Manfreda v.; M. tigrina.*)

FAMILY **SMILACACEAE**, the Catbrier Family

Fls trimerous, regular, unisexual or in some extralimital genera perfect; tep in 2 petaloid cycles, distinct or connate below; filaments variously free and distinct or borne on the perianth and sometimes connate into a column; ovary superior, trilocular or sometimes unilocular, with accordingly axile or parietal placentation; no septal nectaries; ovules 1–many in each locule or on each placenta; fr a berry with mostly 1–6 seeds; embryo mostly small, with a terminal cotyledon, axially embedded in the very hard endosperm; climbing, herbaceous or slenderly woody vines, or less often erect perennial herbs or branching shrubs, arising from creeping, often tuber-bearing rhizomes; lvs simple, mostly alternate, usually with a pair of tendrils arising from the petiole near its junction with the short stipular flange or open sheath; blade well developed and expanded, commonly with 3–7 "parallel," curved-convergent main veins connected by an evident network of smaller veins; vessels present in all vegetative organs; fls commonly in umbels or sets of umbels. 12 genera, all but *Smilax* small.

1. SMILAX L.   Dioecious; fls in axillary pedunculate umbels, rather small, the staminate often a little larger than the pistillate; tep spreading, greenish or yellowish; filaments borne at the base of the tep; anthers basifixed; style none or very short; stigmas solitary or 3, oblong, recurved; ovules orthotropous, 1 or 2 per carpel. 300, cosmop.

1 Stem herbaceous, annual, unarmed; ovules 2 per carpel; fls foetid (*Nemexia*; carrion-fls.)
  2 Peduncles arising from the axis of foliage-lvs; tendrils present.
    3 Tep 1.5–2.5 mm; lateral lf-margins straight or concave .... 1. *S. pseudochina.*
    3 Tep 3.5–5 mm; lateral lf-margins convex .... 2. *S. herbacea.*
  2 Peduncles axillary to bracts below the lowest foliage-lf; tendrils often none .... 3. *S. ecirrhata.*
1 Stem woody, perennial, often prickly; ovules 1 per carpel (true *Smilax*; greenbriers or catbriers).
  4 Peduncles notably longer than the petiole of the subtending lf.
    5 Lvs strongly glaucous beneath, entire .... 4. *S. glauca.*
    5 Lvs green beneath, often serrulate or ciliate.
      6 Lf-margin thin, minutely serrulate (visibly so at 10×) .... 5. *S. hispida.*
      6 Lf-margin conspicuously thickened, about as thick as the outermost longitudinal vein, entire
        or sparsely spinulose-ciliate .... 6. *S. bona-nox.*
  4 Peduncles up to about as long as (or only slightly longer than) the petiole.
    7 Lvs evergreen, broadly linear to oblong or lance-oblong, obtuse or cuneate at base .... 7. *S. laurifolia.*
    7 Lvs deciduous, lance-ovate to suborbicular, with truncate to cordate base.
      8 Fr black; widespread .... 8. *S. rotundifolia.*
      8 Fr red; southern .... 9. *S. walteri.*

1. Similax pseudochina L.   Herbaceous, unarmed, climbing to 2 m, occasionally branched; lvs numerous, mostly with tendrils, the blade 5–12 cm, thin, glabrous, the larger ones hastate, with concave margins, truncate to subcordate at base, rounded and cuspidate at the tip; peduncles axillary to foliage-lvs, often 2 or 3 from the same axil, terete even at maturity, surpassing the petiole and often nearly equaling the blade, with 10–30 fls; tep 1.5–2.5 mm; anthers 0.5–1 mm; fr black, glaucous, 4–6 mm, with 1–3 seeds; $2n=30$. Moist or wet low places on the coastal plain; N.J. and Del. to Ga. May. (*S. tamnifolia.*)

2. Similax herbacea L.   Herbaceous, unarmed, climbing to 2.5 m, often freely branched; lvs numerous (more than 25), most of them with tendrils, at base cordate to rounded, at apex acuminate to cuspidate or broadly rounded, always with convex lateral margins; peduncles numerous, individually axillary to the foliage-lvs, flattened, with numerous (mostly more than 25) fls; tep 3.5–5 mm; anthers 1–2 mm; fr 8–10 mm; seeds 3–6; $2n=26$. Moist soil of open woods, roadsides, and thickets. May, June. Three vars.:

a Lvs glabrous beneath, the lower side pale and somewhat glaucous; lateral lf-margins convex to the
  obtuse or cuspidate tip; peduncles 5–8 times as long as the subtending petioles; fr dark blue,
  glaucous; mainly Appalachian, from Que. and Me. to Ga. and Ala. .... var. *herbacea.*
a Lvs puberulent on the veins beneath.
  b Lvs bright green and shining beneath, distinctly short-acuminate; peduncles mostly 5–10 times as
    long as the subtending petioles, sometimes less; fr black, not glaucous; se. N.Y. to Va. and w.
    N.C., w. irregularly to Ind., Mo., e. Kans., and Ark. (*S. p.*) .... var. *pulverulenta* (Michx.) A. Gray.
  b Lvs paler beneath than above, rounded, blunt, or short-cuspidate at the tip; peduncles seldom over
    twice as long as the subtending petioles; fr dark blue, glaucous; mainly w. of the Appalachian
    Mts., w. to Mont., Wyo., Colo., and Okla., and s. to w. Fla. (*S. l.*) .... var. *lasioneura* (Small) Rydb.

3. Smilax ecirrhata (Engelm.) S. Wats.   Herbaceous, unarmed, ± erect, unbranched, to 8 dm, without tendrils, or producing a few from the upper lvs only; lvs few, mostly less than 20, often only 7–9, ± crowded on the upper part of the stem only, narrowly to broadly ovate, truncate to cordate at base, convexly narrowed to a short cusp, hairy beneath; peduncles axillary to lance-linear bracts on the lfless lower part of the stem, or rarely also from the axil of the lowest lf, ascending, 5–10 cm, with few (seldom more than 25) fls; tep 3.5–5 mm; seeds 3–5; $2n=26$. Rich woods and thickets; Mich. to Minn., s. to Ky., Mo., and e. Okla. May.

S. illinoensis Mangaly, found occasionally throughout most of the range of *S. ecirrhata*, may reflect hybridization with *S. herbacea* var. *lasioneura*. It is erect like *S. ecirrhata*, with the peduncles mainly below the foliage-lvs, but it averages taller, with more numerous lvs (many of them tendriliferous), and has more numerous (mostly more than 25) fls on longer peduncles.

4. **Smilax glauca** Walter. Slender woody vine, seldom climbing very high; stems green and glaucous the first year, beset with stout prickles, the lower generally straight, the upper nodal and recurved; lvs often ± persistent, mostly ovate, often with cordate base, varying to subrotund or nearly triangular, 5–9 cm, half to three-fourths as wide, glaucous (and often long-papillate) beneath, at maturity subcoriaceous and shining above, thin and entire at the margin, 3- or 5(7)-nerved, the reticulate veins not prominently elevated; peduncles flattened, 1.5–3 times as long as the petioles; fr black, glaucous, 8–10 mm, mostly 2- or 3-seeded; $2n=32$. Upland woods, roadsides, and thickets; Conn. to Fla., w. to O., s. Ill., se. Mo., Ark., and Tex. May, June.

5. **Smilax hispida** Muhl. Bristly g. Slender woody vine, often climbing high, beset (at least below) with unequal needle-like prickles; lvs deciduous, shiny-green, ovate to rotund; 8–12 × 6–10 cm at maturity; acute to rounded or cuspidate; at base rounded to truncate or cordate, not thickened at the margin, minutely serrulate (visibly so at 10×), 5- or 7-nerved, the reticulate veinlets not much elevated; peduncles flattened, mostly 2–6 cm, generally at least twice as long as the subtending petiole; fr black, not glaucous, 6–8 mm, with 1(2) seeds; $2n=32$. Moist woods and thickets; Conn. and N.Y. to n. Fla., w. to Mich., s. Minn., Nebr., and Tex., the commonest greenbrier in the w. part of our range. May, June. (*S. pseudochina*, misapplied; *S. tamnoides*, perhaps misapplied.)

6. **Smilax bona-nox** L. Slender woody vine; stems usually 4-angled, diffusely branched and often climbing high, armed with large, stout, flattened prickles especially on the angles, or distally unarmed; lvs subcoriaceous and ± persistent, ovate to deltoid or hastate, 4–8 cm, green beneath, conspicuously thickened at the entire or spinulose margins; veinlets prominent; peduncles flattened, much longer than the subtending petiole; fr black, 6–8 mm, 1-seeded; $2n=32$. Dry woods, thickets, abandoned fields, and roadsides; s. Md. to Mo. and se. Kans., s. to Fla. and Mex. May–July.

7. **Smilax laurifolia** L. Slender woody vine, climbing high; stems glaucous at first, beset with numerous straight, stout prickles below, generally innocuous above; lvs evergreen, leathery, oblong or lanceolate to ovate, 6–10 cm, obtuse or subacute, entire, cuneate at base, 3-nerved, the outer nerves much nearer to the margin than to the midnerve; peduncles terete or angled, about equaling the subtending petioles; fr purple to black, mostly 1-seeded, ripening the second year; $2n=32$. Wet woods and margins of swamps; N.J. to Fla. and Tex., and inland to e. Tenn. Aug., Sept.

8. **Smilax rotundifolia** L. Slender woody vine; stems usually quadrangular, diffusely branched and often climbing high, with scattered stout, flattened prickles below; lvs deciduous, thin, green, shining, ovate to rotund or triangular-ovate, 5–10 × 4–9 cm, or smaller in dry soils, acute to cuspidate or obtuse, entire or rarely sparsely roughened on the margins, at base broadly rounded to truncate or cordate, 5- or 7-nerved, reticulate beneath at maturity; peduncles flattened, about as long as the subtending petioles; staminate tep recurved only above the middle; fr black, usually glaucous, mostly 2–3-seeded; $2n=32$. Open woods, thickets, and roadsides; N.S. to n. Fla., w. to Mich., se. Mo., e. Okla., and e. Tex., the commonest greenbrier in the ne. part of our range. May, June.

9. **Smilax walteri** Pursh. Slender woody vine, unarmed except for a few small, flat, slightly recurved prickles at base; lvs deciduous, firm, shining above, ovate or triangular-ovate or lance-triangular, 7–12 cm, acute or obtuse, entire, 3- or 5-nerved, prominently reticulate; peduncles flattened, about equaling the subtending petioles; staminate tep 6 mm, broadly recurved; frs red, few in a cluster, 7–9 mm, 1–3-seeded; $2n=32$. Swamps and wet woods; Md. to n. Fla., w. to La. May.

FAMILY **DIOSCOREACEAE**, the Yam Family

Fls trimerous, regular, epigynous, unisexual (the plants dioecious) or seldom perfect; tep in 2 similar sets of 3, petaloid or chaffy, mostly connate at base; both septal and tepalar nectaries commonly present; stamens usually 6 in 2 cycles, or the inner set staminodial or obsolete; filaments borne on the base of the perianth; ovary inferior, trilocular with axile placentation; styles or stigmas distinct; ovules mostly 2 per locule; fr mostly capsular; embryo small, with subterminal plumule and broad, lateral cotyledon, axially embedded in the very hard endosperm; twining-climbing or seldom erect herbs from a fleshy-thickened rhizome, or much more often from a large basal "tuber" of complex structure; lvs alternate or sometimes opposite or whorled, generally with a distinct blade and petiole, the blade broad, entire or less often palmately lobed or cleft or even compound, commonly with 3–13 "parallel," curved-convergent main veins and a network of smaller veins; vessels generally present in all vegetative organs; fls small, variously in racemes, spikes, or panicles. 6 genera, all but *Dioscorea* small.

1. **DIOSCOREA** L.   Yam.   Dioecious; infls axillary; fls white to greenish-yellow, the pistillate solitary at each node of the short, interrupted spikes, the staminate in small glomerules or solitary at each node of the panicle; stamens 6; fr a 3-winged loculicidal capsule; seeds very flat, broadly winged. 600, mainly trop. and subtrop.

1 Lvs cordate-ovate, with convex sides; perennial from rhizomes.
  2 Lvs all or nearly all alternate, only the lowermost occasionally ± whorled; rhizomes relatively
    slender, 5–10 mm thick, comparatively straight and not much branched . . . . 1. *D. villosa.*
  2 Many or most of the lvs, at least the lower ones, in whorls of 4–7; rhizomes stout, mostly 10–15
    mm thick, often very irregularly contorted or with many short branches . . . . 2. *D. quaternata.*
1 Lvs halberd-shaped, the sides concave above the conspicuously widened base; perennial from large,
  vertical, root-like tubers. . . . . 3. *D. batatas.*

1. **Dioscorea villosa** L.   Colic-root. Rhizomes relatively slender; stems twining counterclockwise, to 5 m, usually glabrous; lvs all or nearly all alternate, the blades glabrous or hairy beneath, cordate-ovate, abruptly acuminate, 5–10 cm, 7–11-nerved; staminate infl widely branched, with 1–4 fls per node; pistillate spikes 5–10 cm, with solitary fls; ovary fusiform, 5–7 mm; fr 16–26 mm, each of the 3 thin valves semi-orbicular or half-ovate; seeds 8–18 mm; $2n=60$. Moist open woods, thickets, and roadsides; Conn. and N.Y. to Minn., s. to Fla. and Tex. June, July. Some coastal-plain plants with hairy stem and relatively few-fld infls have been segregated as var. *hirticaulis* (Bartlett) Ahles. (*D. hirticaulis.*)

2. **Dioscorea quaternata** (Walter) J. F. Gmelin.   Much like no. 1, differing as indicated in the key; frs and seeds avg larger, the valves half-obovate; $2n=36, 54$. Moist open woods, thickets, and roadsides; Pa. to Ind. and Mo., s. to Fla. and La., chiefly in the Appalachian region. May. Plants with the lvs glaucous beneath have been called var. *glauca* (Muhl.) Fern. (*D. g.*)

3. **Dioscorea batatas** Decne.   Cinnamon-vine. Perennial from deep-seated tubers to 1 m, the slender stems climbing 1–5 m and twining clockwise; lvs alternate, opposite, or ternate, usually with a small axillary tuber, the blade about as wide as long, halberd-shaped, deeply cordate at base; fr probably never produced in our range; $2n=$ca 140–144. Native of China, often escaped from cult. in U.S. and occasional in our range.

FAMILY **IRIDACEAE**, the Iris Family

Fls perfect, regular or irregular; tep 6, mostly bicyclic, petaloid, all alike, or the inner and outer sets unlike, often all connate below into a tube; stamens 3, opposite the sep (outer tep), the filaments often connate below; ovary inferior, trilocular with axile placentation and mostly ± numerous ovules; style 3-lobed, its branches sometimes again divided; fr a loculicidal capsule, or sometimes indehiscent; embryo linear, with a terminal cotyledon; endosperm fleshy, not starchy; mostly geophytic perennial herbs from rhizomes, bulbs, or corms; lvs simple, parallel-veined, narrow, mostly distichous, with an open sheathing base and an often equitant or gladiate, unifacial blade; infl terminal, often subtended by 1 or 2 expanded, bladeless sheaths forming a spathe. 80/1500, especially diversified in S. Afr.

Spp. of *Crocus* and *Gladiolus,* not treated below, rarely escape from cult.

1 Style-branches expanded and petal-like, concealing the stamens . . . . 3. *Iris.*
1 Style-branches neither expanded nor petal-like, the stamens clearly visible.
  2 Filaments distinct; fls 3–5 cm wide . . . . 1. *Belamcanda.*
  2 Filaments connate; fls under 2 cm wide . . . . 2. *Sisyrinchium.*

1. **BELAMCANDA** Adans.   Blackberry-lily. Tep distinct, subequal, widely spreading; filaments distinct, inserted on the base of the tep; style clavate, 3-cleft; capsule oblong to pyriform, 3-lobed, its 3 valves recurving sharply at maturity and soon deciduous, exposing the mass of round, black, fleshy seeds attached to the central axis, the whole resembling a large blackberry; perennial herbs with horizontal rhizomes and ensiform lvs; infl widely branched, cymose-paniculate. 2, e. Asia.

1. **Belamcanda chinensis** (L.) DC.   Stem 3–6 dm, the fls cymose at the ends of the branches, lasting but a day; fls orange with crimson or purple spots and markings, 3–5 cm wide; capsule 2.5–3 × 1–2 cm; $2n=32$. Native of Asia, well established as an escape from cult. in pastures, roadsides, thickets, and hillsides from Conn. to Nebr. and Ga. June, July. (*Gemmingia c.*)

2. SISYRINCHIUM L.   Blue-eyed grass.

Perianth subrotate, the tep alike and spreading, retuse or abruptly aristulate, distinct or very shortly connate at base; filaments connate at least to the middle (to the top in our spp.); style-branches filiform, alternating with the stamens; capsule globose to obovoid, loculicidal; seeds black, globular, ± pitted; low, perennial, grass-like herbs, usually tufted from fibrous roots; lvs equitant, linear and mostly basal; scape-like stems 2-edged or 2-winged, simple or branched from the axils of lf-like bracts; fls delicate and fugacious, in umbel-like clusters from 2-valved spathes, blue to violet or pink, seldom white, rarely yellow, our spp. regularly with a yellow or greenish eye. Our spp. fl chiefly in May, but continue until Sept. 60±, New World.

1 Spathes essentially sessile, solitary or paired at the top of the simple stem.
  2 Outer bract of the spathe with the margins distinct to the base.
    3 Stem very slender, scarcely winged, less than 1 mm wide; se. coastal plain .... 1. *S. capillare.*
    3 Stem wider, distinctly winged, 1–4 mm wide; widespread spp., mostly inland.
      4 Spathes mostly 2 .... 2. *S. albidum.*
      4 Spathe solitary .... 3. *S. campestre.*
  2 Outer bract with its margins united for 2–5 mm at the base.
    5 Stem very slender, up to 1(–1.5) mm wide, barely or scarcely winged; largest lvs up to 1.5 mm
      wide .... 4. *S. mucronatum.*
    5 Stem flattened and winged, 1.5–3+ mm wide; largest lvs 2–3 mm wide .... 5. *S. montanum.*
1 Spathes evidently pedunculate, the peduncles arising from the axils of lf-like bracts, usually more than
  one, the upper part of the stem thus appearing branched.
  6 Stem narrowly or scarcely winged, 0.5–2 mm wide.
    7 Old lf-bases fibrous and persistent; herbage drying dark .... 6. *S. fuscatum.*
    7 Old lf-bases not long-persistent; herbage mostly drying light green .... 7. *S. atlanticum.*
  6 Stem broadly winged, (2–)2.5–4 mm wide .... 8. *S. angustifolium.*

1. **Sisyrinchium capillare** E. Bickn.   Very slender and delicate, 2–4.5 dm, erect, glaucescent, drying pale; stems 0.5 mm wide or less; lvs less than 1 mm wide, the bases persisting as stiff fibers; spathes (1)2(3 at the top of the simple stem, the outer (lower) ones with an erect, setaceous outer bract 2–8.5 cm, the other bracts 1–1.5 cm, subequal, often purplish-tinged, hyaline-margined; tep 6–8 mm, light blue-violet; fr 2–3 mm. Flatwoods and pinelands; se. Va. to Fla.

2. **Sisyrinchium albidum** Raf.   Erect, 1–4 dm, pale green and somewhat glaucous, drying pale; stems 1.5–4 mm wide, distinctly winged; lvs about as wide as the stems; spathes mostly paired at the top of the simple stem, the lower (outer) one with a foliaceous outer bract commonly 3.5–7 cm, the other bracts shorter, 1.5–2.5 cm, often purplish-tinged; tep 8–12 mm, white to violet, fr 2–4 mm; $2n=32, 64$. Prairies, meadows, grassy or sandy places, and open woods; N.Y. to N.D., s. to Ga. and Okla.

3. **Sisyrinchium campestre** E. Bickn.   Much like no. 2, but the spathe mostly solitary, its outer bract elongate and foliaceous, 2.5–4.5 cm, the inner ones 1–2.5 cm; tep white to pale blue, rarely yellow; $2n=32$. Prairies, meadows, sandy places, and open woods; Wis. and Ill. to Minn., S.D., Okla., and Ark. (*S. kansanum.*)

4. **Sisyrinchium mucronatum** Michx.   Erect, 1–4 dm, delicate and wiry, dark green, drying dark; stems very slender, up to 1(–1.5) mm wide, barely or scarcely winged, entire-margined; lvs entire, very narrow, the larger ones up to 1.5 mm wide; spathes mostly solitary at the summit of the simple stem, often geniculate (set at an angle to the stem), anthocyanic or rarely green, the outer bract somewhat foliaceous and with its margins connate near the base, the inner bract shorter, 1.2–2 cm; tep 8–10 mm, bright violet (white); fr 2–4 mm; $2n=32$. Meadows, fields, sandy places, and woods; Me. and w. Mass. to Ont. and Man., s. to N.C., Mich., and N.D.

5. **Sisyrinchium montanum** Greene.   Erect, 1–5 dm, light green and glaucescent, drying pale, or deep green and drying dark; stems usually stout, flattened and winged; 1.5–3+ mm wide, the margins minutely denticulate, as also the lvs; largest lvs 2–3 mm wide; spathes mostly solitary at the summit of the simple stem, mostly not geniculate, usually green, the outer bract 3–7 cm, its margins connate for 2–5 mm near the base, the inner bract much shorter, 1.5–3 cm; tep 8–12 mm, bright violet; fr 4–6 mm; $2n=32, 96$. Sandy open ground and meadows; Que. to Alta., s. to N.Y., N.J., Pa., and Nebr., and along the mts. to Va. and N.C. (*S. angustifolium*, misapplied.) The phase with the herbage blackening in drying has been called var. *crebrum* Fern.

6. **Sisyrinchium fuscatum** E. Bickn.   Light green, but blackening in drying, ± erect, 1–5 dm; old lf-bases persistent as prominent tufts of light brown, fragile fibers often several cm long; stem only narrowly winged, minutely denticulate along the margins, usually not geniculate at the nodes; lvs 1–3.5 mm wide, spathes evidently peduncled from the axils of lf-like bracts, generally not geniculate; outer bract of the spathe mostly

1.5–2 cm, its margins connate for 2.5–4 mm at base; inner bract not much shorter than the outer; tep 5–8 mm, violet; fr 3–5 mm; 2n=32. Sandy areas, mainly near the coast, from Mass. to Fla. and Miss., and irregularly inland to Mich. and Mo. (*S. arenicola; S. farwellii*.)

7. **Sisyrinchium atlanticum** E. Bickn. Light green, somewhat glaucous, usually drying pale, ± spreading and often forming small tussocks; stem slender and wiry, 0.5–2 mm wide, the smooth-margined wings distinctly narrower than the central portion; lvs 1–3 mm wide; spathes evidently peduncled from the axils of lf-like bracts, the stem tending to be geniculate at the nodes and the spathes likewise often geniculate on the peduncles; outer bract of the spathe mostly 1.5–2 cm, its margins connate for 2.5–4 mm at base; inner bract not much shorter than the outer; tep 8–12 mm, blue-violet; fr 3–5 mm; 2n=16, 32, 96. Fields, meadows, open woods, and edges of salt marshes; N.S. and Me., s. in the coastal states to Fla. and Miss., and locally inland to Mich., Ind., Minn., and Mo. (*S. strictum*.)

8. **Sisyrinchium angustifolium** Miller. Bright green, turning darker in drying; stems 1.5–5 dm, somewhat spreading, mostly not geniculate, thin and flattened, broadly winged, (2–)2.5–4 mm wide, the wings individually wider than the central part and minutely denticulate on the margin; lvs mostly 2–6 mm wide; spathes evidently peduncled from the axils of lf-like bracts, mostly not geniculate, the outer bract commonly 2–4 cm, its margins connate for (2.5–)3.5–6 mm at base, the inner bract evidently shorter, commonly 1.3–2.2 cm; tep blue, 7–10 mm; fr 4–6 mm; 2n=48. Meadows, grassy places, and damp woods; Nf. to Minn., s. to Fla. and Tex. (*S. graminoides; S. gramineum*.)

3. **IRIS** L. Iris; flag; fleur-de-lys. Sep spreading or reflexed; pet erect or arching, in our spp. narrower and shorter than the petaloid sep; tep all united below into a perianth-tube; stamens inserted at the base of the sep; ovary 3- or 6-angled or -lobed; style divided distally into 3 petaloid branches arching over the stamens, each 2-lobed at the tip; stigma a thin plate or lip at the outer base of the 2 lobes; fr coriaceous or chartaceous, loculicidal or indehiscent; seeds in 1 or 2 rows per locule; perennial herbs with ensiform or linear lvs, our spp. with horizontal rhizomes and usually erect fl-stalks bearing 1–many fls. 200, N. Temp. Much hybridized in cult.

1 Sep bearded .... 11. *I. germanica.*
1 Sep not bearded.
  2 Plants stemless or nearly so; lvs at anthesis and fl-stalk less than 15 cm.
    3 Sep not crested, but with a pubescent line; lvs up to 1 cm wide .... 1. *I. verna.*
    3 Sep strongly crested, not pubescent; lvs more than 1 cm wide.
      4 Perianth-tube 4–6 cm, much exceeding the spathe and longer than the sep .... 2. *I. cristata.*
      4 Perianth-tube 1–2 cm, little if at all exceeding the spathe and shorter than the sep .... 3. *I. lacustris.*
  2 Plants with well developed stems; lvs and fl-stalks over 15 cm.
    5 Fls yellow, orange, brownish, or copper-color.
      6 Pet about equaling the sep; fls coppery or brown-orange (yellow) .... 4. *I. fulva.*
      6 Pet much shorter than the sep; fls bright yellow (cream) .... 5. *I. peudacorus.*
    5 Fls blue or violet, variegated with white, brown, and/or green (rarely all white).
      7 Ovary and capsule 6-angled .... 6. *I. brevicaulis.*
      7 Ovary and capsule 3-angled.
        8 Lvs up to 1 cm wide; ovary and capsule sharply angled .... 7. *I. prismatica.*
        8 Lvs over 1 cm wide; ovary and capsule bluntly angled.
          9 Pet and sep well developed. .... 9. *I. versicolor.*
            10 Sep-blade oblong-ovate, with a hairy yellow basal spot; ovary 1.8–3.8 cm at anthesis;
                style branches auriculate at base .... 8. *I. virginica.*
            10 Sep-blade ovate to reniform, the base merely papillate, unspotted or with a green-
                ish or greenish-yellow spot; ovary 1–2 cm at anthesis, style-branches not auric-
                ulate .... 9. *I. versicolor.*
          9 Pet much reduced; sep well developed .... 10. *I. setosa.*

1. **Iris verna** L. Dwarf i. Lvs grass-like, bright green, glaucous on one side, 3–4 cm × 3–10 mm at anthesis, later much elongate; spathe 1–2-fld, the valves pale green to whitish or pink-tinged, rather closely sheathing each other; perianth-tube dull green, ± exceeding the spathe-valves, the limb 5–6 cm wide, bright violet (pale or white); sep spreading or recurved; with a broad, yellow, papillose-hairy band extending from the middle of the blade into the claw and frequently surrounded by white markings; pet erect, with long slender claw and obovate blade; fr obtusely 3-angled, slender-beaked, 1.5–2.5 cm; seeds arillate; 2n=42. Sandy open woods and pine-lands. Pa. to Ky., Miss. and Ga. Apr., May. (*Neubeckia v.*)

2. **Iris cristata** Aiton. Dwarf crested i. Lvs broadly linear, usually somewhat curved-arching, light green, slightly glaucescent on one side, well developed at anthesis, 1–2 dm × 9–25 mm, later elongating slightly; spathe 1–2-fld, the valves rather loosely sheathing each other, the lower ones often spaced on a scape-like stalk; perianth-tube greenish, often streaked with purple or brown, 4–6 cm, much exceeding the spathe, the limb 6–8 cm wide, light violet to lilac or purple, rarely pale or white; sep with a strong, several-ridged, yellow or whitish crest bordered by a white zone outlined in violet or purple, not at all pubescent; pet spreading, oblanceolate, slightly emarginate or blunt, with broad claw; capsule ellipsoid, ovoid, or obovoid, sharply

3-angled, minutely beaked; seeds orange-brown, arillate; $2n=24$, 32. Rich woods, banks, and cliffs, in acid soil; Md. to Okla. and Ga. Apr., May.

3. Iris lacustris Nutt.   Dwarf lake-i. Like a smaller form of no. 2; lvs broadly linear, curved-arching, 4–6 cm at anthesis, later to 18 cm × 5–10 mm; spathe-valves closely sheathing each other, the upper and lower equally spaced; perianth-tube dull yellow, 1–2 cm, dilated upwards, not or scarcely exceeding the spathe-valves and shorter than the sep and pet; fls 5–6 cm wide, the pet emarginate; seeds dark brown, arillate; $2n=42$. Gravelly shores and cliffs in calcareous soil around lakes Superior, Michigan, and Huron; now rare. May.

4. Iris fulva Ker Gawler.   Copper-i. Lvs ensiform from a horizontal rhizome, 0.5–1 m; stem 0.5–1.5 m, surpassing the lvs; fls 7–9 cm wide, copper-color to brown-orange, reddish-brown, or rarely yellow; tep widely spreading, the claw scarcely a fourth the length of the blade; ovary 6-angled; mature fr 4.5–8 cm, indehiscent; $2n=42$. Swamps; s. Ill. and Mo. to Ga. and Pa. Apr., May.

5. Iris pseudacorus L.   Water-f; yellow f. Lvs stiff and erect, broadly ensiform from densely crowded rhizomes; stem 0.5–1 m, shorter than or equaling the lvs; fls 7–9 cm wide, bright yellow or cream-color; sep spreading, the crest area outlined by an irregular series of brown markings; pet erect, ligulate, constricted at the middle, 1–2.5 cm, unmarked; capsule 6-angled, cylindric-prismatic to ellipsoid, 5–8.5 cm, the valves widely spreading at maturity; $2n=24$–34. Swamps and shallow water along streams and ponds; native of Europe, and widely established in our range. Apr–June.

6. Iris brevicaulis Raf.   Zigzag i. Lvs numerous, ensiform, to 7 dm; stem strongly zigzag, often decumbent, 2–4 dm, conspicuously shorter than the lvs; fls 8–10 cm wide, light violet to lavender; sep with a white-variegated area at the base of the ovate blade around the yellow papillose crest, the claw as long as the blade, greenish-white with green veins; pet oblanceolate to spatulate; fr ovoid to ellipsoid-ovoid, indehiscent; $2n=42$. Swamps, low woods, and shores; O. and e. Kans. to La. and Ky. May, June.

7. Iris prismatica Pursh.   Slender blue f. Lvs linear, erect, 5–7 dm × 3–7 mm, from slender, widely creeping rhizomes; stems slender, 3–10 dm; fls on slender pedicels much exceeding the spathes, 6–8 cm wide, violet to blue-violet; pet erect, obovate, with a slender claw; sep spreading-reflexed, veined with dark violet and variegated with white, the claw greenish and purple-veined; fr sharply trigonous, 3–5 cm, apically dehiscent; $2n=42$. Marshes, swamps, and damp meadows; near the coast from N.S. to Ga., and in the s. Appalachians. June, July.

8. Iris virginica L.   Southern blue f. Plants softer than no. 9; lvs broadly linear to broadly ensiform, erect or arching; rhizomes thick, creeping, often forming extensive colonies; stem to 1 m; fls short-pediceled, 6–8 cm wide, lavender or light violet to blue-violet or purple, often dark-veined, rarely red-purple or white; sep spreading-recurved, with a bright yellow hairy blotch at the base of the blade; pet somewhat shorter than the sep; ovary 1.8–3.8 cm at anthesis; fr ovoid to ellipsoid-ovate, 4–7 cm, obtusely 3-angled, the valves strongly reflexed after dehiscence; seeds with a shallowly and irregularly pitted surface; $2n$ mostly = 70–72. Swamps, marshes, meadows, and ditches; coastal plain from Md. to Tex., and inland to Ont., Minn., and Okla. May–July. Var. *virginica,* mainly on the coastal plain, is up to 6 dm, unbranched or with a few very short branches; its frs are 4–7 cm and nearly as thick. (*I. caroliniana; I. georgiana.*) Var. *shrevei* (Small) E. Anderson, the inland phase, is up to 1 m and more branched, usually with 1 or 2 widely spreading branches; its frs are 7–11 cm, half as thick, and the fls avg darker in shade than var. *virginica.* (*I. shrevei.*)

9. Iris versicolor L.   Northern blue f. Lvs broadly linear to ensiform, erect or arching from thick creeping rhizomes and forming large clumps, stems 2–8 dm, equaling or usually slightly exceeding the lvs; bracts with vernicose dark margins; fls on short pedicels, 6–8 cm wide, violet or blue-violet to red-purple, lavender, or rarely white; sep spreading, unspotted or with a minutely papillate, greenish-yellow blotch at the base of the blade surrounded by white variegations and purple veins, the veins extending into the claw; pet erect, half to two-thirds as long as the sep, the claw pale-streaked; ovary 1–2 cm at anthesis; fr bluntly 3-angled, prismatic-cylindric, 3.5–5.5 cm, indehiscent; seeds with a ± regularly pebbled surface; $2n$ mostly = 108; perhaps an allopolyploid derived from nos. 8 and 10. Marshes, swamps, meadows, and shores; Nf. and Lab. to Man., s. to Va. and Minn. May–July (Aug.)

10 Iris setosa Pallas.   Arctic blue f. Plants coarse and firm, with lignified stems and frs; lvs linear to ensiform, from a stout creeping rhizome; stems 1–5 dm, unbranched or with 1 or 2 short branches; fls 6–8 cm wide; sep dark blue-violet with lighter streaks and a white blotch at the base of the blade; pet blue-violet, involute, oblanceolate and setose-tipped, 1–1.5 cm; fr short-cylindric to ovoid, 2.5–4 cm, bluntly 3-angled; $2n=34$–38. River shores, sea beaches, and rocky headlands; ne. Asia and Alaska; Nf. and Lab. to Que. and Me. June, July. The Atlantic plants are var. *canadensis* M. E. Foster. (*I. hookeri.*)

11. Iris germanica L.   German i. Lvs broadly ensiform, glaucous; stems stout, to nearly 1 m, the fls nearly sessile in the spathe, 7–10 cm wide; sep broadly ovate, recurved, deep violet with yellow, white, and brown veins at the base of the blade, the median line long-bearded; pet light violet, erect-arching, slightly smaller than the sep; fr trigonous, 4–7 cm, seldom produced; $2n=36$–48. A European cultigen, persisting after cultivation and sometimes spreading into waste places or roadsides; one of the parents of the tall bearded irises of horticulture, which also occasionally persist.

FAMILY **BURMANNIACEAE,** the Burmannia Family

Fls trimerous, regular (ours) or somewhat irregular, perfect, epigynous; perianth corolloid, tubular below, (3- or) 6-lobed, the outer lobes valvate, the inner mostly smaller and induplicate-valvate (seldom wanting); stamens 6, or often only 3 and then alternate with the sep, borne on the perianth-tube, sessile or on short filaments, the anthers sometimes connate into a tube around (but free from) the style; ovary inferior, trilocular or unilocular, the placentation accordingly axile or parietal; fr capsular, often winged; seeds very numerous and tiny, with scanty or virutally no endosperm, the embryo minute and undifferentiated; small, mycotrophic herbs with alternate green lvs, or the lvs more often reduced to scales and the plant without chlorophyll; fls in terminal cymes or racemes, or solitary and terminal. 20/130.

1 Ovary and perianth-tube broadly winged .... 1. *Burmannia.*
1 Ovary and perianth-tube wingless .... 2. *Thismia.*

1. BURMANNIA L. Ovary and perianth-tube 3-angled or 3-winged, the 3 outer perianth-lobes well developed, the inner minute or lacking; anthers 3, sessile in the throat of the perianth-tube; style slender, 3-lobed; ovary trilocular; capsule irregularly dehiscent, crowned by the persistent perianth; mycotrophic, chlorophyllous or (ours) achlorophyllous herbs with scale-like or linear lvs and solitary, capitate, or cymose fls. 60, warm reg.

1. **Burmannia biflora** L. Stem almost capillary, 5–15 cm, with scattered minute scale-lvs; fls 5 mm, solitary and terminal, or in luxuriant specimens several in a bifurcate cyme; ovary-wings porcelain-blue, semi-obovate; outer perianth-lobes 2 mm, the inner smaller. Bogs on the coastal plain; se. Va. to Fla. and Tex. Aug.

2. THISMIA Griffith. Fls regular, wingless; perianth-lobes equal or the inner the larger, the throat partly closed by an annulus; stamens 6, pendent from the annulus within the tube; fr fleshy, crowned by the perianth-tube; fleshy, achlorophyllous herbs, mostly small, with a few minute cauline lvs and 1–few fls. 25, mainly trop.

1. **Thismia americana** Pfeiffer. Stem subterranean, 3–10 mm, with a few minute scales; fl solitary, white or greenish, partly subterranean, obovoid-oblong, 8–15 mm, inner perianth-segments curved-ascending and connate at the tip. Known only from a wet prairie near Chicago, Ill., and not seen since 1913, the site now destroyed. Late summer. Very similar to *T. rodwayi* F. Muell., of Tasmania and N.Z.

FAMILY **ORCHIDACEAE,** the Orchid Family

Fls perfect, irregular, usually resupinate (twisted in ontogeny so that the morphologically adaxial side appears to be abaxial); sep 3 (or 2 by fusion), green or colored, often resembling the lateral pet; pet 3, usually colored or white; the 2 lateral ones consimilar and evidently different from the third (typically the lowest) one, called the *lip*; stamen typically one, adnate to the style on the opposite side from the lip, forming a usually stout stylar *column* with the bilocular anther terminal or subterminal and separated from the proper stigmatic surface by an enlarged *rostellum* derived from the adjacent stigma-lobe, the 2 functional stigma-lobes often connate (in *Cypripedium* 2 anthers and an expanded staminode borne on the column, which lacks a rostellum); pollen monadinous and only loosely coherent in *Cypripedium*, in our other genera tetradinous and organized into 1–6 *pollinia* in each locule of the anther; one end of a pollinium often prolonged into a slender tip attached to a sticky pad, the *viscidium* (a detachable portion of the rostellum), the viscidium and its attached pollinium or pollinia collectively forming a pollinarium; ovary inferior, unilocular, with very numerous, late-developing ovules on 3 expanded parietal placentas; fr mostly capsular, opening by 3(6) longitudinal slits but remaining closed top and bottom; seeds countless, minute; embryo mostly undifferentiated, only seldom with a barely recognizable cotyledon; endosperm wanting; strongly mycotrophic (sometimes nongreen) perennial herbs (many of the tropical ones epiphytic) with alternate (seldom opposite or whorled), parallel-veined, often somewhat fleshy lvs sheathing at base, the fls solitary or more often in racemes, spikes, or panicles; generally individually subtended by a bract. 600/15,000+.

1 Flowering plants lfless, essentially without chlorophyll.
  2 Fl solitary; lip bearded on its face .... 14. *Arethusa*.
  2 Fls in spikes or racemes; lip not bearded.
    3 Fls with a conspicuous, slender, free spur .... 20. *Tipularia*.
    3 Fls without a free spur.
      4 Lip longitudinally ridged; fls not white, or only the lip white.
        5 Lip with 1–3 low longitudinal ridges; pollinia 4.
          6 Stem arising from a cluster of coralloid roots; lvs none .... 21. *Corallorhiza*.
          6 Stem arising from a globose corm; plants with a solitary winter-lf .... 19. *Aplectrum*.
        5 Lip with 7 conspicuous longitudinal-ridges; pollinia 8 .... 15. *Hexalectris*.
      4 Lip not ridged; fls white or nearly so .... 4. *Spiranthes*.
1 Flowering plants at least partly green, normally with one or more lvs.
  7 Lip inflated and pouch-like, saccate, or boat-shaped, at least at base.
    8 Fls 1–3, large, brightly colored.
      9 Lip smooth; anthers 2; lvs 2–several .... 1. *Cypripedium*.
      9 Lip bearded; anther 1; lf 1 .... 18. *Calypso*.
    8 Fls numerous in spikes or spike-like racemes.
      10 Foliage lvs cauline, alternate; fls greenish-purple .... 2. *Epipactis*.
      10 Foliage lvs basal, the cauline reduced to scales; fls white and green .... 5. *Goodyera*.
  7 Lip not inflated or saccate or boat-shaped.
    11 Lip prolonged backward or downward at base into an evident spur.
      12 Lvs 1 or 2, basal or near-basal.
        13 Fls showy, anthocyanic at least in part .... 7. *Orchis*.
        13 Fls inconspicuous, white to greenish-white or yellowish-green .... 8. *Habenaria*.
      12 Lvs 3 or more, usually cauline .... 8. *Habenaria*.
    11 Spur wanting.
      14 Lip uppermost, the fls not resupinate.
        15 Lf solitary; fls large, 9–25 mm; lip bearded .... 13. *Calopogon*.
        15 Lvs 2–several; fls smaller, 2–6 mm; lip not bearded.
          16 Stem bearing scale-lvs between the foliage lvs and the raceme .... 6. *Ponthieva*.
          16 Stem naked between the foliage lvs and the raceme .... 16. *Malaxis*.
      14 Lip lowermost, the fls resupinate.
        17 Lip bearded or ridged or crested on its upper (inner) side.
          18 Cauline lvs whorled .... 11. *Isotria*.
          18 Cauline lvs alternate or solitary.
            19 Sep and lateral pet distinctly different in size, shape, and color .... 10. *Cleistes*.
            19 Sep and lateral pet essentially similar.
              20 Lip prominently bearded.
                21 Lf flat, well developed at anthesis .... 9. *Pogonia*.
                21 Lf plicate, absent or immature at anthesis .... 14. *Arethusa*.
              20 Lip not bearded.
                22 Fls few, from the axils of foliage lvs; lip white to pale pink .... 12. *Triphora*.
                22 Fls numerous in an elongate raceme; lip greenish and purple .... 2. *Epipactis*.
        17 Lip not bearded or ridged or crested.
           23 Foliage lvs in a single pair and apparently opposite.
            24 Lvs near the middle of the stem; lip bilobed or bifid at the tip .... 3. *Listera*.
            24 Lvs near the base of the stem; lip not lobed .... 17. *Liparis*.
           23 Foliage lf or lvs basal or cauline, but not appearing opposite.
            25 Stem bearing scale-lvs between the foliage and the raceme .... 4. *Spiranthes*.
            25 Stem naked between the foliage lvs and the raceme .... 16. *Malaxis*.

1. CYPRIPEDIUM L. Lady-slipper. Sep and lateral pet somewhat differentiated, widely spreading, the 2 lower sep usually connate; lip a large inflated pouch, its margins ± inrolled around the orifice; column declined over the orifice of the lip, bearing a fertile stamen (with granular, irregularly coherent pollen) on each side and dilated staminode above; perennial from fibrous roots, the erect stem bearing 2–several basal or cauline lvs and 1 or 2 large fls. 50, N. Temp.

1 Lower 2 sep separate and spreading .... 1. *C. arietinum*.
1 Lower 2 sep connate into one, this usually 2-nerved or 2-toothed at the tip.
  2 Sep and lateral pet white, broadly oval, obtuse .... 2. *C. reginae*.
  2 Sep and lateral pet yellowish to greenish or brown, ovate to linear, pointed.
    3 Lvs 2–several, cauline; lip yellow to white.
      4 Lip white, 1.5–2.5 cm .... 3. *C. candidum*.
      4 Lip white to yellow, if white then 5–6 cm.

5 Lip bright yellow, usually veined with purple, 2–5 cm; widespread . . . . 4. *C. calceolus.*
5 Lip white to dull yellow, 5–6 cm; e. Ky. . . . . 5. *C. kentuckiense.*
3 Lvs 2, basal; lip pink, varying to rarely white . . . . 6. *C. acaule.*

1. Cypripedium arietinum R. Br.   Ram's head l.-s. Stem slender, 1–4 dm, thinly hairy, with 2–3 sheathing scales below and 3–5 sessile lvs above the middle; lf-blades lanceolate to elliptic, often folded, 5–10 cm, finely ciliate, otherwise glabrous; fl solitary; sep and lateral pet 1.5–2.5 cm, greenish-brown, the upper sep lanceolate, the lower 2 sep separate, linear, the lateral pet lance-linear; lip whitish, strongly red-veined, 1.5–2.5 cm, prolonged downward to form a conical pouch. Moist, usually acid soils in coniferous woods; Que. to Man., s. to Mass., N.Y., Mich. and Minn. May, June.

2. Cypripedium reginae Walter.   Showy l.-s. Stem leafy, 4–10 dm, hirsute; lvs elliptic-oval, 1–2 dm, half as wide, clasping, strongly ribbed, hairy; fls 1–3; sep white, round-oval, obtuse, 2.5–4 × 1–3 cm, the lower 2 completely united; lateral pet white, flat, oblong, nearly as long as the sep but much narrower; lip 3–4 cm, white or suffused with pink, irregularly streaked with rose or purple outside and marked with purple inside; $2n=20$. Swamps, bogs, and wet woods; Nf. and Que. to N.D., s. to N.C., Ga., and Mo. June, July. (*C. hirsutum.*)

3. Cypripedium candidum Muhl.   White l.-s. Stem 1.5–4 dm, with a few sheathing scales below and 3–5 lvs; lf-blades lanceolate to narrowly elliptic, often overlapping at base, 8–15 cm, the uppermost smaller, erect, and subtending the solitary fl; sep and lateral pet lance-linear, greenish, often red-striped, 2–4 cm, the pet often closely twisted; lip white, veined with violet within, 1.5–2.5 cm; $2n=20$. In calcareous soils of marly bogs, open swamps, and wet prairies; N.Y. and N.J. to O., N.D., Nebr., and Mo. May, June. A hybrid with no. 4 is *C. × andrewsii* A. Fuller.

4. Cypripedium calceolus L.   Yellow l.-s. Stem leafy, 2–8 dm; lvs ± sheathing, oval to lance-ovate, 6–20 cm, half as wide; fls 1 or 2, each subtended by an erect foliaceous bract; sep and lateral pet greenish-yellow to purplish-brown; upper sep ovate to lance-ovate, sharply acute, 3–8 cm; united lower sep somewhat narrower, 2-toothed at the tip; lateral pet lanceolate, 3–8 cm, usually twisted; lip yellow, usually veined with purple, 2–5 cm; $2n=20$. Circumboreal, s. in Amer. to S.C., La., and N.M. May–July. Two vars. in N. Amer., partly isolated by differences in pollinators associated with fl-size.

Var. *pubescens* (Willd.) Correll, relatively robust, with large fls, the pet 5–8 cm, the lip 3–5 cm, occurs in mesic woods nearly throughout our range. (*C. pubescens.*)
Var. *parviflorum* (Salisb.) Fern., smaller, with smaller, darker fls, the pet 3–5 cm, the lip 2–3 cm, chiefly of moist or wet low places, is northern and cordilleran, s. in our range to N.Engl., the Lake states, n. N.J., Pa., O., and at progressively higher elev. in the S. Appalachian Mts. to N.C. (*C. parviflorum.*)

5. Cypripedium kentuckiense C. Reed.   Kentucky l.-s. Like a robust, pale-fld *C. calceolus*; plants 6–8 dm; lvs 15–20 cm; fls very large, the upper sep 7–8 cm, the lower connate pair 5–7 cm, the lateral pet 8–9 cm; lip white to creamy or dull yellow, 5–6 cm. Ravines and low woods; irregularly from e. Ky. and e. Tenn. to Ark., Okla., and La. May.

6. Cypripedium acaule Aiton.   Moccasin-fl; pink l.-s. Lvs 2, basal, subopposite, narrowly elliptic, 1–2 dm, thinly hairy, pale beneath; scape 2–4 dm, hairy, with a single lanceolate bract arching forward over the solitary fl; sep and lateral pet yellowish-green to greenish-brown, lanceolate, 3–5 cm, the 2 lower sep united; lip drooping, pink with red veins, 3–6 cm, hairy within, cleft along the upper side from the basal orifice to the summit, the inturned margins in contact; $2n=20$. In acid soil, from swamps and bogs to dry woods and sand-dunes; Nf. and Que. to Alta., s. to N.J. and n. Ind., and along the mts. and coastal plain to S.C. and Ala. Apr.–June. (*Fissipes a.*)

2. EPIPACTIS Zinn, nom. conserv.   Sep and lateral pet similar; lip strongly saccate in the basal half, the terminal lobe broadly ovate, not hinged to the basal part, crested at its base with 2 evident swellings; column very short and broad, the large sessile anther hinged to its back below the middle and somewhat projecting; pollinia 4; rhizomatous or stoloniferous plants with numerous alternate lvs and a terminal raceme. 25, N. Hemis.

1. Epipactis helleborine (L.) Crantz.   Helleborine. Erect from a short, praemorse rhizome, to 8 dm; lvs sessile and clasping, ovate to lanceolate, the lower to 10 cm, the upper progressively smaller; raceme 1–3 dm, many-fld, its bracts linear or narrowly lanceolate, the lower surpassing the fls; sep and lateral pet lance-ovate, 10–14 mm, acute, dull-green, strongly veined with purple; lip greenish and purple; $2n=36–44$. Native of Europe, now widespread in our range along roadsides and in woods. July, Aug. (*E. latifolia; Serapias h.*)

3. LISTERA R. Br., nom. conserv.   Twayblade. Sep and lateral pet much alike, spreading or reflexed; lip declined or projecting horizontally, in our spp. much longer than the lateral pet, 2-lobed or 2-cleft; column erect, shorter than the pet, the anther borne on its back near the tip; pollinia 2; stem bearing a single pair of broad, opposite, sessile lvs near its middle and a terminal raceme of small, green to purple or red fls; our spp. glabrous below the lvs, ± pubescent above them. 25, N. Hemis.

1 Stout, 2–6 dm; lvs 5–15 cm; fls ca 25–60+; locally intr. in Ont. .... 4. *L. ovata.*
1 Slender, 1–3.5 dm; lvs 1–5(–7) cm; fls seldom more than 25; native.
  2 Lip cleft to about the middle or below into linear segments; column under 1 mm.
    3 Lip 3–5 mm, twice as long as the lateral pet; pedicels and axis glabrous .... 5. *L. cordata.*
    3 Lip 6–10 mm, 4 times as long as the pet; pedicels and axis glandular .... 6. *L. australis.*
  2 Lip not cleft to the middle, its lobes oblong or ovate; column 1.5–4 mm.
    4 Lip broad-based, with retrorse auricles but without lateral teeth .... 1. *L. auriculata.*
    4 Lip narrowed to the base, and often with lateral teeth near the base.
      5 Pedicels glandular; lateral teeth of the lip inconspicuous or none .... 2. *L. convallarioides.*
      5 Pedicels glabrous; lateral teeth conspicuously projecting .... 3. *L. smallii.*

1. Listera auriculata Wieg. Auricled t. Stem 1–2.5 dm; lvs ovate or round-ovate, 2–5 cm, longer than the peduncle, broadly rounded at base, obtuse or subacute; axis glandular; pedicels and ovary glabrous; bracts 2 mm; fls up to ca 20; sep and lateral pet lance-ovate to oblong, 3–4.5 mm; lip pale green oblong in outline, 6–11 mm, maintaining its width to the rounded base and there bearing a pair of minute retrorse auricles, slightly constricted at the middle, cleft at the summit for about a fourth its length; column 2.5–3 mm. Wet woods and thickets; Nf. and Que. to N.H. and N.Y.; Isle Royale, Mich. July, Aug. (*Ophrys a.*)

2. Listera convallarioides (Swartz) Torr. Broad-lvd t. Stem 1–3 dm; lvs broadly ovate, 3–5(–7) cm, longer than the peduncle, obtuse or subacute; axis, pedicels, and ovary glandular-puberulent; bracts 3–5 mm; fls up to ca 20; sep and lateral pet 4–5 mm, reflexed; lip greenish, translucent, cuneate, 9–10 mm, deeply retuse and ca 5 mm wide at the summit, usually with a short (often obscure) tooth on each side near the narrow base; column 2.5–4 mm; $2n=36$. Wet woods, usually in deep shade; Nf. to Alas., s. to Mass., N.Y., Mich., Minn., and Ariz. July, Aug. (*Ophrys c.*) A hybrid with no. 1 is *L.* × *veltmanii* Case.

3. Listera smallii Wieg. Appalachian t. Stem 1–3 dm; lvs very broadly ovate to reniform, 1.5–4 cm, often wider than long, usually acute, mostly shorter than the peduncle; axis glandular, pedicels and ovary glabrous; bracts 2–3 mm; fls up to ca 15; sep and lateral pet lance-oblong, 3–4 mm, spreading or reflexed; lip whitish, broadly cuneate-obovate, 7–8 mm, nearly as wide, deeply retuse with a minute tooth in the sinus, bearing a conspicuous pair of spreading lateral teeth near the base; column 1.5–3 mm. Wooded slopes in the mts., s. Pa. to e. Ky. and S.C. July, Aug. (*Ophrys s.*)

4. Listera ovata (L.) R. Br. Big t. Stout and coarse, 2–6 dm; lvs 5–15 cm, ovate to elliptic, usually shorter than the peduncle; infl elongate, with numerous (ca 25–60+) fls, its axis glandular-puberulent; sep ovate, 4–5 mm; lateral pet linear, 4 mm; lip yellow-green, 7–10 mm, cleft nearly to the middle into slender lobes and with a minute tooth in the sinus, acutely angled downward from the narrow base; column 2 mm; $2n=34$–38. Disturbed wet woods; widespread in Eurasia, locally intr. on the Bruce Peninsula of Ont. June, July.

5. Listera cordata (L.) R. Br. Heart-lvd t. Stems 1–2.5 dm; lvs 1–3 cm, shorter than the peduncle, truncate or subcordate at base, but also abruptly contracted at the point of insertion; axis, pedicels, and ovary glabrous; bracts minute, 1 mm; fls up to ca 25, green to purple or red; sep and lateral pet 2–2.5 mm, ovate oblong; lip 3–5 mm, with 2 projecting lateral teeth near the base, cleft to the middle or below into divergent linear lobes; column 0.5 mm. Deep wet woods and sphagnum-bogs; circumboreal, s. in Amer. to N.Y., Pa., Mich., and Minn., and in the mts. to N.C. and N.M. June, July. (*Ophrys c.*)

6. Listera australis Lindl. Southern t. Stem 1.5–3 dm; lvs ovate, 1–3.5 cm, shorter than the peduncle, rounded at base, abruptly apiculate; axis and pedicels glandular; ovaries glabrous; bracts 1 mm; fls up to ca 25, greenish-purple to dull red, the sep and lateral pet ovate-oblong, reflexed, 2 mm; lip 6–10 mm, narrowly oblong, cleft half its length into 2 slender lobes, and with a pair of narrow, retrorse auricles at the base; column under 1 mm. Shaded bogs and wet woods, mainly on the coastal plain from N.J. to Fla. and e. Tex.; irregularly in Ky., Tenn., N.Y., Vt., e. Ont., s. Que., N.B. and N.S. June, July. (*Ophrys a.*)

4. SPIRANTHES Rich. Ladies' tresses. Sep and lateral pet similar, the pet connivent with the upper sep (or with all the sep) and projecting forward over the lip and column; lip oblong or ovate to pandurate in outline, bearing 2 small callosities near the basal angles, its margins below the middle upturned and embracing the short column, the more distal part ± decurved, not crested or ridged, but often with a median area of different color; anther borne on the back of the column, its 4 pollinia attached to the viscidium and commonly surpassed by the bidentate or bifid rostellar beak; slender herbs with narrow (seldom broader), chiefly basal lvs, the erect stem bearing several reduced lvs (or mere sheaths) and a terminal, spirally twisted, spike-like raceme; fls mostly white or whitish at least in large part. (*Ibidium.*) 300, cosmop.

The fls decrease in size progressively from the bottom to the top of the infl. Measurements given here are for fls in the lower third. In spp. 1–5 the infl is ordinarily such a close spiral that the fls appear to be in 3 or 4 vertical, straight or somewhat spiral ranks.

In spp. 6–11 the infl is typically more lax and open, with the fls obviously in a single long spiral, but some of these spp. vary to forms with the infl approaching that of the first group. A number of hybrids have been documented.

1 Lip pandurate, constricted just above the middle and strongly expanded above; sep and lateral pet all connivent to form an open-tubular hood .... 1. *S. romanzoffiana.*
1 Lip otherwise, sometimes narrowed near the middle, but then not much if at all expanded above; lateral sep seldom connivent with the pet and upper sep.
  2 Sep connate for 0.6–0.8 mm at base; viscidium oval; lip largely bright yellow .... 2. *S. lucida.*
  2 Sep distinct or nearly so; viscidium linear or none; lip usually paler.
    3 Root single (2); axis of the infl glabrous .... 11. *S. tuberosa.*
    3 Roots several; axis of the infl sparsely to densely short-hairy or glandular.
      4 Axis of the infl densely beset with short, pointed, eglandular hairs .... 7. *S. vernalis.*
      4 Axis of the infl sparsely to often densely (but shortly) glandular-hairy.
        5 Lip with conspicuous, distally widened green veins .... 9. *S. praecox.*
        5 Lip without such veins.
          6 Lip with a lacerate-dentate distal margin; lvs linear .... 8. *S. laciniata.*
          6 Lip with a wavy or crisped or erose distal margin; lvs often wider.
            7 Fls larger, the upper sep usually over 7.5 mm.
              8 Lvs present at anthesis; lateral sep loose but scarcely spreading .... 4. *S. cernua.*
              8 Lvs wanting at anthesis; lateral sep ± spreading .... 5. *S. magnicamporum.*
            7 Fls smaller, the upper sep usually less than 7.5 mm.
              9 Infl dense, the fls in 3 or 4 vertical files; rostellum and viscidium wanting (unique among our spp.) .... 3. *S. ovalis.*
              9 Infl mostly looser, the fls ± evidently in a single spiral.
                10 Basal lvs ± erect, 5–10 times as long as wide; lip yellowish .... 6. *S. casei.*
                10 Basal lvs spreading, 1.5–3.5 times as long as wide (or wanting at anthesis); lip largely green .... 10. *S. lacera.*

    1. Spiranthes romanzoffiana Cham. Hooded l.-t. Plants 1–4 dm; basal lvs narrow, up to 20(–25) cm, 5–15 mm wide; cauline sheaths 3–5, the uppermost ± bladeless; spikes 3–10 cm, dense (especially above), the fls in 3 or 4 vertical ranks, individually up- and outcurved, white or ochroleucous to chloroleucous; sep basally connate, rather prominently veined, sparsely viscid-hairy, connivent with the lateral pet to form an open-tubular hood 7–12 mm, the lateral sep often with reflexed tip; lip sharply deflexed, about as long as the sep, pandurate with an expanded, erose tip, with prominent divergent veins below the constriction, generally only the midvein continued beyond; basal callosities of the lip inconspicuous, under 1 mm; viscidium 1.3–2 mm; $2n=(30)44(60)$. Swamps, bogs, and wet shores; Nf. and Lab. to Alas., s. to N.Y., n. Pa., n. O., Mich., Io., Nebr., and Calif.; Ireland. July–Sept. (*S. stricta.*)

    2. Spiranthes lucida (H. Eaton) Ames. Shining l.-t. Slender, 1–2.5 dm; basal lvs lance-oblong, to 12 cm, 7–15 mm wide; cauline sheaths 2(3), much reduced, the upper scale-like; axis of the spike minutely glandular-hairy; fls widely spreading; sep 5–5.5 mm, basally connate for 0.6–0.8 mm, directed forward (as also the lateral pet), somewhat spreading only near the tip; lip 5–6 mm, broadly oblong, rounded-truncate and erose distally, the distal two-thirds bright yellow or orange-yellow with white margins; basal callosities of the lip inconspicuous, under 1 mm; viscidium oval, 0.3–0.6 mm; seed-coat netted; $2n=44$. Damp woods, marshes, and wet shores, calciphile; N.S. and N.B. to s. Ont. and Mich., s. to Ky. and w. Va. May–July. (*S. plantaginea,* a preoccupied name.)

    3. Spiranthes ovalis Lindl. Oval l.-t. Plants 1.5–3 dm; basal lvs 2 or 3, to 12 cm, 5–10(–15) mm wide; cauline sheaths 3–4, the lower with divergent recurved blade 2–5 cm; infl 2–6 cm, with glandular-puberulent axis, dense, the fls in 3 or 4 vertical ranks, individually widely spreading or somewhat deflexed, urceolate-cylindric, white or the lip creamy at the middle; sep and lateral pet 3.5–5 mm, the pet usually widest in the distal third and with one vein; lip 4–5 mm, ovate-oblong, rounded distally, often somewhat constricted near the middle, its basal callosities short, slender, less than 1 mm. Moist woods and bottomlands, and also in old fields and pastures; s. Pa. to sw. Mich. and Ill., s. to Fla. and Tex. Sept., Oct. Our plants, mainly autogamous and lacking a viscidium and rostellum (unique among our spp. in these regards) have been described as var. *erostellata* Catling, in contrast to var. *ovalis,* of Ark., La., and Tex., with slightly larger fls of normal structure.

    4. Spiranthes cernua (L.) Rich. Nodding l.-t. Plants 1–4(–8) dm; basal lvs narrow, to 30 cm, 5–50 mm wide; cauline sheaths 3–8, sometimes with a spreading-recurved blade; infl 2–18 cm, with shortly glandular-hairy axis, dense, the fls ± in 3 or 4 vertical ranks, 8–11(–15) mm, individually widely spreading or somewhat declined, urceolate-cylindric, only the upper sep and lateral pet connivent, the lateral sep looser but scarcely spreading; lip moderately to sometimes (var. *ochroleuca*) strongly arcuate-recurved, ovate or ovate-oblong or obcuneate, often dilated and cordate at base, erose or crisped around the broad summit, somewhat yellowish-green in the center, the fls otherwise white or whitish (or more ochroleucous in var. *ochroleuca*); callosities at base of lip often 1 mm or more, higher than thick; viscidium 1–2 mm, narrow; $2n=30, 45, 60, 61$, the plants often asexual. Open, moist, often sandy places; N.S. and Que. to se. N.D., s. to Fla. and Tex. Aug., Sept. Common, variable, and imperfectly divisible into an indefinite number of ecogeographic phases. Robust, diploid plants, to 8 dm, with fls 10–14 mm, often spreading by stolon-like roots, and with the cauline sheaths bearing spreading, ± recurved blades, mainly on the se. coastal plain, often in estuaries, form the ill-defined var. *odorata* (Nutt.) Correll; the fls are said to have the odor of vanilla, in contrast to the inodorous fls of others vars. (*S. odorata.*) Plants with slightly more yellowish fls and a strongly decurved lip with a relatively long claw (0.8–1.5 mm, in contrast to 0.3–0.8 mm in var. *cernua*) thought to be mainly diploid but not always sexual, occurring from P.E.I. and s. N.S. to Mich., s. to Ky. and the mts. of Va. and N.C., have been distinguished as var. *ochroleuca* (Rydb.) Ames. (*S. ochroleuca; S. steigeri.*)

5. **Spiranthes magnicamporum** Sheviak. Great Plains l.-t. Much like no. 4; plants with a strong almond-odor; lvs withering before anthesis; cauline sheaths 4–8, with erect, sheathing, often overlapping blades; lateral sep more spreading, often over the top of the fl; lip more yellowish, its basal callosities short and conical, usually less than 1 mm, as wide as high; $2n=30$, but sometimes asexual. Open, calcareous prairies; Great Plains from N.D. to Tex., e. to nw. Ind. and irregularly to Ky., Mich., sw. Ont., and Pa. Sept., Oct.

6. **Spiranthes casei** Catling & Cruise. Plants 2–4 dm; basal lvs lanceolate or lance-ovate to broadly lance-oblong, 5–15 cm × 8–15 mm; cauline sheaths (2)3 or 4, the lower with short blade; infl 2–16 cm, loosely to fairly densely fld; fls ochroleucous or greenish-yellow, more greenish basally, 5–7.5 mm, stocky, the sep 2–3.7 mm wide, the pet 1.7–2.8 mm wide; lip fleshy, obovate or elliptic-ovate, papillate below, the basal callosities ca 1 mm, higher than wide; $2n=60, 75$. Sandy, acid soil, often with *Polytrichum*; N.S.; Me. to n. Pa., w. to Mich. and Wis. Aug., Sept.

7. **Spiranthes vernalis** Engelm. & A. Gray. Spring l.-t. Plants 2–8 dm; basal lvs narrow, 5–25 cm × 4–12 mm; cauline sheaths usually 5–7, the lower with blades to 15 cm; infl 3–15+ cm, its axis densely beset with pointed hairs 0.2–0.3 mm, the fls mostly in a single long spiral (seldom more condensed), spreading or somewhat deflexed, urceolate-cylindric, white or ivory with the lip more yellowish centrally; sep and lateral pet 5–9.5 mm, the upper sep a bit longer than the spreading-divergent lateral ones; lip 5–8 mm, ovate or broadly ovate, basally cuneate, distally ± obtuse and crisped-erose, papillate on both sides especially distally, its basal callosities 0.5–1 mm, higher than thick; $2n=30$. Open, sandy, moist or dry acid soil; Mass. to Fla., w. to O., se. S.D., Kans., and Tex.; Mex. and C. Amer. June–Aug.

8. **Spiranthes laciniata** (Small) Ames. Lace-lipped l.-t. Plants 2–10 dm; basal lvs narrow, to 30 cm, 3–10(–15) mm wide; cauline sheaths 4–6, the lower with well developed blade; infl 5–16 cm, its axis minutely glandular-hairy, the fls closely spaced in a single straight row or an evident spiral, spreading; urceolate-cylindric, 6–10 mm, mainly white; lip recurved-deflexed, with a yellow center, rather shallowly lacerate-fringed around the end, papillate on the lower surface, its basal callosities 0.7–1.5 mm, higher than thick. Open wet places on the coastal plain from N.J. to Fla. and Tex. July–Sept.

9. **Spiranthes praecox** (Walter) S. Wats. Giant l.-t. Plants 2–8 dm; basal lvs narrow, to 20 cm, 3–10 mm wide; cauline sheaths 4–6, with blades to 10 cm; infl elongate, its axis minutely glandular-hairy, the fls closely spaced in a single spiral or the spiral sometimes so tight as to create 3 or 4 vertical files of fls, these spreading or slightly deflexed, urceolate-cylindric, 5–10 mm, mainly white, the oblong lip with a yellowish center and prominent divergent green veins, glabrous below and distally so above, its basal callosities 0.5–1.2 mm, higher than thick. Moist or wet open places on the coastal plain from N.J. to Fla. and Tex. June, July.

10. **Spiranthes lacera** (Raf.) Raf. Slender l.-t. Slender, 1–4 dm; basal lvs ovate-elliptic or ovate-lanceolate, petiolate, spreading to form a rosette (but sometimes withering early), 1–6 × 0.6–3 cm; cauline sheaths (3)4–7, essentially bladeless; infl 3–15+ cm, its axis sparsely and minutely glandular-hairy especially upwards (hairs 0.1–0.2 mm), the fls in a single loose or compact spiral, spreading or somewhat declined, subcylindric, small, 3.5–6.5 mm, white except for the large green central spot on the lip; lip broadly oblong, erose at the rounded-subtruncate summit, its basal callosities 0.5–1 mm, higher than thick; $2n=30$. Open, sandy places, often with pines; N.S. and N.B. to Sask., s. to Fla. and Tex. Two ill-defined vars. with broadly overlapping range: Var. *lacera*, with loose infl, lvs persistent at anthesis, and fl early (July), is northern, s. to Va., Tenn., and Mo. Var. *gracilis* (Bigelow) Luer, with denser infl, lvs withered by the time of anthesis, and fl late (Aug., Sept.) is southern, n. to s. N.H., c. Vt., s. Wis., and Nebr. (*S. gracilis.*)

11. **Spiranthes tuberosa** Raf. Little l.-t. Slender, 1.5–3 dm; root single (unique among our spp.), turbinate, to 6(–10) cm, 4–12 mm thick, sometimes accompanied by the partly decomposed root of the previous year; basal lvs ephemeral, usually ovate, short-petiolate, 2–4(–5) cm, 6–15(–20) mm wide; usually 4–6 bladeless sheaths on the stem; infl 2–8+ cm, its rachis (as well as the whole herbage) glabrous, the ± cylindric fls rather closely set in a single long spiral, 3–5.5 mm, pure white, the lip broadly ovate, distally truncate or sometimes rounded and crisped or slightly erose, puberulent with minute hairs 0.2–0.4 mm on the upper surface, minutely papillate below, the basal callosities less than 1 mm. Acid, usually rather dry soil; Mass. to Fla., w. to Ill., Mo., and Tex. Aug., Sept. (*S. grayi; S. beckii,* misapplied.)

5. **GOODYERA** R. Br. Rattlesnake-plantain. Upper sep and lateral pet adherent by their margins to form a concave galea extending forward over the lip; lateral sep free, scarcely spreading except at the tip; lip shorter than the galea, ± strongly saccate or pouch-like at base, prolonged distally into a horizontal or deflexed beak; column short, bearing the anther on its back below the usually bifid rostellar tip; pollinia 2; perennial, forming new rosettes by budding from creeping rhizomes; lvs all basal, often white-reticulate, in ours narrowed to a broadly petiolar base; scape bracteate, terminating in a raceme of white or greenish fls; our spp. all with glandular hairy scape and infl, the lateral pet thin, white, differing in texture from the dorsal sep to which they adhere. 25+, cosmop.

1 Beak of the lip less than half as long as the body; infl dense, cylindric .... 1. *G. pubescens.*
1 Beak half to fully as long as the body; infl looser, spirally secund.
  2 Lvs typically white only on the midstrip; fls larger (see description) .... 2. *G. oblongifolia.*
  2 Lvs typically patterned green-and-white; fls smaller (see description).
    3 Lip shallowly saccate, the pouch longer than deep; anther acuminate .... 3. *G. tesselata.*
    3 Lip deeply saccate, the pouch about as deep as long, anther blunt .... 4. *G. repens.*

1. Goodyera pubescens (Willd.) R. Br.   Downy r.-p. Stout, 2–4 dm; scape with 4–14 (avg 7) bracts; lf-blades ovate or lance-ovate, 3–6 cm, the dark green midrib flanked by a pair of broad white stripes, and the primary and secondary lateral veins forming a white reticulum; infl dense, cylindric, 4–10 cm; galea broadly elliptic, very convex, 4–5.5 mm, upturned at the tip; lateral sep broadly ovate to obovate, 3.5–5 mm, abruptly short-acuminate; lip subglobose, 3.5–5 mm, its straight beak less than 1 mm, scarcely projecting beyond the ventricose body; rostellar beak obsolete; $2n=26$. Dry woods; s. Me. and s. Que. to Minn., s. to S.C., Ga., Ala., and Ark. July, Aug. (*Epipactis p.; Peramium p.*)

2. Goodyera oblongifolia Raf.   Western r.-p. Stout, mostly 2–4 dm; scape with 4–7 bracts; lf-blades lance-ovate to narrowly elliptic, 3–6 cm, usually white only on the midstrip, seldom also on some of the other veins; infl a loose to often rather tight spiral; galea 5–10 (avg 7.5) mm; lip 5–8 (avg 6) mm, with a deeply concave base tapering to a spreading or slightly recurved boat-shaped tip with involute or upright margins; anther acuminate, evidently surpassed by the 2.3–3.6 mm rostellar beak; $2n=30$. Dry or moist, hardwood or coniferous forests; P.E.I. and the Gaspé Peninsula to n. Vt.; vicinity of lakes Superior and Huron and n. Lake Michigan; widespread in the w. cordillera. July–Sept. (*G. decipiens; Epipactis d.; Peramium d.*)

3. Goodyera tesselata Lodd.   Alloploid r.-p. Intermediate in most respects between nos. 2 and 4, and very probably of alloploid origin; plants 1.5–3.5 dm; scape with 3–5 bracts; lf-blades mostly 2–5.5 cm, variably white-reticulate, the extremes approaching the putatively ancestral spp.; galea 4–7 (avg 5) mm; lip 3–5.5 mm, shallowly saccate, the pouch longer than deep; anther acuminate, somewhat surpassed by the 0.6–1.7 mm rostellar beak; $2n=60$. Rich coniferous or hardwood forests; Nf. to se. Man., s. to N.J., Md., O., Mich., Wis., and Minn. July, Aug. (*Epipactis t.; Peramium t.*)

4. Goodyera repens (L.) R. Br.   Lesser r.-p. Slender, mostly 1–2 dm; scape with 2–4(5) bracts; lf-blades ovate or oblong, 1–3 cm, dark bluish-green, in our plants usually white-reticulate, the midrib and veins typically green, but the lateral and cross-veins bordered with white (or pale green); infl loose, spirally secund, 3–6 cm; galea 3–5.5 (avg 4) mm; lip 3–4 mm, deeply saccate, the prominent, slender, abruptly deflexed beak-tip with spreading margins; rostellar beak 0.2–0.6 mm, scarcely prolonged beyond the blunt anther; $2n=30$. Dry or moist cold woods, especially under conifers; circumboreal, in our range s. to N.Y., Mich., and Minn., and in the mts. to N.C. July, Aug. Our plants, with white-veiny lvs, belong to var. *ophioides* Fern., mainly of e. N. Amer. (*G. o.; Peramium o.*)

6. PONTHIEVA R. Br.   Fls not resupinate, the lip uppermost; lateral sep inequilateral; lateral pet arising from near the middle of the column, strongly inequilateral, directed forward, cohering at the tip; lip clawed, the broad blade folded and at first embracing the column; anther subterminal, with 4 pollinia; rhizomatous perennials with a cluster of basal lvs and an erect scape bearing a few narrow scales and a terminal raceme. 25, warm Amer.

1. Ponthieva racemosa (Walter) C. Mohr.   Shadow-witch. Lvs very thin, spreading on the ground, elliptic, 5–10 cm, acute, tapering to the base; scape 2–4 dm, glandular-hairy, as also the axis and pedicels; sep white with green stripes, 5–6 mm, ovate to oblong; lateral pet white with greenish-yellow veins, broadly falcate-ovate, 5 mm, nearly as wide; lip arising near the middle of the column, rotund, white, veined with green, 6 mm, abruptly pointed. Bogs and wet woods on the coastal plain; se. Va. to Fla. and trop. Amer. Sept., Oct.

7. ORCHIS L.   Sep and lateral pet similar in color but differing in size, the pet connivent with the upper sep or with all the sep to form a concave hood over the column; lip large, prolonged at its base into a conspicuous spur; anther broadly sessile atop the column, pollen-sacs contiguous; pollinia 2; viscidia 2, enclosed in a single pouch; perennial from short rhizomes and thickened roots, the scape bearing 1 or 2 lvs near the base and a loose, terminal, few-fld raceme. 100, N. Temp.

1 Lvs 2; lip unspotted, not lobed or notched .... 1. *O. spectabilis.*
1 Lf 1; lip purple-spotted, 3-lobed, notched at the summit .... 2. *O. rotundifolia.*

1. Orchis spectabilis L.   Showy orchis. Lvs 2, rather fleshy, narrowly obovate to broadly elliptic, 8–15 cm; scape 1–2 dm, stout; bracts foliaceous, lance-oblong, 15–50 mm; sep and lateral pet pink to pale purple, 13–18 mm, all connivent; lip white, 15–20 mm, rhombic-ovate; spur stout, about equaling the lip; $2n=42$. Rich woods; N.B. to Minn. and Nebr., s. to Ga. and Ark. May, June. (*Galeorchis s.; Galearis s.*)

2. Orchis rotundifolia Banks.   One-lf orchis. Lf 1, elliptic to ovate or obovate, 4–11 cm; scape 1–2.5 dm, slender; bracts lance-linear, 6–15 mm; sep and lateral pet pale purple to white, oblong-obovate, 5–8 mm, the pet slightly narrower than the sep; 2 lateral pet and upper sep connivent, the lateral sep spreading; lip white, spotted with pale purple, 6–9 mm, 3-lobed, the lateral lobes short-ovate, ascending, the terminal one triangular, notched at the summit; spur much shorter than the lip; 2n=42. Wet woods and swamps; Greenl. to Alas., s. to Que., n. N.Y., Mich., Minn., and B.C. June, July. (*Amerorchis r.*)

8. HABENARIA Willd.   Rein-orchid. Sep and lateral pet ± alike in form and color or the pet smaller, in our spp. the pet usually connivent with the dorsal sep to form a hood over the column, the lateral sep spreading or recurved; lip linear to ovate or obovate or 3-lobed, entire to toothed or fringed, prolonged backward at base into a spur; column short, capped by the broadly sessile anther; pollen-sacs 2, commonly separated by connective tissue or by part of the stigma, each containing a clavate (seldom bifid) pollinium attached at its narrow end to a usually ± exposed viscidium; perennial herbs, often from a fascicle of fleshy roots, the erect stem with basal or cauline, alternate lvs and a terminal spike or raceme. 500+, cosmop. (*Coeloglossum, Gymnadeniopsis, Limnorchis, Piperia, Platanthera.*) Hybrids are fairly frequent, in spite of modal differences in pollinators.

1 Lip entire or minutely toothed or erose, but neither fringed nor tripartite.
  2 Foliage lvs cauline.
    3 Lip over 1 cm; spur elongate, mostly 4–5 cm .... 14. *H. correlliana.*
    3 Lip up to ca 1 cm; spur shorter, up to ca 2.5 cm.
      4 Lip with 2 or 3 teeth or short lobes at the tip.
        5 Spur pouch-like, 2–3 mm, shorter than the lip .... 1. *H. viridis.*
        5 Spur elongate, slender, 7–12 mm, evidently longer than the lip .... 2. *H. clavellata.*
      4 Lip entire to minutely denticulate or erose, but not definitely toothed.
        6 Lip with an erect tubercle on the upper side below the middle .... 3. *H. flava.*
        6 Lip not tuberculate.
          7 Fls strongly ascending or appressed, at least the lower conspicuously exceeded by their
            scabrellate-margined bracts.
            8 Lip bluntly lanceolate, gradually widened below; fls greenish .... 4. *H. hyperborea.*
            8 Lip abruptly widened at base; fls white .... 5. *H. dilatata.*
          7 Fls spreading, even the lowest ones surpassing their glabrous bracts.
            9 Spur 4–8 mm; fls golden yellow .... 6. *H. integra.*
            9 Spur 11–23 mm; fls white .... 7. *H. nivea.*
  2 Foliage lvs basal (sometimes with 1 or 2 bladeless sheaths below them); stem naked or with a few
    bract-like lvs only.
    10 Spur mostly 3–9 mm; lvs mostly ca twice as long as wide, or longer.
      11 Lf solitary, persistent through anthesis; lateral sep 3.5–6 mm .... 8. *H. obtusata.*
      11 Lvs mostly 2 or 3, soon withering; lateral sep 1.5–3(–3.5) mm .... 9. *H. unalascensis.*
    10 Spur mostly 13–45 mm; lvs usually broader, nearly or fully as wide as long.
      12 Stem bractless; ovary sessile .... 10. *H. hookeri.*
      12 Stem bracteate; ovary on a pedicel 5–10 mm .... 11. *H. orbiculata.*
1 Lip tripartite or fringed or both.
  13 Body of the lip oblong to oblanceolate, fringed, not tripartite.
    14 Fls orange (rarely light yellow).
      15 Spur 5–9 mm .... 12. *H. cristata.*
      15 Spur 18–28 mm .... 13. *H. ciliaris.*
    14 Fls white .... 15. *H. blephariglottis.*
  13 Body of the lip deeply tripartite.
    16 Fls white or greenish-white or pale yellowish-green; lip deeply fringed.
      17 Pet linear-spatulate, blunt, entire; lateral sep deflexed .... 16. *H. lacera.*
      17 Pet cuneate to broadly obovate, toothed; lateral sep divergent .... 17. *H. leucophaea.*
    16 Fls magenta (white forms rarely occur); lip relatively shallowly fringed, or merely erose.
      18 Lip merely erose, the teeth rarely more than 1 mm deep .... 18. *H. peramoena.*
      18 Lip distinctly fringed, most or all of the segments over 1 mm .... 19. *H. psycodes.*

1. Habenaria viridis (L.) R. Br.   Bracted orchid. Slender or stout, 2–5 dm; fleshy-forked roots renewed annually; lowest 1 or 2 lvs reduced to bladeless sheaths; main foliage lvs 3 or more, obovate to oblanceolate, 5–12 cm, up to 5 cm wide, the upper progressively reduced and passing into bracts; infl 5–20 cm, the lanceolate, foliaceous bracts (especially the lower) surpassing the fls; fls greenish, often tinged with purple; lip 6–10 mm, its margins upcurved at the base (nectariferous within the minute lateral pouches so produced), the body obcuneately widened distally, terminating in 3 teeth, the central one the shortest; lateral pet lanceolate, nearly concealed by the incurved sep; spur 2–3 mm, pouch-like, with a minute orifice; 2n=40. Moist woods; circumboreal, in Amer. s. to N.J., Md., W.Va., O., Io., and Nebr., and in the mts. to N.C. and Colo. May–Aug. Our plants belong to the American and east Asian var. *bracteata* (Muhl.) A. Gray. (*Coeloglossum v.,* and var. *virescens.*)

2. Habenaria clavellata (Michx.) Sprengel. Club-spur orchid. Slender, 1–4 dm; foliage lf 1, near or just below (seldom well below) the middle of the stem, linear-oblong to narrowly elliptic or oblanceolate, 7–16 cm, up to 3 cm wide; upper lvs 1–few, much reduced; infl open, 2–6 cm, the narrowly lanceolate bracts shorter than the 5–15 fls; fls divergent, white or tinged with green or yellow, twisted to one side so that the spur is lateral; lip broadly cuneate, 3–5 mm, shallowly (sometimes obscurely) 3-lobed at the summit; sep and lateral pet broadly ovate, 4–5 mm; spur 7–12 mm, strongly curved, dilated at the tip; $2n=42$. Acid bogs and wet soils, especially in sphagnum; Nf. to Ont. and Minn., s. to Fla. and Tex. June–Aug. (*Platanthera c.*)

3. Habenaria flava (L.) R. Br. Tubercled orchid. Stem 3–7 dm; well developed lvs lance-linear to lanceolate or lance-elliptic, to 20 × 5 cm, the upper much reduced or bract-like; spike loose or compact, 5–20 cm, 1.5 cm thick; bracts lance-linear, shorter to much longer than the fls; fls sessile, greenish-yellow or green, 5–6 mm wide; lip deflexed, 4–6 mm, often irregular on the margin, bearing a conspicuous, fin-like protuberance on the upper side just below the middle, and usually with a small lateral lobe on each side at the base; spur 3–6 mm; $2n=42$. Boggy or swampy ground and flood-plains; N.S. and s. Que. to Minn., s. to Fla. and Tex. June–Sept. (*Platanthera f.; Perularia f.; Perularia scutellata.*) Variable in aspect. Northern plants (including most of those in our range) usually have a leafier stem and a congested spike with elongate bracts. These have been distinguished as var. *herbiola* (R. Br.) Ames & Correll, in contrast to the more southern var. *flava*, extending n. into our s. margin, with loose spike, shorter bracts and broader lip.

4. Habenaria hyperborea (L.) R. Br. Tall northern bog-orchid. Stout, 3–10 dm; lowest lf bladeless; principal foliage lvs lanceolate to oblanceolate, to 25 × 3 cm, the upper smaller and passing into the bracts; spike compact, 6–20 cm, many-fld; bracts lanceolate, the lower 1.5–4 cm long; fls erect or appressed, green or greenish-white; lip lance-ovate, 4–7 mm, blunt, gradually and uniformly widened toward the base; lateral pet lanceolate, directed forward and incurved under the upper sep; spur a little shorter than or as long as the lip; $2n=42$, 84. Bogs and wet woods; Greenl. and Iceland to Alas., s. to N.J., Pa., Ind., Nebr., and N.M. June–Aug. (*Platanthera h.; Limnorchis h.; L. huronensis.*) A frequent hybrid with no. 5 is *H.* × *media* (Rydb.) Niles.

5. Habenaria dilatata (Pursh) Hook. Tall white bog-orchid. Stout or slender, to 1 m; lowest lf bladeless; principal foliage lvs lanceolate or lance-linear, to 20 × 4 cm, the upper smaller and passing into the bracts; spike 1–3 dm, dense or open; bracts narrowly lanceolate, the lowest 1.5–4 cm; fls erect or appressed, white; lip 6–8 mm, blunt, lance-ovate, conspicuously widened at base; lateral pet lance-ovate, falcate, directed forward and incurved under the upper sep; spur slender, as long as the lip; $2n=42$. Bogs and wet woods; Iceland and Greenl. to Alas. and ne. Asia, s. to N.Y., Pa., O., Io., N.M., and Calif. (*Platanthera d.*)

6. Habenaria integra (Nutt.) Sprengel. Yellow orchid. Slender, 3–6 dm; foliage lf 1(2) narrowly lance-linear, folded, recurved, to 25 cm, the second much smaller; upper lvs several, erect, much reduced and bract-like; infl ovoid or cylindric, dense, 3–9 × 2.5 cm; bracts lance-linear, even the lowest shorter than the ovary; fls golden-yellow, divergent; lip ovate, 5 mm, crenulate or erose; spur 4–8 mm, tapering to the tip; pollen-sacs adjacent and parallel, the viscid ends of the caudicles of the pollinia adjacent and immediately above the orifice to the nectary. Open acid bogs and pine-barrens on the coastal plain; s. N.J. and Del. to Fla. and Tex., n. in the interior to Tenn. (*Gymnadeniopsis i.; Platanthera i.*)

7. Habenaria nivea (Nutt.) Sprengel. Snowy orchid. Slender, 3–6 dm; foliage lvs (1)2–3, linear, 5–25 cm, folded; upper lvs several, much reduced, narrowly linear; infl compact, cylindric, 4–10 × 2–2.5 cm; bracts lance-subulate, shorter than the ovary; fls numerous, white, the lip uppermost, narrowly lanceolate or oblong, 6–8 mm, entire; spur slender, directed horizontally, 11–23 mm, usually abruptly bent near the tip; pollen-sacs adjacent, the viscidia in contact. Open acid bogs on the coastal plain; s. N.J. and Del. to Fla. and Tex. Aug., Sept. (*Gymnadeniopsis n.; Platanthera n.*)

8. Habenaria obtusata (Banks) Richardson. Blunt-lf orchid. Scape slender, 1–3 dm, from a slender rhizome, usually bractless; lf solitary, basal, ascending, persistent through anthesis, oblanceolate to obovate, 5–13 × 1–4 cm, blunt, long-tapering to a petiole-like base; infl 2–10 cm; fls few, short-pediceled, greenish-white; lip deflexed, 5–9 mm, lanceolate; lateral pet dilated below the middle, directed obliquely upward to the margins of the upper sep; lateral sep widely spreading; spur 5–9 mm, tapering to the tip; $2n=42$. Moist woods and bogs, especially under evergreens; circumboreal, in Amer. s. to Mass., N.Y., Mich., Minn., and Colo. June–Aug. (*Lysiella o.; Platanthera o.*)

9. Habenaria unalascensis (Sprengel) S. Wats. Alaska orchid. Scapose, 3–5 dm from a pair of tuberous roots, with 1 or 2 sheathing bracts below the 2 or 3(4) ascending, nearly basal lvs, these narrowly lanceolate to broadly oblanceolate, 8–12 cm, acute to rounded, beginning to wither by early anthesis; scape with scattered small bracts; infl 10–30 cm; fls numerous, small, sessile or nearly so, greenish; lip narrowly lanceolate to triangular-ovate, 2.5–4 mm; sep and lateral pet 2–4 mm; spur cylindric to clavellate, 3–5 mm; pollen-sacs adjacent, each with a deeply bifid pollinium, the viscidia close together in the roof of the orifice to the spur. Thin soil of clearings and open woods; cordilleran, disjunct to the e. part of the Upper Peninsula of Mich. and the n. shore of Lake Huron and to Anticosti I. June–Aug. (*Piperia u.*)

10. Habenaria hookeri Torr. Hooker's orchid. Scape arising from a few fleshy roots, 2–4 dm, bractless; lvs 2, basal, prostrate, 6–12 cm, broadly elliptic to rotund, seldom narrower; fls sessile, yellowish-green, ascending; lip lance-triangular, directed outward, upcurved, 8–12 mm; lateral sep widely reflexed behind the fl, alongside the ovary; lateral pet lanceolate, incurved and ± adjacent to the upper sep; spur 13–25 mm, directed downward, tapering to the tip; $2n=42$. rich moist woods; N.S. and Nf. to N.J. and Pa., w. to Man., Minn., and Io. June, July. (*Lysias h.; Platanthera h.*)

11. Habenaria orbiculata (Pursh) Torr. Round-lvd orchid. Scape arising from fleshy roots, 3–6 dm, with a few narrow bracts; lvs 2, basal, prostrate, 10–15(–25) cm, broadly elliptic to rotund, seldom narrower; infl loose, 5–20 × 4–5 cm; fls white or greenish-white, on pedicels 5–10 mm; lip oblong-linear, directed downward, 10–20(–25) mm; sep and lateral pet widely spreading, the pet directed obliquely upward; spur 15–25(–45) mm, usually directed transversely and somewhat dilated distally; tips of the pollen-sacs projecting in front of the nectary and widely spreading; $2n=42$. Moist woods; Lab. to Alas., s. to Pa., Ind., Minn., and Oreg., and in the mts. to N.C. and Tenn. July, Aug. (*Lysias o.; Platanthera o.*) Robust plants with large lvs and large fls with a notably long spur have been distinguished as var. *macrophylla* (Goldie) B. Boivin. (*H. macrophylla.*)

12. Habenaria cristata (Michx.) R. Br. Crested fringed orchid. Stem 3–8 dm; lower 1–3 lvs narrowly lanceolate to linear-oblong, to 20 cm, rarely over 2 cm wide; upper several lvs much reduced; infl compact, narrow, cylindric, 3–12 × 3 cm; fls numerous, orange; sep broadly ovate to obovate, 3–5 mm; lateral pet oblong, shorter than the sep, lacerate or shortly fringed at the summit; lip oblong, long-fringed, 7–10 mm; spur 5–9 mm. Low moist meadows and damp pine woods, especially along the coastal plain, but also in the mts. southward; se. Mass.; L.I.; N.J. to Fla., La., and Tex., and inland to Ky., Tenn., and Ark. July, Aug. (*Blephariglottis c.; Platanthera c.*) A hybrid with no. 15 is *H.* × *canbyi* Ames. The Long Island plants differ in their light yellow fls.

13. Habenaria ciliaris (L.) R. Br. Yellow fringed orchid. Stem 4–10 dm; lower 1–3 lvs lanceolate, to 15 × 4 cm; upper several lvs abruptly reduced; infl compact, cylindric, 5–15 × 5 cm; fls many, orange; sep broadly oval to obovate, 6–8 mm, the lateral ones spreading like ears; lateral pet linear-oblong, shorter than the sep, lacerate at the summit; body of the lip linear-oblong, 10–16 mm, long-fringed; spur 18–30+ mm, usually directed down or back. Bogs, fields, and woods; Mass. to Mich., Wis., and Mo., s. to Fla. and Tex. July, Aug. (*Blephariglottis c.; Platanthera c.*) A hybrid with no. 12 is *H.* × *chapmanii* (Small) Ames.

14. Habenaria correlliana Cronq. Much like no. 15; relatively small, with only 1 or 2 well developed foliage lvs, the others reduced to a few distant bracts; racemes short, broad, and loosely fld; lateral sep reflexed against the ovary; lip narrowly elliptic, ca 13 × 3 mm, not fringed, merely serrate distally; spur curved, mostly 4–5 cm. Deep shade of moist deciduous forests; Cumberland Plateau region in Ky. and Tenn., and irregularly to the Great Smoky Mts. in Tenn. and the Piedmont and inner Coastal Plain in Ga., Ala., and Miss. Aug., Sept.

15. Habenaria blephariglottis (Willd.) Hook. White fringed orchid. Stem 4–10 dm; lower 1–3 lvs linear or lance-linear, to 20 × 2 cm; upper several lvs much reduced; infl compact, ovoid, 5–15 × 4–5 cm; fls white; sep subrotund or broadly ovate, 5–11 mm, the lateral ones inequilateral, reflexed against the ovary; lateral pet shorter than the sep, linear-oblong, erose or lacerate at the summit; body of the lip narrowly oblanceolate, 8–11 mm, fringed except at the slender, claw-like base; spur slender and elongate; $2n=42$. Acid swamps and bogs; Nf. and Ont. to Mich., s. to Fla. and Tex., but missing from most of the Ohio drainage. July, Aug. (*Blephariglottis b.; B. albiflora; Platanthera b.*) Two geographic vars.:

Var. *blephariglottis.* Relatively small, to 6 dm; spur 18–27 mm; fringe of the lip 3–6 mm. Northern, s. to N.J., Pa., and locally in Ill.

Var. *conspicua* (Nash) Ames. Larger and coarser, to 1 m; spur 28–40 mm; fringe of the lip nearly 1 cm. Southern, n. along the coastal plain to s. N.J.

16. Habenaria lacera (Michx.) Lodd. Ragged fringed orchid. Stem 3–8 dm; lower lvs lanceolate to oval, to 15 × 4 cm, acute or blunt; upper lvs much reduced, nearly linear; infl usually compact and many-fld, 5–15(–20) × 3–5 cm; fls pale yellow-green to greenish-white; sep broadly oval, 4.5–7 mm, the lateral ones deflexed behind the lip; lateral pet linear-spatulate, about equaling the sep, blunt, entire; lip 11–16 mm, deeply 3-lobed, each lobe cuneate, fringed with a few long segments at the end only, the terminal lobe long-clawed, the lateral ones cut to below the middle or nearly to the base; spur 14–21 mm, about equaling the ovary; $2n=42$. Open, wet, sunny places; Nf. to se. Man., s. to S.C., Ga., Ala., Ark., and ne. Okla. June–Aug. (*Blephariglottis l.; Platanthera l.*) A hybrid with no. 19 is *H.* × *andrewsii* M. White.

17. Habenaria leucophaea (Nutt.) A. Gray. Prairie fringed orchid. Stem 4–10 dm; lower lvs lanceolate to broadly linear, 10–20 cm, commonly blunt; upper lvs much reduced; spike cylindric, 8–20 × 5–7 cm; lip and lateral pet white or creamy, the sep green or greenish-white; sep broadly oval to obovate, 7–13 mm; lateral pet broadly obovate-cuneate, toothed, slightly longer than the sep; lip deeply 3-lobed, the terminal lobe short-clawed, very broadly cuneate, usually deeply notched in the center, long-fringed, the lateral lobes fringed to the middle or below; spur slender and elongate; $2n=42$. Wet prairies from Mich. and O. to N.D., Nebr., Kans. and La., and rare in bogs and marshes from N.S. and Ont. to O. June, July. Two well marked, largely allopatric, but very similar vars., partly isolated by differences in pollinators:

Var. *leucophaea.* Fls smaller, the lateral pet mostly 8–11 × 4.5–7 mm, the lip 15–21 mm, the spur 30–40 mm; column comparatively small and rounded, the pollinaria closely spaced, their caudicles parallel, the viscidia directly below the pollinia; mainly eastern, extending w. to Ill., Wis., se. Io., and rarely to e. Kans. and e. Okla. (*Blephariglottis l.; Platanthera l.*)

Var. *praeclara* (Sheviak & Bowles) Cronq. Infl avg shorter and fewer-fld; fls larger, the lateral pet mostly 11–15 × 7–11 mm, the lip 22–30 mm, the spur 40–50 mm, column larger, somewhat angular, the pollinaria well spaced, their caudicles widely diverging and directed somewhat forward; more western, extending e. to Minn., Io. and e. Mo.

18. Habenaria peramoena A. Gray. Purple fringeless orchid. Stem 3–10 dm; lower 2–4 lvs lanceolate, 10–20 cm; upper lvs abruptly reduced, lance-linear; infl 6–18 × 4–6 cm, rather loose; fls reddish-purple; sep broadly oval or obovate, 6–10 mm; lateral pet ca equaling the sep, obovate-spatulate above a narrow base, erose; lip 13–22 mm, deeply tripartite, the terminal lobe deeply notched, each lobe broadly cuneate-flabellate, with conspicuous radiating veins, erose at the margin, the teeth only 0.5–1 mm deep; spur 22–31 mm. Open, swampy or vernally wet places, often in acid soil; N.J. to O. and Mo., s. to S.C., Miss., and Ark. July, Aug. (*Blephariglottis p.; Platanthera p.*)

19. Habenaria psycodes (L.) Sprengel. Purple fringed orchid. Stem stout, 3–15 dm; lower lvs lanceolate or oval, the upper much reduced and narrow; infl cylindric, dense; fls rose-purple (white); sep broadly oval to obovate; lateral pet oblong-spatulate, finely toothed, about equaling the sep; lip very broad, deeply 3-parted, each lobe broadly flabellate, deeply toothed or shortly fringed, the teeth shorter than the body of the lobe, the terminal lobe scarcely clawed, not notched at the center. Wet meadows and wet open woods; Nf. to n. Man., s. to Del., Md., Ky., Ind., and Nebr., and along the mts. to N.C. and Tenn. June–Aug. Two well marked but intergradient vars. of nearly coextensive range, partly isolated by differences in pollinators:

Var. *psycodes.* Fls avg smaller, the lateral sep 5–7 × 3–4 mm, the lateral pet 5–7 × 3–6 mm, the lip 7–12 × 8–15 mm, its segments cleft less than one third of their length; column relatively narrow, its pollen-sacs diverging but slightly, so that the viscidia terminate close to and above the entrance to the nectary, the entrance to which is narrowed at the middle and thus dumbbell shaped; $2n=42$.

Var. *grandiflora* (Bigelow) A. Gray. Fls avg larger, the lateral sep 6–10 × 4–6 mm, the lateral pet 6–10 × 5–6 mm, the lip up to 18 × 25 mm, its margins deeply fringed (a third or more of the length of the segments); column larger and broader, the pollen-sacs spread widely apart at the edges of the widely rounded opening to the nectary, the viscidia exsert on columnar processes; plants avg more robust and often with a more elongate and open raceme. (*H. grandiflora; Platanthera g.*)

9. POGONIA Juss. Sep and lateral pet separate, spreading, similar in color, the pet somewhat wider and shorter; lip declined, not lobed, narrow below the middle, distally expanded, fringed on the margin and bearded on the face; column projecting forward over the lip, its terminal anther deflexed; pollinia 2; perennial from a short rhizome with a cluster of fibrous roots, the erect stem bearing a single lf near the middle, often a long-petioled basal lf, and a terminal bracteal lf subtending the solitary, short-peduncled fl; fls rarely 2. 2–several, irregularly cosmop.

1. Pogonia ophioglossoides (L.) Ker Gawler. Rose pogonia; snake-mouth. Slender, (1–)2–4 dm; cauline lf lanceolate or narrowly elliptic, 3–9 × 1–2.5 cm; bracteal lf similar but smaller; sep and lateral pet rose-pink to seldom white, 17–22 mm; sep lanceolate, widely spread, the dorsal one nearly erect; lateral pet elliptic, hovering over the column; lip about as long as the sep, usually pink, veined with red, bearded with yellow hairs; $2n=18$. Open, wet meadows and sphagnum-bogs; Nf. to Minn., s. to Fla. and Tex. May–July.

10. CLEISTES Rich. Sep and lateral pet evidently unlike, the sep slender, widely spreading or distally recurved, the pet directed forward, close together over the lip; lip rather broad, ± ovate in outline, obscurely 3-lobed, the long low lateral lobes enfolding the column, the terminal lobe short, triangular; lip crested with a median ridge and with a ridged and papillate summit; column with a deflexed terminal anther; pollinia 2; perennial from a cluster of fleshy-fibrous roots, the erect stem-bearing 1 or 2 sheathing scales toward the base and a single (2) foliage lf near or above the middle; fl solitary, rarely 2. 25; W. Hemisphere, mainly trop.

1. Cleistes divaricata (L.) Ames. Spreading pogonia. Slender, 2–6 dm; lf long-sheathing, the blade sessile, oblong-linear, obtuse, 5–11 cm × 8–20 mm, bracteal lf similar but smaller; fl peduncled, somewhat nodding; sep linear, purple-brown, 3–6 cm; lateral pet pink to rose-purple, oblanceolate, 3–5 cm, recurved at the tip; lip slightly exceeding the pet, greenish, purple-veined. Swamps and wet woods, mostly on the coastal plain; N.J. to Fla. and Tex.; also inland to e. Ky. and e. Tenn. June, July. (*Pogonia d.*)

11. ISOTRIA Raf. Whorled pogonia; five-lvs. Sep and lateral pet evidently unlike, the sep linear, elongate, the pet shorter and relatively wider; lip not declined, bearing a prominent median ridge from the base to beyond the middle, 3-lobed, the lateral lobes ⅔ as long as the whole lip, or longer, upturned at the margin, the terminal lobe short, broad, inconspicuously erose; column extending forward over the lip, its terminal anther deflexed; pollinia 2; nectar wanting; perennial from a cluster of fleshy-fibrous roots, the erect glabrous, hollow stem bearing a single whorl of usually 5 lvs near the summit and 1(2) large terminal fl. Only 2 spp.

1 Sep 3.5–6.5 cm; peduncle at least as long as the ovary . . . . 1. *I verticillata.*
1 Sep 1.2–2.5 cm; peduncle shorter than the ovary . . . . 2. *I. medeoloides.*

1. **Isotria verticillata** (Willd.) Raf.   Larger w.p.; larger five-lvs. Colonial; not or scarcely glaucous; stems slender, 2–4 dm; lvs oblong-obovate, 3–5 cm at anthesis, later to 8 cm; peduncle 2–5.5 cm; fls outcrossing, fragrant at first, multicolored, with guide lines; sep projecting forward or somewhat spreading, linear, 3.5–6.5 cm, greenish-brown to purple-brown, more than twice as long as the lateral pet, these 1.5–2.5 cm, yellowish-green, connivent over the column and lip; lip 2–2.5 cm, its lateral lobes purple or strongly purple-veined, the body and the terminal lobe greenish-yellow; $2n=18$. Acid soil in woods; s. Me. to Ont. and Mich., s. to n. Fla. and e. Tex. May, June. (*Pogonia v.*)

2. **Isotria medeoloides** (Pursh) Raf.   Little w.p.; little five-lvs. Not colonial, the individuals few and scattered; herbage glaucous; stem 1.5–2.5 cm; lvs as in no. 1; peduncle 1–1.5 cm, fls selfed, inodorous, light green or yellow-green or partly whitish, without guide-lines; sep 1.2–2.5 cm; pet 1.3–1.7 cm; lip 1–1.5 cm, less emarginate and ornate; $2n=18$. Open stands of second-growth hardwoods or pine-hardwoods; irregularly at widely scattered stations from s. Me. to N.C., w. to s. Ont., Mich. and Mo. May, June. Our rarest orchid.

12. TRIPHORA Nutt.   Sep and lateral pet similar, ascending; lip obovate in outline, erect at base, thence decurved, 3-lobed, traversed by 3 elevated nerves, the lateral lobes triangular, the terminal one broadly rounded; anther erect at the end of the column; pollinia 2; delicate herbs, the stem bearing several alternate lvs, arising from a solid cormose base or a cluster of short, fleshy roots; fls few, axillary, ephemeral, white to pale pink, with green veins. 10, New World.

1. **Triphora trianthophora** (Swartz) Rydb.   Three-birds orchid. Stem 1–3 dm, nodding, straightening as it grows; lvs sessile, ovate, 1–2 cm; fls ephemeral; sep and lateral pet lanceolate, 1.5–2 cm; lip 1.5–2 cm, with 3 prominent green ridges, crisped at the tip, the lateral lobes upturned. Rich moist woods, often on rotten logs; s. Me. to s. Wis., s. to Fla., Tex., and C. Amer. Aug., Sept. (*Pogonia t.*)

13. CALOPOGON R. Br.   Grass-pink. Fls nectarless, not resupinate, the lip upper-most; sep and lateral pet nearly alike, spreading, the lateral sep inequilateral, wider than the lateral pet; lip spreading, long-stalked, dilated at the summit and densely bearded along the inner side with long clavate hairs; column elongate, curved, 2-winged distally; anther terminal, with 4 pollinia; perennial from a corm, the slender stem with 1 or 2 basal sheathing scales, a single long grass-like lf near the base, and a short raceme of 2–several fls. 4, e. N. Amer.

1 Fls larger, the lip 15–20 mm, ca 10 mm wide near the summit . . . . 1. *C. tuberosus.*
1 Fls smaller, the lip 8–12 mm, ca 5 mm wide near the summit . . . . .2. *C. pallidus.*

1. **Calopogon tuberosus** (L.) BSP.   Lf linear, to 5 dm × 4 cm, long-sheathing; scape 3–7 dm, with a loose raceme of 3–15 fls; tep rose-purple, 15–25 mm, acute; lip 15–20 mm, with a narrow basal part, the summit broadly flabellate, rose purple, crested on its face with stout hairs tipped with magenta and yellow; winged terminal portion of the column suborbicular (when flattened out), with tapering base. Acid bogs and swamps; Nf. to Minn. and se. Man. s. to Fla., Tex. and Cuba. June, July. (*C. pulchellus; Limnodorum t.*)

2. **Calopogon pallidus** Chapman.   Much like no. 1; fls white to pink, avg smaller, the tep 9–18 mm; lip 8–12 mm, with a dense tuft of hairs at the base of the terminal expanded part; winged terminal portion of the column (when flattened out) triangular-elliptic, with subtruncate base. Wet pinelands and savannas on the coastal plain; se. Va. to Fla. and La. June. (*Limnodorum p.*)

14. ARETHUSA L.   Sep and lateral pet similar, united at base, erect or arched-as-cending over the column; lip ascending at base, thence curved forward and decurved at the widened summit, crested on the face with 3 fimbriate ridges; column erect, petaloid, dilated at the summit; anther subterminal, with 4 pollinia; cormose perennial, the scape with 1–3 loose blunt bracts toward the base; lf 1, grass-like, arising after anthesis from within the upper bract. 2, the other Japanese.

1. **Arethusa bulbosa** L.   Dragon's mouth. Scape 1–3 dm; lf at maturity nearly equaling the scape, 2–4 mm wide; fls 3–6 cm high, subtended by a pair of small bracts; sep and pet lanceolate, magenta; lip pinkish-white, spotted and streaked with purple and yellow, about as long as the lateral pet. Sphagnum-bogs and swampy meadows, usually rare; Nf. to Minn., s. to N.J. and n. Ind., and in the mts. to N.C. and S.C. May, June.

15. HEXALECTRIS Raf.   Sep and lateral pet similar, 5–7-veined; lip 3-lobed, usually with 7 longitudinal ridges convergent below and extending nearly to the tip; lateral lobes

of the lip upturned and partly embracing the fleshy, distally dilated column; anther terminal, with 8 pollinia, yellowish or brownish to purplish, nongreen mycotrophs arising from a cluster of coralloid rhizomes; stem bearing a few sheathing scales and a loose terminal raceme of dull yellow fls striped with purple. 7, N. Amer., esp. Mex.

1. **Hexalectris spicata** (Walter) Barnhart. Crested coral-root. Stems 3–8 dm, purplish, glaucous; raceme 1–3 dm, with 5–15(–25) short-pedicellate, spreading or drooping fls from the axils of lance-ovate bracts; sep oblanceolate, 2 cm, spreading and distally recurved, slightly exceeding the lateral pet, which are often held closely over the column; lip 1.5 cm, its lateral lobes with 3 or 4 curved purple veins, the terminal lobes with a median purple and white vein and ca 6 purple lateral ridges; column white, ca 15 mm. Woods; Md. to O. and Mo., s. to Fla. and Mex. July, Aug. (*H. aphylla.*)

16. **MALAXIS** Sol. Adder's mouth. Sep and pet unlike, the lateral sep ± parallel behind the lip, the pet much smaller, spreading or recurved; lip relatively broad, variously shaped, often auriculate at base; column very short, in our spp. ca 1 mm or less; pollinia 4; small orchids from a corm, with 1–5 lvs at the base or near the middle of the low scape, and a terminal raceme of small fls. 200, irregularly widespread.

1 Foliage lf 1; lip lowermost, in fl, deflexed.
  2 Lip narrowly pointed, not at all bifid . . . . 1. *M. monophyllos.*
  2 Lip strongly bifid at the summit, and with a short central tooth . . . . 2. *M. unifolia.*
1 Foliage lvs 2 or more, lip uppermost in fl, erect.
  3 Pedicels elongate, ca 4–8 mm; southern . . . . 3. *M. spicata.*
  3 Pedicels short and inconspicuous, 1–3 mm; northern . . . . 4. *M. paludosa.*

1. **Malaxis monophyllos** (L.) Swartz. White a.-m. Scape 1–2.5 dm; lf solitary, clasping, oval to elliptic, 3–6 cm, usually at least half as wide; infl very slender, tapering to the tip; pedicels 1–2 mm; sep ovate to lance-ovate, 2–2.5 mm, the median erect, the lower deflexed beyond the lip; lateral pet oblanceolate, 1 mm, spreading horizontally; lip greenish-white, deflexed, 2–3 mm, broadly cordate, auriculate-lobed at base, tapering to a long point, minutely erose. Damp woods and bogs; circumboreal, in Amer. s. to N.J., Pa., Ill., Minn., Colo., and Calif. Most of the Amer. plants, including all ours, belong to var. *brachypoda* (A. Gray) F. Morris, with the fls merely resupinate, the lip lowermost. (*M. brachypoda.*) Typical Eurasian var. *monophyllos* has the fls twisted a full circle at base, so that the lip is uppermost. (*Microstylis m.*)

2. **Malaxis unifolia** Michx. Green a.-m. Scape 1–3 dm; lf solitary, sessile, oval or elliptic, 3–6 cm, a third to half as wide; infl solitary or broadly rounded at the summit; pedicels 4–8 mm; upper sep elliptic, obtuse, erect, 1.3–1.5 mm; lower sep similar, spreading or deflexed; lateral pet linear, 1 mm, recurved behind the fl; lip lowermost, greenish, oblong in general outline, 2–2.5 mm, its basal angles prolonged into auricles behind the fl, and cleft deeply 2-lobed, with a small central tooth in the sinus. Damp woods and bogs; Nf. and Que. to Man., s. to Fla., Tex., W.I., and C. Amer. (*Microstylis u.*)

3. **Malaxis spicata** Swartz. Florida a.-m. Scape 1.5–3 dm; lvs 2, broadly oval, 3–10 cm, usually over half as wide; pedicels evident, slender, 4–8 mm; lip uppermost in fl, broadly cordate, pointed at the tip, with conspicuous basal auricles extending below the column and nearly meeting around it, 4 mm, including the auricles, orange toward the center, paler toward the margins. Damp calcareous soil; se. Va. to Fla. and W.I. July–Aug.

4. **Malaxis paludosa** (L.) Swartz. Bog a.-m. Scape 5–15 cm; lvs 2–5, lance-ovate, 1–3 cm; pedicels short, 1–3 mm; fls yellowish-green, the lip uppermost, 1.5 mm, surpassed by the lateral sep, which stand erect behind it; $2n=28$. Deep bogs; nearly circumboreal, entering our range in n. Minn. July, Aug. Our smallest orchid.

17. **LIPARIS** Rich. Twayblade. Sep spreading, narrow, with incurved margins; lateral pet narrowly linear, usually turned forward under the lip, rolled and twisted, appearing thread-like; lip broad above a narrow base, declined; column longer than in *Malaxis* (in our spp. ca 3–4 mm), strongly curved, with narrow lateral wings above; pollinia 4; low orchids from a corm or pseudobulb, producing a few basal scale-lvs, 2 large, shining foliage lvs near the base, and a naked scape bearing a loose raceme. 250, ± cosmop.

1 Lip ca 1 cm, pale purple . . . . 1. *L. liliifolia.*
1 Lip ca 5 mm, yellowish-green . . . . 2. *L. loeselii.*

1. **Liparis liliifolia** (L.) Rich. Large t.; mauve sleekwort. Scape 1–2.5 dm, with a loose raceme of 5–30 fls; lvs oval to elliptic, 5–15 × 2–6 cm; pedicels and slender ovary together ca 8–15 m, widely spreading;

sep greenish-white, 1 cm; lateral pet greenish to pale purple, 1 cm, projecting forward (or deflexed) under the lip and often crossed; lip pale purple, 1 cm, nearly as wide, rhomboid-obovate, subtruncate at apiculate summit, minutely erose, nearly flat; column 4 mm. Rich woods; Me. to Minn., s. to Ga. and Ark.; China. June, July.

2. **Liparis loeselii** (L.) Rich. Loesel's t.; fen-orchid. Scape 1–2.5 dm, with a loose raceme of 2–12 fls; lvs lanceolate to elliptic, 5–15 × 2–3(–5) cm, somewhat folded along the midrib; pedicels 3–5 mm; sep yellowish-green, 5–6 mm; pet yellowish-green, 5 mm, somewhat reflexed; lip 5 mm; ascending at the narrowed base, the blade yellowish-green, broadly ovate with upturned margins; column 3 mm; $2n=32$. Damp or wet woods; N.S. and Que. to Man., s. to N.J., O., Nebr., Kans., Mo., and in the mts. to Ala.; Europe. June, July.

18. **CALYPSO** Salisb., nom. conserv. Sep and lateral pet lance-linear, similar, ascending over the lip; lip saccate, shoe-shaped, declined, narrowed to the shortly bifid summit, the distal part covered by a delicate transparent apron; column broadly oval, petaloid, overhanging the lip; anther sessile on the lower side of the column just beneath the tip, each locule with a bifid pollinium; perennial from a corm with coralloid roots, producing in autumn a single basal lf that persists through the next anthesis, and in the spring sending up a lfless scape bearing 2 or 3 sheathing bracts and at the summit a solitary, nodding, nectarless fl subtended by a linear bract. Monotypic.

1. **Calypso bulbosa** (L.) Oakes. Calypso. Scape 1–2 dm; lf-blade round-ovate, 3–5 cm, often basally cordate, the petiole about as long; sep and lateral pet pale purple (white), 1–2 cm; lip 1.5–2.3 cm, whitish, becoming yellowish toward the tip, marked with red-brown inside, the apron white, spotted with purple and crested with 3 rows of yellow hairs; column 8–12 mm. Moist coniferous woods; circumboreal, s. in Amer. to N.Y., Mich., Minn., N.M., Ariz., and Calif. May, June. (*Cytherea b.*) Our plants have been segregated as var. *americana* (R. Br.) Luer.

19. **APLECTRUM** Torr. Sep and lateral pet similar, spreading, the pet projecting forward over the column; lip broadly obovate, with 3 low parallel ridges near the center, obliquely 3-cleft, its lateral lobes upcurved, the terminal one dilated and upcurved at the margin; column slender, flattened, bearing a terminal anther with 4 pollinia; perennial from stout globose corms that produce in late summer a single lf that lasts ± throughout the winter, the same corm (in some individuals) producing an achlorophyllous, bracteate scape the following spring, after the lf has withered; fls in a loose terminal raceme. 2, the other Japanese.

1. **Aplectrum hyemale** (Muhl.) Torr. Adam and Eve; putty-root. Corms usually 2, connected by a slender 3 cm rhizome; lf basal, its blade elliptic, 10–20 cm, dark green with whitish veins; scape 3–6 dm, with a few linear-oblong, sheathing nongreen bracts; fls 7–15; sep and lateral pet 10–15 mm, purplish proximally, brownish distally; lip 10–15 mm, white, marked with violet; column 7 mm. Rich woods; Mass., Vt. and s. Que. to N.C. and n. Ga., w. to Sask. and Kans. May, June.

20. **TIPULARIA** Nutt. Sep and lateral pet similar, narrowly oblong or oblanceolate; lip 3-lobed, the lateral lobes basal, short and rounded, upturned, the terminal lobe long and slender; lip prolonged at base into a long, slender, horizontal spur; anther terminal on the column, with 2 pollinia; perennial herbs with a series of corms connected by short rhizomes, the youngest corm producing a single lf in autumn and then the next summer (after the lf has withered) an achlorophyllous, lfless scape bearing a few sheathing scales toward the base and a long, loose, bractless terminal raceme. 3, the others in e. Asia.

1. **Tipularia discolor** (Pursh) Nutt. Crane-fly orchid. Lf basal, 5–10 × 2.5–7 cm, somewhat plicate, green above, purplish beneath, on a 5 cm petiole, withering in the spring; scape produced in midsummer, 2–5 dm, very slender; infl 1–2 dm; fls many; sep and lateral pet 4–8 mm, pale greenish-purple with purple veins, spreading, one pet overlapping the dorsal sep; lip pale purple, 4–8 mm; its basal lobes nearly semicircular, erose, with forwardly directed tip, the terminal lobe linear, arcuate, its margins revolute, the apex minutely notched; spur 15–22 mm; column 3–4 mm; fr pendulous. Rich damp woods; se. Mass. and s. N.Y. to s. Mich., s. to Fla. and e. Tex. June–Aug. (*T. unifolia.*)

21. **CORALLORHIZA** Gagnebin. Coral-root. Sep and lateral pet narrow, similar, usually spreading-ascending or projecting over the column; lateral sep united with the base of the column, often forming a low protuberance or short spur at the summit of the

ovary; lip deflexed, oblong to rotund, often with 2 lateral lobes or teeth, the margins usually upturned, and the face bearing one or 2 short longitudinal ridges; column shorter than the perianth, rather broad or boat-shaped, slightly incurved, its terminal anther with 4 pollinia; fr pendulous; yellow, brown or purplish mycotrophic plants; stem arising from a cluster of coralloid roots, bearing a few sheathing scales and a terminal raceme of small, usually bicolored fls. 10, N. Amer.

1 Lip with a pair of evident lateral lobes or teeth.
  2 Sep and pet 3(5)-nerved; spur generally present (though small) .... 1. *C. maculata.*
  2 Sep and pet 1-nerved, or the pet weakly 3-nerved; spur none .... 2. *C. trifida.*
1 Lip neither lobed nor toothed, but sometimes with erose or wavy margins.
  3 Sep and lateral pet 8–15 mm, with 3–5 strong purple stripes .... 3. *C. striata.*
  3 Sep and lateral pet 3–7.5 mm, not striped.
    4 Lip and lateral pet 5–7.5 mm; fl spring (to midsummer) .... 4. *C. wisteriana.*
    4 Lip and lateral pet 3–4.5 mm; fl late summer and fall .... 5. *C. odontorhiza.*

    1. Corallorhiza maculata (Raf.) Raf. Spotted c.-r. Stem 2–6 dm, often pinkish-purple; infl 5–15 cm, fls 10–40; sep and lateral pet narrowly oblong to oblanceolate, 6–8 mm, 3(5)-nerved, the lateral sep somewhat divergent, the lateral pet ± spotted or suffused with purple; spur a prominent, sometimes divergent swelling near the ovary-tip; lip 6–8 mm, white, generally purple-spotted, with 2 rather small but evident lateral lobes below the middle and 2 short parallel ridges on the face, its terminal lobe rounded and deflexed, 3–4 mm wide. Woods; Nf. and Lab. to B.C., s. to Md. and Ind. and in the mts. to N.C., in the west to Mex. and C. Amer. July–Sept.

    2. Corallorhiza trifida Chatel. Northern c.-r. Stem 1–3 dm, yellowish to greenish; infl 2–8 cm; fls 5–15; sep and lateral pet yellowish-green, to light purplish, lanceolate, 4–5 mm, 1-nerved (or the pet weakly 3-nerved), the lateral sep extending forward under or alongside the lip; spur none; lip 3.5–5 mm, white, sometimes purple-spotted, with 2 short lateral lobes below the middle and 2 short parallel ridges on the face, the terminal lobe truncate and deflexed, 1.5–2 mm wide; 2*n*=42. Wet woods; circumboreal, s. in Amer. to N.J., Pa., Ill., Mo., and N.M. May–July. (*C. corallorhiza.*) The more southern phase of the sp. (and the common form in our range), with unspotted lip, has been segregated as var. *verna* (Nutt.) Fern. from the more northern and circumboreal form with spotted lip.

    3. Corallorhiza striata Lindl. Striped c.-r. Stem 2–4.5 dm, rather stout, purple or magenta; infl 5–20 cm, with 7–25 fls; sep and lateral pet arching forward, lance-oblong, 8–15 mm, yellowish-white with 3–5 conspicuous longitudinal purple stripes, also purple-margined; lip declined, 8–12 mm, white, heavily striped with purple, or wholly purple. Woods; N.B. and s. Que. to B.C., s. Mich., Wis., Minn., N.D., and in the w. cordillera to Mex. May–July.

    4. Corallorhiza wisteriana Conrad. Spring c.-r. Stem 1–4 dm, purple or reddish; infl 3–7 cm, with 10–15 fls; sep and lateral pet extending forward over the column, scarcely spreading, narrowly lanceolate, 5–7.5 mm, greenish-yellowish, tinged with purple and marked with short purple lines; lip deflexed, about as long as the sep, narrowed below into a slender claw more than half as long as the blade, the latter broadly oval, crenulate, notched at the rounded summit, white, dotted with purple. Damp woods; Pa. and s. N.J. to Fla., w. to se. Nebr., Okla., and Tex., and in the w. cordillera. Apr.–June.

    5. Corallorhiza odontorhiza (Willd.) Nutt. Autumn c.-r. Stem 1–2 dm, with a bulbous-thickened white base, otherwise purple to brown, or greenish above; infl 3–6 cm; fls 5–15; sep and lateral pet extending forward over the column, oblong, 3–4.5 mm, yellowish to purplish-green or dark purple; lip declined, shortclawed or nearly sessile, expanded above into a rotund blade 3–4 mm, white, usually purple-margined with 2 purple spots, entire or finely erose, often emarginate. Open woods; Me. and Vt. to N.Y., s. Mich., and s. Minn., s. to Fla. and Mex. Aug.–Sept.

Nomenclatural Innovations

Antennaria neglecta Greene var. **neodioica** (Greene) Cronq.
  Antennaria neodioica Greene, Pittonia 3: 184. 20 July 1897.
Antennaria neglecta Greene var. **petaloidea** (Fern.) Cronq.
  Antennaria neodioica var. *petaloidea* Fern., Proc. Boston Soc. Nat. Hist. **28:** 245. 1898.
Carex atlantica L. Bailey var. **capillacea** (L. Bailey) Cronq.
  Carex interior L. Bailey var. *capillacea* L. Bailey, Bull. Torrey Bot. Club **20:** 426. 1893.
Chenopodium berlandieri Moq. var. **bushianum** (Aellen) Cronq.
  Chenopodium bushianum Aellen, Feddes Repert. **26:** 63. 1929.
Chenopodium berlandieri Moq. var. **macrocalycium** (Aellen) Cronq.
  Chenopodium macrocalycium Aellen, Feddes Repert. **26:** 123. 1929.
Chrysopsis camporum Greene var. **glandulissima** (Semple) Cronq.
  Heterotheca camporum var. *glandulissima* Semple, Brittonia **35:** 146. 1983.

Elytrigia dasystachya (Hook.) A. & D. Löve var **psammophila** (J. M. Gillett & Senn) Cronq.
Agropyron psammophilus J. M. Gillett & Senn, Canad. J. Bot. **39**: 1170. 1961.
Empetrum rubrum L. var. **eamesii** (Fern. & Wieg.) Cronq.
Empetrum eamesii Fern. & Wieg. Rhodora **15**: 214. 1913.
Festuca brachyphylla J. A. Schultes var. **rydbergii** (St.-Yves) Cronq.
Festuca ovina L. subsp. *saximontana* (Rydb.) St.-Yves var. *rydbergii* St.-Yves, Candollea **2**: 245. 1925.
Habenaria correllii Cronq.
Habenaria blephariglottis (Willd.) Hook. var. *integrilabia* Correll, Bot. Mus. Leafl. Harvard Univ. **9**: 153. 1941. Not *H. integrilabris* J. J. Smith, 1909.
Habenaria leucophaea (Nutt.) A. Gray var. **praeclara** (Sheviak & Bowles) Cronq.
Platanthera praeclara Sheviak & Bowles, Rhodora **88**: 278. 1986.
Helianthus pauciflorus Nutt. var. **subrhomboideus** (Rydb.) Cronq.
Helianthus subrhomboideus Rydb. Mem. N. Y. Bot. Gard. **1**: 419. 1900.
Hymenoxys herbacea (Greene) Cronq.
Tetraneuris herbacea Greene, Pittonia **3**: 268. 1898.
Lespedeza hirta (L.) Hornem. var. **curtissii** (Clewell) Cronq.
Lespedeza hirta (L.) Hornem. ssp. *curtissii* Clewell, Brittonia **16**: 25. 1964.
Lithospermum caroliniense (Walter) MacMillan var. **croceum** (Fern.) Cronq.
Lithospermum croceum Fern. Rhodora **37**: 329. 1935.
Oenothera biennis L. var. **austromontana** (Munz) Cronq.
O. biennis L. ssp. *austromontana* Munz, N. Amer. Fl. II. **5**: 134. 1965.
Sagina nodosa (L.) Fenzl var. **borealis** (Crow) Cronq.
Sagina nodosa (L.) Fenzl ssp. *borealis* Crow, Rhodora **80**: 28. 1978.
Viola palmata L. var **pedatifida** (G. Don) Cronq.
Viola pedatifida G. Don, Gen. Syst. Dichlam. Pls. **1**: 320. 1831.
Zigadenus elegans Pursh var. **glaucus** (Nutt.) Preece, an herbarium name here validated with the permission of Sherman Preece.
Melanthium glaucum Nutt. Gen. N. Amer. Pls. **1**: 232. 1818

INDEX

Note: The accepted scientific and common names for plants and plants groups that are considered to be regular members of our flora are given in Roman type in this Index, but it has not been considered necessary to index separately the species in genera represented by less than 10 species in our area. Names listed in synonymy, or casually mentioned as representing hybrids or introduced plants that may not be fully established in our range, are given in *italic* type.